Electrophysiological Disorders of the Heart

Electrophysiological Disorders of the Heart

Sanjeev Saksena, M.B.B.S., M.D., F.A.C.C., F.E.S.C., F.A.H.A., F.H.R.S.
Clinical Professor of Medicine
Robert Wood Johnson School of Medicine
Piscataway, New Jersey
Director, Cardiovascular Institute and Arrhythmia/Pacemaker Service
PBI Regional Medical Center
Passaic, New Jersey

A. John Camm, M.D., F.R.C.P., F.E.S.C., F.A.C.C.
Professor of Clinical Cardiology
St. George's Hospital Medical School
London, United Kingdom

Penelope A. Boyden, Ph.D.
Professor, Department of Pharmacology and the Center for Molecular Therapeutics
Columbia University
New York, New York

Paul Dorian, M.D., M.Sc., F.R.C.P.C.
Professor of Medicine
University of Toronto Faculty of Medicine
Director of Electrophysiology
St. Michael's Hospital
Toronto, Ontario, Canada

Nora Goldschlager, M.D., F.A.C.P., F.A.C.C.
Professor of Clinical Medicine
University of California, San Francisco, School of Medicine
Associate Director, Cardiology Division
Director, Coronary Care Unit, ECG Laboratory and Pacemaker Clinic
San Francisco General Hospital
San Francisco, California

ELSEVIER
CHURCHILL LIVINGSTONE

ELSEVIER
CHURCHILL
LIVINGSTONE

The Curtis Center
170 S Independence Mall W 300E
Philadelphia, Pennsylvania 19106-3399

ELECTROPHYSIOLOGICAL DISORDERS OF THE HEART 0-443-06570-5

Notice

Medicine is an ever-changing field. Standard safety precautions must be followed, but as new research and clinical experience broaden our knowledge, changes in treatment and drug therapy may become necessary or appropriate. Readers are advised to check the most current product information provided by the manufacturer of each drug to be administered to verify the recommended dose, the method and duration of administration, and contraindications. It is the responsibility of the treating physician, relying on experience and knowledge of the patient, to determine dosages and the best treatment for each individual patient. Neither the Publisher nor the editor assumes any liability for any injury and/or damage to persons or property arising from this publication.

The Publisher

Library of Congress Cataloging-in-Publication Data

Electrophysiological disorders of the heart/editors, Sanjeev Saksena, A. John Camm; associate editors, Penelope A. Boyden, Paul Dorian, Nora Goldschlager. — 1st ed.
p. ; cm.
Includes bibliographical references.
ISBN 0-443-06570-5
1. Arrhythmia. I. Saksena, Sanjeev.
[DNLM: 1. Arrhythmia—diagnosis. 2. Arrhythmia —physiopathology. 3. Arrhythmia—therapy. 4. Eletrocardiography. 5. Electrophysiologic Techniques, Cardiac.
WG 330 E383 2004]
RC685.A65E455 2004
616.1′28—dc22

2004043905

Acquisitions Editor: Anne Lenehan
Developmental Editor: Jennifer Shreiner
Project Manager: Jeff Gunning
Book Designer: Gene Harris

Printed in the United States of America

Last digit is the print number: 9 8 7 6 5 4 3 2 1

To Diane, Joy, and our parents and families, whose unfailing support and understanding made this work feasible.

This book is dedicated to the pioneers in our field, who made such progress possible, and to our mentors, who imbued us with the desire to help advance this science.

Contributors

Masood Akhtar, M.D.
Clinical Professor of Medicine, University of Wisconsin Medical School—Milwaukee Clinical Campus; Attending, Aurora Sinai/St. Luke's Medical Centers, Milwaukee, Wisconsin
Ventricular Tachycardia

Kelley Anderson, M.D.
Clinical Associate Professor of Medicine, University of Wisconsin, Medical School, Madison; Cardiologist, Marshfield Clinic, Marshfield, Wisconsin
Ventricular Tachycardia and Ventricular Fibrillation without Structural Heart Disease

Angelo Auricchio, M.D., Ph.D.
Associate Professor of Cardiology, Otto von Guericke University School of Medicine; Director, Cardiac Catheterization Laboratories, Division of Cardiology, University Hospital, Magdeburg, Germany
Device Technology for Congestive Heart Failure

Rabih Azar, M.D., M.Sc., F.A.C.C.
Saint-Joseph University School of Medicine; Attending, Division of Cardiology, Hôtel Dieu de France Hospital, Beirut, Lebanon
Sinus Node Dysfunction; Atrioventricular Block

David G. Benditt, M.D.
Professor of Medicine, University of Minnesota Medical School; Cardiac Arrhythmia Center, University Hospital, Minneapolis, Minnesota
Autonomic Nervous System and Cardiac Arrhythmias

Marcie Berger, M.D.
Clinical Assistant Professor of Medicine, University of Wisconsin Medical School—Milwaukee Clinical Campus; Attending, St. Luke's/Aurora Sinai Medical Centers, Milwaukee, Wisconsin
Ventricular Tachycardia

Tim Betts, M.D., M.B.Ch.B., M.R.C.P.
Consultant Cardiologist and Electrophysiologist, Department of Cardiology, John Radcliffe Hospital, Oxford, United Kingdom
Arrhythmias in Coronary Artery Disease

Saroja Bharati, M.D.
Professor of Pathology, Rush Medical College; Rush–Presbyterian–St. Luke's Medical Center, Chicago; Director, The Maurice Lev Congenital Heart and Conduction System Center, The Heart Institute for Children; Advocate Hope Children's Hospital and Advocate Christ Medical Center, Oak Lawn, Illinois
Sinus Node Dysfunction; Atrioventricular Block; Paroxysmal Supraventricular Tachycardias and the Preexcitation Syndromes; Atrial Tachycardias, Flutter, and Fibrillation; Sustained Ventricular Tachycardia with Heart Disease; Ventricular Fibrillation

David B. Bharucha, M.D., Ph.D.
Assistant Professor of Medicine–Cardiac Physiology, Mount Sinai School of Medicine; Attending Electrophysiologist, Cardiovascular Institute, Mount Sinai Medical Center; Director, Arrhythmia and Cardiac Device Services, Queens Health Network, New York, New York
Postoperative Arrhythmias after Cardiac Surgery

Zalmen Blanck, M.D.
Clinical Associate Professor of Medicine, University of Wisconsin Medical School—Milwaukee Clinical Campus; Attending, St. Luke's/Aurora Sinai Medical Centers, Milwaukee, Wisconsin
Ventricular Tachycardia

Neil E. Bowles, Ph.D.
Assistant Professor of Pediatrics, Baylor College of Medicine; Pediatric Cardiologist, Texas Children's Hospital, Houston, Texas
Genetics and Cardiac Arrhythmias

Josep Brugada, M.D.
Associate Professor of Medicine, University of Barcelona School of Medicine; Director, Arrhythmia Unit, Hospital Clinic, Barcelona, Spain
The Brugada Syndrome

Pedro Brugada, M.D.
Professor of Cardiology, Cardiovascular Research and Teaching Institute; Olv Hospital, Aalst, Belgium
The Brugada Syndrome

Ramon Brugada, M.D.
Assistant Professor of Medicine; Director, Molecular Genetics, Massonic Medical Research Laboratory, Utica, New York
The Brugada Syndrome

Hugh Calkins, M.D.
Professor of Medicine, Johns Hopkins University School of Medicine; Director, Electrophysiology Laboratory, and Director, Arrhythmia Service, Johns Hopkins Hospital, Baltimore, Maryland
Syncope

A. John Camm, M.D., F.R.C.P., F.E.S.C., F.A.C.C.
Professor of Clinical Cardiology, Department of Cardiac and Vascular Sciences, St. George's Hospital Medical School, London, United Kingdom
Atrial Tachycardia, Flutter, and Fibrillation; Nonsustained Ventricular Tachycardia; Noninvasive Electrophysiology

Franco Cecchi, M.D.
Chief, Referral Center for Cardiomyopathies, Department of Cardiology, Azienda Ospedaliera Universitaria Careggi, Florence, Italy
Arrhythmias Associated with Hypertrophic Cardiomyopathy

Nipon Chattipakorn, M.D., Ph.D.
Director, Cardiac Electrophysiology Unit, Department of Physiology, Chiangmai University Faculty of Medicine, Chiangmai, Thailand
Fundamental Concepts and Advances in Defibrillation

Shih-Ann Chen, M.D.
Professor of Medicine, National Yang-Ming University School of Medicine; Director, Cardiac Electrophysiology Laboratory, Taipei Veterans General Hospital, Taipei, Taiwan
Paroxysmal Supraventricular Tachycardias and the Preexcitation Syndromes; Atrial Tachycardia, Flutter, and Fibrillation

Yongkeun Cho, M.D., Ph.D.
Associate Professor of Internal Medicine, Kyungpook National University Medical School and Hospital, Taegu, Korea
Nonsustained Ventricular Tachycardia

Anthony W.C. Chow, M.D., M.R.C.P.
Consultant Electrophysiologist, Department of Cardiology, University College London Hospitals NHS Trust; Honorary Consultant Cardiologist, Department of Cardiology, St. Mary's Hospital NHS Trust, London, United Kingdom
Catheter Mapping Techniques

Jamie Beth Conti, M.D.
Associate Professor of Medicine and Training Program Director, Cardiovascular Diseases, University of Florida College of Medicine; Assistant Director, Clinical Electrophysiology, Shands at the University of Florida, Gainesville, Florida
Arrhythmias during Pregnancy

Ryan Cooley, M.D.
Clinical Assistant Professor of Medicine, University of Wisconsin Medical School—Milwaukee Clinical Campus; Attending, Aurora Sinai/St. Luke's Medical Centers, Milwaukee, Wisconsin
Ventricular Tachycardia

Anne B. Curtis, M.D.
Professor of Medicine, University of Florida College of Medicine; Director, Clinical Electrophysiology, Shands at the University of Florida, Gainesville, Florida
Arrhythmias during Pregnancy

D. Wyn Davies, M.D., F.E.S.C.
Professor of Cardiology, University of London; Consultant in Cardiology, St. Mary's Hospital, London, United Kingdom
Catheter Mapping Techniques

Sanjay Deshpande, M.D.
Clinical Associate Professor of Medicine, University of Wisconsin Medical School—Milwaukee Clinical Campus; Attending, Aurora Sinai/St. Luke's Medical Centers, Milwaukee, Wisconsin
Ventricular Tachycardia

Anwer Dhala, M.D.
Clinical Associate Professor of Medicine, University of Wisconsin Medical School—Milwaukee Clinical Campus; Clinical Associate Professor of Pediatrics, Medical College of Wisconsin; Attending, St. Luke's/Aurora Sinai Medical Centers and Children's Hospital of Wisconsin, Milwaukee, Wisconsin
Ventricular Tachycardia

Michael Domanski, B.S. (Aerospace Engineering), M.D.
Head, Clinical Trials Group, National Heart, Lung, and Blood Institute; Warren G. Magnusson Clinical Center, National Institutes of Health, Bethesda, Maryland
Ventricular Fibrillation

Paul Dorian, M.D., M.Sc., F.R.C.P.C.
Professor of Medicine, University of Toronto Faculty of Medicine; Director, Electrophysiology Service, Department of Cardiology, St. Michael's Hospital, Toronto, Ontario, Canada
Principles of Clinical Pharmacology; Sustained Ventricular Tachycardia with Heart Disease; Ventricular Fibrillation

Nabil El-Sherif, M.D.
Professor of Medicine and Physiology, SUNY Downstate Medical Center College of Medicine; Director, Clinical Cardiac Electrophysiology program, SUNY Downstate Medical Center; Director, Division of Cardiology, VA Medical Center, Brooklyn, New York
Arrhythmias and Electrolyte Disorders

N.A. Mark Estes III, M.D.
Professor of Medicine, Tufts University School of Medicine; Director, Cardiac Electrophysiology Laboratory, New England Medical Center, Boston, Massachusetts
Principles of Catheter Ablation

Marjaneh Fatemi, M.D.
Assistant Professor, Department of Cardiology, University of Brest Faculty of Medicine; Attending Physician, Brest University Hospital, Brest, France
Arrhythmogenic Right Ventricular Cardiomyopathy

Sami Firoozi, M.D., M.R.C.P.
Clinical Research Fellow, St. George's Hospital Medical School; Cardiology Specialist Registrar, St. George's Hospital, London, United Kingdom
Evaluation and Management of Arrhythmias in Athletes

John D. Fisher, M.D.
Professor of Medicine, Department of Medicine–Cardiology, Albert Einstein College of Medicine of Yeshiva University; Director, Arrhythmia Service/CCEP Program Director, Montefiore Medical Center, Bronx, New York
Clinical Electrophysiology Techniques

Anne M. Gillis, M.D.
Professor of Medicine, Department of Cardiac Sciences, University of Calgary Faculty of Medicine; Director of Pacing and Electrophysiology, Department of Cardiac Sciences, Calgary Health Region, Calgary, Alberta, Canada
Proarrhythmia Syndromes

Nora Goldschlager, M.D., F.A.C.P., F.A.C.C.
Professor of Clinical Medicine, University of California, San Francisco, School of Medicine; Associate Director, Cardiology Division, and Director, Coronary Care Unit, ECG Laboratory and Pacemaker Clinic, San Francisco General Hospital, San Francisco, California
Sinus Node Dysfunction; Atrioventricular Block

David E. Haines, M.D.
Director, Heart Rhythm Center, William Beaumont Hospital, Royal Oak, Michigan
Ablation Technology

Michel Haissaguerre, M.D., F.E.S.C.
Professor of Cardiology, University of Bordeaux, Bordeaux; Director, Electrophysiology, University Hospital, Pessac, France
Curative Catheter Ablation for Supraventricular Tachycardia: Techniques and Indications

Stephen Hammill, M.D.
Professor of Medicine, Mayo Medical School; Director, Heart Rhythm Services, Division of Cardiovascular Diseases, Mayo Clinic, Rochester, Minnesota
Sinus Node Dysfunction; Atrioventricular Block

Meleze Hocini, M.D.
University of Bordeaux II, Bordeaux; Research Associate, Department of Cardiology, Hôpital Cardiologique du Haut Lévêque, Bordeaux-Pessac, France
Curative Catheter Ablation for Supraventricular Tachycardia: Techniques and Indications

Stefan H. Hohnloser, M.D.
Professor of Medicine, Department of Cardiology, J.W. Goethe University Faculty of Medicine, Frankfurt, Germany
Evaluation and Management of Arrhythmias in Dilated Cardiomyopathy and Congestive Heart Failure

Munther K. Homoud, M.D.
Assistant Professor of Medicine, Tufts University School of Medicine; Co-Director, Cardiac Electrophysiology Laboratory, Tufts–New England Medical Center, Boston, Massachusetts
Principles of Catheter Ablation

Raymond E. Ideker, M.D., Ph.D.
Jeanne V. Marks Professor of Medicine, Department of Medicine, Division of Cardiovascular Disease; Professor of Biomedical Engineering; and Professor of Physiology, University of Alabama–Birmingham School of Medicine, Birmingham, Alabama
Fundamental Concepts and Advances in Defibrillation

Demosthenes Iskos, M.D.
Assistant Professor, University of Minnesota Medical School; Cardiac Arrhythmia Center, University Hospital, Minneapolis, Minnesota
Autonomic Nervous System and Cardiac Arrhythmias

Pierre Jais, M.D.
University Bordeaux II Victor Ségalen; Electrophysiology, Hopital Haut Lévêque, Bordeaux, France
Curative Catheter Ablation for Supraventricular Tachycardia: Techniques and Indications

José Jalife, M.D.
Professor and Chairman, Department of Pharmacology, and Professor of Medicine and Pediatrics, SUNY Upstate Medical University; Director, Institute for Cardiovascular Research, University Hospital, Syracuse, New York
Mechanisms of Reentrant Arrhythmias

Michiel Janse, M.D., Ph.D.
Emeritus Professor of Experimental Cardiology, University of Amsterdam Faculty of Medicine; Laboratory of Experimental Cardiology, Academic Medical Center, Amsterdam, The Netherlands
Sustained Ventricular Tachycardia with Heart Disease

Werner Jung, M.D., F.E.S.C.
Professor of Medicine and Head, Department of Cardiology, University of Villingen Faculty of Medicine; Attending, Academic Hospital Villingen, Villingen-Schwenningen, Germany
Devices for the Management of Atrial Fibrillation

Alan Kadish, M.D.
Chester and Deborah C. Cooley Professor of Medicine, Northwestern University Feibberg School of Medicine; Senior Associate Chief, Division of Cardiology, Department of Medicine, Northwestern Memorial Faculty Foundation, Chicago, Illinois
Arrhythmias in Coronary Artery Disease

Demosthenes G. Katritsis, M.D., Ph.D., F.R.C.P., F.A.C.C.
Director, Cardiology Service, Athens Euroclinic, Athens, Greece; Honorary Consultant Cardiologist, Cardiothoracic Centre, St. Thomas' Hospital, London, United Kingdom
Nonsustained Ventricular Tachycardia

George J. Klein, M.D., F.A.C.C., F.R.C.P.C.
Professor of Medicine and Chair, Cardiology Division, Department of Medicine, University of Western Ontario Faculty of Medicine; Chief of Cardiology, Department of Medicine, London Health Sciences Centre, London, Ontario, Canada
Asymptomatic ECG Abnormalities

Helmut Klein, M.D.
Professor of Medicine, Otto von Guericke University School of Medicine; Chief, Department of Cardiology, University Hospital, Magdeburg, Germany
Device Technology for Congestive Heart Failure

Peter R. Kowey, M.D.
Professor of Medicine, Thomas Jefferson University, Jefferson Medical College, Philadelphia; Chief, Cardiovascular Services, Main Line Health System, Lankenau Hospital, Wynnewood, Pennsylvania
Postoperative Arrhythmias after Cardiac Surgery

Fred Kusumoto, M.D.
Associate Clinical Professor of Medicine, University of New Mexico College of Medicine, Albuquerque, New Mexico
Sinus Node Dysfunction; Atrioventricular Block

Chu-Pak Lau, M.D.
Chair Professor, University of Hong Kong School of Medicine; Chief of Cardiology, Queen Mary Hospital, Hong Kong, China
Pacing Technology and Its Indications: Advances in Threshold Management, Automatic Mode Switching, and Sensors

Ralph Lazzara, M.D.
Regent's Professor, Department of Medicine, University of Oklahoma College of Medicine; Medical Director, Cardiac Arrhythmia Research Institute, University of Oklahoma Health Science Center, Oklahoma City, Oklahoma
Sinus Node Dysfunction; Atrioventricular Block

Paul LeLorier, M.D., F.A.C.C.
Assistant Professor of Medicine, Department of Medicine, Division of Cardiology, Boston University School of Medicine; Director, Implantable Cardiac Device Center, Arrhythmia Service, Section of Cardiology, Boston Medical Center, Boston, Massachusetts
Asymptomatic ECG Abnormalities

Samuel Levy, M.D., F.E.S.C., F.A.C.C.
Professor of Cardiology, University of Marseille; Head, Department of Cardiology, Hopital Nord, Marseille, France
Paroxysmal Supraventricular Tachycardias and the Preexcitation Syndromes

Hua Li, Ph.D.
Instructor in Pediatrics, Baylor College of Medicine; Pediatric Cardiology, Texas Children's Hospital, Houston, Texas
Genetics and Cardiac Arrhythmias

Bruce D. Lindsay, M.D.
Associate Professor of Medicine, Washington University School of Medicine; Director, Clinical Electrophysiology Laboratory, Barnes-Jewish Hospital, St. Louis, Missouri
Paroxysmal Supraventricular Tachycardias and the Preexcitation Syndromes; Atrial Tachycardia, Flutter, and Fibrillation; Sustained Ventricular Tachycardia with Heart Disease; Ventricular Tachycardia and Ventricular Fibrillation without Structural Heart Disease; Ventricular Fibrillation

Mark S. Link, M.D.
Associate Professor of Medicine, Tufts University School of Medicine; Co-Director, Cardiac Electrophysiology Laboratory, Tufts–New England Medical Center, Boston, Massachusetts
Principles of Catheter Ablation

Berndt Lüderitz, M.D., F.E.S.C., F.A.C.C., F.A.H.A.
Professor and Chairman, Department of Medicine–Cardiology, University of Bonn Faculty of Medicine, Bonn, Germany
Devices for the Management of Atrial Fibrillation

Nandini Madan, M.B.B.S., M.D.
Associate Professor of Pediatrics, Drexel College of Medicine; Attending Cardiologist, St. Christopher's Hospital for Children, Philadelphia, Pennsylavania
Implantable Cardioverter Defibrillators: Technology, Indications, Implantation Techniques, and Follow-up

Yousuf Mahomed, M.D.
Professor of Surgery, Indiana University School of Medicine; Chief of Adult Cardiothoracic Surgery, Section of Cardiothoracic Surgery, Methodist Hospital, Indianapolis, Indiana
Antiarrhythmic Surgery

Vias Markides, M.D., M.B. (Hons.), B.S. (Hons.), M.R.C.P.
Hon. Senior Lecturer, National Heart and Lung Institute, Imperial College; Consultant, Waller Cardiac Department, St. Mary's Hospital, and Department of Cardiology, Royal Brompton and Harefield NHS Trust, London, United Kingdom
Catheter Mapping Techniques

Barry J. Maron, M.D.
Director, Hypertrophic Cardiomyopathy Center, Minneapolis Heart Institute Foundation, Minneapolis, Minneapolis; Adjunct Professor of Medicine, Tufts University School of Medicine, Boston, Massachusetts
Arrhythmias Associated with Hypertrophic Cardiomyopathy

William J. McKenna, M.D., F.R.C.P., F.A.C.C., F.E.S.C.
BHF Professor of Cardiology and Professor of Inherited Cardiovascular Disorders and Cardiology, The Heart Hospital, University College Hospital, London, United Kingdom
Evaluation and Management of Arrhythmias in Athletes

Rahul Mehra, Ph.D.
Senior Director of Arrhythmia Research, Medtronic Inc., Minneapolis, Minnesota
Fundamentals of Cardiac Stimulation

John M. Miller, M.D.
Professor of Medicine, Indiana University School of Medicine; Director, Cardiac Electrophysiology Services, and Director, Clinical Cardiac Electrophysiology Training Program, Clarion Health System, Indianapolis, Indiana
Antiarrhythmic Surgery

L. Brent Mitchell, M.D., F.R.C.P.C.
Professor and Head, Department of Cardiac Sciences, University of Calgary Faculty of Medicine; Director, Libin Cardiovascular Institute of Alberta, Calgary Health Region, Calgary, Alberta, Canada
Ventricular Fibrillation

Arthur J. Moss, M.D.
Professor of Medicine (Cardiology), University of Rochester School of Medicine and Dentistry; Attending Physician, Department of Medicine, University of Rochester Medical Center, Rochester, New York
Nonsustained Ventricular Tachycardia

Robert J. Myerburg, M.D.
Professor of Medicine and Physiology, Department of Medicine, Division of Cardiology, University of Miami School of Medicine; Attending, Jackson Memorial Hospital, Miami, Florida
Sustained Ventricular Tachycardia with Heart Disease; Ventricular Fibrillation

Gerald Naccarelli, M.D.
Bernard Trabin Chair of Cardiology and Professor of Medicine, Pennsylvania State University College of Medicine; Director, Cardiovascular Center, Milton S. Hershey Medical Center, Hershey, Pennsylvania
Sinus Node Dysfunction; Atrioventricular Block

Stanley Nattel, M.D.
Professor of Medicine, Paul-David Chair in Cardiovascular Electrophysiology, University of Montreal Faculty of Medicine; Cardiologist, Montreal Heart Institute, Montreal, Quebec, Canada
Atrial Tachycardia, Flutter, and Fibrillation

Iacopo Olivotto, M.D.
Staff Physician, Department of Cardiology, Azienda Ospedaliera Universitaria Careggi, Florence, Italy
Arrhythmias Associated with Hypertrophic Cardiomyopathy

Craig M. Pratt, M.D.
Professor of Medicine, Baylor College of Medicine; Director of Research, DeBakey Heart Center, and Director of Coronary Care Unit, The Methodist Hospital, Houston, Texas
Interpretation of Clinical Trials: How Mortality Trials Relate to the Therapy of Atrial Fibrillation

Mark Preminger, M.D.
Associate Professor of Medicine, UMDNJ Robert Wood Johnson Medical School; Director, Electrophysiology Laboratory, Robert Wood Johnson University Hospital, New Brunswick, New Jersy
Implantable Cardioverter Defibrillators: Technology, Indications, Implantation Techniques, and Follow-up

Kara J. Quan, M.D.
Assistant Professor of Medicine, Case Western Reserve University School of Medicine; Director, Electrophysiology Laboratory, Heart and Vascular Research Center, MetroHealth Campus, Cleveland, Ohio
Ventricular Fibrillation

Vivek Y. Reddy, M.D.
Director, Experimental Electrophysiology Laboratory, Cardiac Arrhythmia Service, Massachusetts General Hospital, Boston, Massachusetts
Sudden Cardiac Death

Larry A. Rhodes, M.D.
Associate Professor of Pediatrics, University of Pennsylvania School of Medicine; Director, Electrophysiology Unit, The Children's Hospital of Philadelphia, Philadelphia, Pennsylvania
Evaluation and Management of Arrhythmias in a Pediatric Population

Hygriv B. Rao, M.D., D.M.
Research Fellow, Electrophysiology Research Foundation, Warren; PBI Regional Medical Center, Passaic, New Jersey
Devices for the Management of Atrial Fibrillation

Dionyssios A. Robotis, M.D.
Assistant Professor of Medicine, SUNY Downstate Medical Center College of Medicine; Director, Electrophysiology Laboratory, VA Medical Center, Brooklyn, New York
Arrhythmias and Electrolyte Disorders

Dan M. Roden, M.D.
Professor of Medicine and Pharmacology, Department of Clinical Pharmacology, Vanderbilt University School of Medicine; Director, Division of Clinical Pharmacology, Vanderbilt University Hospital, Nashville, Tennessee
Molecular and Cellular Basis of Cardiac Electrophysiology

Micheal R. Rosen, M.D.
Gustavus A. Pfeiffer Professor of Pharmacology and Professor of Pediatrics, Columbia University College of Physicians and Surgeons; Director, Center for Molecular Therapeutics, New York, New York
Principles of Electropharmacology

David Rosenbaum, M.D.
Associate Professor of Medicine, Biomedical Engineering, Physiology, and Biophysics, Case Western Reserve University School of Medicine; Director, Heart and Vascular Research Center, MetroHealth Campus, Case Western Reserve University, Cleveland, Ohio
Ventricular Fibrillation

Jeremy N. Ruskin, M.D.
Associate Professor of Medicine, Harvard Medical School; Director, Cardiac Arrhythmia Service, Massachusetts General Hospital, Boston, Massachusetts
Sudden Cardiac Death

Mohammad Saeed, M.D.
Assistant Professor of Medicine, University of Texas Medical Branch; Director, Cardiac Electrophysiology Laboratory, University of Texas Medical Branch, Galveston, Texas
Principles of Catheter Ablation

Scott Sakaguchi, M.D.
Associate Professor of Medicine, University of Minnesota Medical School; Cardiac Arrhythmia Center, University Hospital, Minneapolis, Minnesota
Autonomic Nervous System and Cardiac Arrhythmias

Sanjeev Saksena, M.B.B.S., M.D., F.A.C.C., F.E.S.C., F.A.H.A., F.H.R.S.
Clinical Professor of Medicine, Robert Wood Johnson School of Medicine, Piscataway; Director, Cardiovascular Institute and Arrhythmia/Pacemaker Service, PBI Regional Medical Center, Passaic, New Jersey
Atrioventricular Block; Paroxysmal Supraventricular Tachycardias and the Preexcitation Syndromes; Sustained Ventricular Tachycardia with Heart Disease; Ventricular Tachycardia and Ventricular Fibrillation without Structural Heart Disease, Implantable Cardioverter Defibrillators: Technology, Indications, Implantation Techniques, and Follow-up; Devices for the Management of Atrial Fibrillation; Device Technology for Congestive Heart Failure

Faramarz H. Samie, M.D.
Resident, Department of Dermatology, University of Rochester Medical Center, Rochester, New York
Mechanisms of Reentrant Arrhythmias

Irina Savelieva, M.D.
Clinical Research Fellow, Department of Cardiac and Vascular Sciences, St. George's Hospital Medical School, London, United Kingdom
Atrial Tachycardia, Flutter, and Fibrillation; Noninvasive Electrophysiology

Richard J. Schilling, M.D.
Honorary Senior Lecturer, Queen Mary University of London; Consultant Cardiologist, St. Bartholomew's Hospital, London, United Kingdom
Catheter Mapping Techniques

Mark H. Schoenfeld, M.D., F.A.C.C.
Clinical Professor of Medicine, Yale University School of Medicine; Director, Cardiac Electrophysiology and Pacemaker Laboratory, Hospital of Saint Raphael, New Haven, Connecticut
Pacemaker Insertion, Revision, Extraction, and Follow-up

Peter J. Schwartz, M.D.
Professor and Chairman, Department of Cardiology, University of Pavia School of Medicine; Chief, Coronary Care Unit, IRCCS Policlinico S. Matteo, Pavia, Italy
The Long QT Syndrome

David Schwartzman, M.D.
Associate Professor of Medicine–Cardiac Electrophysiology, University of Pittsburgh School of Medicine; Director, Atriology, University of Pittsburgh Medical Center–Presbyterian, Pittsburgh, Pennsylvania
Imaging Techniques in Interventional Electrophysiology

Dipen Shah, M.D.
Associate Physician, Cardiology Service, Canton Hospital of the University of Geneva, Geneva, Switzerland
Curative Catheter Ablation for Supraventricular Tachycardia: Techniques and Indications

Sanjay Sharma, M.D., M.R.C.P.
Honorary Clinical Lecturer, University Hospital Lewisham; Consultant Cardiologist, University Hospital Lewisham, London, United Kingdom
Evaluation and Management of Arrhythmias in Athletes

Bramah N. Singh, M.D., D.Phil., D.Sc.
Professor of Medicine, David Geffen School of Medicine at UCLA; Staff Cardiologist, VA Greater Los Angeles Healthcare System, Los Angeles, California
Antiarrhythmic Drugs

Kaori Shinagawa, M.D., Ph.D.
Instructor in Medicine; Keio University School of Medicine; Cardiologist, Eiju-Sogo Hospital, Tokyo, Japan
Atrial Tachycardia, Flutter, and Fibrillation

Allan C. Skanes, M.D., F.R.C.P.C.
Associate Professor, Department of Medicine, University of Western Ontario Faculty of Medicine; Director of Electrophysiology Laboratory, Arrhythmia Service, Division of Cardiology, London Health Sciences Centre, London, Ontario, Canada
Asymptomatic ECG Abnormalities

Jasbir Sra, M.D.
Clinical Professor of Medicine, University of Wisconsin Medical School—Milwaukee Clinical Campus; Attending, St. Luke's/Aurora Sinai Medical Centers, Milwaukee, Wisconsin
Ventricular Tachycardia

William G. Stevenson, M.D.
Associate Professor of Medicine, Harvard Medical School; Director, Clinical Cardiac Electrophysiology Program, Brigham and Women's Hospital, Boston, Massachusetts
Sustained Ventricular Tachycardia with Heart Disease; Ventricular Fibrillation

Gordon Tomaselli, M.D.
Professor of Medicine and Molecular Medicine, Johns Hopkins University School of Medicine; Attending, Johns Hopkins Hospital, Baltimore, Maryland
Molecular and Cellular Basis of Cardiac Electrophysiology

Paul Touboul, M.D.
Professor of Cardiology, University of Lyon, Faculty of Medicine; Head, Cardiovascular Section, Hôpital Cardiovasculaire Louis Pradel, Lyon, France
Arrhythmogenic Right Ventricular Cardiomyopathy

Jeffrey A. Towbin, M.D.
Professor of Pediatrics, Molecular and Human Genetics, Baylor College of Medicine; Chief, Pediatric Cardiology, Texas Children's Hospital, Houston,Texas
Genetics and Cardiac Arrhythmias

Jacques Turgeon, Ph.D., B.Pharm.
Dean, Faculty of Pharmacy, Université de Montréal, Montreal, Quebec, Canada
Principles of Clinical Pharmacology

Gioia Turitto, M.D.
Associate Professor of Medicine, SUNY Downstate Medical Center College of Medicine; Director, Coronary Care Unit and Cardiac Electrophysiology Laboratory, University Hospital of Brooklyn, Brooklyn, New York
Arrhythmias and Electrolyte Disorders

George F. Van Hare, M.D.
Professor of Pediatrics, Stanford University School of Medicine, Stanford; Director, Pediatric Arrhythmia Center at UCSF and Stanford–UCSF Children's Hospital, San Francisco, and Lucile Packard Children's Hospital, Palo Alto, California
Arrhythmias Associated with Congenital Heart Disease

Mattes Vatta, Ph.D.
Assistant Professor of Pediatrics, Baylor College of Medicine; Pediatric Cardiologist, Texas Children's Hospital, Houston, Texas
Genetics and Cardiac Arrhythmias

Victoria L. Vetter, M.D.
Professor of Pediatrics, University of Pennsylvania School of Medicine; Chief, Division of Cardiology, The Children's Hospital of Philadelphia, Philadelphia, Pennsylvania
Evaluation and Management of Arrhythmias in a Pediatric Population

Galen Wagner, M.D.
Associate Professor, Duke University School of Medicine, Durham, North Carolina
Basic Electrocardiography

Albert L. Waldo, M.D.

Walter H. Pritchard Professor of Cardiology, Professor of Medicine, and Professor of Biomedical Engineering, Case Western Reserve University School of Medicine; Director, Clinical Cardiac Electrophysiology Program, University Hospitals of Cleveland, Cleveland, Ohio

Interpretation of Clinical Trials: How Mortality Trials Relate to the Therapy of Atrial Fibrillation

Bruce Walker, M.B.B.S., Ph.D.

Lecturer in Medicine, University of New South Wales Faculty of Medicine; Visiting Medical Officer, Department of Cardiology, Sydney, New South Wales, Australia

Asymptomatic ECG Abnormalities

Mariah L. Walker, Ph.D.

Visiting Scientist, Heart and Vascular Research Center, MetroHealth Campus, Case Western Reserve University, Cleveland, Ohio

Ventricular Fibrillation

Paul J. Wang, M.D.

Professor of Medicine, Stanford University School of Medicine; Director, Cardiac Arrhythmia Service and Cardiac Electrophysiology Laboratory, Stanford Hospital and Clinics, Stanford, California

Principles of Catheter Ablation

Wojciech Zareba, M.D., Ph.D.

Associate Professor of Medicine (Cardiology), University of Rochester Medical School, Rochester, New York

Nonsustained Ventricular Tachycardia

Foreword

The knowledge base in the fields of basic and clinical electrophysiology of cardiac arrhythmias, and the translation from the former to the latter, have increased exponentially since the end of the last century. The rate of change from the 1970s, through the 1980s and 1990s, to the present has been especially dramatic. During the last decade, there have been a great number of publications summarizing the progress in this expanding field. For clinical cardiologists who are close to or entering this "dimension," the amount of new information can be overwhelming. Added to the cognitive base is the fact that electrophysiological techniques have evolved from their initial goals of clarifying diagnoses and mechanisms of arrhythmias to therapeutic methods that can abolish permanently certain disorders of heart rhythm and thus become curative interventional clinical strategies. Finally, new drugs that have contributed to our armamentarium have been introduced for clinical use.

Most of the currently available textbooks that are intended to bring diverse information together into a single source have used a classical style format, in which various elements of knowledge are handled as individual disciplines. Although this approach is useful for the reader interested in a focused topic, it becomes somewhat cumbersome for the person trying to acquire a broad overview. Specifically, the burden placed on the student of the field—integrating multiple sources and orientations of information—can be onerous. For these reasons, the editors of this work have developed a unique communication strategy, intended to integrate multiple elements of a topic into comprehensive chapters. Specifically, in contrast with the common multi-authored *textbook*, the editors have provided a multi-authored *chapter* approach. This brings expertise to the multiple elements covered in single chapters, providing the reader with a high-level overview of the topic. This experiment in scientific communication is particularly effective in Section II of the book, in which individual cardiac arrhythmias are discussed by multiple authors.

The primary editors, Drs. Saksena and Camm, as well as the associate editors, Drs. Boyden, Dorian, and Goldschlager, and the more than 100 contributors, should be congratulated for having presented us with such a valuable and original publication.

Agustin Castellanos, M.D.

Robert J. Myerburg, M.D.

Division of Cardiology
University of Miami School of Medicine
Miami, Florida

Foreword

There is a golden rule in cardiology that every five years, 50% of our knowledge is replaced by new information. This especially holds for cardiac arrhythmology, an area in which the three important contributors—basic science, clinical knowledge, and medical technology—continuously change our insights and understanding.

To stay well informed, it is essential, therefore, to have a comprehensive review of all of the new developments at regular intervals. To successfully accomplish the compilation of such a review is not easy. To produce a well-balanced book, the selection and cooperation of and input from high-level experts in the different areas are required, guided by experienced and well-known editors.

In going through *Electrophysiological Disorders of the Heart,* I came to the conclusion that the editors and authors have reached that goal perfectly. The book is an excellent overview of our current knowledge on mechanisms and management of cardiac arrhythmias and should be read by any cardiologist active in that area.

Hein J. Wellens, M.D.

Preface

The extraordinary development of cardiovascular medicine in the final decades of the 20th century, combined with new information technology for its delivery, has led to an amassing of scientific information that has never previously been witnessed in the medical sciences. A wealth of new information has accumulated on the fundamental mechanisms of cardiac arrhythmias, new investigative and therapeutic techniques, and impact of these therapies in clinical trials. At the dawn of the 21st century, students of cardiac arrhythmias are inundated with information but have few avenues for synthesizing this new information with more traditional teaching tools that underlie medical school curricula in cardiac arrhythmias.

This textbook arose from a perceived need for analyzing and distilling the most recent information and melding it with classical concepts in cardiac arrhythmology. *Electrophysiological Disorders of the Heart* approaches the subject from the vantage point of the practicing clinician (cardiologist or electrophysiologist) or the clinical trainee who is involved in the care of patients with arrhythmias. In this effort, we have tried to break away from some of the common formats that actually have further segmented the field of arrhythmology. This book was designed to be a clinical reference that presents a complete view of the entire field for the practicing clinician (whether cardiologist or arrhythmia specialist), trainee, or investigator and utilize this new information in daily medical practice. The book is sufficiently detailed, however, to serve as a comprehensive reference in all of its sections.

To achieve these ambitious goals, we have had to contend with important developments in information presentation and dissemination that presage future trends in medical education in the 21st century. There has been extraordinary specialization in investigation of cardiac arrhythmias. Yet both of us fervently believe in the need for traditional arrhythmia teaching tools such as electrocardiography, which need to be mined to their fullest extent by the art of deductive analysis. Simplification of physical and chemical data, three-dimensional imaging, and complex intracardiac recordings and relating them to standard classical techniques of electrocardiography became necessary. The pervading theme has been to provide and translate the detail of new knowledge from basic science to medical technology to clinical management for the practicing clinician and investigator. We have fostered a significant use of illustrations, tables, and algorithms to simplify and reach this goal. To be consistent with current medical practice, we have formalized the inclusion of evidence-based medical therapy as a specific section in many of the chapters.

Each of the five sections can stand alone as a focused monograph for different educational needs, yet each section complements other discussions elsewhere in the book. Overlap between sections and chapters has been limited, and links between chapters to avoid repetitive discussion have been included. The use of individually authored sections in the chapters of Section II reflects our bias that the information base needed for such synthesis is vast, requiring experts in each area of study to provide the core knowledge needed for that topic, and we have sought to develop a common theme and progression in these chapters. This unique feature, in our view, makes for an particularly compelling and authoritative review of the subject.

We have been extraordinarily fortunate to assemble a team of editors who have the knowledge and experience to provide a bridge between classical concepts and the new learning. We have all learned much from each other during the travails of this experience since the conception of this textbook in the late 1990s. The contributions of our 115 authors from 15 countries are the heart of this book. These individual contributions reflect the global interest in this science. This unique worldwide effort has produced a truly international textbook. To our contributors, we can only express our deepest gratitude and hope that the final product is, for them, in small measure, a worthwhile outcome of their efforts. In shepherding this project, we could not have arrived at our destination without the continuous support of our staff, colleagues, and the editorial and production staff at Elsevier. We would particularly like to acknowledge the contributions of Ms. Celeste Simmons, Ina Ellen Wendler, Irina Savelieva, Jennifer Shreiner, and Anne Lenehan.

As senior editors, we have had the opportunity to shape this book, and this has in turn helped define our own ideas on the educational needs in cardiac arrhythmology in our time. It is our sincere hope that this textbook will fulfill the expectations of our readers and advance their knowledge and interest in cardiac arrhythmias. Should it do so, this achievement will give this compelling and challenging effort its *raison d'être.*

Sanjeev Saksena

Alan John Camm

Contents

Color plate section follows page xxiv.

Electrophysiological Disorders of the Heart

Color Plates

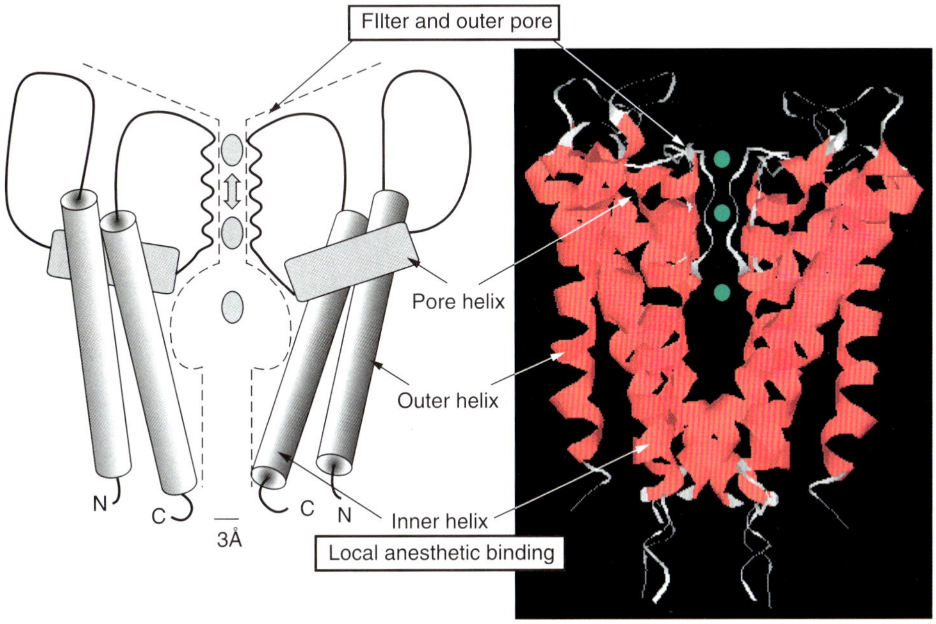

COLOR PLATE 1-10 Crystal structure of the KcsA bacterial inward rectifier K channel.[138] **Left,** Cartoon representation of the major features of the structure of the bacterial channel, KcsA. Each of the four channel subunits (only two are shown) contain two α-helical membrane-spanning repeats, a pore helix, and the K channel signature sequence that forms the K^+-selective pore. **Right,** The structure of KcsA, the peptide backbone is rendered in a ribbon format. The features of the permeation pathway include the presence of two to three K^+ ions in the pore with ion–ion repulsion that facilitates high rates of ion transport and large inner vestibule composed of the carboxyl terminal portion of the outer or M2 helix. The M2 helix corresponds to the S6 domains of voltage-dependent K channels, which mediate antiarrhythmic drug binding.

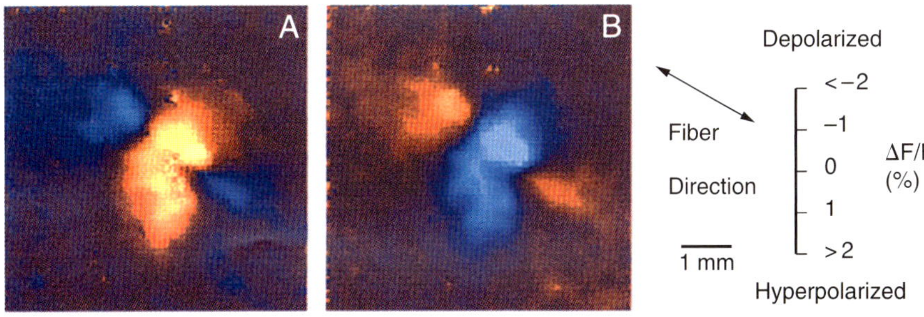

COLOR PLATE 8-14 Images of the transmembrane potential associated with injection of current into refractory cardiac tissue. **A,** The image for a 10 mA, 2 ms cathodal stimulus applied at a point electrode. Note the dog bone–shaped depolarized region (*orange*) and a pair of adjacent hyperpolarized regions (*blue*). The fiber orientation is from lower right to upper left. The *color bar* shows the fractional change in fluorescence. **B,** The complementary image for a 10 mA, 2 ms anodal stimulus at the same location of the heart. Note that the "dog bone" is now hyperpolarized region (*blue*), whereas the adjacent areas are depolarized (*orange*). (Reproduced from Wikswo JP, Lin SF, Abbas RA: Virtual electrodes in cardiac tissue: A common mechanism for anodal and cathodal stimulation. Biophys J 1995;69:2195-210.)

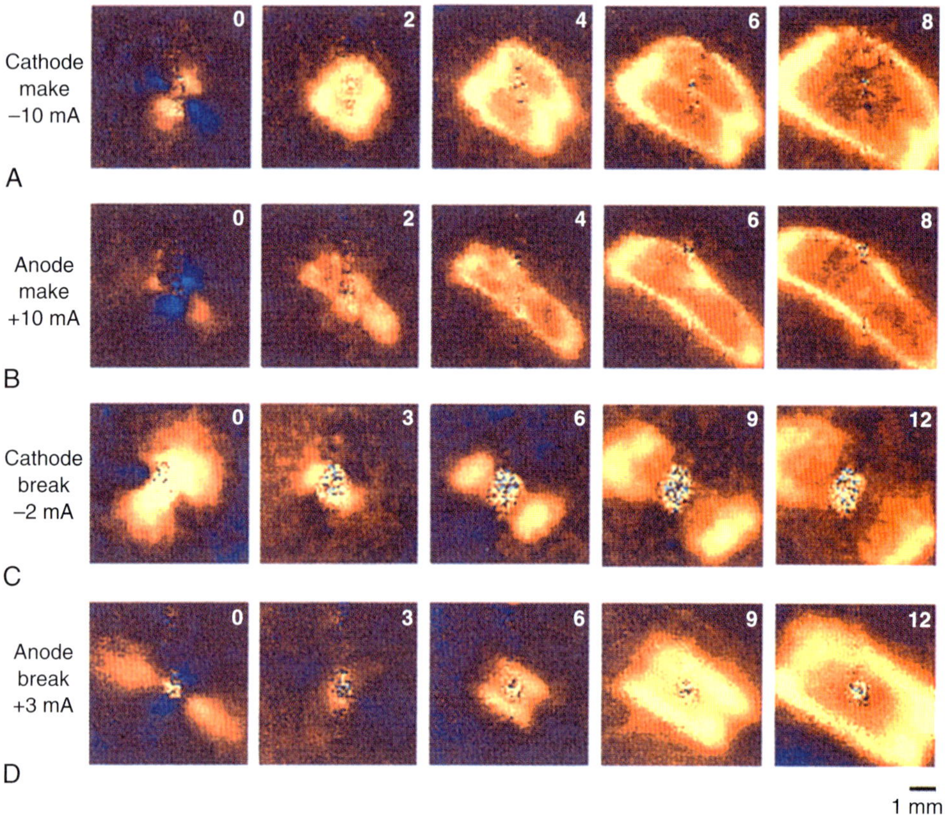

COLOR PLATE 8-16 False-color images of the transmembrane potential associated with injection of current into fully repolarized excitable cardiac tissue. The number inside each frame is time in ms. Upper two rows: make excitation. **A,** Cathodal make excitation with 1 ms 10 mA stimulus; **B,** 1 ms, 10 mA anodal make stimulation in the same heart. Lower two rows: break stimulation. **C,** 180 ms 2 mA cathodal break stimulation with a 2 mA, 180 ms long duration stimulus in another heart. **D,** Anodal break stimulation with a 3 mA 150 ms long-duration stimulus. The direction of the epicardial fibers is from lower right to upper left. The color scale is the same as in Figure 8-14. (Reproduced from Wikswo JP, Lin SF, Abbas RA: Virtual electrodes in cardiac tissue: A common mechanism for anodal and cathodal stimulation. Biophys J 1995;69:2195-210.)

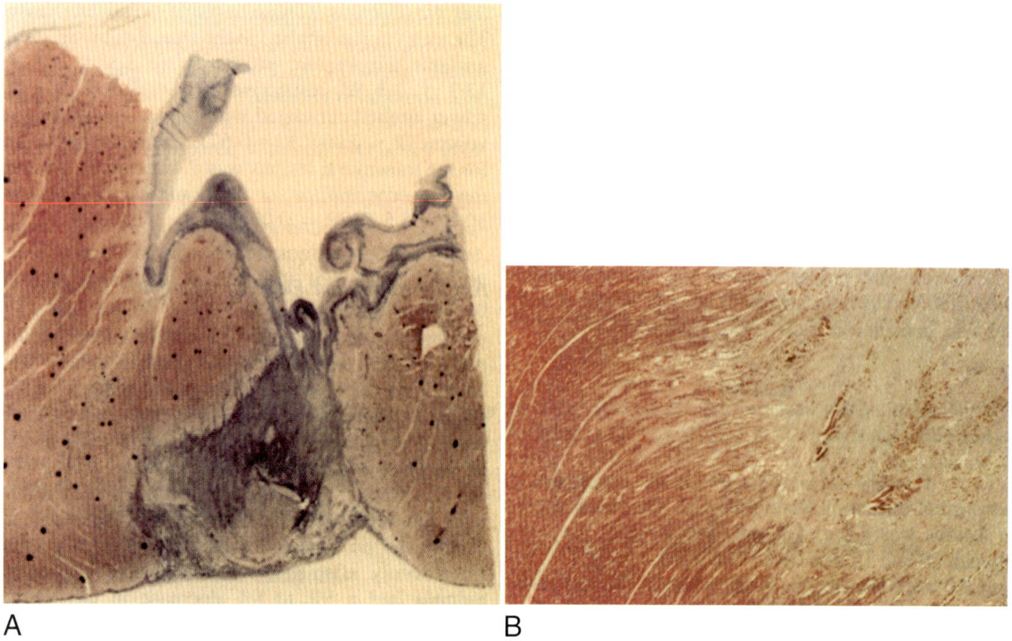

COLOR PLATE 10-7 Histology of microwave lesion. **A,** Low-power magnification of a transmural ventricular lesion created using microwave energy. The lesion is transmural and extends from the endocardial surface to the epicardium. **B,** High-power magnification of a ventricular lesion. Note the extensive fibrosis and well-demarcated lesion border zone. Masson's trichome stain. (Adapted with permission from Vanderbrink B, Gilbride C, Aronovitz M, et al: Safety and efficacy of a steerable temperature monitoring microwave catheter system for ventricular myocardial ablation. J Cardiovasc Electrophysiol 2000;11:305-10.)

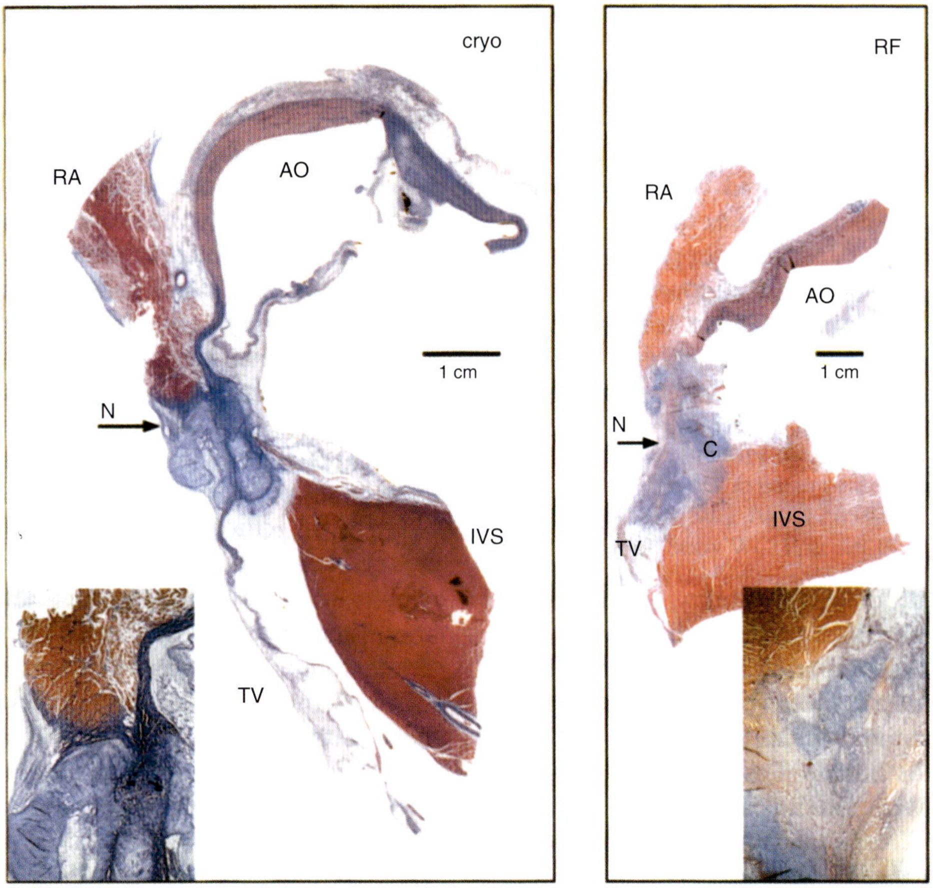

COLOR PLATE 10-8 **Left,** Chronic cryoablation lesion of an atrioventricular (AV) node in a dog (Azan stain). There is homogenous fibrotic tissue extending from the right atrial wall across the central fibrous body to the muscular interventricular septum. **Inset,** Higher magnification: absence of viable myocardium within the fibrous strands. *Blue color* indicates fibrotic tissue, and *red color* indicates normal tissue. **Right,** Chronic radiofrequency lesion of AV node in dog (Azan stain). Note the presence of fibrotic tissue with scattered strands of viable myocardium. **Inset,** Higher magnification: The lesion contains inhomogeneous fibrotic tissue with strands of viable myocytes and cartilage formation. The borders are not as well demarcated as the cryoablation lesion. *AO,* aorta; *C,* cartilage; *IVS,* interventricular septum; *N,* compact AV node; *RA,* right atrium; *TV,* tricuspid valve. (Adapted with permission from Rodriguez LM, Leunissen J, Hoekstra A, et al: Transvenous cold mapping and cryoablation of the AV node in dogs: Observations of chronic lesions and comparison to those obtained using radiofrequency ablation. J Cardiovasc Electrophysiol 1998;9:1055-61.)

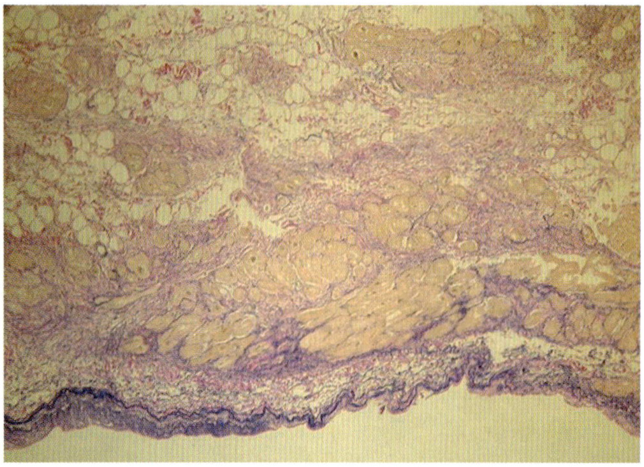

COLOR PLATE 15-9 Light microscopy of atrial myocardium showing accumulation of glycogen and fat in vacuolated, oversized myocytes (trichrome stain) in a patient with sustained atrial fibrillation.

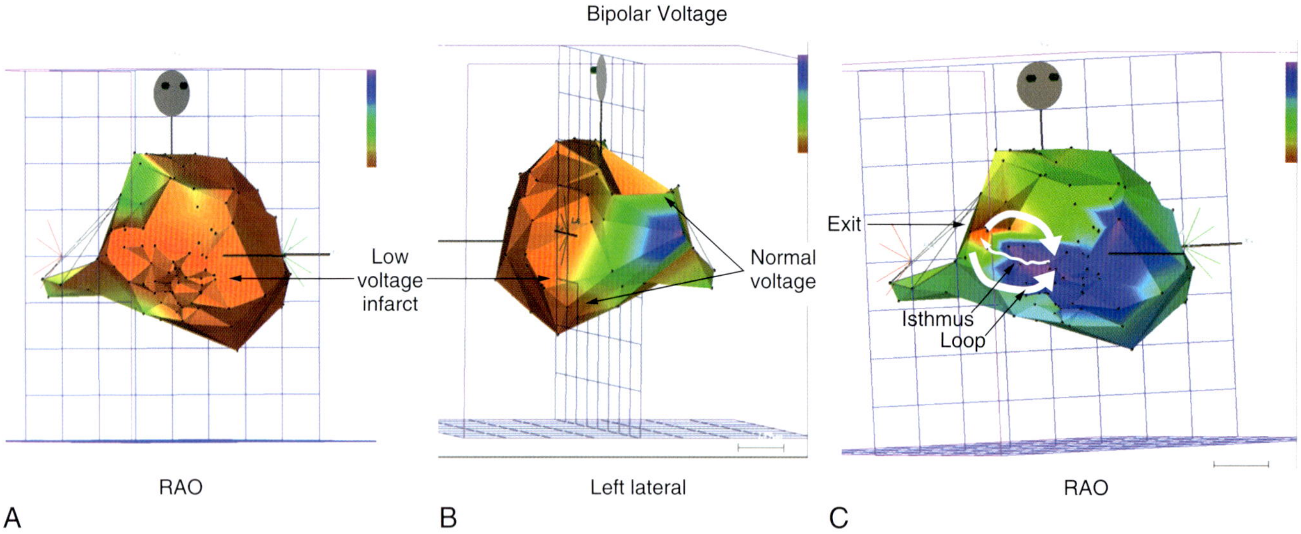

COLOR PLATE 17-17 Mapping data from a patient with prior anterior wall infarction and ventricular tachycardia. **A** and **B,** Map of the left ventricle in right anterior oblique (AO) and left anterior oblique (LAO) projections obtained by moving a mapping catheter from point to point. The color-coding indicates electrogram voltage; low voltage areas are *red, yellow, orange*; normal voltage is *green, blue, purple*. A very large area of low voltage, identifying the infarct region, occupies the septum, apex, and anterior wall. **C,** Ventricular activation sequence map obtained during ventricular tachycardia. The color-coding indicates the sequence of activation, and the circuit is indicated by the *white arrows*. The VT circuit isthmus is located on the septum. The reentry wavefront travels from apical to basal along the septum exiting near the base (*red*), then divides into two wavefronts that propagate superiorly and inferiorly and back to the isthmus.

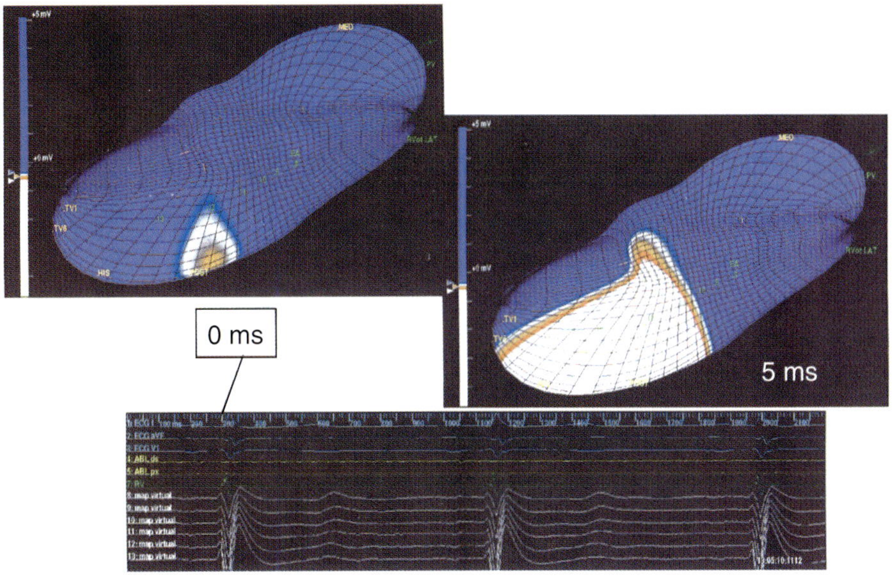

A

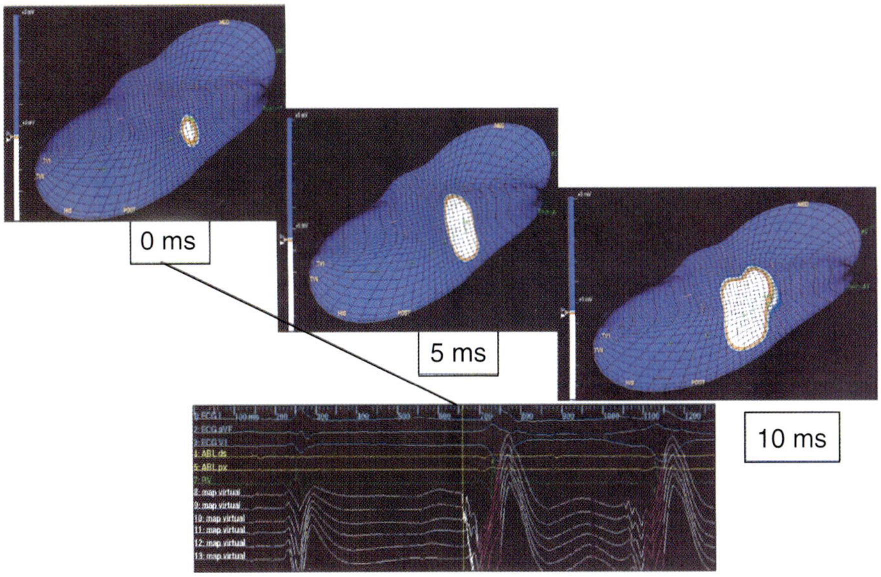

B

COLOR PLATE 18-4 Three-dimensional noncontact mapping of ventricular tachycardia (VT) arising in the right ventricular outflow tract (RVOT). **A,** Sinus rhythm. **B,** VT, arising from RVOT. The noncontact map of the outflow tract is seen as an elliptical oblong chamber with the VT wavefront shown in sequential images. Note the focal origin with centrifugal propagation. RV, right ventricle. Timing is shown in milliseconds after onset. Standard ECG leads, aVF and V, as well as virtual electrograms, are displayed below on three-dimensional map.

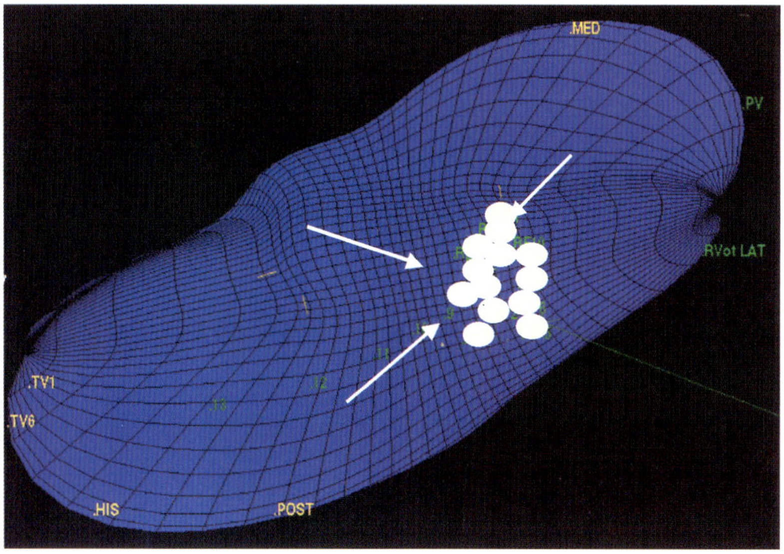

COLOR PLATE 18-5 Catheter ablation guided by three-dimensional noncontact mapping in the same patient with ventricular tachycardia (VT) arising from the right ventricular outflow tract shown in Figure 18-4. Ablation lesions (*circular markers*) are shown individually at the focal origin and in the surrounding region. Successful VT ablation was achieved.

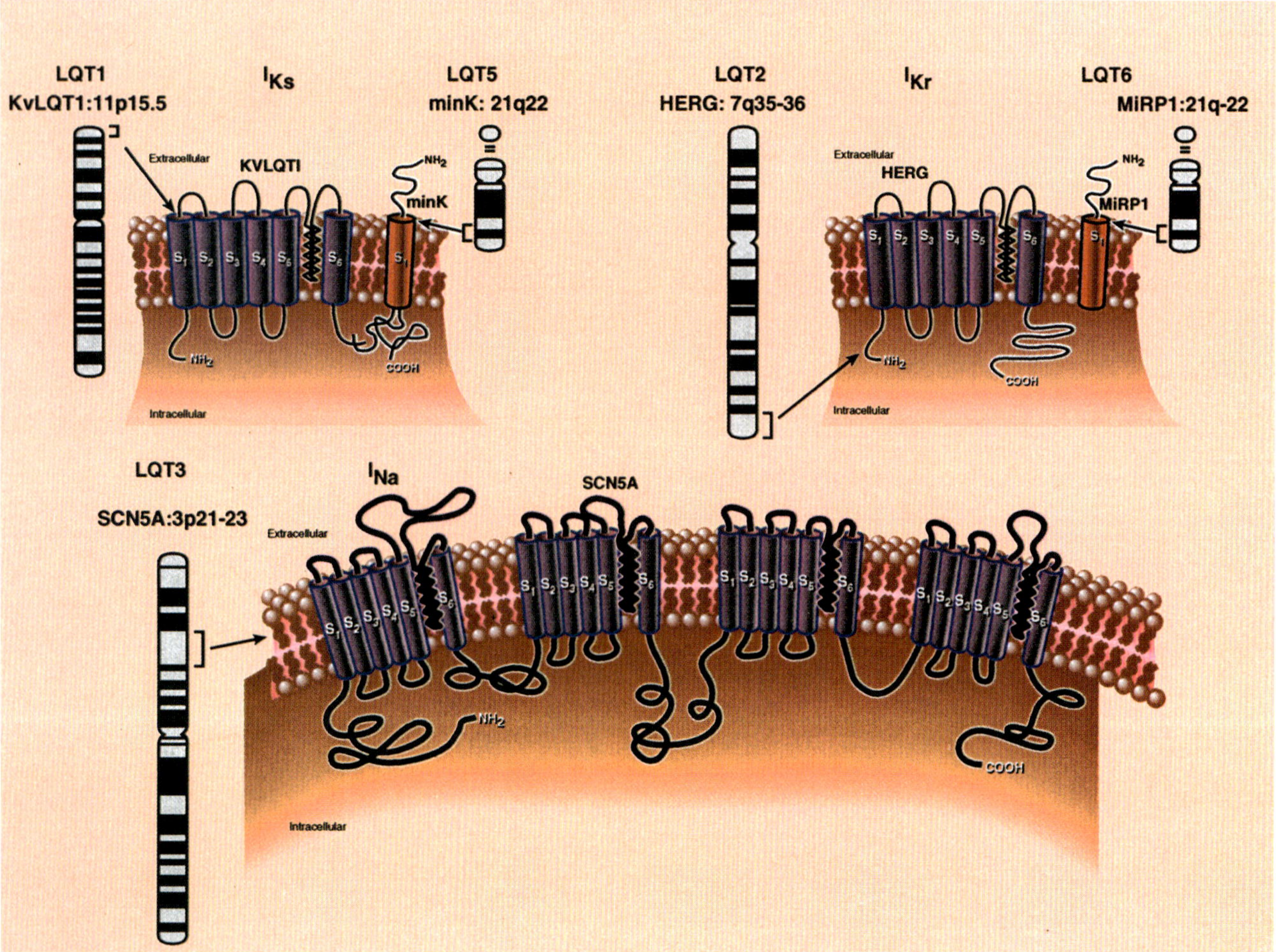

COLOR PLATE 27-3 Genetics of ventricular arrhythmias. The genetic loci and known genes identified for long QT syndrome are shown along with the ion channel protein structure. Note that the potassium channel α-subunits *KvLQT1* and *HERG* require association with β-subunits (*minK* and *KVLQT1*=IKs; *MiRP1* and *HERG*=IKr for normal function).

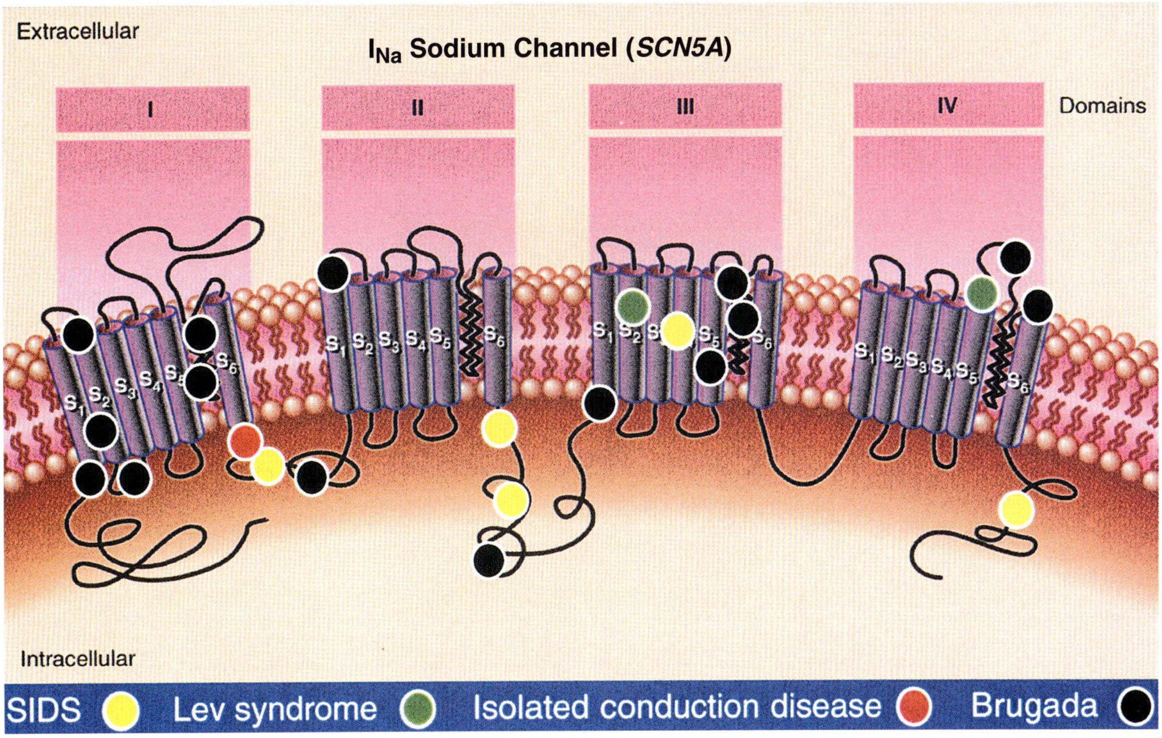

COLOR PLATE 27-7 Cardiac sodium channel (*SCN5A*) gene mutations associated with cardiac arrhythmias and conduction system diseases. Cardiac arrhythmias: SIDS, sudden infant death syndrome (*yellow*); Brugada syndrome (*black*); conduction system disease: Lev syndrome (*green*); isolated conduction disease (*red*). Note that mutations for all the disorders are scattered throughout the channel protein domains. They are found within the transmembrane portions and pore regions of the channel, as well as within intracellular and extracellular regions of the protein.

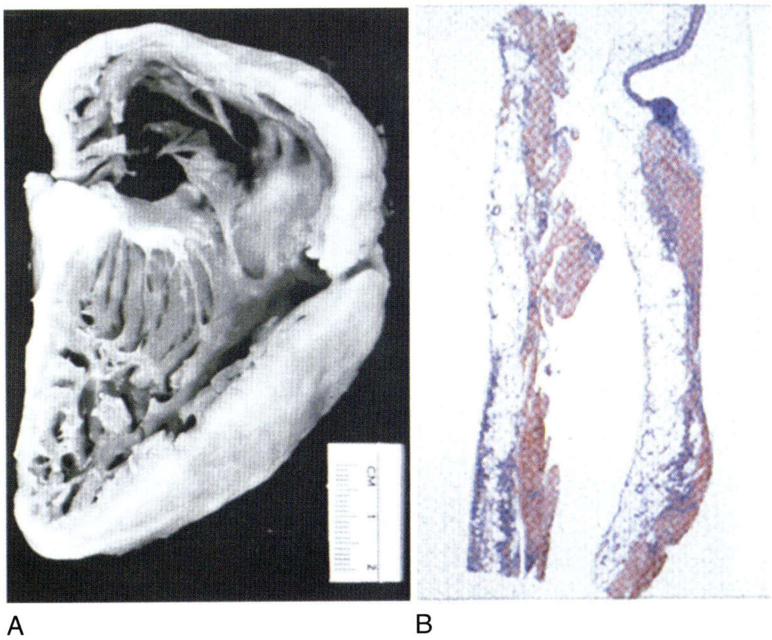

COLOR PLATE 27-8 Arrhythmogenic right ventricular dysplasia/cardiomyopathy (ARVD/C). **A,** Gross anatomy with thinned right ventricle and fatty infiltration. **B,** Histology identifying fibrofatty infiltration of right ventricle.

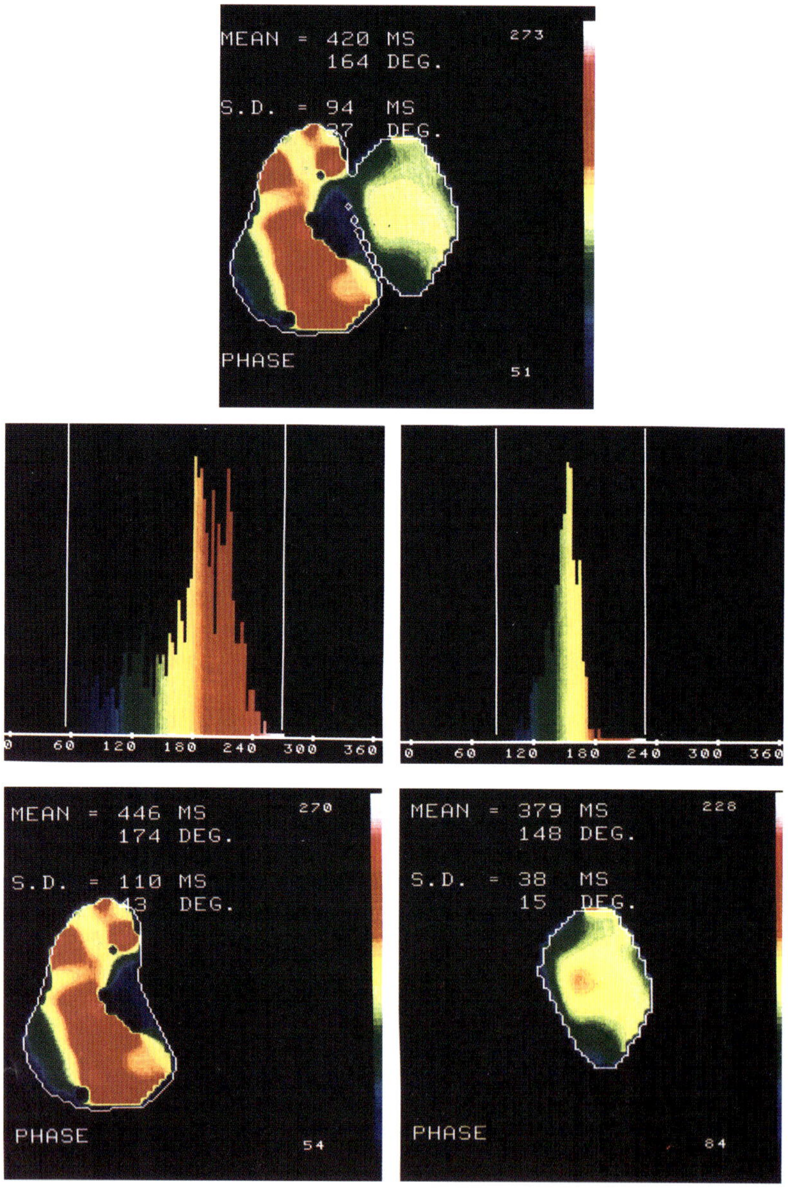

COLOR PLATE 31-13 Gated blood pool tomographic data (short axis) obtained from a patient diagnosed with ARVC with no left ventricular involvement. **Top:** Phase analysis of the right and left ventricles showing significant contraction delay in the right ventricle relative to the left ventricle. **Middle:** Phase histograms of the right ventricle **(left)** and the left ventricle **(right):** the histogram corresponding to the left ventricle shows a homogeneous distribution of contraction phases, whereas, the right ventricular histogram is much more dispersed, due to heterogeneity of contraction. **Bottom:** Phase dispersion more pronounced in the right ventricle **(left)** as compared with the left ventricle **(right),** expressed as mean phase and phase standard deviation.

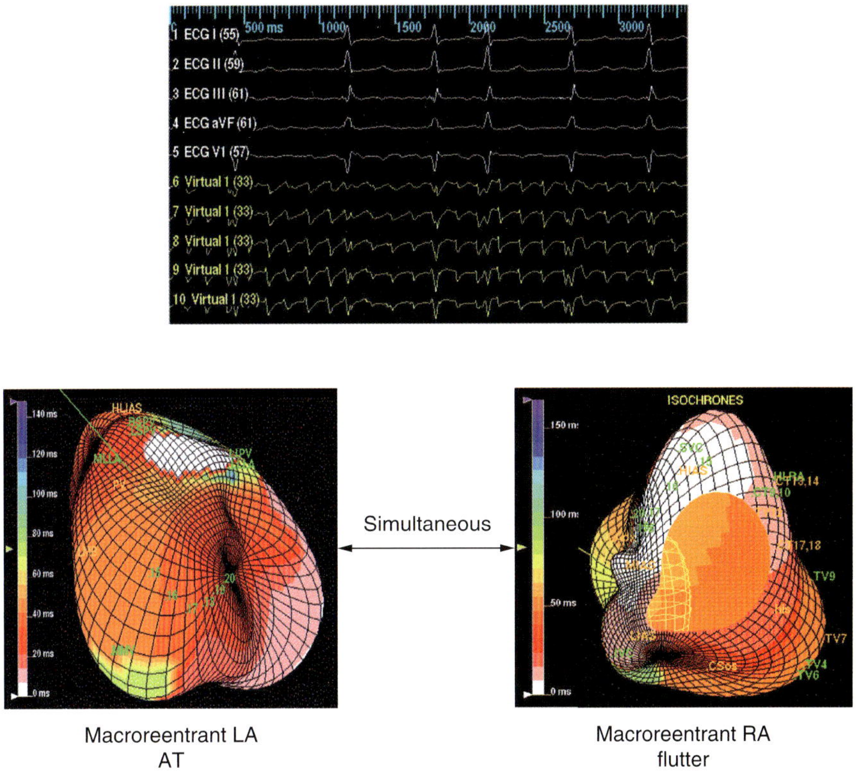

COLOR PLATE 41-1 Three-dimensional map showing focal left atrial tachycardia and macro-reentrant right atrial flutter coexisting simultaneously in a patient with permanent atrial fibrillation. The **upper panel** shows surface ECG leads and virtual electrograms at numbered sites in the atrial maps. The color coding shows activation times at different right and left atrial regions during the cycle.

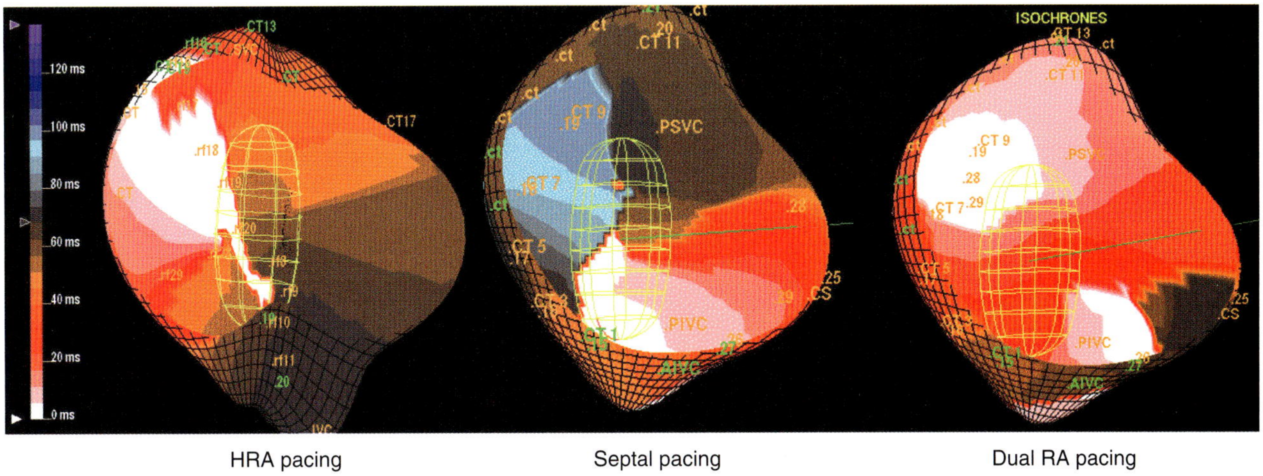

COLOR PLATE 41-4 Right atrial three-dimensional activation maps during high right atrial, septal, and dual site right atrial pacing. Note the shorter activation times and two simultaneous activation wavefronts in the dual site pacing mode that abbreviate and synchronize atrial activation.

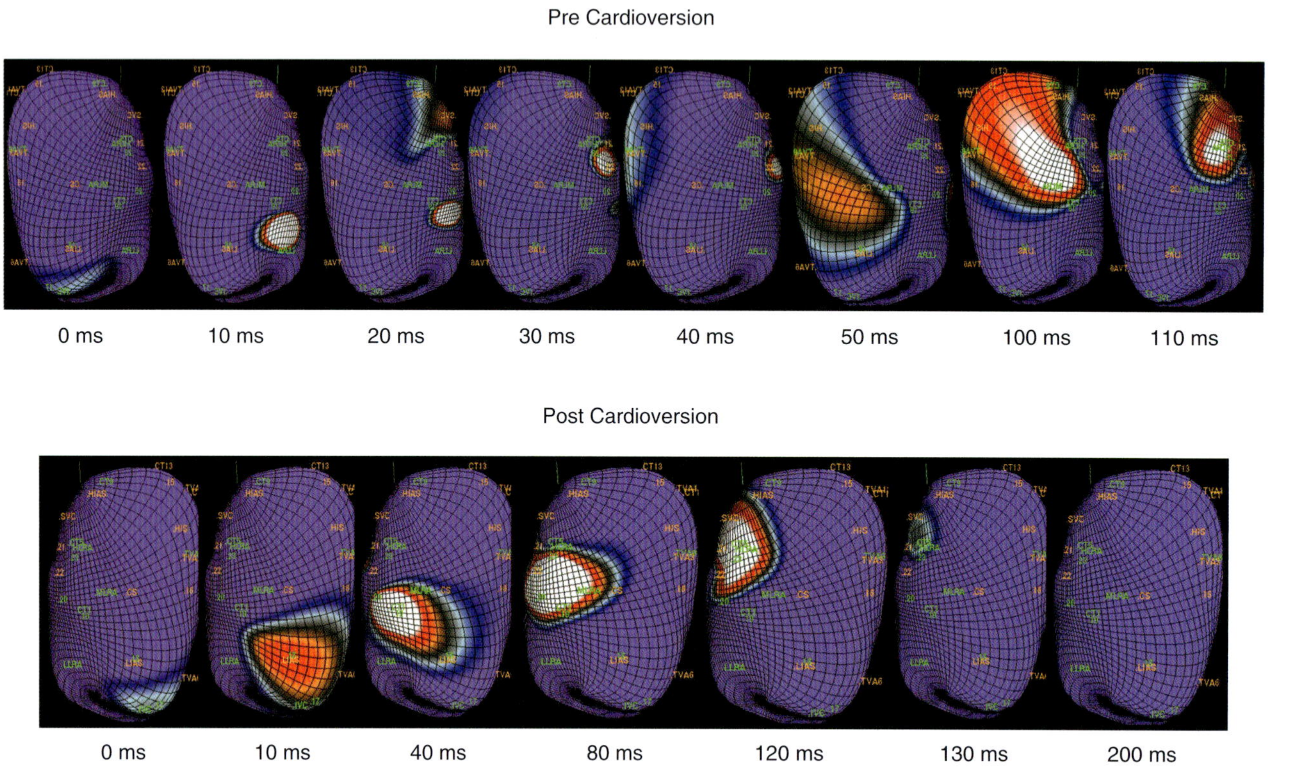

COLOR PLATE 41-7 Three-dimensional non-contact right atrial maps of atrial activation patterns before and after delivery of a cardioversion shock. Before cardioversion more than one wavefront is present and after the shock delivery, only a single macro-reentrant wavefront is left. Although classified as a failed cardioversion, delivery of the shock fundamentally altered the tachyarrhythmias. Individual frames of the activation wavefront and their timing in ms relative to shock delivery are shown.

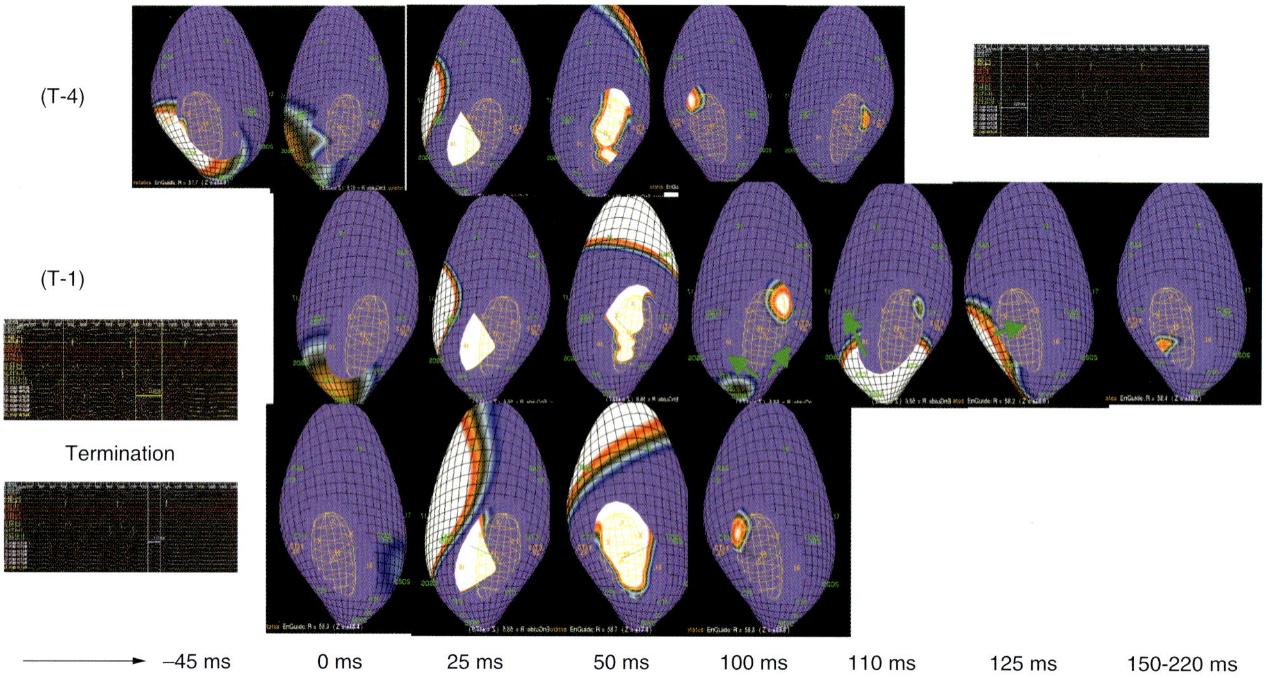

COLOR PLATE 41-8 Three-dimensional noncontact right atrial maps of atrial activation patterns before spontaneous termination of an AF event. Select individual frames from the last four cycles before termination are shown and their timing relative to each other in ms are presented. Before termination, only a single wavefront is present in the fourth cycle before cessation. In the cycle immediately before termination, a concealed atrial premature beat invades the excitable gap in the single circuit and gives rise to antidromic and orthodromic wavefronts. The former collides and terminates the original tachycardia and the latter circulates for one cycle and spontaneously stops propagating, resulting in atrial fibrillation episode termination.

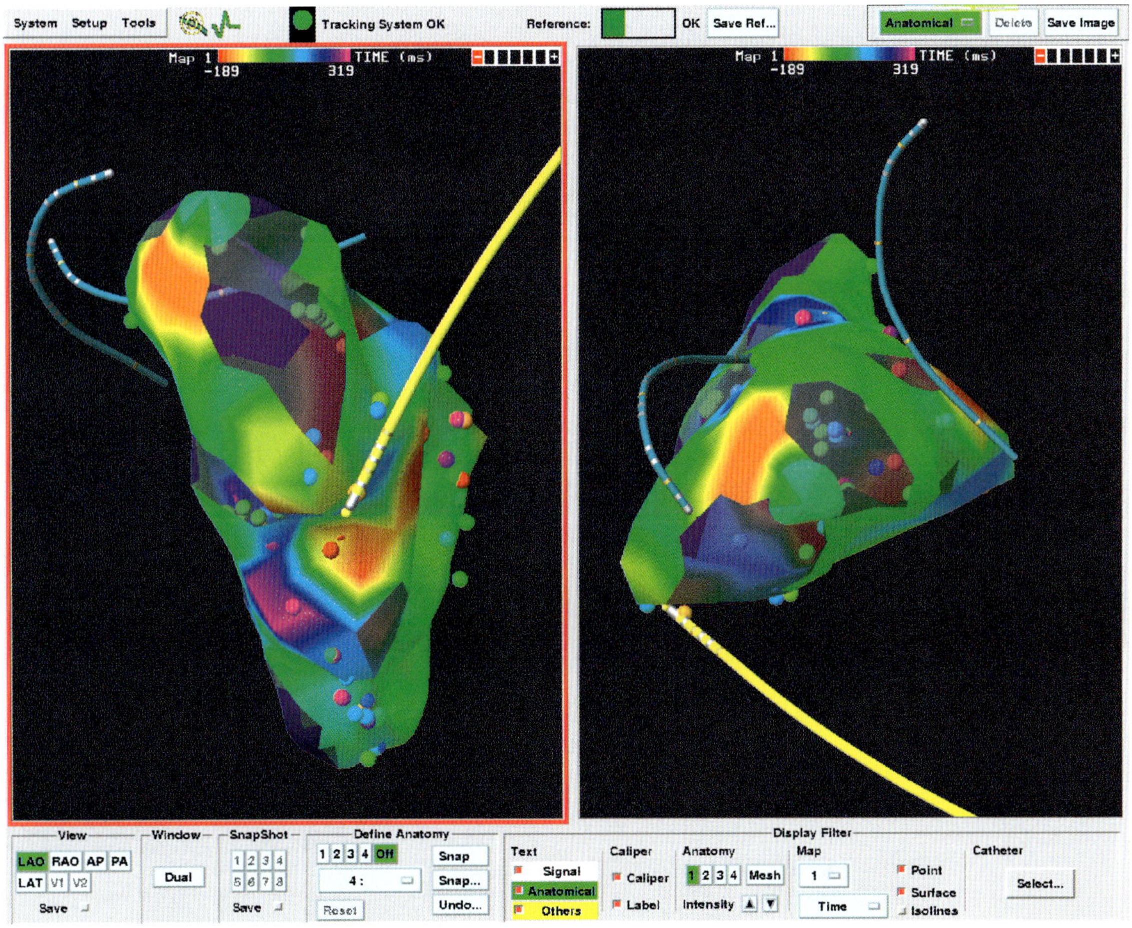

COLOR PLATE 42-14 Realtime positioning management (RPM) system showing three-dimensional spatial navigation and catheter position.

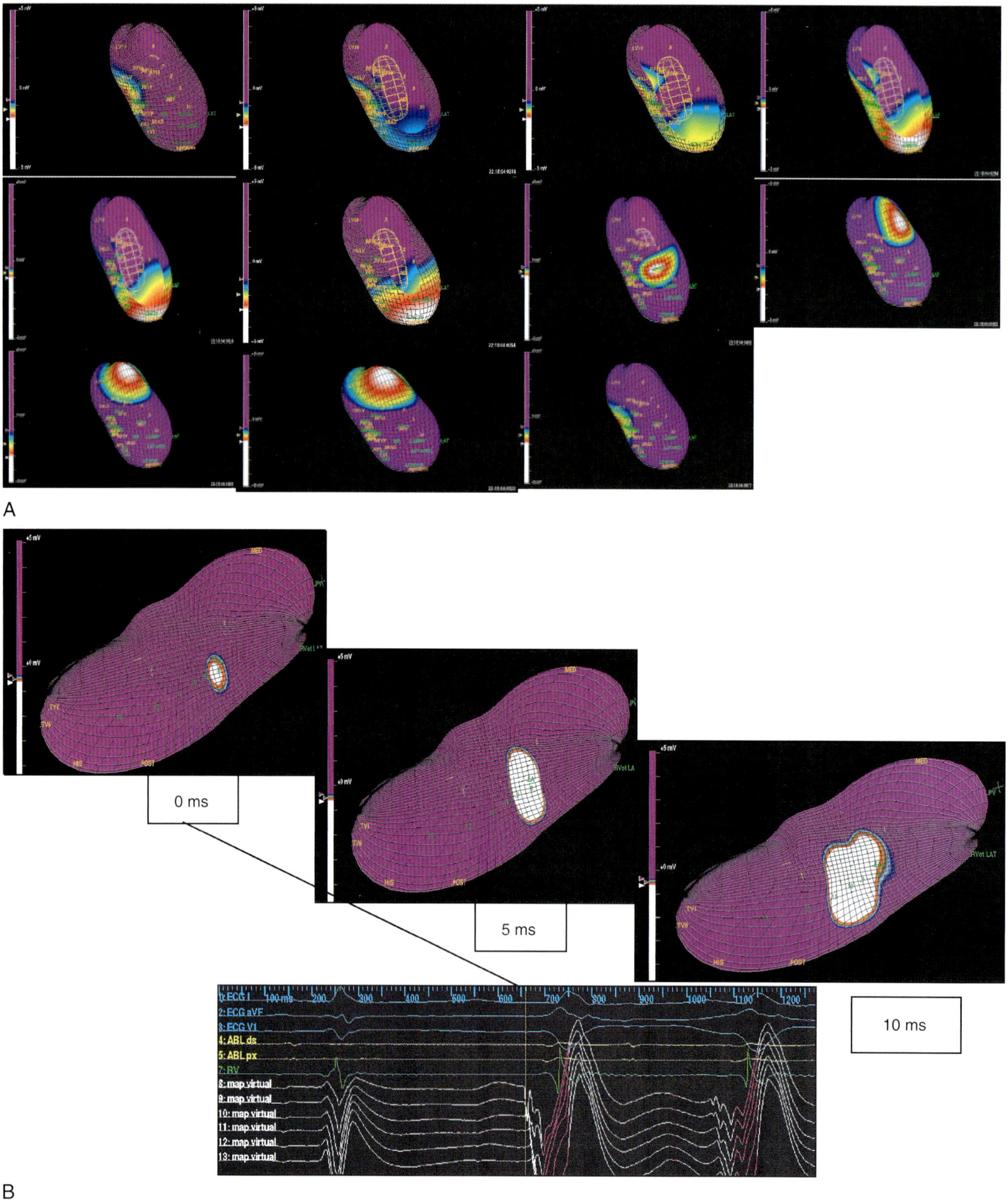

COLOR PLATE 42-15 **A,** Sequence of three-dimensional mapping images during ischemic ventricular tachycardia cycle. The wavefront progression is shown on the left ventricular endocardial surface showing the head to tail relationship seen in a reentrant circuit. Slower and more rapid conduction and spatial locations of regions involved in the tachycardia circuit can be visualized. **B,** Sequence of three-dimensional mapping images during idiopathic ventricular tachycardia. Focal origin is demonstrated, and wavefront progression in centrifugal fashion on the right ventricular endocardial surface can be visualized. The lower panel shows surface ECG leads, intracardiac, and virtual electrograms seen during the mapping process.

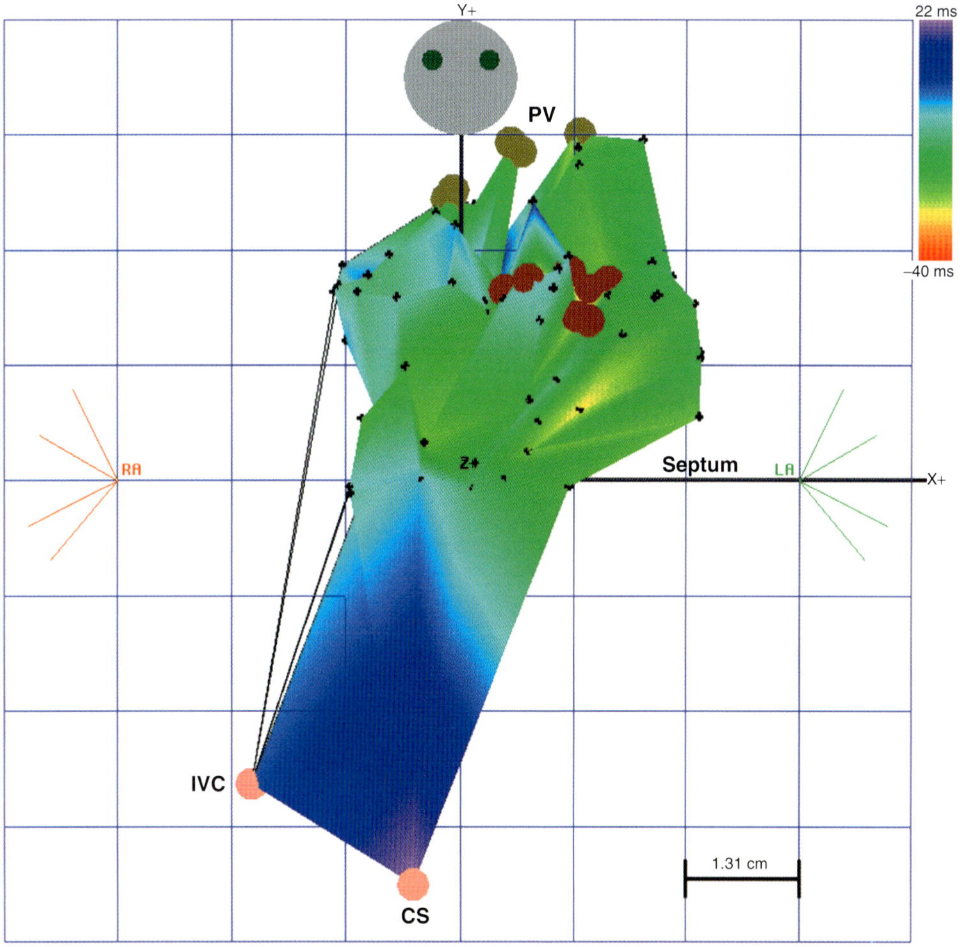

COLOR PLATE 45-2 Activation mapping of idiopathic right ventricular outflow tract (RVOT) tachycardia. The CARTO electroanatomic map of the lower right atrium and RVOT is shown during tachycardia. Red indicates areas with earliest endocardial activation and orange, yellow, green, blue, and purple indicate progressively delayed activation sequence. Shown are locations of inferior vena cava (IVC), coronary sinus (CS), and the pulmonic valve (PV). (From Varanasi S, Dhala A, Blanck Z, et al: Electroanatomic mapping for radiofrequency ablation of cardiac arrhythmias. J CV Electrophys 1999;10:538-544.)

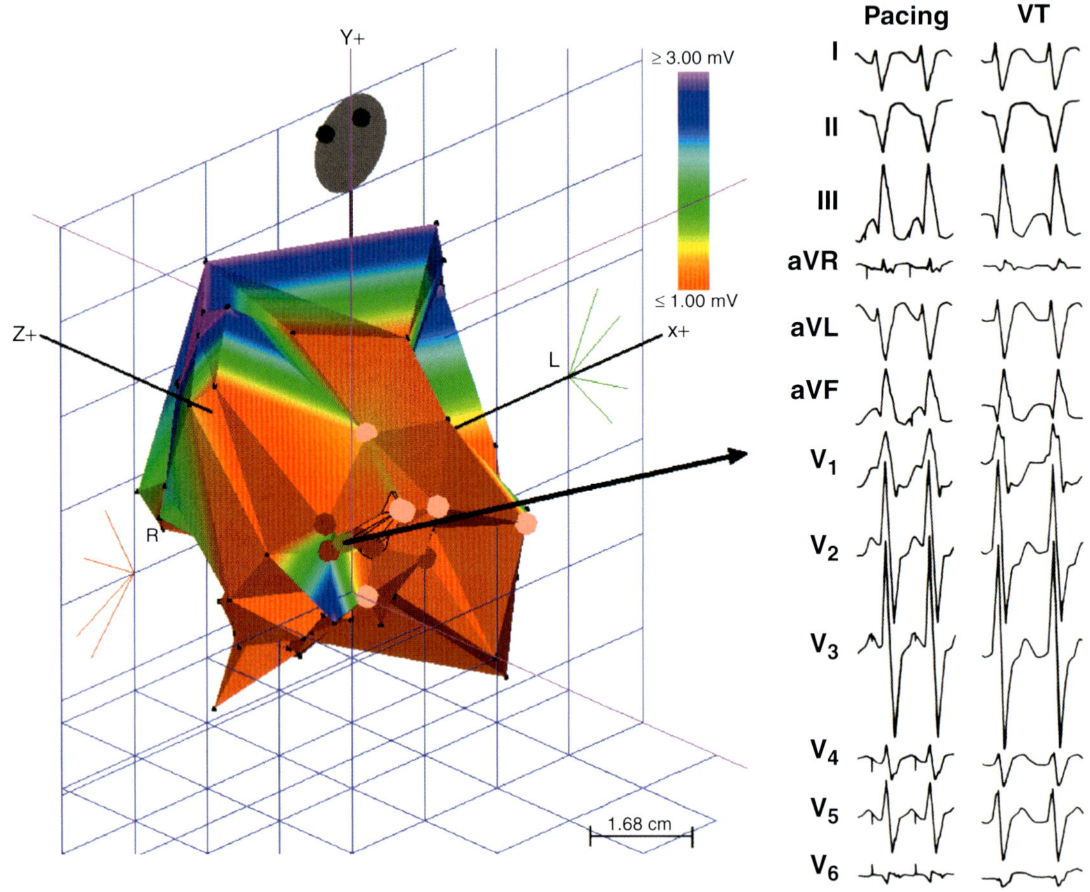

COLOR PLATE 45-5 Voltage map of the LV from patient with VT and coronary disease. Purple represents normal endocardium (amplitude ≥3.0 mV); red, dense scar (amplitude ≤1.0 mV); and ranges between purple and red, border zone. Electrograms of pace map and VT are shown to the right of the voltage map, identifying appropriate site for linear ablation at the edge of the dense scar. Linear lesions are identified by adjacent red circles.

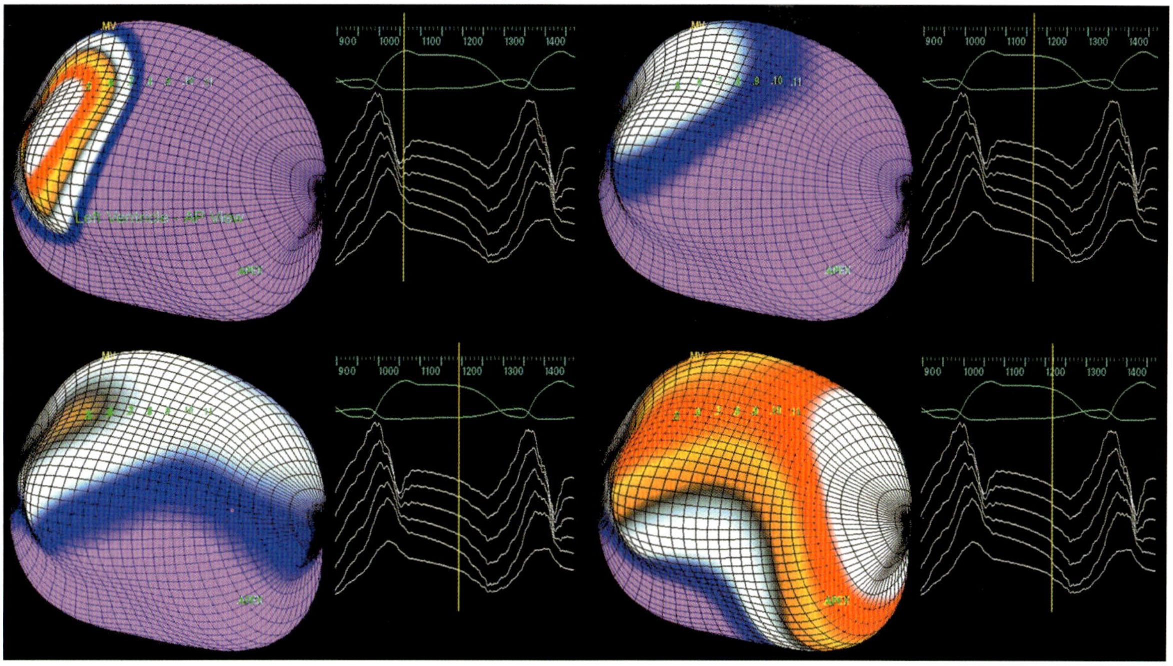

COLOR PLATE 45-6 Surface electrograms, virtual electrograms, and instantaneous isopotential maps at four time points, depicted by vertical lines through the electrograms, trace the path of a VT reentrant circuit during diastole and at the onset of the QRS complex. The VT shown in this example had a cycle length of 370 ms and was associated with marked hypotension. The **top panel on the left** shows endocardial activation 190 ms preceding the onset of the QRS complex during VT. The **top panel on the right** and the **bottom left panel** depict the slow spread of presystolic endocardial activation. The **bottom right panel** shows the VT exit site, corresponding to the onset of the QRS interval on the surface ECG. Linear radiofrequency lesions were delivered to transect the region of slow diastolic activation, as guided by the locator signal. This VT was no longer inducible and did not recur spontaneously following ablation.

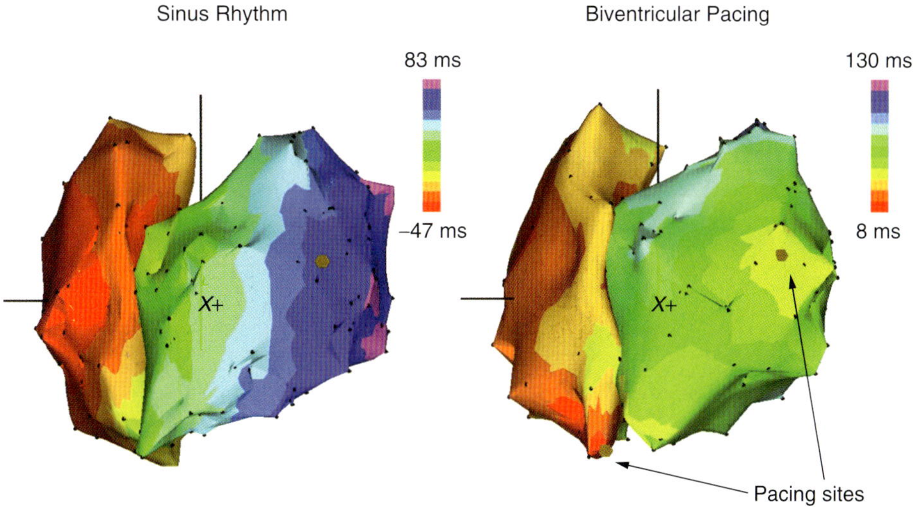

COLOR PLATE 49-1 Three-dimensional electroanatomic, nonfluoroscopic mapping in a patient with dilated cardiomyopathy during sinus rhythm and during biventricular stimulation. In sinus rhythm **(left panel)** the earliest ventricular activation (*red*) is located at the anterolateral wall of the right ventricle. After about 60 ms, the activation breaks through into the left ventricle and slowly proceeds (cell-to-cell conduction) from the septum to the lateral and posterolateral wall. The simultaneous pacing from the apex of the right ventricle and lateral wall restored a more homogeneous electrical activation of both ventricles.

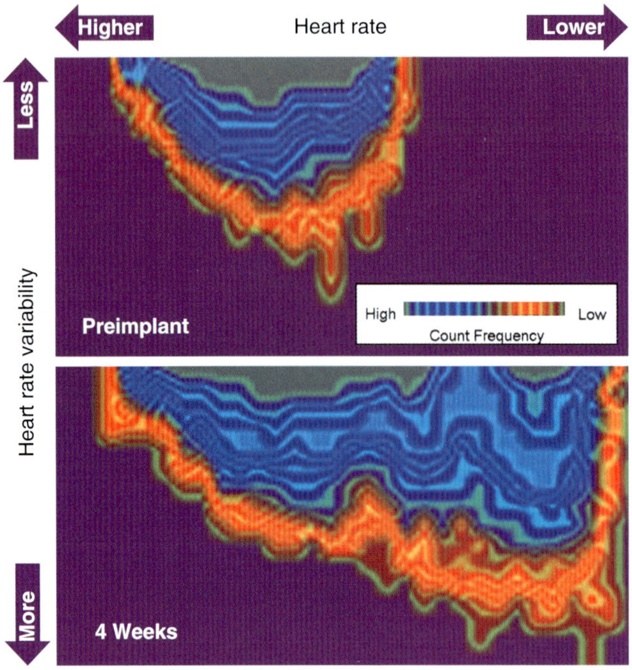

COLOR PLATE 49-7 Three-dimensional plot of heart rate variability and heart rate before implantation of a CRT device and after 12 weeks of continuous pacing. At baseline, a markedly depressed heart rate variability was noted as well as a high resting heart rate. After CRT, the heart rate variability largely increased and the heart rate was greatly reduced, thus leading to a rightward shift of the plot. CRT, Cardiac resynchronization therapy.

SECTION I

Conceptual Basis for Cardiac Arrhythmology

Chapter 1 Molecular and Cellular Basis of Cardiac Electrophysiology

GORDON TOMASELLI and DAN M. RODEN

Many volumes have been dedicated exclusively to the molecular and cellular basis of cardiac electrophysiology; to cover the topic exhaustively is not the aim of this chapter. Our goal is to present instead what is known about the fundamental basis of excitability in the heart starting from individual molecules and incorporating increasingly complex levels of integration from DNA to receptors, channels, and transporters, and from cells to tissues. We illustrate cellular and molecular fundamentals using clinically relevant examples.

Basic Concepts

CELLULAR STRUCTURE OF THE HEART

The myocardium is composed of cardiac myocytes, highly differentiated and specialized cells responsible for conduction of the electrical impulse and the heart's contractile behavior, and nonmyocyte cells. Myocytes occupy two thirds of the structural space of the heart; however, they represent but one third of all cells. Nonmyocytes include fibroblasts responsible for turnover of extracellular matrix that consists predominantly of fibrillar collagen types I and III. The collagen scaffolding provides for myocyte alignment and coordinated transmission of contractile force to the ventricular chamber. Other nonmyocyte cells include endothelial and smooth muscle cells of the intramural vasculature; neuronal elements (such as ganglia); and, under some conditions, inflammatory cells. The gross anatomic features of the heart, the extracellular matrix, and intramural vasculature create both macro- and microanatomic barriers that are central to both the normal electrophysiology of the heart and clinically important arrhythmias.

Cardiac myocytes are a family of structurally distinct cells, with a design commensurate with their function. Pacemaking cells such as those in the sinoatrial and atrioventricular nodes underlie the spontaneous electrical activity of the heart and contain relatively few contractile elements. In contrast, muscle cells are packed with actin and myosin filaments that serve the main function of the heart, propulsion of blood through the vasculature. Contractile myocytes are rod-shaped cells of approximately 100 μm by 20 μm. The myocyte is enveloped by the cell membrane, a lipid bilayer 80 to 100 Å in thickness. This insulating bilayer permits little to no transport of ions and maintains a separation of charge established by active transporters that reside in the cell membrane. Ion channels are transmembrane proteins that serve as a conductive pathway between the inside and outside of the cell, allowing the flow of ions and thus charge (i.e., current). There is also current flow among myocytes. However, unlike skeletal muscle, cardiac tissue is not a true syncytium—cells are connected to one another by low resistance communications called *gap junctions,* which contain intercellular ion channels.

THE MEMBRANE POTENTIAL AND CONDUCTION

Ion concentration and charge gradients across the cell membrane are responsible for the membrane potential of the cardiac myocyte. Transmembrane ionic and electrical gradients are maintained by a series of energy-requiring ion pumps and exchangers that: (1) concentrate K^+ inside the cell, (2) keep the intracellular Na^+ low (<10 mM), and (3) tightly regulate intracellular Ca^{2+} concentrations (10 to 200 nM). The pump responsible for establishing and maintaining most of the monovalent cation gradients is the Na^+K^+ ATPase, although other pumps and exchangers that transport Na^+, Ca^{2+}, and H^+ play significant roles in the genesis of the ionic concentration gradients and membrane potential. Ion channels are passive but selective conduits

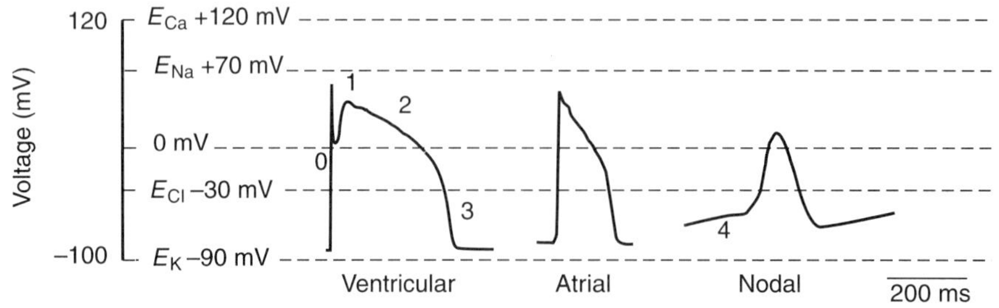

FIGURE 1-1 Schematic of action potentials in different regions of the heart. The permeability of the cell membrane determines membrane voltage. At rest, cells in all regions of the heart are more permeable to potassium than any other cation; hence, the negative membrane potential, near the Nernst potential for K^+. The characteristic shapes of the action potentials are determined by the ionic currents that are active in each cell type during the cardiac cycle (see text for details).

for the flow of ions along electrical and chemical gradients established by active ion transport systems. The physicochemical basis of the membrane potential depends on the Nernst potential (E_x), described as follows:

$$E_x = \frac{RT}{zF}\ \ln\left(\frac{[X]_0}{[X]_i}\right) =\approx 27\ \ln\left(\frac{[X]_0}{[X]_i}\right) \qquad \{\text{EQ. } 1.1\}$$

where R is the gas constant, T is temperature, F is Faraday's constant, and z is the valence of the ionic species. If one assumes that the resting cardiac myocyte has an intracellular $[K^+]$ of approximately 150 mM and the extracellular $[K^+]$ is 4 mM, then the Nernst potential for K^+ (E_K) is roughly −90 mV. The Nernst potential represents the voltage at which the osmotic tendency for K^+ to flow across the membrane is exactly balanced by the electrical tendency to flow in the opposite direction, resulting in a net zero ionic flux and thus no net current flow. Accordingly the K^+ Nernst potential is sometimes referred to as the *zero current potential* and represents the potential at which no K^+ current would flow through open K channels. By contrast, the Nernst potential for Na^+ in the resting cell is approximately +60 mV, indicating that if a pathway for Na^+ to enter the cell were present, Na^+ would enter the cell to move the membrane voltage toward the Nernst potential for Na^+. Indeed, this is precisely what happens when Na channels are open and initiate phase 0 of the action potential (AP). The Nernst potential for potassium is, in fact, very close to the resting membrane potential of the ventricular myocyte, indicating that the resting heart cell membrane is highly permeable to K^+. Actual resting potentials are less negative than E_K, due to small conductances of other ionic species with less negative Nernst potentials. Over the cardiac cycle the cell membrane becomes permeable to different ionic species, and these changes in permeability determine time-dependent changes in membrane potential, with each ion striving to move the membrane voltage to its Nernst potential and inscribing regionally specific APs (Fig. 1-1).

PASSIVE MEMBRANE PROPERTIES AND CABLE THEORY

The cardiac cell membrane can be modeled as a circuit composed of variable resistors (ion channels) in parallel with a capacitor (lipid bilayer), a "resistor capacitor circuit" (Fig. 1-2). The flow of current across the membrane will alter the charge on the capacitor (and therefore the membrane potential) and change the membrane resistance. The flow of current occurs not only across but obviously along the inside and outside of the cell membrane from cell to cell (i.e., current propagates).

Cable theory, originally developed to understand current flow in transoceanic telegraphic cables, can be used to model passive current flow and propagation in a cardiac muscle fiber. In their simplest formulation, the cable equations define the distribution of voltage along a continuous, uniform cable of infinite length stimulated by a point source. The predictions of cable theory are (1) a change in voltage exhibits a characteristic decay along the cable defined by the space constant (distance over which voltage decays to 1/e of the value at the site of injection) (Fig. 1-3*A*) and (2) there

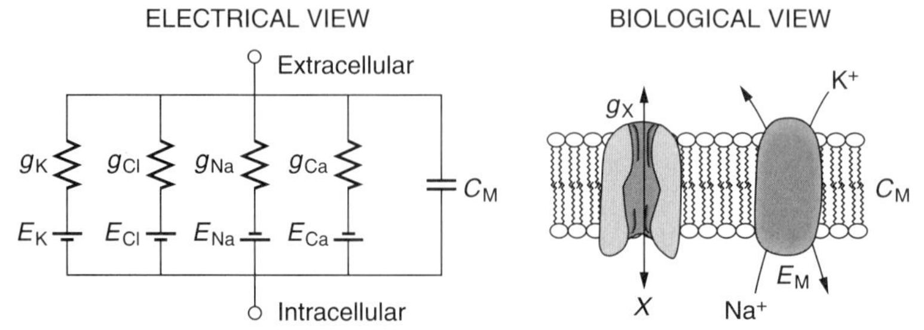

FIGURE 1-2 Electrical and biological representations of the cardiac cell membrane. The cardiac (or any excitable) membrane can be modeled as a resistor capacitor circuit with variable resistors (ion channels) in parallel with a capacitor (cell membrane). The transmembrane voltage is established by a series of energy-requiring pumps that maintain ionic gradients across the membrane.

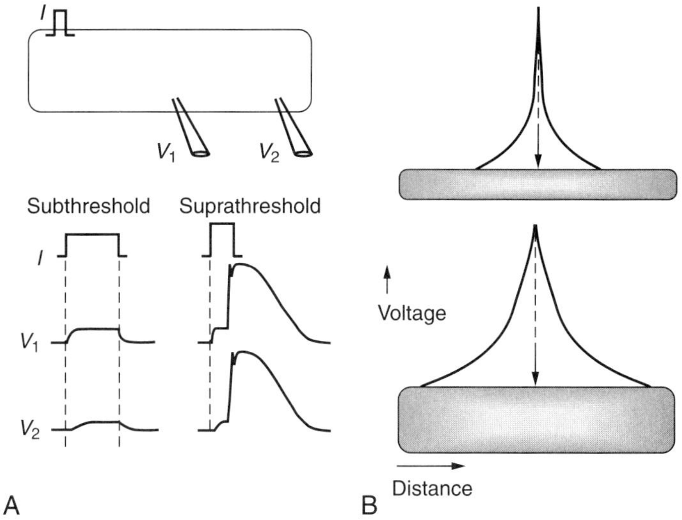

FIGURE 1-3 Spread of current in an idealized cable. **A,** With injection of current into the cable at site I, a transmembrane voltage change of smaller amplitude and slower kinetics is recorded at more distal sites in the cable. In excitable tissue such as in the heart, a stimulus of sufficient amplitude elicits a regenerative response (i.e., an action potential) that can propagate along the length of the preparation. Transmission of the action potential is associated with current flow across and along the membrane. **B,** The characteristics of the cable determine the distance over which the voltage difference induced by the current injection decays (space constant). In a cable of larger diameter, the voltage difference falls over a larger distance (i.e., the space constant, or distance over which the voltage difference falls to 1/e of the value at the site of injection, is proportional to the radius of the cable).

is an inverse relationship between resistance to current flow (both transmembrane and intercellular) and the cable diameter (see Fig. 1-3*B*). That is, the space constant is directly proportional to the cable diameter, thus greater lengths of the cable are influenced by the same current injection into a thick rather than thin cable.

There are substantial anatomic and biophysical limitations when applying the cable theory description of conduction to cardiac muscle. Anatomically, the shape of the heart is complex, and, at any level of integration above a single muscle fiber, it does not resemble a cable. Conduction through the myocardium is not continuous; myocardial cells are connected instead by gap junction channels that create a nonuniformity of intercellular resistance. Macroscopic discontinuities such as fibrous tissue and blood vessels also significantly perturb the cable view of conduction in the myocardium. Finally, the cardiac cell membrane is composed of resistor capacitor circuits that, when stimulated to threshold, will generate APs (see Fig. 1-3*A*). Despite the limitations of cable theory, it serves as the foundation for several important concepts in impulse propagation (see Chapter 2, Mechanisms of Reentrant Arrhythmias) and has been used to demonstrate the electrical nature of conduction in the heart.[1]

Depolarization of cardiac muscle results in the generation of an AP at the site of excitation and, in doing so, sets up a voltage gradient between the excited cells and their nearest neighbors. The current generated by the AP serves as an excitatory current for neighboring cells (source). The neighboring cells at their resting membrane potential (sink) are activated by the source. Impulse propagation depends on the balance between the magnitude of the currents in the source and sink and the resistance along the fiber. Failure of conduction may result from alterations in the source current; examples include reduction of the source by drug blockade of Na (atrial or ventricular muscle) or Ca channels (nodal cells) or from changes in the characteristics of the sink, such as ischemia and activation of ATP-dependent potassium channels ($I_{K\text{-}ATP}$). In the latter case, propagation fails because the tissue with greatly increased repolarizing K current, due to activated I_{KATP}, acts as an infinite sink and cannot be sufficiently depolarized to reach threshold for generation of an AP. Myocardial ischemia will produce changes in the intracellular environment such as decreased pH and increased intracellular Ca^{2+} that will serve to reduce gap junctional conductance and functionally uncouple myocardial cells altering the relationship between source and sink, hindering impulse propagation.

The safety factor for conduction is the magnitude of the current provided by the source that is in excess of that required to activate the sink. The main factors influencing source current are the rate of rise of the upstroke and amplitude of the AP, metrics reflecting the magnitude of inward currents. The factors that influence the current requirements of the sink are the membrane resistance and the difference between the resting and threshold potentials. One major reason for mismatch between the source and sink is an abrupt anatomic change, such as that which occurs at the Purkinje fiber–ventricular muscle junction. Orthodromic conduction over the Purkinje system results in activation of a broad band of ventricular muscle by narrow strands of Purkinje fibers. Such an abrupt transition from an anatomically narrow source to a massive sink makes propagation tenuous, such that small changes in the characteristics of the source or sink are likely to produce failure of conduction. Antidromic conduction from ventricular muscle to Purkinje fiber produces just the opposite source-sink relationship and thus a higher safety factor for conduction.

While this discussion implicitly treats the activation wave in one dimension, the behavior of propagating waves in the heart is more complex. The complexity can be appreciated if one considers propagation in two dimensions. In this circumstance the shape of the wavefront is a major determinant of efficiency of propagation. A convex wavefront, as might be observed after point stimulation, creates a large sink around a smaller activating source. This mismatch reduces conduction velocity and the safety factor for propagation. Conversely, a concave activation front produces a source-sink

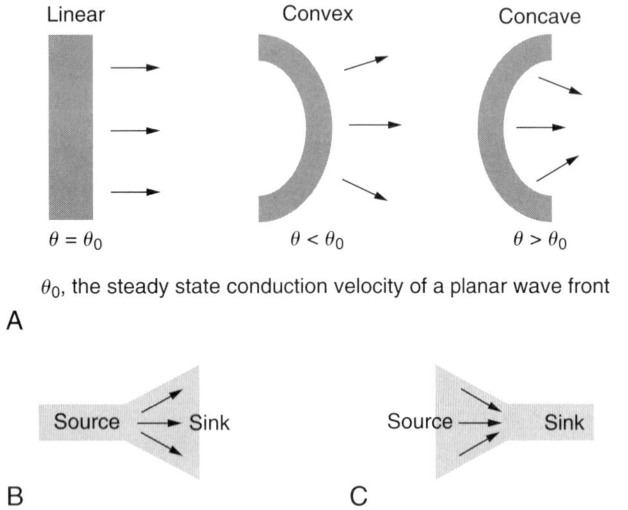

FIGURE 1-4 Relationship of current sources and sinks. **A,** Compared to the steady-state conduction velocity for planar wavefront in homogeneous medium (θ_0), the conduction velocity of a convex wavefront is slower, and a concave wavefront is faster with lower and higher safety margins for conduction, respectively. In situations of structural inhomogeneity, such as a Purkinje fiber activating ventricular muscle (**B**), the conduction velocity is slower with a lower safety margin at the structural discontinuity. **C,** In the case of a larger mass of tissue activating a smaller mass, the conduction velocity is fast with a high safety margin.

mismatch that favors the source, resulting in a high safety factor and more rapid impulse transmission. Thus, source-sink characteristics not only influence propagation but also the curvature of the wavefront (Fig. 1-4).

Directionally different conduction velocity is a characteristic feature of cardiac muscle known as anisotropic conduction. Anisotropic conduction has its basis in the structure of the myocyte and cardiac tissue; myocytes are rod shaped and organized in bundles oriented along the long axis of the cell. Transmurally, the axis of these bundles undergo significant changes in orientation through the ventricular wall (≈120° maximal deviation[2]). The communication among myocytes occurs via gap junction channels that are distributed nonuniformly over the surface of the heart cell with larger numbers of channels poised to propagate the impulse longitudinally rather than transversely to the long axis of the muscle fiber.[3] The implications of anisotropy for conduction in the longitudinal and transverse direction under pathologic conditions are controversial. In the context of uniform depression of conduction, as might exist with antiarrhythmic drug treatment, propagation in the transverse direction is preserved compared to conduction in the longitudinal direction.[4] However, when cellular uncoupling occurs, such as in ischemia, longitudinal propagation may exhibit a higher safety factor than transverse conduction.

MAJOR BREAKTHROUGHS—VOLTAGE CLAMP, MOLECULAR CLONING

A stimulus of sufficient magnitude applied to a myocyte (or any excitable cell) elicits a stereotypical change in membrane potential, an AP. The ionic current basis of the AP was confirmed and quantitatively studied using the voltage clamp developed in the middle of the 20th century.[5-7] Voltage clamping is a technique whereby the experimenter controls the transmembrane voltage and measures the current at that defined voltage. Much of what we know about ionic currents in myocytes comes from voltage clamp experiments and a more recently developed type of voltage clamp called the *patch clamp*.[8] A variant of the patch clamp technique permits the measurement of ionic currents through *single* ion channels.

Typically, in a voltage clamp experiment, the membrane voltage (V) is held near the resting membrane potential, (≈−80 mV for ventricular myocytes) and then stepped to more positive voltages. This voltage step induces two components of membrane current (I_M) flow; at the instant of the voltage change, ions (I_C) flow to charge the membrane capacitance (C_M), after which current reflects the movement of ions through the ion channels (I_i).

$$I_M = I_c + I_i = C_M \times \frac{dV}{dt} + I_i \qquad \{\text{EQ. } 1.2\}$$

Capacitive current is generally small and transient and can usually be electronically compensated for, so the voltage clamp provides a robust measure of current flow through ion channels, thereby permitting the study of the detailed biophysics and pharmacology of ionic currents and channels. The major limitation of such experiments in myocytes is the existence of many currents that are simultaneously active in response to the voltage step. The experimental conditions can be altered to isolate a current of interest; however, this often requires the presence of drugs, toxins, or highly unphysiologic conditions. An alternative to the study of ionic currents in native cells was afforded with the molecular cloning and heterologous expression of ion channel genes. Expression of an ion channel gene in a nonexcitable cell without other overlapping currents permits the study of the ionic current of interest under more physiologic conditions. The fundamental limitation of heterologous expression is that the ion channel is removed from its native cellular background, which may change the behavior of the channel.

The combination of highly sensitive electrophysiological methods such as patch-clamp recording and DNA cloning heralded the era of molecular understanding of the basis of cardiac excitability.

The Molecular Basis of Cardiac Action Potentials

Cardiac myocytes possess a characteristically long AP (200 to 400 milliseconds [ms], see Fig. 1-1) compared to neurons or skeletal muscle cells (1 to 5 ms). The action potential profile is sculpted by the orchestrated activity of multiple ionic currents, each with its distinctive time- and voltage-dependent amplitudes. The currents in turn are carried by complex transmembrane proteins that passively conduct ions down their electrochemical gradients through selective pores (ion channels), actively transport ions against their electrochemical gradient (pumps, transporters), or exchange electrogenically ionic species (exchangers).

APs in the heart are regionally distinct. The regional variability in cardiac APs is the result of differences in the number and type of ion channel proteins expressed by different cell types in the heart. Further, unique sets of ionic currents are active in pacemaking and muscle cells, and the relative contributions of these currents may vary in the same cell type in different regions of the heart.[9,10]

ION CHANNELS AND TRANSPORTERS ARE THE MOLECULAR BUILDING BLOCKS OF THE ACTION POTENTIAL

The currents that underlie the AP are carried by complex, multisubunit transmembrane glycoproteins called *ion channels* (Table 1-1). These channels open and close in response to a number of biological stimuli including a change in voltage, ligand binding (directly to the channel or to a G-protein coupled receptor), and mechanical deformation. Other ion motive transmembrane proteins such as exchangers and transporters make important contributions to cellular excitability in the heart. Ion pumps establish and maintain the ionic gradients across the cell membrane that permit current flow through ion channels. If pumps, transporters, or exchangers are not electrically neutral (e.g., three Na^+ for one Ca^{2+}), they are termed *electrogenic* and can further influence electrical signaling in the heart.

The most abundant superfamily of ion channels expressed in the heart are voltage gated. Various structural themes are common to all voltage-dependent ion channels. First, the architecture is modular, consisting of either four homologous subunits or four internally homologous domains (in Na and Ca channels). Secondly, the proteins wrap around a central pore. The pore-lining ("P segment") regions exhibit exquisite conservation

TABLE 1-1 Human Ion Channel, Exchanger, and Transporter Genes

Channel	Gene	Chromosome	Accession #	Locus Link	Ref
K Channels—α Subunits					
HERG	*KCNH2*	7q35-q36	NM_000238	3757	237
KvLQT1	*KCNQ1*	11p15.5	NM_000218	3784	132
Kv1.4	*KCNA4*	11p14	NM_002233	3739	238
Kv1.5	*KCNA5*	12p13	NM_002234	3741	239
Kv4.3	*KCND3*	1p13	NM_004980	3752	240
Kir2.1	*KCNJ2*	17q23.1-q24.2	NM_000891	3759	141
GIRK4 (Kir3.4, CIR)	*KCNJ5*	11q24	NM_000890	3762	241
GIRK1 (Kir3.1)	*KCNJ3*	2q24.1	NM_002239	3760	242
Kir6.2	*KCNJ11*	11p15.1	NM_000525	3767	148
HA-HCN2	*HCN2*	19p13.3	NM_001194	610	243
hHCN4	*HCN4*	15q24-q25	NM_005477	10021	154
Ancillary Subunits—K Channels					
Mink	*KCNE1*	21q22.1-q22.2	NM_000219	3753	106
MiRP-1	*KCNE2*	21q22.12	NM_005136	9992	109
KChIP2	*KCNIP2*	10	NM_014591	30819	104
SUR2A	*ABCC9*	12p12.1	NM_005691	10060	244
Ca Channels					
Cavα1C	*CACNA1C*	12p13.3	NM_000719	775	245
Cavα1H	*CACNA1H*	16p13.3	NM_021098	8912	94
Cavβ1	*CACNB1*	17q21-q22	NM_000723	782	246
Cavβ2	*CACNB2*	10p12	NM_000724	783	247
Cavα2δ	*CACNA2D*	3p21.3	NM_006030	9254	248
Na Channels					
hH1	*SCN5A*	3p21	NM_000335	6331	16
HNavβ1	*SCN1B*	19q13.1	NM_001037	6324	249
HNavβ2	*SCN2B*	11q23	NM_004588	6327	250
Gap Junction Channels					
Cx-43	*GJA1*	6q21-q23.2	NM_000165	2697	251
Cx-40	*GJA5*	1q21.1	NM_005266	2702	252
Cx-45	*GJA7*	17	NM_005497	10052	252
Transporters and Exchangers					
NaCa3	*NCX1*	2p22-p23	NM_021097	6546	157
Na-K ATPase					
α1	*ATP1A1*	1p13	NM_000701	476	253
α2	*ATP1A2*	1q21-23	NM_000702	477	254
α3	*ATP1A3*	19q13.2	NM_000703	478	255
β1	*ATP1B1*	1q22-25	NM_001677	481	256
β2	*ATP1B2*	17p13.1	NM_001678	482	257

within a given channel family of like selectivity (e.g., jellyfish, eel, fruit fly, and human Na channels have very similar P segments), but not among families with different selectivity. Third, the general strategy for activation gating (opening and closing in response to changes in membrane voltage) is highly conserved: The fourth transmembrane segment (S4), stereotypically studded with positively charged residues, lies within the membrane field and moves in response to depolarization, opening the channel.[11] Fourth, most ion channel complexes include not only the pore-forming proteins (α-subunits) but also auxiliary subunits (e.g., β-subunits) that modify channel function.

SODIUM CHANNELS

Na channels are highly conserved through evolution, existing in all species from the jellyfish to humans. They are nature's solution to the conundrum of coordination and communication within large organisms, particularly when speed is of the essence. Thus, Na channels are richly concentrated in axons and muscle, where they are often the most plentiful ion channels. A mammalian heart cell, for example, typically expresses more than 100,000 Na channels,[12] but only 20,000 or so "large and long-lasting" type (L-type) Ca channels[13] and fewer copies of each family of voltage-dependent K channels.

Na channels were the first ion channels to be cloned and have their sequence determined.[14] In humans more than 10 distinct Na channel genes have been cloned from excitable tissues with striking homology to the cDNA cloned from the eel electroplax.[15-17] The cardiac Na channel gene (*SCN5A*) resides on the short arm of chromosome 3 (3p21) (see Table 1-1). The Na channel complex is composed of several subunits, but only the α subunit is required for function. Figure 1-5 shows that the α subunit consists of four internally homologous domains (labeled I to IV), each of which contains six transmembrane segments. The four domains fold together so as to create a central pore whose structural constituents determine the selectivity and conductance properties of the Na channel.

The peptide linkers between the fifth (S5) and sixth (S6) membrane spanning repeats or P segments of each domain come together to form the pore.[18] The primary structure of the S5-S6 linkers of Na channels in each domain is unique. Thus, the structural basis of permeation differs fundamentally from that of K channels, in which four identical P segments can come together to form a K^+ selective pore (discussed later). Indeed, accessibility mapping studies in Na channels have revealed marked asymmetries in the contributions of each domain to the permeation pathway;[19] particularly prominent are domain III, in which a lysine (K1418 in the human cardiac Na channel sequence) is critical for discrimination for Na^+ over Ca^{2+},[20] and domain IV, in which mutations of various contiguous residues render the channel nonselective among monovalent cations.[19]

One of the seminal contributions of Hodgkin and Huxley was the notion that Na channels occupy several "states" (which we now view as different conformations of the protein) in the process of opening ("activation"); yet another set of conformations is entered when the channels close during maintained depolarization ("inactivation"). The *m* gates that underlie activation and the *h* gate that mediates inactivation were postulated to have intrinsic voltage dependence and function independently.[21] While some of the implicit structural predictions of that formulation have withstood the test of time, others have not. For example, the four S4 segments are now widely acknowledged to serve as the activation voltage "sensors." In the process of activation, several charged residues in each S4 segment physically traverse the membrane through a narrow canaliculus formed by other, as-yet-unidentified regions of the channel (see Fig. 1-5, *bottom panel*).[22-24] However, the idea that the sensors are equivalent and independent turns out to be incorrect. The contributions of each S4 segment to activation are markedly asymmetrical; some of the charged residues play a much more prominent role than others in "homologous" positions.[11,25,26] Other studies have revealed that activation is coupled to inactivation.[11,25] Indeed, the time course of current decay during maintained depolarization predominantly reflects the voltage dependence of activation,[27] although single-channel inactivation itself does vary with voltage (particularly in cardiac Na channels).[28] If the S4s are the sensors, where are the activation "gates" themselves? This crucial question remains unresolved. However, based on experimental evidence, S6 emerges as the leading contender for the physical activation gate. It is notable that local anesthetic antiarrhythmic drugs bind to the S6 transmembrane segment in the fourth homologous domain (see Fig. 1-5, *top*). The homologous domains on calcium and potassium channels are also loci for drug binding.

Inactivation of Na channels is as arcane a process as activation. There is not only loose coupling to activation but there are multiple inactivation processes. One common approach to distinguishing among inactivated states is to determine the rate at which they recover the ability to activate: Repriming from the traditional "fast" inactivation occurs over tens of ms, while recovery from "slow" inactivation can require tens of seconds or longer.[29,30] Fast inactivation is at least partly mediated by the cytoplasmic linker between domains III and IV (the crucial residues are labeled *IFM* for isoleucine, phenylalanine, and methionine, in Fig. 1-5),[11,31-33] which may function as a hinged lid,[33] docking onto a receptor formed by amino acids in the S4-S5 linkers of domains III and IV.[34] This notion is consistent with observations that fast inactivation can be disrupted by internal proteases. Nevertheless, it is increasingly clear that mutations scattered widely throughout the channel affect inactivation gating, undermining somewhat the primacy of the IFM residues of the III-IV linker. The structural determinants of slow inactivation are less well localized than those of fast inactivation. Mutations in the P region of domain I affect both activation gating and slow inactivation,[35,36] while various widely scattered disease mutations identified in paramyotonia congenita and other skeletal myopathies suppress slow inactivation of the Na channel.[37]

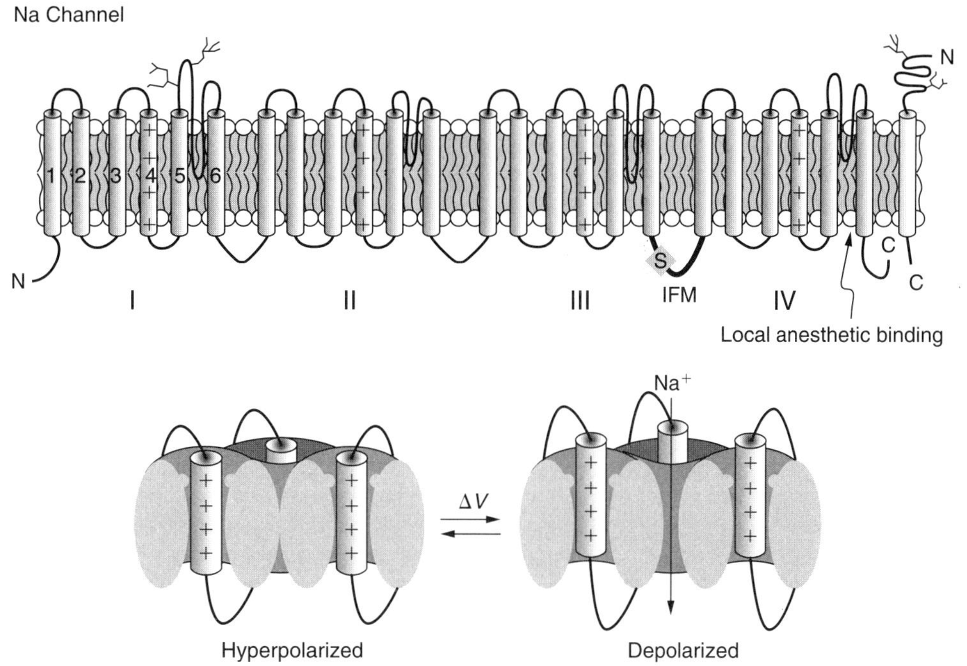

FIGURE 1-5 Models of the Na channel. **(Top)** Topology cartoon of the Na channel α subunit with four pseudohomologous domains (*I-IV*). One or more transmembrane β subunits coassemble with the α subunit to form the intact channel. The fourth membrane-spanning repeat (S4) is charged and serves as the voltage sensor for channel activation. The segments or linkers between S5 and S6 in each domain, called the *permeation* or *P segments*, form the outer pore mouth and the selectivity filter. The linker between the third and fourth domains underlies fast inactivation and contains one of the mutations that underlies the chromosome 3-linked form of the long QT syndrome. The S6 segment of the fourth domain contains residues critical for local anesthetic binding to the channel. **(Bottom)** Cartoon of the canaliculus through which the S4 segment slides during channel activation. An outward movement of the S4 during channel activation is proposed.

The S6 segment of domain IV has been proposed to contain the receptor for local anesthetics that block Na channels in a voltage-dependent manner.[38] Block is enhanced at depolarized potentials or with repetitive pulsing, or both. These observations are consistent with the idea that local anesthetics act as allosteric effectors of the inactivation gating mechanism: When they bind to the channel, they facilitate inactivation.[39] It is clear that gating interacts with local anesthetic block so profoundly that it is difficult to interpret the localization of a "receptor" to S6. Mutations in S6, at the putative receptor sites, alter gating independent of superimposed drug effects.[38,40,41] Further, mutations in distant parts of the molecule can also dramatically alter the phenotype of local anesthetic block. Despite these caveats, the S6 segments appear to play a special role in the effects of drugs in all of the voltage-gated ion channels.

Pharmacologic competition studies and mutagenesis have defined a number of neurotoxin-binding sites on the Na channel. Among these, tetrodotoxin (TTX), a guanidinium-containing blocker, has contributed the most to our understanding of Na channel structure and function. Externally applied TTX blocks neural and skeletal muscle Na channel isoforms potently (in the nM range), but block of cardiac channels requires much higher concentrations ($\approx 10^{-5}$ M). The identity of one particular residue in the P region of domain I accounts for most of the isoform-specific TTX sensitivity: An aromatic residue at this position (373 in the human heart sodium channel) confers high affinity, while its absence renders the channel TTX resistant.[42,43] Many other residues in the outer mouth of the channel contribute to the binding of TTX and the related divalent guanidinium toxin saxitoxin (STX), suggesting that the toxin has a large footprint on the external surface of the channel.[44]

Biochemical studies reveal the existence of two distinct ancillary subunits associated with the brain Na^+ channel, designated β1 and β2.[45] Antibodies directed to the β1 or β2 subunit will immunoprecipitate the entire brain Na^+ channel complex with a subunit stoichiometry of 1:1:1. The β1 subunit is noncovalently associated and β2 is linked by a disulfide bond to the α subunit. Each of the β subunits is heavily glycosylated; up to 36% of the mass of each is carbohydrate.[46,47]

The β1 and β2 subunits have been cloned and the deduced primary structures indicate that they are unrelated proteins of molecular weight 23 and 21 kDa, respectively. The predicted transmembrane topology of the β subunits is similar, each containing a small carboxyl terminal cytoplasmic domain, a single membrane spanning segment, and a large amino terminal extracellular domain with several consensus sites for N-linked glycosylation.[48,49] The β2 subunit has several distinctive features, including an extracellular immunoglobulin-like fold with homology to the neural cell adhesion molecule contactin. Expression of β2 with neuronal α subunits in *Xenopus* oocytes increases the current amplitude, modulates gating, and increases the cell membrane capacitance.[49] Coexpression of β1 subunits with either neuronal or skeletal muscle α subunits in oocytes also produces clear-cut effects on Na channel function. Current density increases, both activation and inactivation gating are hastened, and the steady-state inactivation curves are shifted in the hyperpolarizing direction.[48,50-52]

The mRNA encoding the β1 subunit appears to be widely expressed and is clearly an important component of the neuronal and skeletal muscle Na channels. However, the functional role of this subunit in the heart Na channel is uncertain. Despite expression of β1 mRNA using subtype-specific antisera, no β1 subunit is found in association with the α subunit protein from rat heart. The results of heterologous expression studies are conflicting, with some studies suggesting no effect of β1 while in others changes in level of expression, function, and pharmacology of the cardiac Na channel have been observed. No specific role for the β2 subunit has been defined in the cardiac Na channel.

Regulation of the Na channel by phosphorylation is complex. Isoforms of the Na channel α subunit fall into one of two groups, long (neuronal and cardiac) and short (skeletal muscle and eel). The neuronal isoforms have a substantially larger intracellular linker between domains I and II. The linker contains five consensus sites for cyclic adenosine monophosphate (AMP)-dependent protein kinase (PKA) phosphorylation. In fact, PKA modulates the function of expressed neuronal and cardiac Na channels. Phosphorylation of sites in the I-II linker of the brain channel reduces current amplitude without significantly affecting gating.[53] The cardiac Na channel has eight candidate consensus PKA phosphorylation sites in the I-II linker, all of which are distinct from the neuronal channels. In vitro studies of the expressed cardiac Na channel demonstrate cyclic AMP-dependent phosphorylation on only two of these serines.[54] Interestingly, when the cardiac channel is phosphorylated by PKA, the whole-cell conductance increases, suggesting the specific pattern of phosphorylation is responsible for the functional effect.[54,55]

In contrast to PKA, protein kinase C (PKC) alters the function of all of the mammalian Na channel isoforms. The PKC effect is largely attributable to phosphorylation of a highly conserved serine in the III-IV linker (see Fig. 1-5). PKC reduces the maximal conductance of the channels and alters gating in an isoform-specific fashion. The macroscopic current decay of neuronal channels is uniformly slowed by PKC, suggesting a destabilization of the inactivated state.[56] Cardiac channels exhibit a hyperpolarizing shift in the steady-state availability curve, suggesting an enhancement of inactivation from closed states.[57]

All subunits of the Na channel are modified by glycosylation. The β1, β2, and brain and muscle α subunits are heavily glycosylated, with up to 40% of the mass being carbohydrate.[46,47] In contrast, the cardiac α subunit is only 5% sugar by weight.[58] Sialic acid is a prominent component of the N-linked carbohydrate of the Na channel. The addition of such a highly charged carbohydrate has predictable effects on the voltage dependence of gating through alteration of the surface charge of the channel protein.

Alteration of ion channel function is an important pathophysiologic mechanism of various familial diseases of muscle,[59] brain,[60] and inherited arrhythmias (see Chapter 27).[61] Na channel mutations underlie the aberrant excitability characteristic of some epilepsies,[62] skeletal muscle myotonias and paralysis, as well as the chromosome-3–linked long QT syndrome (LQT3). In general, LQT3-linked mutations in *SCN5A* generally disable fast inactivation of the Na channel (gain-of-function mutation), producing AP prolongation and a predisposition to repetitive electrical activity (polymorphic ventricular tachycardia). LQT3-linked mutations, and the mechanisms by which they generate arrhythmias, are the subject of intensive study. Some forms of idiopathic ventricular tachycardia have also been linked to mutations in *SCN5A*.[63] In contrast to LQT3, these mutations tend to reduce or eliminate channel function. Disruption of Na channel function has been suggested to create an imbalance in early repolarization, particularly in the epicardium of the ventricle that predisposes to functional reentry and ventricular fibrillation.[64]

CALCIUM CHANNELS—L-TYPE

The pore-forming subunit (α1) of the Ca channel is built on the same structural framework as the Na channel.[65] As is the case with the Na channel, there are a number of genes that encode surface membrane Ca channel α1 subunits. The predominant sarcolemmal Ca channels in the heart are the L-type and "tiny and transient" type (T-type) (Table 1-2). The cardiac L-type Ca channel ($Ca_V1.2$) is a multisubunit transmembrane protein composed of α1C ($\alpha_1 1.2$)[66] (165 kDa), β (55 kDa), and α2 (130 kDa)-δ (32 kDa) subunits. Three genes are known to encode L-type Ca channel α1 subunits ($\alpha_1 1.x$-$\alpha_1 3.x$), and the $\alpha_1 1.2$ is the gene expressed in the heart (see Table 1-1). Distinct splice variants of the $\alpha_1 1.2$

TABLE 1-2 Properties of Ca Channels

	L-type	T-type
Pore-forming α subunit	α1C	α1H
Auxiliary subunits	β, α2-δ	?
Permeability	$Ba^{2+} > Ca^{2+}$	$Ba^{2+} \approx = Ca^{2+}$
Activation threshold	>−30 mV	>−60 mV
Inactivation threshold	>−40 mV	>−90 mV
Inactivation		
Rate	Slow	Fast
Ca-dependent	Yes	No
Voltage sensitive	Yes	Yes
Recovery	Fast	Slow
Localization in heart	All	Nodal > Purkinje > atria
Blocker sensitivity		
Dihydropyridines	+++	No
Phenylalkylamines	+++	No
Benzothiazepines	+++	No
Tetralols	++	+++
Ni^{2+}	+	+++
Cd^{2+}	+++	+

gene have been described and contribute to the diversity of the cardiac L-type Ca channel function.[67-70] Similar to the α subunit of the Na channel, the S5-S6 linkers (P segments) of the α1 subunit of the Ca channel form the ion selective pore (Fig. 1-6). However, unlike Na channels, each P segment contributes a glutamic acid to a cluster that serves to bind Ca^{2+} in the channel pore.[71] The β subunit is completely cytoplasmic and noncovalently binds to the α1C subunit, modifying its function[72-76] and contributing to appropriate membrane trafficking of the channel complex.[77] Although β_{2a} has been proposed to be the major L-type Ca channel β subunit, β_{1b}, β_{1c}, and β_3 have also been detected in the heart.[78] As many as five genes have been suggested to encode the α2-δ subunit[79]: These gene products undergo post-translational processing to produce the mature extracellular α2 subunit linked by a disulfide bond to the transmembrane δ subunit.[80,81] In heterologous expression systems, α2-δ subunits enhance expression of Ca channels and hasten current activation and deactivation in the presence of α1 and β subunits.[72,82]

The molecular basis of gating of Ca channels is less well understood than Na or K channels. The α1 subunit of the Ca channel contains a highly basic S4 transmembrane segment in each homologous domain. These segments are thought to be the voltage sensors for channel activation. Activation of skeletal muscle and cardiac L-type Ca channels is distinct; skeletal muscle channels activate much more slowly than their cardiac counterparts. Based on the properties of chimeric channels constructed from cardiac ($\alpha_1$1.2) and skeletal muscle ($\alpha_1$1.1) α1 subunits, the difference in activation gating resides in the first homologous domain.[83] However, it is unclear if the S4 membrane-spanning segment is the crucial structural motif.

Inactivation in Ca channels is complicated by the presence of both voltage-dependent and Ca^{2+}-dependent processes. The structural motifs that underlie voltage-dependent inactivation are uncertain. Ca^{2+}-dependent inactivation has recently been demonstrated to depend on the binding of calmodulin to the channel. Calmodulin is permanently tethered to the channel complex and upon entry of Ca^{2+} into the cell through the L-type channel pore, calmodulin binds Ca^{2+} The Ca^{2+} calmodulin complex then facilitates the interaction of calmodulin with a consensus-binding motif, called the *IQ domain in the carboxyl terminus of the channel.* The binding of calmodulin to the IQ motif occludes the inner mouth of the Ca channel pore and terminates inward Ca^{2+} flux despite continued depolarization. As cytoplasmic Ca^{2+} concentration falls, calmodulin unbinds Ca^{2+} and the IQ motif, relieving Ca^{2+}-dependent inactivation.[84,85] A Ca^{2+}-binding EF hand motif is also present in the carboxyl terminus of α1 subunits that exhibit Ca^{2+}-dependent inactivation (e.g., $\alpha_1$1.2). The EF hand also appears to be necessary to confer Ca^{2+}-dependent inactivation to the $\alpha_1$1.2 subunit, although not through a direct binding of Ca^{2+} (see Fig. 1-6, *bottom*).[86]

L-type Ca channels are found in all myocytes of the mammalian heart and have several important electrophysiological functions. In sinoatrial (SA) nodal tissue, both L- and T-type channels contribute to diastolic depolarization and therefore impulse formation.[87] Modulation of L-type current by the autonomic nervous system is important in controlling the rate of sinus node discharge. Blockade of the L-type channel underlies

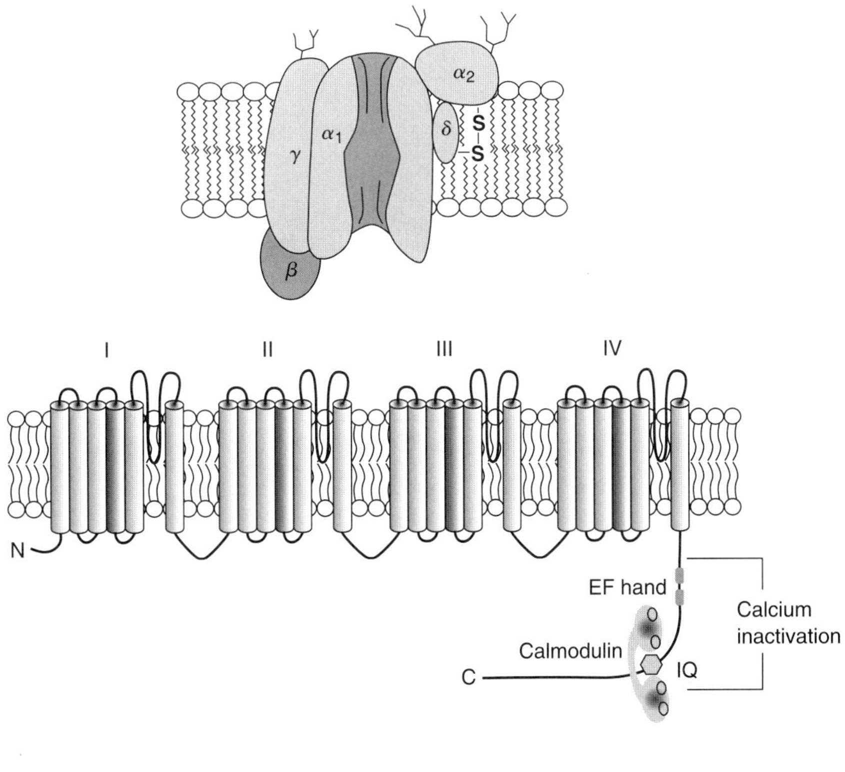

FIGURE 1-6 Top, Subunit structure of the cardiac Ca channel. The $\alpha_1$1.2 (α1C) subunit forms the pore and contains drug-binding sites. β2 and α2-δ subunits coassemble with the α1 subunit. The intact cardiac L-type Ca channel containing the $\alpha_1$1.2 subunit is referred to as Ca_V1.2 (see text for details). **Bottom,** The transmembrane topology of the Ca channel is similar to the Na channel with four homologous domains, each containing six membrane-spanning repeats. The S4 segments *(dark gray)* are generally conserved, but the molecular basis of inactivation is distinct from Na channels. The Ca channel exhibits both voltage- and Ca^{2+}-dependent inactivation. Calmodulin (CAM) is required for Ca^{2+}-induced inactivation of the channel and is tethered to the channel in the carboxyl terminus; upon binding of Ca^{2+} there is a conformation change in this region of the channel that requires CAM, the CAM binding motif (IQ), and the EF hand motif. Together these constitute the Ca^{2+} inactivation (CI) region of the channel.

the sinus node slowing observed with some calcium channel antagonists. The atrioventricular (AV) node is the only place in the body where Ca channels (L-type) normally conduct excitatory impulses. Consequently, it is not surprising that modulators of the L-type current have profound effects on SA and AV conduction. In muscle tissue the L-type Ca current is the major depolarizing current during the AP plateau, and inhibition of this current reduces the voltage of the plateau and shortens the AP duration. Down regulation of the L-type current in atrial myocytes isolated from patients with a history of atrial fibrillation is thought to promote the maintenance of fibrillation.[88]

CALCIUM CHANNELS—T-TYPE

The other major Ca channel present in the sarcolemma of heart (and prominently in vascular smooth muscle) cells is the T-type channel. The T-type channel has a biophysical fingerprint that is distinct from the L-type channel in that it opens at more negative voltages, inactivates more rapidly, and has a lower conductance than the L-type channel (for review see Vassort,[89] Bers[90]). The distribution of the T-type current is more restricted in the heart than the L-type current. The T-type current has been recorded in SA node, AV node, atrium, and Purkinje cells, but not in the adult ventricle (see Table 1-2).[89] The T-type current plays a prominent role in phase 4 diastolic depolarization and the AP upstroke of pace-making cells, but it has not been detected in normal or diseased human ventricular myocytes.[91,92] The genes presumed to encode the T-type Ca current α1 subunits have been cloned. The $\alpha_1 3.1$ (α1G) cDNA was isolated from neuronal tissue and was the first of this new class of Ca channels cloned. The gene resides on human chromosome 17q22 (see Table 1-1) and has a predicted topology similar to that of other Ca channel cDNAs.[93] Heterologous expression of this channel subunit produced currents very much like the T-type current. Shortly after cloning α1G, the same group cloned another cDNA from human heart, $\alpha_1 3.2$ (a1H). This cDNA had a predicted amino acid sequence that was topologically consistent with other Ca channels and resided on human chromosome 16p13.3. Again, expression of $\alpha_1 3.2$ produced a current with the biophysical hallmarks of the T-type Ca current.[94] Interestingly, $\alpha_1 3.2$ lacks a consensus β subunit-binding motif in the I-II linker of the channel and does not have an EF hand or IQ motif suggesting modes of inactivation distinct from the L-type channel.

Four chemical classes of compounds have been used to block Ca currents: dihydropyridines, phenylalkylamines, benzothiazepines, and tetralols. Calcium channel blockers exhibit significant pharmacodynamic heterogeneity across classes and even within a given chemical class. Drugs of the phenylalkylamine (verapamil) and benzothiazepine (diltiazem) classes are effective antiarrhythmics primarily used in termination of some supraventricular arrhythmias, in control of the ventricular response in others and in some forms of idiopathic ventricular tachycardia (see Chapter 18). Dihydropyridines are more potent vasodilators and are not useful as antiarrhythmic compounds. There are a number of mechanisms to explain these clinical differences. The classes of drugs (dihydropyridines, phenylalkylamines, benzothiazepines) that block the L-type channel have distinct but overlapping binding sites on the $\alpha_1 1.2$ subunit.[95] Vascular smooth muscle and cardiac muscle express different splice variants of $\alpha_1 1.2$, and the vascular variant is more sensitive to block by dihydropyridines.[96] Perhaps more important than the intrinsic sensitivity of the specific $\alpha_1 1.2$ variant to a blocking compound is the voltage-dependence and kinetics of block. Like Na channel-blocking local anesthetic antiarrhythmic drugs, Ca channel antagonists exhibit use- and voltage-dependent block as a result of the preference of the drugs to bind to inactivated states of the channel. The enhanced sensitivity of vascular Ca channels to block by dihydropyridines is predominantly due to the depolarized resting membrane potential of vascular smooth muscle cells (Vm ≈−40 mV) compared to cardiac myocytes and the greater occupancy of the inactivated state. Differences among dihydropyridines, phenylalkylamines, and benzothiazepines in blocking cardiac Ca channels is significantly related to the kinetic interaction of the drug and channel. Phenylalkylamines dissociate from the Ca channel very slowly, dihydropyridines do so rapidly and benzothiazepines recover with intermediate kinetics. Mibefradil is a tetralol Ca channel blocker that is relatively selective for T-type over L-type Ca channels. It was briefly marketed for hypertension, but was withdrawn because of a high incidence of adverse effects, often occurring as a result of drug interactions. Whether T-type channels in cardiac cells represent a target of opportunity in cardiovascular therapeutics thus remains an open question.

POTASSIUM CHANNELS

Currents through potassium channels are the major repolarizing currents in the heart, but the relative importance of any specific channel varies regionally. Potassium channels are the most diverse subfamily of channel proteins, being composed of molecules with three distinct molecular architectures (Fig. 1-7). The inward rectifier currents (I_{K1}, I_{KACh}, I_{KAdo}), designated *Kir*, are encoded by a K channel that is evolutionarily the most primitive and is composed of only two membrane-spanning repeats (analogous to S5 and S6) and a pore or P segment. The latter contains the K channel signature sequence (TVGYGDM) that underlies K^+ selective permeability of the channel.[97] The first potassium channel gene isolated was from a mutant fruit fly (*Drosophilia melongaster*), which was called *Shaker* because of its response to ether anesthesia. The gene that caused the *Shaker* phenotype was isolated by positional cloning and encoded a voltage-dependent K (Kv) channel.[98] Since the original cloning of the *Shaker* potassium channel (Kv1.x), a number of potassium channel genes in the same or closely related gene families have been isolated (Kv1.x-Kv9.x). The voltage-dependent K channels that have been identified in the mammalian heart are shown in Figure 1-7. The voltage-dependent K channels, which are structurally similar to

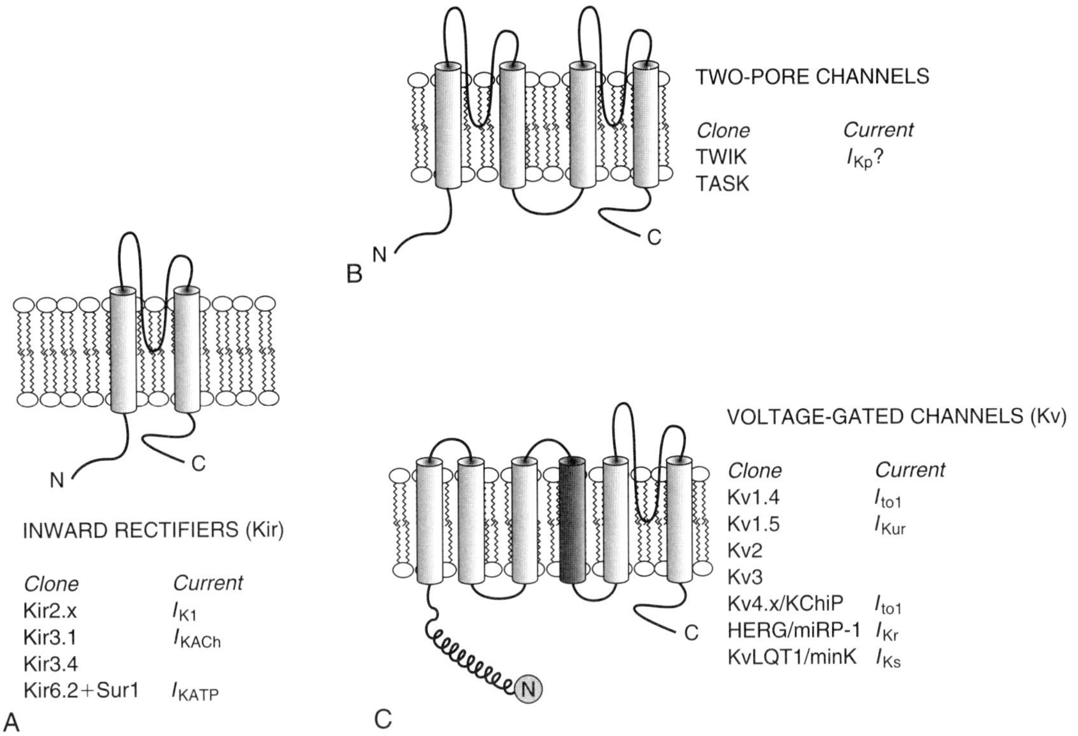

FIGURE 1-7 Families of K channels in the mammalian heart. The predicted transmembrane topology of the three subclasses consists of: **A,** inward rectifier (Kir); **B,** twin pore K; and **C,** voltage-gated (Kv) channels. Examples of specific channels in each class that are expressed in the human heart are provided.

a single domain of the Na or Ca channel, are composed of six membrane-spanning segments, including a highly basic S4 segment. The cytoplasmic half of the S6 membrane-spanning repeat appears to mediate drug block of voltage-gated K channels,[99] analogous to regions of the Na channel that bind local anesthetics.[38] Similar to Kir channels, Kv channels must tetramerize to form the intact channel and are typically associated with ancillary subunits. Within a subfamily of K channels (e.g., Kv1.x) subunits may heteromultimerize, but it is believed that assembly does not occur across subfamilies. It seems likely that two rounds of gene duplication generated Ca and Na channels from the less complex Kv structure. It is possible that a more straightforward gene duplication of an inward rectifier channel produced the third type of K channel, the two-pore K^+ selective channel (see Fig. 1-7).

The cDNAs encoding the α subunits of the K channel are sufficient to generate K^+ selective currents, but a number of ancillary subunits that modify channel function (Kvβ, minK, MiRP-1, KChIP) have been identified. A family of related proteins (Kvβ1-Kvβ3) modulate the function of Kv channels. The β subunits bind to the amino terminus of Kv α subunits to modify function in an isoform-specific fashion (for review see Snyders[100]). The structure of Kvβ2 complexed with the amino terminus of Kv1.1 has recently been solved[101] and has been suggested to function as an oxidoreductase.[102] Indeed, Kvβ1.2 has been shown to confer oxygen sensitivity to Kv4.2 channels.[103] A recently described unrelated family of proteins, KChIPs that contain Ca^{2+} binding EF hand motifs, modulate the function of members of the Kv4 family,[104] suggesting the possibility that more than one type of ancillary subunit can interact with Kv4 channels. The molecular details of the interaction of KChIP and Kv4 subunits remain under study (Fig. 1-8, *left*).

Still other K channel ancillary subunits are predicted to be transmembrane proteins that alter not only channel gating but in some cases may influence channel pore properties.[105] Most relevant to the heart are the gene products of *KCNE1* (minK)[106] and *KCNE2* (MiRP-1), which are thought to coassemble with KvLQT1[107,108] and HERG[109] to form the two components of the delayed rectifier current I_{Ks} and I_{Kr}, respectively (see Fig. 1-8, *right* and later discussion). Recently it has been recognized that α subunits, which by themselves are not functional (e.g., Kv9.x), may modulate the function of other Kv α-encoded channels.[110]

In response to a depolarizing voltage pulse, K channels, like other voltage-gated ion channels, undergo a series of conformational changes that alter function. The S4 membrane spanning repeats are critical components of the activation gating machinery. Many K channels (like Na and Ca channels) close in the face of continued depolarization (i.e., inactivate). The molecular basis of inactivation, however, is mechanistically heterogeneous. The first type of inactivation to be understood in molecular detail in K channels validated a scheme proposed by Armstrong that was referred to as the "ball-and-chain" mechanism.[111] In an elegant series of experiments, Aldrich and coworkers demonstrated the "ball" role of the amino terminus of the *Shaker* K channel in the inactivation process they called

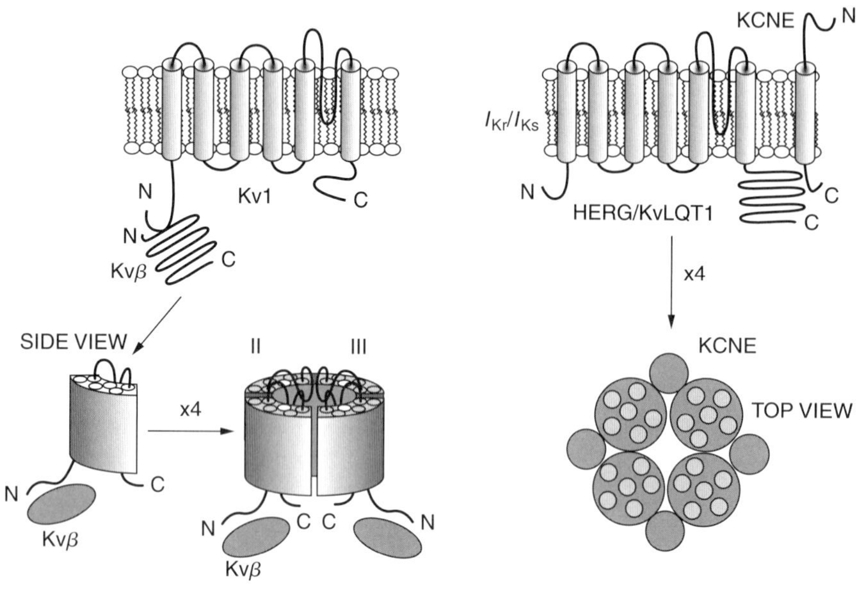

FIGURE 1-8 K channels are multisubunit complexes. At least three types of ancillary subunits are found in the heart. Kvβ associates with the amino terminal section of Kv1α subunits and the carboxyl terminus of Kv4α subunits. KChIPs are Ca^{2+}-binding ancillary subunits that associate with the amino terminus of Kv4α subunits. Kvβ and KChIP subunits increase the current density when coexpressed with Kvα subunits and modify gating. The gene products of *KCNE1* (minK) and *KCNE2* (MiRP-1) are believed to be transmembrane proteins. MinK has been shown to influence the ion conductive pathway of the I_{Ks} channel.

N-type (because it involves the amino terminus).[112,113] After channel activation by a depolarizing stimulus, the amino terminus binds to and plugs the cytoplasmic mouth of the channel pore terminating K^+ flux. Channels that have the amino terminus removed fail to undergo this type of inactivation, but it can be restored if a peptide that resembles the amino terminal ball is added to the cytoplasm of the cell.[113] By examining the sequence of the amino terminus of voltage-dependent K channels, one can easily identify channels that undergo N-type inactivation. A second form of inactivation (C-type for carboxyl-terminal) involves the outer mouth of the channel pore and amino acid residues in S6 and the P segment.[114] It has been suggested that C-type inactivation of the channel protein resembles the closing of a camera shutter (i.e., it involves constriction of the outer pore of the channel).

Potassium channels subserve multiple roles in maintenance of normal cardiac electrophysiology. Of the multiple subtypes (voltage-gated, inward rectifier, twin pore), voltage-dependent K channels underlie both the transient outward (sometimes called *A-type*) and delayed rectifier currents in the heart. The transient outward potassium current activates and inactivates rapidly and is a critical determinant of phase 1 repolarization of the ventricular AP (see Fig. 1-1). There are two components of the transient outward current in the heart, a Ca^{2+} independent K current (I_{to1}) and a Ca^{2+} dependent current (I_{to2}). The latter is a K current in some species and a chloride current in others.[115,116] The channels that encode cardiac I_{to1} vary among species and may vary regionally in the ventricle. Kv1.4 is an important but minor component of I_{to1} in some species, including humans.[117] However, in the human ventricle, I_{to1} is primarily encoded by Kv4.3,[118,119] which recovers from inactivation much faster than homomeric Kv1.4 channels. Indeed, Kv1.4 channels recover so slowly (2 to 3 seconds) that it cannot be a significant component of the cardiac I_{to1} at physiologic heart rates. However, it is possible that heteromultimerization of Kv1.4 with other Kv1 family genes[120] or coassembly with β subunits, or both, could alter the kinetics of the current. Another argument against Kv1.4 as the major component of cardiac I_{to1} is the insensitivity of expressed Kv1.4 to block by low-dose flecainide (10 μM), whereas expressed Kv4 and native cardiac I_{to1} are both flecainide sensitive.[121] The Kv4 family of genes is expressed in relative abundance in the mammalian heart, Kv4.3 in larger mammalian ventricles such as dog and human[118,119] and Kv4.2 in rodent ventricle.[118,122] Recent data suggest that Kv1.4 mRNA and protein are also present in mammalian ventricular myocytes.[123] A physiologic correlate of Kv1.4-based I_{to1} may be a slowly recovering transient outward current in the subendocardium of the human ventricle.[124,125]

The ultrarapid activating delayed rectifier current (I_{Kur}) that is primarily found in the atrium in humans (and throughout the heart in rodents) is generated by Kv1.5,[126] although other rapidly activating delayed rectifiers may be encoded by genes in the Kv3 family in the atria of some species.[127] There is a close correspondence between the biophysical and pharmacologic properties of I_{Kur} in human atrial myocytes and Kv1.5.[128] Furthermore, Kv1.5 protein and mRNA have been observed in human atrial and ventricular tissue,[129] and Kv1.5-specific antisense oligonucleotides suppress I_{Kur} in atrial myocytes.[126] The restricted expression of Kv1.5 in atrium makes it an attractive pharmacologic target for the treatment of supraventricular arrhythmias.

The delayed rectifier K current (I_K) plays a major role in terminating repolarization in cells of large mammalian hearts. I_K is a composite current made up of a rapid component (I_{Kr}) and a slow component (I_{Ks}).[130] Defining the genetics of the LQTS clarified the molecular basis of both components of the delayed rectifier. MinK (*KCNE1*) was initially considered a "minimal K channel" that encoded a current resembling I_{Ks}.[106,131] Subsequently, positional cloning identified the disease gene in chromosome 11–linked LQTS as KvLQT1,[132] but the current encoded by KvLQT1 was a functional orphan, not resembling any known cardiac K current.[107,108] However, the coexpression of KvLQT1 and minK generated a current with a much closer resemblance to native I_{Ks} than either of the subunits expressed alone.[107,108] An alternatively spliced variant of KvLQT1 is expressed in the heart and exerts a dominant negative effect on I_{Ks} in vitro; thus native I_{Ks} may be regulated in part by the extent of such alternative splicing.[133]

Long QT genetics identified *KCNQ1* (KvLQT1) as the gene underlying I_{Ks} and *KCNHZ* (HERG) as the gene underlying I_{Kr}.[134,132] I_{Kr} exhibits a number of unusual physiologic properties whose disruption (by mutations in HERG, by hypokalemia, or by drug block) disrupt normal repolarization. With depolarizations to progressively more positive potentials, activating I_{Kr} actually decreases. This "inward rectification" is a manifestation of the very rapid inactivation that HERG channels undergo once open. The extent of this fast inactivation increases at positive potentials and with lower extracellular K^+. The latter explains the decrease in I_{Kr} (causing AP and QT prolongation) observed with hypokalemia. Further, when the AP enters phase 3, channels recover from inactivation, transitioning rapidly to an open (conducting) state before closing relatively slowly.[135] Thus, as the AP begins to repolarize, I_{Kr} increases markedly, further accelerating repolarization. HERG channels are blocked by methansulfonanilide drugs such as dofetilide and sotalol. As with KvLQT1, HERG may coassemble with another protein to produce native I_{Kr}. Database mining for homologs of minK uncovered a related gene, MiRP-1 (*KCNE2*), in the same locus on chromosome 21 that encodes a topologically similar, small polypeptide with an extracellular amino terminus, a single transmembrane domain, and a cytoplasmic carboxy tail (see Fig. 1-8). When MiRP-1 is coexpressed with HERG voltage-dependent gating, single-channel conductance, regulation by K^+, and biphasic block by methansulfonanilides are all modified.[109] However, MiRP-1's role in native cardiac I_{Kr} remains uncertain. HERG exists in alternatively spliced forms,[136,137] but the role that different splice variants play in generating the native current is unknown. As with HERG and KvLQT1, mutations in minK and MiRP-1 have been linked to LQTS (see Chapter 27).

Another major class of potassium channel genes expressed in the heart encodes inwardly rectifying currents. The term "inward rectification" is used to describe the fact that these channels pass current more readily *into* than *out of* cells (Fig. 1-9). Inward rectifiers all share a similar topology with only two membrane spanning repeats and a pore loop, and they must tetramerize to form the intact channel. In 1998 a major advance in ion channel biology occurred with the determination of the structure of a bacterial inward rectifier channel from *Streptomyces lividans* called *KcsA*.[138] The structure was remarkable in that it accounted for a number of the physical principles that underlie K^+ selective permeation.[139] The crystal structure demonstrated that the linker between the two membrane-spanning domains (P segments) forms the outer mouth of the channel, and the K channel signature sequence forms the selectivity filter. High rates of ion flux are maintained despite relatively avid binding of K^+ due to the presence of 2 K^+ ions in the selectivity region that repel each other. The second membrane-spanning repeat, analogous to the S6 of Kv channels, forms much of the inner mouth of channel, where antiarrhythmic drug binding is predicted to occur (Fig. 1-10).

The inward rectifier family of cDNAs is designated *Kir*. I_{K1}, the current that is important in maintaining the resting membrane potential and facilitating terminal repolarization, is encoded by the Kir2.x subfamily. It is likely that Kir2.1[140,141] encodes I_{K1} in human ventricles, but other Kir2 isoforms have been detected in the heart.[142]

The other inward rectifiers in the heart exhibit specialized functions such as in response to neurohormones or metabolic stress. The Kir3 family of inward rectifier channels underlies the K current that is coupled with the M2 muscarinic (I_{KACh}) or A1 adenosine receptors (I_{KAdo}) in nodal cells and atria.[143] I_{KACh} (I_{KAdo}) is a heteromultimer of the products of two different genes in the Kir3 family initially referred to as GIRK (G-protein inwardly rectifying K channel, Kir3.1) and CIR (cardiac inward rectifier, Kir3.4).[144] Kir3.1 and Kir3.4 tetramerize in a 2:2 ratio[145] to form I_{KACh} channel protein, which encodes a current that is directly activated by the βγ subunits of an inhibitory G protein (Fig. 1-11, *left*).[146] I_{KACh} is the primary mediator of the negative chronotropic and dromotropic effects of parasympathetic activation in the heart.

I_{KATP}, another inward rectifier, links electrical signaling to the metabolic state of the myocyte. Changes in the activity of I_{KATP} profoundly influence the electrophysiology of the heart in ischemia and play a key role in the endogenous cellular mechanism that limits the injurious effect of myocardial ischemia known as *ischemic preconditioning*.[147] I_{KATP} is believed to be a heteromultimeric channel complex composed of a tetrameric assembly of Kir6.2 channels at its core, surrounded by four sulfonylurea receptor subunits (SUR2A). SUR2A is an ATP-binding cassette (ABC) protein[148] that imparts sensitivity to sulfonylureas and K channel openers such as pinacidil and chromakalim to the channel complex.[149]

A third structural class of K channels has been observed in the heart. These channels are composed of four transmembrane segments and two pore loops. Twin-pore acid sensitive K channel (TASK) is a member of the twin pore family of K channel genes that is highly expressed in the heart.[150] The TASK channel exhibits little intrinsic voltage or time dependence and therefore most resembles a background current. The precise role for this channel and other members of the twin pore family in cardiac myocytes is unknown.

I_F "FUNNY" OR PACEMAKER CURRENT

I_f is a current that contributes to diastolic depolarization in pacemaking cells in the heart. The current is found in many cell types, but its features are variable.

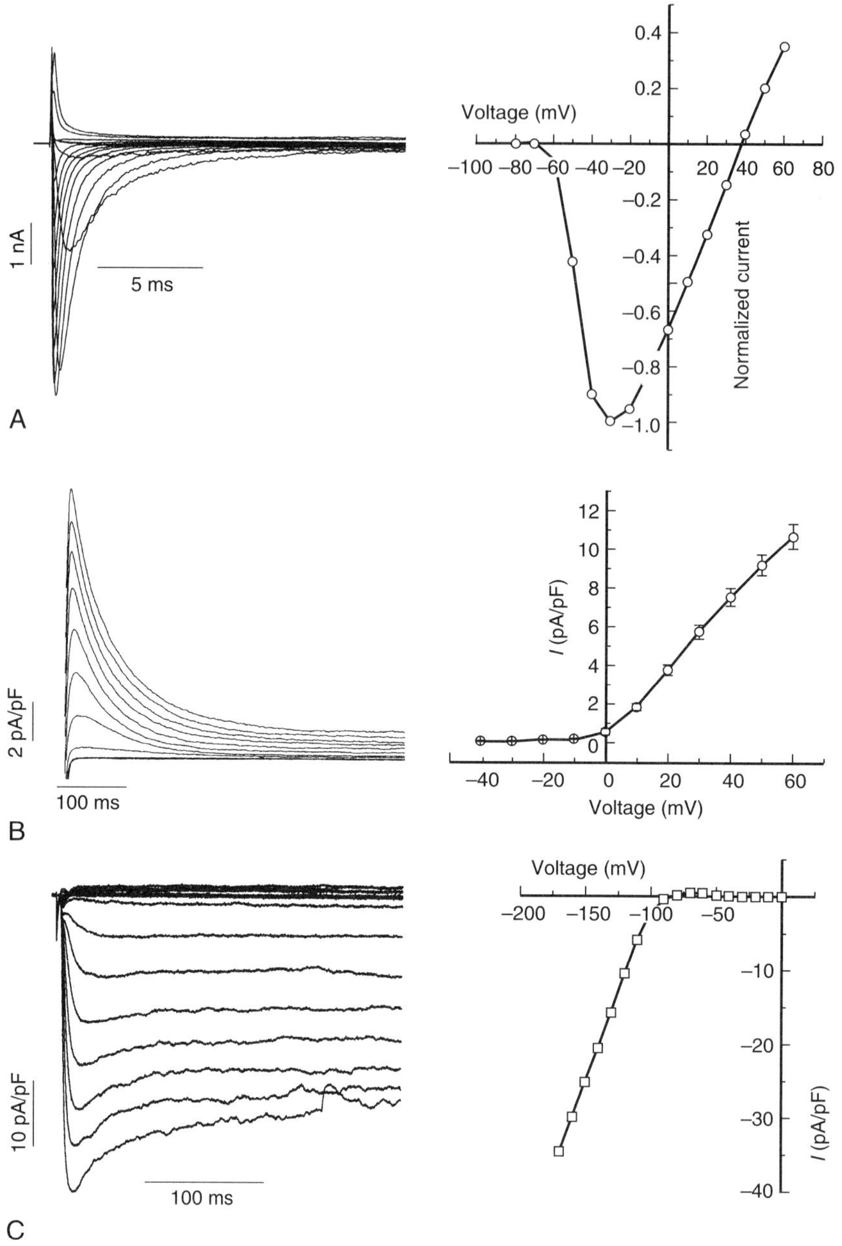

FIGURE 1-9 Properties of ionic currents in the heart. **A,** Whole-cell Na currents recorded from mammalian tissue culture cells transfected with the cDNA encoding the human cardiac Na channel (hH1) (Nav1.5). By convention the current is inward (Na^+ ions flowing into the cell) and therefore negative. The current activates rapidly upon depolarization of the cell membrane and rapidly closes in the face of a maintained depolarization of the cell membrane, a gating process referred to as inactivation. This channel passes current in both the inward and outward direction depending on the transmembrane voltage. **B,** Whole-cell recording of the transient outward K current (I_{to1}) recorded from a human ventricular myocyte. The current activates rapidly with depolarization and inactivates. The current flow is preferentially in the outward direction (positive current). This is referred to as *outward rectification.* **C,** Whole-cell current flow through the inward rectifier K current (I_{K1}). I_{K1} channels are activated at rest and close with membrane depolarization. At voltages where the channel prefers to open (voltages negative to the Nernst potential for K^+), there is little time-dependent current decay or inactivation. Current preferentially flows in the inward (negative) direction, thus this channel and all channels in the Kir family are referred to as *inward rectifiers.*

For example, I_f is present in ventricular myocytes, but its activation voltage is so negative that it is not likely to be of physiologic significance.[151] I_f activates slowly on hyperpolarization and deactivates rapidly with depolarization. I_f supports a mixed monovalent cation (Na^+ and K^+ current with a reversal potential of −20 to −30 mV. The current is highly regulated: β-Adrenergic stimulation increases I_f and hastens diastolic depolarization. A family of genes topologically similar to voltage-dependent K channels and related to cyclic nucleotide-gated channels in photoreceptors in the retina appears to encode I_f.[152,153] A number of

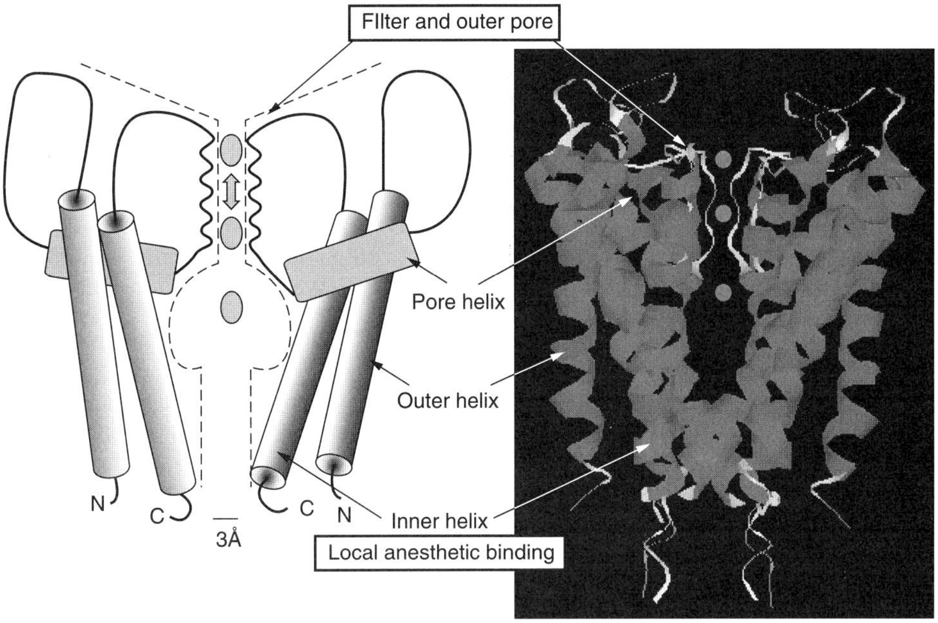

FIGURE 1-10 (See also Color Plate 1-10.) Crystal structure of the KcsA bacterial inward rectifier K channel.[138] **Left,** Cartoon representation of the major features of the structure of the bacterial channel, KcsA. Each of the four channel subunits (only two are shown) contain two α-helical membrane-spanning repeats, a pore helix, and the K channel signature sequence that forms the K^+-selective pore. **Right,** The structure of KcsA, the peptide backbone is rendered in a ribbon format. The features of the permeation pathway include the presence of two to three K^+ ions in the pore with ion–ion repulsion that facilitates high rates of ion transport and large inner vestibule composed of the carboxyl terminal portion of the outer or M2 helix. The M2 helix corresponds to the S6 domains of voltage-dependent K channels, which mediate antiarrhythmic drug binding.

hyperpolarization-activated cyclic nucleotide-gated channels (HA-CNG) have been cloned from the heart, and several exhibit the general features of I_f in cardiac pace-making cells. It has been suggested that I_f itself is a composite current with fast and slow components encoded by HA-HCN2 and HA-HCN4, respectively.[154] Support for I_f as the pacemaker current in the heart also comes from a genetic model of bradycardia in zebra fish with a dramatically reduced I_f.[155]

ELECTROGENIC TRANSPORTERS

Na^+-Ca^{2+} Exchanger

The Na^+-Ca^{2+} exchanger is an electrogenic ion transporter that exchanges three Na^+ ions for one Ca^{2+}. The highest levels of exchange activity have been observed in the heart. The cardiac NCX is a transmembrane glycoprotein originally proposed to have 11 or 12 transmembrane repeats based on hydropathy analysis.[156,157]

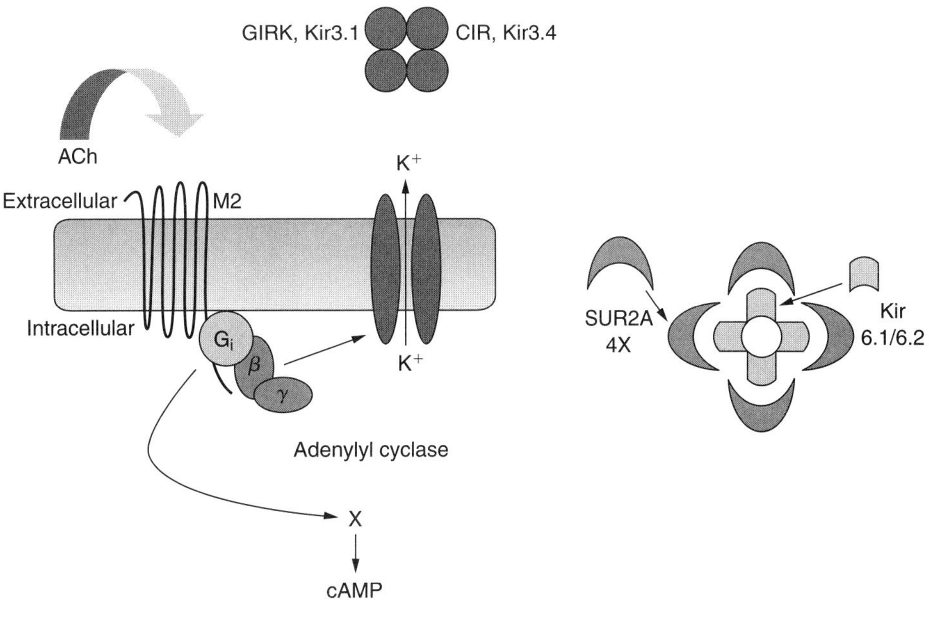

FIGURE 1-11 Subunit structure of I_{KACh} and I_{KATP}. **Left,** I_{KACh} is a G-protein coupled channel that is activated by ACh binding to M2 cholinergic receptors in the heart. The channel is separate from the receptor and is formed by the heterotetramerization of two inward rectifier K channel subunits, Kir3.1 (GIRK1) and Kir3.4 (CIR), in a 1:1 stoichiometry. **Right,** I_{KATP} is formed by the heterooctomeric assembly of the ATP-binding cassette protein, the sulfonylurea receptor (SUR2A) and the inward rectifier Kir6.2.

More recent mutagenesis data challenge the original topological models and suggest instead that there may be only nine transmembrane segments[158] (Fig. 1-12). The NCX contains two membrane-spanning domains with the first five transmembrane segments being separated from the remainder by a large cytoplasmic loop that comprises about half of the molecule. The intracellular loop contains domains that bind Ca^{2+} and the endogenous NCX inhibitory domain, XIP.[159]

Sodium/calcium exchange is an electrochemical process during which three sodium ions are exchanged for one calcium. The exchange is thus electrogenic (i.e., generates a current). Ion exchange can occur in either direction. With each heartbeat, cytosolic $[Ca^{2+}]$ is released from sarcoplasmic reticulum (SR) stores primarily by the ryanodine release channel RYR2. $[Ca^{2+}]_i$ rises from the resting level of less than 100 nM to approximately 1 μM with each cardiac cycle. Under normal physiologic conditions, outward Ca^{2+} flux through the NCX (generating an inward current) along with Ca^{2+} reuptake into the SR by the SR Ca^{2+} ATPase (SERCA) are the major mechanisms of restoration of normal diastolic $[Ca^{2+}]$. NCX is sensitive to the cytoplasmic concentrations of Ca^{2+} and Na^+, which determine the exchanger activity and the potential at which exchange reverses direction. NCX current is time independent and largely reflects changes in intracellular $[Ca^{2+}]$ during the AP. Thus, NCX has an important effect on membrane voltage both at rest and during activation of the myocyte. At very depolarized potentials, reverse mode Na^+Ca^{2+} exchange (Ca^{2+} influx, net outward current) can occur; however, the role of reverse mode exchange in initiating SR Ca^{2+} release and contraction is uncertain.

Increases in intracellular $[Ca^{2+}]$ shift the reversal potential of the NCX in the positive direction and therefore increase the driving force for inward exchanger current. Inward NCX current can depolarize the membrane toward the threshold for firing an AP and thus is potentially arrhythmogenic. NCX current is one important component of the inward current (transient inward current, I_{TI}) that underlies delayed afterdepolarizations (DADs). DADs are spontaneous membrane depolarizations from rest after complete repolarization of the AP.[160] DADs are usually not present under physiologic conditions but are favored by conditions that increase SR Ca^{2+} load such as rapid firing rates,[160] digitalis intoxication,[161] or ischemia/reperfusion.[162] Under these conditions spontaneous SR Ca^{2+} release occurs, which then increases NCX and probably other Ca^{2+}-dependent currents, resulting in membrane depolarization. DADs may produce arrhythmias in two ways. First, if DADs are of sufficient amplitude, they may trigger an AP. Second, even if DADs are subthreshold, they may affect the excitability of the cell, slowing conduction in the myocardium.[163]

Na^+-K^+ ATPase

The Na^+-K^+ ATPase or Na pump is responsible for establishing and maintaining the major ionic gradients across the cell membrane. The Na pump belongs to the widely distributed class of P-type ATPases that are responsible for transporting a number of cations. The P-type

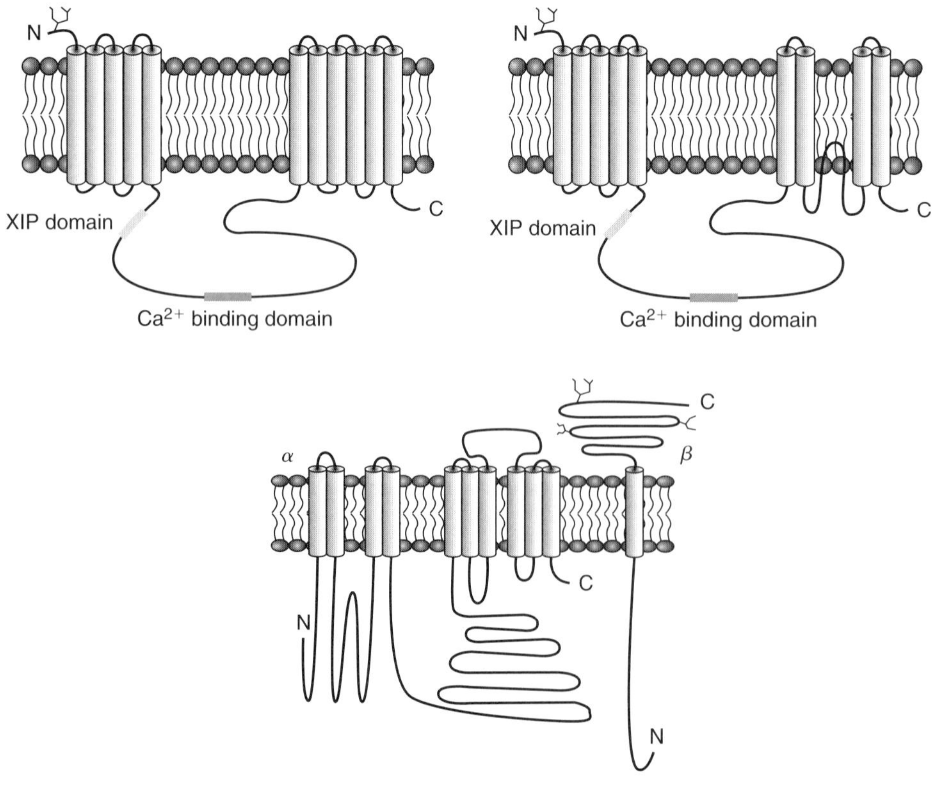

FIGURE 1-12 Subunit structure and transmembrane topology of the Na^+-Ca^{2+} exchanger (NCX) and the Na^+-K^+ ATPase (Na pump). **Top,** Two alternative transmembrane topologies for the NCX. A large cytoplasmic loop is crucial to physiologic regulation of the exchanger and contains Ca^{2+} and inhibitory peptide (XIP) binding domains. **Bottom,** The Na^+-K^+ ATPase is a heteromeric assembly of a large α and smaller single membrane-spanning repeat β subunit (see text for details).

designation of this family of enzymes refers to the formation of a phosphorylated aspartyl intermediate during the catalytic cycle. The Na^+-K^+ ATPase hydrolyzes a molecule of ATP to transport two K^+s into the cell and three Na^+s out and is thereby electrogenic, generating a time-independent outward current. The Na^+-K^+ ATPase is oligomeric, consisting of α, β, and possibly γ subunits. There are four different α and three distinct β isoforms (for review see Blanco[164]). Evidence that the γ subunit is part of the complex comes from photoaffinity labeling with ouabain derivatives[165] and immunoprecipitation studies.[166] The γ subunit belongs to a family of small membrane-spanning proteins including phospholemman[167] that support ionic fluxes.

Na^+-K^+ ATPase isoforms exhibit tissue-specific distributions. The α1β1 isoform is broadly distributed—α2-containing isoforms are preferentially expressed in the heart, skeletal muscle, adipocytes, and brain; α3 is predominantly a brain isoform; and α4 is found in abundance in the testis.[164] The structural diversity of the Na^+-K^+ ATPase comes from variations in α and β genes, splice variants of the α subunits and promiscuity of subunit associations, themes that also underlie the diversity of ion channels, particularly K channels. The α subunit is catalytic and binds digitalis glycosides in the extracellular linker between the first and second membrane-spanning region (see Fig. 1-12, *bottom*). The human heart contains α1, α2, and α3 subunits.[168] In rats, α3 subunits bind glycosides with three orders of magnitude greater affinity than α1-containing pumps. However, in humans the binding affinities of the α subunits are far less variable. The β subunits are essential for normal pump function and influence K^+ and Na^+ affinity of the α subunits as well as serving as chaperones ensuring the proper trafficking of the α subunit to the sarcolemma.[164] Only β2 appears to be present in significant quantities in human heart.[168]

In heart failure, the density of the Na^+-K^+ ATPase decreases as assessed by ^{3}[H]-ouabain binding. The decrease occurs without a significant impact on the inotropic effect of digitalis glycosides in human ventricular myocardium.[169] However, the reduction in the density of the Na pump may influence the electrophysiology of cardiac myocytes and their response to an extracellular K^+ load, as might occur in ischemia.

INTERCELLULAR ION CHANNELS—CONNEXINS AND CONNEXONS

Gap junctions are specialized membrane structures composed of multiple intercellular ion channels that facilitate electrical and chemical communication among cells. Mammalian gap junction channels are built by the oligomerization of a family of closely related genes encoding connexins. Connexins are transmembrane proteins consisting of four highly conserved membrane-spanning α-helices, two extracellular loops, and one intracellular loop. The intracellular amino and carboxy termini are less well conserved among connexins (Fig. 1-13).[170] Three different connexins are prominently expressed in the mammalian heart: connexin-40 (Cx-40), connexin-43 (Cx-43), and connexin-45 (Cx-45), named for their molecular masses (see Table 1-1). Connexin-37 (Cx-37) is found in vascular smooth muscle and atrial and ventricular endocardium.[171] The connexins are hexagonally arranged around a central aqueous pore to form a hemi-channel in the cell membrane. Two hemi-channels from neighboring cells are docked head-to-head and span the intercellular gap to form the gap junction channel or *connexon.* The interaction between hemi-channels is mediated by the relatively well conserved extracellular loops, so it is not surprising that hemi-channels composed of different connexins can form gap junction channels (see Fig. 1-13).

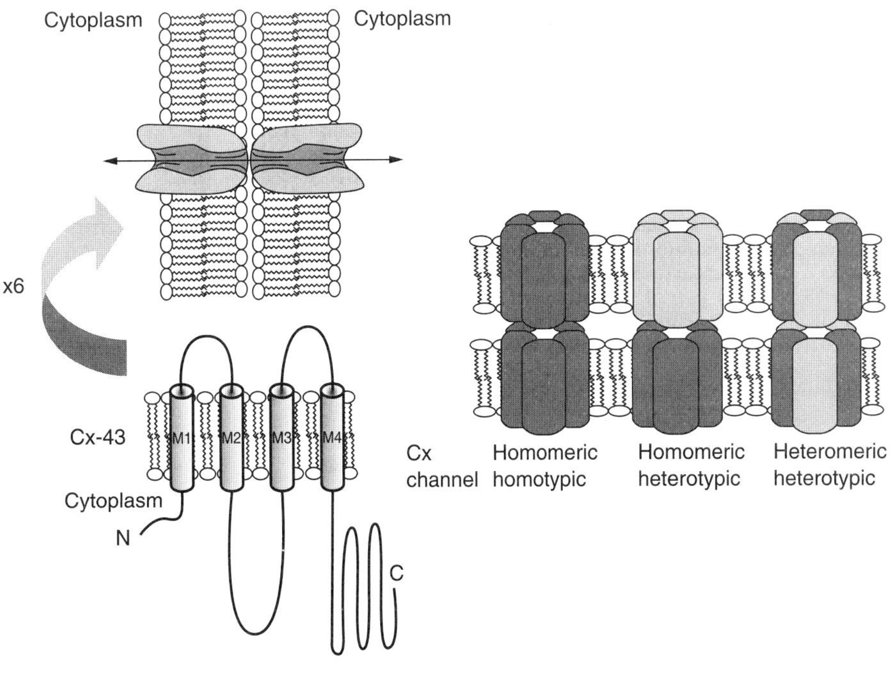

FIGURE 1-13 Subunit structure of the gap junction channel. **Left,** Gap junction channels are intercellular ion channels composed of two hemi-channels or connexons in adjacent cells. Each hemi-channel is composed of six subunits or connexins composed of four highly conserved membrane-spanning repeats, two conserved extracellular loops and more divergent cytoplasmic loop, and amino and carboxyl termini. **Right,** Different subtypes of connexins may assemble to form channels that are homomeric (single type of connexin in each connexon), heterotypic (different connexins in each connexon), or heteromeric heterotypic where more than one type of connexin is present in each connexon.

The structure of Cx-43 has recently been solved at 7.5 angstrom resolution.[172]

The diversity of gap junction channels may be amplified by the existence of channels composed of different connexins. Connexons composed of a single connexin are termed *homomeric*, those composed of a different connexin in each hemi-channel are *heterotypic*, and those with more than one connexin isoform in the hemi-channel are *heteromeric* (see Fig. 1-13). Different connexins colocalize in gap junctions, but it is not known whether the channels are heterotypic or heteromeric. Homomeric channels probably exist in vivo since some cells express only one type of connexin and homomeric recombinant connexons recapitulate the function of native gap junction channels. However, many cells express a number of different connexins and heteromultimerization has been demonstrated by immunofluorescence[173] and Western blotting[174,175] in native tissues and in heterologous expression systems.[176] The ability to form heterotypic channels depends on the extracellular loops and to a lesser extent may be influenced by the C-terminus and the cytoplasmic loop.[177] Two of the major connexins in the heart, Cx-43 and Cx-40, apparently do not form heterotypic gap junctions.

Gap junction channels are permeable to ions and small molecules with molecular weights up to approximately 1 kDa. The permeability of a given molecule depends on both size and charge but includes second messengers in the heart such as Ca^{2+}, cyclic AMP, and inositol triphosphate. Different cardiac tissues are connected by gap junctions that differ in both spatial distribution and constituent connexins.[178-180] The connexin composition of a gap junction channel determines its ion selectivity, conductance, voltage sensitivity, and regulation. Ventricular muscle expresses predominantly Cx-43 and Cx-45, whereas atrial muscle and Purkinje fibers express all three cardiac connexins. Cx-40 gap junction channels exhibit the largest conductance and Cx-45 the smallest. Both Cx-40 and Cx-45 are highly cation selective and their conductance is voltage dependent. Cx-43 has an intermediate conductance is nonselective and freely passes Lucifer yellow.[181]

The distribution of gap junctions varies in different regions of the heart.[182,183] For example, in the mammalian ventricle, myocytes are connected to an average of 11 neighboring cells through gap junctions equally distributed between junctions connecting cells end to end and side to side. A propagating wavefront encounters a greater number of gap junctions per unit length in the transverse compared to the longitudinal direction, producing slower transverse than longitudinal conduction in normal myocardium (i.e., anisotropic conduction). Ventricular muscle exhibits rapid conduction velocities of up to approximately 0.75 m/sec in the longitudinal direction and slower conduction velocities (0.24 m/sec) in the transverse direction for an anisotropy ratio of 3 to 4:1.[184] An activation wavefront in the ventricle will therefore encounter a larger number of gap junctions and therefore larger resistance traveling along the transverse axis of the cell compared to the longitudinal axis. The shape and size of cardiac myocytes is a major determinant of anisotropic conductance (see later discussion).

The crista terminalis is a right atrial structure. It generates functional block that is important in the maintenance of typical atrial flutter (see Chapter 15). The crista's role in atrial flutter likely reflects its especially high conduction velocity (≈1 m/sec) and anisotropy ratio (10:1).[185] The basis for both the high conduction velocity and anisotropy ratio are cell shape, gap junction type, and distribution. Cells of the crista terminalis are cylindrically shaped and connected by fewer gap junctions per unit of cell surface area, most of which are end to end in orientation.[182] The larger longitudinal conduction velocity in the crista and Purkinje fibers may also reflect the expression of Cx-40, which has a larger conductance than Cx-43.

MOLECULAR BASIS OF ACTIVATION AND RECOVERY OF THE HEART

In normal sinus rhythm, cardiac activation begins in the SA node, the specialized collection of pacemaking cells in the roof of the right atrium between the crista terminalis and the right atrial-superior vena cava (RA-SVC) junction. SA nodal cells undergo spontaneous depolarization, repetitively activating the rest of the heart. The resting membrane potentials of SA and AV nodal cells are considerably less negative than those of atrial or ventricular muscle cells, as a result of a lower density of inwardly rectifying K current (I_{K1}) and the presence of a hyperpolarization-activated pacemaker current, I_f. The result is a continuous, slow depolarization of the membrane potential; thus, nodal cells do not have a true resting potential, but the maximum diastolic potential is never more negative than −60 mV.

The aggregate activity of I_f and diminished I_{K1} slowly depolarizes the nodal cell until approximately −40 mV, when Ca currents are activated, hastening the rate of rise of the AP. First the transient T-type Ca current ($I_{Ca,T}$) is activated, driving the membrane potential toward E_{Ca}, followed by activation of the longer lasting, dihydropyridine-sensitive L-type Ca current ($I_{Ca,L}$). Simultaneously more slowly activating outward potassium currents (delayed rectifier, I_K) are activated, hindering the movement of the membrane potential toward E_{Ca}. Ultimately the Ca currents inactivate and the membrane potential moves back toward E_K, turning off I_K and activating I_f, starting the cycle again. Current flows through the electrogenic Na^+-Ca^{2+} exchanger throughout the cycle. The magnitude and direction of this current depends on the membrane potential and intracellular Ca^{2+} and Na^+ concentrations.

The synchronization of the somewhat diffuse pacemaking cells that comprise the sinus node is through gap junction channels composed of Cx-40 and Cx-43.[186] The activity of pace-making cells is synchronized by a process of mutual entrainment whereby each cell in the nodal syncytium constantly modulates the discharge frequency of the other cells.

As in the case of the nodal cell AP, the highly orchestrated activity of a number of ionic currents inscribe the muscle cell AP. A prototypic AP from a ventricular

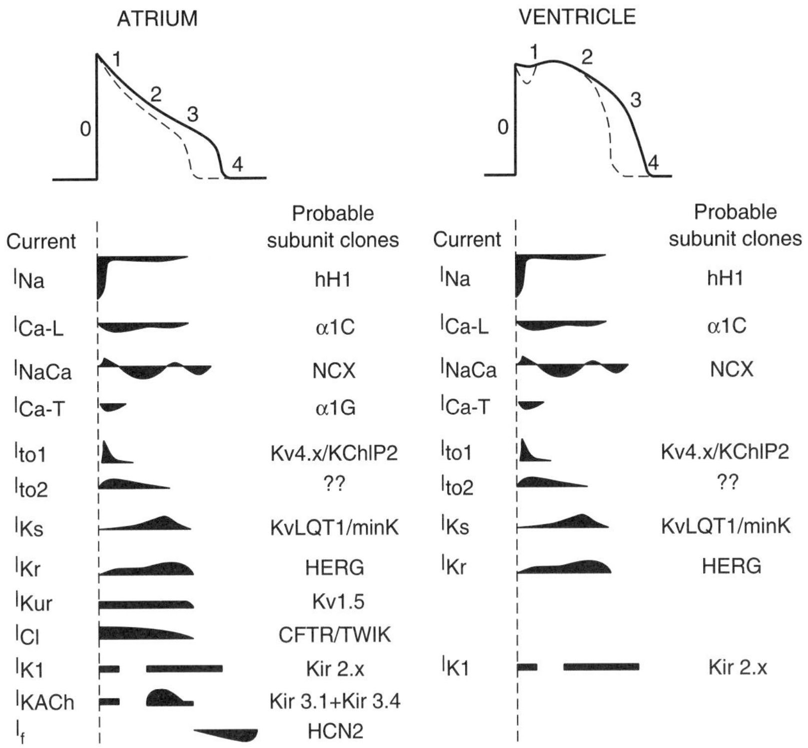

FIGURE 1-14 Action potential and membrane currents active during a ventricular action potential. (Courtesy of Dr. Fadi Akar).

myocyte with a schematic of the trajectory of the underlying ionic currents is shown in Figure 1-14. The AP is divided into five phases: Phase 0 is the rapid upstroke; phase 1 is early repolarization; phase 2 is the plateau; phase 3 is late repolarization; and phase 4 is the resting potential or, in the case of a nodal AP, the diastolic depolarization. Unlike the nodal cells, a true resting potential can be defined in cardiac muscle cells and it is approximately 90 mV, close to E_K; thus, at rest cardiac muscle cells are mostly permeable to K^+ due to the activity of I_{K1}.

Under normal conditions muscle cells are stimulated by spontaneously occurring impulses generated in pacemaking tissue. When this stimulus moves the membrane voltage positive to threshold (≈−65 mV), an AP is initiated. Depolarization beyond threshold explosively activates Na channels, producing an enormous (≈400 pA/pF) but transient (1-2 m/sec) current driving the membrane voltage toward E_{Na} (+65-+70 mV). Although Na channels are by far the most numerous in the myocyte cell membrane, their activity is fortunately short-lived or the transmembrane Na^+ gradient would quickly be erased. The Na current quickly dissipates by inactivation, and the membrane must repolarize to its resting potential before Na channels recover from inactivation and again become available to activate. Thus, the time and voltage dependence of availability of the Na current is the basis of refractoriness in cardiac muscle.

The upstroke of an AP falls short of E_{Na} because of inactivation of Na current and activation of a K current and possibly a Ca^{2+}-dependent Cl^- current (I_{to2}), which produce rapid membrane repolarization to approximately +10 mV (phase 1). The Ca^{2+}-independent transient outward K current (I_{to1}) activates as Na channels inactivate. Activation of I_{to1} is rapid (≈10 ms), and decay of this current occurs over 30 to 40 ms at physiologic temperatures. The density of I_{to1} is less than 5% of the Na current, thus inactivation of Na current is the main reason for early repolarization while I_{to1} is an important determinant of the membrane voltage at the end of phase 1. In canine ventricular myocytes, I_{to2} is a prominent current during phase 1 of the AP[187]; however, its role in human myocytes is uncertain.

Depolarization of the membrane potential activates a number of other currents, albeit more slowly than the Na current and I_{to1}. In ventricular myocytes, $I_{Ca,L}$ is activated and accounts for the major depolarizing current during the AP plateau or phase 2. This current is the main route for Ca^{2+} influx and triggers Ca^{2+}-induced Ca^{2+} release (CICR) from the SR to initiate contraction. $I_{Ca,L}$ tends to depolarize the cell membrane, and delayed rectifier repolarizing K currents active during the plateau and phase 3 oppose this action. Activation of delayed rectifier K currents and inactivation of Ca currents serve to terminate the plateau phase and begin phase 3 or late repolarization. In atrial tissue I_{Kur} is a prominent delayed rectifier that is an important determinant of the plateau height and aids in the termination of the plateau.[188] Delayed rectifiers (especially I_{Kr}) are important in terminating the plateau but are limited in their ability to restore the normal resting potential because they deactivate at voltages less than −40 mV.[135] Final repolarization is mediated by the outward component of I_{K1} even in atrial cells where the density of I_{K1} is small compared to that of ventricular myocytes.

Cellular and Molecular Basis of Cardiac Electrophysiology

EXCITABILITY AND PROPAGATION

Many electrophysiological properties of the heart are direct consequences of ionic current activity during the AP. Cardiac cells are excitable because a stereotypic, regenerative response, the AP, is elicited if the membrane potential exceeds a critical threshold. APs are regenerative because they can be conducted over large distances without attenuation. APs generated in the sinus node serve to excite adjacent atrial muscle and thus the remainder of the heart under normal conditions.

In atrial and ventricular muscle at rest the membrane is most permeable to K^+, the result of activity of I_{K1}. Excitability in cardiac muscle is primarily determined by the availability of the Na current. In response to an external stimulus either from adjacent cells or an artificial pacemaker, depolarization of muscle cells occurs. If the depolarization is sufficient and raises the membrane potential above a critical value known as the *threshold potential*, Na channels open, depolarize the membrane, and initiate an AP. In pacemaking tissues such as the sinus node (SA) or AV node, Na current is absent, and excitability is mediated by activation of Ca currents. The consequence is a higher threshold for activation and slower rate of rise (≈1 to 10 V/sec versus hundreds of V/sec in muscle) of phase 0 of nodal cell APs.

Propagation of a wave of excitation in a homogeneous cable-like medium is continuous and obeys the laws of cable theory (see "Passive Membrane Properties and Cable Theory" earlier). In such a preparation the maximal upstroke velocity of the AP $(dV/dt)_{max}$ is an indirect measure of depolarizing ionic current and conduction velocity.[189] A continuous cable model is a structural oversimplification of all cardiac tissue with the possible exception of normal papillary muscles.[190] Continuous propagation of excitation waves is not characteristic of cardiac tissue. Due to the structural and functional complexities of the myocardium, discontinuous conduction (see following discussion) is the rule.

There is feedback between network properties (cell-to-cell coupling via gap junctions) and active membrane properties (ionic currents) in propagation in cardiac tissue preparations.[3] Under conditions of normal cellular coupling, fluctuations in local conduction velocity, AP shape, and ionic current flow are small. However, with cellular uncoupling such as accompanies ischemia, the interaction between intercellular conduction and active membrane properties assumes greater significance. In cardiac muscle the Na current is the main determinant of membrane depolarization and local circuit current. When cells are uncoupled, discontinuity of conduction increases, the delay between activation of cells increases, and the Na current may sufficiently inactivate such that the currents active during the plateau (i.e., L-type Ca current) of the AP become essential for driving excitatory current through gap junctions.[191] In both experimental models and computer simulations, blocking the L-type Ca current reduced the safety factor for conduction and lowered the intercellular resistance that produced conduction block.[191,192] Cellular uncoupling and discontinuous conduction have important implications for safety factors for propagation of the impulse. With moderate cell-to-cell uncoupling in simple models of propagation, conduction is slower but with a higher safety factor. However, with more significant uncoupling, the transmitted current is so small that insufficient Na current is recruited to initiate an AP.

The most important causes of discontinuous conduction in the heart are macroscopic discontinuities in cardiac tissue. Such anatomic discontinuities exist in all regions of the heart and are especially prominent in trabeculated portions of the atria and ventricles,[193] the layers of the left ventricular wall,[2] and Purkinje-muscle junction.[194] Two-dimensional models of macroscopic discontinuities highlight the importance of the change in geometry and consequently the dispersion of the local circuit current in the characteristics of propagation and block at such sites.[195] Analogous to the feedback between cellular coupling and ionic currents, there is feedback between the current to load mismatch produced by the tissue architecture and ionic current flow. Small current to load mismatches (larger strand to sheet ratio) are associated with minor conduction delays across the tissue discontinuity. In contrast, tissue architecture characterized by a large current-to-load mismatch (narrow strand into a large sheet) is associated with significant conduction delay, and block across the discontinuity can be produced by either Na or Ca channel blockers.[196] Thus, the L-type Ca current is essential for impulse propagation through cardiac tissue with structural discontinuities. Such structural discontinuities are present in the normal heart but may be much more prominent in the aged or diseased (e.g., hypertrophied or infarcted) myocardium.

REPOLARIZATION AND REFRACTORY PERIODS

Refractoriness of tissue is a consequence of the long duration of the cardiac AP allowing only gradual recovery of excitability. Refractoriness is essential to the normal mechanical function of the heart permitting relaxation of cardiac muscle before subsequent activation. Refractoriness of cardiac muscle is classified as either absolute or relative: The former occurs immediately after phase 0 and during the plateau. No stimulus, regardless of its strength, can re-excite the cell. The latter occurs during phase 3 when the cell is excitable but the stimulus strength for activation exceeds that at rest (Fig. 1-15). The molecular basis of refractoriness is the lack of availability of depolarizing current (Na current in muscle) because repolarization to negative potentials is required for channels to recover from fast inactivation and thus be available to pass excitatory current. The duration of refractoriness of any cardiac tissue thus depends upon the complement of ion channels (and in particular depolarizing currents) expressed. When the depolarizing current becomes available to

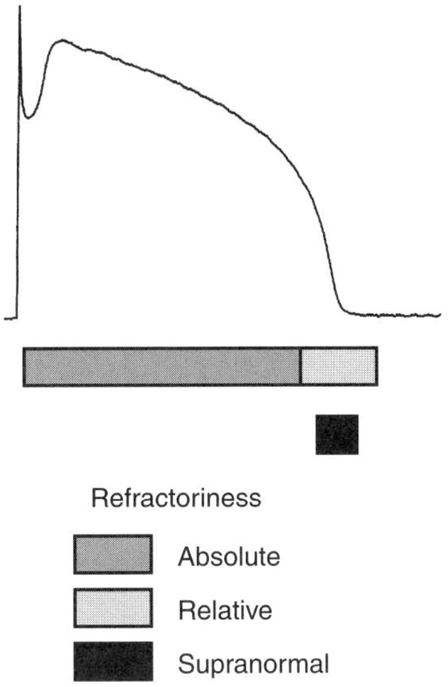

FIGURE 1-15 Absolute and relative refractory periods in the ventricle. Action potential recorded from a ventricular myocyte. The bars beneath the action potential delineate the periods of absolute refractoriness where no stimulus, regardless of amplitude, can elicit another action potential, and relative refractoriness where a subsequent action potential can be initiated with a high strength stimulus. Under the appropriate circumstances during the period of relative refractoriness, the cell may exhibit supranormal excitability (i.e., a stimulus that is normally subthreshold will elicit an action potential).

activate, outward currents (typically delayed rectifier K currents) increase the stimulus strength required to reach threshold, making the tissue relatively refractory (compared to the rested state).

Under some conditions, some tissues, particularly Purkinje fibers, may exhibit supranormal excitability.[197] This phenomenon occurs at the end of repolarization and is the result of reactivation of Na currents when the membrane potential of the heart cell is closer to the threshold for reactivation than when the cell has fully returned to rest. Supranormal excitability is one contributor to the vulnerable period of the cardiac cycle, by increasing the likelihood of re-excitation during terminal repolarization (when heterogeneity of AP durations is most likely to support reentry).

Cellular and Molecular Mechanisms Contributing to Cardiac Arrhythmias

Cardiac arrhythmias result from abnormalities of impulse generation, conduction, or both. It is difficult, however, to establish an underlying mechanism for many clinical arrhythmias. Criteria such as initiation and termination with pacing and entrainment are used in the clinical electrophysiology laboratory to make the diagnosis of reentry in some cases. There are even fewer specific tools available to diagnose nonreentrant arrhythmias. It is clear that molecular changes in the heart predispose to the development of abnormalities of cardiac rhythm. However, an exclusively molecular approach to understanding arrhythmia mechanisms is limited by failure to include cellular and network properties of the heart. We attempt to place in context the role of cellular and molecular changes in the development of clinically significant rhythm disturbances. A summary of the cellular and molecular changes that underlie prototypic arrhythmias and their putative mechanisms is shown in Table 1-3.

ALTERATIONS IN IMPULSE INITIATION: AUTOMATICITY

Spontaneous (phase 4) diastolic depolarization underlies the property of automaticity characteristic of cells in the SA and AV nodes, His-Purkinje system, coronary sinus, and possibly the pulmonary veins. Phase 4 depolarization results from the concerted action of a number of ionic currents, but the relative importance of these currents remains controversial (Fig. 1-16).[151] The inwardly rectifying K current (I_{K1}) maintains the resting membrane potential and resists depolarization, thus the activity of other currents (e.g., Ca currents) or a reduction of I_{K1} (and other K conductances) must occur to permit the cell to reach threshold for firing of an AP. I_f may play a particularly prominent role in the normal automaticity of Purkinje fibers,[198] although this hypothesis is not without controversy.[199] Deactivation of I_K is another mechanism allowing depolarizing currents to move the membrane potential toward threshold. Calcium currents, both the T-type and the L-type, figure prominently in diastolic depolarization and in the upstroke of the AP in nodal tissue and latent atrial pacemakers. Numerous other time-independent currents that may play a role in diastolic depolarization and pace-making activity including currents through the electrogenic Na-K ATPase and the Na-Ca exchanger and background currents.

The rate of phase 4 depolarization and therefore firing rate of pacemaker cells is dynamically regulated. Prominent among the factors that modulate phase 4 is autonomic nervous system tone. The negative chronotropic effect of activation of the parasympathetic nervous systems results from release of acetylcholine that binds to muscarinic receptors, releasing G-protein $\beta\gamma$ subunits that activate a potassium current (I_{KACh}) in nodal and atrial cells (see Fig. 1-11).[200] The resultant increase K^+ conductance opposes membrane depolarization, slowing the rate of rise of phase 4 of the AP. Agonist activation of muscarinic receptors also antagonizes sympathetic nervous system activation through inhibition of adenylyl cyclase, reducing cAMP and inhibiting protein kinase A. Conversely, augmentation of sympathetic nervous system tone increases myocardial catecholamine concentrations, which activate both α and β receptors. The effect of β1-adrenergic stimulation predominates in pace-making cells, increasing the L-type Ca current and shifting the voltage dependence of I_f to more positive potentials, thus augmenting the slope of phase 4 and increasing the rate of SA node firing. L-type Ca current density is increased by PKA-mediated phosphorylation,[201] resulting in an increase

TABLE 1-3 Arrhythmia Mechanisms

	Molecular Components	Mechanism	Prototypic Arrhythmias
Impulse Initiation			
Automaticity	I_f, $I_{Ca\text{-}L}$, $I_{Ca\text{-}T}$, I_K, I_{K1}	Suppression/Acceleration	Sinus bradycardia Sinus tachycardia
Triggered Automaticity	I_{TI}	DADs	Digitalis toxicity Reperfusion VT, IVR Idiopathic VT
	$I_{Ca\text{-}L}$, I_K, I_{Na} I_{Na}, $I_{K\text{-}ATP}$	EADs	Torsades de pointes Ischemic VF
Excitation			
	$I_{Ca\text{-}L}$		AV conduction block
	I_{Na}, $I_{Ca\text{-}L}$, K channels	AP prolongation	Polymorphic VT
Repolarization	Ca homeostasis	EADs	(Hypertrophy, HF)
	I_{Na}, $I_{Ca\text{-}L}$, K channels Ca homeostasis	AP shortening	Atrial fibrillation
Multicellular			
	Connexins I_{Na}, $I_{K\text{-}ATP}$	Uncoupling Conduction delay or block, functional re-entry	Ischemic VT/VF
Cellular coupling			
	Extracellular matrix collagen	Re-entry with an excitable gap	Monomorphic VT

AP, action potential; DADs, delayed afterpolarizations; EADs, early afterdepolarizations; HF, heart failure; IVR, idioventricular rhythm; VF, ventricular fibrillation; VT, ventricular tachycardia.

in the rate of rise of phase 4 and the upstroke velocity of the AP in nodal cells. Enhanced sympathetic nervous system activity can dramatically increase the rate of firing of SA nodal cells producing sinus tachycardia with rates in excess of 200 beats per minute (bpm). On the other hand, the increased rate of firing of Purkinje cells is more limited, rarely producing ventricular tachyarrhythmia in excess of 120 bpm.

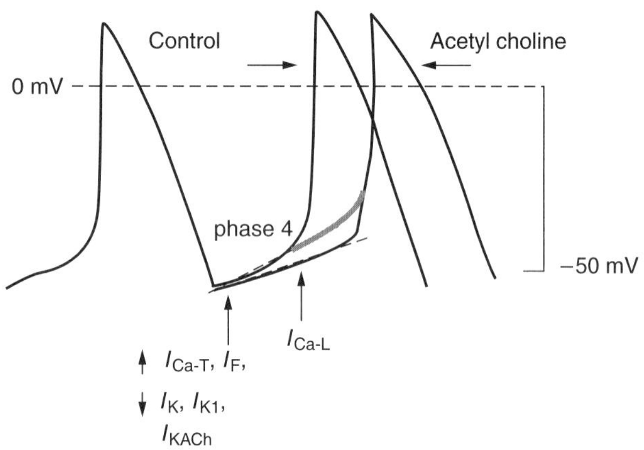

FIGURE 1-16 Nodal action potential and the currents that underlie phase 4 diastolic depolarization. Nodal cells exhibit phase 4 diastolic depolarization that spontaneously brings the cell to threshold, resulting in the production of an action potential. Several currents that play a role in phase 4 include Ca currents (T- and L-type), I_f or the pacemaker current, and a reduction in current flow through several K channels including I_K, I_{K1}, and I_{KACh}. The rate of phase 4 diastolic depolarization is highly sensitive to autonomic nervous system tone. Cholinergic agonists slow phase 4 and sympathomimetics hasten phase 4.

Normal automaticity may be affected by a number of other factors associated with heart disease. Hypokalemia and ischemia may reduce the activity of the Na-K ATPase, thereby reducing the background repolarizing current and enhancing phase 4 diastolic depolarization. The end result would be an increase in the firing rate of pacemaking cells. Slightly increased extracellular potassium may render the maximum diastolic potential more positive, thus also increasing the firing rate of pacemaking cells. A greater increase in $[K^+]_o$ however, renders the heart inexcitable by depolarizing the membrane potential and inactivating the Na current.

Sympathetic stimulation explains the normal response of the sinus node to stress such as exercise, fever, and thyroid hormone excess. Normal or enhanced automaticity of subsidiary latent pacemakers produces escape rhythms in the setting of failure of more dominant pacemakers. Suppression of a pacemaker cell by a faster rhythm leads to an increased intracellular Na^+ load (particularly in cells with an Na^+-dependent AP), and extrusion of Na^+ from the cell by the Na-K ATPase produces an increased background repolarizing current that slows phase 4 diastolic depolarization.[202,203] At slower rates the Na^+ load is decreased, as is the activity of the Na-K ATPase, resulting in a progressively rapid diastolic depolarization and warm-up. Overdrive suppression and warm-up may not be observed in all automatic tachycardias. For example, functional isolation of the pacemaker tissue from the rest of the heart (entrance block) may blunt or eliminate the phenomena of overdrive suppression and warm-up of automatic tissue.

Myocytes in the atrium and ventricle may exhibit spontaneous activity under pathologic conditions associated with depolarization of the resting membrane potential to levels more positive than −60 mV.[204] The mechanism of spontaneous depolarization in contractile cells is uncertain but is likely to involve the activity of numerous depolarizing and repolarizing currents that on balance favor membrane depolarization. Ventricular myocytes do express I_f, although the threshold for activation is well below the resting potential of the cell, and so the functional significance of this current is uncertain. Currents that mediate the upstroke of the AP of abnormally automatic cells depend on the diastolic potential. At more negative diastolic potentials, abnormal automaticity can be suppressed by Na channel blocking drugs. At more positive diastolic potentials (>−50 mV), Na channel blockers are ineffective while calcium channel blockers suppress abnormal automaticity, implicating the L-type Ca channel in the upstroke in this setting.[205]

Abnormally automatic cells and tissues are less sensitive to overdrive suppression than cells and tissues that are fully polarized with enhanced normal automaticity.[206] However, in situations where cells may be sufficiently depolarized to inactivate Na current and limit intracellular Na^+ load, overdrive suppression may still be observed due to increased intracellular Ca^{2+} loading. Such Ca^{2+} loading may activate Ca^{2+}-dependent K conductances (favoring repolarization) and promote Ca^{2+} extrusion through the Na-Ca exchanger, and Ca channel phosphorylation, increasing Na^+ load and thus Na-K ATPase activity. The increase in intracellular Ca^{2+} load may also reduce depolarizing L-type I_{Ca} by promoting Ca^{2+}-induced inactivation of the Ca current.[207]

Abnormal automaticity may underlie atrial tachycardia, accelerated idioventricular rhythms, and ventricular tachycardia, particularly associated with ischemia and reperfusion. It has also been suggested that injury currents at the borders of ischemic zones may depolarize adjacent nonischemic tissue predisposing to automatic ventricular tachycardia.[205]

AFTERDEPOLARIZATIONS AND TRIGGERED AUTOMATICITY

Triggered automaticity or activity refers to impulse initiation that is dependent on afterdepolarizations (Fig. 1-17). Afterdepolarizations are membrane voltage oscillations that occur during (early afterdepolarizations, EAD) or following (delayed afterdepolarization, DAD) an AP.[208]

In the early 1970s DADs were experimentally observed in Purkinje fibers exposed to toxic concentrations of digitalis glycosides.[209,210] The cellular feature common to the induction of DADs is the presence of increased Ca^{2+} load in the cytosol and SR.[211] Inhibition of the Na-K ATPase by digitalis glycosides will increase Ca^{2+} load by increasing intracellular Na^+, which is exchanged for Ca^{2+} by the Na-Ca exchanger. Increased $[Ca^{2+}]_i$ activates a transient inward current, I_{TI},

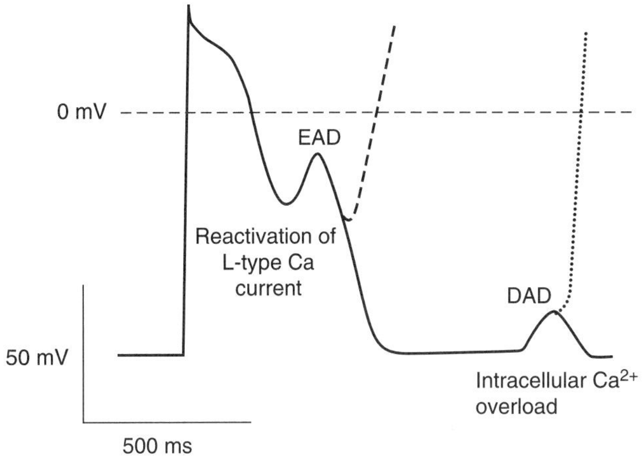

FIGURE 1-17 Afterdepolarizations early and delayed. Interruptions of repolarization before its completion are referred to as early afterdepolarizations (EADs). Most EADs, especially phase 2 and early phase 3, are believed to result from reactivation of the L-type Ca current and perhaps Na-Ca exchanger current. Later phase 3 EADs may also involve reactivation of Na currents (not shown). Afterdepolarizations that occur after the completion of repolarization are referred to as delayed afterdepolarizations (DADs). The mechanism of DAD involves intracellular Ca^{2+} overload and oscillatory release of Ca^{2+} from the SR, activating a number of Ca^{2+}-dependent conductances.

that depolarizes the cell.[212] The ionic basis of I_{TI} is still controversial and may result from electrogenic current through the Na-Ca exchanger or Ca^{2+}-activated depolarizing currents.[161]

Inhibition of the Na-K ATPase by digitalis glycosides facilitates, but is not necessary for creating, the Ca^{2+} overload that predisposes to DADs. Catecholamines and ischemia sufficiently enhance Ca^{2+} loading to produce DADs. The presumed mechanism of cytosolic Ca^{2+} increase and DADs with catecholamine stimulation is an increase in transmembrane Ca^{2+} flux through L-type Ca channels. Catecholamines may also enhance the activity of the Na-Ca exchanger, thus increasing the likelihood of DAD-mediated triggered activity.[213] Elevations in intracellular Ca^{2+} in ischemic myocardium are also associated with DADs and triggered arrhythmias. Accumulation of lysophosphoglycerides in ischemic myocardium with consequent Na^+ and Ca^{2+} overload has been suggested as a mechanism for DADs and triggered automaticity.[214] Cells from damaged areas or surviving the infarction may display spontaneous release of calcium from SR, and this may generate "waves" of intracellular calcium elevation and arrhythmias.[215]

The duration of the AP is a critical determinant of the presence of DADs. Longer APs associated with more transarcolemmal Ca^{2+} influx are more likely to be associated with DADs. If I_{TI} underlies at least part of the DAD, then the voltage dependence of the transient inward current should be reflected in the voltage dependence of DADs. Indeed, at membrane voltages where I_{TI} is near its maximum, DADs exhibit the largest amplitude.[216] Importantly, stimulation of the experimental preparation at fast rates increases the size of the DAD and the presence of triggered activity,[209] likely a function of frequency dependent loading of the SR with Ca^{2+}.

Mutations in the cardiac ryanodine receptor (RYR2), the SR calcium release channel in the heart, have been identified in kindreds with the syndrome of catecholamine-stimulated polymorphic ventricular tachycardia and ventricular fibrillation with short QT intervals.[217] It seems likely that perturbed $[Ca^{2+}]_i$ and thus perhaps DADs will be found to underlie arrhythmias in this syndrome. However, at this time, there are no clinical arrhythmias in humans that are proven to be due to DAD-mediated triggered activity. It is likely that some ventricular tachycardias that complicate digitalis intoxication are initiated by triggered activity. It has also been suggested that DADs underlie some forms of idiopathic ventricular tachycardia, particularly from the right ventricular outflow tract (see Table 1-3).

The other type of afterdepolarizations, EADs, occur during the AP and interrupt the orderly repolarization of the myocyte. They have been classified as phase 2 and phase 3 depending on when they occur, and the subclassification may have mechanistic implications. It has been traditionally held that, unlike DADs, EADs do not depend on a rise in intracellular Ca^{2+}, instead, AP prolongation and reactivation of depolarizing currents are fundamental to their production.[211] More recent experimental evidence suggests a previously unappreciated interrelationship between intracellular calcium loading and EADs. Cytosolic calcium may rise when APs are prolonged. This in turn appears to enhance L-type Ca current (possibly via calcium-calmodulin kinase activation), further prolonging AP duration as well as providing the inward current driving EADs.[218] Intracellular calcium loading by AP prolongation may also enhance the likelihood of DADs. The interrelationship among intracellular $[Ca^{2+}]$ and delayed and early afterdepolarizations may be one explanation for the susceptibility of hearts that are calcium loaded (e.g., in ischemia or congestive heart failure) to develop arrhythmias, particularly on exposure to AP-prolonging drugs.

The plateau of the AP is a time of high membrane resistance when there is little current flow. Consequently, small changes in either repolarizing or depolarizing currents can have profound effects on the AP duration and profile. The ionic mechanisms of phase 2 and 3 EADs and the upstrokes of the APs they elicit may differ. At the depolarized membrane voltages of phase 2, the Na current is inactivated and EADs can result from reactivation of the L-type Ca current.[211,219] Despite less data, it has been suggested that current through the Na-Ca exchanger and possibly the Na current may also participate in the inscription of phase 3 EADs.[220] The upstrokes of the APs elicited by phase 2 and 3 EADs also differ. Phase 2 EAD-triggered AP upstrokes are exclusively mediated by Ca currents[211,219]; these may or may not propagate, but they can substantially exaggerate heterogeneity of the time course of repolarization of AP (a key substrate for reentry), since EADs occur more readily in some regions (e.g., Purkinje, mid-myocardium) than others (e.g., epicardium, endocardium). APs triggered by phase 3 EADs arise from more negative membrane voltages. The upstrokes may be due to both Na and Ca currents and are more likely to propagate.

EAD-triggered arrhythmias exhibit rate dependence. In general, the amplitude of an EAD is augmented at slow rates when APs are longer. Pacing-induced increases in rate shorten the AP duration and reduce EAD amplitude.[221] AP shortening and suppression of EADs with increased stimulation rate is likely the result of augmentation of delayed rectifier K currents and perhaps hastening of Ca^{2+}-induced inactivation of L-type Ca currents. Similarly, catecholamines increase heart rate and decrease AP duration and EAD amplitude, despite the well-described effect of β adrenergic stimulation to increase L-type Ca current.[201]

A fundamental condition that underlies the development of EADs is AP prolongation, which is manifest on the surface electrocardiogram by QT prolongation. Hypokalemia, hypomagnesemia, bradycardia, and drugs can predispose to the formation of EADs, invariably in the context of prolonging the AP; drugs are the most common cause.[222] Antiarrhythmics with class IA and III action produce AP and QT prolongation intended to be therapeutic but frequently causing proarrhythmia. Noncardiac drugs such as some phenothiazines, some nonsedating antihistamines, and some antibiotics can also prolong the AP duration and predispose to EAD-mediated triggered arrhythmias. Decreased $[K^+]_o$ paradoxically decreases some membrane potassium currents (particularly I_{Kr}) in the ventricular myocyte, explaining why hypokalemia causes AP prolongation and EADs.[223,224] Indeed, potassium infusions in patients with the congenital LQTS[225] and with drug-induced QT prolongation reduced the Q–T interval.[226]

EAD-mediated triggered activity likely underlies initiation of the characteristic polymorphic ventricular tachycardia, Torsades de Pointes, seen in patients with congenital and acquired forms of LQTS (see Chapter 27). Acquired prolongation of the Q–T interval most often is the result of drug therapy or electrolyte disturbances as noted previously. However, structural heart disease such as cardiac hypertrophy and failure may also delay ventricular repolarization (so-called *electrical remodeling*) and predispose arrhythmias related to abnormalities of repolarization.[227] The abnormalities of repolarization in hypertrophy and failure are often magnified by concomitant drug therapy or electrolyte disturbances.

ABNORMAL IMPULSE CONDUCTION: REENTRY

The most common arrhythmia mechanism is reentry. Reentry is as much a property of networks of myocytes as it is a property of individual heart cells. Fundamentally, reentry is circulation of an activation wave around an inexcitable obstacle. Thus, the requirements for reentry are two electrophysiologically dissimilar pathways for impulse propagation around an inexcitable region such that unidirectional block occurs in one of the pathways and a region of excitable tissue exists at the head of the propagating wavefront.[228] Structural and electrophysiological properties of the heart may contribute to the development of the inexcitable obstacle and of unidirectional block. The complex geometry of muscle bundles in the heart and spatial heterogeneity of cellular coupling[229] or other active membrane properties (i.e., ionic currents) appear to be critical.

At the macroscopic level conduction through normal myocardial tissue is uniformly anisotropic; that is, propagation is continuous or "smooth" but faster longitudinally than transversely. However, at higher spatial resolution, anisotropy is always nonuniform due to the irregularities of cell shape and gap junction distribution.[3,4] The conversion of macroscopic anisotropy from uniform to nonuniform is correlated with an increased predilection to arrhythmias. One well-studied example is the aged human atrial myocardium, in which nonuniform anisotropy, manifest as highly fractionated electrograms, is associated with lateral uncoupling of myocytes and profound slowing of macroscopic transverse conduction. This produces an ideal substrate for the reentry that may underlie the very common development of atrial fibrillation in the elderly.[230]

Anatomically determined, excitable gap reentry can explain several clinically important tachycardias such as atrioventricular reentry, atrial flutter, and bundle branch reentry tachycardia (see Chapter 2). Strong evidence suggests that arrhythmias such as atrial and ventricular fibrillation, which are associated with more complex activation of the heart, are reentrant. However, this type of reentry ("functional") is mechanistically distinct from excitable gap reentry.

Reflection is a type of reentry that occurs in a linear segment of tissue (e.g., trabecula or Purkinje fiber) containing an area of conduction block with re-excitation occurring over the same segment of tissue. If the region of the segment proximal to the area of block is excited, the wave will propagate and generate APs up to the area of conduction block. Assuming that the area of conduction block remains connected to the remainder of the tissue (by gap junctions), it can be electrotonically activated (i.e., by current flow without AP induction). If the area of conduction block is short and the magnitude of the electrotonic current (source) is sufficiently large, the segment of tissue distal to the blocked area (sink) will be excited but with a significant delay. With the appropriate relationship of the electronic current transmitted through the inexcitable segment and distal excitable tissue, the distal segment can not only be activated but it can reactivate the proximal segment of muscle by electronic current flow from distal to proximal segments.

A key feature in classifying reentrant arrhythmias, particularly for therapy, is the presence and size of an excitable gap (see Chapter 2). An excitable gap exists when the tachycardia circuit is longer than the tachycardia wavelength (λ = conduction velocity × refractory period), allowing appropriately timed stimuli to reset propagation in the circuit. Reentrant arrhythmias may exist in the heart in the absence of an excitable gap and with a tachycardia wavelength nearly the same size as the pathlength. In this case, the wavefront propagates through partially refractory tissue with no anatomic obstacle and no fully excitable gap. This is referred to as leading circle reentry,[231] a form of functional reentry (reentry that depends on functional properties of the tissue). Unlike excitable gap reentry, there is no fixed anatomic circuit in leading circle reentry, and it may therefore not be possible to disrupt the tachycardia with pacing or destruction of a part of the circuit. Furthermore, the circuit in leading circle reentry tends to be less stable than that in excitable gap reentrant arrhythmias, with large variations in cycle length and predilection to termination. Atrial flutter represents an example of a reentrant tachycardia with a large excitable gap not always due to an anatomic constraint, but to functional block (reflecting the special properties of the crista terminalis discussed earlier). Experimental data and computer simulations have highlighted shortcomings of the tenets leading circle reentry and suggest that spiral waves may better explain some forms of functional reentry (see Chapter 2).

Tissue anisotropy is another important determinant of functional reentrant arrhythmias in ischemic heart disease. Changes in functional and anatomic anisotropy are characteristics of both acute and chronic ischemic heart disease. Within 30 minutes of the onset of myocardial ischemia, significant increases in gap junction channel resistance and packing are observed. Further cellular uncoupling and a significant reduction in gap junction protein is observed with 60 minutes of ischemia[232]; this coincides with irreversible cellular damage. These changes exaggerate anisotropic conduction in the ischemic zone.

Chronically ischemic but not infarcted myocardium also exhibits an approximate 50% down regulation of gap junction protein (Cx-43) with a significant change in the pattern or number of intercalated disks.[233] The suggestion that a 50% reduction in gap junction protein influences anisotropic conduction is supported by measurements of conduction velocities in heterozygous Cx-43 knockout mice.[234] The border zones of infarcted myocardium exhibit not only functional alterations of ionic currents but remodeling of tissue and altered distribution of gap junctions in human ventricle[235] and canine infarction.[236] The alterations in gap junction expression in context of macroscopic tissue alterations supports a role for anisotropic conduction in reentrant arrhythmias that complicated coronary artery disease.

SUMMARY

The science of cardiac electrophysiology has its roots in clinical medicine. It began and continues with descriptions of specific arrhythmia syndromes. Understanding normal and abnormal mechanisms underlying such well-defined syndromes has been a key to development and widespread implementation of modern therapies such as targeted ablation for focal or reentrant arrhythmias. Advances in understanding the role of individual current components and their underlying molecular bases, in normal and abnormal electrogenesis, presents us with a further opportunity in this direction.

Indeed, delineation of specific syndromes such as LQTS or interventricular foramen syndrome, followed by an understanding of their molecular underpinnings, is now poised to further revolutionize arrhythmia therapy: Identification of patients with genetic risk factors for arrhythmias may open the way to effective therapies in these groups. Moreover, further understanding of the molecular mechanisms underlying initiation and maintenance of complex and common arrhythmia syndromes such as atrial or ventricular fibrillation may allow development of entirely new drug or nonpharmacologic therapies.

REFERENCES

1. Weidmann S: Effect of current flow on the membrane potential of cardiac muscle. J Physiol (Lond) 1951;115:227-36.
2. LeGrice IJ, Smaill BH, Chai LZ, et al: Laminar structure of the heart: Ventricular myocyte arrangement and connective tissue architecture in the dog. Am J Physiol 1995;269:H571-82.
3. Spach MS, Heidlage JF: The stochastic nature of cardiac propagation at a microscopic level. Electrical description of myocardial architecture and its application to conduction. Circ Res 1995;76:366-80.
4. Spach MS, Kootsey JM, Sloan JD: Active modulation of electrical coupling between cardiac cells of the dog. A mechanism for transient and steady state variations in conduction velocity. Circ Res 1982;51:347-62.
5. Marmont G: Studies on the axon membrane. I. A new method. J Cell Comp Physiol 1949;34:351-82.
6. Cole KS: Dynamic electrical characteristics of the squid axon membrane. Arch Sci Physiol 1949;3:253-8.
7. Hodgkin AL, Huxley AF, Katz B: Ionic currents underlying activity in the axon of the squid. Arch Sci Physiol 1949;3:129-50.
8. Hamill OP, Marty A, Neher E, et al: Improved patch-clamp techniques for high-resolution current recording from cells and cell-free membrane patches. Pflugers Arch 1981;391:85-100.
9. Antzelevitch C, Sicouri S, Litovsky SH, et al: Heterogeneity within the ventricular wall. Electrophysiology and pharmacology of epicardial, endocardial, and M cells. Circ Res 1991;69: 1427-49.
10. Drouin E, Charpentier F, Gauthier C, et al: Electrophysiologic characteristics of cells spanning the left ventricular wall of human heart: Evidence for presence of M cells [see comments]. J Am Coll Cardiol 1995;26:185-92.
11. Stühmer W, Conti F, Suzuki H, et al: Structural parts involved in activation and inactivation of the sodium channel. Nature 1989;339:597-603.
12. Makielski J, Sheets M, Hanck D, et al: Sodium current in voltage clamped internally perfused canine cardiac Purkinje cells. Biophys J 1987;52:1-11.
13. Rose W, Balke C, Wier W, Marban E: Macroscopic and unitary properties of physiological ion flux through L-type Ca^{2+} channels in guinea-pig heart cells. J Physiol (Lond) 1992;456:267-84.
14. Noda M, Shimizu S, Tanabe T, et al: Primary structure of *Electrophorus electricus* sodium channel deduced from cDNA sequence. Nature 1984;312:121-7.
15. Ahmed CM, Ware DH, Lee SC, et al: Primary structure, chromosomal localization, and functional expression of a voltage-gated sodium channel from human brain. Proc Natl Acad Sci U S A 1992;89:8220-4.
16. Gellens ME, George AL Jr, Chen LQ, et al: Primary structure and functional expression of the human cardiac tetrodotoxin-insensitive voltage-dependent sodium channel. Proc Natl Acad Sci U S A 1992;89:554-8.
17. George AL, Knittle TJ, Tamkun MM: Molecular cloning of an atypical voltage-gated sodium channel expressed in human heart and uterus: Evidence for a distinct gene family. Proc Natl Acad Sci U S A 1992;89:4893-7.
18. Yellen G, Jurman ME, Abramson T, MacKinnon R: Mutations affecting internal TEA blockade identify the probable pore-forming region of a K^+ channel. Science 1991;251:939-42.
19. Chiamvimonvat N, Perez-Garcia M, Ranjan R, et al: Depth asymmetries of the pore-lining segments of the Na^+ channel revealed by cysteine mutagenesis. Neuron 1996;16:1037-47.
20. Heinemann S, Terlau H, Stuhmer W, et al: Calcium channel characteristics conferred on the sodium channel by single mutations. Nature 1992;356:441-3.
21. Hodgkin AL, Huxley AF: A quantitative description of membrane current and its application to conduction and excitation in nerve. J Physiol (Lond) 1952;117:500-44.
22. Yang N, Horn R: Evidence for voltage-dependent S4 movement in sodium channels. Neuron 1995;15:213-18.
23. Yang N, George A Jr, Horn R: Molecular basis of charge movement in voltage-gated sodium channels. Neuron 1996;16:113-22.
24. Yang N, George A, Horn R: Probing the outer vestibule of a sodium channel voltage sensor. Biophys J 1997;73:2260-8.
25. Chen L, Santarelli V, Horn R, Kallen R: A unique role for the S4 segment of domain 4 in the inactivation of sodium channels. J Gen Physiol 1996;108:549-56.
26. Kontis K, Rounaghi A, Goldin A: Sodium channel activation gating is affected by substitutions of voltage sensor positive charges in all four domains. J Gen Physiol 1997;110:391-401.
27. Aldrich RW, Corey DP, Stevens CF: A reinterpretation of mammalian sodium channel gating based on single channel recording. Nature 1983;306:436-41.
28. Yue D, Lawrence J, Marban E: Two molecular transitions influence cardiac sodium channel gating. Science 1989;244:349-52.
29. Adelman WJ, Palti Y: The effects of external potassium and long duration voltage conditioning on the amplitude of sodium currents in the giant axon of the squid, *Loligo pealei*. J Gen Physiol 1969;54:589-606.
30. Chandler WK, Meves H: Slow changes in membrane permeability and long lasting action potentials in axons perfused with fluoride solutions. J Physiol 1970;211:707-28.
31. Moorman JR, Kirsch GE, Brown AM, Joho RH: Changes in sodium channel gating produced by point mutations in a cytoplasmic linker. Science 1990;250:688-91.
32. Patton D, West J, Catterall W, Goldin A: Amino acid residues required for fast Na^+-channel inactivation: Charge neutralizations and deletions in the III-IV linker. Proc Natl Acad Sci U S A 1992;89:10905-9.
33. West J, Patton D, Scheuer T, et al: A cluster of hydrophobic amino acid residues required for fast Na^+-channel inactivation. Proc Natl Acad Sci U S A 1992;89:10910-4.
34. McPhee JC, Ragsdale DS, Scheuer T, Catterall WA: A role for intracellular loop IVS4-S5 of the Na^+ channel α subunit in fast inactivation. Biophys J 1996;70:318a.
35. Tomaselli G, Chiamvimonvat N, Nuss H, Balser J, Perez-Garcia M, Xu R, Orias D, Backx P, Marban E. A mutation in the pore of the sodium channel alters gating. Biophys J 1995;68:1814-27.
36. Balser J, Nuss H, Chiamvimonvat N, et al: External pore residue mediates slow inactivation in mu 1 rat skeletal muscle sodium channels. J Physiol 1996;494:431-42.
37. Hayward L, Brown R Jr, Cannon S: Slow inactivation differs among mutant Na channels associated with myotonia and periodic paralysis. Biophys J 1997;72:1204-19.
38. Ragsdale D, McPhee J, Scheuer T, Catterall W: Molecular determinants of state-dependent block of Na+ channels by local anesthetics. Science 1994;265:1724-8.
39. Balser J, Nuss H, Orias D, et al: Local anesthetics as effectors of allosteric gating. Lidocaine effects on inactivation-deficient rat skeletal muscle Na channels. J Clin Invest 1996;98:2874-86.
40. McPhee J, Ragsdale D, Scheuer T, Catterall W: A mutation in segment IVS6 disrupts fast inactivation of sodium channels. Proc Natl Acad Sci U S A 1994;91:12346-50.
41. McPhee J, Ragsdale D, Scheuer T, Catterall W: A critical role for transmembrane segment IVS6 of the sodium channel alpha subunit in fast inactivation. J Biol Chem 1995;270:12025-34.
42. Satin J, Kyle J, Chen M, et al: A mutant of TTX-resistant cardiac sodium channels with TTX-sensitive properties. Science 1992; 256:1202-5.
43. Backx P, Yue D, Lawrence J, et al: Molecular localization of an ion-binding site within the pore of mammalian sodium channels. Science 1992;257:248-51.
44. Lipkind G, Fozzard H: A structural model of the tetrodotoxin and saxitoxin binding site of the Na+ channel. Biophys J 1994;66:1-13.

45. Hartshorne R, Messner D, Coppersmith J, Catterall W: The saxitoxin receptor of the sodium channel from rat brain. Evidence for two nonidentical beta subunits. J Biol Chem 1982; 257:13888-91.
46. Messner D, Catterall W: The sodium channel from rat brain. Separation and characterization of subunits. J Biol Chem 1985;260:10597-604.
47. Roberts R, Barchi R: The voltage-sensitive sodium channel from rabbit skeletal muscle. Chemical characterization of subunits. J Biol Chem 1987;262:2298-303.
48. Isom L, De Jongh K, Patton D, et al: Primary structure and functional expression of the beta 1 subunit of the rat brain sodium channel. Science 1992;256:839-42.
49. Isom L, Ragsdale D, De Jongh K, et al: Structure and function of the beta 2 subunit of brain sodium channels, a transmembrane glycoprotein with a CAM motif. Cell 1995;83:433-42.
50. Bennett P Jr, Makita N, George A Jr: A molecular basis for gating mode transitions in human skeletal muscle Na+ channels. FEBS Letters 1993;326:21-4.
51. Cannon S, McClatchey A, Gusella J: Modification of the Na+ current conducted by the rat skeletal muscle alpha subunit by coexpression with a human brain beta subunit. Pflugers Arch 1993;423:155-7.
52. Patton D, Isom L, Catterall W, Goldin A: The adult rat brain beta 1 subunit modifies activation and inactivation gating of multiple sodium channel alpha subunits. J Biol Chem 1994;269:17649-55.
53. Li M, West J, Numann R, et al: Convergent regulation of sodium channels by protein kinase C and cAMP-dependent protein kinase. Science 1993;261:1439-42.
54. Murphy B, Rogers J, Perdichizzi A, et al: cAmp-Dependent phosphorylation of two sites in the alpha subunit of the cardiac sodium channel. J Biol Chem 1996;271:28837-43.
55. Frohnwieser B, Chen L, Schreibmayer W, Kallen R: Modulation of the human cardiac sodium channel alpha-subunit by cAMP-dependent protein kinase and the responsible sequence domain. J Physiol 1997;498:309-18.
56. West J, Numann R, Murphy B, et al: A phosphorylation site in the Na^+ channel required for modulation by protein kinase C. Science 1991;254:866-8.
57. Qu Y, Rogers J, Tanada T, et al: Phosphorylation of S1505 in the cardiac Na^+ channel inactivation gate is required for modulation by protein kinase C. J Gen Physiology 1996;108:375-9.
58. Cohen S, Levitt L: Partial characterization of the rH1 sodium channel protein from rat heart using subtype-specific antibodies. Circ Res 1993;73:735-42.
59. Cannon S: Sodium channel defects in myotonia and periodic paralysis. Annu Rev Neurosci 1996;19:141-64.
60. Celesia GG: Disorders of membrane channels or channelopathies. Clin Neurophysiol 2001;112:2-18.
61. Wang Q, Shen J, Splawski I, et al: SCN5A mutations associated with an inherited cardiac arrhythmia, long QT syndrome. Cell 1995;80:805-11.
62. Haug K, Sander T, Hallmann K, et al: The voltage-gated sodium channel beta2-subunit gene and idiopathic generalized epilepsy. Neuroreport 2000;11:2687-9.
63. Chen Q, Kirsch GE, Zhang D, et al: Genetic basis and molecular mechanism for idiopathic ventricular fibrillation. Nature 1998;392:293-6.
64. Antzelevitch C, Yan GX, Shimizu W: Transmural dispersion of repolarization and arrhythmogenicity: The Brugada syndrome versus the long QT syndrome. J Electrocardiol 1999;32:158-65.
65. Tanabe T, Takeshima H, Mikami A, et al: Primary structure of the receptor for calcium channel blockers from skeletal muscle. Nature 1987;328:313-8.
66. Ertel EA, Campbell KP, Harpold MM, et al: Nomenclature of voltage-gated calcium channels [letter]. Neuron 2000;25:533-5.
67. Koch WJ, Ellinor PT, Schwartz A: cDNA cloning of a dihydropyridine-sensitive calcium channel from rat aorta. Evidence for the existence of alternatively spliced forms. J Biol Chem 1990;265: 17786-91.
68. Biel M, Hullin R, Freundner S, et al: Tissue-specific expression of high-voltage-activated dihydropyridine-sensitive L-type calcium channels. Eur J Biochem 1991;200:81-8.
69. Diebold RJ, Koch WJ, Ellinor PT, et al: Mutually exclusive exon splicing of the cardiac calcium channel alpha 1 subunit gene generates developmentally regulated isoforms in the rat heart. Proc Natl Acad Sci U S A 1992;89:1497-501.
70. Klockner U, Mikala G, Eisfeld J, et al: Properties of three COOH-terminal splice variants of a human cardiac L-type Ca^{2+}-channel alpha1-subunit. Am J Physiol Cell Physiol 1997;272:H1372-81.
71. Yang J, Ellinor PT, Sather WA, et al: Molecular determinants of Ca^{2+} selectivity and ion permeation in L-type Ca^{2+} channels [see comments]. Nature 1993;366:158-61.
72. Singer D, Biel M, Lotan I, et al: The roles of the subunits in the function of the calcium channel. Science 1991;253:1553-7.
73. Wei XY, Perez-Reyes E, Lacerda AE, et al: Heterologous regulation of the cardiac Ca^{2+} channel alpha 1 subunit by skeletal muscle beta and gamma subunits. Implications for the structure of cardiac L-type Ca^{2+} channels. J Biol Chem 1991;266:21943-7.
74. Lory P, Varadi G, Slish DF, et al: Characterization of beta subunit modulation of a rabbit cardiac L-type $Ca2^+$ channel alpha 1 subunit as expressed in mouse L cells. FEBS Lett 1993;315:167-72.
75. Perez-Reyes E, Castellano A, Kim HS, et al: Cloning and expression of a cardiac/brain beta subunit of the L-type calcium channel. J Biol Chem 1992;267:1792-7.
76. Kamp TJ, Perez-Garcia MT, Marban E: Enhancement of ionic current and charge movement by coexpression of calcium channel beta 1A subunit with alpha 1C subunit in a human embryonic kidney cell line. J Physiol (Lond) 1996;492:89-96.
77. Chien AJ, Zhao X, Shirokov RE, et al: Roles of a membrane-localized beta subunit in the formation and targeting of functional L-type Ca^{2+} channels. J Biol Chem 1995;270:30036-44.
78. Haase H, Kresse A, Hohaus A, et al: Expression of calcium channel subunits in the normal and diseased human myocardium. J Mol Med 1996;74:99-104.
79. Klugbauer N, Dai S, Specht V, et al: A family of gamma-like calcium channel subunits. FEBS Lett 2000;470:189-97.
80. Wiser O, Trus M, Tobi D, et al: The alpha 2/delta subunit of voltage sensitive Ca^{2+} channels is a single transmembrane extracellular protein which is involved in regulated secretion. FEBS Lett 1996;379:15-20.
81. Gurnett CA, De Waard M, Campbell KP: Dual function of the voltage-dependent Ca^{2+} channel alpha 2 delta subunit in current stimulation and subunit interaction. Neuron 1996;16:431-40.
82. Bangalore R, Mehrke G, Gingrich K, et al: Influence of L-type Ca channel alpha 2/delta-subunit on ionic and gating current in transiently transfected HEK 293 cells. Am J Physiol 1996;270: H1521-8.
83. Tanabe T, Adams BA, Numa S, Beam KG: Repeat I of the dihydropyridine receptor is critical in determining calcium channel activation kinetics. Nature 1991;352:800-3.
84. Zuhlke RD, Pitt GS, Deisseroth K, et al: Calmodulin supports both inactivation and facilitation of L-type calcium channels [see comments]. Nature 1999;399:159-62.
85. Peterson BZ, DeMaria CD, Adelman JP, Yue DT: Calmodulin is the Ca^{2+} sensor for Ca^{2+}-dependent inactivation of L-type calcium channels [published erratum appears in Neuron 1999 Apr;22:following 893]. Neuron 1999;22:549-58.
86. Peterson BZ, Lee JS, Mulle JG, et al: Critical determinants of Ca^{2+}-dependent inactivation within an EF-hand motif of L-type $Ca2^+$ channels [in process citation]. Biophys J 2000;78: 1906-20.
87. Irisawa H, Brown HF, Giles W: Cardiac pacemaking in the sinoatrial node. Physiol Rev 1993;73:197-227.
88. Van Wagoner DR, Pond AL, Lamorgese M, et al: Atrial L-type Ca^{2+} currents and human atrial fibrillation [see comments]. Circ Res 1999;85:428-36.
89. Vassort G, Alvarez J: Cardiac T-type calcium current: Pharmacology and roles in cardiac tissues. J Cardiovasc Electrophysiol 1994;5:376-93.
90. Bers DM, Perez-Reyes E: Ca channels in cardiac myocytes: Structure and function in Ca influx and intracellular Ca release. Cardiovasc Res 1999;42:339-60.
91. Beuckelmann DJ, Nabauer M, Erdmann E: Characteristics of calcium-current in isolated human ventricular myocytes from patients with terminal heart failure. J Mol Cell Cardiol 1991;23:929-37.
92. Beuckelmann DJ, Nabauer M, Erdmann E: Intracellular calcium handling in isolated ventricular myocytes from patients with terminal heart failure. Circulation 1992;85:1046-55.

93. Perez-Reyes E, Cribbs LL, Daud A, et al: Molecular characterization of a neuronal low-voltage-activated T-type calcium channel [see comments]. Nature 1998;391:896-900.
94. Cribbs LL, Lee JH, Yang J, et al: Cloning and characterization of alpha1H from human heart, a member of the T-type Ca^{2+} channel gene family. Circ Res 1998;83:103-9.
95. Hockerman GH, Peterson BZ, Johnson BD, Catterall WA: Molecular determinants of drug binding and action on L-type calcium channels. Annu Rev Pharmacol Toxicol 1997;37:361-96.
96. Welling A, Kwan YW, Bosse E, et al: Subunit-dependent modulation of recombinant L-type calcium channels. Molecular basis for dihydropyridine tissue selectivity. Circ Res 1993;73:974-80.
97. Heginbotham L, Abramson T, MacKinnon R: A functional connection between the pores of distantly related ion channels as revealed by mutant K+ channels. Science 1992;258:1152-5.
98. Tempel BL, Papazian DM, Schwarz TL, et al: Sequence of a probable potassium channel component encoded at Shaker locus of Drosophila. Science 1987;237:770-5.
99. Yeola SW, Rich TC, Uebele VN, et al: Molecular analysis of a binding site for quinidine in a human cardiac delayed rectifier K+ channel. Role of S6 in antiarrhythmic drug binding [see comments]. Circ Res 1996;78:1105-14.
100. Snyders DJ: Structure and function of cardiac potassium channels. Cardiovasc Res 1999;42:377-90.
101. Gulbis JM, Zhou M, Mann S, MacKinnon R: Structure of the cytoplasmic beta subunit-T1 assembly of voltage-dependent K+ channels. Science 2000;289:123-7.
102. Gulbis JM, Mann S, MacKinnon R: Structure of a voltage-dependent K+ channel beta subunit. Cell 1999;97:943-52.
103. Perez-Garcia MT, Lopez-Lopez JR, Gonzalez C: Kvbeta1.2 subunit coexpression in HEK293 cells confers O2 sensitivity to kv4.2 but not to Shaker channels. J Gen Physiol 1999;113: 897-907.
104. An WF, Bowlby MR, Betty M, et al: Modulation of A-type potassium channels by a family of calcium sensors. Nature 2000;403:553-6.
105. Tai KK, Goldstein SA: The conduction pore of a cardiac potassium channel. Nature 1998;391:605-8.
106. Murai T, Kakizuka A, Takumi T, et al: Molecular cloning and sequence analysis of human genomic DNA encoding a novel membrane protein which exhibits a slowly activating potassium channel activity. Biochem Biophys Res Commun 1989;161:176-81.
107. Sanguinetti MC, Curran ME, Zou A, et al: Coassembly of K(V)LQT1 and minK (IsK) proteins to form cardiac I(Ks) potassium channel [see comments]. Nature 1996;384:80-3.
108. Barhanin J, Lesage F, Guillemare E, et al: K(V)LQT1 and lsK (minK) proteins associate to form the I(Ks) cardiac potassium current [see comments]. Nature 1996;384:78-80.
109. Abbott GW, Sesti F, Splawski I, et al: MiRP1 forms IKr potassium channels with HERG and is associated with cardiac arrhythmia. Cell 1999;97:175-87.
110. Salinas M, Duprat F, Heurteaux C, et al: New modulatory alpha subunits for mammalian Shab K+ channels. J Biol Chem 1997;272:24371-9.
111. Armstrong CM: Inactivation of the potassium conductance and related phenomena caused by quaternary ammonium ion injection in squid axons. J Gen Physiol 1969;54:553-75.
112. Hoshi T, Zagotta WN, Aldrich RW: Biophysical and molecular mechanisms of Shaker potassium channel inactivation [see comments]. Science 1990;250:533-8.
113. Zagotta WN, Hoshi T, Aldrich RW: Restoration of inactivation in mutants of Shaker potassium channels by a peptide derived from ShB [see comments]. Science 1990;250:568-71.
114. Lopez-Barneo J, Hoshi T, Heinemann SH, Aldrich RW: Effects of external cations and mutations in the pore region on C-type inactivation of Shaker potassium channels. Receptors Channels 1993;1:61-71.
115. Tseng GN, Hoffman BF: Two components of transient outward current in canine ventricular myocytes. Circ Res 1989;64:633-47.
116. Zygmunt AC, Gibbons WR: Calcium-activated chloride current in rabbit ventricular myocytes. Circ Res 1991;68:424-37.
117. Nabauer M, Kaab S: Potassium channel down-regulation in heart failure. Cardiovasc Res 1998;37:324-34.
118. Dixon JE, Shi W, Wang HS, et al: Role of the Kv4.3 K+ channel in ventricular muscle. A molecular correlate for the transient outward current [published erratum appears in Circ Res 1997 Jan;80:147]. Circ Res 1996;79:659-68.
119. Kaab S, Dixon J, Duc J, et al: Molecular basis of transient outward potassium current downregulation in human heart failure: A decrease in Kv4.3 mRNA correlates with a reduction in current density. Circulation 1998;98:1383-93.
120. Po S, Roberds S, Snyders DJ, et al: Heteromultimeric assembly of human potassium channels. Molecular basis of a transient outward current? Circ Res 1993;72:1326-36.
121. Yeola SW, Snyders DJ: Electrophysiological and pharmacological correspondence between Kv4.2 current and rat cardiac transient outward current. Cardiovasc Res 1997;33:540-7.
122. Dixon JE, McKinnon D: Quantitative analysis of potassium channel mRNA expression in atrial and ventricular muscle of rats. Circ Res 1994;75:252-60.
123. Juang G, Burysek M, Ohler A, et al: Regional alterations of K channel subunit expression in canine pacing-tachycardia heart failure. Circulation 2000;102:1281a.
124. Wettwer E, Amos GJ, Posival H, Ravens U: Transient outward current in human ventricular myocytes of subepicardial and subendocardial origin. Circ Res 1994;75:473-82.
125. Nabauer M, Beuckelmann DJ, Uberfuhr P, Steinbeck G: Regional differences in current density and rate-dependent properties of the transient outward current in subepicardial and subendocardial myocytes of human left ventricle. Circulation 1996;93:168-77.
126. Feng J, Wible B, Li GR, et al: Antisense oligodeoxynucleotides directed against Kv1.5 mRNA specifically inhibit ultrarapid delayed rectifier K+ current in cultured adult human atrial myocytes. Circ Res 1997;80:572-9.
127. Yue L, Feng J, Li GR, Nattel S: Characterization of an ultrarapid delayed rectifier potassium channel involved in canine atrial repolarization. J Physiol (Lond) 1996;496:647-62.
128. Wang Z, Fermini B, Nattel S: Sustained depolarization-induced outward current in human atrial myocytes. Evidence for a novel delayed rectifier K^+ current similar to Kv1.5 cloned channel currents. Circ Res 1993;73:1061-76.
129. Mays DJ, Foose JM, Philipson LH, Tamkun MM: Localization of the Kv1.5 K+ channel protein in explanted cardiac tissue. J Clin Invest 1995;96:282-92.
130. Sanguinetti MC, Jurkiewicz NK: Two components of cardiac delayed rectifier K+ current. Differential sensitivity to block by class III antiarrhythmic agents. J Gen Physiol 1990;96: 195-215.
131. Varnum MD, Busch AE, Bond CT, et al: The min K channel underlies the cardiac potassium current IKs and mediates species-specific responses to protein kinase C. Proc Natl Acad Sci U S A 1993;90:11528-32.
132. Wang Q, Curran ME, Splawski I, et al: Positional cloning of a novel potassium channel gene: KVLQT1 mutations cause cardiac arrhythmias. Nat Genet 1996;12:17-23.
133. Demolombe S, Baro I, Pereon Y, et al: A dominant negative isoform of the long QT syndrome 1 gene product. J Biol Chem 1998;273:6837-43.
134. Curran ME, Splawski I, Timothy KW, et al: A molecular basis for cardiac arrhythmia: HERG mutations cause long QT syndrome. Cell 1995;80:795-803.
135. Gintant GA: Characterization and functional consequences of delayed rectifier current transient in ventricular repolarization. Am J Physiol Heart Circ Physiol 2000;278:H806-17.
136. Lees-Miller JP, Kondo C, Wang L, Duff HJ: Electrophysiological characterization of an alternatively processed ERG K+ channel in mouse and human hearts. Circ Res 1997;81:719-26.
137. London B, Trudeau MC, Newton KP, et al: Two isoforms of the mouse ether-a-go-go-related gene coassemble to form channels with properties similar to the rapidly activating component of the cardiac delayed rectifier K+ current. Circ Res 1997;81:870-8.
138. Doyle DA, Cabral JM, Pfuetzner RA, et al: The structure of the potassium channel: molecular basis of K+ conduction and selectivity [see comments]. Science 1998;280:69-77.
139. Hille B, Armstrong CM, MacKinnon R: Ion channels: From idea to reality. Nat Med 1999;5:1105-9.
140. Kubo Y, Baldwin TJ, Jan YN, Jan LY: Primary structure and functional expression of a mouse inward rectifier potassium channel [see comments]. Nature 1993;362:127-33.

141. Ashen MD, O'Rourke B, Kluge KA, et al: Inward rectifier K+ channel from human heart and brain: Cloning and stable expression in a human cell line. Am J Physiol 1995;268:H506-11.
142. Wible BA, De Biasi M, Majumder K, et al: Cloning and functional expression of an inwardly rectifying K+ channel from human atrium. Circ Res 1995;76:343-50.
143. Kubo Y, Reuveny E, Slesinger PA, et al: Primary structure and functional expression of a rat G-protein-coupled muscarinic potassium channel [see comments]. Nature 1993;364:802-6.
144. Krapivinsky G, Gordon EA, Wickman K, et al: The G-protein-gated atrial K+ channel IKACh is a heteromultimer of two inwardly rectifying K(+)-channel proteins. Nature 1995;374:135-41.
145. Corey S, Krapivinsky G, Krapivinsky L, Clapham DE: Number and stoichiometry of subunits in the native atrial G-protein-gated K+ channel, IKACh. J Biol Chem 1998;273:5271-8.
146. Logothetis DE, Kurachi Y, Galper J, et al: The beta gamma subunits of GTP-binding proteins activate the muscarinic K+ channel in heart. Nature 1987;325:321-6.
147. O'Rourke B: Myocardial K(ATP) channels in preconditioning [in process citation]. Circ Res 2000;87:845-55.
148. Inagaki N, Gonoi T, Clement JP, et al: Reconstitution of IKATP: An inward rectifier subunit plus the sulfonylurea receptor [see comments]. Science 1995;270:1166-70.
149. Seino S: ATP-sensitive potassium channels: A model of heteromultimeric potassium channel/receptor assemblies. Annu Rev Physiol 1999;61:337-62.
150. Duprat F, Lesage F, Fink M, et al: TASK, a human background K+ channel to sense external pH variations near physiological pH. EMBO J 1997;16:5464-71.
151. DiFrancesco D: Cardiac pacemaker: 15 years of "new" interpretation. Acta Cardiol 1995;50:413-27.
152. Ludwig A, Zong X, Jeglitsch M, et al: A family of hyperpolarization-activated mammalian cation channels. Nature 1998;393: 587-91.
153. Santoro B, Tibbs GR: The HCN gene family: Molecular basis of the hyperpolarization-activated pacemaker channels. Ann N Y Acad Sci 1999;868:741-64.
154. Ludwig A, Zong X, Stieber J, et al: Two pacemaker channels from human heart with profoundly different activation kinetics. EMBO J 1999;18:2323-9.
155. Baker K, Warren KS, Yellen G, Fishman MC: Defective "pacemaker" current (Ih) in a zebra fish mutant with a slow heart rate. Proc Natl Acad Sci U S A 1997;94:4554-9.
156. Nicoll DA, Longoni S, Philipson KD: Molecular cloning and functional expression of the cardiac sarcolemmal Na^+-Ca^{2+} exchanger. Science 1990;250:562-5.
157. Komuro I, Wenninger KE, Philipson KD, Izumo S: Molecular cloning and characterization of the human cardiac Na+/Ca^{2+} exchanger cDNA. Proc Natl Acad Sci U S A 1992;89:4769-73.
158. Nicoll DA, Ottolia M, Lu L, et al: A new topological model of the cardiac sarcolemmal Na+-Ca^{2+} exchanger. J Biol Chem 1999;274:910-7.
159. Li Z, Nicoll DA, Collins A, et al: Identification of a peptide inhibitor of the cardiac sarcolemmal Na^+-Ca^{2+} exchanger. J Biol Chem 1991;266:1014-20.
160. Fozzard HA: Afterdepolarizations and triggered activity. Basic Res Cardiol 1992;87:105-13.
161. Kass RS, Tsien RW, Weingart R: Ionic basis of transient inward current induced by strophanthidin in cardiac Purkinje fibres. J Physiol (Lond) 1978;281:209-26.
162. Benndorf K, Friedrich M, Hirche H: Reoxygenation-induced arrhythmogenic transient inward currents in isolated cells of the guinea-pig heart. Pflugers Arch 1991;418:248-60.
163. January CT, Fozzard HA: Delayed afterdepolarizations in heart muscle: Mechanisms and relevance. Pharmacol Rev 1988;40: 219-27.
164. Blanco G, Mercer RW: Isozymes of the Na-K-ATPase: Heterogeneity in structure, diversity in function. Am J Physiol 1998;275:F633-50.
165. Forbush Bd, Kaplan JH, Hoffman JF: Characterization of a new photoaffinity derivative of ouabain: Labeling of the large polypeptide and of a proteolipid component of the Na, K-ATPase. Biochemistry 1978;17:3667-76.
166. Mercer RW, Biemesderfer D, Bliss DP Jr, et al: Molecular cloning and immunological characterization of the gamma polypeptide, a small protein associated with the Na,K-ATPase. J Cell Biol 1993;121:579-86.
167. Palmer CJ, Scott BT, Jones LR: Purification and complete sequence determination of the major plasma membrane substrate for cAMP-dependent protein kinase and protein kinase C in myocardium. J Biol Chem 1991;266:11126-30.
168. Wang J, Schwinger RH, Frank K, et al: Regional expression of sodium pump subunits isoforms and Na^+-Ca^{++} exchanger in the human heart. J Clin Invest 1996;98:1650-8.
169. Schwinger RH, Bohm M, Erdmann E: Effectiveness of cardiac glycosides in human myocardium with and without "downregulated" beta-adrenoceptors. J Cardiovasc Pharmacol 1990;15:692-7.
170. Kumar NM, Gilula NB: The gap junction communication channel. Cell 1996;84:381-8.
171. Dhein S: Gap junction channels in the cardiovascular system: Pharmacological and physiological modulation. Trends Pharmacol Sci 1998;19:229-41.
172. Unger VM, Kumar NM, Gilula NB, Yeager M: Three-dimensional structure of a recombinant gap junction membrane channel. Science 1999;283:1176-80.
173. Risek B, Klier FG, Gilula NB: Developmental regulation and structural organization of connexins in epidermal gap junctions. Dev Biol 1994;164:183-96.
174. Jiang JX, Goodenough DA: Heteromeric connexons in lens gap junction channels. Proc Natl Acad Sci U S A 1996;93:1287-91.
175. Bevans CG, Kordel M, Rhee SK, Harris AL: Isoform composition of connexin channels determines selectivity among second messengers and uncharged molecules. J Biol Chem 1998;273: 2808-16.
176. Brink PR, Cronin K, Banach K, Peterson E, et al: Evidence for heteromeric gap junction channels formed from rat connexin43 and human connexin37. Am J Physiol 1997;273:C1386-96.
177. White TW, Paul DL, Goodenough DA, Bruzzone R: Functional analysis of selective interactions among rodent connexins. Mol Biol Cell 1995;6:459-70.
178. Kanter HL, Saffitz JE, Beyer EC: Cardiac myocytes express multiple gap junction proteins. Circ Res 1992;70:438-44.
179. Kanter HL, Laing JG, Beyer EC, et al: Multiple connexins colocalize in canine ventricular myocyte gap junctions. Circ Res 1993;73:344-50.
180. Gourdie RG, Severs NJ, Green CR, et al: The spatial distribution and relative abundance of gap-junctional connexin40 and connexin43 correlate to functional properties of components of the cardiac atrioventricular conduction system. J Cell Sci 1993; 105:985-91.
181. Veenstra RD: Size and selectivity of gap junction channels formed from different connexins. J Bioenerg Biomembr 1996; 28:327-37.
182. Saffitz JE, Kanter HL, Green KG, et al: Tissue-specific determinants of anisotropic conduction velocity in canine atrial and ventricular myocardium. Circ Res 1994;74:1065-70.
183. Saffitz JE, Green KG, Schuessler RB: Structural determinants of slow conduction in the canine sinus node. J Cardiovasc Electrophysiol 1997;8:738-44.
184. Kadish A, Shinnar M, Moore EN, et al: Interaction of fiber orientation and direction of impulse propagation with anatomic barriers in anisotropic canine myocardium. Circulation 1988; 78:1478-94.
185. Spach MS, Miller WT, Geselowitz DB, et al: The discontinuous nature of propagation in normal canine cardiac muscle. Evidence for recurrent discontinuities of intracellular resistance that affect the membrane currents. Circ Res 1981;48:39-54.
186. Gourdie RG: A map of the heart: Gap junctions, connexin diversity and retroviral studies of conduction myocyte lineage. Clin Sci (Colch) 1995;88:257-62.
187. Zygmunt AC, Gibbons WR: Properties of the calcium-activated chloride current in heart. J Gen Physiol 1992;99:391-414.
188. Courtemanche M, Ramirez RJ, Nattel S: Ionic targets for drug therapy and atrial fibrillation-induced electrical remodeling: Insights from a mathematical model. Cardiovasc Res 1999;42: 477-89.
189. Hodgkin AL: A note on conduction velocity. J Physiol (Lond) 1954;125:221-4.
190. Buchanan JW Jr, Saito T, Gettes LS: The effects of antiarrhythmic drugs, stimulation frequency, and potassium-induced resting

membrane potential changes on conduction velocity and dV/dtmax in guinea pig myocardium. Circ Res 1985;56:696-703.
191. Sugiura H, Joyner RW: Action potential conduction between guinea pig ventricular cells can be modulated by calcium current. Am J Physiol 1992;263:H1591-604.
192. Shaw RM, Rudy Y: Ionic mechanisms of propagation in cardiac tissue. Roles of the sodium and L-type calcium currents during reduced excitability and decreased gap junction coupling. Circ Res 1997;81:727-41.
193. Spach MS, Miller WT, Dolber PC, et al: The functional role of structural complexities in the propagation of depolarization in the atrium of the dog. Cardiac conduction disturbances due to discontinuities of effective axial resistivity. Circ Res 1982;50:175-91.
194. Mendez C, Mueller WJ, Urguiaga X: Propagation of impulses across the Purkinje fiber-muscle junctions in the dog heart. Circ Res 1970;26:135-50.
195. Fast VG, Kleber AG: Role of wavefront curvature in propagation of cardiac impulse. Cardiovasc Res 1997;33:258-71.
196. Rohr S, Kucera JP: Involvement of the calcium inward current in cardiac impulse propagation: Induction of unidirectional conduction block by nifedipine and reversal by Bay K 8644. Biophys J 1997;72:754-66.
197. Spear JF, Moore EN: Supernormal excitability and conduction in the His-Purkinje system of the dog. Circ Res 1974;35:782-92.
198. DiFrancesco D: A new interpretation of the pace-maker current in calf Purkinje fibres. J Physiol (Lond) 1981;314:359-76.
199. Yu H, Chang F, Cohen IS: Pacemaker current exists in ventricular myocytes. Circ Res 1993;72:232-6.
200. Sakmann B, Noma A, Trautwein W: Acetylcholine activation of single muscarinic K+ channels in isolated pacemaker cells of the mammalian heart. Nature 1983;303:250-3.
201. Yue DT, Herzig S, Marban E: Beta-adrenergic stimulation of calcium channels occurs by potentiation of high-activity gating modes. Proc Natl Acad Sci U S A 1990;87:753-7.
202. Courtney KR, Sokolove PG: Importance of electrogenic sodium pump in normal and overdriven sinoatrial pacemaker. J Mol Cell Cardiol 1979;11:787-94.
203. Boyett MR, Fedida D: Changes in the electrical activity of dog cardiac Purkinje fibres at high heart rates. J Physiol (Lond) 1984;350:361-91.
204. Katzung BG, Morgenstern JA: Effects of extracellular potassium on ventricular automaticity and evidence for a pacemaker current in mammalian ventricular myocardium. Circ Res 1977;40:105-11.
205. Janse MJ, Wit AL: Electrophysiological mechanisms of ventricular arrhythmias resulting from myocardial ischemia and infarction. Physiol Rev 1989;69:1049-169.
206. Dangman KH, Hoffman BF: Studies on overdrive stimulation of canine cardiac Purkinje fibers: Maximal diastolic potential as a determinant of the response. J Am Coll Cardiol 1983;2:1183-90.
207. Watanabe EI, Honjo H, Boyett MR, et al: Inactivation of the calcium current is involved in overdrive suppression of rabbit sinoatrial node cells. Am J Physiol 1996;271:H2097-107.
208. Cranefield PF: Action potentials, afterpotentials, and arrhythmias. Circ Res 1977;41:415-23.
209. Rosen MR, Gelband H, Merker C, Hoffman BF: Mechanisms of digitalis toxicity. Effects of ouabain on phase four of canine Purkinje fiber transmembrane potentials. Circulation 1973;47:681-9.
210. Ferrier GR, Moe GK: Effect of calcium on acetylstrophanthidin-induced transient depolarizations in canine Purkinje tissue. Circ Res 1973;33:508-15.
211. Marban E, Robinson SW, Wier WG: Mechanisms of arrhythmogenic delayed and early afterdepolarizations in ferret ventricular muscle. J Clin Invest 1986;78:1185-92.
212. Lederer WJ, Tsien RW: Transient inward current underlying arrhythmogenic effects of cardiotonic steroids in Purkinje fibres. J Physiol (Lond) 1976;263:73-100.
213. Han X, Ferrier GR: Contribution of Na^+-Ca^{2+} exchange to stimulation of transient inward current by isoproterenol in rabbit cardiac Purkinje fibers. Circ Res 1995;76:664-74.
214. Pogwizd SM, Onufer JR, Kramer JB, et al: Induction of delayed afterdepolarizations and triggered activity in canine Purkinje fibers by lysophosphoglycerides. Circ Res 1986;59:416-26.
215. Ter Keurs HEDJ, Boyden PA: Ca^{2+} and arrhythmias. In Spooner P, Rosen MR (ed): Foundations of Cardiac Arrhythmias—Basic Concepts and Clinical Approaches. New York, Marcel Dekker, 2001, pp 287-317.
216. Wasserstrom JA, Ferrier GR: Voltage dependence of digitalis afterpotentials, aftercontractions, and inotropy. Am J Physiol 1981;241:H646-53.
217. Priori SG, Napolitano C, Tiso N, et al: Mutations in the cardiac ryanodine receptor gene (hRyR2) underlie catecholaminergic polymorphic ventricular tachycardia. Circulation 2001;103:196-200.
218. Wu Y, MacMillan LB, McNeill RB, et al: CaM kinase augments cardiac L-type Ca^{2+} current: A cellular mechanism for long Q-T arrhythmias. Am J Physiol 1999;276:H2168-78.
219. January CT, Riddle JM: Early afterdepolarizations: Mechanism of induction and block. A role for L-type Ca^{2+} current. Circ Res 1989;64:977-90.
220. Szabo B, Sweidan R, Rajagopalan CV, Lazzara R: Role of Na+:Ca^{2+} exchange current in Cs^+-induced early afterdepolarizations in Purkinje fibers. J Cardiovasc Electrophysiol 1994;5:933-44.
221. Damiano BP, Rosen MR: Effects of pacing on triggered activity induced by early afterdepolarizations. Circulation 1984;69:1013-25.
222. Roden D, Lazzara R, Rosen M, et al: Multiple mechanisms in the long-QT syndrome. Current knowledge, gaps, and future directions. Circulation 1996;94:1996-1202.
223. Shibasaki T: Conductance and kinetics of delayed rectifier potassium channels in nodal cells of the rabbit heart. J Physiol (Lond) 1987;387:227-50.
224. Sanguinetti MC, Jurkiewicz NK: Role of external Ca^{2+} and K+ in gating of cardiac delayed rectifier K+ currents. Pflugers Arch 1992;420:180-6.
225. Compton SJ, Lux RL, Ramsey MR, et al: Genetically defined therapy of inherited long-QT syndrome. Correction of abnormal repolarization by potassium [see comments]. Circulation 1996;94:1018-22.
226. Choy AM, Lang CC, Chomsky DM, et al: Normalization of acquired QT prolongation in humans by intravenous potassium. Circulation 1997;96:2149-54.
227. Tomaselli GF, Beuckelmann DJ, Calkins HG, et al: Sudden cardiac death in heart failure. The role of abnormal repolarization [see comments]. Circulation 1994;90:2534-9.
228. Mines GR: On circulating excitations in heart muscles and their possible relation to tachycardia and fibrillation. Trans R Soc Can 1914;IV:43-52.
229. Rohr S, Kucera JP, Fast VG, Kleber AG: Paradoxical improvement of impulse conduction in cardiac tissue by partial cellular uncoupling. Science 1997;275:841-4.
230. Spach MS, Dolber PC: Relating extracellular potentials and their derivatives to anisotropic propagation at a microscopic level in human cardiac muscle. Evidence for electrical uncoupling of side-to-side fiber connections with increasing age. Circ Res 1986;58:356-71.
231. Allessie MA, Bonke FI, Schopman FJ: Circus movement in rabbit atrial muscle as a mechanism of trachycardia. Circ Res 1973;33:54-62.
232. Hoyt RH, Cohen ML, Corr PB, Saffitz JE: Alterations of intercellular junctions induced by hypoxia in canine myocardium. Am J Physiol 1990;258:H1439-48.
233. Peters NS, Green CR, Poole-Wilson PA, Severs NJ: Reduced content of connexin43 gap junctions in ventricular myocardium from hypertrophied and ischemic human hearts. Circulation 1993;88:864-75.
234. Guerrero PA, Schuessler RB, Davis LM, et al: Slow ventricular conduction in mice heterozygous for a connexin43 null mutation. J Clin Invest 1997;99:1991-8.
235. Smith JH, Green CR, Peters NS, et al: Altered patterns of gap junction distribution in ischemic heart disease. An immunohistochemical study of human myocardium using laser scanning confocal microscopy. Am J Pathol 1991;139:801-21.
236. Peters NS, Coromilas J, Severs NJ, Wit AL: Disturbed connexin43 gap junction distribution correlates with the location of reentrant circuits in the epicardial border zone of healing canine infarcts that cause ventricular tachycardia. Circulation 1997;95:988-96.
237. Warmke JW, Ganetzky B: A family of potassium channel genes related to eag in Drosophila and mammals. Proc Natl Acad Sci U S A 1994;91:3438-42.
238. Philipson LH, Schaefer K, LaMendola J, et al: Sequence of a human fetal skeletal muscle potassium channel cDNA related to RCK4. Nucleic Acids Res 1990;18:7160.

239. Tamkun MM, Knoth KM, Walbridge JA, et al: Molecular cloning and characterization of two voltage-gated K+ channel cDNAs from human ventricle. FASEB J 1991;5:331-7.
240. Kong W, Po S, Yamagishi T, et al: Isolation and characterization of the human gene encoding the transient outward potassium current: Further diversity by alternative mRNA splicing. Am J Physiol 1998;275:H1963-70.
241. Ashford ML, Bond CT, Blair TA, Adelman JP: Cloning and functional expression of a rat heart KATP channel. Nature 1994; 370:456-9.
242. Chan KW, Langan MN, Sui JL, et al: A recombinant inwardly rectifying potassium channel coupled to GTP-binding proteins. J Gen Physiol 1996;107:381-97.
243. Vaccari T, Moroni A, Rocchi M, et al: The human gene coding for HCN2, a pacemaker channel of the heart. Biochim Biophys Acta 1999;1446:419-25.
244. Inagaki N, Gonoi T, Clement JP, et al: A family of sulfonylurea receptors determines the pharmacological properties of ATP-sensitive K+ channels. Neuron 1996;16:1011-7.
245. Powers PA, Gregg RG, Lalley PA, et al: Assignment of the human gene for the alpha 1 subunit of the cardiac DHP-sensitive Ca^{2+} channel (CCHL1A1) to chromosome 12p12-pter. Genomics 1991;10:835-9.
246. Powers PA, Liu S, Hogan K, Gregg RG: Skeletal muscle and brain isoforms of a beta-subunit of human voltage-dependent calcium channels are encoded by a single gene. J Biol Chem 1992;267:22967-72.
247. Rosenfeld MR, Wong E, Dalmau J, et al: Cloning and characterization of a Lambert-Eaton myasthenic syndrome antigen. Ann Neurol 1993;33:113-20.
248. Nagase T, Ishikawa K, Miyajima N, et al: Prediction of the coding sequences of unidentified human genes. IX. The complete sequences of 100 new cDNA clones from brain which can code for large proteins in vitro. DNA Res 1998;5:31-9.
249. Makita N, Bennett PB Jr, George AL Jr: Voltage-gated Na+ channel beta 1 subunit mRNA expressed in adult human skeletal muscle, heart, and brain is encoded by a single gene. J Biol Chem 1994;269:7571-8.
250. Eubanks J, Srinivasan J, Dinulos MB, et al: Structure and chromosomal localization of the beta2 subunit of the human brain sodium channel. Neuroreport 1997;8:2775-9.
251. Fishman GI, Spray DC, Leinwand LA: Molecular characterization and functional expression of the human cardiac gap junction channel. J Cell Biol 1990;111:589-98.
252. Kanter HL, Saffitz JE, Beyer EC: Molecular cloning of two human cardiac gap junction proteins, connexin40 and connexin45. J Mol Cell Cardiol 1994;26:861-8.
253. Kawakami K, Ohta T, Nojima H, Nagano K: Primary structure of the alpha-subunit of human Na,K-ATPase deduced from cDNA sequence. J Biochem (Tokyo) 1986;100:389-97.
254. Yang-Feng TL, Schneider JW, Lindgren V, et al: Chromosomal localization of human Na+, K+-ATPase alpha- and beta-subunit genes. Genomics 1988;2:128-38.
255. Ovchinnikov Yu A, Monastyrskaya GS, Broude NE, et al. Family of human Na+, K+-ATPase genes. Structure of the gene for the catalytic subunit (alpha III-form) and its relationship with structural features of the protein. FEBS Lett 1988;233:87-94.
256. Kawakami K, Nojima H, Ohta T, Nagano K: Molecular cloning and sequence analysis of human Na,K-ATPase beta-subunit. Nucleic Acids Res 1986;14:2833-44.
257. Martin-Vasallo P, Dackowski W, Emanuel JR, Levenson R: Identification of a putative isoform of the Na,K-ATPase beta subunit. Primary structure and tissue-specific expression. J Biol Chem 1989;264:4613-8.

Chapter 2 Mechanisms of Reentrant Arrhythmias

JOSÉ JALIFE and FARAMARZ H. SAMIE

Today it is generally believed that the most deadly cardiac arrhythmias result from electrical waves that rotate at a high frequency, in a self-sustaining manner, and give rise to electrical activity that propagates throughout the ventricles in a complex fashion.[1,2] It is thought that the responsible agent for the spontaneous formation of reentry is often the formation of circuits[3] or vortices[4,5] produced by the interaction of a propagating wavefront with an obstacle. It should be noted, however, that reentry could also be initiated by automatic pacemaker discharge as well as triggered activity (i.e., early or delayed afterdepolarizations).[6,7] The objective of this chapter is to provide a brief historical review of the literature on the mechanisms of initiation and maintenance of reentrant arrhythmias, including those mechanisms thought to underlie tachycardia and fibrillation. Emphasis is placed on the most recent concepts derived from the theory of nonlinear wave of propagation in generic excitable media and their contribution to our understanding of reentry and ventricular fibrillation (VF).

What Is Reentry?

In its simplest form, reentry is the circulation of the cardiac impulse around an obstacle, leading to repetitive excitation of the heart at a frequency that depends on the conduction velocity and the perimeter of the obstacle (Fig. 2-1). According to the original description of George Mines in 1913,[8] reentry occurs around a fixed anatomic obstacle, and the physical disruption of the surrounding circuit will interrupt the activity. As illustrated in Figure 2-1, the initiation of the reentrant activity depends on the occurrence of unidirectional block so that activation occurs only in one direction within the circuit.

It is clear from Figure 2-1 that the rotation time around the circuit should be longer than the recovery period of all segments of the circuit. This excess time required for the impulse to successfully complete a rotation may result from a relatively large circuit, a relatively slow conduction velocity of the impulse, or the relatively short duration of the refractory period. Hence, the "wavelength," which may be calculated roughly as the product of the refractory period times the conduction velocity,[9] must be shorter than the perimeter of the circuit. An excitable region will separate the front of the impulse from its own refractory tail (i.e., excitable

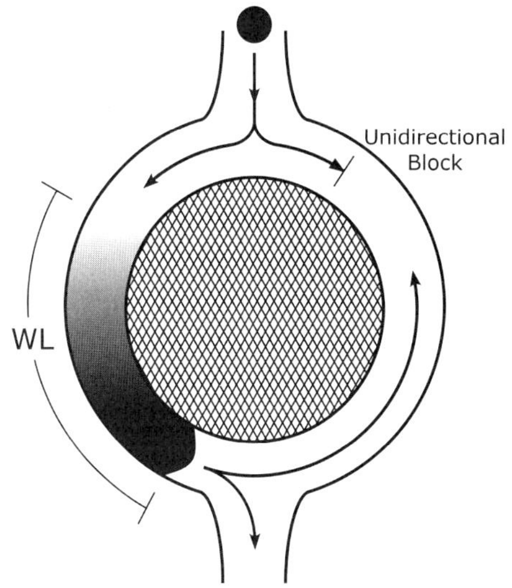

FIGURE 2-1 Circus movement reentry around a ring of heterogeneous tissue surrounding an anatomic obstacle. Reentry is initiated by the application of a premature stimulus (black dot) to the upper branch. As the impulse enters the ring, it encounters tissue recovered on the left side. However, the tissue on the right has not yet recovered from previous excitation (not shown) and unidirectional block occurs. As a result, the wavefront begins to rotate around the obstacle. If the pathlength is long enough or the conduction velocity is slow enough, there will be sufficient time for recovery on the upper right side of the ring and sustained reentry will be initiated. Note that in this hypothetical example, the wavelength *(WL)*, which is equal to the product of conduction velocity *(CV)* times refractory period *(RP)*, is much shorter than the pathlength.

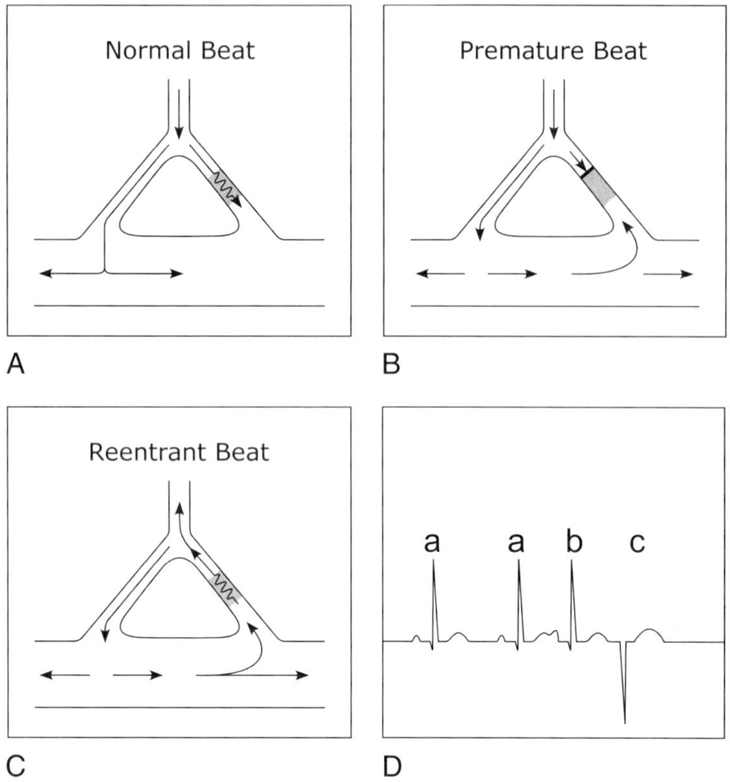

FIGURE 2-2 Schematic representation of circus movement reentry at the Purkinje-muscle junction. A small Purkinje fiber gives rise to two terminal branches that connect directly to ventricular muscle. **A,** Region of low excitability at the right branch (*shaded area*) leads to slowing of conduction, while on the left branch normal propagation rapidly leads to activation of the ventricle. **B,** The region of low excitability prevents penetration of the impulse coming from the top bundle and results in unidirectional block. The impulse that makes it through the left branch and the muscle begins to enter the right branch in a retrograde manner. **C,** Slow retrograde propagation across the area of low excitability permits expiration of the refractory period at the junction with the main branch, and a reentrant beat ensues. **D,** Electrocardiogram manifestation of events occurring in **A-C;** *a,* normal beats; *b,* premature beat; *c,* reentrant beat.

gap) and re-excitation will ensue. The traditional scheme used to describe reentrant activation[10] is displayed in Figure 2-2, where a Purkinje fiber is shown attached to the ventricular myocardium by two terminal branches. In such a scheme, the first prerequisite for reentry is met (i.e., the presence of a predetermined circuit). There is a region of impaired (slow) conduction in one of the terminal branches *(shaded area in the right branch),* which also recovers slowly from previous excitation (this region may represent an area of ischemia where the conditions for both slow conduction and excessively long recovery time are usually encountered). This region may provide the stage for the unidirectional block that is needed for the initiation, as well as the slow conduction that is appropriate for the maintenance reentry. *Panel A* represents activation of the tissue when the Purkinje fiber branches are excited by a beat of sinus origin at a relatively slow frequency. Because of the area of impairment, there is some delay in the activation of the right branch. Yet the impulse moves slowly through that area and eventually reaches the attached myocardium. The result, as recorded by an ECG *(Panel D),* would be a normal QRS complex *(a). Panel B* illustrates the dynamics of propagation across the same anatomic circuit during premature activation (e.g., as a result of sinus nodal tachycardia, an atrial extrasystole, or an electrical stimulus that follows the previous impulse very closely in time). This is represented on ECG as a premature P wave leading to early activation of the ventricles (complex labeled *b*). As shown in *Panel B,* under these conditions, the tissue in the slow conduction region is not yet fully recovered from previous excitation, and anterograde block ensues. Over the left branch, however, the impulse moves unimpaired to activate the ventricular myocardium, and then continues retrogradely to excite the already recovered right branch. *Panel C* shows the impulse reaching the initial site of activation after exciting the entire right branch. The ECG manifestation of this process would be in the form of a ventricular extrasystole *(complex c in panel D).*

Reentry is responsible for various arrhythmias including supraventricular and ventricular extrasystoles,[11] atrial flutter,[12] atrioventricular (AV) nodal reciprocating tachycardias,[13] supraventricular tachycardias associated with accessory AV pathways,[14] bundle branch ventricular tachycardias,[15] and monomorphic ventricular tachycardias associated with myocardial infarction.[16] There is also strong evidence to suggest than more complex arrhythmias such as atrial fibrillation,[17] polymorphic ventricular tachycardia,[18] and VF[19] are the result of reentrant mechanisms, which are somewhat different from those depicted in Figures 2-1 and 2-2.

The classic model of anatomically determined reentry is directly applicable to specific cases of tachyarrhythmias. These include supraventricular tachycardias occurring within the AV node[13] or those using accessory pathways,[14] and bundle branch reentrant tachycardia.[15] However, other types of reentrant arrhythmias require somewhat different mechanistic explanations. For example, the cellular basis of closely coupled ventricular extrasystoles initiated somewhere in the Purkinje fiber network can be explained by the so-called "reflection" mechanism[20] (Fig. 2-3*A*). On the other hand, many

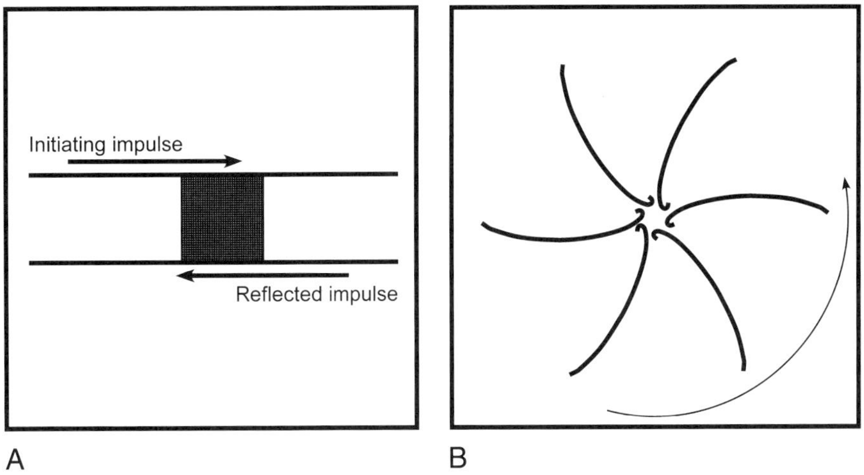

FIGURE 2-3 Two different forms of functional reentry (i.e., in the absence of an anatomic obstacle). **A,** Reflection, where reentry occurs over a single pathway in a linear bundle (e.g., a Purkinje fiber) across an area of depressed excitability *(shaded).* **B,** Functional reentry in two-dimensional myocardium. *Curved lines* are isochrone lines showing consecutive positions of the wavefront. The *curved arrow* indicates the direction of rotation.

tachyarrhythmias that originate in the myocardium (atrial or ventricular) require mechanisms whereby reentrant activation may occur as vortices of electrical excitation rotating over an area of myocardium, in the absence of a predetermined obstacle or circuit (Fig. 2-3*B*). Accordingly, the impulse must circulate around a region of quiescence. The most widely accepted hypothesis used to explain such functionally determined reentry is the so-called "leading-circle" hypothesis,[21] with its two variants of "anisotropic"[22] and "figure-of-8" reentry.[23] A somewhat different postulate for vortex-like reentry, the "spiral wave reentry" hypothesis,[24] is derived from the theory of wave propagation in excitable media[25] and attempts to provide a unifying explanation for the mechanisms of monomorphic and polymorphic ventricular tachycardias, as well as fibrillation.

Let us now focus on the characteristics of reflection and circus movement reentry. Then we will review briefly the concept of the leading circle and contrast it with the more recent ideas on spiral wave reentry. The last sections of this chapter are devoted to the most recent work supporting the applicability of the spiral wave concept to the mechanism of cardiac fibrillation.

Reflection

The presence of a region of severely impaired conduction (but not complete block) in a linear pathway (e.g., a Purkinje fiber or a thin muscle trabecula) may give rise to reentrant excitation even in the absence of an anatomic circuit (Fig. 2-4). In the reflection model of reentry, back and forth activation occurs over the same pathway.[20,26] Because of the simplicity of the experimental model, reflection has been used to analyze in detail the effects of various conditions (e.g., stimulation rate, antiarrhythmic agents, ischemia) on the manifestation of reentrant activity, specifically single reentrant excitation or extrasystoles.[20,26-28]

A convenient approach to study reflection is the sucrose gap preparation.[20,26] As shown in Figure 2-4, the sucrose gap preparation consists of a linear Purkinje fiber bundle that is excised from the endocardium of a dog or sheep ventricle and placed in a three-compartment tissue bath. Each compartment is perfused independently. The central chamber is perfused with an ion-free solution containing isotonic (≈300 mM) sucrose. The cells in the central segment thus become unexcitable, even though they remain connected to the cells of the two outer segments. As illustrated in Figure 2-4, an external bridge (a silver chloride wire with variable resistance determined by a potentiometer) is used to connect the extracellular fluid of the outer chambers. Such a bridge is used to modulate the degree of conduction block across the central compartment (for details see Antzelevitch,[20] Jalife and Moe[26]).

In Figure 2-4, when the proximal *(P)* end of the fiber is stimulated, an action potential *(AP)* propagates toward the central segment. Since cells in the central segment (the gap) are not excitable (because of the lack of sodium ions in the extracellular space that bathes them), the AP is unable to propagate through

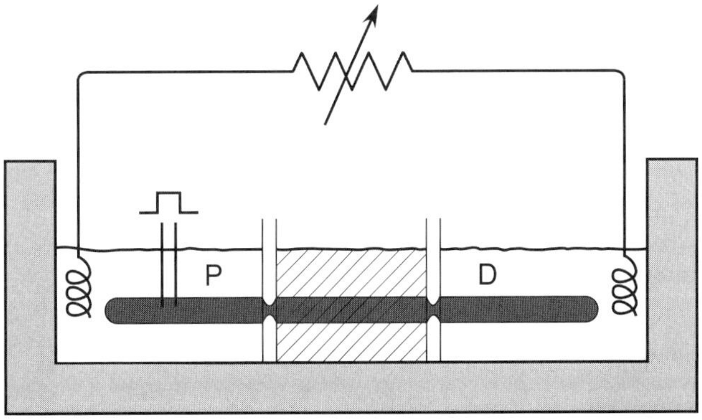

FIGURE 2-4 The sucrose gap preparation used to demonstrate reflection. A Purkinje fiber, depicted in *dark gray,* is placed in a three-compartment tissue bath. The central compartment is perfused with an ion-free sucrose solution. The outer compartments are perfused with normal Tyrode's solution. Stimuli are delivered to the proximal side *(P),* and recordings are obtained from both here and the distal side *(D).* An external wire with a variable resistance is used to shunt the sucrose compartment.

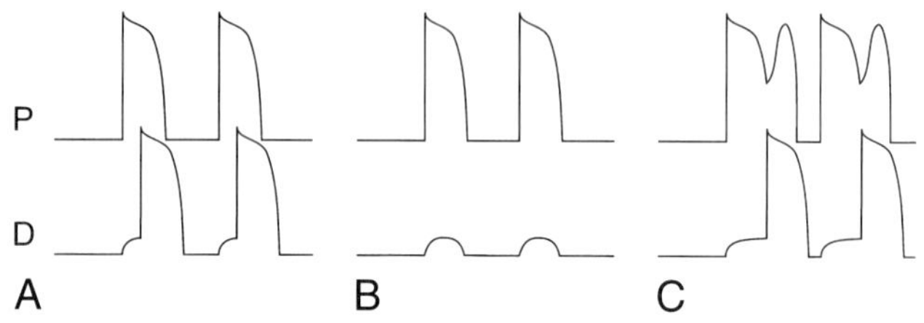

FIGURE 2-5 Conduction delay, block, and reflection across an area of depressed excitability. Action potentials were recorded by two microelectrodes, one located in the proximal *(P)* and the other in the distal *(D)* segments of a sucrose gap preparation (see Fig. 2-4). **A,** When the shunt resistance is relatively low, propagation across the gap is successful but slow. **B,** When the shunt resistance is extremely high, there is complete failure of propagation with only subthreshold depolarizations manifested in the *D* segment. **C,** At an intermediate level of shunt resistance, propagation is again successful, but the delay from *P* to *D* is much longer than in panel *A*. Consequently, the action potential generated in *D* occurs with enough delay to allow expiration of the refractory period in *P*, and reflection will occur. In this example each *P* discharge is followed by a reflected (R) discharge.

these cells. Yet the local circuit current generated at the site where the AP stops is sufficient to depolarize passively (i.e., electrotonically) the inexcitable cells near the boundary between the proximal and distal *(D)* segments. Electrotonic current decays rapidly with distance.[29] Thus, if the length of the gap is relatively small (≈1 mm), enough current may reach the distal segment to bring the membrane of those cells to threshold and initiate an AP after an appreciable delay.[20,26] In this case, propagation from proximal to distal segments is successful. The amount of current reaching the distal segment (i.e., the source) and the requirement of current for excitation of the distal segment (i.e., the sink) will determine whether propagation is successful, as well as the time required for excitation of the distal segment.[20,26] As shown in Figure 2-5*A*, when the balance between source and sink is appropriate, proximal to distal conduction is not significantly affected. When the source current is decreased appreciably or the sink requirements are too high (i.e., low distal excitability), or both, there may be complete block. This is shown in Figure 2-5*B*, whereby only a small (local) depolarization is observed in the distal segment following each proximal AP. Finally, Figure 2-5*C* shows a situation in which the balance between source and sink is such that slow conduction occurs. Under these conditions, proximal to distal propagation may be so slow that the distal AP occurs when the proximal segment has already recovered from activation. Thus, the distal segment becomes the source and reactivates the proximal segment (i.e., there is reflection).

Circus Movement Reentry

Undoubtedly, the concept of circus movement reentry, in which a cardiac impulse travels around a predetermined circuit or around an anatomic obstacle, can be applied successfully to various clinical situations. Two clear examples of reentrant arrhythmias based on the circus movement mechanisms are: supraventricular tachycardias observed in patients with Wolff-Parkinson-White (WPW) syndrome[14] and bundle branch reentrant ventricular tachycardia,[15] which is more commonly seen in patients with idiopathic dilated cardiomyopathy. All conditions required by the original idea[8] of circus movement reentry may be found in these two types arrhythmias, as follows:

1. There is a need for an intact predetermined anatomic circuit. As shown schematically in Figure 2-6*A*, in the case of the Wolff-Parkinson-White (WPW) syndrome, various types of structures, including the AV node, the His-Purkinje system, ventricular muscle, and an accessory atrioventricular pathway, form the circuit. In the case of bundle branch reentry (Fig. 2-6*B*), the circuit is composed of the main bundle branches and the interventricular septum. The need for the integrity of the circuit is demonstrated by the fact that physical interruption of the circuit at any point leads to the interruption of the arrhythmia.
2. There must be unidirectional block before the onset of the reentrant activity. In most cases, unidirectional block occurs in the region of longest refractory period and is the result of an increase in heart rate. Unidirectional block may occur as a result of various conditions including: (1) increase in sinus rate; (2) rapid or premature atrial pacing; (3) retrograde activation from a ventricular extrasystole; (4) autonomic influences; (5) antiarrhythmic drugs; and (6) ischemia.
3. Slow conduction in part of the circuit facilitates reentry. In the case of WPW syndrome, the arrhythmia may begin after significant prolongation of the anterograde AV nodal conduction time. The activation of the ventricles occurs when both accessory pathway and atria are recovered. This leads to retrograde activation of the accessory pathway and initiation of reentrant arrhythmia.
4. The wavelength of the impulse must be shorter than the length of the circuit. As shown in Figure 2-7, there is a segment within the circuit that remains excitable during reentrant activity. The presence of an excitable gap has major significance for various reasons: (a) The reentrant activity will likely be stable in the presence of an excitable gap because the reentrant wavefront will find only fully recovered tissue in its path; (b) the activity may be entrained or interrupted by means of external stimulation, or both (see later). An externally initiated impulse may invade the circuit during the excitable gap and thus advance the activation front. Depending on the timing or the rate of external stimulation, the wavefront may be premature enough to collide with the repolarizing tail and thus terminate the activity; and (c) agents that prolong the refractory period may not affect the reentrant process unless the prolongation of refractoriness totally obliterates the excitable gap.

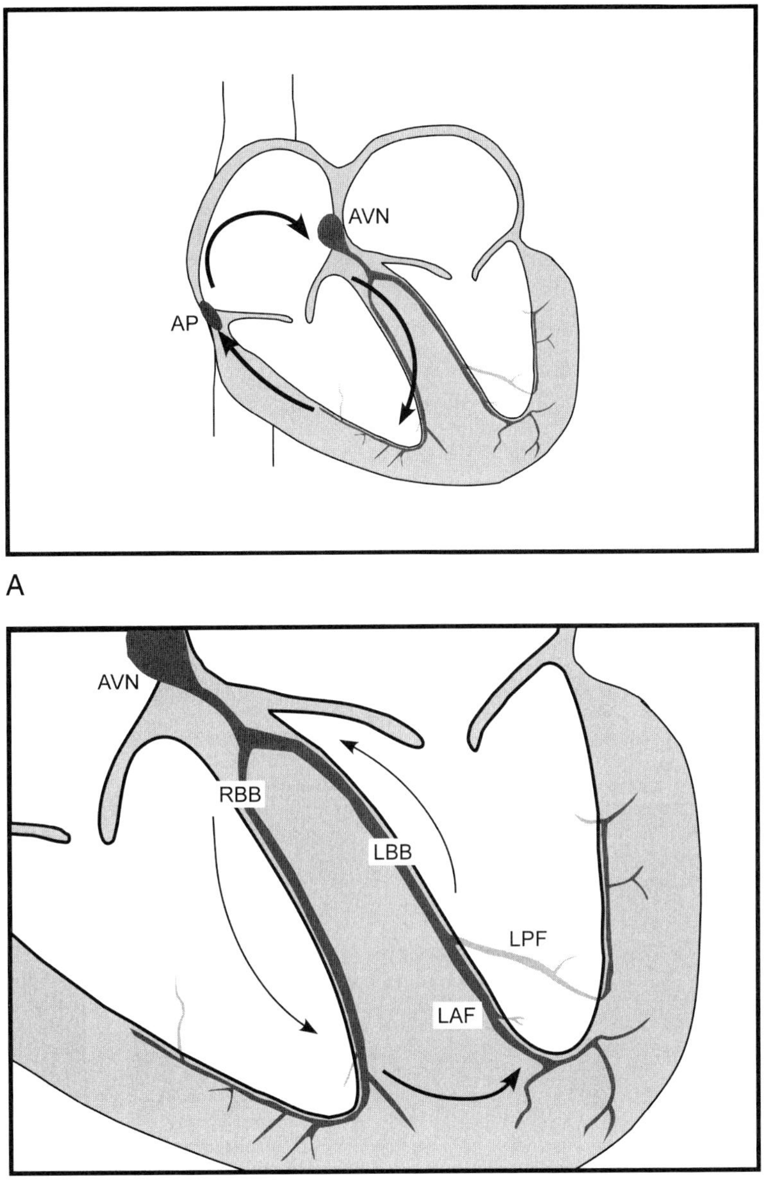

FIGURE 2-6 **A,** Atrioventricular reentry in the presence of an accessory pathway *(AP).* **B,** Bundle branch reentry using the right bundle branch *(RBB)* and the left anterior fascicle *(LAF)* as the two major components of the circuit. *AVN,* Atrioventricular node; *LBB,* left bundle branch; *LPF,* left posterior fascicle.

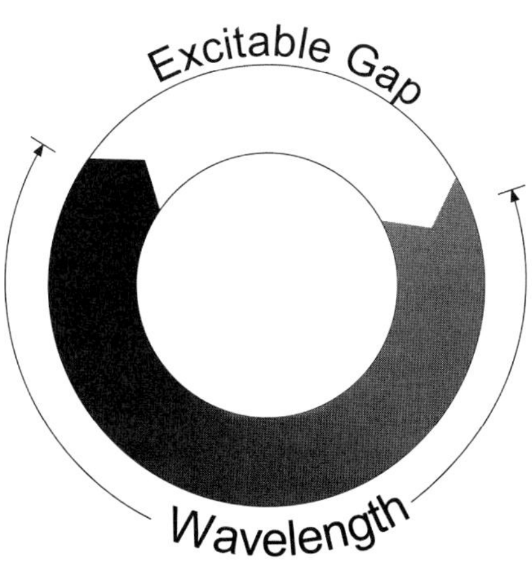

FIGURE 2-7 Ring-type model of reentry, which requires a fully excitable gap. See text for further discussion.

Functionally Determined Reentry

Circus movement reentry results when an electrical impulse propagates around a one-dimensional circuit or ring-like structure.[8] Although the model is entirely applicable to arrhythmias such as those observed in the presence of atrioventricular accessory pathways, it may not represent a realistic model for reentrant arrhythmias occurring in the atria or ventricles. Reentrant activity may indeed occur in the absence of a predetermined circuit.[21,30,31] Furthermore, the electrical impulse may rotate around a region that is anatomically normal and uniform but functionally discontinuous.[25]

In 1924, Garrey[32] presented the first description of reentrant excitation in the absence of anatomic obstacles in experimental studies on circus movement in the turtle heart. Garrey's observations suggested that point stimulation of the atrium was sufficient to initiate a regular wave of rotation around the stimulus site. Subsequently, in 1946, Wiener and Rosenblueth[9] developed the first mathematical model of circus movement reentry, which supported waves of rotation around a sufficiently large barrier, but they could not demonstrate reentry in the absence of an obstacle. This prompted Wiener and Rosenblueth to suggest that perhaps Garrey may have unwittingly produced a transient artificial obstacle near the stimulation site.

The "Leading Circle" Model

In 1973 Maurits Allessie and his associates at the University of Limburg in Maastricht, the Netherlands provided the first direct experimental demonstration that the presence of an anatomic obstacle is not essential for the initiation or maintenance of reentry.[30] These authors studied the mechanism of tachycardia in small pieces of isolated rabbit left atrium by the application of single premature stimuli. Through multiple electrode mapping techniques they demonstrated that the tachycardias were based on rotating waves (Fig. 2-8) and suggested that such waves were initiated as a result of unidirectional block of the triggering premature input.[21,30,31] Transmembrane potential recordings demonstrated that cells at the center of the vortex were not excited but developed local responses.[30] It was hypothesized that such depolarizations led to some degree of refractoriness and served as a functional obstacle around which the impulse rotated. These observations were the basis for the development of the "leading-circle" concept of functional reentry.[21]

According to the leading-circle concept, in the absence of an anatomic obstacle, the dynamics of reentry are determined by the smallest possible loop in which the impulse can continue to circulate.[21,30,31] As depicted in Figure 2-8, under these conditions, the wavefront must propagate through relatively refractory tissue, in which case there will be no "fully excitable gap" and the wavelength will be very close to the length of the circuit. Thus, there are several differences between circus movement reentry occurring around fixed anatomic obstacle and leading-circle reentry: (1) Because in leading-circle reentry there is no anatomically determined circuit, there is no theoretical possibility of interrupting the arrhythmia by disrupting the circuit; (2) the absence of an excitable gap makes the arrhythmia unstable. That is, relatively small variations in the electrophysiologic characteristics of the tissues involved (e.g., a small increase in the refractory period) may result in a change in the cycle length of the arrhythmia or, eventually, termination of the activity. Also as a result of the absence of an excitable gap, leading-circle reentry would be expected to be insensitive to electrical stimulation. Thus, entrainment[33] and annihilation of the arrhythmia by externally applied stimuli is theoretically very unlikely; (3) finally, when compared with reentrant activity around an anatomic obstacle, leading-circle type reentry is expected to have a shorter cycle length.

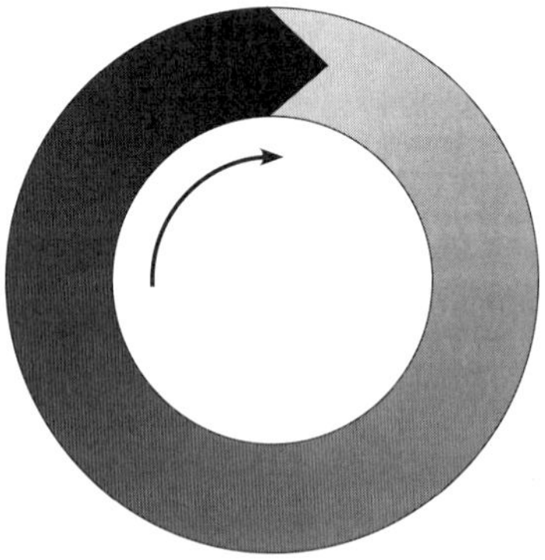

FIGURE 2-8 Leading-circle–type reentry, where there is no fully excitable gap and the wavefront *(black)* "bites" its tail of refractoriness *(light gray)*. The *arrow* shows the direction of propagation.

As discussed in detail later, while the leading circle idea has paved the way for major advances in our understanding of functional reentry, over the past several years, many of its original predictions were proven inaccurate by multiple experimental observations (for review see Jalife et al.[34]). In addition, the model of the leading circle seems incompatible with some of the major properties of functionally determined reentry that are commonly observed experimentally in normal cardiac muscle, including the phenomenon of reentry "drift,"[4,5] which results in beat-to-beat changes in the location of the rotation center (see Drifting Vortices and VF).

Anisotropic Reentry

Work in the 1980s implicated microscopic structural complexities of the cardiac muscle in the mechanism of reentrant activation in both atria and ventricles, particularly in relation to the orientation of the myocardial fibers, the manner in which the fibers and fiber bundles are connected to each other, and the effective electrical resistivities that depend on the fiber orientation.[35-38] Indeed, it is well known that AP propagation in the heart is determined not only by the electrical properties associated with cell excitability and refractoriness but also by the high degree of anisotropy in cell-to-cell communication resulting from the specific parallel arrangement of the fiber bundles[35,36,39] and the paucity of the transverse electrical connections between them.[35,36] Consequently, propagation velocity in the cardiac muscle is three to five times faster in the longitudinal axis of the cells than along the transverse axis.[35] In addition, asymmetry in the safety factor for propagation may result in conduction block occurring first in one direction (i.e., unidirectional block). Thus, it has been suggested that structural anisotropy may set the stage for heterogeneity of functional properties and therefore lead to the initiation and maintenance of reentry.[35,36] Mapping studies using multiple extracellular electrodes have shown that, in the setting of myocardial infarction, reentry may occur in the survival epicardial rim of tissue.[22,38,39] Under such conditions, the wave circulates

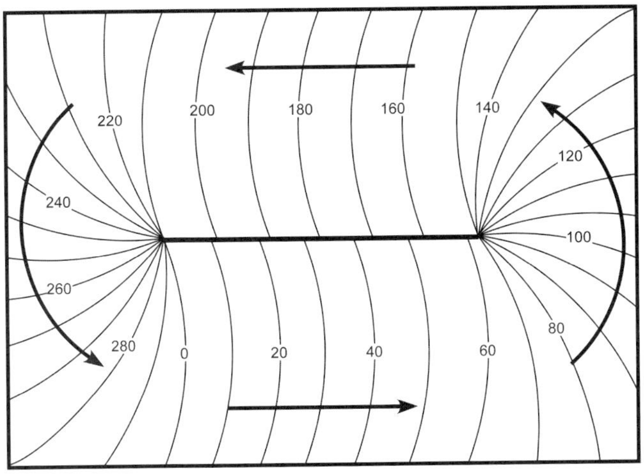

FIGURE 2-9 Anisotropic reentry around a line of block. Curved lines are isochrones. The distance between lines denotes velocity of propagation. Velocity is faster in the horizontal direction than around the pivot points.

around a functionally determined elongated region of block, the so-called *line of conduction block*. Based on the orientation of such a line of block, it was thought that anisotropic propagation played a major role both in the initiation as well as in the maintenance of reentry in ventricular tissue surviving a myocardial infarction (Fig. 2-9). In addition, propagation velocity is exceedingly slow at the edges of the lines of block, which has also been attributed to anisotropic propagation.[22,38,39]

Thus, anisotropic reentry is seen as a model of functionally determined reentry, in which both initiation and maintenance of the activity is based on the histological properties of the tissue.[37,38] However, the true role of anisotropic propagation in determining the reentrant circuit remains unclear. First, the line of block does not follow strictly the direction of the fibers.[40] Furthermore, in most published examples of so-called *anisotropic reentry*, there are beat-to-beat changes in the direction of the line of block, which cannot be attributed to transient changes in anisotropy.[38,40] Most importantly, recent experimental studies have demonstrated that in reentrant circuits occurring around a thin linear anatomic obstacle (i.e., a linear lesion produced by a laser beam), the slowest propagation is observed at the pivot points regardless of the actual orientation of the fibers.[41] Finally, the anisotropic properties of the tissue have also been implicated in the establishment of an excitable gap. This has recently been challenged by computer simulations in which the rotation period and the excitable gap were not significantly modified by the "addition" of anisotropy to the circuit.[5] Therefore, we believe that although anisotropic propagation may play a role in arrhythmogenesis, there are other aspects of wave propagation, such as wavefront curvature (see later discussion) and macroscopic tissue structure, which must be taken into consideration to better understand two- and three-dimensional (3-D) reentry in the myocardium.

"Figure-of-8" Reentry

Figure-of-8 reentry has been recognized as an important pattern of reentry in the late stages of myocardial infarction.[42,43] In most cases, two counter-rotating waves coexist at a relatively short distance from each other (Fig. 2-10). As described for the case of single reentrant circuits, each wave of the figure-of-8 reentry circulates around a thin line or arc of block. The region separating the lines of block is called the "common pathway." A detailed description of the common pathway is of great practical importance since there is evidence that it could be a strategic region for surgical or catheter ablation in this type of reentry. In fact, unlike other forms of functionally determined reentry, figure-of-8 reentry may indeed be interrupted by physical disruption of the circuit. Several studies have attempted to describe the characteristics of propagation in the common pathway.[44-46] However, the properties of the common pathway are still not clearly defined. The common pathway effectively behaves like an isthmus limited by two functionally determined barriers. In addition, there are two wavefronts that interact in the common pathway. As a result, propagation may be determined by a combination of factors other than those analyzed in most experimental studies such as anisotropy. The study of propagation across an isthmus and the influence of wavefront curvature may have significant implications in understanding the properties of the common pathway, as discussed in the next section.

Spiral Wave Reentry

Computer simulations of wave propagation in two-dimensional (2-D) excitable media[47,48] have shown that waves can rotate in such media with periods much higher than the refractory period measured with conventional

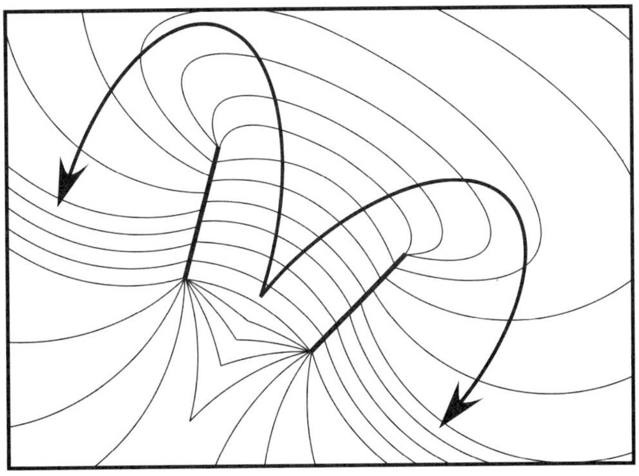

FIGURE 2-10 Figure-of-8 reentry. It consists of two counter-rotating wavefronts, one in the clockwise direction and the other counterclockwise, around their respective lines of block. The wavefronts coalesce in the lower part of the circuits, and they move at varying velocities across the common pathway.

premature stimulation protocols. Moreover, the rotating waves may have a large excitable gap between the wavefront and the repolarizing tail of the previous excitation. Such observations led to the development of a slightly different approach to the problem of ventricular arrhythmias, which is based on the predictions of a few theoretical biologists, mathematicians, and physicists.[48-50] This approach takes advantage of the knowledge gained recently regarding the ubiquitous formation of spiral waves in 2-D media. Details regarding the formation of rotating waves are described later in "Spontaneous Formation of Rotors." Spiral wave reentry differs from the more traditional concept of functional reentry (i.e., the leading circle) in two major aspects: (1) initiation of reentry and (2) circulation of the activity.[34]

According to traditional concepts, circus movement reentry may be initiated in the heart because block is predetermined by the inhomogeneous functional characteristics of the tissue,[51] whereas spiral waves could be formed in the heart even if cardiac muscle was completely homogenous in its functional properties.[48,50,52] This is because the initiation of rotating activity may depend solely on transient local conditions[34] (e.g., the conditions created by cross-field stimulation). Moreover, according to the traditional concept of reentry, the circulation of the activity occurs around an anatomically or functionally predetermined circuit, and the rotating activity cannot drift. In other words, the circuit gives rise to and maintains the rotation.[34] However, spiral waves occur due to initial curling of the wavefront, and in fact, the curvature of the wavefront determines the size and shape of the region, called the *core*, around which activity rotates (Fig. 2-11). Importantly, the core remains unexcited by the extremely curved activation front and it is readily excitable.[53] This explains the mechanism underlying the drift of spirals.[4,5]

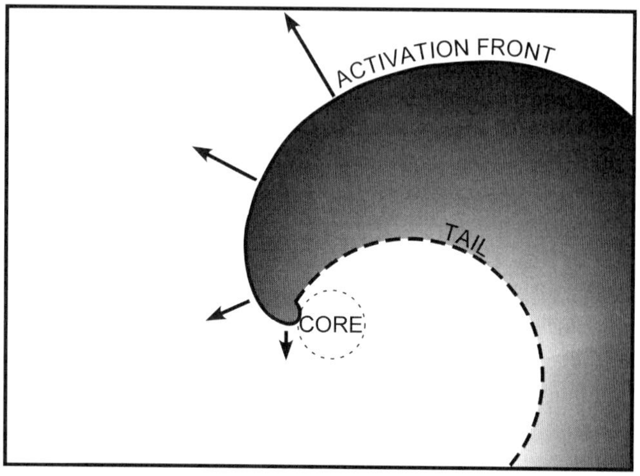

FIGURE 2-11 Spiral wave reentry. The activation front has increasing curvature from the periphery to the center. At the tip, curvature is so extreme that it cannot propagate into the core. Note that the activation front meets its tail of refractoriness (*dotted line*) at a specific point.

Spiral Wave Theory and Ventricular Fibrillation

Today the question still remains whether VF is the result of the random propagation of multiple independent wavelets or multiple wavelets are the consequence of the sustained activity of a single or a small number of reentrant sources activating the ventricles at high frequencies. This area is the subject of great interest for cardiac electrophysiologists and researchers in the field of nonlinear dynamics.

Modes of Initiation of Spiral Wave Reentry

In 1923 De Boer[54] demonstrated that a single electrical shock applied during the late systole to the frog's ventricles induced VF. In 1940 Wiggers and Wégria[55] confirmed De Boer's experimental observation. Furthermore, they demonstrated that the application of a shock to normal hearts of young and old dogs induced fibrillation only when the shock was applied during the late systolic phase, which they called the *vulnerable period*. In the following year Wégria and colleagues demonstrated that the vulnerable period of premature beats is extended, although the fibrillation threshold is not significantly altered in comparison to normal beats.[56] Subsequently, Moe et al. performed a detailed analysis of the initiation of fibrillation during the vulnerable period by electrocardiographic studies.[57] They concluded that repetitive discharges from a center or centers, which are accompanied by a progressive decrease in the refractory period and combined with an increase in the conduction time, are essential to the initiation of VF after a strong electrical shock.[57] Today it is widely known that stimulating the ventricles during the vulnerable period induces VF.

In 1946 Weiner and Rosenblueth published a theoretical description of the mechanisms of initiation of flutter and fibrillation in cardiac muscle in the presence and the absence of anatomic obstacles.[9] They proposed that wave rotation around single or multiple obstacles was required for the initiation and maintenance of both types of arrhythmias, which they assumed to result from a single reentrant mechanism.

More than 3 decades later, another theory of initiation of vortices in two dimensions was suggested, and it has been supported by experiments in a number of different excitable media. It is based on Winfree's "pinwheel experiment" protocol.[49,58] As shown in Figure 2-12, this protocol involves crossing a spatial gradient of momentary stimulus with a spatial gradient of phase (i.e., refractoriness, established by prior passage of an activation front through the medium).[58] In accordance with this theory, when a stimulus of the right size (S*) is given at the proper time, mirror image vortices begin to rotate around crossings of critical contours of transverse gradients of phase and stimulus intensity. The critical phase is roughly similar to the vulnerable phase described by Wiggers and Wégria.[55] Based on this theory,

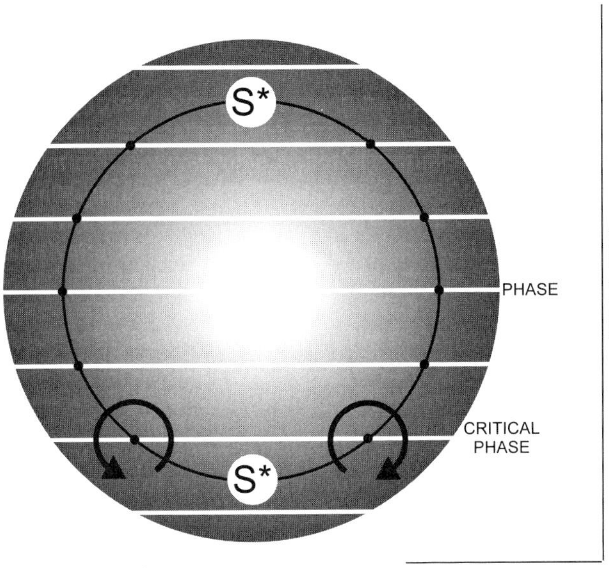

FIGURE 2-12 Winfree's pinwheel experiment.[25,49,58] The *circular surface* represents a two-dimensional sheet of cardiac muscle. The *horizontal white lines* indicate different phases of the action potential. The *white circles* represent the critical magnitude *(S*)* of a stimulus applied at the very center of the tissue. The *black dots* represent different stimuli occurring at the indicated phases. At the crossing of S* with the critical phase, two counter-rotating vortices emerge.

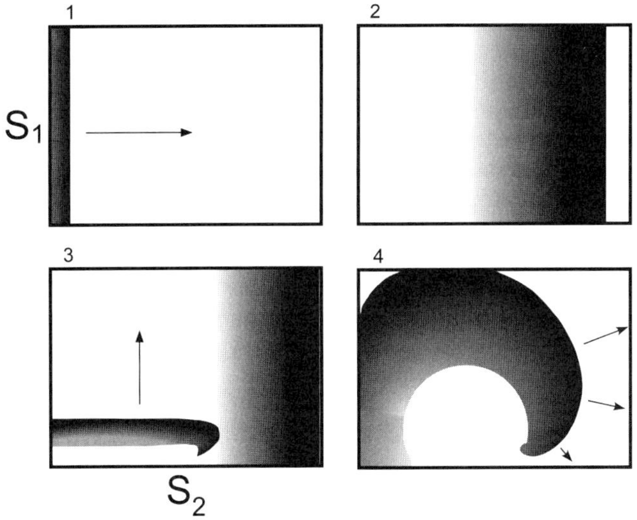

FIGURE 2-13 Cartoon illustrating the cross-field stimulation protocol used to initiate spiral wave (vortex-like) activity in a rectangular sheet of cardiac muscle. At time 1, an S_1 stimulus is applied to the entire left border of the sheet. At time 2, a planar wavefront *(black)* is reaching the right border followed by its tail of refractoriness *(fading gray)*. At time 3, an S_2 stimulus is applied perpendicularly to S_1 when the right border has not yet fully recovered from previous excitation. Consequently, the S_2 wavefront breaks into the refractory tail of S_1 and develops a pronounced curvature. At time 4, the wavefront has curled sufficiently to initiate sustained spiral wave activity.

a vulnerable domain was described. Its timing occurred just before the complete recovery from previous excitation. Thus, with its limits of timing and stimulus intensity, the idea of vulnerable domain was similar to the empirical concept of the vulnerable period. Shibata et al.[59] experimentally demonstrated the application of Winfree's theory[49,58] to the induction of VF in the heart. They concluded that the response to administered shocks during the vulnerable period is complex. However, in accordance with theory, during pacing of the ventricles, if a shock of the proper amplitude and delay is applied during the vulnerable period, two counter rotating vortices can be formed.[59] Thus, as predicted by theory, vortices can be formed even in the normal myocardium.[58,60] Subsequently, Frazier et al.[52] used an extracellular recording array with a modification of the pinwheel experiment, the so called *twin-pulse protocol*, to demonstrate the mechanism of reentry and fibrillation in the dog heart. They used the term *critical point* to refer to a phase singularity and provided strong support for what is referred to as the *critical point hypothesis for the initiation of vortex-like reentry and fibrillation.* They also demonstrated that there is an upper limit of vulnerability for VF such that during the vulnerable period, if shock is applied but its strength is larger than a certain limit, then most likely VF will not be induced.[52]

Another approach for initiating vortices is the cross-field stimulation protocol.[4,5] This method is different from the pinwheel protocol in that it does not require a large stimulus. As shown in Figure 2-13, in cross-field stimulation, a conditioning stimulus (S1) is used to initiate a plane wave propagating in one direction. Subsequently, a second stimulus, S2, is applied perpendicular to S1 and timed in such a way to allow interaction of the S2 wavefront with the recovering tail of the S1 wave. The S2 wavefront cannot invade the refractory tissue at the site of the interaction with the S1 wave tail; consequently, a wave break or phase singularity is formed at the end of the S2 wave, and rotation about this point occurs. Computer simulations and experiments have verified the ability of cross-field stimulation to induce reentry.[61,62] Finally, both the pinwheel and the cross-field stimulation protocols require two different stimuli at different locations.

Reentry and fibrillation can also be induced by rapid stimulation through a single unipolar[63] or bipolar electrode.[64] As suggested by Keener, the discrete nature of cardiac tissue and its structural anisotropy may play a crucial role.[65] More recently, Cheng et al.[66] demonstrated that in rabbit hearts the application of shocks with implantable defibrillator electrodes during the refractory period produced virtual electrode polarization (VEP), with both positive and negative values. They found that after a shock, a new propagated wavefront emerged at the boundary between the two regions and reexcited negatively charged polarized regions.[66] Moreover, wavebreaks were produced, and they degenerated into arrhythmias under appropriate conditions (slow conduction). Thus, the formation of virtual electrodes represents another method for the induction of rotors.

Spontaneous Formation of Rotors

A major contribution of wave propagation theory in excitable media to the understanding of the mechanisms of initiation reentrant arrhythmias is the concept of wavebreak.[67-69] As previously demonstrated in computer simulations and in the Ce-catalyzed Belousov-Zhabotinsky (BZ) reaction,[70] the interaction of a wavefront with an obstacle can lead to wavefront fragmentation and rotor formation. The reentrant wave can begin either as a single vortex,[4,21] as a pair of counter-rotating vortices,[42] or as two pairs of counter-rotating vortices (quatrefoil reentry; for review see Lin et al.[71]).

The concept of wavebreak is illustrated schematically in Figure 2-14, which shows the dynamics of the interaction of a wavefront with an anatomic obstacle in a 2-D sheet of cardiac tissue with two different excitability conditions. In *panel A*, when the tissue excitability is normal, upon circumnavigating the obstacle, the broken ends of the wave join and recover the previous shape of the wavefront and continue. However, in *panel B*, when the excitability is low, the broken ends of the wave do not fuse. Instead, the broken ends rotate in the opposite direction. As illustrated by the diagrams in Figure 2-15, during "normal" propagation, initiated by a linear source (planar wave, *panel A*) or a point source (circular wave, *panel B*), the wavefront is always followed by a recovery band or wave tail. Under these conditions, the front and tail never meet and the distance between them corresponds to the wavelength of excitation. In contrast, as shown in *panel C*, broken waves demonstrate a unique feature whereby the front and tail meet one another at the wavebreak.[67,69] In this situation, the wavefront curls and its velocity decreases toward the wavebreak. In fact, at the wavebreak the curvature is so pronounced that the wavefront fails to activate the tissue ahead. Consequently, the wavebreak effectively serves as a pivoting point, which forces the wavefront to acquire a spiral shape as it rotates around a small central region called the *core*.[72]

Using a generic model of excitable 2-D medium, Pertsov et al.[73] studied the conditions in which a wavebreak forms after collision with an obstacle. They concluded that a wavebreak leads to lateral instabilities, the dynamics of which depend on the existence of a critical curvature for the medium. If the curvature of the front in the region of the break was higher than the critical curvature, then the wave would shrink and result in decremental propagation.[74] However, if the curvature of the front was lower than the critical curvature, the wave would expand and under the proper conditions (i.e., slow expansion and a sufficiently large obstacle), the broken end would curve and give rise to reentrant activity.[73,74]

In cardiac tissue, multitudes of obstacles, both anatomic and functional, are present. However, the excitation of the heart, which is triggered by signals that originate in the sinus node and subsequently propagate throughout the atria and the ventricles, occurs repeatedly in a rhythmic manner. This process occurs without the induction of arrhythmias because the normal sequence of activation through the His-Purkinje system prevents the formation of wavebreaks. Consequently, the presence of obstacles is not a sufficient condition for the establishment of reentry. Cabo et al.,[75] using a voltage-sensitive dye in conjunction with a high resolution video imaging system, demonstrated that certain critical conditions must be met in order for unexcitable obstacles to destabilize propagation and produce self-sustained vortices that result in uncontrolled high-frequency stimulation of the heart. They demonstrated that the critical condition was the excitability of the tissue such that, when the tissue excitability was low, a broken wave would contract and vanish (i.e., conduction would be blocked). However, at an intermediate level of excitability, the broken wave detached from the barrier and formed a vortex in a manner visually similar to the separation of the main stream from a body in a hydrodynamic system, where there is subsequent eddy formation during turbulence.[76] Moreover, Cabo et al.[75] demonstrated

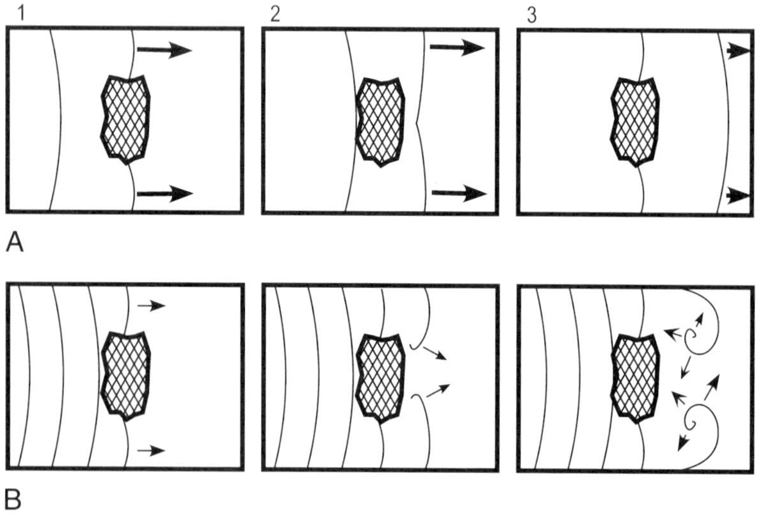

FIGURE 2-14 Initiation of functional reentry by the interaction of a wavefront with an anatomic obstacle in a rectangular sheet of cardiac muscle. Two conditions of tissue excitability are represented. **A,** Under conditions of high excitability, quasiplanar wavefronts initiated at the left border move rapidly toward the obstacle, break and circumnavigate it, and then fuse again to continue propagating toward the right border. **B,** When the excitability is lower, conduction velocity is slower. Upon reaching the obstacle, the wavefront again breaks. However, in this case, the newly formed wavebreaks detach from the obstacle as they move toward the right border and begin to curl, giving rise to two counter-rotating spirals.

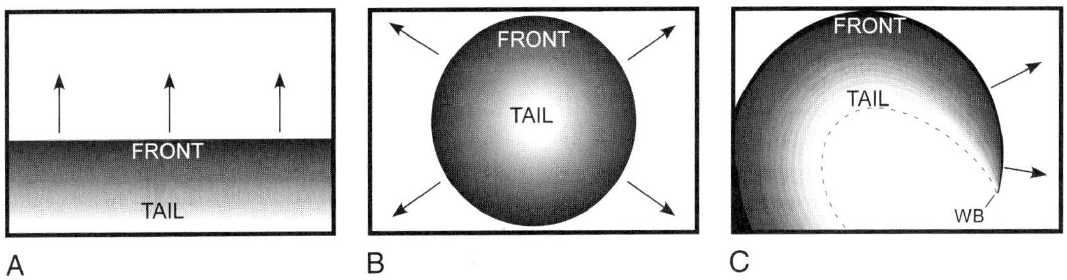

FIGURE 2-15 Expected conditions of propagation of different types of waves in a homogeneous and isotropic sheet of cardiac muscle. **A,** Planar wave initiated by stimulation of the entire bottom border of the sheet. **B,** Circular wave initiated by point stimulation in the center of the sheet. **C,** Spiral wave initiated by cross-field stimulation. Note that for both planar and circular waves, the wavefront never meets the refractory tail. In contrast, during spiral wave activity, the wavefront and the wave tail meet at the wavebreak *(WB)*.

that high-frequency stimulation, which decreases excitability, in the presence of anatomic obstacles also resulted in the detachment of the broken wave and the generation of vortices. This phenomenon has been termed *vortex shedding*. In summary, the dynamics of wavebreaks are determined by: (1) the critical curvature of the wavefront (i.e., the curvature at which propagation fails), (2) the excitability of medium, and (3) the frequency of stimulation or wave succession.

The characteristics of the obstacle (including its size and shape) with which the wavefront interacts also play a role in the formation of wavebreaks and vortex-like activity. Agladze and colleagues[70] used a chemical excitable medium, the Belousov-Zhabotinsky (BZ) reaction, to show that for rotating waves to be initiated in the presence of an unexcitable barrier, the barrier must possess sharp corners, since a wavebreak does not detach from a slowly curving barrier. Moreover, they showed that if the size of the obstacle is small, or the stimulation frequency is too low, a planar wave initiated proximally to the obstacle will separate at the obstacle into two waves with free ends. Subsequently, each will circumnavigate the obstacle, and the broken ends will fuse on the distal end of the obstacle, thus forming a single wave again.[70] In contrast, if the obstacle is large or the stimulation frequency is higher than a critical value, then the wave splits and the ends remain separated from each other and the obstacle. Depending on the excitability of the medium, a pair of counter rotating waves may be initiated.

Role of Wavebreaks in VF

Based on investigations in other excitable media[70,73,74,77] it is known that under normal conditions of excitability and stimulation, the interaction of the wavefront with an obstacle does not produce a wavebreak. However, when the excitability is lowered, wavebreaks may be initiated and persist after the collision of the front with the appropriate obstacles, anatomic or functional. However, what is an appropriate obstacle? As predicted by theory, the obstacle size must be equal to or greater than the width of the wavefront for perturbation of propagation to occur. The width of the wavefront is equal to the product of the AP duration (APD) and the conduction velocity (CV). At a propagation speed of 0.5 cm/sec, the wavefront width in normal cardiac muscle is approximately 1 mm.[78] Consequently, obstacles of 1 mm or larger have the potential for generating wavebreaks in the propagating waves and producing vortex-like reentry.

The complicated anatomic structure of the ventricles may provide the necessary substrate for the establishment of reentry. For example, the complicated and uneven endocardial structure (*trabeculae carneae* and papillary muscles) and the highly heterogeneous geometric arrangement of cardiac cells may play a role in the initiation of reentry and add to the complexity or stability, or both, of reentrant rhythms. However, as noted previously, arrhythmias do not occur under normal circumstances and, in fact, they only occur in the presence of premature or high-frequency stimulation or pathologic conditions such as congestive heart failure, myocardial ischemia, and infarction.[79] In diseased and elderly hearts, fibrotic patches of significant size (i.e., >1 mm) are often found. Thus, the presence of such obstacles and decreased excitability (due to ischemia) may make elderly individuals more susceptible to reentrant arrhythmias.[80] In addition, the dispersion of refractoriness (i.e., temporal differences in the duration of the refractory period between cells), secondary to uneven chamber enlargement, has also been implicated in the initiation of reentry.[80] Such dispersion could set the stage for functionally determined obstacles, which would interfere with the normal propagation of the waves in the heart. Hence, in the heart wavebreaks may be generated by any of the previously mentioned causes because they can interfere with the normal propagation of the wavefront.[81] Moreover, during reentrant activity itself, additional wavebreaks could also be produced by the interaction of the wavefront with the wave tail.[81] It should be noted, however, that wavebreaks do not always lead to reentrant activity.

Clearly, numerous possibilities and conditions can lead to the formation of wavebreaks. However, what is the relationship between wavebreaks and VF? It is the hypothesis of our laboratory that the numerous fragmented wavefronts observed during VF form as the result of the interaction of waves emanating from a high-frequency source with the obstacles present in cardiac tissue.[80] Because of their lateral instability, some waves may shrink and undergo decremental conduction, but other waves may continue unchanged until annihilated by other waves. Still others may undergo curling and form new rotors. The final result is the fragmentation of the mother waves away from the source into multiple short-lived daughter waves that produce a complex pattern of propagation during VF.

DRIFT AND ANCHORING OF ROTORS

Studies based on the theory of spiral waves in excitable media, together with experiments using high-resolution optical mapping,[5,62,82] have suggested that the behavior of the core of the vortex plays an important role in determining the ECG manifestation of the arrhythmia. In fact, it has been demonstrated that a stationary position of the core produces a monomorphic pattern of excitation (i.e., ventricular tachycardia). In contrast, a drifting core (i.e., its position demonstrates beat-to-beat changes) leads to an irregular pattern of activation.[62] Davidenko[62] proposed that when the core drifts in one direction, it produces a Doppler shift in the excitation period in a manner that two coexisting frequencies can be observed, one ahead and one behind the drifting core. Under this condition, the activation frequency behind the core is always slower than the frequency ahead of the core.[62] Simulated ECGs obtained during reentrant activity showed that in the presence of a unidirectional drifting rotor, there is undulating pattern (i.e., waxing and waning, whereby the axis of the depolarization complex demonstrates a gradual torsion much like torsades de pointes).[5] However, when the core drifts at higher speed and in many directions, the ECG pattern is more complicated and resembles VF.[5,62,82] Thus, in this manner the dynamics of the rotor can determine the arrhythmia manifestation.

According to excitable media theory, drift may be the result of spatial gradients in parameters such as refractory period,[83] fiber orientation,[5] or in a bounded medium as the result of interaction of the rotor with a border.[84] It is well known that in the healthy myocardium there are nonuniformities in refractoriness, excitability, and fiber orientation, which allow for spiral drift. However, it should be noted that spiral drift may be short-lived since spirals may spontaneously terminate or they may anchor to regions of low excitability or small discontinuities (e.g., patches of fibrosis or small vessels).[4,5]

Mechanisms of VF Maintenance

IS VF RANDOM OR ORGANIZED?

Based on his cinematographic studies, Wiggers concluded in 1940 that VF could not be adequately described as an asynchronous contraction of myocardial fibers.[85] Wiggers observed that the lack of coordination and asynchrony first involved comparatively large sections of the myocardium, which progressively multiplied and decreased in size as fibrillation continued; however, even in the latter stage of fibrillation, asynchronic contraction of adjacent fibers did not seem to occur.[85] These observations are in agreement with the notion that VF arises from wandering wavefronts that are ever changing in direction and number.[86] Furthermore, it is possible to suggest that the fragmentation of the wavefront into multiple independent wavelets may arise from the interaction with obstacles and refractory tails of other waves.[87] As the front breaks, some waves may shrink and cease to exist (i.e., decremental propagation), while others may propagate until terminated by the collision with other waves or boundaries, and still others may give rise to new vortices.[80] The product of such phenomena may be the complex patterns of propagation that characterize VF. However, today there is ample evidence in the literature suggesting that VF is not entirely a random phenomenon. A summary of the studies that have documented "organization" during VF follows.

In 1981 Ideker and colleagues[88] documented that ventricular activation during the transition to VF arose near the border of the ischemic-reperfused region of the dog heart and was organized as it passed across the nonischemic tissue, but the body surface ECG appeared disorganized as judged by the variable spacing between successive, coexistent activation fronts. More recently, Damle et al.[89] demonstrated that epicardial activation during VF in a canine model of healing infarction is not random. Moreover, they showed that during VF there is both spatial and temporal "linking" of activation, in which the same path of conduction is traversed by several consecutive wavefronts in a relatively rhythmic manner. However, the strength of spatial linking is relatively greater.[89]

Other techniques, such as spectral correlation and coherence analysis,[90,91] as well as nonlinear dynamics approaches, have also been used to study VF organization.[92] Ropella et al.[90] measured the coherence spectrum of bipolar electrograms in patients during sinus rhythm, paroxysmal supraventricular tachycardia, monomorphic ventricular tachycardia, and VF and determined that fibrillatory rhythms exhibited significantly less coherence than nonfibrillatory events; nevertheless, they documented a certain degree of coherence even during VF. Another attempt to measure spatio-temporal correlations during VF was carried out by Damle et al.,[89] who used vector loops, calculated by summing two orthogonal bipolar electrograms obtained with a recording plaque at multiple sites in the heart. In this manner, vector mapping was used to analyze epicardial activation directions in an attempt to detect and quantify underlying organization at adjacent sites in the heart and during different cycles of VF. Moreover, they developed a linear regression model and showed a predictable relation between activation at adjacent regions during a given beat of VF, suggesting that the activation of the myocardium proceeds as a wavefront rather than as a random localized event.[89] This resulted in strong evidence for the presence of spatio-temporal organization during VF.

Garfinkel et al.[92] used nonlinear dynamics theory to study fibrillation in a computer model and three stationary forms of arrhythmias: in human chronic atrial fibrillation, in a stabilized form of canine VF, and in fibrillation-like activity in thin sheets of canine and human ventricular tissue. They found that fibrillation arose through a quasiperiodic stage of period and amplitude modulation; thus, they concluded that fibrillation is a form of spatio-temporal chaos. More recently, however, Bayly et al.[93] explored several techniques to quantify spatial organization during VF. They used epicardial electrograms recorded from pig hearts using rectangular arrays of unipolar extracellular electrodes and concluded that VF is neither "low-dimensional chaos" nor "random" behavior but rather a high-dimensional response with a degree of spatial coherence.[93]

Recently, the development of an analytic technique by Gray et al.,[94] which markedly reduces the amount of data required to depict the complex patterns of fibrillation, has enabled investigators to study the detailed dynamics of wavelets and rotors, including their initiation, life span, and termination. Using a fluorescent potentiometric dye and video imaging, Gray et al. recorded the dynamics of transmembrane potentials from a large region of the heart and determined that transmembrane signals at many sites exhibit a strong periodic component. With this analysis, the periodicity is seen as an attractor in 2-D phase space, and each site can be represented by its phase around the attractor. Using spatial phase maps at each instant in time, Gray et al.[94] revealed the "sources" of fibrillation in the form of topologic defects, or phase singularities,[49] at several sites. Thus, they demonstrated that a substantial amount of spatial and temporal organization underlies cardiac fibrillation in the whole heart.[94]

Work from our laboratory[95] in the isolated Langendorff-perfused rabbit heart has also demonstrated organization during VF in the form of sequences of wave propagation that activated the ventricles in a spatially and temporally similar fashion. Furthermore, the frequency of the periodic activity was shown to correspond to the dominant peak in both the global bipolar electrogram and the optical pseudo-ECG, which suggests that the sources of the periodic activity are the dominant sources that maintain VF in this model. Moreover, quantification of wavelets revealed that during VF, wavebreaks underlie wavelet formation; however, breakup of rotor waves was not a robust mechanism for the maintenance of VF. Overall, the results suggested that the organized activity of periodic sources was responsible for most of the frequency content of VF and was therefore important for the maintenance of this arrhythmia.[95]

ROTORS AND VF

About 3 decades ago, a new idea emerged based on theoretical[96] and experimental[30] findings, demonstrating that the heart could sustain electrical activity that rotated about a functional obstacle. These "rotors" were thought to be the major organizing centers of fibrillation. Since then, much work has focused on rotors as the underlying mechanism for VF in the heart. However, two schools of thought have emerged. On one hand, many recently proposed mechanisms for fibrillation have focused on transience and instability of rotors.[97,98] These mechanisms suggest that the breakup of rotors results in the "turbulent" nature of fibrillation. One such mechanism, the *restitution hypothesis*, suggests that fractionation of the rotor ensues when the oscillation of the APD is of sufficiently large amplitude to block conduction along the wavefront.[92,99,100] Another mechanism for breakup focuses on the fact that propagation within the 3-D myocardium is highly anisotropic due to the intramural rotation of the fibers, thus producing twisting and instability of the organizing center (filament), which results in multiplication following repeated collisions with boundaries in the heart.[97]

Studies from our laboratory have also focused on rotors as the primary engines of fibrillation.[4,5,101] However, here the breakup of the rotor is not regarded as the underlying mechanism of VF. Rather, it is proposed that VF is a problem of self-organization of nonlinear electrical waves with both deterministic and stochastic components.[72,94,101,102] This has led to the hypothesis that there is both spatial and temporal organization during VF in the structurally normal heart, although there is a wide spectrum of behavior during fibrillation. On one end, it has been demonstrated that a single drifting rotor can give rise to a complex pattern of excitation that is reminiscent of VF.[101] On the other end, it has been suggested that VF is the result of a high-frequency stable source and the complex patterns of activation are the result of the fragmentation of emanating electrical activity from that source (i.e., fibrillatory conduction).[103] In the following sections, these two extremes are examined.

DRIFTING VORTICES AND VF

Gray et al.,[94,101] using novel techniques of high-resolution video imaging of voltage-sensitive fluorescent dye in the structurally normal isolated Langendorff-perfused rabbit heart, studied the applicability of spiral wave theory to VF. In that study, they demonstrated the presence of a drifting rotor on the epicardial surface of the heart. Simultaneous recording of a volume-conducted ECG and fluorescence imaging demonstrated that a single rapidly moving rotor was associated with turbulent polymorphic electrical activity, which was indistinguishable from VF. It was assumed that rotors were the 2-D epicardial representation of a 3-D scroll wave. In addition, computer simulations incorporating a realistic 3-D heart geometry and appropriate model parameters demonstrated the ability to form a rapidly drifting rotor similar to that observed in the experiments.[101,102] Frequency analysis of the irregular ECGs for both the experiments and simulations demonstrated spectra that were consistent with previously published data.[104,105] Furthermore, they showed, through the Doppler relationship, that the width of the frequency spectrum can be related to the frequency of the rotation of the rotor, the speed of its motion, and the wave speed.[102]

FIBRILLATORY CONDUCTION

Gray et al.[101] have demonstrated unequivocally that, in the rabbit heart, even a single drifting rotor can produce an ECG that is indistinguishable from VF. However, it has been demonstrated that in other hearts a more complex spatio-temporal organization may prevail. This led Jalife and colleagues[103] to suggest that some forms of fibrillation depend on the uninterrupted periodic activity of discrete reentrant circuits. The faster rotors act as dominant frequency sources that maintain the overall activity. The rapidly succeeding wavefronts emanating from these sources propagate throughout the ventricles and interact with tissue heterogeneities, both functional and anatomic, leading to fragmentation and wavelet formation.[103] The newly formed wavelets may undergo decremental conduction or they may be annihilated by collision with another wavelet or a boundary, and still others may form new sustained rotors.[103] Thus, the result would be fibrillatory conduction or the frequency-dependent fragmentation of wavefronts, emanating from high-frequency reentrant circuits into multiple short-lived wavelets.[64,95]

Zaitsev et al.,[106] using spectral analysis of optical epicardial and endocardial signals for sheep ventricular slabs, have provided additional evidence suggesting that fibrillatory conduction may be the underlying mechanism of VF. Zaitsev and colleagues[106] present data showing that the dominant frequencies of excitation do not change continuously on the ventricular surfaces of slabs. Rather, the frequencies are constant over regions termed domains; moreover, there are only a small number of discrete domains found on the ventricular surfaces. They also demonstrated that the dominant frequency of excitation in the adjacent domains was often related to the fastest dominant frequency domain in 1:2, 3:4, or 4:5 ratios, and this was suggested to be the result of intermittent, Wenckebach-like conduction block occurrences at the boundaries between domains.[106] Thus, they concluded that, in their model, VF may have resulted from a sustained high-frequency 3-D intramural scroll wave, which created complex patterns of propagation as the result of fragmentation when waves emanating from a high frequency scroll interacted with tissue heterogeneities.[106]

Samie et al.[107] presented new evidence in the isolated Langendorff-perfused guinea pig heart that strongly supports the hypothesis that fibrillatory conduction from a stable high-frequency reentrant source is the underlying mechanism of VF. Optical recordings of potentiometric dye fluorescence from the epicardial ventricular surface were obtained along with a volume-conducted "global" ECG. Spectral analysis of optical signals (pixel by pixel) was performed, and the dominant frequency (DF) (peak with maximal power) from each pixel was used to generate a DF map. Pixel-by-pixel Fast Fourier Transform (FFT) analysis revealed that DFs were distributed throughout the ventricles in clearly demarcated domains. The highest frequency domains were always found on the anterior wall of the left ventricle. Correlation of rotation frequency of rotors with the fastest DF domain strongly suggests that rotors are the underlying mechanism of the fastest frequencies. Further analysis of optical recordings demonstrates that fragmentation of wavefronts emanating from high-frequency rotors occurs near the boundaries of the DF domains. Thus, the results demonstrate that in the isolated guinea pig heart, a high-frequency reentrant source that remains stationary in the LV is the mechanism that sustains VF.

ACKNOWLEDGEMENTS

This work is supported in part by grants PO1 HL39707 and RO1 HL60843 from the National Heart and Blood Institute, National Institutes of Health.

REFERENCES

1. Zipes DP, Wellens HJ: Sudden cardiac death. Circulation. 1998;98:2334-51.
2. Zipes DP, Jalife J, eds. Cardiac Electrophysiology from Cell to Bedside. Philadelphia, WB Saunders, 2000.
3. Frazier DW, Wharton JM, Wolf PD, et al: Mapping the electrical initiation of ventricular fibrillation. J Electrocardiol 1989; 22(Suppl):198-9.
4. Davidenko JM, Pertsov AV, Salomonsz R, et al: Stationary and drifting spiral waves of excitation in isolated cardiac muscle. Nature 1992;355:349-51.
5. Pertsov AM, Davidenko JM, Salomonsz R, et al: Spiral waves of excitation underlie reentrant activity in isolated cardiac muscle. Circ Res 1993;72:631-50.
6. Marban E, Robinson SW, Wier WG: Mechanisms of arrhythmogenic delayed and early afterdepolarizations in ferret ventricular muscle. J Clin Invest 1986;78:1185-92.
7. Asano Y, Davidenko JM, Baxter WT, et al: Optical mapping of drug-induced polymorphic arrhythmias and torsades de pointes in the isolated rabbit heart [see comments]. J Am Coll Cardiol 1997;29:831-42.
8. Mines GR: On dynamic equilibrium in the heart. J Physiol 1913;46:349-83.
9. Weiner N, Rosenblueth A: The mathematical formulation of the problem of conduction of impulses in a network of connected excitable elements, specifically in cardiac muscle. Arch Inst Cardiol Mex 1946;16:205-65.
10. Schmitt FO, Erlanger J: Directional differences in the conduction of the impulse through heart muscle and their possible relation to extrasystolic and fibrillary contractions. Am J Physiol 1928-1929;87:326-47.
11. Cranefield PF: Conduction of the Cardiac Impulse. Mount Kisko, NY, Futura, 1975.
12. Puech P, Latour H, Grolleau R: Le flutter et ses limites. Arch Mal Coeur Vaiss 1970:116-44.
13. Josephson ME: Paroxysmal supraventricular tachycardia: An electrophysiologic approach. Am J Cardiol 1978;41:1123-6.
14. Gallagher JJ: Variants of preexcitation: Update 1984. In DP Zipes, J Jalife (eds): Cardiac Electrophysiology and Arrhythmias. Orlando, Fla., Grune & Stratton, 1985, pp 419-33.
15. Akhtar M, Gilbert C, Wolf FG, Schmidt DH: Reentry within the His-Purkinje system. Elucidation of reentrant circuit utilizing right bundle branch and His bundle recordings. Circulation 1978;58:295-304.
16. Josephson ME, Buxton AE, Marchlinski FE, et al: Sustained ventricular tachycardia in coronary artery disease—evidence for reentrant mechanisms. In DP Zipes, J Jalife (eds): Cardiac Electrophysiology and Arrhythmias. Orlando, Fla., Grune & Stratton, 1985, pp 409-18.
17. Allessie MA, Lammers WEJEP, Bonke FIM, Hollen J: Experimental evaluation of Moe's multiple wavelet hypothesis of atrial fibrillation. In Zipes DP, Jalife J (eds): Cardiac Electrophysiology and Arrhythmias. Orlando, Fla., Grune & Stratton, 1985, pp 265-75.
18. Gray RA, Jalife J, Panfilov A, et al: Nonstationary vortexlike reentrant activity as a mechanism of polymorphic ventricular tachycardia in the isolated rabbit heart. Circulation 1995;91:2454-69.

19. Mines GR: On circulating excitation on heart muscles and their possible relation to tachycardia and fibrillation. Trans R Soc Can 1914;4:43-53.
20. Antzelevitch C, Jalife J, Moe GK: Characteristics of reflection as a mechanism of reentrant arrhythmias and its relationship to parasystole. Circulation 1980;61:182-91.
21. Allessie MA, Bonke FI, Schopman FJ: Circus movement in rabbit atrial muscle as a mechanism of tachycardia. III. The "leading circle" concept: A new model of circus movement in cardiac tissue without the involvement of an anatomical obstacle. Circ Res 1977;41:9-18.
22. Dillon S, Allessie MA, Ursell PC, Wit AL: Influence of anisotropic tissue structure on reentrant circuits in the subepicardial border zone of subacute canine infarcts. Circ Res 1988;63:182-206.
23. El-Sherif N: The Figure 8 model of reentrant excitation in the canine post-infarction heart. In DP Zipes, J Jalife (eds): Cardiac Electrophysiology and Arrhythmias. Orlando, Fla., Grune & Stratton, 1985, pp 363-78.
24. Winfree AT: Electrical turbulence in three-dimensional heart muscle. Science 1994;266:1003-6.
25. Winfree AT: When Time Breaks Down. Princeton, N.J., Princeton University Press, 1987.
26. Jalife J, Moe GK: Excitation, conduction and reflection of impulses in isolated bovine and canine Purkinje fibers. Circ Res 1981;49:233-47.
27. Davidenko JM, Antzelevitch C: The effects of milrinone on conduction, reflection, and automaticity in canine Purkinje fibers. Circulation 1984;69:1026-35.
28. Antzelevitch C, Bernstein MJ, Feldman HN, Moe GK: Parasystole, reentry, and tachycardia: A canine preparation of cardiac arrhythmias occurring across inexcitable segments of tissue. Circulation 1983;68:1101-115.
29. Weidmann S: The electrical constants of Purkinje fibers. J Physiol (Lond) 1952;118:348-60.
30. Allessie MA, Bonke FI, Schopman FJ: Circus movement in rabbit atrial muscle as a mechanism of tachycardia. Circ Res 1973; 33:54-62.
31. Allessie MA, Bonke FI, Schopman FJ: Circus movement in rabbit atrial muscle as a mechanism of tachycardia. II. The role of nonuniform recovery of excitability in the occurrence of unidirectional block, as studied with multiple microelectrodes. Circ Res 1976;39:168-77.
32. Garrey WE: Auricular fibrillation. Physiol Rev 1924;4:215-50.
33. Waldo AL, Maclean WAH, Karp RB, et al: Entrainment and interruption of atrial flutter with atrial pacing. Studies in man following open heart surgery. Circulation 1977;56:737-54.
34. Jalife J, Davidenko J, Michaels D: A new perspective on the mechanisms of arrhythmias and sudden cardiac death: Spiral waves of excitation in heart muscle. J Cardiovasc Electrophysiol 1991;2: S133-52.
35. Spach MS, Dolber PC: Relating extracellular potentials and their derivatives to anisotropic propagation at a microscopic level in human cardiac muscle. Evidence for electrical uncoupling of side-to-side fiber connections with increasing age. Circ Res. 1986;58:356-71.
36. Spach MS, Dolber PC, Heidlage JF, et al: Propagating depolarization in anisotropic human and canine cardiac muscle: Apparent directional differences in membrane capacitance. A simplified model for selective directional effects of modifying the sodium conductance on Vmax, tau foot, and the propagation safety factor. Circ Res 1987;60:206-19.
37. Spach MS, Dolber PC, Heidlage JF: Influence of the passive anisotropic properties on directional differences in propagation following modification of the sodium conductance in human atrial muscle. A model of reentry based on anisotropic discontinuous propagation. Circ Res 1988;62:811-32.
38. Dillon SM, Coromilas J, Waldecker B, Wit AL: Effects of overdrive stimulation on functional reentrant circuits causing ventricular tachycardia in the canine heart: Mechanisms for resumption or alteration of tachycardia. J Cardiovasc Electrophysiol 1993;4: 393-411.
39. Peters NS, Wit AL: Myocardial architecture and ventricular arrhythmogenesis. Circulation 1998;97:1746-54.
40. Ciaccio EJ, Scheinman MM, Fridman V, et al: Dynamic changes in electrogram morphology at functional lines of block in reentrant circuits during ventricular tachycardia in the infarcted canine heart: A new method to localize reentrant circuits from electrogram features using adaptive template matching. J Cardiovasc Electrophysiol 1999;10:194-213.
41. Girouard SD, Pastore JM, Laurita KR, et al: Optical mapping in a new guinea pig model of ventricular tachycardia reveals mechanisms for multiple wavelengths in a single reentrant circuit. Circulation 1996;93:603-13.
42. El Sherif N, Smith RA, Evans K: Canine ventricular arrhythmias in the late myocardial infarction period. 8. Epicardial mapping of reentrant circuits. Circ Res 1981;49:255-65.
43. El-Sherif N, Mehra R, Gough WB, Zeiler RH: Ventricular activation patterns of spontaneous and induced ventricular rhythms in canine one-day-old myocardial infarction. Evidence for focal and reentrant mechanisms. Circ Res 1982;51:152-66.
44. Harada T, Stevenson WG, Kocovic DZ, Friedman PL: Catheter ablation of ventricular tachycardia after myocardial infarction: Relationship of endocardial sinus rhythm potentials to the reentry circuit. J Am Coll Cardiol 1997;30:1015-23.
45. Stevenson WG, Friedman PL, Sager PT: Exploring post infarct reentrant ventricular tachycardia with entrainment mapping. J Am Coll Cardiol 1997;29:1180-9.
46. Stevenson WG, Friedman P: Catheter ablation of ventricular tachycardia. In Zipes DP, Jalife J (eds): Cardiac Electrophysiology from Cell to Bedside. Philadelphia, WB Saunders, 2000, pp 1049-56.
47. Mikhailov AS, Krinsky VI: Rotating spiral waves in excitable media: The analytic results. Physica D 1983;90:346-71.
48. Pertsov AM, Emarkova EA, Panfilov AV: Rotating spiral waves in modified Fitz Hugh-Nagumo model. Physica D 1984;14:117-24.
49. Winfree AT: Chemical waves and fibrillating hearts: Discovery by computation. J Biosci 2002;27:465-473.
50. Zykov VS: Simulation of Wave Process in Excitable Media. New York, Manchester University Press, 1987.
51. Allessie MA, Schalij MJ, Kirchhof CJ, et al: Electrophysiology of spiral waves in two dimensions: The role of anisotropy. Ann N Y Acad Sci 1990;591:247-56.
52. Frazier DW, Wolf PD, Wharton JM, et al: Stimulus-induced critical point. Mechanism for electrical initiation of reentry in normal canine myocardium. J Clin Investig 1989;83:1039-52.
53. Gray RA, Pertsov AM, Jalife J: Incomplete reentry and epicardial breakthrough patterns during atrial fibrillation in the sheep heart. Circulation 1996;94:2649-61.
54. De Boer S: Die physiologie und pharamkologie des Flimmers. Ergeb Physiol 1923;21:1-20.
55. Wiggers CJ, Wegria R: Ventricular fibrillation due to single, localized induction and condenser shocks applied during the vulnerable phase of ventricular systole. Am J Physiol 1940;128:500-5.
56. Wegria R, Moe GK, Wiggers CJ: Comparison of the vulnerable periods and fibrillation thresholds of normal and idioventricular beats. Am J Physiol 1941;133:651-57.
57. Moe GK, Harris AS, Wiggers CJ: Analysis of the initiation of fibrillation by electrographic studies. Am J Physiol 1941;134: 473-92.
58. Winfree AT: Vortex action potentials in normal ventricular muscle. Ann N Y Acad Sci 1990;591:190-207.
59. Shibata N, Chen PS, Dixon EG, et al: Influence of shock strength and timing on induction of ventricular arrhythmias in dogs. Am J Physiol 1988;255:H891-901.
60. Winfree AT: Electrical instability in cardiac muscle: phase singularities and rotors. Journal of Theoretical Biology 1989; 138:353-405.
61. Beaumont J, Davidenko N, Davidenko JM, Jalife J: Spiral waves in two-dimensional models of ventricular muscle: Formation of a stationary core. Biophys J 1998;75:1-14.
62. Davidenko JM: Spiral wave activity: A possible common mechanism for polymorphic and monomorphic ventricular tachycardias. [Review] [63 refs]. J Cardiovasc Electrophysiol 1993;4: 730-46.
63. Matta RJ, Verrier RL, Lown B: Repetitive extrasystole as an index of vulnerability to ventricular fibrillation. Am J Physiol 1976;230:1469-73.
64. Samie FH, Mandapati R, Gray RA, et al: A mechanism of transition from ventricular fibrillation to tachycardia: Effect of calcium channel blockade on the dynamics of rotating waves. Circ Res 2000;86:684-91.

65. Keener JP: On the formation of circulating patterns of excitation in anisotropic excitable media. J Math Biol 1988;26:41-56.
66. Cheng Y, Mowrey KA, Van Wagoner DR, et al: Virtual electrode-induced reexcitation: A mechanism of defibrillation. Circ Res 1999;85:1056-66.
67. Krinsky VI: Mathematical models of cardiac arrhythmias (spiral waves). Pharmacol Ther Part B: General & Systematic Pharmacology 1978;3:539-55.
68. Winfree AT: Evolving perspectives during 12 years of electrical turbulence. Chaos 1998;8:1-19.
69. Krinsky VI: Self-Organization: Autowaves and Structures Far from Equilibrium. Berlin, Springer, 1984.
70. Agladze K, Keener JP, Muller SC, Panfilov A: Rotating spiral waves created by geometry. Science 1994;264:1746-8.
71. Lin SF, Roth BJ, Wikswo JP Jr: Quatrefoil reentry in myocardium: An optical imaging study of the induction mechanism. J Cardiovasc Electrophysiol 1999;10:574-86.
72. Jalife J, Gray RA, Morley G, Davidenko J: Self-organization and the dynamical nature of ventricular fibrillation. Chaos 1998; 8:79-93.
73. Pertsov AM, Panfilov AV, Medvedeva FU: [Instabilities of autowaves in excitable media associated with critical curvature phenomena]. Biofizika 1983;28:100-102.
74. Nagy-Ungvarai Z, Pertsov AM, Hess B, Muller SC: Lateral instabilities of a wave front in the Ce-catalyzed Belousov-Zhabotinsky reaction. Physica D 1992;61:205-12.
75. Cabo C, Pertsov AM, Davidenko JM, et al: Vortex shedding as a precursor of turbulent electrical activity in cardiac muscle. Biophys J 1996;70:1105-11.
76. Tritton DJ: Physical Fluid Dynamics. Berkshire, UK, Van Nostrand Reinhold, 1977.
77. Starobin JM, Zilberter YI, Rusnak EM, Starmer CF: Wavelet formation in excitable cardiac tissue: The role of wavefront-obstacle interactions in initiating high-frequency fibrillatory-like arrhythmias. Biophys J 1996;70:581-94.
78. Pertsov AM: Scale of geometric structures. In Spooner PM, Joyner RW, Jalife J (ed): Discontinuous Conduction in the Heart. Armonk, N.Y., Futura Publishing Company, 1997, pp 273-93.
79. Myerburg RJCA: Cardiac arrest and sudden cardiac death. In Braunwald E (ed): Heart Disease: A Textbook of Cardiovascular Medicine. Philadelphia, WB Saunders, 1997, pp 742-79.
80. Jalife J: Ventricular fibrillation: mechanisms of initiation and maintenance. Ann Rev Physiol 2000;62:25-50.
81. Winfree AT: Varieties of spiral wave behavior: An experimentalist's approach to the theory of excitable media. 1991;1:303-334.
82. Gray RA, Jalife J, Panfilov A, et al: Nonstationary vortexlike reentrant activity as a mechanism of polymorphic ventricular tachycardia in the isolated rabbit heart. Circulation 1995; 91:2454-69.
83. Fast VG, Pertsov AM: Drift of a vortex in the myocardium. Biophysics 1990;35:489-94.
84. Yermakova YA, Pertsov AM: Interaction of rotating spiral waves with a boundary. Biophysics 1986; 31:932-940.
85. Wiggers CJ: The mechanism and nature of ventricular fibrillation. Am Heart J 1940;20:399-412.
86. Moe GK, Rheinboldt WC, Abildskov JA: A computer model of atrial fibrillation. Am Heart J 1964;67:200-220.
87. Panfilov AV: Spiral breakup as a model of ventricular fibrillation. Chaos 1998;8:57-64.
88. Ideker RE, Klein GJ, Harrison L, et al: The transition to ventricular fibrillation induced by reperfusion after acute ischemia in the dog: A period of organized epicardial activation. Circulation 1981;63:1371-9.
89. Damle RS, Kanaan NM, Robinson NS, et al: Spatial and temporal linking of epicardial activation directions during ventricular fibrillation in dogs. Evidence for underlying organization. Circulation 1992;86:1547-58.
90. Ropella KM, Sahakian AV, Baerman JM, Swiryn S: The coherence spectrum. A quantitative discriminator of fibrillatory and nonfibrillatory cardiac rhythms. Circulation 1989;80:112-9.
91. Bayly PV, Johnson EE, Wolf PD, et al: A quantitative measurement of spatial order in ventricular fibrillation. J Cardiovasc Electrophysiol 1993;4:533-46.
92. Garfinkel A, Chen PS, Walter DO, et al: Quasiperiodicity and chaos in cardiac fibrillation. J Clin Invest 1997;99:305-14.
93. Bayly PV, KenKnight BH, Rogers JM, et al: Spatial organization, predictability, and determinism in ventricular fibrillation. Chaos 1998;8:103-15.
94. Gray RA, Pertsov AM, Jalife J: Spatial and temporal organization during cardiac fibrillation. Nature 1998;392:75-8.
95. Chen J, Mandapati R, Berenfeld O, et al: High-frequency periodic sources underlie ventricular fibrillation in the isolated rabbit heart. Circ Res 2000;86:86-93.
96. Krinskii VI: [Excitation propagation in nonhomogenous medium (actions analogous to heart fibrillation)]. Biofizika 1966;11:676-83.
97. Fenton F, Karma A: Vortex dynamics in three-dimensional continuous myocardium with fiber rotation: Filament instability and fibrillation. Chaos 1998;8:20-47.
98. Riccio ML, Koller ML, Gilmour RF Jr: Electrical restitution and spatiotemporal organization during ventricular fibrillation. Circ Res 1999;84:955-63.
99. Karma A: Electrical alternans and spiral wave breakup in cardiac tissue. Chaos 1994;4:461-72.
100. Weiss JN, Garfinkel A, Karagueuzian HS, et al: Chaos and the transition to ventricular fibrillation: A new approach to antiarrhythmic drug evaluation. Circulation 1999;99:2819-26.
101. Gray RA, Jalife J, Panfilov AV, et al: Mechanisms of cardiac fibrillation. Science 1995;270:1222-3.
102. Jalife J, Gray R: Drifting vortices of electrical waves underlie ventricular fibrillation in the rabbit heart. Acta Physiol Scand 1996;157:123-31.
103. Jalife J, Berenfeld O, Skanes A, Mandapati R: Mechanisms of atrial fibrillation: Mother rotors or multiple daughter wavelets, or both? J Cardiovasc Electrophysiol 1998;9(8 Suppl):S2-12.
104. Herbshlef JN, Heethaar RM, van der Tweel I, et al: Signal analysis of ventricular fibrillation. IEEE Comp Cardiol 1979; 49-52.
105. Herbshlef JN, Heethaar RM, van der Tweel I, Meijler FL: Frequency analysis of the ECG before and during ventricular fibrillation. IEEE Comp Cardiol 1980;365-368.
106. Zaitsev AV, Berenfeld O, Mironov SF, et al: Distribution of excitation frequencies on the epicardial and endocardial surfaces of fibrillating ventricular wall of the sheep heart. Circ Res 2000;86:408-17.
107. Samie FH, Berenfeld O, Mironov SF, et al: An ionic mechanism for ventricular fibrillation in the Langendorff-perfused guinea pig heart. Circulation 2000;102:II-341-2 (Abstract).

Chapter 3 Autonomic Nervous System and Cardiac Arrhythmias

DAVID G. BENDITT, DEMOSTHENES ISKOS, and SCOTT SAKAGUCHI

Supervision of the status and operation of the entire cardiovascular system is one of the principal responsibilities of the autonomic nervous system (ANS). The ANS continuously monitors afferent neural signals from multiple organ systems and coordinates efferent neural traffic to the heart and blood vessels in response to ever-changing physiologic and metabolic requirements. The sympathetic and parasympathetic components of the ANS are the dominant players in cardiovascular control.[1,2] However, the ANS also incorporates the actions of cardiac and extra-cardiac neurohumoral agents, intracardiac reflex arcs, and a number of less well understood agents such as vasoactive intestinal peptide (VIP),[3] neuropeptide Y,[4] transmitters released by so-called *purinergic* nerve endings,[5-8] and the ubiquitous nitric oxide.[9,10]

The ANS oversees and modulates cardiac chronotropic, dromotropic, and inotropic properties while monitoring and modifying reflexes governing systemic arterial pressure, venous return, and respiratory rate and tidal volume. By virtue of this critical "oversight" responsibility, ANS activity becomes in its own right a potentially interesting "barometer" of cardiovascular stability. The latter has been of particular interest in regard to predicting arrhythmia susceptibility. Specifically, there is increasing interest in using quantitative albeit indirect markers of ANS activity (e.g., baroreceptor sensitivity, heart rate variability (HRV), cardiac output [Q–T] interval dispersion) to determine future sudden death risk in individual patients.[11,12]

Given the pivotal position played by ANS activity in cardiovascular control, it is not unexpected that any disturbance of ANS operation may lead to clinically important consequences. In terms of cardiac arrhythmias and related disorders, two of the most common circumstances in which this occurs are: (1) in the setting of acute myocardial ischemia and (2) during the neurally mediated vasovagal faint. In the former instance the outcome can be life threatening, whereas in the latter the ANS abnormality is self-limited, functional, and generally benign in nature. Other circumstances in which serious and occasionally life-threatening outcomes may be encountered when ANS function is perturbed include structural central nervous system catastrophes (e.g., subarachnoid hemorrhage, infections, seizures), congestive heart failure, metabolic derangements, or drug toxicity (e.g., proarrhythmic actions of certain drugs).

The objective of this chapter is to provide an overview of the anatomy and physiology of autonomic innervation as it pertains to cardiac arrhythmias, conduction system disturbances, and related disorders. Additionally, certain of the more important clinical scenarios in which the ANS plays a key role in arrhythmogenesis or the development of symptoms associated with cardiac arrhythmias, or both, are examined.

Anatomy and Physiology of Cardiac Autonomic Nervous System Innervation and Influence

PARASYMPATHETIC AND SYMPATHETIC ACTIVITY

The ANS, operating primarily through sympathetic and parasympathetic nerves and to a lesser extent through release of certain neurohumoral agents, regulates the two key electrophysiological properties of the human heart: chronotropism and dromotropism (Fig. 3-1).[1] In this regard, the neural connections at the level of the heart have been the subject of considerable study. However, the precise anatomy of the operative neural sites within the brain and the cellular connections at the level of the myocardium are known only incompletely (Fig. 3-2), and principally as a result of studies in animal models.[1,2,13-14]

In the canine heart, efferent nerves and their effects on sinus node, atrioventricular (AV) node, and contractile

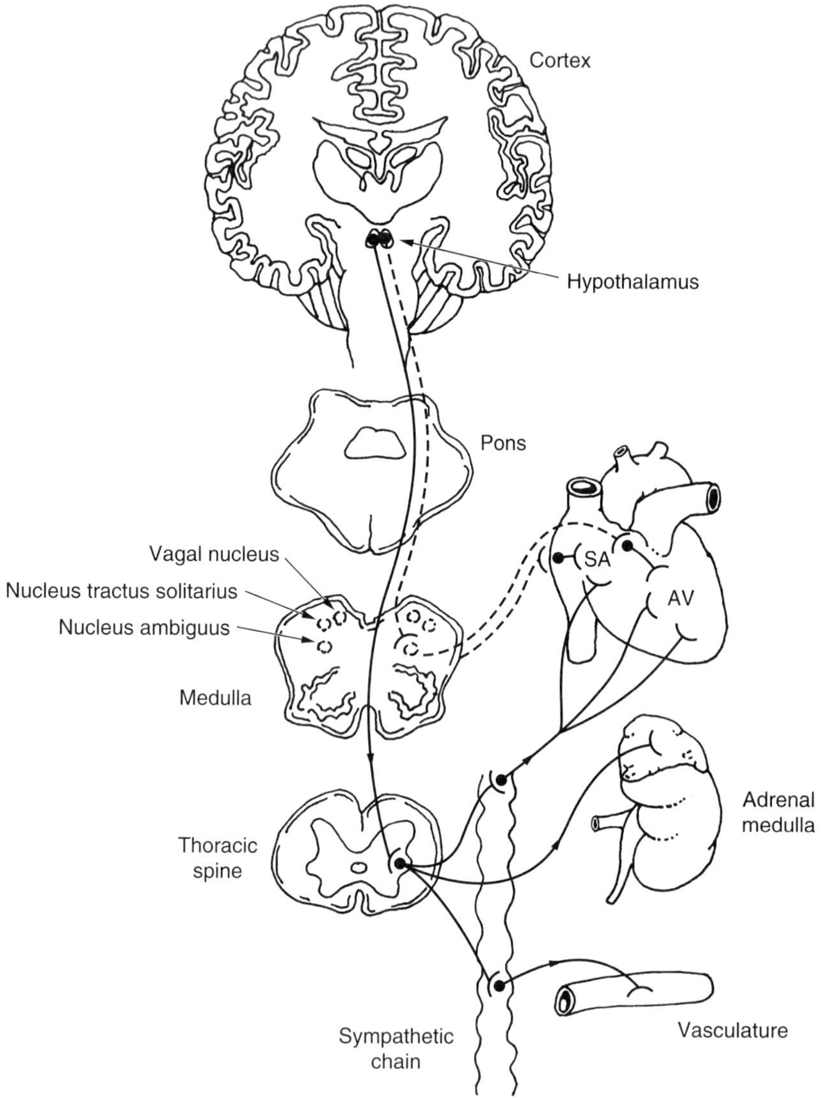

FIGURE 3-1 Schematic illustrating the course of sympathetic *(dashed line)* and parasympathetic *(dotted line)* nerve pathways to key cardiac and vascular structures.

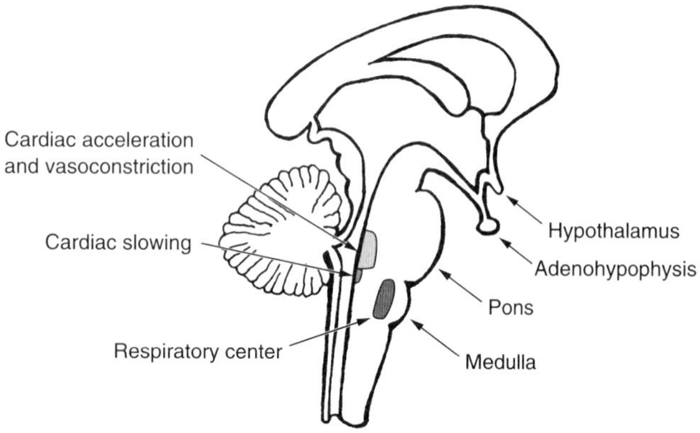

FIGURE 3-2 The approximate midbrain sites considered important for basic autonomic nervous system control of cardiovascular function.

function have been the subject of detailed study.[2] Findings suggest that vagal fibers to the region of the sinus node follow the right pulmonary veins, and interruption at the pulmonary vein-cardiac site has little or no effect on AV nodal function. Conversely, vagal control to the AV node region appears to arrive at the level of the inferior vena cava. Vagal nerve interruption at this site does not affect sinus node parasympathetic control. Further, neither of these sites affect sympathetic input to either node.

Sympathetic inotropic nerve traffic to the canine heart follows the ventrolateral cardiac nerve as well as nerves coursing along the main pulmonary artery. Chronotropic influence to the sinus node appears in part to follow the ventrolateral cardiac nerve and the pulmonary veins (left-sided sympathetics), and from the pulmonary artery region for right-sided sympathetic nerve access. The right-sided inputs seem to be the more important in terms of their activity focusing on sinus node chronotropism (i.e., they have less impact on ventricular inotropic state). AV node sympathetic input is derived from fibers arriving from the left and right sides and coursing within the ventrolateral cardiac nerve and pulmonary artery region.

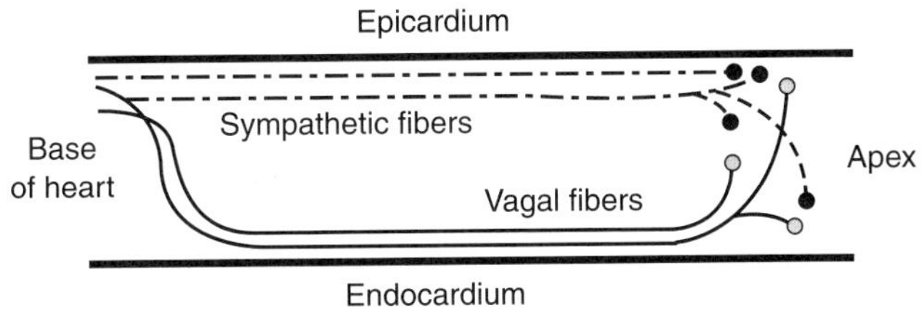

FIGURE 3-3 Epicardial and endocardial locations of ventricular sympathetic nerves and parasympathetic nerves, respectively. (Adapted from Zipes DP, Inoue H: Autonomic neural control of cardiac excitable properties. In Kulbertus HE, Franck G (eds): Neurocardiology. Mount Kisco, NY, Futura Publishing, 1988, pp 59-84.)

Current evidence suggests that, within the heart, sympathetic nerves course predominantly on the epicardial surface, following vascular structures. Endocardial sympathetic activation occurs as the fibers penetrate inward over the ventricular surface.[15-18] Parasympathetic nerve fibers seem, at the level of the AV groove, to cross onto the endocardial ventricular surface (Fig. 3-3).[15-18] In the past, the penetration of parasympathetic nerves into the ventricular conduction system and myocardium was thought to be limited. In recent years this concept has been reconsidered; increasing evidence favors a more important parasympathetic role in the ventricles than had previously been thought.

Within the central nervous system, in experimental canine studies, vagal efferents targeting both the sinus node and the AV node appear to arise predominantly within the *nucleus ambiguus* situated within the medulla oblongata (see Fig. 3-2). In humans the dorsal nucleus of the vagus and the nucleus ambiguus lie in close proximity to both the fourth ventricle and the nucleus tractus solitarii (the key center associated with afferent nerve traffic from peripheral baro/chemo receptors). With respect to sinus node and AV node function, separate cells within the nucleus ambiguus appear to be involved. The manner in which coordination occurs is uncertain, but it seems evident that neither chronotropic state nor dromotropic state can be used as surrogate measures for the other. The same, of course, applies to inotropic state.

The physiologic effects of cardiac innervation are not readily apparent by anatomic assessment alone. In humans at rest, parasympathetic influence appears to predominate in the case of the sinus node chronotropic state,[19] whereas the parasympathetic and sympathetic effects seem to be more balanced in regard to the dromotropic capacity of the AV node.[20] Of course, multiple factors alter this situation throughout the day; the most obvious of these are physical exercise and emotional state. Over the longer term, other factors such as the ageing process and drug therapy impact neural balance.

Detailed review of cardiovascular neuropathology is beyond the scope of this chapter and, in certain respects, is not yet pertinent given current inability to infer the functional implications of observed neuroanatomic disturbances. Nevertheless, it appears safe to conclude that universal cardiac denervation, as in the transplanted heart, does not create a critical electrophysiologic problem. However, the outcome may not be as benign for localized or regional disturbances of autonomic control. For instance, at least experimentally, selective parasympathetic denervation may result in persistent sinus tachycardia.[21] This mechanism may account for certain cases of the syndrome of inappropriate sinus tachycardia. By way of further example, it has been suggested that regional denervation accompanying myocardial infarction (MI) increases arrhythmogenic susceptibility (see later discussion).

PURINERGIC NEURAL INFLUENCES

The relationship of purinergic agonists to overall ANS control of cardiac conduction and arrhythmias is not well understood. Nevertheless, purinergic agonists appear to have a role in regulation of certain conduction system properties.[5-8] The most important purinergic effectors are adenosine activation of so-called *P1 receptors* and adenosine triphosphate (ATP) activation of *P2 receptors*.

Four principal subclasses of P1 receptors are expressed in the heart (A1, A2a, A2b, and A3), with A1 being the most important from a conduction system viewpoint.[8,22] Adenosine action at this site causes negative chronotropic, dromotropic, and inotropic effects, as well as an antiadrenergic action. The A2a site mediates coronary artery vasodilation. ATP acts at both P2 sites as well as P1 sites due to its rapid degradation to adenosine.

In terms of specific transmembrane currents, both purinergic and cholinergic agonists activate I_k (Ach, ado), an inwardly rectifying potassium channel in atrial tissues. They also inhibit I_f, which is currently thought to be important in sinus node diastolic depolarization. It may be in part through these channel activities that acetylcholine and adenosine exhibit their negative chronotropic effects on the sinus node and negative dromotropic effects on the AV node. In particular, it is by means of the latter (i.e., negative dromotropic) effect that vagomimetic agents (e.g., digitalis) and adenosine slow and often abruptly terminate AV node–dependent reentry supraventricular tachycardias. On the other hand, as a rule, neither acetylcholine nor adenosine exhibits prominent direct electrophysiological effects on the ventricle; the exception is an anti-adrenergic effect, and in particular block of catecholamine-stimulated L-type Ca++ channel current (L-type I_{Ca}).[8,23] The manner in which adenosine often interrupts certain ventricular tachycardias (e.g., right ventricular outflow tract, left ventricular fascicular ventricular tachycardia) is as yet unclear. However, to the extent that these are due to adrenergically driven triggered activity, the benefit may be due to attenuating β-adrenergic accentuated L-type I_{Ca}.

Autonomic Nervous System and Cardiac Conduction System Physiology

SINUS NODE

Normal electrical activation of the heart is conventionally described as being initiated by spontaneously depolarizing cells within an anatomic region (the sinus node) situated laterally in the epicardial groove of the sulcus terminalis near the right atrium–superior vena cava junction. However, sinus node anatomy is more complex than that provided by conventional description. For instance, pacemaker tissue extends more caudally than usually depicted. Further, the normal sinus node is composed of multiple pacemaker cell "nests" residing in a fibrous tissue matrix. As a rule, the more rapidly depolarizing cells predominate with others serving as subsidiary back-up pacemakers. The subsidiary pacemaker cells or cell groups may take over under a variety of physiologic conditions, usually based on ANS direction.[24,25]

ANS influence is the most important of the many extrinsic factors (e.g., drugs, hormones) affecting sinus node function. The sinus node region has an abundance of parasympathetic and sympathetic nerve endings.[24,25] Acetylcholine (derived from parasympathetic nerve endings), norepinephrine (predominantly from sympathetic nerves), and epinephrine (adrenal origin) alter depolarization rates of sinus node cells and influence the site of the principal pacemaker within the node. For example, acetylcholine increases transmembrane resting potential (i.e., makes the resting potential more negative) and reduces spontaneous phase 4 (diastolic) slope (the portion of the action potential during which pacemaker cells spontaneously depolarize toward threshold for "firing"), thereby tending to reduce the rate at which depolarization occurs. Acetylcholine also tends to prolong refractoriness of sinus node cells. Conversely, catecholamines (norepinephrine and epinephrine) increase the rate of phase 4 depolarization, thereby increasing sinus rate. Thus, excessive parasympathetic influence may induce marked sinus bradycardia, sinus arrest, and sinoatrial exit block, while catecholamines typically increase heart rate and may reverse sinus arrest and sinoatrial exit block. In disease states, these same neurotransmitters may facilitate the development of "ectopic" pacemaker activity, leading to abnormal sinoatrial and atrial arrhythmias.

In the healthy heart, fluctuation of ANS influence on sinus node function results in a normal respiratory-induced variation of sinus cycle length (i.e., respiratory sinus arrhythmia). In the case of respiratory sinus arrhythmia, the variations may be substantial (suggesting sinus pauses) or more subtle. The same may be the case for ventriculophasic sinus arrhythmia, in which the P–P interval (i.e., atrial cycle length) surrounding a QRS complex is relatively short compared to baroreceptor-mediated prolongation of the subsequent sinus cycle. Absence of sinus arrhythmia variation has become recognized as a sign of cardiac disease (see later discussion of HRV).

Age-related changes of ANS effects on sinoatrial function have been the subject of recent study.[26,27] In general terms, parasympathetic influence on sinus node chronotropism progressively diminishes with increasing age. However, at the same time there is an age-related decrease of intrinsic heart rate (i.e., the heart rate in the absence of autonomic influences—see later discussion).[28] Thus, maintenance of an appropriate heart rate and chronotropic responsiveness in older individuals is increasingly dependent on the integrity of ANS sympathetic tone.

ATRIOVENTRICULAR NODE AND CARDIAC CONDUCTION SYSTEM

Normally, the AV node and specialized cardiac conduction system provide the only connection for transmission of electrical impulses to the ventricles. In regard to AV nodal function, a number of factors favor both slow and decremental conduction. The cells are small and dispersed in a complex fibrous tissue matrix with relatively large extracellular space. Action potentials in the central regions have relatively low resting potentials, slow upstrokes (Ca^{++}-dependent), and properties of refractoriness that persist well after repolarization has been completed (i.e., time-dependent refractoriness). Further, the AV nodal region is heavily infiltrated with neural connections; these are derived not only from the sympathetic and parasympathetic elements of the ANS but also from nerves considered to be purinergic in nature (i.e., adenosine-mediated effects).[1,2,13,29]

As a rule, AV nodal dromotropic responsiveness in the resting patient is under relatively balanced sympathetic and parasympathetic neural influence.[20,30,31] However, this situation may be readily altered by physiologic events (e.g., exercise, sleep), the impact of disease states, or drug effects. In such cases, any tendency toward parasympathetic predominance markedly enhances AV nodal decremental properties; in the extreme this can be associated with transient complete AV nodal block. The latter is, in fact, a relatively common finding in sleeping patients and in very fit resting subjects.

The relationship between ANS control of sinus node rate and AV conduction properties appears to foster both maintenance of 1:1 AV conduction and a relatively optimal AV conduction interval. Malik et al[32] observed that during sinus rhythm, spectral high and low frequency (HF, LF) patterns were similar at various levels of the conduction system (i.e., sinus, atrial-His [A–H] interval). However, when atrial rate was fixed by pacing, parasympathetic predominance appeared at the AV node level. Thus, there are important interactions among autonomic effects at various levels of the conduction system. In sinus rhythm, sympathovagal balance can be estimated by measuring HRV at the ventricular level. However, if atrial cycle length is fixed, the internal balance is lost and the impact of ANS influence on AV conduction becomes evident.

The His bundle and bundle branches are composed of cells with larger surface area, more negative resting membrane potentials, and faster (Na^{+}-dependent) action potentials than those of the AV node. Furthermore, cells comprising the cardiac conduction system have abundant intercellular connections and are physically arranged so as to promote longitudinal conduction. Consequently, decremental conduction (at least to the extent that it can be appreciated with conventional recording techniques) is absent, except in the setting of relatively severe conduction system disease.

Sympathetic nerve endings are generally better represented in the distal aspects of the specialized conduction system than are parasympathetic nerves. Presumably sympathetic activation facilitates antegrade conduction and modulates the rate of junctional pacemaker sites (which are occasionally necessary "back-up" pacemakers for the heart). However, it has become evident that parasympathetic influence penetrates farther (although perhaps variably) into the specialized cardiac conduction system than had previously been thought. The role the parasympathetic effects play, apart from modulating automaticity of junctional subsidiary pacemakers, is not clear.

VENTRICULAR MYOCARDIUM

Studies of the innervation of the ventricular myocardium have relied largely on the canine model (see Fig. 3-3). In this setting, the left ventricular sympathetics tend to lie within the subepicardial layer and follow the large coronary vessels as they spread out over the myocardium.[15-18] The parasympathetics, on the other hand, tend to penetrate the myocardium after crossing the AV groove and are thereafter subendocardial in location. The parasympathetic vagal efferents to the myocardium terminate not on the muscle cells themselves but on intracardiac ganglia. Evidence suggests that these ganglia form not only relay stations but also subserve certain local integrative functions including intracardiac reflex activity.[33]

In the right ventricle the sympathetic nerves are epicardial once again, radiating perpendicularly from the right lateral AV groove or similarly from the left anterior descending artery.[17,18] The region of the right ventricular outflow tract appears to differ somewhat from the remainder of the right ventricle inasmuch as some of the sympathetic nerves are more intramural in location. Vagal nerve fibers in all regions appear to be intramural.

Heightened adrenergic activation in ventricular myocardium is potentially arrhythmogenic in several ways. First, increased sympathetic tone enhances pacemaker activity and is known to increase the frequency and rate of automaticity. While this alone is unlikely to be the basis for a life-threatening ventricular arrhythmia, it may increase the chance of a reentry rhythm being triggered. Secondly, elevated adrenergic tone is known to increase the likelihood of the generation of early and delayed afterdepolarizations (EADs, DADs).[34,35] EADs have become closely associated with the polymorphic ventricular tachycardias in abnormal repolarization syndromes (e.g., torsades de pointes) and may be enhanced by adrenergic activity. DADs are associated with cardiac glycoside toxicity but can also be generated by adrenergic stimulation alone. Further, their amplitude and frequency are directly correlated with heart rate (i.e., in the setting of sinus tachycardia both the propensity for DADs, and the resulting DAD tachycardia rate are increased).

Compared to the atria, parasympathetic cholinergic effects are much less prominent at the ventricular level under most conditions. In canines, the parasympathetic effect is mainly observed as an antiadrenergic action in the setting of increased adrenergic tone. However, in humans, evidence favors a greater ventricular parasympathetic effect whether there is elevated adrenergic activity (e.g., acute ischemia) or not.[36,37] The outcome of this activity may be to diminish production of adrenergically induced EADs and DADs and thereby reduce arrhythmic risk.

Autonomic Nervous System Imaging

The ability to assess ANS status in humans is largely limited to noninvasive indirect techniques such as HRV and baroreceptor sensitivity (BRS) measures. Radionuclide imaging techniques are evolving, however. Ultimately, it may be possible to classify ANS disturbances more precisely not only in terms of their impact on the cardiovascular system, but perhaps even the converse (i.e., the magnitude of disease-effects such as MI on ANS neural distributions in individual patients).

To date, the imaging agents $_{123}$I-metaiodobenzylguanidine (MIBG) and $_{11}$C-m-hydroxyephedrine (HED) have been most studied in humans. Both are pseudotransmitters used to evaluate sympathetic presynaptic function. Postsynaptic markers, such as labeled β-adrenergic blockers for β-adrenergic sites and others for muscarinic cholinergic sites, have also been developed but are as yet more problematic in terms of interpretation of results.

MIBG and HED studies have been carried out in various conditions in which cardiac denervation is expected (e.g., transplanted heart, acute MI) with some success.[38-40] Findings support the notions first derived in experimental work that zones of denervation and areas of under-perfusion are often mismatched.[40-43] Further, regeneration of neural elements has been observed with time. The latter may in fact tend to be an arrhythmogenic element as reinnervation is nonuniform at least for a time.[44]

Currently, ANS imaging techniques are primarily of research interest. However, they offer the potential for improving understanding of the relationship between structural heart disease and the propensity for periodic rhythm disturbances.

Autonomic Nervous System and Specific Bradyarrhythmias and Cardiac Conduction System Disturbances

SINUS NODE DYSFUNCTION

Sinus node dysfunction (sick sinus syndrome) encompasses a wide range of electrocardiographic and electrophysiological phenomena. These include abnormalities of sinus node impulse generation, disturbances of impulse emergence into the atrium, abnormal impulse transmission within the atria (and in some cases from the atria to the ventricles), increased susceptibility to atrial tachycardias (particularly atrial fibrillation), chronotropic incompetence, and inappropriate sinus tachycardia.[24,25] The clinical manifestations may vary from seemingly asymptomatic electrocardiographic findings to a wide range of complaints including syncope,

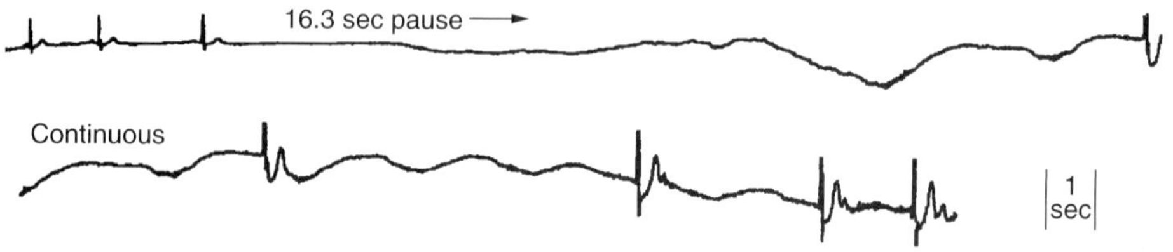

FIGURE 3-4 Electrocardiographic monitor recording illustrating a prolonged asystolic spell with spontaneous heart rhythm recovery in a patient with a spontaneous vasovagal faint during a medical procedure.

dizziness, shortness of breath, palpitations, fatigue, lethargy, stroke, and premature mental incapacity.

The causes of sinus node dysfunction are numerous but may be conveniently categorized as: (1) conditions that alter either sinus node or sinoatrial structure directly (so-called *intrinsic sinus node disease*), or both, or (2) operate indirectly to impair sinoatrial function (i.e., caused by extrinsic factors such as autonomic disturbances or drug effects).

Idiopathic degenerative and fibrotic changes associated with the ageing process are probably the findings most closely associated with "intrinsic" sinus node dysfunction. In regard to "extrinsic" sinus node dysfunction, drugs are the most important contributors. β-Adrenergic blockers, calcium channel blockers, membrane-active antiarrhythmics, and, to a lesser extent, digitalis, are the most frequently implicated. Each may alter sinus node function as a result of direct pharmacologic effects (e.g., flecainide, d-sotalol, verapamil) or indirectly via the ANS (e.g., β-adrenergic blockers), or both (e.g., quinidine, disopyramide, propafenone, amiodarone, digitalis).[24,45] In terms of clinical outcomes, cardioactive drugs may initiate or aggravate sinus bradyarrhythmias or induce chronotropic incompetence.

Apart from drug-induced autonomic disturbances, the ANS may also contribute directly to "apparent" disturbances of sinus node function. Sinus bradycardia, sinus pauses, sinoatrial exit block, and slow ventricular responses in atrial fibrillation may occur in the setting of parasympathetic predominance despite apparently normal underlying intrinsic sinus node or atrial function. In some cases the bradyarrhythmias are, in fact, extreme forms of sinus arrhythmia. Perhaps the best example of the latter is the physically fit individual in whom parasympathetic predominance at both the sinus node and AV node levels may be present on a chronic basis. In such cases, sinus pauses and various degrees of AV block have been reported during sleep or rest. Generally, these are asymptomatic and of little clinical consequence. Nonetheless, their occurrence (often detected inadvertently) may cause alarm. Carotid sinus syndrome and related conditions, in which excessive hypervagotonia is transient, are other instances in which intrinsic conduction system function is usually relatively normal yet manifests clinically important ANS-induced disturbances. It has even been suggested that in rare cases, a vagally mediated extended period of asystole occurring on a neurally mediated basis may have precipitated sudden cardiac death syndrome (most likely by triggering ventricular fibrillation) in otherwise well individuals.[46,47] Fortunately, even in the setting of an apparently prolonged asystolic event, spontaneous restoration of cardiac rhythm occurs in by far the vast majority of cases (Fig. 3-4).

Another example of a clinical circumstance in which the ANS appears to play a primary role in "arrhythmogenesis" is the syndrome of persistent or inappropriate sinus tachycardia. The basis for the tachycardia is believed to be abnormal enhanced automaticity within the sinus node or nearby atrial regions. The specific cause is for the most part unknown. However, current concepts point toward diminished parasympathetic control of sinus node function. In particular, given the frequent association with recent radiofrequency ablation of cardiac structures (or in former times to surgical ablation of accessory connections), an iatrogenic disturbance of intracardiac vagal reflexes has been proposed. Fortunately, the problem is often short-lived (weeks to months in duration) when it occurs as a consequence of a recent surgical or ablation procedure or a self-limited illness. In such cases, reassurance and/or β-blockade therapy may suffice. On occasion, however, refractory inappropriate sinus tachycardia necessitates more invasive measures, including attempted transcatheter ablation/modification of the sinoatrial region. Unfortunately, long-term results have been less than satisfactory to date.

The coexistence in the same patient of periods of bradyarrhythmia interspersed with bouts of atrial fibrillation, or less commonly other paroxysmal primary atrial tachycardias, is a common manifestation of sinus node dysfunction (so-called *bradycardia-tachycardia syndrome*). In regard to the onset of atrial fibrillation, an excess of either cholinergic or purinergic agonists may increase arrhythmogenic susceptibility. Both acetylcholine and adenosine shorten atrial refractoriness, thereby facilitating maintenance of multiple intra-atrial reentry pathways. Adrenergic agonists may have the same proarrhythmic effect (Fig. 3-5).

In bradycardia-tachycardia syndrome, symptoms may be the result of either the rapid heart beat or the bradycardic component, or both.[24,25] ANS influences are rarely entirely to blame for this manifestation of sinus node dysfunction, but it is likely that they often play a facilitating role. Similarly, true chronotropic incompetence is not usually attributable to ANS effects alone. As a rule, patients with parasympathetic predominance may exhibit low resting heart rates but ultimately manifest normal chronotropic responses to physical exertion. True chronotropic incompetence (i.e., inability of the heart to adjust its rate appropriately in response to metabolic need) most often implies intrinsic sinus node dysfunction or the undesirable effect of concomitant

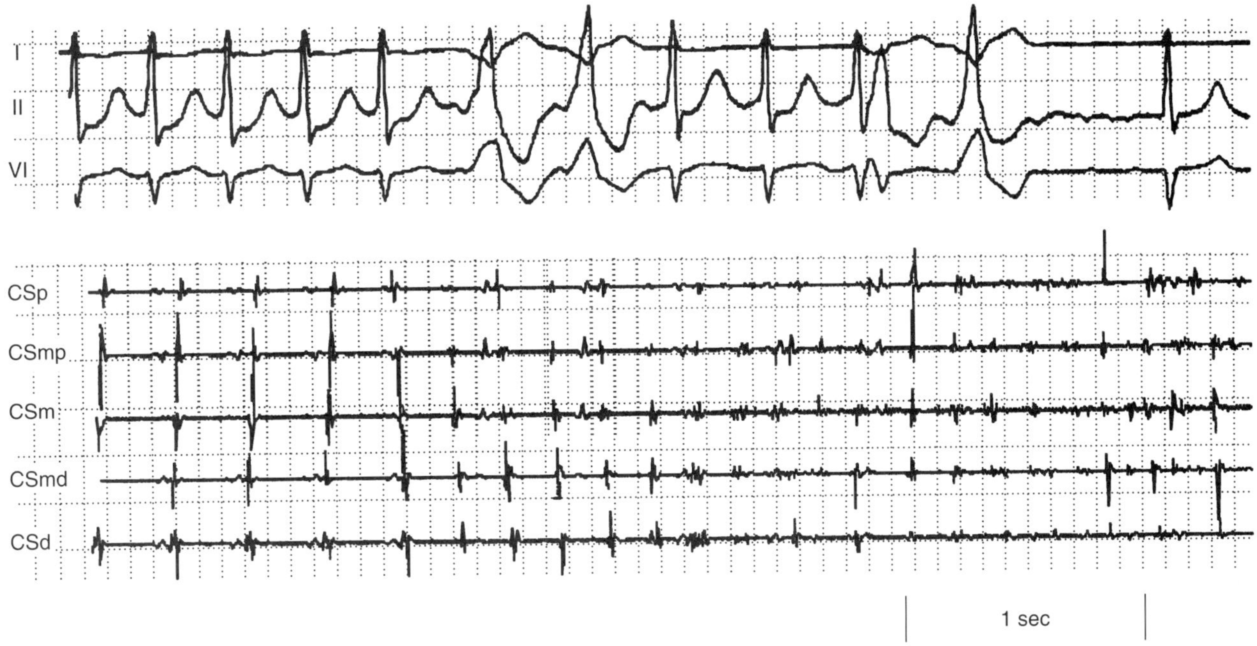

FIGURE 3-5 Spontaneous triggering of atrial fibrillation in association with an atrial ectopic beat. Electrocardiographic recordings I, II, and VI are provided along with a series of atrial electrograms recorded within the coronary sinus (proximal [CSp] to distal [CSd]). The patient had been resting, and isoproterenol infusion had just been initiated approximately 1 to 2 minutes earlier. The origin of the first atrial ectopic beat cannot be stated with certainty, but the earliest recording is noted on the most proximal CS electrogram.

drug treatment, or both. In this regard, although conventional exercise testing is not generally useful in identifying most forms of sinus node dysfunction, such testing may be helpful in differentiating those patients with resting sinus bradycardia but essentially normal exercise heart rate responses (e.g., physically trained individuals) from patients with true inadequate chronotropic responsiveness.

Assessment of sinus node responses to pharmacologic interventions (e.g., autonomic blockade), neural reflexes (e.g., carotid sinus massage, Valsalva's maneuver, heart rate response to upright tilt), or induced hypotension (e.g., by administration of amyl nitrite) are important elements of the diagnostic assessment of sinus node function. For example, pharmacologic interventions may assess sinus node response to β-adrenergic blockade, β-adrenergic stimulation, or parasympathetic muscarinic blockade (i.e., atropine infusion). The most important of these tests is assessment of intrinsic heart rate (IHR, sinus node rate in the "absence" of neural control) by pharmacologic autonomic blockade with combined administration of a β-adrenergic blocker and atropine. The observed sinus rate following drug administration is essentially independent of ANS input to the node and can be characterized as the observed IHR (IHRo). Normal values for IHR can be predicted approximately (IHRp) from the linear regression: IHRp = 118.1 − (0.57 × age).[28] Thus, a normal IHRo in a patient with resting bradycardia or sinus pauses may, in the absence of drug effects, be used as evidence suggesting an extrinsic form of "ANS-mediated" sinus node dysfunction.[24,48] Autonomic blockade can also be used to ascertain the impact of ANS on sinoatrial conduction time (SACT) and sinus node recovery time (SNRT) measurements.[24] Although the clinical implications of such measurements are not well established, it does provide further evidence regarding the role the ANS might play in inducing disturbances of sinus node function in certain individuals.

ATRIOVENTRICULAR CONDUCTION DISTURBANCE

As noted earlier, in the normal resting state sympathetic and parasympathetic influences tend to exert approximately equal influence on AV nodal function (in contrast to the sinus node, where parasympathetic influences usually dominate).[20,30,31] However, in some situations (e.g., well-trained athletes, cardiac glycoside excess, hypervagotonia associated with conditions such as carotid sinus syndrome or other forms of neurally mediated syncope) parasympathetic effects become dominant, at least transiently. The result may be development of first-degree, second-degree Mobitz type 1 (Wenckebach), or even higher degrees of AV block. The autonomic etiology may be suggested by both clinical and electrocardiographic clues. Most importantly, in terms of electrocardiographic evidence, the autonomic basis may be suspected by virtue of AV block being associated with a real or relative sinus bradycardia, rather than the sinus tachycardia that would be expected if the hemodynamic embarrassment were due to structural conduction system disease.

First- and second-degree type 1 AV blocks are most often the result of conduction disturbances at the level of the AV node (i.e., prolonged A–H interval) and are frequently attributable to ANS influences. This is especially the case when there is no evidence of underlying cardiac disease, when the QRS morphology is normal, and when the individual is young or physically fit, or both.[49-52] Of course, drug-induced AV block must also be excluded. In the presence of a narrow QRS complex, first-degree AV block is due to AV nodal delay in more than 85% of patients and due to delay within the His bundle in less than 15%. Similarly, type I second-degree AV block is AV nodal in origin in the large majority of cases in which the QRS is narrow. Invasive electrophysiologic recordings long ago documented the progressive prolongation of the A–H interval during classic Wenckebach periodicity and its reversibility following muscarinic blockade with atropine.[49]

ANS-mediated higher degrees of AV block may also be observed. These episodes of "paroxysmal AV block" are generally benign from a mortality perspective, although they may be associated with dizziness and syncope (e.g., vasovagal faint) and risk of physical injury. Sustained third-degree AV block is, however, not usually attributable to ANS effects. In adults acquired complete heart block is almost always associated with structural heart disease and, more often than not, is associated with a wide QRS morphology. However, ANS effects may also contribute to the electrocardiographic findings.

In the setting of acute anterior MI, transient or fixed complete AV block is reported to occur in 5% of cases and is typically infranodal.[49] The ultimate poor prognosis in these patients is related to the magnitude of ventricular damage. By contrast, complete AV block occurs more frequently (10% to 15% of patients) after inferior wall MI, but in these instances ANS effects are often the prime cause. Indeed, the block may progress through stages beginning with P–R interval prolongation or type 1 second-degree AV block, or both; the site of block is within the AV node. Apart from ANS-mediated parasympathetic effects, the mechanisms eliciting this form of AV block also include nodal ischemia and adenosine release. Nevertheless, the block can often be reversed (at least temporarily) by atropine administration, thereby supporting the importance of the parasympathetic autonomic etiology.

Drug effects are a common cause of AV nodal conduction disturbances. Various cardioactive drugs affect the AV node, either directly by cellular action or indirectly as a result of their actions on the autonomic nervous system, or both. Cardiac glycosides are widely known to affect the AV node by ANS-mediated effects; first- or second-degree type 1 AV block occurs as a result of glycoside-induced enhanced vagal tone at the AV node. β-Adrenergic blockers result in AV nodal conduction slowing or block, or both, by diminishing sympathetic neural effects on the AV junction.

For the most part, calcium channel blockers (particularly verapamil and diltiazem) and most antiarrhythmic drugs (especially class 1C drugs) act directly to slow conduction in the AV node. However, in some cases noncardiac calcium channel blocking actions (e.g., vasodilation) initiate neural reflex effects (i.e., increased sympathetic activity) that enhance AV conduction. Nifedipine's effects are well known in this regard.

Certain antiarrhythmic drugs have important ANS effects that must be accounted for when they are prescribed. Both quinidine and disopyramide manifest prominent vagolytic actions, which tend to counterbalance their negative dromotropic direct effects. This vagolytic effect can lead to apparently "paradoxical" increases of ventricular rate when these drugs are used to treat patients with certain primary atrial tachycardias, especially atrial flutter. This outcome occurs by virtue of the tendency of the drugs to slow atrial rate while enhancing AV conduction; the net effect is less block at the level of the AV node and a more rapid ventricular response. For some other antiarrhythmic drugs (e.g., amiodarone, sotalol), negative dromotropic actions result from not only direct but also indirect ANS effects; in such cases, these synergistic effects can lead to unexpectedly severe bradycardia.

Autonomic Nervous System and Specific Tachyarrhythmias

ANS activity may be implicated to some extent in the initiation, maintenance, and clinical impact of virtually all tachyarrhythmias. For instance, sympathetic, parasympathetic, and purinergic neural input at the AV node may in large part determine whether AV node reentry or AV reentry supraventricular tachyarrhythmias can be triggered or sustained at a particular time in patients known to be susceptible to these arrhythmias. In essence, the ability of a premature atrial or ventricular beat to dissociate conduction pathways and thereby permit reentry may vary from moment to moment depending on neural influences. In a few instances, ANS effects appear to exhibit even greater responsibility for rhythm disturbances. The best examples are the vagally mediated and adrenergically mediated forms of atrial fibrillation. Torsades de pointes ventricular tachycardia in the setting of certain ventricular repolarization disturbances may also be included in this category. Finally, the importance of ANS influences on arrhythmia susceptibility in the setting of myocardial ischemia is well known although far from well understood.

ATRIAL FIBRILLATION

The ANS may play a role both in setting the electrophysiological stage as well as triggering certain forms of atrial fibrillation. For instance, although catecholamine-induced atrial fibrillation is not common in the clinic, it is observed relatively commonly in the electrophysiology laboratory (see Fig. 3-5).[53-60] In addition, the relative balance of ANS input to the cardiac conduction system is a crucial determinant of the ventricular response associated with an episode of atrial fibrillation. Little is known regarding the possibility that ANS elements may also participate in termination of periods of atrial fibrillation.

There exists considerable experimental evidence implicating acetylcholine, and more recently adenosine, in promoting susceptibility to atrial fibrillation. Both of these agents are known to shorten atrial refractoriness, an electrophysiologic finding associated with increased susceptibility to atrial fibrillation. The cellular basis for this effect could be increased repolarization current via acetylcholine-activated potassium channels. Reduced inward calcium current, due to diminished intracellular adenyl cyclase activity, may also be relevant. These effects, especially if anatomically inhomogeneous due to variability of ANS atrial innervation, may enhance the potential for coexistence of simultaneous multiple "functional" reentry pathways. The outcome would facilitate both triggering and maintenance of an atrial fibrillation episode.

ANS influences may also be associated with the triggering of atrial fibrillation events, either by facilitating triggered activity as a basis for "focal origin" atrial fibrillation, or by "supplying" the necessary ectopic activity required to permit maintenance of atrial fibrillation. In this regard, relatively little is known beyond preliminary findings suggesting that the factors leading to onset of spontaneous events are not predictable even in a single patient.[60] Consequently, although the ANS may be an important contributor, the overall susceptibility to arrhythmia onset is almost certainly multifactorial.

The possibility that reduction of susceptibility to atrial fibrillation, or even atrial fibrillation termination, may be facilitated by ANS manipulations has received relatively little attention. However, Elvan et al.[61,62] found that the placement of radiofrequency linear lesions in canine atria reduced susceptibility to atrial fibrillation induction by pacing and vagal stimulation, while atrial fibrillation remained inducible with high-dose methacholine. These findings suggest that modification of atrial innervation may impact atrial fibrillation induction, and perhaps part of the benefit of so-called *maze* procedures may be due partly to alteration of ANS influences on the atria.

Atrial Fibrillation Triggering (Vagally and Adrenergically Mediated)

Vagally mediated (bradycardia- or pause-dependent) atrial fibrillation is relatively uncommon.[53,54] It tends to occur more commonly in men than women (≈ 4:1 ratio), and the episodes begin at night or during the early morning hours when vagal predominance is greatest. The same individuals may experience postprandial atrial fibrillation. Clinical recognition of vagally mediated atrial fibrillation is important given the fact that cardiac glycosides and β-adrenergic blockers are contraindicated in treatment. In this regard, a recent study comparing disopyramide and metoprolol in this population revealed a markedly greater recurrence rate with β-adrenergic blocker therapy than with disopyramide.[56] Vagolytic antiarrhythmics (e.g., quinidine, disopyramide) may be helpful but are often poorly tolerated or generally unwanted by these otherwise healthy young patients. Atrial-based cardiac pacing may also have a useful role to play in this setting (i.e., prevention of the vagally mediated bradycardia associated with onset of events) but is often considered "unacceptable" in this relatively young and otherwise generally healthy patient population.[63]

Excluding the probable role played by heightened adrenergic tone in postoperative arrhythmias (see later discussion), true adrenergically mediated forms of paroxysmal atrial fibrillation are, perhaps somewhat surprisingly, less common than the vagally mediated form. Rarely, adrenergically mediated atrial fibrillation is secondary to a noncardiac disease process such as hyperthyroidism or pheochromocytoma. Most often the medical history suggests onset during the waking hours (usually in the morning) in association with stress or physical exertion.[53,54,63] Underlying cardiac disease may or may not be present. Some individuals seem to manifest this problem as a form of "lone atrial fibrillation." In others, recent cardiac or chest surgery seems to play a role. The possibility that this form of atrial fibrillation is largely the result of triggered activity (i.e., a form of so-called *focal* atrial fibrillation) needs further consideration. Antiadrenergic therapy is the first choice in these cases.

As alluded to earlier, the ANS appears to play an important role in postoperative atrial fibrillation; however, the relationship is largely inferential. In this regard, Dimmer et al.,[59] using HRV measures, observed diminished vagal tone and enhanced sympathetic tone in individuals developing postoperative atrial fibrillation. At the cellular level, one study indicated that patients manifesting postoperative atrial fibrillation had had a higher density of β-adrenergic receptors preoperatively (while still in sinus rhythm) than did individuals who had no postoperative atrial fibrillation.[64] Further, several reports suggest that prophylactic administration of β-adrenergic blockers reduces the frequency of postoperative atrial fibrillation in patients undergoing coronary artery bypass surgery.[65-67]

Control of Ventricular Rate

The manner by which the AV node responds to high rate "bombardment" from a fibrillating atrium and generates an irregular ventricular response remains uncertain. However, whatever the mechanisms at play, it seems clear that the ANS contributes importantly to modulating the ventricular rhythm. Increased vagal tone or diminished sympathetic tone, or both, are associated with a decrease in average ventricular rate. The converse clearly increases the average ventricular rate. Whether the degree of R-R variability is diminished by these maneuvers is less clear. Nevertheless, pharmacologic manipulation of ANS activity (e.g., cardiac glycosides, β-adrenergic blockers) is a crucial part of everyday clinical practice in the management of heart rate in atrial fibrillation patients.[68]

SUPRAVENTRICULAR TACHYCARDIAS (OTHER THAN ATRIAL FIBRILLATION)

For purposes of this discussion, the supraventricular tachycardias are categorized into those that are dependent on AV node conduction and those that

TABLE 3-1 Supraventricular Tachycardias

Atrioventricular (AV) Node–Dependent (i.e., maintenance dependent on autonomic nervous system [ANS] effects on AV node)
AV node reentry
AV reentry
• Orthodromic reciprocating tachycardia
• Antidromic reciprocating tachycardia using the AV node in the retrograde direction
AV Node–Independent (i.e., ANS effects primarily affect ventricular response, not arrhythmia maintenance)
Atrial fibrillation
Atrial flutter
Sinus node reentry tachycardia
"Multifocal" atrial tachycardia
Other primary atrial tachycardias

are not (Table 3-1). The first group is exemplified by AV nodal reentry tachycardia (AVNRT) and AV reentrant tachycardia (AVRT) using accessory AV connections. The AV nodal–independent tachycardias include both reentrant arrhythmias as well as those thought to be automatic or "triggered" in origin. Most prominent among the reentry forms are so-called *sinus node reentry tachycardia* and other *intra-atrial reentry tachycardias* (including the various forms of atrial flutter). The nonreentrant tachycardias in this category include "inappropriate sinus tachycardia" (see also the earlier discussion of Sinus Node Dysfunction), and various "focal" atrial tachycardias.

With respect to ANS influences, the rate and stability of AV nodal–dependent tachycardias are critically determined by the important effects that the ANS exerts on the AV conduction system (particularly the AV node, see earlier discussion). Thus, initiation, maintenance, and termination of these arrhythmias may be influenced by ANS action. By contrast, AV node "independent" tachycardias are less influenced by ANS effects; their atrial rate is determined principally by the nature of the substrate (although the atrial cycle length can be influenced to some extent by sympathetic ANS effects). On the other hand, the ventricular rate response in "AV node independent" tachycardias is very much influenced by autonomic effects at the AV nodal level. In this regard, the ANS impact is essentially the same as that described earlier for atrial fibrillation.

Atrioventricular Node–"Dependent" Tachycardias

Tachycardias in which the AV node is an essential component of the reentry circuit are termed *AV node "dependent."* In these cases ANS influences determine AV nodal conduction properties and thereby play a crucial role in tachycardia initiation, its stability and cycle length, and its manner of termination. By way of example, Waxman and colleagues pointed out that a complex sequence of events may occur following onset of a supraventricular tachycardia and may result in self-termination of the arrhythmia.[69-70] Hypotension at onset of the arrhythmia may initiate a reflex sympathetic drive to permit blood pressure to stabilize. As blood pressure recovers, reflex-enhanced vagal tone may slow and terminate the arrhythmia (Fig. 3-6). In other cases, especially in patients with heart disease in whom hemodynamic recovery to tachycardia stress is limited, it is not uncommon for sympathetic drive triggered by hypotension to cause a further shortening of the cardiac cycle length and aggravation of hypotension. In these cases, a "therapeutic vagal rebound" doesn't occur.

Alteration of venous return, such as occurs with deep inspiration or cough, may also act to terminate a reentrant tachycardia. In this case, the physical effects of stretch on atrial tissue may play a role. Alternatively, the triggering of atrial-based neural reflexes by activation of mechanically sensitive receptor sites could result in alteration of ANS influence on atrial and AV nodal conduction properties. Thus, Valsalva's maneuver, cough,

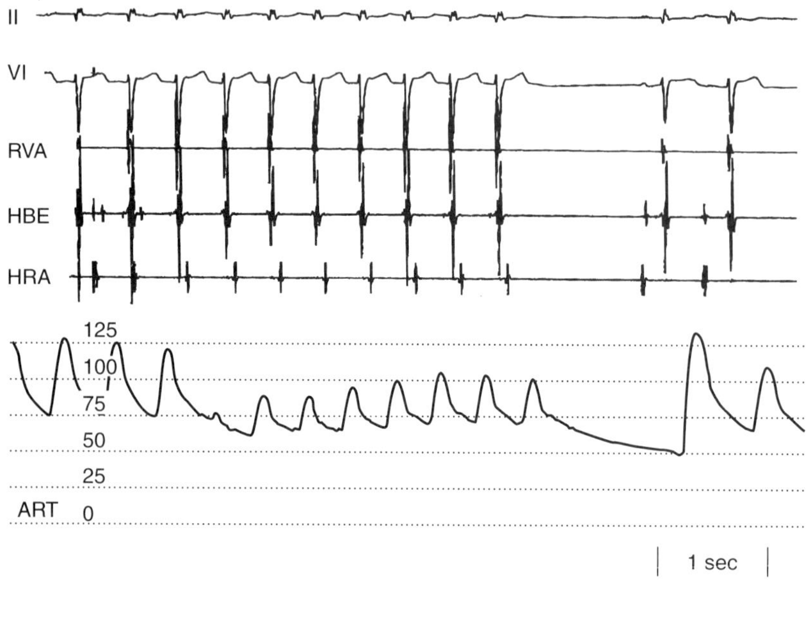

FIGURE 3-6 Electrocardiographic, intracardiac, and arterial pressure traces during a brief episode of reentry supraventricular tachycardia. Note the initial drop of pressure at onset of the arrhythmia. Thereafter the pressure slowly rebuilds. The pressure recovery may play a role in tachycardia termination (see text).

head-down position, and other techniques that augment venous return presumably act similarly to terminate tachycardias through both their mechanical effects as well as ANS reflex pathways. The latter principally affects the AV node since it is the most neurally sensitive structure essential to maintaining many of these arrhythmias.

Atrioventricular Node–"Independent" Tachycardia

Excluding the ANS impact on ventricular rate control, AV node–independent supraventricular tachycardias vary widely in terms of their susceptibility to ANS effects. As in the case of other tachycardias, it is reasonable to assume that ANS influences may play a role in triggering seemingly sporadic events in patients who presumably have the substrate present at all times. However, it is usually only possible to speculate on the ANS contribution to the onset of an episode. Potential mechanisms include facilitation of the occurrence of ectopic "trigger" beats through direct neurohumoral action on atrial foci, or by indirect effects such as stretch. ANS-induced atrial cycle length changes (e.g., sinus arrhythmia), or alteration of conduction properties within a reentry circuit may also account for de novo appearance of a tachycardia in an individual who had been arrhythmia free for some time.

Sinus node reentry tachycardia[24,71,72] and inappropriate sinus tachycardia[73,74] are infrequent clinically, but nevertheless are examples of AV node–independent supraventricular tachycardias in which ANS effects appear to be important. Aside from the triggering issue alluded to earlier, both of these arrhythmias are associated with wide heart rate excursions, suggesting significant ANS effect. In the case of sinus node reentry, treatment with a β-adrenergic blocker or calcium channel blockers, or both, is often effective. Unfortunately, this same approach is generally much less reliable in the case of inappropriate sinus tachycardia.

Atrial flutter and other forms of intra-atrial reentry occur relatively commonly in the setting of heart disease. As a rule, and again putting aside the "trigger" issue discussed earlier, these arrhythmias are less susceptible to ANS effects than are sinus node reentry or inappropriate sinus tachycardia. However, ANS-induced changes of atrial volume or stretch may act secondarily to impact tachycardia cycle length and stability. Further, tachycardia-induced hypotension, if sustained, may increase circulating catecholamines and thereby further increase both tachycardia rate as well as the ventricular response through effects on AV nodal conduction.

Iatrogenic Factors

Nonpharmacologic and pharmacologic iatrogenic effects, acting at least in part via the ANS, are important considerations when discussing the supraventricular tachycardias. In terms of nonpharmacologic effects, maneuvers that transiently modify ANS tone (sometimes in a complex fashion) are widely used in an attempt to terminate certain supraventricular tachycardias (usually AV node dependent) or slow the ventricular response in others (typically AV node independent). Thus, carotid sinus massage, facial immersion (usually only effective in children), and the injection of drugs such as phenylephrine hydrochloride (Neo-Synephrine) to transiently increase blood pressure (a technique now largely abandoned) are known to enhance vagal tone to the heart and the AV node in particular. Optimally, the induced conduction slowing alters the stability of the reentry circuit in AV-dependent tachycardias and terminates the tachycardia (usually by means of A–H interval block). Valsalva's maneuver, cough, and head-down posture are other techniques employed for similar purposes. In these cases the mechanism of action may be only partly via ANS reflex, since induced chamber volume changes resulting from altered preload and afterload may contribute by means of mechanical impact on electrophysiology properties of the circuit.

In terms of pharmacologic contributors to altered ANS state, the vagotonic effects of cardiac glycosides have long been used to diminish susceptibility to AV node–dependent supraventricular tachycardias and to reduce the ventricular rate of all types of supraventricular tachycardias.[68] The potential adverse effects of cardiac glycosides in the setting of pre-excitation syndromes (i.e., reducing accessory pathway antegrade refractory periods and increasing rate response of pre-excited beats in atrial fibrillation) was only recognized much later.[75] The mechanism for this effect is thought to be a direct drug effect on bypass pathways, but indirect ANS-mediated actions (e.g., enhanced central sympathetic tone, reduced retrograde concealment due to increased AV nodal block) cannot be discounted. A similar argument pertains to acceleration of pre-excited atrial fibrillation responses following verapamil administration.[76,77] Once again drug-induced hypotension (with increased reflex sympathetic drive via the ANS) or increased AV nodal block (with consequent diminished retrograde concealment into the bypass connection), or both, may in part account for this observation.

Perhaps the most frequent adverse iatrogenic effect associated with the supraventricular tachycardias is that associated with use of certain antiarrhythmic agents (e.g., quinidine, disopyramide, procainamide) in the treatment of primary atrial tachycardias. All of these drugs happen to exhibit a measure of vagolytic effect. This action may lead not only to certain undesirable side effects (e.g., dry eyes, dry mouth, constipation) but also to potentially hazardous electrophysiological consequences. Thus, in the case of the treatment of primary atrial tachycardias, the direct antiarrhythmic effects of these agents prolong atrial cycle length, while at the same time the vagolytic action enhances AV nodal conduction. The net result may be a dangerously rapid increase in ventricular rate due to 1:1 AV conduction of the somewhat slower tachycardia. Thus, concomitant use of a β-adrenergic blocker or calcium channel blocker is a mandatory precaution at the initiation of antiarrhythmic drug therapy in this setting.[78]

VENTRICULAR TACHYCARDIA

The importance of the ANS in determining susceptibility to ventricular tachyarrhythmias in certain disease states is well established. Acute ischemic heart disease is the best example.[79] However, ANS influences may also be instrumental in triggering tachycardia events in patients with a well-established, long-standing substrate, such as those with preexisting fibrotic areas as a consequence of prior MI or remote cardiac surgery (e.g., childhood ventricular septal defect repairs). ANS participation in the triggering of arrhythmias is almost certainly pertinent in other chronic states in which the arrhythmia substrate is present all the time, yet rhythm disturbances occur only sporadically. Among the best examples of the latter scenario are the abnormal ventricular repolarization syndromes (e.g., long QT syndromes [LQTS], Brugada syndrome).[80-83] In these cases, it is usually impossible to judge the exact importance, or precise nature, of any contributing transient ANS disturbance to a sudden arrhythmic event. Nevertheless, in some cases the association of life-threatening ventricular arrhythmias with "startle" reactions (e.g., to the ringing of an alarm clock) strongly supports the potential importance of ANS effects.

Among the better known effects of the ANS on ventricular arrhythmias is the evident, although variable, impact of heart rate on ambient ventricular ectopy. In this case it is difficult to distinguish the effect of heart rate from other effects of ANS mediators (e.g., increases in blood pressure). Thus, in some cases exercise results in diminution of ventricular ectopy frequency. In this setting of increased heart rate, enhanced sympathetic stimulation, and parasympathetic withdrawal, it would seem intuitively that the rate effect predominates since the others would tend to be proarrhythmic. Conversely, in some patients heart rate slowing is associated with diminished ectopy. In such case one might assume that the basis is decreased sympathetic drive or increased parasympathetic influence, or both.

Understanding mechanisms by which ANS effects may alter susceptibility to ventricular arrhythmia is complicated by the indirect effects associated with changes of sympathetic and parasympathetic tone.[84] Thus, in ischemic heart disease coronary flow may be sufficiently disturbed by ANS effects to promote arrhythmias. Similarly, increases in systemic pressure may alter myocardial metabolism (afterload effect) or directly induce ectopy through a stretch mechanism. The latter may be particularly important in the case of severe mitral valve prolapse, in which tension on the papillary muscle apparatus has been purported to be a basis for ventricular ectopy. Further, Waxman et al.[85] have observed termination of ventricular tachycardia by certain maneuvers that alter ANS output to the cardiovascular system (e.g., Valsalva's maneuver). In this case the effect appeared to be secondary to ventricular volume changes (potentially a stretch effect), rather than the direct effect of the change in autonomic balance at the level of the myocardium.[85]

Ischemic Heart Disease

The ANS contributes importantly to arrhythmogenesis in acute myocardial ischemia.[86,93] In brief, the risk of potentially life-threatening arrhythmias and sudden death increases in response to ischemia-associated increased sympathetic activity and is diminished by sympathetic blockade or parasympathetic enhancement, or both. A detailed survey of the literature supporting this statement lies beyond the scope of this chapter, but certain key observations are touched upon.

1. Sympathetic Neural Influences

Clinically, acute myocardial ischemia is often associated with findings consistent with elevated adrenergic tone, most importantly sinus tachycardia and hypertension (although heightened vagal tone may also occur, especially in the case of inferior wall ischemia). Additionally, it appears that this increased sympathetic activity is associated with arrhythmogenic electrophysiologic effects.[87-91] In this regard, perhaps the most convincing clinical evidence supporting the close relationship between increased sympathetic neural activity and greater arrhythmogenicity in patients with ischemic heart disease is multiple published reports indicating that β-adrenergic blockade reduces sudden death risk.[86,87,92-94] A summary of these findings has been provided in a meta-analysis by Yusuf and Teo comprising more than 52,000 individuals.[94] Overall, there was a 17 percent mortality reduction when β-blocker therapy was compared to placebo. A comparable result has yet to be achieved by any other pharmacologic agent.

Acute myocardial ischemia and infarction result in central ANS alterations, presumably on the basis of the initiation of neural activity in cardiac afferent nerves; central effects resulting from pain, fear, and so on; and disturbances of peripheral neural distribution to the ventricular myocardium. In the central nervous system, myocardial ischemia may trigger varying levels of sympathetic neural efferent activity with both increases and decreases being observed. At the level of the myocardium, acute infarction appears to disrupt sympathetic neural distribution to the ventricles in excess of the actual muscle damage.[40] As time passes, it appears that a greater concordance develops. Nevertheless, for some considerable time centrally mediated as well as peripheral discrepancies in sympathetic influence on infarcted and noninfarcted myocardium may facilitate arrhythmia susceptibility (i.e., inhomogeneity of ANS sympathetic input to the myocardium).

β-Adrenergic effects seem to be the predominant pathway by which abnormal sympathetic neural activity adversely alters cardiac electrophysiology. In the ischemic partially depolarized myocardium, enhanced β-adrenergic activity may facilitate development and maintenance of reentry circuits by enhancing inward calcium currents; in the absence of sympathetic drive, conduction block may prevent sustained arrhythmias from becoming manifest. Additionally, β-adrenergic drive may promote triggered activity, again setting the stage for sustained arrhythmias on either a triggered or reentrant basis.

The role of α-adrenergic stimulation in the ischemic heart has not been as clear as that of β-adrenergic stimulation. There is evidence both for and against a protective value of α-adrenergic blockade in acute ischemia.

For the most part, current thought based on canine studies does not support a protective effect of α-adrenergic blockade in most acute ischemic syndromes. However, it is still possible that α-adrenergic blockade, operating by reversing alpha-agonist–mediated calcium ion mobilization within myocytes, may have some value in diminishing reperfusion-associated cardiac arrhythmias.

2. Parasympathetic Neural Influences

Not too many years ago, increased parasympathetic tone was thought to be associated with increased arrhythmia and mortality risk in ischemic heart disease. This is no longer the case. Both experimental and clinical evidence supports the view that enhanced parasympathetic tone diminishes arrhythmic risk in the setting of acute ischemia. In this regard recent studies suggest that both neurally mediated heart rate slowing effects and direct parasympathetic agonist effects contribute comparably to the overall benefit.[95] Further, a number of experimental studies suggest that while muscarinic agonists are not as effective in terms of antiarrhythmic action in the setting of acute ischemia as are β-adrenergic blockers, they nonetheless could offer some additional advantage were it not for their troublesome adverse side effects (i.e., dry eyes, dry mouth, constipation).

Experimental and clinical studies of baroreceptor sensitivity (BRS, a measure of vagal influence on the heart) offer important insight into the potentially protective role played by the parasympathetic nervous system in patients with ischemic heart disease.[96-98] First, BRS was noted to be lower immediately following MI. Thereafter, BRS values were found to be even lower in a subset of those post–MI patients who had experienced an episode of ventricular fibrillation. In a prospective trial of relatively low-risk patients, BRS values again proved to be lower in those individuals who failed to survive the 2-year follow-up, and the effect appeared to be independent of left ventricular function as assessed by ejection fraction measurement.[98] Most recently a multicenter trial (Autonomic Tone and Reflexes after Acute Myocardial Infarction, ATRAMI) provided convincing additional evidence.[99] ATRAMI enrolled 1284 patients who had had a recent MI, could exercise, and did not require surgical revascularization. Patients were followed for 21 ± 8 months. Cardiac mortality was higher (9% versus 2%; 10% versus 2%) among individuals with low BRS (<3 ms/mmHg) or low SDNN (<70 ms) than those with normal BRS or SDNN (>6.1 ms/mmHg, >105 ms). Combining both indices resulted in even greater risk recognition. Once again, the effect appeared to be independent of ejection fraction.

The observations related to BRS lead to examination of the concept that heart rate variation (HRV) may provide a means to stratify patients with ischemic heart disease at risk of lethal arrhythmias.[100-102] As was the case with SDNN in the ATRAMI study, findings suggest that diminished HRV is associated with a much greater mortality risk in post–MI patients.[99] Thus, while HRV may be more a mixture of both sympathetic and parasympathetic influences to a greater extent than BRS is, the finding that parasympathetic predominance is protective while diminished parasympathetic influence is detrimental remains consistent.

Given the experimental observations suggesting the benefit of parasympathetic predominance, carefully supervised exercise training would appear to be an advisable approach to improve mortality in ischemic heart disease patients. Prospective studies examining this hypothesis are currently in progress.

Long QT Syndromes (LQTS) and Brugada Syndrome

Disturbances of ventricular repolarization have been the subject of considerable interest in recent years.[80-83,103-105] The arrhythmias associated with these conditions (primarily torsades de pointes) are most often iatrogenic, almost always life-threatening, and usually treatable/preventable if the underlying problem is recognized promptly.

1. Acquired LQTS

The acquired form of LQTS is by far the most common form of LQTS and is most frequently the result of medical treatment with Q–T interval–prolonging drugs. Torsades in this setting is typically seen during periods of bradycardia (e.g., sleep) or following pauses in the cardiac rhythm (e.g., post-PVC) that accentuate the Q–T interval.[104-106] Some of the best known offending drugs are listed in Table 3-2. The majority of these act

TABLE 3-2 Pharmacologic Agents Associated with Q–T Interval Prolongation

Antiarrhythmic Agents
CLASS IA
Disopyramide
Procainamide
Quinidine
CLASS III
Amiodarone
Dofetilide
D-sotalol
Ibutilide
N-acetyl procainamide (NAPA)
Sotalol
Antianginal Agents
Bepridil
Psychoactive Agents
Phenothiazines
Thioridazine
Tricyclic Antidepressants
Amitriptyline
Imipramine
Antibiotics
Erythromycin
Pentamidine
Fluconazole
Nonsedating Antihistamines
Terfenadine
Astemizole
Others
Cisapride

by antagonizing outward (i.e., repolarizing) potassium currents (e.g., class 1A and class 3 antiarrhythmic drugs). Others in this list are reported to interfere with the metabolism of drugs that directly prolong the Q–T interval. In general the risk of torsades increases in proportion to the duration of the Q–T interval.

The impact of the ANS on initiating torsades in drug-induced LQTS is not well documented. Nevertheless, circumstantial evidence strongly suggests an important link. Thus, in some cases, bradycardia (not infrequently of ANS origin) may be a contributory factor. In this regard, cardiac pacing has been considered by some to be an effective treatment strategy, although its benefits are often obscured by the concomitant use of anti-adrenergic therapy in most reports.[107,108] Schwartz and Priori[108] examined Long QT Registry data and concluded that pacing does offer a benefit, but the magnitude of the effect may not be great enough to warrant its use as an initial treatment approach.

2. Congenital LQTS

Congenital, idiopathic, or familial LQTS is caused by mutations in cardiac ion channels that contribute to the action potential repolarization process. Congenital LQTS is infrequent, but its identification can be life saving. Affected individuals have QT prolongation and a high risk of recurrent syncope and sudden cardiac death due to torsades de pointes. The role of the ANS in triggering torsades in LQTS patients is highly suspected even if not well understood. Syncope and sudden death in this setting are frequently associated with emotional or physical arousal (e.g., fear, loud noises, exertion).[105] Heterogeneity in clinical presentation exists, however, so that in other individuals torsades de pointes occurs due to bradycardia or during sleep in conjunction with rate-dependent Q–T interval prolongation.

3. Brugada Syndrome

Brugada syndrome is a relatively recently recognized genetic defect of the cardiac sodium channel gene leading to susceptibility to life-threatening ventricular arrhythmias. The relationship between Brugada syndrome and ANS effects is suggested by the observation that sudden death episodes in this setting have often been during sleeping hours,[82] possibly implicating sleep-related bradycardia as a trigger factor.

Other Forms of Idiopathic Ventricular Tachycardia

The role of ANS activity in triggering or sustaining arrhythmic events in patients with other forms of idiopathic ventricular tachycardia is suspected but has been the subject of only infrequent study. For instance, in the electrophysiology laboratory, parenterally administered β-adrenergic agonists are often needed to "induce," and β-adrenergic blockade has been used to terminate both ventricular tachycardia of right ventricular outflow tract origin and ventricular tachycardia considered to be of left ventricular fascicular origin.

In one study of patients with idiopathic ventricular tachycardia in the setting of normal left ventricular function, Engelstein et al[109] used imaging techniques to infer the presence of myocardial glucose metabolism abnormalities and disturbances of sympathetic innervation in excess of any demonstrable perfusion defects. The implication is that apparently minor heterogeneities of myocardial metabolism or innervation may translate into functionally important arrhythmogenic substrates. Clearly, this is a topic in need of more detailed study.

Autonomic Nervous System and Syncope

Syncope is best viewed as a syndrome characterized by transient loss of consciousness, usually associated with concomitant loss of postural tone, and subsequent spontaneous recovery. In this context, it is important to distinguish true syncope (Table 3-3) from other nonsyncopal conditions or apparent disturbances of consciousness (e.g., seizures, sleep disorders, "drop attacks"). These other conditions are not the subject of this discussion. Comprehensive discussions of the causes of syncope and its treatment are provided elsewhere.[110-113]

Mechanistically, syncope is most often the result of transient disturbances of cerebral blood flow. In this regard, maintenance of cerebral blood flow is normally facilitated by several factors, all of which are to some extent importantly influenced by the ANS. Certain of these factors include: (1) cardiac output; (2) baroreceptor-induced adjustments of heart rate and systemic vascular resistance; (3) cerebrovascular autoregulation (which is contributed to in part by the status of systemic arterial pressure as well as local metabolic

TABLE 3-3 Diagnostic Classification of Syncope

1. Neurally Mediated Reflex Disturbances of Blood Pressure Control
- Vasovagal faint
- Carotid sinus syncope
- Others

2. Orthostatic

3. Primary Cardiac Arrhythmias
- Sinus node dysfunction (including bradycardia/tachycardia syndrome)
- Atrioventricular conduction system disease
- Paroxysmal supraventricular tachycardias
- Paroxysmal ventricular tachycardia (including torsades de pointes)

4. Structural cardiovascular or cardiopulmonary disease
- Cardiac valvular disease/ischemia
- Acute myocardial infarction
- Obstructive cardiomyopathy
- Primary pulmonary hypertension

5. Noncardiovascular
- Metabolic/endocrine disturbances
- Psychogenic "syncope"

factors, particularly PCO_2); and (4) regulation of vascular volume by the kidneys and hormonal influences.

Failure of physiologic mechanisms designed to protect cerebrovascular blood flow or the intervention of factors that impair O_2 delivery to excessively low levels for 8 to 10 seconds or longer may compromise cerebral function sufficiently to induce a syncopal episode. In this regard, the older or ill patient is obviously at greater risk than the young or healthy patient. Ageing alone has been associated with diminution of cerebral blood flow, while compensatory mechanisms that rely on neural reflexes, such as the carotid baroreceptors, may become functionally less reliable.

Of the many causes of syncope, ANS effects are of greatest importance in the various forms of neurally mediated syncope.[111-113] The vasovagal faint and carotid sinus syndrome are the most common of these. Other conditions in this group (e.g., postmicturition syncope, cough syncope, swallow syncope) are relatively uncommon. However, ANS effects are crucial contributors to syncope associated with orthostatic stress and may even play an important contributory role in certain tachyarrhythmias and cases of valvular heart disease.

NEURALLY MEDIATED SYNCOPE

Current understanding suggests that, as a group, the neurally mediated syncopal syndromes exhibit a number of common pathophysiologic elements. Differences among the various forms of neurally mediated syncope are primarily due to the "trigger factors" associated with each and possibly the manner in which the ANS processes incoming signals.[111-115] In general, the signals that trigger these forms of syncope are believed to originate from any of various receptors that can respond to mechanical or chemical stimuli, pain, or, less commonly, temperature change. Posture, circulating volume, and emotional state play important contributory roles. In the case of carotid sinus syndrome, carotid artery mechanoreceptors (baroreceptors) are the presumed origin of the afferent neural signals. However, recent studies suggest that disturbances of signals from other neck structures (particularly neck muscles) may play a crucial facilitatory role.[116]

In the vasovagal faint, and especially faints associated with stress or emotional upset, primary central nervous system stimuli are believed to be responsible for the "trigger" signals. However, receptors in any of various organ systems may contribute. For instance, mechanoreceptors and to some extent chemoreceptors located in atrial and ventricular myocardium may participate in certain neurally mediated events by initiating afferent neural signals if subjected to increased wall tension or changes in the chemical environment (e.g., myocardial ischemia).[111,116] Similarly, mechanoreceptors and chemoreceptors in the central great vessels and lungs may contribute, thereby accounting for the reported occurrence of vasovagal faints in heart transplant recipients.[116] The basis for apparent variations in vasovagal syncope susceptibility among seemingly otherwise well individuals and the factors causing a faint to occur at a certain time are unknown.

Bradycardia in neurally mediated syncope is primarily the result of increased efferent parasympathetic tone mediated via the vagus nerve (see Figs. 3-4 and 3-7). It may manifest as asystole, sinus bradycardia, or even paroxysmal AV block (see Fig. 3-7). If the bradyarrhythmia is sufficiently severe, it may be the principal cause of the faint (i.e., "cardioinhibitory" syncope). However, most patients also exhibit a "vasodepressor" picture comprising inappropriate ANS-induced vasodilation.[115,117,118] The mechanism of the vasodilation is

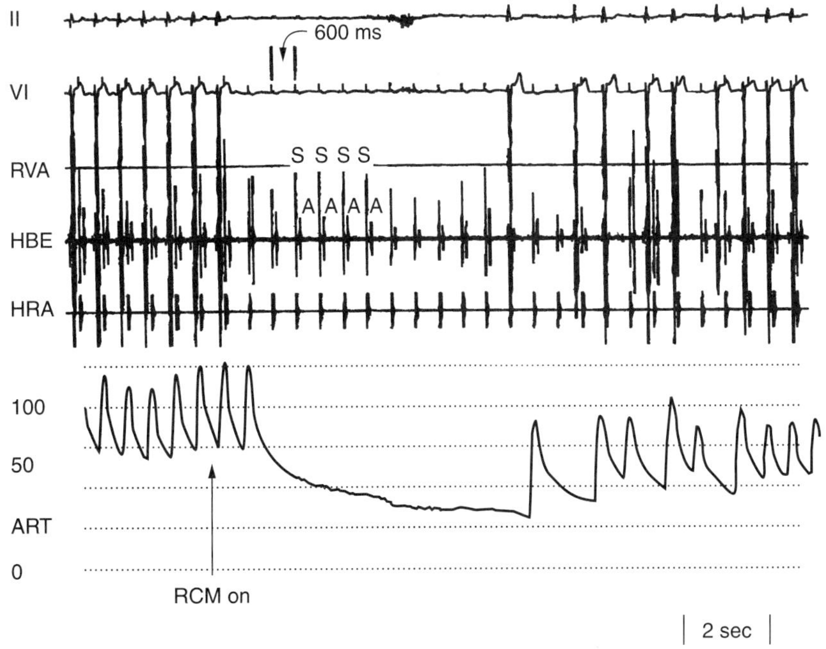

FIGURE 3-7 Electrocardiographic, intracardiac, and blood pressure tracings illustrating the development of paroxysmal atrioventricular (AV) block during right-sided carotid sinus massage (RCM of ≈5 seconds duration). In this case the atria (*A*, atrial electrogram) are being paced (*S*, stimulus) to prevent atrial bradycardia and thereby "unmask" the AV block. Note that following return to conducted rhythm, the blood pressure remains relatively low. The latter implies the concomitant presence of a clinically significant vasodepressor component to the reflex in this patient.

believed to be mainly the result of abrupt peripheral sympathetic neural "withdrawal," although potential contributions of excess β-adrenergic tone due to frequently associated elevated circulating epinephrine levels or altered epinephrine/norepinephrine balance are as yet uncertain. Individuals in whom vasodilation is the primary cause of symptomatic hypotension may be classified as having a "vasodepressor" faint. Most individuals, however, exhibit a mixed response as the cause of the faint (i.e., both "vasodepressor" and "cardioinhibitory" components participate). The type of response may vary from episode to episode.

ORTHOSTATIC SYNCOPE

The ANS participates importantly in the ubiquitous presyncopal or syncopal symptoms associated with abrupt postural changes. For the most part, these symptoms result from actual or relative central vascular volume depletion due to inadequate or delayed peripheral vascular compensation in the presence of a change in gravitational stress (e.g., moving to upright posture).[119,120] The outcome is posture-related symptomatic hypotension. Iatrogenic factors such as excessive diuresis or overly aggressive use of antihypertensive agents are important contributors. The former deplete central volume, while the latter diminish desirable ANS-induced vasoconstriction. In the older or infirm patient, environmental factors (e.g., excessive heat), impaired mobility, and a reduced appetite may similarly aggravate the problem by both reducing circulating fluid volume and diminishing responsiveness of vascular constriction.

Primary ANS disturbances are relatively rare, but increasingly recognized, causes of abnormal vascular control leading to syncope.[120-123] On occasion these occur in the absence of other neurologic disturbances, and subtle forms may be easily overlooked. ANS dysfunction may also occur in association with multiple system involvement (formerly termed Shy-Drager syndrome). However, far more common clinically than any of these primary ANS disease disturbances are those that are secondary in nature. Examples include neuropathies of alcoholic or diabetic origin, dysautonomias occurring in conjunction with certain inflammatory conditions (e.g., Guillain-Barré) or paraneoplastic syndromes.

PRIMARY CARDIAC ARRHYTHMIAS

Primary cardiac arrhythmias imply rhythm disturbances associated with intrinsic cardiac disease or other structural anomalies (e.g., accessory conduction pathways) and are among the most frequent causes of syncope. The role played by the ANS in sinus node dysfunction, conduction system disturbances, and certain tachyarrhythmias were discussed earlier.[19,26,27,31,53,54,73,124-126] However, when syncope occurs in these settings, the basis is multifactorial, including not only the type and rate of the arrhythmia but also the status of left ventricular function and the appropriateness of vascular compensation to the hemodynamic stress.

Recent studies have implicated neural reflex vasodepression as a potential cause of syncope in patients with sinus node dysfunction, particularly those with paroxysmal atrial fibrillation. The same seems to be the case for other paroxysmal supraventricular tachycardias, and possibly even ventricular tachyarrhythmias.[124-126] In essence, either inappropriate vasodilation at onset of a tachycardia or excessively slow or incomplete vasoconstriction may cause a period of sufficient hypotension as to provoke a syncopal episode.

STRUCTURAL CARDIOVASCULAR OR CARDIOPULMONARY DISEASE

The most common cause of syncope attributable to left ventricular disease is that which occurs in conjunction with acute myocardial ischemia or infarction.[127] In such cases, the contributory factors are multiple, including not only transient reduction of cardiac output, and cardiac arrhythmias, but also important neural reflex effects (discussed earlier).[127-131] Other acute medical conditions occasionally associated with syncope include pulmonary embolism and pericardial tamponade. Again, the basis of syncope is multifactorial with neural-reflex contributions probably playing an important role.

Syncope as a result of obstruction to left ventricular outflow is infrequent but carries a poor prognosis if the underlying problem is not recognized and addressed promptly (e.g., aortic stenosis, hypertrophic obstructive cardiomyopathy).[129-131] The basis for the faint may be in part inadequate cerebrovascular blood flow due to mechanical obstruction, but once again (especially in the case of valvular aortic stenosis) ventricular mechanoreceptor-mediated reflex bradycardia and vasodilation are thought to contribute importantly.

Syncope and lightheadedness may also occur in the setting of AV dissociation or in any circumstance in which atrial contraction occurs too closely after the preceding QRS (e.g., retrograde atrial activation, exceedingly long P–R interval). Symptoms in this case are due to both loss of atrial "kick" as well as neural reflex effects. The latter, in fact, seems to be the more important contributor. Atrial contraction against a closed AV valve elicits a number of potential neural (e.g., via mechanoreceptor stretch) and neurohumoral (e.g., atrial peptides) effects; the outcome is inappropriate vasodilatation and hypotension, as well as other less well understood systemic effects. One of the more common and usually readily avoidable circumstances in which this scenario plays out is so-called *pacemaker syndrome*.[132,133] In this case, the most principal cause is single-chamber ventricular pacing with either AV dissociation, or (even worse) 1:1 ventriculoatrial conduction with retrograde atrial "capture."

SYNCOPE OF NONCARDIOVASCULAR ORIGIN

Most often, noncardiovascular causes result in "syncope-mimics" rather than true syncope. However, in some patients temporal lobe seizures may closely mimic or even induce neurally mediated reflex bradycardia and hypotension. Similarly, metabolic/endocrine disturbances do not often cause true syncope. Acute hyperventilation provoked by or associated with panic/anxiety attacks and thus perhaps ANS related) is the most important exception. In these cases, abrupt reduction of P_{CO_2} levels may result in sufficient cerebral vasoconstriction to cause syncope.

The role of the ANS in so-called *chronic fatigue syndrome* has been the source of some controversy since publication of findings suggesting an overlap with tilt-induced hypotension-bradycardia.[134] It is most likely that ANS effects do play a role, but the magnitude of the impact is probably quite variable, and the evidence supporting a close connection with the neurally mediated reflex syncopal syndromes is far from convincing at this stage.

SUMMARY

The ANS impacts cardiac electrophysiology and arrhythmic risk through various direct and indirect effects. For the most part, our understanding of these effects remains relatively superficial. This chapter has provided a limited survey of some of the more important known relationships between cardiac arrhythmias and ANS effects. Of necessity, the work of many key investigators in the field has been omitted. Hopefully, however, this brief examination will prompt some readers to delve further into this vast field. Meanwhile, as technologic innovation progresses, we will inevitably become even more impressed with the importance of the ANS in establishing arrhythmic risks and determining the ultimate clinical outcomes in individuals with a wide variety of heart rhythm disturbances.

ACKNOWLEDGEMENT

The authors would like to acknowledge the valuable assistance of Wendy Markuson and Barry LS Detloff in the preparation of this manuscript.

REFERENCES

1. Levy MN, Martin PJ: Neural control of the heart. In Berne RM, Sperelakis N (eds): Handbook of Physiology, Section 2: The Cardiovascular System, vol 1, The Heart, Bethesda, Md, American Physiological Society, 1979, pp 581-620.
2. Randall WC, Wurster RD: Peripheral innervation of the heart. In Levy MN, Schwartz PJ (eds): Vagal Control of the Heart: Experimental Basis and Clinical Implications, Armonk, NY, Futura Publishing, 1994, pp 21-32.
3. Hill MRS, Wallick DW, Martin PJ, Levy MN: The effects of repetitive vagal stimulation on heart rate and on cardiac vasoactive intestinal polypeptide efflux. Am J Physiol 1995;268: H1939-46.
4. Warner MR, Levy MN: Neuropeptide Y as a putative modulator of the vagal effects on heart rate. Circ Res 1989;64:882-9.
5. Burnstock G: Purinergic receptors in the heart. Circ Res 1980;46:I175-82.
6. Burnstock G, Wood JN: Purinergic receptors: Their role in nociception and primary afferent neurotransmission. Curr Opin Biol 1996;6:526-32.
7. Shyrock JC, Belardinelli L: Adenosine and adenosine receptors in the cardiovascular system: Biochemistry, physiology, and pharmacology. Am J Cardiol 1997;79:2-10.
8. Belardinelli L, Song Y, Shyrock JC: Cholinergic and purinergic control of cardiac electrical activity. In Zipes DP, Jalife J (eds): Cardiac Electrophysiology: From Cell to Bedside, 3rd ed. Philadelphia, WB Saunders, 2000, pp 294-300.
9. Elvan A, Rubart M, Zipes DP: NO modulates autonomic effects on sinus discharge rate and AV nodal conduction in open-chest dogs. Am J Physiol 1997;272:H263-71.
10. Yabe M, Nishikawa K, Terai T: The effects of intrinsic nitric oxide on neural regulation in cats. Anesth Analg 1998;86:1194-1200.
11. Camm AJ, Katritis DG: Risk stratification of patients with ventricular arrhythmias. In Zipes DP, Jalife J (eds): Cardiac Electrophysiology: From Cell to Bedside, 3rd ed. Philadelphia, WB Saunders, 2000, pp 808-28.
12. Verrier RL, Cohen RJ: Risk identification by noninvasive markers of cardiac vulnerability. In Spooner PM, Rosen MR (eds): Foundations of Cardiac Arrhythmias: Basic Concepts and Clinical Approaches. New York, Marcel Dekker, 2001, pp 745-77.
13. Randall WC: Efferent sympathetic innervation of the heart. In Armour JA, Ardell JL (eds): Neurocardiolog. New York, Oxford University Press, 1994, pp 77-94.
14. Rossi L: Histology of vagal cardiac innervation in man. In Levy MN, Schwartz PJ (eds): Vagal Control of the Heart: Experimental Basis and Clinical Implications. Armonk, NY, Futura Publishing, 1994, pp 3-20.
15. Barber MJ, Mueller TM, Henry DP, et al: Transmural infarction in the dog produces sympathectomy in noninfarcted myocardium. Circulation 1983;64:787-96.
16. Barber MJ, Mueller TM, Davies BG, Zipes DP: Phenol topically applied to canine left ventricular epicardium interrupts sympathetic but not vagal afferents. Circ Res 1984;55:532-44.
17. Takahashi N, Barber MJ, Zipes DP: Efferent vagal innervation of the canine ventricle. Am J Physiol 1985;248:H80-97.
18. Ito M, Zipes DP: Efferent sympathetic and vagal innervation of the canine ventricle. Circulation 1994;90:1459-68.
19. Desai JM, Scheinman MM, Strauss HC, et al: Electrophysiologic effects of combined autonomic blockade in patients with sinus node disease. Circulation 1981;63:953-60.
20. Levy MN, Zieske H: Autonomic control of cardiac pacemaker activity and atrioventricular nodal transmission. J Appl Physiol 1969;27:465-70.
21. Pappone C, Stabile G, Oreto G, et al: Inappropriate sinus tachycardia after radiofrequency ablation of para-Hisian accessory pathways. J Cardiovasc Electrophysiol 1997;8:1357-65.
22. Fredholm BB, Abbrachio MP, Burnstock G, et al: Nomenclature and classification of purinoceptors. Pharmacol Rev 1994;46:143-56.
23. Isenberg G, Belardinelli L: Ionic basis for the antagonism between adenosine and isoproterenol on isolated mammalian ventricular myocytes. Circ Res 1984;55:309-25.
24. Benditt DG, Sakaguchi S, Goldstein MA, et al: Sinus node dysfunction: Pathophysiology, clinical features, evaluation and treatment. In Zipes DP, Jalife J (eds): Cardiac Electrophysiology: From Cell to Bedside, 2nd ed. Philadelphia, WB Saunders, 1995, pp 1215-47.
25. Schuessler RB, Boineau JP, Saffitz JE, et al: Cellular mechanisms of sinoatrial activity. In Zipes DP, Jalife J (eds): Cardiac Electrophysiology: From Cell to Bedside, 3rd ed. Philadelphia, WB Saunders, 2000, pp 187-95.
26. de Marneffe M, Jacobs P, Haardt R, et al: Variations of normal sinus function in relation to age: Role of autonomic influence. Eur Heart J 1986;7:662-72.
27. de Marneffe M, Gregoire JM, Waterschoot P, et al: Autonomic nervous system in relation to age in patients with and without sinus node disease. Eur J Cardiac Pacing Electrophysiol 1992;2:44-52.
28. Jose AD, Collison D: The normal range and the determinants of the intrinsic heart rate in man. Cardiovasc Res 1970;4:160-7.
29. Randall WC: Selective autonomic innervation of the heart. In Randall WC (ed): Nervous Control of Cardiovascular Function. New York, Oxford University Press, 1984, pp 46-67.
30. Warner MR, deTarnowsky JM, Whitson CC, Loeb JM: Beat-by-beat modulation of AV conduction II. Autonomic neural mechanisms. Am J Physiol 1986;251:H1134-42.
31. Benditt DG, Klein GH, Kriett JM, et al: Enhanced atrioventricular nodal conduction in man: Electrophysiologic effects of pharmacologic autonomic blockade. Circulation 1984;69:1088-95.
32. Malik M, Kautzner J, Hnatkova K, Camm AJ: Identification of electrocardiographic patterns. PACE 1996;19:245-51.
33. Xi X, Randall WC, Wurster RD: Intracellular recording of spontaneous activity of canine intracardiac ganglion cells. Neurosci Lett 1991;128:129.
34. Cranefield PF: Action potentials, afterpotentials, and arrhythmias. Circ Res 1977;41:415-23.

35. Ben-David J, Zipes DP: Differential response to right and left ansae subclaviae stimulation of early afterdepolarizations and ventricular tachycardia induced by cesium in dogs. Circulation 1988;78:1241-50.
36. Morady F, Kou WH, Nelson SD, et al: Accentuated antagonism between beta-adrenergic and vagal effects on ventricular refractoriness in humans. Circulation 1988;77:289-97.
37. Prystowsky EN, Jackman WM, Rinkenberger RL, et al: Effect of autonomic blockade on ventricular refractoriness and atrioventricular nodal conduction in humans. Evidence supporting a direct cholinergic action on ventricular muscle refractoriness. Circ Res 1981;49:511-8.
38. Rosenpire K, Haka MS, Van Dort ME, et al: Synthesis and preliminary evaluation of 11C-metahydroxyephedrine: A false transmitter for neuronal imaging. J Nucl Med 1990;31:1328-34.
39. Wieland D, Swanson DP, Brown LE, Beierwaltes WH: Imaging the adrenal medulla with an I131 labeled antiadrenergic agent. J Nucl Med 1979;20:155-8.
40. Stanton MS, Tuli MM, Radtke NL, et al: Regional sympathetic denervation after myocardial infarction in humans detected noninvasively using I123 Metaiodobenzylguanidine. J Am Coll Cardiol 1989;14:1519-26.
41. Wharton JM, Friedman JM, Greenfield RA: Quantitative perfusion and sympathetic nerve defect size after myocardial infarction in humans. J Am Coll Cardiol 1992;19:264A.
42. Spinnler MT, Lombardi F, Moretti C, et al: Evidence of functional alterations in sympathetic activity after a myocardial infarction. Eur Heart J 1993;14:1334-43.
43. Katoh K, Nishimura S, Nakanishi S, et al: Stunned myocardium and sympathetic denervation: Clinical assessment using MIBG scintigraphy. Jpn Circ J 1991;55:919-22.
44. Lai AC, Wallner K, Cao J-M: Colocalization of tenascin and sympathetic nerves in a canine model of nerve sprouting and sudden cardiac death. J Cardiovasc Electrophysiol 2000;11:1345-51.
45. Scheinman MM, Strauss HC, Evans GT, et al: Adverse effects of sympatholytic agents in patients with hypertension and sinus node dysfunction. Am J Med 1978;64:1013-20.
46. Engel GL: Psychologic stress, vasodepressor (vasovagal) syncope, and sudden death. Ann Intern Med 1978;89:403-12.
47. Milstein S, Buetikofer J, Lesser J, et al: Cardiac asystole in patients with neurally mediated hypotension-bradycardia syndrome. J Am Coll Cardiol 1989;14:1626-32.
48. Jordan JL, Yamaguchi I, Mandel WJ: Studies on the mechanism of sinus node dysfunction in the sick sinus syndrome. Circulation 1978;57:217-23.
49. Puesch P, Grolleau R, Guimond C: Incidence of different types of A-V block and their localization by His bundle recordings. In Wellens HJJ, Lie KI, Janse MJ (eds): The Conduction System of the Heart. Philadelphia, Lea & Febiger, 1976, pp 467-84.
50. Coumel PH: Autonomic influences in atrial tachyarrhythmias. J Cardiovasc Electrophysiol 1996;7: 999-1007.
51. Narula OS, Scherlag BJ, Javier RP, et al: Analysis of the A-V conduction defect in complete heart block utilizing His bundle electrograms. Circulation 1970;41:437-48.
52. Smith ML, Carlson MD, Thames MD: Reflex control of the heart and circulation: Implications for cardiovascular electrophysiology. J Cardiovasc Electrophysiol 1991;2:441-9.
53. Coumel PH: Autonomic arrhythmogenic factors in paroxysmal atrial fibrillation. In Olsson SB, Allessie MA, Campbell RWF (eds): Atrial Fibrillation: Mechanisms and Therapeutic Strategies. Armonk, NY, Futura Publishing, 1994, pp 171-85.
54. Rosen KM, Rahimtoola SH, Chuquimia R, et al: Electrophysiological significance of first degree atrioventricular block with intraventricular conduction disturbance. Circulation 1971; 43:491-502.
55. Sharifov OF, Fedorov VV, Rosenshtraukh LV, Yusshmanova AV: Effects of isoproterenol on acetylcholine mediated atrial fibrillation in the dog (abstract). Eur Heart J 2000;21:325.
56. Lokshyn S, Pravosudovich, Bondarenko O: Night atrial fibrillation: Disopyramide versus metoprolol (abstract). Eur Heart J 2000;21:326.
57. Hwang C, Peter C, Chen P-S: Mechanisms of adrenergic atrial fibrillation (abstract). Circulation 1998;98:I-282.
58. Alessi R, Nusynowitz M, Abildskov JA, Moe GK: Non-uniform distribution of vagal effects on the atrial refractory period. Am J Physiol 1958;194:406-10.
59. Dimmer C, Tavernier R, Gjorgov N, et al: Variations of autonomic tone preceding onset of atrial fibrillation after coronary artery bypass grafting. Am J Cardiol 1998;82:22-5.
60. Hoffmann E, Janko S, Dorwath U, Steinbeck G: Is the onset mechanism of atrial fibrillation always the same? In Raviele A (ed): Cardiac Arrhythmias 1999, vol 2. Milan, Springer-Verlag Italia, 2000, pp 211-3.
61. Elvan A, Pride HP, Eble JN, Zipes DP: Radiofrequency catheter ablation of the atria reduces the inducibility and duration of atrial fibrillation in dogs. Circulation 1995;91:2235-44.
62. Elvan A, Huang X, Pressler M, Zipes DP: Radiofrequency catheter ablation of the atria eliminates pacing induced sustained atrial fibrillation and reduces connexin 43 in dogs. Circulation 1997;96:1675-85.
63. Benditt DG, Samniah N: Antiarrhythmic drug treatment of atrial fibrillation: Relation to atrial fibrillation "type," and nature and severity of underlying heart disease. In Santini M (ed): Progress in Clinical Pacing. Rome, CEPI Srl, 2000, pp 397-406.
64. Kempf FC, Hedberg A, Molinoff P, et al: The relation of atrial beta receptor density to postoperative arrhythmias (abstract). J Am Coll Cardiol 1984;3:487.
65. Mantangi MF, Eutze JM, Graham IC, et al: Arrhythmia prophylaxis after aorta-coronary bypass: The effect of minidose propranolol. J Thorac Cardiovasc Surg 1985;89:439-43.
66. Stephanson LW, MacVaugh H, Tomasello DN, et al: Propranolol for prevention of postoperative cardiac arrhythmias: A randomized study. Ann Thorac Surg 1980;29:113-6.
67. White HD, Antman GM, Glynn MA, et al: Efficacy and safety of timolol for prevention of supraventricular tachyarrhythmias after coronary artery bypass surgery. Circulation 1984;70:479-84.
68. DiMarco JP, Drucker M: Clinical pharmacology of AV nodal conduction. In Mazgalev TN, Tchou PJ (eds): Atrial-AV Nodal Electrophysiology. Armonk, NY, Futura Publishing, 2000, pp 323-33.
69. Waxman MB, Cameron DA: The reflex effects of tachycardias on autonomic tone. Ann N Y Acad Sci 1990;601:378-93.
70. Waxman MB, Sharma AD, Cameron DA, et al: Reflex mechanisms responsible for early spontaneous termination of paroxysmal supraventricular tachycardia. Am J Cardiol 1982;49: 259-72.
71. Narula OS: Sinus node re-entry. A mechanism for supraventricular tachycardia. Circulation 1974;50:1114-28.
72. Allessie MA, Bonke FIM: Direct demonstration of sinus node reentry in the rabbit heart. Circ Res 1979;44:557-68.
73. Morillo CA, Klein GJ, Thakur RK, et al: Mechanism of "inappropriate" sinus tachycardia: Role of sympathovagal balance. Circulation 1994;90:873-7.
74. Olsovsky MR, Ellenbogen KA: Autonomic effects of radiofrequency catheter ablation. In Mazgalev TN, Tchou PJ (eds): Atrial-AV Nodal Electrophysiology. Armonk, NY, Futura Publishing, 2000, pp 479-91.
75. Sellers TD, Bashore TM, Gallagher JJ: Digitalis in the pre-excitation syndrome: Analysis during atrial fibrillation. Circulation 1977; 56:260-7.
76. Gulamhusein S, Ko P, Klein GJ: Ventricular fibrillation following verapamil in the Wolff-Parkinson-White syndrome. Am Heart J 1983;106:145-7.
77. Klein GJ, Gulamhusein S, Prystowsky EN: Comparison of electrophysiologic effects of intravenous and oral verapamil in patients with paroxysmal supraventricular tachycardia. Am J Cardiol 1982;49:117-24.
78. Waldo AL, Mackall JA, Biblo LA: Mechanisms and medical management of patients with atrial flutter. Cardiol Clinics 1997;15:661-76.
79. Shusterman V, Aysin B, Gottipaty V, et al: Autonomic nervous system activity and the spontaneous initiation of ventricular tachycardia. J Am Coll Cardiol 1998;32:1891-9.
80. Moss AJ, Zareba W, Benhorin J, et al: ECG T-wave patterns in genetically distinct forms of the hereditary long QT syndrome. Circulation 1995;92:2929-34.
81. Zareba W, Moss AJ, Schwartz PJ, et al: Influence of genotype on the clinical course of the long QT syndrome. International Long QT Registry Research Group. N Engl J Med 1998;339:960-5.
82. Brugada J, Brugada R, Brugada P: Right bundle-branch block and ST segment elevation in leads V1 through V3: A marker for sudden death in patients without demonstrable structural heart disease. Circulation 1998;97:457-60.

83. Priori S: Long QT and Brugada syndromes: From genetics to clinical management. J Cardiovasc Electrophysiol 2000;11:1174-80.
84. De Ferrari GM, Salvati P, Grossoni M, et al: Pharmacologic modulation of the autonomic nervous system in the prevention of sudden cardiac death. J Am Coll Cardiol 1993;22:283-90.
85. Waxman MB, Wald RW: Termination of ventricular tachycardia by an increase in cardiac vagal drive. Circulation 1977;56:385-91.
86. Zipes DP, Wellens HJJ: Sudden cardiac death. Circulation 1998;98:2334-51.
87. Schwartz PJ, Priori SG: Sympathetic nervous system and cardiac arrhythmias. In Zipes DP, Jalife J (eds): Cardiac Electrophysiology: From Cell to Bedside. Philadelphia, WB Saunders, 1990, pp 330-43.
88. Schomig A, Richardt G, Kurz T: Sympatho-adrenergic activation of the ischemic myocardium and its arrhythmogenic impact. Herz 1995;29:169-86.
89. Du XJ, Cox HS, Dart AM, Esler MD: Sympathetic activation triggers ventricular arrhythmias in rat heart with chronic infarction and failure. Cardiovasc Res 1999;43:919-29.
90. Podrid PJ, Fuchs T, Candinas R: Role of the sympathetic nervous system in the genesis of ventricular arrhythmia. Circulation 1990;82(suppl):I103-13.
91. Neely BH, Hageman GR: Differential cardiac sympathetic activity during myocardial ischemia. Am J Physiol 1990; 258:H1534-41.
92. Beta Blocker Heart Attack Trial Research Group: A randomized trial of propranolol in patients with acute myocardial infarction. JAMA 1992;247:1707-14.
93. Armour JA: Myocardial ischemia and the cardiac nervous system. Cardiovasc Res 1999;41:41-54.
94. Yusuf S, Teo KK: Approaches to prevention of sudden death: Need for fundamental reevaluation. J Cardiovasc Electrophysiol 1991;2:S233-9.
95. de Ferrari GM, Vanoli E, Schwartx PJ: Vagal activity and ventricular fibrillation. In Levy MN, Schwartz PJ (ed): Vagal Control of the Heart: Experimental Basis and Clinical Implications. Armonk, NY, Futura Publishing, 1994, pp 613-35.
96. Schwartz PJ, Vanoli E, Stramba-Badiale M, et al: Autonomic mechanisms and sudden death. New insight from the analysis of baroreceptor reflexes in conscious dogs with and without a myocardial infarction. Circulation 1088;78:969-79.
97. La Rovere MT, Mortara A, Specchia G, Schwartz PJ: Baroreflex sensitivity, clinical correlates and cardiovascular mortality among patients with a first myocardial infarction: A prospective study. Circulation 1988;78:816-24.
98. Farrell TG, Odemuyiwa O, Bashir Y, et al: Prognostic value of baroreceptor sensitivity testing after acute myocardial infarction. Br Heart J 1992;67:129-37.
99. La Rovere MT, Bigger JT Jr, Marcus FI, et al: Baroreflex sensitivity and heart rate variability in prediction of total cardiac mortality after myocardial infarction. Lancet 1998;351:487-494.
100. Kleiger RE, Miller JP, Bigger JT Jr, et al: Decreased heart rate variability and its association with increased mortality after acute myocardial infarction. Am J Cardiol 1987;59:256-62.
101. Lanza GA, Guido V, Galeazzi N, et al: Prognostic role of heart rate variability in patients with a recent myocardial infarction. Am J Cardiol 1998;82:1323-8.
102. Task Force of the European Society of Cardiology and the North American Society of Pacing and Electrophysiology: Heart rate variability. Standards of measurement, physiological interpretation, and clinical use. Circulation 1996;93:1043-65.
103. Jackman WM, Friday KJ, Anderson JL, et al: The long QT syndromes: A critical review, new clinical observations and a unifying hypothesis. Prog Cardiovasc Dis 1988;31:115-72.
104. Moss AJ, Schwartz PJ, Crampton RS, et al: The long QT syndrome: Prospective longitudinal study of 328 families. Circulation 1991;84:1136-44.
105. Schwartz PJ, Zaza A, Locati E, Moss AJ: Stress and sudden death: The case of the long QT syndrome. Circulation 1991;83(suppl II):71-80.
106. Tobe TJM, de Langen CD, Bink-Boelkens Mt, et al: Late potentials in bradycardia-dependent long QT syndrome associated with sudden death during sleep. J Am Coll Cardiol 1992;19:541-9.
107. Eldar M, Griffin JC, Abbott JA, et al: Permanent cardiac pacing in patients with the long QT syndrome. J Am Coll Cardiol 1987;10:600-7.
108. The long QT syndrome. In Zipes DP, Jalife J (eds): Cardiac Electrophysiology. From Cell to Bedside, 3rd ed. Philadelphia, WB Saunders, 2000, p 611.
109. Engelstein ED, Sawada S, Hutchins GD, et al: "Idiopathic" ventricular tachycardia is not really idiopathic. Structural ventricular abnormalities detected by positron emission tomography. J Am Coll Cardiol 1998; 31:180A.
110. Kapoor W: Evaluation and outcome of patients with syncope. Medicine 1990;69:160-75.
111. Hainsworth R: Syncope and fainting: Classification and pathophysiological basis. In Mathias CJ, Bannister R (eds): Autonomic Failure. A textbook of clinical disorders of the autonomic nervous system, 4th ed. Oxford, England, Oxford University Press, 1999, pp 428-36.
112. Benditt DG: Syncope. In Evans RW (ed): Diagnostic Testing in Neurology. Philadelphia, WB Saunders, 1999, pp 391-404.
113. Benditt DG: Sinus node dysfunction. In Willerson JT and Cohn JN (eds): Cardiovascular Medicine. New York, Churchill Livingston, 1994, pp 1296-1316.
114. Almquist A, Gornick C, Benson DW Jr, et al: Carotid sinus hypersensitivity: Evaluation of the vasodepressor component. Circulation 1985;71:927-36.
115. Benditt DG, Lurie KG, Adler SW, Sakaguchi SW: Rationale and methodology of head-up tilt table testing for evaluation of neurally mediated (cardioneurogenic) syncope. In Zipes DP, Jalife J (ed): Cardiac Electrophysiology. From Cell to Bedside, 2nd ed. Philadelphia, WB Saunders, 1995, pp 1115-28.
116. Tea SH, Mansourati J, L'Heveder G, et al: New insights into the pathophysiology of carotid sinus syndrome. Circulation 1996;93:1411-6.
117. Sutton R, Petersen M, Brignole M, et al: Proposed classification for tilt induced vasovagal syncope. Eur J Cardiac Pacing Electrophysiol 1992;2:180-3.
118. Benditt DG, Ferguson DW, Grubb BP, et al: ACC Expert Consensus Document: Tilt Table Testing for Assessing Syncope. JACC 1996;28:263-75.
119. Bannister R: Chronic autonomic failure with postural hypotension. Lancet 1979;ii:404-6.
120. Low PA: Autonomic nervous system function. J Clin Neurophys 1993;10:14-27.
121. Low PA, Opfer-Gherking TL, McPhee BR, et al: Prospective evaluation of clinical characteristics of orthostatic hypotension. Mayo Clin Proc 1995;70:617-22.
122. Weiling W, van Lieshout JJ: Investigation and treatment of autonomic circulatory failure. Curr Opinion Neurology Neurosurg 1993;6:537-43.
123. Edmonds ME, Sturrock RD: Autonomic neuropathy in the Guillain-Barré syndrome. BMJ 1979;2:668-70.
124. Leitch JW, Klein GJ, Yee R, et al: Syncope associated with supraventricular tachycardia: An expression of tachycardia or vasomotor response. Circulation 1992;85:1064-71.
125. Brignole M, Gianfranchi L, Menozzi C, et al: Role of autonomic reflexes in syncope associated with paroxysmal atrial fibrillation. J Am Coll Cardiol 1993;22:1123-9.
126. Alboni P, Menozzi C, Brignole M, et al: An abnormal neural reflex plays a role in causing syncope in sinus bradycardia. J Am Coll Cardiol 1993;22:1130-4.
127. Pathy MS: Clinical presentation of myocardial infarction in the elderly. Br Heart J 1967;29:190-9.
128. Dixon MS, Thomas P, Sheridon DJ: Syncope is the presentation of unstable angina. Int J Cardiol 1988;19:125-9.
129. Johnson AM: Aortic stenosis, sudden death, and the left ventricular baroreceptors. Br Heart J 1971;33:1-5.
130. Lombard JT, Selzer A: Valvular aortic stenosis. Ann Intern Med 1987;106:292-8.
131. Atwood JE, Kawanishi S, Myers J, et al. Exercise testing in patients with aortic stenosis. Chest 1988;93:1083-7.
132. Ausubel K, Furman S: The pacemaker syndrome. Ann Intern Med 1985;103:420-9.
133. Ellenbogen K, Wood MA, Stambler B: Pacemaker syndrome: Clinical, hemodynamic, and neurohumoral features. In Barold SS, Mugica J (eds): New Perspectives in Cardiac Pacing, vol 3. Mount Kisco, NY, Futura Publishing, 1993, pp 85-112.
134. Rowe P, Bou-Holaigah I, Kan J, Calkins H: Is neurally mediated hypotension an unrecognized cause of chronic fatigue? Lancet 1995;345:623-4.

Chapter 4 Fundamental Concepts and Advances in Defibrillation

NIPON CHATTIPAKORN and RAYMOND E. IDEKER

Sudden cardiac death is a major health problem in industrialized countries.[1] Most deaths are believed to be caused by ventricular fibrillation (VF).[2] Currently, electrical defibrillation is the only effective means for terminating this fatal arrhythmia. The mortality rate from sudden cardiac death has decreased in the past decade, partly due to better understanding of the nature of this fatality and the development of defibrillation devices. Recent advances in external defibrillators have led to the introduction of public access defibrillation, which promises to significantly reduce the mortality rate due to sudden cardiac death. In addition, recent advances in implantable defibrillators, such as the use of a biphasic waveform, have led to smaller intravenous devices that have been shown to significantly benefit certain groups of patients.[2-4] Despite these wide applications of transthoracic and intracardiac defibrillators, there is still a great need to improve defibrillation. The better we understand the fundamental mechanisms of defibrillation, the more likely we can devise strategies to improve defibrillation. In this chapter, we present factors that are believed to be important and, perhaps, *crucial* in determining the outcome of a defibrillation shock.

Potential Gradient Distribution Created by the Shock

Defibrillation success depends on the strength of the shock and is thought to be achieved by the shock changing the transmembrane potential of the myocardial fibers. This transmembrane potential change is caused by current flow between the extracellular and intracellular space generated by the electrical shock. When the shock is delivered, different amounts of current flow through different parts of the heart. The distribution of this current flow is directly related to the potential gradient, the change in shock potential over space, and the spatial derivative of the potential gradient generated by the shock across the heart.[5,6] For shocks delivered from intracardiac electrodes, the potential gradient distribution is markedly nonuniform in that high potential gradients are located near the shocking electrodes, and low potential gradients exist at some distance away from the electrodes.[7] Other factors such as myocardial fiber curvature and orientation,[8] myocardial connective tissue barriers,[9,10] blood vessels, and scar tissue[11] also have been shown to directly influence the transmembrane potential.

When a shock is delivered to the heart, the pattern of potential and potential gradient distributions created by the shock depends on the configuration of the shocking electrodes. For a 1-V shock delivered from electrodes on the right atrium (anode) and left ventricular (LV) apex (cathode), the largest negative potentials were created by the shock at the apex. The voltage drop was also marked at the ventricular apex, the region where the defibrillation electrode was located (Fig. 4-1*A*).[12] Figure 4-1*B* shows the potential gradients calculated from this potential distribution. For this shocking electrode configuration, the potential gradient distribution is markedly uneven in that the potential gradient is much larger and changes faster (as indicated by narrow spacing between isogradient lines) in the apical portion than in the basal portion of the ventricles. For a 1.5-V shock delivered from electrodes at the lateral base of the right (anode) and left (cathode) ventricles (Fig. 4-2), the voltage drop was marked at the regions close to the shocking electrodes (see Fig. 4-2*A*). The strongest potential gradient regions were in the basal portion of the ventricles near the electrodes and the weakest gradient regions were near the apex, far from the shocking electrodes (see Fig. 4-2*B*).[12] Thus, regardless of the shocking electrode configuration, the distribution of the potential gradient created by shocks has a similar pattern. The high potential gradient region with rapidly changing gradients is always near the defibrillation electrodes, and the low potential

Supported by TRF-RSA4680010.

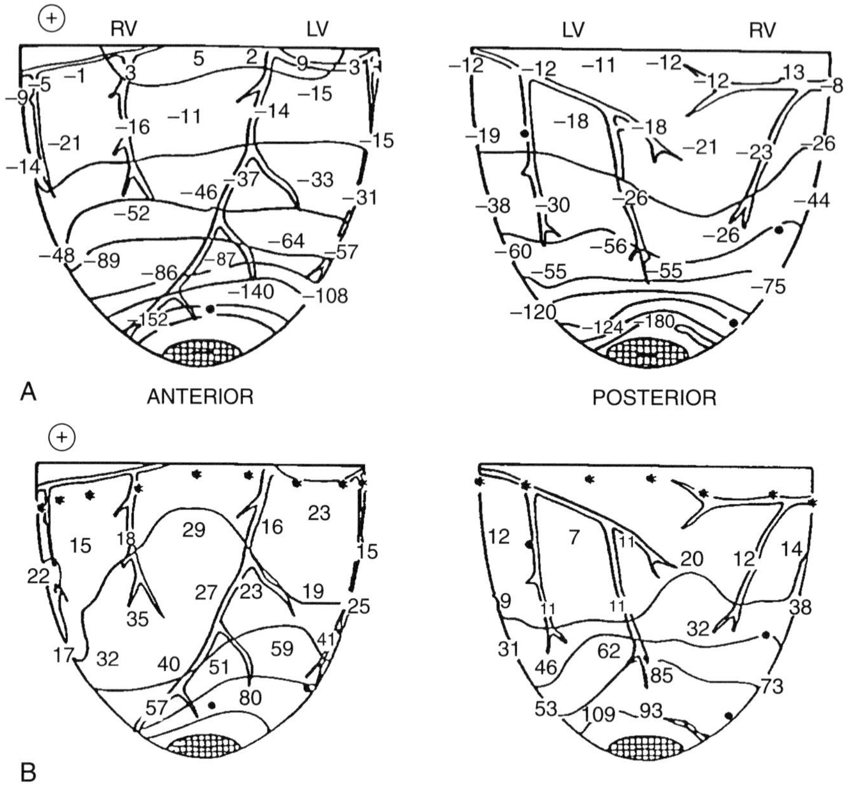

FIGURE 4-1 The epicardial maps of potential and potential gradient distribution created by a 1-V shock delivered from electrodes *(cross-hatched circles)* at the apex (cathode[−]) and right atrium (anode [+]). The maps are displayed as two complementary projections of the anterior *(right map)* and posterior ventricles *(left map)*. Numbers represent the potential (**Panel A,** *mV*) and potential gradient (**Panel B,** *mV/cm*) at their sites. *Closed circles* indicate inadequate recording sites. *Asterisks* indicate electrode sites near the ventricular border for which the gradient could not be calculated. The isopotential lines are 25 mV per shock volt (25 mV/V) apart; the isogradient lines are 25 mV/cm/V apart. **A,** The isopotential map. The isopotential lines were close together at the apex, where the shocking electrode was located, and gradually became farther apart toward the base, indicating an uneven distribution of the potentials across the ventricles. **B,** The isogradient map calculated from **Panel A**. An uneven gradient distribution caused by the shock is shown with the high gradient area close to the apex, where the shocking electrode was located, and the low gradient area at the base far from the electrode. (Reproduced with permission from Chen P-S, Wolf PD, Claydon FJ III, et al: The potential gradient field created by epicardial defibrillation electrodes in dogs. Circulation 1986;74:626-36.)

gradient regions frequently occur distant from the shocking electrodes.[7,12]

Myocardial Responses to Electrical Stimuli

When an electrical stimulus is delivered to a myocardial fiber, several responses can be observed depending on the stimulation strength (i.e., potential gradient created by a stimulus) and the phase of the action potential (AP) of the fiber at the time of stimulation.[13] If the stimulus is strong (above threshold) and is delivered to a cell that is in its resting state or relatively refractory, a new AP will be generated (Fig. 4-3*A*).[14] If the stimulus is weak (below the threshold) and is delivered to the cell at its resting state or refractory state, no response will be observed. However, if the stimulus is very strong (much above the threshold) and is delivered to a cell even at its highly refractory state, a graded response occurs (Fig. 4-3*B*). The size of the graded response increases as the stimulus magnitude or coupling interval, or both, increase.[15] This graded response prolongs the AP duration as well as the refractory period of the cardiac cell.[16,17] This type of cardiac response has been proposed to be a possible defibrillation mechanism known as the *refractory period extension hypothesis.*[18,19] This hypothesis states that successful shocks must be sufficiently strong to prolong the refractory period of cardiac tissue across the heart so that ectopic activation occurring after the shock, if any, will be prevented from propagation that could lead to reentry and VF. However, potential gradients are distributed unevenly. Therefore, potential gradients generated by the shock are stronger than needed in most regions in order to achieve the minimum potential gradient where the shock field is weakest.

Shocks Delivered During Ventricular Fibrillation and Myocardial Responses

During VF, many wandering activation fronts are present at all times on the heart.[20] At different times during the

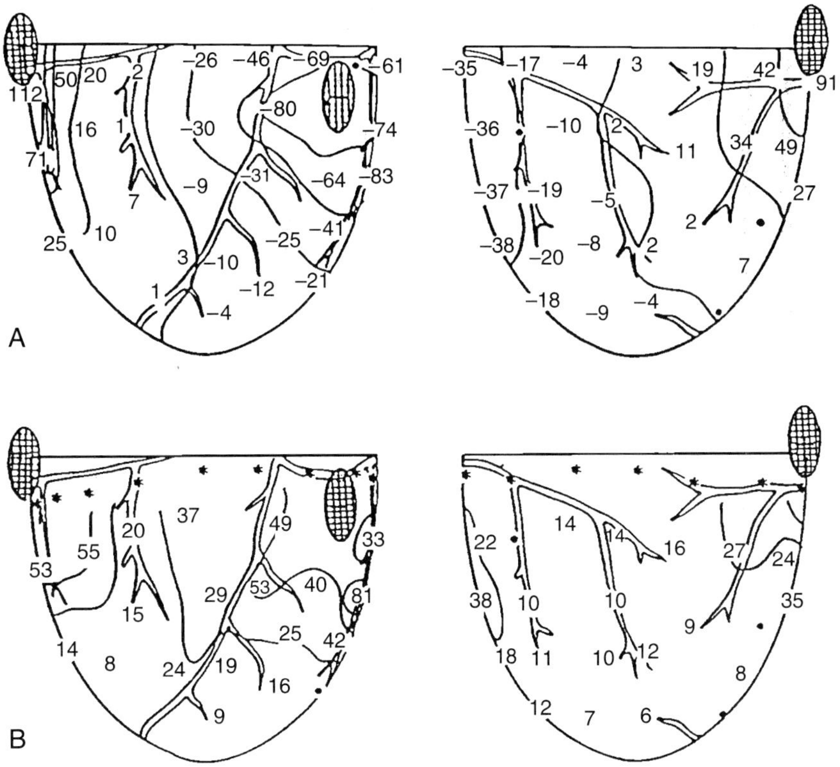

FIGURE 4-2 The epicardial maps of potential and potential gradient distribution created by a 1.5-V shock delivered from electrodes *(cross-hatched circles)* at the right (anode [+]) and left (cathode[−]) ventricular bases. **A,** The isopotential lines were close together near the two shocking electrodes. **B,** The isogradient map calculated from *Panel A*. Regions of high potential gradient were near the two defibrillation electrodes. (Reproduced with permission from Chen P-S, Wolf PD, Claydon FJ III, et al: The potential gradient field created by epicardial defibrillation electrodes in dogs. Circulation 1986;74:626-36.)

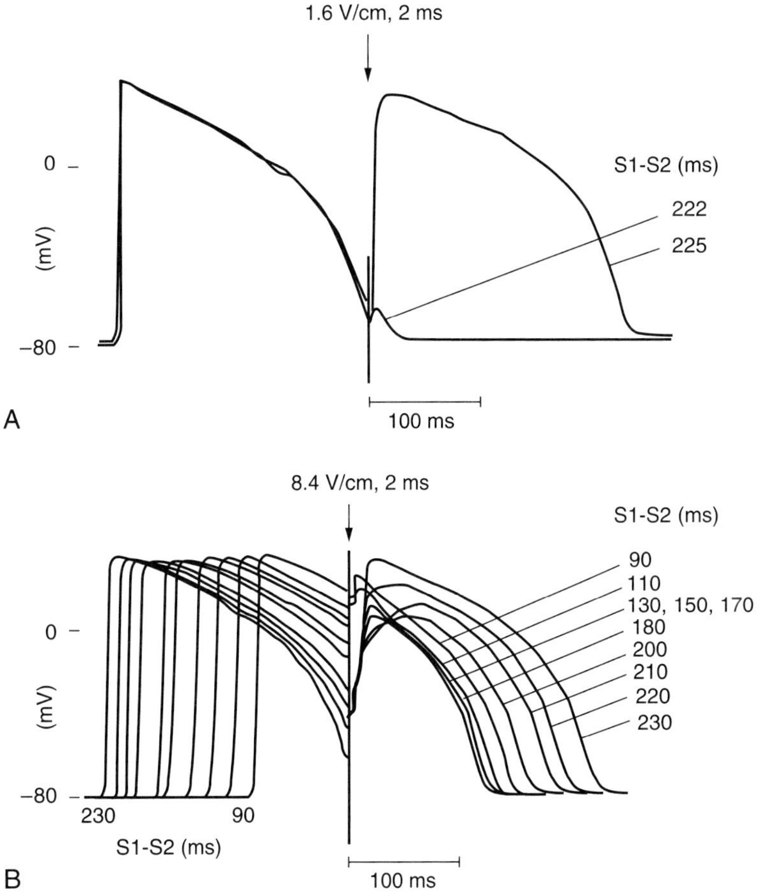

FIGURE 4-3 A, All or no cellular responses to an S2 shock field stimulus (1.6 V/cm at the cell) taken from single cell recordings. There was almost no cellular response when S2 was delivered at an S1-S2 interval of 222 milliseconds (ms). However, a new action potential was generated when an S2 shock was delivered only 3 ms later than the first one that had no response. **B,** Recordings illustrating a range of action potential prolongation caused by an S2 field stimulus (8.4 V/cm at the cell). The recordings are taken from the same cell and are aligned with the S2 time. The coupling intervals of the S1-S2 are indicated at the bottom of the recordings before S2 is given. The S1-S2 intervals for each response after S2 are indicated to the right of the recordings. The degree of action potential prolongation increased as the S1-S2 interval increased. (Reproduced with permission from Knisley SB, Smith WM, Ideker RE: Effect of field stimulation on cellular repolarization in rabbit myocardium: Implications for reentry induction. Circ Res 1992;70:707-15.)

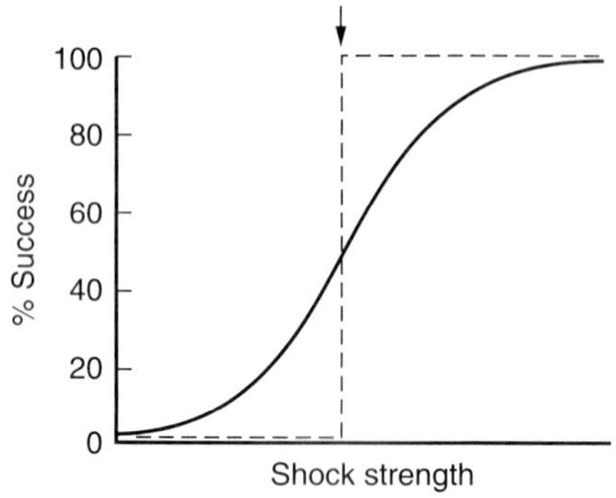

FIGURE 4-4 The probability of defibrillation success curve. The defibrillation threshold is not a discrete value *(dashed line)*. The relationship between shock strengths and defibrillation success is characterized as a sigmoidal-shaped dose-response curve *(solid line)*. High-strength shocks have a greater chance of defibrillation success than low-strength shocks. (Modified by permission from Davy JM, Fain ES, Dorian P, et al: The relationship between successful defibrillation and delivered energy in open-chest dogs: Reappraisal of the "defibrillation threshold" concept. Am Heart J 1987; 113:77-84.)

same or different VF episodes, activation sequences are not constant and can differ markedly.[21,22] When shocks of the same or different strengths are delivered during VF, myocardial responses to each shock can differ from one shock to the next, depending on the state of the ventricles when the shock is given. Since these responses are thought to be crucial in determining defibrillation success, shocks of the same strength delivered to a fibrillating heart can sometimes succeed and other times fail to defibrillate. As a result, there is no definite threshold in shock strength that demarcates successful from failed defibrillation. The relationship between shock strength and defibrillation success can, therefore, be characterized as probabilistic. Other factors that may contribute to the probabilistic nature of defibrillation success include changes in autonomic tone and changes in heart volume during VF.[23,24] Although defibrillation success is probabilistic, as shock strength becomes stronger, the chance of defibrillation success becomes greater (Fig. 4-4) for both transthoracic and intracardiac defibrillation.[25,26]

Regions of Immediate Postshock Activation

Defibrillation studies demonstrate that the relationship between regions where activation appears on the heart soon after the shock and the extracellular potential gradient distribution created by the shock are well correlated. For defibrillation to be achieved, it has been proposed that it is necessary to raise the potential gradient throughout all or almost all of the ventricular myocardium to a certain minimum level.[27] This statement is supported by the findings that, following weak shocks that failed to defibrillate, the immediate postshock activations arose at multiple sites throughout the ventricles.[28] As shock strength was stronger, the numbers of sites of immediate postshock activation were decreased and were no longer found in the high potential gradient regions. Figure 4-5 demonstrates the sites of postshock activation recorded from the same animal as shown in Figures 4-1 and 4-2. For shocks delivered from two different electrode configurations, the site of the earliest recorded postshock activation was at the base of the ventricles for a shock given from the right atrial and ventricular apical electrodes (see Fig. 4-5*A*) and was at the apex of the ventricles for the shock given from the right and LV basal electrodes (see Fig. 4-5*B*). Both regions correspond to the weak potential gradient area created by shocks delivered from each shocking electrode configuration (see Figs. 4-1*B* and 4-2*B*). These results suggest that the potential gradient field created by the shock is important in determining the immediate response of the myocardium to defibrillation shocks. Since the potential gradient field created by the shock is markedly uneven, a strong shock is normally required to create a potential gradient that reaches the optimal level at the region where the gradient field is weakest in the ventricles. However, this strong shock can be detrimental since it can create an excessively high gradient near the shocking electrodes and may damage the myocardium.[29,30] Several studies have shown that this detrimental effect can lead to postshock conduction block and arrhythmias.[31]

Why do Shocks Fail to Defibrillate?

Following the near-threshold shocks that fail to defibrillate, the earliest recorded postshock activation that propagates throughout all or almost all of the myocardium always arises in the low potential gradient regions. After several such organized cycles in rapid succession, activation becomes more disorganized, allowing fibrillation to resume. Currently, two possible mechanisms are thought to explain the origin of these early postshock activation cycles in the low potential gradient regions. The first one is known as the *critical mass hypothesis*. According to this hypothesis, the gradient field created by the shock is too weak to halt the fibrillatory wavefronts present in those regions, allowing the fibrillation to continue propagation after the shock.[32-34] The second hypothesis is known as the *upper limit of vulnerability hypothesis for defibrillation.* This hypothesis suggests that a shock of near-threshold strength is already strong enough to terminate all fibrillatory wavefronts including those in the low gradient regions. However, this shock fails to defibrillate because it creates new activation fronts in these regions.[28,35,36] These activations then spread out, eventually causing block and disorganized activations across the ventricles, degenerating back into VF. These two hypotheses continue to be debated.[37-39] Whatever the mechanism is, since the direct effect of the shock at each myocardial region depends on both the strength of the shock (i.e., the potential gradient) and the phase of the cardiac cycle at the time the shock is delivered, these two hypotheses agree that the shock potential gradient or its derivative must be sufficiently high to stop fibrillatory

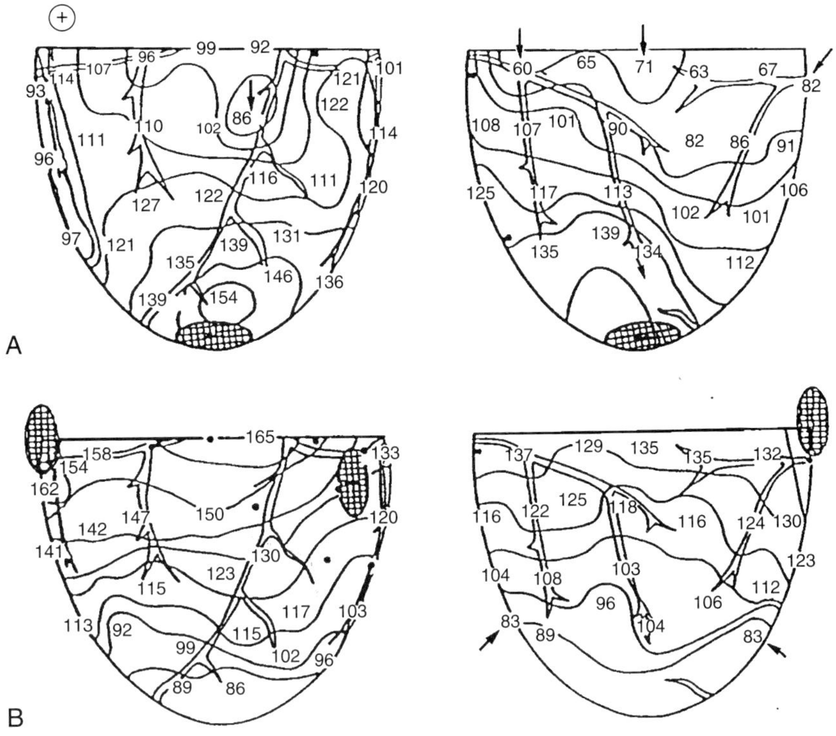

FIGURE 4-5 Isochronal maps of the first postshock activation. The thin solid lines are isochrones spaced 10 milliseconds (ms) apart. Numbers represent activation times at each recording electrode in ms relative to the shock onset. **A,** A 4.9-J failed defibrillation shock given via the electrodes placed at the right atrium and the apex during ventricular fibrillation (VF). The sites of earliest postshock activation were located at the base of the ventricles *(arrows).* The weak potential gradient region created by the shock delivered from this electrode configuration is indicated by the gradient map in Figure 4-1*B.* **B,** A 23.2-J failed defibrillation shock given via the electrodes placed at the right and left ventricular bases during VF. The sites of earliest activation were located at the posterior and apical aspects of the ventricles *(arrows).* The weak potential gradient region for this shocking electrode configuration is indicated by the gradient map in Figure 4-2*B.* (Reproduced with permission from Chen P-S, Wolf PD, Claydon FJ III, et al: The potential gradient field created by epicardial defibrillation electrodes in dogs. Circulation 1986;74:626-36.)

fronts on the ventricles or not create new activations that allow fibrillation to resume, or both.[28,35,36]

The Critical Point Hypothesis: Classic Interpretation

Previous cardiac mapping studies demonstrated that following failed defibrillation with shocks near the defibrillation threshold in strength, the pattern of activation after the shock was different from the VF activation pattern immediately before the shock.[28,35] These findings suggest that the postshock activation was not the unaltered activation continuing from VF activation before the shock in that region. Rather, the shock terminated all VF activation fronts but failed to defibrillate because it generated a new activation in the weak gradient region, which degenerated into VF. The *critical point hypothesis* has been proposed to explain how a shock generates a new activation in this weak gradient area that leads to fibrillation. The concept of this hypothesis is based on the relationships among the distribution of potential gradients created by the shock, the state of the myocardium at the time of the shock, and the myocardial response to the shock. This hypothesis was proposed theoretically by Winfree[40] and later demonstrated experimentally by Frazier et al.[41] During the shock, different responses of cardiac tissue to an electrical stimulation can be observed. As a result, depending on the state of the cardiac tissue at the time of the shock, some regions of the myocardium can be directly activated by the shock field to undergo a new AP while other regions can undergo refractory period extension caused by a graded response of the AP.[14,16,42,43] Thus, an activation front arises after the shock that terminates at a critical point on the boundary between these two types of regions. This blindly ending activation front propagates to form a functional reentrant circuit.

Frazier et al.[41] tested this hypothesis by delivering a shock to the dog heart during paced rhythm (Fig. 4-6). A row of epicardial stimulating wires on the right of the recording region was used to deliver S1 pacing. Figure 4-6*A* shows activation times and recovery times distributed across the mapped region during the last S1 pacing beat. *Solid lines* represent the spread of the activation front away from the S1 electrodes and *dashed lines* represent the recovery times estimated from the refractory period to a local 2-mA stimulus. The S2 shock was delivered through a long narrow electrode placed near the bottom of the mapped region that was perpendicular to the activation front arising from the S1 pacing stimulus. The potential gradients created by a large premature S2 shock (see Fig. 4-6*B*) demonstrate that the highest potential gradient was located in the region close to the S2 electrode and weakened with distance away from the S2 electrode. S2 shocks were delivered to scan the vulnerable period following the last S1.

Figure 4-6*C* demonstrates the initial activation pattern when a reentrant circuit was formed after the S2 shock was delivered at an S1-S2 coupling interval when a dispersion of refractoriness was present across the mapped region. Following the strong S2 stimulation, an activation front first appeared a few centimeters away from the S2 electrode with one end terminating blindly

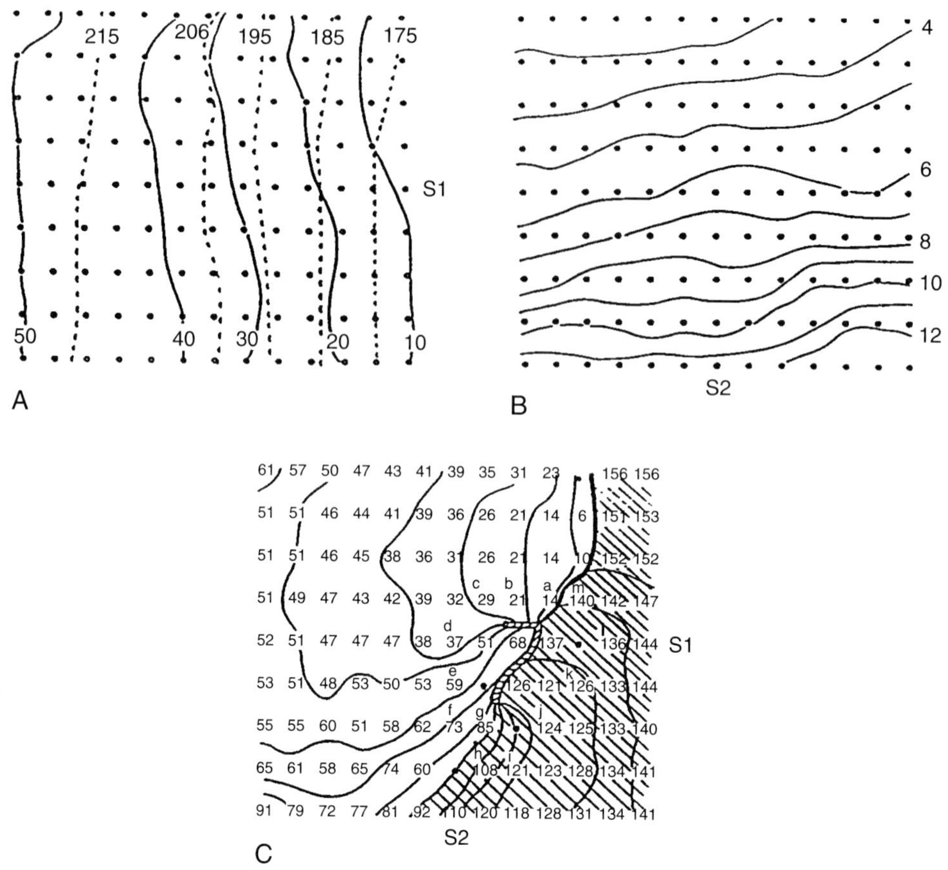

FIGURE 4-6 Reentrant circuit formation by a shock. **A,** Activation times during the last S1 beat *(solid lines)* and recovery times to a local 2-mA stimulus *(dashed lines)* in milliseconds (ms). **B,** Distribution of potential gradient created by the S2 stimulus in V/cm. **C,** The initial activation pattern immediately after the S2 is delivered. Numbers give activation times at each recording electrode in ms timed from the S2 stimulus. The solid lines portray isochronal lines spaced at 10-ms intervals. The hatched area indicates portions of the mapped region thought to be directly activated by the S2 stimulus field. A frame line *(heavy solid line)* represents the origin of the activation front propagating away from the directly activated region and also indicates the transition between the map for this activation cycle and the map for the next cycle (not shown). The *hatched line* indicates a zone of functional conduction block at the center of the reentrant circuit. (Reproduced with permission from Frazier DW, Wolf PD, Wharton JM, et al: Stimulus-induced critical point: Mechanism for electrical initiation of reentry in normal canine myocardium. J Clin Invest 1989;83:1039-52.)

at a point in the center of the mapped region where the S2 potential gradient was approximately 6 V/cm and where the tissue was just passing out of its absolute refractory period.[41] This point is called the *critical point for reentry*. This activation front then propagated away from the S1 electrode, pivoted around the critical point, later spread through the lower left quadrant, and formed a reentrant circuit as it entered the right lower quadrant that continued for more than 10 cycles before degenerating into VF.

The formation of the reentrant circuit is proposed to be due to different cardiac tissue responses to the S2 shock field in different cardiac regions at the time the shock is delivered. According to the cardiac tissue responses to the S2 shock, the mapped region can be divided into four zones that roughly form quadrants (centered at the critical point). Myocardium in the hatched region (top and bottom right quadrants) had recovered sufficiently at the time of the shock that it was directly activated. Myocardium in the top left quadrant was not directly activated because it was more refractory than tissue in the hatched region. Since the shock potential gradient was weaker than the critical value needed to cause a graded response in the top quadrant, the myocardial refractoriness in this quadrant was not extended. However, the potential gradient created by the shock in the bottom quadrant was stronger than the critical value. Therefore, the refractoriness of cardiac tissue in the bottom left quadrant close to the S1 pacing electrode was prolonged by a graded response, since myocardium in this region was less refractory than that far from the S1 pacing electrode. The majority of myocardium in the bottom left quadrant was too refractory to be affected by the shock, although it was exposed to a strong potential gradient. As a result, the activation front forming in the directly activated region (hatched) could only propagate from the top right to the top left quadrant. Directly excited activation in the bottom right quadrant could not propagate to the left since it was blocked by myocardium in a prolonged refractory state. By the time the activation front from the top left quadrant entered the bottom left quadrant, cardiac tissue in the bottom left quadrant had already recovered. This allowed activation to propagate through and reenter the directly activated tissue in the lower right quadrant, which had by this time also recovered excitability. As a result, a counterclockwise reentrant circuit was formed around the critical point.

The Critical Point Hypothesis: New Interpretation

Optical mapping studies have shown that when an electrical stimulus was applied to the myocardium, different polarities of transmembrane potential changes were observed near the stimulating electrode.[6,9,44-46] During a defibrillation shock, it has been proposed that depolarized and hyperpolarized regions caused by the shock are interspersed throughout the heart.[9,47,48] By using an optical mapping technique to investigate the mechanism of failed defibrillation, Efimov and colleagues demonstrated a different type of reentrant circuit formation than that by the classic critical point.[47] Their results showed that the formation of a critical point where a reentrant circuit was observed depends on the magnitude and distribution of depolarization and hyperpolarization of the transmembrane potential created by the shock. A critical point is formed on the boundary between depolarized and hyperpolarized regions when the magnitude and rate of change of transmembrane polarization across this boundary are sufficiently large. In the region where the tissue is hyperpolarized by the shock, this hyperpolarization can "de-excite" tissue that was depolarized just before the shock, thus restoring excitability of myocardium in that region.[6,47] According to these results, the potential gradient and refractoriness are not critical for reentrant circuit formation. Figure 4-7 illustrates reentrant circuit formation based on the classic and new forms of the critical point hypothesis.

Figure 4-7*A* illustrates the classic type of critical point formation as shown in the study by Frazier and colleagues.[41] The S1 pacing electrode is located on the left and the S2 shocking electrode is at the bottom of the mapped region. When the shock is delivered, myocardium that is directly activated (DA) by the S2 shock is located near the S1 pacing electrode. Myocardium that has refractory period extension (RPE) is in the region near the S2 electrode. As a result, the activation arising in the DA region can only propagate unidirectionally in a clockwise manner around the critical point.

An idealized diagram illustrating the new type of critical point formation is shown in Figure 4-7*B*. When a shock is delivered, it creates regions of depolarization adjacent to regions of hyperpolarization.[9,45]

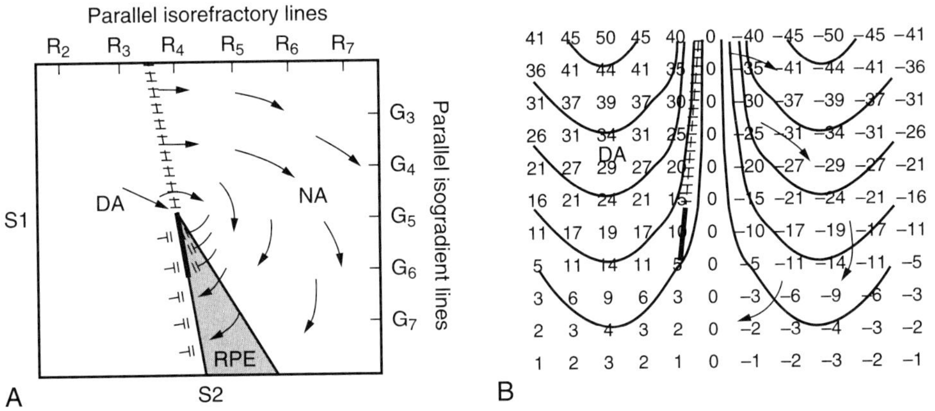

FIGURE 4-7 Two types of hypothesized critical points. **A,** Idealized diagram corresponding to the experiment shown in Figure 4-6 is shown with a critical point formed at the intersection of a critical shock potential gradient of G5 and a critical tissue refractoriness of R4. S1 pacing is performed from the left to cause a dispersion of refractoriness at the time of the S2 shock, with *R2* representing *less* refractoriness and *R7, more.* The S2 shock is given during the vulnerable period from the bottom of the region with large gradient *G7* at the bottom and small gradient *G3* at the top. The region labeled *DA* is sufficiently recovered so that it is directly activated by the gradient field. The area in the stippled region, although more refractory, is exposed to a higher gradient and undergoes refractory period extension *(RPE)* so that activation in the DA tissue cannot propagate through this region. The region *NA* is too refractory to be affected even with a large gradient. Thus, propagation conducts unidirectionally from the DA to NA region at the top, encircling the critical point, and then reentering the DA region to create a reentrant circuit. **B,** An idealized diagram is shown of a critical point caused by adjacent regions of depolarized and hyperpolarized transmembrane potential changes. Numbers represent transmembrane changes with isolines spaced every 10 mV beginning at −45 mV. DA occurs to the left of the frame line, where depolarized transmembrane potential changes are suprathreshold. Where the gradient in transmembrane potential is high, as indicated by the closely spaced isolines at the top center of the panel, conduction can occur into the hyperpolarized region. Below, where the gradient in transmembrane potential is smaller, propagation cannot occur. A critical point is formed at the intersection of the frame and block lines, where one end of the propagating activation front terminates in both panels. (Reproduced with permission from Chattipakorn N, Ideker RE: Mechanism of defibrillation. In Aliot E, Clémenty J, Prystowsky EN (eds): Fighting Sudden Cardiac Death: A Worldwide Challenge. Armonk, NY, Futura Publishing, 2000.)

The magnitude of depolarization caused by the shock in the DA region is high at the top *(large positive numbers)* and gradually decreases toward the bottom *(small positive numbers)*. Adjacent to the depolarized region is a region of hyperpolarization. The magnitude of hyperpolarization caused by the shock is also high *(large negative numbers)* and gradually decreases from top to bottom *(small negative numbers)*. The pattern of transmembrane potential distribution produced by the shock creates a large gradient between the depolarized and hyperpolarized regions as indicated by the closely spaced *isolines* at the top center of the panel. The large gradient between the two adjacent transmembrane polarities allows an activation front to propagate from the depolarized region into the hyperpolarized region *(arrow at the top)*. There is no propagation in the bottom half of the panel because the gradient in transmembrane potentials is too small. As a result, a critical point is formed at the intersection of the frame and block lines (indicated by *hatched and solid lines*, respectively) where one end of the propagating activation front terminates. Thus, activation arising in the DA region propagates unidirectionally from top to bottom *(arrows)* and reenters the DA region later, forming a clockwise reentrant circuit.

Although these two interpretations suggest different mechanisms of critical point formation, they both indicate that failed defibrillation is due to reentrant activation caused by the shock, which later degenerates into VF. Most cardiac mapping studies using a large-animal model have shown that a reentrant activation pattern is rarely observed after a shock that fails to defibrillate and that is near the defibrillation threshold in strength.[28,34,49-52] Epicardial focal activation patterns are commonly observed in those studies. Transmural or Purkinje-myocardial reentry has been proposed as a possible mechanism giving rise to activation fronts. In those studies, the epicardial focal activation pattern was caused by epicardial breakthrough instead of having a true focal origin.[28,34,50-52] To test this hypothesis, three-dimensional mapping is needed. Currently, only a few three-dimensional studies have been performed to investigate the defibrillation mechanism. Chen and colleagues performed a transmural cardiac mapping study and demonstrated that only a few episodes of fibrillation following a failed defibrillation shock were initiated by a reentrant circuit.[49] Many appeared to arise from a focus. These findings suggest that the current interpretations of critical point formation may only partially explain defibrillation mechanisms and that the relationship between the shock delivered to a fibrillating heart and the cardiac responses to the shock is complex.

The Upper Limit of Vulnerability and Defibrillation Mechanism

VF can be induced when an electrical stimulus within a certain range of strengths is delivered to the myocardium during the vulnerable period of the cardiac cycle in normal sinus or paced rhythm.[53] The lowest stimulation strength that can induce VF is known as the VF threshold (Fig. 4-8). As the stimulation strength is increased, VF can still be induced until the stimulus strength reaches a value above which VF again can no longer be induced. This strong stimulation strength that no longer induces VF, no matter when this stimulus is delivered during the vulnerable period of repolarization, is known as the upper limit of vulnerability (ULV) (see Fig. 4-8).[54] Table 4-1 shows the estimated threshold of the stimulus current and the voltage gradient required for different myocardial responses.[27]

The existence of the ULV has been linked to the defibrillation mechanism.[35,36] Since fibrillation is thought to be maintained by reentry, activation fronts should be continuously present. If so, then repolarization should occur continuously also. When a shock is delivered to defibrillate the heart, it is probable that

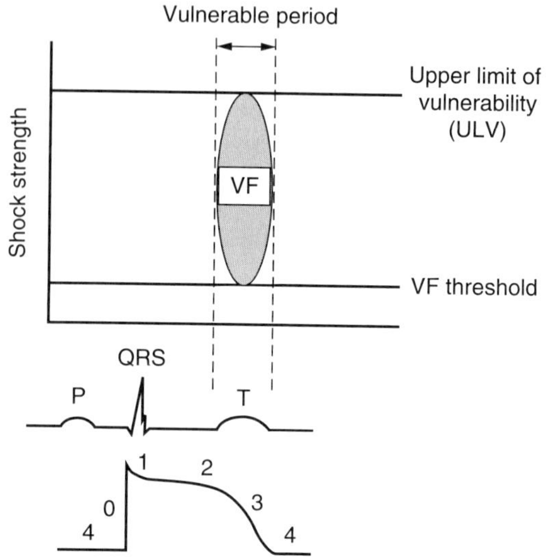

FIGURE 4-8 The relationship among the shock strength, vulnerable period, and ventricular fibrillation *(VF)*. Shocks of a strength at or above the VF threshold induce VF *(filled oval)* when delivered at an appropriate time during the vulnerable period (corresponding to a portion of the T-wave on the ECG or the repolarization phase of the action potential). Shocks stronger than the upper limit of vulnerability *(ULV)*, however, no longer induce VF when given at any time during the cardiac cycle. (Reproduced with permission from Chattipakorn N, Ideker RE: Mechanism of defibrillation. In Aliot E, Clémenty J, Prystowsky EN (eds): Fighting Sudden Cardiac Death: A Worldwide Challenge. Armonk, NY, Futura Publishing, 2000.)

TABLE 4-1 Threshold

	mA	V/cm
Diastolic pacing	0.5	1
Ventricular fibrillation	20	6
Defibrillation	10,000	6

Reproduced with permission from Ideker RE, Zhou X, Knisley SB: Correlation among fibrillation, defibrillation, and cardiac pacing. Pacing Clin Electrophysiol 1995;18:512-25.

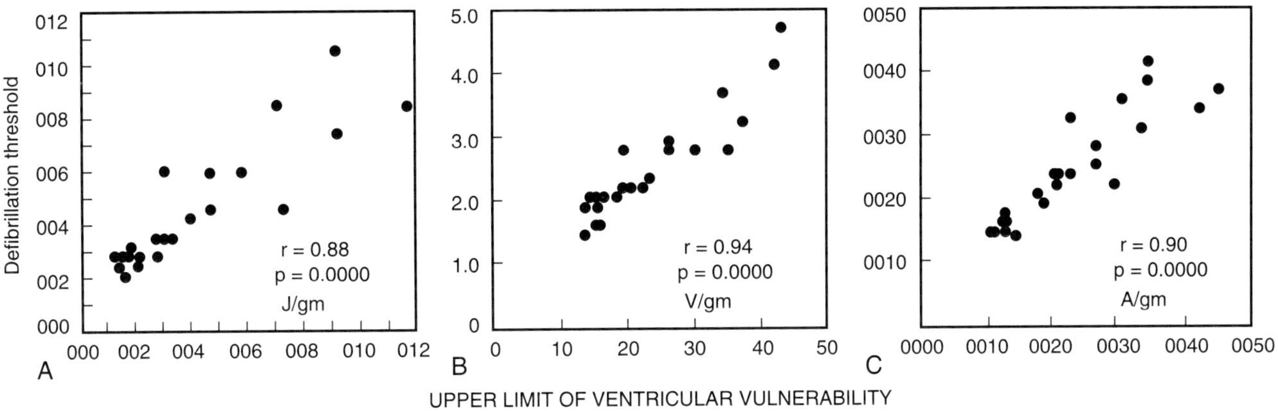

FIGURE 4-9 Correlation of the defibrillation threshold and the upper limit of vulnerability (ULV) for electrodes on the right atrium (anode) and the left ventricular apex (cathode). Results are obtained from 22 dogs and expressed in units of energy (**A**), voltage (**B**), and current (**C**). All units are expressed per gram of heart weight. (Reproduced with permission from Chen P-S, Shibata N, Dixon EG, et al: Comparison of the defibrillation threshold and the upper limit of ventricular vulnerability. Circulation 1986;73:1022-8.)

some portions of myocardium are in the vulnerable period while they are exposed to the shock. Therefore, defibrillation shock can induce VF in this region if the shock gradient there is stronger than the VF threshold but weaker than the ULV, resulting in failed defibrillation. However, if the shock is sufficiently strong that it surpasses the ULV across the whole heart, VF will not be induced, resulting in successful defibrillation. Therefore, according to this concept, to successfully defibrillate, the shock must be sufficiently strong that it will not induce new activations that can lead to refibrillation after the shock. Indeed, it has been shown that the ULV and defibrillation thresholds are well correlated (Fig. 4-9),[55,56] suggesting that the existence of the ULV is a possible explanation for a defibrillation mechanism.[28,35,54,57] This hypothesis is known as the *ULV hypothesis for defibrillation*, which states that a successful shock must terminate all VF activation fronts and, at the same time, not generate new postshock activation that can reinitiate fibrillation.[58,59]

The critical point hypothesis has been proposed to explain the relationship between the defibrillation threshold and the ULV (Fig. 4-10).[27] When a shock is delivered, the potential gradients created by the shock are high in the region closest to the electrode and progressively decrease with distance from the electrode.

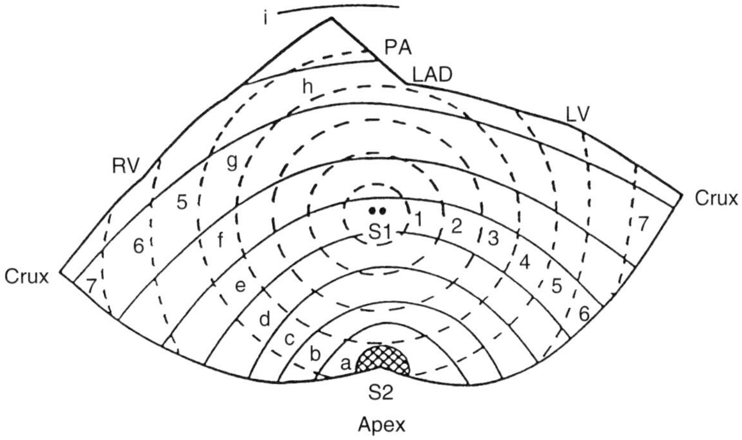

FIGURE 4-10 A hypothesized critical point formation when a shock is given to the ventricles. The epicardial surface of the canine heart is depicted as if the ventricles were folded out after an imaginary cut was made from the crux to the apex. Isorecovery lines *(dashed lines 1-7)*, representing different degrees of refractoriness, are concentric about the pacing site labeled *S1*. Large premature stimuli are delivered from the apex of the heart through the electrode labeled *S2* with the return electrode located elsewhere in the body away from the heart. Isogradient lines *(solid lines a-i)*, representing different levels of extracellular potential gradient, are concentric about the S2 electrode. The smallest values in the ventricles, occurring in the small region at the top of the ventricles, represent the pulmonary outflow tract. LAD, left anterior descending coronary artery; LV, left ventricle; PA, pulmonary artery; and RV, right ventricle. (Reproduced with permission from Ideker RE, Tang ASL, Frazier DW, et al: Ventricular defibrillation: Basic concepts. In El-Sherif N, Samet P (eds): Cardiac Pacing and Electrophysiology. Orlando, Fla., WB Saunders, 1991, pp 713-26.)

Since a certain critical potential gradient is required to form a critical point, a weak shock (i.e., at the VF threshold) can create the critical potential gradient only in the region closest to the shocking electrode *(line a)*, resulting in critical point formation near the S2 electrode. As the shock strength increases, the critical gradient will move further from the shocking electrode *(lines b to h)*. Thus, the formation of critical points and reentrant circuits will move farther away from the shocking electrode. When the shock is strong enough that the potential gradients created by the shock are greater than the critical value throughout the entire ventricular myocardium *(line i)*, no critical point will be created in the ventricles and no reentrant circuit will be formed. This shock strength therefore reaches the level of the ULV when the shock is delivered during the vulnerable period of normal sinus or paced rhythm and will successfully defibrillate when the shock is delivered during VF. This concept was tested and supported experimentally by a study from Idriss and colleagues.[60]

Although reentrant circuit formation around the critical point has been proposed to be responsible for defibrillation failure and VF induction during a T-wave shock, the reentrant pattern is not frequently observed in most studies.[34,50-52,61] These results suggest that the critical point hypotheses may only partially explain defibrillation and VF induction mechanisms. For example, recent VF induction and defibrillation studies in pigs, using near-threshold strength shocks,[52,61,62] demonstrated that rapid repetitive postshock activations arising focally from the weak potential gradient region are responsible for VF reinduction in failed defibrillation. No epicardial reentry was found in those studies. While intramural reentry may be responsible, several studies suggest that shock-induced automaticity or triggered activity may be responsible for these rapid repetitive activations arising after the shock.[63-66] These findings suggest that other mechanisms (e.g., focal activity) may be responsible for defibrillation failure. It is important to note that most reentrant activity has been reported in studies that applied weak shocks (i.e., well below the defibrillation threshold) in a small heart model such as a guinea pig or rabbit.[47,67,68] Focal activity, however, has been demonstrated in studies that applied strong shocks (i.e., near the defibrillation threshold) to induce VF or to defibrillate the heart in dogs and pigs.[35,50,52,61]

Near-Threshold Shocks and Mechanism of Defibrillation

The answer to a simple question "Why do some shocks succeed and others fail in terminating VF?" remains surprisingly elusive. Although defibrillation has been

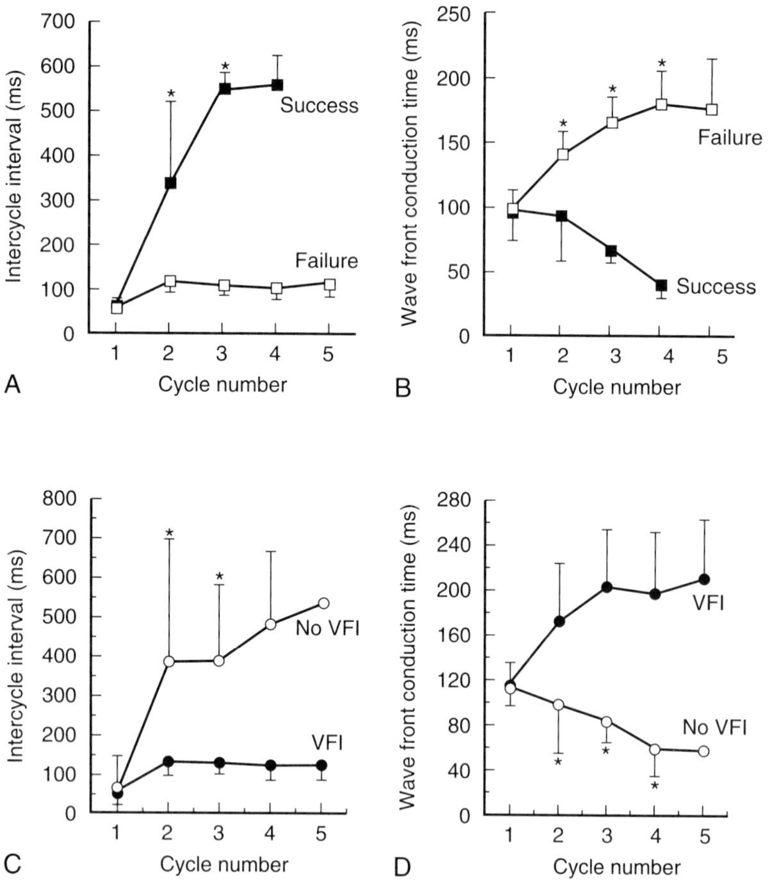

FIGURE 4-11 The intercycle interval (the interval between the onset of two successive postshock cycles) and the wavefront conduction time (the time the cycle needs to traverse across the ventricles) of the first five postshock cycles following defibrillation shocks, all of the same strength, that successfully defibrillate 50% of the time, DFT_{50} **(Panels A and B),** and shocks during the vulnerable period of paced rhythm that induced ventricular fibrillation (VF) 50% of the time, ULV_{50} **(Panels C and D)**. An *asterisk* signifies a significant difference between the two outcomes for that cycle. Failure, failed defibrillation; NoVFI, failed VF induction by ULV shocks; Success, successful defibrillation; VFI, successful VF induction by ULV shocks.

studied extensively for many years, its mechanisms continue to be debated.[37-39] The inconsistent results obtained from those studies could be influenced by the differences in shock strengths and species used in those studies. To minimize the effect of shock strength and species differences on the shock outcome, studies with whole pig hearts have investigated the mechanisms of defibrillation by using only the near–defibrillation threshold strength shocks. A large animal model is used because it has physiologic and anatomic similarities to the human heart.[50-52,69]

Following near-threshold defibrillation shocks that successfully defibrillated 50% of the time (DFT_{50}), Chattipakorn and colleagues demonstrated that the patterns of the first postshock activation cycle were indistinguishable between successful and failed shocks.[52] However, starting at cycle 2, activation cycles arose on the epicardium progressively faster and the time to traverse the ventricles was progressively slower in failed shocks than in successful shocks (Figs. 4-11*A* and 4-11*B*). These first few postshock activations always arose focally at the LV apex, the region where the shock potential gradient field was weak for the shocking electrode configuration (right ventricular [RV] apex–superior vena cava [SVC]) used in this study (Fig. 4-12).[7] These findings suggested that it was the number and rapidity of the postshock activation cycles, not the immediate postshock activation, that determined the shock outcome. To test this hypothesis, a subsequent study was performed by delivering one to five pacing stimuli at the LV apex following a shock above the defibrillation threshold.[70] Thus, the shock itself always successfully defibrillated, and the five pacing stimuli mimicked the rapid successive postshock cycles observed in the DFT_{50} shock study. The results from this study demonstrated that to reinitiate VF, three rapid successive postshock pacing stimuli were always required; one or two pacing stimuli never reinitiated VF even at the shortest coupling intervals that captured the myocardium. These findings were consistent with the DFT_{50} shock study and support the hypothesis that the number and rapidity of postshock activations determine shock outcome.[52,70]

Since the ULV hypothesis states that failed defibrillation has a mechanism similar to that of VF induction caused by a shock delivered during the vulnerable period, one study used shocks of the same strength near the ULV that induced VF 50% of the time (ULV_{50}) to see if the results were similar to those when DFT_{50} shocks were given during fibrillation.[61] The patterns of the first postshock activation were indistinguishable between shocks that induced VF and shocks that did not (Fig. 4-13). In VF induction episodes, the subsequent cycles arose progressively faster, and the time to traverse the ventricles was progressively longer than in the episodes in which shocks did not induce VF (see Figs. 4-11*C* and 4-11*D*). These results were similar to the findings obtained from the DFT_{50} shock study,[52] supporting the ULV hypothesis for defibrillation. To test whether the number and rapidity of postshock cycles determined the shock outcome for VF induction as suggested by the defibrillation pacing study

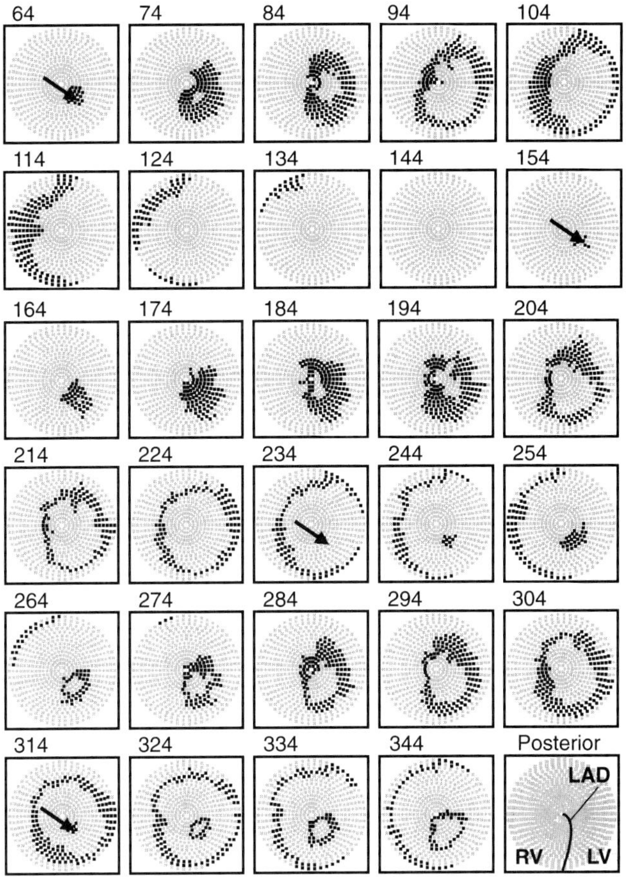

FIGURE 4-12 Example of postshock cycles following a failed DFT_{50} defibrillation shock. The orientation of the recording electrodes *(gray or black squares)* relative to the ventricles is shown in the bottom right map. Each panel shows in *black* the electrode sites at which $dV/dt \leq -0.5$ V/sec at any time during a 10 millisecond (ms) interval, indicating activation. *Numbers* above the frames indicate the start of each interval in ms relative to the shock onset. *Arrows* indicate the site of earliest recorded activation for each cycle. The first cycle appeared on the epicardium 64 ms after the shock at the antero-apical LV and propagated toward the antero-basal LV. The second cycle (154 ms) arose on the epicardium in the same region as the first cycle and also propagated away in a focal pattern. The third (235 ms) and the fourth (315 ms) cycles arose before the activation front from the previous cycle disappeared. DFT_{50}, defibrillation 50% of the time; dV/dt, first derivative of voltage with respect to time; LV, left ventricle. (Reproduced with permission from Chattipakorn N, Fotuhi PC, Ideker RE: Prediction of defibrillation outcome by epicardial activation patterns following shocks near the defibrillation threshold. J Cardiovasc Electrophysiol 2000;11: 1014-21.)

described earlier, a similar study delivered one to five pacing stimuli at the LV apex following a shock stronger than the ULV.[62] Thus, the shock itself never induced VF, and the five pacing stimuli mimicked the rapid successive postshock cycles observed in the near-threshold shock study. The results demonstrated that three rapid successive postshock pacing stimuli were always required to induce VF; one or two pacing stimuli never reinitiated VF even at the shortest coupling intervals that captured the myocardium. All these findings

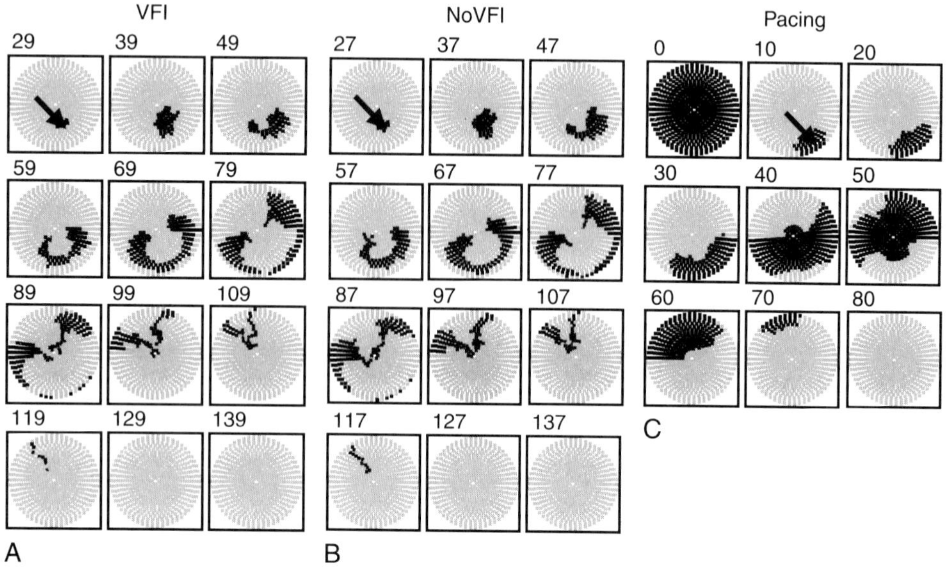

FIGURE 4-13 Examples of postshock cycle 1 following ULV_{50} shocks for VFI (**A**) and NoVFI (**B**) and of a paced cycle (**C**) from the same animal. Map orientation is similar to Figure 4-12. *Arrows* indicate the early site for each cycle. **A,** Cycle 1 arose at anteroapical LV, propagated toward anterobasal LV, and blocked over RV apex. **B,** Cycle 1 arose in the same region as in *A* and propagated similarly. **C,** Activation initiated by pacing from anterobasal epicardial LV propagated without slowing across the apex, suggesting that there was no anatomic block at the apex. LV, left ventricle; NoVFI, failed VF induction by ULV shocks; RV, right ventricle; ULV_{50}, upper limit vulnerability half the time VF induced by shock; VFI, successful ventricular fibrillation induction by ULV shocks. (Reproduced with permission from Chattipakorn N, Rogers JM, Ideker RE: Influence of postshock epicardial activation patterns on initiation of ventricular fibrillation by upper limit of vulnerability shocks. Circulation 2000;101:1329-36.)

were consistent with the defibrillation pacing study and confirmed that the number and rapidity of postshock activations determined shock outcome.[52,61,62,70]

Small Arrhythmogenic Region After Near-Threshold Shocks

Results from the near-threshold studies for both defibrillation and VF induction are all consistent in that, following the shock, the sites of earliest activation always arose at the LV apex, the low shock potential gradient region for the RV apex–SVC shocking electrode configuration.[7] Activations arose repeatedly faster but in an organized pattern from this region for at least five cycles before degenerating into VF, as observed in both defibrillation and VF induction studies.[50,52,61,62,70] It is not known what produces these postshock activations and why this postshock activity spontaneously stops after a few cycles, leading to a successful defibrillation/failed VF induction in some cases, yet continues and generates VF, leading to failed defibrillation/successful VF induction in others. However, these findings suggest the extreme importance of this small arrhythmogenic region after the shock. The similarity of the immediate postshock activation pattern between successful and failed shocks suggests that the global dispersion of refractoriness following the shock may not be the key determinant for the success or failure of defibrillation in these normal hearts. Rather, the state of the small arrhythmogenic region from which the postshock cycles arise is the crucial determinant of shock outcome.

Several studies have strongly supported this hypothesis. When a tiny shock, 50 to 100 V, was given to a small electrode on the epicardium at the LV apex, the site of weakest potential gradient where the early postshock cycles arise, just before or after the standard defibrillation shock was given from electrodes at the RV apex and SVC, the total defibrillation threshold energy was decreased by 60% compared with the defibrillation threshold for shocks through the RV-SVC electrodes alone.[71-73] When the small electrode was placed elsewhere on the epicardium, it had little or no effect on the defibrillation threshold. One VF induction study showed that the same general phenomenon occurred after the initiation of VF by a stimulus slightly larger than the VF threshold.[74] Defibrillation shocks delivered from an electrode immediately adjacent to the electrode from which VF was initiated could significantly lower the defibrillation threshold for shocks delivered for the first three postinduction cycles compared with the defibrillation threshold for defibrillation shocks delivered after 10 seconds of VF when activations no longer arose solely from the area of original initiation (Fig. 4-14). In the most recent study, subendocardial radiofrequency ablation was performed at the site where the early postshock activation arose after VF induction by the near-ULV shocks.[75] Ablation of this arrhythmogenic site resulted in a marked decrease in the ULV, but when ablation was performed elsewhere,

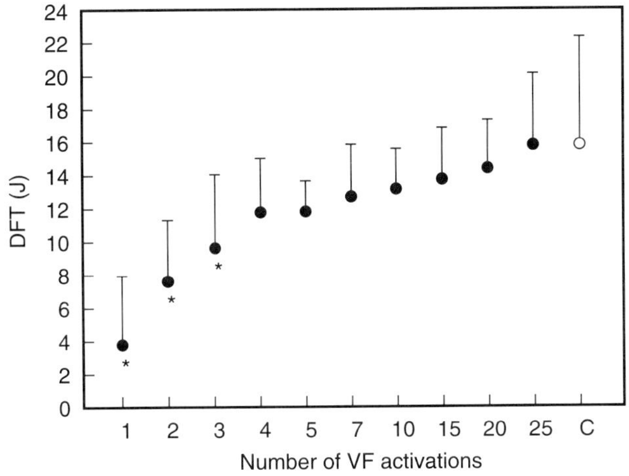

FIGURE 4-14 Increase in DFT with time demonstrating that shocks given from defibrillation electrodes located near the site of initiation of VF have a low DFT for the first few cycles following VF induction, implying VF is maintained for the first few cycles by activation fronts arising in this localized region. VF was initiated from and defibrillation shocks were given from the RV in seven pigs. Defibrillation shocks were timed to be given after 1 to 5 cycles and after 10 seconds of VF (control, *C*). The mean ± SD DFT rises sharply for shocks given after 1 to 4 VF cycles and then rises gradually for shocks given after up to 25 VF activations. An *asterisk* signifies the DFT is significantly less at that number of VF cycles than for *C*. DFT, defibrillation threshold; RV, right ventricle; SD, standard deviation; VF, ventricular fibrillation. (Reproduced with permission from Strobel JS, Kenknight BH, Rollins DL, et al: The effects of ventricular fibrillation duration and site of initiation on the defibrillation threshold during early ventricular fibrillation. J Am Coll Cardiol 1998;32:521-7.)

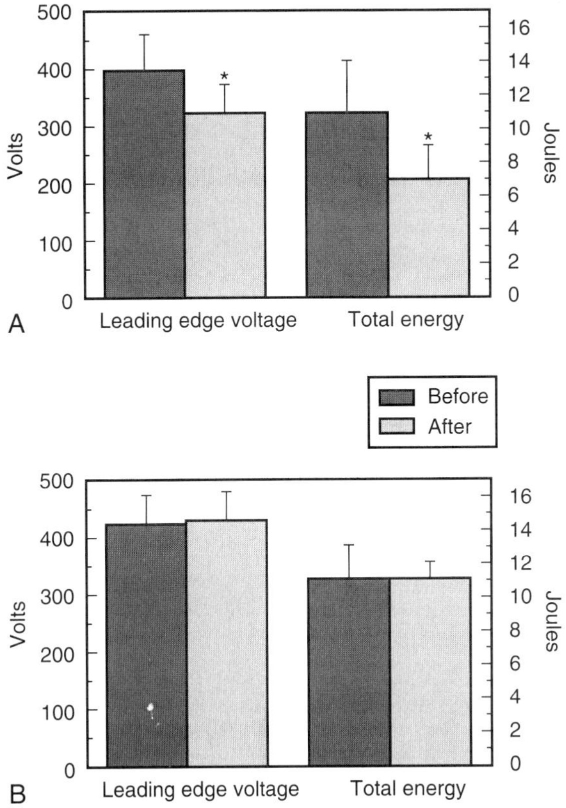

FIGURE 4-15 ULV shocks determined before and after ablation was performed at the LV apex (**A**) and LV base (**B**). Delivered voltage and energy at the ULV were significantly decreased after the LV apex ablation by 19% and 34%, respectively. However, there was no difference in the ULV shocks required before and after the LV base ablation. An *asterisk* signifies the ULV is significantly less after ablation than before ablation. LV, left ventricular; ULV, upper limit of vulnerability. (Reproduced with permission from Chattipakorn N, Fotuhi PC, Zheng X, et al: Left ventricular apex ablation decreases the upper limit of vulnerability. Circulation 2000;101:2458-60.)

it had little or no effect on the ULV (Fig. 4-15). All of these studies indicate the crucial importance of the small region that gives rise to activations after the shock in determining defibrillation outcome.

Postshock Isoelectric Window: Is It Truly Electrically Silent?

Previous cardiac mapping studies have demonstrated that, following defibrillation shocks, the postshock interval (the interval between the shock and the first postshock activation that propagated globally across the ventricles) was longer for successful than for failed defibrillation.[28,35] This postshock interval was believed to be electrically silent and was known as the "isoelectric window."[35] The existence of the isoelectric window is thought to be due to the refractory period prolongation caused by the shock. Several studies showed that when compared to failed defibrillation shocks, successful defibrillation shocks caused a larger degree of refractory period extension that occurred over a larger area.[18,76]

By using an optical mapping technique, Dillon demonstrated that, during fibrillation, refractory period extension caused by the shock could be observed at any time if the shock was delivered to the myocardium when the cardiac tissue just passed its upstroke of the fibrillatory AP.[77] Regardless of the electrical state of the myocardium immediately before the shock, he found that all myocardium in the mapped region repolarized and came back to the resting state at the same time following successful defibrillation. Dillon assumed that failed defibrillation occurs by reentry after the shock caused by the nonuniform dispersion of refractory period across the heart and hypothesized that successful shocks require the immediate postshock repolarization time to be constant throughout the myocardium to decrease the dispersion of refractoriness. This is known as the *synchronization of repolarization hypothesis*. Since defibrillation success is shock strength dependent, the concept of synchronization of repolarization has been extended to a new hypothesis for defibrillation known as the *progressive depolarization hypothesis*.[38] This hypothesis was proposed to unify the mechanism by which shocks terminate or induce fibrillation. As the shock strength progressively increases, the depolarized regions as well as the degree of refractory period extension progressively increase across the heart, resulting in less dispersion of refractoriness due to synchronization of repolarization time. Similar to the critical point hypothesis, the progressive depolarization hypothesis suggests that the immediate myocardial responses following the shock are crucial in determining defibrillation or VF induction outcome.

Electrical[51] and optical cardiac mapping studies[47,67] have demonstrated that the postshock interval is not totally electrically silent. While most shocks were well below the defibrillation threshold in optical mapping studies, only shocks near the defibrillation threshold were used in the electrical mapping study. Chattipakorn and colleagues reported that, before the first postshock activation that propagates globally across the heart is observed, activations occur immediately after the shock but only propagate locally for a short distance and then disappear.[51] These locally propagated activations were observed in both successful and failed defibrillation (Fig. 4-16). Thus, the isoelectric window is not truly electrically silent. The existence of these immediate postshock locally propagated activations suggest that a uniform refractoriness distribution across the ventricles may not be necessary for successful defibrillation since such refractoriness should have prevented this activation. Results from recent defibrillation and VF induction studies using shocks of similar strength near the threshold, so that variations in shock strength would not influence the results obtained, strongly support this idea.[34,50,51,61,62] These recent findings suggest that the progressive depolarization observed in previous studies is shock strength dependent and may not have a direct cause and effect relationship on defibrillation outcome.

Harmful Effects of Strong Shocks

Although defibrillation success is shock strength dependent, successful defibrillation is not always observed with a very strong shock. Figure 4-17 demonstrates the detrimental effect when a very high-strength shock is delivered to the heart, since the probability of success is decreased.[78] It has been shown that when a shock much stronger than the ULV is given to the heart, it can induce VF regardless of the state of the AP of that cardiac tissue.[56,79] For defibrillation using an intracardiac electrode configuration, the region exposed to the high potential gradient is close to the shocking electrode. For shocks near the defibrillation threshold delivered from electrodes placed at the SVC and RV apex, Walker et al.[80] demonstrated that the immediate postshock activation often arose at the LV apex, where the potential gradient was weak. When the shock strength was increased to a few hundred volts above the defibrillation threshold, the immediate postshock activation arose near the RV electrode, where the potential gradient was high, but VF was not induced. With a shock a few hundred volts higher, tachyarrhythmic-like activation arose from the high gradient region immediately following the shock. Jones and colleagues discovered electroporation and myocardial damage when an excessively high-strength shock (above ≈50 V/cm) was given to myocardium.[81] This could be responsible for the low probability of defibrillation success as well as the high chance of VF induction with very strong shocks.[18,81,82]

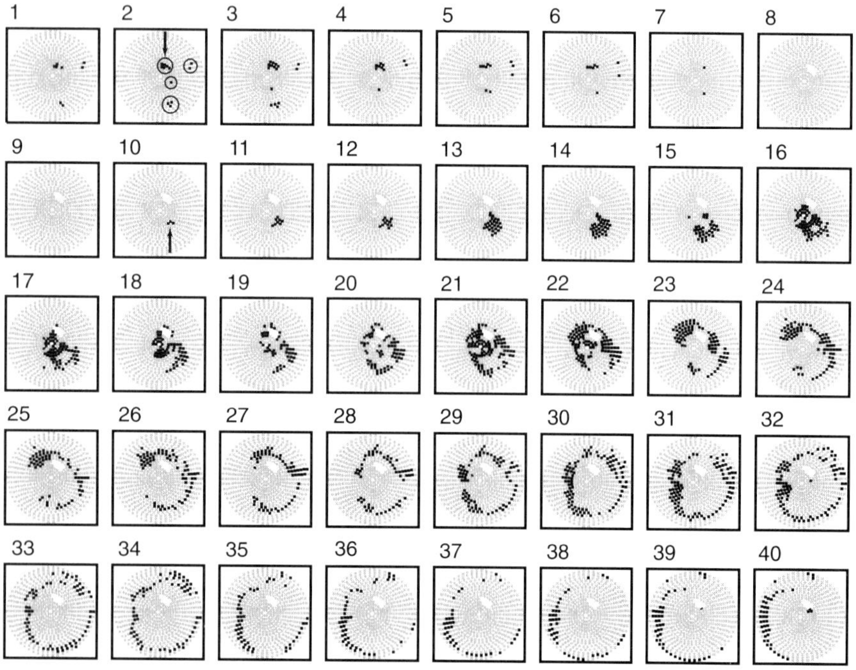

FIGURE 4-16 The presence of locally propagated activation (LPA) following a failed defibrillation shock. Each map represents a polar view of the ventricles. The interval between consecutive maps is 2 milliseconds (ms). Each black dot represents local activation at 1 of 504 epicardial electrodes. LPAs were detected 48 ms after the shock *(circles, frame 1)*. They propagated locally and disappeared, after which a first globally propagated activation (GPA) was observed *(arrow, frame 10)*. The GPA wavefront blocked without propagating through one LPA region *(circle with arrow in frame 2)*. *White dots* from frames 8 through 40 indicate the LPA region in which this block occurred. (Reproduced with permission from Chattipakorn N, KenKnight BH, Rogers JM, et al: Locally propagated activation immediately after internal defibrillation. Circulation 1998;97:1401-10.)

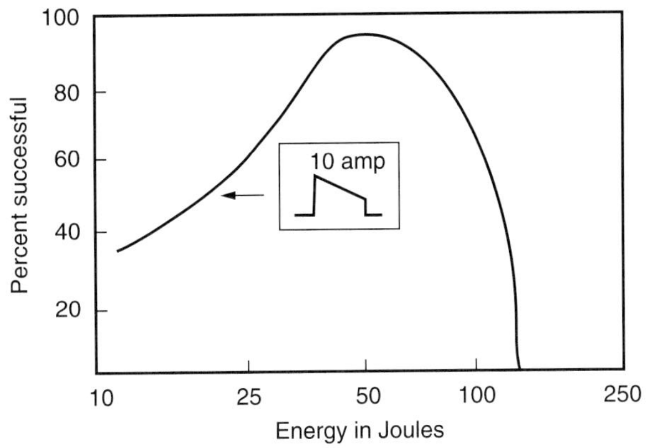

FIGURE 4-17 Relationship between success rate of transthoracic ventricular defibrillation and energy content of a trapezoidal shock in dogs. The energy of the 10-ampere leading edge shock is increased by increasing the duration of the shock. (Modified with permission from Schuder JC, Rahmoeller GA, Stoeckle H: Transthoracic ventricular defibrillation with triangular and trapezoidal waveforms. Circ Res 1966;19:689-94.)

Defibrillation Mechanisms Continue to be Debated

Although defibrillation has been extensively investigated over the past few decades and much has been discovered, defibrillation mechanisms continue to be debated. The immediate postshock myocardial responses to the shock as well as the immediate postshock activation pattern (i.e., reentry) have been proposed to be crucial in defibrillation success. However, recent studies suggest that the immediate postshock myocardial responses can have a focal origin and are not absolutely crucial in determining shock outcome.[51,61,83] Instead, for shocks near the defibrillation threshold, the number and rapidity of repetitive postshock activations arising from the small arrhythmogenic region located in the low potential gradient area appear to be the key determinants for defibrillation outcome.[61,62] These different findings suggest that the mechanism of defibrillation is complex. It is hoped that the development of high resolution three-dimensional recording techniques that do not alter the electrophysiologic properties of myocardium will allow the inconsistencies of results from previous studies to be resolved.

REFERENCES

1. Zipes DP, Wellens HJJ: Sudden cardiac death. Circulation 1998;98:2334-51.
2. Maron BJ, Shen WK, Link MS, et al: Efficacy of implantable cardioverter-defibrillators for the prevention of sudden death in patients with hypertrophic cardiomyopathy. N Engl J Med 2000;342:365-73.
3. Moss AJ, Hill WJ, Cannom DS, et al: Improved survival with an implanted defibrillator in patients with coronary disease at high risk for ventricular arrhythmia. N Engl J Med 1996;335:1933-40.
4. Zipes DP, Roberts D: Results of the international study of the implantable pacemaker cardioverter-defibrillator: A comparison of epicardial and endocardial lead systems. Circulation 1995;92:59-65.
5. Knisley SB: Transmembrane voltage changes during unipolar stimulation of rabbit ventricle. Circ Res 1995;77:1229-39.
6. Wikswo JP Jr, Lin S-F, Abbas RA: Virtual electrodes in cardiac tissue: A common mechanism for anodal and cathodal stimulation. Biophys J 1995;69:2195-210.
7. Tang ASL, Wolf PD, Claydon FJ III, et al: Measurement of defibrillation shock potential distributions and activation sequences of the heart in three-dimensions. Proc IEEE 1988;76:1176-86.
8. Eason J, Trayanova N: The effects of fiber curvature in a bidomain tissue with irregular boundaries. Proc 15th Annu Intl Conf IEEE Engineering in Medicine and Biology Society 1993:744-5.
9. Gillis AM, Fast VG, Rohr S, et al: Spatial changes in transmembrane potential during extracellular electrical shocks in cultured monolayers of neonatal rat ventricular myocytes. Circ Res 1996;79:676-90.
10. Fast VG, Rohr S, Gillis AM, et al: Activation of cardiac tissue by extracellular electrical shocks. Formation of "secondary sources" at intercellular clefts in monolayers of cultured myocytes. Circ Res 1998;82:375-85.
11. White JB, Walcott GP, Pollard AE, et al: Myocardial discontinuities: A substrate for producing virtual electrodes to increase directly excited areas of the myocardium by shocks. Circulation 1998;97:1738-45.
12. Chen P-S, Wolf PD, Claydon FJ III, et al: The potential gradient field created by epicardial defibrillation electrodes in dogs. Circulation 1986;74:626-36.
13. Katz AM: The cardiac action potential. In Katz: Physiology of the Heart. New York, Raven Press, 1992, pp 438-72.
14. Knisley SB, Smith WM, Ideker RE: Effect of field stimulation on cellular repolarization in rabbit myocardium: Implications for reentry induction. Circ Res 1992;70:707-15.
15. Kao CY, Hoffman BF: Graded and decremental response in heart muscle fibers. Am J Physiol 1958;194:187-96.
16. Dillon SM: Optical recordings in the rabbit heart show that defibrillation strength shocks prolong the duration of depolarization and the refractory period. Circ Res 1991;69:842-56.
17. Knisley SB, Hill BC: Optical recordings of the effect of electrical stimulation on action potential repolarization and the induction of reentry in two-dimensional perfused rabbit epicardium. Circulation 1993;88:I:2402-14.
18. Jones JL, Tovar OH: The mechanism of defibrillation and cardioversion. Proc IEEE 1996;84:392-403.
19. Sweeney RJ, Gill RM, Steinberg MI, et al: Ventricular refractory period extension caused by defibrillation shocks. Circulation 1990;82:965-72.
20. Moe GK, Rheinboldt WC, Abildskov JA: A computer model of atrial fibrillation. Am Heart J 1964;67:200-20.
21. Janse MJ, Wilms-Schopman FJG, Coronel R: Ventricular fibrillation is not always due to multiple wavelet reentry. J Cardiovasc Electrophysiol 1995;6:512-21.
22. Huang J, Rogers JM, KenKnight BH, et al: Evolution of the organization of epicardial activation patterns during ventricular fibrillation. J Cardiovasc Electrophysiol 1998;9:1291-1304.
23. Idriss SF, Anstadt MP, Anstadt GL, et al: The effect of cardiac compression on defibrillation efficacy and the upper limit of vulnerability. J Cardiovasc Electrophysiol 1995;6:368-78.
24. Strobel JS, Kay GN, Walcott GP, et al: Defibrillation efficacy with endocardial electrodes is influenced by reductions in cardiac preload. J Intervent Cardiac Electrophys 1997;1:95-102.
25. Gold JH, Schuder JC, Stoeckle H: Contour graph for relating percent success in achieving ventricular defibrillation to duration, current, and energy content of shock. Am Heart J 1979;98:207-12.
26. Davy JM, Fain ES, Dorian P, et al: The relationship between successful defibrillation and delivered energy in open-chest dogs: Reappraisal of the "defibrillation threshold" concept. Am Heart J 1987;113:77-84.
27. Ideker RE, Chen P-S, Zhou X-H: Basic mechanisms of defibrillation. J Electrocardiol 1991;23(suppl):36-8.
28. Shibata N, Chen P-S, Dixon EG, et al: Epicardial activation following unsuccessful defibrillation shocks in dogs. Am J Physiol 1988;255:H902-9.
29. Dahl CF, Ewy GA, Warner ED, et al: Myocardial necrosis from direct current countershock: Effect of paddle size and time interval between discharge. Circulation 1974;50:956-61.

30. Jones JL, Lepeschkin E, Jones RE, et al: Response of cultured myocardial cells to countershock-type electric field stimulation. Am J Physiol 1978;235:H214-22.
31. Jones JL, Jones RE: Postshock arrhythmias: A possible cause of unsuccessful defibrillation. Crit Care Med 1980;8:167-71.
32. Zipes DP, Fischer J, King RM, et al: Termination of ventricular fibrillation in dogs by depolarizing a critical amount of myocardium. Am J Cardiol 1975;36:37-44.
33. Witkowski FX, Penkoske PA, Plonsey R: Mechanism of cardiac defibrillation in open-chest dogs with unipolar DC-coupled simultaneous activation and shock potential recordings. Circulation 1990;82:244-60.
34. Zhou X, Daubert JP, Wolf PD, et al: Epicardial mapping of ventricular defibrillation with monophasic and biphasic shocks in dogs. Circ Res 1993;72:145-60.
35. Chen P-S, Shibata N, Dixon EG, et al: Activation during ventricular defibrillation in open-chest dogs: Evidence of complete cessation and regeneration of ventricular fibrillation after unsuccessful shocks. J Clin Invest 1986;77:810-23.
36. Chen P-S, Wolf PD, Ideker RE: Mechanism of cardiac defibrillation: A different point of view. Circulation 1991;84:913-19.
37. Chen PS, Swerdlow CD, Hwang C, et al: Current concepts of ventricular defibrillation. J Cardiovasc Electrophysiol 1998;9:553-62.
38. Dillon SM, Kwaku KF: Progressive depolarization: A unified hypothesis for defibrillation and fibrillation induction by shocks. J Cardiovasc Electrophysiol 1998;9:529-52.
39. Efimov IR, Gray RA, Roth BJ: Virtual electrodes and deexcitation: New insights into fibrillation induction and defibrillation. J Cardiovasc Electrophysiol 2000;11:339-53.
40. Winfree AT: When time breaks down: The three-dimensional dynamics of electrochemical waves and cardiac arrhythmias. Princeton, NJ, Princeton University Press, 1987, pp 1-153.
41. Frazier DW, Wolf PD, Wharton JM, et al: Stimulus-induced critical point: Mechanism for electrical initiation of reentry in normal canine myocardium. J Clin Invest 1989;83:1039-52.
42. Zhou X, Knisley SB, Wolf PD, et al: Prolongation of repolarization time by electric field stimulation with monophasic and biphasic shocks in open chest dogs. Circ Res 1991;68:1761-7.
43. Jones JL, Jones RE, Milne KB: Refractory period prolongation by biphasic defibrillator waveforms is associated with enhanced sodium current in a computer model of the ventricular action potential. IEEE Trans Biomed Eng 1994;41:60-8.
44. Clark DM, Pollard AE, Ideker RE, et al: Optical transmembrane potential recordings during intracardiac defibrillation-strength shocks. J Interv Card Electrophysiol 1999;3:109-20.
45. Efimov IR, Cheng YN, Biermann M, et al: Transmembrane voltage changes produced by real and virtual electrodes during monophasic defibrillation shocks delivered by an implantable electrode. J Cardiovasc Electrophysiol 1997;8:1031-45.
46. Lindblom AE, Roth BJ, Trayanova NA: Role of virtual electrodes in arrhythmogenesis: Pinwheel experiment revisited. J Cardiovasc Electrophysiol 2000;11:274-85.
47. Efimov IR, Cheng Y, Van Wagoner DR, et al: Virtual electrode-induced phase singularity: A basic mechanism of defibrillation failure. Circ Res 1998;82:918-25.
48. Lin SF, Roth BJ, Wikswo JP Jr: Quatrefoil reentry in myocardium: An optical imaging study of the induction mechanism. J Cardiovasc Electrophysiol 1999;10:574-86.
49. Chen P-S, Wolf PD, Melnick SD, et al: Comparison of activation during ventricular fibrillation and following unsuccessful defibrillation shocks in open chest dogs. Circ Res 1990;66:1544-60.
50. Usui M, Callihan RL, Walker RG, et al: Epicardial sock mapping following monophasic and biphasic shocks of equal voltage with an endocardial lead system. J Cardiovasc Electrophysiol 1996;7:322-34.
51. Chattipakorn N, KenKnight BH, Rogers JM, et al: Locally propagated activation immediately after internal defibrillation. Circulation 1998;97:1401-10.
52. Chattipakorn N, Fotuhi PC, Ideker RE: Prediction of defibrillation outcome by epicardial activation patterns following shocks near the defibrillation threshold. J Cardiovasc Electrophysiol 2000;11:1014-21.
53. Wiggers CJ, Wégria R: Ventricular fibrillation due to single, localized induction and condenser shocks applied during the vulnerable phase of ventricular systole. Am J Physiol 1940; 128:500-5.
54. Shibata N, Chen P-S, Dixon EG, et al: Influence of shock strength and timing on induction of ventricular arrhythmias in dogs. Am J Physiol 1988;255:H891-901.
55. Chen P-S, Shibata N, Dixon EG, et al: Comparison of the defibrillation threshold and the upper limit of ventricular vulnerability. Circulation 1986;73:1022-8.
56. Lesigne C, Levy B, Saumont R, et al: An energy-time analysis of ventricular fibrillation and defibrillation thresholds with internal electrodes. Med Biol Eng 1976;14:617-22.
57. Chen P-S, Wolf PD, Dixon EG, et al: Mechanism of ventricular vulnerability to single premature stimuli in open-chest dogs. Circ Res 1988;62:1191-1209.
58. Walcott GP, Walcott KT, Ideker RE: Mechanisms of defibrillation. J Electrocardiol 1995;28:1-6.
59. Ideker RE, Tang ASL, Frazier DW, et al: Ventricular defibrillation: Basic concepts. In El-Sherif N, Samet P (eds): Cardiac Pacing and Electrophysiology. Orland, Fla., WB Saunders, 1991, pp 713-26.
60. Idriss SF, Wolf PD, Smith WM, et al: Effect of pacing site on ventricular fibrillation initiation by shocks during the vulnerable period. Am J Physiol (Heart Circ Physiol 46) 1999;277:H2065-82.
61. Chattipakorn N, Rogers JM, Ideker RE: Influence of postshock epicardial activation patterns on initiation of ventricular fibrillation by upper limit of vulnerability shocks. Circulation 2000;101:1329-36.
62. Chattipakorn N, Fotuhi PC, Sreenan KM, et al: Pacing after shocks stronger than the upper limit of vulnerability: Impact on fibrillation induction. Circulation 2000;101:1337-43.
63. Li HG, Jones DL, Yee R, et al: Defibrillation shocks produce different effects on Purkinje fibers and ventricular muscle: Implications for successful defibrillation, refibrillation and postshock arrhythmia. J Am Coll Cardiol 1993;22:607-14.
64. Sano T, Sawanobori T: Mechanism initiating ventricular fibrillation demonstrated in cultured ventricular muscle tissue. Circ Res 1970;26:201-10.
65. Antoni H, Tagtmeyer H: Die Wirkung starker Ströme auf Errgungsbildung und Kontraktion des Herzmuskels. Beitr. Ersten Hilfe Bei Unfaellen Durch Elekt. Strom 1965;4:1.
66. Antoni H, Berg W: Wirkungen des Wechselstroms auf Erregungsbildung und Kontraktion des Saugetiermyocards. Beitr. Ersten Hilfe Bei Unfaellen Durch Elekt. Strom 1967;5:3.
67. Kwaku KF, Dillon SM: Shock-induced depolarization of refractory myocardium prevents wave-front propagation in defibrillation. Circ Res 1996;79:957-73.
68. Girouard SD, Patore JM, Laurita KR, et al: Optical mapping in a new guinea pig model of ventricular tachycardia reveals mechanisms for multiple wavelengths in a single reentrant circuit. Circulation 1996;93:603-13.
69. Huang J, KenKnight BH, Walcott GP, et al: Effects of transvenous electrode polarity and waveform duration on the relationship between defibrillation threshold and upper limit of vulnerability. Circulation 1997;96:1351-9.
70. Chattipakorn N, Fotuhi PC, Ideker RE: Pacing following shocks stronger than the defibrillation threshold: Impact on defibrillation outcome. J Cardiovasc Electrophysiol 2000;11:1022-8.
71. KenKnight BH, Walker RG, Ideker RE: Marked reduction of ventricular defibrillation threshold by application of an auxiliary shock to a catheter electrode in the left posterior coronary vein of dogs. J Cardiovasc Electrophysiol 2000;11:900-906.
72. Walker RG, KenKnight BH, Ideker RE: Impact of low-amplitude auxiliary shock strength on endocardial defibrillation threshold reductions with novel dual-shock therapy (abstract). Pacing Clin Electrophysiol 1998;21:900.
73. Walker RG, KenKnight BH, Ideker RE: Reduction of defibrillation threshold by 50% with a low-amplitude auxiliary shock (abstract). Pacing Clin Electrophysiol 1998;21:853.
74. Strobel JS, Kenknight BH, Rollins DL, et al: The effects of ventricular fibrillation duration and site of initiation on the defibrillation threshold during early ventricular fibrillation. J Am Coll Cardiol 1998;32:521-7.
75. Chattipakorn N, Fotuhi PC, Zheng X, et al: Left ventricular apex ablation decreases the upper limit of vulnerability. Circulation 2000;101:2458-60.
76. Tovar OH, Jones JL: Relationship between "extension of refractoriness" and probability of successful defibrillation. Am J Physiol 1997;272:H1011-9.

77. Dillon SM: Synchronized repolarization after defibrillation shocks: A possible component of the defibrillation process demonstrated by optical recordings in rabbit heart. Circulation 1992;85:1865-78.
78. Schuder JC, Rahmoeller GA, Stoeckle H: Transthoracic ventricular defibrillation with triangular and trapezoidal waveforms. Circ Res 1966;19:689-94.
79. Fabiato A, Coumel P, Gourgon R, et al: Le seuil de réponse synchrone des fibres myocardiques. Application à la comparaison expérimentale de l'efficacité des différentes formes de chocs électriques de défibrillation. Arch Mal Cœur 1967;60:527-44.
80. Walker RG, Walcott GP, Smith WM, et al: Sites of earliest activation following transvenous defibrillation (abstract). Circulation 1994;90:Abstract.
81. Jones JL, Proskauer CC, Paul WK, et al: Ultrastructural injury to chick myocardial cells in vitro following "electric countershock." Circ Res 1980;46:387-94.
82. Tung L: Detrimental effects of electrical fields on cardiac muscle. Proc IEEE 1996;84:366-78.
83. Ujhelyi MR, Sims JJ, Miller AW: Induction of electrical heterogeneity impairs ventricular defibrillation: an effect specific to regional conduction velocity slowing. Circulation 1999;100: 2534-40.
84. Chattipakorn N, Ideker RE: Mechanism of defibrillation. In Aliot E, Clémenty J, Prystowsky EN (eds): Fighting Sudden Cardiac Death: A Worldwide Challenge. Armonk, NY, Futura Publishing, 2000.
85. Ideker RE, Zhou X, Knisley SB: Correlation among fibrillation, defibrillation, and cardiac pacing. Pacing Clin Electrophysiol 1995;18:512-25.

Chapter 5 Principles of Clinical Pharmacology

JACQUES TURGEON and PAUL DORIAN

Most antiarrhythmic drugs are administered in a relatively fixed dose, without taking into account the many and various sources of variability in effect produced by a given dose. Although the extent of this variability is difficult to quantify in individual patients, and the relationship between drug dose and clinical outcome in individual patients may be impossible to predict, knowledge of pharmacokinetic and pharmacodynamic principles can be very useful for the clinician to enhance efficacy and decrease toxicity of antiarrhythmic drugs.

It cannot be overemphasized that standard dose recommendations for antiarrhythmic drugs apply to the hypothetical "average patient," and that marked interindividual variability in drug concentration for a particular dose can occur. In addition, the relationship between drug dose and drug concentration is not linear over the entire dosage range usually employed, and thus a given dose increment may result in differential relative increases in drug effect at the lower versus the upper end of the dosage range. Given the marked variability and unpredictability of drug effect, the clinician needs to be alert to the possibility of a greater than or less than expected effect for a "standard" dose of a given drug; a useful general approach is to identify a priori some target clinical effect before drug administration and to carefully observe patients for toxicity during the initial phases of drug treatment. If the desired effect (e.g., a given amount of refractoriness or cardiac repolarization [QT] prolongation, heart rate slowing, blood pressure reduction) is not achieved and toxicity is absent, doses may be increased until some predefined effect threshold is encountered, or the maximum recommended dose of the drug is administered. Although some patients could potentially receive additional benefit from using larger than recommended doses of a given drug, increasing doses in this situation is not recommended, given the paucity of data from clinical trials regarding the safety of such an approach.

Basic Concepts in Pharmacokinetics

Pharmacokinetics is the science that describes the relationship between the dose of a drug administered and the concentrations observed in biologic fluids. Two parameters are of major importance in order to understand pharmacokinetics: the clearance (CL) and the volume of distribution (Vd). These parameters are independent but constitute major determinants of drug disposition; in other words, they will not influence each other, but both of them will dictate the time that a drug resides within the organism: the elimination half-life ($t_{1/2}$). From this concept, the following equation is derived:

$$CL = \frac{Vd \times 0.693}{t_{1/2}} \qquad \text{\{EQ. 5-1\}}$$

Thus, the greater the clearance, the shorter the elimination half-life. The larger the volume of distribution, the longer the elimination half-life.

"Clearance" reflects the ability of an organ or of the entire body to get rid of ("clear"), in an irreversible manner, the drug. This ability to clear the drug will dictate the mean plasma concentrations observed after a given dose:

$$CL = \frac{\text{Dose}}{\text{Average Concentration}} \qquad \text{\{EQ. 5-2\}}$$

Thus, conditions that increase the clearance of a drug (such as enzyme induction) will tend to decrease the mean plasma concentrations; the elimination half-life will also become shorter. Conversely, conditions that decrease the clearance of a drug (such as enzyme inhibition) will increase the mean plasma concentrations of the drug; its elimination half-life will become longer. Finally, the total body clearance of a drug reflects the ability of each organ to clear this drug.

$$CL = CL_{kidneys} + CL_{liver} + CL_{intestine} + CL_{skin} + \cdots \quad \{EQ. 5\text{-}3\}$$

$$CL = CL_{renal} + CL_{metabolic} \quad \{EQ. 5\text{-}4\}$$

When the metabolic or renal clearance of a drug is decreased, the total clearance becomes smaller, the plasma concentrations rise, and the elimination half-life becomes longer.

The volume of distribution reflects the apparent volume of liquid in which the drug is dissolved (distributed) in the organism. The larger the volume of distribution, the lower the observed plasma concentrations are and the less available the drug is for being eliminated by specific organs (the elimination half-life is then longer). For example, the distribution of antiarrhythmic drugs into body tissues will yield, for some drugs such as amiodarone, a very large volume of distribution that results in extremely long half-lives. Conversely, digoxin is distributed in lean body tissues, and the volume of distribution is lower in patients with renal failure, thus compounding the effects of decreased renal excretion of digoxin and increasing the likelihood of digoxin toxicity in these patients. Changes in volume of distribution are also important in older patients receiving drugs such as sotalol, in whom the creatinine concentration in the blood is only one measure of renal elimination; thus, elderly patients with low muscle mass can have substantial renal impairment and thus be predisposed to the risks of sotalol toxicity, even with creatinine concentrations near the normal range.

Intersubject Variability in Drug Action

It seems self-evident that each of us is an individual at birth and that our physiologic characteristics are unique. On the other hand, we are always disconcerted when unexpected effects are observed in a particular patient following administration of a drug. These effects are labeled as *unexpected* on the basis of "usual" response observed in the "normal" population. The so-called *expected* response (which, in fact, reflects the *average* response) is often derived from selected patients enrolled in clinical trials during drug development under well-controlled conditions. This may not always represent the "real world" situation. In everyday practice, patients are treated in the setting of multiple drugs administered, concomitant diseases, and varying physiologic and pathologic conditions.

Several factors can modulate the response obtained following administration of a particular drug to a particular patient at a particular time. This statement argues against the "one size fits all" concept and clearly defines the need for individualized drug therapy. To fully integrate the basic principles underlying clinical pharmacology, the prescriber needs to fully understand the principles of pharmacokinetics, pharmacodynamics, and drug efficacy. Figure 5-1 depicts the three major principles that define the relationship between drug dose and clinical outcome.

As discussed earlier, pharmacokinetics describes the relationship between the dose administered and the observed concentrations of a drug or its metabolites in selected biologic fluids. Concentrations of active or toxic substances at their effector or toxic sites are often of the greatest interest. Pharmacodynamics describes the relationship between the concentration of an active substance at its effector site and the physiologic effects observed. Currently, most drugs are aimed at either direct or indirect modulation of a protein function. For most of them, there is a range of concentrations for which changes in protein function are linearly related to drug concentration. Finally, drug efficacy links the physiologic effects observed following administration of a drug to clinical outcome. Several major clinical trials in recent years, such as the Cardiac Arrhythmia Suppression Trial, have taught us that achievement of expected pharmacodynamic response is not necessarily related to a desirable clinical outcome (i.e., drug effectiveness).[1,2]

Narrow Therapeutic Index Drugs: Antiarrhythmic Agents

The notion that monitoring plasma drug concentrations could provide a method for adjusting doses to reduce interindividual variability in response arose during the development of new antimalarial drugs during World War II. Shortly thereafter, this notion was applied to quinidine therapeutics.[3] This concept was derived from the well-recognized relationships between "normal" plasma ion concentrations or hormonal levels and a "normal" physiologic state. Using such a framework, it was observed in initial trials that plasma concentrations of quinidine below 3 μg/mL were rarely associated with an antiarrhythmic response, while concentrations above 8 μg/mL were frequently associated with QRS widening, cinchonism, and hypotension.[4] Thereby, a tentative therapeutic range of 3 to 8 μg/mL was defined.

Using the same approach, relatively well-defined therapeutic ranges were also established for lidocaine (4 to 8 μg/mL), mexiletine (500 to 1000 ng/mL) and procainamide (4 to 8 μg/mL) for patients presenting with ventricular arrhythmias.[5-8] However, as drug assays developed further and experience accumulated, it became evident that the therapeutic concentration window was very wide with these antiarrhythmic agents and that wide intersubject variability existed. Therapeutic ranges had to be redefined, such as the one of quinidine (2 to 5 μg/mL), due to impurities and metabolites interfering with early fluorometric methods.[9] Also, there was significant overlap between effective and toxic concentrations (narrow therapeutic-toxic window) in different patients, and it became almost impossible to predict, for a specific patient, plasma levels associated with efficacy or toxicity.

Subsequently, another important source of intersubject variability was identified in patients treated with the potent class Ic antiarrhythmic agent encainide.[10] Ten out of 11 patients with ventricular arrhythmias in a small clinical study responded to the drug with arrhythmia suppression and QRS widening, while the eleventh had no response. In the 10 responders, peak plasma

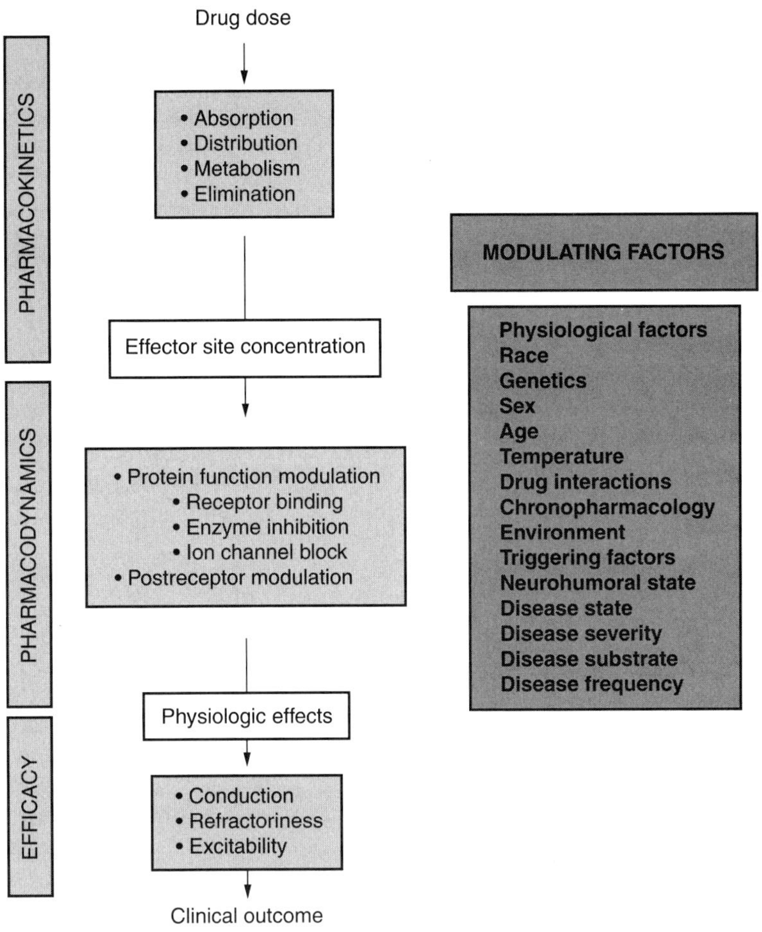

FIGURE 5-1 Principles of clinical pharmacology: factors that affect the relationship between drug dose and clinical outcome for antiarrhythmic drugs. Note that this illustration does not take into account extracardiac (e.g., autonomic) effects of drugs, which further complicate the relationship between physiologic state and drug effect.

encainide ranged from 3 to 200 ng/mL. In the single nonresponder, peak plasma encainide was the highest (300 ng/mL). Further studies demonstrated the importance of active metabolites (ODE and MODE) in accounting for encainide action, and a simple therapeutic range based solely on the plasma concentrations of the parent compound, or in combination with the metabolites, could not be defined.[11]

Propafenone is another class Ic antiarrhythmic agent that shows wide intersubject variability in its response and in the formation of active metabolites.[12] In addition, the drug exhibits varying electrophysiological (sodium, calcium, and potassium channel block) and pharmacologic (β-blocking) effects depending on the route of administration, on the metabolism status, and on the plasma concentrations of its enantiomers.[12,13] Several investigators have tried to derive combined therapeutic ranges for the metabolites, the enantiomers, and for the combinations of parent drug plus metabolites, without success.

The situation observed with antiarrhythmic agents is not unique and is observed with other drugs that have a narrow therapeutic index. For example, doses and plasma concentrations of warfarin required to maintain the International Normalized Ratio (INR) within acceptable limits (two to three) vary widely among individuals.[14-16] There is no rationale to use the plasma concentrations of each warfarin enantiomer rather than INR values to adjust warfarin doses.

The notion that plasma concentrations of a drug should be maintained within a range of concentrations in order to guarantee drug response and prevent toxicity is appealing. The problem is that this range most likely needs to be defined for each individual. Several factors must then be considered in addition to the plasma concentrations of the parent compound. A better understanding of the clinical pharmacology of drugs with cardiac electrophysiological effects, including antiarrhythmic and nonantiarrhythmic agents, can be useful to allow optimal prescribing.

Pharmacogenetics

As discussed earlier, at the same dose, not every individual will have the same plasma concentrations (pharmacokinetics). As well, at the same plasma concentration of a drug, not every individual will exhibit the same physiologic response (pharmacodynamics). And with the same physiologic response, not every individual will have the same clinical outcome (drug efficacy). Part of this variability can be explained by genetic factors: The study of interindividual variability in drug response due to genetic factors defines pharmacogenetics.

GENETICALLY DETERMINED PHARMACOKINETIC FACTORS

Genetically determined abnormalities in the ability to biotransform drugs range from apparently benign conditions such as Gilbert's syndrome (a deficiency in glucuronyl transferase activity) to the rare but potentially fatal syndrome of pseudocholinesterase deficiency. The most widely studied polymorphic drug oxidation trait is a deficiency in the cytochrome P450 isozyme (CYP2D6) responsible, among others, for the biotransformation of the antihypertensive drug debrisoquine to its inactive 4-hydroxy metabolite.[17,18] Following the oral administration of a single 10 mg dose of debrisoquine, a metabolic ratio (debrisoquine/4-hydroxydebrisoquine) established from an 8-hour urinary excretion profile can discriminate between two distinct phenotypes.[19] Individuals with a ratio greater than 12.6 are defined as poor metabolizers (PMs) whereas a value less than this antimode reflects the ability to extensively metabolize (EM) the probe drug. Family studies indicated that the deficient trait is inherited as an autosomal recessive character.[17] Regardless of geographic location, about 5% to 10% of Caucasians are PMs.

The *CYP2D6* gene is located on the long arm of chromosome 22 (q11.2-qter).[20] Deletion or transition mutations in the gene lead to splicing errors during mRNA processing and result in unstable proteins.[21,22] Therefore, the CYP2D6 protein is functionally absent in PMs. DNA assays based on allele-specific amplification with the polymerase chain reaction (PCR) allow identification of approximately 95% of all PMs.[22-24]

CYP2D6 can metabolize substances via various C-oxidations including aromatic, alicyclic and aliphatic hydroxylation, N- and S-oxidation, as well as O-dealkylation. For example, the metabolism of several classes of cardiovascular drugs such as β-blockers and class I antiarrhythmic drugs, as well as the metabolism of neuroleptics and antidepressants, cosegregates with the debrisoquine 4-hydroxylase polymorphism.[25] The clinical consequences of genetically determined polymorphic drug metabolism depends on the pharmacologic activity or toxicity of the parent compound compared to that of the metabolites formed by CYP2D6. Four situations where such variation can be clinically important can be encountered:

1. Pharmacologic effects are mediated by the parent compound alone.
2. A metabolite is more active than the parent compound.
3. The parent compound and the metabolite have different pharmacologic effects.
4. Toxicity resides within the metabolite.

The following examples are illustrative since some drugs are no longer or rarely used. The principles nevertheless are important to consider in the prescribing of antiarrhythmic agents.

Pharmacologic Effects Are Mediated by the Parent Compound Alone

Mexiletine is a class Ib antiarrhythmic agent that undergoes stereoselective disposition due to an extensive metabolism; less than 10% of an administered oral dose is recovered unchanged in urine.[26,27] The major metabolites formed by carbon and nitrogen oxidation are hydroxymethylmexiletine, p-hydroxymexiletine, m-hydroxymexiletine, and N-hydroxymexiletine.[26-29] Antiarrhythmic activity resides solely in mexiletine, and all metabolites are inactive. Formation of hydroxymethylmexiletine, p-hydroxymexiletine, and m-hydroxymexiletine is genetically determined and cosegregates with polymorphic debrisoquine 4-hydroxylase (CYP2D6) activity.[30] Hence, subjects with the extensive metabolizer (EM) phenotype form large amounts of these metabolites. In converse, clearance of mexiletine is twofold smaller and elimination half-life is longer in subjects with the PM phenotype. Consequently, at the same dose, mean plasma concentrations of mexiletine are higher, and drug accumulation is expected to occur in PM patients during chronic therapy.[30]

Combined administration of low-dose quinidine, a selective and potent inhibitor of CYP2D6, inhibits mexiletine metabolism through its three CYP2D6 major oxidative pathways and alters mexiletine disposition to such an extent that pharmacokinetic parameters of the drug are no longer different between EMs and PMs.[30] Mexiletine and quinidine have been used in combination to improve antiarrhythmic efficacy and to decrease the incidence of gastrointestinal side effects.[31] Because of a decreased clearance and an increased elimination half-life during quinidine coadministration, EM patients undergoing combined therapy should exhibit higher trough concentrations and lesser peak-to-trough fluctuations in mexiletine plasma concentrations. Drug accumulation and long-term side effects remain a risk if dosage adjustments are not made.

A Metabolite Is More Active Than the Parent Compound

Initial clinical trials with encainide illustrate the series of observations that can lead to important conclusions with regard to the potential role of active metabolites in mediating drug effects. In the study of encainide effects as related to metabolite concentrations, O-desmethyl encainide (ODE) and 3-methoxy ODE (MODE) were found in urine in all 10 responders with respect to clinical effects but were not detected in the nonresponder.[10] Electrophysiological studies demonstrated that ODE is approximately 10-fold more potent a sodium channel blocker than the parent drug, while MODE is approximately 3-fold more potent; the metabolites had refractoriness-prolonging properties while the parent drug had minor effects.[11,32-34]

Drug metabolism studies clearly demonstrated that CYP2D6 is involved in the sequential metabolism of encainide into ODE and into MODE.[11] Patients unable to form ODE or MODE are therefore PMs with low CYP2D6 activity. In normal volunteers with the EM phenotype, pretreatment with low-dose quinidine decreased encainide systemic clearance 5-fold and decreased the partial metabolic clearance of encainide to ODE + MODE 13-fold.[35] These data are compatible with inhibition of encainide biotransformation by quinidine (inhibition of CYP2D6). Coadministration of quinidine to volunteers having EM properties blunted encainide-induced QRS prolongation.[35]

The Parent Compound and the Metabolite Have Different Pharmacologic Effects

Systematic evaluation of the dose- and concentration-response relations for propafenone demonstrated substantial interindividual variability in extent of QRS prolongation and in minimal effective plasma concentrations required for arrhythmia suppression. Follow-up studies have shown that propafenone biotransformation to 5-hydroxy propafenone is catalyzed by CYP2D6, and that 5-hydroxy propafenone exerts sodium channel–blocking action in vitro similar to those of the parent drug, while a second metabolite, N-desalkyl propafenone, is somewhat less potent.[36-38] Administration of low-dose quinidine for a short period to a group of patients receiving chronic propafenone therapy resulted in a 2.5-fold increase in plasma propafenone with a commensurate decrease in 5-hydroxy propafenone concentrations.[39]

Although propafenone and 5-hydroxy propafenone are roughly equipotent as sodium channel blockers, the parent drug is substantially more potent as a β-blocker.[13] High concentrations of propafenone that can be observed in PMs can produce clinically detectable β-blockade similar to approximately 20 mg of propranolol every 8 hours. Propafenone metabolism is known to be saturable in EMs; that is, doubling the daily dosage from 450 to 900 mg/day results in a disproportionate sixfold increase in mean plasma propafenone concentrations.[36] Thus, β-blocking effects are expected in patients with the PM phenotype or in EMs receiving high dosages of the drug.[40]

Combined administration of propafenone and quinidine was also tested over a 1-year period in patients with atrial fibrillation.[41] The objective of the study was to demonstrate that combined administration of propafenone and quinidine would be superior to propafenone alone to prevent recurrence of atrial fibrillation (it was called the *CAQ-PAF study*). The rationale was that increased plasma propafenone concentrations due to combined quinidine administration would be associated with additional electrophysiologic (sodium, potassium, and calcium channel block) and pharmacologic (β-blocking) effects that are mediated mostly by propafenone itself compared to the effects that can be observed from propafenone and its 5-hydroxy metabolite. Results demonstrated that chronic administration of quinidine was able to inhibit CYP2D6 and propafenone metabolism over a 1-year period. Recurrence of atrial fibrillation was very low in genetically determined PMs (1/11) and in patients with propafenone plasma levels greater than 1500 ng/mL but very high in patients with propafenone plasma concentrations lower than 1000 ng/mL. This example illustrates that combined drug administration to alter patient phenotype can be associated with improved efficacy of a drug.

Venlafaxine is another example of a drug and its metabolite with different pharmacologic effects between EMs and PMs. Venlafaxine is a new-generation drug considered a first-line agent for the treatment of depressive disorders. It strongly inhibits presynaptic reuptake of noradrenaline and serotonin and weakly inhibits presynaptic reuptake of noradrenaline and serotonin. It also weakly inhibits dopamine reuptake.[42] Following oral administration, venlafaxine undergoes extensive first-pass metabolism.[43,44] It is metabolized to several metabolites including O-desmethyl venlafaxine, a pharmacologically active metabolite that inhibits noradrenaline and serotonin reuptake with similar potencies to those of venlafaxine.[45] The disposition of venlafaxine is genetically determined and cosegregates with CYP2D6 activity in man.[46] Subjects with the PM phenotype have 4- to 8-fold higher plasma concentrations of venlafaxine and a 20-fold lower capability to form the O-desmethyl metabolite. Since the O-desmethyl metabolite and venlafaxine have a similar potency for serotonin reuptake, no difference in antidepressant activity is expected between EMs and PMs of CYP2D6. However, case studies suggested that higher plasma concentrations of venlafaxine due to low CYP2D6 activity could increase the risk of cardiovascular toxicity since venlafaxine (and possibly not the metabolite) is a potent blocker of the cardiac sodium channel.[47] Venlafaxine has weak affinity for CYP2D6 and low propensity for causing drug interaction. However, several other CYP2D6 substrates, such as the first-generation histamine H1 antagonist diphenhydramine, can inhibit the metabolism of venlafaxine, increase the plasma concentrations of the parent compound up to fourfold, and potentially predispose patients to increased risk of cardiac toxicity.[48]

Toxicity Resides Within the Metabolite

A major form of toxicity-limiting chronic procainamide therapy is the drug-induced lupus syndrome.[49] The exact mechanism whereby procainamide is capable of initiating this autoimmune syndrome is unclear. Preliminary metabolic studies have indicated that incubation of procainamide with mouse hepatic microsomes produced a reactive metabolite.[50] Comparison with microsomal incubations of compounds modified at the site of the aromatic amine (N-acetyl procainamide [NAPA], p-hydroxyprocainamide, or desaminoprocainamide) led to the conclusion that oxidation of the primary aromatic amine of procainamide is involved in the production of such a reactive metabolite.[49,51] The formation of N-hydroxyprocainamide was confirmed in both rat and human hepatic microsomes, and characterization of the reaction showed that it was cytochrome P450 mediated.[52,53] Moreover, in vitro studies with genetically engineered microsomes expressing high levels of CYP2D6 exhibited the highest activity for the formation of N-hydroxyprocainamide.[54] In vitro results were corroborated by the clinical observations that formation of the potentially stable end-product of N-hydroxyprocainamide, nitroprocainamide, was absent in PMs of CYP2D6 but present in subjects with high CYP2D6 activity.[46] Finally, formation of N-hydroxyprocainamide was prevented in EMs during the combined administration of quinidine, a potent CYP2D6 inhibitor.[46] These results indicate that CYP2D6 becomes the key enzyme involved in the formation of the toxic metabolite. Subjects with functionally deficient CYP2D6 activity (PMs) may therefore be at lower risk of procainamide-induced lupus erythematosus.

GENETICALLY DETERMINED PHARMACODYNAMIC FACTORS

Over the past decade, great advances in the field of molecular biology have made it possible to elucidate genetic causes of the **inherited** forms of the long QT syndrome (LQTS).[55-58] These exciting discoveries have important implications for the comprehension and therapy of this condition, and have led to a better understanding of cardiac repolarization and arrhythmias in general. However, prevalence of the inherited LQTS is low. In counterpart, it is more and more recognized that the concomitant use of older and recently introduced agents, from new or previously believed safe therapeutic classes (such that they were made available over the counter) put patients at increased risk for cardiac toxicity. Indeed, the list of drugs associated with the **acquired** form of LQTS is still growing.

Genetic markers associated with an increased risk of drug-induced LQTS have also been identified.[59] That is, mutations in genes encoding for specific ion channel proteins predispose patients, otherwise apparently "normal," to excessive response to drugs causing prolongation of cardiac repolarization and increased risk of torsades de pointes. In 1998 Priori et al. demonstrated for the first time that a recessive variant of the Romano-Ward LQTS is present in the population.[60] A homozygous missense mutation in the pore region of KvLQT1 was found in a 9-year-old boy with normal hearing, a prolonged Q–T interval, and syncopal episodes during exercise. However, the parents of the proband were heterozygous for the mutation and had a normal Q–T interval. In 1997 Donger et al. identified a missense mutation in the C-terminal domain of KvLQT1 that was not associated with significant prolongation of the Q–T interval but predisposed patients to torsades de pointes upon administration of QT-prolonging drugs.[61] These recent observations suggest that mutations in cardiac potassium channel genes (and possibly other genes encoding for proteins involved in cardiac repolarization) may predispose patients with normal Q–T intervals to the acquired LQTS during treatment with drugs modulating cardiac repolarization.

Drug Interactions

Clinicians and regulatory agencies have recently been concerned about the risk of drugs other than antiarrhythmic drugs causing prolongation of cardiac repolarization. This concern is well placed since ECG monitoring is not routinely employed in therapy with several of these agents. Such undesirable drug actions were first reported as proarrhythmic events following the administration of the histamine-H1 antagonist terfenadine.[62,63] The underlying mechanism of QT prolongation and torsades de pointes during terfenadine therapy was shown to be related to I_{Kr} block.[64,65] Block of I_{Kr} was also demonstrated for several other agents such as astemizole, cisapride, pimozide, thioridazine, droperidol, domperidone, macrolide antibiotics (erythromycin, clarithromycin), imidazole antifungals, and sildenafil, which have all been associated with proarrhythmic events and deaths in some patients.[66-74]

Proarrhythmia with these drugs is almost always observed during combined drug administration. Therefore, some authors have concluded that concomitant treatment with I_{Kr} blockers may predispose patients to proarrhythmia. However, this hypothesis has not been proven. Competitive antagonism at the receptor level would predict that combined use of I_{Kr} blockers should lead to a decrease in drug effects rather than synergistic activity. Indeed, data from our laboratory indicate that combined use of dofetilide and NAPA, or NAPA and diphenhydramine, is associated with a decrease in action potential prolongation when the drugs were used together compared to when the drugs were used alone. Similarly, concomitant administration of dofetilide and erythromycin was associated with a decrease in overall action potential prolongation compared to dofetilide alone.[75] Thus, proarrhythmia observed during the concomitant administration of I_{Kr} blockers in patients cannot be related solely to their electrophysiological properties on I_{Kr}.

Proarrhythmia with combined use of I_{Kr} blockers is usually observed under conditions of decreased metabolic capacity. For example, the induction of torsades de pointes during concomitant therapy with terfenadine and erythromycin or ketoconazole has been explained mainly on the basis of a specific cytochrome P450 enzyme inhibition.[76,77] Terfenadine is known to be metabolized by CYP3A4.[78] Erythromycin and imidazole oral antifungals are known inhibitors of CYP3A4; in subjects receiving the combination of terfenadine and erythromycin, erythromycin causes a decrease in the formation of the inactive acid metabolite, and accumulation of terfenadine that may lead to prolongation of cardiac repolarization (QT) and torsades de pointes. A similar mechanism can be described for other agents. Thus, combined administration of CYP3A4 substrates leads to the accumulation of one of these drugs; if the drug exhibits potent I_{Kr} block properties, proarrhythmia (torsades de pointes) due to prolonged repolarization may be observed.

A third factor may also play a major role in drug-induced LQTS. P-glycoprotein (P-gp) is a versatile transporter that is able to pump a wide variety of xenobiotics outside a cell.[79] P-gp is located primarily in the villous columnar epithelial cells of the small intestine and in hepatocytes, but it can also be found in cardiac myocytes.[80] CYP3As and P-gp can function together by preventing cellular entry of lipophilic toxic compounds or by decreasing intracellular concentration of drugs. P-gp and CYP3As share tremendous substrate or inhibitor specificity, or both, so that substrates/inhibitors of CYP3A4 can also simultaneously inhibit P-glycoprotein. Under conditions of combined

I_{Kr}/CYP3A4/P-gp substrates treatment, not only plasma concentrations but also intracellular cardiac concentrations of I_{Kr} blockers can be increased.

Finally, as with CYP3A4, there is significant interindividual variation in the expression of P-gp, and genetic polymorphisms have been described for both *CYP3As* and *MDR1* (P-gp).[81,82] We have found that 29% of Canadians from French extraction possess two mutated alleles (exon 26) of *MDR1*, which have recently been associated with altered drug concentrations.[82] Thus, some patients may be at increased of proarrhythmia due to mutations in these genes as well as mutations in genes associated with LQTS.

SUMMARY

Individualized therapy is slowly emerging as a favored approach to improve efficacy and limit toxicity. This approach will be even more important in the near future as we march toward the use of treatments derived from biotechnologies. In fact, these treatments may be aimed at correction of a specific gene defect (specific mutation or gene deletion) in a specific patient and will require fully individualized therapy. Obtaining optimal response to conventional drug treatment may also be achieved by individualized drug therapy. This is partly explained by genetic factors or concomitant drug interactions that modulate response to drugs between patients and within patients with time.

REFERENCES

1. Ruskin JN: The Cardiac Arrhythmia Suppression Trial (CAST). N Engl J Med 1989;321:386-8.
2. The Cardiac Arrhythmia Suppression Trial (CAST) Investigators: Preliminary report: Effect of encainide and flecainide on mortality in a randomized trial of arrhythmia suppression after myocardial infarction. N Engl J Med 1989; 321:406-10.
3. Edgar AL, Sokolow M: Experiences with the photofluorometric determination of quinidine in blood. J Lab Clin Med 1950;36: 478-84.
4. Sokolow M, Edgar AL: Blood quinidine concentrations as a guide in the treatment of cardiac arrhythmias. Circulation 1950;1: 576-92.
5. Gianelly R, von der Gruben JO, Spivack AP, Harrison DC: Effect of lidocaine on ventricular arrhythmias in patients with coronary heart disease. N Engl J Med 1967;277:1215-9.
6. Campbell NPS, Kelly JG, Shanks RG, et al: Mexiletine (Ko 1173) in the management of ventricular dysrhythmias. Lancet 1973; August:404-7.
7. Koch-Weser J: Serum drug concentrations as therapeutic guides. N Engl J Med 1972;287:227-31.
8. Koch-Weser J: Correlation of serum concentrations and pharmacologic effects of antiarrhythmic drugs. In Acheson GH, Maxwell RA (eds): Pharmacology and the Future of Man. Basel, Karger, 1973, pp 69-85.
9. Kessler KM, Lowenthal DT, Warner H, et al: Quinidine elimination in patients with congestive heart failure or poor renal function. N Engl J Med 1974;290:706-9.
10. Roden DM, Reele SB, Higgins SB, et al: Total suppression of ventricular arrhythmias by encainide. Pharmacokinetic and electrocardiographic characteristics. N Engl J Med 1980;302: 877-82.
11. Barbey JT, Thompson KA, Echt DS, et al: Antiarrhythmic activity, electrocardiographic effects and pharmacokinetics of the encainide metabolites *O*-desmethyl encainide and 3-methoxy-*O*-desmethyl encainide in man. Circulation 1988;77:380-91.
12. Funck-Brentano C, Kroemer HK, Lee JT, Roden DM: Propafenone. N Engl J Med 1990;322:518-25.
13. Kroemer HK, Funck-Brentano C, Silberstein DJ, et al: Stereoselective disposition and pharmacologic activity of propafenone enantiomers. Circulation 1989;79:1068-76.
14. Breckenridge A, Orme M, Wesseling H, et al: Pharmacokinetics and pharmacodynamics of the enantiomers of warfarin in man. Clin Pharmacol Ther 1974;15:424-30.
15. Chan E, McLachlan A, O'Reilly R, Rowland M: Stereochemical aspects of warfarin drug interactions: Use of combined pharmacokinetic-pharmacodynamic model. Clin Pharmacol Ther 1994;56:286-94.
16. Furuya H, Fernandez-Salguero P, Gregory W, et al: Genetic polymorphism of CYP2C9 and its effect on warfarin maintenance dose requirement in patients undergoing anticoagulation therapy. Pharmacogenetics 1995;5:389-92.
17. Price Evans DA, Mahgoub A, Sloan TP, et al: A family and population study of the genetic polymorphism of debrisoquine oxidation in a white British population. J Med Genet 1980;17:102-5.
18. Idle JR, Mahgoub A, Angelo MM, et al: The metabolism of [^{14}C]-debrisoquine in man. Br J Clin Pharmacol 1979;7:257-66.
19. Daily AK, Armstrong M, Monkman SC, et al: Genetic and metabolic criteria for the assignment of debrisoquine 4-hydroxylation (cytochrome P4502D6) phenotypes. Pharmacogenetics 1991;1:33-41.
20. Kimura S, Umeno M, Skoda RC, et al: The human debrisoquine 4-hydroxylase (CYP2D) locus: Sequence and identification of the polymorphic CYP2D6 gene, a related gene, and a pseudogene. Am J Hum Genet 1989;45:889-904.
21. Gonzalez FJ, Skoda RC, Kimura S, et al: Characterization of the common genetic defect in humans deficient in debrisoquine metabolism. Nature 1988;331:442-6.
22. Gonzalez FJ, Meyer UA: Molecular genetics of the debrisoquin-sparteine polymorphism. Clin Pharmacol Ther 1991;50:233-8.
23. Broly F, Gaedigk A, Heim M, et al: Debrisoquine/sparteine hydroxylation genotype and phenotype: Analysis of common mutations and alleles of CYP2D6 in a European population. DNA Cell Biol 1991;10:545-58.
24. Broly F, Marez D, Sabbagh N, et al: An efficient strategy for detection of known and new mutations of the CYP2D6 gene using single strand conformation polymorphism analysis. Pharmacogenetics 1995;5:373-84.
25. Eichelbaum M, Gross AS: The genetic polymorphism of debrisoquine/sparteine metabolism. Clinical aspects. Pharmacol Ther 1990;46:377-94.
26. Beckett AH, Chidomere EC: The identification and analysis of mexiletine and its metabolic products in man. J Pharm Pharmacol 1977;29:281-5.
27. Beckett AH, Chidomere EC: The distribution, metabolism and excretion of mexiletine in man. Postgrad Med J 1977; 53(suppl 1):60-6.
28. Grech-Bélanger O, Turgeon J, Lalande M, Bélanger PM: Metahydroxymexiletine, a new metabolite of mexiletine. Isolation, characterization, and species differences in its formation. Drug Metab Dispos 1991;19:458-61.
29. Turgeon J, Paré JRJ, Lalande M, et al: Isolation and structural characterization by spectroscopic methods of two glucuronide metabolites of mexiletine after N-oxidation and deamination. Drug Metab Dispos 1992;20:762-9.
30. Turgeon J, Fiset C, Giguère R, et al: Influence of debrisoquine phenotype and of quinidine on mexiletine disposition in man. J Pharmacol Exp Ther 1991;259:789-98.
31. Duff HJ, Roden DM, Primm RK, et al: Mexiletine in the treatment of resistant ventricular tachycardia: Enhancement of efficacy and reduction of dose-related side effects by combination with quinidine. Circulation 1983;67:1124-8.
32. Carey EL Jr, Duff HJ, Roden DM, et al: Encainide and its metabolites. Comparative effects in man on ventricular arrhythmia and electrocardiographic intervals. J Clin Invest 1984;73:539-47.
33. Dawson AK, Roden DM, Duff HJ, et al: Differential effects of O-dimethyl encainide on induced and spontaneous arrhythmias in the conscious dog. Am J Cardiol 1984;54:654-8.
34. Duff HJ, Dawson AK, Carey EL, et al: The electrophysiologic actions of O-demethyl encainide: An active metabolite [abstract]. Clin Res 1981;29:270A.

35. Funck-Brentano C, Turgeon J, Woosley RL, Roden DM: Effect of low dose quinidine on encainide pharmacokinetics and pharmacodynamics: Influence of genetic polymorphism. J Pharmacol Exp Ther 1989;249:134-42.
36. Siddoway LA, Roden DM, Woosley RL: Clinical pharmacology of propafenone: pharmacokinetics, metabolism and concentration-response relations. Am J Cardiol 1984;54:9D-12D.
37. Malfatto G, Zaza A, Forster M, et al: Electrophysiologic, inotropic and antiarrhythmic effects of propafenone, 5-hydroxypropafenone and *N*-depropylpropafenone. J Pharmacol Exp Ther 1988; 246:419-26.
38. Thompson KA, Iansmith DHS, Siddoway LA, et al: Potent electrophysiologic effects of the major metabolites of propaferone in canine Purkinje fibers. J Pharmacol Exp Ther 1988;244:950-5.
39. Funck-Brentano C, Kroemer HK, Pavlou H, et al: Genetically-determined interaction between propafenone and low dose quinidine: Role of active metabolites in modulating net drug effect. Br J Clin Pharmacol 1989;27:435-44.
40. Lee JT, Kroemer HK, Silberstein DJ, et al: The role of genetically determined polymorphic drug metabolism in the beta-blockade produced by propafenone. N Engl J Med 1990; 322:1764-68.
41. O'Hara GE, Philippon F, Gilbert M, et al: Combined administration of quinidine and propafenone for atrial fibrillation: The CAQ-PAF pilot study. Eur Heart J 1999; 20:104 (Abstract).
42. Bolden-Watson C, Richelson E: Blockade by newly-developed antidepressants of biogenic amine uptake into rat brain synaptosomes. Life Sci 1993;52:1023-29.
43. Howell SR, Husbands GEM, Scatina JA, Sisenwine SF: Metabolic disposition of ^{14}C-venlafaxine in mouse, rat, dog, rhesus monkey and man. Xenobiotica 1993;23:349-59.
44. Wang CP, Howell SR, Scantina J, Sisenwine SF: The disposition of venlafaxine enantiomers in dogs, rats, and human receiving venlafaxine. Chirality 1992;4:84-90.
45. Muth EA, Moyer JA, Haskins JT, et al: Biochemical, neurophysiological, and behavioral effects of Wy-45,030, an ethyl cyclohexanol derivative. Drug Dev Res 1991;23:191-3.
46. Lessard E, Yessine MA, Hamelin BA, et al: Influence of CYP2D6 activity on the disposition of the antidepressant agent venlafaxine in humans. Pharmacogenetics 1999;9:435-43.
47. Khalifa M, Daleau P, Turgeon J: Mechanism of sodium channel block by venlafaxine in guinea pig ventricular myocytes. J Pharmacol Exp Ther 1999;291:280-4.
48. Lessard E, Yessine MA, Hamelin BA, et al: Diphenhydramine alters the disposition of venlafaxine trough inhibition of CYP2D6 activity in humans. J Clin Psychopharmacol 2001;21:175-84.
49. Uetrecht JP, Freeman RW, Woosley RL: The implications of procainamide metabolism to its induction of lupus. Arthritis Rheum 1981;24:994-9.
50. Freeman RW, Uetrecht JP, Woosley RL, et al: Covalent binding of procainamide in vitro and in vivo to hepatic protein in mice. Drug Metab Dispos 1981;9:188-92.
51. Freeman RW, Woosley RL, Oates JA, Harbison RL: Evidence for the biotransformation of procainamide to a reactive metabolite. Toxicol Appl Pharmacol 1979;50:9-16.
52. Uetrecht JP, Sweetman BJ, Woosley RL, Oates JA: Metabolism of procainamide to a hydroxylamine by rat and human hepatic microsomes. Drug Metab Dispos 1984;12:77-81.
53. Budinsky RA, Roberts SM, Coats EA, et al: The formation of procainamide hydroxylamine by rat and human liver microsomes. Drug Metab Dispos 1987;15:37-43.
54. Lessard E, Fortin A, Bélanger PM, et al: Role of CYP2D6 in the N-hydroxylation of procainamide. Pharmacogenetics 1997;7: 381-90.
55. Roden DM, Spooner PM: Inherited long QT syndromes: A paradigm for understanding arrhythmogenesis. J Cardiovasc Electrophysiol 1999;10:1664-83.
56. Keating MT: The long QT syndrome. A review of recent molecular genetic and physiologic discoveries. Medicine 1996;75:1-5.
57. Splawski I, Shen J, Timothy KW, et al: Genomic structure of three long QT syndrome genes: KVLQT1, HERG, and KCNE1. Genomics 1998;51:86-97.
58. Wang Q, Shen J, Splawski I, et al: SCN5A mutations associated with an inherited cardiac arrhythmia, long QT syndrome. Cell 1995;80:805-11.
59. Mitcheson JS, Chen J, Lin M, et al: A structural basis for drug-induced long QT syndrome. Proc Natl Acad Sci U S A 2000;97:12329-33.
60. Priori SG, Schwartz PJ, Napolitano C, et al: A recessive variant of the Romano-Ward long-QT syndrome? Circulation 1998;97:2420-5.
61. Donger C, Denjoy I, Berthet M, et al: KVLQT1 C-terminal missense mutation causes a forme fruste long-QT syndrome. Circulation 1997;96:2778-81.
62. MacConnell TJ, Stanners AJ: Torsades de pointes complicating treatment with terfenadine. BMJ 1991;302:1469.
63. Mathews DR, McNutt B, Okerholm R, et al: Torsades de pointes occurring in association with terfenadine use. JAMA 1991;266:2375-6.
64. Rampe D, Wible B, Brown AM, Dage RC: Effects of terfenadine and its metabolites on a delayed rectifier K^+ channel cloned from human heart. Mol Pharmacol 1993;44:1240-5.
65. Roy ML, Dumaine R, Brown AM: HERG, a primary human ventricular target of the nonsedating antihistamine terfenadine. Circulation 1996;94:817-23.
66. Salata JJ, Jurkiewicz NK, Wallace AA, et al: Cardiac electrophysiological actions of the histamine H_1-receptor antagonists astemizole and terfenadine compared with chlorpheniramine and pyrilamine. Circ Res 1995;76:110-9.
67. Drolet B, Khalifa M, Daleau P, et al: Block of the rapid component of the delayed rectifier potassium current by the prokinetic agent cisapride underlies drug-related lengthening of the QT interval. Circulation 1998;97:204-10.
68. Kang J, Wang L, Cai F, Rampe D: High affinity blockade of the HERG cardiac K(+) channel by the neuroleptic pimozide. Eur J Pharmacol 2000;392:137-40.
69. Drolet B, Vincent F, Rail J, et al: Thioridazine lengthens repolarization of cardiac ventricular myocytes by block of the delayed rectifier potassium current. J Pharmacol Exp Ther 1999;288:1261-8.
70. Drolet B, Zhang S, Deschênes D, et al: Droperidol lengthens cardiac repolarization due to block of the rapid component of the delayed rectifier potassium current. J Cardiovasc Electrophysiol 1999;10:1597-1604.
71. Drolet B, Rousseau G, Daleau P, et al: Domperidone should not be considered a no-risk alternative to cisapride in the treatment of gastrointestinal motility disorders. Circulation 2000;102:1883-5.
72. Daleau P, Lessard E, Groleau MF, Turgeon J: Erythromycin blocks the rapid component of the delayed rectifier potassium current and lengthens repolarization of guinea pig ventricular myocytes. Circulation 1995;91:3010-6.
73. Dumaine R, Roy ML, Brown AM: Blockade of HERG and Kv1.5 by ketoconazole. J Pharmacol Exp Ther 1998;286:727-35.
74. Geelen P, Drolet B, Rail J, et al: Sildenafil (Viagra) prolongs cardiac repolarization by blocking the rapid component of the delayed rectifier potassium current. Circulation 2000;102:275-7.
75. Antzelevitch C, Sun ZQ, Zhang ZQ, Yan GX: Cellular and ionic mechanisms underlying erythromycin-induced long QT intervals and torsades de pointes. J Am Coll Cardiol 1996;28:1836-48.
76. Honig PK, Wortham DC, Zamani K, et al: Terfenadine-ketoconazole interaction. Pharmacokinetic and electrocardiographic consequences. JAMA 1993;269:1513-8.
77. Honig PK, Woosley RL, Zamani K, et al: Changes in the pharmacokinetics and electrocardiographic pharmacodynamics of terfenadine with concomitant administration of erythromycin. Clin Pharmacol Ther 1992;52:231-8.
78. Kuang TY, Morgan A, Lazarev A, Cantilena LR: Human CYP3A4 as a potential in vitro screening system for terfenadine drug interactions [abstract]. Clin Pharmacol Ther 1994;55:139.
79. Bellamy WT: P-glycoproteins and multidrug resistance. Annu Rev Pharmacol Toxicol 1996;36:161-83.
80. Cayre A, Moins N, Finat-Duclos F, et al: In vitro detection of the MDR phenotype in rat myocardium: Use of PCR, [^{3}H]daunomycin and MDR reversing agents. Anticancer Drugs 1996;7:833-7.
81. Paulussen A, Lavrijsen K, Bohets H, et al: Two linked mutations in transcriptional regulatory elements of the CYP3A5 gene constitute the major genetic determinant of polymorphic activity in human. Pharmacogenetics 2000;10:415-24.
82. Kim RB, Leake BF, Choo EF, et al: Identification of functionally variant MDR1 alleles among European Americans and African Americans. Clin Pharmacol Ther 2001;70:189-99.

Chapter 6 Basic Electrocardiography

GALEN WAGNER

Historical Perspective

We have reached the 100th anniversary of the introduction of the clinical ECG. The initial 3 leads have been expanded to 12 to provide 6 views of cardiac electrical activity in the frontal and 6 in the horizontal (transverse) planes. During this century of development of more sophisticated and expensive cardiac diagnostic tests, the standard 12-lead ECG has had increasingly expanded clinical importance, particularly in the evaluation of patients with ischemic heart disease.

In the early 1900s, Einthoven and colleagues[1] placed recording electrodes on the right and left arms and the left leg and an additional electrode on the right leg to ground the "elektrokardiogramme" (EKG). Three leads (I, II, and III) were produced; each using a pair of the limb electrodes, with one serving as the positive and one as the negative pole. Each lead can be considered to provide two "views" of the cardiac electrical activity: one from the positive pole and an inverted or "reciprocal" view from the negative pole. The positive poles of these leads are located to the left or inferiorly so that "normal" cardiac waveforms typically appear primarily upright on the recording. For lead I, the left arm electrode is the positive pole, and the right arm electrode is the negative pole. Lead II, with its positive pole on the left leg and its negative pole on the right arm, provides a view of the electrical activity along the long (base to apex) axis of the heart. Finally, lead III has its positive pole on the left leg and its negative pole on the left arm (Fig 6-1*A*).

These three leads form the Einthoven triangle, a simplified model of the true orientation of the leads in the frontal plane. Consideration of these three leads so that they intersect in the center of the frontal plane, but retain their original orientation, provides a triaxial reference system for viewing cardiac electrical activity (see Fig. 6-1*B*).

The 60-degree angles among leads I, II, and III create wide gaps among the three views of cardiac electrical activity. Wilson and coworkers[2] developed a method for filling these gaps by creating a central terminal, connecting all three limb electrodes through a 5000-ohm resistor. A lead using this central terminal as its negative pole and an exploring electrode at any site on the body surface as its positive pole is termed a *V lead.* When the central terminal is connected to an exploring electrode on an extremity, the electrical signals are small. The amplitude of these signals in the frontal plane may be increased or augmented by disconnecting the attachment of the central terminal to the explored limb. Such an augmented V lead is termed *aV.* For example, aVF measures the potential difference between the left leg and the average of the potentials at the right and left arms. The gap between leads I and II is filled by lead aVR, between leads II and III by lead aVF, and between leads III and I by lead aVL. Leads aVR, aVL, and aVF were introduced in 1932 by Goldberger and colleagues. The positive poles of aVL and aVF are located to the left or inferiorly so that "normal" cardiac waveforms typically appear primarily upright on the recording; however, the positive pole of lead aVR is located to the right and superiorly so that "normal" cardiac waveforms typically appear primarily downward.

Addition of these three aV leads to the triaxial reference system produces a hexaxial system for viewing the cardiac electrical activity in the frontal plane with the six leads separated by angles of only 30 degrees. This provides a perspective of the frontal plane similar to the face of a clock, as discussed later in section III and illustrated in Figure 6-2. Using lead I (located at 0 degrees) as the reference, positive designations increase at 30-degree increments in a clockwise direction to +180 degrees, and negative designations increase at the same increments in a counterclockwise direction up to −180 degrees. Lead II appears at +60 degrees, aVF at +90 degrees, and III at +120 degrees, respectively. Leads aVL and aVR have designations of −30 degrees and −150 degrees, respectively. The negative poles of each of these leads complete the "clock face." Most modern electrocardiographs use digital technology. They record leads I and II only and then calculate the remaining limb leads in real time based on Einthoven's law: I + III = II.[1] The algebraic outcome of the formulas for calculating the aV leads from leads I, II, and III are:

$$aVR = -1/2\ (I + II)$$

$$aVL = I - 1/2\ (II)$$

$$aVF = II - 1/2\ (I)$$

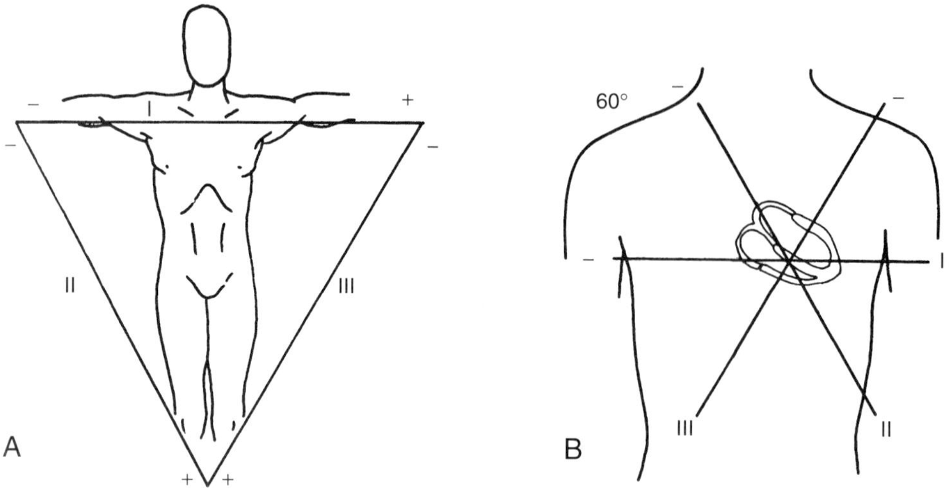

FIGURE 6-1 **A,** The equiangular (60-degree) Einthoven triangle formed by leads I, II, and III is shown with positive (I, II, III) and negative poles (–) of each of the leads indicated. **B,** The Einthoven triangle is shown in relation to the schematic view of the heart, and the three leads are shown to intersect at the center of the cardiac electrical activity.

thus,

$$aVR + aVL + aVF = 0$$

Today's standard 12-lead ECG includes these 6 frontal plane leads and also 6 leads relating to the transverse plane of the body. These leads, introduced by Wilson,[3-7] are produced by connecting the central terminal to an exploring electrode placed at various positions across the chest wall. Since the sites of these leads are close to the heart, they are termed *precordial*, and the electrical signals have sufficient amplitude so that no augmentation is necessary. The six leads are labeled V1 through V6, because the central terminal connected to all three of the limb electrodes provides their negative poles (Fig. 6-3). Lead V1, with its positive pole on the right anterior precordium and its negative pole on the left posterior thorax, provides the view of cardiac electrical activity that best distinguishes left versus right cardiac activity (Fig. 6-4).

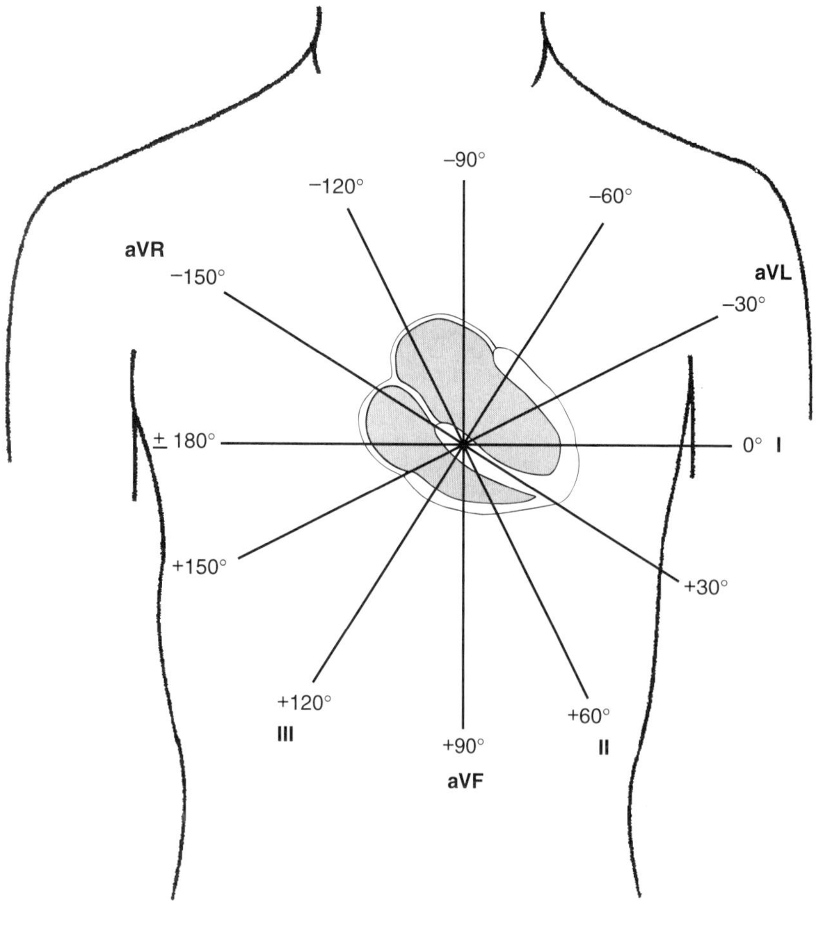

FIGURE 6-2 The locations of the positive and negative poles of each lead around the 360 degrees of the "clock face" are indicated, with the names of the six leads appearing at their positive poles.

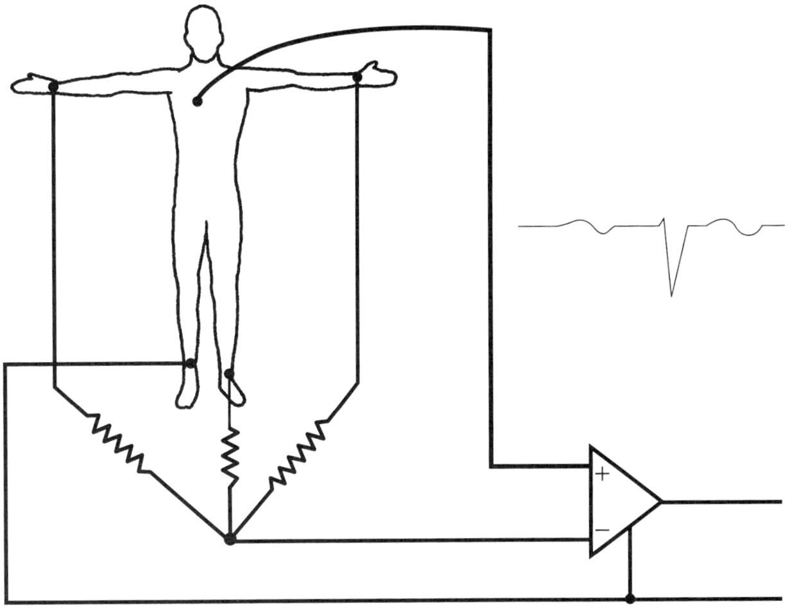

FIGURE 6-3 The method of ECG recording of the precordial leads is illustrated, along with an example of lead VI. The *wavelike lines* indicate resistors in the connections between the recording electrodes on the three limb leads that produce the negative poles for each of the V leads. (Modified from Netter FH: The Ciba collection of medical illustrations, vol 5. Heart. Summit, NJ, Ciba-Geigy, 1978, p 51.)

The sites of the exploring electrode are determined by bony landmarks on the anterior and left lateral aspects of the precordium, and the angles between the six transverse plane leads are approximately 30 degrees, the same as the angles between the six frontal plane leads.

The views of the cardiac electrical activity from the positive poles of these 12 standard ECG leads are presented in the typical displays provided by electrocardiographic recorders. However, the additional 12 views from the negative poles could also be presented to provide a "24-view ECG."

Basic Principles

ANATOMIC ORIENTATION OF THE HEART WITHIN THE BODY

The position of the heart within the body determines the "view" of the cardiac electrical activity that can be observed from any ECG recording electrode site on the body surface. The atria are located in the top or base of the heart, and the ventricles taper toward the bottom or apex. However, the right and left sides of the heart are not

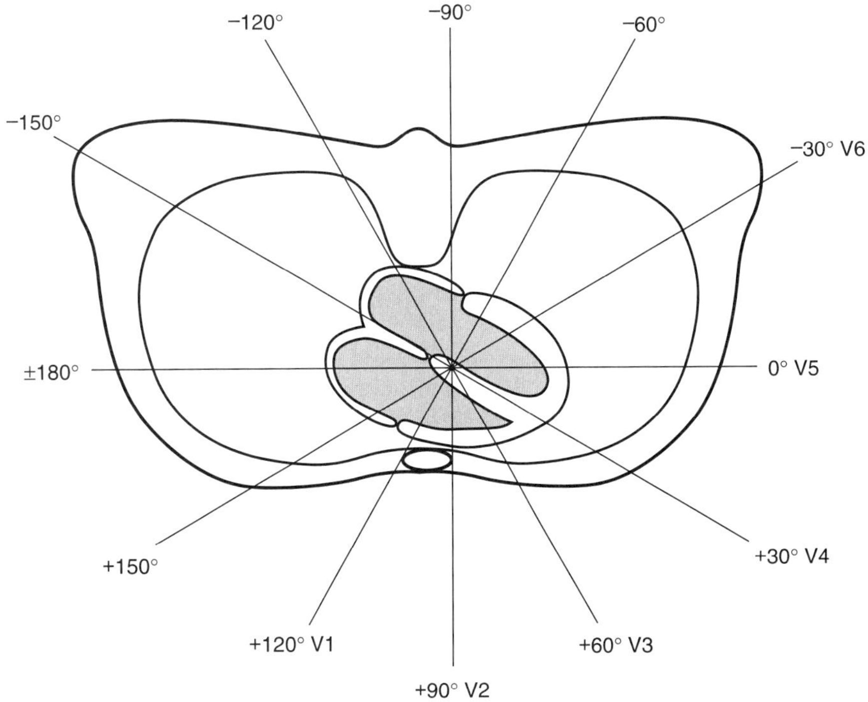

FIGURE 6-4 The orientation of the six precordial leads is indicated by *solid lines* from each of their recording sites through the approximate center of cardiac electrical activity. Extension of these lines through the chest indicates the opposite positions, which can be considered the locations of the negative poles of the six precordial leads.

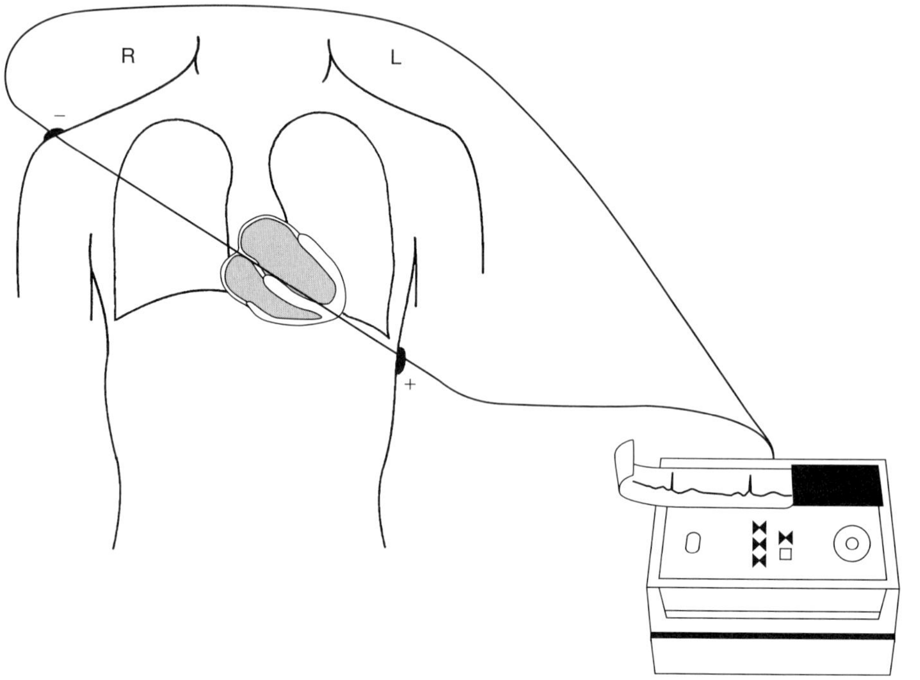

FIGURE 6-5 The schematic frontal-plane view of the heart in the thorax with electrodes where the long axis intersects with the body surface. The positive electrode on the left lower thoracic wall and the negative electrode on the right shoulder are aligned from the cardiac base to apex parallel to the interatrial and interventricular septa and are attached to a single-channel ECG recorder. The ventricular repolarization wave is positively oriented.

directly aligned with the right and left sides of the body. The long axis of the heart, which extends from base to apex, is tilted to the left and anteriorly at its apical end (Fig. 6-5). Also, the heart is rotated so that the right atrium and ventricle are more anterior than the left atrium and ventricle.[8,9] These anatomic relationships dictate that an ECG lead providing a right anterior to left posterior view (such as V1) provides better differentiation of right versus left cardiac activity than does a lead providing a right lateral to left lateral view such as lead I (Fig. 6-6).

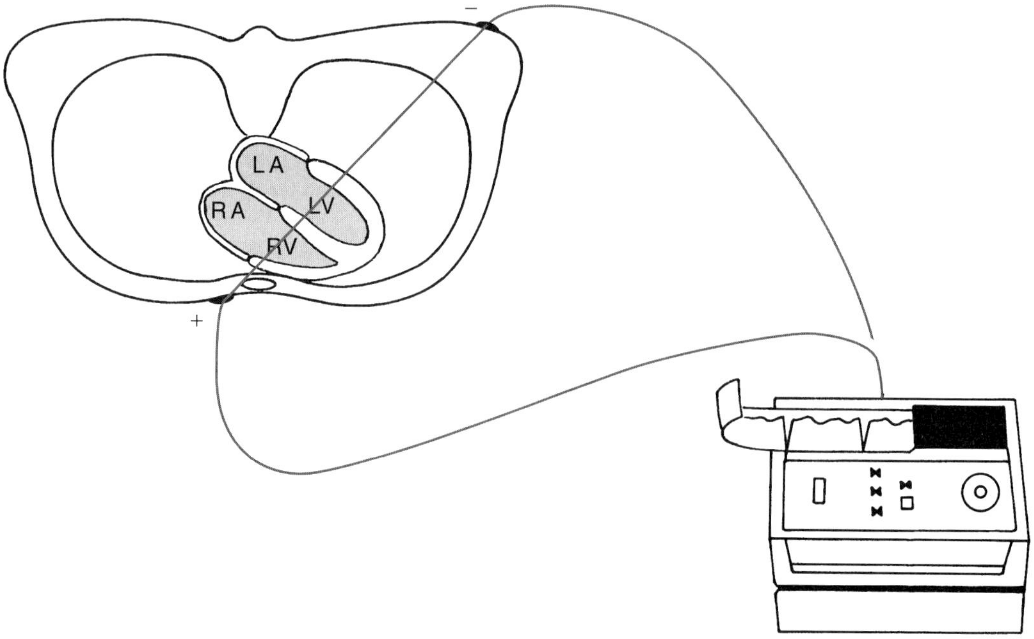

FIGURE 6-6 The schematic transverse-plane view of the heart in the thorax with electrodes where the short axis intersects with the body surface. The positive electrode to the right of the sternum and the negative electrode on the back are aligned perpendicular to the interatrial and interventricular septa and are attached to a single-channel ECG recorder. The typically diphasic P and T waves and the predominately negative QRS complex recorded by electrodes at these positions are indicated on the ECG.

THE CARDIAC CYCLE

The timing and synchronization of contraction of myocardial cells are controlled by cells of the pacemaking and conduction system. Impulses generated within these cells create a rhythmic repetition of events called cardiac cycles. Each cycle is composed of electrical and mechanical activation (systole) and recovery (diastole). Since the electrical events initiate the mechanical events, there is a brief delay between the onsets of electrical and mechanical systole and of electrical and mechanical diastole.

The electrical recording from inside a single myocardial cell as it progresses through a cardiac cycle is illustrated in the top panel of Figure 6-7. During electrical diastole, the cell has a baseline negative electrical potential and is also in mechanical diastole with separation of its contractile proteins. An electrical impulse arriving at the cell allows positively charged ions to cross the cell membrane, causing its depolarization. This movement of ions initiates electrical systole, which is characterized by an action potential *(middle panel).* This electrical event then initiates mechanical systole in which the contractile proteins slide over each other, thereby shortening the cell. Electrical systole continues until the positively charged ions are pumped out, causing repolarization of the cell. The electrical potential returns to its negative resting level. This return of electrical diastole causes the contractile proteins to separate again. The cell is then capable of being reactivated if another electrical impulse arrives at its membrane.

The ECG recording is formed by the summation of electrical signals from all of the myocardial cells (see Fig. 6-7, *lower panel*). When the cells are in their resting state, the ECG recording produces a flat baseline. The onset of depolarization of the cells produces a relatively high frequency ECG waveform. Then, while depolarization persists, the ECG returns to the baseline. Repolarization of the myocardial cells is represented on the ECG by a lower frequency waveform in the opposite direction from that representing depolarization.

CARDIAC IMPULSE FORMATION AND CONDUCTION

The electrical activation of a single cardiac cell or even a small group of cells does not produce enough current to be recorded on the body surface. Clinical electrocardiography is made possible by the activation of atrial and ventricular myocardial masses that are of sufficient magnitude for their electrical activity to be recorded on the body surface.

Myocardial cells normally lack the ability for either spontaneous formation or rapid conduction of an electrical impulse. They are dependent for these functions on special cells of the cardiac pacemaking and

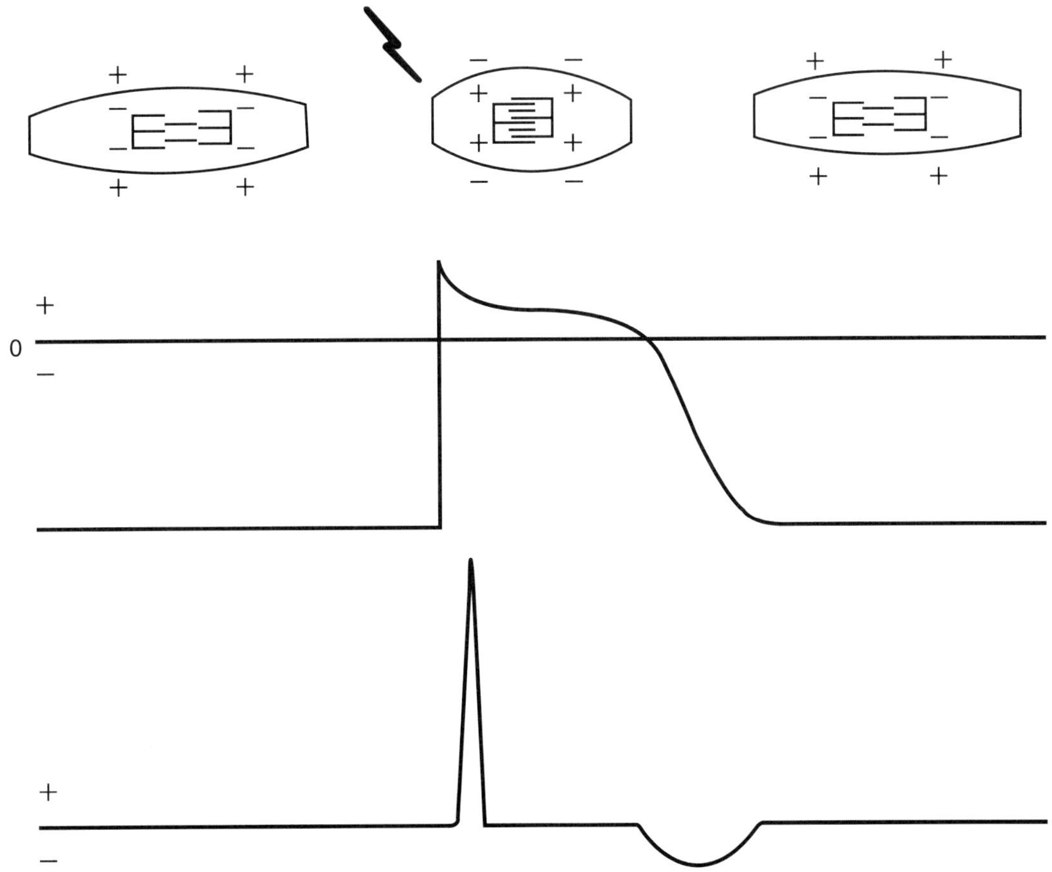

FIGURE 6-7 The schematic ECG recording beneath a cardiac cellular action potential. (Modified from Thaler MS: The Only EKG Book You'll Ever Need. Philadelphia, JB Lippincott, 1988, p 11.)

conduction system placed strategically through the heart (Fig. 6-8). These cells are arranged in nodes, bundles, bundle branches, and branching networks of fascicles. They lack contractile capability but are able to achieve spontaneous electrical impulse formation (act as pacemakers) and to alter the speed of electrical conduction. The intrinsic pacemaking rate is most rapid in the specialized cells in the sinus node and slowest in the specialized cells in the ventricles. The intrinsic pacing rate is altered by the balance between the sympathetic and parasympathetic components of the autonomic nervous system.[10-13]

A

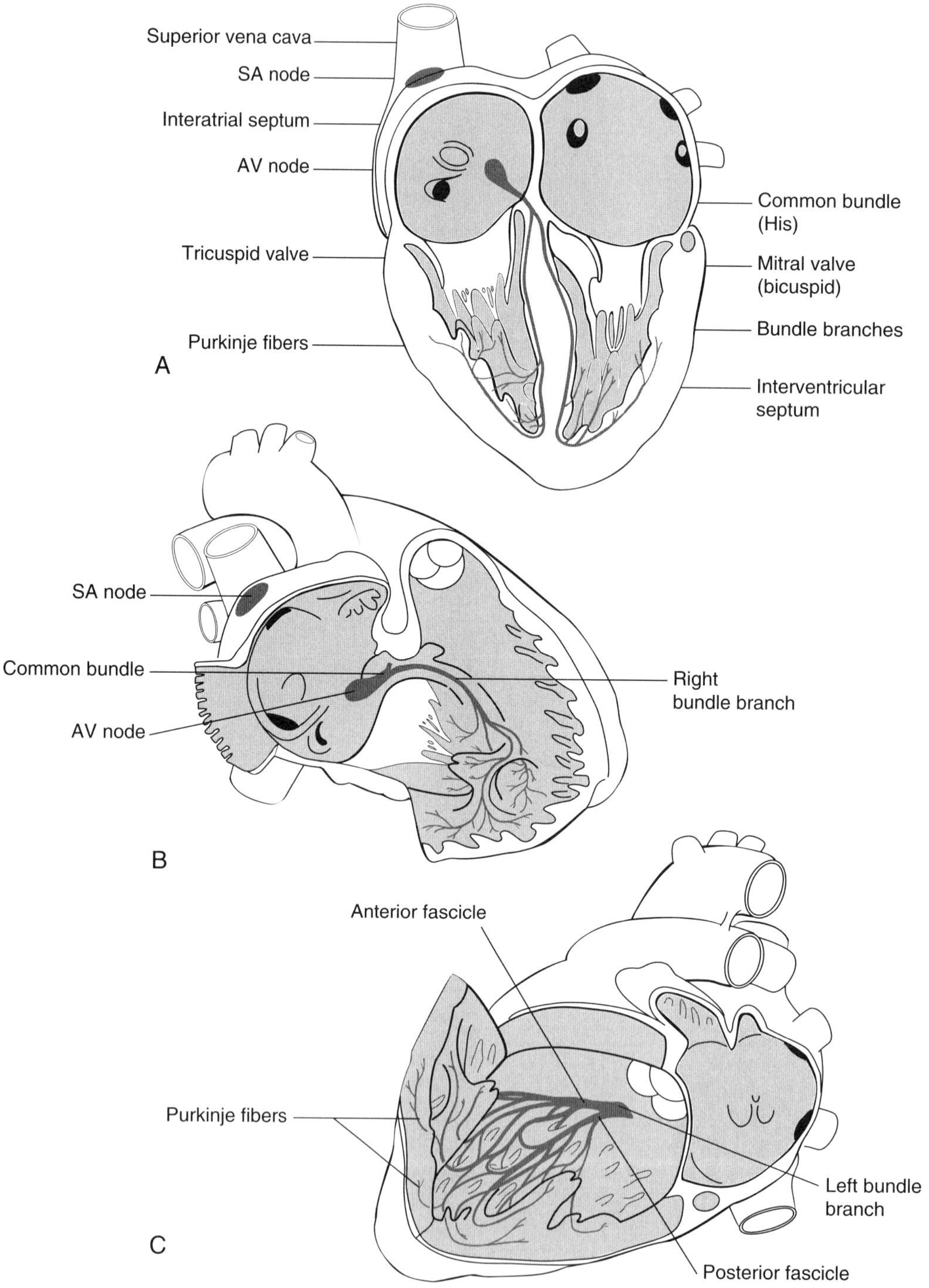

B

C

FIGURE 6-8 Three views of the anatomic relationships between the cardiac pumping chambers and the structures of the pacemaking and conduction system. **A,** From the anterior precordium. **B,** From the right anterior precordium looking onto the interatrial and interventricular septa through the right atrium and ventricle. **C,** From the left posterior thorax looking onto the septa through the left atrium and ventricle. (Modified from Netter FH: The Ciba collection of medical illustrations, vol 5. In Yonkman FF (ed): Heart. Summit, NJ, Ciba-Geigy, 1978, pp 13, 49.)

The intraventricular conduction pathways include a common bundle (bundle of His), leading from the atrioventricular (AV) node to the summit of the interventricular septum, and its right and left bundle branches, proceeding along the septal surfaces to their respective ventricles (see Fig. 6-8*A*). The left bundle branch fans into fascicles that proceed along the left septal surface and toward the two papillary muscles of the mitral valve (see Fig. 6-8*B*). The right bundle branch remains compact until it reaches the right distal septal surface, where it branches into the distal interventricular septum and toward the lateral wall of the right ventricle (Fig. 6-8*C*). These intraventricular conduction pathways are composed of fibers of Purkinje cells with specialized capabilities for both pacemaking and rapid conduction of electrical impulses. Fascicles composed of Purkinje fibers form networks that extend just beneath the surface of the right and left ventricular (LV) endocardium. The impulses then proceed slowly from endocardium to epicardium throughout the right and left ventricles.[14-16]

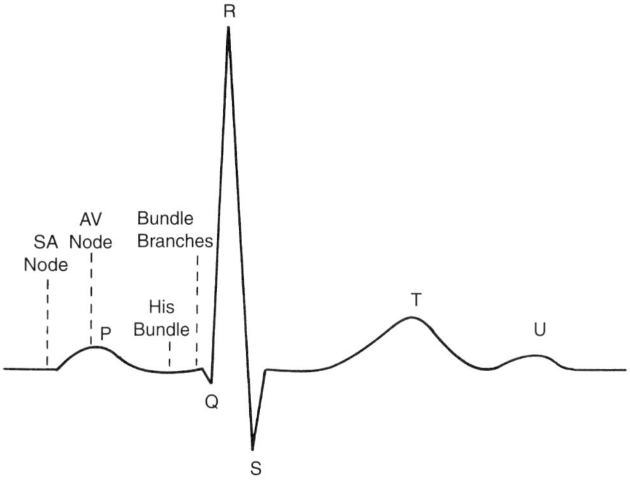

FIGURE 6-9 The visible waveforms represent activation of the atria (*P*), ventricles (*Q*, *R*, and *S*), and recovery of the ventricles (*T* and *U*). The timing of activation of the structures of the pacemaking and conduction system is also indicated.

THE ELECTROCARDIOGRAM WAVEFORMS

The initial electrical wave of a cardiac cycle represents activation of the atria and is called the P wave (Fig. 6-9). Since the sinus node is located in the right atrium, the first part of the P wave represents the activation of this chamber. The middle section of the P wave represents completion of right atrial activation and initiation of left atrial activation. The final section of the P wave represents completion of left atrial activation. The AV node is activated during the inscription of the P wave. The wave representing electrical recovery of the atria is usually obscured by the larger QRS complex, representing the activation of the ventricles. From ECG leads II oriented from cardiac base to apex, the P wave is entirely positive, and the QRS complex is predominantly positive. Minor portions at the beginning and end of the QRS complex may appear as downward or negative waves. The QRS complex may normally appear as one (monophasic), two (diphasic), or three (triphasic) individual waveforms. By convention, a negative wave at the onset of the QRS complex is called a Q wave. The first positive wave is called the R wave, regardless of whether or not it is preceded by a Q wave. A negative deflection following aVR wave is called an S wave. When a second positive deflection occurs, it is termed *R′*. A monophasic negative QRS complex should be termed a QS wave.

The wave in the cardiac cycle that represents recovery of the ventricles is called the T wave. Since recovery of the ventricular cells (repolarization) causes a countercurrent to that of depolarization, one might expect the T wave to be inverted in relation to the QRS complex. However, epicardial cells repolarize earlier than endocardial cells, thereby causing the wave of repolarization to spread in the direction opposite that of depolarization. This results in a T wave deflected in a similar direction as the QRS complex (Fig. 6-10). The T wave is sometimes followed by another small upright wave (the source of which is uncertain) called the U wave.

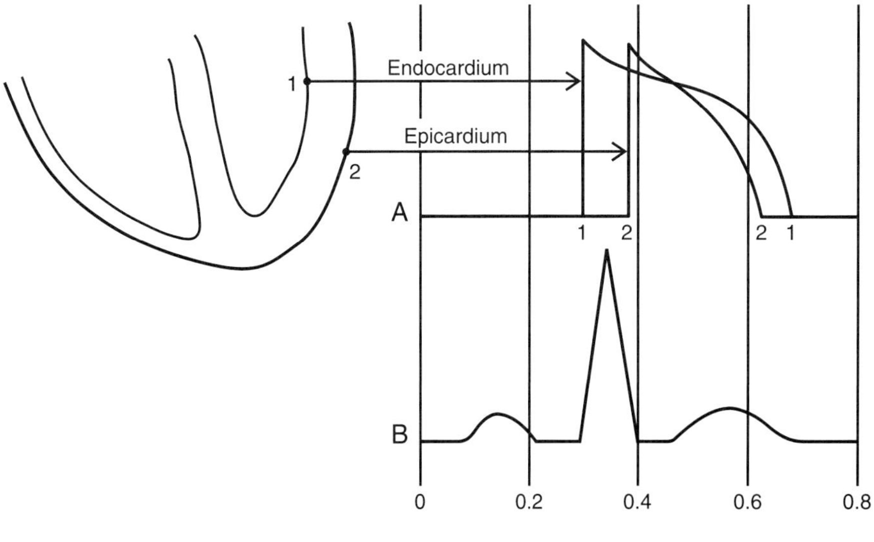

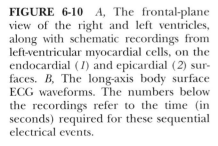

FIGURE 6-10 *A,* The frontal-plane view of the right and left ventricles, along with schematic recordings from left-ventricular myocardial cells, on the endocardial (*1*) and epicardial (*2*) surfaces. *B,* The long-axis body surface ECG waveforms. The numbers below the recordings refer to the time (in seconds) required for these sequential electrical events.

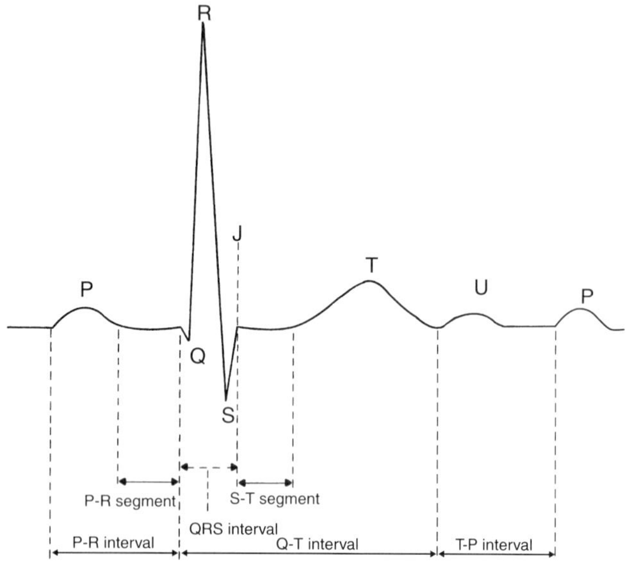

FIGURE 6-11 The magnified recording from the cardiac long-axis viewpoint is presented, with the principal ECG segments (*P-R* and *S-T*) and time intervals *(P–R, QRS, Q–T, and T–P)* indicated.

The time from the onset of the P wave to the onset of the QRS complex is called the P–R interval, whether the first wave in this complex is a Q wave or an R wave (Fig. 6-11). This interval measures the time between the onsets of activation of the atrial and ventricular myocardium. The designation PR segment refers to the time from the end of the P wave to the onset of the QRS complex. The QRS interval measures the time from beginning to end of ventricular activation. Since activation of the thicker left ventricle requires more time than the right ventricle, the terminal portion of the QRS complex represents only LV activation.

The ST segment is the interval between the end of ventricular activation and the beginning of ventricular recovery. The term ST segment is used regardless of whether the final wave of the QRS complex is an R or an S wave. The junction of the QRS complex and ST segment is called the J point. The interval from the onset of ventricular activation to the end of ventricular recovery is called the Q–T interval. This term is used whether the QRS complex begins with a Q or an R wave.

At low heart rates in a healthy person, the PR, ST, and TP segments are at the same horizontal level and form the isoelectric line. This line is considered as the baseline for measuring the amplitudes of the various waveforms. The TP segment disappears at higher heart rates when the T wave merges with the following P wave.[17-19]

DETERMINING LEFT VERSUS RIGHT CARDIAC ELECTRICAL ACTIVITY

It is often important to determine if an abnormality originates from the left or the right side of the heart. The optimal site for recording left versus right cardiac electrical activity is located where the extension of the short axis of the heart (perpendicular to the interatrial and interventricular septa) intersects with the precordial body surface (lead V1) (Fig. 6-12).

The initial part of the P wave representing right atrial activation appears positive in lead V1 because of progression of the electrical activity from the interatrial septum toward the right atrial lateral wall. The terminal part of the P wave representing left atrial activation appears negative because of progression from the interatrial septum toward the left atrial lateral wall. This activation sequence produces a diphasic P wave (Fig. 6-13).

The initial part of the QRS complex represents the progression of activation in the interventricular septum. This movement is predominantly from left to right, producing a positive (R wave) deflection at this left- versus right-sided recording site. The midportion of the QRS complex represents progression of electrical activation through the right and LV myocardium. Since the posteriorly positioned left ventricle is much thicker, its activation predominates over that of the anteriorly placed right ventricle, resulting in a deeply negative deflection (S wave). The final portion of the QRS complex represents the completion of activation of the left ventricle. This posteriorly directed excitation is represented by the completion of the S wave.

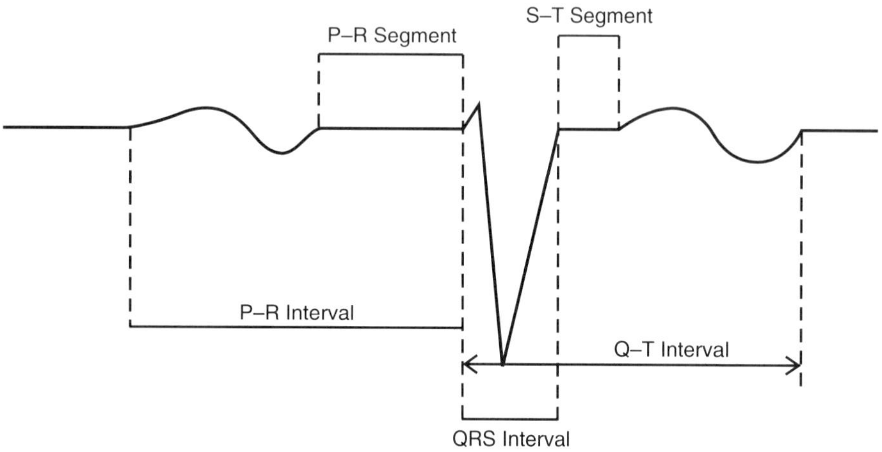

FIGURE 6-12 Magnified view of recording from the cardiac short-axis viewpoint with the principal ECG segments and time intervals indicated for the long-axis view.

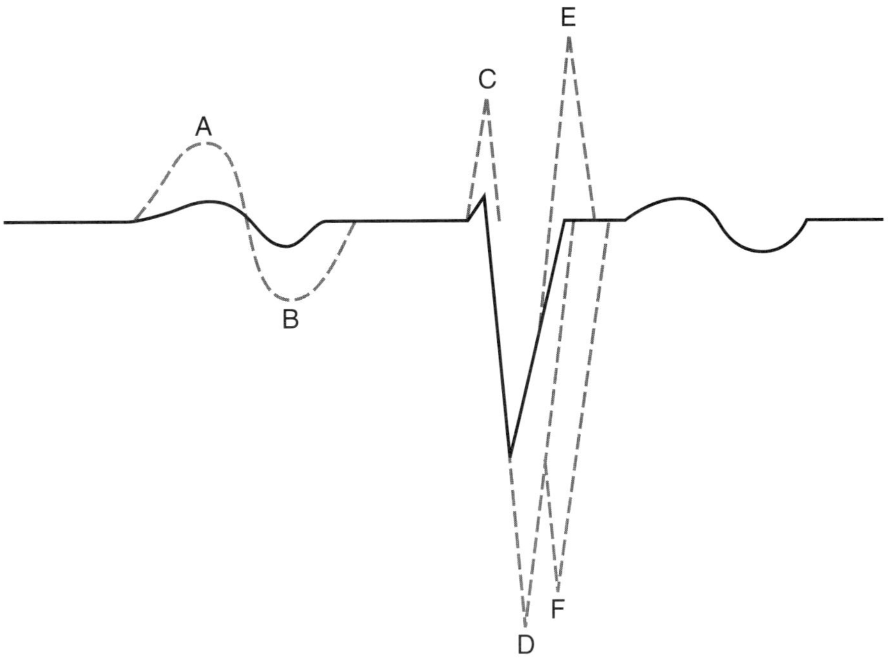

FIGURE 6-13 The ECG waveforms are reproduced with the alterations, indicated by *dashed lines,* that would typically result from enlargements of the right (*A*) and left (*B*) atrial chambers and the right (*C*) and left (*D*) ventricular chambers and from right-sided (*E*) and left-sided (*F*) intraventricular conduction delays.

The left versus right recording site is the key ECG view for identifying enlargement of one of the four cardiac chambers and localizing the site of a delay in ventricular activation (Fig. 6-14). Right atrial enlargement produces an abnormally prominent initial part of the P wave, while left atrial enlargement produces an abnormally prominent terminal part of the P wave. Right ventricular (RV) enlargement produces an abnormally prominent R wave, whereas LV enlargement produces abnormally prominent S wave. A delay in the right bundle branch causes RV activation to occur after LV activation is completed, producing an R′ deflection. A delay in the left bundle branch markedly postpones LV activation, resulting in an abnormally prominently S wave.

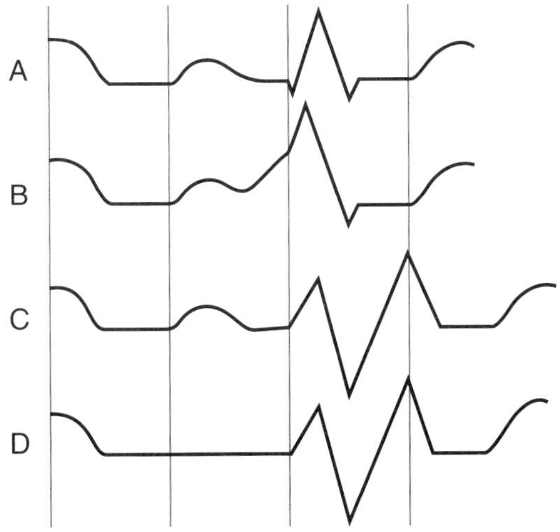

FIGURE 6-14 The normal P-to-QRS relationship and appearances (*A*) are contrasted with various conditions that produce obvious abnormalities (*B-D*). Each example begins and ends during a T wave.

Interpretation of the Normal Electrocardiogram

In interpreting every ECG there are nine features that should be examined systematically:

1. Rate and regularity
2. Rhythm
3. P wave morphology
4. P–R interval
5. QRS complex morphology
6. ST segment morphology
7. T wave morphology
8. U wave morphology
9. Q–Tc interval

Rate, regularity, and rhythm (1 and 2) are considered elsewhere in this volume, and P wave morphology (3) has been discussed earlier.

P–R INTERVAL

The P–R interval measures the time required for the impulse to travel from the atrial myocardium adjacent to the sinus node to the ventricular myocardium adjacent to the fibers of the Purkinje network. This duration is normally from 0.10 to 0.22 seconds. A major portion of the P–R interval is due to the slow conduction through the AV node, and this is controlled by the sympathetic-parasympathetic balance of the autonomic nervous system. Therefore, the PR interval varies with the heart rate, being shorter at faster rates when the sympathetic component predominates, and vice versa. The P–R interval tends to increase with age.[20]

An abnormal P wave direction is often accompanied by an abnormally short P–R interval since the site of impulse formation has moved from the sinus node to a position closer to the AV node (Fig. 6-15). However, a

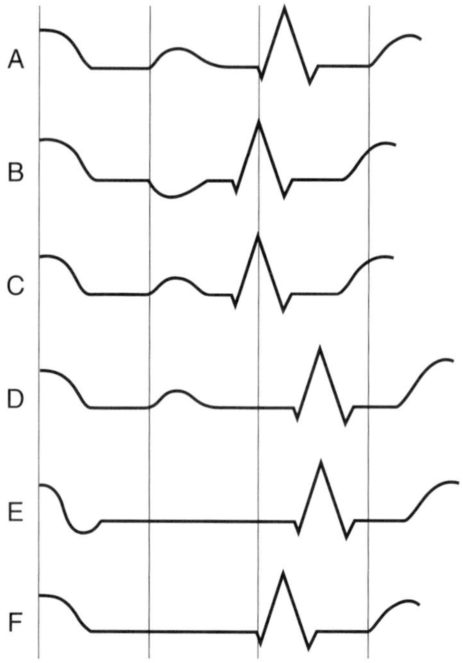

FIGURE 6-15 The normal P-to-QRS relationship (*A*) is contrasted with various abnormal relationships (*B-F*). Each example begins with the completion of a T wave and ends with the initiation of the T wave of the following cardiac cycle. The vertical time lines are at 0.2-sec intervals. The P–R interval in (*A*) is therefore 0.2 sec, which is near the upper limit of normal.

short P–R interval in the presence of a normal P wave axis suggests either an abnormally rapid conduction pathway within the AV node or the presence of an abnormal bundle of cardiac muscle (bundle of Kent) connecting the atria and ventricles and bypassing the AV node. This earlier than normal activation of the ventricular myocardium (ventricular preexcitation) creates the potential for electrical reactivation or reentry into the atria to produce a tachyarrhythmia (the Wolff-Parkinson-White syndrome).

A longer than normal P–R interval in the presence of a normal P wave axis indicates delay in impulse transmission at some point along the pathway between the atrial and ventricular myocardium (Fig. 6-16). When a prolonged P–R interval is accompanied by an abnormal P wave direction, the possibility that the P wave is actually associated with the preceding rather than the following QRS complex should be considered. When such retrograde activation from ventricles to atria occurs, the P–R interval is usually even longer than the preceding QRS to P (R–P) interval. When the P–R interval cannot be

TABLE 6-1 Wave Duration Limits

Limb Leads		Precordial Leads	
Lead	**Upper Limit**	**Lead**	**Upper Limit**
I	<0.03 sec	V1	Any Q
II	<0.03 sec	V2	Any Q
III	None	V3	Any Q
aVR	None	V4	<0.02 sec
aVL	<0.03 sec	V5	<0.03 sec
aVF	<0.03 sec	V6	<0.03 sec

Modified from Wagner GS, Freye CJ, Palmeri ST, et al: Evaluation of a QRS scoring system for estimating myocardial infarct size. I. Specificity and observer agreement. Circulation 1982;65:345.

determined, because of absence of any visible P wave, there is an obvious abnormality of the cardiac rhythm.

QRS COMPLEX

The QRS complex is composed of higher frequency signals than are the P and T waves, thereby causing its contour to be peaked rather than rounded. Positive and negative components of the P and T waves are simply termed positive and negative deflections, while those of the QRS complex are assigned specific labels such as Q wave.

Q Waves. In some leads—V1, V2, and V3—the presence of a Q wave should be considered abnormal, and in all other leads (except III and aVR), a "normal" Q wave would be very small. The "upper limit of normal" for such Q waves in each lead is indicated in Table 6-1.[21]

The absence of small Q waves in leads V5 and V6 should be considered abnormal. A Q wave of any size is normal in III and aVR because of their rightward orientations. Q waves may be enlarged by conditions such as local loss of myocardial tissue (infarction), hypertrophy or dilatation of the ventricular myocardium, or abnormalities of ventricular conduction.

R Waves. Since the precordial leads provide a panoramic view of the cardiac electrical activity progressing from the thinner right ventricle across the thicker left ventricle, the positive R wave normally increases in amplitude and duration from V1 to V4 or V5 (Fig. 6-17). Reversal of this sequence with larger R waves in V1 and V2 can be produced by RV enlargement, and accentuation of the normal sequence with larger R waves in V5 and V6 can be produced by LV enlargement. Loss of normal R wave progression from

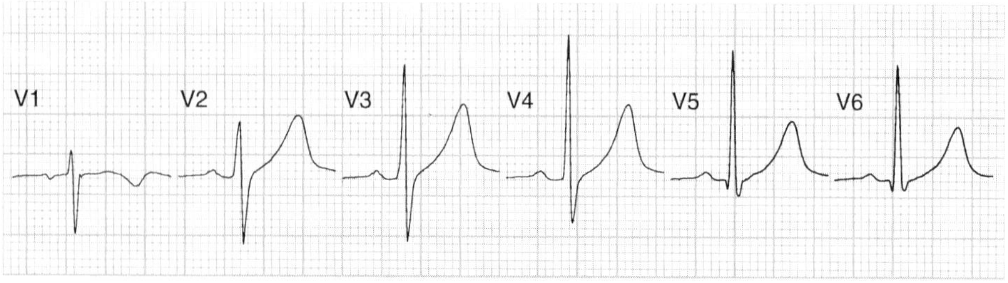

FIGURE 6-16 The typical panoramic display of the six precordial leads of the ECG, illustrating the normal progression and regression of R- and S-wave amplitudes.

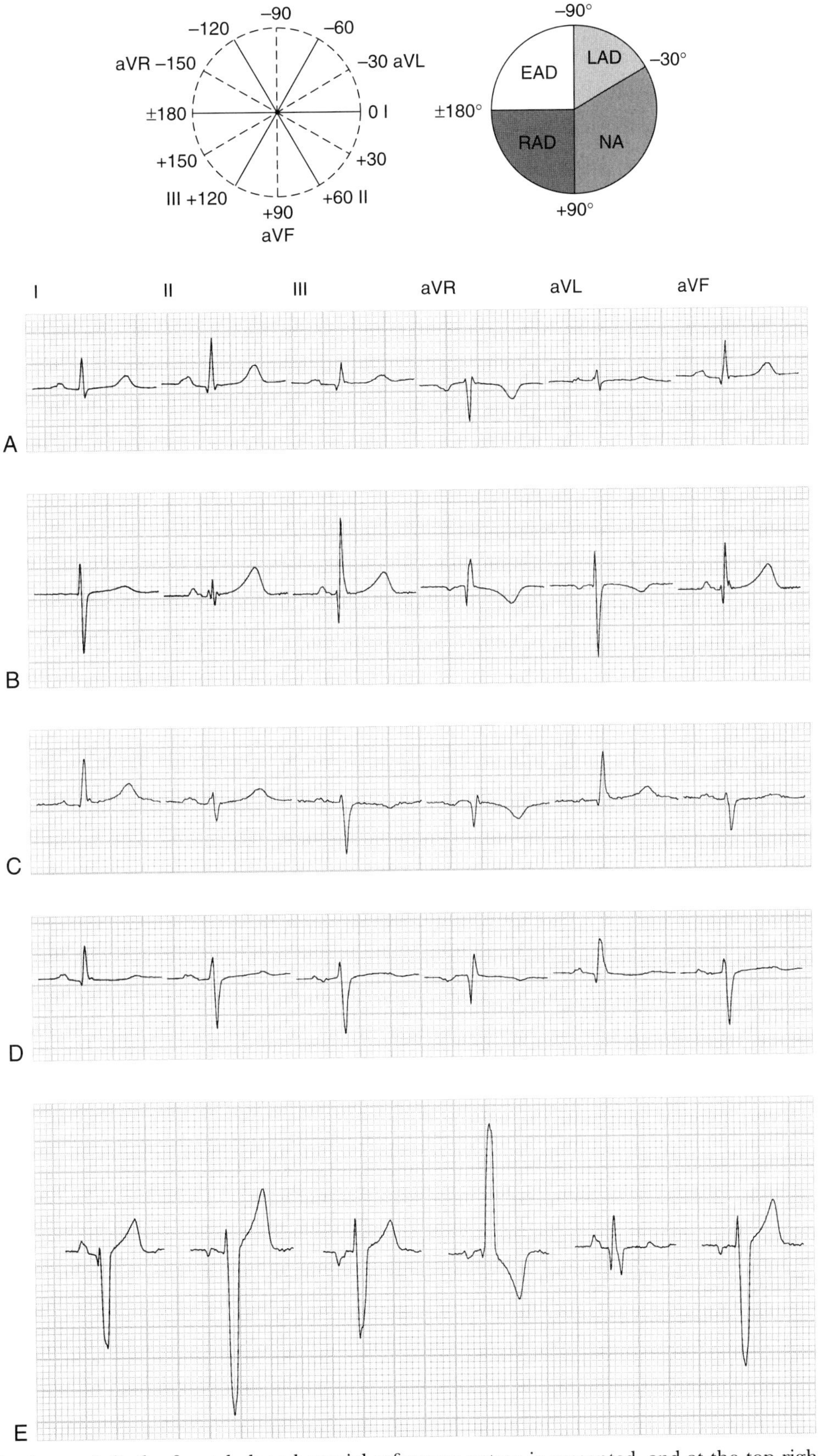

FIGURE 6-17 At the top left, the frontal-plane hexaxial reference system is presented, and at the top right the sectors indicating the various designations of the frontal-plane QRS axis in the adults are identified: normal axis (*NA*), right-axis deviation (*RAD*), left-axis deviation (*LAD*), and extreme axis deviation (*EAD*). At the bottom, examples of various frontal plane QRS axes are shown: **A,** +60 degrees; **B,** +150 degrees; **C,** −30 degrees; **D,** −60 degrees; and **E,** −120 degrees.

V1 to V5 may indicate loss of myocardium in the LV wall, due to myocardial infarction (MI).

S Waves. The S wave also has a normal sequence of progression in the precordial leads. It should be large in V1, larger in V2, and then progressively smaller from V3 through V6 (see Fig. 6-17). As with the R wave, alteration of this sequence could be produced by enlargement of one of the ventricles.

The duration of the QRS complex is termed the *QRS interval,* normally ranging from 0.07 to 0.10 seconds. It tends to be slightly longer in males than in females. The QRS interval is measured from the beginning of the first appearing Q or R wave to the end of the last appearing R, S, or R' wave. Multilead comparison is useful since either the beginning or end of the QRS complex may be isoelectric in any single lead, causing underestimation of QRS duration. The onset of the QRS complex is usually quite apparent in all leads, but its offset at the junction with the ST segment is often indistinct, particularly in the precordial leads. The QRS interval has no lower limit that indicates abnormality. QRS prolongation may be caused by LV enlargement, an abnormality in impulse conduction, or a ventricular site of origin of the QRS complex.

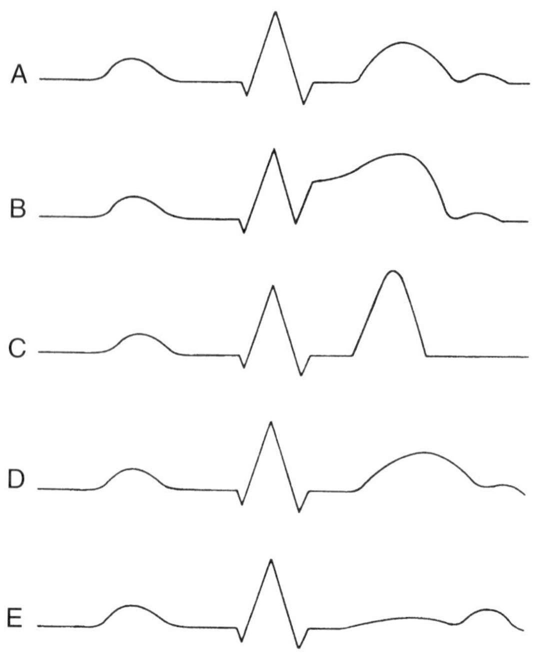

FIGURE 6-18 The normal QRS-to-T relationship **(A)** is contrasted with various abnormalities that may be associated with cardiac arrhythmias **(B-E)**. Each example begins with the completion of a TP segment and ends with the initiation of the following TP segment.

QRS AXIS DETERMINATION

When the typical ECG display is used, there is a three-step method for determining the frontal plane QRS direction (termed *axis*):

1. Identify the transitional lead, as defined by positive and negative components of the QRS complex of approximately equal amplitudes.
2. Identify the lead that is oriented perpendicular to the transitional lead by using the hexaxial reference system (Fig. 6-18).
3. Consider the predominant direction of the QRS complex in the lead identified in step 2. If the direction is positive, the axis is equal to the positive pole of that lead. If the direction is negative, the axis is equal to the negative pole of that lead.

The frontal plane axis is normally directed leftward and either slightly superiorly or inferiorly: between −30 degrees and +90 degrees (see Fig. 6-18). Therefore, the QRS complex is normally predominately positive in both leads I (with its positive pole at 0 degrees) and II (with its positive pole at +60 degrees). If the QRS is positive in lead I but negative in II, the axis is deviated leftward between −30 and −120 degrees. However, if the QRS is negative in I but positive in II, the axis is deviated rightward between +90 and +180 degrees. The axis is rarely directed entirely opposite to normal with predominately negative QRS orientation in both leads I and II.

The frontal plane axis is typically rounded to the nearest multiple of 15 degrees. If it is directly aligned with one of the limb leads, the axis is designated as −30 degrees, 0 degrees, +30 degrees, +60 degrees, and so on. If it is located midway between two of the limb leads, it is designated as −15 degrees, +15 degrees, +45 degrees, +75 degrees, and so on. Examples of patients with various frontal plane QRS axes are presented in panels *A* through *E.*

The normal frontal plane QRS axis is rightward in the neonate, moves to a vertical position during childhood, and then moves to a more horizontal position during adulthood. In normal adults the electrical axis is almost parallel to the anatomic base to apex axis of the heart, in the direction of lead II. However, the axis is more vertical in thin individuals and more horizontal in heavy individuals. A QRS axis more positive than +90 degrees in an adult should be designated *right axis deviation (RAD) (panel B)*; and an axis more negative than −30 degrees at any age should be *designated left axis deviation (LAD) (panel C).* RV hypertrophy may produce RAD and LV hypertrophy LAD. An axis between −90 and +180 degrees should be considered extreme axis deviation (EAD) without designating it as either rightward or leftward *(panel E).*

ST SEGMENT MORPHOLOGY

The ST segment represents the period of time when the ventricular myocardium remains in an activated or depolarized state. At its junction with the QRS (J point), it typically forms a nearly 90-degree angle and then proceeds horizontally until it curves gently into the T wave. The length of the ST segment is influenced by factors that alter the duration of ventricular activation. Points along the ST segment are designated with reference to the number of milliseconds beyond the J point, such as "J + 20," "J + 40," "J + 60."

The first section of the ST segment is normally located at the same horizontal level as the baseline formed by the PR segment and the TP segment that fills in the space between electrical cardiac cycles. Slight up-sloping, down-sloping, or horizontal depression of the ST segment may occur as a normal variant. Another normal variant appears when there is early repolarization in epicardial areas within the ventricles.[22] This causes displacement of the ST segment in the direction of the following T wave. Occasionally, there may be as much as a 4-mm ST elevation in leads V1 to V3 in normal young males.[23] The appearance of the ST segment may also be altered during exercise or when there is an altered sequence of activation of the ventricular myocardium.

T WAVE MORPHOLOGY

The smooth, rounded shape of the T wave resembles that of the P wave. However, there is greater normal variation of monophasic versus diphasic appearance in the various leads. The initial deflection of the T wave is typically longer than the terminal deflection, producing a slightly asymmetrical shape. Slight "peaking" of the T wave may occur as a normal variant, and notching of the T waves is common in children. The duration of the T wave itself is not usually measured but is instead included in the Q–T interval discussed later. The amplitude of the T wave, like that of the QRS complex, has wide normal limits. It tends to diminish with age and is larger in males than in females. T wave amplitude tends to vary with QRS amplitude and should always be greater than that of any U wave that is present. T waves do not normally exceed 5 mm in any limb lead or 10 mm in any precordial lead. The T wave amplitude tends to be lower in the leads providing the extreme views of both the frontal and transverse planes: T waves do not normally exceed 3 mm in leads aVL and III, or 5 mm in leads V1 and V6.[24]

The direction of the T wave should be evaluated in relation to that of the QRS complex. The rationale for similar directions of these waveforms that represent the opposite myocardial events of activation and recovery has been presented earlier in "Basic Principles." The method presented earlier for determining the direction of the QRS complex in the frontal plane should be applied for determining the direction of the T wave. The term *QRS-T angle* is used to indicate the number of degrees between the QRS complex and T wave axes in the frontal plane.[25] A similar method can be applied in the transverse plane.

U WAVE MORPHOLOGY

The U wave is normally either absent or present as a small, rounded wave following the T wave. It is normally in the same directions as the T wave, but approximately 10% of its amplitude. It is usually most prominent in leads V2 or V3.

Q–Tc INTERVAL

The Q–T interval measures the duration of activation and recovery of the ventricular myocardium. It varies inversely with the heart rate. To assure that there is complete recovery from one cardiac cycle before the following cycle begins, the duration of recovery must decrease as the rate of activation increases. Therefore, the "normality" of the Q–T interval can be determined only by correcting for the heart rate. The corrected Q–T interval (Q–Tc interval) rather than the measured Q–T interval is included in the ECG analysis. Bazett developed a formula for performing this correction,[20] which has since been modified by Hodges and coworkers[21] and MacFarlane and Veitch Lawrie.[22]

$$Q\text{–}Tc = Q\text{–}T + 1.75\ (\text{ventricular rate} - 60)$$

The normal value of Q–Tc is approximately 0.41 seconds. The Q–Tc is slightly longer in females than in males and increases slightly with age. The accommodations of the duration of electrical recovery to the rate of electrical activation does not occur immediately but requires several cardiac cycles. Thus, an accurate Q–Tc can be calculated only after a series of regular, equal cardiac cycles.

The diagnostic value of the Q–Tc interval is seriously limited by the difficulty of identifying the completion of ventricular recovery.

1. There is commonly a variation in the Q–T interval among the various leads. This occurs when the terminal portion of the T wave is isoelectric in some of the leads.[23] The longest Q–T interval measured in multiple leads should, therefore, be considered the true Q–T interval.
2. The U wave may merge with the T wave, creating a TU junction, which is not on the baseline. In this instance, the onset of the U wave should be considered the approximate end of the Q–T interval.
3. At faster heart rates, the P wave may merge with the T wave, creating a TP junction, which is not on the baseline. In this instance, the onset of the P wave should be considered the approximate end of the Q–T interval.

Marked elevation of the ST segment, increase or decrease in T wave amplitude, prolongation of the Q–Tc interval, or increase in U wave amplitude maybe indications of underlying cardiac conditions that may produce serious cardiac rhythm abnormalities (Fig. 6-19).

Chamber Enlargement

RIGHT VENTRICULAR DILATION

The right ventricle dilates either during compensation for a volume overload or after its hypertrophy eventually fails to compensate for a pressure overload. This dilation causes stretching of the right bundle branch, which courses from base to apex on the endocardial surface of the right side of the interventricular septum as illustrated in Figure 6-8. Conduction of impulses within these right bundle Purkinje fibers is slowed so much that electrical activation arrives at the RV myocardium only after it has already been activated by spread of impulses from the left ventricle.

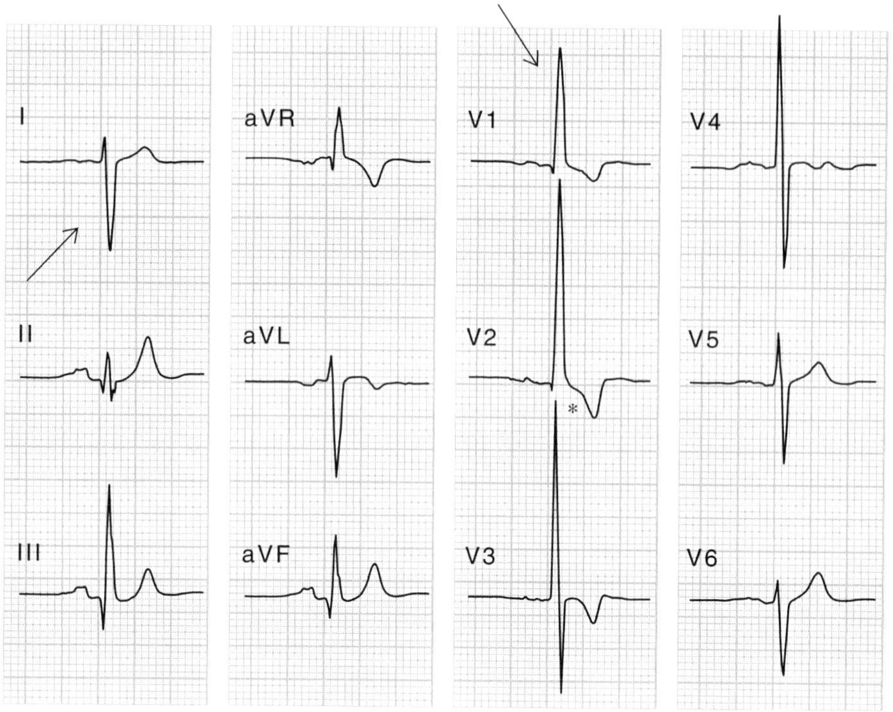

FIGURE 6-19 An 18-year-old man with congenital heart disease and pulmonary hypertension. *Arrows* indicate the changes of right ventricular hypertrophy in the QRS waveforms, and an *asterisk* indicates the ST- and T-wave changes of right ventricular strain.

This phenomenon referred to as *right bundle branch block* (RBBB) is discussed later in "Intraventricular Conduction Abnormalities." This RV conduction abnormality may appear suddenly during the early or compensatory phase of a volume overload or during the advanced or failing phase of a pressure overload.

RIGHT VENTRICULAR HYPERTROPHY

The right ventricle hypertrophies because of compensation for pressure overload. In the neonate, the right ventricle is more hypertrophied than the left, because there is greater resistance in the pulmonary circulation than in the systemic circulation during fetal development. Right-sided resistance is greatly increased when the placenta is removed.[26] From this time onward, the ECG evidence of RV predominance is gradually lost as the left ventricle becomes hypertrophied in relation to the right. Therefore, hypertrophy, like dilation, may be a compensatory rather than a pathologic condition.[27] A pressure overload of the right ventricle may recur in later years because of increased resistance to the flow of blood through either the pulmonary valve, the pulmonary circulation, or the left side of the heart.

The normal QRS complex in the adult is predominately negative in lead V1 with a small R wave followed by a prominent S wave. When the right ventricle hypertrophies in response to a pressure overload, this negative predominance may be lost. In milder forms, a late positive R′ wave appears. With moderate hypertrophy, the initial QRS forces move anteriorly (increased lead V1 R wave), and the terminal QRS forces move rightward (increased lead I S wave). With marked hypertrophy, the QRS complex may even become predominately positive (Fig. 6-20). This severe pressure overload causes sustained delayed repolarization of the RV myocardium, producing negativity of the ST segment and the T wave, which has been termed RV strain.

LEFT VENTRICULAR DILATION

The left ventricle dilates for the same reasons indicated previously for the right ventricle. However, the dilation does not stretch the left bundle enough to cause complete *left bundle branch block* (LBBB). This is most likely due to differences between the anatomy of the right and left bundles. The right bundle continues as a single bundle along its septal surface, but the left bundle divides almost immediately into multiple fascicles. LV dilation may produce only a partial or incomplete LBBB.

Dilation enlarges the surface area of the left ventricle and moves the myocardium closer to the precordial electrodes, which increases the amplitudes of leftward and posteriorly directed QRS waveforms.[28] The S wave amplitude is increased in leads V2 and V3, and the R wave amplitude is increased in leads V5 and V6.

LEFT VENTRICULAR HYPERTROPHY

As discussed earlier, the left ventricle normally becomes hypertrophied relative to the right ventricle following the neonatal period. Abnormal hypertrophy, which occurs in response to a pressure overload, produces exaggeration of the normal pattern of LV predominance on the ECG. Like dilation, hypertrophy enlarges the surface area of the left ventricle, which increases the voltages of leftward and posteriorly directed QRS waveforms, thereby causing similar shifts in the frontal plane axis and transverse plane transitional zone.

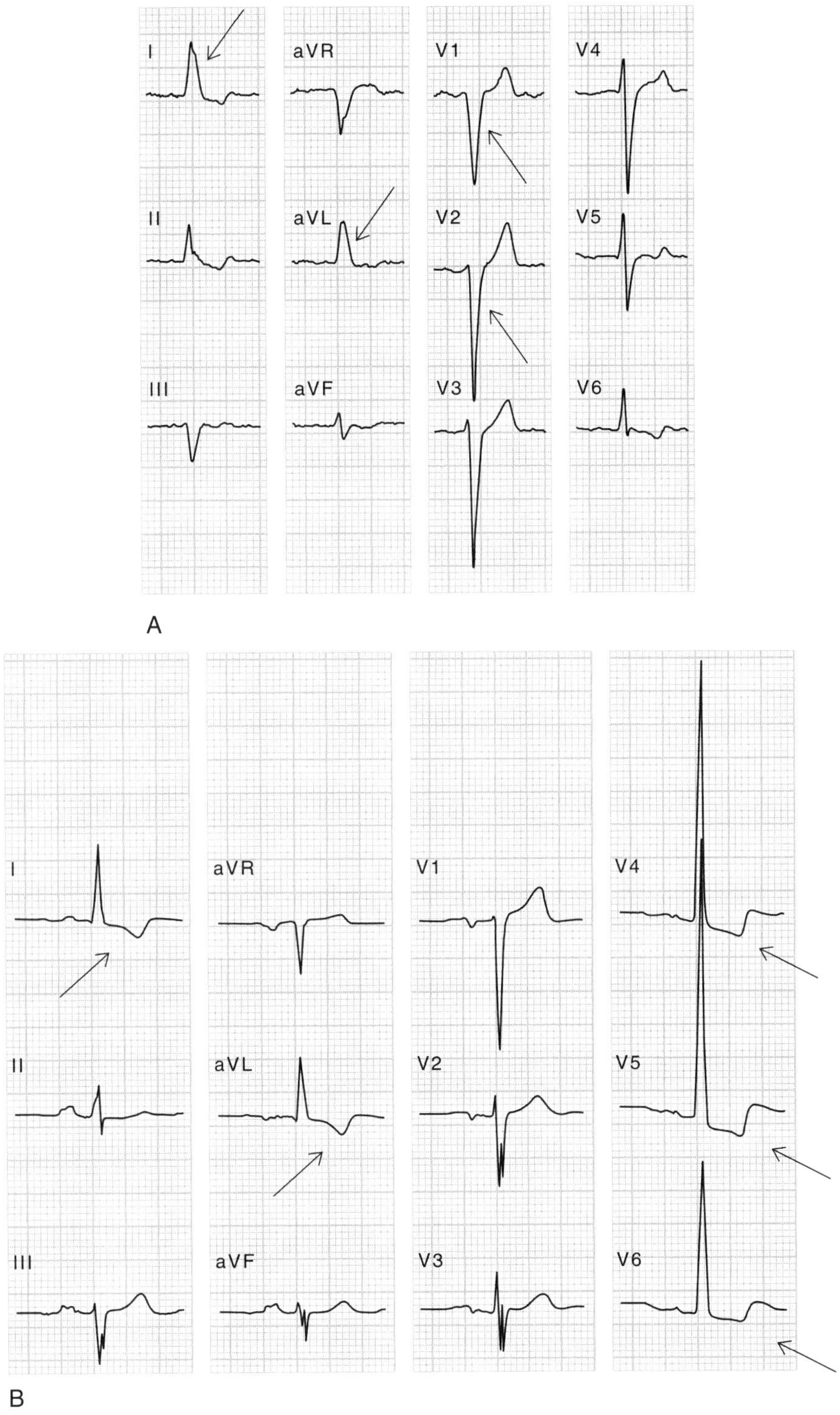

FIGURE 6-20 Twelve-lead ECGs from a 75-year-old woman with symptoms of heart failure caused by longstanding hypertension (**A**) and a 70-year-old man with severe aortic valve stenosis just before surgical replacement (**B**). *Arrows* indicate the intraventricular conduction delay in **A** and the ST-segment depression and T-wave inversion in **B**.

A longer time is required for spread of electrical activation from the endocardial to epicardial surface, prolonging both the time to peak R wave (intrinsicoid deflection) and the overall QRS duration. These conduction delays, induced by hypertrophy, may mimic incomplete or even complete LBBB (see "Intraventricular Conduction Abnormalities" later) (Fig. 6-21).

Pressure overload leads to sustained delayed repolarization of the left ventricle, which produces negativity of both the ST segment and the T wave in leads with leftward or posterior orientation. This is referred to as *LV strain.*[29] The epicardial cells no longer repolarize early, causing the spread of recovery to proceed from endocardium to epicardium. This leads to deflection of the T wave in the opposite direction of the QRS complex. The mechanism that produces the strain is uncertain, but there are several factors believed to contribute. The development of strain correlates well with increasing *LV* mass as determined by echocardiography.[30] Myocardial ischemia and slowing of intraventricular conduction are factors that may also contribute to strain.

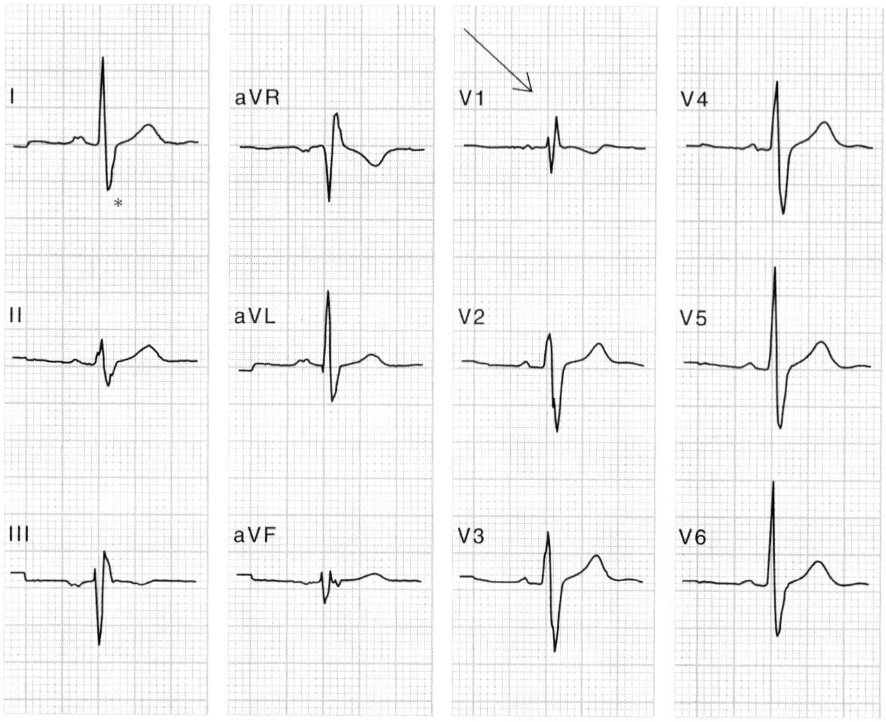

FIGURE 6-21 Twelve-lead ECGs from a 17-year-old girl with an ostium secundum atrial septal defect. *Arrows* indicate the prominent terminal R′ wave in V1, and *asterisks* indicate the rightward and leftward axis shifts, respectively.

Intraventricular Conduction Abnormalities

BUNDLE BRANCH AND FASCICULAR BLOCK

Since the activation of the ventricular Purkinje system (see Fig. 6-8) is not represented on the surface ECG, abnormalities of its conduction must be detected indirectly by their effects on myocardial activation and recovery. The most specific changes occur within the QRS complex. A conduction disturbance within the right bundle branch, left bundle branch, or left bundle fascicles or between the Purkinje fibers and the adjacent myocardium may alter the QRS complex and T wave. A conduction disturbance in the common or His bundle has a similar effect on activation of both ventricles and, therefore, does not alter the appearance of the QRS complex or T wave.

UNIFASCICULAR BLOCKS

This term is used when there is ECG evidence of blockage of only one of the fascicles. Isolated RBBB or left anterior fascicular block (LAFB) commonly occurs, while left posterior fascicular block (LPFB) is rare. Rosenbaum et al. identified only 30 patients with LPFB as compared with 900 patients with LAFB.[31]

Right Bundle Branch Block

Since the right ventricle contributes minimally to the normal QRS complex, RBBB produces little distortion during the time required for LV activation. Figure 6-12*A* illustrates the minimal distortion of the early portion and marked distortion of the late portion of the QRS complex that typically occur with RBBB. The minimal contribution of the normal RV myocardium is completely subtracted from the early portion of the QRS complex and then added later when the right ventricle is activated via the spread of impulses from the left ventricle. This produces a late prominent positive wave in lead V1 termed *R′*, because it follows the earlier positive R wave produced by normal left to right spread of activation through the interventricular septum (Fig. 6-22).

Left Anterior-Superior Fascicular Block

If the LA fascicles of the LBB are blocked, the initial activation of the LV free wall occurs via the LP fascicles (Fig. 6-23). Activation spreading from endocardium to epicardium in this region is directed inferiorly and rightward. Since the block in the LA fascicles has removed the competition from activation directed superiorly and leftward, Q waves appear in leads with their positive electrode on the left arm (leads I and aVL). Following this initial period, the activation wave spreads over the remainder of the LV free wall in a superior and leftward direction. This produces prominent R waves in leads I and aVL and prominent S waves in leads II, III, and aVF, causing a leftward shift of the QRS axis to at least −45 degrees. The overall QRS duration is prolonged by 0.10 to 0.20 seconds.[32]

LAFB is by far the most commonly occurring conduction abnormality involving the LBB. Its presence was detected in 1.5% of a population of 8000 men 45 to 69 years of age.[33]

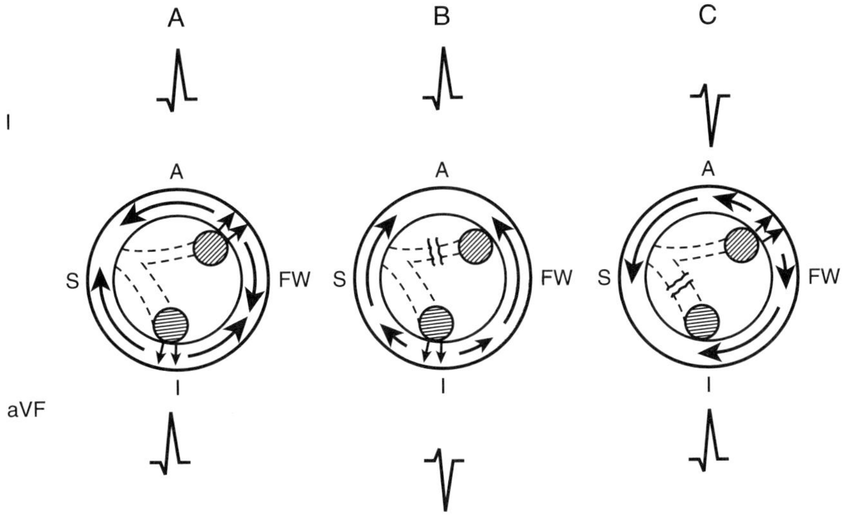

FIGURE 6-22 Schematic left ventricle viewed from its apex upward toward its base. The interventricular septum (*S*), left-ventricular free wall (*FW*), and anterior (*A*) and inferior (*I*) regions of the left ventricle are indicated. The typical appearances of the QRS complexes in leads 1 (*top*) and aVF (*bottom*) are presented for normal (**A** *at top left*), left anterior fascicular block (**B**), and left posterior fascicular block left-ventricular activation (**C**). *Dashed lines* within the inner circles represent the fascicles; the two *wavy lines* crossing a fascicle indicate the sites of block. *Small crosshatched circles* represent the papillary muscles; *outer rings* represent the endocardial and epicardial surface of the left ventricular myocardium. *Arrows* within the outer rings indicate the directions of the wavefronts of activation as they spread from the unblocked fascicles through the myocardium.

Left Posterior-Inferior Fascicular Block

If the LP fascicles of the LBB are blocked, the situation is reversed (see Fig. 6-23). The initial LV free wall activation occurs via the LA fascicles. Activation spreading from endocardium to epicardium in this region is directed superiorly and leftward. Since the block in the LP fascicles has removed the competition provided by activation directed inferiorly and rightward, Q waves appear in leads with their positive electrode on the left leg (leads II, III, and aVF). Following this initial period, the activation wave spreads over the remainder of the LV free wall in an inferior and rightward direction. This produces prominent R waves in leads II, III, and aVF and prominent S waves in leads I and aVL, causing a rightward shift of the QRS axis to at least +90 degrees.[33] The QRS duration is slightly prolonged as in LAFB. The diagnosis of LPFB requires that there be no evidence of RV hypertrophy (RVH), because RVH itself can produce the same ECG pattern as LPFB.

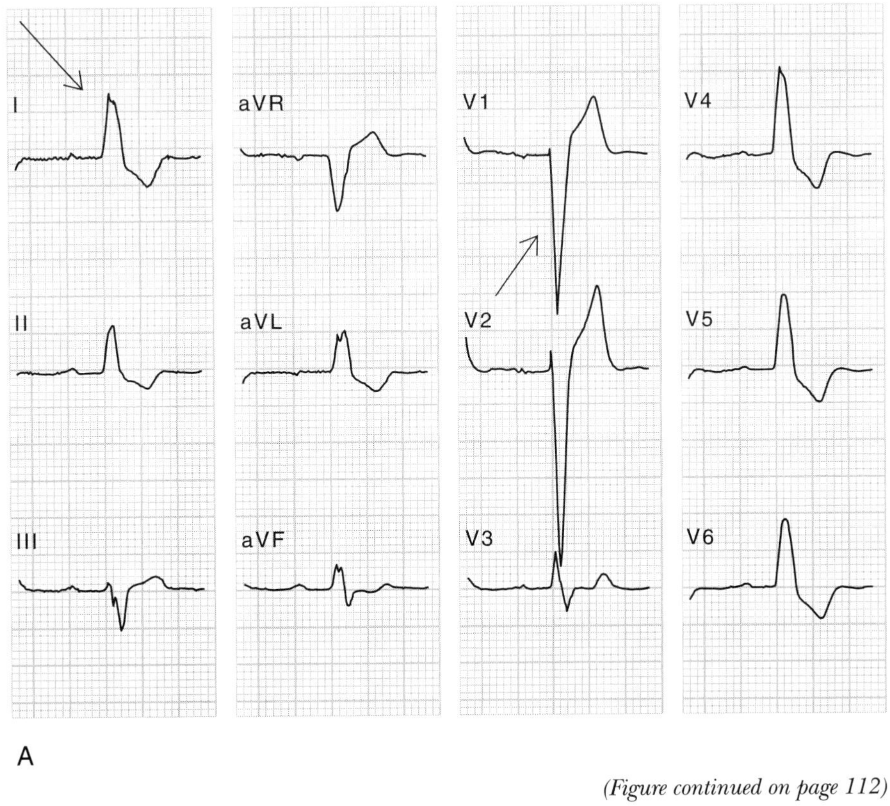

FIGURE 6-23 Twelve-lead ECGs from an 82-year-old woman with no medical problems (**A**), a 71-year-old man with chronic heart failure (**B**), and a 74-year-old man with a long history of hypertension (**C**). *Arrows* in **A** and **C** indicate the typical characteristics of left bundle branch block in leads I and V1, and *arrows* in **B** indicate the deep S waves in leads II, III, and aVF and decreased R waves in leads V2 to V4.

(Figure continued on page 112)

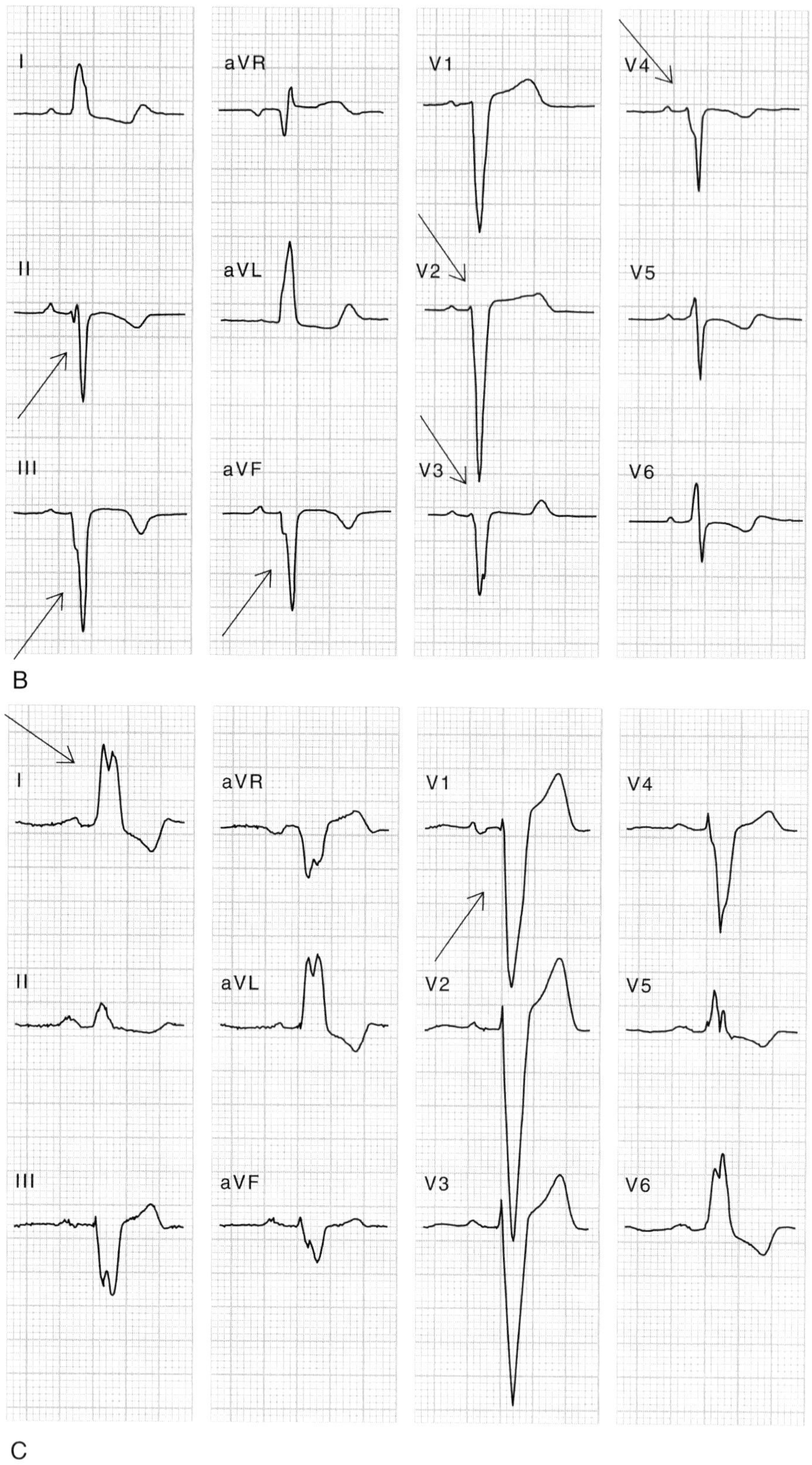

FIGURE 6-23 cont'd.

BIFASCICULAR BLOCKS

This term is used when there is ECG evidence of involvement of two of the major Purkinje fascicles. Such evidence may appear at different times or may coexist on the same ECG. The term bifascicular block is sometimes applied to complete LBBB but is more commonly applied to the combination of RBBB with either LAFB or LPFB. The term *bilateral bundle branch block* is also appropriate when RBBB and either LAFB or LPFB are present.[34] When there is bifascicular block, the QRS duration is prolonged to at least 0.12 seconds.

Left Bundle Branch Block

Figure 6-24 illustrates the marked distortion of the entire QRS complex produced by LBBB. Complete LBBB may be caused either by disease in the main LBB (predivisional) or in all of its fascicles (postdivisional). When the impulse cannot progress along the LBB, it must first enter the right ventricle and then travel through the interventricular septum to the left ventricle.

Normally, the interventricular septum is activated from left to right, producing an initial R wave in the right precordial leads and a Q wave in leads I, aVL, and the left precordial leads. When complete LBBB is present, the septum is activated from right to left. This produces initial Q waves in the right precordial leads and eliminates the normal Q waves in the leftward oriented leads.[35] The activation of the left ventricle then proceeds sequentially from the interventricular septum to the adjacent anterior-superior and inferior walls to the posterior-lateral free wall. This sequence of ventricular activation in complete LBBB tends to produce monophasic QRS complexes: QS in lead V1 and R in leads I, aVL, and V6.

Right Bundle Branch Block with Left Anterior-Superior Fascicular Block (RBBB + LAFB)

Just as LAFB appears as a unifascicular block much more commonly than LPFB, it more commonly accompanies RBBB as a bifascicular block. The diagnosis is made by observing the late prominent R or R′ wave in precordial lead V1 of RBBB and the initial R waves and the prominent S waves in limb leads II, III, and aVF of LAFB. The QRS duration should be at least 0.12 seconds, and the frontal plane axis should be between −45 and −120 degrees (Fig. 6-25).[33]

Right Bundle Branch Block with Left Posterior-Inferior Fascicular Block (RBBB + LPFB)

This example of bifascicular block rarely occurs. Even when the ECG changes are entirely typical, the diagnosis should be made only if there is no clinical evidence of RVH. The diagnosis of RBBB with LPFB should be considered when there are typical changes in precordial lead V1 of RBBB and the initial R waves and prominent S waves in limb leads I and aVL of LPFB. The QRS duration should be at least 0.12 seconds and the frontal plane axis at least +90 degrees (Fig. 6-26).[36,37]

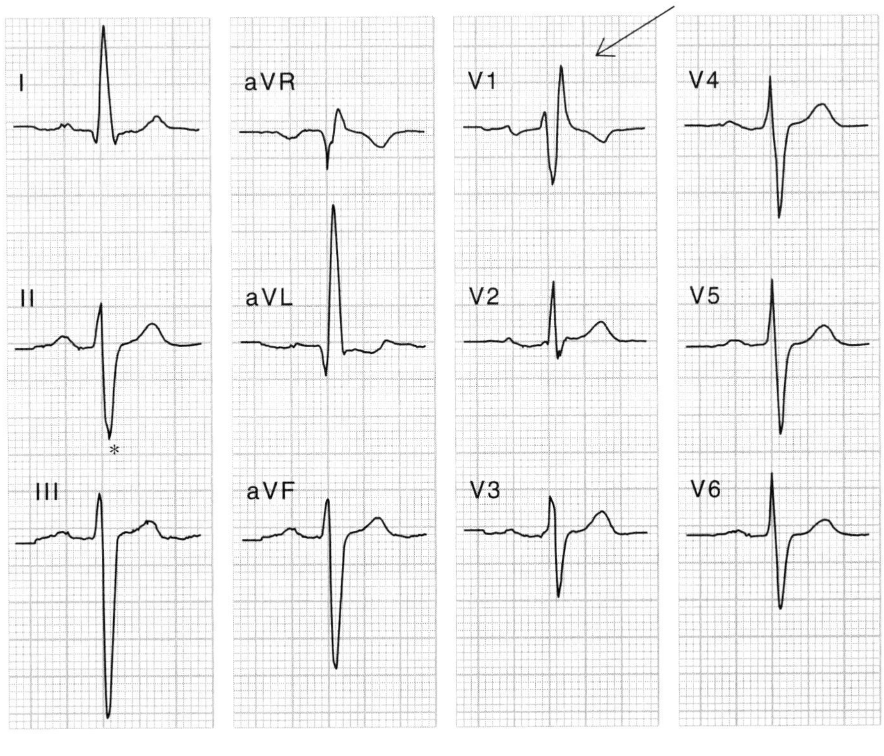

FIGURE 6-24 An 82-year-old man with fibrosis of both the right bundle branch and the anterior fascicle of the left bundle branch. *Arrows* indicate the prominent terminal R′ wave in V1, and *asterisks* indicate the rightward and leftward axis shifts, respectively.

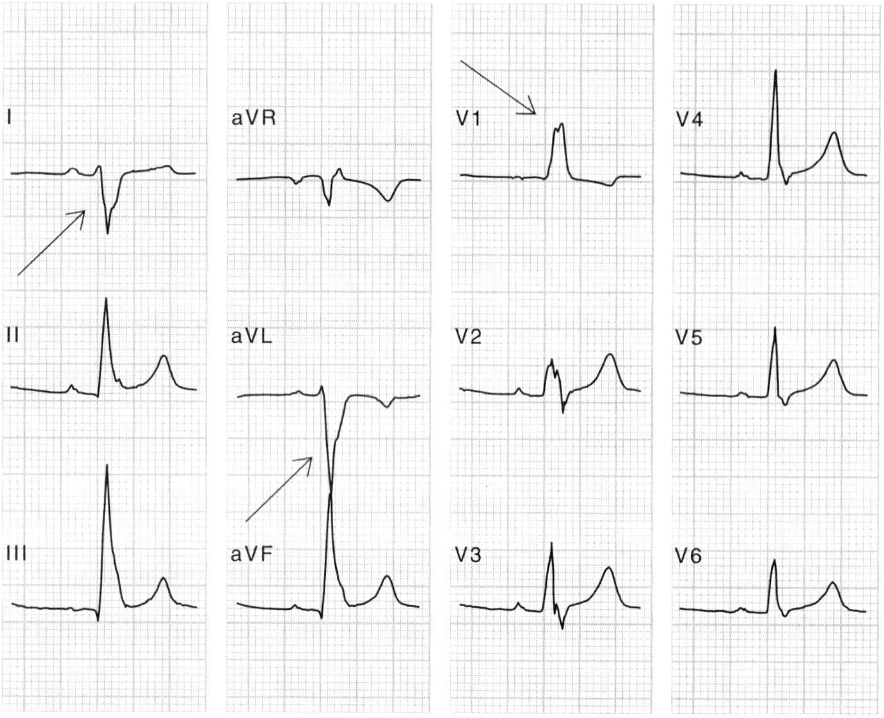

FIGURE 6-25 Twelve-lead ECGs from an 82-year-old woman with no complaints and no other evidence of heart disease. *Arrows* indicate the prominent S waves in I and aVL and RR′ complex in V1.

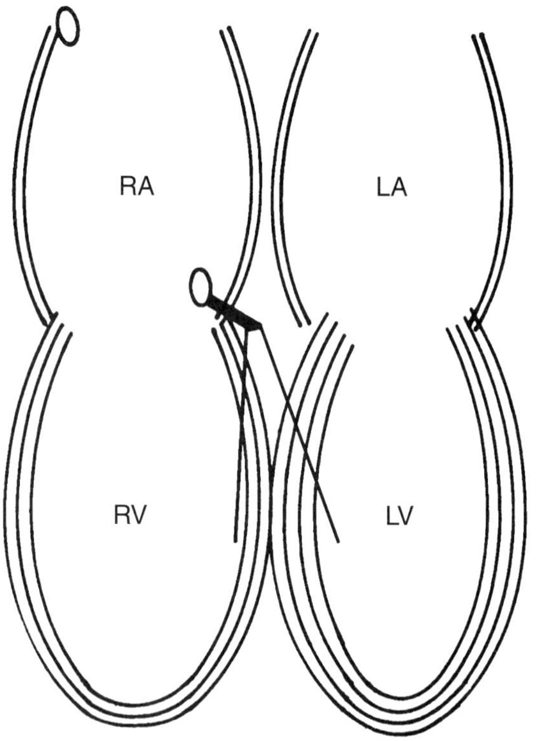

FIGURE 6-26 Schematic comparison of the relative thicknesses of the myocardium in the four cardiac chambers. The *ovals* indicate the locations of the sinoatrial (SA) and atrioventricular (AV) nodes; the His bundle (*thick short line*) and right and left bundle branches (*thin longer lines*) descend from the AV node into the interventricular septum. *LA*, left atrium; *LV*, left ventricle; *RA*, right atrium; *RV*, right ventricle. (Modified from Wagner GS, Waugh RA, Ramo BW: Cardiac Arrhythmias. New York, Churchill Livingstone, 1983, p 2.)

Acute and Chronic Ischemic Heart Disease

GENERAL ELECTROPHYSIOLOGICAL PRINCIPLES

The process of electrical recovery of myocardial cells is more susceptible to ischemia than is that of electrical activation. Since ischemia caused by an increase in myocardial demand is not as profound as that caused by a complete cessation of coronary blood flow, it is manifested on the ECG only by changes in the waveforms representing the recovery process—the ST segments and T waves. The more profound ischemia that occurs with acute coronary occlusion produces a different array of changes in the ST segments and T waves, and may even alter the QRS complexes. When blood flow is not rapidly restored, the more permanent changes in the QRS waveforms of MI sequentially evolve.[38,39]

The ECG changes caused by a potentially reversible increase in myocardial metabolic demand or decrease in coronary blood flow are typically termed *ischemia* when the direction of the T wave is altered, and *injury* when the level of the ST segment is deviated from those of the TP and PR segments of the baseline. However, the term *ischemia* is used more generally in this chapter in reference to the condition of a pathologic imbalance between supply and demand.

The thicker-walled left ventricle is much more susceptible to ischemia than is the right ventricle, and the endocardial linings of both are supplied by the cavitary blood. The subendocardial layer of LV cells is that most likely to become ischemic, because it

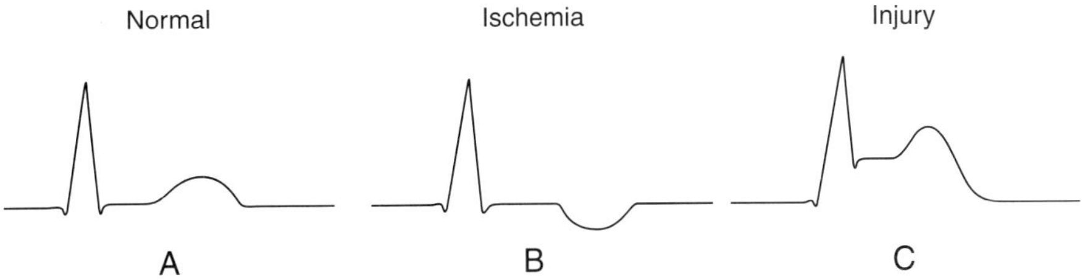

FIGURE 6-27 Schematic single ventricular cycles from an ECG lead oriented to the cardiac long axis are shown for normal (**A**), ischemic (**B**), and injury conditions (**C**).

is located "at the end of the supply line" (Fig. 6-27). Ischemia occurs when, in the presence of underlying coronary atherosclerosis, there is either an increase in myocardial demand or a decrease in coronary blood flow.[40,41]

The typical ECG manifestation of ischemia caused by an increase in myocardial demand is deviation of the ST segment away from the involved left ventricle. The T wave may or may not be altered in the same direction as the ST segment. Because of the location of the involved layer of the myocardium and the characteristic deviation of the ST segment of the ECG baseline, the term used is *subendocardial ischemia (SEI).*

The typical ECG manifestation of ischemia caused by acute occlusion of myocardial blood flow is deviation of the ST segment toward the specifically involved area. Again, the T wave may or may not be altered in the same direction as the ST segment. Because of the involvement of all layers of the myocardium (including the epicardium), the term that most accurately describes the ECG changes is *epicardial injury (EI).* Note that the term *injury* replaces *ischemia,* because the presence of these ECG changes typically indicates at least a portion of the involved myocardium is being irreversibly changed—infarcted. The process of infarction deviates both the QRS complex and T wave away from the involved area.[42] Unless poor myocardial remodeling results in wall thinning or even aneurysmal dilatation, the ST segment soon returns to a position isoelectric with the remainder of the ECG baseline.

MYOCARDIAL ISCHEMIA

Normally, the directions of the QRS complex and T wave are similar rather than opposite because of prolonged maintenance of the activated condition in the endocardial layer of myocardium. The ischemic subendocardial cells are unable to maintain prolonged activation, thereby causing the T wave on the ECG to become "inverted" in relation to the QRS complexes (Fig. 6-28). There is normally an angle of less than 45 degrees between the directions of the QRS complexes and the T waves in the frontal plane and less than 60 degrees in the transverse plane. When the angles exceed these limits, in the absence of other abnormal conditions such as ventricular hypertrophy or bundle branch block, the presence of myocardial ischemia should be considered. The location of the ECG leads demonstrating these inverted T waves may be indicative of the specific location of the ischemic area within the

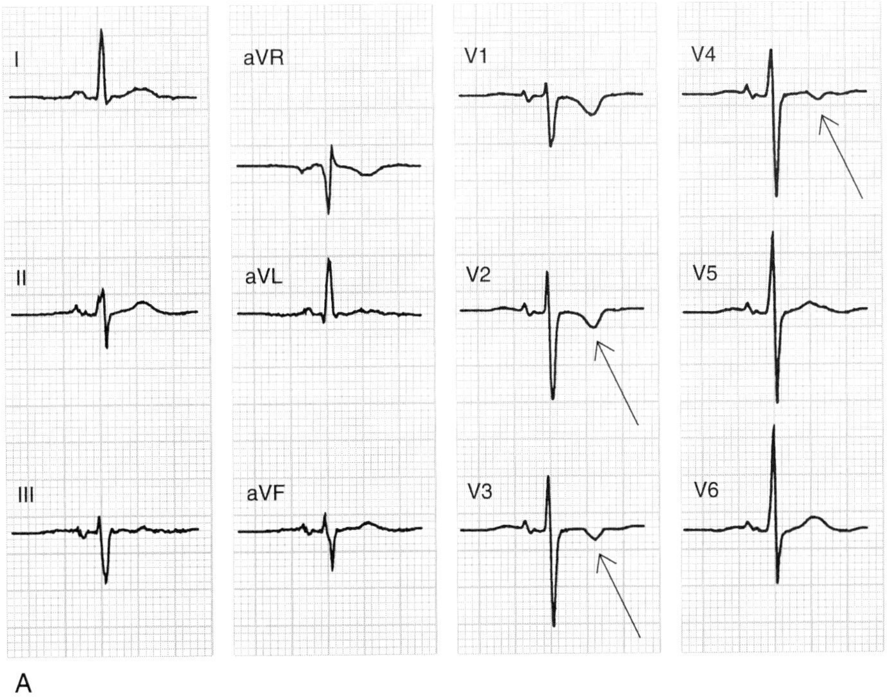

FIGURE 6-28 Twelve-lead ECGs from a 63-year-old woman presenting to the emergency department with 2 hours of substernal chest pain (**A**), a 78-year-old man with an occluded right coronary artery vein graft 3 days after coronary bypass surgery (**B**), and an 83-year-old man with a previous anterior infarct and recurrent resting chest pain following abdominal surgery (**C**). Coronary angiography revealed high-grade stenosis of the proximal left circumflex artery. *Arrows* in **A, B,** and **C** indicate abnormally directed T waves.

(Figure continued on page 116)

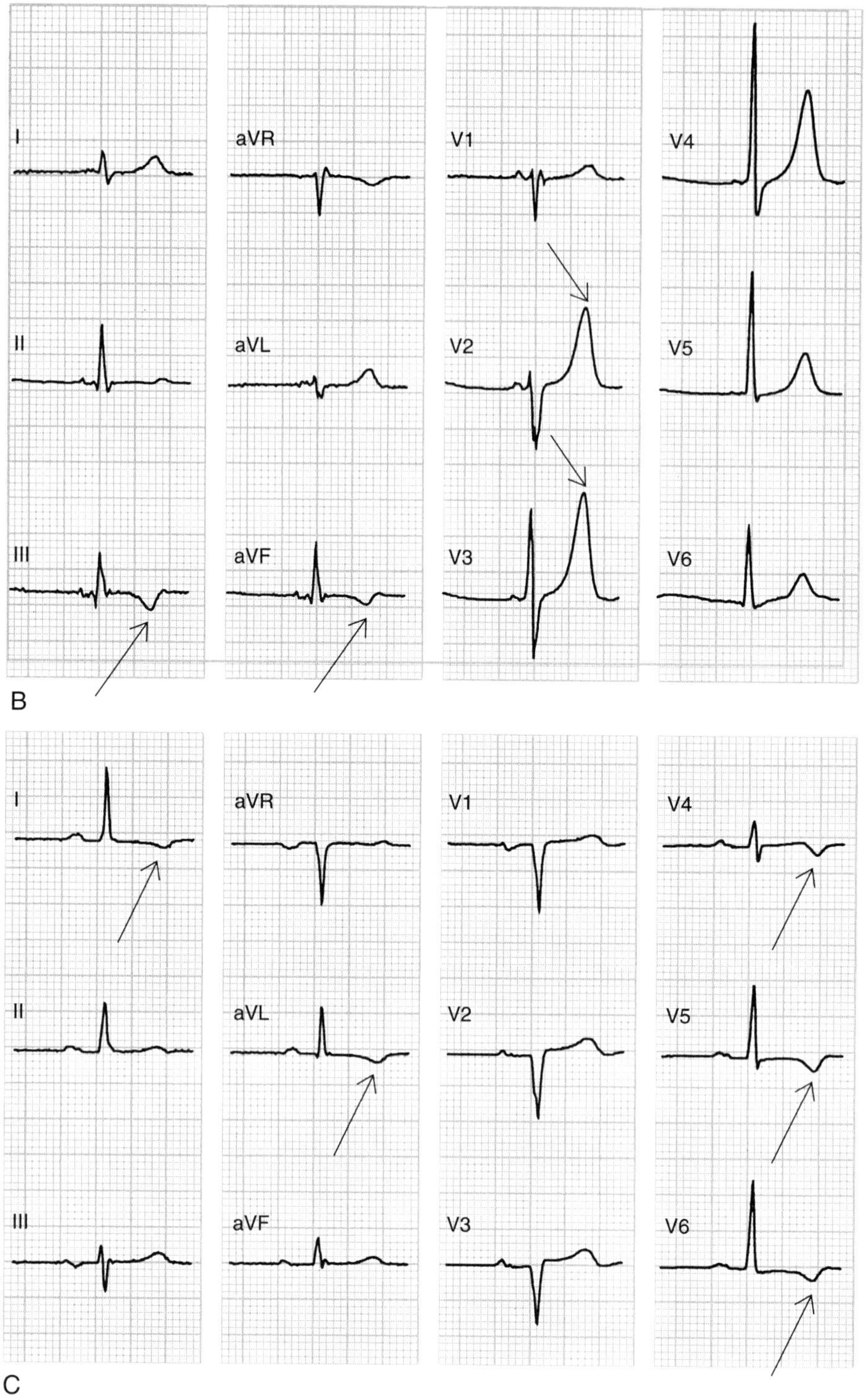

FIGURE 6-28 cont'd.

LV myocardium. The lead groups that typically localize the ischemia in the distributions of the three major coronary arteries are indicated in Figure 6-29.

T wave changes are not as specific as are the ST segment changes discussed in the following section for establishing the diagnosis of myocardial ischemia due to increased demand. Inversion from the direction of the QRS commonly occurs as a normal variant or with other cardiac or noncardiac conditions. T wave inversion is also not a sensitive sign of LV ischemia. The typical ST segment changes of SEI occur in both the presence and absence of T wave inversion during periods of increased metabolic demand (Fig. 6-30).[43]

Like ST segment depression, T wave inversion usually resolves when the increased LV workload is removed. Unlike ST depression, however, T wave inversion is

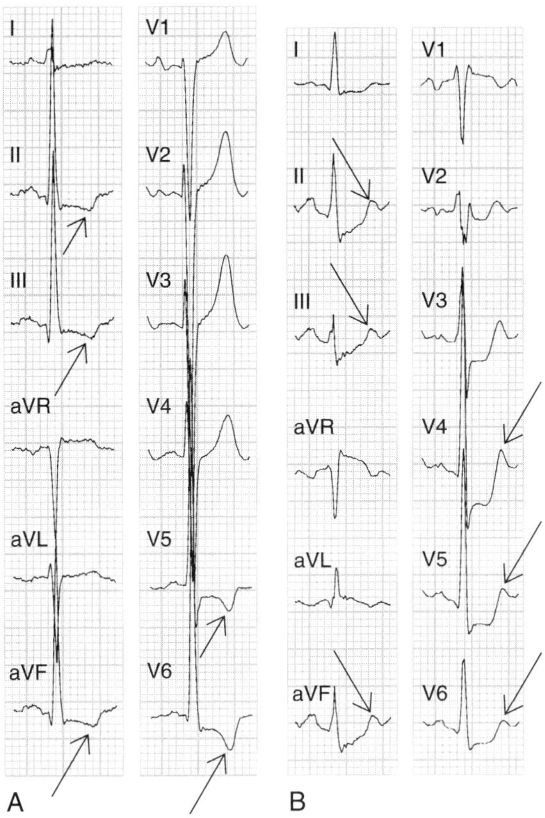

FIGURE 6-29 Resting and exercise 12-lead ECGs from a 47-year-old man (**A**) and a 72-year-old man with sustained chest pain during exercise stress testing (**B**). *Arrows* indicate negative T waves in **A** and positive T waves in **B** following the diagnostically depressed ST segments.

typically present for a prolonged period following the acute phase of MI. This chronic T wave inversion should not be considered evidence for persistent ischemia. It represents an alteration in electrical recovery secondary to the infarction-induced changes in electrical activation in the same manner that T wave inversion is an expected secondary occurrence with LBBB.

SUBENDOCARDIAL INJURY

Normally, the ST segment is located at the same level as the remainder of the ECG baseline: isoelectric with the PR and TP segments. Observation of the changes in the appearance of the ST segment of a patient with a positive exercise stress test provides the pattern of the ECG changes termed *subendocardial ischemia (SEI).*[44]

When partial obstruction of a coronary artery prevents the blood flow from increasing sufficiently during a time of increased metabolic demand, a "current of injury" is produced by the electrophysiological imbalance between the involved subendocardial layer and the noninvolved mid and epicardial layers of the LV myocardium.[45,46] The ST segment changes typically disappear when the myocardial demands are returned to baseline by reducing the metabolic demand, indicating that the myocardial cells have been only reversibly "ischemic." Such changes may occur during psychologically as well as physiologically induced episodes of increased myocardial metabolic demand.

A combination of two diagnostic criteria have been typically required for the diagnosis of SEI:

1. At least 0.10 mV depression at the J point of the ST segment
2. Either a horizontal or downward sloping of the ST segment toward the T wave

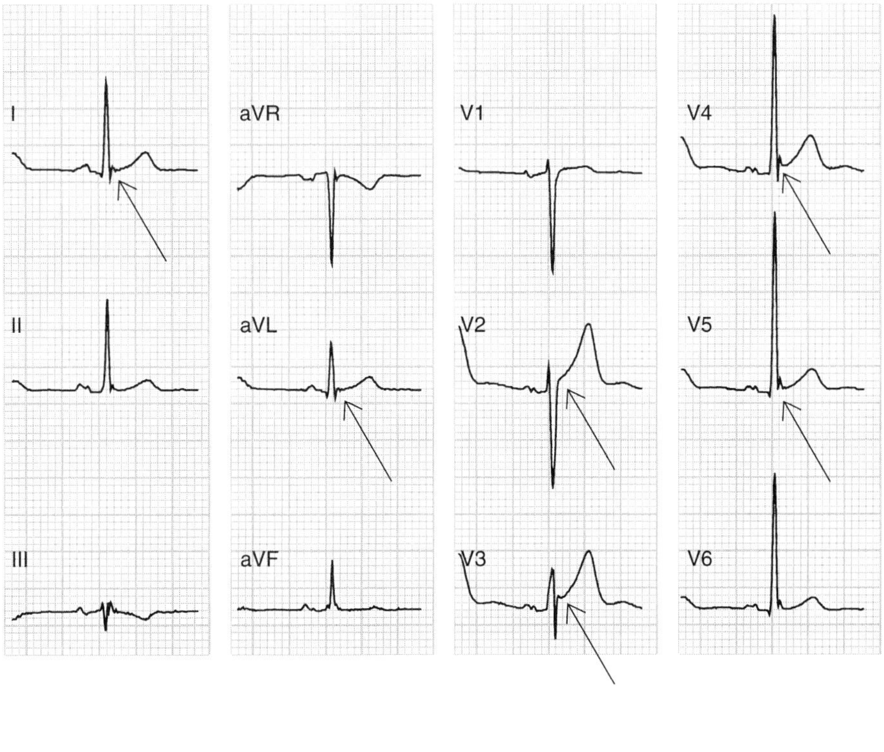

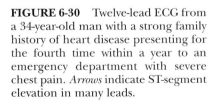

FIGURE 6-30 Twelve-lead ECG from a 34-year-old man with a strong family history of heart disease presenting for the fourth time within a year to an emergency department with severe chest pain. *Arrows* indicate ST-segment elevation in many leads.

Lesser deviation of the ST segments could be caused by SEI or could be a variation of normal. Even the "diagnostic" ST segment changes could be due to an extreme variation of normal. When these ECG changes appear, they should be considered in the context of other manifestations of ischemia such as precordial pain, decreased blood pressure, or cardiac arrhythmias.[45,47]

ST-J point deviation followed by an up-sloping ST segment may also be abnormal. A 0.1 to 0.2 mV depression of the J point followed by an up-sloping ST segment that remains 0.1 mV depressed for 0.08 seconds, or a 0.2 mV depression of the J point followed by an up-sloping ST segment that remains 0.2 mV depressed for 0.08 seconds may also be considered "diagnostic" of SEI.[48,49]

The positive poles of most of the standard limb and precordial ECG leads are directed toward the left ventricle. In deviating away from the left ventricle, the ST segment changes of SEI appear negative or depressed in groups of either leftward (I, aVL, or V4 to V6) or inferiorly (II, III, aVF) oriented leads. The location of the ECG leads with the ST segment depression is not indicative of the location of the ischemic area of the LV subendocardium.

The appearance of the ST segments with LV SEI is similar to that occurring with severe LV hypertrophy, referred to as *strain* (see "Chamber Enlargement" earlier). The ST segment depression of LV strain appears chronically as one of the manifestations of LV hypertrophy.

The maximal ST depression of SEI is almost never seen in leads V1 to V3. (When the maximal ST depression is located in these leads, the cause is either RV strain or posterior EI.) The ST segment depression of LV SEI usually resolves immediately following removal of the excessive cardiovascular stress. When the ST segment depression continues in the absence of an increased LV workload, the presence of subendocardial infarction should be considered. However, infarction is only confirmed if there is accompanying abnormal elevation of serum biochemical markers.[50]

EPICARDIAL INJURY

Just as ST segment changes are reliable indicators of ischemia due to increased myocardial demand, they are also reliable indicators of ischemia due to insufficient coronary blood flow. Observation of the position of the ST segments (relative to the PR and TP segments) in a patient experiencing acute precordial pain provides clinical evidence regarding the presence or absence of severe myocardial ischemia or developing infarction. However, there are many normal variations in the appearance of ST segments (Fig. 6-31).[51-53]

It may be difficult to differentiate the abnormal ST segment changes of EI from variations of normal when the ST segment deviation is minimal. Presence of one of the following criteria is typically required for diagnosis of EI:

1. Elevation of the origin of the ST segment at its junction (J point) with the QRS of:
   a. Greater than 0.10 mV in two or more limb leads or precordial leads V4 to V6

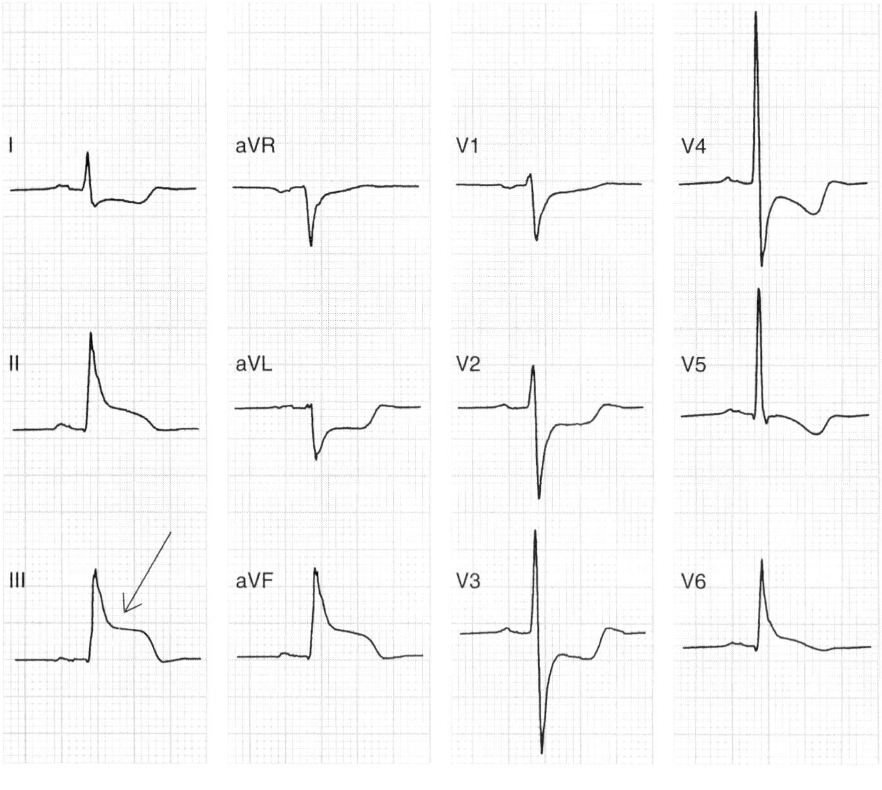

FIGURE 6-31 Twelve-lead ECGs illustrating acute epicardial injury after 1 minute of balloon occlusion in the mid–right coronary artery of a 47-year-old man with symptoms of unstable angina. *Arrow* indicates the maximal ST-segment deviation directed toward the involved regions.

b. Greater than 0.20 mV in two or more precordial leads V1 to V3
2. Depression of the origin of the ST segment at the J point of greater than 0.10 mV in two or more of precordial leads V1 to V3

The greater threshold is required for ST elevation in leads V1 to V3, because there is often a normal slight ST elevation present.

The deviated ST segments typically are either horizontal or slope toward the direction of the T waves. The sloping produces greater deviation of the ST segment as it moves farther from the J point toward the T wave. Various positions along the ST segment are sometimes selected for measurement of ST segment deviation, either for establishing the diagnosis of EI or for estimating its extent. The "J," "J + 0.02 seconds," and "J + 0.06 seconds" points of the ST segment have all been considered.

Since the ST segment changes of EI deviate toward the involved area of myocardium, they appear positive or elevated in the ECG leads with their positive poles pointing toward the lateral, inferior, or anterior aspects of the left ventricle. The ST segments appear negative or depressed in the ECG leads with their positive poles pointing away from the posterior aspect of the left ventricle.

Often, both ST segment elevation and depression appear on different leads of a standard 12-lead ECG. Usually, the direction of the greater deviation should be considered primary, and the direction of the lesser deviation considered secondary or reciprocal. However, there are exceptions to this rule. When EI involves both inferior and posterior aspects of the left ventricle, the ST depression in leads V1 to V3 may equal or exceed the elevation in II, III, and aVF (Fig. 6-32).

EI most commonly occurs in the distal aspect of the area of the LV myocardium supplied by one of the three major coronary arteries (Fig. 6-33).[54] The relationships among the coronary artery, LV quadrant, sectors of the quadrant, and diagnostic ECG leads are indicated in Table 6-2.

In about 90% of individuals, the posterior descending artery originates from the right coronary artery (RCA), and the left circumflex artery (LCX) supplies only part of a single LV quadrant. This has been termed *right coronary dominance.* In the other 10% with left coronary dominance, the posterior descending artery originates from the LCX, and the RCA supplies only the right ventricle.

EI may also involve the thinner-walled RV myocardium when its blood supply via the RCA becomes insufficient. RV EI is represented on the standard ECG as ST segment elevation in leads V1 and V2, with greater elevation in lead V1 than in V2, and with even greater elevation in the more rightward additional leads V3R and V4R (Fig. 6-34).

The entire ST segment elevation disappears abruptly when EI persists for only the 1 to 2 minutes required for percutaneous transluminal coronary angioplasty (PTCA). However, EI produced by coronary thrombosis typically persists throughout the minutes to hours required for clinical administration of some form of reperfusion and then only resolves gradually following

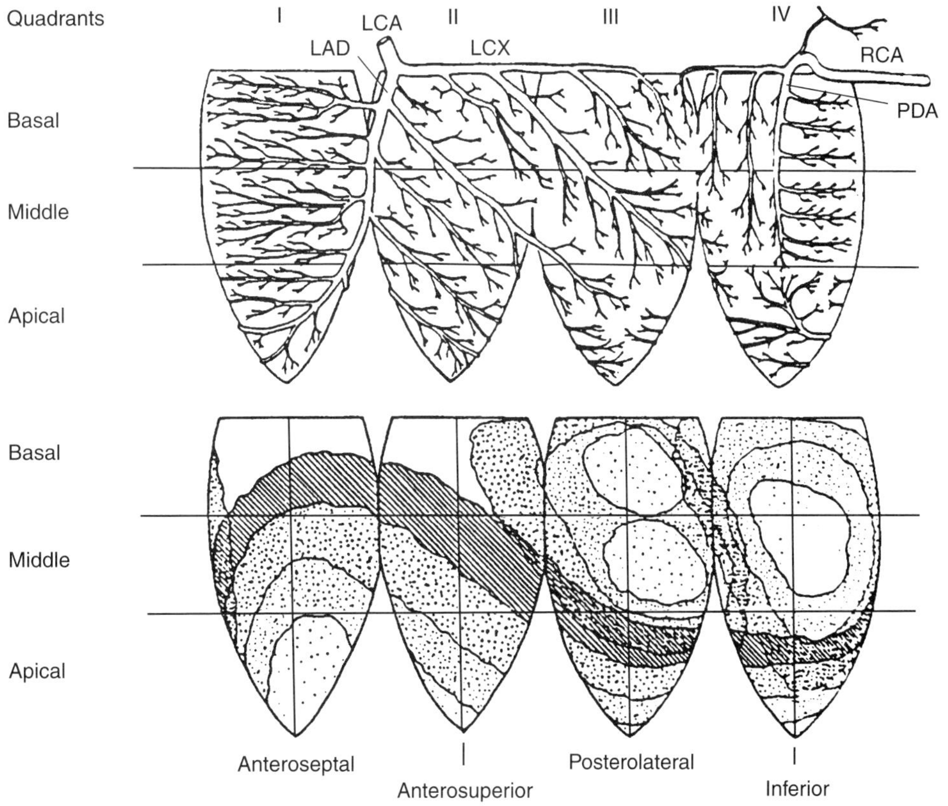

FIGURE 6-32 The 12 sectors of the left ventricular myocardium defined by the four quadrants and the three levels. The distributions of the coronary arteries (*LCA, LAD, LCX, RCA,* and posterior descending [*PDA*]) (*top*) are related to the distributions of insufficient blood supply resulting from occlusions of the respective arteries (*bottom*). The four grades of *shading* from light to dark indicate the size of the involved region as small, medium, large, and very large, respectively. (From Califf RM, Mark DB, Wagner GS: Acute coronary care in the thrombolytic era, 1st ed. Chicago, Year Book, 1988, pp 20-1.)

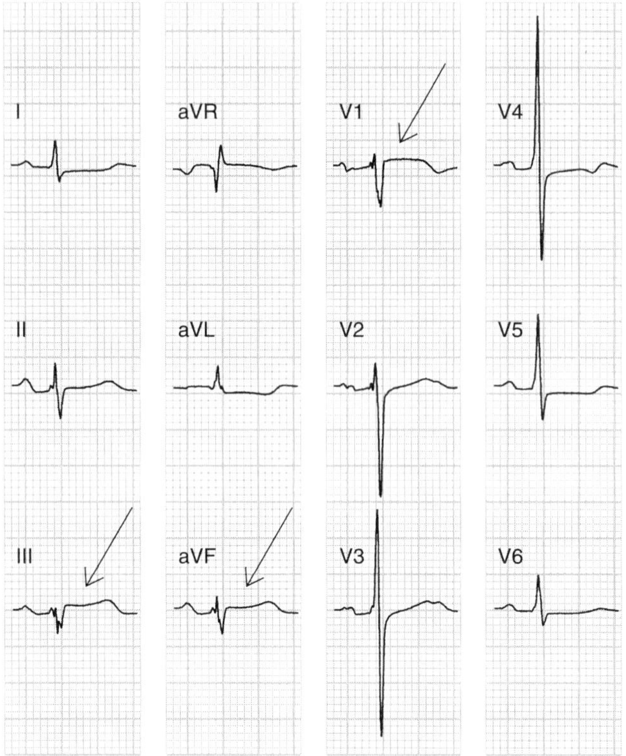

FIGURE 6-33 A 12-lead ECG after 1 minute of balloon occlusion in the proximal right coronary artery in a 65-year-old woman presenting with acute precordial pain of sudden onset. *Arrows* indicate the inferior epicardial injury appearing as ST-segment elevation in leads III and aVF, and right-ventricular epicardial injury appearing as ST-segment elevation in lead V1.

restoration of flow. The disappearance of EI may reveal the already developed QRS changes of infarction that have previously been obscured by the injury current. In some patients, multiple episodes of ST segment elevation and resolution have been documented by continuous monitoring following the initiation of IV thrombolytic therapy. In the absence of successful thrombolytic therapy, there is eventual gradual resolution of the ST segment elevation as the area with EI becomes infarcted.[55,56]

Sometimes the ST segment deviation of EI is accompanied by marked increase in T wave amplitude produced by local hyperkalemia. These primary T wave elevations have been termed *hyperacute T waves* and persist for only a brief period of time after the acute coronary thrombosis[57](Fig. 6-35). Hyperacute T waves may, therefore, be useful in timing the duration of the EI when a patient presents with acute precordial pain.

Definition of the amplitude of the T wave required to identify hyperacute changes during EI requires reference to the upper limits of T wave amplitudes in the various ECG leads of normal subjects. Table 6-3 presents the upper limits of T wave amplitudes in each of the 12 standard leads for females and males in the older than 40-year-old age groups in the Glasgow, Scotland, normal database.[58] A rough estimate of the normal upper limits of T wave amplitude would be: at least 0.5 mV in the limb leads and at least 1.00 mV in the precordial leads.

EI begins during the QRS complex. This may result in secondary deviation of the QRS waveforms in the same direction as that of the ST segments. This distortion affects the amplitudes but not the durations of the QRS waveforms and in their later more than their earlier waveforms.[59]

The deviation of the ST segments confounds the capability of measuring the amplitudes of the QRS waveforms. As illustrated in Figure 6-36, the PR segment baseline remains as the reference for the initial QRS waveform, but the terminal waveform maintains its relationship with the ST segment baseline.

There may also be primary changes in the QRS complexes immediately following the onset of EI as seen in Figure 6-37 during PTCA. The deviation of the QRS waveforms toward the area of the EI is considered primary, because its increase in amplitude is greater than that of the ST segment, and its duration is also prolonged. The most likely cause of the primary QRS deviation is ischemia-induced delay in subendocardial electrical activation. The mid and epicardial layers of the area with EI are activated late, thereby producing an unopposed positive QRS waveform. The QRS would deviate in the negative direction in leads V1 to V3 with EI in the posterior-lateral left ventricle.[60]

MYOCARDIAL INFARCTION

When insufficient coronary blood flow persists after the myocardial glycogen reserves have been depleted, the cells become irreversibly ischemic, and the process of necrosis or MI begins.[61,62] The QRS complex is the most useful aspect of the ECG for the evaluation of the presence, location, and extent of MI. As indicated previously, the QRS waveforms deviate toward the area of potentially reversible EI; secondarily due to the current of injury and primarily due to myocardial activation delay. The process of infarction begins in the

TABLE 6-2 Injury Terminology Relationships

Coronary Artery	LV Quadrant	Sectors	Diagnostic Leads	Common Terms
Left anterior descending	Anteroseptal	All	V1-V3 (elevation)	Anterior
	Anterosuperior	All	I, aVL (elevation)	Lateral
	Inferior	Apical	V4-V6 (elevation)	Lateral
	Posterolateral	Apical	V4-V6 (elevation)	Lateral
Posterior descending	Inferior	Basal, middle	II, III, aVF (elevation)	Inferior
Left circumflex	Posterolateral	Basal, middle	V1-V3 (depression)	Posterior

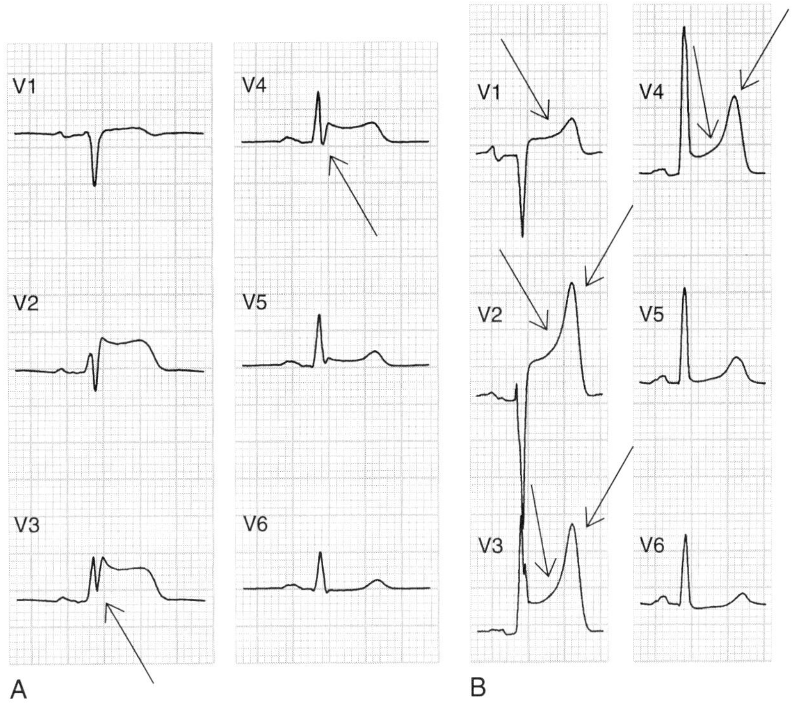

FIGURE 6-34 The six precordial leads of the ECG after 1 minute of balloon occlusion of the left anterior descending coronary artery in a 74-year-old woman with a 5-year history of exertional angina (**A**) and a 51-year-old man with an initial episode of precordial pain (**B**). *Arrows* indicate ST-segment elevation and hyperacute T waves in **A** and disappearance of the S wave from below the TP-PR segment baseline in **B**.

most severely ischemic subendocardial layer. QRS deviation toward the ischemic area is replaced by QRS deviation away from the infarcted area.[63] Since there is no electrical activation of the infarcted myocardium, the summation of activation spread is away from the involved area.

The rapid appearance of the abnormalities of the QRS complex produced by an anterior infarction during continuous ischemia monitoring is illustrated in Figure 6-38. Myocardial reperfusion is accompanied by rapid resolution of the EI and a shift of the QRS waveforms away from the anterior LV wall. Though it may appear that the therapy has caused the infarction, it is much more likely that the infarction had already occurred before the initiation of the therapy, but its detection on the ECG was obscured by the secondary QRS changes of the EI.

As previously mentioned, EI involving the thin RV free wall may be manifested on the ECG by ST segment deviation, but RV infarction is not manifested by significant alteration of the QRS complex. RV free wall activation is insignificant in comparison with activation of the thicker interventricular septum and LV free wall.

MI evolves from EI in the distal aspects of the areas of LV myocardium supplied by one of the three major coronary arteries (see Fig. 6-33 and Table 6-2).[64] The basal and middle sectors of the posterior-lateral quadrant

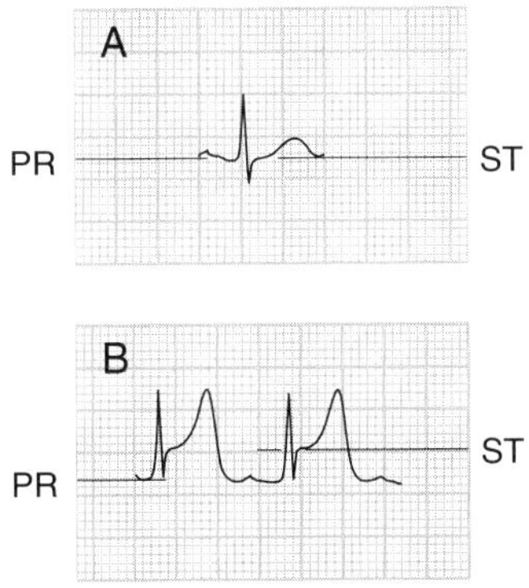

FIGURE 6-35 **A,** An ECG recording made before balloon inflation, in which the PR and ST baselines are at the same level and a 0.03-mV S wave is present. **B,** During balloon occlusion, the ST segment is elevated by 0.03 mV by the epicardial injury current, and the S wave also deviates upward so that it just reaches the PR-segment baseline.

TABLE 6-3 T Wave Amplitude Limits*

Lead	Males 40–49	Females 40–49	Males 50+	Females 50+
aVL	0.30	0.30	0.30	0.30
I	0.55	0.45	0.45	0.45
-aVR	0.55	0.45	0.45	0.45
II	0.65	0.55	0.55	0.45
aVF	0.50	0.40	0.45	0.35
III	0.35	0.30	0.35	0.30
V1	0.65	0.20	0.50	0.35
V2	1.45	0.85	1.40	0.70
V3	1.35	0.85	1.35	0.85
V4	1.15	0.85	1.10	0.75
V5	0.90	0.70	0.95	0.70
V6	0.65	0.55	0.65	0.50

*Presentation of upper-limit T-wave amplitudes (in millivolts rounded to the nearest 0.05) in each lead by gender and age for normal subjects from Glasgow, Scotland. The leads are arranged in the panoramic sequence.

Modified from Macfarlane PW, Lawrie TDV: In *Comprehensive electrocardiology*, vol. 3. New York, Pergamon Press, 1989, pp 1446–57.

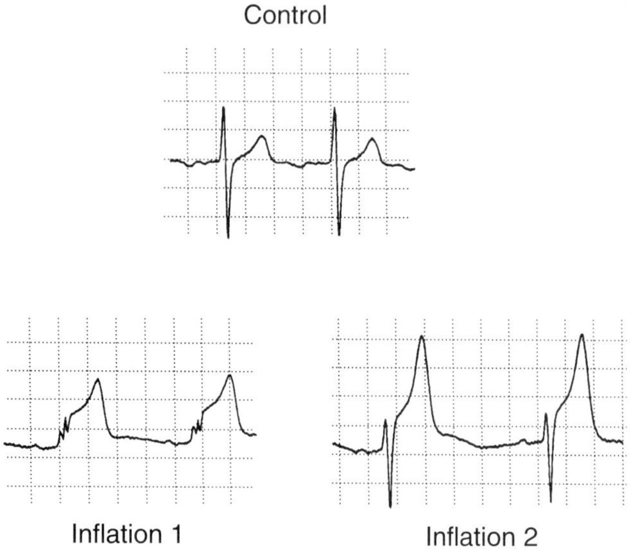

FIGURE 6-36 Recordings of two cardiac cycles in lead V2 from baseline (control) and after 2 minutes each of two different periods of balloon occlusion of the left anterior descending coronary artery. (From Wagner NB, Sevilla DC, Drucoff MW, et al: Transient alterations of the QRS complex and the ST segment during percutaneous transluminal balloon angioplasty of the left anterior descending coronary artery. Am J Cardiol 1988;62:1038-42.)

of the left ventricle are located away from the positive poles of all 12 of the standard ECG leads. Therefore, posterior infarction is indicated by a positive rather than negative deviation of the QRS complex. Additional leads on the posterior-lateral thorax would be required to record the ST segment elevation due to EI and the negative QRS deviation due to MI in this area.[65]

The initial portion of the QRS complex deviates most prominently away from the area of infarction and is represented on the ECG by prolonged Q wave duration. The initial QRS waveform may normally be negative (a <30 millisecond [ms] Q wave) in all leads except V1 to V3. The presence of any Q wave is considered abnormal in only these 3 of the 12 standard leads. Table 6-1 indicates the upper limits of normal of the Q wave duration in the various ECG leads.[66] Instead of amplitude, duration of the Q wave should be used in the definition of abnormality, because the amplitudes of the individual QRS waveforms vary with the overall QRS amplitude. As discussed later, Q wave amplitudes may be considered abnormal only in relation to R wave amplitudes.

Many cardiac conditions other than MI are capable of producing abnormal initial QRS waveforms. Ventricular hypertrophy, intraventricular conduction abnormalities, and ventricular preexcitation commonly cause prolongation of Q wave duration. The term *Q wave* as used here also refers to the Q wave equivalent of abnormal R waves in leads V1 and V2. Therefore, the following steps should be considered in the evaluation of Q waves regarding the presence of MI:

1. Are abnormal Q waves present in any lead?
2. Are criteria present for other cardiac conditions that are capable of producing abnormal Q waves?
3. Does the extent of Q wave abnormality exceed that which could have been produced by that other cardiac condition?

The deviation of the QRS complex away from the area of the MI may, in the absence of abnormal Q waves, be represented by diminished R waves. Table 6-4 indicates the leads in which R waves of less than a certain amplitude or duration may be indicative of MI.[67]

An infarct produced by insufficient blood flow via the LAD might be limited to the anterior-septal quadrant (see Fig. 6-33). It might also extend into the anterior-superior quadrant or into the apical sectors of other quadrants commonly referred to as anterior, anterior-lateral, or anterior-apical infarction, respectively.

When the RCA is dominant, its sudden complete obstruction typically produces an inferior infarction in the basal and middle sector of the inferior quadrant. Also, with this anatomy, the typical distribution of the LCX is limited to the LV free wall between the distributions of the anterior and posterior descending arteries. Sudden complete occlusion produces only a posterior infarction, as illustrated by QRS deviation away from that region in Figure 6-39.

When the left coronary artery is dominant, a sudden complete obstruction of the RCA can only produce

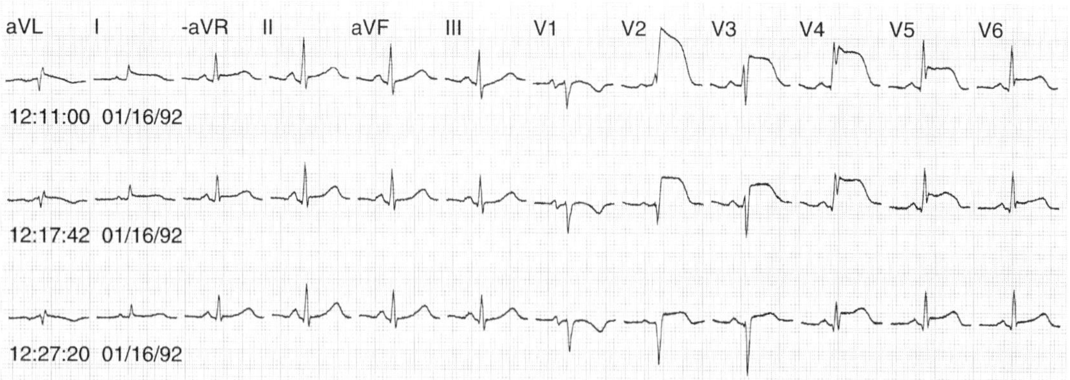

FIGURE 6-37 Continuous ECG monitoring during the first 27 minutes of IV thrombolytic therapy (begun at 12:00:00) in a 69-year-old man with acute thrombotic occlusion of the left anterior descending coronary artery. The 12 standard leads of the ECG are presented in the panoramic format after 11, 17, and 27 minutes of therapy.

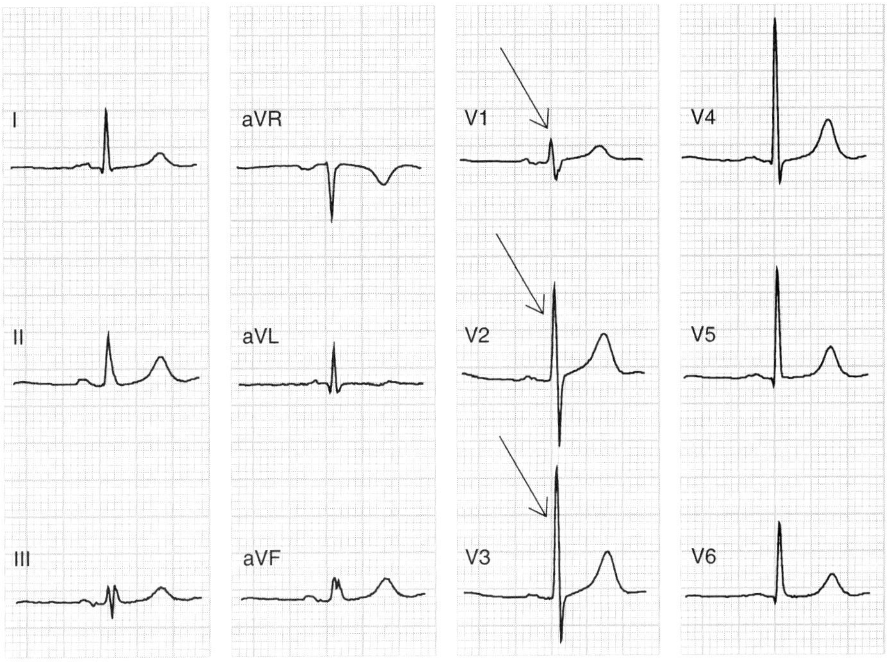

FIGURE 6-38 A 12-lead ECG from a 70-year-old man 1 year after an acute posterior-wall myocardial infarction. Coronary angiography showed complete occlusion of a nondominant left circumflex coronary artery (the right coronary artery supplied the posterior descending artery). *Arrows* indicate the increased R waves in leads V1 to V3.

infarction in the right ventricle, which would not likely produce changes in the QRS complex. The LCX then supplies the middle and basal sectors of both the posterior-lateral and inferior quadrants, and its obstruction can produce an inferior-posterior infarction. This same combination of LV locations can be involved when there is RCA dominance and one of its branches extends into the typical LCX distribution. The ECG, therefore, indicates the region infarcted, but not whether the RCA or LCX is the "culprit artery."

There are variations among individuals regarding the areas of LV myocardium supplied by the three major coronary arteries. These variations may occur either congenitally or because atherosclerotic obstruction in one artery results in collateral blood supply from another artery. For example, the posterior descending artery may extend its supply to include the apical sector of the inferior quadrant. In this instance, its sudden complete obstruction could result in QRS deviation away from leads V4 to V6 in addition to leads II, III, and aVF, causing an inferior-apical infarction. Similarly, the LCX could supply the apical sector of the posterior-lateral quadrant, causing a posterior-apical infarction. A marginal branch of the LCX may supply a portion of the anterior-superior quadrant and be responsible for a posterior-lateral infarction.

The posterior aspect of the apex may be involved when either a dominant RCA or LCX is acutely obstructed, and inferior, posterior, and apical locations are apparent on the ECG (Fig. 6-40).

TABLE 6-4 R Wave Lower Limits

Limb Leads		Precordial Leads	
Lead	**Criteria for Abnormal**	**Lead**	**Criteria for Abnormal**
I	R amp ≤ 0.20 mV	V1	None
II	None	V2	R dur ≤ 0.01 sec or amp ≤ 0.10 mV
III	None	V3	R dur ≤ 0.02 sec or amp ≤ 0.20 mV
aVR	None	V4	R amp ≤ 0.70 mV or ≤ Q amp
aVL	R amp ≤ Q amp	V5	R amp ≤ 0.70 mV or ≤ 2 × Q amp
aVF	R amp ≤ 2 × Q amp	V6	R amp ≤ 0.60 mV or ≤ 3 × Q amp

amp, amplitude; dur, duration.

ESTIMATING INFARCT SIZE

An individual patient may have single infarcts varying in size in the distributions of any of the three major coronary arteries or may have multiple infarcts. Selvester and coworkers developed a method for estimating the total percentage of the left ventricle infarcted using a weighted scoring system.[68] Computerized simulation of the sequence of electrical activation of the normal human left ventricle formed the basis for the 31-point scoring system, with each point accounting for 3% of the left ventricle.[69] The Selvester QRS scoring system includes 50 criteria from 10 of the 12 standard leads with weights ranging from one to three points per criteria (Fig. 6-41). There are criteria in precordial leads V1 and V2 for both anterior and posterior infarct locations. In addition to the Q wave and decreased R wave criteria typically used for diagnosis and localization of infarcts, this system for size estimation also contains criteria relating to the S wave.[67]

In the Selvester scoring system, a very important consideration is the Q wave duration. This measurement is easy when the QRS complex has discrete Q, R, and S waves.[67] Figure 6-42 presents sequentially smaller

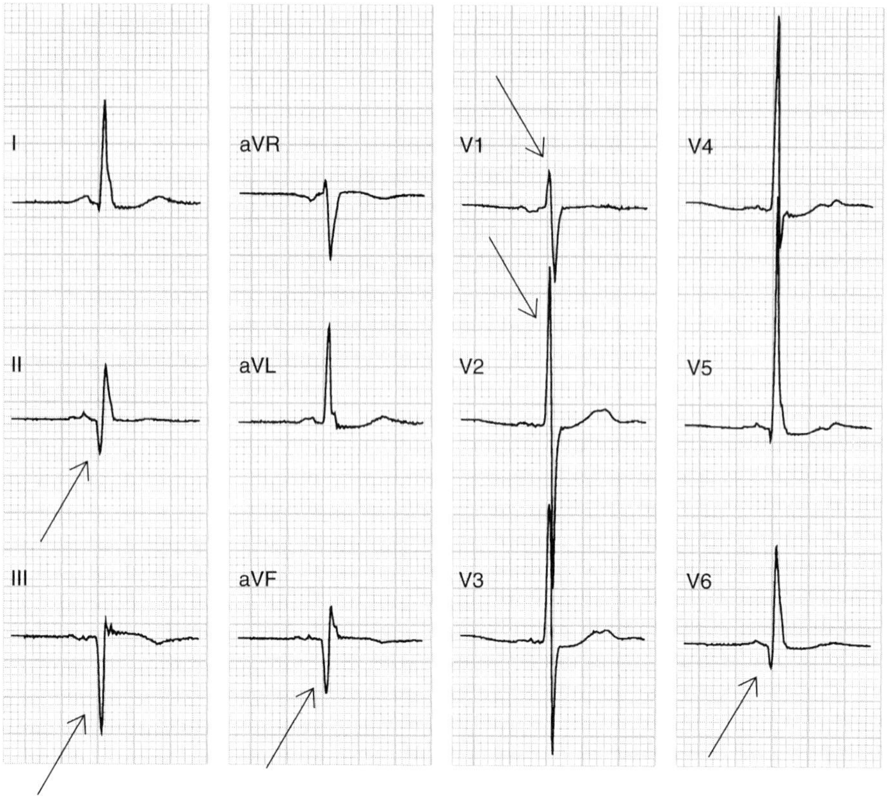

FIGURE 6-39 Serial 12-lead ECGs from a previous routine examination. *Arrows* indicate abnormal initial QRS forces.

Complete 50-Criteria, 31-Point QRS Scoring System*

Lead	Maximum Lead Points	Criteria	Points
I	(2)	Q ≥ 30 ms	(1)
		R/Q ≤ 1	(1)
		R ≤ 0.2 mV	(1)
II	(2)	Q ≥ 40 ms	(2)
		Q ≥ 30 ms	(1)
aVL	(2)	Q ≥ 30 ms	(1)
		R/Q ≤ 1	(1)
aVF	(5)	Q ≥ 50 ms	(3)
		Q ≥ 40 ms	(2)
		Q ≥ 30 ms	(1)
		R/Q ≤ 1	(2)
		R/Q ≤ 2	(1)
V_1 Anterior	(1)	Any Q	(1)
V_1 Posterior	(4)	R/S ≥ 1	(1)
		R ≥ 50 ms	(2)
		R ≥ 1.0 ms	(2)
		R ≥ 40 ms	(1)
		R ≥ 0.6 mV	(1)
		Q and S ≤ 0.3 mV	(1)
V_2 Anterior	(1)	Any Q	(1)
		R ≤ 10 ms	(1)
		R ≤ 0.1 mV	(1)
		R ≤ RV_1 mV	(1)
V_2 Posterior	(4)	R/S ≥ 1.5	(1)
		R ≥ 60 ms	(2)
		R ≥ 2.0 mV	(2)
		R ≥ 50 ms	(1)
		R ≥ 1.5 mV	(1)
		Q and S ≤ 0.4 mV	(1)
V_3	(1)	Any Q	(1)
		R ≤ 20 ms	(1)
		R ≤ 0.2 mV	(1)
V_4	(3)	Q ≥ 20 ms	(1)
		R/S ≤ 0.5	(2)
		R/Q ≤ 0.5	(2)
		R/S ≤ 1	(1)
		R/Q ≤ 1	(1)
		R ≤ 0.7 mV	(1)
V_5	(3)	Q ≥ 30 ms	(1)
		R/S ≤ 1	(2)
		R/Q ≤ 1	(2)
		R/S ≤ 2	(1)
		R/Q ≤ 2	(1)
		R ≤ 0.7 mV	(1)
V_6	(3)	Q ≥ 30 ms	(1)
		R/S ≤ 1	(2)
		R/Q ≤ 1	(2)
		R/S ≤ 3	(1)
		R/Q ≤ 3	(1)
		R ≤ 0.6 mV	(1)

FIGURE 6-40 The maximal number of points that can be awarded for each lead is shown in parentheses following each lead name (or left-ventricular region within a lead for leads V1 and V2), and the number of points awarded for each criterion is indicated in parentheses after each criterion name. The QRS criteria from 10 of the 12 standard ECG leads are indicated. Only one criterion can be selected from each group of criteria within a bracket. All criteria involving R/Q or R/S ratios consider the relative amplitudes of these waves. (Modified from Selvester RH, Wagner GS, Hindman NB: The Selvester QRS scoring system for estimating myocardial infarct size. The development and application of the system. Arch Intern Med 1985;145:1877–81. Copyright 1985 American Medical Association.)

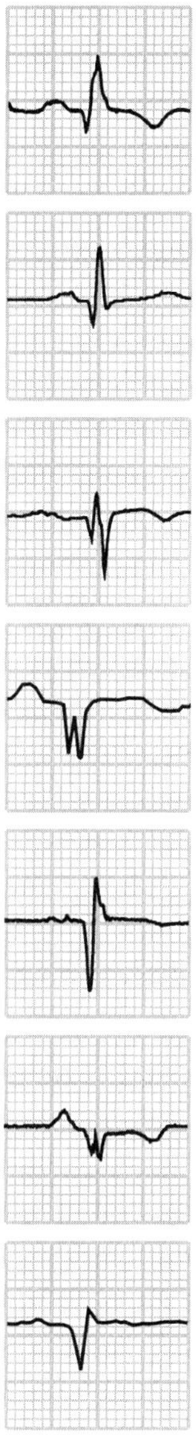

LEAD aVF	Q DURATION	R/Q RATIO	TOTAL POINTS
A	.03 sec (1)		1
B	.03 sec (1)	≤2:1 (1)	2
C	.03 sec (1)	≤1:1 (2)	3
D	.03 sec (1)	≤1:1 (2)	3
E	.04 sec (2)	≤1:1 (2)	4
F	.04 sec (2)	≤1:1 (2)	4
G	≥.05 sec (3)	≤1:1 (2)	5

FIGURE 6-41 A-G, Variations in the appearance of the QRS complex in lead aVF representing the changes of inferior infarction. The numbers of QRS points awarded for the Q-wave duration and the R/Q amplitude ratio criteria met in the various ECGs given as examples are indicated in parentheses. The total number of QRS points awarded for lead aVF is indicated for each example in the final column. (Modified from Wagner GS, Freye CJ, Palmeri ST, et al: Evaluation of a QRS scoring system for estimating myocardial infarct size. I. Specificity and observer agreement. Circulation 1982;65:342–7.)

positive deflections located between the initial abnormal negative deflection *(Q wave)* and the terminal normal negative deflection *(S wave).* The true Q wave duration should be measured along the ECG baseline from the onset of the initial negative reflection to either the onset of the positive deflection or to the point directly above the peak of the notch in the negative deflection. Satisfaction of only a single Selvester scoring criterion may represent either a normal variant or an extremely small infarct. This system may be confounded by two infarcts located in opposite sectors of the left ventricle. The opposing effects on the summation of the electrical forces may cancel each other, producing falsely negative ECG changes. For example, the coexistence of both anterior and posterior infarcts creates the potential for underestimation of the total percentage of left ventricle infarcted.

The ST segment changes that are prominent during EI typically disappear when the ischemic myocardium either infarcts or regains sufficient blood supply. Their time course of resolution is accelerated by reperfusion via the culprit artery. When re-elevation of the ST segments is observed, further EI or a disturbance in the pericardium is suggested. EI is typically limited to a particular area of the left ventricle. When the ST elevation occurs in leads representing multiple LV areas, acute bleeding into the pericardium should be considered.[70] This may be the first indication that the infarct has caused a myocardial rupture with leakage of blood into the pericardial sac. If this process remains undetected, cardiac arrest may result from pericardial tamponade, in which myocardial relaxation is restricted by the blood in the enclosed pericardial space.

In some patients, the ST segment elevation does not completely resolve during the acute phase of the MI. This more commonly occurs with anterior infarcts than with those in the other locations. This lack of ST segment resolution has been associated with the thinning

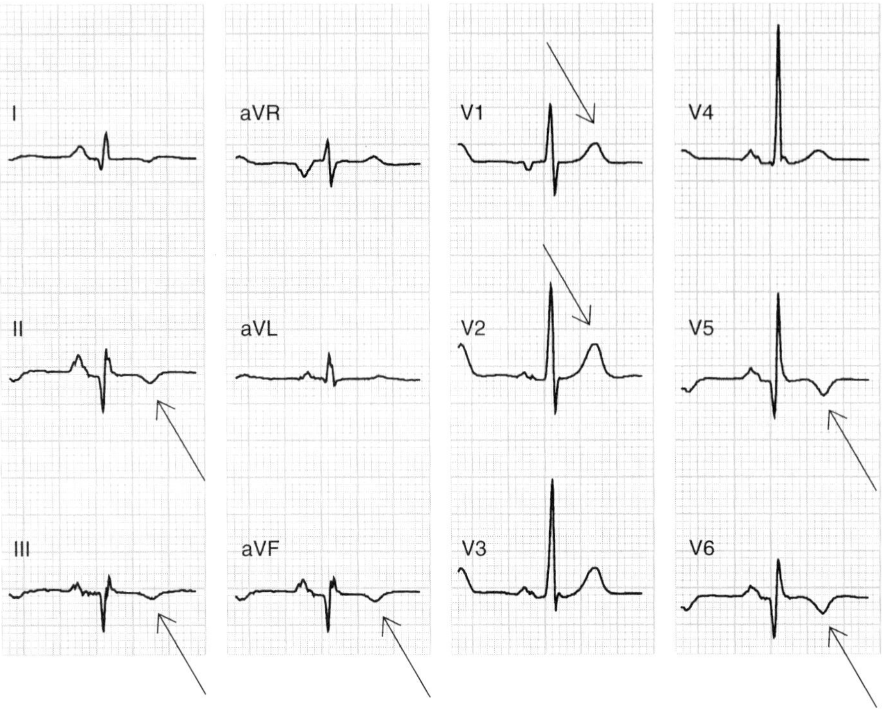

FIGURE 6-42 A 12-lead ECG from a 53-year-old man 5 days after an inferior-posterior-lateral infarction. *Arrows* indicate negative T waves in leads with abnormal Q waves but positive T waves in leads with abnormal R waves.

of the LV wall caused by infarct expansion.[71,72] The extreme manifestation of infarct expansion is the formation of a ventricular aneurysm. The incidence of such extreme infarct expansion is reduced by successful thrombolytic therapy.

The movement of the T waves toward the area of the EI, like that of the ST segments, resolves as the ischemic myocardium either recovers or infarcts. Unlike the ST segments, however, the T waves do not typically return to their normal positions as the process of infarction evolves. The T waves move past the isoelectric position until they are directed away from the area of infarction.[73] They assume an appearance identical to that of "ischemic T waves," even though there is no ongoing myocardial ischemia. Typically, the terminal portion of the T wave is the first to become inverted, followed by the middle and initial portions.

Similarly, when the posterior-lateral quadrant of the LV is involved, the T waves eventually become markedly positive. Figure 6-42 illustrates the tall positive T waves in leads V1 and V2 that accompany the negative T waves in other leads during the chronic phase of an inferior-posterior-apical infarction.

FUTURE ELECTROCARDIOGRAM ROLES

Many of these considerations of practical information provided by careful quantitative interpretation of the standard 12-lead ECG were unknown 5 years ago. As this "diagnostic" method enters its second century, there remain many areas for future elucidation. Some examples are:

How can ECG indicators of timing of the acute infarction process add to historical timing to predict myocardial salvagability?

How well does the amount of changes in the various waveforms reflect the size of the acutely ischemically threatened area?

Can infarct sizing methods be improved to more accurately quantify multiple infarcts?

As large multicenter clinical trials provide the ECG data and non–ECG standards for definitions of criteria, these and many other new clinical diagnostic and prognostic methods will emerge in the second century of electrocardiography.

REFERENCES

1. Einthoven W, Fahr G, de Waart A: Uber die richtung und die manifeste grosse der potentialschwangkungen im menschilichen herzen und uber dem einfluss der herzlage auf die from des elektrokardiogramms. Pfluegers Arch 1913;150:275-315. (Translation: Hoff HE, Sekelj P: Am Heart J 1950;40:163-94.)
2. Wilson FN, Macloed AG, Barker PS: The interpretation of the initial deflections of the ventricular complex of the electrocardiogram. Am Heart J 1931;6:637-64.
3. Wilson FN, Johnston FD, Macloed AG, Barker PS: Electrocardiograms that represent the potential variations of a single electrode. Am Heart J 1934;9:447-71.
4. Kossmann CE, Johnston FD: The precordial electrocardiogram. I. The potential variations of the precordium and of the extremities in normal subjects. Am Heart J 1935;19:925-41.
5. Joint recommendations of the American Heart Association and the Cardiac Society of Great Britain and Ireland. Standardization of Precordial Leads. Am Heart J 1938;15:107-8.
6. Committee of the American Heart Association for the Standardization of Precordial Lead. Supplementary report. Am Heart J 1938;15:235-9.
7. Committee of the American Heart Association for the Standardization of Precordial Leads. Second supplementary report. JAMA 1943;121:1329-51.
8. De Vries PA: Development of the ventricles and spiral outflow tract of the human heart. Contrib Embryol 1962;37:87.
9. Mall FP: On the development of the human heart. Am J Anat 1912;13:249.

10. Rushmer RF: Functional anatomy and the control of the heart, part I. In Rushmer RF (ed): Cardiovascular Dynamics. Philadelphia, WB Saunders, 1976;76-104.
11. Langer GA: Heart: Excitation-contraction coupling. Ann Rev Physiol 1973;35:55-85.
12. Weidmann S: Resting and action potentials of cardiac muscle. Ann N Y Acad Sci 1957;65:663.
13. Rushmer RF, Guntheroth WG: Electrical activity of the heart, part I. In Rushmer RF (ed): Cardiovascular Dynamics. Philadelphia, WB Saunders, 1976.
14. Meyerburg RJ, Gelband H, Castellanos A, et al: Electrophysiology of endocardial intraventricular conduction: The role and function of the specialized conducting system. In Wellens HJJ, Lie KL, Janse MJ (eds): The conduction system of the heart. The Hague, Martinus Nijhoff, 1978.
15. Guyton AC: Rhythmic excitation of the heart. In Guyton AC (ed): Textbook of Medical Physiology. Philadelphia, WB Saunders, 1991.
16. Scher AM: The sequence of ventricular excitation. Am J Cardiol 1964;14:287.
17. Graybiel A, White PD, Wheeler L, Williams C: The typical normal electrocardiogram and its variations. In Graybiel A, White PD, Wheeler L, Williams C (eds): Electrocardiography in Practice. Philadelphia, WB Saunders, 1952.
18. Netter FH: Section II, the electrocardiogram. The CIBA Collection of Medical Illustrations, vol 5. New York, CIBA, 1978.
19. Barr, RC: Genesis of the electrocardiogram. In MacFarlane PW, Veitch Lawrie TD (eds): Comprehensive Electrocardiology, vol I. New York, Pergamon Press, 1989, pp 139-47.
20. Bazett HC: An analysis of the time relations of electrocardiograms. Heart 1920;7:353-70.
21. Hodges M, Salerno D, Erlien D: Bazett's QT correction reviewed. Evidence that a linear QT correction for heart is better [abstract]. J Am Coll Cardiol 1983;1:694.
22. Macfarlane PW, Veitch Lawrie TD: The normal electrocardiogram and vectorcardiogram. In Macfarlane PW, Veitch Lawrie TD (eds): Comprehensive electrocardiology, vol I. New York, Pergamon Press, 1989, pp 451-2.
23. Day CP, McComb JM, Campbell RW: QT dispersion in sinus beats and ventricular extrasystoles in normal hearts. Br Heart J 1992; 67:39-41.
24. Kleiger RE, Miller JP, Bigger JT, Moss AJ: The Multi-Center Post-Infarction Research Group. Decreased heart rate variability and its association with increased mortality after acute myocardial infarction. Am J Cardiol 1987;59:256-62.
25. Rushmer RF: Cardiovascular Dynamics. Philadelphia, WB Saunders, 1991, pp 452-56.
26. Rushmer RF: Cardiac compensation, hypertrophy, myopathy and congestive heart failure. In Rushmer RF (ed): Cardiovascular Dynamics. Philadelphia, WB Saunders, 1976, pp 532-65.
27. Cabrera E, Monroy JR: Systolic and diastolic loading of the heart. II. Electrocardiographic data. Am Heart J 1952;43:661.
28. Devereux RB, Reichek N: Repolarization abnormalities of left ventricular hypertrophy. J Electrocardiol 1982;15:47.
29. Casale PN, Devereux RB, Kligfield P, et al: Electrocardiographic detection of left ventricular hypertrophy: Development and prospective validation of improved criteria. J Am Coll Cardiol 1985;6:572.
30. Rosenbaum MB, Elizari MV, Lazzari JO: The Hemiblocks. Oldsmar, FL, Tampa Tracings, 1970.
31. Rosenbaum MB: Types of left bundle branch block and their clinical significance. J Electrocardiol 1969;2:197.
32. Yano K, Peskoe SM, Rhoads GG, et al: Left axis deviation and left anterior hemiblock among 8000 Japanese-American men. Am J Cardiol 1975;35:809.
33. Rosenbaum MB: The Hemiblocks: Diagnostic criteria and clinical significance. Mod Concepts Cardiovasc Dis 1970;39:141-6.
34. Hindman MC, Wagner GS, JaRo M, et al: The clinical side of bundle branch block complicating acute myocardial infarction. II. Indications for temporary and permanent insertion. Circulation 1978;58:689-99.
35. Scher AM, Young AC, Malmgren AL, Erickson RV: Activation of the interventricular septum. Circ Res 1963;3:56-64.
36. Willems JL, Robles De Medina EO, Bernard R, et al: Criteria for intraventricular-conduction disturbances and pre-excitation. J Am Coll Cardiol 1985;6:1261-75.
37. Hassett MA, Williams RR, Wagner GS: Transient QRS changes simulating myocardial infarction. Circulation 1980;62:975-9.
38. Ekmekci A, Toyoshima HJ, Kwoczynski JK, et al: Angina pectoris: Giant R and receding S wave in myocardial ischemia and certain non-ischemic conditions. Am J Cardiol 1961;7:521-32.
39. Wagner NB, Sevilla DC, Krucoff MW, et al: Transient alterations of the QRS complex and the ST segment during percutaneous transluminal balloon angioplasty of the left anterior descending artery. Am J Cardiol 1988;62:1038-42.
40. Reimer KA, Lowe JE, Ramussen MM, Jennings RB: The wavefront phenomenon of ischemic cell death: I. Myocardial infarct size vs. duration of coronary occlusion in dogs. Circulation 1977; 56:786-94.
41. Reimer KA, Jennings RB: The "wavefront phenomenon" of myocardial ischemic cell death: II. Transmural progression of necrosis within the framework of ischemic bed size (myocardium at risk) and collateral flow. Lab Invest 1979;40:633-44.
42. Wagner NB, Wagner GS, White RD: The twelve lead ECG and the extent of myocardium at risk of acute infarction: Cardiac anatomy and lead locations, and the phases of serial changes during acute occlusion. In Califf RM, Mark DB, Wagner GS (eds): Acute coronary care in the thrombolytic era. Chicago, Year Book, 1988, pp 31-45.
43. Stuart RJ, Ellestad MH: Upsloping ST segments in exercise stress testing. Am J Cardiol 1976;37:19.
44. Sheffield LT, Holt JH, Reeves TJ: Exercise graded by heart rate in electrocardiographic testing for angina pectoris. Circulation 1965;32:622.
45. Ellestad MH, Cooke BM, Greenberg PS: Stress testing; clinical application and predictive capacity. Prog Cardiovasc Dis 1979; 21:431-60.
46. Lepeschkin E: Modern electrocardiography, vol I. Baltimore, Williams & Wilkins, 1951, p 798.
47. Ellestad MH, Savitz S, Bergdall D, Teske JE: The false positive stress test: Multivariate analysis of 215 subjects with hemodynamic, angiographic and clinical data. Am J Cardiol 1977; 40:681.
48. Rijneki RD, Ascoop CA, Talmon JL: Clinical significance of upsloping ST segments in exercise electrocardiography. Circulation 1980;61:671.
49. Kurita A, Chaitman BR, Bourassa MG: Significance of exercise-induced junctional ST depression in evaluation of coronary artery disease. Am J Cardiol 1977;40:492-7.
50. Hurst JW, Logue RB: The heart: Arteries and veins. New York, McGraw-Hill, 1966, p 147.
51. Prinzmetal M, Goldman A, Massumi RA, et al: Clinical implications of errors in electrocardiographic interpretation: Heart disease of electrocardiographic origin. JAMA 1956;161:138.
52. Levine HD: Non-specificity of the electrocardiogram associated with coronary artery disease. Am J Med 1953;15:344.
53. Marriott HJL: Coronary mimicry: Normal variants, and physiologic, pharmacologic and pathologic influences that stimulate coronary patterns in the electrocardiogram. Ann Intern Med 1960;52:411.
54. Wagner GS, Wagner NB: The 12-lead ECG and the extent of myocardium at risk of acute infarction: Anatomic relationships among coronary, Purkinje, and myocardial anatomy. In Acute Coronary Care in the Thrombolytic Era. Chicago, Year Book, 1988 pp 20-1.
55. Krucoff MW, Croll MA, Pope JE, et al: Continuously updates 12-lead ST-segment recovery analysis for myocardial infarct artery patency assessment and its correlation with multiple simultaneous early angiographic observations. Am J Cardiol 1993;71: 145-51.
56. Kondo M, Tamura K, Tanio H, Shimono Y: Is ST segment re-elevation associated with reperfusion an indicator of marked myocardial damage after thrombolysis? J Am Coll Cardiol 1993;21:62-7.
57. Dressler W, Roesler H: High T waves in the earliest stage of myocardial infarction. Am Heart J 1947;34:627-45.
58. Macfarlane PW, Lawrie TDV (eds): Comprehensive electrocardiography. New York, Pergamon Press, 1989, pp 1446-57.
59. Wagner NB, Sevilla DC, Krucoff MW, et al: Transient alterations of the QRS complex and ST segment during percutaneous transluminal balloon angioplasty of the left anterior descending artery. Am J Cardiol 1988;62:1038-42.

60. Selvester RH, Wagner NB, Wagner GS: Ventricular excitation during percutaneous transluminal angioplasty of the left anterior descending coronary artery. Am J Cardiol 1988;62:1116-21.
61. Reimer KA, Lowe JE, Rasmussen MM, Jennings RB: The wavefront phenomenon of ischemic cell death: I. Myocardial infarct size vs. duration of coronary occlusion in dogs. Circulation 1977;57:786-94.
62. Reiner KA, Jennings RB: The "wavefront phenomenon" of myocardial ischemic cell death: II. Transmural progression of necrosis within the framework of ischemic bed size (myocardium at risk) and collateral flow. Lab Invest 1979;40:633-44.
63. Wagner NB, White RD, Wagner GS: The 12-lead ECG and the extent of myocardium at risk of acute infarction: Cardiac anatomy and lead locations, and the phases of serial changes during acute occlusion. In Califf RM, Mark DB, Wagner GS (eds): Acute coronary care in the thrombolytic era. Chicago, Year Book, 1988, pp 36-41.
64. Wagner GS, Wagner NB: The 12-lead ECG and the extent of myocardium at risk of acute infarction: Anatomic relationships among coronary, Purkinje, and myocardial anatomy. In Califf RM, Mark DB, Wagner GS (eds): Acute coronary care in the thrombolytic era. Chicago, Year Book, 1988, pp 16-30.
65. Flowers NC, Horan LG, Sohi GS, et al: New evidence for inferior-posterior myocardial infarction on surface potential maps. Am J Cardiol 1976;38:576.
66. Wagner GS, Freye CJ, Palmeri ST, et al: Evaluation of a QRS scoring system for estimating myocardial infarct size. I. Specificity and observer agreement. Circulation 1982;65:342-7.
67. Hindman NB, Schocken DD, Widmann M, et al: Evaluation of a QRS scoring system for estimating myocardial infarct size. V. Specificity and method of application of the complete system. Am J Cardiol 1985;55:1485-90.
68. Selvester RH, Wagner JO, Rubin HB: Quantitation of myocardial infarct size and location by electrocardiogram and vectrocardiogram. In Snellen HA, Hemker HC, Hugenholtz PG, von Bemmel JH: Boerhave Course in Quantitation in Cardiology. The Netherlands, Leiden University Press, 1972, p. 31.
69. Selvester RH, Soloman J, Sapoznikov D: Computer simulation of the electrocardiogram. In Cady L (eds): Computer techniques in cardiology. New York, Marcel Dekker, 1979, p 417.
70. Olivia PB, Hammill SC, Edwards WD: Electrocardiographic diagnosis of post infarction regional pericarditis: Ancillary observations regarding the effect of reperfusion on the rapidity and amplitude of T wave inversion after acute myocardial infarction. Circulation 1993;88:896-904.
71. Lindsay J Jr, Dewey RC, Talesnick BS, Nolan NG: Relation of ST segment elevation after healing of acute myocardial infarction to the presence of left ventricular aneurysm. Am J Cardiol 1984;54:84-6.
72. Arvan S, Varat MA: Persistent ST-segment elevation and left ventricular wall abnormalities: Two-dimensional echocardiographic study. Am J Cardiol 1984;38:178-88.
73. Mandel WJ, Burgess MJ, Neville J Jr, Abidskov JA: Analysis of T wave abnormalities associated with myocardial infarction using a theoretic model. Circulation 1968;38:178-88.

Chapter 7 Principles of Electropharmacology

MICHAEL R. ROSEN

This paper is dedicated to—and should have been written by—the late Ronnie Campbell, whose ideas and humor swayed many a meeting on electropharmacology and whose influence and presence are sorely missed.

Not long ago, electropharmacology appeared to be evolving into a simple, straightforward field. Basic research on cardiac action potentials (APs) demonstrated antiarrhythmic drug effects that were consistently predictable when interpreted in light of the species of animal and the site in the heart from which they were recorded. Clinical investigation of antiarrhythmic drugs had sufficient success (albeit somewhat anecdotal) to suggest that modification of known molecules to enhance duration of action or reduce certain unwanted side effects would lead to optimization of therapy. However, it was the very limitations of our knowledge that made the future of antiarrhythmic therapy appear predictable and perhaps a bit unchallenging, not auguring the clinical and experimental studies that were to so complicate the future.

In this chapter I first review the status of the field 34 years ago, then reconstruct the train of information that revealed our limited understanding of electropharmacology, and, finally, explore the possibilities that currently confront the clinician and scientist who contemplate the possibility of future electropharmacologic approaches to cardiac arrhythmias.

Electropharmacology in 1970: A Field Whose Future Had Arrived

A key moment in electropharmacology occurred in 1970, when Miles Vaughan Williams reported a classification of antiarrhythmic drugs at an Astra symposium.[1] The initial important update to that classification appeared in 1972, when Bramah Singh, Vaughan Williams' student and research partner, published a joint paper with his mentor, describing a fourth class of antiarrhythmic action.[2] Hence, the four so-called *Vaughan Williams classes* were born. These had the following actions: "[class 1] direct interference with depolarization (local anesthetic type); [class 2] antisympathetic action, by neurone blockade or transmitter competition; [class 3] delay of repolarization; ... and [class 4] interference with calcium conductance."[2]

Although subsequent important updates appeared,[3,4] the Vaughan Williams classification in 1972 not only summarized the art and science of electropharmacology to that date, but effectively determined much of the direction of antiarrhythmic drug development and therapy for the rest of the 20th century.

The beauty of the Vaughan Williams classification lay in the simplicity with which it permitted information about diverse drugs to be categorized in terms of commonalities of action. The approach facilitated the tasks of teaching and learning about antiarrhythmic drugs. Moreover, many of the advances in antiarrhythmic research that occurred in the 1960s to 1990s were interpreted in light of the Vaughan Williams classification in a way that provided further enlightenment regarding drug action. One example is seen in the local anesthetic drugs of class 1, for which knowledge regarding effects on conduction and on repolarization permitted the class to be subdivided. Hence, in 1981 Harrison et al. proposed that "there appear to be three groups included within class 1. Quinidine, procainamide, and disopyramide could be classed 1A—these drugs increase both His-Purkinje conduction and the duration of repolarization. . . . Lidocaine, tocainide, and mexiletine are class 1B agents. They have minimal effect on phase 0 of normal tissue, but decrease it in ischemic tissue and shorten the duration of the AP. . . . encainide, flecainide, lorcainide . . . slow phase 0 but have little or no effect on . . . repolarization. They are class 1C agents."[3]

In subsequent years, several groups published information regarding the kinetics of the effect of sodium channel-blocking local anesthetics; in part, this helped explain the different clinical effects of class 1 drugs. The kinetics were expressed in terms of drug actions on conduction velocity, the $\dot{V}_{max}$ of phase 0 of the AP, or Na current per se. These effects derived from drug blockade of the sodium channel (decreasing $\dot{V}_{max}$ and slowing conduction), which was augmented by increases in stimulation (or heart) rate, a phenomenon referred to as *use dependence*.[5-7] Use dependence meant that the more frequently an Na^+ channel opened, was inactivated, and closed (as would occur with faster heart rates), the more readily a drug could reach its channel-binding site, equilibrate with it, and block the channel. Whereas some highly unionized and lipid-soluble drugs like benzocaine show little to no use dependence (and therefore, only a "tonic" block), most local anesthetics are, to some degree, ionized and show varying degrees of use dependence.[6,7]

What may be most striking in this evolution of knowledge is the internal consistency of the research and its interpretation. However, as the next section explores, there were signs that many clinical and basic research studies had not been designed appropriately to reflect the complexity that characterizes the interaction of drugs, diseased hearts, and arrhythmias.

Electropharmacology Since 1970: The Problems That Arose

For those of us who believed that electrophysiology had a bright future, one of the high points was the Cardiac Arrhythmia Pilot Study (CAPS), reported in 1988.[8] It demonstrated that three Vaughan Williams class 1 drugs—flecainide, encainide, and moricizine (Ethmozine)—suppressed ventricular premature depolarizations in patients following myocardial infarction. Since many individuals then believed that the self-same ectopy was responsible for ventricular tachycardia and fibrillation, it was logical to test the ability of these drugs to reduce mortality. The resultant Cardiac Arrhythmia Suppression Trial (CAST)[9,10] was a major blow to the use of local anesthetic antiarrhythmics. The design of the study and the selection of drugs had the following rationale: (1) flecainide, encainide, and moricizine (the first two, Vaughan Williams class 1C, and the third having mixed class 1 characteristics) suppressed ventricular premature depolarizations in the CAPS Trial; (2) suppression of this type of ventricular ectopy postinfarction was expected to reduce the incidence of lethal arrhythmias; (3) the placebo-treated subjects enrolled in CAST were expected to have a sudden death mortality of 11% in the 3 years after myocardial infarction.[11]

Regrettably, the latter two assumptions were invalid, and CAST, rather than demonstrate the expected reduction in sudden death, provided an important—albeit negative—lesson. First, we learned that drug-induced suppression of ventricular ectopy is not necessarily associated with a reduction in sudden death. Second, we found that the mortality rate in the placebo-treated population studied was not as high as projected. Most importantly, the flecainide/encainide group had a total 10-month mortality of 7.76% versus 3.0% in the placebo group. Looking at the subcategory of nonfatal cardiac arrest or sudden death, the respective percentages were 4.5 and 1.2.[11] This series of findings seriously curtailed the clinical use of flecainide, stopped the clinical development of encainide and moricizine altogether, and put research and development of class 1 drugs, regardless of their actions, into a tailspin from which they have not recovered.

Well before CAST, interest in "class 3" drugs had been increasing.[1] In fact, while various antiarrhythmics were developed in the 1970s through the 1990s, the greatest focus has been on those drugs that prolong repolarization.[12] This effort has been fueled by results with amiodarone, initially considered a class 3 drug.[1] The demonstration of amiodarone's therapeutic efficacy led investigators and pharmaceutic houses alike to stress the importance of developing similar drugs, with the word "similar" interpreted in light of amiodarone's action to prolong repolarization. In other words, it was assumed that the totality of amiodarone's antiarrhythmic effect derived from the prolongation of repolarization. Subsequently, we have learned that amiodarone has many effects—so many that its manufacturer has advertised it as the only drug manifesting all four Vaughan Williams classes of action.

None of the drugs developed as class 3 or "pure class 3" agents has yet proven more effective than, or even as effective as, amiodarone for the treatment of various arrhythmias. Moreover, the Survival with ORal-D Sotalol (SWORD) trial (in which d-sotalol was associated with more deaths than placebo in a postmyocardial infarction population, having somewhat different selection criteria than CAST[13]) demonstrated that a so-called *class 3 effect* may be arrhythmogenic and lethal. In part, this result has been associated with the "reverse use dependence" of many of these drugs. Reverse use dependence refers to the property to prolong repolarization more at low than at rapid heart rates—exactly the opposite of what one would hope for in a drug developed to suppress tachycardias by prolonging repolarization and refractoriness. This prolongation of repolarization at low heart rates facilitates induction of an arrhythmia, torsades de pointes, that is characteristically associated with long Q–T intervals, whether congenital or drug induced.[14]

Given this history, antiarrhythmic drug development is not a high-priority issue for many drug companies now. Despite the fact that sudden death kills over 300,000 persons annually in the United States alone,[15] and that atrial fibrillation (AF) affects 5% of the population older than age 65,[16] the problem of proarrhythmia considered in light of the effectiveness of devices and ablation has markedly reduced enthusiasm for the development of new drugs as well as for the administration of antiarrhythmic drugs to patients.

Why did this situation come about? We assumed electropharmacology was not complex; we assumed we knew more than we did about the pathogenesis of arrhythmias and the actions of antiarrhythmic drugs; in brief, we oversimplified and overstepped. Based on the realization of our ignorance, we have sought new ways to understand the electropharmacology of arrhythmias in light of the electrophysiological mechanisms of the arrhythmias themselves. These considerations are the subject of the rest of this chapter.

Electropharmacology in Light of the Electrophysiological Mechanisms of Arrhythmias

Following the CAST trial, a meeting was convened to consider the classification of antiarrhythmics in the "post–CAST era." The resultant report, referred to as the "Sicilian Gambit,"[17] is often misinterpreted as proposing a new antiarrhythmic drug classification. In fact, it argued against the classification of antiarrhythmics. It did so because a major quandary facing those who study antiarrhythmic therapy and the mechanisms of action of antiarrhythmic drugs resides in the complexities of drug action. To demonstrate, Table 7-1 groups drugs in a fashion not dissimilar from the Vaughan Williams classes: "Class 1A" drugs are grouped together, "class 1B drugs" are grouped together, and so on, although they are not identified as such. As an example of complexity, consider just the class 1A compounds: quinidine, procainamide, and disopyramide. Although their effects on the fast sodium channel are similar, their effects on repolarizing currents and on autonomic function differ importantly, in ways that likely contribute to the variability of effect of these drugs clinically.

TABLE 7-1 Sicilian Gambit Listing of Antiarrhythmic Drugs

Drug	Channels						Receptors				Pumps	Clinical Effects			ECG Effects		
	Na Fast	Na Med	Na Slow	Ca	K	I_f	α	β	M_2	A1	Na-K ATPas	Left ventricular function	Sinus Rate	Extracardiac	P–R interval	QRS width	J–T interval
Lidocaine	○											→	→	⊘			↓
Mexiletine	○											→	→	⊘			↓
Tocainide	○											→	→	●			↓
Moricizine	●I											↓	→	○		↑	
Procainamide		●A			⊘							↓	→	●	↑	↑	↑
Disopyramide		●A			⊘				○			↓	→	⊘	↑↓	↑	↑
Quinidine		●A			⊘		○		○			→	↑	⊘	↑↓	↑	↑
Propafenone		●A						⊘				↓	↓	○	↑	↑	
Aprinidine		●I		○	○	○						→	→	⊘	↑	↑	→
Cibenzoline			●A	○	⊘				○			↓	→	○	↑	↑	→
Pirmenol			●A		⊘				○			↓	↑	○	↑	↑	↓→
Flecainide			●A		○							↓	→	○	↑	↑	
Pilsicainide			●A									↓→	→	○	↑	↑	
Encainide			●A									↓	→	○	↑	↑	
Bepridil	○			●	⊘							?	↓	○			↑
Verapamil	○			●			⊘					↓	↓	○	↑		
Diltiazem				⊘								↓	↓	○	↑		
Bretylium					●		◪	◪				→	↓	○			↑
Sotalol					●			●				↓	↓	○	↑		↑
Amiodarone	○			○	●		⊘	⊘				→	↓	●	↑		↑
Alinidine					⊘	●						?	↓	●			
Nadolol								●				↓	↓	○	↑		
Propranolol	○							●				↓	↓	○	↑		
Atropine									●			→	↑	⊘	↓		
Adenosine										□		?	↓	○	↑		
Digoxin									□		●	↑	↓	●	↑		↓

Relative potency of block: ○ Low ⊘ Moderate ● High
□ = Agonist ◪ = Agonist/Antagonist
A = Activated state blocker
I = Inactivated state blocker

(Modified from Members of the Sicilian Gambit: Antiarrhythmic Therapy: A Pathophysiologic Approach. Futura Publishing Co, Inc, Armonk, NY, 1994.)

More importantly, consider the so-called *class 3 drugs.* Amiodarone, thought to be prototypical of the group, not only prolongs repolarization by blocking repolarizing potassium channels but also blocks inward sodium and calcium currents and is antiadrenergic. In contrast, dofetilide prolongs repolarization via block of the potassium current, I_{Kr}, alone[18,19]; azimilide blocks both the potassium currents I_{Kr} and I_{Ks}; and ibutilide blocks I_{Kr} and increases I_{Na}.[20] Hence, each of these four class 3 drugs has a very different mechanism of action. In fact, any drug that increases inward current or decreases outward current, or both, during repolarization would have a class 3 effect.

How, then, do we classify drugs? In fact, can we classify drugs; or should we be thinking more about arrhythmias per se and the effects of drugs in light of the electrophysiology of arrhythmias?

Originally, the Sicilian Gambit proposed that understanding the electrophysiology of arrhythmias and the ion channel mechanisms that determined normal and abnormal electrical activity might impose an adequate formalism for understanding drug actions. In part, this was based on the information in Figure 7-1. This demonstrates two points: (1) even in the normal heart, different tissues have different APs to which the contributions of specific ion channels vary importantly and (2) for those ion channels contributing to the AP, we have a growing body of information about the molecular determinants. A third factor, not presented in the figure, is that important alterations in ion channels occur in specific disease entities and these alterations complicate drug-channel interactions and the expression of drug effect.[21]

Ion Channels and Their Molecular Determinants

Consider the ventricular myocardium as an example: Its AP is initiated by the entry of sodium ion during phase 0. The gene determining the fast inward sodium current in human subjects is *SCN5A*. The sodium channel operates not only to depolarize the cell during phase 0 but to maintain a small inward current during the plateau of the AP. Using this channel alone as an example, we can begin to explore some understanding of pathophysiology. For example, if the inward sodium

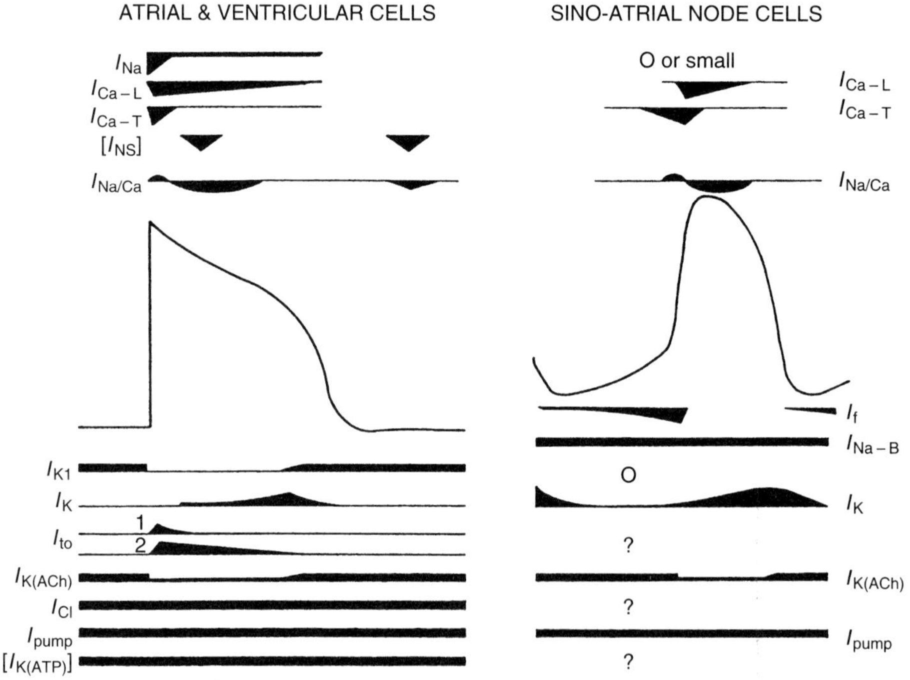

FIGURE 7-1 Currents and channels involved in generating the resting and action potential. The time course of a stylized action potential of atrial and ventricular cells is shown on the **left** and of sinoatrial node cells on the **right. Above** and **below** are the various channels and pumps that contribute currents underlying the electrical events. Where possible, the approximate time courses of the currents associated with the channels or pumps are shown symbolically without effort to represent their magnitudes relative to each other. I_K incorporates at least two currents, I_{Kr} and I_{Ks}. There appears to be an ultrarapid component, as well, designated I_{Kur}. The *heavy bars* for I_{Cl}, I_{pump}, and $I_{K(ATP)}$ only indicate the presence of these channels or pump, without implying magnitude of currents, since that would vary with physiologic and pathophysiologic conditions. The channels identified by *brackets* (I_{Ns} and $I_{K(ATP)}$) imply that they are active only under pathologic conditions. For the sinoatrial node cells, I_{Na} and I_{K1} are small or absent. *Question marks* indicate that experimental evidence is not yet available to determine the presence of these channels in sinoatrial cell membranes. Although it is likely that other ionic current mechanisms exist, they are not shown here because their roles in electrogenesis are not sufficiently well defined. (Modified from Members of the Sicilian Gambit: Antiarrhythmic Therapy: A Pathophysiologic Approach. Armonk, NY, Futura Publishing, 1994.)

current were sustained longer or were of greater magnitude than normal, or both, during the plateau, then the AP duration would increase. It is precisely this that occurs in one variant of the congenital long QT syndrome (LQTS). Here, the pathologically prolonged Q–T interval results from any one of several anomalies of the structure of *SCN5A*. Yet other variations in structure of the same sodium channel result in its faster recovery from inactivation following the upstroke of an AP. The result is elevation of the J point and early ST segment on selected ECG leads and an increased likelihoodof arrhythmias in the absence of QT prolongation. These variants on *SCN5A* structure are responsiblefor another type of inherited arrhythmia, Brugada syndrome.

The complexity of drug-channel interactions is compounded by the contributions of several inward and outward currents to the AP plateau. In addition to the earlier-mentioned *SCN5A*, an inward current carried by calcium ($I_{Ca,L}$) sustains the plateau.[17] This channel has been defined as a receptor for dihydropyridines, a number of which are used clinically as calcium current blockers.

The outward currents determining AP repolarization are carried by potassium. The first of these, the transient outward potassium current, I_{to}, is determined in large part in the human and the dog by the gene encoding its α subunit, *Kv4.3*.[22] Very importantly, with respect to our understanding of transmural differences in AP characteristics is that there is a gradient for Kv4.3 and for I_{to} across the ventricular wall, with high current density in the epicardium and minimal current density subendocardially. This transmural gradient in current is accompanied by a difference in the phase 1 notch of repolarization, which is determined by I_{to} and is large epicardially and disappearingly small in the subendocardium. Although I_{to} has not been targeted specifically in antiarrhythmic therapy, a number of local anesthetics (e.g., quinidine, flecainide) and AP-prolonging drugs (e.g., azimilide) also block this current, adding to the complexity of their action.[17]

The major outward current responsible for repolarization is the delayed rectifier, I_K. This has three components; one, ultra-rapidly activating and referred to as I_{Kur}, is present more in atrium than ventricle.[23,24] I_{Kr}, a rapidly activating component, and I_{Ks}, a slowly activating component, are present in atrium and ventricle.[25] The genes for the α subunits of I_{Kr} and I_{Ks}, respectively *HERG* (the **H**uman **E**ther-a-go-go-**R**elated **G**ene) and *KvLQT1*, are important not only physiologically but pathophysiologically, as abnormalities in both have been implicated in different variants on the congenital LQTS.[14] Specifically, these abnormalities reduce repolarizing potassium current, prolonging the AP duration and predisposing to torsades de pointes. Moreover, the *HERG* channel is bound to by various cardiac and noncardiac drugs that can induce an acquired LQTS.[26] These drugs (partially listed in Table 7-2) all have in common a propensity to block I_{Kr}, such that when given individually and—more frequently in combination (e.g., a nonsedating antihistamine and a macrolide antibiotic)—they pose the risk of pathologically prolonging the Q–T interval and inducing torsades de pointes.

TABLE 7-2 Examples of Drugs That Can Prolong the Q–T Interval and Induce Torsades De Pointes

Class	Examples
Antiarrhythmic drugs:	Amiodarone Azimilide Bretylium Clofilium Dofetilide Disopyramide Ibutilide N-acetyl-procainamide Procainamide Propafenone Quinidine Sematilide D,L-sotalol, D-sotalol
Vasodilators/anti-ischemic agents:	Bepridil Lidoflazine Prenylamine
Psychiatric drugs:	Amitriptyline Chlorpromazine Citalopram Desipramine Doxepin Droperidol Haloperidol Imipramine Lithium Maprotiline Pimozide Sertindole Sultopride Thioridazine Timiperone Zimelidine
Antimicrobial and antimalarial drugs:	Amantadine Clarithromycin Chloroquine Cotrimoxazole Erythromycin Grepafloxacin Halofantrine Ketoconazole Pentamidine Quinine Spiramycin Sparfloxacin
Antihistaminics:	Astemizole Diphenhydramine Terfenadine
Miscellaneous drugs:	Budipine Cisapride Probucol Terodiline Vasopressin

Modified from Haverkamp W, Breithardt G, Camm AJ, et al: The potential for QT prolongation and proarrhythmia by non-antiarrhythmic drugs: Clinical and regulatory implications. Report on a policy conference of the European Society of Cardiology. Eur Heart J 2000;21:1216-1231.

Another ion current of importance to repolarization is the inward rectifier, I_{K1}, which is responsible for the terminal portion of phase 3 as well as for the maintenance of a negative transmembrane potential.[17] The gene family determining this is *KIR*.

Distribution of Ion Channels is Reflected in Regional Dispersion of Action Potentials

Critically important to our understanding of ion channels is how their distribution varies across sites in the myocardium, thereby determining differences in local electrophysiological properties. For example, I_{to} is of high density in the ventricular epicardium, intermediate density in the midmyocardium, and low density in the endocardium.[27,28] I_{Na} is largest in the midmyocardium and I_{Ks} is smallest in the midmyocardium.[29] The net expression of these different densities electrophysiologically is as a longer AP duration in midmyocardium than in epi- and endocardium (the result of a larger I_{Na} and a smaller I_{Ks} in midmyocardium) and a large phase 1 notch in epi- and midmyocardium but not in endocardium (reflecting the transmural gradient for I_{to}).[30-32]

There also are different densities of I_{K1} at different sites. In ventricular myocardium, I_{K1} is large, providing a net outward current during the termination of phase 3 and during phase 4.[17] This net outward current "clamps" the membrane at a negative potential and maintains it as such. In pacemaker tissues, such as Purkinje fiber or sinus node, I_{K1} is of lesser density. The result is that the magnitude of outward current is less able to counteract the inward current carried by the pacemaker current, I_f, thereby permitting spontaneous activity to appear. These are just two examples of the profound heterogeneity in repolarizing currents and in AP characteristics that occur regionally in the heart. Additional modifiers having significant importance but beyond the scope of this discussion are changes in currents with growth, development and aging, and differences in currents relating to gender and to gonadal steroids.[33,34]

Impulse Initiation and Impulse Conduction

The process of pace-making is critical to normal impulse initiation in the heart. The major pacemaker current, I_f, is inward, carried by Na, and activated on hyperpolarization of the cell membrane.[35,36] I_f determines the initiation and initial slope of the pacemaker potential. It decreases relatively rapidly and the phase 4 depolarization is then maintained by a diminishing outward potassium current.[37] Both the inward current at the start of phase 4 and the diminishing potassium current during phase 4 have the net effect of accumulating positive charge in the cell, such that the membrane depolarizes. Finally, during the latter portion of phase 4, calcium carried via I_{CaT} provides additional inward current.[38]

Also under intense electrophysiological study at present are the characteristics of cell–cell interactions in determining both activation and repolarization. When studied in isolation, cells in different regions of the heart have dissimilar action potentials, as described earlier.[30,31] Despite the different populations of ion channels in these regions, in a normal heart, cells that are near one another have very similar AP contours. The imposition of a homogenous repolarization pattern in the normal heart is likely an important byproduct of the gap junctions. Junctions that are concentrated at the ends (i.e., along the long axes) of cardiac myocytes provide a low resistance site for current flow from cell to cell.[39] There are also gap junctions at the lateral margins of cells, but these are of far lower density.[39] Each cell contributes a hemi-channel—or connexon—to the pairing, and two hemi-channels together constitute the gap junction that links two cells. Assembly of a hemichannel requires the incorporation of six gap junctional proteins, or connexins, the major one in the ventricle being connexin43.[40,41] The electrotonic flow of current from cell to cell results in the similarity of AP contours in nearby cells. In ischemia, infarction, or other pathologic states, gap junctions can be uncoupled to varying degrees. In these situations, the expression of different ion channel properties in different regions can come to the fore in promoting inhomogeneity of repolarization.

There are two other aspects of connexins that are important here. First is their property to transport rather large molecules important in cell–cell communication and signaling processes from cell to cell.[42,43] Second is the density of low-resistance junctions at the ends of the myocytes. This pattern determines the direction and rapidity of propagation of the cardiac impulse. Hence, along with the upstroke of an AP, which provides the depolarizing signal for the next cell in a conducting pathway, the gap junction that determines the resistance encountered by an advancing wavefront is a major factor in impulse propagation.

Ion Channel Changes in Pathologic States

In the early days of electropharmacology, most investigators focused on the study of normal cells alone: In fact, the entire Vaughan Williams classification is based on this. However, to consider normal cells uniquely is to avoid the picture characteristic of most diseased hearts and most arrhythmic hearts. Using congestive failure as an example, there tends to be a depolarization of membrane potential, a loss of the AP notch, and a prolongation of AP duration.[44] These changes are attributed to reduced densities of I_{to} and I_{K1} and possibly the delayed rectifier, I_K. In addition, there is often triggered activity that may result from early or delayed after depolarizations.[44]

Another example is myocardial infarction, in which significant changes occur in cells in the infarct itself. These cells are dying or dead and depolarized to varying degrees during evolution of the infarct. There are also important changes in surviving cells of the epicardial border zones (i.e., the tissues surrounding the infarct).

They become hypertrophic and with this show prolongation of repolarization and associated reduction in density of I_{to} and I_{K1}.[45]

Both conditions, congestive failure and infarction, are importantly complicated by arrhythmias. How can an understanding of drug effect on normal cells help us consider how these drugs might act in arrhythmic settings? Indeed, it is likely that in the setting of congestive failure or infarction, or both, altered ion channels are important to the genesis of arrhythmias and are potential targets for antiarrhythmic drugs. Recent attempts to understand the electrophysiological mechanisms for arrhythmias in these settings are leading to exploration of different types of effects of existing drugs; based on the differences in the individual channel characteristics and the overall channel populations.[21]

Gap junctions are altered in various pathologic situations such that their density and ability to pass current may decrease, and their distribution can be changed such that they are of increased density on the lateral margins of the cells. Such alteration of gap junctional distribution and density occurs in the setting of myocardial infarction, especially in the epicardial border zone,[46] where it can facilitate altered pathways of activation and promote reentrant arrhythmias. Hence, the gap junctions have become an important target for research, with the thought that their pharmacologic modulation might be antiarrhythmic.

Electropharmacology in Light of the Pathophysiology of Arrhythmias

I shift focus to the more general descriptor "pathophysiology of arrhythmias" rather than "electrophysiological mechanisms of arrhythmias" to make the following point: Too much of our research and our thought has gone into thinking of arrhythmias in general and electropharmacology in particular with blinders that have limited us to electrophysiological mechanisms. This is not to imply that the relatively narrow focus on electrophysiology is flawed: indeed, it is this focus that has taught us so much about the determinants of cardiac rhythm in various settings. It becomes flawed as a focus, however, when we want to understand and influence the spectrum of events that determine arrhythmias in diseased hearts. In other words, not enough effort has been expended on the overall pathophysiology of the arrhythmia: the real blend of factors that determines an arrhythmic event. This distinction can be understood by reviewing Figure 7-2, which emphasizes the various determinants of cardiac rhythm and arrhythmias. Constitutional factors such as an individual's genetic makeup and gender are central determinants. Important environmental modifiers include the autonomic nervous system, hemodynamic load, myocardial perfusion, and age.

In the state of health, we can envisage an active gene program and normal phenotype of cardiac function

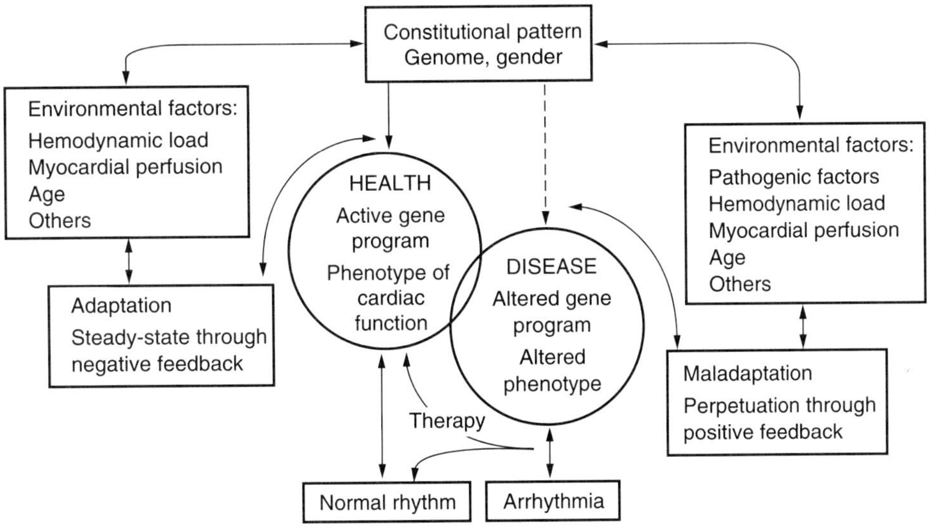

FIGURE 7-2 A schematic relating the occurrence of sinus rhythm and of arrhythmias to the general cardiac homeostasis. A complex series of events contributes to the arrhythmia, as depicted here. Some are immediately related to the arrhythmia, and some more distantly related. Hence, we may consider both the genome and the gender of an individual as prime determinants of normal function, modified importantly by environmental factors such as those listed. All these factors will condition the response of the individual to the intervention of a disease process, which displaces the homeostasis. To the extent that therapy focuses on a target immediately associated with the arrhythmia (e.g., an ion channel in the patient with ischemic heart disease and ventricular tachycardia), its beneficial effects may be counterbalanced by its deleterious effects on, for example, contractility of the myocardium. The ability to identify targets that are "upstream" (i.e., that modify the activity of molecules that are prime determinants of the disease process) should improve significantly the extent to which successful prevention is effected. This approach identifies prevention of arrhythmias as being of more benefit to the patient than their treatment, while recognizing the need to search for better treatment in settings where prevention has failed. (Modified from Members of the Sicilian Gambit: Antiarrhythmic Therapy: A Pathophysiologic Approach. Armonk, NY, Futura Publishing, 1994.)

being maintained in a steady state of adaptation to the continual demands put on the heart, via a negative feedback pattern. In contrast, in the setting of disease, variations on the constitutional or environmental factors mentioned can be construed to operate via positive feedback to produce both an altered gene program and altered phenotype, resulting in an arrhythmia (as well as other expressions of the disease process). Figure 7-3 focuses us on a major reason for the frequent failure of antiarrhythmic drug therapy; that is, it has been directed entirely toward returning the heart to sinus rhythm. Far more important as we understand more about the processes that cause arrhythmias would be to modify the underlying disease itself, thereby returning to a state of health.

An instructive example of how this approach can be applied to therapy is seen with the congenital LQTS. As mentioned earlier, lesions in a sodium channel (*SCN5A*) and in two potassium channels (*HERG* and *KvLQT1*) account for many of the LQTS cases studied genetically to date.[14] Yet there are complicating factors: For example, one might expect that in this autosomal dominant disease, the occurrences of arrhythmias and death would be pretty much determined by the expression of the genetic abnormality and of the prolonged Q–T interval on ECG, and that pathology would be equally distributed in males and females. However, there is female predominance in the risk of sudden death.[14] Moreover, some individuals with long Q–T intervals do not have arrhythmias or die and others with relatively normal Q–T intervals have arrhythmias and die. It is clearly the ion channel abnormalities that induce the QT prolongation (*SCN5A* by increasing inward plateau current and *HERG* and *KvLQT1* by decreasing repolarizing K current). Yet some other factor—a trigger—is required if arrhythmias and death are to occur. For *HERG* and especially *KvLQT1,* this extraneous factor appears to be catecholamine via the β-adrenergic receptor. Interestingly, it had been proposed initially that LQTS results from an imbalance of sympathetic innervation to the heart.[47,48] This viewpoint supported the use of β-blockade as primary therapy, with rather favorable results overall,[14] despite the fact that the primary lesions reside in specific cardiac ion channels.[14]

If we consider LQTS as a paradigm for thinking about the causes of arrhythmias, what does it teach us? First, for targeting therapy, we may want to identify targets in the myocardial substrate (e.g., the ion channels) or trigger targets (e.g., the β-adrenergic receptor), or both. Second, even if we identify primary pathologic events, seen in the focus on substrate and trigger, there may be other important modifiers (e.g., gender). Third, if we can focus on a specific target, then highly selective therapy may be designed that might be expected to impact upon the arrhythmia (e.g., the use of mexiletine to treat *SCN5A* patients[49] based on mexiletine's propensity to block the inward plateau sodium current). Fourth, with the best of therapeutic intentions and the most knowledge, superior therapy may be arrived at based on wrong or incomplete information (e.g., the success of β-blockade as therapy). Finally, given the association of gene with lesion, it is likely to become possible to identify disease-specific, and even patient-specific, therapy.

Arrhythmias and Remodeling

The problem with LQTS as an example is that despite its complexity, it is ultimately simple in comparison to garden-variety diseases such as ischemia- or failure-induced arrhythmogenesis, or both. Yet it is in the latter

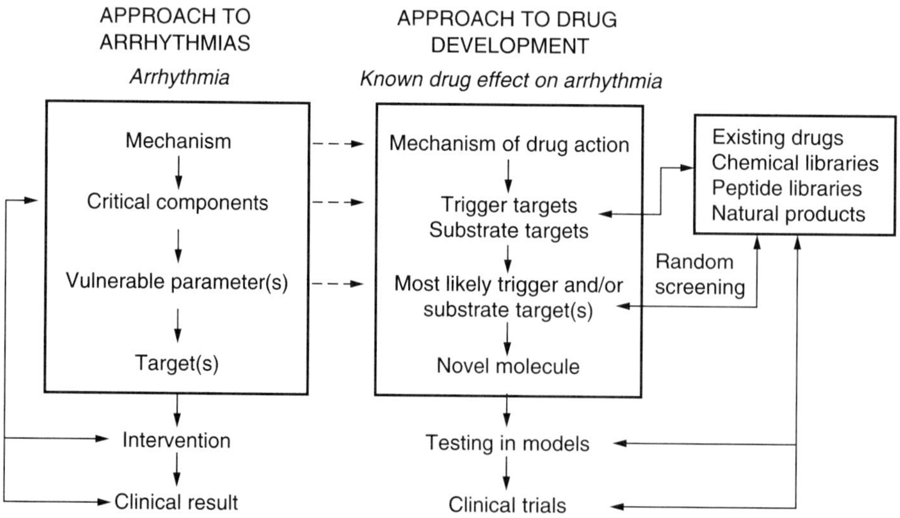

FIGURE 7-3 The relationship of the Sicilian Gambit approach to arrhythmia diagnosis and the development of antiarrhythmic therapies. For every critical component of an arrhythmia, there are potential substrate targets or trigger targets, or both. Various sources provide molecules to be screened with relation to the potential targets. With random (or, when possible) focused screening of both targets and molecules, one can identify a novel and effective molecule and can use this to intervene in arrhythmia models and eventually in patients. See text for further discussion. (Modified from Members of the Sicilian Gambit: Antiarrhythmic Therapy: A Pathophysiologic Approach. Armonk, NY, Futura Publishing, 1994.)

that the major challenge to and demand for prevention and therapy reside. A great part of the challenge lies in the capacity of the heart to remodel. Remodeling is a complex process that can be physiologic or pathologic.[21,50] An example of the former is the left ventricular hypertrophy that occurs postnatally in response to the closure of the patent ductus arteriosus and the demands the systemic circulation places on the left ventricle. In this example, there is appreciable hypertrophy, but the entire system remains compensated. An obvious—and parallel—example of pathologic remodeling is the left ventricular hypertrophy that occurs in systemic hypertension. If untreated, this progresses to the point of decompensation.

With respect to arrhythmias, remodeling is generally thought of as electrophysiological or structural. A well-known example of electrophysiological remodeling is the shortening of the effective refractory period and of AP duration that occurs when the atria are paced rapidly.[51] Moreover, this change is associated with an increased ease of inducing AF, which, itself, can accelerate repolarization and shorten the effective refractory period.[51] This behavior led Wijffels et al to state that "AF begets AF," a quote that has become the byword for electrical remodeling.[51] The problem in considering this type of electrophysiological remodeling is that it does not occur in isolation: Rather, not only does pacing induce electrophysiological remodeling and AF, but it induces structural change in the atria as well.[52] Hence, what commences as a simple intervention—a change in heart rate with an altered activation pathway (depending on the site from which the atrium is paced) such that repolarization and refractoriness are altered rapidly—becomes a widespread lesion in the atrium involving its structure.

Not only do changes in the AP that accompany rapid pacing induce AF, but there are also changes in ion channel properties. In brief, as documented in human atrium, there are roughly 60% reductions in density of $I_{Ca,L}$, I_{to},[53,54] and—in one report—I_{Kur}.[55] There are also reports of increases in I_{K1}.[55,54] The net result of all these changes is the decreased AP duration and lower plateau that characterizes the fibrillating atrium. Interestingly, these alterations in the AP—which are largely consistent with those identified in rapidly paced canine atria—are seen even when rapid pacing does *not* cause fibrillation in the canine heart. This leads to the inescapable conclusion that the AP changes and the ion channel changes that characterize AF cannot or do not in themselves cause fibrillation but rather are determined by atrial rate or activation, or both, and are the byproduct of some causative event.

What is the causative event? This very important question has not been answered. Indeed, it is likely that there is no single causative event, given the multiple conditions associated with susceptibility to AF. Representative of these are age, structural heart disease, and cardiac failure.[16] What all have in common is an alteration in the structural substrate of the atrium, such that stretch on myocytes is altered.[16] Hence, it is possible that seeking means to normalize stress-strain relationships may identify new targets. In the meantime, the example of AF shows us that remodeling is a powerful and arrhythmogenic result of and determinant of altered cardiac structure and function.

Future Strategies for Electropharmacology

How do we derive an electropharmacologic strategy in light of the complexities that characterize the arrhythmic heart? Insights may be derived from a recent workshop[16] as follows: First, the basis for the initial success and ultimate failure of antiarrhythmic drugs and their inconsistencies of action[56,57] must be understood. We must learn whether ion channel changes evolve to a point beyond the range at which antiarrhythmic drugs can be expected to be effective, or if there is excess structural change such as fibrosis or uncoupling of gap junctions, or both. It is possible that arrhythmia prevention may be achieved with better ion channel–targeted drugs. For example, given the important downregulation of I_{Kur} and $I_{Ca,L}$ in remodeled atria,[53,55] development of atrial-selective drugs that upregulate or open these channels might be appropriate.

If designing a better ion channel–blocking drug is not an optimal antiarrhythmic goal, better answers may lie in considering the determinants of remodeling and incorporating them in a wider antiarrhythmic strategy. AF provides a good example here. Clinical trials have already suggested that suppressing angiotensin II synthesis or preventing its receptor binding reduces recurrences of AF.[58,59] Angiotensin II is a powerful remodeling agent, causing cardiac hypertrophy and increasing the L-type Ca current.[60,61] Moreover, its synthesis by cardiac cells is induced by altering stress-strain relationships,[60] precisely the types of events that may occur in the setting of cardiac ischemia/failure or valvular heart disease. Hence, there is a pathophysiologic basis for understanding why interference with angiotensin II synthesis or binding may be a good antiarrhythmic strategy. If angiotensin-converting enzyme (ACE) inhibitors slow or reverse mechanical remodeling and create more homogeneous contraction patterns, this would promote homogeneous mechanoelectric coupling and reduce electrophysiological dispersion.

Other directions that have been suggested for prevention of AF relate to the complexity of changes in Ca handling that appear involved in arrhythmogenesis as well as the metabolic status of the heart.[16] The latter may influence the propensity of atrium to undergo electrophysiological remodeling once an arrhythmia begins. Interventions altering the redox state of fibrillating atria may prevent the short-term electrophysiologic remodeling that accompanies AF initiation and decrease the propensity of AF to reinitiate after cardioversion.[16] Moreover, the time required to recover normal electrophysiological function upon termination of AF may relate to its metabolic state at that time as well as to factors such as autonomic remodeling.[62]

To summarize, then, while it is still useful to have available specific antiarrhythmic drugs as part of the therapeutic armamentarium, the understanding of pathophysiology and exploration of the interactions among electrophysiological, structural, humoral, genetic, and other components should permit us to devise more effective electropharmacologic approaches to therapy than now is the case.

Given our expectations for the future, how can the clinician and the pharmacologist approach the questions of antiarrhythmic therapy and of drug design? There is no easy answer to this question. To date, empiricism has driven most of the work of clinicians and drug developers alike. However, the development of enlightened and individualized therapy and incorporation of the needs of patient populations into drug design can be facilitated by understanding pathophysiology in general, and by trying to understand the pathophysiology in each patient. Some years ago, the Sicilian Gambit suggested the approach described in Figure 7-2 as a means to understand the pathology in each patient, to individualize therapy, and to try to take the lessons of success or failure into account, thereby improving matters for the next generation of patients.

In brief, for any arrhythmia for which we know or surmise a mechanism, certain critical components can be identified that constitute triggers and substrates. For example, in AV nodal reentry the substrate includes not only atrial and ventricular myocardium but the AV node and the pathways for reentry to the right atrium. Vulnerable parameters that are the most likely targets for intervention are identified. For drug therapy, the target is the AV nodal AP, and its suppression by calcium channel antagonists or β-blocking drugs. For ablation, the reentrant site of the autonomic pathway into the atrium is the preferable target. Intervention gives rise to a clinical result that—at best—is curative and, in any case, instructive. The challenge is to attain the success seen in relatively straightforward settings like AV nodal reentry in more complex arrhythmias such as ischemia-induced ventricular tachycardia or fibrillation.

A modification of this approach has been used to recommend linkages between clinical needs and drug development (see Fig. 7-3). Targets can be identified at the level of trigger or substrate, or both. By screening existing drugs, libraries, and natural products, one can attempt to optimize target identification and understand drug-target intervention. Novel molecules can then be synthesized to interact with this target, or existing molecules can be altered to optimize benefit over risk of therapy, or both.

SUMMARY

Although the field of electropharmacology has "come a long way," much antiarrhythmic therapy has been based more on empiricism than strict scientific knowledge, and some has been harmful. It is likely that continuing to consider antiarrhythmic therapy and drug development in the microcosm of the management and prevention of an electrical event is doomed to be a holding operation, palliating at best, but not curing. By opening the vista of electropharmacology to information that is molecular, genetic, structural, and pathophysiologically based, we find room for growth in our knowledge and in the possibilities offered for design and delivery of therapy. It is in these directions that our most exciting opportunities lie and to which our attention must increasingly turn.

ACKNOWLEDGEMENTS

Certain referenced studies were supported by USPHS-NHLBI grant HL-28958, HL-53956, and HL-67449 and by the Wild Wings Foundation.

The author expresses his gratitude to Ms. Eileen Franey for her careful attention to the preparation of this manuscript.

REFERENCES

1. Vaughan Williams EM: Classification of antiarrhythmic drugs. In Sandoe E, Flensted-Jensen E, Olsen KH (eds): Cardiac Arrhythmias. Sodertalje, Sweden, Astra, 1971, pp 449-72.
2. Singh BN, Vaughan Williams EM: A fourth class of anti-dysrhythmic action? Effect of verapamil on ouabain toxicity, on atrial and ventricular intracellular potentials, and on other features of cardiac function. Cardiovasc Res 1972;6:109-19.
3. Harrison DC, Winkle RA, Sami M, et al: Encainide: A new and potent antiarrhythmic agent. In Harrison DC, Hall GK (eds): Cardiac Arrhythmias, a Decade of Progress. Boston, Medical Publishers, 1981, pp 315-30.
4. Campbell TJ: Kinetics of onset of rate-dependent effects of class I antiarrhythmic drugs are important in determining their effects on refractoriness in guinea-pig ventricle, and provide a theoretical basis for their subclassification. Cardiovasc Res 1983;17:344-52.
5. Courtney K: Mechanisms of frequency-dependent inhibition of sodium currents in frog myelinated nerve by the lidocaine derivative GEA-968. J Pharmacol Exp Ther 1975;195:225-36.
6. Hondeghem LM, Katzung BG: Time- and voltage-dependent interactions of antiarrhythmic drugs with cardiac sodium channels. Biochim Biophys Acta 1977;472:373-98.
7. Hondeghem LM, Katzung BG: Antiarrhythmic agents: The modulated receptor mechanism of action of sodium and calcium blocking drugs. Annu Rev Pharmacol Toxicol 1984;24:387-423.
8. The Cardiac Arrhythmia Pilot Study (CAPS) Investigators: Effects of encainide, flecainide, imipramine and morizine on ventricular arrhythmias during the year after acute myocardial infarction. Am J Cardiol 1988;61:501-9.
9. CAST Investigators: Preliminary Report: Effects of encainide and flecainide on mortality in a randomized trial of arrhythmia suppression. N Engl J Med 1989;321:406-12.
10. The Cardiac Arrhythmia Suppression Trial-II (CAST) Investigators: Effect of the antiarrhythmic agent moricizine on survival after myocardial infarction. N Engl J Med 1992;327:227-33.
11. Task Force of the Working Group on Arrhythmias of the European Society of Cardiology: CAST and beyond. Implications of the cardiac arrhythmia suppression trials. Eur Heart J 1990;11:194-9.
12. Hondeghem LM, Snyders DJ: Class III antiarrhythmic agents have a lot of potential but a long way to go. Reduced effectiveness and dangers of reversed use dependence. Circulation 1990;81:676-90.
13. Waldo AL, Camm AJ, Deruyter H, et al: Effect of d-sotalol on mortality in patients with left ventricular dysfunction after recent and remote myocardial infarction. Lancet 1996;348:7-12.
14. Roden DM, Lazzara R, Rosen MR, et al: Multiple mechanisms in the long-QT syndrome: Current knowledge, gaps, and future directions. Circulation 1996;94:1996-2012.
15. Spooner PM, Rosen MR: Perspectives on arrhythmogenesis, antiarrhythmic strategies and sudden cardiac death. In Spooner PS, Rosen MR (ed): Foundations of Cardiac Arrhythmias. New York, Marcel Dekker, 2000.
16. Allessie MA, Boyden PA, Camm AJ, et al: Pathophysiology and prevention of atrial fibrillation. Circulation 2001;103:769-77.
17. Task Force of the Working Group on Arrhythmias of the European Society of Cardiology: The Sicilian Gambit: A new approach to the

classification of antiarrhythmic drugs based on their action on arrhythmogenic mechanisms. Circulation 1991;84:1831-51.
18. Carmeliet E: Voltage- and time-dependent block of the delayed K^+ current in cardiac myocytes by dofetilide. J Pharmacol Exp Ther 1993;262:809-17.
19. Jurkiewicz NK, Sanguinetti MC: Rate-dependent prolongation of cardiac action potentials by a methansulfonanilide class III antiarrhythmic agent. Circ Res 1993;72:75-83.
20. Wesley RC Jr, Farkhani F, Morgan D, et al: Ibutilide: Enhanced defibrillation via plateau sodium current activation. Am J Physiol 1993;264:H1269-74.
21. Members of the Sicilian Gambit: The search for novel antiarrhythmic strategies. Eur Heart J 1998;19:1178-96.
22. Dixon JE, Shi W, Wang H-S, et al: Role of the Kv4.3 K^+ channel in ventricular muscle. A molecular correlate for the transient outward current. Circ Res 1996;79:659-68.
23. Backx PH, Marban E: Background potassium current active during the plateau of the action potential in guinea pig ventricular myocytes. Circ Res 1993;72:890-900.
24. Wang Z, Fermini B, Nattel S: Delayed rectifier outward current and repolarization in human atrial myocytes. Circ Res 1993;73:276-85.
25. Sanguinetti MC, Jurkiewicz NK: Two components of cardiac delayed rectifier K^+ currents. Differential sensitivity to block by class III antiarrhythmic agents. J Gen Physiol 1990;96:195-215.
26. Haverkamp W, Breithardt G, Camm AJ, et al: The potential for QT prolongation and proarrhythmia by non-antiarrhythmic drugs: Clinical and regulatory implications. Report on a policy conference of the European Society of Cardiology. Eur Heart J 2000;21:1216-31.
27. Litovsky SH, Antzelevitch C: Transient outward current prominent in canine ventricular epicardium but not endocardium. Circ Res 1988;62:116-26.
28. Litovsky SH, Antzelevitch C: Rate dependence of action potential duration and refractoriness in canine ventricular endocardium differs from that of epicardium: Role of the transient outward current. J Am Coll Cardiol 1989;14:1053-66.
29. Liu DW, Gintant GA, Antzelevitch C: Ionic bases for electrophysiological distinctions among epicardial, midmyocardial, and endocardial myocytes from the free wall of the canine left ventricle. Circ Res 1993;72:671-87.
30. Sicouri S, Antzelevitch C: A subpopulation of cells with unique electrophysiological properties in the deep subepicardium of the canine ventricle: The M cell. Circ Res 1991;68:1729-41.
31. Sicouri S, Antzelevitch C: Electrophysiologic characteristics of M cells in the canine left ventricular wall. J Cardiovasc Electrophysiol 1995;6:591-603.
32. Anyukhovsky EP, Sosunov EA, Gainullin RZ, et al: The controversial M cell. J Cardiovasc Electrophysiol 1999;10:244-60.
33. Drici MD, Burklow TR, Haridasse V, et al: Sex hormones prolong the QT interval and downregulate potassium channel expression in the rabbit heart. Circulation 2000;94:1471-4.
34. Hara M, Danilo PJ, Rosen MR: Effects of gonadal steroids on ventricular repolarization and on the response to E4031. J Pharmacol Exp Ther 1998;285:1068-72.
35. DiFrancesco D: A new interpretation of the pacemaker current in calf Purkinje fibres. J Physiol 1981;314:359-76.
36. DiFrancesco D: A study of the ionic nature of the pacemaker current in calf Purkinje fibres. J Physiol 1981;314:377-93.
37. Vassalle M, Yu H, Cohen I: A time-dependent decay of a K^+ current generates a "pacemaker" current at diastolic potentials in Purkinje myocytes. Circulation 1994;90:14.
38. Hagiwara N, Irisawa H, Kameyama M: Contribution of two types of calcium currents to the pacemaker potentials of rabbit sinoatrial node cells. J Physiol (Lond) 1988;395:233-53.
39. Hoyt RH, Cohen ML, Saffitz JE: Distribution and three-dimensional structure of intercellular junctions in canine myocardium. Circ Res 1989;64:563-74.
40. Kumar NM, Gilula NB: The gap junction communication channel. Cell 1996;84:381-8.
41. Gros DB, Jongsma HJ: Connexins in mammalian heart function. BioEssays 1996;18:719.
42. Imanaga I: Cell-to-cell diffusion of procion yellow in sheep and calf Purkinje fibers. J Membr Biol 1974;16:381-8.
43. Imanaga I, Kameyama M, Irisawa H: Cell-to-cell diffusion of fluorescent dyes in paired ventricular cells. Am J Physiol 1987;252:H223-32.
44. Kääb S, Nuss B, Chiamvimonvat N, et al: Ionic mechanism of action potential prolongation in ventricular myocytes from dogs with pacing-induced heart failure. Circ Res 1996;78:262-73.
45. Pinto JMB, Boyden PA: Electrophysiologic abnormalities in hypertrophied, failed or infarcted hearts. In Zaza A, Rosen MR (eds): An Introduction to Cardiac Electrophysiology, Amsterdam, Harwood Academic Publishers, 2000, pp 179-97.
46. Peters NS, Coromilas J, Severs NJ, et al: Disturbed connexin43 gap junction distribution correlates with the location of reentrant circuits in the epicardial border zone of healing canine infarcts that cause ventricular tachycardia. Circulation 1997;95:988-96.
47. Schwartz PJ: Sympathetic imbalance and cardiac arrhythmias. In Randall WC (ed): Nervous Control of Cardiovascular Function. New York, Oxford University Press, 1984, pp 225-52.
48. Schwartz PJ, Locati E: The idiopathic long QT syndrome. Pathogenetic mechanisms and therapy. Eur Heart J 1985;6:103-14.
49. Schwartz PJ, Priori SG, Locati EH, et al: Long QT syndrome patients with mutations of the SCN5A and HERG genes have differential responses to Na^+ channel blockade and to increases in heart rate. Implications for gene-specific therapy. Circulation 1996;92:3381-6.
50. Weber KT: Targeting pathological remodeling. Concepts of cardioprotection and reparation. Circulation 2000;102:1342-5.
51. Wijffels MCEF, Kirchof CJHJ, Dorland R, et al: Atrial fibrillation begets atrial fibrillation: A study in awake chronically instrumented goats. Circulation 1995;92:1954-68.
52. Morillo CA, Klein GJ, Jones DL, et al: Chronic rapid atrial pacing: Structural, functional, and electrophysiologic characteristics of a new model of sustained atrial fibrillation. Circulation 1995;91:1588-95.
53. Van Wagoner DR, Pond AL, Lamorgese M, et al: Atrial L-type Ca^{2+} currents and human atrial fibrillation. Circ Res 1999;85:428-36.
54. Bosch RF, Zeng X, Grammar JB, et al: Ionic mechanisms of electrical remodeling in human atrial fibrillation. Cardiovasc Res 1999;44:121-31.
55. Van Wagoner DR, Pond AL, McCarthy PM, et al: Outward K^+ current densities and Kv1.5 expression are reduced in chronic human atrial fibrillation. Circ Res 1997;80:772-81.
56. Daoud EG, Knight BP, Weiss R, et al: Effect of verapamil and procainamide on atrial fibrillation-induced electrical remodeling in humans. Circulation 1997;96:1542-50.
57. Yu WC, Chen SA, Lee SH, et al: Tachycardia-induced change of atrial refractory period in humans: Rate dependency and effects of antiarrhythmic drugs. Circulation 1998;97:2331-7.
58. Pedersen OD, Bagger H, Kober L, et al: Trandolapril reduces the incidence of atrial fibrillation after acute myocardial infarction in patients with left ventricular dysfunction. Circulation 1999;100:376-80.
59. Crijns HJ, van den Berg MP, Van Gelder IC, et al: Management of atrial fibrillation in the setting of heart failure. Eur Heart J 1997;18:C45-49.
60. Sadoshima JI, Izumo S: Molecular characterization of angiotensin II-induced hypertrophy of cardiac myocytes and hyperplasia of cardiac fibroblasts. Circ Res 1993;73:413-23.
61. Kass RS, Blair ML: Effects of angiotensin II on membrane current in cardiac Purkinje fibers. J Mol Cell Cardiol 1981;13:797-809.
62. Jayachandran JV, Sih HJ, Winkle N, et al: Atrial fibrillation produced by prolonged rapid atrial pacing is associated with heterogeneous changes in atrial sympathetic innervation. Circulation 2000;101:1185-91.

Chapter 8 Fundamentals of Cardiac Stimulation

RAHUL MEHRA

The response of cardiac tissue by electrical stimulation was first recognized by early pioneers such as Bichat and Galvani. It was not until 1927 that Marmorstein demonstrated that the heart rate of a dog could be increased by pacing with the help of an external pulse generator.[1] Mark Lidwill and Albert Hyman took these concepts into the clinical arena and conducted electrical stimulation of the human heart in the early 1930s.[2,3] Over the past few decades, the scientific basis and the technology of electrical stimulation of the heart has advanced dramatically and spurred the development of implantable pacemakers and defibrillators, providing clinical benefit to millions of patients. Concurrently, scientists have been challenged to understand the fundamental mechanisms of electrical stimulation armed with sophisticated experimental tools and computational power. Understanding the mechanisms will accelerate development of innovative applications as well as increase the efficiency of electrical stimulation, thereby increasing longevity and decreasing the size of the implantable devices.

The fundamental concepts of electrical stimulation can be divided into three aspects: extracellular stimulation, intracellular stimulation, and the relationship between the two. Extracellular stimulation occurs when an electrode is placed outside the cardiac cell. Several characteristics of the stimulus, electrode, and tissue affect excitation of the cardiac tissue. Intracellular stimulation occurs when one of the electrodes is inside the cell, and this helps determine the fundamental transmembrane changes required to excite cardiac cells. One needs to understand how the extracellular stimulus sets up the necessary conditions for intracellular stimulation. Recent experimental and modeling data have provided unique insights into this complex relationship.

Extracellular Stimulation

STIMULUS VARIABLES

Decades of experimental observations indicate that excitation occurs when a minimum "strength" of a stimulus initiates a propagated response in cardiac tissue. The stimulus is applied through an electrode and its strength is measured as a voltage or current amplitude (Voltage = Current × Resistance). Physiologically, a minimum current must be delivered through an electrode to excite the tissue, irrespective of the resistance of the electrode tissue interface. If the resistance is high, the voltage threshold will increase, as it is equal to the current threshold times the resistance of the tissue. Studies on atrial and ventricular excitation show that the threshold of excitation is not only different at various intervals within the cardiac cycle, but it varies with stimulus duration; polarity of the electrode; electrode shape, area, and material; as well as its distance from the tissue. The polarity of the stimulus is referred to as *cathodal* when the electrode is negative with an indifferent positive electrode and as *anodal* when the electrode is positive. The effect of these variables on cardiac excitation is discussed as follows.

Strength-Interval Curve

The stimulus threshold required at various intervals within the cardiac cycle is represented by the strength-interval curve. During cardiac diastole, the threshold is lower and tends to increase as the stimulus coupling is shortened and systole is approached. This indicates that it is more difficult to re-excite the cardiac tissue soon after it has been excited.[4-6] At very short coupling interval, the tissue cannot be re-excited even with a large stimulus. Figure 8-1 shows a set of strength-interval curves that were determined in a patient with a constant current stimulus when the ventricle was paced at 90 beats/min. With a cathodal stimulus, the threshold is relatively low (0.4 milliamperes [mA]) at intervals greater than 310 milliseconds (ms) (i.e., after the end of the T wave of the paced beat). The tissue cannot be re-excited any earlier than 288 ms even with an 8 mA stimulus; this is its *relative refractory period.* The relative refractory period is a fundamental property of excitable tissue, and it determines the maximum rate at which it can be stimulated. When the polarity of the electrode is reversed and the stimulus polarity is anodal, the shape of the strength-interval curve changes significantly. The anodal threshold is higher during diastole, and the

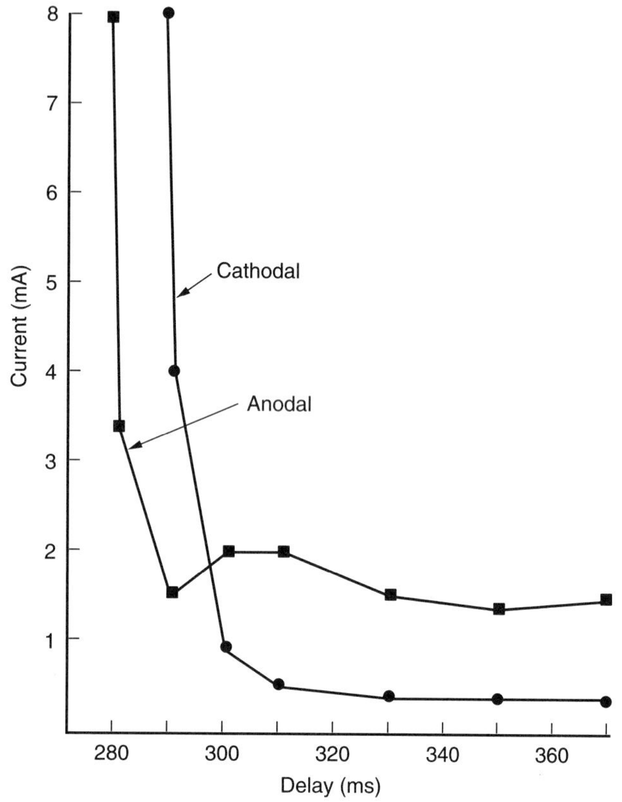

FIGURE 8-1 Typical unipolar cathodal and anodal strength-interval curves obtained in a patient with an acute unipolar electrode (11 mm^2 in area). The ordinate represents the excitation current threshold in mA and the abscissa, the delay of the stimulus from the stimulus pacing the ventricle at 90 beats/min. Note that the anodal refractory periods are shorter than the cathodal periods for most stimulus currents. (Reproduced from Mehra R, Furman S: Comparison of cathodal, anodal and bipolar strength-interval curves with temporary and permanent electrodes. Br Heart J 1979;41:468-76.)

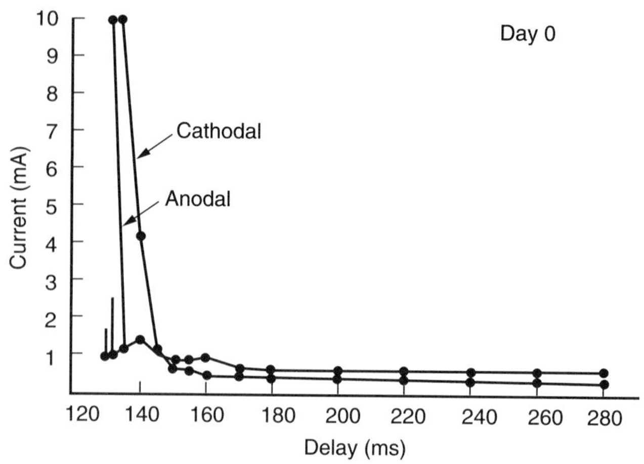

FIGURE 8-2 Unipolar cathodal and anodal strength-interval curves obtained in a canine ventricle immediately following endocardial electrode insertion. In the anodal strength-interval curves, the *dotted lines* represent the range of stimulus currents, which were effective in initiating excitation. For example, at a 130-ms delay, only a stimulus between 1.0 and 1.7 mA caused excitation. Larger stimuli were ineffective. This is referred to as the *no-response phenomena* and has only been observed with anodal stimuli and at short coupling intervals. (Reproduced from Mehra R, McMullen M, Furman S: Time dependence of unipolar cathodal and anodal strength interval curves. Pacing Clin Electrophysiol 1980;3:526-30.)

relatively refractory period is observed to be shorter. This implies that an anodal stimulus can initiate excitation earlier on the T wave than a cathodal stimulus. In Figure 8-1, with an 8 mA stimulus, the anodal refractory period is about 10 ms shorter than with a cathodal stimulus. Another key difference is that as the coupling interval of the anodal stimulus is decreased, the threshold increases, then decreases, and is followed again by a significant increase in threshold at the shortest coupling. This reduction in threshold at the short coupling interval of the stimulus is referred to as the *anodal dip.*

A very unusual property of anodal stimulation in the relative refractory period is the "no-response" phenomenon.[7-9] Typically, the term *threshold* implies that any stimulus that is of greater magnitude results in excitation of the tissue. Within the relative refractory period, when the current strength of an anodal stimulus is increased above threshold, excitation is initiated. Paradoxically, when the stimulus amplitude of the anodal stimulus is increased further, the stimulus becomes ineffective and is no longer able to excite the tissue. For example, in the strength interval curve of Figure 8-2, an anodal stimulus at 130 ms delay can only initiate excitation when its magnitude is between 1.0 and 1.7 mA. When the stimulus is increased further, excitation does not occur and this is termed the *no-response zone.* At a much larger stimulus, excitation may again be re-established. The time interval in the relative refractory period during which this "no-response" phenomena occurs is usually very short. Although this phenomenon has been observed in canine hearts, it has not been reported in the clinical literature.

Strength-Duration Curve

Experimental studies on excitability indicate that it is much easier to excite tissue when the stimulus duration is long rather than short. The strength-duration curve refers to the relationship between the strength of the stimulus required to elicit excitation and the duration of the stimulus. With increasing duration, the thresholds always decrease, although the anodal threshold is observed to be higher than cathodal. Figure 8-3 shows the cathodal and anodal strength-duration curves measured in a canine ventricle. Both curves approach a minimum threshold at long pulse durations, which is called the *rheobase.* The rheobase is the minimum stimulus strength required to excite the cardiac muscle with an extremely long duration and is around 0.2 mA for the cathodal and 0.5 mA for an anodal stimulus in Figure 8-3. The term *chronaxie* defines the pulse duration required to cause excitation with stimulus amplitude of twice the rheobase. In Figure 8-3, the chronaxie for anodal stimulation would be around 0.8 ms. Knowing the rheobase (Io) and chronaxie (T),

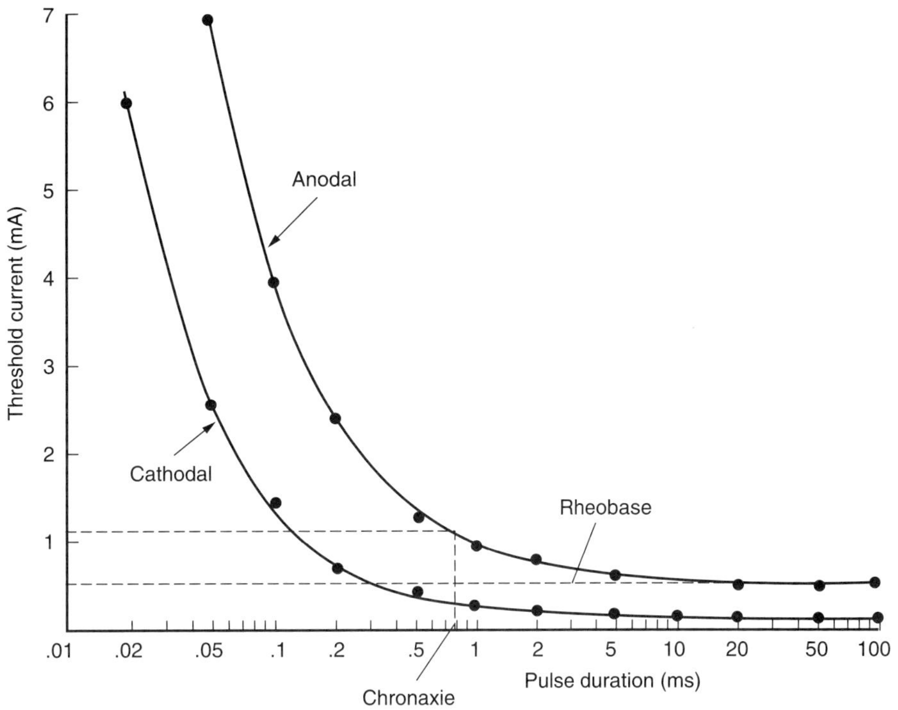

FIGURE 8-3 Strength-duration curve measured with cathodal and anodal stimuli in a canine ventricle. The pulse durations are plotted on a logarithmic scale. Note that the threshold becomes stable at pulse durations greater than 2 to 5 ms. This threshold is called the *rheobase.* Chronaxie is the stimulus duration required to excite tissue at twice rheobase. (Reproduced from Mehra R: Myocardial Vulnerability to Arrhythmias with Cathodal, Anodal and Bipolar Stimulation. [PhD Dissertation #75-21384]. University of Michigan Microfilms, Ann Arbor, Mich, 1975.)

the equation It = Io (1 + t/T) defines the approximate threshold current at any pulse duration "t".[11] The basic electrophysiologic mechanism that is responsible for lower thresholds at longer pulse duration is discussed later.

For an implantable device, one needs to maintain consistent pacing and use minimum energy of the stimulus to maximize longevity of the device. Optimization of electrode design and pulse duration can help minimize the energy dissipated per pulse. The energy dissipated per pulse can shown to be a product of $I^2 \times R \times t$ where I is the current threshold, R the resistance of the pacing circuit, and t the pulse duration. Unlike the current strength-duration curves, the energy strength-duration curve typically has a minimum, indicating there is a duration at which the energy dissipated per pulse would be minimum, as illustrated in Figure 8-4. Pacing at this unique pulse duration can therefore maximize device longevity. The duration at which the energy dissipated is minimum is not always the same

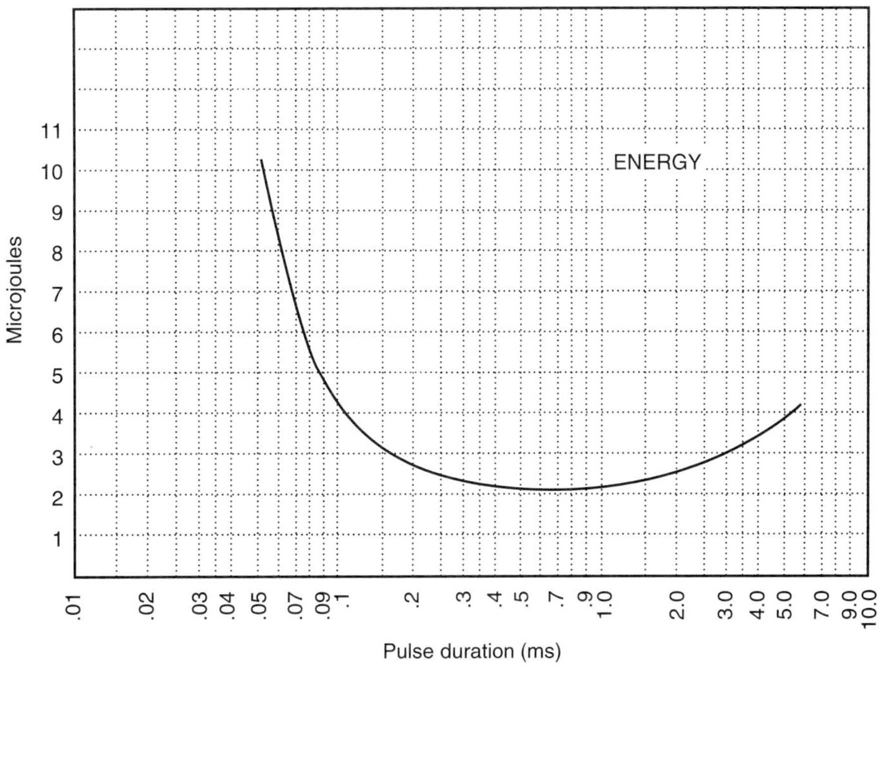

FIGURE 8-4 Energy required to pace the ventricle at various pulse durations with a chronic endocardial lead in a canine ventricle. In this case, the minimum energy is consumed at durations ranging from .25 to 1.0 ms. Energy consumption increases at shorter and longer durations. (Reproduced from Furman S, Hurzeler P, Mehra R: Cardiac pacing and pacemakers IV: Threshold of cardiac stimulation. Am Heart J 1977;94:115-24.)

and will depend upon the electrode material and its design.

Current versus Voltage Threshold

As discussed previously, the amplitude of the stimulus can be measured in current (amperes) or voltage (volts), and these two variables are interrelated by Ohm's law (Voltage = Current × Resistance). The current threshold is a more physiologic measurement of excitation as it is independent of the resistance of the electrode or the tissue. As a first approximation, the current density (current/unit area) is even a more accurate determinant of the basic requirement for excitation. It is postulated that in order to elicit excitation, a certain minimum number of cells must be excited. Since the heart is a sanctum (i.e., an aggregate of interconnected cardiac cells), once this critical volume of cells is excited, the conduction propagates to the rest of the heart. Therefore, a minimum amount of current per unit area must exist in order to excite this aggregate of cardiac cells. This is called the *current density threshold*. The current density required for cardiac excitation is typically between 2 and 2.5 mA/cm^2 at the electrode tissue interface.[13,14]

Cardiac stimulators can be designed to provide an output pulse that maintains its voltage or current throughout the duration of the pulse. These stimulators are referred to as *constant voltage* or *constant current stimuli*. Physiologic stimulators used in laboratories typically provide constant voltage or constant current stimuli, whereas most pacemakers are neither and typically have an exponentially decaying voltage output profile (Fig. 8-5*A* and *B*). Figure 8-5*A* illustrates a typical constant current stimulus, where the output is constant in amplitude from the onset to the offset of the stimulus. Due to the polarization characteristics of the electrode tissue interface, the resistance increases during the pulse and an increasing voltage is required to maintain the constant current stimulus. This is illustrated by a significant increase from *Vo* at the beginning of the stimulus to voltage *Vt* at the end of the stimulus. Figure 8-5*B* illustrates the typical waveform of a pacemaker stimulus where the output is an exponentially decaying

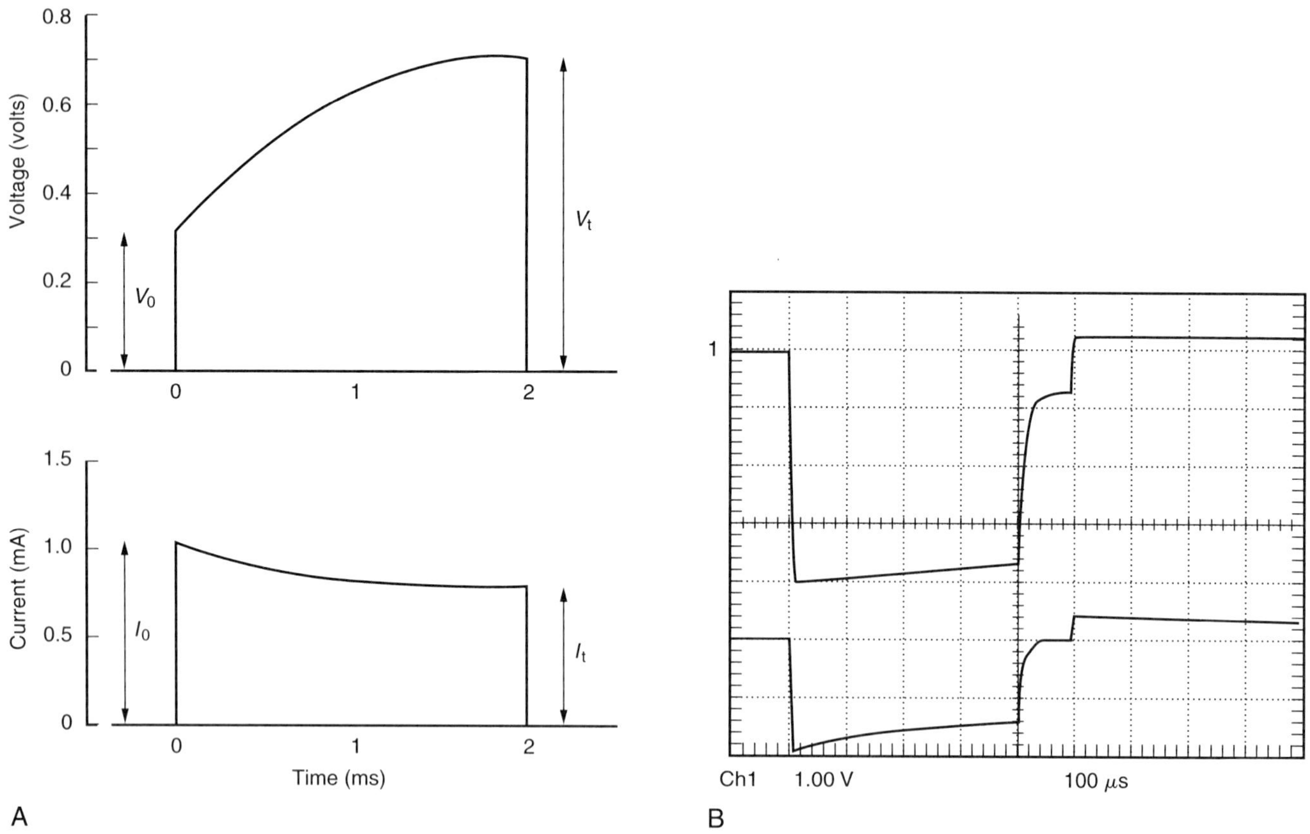

FIGURE 8-5 **A,** Voltage waveform measured at the electrode during a constant current stimulus. The current amplitude is almost constant at 1.0 mA through the stimulus duration of 2 ms. Significantly increasing stimulus voltage from 0.3 volts at the beginning of the stimulus to 0.7 volts at its end is required to maintain this constant current through the tissue. This is due to the increasing polarization resistance during the pulse. (Reproduced from Furman S, Parker B, Escher DJW, Solomon N: Endocardial threshold of cardiac response as a function of electrode surface area. J Surg Res 1968;8:161-6). **B,** Voltage (*top*) and current (*bottom*) waveform of a 4 volt 0.4 ms stimulus measured at the output of a typical pacemaker. The voltage is decaying exponentially during the pulse and is followed by a "recharge" of an opposite polarity. The current waveform decays at a greater rate due to increasing resistance from the beginning to the end of the stimulus. After the end of the 0.4 ms stimulus, the current returns to zero and is followed by a "recharge" stimulus of a small magnitude but longer duration.

voltage stimulus, and there is a slight decrease in voltage output from the beginning to the end of the stimulus. The exponentially decaying voltage output results from a capacitor at the output of the pacemaker that is used to deliver the stimulus. A capacitor is used to maintain low energy drain for the implantable device and ensure safe performance. The current waveform that is generated has an even higher rate of decay because of polarization resulting in increasingly higher resistance during the pulse. It is important to note that the general shape of the strength-duration or strength-interval curves is not significantly altered with either of these stimulus waveforms.

ELECTRODE VARIABLES

Unipolar versus Bipolar Electrodes

In the previous sections, cardiac stimulation properties were evaluated when the electrode was a cathode or an anode. In each circumstance, the other indifferent electrode was assumed to be relatively large and far from cardiac tissue. Frequently, cardiac pacing is conducted with bipolar electrodes, where the cathode and anode are close to each other and relatively small in size. Figure 8-6 illustrates such an electrode. With such endocardial electrodes, the tip electrode is made the cathode, as it will typically contact the myocardium and therefore provide lower thresholds. The anodal proximal electrode may not always contact cardiac tissue. If it does, it could elicit an excitation depending on the stimulus output and the anodal threshold. The bipolar

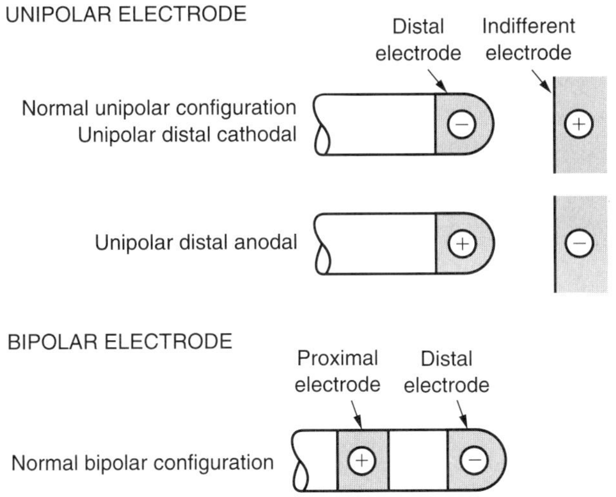

FIGURE 8-6 Unipolar and bipolar electrode configurations. The normal unipolar pacing configuration is one in which the tip of the electrode is negative (distal cathodal) with an indifferent positive electrode. The unipolar distal anodal configuration reverses the polarities and is not used because of higher anodal thresholds. The normal bipolar configuration uses the tip (distal) electrode as the cathode and the ring (proximal) electrode as the anode. (Reproduced from Mehra R, Furman S: Comparison of cathodal, anodal and bipolar strength-interval curves with temporary and permanent electrodes. Br Heart J 1979;41:468-76.)

excitation threshold at any time in the cardiac cycle would be equal to the lower of the cathodal and anodal thresholds determined at the two electrode sites.[5,6] Therefore, the bipolar strength-interval curve would follow the lower of the cathodal and anodal curves as illustrated in Figure 8-7*A*. This indicates that with a bipolar stimulus, excitation would arise from *both* electrode sites if the stimulus output were above the anodal and cathodal thresholds and *selectively* from one electrode if the stimulus output were between the cathodal and anodal thresholds. For example, in Figure 8-7*A*, a 2 ms stimulus at 380 ms coupling would only excite the tissue at the cathode, whereas a 7 mA stimulus would result in excitation from the cathode as well as the anode.

Electrode Size

Since a minimum current density is a measure of the requirement for excitation, the threshold of excitation would also be dependent on electrode size. Studies have demonstrated that during diastole, the current threshold for cardiac pacing is directly proportional to the electrode surface area with the areas ranging from about 10 to 50 mm^2 (Fig. 8-8). As indicated in Figure 8-8, smaller electrodes would require less current for excitation. At the same time, the resistance of a smaller electrode is higher since the resistance of a spherical electrode is inversely proportional to its radius. Therefore, the question arises whether the energy required for excitation would increase or decrease with smaller electrodes. The energy of a stimulus is equal to I^2Rt, where I is the current threshold, R the resistance, and t the stimulus duration. With small electrodes, even though the resistance increases due to a smaller radius, the reduction in current threshold is much greater, resulting in lower energy requirements. For this reason, the surface area of commercial implantable electrodes has become smaller over the years. Making the electrodes extremely small has some disadvantages. Extremely small electrodes are susceptible to being dislocated easily, and they also tend to show greater rise in chronic thresholds due to the formation of the fibrotic capsule.

Tissue Fibrosis

It has been observed that due to the trauma caused by the electrode, tissue damage and edema occurs acutely. This is replaced in time by a fibrotic capsule (Fig. 8-9).[16,17] The fibrotic capsule tends to increase the separation between the electrode and the excitable tissue and therefore decreases the current density (current per unit area) at the excitable tissue. As the electrode matures from acute to chronic, the radius of the virtual electrode increases from "r" to "r + Δ," as illustrated in Figure 8-9. Therefore, greater stimulus current is needed to reach the threshold current density in the excitable tissue. This tends to increase the pacing thresholds after electrode implantation, as shown in Figure 8-10. There is a significant increase in thresholds in the 4- to 8-week period due to edema developed

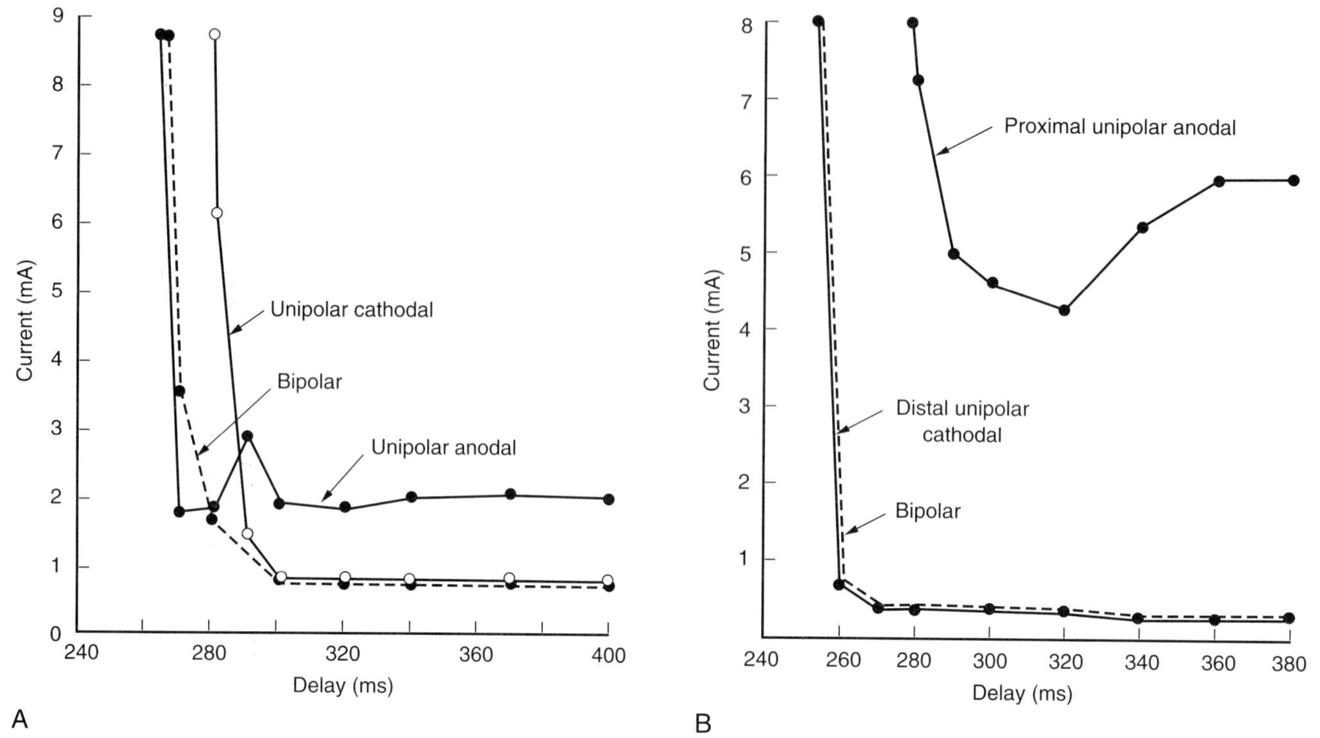

FIGURE 8-7 **A,** Relationship among distal unipolar cathodal, proximal unipolar anodal, and bipolar strength-interval curves in a patient with a temporary lead. For the proximal unipolar anodal thresholds, the ring is the anode and the indifferent electrode is the cathode. For the distal unipolar cathodal threshold, the lead tip is the cathode. The bipolar threshold at any delay is equal to the lower of the cathodal and anodal thresholds. (Reproduced from Mehra R, Furman S: Comparison of cathodal, anodal and bipolar strength-interval curves with temporary and permanent electrodes. Br Heart J 1979;41:468-76). **B,** In this patient the proximal unipolar anodal thresholds are relatively high, and therefore the distal unipolar cathodal and bipolar strength-interval curves are the same. (Reproduced from Mehra R, Furman S: Comparison of cathodal, anodal and bipolar strength-interval curves with temporary and permanent electrodes. Br Heart J 1979;41:468-76.)

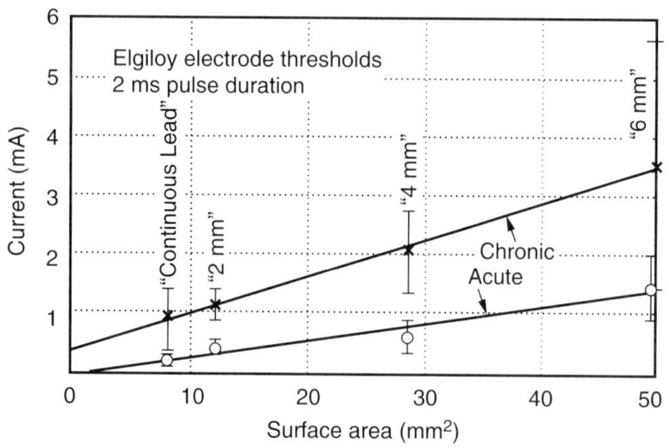

FIGURE 8-8 Acute and chronic current thresholds of four electrodes with increasing surface area. Each threshold is measured with a 2 ms duration pulse. The terms *"Continuous Lead," "2 mm," "4 mm,"* and *"6 mm"* denote electrodes of increasing surface area. Note that both the acute and chronic thresholds increase with increasing area, but the slope of the lines is different. (Reproduced from Furman S, Hurzler P, Parker B: Clinical thresholds of endocardial cardiac stimulation: A long-term study. J Surg Res 1975;19:149-55.)

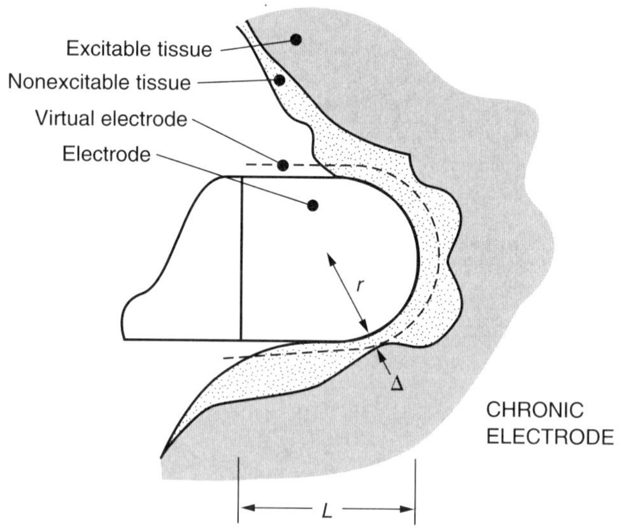

FIGURE 8-9 Schematic diagram of a chronically implanted electrode. An irregularly shaped fibrous capsule of nonexcitable tissue forms around the electrode, which increases the separation between the electrode and the excitable tissue. The *dotted line* indicates that the radius of the electrode has increased from r to a virtual electrode of radius $r + \Delta$. This is responsible for the chronic increase in pacing thresholds. (Reproduced from Furman S, Hurzler P, Parker B: Clinical thresholds of endocardial cardiac stimulation: A long-term study. J Surg Res 1975;19:149-55.)

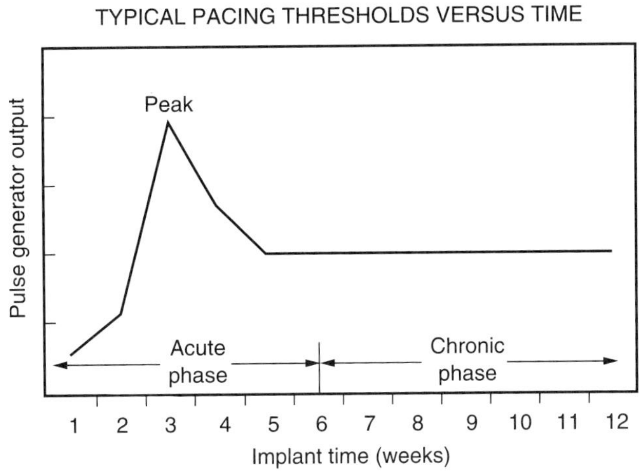

FIGURE 8-10 Typical change in pacing threshold as a function of implant time with a nonsteroid pacing electrode. Between 2 and 3 weeks, the threshold increases significantly but stabilizes to a lower value afterwards.

under the electrode. With increasing time, the size of this edematous capsule shrinks and is replaced by a fibrotic layer. This causes the excitation threshold to decrease again and stabilize chronically.

A major advance in electrode technology has been the development of electrodes that elute small quantities of corticosteroid dexamethasone sodium phosphate.[18] These steroid electrodes are characterized by a minimal change in threshold from implantation to a follow-up period of several years. The acute rise in stimulation threshold that occurs due to formation of inflammation and edema adjacent to the pacing electrode is reduced dramatically. Although it has been hypothesized that the steroid reduces the size of the fibrotic layer and therefore minimizes the rise in threshold, the exact mechanism by which steroid leads improve pacing efficiency is not well understood.

Electrode Polarization

For pacing efficiency to be high, minimal energy should be lost at the electrode tissue interface, and all the energy of the stimulus should be dissipated in the cardiac tissue. Unfortunately, energy is lost with most electrodes due to the electrochemical reactions at the electrode tissue interface, referred to as *polarization*.[19,20] In the electrode, the flow of current is due to the presence of electrons, whereas in the tissue it is ionic. The chemical reactions that occur at the electrode tissue interface to convert electronic to ionic flow are complex and vary with the metal and electrode geometry being used. This polarization tends to dissipate and waste energy. If the electrode was nonpolarizable, the voltage and current waveforms would be identical, and no energy would be wasted at the interface. Figure 8-5*A* illustrates a typical voltage waveform with a polarizable electrode when a constant current stimulus is applied. Initially the voltage is low, but it increases due to increasing impedance as a result of the polarization. The electrochemical reactions are different at the anode; therefore, they would give rise to a different shape of the current waveform. New materials that tend to reduce polarization have been developed. Among these are the porous and platinized electrodes, which have voltage/current characteristics that approach those of nonpolarizable electrodes.

GOALS OF CARDIAC STIMULATION IN IMPLANTABLE DEVICES

The pacing thresholds in a physiologic situation are dynamic and change with alterations in metabolic activity, pH, lead maturity, and so on. If one were to set the magnitude of the pacemaker output at the pacing threshold measured at the time of implant, pacing capture could be lost frequently. There are two methods of resolving this issue: (1) pace with a higher stimulus (i.e., incorporate a "safety factor") and (2) develop systems that automatically measure the threshold and adjust the stimulus output accordingly. The appropriate "safety factor" setting would be based on the type of lead and the expected changes in the threshold. Clearly, the higher the safety factor, the higher the pacing output, which results in greater battery drain and shorter device longevity. The systems that automatically adjust the output based on the pacing threshold require the ability to assess capture from the same lead as the one from which pacing is being conducted. This requires the use of special low polarization leads or unique algorithms and has been shown to be effective in the ventricle but continues to be a challenge in the atrium.

Intracellular Stimulation

A cardiac cell can be excited when it is depolarized and, paradoxically, even when it is hyperpolarized to a critical value. In normal cardiac cells, the transmembrane potential is between 70 and 90 millivolts (mV) with the inside of the cell negative. When a depolarizing current is applied, the transmembrane potential is reduced. When it reaches a critical value, an action potential is elicited. Figure 8-11 illustrates the situation where subthreshold stimuli are ineffective until a threshold stimulus brings the transmembrane potential to a critical value of minus 65 mV and excitation occurs. At this critical membrane potential (CMP), the changes in membrane permeability cause large movements of sodium ions into the cells. A reversible sequence of changes in permeability to various ions is triggered, giving rise to an action potential.

The transmembrane potential across the cell does not reach its stable value instantaneously when a rectangular stimulus is turned on, as illustrated in Figure 8-11. Instead, the transmembrane potential increases slowly due to the ability of the cell membrane to store charge (capacitance). Similarly, when the stimulus is turned off, the transmembrane potential recovers slowly. As the stimulus amplitude is increased, a threshold transmembrane potential of minus 65 mV is reached and the active transport mechanisms responsible for the action

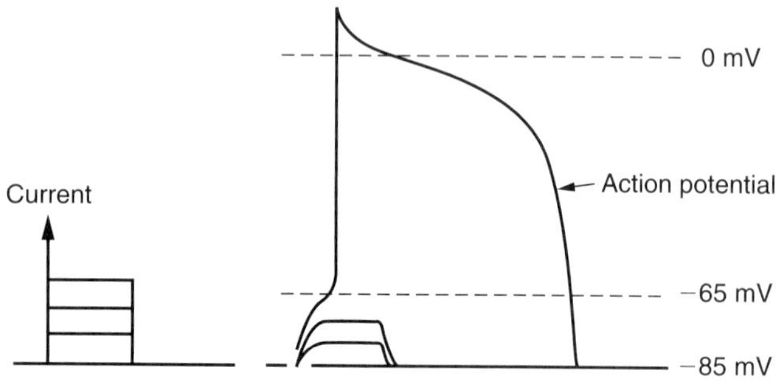

FIGURE 8-11 Schematic representation of the effect of three increasing current stimuli on the transmembrane potential of a cardiac cell. With the smallest stimulus, the membrane potential changes from −85 to about −80 mV and recovers back to −85 mV after the stimulus is turned off. With the largest stimulus, the cell transmembrane potential reaches the critical threshold value of −65 mV, and an action potential is initiated.

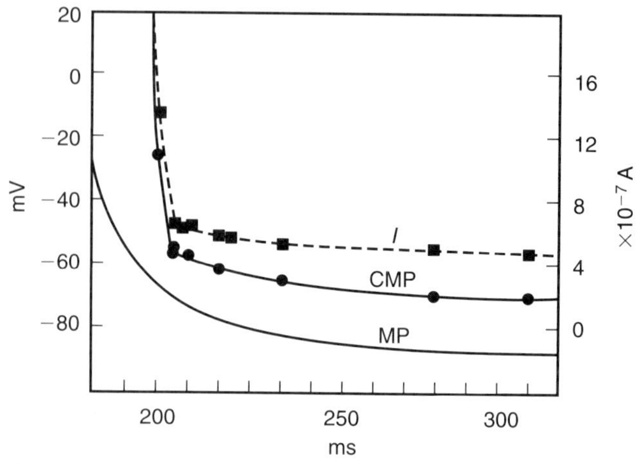

FIGURE 8-12 Critical membrane potential (*CMP*), depolarizing current threshold (*I*), and transmembrane potential (*MP*) as a function of delay from the onset of an action potential. The ordinate is the potential in mV or current threshold and the abscissa is in ms. The *CMP* is almost constant at intervals longer than 200 ms, and *I* represents the intracellular strength-interval curve with a depolarizing stimulus. (Reproduced with permission from Hoshi T, Matsuda K: Excitability cycle of cardiac muscle examined by intracellular stimulation. Jpn J Physiol 1962;12:433.)

potential are initiated. If excitation occurs after the stimulus is turned on and before it is turned off, it is referred to as *make excitation.* Due to the membrane capacitance, if the stimulus were to be turned off very quickly (i.e., after a very short pulse duration), the threshold transmembrane potential would not be reached, and the cell would not elicit an action potential. Significantly higher stimulus amplitude would be required to reach this threshold transmembrane potential at short pulse duration. This property of cell capacitance gives rise to a relationship between the stimulus duration and the amplitude required to elicit excitation (i.e., the strength-duration curve) (see Fig. 8-3).

The CMP required to elicit excitation with a stimulus also depends on its timing within the cardiac cycle. With a depolarizing cathodal stimulus, it is lowest in diastole and increases as the stimulus approaches systole. The CMP and the transmembrane current required to excite tissue with intracellular stimulation at various intervals in the cardiac cycle were measured by Hoshi et al. in canine Purkinje fibers (Fig. 8-12).[21] The figure shows that the CMP was between 60 and 65 mV and relatively constant throughout the diastolic phase and the terminal phase of repolarization of the membrane potential (MP). In the rapid phase of repolarization, the CMP increased sharply and so did the current threshold (*I*). This results in a hyperbolic-shaped curve that describes the relationship between the current threshold for excitation and the time delay from the onset of the action potential of a depolarizing cathodal stimulus. This curve is referred to as the *intracellular strength-interval curve* and has a shape very similar to the strength-interval curve with an extracellular cathodal electrode (see Fig. 8-1).

The second type of excitation occurs when the cardiac cell is hyperpolarized to a critical value (ΔVm is negative) and after the stimulus is turned off. This is called *break excitation.* It was initially observed in nerve fibers and associated with termination of an anodal stimulus and was called *anodal break stimulation.*[22,23] In cardiac tissue, break excitation was first demonstrated by Dekker et al.[24] He measured the excitation thresholds during the "make" (following the stimulus onset and before its termination) and "break" (following the end of the stimulus) of the pulses and demonstrated that these thresholds varied with the polarity of the stimulus as well as the timing of the stimulus within the cardiac cycle (Fig. 8-13). The make and break thresholds were measured when anodal and cathodal stimuli were delivered through epicardial electrodes of the canine ventricle. Figure 8-13 illustrates the make and break thresholds with long duration stimuli. For break threshold measurements, the stimulus was initiated in the absolute refractory period and ended at different intervals within the cardiac cycle. Excitation was observed after the termination of the stimulus. The thresholds for break excitation had the characteristic "anodal dip" with higher thresholds during diastole, and the make excitation thresholds had the characteristic hyperbolic shape. Break excitation has been observed experimentally with long duration as well as short duration stimuli.[24-26] With optical mapping techniques, Wikswo et al. demonstrated that the break excitation occurred after the end of the stimulus from the region that was hyperpolarized.[25] This was observed during anodal and cathodal break excitation. During the hyperpolarized state, the tissue is inexcitable because its sodium channels are inactivated. Excitation is only observed after the hyperpolarizing stimulus is terminated.

The mechanism responsible for break excitation is controversial. Two concepts have been proposed. In the mechanism proposed by Roth et al., break excitation from the hyperpolarized region occurs only when there are adjacent depolarized areas, and it does not occur in tissue that is uniformly hyperpolarized.[27] At the end of

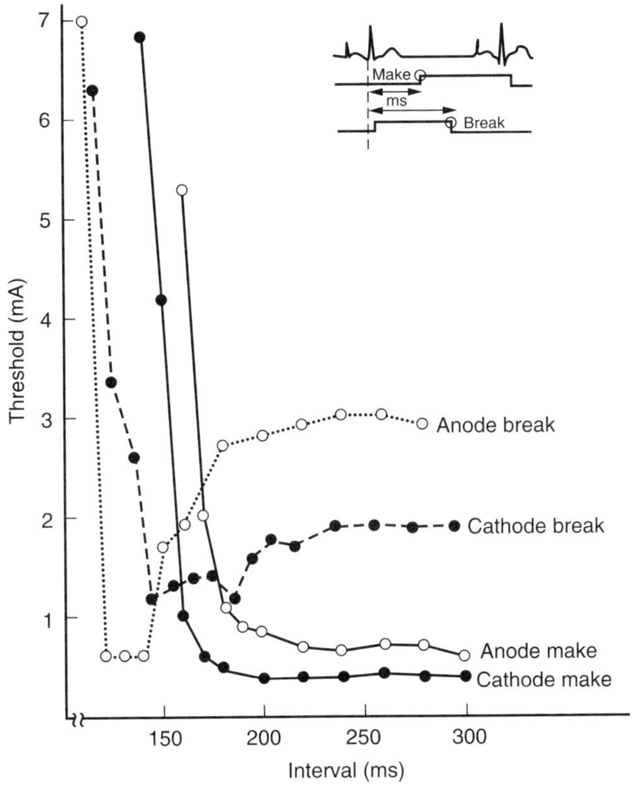

FIGURE 8-13 Typical ventricular thresholds plotted as a function of the interval in ms after the preceding normally conducted QRS complex for each mode of stimulation—anodal make, anodal break, cathodal make, and cathodal break for a long-duration stimulus. "Make" excitation occurs after the stimulus onset and before it is offset, whereas "break" excitation occurs after the stimulus has been terminated. (Reproduced from Dekker E: Direct current make and break thresholds for pacemaker electrodes on the canine ventricle. Circ Res 1970;27:811.)

the stimulus, current diffuses from the depolarized to hyperpolarized areas to initiate excitation. Based on this hypothesis, break excitation is a result of the syncytial properties of the tissue and not a property of the cell membrane alone. On the other hand, simulation conducted by other investigators indicates that break excitation in the hyperpolarized tissue may be possible by itself without the diffusion of current from adjacent tissue.[28] This controversy has not been resolved as yet.

Relationship Between Intracellular and Extracellular Stimulation

The previous section discussed the observation that in order to elicit excitation, the transmembrane potential (ΔVm) must be altered. The question then arises: How does an extracellular stimulus change the transmembrane potential? Several investigators have used optical mapping techniques to measure the transmembrane potentials around an extracellular unipolar electrode.[25,26,29] Their key observation was that adjacent areas of hyperpolarization and depolarization are observed; this pattern has been referred to as the *"dog bone" shape* (Fig. 8-14). For example, with cathodal stimulation (Fig. 8-14*A*), the tissue directly underneath the electrode is depolarized and it has a "dog bone" shape with the long axis of the dog bone perpendicular to the axis of the fibers. In the axis perpendicular to the fibers, the tissue is only depolarized, and its amplitude decreases with increasing distance. However, in the axis parallel to the fibers, the tissue directly underneath the electrode is depolarized but changes polarity and is hyperpolarized about a millimeter away from the electrode, as illustrated by the blue image in Figure 8-14*A*. If the polarity of the extracellular stimulus is reversed to an anodal stimulus, the shape of the "dog bone" remains quite similar with reversal of hyperpolarized and depolarized regions (Fig. 8-14*B*).

These observations are quite compatible with the predictions of the bidomain model. In the bidomain model, the intracellular and extracellular spaces are considered as two distinct domains and separated by a high resistance cell membrane. It is assumed that the intracellular space of each cell is coupled to its neighbors through low-resistance intracellular channels. The intracellular and extracellular spaces are also assumed

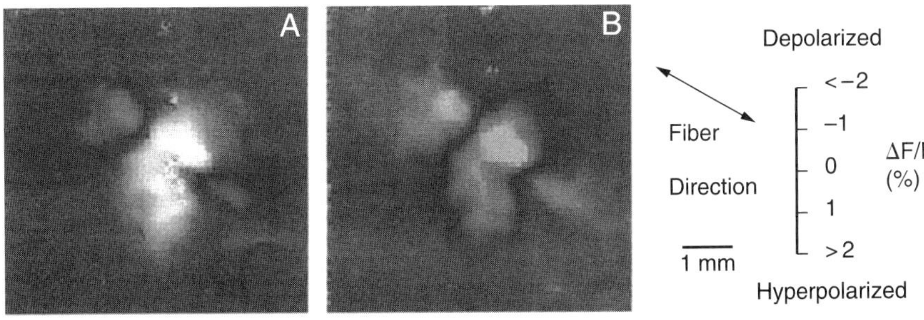

FIGURE 8-14 (See also Color Plate 8-14.) Images of the transmembrane potential associated with injection of current into refractory cardiac tissue. **A,** The image for a 10 mA, 2 ms cathodal stimulus applied at a point electrode. Note the dog bone–shaped depolarized region (*orange*) and a pair of adjacent hyperpolarized regions (*blue*). The fiber orientation is from lower right to upper left. The *color bar* shows the fractional change in fluorescence. **B,** The complementary image for a 10 mA, 2 ms anodal stimulus at the same location of the heart. Note that the "dog bone" is now hyperpolarized region (*blue*), whereas the adjacent areas are depolarized (*orange*). (Reproduced from Wikswo JP, Lin SF, Abbas RA: Virtual electrodes in cardiac tissue: A common mechanism for anodal and cathodal stimulation. Biophys J 1995;69:2195-210.)

to be anisotropic (i.e., the tissue's electrical conductivity is different when it is measured parallel versus perpendicular to the orientation of the myocytes). Early theoretical work with bidomain models assumed that the anisotropic conductivity ratios for the intracellular and extracellular spaces were equal. If equal anisotropic ratios are assumed for intracellular and extracellular spaces, and an extracellular stimulus is applied from a point source, the shape of the transmembrane potential contours is ellipsoid with the transmembrane potential decreasing with increasing distance.[25] Therefore, the tissue would be either depolarized or hyperpolarized. This is incompatible with the experimental observations, which demonstrate adjacent areas of depolarization and hyperpolarization during extracellular stimulation (see Fig. 8-14). Recent simulations with the bidomain model have assumed that the intracellular and extracellular space are both anisotropic, but not to the same extent. The anisotropic conductivity ratio in the extracellular space is typically about 2:1 whereas it is 10:1 in the intracellular space. This property of cardiac tissue seems to be responsible for many of its interesting electrical features.[25] With simulation of unequal anisotropic properties, adjacent areas of depolarization and hyperpolarization are observed with a "dog bone" shape, which are very similar to the experimental observations.[25,30] Adjacent areas of depolarization and hyperpolarization are produced because the rate of change of the intracellular and extracellular potentials is nonuniform. The size and shape of the depolarized and hyperpolarized regions changes with the anisotropy ratios. This is demonstrated in the two-dimensional schematic of Figure 8-15. The "dog bone" shape represents the depolarized region; the shaded regions are hyperpolarized. Depolarization occurs underneath the cathodic electrode where the current enters a syncytium of cells, and

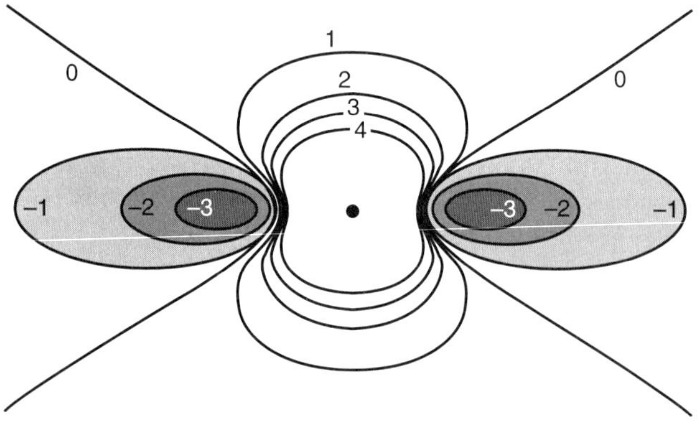

FIGURE 8-15 Transmembrane potentials induced by an extracellular unipolar electrode in a two-dimensional tissue based on the bidomain model with unequal anisotropic ratios in the intracellular and extracellular space. The central dot denotes an electrode through which the cathodal stimulus is delivered. The fibers are arranged horizontally and the hyperpolarized regions of the tissue are shaded. The units of transmembrane potential are arbitrary. (Reproduced from Roth BJ: The pinwheel experiment revisited. J Theor Biol 1998;190: 389-93.)

hyperpolarization occurs where the current exits the cells. If the polarity of the stimulus were anodal, the "dog bone" region would be hyperpolarized, and depolarization would be observed in the adjacent region. These voltage nonuniformities can be further accentuated by electrical uncoupling of cells by ischemia, presence of fibrotic tissue, blood vessels, fiber curvature, and so on. All these variables may also play a pivotal role in establishing ΔVm during extracellular stimulation.

Apart from the shape of the depolarized and hyperpolarized regions, the optical mapping studies and the bidomain model define the region where excitation is initiated with extracellular stimulation. Depending on which threshold is lower, make excitation can occur from the depolarized area or break excitation from the hyperpolarized area. For example, during cathodal stimulation, the region directly beneath the electrode is depolarized. If it reaches the critical threshold, the excitation wavefront is initiated. This is illustrated in the optical maps of Figure 8-16 and schematically in Figure 8-17.[30] During a cathodal stimulus of 10 mA, excitation is typically initiated from the depolarized "dog bone" region (see Figs. 8-16*A* and 17*A*) following the make of the stimulus. If the stimulus were anodal, make excitation could occur in the depolarized region a few millimeters from the electrode, provided the make threshold is lower than the break threshold in the hyperpolarized region. In Figure 8-16*B* the make threshold is lower, and excitation occurs from the two yellow regions.

As discussed previously, excitation can also be initiated from the hyperpolarized region after the stimulus is turned off (i.e., "break stimulation"). "Break" excitation occurs from the hyperpolarized region, and its mechanism has been discussed. For example, during a cathodal pulse, the tissue underneath the electrode is depolarized, and the distal tissue is hyperpolarized (Figs. 8-16*C* and 8-17*C*). If the stimulus amplitude is not enough to elicit make excitation from the depolarized regions, then after the end of the pulse, "break" excitation could begin from the hyperpolarized region as in Figure 8-16*C*. This will propagate rapidly along the fiber axis. In a similar manner, if the anodal make excitation does not occur from the depolarized region, "break" excitation will occur from the hyperpolarized regions as illustrated in Figures 8-16*D* and 8-17*D*.

The bidomain model not only predicts the shape of the depolarized and hyperpolarized regions and explains "make" and "break" excitation, but it has also been able to simulate the shape of the cathodal and anodal strength-interval curves, presence of the "anodal dip," the "no-response phenomena,"[32] and the ratio of anodal and cathodal thresholds with large surface area electrodes.[33] Roth et al. modeled the cathodal and anodal strength-interval curves using the bidomain model and simulated some of its key characteristics. The model predicts that with short duration stimuli, the shape of the cathodal strength-interval curve is primarily determined by make excitation of the depolarized tissue. However, with an anodal stimulus at long coupling intervals, stimulation occurs due to make excitation distal to the electrode, as it has the lowest threshold.

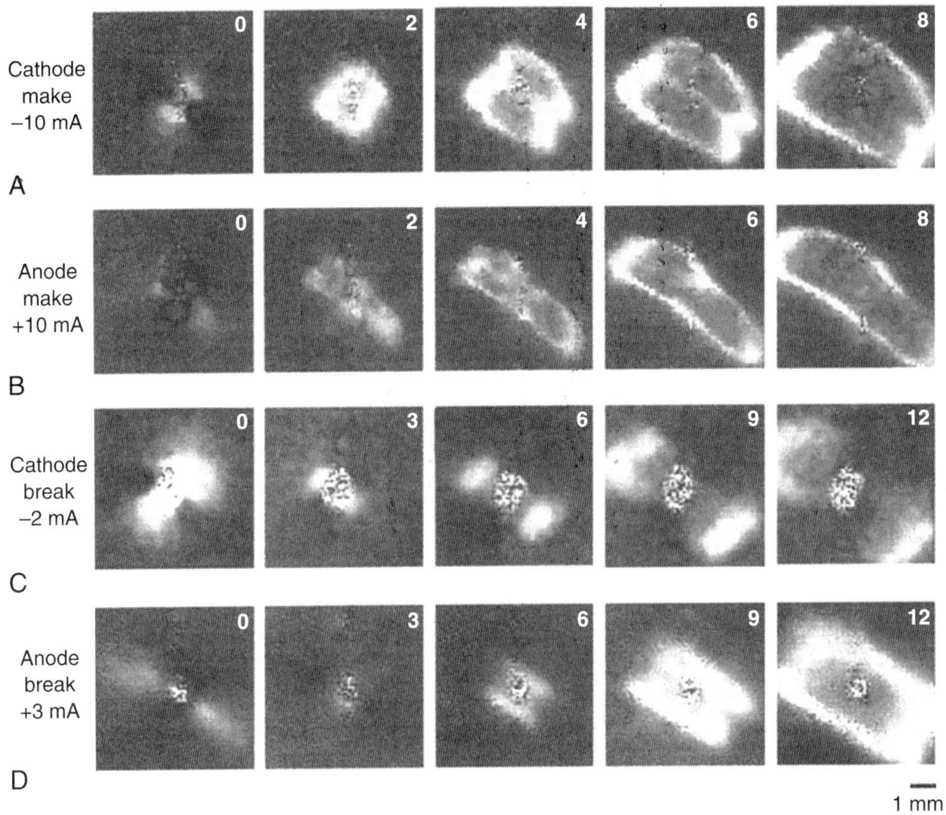

FIGURE 8-16 (See also Color Plate 8-16.) False-color images of the transmembrane potential associated with injection of current into fully repolarized excitable cardiac tissue. The number inside each frame is time in ms. Upper two rows: make excitation. **A,** Cathodal make excitation with 1 ms 10 mA stimulus; **B,** 1 ms, 10 mA anodal make stimulation in the same heart. Lower two rows: break stimulation. **C,** 180 ms 2 mA cathodal break stimulation with a 2 mA, 180 ms long duration stimulus in another heart. **D,** Anodal break stimulation with a 3 mA 150 ms long-duration stimulus. The direction of the epicardial fibers is from lower right to upper left. The color scale is the same as in Figure 8-14. (Reproduced from Wikswo JP, Lin SF, Abbas RA: Virtual electrodes in cardiac tissue: A common mechanism for anodal and cathodal stimulation. Biophys J 1995;69:2195-210.)

However, at short coupling intervals, the "break" threshold in the hyperpolarized regions becomes lower and gives rise to the "dip" in the anodal strength-interval curve. These observations are very similar to the experimental observations of Dekker et al. 30 years ago.[24] At long coupling intervals, the make thresholds were lower than break thresholds, with the cathodal make threshold being the lowest (see Fig. 8-13). The cathodal and anodal make strength-interval curves were hyperbolic in shape as compared to the break-excitation curves, which had shorter refractory periods and a characteristic "dip." At any coupling interval within the cardiac cycle, the lower of the "make" or "break" thresholds determines the excitation threshold. For example, with an anodal stimulus, excitation would be due to make excitation at long coupling and due to break excitation at shorter coupling (see Fig. 8-13). All these experimental observations are compatible with the bidomain model simulations.

In experimental clinical studies, the ratio of anodal and cathodal thresholds is observed to be 2.8 with acute electrodes and 1.5 with chronic electrodes.[10] This has also been modeled using the bidomain model, which indicates that if the distance between the electrode and the excitable tissue is increased due to a perfusion bath or fibrotic tissue, the ratio of the excitation thresholds decreases.[34] The amplitude of depolarization potential change induced by a cathodal stimulus is reduced dramatically close to the electrode, whereas with an anodal stimulus, the change in the depolarization potential farther away from the electrode is reduced much less. The ratio of the maximum depolarization to maximum hyperpolarization also decreased with increasing electrode tissue separation, thus reducing the ratio of anodal and cathodal excitation thresholds.

Future of Cardiac Excitation

Our understanding of cardiac stimulation has increased dramatically over the past decade with the availability of optical mapping techniques and computational efficiencies. The focus of this chapter has been primarily on the fundamentals aspects of electrical stimulation related to "near-field" stimulation (i.e., stimulation close to the electrode) rather than "far-field" stimulation. "Far-field" stimulation plays an important role in cardiac defibrillation. Significant strides have been made

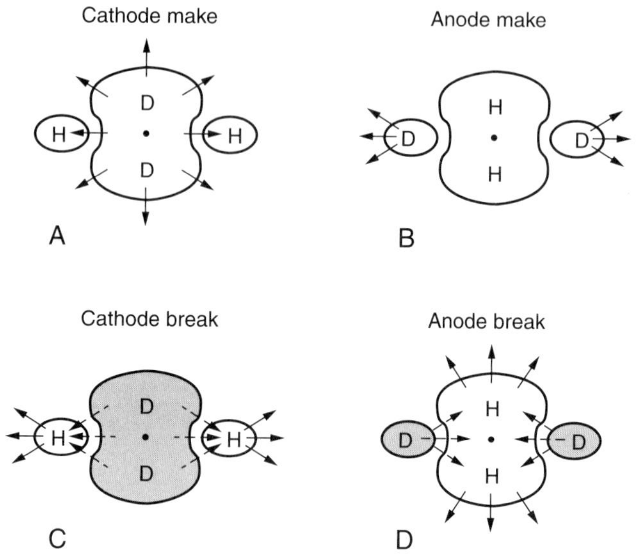

FIGURE 8-17 A schematic representation of the four mechanisms of stimulation based on the results of the bidomain model: **A,** Cathode make; **B,** Anode make; **C,** Cathode break, and **D,** Anode break. The myocardial fibers are oriented horizontally and the electrode site is indicated horizontally. *"D"* indicates depolarized tissue and *"H"* indicates hyperpolarized tissue. *Solid arrows* show the initial direction of wavefront propagation; *dashed arrows* indicate diffusion of depolarization into the adjacent hyperpolarized tissue after the end of the stimulus. *Shading* denotes the tissue that is rendered inexcitable by a prolonged depolarization. (Reproduced from Roth BJ: Strength-interval curves for cardiac tissue predicted using the bidomain model. J Cardiovasc Electrophysiol 1996;7:722.)

in our understanding of the mechanisms of far-field stimulation and are still being evaluated. Several other areas of cardiac stimulation, such as its statistical nature and the effect of stimulation waveforms, are not discussed. Their fundamental understanding will also be enhanced by mapping and simulation techniques in the future.

REFERENCES

1. Marmorstein M: Contribution a l'etude des axcitations electriques localisees sur le Coeur en rapport avec al topographie de l'innervation du Coeur chez le chien. J Physiol Pathol 1927;25:617-25.
2. Lidwill M: Cardiac disease in relation to anesthesia. Transactions of the third session, Australasian Medical Congress (British Medical Association). Sydney, Australia, 1929.
3. Hyman A: Resuscitation of the stopped heart by intracardial therapy. II. Experimental use of artificial pacemaker. Arch Intern Med 1932;50:283.
4. Brooks C. McC, Hoffman BF, Suckling EE, et al: Chapters III and IV. In: Excitability of the Heart. New York/London, Grune & Stratton, 1955.
5. Van Dam RT, Durrer D, Strackee J, et al: The excitability cycle of the dog's ventricle determined by anodal, cathodal and bipolar stimulation. Circ Res 1956;4:196.
6. Mehra R, Furman S: Comparison of cathodal, anodal and bipolar strength-interval curves with temporary and permanent electrodes. Br Heart J 1979;41:468-76.
7. Brooks C. McC, Orias O, Gilbert JL, et al: Excitability of the mammalian heart during the cardiac cycle: The auricle. Fed Proc 1950;9:18.
8. Cranefield PF, Hoffman BF, Siebens AA: Anodal excitation of cardiac muscle. Am J Physiol 1957;190:383.
9. Mehra R, McMullen M, Furman S: Time dependence of unipolar cathodal and anodal strength interval curves. Pacing Clin Electrophysiol 1980;3:526-30.
10. Mehra R: Myocardial Vulnerability to Arrhythmias with Cathodal, Anodal and Bipolar Stimulation. [PhD Dissertation #75-21384]. University of Michigan Microfilms, Ann Arbor, Mich, 1975.
11. Lapicque L: Recherches quantitative sur l'excitation electriques des nerfs traitee' comme une polarisation. J Physiol Paris 1907; 9:620.
12. Furman S, Hurzeler P, Mehra R: Cardiac pacing and pacemakers IV: Threshold of cardiac stimulation. Am Heart J 1977;94:115-24.
13. Furman S, Parker B, Escher DJW, Solomon N: Endocardial threshold of cardiac response as a function of electrode surface area. J Surg Res 1968;8:161-6.
14. Irnich W: Engineering concepts of pacemaker electrodes. In Schaldach M, Furman S (eds): Engineering in Medicine I. Advances in Pacemaker Technology. New York, Heidelberg, Berlin, Springer-Verlag, 1975, pp 241-72.
15. Furman S, Hurzler P, Parker B: Clinical thresholds of endocardial cardiac stimulation: A long-term study. J Surg Res 1975;19:149-55.
16. Beyersdorf F, Schneider M, Kreuzer J, et al: Studies of the tissue reaction induced by transvenous pacemaker electrodes. I. Microscopic examination of the extent of connective tissue around the electrode in the human heart ventricle. Pacing Clin Electrophysiol 1988;11:1753-9.
17. Kay G: Basic aspects of cardiac pacing. In Ellenbogen K (ed): Cardiac Pacing. Boston, Blackwell Scientific Publications, 1992, pp 32-119.
18. Stokes K, Graf J, Wiebusch W: Drug eluting electrodes improved pacemaker performance. New York, Fourth Annu Conf IEEE Eng Med Biol Soc, 1982, p 499.
19. Schwan H, Kay C, Bothwell T: Electrical resistivity of living body tissues at low frequencies. Proc Fed Biol Sci 1954;13:1.
20. Mansfield PB: Myocardial stimulation: The electrochemistry of electrode-tissue coupling. Am J Physiol 1967;212:1475.
21. Hoshi T, Matsuda K: Excitability cycle of cardiac muscle examined by intracellular stimulation. Jpn J Physiol 1962;12:433.
22. Frankenhaenser B, Widen L: Anode break excitation in desheathed frog nerve. J Physiol 1956;131:243.
23. Ooyama H, Wright EB: Anode break excitation on single Ranvier node of frog nerve. Am J Physiol 1961;200:209.
24. Dekker E: Direct current make and break thresholds for pacemaker electrodes on the canine ventricle. Circ Res 1970; 27:811.
25. Wikswo JP, Lin SF, Abbas RA: Virtual electrodes in cardiac tissue: A common mechanism for anodal and cathodal stimulation. Biophys J 1995;69:2195-210.
26. Knisley SB: Transmembrane voltage changes during unipolar stimulation of rabbit ventricle. Circ Res 1995;77:1229-39.
27. Roth BJ: A mathematical model of make and break electrical stimulation of cardiac tissue by a unipolar anode or cathode. IEEE Trans Biomed Eng 1995;42:1174-84.
28. Ranjan A: A novel mechanism of anode-break stimulation predicted by bidomain modeling. Circ Res 1999;84:153-6.
29. Neunlist M, Tung L: Spatial distribution of cardiac transmembrane potentials around an extracellular electrode: Dependence of fiber orientation. Biophysical J 1995;68:2310-22.
30. Sepulveda NG, Roth BJ, Wikswo JP: Current injection into a two-dimensional anisotropic bidomain. Biophys J 1989;55:987-99.
31. Roth BJ: Strength-interval curves for cardiac tissue predicted using the bidomain model. J Cardiovasc Electrophysiol 1996;7:722.
32. Roth BJ: Nonsustained reentry following successive stimulation of cardiac tissue through a unipolar electrode. J Cardiovasc Electrophysiol 1997;8:768.
33. Patel GP, Roth BJ: How electrode area affects the electric potential distribution in cardiac tissue. IEEE Trans Biomed Eng 2000;47:1284-7.
34. Latimer DC, Roth BJ: Electrical stimulation of cardiac tissue by a bipolar electrode in a conductive bath. IEEE Trans Biomed Eng 1998;45:1449-58.
35. Roth BJ: The pinwheel experiment revisited. J Theor Biol 1998;190:389-93.

Chapter 9 Clinical Electrophysiology Techniques

JOHN D. FISHER

The term "technique" implies an objective. The electrophysiological techniques described in this chapter have the objective of determining whether a patient is electrophysiologically normal, adequate, or abnormal. Tests have been developed to assess the multiple electrophysiological levels of the heart ranging from the sinus node to the atrium, atrioventricular node (AVN), His-Purkinje system, ventricle, and associated structures such as the pulmonary veins. At any of these levels, abnormalities that can be associated with bradycardias or tachycardias are sought. Other chapters in this textbook detail the various types of abnormalities found at each of these levels and appropriate therapies. This chapter covers the elements of a "complete electrophysiological study" (EPS).

Indications for Electrophysiological Studies

Not everyone with a known or suspected arrhythmia needs an EPS. A simple ECG or one of the many noninvasive tests may provide definitive information. Guidelines on indications for EPS and ablation are published from time to time by the American College of Cardiology and the American Heart Association (ACC/AHA Guidelines)[1] with input from the North American Society of Pacing and Electrophysiology (NASPE), now Heart Rhythm Society. As the international community has expanded, there is also growing input from organizations such as the European Society of Cardiology. The Guidelines are likely to be updated in the near future.

Preparing for an Electrophysiological Study

THE ELECTROPHYSIOLOGIST

Performing an EPS is not a routine procedure; it must be tailored to the individual patient. As the study proceeds, the electrophysiologist must recognize the implications of each finding and adjust the remainder of the study accordingly. All of this implies an intimate knowledge of the patient and the indications for this specific EPS, as well as knowledge of clinical electrophysiology (EP) and testing techniques, including the risks and benefits associated with each test or maneuver. Clinical electrophysiology is recognized as a subspecialty of cardiology requiring substantial training and experience. Diagnostic and therapeutic (interventional or device) EPS should be performed only by physicians who have such a background.[2,3]

THE ELECTROPHYSIOLOGY LABORATORY

A detailed list of the requirements for an adequate EP laboratory are beyond the scope of this chapter. Analogous to the situation in cardiac catheterization, some laboratories limit themselves to diagnostic studies, whereas others perform both diagnostic and therapeutic/interventional procedures. The EP laboratory can be dedicated, or may be shared with catheterization/angiography. A summary of the needs of an adequately equipped EP laboratory includes: (1) an imaging/fluoroscopy system with pulsed fluoroscopy or other means of limiting radiation exposure, especially for laboratories performing ablation; (2) a suitable amplifier/recording system; (3) a programmable electrophysiological stimulator; (4) a selection of EP catheters, introducers, connectors, and so on; (5) an ablation system (currently a radiofrequency generator); (6) a selection of antiarrhythmic medications that can be delivered intravenously; and (7) equipment for defibrillation and advanced cardiac life support (ACLS). There must be provisions for blood pressure and oxygen saturation monitoring, and it is desirable to monitor expired CO_2.

STAFFING[4]

Requirements for staffing of an EP laboratory depend on the types of procedures done. In all cases, arrangements should optimize patient care and safety and minimize risks or complications including infections. The minimal staffing for a simple case is two: the electrophysiologist and assistant, usually a nurse or preferably a nurse/technician. When conscious sedation is used, legalities require that providing conscious sedation is the nurse's *only* major responsibility.[5] In a long case involving multiple and frequently moved catheters, it is difficult for the electrophysiologist to both maneuver the catheters and operate the stimulator and recording apparatus. Sometimes there is a temptation to move quickly between the patient on the procedure table and the recording/stimulator area, with just a change in gloves to serve the interest of infection control; this is *not* optimal. It is far preferable for a laboratory contemplating anything beyond simple and occasional studies to have a dedicated team of professionals who can operate the equipment and provide nursing support including conscious sedation, defibrillation, and so on while the physician remains scrubbed. For complex procedures it is sometimes invaluable to have involvement by a second electrophysiologist or cardiology/EP trainee or a trained physician assistant or nurse.

PATIENT PREPARATION

When an elective and scheduled EPS is planned, early patient preparation includes the information provided by the referring physician, and it is important that this be as accurate as possible. Some patients arrive at the electrophysiologist's consulting office under the impression that they will be having a 15- to 20-minute procedure that is virtually risk free. In fact, it is often difficult to predict how long a procedure will take. It is better for the patient to hear from the beginning that the procedure may take several hours, but that through the use of conscious sedation, time will pass quickly. Patients must understand that there are risks "including, but not limited to," bleeding, infection, perforation, stroke, heart attack, and death, and that a pacemaker may need to be implanted as a result of the study itself. It is equally important that this information be put in context and proportion. Most patients can understand analogies well, such as "imagining all the things that could conceivably happen to you when walking across the street, but of course they are very unlikely if one is careful." In the days before the scheduled procedure, plans must be made regarding continuation or discontinuation of anticoagulants or antiarrhythmic medications, if any.

When the patient arrives in the EP laboratory, a sympathetic staff can do much to allay the patient's anxieties. A professional but friendly and reassuring attitude can work wonders. For patients with a history of various aches and pains, strategic cushioning of the procedure table can lead to a smoother postprocedure course. When used by trained staff,[5] conscious sedation is a major contributor to toleration of longer procedures and implanted cardioverter defibrillator (ICD) testing and is safe. For longer procedures, especially with significant conscious sedation, urinary difficulties are common. Some patients are unable to urinate, whereas others may have frequency problems or a need to urinate at a point when bedpans or urinals cannot be placed. When this is anticipated, both patient comfort and monitoring are improved by the insertion of a Foley catheter before the procedure. Discomfort associated with catheter insertion can be substantially decreased in males by using a commercially available lidocaine jelly preparation injected into the urethra for both lubrication and topical anesthesia.

Catheter Electrode Insertion and Positioning

There is no single "correct rig." Only the use of the Seldinger technique is virtually universal. Sites of insertion include the brachial, subclavian, internal and external jugular, and femoral veins. The number of catheters used and the insertion sites also depend on the purpose of the EPS and whether there is any intention of retaining one or more catheter electrodes ("wires") for subsequent use another day. Common variations are shown in Figure 9-1.

In our own laboratory, we tend to favor the femoral vein approach for most diagnostic studies. Some electrophysiologists insert multiple catheters into the same vein, and others prefer to distribute the wires between the right and left femoral veins, believing that there may be less risk of laceration or thrombus. In a related matter, some electrophysiologists routinely heparinize patients during an EPS, and others do not unless catheters will be placed in the arterial circulation or the left atrium or left ventricle.

Subclavian and internal jugular (IJ) approaches: Both have their advocates. The IJ approach has a lower risk of pneumothorax, but the wires may be more difficult to maneuver. With the IJ approach, sterility may be more difficult to maintain during a longer procedure or with a retained wire because of proximity to the hair and turning of the patient's head. The subclavian, particularly on the left, permits the wires to follow a smooth arc toward the heart. In our laboratory we tend to prefer the subclavian, in part because we often retain the wires overnight after certain ablation procedures for confirmation testing, before discharge on the morning after an ablation. The "single stick–double wire" technique works well. Once subclavian access is obtained, a 6 French (6F) nonsidearm sheath with a diaphragm is introduced. Guidewires for 6F and 4F introducers are passed through the diaphragm and into the subclavian, and the 6F sheath is removed. A 4F sheath is then introduced over one of the guidewires, and a bipolar wire advanced to the right ventricular apex (RVA). A 6F or 7F sheath is then introduced using the other guidewire, and an octopolar, decopolar, or 20-pole catheter is then introduced. In most cases this goes into the coronary sinus. The 20-pole catheter can be maneuvered to have poles in the coronary sinus (CS), the isthmus between the coronary sinus and the inferior vena cava, and then along the lateral walls of the right atrium. This serves much the same purpose as the multipolar catheters usually inserted from the femoral approach during atrial flutter ablations, but

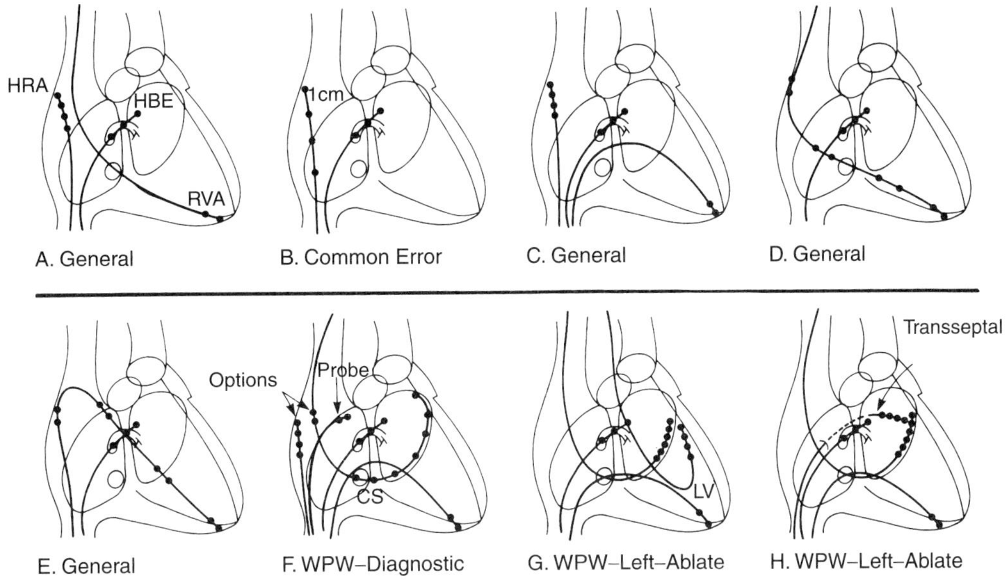

FIGURE 9-1 Common electrophysiology wire rigs. The number and type of electrode catheters used for an electrophysiology study (EPS) are determined by the type of study, the preferences of the physician and laboratory, and patient-specific issues. **Panel A** shows a setup for a standard diagnostic EPS. Note that the quadripolar catheter in the high right atrium (*HRA*) has closely spaced poles. This allows stimulation through the distal role and recording of the resulting electrograms through the proximal pair of poles, which are still in or near the HRA. Use of a catheter with 1 cm or more interelectrode distance for this purpose no longer allows recordings from the HRA, as illustrated in **panel B**. In **panel A,** one catheter was inserted "from above" (i.e., from an arm, subclavian, or jugular vein). **Panel C** shows a comparable rig with all catheters inserted from below. In **panels D** and **E,** somewhat comparable effects are obtained using a multipolar catheter, eliminating the need for separate HRA and RVA leads. As shown, both stimulation and recording in the HRA would need to be done through the same catheter. This is possible with some, but not all, recording systems and often results in the ability to record only those beats that are not directly stimulated. Options for more complex studies are shown in **panels F, G,** and **H.**

the latter technique also allows retention of the wires if next-day confirmation testing is contemplated. Optimal transisthmus positioning is facilitated by using a commercially available lead with an "offset" curve to achieve the necessary, rather anterior position. This rig, which allows for ventricular stimulation, and atrial stimulation and recording from several points, allows accurate next-morning confirmation testing after many types of ablation, all without the need for reinstrumenting the patient.

The femoral approach: This is the most common approach for all EPSs, and some electrophysiologists use it almost exclusively. It is possible to access the CS from the femoral approach, especially if deflectable leads are used. Although the femoral veins are quite forgiving, there is always a danger of laceration from excessive punctures, thrombus formation, retroperitoneal bleed. Multipolar catheters can help economize on the number of punctures needed. For example, octopolar and decopolar catheters with appropriately spaced poles can be positioned so that a single wire can perform pacing and recording from the right atrium and right ventricular apex. Such a wire, together with another for recording the His bundle potential, is all that is needed for most comprehensive diagnostic EPSs.

Guiding sheaths: In recent years, large numbers of specially shaped sheaths have become available. They are designed to help position or stabilize an electrode catheter in a certain area or site in the heart. These are usually long sheaths, with the distal portions in the heart and only a few centimeters of the electrode catheter protruding. Such sheaths are often invaluable during ablation procedures. Because the sheaths are designed to direct the catheters to a certain part of the heart, some flexibility is lost, and the sheaths may need to be replaced from time to time with other shapes or more standard shorter sheaths to facilitate wire placement in other positions. Guiding sheaths play a particular role during transseptal catheterization.

Transseptal puncture and catherization:[6-9] This technique dates back to the 1950s but has become important to the electrophysiologist for its role in the ablation of arrhythmias on the left side of the heart. The classic procedure uses a Brockenbrough needle, which is a long hollow instrument curving near its tip, with stiffness that allows precise torquing and positioning, and a lumen that can accommodate a stylet, guidewire, or fluid/contrast injection. The Brockenbrough needle is advanced through a special sheath. For electrophysiology purposes, the classic Mullins sheath with a distal curve like an umbrella handle has largely been superceded by sheaths with special shapes to facilitate placement in desired portions of the left atrium. The biggest danger associated with the transseptal technique is tamponade due to perforation of the left atrium, usually posteriorally.

Several techniques have been used to diminish this risk. When it is thought that the tip of the Brockenbrough needle/sheath is engaged in the *fossa ovalis* of the interatrial septum, the tip of the needle is advanced. Some electrophysiologists inject a small amount of dye to "mark" the septum. With the needle through the septum into the presumed left atrium, the position can be confirmed by advancing a small (e.g., 0.015 inch) guidewire and determining that it crosses the left atrium, ideally advancing into a left pulmonary vein. Alternatively, pressures can be measured as the needle is advanced from the right atrium, to the fossa (pressure damped), and then into the left atrium (looking for a typical atrial pressure curve versus a much more damped or absent curve for the pericardium, and much higher pressures in the event that the needle has pierced the aorta). Dye can also be injected at this point to confirm placement in the left atrium. Some physicians prefer the concurrent use of intracardiac or transesophageal echo to assist placement.[7] The literature suggests a comparably low incidence of complications with all of these techniques,[6-9] so much is left to the individual electrophysiologist's discretion. It must be underscored that the transseptal technique, however, is fraught with many subtleties and potential complications. The details are far beyond the scope of this section, and no one should attempt the technique based only on general experience and reading this paragraph.

For certain types of ablations, either the transseptal or retrograde (femoral artery to aortic valve to left ventricle) approach may be used. The Mullins sheath was originally designed to place a catheter in the left ventricle, and technique can still be used for ablation or mapping procedures involving the left ventricle. Approaches to the pulmonary veins, presently a matter of interest of atrial fibrillation ablation, are almost exclusively via the transseptal route, often with several punctures. Ablation of the Wolff-Parkinson-White Syndrome (WPW) can be accomplished using either the transseptal or retrograde approach.

Stimulation Techniques

INCREMENTAL/DECREMENTAL: HOW FAST IS THAT?

Unfortunately, mutually contradictory terminology is widespread and in general use. Most physicians describe the heartbeat in terms of the "pulse" *rate* in *beats per minute* (bpm). Temporary pacemakers are also usually calibrated in bpm. "Incremental pacing" may therefore describe progressively faster stimulation reported in bpm. Such terminology is satisfactory when the heart rate is regular during the period being measured, or when one is interested in an average overall rate, such as in atrial fibrillation. Problems arise, however, when one is interested in the effects of atrial or ventricular premature beats, whether spontaneous or stimulated.

The use of *intervals* or *cycle lengths* (CL) in milliseconds (ms) between successive beats or stimuli allows a more precise description of specific events and their consequences. However, there is an inverse relationship between bpm and CL. This leads to descriptions of *incremental* (progressively faster) pacing coupled with extrastimuli delivered at *decremental* (faster/shorter) intervals, with the electrophysiologist observing for *decremental* conduction (longer/slower CLs). Usually the meaning is clear from the context, but it is probably wise to provide clarification through terms such as "rate incremental pacing."

Rate and frequency: The most common English usage is for bpm to be described as "rate." "Frequency" usually refers to: (a) how often an arrhythmia occurs spontaneously or (b) the signal filtering settings in Hz. In other languages, "frequency" often means "rate in bpm," and this can introduce confusion in some communications.

Straight pacing: Pacing stimuli are delivered at a constant rate or cycle length. The duration may be indefinite as with temporary pacing, but for electrophysiological studies is usually for a specified number of stimuli or seconds. As indicated in the previous paragraph, progressively faster pacing can be described as incremental (bpm) or decremental (CL).

Ramps:[7] With pacing, a series of stimuli is delivered with each interstimulus interval successively differing from its predecessor. Most often the interval decreases, resulting in progressively faster pacing during the ramp. For example, stimulation could begin at cycle length 400 ms (150 bpm) with each of 10 successive intervals shortening by 10 ms so that at the end of the ramp, the cycle length would be 300 ms (200 bpm). Clinically, ramps are used for assessment of conduction and both initiation and termination of tachycardias. Ramps that start fast and then slow down are occasionally used in the treatment of tachycardias. The various uses of ramp pacing are detailed in the following sections.

Extrastimulus technique: After a series of spontaneous or paced beats at a constant cycle length (a type of straight pacing, designated S1S1), an extrastimulus is introduced at a somewhat shorter cycle length that is designated S1S2. After observing the response, the process is repeated, with the S1S2 intervals progressively shortened. Sometimes double or triple extrastimuli (S3S4) are introduced. The extrastimulus technique is used for assessment of refractory periods (see later), for initiation and termination of tachycardias, and as a diagnostic tool during ongoing tachycardias.

Stimulus amplitude and pulse duration: These are particularly important during extrastimulus testing. Higher amplitudes and longer pulse durations permit more closely coupled stimuli to "capture" (depolarize) the heart.[10,11] Excessively large stimuli may cause fibrillation. For these reasons, most EPS involves stimuli at two to four times diastolic threshold (in milliamperes [mA] or volts [V]) and 1 to 2 ms pulse duration.[12]

Ultra-rapid train stimulation:[13,14] A series of stimuli is delivered at such a rapid rate (typically, cycle lengths are 10 to 60 ms) that it is not expected that each of these will capture the tissue being stimulated. Using progressively higher stimulus amplitudes with successive trains, the method has been used to assess vulnerability to inducible ventricular fibrillation (VF). With stimulus amplitudes in conventional pacing ranges, trains have been used for both initiation and termination of regular monomorphic tachycardias. At very low (subthreshold) amplitudes, trains have been used as tools to alter local tissue responses, which in turn can affect certain tachycardias.

Stimulation Protocols

Clinical electrophysiology protocols did not spring fully formed from the head of Zeus. Different centers around the world evolved their own EP protocols. In general terms, there are broad differences in what constitutes a "complete" EPS. In more specific terms, there are differences in the sequence in which extrastimuli are delivered and in stimulus amplitudes and pulse durations used. These differences can affect the sensitivity and the specificity of the protocols as well as the time required to complete a protocol.

Over the years there have been calls for a "universal protocol," but this is not likely to happen. Individual centers have established databases that depend on using their own protocols. New methods that improve efficiency while maintaining sensitivity and specificity continue to be introduced and evaluated, and there is concern that imposition of a universal protocol, although desirable in many ways, would lead to petrification and stagnation. To deal with this, outcomes-based guidelines are generally accepted. For example, the North American Society of Pacing and Electrophysiology (NASPE) states that for patients with coronary artery disease and sustained ventricular tachycardia (VT), the stimulation protocol must be able to reproduce the clinical arrhythmia in at least 90% of cases.[12]

EFFECT OF SIGNAL FILTRATION AND INTERELECTRODE SPACING (FIG. 9-2)

During electrophysiological studies, the scale of interest ranges from macro (when does the QRS complex start,

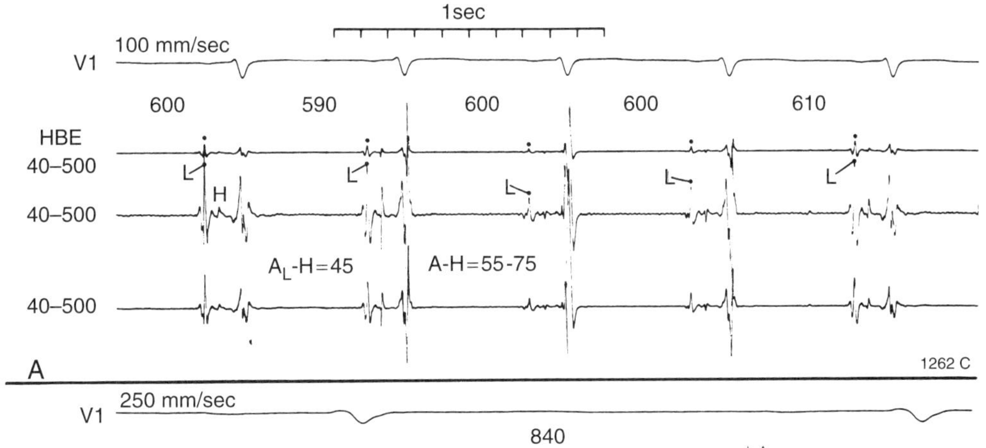

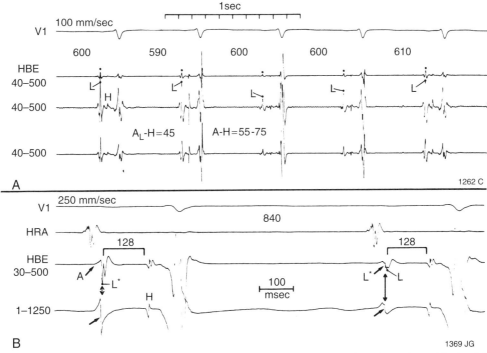

FIGURE 9-2 Beat-to-beat variation during normal sinus rhythm. ECG lead *V1* is displayed together with the *HBE lead* filtered at 40 to 500 Hz and recorded simultaneously at three gain settings. Paper speed is 100 mm/sec. Beat-to-beat variation can materially affect the slope of "first rapid deflection." In the five beats shown, these variations resulted in a 20-second spread in traditionally measured A–H intervals. The A_L is designated by an *L* connected by a line to a dot at the point in the upper low-gain HBE tracing. The A_L was easy to measure throughout the changes in amplitude and remained in constant relationship with the H deflection. **Panel B,** Atrial electrograms during normal sinus rhythm at paper speed 250 mm/sec. The onset of the "first rapid deflection" changes from beat to beat. Measurement of A_L-*H* is constant at 128 ms. Local atrial activity *L* and *L** on the HBE at 30 to 500 Hz was determined by using a simultaneously unfiltered tracing at 1 to 1250 Hz. *A* = onset of atrial activity as conventionally measured and designated by an *arrow* from the letter *A* to the site measured; *L* = timing of local activity on electrograms filtered at 30 to 40 to 500 Hz. The letters *L*, *L** are connected by a line to a dot at the designated site of the tracings; *L** = timing of the electrogram that corresponds most closely to the local electrogram recorded on the unfiltered lead; on the tracings, the timing of the local deflection is indicated by an *arrow*. (With permission from Fisher JD, Baker J, Ferrick KJ, et al: The atrial electrogram during clinical electrophysiologic studies: Onset versus the local/intrinsic deflection. J Cardiovascular Electrophysiol 1991;2:398-407.)

or what is the QRS duration?) to micro (how many ms does it take for a wavefront to move from the His bundle to the right bundle branch?). The ability to distinguish between macro and micro events is based on the effects of signal filtration and interelectrode spacing.

Low and wide: Macro events such as the QRS duration are best measured with electrograms filtered at about 0.5 to 100 Hz. Most electrical energy of the heartbeat occurs in these low frequency ranges, and low frequency signals also propagate over greater distances than high frequency signals. The archetype of these recordings is the surface ECG. The recommended filtration is 0.5 to 100 Hz, but very little difference is noted in the appearance of the QRS complex with filtration set as low as 0.5 to 20 Hz. Interelectrode distances are also relatively great (e.g., arm to arm, or arm or chest to leg). With intracardiac catheter electrodes, inclusion of the lower frequencies (starting at 0.5 Hz) and relatively wide interelectrode spacing will maximize inclusion of "far field" signals from parts of the heart that are not in close proximity to the recording electrodes. For example, with filtering at 0.5 to 100 Hz, the lead in the ventricle will record a relatively broad electrogram, possibly with a low amplitude atrial wave followed by a higher amplitude ventricular signal, followed in turn by a broad, gently rounded T wave. The ventricular deflection will usually show several components: 1) As the wavefront approaches the electrodes from afar, there will be a progressively steeper deflection that culminates in a point. 2) At this point, as the wavefront passes by the recording electrodes, there is a very rapid reversal in slope, making a near-vertical deflection, which has the highest dV/dt of the entire electrogram and is known as the *intrinsic deflection.* 3) This culminates in another reversal of direction as the wavefront continues to move away from the recording electrodes and inscribes a deflection that is a mirror image of the curve inscribed by the approaching wavefront.

Higher and closer: "Local" events, such as precise timing of a His bundle deflection, are best accomplished by filtration settings beginning at 30 or 40 Hz and going up to 500 Hz or higher. This takes advantage of the fact that higher frequencies do not propagate as well, and the fact that many of the structures of interest (such as the His bundle) contain relatively few cardiac fibers, and thus relatively low amounts of energy. Close interelectrode spacing (2 to 10 mm) together with filtering at 30 to 500 Hz combine to maximize recording of local "micro" signals, and exclude far field or macro signals.

Theoretically, it might be possible to focus even more closely on local events by filtering at 100 to 1000 Hz. However, there is so little energy at these frequencies that the recorder/amplifier system must be used at high gains, where the signal-to-noise ratio tends to eliminate any potential benefit.

In the world of signal filtration, the terminology again becomes somewhat confusing, rather like the alternative meanings for incremental and decremental pacing. With filtration, the "high pass" level is *not* the higher end of the frequency range but rather the hurdle above which frequencies are recorded. Similarly, the "low pass" level is the frequency below which a signal will be recorded. Thus, for a typical 30 to 500 Hz filtration range, the high pass is 30 Hz and the low pass is 500 Hz.

TIMING OF ELECTRICAL EVENTS[15]

Conduction in the heart is ionic rather than electronic or photonic. This means that conduction does not proceed at the speed of light, but at millimeters to meters per second. This in turn means that it is possible to measure sequences of depolarization rather easily using simple calipers or rulers. As indicated in the previous section, filtration and interelectrode distance affect the ability to record events at varying distances from a given point within the heart. All recorded signals will have characteristics such as duration and amplitude. As a general rule, if one is looking for the first evidence of an electrical event, several simultaneously recorded leads are observed for *onset* of a deflection that is used for the relevant measurement, and far field signals are welcome in some instances. The *local timing* of an event is important during mapping studies, when one is interested in the timing of an event at the site where the mapping electrode or probe is located. Here, far field signals are unwelcome, and closely spaced electrodes filtered at 30 to 500 Hz are critical. However, for some measurements (e.g., the atrial deflection in the His bundle region) the *local* timing at 40 to 500 Hz corresponds to the intrinsic deflection of "less filtered" recordings (see Fig. 9-2).[15]

Several factors can alter the shape of the wavefront. If the recording electrodes are in close proximity to the initial site of ventricular repolarization (e.g., near the site of initial depolarization during sinus rhythm, at the site of a tachycardia focus), the intrinsic deflection may come very early in the overall complex. This is particularly notable if *unipolar* recordings are made; an initial rapid negative intrinsic deflection is evidence that a recording electrode is *at* the initial site of depolarization. Unipolar stimulation is generally undesirable because the larger field (usually intracardiac—the "unipole"—to a surface electrode) creates a large stimulus artifact that can obscure the recordings.

When filtered at 30 to 500 Hz, the recorded electrogram takes on a more jagged appearance, often with several sharp changes in direction. The total duration of the electrogram is less than that recorded at 0.5 to 100 Hz because less far field information is included. However, as previously indicated, the timing of the maximum or peak deflection can be similar for electrograms recorded at 0.5 to 100 and 30 to 500 Hz.[15] Similarly, if recordings are made in areas that are scarred or damaged, overall signal amplitude may be low (<1.0 mv), and there may be a series of low amplitude deflections, none of which fulfills the criteria for local or intrinsic deflection. Usually, the first of the relatively larger deflections (this can be quite subjective) is used for local timing.

Computerized and automated mapping systems: These use algorithms similar to those used in the sensing amplifiers of pacemakers. Once a signal achieves a certain amplitude, the timing relative to a reference is recorded. In some cases there are dV/dt criteria that must be met to eliminate T wave sensing and far field information. The signal amplitudes vary considerably, so that amplitude criteria may be met relatively earlier at a site with a large amplitude than at another site that only marginally meets sensing criteria. Thus, the signals may be consistent in marking times as they reach the threshold for sensing; but these timings may differ from those that the electrophysiologist might choose as meeting onset or peak/local criteria. Computerized sensing algorithms are improving, but it is still wise for the electrophysiologist to review the point that the computer has chosen for measurement.

Clipping and limiting: In some instances, very large gains are necessary to focus on an area of interest. For example, recordings in the His bundle region typically include the low medial right atrium, the His bundle deflection, and the right ventricular inflow region. If only a very small His bundle deflection can be recorded, the gain may be increased on the recording system. This may make the atrial and right ventricular inflow deflections so large that they would cover much of the screen, overlapping with other simultaneous recordings. In such instances, "clippers" or "limiters" are applied; they electrically chop the maximum amplitude of signals so that they remain within designated bounds. In the case of the recordings from the His region, such limiting is not usually a problem. However, when measuring from areas where timing of the peak/local deflection is of interest, the use of a limiter/clipper will make such measurements impossible.

Notch filters: During the construction of an electrophysiology laboratory, it is important to have significant bioengineering input to ensure that electromagnetic interference with the recording apparatus is minimized. Nevertheless, some AC interference is virtually unavoidable, creating the telltale 50 to 60 Hz pattern that can render recordings almost unreadable. Notch filters that are optimized for 50 or 60 Hz can go far to mitigate the interference problem. Although the notch is optimized at 50 or 60 Hz, there is some attenuation of signals at nearby frequencies. In most instances this does not have a clinically important effect on the recorded electrograms.

The Electrophysiology Study

CHOICE OF SURFACE ECG AND INTRACARDIAC RECORDINGS

It is cumbersome to display all 12 leads of the regular surface ECG. One option is to use mutually perpendicular leads (I, AVF, and V1), often supplemented by lead II, which gives an indication as to the presence of abnormal left axis deviation. Many electrophysiologists have their own personal favorite lead selections. Orthogonal lead systems (X, Y, Z) are logical but not commonly used.

Intracardiac leads are placed strategically at various positions within the heart to record *local* events in the region of the lead, rather than far field events. This is accomplished by filtering intracardiac electrograms.

STIMULATION PROTOCOL AND SEQUENCE

Essential tools to use during an EPS are the extrastimulus technique, (rate) incremental pacing, and pharmacologic probes. Variations are used for assessment of the entire heart, from sinus node to ventricle, as shown later. The protocol must meet the objectives of the study. There is no "universal protocol," but general outcome standards have been established for some indications (e.g., induction of monomorphic VT in >90% of patients with coronary artery disease and a history of such VT).[12]

The *extrastimulus technique* (briefly introduced earlier) is the heart of an EPS used primarily for tests of refractoriness (see later) and tachycardia induction. Typically there is a baseline drive of eight beats, which may be sinus rhythm but is usually delivered at a constant cycle length (S1S1) by a stimulator. The drive is then followed successively by single (S2), double (S2 and S3), and triple extrastimuli (S2, S3, and S4), which are designated S2, S3, and S4, respectively. Generally the extrastimuli are initiated at 80% to 90% of the drive CL. All these stimuli are delivered with uniform specifications, usually 1- to 2-ms pulse duration and an amplitude two to four times diastolic threshold. Of the various methods for shortening (decrementing) the S1S2 intervals and subsequently S2S3 and S3S4, most are variations of either the "tandem" method or the "simple sequential" method (Fig. 9-3).[16]

Tandem method: The S1S2 is decremented in 10 ms steps until S2 does not capture, and then the S1S2 interval is increased by 40 to 50 ms. A late S3 is then introduced, and the S2S3 interval is decremented until S3 fails to capture. Then, in alternating fashion, the S1S2 and S2S3 intervals are alternating (tandem) until S3 fails to capture. S3 is then moved out 40 to 50 ms, and the process repeated with S4. Some laboratories do use an S5 or S6 as well, but stopping with triple extrastimuli (S4) is more typical.

Simple sequential method: The S1S2 interval is decremented in 10-ms steps until S2 fails to capture. It is then moved 10 ms later or until reliable capture is attained. At that point the second extrastimulus (S3) is decremented similarly, and finally a third (S4). A prospective randomized trial has shown that the tandem and simple sequential methods produced comparable results in terms of inducibility of the clinical arrhythmia, inducibility of "nonclinical" arrhythmias, and noninducibility.[16] The simple sequential method takes significantly less time to perform.[16]

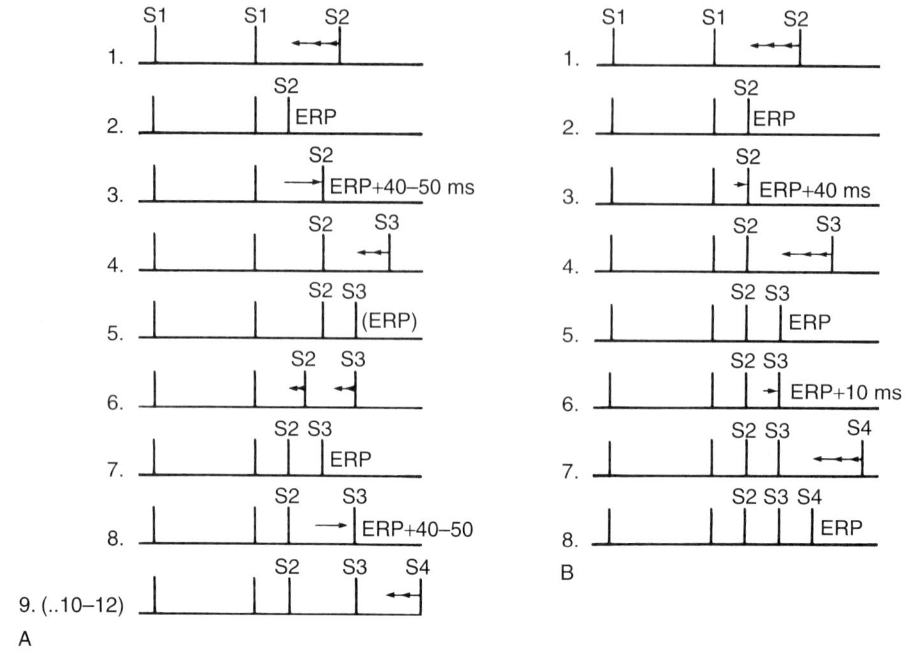

FIGURE 9-3 Method of introducing extrastimuli. **Panel A,** The tandem method; after the first extrastimulus (*S2*) reaches the effective refractory period (*ERP*), it is moved out 40 to 50 ms. The second extrastimulus (*S3*) is introduced and the S2S3 interval is decremented until S3 fails to capture. The S1S2 interval is then decremented until S3 again captures. Decrements of S2 and S3 continue in this tandem fashion until reaching the shortest intervals where S2 and S3 capture. For the third extrastimulus (*S4*), S3 is moved out 40 to 50 ms, and the process is repeated for S3 and S4. **Panel B,** Simple sequential method. The S1S2 interval is decremented until the ERP of S2 is reached. S2 is then placed 10 to 20 ms later to assure capture, and the process is repeated with S3 and then S4. Clinical results are comparable with the two methods. (With permission from Roelke M, Smith AJ, Palacios IF: The technique and safety of transseptal left heart catheterization: The Massachusetts General Hospital experience with 1,279 procedures. Catheter Cardiovasc Diag 1994;32:332-9.)

REFRACTORY PERIODS AND CONDUCTION INTERVALS

These closely interwoven concepts often produce some level of confusion. At the simplest level refractory periods are established by responses to extrastimulus testing, and conduction is accessed by rate incremental pacing. Details follow.

Refractory Periods[17-25]

The classic technique is to deliver eight drive stimuli all designated as *S1*. After the last S1, an extrastimulus (S2) is delivered at an interval somewhat shorter than the S1S1 interval. The process is repeated with decrements in the S1S2 interval, usually until S2 reaches refractoriness (i.e., fails to capture). The S1S2 interval is usually decremented in 10-ms steps, although 20- or even 30-ms steps may be used in clinical laboratories for very long S1S2 intervals. Refractory periods tend to be shorter with shorter S1S1 intervals. An exception is sometimes seen in the AVN where refractory periods may prolong as the S1S1 is decreased from 1000 to 600 ms, but thereafter usually tend to shorten as the S1S1 is further decreased (Fig. 9-4). Depending on the objective of the EPS, refractory period testing may be carried out with stimulation at several atrial or ventricular sites and at two or more drive (S1S1) cycle lengths. Several different "refractory periods" require description.

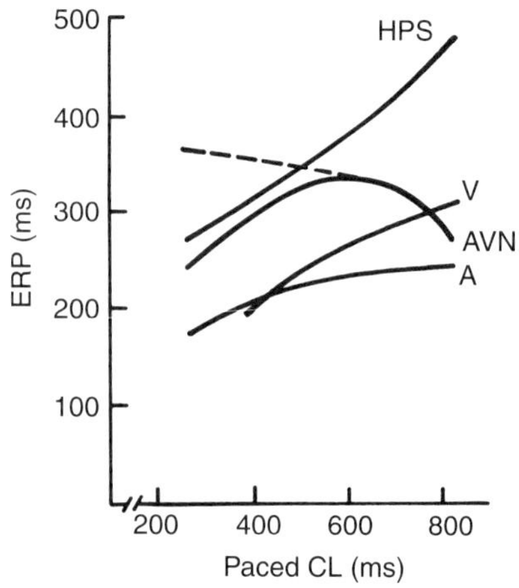

FIGURE 9-4 Effective refractory period (*ERP*) variability in different parts of the atrial ventricular conduction system. In the case of the His-Purkinje system (*HPS*), atrium (*A*), and ventricle (*V*), the ERP consistently shortens with shorter-paced cycle lengths (*CL*). The atrial ventricular nodal (*AVN*) ERP may initially prolong with shorter-paced CLs, and this trend can continue (*dashed line*) or the AVN ERP can shorten as the paced CL is further shortened. (With permission from Fisher JD: Primer of Cardiac Electrophysiology. Armonk, NY, Futura, In press.)

Absolute refractory period (ARP): The shortest S1, S2 interval that fails to capture, even at the maximum available S2 stimulus amplitude.

Relative refractory period (RRP): Just longer than the absolute refractory period, the stimulated tissue has repolarized enough to permit depolarization in response to a stimulus. However, because repolarization is not yet complete, the action potential amplitude and dV/dt or slew rate are also low. This results in a slowly conducting wavefront that has a low "safety factor" for propagating to nearby cells.

Effective refractory period (ERP): The *longest* S1S2 interval that *fails* to capture or depolarize the tissue of interest. Note that the ERP is defined by stimulus–stimulus (S1S2) intervals. The definition becomes somewhat problematic when double (S3) or triple (S4) extrastimuli are introduced and ERPs are defined for them. In this case, the definition is altered slightly, and the ERP of the first extrastimulus is set as the *shortest* coupling interval that *does* result in capture. Such an ERP should thus be 10 ms longer than the longest S1S2 that *fails* to capture; but if the earlier extrastimuli failed to capture, then responses to the later extrastimuli would have little meaning.

There is a downstream effect with ERPs: If measured at the site of pacer stimulation, the interval set in the programmable stimulator can be read directly to determine the ERP. Suppose, however, one is interested in the ERP of the His bundle. If stimulation is delivered to the high right atrium (HRA), as is conventional, it may be necessary to reduce the S1S2 interval to the point where the RRP of the atrium is reached, resulting in a prolonged conduction time from the HRA to the atrioventricular node (AVN)/His bundle region. The impulse may also encounter the RRP of the AVN, further delaying the arrival of the wavefront generated by S2 at the His bundle (H). Thus, the H1H2 interval will be substantially longer than the S1S2 interval. Because in this case the tissue of interest is the His bundle, its ERP is defined as the longest H1H2 interval that fails to propagate. As indicated, this may differ substantially from the S1S2 delivered in the HRA. Often, the ERPs at a proximal site (e.g., the HRA) are longer than at distal sites, making it impossible to determine the ERP distally.[17-25] The importance of taking the extra steps to measure correctly depends on the nature of the EPS.

Functional refractory period (FRP): This is defined as the *shortest* coupled response—response time at the tissue of interest as a result of an S1S2. Again, depending on the tissue of interest, the shortest stimulation intervals may not produce the shortest response intervals. For example, when assessing the AVN, 10-ms decrements in the S1S2 stimuli delivered to the HRA may produce substantially longer increments in AV nodal conduction as accessed by the resulting H1H2 intervals. Thus, the FRP of the AVN may occur at S1S2 intervals that are longer than the S1S2 intervals needed to reach the ERP of the AVN.[18] Normal variations in these relationships, identified earlier, are summarized in Figures 9-4 to 9-6.

SPECIFIC REFRACTORY PERIOD DEFINITIONS[23-25]

Atrial ERP: The longest S1S2 resulting in capture of the atrium at the stimulation site.

Atrial FRP: The shortest A1A2 in response to any S1S1. As with all FRPs, the site of stimulation and response must be defined. The term *A1A2* is often used to indicate that atrial depolarization near the AVN region should be used for this measurement, rather than the stimuli delivered to a different site such as the HRA. Usually for the atrium this means stimulation in the HRA and response in the low medial right atrium in the region recording the His bundle. However, one might choose interatrial testing (e.g., from HRA to left atrium or other designated sites).

Atrioventricular nodal (AVN) ERP: The longest A1A2 (in the low right atrium) that fails to conduct to the His bundle (i.e., to produce an H2).

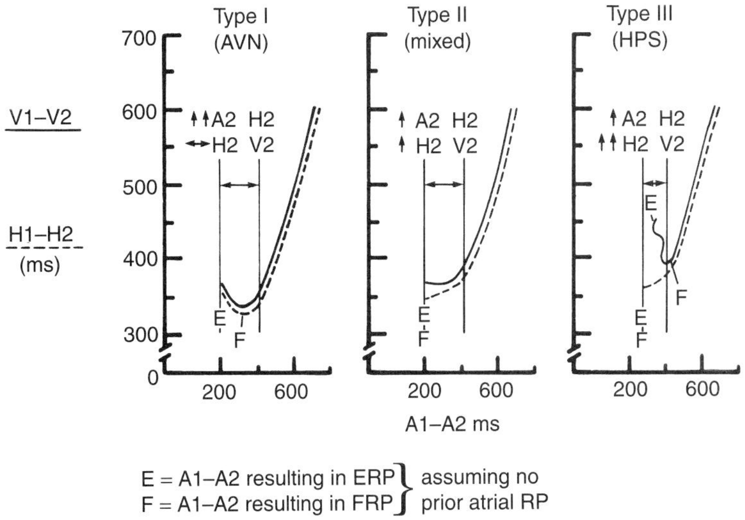

FIGURE 9-5 Patterns of atrioventricular (AV) conduction and refractoriness. Abbreviations are the same as for other figures. In response to decremental extrastimulus testing, conduction over the AV node (*A2H2*) and His-Purkinje system (*H2V2*) may differ substantially from each other. Major delays may be encountered at the A2H2 level with minimal H2V2 prolongation (*Type I*); at the H2V2 level (*Type III*); or at both levels (*Type II*). (With permission from Fisher JD: Primer of cardiac electrophysiology. Armonk, NY, Futura, In press.)

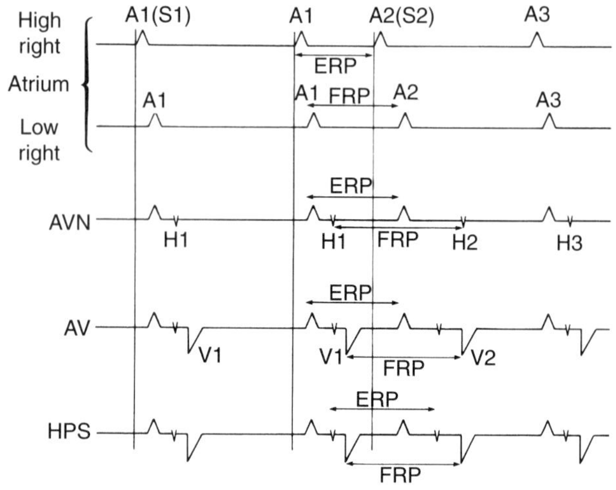

FIGURE 9-6 Effective and functional refractory periods (*EPR, FRP*) in the atrioventricular conduction system. *A1 (S1), H1,* and *V1* represent baseline depolarizations. *A2 (S2)* in the high right atrium represent an extrastimulus with the responses *A2* in the low right atrium, *H2* in the His-Purkinje system, and *V2* in the ventricle. *A3* is the recovery beat. See text for details. At each level of the conduction system, the ERP is the longest stimulus-to-stimulus interval that fails to capture or conduct, and the FRP is the shortest result of such conduction in the specified tissue. Note that the ERPs are therefore specified not by the timing of the stimulus in the high right atrium, but by the component of the conduction system that is acting as the stimulus for the next step. Hence, at the bottom of the cascade, the ERP of the HPS is defined in terms of the H1H2 interval. Note: In the AVN line, the H1H2 interval should be shorter, identical with that of the following two lines. (With permission from Fisher JD: Primer of Cardiac Electrophysiology. Armonk, NY, Futura, In press.)

Atrioventricular nodal (AVN) FRP: The shortest H1H2 in response to any A1A2.

His bundle (or His-Purkinje) ERP: The longest H1H2 that fails to propagate to the ventricle (i.e., no V2). Sometimes it is also possible to determine that conduction has gone beyond the His (e.g., in the production of a right bundle branch potential, or perhaps a beat that is conducted with a bundle branch block morphology). In such cases the ERPs and FRPs of the bundle branches can be determined in a similar manner.

His bundle FRP: The shortest V1V2 in response to any H1H2.

Ventricular ERP: The longest S1S2 (delivered in the ventricle) that fails to capture.

Ventricular FRP: The shortest V1V2 interval at a designated ventricular site in response to any S1S2 delivered at a ventricular site that may or may not be the same site designated as V1V2 used for FRP.

Accessory pathway ERP: The longest S1S2 that fails to propagate across an accessory pathway. It is important that this should be delivered or measured near the site of the accessory pathway itself. If the HRA is stimulated and the accessory pathway is located in the left lateral mitral annulus, there may be a great difference in the timing between the S1S1 in the HRA and the A1A2 arriving in the lateral left atrium.

Accessory pathway FRP: The shortest pre-excited V1V2 in response to any A1A2 interval. As with determination of the ERP, the site of stimulation needs to be taken into account.

Retrograde refractory period: These are determined using ventricular S1S2, with responses to the reverse of those described previously.

CONDUCTION[18,19,20,27-31]

Conduction is usually defined as the ability of (a string of) tissue to propagate a wavefront in response to increasingly faster pacing rates. No extrastimuli are involved. Straight pacing and slow ramp techniques can both be used for this purpose.[18] With straight pacing, the first several beats involve stimuli at cycle lengths substantially shorter than the resting cycle lengths, so that a period of "accommodation" occurs that, in some instances, takes as long as 45 seconds to resolve, although 10 to 15 seconds is usually sufficient.[30,31] The ability of an impulse to conduct from the HRA to the ventricle is often tested with straight incremental pacing, often simultaneously with the same type of pacing used for assessment of sinus node recovery times (SNRTs) (see page 164). A slow ramp (rate increasing by 2 to 4 bpm/sec) results in accommodation on a beat–beat basis and allows assessment of AV and VA conduction in a total of 10 to 15 seconds.[18,26-29] The ramp technique is therefore useful as a screening tool, and as a means of assessment after interventions such as catheter ablation. Some phenomena can be seen more readily with the ramp technique. For example, in patients with dual AVN physiology, a sudden discontinuity or "jump" in the A–H interval can often be observed during a ramp, but much less often with straight pacing because the stepped series of cycle lengths used in straight pacing are not close enough together to identify a jump.

Conduction intervals:[18,26] It is routine during EPS to indicate the cycle lengths at which AV or VA Wenckebach or other block occurs. These intervals are usually defined in terms of the stimulation cycle lengths and thus are analogous in some ways to ERPs. The timings of depolarizations, which are downstream from the site of stimulation, are analogous to FRPs. Because of accommodation with straight pacing, and near-elimination of this factor with ramp pacing, effective and functional *conduction* periods are usually much nearer to each other than the comparable *refractory* periods (Fig. 9-7).

WHAT PARTS OF THE HEART TO STIMULATE, AND IN WHAT ORDER?

It is traditional to begin atrial stimulation in the HRA and then proceed to other sites such as the coronary sinus, if clinically indicated. In the ventricle, stimulation usually begins at the right ventricular apex (RVA), and, if further stimulation is necessary, proceeds to the RV outflow tract (RVOT). Sometimes other sites are used, such as the RV inflow tract or septum or the left ventricle.

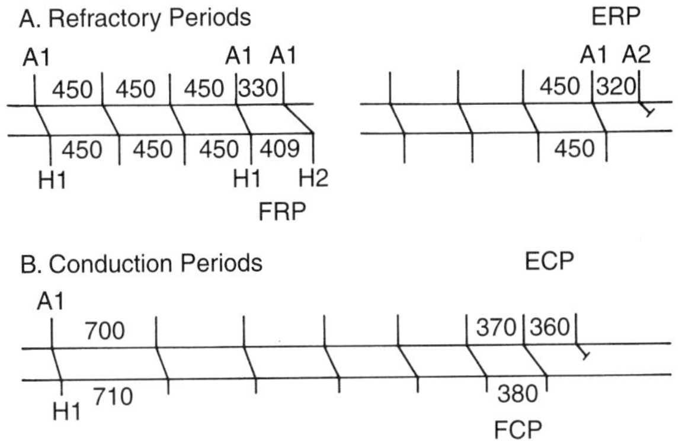

FIGURE 9-7 Schematic comparison of refractory and conduction periods. *A*, Atrioventricular nodal refractory periods. Impulses in the low right atrium at regular intervals of 450 ms (*A1A1*) are followed by an extrastimulus (*A2*) with an *A1A2* interval of 330 ms. The resulting *H1H2* interval of 409 ms is the shortest achieved and represents the atrioventricular nodal function refractory period (FRP). To the right in *A*, an extrastimulus at *A1A2 = 320 ms* fails to conduct and represents the effective refractory period (*ERP*). The A1A2 intervals resulting in the FRP and ERP are not always juxtaposed. The FRP is typically much longer than the ERP. *B*, Atrioventricular nodal conduction periods. Pacing at progressively shorter cycle lengths ultimately results in block. The shortest resulting HH interval is the functional conduction period (*FCP*), and the longest AA failing to conduct is the effective conduction period (*ECP*). The intervals are typical, with relatively little difference between ECP and FCP. (With permission from Fisher JD, Zhang X, Waspe LE, et al: Tests of refractoriness and conduction during clinical electrophysiologic studies: Yields and roles. J Electrophysiol 1988;2:175.)

As a general rule, more accurate refractory periods can be determined, and arrhythmias are more easily induced if the site of stimulation is in proximity to the tissues being assessed. Therefore, the clinical situation dictates the chosen site and sequence of stimulation.

It is debatable whether it is wise to complete double extrastimulus testing at multiple drive rates at all selected sites of stimulation before moving on to triple extrastimuli.[34-37] It is known that triple (or quadruple) extrastimuli, particularly at more rapid drive rates, induce a higher percentage of nonclinical arrhythmias.[38] However, completion of double extrastimuli at all sites before returning and repeating the process with triple extrastimuli has several drawbacks. It may require multiple moves of the catheter; and refractory periods often change over a period of several minutes[39] so that the rather tedious process of establishing the refractory periods for the first and second extrastimuli must be repeated before delivering the third. At any given site, triple extrastimuli delivered during sinus rhythm and an initial drive pacing rate have a high likelihood of inducing clinical arrhythmias and a low incidence of induction of nonclinical arrhythmias.[34,36,38] Indeed, this "yield ratio" for triple extrastimuli delivered at the slower drive rates is better than for double extrastimuli and certainly for triple extrastimuli at the faster drive rates (Fig. 9-8).[36] Thus, if the objective is to minimize the induction of nonclinical arrhythmias, it would be best to deliver single, double, and triple extrastimuli during sinus rhythm and a slower drive rate and *single* extrastimulus at the faster drive rate, all at the same site

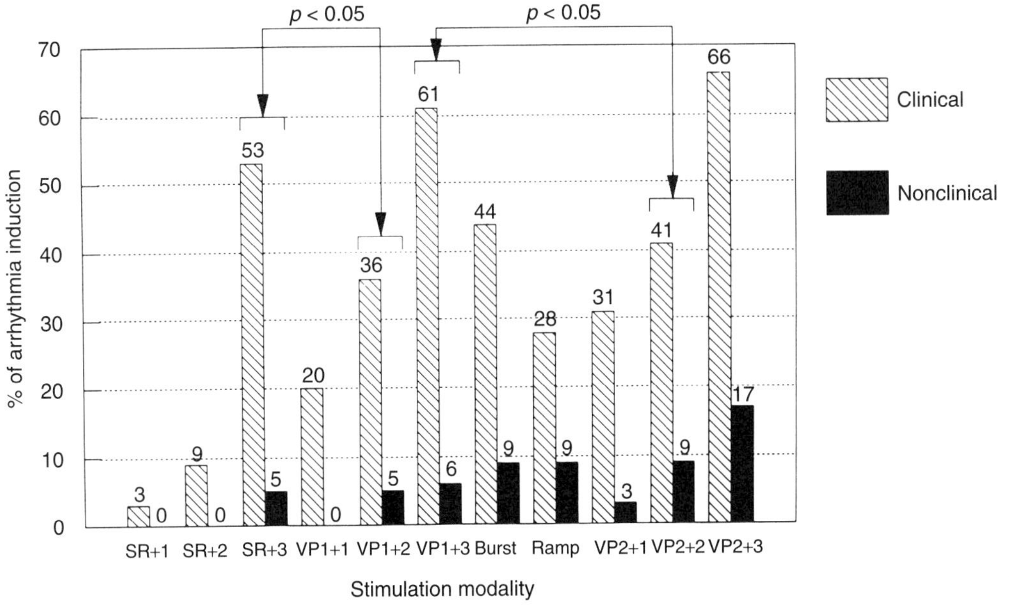

FIGURE 9-8 Frequency of induced clinical and nonclinical arrhythmias with each stimulating modality (protocol step) in a consecutive series of patients with inducible ventricular tachycardia. All patients received *all* steps. The sequence actually used is displayed on the abscissa. The ordinate indicates the percent of uses of each modality that resulted in a clinical (hatched) or nonclinical (solid) arrhythmia. Each bar graph is capped by a number indicating more precisely the percent of arrhythmias induced. *SR* = sinus rhythm; *VP1* = the slower-paced drive rate; and *VP2* = the faster rate. Numbers *1, 2,* and *3* indicate the number of extrastimuli. Note that *SR + 3* was more efficacious than either *VP1 + 2* or *VP2 + 2*, and that *VP1 + 3* was superior to *VP2 + 2*. In addition to the *P* values indicated, *VP2 + 3* was more likely to induce nonclinical arrhythmias than *SR + 3* or *VP1 + 3* ($P < 0.05$). (With permission from Artoul SG, Fisher JD, Kim SG, et al: Stimulation hierarchy: Optimal sequence for double and triple extrastimuli during electrophysiological studies. Pacing Clin Electrophysiol 1992;15:790-800.)

before moving to a second site and returning to complete the protocol. On a practical basis, it seems most reasonable to complete the protocol at one site before moving to a second.[34] However, many laboratories have established databases and outcomes that are related to their historic sequence of delivery extrastimuli, and this too may be important in determining a laboratory's choice of stimulation sequence.

"PROBLEM-ORIENTED" VERSUS "COMPLETE" ELECTROPHYSIOLOGICAL STUDY

Sometimes there are limited or problem-oriented indications for EPS. Typical examples are assessment of the effects of an antiarrhythmic drug on inducibility of any arrhythmia or on defibrillation threshold in a patient with an ICD. Sometimes scheduling or workload issues dictate a limited study, with the option for a more complete study later if the clinical situation warrants.

Generally it is preferable to plan for a "complete diagnostic EPS" with or without ablation of known or suspected arrhythmias. The reason is that patients with indications for EPS are often found to have multiple abnormalities. Thus, patients being studied for syncope of undetermined origin (SUO) may have abnormalities of the sinus node, AVN, or His-Purkinje system; inducible arrhythmias; or some combination of these abnormalities that could affect the choice of treatment. For example, sinus node abnormalities identified in a patient with VT would suggest the use of a dual-chamber ICD as part of the therapy. As an additional example, in the case of supraventricular tachycardias (SVTs), about 15% of patients have more than one reproducibly inducible mechanism; hence, it is not uncommon to find arrhythmias related to accessory pathways and dual–AVN physiology in the same patient. Often these arrhythmias have somewhat similar rates, and previous "documentation" is often unable to definitively rule out both arrhythmias as sources of clinical tachycardias. Thus, many electrophysiologists believe that after ablating what they think is the main arrhythmia, it is wise to proceed and ablate other inducible arrhythmias so that the patient will not need to return to the laboratory for another invasive procedure. Except for induction of heart block, the major risk of additional procedures is related to the instrumentation process; therefore, eliminating arrhythmias that seem likely to be clinically relevant should be done, if possible, during a single EPS.

THE COMPLETE ELECTROPHYSIOLOGICAL STUDY

The term "complete" is necessarily relative. For example, it may be impossible to use each of the available pharmacologic probes during a single EPS, so the electrophysiologist must select the most important components that will satisfy the general criteria of "complete."

A complete diagnostic EPS should include assessments of the sinus node, atrium, AVN, His-Purkinje system, retrograde conduction, and inducibility of supraventricular and ventricular arrhythmias. In many cases, a single set of stimulation sequences provides information about a broad range of cardiac chambers and conduction system components. For example, rate-incremental atrial pacing can be used to assess SNRTs, as well as conduction at the AVN and His-Purkinje levels. Atrial extrastimulus testing can provide information on the sinoatrial conduction time (SACT), vulnerability to atrial tachycardias, evidence of normal or dual AVN physiology, and vulnerability to induced AVN reentry tachycardia (AVNRT), as well as refractory period data on the atrium, AVN, and His-Purkinje system. Similarly, ventricular pacing can provide information on retrograde conduction as well as vulnerability to ventricular arrhythmias.

A TOP-TO-BOTTOM REPORT

If one conceptualizes the heart as starting at the sinus node and ending at the ventricle, with tests of the various structures in between, organization becomes rational and simple. An EPS report should comment on each of these steps or levels, indicating normal, borderline, or abnormal findings. Many of the same stimulation processes accomplish several tests at once. Other chapters in this textbook provide details on some of the testing methods mentioned only briefly in the present section.

1. **Baseline intervals:** These are obtained from the "passive His recording" (i.e., from the combination of the surface ECGs and catheters that allow timing of the P–A, A–H, H, and H–V intervals in the resting state without simulation or additional medications) (Fig. 9-9). Normal intervals appear in Table 9-1.[15,23,39-47] The electrophysiologist should indicate whether the intervals are normal, borderline, or abnormal.
2. **Sinus node function tests:** These include sinus node recovery times (SNRTs)[22,23,48-75] and sinoatrial conduction times (SACTs) using the Narula,[72,76] Strauss,[55,71,72,74] or direct[77-81] methods. Sinus node tests are discussed in greater detail in later chapters, as are tests of other structures. It is important to understand that there is no single "SNRT value." For normal patients, pacing at 120 bpm (500 ms CL) for anywhere from 15 to 120 seconds will result in SNRTs with a maximal spread of 250 ms (±2 SD). Slower or shorter drive rates may produce insufficient overdrive suppression of this archetype of an automatic focus. Faster pacing may produce periods of block or physical sensations. These sensations produce reflexes resulting in neurohumoral changes that affect the SNRTs.

Chronotropic incompetence: A patient may have a normal resting sinus rate, but fail to respond appropriately to stress, resulting in a symptomatic relative bradycardia. Exercise testing and isoproterenol or atropine infusion are among the methods to unmask chronotropic incompetence.

Intrinsic heart rate (IHR): The sinus node is the archetype of the automatic focus, but is heavily influenced

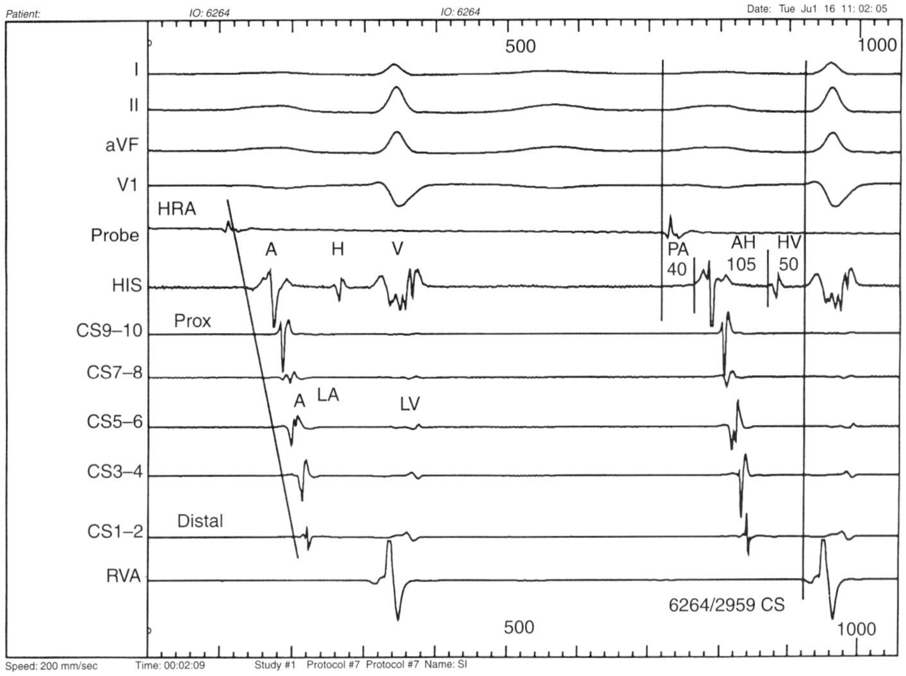

FIGURE 9-9 Normal intracardiac electrograms. This is a "passive HBE" (i.e., a His bundle electrogram and accompanying recordings during sinus rhythm with no ongoing manipulations, stimuli, or intentional stresses). Semi-orthogonal leads *II, AVF,* and *V1* are used together with lead 2, which is helpful in determining axis. The measurement of intervals is demonstrated. The coronary sinus *(CS) poles 1-2 (distal)* to *9-10 (proximal)* represent left atrial (*LA*) and left ventricular (*LV*) events. Definitions are provided in the text.

by neurohumoral tone. It is sometimes important to distinguish whether inappropriate sinus rates or responses to SNRT or SACT testing are due to intrinsic sinus node dysfunction or to extrinsic influences. To achieve autonomic blockade, divided doses of medications are given over a few minutes, including propranolol 0.2 mg per kilogram and atropine 0.04 mg/per kilogram. After these medicines, the IHR = 118.1 − (0.57 × age); normal values are ±14% for ages younger than 45 and ±18% for ages older than 45. Interpretations of SNRTs and SACTs after neurohumoral blockade are uncertain.

3. **Carotid sinus massage (CSM):**[82-88] The effects of CSM can be assessed particularly well in the EP laboratory. Pauses of greater than 3 seconds, (cardioinhibitory response) are abnormal due to sinus node arrest or atrioventricular block (the latter at the A-H level). If there is sinus arrest, atrial pacing will give some inkling as to whether the AVN is also affected. Blood pressure measurement during CSM may identify a *vasodepressor* effect.
4. **Atrium:**[42-46,89-103] The atrium is assessed for conduction, refractory periods, and induction of arrhythmias. Intra-atrial conduction is determined from the P–A interval (see Fig. 9-9). The P–A interval is measured from the first onset of evidence of sinus node depolarization (either a normal-appearing P wave or a deflection on the high right atrial intracardiac lead) to the first rapid deflection of the atrial depolarization seen on the intracardiac electrodes recording the His bundle deflection. With atrial pacing, the P–A interval may prolong for several reasons: "latency" at the site of stimulation; initial slow conduction out of the site of stimulation, perhaps due to anisotropic conduction; or stimulation at a site that does not directly engage the preferred pathways between the sinus node and the region of the AVN. The P–A interval may also prolong as a result of atrial disease, postoperative scarring, or other abnormalities. The interatrial conduction time between the right atrium and the left atrium is generally measured between the HRA and a mid-distal coronary sinus lead, and it should not exceed 130 ms. The routes of interatrial conduction are complex; they are still a subject of investigation in the human.

Atrial extrastimulus testing is used for refractory period assessment and induction of arrhythmias.

Atrial fibrillation can be induced in virtually any patient with atrial stimulation that is both very prolonged and rapid, and certainly if enhanced

TABLE 9-1 Normal Conduction Intervals in ms (±2 SD), Based on a Compilation of Sources[15,23,39-47]

	P-A Intra-atrial	A-H AV Node	H His	H-V His to V	P-LA Inter-atrial
Passive/Baseline	10-45	55-130	<25	30-55	40-130
Atrial Paced	10-75	*	<25	30-55	65-150

*Progressive prolongation; see Figure 9-10.

See text for explanation and definition of abbreviations.

with vagal tone, either directly or indirectly through the use of agents such as adenosine. Interpretation of such a finding again requires clinical wisdom.

Induction of an organized regular atrial tachycardia or flutter is much less likely to be a normal variation, particularly if the arrhythmia is sustained.

5. **Atrioventricular node (AVN) function:**[23,89-122] The AVN competes with the sinoatrial node (SAN) for honors as the part of the heart most influenced by neurohumoral tone. The AVN is the chief arbiter of the P–R interval on the surface ECG, which prolongs or shortens in response to physiologic needs. As a general rule, sympathetic tone as seen in exercise, anxiety, and similar tends to shorten the A–H interval, which is the reflection of AVN conduction. Parasympathetic (vagal) tone tends to prolong AV conduction at the AV level and may even cause nocturnal block in normal patients or block during low-level activities in highly trained athletes. Atropine and adenosine are commonly used "pharmacologic probes" in assessment of both sinus and AVN function.

With rate incremental pacing, the A–H interval prolongs in a typical complex but smooth curve, with large increases coming just before block (Fig. 9-10). Similar effects on the A–H interval are observed with extrastimulus testing in the atrium. In essence, the AVN is "set" to conduct at an A–H interval that is appropriate for the patient's physiologic condition at the time of the EPS. In general this means a sedentary state, often with mild sedation.

Wenckebach periodicity is most common at the AVN level in response to pacing, parasympathetic tone, medications, and disease. The mechanisms are complex and debated.[112-116]

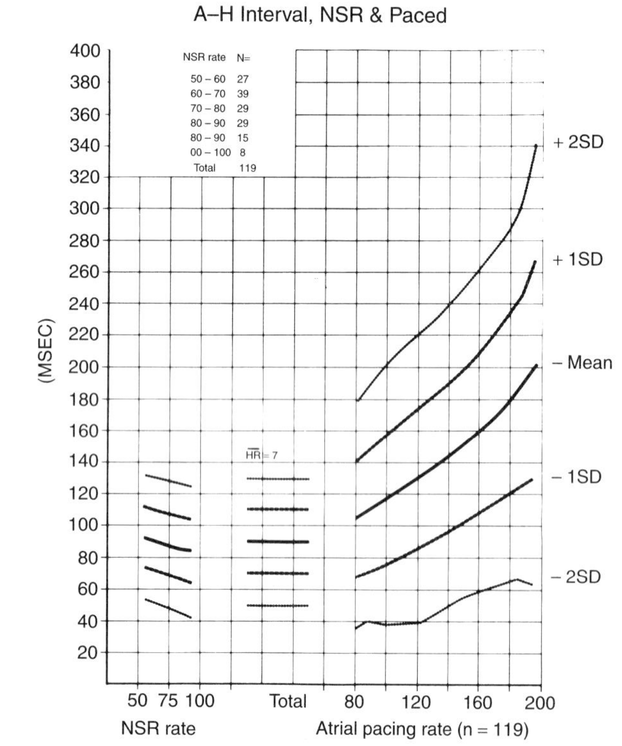

FIGURE 9-10 The A–H interval during normal sinus rhythm (*NSR*) and during high right atrial pacing. Mean and standard deviations are shown for subjects without evidence of AV nodal disease. (With permission from Fisher JD: Primer of Cardiac Electrophysiology. Armonk, NY, Futura, In press.)

Dual or multiple AVN pathways: Humans do not have the very discreet internodal pathways that can be identified in some animals. However, remains of these pathways may exist in the "slow posterior," "fast anterior," and "left atrial input" fibers entering the AVN. Even this is a simplification of very complex anatomy, but it means there is a possibility that the electrophysiology of all these pathways may differ somewhat, so that entrance into the AVN may be blocked by refractoriness at the same interval which allows entrance through an alternate pathway. Among the possible effects of such alternate entryways into the AVN is a discontinuity in the A–H interval in response to rate incremental pacing or extrastimulus testing. Such discontinuities or "jumps" are defined as increases in the usually smooth prolongation of the A–H interval by 50 ms or more in response to a decrease of 10 ms in the pacing stimulus interval (e.g., during a ramp) or extrastimulus interval. Such jumps are almost routinely observed in patients with AVNRT. However, jumps may also be seen in patients without a history of documented tachycardia, and this may present the electrophysiologist with a dilemma. If the patient has had unexplained syncope or palpitations, it may be elected to proceed with ablation of the slow pathway, thus avoiding a future trip back to the EP laboratory, but at the same time accepting a small chance that the patient will require a pacemaker.

6. **His-Purkinje system:**[22,123-152] His bundle responses are also observed during passive recordings, rate incremental pacing, and extrastimulus testing. They also occur in response to pharmacologic probes.

Validation of the His bundle: The deflection accepted as a His bundle potential might actually include some tail-end deflections from the atrium or perhaps right bundle branch (RBB) components as well. "Validation" of the His bundle is accomplished by pacing through the lead recording the presumed His bundle, and confirming that the stimulus-V interval is the same as the presumed H–V interval and that the QRS morphology is unchanged from that seen during sinus rhythm (allowing for the effect of a superimposed retro-conducted atrial deflection). It is often difficult to stimulate the His bundle alone without engaging some of the surrounding right ventricular inflow muscle. In some individuals, it seems preferable to change the stimulation settings somwhat, moving to a wider pulse duration (up to 5 ms) rather than increasing the current or voltage amplitude.

Longitudinal dissociation and "normalization" of left bundle branch block (LBBB):[144,145] There is some anisotropic conduction between the longitudinal fibers in the His bundle that are destined to become the right and left bundle branches (RBB, LBB). However, in some patients with LBBB pacing, the His bundle will normalize the QRS complex (Figs. 9-11 and 9-12). If the stimulus-to-ventricle (ST-V) time is longer than the baseline HV, normalization of the QRS may be due to balanced first-degree block in the bundle branches, with prior first-degree LBBB now matched by pacing-induced first-degree RBB block (RBBB). When the ST-V is equal to the HV, this may indicate disease in the His bundle, with block of fibers committed to become the LBB—pacing depolarizes the His bundle below the level of block. This is potentially the more serious scenario because progression of the lesion could cause complete heart block at the HV level.

Catheter bumping: The HV can be affected by trauma during placement of EP catheters. Most often RBBB is seen and rarely persists more than a few hours. Complete heart block is uncommon. A notable exception is seen in patients with preexisting LBBB; in these, trauma to the His or RBB can produce complete AV block.

Response to stimulation: Rate incremental atrial pacing and atrial extrastimuli most often have negligible effects on the H–V interval. If the patient has a good AVN, then sequential impulses may reach some part of the His-Purkinje system within its ERP. The most typical example is RBBB observed with increasingly premature atrial extrastimuli; it is often seen with short cycles during atrial fibrillation, known as the *Ashman phenomenon.* Further prematurity of the atrial extrastimuli may result in additional increases in the A–H interval, so that the interval between subsequent H–H intervals actually increases, leading to a resolution of the RBBB. This gap in conduction is a normal finding. See the sections on "gap" phenomena and refractory periods in this and other chapters.

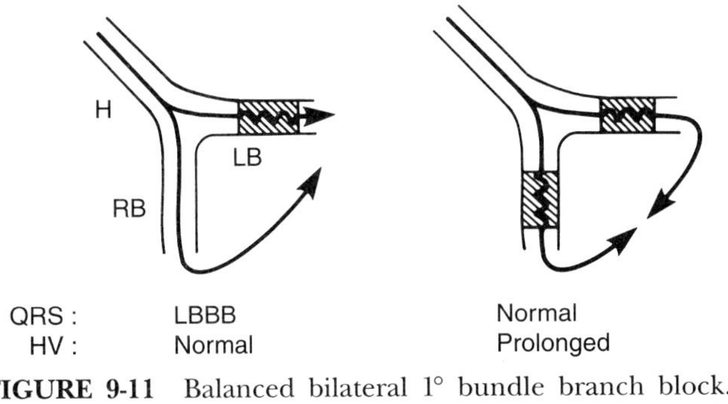

FIGURE 9-11 Balanced bilateral 1° bundle branch block. In the *left hand panel,* there is delay in the left bundle (*LB*), while the right bundle (*RB*) conducts normally. This results in a left bundle branch block morphology on the surface ECG and a normal H–V interval. The *right panel* shows a patient with balanced 1° bundle branch block. Equal delays in both the right and the left bundle branch result in a prolonged H–V interval but a normal QRS morphology. (With permission from Fisher JD: Primer of Cardiac Electrophysiology. Armonk, NY, Futura, In press.)

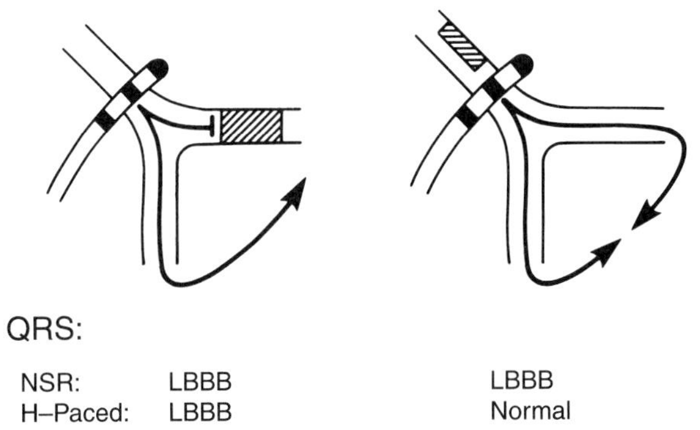

FIGURE 9-12 Proximal versus distal disease causing left bundle branch block (LBBB) pattern. See text. In the **left hand panel,** disease in the left bundle branch created an LBBB pattern on the surface ECG, and also during pacing of the His bundle, which is proximal to the site of block. In the **right hand panel,** there is disease in those fibers of the His bundle destined to become the left bundle branch. In patients with such longitudinal dissociation, the surface ECG will also show an LBBB pattern. However, during His pacing at a site distal to the block, the paced wavefront will propagate down both the right and left bundle branches with a normal stimulus-V time (equivalent to the HV) and a normal QRS morphology. (With permission from Giusti RP, Fisher JD: Prolongation of intra-atrial conduction time in response to atrial pacing. Clin Res 1976;24:218A.)

As a rule of thumb, block at the HV level is uncommon in response to incremental pacing or extrastimulus testing with H–H interval 400 ms or longer. Block at the H–V interval during straight pacing, at cycle lengths not producing Wenckebach at the AVN level, is distinctly abnormal and an indication for pacing.[127]

Pharmacologic probes: Isoproterenol has a minimally positive effect on conduction to the His-Purkinje system. If used in a patient with profound spontaneous bradycardia, it must be understood that isoproterenol is more likely to increase the idioventricular rhythm rate and not to promote A-V conduction unless that delay was at the AVN rather than the HV level.

Lidocaine and procainamide, as well as ajmaline,[130-135] have been used as part of the assessment of "suspect" His-Purkinje conduction. Block or prolongation of the H–V interval by more than 30% in response to these agents is abnormal. In the case of lidocaine, typical doses are 75 mg delivered twice over a period of 30 seconds; procainamide is usually given as 10 mg per kilogram with a delivery rate of about 25 to 50 mg per minute. At the end of the drug infusions, rate incremental pacing may help by precipitating block, or otherwise document the tenuousness of HV conduction.

7. **Retrograde ventriculoatrial (VA) conduction:**[153-160] During clinical EPS's, 20% to 50% of patients with apparently normal anterograde conduction have no retrograde/V-A conduction. This is a normal variation. In many of these, administration of isoproterenol results in a normal pattern of retrograde conduction.

For those with normal retrograde conduction, earliest atrial activation occurs in the atrium, specifically the relatively anterior septal portion that records the His bundle deflection as well. Occasionally the earliest activation will occur more posteriorly, presumably due to use of the "slow pathway." If the earliest atrial depolarization occurs elsewhere, this is sometimes referred to as an *eccentric pattern* and may imply the presence of an accessory pathway.

With stepwise or ramp rate incremental pacing, prolongation in V-A conduction times may be very subtle in some patients but easily identified with the extrastimulus technique. This is important because retrograde conduction is normally "decremental" (i.e., there is a gradual prolongation of the V-A conduction time in response to both pacing and extrastimulus testing). Absence of such prolongation may imply the presence of an accessory pathway.

Retrograde block can occur at any level in the conduction system. The distal Purkinje fibers of the bundle branches have particularly long ERPs (the "distal gate")[159-160]; this may contribute to some of the phenomena mentioned in the following section.

As with other portions of the EPS, note that basic technique is a combination of rate incremental pacing and extrastimulus testing, this time from the ventricle.

8. **Unusual phenomena:** Various types of *gap phenomena* (see other chapters)[98,161-170] are described with retrograde as well as anterograde conduction. Keep in mind that these are unusually indeed phenomena rather than abnormalities. *Retrograde "jumps"* during ventricular stimulation may also occur for several reasons, among them: (a) dual AVN physiology that may only be apparent during retrograde conduction; (b) conduction using an accessory pathway, which then blocks, with subsequent conduction switching to the normal pathway, or the reverse of this process; (c) block of VA conduction in the right bundle branch (RBB), resulting in transventricular septal conduction and retrograde conduction up the left bundle branch (LBB). In some instances this latter phenomenon results in a *bundle branch reentry*[171-172] in which the impulse traveling retrograde in the LBB continues retrograde to the atrium via the His, but also reaches a turnaround point at the bundle branch bifurcation region and travels anterogradely down the RBB to the ventricle. Demonstration of a few beats of bundle branch reentry is classified as a normal phenomenon, to be distinguished from sustained bundle branch reentry tachycardia (BBRT, as described elsewhere in this textbook). *Supernormal conduction*[173] occurs when a precisely timed beat results in conduction in a previously blocked area. Also see elsewhere in the textbook for other phenomena such as phase 3 and phase 4 block and supernormal excitability.

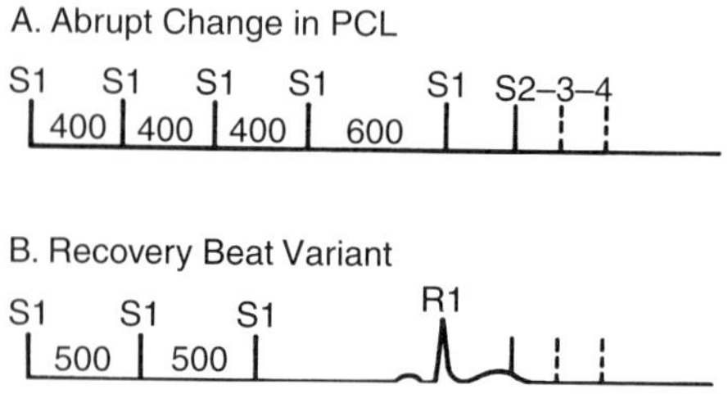

FIGURE 9-13 Dispersion protocols. **Panel A,** Abrupt change in pacing cycle length (*PCL*). **Panel B,** Recovery beat variant in which the delayed extrastimuli (*S2-3-4*) follow the first spontaneous beat (*R1*) after cessation of the drive pacing *S1*. (With permission from Fisher JD: Primer of Cardiac Electrophysiology. Armonk, NY, Futura, In press.)

9. **Ventricle:** In the ventricle the most important question is whether sustained monomorphic VT is inducible, particularly *reproducibly* inducible. Given enough stimulation and enough electricity, it is possible to fibrillate anybody so that the induction of ventricular fibrillation is considered to be nonspecific except in special circumstances detailed elsewhere in the textbook.

   As indicated earlier, there is no "universal stimulation protocol" (see Figs. 9-3, 9-8; Figs. 9-13 and 9-14).[12,14,32-36] Virtually all techniques in common clinical use involve the serial introduction of single, double, triple, and occasionally quadruple extrastimuli during at least two drive rates and at

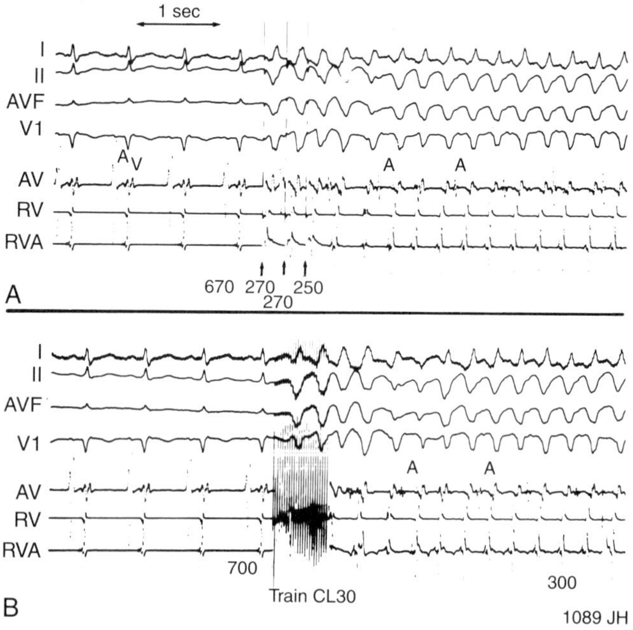

FIGURE 9-14 Programmed stimulation and ultrarapid trains for induction of ventricular tachycardia (VT). **Panel A,** After a series of baseline (S1) beats in sinus rhythm, triple extrastimuli are introduced at the coupling intervals shown, and VT is induced. **Panel B,** In the same patient on the same day, an ultrarapid train at cycle length 30 ms produces three captures and the same VT as in **panel A**. *AV*, intracardiac electrogram recording atrial (A) and ventricular (V) signals. *RV*, right ventricle; *RVA*, RV apex.

TABLE 9-2 Description of Ventricular Tachycardia Morphologies in Terms of Similarity to Bundle Branch Block (BBB) Morphologies

	2-Choice Axis	QRS in Lead			
BBB	*Examples*	*I*	*II*	*AVF*	*VI*
RBBB	Superior	+,−	−	−	+
RBBB	Inferior	+,−	+	+	+
LBBB	Superior	+,−	−	−	−
LBBB	Inferior	+,−	+	+	−
	4-Choice Axis				
RBBB	RBNA	+	+	+	+
RBBB	RBLA	+	−	−	+
RBBB	RBRA	−	+	+	+
RBBB	RBXA	−	−	−	+
LBBB	LBNA	+	+	+	−
LBBB	LBLA	+	−	−	−
LBBB	LBRA	−	+	+	−
LBBB	LBXA	−	−	−	−

LA, left axis; LBBB, LB, left bundle branch block; NA, normal axis; RA, right axis; RBBB, RB, right bundle branch block; XA, extreme axis ("Northwest"); +, positive; –, negative.

least two ventricular sites. Rapid pacing in the form of bursts and rapidly increasing ramps are also used in many centers. In certain conditions and in patients strongly suspected of having ventricular arrhythmias, the protocol may be repeated in the presence of isoproterenol, which is usually delivered according to one of several local formulas: (a) sufficient to raise the sinus or baseline rate to about 120 bpm or (b) sufficient to increase the baseline rate by 25% (or some other percentage). Provocation of automatic arrhythmias (in the atrium as well as in the ventricles) may require sequential increases in isoproterenol dosing. An example pattern would be to start at 1 or 2 micrograms (mcg) per minute for 5 minutes and then increase the infusion rate every 5 minutes to 4, 6, 8, and 10 mcg per minute. Most patients have no further response after 6 to 8 mcg per minute, indicating that the β-receptors are saturated. Certainly the infusion of isoproterenol should be avoided or limited strictly in patients with ischemic heart disease, particularly soon after an acute infarction. Induction of 10 to 15 beats of (N-"fix") monomorphic VT, polymorphic VT (PVT), or ventricular flutter (VFL) is probably normal, particularly at the extremes of the protocol such as tightly coupled triple extrastimuli during the second drive rate or very rapid pacing.[34] The definition of "sustained" tachycardia varies somewhat, with some laboratories using 15 seconds, others 30 seconds, and most accepting the need for rescue cardioversion or termination. Sustained monomorphic VT implies that the rhythm has a shape or morphology that can be described in terms of bundle branch blocks (with the understanding that these are not truly related to AV conduction difficulties). Examples of such descriptions are given in Table 9-2.

SUMMARY

The techniques described in this chapter for setting up and performing a "complete diagnostic EPS" contain the components that are essential to all EPSs: selection of stimulation sites and recordings; methods of stimulation including stimulus characteristics and patterns (straight, ramp, extrastimulus); timing of overall sequence and local events; and the use of pharmacologic probes. Other chapters will expand on these basics for the differential diagnosis of complex arrhythmias and for therapeutic/interventional electrophysiology.

REFERENCES

1. Zipes DP, DiMarco JP, Gillette PC, et al: ACC/AHA task force report—guidelines for clinical intracardiac electrophysiological and catheter ablation procedures. A report of the ACC/AHA task force on practice guidelines (committee on clinical intracardiac electrophysiological and catheter ablation procedures), developed in collaboration with NASPE. J Am Coll Cardiol 1995; 26:555-73.
2. Tracy CM, Akhtar M, DiMarco JP, et al: ACC/AHA clinical competence statement on invasive electrophysiology studies, catheter ablation, and cardioversion. A report of the American College of Cardiology/American Heart Association/American College of Physicians-American Society of Internal Medicine Task Force on Clinical Competence. Developed in collaboration with NASPE. Circulation 2000;2309-20.
3. Curtis AB, Landberg JJ, Tracy CM: Clinical competency statement: Implantation and follow-up of cardioverter defibrillators. J Cardiovasc Electrophysiol 2001;12:280-4.
4. Fisher JD et al: Catheter ablation for cardiac arrhythmias: Clinical applications, personnel and facilities. ACC position statement—J Am Coll Cardiol 1994;24:828-33.
5. Bubien RS, Fisher JD, Gentzel JA, et al: NASPE expert consensus document: Use of IV (conscious) sedation/analgesia by nonanesthesia personnel in patients undergoing arrhythmia specific diagnostic, therapeutic, and surgical procedures. Pacing Clin Electrophysiol 1998;21:375-85.
6. Fisher JD, Kim SG, Ferrick KJ, et al: Internal transcardiac pericardiocentesis for acute tamponade. J Interv Card Electrophysiol 2001;5:173-6.
7. Hahn K, Gal R, Sarnoski J, et al: Transesophageal echocardiographically guided atrial transseptal catheterization in patient with normal-sized atria; incidence of complications. Clin Cardiol 1995;18:217-20.
8. Roelke M, Smith AJ, Palacios IF: The technique and safety of transseptal left heart catheterization: The Massachusetts General Hospital experience with 1,279 procedures. Catheter Cardiovasc Diag 1994;32:332-9.
9. Clugston R, Lau FY, Ruiz C: Transseptal catheterization update. Cathet Cardiovasc Diagn 1992;26:266-74.
10. Mehra R, Furman S: Comparison of cathodal, anodal, and bipolar strength-interval curves with temporary and permanent pacing electrodes. Br Heart J 1979;41:468-76.
11. Roth BJ: Strength-interval curves for cardiac tissue predicted using the bidomain model. J Cardiovasc Electrophysiol 1996;7: 722-37.
12. Waldo AL, Akhtar M, Brugada P, et al: NASPE policy statement: The minimally appropriate electrophysiologic study for the initial assessment of patients with documented sustained monomorphic ventricular tachycardia. Pacing Clin Electrophysiol 1985;8:918.
13. Fisher JD, Ostrow E, Kim SG, Matos JA: Ultrarapid single-capture train stimulation for termination of ventricular tachycardia. Am J Cardiol 1983;51:1334-8.
14. Fisher JD, Platt SB, Cua MC, et al: Ultrarapid train stimulation: An efficient alternative to conventional programmed electrical stimulation for induction of ventricular arrhythmias. J Interv Card Electrophysiol 1997;1:15-21.
15. Fisher JD, Baker J, Ferrick KJ, et al: The atrial electrogram during clinical electrophysiologic studies: Onset versus the

local/intrinsic deflection. J Cardiovascular Electrophysiol 1991;2: 398-407.
16. Fisher JD, Kim SG, Ferrick KJ, Roth JA: Programmed ventricular stimulation using tandem vs. simple sequential protocols. Pacing Clin Electrophysiol 1994;17:286-94.
17. Dubrow IW, Fisher EA, Amat-Y-Leon F, et al: Comparison of cardiac refractory periods in children and adults. Circulation 1975;51:485.
18. Fisher JD, Zhang X, Waspe LE, et al: Tests of refractoriness and conduction during clinical electrophysiologic studies: Yields and roles. J Electrophysiol 1988;2:175.
19. Wit AL, Weiss MB, Berkowitz WD, et al: Patterns of atrioventricular conduction in the human heart. Circ Res 1970;27:345.
20. Akhtar M, Damato AN, Batsford WP, et al: A comparative analysis of antegrade conduction patterns in man. Circulation 1975; 52:766.
21. Bisset JK, Kane JJ, DeSoyza N, et al: Electrophysiological significance of rapid atrial pacing as a test of atrioventricular conduction. Cardiovasc Res 1975;9:593.
22. Fisher JD: Primer of Cardiac Electrophysiology. Armonk, NY, Futura, In press.
23. Fisher JD: Role of electrophysiologic testing in the diagnosis and treatment of patients with known and suspected bradycardias and tachycardias. Prog Cardiovasc Dis 1981;24:25-90.
24. Josephson ME: Clinical Cardiac Electrophysiology: Techniques and Interpretations, 3rd ed. Philadelphia, Lippincott Williams & Wilkins, 2002, p 39.
25. Denes P, Wu D, Dhingra R, et al: The effects of cycle length on cardiac refractory periods in man. Circulation 1974;49:32.
26. Fisher JD, Brodman R, Kim SG, et al: Attempted nonsurgical electrical ablation of accessory pathways via the coronary sinus in the Wolff-Parkinson-White syndrome. J Am Coll Cardiol 1984;4:685-94.
27. Loeb JM, deTarnowsky JM, Warner MR, et al: Dynamic interactions between heart rate and atrioventricular conduction. Am J Physiol 1985;249:H505.
28. Warner MR, Loeb JM: Beat-by-beat modulation of AV conduction. I. Heart rate and respiratory influences. Am J Physiol 1986;251:H1126.
29. Warner MR, Loeb JM: Beat-by-beat modulation of AV conduction. II. Autonomic neural mechanisms. Am J Physiol 1986;251:H1134.
30. Narula OS: Disorders of sinus node function. In Narula OS (ed): Electrophysiologic Evaluation. His Bundle Electrocardiography and Clinical Electrophysiology. Philadelphia, FA Davis Publishers, 1975:145-6.
31. Lehmann MH, Denker S, Mahmud R, Akhtar M: Patterns of human atrioventricular nodal accommodation to a sudden acceleration of atrial rate. Am J Cardiol 1984;53:71.
32. Denker S, Lehmann M, Mahmud R, et al: Facilitation of ventricular tachycardia induction with abrupt changes in ventricular cycle length. Am J Cardiol 1984;53:508-15.
33. Simpson JS, Gang ES, Mandel W, et al: Increasing the yield of ventricular tachycardia induction: A prospective, randomized comparative study of the standard, ventricular stimulation protocol to a short-to-long protocol and a new two-site protocol. Am Heart J 1991;121:68-76.
34. Fisher JD, Kim SG, Ferrick KJ, et al: Programmed electrical stimulation of the ventricle: An efficient, sensitive, and specific protocol. Pacing Clin Electrophysiol 1992;15:435-50.
35. Fisher JD, Kim SG, Ferrick, Roth JA: Programmed electrical stimulation protocols: Variations on a theme. Pacing Clin Electrophysiol 1992;15:2180-7.
36. Artoul SG, Fisher JD, Kim SG, et al: Stimulation hierarchy: Optimal sequence for double and triple extrastimuli during electrophysiological studies. Pacing Clin Electrophysiol 1992;15: 790-800.
37. Josephson ME: Clinical Cardiac Electrophysiology: Techniques and Interpretations, 3rd ed. Philadelphia, Lippincott Williams & Wilkins, 2002, p 452.
38. Josephson ME: Clinical Cardiac Electrophysiology: Techniques and Interpretations, 3rd ed. Philadelphia, Lippincott Williams & Wilkins, 2002, p 446.
39. Damato AN, Lau SH, Helfant RH, et al: Study of atrioventricular conduction in man using electrode catheter recordings of His bundle activity. Circulation 1969;39:287-96.
40. Narula OS, Cohen LS, Samet P, et al: Localization of A-V conduction defects in man by recording of the His bundle electrogram. Am J Cardiol 1970;25:228.
41. Josephson ME, Seides SF: Electrophysiologic investigation: Technical aspects. In Josephson ME, Seides SF (eds): Clinical Cardiac Electrophysiology: Techniques and Interpretations. Philadelphia, Lea & Febiger Publishers, 1979, pp 17-18.
42. Ausubel K, Klementowicz P, Furman S: Interatrial conduction during cardiac pacing. Pacing Clin Electrophysiol 1986;9:1026.
43. Buxton AE, Waxman HL, Marchlinski FE, Josephson ME: Atrial conduction: Effects of extrastimuli with and without atrial dysrhythmias. Am J Cardiol 1984;54:755.
44. Camous J-P, Raybaud F, Dolisi C, et al: Interatrial conduction in patients undergoing AV stimulation: Effects of increasing right atrial stimulation rate. Pacing Clin Electrophysiol 1993; 16:2082.
45. Plumb VJ, Karp RB, James TN, Waldo AL: Atrial excitability and conduction during rapid atrial pacing. Circulation 1981;63:1140.
46. Dhingra RC, Rosen KM, Rahimtoola SH: Normal conduction intervals and responses in sixty-one patients using His bundle recording and atrial pacing. Chest 1973;64:55-9.
47. Josephson ME: Clinical Cardiac Electrophysiology: Techniques and Interpretations, 3rd ed. Philadelphia, Lippincott Williams & Wilkins, 2002.
48. Narula OS: Disorders of sinus node function. In Narula OS (ed): Electrophysiologic Evaluation. His Bundle Electrocardiography and Clinical Electrophysiology. Philadelphia, FA Davis Publishers, 1975:275-311.
49. Narula OS: Sick sinus syndrome: Key references. Circulation 1979;60:1422-4.
50. Scarpa WJ: The sick sinus syndrome. Am Heart J 1976;92:648.
51. Seinfeld D, Altschuler H, Yipintsoi T, et al: Pathogenetic locus of the sick sinus syndrome. Clin Res 1978;26:270A.
52. Jordan JL, Yamaguchi I, Mandel WJ: Studies on the mechanism of sinus node dysfunction in the sick sinus syndrome. Circulation 1978;57:217-22.
53. Mandel W, Hayakawa H, Danzig R, et al: Evaluation of sinoatrial node function in man by overdrive suppression. Circulation 1971;44:59-66.
54. Mandel WJ, Laks MM, Orayashi K: Sinus node function: Evaluation in patients with and without sinus node disease. Arch Intern Med 1975;135:388-94.
55. Strauss HC, Bigger JT Jr, Saroff AL, Giardina EG: Electrophysiologic evaluation of sinus node function in patients with sinus node dysfunction. Circulation 1976;53:763-76.
56. Kulbertus HE, DeLeval-Rutten D, Mary L, Casters P: Sinus node recovery time in the elderly. Br Heart J 1975;37:420.
57. Breithardt G, Seipel L, Loogen F: Sinus node recovery time and calculated sinoatrial conduction time in normal subjects and patients with sinus node dysfunction. Circulation 1977;56: 43-50.
58. Gupta PK, Lichstein F, Chadda KD, Badui E: Appraisal of sinus nodal recovery time in patients with sick sinus syndrome. Am J Cardiol 1974;34:265-70.
59. Altschuler H, Fisher JD: Increasing the yield of the sinus recovery time in the sick sinus syndrome. Clin Res 1976;24:205A.
60. Benditt DG, Strauss HC, Scheinman MM, et al: Analysis of secondary pauses following termination of rapid pacing in man. Circulation 1976;54:436-41.
61. Dhingra RC, Amat-Y-Leon F, Wyndham C, et al: Electrophysiologic effects of atropine on sinus node and atrium in patients with sinus node dysfunction. Am J Cardiol 1976;38:848.
62. Reiffel JA, Bigger JT Jr, Giardina EGV: "Paradoxical" prolongation of sinus nodal recovery time after atropine in the sick sinus syndrome. Am J Cardiol 1975;36:98.
63. Dhingra RC, Deedwania PC, Cummings JM, et al: Electrophysiologic effects of lidocaine on sinus node and atrium in patients with and without sinoatrial dysfunction. Circulation 1978;57:448.
64. Evans TR, Callowhill EA, Krikler DM: Clinical value of tests of sino-atrial function. Pacing Clin Electrophysiol 1978;1:2.
65. Alboni P, Malcarne C, Pedroni P, et al: Electrophysiology of normal sinus node with and without autonomic blockade. Circulation 1982;65:1236-42.
66. Nalos PC, Deng Z, Rosenthal ME, et al: Hemodynamic influences on sinus node recovery time: Effects of autonomic blockade. J Am Coll Cardiol 1986;7:1079.

67. Karagueuzian HS, Jordan JL, Sugi K, et al: Appropriate diagnostic studies for sinus node dysfunction. Pacing Clin Electrophysiol 1985;8:242-54.
68. Tonkin AM, Heddle WF: Electrophysiological testing of sinus node function. Pacing Clin Electrophysiol 1984;7:735-48.
69. De Marneffe M, Jacobs P, Englert M: Reproducibility of electrophysiologic parameters of extrinsic sinus node function in patients with and without sick sinus syndrome. Pacing Clin Electrophysiol 1986;9:482-9.
70. De Marneffe M, Waterschoot P, Melot C, Englert M: Electrophysiology of the normal sinus node. J Electrophysiol 1988;2: 155-74.
71. Reiffel JA, Kuehnert MJ: Electrophysiological testing of sinus node function: Diagnostic and prognostic application including updated information from sinus node electrograms. Pacing Clin Electrophysiol 1994;17:349-65.
72. Narula OS, Shantha N, Narula LK, Alboni P: Clinical and electrophysiological evaluation of sinus node function. In Narula OS (ed): Cardiac Arrhythmias Electrophysiology: Diagnosis and Management. Baltimore/London, Williams & Wilkins, 1979:176-206.
73. Engel TR, Schaal SF: Digitalis in the sick sinus syndrome: The effects of digitalis on sinoatrial automaticity and atrioventricular conduction. Circulation 1973;48:1201-7.
74. Dhingra RC, Amat-Y-Leon F, Wyndham C, et al: Clinical significance of prolonged sinoatrial conduction time. Circulation 1977; 55:8-15.
75. Josephson ME: Clinical Cardiac Electrophysiology: Techniques and Interpretations, 3rd ed. Philadelphia, Lippincott Williams & Wilkins, 2002, pp 68-91.
76. Narula OS, Shantha N, Vasquez M, et al: A new method for measurement of sinoatrial conduction time. Circulation 1978;58: 706-14.
77. Reiffel JA, Zimmerman G: The duration of the sinus node depolarization on transvenous sinus node electrograms can identify sinus node dysfunction and can suggest its severity. Pacing Clin Electrophysiol 1989;12:1746-56.
78. Juillard A, Guillerm F, Van Chuong H, et al: Sinus node electrogram recording in 59 patients. Comparison with simultaneous estimation of sinoatrial conduction using premature atrial stimulation. Br Heart J 1983;50:75.
79. Gomes JAC, Kang PS, El-Sherif N: The sinus node electrogram in patients with and without sick sinus syndrome: Techniques and correlation between directly measured and indirectly estimated sinoatrial conduction time. Circulation 1982;66:864-73.
80. Reiffel JA, Gang E, Gliklich J, et al: The human sinus node electrogram: A transvenous catheter technique and a comparison of directly measured and indirectly estimated sinoatrial conduction time in adults. Circulation 1980;62:1324-34.
81. Reiffel JA, Bigger JT Jr: The relationship between sinoatrial conduction time and sinus cycle length revisited. J Electrophysiol 1987;1:290-9.
82. Moore EN, Spear JF: Effect of autonomic activity on pacemaker function and conduction. In Wellens HJ, Lie KL, Janse MJ (eds): The Conduction System of the Heart. Philadelphia, Lea & Febiger, 1976, pp 100-110.
83. Martin P: The influence of the parasympathetic nervous system on atrioventricular conduction. Circ Res 1977;41:593.
84. Sigler LH: Clinical observations on the carotid sinus reflex I. The frequency and the degree of response to carotid sinus pressure under various diseased states. Am J Med Sci 1933;186:110.
85. Sigler LH: Further observation on the carotid sinus reflex. Ann Intern Med 1936;9:1380.
86. Heidorn GH, McNamara AP: Effect of carotid sinus stimulation on the electrocardiograms of clinically normal individuals. Circulation 1956;14:1104.
87. Hartzler GO, Maloney JD: Cardioinhibitory carotid sinus hypersensitivity. Intracardiac recordings and clinical assessment. Arch Intern Med 1977;137:727.
88. Fisher JD, Katz G, Furman S, Rubin I: Differential responses to carotid sinus massage in cardiac patients with and without syncope. In Feruglio GA (ed): Cardiac Pacing, Electrophysiology and Pacemaker Technology. Florence, Italy, Piccin Medical Books Publishers, 1982, pp 521-2.
89. Josephson ME, Scharf DL, Kastor JA, et al: Atrial endocardial activation in man: Electrode catheter technique for endocardial mapping. Am J Cardiol 1977;39:972.
90. James TN: The connecting pathways between the sinus node and AV node and between the right and left atrium in the human heart. Am Heart J 1963;66:498.
91. Sherf L: The atrial conduction system: Clinical implications. Am J Cardiol 1976;37:814.
92. Ross AM, Proper MC, Aronson AL: Sinoventricular conduction in atrial standstill. J Electrocardiol 1976;9:161.
93. Giusti RP, Fisher JD: Prolongation of intra-atrial conduction time in response to atrial pacing. Clin Res 1976;24:218A.
94. Millar RNS, Mauer BJ, Rei DS, et al: Studies of intra-atrial conduction with bipolar atrial and His electrograms. Br Heart J 1973;35:604.
95. Fisher JD, Lehmann MH: Marked intra-atrial conduction delay with split atrial electrograms: Substrate for reentrant supraventricular tachycardia. Am Heart J 1986;111:781.
96. Simpson RJ Jr, Foster JR, Gettes LS: Atrial excitability and conduction in patients with interatrial conduction defects. Am J Cardiol 1982;50:1331.
97. Simpson RJ Jr, Amara I, Foster JR, et al: Thresholds, refractory periods, and conduction times of the normal and diseased human atrium. Am Heart J 1988;116:1080.
98. Akhtar M, Caracta AR, Lau SH, et al: Demonstration of intra-atrial conduction delay, block, gap and reentry: A report of two cases. Circulation 1978;58:947.
99. Narula OS, Runge M, Samet P: Second-degree Wenckebach type AV block due to block within the atrium. Br Heart J 1972; 34:1127.
100. Castellanos A Jr, Iyengar R, Agha AS, Castillo CA: Wenckebach phenomenon within the atria. Br Heart J 1972;34:1121.
101. Calkins H, El-Atassi R, Kalbfleisch S, et al: Effects of an acute increase in atrial pressure on atrial refractoriness in humans. Pacing Clin Electrophysiol 1992;15:1674.
102. Niwano S, Aizawa Y: Fragmented atrial activity in patients with transient atrial fibrillation. Am Heart J 1991;121:62.
103. Tanigawa M, Fukatani M, Konow A, et al: Prolonged and fractionated right atrial electrograms during sinus rhythm in patients with paroxysmal atrial fibrillation and sick sinus node syndrome. J Am Coll Cardiol 1991;17:403.
104. Narula OS, Narula JT: Junctional pacemakers in man: Response to overdrive suppression with and without parasympathetic blockade. Circulation 1978;57:880.
105. Narula OS, Narula JT: Junctional pacemakers in man. Response to overdrive suppression with and without parasympathetic blockade. Circulation 1978;57:880.
106. Mazgalev TN, Tchou PF (eds): Atrial—AV Nodal Electrophysiology. Armonk, NY, Futura 2000.
107. Jackman WM, Beckman KJ, McClelland JH, et al: Treatment of supraventricular tachycardia due to atrioventricular nodal reentry by radiofrequency catheter ablation of slow-pathway conduction. N Engl J Med 1992;327:313-8.
108. Mitrani RD, Klein LS, Hackett FK, et al: Radiofrequency ablation for atrioventricular node reentrant tachycardia: Comparison between fast (anterior) and slow (posterior) pathway ablation. J Am Coll Cardiol 1993;21:432-41.
109. Atrioventricular node of the isolated rabbit heart. In Wellens HJJ, Lie KI, Janse MJ (eds): The Conduction System of the Heart: Structure, Function and Clinical Implications. Philadelphia, Lea & Febiger Publishers, 1976.
110. Jackman WM, Prystowsky EN, Naccarelli GV, et al: Reevaluation of enhanced atrioventricular nodal conduction: Evidence to suggest a continuum of normal atrioventricular nodal physiology. Circulation 1983;67:441.
111. Billette J, Bonin JP: Rate-induced shortenings in refractory periods and conduction time in the dog atrioventricular node. In Meere C (ed): Proceedings of the 6th World Symposium on Cardiac Pacing. Montreal, Pacing Clin Electrophysiol Symposium, 1979, chap 11-4.
112. Billette J, Nattel S: Dynamic behavior of the atrioventricular node: A functional model of interaction between recovery, facilitation, and fatigue. J Cardiovasc Electrophysiol 1994;5:90.
113. Rosenblueth A: Mechanism of Wenckebach-Luciani cycles. Am J Physiol 1958;194:491-4.
114. Young M-L, Wolff GS, Castellanos A, Gelband H: Application of the Rosenblueth hypothesis to assess cycle length effects on the refractoriness of the atrioventricular node. Am J Cardiol 1986;57:142.

115. Malik M, Ward D, Camm AJ: Theoretical evaluation of the Rosenblueth hypothesis. Pacing Clin Electrophysiol 1988; 11:1250.
116. Aranda J, Castellanos A, Moleiro F, Befeler B: Effects of the pacing site on A-H conduction and refractoriness in patients with short P-R intervals. Circulation 1976;53:33.
117. Batsford WP, Akhtar M, Caracta AR, et al: Effect of atrial stimulation site on the electrophysiological properties of the atrioventricular node in man. Circulation 1974;50:283.
118. Iinuma H, Dreifus LS, Price R, Michelson EL: Influence of the site of stimulation on atrioventricular nodal refractory periods and the effect of Verapamil. Am J Cardiol 1986;57:1167.
119. Amat y Leon F, Denes P, Wu D, et al: Effects of atrial pacing site on atrial and atrioventricular nodal function. Br Heart J 1975; 37:576.
120. Mae GK, Preston JD, Burlington H: Physiologic evidence of a dual AV transmission system. Circ Res 1956;4:357-75.
121. Prystowsky EN, Klein GJ: Cardiac Arrhythmias: An Integrated Approach for the Clinician. New York, McGraw-Hill Publishers, 1994, p 120.
122. Dhingra RC, Wyndham C, Amat-Y-Leon F, et al: Significance of A-H interval in patients with chronic bundle branch block: Clinical, electrophysiologic and follow-up observations. Am J Cardiol 1976;37:231.
123. Damato AN, Lau SH, Helfant RH, et al: A study of heart block in man using His bundle recordings. Circulation 1969;39:297-305.
124. Narula OS, Samet P: Predilection of elderly females for intra-His bundle (BH) blocks. Circulation 1974;49-50(suppl III):195.
125. Narula OS, Samet P: Wenckebach and Mobitz type II A-V block due to block within the His bundle and bundle branches. Circulation 1970;41:947-65.
126. Dhingra RC, Wyndham C, Bauernfeind R, et al: Significance of block distal to the His bundle induced by atrial pacing in patients with chronic bifascicular block. Circulation 1979;60:1455-64.
127. Gregoratos G, Cheitlin MD, Conill A, et al: ACC/AHA Guidelines for Implantation of Cardiac Pacemakers and Antiarrhythmia Devices. A report of the American College of Cardiology/American Heart Association Task Force on Practice Guidelines (Committee on Pacemaker Implantation). J Am Coll Cardiol 1998;1175-209.
128. Josephson ME: Clinical Cardiac Electrophysiology: Techniques and Interpretations, 3rd ed. Philadelphia, Lippincott Williams & Wilkins, 2002, pp 110-39.
129. Prystowsky EN, Klein GJ: Cardiac Arrhythmias: An Integrated Approach for the Clinician. New York, McGraw-Hill Publishers, 1994, pp 47-70.
130. Gupta PK, Lichstein E, Chauda KD: Lidocaine-induced heart block in patients with bundle branch block. Am J Cardiol 1974:33:487-92.
131. Fisher JD, Zilo P, Kim SG: Procainamide and lidocaine as electrophysiologic stress tests. In Feruglio GA (ed): Cardiac Pacing: EP and Pacemaker Technology. Florence, Italy, Piccin Medical Books Publishers, 1982, pp 125-6.
132. Josephson ME, Caracta AR, Ricciutti MA, et al: Electrophysiologic properties of procainamide in man. Am J Cardiol 1974;33: 596-603.
133. Cannom DS: The effects of procainamide in left bundle branch block. Circulation 1975;51-52(suppl II):137.
134. McKenna WJ, Rowland E, Davies J, Krikler DM: Failure to predict development atrioventricular block with electrophysiological testing supplemented by ajmaline. Pacing Clin Electrophysiol 1980;3:666-9.
135. Kaul U, Dev V, Narula J, et al: Evaluation of patients with bundle branch block and "unexplained" syncope: A study based on comprehensive electrophysiologic testing and ajmaline stress. Pacing Clin Electrophysiol 1988;11:289-97.
136. Dhingra RC, Denes P, Wu D, et al: Prospective observations in patients with chronic bundle branch block and marked H-V prolongation. Circulation 1976;53:600-4.
137. Rosen KM, Dhingra RC, Wyndham C, et al: Significance of HV interval in 515 patients with chronic bifascicular block. Am J Cardiol 1980;45:405.
138. Scheinman MM, Peters RW, Modin G, et al: Prognostic value of infranodal conduction time in patients with chronic bundle branch block. Circulation 1977;56:240-4.
139. Gupta PK, Lichstein E, Chadda KD: Follow-up studies in patients with right bundle branch block and left anterior hemiblock: Significance of H-V interval. J Electrocardiol 1977;10:221-4.
140. McAnulty JH, Rahimtoola SH, Murphy ES, et al: A prospective study of sudden death in "high-risk" bundle branch block. N Engl J Med 1978;299:209-15.
141. Vera Z, Mason DT, Fletcher RD, et al: Prolonged His-Q interval in chronic bifascicular block. Relation to impending complete heart block. Circulation 1976;53:47-55.
142. Altschuler H, Fisher JD, Furman S: Significance of isolated H-V interval prolongation in symptomatic patients without documented heart block. Am Heart J 1979;97:19-26.
143. Castellanos A Jr: Letter to the editor: H-V intervals in LBBB. Circulation 1973;47:1133.
144. Narula OS: Longitudinal dissociation in the His bundle: Bundle branch block due to asynchronous conduction within the His bundle in man. Circulation 1977;56:996-1006.
145. El-Sherif N, Amat-Y-Leon F, Schonfield C: Normalization of BBB patterns by distal His bundle pacing: Clinical and experimental evidence of longitudinal dissociation in the pathologic His bundle. Circulation 1978;57:473-83.
146. Hindham MC, Wagner GS, Jaro M: The clinical significance of bundle branch block complicating acute myocardial infarction. I. Clinical characteristics, hospital mortality, and one-year follow-up. Circulation 1978;58:679.
147. Hindham MC, Wagner GS, Jaro M: The clinical significance of bundle branch block complicating acute myocardial infarction. II. Indications for temporary and permanent pacemaker insertion. Circulation 1978;58:689.
148. Ritter WS, Atkins JM, Blumqvist CG, Mullins CB: Permanent pacing in patient with transient trifascicular block during acute myocardial infarction. Am J Cardiol 1976;38:205.
149. Lichstein E, Gupta PK, Chadda KD: Long-term survival of patients with incomplete bundle-branch block complicating acute myocardial infarction. Br Heart J 1975;83:924.
150. Lie KI, Wellens HJ, Schuilenburg RM, et al: Factors influencing prognosis of bundle branch block complicating acute anteroseptal infarction: The value of His bundle recordings. Circulation 1974;50:935.
151. Harper R, Hunt D, Vohra J, et al: His bundle electrogram in patients with acute myocardial infarction complicated by atrioventricular or intraventricular conduction disturbances. Br Heart J 1975;37:705.
152. Seinfeld D, Fisher JD: Unpredictability of site of conduction system injury due to acute myocardial infarction. Clin Progress EP & Pacing 1985;3:202-6.
153. Akhtar M, Gilbert C, Wolf FG, Schmidt DH: Retrograde conduction in the His-Purkinje system: Analysis of the routes of impulse propagation using His and right bundle branch recordings. Circulation 1979;59:1252.
154. Mahmud R, Lehmann M, Denker S, et al: Atrioventricular sequential pacing: Differential effect on retrograde conduction related to level of impulse collision. Circulation 1983;68:23-32.
155. Akhtar M: Retrograde conduction in man. Pacing Clin Electrophysiol 1981;4:548.
156. Josephson ME, Kastor JA: His-Purkinje conduction during retrograde stress. J Clin Invest 1978;61:171.
157. Mann DE, Sensecqua JE, Easley AR, Reiter MJ: Effects of upright posture on anterograde and retrograde atrioventricular conduction in patients with coronary artery disease, mitral valve prolapse or no structural heart disease. Am J Cardiol 1987; 60:625-9.
158. Akhtar M, Gilbert C, Wolf FG, et al: Reentry within the His-Purkinje system: Elucidation of reentrant circuit utilizing right bundle branch and His bundle recordings. Circulation 1978; 58:295.
159. Gallagher JJ, Damato AN, Varghese PJ, Lau SH: Localization of an area of maximum refractoriness or "gate" in the ventricular specialized conduction system in man. Am Heart J 1972; 84:310.
160. Akhtar M, Denker S, Lehmann MH, Mahmud R: Macro-reentry within the His-Purkinje system. Pacing Clin Electrophysiol 1983;6:1010.
161. Wu D, Denes P, Dhingra R: Nature of the gap phenomenon in man. Circ Res 1974;34:682.

162. Akhtar M, Damato AN, Batsford WP, et al: Unmasking and conversion of gap phenomenon in the human heart. Circulation 1974;49:624.
163. Akhtar M, Damato AN, Caracta AR, et al: The gap phenomena during retrograde conduction in man. Circulation 1974;49:811.
164. Waleffe A, Bruninx P, Demoulin J-C, Kulbertus HE: Two unusual cases of intra-atrial gap. Br Heart J 1977;39:451-5.
165. Mazgalev T, Dreifus LS, Michelson EL: A new mechanism for atrioventricular nodal—vagal modulation of conduction. Circulation 1989;79:417.
166. Mazgalev T, Tchou P: Atrioventricular nodal conduction gap and dual pathway electrophysiology. Circulation 1995;92:2705.
167. Reddy CP, Harris B: Gap phenomenon in "the right and left bundle branch systems" during retrograde conduction in man. Am Heart J 1979;97:216.
168. Bexton RS, Hellestrand KJ, Nathan AW, et al: Retrograde gap in fast pathway conduction accentuated by the class I antiarrhythmic agent, flecainide. Pacing Clin Electrophysiol 1983;6:1273.
169. Weiss J, Stevenson WG: A new type of ventriculo-atrial gap phenomenon in man. Pacing Clin Electrophysiol 1984;7:46.
170. Agha AS, Castellanos A Jr, Wells D, et al: Type I, type II, and type III gaps in bundle-branch conduction. Circulation 1973; 47:325.
171. Fisher JD: Bundle branch reentry tachycardia: Why is the HV interval often longer than in sinus rhythm? J Interv Card Electrophysiol 2001;5:173-6.
172. Casceres J, Jazayeri M, McKinnie J, et al: Sustained bundle branch reentry as a mechanism for clinical tachycardia. Circulation 1989;79:256.
173. Puech P, Guimond C, Nadeau R, et al: Supernormal conduction in the intact heart. In Narula OS (ed): Cardiac Arrhythmias: Electrophysiology, Diagnosis and Management. Baltimore, Williams & Wilkins Publishers, 1979, pp 40-56.

Chapter 10 Principles of Catheter Ablation

MOHAMMAD SAEED, PAUL J. WANG,
MARK S. LINK, MUNTHER K. HOMOUD, and
N.A. MARK ESTES III

Ablative therapy for cardiac arrhythmias is based on the rationale that for every arrhythmia, there is a critical anatomic region of abnormal pulse generation or propagation required for that arrhythmia to be initiated and sustained. Selective destruction of that myocardial tissue will result in elimination of the arrhythmia. This principle was first demonstrated in 1968 in successful surgical division of a bundle of Kent in a patient with Wolff-Parkinson-White syndrome.[1] Since then, catheter-mediated ablative techniques have nearly completely replaced surgical procedures because of high clinical success rates and minimal morbidity associated with catheter techniques. Presently, the conventional energy source used for nearly all clinical catheter ablations being performed is radiofrequency (RF) energy. Alternative methods for catheter ablation include microwave, laser, cryoablation, ultrasound, and chemical ablation (Table 10-1).

Subsequent chapters in this book deal with the specific mapping and ablation techniques for supraventricular tachycardia, atrial arrhythmias, atrial fibrillation, atrioventricular (AV) nodal ablation, and ventricular tachycardia ablation; this chapter deals specifically with conceptual basis and principles of catheter ablation including the biophysics and pathology of catheter ablation with specific emphasis on RF ablation techniques.

Radiofrequency Ablation

HISTORICAL BACKGROUND

Catheter ablation using RF current represents a conceptual revolution of therapy for cardiac arrhythmias. Although the initial clinical use of RF energy for ablation of cardiac arrhythmias was reported in 1987, the historical background for this approach goes back more than a century. The principle of using alternating current for therapeutic purposes while avoiding the undesirable effect of neuromuscular stimulation during surgical procedures was initially introduced in 1891.[2] The first clinical application of RF current occurred when the famous neurosurgeon Harvey Cushing, working with W.T. Bovie, used high frequency current for the purpose of electrocoagulation.[3] Subsequent detailed investigations, including pivotal pathologic observations indicating that high frequency lesions made in the RF range were very reproducible and had sharp, well-defined borders, led to its widespread use.

Although RF current had a long history, the first catheter techniques of ablation used direct current. Scheinmann and Gallagher reported use of catheter ablation for elimination of AV conduction in humans in 1982.[4,5] These initial reports used direct current shock

TABLE 10-1 Alternative Energy Sources for Catheter Ablation

	Radiofrequency	Microwave	Laser	Cryoablation	Ultrasound	Chemical
Catheter-based clinical experience	Extensive	Minimal	Minimal	Minimal	Minimal	Minimal
Lesion size	Small	Med–Large	Med–Large	Med–Large	Med–Large	Large
Ability to create transmural lesion	Limited	Low–High	High	High	Limited	High
Thrombogenicity	Medium	Medium	Med–High	Low	Medium	Low
Lesion dependent on good contact	Yes	No	No	Yes	No	No
Risk of perforation	Low	Low	High	Low	Low	High
Ability to reversibly map	No	No	No	Excellent	No	No

using a conventional external defibrillator to deliver energy between the catheter placed adjacent to the cardiac structure targeted and a large-surface-area adhesive electrode applied to the skin. Although direct current shock was effective for ablation of cardiac tissue and was used therapeutically for creation of complete heart block and cure of Wolff-Parkinson-White syndrome, as well as selected ventricular tachycardias, there were significant complications related to intracardiac direct current shocks.[6] Accordingly, catheter-based ablative techniques did not gain widespread acceptance for arrhythmia therapy until the advent of RF ablation. Initial descriptions of RF energy used in a closed chest canine model in 1985 reported that ablation of the AV junction with a conventional 2-mm–tipped electrode catheter was feasible using RF energy.[7,8] Multiple in vitro and in vivo animal studies evaluated the influence of electrode size, electrode-tissue contact pressure, pulse power, duration, impedance, and temperature.[9-11] In vivo observations were made in animal models with the use of RF energy in the ventricles,[12-16] the atrium,[17-21] the coronary sinus,[21-23] and the tricuspid annulus.[17-21] These in vitro and in vivo observations were critical to the successful application for human arrhythmias reported in the early 1990s.[21,24,25]

BIOPHYSICAL ASPECTS OF RADIOFREQUENCY ABLATION

RF energy spans a wide range of frequencies, exhibiting a varying combination of resistive and dielectric properties. The RF band of 300 kHz to 30,000 kHz is used for coagulation, ablation, and cauterizing tissues in medicine. Frequencies below 300 kHz can also be used for ablation, but frequencies lower than 10 kHz stimulate excitable muscular and cardiac tissues. Frequencies greater than 1000 kHz are effective in generating tissue heating, but as frequencies increase, the losses along the transmission line increase and the mode of heating changes from resistive heating to dielectric heating.[26] Therefore, for ease of application, safety, and patient comfort, most standard RF generators produce signals in the range of 300 to 750 kHz.[27]

Application of RF current to the myocardium results in generation of heat due to resistive losses in a dielectric medium (Fig. 10-1). The magnitude of heating is proportional to the power density within the tissue, which diminishes in proportion to the radius to the fourth power. Because of this, only a small volume of myocardial tissue is heated directly by the RF energy (<2 mm in most cases), while the remaining tissue heating occurs via conduction.[11] Haines and Watson have developed a simplified thermodynamic model to study the dynamics of heat transfer within a uniform medium at steady state. In this model, the tissue temperature falls in an inverse proportion to the distance from the heat source (Fig. 10-2). It also predicts that when the temperature at electrode-tissue interface is maintained constant by varying power, the lesion size should be proportional to the temperature and to the electrode radius.[11] Clinical studies have demonstrated the association of higher electrode-tissue temperature and

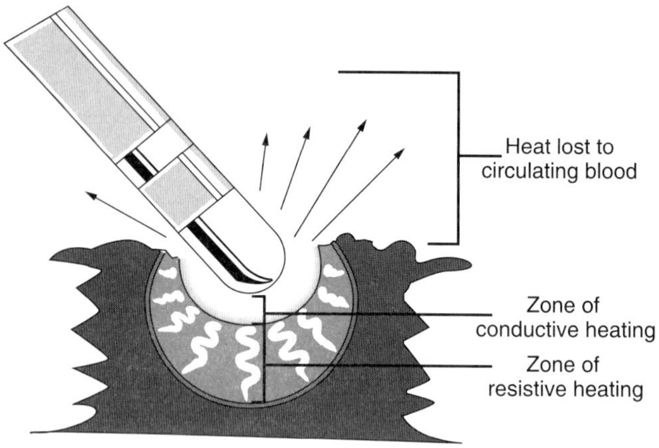

FIGURE 10-1 Tissue heating with radiofrequency energy. Superficial tissue close to the electrode-tissue interface is ablated via resistive heating, while deeper myocardium is heated via conductive heating. Convective cooling occurs because of blood flow near the electrode-tissue interface and in the coronary circulation. (Adapted with permission from Avitall B, Helms R: Determinants of Radiofrequency-Induced Lesion Size, 2nd ed. Armonk, NY, Futura, 2000.)

larger-tip electrodes with larger RF lesion size and procedure efficacy.[24,28] However, temperatures above 100°C are associated with coagulum formation on the RF ablation electrode and electrical impedance rise, which potentially limits lesion size. When the catheter tip is cooled at the endocardial surface, very high power can be delivered without the risk of coagulum formation and a sudden rise in impedance. Thus, ablation with high power and convective cooling of the catheter tip can produce lesions greater than 10 mm in depth. Convective cooling of the catheter tip may be accomplished by cooling the electrode tip with perfused or circulating saline. Because the peak temperature with this technique is below the endocardial surface, very high tissue temperature may be achieved

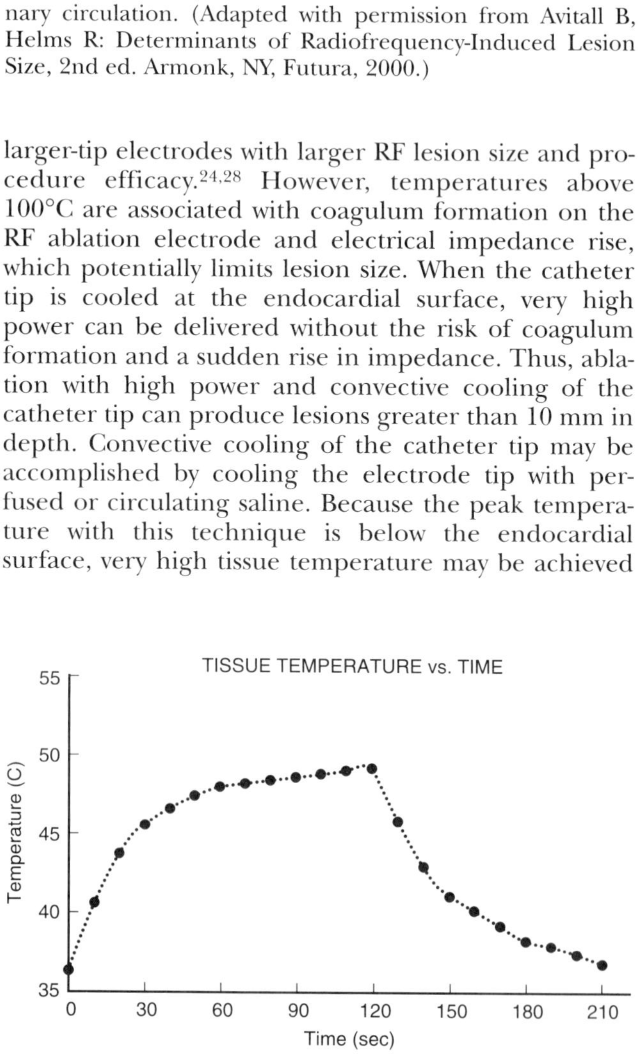

FIGURE 10-2 Radiofrequency energy was delivered to isolated perfused and superfused canine right ventricular free wall. The tissue temperature (°C) was measured using a thermistor. The steady-state temperature as a function of distance is shown. (Adapted with permission from Wang P, et al: Physics and Biology of Catheter Ablation. Armonk, NY, Futura, 1998.)

despite low temperatures measured by catheter tip thermistors. In the worst case, the temperature of the subendocardial tissue can exceed 100°C and result in rapid steam expansion and crater formation, often accompanied by an audible "pop."[10] These so-called *pop lesions* may be thrombogenic or may result in rupture of the thin-walled structures like the atria. However, the increased lesion depth and width achieved with this technique may be useful in the ablation of difficult arrhythmia substrates such as those found with reentrant ventricular tachycardia.

Generally, RF lesions are created by application in a unipolar fashion. Current passes between the tip of an ablating electrode in contact with myocardium and a grounded reference patch electrode placed externally on the patient's skin.[29] Because RF current is alternating current, the polarity of connections from the electrodes to the generator is unimportant. The surface area of the standard ablating electrode (≈12 mm^2) is significantly smaller than the surface area of the dispersive electrode (100 to 250 cm^2), and thus the power density and resultant tissue heating are highest at the electrode-tissue contact point, and no substantive heating occurs at the dispersive skin electrode.

The temperature rise at the electrode-tissue interface is rapid ($t^{1/2}$, 7 to 10 seconds). However, thermal conduction to deeper tissue layers occurs more slowly, resulting in a steep tissue gradient. Studies have shown that steady-state lesion size is not achieved until after 40 seconds of ablation and intramyocardial temperature rises steeply until it reaches a plateau at 60 seconds, with the majority of heating occurring within the first 30 seconds (Fig. 10-3).[30] However, because the heating of deeper layers is delayed, the tissue temperature will continue to rise in deeper layers despite termination of energy delivery.[31] This thermal latency phenomenon may be responsible for the fact that undesirable effects of

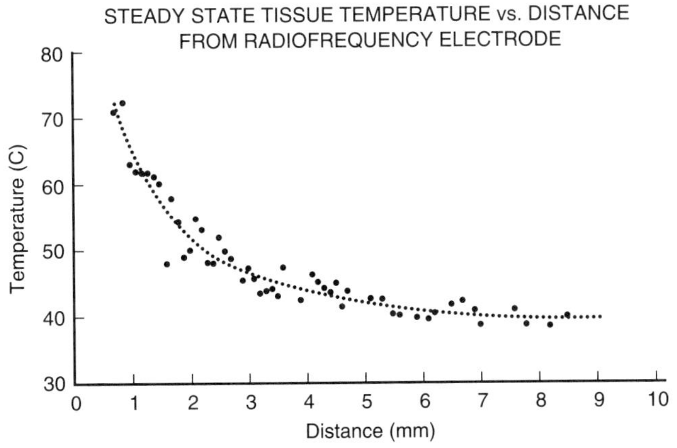

FIGURE 10-3 Radiofrequency (RF) energy was delivered to isolated perfused and superfused canine right ventricular free wall. The tissue temperature (°C) was measured using a thermistor at a distance of 2.5 ± 0.2 mm from the RF electrode and after the 120 seconds of RF energy delivery. The electrode tip temperature was 80°C. (Adapted with permission from Wang P, et al: Physics and Biology of Catheter Ablation. Armonk, NY, Futura, 1998.)

ablation-like AV nodal block can progress despite prompt cessation of energy delivery in clinical situations.

Clinically, the efficacy of RF ablation depends on precise target site mapping and adequate lesion formation.[28,32,33] A wide range of experimental and clinical observations support the notion that applied energy, power, and current are not precise indicators of the extent of lesion formation.[34] By contrast, actual electrode-tissue interface temperature remains the best predictor of actual lesion volume.[30] The range of tissue temperatures effective for RF catheter ablation is between 50°C and 90°C. Typically, 55°C to 65°C is the target or optimal temperature for many temperature control catheters used clinically.[35-38] If the temperature is lower than 50°C, none or minimal tissue necrosis is expected. On the other hand, excessive heating leads to coagulum formation and limitation of lesion size. A number of clinically available RF ablation systems use temperature monitoring during ablation to prevent coagulum formation and achieve adequate electrode-myocardial tissue temperature. They either use a thermocouple or a thermistor placed in the catheter tip.[39] A thermocouple is composed of two metals in contact with each other that generate a small current proportional to ambient temperature. Typically thermocouples are placed within the catheter tip in contact with the distal electrode. A thermistor is essentially a semi-conductor, which has an intrinsic resistance that changes in a predictable fashion with temperature. Small thermistors can be incorporated into a catheter tip and are thermally isolated from the shredding tissue with an insulating sleeve.[39] Currently such catheters are used as part of a temperature monitoring, closed-loop feedback system in which the RF generator automatically adjusts the power to maintain a user-program temperature. Clinical use of temperature monitoring has resulted in a decreased incidence of coagulum formation during ablation.[32]

Numerous limitations may result in discrepancies between the recorded temperature and the true electrode-myocardial tissue temperature. The catheter electrode may not be in intimate contact with the myocardium at the site of the thermistor or thermocouple or at all parts of the electrode. This may underestimate the myocardial interface temperature.[36] This is particularly true at locations in which the catheter electrode may be parallel rather than perpendicular to the endocardial surface, such as occurs with accessory pathway ablation via the transseptal approach. In addition, temperature monitoring becomes difficult as the electrode size increases or the electrode geometry changes. As the electrode size increases, the maximal temperature may be greatly underestimated by measuring just the tip temperature by a single point thermocouple. This may be particularly true in longer electrodes that are being used in linear catheter ablation for atrial fibrillation. Also, in longer electrodes being employed for linear ablation, the temperature at the edges may exceed the temperature in the electrode body, a phenomenon called *edge effect*.[40] Therefore, measuring temperatures in the midpoint will also underestimate the peak temperatures.

PATHOLOGIC ASPECTS OF RADIOFREQUENCY ABLATION

Cellular injury during RF ablation is primarily mediated via thermal mechanisms. In addition, there might be some direct effects of electrical energy on the myocyte, particularly the sarcolemmal membrane. Thermal injury seems to be the dominant cause of coagulation necrosis in the central zone of RF lesion, while at the border zone both thermal and electrical energy probably contribute to the lesion formation.

Research using mammalian cell culture lines has shown that tissue survival during hyperthermia is a function of both temperature and time.[41] Higher temperatures lead to a more rapid and greater degree of cell death. In vitro studies have shown that irreversible myocardial cell death during conventional ablation likely occurs between 52°C and 55°C.[42] Typically RF catheter ablation results in a high temperature (55°C to 60°C) for a relatively short time (30 to 60 seconds). In a tissue immediately adjacent to the electrode, there is rapid tissue injury, but in the tissue heated by conductive heating, there is a relatively delayed myocardial injury with increasing distance from the RF electrode.[10] Clinical outcome during catheter ablation has been shown to correlate with maximum tissue temperature achieved, with a temperature of 62 ± 15°C associated with *irreversible* electrophysiological effects and a temperature of 50 ± 15°C associated with *reversible* electrophysiological effects.[28] In a study of patients undergoing ablation of AV junction, an accelerated junctional rhythm without heart block was observed at 51 ± 4°C, whereas heart block was seen at 58 ± 6°C.[43] Thus, a physiologic effect of hyperthermia was observed at a lower temperature, but higher temperatures were needed to sustain irreversible tissue damage.

The hyperthermia induced by RF energy has immediate and profound effects on cellular electrophysiology, structure, and function. These include the cellular effects on plasma membrane, the cytoskeleton, the nucleus, and cellular metabolism. Exposure to high temperatures results in inhibition of membrane transport proteins[44] and structural changes to ion channels and ion pumps.[45-49] The thermal effects on the cytoskeleton include disruption of stress elements,[50] loss of plasma membrane support, and membrane blebbing.[51] There is loss of nuclear structure and function.[52] With denaturation of metabolic proteins there is inhibition of cellular metabolism.[53] There is damage to the microvasculature and reduced microvascular blood flow.[54] At the level of the plasma membrane there is formation of transient nonspecific ion-permanent pores in a plasma membrane.[55] These effects cause an initial reversible loss of conduction immediately after the initiation of RF energy, which is likely caused by an electrotonic- and heat-induced cellular depolarization. There is irreversible loss of electrophysiological function immediately after successful ablation, and this is likely caused by thermal tissue injury, which results in focal coagulation necrosis. There is a secondary inflammatory response and ischemia as a consequence of microvascular damage, which may cause progression of tissue injury within the border zone.[54] Progressive tissue injury may result in RF lesion extension over time and may be the pathophysiologic mechanism for the late loss of electrophysiological function observed after RF ablation. Alternatively, in approximately 5% to 10% of cases, there is resolution of the RF-induced tissue injury within the border zone leading to late recovery of electrophysiology function and arrhythmia recurrence.

RF energy produces lesions that are homogenous and well demarcated from the surrounding tissue (Fig. 10-4). A potential advantage of RF ablation is the relatively small lesion size. With alternate energy sources such as microwave or laser, in which a larger lesion would be preferable because of the deeper location of the arrhythmia focus, there may be a higher risk of cardiac perforation or other complications.

DETERMINANTS OF LESION SIZE

The lesion size in RF ablation is affected by several factors including electrode configuration and size, power,

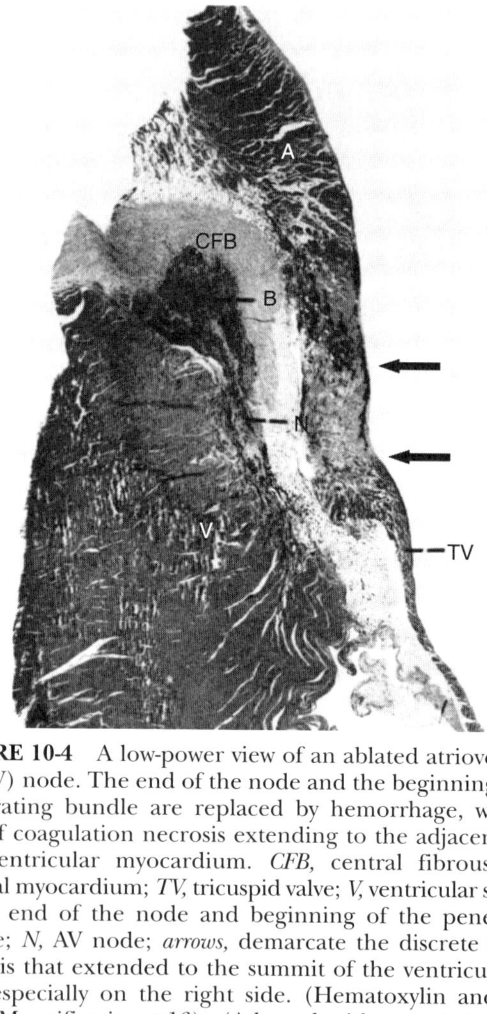

FIGURE 10-4 A low-power view of an ablated atrioventricular (AV) node. The end of the node and the beginning of the penetrating bundle are replaced by hemorrhage, with the area of coagulation necrosis extending to the adjacent atrial and ventricular myocardium. *CFB,* central fibrous body; *A,* atrial myocardium; *TV,* tricuspid valve; *V,* ventricular septum; *B,* the end of the node and beginning of the penetrating bundle; *N,* AV node; *arrows,* demarcate the discrete area of necrosis that extended to the summit of the ventricular septum, especially on the right side. (Hematoxylin and eosin stain. Magnification ×16). (Adapted with permission from Huang S, Bharti S, Graham A, et al: Closed chest catheter dessication of the atrioventricular junction using radiofrequency energy—a new method of catheter ablation. J Am Coll Cardiol 1987;9:349-58.)

duration, temperature, and myocardial tissue properties.[56] The shape of the electrode by affecting the current density can affect the size of the RF lesion. Although needle-shaped electrodes have yielded the largest lesion size, dome-shaped electrodes are in use clinically because of their practical design.[57] Electrode size has also been shown to be an important determinant of RF ablation lesion size. With increasing electrode size, as long as electrode surface maintains contact with endocardial surface and power is used to maintain the current density, the lesion size increases.[58] Increasing the electrode size at some point may decrease the lesion size when a significant proportion of the electrode is not in contact with the myocardium, and thus energy is dissipated into the blood pool. Studies have shown that lesion size is nearly double with 4 mm electrodes than with 2 mm ones. Similarly, lesion size continues to increase as tip electrode size is increased from 4 mm to 8 mm but decreases with larger electrode sizes.[59] In addition, significant char and coagulation formation is observed with electrode size greater than 8 mm.

With increasing duration of energy delivery, the lesion size increases but then reaches a plateau. In vitro studies have shown that RF ablation lesion size increases within the first 30 seconds, but the rate of increase significantly diminishes with extended power delivery.[42] Similarly, lesion size increases as the temperature increases. However, as discussed earlier, when the temperature becomes excessive, the incidence of coagulation formation increases, thus impairing the ability to increase lesion size further. The coupling of RF energy to the endocardial surface is enhanced by improved contact of the electrode with the myocardium.[30] Instability of catheter electrode will result in fluctuating temperatures due to convective loss of energy resulting in poor tissue heating. The efficiency of energy transfer between the RF catheter tip and the endocardium is also dependent on catheter-tissue orientation and the angle of the electrode contact. With perpendicular or oblique placement, lower temperatures are achieved compared to parallel electrode orientation.[60,61] Recent investigations have determined that lesion dimensions can be influenced using higher frequencies. A comparison of lesion dimension for frequency of 100, 600, 1200, and 2000 kHz using a constant electrode size (1 mm) and constant temperature control with a setting of 90°C lesion width was not statistically different for the various frequencies but the depth increased with higher frequency.[62] However, there are few data on the use of RF generators with frequencies other than 500 to 750 kHz with standard catheter technology.

Energy can be delivered in a bipolar mode between two endocardial electrodes and may have applications for treating long linear lesions for the ablation of atrial fibrillation. In comparative studies of unipolar versus bipolar RF energy, the unipolar lesions were longer (14.9 + 1.6 mm versus 9.2 + 1.5 mm) and wider (8.1 + 1.8 versus 5.6 + 1.5) than bipolar lesions.[63] Other studies comparing the two techniques found that the lesions generated with bipolar RF energy were longer and wider than those created with unipolar delivery, although the depth of the lesions did not differ between the two methods.[64] In vitro observations indicate that the bipolar energy delivery resulted in a greater lesion depth than unipolar delivery, but the width and depth were not statistically significantly different between the two methods.[65]

Cooled Radiofrequency Ablation

As mentioned previously, the electrode-tissue interface temperature is an important determinant of lesion size. Cooling of the ablation electrode tip has been shown to prevent both overheating of the electrode-tissue interface and impedance rise during RF delivery, allowing greater power delivery and larger/deeper lesions.[66] During cooled RF ablation, maximum temperature is achieved just below the endocardial surface. The electrode-myocardial interface temperature is usually lower but may be close to the maximal temperature. Because the tip is being constantly cooled and may not accurately reflect the interface temperature, improved methods are needed to measure the electrode-myocardial interface temperature during ablation.

The tip of the catheter can be cooled either by using a closed-loop design in which saline is circulated through the catheter lumen and the ablation electrode or by an open design in which saline is irrigated via small holes at the tip of the electrode. Both these designs have been shown to maintain a low electrode-tissue interface temperature, allowing higher power application and producing significantly larger lesions.[67,68] Cooled RF ablation has been shown to be effective and safe in patients with de novo and refractory atrial flutter and has been used with a high degree of success in patients with ventricular arrhythmia and structural heart disease.[69-71]

Microwave Ablation

Microwave ablation acts mainly via dielectric heating of the myocardium. The microwave energy generates heat by producing oscillation of the water molecule dipoles within the myocardium. The microwave antenna placed at the end of a catheter radiates an electromagnetic field into the surrounding tissue and does not depend on current flow from ablation catheter to the tissue as with an RF electrode.[72] Therefore, as opposed to RF ablation, contact with the myocardium is not required in order to heat myocardial tissue.[29]

The microwave ablation system consists of a microwave generator and a microwave antenna incorporated into the ablation catheter. Microwave generators operate at either 915 MHz or 2450 MHz. Several measurements are used to characterize the properties of microwave systems used for myocardial ablation. A specific absorption rate (SAR) pattern provides a three-dimensional picture of the electromagnetic field created during microwave ablation. SAR is measured by giving a brief energy pulse and measuring the temperature rise in three dimensions around the antenna. The ability of the energy source to deliver energy to the medium is termed as "matching" of the antenna to the dielectric properties of the medium and can be measured either in terms of efficiency of energy transfer or ratio of reflected power to delivered power. When the antenna and tissue are optimally matched, there is high efficiency of energy transfer, and very little power is reflected. Most ablation systems are optimized for specific load or impedance, although some ablation systems can vary the impedance of the system, compensating for the differences in myocardial tissue properties.

The heating pattern and efficiency of microwave ablation system depends on the antenna design (Fig. 10-5). The helical coil antenna generates a circumferential heating pattern, making placement parallel to the endocardial surface optimal, whereas the dipole antenna creates a heating pattern that projects the ablation energy forward. This makes the use of helical coil antennas more suitable for making long linear lesions such as for atrial fibrillation ablation, while a forward-projecting antenna might be ideal for accessory pathway ablation. A spiral microwave antenna was shown in vivo to create lesions that were significantly larger and deeper than the other microwave antenna systems.[73] This may prove useful in ablation of ventricular arrhythmia in structurally abnormal hearts, where the arrhythmogenic focus is located very deep within the ventricular myocardium or is subepicardial.

Microwave ablation can achieve greater temperature rise as a function of depth as compared to RF ablation.[72] As the power and duration of microwave energy is increased, the lesion size also increases. Unlike RF, microwave energy is not limited by impedance rise. Microwave ablation lesion growth is considerably slower than RF ablation.[74] Whayne et al. demonstrated that microwave lesion size using a 915 MHz monopolar antenna system continued to increase over 300 seconds of ablation (Fig. 10-6).[42] Pathologically, microwave lesions are comparable to RF lesions. Like RF lesions, the lesions produced by microwave ablation have greater width than depth for a given power. Acutely, they demonstrate considerable hemorrhage, but chronically, they become well-circumscribed dense fibrous scars (Fig. 10-7).

Microwave catheter ablation using a 2450 MHz generator has been demonstrated to be effective in creating chronic AV block in open- and closed-chest animals.[75-77] Myocardial ablation has also been used to ablate atrial myocardial tissue. Microwave energy successfully ablated aconitine-induced atrial tachycardia.[78] In another experiment, a deflectable microwave catheter was used to create bidirectional block at the isthmus in seven canines with an average of 2.7 ± 1.3 ablations, suggesting the potential benefit of microwave energy in creating long linear lesions.[79] Microwave ablation may also be used for ventricular myocardial ablation at either 2450 MHz or 915 MHz. In vivo experiments with canine left ventricular myocardium showed that the size of the microwave lesion increased with increasing power and duration.[76] Microwave energy, unlike RF energy, may be delivered despite tissue coagulation. Thus, temperature control of the ablation electrode may be even more important for microwave than for RF ablations to avoid thromboembolic risks. A steerable microwave catheter with temperature feedback control was tested in 11 dogs.[80] Delivering three microwave lesions for 60 seconds each, with mean power of 75 watts (W), created large transmural lesions. Although there were no thrombi noted, there was a high incidence of ventricular tachycardia within the first week after ablation.

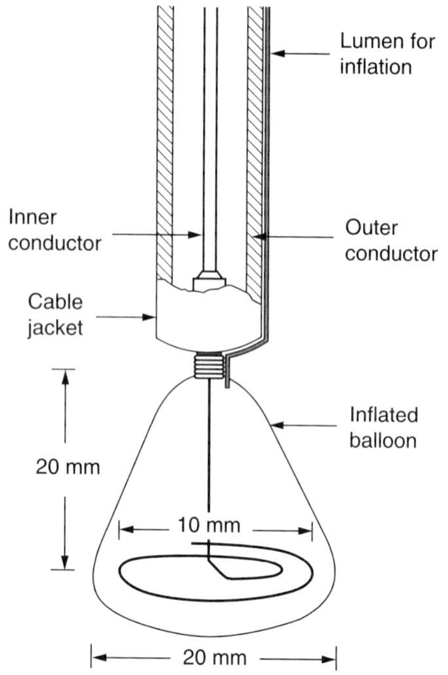

FIGURE 10-5 Schematic diagram of the spiral microwave antenna design. (Adapted with permission from Vanderbrink B: Microwave ablation using a spiral antenna design in a porcine thigh muscle preparation: In vivo assessment of temperature profile and lesion geometry. J Cardiovasc Electrophysiol 2000; 11:193-8.)

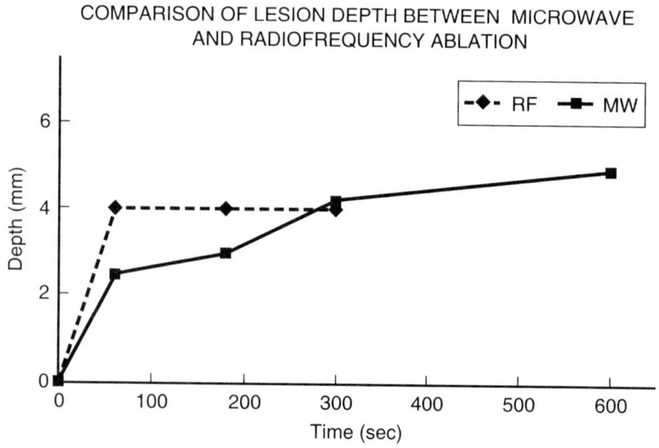

FIGURE 10-6 Comparison of lesion depth between microwave and radiofrequency ablation. Although microwave heating is considerably slower, if more energy is used, deeper lesions are created. (Adapted with permission from Wang P, et al: Physics and Biology of Catheter Ablation. Armonk, NY, Futura, 1998.)

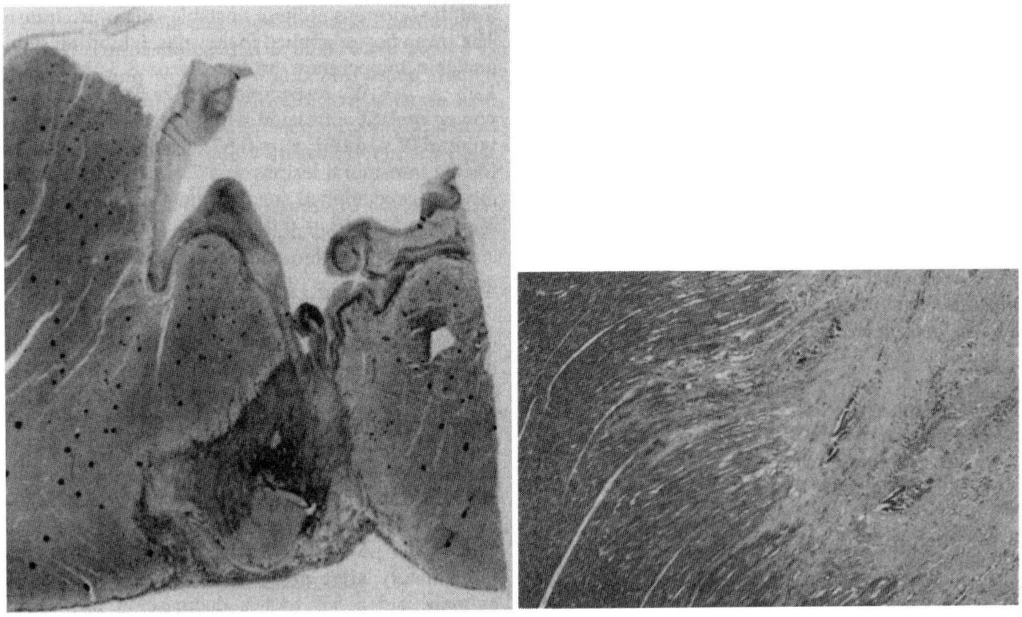

A B

FIGURE 10-7 (See also Color Plate 10-7.) Histology of microwave lesion. **A,** Low-power magnification of a transmural ventricular lesion created using microwave energy. The lesion is transmural and extends from the endocardial surface to the epicardium. **B,** High-power magnification of a ventricular lesion. Note the extensive fibrosis and well-demarcated lesion border zone. Masson's trichome stain. (Adapted with permission from Vanderbrink B, Gilbride C, Aronovitz M, et al: Safety and efficacy of a steerable temperature monitoring microwave catheter system for ventricular myocardial ablation. J Cardiovasc Electrophysiol 2000;11:305-10.)

Laser Ablation

The mechanism of laser myocardial ablation is predominantly via thermal heating. Laser energy wavelength can range from 250 nanometers (nm) to 10,600 nm. The tissue response to the laser energy depends on the wavelength delivered. Wavelengths in the infrared region such as neodymium-yttrium-aluminum-garnet (Nd:YAG) result in scattering of photons causing tissue destruction via photocoagulation without vaporization. Due to its low absorption in normal and coagulated myocardium, Nd:YAG laser light has the potential to reach deep intramural sites from the endo- or epicardial surfaces. This results in deeper lesions and has been extensively applied to the intraoperative ablation of ventricular tachycardia. In contrast, the CO_2 laser produces predominantly absorption compared to scatter, resulting in tissue vaporization and cutting. Other laser lights like Xenon and Fluoride excimer produce vaporization of tissue due to absorption of energy concentrated in a thin layer.[29,81]

The laser ablation system consists of a flexible optical fiber cable housed inside a delivery catheter coupled to a laser energy source. Because a considerable amount of heat can be generated at the ablation site, the catheter must be continuously flushed with saline for cooling. This also helps to clear the blood in front of the catheter tip. Laser, like microwave energy is projected onto the myocardium from the fiberoptic tip. Thus, unlike RF ablation, contact is not required between the fiberoptic tip and myocardium. Continuous-wave laser irradiation produces gradual, dose-dependent coagulation of the irradiated myocardium.[82] Increased energy may be required to create a comparable lesion in a diseased ventricular tissue compared to normal ventricular tissue.[83] Lesion size increases with greater energy. However, laser ablation in situ results in lesions that are considerably larger than in vitro.[84] Temperature monitoring may be important to prevent excessive heating of the tissue. Pulsing of the laser energy at low repetition rates has also been investigated to decrease the degree of tissue charring.

Lesions induced by laser energy can demonstrate varying pathology depending on the type of laser used. With an Nd:YAG laser, the predominant finding is coagulation necrosis with minimal vacuolization and crater formation.[85-87] Lesions are typically well demarcated. When central vaporized crater is present, it is surrounded by a rim of necrotic tissue. Chronically the lesions are healed with a homogeneous region of fibrosis. Water-soluble gaseous substances are produced as a byproduct of laser ablation. Cardiac perforation is also a potential risk with laser ablation.

Laser ablation has been used successfully to create permanent AV block.[88,89] Laser catheter ablation using a preformed catheter with a pin electrode was performed in 10 patients for AV nodal reentrant tachycardia with successful outcome in all 10 patients and no adverse outcomes after a mean follow-up of 27 months.[90] Laser ablation has also been used extensively for treatment of ventricular tachycardia intraoperatively. In a study of 57 patients, an Nd:YAG laser system was used to treat ventricular tachycardia. The operative mortality was 14%. Of the 49 survivors, only 3 remained inducible for ventricular tachycardia.[91] More recently, slow intramural heating with diffuse laser light has been proposed as a safer and more effective way of myocardial ablation.[82] A steerable laser ablation catheter is currently being developed, and initial clinical studies for treatment of ventricular tachycardia have begun.

Cryoablation

All the ablative modalities discussed thus far cause tissue destruction by hyperthermia. In contrast, cryoablation causes tissue injury by freezing. This results in minimal tissue destruction and preserves basic underlying tissue architecture. There is a large body of clinical experience in the use of cryothermy in the surgical treatment of tachyarrhythmias, in which it has been proven to be safe and effective. Cryoablation has been used extensively intraoperatively for the creation of AV block or ablation of atrial and ventricular tachycardias.[92-94]

Application of a cryoprobe to a tissue surface results in the formation of a well-demarcated hemispherical block of frozen tissue. Initially at the cellular level there is formation of intracellular and extracellular ice crystals followed by thawing. Within 24 hours, coagulation necrosis results with hemorrhage, edema, and inflammation. Over time, the lesion becomes sharply demarcated from surrounding tissue and is finally replaced with fatty and fibrous tissue by 2 to 4 weeks (Fig. 10-8).[95] Acutely after cryoablation, frequent ventricular ectopic beats occur, particularly within the first 24 hours. The frequent ventricular activity is completely resolved by 1 week or more after cryoablation. The mechanism of these acute arrhythmias is thought to be increased automaticity.[96,97]

Catheter-based techniques that are capable of freezing myocardial tissue have been developed for the cryoablation of arrhythmias.[98,99] Cryoablation of myocardial tissue is usually accomplished by achieving temperatures of −60°C for 2 or more minutes. Several designs

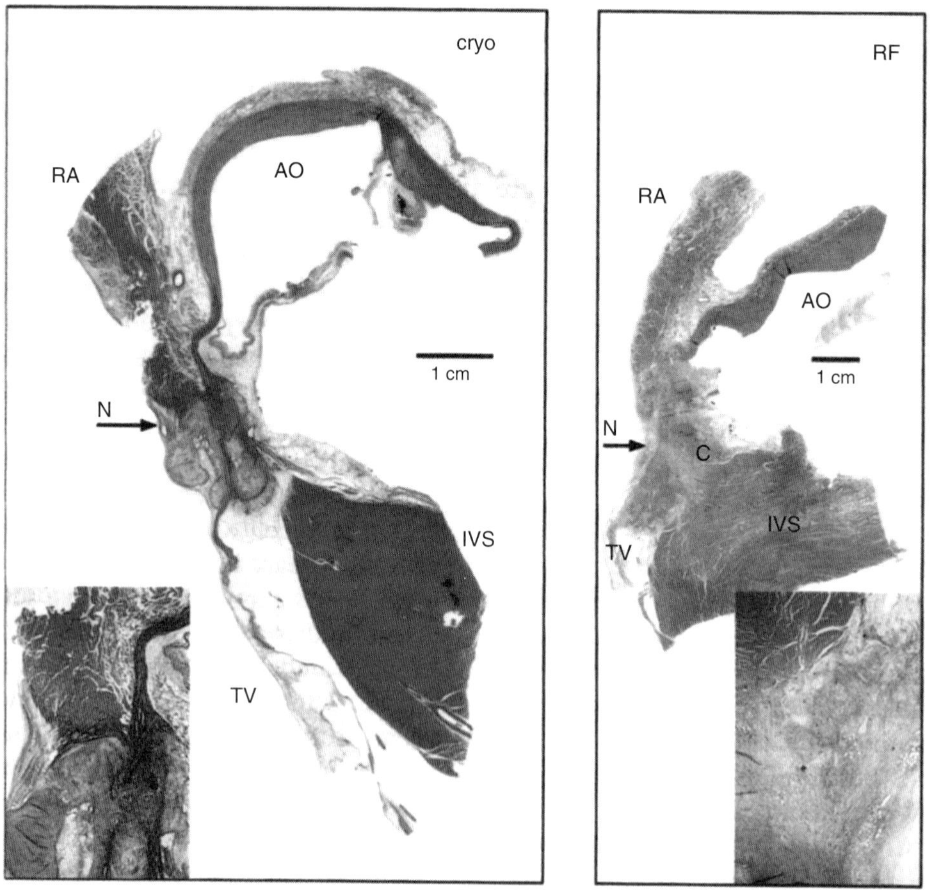

FIGURE 10-8 (See also Color Plate 10-8.) **Left,** Chronic cryoablation lesion of an atrioventricular (AV) node in a dog (Azan stain). There is homogenous fibrotic tissue extending from the right atrial wall across the central fibrous body to the muscular interventricular septum. **Inset,** Higher magnification: absence of viable myocardium within the fibrous strands. *Blue color* indicates fibrotic tissue, and *red color* indicates normal tissue. **Right,** Chronic radiofrequency lesion of AV node in dog (Azan stain). Note the presence of fibrotic tissue with scattered strands of viable myocardium. **Inset,** Higher magnification: The lesion contains inhomogeneous fibrotic tissue with strands of viable myocytes and cartilage formation. The borders are not as well demarcated as the cryoablation lesion. *AO,* aorta; *C,* cartilage; *IVS,* interventricular septum; *N,* compact AV node; *RA,* right atrium; *TV,* tricuspid valve. (Adapted with permission from Rodriguez LM, Leunissen J, Hoekstra A, et al: Transvenous cold mapping and cryoablation of the AV node in dogs: Observations of chronic lesions and comparison to those obtained using radiofrequency ablation. J Cardiovasc Electrophysiol 1998;9:1055-61.)

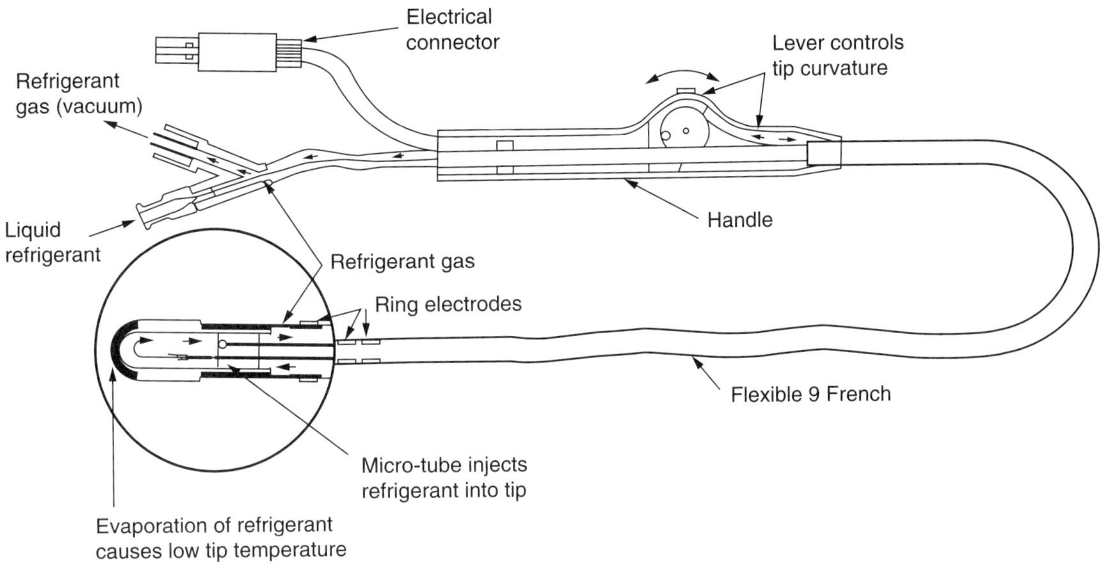

FIGURE 10-9 Schematic of a cryoablation catheter. (Adapted with permission from Dubuc M, Roy D, Thibault B, et al: Transvenous catheter ice mapping and cryoablation of the atrioventricular node in dogs. Pacing Clin Electrophysiol 1999;22:1488-98.)

have been employed to achieve cryoablation using a catheter-based system (Fig. 10-9). They differ among the techniques employed to achieve freezing temperatures. Some designs are based on the Joule Thompson effect, in which a pressure drop results in significant cooling or freezing. Gases such as nitrous oxide or halocarbon are used as refrigerants. The size of the cryoablative lesion depends on various factors such as temperature, duration, surface area of the cooling probe, and the number of applications. Increased contact pressure and longer duration of cryoablation achieve significantly larger lesion size.[100,101] In fact, studies in warm blood–filled space showed that cryoablation lesion size continues to increase up to 10 minutes of application time. The size of the cryoablative probe is limited greatly by the volume of heating provided by the circulating blood flow in the ventricles.[102] Experimental studies have demonstrated that cryoablative catheter techniques can create permanent AV block.[99] Transvenous cryoablation of ventricular tachycardia has been performed successfully in animals. Early results are encouraging, showing an ability to produce transmural lesions. Clinical studies examining the role of catheter cryoablation in ventricular tachycardia are in early stages.

Cryothermal energy also allows for reversible "ice mapping," in which the area likely responsible for the arrhythmia can be evaluated by suppressing its electrophysiological properties before the creation of an irreversible state.[103] Analysis of canine ventricular myocardium exposed to temperatures decreasing from 37°C to 30°C showed progressive slowing of conduction to the point of complete block.[104] Intraoperative cryomapping has been used in the characterization and treatment of supraventricular and ventricular arrhythmias. Catheter-based transvenous cryomapping of AV node has been performed in dogs.[105,106] Friedman et al. successfully used cryomapping in 16 out of 18 patients to identify slow pathway during cryoablation for AV nodal reentrant tachycardia.[107] Clinical studies using transvenous catheter cryoablation are in early stages of implementation, and results will be available in the near future.

Ultrasound Ablation

Ultrasound ablation is an evolving technology with potential therapeutic applications in the field of cardiac electrophysiology. Using frequencies of 500 kHz to 10 MHz, ultrasound energy causes localized hyperthermia at predictable depths without injuring intervening tissue. Applications of cardiac ultrasound in selective cardiac tissue ablation have been promising. Using 10 MHz ultrasound transducers mounted on catheters, He et al. created lesions up to 8.7 ± 2.9 mm in depth in canine left ventricle.[108] In another study, lesions up to 6.4 ± 2.5 mm deep were reported using cylindrical ultrasound transducers mounted on a catheter.[109] There was a linear relationship between increasing power and depth of lesion. Temperature also increased while maintaining constant power increased lesion size. While ultrasound ablation appears to be free of the "pop" phenomenon seen in RF ablation, blood coagulum formation at the tip of the catheter was reported in one study.[110] High intensity–focused ultrasound ablation has been applied from outside the beating heart to successfully create AV block in 10 open chest dogs without damaging the overlying or underlying tissues.[111] Thus, potentially ultrasound energy may allow for noninvasive ablation of cardiac arrhythmias in the future. Combining ultrasound imaging with ultrasound ablation is also a potential advantage of this technique.[112,113] Ultrasound ablation has also been used in electrically isolating pulmonary veins for the treatment of atrial fibrillation. Early clinical experience suggests that circumferential ultrasound ablation of pulmonary veins is feasible in man. Wilber et al. evaluated the performance of a circumferential ultrasound ablation system in 10 patients with recurrent symptomatic atrial fibrillation.[114] After 12 weeks of follow-up, nearly half the patients remained free of atrial fibrillation. The procedure was complicated by stroke in one patient.

Chemical Ablation

Catheter chemical ablation is based on the rationale that application of cytotoxic agents through the coronary circulation to selectively destroy myocardial tissue can be effective in treating various cardiac arrhythmias. In this technique, using angioplasty catheters, branches of coronary circulation are selectively cannulated. Cytotoxic agents such as ethanol are then delivered via an infusion catheter into these small branches. Cardiac myocytes *only* within the supply territory of that particular coronary artery are selectively destroyed. Various experimental and clinical arrhythmias have been treated based on this technique. Increasing ethanol concentration seems to be more effective in creating large lesions but is associated with more impairment of left ventricular function and a greater likelihood of acute and early chronic arrhythmia aggravation in animal studies.[115]

Clinically, ventricular tachycardia has been successfully ablated using chemical ablation. In a study by Nellens et al., 10 out of 12 patients with incessant ventricular tachycardia following myocardial infarction were successfully treated with intracoronary ethanol infusion.[116] In follow-up, six out of seven patients still alive remained free of recurrences. In another study, 10 out of 23 patients with sustained monomorphic ventricular tachycardia were injected with ethanol. In nine patients ventricular tachycardia was no longer inducible acutely but recurred in two patients by 5 to 7 days.[117] Four patients developed complete heart block, and one patient developed pericarditis.

Chemical ablation has also been used for ablation of AV nodes by selective delivery of ethanol into the AV nodal artery.[118] In a study of 12 patients who had medically refractory atrial arrhythmias and had failed RF ablation of AV node, infusion of 2 mL of 96% ethanol induced complete heart block in 10 patients.[119] Two patients developed transient ST segment elevation inferiorly, but none had a decrease in left ventricular ejection fraction or new wall motion abnormality. Another study reported successful chemical ablation of AV node in 13 out of 19 patients without any major complications.[116] The cardiac venous system has also been used for delivery of cytotoxic agents in experimental animals.[120] Use of the cardiac venous system for chemical ablation may have the advantage that unlike coronary arteries, the cardiac veins are rarely stenosed.

Methods of controlling the size of myocardial necrosis remain an issue in chemical ablation technology. Haines et al. showed that although lesions are produced within the targeted coronary bed with ethanol infusion, reflux into other vascular territories is frequently seen.[121] Injection of contrast agent before delivery of a cytotoxic agent may be used to estimate the region of necrosis. Preventing reflux of the cytotoxic agents into adjacent branches has also been the focus of efforts to improve these techniques. Improved catheter design in the future will probably permit delivery into very small and distal coronary arteries or cardiac venous branches without reflux into adjacent vascular territories.

SUMMARY

Various catheter ablation techniques have been developed for the selective ablation of myocardial cells for abolition of cardiac arrhythmias. The myocardial cells are destroyed by absorption of energy or heating when RF, laser, microwave, or ultrasound is used as the energy source. Cryoablation is unique in that freezing of cells results in myocyte death. Although RF has emerged as the dominant energy source in clinical practice, it has some limitations. Future studies are needed to determine the relative use of these techniques in various clinical situations.

REFERENCES

1. Cobb R, Blumenschein S, Sealy W: Successful surgical interruption of the bundle of Kent in a patient with Wolff-Parkinson-White syndrome. Circulation 1968;38:1018-29.
2. D'arsonval M: Action physiologique des courants alternatifs. Comp Rend Soc Biol 1891;43:283-7.
3. McLean A: The Bovie electrosurgical current generator. Arch Surg 1929;18:1863-70.
4. Scheinman M, Morady F, Hess D, et al: Catheter-induced ablation of the atrioventricular junction to control refractory supraventricular arrhythmias. JAMA 1982;248:194-200.
5. Gallagher J, Svenson R, Kasell J, et al: Catheter technique for closed-chest ablation of the atrioventricular conduction system: A therapeutic alternative for the treatment of refractory supraventricular tachycardia. N Engl J Med 1982;306:194-200.
6. Bardy G, Ivey T, Coltorti F, et al: Developments, complications and limitations of catheter-mediated electrical ablation of posterior accessory atrioventricular pathways. Am J Cardiol 1988;61:309-16.
7. Huang S, Jordan N, Graham A, et al: Closed-chest catheter dessication of atrioventricular junction using radiofrequency energy—a new method of catheter ablation. Circulation 1985;72:III389.
8. Huang S, Bharati S, Graham A, et al: Closed chest catheter desiccation of the atrioventricular junction using radiofrequency energy—a new method of catheter ablation. J Am Coll Cardiol 1987;9:349-58.
9. Ring M, Huang S, Gorman G, et al. Determinants of impedance rise during catheter ablation of bovine myocardium with radiofrequency energy. Pacing Clin Electrophysiol 1989;12:1502-13.
10. Haines D, Verow A. Observations on electrode-tissue interface temperature and effect on electrical impedance during radiofrequency ablation of ventricular myocardium. Circulation 1990;82:1034-8.
11. Haines D, Watson D. Tissue heating during radiofrequency catheter ablation: A thermodynamic model and observations in isolated perfused and superfused canine right ventricular free wall. Pacing Clin Electrophysiol 1989;12:962-76.
12. Huang S, Graham A, Hoyt R, et al: Transcatheter desiccation of the canine left ventricle using radiofrequency energy: A pilot study. Am Heart J 1987;114:42-8.
13. Huang S, Graham A, Wharton K: Radiofrequency catheter ablation of the left and right ventricles: Anatomic and electrophysiologic observations. Pacing Clin Electrophysiol 1988;11:449-59.

14. Ring M, Huang S, Graham A, et al: Catheter ablation of the ventricular septum with radiofrequency energy. Am Heart J 1989;117:1233-40.
15. Haverkamp W, Hindricks G, Gulker H, et al: Coagulation of ventricular myocardium using radiofrequency alternating current: Bio-physical aspects and experimental findings. Pacing Clin Electrophysiol 1989;12:187-95.
16. Oeff M, Langberg J, Franklin J, et al: Effects of multipolar electrode radiofrequency energy delivery on ventricular endocardium. Am Heart J 1990;119:599-607.
17. Lee M, Huang S, Graham A, et al: Radiofrequency catheter ablation of the canine atrium and tricuspid annulus. Circulation 1987;76:IV-405.
18. Chauvin M, Wihelm J-M, Dumont P, et al: The ablation of canine atrial tissue by high-frequency currents: Anatomical and histological findings. J Electrophysiol 1988;2.
19. Lavergne T, Prunier L, Guize L, et al: Transcatheter radiofrequency ablation of atrial tissue using a suction catheter. Pacing Clin Electrophysiol 1989;12:177-86.
20. Lee M, Huang S, Graham A, et al: Transcatheter radiofrequency ablation in the canine right atrium. J Intervent Cardiol 1991; 4:125-33
21. Jackman W, Kuck K-H, Naccarelli G, et al: Catheter ablation at the tricuspid annulus using radiofrequency current in canines [abstract]. J Am Coll Cardiol 1987;9:99A.
22. Huang S, Graham A, Bharati S, et al: Short- and long-term effects of transcatheter ablation of the coronary sinus by radiofrequency energy. Circulation 1988;78:416-27.
23. Langberg J, Griffin J, Herre J, et al: Catheter ablation of accessory pathways using radiofrequency energy in the canine coronary sinus. J Am Coll Cardiol 1989;13:491-6.
24. Jackman W, Wang X, Friday K, et al: Catheter ablation of accessory atrioventricular pathways (Wolff-Parkinson-White syndrome) by radiofrequency current. N Engl J Med 1991;324:1605-11.
25. Calkins H, Sousa J, El-Atassi R, et al: Diagnosis and cure of Wolff-Parkinson-White syndrome or paroxysmal supraventricular tachycardia during a single electrophysiologic test. N Engl J Med 1991;324:1612-18.
26. Wang P, et al: Physics and Biology of Catheter Ablation. Armonk, NY, Futura, 1998.
27. Lin J: Engineering and Biophysical Aspects of Microwave and Radiofrequency Radiation. Glasgow, Scotland, Blackie, 1986.
28. Langberg J, Calkins H, el-Atassi R, et al: Temperature monitoring during radiofrequency catheter ablation of accessory pathways. Circulation 1992;86:1469-74.
29. Avitall B, Khan M, Krum D: Physics and engineering of transcatheter cardiac tissue ablation. J Am Coll Cardiol 1993;22:932.
30. Haines DE: Determinants of lesion size during radiofrequency catheter ablation: The role of electrode-tissue contact pressure and duration of energy delivery. J Cardiovasc Electrophysiol 1991;2:509-15.
31. Wittkampf F, Nakagawa H, Yamanashi W, et al: Thermal latency in radiofrequency ablation. Circulation 1996;93:1083-6.
32. Calkins H, Prystowsky E, Carlson M, et al: Temperature monitoring during radiofrequency catheter ablation procedures using closed loop control. Circulation 1994;90:1279-86.
33. Strickberger S, Hummel J, Gallagher M, et al: Effect of accessory pathway location on the efficiency of heating during RF catheter ablation. Am Heart J 1995;129:54-8.
34. Nath S, Di MJ, Haines D: Basic aspects of radiofrequency catheter ablation. J Cardiovasc Electrophysiol 1994;5:863-76.
35. Langberg J, Harvey M, Calkins H, et al: Titration of power output during radiofrequency catheter ablation of atrioventricular nodal reentrant tachycardia. Pacing Clin Electrophysiol 1993;16:465-70.
36. McRury ID, Whayne JG, Haines DE: Temperature measurement as a determinant of tissue heating during radiofrequency catheter ablation: An examination of electrode thermistor positioning for measurement accuracy. J Cardiovasc Electrophysiol 1995;6: 268-78.
37. Pires L, Huang S, Wagshal A, et al: Temperature-guided radiofrequency catheter ablation of closed-chest ventricular myocardium with a novel thermistor-tipped catheter. Am Heart J 1996;127: 163-73.
38. Dinnerman J, Berger R, Calkins H, et al: Temperature monitoring during radiofrequency ablation. J Cardiovasc Electrophysiol 1996;7:163-73.
39. Nath S, Haines DE: Biophysics and pathology of catheter energy delivery systems. Prog Cardiovasc Dis 1995;37:185-204.
40. McRury ID, Panescu D, Mitchell MA, et al: Nonuniform heating during radiofrequency catheter ablation with long electrodes: Monitoring the edge effect. Circulation 1997;96:4057-64.
41. Bauer K, Henle K: Arrhenius analysis of heat survival curves from normal and thermotolerant CHO cells. Radiat Res 1979;78: 251-63.
42. Whayne JG, Nath S, Haines DE: Microwave catheter ablation of myocardium in vitro. Assessment of the characteristics of tissue heating and injury. Circulation 1994;89:2390-5.
43. Nath S, Di MJ, Mounsey J, et al: Correlation of temperature and pathophysiological effect during radiofrequency catheter ablation of the AV junction. Circulation 1995;92:1188-92.
44. Kapiszewska M, Hopwood LE: Changes in bleb formation following hyperthermia treatment of Chinese hamster ovary cells. Radiat Res 1986;105:405-12.
45. Stevenson A, Galey W, Tobey R: Hyperthermia-induced increase in potassium transport in Chinese hampster cells. J Cell Physiol 1983;115:75-86.
46. Yi P: Cellular ion content changes during and after hyperthermia. Biochem Biophys Res Commun 1979;91:171-82.
47. Vidair C, Dewey W: Evaluation of a role for intracellular Na, K, Ca and Mg in hyperthermic cell killing. Radiat Res 1986;105(2): 187-200.
48. Boonstra J, Schamhart D, Delaat S, et al: Analysis of K and Na transport and intracellular contents during and after heat shock and their role in protein synthesis in rat hepatoma cells. Cancer Res 1984;44:955-60.
49. Borelli M, Carlini W, Ransom B, et al: Ion-selective microelectrode measurements of free intracellular chloride and potassium concentrations in hyperthermia-treated neuroblastoma cells. J Cell Physiol 1986;129:175-84.
50. Glass J, DeWitt R, Cress A: Rapid loss of stress fibres Chinese hamsters ovary cells after hyperthermia. Cancer Res 1985;11:258-62.
51. Borelli M, Wong R, Dewey W: A direct correlation between hyperthermia-induced membrane blebbing and survival in synchronous G1 CHO cells. J Cell Physiol 1986;126:181-90.
52. Warters R, Rori J: Hyperthermia and cell nucleus. Radiat Res 1982;92:458-62.
53. Haines DE, Whayne J, Walker J, et al: The effect of radiofrequency catheter ablation on myocardial creatine kinase activity. J Cardiovasc Electrophysiol 1995;6:79-88.
54. Nath S, Whayne J, Kaul S, et al: Effects of radiofrequency catheter ablation on regional myocardial blood flow: Possible mechanism for late electrophysiologic outcome. Circulation 1994;89:2667-72.
55. Chang D: Cell poration and cell fusion using an oscillating electric field. Biophys J 1989;56:641-52.
56. Avitall B, Helms R: Determinants of Radiofrequency-Induced Lesion Size, 2nd ed. Armonk, NY, Futura, 2000.
57. Blouin L, Marcus F: The effect of electrode design on the efficiency of delivery of radiofrequency energy to cardiac tissue in vitro. Pacing Clin Electrophysiol 1989;12:136-43.
58. Haines D, Watson D, Verow A: Electrode radius predicts lesion radius during radiofrequency energy heating. Validation of a proposed thermodynamic model. Circ Res 1990;67:124-9.
59. Langberg J, Lee Michael A, Chin M, et al: Radiofrequency catheter ablation: The effect of electrode size on lesion volume in vivo. Pacing Clin Electrophysiol 1990;13:1242-8.
60. Panescu D, Whayne J, Fleischman S, et al: Three-dimensional finite element analysis of current density and temperature distributions during radiofrequency ablation. IEEE Trans Biomed Eng 1995;42:879-90.
61. Kongsgaard E, Steen T, Jensen O, et al: Temperature guided radiofrequency catheter ablation of myocardium: Comparison of catheter tip and tissue temperature in vitro. Pacing Clin Electrophysiol 1997;20:1252-60.
62. Kovoor P, Eipper V, Dewsnap B, et al: The effect of different frequencies on lesion size during radiofrequency ablation. Circulation 1996;94:3953.
63. Haines DE, McRury ID, Whayne, et al: Radiofrequency ablation at the tricuspid-inferior vena cava isthmus: Unipolar versus bipolar delivery. Circulation 1994;90:3202;I-594.
64. Anfinsen O, Kongsgaard E, Foerster A, et al: Radiofrequency catheter ablation of pig right atrial free wall: Larger lesions but

similar incidence of lung injury and diaphragmatic paresis with two-catheter bipolar compared to unipolar electrode configuration. Circulation 1996;92:3260;I-557.
65. Anfinsen O, Kongsgaard E, Aass H, et al: Radiofrequency current ablation of thin-walled structures; an in vitro study comparing unipolar and bipolar electrode configuration. Pacing Clin Electrophysiol 1996;19:714.
66. Ruffy R, Imran MA, Santel DJ, et al: Radiofrequency delivery through a cooled catheter tip allows the creation of larger endomyocardial lesions in the ovine heart. J Cardiovasc Electrophysiol 1995;6:1089-96.
67. Nakagawa H, Yamanashi W, Pitha J, et al: Comparison of in vivo tissue temperature profile and lesion geometry for radiofrequency ablation with a saline-irrigated electrode versus temperature control in a canine thigh muscle preparation. Circulation 1995;91:2264-73.
68. Mittleman R, Huang S, de Guzman, et al: Use of saline infusion electrode catheter for improved energy delivery and increased lesion size in radiofrequency catheter ablation. Pacing Clin Electrophysiol 1995;18:1022-7.
69. Jais P, Haissaguerre M, Shah D, et al: Successful irrigated-tip catheter ablation of atrial flutter resistant to conventional radiofrequency ablation. Circulation 1998;98:835-8.
70. Jais P, Shah DC, Haissaguerre M, et al: Prospective randomized comparison of irrigated-tip versus conventional-tip catheters for ablation of common flutter. Circulation 2000;101:772-6.
71. Calkins H, Epstein A, Packer D, et al: Catheter ablation of ventricular tachycardia in patients with structural heart disease using cooled radiofrequency energy: Results of a prospective multicenter study. Cooled RF Multi Center Investigators Group. J Am Coll Cardiol 2000;35:1905-14.
72. Wonnell TL, Stauffer PR, Langberg JJ: Evaluation of microwave and radio frequency catheter ablation in a myocardium-equivalent phantom model. IEEE Trans Biomed Eng 1992;39:1086-95.
73. Vanderbrink BA, Gu Z, Rodriguez V, et al: Microwave ablation using a spiral antenna design in a porcine thigh muscle preparation: In vivo assessment of temperature profile and lesion geometry. J Cardiovasc Electrophysiol 2000;11:193-8.
74. Thomas S, Clout R, Deery C, et al: Microwave ablation of myocardial tissue: The effect of element design, tissue coupling, blood flow, power, and duration of exposure on lesion size. J Cardiovasc Electrophysiol 1999;10:72-8.
75. Langberg J, Wonnell T, Chin M, et al: Catheter ablation of the atrioventricular junction using a helical microwave antenna: A novel means of coupling energy to the endocardium. Pacing Clin Electrophysiol 1991;14:2105-13.
76. Yang X, Watanabe I, Kojima T, et al: Microwave ablation of the atrioventricular junction in vivo and ventricular myocardium in vitro and in vivo. Effects of varying power and duration on lesion volume. Jpn Heart J 1994;35:175-91.
77. Lin J, Beckman K, Hariman R, et al: Microwave ablation of the atrioventricular junction in open-chest dogs. Bioelectromagnetics 1995;16:97-105.
78. Rho T, Ito M, Pride H, et al: Microwave ablation of canine atrial tachycardia induced by aconitine. Am Heart J 1995;129:1021-5.
79. Liem L, Mead R: Microwave linear ablation of the isthmus between the inferior vena cava and tricuspid annulus. Pacing Clin Electrophysiol 1998;21:2079-86.
80. Vanderbrink B, Gilbride C, Aronovitz M, et al: Safety and efficacy of a steerable temperature monitoring microwave catheter system for ventricular myocardial ablation. J Cardiovasc Electrophysiol 2000;11:305-10.
81. Isner JM, Donaldson RF, Deckelbaum LI, et al: The excimer laser: Gross, light microscopic and ultrastructural analysis of potential advantages for use in laser therapy of cardiovascular disease. J Am Coll Cardiol 1985;6:1102.
82. Ware D, Boor P, Yang C, et al: Slow intramural heating with diffused laser light: A unique method for deep myocardial coagulation. Circulation 1999;99:1630-6.
83. Weber HP, Heinze A, Enders S, et al: Laser catheter coagulation of normal and scarred ventricular myocardium in dogs. Lasers Surg Med 1998;22:109-19.
84. Lee BI, Gottdiener JS, Fletcher RD, et al: Transcatheter ablation: Comparison between laser photoablation and electrode shock ablation in the dog. Circulation 1985;71:579-86.
85. Weber HP, Heinze A, Enders S, et al: Catheter-directed laser coagulation of atrial myocardium in dogs. Eur Heart J 1994;15:971-80.
86. Weber H, Heinze A, Enders S, et al: Mapping guided laser catheter ablation of the atrioventricular conduction in dogs. Pacing Clin Electrophysiol 1996;19:176-87.
87. Weber HP, Heinze A, Enders S, et al: Laser versus radiofrequency catheter ablation of ventricular myocardium in dogs: A comparative test. Cardiology 1997;88:346-52.
88. Curtis A, Abela G, Griffin JC, et al: Transvascular argon laser ablation of atrioventricular conduction in dogs: Feasibility and morphological results. Pacing Clin Electrophysiol 1989;12:347-57.
89. Curtis AB, Mansour M, Friedl SE, et al: Modification of atrioventricular conduction using a combined laser-electrode catheter. Pacing Clin Electrophysiol 1994;17:337-48.
90. Weber H, Kaltenbrunner W, Heinze A, et al: Laser catheter coagulation of atrial myocardium for ablation of atrioventricular nodal reentrant tachycardia. First clinical experience [see comments]. Eur Heart J 1997;18:487-95.
91. Svenson R, Littman L, Splinter R, et al: Current status of lasers for arrhythmia ablation. J Cardiovasc Electrophysiol 1992;3:345-53.
92. Harrison L, Gallagher J, Kasell J: Cryosurgical ablation of the AV node. Circulation 1977;55:463-70.
93. Gallagher L, Selay W, Anderson W: Cryo-surgical ablation of accessory atrioventricular connections: A method for correction of pre-excitation syndrome. Cardiology 1977;55:471-8.
94. Otto D, Garson A, Cooley D: Cryoablative techniques in the treatment of cardiac tachyarrhythmias. Ann Thorac Surg 1987;26:438-41.
95. Whittaker D: Mechanisms of tissue destruction following cryosurgery. Ann R Coll Surg Engl 1984;66:313-8.
96. Klein G, Harrison L, Ideker R, et al: Reaction of myocardium to cryosurgery: Electrophysiology and arrhythmogenic potential. Circulation 1979;59:364-72.
97. Hunt G, Chard R, Johnson D, et al: Comparison of early and late dimensions and arrhythmogenicity of cryolesions in the normothermic canine heart. J Thorac Cardiovasc Surg 1989;97:313-8.
98. Dubuc M, Friedman PL, Roy D: Reversible electrophysiologic effects using ice mapping with a cryoablation catheter. Pacing Clin Electrophysiol 1997;20:1203.
99. Dubuc M, Talajic M, Roy D, et al: Feasibility of cardiac cryoablation using a transvenous steerable electrode catheter. J Interv Card Electrophysiol 1998;2:285-92.
100. Matsudaira K, Nakagawa H, Yamanashi W, et al: Effect of contact pressure on lesion size during catheter cryoablation [abstract]. Circulation 2000;102:II-526.
101. Matsudaira K, Nakagawa H, Yamanashi W, et al: Effect of application time on catheter cryoablation lesion size [abstract]. Circulation 2000;102:II-527.
102. Imae S, Nakagawa H, Yamanashi W: Cryo-catheter ablation: Effect of blood flow on lesion formation. Pacing Clin Electrophysiol 1997;20:1204.
103. Gallagher J, Del Rossi A, Fernandes J: Cryothermal mapping of recurrent ventricular tachycardia in man. Circulation 1985;71:732-9.
104. Gessman J, Agarwal J, Endo T: Localization and mechanism of ventricular tachycardia by uce mapping 1 week after the onset of myocardial infarction in dogs. Circulation 1983;68:657-66.
105. Rodriguez LM, Leunissen J, Hoekstra A, et al: Transvenous cold mapping and cryoablation of the AV node in dogs: Observations of chronic lesions and comparison to those obtained using radiofrequency ablation. J Cardiovasc Electrophysiol 1998;9:1055-61.
106. Dubuc M, Roy D, Thibault B, et al: Transvenous catheter ice mapping and cryoablation of the atrioventricular node in dogs. Pacing Clin Electrophysiol 1999;22:1488-98.
107. Friedman PL, Stevenson W, Tchou P, et al: Reversible cryomapping of slow pathway during catheter cryoablation of atrioventricular nodal reentrant supraventricular tachycardia [abstract]. Circulation 2000;102:II-368.
108. He DS, Zimmer JE, Hynynen K, et al: Application of ultrasound energy for intracardiac ablation of arrhythmias [see comments]. Eur Heart J 1995;16:961-6.

109. Hynynen K, Dennie J, Zimmer J, et al: Cylindrical ultrasonic transducers for cardiac catheter ablation. IEEE Trans Biomed Eng 1997;44:144-51.
110. Ohkubo T, Okishige K, Goseki Y, et al: Experimental study of catheter ablation using ultrasound energy in canine and porcine hearts. Jpn Heart J 1998;39:399-409.
111. Strickberger S, Tokano T, Kluiwstra J, et al: Extracardiac ablation of the canine atrioventricular junction by use of high-intensity focused ultrasound. Circulation 1999;100:203-8.
112. Packer D, Chan R, Johnson S: Ultrasound cardioscopy: Initial experience with a new resolution combined intracardiac ultrasound/ablation system. Pacing Clin Electrophysiol 1994;17:863.
113. Packer DL SJ, Chan RC: The utility of a new integrated high resolution intracardiac ultrasound/ablation system in a canine model [abstract]. J Am Coll Cardiol 1995;353A.
114. Wilber D, Packer D, Natale A, et al: Initial US clinical experience with circumferential ultrasound ablation of the pulmonary veins for treatment of atrial fibrillation [abstract]. Circulation 2000;102:II-443.
115. Haines DE, Whayne JG, DiMarco JP: Intracoronary ethanol ablation in swine: Effects of ethanol concentration on lesion formation and response to programmed ventricular stimulation. J Cardiovasc Electrophysiol 1994;5:422-31.
116. Nellens P, Gursoy S, Andries E, et al: Transcoronary chemical ablation of arrhythmias. Pacing Clin Electrophysiol 1992;15:1368-73.
117. Kay G, Epstein A, Bubien R, et al: Intracoronary ethanol ablation for the treatment of recurrent sustained ventricular tachycardia. J Am Coll Cardiol 1992;19:159-68.
118. Wang P, Ursell P, Sosa-Suarez G, et al: Permanent AV block or modification of AV nodal function by selective AV nodal artery ethanol infusion. Pacing Clin Electrophysiol 1992;15:779-89.
119. Kay G, Bubien R, Dailey S, et al: A prospective evaluation of intracoronary ethanol ablation of the atrioventricular conduction system. J Am Coll Cardiol 1991;17:1634-40.
120. Wright KN, Morley T, Bicknell J, et al: Retrograde coronary venous infusion of ethanol for ablation of canine ventricular myocardium [see comments]. J Cardiovasc Electrophysiol 1998;9:976-84.
121. Haines D, Verow A, Sinusas A, et al: Intracoronary ethanol ablation in swine: Characterization of myocardial injury in target and remote vascular beds. J Cardiovasc Electrophysiol 1994;5:41-9.

Chapter 11 Interpretation of Clinical Trials: How Mortality Trials Relate to the Therapy of Atrial Fibrillation

ALBERT L. WALDO and CRAIG M. PRATT

Clinical trials have become central to testing the efficacy and safety of therapy of all sorts. This was not always true. For a long time, there were few trials that examined both the safety and efficacy of cardiac arrhythmia treatment modalities. Fortunately, this changed, influenced particularly by the Cardiac Arrhythmia Suppression Trial (CAST)[1,2] and reinforced by subsequent clinical trials as well as regulatory policies. This chapter discusses several important, in fact, *landmark* trials of the treatment of cardiac arrhythmias, thereby examining not only the lessons learned but also how they have been clinically applied. This discussion serves as a paradigm for the interpretation of clinical trials.

The Cardiac Arrhythmia Suppression Trials I[1] and II[2]

BACKGROUND

Based on a series of prior studies,[3-9] it was recognized that the frequency or complexity of ventricular premature beats, or both, after myocardial infarction were a risk factor for sudden cardiac death. Furthermore, it was widely assumed that suppression of ventricular ectopy in this patient population would suppress spontaneous induction of ventricular tachyarrhythmias, the presumed mode of sudden cardiac death. In fact, many physicians had been treating these patients with antiarrhythmic agents in order to suppress their premature ventricular beats.[10,11] Furthermore, a whole lingo (spin) developed in which ventricular arrhythmias were described as *benign, potentially malignant,* or *malignant,* with premature ventricular beats included under the categories of benign or potentially malignant.[12-15]

In this context, and because of the important incidence of sudden cardiac death after myocardial infarction, the National Heart, Lung, and Blood Institute initiated the Cardiac Arrhythmia Pilot Study (CAPS), a feasibility study to test the hypothesis that long-term antiarrhythmic drug suppression of ventricular arrhythmias after an acute myocardial infarction would improve survival.[16] When CAPS ended, it had enrolled 502 patients with an acute myocardial infarction between 6 and 60 days before entry into the study. These patients met the following criteria: a left ventricular ejection fraction (≥0.20); younger than 75 years of age; and demonstration of either an average of 10 or more premature ventricular complexes (PVCs) per hour or 5 or more episodes of nonsustained ventricular tachycardia in a qualifying 24-hour ambulatory electrocardiogram (Holter monitor) recording. Nonsustained ventricular tachycardia was defined as three to nine PVCs in a row at a rate greater than or equal to 100 beats per minute (bpm). Four drugs were used in this placebo-controlled pilot study—encainide, flecainide, moricizine, and imipramine, the latter a "quinidine equivalent" thought to be better tolerated than quinidine. The then investigational antiarrhythmic drugs encainide, flecainide, and moricizine showed good to outstanding suppression of ventricular premature beats, and were well tolerated compared to placebo.[17] It was then elected to proceed with a full study.

CARDIAC ARRHYTHMIA SUPPRESSION TRIAL I

The Protocol

CAST was a multicenter, randomized, placebo-controlled study designed to test whether the suppression of asymptomatic or mildly symptomatic ventricular arrhythmias after myocardial infarction would reduce the rate of death from arrhythmia.[1,2,18] Because in CAPS, encainide, flecainide, and moricizine were shown to suppress these arrhythmias sufficiently in the target population[17] and were well tolerated,[17] these drugs were selected for use in CAST. Also, the protocol was changed in important ways from the CAPS trial. In CAST, patients eligible for the study had to be (1) between 6 days and 2 years after a documented myocardial infarction; (2) have six or more premature ventricular beats per hour on a Holter monitor recording that had a minimum 18 hours' analyzable data; (3) have a left ventricular ejection fraction (LVEF) of 0.55 or less if the qualifying Holter recording was obtained within 90 days of the myocardial infarction; and (4) have an LVEF of 0.40 or less if the qualifying Holter recording was performed between 90 days and 2 years after the myocardial infarction. The reason for making patients eligible beyond 90 days of their qualifying myocardial infarction if the LVEF was 40 or less was that this group with ventricular ectopy had a 10% incidence of sudden cardiac death per a then "current" (actually a few years behind) Duke University database of post–myocardial infarction patients. Finally, patients could not have nonsustained ventricular tachycardia greater than or equal to 15 consecutive ventricular beats at a rate of 120 or greater bpm.

Eligible patients first underwent an open label titration phase, during which up to three drugs (encainide, flecainide, and moricizine) at two oral doses each were evaluated. The titration was terminated when a drug and dose that suppressed the ventricular ectopy were found. The criteria for suppression were 80% or greater reduction of premature ventricular beats and 90% or greater reduction of runs of nonsustained ventricular tachycardia. Patients in whom suppression of the arrhythmia occurred were then randomized to either the effective drug or placebo. Patients who did not meet the effective suppression criteria were randomized to receive "the best drug" or placebo. Patients with LVEFs of 0.30 or greater were randomly assigned to one of two titration sequences: encainide–moricizine–flecainide or flecainide–moricizine–encainide. Patients whose LVEFs were less than 0.30 were randomly assigned to receive either encainide followed by moricizine or moricizine followed by encainide, as flecainide was not given to this group of patients because of concern about its negative inotropic properties.

About 22 months into the study, the CAST Data and Safety Monitoring Board recommended discontinuation of the encainide and flecainide limbs because of excessive mortality compared to placebo.[1] This was true both for the primary end point of arrhythmic death or resuscitated cardiac arrest (Fig. 11-1) or the secondary end point of total mortality. After a mean follow-up of 10 months, of the 1498 patients in these two limbs of the study, 89 had died (59 of arrhythmia, 43 receiving drug versus 16 receiving placebo; $P = .0004$), 22 of nonarrhythmic causes (17 receiving drug versus 5 receiving placebo; $P = .01$), and 8 of noncardiac causes (3 receiving drug versus 5 receiving placebo). At that time, 857 patients had been assigned to receive encainide or its placebo (432 to active drug and 425 to placebo), and 641 to receive flecainide or its placebo (323 to active drug and 318 to placebo). The baseline characteristics of the patients randomly assigned to receive flecainide or encainide as compared with those receiving placebo were similar with respect to all characteristics examined.

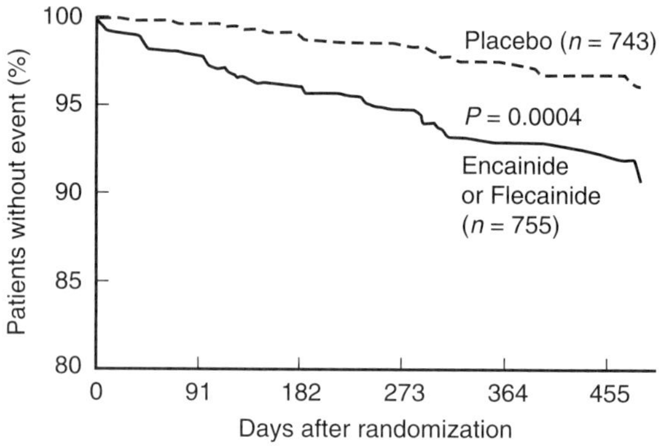

Placebo	743	632	516	412	292	201
Active drug	755	631	507	392	286	198

FIGURE 11-1 Actuarial probabilities of freedom from death or cardiac arrest due to arrhythmia in 1948 patients receiving encainide, flecainide, or corresponding placebo. The number of patients at risk of an event is shown along the bottom of the figure. (From Anderson JL, Platia EV, Hallstrom RW, et al: Interaction of baseline characteristics with hazard of encainide, flecainide, and moricizine therapy in patients with myocardial infarction; possible explantation for increased mortality in the Cardiac Arrhythmia Suppression Trial (CAST). Circulation 1994;90:2843-52.)

Analysis of Cardiac Arrhythmia Suppression Trial I

Unexpected Results

There were several striking aspects to CAST I. First and foremost was the unexpected harm due to the encainide and flecainide therapy, presumably due to proarrhythmia. Second was the unexpectedly low use of β-blockers (27% for both drug treatment arms and placebo) despite the fact that this was the one category of drug that had already been shown unequivocally to have a favorable effect on outcome in this patient population.[19-21] Third was the unexpected and remarkably low arrhythmic death rate in the CAST placebo group. With regard to the last point, a comparison of the baseline characteristics and mortality results in CAPS and CAST I are most revealing.[22] The total mortality at 1 year and the sudden death mortality at 1 year for the CAPS placebo groups were 9.0% and 4.6%, respectively, but only 3.6% and 1.5%, respectively, for the CAST placebo group. Intuitively, one would have expected that the CAST placebo patients should have had a higher incidence of arrhythmic death than that observed in CAPS because more patients in CAST I had lower LVEFs, and there was also a higher incidence of congestive heart failure. Not only was the incidence of arrhythmic death in the CAST I placebo group remarkably low compared with the CAPS placebo group, but it was also low compared with the published experience at that time for patients surviving an acute myocardial infarction stratified for the extent of ventricular arrhythmia on Holter monitor recordings.[4,5,9,17,19]

The Cardiac Arrhythmia Suppression Trial I Placebo Mortality Data

One potential explanation for the unexpectedly low CAST I arrhythmic death rate in placebo patients is that during the open label ventricular arrhythmia suppression phase of CAST I, which took approximately 2 weeks, and which was not present in CAPS, 62 deaths that were not included in the mortality analysis occurred. Had these deaths occurred after randomization, they likely would have been equally distributed across treatment groups. Although this would have increased the placebo arrhythmic death rate, thereby more closely approximating previously published experience, it is quite unlikely that this would have led to a different conclusion. A second potential partial explanation lies in the subsequent demonstration that in CAST I, suppression of ventricular ectopy by antiarrhythmic drug therapy (N.B.: the study enrolled drug responders) selected a population with a lower risk of arrhythmic death than those who were nonresponders.[23]

A third very important partial explanation of the low placebo mortality rate in CAST I compared to CAPS and other previous studies seems related to advances in medical management of the CAST I patients compared with the CAPS patients. The CAPS patients were randomized between 1983 and 1985. The CAST I patients were randomized from June 1987 through mid-April 1989. During that period, advances in the care of patients with myocardial infarction were such that a much larger percentage of patients received thrombolytic therapy, percutaneous transluminal coronary angioplasty (PTCA), and coronary artery bypass graft surgery (CABG) in CAST I than in CAPS (thrombolytic therapy: 24.3% versus 17%; CABG/PTCA: 54.1% versus 18.0%, respectively). Furthermore, 32% of the CAPS placebo group versus 20.1% of the CAST I placebo group had nonsustained ventricular tachycardia, an important risk stratifier for sudden cardiac death.[3,8,13,16] And finally, none of the patients in the CAST I placebo group whose index ventricular arrhythmia was beyond 90 days of their myocardial infarction died, contrary to the expectation from the "old" Duke University database.

Proarrhythmia

It seemed apparent that the excessive mortality in the encainide and flecainide treatment arms were due to their proarrhythmic effects in this patient population. Moreover, CAST I demonstrated that, contrary to the clinical mindset up to that time, proarrhythmic effects of antiarrhythmic drugs are not just limited to the period soon after initiation of therapy. In CAST I, these adverse effects were present continuously over the length of the study (mean 10 months).[1]

The mechanism of this proarrhythmia is suggested by two studies in canine models of acute ischemia and infarction, one by Nattel et al.[24] using aprindine (a class IC agent, and in that sense, similar to encainide and flecainide), and one by Kou et al.[25] using flecainide. Both studies demonstrated that pretreatment with the respective IC antiarrhythmic agent increased the incidence of ventricular tachycardia/ventricular fibrillation compared with controls. Extrapolating from these studies, a possible explanation of the excessive mortality in the CAST I encainide/flecainide patient groups is that acute ischemia is not good for patients, but in patients pretreated with an IC agent, acute ischemia is worse. And in this context, it appears that the excessive mortality associated with encainide and flecainide in CAST I was in patients with non–q-wave myocardial infarcts (i.e., patients with myocardium still at risk).[26,27] Additional support for this perspective comes from the fact that in CAST, the single most important predictor of a favorable outcome was having had CABG.[28] It follows logically that coronary artery disease and IC agents do not mix.

Importance, Impact, and Lessons of Cardiac Arrhythmia Suppression Trial I

The reverberations of CAST I data were and still are considerable. Encainide was ultimately withdrawn from the market by Bristol-Myers Squibb, in large measure because of the results of CAST I. However, 3M Pharmaceuticals subsequently demonstrated that flecainide was both safe and effective treatment in patients with paroxysmal atrial fibrillation in the absence of structural heart disease, and as a result, flecainide received U. S. Food and Drug Administration (FDA) approval for this indication after the CAST I results were well understood and published. However, it is now widely accepted, and, in fact, a community standard, that IC antiarrhythmic agents are contraindicated for use in patients with ischemic heart disease. Furthermore, the publication of the CAST I results led to the performance of a meta-analysis by Teo et al.[29] that showed that use of class I antiarrhythmic agents in CAST I type populations demonstrated an excessive mortality in the treatment arm compared with the placebo arm ($P = .05$). Thus, at least on the basis of that meta-analysis, not only are the IC agents inappropriate to use in this patient population, but also, all of the class I antiarrhythmic drugs as well. Importantly, these data have been extrapolated to the treatment of atrial fibrillation with class I antiarrhythmic drugs in patients with structural heart disease, particularly with coronary artery disease.[30] In fact, such extrapolation has become the paradigm for antiarrhythmic drugs that were studied in patients with ventricular arrhythmias, but which have efficacy in the treatment of atrial fibrillation therapy.

In sum, many lessons were learned from CAST I: (1) the importance of placebo-controlled trials (had historical controls been used, the data would have suggested neutral results in CAST I patients receiving encainide and flecainide); (2) the inadequate use (only in 27%) of β-blockers in the CAST I patients despite the fact that three trials clearly demonstrated their efficacy in favorably affecting the incidence of sudden cardiac death in this patient population; (3) the real possibility of an adverse outcome when a patient population with low risk is exposed to drugs that have the potential for adverse effects, particularly proarrhythmia (it would have been far better in CAST I to have selected a higher-risk patient population, but clearly that was not appreciated until analysis of the study data was performed); (4) the recognition that protection against arrhythmic death, sudden cardiac death, or mortality in the CAST I patient population is not, per se, conferred by the suppression of putative arrhythmic triggers for ventricular tachyarrhythmia; (5) the recognition that proarrhythmia is not necessarily limited to the early period after initiation of an antiarrhythmic drug; and (6) the recognition that the risk for arrhythmic or sudden cardiac death is very low in patients who are 90 days or beyond post myocardial infarction with an LVEF greater than or equal to 40 despite the presence of CAST I ventricular ectopy.

CARDIAC ARRHYTHMIA SUPPRESSION TRIAL II

Protocol Changes

The results of CAST I had indicated that the ventricular ectopy suppression hypothesis in patients with a recent myocardial infarction and ventricular dysfunction was either wrong or might be drug specific. Despite the results of CAST I, the Data and Safety Monitoring Board recommended continuation of the study, as it was later learned, because the preliminary data suggested moricizine might have favorable effects on outcome.[2] Although consideration was given to adding a drug or drugs to the study, no other drugs met the original criteria for acceptability, and it would have been logistically difficult to add another drug in any event. Therefore, CAST II continued as moricizine versus placebo.

There were other changes in the protocol[2] as a result of the analysis of CAST I. Because no sudden arrhythmic deaths occurred in the patients who were receiving placebos and were enrolled beyond 90 days of myocardial infarction, such patients were no longer considered for enrollment. In addition, the qualifying LVEF was lowered to 0.40 or less. An additional change was that disqualifying nonsustained ventricular tachycardia was redefined to exclude from the trial patients with any runs lasting 30 seconds or more at a rate greater than or equal to 120 bpm. Yet patients with repetitive ventricular beats of 15 or more lasting up to 30 seconds without symptoms were allowed to enroll (such patients were excluded from CAST I). A third and higher dose of moricizine (300 mg every 8 hours) was allowed, and patients formerly treated with encainide, flecainide, or their placebo equivalents could be rerandomized if they met the new entry criteria (of course, except for the criterion of time since the index myocardial infarction). In addition, the titration phase of CAST II was now placebo controlled (recall that in CAST I, the titration phase was open label) to determine if there was a mortality benefit or risk during the first 2 weeks of exposure to the drug.

Analysis of Cardiac Arrhythmia Suppression Trial II

After continuing CAST II for 19½ more months beyond the termination of CAST I, the Data and Safety Monitoring Board recommended early termination of the study for two reasons.[2] First, when examining the data for the 2-week, placebo-controlled, titration-phase trial of the low-dose moricizine (200 mg every 8 hours), there was an increased mortality among patients treated with moricizine compared with patients who received placebo (17 versus 4 deaths or cardiac arrests; $P = <.02$) (Fig. 11-2). Second, because the long-term data comparing moricizine and placebos were a wash ($P = .40$, the end point being death or nonfatal cardiac arrest due to arrhythmia) (Fig. 11-3), it was unlikely that the long-term study had any chance of showing improved survival of patients treated with moricizine compared with placebos. Thus, the combination of early adverse effects of the drug and long-term neutral effects of the drug led to termination of the study.

There were 49 deaths or cardiac arrests due to arrhythmia among the 581 patients in the moricizine group compared with 42 such events among the 574 patients in the placebo group ($P = .40$) (see Fig. 11-3). In the placebo group, 4.8% of the patients died of arrhythmia or had a cardiac arrest due to arrhythmia during the first year. This compared virtually *exactly* with the sudden death mortality at 1 year of 4.6% in the CAPS placebo group,[17] so that the changes in the eligibility criteria for CAST II *did* significantly change the risk to that originally expected. In addition, there were 87 deaths or cardiac arrests in the moricizine group, and 71 in the placebo group, with a 2-year survival rate of 81.7% and 85.6%, respectively. The number of deaths or cardiac arrests from cardiac causes other than arrhythmias was similar in the moricizine and placebo groups (23 versus 20). Interestingly, the substudy of

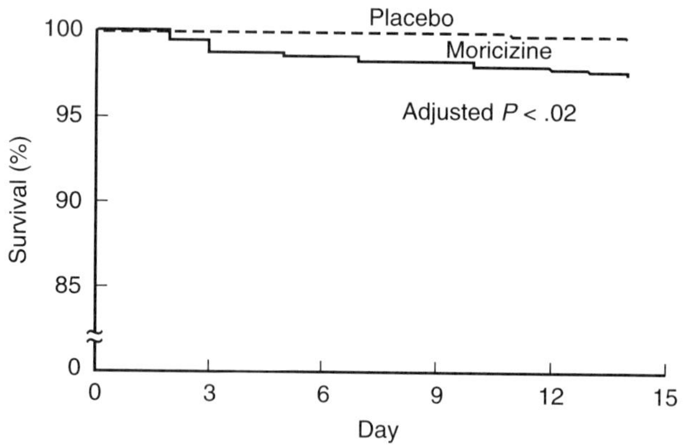

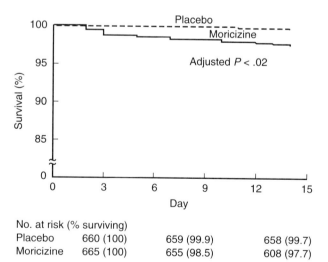

No. at risk (% surviving)			
Placebo	660 (100)	659 (99.9)	658 (99.7)
Moricizine	665 (100)	655 (98.5)	608 (97.7)

FIGURE 11-2 Survival of patients during the first 14 days of treatment with moricizine or placebo. The end point was death or nonfatal cardiac arrest from any cause. The adjusted *P* value is based on the log-rank statistic and adjusted for sequential monitoring. Fifty patients who began immediate titration with moricizine completed titration, and their data were censored before 14 days. (From The Cardiac Arrhythmia Suppression Trial-II (CAST) Investigators: Effect of the antiarrhythmic agent moricizine on survival after myocardial infarction. N Engl J Med 1992;327:227-33.)

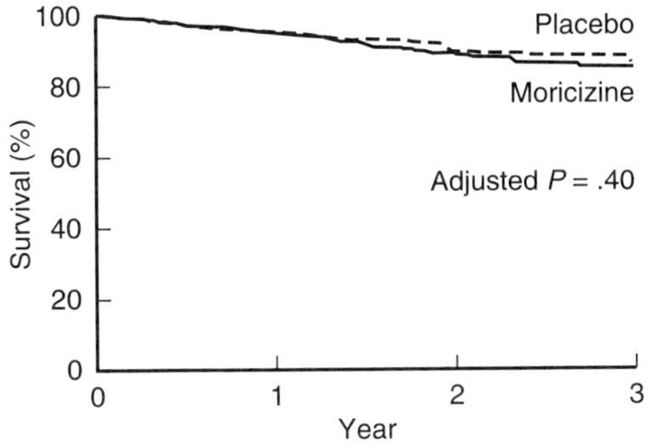

No. at risk (% surviving)

Placebo	574 (100)	351 (95.2)	175 (89.6)	56 (88.0)
Moricizine	581 (100)	350 (94.9)	181 (88.8)	56 (85.0)

FIGURE 11-3 Survival of patients in the long-term main study during treatment with moricizine or placebo after adequate suppression of ventricular premature complexes with moricizine. The end point was death or nonfatal cardiac arrest due to arrhythmias. The adjusted *P* value is two-tailed. (From The Cardiac Arrhythmia Suppression Trial-II (CAST) Investigators: Effect of the antiarrhythmic agent moricizine on survival after myocardial infarction. N Engl J Med 1992;327: 227-33.)

patients with partial suppression of ventricular ectopy included 219 patients, of whom 110 were randomized to receive moricizine, and 109 to receive placebo. There were 17 deaths in the moricizine group (10 due to arrhythmias) compared with 15 deaths in the placebo group (9 due to arrhythmias).

Lessons from Cardiac Arrhythmia Suppression Trial II

These data support the CAST I conclusion that suppression of ventricular ectopy in the CAST patient population either has no favorable effect or adverse effects. In fact, a stronger statement is in order—CAST II put an end to the hypothesis that suppression of ventricular ectopy is a useful treatment in preventing sudden cardiac death.

The data from the placebo-controlled moricizine titration phase demonstrating that moricizine was harmful not only again emphasized the potential risk of giving antiarrhythmic drugs prophylactically to this group of patients, but also supported the now widely established practice of hospitalizing patients with structural heart disease for initiation of antiarrhythmic drug therapy. Whether the risk associated with the initiation of moricizine therapy could have been avoided by hospitalizing these apparently high-risk patients at the beginning of therapy cannot be answered on the basis of CAST II, although 12 of the 17 deaths in the moricizine group occurred during the first week of therapy and 10 of the 17 occurred in patients with an LVEF of less than 0.30.[2] Conversely, 5 of the 17 deaths (29%) occurred after the first week of therapy, indicating an important ongoing risk even if hospitalization had been provided for a week.[2]

Despite the recognition that in CAST I, patients with a clear indication for β-blocker therapy only received it about 27% of the time, this did not change in CAST II. In fact, only 28.1% of patients on moricizine and 29.9% of patients on placebo in the long-term arm of CAST II received a β-blocker, and only 30.3% of patients on moricizine and 27.1% of patients on placebo received a β-blocker during the 2-week, double-blind, placebo-controlled titration phase.[2]

Extrapolation of the CAST II results to treatment of patients with atrial fibrillation is rather tantalizing. Because moricizine is an effective drug in the treatment of atrial fibrillation,[31] because in the long term, it was a wash versus placebo, and because 42% of the patients in CAST II receiving moricizine had an ejection fraction of less than or equal to 0.30,[2] one might say that perhaps moricizine can be used safely to treat CAST II–type patients with atrial fibrillation who are beyond a certain period (at least 90 days, and more conservatively, 6 months) of their myocardial infarction. This is perhaps most tantalizing because the drug seemed safe in the long term follow-up despite a large percentage of the patients having serious left ventricular dysfunction.

The Survival With ORal D-sotalol Trial

BACKGROUND

At the start of the Survival With ORal D-sotalol (SWORD) trial in August 1992, it was recognized that depressed ventricular function,[3,4,8] ventricular ectopic activity,[3-9] signal-averaged late potentials,[32] low heart rate variability,[33] and low baroreflex sensitivity[34] identified patients at highest risk after a recent myocardial infarction. However, when a myocardial infarction was remote, although the risk of sudden cardiac death remained substantial,[35] identification of high-risk patients with ventricular dysfunction was still not well characterized. For these "remote" myocardial infarction patients, left ventricular dysfunction was regarded as the best predictor of total mortality. However, because of the increased use of improved treatment modalities (e.g., CABG, afterload reduction, thrombolysis, angioplasty), the actual incidence of sudden arrhythmic death in this patient population with depressed left ventricular function was uncertain and a moving target. In addition, the actual incidence of sudden arrhythmic death in patients with a "remote" myocardial infarction and left ventricular dysfunction still had not been well characterized.

Also at this time (August 1992), as discussed earlier, it was well established that the class I antiarrhythmic agents were associated with an increased rather than a decreased all-cause mortality and sudden cardiac death rate in patients with coronary artery disease, ventricular dysfunction, and ventricular ectopy despite suppression of frequent and complex ventricular ectopy. However, results with class III antiarrhythmic drugs were expected to be different, although it was recognized that the pure class III agents had an important potential for ventricular proarrhythmia in the form of torsades de pointes ventricular tachycardia (torsades VT). Investigation of new class III agents, therefore, seemed warranted.[36]

D-sotalol, an Ik_r blocker, had been the most widely used of the new class III agents, and was expected to be well tolerated by patients with severe left ventricular dysfunction because it was a mildly positive inotropic agent. Unlike dl-sotalol, d-sotalol had no clinically significant β-blocking activity in humans, appeared to be well tolerated, and had a low incidence of reported adverse effects, the latter including an approximately 1.2% incidence of torsades VT.[36,37] It was in this context that the SWORD trial was launched as a multinational, multicenter, randomized, double-blind, placebo-controlled trial to test the hypothesis that d-sotalol, a drug with pure potassium channel–blocking action, reduced all-cause mortality in patients with a previous myocardial infarction and left ventricular dysfunction.[36,37]

THE PROTOCOL[36,37]

SWORD enrolled eligible patients 18 years of age and older with LVEFs less than or equal to 0.40 determined within 6 months of randomization. Patients also had to have evidence of a prior myocardial infarction. Two thirds of the patients were to be drawn from those with a myocardial infarction 6 to 42 days before randomization with or without overt heart failure, and one third were to be drawn from patients with both a myocardial infarction occurring more than 42 days before randomization and a history of overt heart failure (New York Heart Association [NYHA] functional class II or III).

Major exclusions were designed, particularly because of the potential for torsades VT, and included (1) corrected QT (Q-Tc) interval greater than 460 milliseconds (ms); (2) concomitant antiarrhythmic agents or drugs that prolong the Q–T interval; (3) specific electrolyte abnormalities (serum potassium <4.0 mEq/L or serum magnesium <1.5 mEq/L); (4) renal dysfunction; (5) recent (within 14 days) coronary angioplasty or CABG; (6) unstable angina; (7) a history of life-threatening arrhythmia unrelated to myocardial infarction; (8) nonischemic or severe (NYHA class IV) heart failure; (9) sick sinus syndrome or high-degree AV block without a pacemaker; and (10) liver abnormalities.

Eligible patients were randomly assigned to receive a titrated dose of either d-sotalol or placebo. Doses could not exceed 200 mg twice daily for the duration of the study. Patients were initially assigned oral d-sotalol 100 mg twice daily or matching placebo for 1 week. If this dose was tolerated with a Q-Tc of <520 ms, the dose was increased to 200 mg twice daily, or matching placebo for 1 more week. If tolerated with a Q-Tc of <560 ms, this dose was given for the duration of the study. If the Q-Tc exceeded 560 ms at any time during the follow-up, the dose was reduced. If the Q-Tc was still above this value at the 100 mg twice daily dose, the study drug was discontinued. Concomitant treatment with β-blockers, digoxin, angiotensin-converting enzyme inhibitors, and calcium antagonists was allowed. The study consisted of a screening and a double-blind phase. During the screening phase, a 24-hour ambulatory ECG (Holter) recording identified baseline characteristics, but randomization was not contingent upon the results of the ambulatory ECG. There was, however, a run-in phase in which 500 patients had a Holter monitor performed at 3 months after randomization. The results were only reviewed by the Data and Safety Monitoring Board to screen for the possibility of proarrhythmia due to torsades VT.

The primary efficacy end point of SWORD was all-cause mortality. Cardiac mortality was a secondary end point. Tertiary end points were cardiovascular mortality, presumed arrhythmic death, nonfatal severe arrhythmic events, hospitalizations for cardiovascular causes, and composites of these end points. The goal was to enroll 6400 patients.

RESULTS[37]

By November 1, 1994, after 2.3 years of patient recruitment, and after the SWORD trial had enrolled 3121 patients with a mean follow-up of 148 days, the Data and Safety Monitoring Board recommended termination of the trial. An interim analysis had shown an increased mortality in patients assigned to d-sotalol that crossed the prespecified advisory statistical boundary. The trial originally intended to recruit two patients with a recent myocardial infarction for each one with a remote myocardial infarction. This was because of concerns over potential differences in cause-specific mortality between patients in each group. However, and *critically important,* at the time of the study's termination, this ratio was reversed. There were 915 patients with a recent myocardial infarction and 2206 with a remote myocardial infarction (ratio 1:2.04). The consequences will be seen below.

When the study was terminated, there had been 30 excess deaths in the d-sotalol–assigned patients compared with the placebo-assigned patients (78/1549 [5.0%] versus 48/1572 [3.1%]; relative risk 1.65 [95% CI 1.15 to 2.36]; $P = .006$).[37] Survival curves are shown in Figure 11-4. Significantly greater numbers of cardiac and arrhythmic deaths accounted for this increased mortality. The reported incidence of torsades VT in these patients was low (3/1549 [0.2%]). Exploratory subgroup analyses generally showed a consistency of drug effect on mortality (Table 11-1).[37] In all subgroups of interest, patients assigned to d-sotalol had a higher death rate than patients assigned to placebo. The observed risk with d-sotalol was present regardless of age, sex, time from index myocardial infarction, LVEF, or concomitant therapy. However, the adverse effect associated with d-sotalol was remarkably more pronounced in patients with relatively better ventricular function (LVEF: 0.31 to 0.40) than those with a lower LVEF, and in women rather than in men, although few women were studied.[37] Concomitant therapy with β-blockers, diuretics, or calcium channel blockers did not alter this adverse effect.

ANALYSIS OF THE DATA

One of the obvious possible explanations of the SWORD trial data was that the excess deaths on d-sotalol were due to ventricular proarrhythmia, particularly torsades VT.

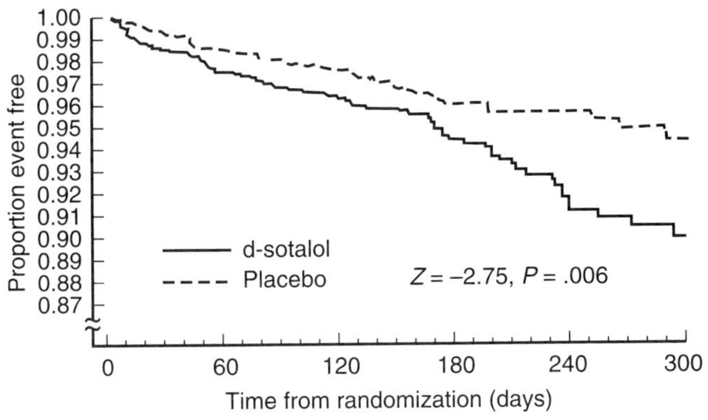

Patients at risk					
Placebo	1572	1170	874	551	330
d-sotalol	1549	1150	844	544	323

FIGURE 11-4 Survival among 3121 patients randomly assigned d-sotalol or placebo. (From Waldo AL, Camm AJ, deRuyter H, et al: Effect of *d*-sotalol on mortality in patients with left ventricular dysfunction after recent and remote myocardial infarction. Lancet 1996;348:7-12.)

However, starting with the initial analysis, this was not at all obvious. First, during the run-in phase, there were only three cases of torsades VT reported, and with the exception of the more pronounced risk of death in women on d-sotalol (N.B.: only 14% of the population was female), other baseline characteristics likely to favor torsades VT were not associated with the increased mortality. However, additional analysis was quite revealing. The two original SWORD myocardial infarction groups, recent (6 to 42 days) and remote (>42 days), were further stratified into those with poor LVEF (0.31 to 0.40) and very poor (LVEF ≤0.30) ventricular function.[38] With these two LVEF groups and the two index myocardial infarction groups becoming four major subgroups, variables of known prognostic importance (age, LVEF, heart failure class),[3-9,13,16] as well as variables known to be related to torsades VT (gender, temporal relationship to initiation of trial medication, Q-Tc bradycardia, serum potassium and magnesium, and diuretic use),[39] were assessed. Treatment-specific Kaplan-Meier survival curves were constructed within the four major subgroups defined by the myocardial infarction

TABLE 11-1 Effect of d-Sotalol on Mortality in Subgroups

Prerandomization characteristic		n	Number of deaths: d-Sotalol	Number of deaths: Placebo	Relative risk (95% CI)
All patients		3121	78	48	
Sex	Male	2684	64	45	
	Female	437	14	3	
Age (years)	<60	1368	23	17	
	≥60	1753	55	31	
Days from MI	6–42	915	23	13	
	>42	2206	55	35	
LVEF (%)	≤30	1361	45	40	
	31–40	1760	33	8	
VPD per h	<6	1325	13	8	
	≥6	1598	55	35	
QTc (ms)	<420	1699	38	22	
	420–460	1334	37	20	
Potassium (mmol/L)	≥4.5	1767	40	29	
	>4.5	1300	37	17	
Creatinine (μmol/L)	≤97.2	1397	31	16	
	>97.2	1670	46	30	
Diuretic use	Yes	1487	55	43	
	No	1560	18	5	
Digoxin use	Yes	817	37	25	
	No	2230	36	23	
β-blocker use	Yes	983	16	10	
	No	2064	57	38	
Calcium-blocker use	Yes	575	15	12	
	No	2472	58	36	
ACE inhibitor use	Yes	2171	55	38	
	No	876	18	10	
Aspirin use	Yes	1980	37	26	
	No	1067	36	22	

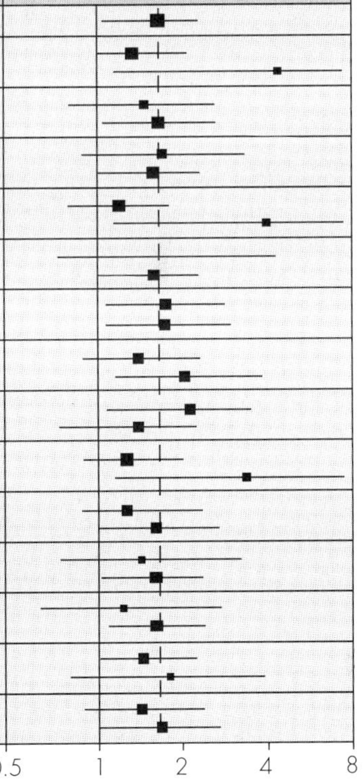

**Squares* indicate the relative risk in that subgroup: the area of the square is proportional to the number of deaths. *Solid vertical line* indicates a finding of no effect. *Dashed vertical line* indicates the effect observed in a full sample. Deviations of subgroup relative risks from overall relative risk were assessed by testing for interaction by proportional hazards regression. $\chi^2 = 7.27$ ($P = .007$) for interaction of the effect of d-sotalol on mortality over subgroups of ejection fraction. Tests for interaction with other characteristics were not significant.

ACE, angiotensin-converting enzyme; MI, myocardial infarction; QTc, corrected QT [interval]; VPD, ventricular premature depolarization.

From: Waldo AL, Camm AJ, deRuyter H, et al: Effects of *d*-sotalol on mortality in patients with left ventricular dysfunction after recent and remote myocardial infarction. Lancet 1996;348:7-12.

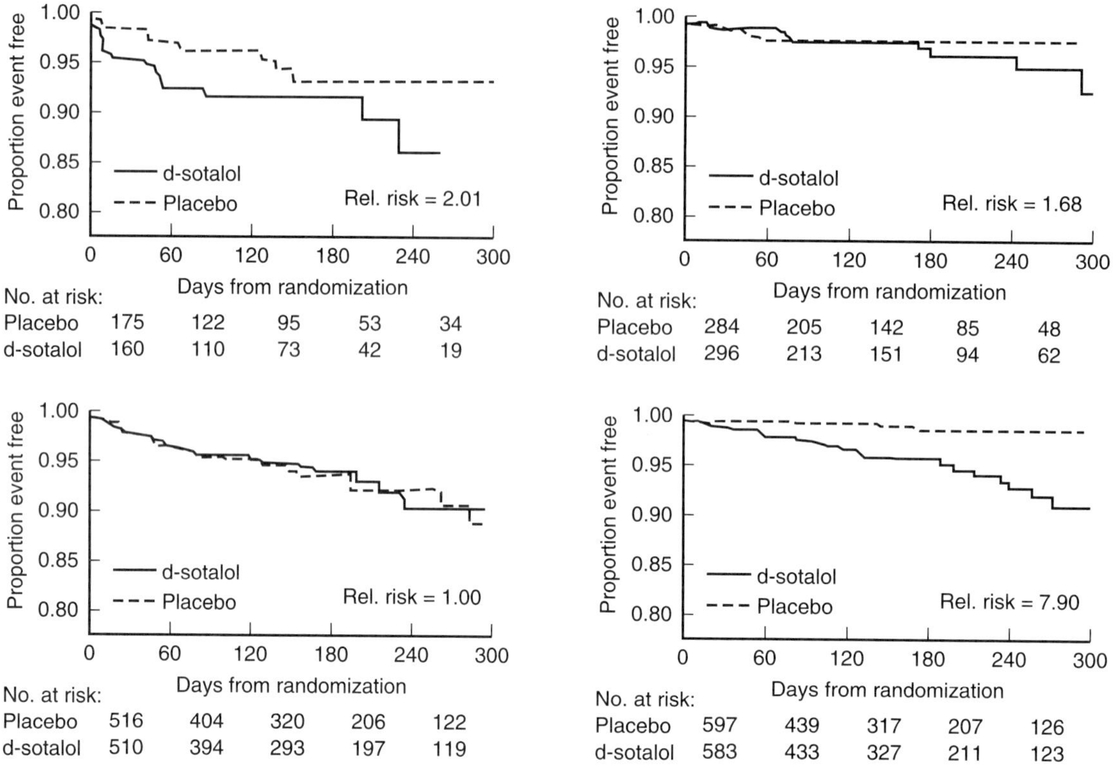

FIGURE 11-5 Survival curves by treatment group for *total mortality*. The four major subgroups are represented. Recent myocardial infarction (MI), left ventricular ejection fraction (LVEF) less than or equal to 30% *(top left)*; recent MI, LVEF 31% to 40% *(top right)*; remote MI, LVEF less than or equal to 30% *(bottom left)*; remote MI, LVEF 31% to 40% *(bottom right)*. The d-sotalol–treated patients in this latter group exhibit a progressive increased mortality risk compared with a nearly event-free, placebo-assigned group (relative risk = 7.9). (From Pratt CM, Camm AJ, Cooper W, et al: Mortality in the Survival With Oral d-Sotalol (SWORD) Trial: Why did patients die? Am J Cardiol 1998;81:869-76.)

group (recent or remote) and the degree of ventricular dysfunction (LVEF ≥0.30 or 0.31 to 0.40) (Fig. 11-5).[38]

Although there were no significant differences in baseline characteristics known to be of prognostic importance between d-sotalol and placebo-assigned patients as previously described,[36] there were many differences in the baseline characteristics of the two distinct, prospectively defined index myocardial infarction groups, recent and remote (Table 11-2).[38] Significant differences were consistent with a greater degree of systolic ventricular dysfunction and symptomatic heart failure in the remote myocardial infarction group. Furthermore, and *critically important*, the d-sotalol–associated total mortality risk and presumed arrhythmia death varied widely (1.0 to 7.9 and 1.2 to 20.7, respectively) among the groups, and was greatest in the remote myocardial infarction group with the LVEF of 0.31 to 0.40.[38] Treatment-specific survival curves for the total mortality in these four subgroups are shown in Figure 11-5. It is striking that in the remote myocardial infarction, LVEF 0.31 to 0.40 group, there was a very low placebo mortality, with only three deaths, only one of which was arrhythmic. It is difficult to see how the d-sotalol could have helped these patients. Moreover, it is clear that these patients were only subjected to potential risk, with essentially no hope of potential benefit. In addition, the mean time from the index myocardial infarction in the remote group was 4.2 years, whereas the mean time in the recent myocardial infarction group was 22.4 days. So, not only was the study skewed in the wrong direction in the 2:1 ratio for enrolling patients with recent compared to remote infarcts, but also there was no cap on the time from the myocardial infarct in the remote group. Because so many patients were enrolled so late after their index myocardial infarction, in essence, they had already demonstrated that they were going to survive.

Considering gender influence in the results, although men and women were at similar risk of death in SWORD, within the d-sotalol group, women tended to be at higher risk than men, whereas in the placebo group, the reverse was true. Relative to placebo, d-sotalol was associated with a 4.7-fold increased risk of death in women and a 1.4-fold increased risk in men.[37,38] In women, the total number of d-sotalol–associated deaths was small (14 of 219 [6.4%]), and only 2 of 14 d-sotalol–associated deaths in women occurred in the first 2 weeks. Most (10 of 14 [71%]) occurred in the recent myocardial infarction group despite the fact that most (133 of 219 [61%]) of the women receiving d-sotalol were in the remote index myocardial infarction group (P = .005). Exploratory analyses of other selected baseline variables failed to reveal any additional significant interactions.[37,38]

TABLE 11–2 Baseline Characteristics by Index Myocardial Infarction

Characteristic	Index MI: Recent (*n* = 915)	Index MI: Remote (*n* = 2206)	*p* Value
Men (%)	83	87	0.007
Caucasian (%)	91	93	0.01
Age (yrs) (mean, SD)	59.5 (10.6)	60.4 (9.7)	0.01
Medical history (%)			
MI, before index MI	29	35	<0.001
Hypertension	37	36	0.51
CABG	9	28	<0.001
PTCA	15	15	0.97
Chronic AF	2	4	<0.001
Weeks from index MI (mean, SD)	3.2 (1.6)	219 (255)	—
LVEF (%)			
≤30%	37	47	
31%–40%	63	54	
Mean (SD)	31.9 (6.4)	30.5 (7.0)	<0.001
NYHA class (%)			
I	25	0	
II	56	78	<0.001
III	19	22	
12-Lead ECG			
Heart rate (mean, SD)	73.9 (13.7)	72.8 (13.6)	0.03
QT (ms) (mean, SD)	385 (43)	386 (40)	0.42
QTc (ms) (mean, SD)	415 (32)	416 (31)	0.44
Abnormal Q waves (%)	70	81	<0.001
LVH (%)	7	16	<0.001
Pacemaker (%)	1	1	0.34
LBBB (%)	4	7	0.01
RBBB (%)	5	6	0.08
24-h ECG			
Number of patients	875	2057	
PVC/h (%)			
None	4	2	
Present (<6)	58	37	
≥6	39	61	
Mean (SD)	35 (101)	65 (152)	<0.001
Paired PVCs	47	67	<0.001
VT runs (%)	27	40	<0.001
Concomitant medications at randomization (%)			
Diuretics	36	54	<0.001
Digoxin	17	31	<0.001
ACE inhibitors	69	72	0.11
Nitrates	52	54	0.43
β-blockers	41	29	<0.001
Calcium blockers	12	22	<0.001
Aspirin	71	62	<0.001

ACE, angiotensin-converting enzyme; AF, atrial fibrillation; CABG, coronary artery bypass graft surgery; F, atrial fibrillation; LBBB, left bundle branch block; LVH, left ventricular hypertrophy; MI, myocardial infarction; NYHA, New York Heart Association; PTCA, percutaneous transluminal coronary angioplasty; PVC, premature ventricular complex; QTc, corrected QT [interval]; RBBB, right bundle branch block; SD, standard deviation; VT, ventricular tachycardia.

From Pratt CM, Camm AJ, Cooper W, et al: Mortality in the Survival With ORal d-Sotalol (SWORD) Trial: Why did patients die? Am J Cardiol 1998;81:869-76.

The subset analyses of SWORD demonstrated remarkable differences in mortality rates among the subgroups (see Fig. 11-5). Of special relevance are the patients remote (mean 4.5 years) from myocardial infarction with an LVEF of 31-40%. Remarkably, the placebo mortality was less than 2%! The recent clinical trials utilizing an implantable cardioverter defibrillator identify patients post-myocardial infarction with cutoffs for LVEF. The Multicenter Automatic Defibrillator Implantation Trial (MADIT II) required an LVEF <30% in post-myocardial infarction patients.[40] The SWORD data support this population as having considerable risk. However, the Multicenter Unsustained Tachycardia Trial (MUSTT) included CAD patients with unsustained VT with an LVEF of up to 40%.[41] The SWORD data do not support a likely benefit of ICD therapy in the MUSTT patients whose LVEFs were above 30% because the mortality was so low.

Primary observations from analysis of the SWORD data included: (1) most of the d-sotalol–associated mortality risk was greatest in patients very remote from their myocardial infarction (mean = 4.1 years) who had an LVEF of 0.31 to 0.40, whereas comparable placebo-treated patients had an extremely low mortality risk; (2) there was a relation between d-sotalol–associated mortality risk and gender in that women with a recent myocardial infarction tended to have a higher d-sotalol–associated mortality risk than men; (3) the d-sotalol–associated mortality was not related to the severity of heart failure, baseline measures of heart rate, or indirect measures of ischemia; (4) and the increased d-sotalol–associated mortality risk was not accentuated in the first weeks of drug administration; rather, it was evenly distributed throughout the entire trial period for both men and women. Based on this detailed analysis, data supporting the hypothesis that d-sotalol–associated mortality in the SWORD trial is due to torsades VT is limited. In fact, the actual data provide virtually no support for any proarrhythmic mechanism.

LESSON FROM THE SWORD TRIAL

The SWORD trial has, once again, demonstrated the critical importance of placebo-controlled, randomized trials. In the SWORD trial, patients receiving placebos who were remote from their myocardial infarction and had an LVEF of 0.31 to 0.40 had an excellent prognosis despite having clinical heart failure. This population of patients receiving d-sotalol had most of the d-sotalol–associated risk seen in the total trial. Clearly, identifying the patient population at risk to be studied must be done carefully. Patients with little need for the potential benefits of a medication should not be treated because this simply and needlessly exposes them to the potential risks of the medication. And patient populations at risk are, not uncommonly, moving targets. We previously saw the latter when comparing CAPS and CAST I. In the case of the SWORD patient population, use of new treatment measures likely improved patient outcome, particularly in the remote myocardial infarction group. In this light, the results of the SWORD trial indicate that the predictive accuracy of classical risk factors such as poor LV function derived from retrospective data collected more than a decade ago in the prethrombolytic, preafterload reduction era must be re-evaluated.

With regard to implications for treatment of atrial fibrillation, because of the results of the SWORD trial and one subsequent trial comparing d-sotalol with amiodarone in the treatment of atrial fibrillation, d-sotalol was withdrawn from the market. Nevertheless, another pure Ik_r blocker, dofetilide, has been FDA approved for treatment of atrial fibrillation in patients based on data from a series of well-planned studies.[42-46] The question to be asked is *what is the difference between d-sotalol and dofetilide?* It seems clear that one of the differences is study design. We may have lost a perfectly good drug in d-sotalol because it was used inappropriately in a population that was inappropriate.

In addition, it seems clear that the patients at biggest risk after myocardial infarction with left ventricular dysfunction are those within the first 6 months or less of their myocardial infarct. And all the continuing, excellent advances in the treatment of acute myocardial infarction and its after effects, including left ventricular dysfunction, clearly will impact morbidity and mortality. Thus, in clinical trials of sudden cardiac death, selection of patients at risk is especially critical, because one could easily select patients who are unlikely to benefit much, if at all, from the study drug(s), subjecting them needlessly to risk and adversity. That seems to have been the case with d-sotalol and SWORD, but not the case with dofetilide and the Danish Investigation of Arrhythmia and Mortality ON Dofetilide (DIAMOND) trials, discussed next.

Danish Investigation of Arrhythmia and Mortality ON Dofetilide Trials

BACKGROUND

Dofetilide, an Ik_r blocker, is FDA approved as a drug for cardioversion and prevention of the recurrence of atrial fibrillation.[44,47] The effect of dofetilide on mortality has also been evaluated, both in populations with previous myocardial infarction and clinical congestive heart failure (CHF). Both trials required that patients have left ventricular systolic dysfunction.[43,45] The DIAMOND trials were designed to assess the potential of dofetilide to reduce mortality. In the United States, convincing mortality data are generally considered necessary to support safety (neutral mortality versus placebo) for registration of new drugs for treatment of atrial fibrillation. For example, sotalol is approved to treat atrial fibrillation, and was tested in a previous mortality trial supporting safety.[48] Multiple mortality trials have examined the mortality effect of amiodarone in patients with myocardial infarction and clinical CHF.[49-51] They are discussed in a subsequent section. The emphasis in this section is on the lessons learned from DIAMOND regarding safe use of dofetilide.

THE PROTOCOL

DIAMOND consisted of two clinical trials with identical designs, but different patient populations.[43,45] The hypothesis was that the treatment with dofetilide would reduce total mortality and morbidity in high-risk patients with left ventricular dysfunction in association with CHF or recent myocardial infarction. Inclusion required a wall motion index on echocardiogram approximating an LVEF less than or equal to 35%, in association with either a recent myocardial infarction (DIAMOND-MI; n = 1510) or clinical CHF (DIAMOND-CHF; n = 1518).[43,45] The design was randomized, parallel, placebo controlled, and double blind. The primary end point of each trial was total mortality.

The population in DIAMOND-MI consisted of patients with an LVEF of less than or equal to 35% who were hospitalized for myocardial infarction within 2 to 7 days of their acute event. The population in the DIAMOND-CHF trial consisted of patients with an LVEF of less than or equal to 35% who were hospitalized for new or worsening CHF within 7 days.

The exclusion criteria for these studies included Q-Tc prolongation (≥460 ms), bradycardia (<50 bpm), high-degree heart block, hypokalemia, or a history of polymorphic ventricular tachycardia. Other more general exclusion criteria included underlying conditions likely to lead to death within the course of the study and those with significant liver dysfunction or severely impaired renal dysfunction. Concomitant therapy with class I or III antiarrhythmic drugs was prohibited, as was the use of other drugs known to be associated with the genesis of torsades VT. Patients with an implantable cardioverter defibrillator and those with CHF likely to require cardiac transplantation were also excluded.

Patients received dofetilide or placebo at a dose of 500 micrograms (μg) twice a day. Because dofetilide is primarily eliminated via renal excretion, the dose of dofetilide was adjusted downward for those patients with an abnormal creatinine clearance. Creatinine clearance was calculated using the Cockcroft-Gault formula.[52] Those with a creatinine clearance of greater than or equal to 60 mL/min required no dosage adjustment. Dosage was adjusted downward incrementally for lower clearances. Patients with a creatinine clearance less than or equal to 20 mL/min were excluded. There was one exception to this dosing regimen. Patients enrolled in the DIAMOND trial who had atrial fibrillation at baseline were treated with dofetilide, 250 μg twice a day (500 μg twice a day was not an option) or placebo. As with the other patients, the dose could be adjusted downward if the calculated creatinine clearance was sufficiently reduced.

As with all dofetilide protocols, dosing was mandated to begin in the hospital with 3 days of telemetry monitoring. The purpose of this inpatient surveillance was threefold: (1) to monitor for arrhythmias, (2) to alter drug dose for reduced creatinine clearance, and (3) to monitor Q–T interval prolongation. In addition to excluding patients with a Q–Tc interval greater than or equal to 460 ms (500 ms with bundle branch block) at baseline, the study drug was reduced if the Q–T or Q–Tc interval prolonged greater than or equal to 20% of baseline or exceeded 500 ms but was discontinued if the Q–T or Q–Tc interval became greater than or equal to 550 ms during the trial. The development of polymorphic ventricular tachycardia resembling torsades VT was also an indication for discontinuation of study drug.

The efficacy end point of the two DIAMOND trials was total mortality. Arrhythmic mortality was an important secondary end point, as were total hospitalizations and hospitalization for CHF. In addition, it was prespecified to evaluate all patients who had atrial fibrillation at baseline in both the DIAMOND trials. These analyses of atrial fibrillation support the efficacy and safety of dofetilide for use in treatment of atrial fibrillation, and are discussed later in "Atrial Fibrillation and Congestive Heart Failure: Lessons from the Amiodarone and Dofetilide Trials."

RESULTS

A total of 3028 patients were randomized in DIAMOND, and 1511 were assigned to dofetilide.[43,45] Table 11-3 summarizes the baseline demographic characteristics.[43,45,53] Compared with the DIAMOND-MI population, the DIAMOND-CHF population was slightly older, with a greater preponderance of NYHA class III CHF. The DIAMOND-MI population had a higher β-blocker use, while the DIAMOND-CHF population had a larger proportion of patients on ACE inhibitors and calcium channel blockers. Mortality risk was evenly distributed between placebo and dofetilide in each trial.

The mortality results of the two trials are presented in Table 11-4. The study populations were both high risk (placebo mortality of 31% in DIAMOND-MI and 42% in DIAMOND-CHF).[43,45] The higher mortality in the population with both low LVEF and clinical CHF is not surprising. There was no mortality difference between the dofetilide-assigned and placebo-assigned patients. In both trials, an independent, blinded committee classified the cause of death. Approximately 60% of all deaths were classified as arrhythmic. There was no significant difference in arrhythmic mortality between dofetilide-assigned and placebo-assigned patients.

Additional subgroup analyses were carried out for clinically relevant populations. There was no difference in treatment effect for the following variables: NYHA class, gender, age, previous myocardial infarction, concomitant use of β-blockers, ACE inhibitors, calcium channel blockers, or the presence/absence of atrial fibrillation or atrial flutter at baseline.

TABLE 11-4 Mortality in the DIAMOND Trials

	DIAMOND MI*		DIAMOND CHF†	
	Dofetilide	*Placebo*	*Dofetilide*	*Placebo*
Total Population	749	761	762	756
Deaths	230	243	311	317

*$P = .226$

†$P = .557$

CHF, congestive heart failure; DIAMOND, Danish Investigation of Arrhythmia and Mortality on Dofetilide; MI, myocardial infarction.

ANALYSIS OF THE DATA: FOCUS ON HOSPITALIZATION

Since the primary mortality end point was neutral, other secondary end points and subset analyses should be considered exploratory in nature. However, there is fascinating information on hospitalization rates from the DIAMOND trials. The preponderance of these data come from the DIAMOND CHF population. Such patients would be expected to be hospitalized more

TABLE 11-3 Demographic Characteristics of Patients in DIAMOND CHF and MI Studies

	DIAMOND CHF		DIAMOND MI	
Characteristics	*Dofetilide n = 762*	*Placebo n = 756*	*Dofetilide n = 749*	*Placebo n = 761*
Median duration of heart failure—*n*	12	12	N/A	N/A
Day of randomization following MI (day [range])	N/A	N/A	3 (1-11)	3 (0-13)
Mean age—years (range)	70 (26-94)	70 (32- 92)	68 (34-89)	69 (33-92)
Male—*n* (%)	546 (72)	568 (75)	542 (72)	569 (75)
History—*n* (%)				
MI	389 (51)	390 (52)	269 (36)	278 (37)
Ischemic heart disease	509 (67)	508 (67)	749 (100)	761 (100)
Diabetes	152 (20)	140 (19)	97 (13)	96 (13)
Hypertension	111 (15)	115 (15)	120 (16)	130 (17)
AF/AFl at randomization—*n* (%)	190 (25)	201 (27)	59 (8)	56 (7)
Median wall-motion index (range)	0.9 (0.3-1.2)	0.9 (0.3-1.2)	1.1 (0.4-1.4)	1.1 (0.4-1.9)
Treatment at randomization—*n* (%)				
β-blocker	72 (9)	80 (11)	266 (36)	282 (37)
ACE inhibitor	552 (72)	541 (76)	445 (59)	431 (57)
Calcium channel blocker	153 (20)	170 (23)	129 (17)	130 (17)
NYHA function class—*n* (%)				
I	16 (2)	17 (2)	80 (11)	82 (11)
II	268 (35)	297 (40)	383 (54)	390 (53)
III	423 (56)	385 (51)	215 (30)	233 (32)
IV	49 (7)	52 (7)	34 (5)	27 (4)
Not available	6 (<1)	5 (<1)	37 (5)	29 (4)

ACE, angiotensin-converting enzyme; AF, atrial fibrillation; AFl, atrial flutter; CHF, congestive heart failure; DIAMOND, Danish Investigation of Arrhythmia and Mortality on Dofetilide; MI, myocardial infarction; N/A, not applicable; NYHA, New York Heart Association.

often for worsening clinical heart failure. In fact, over the course of the study, there were 229/762 (30%) dofetilide-assigned patients as compared with 290/756 (38%) placebo-assigned patients hospitalized at some time during the trial with worsening CHF.[43,45,53] When the end point of prolongation of the time-to-first event of hospitalization for worsening heart failure is considered, dofetilide-assigned patients had a better outcome than placebo-assigned patients (P = <.001, relative risk 0.75, 95% CI = 0.63 to 0.89). This observation was not explained by differences in the risk of hospitalization for heart failure at baseline or concomitant therapy during the trial. During the trial, there was no significant change in NYHA class between dofetilide-assigned and placebo-assigned patients, although there was a trend in favor of dofetilide-assigned patients.

LESSONS FROM THE DIAMOND TRIALS

We focus on three major lessons from the DIAMOND trials. First, dofetilide is safe when used according to the inpatient algorithm of telemetry monitoring with dosage adjustments for both renal function and Q-Tc changes during therapy. Second, when used appropriately, dofetilide is safe even in high-risk subsets (female patients, as well as patients with recent myocardial infarction, left ventricular systolic dysfunction, and compensated CHF). Third, DIAMOND provides randomized, placebo-controlled observations of 506 patients with atrial fibrillation at baseline in a patient population with severe structural heart disease. This third issue is the focus of "Atrial Fibrillation and Congestive Heart Failure: Lessons from the Amiodarone and Dofetilide Trials." The focus here is on the treatment algorithm for dofetilide, allowing its safe use. The dofetilide development program required inpatient initiation and 3 days of telemetry ECG monitoring with dosage adjustment based on creatinine clearance and QTc change. What was the result of this conservative approach?

First, consider that there was no excessive mortality in dofetilide-assigned patients compared with placebo-assigned patients in the DIAMOND trials. In fact, there were 20 fewer deaths in dofetilide-assigned patients than in placebo-assigned patients, despite 32 cases of torsades VT. Table 11-5 summarizes the torsades VT cases. Dofetilide use in patients with CHF was associated with a substantially higher torsades VT risk than in the post–myocardial infarction population. Only two patients died of torsades VT. The vast majority (25/32) of torsades VT events occurred in the initial 3-day inpatient period.[43,45]

TABLE 11-5 Torsades de Pointes Ventricular Tachycardia in DIAMOND

	DIAMOND CHF	DIAMOND MI
No. patients	762	749
No. (%) with torsades VT	25 (3.3%)	7 (0.9%)
Required intervention	16/25	7/7
No. died of torsades VT	2	0

CHF, congestive heart failure; DIAMOND, Danish Investigation of Arrhythmia and Mortality on Dofetilide; MI, myocardial infarction; torsades VT, torsades de pointes ventricular tachycardia.

The data contained in Table 11-6 provide an explanation for the value of the treatment algorithm. This table lists the frequency of adjustment of dofetilide dosage using the inpatient treatment algorithm in both trials. Dofetilide was either discontinued or the dosage adjusted based on the measurements of creatinine clearance as well as QTc changes.[43,45,53] A large proportion (≥40%) of DIAMOND patients had dosage reduction based on the calculated creatinine clearance.[52] An additional 82 dofetilide-assigned patients, but only 4 placebo-assigned patients, had dosage reductions due to QT prolongation during the 3-day inpatient titration. Overall, nearly 50% of dofetilide-assigned (versus 42% of placebo-assigned) patients had the dofetilide dosage reduced or discontinued during therapy. Presumably, many patients at risk for torsades VT were identified and had dofetilide dosage reduced, thereby avoiding a clinical event. This provides an explanation for the observed lack of an adverse mortality outcome with dofetilide, supporting the use of inpatient drug initiation. An outpatient setting probably would *not* guarantee the same safety margin compared with this conservative inpatient algorithm.

TABLE 11-6 Dofetilide Dosage Adjustment Using Inpatient Treatment Algorithm

	DIAMOND CHF		DIAMOND MI	
	Dofetilide No. (%) Patients	*Placebo No. (%) Patients*	*Dofetilide No. (%) Patients*	*Placebo No. (%) Patients*
Total	762	756	749	761
Discontinuation				
Torsades VT	15 (2.0)	0	4 (0.5)	0
QTc prolongation	2 (0.3)	0	2 (0.3)	0
Total patients discontinued	17 (2.3)	0	6 (0.8)	0
Dose adjustment				
Creatinine clearance	342 (44.9)	314 (41.5)	312 (41.7)	323 (42.4)
QTc prolongation	32 (4.2)	2 (0.3)	46 (6.1)	2 (0.3)
Total patients dose adjusted	374 (49.1)	316 (41.8)	358 (47.6)	325 (42.7)

CHF, congestive heart failure; DIAMOND, Danish Investigation of Arrhythmia and Mortality on Dofetilide; MI, myocardial infarction; QTc, corrected QT [interval]; torsades VT, torsades de pointes ventricular tachycardia.

Clinical Trials of Amiodarone

BACKGROUND

Amiodarone is a commonly used antiarrhythmic drug for both ventricular as well as supraventricular arrhythmias. In the United States, amiodarone is FDA approved solely for the treatment of life-threatening ventricular tachycardia, even though it is used frequently to treat atrial fibrillation.[54-56] More than 3300 patients were enrolled in three placebo-controlled clinical trials evaluating the efficacy of amiodarone in patients surviving myocardial infarction or in patients with clinical CHF, or both. These include the Survival Trial of Antiarrhythmic Therapy in Congestive Heart Failure (CHF-STAT) trial,[49] the European Myocardial Infarction Amiodarone Trial (EMIAT),[50] and the Canadian Amiodarone Myocardial Infarction Arrhythmia Trial (CAMIAT).[51]

PROTOCOLS

A brief review of essential design features of the three amiodarone clinical trials are presented in Table 11-7. The CHF-STAT trial consisted of patients with a low ejection fraction (≤40%), clinical CHF, and ventricular arrhythmia. EMIAT and CAMIAT both featured patients surviving myocardial infarction. The latter two differ in that EMIAT required a depressed LVEF, whereas the CAMIAT required the presence of ventricular arrhythmia on an ambulatory electrocardiographic (Holter monitor) recording (see Table 11-7). Both the CHF-STAT and EMIAT had patients with low LVEFs, an important population for establishing the range of safety of amiodarone. While the primary end point of both CHF-STAT and EMIAT was total mortality, the primary end point of CAMIAT was arrhythmic death and resuscitated ventricular fibrillation. Exclusion criteria in the three trials were quite standard and were enumerated in previous publications.[49-51] Baseline comparison of risk in each trial for amiodarone and placebo-assigned patients did not differ.

RESULTS

The mortality results are presented in Table 11-8. For each of the three clinical trials, there was no difference in total mortality between placebo and amiodarone. However, sudden death mortality was marginally reduced in amiodarone-assigned compared with placebo-assigned patients in EMIAT and CAMIAT. What is interesting about these mortality figures is that in each of the three clinical trials, there was at least a trend to a reduction in sudden cardiac death, whereas total mortality was comparable to placebo. This implies either a problem with the death classification system or the possibility that amiodarone was related to an increase in noncardiac deaths, or both.[49-51]

A review of the toxicities of amiodarone in the three placebo-controlled clinical trials is very instructive. Perhaps one of the most interesting observations is that there were no reported cases of torsades VT in any of the clinical trials. The lack of amiodarone-associated proarrhythmia is a positive aspect, but substantial evidence exists from these trials to suggest the present of severe amiodarone-associated noncardiac toxicity. Table 11-9 presents the incidence of severe noncardiac toxicities, which usually required withdrawal from the study. In four categories consisting of pulmonary fibrosis, thyroid function abnormalities, neurologic manifestations such as ataxia, and cutaneous manifestations, amiodarone-assigned patients had a notably higher incidence of withdrawals than placebo-assigned patients.[47-49]

ANALYSIS OF DATA

While each previously mentioned toxicity is concerning, the data supporting the cardiac safety profile of amiodarone are quite strong. Consider that outpatient initiation of amiodarone was frequent in these trials, yet no torsades VT cases were seen in the three amiodarone trials. Amiodarone safety, represented by total mortality, was neutral in all trials. For comparison, there were 20 fewer deaths in dofetilide-assigned than placebo-assigned patients in the DIAMOND trials.

TABLE 11-7 Selected Design Aspects of Amiodarone Trials

	CHF-STAT	EMIAT	CAMIAT
No. patients	674	1486	1202
Primary inclusion criteria	LVEF ≤40% ≥10 VPCs/hr Clinical CHF	LVEF ≤40% Acute MI (5-21 days)	Acute MI (6-45 days) and ≥10 VPCs/hr or NSVT
Primary end point	Total mortality	Total mortality	Arrhythmic death plus resuscitated VF
Duration of follow-up	45 mos (median)	21 mos (median)	1.79 yr (mean)
Mean LVEF	≈2/3 LVEF ≤30%	25%	N/A
Amiodarone dose regimen	800 mg × 14 days; 400 mg × 50 weeks; then 300 mg/day until completion	800 mg × 14 days; 400 mg × 14 weeks; 200 mg until completion	10 mg/kg × 2 wk; 400 mg/day; (300 mg >75 yo, <60 yr) Further reduced to 200 mg with VPC suppression

CAMIAT, Canadian Amiodarone Myocardial Infarction Arrhythmia Trial; CHF, congestive heart failure; CHF-STAT, Survival Trial of Antiarrhythmic Therapy in Congestive Heart Failure; EMIAT, European Myocardial Infarction Amiodarone Trial; LVEF, left ventricular ejection fraction; MI, myocardial infarction; N/A, not available; NSVT, nonsustained ventricular tachycardia; VF, ventricular fibrillation; VPC, ventricular premature complex.

TABLE 11-8 Mortality Results of Amiodarone Trials

	CHF-STAT		EMIAT		CAMIAT	
	Placebo	*Amiodarone*	*Placebo*	*Amiodarone*	*Placebo*	*Amiodarone*
Total mortality	143	131	(102)	103	68	57
Sudden death	75	64	50	33*	31	15†

*P = .05

†P = .02

CAMIAT, Canadian Amiodarone Myocardial Infarction Arrhythmia Trial; CHF-STAT, Survival Trial of Antiarrhythmic Therapy in Congestive Heart Failure; EMIAT, European Myocardial Infarction Amiodarone Trial.

An almost identical result was seen in the three combined amiodarone trials (22 fewer amiodarone-associated than placebo-associated deaths). The latter was achieved without a mandated rigorous inpatient initiation algorithm.

By combining the three trials, there is placebo-controlled safety information in more than 3300 patients, from which we can accurately estimate the placebo-subtracted incidence of severe noncardiac toxicities involving amiodarone. Of the 1685 patients assigned to amiodarone, the incidence of severe pulmonary, thyroid, neurologic, and cutaneous problems in amiodarone-assigned patients is contained in Table 11-10. The placebo-subtracted incidence of severe amiodarone-associated reactions ranges from 0.7% to 3%.

LESSONS FROM THE AMIODARONE TRIALS

Many patients in these three clinical trials of amiodarone had outpatient initiation of drug. The total lack of cases of torsades VT is reassuring when considering outpatient initiation of amiodarone for atrial fibrillation. In addition, all three clinical trials show no increased mortality risk compared with placebo. Two of the three clinical trials show a trend or statistical difference in the rate of arrhythmic death favoring amiodarone-assigned patients. At a minimum, one can conclude that amiodarone appears safe for use in severe structural heart disease in patients with atrial fibrillation. *It should be emphasized that treatment of atrial fibrillation is not an FDA-approved indication for amiodarone.*

Atrial Fibrillation and Congestive Heart Failure: Lessons from the Amiodarone and Dofetilide Trials

BACKGROUND

This discussion focuses on patients who had atrial fibrillation at baseline in the DIAMOND and CHF-STAT trials. The focus is on the use of dofetilide and amiodarone for treatment of atrial fibrillation in the setting of severe left ventricular systolic dysfunction. This is an important patient subgroup, since atrial fibrillation substantially contributes to the morbidity and mortality of clinical heart failure.[57]

The DIAMOND and CHF-STAT trials provide us with the first large cohorts of patients with heart failure and atrial fibrillation (DIAMOND—506 patients, CHF-STAT—103 patients). These studies provide placebo-controlled information, allowing an accurate reflection of the efficacy of dofetilide and amiodarone. There are disadvantages and limitations to these analyses. First, it is recognized that in both trials these are subset analyses of neutral mortality trials. The data presented represent a selected population (patients with

TABLE 11-9 Selected Aspects of Amiodarone Noncardiac Toxicity

	CHF-STAT* (*n* = 674)		EMIAT† (*n* = 1486)		CAMIAT† (*n* = 1202)	
	Placebo	*Amiodarone*	*Placebo*	*Amiodarone*	*Placebo*	*Amiodarone*
Pulmonary fibrosis/toxicity	4	10	3	6	7	23
Thyroid function abnormalities	2	4	12	44	5	24
Neurologic/ataxia	6	16	1	4	5	19
Cutaneous reactions	2	1	1	9	8	12

*Classified as severe reactions.

†Reasons for discontinuation.

CAMIAT, Canadian Amiodarone Myocardial Infarction Arrhythmia Trial; CHF-STAT, Survival Trial of Antiarrhythmic Therapy in Congestive Heart Failure; EMIAT, European Myocardial Infarction Amiodarone Trial.

TABLE 11-10 Incidence and Placebo-Subtracted Incidence of Severe Noncardiac Toxicities Associated with Amiodarone Therapy (n = 1685 patients treated)

Disease Category	Incidence*	Placebo-Subtracted Incidence†
Pulmonary	39/1685 = 2.3%	25/1685 = 1.5%
Thyroid	72/1685 = 4.3%	51/1685 = 3.0%
Neurologic	39/1685 = 2.3%	27/1685 = 1.6%
Cutaneous	27/1685 = 1.6%	12/1685 = 0.7%

Includes only "severe" cases or drug discontinuations.

*Represents all amiodarone-assigned patients in CHF-STAT, EMIAT, and CAMIAT.

†Represents all amiodarone-assigned patients in CHF-STAT, EMIAT, and CAMIAT minus the placebo event rate.

CAMIAT, Canadian Amiodarone Myocardial Infarction Arrhythmia Trial; CHF-STAT, Survival Trial of Antiarrhythmic Therapy in Congestive Heart Failure; EMIAT, European Myocardial Infarction Amiodarone Trial.

atrial fibrillation) from within a randomized, placebo-controlled population. Since this is a subset analysis, there is less reliable information as to the equality of risk of the populations at baseline or during the trial.[58] There is also limited information available regarding cotherapies at baseline or during the trials. However, these two trials represent the largest antiarrhythmic experiences in patients with atrial fibrillation and CHF.

RESULTS

The distribution of atrial fibrillation patients (n = 506) in the DIAMOND trials is contained in Figure 11-6. A total of 506 patients had atrial fibrillation at baseline in the DIAMOND trials.[46] More than three times as many patients in DIAMOND-CHF had atrial fibrillation at baseline than in DIAMOND-MI. Thus, the association of atrial fibrillation with CHF is strong, since both trials required an LVEF of less than or equal to 35% for enrollment, but they were differentiated by the presence of clinical CHF in the DIAMOND-CHF trial.

TABLE 11-11 Baseline Characteristics: DIAMOND AF Population (n = 506)

	Dofetilide (n = 249)	Placebo (n = 257)
Mean age (yr)	72	72
Gender (% male)	76/24	78/22
LVEF (<25%)	21%	24%
NYHA		
I	44%	40%
II	52%	49%
III	7%	7%
Medication use %		
Digoxin	92%	89%
β-Blockers	12%	12%
ACE inhibitors	67%	63%

ACE, angiotensin-converting enzyme; AF, atrial fibrillation; DIAMOND, Danish Investigation of Arrhythmia and Mortality on Dofetilide; LVEF, left ventricular ejection fraction; NYHA, New York Heart Association.

Although this is not a separately randomized cohort, the 506 patients with atrial fibrillation at baseline are presented for their baseline characteristics (Table 11-11).[46] This population was characterized by a very low LVEF, consisting of patients with class II and class III NYHA heart failure. The majority of patients were on digoxin, and two thirds of the patients were on an ACE inhibitor. Of note is the fact that very few of these patients were on a β-blocker.

The distribution of patients with atrial fibrillation in the CHF-STAT population (n = 103) is highlighted in Figure 11-7.[58] Overall, 15.4% of the population had atrial fibrillation at baseline, quite close to the overall incidence of atrial fibrillation in the DIAMOND studies (16.7%). These 103 patients were equally distributed between amiodarone-assigned and placebo-assigned patients. The baseline characteristics are listed in Table 11-12. As compared with the dofetilide patients, amiodarone patients were slightly younger with a fairly comparable degree of left ventricular systolic dysfunction.

ANALYSIS OF DATA

Patients with atrial fibrillation at baseline in the DIAMOND-AF population are summarized in Table 11-13.

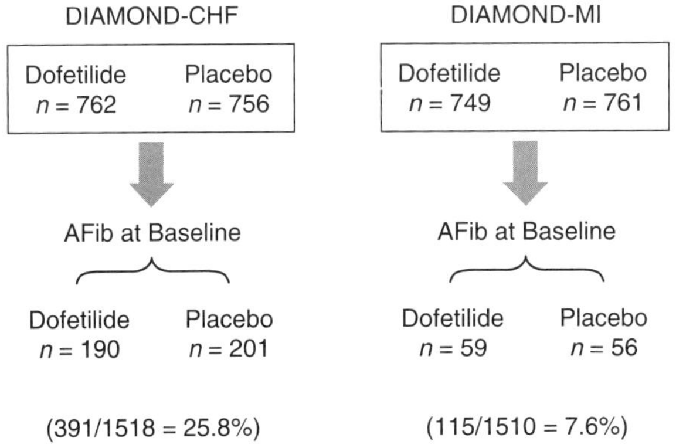

FIGURE 11-6 Distribution of patients with atrial fibrillation (*AFib*) in the Danish Investigation of Arrhythmia and Mortality on Dofetilide (DIAMOND) trials.

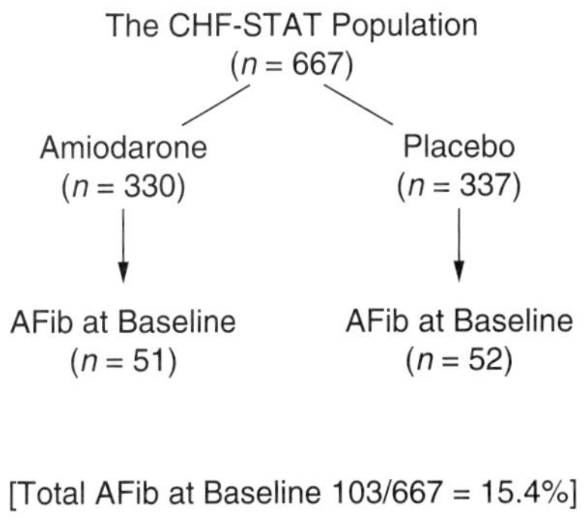

FIGURE 11-7 Distribution of patients with atrial fibrillation in the Survival Trial of Antiarrhythmic Therapy in Congestive Heart Failure (CHF-STAT) study.

TABLE 11-12 Baseline Clinical Variables in Patients with Atrial Fibrillation Randomized to Amiodarone versus Placebo

	Amiodarone (*n* = 51)	Placebo (*n* = 52)
Mean age, yr	67 ± 7	67 ± 8
Hypertension %	61	57
Diabetes %	27	33
Active tobacco use, %	18	23
Current alcohol use, %	27	27
Prior MI, %	45	50
Cause of CHF, %		
Ischemic	61	58
Nonischemic	39	42
Mean LVEF, %	25 ± 8.4	26 ± 8.1

CHF, congestive heart failure; LVEF, left ventricular ejection fraction; MI, myocardial infarction.

Used with permission from: Deedwania PC, Singh BN, Ellenbogen K, et al: Spontaneous conversion and maintenance of sinus rhythm by amiodarone in patients with heart failure and atrial fibrillation. Circulation 1998;98:2574-9.

In a total of 148 dofetilide-assigned patients and 86 in placebo-assigned patients, atrial fibrillation converted to sinus rhythm at some point during the trial. These conversions are a combination of electrical cardioversion and spontaneous cardioversion. A total of 55 placebo-assigned patients developed new onset atrial fibrillation as compared with 21 dofetilide-assigned patients (P = <.01). These data support the fact that dofetilide was effective in both the conversion of atrial fibrillation to sinus rhythm as well as preventing the recurrence of atrial fibrillation during the trial.[46,53]

The most impressive feature of the DIAMOND-AF population is that there was a 31% reduction (P = .014) in both total hospitalizations and hospitalizations for clinical CHF in dofetilide-assigned as compared with placebo-assigned patients. Importantly, mortality was nearly identical in the dofetilide-assigned and placebo-assigned atrial fibrillation cohorts (hazard ratio 1.03; 95% CI, 0.79 to 1.35).

Analysis of data from the CHF-STAT atrial fibrillation population is provided in Table 11-14. The percentage of patients in whom atrial fibrillation converted to sinus rhythm in the placebo-assigned group is markedly lower (8%) than the amiodarone-assigned (31%) group (P = .002). Because of the relatively smaller number of patients, evaluation of hospitalization data was not feasible. A total of 11 of the 206 amiodarone-assigned patients who were in sinus rhythm at baseline compared with 22 of the 263 placebo-assigned patients who were in sinus rhythm developed atrial fibrillation (P = .005).[59]

In summary, both amiodarone and dofetilide exhibit modest efficacy in maintaining sinus rhythm in patients with atrial fibrillation. They also prevent the new development of atrial fibrillation in a population with significant systolic dysfunction and clinical CHF. Unique data from DIAMOND suggest that dofetilide therapy in such a patient population may result in decreased hospitalization.

Table 11-15 summaries the results. Atrial fibrillation was a common concomitant of clinical CHF in both trials, as high as 25% of patients in the DIAMOND-CHF trial. The percentage of patients in whom sinus rhythm was achieved was dramatically higher in the DIAMOND trial than the CHF-STAT trial. This is partly explained by differences in the use of electrical cardioversion. The populations may also be quite different, given the comparative placebo efficacy rates (see Table 11-15).

As opposed to the uncorrected efficacy numbers, the placebo-subtracted efficacy of the two drugs is very similar, both showing modest ability to maintain sinus rhythm (placebo-subtracted efficacy of 25% and 23% for dofetilide and amiodarone, respectively; see Table 11-15). An enduring lesson is the importance of requiring a placebo group to permit an accurate comparison of relative efficacy. Direct comparison of the efficacy of these two drugs is inappropriate, since the patient populations and study designs are not comparable, and this was a subgroup analysis.[58]

TABLE 11-13 DIAMOND AF Population (Converted to Sinus Rhythm)

	Dofetilide	Placebo
Total no. patients in AF at baseline	249	257
No. converted to normal sinus rhythm	148	86
No. (%) hospitalized for CHF*	73 (29%)	102 (40%)

*Hazard ratio = 0.69; 95%; CI = 0.51-0.93; P = .02.

AF, atrial fibrillation; CHF, congestive heart failure; DIAMOND, Danish Investigation of Arrhythmia and Mortality on Dofetilide.

TABLE 11-14 CHF-STAT AF Population (Converted to Sinus Rhythm)

	Amiodarone	Placebo
Total no. patients in AF at baseline	51	52
No. converted to NSR	16	4*
Percent converted to NSR	31%	8%

*P = .002

AF, atrial fibrillation; CHF-STAT, Survival Trial of Antiarrhythmic Therapy in Congestive Heart Failure; NSR, normal sinus rhythm.

TABLE 11-15 Achievement of Sinus Rhythm During the Trials: Patients with Atrial Fibrillation at Baseline

	DIAMOND Trials (Dofetilide)	CHF-STAT Trial (Amiodarone)
Total no. patients with AF	506	103
Percent of patients with AF	16.7%	15.4%
Efficacy		
Active	59%	31%
Placebo	34%	8%
Placebo-subtracted efficacy	25%	23%

AF, atrial fibrillation; CHF-STAT, Survival Trial of Antiarrhythmic Therapy in Congestive Heart Failure; DIAMOND, Danish Investigation of Arrhythmia and Mortality on Dofetilide.

LESSONS RELATIVE TO ATRIAL FIBRILLATION

The observations of an improvement in CHF hospitalization on dofetilide are promising, and presumably are related to improved therapy of atrial fibrillation. Hopefully, future trials of antiarrhythmics will target atrial fibrillation and low LVEF to evaluate prospectively the ability of antiarrhythmic drug therapy to improve heart failure outcome. For now, the possibility that dofetilide reduces this clinically important morbidity remains an intriguing and important hypothesis. Amiodarone and dofetilide are the only two antiarrhythmic drugs with substantial data regarding maintenance of sinus rhythm in such high risk patients.

ACKNOWLEDGEMENTS

This work was supported in part by Grant RO1 HL38408 from the U.S. Public Health Service, National Institutes of Health, National Heart, Lung, and Blood Institute, Bethesda, Md.

REFERENCES

1. The Cardiac Arrhythmia Suppression Trial (CAST) Investigators: Effect of encainide and flecainide on mortality in a randomized trial of arrhythmia suppression after myocardial infarction. N Engl J Med 1989;321:406-12.
2. The Cardiac Arrhythmia Suppression Trial-II (CAST) Investigators: Effect of the antiarrhythmic agent moricizine on survival after myocardial infarction. N Engl J Med 1992;327:227-33.
3. Bigger JT Jr, Fleiss JL, Kleiger R, et al: The relationships among ventricular arrhythmias, left ventricular dysfunction and mortality in the 2 years after myocardial infarction. Circulation 1984;69:250-8.
4. The Multicenter Postinfarction Research Group: Risk stratification and survival after myocardial infarction. N Engl J Med 1983; 309:331–6.
5. The Multicenter Diltiazem Postinfarction Trial Research Group: The effect of diltiazem on mortality and reinfarction after myocardial infarction. N Engl J Med 1988;319:385-92.
6. Ruberman W, Weinblatt E, Goldberg JD, et al: Ventricular premature complexes and sudden death after myocardial infarction. Circulation 1981;64:297-305.
7. The Miami Trial Research Group: Metoprolol in acute myocardial infarction (MIAMI): A randomized placebo-controlled international trial. Eur Heart J 1985;6:199-226.
8. Mukharji J, Rode RE, Poole K, et al: Late sudden death following acute myocardial infarction: Importance of combined presence of repetitive ectopy and left ventricular dysfunction. Clin Res 1982;30:108A.
9. Moss AJ, Davis HT, DeCamilla J, Bayer LW: Ventricular ectopic beats and their relation to sudden and non-sudden cardiac death after myocardial infarction. Circulation 1979;60:998-1003.
10. Vlay SC: How the university cardiologist treats ventricular premature beats: A nationwide survey of 65 University Medical Centers. Am Heart J 1985;110:904-12.
11. Morganroth J, Bigger JT Jr, Anderson JL: Treatment of ventricular arrhythmias by United States cardiologists: A survey before the Cardiac Arrhythmia Suppression Trial results were available. Am J Cardiol 1990;65:40-8.
12. Morganroth J: Premature ventricular complexes. JAMA 1984; 252:673-6.
13. Bigger JT Jr: Identification of patients at high risk for sudden cardiac death. Am J Cardiol 1984;54:3D-8D.
14. Morganroth J, Oshrain C, Steele PP: Comparative efficacy and safety of oral tocainide and quinidine for benign and potentially lethal arrhythmias. Am J Cardiol 1985;56:581-5.
15. Morganroth J: Comparative efficacy and safety of mexiletine and quinidine in benign or potentially lethal ventricular arrhythmias. Am J Cardiol 1987;60:1276-81.
16. The CAPS Investigators: The Cardiac Arrhythmia Pilot Study. Am J Cardiol 1986;57:91-5.
17. The Cardiac Arrhythmia Pilot Study (CAPS) Investigators: Effects of encainide, flecainide, imipramine and moricizine on ventricular arrhythmias during the year after acute myocardial infarction: The CAPS. Am J Cardiol 1988;61:501-9.
18. Greene HL, Roden DM, Katz RJ, et al: The Cardiac Arrhythmia Suppression Trial: First CAST ... then CAST-II. J Am Coll Cardiol 1992;19:894-8.
19. Beta-blocker Heart Attack Trial Research Group: A randomized trial of propranolol in patients with myocardial infarction: I. Mortality results. JAMA 1982;247:1707-17.
20. Norwegian Multicenter Study Group: Timolol-induced reduction in mortality and reinfarction in patients surviving acute myocardial infarction. N Engl J Med 1981;304:801-7.
21. Hjalmerson A, Elmfeldt D, Herlitz J, et al: Effect on mortality of metoprolol in acute myocardial infarction. A double-blind randomized trial. Lancet 1981;iv:823-7.
22. Pratt CM, Moye L: The Cardiac Arrhythmia Suppression Trial: Background, interim results, and implications. Am J Cardiol 1990;65:20B-9B.
23. Goldstein S, Brooks MM, Ledingham R, et al: Association between ease of suppression of ventricular arrhythmia and survival. Circulation 1995;91:79-83.
24. Nattel S, Pederson DH, Zipes DP: Alterations in regional myocardial distribution and arrhythmogenic effects of aprindine produced by coronary artery occlusion in the dog. Cardiovasc Res 1981;15:80-5.
25. Kou WH, Nelson SD, Lynch JJ, et al: Effect of flecainide acetate on prevention of electrical induction of ventricular tachycardia and occurrence of ischemic ventricular fibrillation during the early post myocardial infarction period: Evaluation of a conscious canine model of sudden death. J Am Coll Cardiol 1987;9:359-65.
26. Echt DS, Liebson PR, Mitchell LB, et al: Mortality and morbidity in patients receiving encainide, flecainide, or placebo. The Cardiac Arrhythmia Suppression Trial. N Engl J Med 1991;324:781-8.
27. Anderson JL, Platia EV, Hallstrom RW, et al: Interaction of baseline characteristics with hazard of encainide, flecainide, and moricizine therapy in patients with myocardial infarction; possible explantation for increased mortality in the Cardiac Arrhythmia Suppression Trial (CAST). Circulation 1994;90:2843-52.
28. CAST database.
29. Teo KK, Yusuf S, Furberg CD: Effects of prophylactic antiarrhythmic drug therapy in acute myocardial infarction: An overview of results from randomized control trials. JAMA 1993;270:1589-95.
30. Fuster V, Ryden LE, Asinger RW, et al: ACC/AHA/ESC guidelines for the management of patients with atrial fibrillation: Executive summary. A report of the American College of Cardiology/ American Heart Association Task Force on Practice Guidelines and the European Society of Cardiology Committee for Practice Guidelines and Policy Conferences (Committee to develop guidelines for the management of patients with atrial fibrillation). Developed in collaboration with the North American Society of Pacing and Electrophysiology. Circulation 2001;104:2118-50.
31. Geller JC, Geller M, Carlson MD, Waldo AL: Efficacy and safety of moricizine in the maintenance of sinus rhythm in patients with recurrent atrial fibrillation. Am J Cardiol 2001;87:172-7.
32. Steinberg J, Regan A, Sciacca R, et al: Predicting arrhythmic events after acute MI using the signal-averaged electrocardiogram. Am J Cardiol 1992;69:13-21.
33. Odemuyiwa O, Malik M, Farrell TG, et al: Comparison of the predictive characteristics of heart rate variability index and left ventricular ejection fraction for all-cause mortality, arrhythmic events and sudden death after acute myocardial infarction. Am J Cardiol 1991;68:434-9.
34. Schwartz PJ, LaRovere MT, Vanoli E: Autonomic nervous system and sudden cardiac death. Circulation 1992;85(suppl I):78-91.
35. SOLVD Investigators: Effect of enalapril on survival in patients with reduced left ventricular ejection fractions and congestive heart failure. N Engl J Med 1991;325:293-302.
36. Waldo AL, Camm AJ, deRuyter H, et al: Survival with ORal d-Sotalol in patients with left ventricular dysfunction after myocardial infarction: Rationale, design, and methods (the SWORD Trial). Am J Cardiol 1995;75:1023-7.
37. Waldo AL, Camm AJ, deRuyter H, et al: Effect of *d*-sotalol on mortality in patients with left ventricular dysfunction after recent and remote myocardial infarction. Lancet 1996;348:7-12.

38. Pratt CM, Camm AJ, Cooper W, et al: Mortality in the Survival With ORal d-Sotalol (SWORD) Trial: Why did patients die? Am J Cardiol 1998;81:869-76.
39. Priori SG, Diehl L, Schwartz PJ: Torsades de pointes. In Podrid PJ, Kowey PR (ed): Cardiac Arrhythmia; Mechanisms, Diagnosis and Management. Baltimore, Williams & Wilkins, 1995, pp 951-63.
40. Moss AJ, Zareba W, Hall J, et al: Multicenter Automatic Defibrillator Implantation Trial II Investigators. Prophylactic implantation of a defibrillator in patients with myocardial infarction and reduced ejection fraction. N Engl J Med 2002;346:877-83.
41. Buxton AE, Lee KL, Fisher JD, et al: Multicenter Unsustained Tachycardia Trial Investigators. A randomized study of the prevention of sudden death in patients with coronary artery disease. Multicenter Unsustained Tachycardia Trial Investigators. N Engl J Med 1999;341:1882-90.
42. DIAMOND Study Group: Dofetilide in patients with left ventricular dysfunction and either heart failure or acute myocardial infarction: Rationale, design, and patient characteristics of DIAMOND study. Clin Cardiol 1997;20:704-10.
43. Torp-Pedersen C, Moller M, Bloch-Thomsen PE, et al: Dofetilide in patients with congestive heart failure and left ventricular dysfunction. N Engl J Med 1999;341:857-65.
44. Singh S, Zoble RG, Yellen L, et al: Efficacy and safety of oral dofetilide in converting to and maintaining sinus rhythm in patients with chronic atrial fibrillation and flutter. The Symptomatic Atrial Fibrillation Investigation Research on Dofetilide (SAFIRE-D) Study. Circulation 2000;2385-90.
45. Kober L, Bloch Thomsen PE, Moller M, et al: Effect of dofetilide in patients with recent myocardial infarction and left ventricular dysfunction: A randomized trial. Lancet 2000;356:2052-8.
46. Pedersen OD, Bagger H, Keller N, et al: Efficacy of dofetilide in the treatment of atrial fibrillation-flutter in patients with reduced ventricular function: A Danish Investigation of Arrhythmia and Mortality ON Dofetilide (DIAMOND) substudy. Circulation 2001;104:292-6.
47. Sedgwick ML, Rasmussen HS, Cobbe SM: Clinical and electrophysiologic effects of intravenous dofetilide (UK-68,798), a new class III antiarrhythmic drug, in patients with angina pectoris. Am J Cardiol 1992;69:513-7.
48. Julian DG, Prescott RJ, Jackson FS, Szekely P: Controlled trial of sotalol for one year after myocardial infarction. Lancet 1982;1:1142-7.
49. Singh SH, Fletcher RD, Fisher SG, et al, for the Survival Trial of Antiarrhythmic Therapy in Congestive Heart Failure: Amiodarone in patients with congestive heart failure and asymptomatic ventricular arrhythmia. N Engl J Med 1995;333:77-82.
50. Julian DG, Camm AJ, Frangin G, et al, for the European Myocardial Infarct Amiodarone Trial investigators: Randomised trial of effect of amiodarone on mortality in patients with left-ventricular dysfunction after recent myocardial infarction: EMIAT. Lancet 1997;349:667-74.
51. Cairns JA, Connolly SJ, Robert R, Gent M, for the Canadian Amiodarone Myocardial Infarction Arrhythmia Trial investigators: Randomised trial of outcome after myocardial infarction in patients with frequent or repetitive ventricular premature depolarisations: CAMIAT. Lancet 1997;349:675-82.
52. Cockcroft DW, Gault MH: Prediction of creatinine clearance from serum creatinine. Nephron 1976;16:31-41.
53. Dofetilide (Tikosyn) presentation at the Cardiorenal Advisory Board meeting of the Food and Drug Administration, Bethesda, MD, Jan. 14, 1999.
54. Ceremuzynski L, Kleczar E, Krzeminska-Pakula M, et al: Effect of amiodarone on mortality after myocardial infarction: A double-blind, placebo-controlled, pilot study. J Am Coll Cardiol 1992;20:1056-62.
55. Roy D, Taljic M, Dorian P, et al, for the Canadian Trial of Atrial Fibrillation investigators: Amiodarone to prevent recurrence of atrial fibrillation. N Engl J Med 2000;342:913-20.
56. Herre JM, Sauve MJ, Malone P, et al: Long-term results of amiodarone therapy in patients with recurrent sustained ventricular tachycardia or ventricular fibrillation. J Am Coll Cardiol 1989;13:442-9.
57. Dries DL, Exner DV, Gersh BJ, et al: Atrial fibrillation is associated with an increased risk for mortality and heart failure progression in patients with asymptomatic and symptomatic left ventricular systolic dysfunction: A retrospective analysis of the SOLVD trials. J Am Coll Cardiol 1998;32:695-703.
58. Assmann SF, Pocock SJ, Enos LE, Kasten LE: Subgroup analysis and other (mis)uses of baseline data in clinical trials. Lancet 2000;355:1064-9.
59. Deedwania PC, Singh BN, Ellenbogen K, et al, for the Department of Veterans Affairs CHF-STAT investigators: Spontaneous conversion and maintenance of sinus rhythm by amiodarone in patients with heart failure and atrial fibrillation. Circulation 1998;98:2574-9.

SECTION II

Cardiac Rhythms and Arrhythmias

Chapter 12 Sinus Node Dysfunction

SAROJA BHARATI, NORA GOLDSCHLAGER, FRED KUSUMOTO, RALPH LAZZARA, RABIH AZAR, STEPHEN HAMMILL, and GERALD NACCARELLI

Epidemiology: Nora Goldschlager, Fred Kusumoto, and Rabih Azar
Anatomy and Pathology: Saroja Bharati
Basic Electrophysiology: Ralph Lazzara
Etiology: Nora Goldschlager, Fred Kusumoto, and Rabih Azar
Diagnostic Evaluation: Gerald Naccarelli
Clinical Electrophysiology: Stephen Hammill
Evidence-Based Therapy: Nora Goldschlager, Fred Kusumoto, and Rabih Azar

In the early 1900s Keith and Flack identified the sinus node as the region responsible for activation of the heart.[1] Laslett first suggested sinus node dysfunction as a cause of bradycardia in 1909, and during the 1950s and 1960s Short, Ferrer, Lown, and others described the clinical spectrum of sinus node dysfunction that is commonly called the "sick sinus syndrome resting sinus bradycardia."[2-5] Sinus node dysfunction can have multiple electrocardiographic manifestations including sinus pauses, the bradycardia-tachycardia syndrome, and inappropriate sinus node response to exercise (chronotropic incompetence).

Epidemiology

It can be difficult to differentiate sinus node dysfunction from physiologic sinus bradycardia in a specific population. In a study of 50 young adult males, 24-hour ambulatory electrocardiographic monitoring revealed that 24% of the study group had transient heart rates less than 40 beats per minute (bpm), pauses up to 1.7 seconds while awake, and 2.1-second pauses while asleep.[6] Similarly, in a study of 50 young adult women, pauses from 1.6 to 1.9 seconds were observed.[7] In older asymptomatic individuals, transient heart rates less than 40 bpm and pauses of 1.5 to 2.0 seconds were observed in less than 2%.[8] The decreased incidence of nocturnal bradycardia in this normal elderly population is probably due to the decrease in vagal tone that occurs with increasing age.

Sinus node dysfunction should be suspected when a patient describes symptoms of fatigue, syncope or presyncope, or exercise intolerance and is noted to have sinus bradycardia or pauses on the 12-lead ECG or during Holter monitoring. In general, symptomatic sinus node dysfunction increases with age, with the incidence doubling between the fifth and sixth decades of life. In one study of approximately 9000 patients visiting a regional cardiac center in Belgium, Kulbertus et al. estimated that the incidence of sinus node dysfunction is less than 5 per 3000 people older than 50 years of age.[9] However, sinus node dysfunction is more commonly identified today because of an increased elderly population and increased physician awareness. In a 1997 survey, sinus node dysfunction accounted for approximately 50% of the 150,000 new pacemakers implanted in the United States and 20% to 30% of the 150,000 new pacemakers implanted in Europe.[10,11]

Although more common in elderly patients, it is important to note that several specific younger patient groups can also have sinus node dysfunction. First, patients with congenital heart disease can have sinus node dysfunction. In 39 patients younger than 40 years of age who underwent pacemaker implant for sinus node dysfunction at the Mayo Clinic, 64% had associated congenital heart disease.[12] The most common condition was transposition of the great arteries corrected by a

Mustard operation, since this procedure requires extensive atriotomies. Second, several familial forms of sinus node dysfunction have been identified and account for approximately 2% of patients who present with sinus node dysfunction.[13-15] Bharati et al. described a family with congenital absence of sinus rhythm. Several other investigators have described different forms of sinus node dysfunction that appeared to be genetically transmitted.[13] Finally, sinus node dysfunction is observed in approximately 4% of patients after cardiac transplant.[16,17] Some, but not all, studies have found that a heart from a donor older than 40 years of age is associated with a higher incidence of sinus node dysfunction requiring permanent cardiac pacing.[16,17]

Anatomy

The sinoatrial (SA) node is situated in the sulcus terminalis between the superior vena cava and the right atrial appendage in a somewhat curved fashion and extends from the hump of the atrial appendage, tapering toward the inferior vena cava (Fig. 12-1*A*). The SA node is a sizable structure, and although it is situated epicardially, some of the fibers may extend deep intramyocardially. At the gross level, the SA node cannot be dissected. It can be identified at the light microscopic level; it consists of small fusiform cells, distinctly smaller than the surrounding atrial myocardial cells, arranged to a considerable extent along the line of the sulcus terminalis but also serpiginously surrounding the SA nodal artery. The cytoplasm of the cells stain lighter than do the atrial myocardial cells (see Fig. 12-1*B*). The myofibrils are distinctly less prominent than the surrounding atrial myocardial cells, and the striations are scarce but do increase with age. There are no intercalated discs at the light microscopic level. The cells lie in a massive amount of thick collagenous and thinner elastic fibers.[18-22]

There are fewer myofibrils at the electron microscopic level as compared with the atrial cells. The myofibrils are arranged in a disorganized pattern with a lesser number of mitochondria, which are also disorganized. The sarcoplasmic reticulum is poorly developed, and there is no transverse tubular system. There are very few fascia adherents and scant gap junctions, and toward the periphery there are transitional cells, which finally end in atrial cells.[18-20]

BLOOD SUPPLY

The SA node is supplied by way of ramus ostii cavae superioris (SA nodal artery) and reinforced by other atrial branches, and branches from the bronchial arteries. In addition, Kugel's artery, a branch from the left coronary artery, also supplies the SA node. The ramus ostii cavae superioris originates in about 55% of cases from the right coronary artery and in 45% of cases from the left coronary artery.[18-22] There are nerve cells and fibers evident in the periphery of the node. Nerve fibers are present in the midst of the head of the SA node.[18-23]

PATHOLOGY

The anatomic base of supraventricular arrhythmias may lie in the SA node and its approaches, as well as in the atrial preferential pathways, the approaches to the atrioventricular (AV) node, and the AV node. Although at

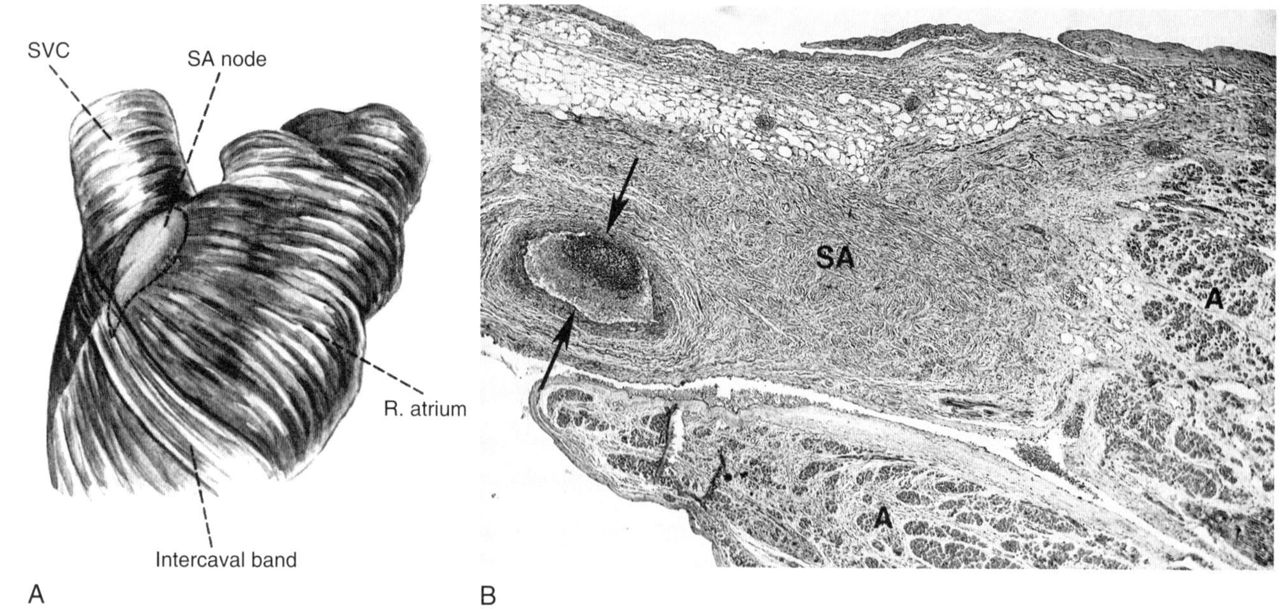

FIGURE 12-1 **A,** Diagrammatic sketch of location of sinoatrial (SA) node. R. atrium, right atrium; SA node, sinoatrial node; SVC, superior vena cava. **B,** Transverse section through the SA node and its approaches. Hematoxylin-eosin stain ×50. *Arrows* point to the SA nodal artery. A, atrial myocardium; SA, sinoatrial node. (From Lev M: The conduction system. In Gould SE (ed): Pathology of the Heart and Blood Vessels, 3rd ed. Springfield, Ill, Charles C Thomas, 1968.)

the moment there is no pathologic specificity for any one of the atrial arrhythmias, it would appear that the same type of pathology may be responsible for both bradyarrhythmias and tachyarrhythmias on different occasions.[23] Rarely, sinus node dysfunction can be familial[13] (Fig. 12-2).

SICK SINUS SYNDROME IN THE YOUNG

Although sick sinus syndrome is usually seen in the elderly, it may be seen in the young as well. Pathologically, a marked increase in connective tissue is found, with fat in the approaches to the SA node, the atrial preferential pathways, and approaches to the AV node. The penetrating AV bundle may be septated. In other cases an increased amount of fat in other parts of the atria, as well as around the SA and AV nodes, is seen.[24]

SICK SINUS SYNDROME IN THE ELDERLY-AGING

In older patients with sick sinus syndrome, significant coronary artery disease is usually present; however, the basic pathology is variable. An abnormal formation of the atrial septum, resulting in fibrosis in the approaches to the SA and AV nodes and atrial preferential pathways, may be seen, or there may be similar findings within these structures along with fat. With advancing age, there may be loss of cells to a varying extent in the SA and AV nodes and their approaches and in the surrounding atrial myocardium, with replacement by fat and fibrosis, associated with varying degrees of focal hypertrophy or atrophy of myocardium (Fig. 12-3). In addition, arteriolosclerosis, necrosis, or chronic inflammation may be present in the atria.[25-27]

SENILE AMYLOIDOSIS

This is a type of primary amyloidosis seen in patients older than 80 years of age. It affects the atrial

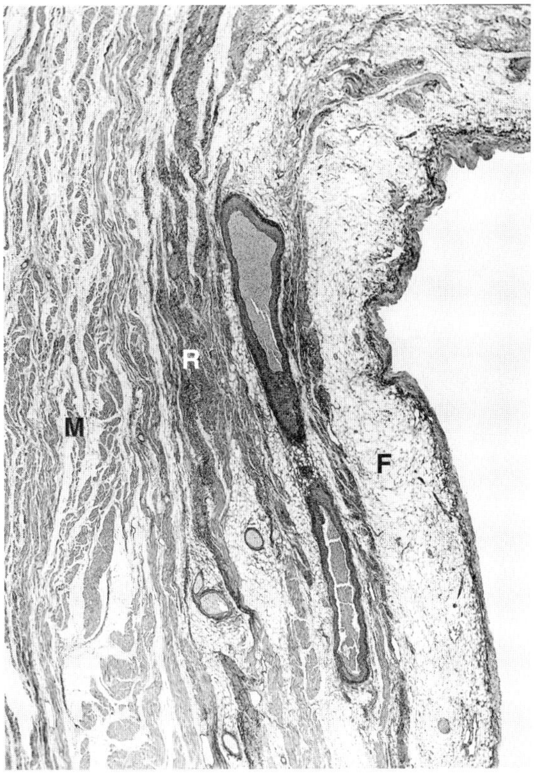

FIGURE 12-2 Histologic characteristics of the sinoatrial (SA) nodal area in a 72-year-old male member of a family with congenital absence of sinus rhythm and a tendency to develop atrial fibrillation at an early age who collapsed rising from a chair and could not be resuscitated. He had a history of atrial fibrillation, at first paroxysmal and later chronic for 18 years, but was only mildly incapacitated. Note the fragmented SA node. Weigert-van Gieson stain ×30. F, fat; M, atrial myocardium; R, remnant of SA node. (From Bharati S, Surawicz B, Vidaillet, HJ, Lev M: Familial congenital sinus rhythm anomalies, clinical and pathological correlations. Pacing Clin Electrophysiol 1993;15:1720-9).

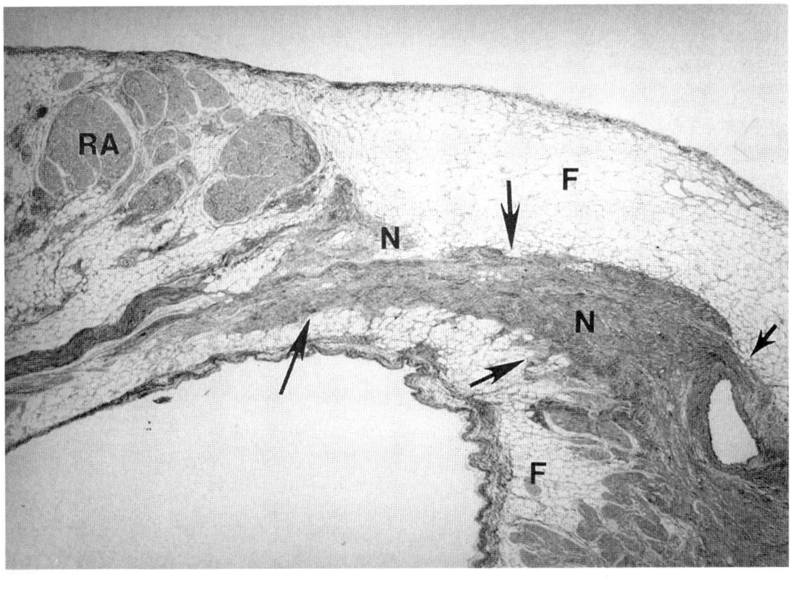

FIGURE 12-3 Fat accumulation around the SA node and replacement of SA node by fat with almost total separation of SA node from its approaches in an elderly patient with sick sinus syndrome. Weigert-van Gieson stain ×45. F, fat; N, SA node; RA, right atrium. *Arrows* point to the isolated SA node with marked fat in and around the node. (From Lev M, Bharati S: Age related changes in the cardiac conduction system. Intern Med 1981;2:19-21.)

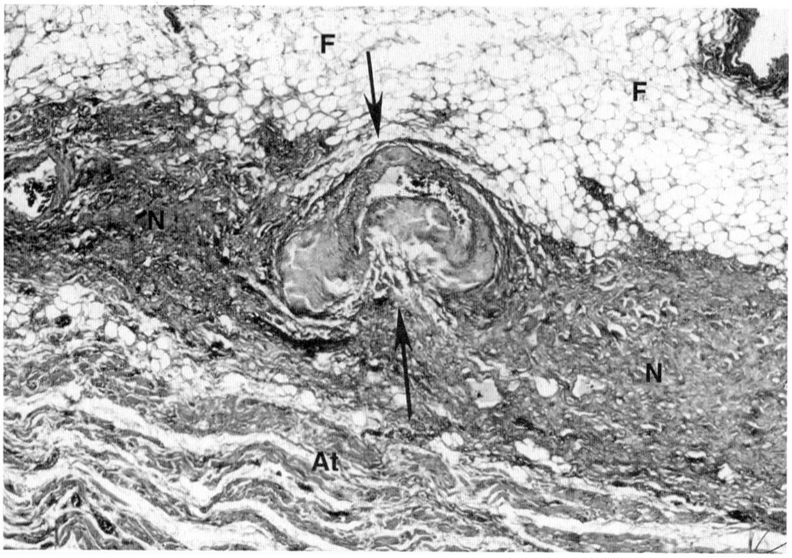

FIGURE 12-4 Amyloid infiltration of large arteriole in sinoatrial (SA) node in an 83-year-old man with syncope, left bundle branch block, prolonged H–V interval, and a sinus echo zone with repetitive sinus nodal reentrance. Weigert-van Gieson stain ×45. At, atrial approaches with amyloid; F, fat; N, SA node with marked fat in and around with amyloid. *Arrows* point to amyloid in large arteriole. (From Bharati S, Lev M, Dhingra R, et al: Pathologic correlations in three cases of bilateral bundle branch disease with unusual electrophysiologic manifestations in two cases. Am J Cardiol 1976; 38:508-18.)

myocardium including the SA node and its approaches, SA nodal artery, and the arterioles of the atria (Fig 12-4). Senile amyloidosis may be associated with sinus bradycardia, the bradycardia-tachycardia syndrome, and sinus echo phenomena.[27-30]

TRANSIENT AND PERSISTENT ATRIAL STANDSTILL

Here one may find arteriolosclerosis of the SA and AV nodes with fibrosis and fibroelastosis in the approaches to the AV node and within the AV node.[29]

IATROGENIC CONSIDERATIONS

Sinus node dysfunction has been documented following cardiac surgery in both congenital and acquired heart disease.[30-31] Radiation therapy for Hodgkin's disease and blunt trauma to the chest wall may result in SA nodal dysfunction.[18]

In summary, pathologic findings in and around the SA node, atria, approaches to the AV node, the AV node, the AV bundle and the ventricular myocardium are almost always present to varying degrees in chronic or permanent SA node dysfunction. However, isolated pathology of the SA node or the atria, or both, may be seen. Any pathologic state, such as myotonia dystrophica, Kearns-Sayre syndrome, primary or secondary tumor, primary or secondary amyloidosis, and numerous other diseases that affect the atria or SA node and its approaches and the AV node and its approaches may produce varying clinical forms of SA nodal dysfunction.[22]

BASIC ELECTROPHYSIOLOGY OF THE SINUS NODE

The structural and functional organization of the sinus node is quite complex, and there are significant differences in the organization of the sinus node among species.[32,33] The most extensive studies of the sinus node have been performed in rabbits. The node contains prototypical, small, structurally primitive pacemaker cells concentrated in the center, larger transitional latent pacemaker cells concentrated more peripherally; and intermingled, nonpacing atrial cells extending into the node from the atrial margins of the node in strands that are more prominent in the periphery. The cells within the node are relatively poorly coupled by gap junctions, and there is substantial interstitial tissue interspersed among the fascicles of nodal cells.[34] The resultant relatively poor intercellular communication slows the propagation of the impulse from the central pacemaking regions toward the periphery of the node. In addition, the coupling of the transitional cells near the margins of the node with the atrial myocardial cells is not a smooth continuum but consists of irregular junctions of interweaving strands of transitional cells and atrial cells extending into the interior of the node. Mapping of electrograms in and around the node has disclosed an apparent "multicentric" initiation of activation that more likely represents irregular propagation to atrial myocardium than simultaneous generation of impulses at separated sites. The node preferentially connects to the atrium in the superior aspects of the crista terminalis, and there appears to be conduction block exiting the node in the direction of the atrial septum.

A dense representation of sympathetic and parasympathetic nerves and ganglia in the node ensures a sensitive autonomic responsiveness.[35] With increasing vagal influence, the primary pacemaking site near the superior aspect of the node tends to migrate inferiorly, whereas an increasing adrenergic influence produces return to the primary dominant pacemaker sites in the superior region of the node. Detailed and sensitive analyses of P wave morphology, as well as mapping of the sinus node, indicate a dynamic shifting of pacemaker sites with changes in heart rate.

Action Potentials

Small, primitive pacemaker cells in the interior of the node generate the dominant pacemaker potentials

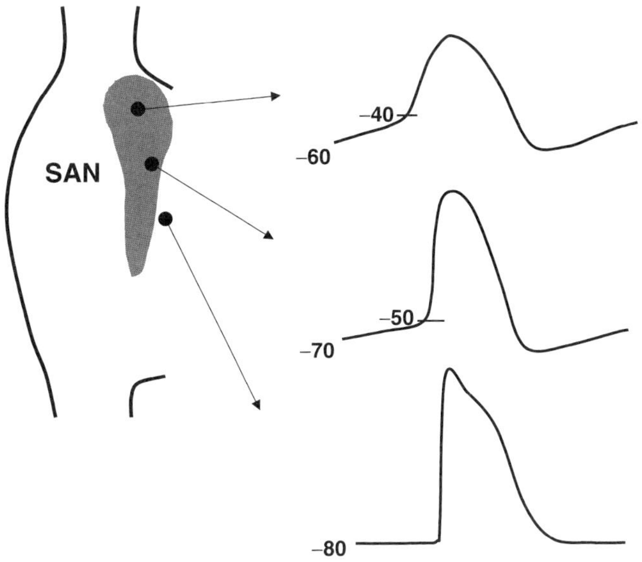

Figure 12-5 Samples of action potentials recorded from different regions of the sinus node (SAN) from the center to the periphery of the node.

and show the least polarized maximum diastolic potentials (−60 to −40 mV), the most rapid rates of diastolic depolarization with smooth transition from end-diastole to the upstroke, and the slowest upstroke velocities (≈1-10 V/sec), as shown in Figure 12-5.[36] Latent pacemakers are concentrated more peripherally in the node. These cells are more polarized, show less rapid diastolic depolarization, a more abrupt transition from diastole to upstroke, and more rapid upstrokes. There is a continuous gradation of the properties of transitional cells from the characteristics of primary pacemaker cells to the properties of highly polarized surrounding atrial cells with stable resting potentials close to −80 mV and upstroke velocities approaching 100 V/sec. This gradation of properties is conditioned by cell coupling and electrotonic interaction, as well as by the differing intrinsic properties of myocytes within the node. Due to the electrotonic influence of atrial cells, the pacemaker capabilities of transitional cells are muted, whereas the more remotely connected interior cells are shielded from the nonpacing atrial myocytes, allowing them to maintain pacemaker dominance.[37] Separation of latent pacemaker cells near the atrial margins from the atrium and from the centrally located dominant pacemakers result in a faster intrinsic rate in latent pacemakers freed from the influence of the nonpacemaking atrial cells. Cells in other parts of the atrium can also act as backup pacemakers, especially in the inferior portions of the node near the coronary sinus. These pacemaker cells can respond appropriately to autonomic influence and under abnormal conditions can usurp control of the heart. The electrophysiological properties of these subsidiary pacemakers have not been as well characterized as those of the sinus node.

Currents

Sinus node pacemaker cells are relatively depolarized because of an absence of paucity of channels for the current I_{K1}.[38] These channels are plentiful and open at negative membrane potentials in atrial and ventricular myocytes. They establish a dominance of K^+ permeability in the resting state, thereby determining a resting potential approximating the K^+ equilibrium potential (≈−90 mV). The absence of these channels is most complete in the small, central pacemaker cells operating at diastolic potentials between −60 mV and −30 mV. In larger, more polarized transitional cells, I_{K1} may be present but reduced to varying degrees. The low upstroke velocities of these cells are related to a lack of operating Na channels, those channels that transmit the intense excitatory Na current in atrial and ventricular cells.[39] The absence or paucity of this excitatory current is due to a deficiency of the channels in individual myocytes and the depolarization of the myocytes during diastole to levels of membrane potential at which Na channels, if present, would be inactivated. The smallest cells may lack Na channels entirely. Larger transitional cells may contain Na channels and may operate at diastolic potentials at which Na channels can be activated to provide some excitatory Na current and more rapid upstrokes. The slow and diminutive upstrokes in the primary pacemaker cells are generated by the L-type Ca^{2+} current (I_{CaL}), which serves as the trigger for release of Ca^{2+} by the sarcoplasmic reticulum and therefore the trigger for contraction in all cardiac cells but is the primary excitatory current in depolarized sinus nodal cells. This current is slower and far less intense than the Na current, accounting for the poor upstrokes and slow conduction within the node. The currents producing diastolic depolarization, the fundamental pacemaker potential, comprise a multitude of candidates about which there is no uniform consensus.[40,41] The "funny" current I_f is a nonspecific cation current that is mainly an inward Na current activating relatively slowly at negative membrane potentials in the range of diastolic potentials.[40] It becomes more intense at more negative membrane potentials. This current is well expressed in sinus nodal cells, responds appropriately to adrenergic and cholinergic stimulation, and thus is a plausible candidate as an important pacemaker current. However, some studies have shown activation at more negative levels than the diastolic potentials of the primary pacemakers (below −60 mV) and a greater representation in peripheral latent pacemakers. This has led some to suggest that I_f is more active in latent pacemaker cells with a role to maintain pacemaker activity and counter the electrotonic influence of stable, well-polarized diastolic potentials of atrial cells. The I_f current is activated by the second messenger adenylate cyclase (cAMP), which shifts the voltage activation curve into a more positive range. It is argued that at physiologic levels of cAMP, the voltage activation curve is shifted into the range of the diastolic potentials of primary pacemaker cells.

In the absence of I_{K1}, the delayed rectifier K^+ currents I_{Kr} and I_{Ks}, which are activated during the action potential, can play a role in the attainment of the maximum diastolic potential by providing the K^+ permeability that would bring the transmembrane potential close to the K^+ equilibrium potential at the end of the action potential when they are fully activated. As these currents deactivate in diastole, the membrane potential would drift positive toward the more positive equilibrium potentials of other major ions such a Na, Ca^{2+}, and Cl. It has been argued that one or another of the delayed rectifier currents is the major pacemaker current.[41] However, although I_{Ks} has an appropriate autonomic sensitivity, I_{Kr} does not. I_{Ks} is not prominent in the sinus nodes of all species.

Ca^{2+} currents, both I_{CaL} and the T-type Ca^{2+} current (I_{CaT}), have also been implicated in pacemaker function. I_{CaL}, which has an activation threshold more positive than −40 mV in most cardiac cells, could be active toward the end of diastolic depolarization in the depolarized primary pacemaker cells whereas I_{CaT}, with an activation threshold more negative, might be active in earlier phases of diastolic depolarization and in latent pacemaker cells. I_{CaL} responds to autonomic influence in a manner like the sinus node, and recent studies have suggested that the activation threshold may be more negative in sinus nodal cells. The role of I_{CaT} remains uncertain. Other currents that have been nominated include the Na:Ca^{2+} exchange current generating inward current as Ca^{2+} entering the cells during the action potential is extruded in diastole and a newly described sustained inward current. It is possible that multiple currents can be involved in pacemaking with different roles in primary versus latent pacemaker cells and under different conditions.

Pacemaker activity can be notably influenced by the acetylcholine-activated K^+ current, $I_{K,Ach}$, which markedly increases K^+ permeability throughout the cycle, speeding repolarization, hyperpolarizing the cell, and reducing the rate of diastolic depolarization. This current appears to be less sensitive to acetylcholine than the I_f current in the sinus node. The complexities of pacemaker function remain to be fully clarified. Contemporary molecular biologic techniques will be powerful tools to clarify the location and function of channels in the SA node and their roles in SA nodal pacemaking.

Etiology

Sinus node dysfunction results from various conditions that have in common the capability to depress automaticity in and electrical conduction from the sinus node and perinodal and atrial tissue. These conditions can be intrinsic, resulting from structural damage to the sinus node, or extrinsic, caused by medications or systemic illnesses. The manifestation of sinus node dysfunction is inappropriate sinus or atrial bradycardia.[42-45] Some patients may also experience episodes of supraventricular tachycardias (tachycardia-bradycardia syndrome). Because the clinical manifestations of sinus node dysfunction can mimic normal physiologic conditions (bradycardia) or can be caused by diseases that do not affect the sinus node (supraventricular tachycardias), the assessment and management of patients with suspected sinus node dysfunction can be challenging.[6,46-48]

EXTRINSIC CAUSES OF SINUS NODE DYSFUNCTION

Sinus bradycardia can be caused by medications that suppress automaticity; these include β-blockers, some calcium channel blockers (diltiazem and verapamil), digoxin (especially in the presence of high vagal tone), class I and III antiarrhythmic medications, and sympatholytic drugs such as clonidine. Sinus bradycardia in these cases is frequently transient and reversible once the offending agent is withdrawn.

Sinus bradycardias can also be a manifestation of systemic illnesses or other extrinsic conditions such as hypothyroidism; hypoxemia caused by sleep apnea; increased intracranial pressure; or increased vagal tone such as occurs during endotracheal suctioning, vomiting, and Valsalva's maneuver. These conditions are important to diagnose, because their appropriate treatment often results in resumption of normal sinus node function.

INTRINSIC CAUSES OF SINUS NODE DYSFUNCTION

Intrinsic sinus node dysfunction is usually caused by degenerative processes involving the sinus node and SA area. The syndrome is usually acquired but can rarely be familial.[8] Sinus node dysfunction is present when inappropriate sinus bradycardia, pauses in sinus rhythm (sinus arrest), SA block, or a combination of these exist.[49] The degenerative process and associated fibrosis may also involve the AV node and intraventricular conduction system; as many as 17% of patients with sick sinus syndrome have evidence of AV block and bundle branch block.[50]

NATURAL HISTORY

The natural history of sick sinus syndrome is one of spontaneous clinical improvement alternating with periods of clinical deterioration. Patients generally seek medical attention when they are symptomatic from bradycardia. In the majority of cases, the heart rate spontaneously increases, and symptoms diminish.[51,52] However, the clinical course is not predictable. Even patients with more severe symptoms such as syncope may remain free of symptom recurrence for years, and slightly more than half do not experience another syncopal episode over a 4-year follow-up.[53] There is no clear explanation for this erratic course of the syndrome. The autonomic nervous system may play an important role in the genesis of symptoms, especially in the trigger of syncope.[53] The prevalence of abnormal responses to carotid sinus massage (pauses exceeding 3 seconds) and tilt table testing is significantly higher in patients who experience syncope than in those who do not, highlighting the contribution of abnormal neural reflexes in the pathophysiology (Fig. 12-6).[53,54]

Although symptoms are common in patients with sick sinus syndrome, survival is usually not affected, even in patients who develop syncope.[55] Death related directly to dysfunction of the sinus node occurs in less than 2%

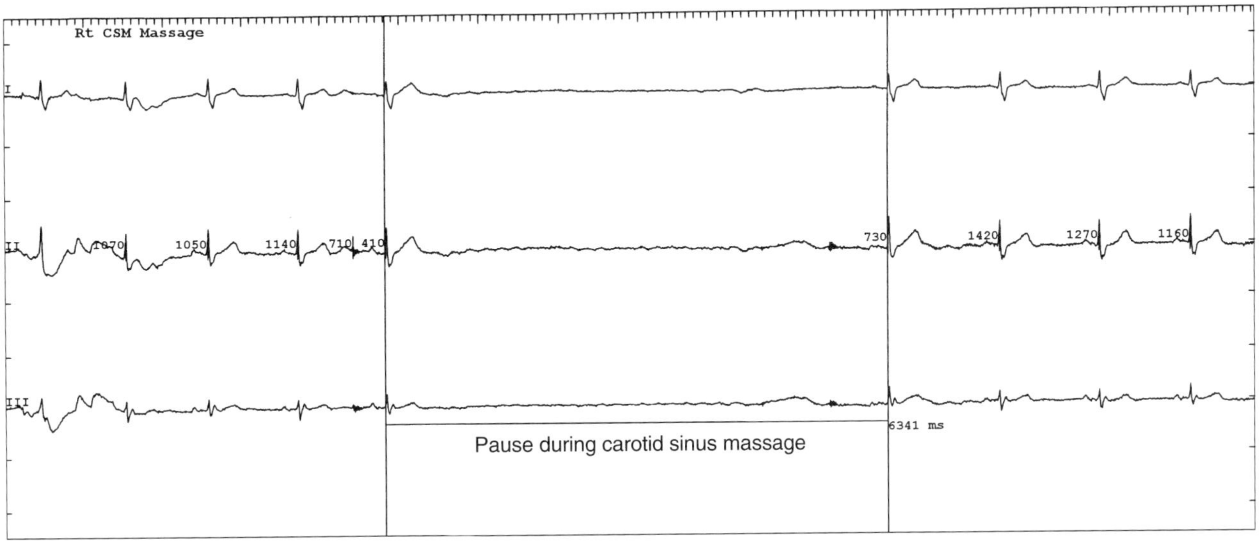

FIGURE 12-6 Sinus arrest during right carotid massage in an elderly female patient with recurrent syncope with simultaneous three-lead ECG rhythm strip. Note that the pause exceeds 6 seconds and persists beyond the duration of the massage. Numbers on ECG leads II and III indicate duration of intervals in milliseconds (ms).

of patients over 6 to 7 years of follow-up.[56,57] However, patients with sick sinus syndrome often have comorbid conditions that can shorten their life span. Coronary atherosclerosis is the most prevalent among these conditions, although myocardial ischemia and infarction, congestive heart failure, and advanced age are also common.[58,59] In one report, patients with sick sinus syndrome had a 4% to 5% excess annual mortality in the first 5 years of follow-up compared with an age and sex-matched population.[60] However, the mortality in patients without other coexisting disease at the time of diagnosis of sinus node dysfunction did not differ significantly from that observed in controls. Thus, although medical intervention and permanent cardiac pacing may be required to improve symptoms, no data exist supporting that these management strategies improve survival.

CLINICAL MANIFESTATIONS

The clinical manifestations of sick sinus syndrome are due to both the bradycardia and tachycardia. In the bradycardia-tachycardia syndrome, patients experience paroxysmal episodes of supraventricular tachycardia, which can be atrial tachycardia, atrial flutter or fibrillation, or reentry tachycardia (Fig. 12-7). More than one type of tachycardia may occur in the same patient. Episodes of rapid heart rate can lead to palpitations, angina, and syncope. Conversely, slowing of the heart rate without a compensatory increase in stroke volume leads to a reduction in cardiac output resulting in fatigue, weakness, lightheadedness, and dizziness.

Failure of the heart rate to increase appropriately with exercise (chronotropic incompetence), is a manifestation of sinus node dysfunction. Proposed definitions for chronotropic incompetence include failure to reach 85% of age-predicted maximum heart rate at peak exercise, failure to achieve a heart rate above 100 bpm, or a maximal heart rate during exercise greater than two standard deviations below that of a control population.[61-63] In patients with chronotropic incompetence, symptoms may be present only during activity.

In about 17% of patients with sick sinus syndrome, overt congestive heart failure may develop,[52] which may be related or contributed to by the slow heart rate, loss of the atrial contribution to left ventricular filling in patients who develop atrial fibrillation, or loss of AV synchrony in patients with implanted ventricular pacing systems.

Syncope or severe presyncope is one of the classical manifestations of sick sinus syndrome and occurs in about 25% of patients. In one prospective trial, the actuarial rates of syncope were 16% and 31% after 1 and 4 years of follow-up, respectively.[51,52] Syncope is usually due to excessive slowing of or transient pauses in sinus rhythm, or sinus exit block with an inadequate escape rhythm (Fig.12-8). In patients with the bradycardia-tachycardia syndrome, overdrive suppression of the sinus node may occur when the tachycardia terminates, resulting in prolonged pauses and syncope. Use of antiarrhythmic medications in these patients, while successful in controlling the tachycardia, frequently leads to worsening of the bradycardia episodes with exacerbation of symptoms, and permanent cardiac pacing is required for rate support.

About one third to one half of patients with sick sinus syndrome experience episodes of supraventricular tachycardia.[51,57] Published reports indicate that chronic atrial fibrillation occurs in about 11% of cases after a mean follow-up of 19 months, increases to 16% at 5 years, and to 28% at 10 years.[51,52,63] Variables that have been associated with the development of chronic atrial fibrillation are age; left ventricular end-diastolic diameter; presence of valvular heart disease; and ventricular, rather

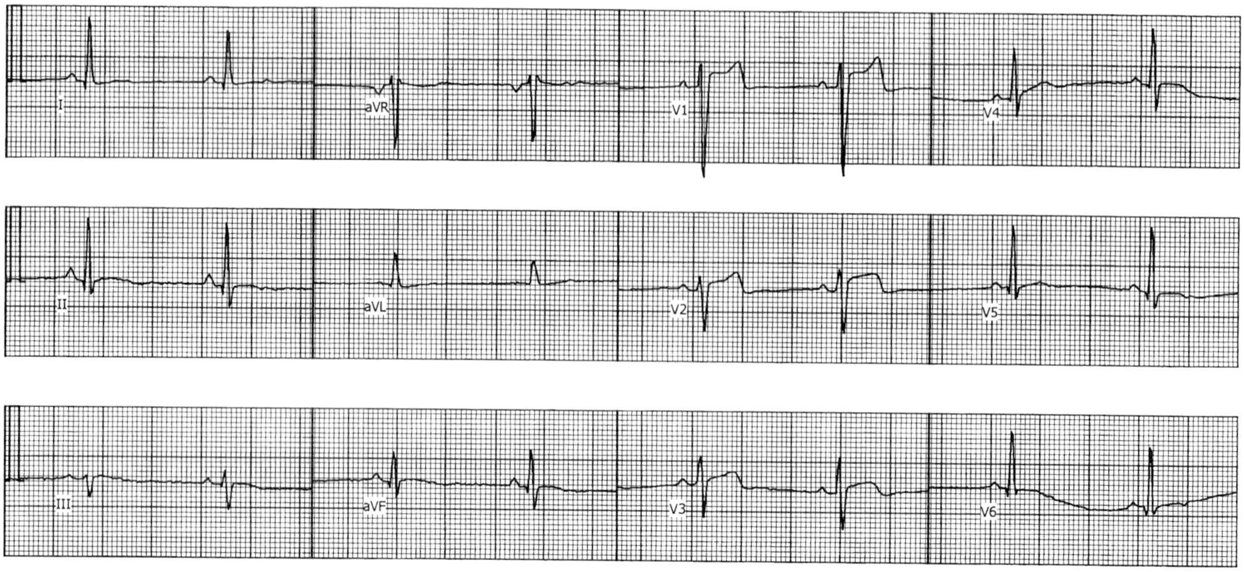

Sinus bradycardia

A

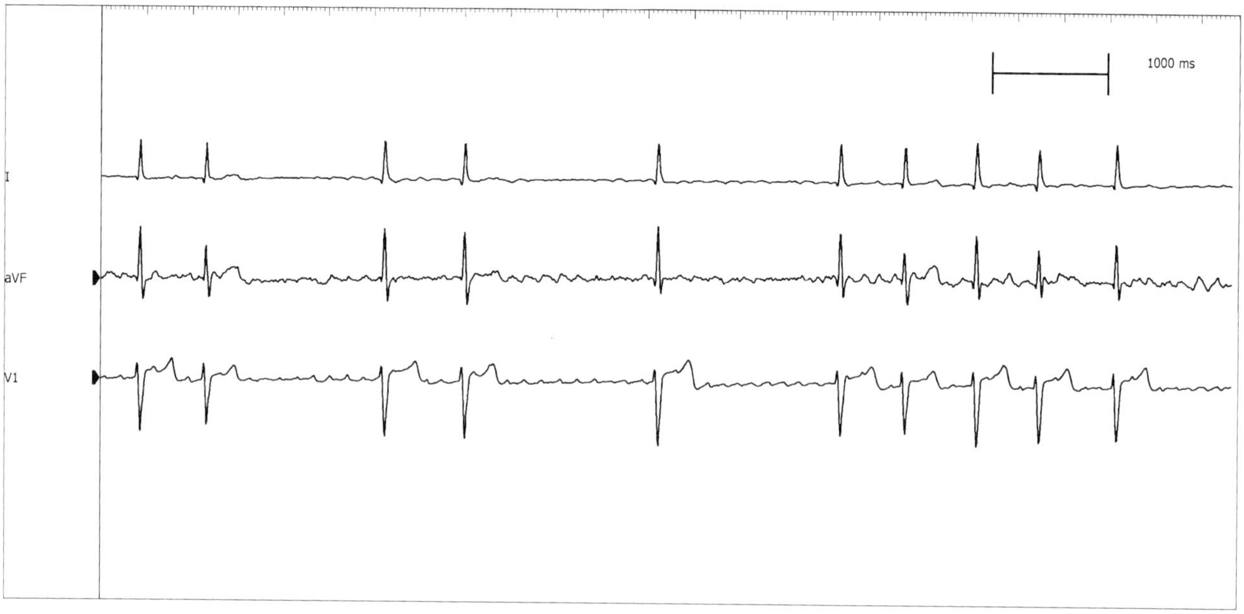

Bradycardia-tachycardia syndrome

B

FIGURE 12-7 **Panel A,** 12-lead ECG during sinus bradycardia in a patient with bradycardia-tachycardia syndrome. **Panel B,** Tachycardia in the same patient in **panel A** with rapid and slow atrioventricular conduction as seen on a three-lead rhythm strip. The patient is in coarse atrial fibrillation, which is paroxysmal in nature in this individual.

than physiologic atrial or dual-chamber, pacing.[63,65] Chronic atrial fibrillation is also more common in patients with the tachycardia-bradycardia syndrome who have had paroxysmal atrial fibrillation. When atrial fibrillation develops, many patients previously symptomatic from bradyarrhythmia experience a substantial improvement, likely due to the increase in ventricular rate.[66]

Thromboembolic events, which occur in 3% to 9% of patients, may be a manifestation of sick sinus syndrome.[51,52] In patients with bradycardia-tachycardia syndrome, the incidence increases to 24%.[56] Ventricular pacing (as opposed to atrial or dual-chamber pacing) and the presence of preexisting cerebrovascular disease are also associated with a higher thromboembolic event rate during follow-up.[63-65] High-risk patients should be carefully identified and placed on long-term anticoagulation therapy. However, not all cerebrovascular accidents in these patients are due to embolic events. Elderly patients with atherosclerotic cerebrovascular disease

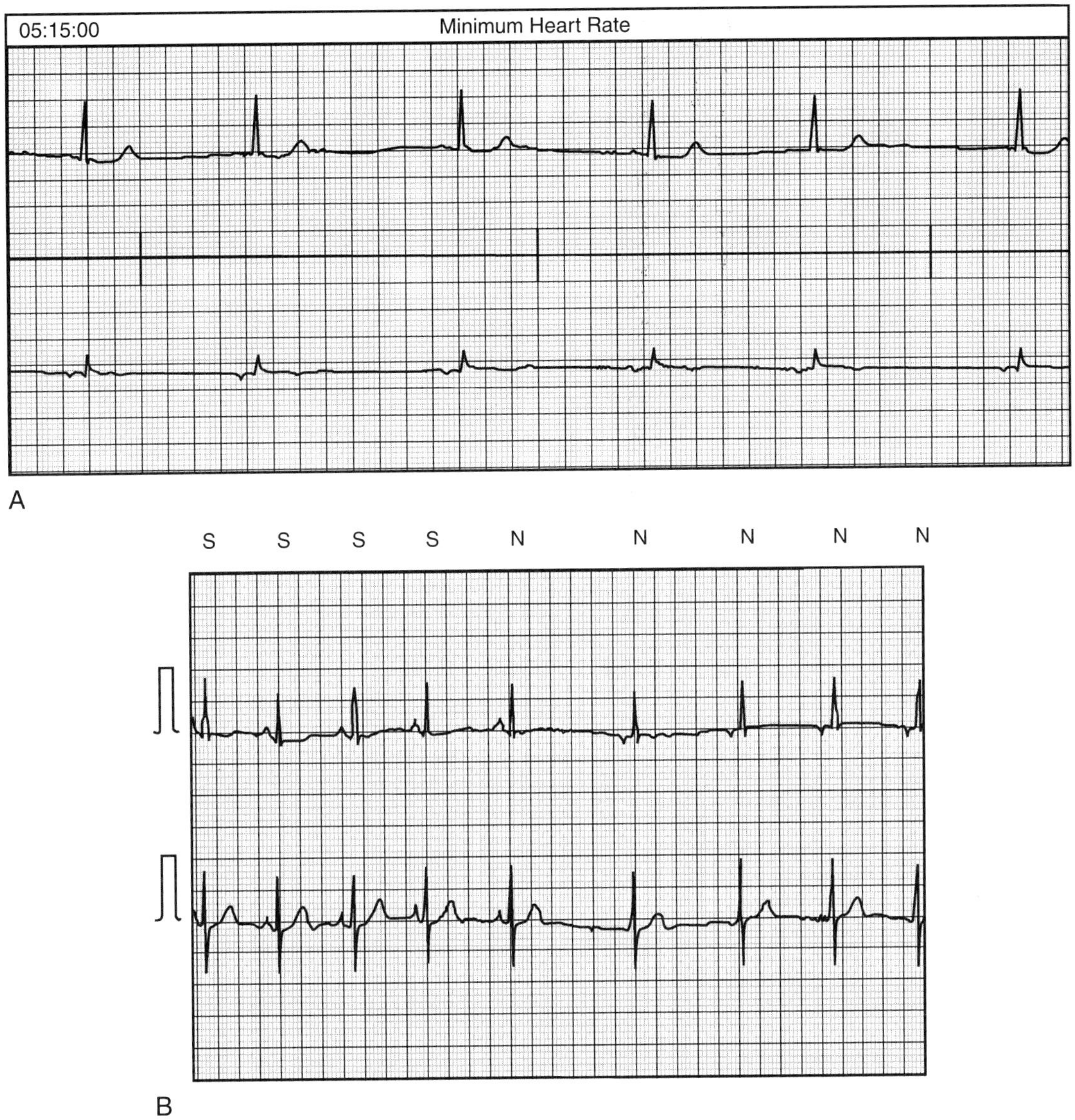

FIGURE 12-8 **Panel A,** Sinus bradycardia in a young patient with primary electrical disease and syncope seen on a dual-channel Holter monitor recording. **Panel B,** Sinus pause in the same patient as in *panel A* with an escape low atrial or junctional rhythm on another dual-channel Holter recording conducted on a different occasion. Escape rhythms may have a rate slow enough to cause symptoms of cerebral hypoperfusion.

may have transient ischemic attacks or frank cerebral infarction if bradyarrhythmia or the tachyarrhythmia is associated with a fall in cardiac output and reduction in cerebrovascular perfusion.

ATRIOVENTRICULAR CONDUCTION IN SINUS NODE DYSFUNCTION

At the time of diagnosis up to 17% of patients with sick sinus syndrome have evidence of AV conduction system disease, although high-degree AV block is unusual, reported in only 5% to 10% of cases.[50,67] The presence of AV conduction disease will affect therapeutic decisions such as safety of concomitant antiarrhythmic drug use and pacemaker mode choice. Electrocardiographic and electrophysiological findings suggestive of significant AV conduction system involvement include a P–R interval greater than 240 milliseconds (ms), complete bundle branch block, development of type I second-degree block during atrial pacing at rates of 120 bpm or less, H–V interval prolongation, and second- or third-degree AV block. During follow-up, AV conduction in patients with sick sinus syndrome usually remains stable.[68-70] In one literature survey of 28 studies of atrial pacing, an annual incidence of second- and third-degree AV block of 0.6% per year was reported.[68] A similarly low incidence (4 of 110 patients) was reported in a prospective trial of atrial versus ventricular pacing in patients with sick sinus syndrome[65] (included were those 70 years of age or younger with P–Q intervals ≤220 ms and those older than 70 years of age with P–Q intervals ≤260 ms); in addition, all patients had normal AV conduction at an atrial pacing rate of 100 bpm at pacemaker implantation.

The progression of AV conduction disease is thus usually slow and can be detected by careful clinical and electrocardiographic monitoring of these patients. Extrinsic influences, such as exposure to antiarrhythmics or drugs that can block conduction in the AV node, are more frequently responsible for worsening of AV conduction than is progressive degeneration within the conduction system.[71,73]

SINUS NODE DYSFUNCTION IN ACUTE MYOCARDIAL INFARCTION

Sinus bradycardia occurs commonly in patients with acute myocardial infarction, especially those with inferior and posterior infarction.[74,75] The bradycardia is usually due to stimulation of the afferent vagus nerve terminals, which are more common in the inferior and posterior ventricular walls. This vagal response can be potentiated by pain and by the use of vagotonic medications such as morphine sulfate and can be associated with a vasodepressor response resulting in systemic hypotension. Intravenous atropine usually reverses the vagal effects associated with myocardial infarction.

In addition to autonomic nervous system influences, sinus bradycardia may also be caused by ischemia of the sinus node or atrial tissue, although this diagnosis is rarely made clinically. Sinus node ischemia is also more common in inferior wall myocardial infarction, since the sinus node is usually supplied by the right coronary artery. The clinical manifestations are sinus bradycardia in the majority of cases; however, bradycardia alternating with episodes of supraventricular tachycardia has been reported in up to 35% of patients.[76] In the majority of cases, the sinus node dysfunction is temporary and normal sinus rhythm returns during the hospitalization.[76,77] Pacemaker implantation is rarely indicated. However, patients who experience alternation of bradycardia and tachycardia occasionally may require long-term antiarrhythmic therapy.[76]

Both noninvasive and invasive means of diagnosing sinus nodal dysfunction are available. Generally the noninvasive methods of electrocardiographic monitoring, exercise testing, and autonomic testing are used first. However, if symptoms are infrequent, invasive electrophysiological testing may be needed.

Diagnostic Evaluation

NONINVASIVE TESTING

The diagnosis of sinus nodal dysfunction is rarely made from a random ECG. If symptoms suggestive of sinus node dysfunction are frequent, Holter monitoring may be useful.[56,78,79] Documentation of symptoms by the patient in a diary, while wearing the Holter monitor, is essential for correlation of symptoms with the heart rhythm recorded at the time. In many cases, a Holter monitor can exclude sinus nodal dysfunction as the cause of symptoms if normal sinus rhythm is documented during dizziness, presyncope, or syncope. However, sinus bradycardia and sinus pauses may be recorded in asymptomatic individuals, reducing the specificity of these findings for the diagnosis.[6,80,81]

Event recorders are more useful than Holter monitors in patients with infrequent symptoms. Patient-activated models exist but are limited to patients who have symptoms prolonged enough to record the rhythm during an event. For patients with little to no warning, a loop recorder event monitor can be used. These recorders can be activated as soon as symptoms occur or after the fact, since the last 45 seconds of ECG recording are "frozen." Newer models that can be automatically triggered by bradycardia or tachycardia are available and useful in some patients. When an ECG diagnosis cannot be recorded by less invasive means, implantable loop recorders are useful in patients with recurrent symptoms suggestive of a bradyarrhythmia.

Exercise testing is of limited value in diagnosing sinus node dysfunction.[49] However, it is useful in differentiating patients with chronotropic incompetence from those with resting bradycardia who are able to demonstrate a normal heart rate increase with exercise. Patients with sinus nodal dysfunction and chronotropic incompetence exhibit abnormal heart rate responses to exercise. The increase in heart rate at each stage of exercise may be less than normal, with a plateau seen below the maximum age-predicted heart rate. Other patients may achieve an appropriate peak heart rate during exercise but have slow heart rate acceleration in the initial stage of exercise or a rapid deceleration of heart rate in the recovery stage. These abnormal chronotropic responses can help identify the cause of exercise intolerance in some patients with sinus nodal dysfunction and help determine their pacemaker prescription.[82]

Autonomic testing of the sinus node includes various pharmacologic interventions and maneuvers to test reflex responses. An abnormal response to carotid sinus massage (pause greater than 3 seconds) indicates carotid sinus hypersensitivity and may suggest the presence of carotid sinus syndrome. This response may also occur in asymptomatic elderly individuals.[83] Heart rate response to Valsalva's maneuver (normally decreased) or upright tilt (normally increased) can also be used to verify that the autonomic nervous system is itself intact.

The most commonly used pharmacologic intervention in the evaluation of sinus node dysfunction is the determination of the intrinsic heart rate.[84]

A low intrinsic heart rate is consistent with abnormal intrinsic sinus nodal function. A normal intrinsic heart rate in a patient with known clinical sinus nodal dysfunction suggests abnormal autonomic regulation.

INVASIVE TESTING

Sinus nodal function can be evaluated invasively with an electrophysiological study. This type of testing is usually reserved for symptomatic patients in whom sinus nodal dysfunction is suspected but cannot be documented in association with symptoms by noninvasive means. The pacing tests most commonly used are the sinus nodal recovery time and SA conduction time.

Pacing the atrium at rates faster than the inherent sinus rate is used to record the sinus node recovery time.

A delay in the return of spontaneous pacemaker activity (overdrive suppression) is a normal finding immediately after cessation of rapid atrial pacing. In patients with sinus nodal dysfunction, however, the sinus node generally takes longer to recover. A better measurement, however, is the corrected sinus nodal recovery time, which is obtained by subtracting the spontaneous sinus cycle length before pacing from the sinus nodal recovery time. Thus, a patient with an abnormally long sinus nodal recovery time could have a normal corrected sinus nodal recovery time if the resting heart rate is slow.[85] The indirect measurement of the corrected sinus nodal recovery time reflects both SA conduction time and sinus automaticity and thus has some limitations. Indirect sinus nodal recovery time measurements can be confounded by sinus nodal entrance block during rapid atrial pacing with resultant shortening of the sinus nodal recovery time and by sinus nodal exit block post pacing, thereby prolonging the measured sinus nodal recovery time. An abnormal sinus nodal recovery time is not found in all patients with sinus nodal dysfunction, partly due to sinus nodal dysfunction not being a homogenous entity from a pathologic standpoint. Despite these limitations, the indirect corrected sinus nodal recovery time is employed frequently in the evaluation of sinus nodal function.

The SA conduction time, another commonly used invasive pacing test, is traditionally measured indirectly from the high right atrium.[86,87] Several assumptions are used in the calculation, including sinus nodal automaticity not being affected by the premature beat, conduction in and out of the sinus node being equal, and the premature atrial beat not causing a shift in the principal pacemaker site. An additional limitation of the sinoatrial conduction time test is the need for a regular cycle length in sinus rhythm. Sinus arrhythmia may make calculation of the SA conduction time by this method impossible.

Clinical Electrophysiology

Electrophysiological testing for sinus node dysfunction is usually reserved for patients with significant symptoms suspicious for sinus node disease in whom a documented association between symptoms and sinus node abnormalities cannot be established by noninvasive testing. The three most common tests to assess sinus node dysfunction are sinus node recovery time, SA conduction time, and assessment of intrinsic heart rate. When used together, these tests have a sensitivity of 65% to 70% and a specificity of approximately 90%.[88] A negative test result does not exclude sinus node dysfunction, whereas a positive test result strongly favors intrinsic sinus node disease.

SINUS NODE RECOVERY TIME

The most commonly used test to assess sinus node function is the sinus node recovery time (SNRT). Cardiac pacemaker activity can be suppressed by pacing the heart at faster rates than that of the intrinsic pacemaker. Sinus node recovery time is assessed by pacing the heart at progressively shorter cycle lengths using a pacing catheter placed in the right atrium near the sinus node and then observing the recovery of sinus rhythm after cessation of pacing. Several protocols have been proposed, but a commonly used protocol is to start pacing the heart at rates slightly faster than sinus rhythm and increase the rate in increments to a cycle length of 300 ms (200 bpm), pacing for 30 seconds at each step and waiting approximately 1 minute between pacing drives. Two common measurements are then made following pacing. The maximum SNRT is measured as the longest pause from the last paced atrial complex to the first sinus return cycle. This measurement is made on the recording electrode closest to the sinus node, which usually is the high right atrial electrogram. The second measure is the corrected sinus node recovery time (CSNRT), which is obtained by subtracting the sinus cycle length from the SNRT. Several authors have reported normals for SNRT and CSNRT[89-93] (Table 12-1). The common upper limit of normal for SNRT is 1500 ms and for CSNRT is 550 ms (Fig. 12-9). The CSNRT is simple to use, takes into account the basic sinus cycle length, and is the most commonly used measurement to assess sinus node function. Importantly, the CSNRT may be normal in the presence of a prolonged SNRT in patients with sinus bradycardia. For example, patients with a resting heart rate of 50 bpm corresponding to a sinus cycle length of 1200 ms may have a "prolonged" SNRT of 1650 ms and yet a normal CSNRT of 450 ms. The CSNRT takes into account the resting bradycardia. Usually, the resting cycle length is achieved within 5 to 6 beats after termination of pacing, and it is not unusual to find oscillations in sinus cycle length shortening as far out as 10 beats following termination of pacing.[94] Frequently, the longest post pacing cycle length occurs in the second or third cycle following cessation of sinus pacing. This is due to SA entrance block during rapid pacing rates, which results in a shortening of the first recovery interval following the termination of pacing. The longest pause within the first six beats upon termination of pacing can be used as the measure of SNRT. These pauses occur most commonly with more rapid atrial pacing rates, making it reasonable to continue atrial pacing to a rate of 200 bpm (300 ms cycle length). Measurements of sinus node function in the electrophysiology laboratory are rarely abnormal in patients without clinical disease of the sinus node.

TABLE 12-1 Sinus Node Dysfunction: Normal Values

Reference	SNRT (ms)	CSNRT (ms)	SACT (ms)
Breithardt[2]	<1480	<508	<120
Mandel[3]	1040 ± 56		
Delius[4]	<1400	<525	
Josephson[5]		<550	45-125
Strauss[6]			68-156
Recommended	<1500	<550	<125

CSNRT, corrected sinus node recovery time; ms, milliseconds; SACT, sinoatrial conduction time; SNRT, sinus node recovery time.

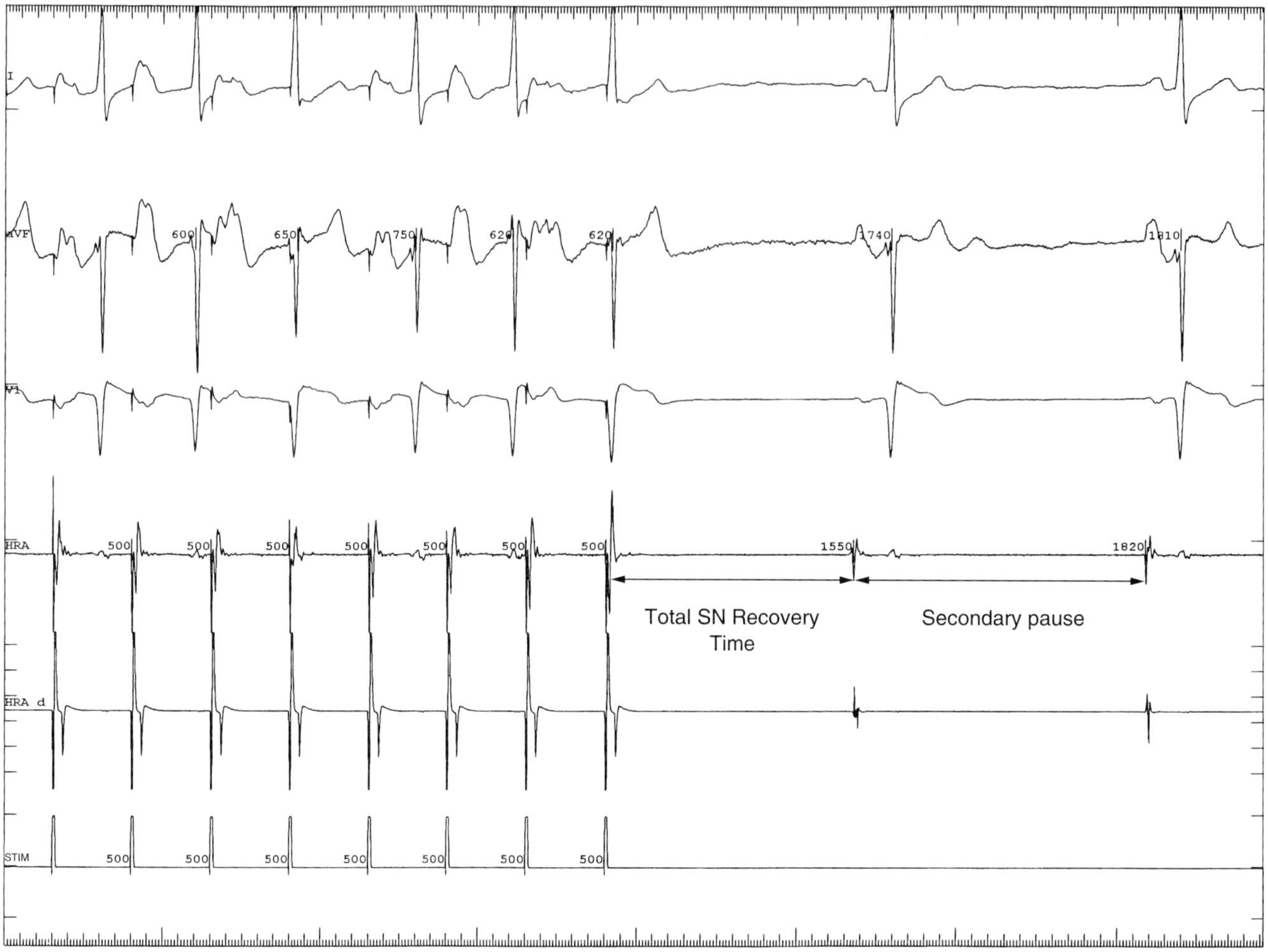

FIGURE 12-9 Sinus node recovery time after rapid atrial pacing in an elderly male with syncope and sinus bradycardia during electrophysiological testing. Note that the total recovery time is slightly prolonged, but a secondary pause is seen after the first recovery cycle. Numbers on high right atrial recording (*HRA*) indicate intervals in milliseconds. SN, sinus node; STIM, stimulation artefact.

However, patients with documented sinus node disease have a high incidence of falsely negative SNRTs during electrophysiological testing. Thus, the test has a low sensitivity but high specificity.

SINOATRIAL CONDUCTION TIME

An indirect method to evaluate SA conduction is the response to induced atrial premature depolarizations (APDs, Figs. 12-10 and 12-11). This test assesses the effect of APDs on the timing and duration of the sinus cycle. The method, devised by Strauss,[93] involves scanning the entire sinus cycle (A1A1) with programmed stimulation of single APDs delivered in the high right atrium near the sinus node. After approximately every eighth to tenth beat of a stable sinus rhythm, the premature beat is delivered with a gradual decrease in the coupling interval (A1A2), and the subsequent response of the next sinus return cycle (A2A3) is measured. The subsequent results are often plotted on a graph (see Fig. 12-10) comparing the timing of the atrial premature beat (A1A2) with the first sinus return cycle (A2A3). The response of the sinus node falls into four different zones (see Fig. 12-11): (1) collision; (2) reset; (3) entrance block or interpolation; and (4) reentry.

Zone 1 is referred to as the zone of collision, interference, or nonreset, since the mechanism of the response is due to a collision of the stimulated atrial impulse (A2) with the spontaneous sinus impulse (A1). In this zone, the return cycle intervals (A2A3) are fully compensatory with A1A2 plus A2A3 being equal to 2 times A1A1. This response usually occurs when APDs fall in the last 20% to 30% of the spontaneous cycle length. The sinus pacemaker is unaffected by A2, and the subsequent beat occurs "on time."

Zone 2 is referred to as the reset zone, in which the spontaneous cycle length (A2A3) following the premature beat remains constant, producing a plateau in the curve. In this zone, the premature beat (A2) causes reset of the sinus pacemaker, resulting in a recovery cycle (A2A3) that exceeds the basic sinus cycle. However, the subsequent return cycle causes less than a compensatory pause with the sum of A1A2 and A2A3 being less than 2 times A1A1. Zone 2 typically occupies 40% to 50% of the cardiac cycle, and the plateau results from A2 entering the sinus node and resetting the pacemaker

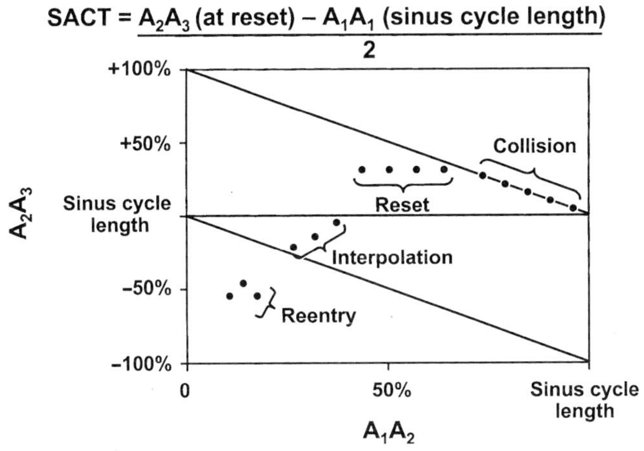

FIGURE 12-10 Assessing sinoatrial conduction time (SACT) using atrial premature depolarizations (APDs). The coupling interval of the APD, A1A2, and the return cycle, A2A3, are expressed as percentages of the sinus cycle length A1A1. The zones of collision, reset, interpolation, and reentry are explained further in Figure 12-11. The SACT is determined by subtracting the sinus cycle length, A1A1, from the A2A3 interval during the plateau or reset phase. This number theoretically represents conduction into and out of the sinus node and is then divided in half to obtain sinoatrial conduction. (Modified from Ward DE, Camm AJ: The current status of ablation of cardiac conduction tissue and ectopic myocardial foci by transvenous electrical discharges. Eur Heart J 1982;3:267.)

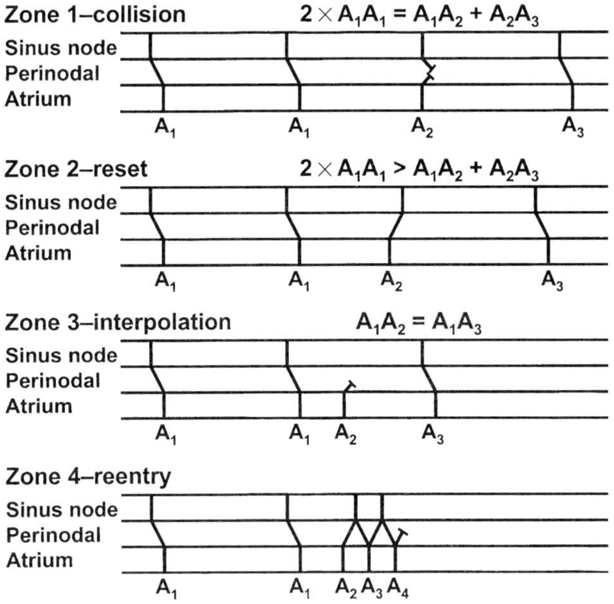

FIGURE 12-11 Schematic ladder diagram demonstrating the effect of atrial premature depolarizations (APDs) delivered during sinus rhythm on the subsequent sinus cycle to measure the sinoatrial conduction time using the Strauss method. The responses of the sinus node fall into four types based on how the APD interacts with the sinus node. In Zone 1, the APD, A2, collides with the sinus impulse and fails to enter the sinus node. This zone is termed the *Zone of collision*, and the sinus cycle length is not affected (2 × A1A1 = A1A2 + A2A3). In Zone 2, the APD is delivered earlier in the sinus cycle at a point where the impulse is able to depolarize the sinus node and the next impulse, A3, is advanced, or reset. This is termed the *Zone of reset* (2 × A1A1 > A1A2 + A2A3). In Zone 3, the APD, A2, is delivered at even earlier coupling intervals and fails to invade or reset the sinus node or affect conduction through the perinodal tissue. The next sinus beat, A3, occurs on time, resulting in interpolation (A1A2 = A1A3). Zone 4 occurs infrequently when an APD results in a sudden transition to short A2A3 and A3A4 intervals that are much shorter than the sinus cycle length, A1A1. This is believed to reflect sinoatrial reentry. (Modified from Ward DE, Camm AJ: The current status of ablation of cardiac conduction tissue and ectopic myocardial foci by transvenous electrical discharges. Eur Heart J 1982;3:267.)

but having no effect on pacemaker automaticity. It has been assumed that conduction into and out of the sinus node is equal. The conduction from the sinus node to the atrium is the SA conduction time and is calculated as the A2A3 interval that results in reset minus the A1A1 interval divided by 2. However, isolated studies have shown that this assumption is not entirely correct and that conduction time out of the sinus node is greater than conduction time into the sinus node.[92] This has also been confirmed by direct recording of the sinus node electrogram.[95] Despite this and other limitations, the SA conduction time is reasonably estimated by the introduction of APDs when a true plateau is present in Zone 2. The measurement of SA conduction time is considered normal if it is less than 125 ms. Errors in the measurement of SA conduction time may occur if sinus arrhythmia is present, a slow increase in the plateau due to a shift to slower sinus node pacemaker cells occurs, or there is an increase in conduction time into the sinus node due to refractoriness of the atrial tissue surrounding the node.

In Zone 3 (interpolation) and Zone 4 (reentry), the return cycle (A2A3) is shorter than the spontaneous sinus cycle (A1A1) in response to the premature beat (A2). The response in Zone 3 is due to interpolation and results from the premature beat encountering refractoriness of the tissue surrounding the sinus node, resulting in sinus node entrance block. SA conduction time is an insensitive indicator of sinus node disease, but a positive test can help establish sinus node dysfunction.

OTHER TESTS TO ASSESS SINUS NODE DYSFUNCTION

Direct recording of sinus node electrograms and the measurement of intrinsic sinus node function following autonomic blockade have both been reported for assessing patients with suspected sinus node dysfunction.[95,96] Sinus node electrograms are recorded using catheters with 0.5 to 1.5 cm interelectrode distance positioned at the junction of the superior vena cava and right atrium in the region of the sinus node or a catheter looped in the right atrium, allowing firm contact in the region of the superior vena cava and junction with the atrium. Sinus node recordings can be obtained in 40% to 90% of patients studied but do take time and experience (Fig. 12-12). The high pass filter is usually between 0.1 and 0.6 Hz, the low pass filter between 20 and 50 Hz,

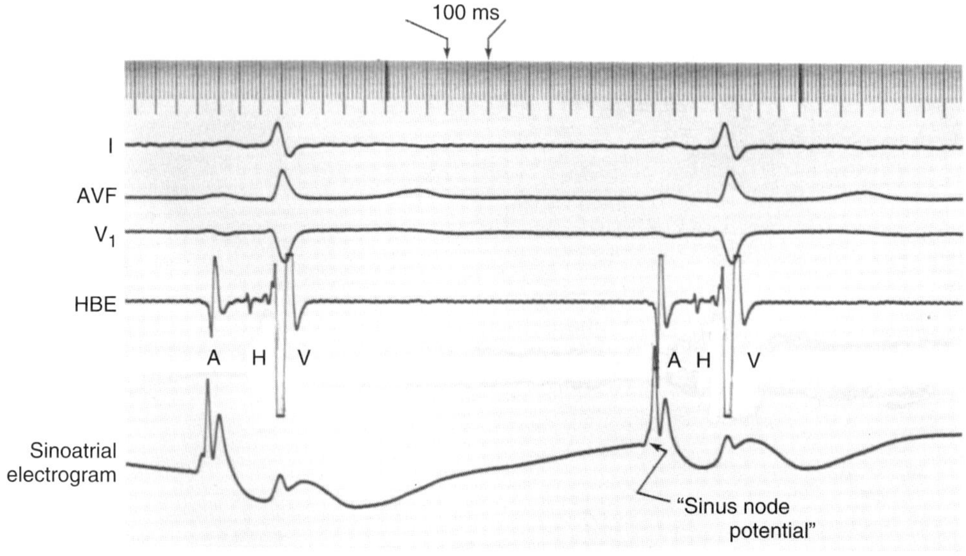

FIGURE 12-12 Sinus node electrogram potential recording at electrophysiological testing. Note the slow upstroke of the prepotential on the unfiltered atrial electrogram on the bipolar recording obtained at the sinoatrial junction region. (Courtesy S. Saksena).

and the signal gained at a range of 50 to 100 microvolt/cm. Using this approach, Josephson reports obtaining stable sinus node electrograms in only 50% of an unselected population of patients.[92] Studies of patients with sinus node dysfunction that compare both the indirect and direct methods[75] show a good correlation between the directly measured SA conduction time and that made by indirect techniques using the introduction of APDs.

Autonomic blockade has been used to determine the intrinsic heart rate (IHR) using a combination of atropine and β-blockers. This testing assumes that autonomic blockade will remove the autonomic effects on sinus node function and unveil unmodulated sinus node activity. A pharmacologic denervation is achieved using propranolol, 0.2 mg/kg, and atropine, 0.04 mg/kg, to identify the IHR. The effects usually peak after 5 minutes, and the measurement of IHR is age dependent and defined by the regression equation:

$$IHR = 117.2 - (0.53 \times age) \text{ bpm}$$ [97,98]

The IHR helps to distinguish patients with true sinus node dysfunction from those with enhanced parasympathetic tone. Additional studies have demonstrated that the response to atropine provides as much information as the combined response to atropine plus β-blockade.[99]

The direct electrogram used to record sinus node activity is difficult to obtain, and the technique is rarely used in the clinical electrophysiology laboratory. Similarly, measurement of IHR following autonomic blockade is infrequently used.

EFFECTIVENESS OF DRUGS ON ASSESSMENT OF SINUS NODE FUNCTION

Digoxin shortens the SNRT in patients with clinical sinus node dysfunction.[92] This is possibly related to an increase in refractoriness in the atrial tissue surrounding the sinus node, creating SA entrance block. In this situation, fewer atrial-paced beats would enter the sinus node to suppress sinus node function. Propranolol has been demonstrated to increase the SNRT, possibly by decreasing sinus node automaticity. Verapamil and diltiazem have been shown to have minimal effects on sinus node function in normal persons. Antiarrhythmic drugs can adversely affect sinus node function in patients with evidence of sinus node disease but, in general, have minimal effects on sinus node function in normal persons. In general, electrophysiological assessment of sinus node function should not be performed in the presence of drugs known to affect sinus node activity, including digoxin, β-blockers, calcium channel antagonists, and antiarrhythmic drugs. Such drugs may exacerbate sinus node function in patients with sinus node disease and may also adversely influence sinus activity in otherwise normal patients.

Management of patients with sinus node disease is most appropriately conducted using clinical, electrocardiographic, and long-term monitoring information. Electrophysiological testing of sinus node function infrequently adds important contributions to the care of patients with sinus node disease. A normal test result does not exclude the diagnosis of sinus node disease, and an abnormal test is not itself an indication for pacing in an asymptomatic patient. Drugs have an important effect on sinus node automaticity and atrial refractoriness and can exacerbate sinus node function in the presence of SA disease.

Evidence-Based Therapy

During the past decade, the importance of making clinical decisions on the basis of *evidence* from clinical trials has been emphasized in clinical medicine. This section reviews the available information from clinical trials on the efficacy of pacing therapy and the effects of pacing mode selection in patients with sinus node dysfunction.

EFFECTIVENESS OF PACING THERAPY

There is only a single published study evaluating the effectiveness of pacing therapy versus no therapy or pharmacologic therapy for preventing symptoms in patients with sinus node dysfunction. In the THEOPACE study, 107 patients with presumed sinus node dysfunction (older than 45 years of age with a mean resting sinus rate < 50 bpm or intermittent sinoatrial block, or both, noted during a diurnal ECG on two separate occasions, and symptoms thought to be secondary to sinus node dysfunction) were randomized to no treatment, oral theophylline, or dual-chamber rate-adaptive pacing.[51] Patients with severe sinus node dysfunction, defined as symptomatic heart rates less than 30 bpm or sinus pauses greater than 3 seconds, were excluded. After an average 18-month follow-up, syncope had occurred in 6% of patients who received pacing therapy and 17% and 23% in the theophylline and control arms, respectively. In all three groups the incidence of atrial tachycardias was similar (26% to 28%). THEOPACE demonstrates that pacing therapy provides symptomatic benefit in patients with sinus node dysfunction. However, it is unlikely that pacing therapy confers a survival benefit, since natural history studies suggest that sinus node dysfunction by itself does not appear to be associated with an increased risk of death.[55]

PACING MODE CHOICE

In patients with sinus node dysfunction, bradycardia can be prevented by single-chamber ventricular pacing (VVI mode), single-chamber atrial pacing (AAI mode), or dual-chamber pacing (DDD mode). Several randomized studies have evaluated the effects of pacing mode in patients with sinus node dysfunction. The first prospective study, initially published in 1994 with follow-up data presented in 1997 and 1998, evaluated 225 patients with sinus node dysfunction who were randomized to single-chamber atrial pacing or single-chamber ventricular pacing.[65] After a mean follow-up of 3.3 years, atrial pacing was associated with a significant decrease in thromboembolic events (atrial pacing: 5.5%; ventricular pacing: 17.4%) and a nonsignificant reduction in atrial fibrillation (atrial pacing: 14%; ventricular pacing: 23%). In addition, progression of heart failure symptoms was observed in 9% of patients in the atrial pacing group and 31% of the ventricular pacing group. In the Pacemaker Selection in the Elderly (PASE) trial 407 elderly patients (mean age, 76 years) were randomized to either the VVIR or DDDR pacing mode with a mean follow-up of 30 months.[100] In the 175 sinus node dysfunction patients enrolled, the DDDR pacing mode was associated with improved cardiovascular functional status and better quality-of-life scores in the role physical, role emotion, and social function categories of the SF-36 questionnaire. The DDDR pacing mode was associated with insignificant reductions in mortality (DDDR, 12%; VVIR, 20%) and incidence of atrial fibrillation (DDDR, 19%; VVIR, 28%). For the entire study group, 26% of patients crossed over from the VVIR pacing mode to the DDDR pacing mode because of symptoms related to pacemaker syndrome.

The Canadian Trial of Physiologic Pacing (CTOPP) evaluated the effects of pacing mode choice in patients with symptomatic bradycardia.[101] The 2568 patients were randomized to single-chamber ventricular pacing or a "physiologic" pacing mode that preserved AV synchrony (single-chamber atrial pacing or dual-chamber pacing). While no significant difference in the annual rate of stroke or death was detected (ventricular pacing, 5.5%; physiologic pacing, 4.9%), the annual rate of atrial fibrillation was significantly lower in the physiologic pacing group (ventricular pacing, 6.6%; physiologic pacing, 5.3%). The reduction in atrial fibrillation became more apparent 2 years after initial randomization. Details on the 40% of patients with sinus node disease have not been published, but subgroup analysis suggested that the indication for permanent pacing (sinus node dysfunction or AV block) did not have a significant effect on the annual rate of stroke or death. An important caveat for this study is that after 5 years of follow-up, while 95% of patients randomized to ventricular pacing were still programmed to ventricular pacing, only 75% of patients randomized to physiologic pacing were actually still in a physiologic pacing mode.

Finally, in the Mode Selection Trial (MOST), 2010 patients with sinus node dysfunction were randomized to dual-chamber pacing or single-chamber ventricular pacing.[102] After a 5-year follow-up, no differences in mortality or stroke were detected, but there was a marked reduction in progression to atrial fibrillation, particularly in patients without a prior history of the arrhythmia; however, the crossover rate to dual chamber pacing was 31% due to symptoms of the pacemaker syndrome, and dual-chamber pacing was associated with improved quality of life.

Thus, available data suggest that pacing modes that preserve AV synchrony are associated with a reduced incidence of atrial fibrillation, particularly after several years. In addition, preservation of AV synchrony is associated with improved quality of life and a reduction in the incidence of pacemaker syndrome.

Management

The first step in the evaluation and management of patients with sinus node dysfunction is to exclude physiologic sinus bradycardia due to extrinsic conditions affecting the sinus node or sinoatrial tissue. β-blockers, some calcium channel blockers (verapamil and diltiazem), digoxin, and other medications having sympatholytic activity can result in sinus node dysfunction. It is therefore important to carefully review all medications taken by the patients, since withdrawal of offending agents usually results in restoration of normal sinus function. Other causes of extrinsic sinus node dysfunction should be investigated and excluded; these include hypothyroidism, sleep apnea, and other systemic diseases. The specific situation in which the bradycardia occurs should be analyzed in order to document bradycardia triggered by an increase in vagal tone, such as occurs during suctioning or vomiting. Whenever possible, treatment should be directed toward correcting the extrinsic condition causing bradycardia.

GUIDELINES FOR MANAGEMENT OF PATIENTS WITH INTRINSIC SINUS NODE DYSFUNCTION

When intrinsic sinus node dysfunction is suspected, it is important to attempt to correlate symptoms with documentation of the arrhythmia, since sinus node dysfunction is common, especially in elderly patients, and may not cause symptoms. The intermittency of symptoms and ECG features characteristic of the syndrome may result in difficulty in establishing a cause-effect relationship. A Holter monitor or event recorder may be useful to establish the diagnosis, but prolonged monitoring may be required. In selected cases, an implantable loop recorder that continuously acquires electrocardiographic signals can be used.[102] Krahn et al. placed implantable loop recorders in 16 patients with syncope and negative electrophysiological and tilt table tests. Fifteen of the patients had recurrent syncope, and sinus arrest was documented in five patients.[102] Invasive electrophysiological studies are usually not required to specifically evaluate sinus node function.

Therapy is aimed at improving symptoms. There is no evidence that medical therapy or pacemaker implantation improves survival; this is in part likely due to the low mortality rate related to the bradyarrhythmia per se.[51]

INDICATIONS FOR PERMANENT PACING

Pharmacologic therapy for bradycardia due to sinus node dysfunction is generally ineffective; pacemaker implantation is therefore the optimum therapy. In the United States, sinus node dysfunction is the most common indication for pacemaker implantation.[10] The benefit to be expected from permanent pacing depends largely on the appropriateness of the indication.

Guidelines for pacemaker implantation in sinus node dysfunction have been published.[104] Indications for permanent pacing in sinus node disease are summarized in Table 12-2. The guidelines emphasize the importance of correlating symptoms and bradycardia whenever possible. The principal benefits of pacing therapy are prevention of syncope, improvement of symptoms due to poor tissue perfusion, and congestive heart failure due to decreased cardiac output caused by slow heart rates. In an observational series of severely symptomatic patients from the 1970s, pacemaker therapy improved symptoms of fatigue, lightheadedness, and near syncope.[105] In a randomized trial from the 1990s conducted to assess the efficacy of pacemakers in patients with sick sinus syndrome, the occurrence of syncope was lower in the paced group over a mean follow-up duration of 18 months (6% versus 23% for controls; $P = .02$).[51] Heart failure also occurred less often in patients assigned to pacemaker therapy (3% versus 17%; $P = .05$), whereas the incidence of sustained paroxysmal tachyarrhythmias, chronic atrial fibrillation, and thromboembolic events were not different between the groups. Compared with observational series and retrospective studies, pacemaker implantation in this randomized trial did not demonstrate different effects on "minor" symptoms such as fatigue, dizziness, palpitation, and New York Heart Association class compared with the "no-treatment" group; this was due to subjective improvement occurring in the placebo group as early as 3 months following randomization.[51]

In some patients, the bradycardia may be iatrogenic and exacerbated by medications used to treat supraventricular tachycardias. If these medications constitute

TABLE 12-2 Indications for Permanent Pacing in Sinus Node Dysfunction

Class	Indications
Class I	Sinus node dysfunction with documented symptomatic bradycardia, including frequent sinus pauses that produce symptoms. In some patients, bradycardia is iatrogenic and will occur as a consequence of essential long-term drug therapy of a type and dose for which there are no acceptable alternatives
	Symptomatic chronotropic incompetence
Class IIa	Sinus node dysfunction occurring spontaneously or as a result of necessary drug therapy with a heart rate <40 bpm when a clear association between symptoms consistent with bradycardia and actual presence of bradycardia has not been documented
	Syncope of unexplained origin when major abnormalities of sinus node function are discovered or provoked in electrophysiological studies
Class IIb	Minimally symptomatic patients, chronic heart rate <40 bpm while awake
Class III	Sinus node dysfunction in asymptomatic patients, including those in whom substantial sinus bradycardia (heart rate <40 bpm) is a consequence of long-term drug treatment
	Sinus node dysfunction in patients with symptoms suggestive of bradycardia that are clearly documented as not associated with a slow heart rate
	Sinus node dysfunction with symptomatic bradycardia due to nonessential drug therapy

Class I: conditions for which there is evidence or general agreement, or both, that a given procedure or treatment is beneficial, useful, and effective.
Class II: conditions for which there is conflicting evidence or a divergence of opinion, or both, about the usefulness/efficacy of a procedure or treatment.
Class IIa: weight of evidence/opinion is in favor of usefulness/efficacy. **Class IIb:** usefulness/efficacy is less well established by evidence/opinion.
Class III: conditions for which there is evidence or general agreement, or both, that a procedure/treatment is not useful/effective and, in some cases, may be harmful.

From Gregoratos G, Cheitlin MD, Conill A, et al: ACC/AHA guidelines for implantation of cardiac pacemakers and antiarrhythmia devices: A report of the American College of Cardiology/American Heart Association Task Force on Practice Guidelines (Committee on Pacemaker Implantation). J Am Coll Cardiol 1998;5:1175-1209; and Gregoratos G, Abrams J, Epstein AE, et al: ACC/AHA/NASPE 2002 guideline update for implantation of cardiac pacemakers and antiarrhythmia devices. J Am Coll Cardiol 2002;40:1703-1719).

the only alternative for management of the tachycardia, and if the patients are symptomatic from the bradycardia, they should receive permanent pacing. However, if the bradycardia does not produce symptoms, the rhythm per se is not an indication for pacing.

The published guidelines also take into consideration the erratic course of the disease and the difficulty in establishing the cause-effect relationship between symptoms and arrhythmia. Accordingly, pacing is considered useful in patients with heart rates less than 40 bpm who have symptoms consistent with bradycardia, but in whom the correlation between the two cannot be clearly established. Pacing is not indicated in asymptomatic patients, patients with bradycardia due to nonessential medical therapy, and when symptoms are clearly documented not to be caused by bradycardia.

Permanent pacing is indicated in patients with chronotropic incompetence who become symptomatic during activity because of inability to increase heart rate and cardiac output. These patients have an improvement in symptoms and exercise tolerance with rate-responsive pacing.[9,96]

PACING MODE SELECTION

Single-chamber atrial pacemakers, single-chamber ventricular pacemakers, and dual-chamber pacemakers will all prevent bradycardia in the patient with sinus node disease. Each pacemaker type is associated with inherent advantages and disadvantages. Single-chamber atrial pacemakers are simple, relatively inexpensive (approximately $3000 to $4000), and maintain AV synchrony. However, they will not prevent ventricular bradycardia if AV block develops. Up to 20% of patients will have abnormal AV conduction at the time of diagnosis of sinus node dysfunction; these patients are not candidates for single-chamber atrial pacing.[50] In a prospective study of 225 patients with 1:1 conduction at heart rates less than 100, Andersen et al. found that AV conduction was unchanged from initial evaluation after a mean 5.5-year follow-up.[69] The annual incidence of second- and third-degree AV block that required implantation of a dual-chamber pacing system was only 0.6% per year.[69]

Single-chamber ventricular pacemakers are also simple and inexpensive. They prevent bradycardia in the presence of AV block but do not maintain AV synchrony. Loss of AV synchrony is associated with a 20% to 30% decrease in cardiac output and is associated with "pacemaker syndrome."[108] Pacemaker syndrome is a constellation of symptoms that can include dizziness, chest pain, weakness, effort intolerance, presyncope, and syncope. The mechanism of pacemaker syndrome is complex but appears to be due to decreased cardiac output from loss of AV contraction and retrograde conduction through the His-Purkinje–AV node axis. Reduced cardiac output leads to reflex sympathetic activation. Atrial contraction when the mitral and tricuspid valves are closed also leads to an increase in atrial pressure and release of atrial natriuretic peptide and peripheral venous and arterial dilation. The reported incidence of pacemaker syndrome has varied widely among studies (1% to 80%), which probably reflects variability in definition rather than a true variability in incidence.[107,109] In the large randomized trials such as PASE and MOST, crossover from single-chamber ventricular pacing to dual-chamber pacing due to pacemaker syndrome was approximately 25% to 30%.[101,110,111]

Dual-chamber pacemakers maintain AV synchrony and prevent bradycardia from all causes. However, dual-chamber pacemakers are more complex and relatively expensive ($5000 to $7000). In addition, since two intracardiac leads are required, the incidence of lead dislodgement is higher for dual-chamber systems (dual-chamber, 6%; single-chamber, 2%).

Currently available pacing systems have a rate-adaption feature. When rate adaption is programmed "on," the pacing system employs a sensor such as body motion, minute ventilation, Q–T interval changes, or combinations of these to estimate metabolic need. The pacemaker will change the pacing rate depending on input from the sensor. This feature is particularly useful for patients with sinus node dysfunction associated with chronotropic incompetence. Pacemakers have monitoring capabilities that allow the clinician to evaluate the range of heart rates a patient has over a specific interval of time. If a blunted range of atrial rates is recorded, the presence of chronotropic incompetence should be suspected.

Sinus Node Dysfunction in Specific Conditions

ACUTE MYOCARDIAL INFARCTION

Sinus bradycardia is common in acute myocardial infarction, especially in inferior and posterior wall infarction, where it is usually due to increased vagal tone or ischemia of the SA tissue.[16,17] Increased vagal tone may also result in transient AV block and hypotension from peripheral vasodilatation. This arrhythmia usually does not require treatment unless the patient is symptomatic (hypotension, ischemia, or bradycardia-related ventricular arrhythmia). It usually responds well to intravenous atropine. In symptomatic patients who are unresponsive to atropine or who have recurrences requiring multiple doses of atropine, temporary transvenous pacing may be required. Pacing is usually performed at the right ventricular apex, but where it is important to maintain AV synchrony (such as refractory hypotension), an additional "J"-shaped electrode can be placed in the right atrial appendage for dual-chamber pacing. Alternatively, atrial pacing can be achieved using a temporary electrode positioned in the proximal coronary sinus. Sinus node dysfunction occurring during acute myocardial infarction is usually temporary, and permanent pacing is rarely required.[76,77]

CAROTID SINUS HYPERSENSITIVITY AND CAROTID SINUS SYNDROME

An abnormal response to carotid sinus massage (>3 seconds of asystole) may occur in asymptomatic patients and does not constitute an indication for therapy; correlation with symptoms is essential. Up to 64% of patients with syncope caused by a hypersensitive carotid syndrome can remain asymptomatic during follow-up; therapy should be reserved for patients with recurrent presyncope or syncope.[112] Drugs that can enhance the hypersensitive response to carotid sinus massage, such as digoxin and sympatholytics (clonidine, methyldopa), should be discontinued if possible.

The type and success of therapy (pacing versus pharmacologic) are based on the mechanism of syncope. Pacing is efficacious in the cardioinhibitory response to carotid sinus massage.[113-116] In patients with a predominant vasodepressor response, neither pacing nor anticholinergic agents prevent the fall in blood pressure, since this is caused by inhibition of sympathetic vasoconstrictor nerves as well as by activation of cholinergic sympathetic vasodilator fibers.[116] Elastic support stockings and volume expansion with sodium-retaining drugs may be useful for ameliorating symptoms.

Recent data suggest that unexplained falls in the elderly might be due to carotid sinus hypersensitivity, and that in some cases pacing therapy may be beneficial.[117] A selected group of 175 patients with a history of nonaccidental falls and significant bradycardia (asystole >3 seconds) in response to carotid sinus massage were randomized into pacing-therapy and no–pacing-therapy groups; after a 1 year follow-up, injurious events were reduced by 70% in the pacing-therapy group.

The indications for permanent pacing in the carotid sinus syndrome are summarized in Table 12-3. Single-chamber atrial pacing is contraindicated, since vagal activation frequently results in AV block and absence of a ventricular escape rhythm. Although single-chamber ventricular pacing prevents bradycardia, it can potentially exacerbate symptoms due to the neurohormonal effects associated with the pacemaker syndrome. Dual-chamber pacing is therefore preferred, since it maintains AV synchrony regardless of the cause of bradycardia. In addition, the current generation of dual-chamber pacing systems allows programming of different heart rates after sensed and paced ventricular beats (hysteresis). By programming a pacemaker to pace at a relatively fast rate on initiation of pacing, symptoms associated with carotid sinus syndrome can be ameliorated even in the presence of a significant vasodepressor response.[118]

VASOVAGAL SYNCOPE

Vasovagal syncope is the most common cause of syncope in young people. Lewis, in 1932, used the phrase "vasovagal" to emphasize the combination of arterial vasodilation and bradycardia associated with this syndrome.[119] While the exact mechanism of vasovagal syncope is not known, it does appear that activation of cardiac mechanoreceptors leads to activation of higher neural centers and reflex withdrawal of sympathetic tone and increased vagal tone.

The most common cause for bradycardia in vasovagal syncope is sinus bradycardia or sinus arrest. For this reason, although they do not affect the vasodepressor component of this syndrome, pacemakers have been used to treat the subset of patients that has particularly severe symptoms unresponsive to drug therapy. Two relatively large randomized studies have evaluated the efficacy of pacing therapy for the treatment of vasovagal syncope. In the North American Vasovagal Pacemaker Study (VPS-I), 54 patients with severe vasovagal syncope and relative bradycardia (trough heart rate <60 bpm during tilt table testing) were randomized to pacing or no pacing.[120] The study was terminated after 2 years when analysis showed that pacing was associated with an 85% decrease in syncopal episodes. Similarly, in the Vasovagal Syncope International Study (VASIS-I), 42 patients with severe drug refractory vasovagal syncope were randomized to pacing or no pacing.[121] Only one patient in the pacing group had syncope, while 14 patients in the no-pacing group had syncope during a mean 3.7-year follow-up. More recently, VPS-II found

TABLE 12-3 Indications for Permanent Pacing In Hypersensitive Carotid Syndrome

Class	Indications
Class I	Recurrent syncope caused by carotid sinus stimulation; minimal carotid sinus pressure induces ventricular asystole of >3 sec duration in the absence of any medication that depresses the sinus node or atrioventricular conduction
Class IIa	Recurrent syncope without clear, provocative events and with a hypersensitive cardioinhibitory response
	Significantly symptomatic and recurrent neurocardiogenic syncope associated with bradycardia documented spontaneously or at the time of tilt table testing
Class IIb	None
Class III	A hyperactive cardioinhibitory response to carotid sinus stimulation in the absence of symptoms
	A hyperactive cardioinhibitory response to carotid sinus stimulation in the presence of vague symptoms such as dizziness, light-headedness, or both
	Recurrent syncope, light-headedness, or dizziness in the absence of a hyperactive cardioinhibitory response

Class I: conditions for which there is evidence or general agreement, or both, that a given procedure or treatment is beneficial, useful, and effective.
Class II: conditions for which there is conflicting evidence or a divergence of opinion, or both, about the usefulness/efficacy of a procedure or treatment.
Class IIa: weight of evidence/opinion is in favor of usefulness/efficacy. **Class IIb:** usefulness/efficacy is less well established by evidence/opinion.
Class III: conditions for which there is evidence or general agreement, or both, that a procedure/treatment is not useful/effective and, in some cases, may be harmful.

From Gregoratos G, Cheitlin MD, Conill A, et al: ACC/AHA guidelines for implantation of cardiac pacemakers and antiarrhythmia devices: A report of the American College of Cardiology/American Heart Association Task Force on Practice Guidelines (Committee on Pacemaker Implantation). J Am Coll Cardiol 1998;5:1175-1209; and Gregoratos G, Abrams J, Epstein AE, et al: ACC/AHA/NASPE 2002 guideline update for implantation of cardiac pacemakers and antiarrhythmia devices. J Am Coll Cardiol 2002;40:1703-1719).

a 30% risk reduction in time to syncope with DDD pacing ($P = .14$), which was considerably lower than in previous studies.[122-125] Patients who require pacing therapy for drug-resistant vasovagal syncope should receive a dual-chamber pacemaker, because transient AV block can be observed during bradycardia episodes. Special programming features that provide an initial higher pacing rate when pacing is initiated have been developed to optimize pacing therapy for patients with vasovagal syncope.[123,126] More recent studies such as VASIS-II are under way to evaluate the effectiveness of these features.

REFERENCES

1. Keith A, Flack M: The form and nature of the muscular connections between the primary divisions of the vertebrate heart. J Anat Physiol 1907;41:172-89.
2. Laslett EE: Syncopal attacks associated with prolonged arrest of the whole heart. QJM 1909;2:347-53.
3. Ferrer MI: The sick sinus syndrome in atrial disease. JAMA 1968;206:645.
4. Lown B: Electrical reversion of cardiac arrhythmias. Br Heart J 1967;29:469-89.
5. Short DS: The syndrome of alternating bradycardia and tachycardia. Br Heart J 1954;16:208-14.
6. Brodsky M, Wu D, Denses P, et al: Arrhythmias documented by 24 hour continuous electrocardiographic monitoring in 50 male medical students without apparent heart disease. Am J Cardiol 1977;39:390-5.
7. Sobotka PA, Mayer JH, Bauernfeind RA, et al: Arrhythmias documented by 24 hour continuous ambulatory electrocardiographic monitoring in young women without apparent heart disease. Am Heart J 1981;101:753-9.
8. Fleg JL, Kennedy HL: Cardiac arrhythmias in a healthy elderly population: Detection by 24-hour ambulatory electrocardiography. Chest 1982;81:301-7.
9. Kulbertus HE, De Leval-Rutten F, Mary L, et al: Sinus node recovery time in the elderly. Br Heart J 1975;37:420-5.
10. Bernstein AD, Parsonnet V: Survey of cardiac pacing and implanted defibrillator practice patterns in the United States in 1997. Pacing Clin Electrophysiol 2001;24:842-55.
11. Ector H, Rickards AF, Kappenberger L, et al: The World Survey of cardiac pacing and implantable cardioverter defibrillators. Pacing Clin Electrophysiol 2001;24:842-55.
12. Albin G, Hayes DL, Holmes DR: Sinus node dysfunction in pediatric and young adult patients: Treatment by implantation of a permanent pacemaker in 39 cases. Mayo Clin Proc 1985;60: 667-72.
13. Bharati S, Surawicz B, Vidaillet HJ, Lew M: Familial congenital sinus rhythm abnormalities: Clinical and pathological correlations. Pacing Clin Electrophysiol 1992;15:1720-9.
14. Mehta AV, Chidambaram B, Garrett A: Familial symptomatic sinus bradycardia: Autosomal dominant inheritance. Pediatr Cardiol 1995;16:231-4.
15. Olson TM, Keating MT: Mapping a cardiomyopathy locus to chromosome 3p22-p25. J Clin Invest 1996;97:528-32.
16. Chau EM, McGregor CG, Rodeheffer RJ, et al: Increased incidence of chronotropic incompetence in older donor hearts. J Heart Lung Transplant 1995;14:743-8.
17. Heinz G, Ohner T, Laufer G, et al: Demographic and perioperative factors associated with initial and prolonged sinus node dysfunction after orthotopic heart transplantation. The impact of ischemic time. Transplantation 1991;51:1217-24.
18. Bharati, S: Pathology of the conduction system. In Silver MD, Gotlieb AI, Schoen FJ (eds): Cardiovascular Pathology, 3rd ed. New York, Churchill Livingston, 2001, pp 607-28.
19. Bharati S, Lev M: The anatomy of the normal conduction system: Disease-related changes and their relationship to arrhythmogenesis. In Podrid PJ, Kowey PR (eds): Cardiac Arrhythmias: Mechanism, Diagnosis and Management. Baltimore, Williams & Wilkins, 1995, pp 1-15.
20. Bharati S, Lev M: Morphology of the sinus and atrioventricular nodes and their innervation. In Mazgalev T, Dreifus LS, Michelson EL (eds): Electrophysiology of the Sinoatrial and Atrioventricular Nodes: Progress in Clinical and Biological Research. New York, Alan R Liss, 1988, 275, pp 3-14.
21. Lev M, Bharati S: Lesions of the conduction system and their functional significance. In Sommers SC (ed): Pathology Annual 1974. New York, Appleton-Century-Crofts, 1974, pp 157-208.
22. Lev M: The conduction system. In Gould SC (ed): Pathology of the Heart and Blood Vessels, 3rd ed. Springfield, Ill, Charles C Thomas, 1968, pp 180-220.
23. Lev M, Bharati S: The pathologic base of supraventricular arrhythmias. In Iwa T, Fontaine G (eds): Cardiac Arrhythmias: Recent Progress in Investigation and Management. Amsterdam, Elsevier, 1988, pp 1-13.
24. Bharati S, Nordenberg A, Bauernfeind R, et al: The anatomic substrate for the sick sinus syndrome in adolescence. Am J Cardiol 1980;46:163-72.
25. Kaplan BM, Langendorf R, Lev M, et al: Tachycardia-bradycardia syndrome (so-called "sick sinus syndrome"). Pathology, mechanisms and treatment. Am J Cardiol 1973;31:497-508.
26. Bharati S, Lev M: The pathologic changes in the conduction system beyond the age of ninety. Am Heart J 1992;124-486.
27. Bharati S, Lev M: Pathologic changes of the conduction system with aging. Cardiol Elderly 1994;2:152-60.
28. Bharati S, Lev M, Dhingra R, et al: Pathologic correlations in three cases of bilateral bundle branch disease with unusual electrophysiologic manifestations in two cases. Am J Cardiol 1976;38:508-18.
29. Rosen KM, Rahimtoola SH, Gunnar RM, et al: Transient and persistent atrial standstill with His bundle lesions, electrophysiologic and pathologic correlations. Circulation 1971;44:220-36.
30. Bharati S, Lev M: Sequelae of atriotomy and ventriculotomy on the endocardium, conduction system and coronary arteries. Am J Cardiol 1982;50:580-7.
31. Bharati S, Molthan ME, Veasy LG, et al: Conduction system in two cases of sudden death two years after the Mustard procedure. J Thorac Cardiovasc Surg 1979;77:101-8.
32. Schuessler RB, Boineau JP, Brombert BI: Origin of the sinus impulse. J Cardiovasc Electrophysiol 1996;7:263-274.
33. Anderson RH, Ho SY: The architecture of the sinus node, the atrioventricular conduction axis, and the internal atrial myocardium. J Cardiovasc Electrophysiol 1998;9:1233-1248.
34. Anumonwo JMB, Jalife J: Cellular and subcellular mechanisms of pacemaker activity initiation and synchronization in the heart. In Zipes DP, Jalife J (eds): Cardiac Electrophysiology: From Cell to Bedside. Philadelphia, WB Saunders, 1995, pp 151-164.
35. Crick SJ, Wharton J, Sheppard MN, et al: Innervation of the human cardiac conduction system. A quantitative immunohistochemical and histochemical study. Circulation 1994;89: 1697-1708.
36. Bleeker WK, MacKaay AJC, Masson-Pevet M, et al: Functional and morphologic organization of the rabbit sinus node. Circ Res 1980;46:11-22.
37. Watanabe E, Honjo H, Anno T, et al: Modulation of pacemaker activity of sinoatrial node cells by electrical load imposed by an atrial cell model. Am J Physiol 1995;269:H1735-1742.
38. Shibata EF, Giles WR: Ionic currents that generate the spontaneous diastolic depolarization in individual cardiac pacemaker cells. Proc Natl Acad Sci USA 1985;82:7796-7800.
39. Honjo H, Boyett MR, Kodama I, Toyama J: J Physiol 1996;496: 795-808.
40. Robinson RB, Difrancesco D: Sinoatrial node and impulse initiation. In Spooner P, Rosen M (eds): Foundations of Cardiac Arrhythmias: Basic Concepts and Clinical Approaches. New York, Marcel Dekker, 2001, pp 151-170.
41. Vassalle M, Yu H, Cohen IS: Pacemaker channels and cardiac automaticity. In Zipes DP, Jalife J (eds): Cardiac Electrophysiology: From Cell to Bedside. Philadelphia, WB Saunders, 1995, pp 94-103.
42. Thiene G (ed): Cardiac Arrhythmias: Recent Progress in Investigation and Management. Amsterdam, Elsevier, 1988, p 1.
43. Spodick DH, Raju P, Bishop RL, Rifkin RD: Operational definition of normal sinus heart rate. Am J Cardiol 1992;69:1245-6.
44. Spodick DH: Survey of selected cardiologists for an operational definition of normal sinus heart rate. Am J Cardiol 1993;72: 487-8.

45. Spodick DH: Normal sinus heart rate: Appropriate rate thresholds for sinus tachycardia and bradycardia. South Med J 1996; 89:666-7.
46. Viitasalo MT, Kala R, Eisalo A: Ambulatory electrocardiographic recording in endurance athletes. Br Heart J 1982;47:213-20.
47. Kantelip JP, Sage E, Duchene-Marullaz P: Findings on ambulatory monitoring in subjects older than 80 years. Am J Cardiol 1986; 57:398-401.
48. Northcote RJ, Canning GP, Ballantyne D: Electrocardiographic findings in male veteran endurance athletes. Br Heart J 1989;61: 155-60.
49. Lorber A, Maisuls E, Palant A: Autosomal dominant inheritance of sinus node disease. Int J Cardiol 1986;15:252-6.
50. Sutton R, Kenny R: The natural history of sick sinus syndrome. Pacing Clin Electrophysiol 1986;9:1110-4.
51. Alboni P, Menozzi C, Brignole M, et al: Effects of permanent pacemaker and oral theophylline in sick sinus syndrome. The THEOPACE Study: A randomized controlled trial. Circulation 1997;96:260-6.
52. Nenozzi C, Brignole M, Albini P, et al: The natural course of untreated sick sinus syndrome and identification of the variables predictive of unfavorable outcome. Am J Cardiol 1998;82:1205-9.
53. Alboni P, Menozzi C, Brignoli M, et al: An abnormal neural reflex plays a role in causing syncope in sinus bradycardia. J Am Coll Cardiol 1993;22:1130-4.
54. Brignole M, Menozzi C, Gianfranchi L, et al: Neurally mediated syncope detected by carotid sinus massage and head-up tilt test in sick sinus syndrome. Am J Cardiol 1991;68:1032-6.
55. Shaw DB, Holman RR, Gowers JI: Survival in sinoatrial disorder (sick sinus syndrome). Br Med J 1980;280:139-41.
56. Rubenstein JJ, Schulman CL, Yurchak PM, DeSanctis RW: Clinical spectrum of the sick sinus syndrome. Circulation 1972; 46:5-13.
57. Lien WP, Lee YS, Chang FZ, et al: The sick sinus syndrome. Natural history of dysfunction of the sinoatrial node. Chest 1977; 72:628-34.
58. Bigger JT Jr, Reiffel JA: Sick sinus syndrome. Annu Rev Med 1979;30:91-118.
59. Chung EK: Sick sinus syndrome: Current views. Mod Concepts Cardiovasc Dis 1980;49:61-6.
60. Skagen K, Hansen JF: The long-term prognosis for patients with sinoatrial block treated with permanent pacemaker. Acta Med Scand 1975;199:13-5.
61. Ellestad MH, Wan MK: Predicted implications of stress testing: Follow-up of 2700 subjects after maximum treadmill stress testing. Circulation 1975;51:363-9.
62. Wiens R, Lafia P, Marder CM, et al: Chronotropic incompetence in clinical exercise testing. Am J Cardiol 1984;54:74-82.
63. Sgarbossa EB, Pinski SL, Maloney JD, et al: Chronic atrial fibrillation and stroke in paced patients with sick sinus syndrome. Relevance of clinical characteristics and pacing modalities. Circulation 1993;88:1045-53.
64. Andersen HR, Thuesen L, Bagger JP, et al: Prospective randomized trial of atrial versus ventricular pacing in sick sinus syndrome. Lancet 1994;344:1523-8.
65. Andersen HR, Nielsen JC, Thomsen PEB, et al: Long-term follow-up of patients from a randomized trial of atrial versus ventricular pacing for sick-sinus syndrome. Lancet 1997;350:1210-6.
66. Vera Z, Mason DT, Awan NA, et al: Improvement of symptoms in patients with sick sinus syndrome by spontaneous development of stable atrial fibrillation. Br Heart J 1977;39:160-7.
67. Gann D, Tolentino R, Samet P: Electrophysiologic evaluation of elderly patients with sinus bradycardia: A long-term follow-up study. Ann Intern Med 1979;24:29.
68. Rosenqvist M, Obel IW: Atrial pacing and the risk for AV block: Is there a time for change in attitude? Pacing Clin Electrophysiol 1989;12:97-101.
69. Narula OS: Atrioventricular conduction defects in patients with sinus bradycardia. Analysis by His bundle recordings. Circulation 1971;44:1096-110.
70. Mortensen PT: Atrioventricular conduction during long-term follow-up of patients with sick sinus syndrome. Circulation 1998;98:1515-1521.
71. Van Mechelen R, Segers A, Hagemeijer F: Serial electrophysiologic studies after single chamber atrial pacemaker implantation in patients with symptomatic sinus node dysfunction. Eur Heart J 1984;5:628-36.
72. Stangl K, Wirtzfeld A, Seitz K, et al: Atrial stimulation: Long-term follow-up of 110 patients. In Belhassen B, Feldman S, Copperman Y (eds): Cardiac Pacing and Electrophysiology. Proceedings of the VIII World Symposium on Cardiac Pacing and Electrophysiology. Jerusalem, R&L Creative Communications, 1987, pp 283-5.
73. Santini M, Alexidou G, Ansalone G, et al: Relation of prognosis in sick sinus syndrome at age, conduction defects and modes of permanent cardiac pacing. Am J Cardiol 1990;565:729-35.
74. Grauer LE, Gershen BJ, Orlando MM, Epstein SE: Bradycardia and its complications in the pre-hospital phase of acute myocardial infarction. Am J Cardiol 1973;32:607-11.
75. Rotman M, Wagner GS, Wallace AGP: Bradyarrhythmias in acute myocardial infarction. Circulation 1972;45:703-22.
76. Parameswaran R, Ohe T, Goldberg H: Sinus node dysfunction in acute myocardial infarction. Br Heart J 1976;38:93-6.
77. Hatle L, Bathen J, Rokseth R: Sinoatrial disease in acute myocardial infarction. Long-term prognosis. Br Heart J 1976;38:410-4.
78. Mangrum JM, Dimarco JP: The evaluation and management of bradycardia. N Engl J Med 2000;342:703-9.
79. Kaplan BM, Langendorf R, Lev M, Pick A: Tachycardia-bradycardia syndrome (so-called "sick sinus syndrome"). Pathology, mechanisms and treatment. Am J Cardiol 1973;31:497-508.
80. Mazuz M, Friedman HS: Significance of prolonged electrocardiographic pauses in sinoatrial disease: Sick sinus syndrome. Am J Cardiol 1983;52:485-9.
81. Hilgard J, Ezri MD, Denes P: Significance of ventricular pauses of three seconds or more detected on twenty-four-hour Holter recordings. Am J Cardiol 1985;55:1005-8.
82. Holden W, McAnulty JH, Rahimtoola SH: Characterization of heart rate response to exercise in the sick sinus syndrome. Br Heart J 1978;40:923-30.
83. Mandel WJ, Hayakawa H, Allen HN, et al: Assessment of sinus node function in patients with sick sinus syndrome. Circulation 1972;46:761-9.
84. Jose AD, Collison D: The normal range and determinants of the intrinsic heart rate in man. Cardiovasc Res 1970;4:160-7.
85. Josephson ME: Sinus Node Function. Clinical Cardiac Electrophysiology, 2nd ed. Philadelphia, Lea & Febiger, 1993, pp 83-4.
86. Strauss HC, Saroff AL, Bigger JT Jr, Giardina EG: Premature atrial stimulation as a key to the understanding of sinoatrial conduction in man. Presentation of data and critical review of literature. Circulation 1973;47:86-93.
87. Narula OS, Shantha N, Vasquez M, et al: A new method for measurement of sinoatrial conduction time. Circulation 1978; 58:706-14.
88. Benditt DG, Milstein S, Goldstein M, et al: Sinus node dysfunction: Pathophysiology, clinical features, evaluation, and treatment. In Zipes DP, Jalife J (eds): Cardiac Electrophysiology: From Cell to Bedside. Philadelphia, WB Saunders, 1990, pp 708-34.
89. Breithardt G, Seipel L, Loogen F: Sinus node recovery time and calculated sinoatrial conduction time in normal subjects and patients with sinus node dysfunction. Circulation 1977;56:43-50.
90. Mandel W, Hayakawa H, Danzig R, et al: Evaluation of sinoatrial node function in man by overdrive suppression. Circulation 1971;44:59-66.
91. Delius W, Wirtzfeld A: The significance of the sinus node recovery time in the sick sinus syndrome. In Luderitz B (ed): Clinical Pacing, Diagnostic and Therapeutic Tools. Berlin, Springer-Verlag, 1976, pp 25-32.
92. Josephson ME: Sinus node function. In : Clinical Cardiac Electrophysiology. Malvern, Pa, Lea & Febiger, 1992, p 83.
93. Strauss HC, Saroff AL, Bigger JT, Giardina EGV: Premature atrial stimulation as a key to the understanding of sinoatrial conduction in man. Circulation 1974;47:86.
94. Benditt DG, Strauss HC, Scheinman MM, et al: Analysis of secondary pauses following termination of rapid atrial pacing in man. Circulation 1976;54:436.
95. Reiffel JA, Gang E, Livelli F, et al: Indirectly estimated sinoatrial conduction time by the atrial premature stimulus technique: Patterns of error and the degree of associated inaccuracy as assessed by direct sinus node electrography. Am Heart J 1983;106:459.

96. Gomes JAC, Kang PS, El-Sherif N: The sinus node electrogram in patients with and without sick sinus syndrome: Techniques and correlation between directly measured and indirectly estimated sinoatrial conduction time. Circulation 1982;66:864.
97. Desai JM, Scheinman MM, Strauss HC, et al: Electrophysiologic effects on combined autonomic blockade in patients with sinus node disease. Circulation 1981;63:953-60.
98. Jose AD: Effect of combined sympathetic and parasympathetic blockade on heart rate and cardiac function in man. Am J Cardiol 1966;18:476.
99. Tonkin AF, Heddle WF: Electrophysiological testing of sinus node function. Pacing Clin Electrophysiol 1984;7:735.
100. Lamas GA, Orav EJ, Stambler BS, et al: Quality of life and clinical outcomes in elderly patients treated with ventricular pacing as compared with dual-chamber pacing. N Engl J Med 1998;338: 1097-104.
101. Connolly SJ, Kerr CR, Gent M, et al: Effects of physiologic pacing versus ventricular pacing on the risk of stroke and death due to cardiovascular causes. N Engl J Med 2000;342:1385-91.
102. Lamas GA, Lee K, Sweeney MO, et al: Ventricular pacing or dual chamber pacing for sinus node dysfunction. N Engl J Med 2002;346:1854-1862.
103. Krahn AD, Klein GJ, Norris C, Yee R: The etiology of syncope in patients with negative tilt table and electrophysiological testing. Circulation 1995;92:1819.
104. Gregoratos G, Cheitlin MD, Conill A, et al: ACC/AHA guidelines for implantation of cardiac pacemakers and antiarrhythmia devices: A report of the American College of Cardiology/ American Heart Association Task Force on Practice Guidelines (Committee on Pacemaker Implantation). J Am Coll Cardiol 1998;5:1175-1209.
105. Breivik K, Ohm OJ, Segadal L: Sick sinus syndrome treated with permanent pacemaker in 109 patients. A follow-up study. Acta Med Scand 1979;206:153-9.
106. Rosenqvist M, Aren C, Kristensson BE, et al: Atrial rate-responsive pacing in sinus node disease. Eur Heart J 1990;11:537-42.
107. Batey L, Sweesy MW, Scala G, Forney RC: Comparison of low rate dual chamber pacing to activity responsive rate variable ventricular pacing. Pacing Clin Electrophysiol 1990;13:646-52.
108. Ausubel K, Furman S: The pacemaker syndrome. Ann Intern Med 1985;103:420-9.
109. Sulke N, Dritsas A, Bostock J, et al: "Subclinical" pacemaker syndrome: A randomized study of symptom free patients with ventricular demand (VVI) pacemakers upgraded to dual chamber devices. Br Heart J 1992;67:57-64.
110. Heldman D, Mulvihill D, Nguyen H, et al: True incidence of pacemaker syndrome. Pacing Clin Electrophysiol 1990;13:1742-50.
111. Link MS, Helkamp AS, Estes NA III, et al: High incidence of pacemaker syndrome in patients with sinus node dysfunction treated with ventricular based pacing in the Mode Selection Trial (MOST). J Am Coll Cardiol 2004;43:2066-71.
112. Strasberg B, Sagie A, Erdman S, et al: Carotid sinus hypersensitivity and the carotid sinus syndrome. Prog Cardiovasc Dis 1989;31:379-91.
113. Richardson DA, Bexton RS, Shaw FE, Kenny RA: Prevalence of cardioinhibitory carotid sinus hypersensitivity in patients 50 years or over presenting to the accident and emergency department with "unexplained" or "recurrent" falls. Pacing Clin Electrophysiol 1997;20:820-3.
114. Huang SK, Ezri MD, Hauser RG, et al: Carotid sinus hypersensitivity in patients with unexplained syncope: Conical, electrophysiologic, and long-term follow-up observations. Am Heart J 1988;116:989-96.
115. Brignole M, Menozzi C, Lolli G, et al: Long-term outcome of paced and non-paced patients with severe carotid sinus syndrome. Am J Cardiol 1992;69:1039-43.
116. Sugrue DD, Gersh BJ, Holmes DR, et al: Symptomatic "isolated" carotid sinus hypersensitivity: Natural history and results of treatment with anticholinergic drugs or pacemakers. J Am Coll Cardiol 1986;7:158-62.
117. Kenny RA, Richardson DA, Steen N, et al: Carotid Sinus Syndrome: A modifiable risk factor for nonaccidental falls in older adults (SAFE PACE). J Am Coll Cardiol 2001;38:1491-6.
118. Katritsis D, Ward DE, Camm AJ: Can we treat carotid sinus syndrome? Pacing Clin Electrophysiol 1991;14:1367-9.
119. Lewis T: Vasovagal syncope and the carotid sinus mechanism. BMJ 1932;1:873-6.
120. Connolly SJ, Sheldon R, Roberts R, et al: The North American vasovagal pacemaker study: A randomized trial of permanent cardiac pacing for the prevention of vasovagal syncope. J Am Coll Cardiol 1999;33:16-20.
121. Sutton R, Brignole M, Menozzi C, et al: Dual chamber pacing in the treatment of neurally-mediated tilt positive cardioinhibitory syncope: Pacemaker versus no therapy: A multicentre randomized study. The Vasovagal Syncope International Study. Circulation 2000;102:294-9.
122. Connolly SJ, Sheldon R, Thorpe KE, et al: The Second Vasovagal Pacemaker Study (VPSII). Pacing Clin Electrophysiol 2003;26:1565.
123. Sutton R: Has cardiac pacing a role in vasovagal syncope? J Intervent Cardiac Electrophysiol 2003;9:145-9.
124. Connolly SJ, Sheldon R, Thorpe KE, et al for VPS Investigators: Pacemaker therapy for prevention of syncope in patients with recurrent severe vasovagal syncope. Second vasovagal pacemaker study (VPS-2). A randomized trial. JAMA 2003;209: 2224-9.
125. Ammirati F, Colivicchi F, Toscano S, et al: DDD pacing with rate drop response function versus DDI pacing with rate hysteresis pacing for cardioinhibitory vasovagal syncope. Pacing Clin Electrophysiol 21:2178-81.
126. Ammirati F, Colivicchi F, Santini M: Permanent cardiac pacing versus medical treatment for the prevention of recurrent vasovagal syncope: A multicenter, randomized, controlled trial. Circulation 2001;104:52-7.
127. Gregoratos G, Abrams J, Epstein AE, et al: ACC/AHA/NASPE 2002 guideline update for implantation of cardiac pacemakers and antiarrhythmia devices. J Am Coll Cardiol 2002;40:1703-19.

Chapter 13 Atrioventricular Block

NORA GOLDSCHLAGER, SANJEEV SAKSENA, SAROJA BHARATI, RALPH LAZZARA, GERALD NACCARELLI, STEPHEN HAMMILL, FRED KUSUMOTO, and RABIH AZAR

Epidemiology: Nora Goldschlager
Anatomy: Sanjeev Saksena
Pathology: Saroja Bharati
Basic Electrophysiology: Ralph Lazzara
Electrocardiography: Sanjeev Saksena
Diagnostic Techniques: Gerald Naccarelli
Clinical Electrophysiology: Stephen Hammill
Evidence-Based Therapy: Fred Kusumoto, Nora Goldschlager, and Rabih Azar

In 1852, Stannius noted that placing a ligature between the atria and ventricles could cause bradycardia in a frog's heart.[1] In the late 1800s Tawara and His identified the atrioventricular (AV) node and His bundle as the normal conduction axis between the atria and ventricles in humans, and Wenckebach suggested blocked AV conduction as a cause for slow and irregular pulses.[2-4]

Epidemiology of Atrioventricular Block

Transient AV block can be observed in children and young adults during sleep[5]; persistent AV block is unusual. This AV block is usually due to increased vagal tone and is often a normal finding. In a continuous monitoring study of 100 healthy teenaged boys, transient first-degree AV block was observed in 12% and second-degree AV block in 11% of the population.[5] In young adults the incidence of transient AV block decreases to about 4% in women and 6% in men.[6,7] In the normal elderly population, transient type I (Wenckebach) second-degree AV block is seen only rarely (1%), and higher-grade AV block is not observed.[8]

Persistent first-degree AV block is rarely seen in young adults. Review of more than 70,000 ECGs from young men entering the Canadian and U.S. military demonstrated a prevalence of first-degree AV block of less than 1%.[9,10] Electrocardiographic studies have shown increased P–R intervals and an increased incidence of first-degree AV block with aging.[9-14] While approximately 2% of adults older than 20 years of age have first-degree AV block, the incidence increases to more than 5% in people older than 50 years of age.[11,13] With increasing age, the development of AV conduction disorders is more common; in one epidemiologic study of 1500 patients older than 65 years of age, AV conduction and intraventricular conduction defects were identified in 30% of patients.[14] Using high-resolution ECG techniques, the increased P–R interval associated with aging is due to delay in conduction in the AV node or the proximal portion of the His bundle.[11] It is uncommon (4%) for persistent first-degree AV block to progress to second-degree or higher-grade AV block in the absence of associated disease in people younger than 60 years of age.[13]

Acquired persistent second-degree and third-degree AV blocks are almost never observed in normal populations regardless of age. The incidence of symptomatic high-grade AV block is currently estimated to be 200/million per year.[15]

Isolated congenital complete (third-degree) AV block is a well-described problem that occurs in approximately 1 in every 20,000 live births.[16] Congenital complete AV block is the most common manifestation of neonatal lupus erythematosus and appears to be associated with the development of autoantibodies in the maternal circulation. Other hereditary conditions associated with AV block are the Kearns-Sayre syndrome (ophthalmoplegia, retinitis pigmentosa) and myotonic dystrophy.[17]

Anatomy of the Atrioventricular Bundle

The anatomy of the AV node, bundle of His, and the atrionodal connections have been substantially redefined.[18] In early studies Truex and Smythe described extensions of the compact AV node that connected to atria, the most prominent of which was directed to the coronary sinus ostium. Anderson and colleagues

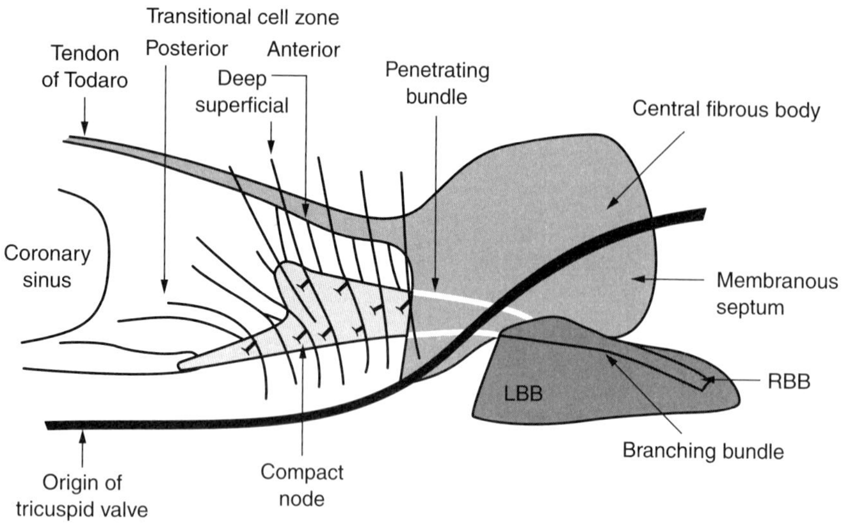

FIGURE 13-1 Schematic diagram of the atrionodal junctional region and the atrioventricular node and surrounding anatomic structures in the human heart. LBB, left bundle branch; RBB, right bundle branch. (Reproduced from Anderson et al: The human atrioventricular junctional area: A morphological study of the AV node and the bundle. Eur J Cardiol 1975;3:11.)

described zones of cellular aggregation at the atrium-AV nodal transition region as superficial, deep, and posterior (Fig. 13-1). Racker described superior, medial, and posterior atrionodal bundles that converged into the compact node. The properties of these atrionodal bundles differed from working atrial myocardium. For a more detailed discussion of this subject, see Chapter 14.

Pathology of Atrioventricular Block

Complete AV block may be viewed from an etiologic standpoint as falling into two major categories: *congenital* and *acquired.* On the other hand, from an electrophysiological standpoint, AV block may be classified according to the location of the block with reference to the AV bundle. Thus, AV block may occur proximal to the bundle of His recording site, within the bundle of His (intra-Hisian), or distal to the bundle of His recording site. Pathologically, AV block can be considered as being due to or occurring in association with various disease states: (1) congenital heart disease; (2) coronary artery disease; (3) hypertensive heart disease; (4) sclerosis of the left side of the cardiac skeleton (i.e., idiopathic or primary AV block); (5) aortic valve disease; (6) collagen diseases; (7) myocarditis; (8) infective endocarditis; (9) iatrogenic; (10) tumors of the heart; and (11) miscellaneous diseases and altered physiologic states (Table 13-1).[19-49]

CONGENITAL ATRIOVENTRICULAR BLOCK

Congenital AV block may occur in any type of a congenital heart disease or it may occur in an otherwise normal heart.[19-25] Pathologically, there are four types of congenital AV block: lack of connection between the atria and the peripheral conduction system; interruption of the AV bundle; bundle branch disease; and abnormal formation and interruption of the AV bundle.

LACK OF CONNECTION BETWEEN THE ATRIA AND THE PERIPHERAL CONDUCTION SYSTEM

This is the most common type of congenital AV block that is seen either with or without an associated congenital cardiac anomaly.[19-25] In this type of block, the

TABLE 13-1 Causes of Complete Heart Block

Congenital
Acquired
Degenerative
Fibrosis and calcification
Ischemia and infarction
Medications
β-blockers
Calcium channel blockers
Clonidine
Lithium
Infectious conditions
Endocarditis
Lyme disease
Diphtheria
Toxoplasmosis
Syphilis
Chagas' disease
Inflammation
Myocarditis
Rheumatic fever
Rheumatic and autoimmune disease
Systemic lupus erythematosus
Ankylosing spondylitis
Reiter's disease
Infiltrative disease
Amyloidosis
Sarcoidosis
Infiltrative malignancies
Increased vagal tone
Trauma to the heart
Neuromyopathic disorders
Other
Hypoxia
Hyperkalemia
Following aortocoronary bypass graft or valve surgery (especially aortic)

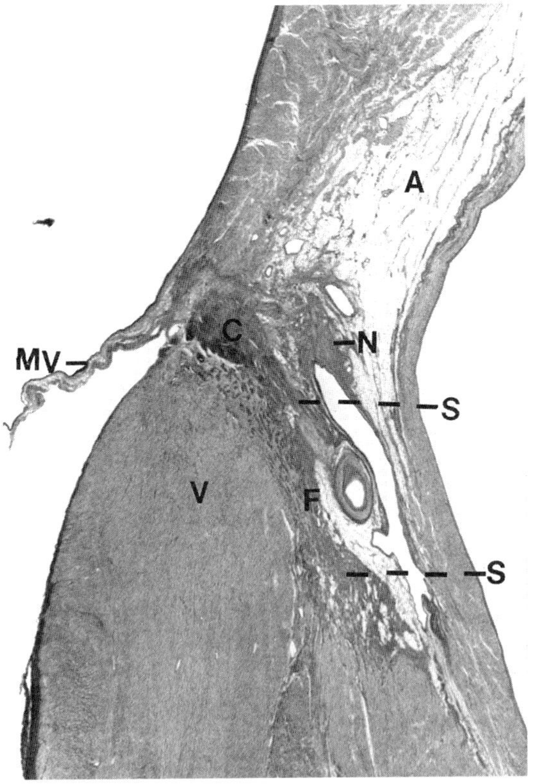

FIGURE 13-2 Congenital atrioventricular block in a 49-year-old man demonstrated marked fatty infiltration in the approaches to the AV node and total isolation of the node. Weigert-van Gieson stain ×17. A, approaches to the AV node almost completely replaced by fat and slight fibrosis; C, central fibrous body; F, fibrosis; MV, mitral valve; N, AV node isolated from the atrial approaches; S, right side of the summit of the ventricular septum; V, ventricular septum. (From Bharati S, Rosen KM, Strassberg B, et al: Anatomic Substrate for Congenital AV Block in Middle Aged Adults. Pacing Clin Electrophysiol 1982;5:860-9.)

myocardium in the distal part of the atria is absent either completely or in part and is replaced by fat, vascular channels, calcification at times, and mononuclear cells with fibrosis (Fig. 13-2). The AV node is usually deficient or abnormally formed and has no continuity with the surrounding atrial myocardium. In extreme forms, the AV node may be totally absent. In still other forms, there may be remnants of nodal cells within the central fibrous body. In some cases, the penetrating AV bundle may be deficient as well. The block is thus proximal to the bundle of His recording site. The remainder of the conduction system is usually normal. Congenital AV block in an otherwise normally developed heart may permit survival to the fourth or fifth decade of life.[25]

INTERRUPTION OF THE ATRIOVENTRICULAR BUNDLE

In this type of less common block, the approaches to the AV node and the AV node are normally formed. However, the penetrating portion of the AV bundle is markedly fragmented. The branching portion of the AV bundle and the beginning of the bundle branches may be normal or deficient as well. Nevertheless, a part of one or both bundle branches is connected to the ventricular myocardium. This type of block may therefore give rise to a narrow or wide QRS complex depending on the extent of deficiency of the branching part of the AV bundle.[19-25]

INTERRUPTION OF THE BEGINNING OF THE BUNDLE BRANCHES

This may be a familial form of AV block with beginning of the bundle branches replaced by empty spaces, fat and fibrous tissue. This is a rare type of a bundle branch disease resulting in complete AV block.[19-21,23,24]

ABNORMAL FORMATION AND INTERRUPTION OF ATRIOVENTRICULAR BUNDLE

Interruption in an aberrant conduction system is seen in corrected transposition of the heart (mixed levocardia with ventricular inversion), in which the atria are situated more or less normally. This may be associated with congenital AV block. The block may be due to either deficiency or absence of the AV node and the surrounding myocardium or interruption within the penetrating or branching bundle, or both. In corrected transposition, in most cases, the AV bundle is situated anterosuperiorly in the morphologically left ventricle in an abnormal location, making it vulnerable to hemodynamic stresses, thereby resulting in interruption.[19-21,23,24]

Congenital AV block may occur in any type of congenital heart disease; however, in corrected transposition there is an increased incidence for its development. Usually, there is absence of the AV node and the surrounding atrial myocardium, with infiltration of mononuclear cells, calcification, fibrosis, and fat.[19,24]

ACUTE MYOCARDIAL INFARCTION—CORONARY ARTERY DISEASE

Complete AV block may occur in association with acute or chronic coronary artery disease. In acute myocardial infarction, a distinction must be made between anterior and posterior wall infarction.[19-25,26-29]

In posteroseptal wall myocardial infarction, there may be infarction of part of the sinoatrial node and its approaches, as well as the approaches to the AV node, with focal necrosis of the node and the bundle. In some cases there may not be any obvious pathologic change within the conduction system, whereas in others, there may or may not be changes in the AV bundle and the bundle branches.

In contrast to posteroseptal wall infarction, in infarction of the anteroseptal wall, the branching bundle and the bundle branches are usually affected by the necrotic process. The AV node and the penetrating part of the AV bundle may be involved in the infarction as well. The infarction is usually extensive in nature; however, if the patient recovers, the heart block usually resolves. Chronic AV block after acute myocardial infarction is decidedly uncommon. Thus, the site of infarction is more important than the site of coronary artery occlusion in producing lesions in the conduction system.

CHRONIC CORONARY INSUFFICIENCY

Chronic coronary insufficiency, with or without previous myocardial infarction, may affect various portions of the conduction system, predominantly the bundle branches that may, over time, progress to complete AV block. In chronic coronary insufficiency, in addition to an ischemic factor, there are aging changes in the conduction system in the form of sclerosis of the left side of the cardiac skeleton; there may be an abnormal formation of the AV bundle itself. In addition, there may be associated small vessel disease (arteriolosclerosis). Thus, in chronic coronary insufficiency, several mechanisms may play a role in the pathogenesis of AV block. In general, in chronic ischemia, the right bundle branch is replaced by fibrous tissue; the left bundle branch fibers are replaced by both the chronic ischemic process and other factors such as a degeneration. For example, calcific mass at the summit of the interventricular septum can impinge upon the branching AV bundle and the origin of the main left bundle branch (Fig. 13-3).

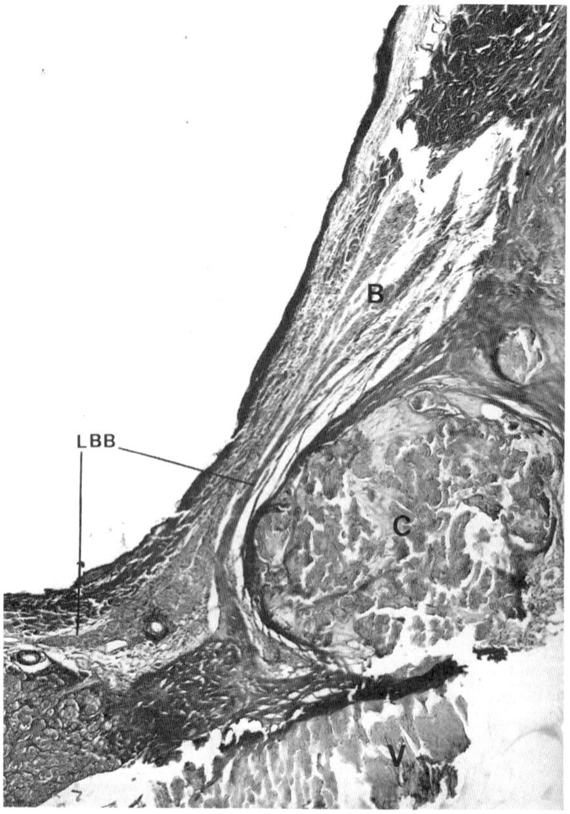

FIGURE 13-3 Chronic coronary insufficiency with atrioventricular (AV) block demonstrating calcification in the summit of the ventricular septum compressing the branching AV bundle and the left bundle branch Weigert-van Gieson stain ×30. B, AV bundle showing fibrosis and loss of fibers; C, calcific mass; LBB, remnant of left bundle branch; V, calcified ventricular septum.

The fibrosis of the left bundle branch is probably related to both mechanical factors produced by the calcific impingement, as well as chronic coronary insufficiency. Clinically, there may be chronic right bundle branch block or left bundle branch block that may progress to chronic complete AV block. In some cases, small vessel disease within the ventricular septum in the absence of major coronary artery disease may cause destruction of the branching bundle and the bundle branches. Chronic coronary insufficiency is the most common cause of bilateral bundle branch block.[19-21,23,24,26-29]

HYPERTENSIVE HEART DISEASE

Arteriolosclerosis (small vessel disease) of the heart is frequently present in hypertensive patients, with consequent conduction disturbances in the form of arteriolosclerosis of the sinoatrial node, AV node, AV bundle, and bundle branches. The pathologic effects of hypertensive heart disease on the conduction system are related to the mechanical forces as well as the arteriolosclerosis process. Since the branching AV bundle is a subendocardial and superficial structure, it is vulnerable to the mechanical stress and strain on the fibrous skeleton of the heart produced by hypertensive heart disease. Since hypertensive heart disease affects the entire heart, the pathologic changes and arteriolosclerosis in the sinoatrial node, AV node, AV bundle, and the bundle branches will result in various types of arrhythmias, such as sinus node dysfunction and right or left bundle branch block or complete AV block.[19-21,23,24,26-29]

AGING CHANGES IN THE SUMMIT OF THE VENTRICULAR SEPTUM

With advancing age, the summit of the ventricular septum, the membranous part of the ventricular septum, central fibrous body, aortic-mitral annulus, mitral annulus, aortic valve base, and the aortic valve undergo degenerative changes. The pathologic changes include fibrosis, calcification, arteriolosclerosis, fatty infiltration, and loss of conduction fibers with replacement of fibrotic strands. This type of degenerative process is referred to as "sclerosis of the left side of the cardiac skeleton." It occurs as a result of "stress and strain" in a high pressure system that may affect the adjacent AV bundle and the bundle branches. Frequently, the branching part of the AV bundle, the beginning of the right and left bundle branches, or the main left bundle branch are compressed by calcium and replaced by fibroelastic process, loss of cells, and space formation with or without fat. The main left bundle branch may be totally disrupted from the branching AV bundle. The second part of the right bundle branch may be completely replaced by fibroelastic process. These changes may result in complete AV block, left bundle branch block, or bilateral bundle branch block. Aging of the summit of the ventricular septum usually begins in the fourth decade of life and is accelerated by hypertensive heart disease, coronary artery disease, and diabetes mellitus. The branching and the bifurcating part of the AV bundle, the beginning of the main left bundle branch, and the beginning of the right bundle branch are quite superficial subendocardial structures and as such are vulnerable to degenerative changes caused by aging.[19-21,23,24,26-29]

COLLAGEN DISEASES

Since there is a considerable amount of collagen connective tissue in the conduction system, varying types of collagen connective tissue disorders may affect the conduction system to varying degrees. Thus, AV block or bundle branch block may occur in lupus erythematosus, dermatomyositis, scleroderma, and other mixed types of connective tissue disorders. Depending on the site of involvement of the conduction system, there may be bundle branch block or complete AV block. For example, complete replacement of the AV node by granulation tissue in lupus, fibrotic disruption of bundle branches to varying degrees in dermatomyositis, and granulomas in various parts of the conduction system in rheumatoid arthritis may be seen histologically.[19-21,23,24,30,31]

MYOCARDITIS

Myocarditis of any type, whether acute or chronic, may cause complete AV block. If the patient survives, the heart block usually resolves; chronic AV block after myocarditis is rare. Pathologically, in the acute phase of the illness, the inflammatory changes in the ventricular myocardium predominate, affecting the distal part of the conduction system, such as the branching bundle and the bundle branches. Rarely, a part of the conduction system may be affected. In general, the conduction system and the surrounding myocardium are involved by the inflammatory process. In rare cases, the conduction system is involved more than the surrounding myocardium.[19-21,23,24]

INFECTIVE ENDOCARDITIS

Infective endocarditis of the aortic or mitral valve, or both, may extend to the aortic-mitral annulus, the membranous part of the AV septum, and the summit of the ventricular septum and thereby affect the AV node and AV bundle to produce AV block.[19-21,23,24]

TUMORS OF THE HEART

Any type of tumor that metastasizes to the heart has the potential to affect various parts of the conduction system to a varying degree, resulting in AV block.[19-21,23,24,32]

One specific type of benign tumor, a mesothelioma or celothelioma of the AV node, has an affinity to affect only the AV node and its approaches (Fig. 13-4). Although rare, the tumor may totally replace the AV node and its approaches and produce complete AV block with a narrow QRS complex in the ECG. In general, this is a slow-growing tumor and may manifest clinically as AV block from birth to old age. The tumor consists of neoplastic cells of benign nature with no anaplasia and tends to form cysts. The tumor is generally seen in young women who may remain totally asymptomatic; sudden death may be the first manifestation.

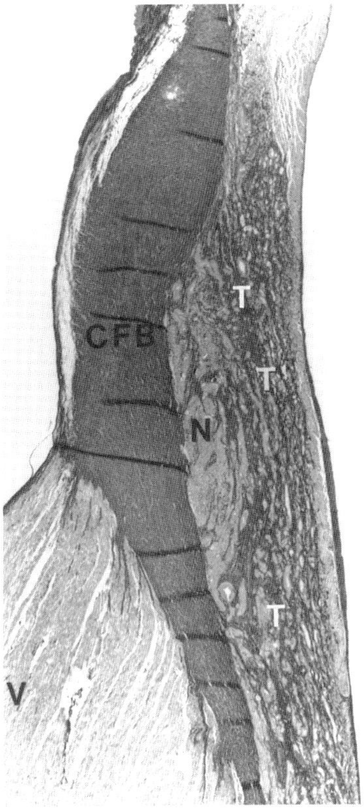

FIGURE 13-4 A 16-year-old asymptomatic girl developed 2:1 atrioventricular (AV) block with intermittent complete AV block immediately after delivering a full-term healthy infant. Six weeks later she developed complete AV block and died suddenly. Photomicrograph showing mesothelioma of the AV node replacing the AV node and its approaches. Weigert-van Gieson stain ×45. CFB, central fibrous body; N, AV node; T, tumor replacing the AV node and its approaches; V, summit of the ventricular septum. (From Bharati S, Lev M: The Cardiac Conduction System in Unexplained Sudden Death. Mt. Kisco, NY, Futura, 1990, p 136.)

It is of interest that mesothelioma, although a benign tumor pathologically, may cause sudden death in otherwise normal, young, healthy persons. It is also of interest that the site of heart block produced by the tumor is proximal to the AV bundle, with a narrow QRS complex in the ECG, which is generally considered to have a favorable clinical course.[19-21,23,24,32]

SURGICAL ATRIOVENTRICULAR BLOCK

Surgical injury to the AV bundle during repair of a ventricular septal defect and ostium primum defect rarely occurs today. On the other hand, AV block occurring during closure of defects in hearts with corrected transposition is seen frequently. Septation of a single-ventricle heart often results in postoperative AV block. In general, when the ventricles are inverted or are in a crisscross position, the AV bundle is situated anterosuperiorly, in a vulnerable position for surgical injury. On the other hand, a right-sided AV bundle in simple, complete transposition of the great vessels or ventricular septal defect may get caught in surgical closure of the defect. The Mustard procedure for complete transposition, in which the atrial septum is removed and a baffle is placed, can lead to distortion of the atrial septum with fibrosis of the approaches to the AV node. This may result in AV block and sudden death. Further, in the Mustard procedure, the sinoatrial node may be affected to a varying degree, resulting in supraventricular arrhythmias.[19-21,23,24,33-36]

MISCELLANEOUS DISEASES AND ALTERED PHYSIOLOGIC STATES

AV block is known to occur in numerous diseases and altered physiologic or metabolic states. These include mitral valve prolapse, Marfan's syndrome, Chagas' disease, Reiter's disease, Lyme disease, Kearns-Sayre syndrome, hemochromatosis, leukemia, alcoholism, radiation therapy,[37] hypothermia, exercise testing, introduction of catheters, prolonged coughing, sleep apnea with or without obesity, scorpion bites, and posture changes.[19]

Basic Electrophysiology

It is generally accepted that the AV junction consists of the AV node with its atrial connections of transitional fibers and the bundle of His.[38] In electrophysiological considerations of AV block, it is useful to define the AV junction more broadly, including the atrial connections to the AV node, the AV node, the bundle of His, and the proximal portions of the bundle branches that are insulated from ventricular myocardium by fibrous matrix. Heart block can occur from interruption of conduction in any of these components of the AV junction as defined broadly. In this array of components, there is great electrophysiological and histologic diversity (Fig. 13-5). The range of conduction velocities, resting and action potential (AP) amplitudes, AP durations, and intercellular conductivity observed within the AV junction is greater than in the remainder of the atrial and ventricular myocardium.

THE COMPACT ATRIOVENTRICULAR NODE

Within the central compact node are cells with distinct electrophysiological characteristics. These characteristics of the prototypical AV nodal cells manifested by intracellular recordings in multicellular preparations have been replicated in cells isolated from the AV node, allowing determination of the activities of specific currents, pumps, and exchangers.[39,40] The amplitudes of resting potentials (−60 mV), APs (85 mV), and upstroke velocities (<10 V/sec) are reduced. These cells manifest diastolic depolarization, a high total transmembrane (input) resistance, and relative insensitivity to changes in extracellular K^+. They have postrepolarization refractoriness; the absolute refractory period outlasts the AP. The relative refractory period, during which relatively diminished, slowly conducting APs are generated, is prolonged during diastole. The amplitudes and upstroke velocities of APs and conduction velocity are strongly modulated by autonomic activity, increasing with adrenergic (sympathetic) stimulation and decreasing with cholinergic (vagal) stimulation. The AV node is richly innervated with vagal and sympathetic fibers. These characteristics of the APs, in large part, account for the slow conduction and spontaneous automaticity of the compact AV node.

Important features of the transmembrane potentials can be explained by the array and activities of specific ionic currents. Like sinoatrial nodal cells, AV nodal cells lack the ionic current I_{K1}. These potassium ion channels in atrial and ventricular cells are open at strongly negative membrane potentials, providing the dominant potassium ion permeability that maintains the transmembrane potential near the potassium ion equilibrium potential. The absence of these channels in AV nodal cells is responsible for the relatively reduced resting potential positive to the potassium ion equilibrium potential, the insensitivity to extracellular potassium ion concentration, and the high membrane resistance of these cells.

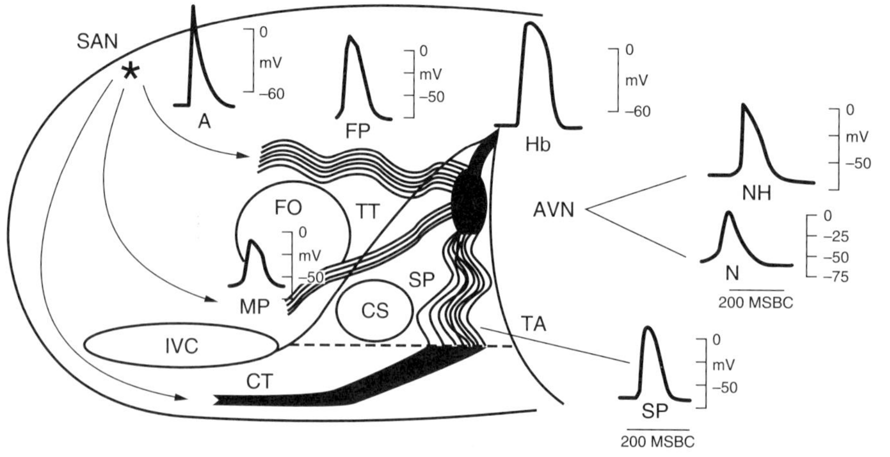

FIGURE 13-5 A representation of the right atrium and action potentials from the atrium and atrioventricular (AV) junction. A, atrial potential; AVN, AV node; BB, bundle branch potential; CS, coronary sinus; CT, crista terminalis; FO, fossa ovalis; FP, fast pathway and potential; SP, slow pathway and potential; Hb, His bundle and potential; IVC, inferior vena cava; MP, mid pathway and potential; N, central nodal potential; NH, distal nodal potential; TT, tendon of Todaro.

AV nodal cells, like sinoatrial nodal cells, lack sodium channels, which provide the excitatory current for atrial and ventricular cells; some cells may have channels that are inactivated at the low resting potentials.[41,42] The excitatory current for AV nodal cells is I_{CaL}, a calcium current that generates a slow upstroke because it is slower and less intense than the Na^+ current. L-type calcium ion channels require a relatively long period to recover from inactivation, hence the postrepolarization refractoriness of the AV node. The potent actions of autonomic stimulation on AV nodal conduction are largely due to the modulation of I_{CaL} by adrenergic and cholinergic stimulation via specific receptors and transduction systems. This current is enhanced and its kinetics of activation, inactivation, and recovery are accelerated by adrenergic stimulation; it is conversely affected by cholinergic stimulation.

AV nodal cells contain the pacemaker current I_f, which may contribute to the automaticity of AV nodal cells. On the other hand, the current I_{to}, a rapidly activating and inactivating potassium current that produces the notch of the AP (phase I) and is prominent in atrial cells, is sparse or absent in AV nodal cells. The repolarizing potassium current I_{Kr} is present and probably a major factor in the repolarization of AV nodal cells. The repolarizing current I_{Ks}, a contributor to repolarization in atrial, ventricular, and His-Purkinje cells, appears to be absent in AV nodal cells.

The potassium ion current $I_{Ach,\,Ado}$ is prominent in sinoatrial nodal and atrial cells but not in ventricular cells. It is also present in AV nodal cells. Activation of this current by acetylcholine or adenosine drives the resting potential to more negative values, suppresses automaticity, and diminishes the AP, thus slowing AV conduction.

Gap junctions appear to be sparse and smaller in the central AV node than in atrial and ventricular myocardium.[43] The gap junctions in the node are composed predominantly of connexin 40 rather than connexin 43, which is dominant in ventricular myocardium. The relatively poor intercellular communication in the AV node is probably the result of these gap junction properties plus a relative increase in the volume of extracellular space surrounding AV nodal cells compared with atrial and ventricular myocardium.

THE TRANSITIONAL ATRIONODAL CONNECTIONS

The transitional fibers connecting the atrium to the AV node aggregate in zones that are functionally relatively discrete, though not insulated from atrial myocardium. Recent observations indicate that anisotropy is a major determinant of directional conduction in these pathways.[44] They extend from the node to and beyond the border of Koch's triangle and connect to the node at relatively discrete sites along its margins.

These atrionodal connections have APs with characteristics that are intermediate between atrial cells and prototypical cells in the compact node. Like the AV node, these transitional connections may show decremental conduction and Wenckebach type of block. Until recently, functional definitions of these connections distinguished a "fast pathway" located at the anterior-superior aspect of the node and connecting to the atrium anterior to the fossa ovalis, and a "slow pathway" connecting to the node inferiorly and posteriorly and extending posterior to the isthmus region inferior to the coronary sinus os. The terms "fast" and "slow" do not necessarily represent different conduction velocities in these pathways but different conduction times determined by access and length of the pathways depending on the site of origin of global atrial activation. In man, the "fast" pathway is usually but not invariably considered to have a longer refractory period than the "slow" pathway.

Recent observations indicate that the atrial connections to the node are more complex.[38] A mid pathway has been described with deep (in relation to the right endocardium) connections to the node in its superior aspect and connections to the atrium in the septum posterior to the fossa ovalis. Left-sided connections have been described in man.[45] The functional interrelationships of the pathways in normal AV conduction are not fully elucidated. It is clear that these pathways, the atria, and the AV node can form various configurations of reentry circuits that cause AV nodal reentrant tachycardia. Interruption of the atrial connections to the AV node can produce AV block.

THE BUNDLE OF HIS AND BUNDLE BRANCHES

The bundle of His and bundle branches, which are insulated by fibrous matrix from ventricular myocardium, constitute the ventricular aspect of the AV conduction axis. They function to rapidly disseminate activation to the ventricular myocardium in a pattern that optimizes ejection of blood by a synchronized and coordinated apex to base contraction.[46] Conduction in these tissues is the most rapid of all cardiac tissue, promoting synchronization of activation of distal sites. The architecture of the bundle branches and their proximal insulation promotes apex-to-base contraction and septal-free wall synchronization. The APs of these fibers in multicellular preparations manifest very rapid upstrokes, prolonged plateaus and repolarization, and automaticity.

Quantitative measurement of individual currents in isolated Purkinje cells has been relatively limited. Sodium current is abundant, accounting in part for the fast upstroke and the rapid conduction velocity. The ionic basis for the prolonged plateau has not been determined, but I_{Ks} appears to be less active in Purkinje than in ventricular myocardial cells.[47]

The basis for the automaticity of the His-Purkinje system is debated. Some favor the decay of a potassium current activated during the AP as the primary basis for automaticity, whereas others favor the current I_f.[48] As in the sinoatrial node and AV node, there may be multiple ionic determinants of automaticity. The automaticity of the His-Purkinje system is more prominent proximally than distally. Automatic firing of the distal Purkinje system is slow and erratic. As a result, heart block due to degeneration and fibrosis of the bundle branches is more malignant than heart block caused by interruption of conduction in the AV junction proximal to the bundle of His.

Gap junctions are abundant in the Purkinje strands of the bundle branches and their distal ramifications. They are uniformly distributed along the lengths and ends of the fibers promoting good intracellular communication throughout the margins of the cells and rapid conduction. However, the bundle of His fibers appear relatively poorly connected in the transverse dimension so that dissociation within the bundle of His has been observed in experimental animals and in man.[49]

Electrocardiography of Atrioventricular Block

The earliest diagnoses of AV block were made by clinicians using irregularities in the pulse rate or a pulse deficit and were later documented by peripheral pulse recordings by Wenckebach and other investigators. While the peripheral pulse and its irregularity, whether periodic or not, can provide the clinician with initial inkling that AV conduction has been compromised, the surface ECG has become the clinical mainstay to diagnose type and potential location of AV block. It is conceptually convenient to consider the three types of AV block as reflecting delayed conduction in the AV conduction system (first-degree AV block), intermittent conduction (second-degree AV block), or failure of conduction (third-degree or complete heart block). First-degree AV block is characterized by prolongation of the PQ or P–R interval beyond 200 milliseconds (ms) (Fig. 13-6). The site of delay may lie in the atrial, AV node, bundle of His, or His-Purkinje system conduction. The associated QRS complex provides insight into the sites of diseased conduction but is not proof of location of the AV block. There may be evidence of distal conduction disease as seen in Figure 13-6 in the form of bundle branch block. Alternatively, a narrow QRS complex would favor conduction delay in the AV node or much less commonly in the bundle of His. However, the latter diagnosis should be suspected in calcific mitral valve disease, typically seen in middle-aged women with syncope or near syncope. Second-degree AV block can be categorized as type 1 (Wenckebach or Mobitz type 1), characterized by gradual prolongation of the P–R interval followed by a nonconducted P wave (Fig. 13-7). The beat following this pause in conduction is characterized by a shorter P–R interval, which may prolong in subsequent beats. Importantly, the last few cycles before the

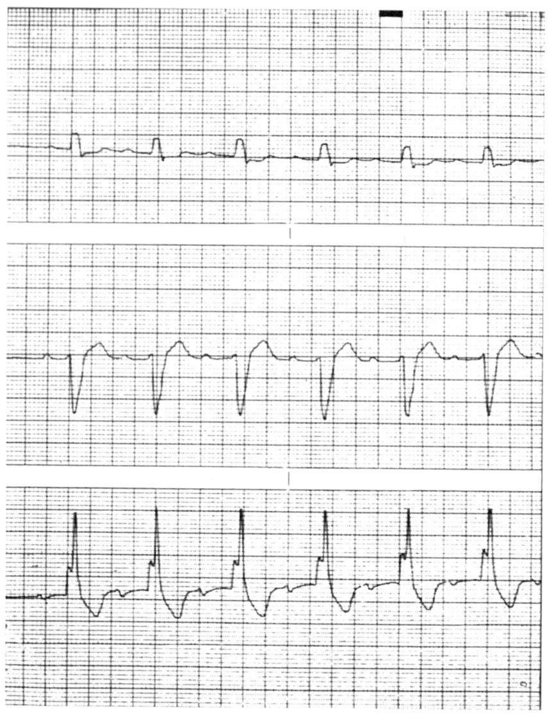

FIGURE 13-6 First-degree atrioventricular block with P–R prolongation and a bundle branch block QRS pattern on surface ECG.

nonconducted P wave may not show significant P–R prolongation, and if the early cycles are not seen, this diagnosis may be missed. This is occasionally referred to as *atypical Wenckebach conduction.* An important differential diagnosis is spontaneously alternating conduction in the slow and fast AV nodal pathways in patients with dual AV nodal conduction pathways (Fig. 13-8). This may masquerade as type 1 block on a cursory ECG examination. Furthermore, this may demonstrate grouped beating during tachycardias such as atrial fibrillation and flutter, further confusing the conduction pattern analysis. Mobitz type 2 second-degree block is characterized by a lack of P–R interval prolongation before the nonconducted P wave (Fig. 13-9). Importantly, the P–R interval of the subsequently conducted beat can be shorter than the subsequent stable P–R interval before the nonconducted P wave. This must be differentiated carefully from the atypical type 1 second-degree block. Type 2 second-degree block can be characterized by an occasional nonconducted P wave or a series of such P waves when it is referred to as *paroxysmal high-degree AV block.* This type of block is typically

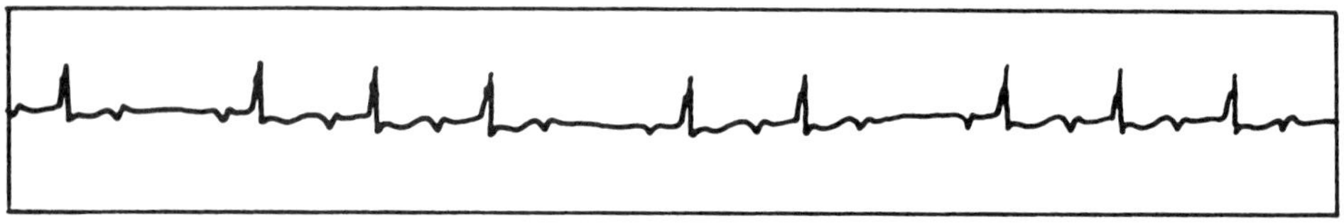

FIGURE 13-7 Second-degree type 1 atrioventricular block (Mobitz type 1 or Wenckebach type) showing gradual P–R prolongation before the nonconducted P wave.

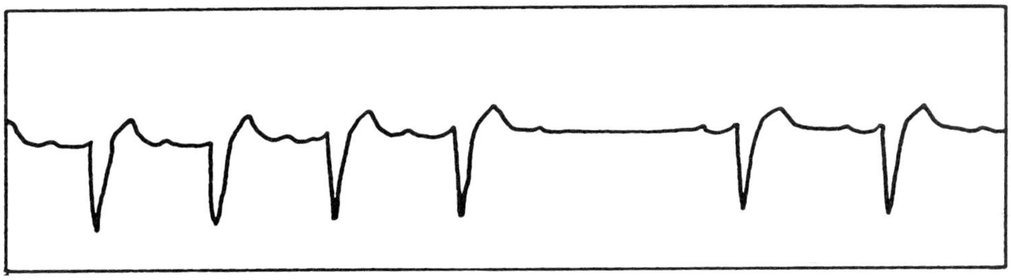

FIGURE 13-8 Second-degree type 2 atrioventricular block (Mobitz type II) showing absence of P–R prolongation before the nonconducted P wave.

seen in the distal or infra-His conduction system and is often associated with bundle branch block on the conducted beats on the surface ECG. This is suggestive of progressive bundle of His and bundle branch involvement in the pathologic process. Complete failure of conduction in third-degree AV block or complete heart block can be localized with some success based on the escape rhythm (Figs. 13-10 and 13-11). A narrow QRS escape rhythm suggests a nodal or bundle of His escape pacemaker and usually localizes the block to the AV node (much less often to intra-His locations). Narrow QRS complexes are typically seen in patients with congenital complete heart block, and a stable junctional pacemaker may be present for many years (see Fig. 13-10). More commonly, complete heart block is associated with a more distal disease process and an idioventricular escape rhythm at rates usually below 40 beats per minute (see Fig. 13-11). The site of AV block can be well directed by careful electrocardiographic analysis and confirmed by clinical electrophysiological investigation as indicated later in this chapter.

Diagnostic Techniques

Since the prognosis and treatment differ in AV block depending on whether block is within the AV node or infranodal, determining the site of block is important. In many cases this can be done noninvasively. The QRS duration, P–R intervals, and ventricular rate on the surface electrogram can provide important clues in localizing the level of block. Several noninvasive interventions may also prove helpful, such as vagal maneuvers, exercise, or administration of atropine. These methods take advantage of the differences in autonomic innervation of the AV node and His-Purkinje system.[61] While the AV node is richly innervated and highly responsive to both sympathetic and vagal stimuli, the His-Purkinje system is influenced minimally by the autonomic nervous system. Carotid sinus massage increases vagal tone and worsens AV nodal block. Exercise or atropine improves AV nodal conduction due to sympathetic stimulation. In contrast, carotid sinus massage improves infranodal block, while exercise and atropine worsens infranodal block, due to the change in the rate of the impulses being conducted through the AV node.

Exercise testing is a useful tool to help confirm the level of block that is already suspected in second- or third-degree block with a narrow or wide QRS complex. Patients with presumed AV nodal block or congenital complete heart block and a normal QRS complex will usually increase their ventricular rate with exercise. However, patients with acquired complete heart block and a wide QRS complex usually show minimal or no increase in ventricular rate.

An electrophysiological study is indicated in a patient with suspected high-grade AV block as the cause of syncope or presyncope when documentation cannot be obtained noninvasively.[62,63] In patients with coronary artery disease, it may be unclear whether symptoms are secondary to AV block or ventricular tachycardia, and

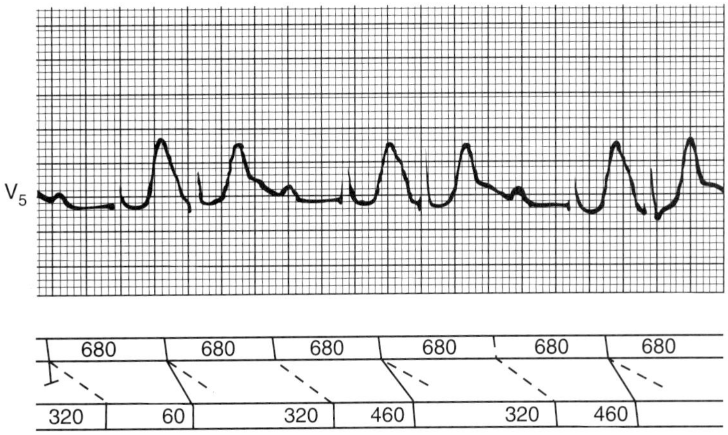

FIGURE 13-9 Dual atrioventricular nodal pathway conduction potentially masquerading as second-degree AV block. The ladder diagram shows the intervals in milliseconds of the P wave and P–R intervals.

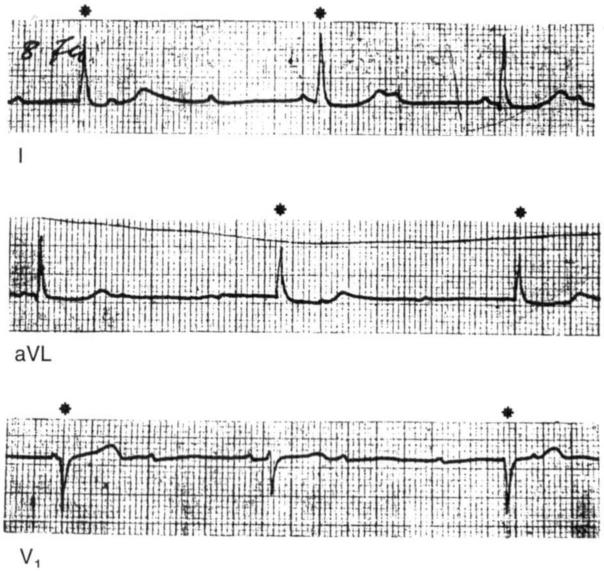

FIGURE 13-10 Congenital complete heart block in a young asymptomatic woman with excellent exercise response of the junctional pacemaker.

an electrophysiological study can be useful in establishing the diagnosis. Some patients with known second- or third-degree block may benefit from an invasive study to localize the site of AV block and help determine therapy or assess prognosis. Once symptoms and AV block are correlated by electrocardiography, further documentation by invasive studies is not typically required. Asymptomatic patients with transient Wenckebach block associated with increased vagal tone should not undergo invasive electrophysiological investigation.

The electrophysiological study allows analysis of the bundle of His–electrogram, as well as atrial and ventricular pacing to look for conduction abnormalities and inducible ventricular tachycardia. The atrio–His (A–H) and His ventricle (H–V) intervals are measured from the bundle of His–electrogram.[64] Atrial pacing techniques are used to help define the site of block and assess AV nodal and His-Purkinje conduction. During decremental atrial pacing, the A–H interval normally will gradually lengthen until AV nodal Wenckebach block is noted. The H–V interval will normally remain consistent despite different pacing rates. Abnormal AV nodal conduction is defined as Wenckebach block occurring at slower atrial-paced rates than what is normally seen (i.e., >500 ms). To determine whether AV nodal disease is truly present or just under the influence of excessive vagal tone, atropine alone or autonomic blockade with atropine and propranolol can be given to differentiate inherently abnormal AV nodal conduction from vagally mediated abnormalities. Infranodal block (Mobitz 2) is present when the atrial deflection is followed by the His electrogram, but no ventricular depolarization is seen. Block below the His is abnormal, unless associated with very short–paced cycle lengths (350 ms or less).[56]

Clinical Electrophysiology

Intracardiac electrophysiological studies in patients with conduction system disease can provide information about the site of AV block, assess possible mechanisms of syncope in affected patients, and stress the conduction system by atrial pacing and pharmacologic

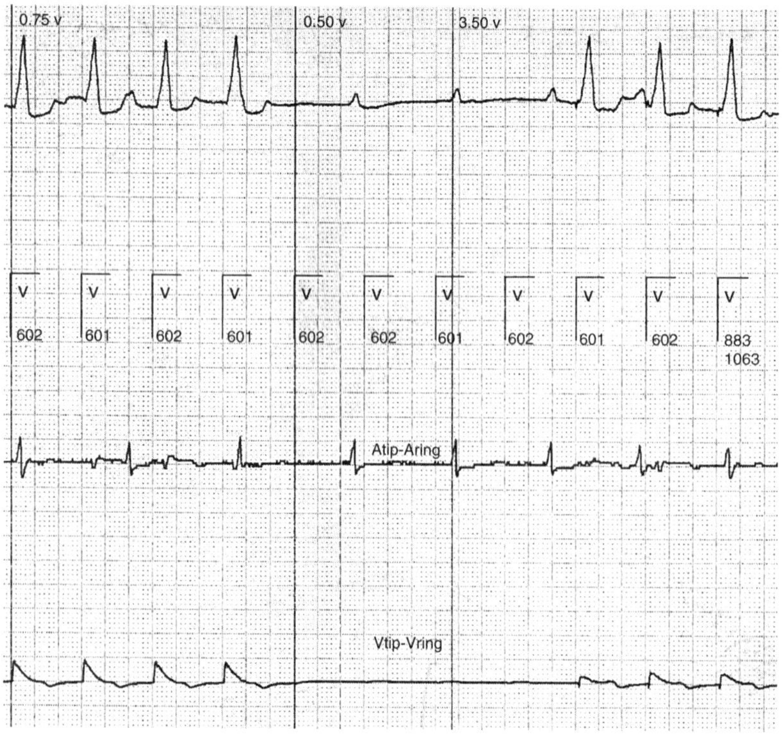

FIGURE 13-11 Complete heart block during threshold testing in a patient with an implanted permanent pacemaker. Note the absence of an escape idioventricular rhythm. Pacing had to be resumed immediately due to symptoms. Atip-A-ring, bipolar atrial electrogram; Vtip-Vring, bipolar ventricular electrogram.

intervention to help determine the risk for developing heart block and the need for permanent pacing.[50] Patients with conduction disease may present with cardiac syncope believed due to intermittent heart block. However, the conduction disease is often due to myocardial disease such as cardiomyopathy, but the syncope can be due to ventricular tachycardia. In this situation, both the conduction disease and propensity to malignant ventricular arrhythmia can be assessed to determine more appropriate therapy.[51] During electrophysiological assessment of patients with significant conduction disease, the presence of intact ventriculoatrial conduction can be assessed to help determine the type of pacemaker to be used. Patients with intact ventriculoatrial conduction should have a dual-chamber pacemaker placed to avoid pacemaker syndrome or pacemaker-mediated tachycardia.

FIRST-DEGREE ATRIOVENTRICULAR NODAL BLOCK

The most common cause of P–R interval prolongation is delay in conduction through the AV node. This can be identified readily at the time of intracardiac study by measuring the A–H interval and assessing the response to pharmacologic testing. The A–H interval represents conduction through the AV node and is measured from onset of the atrial electrogram to onset of the bundle of His–electrogram as recorded by the bundle of His–catheter. The normal A–H interval should not exceed 130 ms (Table 13-2). Most commonly, a prolonged A–H interval is due to either residual drug effect (β-blocker, calcium channel blocker, digoxin, antiarrhythmic drugs) or enhanced vagal tone. Atropine can be administered to reverse vagal tone to help determine whether intrinsic AV nodal conduction disease is present in the absence of AV nodal–blocking drugs. A prolonged A–H interval with AV node Wenckebach during slow atrial pacing (<100 bpm) in the presence of vagal blockade with atropine implies intrinsic AV nodal conduction disease.

SECOND-DEGREE ATRIOVENTRICULAR BLOCK

Type 1 second-degree AV block (Wenckebach) usually occurs within the AV node. Type 2 second-degree AV block is usually infranodal in the His-Purkinje system. However, either type of block may occur within or below the node. The exact level of block can be assessed with electrophysiological testing (Figs. 13-12 and 13-13). Electrophysiological testing can further define if AV block is due to disease in the His-Purkinje system or delayed conduction within the AV node in patients who present with first-degree AV block on their resting ECG and have associated bundle branch block.

Patients with a 2:1 or higher degree of AV block can have the site of block in the AV node or His-Purkinje system. Block in the AV node is often treated by altering the patient's medications, while block in the His-Purkinje system usually requires permanent pacing. The exact level of block in such patients can be determined at the time of electrophysiological testing and appropriate therapy initiated.

TABLE 13-2 Electrophysiological Intervals: Normal Values

Reference	P–A (ms)	A–H (ms)	H–V (ms)
Dhingra[52]	9-45	54-130	31-55
Gallagher[53]	24-45	60-140	30-55
Fisher[54]	10-45	55-130	30-55
Josephson[55]	—	60-125	35-55
Recommended	25-50	50-130	35-55

A–H, interval from the onset of the atrial electrogram to the onset of the His electrogram recorded by the His catheter; H–V, interval from the onset of the His electrogram to earliest ventricular activation; P–A, interval from the onset of atrial activity to the atrial electrogram recorded by the His catheter.

HIS-PURKINJE DISEASE

Abnormal conduction within the distal conduction system is referred to as *infranodal* or *distal-to-His conduction disease.* Conduction delay in the His-Purkinje system is clinically important because of the risk for progressing to complete heart block and need for permanent pacing. Usually, patients who will benefit from a permanent pacemaker can be identified clinically based on noninvasive ECG recording, either short or long term, and correlation of symptoms with ECG abnormalities. Conduction disease in the His-Purkinje system can be assessed by measuring the H–V interval. The H–V interval is determined as the conduction time between the onset of the His electrogram recorded by the His-bundle catheter and earliest ventricular activity recorded by either intracardiac or surface electrograms and usually consists of a sharp single deflection on the recording (see Fig. 13-12*A*). The normal H–V interval is between 35 and 55 ms (see Table 13-2). Shorter intervals are due to ventricular preexcitation or recording the right bundle electrogram instead of the His-bundle electrogram. First-degree intra-His block can be demonstrated by atrial extrastimuli during programmed stimulation (see Fig. 13-12*B*). Typically, split His potentials are seen (indicated as *H1* and *H2* on the figure). Type 1 second-degree block within the His-Purkinje system can occur and is demonstrated by progressive beat-to-beat increases in the H–V interval prior to infra His block (see Fig 13-12*C*). This type of block is uncommon. Type 2 second-degree block within the His-Purkinje block is more common and is usually associated with bundle-branch block, a prolonged H–V interval at rest, and the development of either spontaneous or pacing induced infra His block during electrophysiological testing (Fig. 13-13). This constellation of abnormalities of the His-Purkinje system most often warrants permanent pacing. At times, intracardiac recordings may demonstrate a split His potential with the clear recording of two His deflections separated by approximately 15 to 25 ms. Block can occur between the two His potentials, and, although this problem is rare, it should be treated as second-degree infranodal block with permanent pacing.

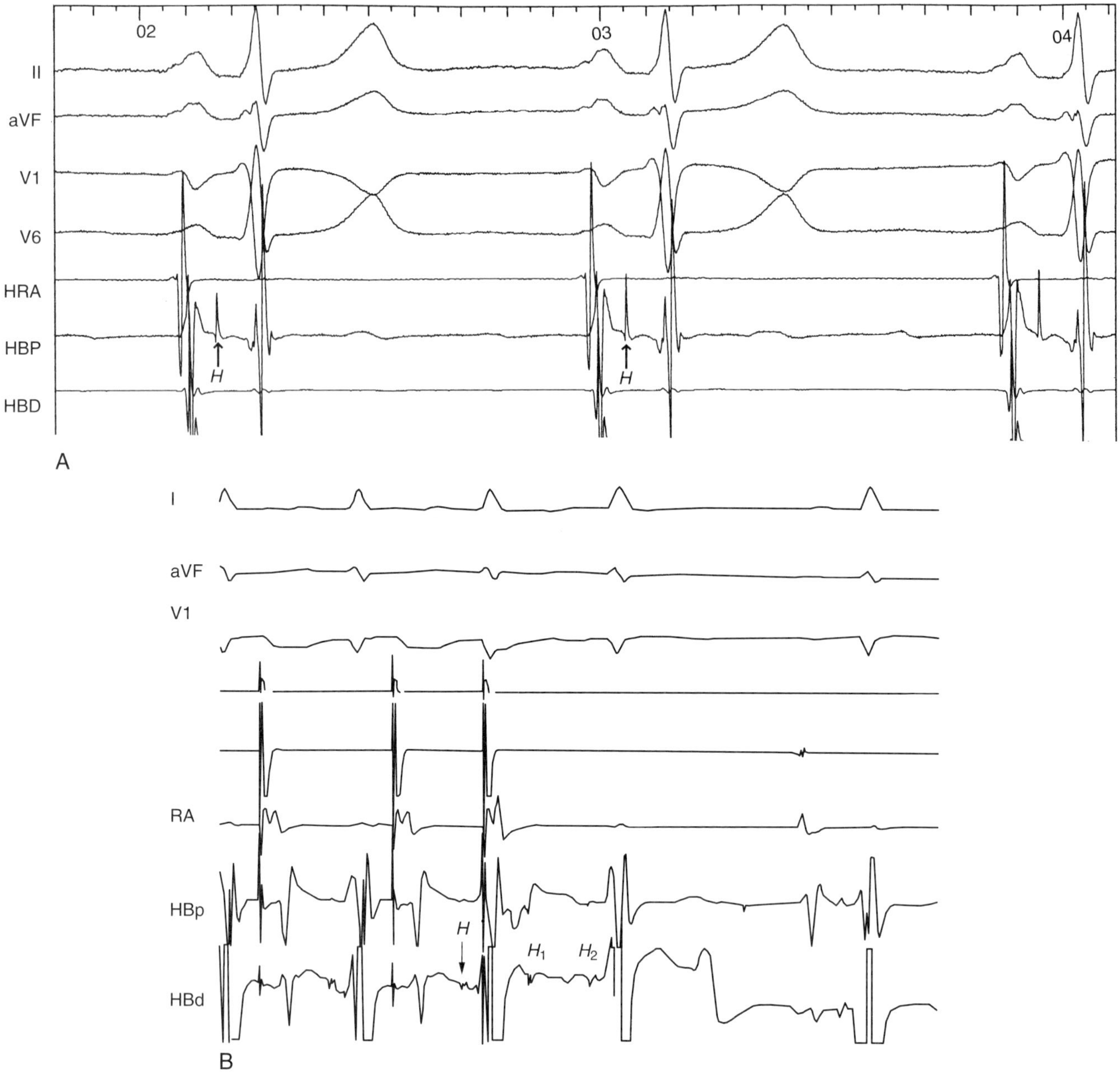

FIGURE 13-12 **Panel A,** Normal atrioventricular conduction with sharp well-defined His bundle (*H*) recording and normal H–V interval. HBD, distal His bundle; HBP, proximal His bundle; HRA, right atrial electrogram. **Panel B,** First-degree intra-His block on atrial extrastimulation. Note the low amplitude His potential (*H*) and the split His potentials (H_1 and H_2) on the distal His bundle recording. While there is a hint of a His bundle electrogram on the proximal recording, H_1 is only seen in the distal recording. HBd, distal His bundle; HBp, proximal His bundle; RA, right atrial electrogram.

BUNDLE BRANCH BLOCK

The H–V interval reflects the shortest conduction time from the His bundle to the ventricle and is often used to assess the state of the remaining conducting fascicles in patients with chronic bundle branch block. For example, patients with chronic left bundle branch block who have a prolonged H–V interval have damage to the remaining right fascicle represented by the H–V interval prolongation. Typically, asymptomatic patients with an H–V interval of 55 to 70 ms have a 1% annual incidence of progression to spontaneous second-degree or third-degree infranodal block, and treatment with permanent pacing should be individualized.[56] Asymptomatic patients with an H–V interval of 70 to 100 ms have a 4% annual incidence of complete heart block and probably should be paced, whereas asymptomatic patients with an H–V interval of greater than 100 ms have an 8% annual incidence of complete heart block and should undergo implantation of a permanent pacemaker.[57-59] Patients with spontaneous pacing-induced distal-to-His block at slow rates have a high incidence of subsequent complete heart block and sudden death and should be paced.[56]

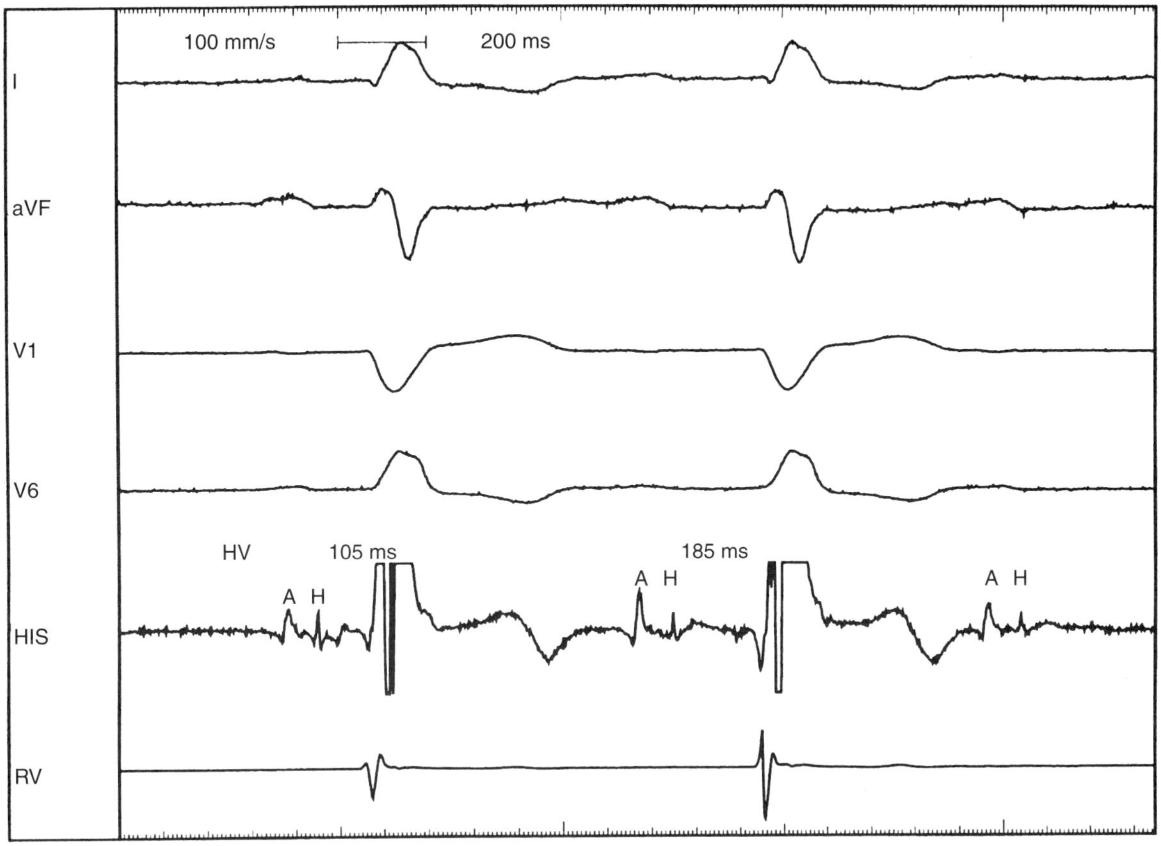

FIGURE 13-12 (cont'd) **Panel C,** Demonstration of spontaneous type 1 second-degree atrioventricular block (Wenckebach) in the His-Purkinje system in a patient with P–R interval prolongation, left axis deviation, and left bundle branch block. During sinus rhythm, the A–H interval remains fixed and the H–V interval prolongs from 105 milliseconds (ms) to 185 ms before block distal to His. I, aVF, V_1, V_6, surface ECG leads; His, electrogram recorded from the His-bundle region; RV, right ventricular electrogram.

The value of stressing the His-Purkinje conduction by using intravenous procainamide was assessed in a study by Girard and associates.[60] Seventy-nine patients who underwent electrophysiological testing received intravenous procainamide. An abnormal response to procainamide (defined as an H–V interval >80 ms) was observed in only 3% of 37 patients with a normal baseline H–V interval of less than 55 ms, 48% of 27 patients with mild H–V prolongation (56 to 70 ms), and in all 15 patients with moderate H–V prolongation (>70 ms). Procainamide induced distal-to-His block in 14% of patients studied for syncope and in 2% of patients studied for ventricular tachycardia. Syncope as the indication for electrophysiological study and left bundle branch block on the ECG were the best predictors of subsequent distal-to-His AV block. There was a strong linear correlation between post drug and baseline H–V intervals. This linear response to procainamide and published prospective studies support pacing patients with syncope who have a baseline H–V interval of greater than 70 ms. This study concluded that procainamide infusion during electrophysiological study of patients with undifferentiated syncope should be reserved for those patients with mild H–V prolongation of 55 to 70 ms in duration. The exact sensitivity and specificity of atrial pacing and IV procainamide to stress His-Purkinje conduction for predicting long-term outcome remains unclear.

COMPLETE HEART BLOCK

Electrophysiological testing has a minimal role in the assessment of patients with complete AV block, especially if they present with a slow escape rhythm and a wide QRS complex, indicating a ventricular escape focus. Such patients require pacing. Patients with complete heart block proximal to the His bundle usually have enhanced vagal tone, inferior myocardial infarction with ischemia of the AV node, or AV nodal block due to drugs such as β-blockers, calcium channel blockers, or digoxin. In these situations, electrophysiological testing is usually not required as the diagnosis and further management is handled clinically. Similarly, the role of electrophysiological testing in patients with congenital complete heart block has not been determined, and the test usually does not provide helpful information on which to base clinical management.

Electrophysiological studies usually do not make a significant contribution to the management of patients with documented AV block and associated symptoms of altered consciousness as the decision to implant a pacemaker is made clinically. However, in selected patients

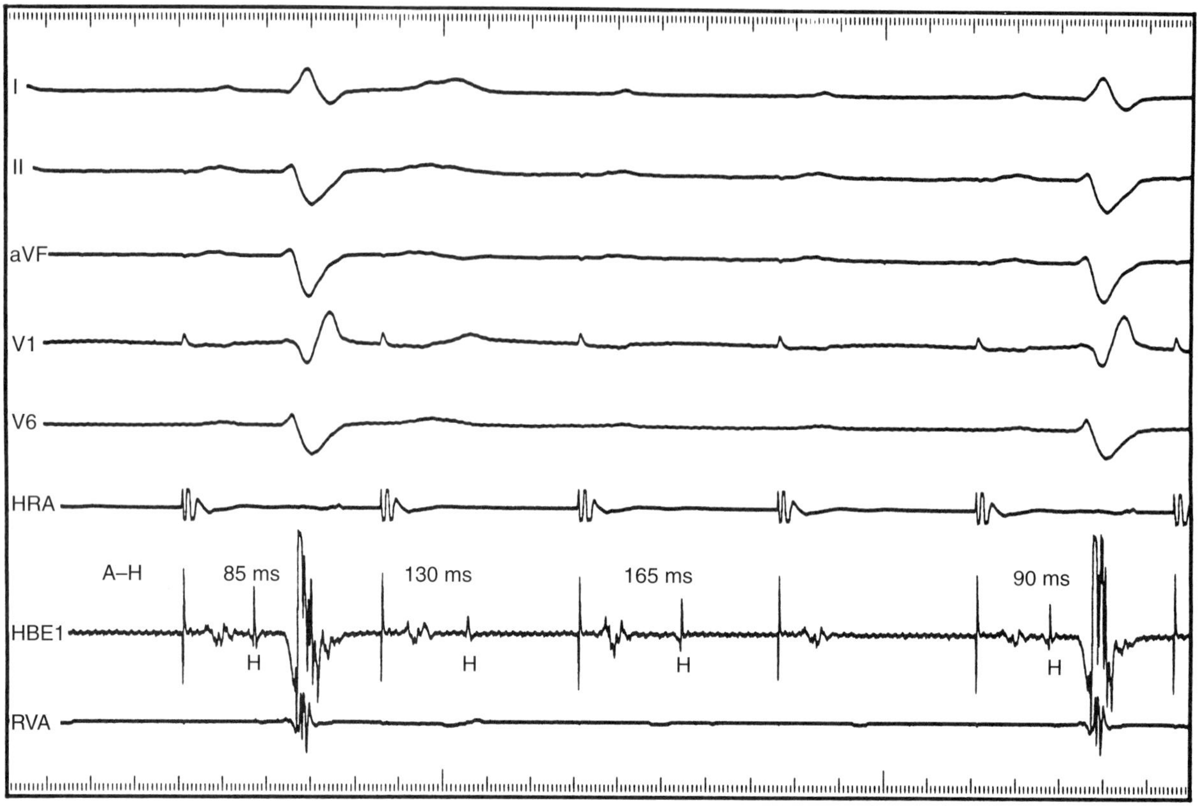

FIGURE 13-13 Demonstration of induced type 1 second-degree atrioventricular (AV) block (Wenckebach) in the AV node in addition to distal-to-His block in a patient with left axis deviation and right bundle branch block. During atrial pacing at 133 bpm, the A–H interval prolongs from 85 milliseconds (ms) to 165 ms before block in the AV node. The second and third beats demonstrate block distal to His. This illustrates multiple levels of conduction system disease in this patient. I, II, aVF, V_1, and V_6, surface ECG leads; HBE1, electrogram recorded from the region of the His bundle; HRA, high right atrial electrogram; RVA, right ventricular apex electrogram.

with conduction disease, including bifascicular block, prognostic information can be gained from invasive electrophysiological testing to assess the integrity of the His-Purkinje system and the risk for developing complete heart block. In addition, assessment of the electrical properties of the AV conduction system and inducibility of ventricular arrhythmias in patients with syncope and heart disease can be extremely helpful in determining appropriate management.

Evidence-Based Therapy

While retrospective studies provide useful information that can often help in hypothesis generation, they frequently exaggerate the benefit of treatment. It is always preferable to base clinical decisions, if possible, on data from prospective randomized trials. However, no prospective randomized trials have evaluated the efficacy of pacing therapy in patients with AV block since there are no alternatives to pacing therapy for the patient with symptomatic AV block (not due to reversible causes). In addition, it is difficult to assemble large groups of asymptomatic patients with specific types of AV block. Recommendations for permanent pacemaker implantation are based on observational studies on the natural history of AV block. In general, permanent pacemakers are implanted in patients with symptomatic AV block and in patients with asymptomatic AV block due to His-Purkinje disease.[65]

Pacing Mode Choice

Prospective data on the effects of pacing mode choice in patients with AV block are limited. Both single-chamber ventricular pacing and dual-chamber pacing will prevent bradycardia in patients with AV block. However, only dual-chamber pacing maintains AV synchrony. Dual-chamber pacing also reduces the incidence of pacemaker syndrome. Despite these potential advantages, the importance of dual-chamber pacing in patients with AV block has not been well established, particularly in the elderly.

In the Pacemaker Selection in the Elderly (PASE) trial, 407 elderly patients (mean age 76 years) were randomized to either the rate-adaptive ventricular inhibited (VVIR) or rate-adaptive dual chamber (AV) inhibited/triggered (DDDR) pacing mode.[66] In the 201 patients who had pacing systems implanted for AV block, no reduction in mortality, stroke, or atrial fibrillation was observed. In contrast to the patients with sinus node dysfunction, dual-chamber pacing was not associated with improvement in quality-of-life indices. In the Canadian Trial of Physiologic Pacing (CTOPP), 2568 patients with symptomatic bradycardia were randomized to single-chamber ventricular pacing or a "physiologic" pacing mode that preserved AV synchrony (single-chamber atrial pacing or dual-chamber pacing).[67] AV block was the indication for pacing in approximately 60% of the patients. Physiologic pacing was not associated with a reduction in stroke or death due to cardiovascular disease, which was the study's primary end point. Interestingly, physiologic pacing was associated with a reduction in the development of chronic atrial fibrillation even in those patients with pacing systems implanted for AV block.[68] A large prospective study recently examined the importance of pacing mode selection in elderly patients with AV block.[69] In the United Kingdom Pacing and Clinical Events (UK-PACE) Trial, patients older than 70 years of age with AV block were randomized to the ventricular inhibited (VVI), VVIR, and DDD pacing modes. Patients were followed for at least 3 years. The primary end point was all-cause mortality, although other secondary outcomes such as atrial fibrillation, heart failure, stroke/transient ischemic attack/thromboembolism, pacing system revision, cardiovascular events such as angina or myocardial infarction, exercise capacity, and quality of life were also evaluated. A composite end point of all outcomes was also examined. Patient recruitment occurred from 1995 to 1999, and the trial terminated in September 2002 with a median follow-up of 4 years.[69] Of the patients, 504 were randomized to VVI pacing, 505 to VVIR pacing, and 1012 to DDD pacing. Some patients remained in the randomized mode: 96.6% of VVI patients, 96.4% of VVIR patients, and 87.7% of DDD patients. There was no difference in all-cause mortality or time to atrial fibrillation or other cardiovascular events. There was an increased stroke risk with VVI pacing (hazard ratio 1.58), but no differences were elicited in any other secondary end points or composite end point. This trial suggests that the elderly patient derives limited and specific benefits and fails to obtain other favorable results compared with the younger patient with DDD pacing. This may reflect more advanced atrial and cardiovascular disease status or other variables.

In summary, it is unlikely that any randomized trial will be performed to evaluate the efficacy of pacing therapy in patients with AV block. Although the evidence for the usefulness of dual-chamber pacing in patients with AV block is incomplete at present, most patients with AV block should receive a dual-chamber pacing system.[65,70]

Management of Atrioventricular Block

When AV block is identified, management requires a search for reversible causes, an assessment of hemodynamic stability to determine whether temporary pacing is required, and, finally, a decision on whether permanent pacing will be necessary.

TEMPORARY PACING

In the hemodynamically unstable patient with AV block, the clinician must identify any rapidly reversible causes. Reversible causes include hyperkalemia, hypoxia, increased vagal tone, and ischemia. While treatment for reversible causes is initiated, it must be decided quickly whether temporary pacing will be required for the hemodynamically unstable patient. Temporary pacing is most quickly initiated by transcutaneous passage of current between two specially designed pads placed on the chest wall. Patch position is the most important factor for determining effectiveness of transcutaneous pacing. The cathode should be placed on the left chest over the cardiac apex, and the anode placed on the back between the spine and scapula or anteriorly just above the right nipple. Currents of 20 to 140 mA are usually required to capture and activate the ventricles. Using the correct technique, transcutaneous pacing is effective in more than 90% of cases and associated with very few complications.[71,72] However, transcutaneous pacing cannot be used for prolonged periods because of patient discomfort and unreliability of capture due to impedance changes.

If temporary pacing must be used for more than 30 minutes, transvenous pacing should be initiated since it is far more stable and better tolerated. With transvenous pacing, intravascular access is obtained, usually through the right internal jugular vein, and a pacing catheter is positioned into the right ventricle. The pacing catheter is connected to a pulse generator; ventricular capture can usually be obtained with currents of 1 to 2 mA. Transvenous pacing can be used for long periods with minimal complications (<2% once venous access is achieved).[71]

PERMANENT PACING

Once the patient is stabilized, the clinician must assess whether the AV block will be permanent. There are several conditions where persistent AV block will gradually resolve. In approximately 10% to 15% of patients with inferior and posterior wall myocardial infarctions, transient second-degree, advanced, or complete AV block will be observed.[75] In almost all cases the AV block will resolve, and permanent pacing is not required. The use of fibrinolytic therapy has not altered the incidence of AV block in inferior wall myocardial infarction. Approximately 8% to 10% of patients with Lyme disease will have transient AV conduction abnormalities due to myocarditis involving the AV nodal region. AV block usually resolves within several weeks, and permanent pacing is almost never required. Acute rheumatic fever can also present with AV block that is expected to resolve after several weeks.[75]

INDICATIONS FOR PERMANENT PACING

Indications for permanent pacing in acquired AV block have been published by a Joint Committee of the American College of Cardiology (ACC) and the American Heart Association (AHA) in 1984, 1991, 1998, and 2002.[65]

TABLE 13-3 Indications for Permanent Pacing in Atrioventricular Block

Asymptomatic	
First-degree	Pacing not indicated
Second-degree	Pacing not indicated if type 1 second-degree AV block with a narrow QRS complex is present Electrophysiological study should be considered if type 1 second-degree AV block with a wide QRS complex is present Pacing if type 2 block is present
Third-degree	Pacing
Symptomatic	Pacing regardless of type

The ACC/AHA/NASPE 2002 guidelines use the standard three-group classification schema. Class I indications are conditions for which there is evidence or general agreement that a given procedure or treatment is beneficial, useful, and effective. Class II indications are conditions for which there is conflicting evidence or a divergence of opinion about the usefulness/efficacy of a procedure or treatment. Class II has been further divided into class IIa, where the weight of evidence/opinion is in favor of usefulness/efficacy, and class IIb, where usefulness/efficacy is less well established by evidence/opinion. Class III indications are conditions for which there is evidence or general agreement that a procedure/treatment is *not* useful/effective and, in some cases, may be harmful. Despite some shortcomings,[76] the ACC/AHA/NASPE guidelines provide a useful framework upon which management is based. The indications for permanent pacing in *symptomatic* patients with second- or third-degree AV block are often straightforward (Table 13-3). Controversial indications involve mostly asymptomatic patients.[76,77] The decision to implant a pacemaker in *asymptomatic* patients is more difficult and requires knowledge of the pathophysiology and natural history of AV block. As a general rule, since escape rhythms from ventricular tissue are unreliable, a pacemaker should be implanted in asymptomatic patients if AV block occurs in His-Purkinje tissue.

THIRD-DEGREE ATRIOVENTRICULAR BLOCK

Symptomatic third-degree AV block is a class I indication for pacing therapy. For asymptomatic patients, the current guidelines use a rate cutoff, with escape ventricular rates less than 40 beats per minute designated as a class I indication and ventricular rates greater than 40 beats per minute designated as class IIa. Although this distinction may seem reasonable, one must confirm whether the patient is truly asymptomatic; in some cases symptoms associated with third-degree AV block are subtle. More importantly, prognosis depends on the stability of the escape rate pacemaker rather than the actual rate; escape rhythms from ventricular tissues (wide QRS complexes) are inherently unstable. In practice, acquired third-degree AV block not associated with any reversible causes is usually considered a class I indication for permanent pacing regardless of whether symptoms are present or absent.

SECOND-DEGREE ATRIOVENTRICULAR BLOCK

Symptomatic second-degree AV block is a class I indication for pacing regardless of type of block. The use of pacing therapy for asymptomatic patients with second-degree AV block is controversial and depends on the type (site) of AV block.

In general, type 1 second-degree AV block associated with a narrow QRS complex (<0.12 sec) is due to block in the AV node, and the current published guidelines do not recommend pacemaker implantation in the asymptomatic patient. In a study of 56 patients with documented chronic second-degree AV block due to AV nodal conduction delay, those without associated cardiac disease had a benign course while those with associated cardiac disease had a poor prognosis due to progression of underlying cardiac disease rather than to the development of sudden bradycardia.[79] However, in another retrospective study of 214 patients with second-degree AV block, survival and requirement for pacing were not different among patients with type 1 and type 2 heart block, and the presence or absence of bundle branch block did not appear to aid in the prediction of survival.[80] In view of these conflicting data, it is prudent to closely monitor patients with type 1 second-degree AV block and a narrow QRS complex for symptoms and for progression of conduction tissue disease (e.g., development of fascicular block or QRS widening). If type 1 second-degree AV block is associated with a wide QRS complex (>0.12 sec) AV block will be located in the AV node in 30% to 40% of patients and in the His-Purkinje system in 60% to 70% of cases.[81,82] In these cases an invasive electrophysiological study is often required to identify the site of block. If intra-Hisian or infra-Hisian block is identified, a pacemaker should be implanted since these conditions usually progress to complete heart block within 5 years.[83]

Type 2 second-degree AV block probably always occurs in His-Purkinje tissue. Asymptomatic patients with type 2 AV block usually do develop symptoms and will require permanent pacing.[84,85]

In 2:1 second-degree AV block, every other P wave conducts, preventing comparison of consecutive P–R intervals. The QRS complex provides a clue as to the site of block: A narrow QRS complex is associated with His-Purkinje block 30% of the time, and a wide QRS complex is associated with His-Purkinje block approximately 80% of the time.[85] In the asymptomatic patient with 2:1 block, maneuvers to alter the conduction ratio between the atria and ventricles (such as exercise, atropine, or continuous ECG monitoring) may allow localization of the site of block. However, in some cases electrophysiological evaluation to determine the site of block will be required.

FIRST-DEGREE ATRIOVENTRICULAR BLOCK

If first-degree AV block is severe, atrial activation and contraction can occur while the ventricles are contracting in response to the previous atrial contraction, which leads to an inappropriate rise in atrial pressures and symptoms similar to the pacemaker syndrome. In this situation symptoms can be significant even with exercise since the P–R interval does not shorten appropriately with adrenergic stimulation. The current guidelines[65] classify symptomatic first-degree AV block as a class IIa indication for pacemaker implantation. Asymptomatic first-degree AV block is a class III indication for pacing. The one exception to this recommendation is the asymptomatic patient with first-degree AV block, abnormal QRS axis due to left anterior or left posterior fascicular block, and neuromuscular disease.[70] Neuromuscular diseases such as myotonic dystrophy and Kearns-Sayre syndrome are associated with progressive AV block; since development of complete heart block can be unpredictable, a permanent pacemaker is justified.

PACING MODE CHOICE

Three pacing modes (DDD, VDD, and VVI) can be used to prevent bradycardia in patients with AV block (Fig. 13-14).

VVI AND VVIR PACING

In the VVI or VVIR pacing mode, bradycardia is prevented, and if rate adaption is programmed "on," the heart rate will increase with exercise. However, in neither of these pacing modes is AV synchrony present. The importance of AV synchrony is controversial in patients with AV block, since the main contribution to increased cardiac output with exercise is heart rate rather than AV synchrony, and the incidence of pacemaker syndrome is lower in patients with AV block compared with patients with sinus node dysfunction. However, it seems intuitively reasonable that in the presence of organized atrial activity, the VDD and DDD pacing modes are most appropriate. In fact, since the early 1990s the British Pacing and Electrophysiology Group guidelines have stated that the only indication for the VVI and VVIR pacing modes is atrial fibrillation/flutter with AV block or slow ventricular response.[70] The VVI and VVIR pacing modes may also be appropriate in patients who are incapacitated and inactive, as well as those with other medical problems associated with a short life expectancy.

DDD AND DDDR MODES

The DDD pacing mode prevents bradycardia and provides AV synchrony in patients with AV block. Although less important for exercise-related increases in cardiac output, several studies have demonstrated that AV synchrony is associated with improved symptoms at baseline levels of activity. A large randomized study (UK-PACE) that compares the VVI and DDD pacing modes in elderly patients (older than 70 years old) with second-degree or third-degree AV block has been reported.[69] Although the DDD pacing mode is the most complex, most guidelines recommend this pacing mode in patients with AV block since it maintains AV synchrony regardless of the cause of the bradycardia.

VDD AND VDDR MODES

Although the DDD pacing mode provides AV synchrony, its use requires the presence of a separate atrial lead. Recently, a single lead that has both ventricular and atrial electrodes has been developed; the ventricular electrodes are in direct contact with the right ventricular endocardium, while the atrial electrodes "float" in the right atrium. When used with a pacemaker in the VDD pacing mode, this lead allows sensing of atrial activity and ventricular pacing. In the presence of normal sinus rates, AV synchrony is maintained using a less complex system. However, if sinus bradycardia develops, AV synchrony is lost since the atria cannot be paced (see Fig. 13-14).

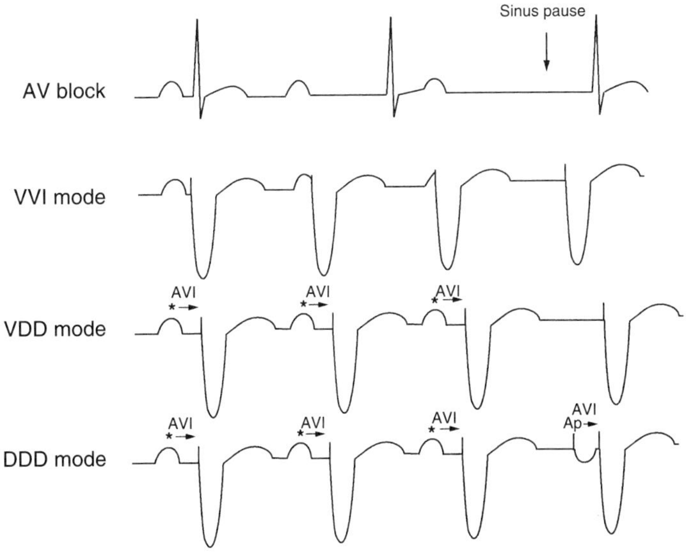

FIGURE 13-14 Responses of various modes in a patient with complete atrioventricular (AV) block and a sinus pause. *First row,* Baseline rhythm strip before pacing shows complete heart block. In addition, there is a sinus pause. *Second row,* Same strip with a pacing system in the VVI pacing mode. Bradycardia is prevented, but AV synchrony is not present. *Third row,* In the VDD pacing mode, atrial activity is sensed by the pacemaker (*), which initiates the A–V interval and maintains AV synchrony. However, when a sinus pause occurs, since the atria are not paced, ventricular pacing occurs, and AV synchrony is lost. *Fourth row,* In the DDD pacing mode AV synchrony is maintained regardless of the cause of bradycardia. When a sinus pause occurs, an atrial stimulus (Ap) is provided.

REFERENCES

1. Stannius HF: Zwei Reihen physiologischer Verusche. Arch Anat Physiol Wis Med 1852;2:85-100.
2. His W Jr: Die Tatigkeit des embryonalen Herzens und deren Bedeutung fur die Lehre von der Herzbewegung beim Erwachsenen. Arb Med Klin Leipzig 1893;14:49.
3. Tawara S: Das Reizleitungssystem des Saugetierherzens. Eine anatomisch histologische Studie uber das Atrioventrikularbundel und ide Purkinjeschen Faden. Mit einem Vorwort von L. Aschoff (Marburg). Fischer, Jena, 1906.
4. Wenckebach KF: Beirage zur Kenntnis der menschlichen Herztatigleit. Arch Anat Physiol 1906;297-354.
5. Dickinson DF, Scott O: Ambulatory electrocardiographic monitoring in 100 teenage boys. Br Heart J 1984;51:179-83.
6. Sobotka PA, Mayer JH, Bauernfeind RA, et al: Arrhythmias documented by 24 hour continuous ambulatory electrocardiographic monitoring in young women without apparent heart disease. Am Heart J 1981;101:753-9.
7. Brodsky M, Wu D, Denes P, et al: Arrhythmias documented by 24 hour continuous electrocardiographic monitoring in 50 male medical students without apparent heart disease. Am J Cardiol 1977;39:390-5.
8. Manning GW: Electrocardiography in the selection of Royal Canadian Air Force crew. Circulation 1954;10:401-12.
9. Johnson RL, Averill KH, Lamb LE: Electrocardiographic findings in 67,375 asymptomatic subjects. Am J Cardiol 1960;6:153-7.
10. Perlman LV, Ostrander LD, Keller JB, Chiang BN: An epidemiologic study of first degree atrioventricular block in Tecumseh, Michigan. Chest 1971;59:40-6.
11. Logue RB, Hanson JF: Heart block. A study of 100 cases with prolonged PR interval. Am J Med Sci 1944;207:765-9.
12. Fleg JL, Kennedy HL: Cardiac arrhythmias in a healthy elderly population: Detection by 24-hour ambulatory electrocardiography. Chest 1982;81:301-7.
13. Mymin D, Mathewson FAL, Tate RB, et al: The natural history of first-degree atrioventricular heart block. N Engl J Med 1986;315: 1183-8.
14. Bhat PK, Watanabe K, Rao DB, Luisado AA: Conduction defects in the aging heart. J Am Geriatr Soc 1974;22:517-20.
15. Morgensen L: Cardiac arrhythmias in the asymptomatic individual without overt cardiac disease. In Kulbertus H (ed): Medical Management of Cardiac Arrhythmias. Edinburgh, Churchill Livingstone, 1986.
16. McCue CM, Mantakas ME, Tingelstad JB, et al: Congenital heart block in newborns of mothers with connective tissue disease. Circulation 1977;56:82-90.
17. Barold SS: Acquired AV block. In Kusumoto FM, Goldschlager N (ed): Cardiac Pacing for the Clinician. Philadelphia, Lippincott Williams & Wilkins, 2000.
18. El-Sherif N, Turitto G, Jain P: Advances in the anatomy and physiology of atrioventricular nodal and atrioventricular pathways. In Saksena S, Luderitz B: Interventional Electrophysiology, 2nd ed. Armonk, NY, Futura, 1995, pp 3-24.
19. Bharati S: Pathology of the conduction system. In Silver MD, Gotlieb AI, Schoen FJ (ed): Cardiovascular Pathology. New York, Churchill Livingston, 2001, 20:607-28.
20. Bharati S, Lev M: The anatomy of the normal conduction system: Disease-related changes and their relationship to arrhythmogenesis. In Podrid P, Kowey P (eds) : Cardiac Arrhythmias: Mechanisms, Diagnosis and Management. Baltimore, Williams & Wilkins, 1995, Chapter 1, pp 1-15.
21. Bharati S, Lev M: Pathology of atrioventricular block. In: Cardiology Clinics. Philadelphia, WB Saunders, 1984, 4:741-51.
22. Bharati S, Swerdlow MA, Vitullo D, et al: Neonatal lupus with congenital atrioventricular block and myocarditis. Pacing Clin Electrophysiol 1987, 10:1058-70.
23. Bharati S, Lev M: The anatomy and pathology of the conduction system. In Samet P, El-Sherif N (eds): Cardiac Pacing. New York, Grune & Stratton, 1980, pp 1-35.
24. Lev M, Bharati S: Lesions of the conduction system and their functional significance. In Sommers SC (ed): Pathology Annual 1974. New York, Appleton-Century Crofts, 1974, 9:157-208.
25. Bharati S, Rosen KM, Strasberg B, et al: Anatomic substrate for congenital atrioventricular block in middle aged adults. Pacing Clin Electrophysiol 1982;5:860-9.
26. Lev M, Bharati S: The conduction system in coronary artery disease. In Donoso, Lipski J: Current Cardiovascular Topics. New York, Stratton Intercontinental Medical Book Corp, 1978, pp 1-16.
27. Rossi L: Histopathologic Features of Cardiac Arrhythmias. Milan, Ambrosiana, 1969.
28. Lenegre J: Etiology and pathology of bilateral bundle branch block in relation to complete heart block. Prog Cardiovasc Dis 1963-1964;6:409.
29. Davies MJ: Pathology of chronic AV block. Acta Cardiol 1976; 21:19.
30. Bharati S, de la Fuente DJ, Kallen RJ, et al: The conduction system in systemic lupus erythematosus with atrioventricular block. Am J Cardiol 1975;35:299-304.
31. Lev M, Bharati S, Hoffman FG, Leight L: The conduction system in rheumatoid arthritis with complete atrioventricular block. Am Heart J 1975;90:78-83.
32. Bharati S, Bicoff JP, Fridman JL, et al: Sudden death caused by benign tumor of the atrioventricular node. Arch Intern Med 1976;136:224-8.
33. Bharati S, Lev M: Sequelae of atriotomy and ventriculotomy on the endocardium, conduction system and coronary arteries. Am J Cardiol 1982;50:580.
34. Bharati S, Molthan ME, Veasy LG, et al: Conduction system in two cases of sudden death two years after the Mustard procedure. J Thorac Cardiovasc Surg 1979;77:101.
35. Bharati S, Lev M: Conduction system in cases of sudden death in congenital heart disease many years after surgical correction. Chest 1986;90:861.
36. Bharati S, Lev M: The Pathology of Congenital Heart Disease: A Personal Experience with More Than 6,300 Congenitally Malformed Hearts. Armonk, NY, Futura, 1996, 2:1445-1498.
37. Kaplan BM, Miller AJ, Bharati S, et al: Complete AV block following mediastinal radiation therapy: Electro-cardiographic and pathologic correlation and review of the world literature. J Interv Card Electrophysiol 1997;1:175-88.
38. Scherlag BJ, Patterson E, Yamanashi W, et al: The AV conjunction: A concept based on ablation techniques in the normal heart. In Mazgalev TN, Tchou PJ, eds: Atrial-AV Nodal Electrophysiology: A View from the Millennium. Armonk, NY, Futura, 2000.
39. Lazzara R, Scherlag BJ, Belardinelli L: Atrioventricular conduction. In Spooner PM, Rosen MR (eds): Foundations of Cardiac Arrhythmias: Basic Concepts and Clinical Approaches. New York, Marcel Dekker, 2001.
40. Petrecca K, Shrier A: Spatial distribution of ion channels, receptors, and innervation in the AV node. In Mazgalev TN, Tchou PJ (eds): Atrial-AV Nodal Electrophysiology: A View from the Millennium. Armonk, NY, Futura, 2000.
41. Petrecca K, Amellal F, Laird DW, et al: Sodium channel distribution within the rabbit atrioventricular node as analyzed by confocal microscopy. J Physiol (Lond) 1997;501:263-74.
42. Zipes DP, Mendez C: Action of manganese ions and tetrodotoxin on atrioventricular nodal transmembrane potentials in isolated rabbit hearts. Circ Res 1973;32:447-54.
43. Saffitz JE, Yamada KA, Schuessler RB: Distribution and function of gap junction proteins in atrial-AV nodal conduction. In Mazgalev TN, Tchou PJ (eds): Atrial-AV Nodal Electrophysiology: A View from the Millennium. Armonk, NY, Futura, 2000.
44. Hocini M, Loh P, Siew Y, et al: Anisotropic conduction in the triangle of Koch of mammalian hearts: Electrophysiologic and anatomic correlations. J Am Coll Cardiol 1989;31:629-36.
45. Otomo K, Wang Z, Lazzara R, et al: Atrioventricular nodal reentrant tachycardia: Electrophysiological characteristics of four forms and implications for the reentrant circuit. In Zipes D, Jalife J (eds): Cardiac Electrophysiology: From Cell to Bedside, 3rd ed. Philadelphia, WB Saunders, 2000.
46. Lazzara R: Electrophysiology of the specialized conduction system: Selected aspects relevant to clinical bradyarrhythmias. In Samet P, El-Sherif N (eds): Cardiac Pacing, 2nd ed. New York, Grune & Stratton, 1980.
47. Han W, Wang Z, Nattel S: Slow delayed rectifier current and repolarization in canine cardiac Purkinje cells. Am J Physiol Heart Circ Physiol 2001;280:H1075-80.
48. Vasalle M, Yu H, Cohen IS: Pacemaker channels and cardiac automaticity. In Zipes D, Jalife J (eds): Cardiac Electrophysiology: From Cell to Bedside, 3rd ed. Philadelphia, WB Saunders, 2000.

49. Narula OS: Longitudinal dissociation in the His bundle. Bundle branch block due to asynchronous conduction within the His bundle in man. Circulation 1977;56:996-1006.
50. Hammill SC, Sugrue DD, Gersh BG, et al: Clinical cardiac electrophysiologic testing: Technique, diagnostic indications, and therapeutic uses. Mayo Clin Proc 1986;61:478-503.
51. Click RL, Gersh BG, Sugrue DD, et al: Role of invasive electrophysiologic testing in patients with symptomatic bundle branch block. Am J Cardiol 1987;59:817-23.
52. Dhingra RC, Rosen KM, Rahimtoola SH: Normal conduction intervals and responses in sixty-one patients using His bundle recording and atrial pacing. Chest 1973;64:55-9.
53. Gallagher JJ, Damato AN: Technique of recording His bundle activity in man. In Grossman W (ed): Cardiac Catheterization and Angiography, 2nd ed. Philadelphia, Lea & Febiger, 1980, pp 283-301.
54. Fisher JD: Role of electrophysiologic testing in the diagnosis and treatment of patients with known and suspected bradycardias and tachycardias. Prog Cardiovasc Dis 1981;24:25-90.
55. Josephson ME: Clinical Cardiac Electrophysiology. Malvern, Pa, Lea & Febiger, 1992, pp 96-116.
56. Dhingra RC, Wyndham C, Bauernfeind R, et al: Significance of block distal to the His bundle induced by atrial pacing in patients with chronic bifascicular block. Circulation 1979;60:1455-64.
57. McAnulty JH, Rahimtoola SH, Murphy E, et al: Natural history of "high-risk" bundle-branch block: Final report of a prospective study. N Engl J Med 1982;307:137-43.
58. Scheinman MM, Peters RW, Modin G, et al: Prognostic value of infranodal conduction time in patients with chronic bundle branch block. Circulation 1977;56:240-44.
59. Dhingra RC, Palileo E. Strasberg B, et al: Significance of the HV interval in 517 patients with chronic bifascicular block. Circulation 1981;64:1265-71.
60. Girard SE, Munger TM, Hammill SC, Shen WK: The effect of intravenous procainamide on the HV interval at electrophysiologic study. J Interv Card Electrophysiol 1999;3:123-37.
61. James TN: Cardiac innervation: Anatomic and pharmacologic relations. Bull NY Acad Sci 1967;43:1041-86.
62. Damato AN, Lau SH, Helfant R, et al: A study of heart block in man using His bundle recordings. Circulation 1969;39:297-305.
63. Rosen KM: The contribution of His bundle recording to the understanding of cardiac conduction in man. Circulation 1971; 43:961-6.
64. Scheinman MM, Peters RW, Sauve MJ, et al: Value of the H-Q interval in patients with bundle branch block and the role of prophylactic permanent pacing. Am J Cardiol 1982;50:1316-22.
65. Gregoratos G, Abrams J, Epstein AE, et al: ACC/AHA/NASPE 2002 guideline update for implantation of cardiac pacemakers and antiarrhythmia devices. J Am Coll Cardiol 2002;640:1703-1719.
66. Lamas GA, Orav EJ, Stambler BS, et al: Quality of life and clinical outcomes in elderly patients treated with ventricular pacing as compared with dual-chamber pacing. N Engl J Med 1998;338: 1097-104.
67. Connolly SJ, Kerr CR, Gent M, et al: Effects of physiologic pacing versus ventricular pacing on the risk of stroke and death due to cardiovascular causes. N Engl J Med 2000;342:1385-91.
68. Skanes AC, Krahn AD, Yee R, et al: Progression to chronic atrial fibrillation after pacing: The Canadian Trial of Physiologic Pacing. J Am Coll Cardiol 2001;38:167-72.
69. Toff WD, Skehan JD, deBono DP, Camm AJ for the UK PACE investigators: The United Kingdom Pacing and Cardiovascular Events (UK-PACE) trial. Presented at the Late Breaking Clinical Trials Session, Annual Sessions of the American College of Cardiology, Chicago, Illinois, March 2003.
70. Clarke M, Sutton R, Ward D, et al: Recommendations for pacemaker prescription for symptomatic bradycardia. Report of a working party of the British Pacing and Electrophysiology Group. Br Heart J 1991;66:185-91.
71. Sheldon M, Kusumoto F, Goldschlager N: Techniques for temporary pacing. In Kusumoto FM, Goldschlager N (ed): Cardiac Pacing for the Clinician. Philadelphia, Lippincott Williams & Wilkins, 2000.
72. Zoll PM: Noninvasive cardiac stimulation revisited. PACE 1990; 13:2014.
73. Behar S, Eldar M, Hod H, et al: Conduction disturbances complicating acute myocardial infarction. Card Electrophysiol Rev 1997;1/2:171-6.
74. Van der Linde MR, Crijns H, Lie KI: Transient complete atrioventricular block in Lyme disease. Chest 1989;96:219-21.
75. Barold SS, Sischy D, Punzi K, et al: Advanced atrioventricular block in a 39-year-old man with acute rheumatic fever. PACE 1998;21:2025-28.
76. Hayes DL, Barold SS, Camm AJ, Goldschlager NF: Evolving indications for permanent cardiac pacing: An appraisal of the 1998 American College of Cardiology/American Heart Association Guidelines [editorial]. Am J Cardiol 1998;82:1082-1086.
77. Barold SS, Barold HS: Pitfalls in the characterization of second-degree AV block. Heartweb on line: Available at http://www. 1997; Article #97040001.
78. Strasberg B, Amat-Y-Leon F, Dhingra RC, et al: Natural history of chronic second-degree atrioventricular nodal block. Circulation 1981;63:1043-9.
79. Shaw DB, Kekwick CA, Veale D, et al: Survival in second-degree atrioventricular block. Br Heart J 1985;53:587-93.
80. Narula OS: Atrioventricular block. In Narula OS (ed): Cardiac Arrhythmias. Electrophysiology, Diagnosis, and Management. Baltimore, Williams & Wilkins, 1979, pp 85-113.
81. Peuch P, Wainwright RJ: Clinical electrophysiology of atrioventricular block. Cardiol Clin 1983;1:209-24.
82. Barold SS, Hayes DL: Second-degree atrioventricular block: A reappraisal. Mayo Clin Proc 2001;76:44-57.
83. Sumiyoshi M, Nakata Y, Yasuda M, et al: Changes in conductivity in patients with second or third degree atrioventricular block after pacemaker implantation. Jpn Circ J 1995;59:284-91.
84. Dhingra RC, Denes P, Wu D, et al: The significance of second-degree atrioventricular block and bundle branch block: Observations regarding site and type of block. Circulation 1974; 49:638-46.

Chapter 14 Paroxysmal Supraventricular Tachycardias and the Preexcitation Syndromes

SANJEEV SAKSENA, SAROJA BHARATI, SHIH-ANN CHEN, SAMUEL LEVY, and BRUCE D. LINDSAY

Epidemiology: Sanjeev Saksena
Anatomy and Pathology: Saroja Bharati
Basic Electrophysiology: Sanjeev Saksena
Clinical Presentation: Sanjeev Saksena
Electrocardiography: Bruce D. Lindsay
Diagnostic Approach: Samuel Levy
Clinical Electrophysiology: Shih-Ann Chen
Principles of Management: Sanjeev Saksena
Evidence-Based Therapy: Sanjeev Saksena

Paroxysmal supraventricular tachycardia (PSVT) has been a well-recognized clinical syndrome and an electrocardiographically defined arrhythmia since the early days of electrocardiography. The clinical syndrome was defined in European literature in the 19th century. In 1867, Cotton reported on an "unusually rapid action of the heart," and this was followed by further observations by French and German scientists.[1,2] Variously referred to as *Bouveret's syndrome* and later as *paroxysmal atrial tachycardia*, it was described classically as "a fully unprovoked tachycardia attack, lasting a few seconds or several days, in patients who as a rule have otherwise healthy hearts."[1] It has been electrocardiographically defined in the 10th Bethesda Conference on Optimal Electrocardiography as "a tachycardia usually characterized by an atrial rate of 140 to 240 beats per minute (bpm) and by an abrupt onset and termination. It may or may not be associated with intact A-V conduction. Specific electrophysiological studies may elicit specific mechanisms such as retrograde and anterograde pathways and sites of reentry."[3]

In the past 25 years, this has been an intensively studied arrhythmia, with extensive definition of its genesis, presentation, subtypes, and electrophysiology. Pharmacologic therapy and, later, nonpharmacologic therapy have been investigated and refined. This is now a classic story in the evolution of clinical cardiac electrophysiology and forms a fundamental cornerstone in the modern treatment of cardiac arrhythmias.

Epidemiology

The epidemiology of PSVT has not been widely investigated in modern times. Early electrocardiographic reports were useful in arrhythmia detection for patients presenting with sustained palpitations and persistent episodes of supraventricular tachycardias (SVT) but had little role for this purpose in the large body of patients with brief SVT events terminating before presentation to the physician.[4,5] These efforts were supplemented by the advent of ambulatory electrocardiography, which documented a large number of cardiac arrhythmias in asymptomatic patients. Conversely, it confirmed that many symptoms experienced by patients with or without heart disease that are suggestive of tachyarrhythmias occur when no arrhythmias or simply premature beats are documented on monitoring. Epidemiologic data are extensively colored by the selection criteria and the electrocardiographic documentation mode used in the study. Brodsky and coworkers reported that ambulatory electrocardiographic recordings in 50 male medical students detected atrial premature beats in 56%, but only one had more than 100 beats in 24 hours.[6] The limited period of observation precludes objective assessment of development of SVT events in these subjects. Thus, Hinkle et al. in a study of 301 men with a median age of 56 years, detected various supraventricular arrhythmias in 76% of these individuals.[7] However, coronary disease was present in 20% of these patients.

Clark et al. studied an apparently normal population ranging from 16 to 65 years old and noted a low incidence of supraventricular arrhythmias.[8,9] In contrast, recent longitudinal studies with telemetric monitoring and even implanted cardiac pacemakers document a very high incidence of asymptomatic and symptomatic atrial arrhythmias, particularly atrial fibrillation in patients with bradyarrhythmias.[10] Thus, it would be appropriate to infer that the true incidence of PSVT in the general population remains unknown due to its evanescent nature and limited methods of detection. It would also be appropriate to surmise that the incidence may be higher than generally believed over a long observation period. Atrial arrhythmias also increase with age.[11] In the Cardiovascular Health Study, short runs of PSVT occurred in 50% of men and 48% of women, doubling in prevalence in octagenarians. Twenty-eight percent of nursing home residents demonstrate PSVT.[11]

More data are available for the preexcitation syndromes, particularly the Wolff-Parkinson-White (WPW) syndrome. Electrocardiographic studies of healthy individuals suggest that the incidence of this condition is 3 in 1000 of the general population.[12] Early studies suggested that the morbidity and mortality in WPW syndrome with tachyarrhythmias were greater in adults with ventricular fibrillation occurring in patients with atrial fibrillation and antegrade preexcitation.[13] However, Klein et al. have reported on the natural history of asymptomatic WPW syndrome; they noted a very low incidence of major morbidity and serious symptomatic arrhythmias including atrial fibrillation with rapid antegrade conduction and mortality.[14]

Supraventricular arrhythmias are seen at all ages and are particularly common in infants, children, and young adults. Age-related behavior of these rhythms has also been the subject of epidemiologic study. In a study of infants younger than 1 year of age, Mantakas et al. noted that associated congenital heart disease was present in 35%, and fully 90% developed SVTs with narrow (35%) or wide QRS (45%) complexes.[15] Most patients (90%) improved with growth or remained stable, but patients with congenital heart disease could have refractory arrhythmias. The quality and duration of life in children with WPW syndrome without clinical dysrhythmias have been reported to be normal.

Anatomy and Pathology

NORMAL ANATOMY OF THE ATRIOVENTRICULAR JUNCTION

In order to understand the pathologic base for atrioventricular (AV) nodal tachycardia and AV reentry tachycardia, the normal anatomy of the AV node and its approaches, including the atria and the AV bundle, is briefly reviewed in the following regions of interest: (1) approaches to the AV node including the atrial septum; (2) AV node; and (3) AV nodo-bundle junction.[16-27]

APPROACHES TO THE ATRIOVENTRICULAR NODE

Approaches to the AV node include the atrial myocardium located in the anterior, superior, midseptal, and inferior regions as they converge to the AV node. The approaches beneath the coronary sinus area may be called the *posterior* or *inferior* approaches. In addition, the approaches also originate from the tricuspid valve. The left-sided approaches include those from the left atrial myocardium and the mitral valve. The superior approaches include the pectinate muscles as they merge from the superior, lateral, and posterior walls of the right atrium toward the AV nodal area, atrial septum, and Todaro's tendon. Approaches to the AV node are formed by different types of myocardial fibers coming from different directions as they merge toward the AV nodal area (Fig. 14-1). Histologically, in general the cells are relatively loosely arranged with lighter staining smaller cells. The size and shape of the myocardial fibers vary considerably from one approach to the other in the vicinity of the node. In general, there is increase in the elastic collagen connective tissue intermingling with the cells, fat, and a large amount of nerve fibers. At the electron microscopic

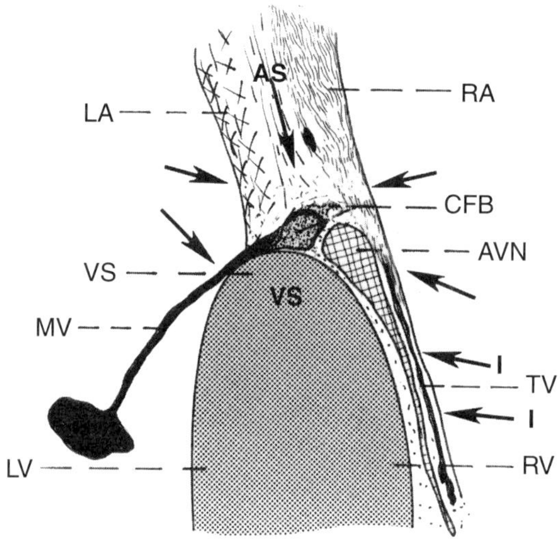

FIGURE 14-1 Schematic representation of the atrioventricular (AV) junction depicting the approaches to the AV node. AVN, AV node; CFB, central fibrous body; LA, left atrial myocardium—left atrial approaches; LV, left ventricular side of the septum; MV, approaches from the mitral valve; RA, right atrial myocardium—superior approaches; AS, atrial septum; S, superior approaches; I, inferior approaches; RV, right ventricular side of the septum—right ventricular approaches; TV, approaches from the tricuspid valve; VS, summit of the ventricular septum. *Arrows* point to the approaches to the AV node from the tricuspid valve area, right atrial aspect, right ventricular aspect, atrial septal aspect, left atrial aspect, and mitral valvular aspect. (With permission from Bharati S, Lev M: The anatomy of the normal conduction system: Disease-related changes and their relationship to arrhythmogenesis. In Podrid PJ, Kowey PR [eds]: Cardiac Arrhythmias, Mechanism, Diagnosis and Management. Baltimore, Williams & Wilkins, 1995, p 1.)

level, the mitochondria and myofibrils of the atrial myocardial cells are not as well organized as the ventricular cells, and some of them do not have a transverse tubular system.[16]

SIGNIFICANCE OF VARIATIONS IN THE SIZE OF APPROACHES TO THE ATRIOVENTRICULAR NODE

The atrial myocardial fibers including the collagen, elastic tissue, and nerve elements in the approaches to the AV nodal area can vary in size, shape, and direction, suggesting that there may be functional differences in the speed of conduction.[16-20] Anatomically, one approach to the AV node may be more dominant than the other. For example, dominant posterior approaches with less prominent superior approaches in the human suggest the possibility of dominant slow pathway conduction. Likewise, dominant superior approaches with practically absent inferior approaches suggest the possibility of dominance of fast pathway conduction or other alterations in AV nodal conduction. Myocardial fibers in the atria may get entrapped within the central fibrous body and join the AV node. In other instances, approaches to the AV node and the AV node may be entrapped within the tricuspid or mitral valve annulus or the base of the aortic valve. During an altered physiologic state, these anomalies may form an anatomic substrate for a reentry circuit resulting in various types of supraventricular or junctional arrhythmias.[16-18]

THE ATRIOVENTRICULAR NODE

The AV node is normally in continuity with its atrial approaches.[16-23,26] It is a sizable structure and is usually located near the annulus of the septal leaflet of the tricuspid valve. In the adult, it measures approximately 5 to 7 mm in length and 2 to 5 mm in width. The node extends closely beneath the endocardium of the right atrium, adjacent to the septal leaflet of the tricuspid valve, and lies very close to the right ventricular aspect of the ventricular septum and the central fibrous body. The size and shape of the node are not uniform in nature and vary considerably from heart to heart. At the light microscopic level, the AV node consists of a meshwork of cells that are approximately the size of atrial cells but are smaller than ventricular cells. Histologically, the AV node may be divided into three layers: (1) superficial or subendocardial, (2) intermediate or midzone, and (3) deep or innermost layer.[16-18]

SUPERFICIAL OR SUBENDOCARDIAL LAYER OF THE ATRIOVENTRICULAR NODE

The approaches that come from various directions merge gradually with the superficial or subendocardial part of the AV node. These fibers are loosely arranged with smaller nodelike cells oriented along the atrial cells, some intermingling with the atrial cells, fat, elastic tissue, collagen, and nerve fibers. A distinct increase in fat occurs with normal aging of the heart.[16-18]

INTERMEDIATE OR MIDLAYER OF THE AV NODE

AV nodal cells are more or less compact; however, the orientation and arrangement of the cells vary considerably. The collagen and elastic tissue content is less than that seen in the superficial layer with fewer nerve fibers. In the older age group, fat may be present in the intermediate layer of the node.[16-18]

DEEP OR INTERMOST LAYER OF THE NODE

AV nodal cells are tightly arranged and may be considered compact. However, the arrangement and orientation of these cells also vary considerably. There is some amount of collagen and elastic tissue, though somewhat less than the intermediate and superficial layers. Fat may be seen intermingling with the nodal cells in the older age group. At the light microscopic level, the nodal fibers vary from the periphery toward the central fibrous body.[16-18] At the electron microscopic level, there are fewer myofibrils and mitochondria, which are randomly arranged. The cytoplasmic reticulum is poorly developed, and there is no transverse tubular system. It is not known whether the AV node contains more glycogen than the surrounding atrial and ventricular myocardium. The gap junctions are scarce, but desmosomes are frequent. Fascia adherens are more than in the sinoatrial nodal cells but not as frequent as in atrial and ventricular myocardial cells.[16-18]

BLOOD SUPPLY AND NERVE SUPPLY TO THE ATRIOVENTRICULAR NODE

In approximately 90% of hearts, the AV node is supplied by ramus *septi fibrosi*, a branch from the right coronary artery reinforced by branches from the anterior descending coronary artery.[16-23] Copious nerve cells surround the AV node, especially in the atrial septum adjacent to the AV node and nerve fibers within the node. The exact distribution and destination of the nerves in the human in the AV nodal area are still unknown. However, the rich autonomic innervation of the sinoatrial node and AV nodal areas in the canine heart indicates that the sinoatrial node is particularly responsive to parasympathetic adrenergic regulation, whereas the AV nodal conduction is preferentially sensitive to sympathetic adrenergic regulation.

VARIATION IN SIZE, SHAPE, AND LOCATION OF THE ATRIOVENTRICULAR NODE

The node lies beneath the septal leaflet of the tricuspid valve in close proximity to the right ventricular aspect of the ventricular septum and the central fibrous body. The AV node may be draped over the central fibrous body within the atrial septum, or some of the fibers may be in part within the tricuspid valve annulus and may be in part within the central fibrous body. In some cases the fibers may be situated toward the left atrial aspect or in part within the mitral valve annulus, or they may be situated close to the base of the aortic valve.[16-23] AV nodal–like cells may also be seen near the tricuspid, mitral, and aortic valve annuli. Some of these nodelike cells may enter the central fibrous body and eventually join the regular posterior AV node. In addition, an accessory AV node may be seen in the parietal wall of the right atrium near the annulus of the tricuspid valve, anterosuperiorly.[16] Note that not all the AV nodal cells eventually form the AV bundle; some remain, such as leftover nodal cells, and lie adjacent to the central fibrous body or near the valves.

MAHAIM FIBERS

Conduction fibers from the AV node, AV bundle, and left bundle branch may join the ventricular myocardium. They have been referred to as *Mahaim fibers* or *paraspecific fibers of Mahaim.* They may form the substrate for a unique variety of ventricular preexcitation. The myocardial fibers resemble the cells of the tissue of origin and gradually take over the characteristics of the ventricular myocardial cells. Mahaim fibers may be present from the AV node to the right, left, or midpart of the ventricular septum.[16-23]

ACCESSORY ATRIOVENTRICULAR BYPASS PATHWAYS—FIBERS OF KENT

Accessory AV bypass pathways bypassing the AV node are seen in normal infants up to 6 months of age.[21] The myocardial cells on the atrial side resemble atrial cells, and those on the ventricular aspect resemble ventricular cells.[16-22] In adults, it has been well documented that such pathways cause preexcitation with varying types of supraventricular arrhythmias.

AV NODO-BUNDLE JUNCTION

The AV node penetrates the central fibrous body to become the penetrating part of the AV bundle. The penetrating AV bundle undergoes or assumes different shapes and contours. The orientation of fibers differs significantly. The nodo-bundle junction is very small, measuring approximately 1 to 1.5 mm in greatest dimension or less in the majority of hearts. The nodo-bundle junction becomes a part of the AV node and may be considered as the most distal part of the AV node or the beginning part of the penetrating AV bundle.[16-18] The AV nodal cells that are close to the tricuspid valve and the posterior approaches are first molded to form the penetrating AV bundle. On the other hand, the nodal fibers from the anterior/superior aspect are the last to lose their continuity with the superior approaches as they enter the central fibrous body to form the AV bundle. Thus, the formation of the AV bundle within the central fibrous body occurs differentially. The posterior part of the AV node penetrates the central fibrous body earlier than the anterior or superior part. The superior part of the AV node includes the nodal fibers closer to the left atrial side, atrial septal, and right atrial aspects.[16-18]

FUNCTIONAL SIGNIFICANCE OF THE NODO-BUNDLE JUNCTION

The variation in sizes of the superior and inferior components of the node, as well as the variations in which they form the AV bundle, may provide a substrate for arrhythmias. Since histologically the nodo-bundle junction has the characteristics of both the AV node and the AV bundle, its functional properties may be intermediate, having characteristics of both the node and the penetrating AV bundle. Likewise, the atrial myocardial approaches from the mitral valve, tricuspid valve, the right ventricular myocardium, and the atrial septum tend to get entrapped within the central fibrous body and may later join the AV node or the nodo-bundle junction, or both. The normal variations in morphology of the AV junction have the potential for supporting AV nodal arrhythmias.[16-18] The variation in the morphology of the AV junctional area also probably predisposes to varying types of physiologic phenomena, including dual AV nodal pathways and other junctional arrhythmias.[15,16] For example, an anterior AV node in the parietal wall of the atrium or near the tricuspid valve annulus or double AV node may alter its conduction velocity. Dual AV nodal pathways may be normal or abnormal. They may be transient or become permanent.[16-18]

ACCESSORY ATRIOVENTRICULAR NODE AND ITS RELATIONSHIP TO PREEXCITATION AND ATRIOVENTRICULAR JUNCTIONAL TACHYCARDIAS

Few pathologic studies document an accessory AV node being responsible for preexcitation or other arrhythmias originating from the AV junction. Figure 14-2 is an example of a 5-month-old child with a history of intractable junctional tachycardia. He died suddenly, and postmortem examination demonstrated a displaced coronary sinus relative to the septal leaflet of the tricuspid valve with a double AV node and double AV bundle (see Figs. 14-2*A* and *B*). In other instances, the AV node has been located in part within the central fibrous body with a left-sided AV bundle, or the right AV node has been completely interrupted by sutures, and a left-sided AV node was connected to the atrial septum. It was recently documented that an accessory AV node located anterior to the AV junction directly communicated with the right atrium and right ventricle in a curved fashion that produced ventricular preexcitation and formed the retrograde limb of typical AV reentrant tachycardia (AVRT). In another patient with typical preexcitation and recurrent AVRT who died suddenly, there were no conventional anomalous pathways on the right side. Instead, an anterior accessory AV node in the right atrium continued as the infundibular right ventricular myocardium.

These types of abnormalities of the AV junction are seen in children as well as adults. Pathologic study in a 13-month-old infant with hypertrophic cardiomyopathy and paroxysmal SVTs, consistent with AV reciprocating tachycardia using a concealed posterior accessory

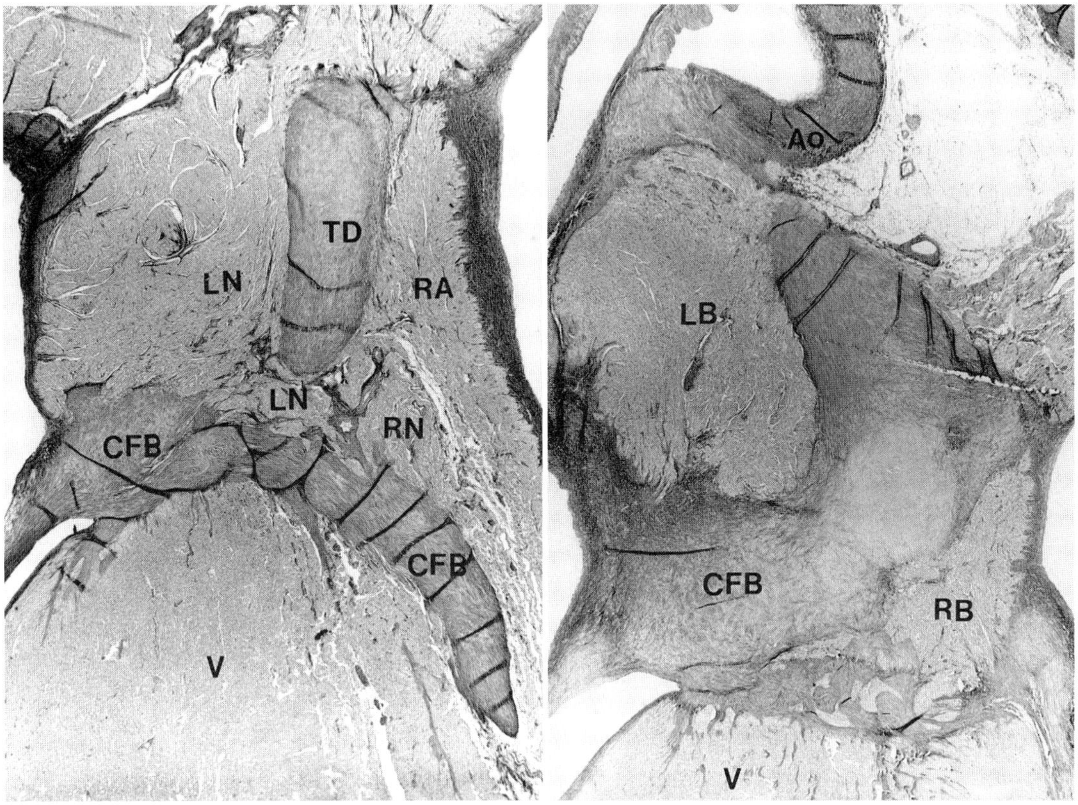

FIGURE 14-2 **A** and **B,** A 5-month-old Hispanic child with a lifelong history of tachycardia had a heart rate of 206 beats per minute. He was diagnosed to have junctional tachycardia. During cardiac catheterization he developed ventricular fibrillation and died. **A,** Pathologic examination revealed a double AV node within the central fibrous body. **B,** Two AV bundles are seen in this section within the central fibrous body. Weigert-van Gieson stain ×19.5. RN, right-sided AV node; LN, left-sided AV node; CFB, central fibrous body; RA, right atrium; LB, large left-sided AV bundle; RB, small right-sided AV bundle; V, summit of the ventricular septum; TD, tendon of Todaro; AO, aorta. (With permission from Bharati S, Moskowich, Sheinman M, et al: Junctional tachycardias: Anatomic substrate and its significance in ablative procedures. J Am Coll Cardiol 1991;15:172-86.)

pathway, revealed that the central fibrous body was abnormally formed with numerous Mahaim fibers (nodoventricular) on both sides of the septum with fibrosis of the left bundle branch (Fig. 14-3).

Concepts and Classification

The fundamental concepts underlying recurrent PSVT have been elucidated by decades of experimental and clinical electrophysiological investigation. Early experimental studies of Moe and associates and the seminal clinical studies of Durrer, Wellens, Castellanos, Rosen, and Denes, among others, helped define the mechanisms and substrates involved in these arrhythmias.[28-31] Extensive pathologic studies as mentioned earlier have further enhanced our understanding of the anatomic basis of these arrhythmias. Fundamental to our understanding is the concept that these arrhythmias may be due to either enhanced automaticity or reentry. While the latter may predominate in certain populations and clinical practice, there is considerable overlap in clinical and electrocardiographic features. Automatic arrhythmias may arise from the sinoatrial region and from working atrial myocardium or the AV junction. Reentrant rhythms may arise in these structures as well and may also involve accessory AV connections or other variants in the preexcitation syndrome. Table 14-1 provides a tabulation of PSVT categories by electrophysiological mechanisms and substrate. A summary of the basic concepts involved in the genesis of these arrhythmias is provided in the following discussion.

Basic Electrophysiology

The anatomic and electrophysiological substrate and physiology of SVTs have been given fresh investigative impetus by the evolution of catheter ablation procedures. Based on anatomic dissections and pathologic study of involved hearts, the accessory bypass tract has been recognized as working myocardial bundles that may cross the AV annulus at any location or bridge elements of the specialized conduction system with atrial or ventricular working myocardium. The reentrant circuit in AVRTs has four components—the atrium, normal AV

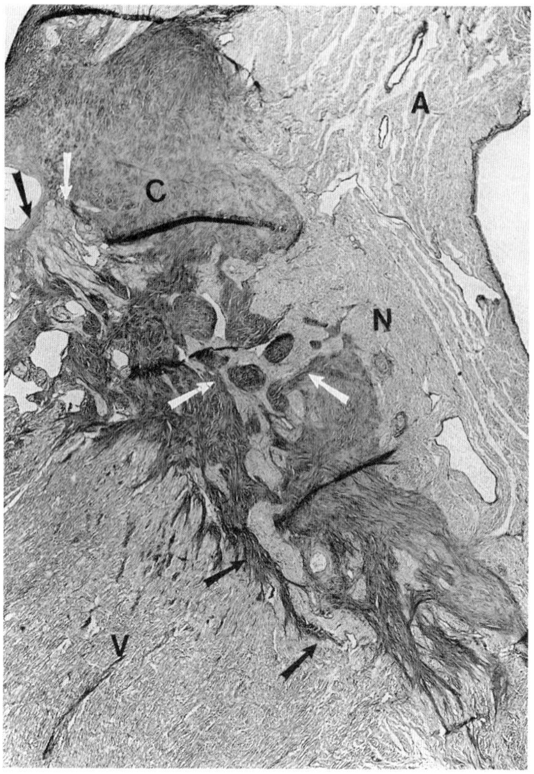

FIGURE 14-3 A 13-month-old boy diagnosed with hypertrophic cardiomyopathy and supraventricular tachycardia died suddenly. Photomicrograph demonstrating the Mahaim fibers from the atrioventricular (AV) node joining the ventricular septum. *Black arrows* point to profuse nodo-ventricular fibers from the AV node joining the ventricular septum on the right side and on the left side. The *white arrows* point to loop formation of AV nodal fibers within the central fibrous body. Weigert-van Gieson stain ×30. A, atrial approaches to the AV node; C, central fibrous body; N, AV node; V, ventricular septum. (With permission from Cantor RJ, Gravatt A, Bharati S: Pathologic findings following sudden death in an infant with hypertrophic cardiomyopathy and supraventricular tachycardia. J Cardiovasc Electrophysiol 1997;8:222-5.)

node, ventricle, and accessory pathway. In some instances, the tachycardia is located in the atrium or AV node and simply uses the accessory pathway for conduction to the ventricles, the so-called *bystander pathway*. In contrast, AV nodal pathways are believed to be located within or in the immediate environs of the AV node. This latter structure has been stratified into transitional regions with the atrium (AN region), compact node (N region), and the His Bundle (NH region). The N region is synonymous with the anatomic descriptions of the compact AV node. The reentrant circuit in AV nodal reentry is confined to the AV node and adjoining atrial myocardium. A concise discussion of these two substrates and their physiology is used as the basic structure to define specific variants of these arrhythmias.

ATRIOVENTRICULAR NODAL REENTRANT TACHYCARDIAS

Pioneering anatomic studies of Tawara and Kent and physiologic demonstration of dual AV nodal physiology

TABLE 14-1 Classification of Regular Supraventricular Tachycardias with Narrow QRS Complexes

Sinus Node Disorders
Paroxysmal sinus tachycardia
Nonparoxysmal sinoatrial tachycardia
Atrioventricular (AV) Nodal Reentrant Tachycardias
Slow-fast (type 1)
Fast-slow (type 2)
Slow-slow (type 3)
Reentrant and Ectopic Atrial Tachycardias
Intra-atrial reentrant tachycardia
Automatic atrial tachycardia (unifocal or multifocal)
Preexcitation Syndrome: Wolff-Parkinson-White Syndrome
Orthodromic atrioventricular reentry
Permanent junctional reciprocating tachycardia
Antidromic atrioventricular reentry
Atrial tachycardia, atrial flutter, or atrial fibrillation with or without accessory pathway conduction
Other Preexcitation Syndromes: Mahaim Conduction
Nodoventricular and nodofascicular reentry
Atrial tachycardia, AV nodal reentry, or atrial fibrillation with nodoventricular or nodofascicular bystander conduction
Other Preexcitation Syndromes: Lown-Ganong-Levine syndrome
Atrial tachycardia, atrial flutter, or atrial fibrillation with enhanced AV nodal conduction
Automatic AV Junctional Tachycardias

in dogs by Moe have laid the foundation for our current understanding of this condition.[28,32,33] Tawara and others examined the compact AV node, but more recent studies have focused on the connections of the node. Moe and coworkers first suggested the presence of functionally and spatially distinct pathways with fast and slow conduction properties in the canine AV node.[30] Subsequently, cellular electrophysiological properties of the node and its environs were studied (see Chapter 13). The AN and N regions lie in the triangle of Koch framed by Todaro's tendon, the tricuspid annulus, and the ostium of the coronary sinus. This anatomic location is critical to understanding the technique of catheter ablation of this arrhythmia. The N region is located at the apex of the triangle where the NH region in the nonbranching part of the AV bundle penetrates the central fibrous body to become the His bundle. Considerable uncertainty existed in our understanding of the AN region and the atrial connections of the AV node. Truex and Smythe demonstrated atrial extensions of the compact AV node, with a prominent posterior tail that extended to the ostium of the coronary sinus.[24] Anderson and coworkers defined superficial, deep, and posterior zones in these connections.[34] The superficial zone extended into the anterior and superior part of the node, the posterior zone into the inferior and posterior aspect of the compact node, and the deep fibers connected the left atrial septum to the node. This latter connection was noted by Anderson to join the distal part of the node, suggesting the possibility of a lesser degree of AV nodal conduction modulation by slow currents and more rapid AV conduction. Definition of these three inputs further refined our clinical techniques of AV nodal modification or ablation. Racker also described discrete superior, medial, and lateral atrionodal bundles.[35] The medial and lateral bundles converged into a single proximal AV bundle leading into the compact AV node. These two bundles are believed to form in part or in entirety the anterior "fast" and posterior "slow" AV nodal physiologic pathways.

Moe and colleagues demonstrated the electrophysiological substrate for AV nodal reentrant tachycardias (AVNRTs) by developing the concept of dual AV nodal pathways as its basis.[28] They noted a marked prolongation of AV nodal conduction when a premature atrial extrastimulus was delivered at a critical coupling interval. This was often associated with an echo beat, postulated to be due to two AV nodal pathways—a β-pathway with a faster conduction and long refractoriness and an α-pathway with shorter refractoriness and slower conduction times. At critical coupling intervals, the premature beat shifts conduction from the β- to the α-pathway and reengages the β-pathway retrogradely to result in an atrial echo beat. More recent, elegant, isolated canine and pig hearts, McGuire, de Bakker, and Janse have shown nodal type action potentials in cells around both mitral and tricuspid valve rings.[36] These cells are separated from atrial cells by a zone of cells with intermediate action potentials. Adenosine reduced amplitude and upstroke velocity of action potentials in nodal-type cells, but their morphologic characteristics were indistinguishable from atrial myocytes. However, they could be distinguished by the absence of connexin 43, which was present in atrium and ventricle. The posterior AV nodal approaches were dissociated by pacing from the atrium and AV node and echo beats used the slow posterior pathway retrogradely and preceded atrial activation. These posterior approaches were not used in fast pathway conduction.

HUMAN CORRELATES OF EXPERIMENTAL OBSERVATIONS

Clinical correlation of these experimental concepts of AV nodal physiology was demonstrated by the early work of Bigger and Goldreyer, as well as Denes, Rosen, and coworkers.[31,37] The critical role of AV nodal conduction delay in the onset of SVT was recognized. Denes et al. demonstrated longitudinal dissociation of the AV nodal conduction in patients with recurrent SVT.[31] The critical link between anatomic and physiologic concepts in humans was provided by a landmark study by Sung et al.[38] They demonstrated the posterior input and output of the "slow" AV nodal pathway and the anterior location of the "fast" AV nodal conduction pathway. Submerged in controversy for some time, this observation formed the basis for the development of fast and slow pathway ablation for the cure of AV nodal reentry. Several years later, this study was directly validated in intraoperative studies in humans. McGuire and coworkers mapped Koch's triangle to define the earliest atrial activation during AVNRT in patients during cardiac surgery. A zone of slow conduction was found in the triangle in 64% of patients.[39] Atrial activation patterns confirmed that the fast pathway was connecting at the apex of the triangle near the AV node–His bundle junction, while the slow pathway did so at the orifice of the coronary sinus near the posterior aspect of the AV node. Two types of AVNRT were distinguished: (1) the common variety or type 1, called the *"slow-fast" form,* which used the slow pathway for anterograde propagation and the fast pathway for retrograde conduction and (2) the uncommon type or type 2, called *"fast-slow"* AVNRT. Contiguous ablation lesions can affect both pathways, confirming their proximity. The presence of multiple "slow" pathways has also been identified, resulting in a third form of reentry, called *"slow-slow" or type 3* AVNRT. These different types and pathways support the concept that these arrhythmias are supported by AV nodal tissue, transitional atrio-nodal inputs, and other perinodal tissues, which have varying electrophysiological properties and functionally simulate distinct electrical pathways.

ATRIOVENTRICULAR REENTRANT TACHYCARDIAS

The classification of preexcitation syndromes has evolved since the original eponyms were given to Kent, Mahaim, and James fibers. The long proposed and accepted classification by the European Study Group is shown in Table 14-2. However, limitations of this classification are being recognized, and correlation with the previous eponyms remains occasionally tangential. The major reason for this observation is the recognition that decremental conduction is a property not solely confined to the AV node but also seen with accessory AV connections. Speculation around the embryologic basis of these connections swirls around the original suggestion by Gallagher that these may be displaced AV nodal and specialized conduction system tissues.[40] Decremental conduction has been observed in posteroseptal AV connections in the permanent form of junctional reciprocating tachycardia, as well as in atriofascicular bypass tracts, which commonly link the parietal atrial myocardium in the right atrium to the right bundle branch at its distal portion and are detected as Mahaim physiology. The tachycardia propagates antegradely over the atriofascicular pathway and retrogradely over the normal AV conduction system and involves the atrium and ventricle as critical elements in the circuit.

TABLE 14-2 Classification of Preexcitation Syndromes

1. Atrioventricular (AV) bypass tracts providing direct connections between the atrium and ventricle
2. Nodoventricular connections between the AV node and ventricular myocardium
3. Fasciculoventricular connections between the fascicles of the specialized conduction system and the ventricular myocardium
4. AV nodal bypass tracts with direct connections between the atrium and His bundle

This is quite distinct from the nodoventricular connections originally described by Mahaim. In this latter instance, the tachycardia has a similar propagation sequence but can exist totally without atrial involvement and has been referred to as a "subjunctional tachycardia."

The anatomic basis for AVRT is the accessory AV pathway, a small band of working myocardium bridging the AV annulus. The location of these pathways is most frequent in the left ventricular free wall, followed by posteroseptal and paraseptal locations and, least commonly, in the right atrial free wall and left anterior AV annulus. Multiple AV connections are seen in approximately 15% of patients. Typical dimensions for these locations and such connections have been elucidated by pathologic studies. In cadaver hearts the posterior septal space was noted to extend from the coronary sinus orifice for 2.3 ± 0.4 cm, and the length of the left ventricular free wall was 5 ± 1 cm. Posteroseptal accessory pathways were located in the proximal 1.5 cm of the coronary sinus in the posterior septum and those between 1.5 and 3 cm could be either in the left free wall or the posterior septal space. Posteroseptal accessory pathways beyond 3 cm were invariably in the free wall.

BASIC ELECTROPHYSIOLOGICAL CONCEPTS IN PREEXCITATION SYNDROMES

AV accessory pathways, as well as AV nodal bypass tracts, generally exhibit "all or none" conduction behavior during electrophysiological evaluation. Rapid nondecremental conduction up to the point of refractoriness is the norm and is exhibited during antegrade and retrograde conduction, especially when competing AV nodal conduction is absent. When decremental conduction in response to progressive premature stimulation is observed, it is usually due to switching of conduction to the AV nodal-His axis, though enhanced AV nodal conduction may occur in the same patient. When the accessory pathway conducts antegradely, the most common manifestation is WPW syndrome with a short P–R interval and δ wave (due to ventricular preexcitation by the pathway) and prolongation of QRS duration (due to abnormal intraventricular conduction patterns). In some instances, when the antegrade refractoriness of the accessory pathway is particularly long or if the pathway fails to permit such conduction, it may remain concealed and only manifest when retrograde propagation occurs over echo beats or AVRT. Patients with concealed accessory pathways have normal surface ECGs. Rarely, decremental conduction is observed in the anomalous pathway. While this is occasionally due to nodoventricular connections, more often this is due to slowly propagating accessory pathways or other causes of Mahaim physiology. The electrophysiological basis for decremental conduction in accessory pathways has been speculative. Suggestions include impedance mismatch at the interface between the pathway and atrial or ventricular myocardium or extreme tortuosity of these fibers in some instances, which has been seen pathologically. Investigations suggest that block in such pathways usually occurs at the ventricular connection and may be related to anisotropy, fiber narrowing, or altered intercellular junctions. Age and autonomic state–related changes in electrophysiological properties may lead to intermittent manifestations of preexcitation syndromes and the arrhythmias supported by them in a patient's lifetime. It is common for incessant tachycardias and manifest preexcitation of infancy to wane during childhood and adolescence or for there to be an initial appearance of the condition in young adults. Aging generally can impair pathway function, and it is uncommon, though not rare, for the condition to manifest itself for the first time in an elderly patient.

SUPRAVENTRICULAR TACHYCARDIAS DUE TO ENHANCED AUTOMATICITY

Accelerated junctional rhythms manifest as supraventricular tachyarrhythmias have been recorded with particular frequency in the era of extensive digitalis use without blood concentration determinations.[41] They have been noted in postoperative patients, after myocardial infarction and during electrolyte abnormalities.[41,42] These arrhythmias are believed to be due to triggered automaticity resulting from delayed afterdepolarizations in most instances. Pacing can enhance the afterdepolarizations and they can be increased in amplitude by increase in extracellular calcium concentrations, which can be seen in digitalis toxicity. Ischemia in experimental models can also result in accelerated, triggered automaticity in the coronary sinus region. Hypokalemia can predispose to afterdepolarizations. Acceleration is often seen in triggered rhythms at initiation and can clinically manifest as nonparoxysmal junctional tachycardias on the ECG. The triggered arrhythmias can often be induced by overdrive pacing and do not necessarily resume after pacing ceases, unlike other automatic rhythms. Triggered rhythms have also been induced in human diseased atrial tissues resected at surgery and in coronary sinus cells during experimental studies.

Clinical Presentation

PSVT presents most commonly as sudden onset of palpitations and may be associated with chest discomfort, dyspnea, near syncope, and syncope.[1,43] It is of variable duration and may last from seconds to days. Termination is usually sudden as well, though in some forms gradual disappearance may be noted. Chest discomfort in children and adults without overt heart disease may be related to the perception of rapid heart action; after an episode, it may persist in a milder form for a period of time. In older patients and in the presence of heart disease, this may be related to myocardial ischemia. Dyspnea may be a prominent symptom, often in patients with preexisting left ventricular dysfunction, when pulmonary congestion may worsen due to poor forward cardiac output. Symptoms suggestive of near syncope and syncope are seen with extremely rapid

PSVT and result from a compromise of cardiac output. In many forms of PSVT, atrial and ventricular activation are not timed sequentially to achieve appropriate ventricular filling.[44] Rapid rates further compromise this, and forward cardiac output can be seriously compromised, especially with heart rates greater than 200 bpm. In some patients, PSVT can be minimally symptomatic or even asymptomatic. In children and infants, incessant PSVT can lead to a tachycardia-induced cardiomyopathy with symptoms of left ventricular failure, failure to thrive, and syncope. Very rapid or hemodynamically unstable episodes of this arrhythmia have been known to precipitate myocardial infarction, ventricular tachyarrhythmias, and cardiac arrest.

Electrocardiography

The distinguishing electrocardiographic features of PSVT reflect the underlying mechanism of the arrhythmia. The major criteria that have been used to separate these mechanisms depend on the features of onset, the position of the P wave in the R–R interval during SVT, the presence or absence of QRS alternans, cycle length variation, the P-wave morphology, effects of bundle branch block (BBB), the presence of a pseudo R′ deflection in V1, and whether there is ventricular preexcitation. Some of these criteria have proven to be particularly useful, but there is considerable overlap among the ECG manifestations of the different mechanisms of SVT.

1. *Onset.*
   AVNRT is usually triggered by a premature atrial beat that differs in morphology from sinus rhythm. As shown in Figure 14-4, initiation of the tachycardia is characterized by sudden prolongation of the P–R interval because conduction from the premature beat blocks in the fast pathway and conducts down the slow pathway. Reentry occurs if the fast pathway has recovered and is capable of conducting retrograde. In contrast, while a triggered or reentrant atrial tachycardia may be initiated by an APD, these arrhythmias are not heralded by marked P-R prolongation with the onset of tachycardia. Automatic atrial tachycardias are characterized by gradual acceleration and a P wave morphology that differs from sinus rhythm. AVRT mediated by accessory pathways may be triggered by either premature atrial or ventricular beats. Ventricular ectopy can induce AVRT or atrial tachycardias, but it is a far less common mode of induction for these arrhythmias.
2. *Position of the P wave.*
   During typical AVNRT the atria are generally activated simultaneously with the ventricles, so the P wave is buried within the QRS complex, though in some cases it may extend into the early portion of the ST segment (Fig. 14-5*A*). In the atypical form of AVNRT, during which antegrade conduction is mediated by the fast pathway and retrograde conduction by the slow pathway, the P wave may fall into the second half of the R–R interval because of slow retrograde conduction to the atria. During orthodromic AVRT mediated by an accessory pathway, retrograde atrial activation generally begins about 70 milliseconds (ms) after the onset of the surface QRS and extends well into the ST segment in the first half of the R–R interval (see Fig. 14-5*B*). In patients with atrial tachycardias the P wave is usually detected in the second half of the R–R interval. The exception is patients with atrial tachycardias in whom AV conduction is delayed because of the effects of antiarrhythmic drugs or intrinsic conduction system disease. Kalbfleisch evaluated ECGs in patients who had undergone electrophysiology studies and found that the P wave was in the first half of the R–R interval in 91% of patients with AVNRT, 87% of patients with AVRT, and 11% of patients with atrial tachycardias.[45]
3. *QRS Alternans.*
   Green observed that ECGs recorded from some patients with SVT exhibited QRS alternans and, when present, it was usually associated with AVRT mediated by an accessory pathway.[46] Subsequent studies by Kay and Kalbfleisch demonstrated that QRS alternans depends on the abrupt onset of SVT and is more common in rapid tachycardias. In their studies, the incidence of QRS alternans was 27% to 38% in orthodromic AVRT and 13% to 23% in patients with AVNRT.[45,47] It was much less common in patients with atrial tachycardias. The differences between the results of these studies may reflect criteria used to identify QRS alternans and the number of leads used for the recordings. For example, the alteration in QRS amplitude may only be apparent in selected leads. Recordings obtained from telemetric monitoring or using

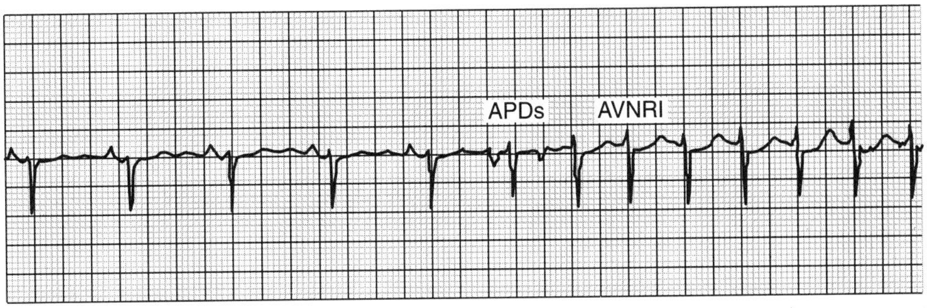

FIGURE 14-4 Sinus rhythm is interrupted by APDs that have a different P-wave morphology. The second APD blocks in the fast pathway and conducts down the slow pathway to induce atrioventricular node reentry. P-waves are not evident during the tachycardia because they are buried in the QRS complex.

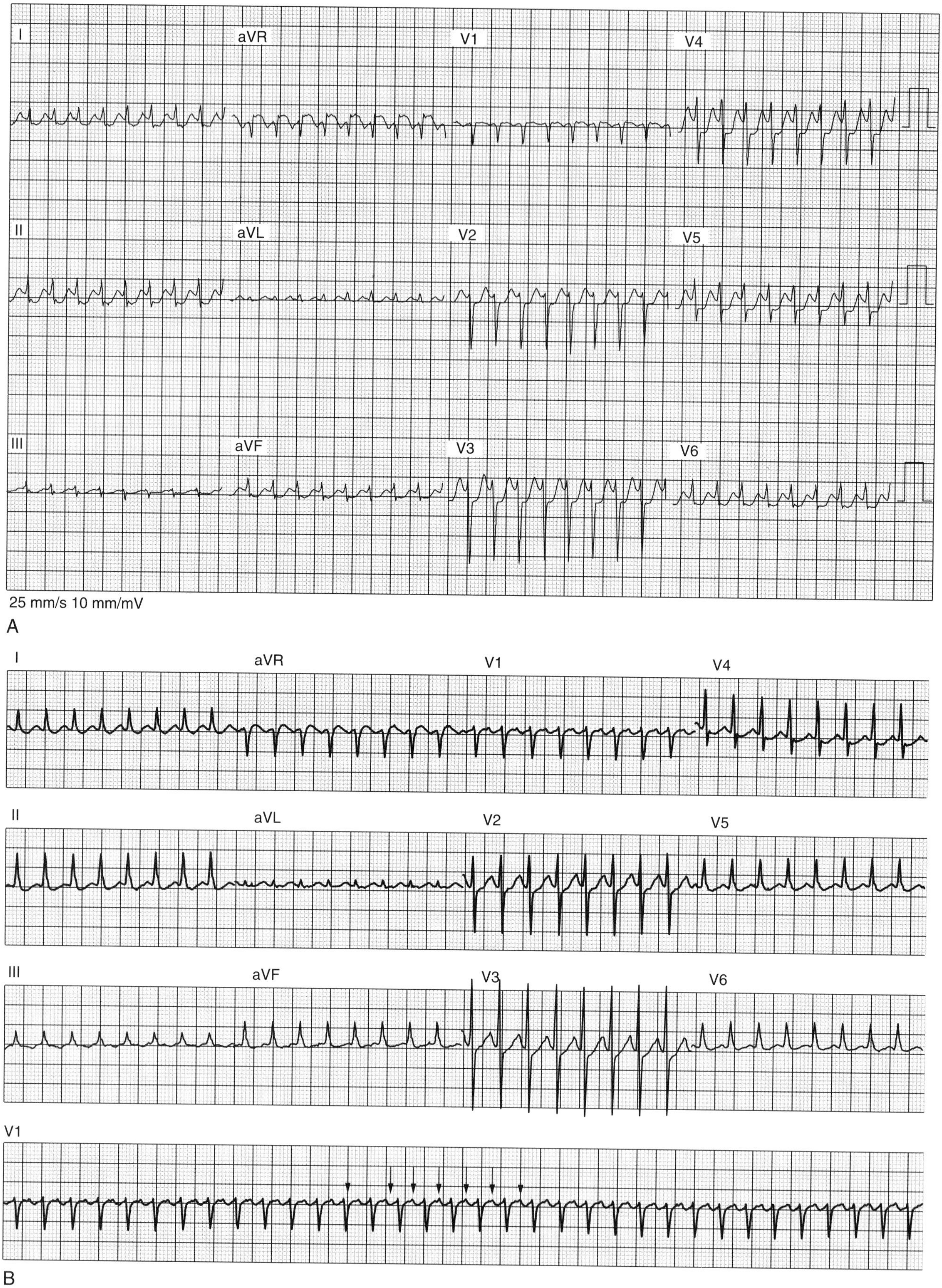

FIGURE 14-5 **A,** Narrow complex supraventricular tachycardia (SVT) at a rate of 240 beats per minute (bpm) with no retrograde or antegrade P wave being visible in the RR cycle. The tachycardia was subsequently confirmed to be type 1 atrioventricular nodal reentrant tachycardia on electrophysiological study. **B,** Narrow complex SVT at a rate of 206 bpm with the retrograde P wave being clearly visible in the mid-RR cycle, especially in lead V1. The tachycardia was subsequently confirmed to be AVRT with a retrogradely conducting postero-septal accessory pathway on electrophysiological study.

only a limited number of surface leads may not demonstrate QRS alternans quite as well as a 12-lead ECG.[48]

4. *Rate and Cycle Length Alternations.* Several studies have evaluated the rate of SVT without demonstrating significant differences that would be useful to discriminate the underlying mechanism.[44-46] Cycle length variation is relatively uncommon in the reentrant tachycardias. One would expect reentrant tachycardias such as AVRT and AVNRT to have relatively constant cycle lengths as they usually do, but sometimes variable conduction in one limb of the circuit may lead to variations in cycle length even during reentry. Figure 14-6 shows unusual and striking variation in the R–R intervals recorded during orthodromic AVRT mediated by a left lateral accessory pathway in a patient who also had dual AV node physiology. The variation in R–R interval is attributable to differences in conduction through the AV node depending on whether antegrade conduction occurred over the fast or slow pathway.
5. *P wave Morphology.* When a P wave can be identified during SVT, it is often difficult to determine the morphology and axis because it may be obscured by ventricular repolarization. Kalbfleisch reported that when the P wave was visible, its axis could be determined in the vertical, horizontal, and anterior-posterior planes in only 32%, 11%, and 9% of patients.[45] One might expect the P wave morphology to be more accurately analyzed during atrial tachycardias, because the P wave falls in the second half of the R–R interval and is less likely to be obscured by the T wave. Tang developed an algorithm to differentiate left atrial from right atrial focal tachycardias based on surface ECGs recorded from patients who underwent ablation of their arrhythmias.[49] In their analysis of 31 patients, there were no differences in P waves arising from the lateral right atrium compared with the right atrial appendage. Leads aVL and V1 were most useful in distinguishing left from right atrial foci. Right atrial foci were associated with a positive or biphasic P wave in AVL, whereas the P wave was negative or isoelectric in aVL in patients with left atrial foci. A positive P wave in V1 was observed in patients' left atrial foci in contrast to negative or biphasic P waves recorded from patients with right atrial foci. They also found that the P wave morphology in the inferior leads distinguished inferior (negative P waves) from superior (positive P waves) foci. Morton analyzed P wave morphologies from nine patients with atrial tachycardias that arose from the tricuspid annulus.[50] In their series the P wave was upright in aVL, inverted in III and V1, and either inverted or biphasic in leads V2 to V6. Tada developed an algorithm for identifying the origin of focal right atrial tachycardias based on ECGs recorded from 32 patients who underwent ablation procedures.[51] They found that negative P waves in the inferior leads differentiated inferior from superior origins, and a negative P wave in aVR was characteristic of tachycardias arising along the crista terminalis. All these investigations are limited in that the number of patients studied is relatively small, and prospective analysis of P wave morphology is often hindered by inability to accurately separate the terminal part of the T wave and P wave during the tachycardia.
6. *Effect of BBB.* The development of BBB during SVT may provide a clue to the diagnosis of AVRT. Coumel and colleagues first observed that the development of BBB ipsilateral to an accessory pathway may result in cycle length prolongation (decrease in rate) because the circuit is prolonged.[52,53] Figure 14-7 was recorded during transition from SVT with left

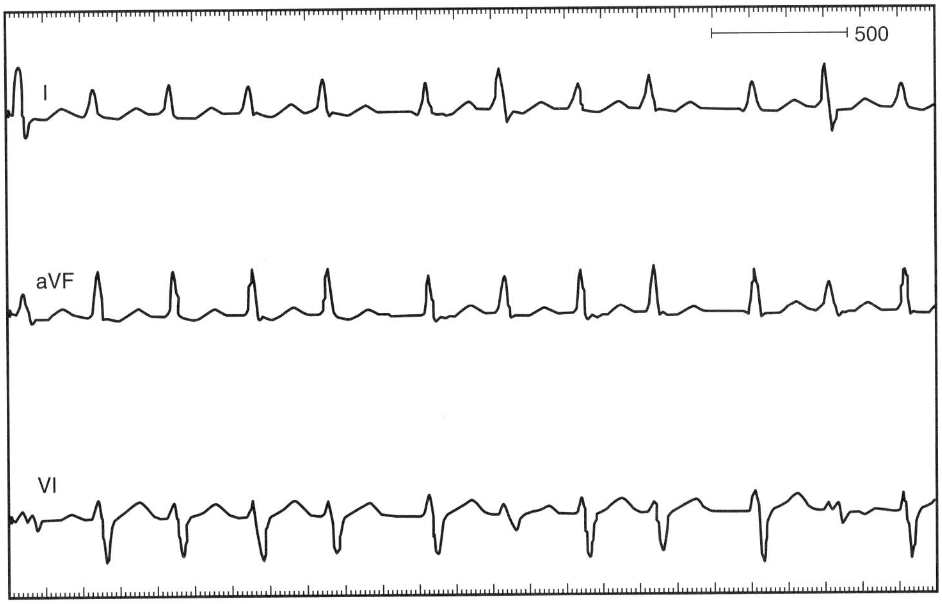

FIGURE 14-6 Surface leads *I*, *aVF*, and *V1* demonstrate marked cycle length variability in a patient with a concealed left lateral accessory pathway and dual AV node physiology. The irregular R–R intervals were recorded during orthodromic atrioventricular (AV) reentry and were attributable to intermittent conduction down the slow AV node pathway.

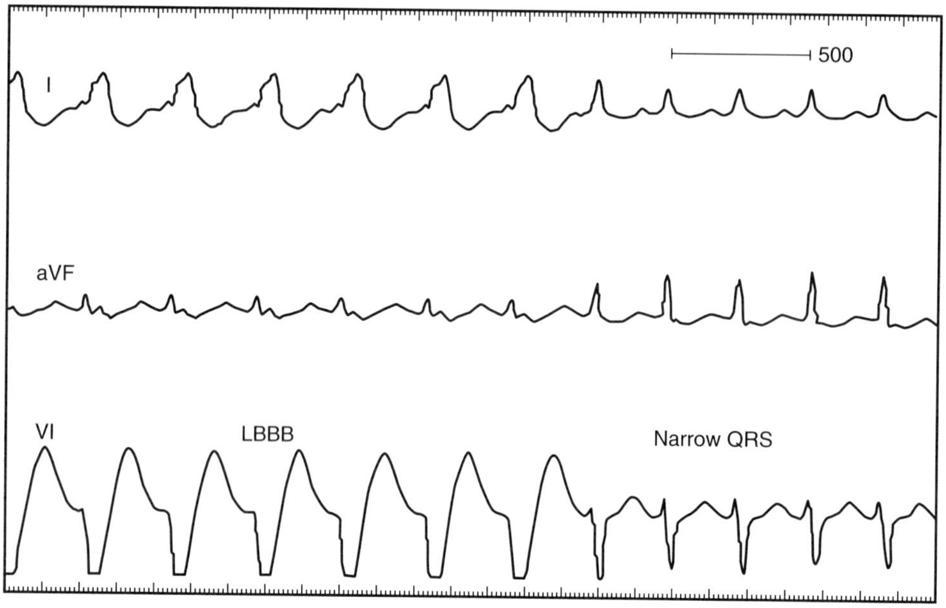

FIGURE 14-7 Surface leads *I*, *aVF*, and *V1* demonstrate orthodromic atrioventricular reentrant tachycardia during the transition from left bundle branch block to a normal QRS. The accessory pathway was located in a left lateral position. Note the increase in rate from 200 to 230 beats per minute when the QRS becomes normal.

BBB to a narrow QRS from a patient with a left lateral accessory pathway. Concomitantly, the cycle length of the tachycardia decreased by 40 ms (300 to 260 ms), resulting in an increase in the rate from 200 to 230 bpm. The rate of tachycardias dependent on right-sided accessory pathways may decrease with the development of right BBB, but tachycardias dependent on accessory pathways located in the septum do not change rate appreciably with BBB because the circuit is not significantly prolonged. Even when BBB develops on the same side as the accessory pathway, the rate does not invariably change because the increase in ventriculo-atrial conduction time may be associated with shortening of antegrade conduction time in the AV node, so that the net effects on the tachycardia cycle length are negated.

7. *Pseudo R′ in V1.*
The development of a pseudo R′ in V1 is observed more frequently in AVNRT than in either AVRT or atrial tachycardias. Although this is not a particularly sensitive criterion (58%), it is relatively specific for AVNRT (91%).[45] It is attributable to distortion of the terminal portion of the QRS by a retrograde P wave.
8. *Preexcited Tachycardias.*
Tachycardias associated with ventricular preexcitation may be attributable to antidromic AVRT, atrial fibrillation or atrial tachycardias that conduct over an accessory pathway, or tachycardias mediated by a Mahaim fiber.[54] As shown in Figure 14-8, maximal preexcitation is present during antidromic AVRT because ventricular activation occurs exclusively through the accessory pathway. ECGs recorded during antidromic SVT show a regular, wide, monomorphic QRS complex that resembles ventricular tachycardia. Retrograde P waves may be detectable during the first half of the R–R interval, but they are extremely difficult to appreciate because they are obscured by the marked repolarization abnormality associated with preexcited complexes. When evident, P waves have a 1:1 relationship with the QRS because block of conduction in either the accessory pathway or AV node would terminate the tachycardia. Sometimes sudden changes in cycle length are observed during antidromic AVRT that reflects which bundle is used during retrograde conduction.[55] During atrial fibrillation with ventricular preexcitation, the ventricular rate may be rapid, the R–R intervals are irregular, and the QRS morphology varies depending on the degree to which the ventricle is activated by the accessory pathway or the AV node. The irregular rate and QRS morphology differentiate this arrhythmia from antidromic AVRT or monomorphic ventricular tachycardia. Patients with preexcited R–R intervals shorter than 220 to 250 ms have accessory pathways with short refractory properties and are at increased risk that atrial fibrillation could induce ventricular fibrillation because of rapid conduction over the accessory pathway.[13,56]
Tachycardias that are mediated by Mahaim fibers, shown in Figure 14-9, exhibit a QRS morphology that resembles left BBB. The QRS axis is typically between 0 and −75 degrees; QRS duration is 0.15 seconds or less; an R wave is present in limb lead I; an rS is seen in precordial lead V1; and there is transition in the precordial leads from a predominantly negative to positive QRS complex in leads V4 to V6.[57]
9. *Patterns of Ventricular Preexcitation.*
The ECG pattern of ventricular preexcitation provides useful information about the location of the accessory pathway and whether more than one pathway may be present. Several authors have studied the ECG manifestations of preexcitation and have developed criteria for localization of the

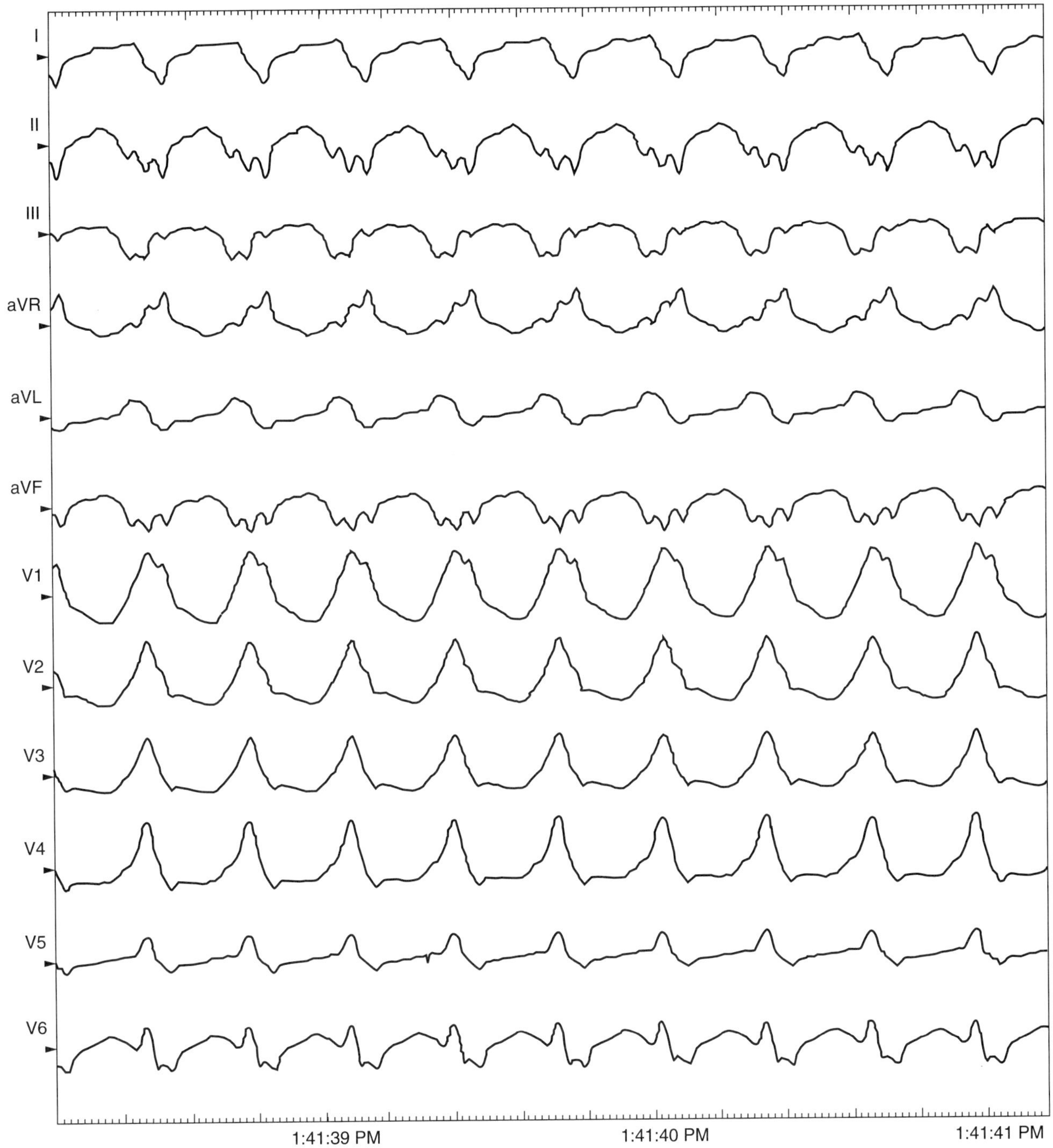

FIGURE 14-8 Twelve-lead ECG recorded during antidromic atrioventricular reentrant tachycardia. The QRS complexes are regular and maximally preexcited with a uniform morphology that is characteristic of a left lateral accessory pathway and matches antegrade preexcited morphology in sinus rhythm or atrial pacing.

accessory pathway.[58-64] The accuracy of these methods depends on the degree of ventricular preexcitation at the time of the recording and whether there is underlying heart disease that modifies the usual patterns of preexcitation. These criteria have evolved over time as additional experience has been acquired. They are based on the concept that δ wave polarity, QRS axis, and R wave transition in the precordial leads reflect the position of the accessory pathway. When preexcitation is pronounced, left-sided pathways have a prominent R wave in V1 with a positive δ wave. If the pathway is located on the posterior aspect of the mitral annulus, the polarity of the δ wave is generally negative in the inferior leads, and a QS complex is present. Pathways that are located more anterior-laterally on the mitral annulus have negative δ waves and QS morphology in aVL, but

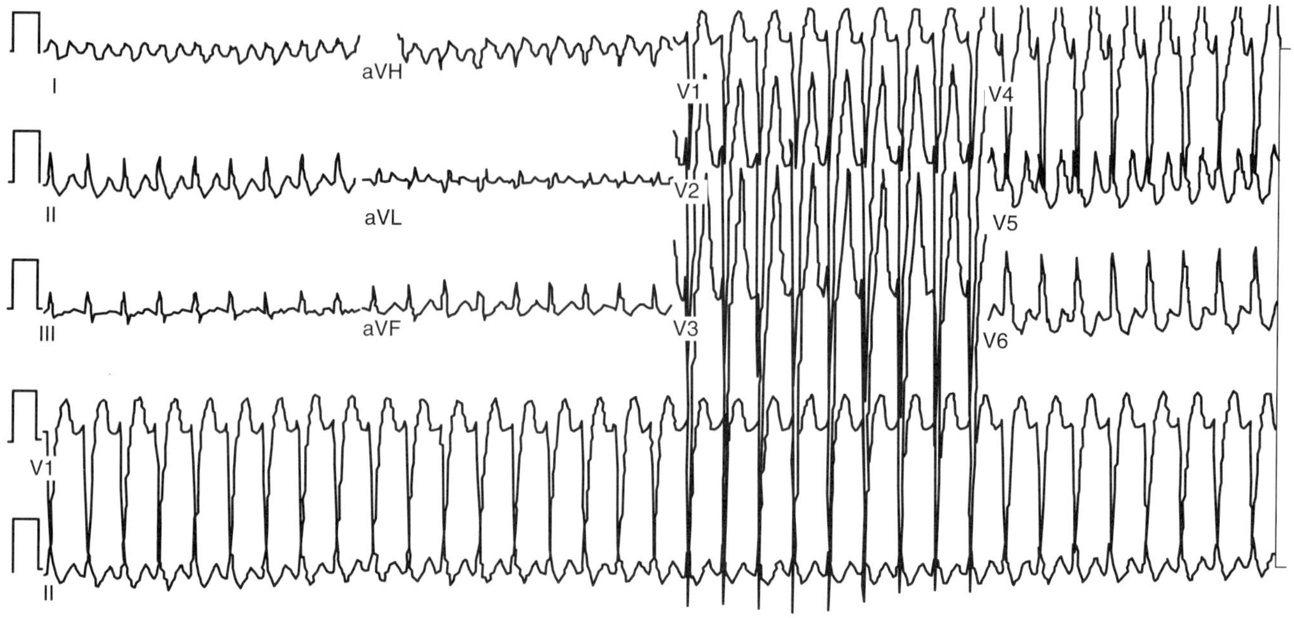

FIGURE 14-9 Characteristic 12-lead ECG recorded during macroreentry mediated by a Mahaim fiber. Note the QRS morphology resembling left bundle branch block with an rS configuration in V1. The differential diagnosis would include aberrancy, antidromic tachycardia, or ventricular tachycardia and requires electrophysiological study for confirmation.

the δ waves are positive in the inferior leads. Posterior septal pathways have negative δ waves in the inferior leads, but the R/S ratio is less than 1 in V1. There is an abrupt transition in to R/S ratio greater than 1 in V2. Arruda found that subepicardial accessory pathways characteristically have clear negative δ waves in lead II in the first 20 ms after the onset of the δ wave.[63] Accessory pathways located on the right side exhibit an R/S less than 1 in V1 and have delayed δ wave progression. The polarity of the δ waves in the inferior limb leads is negative if the pathway is posterior. Positive δ waves in the inferior leads suggest a more anterior position. Pathways located in the middle to anterior septum have positive δ waves in the inferior leads and may have negative δ waves in V1. They are distinguished by the R/S ratio greater than 1 in lead III with anteroseptal pathways and R/S equal to 1 with midseptal pathways.[61]

The assessment of accessory pathway locations is difficult when preexcitation is not pronounced. In one of the larger series, Chiang evaluated a stepwise algorithm that was based on the R/S ratio in V2, the δ wave polarity in lead III (initial 40 ms), the δ wave polarity in V1 (initial 60 ms), and the δ wave polarity in aVF (initial 40 ms).[64] The algorithm correctly predicted the location of the accessory pathway in 93% of the patients. Arruda developed an algorithm base on 135 consecutive patients and prospectively applied it to 121 consecutive patients with an overall sensitivity of 90% and specificity of 99%. Their criteria were based on the initial 20 ms of the δ wave in leads I, II, aVF, and the R/S ratios in III and V1. The algorithm they employed is shown in Table 14-3. The presence of multiple accessory pathways may be recognized by variable patterns of preexcitation with characteristics of more than one pathway or hybrid patterns that do not fit the usual algorithms. Fananapazir found that in patients with multiple pathways, more than one pattern of preexcitation was apparent in only 32% of recordings made during sinus rhythm, but the presence of more than one pathway was recognized in 55% of recordings obtained during atrial fibrillation.[65]

Diagnostic Approach to the Patient with Supraventricular Tachycardia

Investigation of SVT is based on understanding the underlying mechanism of the tachycardia and clinical context in which it occurs. A systematic approach should start with the history and physical examination, which provides two types of information: (1) the presence and type of symptoms and (2) the clinical context, particularly the existence of associated heart disease. Electrocardiographic documentation of the arrhythmia is an essential prerequisite to tachycardia management.[66]

CLINICAL EVALUATION

Careful history taking may provide valuable information. SVT may be symptomatic or asymptomatic. Palpitations due to tachycardia are the most common symptom.

TABLE 14–3 Stepwise Algorithm for Determination of Accessory Pathway Location

Nomenclature Used to Describe Accessory Pathway Location

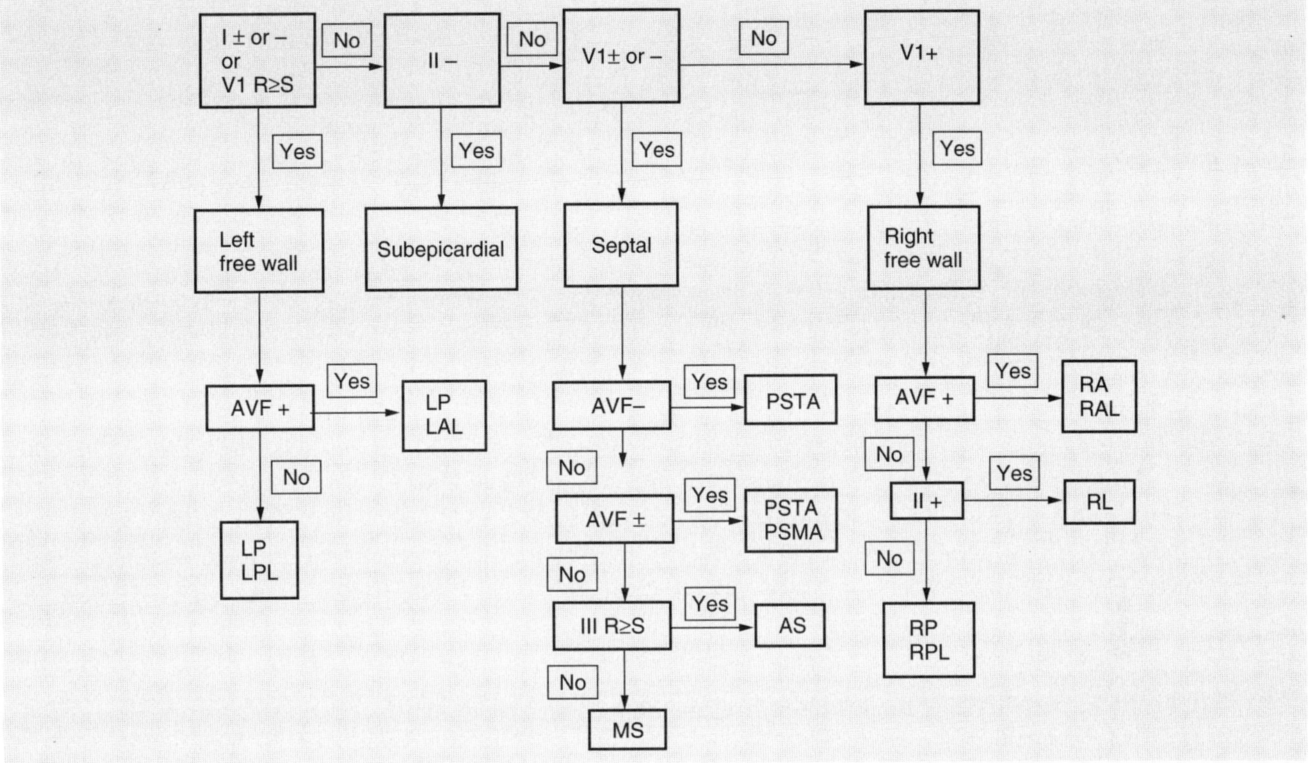

AS, anteroseptal; LAL, left anterolateral; LL, left lateral; LP, left posterior; LPL, left posterolateral; MSTA, mid-septal tricuspid annulus; PSMA, posteroseptal mitral annulus; PSTA, posteroseptal tricuspid annulus; RA, right anterior; RAL, right anterolateral; RAPS, right anterior paraseptal; RL, right lateral; RP, right posterior; RPL, right posterolateral. From Arruda, et al: J Cardiovasc Electrophysiol 1998;9:2-12.

Episodes of palpitations with abrupt onset and termination, followed by polyuria, suggest PSVT. This syndrome, known in French literature as "syndrome de Bouveret," was described in 1888, before the advent of electrocardiography.[2] It is characteristic of paroxysmal junctional tachycardia but may be found in other types of supraventricular or ventricular tachycardias, such as verapamil-sensitive ventricular tachycardia. Occasionally, SVT may be the cause of syncope. A history of palpitations preceding syncope is often reported in this presentation. When syncope or presyncope occurs immediately after termination of fast palpitations, the tachycardia-bradycardia syndrome should be suspected. Other symptoms may be associated with SVT and are indicative of poor tolerance (e.g., syncope, dizzy spells, chest discomfort, dyspnea, or even pulmonary edema). SVT may be mildly symptomatic or asymptomatic and discovered incidentally on recordings performed for another reason. Sometimes, the arrhythmia is specifically suspected due to its complications (e.g., asymptomatic paroxysmal atrial fibrillation in a patient with a cerebrovascular accident suspected to be of embolic origin). Physical examination is focused on any associated heart disease but, in general, paroxysmal SVT occurs in patients without organic heart disease. In contrast, heart disease is present in 70% of patients with atrial fibrillation.

DIFFERENTIAL DIAGNOSIS OF SUPRAVENTRICULAR TACHYCARDIA FROM THE ECG (Table 14-4)

Electrocardiographic documentation of the tachycardia is essential for the proper diagnosis and management of SVT. Recording of the tachycardia episode is easily obtained when tachycardia occurs frequently or is of prolonged duration. SVT by definition arises above the bifurcation of the His bundle, either in the atria or the AV junction and, therefore, is generally associated with narrow QRS complexes. SVT may sometimes present with wide QRS complexes either because the patient had a preexisting BBB or because aberrant conduction is present. Differentiating SVT from ventricular tachycardia may, at times, be difficult, particularly when preexcitation is present.

a. *Tachycardia with narrow QRS complexes* (Table 14-5). Most regular SVTs use the AV node either passively as in atrial tachycardias and atrial flutter or as a critical component of the circuit as in PSVT. The diagnostic approach to tachycardias with narrow QRS complexes should be undertaken in a stepwise fashion.[66-68]

The first step is to assess the regularity of the R–R interval.

TABLE 14-4 Classification of Supraventricular Tachycardias

Atrial Origin	Atrial fibrillation Atrial flutter Atrial tachycardia Enhanced automaticity: atrial focus Reentry: atrial or sinoatrial
AV Junction	Atrioventricular nodal reentrant tachycardia (AVNRT) Common type: slow pathway anterograde–fast pathway retrograde Uncommon type: fast pathway anterograde–slow pathway retrograde "Slow-slow": slow pathway anterograde–slow pathway retrograde Atrioventricular reentrant tachycardia (AVRT) Preexcitation may be overt or concealed on ECG in sinus rhythm (see text) AV junction anterograde–accessory pathway retrograde

1. *If the R–R interval is irregular,* atrial fibrillation or atrial flutter should be considered as the most likely diagnosis. However, when atrial fibrillation is associated with rapid ventricular response, it may seem regular. Permanent junctional reciprocating tachycardia (PJRT) is an SVT in which the R–R interval is often irregular. This tachycardia described by Coumel et al. was found to be related to a concealed accessory connection capable of decremental conduction.[69] The tachycardia uses the AV node in the antegrade conduction and a slow conducting accessory connection in the retrograde direction (Fig. 14-10). This tachycardia may be incessant, accounting for the descriptive term "permanent" in its title. It becomes sustained in clinical situations with increased catecholamine output such as exercise or emotion. It warrants therapy as it may have a deleterious effect on cardiac function in addition to tachycardias related symptom relief. The differential diagnoses include atypical AVNRT using the fast pathway in the antegrade direction and the slow pathway in the retrograde direction, and atrial tachycardia arising from the inferior atrium near the coronary sinus ostium.
2. *When the R–R intervals have been assessed to be regular,* the next step is also to look for the P waves ("cherchez le P") as advocated by Marriott.[66] The presence, morphology, and position as to the QRS complexes are important in the diagnosis of the site of origin and for the suspected mechanism of narrow QRS complex tachycardias. When the QRS complexes are preceded by P waves, which are different in configuration from the sinus P waves and conducted with a P–R interval equal or longer than the P–R interval of the sinus P waves, the most likely diagnosis is atrial tachycardia arising from an ectopic focus. The other mechanism of this type of SVT is intra-atrial or sinoatrial (SA) reentrant tachycardia, a diagnosis that requires an electrophysiological study for substantiation. If the P waves have the same configuration as the sinus P waves, the differential diagnosis includes appropriate or "inappropriate" sinus nodal tachycardia. Inappropriate sinus node tachycardia is a rare arrhythmia that has been recognized recently. This atrial tachycardia is characterized by an inappropriate and exaggerated acceleration of heart rate during physiologic stresses.[70] While its mechanism remains speculative, there are a number of possible hypotheses as to its basis. These include an ectopic atrial focus located in the SA node area, a normal SA node with increased response to the sympathetic tone or failure to respond to vagal stimulation, and an intrinsic anomaly of the SA node. When the P waves are submerged within the QRS complex in SVT and are therefore not identifiable, the most likely diagnosis is type 1 AVNRT (i.e., tachycardia involving the AV node and

TABLE 14-5 Diagnostic Algorithm for Narrow QRS Complex Tachycardia

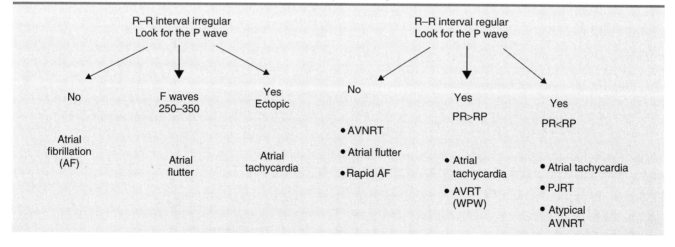

R–R interval irregular Look for the P wave			R–R interval regular Look for the P wave		
No	F waves 250–350	Yes Ectopic	No	Yes PR>RP	Yes PR<RP
Atrial fibrillation (AF)	Atrial flutter	Atrial tachycardia	• AVNRT • Atrial flutter • Rapid AF	• Atrial tachycardia • AVRT (WPW)	• Atrial tachycardia • PJRT • Atypical AVNRT

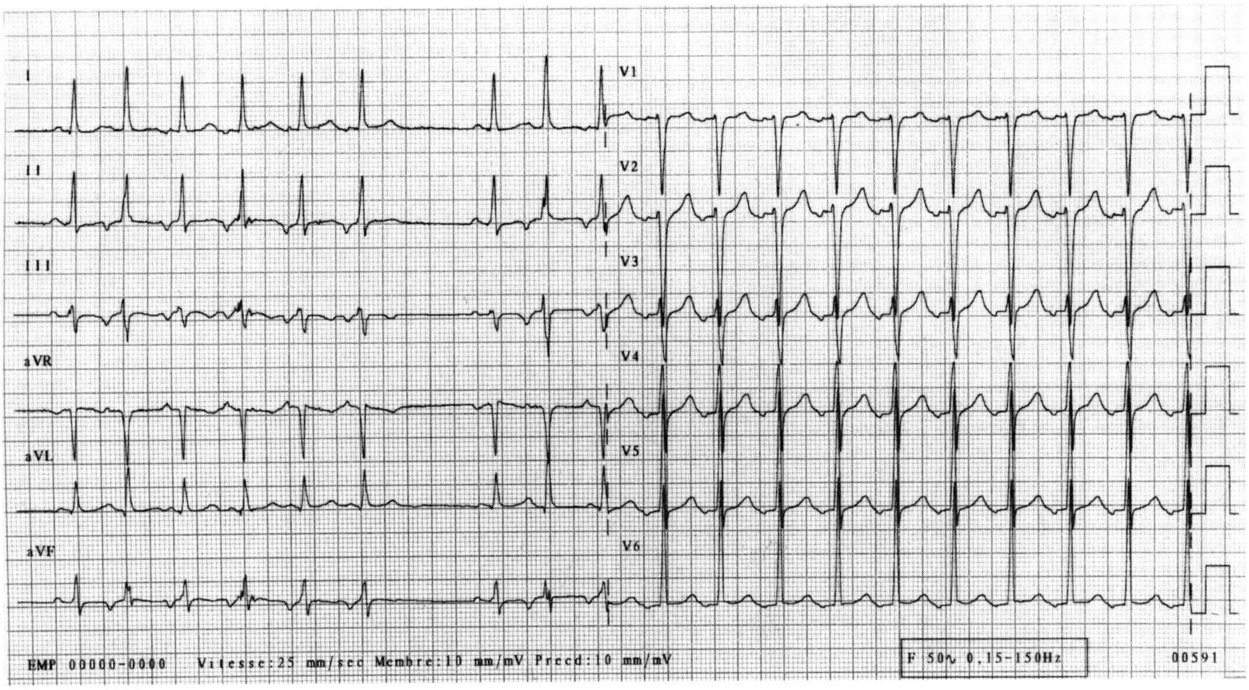

FIGURE 14-10 Twelve-lead ECG of a tachycardia related to a slow conducting accessory pathway used in the retrograde direction. Note in the left panel that the R–R intervals are slightly irregular as the pathway is capable of decremental conduction.

using in the typical form a slow pathway in the anterograde direction and a fast pathway in the retrograde direction, resulting in a P wave within or immediately after the QRS complex).[71,72] Wellens has described an electrocardiographic sign, which is suggestive of AVNRT. It consists of an incomplete BBB pattern (RSR') in lead V1 during the tachycardia that is not present in the 12-lead ECG in sinus rhythm.[72] However, atrial flutter with 2:1 conduction should be suspected as a differential diagnosis if the ventricular rate of the SVT with narrow QRS complexes is around 150 bpm. If the P waves during SVT have been identified and follow the QRS complexes at a significant interval resulting in an R–P interval equal to or greater than the P–R interval, the most likely diagnosis is orthodromic AVRT (i.e., involving the AV node in the antegrade direction and an accessory AV pathway in the retrograde direction). Other helpful electrocardiographic clues that have been reported and shed light on the mechanism of SVT include a negative P wave in leads I and V1, due to left-to-right atrial activation in AVRT using a left-sided AV connection in its retrograde limb.[72] When two recordings of the tachycardia are available, one with narrow QRS complexes and the other with left BBB with a longer cycle length, a finding first described in Paris, the diagnosis of AVRT involving a left-sided accessory connection can be made.[52,73] The presence of QRS alternation is also in favor of AVRT.

The differential diagnosis between PSVT and atrial fibrillation with rapid ventricular response or atrial flutter with 1:1 conduction may at times be difficult. Vagal maneuvers, particularly carotid sinus massage and adenosine injection, may be of great help in clinical diagnosis (Fig. 14-11). In AVRT or AVNRT, the arrhythmia may terminate abruptly or remain unaffected. In contrast, atrial fibrillation or flutter is rarely terminated by these techniques but can be slowed in its ventricular response,

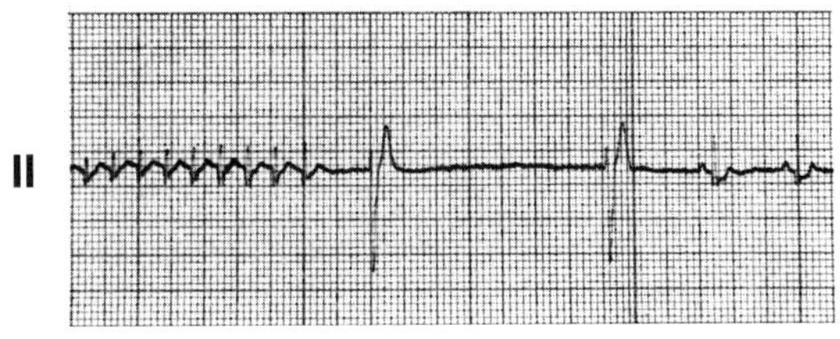

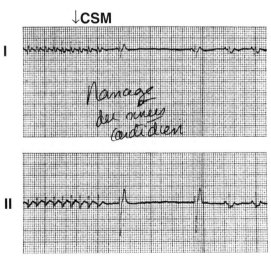

FIGURE 14-11 Termination of narrow QRS complex tachycardia with carotid sinus massage (CSM).

TABLE 14-6 Tachycardias with Wide QRS Complexes

1. Atrial fibrillation with conduction through an atrioventricular (AV) accessory connection (Kent's bundle)
2. Atrial flutter with 1:1 conduction through an AV accessory connection (Kent's bundle)
3. Antidromic tachycardia: Kent's bundle anterograde–AV junction retrograde
4. Tachycardia using 2 accessory connections: overt Kent's bundle in one ventricle anterograde, overt or concealed (on the ECG in sinus rhythm) Kent's bundle in the other ventricle retrograde
5. Tachycardia using a Mahaim fiber (AV-like structure) anterogradely and the AV node retrogradely
6. AV nodal reentrant tachycardia with antegrade conduction over an accessory AV pathway ("bystander Kent")

thus exposing the underlying atrial flutter or fibrillation waves. PSVT and atrial flutter/AF may also coexist in the same patient. Adenosine administration may also be used as a diagnostic test to assess dual AV nodal pathway conduction, efficacy of slow pathway ablation, and detection of concealed accessory pathways.[74]

b. *Tachycardia with regular wide QRS complex* (Table 14-6). Regular SVT may present with wide (>0.12 sec) QRS complex, and differentiating SVT from VT may be difficult. Another etiology of wide QRS complex tachycardia is the preexcitation syndrome, which may be overt (WPW syndrome) or concealed (with the pathway only conducting in the retrograde direction), and this is discussed later. *It bears re-emphasizing that the vast majority of regular SVTs occur in patients without organic heart disease in contrast to patients with VT.* Three clues may be used in the differentiation diagnosis of wide QRS tachycardias.
   1. The age of the patient; in the child and young adult, preexcitation is more common than ventricular tachycardia. In this instance the ECG in sinus rhythm, if available, shows preexcitation, and the ECG during the tachycardias shows an identical morphology.
   2. In making this distinction in SVT diagnosis, it is necessary to locate the P wave during tachycardias. The presence of AV dissociation is generally diagnostic of ventricular tachycardia. However, it is only present in about 40% of ventricular tachycardias. Differentiating SVT from VT may require an electrophysiological study with endocavitary recordings.[66,75-79]
   3. Wellens et al. have described a number of criteria that may be of help differentiating SVT from VT.[79] A few that we have found particularly useful include left axis deviation (beyond −30 degrees), QRS complexes wider than 0.14 sec, and the QRS morphology in V1 and V6 (i.e., a monophasic R wave or a QR pattern in V1, or an rS or a QS pattern in V6), all of which favor a diagnosis of VT. Another clue that might be helpful is the presence of BBB in sinus rhythm. An identical morphology during the tachycardia is suggestive of SVT as well. When the QRS complex is narrow in sinus rhythm and wide during SVT, aberrant conduction is also a consideration. We have found it practical to consider aberrancy only when the typical pattern of right or left BBB is present during tachycardia. This still does not exclude the possible diagnosis of VT. For these reasons, it is a wise rule "*to consider any tachycardia with wide QRS complexes as being VT unless proven otherwise*" (Agustin Castellanos Jr). Detailed electrophysiological studies are indicated in all patients with wide QRS complex tachycardias.

c. *Tachycardia with preexcited QRS complexes* (Fig. 14-12). The differential diagnosis of wide QRS complex tachycardia may be extremely difficult if preexcitation is present. Fortunately, tachycardias with preexcited QRS complexes represent less than 5% of all wide QRS complex tachycardias. A number of mechanisms may account for a tachycardia with preexcited QRS complexes. The most common is atrial fibrillation with conduction over the Kent bundle. The R–R interval is frequently irregular, and the QRS width may change from one complex to

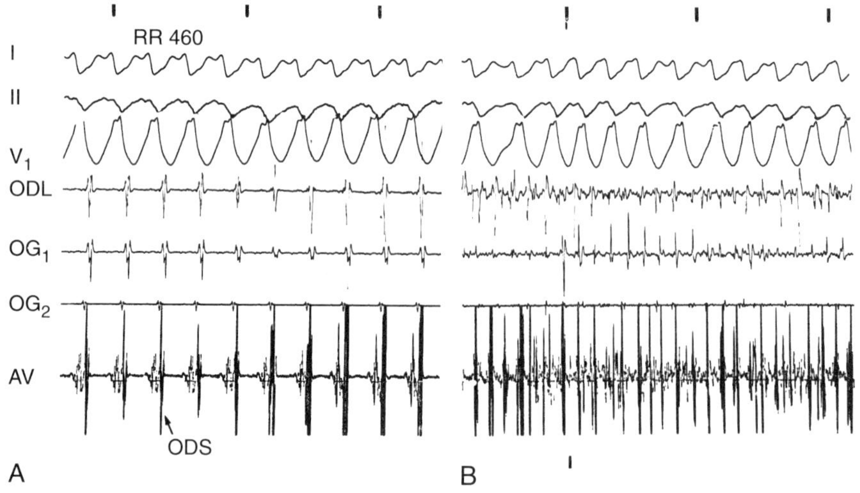

FIGURE 14-12 Tachycardias with preexcited QRS complexes. The same leads as those in Figure 14-3 were used. **Panel A** is consistent with an antidromic reciprocating tachycardia. **Panel B** shows atrial fibrillation conducted over a left-sided accessory AV pathway termination of supraventricular tachycardia with vagal maneuvers.

the other. Atrial flutter with 1:1 conduction over an accessory connection is another possibility; however, this diagnosis should be suspected when the ventricular rate ranges from 250 to 300 bpm. In a young person, a tachycardia with a left BBB and left axis deviation should suggest the possibility of a Mahaim fiber. The other mechanisms may require endocavitary recordings and can only be ascertained by a detailed electrophysiological study.

d. *Noninvasive investigations.*

As previously mentioned, electrocardiographic documentation is an essential step in the diagnostic approach to SVT. This can be achieved by several methods.

1. *Ambulatory ECG Monitoring.*

Ambulatory ECG recordings are only warranted when symptoms are frequent enough to allow documentation within the 24- or 48-hour recordings. Otherwise, the information provided by such monitoring is limited to the possible trigger (extrasystoles). Occasionally, monitoring of the tachycardia shows that it is similar to sinus tachycardia, making ambulatory ECG recordings extremely useful in the diagnosis of "inappropriate" sinus node tachycardias in our experience. Chest pain and dyspnea precipitated by the tachyarrhythmia may be the dominant or stand-alone symptoms, and monitoring may be useful in correlating these symptoms to the tachycardia.

2. *Event Recorders.*

If the episodes of tachycardia are infrequent or of short duration, event recorders are often useful. In recent years, the advances in technology have prompted the development of recorders with transtelephonic transmission and will allow transtelephonic surveillance of selected patients. Daily monitoring and symptomatic transmissions permit the clinician to assess arrhythmia-related symptoms.

3. *Telemetric Monitoring.*

When the tachycardia is severely symptomatic and occurs frequently enough to be recorded during a hospital stay, telemetric monitoring in the hospital offers another diagnostic option. However, this surveillance requires hospitalization and is associated with a high cost, which often makes this technique impractical for any prolonged period.

4. *Exercise Testing.*

Exercise testing is particularly valuable when the tachycardia is precipitated by exercise or is otherwise believed to be catecholamine dependent. For example, atrial tachycardia and atrial fibrillation are not uncommonly induced by exercise. Although junctional tachycardias may occasionally be precipitated by exercise, this test is seldom indicated in patients with paroxysmal SVT. Exercise testing is a valuable tool in patients suspected to have "inappropriate sinus node tachycardia."

5. *Implantable Loop Recorders and Other Devices.*

Implantable loop recorders are subcutaneous ECG recording devices without intracardiac electrodes. They have been developed recently for symptom-ECG correlation. Loop recorders, such as the "Reveal Plus" system, have been used recently in syncope and myocardial infarction populations. While such devices can document tachycardias, large-scale study of the yield of such systems in PSVT is unavailable. Furthermore, these devices cannot reliably differentiate SVT from VT. In patients with previously implanted pacemakers or defibrillators, documentation of arrhythmia and its site of origin is generally easy to achieve.

6. *Electrophysiological Study* (Fig. 14-13).

Electrophysiological studies are used for diagnostic and therapeutic purposes in patients with PSVT. SVT initiation during these

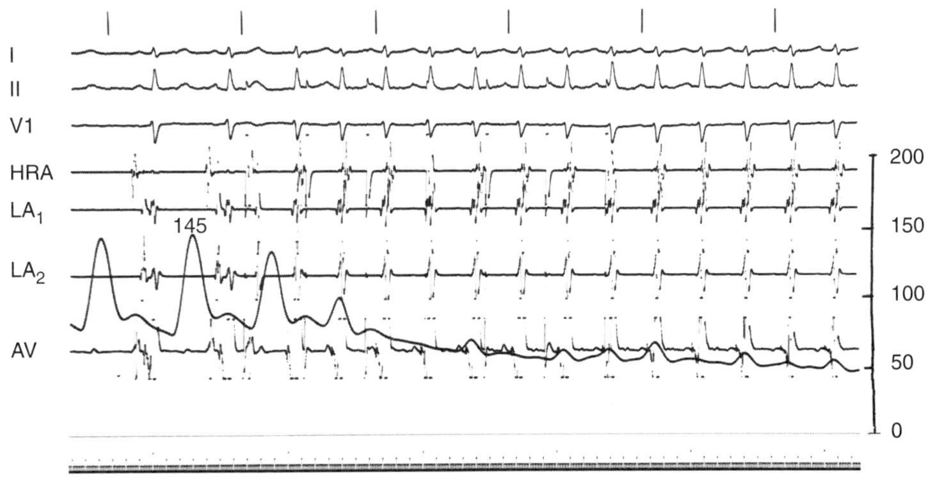

FIGURE 14-13 Recordings in a 76-year-old woman complaining of tachycardia followed by syncope. From top to bottom, ECG leads *I, II,* and *V1,* high right atrium (*HRA*), left atrium (*LA1* and *LA2*) femoral artery pressure and atrioventricular lead (*AV*). Note the fall of blood pressure after AV nodal reentrant tachycardia is induced using overdrive atrial pacing.

procedures allows ECG documentation of the tachycardia, if it has not been previously recorded; definition of the mechanism of the tachycardia and the critical components of the circuit; evaluation of symptoms during the arrhythmia; and, if indicated, performance of radiofrequency ablation or to evaluate the efficacy of antiarrhythmic therapy. Most SVTs are due to a reentrant mechanism and may be induced in the laboratory using programmed electrical stimulation.[75-79] In AV junctional tachycardias, electrophysiological study permits induction of the tachycardia in more than 90% of patients with AVRT or AVNRT. In patients with AVNRT, premature atrial stimulation can demonstrate dual AV nodal conduction and induce the tachycardia in order to determine its mechanism.[75] The arrhythmias associated with the WPW syndrome include reciprocating or circus movement tachycardias and atrial arrhythmias.[71,77] In some instances of AVNRT, there may be concomitant preexcited QRS complexes due to passive conduction over an accessory pathway, which may serve as a passive bystander. Tachycardias related to Mahaim fibers have a particular electrocardiographic presentation with left BBB and left axis. The ECG in sinus rhythm shows the features of the WPW syndrome. The electrophysiological study will often demonstrate decremental conduction over the accessory pathway.

Clinical Electrophysiology

ATRIOVENTRICULAR NODAL REENTRANT TACHYCARDIA

Concept of Dual Atrioventricular Nodal Pathway

During electrophysiological evaluation, typical AVNRT manifests antegrade conduction via a "slow" pathway and retrograde conduction via a "fast" pathway.[31] The electrophysiological characteristic is the presence of dual AV nodal pathway physiology demonstrated by a discontinuous AV node functional curve. In contrast, atypical AVNRT has antegrade conduction through a "fast" or "slow" pathway and retrograde conduction through a "slow" pathway.[80,81] Discontinuous AV nodal conduction is defined as a sudden increment of 50 ms or greater in A–H or H–A interval ("jump") with a decrement in prematurity of the extrastimulus by 10 to 20 ms. In some patients a jump (>50 ms) of any consecutive A–H intervals during incremental atrial pacing would be found, which might be a manifestation of dual AV nodal pathways.

Two important concepts of dual AV nodal pathways have been further clarified by provocative pharmacologic testing and catheter ablation in patients with AV nodal reentrant tachycardia. One major question has been whether dual AV nodal pathways are fully intranodal and due to longitudinal dissociation of AV nodal tissue or extranodal involving the separate atrial inputs into the AV node. From clinical studies of slow pathway potentials, LH (low followed by high) frequency potentials are observed during asynchronous activation of muscle bundles above and below the coronary sinus orifice; HL (high followed by low) frequency potentials are caused by asynchronous activation of atrial cells and a band of nodal-type cells that may represent the substrate of the slow pathway.[82-85] Thus, the slow and fast pathways are likely to be atrionodal approaches or connections rather than discrete intranodal pathways. The results of catheter ablation, indicate that the fast and slow pathways have their origins outside the limits of the compact AV node, and the tissues targeted during successful ablation are composed of ordinary working atrial myocardium surrounding the AV node itself.[83-88] Furthermore, AV or VA conduction block during AVNRT favors the concept that atrial and ventricular tissue is not involved in maintenance of this tachycardia (Fig. 14-14).[90,91]

The electrophysiological characteristics of the retrograde pathway during the tachycardia differ from those of the anterogradely conducting pathway; these characteristics can be demonstrated by the differential response to atrial or ventricular pacing, and response to antiarrhythmic drugs. For example, procainamide prolongs conduction time in the retrogradely conducting, but not in the anterogradely conducting, pathway. Also, pacing at short cycle lengths prolongs anterograde, but not retrograde, AV nodal conduction time in some patients.

Unusual Physiology of Dual Atrioventricular Nodal Pathways

Some patients with AVNRT have multiple antegrade and retrograde AV nodal pathways with multiple discontinuities in the AV node function curve or dual AV nodal pathways with a continuous curve during programmed electrical stimulation.[87,88] Furthermore, variant forms (slow-slow, slow-intermediate, fast-intermediate) of AVNRT have been noted (Figs. 14-15 and 14-16).[86,91] Whether multiple antegrade and retrograde AV nodal pathways originate from anatomically different pathways or represent anisotropic conduction–induced functional pathways is still controversial. Several investigators have demonstrated the marked heterogeneity of the transitional cells surrounding the compact AV nodal pathway. The nonuniform properties of the AV node can produce anisotropic conduction and suggest that the antegrade and retrograde fast pathways are anatomically distant from the multiple "antegrade slow" and "retrograde slow" or intermediate pathways, respectively. Clinical studies have demonstrated that successful ablation or modification of "retrograde slow" and intermediate pathways occurs at different sites from "antegrade fast or slow" pathway, and the possibility of anatomically different antegrade or retrograde multiple pathways should be considered. Furthermore, in the patients who have successful ablation of multiple "antegrade slow" pathways or "retrograde slow" and

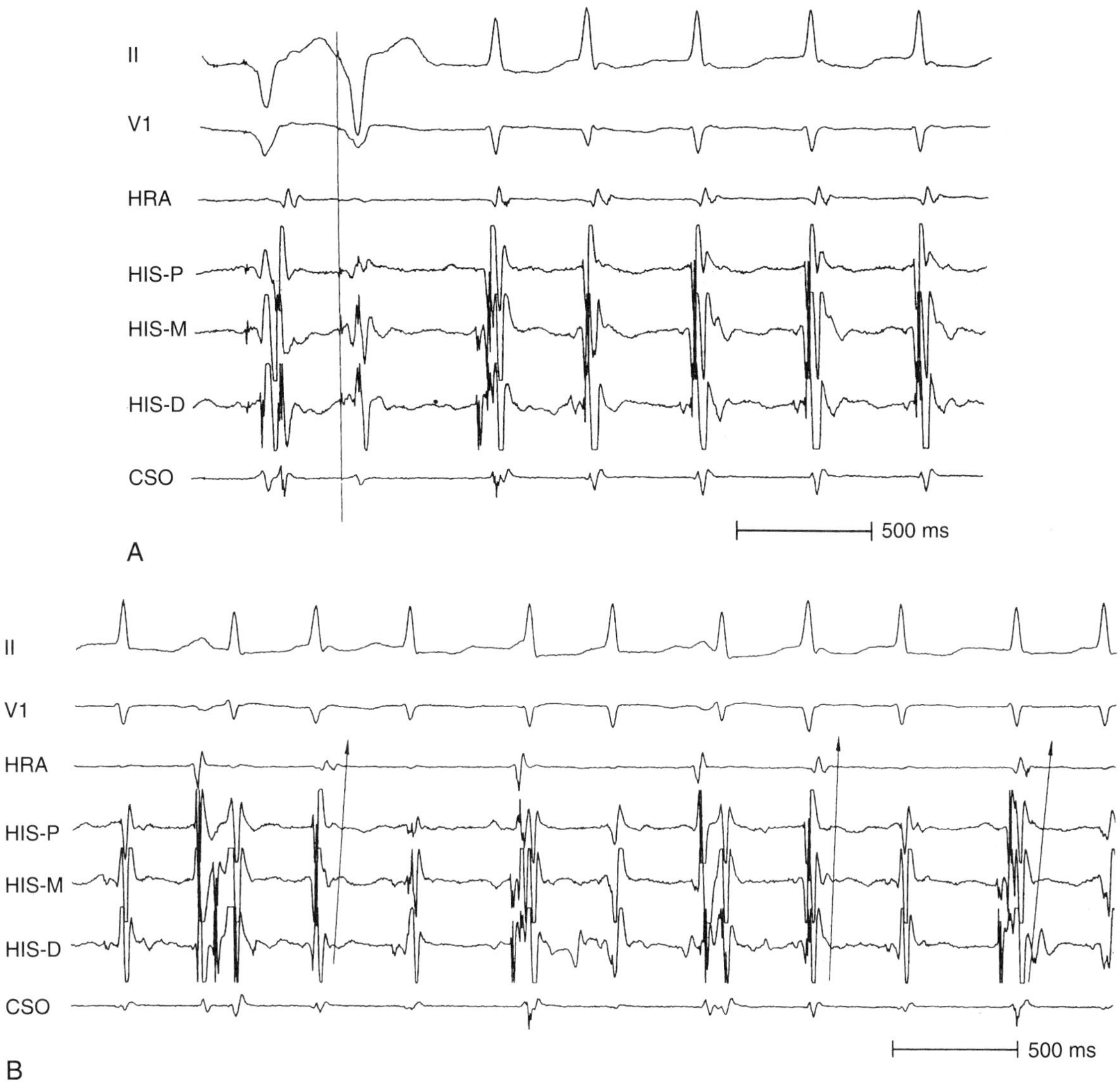

FIGURE 14-14 **A,** Right ventricular stimulus induces atrioventricular nodal reentrant tachycardia (AVNRT). Although the last pacing beat (*with vertical line*) does not conduct to the atrium, AVNRT still happens. **B,** AVNRT with occasional retrograde conduction to the atrium (*oblique arrows*).

intermediate pathways at a single site, anisotropic conduction over the low septal area of the right atrium is a possible explanation for the presence of multiple antegrade and retrograde AV nodal pathways.[86-88,91]

Patients with AVNRT can have continuous AV node conduction curves. These patients do not exhibit an AH jump using two extrastimuli and two drive cycle lengths during atrial pacing from the high right atrium and coronary sinus ostium. The possible mechanisms of the continuous AV node function curves in AVNRT include (1) the functional refractory period of the atrium limits the prematurity with which atrial premature depolarization will encounter the refractoriness in the AV node, which in turn produces inability to dissociate the fast and slow AV node pathways, and (2) fast and slow AV nodal pathways have similar refractory periods and conduction time.[87]

ATRIOVENTRICULAR REENTRY TACHYCARDIA

Anatomy and Electrophysiology of Accessory Pathways

The oblique orientation of most accessory pathways has been demonstrated by detailed endocardial and epicardial mapping techniques.[92-94] The atrial and ventricular insertion sites of accessory pathways can be up to 2 cm disparate in location; furthermore, some accessory pathways have antegrade and retrograde conduction fibers at different locations, and this finding has been proven by different ablation sites for antegrade and retrograde conduction.[95] Thus, the anatomic and functional dissociation of the accessory pathway into atrial and ventricular insertions and antegrade and retrograde components is possible. Approximately 90% of

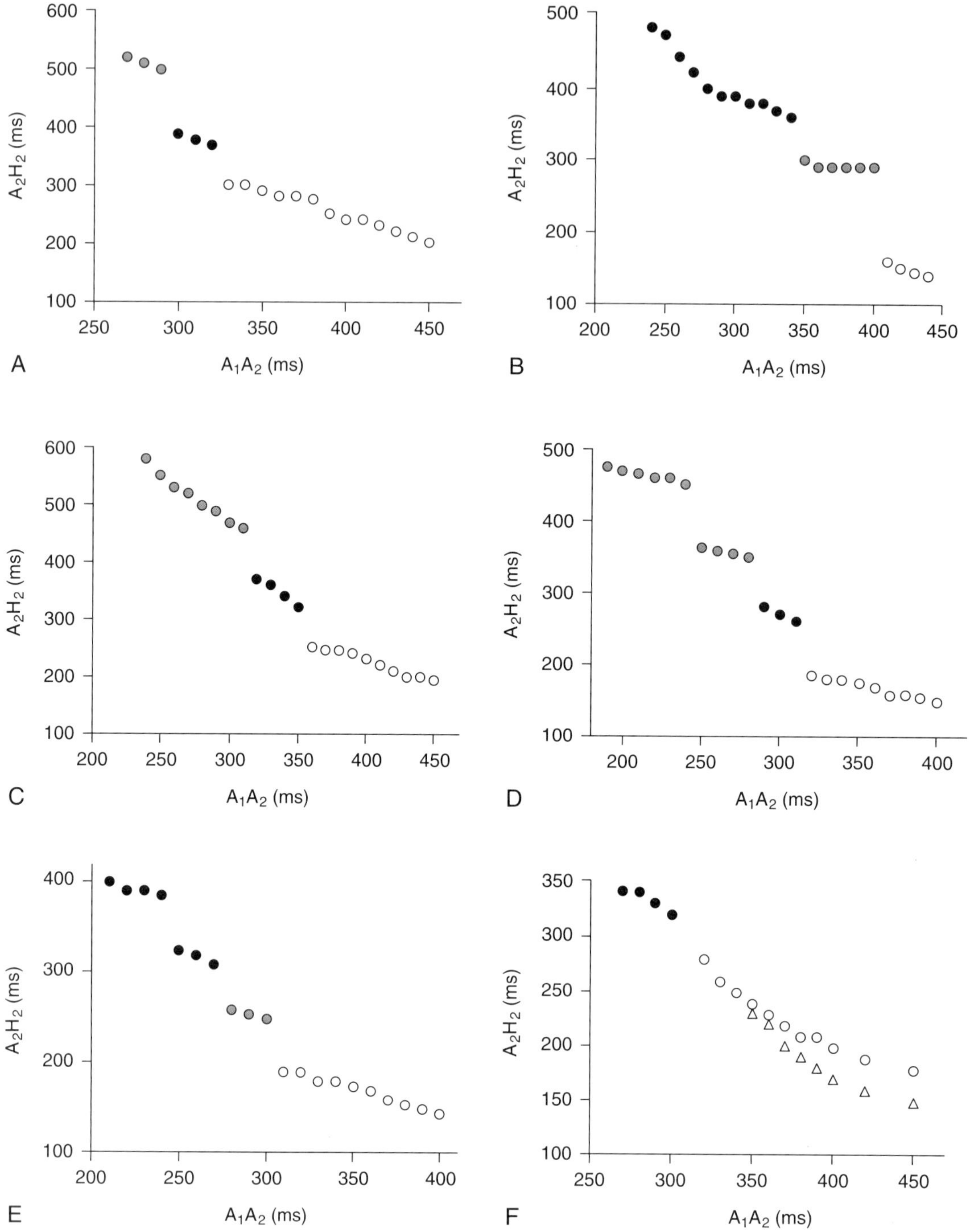

FIGURE 14-15 A–D, Atrioventricular (AV) nodal conduction curve with multiple jumps. Five patterns of tachycardia (slow-fast form) induction demonstrated by AV node conduction curves (A_2H_2 versus A_1A_2). A_2H_2, atrio–His bundle conduction interval in response to atrial extrastimulus; A_1A_2, coupling interval of atrial extrastimulus. Open circles indicate fast pathway conduction; open circles with central cross indicate slow pathway conduction without initiation or maintenance of sustained tachycardia; closed circles indicate slow pathway conduction with initiation and maintenance of sustained tachycardia; closed circles with central cross indicate slow pathway conduction with initiation but without maintenance of sustained tachycardia. **A,** Only the first slow pathway is used for induction and maintenance of sustained tachycardia (*pattern 1*). **B,** Only the second slow pathway is used for induction and maintenance of sustained tachycardia (*pattern 2*). **C,** The first slow pathway is used during sustained tachycardia; either the first or the second slow pathway is used for initiation of tachycardia (*pattern 3*). **D,** The first slow pathway is used during sustained tachycardia; any of the three slow pathways is used for initiation of tachycardia (*pattern 4*). **E,** The third slow pathway is used during sustained tachycardia; either the second or third slow pathway is used for initiation of tachycardia (*pattern 5*). (From Tai CT, et al: J Am Coll Cardiol 1996;28:725-31.) **F,** AV nodal conduction curve without AH jump. Curve relating A_2H_2 interval to prematurity of atrial extrastimuli (A_1A_2). Before ablation, a smooth curve without evidence of dual AV nodal pathway is present (○). After critical AH delay, tachycardia also is induced (•). Ablation in the slow pathway zone eliminates the "tail" of the curve (△). (From Tai CT, et al: Circulation 1997;95:2541-7.)

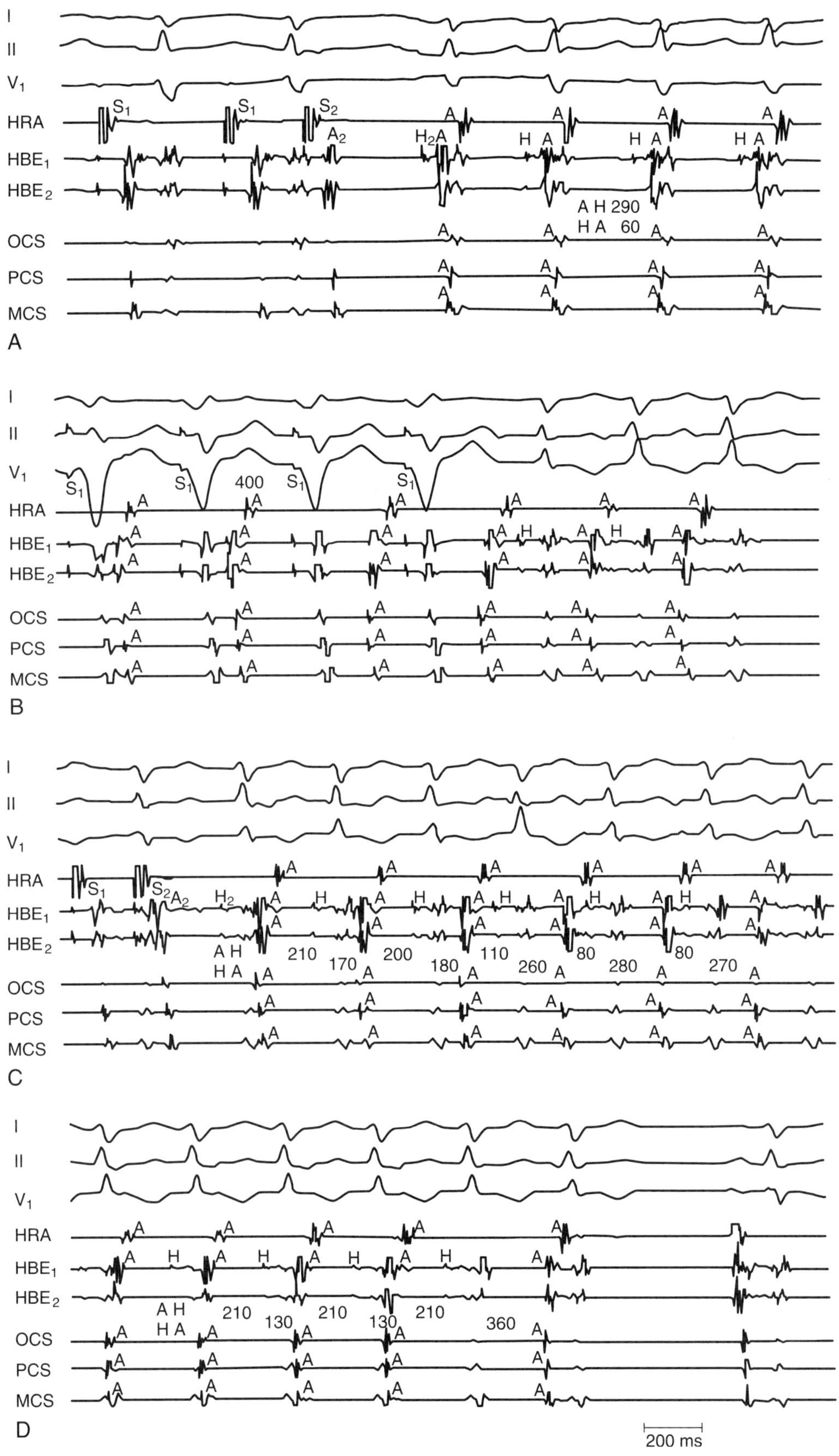

FIGURE 14-16 Recordings show four types of atrioventricular nodal reentrant tachycardia (AVNRT) and echo. **A** and **B,** the baseline state. **C** and **D,** intravenous infusion of isoproterenol. **A,** induction of slow-fast form of AVNRT by atrial extrastimulus, with the earliest atrial activation at the ostium of the coronary sinus (OCS). **D,** a slow-slow form AVNR echo before termination of slow-intermediate form of AVNRT. A_2 and H_2, atrial and His bundle response to the atrial extrastimulus (S_2), respectively; HRA, HBE_1, HBE_2, PCS, and MCS, electrograms recorded from the high right atrium, the distal His-bundle area, proximal His-bundle area, proximal coronary sinus, and middle coronary sinus, respectively; S_1, basic paced beats; S_2, extrastimulus. (From Tai CT, et al: Am J Cardiol 1996;77:52-8.)

AV accessory pathways have fast conduction properties, and the other accessory pathways (including Mahaim fibers) show decremental conduction properties during atrial or ventricular stimuli with shorter coupling intervals.[96-99] These pathways with decremental conduction may be sensitive to several antiarrhythmic drugs, including verapamil and adenosine. Accessory pathways in the right free wall and posteroseptal areas have a higher incidence of decremental conduction properties. When decremental conduction is present, the possibility of Mahaim fiber, such as atriofascicular or nodoventricular pathway, must be considered. Several studies have demonstrated that most of the ventricular insertion sites of these particular bypass tracts are close to the right bundle branch, and the typical Mahaim fiber potential can be recorded along the tricuspid annulus in patients with the atriofascicular pathways (Fig. 14-17).[100-102]

Electrophysiological Findings in Atrioventricular Reentry Tachycardia

The manifest accessory pathway can be diagnosed from the 12-lead surface ECG with typical δ wave and is confirmed by a reduced, absent, or negative H–V interval. During electrophysiological study, recordings should be obtained from the tricuspid and mitral annulus directly or indirectly from the coronary sinus as well as the normal AV nodal His conduction system. Atrial and ventricular pacing and extrastimulation, isoproterenol provocation, and induction of atrial fibrillation to assess antegrade conduction over an accessory pathway are essential elements of electrophysiological study. Ventricular pacing and extrastimulation can define retrograde conduction properties such as refractoriness and conduction time and location of the pathway. Switching of conduction between the accessory pathway and AV nodal–His axis can be demonstrated upon reaching effective refractoriness of one or the other conduction pathway. Atrial pacing or extrastimulation can accentuate antegrade preexcitation up to the refractoriness of the pathway. Tachycardia induction requires unidirectional block in one of the AV conduction pathways (AV node/His or accessory pathway)coupled with critical conduction delay in the circuit.

For the diagnosis of accessory pathway–mediated AV reentry tachycardia, a premature ventricular depolarization can be delivered during the tachycardia at a time when the His bundle is refractory, and the impulse still conducts to the atrium; this indicates that retrograde propagation conducts to the atrium over a pathway other than the normal AV conduction system. The definition of AV reentry tachycardia involves reentry over one or more AV accessory pathways and the AV node, and the classic classification of AV reentry tachycardia includes orthodromic and antidromic tachycardias.[103-105] For the initiation of orthodromic tachycardia, a critical degree of AV or VA delay, which can be in the AV node or His-Purkinje system, is usually necessary.

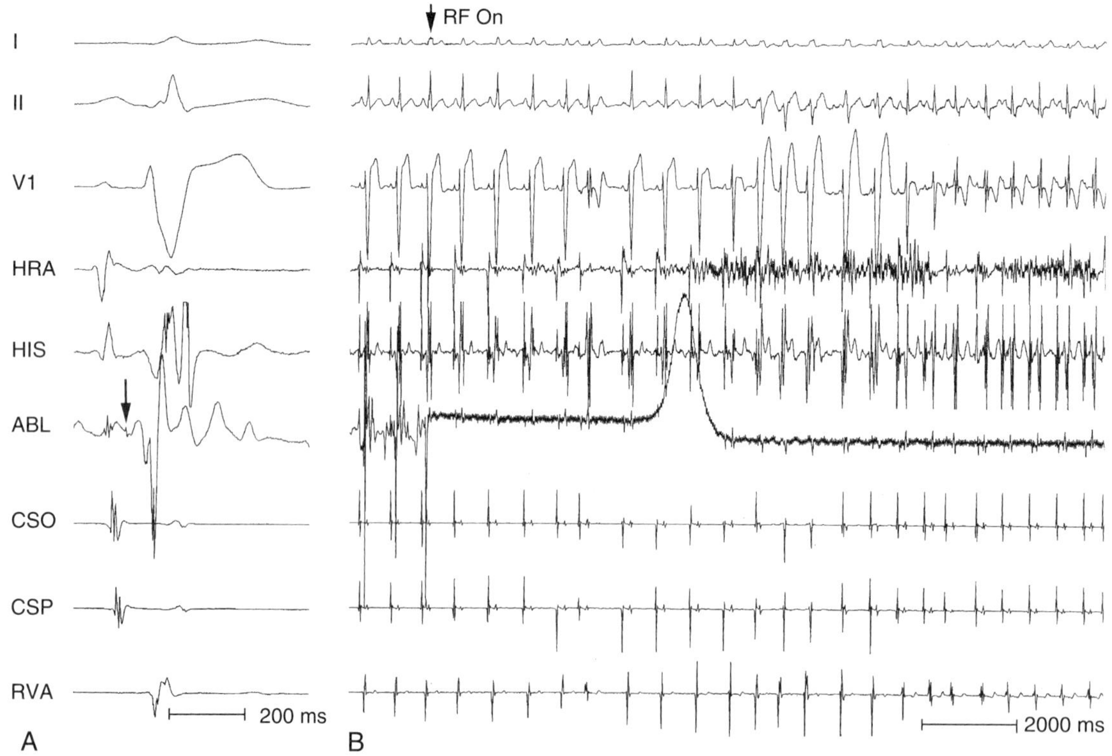

FIGURE 14-17 A mapping catheter along the posterolateral aspect of tricuspid annulus records the Mahaim fiber potential (*arrow in* **A**). Application of radiofrequency energy on this point eliminates Mahaim fiber conduction with disappearance of ventricular preexcitation (**B**).

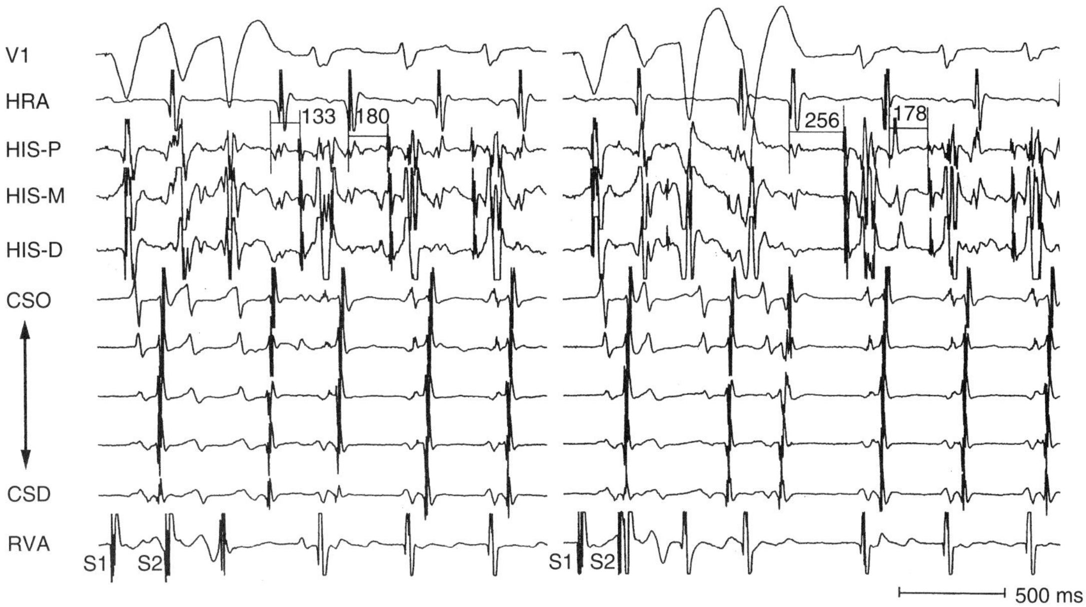

FIGURE 14-18 Left and right panels show right ventricular extrastimulus (S1S2) with V3 phenomenon, and induces AV reentry tachycardia with retrograde conduction through the left lateral accessory pathway. In the left panel, the first and second tachycardia beats show antegrade conduction time (A–H interval) of 133 and 180 milliseconds (ms), respectively. In the right panel, the S1S2 coupling interval is shorter than the left panel, and the antegrade conduction time is much longer in the first tachycardia beat than the second tachycardia beat (256 versus 178 ms), with the difference larger than 50 ms, suggesting the possibility of antegrade conduction through the slow pathway in the first tachycardia beat.

However, dual AV nodal pathway physiology, with or without AV nodal reentrant tachycardia, can be noted in some patients (Fig. 14-18).[106] Ventricular pacing from different sites can provide valuable information about retrograde conduction through the AV node or via a septal pathway (Fig. 14-19).[107] The incidence of antidromic AV reentry tachycardia is much lower than orthodromic AV reentry tachycardia. Rapid conduction in retrograde AV nodal–His axis is necessary for initiation and maintenance of antidromic tachycardia.[106-7] The atrial premature beat usually can advance the next preexcited ventricular complex through the antegrade accessory pathway, or it can terminate the AV reentry tachycardia through collision with the previous retrograde wavefront (Fig. 14-20). The incidence of multiple accessory pathways is about 5% to 20% and antidromic tachycardia is common in this situation.

The most difficult situation for differential diagnosis of AV reentry tachycardia is the so-called *Mahaim tachycardia,* including atriofascicular or nodofascicular (or nodoventricular) reentry tachycardia and AV nodal reentry tachycardia with innocent bystander bypass tract. These arrhythmias often appear as wide complex tachycardias with a left BBB and left axis deviation morphology. However, presence of VA dissociation favors nodofascicular tachycardia.

Principles of Management

The principles of management in many forms of paroxysmal SVT have undergone a sea of change with a better understanding of the natural history and progression of these arrhythmias, as well as the development of more effective ablative and drug therapy for many of these rhythms. Immediate and long-term prophylactic therapy differs significantly in considerations in selection and clinical application. Immediate therapy is directed at resolution of an individual arrhythmic event. Prophylaxis is largely focused on curative approaches, though suppressive and tachycardia termination methods exist and are occasionally applied in specific clinical scenarios. The spontaneous occurrence of a single SVT event may or may not mandate immediate therapy based on symptomatology and certainly does not mandate prophylactic therapy unless recurrence is anticipated, patterns of recurrence and duration become clearer, or the event is potentially malignant or life threatening. Immediate therapy is indicated if a patient develops angina, heart failure, or syncopal symptomatology during SVT. Sustained palpitations can impair functionality and result in serious patient discomfort and require immediate termination. In SVT observed during infancy and childhood, resolution can be observed with growth and development in selected individuals. Diagnosis of these rhythms even in preterm infants has resulted in better characterization of the clinical outcome and therapy selection. While observation with a conservative management approach was more prevalent in this and older populations, the advent of effective curative therapy in the form of catheter ablation has lowered the threshold for intervention in both adult and pediatric populations. As mentioned earlier, a change in pattern of SVT over time is not uncommon and should be considered in prophylaxis.

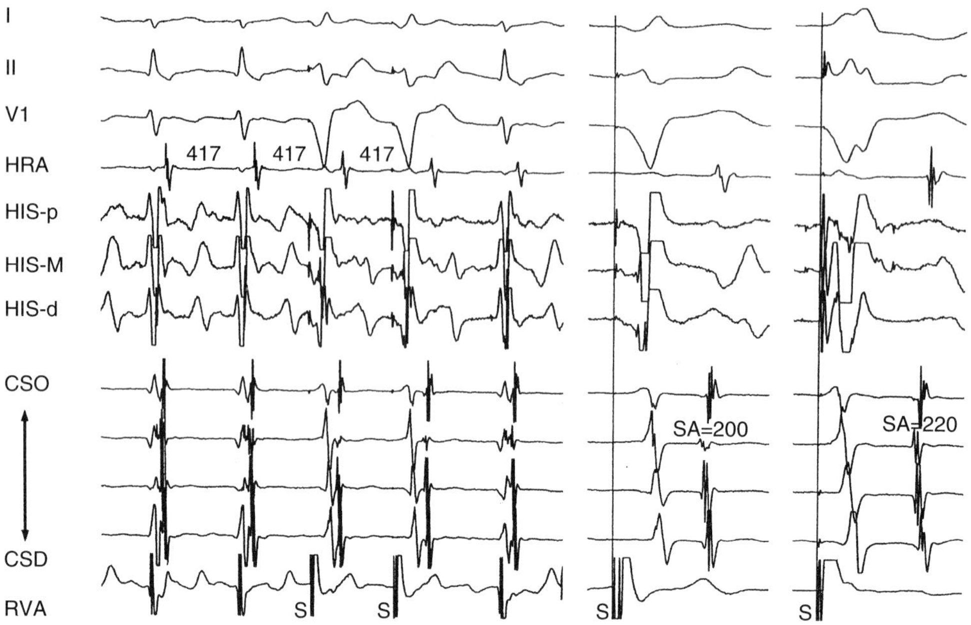

FIGURE 14-19 Left panel shows delivery of two right ventricular apex stimuli during tachycardia, without changing the tachycardia cycle length (417 milliseconds [ms]). Furthermore, the earliest atrial activation is on the CS proximal part. The middle panel shows ventricular stimuli from right ventricular apex with the same pacing cycle length as tachycardia cycle length, and the interval from stimulus to atrial activation on CS is 220 ms. The right panel shows ventricular stimuli from the basal part of the right ventricle with the same pacing cycle length as tachycardia cycle length, and the interval from stimulus to atrial activation on CS is 220 ms. The retrograde His bundle potential is found before atrial activation. The stimulus to the A interval (220 ms) is much longer than the QRS–A interval during tachycardia. These findings suggest atrioventricular nodal reentrant tachycardia (AVNRT) with demonstration of lower common pathway during ventricular pacing and exclude the possibility of retrograde accessory pathway in the posteroseptal area.

The success and safety of ablation of AVNRT, AVRT, and other focal atrial tachycardias has offered promise of cure. The prevalence of these arrhythmias in young populations with the prospect of lifelong recurrent SVT has rapidly motivated physicians to offer catheter ablation as definitive first-line treatment after resolution of the acute episode. Drug therapy is generally reserved for acute management of a symptomatic episode, short-term prophylaxis as a bridge to cure, or rarely in patients in whom ablation is contraindicated or not feasible. While device therapy has had a niche role and can be effective in many of these rhythms, it is now largely relegated to an adjunctive role when implantation of a device is contemplated for other clinical indications in a patient with recurrent SVT. Device therapy is most often initiated for other indications, with activation of antitachycardia pacing for coexisting SVT termination. The use of device therapy in patients with failed ablation and drug therapy is an increasingly rare event.

Evidence-Based Therapy

Spontaneous termination of SVT is common and can occur quite early. Waxman et al. concluded that the spontaneous termination of SVT occurred within the AV node. Further pharmacologic testing indicated that the initial hypotension during SVT onset provokes a sympathetic response to raise blood pressure, which in turn enhances vagal tone that can terminate the arrhythmia.[110] Immediate therapy in this SVT usually consists of vagal or other physical maneuvers to terminate the event or the parenteral administration of a short-acting effective AV nodal blocking agent. Commonly used vagal maneuvers include unilateral carotid sinus massage, diving reflex activation with placement of the face in cold water, or a Valsalva maneuver. These are often quite effective and can be performed by the patient or a health professional. In a comparative clinical study, the efficacy of the initial use of the Valsalva maneuver was 19% and of carotid sinus massage was 11%. However, the sequential use of both techniques improved overall success to 28%.[111]

Due to its extremely short half-life, safety, and efficacy, intravenous adenosine has become the drug of choice in the United States. A single peripheral bolus of 6 or 12 mg followed by saline bolus for rapid transit with minimal dilution is employed. In clinical trials, the efficacy of a single 6 mg intravenous bolus was 63%, rising to 91% at 12 mg.[112] Central administration is not more effective than peripheral bolus injection, but lower doses of 3 mg and 6 mg are more effective.[113] Adenosine may result in complete but very transient AV block or sinus arrest in individual patients; it resolves spontaneously within a few seconds and reinitiation of SVT occurs in less than 10% of patients. In a field study, vagal maneuvers were followed by adenosine administration. There was a 90% conversion of PSVT, consistent with dose-ranging studies, and there was no difference in

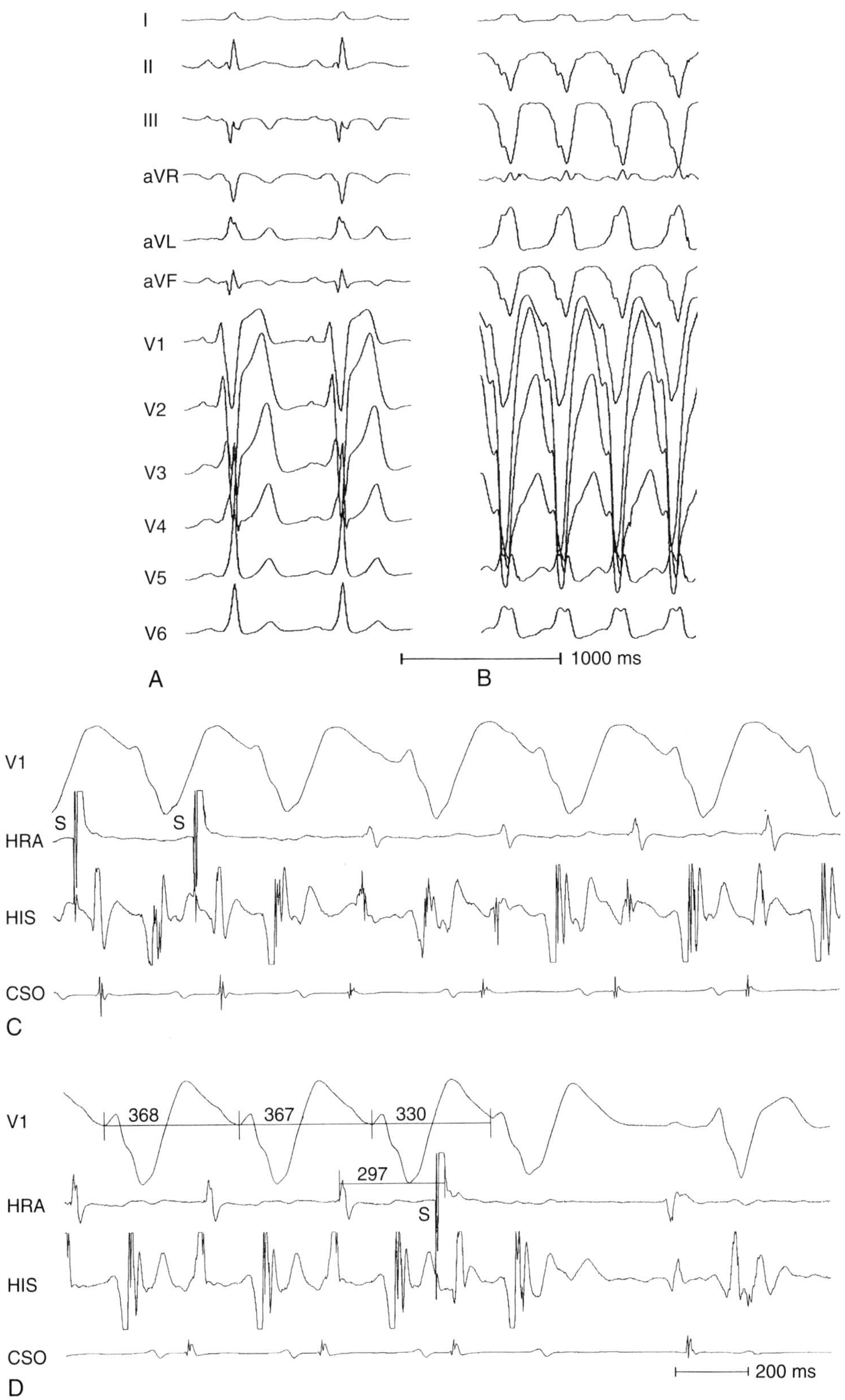

FIGURE 14-20 **A,** A 12-lead ECG during sinus rhythm. **B,** Surface 12-lead ECG of antidromic tachycardia. **C,** Rapid atrial pacing (*S*) induces antidromic atrioventricular (AV) reentry tachycardia with antegrade conduction through right free wall accessory pathway and retrograde conduction through slow AV nodal pathway (earliest retrograde atrial activation in the CS ostium). **D,** One atrial premature beat preexcites the ventricle through the accessory pathway; however, this atrial premature beat (*S*) also collides with retrograde atrial activation, and AV reentry tachycardia is terminated.

the asystolic pause seen in PSVT or AF patients. Adenosine use may also help unmask the underlying mechanism of SVT.[115,116] It is commonly used as a test to block the AV node and elicit accessory pathway conduction during electrophysiological study and may show δ waves transiently in SVT due to AV reentry. In other instances, it may show underlying dual AV nodal physiology in patients with AVNRT.

Adenosine is expensive and other alternative agents that are highly effective include intravenous calcium blockers such as diltiazem or verapamil, type 1 agents or β-blockers. Verapamil is typically administered as a bolus in 5 mg aliquots, and a total of 10 to 15 mg is almost invariably effective. In refractory patients, use of a drug infusion may be helpful. In a direct comparison, the efficacy of intravenous verapamil was 73% in one study and did not differ significantly from adenosine, but hypotension was more common with verapamil.[114] Rarely, fast pathway block with the use of intravenous type 1 agents such as procainamide may be contemplated. Intravenous flecainide has been used in a dose of 2 mg/kg body weight and terminates AVNRT and AVRT with more than 90% efficacy.[117] Electrical reinitiation of PSVT is uncommon and is associated with a markedly slower tachycardia. Comparison of intravenous diltiazem at a dose of 0.2 mg/kg and esmolol at 0.5 mg/kg showed superior efficacy of diltiazem.[118] New intravenous agents include dofetilide, which has been tested in the WPW syndrome in patients with atrial fibrillation and AVRT. The overall efficacy in one study was 71%. Intravenous propafenone was also highly effective in the same patient population in another study.[119]

It is now extremely rare to use electrical antitachycardia therapies for acute conversion of SVT.[76] However, it is important for the treating physician to know that atrial antitachycardia pacing and cardioversion can be used effectively for this purpose. Bursts, programmed extrastimuli, or ramp pacing in the atrium can effectively terminate AVNRT, and this may also be achieved with rapid ventricular pacing in many patients. Thus, in the presence of pacing lead systems on a temporary or permanent basis, any of these modes may effectively terminate an episode of AVNRT without major adverse sequelae in most patients. Induction of atrial fibrillation may occasionally occur with atrial pacing methods, and ventricular tachyarrhythmias may be rarely initiated in a predisposed patient with rapid ventricular pacing. Direct current cardioversion is highly effective but is rarely needed even in emergent circumstances. However, in patients with syncope and very rapid PSVT, this should be considered in emergent circumstances.

Oral single-dose drug therapy to terminate a single arrhythmic event has been gaining currency, particularly in Europe. This is widely promoted in cardioversion of atrial fibrillation. However, this principle can be applied to a single PSVT event that is not highly symptomatic and can be tolerated by the patient. β-Blockers, calcium blockers, and even type 1 drugs can be considered, but in contrast to atrial fibrillation, formal efficacy and safety studies are lacking.

PROPHYLAXIS OF RECURRENT PAROXYSMAL SUPRAVENTRICULAR TACHYCARDIAS

Prophylaxis of recurrent AVNRT has been attempted with drug therapy, ablation, and device therapy. Catheter ablation is the first-line therapy for prophylaxis of PSVT in virtually all populations and age groups, is curative in nature, and uses a variety of techniques.[83,84,105,121,123] It may be avoided in patients who have major contraindications to the procedure such as uncontrolled infections, bleeding diatheses, unstable cardiovascular hemodynamics, and implanted prosthetic heart valves, as well as in very elderly and debilitated patients. However, successful procedures have been performed in our laboratory in patients in their 10th decade of life. In AV nodal tachycardia, the most widely used technique is ablation of the slow AV nodal pathway, while in AVRT, ablation of the accessory bypass tract is performed. The technical details of each technique are described in Section V of this textbook, *Pharmacologic and Interventional Therapies.* Catheter ablation can be performed using map-guided or anatomically based approaches in AV nodal reentry but is invariably performed using map-guided methods for AVRTs, as well as in other preexcitation syndrome substrates. In brief, slow pathway ablation is performed at or above the posterior aspect of the Triangle of Koch just above the coronary sinus ostium. Fast pathway ablation is performed in the superior part of the triangle, just above the AV node–His bundle axis. Specific ablation methods are used for free wall accessory pathways in the left or right atrium, anteroseptal, midseptal, posteroseptal, and paraseptal accessory pathways. Finally, ablation of the nodoventricular and fasciculo-ventricular and atrio-His fibers has also been successfully performed with map-guided methods. The success of these techniques in suppression of recurrent SVT exceeds 90%, with complication rates of 1% or less.[120,121] Mortality is rare with inadvertent complete AV block and myocardial perforation being the most important major complications, estimated at 1% to 2%. Successful cure of AVRT can vary with ablation site, with left free wall and posteroseptal or paraseptal pathways showing higher efficacy rates than right free wall or right anteroseptal pathways. Alternative ablation techniques for AV nodal tachycardia include ablation of the fast AV nodal pathway. Fast pathway ablation can be equally effective in cure of AV nodal reentry but is associated with a higher complication rate with respect to AV nodal block, approaching 5% in some series. Similarly, ablation of intermediate septal pathways in the WPW syndrome can have a similar risk of complete AV block. It is rarely necessary and generally inadvisable to perform complete AV nodal ablation in either arrhythmia. In a comparative clinical trial, catheter ablation provided superior control of symptoms and better quality of life than drug therapy in patients with recurrent PSVT.

Prospective clinical studies have documented a high degree of efficacy with catheter ablation in this condition. Tables 14-7 and 14-8 show the NASPE voluntary registry of 3423 procedures performed at 68 centers in the United States. It reports efficacy rates for AV

TABLE 14-7 Percentage of Successful Ablations and Complications for Teaching Versus Nonteaching Hospitals

Procedure	Teaching Hospital, % Success	Nonteaching Hospital, % Success
Atrioventricular junction	171/176, 97.2	458/470, 97.4
Atrioventricular nodal reentrant tachycardia	456/476, 95.8	705/732, 93.2
AP (total)	255/275, 92.7	372/399, 93.2
Left free wall	160/172, 93.6	232/247, 93.9
Right free wall	26/27, 96.3	54/56, 96.4
Septal	75/83, 90.4	90/103, 87.4

junctional ablation, AVNRT ablation, and accessory pathway ablation by tract location at teaching and nonteaching hospitals in the United States.[120] Note that efficacy rates exceed 90% for all arrhythmias, and there is virtually no significant difference in outcome by location. In addition, comparison of data for large-volume (>100 procedures/yr) and lower-volume centers did not show differences, indicating that the learning curve for the procedure was over. There was a slightly lower success rate with right free wall or septal pathways compared with free wall pathways. Significant complications in this series of patients undergoing AV junction ablation included a very low mortality and a small incidence of sudden death due to polymorphous VT after ablation and cardiac pacing at relatively rapid rates is recommended for several weeks after AV junction ablation. For patients undergoing slow pathway ablation, the risk of second-degree AV block was 0.16% and for complete heart block was 0.74%. Complications included a low incidence of cardiac tamponade, AV block, and rare instances of coronary occlusion and pulmonary embolism. The overall complication rate for this procedure is estimated at less than 1%.

Drug therapy has been largely relegated to temporary, intermittent, or second-line choice. Digoxin therapy, long established for this purpose, has now been supplanted by more effective β-blocker, calcium blocker, and types 1C and III drug therapy in the last 2 decades. In prospective clinical trials, oral flecainide therapy was associated with an actuarial 79% to 82% freedom from symptomatic PSVT events compared with only 15% on placebo at 60 days ($P < .001$).[124] The median time to the first symptomatic PSVT event was greater in the flecainide group, and the interval between attacks was increased by flecainide. Similarly, propafenone reduced recurrent SVT in prospective studies by 80%.[125-126] Oral verapamil was shown to be effective in the prophylaxis of SVT in comparison with placebo, with reduced need for pharmacologic cardioversion and event rates.[127] In a comparison with flecainide therapy, it was equally effective and well tolerated. In one study, both drugs showed a marked reduction in the frequency of attacks of PSVT, with a small advantage for flecainide.[128] Thirty percent of patients on flecainide had resolution of symptomatic attacks versus 13% of the patients on verapamil ($P = .026$). Eleven percent of patients discontinued flecainide, and 19% discontinued verapamil for inefficacy at 1 year. Both drugs were well tolerated; 19% of the flecainide group discontinued therapy due to adverse effects, compared with 24% discontinuing verapamil for this reason. Randomized controlled trials of propafenone with placebo show a sixfold increase in time to first PSVT recurrence at a dose of 600 mg/day.[126] While the higher dose of 900/mg was even more effective if tolerated, there was a significant increase in adverse effects. Comparative studies have shown propafenone and flecainide to have similar efficacy and safety.[126] Newer agents include dofetilide, azimilide, and sotalol.[129-131] While data on dofetilide are limited, at a dose of 500 μg/day, dofetilide is equally effective as propafenone at a relatively low dose of 450 mg/day. The probability of freedom from recurrent PSVT was 50% and 54%, respectively, with a 6% probability on placebo. Thus, these efficacy rates remain well below levels seen with catheter ablation and relegate this approach to a second line of therapy. In addition, the improvement of quality of life with catheter ablation is superior to medical therapy.[132] The improvement in quality of life was seen in patients with moderate or severe symptoms due to PSVT. The safety profile of type 1c agents in the elderly is also of concern due to the risk of proarrhythmia.[133]

TABLE 14-8 Percentage of Successful Ablations and Complications for Medical Centers Handling More or Fewer than 100 Cases in 1998

Procedure	With >100 Cases, % Success	With <100 Cases, % Success
Atrioventricular junction	354/366, 96.7	275/280, 98.2
Atrioventricular nodal reentrant tachycardia	603/627, 96.2	558/581, 95.8
AP (total)	315/339, 92.9	312/335, 93.1
Left free wall	201/216, 93.1	191/202, 94.6
Right free wall	37/39, 94.9	43/44, 97.7
Septal	144/160, 90.0	85/98, 86.7

Surgical Ablation of Atrioventricular Nodal Reentrant Tachycardia

Surgical ablation of PSVT has been well developed since the 1980s. Established and progressively refined techniques have been used since that period. The surgical basis for SVT ablation involves resection or modification of the substrate for the arrhythmia applying and actually pioneering many of the principles that have come in vogue in catheter ablation procedures. Experimental studies have refined surgical techniques since the original description of SVT control by ligature of the His bundle.[132] Since that time, complete AV block has been avoided, and modification of selective pathways with preservation of normal AV conduction has been pursued. In animal studies, McGuire and coworkers have disconnected the anterior part of the

AV node and showed mild impairment of AV nodal conduction, and ventriculo-atrial conduction was destroyed in 50% of the animals.[36] This helped define the contribution of each of these connections to AV conduction patterns in this animal model. It also suggested that the fast pathway is the preferred route of conduction through the AV node, but other inputs can assume these responsibilities with limited loss of efficacy. This became the basis for one surgical approach to cure AVNRT by fast pathway interruption. More recently, slow pathway interruption by surgical or catheter ablation methods has become the procedure of choice, with lesser opportunity of impaired AV conduction and a high degree of efficacy.[83,84] However, in failures, ablation alternatives such as elimination of the fast pathway or left-sided connections should be considered.

Accessory pathway ablation using an endocardial or epicardial approach using linear incision at the mapped pathway location was widely used in the 1980s until it was supplanted by catheter ablation.[103,133] Near-perfect efficacy has been described in the largest series with mortality rates of less than 1% in patients without cardiac disease or other surgical procedures.[134] The details of the mapping and surgical technique are beyond the purview of this chapter and are described elsewhere in this text or in referenced literature. Suffice it to state that the epicardial approach to accessory pathway eliminated the need for cardiopulmonary bypass, and cryoablation of the AV groove resulted in a safe and highly effective procedure. Other energy sources such as laser have also been applied clinically.[135] The usefulness of this technique lies in the ability to perform curative ablation procedures in PSVT patients undergoing cardiac surgery for other indications and in patients with failed catheter ablation attempts, particularly if multiple accessory pathways are present. The role of this approach in sino-atrial node reentry or nonparoxysmal sinus tachycardia or other atrial tachycardias is less well established. While surgical success in select patients has been reported, this is not as effective an approach in these substrates.[136]

Device Therapy

The use of antitachycardia devices for PSVT termination and prevention has become of historical significance except in a few specific situations with newer devices. It is mentioned here for completion, as well as for limited use in future devices. Early studies established the efficacy of antitachycardia pacemakers in recurrent PSVT.[137-140] Atrial burst and ramp pacing have been shown to be highly effective in termination of PSVT and common atrial flutter with implanted pacemaker devices.[141] Prevention of PSVT in different substrates has also been shown. Permanent dual-chamber pacing with a short A–V interval can prevent reentrant PSVT due to collision of the atrial and ventricular wavefronts in the critical elements of the PSVT circuit. This pacing mode has been employed effectively for this purpose. Long-term management of PSVT is feasible with automatic and patient activated antitachycardia pacing.[138-140] Current devices have these pacing modes available in their repertoire. The use of these modes may be considered in patients with implanted pacemakers or pacemaker defibrillators for other arrhythmia indications, who have coexisting PSVT of modest frequency. Incessant or highly frequent PSVT should be considered for ablative therapy.

REFERENCES

1. Cotton RP: Notes and observations of unusually rapid action of the heart. BMJ 1867;I:629.
2. Bouveret 1888, quoted by Froment R: Précis de Cardiologie Clinique. Masson et Cie 1962;238-42.
3. Surawicz B, Uhley H, Borun R, et al: Task Force 1: Standardization of terminology and interpretation. Am J Cardiol 1978;41:130-45.
4. Wolff L, Parkinson W, White PD: Bundle branch block with short P–R interval in healthy young people prone to paroxysmal tachycardia. Am Heart J 1930;5:685-704.
5. Lown B, Ganong WF, Levine SA: The syndrome of short PR interval, normal QRS complex and paroxysmal rapid heart action. Circulation 1952;5:693-706.
6. Brodsky M, Wu D, Denes P, et al: Arrhythmias documented by 24 hour continuous electrocardiographic monitoring in 50 male medical students without apparent heart disease. Am J Cardiol 1977;39:390-5.
7. Hinkle LE Jr, Carver ST, Stevens M: The frequency of symptomatic disturbances of cardiac rhythm. Am J Cardiol 1969;24:629-50.
8. Clark JM, Hamer J, Shelton JR, et al: The rhythm of the normal human heart. Lancet 1976;2:508-12.
9. Harrison DC, Fitzgerald JW, Winle RA: Contribution of ambulatory electrocardiographic monitoring to antiarrhythmic management. Am J Cardiol 1978;41:996-1004.
10. Roche F, Gaspoz JM, Da Costa A, et al: Frequent and prolonged asymptomatic episodes of paroxysmal atrial fibrillation revealed by automatic long-term event recorders in patients with a negative 24-hour Holter. Pacing Clin Electrophysiol 2002;25:1587-93.
11. Manolio TA, Furberg CD, Rautaharju PM, et al: Cardiac arrhythmias on 24-h ambulatory electrocardiography in older women and men: The Cardiovascular Health Study. J Am J Cardiol 1994;23:916-25.
12. Berkman NL, Lamb LE: The Wolff-Parkinson-White electrocardiogram: A follow-up study of 5 to 28 years. N Engl J Med 1968;278:492-4.
13. Dreifus LS, Wellens HJJ, Watanabe Y, et al: Sinus bradycardia and atrial fibrillation associated with the Wolff-Parkinson-White syndrome. Am J Cardiol 1976;38:149-56.
14. Klein GJ, Bashore TM, Sellers TD, et al: Ventricular fibrillation in the Wolff-Parkinson-White syndrome. N Engl J Med 1979;301:1080-5.
15. Mantakas ME, McCue CM, Miller WW: Natural history of Wolff-Parkinson-White syndrome discovered in infancy. Am J Cardiol 1978;41:1097-1103.
16. Bharati S: Pathology of the conduction system. In Silver MD, Gotlieb AI, Schoen FJ (eds): Cardiovascular Pathology. New York, Churchill Livingston, 2001, pp 607-28.
17. Bharati S: Anatomic-morphologic relation between AV nodal structure and function in the normal and diseased heart. In Mazgalev TN, Tchou PJ (eds): Atrial-AV Nodal Electrophysiology: A View from the Millennium. Armonk, NY, Futura, 2000, pp 25-48.
18. Bharati S, Lev M: The anatomy of the normal conduction system: Disease-related changes and their relationship to arrhythmogenesis. In Podrid PJ, Kowey PR (eds): Cardiac Arrhythmias, Mechanism, Diagnosis and Management. Baltimore, Williams & Wilkins, 1995, p 1.
19. Bharati S, Lev M: The anatomy and histology of the conduction system. In Chung EK (ed): Artificial Cardiac Pacing: Practical Approach. Baltimore, Williams & Wilkins, 1984, p 12.
20. Lev M, Bharati S: A method of study of the pathology of the conduction system for electrocardiographic and His bundle electrogram correlations. Anat Rec 1981;201:43.

21. Lev M, Bharati S: Lesions of the conduction system and their functional significance. In Sommers SC (ed): Pathology Annual. New York, Appleton-Century-Crofts, 1974, p 157.
22. Lev M: The conduction system. In Gould SC, Charles C (ed): Pathology of the Heart and Blood Vessels. Springfield, Ill, Charles C Thomas, 1968, p 180.
23. Wildran J, Lev M: The dissection of the human AV node, bundle and branches. Circulation 1951;4:863.
24. Truex RC, Smythe MO: Recent observations on the human cardiac conduction system with special considerations of the atrioventricular node and bundle. In Taccardi B (ed): Electrophysiology of the Heart. New York, Pergamon Press, 1965, pp 177-98.
25. Truex RC, Bishof JK, Hoffman EL: Accessory atrioventricular muscle bundles of the developing human heart. Anat Rec 1958;131:45.
26. Bharati S, Lev M: The morphology of the AV junction and its significance in catheter ablation (editorial). Pacing Clin Electrophysiol 1989;12:879.
27. Randall WC: Differential autonomic control of SAN and AVN regions of the canine heart: Structure and function. In Liss A (ed): Progress in Clinical and Biological Research. Electrophysiology of the Sinoatrial and Atrioventricular Nodes. New York, 1988, 275:15.
28. Moe GK, Preston JB, Burlington H: Physiologic evidence for a dual A-V transfusion symptom. Circ Res 1956;357-75.
29. Durrer D, Schoo L, Schulenberg RM, Wellens HJJ: The role of premature beats in the initiation and termination of supraventricular tachycardia in the Wolff-Parkinson-White syndrome. Circulation 1967;36:644-62.
30. Castellanos A Jr, Chapuroff E, Castello O, et al: His Bundle electrograms in two cases of Wolff-Parkinson-White (pre-excitation) syndrome. Circulation 1970;41:399-411.
31. Denes P, Wu D, Dhingra RC, et al: Demonstration of dual AV nodal pathways in patients with paroxysmal supraventricular tachycardia. Circulation 1973;48:549-55.
32. Tawara S: Das Reizleitungssystem des Saugetierherzens. Eine anatomisch-histologische Studie uber das Atrioventricularbundel und die Purkinjeschen Faden. Mit einerm Vorwrt von L. Aschoff (Marburg), Fischer, Jena.
33. Kent AFS: Researches on the structure and function of the mammalian heart. J Physiol 1893;14:233-54.
34. Anderson R, Becker A, Brechenmacher C, et al: The human atrioventricular junctional area: A morphological study of the AV node and the bundle. Eur J Cardiol 1975;3:11-8.
35. Racker DK: Atrioventricular node and input pathways: A correlated gross anatomical and histological study of the canine atrioventricular junctional region. Anat Rec 1989;224:336.
36. McGuire MA, de Bakker JM, Vermuelen JT, et al: Origin and significance of double potentials near the atrioventricular node: Correlation of extracellular potentials, intracellular potentials and histology. Circulation 1994;89:2351-60.
37. Goldreyer BN, Bigger JT Jr: Site reentry in paroxysmal supraventricular tachycardia in man. Circulation 1971;43:15-26.
38. Sung RJ, Waxman H, Saksena S, et al: Sequence of retrograde atrial activation in patients with dual atrioventricular nodal pathways. Circulation 1981;64:1053-60.
39. McGuire MA, Bourke JP, Robinson MC, et al: High resolution mapping of Koch's triangle using sixty electrodes in humans with atrioventricular junctional (AV nodal) reentrant tachycardia. Circulation 1993;88:2315-28.
40. Gallagher JJ, Pritchett ELC, Sealy WC, et al: The preexcitation syndromes. Prog Cardiovasc Dis 1978;20:285-327.
41. Bigger JT, Sahar DL: Clinical types of proarrhythmic response to antiarrhythmic drugs. Am J Cardiol 1987;59:2E-9E.
42. Bash SE, Shah JJ, Albers WH, Geiss DM: Hypothermia for the treatment of post surgical greatly accelerated junctional ectopic tachycardia. J Am Coll Cardiol 1987;10:1095-9.
43. Brembilla Perot B, Marcon F, Bosser G, et al: Paroxysmal tachycardia in children and teenagers with normal sinus rhythm and without heart disease pacing. Pacing Clin Electrophysiol 2001;24:41-5.
44. Saksena S, An H: Electrophysiologic mechanisms underlying termination and prevention of supraventricular tachyarrhythmias. In Saksena S, Goldschlager N (eds): Electrical Therapy for Cardiac Arrhythmias. Philadelphia, WB Saunders, 1990, pp 384-94.
45. Kalbfleisch SJ, El-Atassi R, Calkins H, et al: Differentiation of paroxysmal narrow QRS complex tachycardias using the 12-lead electrocardiogram. J Am Coll Cardiol 1993;21:85-9.
46. Green M, Heddle B, Dassen W, et al: Value of QRS alternation in determining the site of origin of narrow QRS supraventricular tachycardia. Circulation 1983;68:368-73.
47. Kay GN, Pressley JC, Packer DL, et al: Value of the 12-lead electrocardiogram in discriminating atrioventricular nodal reciprocating tachycardia from circus movement atrioventricular tachycardia utilizing a retrograde accessory pathway. Am J Cardiol 1987;59:296-300.
48. Morady F: Significance of QRS alternans during narrow QRS tachycardias. PACE 1991;14:2193-8.
49. Tang CW, Scheinman MM, Van Hare GF, et al: Use of P wave configuration during atrial tachycardia to predict site of origin. J Am Coll Cardiol 1995;26:1315-24.
50. Morton JB, Sanders P, Das A, et al: Focal atrial tachycardia arising from the tricuspid annulus: Electrophysiologic and electrocardiographic characteristics. J Cardiovasc Electrophysiol 2001;12: 653-9.
51. Tada H, Nogami A, Naito S, et al: Simple electrocardiographic criteria for identifying the site of origin of focal right atrial tachycardia. PACE 1998;21:2431-9.
52. Coumel P, Cabrol C, Fabiato A, et al: Tachycardie permanente par rhythme réciproque. Arch Mal Coeur Vaiss 1967;60:1830-64.
53. Coumel P, Attuel P: Reciprocating tachycardia in overt and latent preexcitation. Influence of functional bundle branch block on the rate of the tachycardia. Eur J Cardiol 1974;423-36.
54. Bardy GH, Packer DL, German LD, Gallagher JJ: Preexcited reciprocating tachycardia in patients with Wolff-Parkinson-White syndrome: Incidence and mechanisms. Circulation 1984;377-91.
55. Kuck KH, Brugada P, Wellens HJJ: Observations on the antidromic type of circus movement tachycardia in the Wolff-Parkinson-White syndrome. J Am Coll Cardiol 1983;2:1003-10.
56. Montoya PT, Brugada P, Smeets J, et al: Ventricular fibrillation in the Wolff-Parkinson-White syndrome. Eur Heart J 1991;12:144-50.
57. Bardy GH, Fedor JM, German LD, et al: Surface electrocardiographic clues suggesting presence of a nodofascicular Mahaim fiber. J Am Coll Cardiol 1984;3:1161-8.
58. Tonkin AM, Wagner GS, Gallagher JJ, et al: Initial forces of ventricular depolarization in the Wolff-Parkinson-White syndrome. Circulation 1975;52:1030-5.
59. Fontaine FG, Guiradon G, Cabrol C, et al: Correlation entr l'orientation de l'onde delta et la topographie de la pre-excitation dans le syndrome de Wolff-Parkinson-White. Arch Mal Coeur Vaiss 1977;5:441-50.
60. Lindsay BD, Crossen KJ, Cain ME: Concordance of distinguishing electrocardiographic features during sinus rhythm with the location of accessory pathways in the Wolff-Parkinson-White syndrome. Am J Cardiol 1987;59:1093-102.
61. Rodriguez LM, Smeets JLMR, deChilou C, et al: The 12-lead electrocardiogram in midseptal, anteroseptal, posteroseptal and right free wall accessory pathways. Am J Cardiol 1993;72:1274-80.
62. Iwa T, Iwase T: Localization of accessory pathway conduction in the Wolff-Parkinson-White syndrome by electrocardiogram. Heart (Japan) 1984;16:225-32.
63. Arruda MS, McCleland JH, Wang X, et al: Development and validation of an ECG algorithm for identifying accessory pathway ablation site in Wolff-Parkinson-White syndrome. J Cardiovasc Electrophysiol 1998;9:2-12.
64. Chiang CE, Chen SA, Tai CT, et al: Prediction of successful ablation site of concealed posteroseptal accessory pathways by a novel algorithm using baseline electrophysiological parameters: Implications for an abbreviated ablation procedure. Circulation 1996;93:982-91.
65. Fananapazir L, German LD, Gallagher JJ, et al: Importance of preexcited QRS morphology during induced atrial fibrillation to the diagnosis and localization of multiple accessory pathways. Circulation 1990;81:578-85.
66. Marriott HJL: Systematic approach to diagnosis of arrhythmias. In Practical Electrocardiography, 7th ed. Baltimore, William & Wilkins, 1984, pp 109-25.
67. Davies DW, Toff WD: Diagnosis of cardiac arrhythmias from the surface electrocardiogram. In Camm AJ, Ward DE (eds): Clinical Aspects of Cardiac Arrhythmias. London, Kluwer Academic, 1988, pp 95-117.

68. Lévy S: Diagnostic approach to cardiac arrhythmias. J Cardiovasc Pharmacol 1991;17(suppl 6):524-31.
69. Coumel P, Attuel P, Motte G, et al: Paroxysmal junctional tachycardia. Determination of the inferior point of junction of the reentry circuit. Dissociation of the intra-nodal reciprocal rhythms. Arch Mal Coeur Vaiss 1975;68:1255-68.
70. Morillo CA, Klein GJ, Thakur RK, et al: Mechanism of inappropriate sinus tachycardia, role of sympathovagal balance. Circulation 1994;90:873-7.
71. Akhtar M, Jazayeri MR, Sra J, et al: Atrioventricular re-entry. Clinical, electrophysiological and therapeutic considerations. Circulation 1993;88:282-95.
72. Farré J, Wellens HJJ: The value of the electrocardiogram in diagnosing the site of origin and mechanism of supraventricular tachycardia. In Wellens HJJ, Kulbertus HG (eds): What's New in Electrocardiography? The Hague, Martinus Nijhoff, 1981, pp 131-71.
73. Slama R, Coumel P, Bouvrain Y: Tachycardies paroxystiques liées à un syndrome de Wolff-Parkinson-White inapparent. Arch Mal Coeur Vaiss 1973;66:639-53.
74. Viskin S, Fish R, Glick A, et al: The adenosine triphosphate test: A bedside diagnostic tool for identifying the mechanism of supraventricular tachycardia inpatients with palpitations. J Am Coll Cardiol 2001;38:173-7.
75. Josephson ME: Paroxysmal supraventricular tachycardia: An electrophysiologic approach. Am J Cardiol 1978;41:1122-26.
76. Wellens HJJ: Contribution of cardiac pacing to our understanding of the Wolff-Parkinson-White syndrome. Br Heart J 1975;37:231-41.
77. Akhtar M: Supraventricular tachycardias: Electrophysiologic mechanisms, diagnosis and pharmacologic therapy. In Josephson ME, Wellens HJJ (eds): Tachycardias: Mechanisms, diagnosis, and treatment. Philadelphia, Lea & Febiger, 1984, p 137.
78. Camm AJ: The recognition and management of tachycardias. In Julian DG, Camm AJ, Fox KM, et al (eds): Diseases of the Heart. London, Balliere Tindal, 1989, pp 509-83.
79. Wellens HJJ, Bar FW, Vanagt EJ, et al: The differentiation between ventricular tachycardia and supraventricular tachycardia with aberrant conduction: The value of the 12-lead electrocardiogram. In Wellens HJJ, Kulbertus HG (eds): What's New in Electrocardiography? The Hague, Martinus Nijhoff, 1981, pp 164-99.
80. Wu D, Denes P, Amat-y-Leon F, et al: An unusual variety of atrioventricular node reentry due to dual atrioventricular nodal pathways. Circulation 1977;56:50-9.
81. Sung RJ, Styperek JL, Myerburg RJ, et al: Initiation of two distinct forms of atrioventricular nodal reentrant tachycardia during programmed ventricular stimulation in man. Am J Cardiol 1978;43:404-15.
82. Ross DL, Johnson DC, Denniss AR, et al: Curative surgery for atrioventricular junctional ("AV nodal") reentrant tachycardia. J Am Coll Cardiol 1985;6:1383-92.
83. Jackman WM, Beckman KJ, McClelland JH, et al: Treatment of supraventricular tachycardia due to atrioventricular nodal reentry by radiofrequency ablation of slow-pathway conduction. N Engl J Med 1992;32:313-16.
84. Haissaguerre M, Gaita F, Fischer B, et al: Elimination of atrioventricular nodal reentrant tachycardia using discrete slow potentials to guide application of radiofrequency energy. Circulation 1992;85:2162-75.
85. McGuire MA, Lau KC, Johnson DC, et al: Patients with two types of atrioventricular junctional (AV nodal) reentrant tachycardia; evidence that a common pathway of nodal tissue is not present above the reentrant circuit. Circulation 1991;83:1232-46.
86. Yeh SJ, Wang CC, Wen MS, et al: Radiofrequency ablation therapy in atypical or multiple atrioventricular node reentry tachycardias. Am Heart J 1994;128:742-58.
87. Tai CT, Chen SA, Chiang CE, et al: Complex electrophysiological characteristics in atrioventricular nodal reentrant tachycardia with continuous atrioventricular node function curves. Circulation 1997;95:2541-7.
88. Tai CT, Chen SA, Chiang CE, et al: Multiple anterograde atrioventricular node pathways in patients with atrioventricular node reentrant tachycardia. J Am Coll Cardiol 1996;28:725-31.
89. Wellens HJJ, Wesdorp JC, Duren DR, et al: Second degree block during reciprocal atrioventricular nodal tachycardia. Circulation 1976;53:595-9.
90. Lee SH, Chen SA, Tai CT, et al: Electrophysiologic characteristics and radiofrequency catheter ablation in atrioventricular node reentrant tachycardia with second-degree atrioventricular block. J Cardiovasc Electrophysiol 1997;8:502-11.
91. Tai CT, Chen SA, Chiang CE, et al: Electrophysiologic characteristics and radiofrequency catheter ablation in patients with multiple atrioventricular nodal reentry tachycardias. Am J Cardiol 1996;77:52-8.
92. Gallagher JJ, Kasell J, Sealy WC, et al: Epicardial mapping in the Wolff-Parkinson-White syndrome. Circulation 1978;57:854-66.
93. Jackman WM, Friday KJ, Yeung LW, et al: New catheter technique for recording left free-wall accessory atrioventricular pathway activation: Identification of pathway fiber orientation. Circulation 1988;78:598-610.
94. Tai CT, Chen SA, Chiang CE, et al: Identification of fiber orientation in left free-wall accessory pathways: Implication for radiofrequency ablation. J Interv Card Electrophysiol 1997;1:235-41.
95. Chen SA, Tai CT, Lee SH, et al: Electrophysiologic characteristics and anatomical complexities of accessory atrioventricular pathways with successful ablation of anterograde and retrograde conduction at different sites. J Cardiovasc Electrophysiol 1996;7:907-15.
96. Gallagher JJ, Sealy WC: The permanent form of junctional reciprocating tachycardia: Further elucidation of the underlying mechanism. Eur Heart J 1978;8:413-20.
97. Farre J, Ross D, Wiener I, et al: Reciprocal tachycardia using accessory pathways with long conduction time. Am J Cardiol 1979;44:1099-109.
98. Tchou P, Lehmann MH, Jazayeri M, et al: Atriofascicular connection or a nodoventricular mahaim fiber? Electrophysiologic elucidation of the pathway and associated reentrant circuit. Circulation 1988;77:837-48.
99. Chen SA, Tai CT, Chiang CE, et al: Electrophysiologic characteristics, electropharmacologic responses and radiofrequency ablation in patients with decremental accessory pathway. J Am Coll Cardiol 1996;28:732-37.
100. Klein LS, Hackett FK, Zipes DP, et al: Radiofrequency catheter ablation of Mahaim fibers at the tricuspid annulus. Circulation 1993;87:738-47.
101. McClelland JH, Wang X, Beckman KJ, et al: Radiofrequency catheter ablation of right atriofascicular (Mahaim) accessory pathways guided by accessory pathway activation potentials. Circulation 1994;89:2655-66.
102. Cappato R, Schluter M, Weiss C, et al: Catheter-induced mechanical conduction block of right-sided accessory fibers with Mahaim-type preexcitation to guide radiofrequency ablation. Circulation 1994;90:282-90.
103. Selle JG, Sealy WC, Gallagher JJ, et al: Technical considerations in the surgical approach to multiple accessory pathways in the Wolff-Parkinson-white syndrome. Ann Thorac Surg 1987;43:579-84.
104. Yeh SJ, Wang CC, Wen MS, et al: Catheter ablation using radiofrequency current in Wolff-Parkinson-White syndrome with multiple accessory pathways. Am J Cardiol 1992;71:1174-80.
105. Huang JL, Chen SA, Tai CT, et al: Long-term results of radiofrequency catheter ablation in patients with multiple accessory pathways. Am J Cardiol 1996;78:1375-9.
106. Csanadi Z, Klein GJ, Yee R, et al: Effect of dual atrioventricular node pathways on atrioventricular reentrant tachycardia. Circulation 1995;91:2614-8.
107. Hirao K, Otomo K, Wang X, et al: Para-Hisian pacing. A new method for differentiating retrograde conduction over an accessory AV pathway from conduction over the AV node. Circulation 1996;94:1027-35.
108. Packer DL, Gallagher JJ, Prystowsky EN: Physiological substrate for antidromic reciprocating tachycardia—prerequisite characteristics of the accessory pathway and atrioventricular conduction system. Circulation 1992;85:574-88.

109. Luria DM, Chugh SS, Munger TM, et al: Electrophysiologic characteristics of diverse accessory pathway locations of antidromic reciprocating tachycardia. Am J Cardiol 2000;86: 1333-8.
110. Waxman MB, Sharma AD, Cameron DA, et al: Reflex mechanisms responsible for early spontaneous termination of paroxysmal supraventricular tachycardia. Am J Cardiol 1982;49: 259-72.
111. Lim SH, Anantharaman V, Teo WS, et al: Comparison of the treatment of supraventricular tachycardia by Valsalva maneuver and carotid sinus massage. Ann Emerg Med 1998;31:30-5.
112. DiMarco JP, Miles W, Akhtar M, et al: Adenosine for paroxysmal supraventricular tachycardia: Dose ranging and comparison with verapamil. Assessment in placebo-controlled, multicenter trials. The Adenosine for PSVT Study Group. Ann Intern Med 1990;113:104-10.
113. McIntosh-Yellin NL, Drew BJ, Scheinman MM: Safety and efficacy of central intravenous bolus administration of adenosine for termination of supraventricular tachycardia. J Am Coll Cardiol 1993;22:741-5.
114. Hood MA, Smith WM: Adenosine versus verapamil in the treatment of supraventricular tachycardia: A randomized double-crossover trial. Am Heart J 1992;123:1543-9.
115. Dierkes S, Vester EG, Dobran LJ, et al: Adenosine in the noninvasive diagnosis of dual AV nodal conduction: Use as a follow-up parameter after slow pathway ablation in AVNRT. Acta Cardiol 2001;56:103-8.
116. Belhassen B, Fish R, Viskin S, et al: Adenosine-5'-triphosphate test for the noninvasive diagnosis of concealed accessory pathway. J Am Coll Cardiol 2000;36:803-10.
117. Gambhir DS, Bhargava M, Arora RK, Khahilullar M: Electrophysiologic effects and therapeutic efficacy of intravenous flecainide for termination of paroxysmal supraventricular tachycardia. Indian Heart J 1995;47:237-43.
118. Gupta A, Nail A, Vora A, Lokhandwala Y: Comparison of the efficacy of intravenous diltiazem and esmolol in terminating supraventricular tachycardia. J Assoc Physicians India 1999;47: 969-72.
119. Reimold SC, Maisel WH, Antman EM: Propafenone for the treatment of supraventricular tachycardia and atrial fibrillation: A meta analysis. Am J Cardiol 1998;82:66N-71N.
120. Melvin Scheinman, Hugh Calkins, Paul Gillette, et al: NASPE policy statement on catheter ablation: Personnel, policy, procedures, and therapeutic recommendations. Pacing Clin Electrophysiol 2003;26:789-99.
121. Kugler JD, Dansford DA, Houson K, Felix G: Radiofrequency catheter ablation for paroxysmal supraventricular tachycardia in children and adolescents without mitral heart disease. Pediatric EP Society Radiofrequency catheter ablation registry. Am J Cardiol 1997;80:1438-43.
122. Tai CT, Chen SA, Chang CE, et al: Accessing atrioventricular pathways and atrioventricular nodal reentrant tachycardia in teenagers. Electrophysiologic characteristics and radiofrequency catheter ablation. Jpn Heart J 1995;36:305-17.
123. Lau CP, Tai YL, Lee PW: The effects of radiofrequency ablation versus medical therapy on the quality of life and exercise capacity in patients with supraventricular tachycardia: A treatment comparison study. Pacing Clin Electrophysiol 1995;18:424-32.
124. Henthorn RW, Waldo AL, Anderson JL, et al: Flecainide acetate prevents recurrence of symptomatic paroxysmal supraventricular tachycardia. The Flecainide Supraventricular Tachycardia Study Group. Circulation 1991;83:119-25.
125. Benditt DG, Dunnigan A, Buetikofer J, Milstein S: Flecainide acetate for long-term prevention of paroxysmal supraventricular tachyarrhythmias. Circulation 1991;83:345-9.
126. Pritchett EL, McCarthy EA, Wilkinson WE: Propafenone treatment of symptomatic paroxysmal supraventricular arrhythmias. A randomized, placebo-controlled, crossover trial in patients tolerating oral therapy. Ann Intern Med 1991;114:539-44.
127. Mauritson DR, Winniford MD, Walker WS, et al: Oral verapamil for paroxysmal supraventricular tachycardia: A long-term, double-blind randomized trial. Ann Intern Med 1982;96:409-12.
128. Dorian P, Naccarelli GV, Coumel P, et al: A randomized comparison of flecainide versus verapamil in paroxysmal supraventricular tachycardia. The Flecainide Multicenter Investigators Group. Am J Cardiol 1996;77:89A-95A.
129. Tendera M, Wuuk-Wojanr AM, Kulakowski P, et al: Efficacy and safety of dofetilide in the prevention of symptomatic episodes of paroxysmal supraventricular tachycardia: A 6 month double blind comparison with propafenone and placebo. Am Heart J 2001;142:93-8.
130. Page RL, Connolly SJ, et al: Antiarrhythmic effects of Azimilide in paroxysmal supraventricular tachycardia: Efficacy and dose response. Am Heart J 2002;143:643-9.
131. Wanless RS, Anderson K, Joy M, Joseph SP: Multicenter comparative study of the efficacy of sotalol in the prophylactic treatment of patients with paroxysmal supraventricular tachycardia. Am Heart J 1997;133:441-6.
132. Dreifus LS, Nichol H, Morse DH, et al: Control of recurrent tachycardia of Wolff-Parkinson-White syndrome by surgical ligature of the A-V bundle. Circulation 1968;38:1030-6.
133. Gallagher JJ, Sealy WC, Anderson RW, et al: Cryosurgical ablation of accessory atrioventricular connections: A method for correction of the pre-excitation syndrome. Circulation 1977; 55:471-479.
134. Guiradon GM, Klein GJ, Sharma AD, et al: Surgery of the Wolff-Parkinson-White syndrome: The epicardial approach. Semin Thorac Cardiovasc Surg 1989;1:21-33.
135. Saksena S, Hussain SM, Gielchinsky I, Pantopoulos D: Intraoperative mapping-guided argon laser ablation of supraventricular tachycardia in Wolff-Parkinson-White syndrome. Am J Cardiol 1987;60:196-9.
136. Yee R, Guiraudon GM, Gardner MJ, et al: Refractory paroxysmal sinus tachycardia management by subtotal right atrial exclusion. JACC 1984;3:400-4.
137. Castellanos A, Waxman HL, Moleiro F, et al: Preliminary studies with an implantable multimode A-V pacemaker for reciprocating atrioventricular tachycardias. Pacing Clin Electrophysiol 1980;3:257-60.
138. Saksena S, Pantopoulos D, Parsonnet V, et al: Usefulness of an implantable antitachycardia pacemaker system for supraventricular or ventricular tachycardia. Am J Cardiol 1986;58:70-4.
139. Jung W, Mletzko R, Manz M, Luderitz B: Clinical results of chronic antitachycardia pacing in supraventricular tachycardia. In Luderitz B, Saksena S (eds): Interventional Electrophysiology. Mt Kisco, NY, Futura, 1992, pp 197-211.
140. Boccadamo R, Toscano S: Prevention and interruption of supraventricular tachycardia by antitachycardia pacing. In Luderitz B, Saksena S (eds): Interventional Electrophysiolog. Mt Kisco, NY, Futura, 1992, pp 213-23.
141. Vollmann D, Stevens J, Buchwald AB, Unterberg C: Automatic atrial anti-tachy pacing for the termination of spontaneous atrial tachyarrhythmias: Clinical experience with a novel dual chamber pacemaker. J Interv Card Electrophysiol 2001;5:477-85.

Chapter 15 Atrial Tachycardia, Flutter, and Fibrillation

A. JOHN CAMM, IRINA SAVELIEVA, SAROJA BHARATI, BRUCE D. LINDSAY, STANLEY NATTEL, KAORI SHINAGAWA, and SHIH-ANN CHEN

Introduction: A. John Camm
Epidemiology and Classification: Irina Savelieva, A. John Camm
Anatomy and Pathology: Saroja Bharati
Clinical Electrocardiography: Bruce D. Lindsay
Basic Electrophysiology: Stanley Nattel, Kaori Shinagawa
Investigations in Atrial Tachyarrhythmias: Irina Savelieva, A. John Camm
Clinical Electrophysiology: Shih-Ann Chen
Principles of Practice: A. John Camm
Evidence-Based Therapy and Management of Atrial Tachyarrhythmias: A. John Camm, Irina Savelieva

Atrial tachyarrhythmias are multifarious: different forms with diverse pathogeneses, electrophysiological mechanisms, clinical presentations, and associated therapies. Among these, atrial fibrillation (AF) is the most common sustained arrhythmia, with an incidence of 19.2 per 1000 person-years in individuals 65 years of age or older,[1] followed by atrial flutter (AFL), which occurs at a rate of 3.17 per 1000 person-years in those older than age 50.[2] The epidemiology of atrial tachycardia (AT) is largely unknown, but its prevalence is believed to be 0.34% to 0.46% in patients with arrhythmias.[3] Affecting mostly the elderly population, atrial tachyarrhythmias, particularly AF, are associated with a considerable increase in morbidity and mortality. In addition to disabling symptoms and impaired quality of life,[4] atrial tachyarrhythmias pose significant risk for ischemic stroke[5] and heart failure,[6,7] resulting in a doubling of hospital admissions in 1996 compared with 1986, according to recent surveys.[8,9] Both AF and AFL confer excess risk for all-cause death and cardiovascular mortality even after adjustment for other risk factors and underlying pathologies.[10,11]

Although our knowledge about the pathogenesis, pathophysiology, and precipitating factors of atrial tachyarrhythmias have increased considerably over the past 2 decades, therapeutic progress in clinical practice has been less encouraging, particularly in the management of AF. A likely explanation for the lack of successful treatment of these arrhythmias is that until recently, there has been a simplistic attitude toward their classification; AF, AFL, and often AT have been considered as single entities demanding similar therapeutic approaches.

The first attempt to rationalize indications for antiarrhythmic drug therapy in these different clinical settings was made more than a decade ago by the Sicilian Gambit group on the basis of their action on arrhythmogenic mechanisms, mainly the width of the excitable gap.[12] This initial approach was further refined based on the new evidence for arrhythmia mechanisms and the development of newer antiarrhythmic agents.[13] These include modified structural analogues of traditional antiarrhythmic drugs with additional novel mechanisms of action and less complex metabolic profiles that may improve their efficacy and safety and the development of innovative antiarrhythmic agents with unconventional antiarrhythmic mechanisms. An attractive prospect is the introduction of agents with high affinity to atrial tissue and ion channels involved in repolarization processes exclusively in the atria, other agents that are capable of reversing ion channel remodeling, and yet others that regulate intracellular calcium homeostasis and cell-to-cell coupling.[14]

Differentiating the mechanisms underlying atrial tachyarrhythmias has led to the rapid development of nonpharmacologic treatment alternatives, including various catheter ablation techniques, which may "cure" many of these arrhythmias and have become a first-line therapy in typical AFL[15] and some forms of AT.[16] Electrophysiological mapping studies in patients with AF have shown that reentrant circuits or ectopic foci responsible for the initiation and perpetuation of the arrhythmia are likely to be located in the posterior wall of the left atrium or in the pulmonary veins rather than the right atrium, suggesting that the left atrium acts as an electrical driving chamber and should be the target for nonpharmacologic treatment options. Although the usefulness of ablation techniques has been marred by the absence of a clear anatomic substrate for AF, considerable progress has been achieved in modification of the susceptibility of the atria to recurrent AF (ablation or isolation of "focal" arrhythmia emanating from the orifices of pulmonary veins or a modified catheter-based maze procedure that reduces the effective size of the atria).[17-19] There is now an increased interest in "hybrid" therapy, which combines two or more therapeutic modalities, providing synergistic effects on rhythm control in atrial tachyarrhythmias.[20] For example, antiarrhythmic drugs that have proven effective in treating AF may have an intermediate effect of organizing the arrhythmia into a fixed circuit reentry arrhythmia such as AFL, which can then be amenable to radiofrequency catheter ablation. Modification of the arrhythmia substrate by atrial linear ablation in combination with antiarrhythmic drugs and atrial pacing has also been shown to be a potentially effective therapeutic approach in selected patients.[21]

The results of several large randomized trials in patients with sinus node dysfunction suggest that atrial or dual-chamber pacing confers a significant benefit in terms of risk reduction for the development of sustained atrial tachyarrhythmias compared with ventricular pacing.[22,23] New devices that combine features of pacemakers and implantable cardioverter-defibrillators (ICDs) can provide continuous surveillance and detection of atrial tachyarrhythmias. They can also be programmed to deliver atrial-tiered therapies, including antitachycardia pacing, to terminate the arrhythmia and atrial preventative therapies to reduce the initiation or perpetuation of the arrhythmia in the presence of proarrhythmic triggers.[24]

"Upstream" therapeutic approaches focused on treatment of the underlying pathology, such as angiotensin-converting enzyme (ACE) inhibitors, angiotensin II type 1 (AT-1) receptor antagonists, and β-blockers, may prevent or delay myocardial changes leading to atrial remodeling by unloading the atria or produce the direct effects on the evolution of the electrophysiological milieu.[25] Thus a broader perspective is being developed regarding the systemic, organ, tissue, myocyte membrane, and intracellular contributions to the genesis of atrial tachyarrhythmias.

Epidemiology and Classification of Atrial Tachyarrhythmias

Much of our knowledge about the natural history and risk factors of atrial tachyarrhythmias in the general population has come from a galaxy of epidemiologic studies in AF.[26-30] Among atrial tachyarrhythmias, AF is the most prevalent sustained arrhythmia, currently affecting approximately 1.5% of the population. Recent advances in the treatment of heart disease have led to a large population of patients who have survived to old age with significant but non–life-threatening underlying heart disease. Two North American studies, the Framingham Heart Study[26] and the Cardiovascular Health Study,[1] have shown that the incidence of AF in subjects younger than 64 years is 3.1 cases in men and 1.9 cases in women per 1000 person-years, rising sharply to about 19.2 per 1000 person-years in those 65 to 74 years old. AF is as high as 31.4 to 38.0 in octogenarians.[1,26] Similar reports have come from the Manitoba Follow-up Study in Canadian citizens[27] and the only European population-based study conducted in Sweden over 25 years (Fig. 15-1).[28] Projected data from a recent cross-sectional AnTicoagulation and Risk Factors in Atrial Fibrillation (ATRIA) study of adults 20 years of age or older who are enrolled in a large health care maintenance organization in California have shown that the number of patients with AF in the United States is likely to increase 2.5-fold from 2.3 million to more than 5.6 million during the next 50 years, with more than 50% of affected individuals 80 years of age or older.[29]

The only reported epidemiologic study of patients with AFL is based on a selected sample of 58,820 residents served by the Marshfield Clinic in Wisconsin (Marshfield Epidemiologic Study Area, MESA).[2]

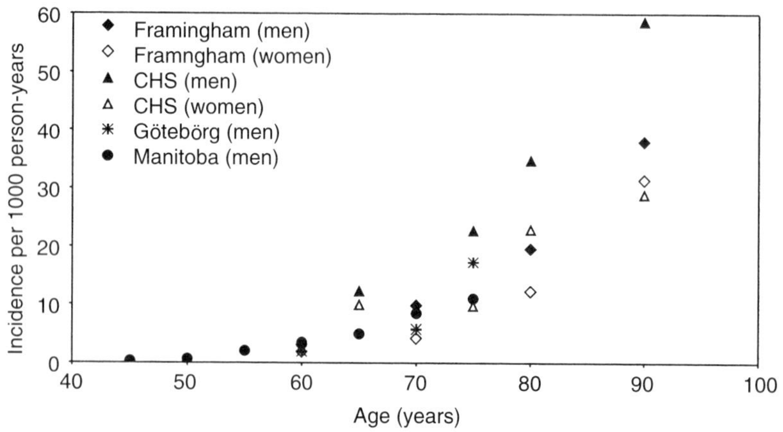

FIGURE 15-1 Incidence of atrial fibrillation in four population-based studies across North America, Canada, and Europe: the Cardiovascular Health Study, the Framingham Heart Study, the Manitoba Follow-up Study, and the Göteborg Multifactor Primary Prevention Study. Data from references 1,26-28.

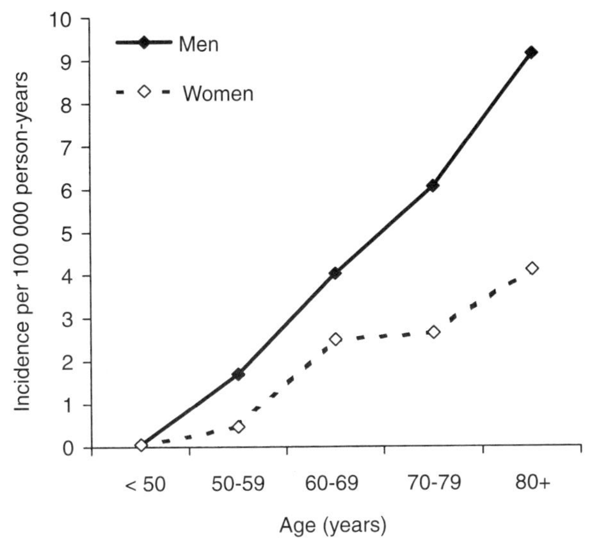

FIGURE 15-2 Incidence of atrial flutter by age and gender in the MESA Study. (Modified from Granada J, Uribe W, Chyou PH, et al: Incidence and predictors of atrial flutter in the general population. J Am Coll Cardiol 2000;36:2242-6.)

The overall incidence of AFL is 0.88 per 1000 person-years, but 58% of these patients also have AF. AFL alone is found in only 0.037%. As with AF, the incidence of AFL increases markedly with age, from 5 per 100,000 in individuals younger than 50 years of age to 587 per 100,000 in those older than 80 years of age (Fig. 15-2). Similar to AF, which is 1.5 times more common in men, AFL is observed 2.5 times more frequently in men than in women. The prevalence of AT in the general population is unknown. It is observed in 0.34% to 0.46% of patients with arrhythmias and constitutes about 10% to 15% of individuals referred for catheter ablation.[3]

Much of our knowledge of the epidemiology of AF is based on predominantly white cohorts. Although African Americans constituted only 5% of the study population, their incidence of AF was lower than the incidence among whites, according to the Cardiovascular Health Study.[1] Among persons 50 years of age or older enrolled in the ATRIA study, the prevalence of AF was also higher in white than in black patients (2.2% versus 1.5%).[29]

ASSOCIATED DISEASE AND RISK FACTORS

Atrial tachyarrhythmias are often found in association with underlying heart disease, such as hypertension and heart failure (Fig. 15-3).[1,26] The prevalence of AF associated with left ventricular dysfunction and congestive heart failure varies from 4% to 50% depending on New York Heart Association (NYHA) class. Although AF is classically caused by mitral stenosis, thyrotoxicosis, and alcohol, these are relatively minor associations. AF is a common complication of acute myocardial infarction and hypertrophic cardiomyopathy. Congenital heart disease and preexcitation syndromes due to accessory pathways are also associated with AF or AFL. Idiopathic or "lone" AF constitutes about half the cases of paroxysmal AF and 20% of persistent AF, particularly in relatively young patients. When studied in detail, some may have evidence of inflammation and atrial myocarditis, mild diastolic ventricular dysfunction, subclinical thyroid disease, autoimmune disorders, or sinus node dysfunction. Contrary to general belief, typical AFL and AT are commonly associated with organic heart disease. In the MESA population, nearly all cases of AFL were linked to comorbid conditions such as heart failure, hypertension, and chronic lung disease or occurred in association with a specific precipitating event (i.e., major surgery, pneumonia, or acute myocardial infarction).[2] Only 1.7% of cases had no structural cardiac disease or precipitating causes ("lone" AFL). Coronary artery bypass and valvular heart surgery are not uncommon causes of postoperative AF in older patients. AT and AFL (so-called *incisional reentry*, not isthmus dependent, AFL) can often occur after repair of congenital heart disease.

AFL and AF probably share risk factors, but much of this evidence has come from the epidemiologic studies of AF, which included a small proportion (10% to 20%) of patients with AFL as a predominant arrhythmia. The Framingham Heart Study, initiated in 1948, introduced the concept of risk assessment and prevention and identified several independent risk factors for AF including advanced age, congestive heart failure, valvular heart disease, hypertension, coronary artery disease (predominantly, myocardial infarction), and diabetes mellitus.[33] These conventional risk factors have been confirmed in many other population surveys and further refined in different models (Table 15-1).[1,26-28,30]

Significant progress in treatment and aggressive strategies of primary and secondary prevention of cardiovascular diseases has resulted in changes in the structure and distribution of risk factors for atrial tachyarrhythmias. Valvular heart disease, particularly of rheumatic etiology, one of the most common causes and a powerful risk factor for AF in the Framingham and other early studies, no longer holds its leading role in more recent surveys but is still important in developing countries or in the very elderly. On the other hand, an increasing number of surviving patients with chronic heart failure, a significant proportion of whom develop atrial tachyarrhythmias, has prompted recognition of congestive heart failure as an extremely important risk factor. The EuroHeart survey conducted in 2000-2001 in 24 countries has reported the overall prevalence of new onset AF in patients hospitalized for heart failure to be 13%, varying from 8% to 36% in different regions.[31] Furthermore, diastolic ventricular dysfunction with subsequent increases in filling pressures mediates atrial remodeling and is associated with a 5.26-fold increased risk for the development of AF compared with normal diastolic function.[32]

CLASSIFICATION OF ATRIAL TACHYARRHYTHMIAS

Current classification of atrial tachyarrhythmias, based on electrocardiographic presentations and electrophysiological mechanisms, include:[33]

- Sinoatrial (SA) nodal reentrant tachycardia
- Focal atrial tachycardia due to automatic, triggered, or microreentrant mechanisms

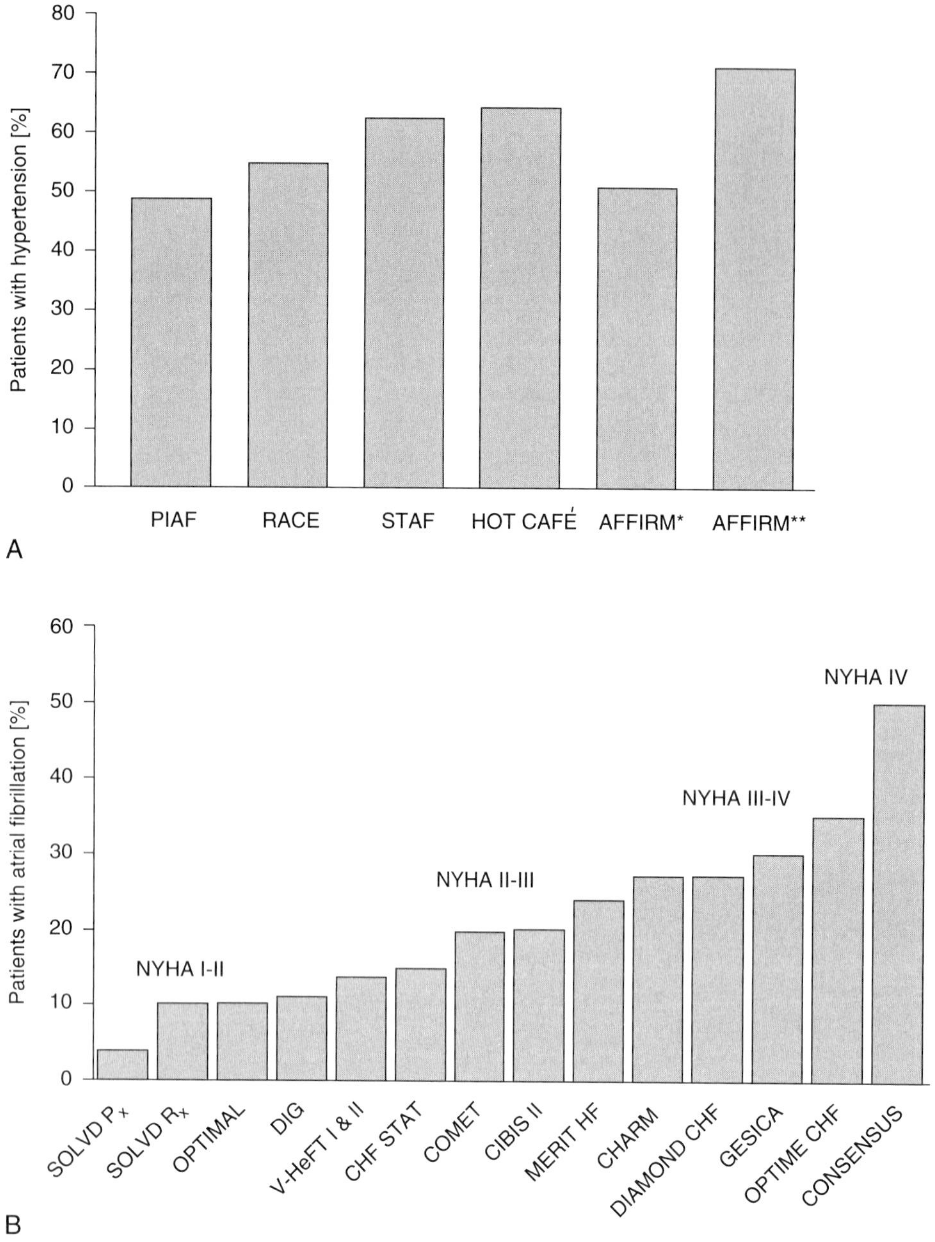

FIGURE 15-3 **A,** Prevalence of hypertension as underlying cardiovascular disorder in patients with atrial fibrillation. **B,** Prevalence of atrial fibrillation in heart failure studies. *, hypertension as a predominant cardiac diagnosis; **, the overall prevalence of hypertension. (Modified from Savelieva I, Camm AJ: Supraventricular and ventricular arrhythmias in heart failure. In Van Veldhuisen DJ, Pitt B (eds): Amsterdam, Benecke N.I., Chronic Heart Failure, 2002, pp 111-43.)

- Typical AFL due to a macroreentrant mechanism
  - Counterclockwise
  - Clockwise (reverse)
- Incisional reentry AFL (or AT)
- Atypical right AFL (isthmus dependent)
- Atypical left atrial flutter (pulmonary vein or mitral valve annulus dependent)
- Atrial fibrillation

In contrast to a relatively straightforward electrophysiological classification, the clinical classification of atrial tachyarrhythmias has caused more controversy as, ideally, it would incorporate the multiple etiologies, risk factors, and precipitating agents; various clinical presentations; modes of onset; and variable temporal patterns of behavior. All of these can have an important influence on the selection of the therapeutic strategy and, ultimately, the effect of treatment.

The most recent clinical classification suggested by the American College of Cardiology/American Heart Association/European Society of Cardiology (ACC/AHA/ESC) Task Force on AF includes first detected, paroxysmal, persistent, and permanent forms of the arrhythmia (Fig. 15-4*A*).[34] First onset AF is the first clinical presentation of the arrhythmia where the patient is still in AF when evaluated, and the episode has been present less than 48 hours. The onset of AF may be asymptomatic and the "first detected episode" should not be regarded as necessarily the true onset. After its first recognition, the arrhythmia may not convert spontaneously and may be refractory to cardioversion, in which case permanent AF is diagnosed. If the physician or patient chooses not to treat the arrhythmia by a cardioversion technique and allows it to remain, the term "accepted" AF is applied. In patients with the

TABLE 15-1 Independent Risk Factors for Atrial Fibrillation in Population Surveys (Odds Ratio, 95% CI)

Risk Factor	Framingham Study (Since 1948 for 38 Yr)[26]	Manitoba Study (Since 1948 for 44 Yr)[27]	CHF (Since 1989 for 3 Yr)[30]	CHF for New Onset of AF (Since 1989 for 3 Yr)[1]	Wilhemsen et al. (Since 1970 for 27 Yr)[28]	MESA Study of Atrial Flutter (Since 1991 for 4 Yr)[2]
Age	2.1 (1.8-2.5)	—	1.03	1.05 (1.03-1.08)	1.1 (1.07-1.16)	Values not stated
Hypertension	1.5 (1.2-2.0)	1.42 (1.10-1.84)	1.39	1.11 (1.05-1.18)*	1.33 (1.07-1.65)	0.9 (0.6-1.4)
CHF	4.5 (3.1-6.6)	3.37 (2.29-4.96)	2.67	1.51 (1.17-1.97)†	6.7 (5.17-8.87)§	3.5 (1.7-7.1)
CAD/MI	1.4 (1.0-2.0)	3.62 (2.59-5.07)	—	1.48 (1.13-1.95)		1.3 (0.7-2.2)
Valvular disease	1.8 (1.2-2.5)	3.15 (1.99-5.00)	3.27	2.42 (1.62-3.60)	—	4.0 (0.4-30)
Diabetes	1.4 (1.0-2.0)	—	—	1.08 (1.03-1.13)	—	1.8 (0.9-3.6)
LVH	1.4 (0.9-2.4)	—	—	—	—	—
Cholesterol	—	—	—	0.86 (0.76-0.98)	—	0.8 (0.5-1.2)
Smoking	1.1 (0.8-1.5)	—	—	—	—	—
Alcohol	1.01 (0.99-1.03)‡	—	—	0.96 (0.93-0.99)	—	—
Body mass index	1.03 (0.99-1.06)‡	1.28 (1.02-1.62)	—	—	1.07 (1.04-1.1)	—
Height	—	—	—	1.03 (1.02-1.05)	1.04 (1.03-1.06)	—
Black race	—	—	—	0.47 (0.22-1.01)	—	—

* Systolic blood pressure, per 10 mm Hg.

†Use of diuretics but not a history of CHF.

‡Adjusted for gender and age but not for cardiac risk factors.

§CHF and MI combined.

AF, atrial fibrillation; CAD, coronary artery disease; CHF, congestive heart failure; CI, confidence intervals; LVH, left ventricular hypertrophy; MI, myocardial infarction.

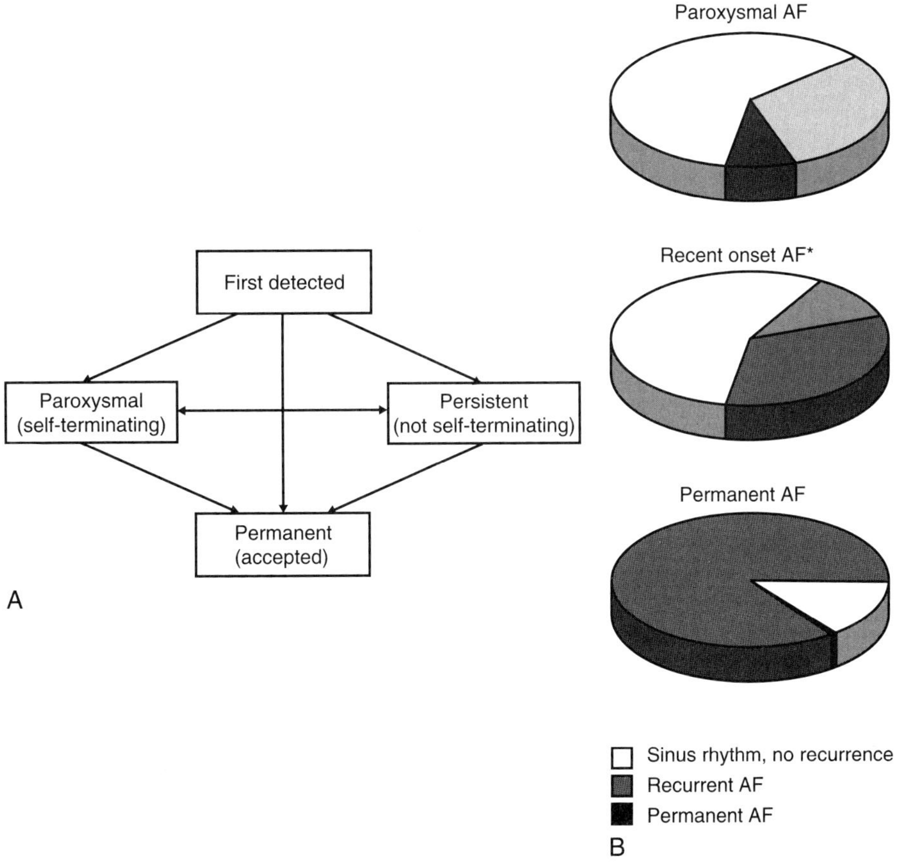

FIGURE 15-4 **A,** Classification of AF suggested by the American College of Cardiology, American Heart Association, and European Society of Cardiology Task Force on AF. (Reprinted from Fuster V, Rydén LE, Asinger RV, et al: Task Force Report: ACC/AHA/ESC guidelines for the management of patients with AF. Eur Heart J 2001;22:1852-1923). **B,** Outcomes of different forms of AF during a mean follow-up of 8.6 months. *, Recent onset AF is defined as persistent (not self-terminating) arrhythmia lasting 7 days or more but less than 1 month. AF, atrial fibrillation. (Reprinted from Lévy S, Maarek M, Coumel P, et al, on behalf of the College of French Cardiologists: Characterization of different subsets of atrial fibrillation in general practice in France: The ALFA Study. Circulation 1999;99:3028-3035.)

paroxysmal variety, most episodes convert back to sinus rhythm spontaneously, whereas the persistent form of the arrhythmia requires an active intervention to restore sinus rhythm. There are mixed forms where the recurrence may or may not cardiovert spontaneously, and there is often, but not always, a progression of the disease from the paroxysmal to the persistent and eventually the permanent (or accepted) form (see Fig. 15-4*B*).[35]

MORBIDITY AND MORTALITY IN ATRIAL TACHYARRHYTHMIAS

The clinical significance of sustained atrial tachyarrhythmias lies in thromboembolic risk and the risk of symptomatic left ventricular dysfunction because of incessant, fast ventricular rates. Both AT and AFL are closely linked to AF, for which these risks and their reduction are well appreciated in epidemiologic studies and large randomized trials.

Risk of Stroke

The presence of AF has been estimated to increase the risk of stroke by about fivefold.[36] In the Framingham Heart Study, the annual risk of stroke attributable to AF among patients 50 to 59 years of age is 1.5% and increases to 23.5% in those older than 80 years of age.[36] When transient ischemic attacks and silent cerebral thromboembolic events are included, the annual risk of ischemic stroke exceeds 7%.[37] In the Veterans Affairs Stroke Prevention in Nonrheumatic Atrial Fibrillation (SPINAF) study, about 15% of neurologically asymptomatic patients had evidence of one or many silent cerebral infarcts.[38] The risk is considerably higher (about 12% per year) for recurrent stroke in patients with previous stroke or transient ischemic attack.[39]

Contrary to popular belief that because of their paroxysmal character and more organized mechanical atrial function, AFL and AT pose a lower risk of thromboembolism, a recent retrospective analysis of 337,428 patients with AF and 17,413 cases of AFL extracted from the Medicare database showed a comparable relative risk for stroke of 1.64 and 1.4, respectively.[40] However, the subgroup analysis has shown that the greatest risk is in individuals with both AFL and AF, whereas isolated AFL poses a minor risk compared with control subjects.

Tachycardia-Induced Cardiomyopathy

In patients with atrial tachyarrhythmias and little or no structural heart disease, symptomatic left ventricular dysfunction may result from poor rate control, irregularity of ventricular response, and loss of atrial contraction. Loss of atrioventricular (AV) synchrony is associated with impaired diastolic filling, reduced stroke volume, and elevated diastolic atrial pressure, resulting in an approximately 20% reduction in cardiac output.[41] An irregular ventricular response decreases cardiac output, increases right atrial pressure and pulmonary capillary wedge pressure independent of rate.[42] Such ventricular dysfunction associated with significant heart dilatation and symptoms of heart failure is termed tachycardia-induced cardiomyopathy.

The rate and duration of the arrhythmia required to cause cardiomyopathy are unknown, but it is generally accepted that sustained ventricular rates of greater than 120 beats per minute (bpm) may pose a risk. Although it is generally applied to persistent forms of atrial tachyarrhythmias, this may also hold true for a paroxysmal variety with frequent occurrences where rate control is more difficult. It is sufficient for tachycardia to be present for 10% to 15% of the day to cause impairment of ventricular function.[43] Tachycardia-induced cardiomyopathy may reverse completely after sinus rhythm is restored or adequate ventricular rate control is achieved either by pharmacologic or nonpharmacologic means (Fig. 15-5).[44]

In patients with compromised ventricular function, atrial tachyarrhythmias can precipitate overt heart failure. Data from the Cardiovascular Health Study and the Digitalis Investigation Group (DIG) have shown that a history of AF carried a 1.65-fold risk of developing congestive heart failure in individuals older than 65 years of age[6] and a threefold risk of worsening heart failure.[45]

Mortality

Data from the Framingham Heart Study suggest that risk of death conferred by AF is nearly doubled even in the absence of identifiable structural heart disease.[46] In the Paris Prospective Study of 6722 men aged 43 to 52 years of age followed up for 23 years, "lone" AF entails a 4.2-fold excess in all-cause and a 1.97-fold excess in cardiovascular mortality.[47] The risk of mortality associated with AF is significantly higher in women than in men (odds ratio 1.9 versus 1.5) after adjustment for age, hypertension, smoking, diabetes, left ventricular hypertrophy, myocardial infarction, congestive heart failure, valvular heart disease, and stroke or transient ischemic attack.[46]

In the MESA population followed up for a mean period of 3.6 years, AFL alone is associated with a 1.7-fold increase in mortality, and AFL coexistent with AF confers a 2.5-fold greater risk of death.[11] At 5 years, the mortality rates conferred by AFL and AF are practically identical (Fig. 15-6).

SPECIFIC FORMS OF ATRIAL TACHYARRHYTHMIAS

Silent Atrial Tachyarrhythmias

Recently, the challenge of asymptomatic, or silent, atrial tachyarrhythmias has been recognized.[44] Usually atrial tachyarrhythmias are associated with various symptoms: palpitations, dyspnea, chest discomfort, fatigue, dizziness, and syncope (Fig. 15-7). Paroxysmal forms are more likely to be symptomatic and frequently present with specific symptoms, while permanent forms are usually associated with less specific symptoms. In the Etude en Activité Liberale sur le Fibrillation Auriculaire (ALFA) study of 756 patients with AF from general practice in France, of 86 participants who reported no symptoms,

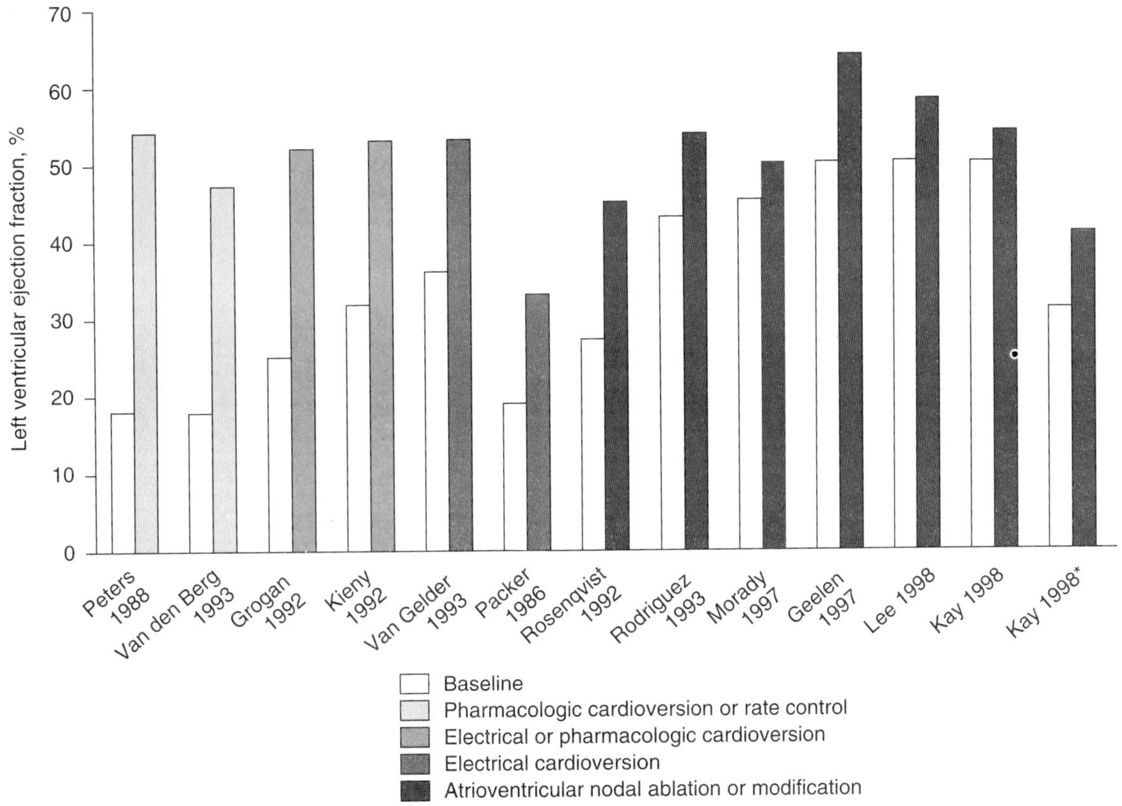

FIGURE 15-5 Improvement of left ventricular function following pharmacologic treatment, electrical cardioversion, and atrioventricular nodal ablation or modification in patients with atrial fibrillation and atrial tachycardia. *, patients with significantly decreased left ventricular function at baseline. (Reprinted from Savelieva I, Camm AJ: Clinical relevance of silent atrial fibrillation: Prevalence, prognosis, quality of life, and management. J Interv Card Electrophysiol 2000;4:369-82.)

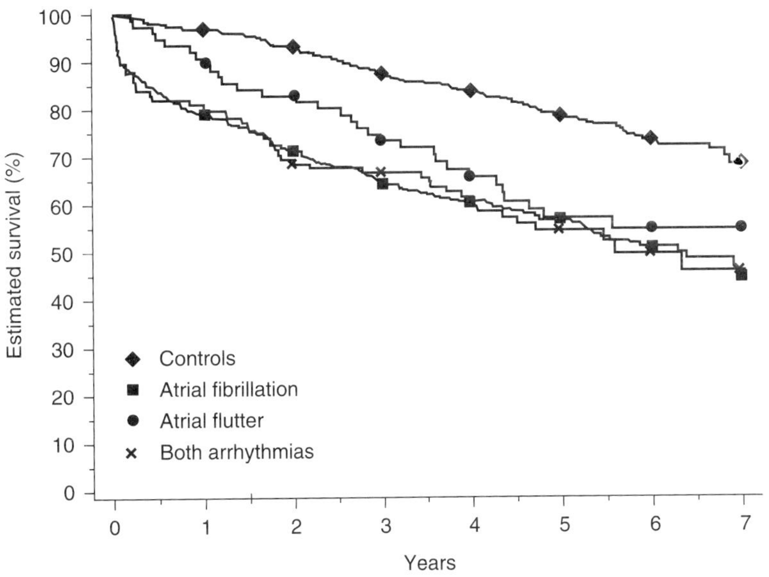

FIGURE 15-6 Kaplan-Meier curves of survival in patients with atrial tachyarrhythmias. (Reprinted from Vidaillet H, Granada JF, Chyou PH, et al: A population-based study on mortality among patients with atrial fibrillation or flutter. Am J Med 2002;113:365-70.)

63 (73%) presented with permanent AF, 14 (16%) were diagnosed with persistent AF, and only 9 (11%) had a paroxysmal form.[35] During transtelephonic monitoring, clinical symptoms correlated with the presence of the arrhythmia in 80.2% patients with AF and 40.6% of those with AFL.[48]

Silent atrial tachyarrhythmias are found incidentally during routine physical examinations, preoperative assessments, occupation assessments, or population surveys. In some cases, they are diagnosed only after complications such as stroke or heart failure have occurred. Usually, the duration of the arrhythmia

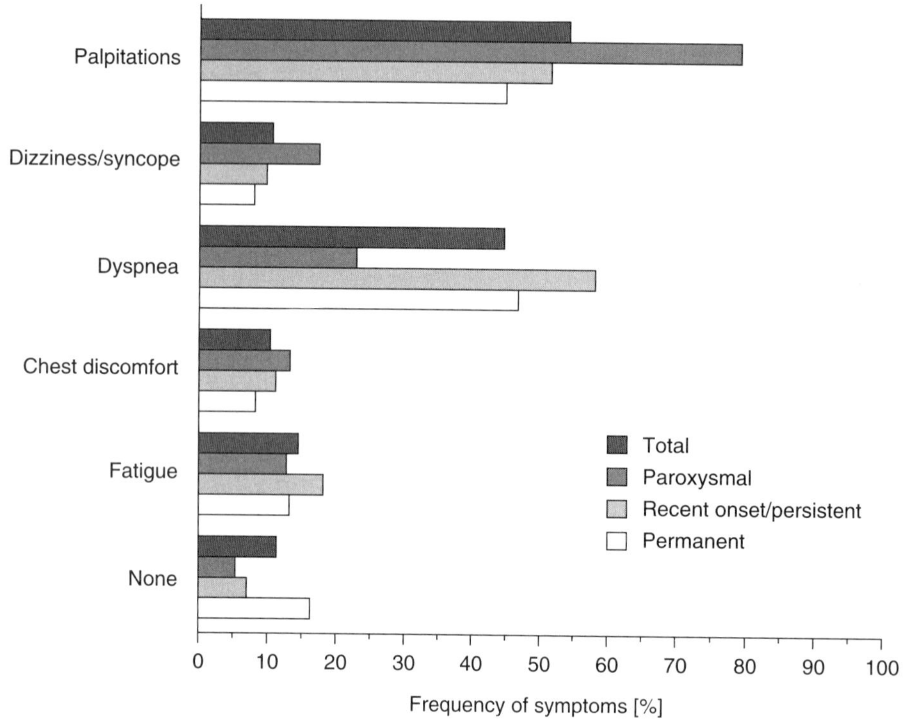

FIGURE 15-7 Distribution of symptoms in different forms of atrial fibrillation. (Modified from Lévy S, Maarek M, Coumel P, et al, on behalf of the College of French Cardiologists: Characterization of different subsets of atrial fibrillation in general practice in France: The ALFA Study. Circulation 1999;99:3028-3035.)

cannot be established. The prevalence of sustained silent tachyarrhythmias is believed to be 25% to 30%, but implantable rhythm control devices, such as pacemakers and ICDs, have revealed that a very large proportion of patients (50% to 100%) may have unsuspected episodes of the arrhythmia.[49] Pharmacologic treatment has long been known to convert a symptomatic form to an entirely asymptomatic variety. In the Prevention of Atrial Fibrillation After Cardioversion (PAFAC) study involving more than 1000 patients with AF, 90% of recurrences were rendered asymptomatic by antiarrhythmic drug therapy and were only detected by routine daily transtelephonic ECG monitoring.[50]

Although symptoms may not stem directly from the arrhythmia, the risk of complications is probably the same for symptomatic and asymptomatic patients. For example, AF is found incidentally in about 25% of admissions for stroke.[51] Silent embolic signals were detected by transcranial Doppler in 13% of patients with symptomatic and 16% of patients with asymptomatic atrial tachyarrhythmias.[52] It is worth noting that AV node ablation or modification, or both, may convert the arrhythmia into a clinically silent variety, but does not reduce the risk of stroke.

Atrial Fibrillation and Heart Failure

National surveys of heart failure have shown that two thirds of patients with this condition are 65 to 70 years of age or older and are likely to have AF as a coexistent disorder.[31] Heart failure may precipitate AF by increasing volume and pressure load, leading to atrial dilatation and stretch and altered atrial electrophysiology, including shortening of the atrial effective refractory period, a major determinant of the arrhythmia. Hypertrophy of atrial myocytes and patchy fibrosis result in numerous areas of conduction delay or block favoring micro-reentry. Calcium overload and activation of stretch-mediated ion channels increase the likelihood of afterdepolarizations and triggered activity in the atria. The onset of AF in patients with heart failure may be associated with overt clinical decompensation, worsening NYHA functional class, a decrease in peak exercise oxygen consumption, a reduction in cardiac index, and an increase in mitral and tricuspid regurgitation, strongly suggesting that AF may be the cause and not just a marker of more severe left ventricular dysfunction.[53]

The prevalence of AF in patients referred for management of mild to moderate heart failure is in the range of 10% to 20% and increases up to 50% in patients with severe decompensation (see Fig. 15-3*B*).[54] The European Heart Failure survey reports that 45% of 10,464 patients present with AF.[31] AF is a marker of increased mortality in heart failure. The Studies of Left Ventricular Dysfunction (SOLVD) Treatment and Prevention trials report a 1.34-fold greater risk of all-cause death conferred by AF, which is largely explained by increased risk of death from progressive pump failure.[7] In the DIG study, the development of atrial tachyarrhythmias entailed a 2.5-times greater risk of subsequent mortality and a 3-times greater risk of hospitalizations for heart failure.[45] Of interest, some investigators suggest that AF is independently associated with increased mortality rates only in patients with relatively preserved ventricular function, while in individuals with advanced disease, the relationship is more complicated and dependent on other variables.[55,56] However, in the Vasodilator in Heart Failure Trials (V-HeFT), which included patients with NYHA class II and III, the presence of AF did not affect survival.[57] A likely explanation may be a relatively short follow-up (2 years) and the absence of the placebo arm in V-HeFT-II.

Atrial Fibrillation and Myocardial Infarction

AF is a common finding in acute myocardial infarction. The Global Use of Strategies To open Occluded Coronary Arteries (GUSTO-I) study (40,891 patients, 4280 with AF)[58] and the Cooperative Cardiovascular Project (106,780 patients, 23,665 with AF)[59] reported that AF was associated with a high incidence of stroke and heart failure and conferred a 1.3-fold greater risk of death at 1 year. In the selected patient population with myocardial infarction and mild to moderate heart failure enrolled in the Optimal Trial in Myocardial Infarction with the Angiotensin II Antagonist Losartan (OPTIMAAL) study, the presence of AF increased risks of stroke and death by 2.5 and 2.6.[60] The results of the Trandolapril Cardiac Evaluation (TRACE) study in 1749 patients with myocardial infarction and left ventricular dysfunction followed up for 5 years suggest that the presence of AF or AFL entails a 30% increased risk of long-term mortality.[61]

AF found in 6.4% patients with acute coronary syndrome without ST segment elevation portended higher rates of mortality and stroke at 6 months after its occurrence in the Platelet IIb/IIIa Underpinning the Receptor for Suppression of Unstable Ischemia Trial (PURSUIT) (hazard ratios 3.0 and 2.9, respectively).[62]

Atrial Fibrillation and Hypertrophic Cardiomyopathy

AF is a common and clinically important complication in hypertrophic cardiomyopathy, with a prevalence of 18% to 28% and an annual incidence of 2%.[63,64] Age older than 50 years, NYHA class higher than II, and the left atrial size 4.5 cm or greater are independent predictors of the development of AF. It is associated with limiting symptoms and increased morbidity and mortality. In 480 patients with hypertrophic cardiomyopathy who had been followed up for 9 years, AF conferred a 3.7-fold increased risk of mortality, predominantly due to excess death from progressive heart failure and stroke.[64] At the time of onset, AF produced new or worsening clinical manifestations in 84% of patients and was associated with substantial functional deterioration during a long-term follow-up, with a nearly threefold increased risk for developing advanced heart failure. Stroke and other thromboembolic events occurred in 6% of patients, with the overall incidence of 0.8% per year and 1.9% per year in those older than 60 years.[64] Thromboembolic complications were substantially more common in the presence of AF (10.2-fold increased risk).[65] Patients with a permanent form of the arrhythmia had a significantly higher probability of combined mortality, functional impairment, and stroke, suggesting that therapy aimed at prevention or delay of the transition from paroxysmal to permanent AF might improve outcome.

Atrial Fibrillation in Preexcitation Syndrome

AF is found in 30% of patients with Wolff-Parkinson-White syndrome and is associated with an increased risk of sudden death, which may be a first manifestation of the disease in younger individuals. Unlike the AV node, accessory pathways usually exhibit rapid, nondecremental conduction, and if the effective refractory period of an accessory pathway is short (<250 milliseconds [ms]), this may result in a rapid ventricular response during AF with subsequent degeneration to ventricular fibrillation. The incidence of sudden death is 0.15% to 0.39% per patient-year and is increased in the presence of multiple accessory pathways, Epstein's anomaly, and familial Wolff-Parkinson-White syndrome.[66]

Atrial Tachyarrhythmias after Heart Surgery

The incidence of postoperative AF is 27% to 37% after coronary bypass surgery and exceeds 50% after valvular surgery.[67,68] Postoperative AF occurs predominantly during the first 4 days and is associated with increased morbidity and mortality, largely due to stroke and circulatory failure, and longer hospital stay. More than 90% patients present with a paroxysmal or first onset form of the arrhythmia. Although AFL and AT are also not uncommon, these are usually linked to corrective surgery for congenital heart disease and occur by the incisional reentry mechanism in which the reentrant circuit travels around the line of block created by incision.[33,69]

Clinical factors that convey a higher risk for the development of postoperative atrial tachyarrhythmias include age, male sex, a previous history of arrhythmias, hypertension, congestive heart failure, valvular heart disease, chronic obstructive pulmonary disease, chronic renal failure, previous cardiac surgery, left atrial enlargement, inadequate cardioprotection and hypothermia, right coronary artery grafting, and a longer bypass time.[68] Recent observations suggest that the incidence of postoperative atrial tachyarrhythmias may be lower with minimally invasive techniques, especially for valvular surgery.[69a]

Familial Atrial Fibrillation

Although AF is commonly associated with underlying heart disease, in a substantial proportion of patients presenting with AF, no cardiovascular abnormality can be detected.[70,71] The pathogenesis of "lone," or "idiopathic," AF is unknown, but genetic predisposition or specific genetically predetermined forms of the arrhythmia have been proposed. Recently, this hypothesis has been confirmed by identification of a gene defect linked to chromosome 10q22-q24 in three Spanish families, 21 of 49 members of which presented with AF with fast ventricular rates at a relatively young age, ranging from 2 to 46 years.[72] Shortly afterward, another locus for AF was mapped to chromosome 6q14-16.[73] The mutated gene has not yet been identified.

A survey from the Mayo Clinic demonstrated that about 5% of all patients with AF and 15% of those with "lone" AF present with a heritable disorder.[74] Fifty probands from four multigeneration families in which AF appeared to segregate as an autosomal dominant trait were identified. Three families presented with symptomatic paroxysmal AF associated with a rapid ventricular response and reversible tachycardia-induced cardiomyopathy. The phenotype of disease in the fourth family was different and included asymptomatic AF, which did not require rate control, progressive AV block, and left ventricular enlargement with low to normal ejection fraction, raising the possibility of diffuse underlying heart disease.[74] Indeed, the genetic interval for "lone" AF defined on chromosome 6q14-16 overlaps with a known locus for dilated cardiomyopathy, suggesting that "lone" AF and some forms of cardiomyopathy may be allelic.

Several other gene mutations associated with an increased incidence of AF have been reported (Table 15-2). A missense mutation in the *lamin A/C* gene in chromosome 1p1-q21 is linked to dilated cardiomyopathy and atrial myopathy accompanied by progressive conduction system disease, AF, stroke, and sudden death.[75,76] The mutation Arg663His in the β-cardiac myosin heavy chain gene has been identified in patients with a specific phenotype of familial hypertrophic cardiomyopathy presenting with moderate left ventricular hypertrophy, predominantly localized in the proximal segment of the interventricular septum, and a 47% prevalence of AF.[77] The association between AF and polymorphism of the *minK* gene has been reported, suggesting that subjects with the 38G allele are more likely to develop AF than those with the 38S allele in the presence of a risk factor.[78] Very recently, the mutation S140G in the *KCNQ1* gene in chromosome 11p15.5 has been reported in Chinese patients with persistent AF.[79] Opposite to the *loss of function* mutations in long QT syndrome, the S140G mutation produces a *gain of function* effect on the cardiac potassium currents and may promote AF due to shortening of action potential (AP) duration in atrial myocytes. These data indicate that the genetic substrate for familial AF is rather diverse. Further exploration may reveal a number of possible targets for medical therapy for prevention or reversal of the arrhythmia.

TABLE 15-2 Genetic Polymorphism in Atrial Fibrillation

Expression	Gene	Chromosome
Lone AF (3 families, Catalonia)	Not identified	10q22-q24
Lone AF (1 family, United States)	Not identified	6q14-16
Lone AF (4 families, United States)	Not identified	Not identified but neither of the above
Lone AF (1 family, China)	*KCNQ1*	11p15.5
AF in a general population (China)	*KCNE1 S38G*	Xx21q22.1-q22.2
AF associated with long QT syndrome	Not identified	4q25-q27
AF and familial WPW syndrome	*PRKAG2*	7q36
AF associated with hypertrophic cardiomyopathy	*MYH7* *β-MHC*	14q12 14q1
AF associated with dilated cardiomyopathy	Lamin A/C Not identified	1p1–q21 3p22-p25
AF associated with mitral valve prolapse	Not identified Not identified	16p11 11p15.4

ATRIAL TACHYARRHYTHMIAS AND QUALITY OF LIFE

Until recently, most of the investigational approaches in patients with atrial tachyarrhythmias have used objective physiologic measures, such as frequency and duration of arrhythmia episodes, ventricular function, exercise tolerance, ventricular rates at rest and during exercise, and maximum oxygen volume. It is a common belief that the concept of quality of life is inherently subjective, definitions vary, and the data are difficult to interpret because of the absence of normative or cardiac disease–related controls.[4] Quality of life can be assessed on the basis of health profiles, including physical condition, psychological well-being, social activities, and everyday activity.

AF-related quality of life in 152 patients with paroxysmal AF has been found to be markedly reduced compared with a healthy population and similar on most scales to health impairment seen in patients with myocardial infarction or congestive heart failure who had more significant structural heart disease.[80] In the Canadian Trial of Atrial Fibrillation (CTAF) study, 264 patients with paroxysmal or persistent AF reported better quality of life and had significantly higher scores in physical functioning, vitality, mental health, and role- emotional when they perceived themselves to be in sinus rhythm.[81] Of interest, the extent of subjective impairment of quality of life is poorly related to conventional objective measures of illness severity, such as the frequency and duration of arrhythmia episodes, NYHA class, and left ventricular ejection fraction. Women are more likely to report poor quality of life than men, despite comparable severity of underlying pathology.

A significant improvement in quality of life has been consistently reported in studies involving AV node ablation. These studies usually address highly symptomatic patients with poorly controlled AF.[80,82,83] However, when quality of life was compared in patients with AF who were randomly assigned to AV node ablation or AV node modification, ablation was shown to result in a greater improvement in quality of life and reduction of symptoms, probably due to better control of rate and regularity of ventricular response (Fig. 15-8).[83] In a small series of patients with permanent AF and a normal ventricular response rate, there also was a significant improvement in symptoms, such as effort and rest dyspnea, exercise tolerance, and lethargy, and in the perception of general well-being following AV node ablation, suggesting that even in patients whose ventricular rate was believed to be well controlled by drug therapy, quality of life might be reduced.[84] Even patients with asymptomatic AF report a significantly poorer

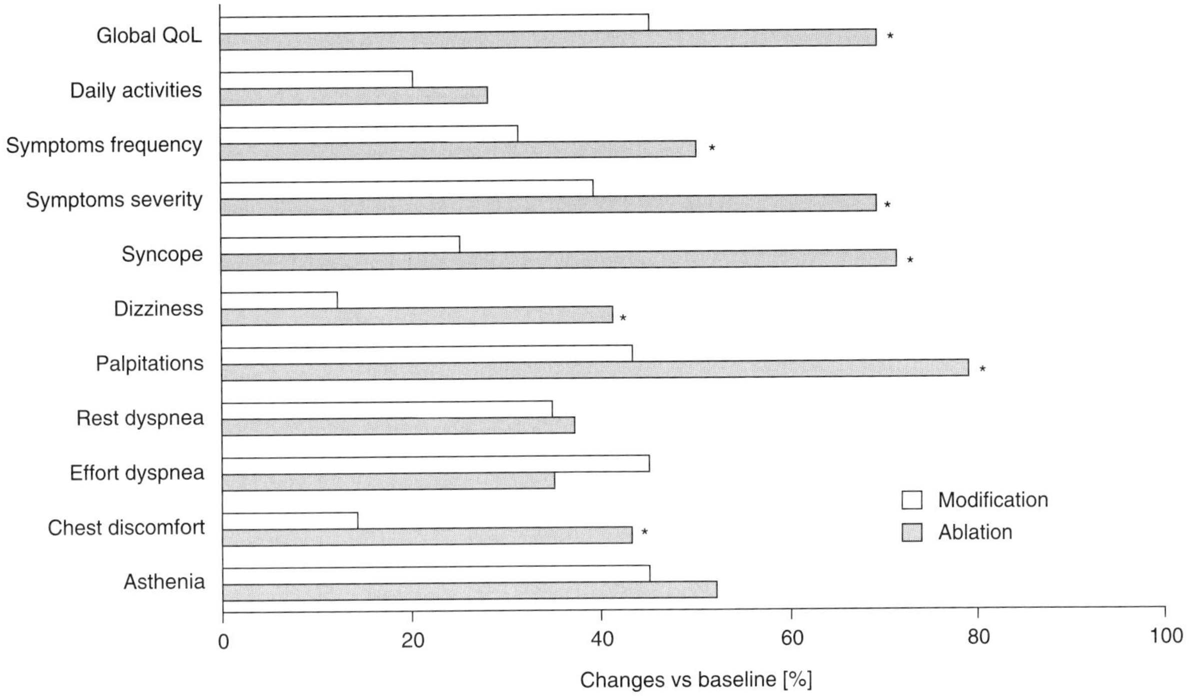

FIGURE 15-8 Comparison of quality of life in patients after atrioventricular node ablation and atrioventricular node modification at 6 months of follow-up. QoL, quality of life; *, $P < .01$ atrioventricular node ablation versus modification. (Modified from Lee SH, Chen SA, Tai CT, et al: Comparisons of quality of life and cardiac performance after complete atrioventricular junction ablation and atrioventricular junction modification in patients with medically refractory atrial fibrillation. J Am Coll Cardiol 1998;31:637-44.)

perception of general health and decreased global life satisfaction compared with their healthy counterparts, despite similar scores on all other scales.[85] This study suggested that the subjective effects of AF on isolated physical aspects or on social and emotional spheres might be subtle in patients with little or no symptoms, but the arrhythmia might significantly decrease the overall perception of well-being in this population.

COSTS OF ATRIAL FIBRILLATION

AF imposes a significant and growing economic burden on health care systems. The majority of AF expenditure constitutes hospital charges due to frequent and often prolonged admissions for AF and particularly its costly complications, such as stroke and heart failure. In addition, the cost of caring for patients with other cardiovascular diseases is significantly higher in the presence of AF. The report from a prospective North-American cohort study during 1991 and 1992 revealed up to 22% higher annual costs of hospitalized Medicare patients with AF compared with their non-AF counterparts.[85a] The cost of caring for patients with AF also continued to be significantly higher during the two and three years after initial hospitalization. In general, hospitalizations account for approximately one half of the total cost, followed by drug prescriptions and consultations which constitute 20% to 25% and 10% to 15%, respectively, of the total expenditure.[85b]

Most economic analyses of AF embraced the costs of cardioversions, antiarrhythmic and rate-controlling drugs, and anticoagulation. With novel effective nonpharmacologic therapies emerging that may cure AF completely, the long-term cost-effectiveness could be potentially improved. The Cost Analysis of Catheter Ablation for paroxysmal Atrial Fibrillation (COCAF) study suggests that projected annual costs of conventional medical management and pulmonary vein ablation break even after three years, and after ten years, the cost-effective ratio was .54 in favor of ablation.[85c]

Anatomy and Pathology

In chronic AT, AFL, and AF the architecture of the atrial myocardium is structurally altered to a varying degree. Various pathologic changes occur in the atria that may alter the normal atrial myocardium, resulting in varying types of atrial arrhythmias. However, in discussing the subject of atrial arrhythmias, it is important to consider that part of the conduction system that is related to the atria, such as the sino-artial (SA) node and its approaches, atrial preferential pathways, approaches to the AV node, the AV node, pectinate muscles, Bachmann's bundle, pulmonary veins, and the coronary sinus. In chronic AF, particularly in the elderly, one may find varying types of pathologic changes not only in the atria but also in the distal conduction system including the ventricular myocardium. Rarely "an isolated atrial disease" without involvement of the ventricles may give rise to AF. However, in the majority of instances, pathologic findings are seen in the entire heart, including the conduction system.[76,86-100]

ANATOMY OF THE ATRIA

Histologically, the atrial myocardium is somewhat more loosely arranged than the ventricular myocardium and has a greater amount of connective tissue. The normal orientation of the myocardial fibers is chaotic in the left atrium. In addition, the left atrial myocardium forms a sleeve around the pulmonary veins to a considerable extent with various atrial cells within it. On the other hand, the orientation of the myocardial fibers in the right atrium is relatively smooth in the limbic margin (or the atrial septum). However, the superior wall of the right atrium has pectinate muscles separated literally by spaces as they proceed from the SA node to the AV node. Fatty metamorphosis is almost constant in the right atrium compared with the left. In both atria, numerous elastic fibers are seen. By middle age, fat replacement in the right atrial myocardium becomes quite prominent. In the fourth and fifth decades, collagen and elastic tissue fibers become abundant in the atrial myocardial interstitium in the right atrium, and there is an increase in interstitial tissue with elastification of fat tissue in both atria. In old age, there is atrophy of the musculature of the atria, vacuolar degeneration, fat, and loss of myocardium to a varying extent.[86-88]

PATHOLOGY OF THE ATRIA

The pathologic changes in the atria include congenital abnormalities of the atria, SA and AV nodes, and the AV junction. The acquired changes are: fatty metamorphosis, fibrosis, infiltrative diseases, coronary artery disease, acute and chronic inflammatory processes, and tumors. Likewise, changes in the mitral and tricuspid valves, the tendon of Todaro, central fibrous body, AV membranous septum, and pulmonary veins may be associated with varying types of atrial arrhythmias including AF.[86]

Once established, AF induces further structural alterations of the atrial myocardium, a process known as *atrial remodeling,* which favors perpetuation of the arrhythmia.[101] Atrial myocytes change to a more fetal phenotype, so-called *dedifferentiation.*[102] Atrial myocytes show increased cellular volume, sarcomere misalignment, loss of contractile elements, and accumulation of glycogen (Fig. 15-9). Further changes involve gap-junctional remodeling with the reduction in expression of connexin Cx40 and Cx43. Experimental evidence suggests that these structural changes may lessen or reverse after restoration of sinus rhythm, although the appearance of small elongated mitochondria and loss of myofilament alignment may persist after several months.[103] Chronic stretch and calcium overload during fast AF are likely to contribute to sustained proteolysis, resulting in slow recovery of contractile elements. More advanced changes include atrial hibernation, myolysis, and hypertrophy, followed by irreversible fibrosis and cell death in long-standing arrhythmia, making restoration and maintenance of sinus rhythm unattainable.

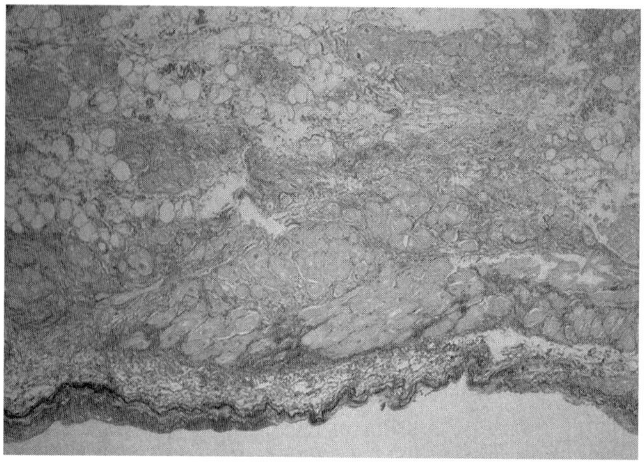

FIGURE 15-9 (See also Color Plate 15-9.) Light microscopy of atrial myocardium showing accumulation of glycogen and fat in vacuolated, oversized myocytes (trichrome stain) in a patient with sustained atrial fibrillation.

Of interest, AF-induced atrial myopathy may also cause remodeling of the sinus node, resulting in depressed sinus node function and the development of tachycardia-bradycardia syndrome.[104]

ATRIAL MYOCARDIUM IN PULMONARY VEINS

Extension of the atrial myocardium into the pulmonary vein wall occurs in normal development of the embryo, and the "sleeves" of myocardial tissue are found in 85% to 95% of the pulmonary veins in asymptomatic adults and in 100% of patients with AF.[105,106] These remnants of atrial tissue may become active as a result of aging metamorphosis and stretch and dilatation of the pulmonary veins in later life and may produce spontaneous, rapid discharge, triggering AT and AF. The average length of "sleeves" is 9.3 mm; they are always external to the smooth muscle layer and are significantly longer in the superior pulmonary veins, which are more frequently involved in the initiation of atrial tachyarrhythmias than the inferior pulmonary veins. Extension of atrial myocardium in patients with AF exhibits a significantly higher degree of hypertrophy, discontinuity, and fibrosis than in control subjects.[106]

Although any chronic disease that affects the atria may produce AF, aging of the heart is the most common cause of AF. In this chapter, we discuss the common causes of AF, such as aging, hypertensive heart disease, coronary artery diseases, diabetes mellitus, postoperative heart disease, infiltrative diseases, genetic causes, and congenital abnormality of the AV junction.[76,86-100]

NORMAL AGING

The normal aging changes of the left atrial myocardium eventually result in a loss of myocardial fibers with an increase in fatty metamorphosis, connective tissue, and fibrous tissue to a varying degree. Some of the atrial myocardial fibers may be atrophied, some hypertrophied, and others in varying stages of degeneration with large spaces between myocardial fibers. This is usually associated with hypertrophy and enlargement of the atria. These changes increase with age, and with time are associated with paroxysmal AF, which eventually results in chronic AF and AFL. It is noteworthy that the normal disarray pattern of the left atrial myocardium and the varying types of myocardial cells within the sleeve around the pulmonary veins form a genetic tendency for the development of AF.[86]

SENILE AMYLOIDOSIS ASSOCIATED WITH AGING

Senile amyloidosis is present in those older than 80 or 85 years of age. A form of primary amyloidosis, it is seen with senility and affects the atrial myocardium more than the ventricular myocardium. Amyloid infiltrates the interstitium and compresses the surrounding myocardial fibers, resulting in degenerative phenomena of the atria (Fig. 15-10). The SA node and its approaches may be infiltrated to a varying degree by amyloid. Amyloid may also involve the SA nodal artery, as well as small blood vessels (arterioles) in the atria.[86,88,89]

HYPERTENSIVE HEART DISEASE

Hypertensive heart disease is associated with the increased incidence of supraventricular arrhythmias and AF. In this disease, there is hypertrophy, and enlargement of both atria associated with moderate to severe coronary artery disease in most hearts. There is pressure hypertrophy of the left ventricle with associated hypertrophy of the heart. This is accompanied by severe small vessel disease (arteriolosclerosis) in various parts of the conduction system, such as SA and AV nodes and their approaches. The SA and AV nodes may be replaced by fibroelastic tissue with or without fat. There is also fibrofatty degeneration of the distal part of the conduction system in many patients with hypertensive heart disease and AF.[86,88,90] Usually, the pathologic process is accelerated with increasing age and coronary artery disease with distinct peripheral conduction system disease. Thus, clinically, left bundle branch block or right bundle branch block may accompany AF in the elderly.[86,88,90]

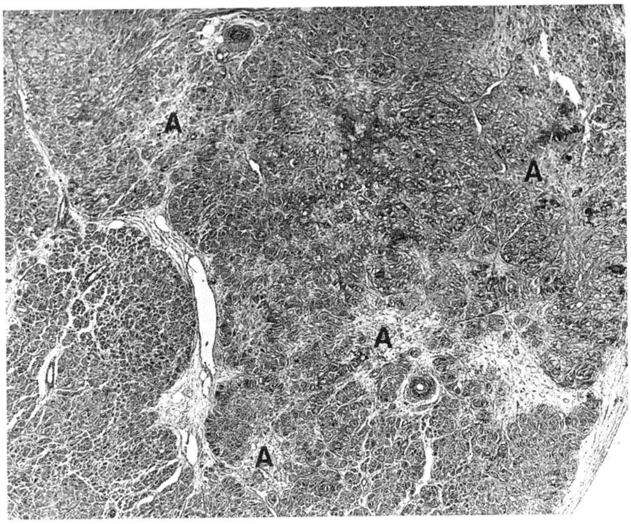

FIGURE 15-10 Atrial myocardium in an 83-year-old man with bundle branch block and atrial fibrillation. Extensive amyloid infiltration of the right atrium with degenerative changes. Hematoxylin-eosin stain ×150. A, amyloid.

CORONARY ARTERY DISEASE

Coronary artery disease either in the acute or chronic form may compromise the blood supply to the proximal and distal conduction system, resulting in varying types of atrial arrhythmias. AF frequently occurs either during acute myocardial infarction or in chronic coronary artery disease. Atrial infarction, with or without infarction of the SA node, with AV block is seen in acute posteroseptal wall infarction with marked narrowing of AV nodal artery.[86,91,92] Chronic coronary insufficiency is usually associated with aging and hypertensive heart disease. The pathologic changes are similar in all of these conditions.[86,91-93]

SICK SINUS SYNDROME IN THE ELDERLY

Sick sinus syndrome in the elderly is usually associated with coronary artery disease and diabetes mellitus. It may be due to arteriolosclerosis with acute or chronic degenerative changes, fibroelastic proliferation, fatty metamorphosis, and focal fibrinoid necrosis of the atrial myocardium including the SA and AV nodes.[86,90,94]

POSTOPERATIVE HEART DISEASE

AF occurs frequently in postoperative cases of various types of congenital cardiac malformations such as closure of an atrial septal defect, the Mustard or Senning procedure for complete transposition, atrial septectomy, and others.[95,96] In the Mustard and Senning procedures the atrial myocardium is replaced by fibrosis, fat, and chronic inflammatory cells. In addition, the AV node practically has no continuity with the surrounding atrial myocardium. In postoperative atrial septal defect, the approaches to the SA node may reveal foreign body reaction to the sutures with fibrosis, neuritis, and fat. The atrial preferential pathways, the approaches to the AV node, may also present similar findings. Thus, fat and fibrosis in the atrial septum and in the SA nodal area with foreign body reaction form the anatomic base for the development of AF following various types of surgery in the atrial septum.[95,96] These changes are usually associated with hypertrophy and enlargement of the atria. In elderly patients, chronic pathologic changes may be the reason for arrhythmias to worsen following closure of the defect.[95]

Although AF is frequently present after thoracotomy or coronary bypass surgery, an anatomic/functional basis has not been clearly defined. The probable causes include surgical injury to the SA node and its approaches, postoperative pericarditis, fluid overload, or other physiologic phenomena. In general, older patients are found to be at greater risk for the development of supraventricular arrhythmias following surgery.[86]

INFILTRATIVE DISEASE OF THE HEART

AF may occur in infiltrative heart diseases, such as sarcoidosis and amyloidosis.[97] As discussed previously, in amyloidosis, the SA node may be infiltrated with amyloid or there may be infiltration in the surrounding atrial myocardium, as well as in the small blood vessels. The atrial myocardium may present marked amyloid infiltration, severe degeneration of muscle cells (see Fig. 15-10), atrophy of some cells, fat, and necrosis.

Likewise, sarcoidosis may affect the atria to a varying degree, resulting in atrial arrhythmias.[97] On the other hand, marked fatty infiltration of the atrial septum, as such, may mimic a lipoma or a tumor and cause AF, particularly in obesity of long duration.[86]

ALCOHOL

Chronic alcohol consumption, especially in the elderly, triggers the onset of AF. Pathologically, there is enlargement of the atria with degenerative changes to a varying degree with a considerable amount of fat extending from the SA node to the AV node.[86]

CONGENITAL ABNORMALITIES

Congenital abnormalities of the AV junction, such as an atrio-Hisian connection, may result in intractable AF in the adult that may be difficult to ablate.[98] Congenital AV block is characterized not only by a lack of connection between the atria and the peripheral conduction system due to extensive fatty metamorphosis of the atria, but the AV node may be absent or hypoplastic. In addition, varying types of degenerative changes are seen in the atria and the peripheral conduction system.[99]

GENETIC ABNORMALITIES OF THE CONDUCTION SYSTEM WITH ATRIAL FIBRILLATION

Familial AF and flutter with advanced or complete AV block may result in sudden death. Pathologically, there is degeneration of atrial myocardium to a varying degree with fat and fibrosis (Fig. 15-11).[100] It was demonstrated recently that missense mutation in the *lamin A/C* gene as a cause of dilated cardiomyopathy was associated with progressive conduction system disease, AF, AV block, congestive heart failure, stroke, and sudden death.[76] Pathologically, the findings in the heart including the conduction system are similar to the findings in AF associated with aging, hypertensive heart disease, and coronary artery disease.[86,88,90-93]

PREEXCITATION

AF is frequently seen in preexcitation. The anomalous AV bypass pathways may conduct in a retrograde fashion and be responsible for the initiation of AF. Pathologically, in addition to the accessory pathway, the atria present similar findings described previously.[86]

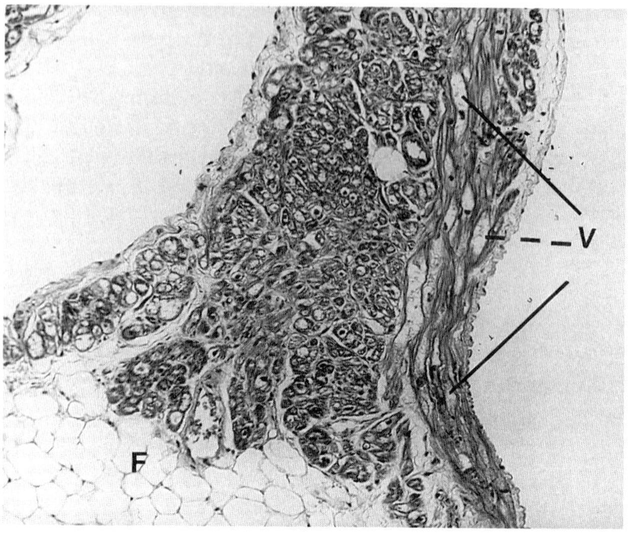

FIGURE 15-11 Atrial biopsy from a 40-year-old woman with a familial history of atrial fibrillation and flutter with advanced or complete atrioventricular block showing vacuolar degeneration, fatty infiltration, hypertrophy of atrial cells and early necrosis. Gomori trichrome stain ×165. F, fat; V, vacuolar degeneration. Arrows point to early necrosis.

MISCELLANEOUS

AF is known to occur in hypertrophic cardiomyopathy, neuromuscular disorders, and many other diseases; however, the pathologic findings are similar to those described previously.[86]

In summary, there are varying types of structural abnormalities of the atrial myocardium in chronic cases of AF. Any chronic disease of the heart has the potential for developing AF. The pathologic changes include loss of myocardial fibers; fat; fibrosis; necrosis; acute and chronic inflammatory changes; atrophy or hypertrophy of myocardium; infiltrative diseases such as sarcoid, amyloid, tumor infiltration of the atrial myocardium; coronary artery disease; and small vessel disease.

These pathologic changes are present to a varying degree in the atrial myocardium on both sides, including the SA node, the AV node, and their approaches. It is also emphasized that there may be distal conduction system disease as well. It is noteworthy that the same kind of pathologic changes responsible for AF may cause other types of atrial arrhythmias.[76,86,100] It is also emphasized that associated metabolic, biochemical, or altered physiologic changes; abnormal neural responses; hormonal imbalances; emotion; and other unknown factors may trigger AF, particularly in patients with preexisting histologic abnormalities of the atria.[86] Genetic abnormalities in ion channel genes may cause a familial type of cardiomyopathy with AF and progressive conduction system disease that may manifest clinically in the fourth or fifth decade of life.[76]

Clinical Electrocardiography

The differentiation of AF from AFL has practical implications when physicians interpret ECGs because it may influence decisions regarding anticoagulation therapy, the use of IV medication for conversion to sinus rhythm, the magnitude of energy used for cardioversion, or referral for ablation procedures. A recent survey showed that cardiologists, cardiology fellows, house officers, and internists have difficulty in distinguishing these arrhythmias.[107] An ECG displaying AF with prominent atrial activity in leads III and V1 was correctly identified by only 31% of physicians. Even among cardiologists and cardiology fellows, less than one third were able to correctly identify the recording as AF. Their accuracy was much better (95%) when they evaluated a recording of AF with prominent atrial activity in V1 that was not apparent in the inferior leads. Cardiologists also identified typical AFL with a high degree of accuracy (92%). The study suggests that both cardiologists and internists have difficulty distinguishing AF from AFL when AF occurs with prominent atrial activity in more than one lead. The ACC/AHA/ESC guidelines and a North American Society for Pacing and Electrophysiology position paper on the classification of atrial flutter and atrial tachycardias have attempted to clarify the electrocardiographic features of these arrhythmias.[33,34]

ATRIAL FLUTTER

AFL is more organized than AF and features a sawtoothed pattern of regular activation that is particularly apparent in leads II, III, and aVF, without an isoelectric baseline between deflections. The rate is typically 240 to 300 bpm. In typical AFL (see Fig. 15-12) that rotates *counterclockwise* in the right atrium, the flutter waves are inverted in leads II, III, aVF, and V6 and upright in lead V1. Biphasic or negative deflections in V1 are less common. Leads I and aVL usually have low amplitude *deflections.*[108,109] When the activation sequence is reversed (*clockwise* rotation), the flutter waves may be upright in leads II, III, and aVF and inverted in lead V1. Wide negative deflections in lead V1 and a positive flutter wave in V6 are characteristic of clockwise rotation in the right atrium.[108,109] Figure 15-13 was recorded in a patient with clockwise AFL. The surface lead characteristics that differentiate clockwise and counterclockwise AFL are summarized in Table 15-3.

TABLE 15-3 Predominant Flutter Wave Morphology in Typical Clockwise and Counterclockwise Right Atrial Flutter

ECG Lead	Clockwise Flutter	Counterclockwise Flutter
II, III, aVF	Positive or negative	Negative
I	Positive	Biphasic/isoelectric
aVL	Biphasic/isoelectric	Positive
V1	Negative or biphasic	Positive
V6	Positive	Negative

Counterclockwise Atrial Flutter

The silent isoelectric zone of the ECG that precedes the negative deflections in the inferior leads corresponds to activation of the low right atrium and isthmus between the tricuspid annulus and inferior vena cava and precedes activation of the left atrium, which begins from the lower septum.[108-114] These studies also show that the left atrial activation sequence is the predominant determinant of the flutter wave morphology. Figure 15-12

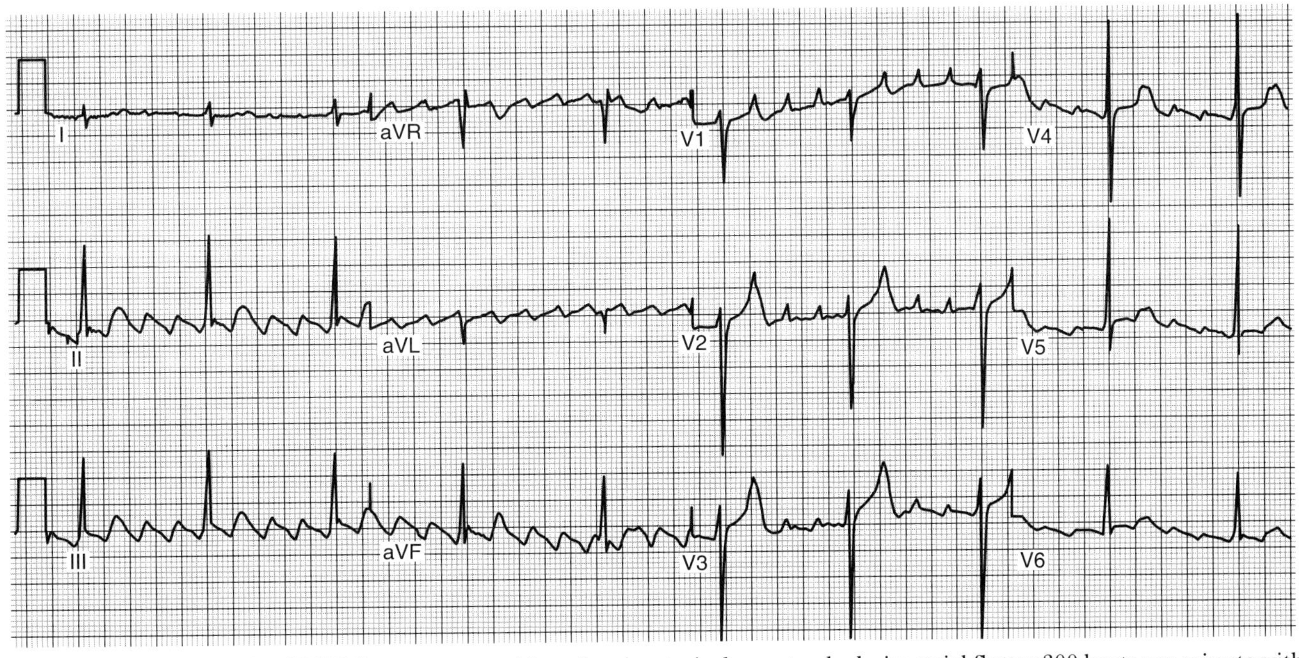

FIGURE 15-12 Twelve-lead ECG from a young athlete showing typical counterclockwise atrial flutter 300 beats per minute with 3:1 atrioventricular block. Note the presence of left ventricular hypertrophy.

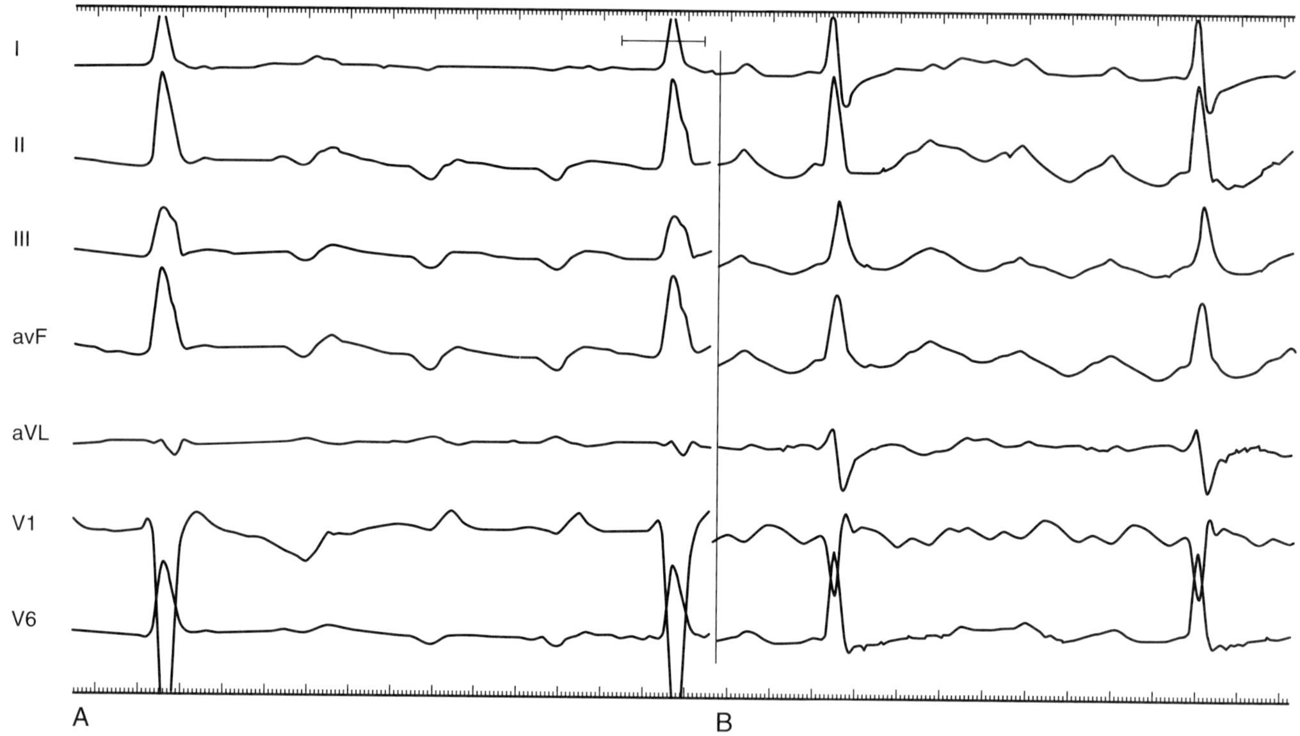

FIGURE 15-13 **A,** Typical counterclockwise atrial flutter. The F waves are negative in leads II, III, avF, and V6 and positive in V1 corresponding to the wavefront traveling down the lateral wall of the right atrium, through the eustachian ridge, and up the interatrial septum. **B,** Reverse (clockwise) atrial flutter. Note the reverse polarity of the F waves: positive in leads I, II, III, avF, and V6 and biphasic in V1.

shows a representative example of counterclockwise AFL. Ndrepepa et al.[110] demonstrated that in typical counterclockwise right AFL the left atrium is activated from both the inferior and superior septum. The posterior wall was activated preferentially from the inferior septum, and the anterior wall was activated from the superior septum. Activation of the lateral wall of the left atrium reflected variable inputs from both regions. Left atrial activation, which required a mean of 133 ± 28 ms, was coincident with the negative component in leads I, II, III, aVF, and V6 and the first flat or slowly rising component in V1. Activation of the lateral wall of the right atrium coincided with the positive deflections in leads I, V1, and V6 and the upstroke component in the inferior leads. The plateau duration in lead III was correlated with the time required for conduction through the isthmus between the tricuspid annulus and the inferior vena cava. Sippens-Groenewegen, et al.[113] correlated body surface mapping with simultaneous endocardial mapping and concluded that the flutter wave cycle length could be divided into three time segments. Caudocranial activation of the right atrial septum occurred in conjunction with proximal to distal activation along the coronary sinus and corresponded to the initial segment of the flutter wave. Craniocaudal activation of the right atrial free wall occurred during the intermediate portion of the flutter wave, and activation of the lateral subeustachian isthmus occurred during the terminal flutter wave.

Clockwise Atrial Flutter

The silent or isoelectric zone of clockwise AFL is shorter compared with counterclockwise AFL.[108,109,113] Saoudi et al.[108] observed a "saw-tooth" pattern in clockwise AFL with a negative deflection in the inferior leads that was interpreted as being very similar to the pattern observed in counterclockwise AFL. A shorter plateau phase was accompanied by widening of the negative component of the flutter wave. Only 3 of the 18 patients in this study exhibited positive flutter waves in the inferior leads. A negative flutter wave in V1 was a constant finding, and flutter waves were predominantly positive in V6 (see Fig. 15-13). Caudal to cranial activation of the lateral wall of the right atrium corresponded to the end of the plateau and the descending part of the negative portion of the flutter wave. The ascending portion of the flutter wave corresponded to the descending activation of the septum and occurred synchronously with proximal to distal activation in the coronary sinus. These results are similar to those reported by Kalman et al.[109] who demonstrated that activation of the lateral right atrium from caudal to cranial corresponded to an inverted component on the inferior leads of variable amplitude just before the development of upright notched flutter waves. In some patients this time period was an electrically silent isoelectric segment. In contrast to work by Saoudi et al.[108] all the patients with clockwise AFL that Kalman et al.[109] studied had prominent upright flutter waves in the inferior leads. The upstroke began when the wavefront of activation reached the superior part of the crista terminalis in the vicinity of Bachmann's bundle. This also corresponded with the onset of the major deflections in the precordial leads. The bulk of the flutter wave was presumably determined by the left atrium.

During clockwise AFL, Ndrepepa et al.[110] observed a dominant breakthrough to the left atrium in the high anteroseptal area in four of five patients. A second breakthrough occurred in the low posterior septal area. Left atrial activation required 130 ± 13 ms and was coincident with positive components on the surface of ECG leads I, II, II, aVF, and V6 and the first negative component in V1. Activation of the lateral wall of the right atrium coincided with the negative components in lead I, inferior leads, and V6. They observed overlap between the initial and terminal activation of the left atrium and right atrial activation. Rodriquez et al.[111] reported similar activation sequences of the left atrium. The body surface maps obtained by Sippens-Groenewegen, et al.[113] attributed the initial segment of the flutter wave to craniocaudal excitation of the right atrial septum. The intermediate segment corresponded to excitation of the isthmus and proximal-to-distal activation along the coronary sinus. The terminal segment corresponded to caudocranial excitation of the right free wall.

Difficulties with ECG Interpretation

The interpretation of AFL morphologies depends on a sufficient degree of AV block to separate the flutter wave from ventricular activation and repolarization. Atypical forms of AFL with diverse flutter wave morphologies that do not have a standard nomenclature complicate ECG assessments. Sometimes the flutter wave morphology is low in amplitude or may be obscured by ventricular repolarization when the ventricular response is rapid. ECGs of atypical right atrial macroreentrant circuits can be difficult to interpret.[114,116] Complex forms of left atrial macroreentry, which may resemble typical right AFL, tend to have predominantly positive flutter waves in V1.[117] Figure 15-14 was recorded from a patient who had undergone a Fontan operation because of transposition of the great vessels. The patient developed AFL, which involved rotation around a scar on the lateral aspect of the right atrium.

ATRIAL TACHYCARDIA

The differentiation of focal AT from AFL may also be confusing. When AFL is treated with antiarrhythmic drugs, the rate may decrease appreciably and overlap with the rate of focal AT that ranges from 130 to 240 bpm (rarely 300 bpm). The isoelectric segment is generally longer, but it may be difficult to distinguish from AFL if the rate is rapid.

The mechanism of AT is attributed to enhanced automaticity, triggered activity, or intra-atrial microreentry. Macroreentrant AT often occurs after surgery for congenital heart disease. Reentrant AT is usually relatively slow (130 to 170 bpm) and can be initiated and terminated by an atrial premature beat. The P–R interval is linked to the rate of tachycardia and is longer than in sinus rhythm at the same rate. A progressive increase in atrial rate with AT onset ("warming up") and progressive decrease before termination ("cooling down") suggest

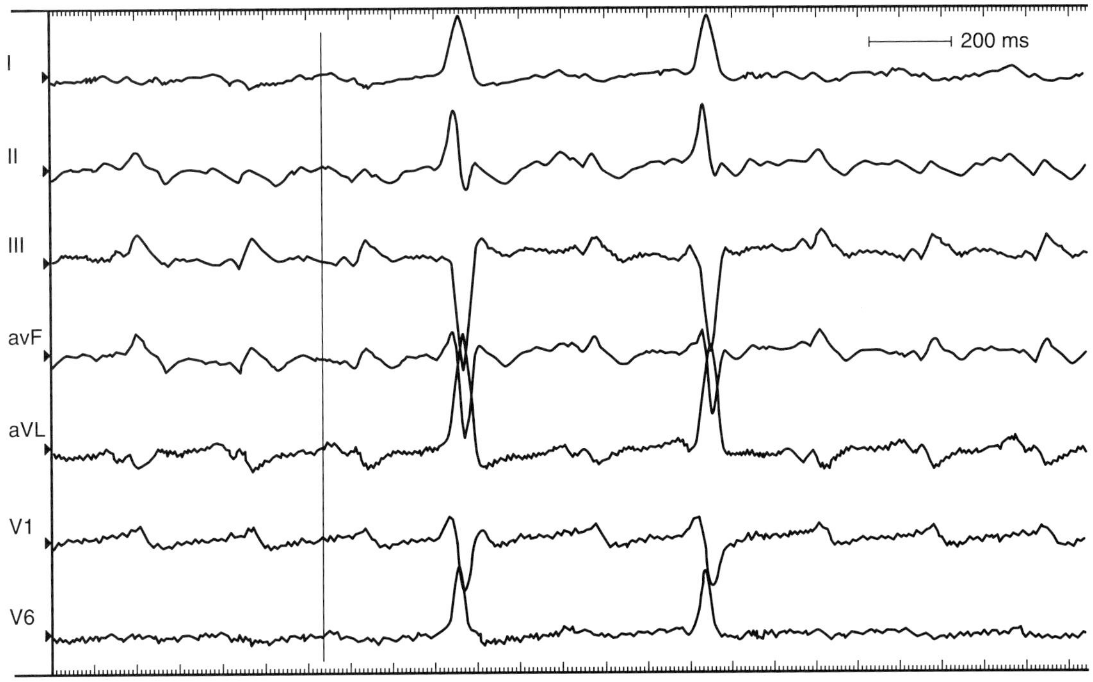

FIGURE 15-14 Atrial flutter after a Fontan operation. Neither typical nor reverse typical pattern of atrial activation can be seen. It is "incisional" atrial flutter in which the wavefront circulates around the scar.

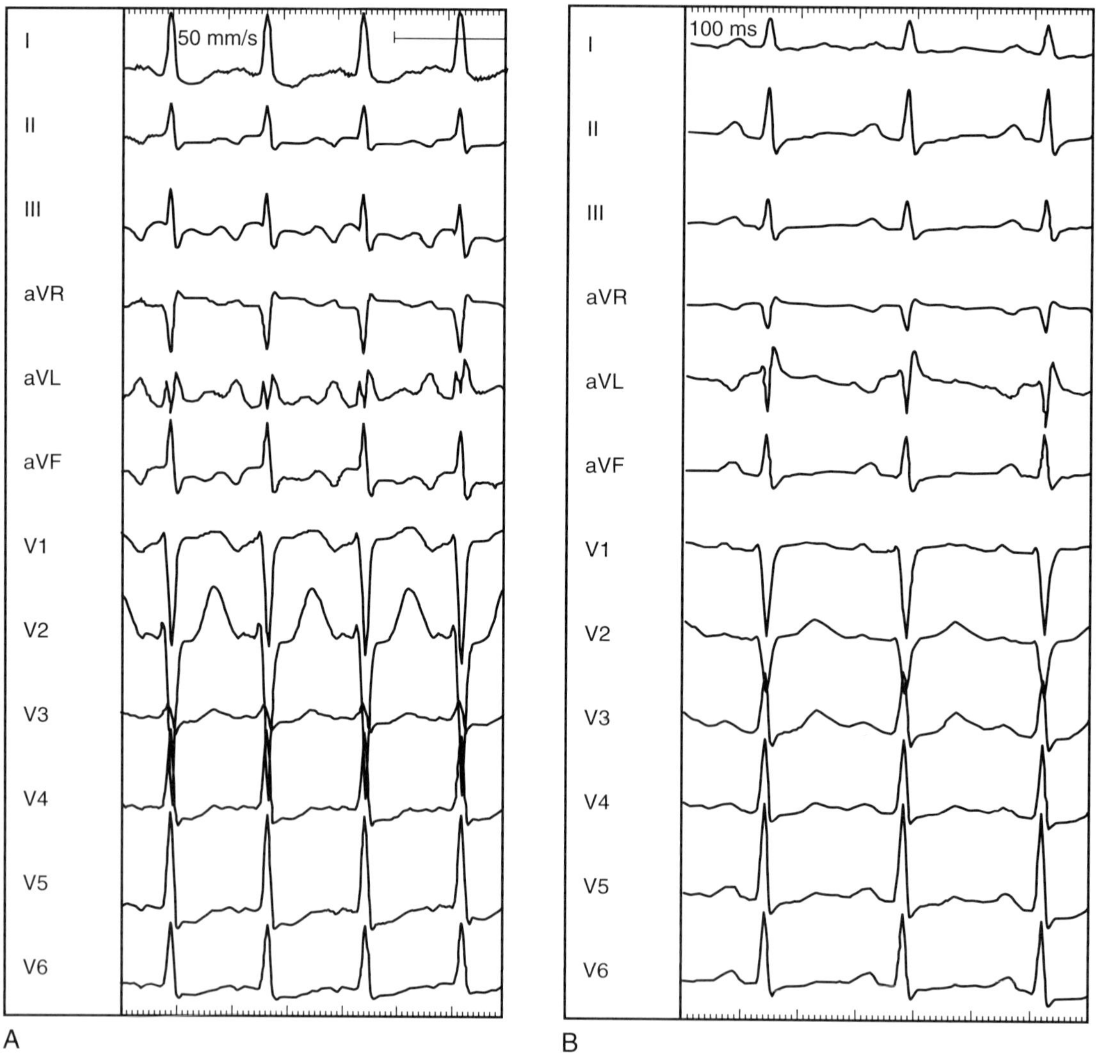

FIGURE 15-15 **A,** Focal atrial tachycardia originating from the right atrium. The P waves are positive in aVL and negative in V_1. **B,** Focal atrial tachycardia originating from the left atrium. Note negative P waves in avL and positive P waves in V_1.

an automatic mechanism. Automatic AT may present as an incessant variety leading to tachycardia-induced cardiomyopathy.

P wave morphology depends on the site of origin. Left atrial AT presents with the negative P waves in leads I, aVL, V5, and V6 (Fig. 15-15). Very fast AT initiated by an atrial premature beat with similar P wave morphology indicates the presence of focal atrial activity, usually in the vicinity of pulmonary veins. P wave morphology similar to that in sinus rhythm suggests SA reentrant tachycardia, which originates from the reentrant circuit within the SA node or involves adjacent atrial myocardium. The average rate is 130 to 160 bpm but may vary from 80 to 200 bpm. Sudden onset and termination helps differentiate between AT and sinus tachycardia. AT with AV block occurs commonly in patients with organic heart disease and in 50% to 75% of cases is due to digitalis toxicity. Multifocal AT presents as rapid, irregular atrial activity with discrete P waves of varying morphology and is considered a transitional rhythm between AT and AF (Fig. 15-16).

RELATIONSHIP BETWEEN ATRIAL FIBRILLATION AND ATRIAL FLUTTER

Although the electrophysiological mechanisms of AFL and AF are distinct, both arrhythmias may occur in the same patient. Multisite endocardial mapping performed by Roithinger et al.[118] demonstrated spontaneous conversion of AF to AFL in 10 of 80 consecutive patients

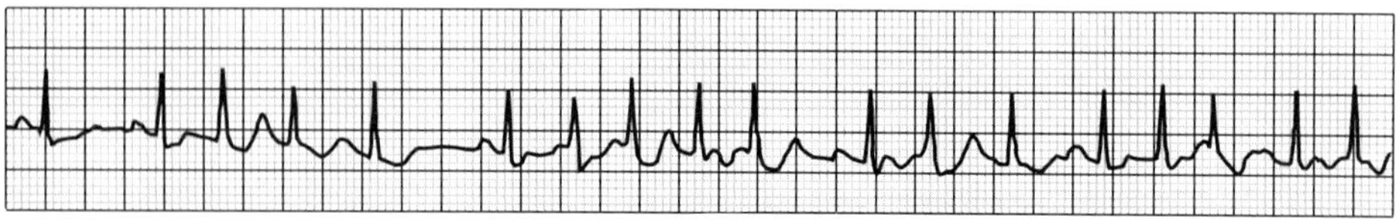

FIGURE 15-16 An ECG strip showing multifocal atrial tachycardia. Note varying morphology of the discrete P waves, suggesting different atrial foci that activate the ventricles at a different rate.

who were referred for ablation of AFL. In these cases gradual organization of AF was observed with changes in the activation sequence on the lateral wall of the right atrium. Counterclockwise AFL was preceded by organization of caudocranial activation on the right free wall, and organization of the right free wall in the opposite direction preceded the onset of clockwise AFL. Just as AF can serve as a trigger for AFL, the observation that 6.4% of AF episodes were preceded by AF demonstrates the interaction between these discrete arrhythmias.[114] The results of these studies fit the observation that ablation of AFL may eliminate AF in some patients.

ATRIAL FIBRILLATION

The ACC/AHA/ESC guidelines[34] have defined AF as consisting of rapid oscillations or fibrillatory waves that vary in size, shape, and timing. The ventricular response to AF depends on the electrophysiological properties of the AV node, the effects of drugs, and the balance between sympathetic and parasympathetic tone. The R–R intervals are irregular unless the patient has AV block or a paced rhythm.

Paroxysmal or intermittent AF appears to be highly dependent on initiation by atrial ectopy. Kolb et al.[114] used 12-lead Holter recordings to characterize the initiation of spontaneous episodes of paroxysmal AF. He observed that 93% were triggered by atrial premature depolarizations, and 6.4% were preceded by typical AFL. The morphology of the initiating P waves was used to estimate the origin of triggering events. Within the limitations of this method, he estimated that 77.5% arose from the left atrium, 2.0% were of right atrial origin, and 13.5% were nonspecific. There was generally an increase in the frequency of atrial ectopy in the 30 seconds that preceded the onset of AF. These results are qualitatively similar to the description by Lu et al.[119] of the electrophysiological characteristics of focal initiation of paroxysmal AF. In this study, 93% of episodes came from the pulmonary veins, and the remainder arose from the superior vena cava. The beats that initiated AF had shorter coupling intervals than those that failed to initiate AF. More than half the episodes were also preceded by cycle length variation with short-long sequences. While there appeared to be qualitative differences in the homogeneity of right atrial activation in paroxysmal compared with chronic AF,[120] no consistent electrocardiographic criteria have been developed to distinguish the standard 12-lead ECG morphology of chronic and paroxysmal AF.

Basic Electrophysiology of Atrial Fibrillation

The basic mechanisms underlying cardiac arrhythmias are discussed in detail in Chapter 3 and are not repeated here. This section deals in detail with the present state of knowledge regarding the basic electrophysiology of AF.

HISTORICAL ASPECTS

In the late 1800s, AF was shown to be the mechanism underlying "delirium cordis," in which the heart was noted to beat without any apparent regularity. With the subsequent development of electrocardiography and of methods to study cardiac electrophysiology, three basic theories emerged regarding the mechanism of AF.[123] These mechanisms are illustrated in Figure 15-17. Mines and Garrey observed the occurrence of regular reentry and fibrillation in cardiac tissue preparations and considered AF to be due to continuous irregular reentrant activity occurring in a dyssynchronous fashion in various atrial regions ("multiple circuit reentry," see Fig. 15-17*A*). Garrey first put forward the idea that fibrillation requires a critical mass of tissue to support a sufficient number of irregular reentrant wavefronts to maintain the arrhythmia. Lewis[124] believed that AF is due to a single rapid macroreentry circuit (see Fig. 15-17*B*), with wavefronts emanating from the primary "driver" circuit breaking against regions of varying and greater refractoriness, producing "fibrillatory conduction" and the irregular global activity characterizing the arrhythmia. Others held that AF is caused by very rapid activity, with either a single source giving rise to fibrillatory conduction (see Fig. 15-17*C*) or multiple ectopic foci producing fibrillation by virtue of dyssynchronous activity and colliding wavefronts. Over subsequent years, various lines of evidence pointed to the relevance of multiple circuit reentry to clinical AF, and from then until relatively recently, multiple circuit reentry (see Fig. 15-17*A*) was widely assumed to be the dominant mechanism underlying clinical AF.

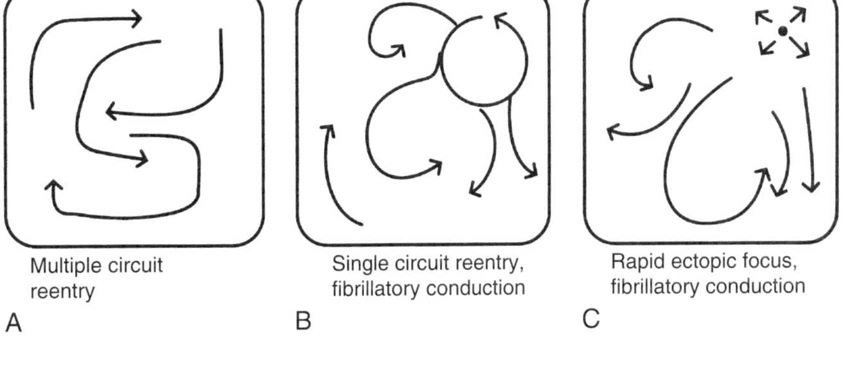

FIGURE 15-17 Theories of atrial fibrillation mechanisms in the early 20th century.

Moe refined the multiple circuit reentry concept by suggesting that activity during AF need not involve complete reentry circuits beginning and ending at the same location, but rather simultaneous wavelets that either extinguish (not contributing to arrhythmia maintenance) or succeed in continuously encountering excitable tissue and maintaining the arrhythmia.[123] Moe viewed the maintenance of AF as a probabilistic function of tissue properties and size, with a minimum temporal density of reentrant wavelets needed to sustain the arrhythmia.[123]

Allessie subsequently added a quantitative element to Moe's reasoning by emphasizing the importance of the wavelength (Fig. 15-18). The wavelength (see Fig. 15-18*A*) is the distance traveled by a cardiac impulse during the refractory period (refractory period × conduction speed) and indicates the shortest path length that can maintain reentry. In circuits shorter than the wavelength, the head of the impulse will impinge on a still-refractory tail after one cycle, and the impulse will die out. According to Allessie's "leading circle" concept of functional reentry, functional reentry circuits naturally form around a perimeter equal to the wavelength.[126] Allessie postulated that shorter wavelengths favor AF maintenance by increasing the number of simultaneous functional reentrant circuits that the atria can accommodate (see Fig. 15-18*C*). A corollary of this notion is that multiple circuit reentry AF can be terminated or prevented by increasing the refractory period and consequently the wavelength, thereby reducing the number of circuits possible and making AF maintenance less likely (see Fig. 15-18*D*).[127] By contrast, factors that reduce the wavelength (like short refractory periods and slow conduction) should favor AF.[128] Atrial dilatation should also favor reentry, by increasing atrial size and thereby increasing the number of circuits that the atria can accommodate (see Fig. 15-18*E*). Heterogeneity in refractory properties should also favor AF by promoting the fractionation of impulses into multiple reentrant wavefronts.

In recent years, more sophisticated models of functional reentry have been developed, such as the "spiral wave" model. A spiral wave involves continuous activity in a spiral pattern, somewhat like a hurricane. It differs from a leading circle in that the center of spiral wave reentry is excitable but not activated (like the eye of a hurricane), and the perpetuation of activity is determined by the angle of curvature of radiating activity. The consequences of the spiral wave activity concept for the initiation and maintenance of AF are poorly understood at the moment; therefore, the leading circle remains the point of reference for understanding multiple circuit reentry.

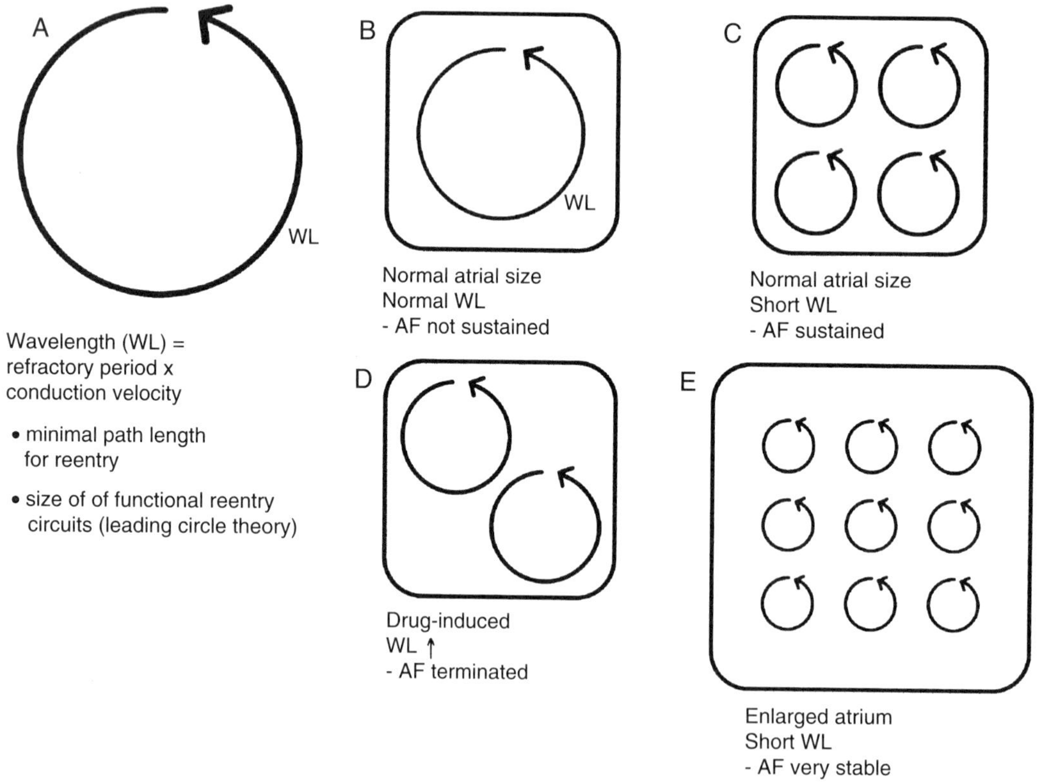

FIGURE 15-18 **A,** The role of the wavelength in determining minimum circuit size for reentry and the size of functional reentry circuits according to the leading circle concept. **B,** In atria of normal size with normal WL values, the atria cannot support multiple circuit reentry, and even if atrial fibrillation is induced, it usually stops spontaneously. **C,** If WL is reduced (e.g., by vagal stimulation or electrical remodeling), enough circuits can be accommodated to maintain AF even in normal atria. **D,** Drugs that increase refractory period increase wavelength and circuit size, thereby making AF harder to sustain, promoting cardioversion and sinus rhythm maintenance. **E,** If atria are enlarged and have short wavelengths, AF tends to be very stable and sinus rhythm very hard to produce and maintain. AF, atrial fibrillation; WL, wavelength.

ELECTRICAL REMODELING

A key concept in understanding AF is that of electrical remodeling. It has become obvious over the past few years that atrial electrical properties are altered by sustained AF, such that the atria become more susceptible to the initiation and maintenance of the arrhythmia.[129] The primary factor driving AF-induced remodeling appears to be atrial tachycardia, and similar changes can be produced by other atrial tachyarrhythmias like AFL and AT. Electrical remodeling occurs in a time-related fashion following the onset of AF and likely involves a series of adaptations stimulated by increased cellular Ca^{2+} loading due to the increased rate of activation, with Ca^{2+} entering the cell with each activation.[130] The electrophysiological changes caused by tachycardia-induced atrial electrical remodeling are summarized in Figure 15-19.

Short-term changes involve functional alterations, primarily Ca^{2+} current (I_{Ca}) inactivation, that reduce AP duration (APD), refractory period, and wavelength and thereby promote AF. Longer-term changes include alterations in the production of ion channel proteins, among which a key alteration appears to be downregulation in the L-type Ca^{2+} current ($I_{Ca.L}$) that maintains the plateau of the AP and triggers cardiac contraction.[131] $I_{Ca.L}$ downregulation reduces APD and attenuates APD adaptation to rate, which is largely due to tachycardia-dependent $I_{Ca.L}$ reduction (that normally reduces APD at fast rates). In addition, Na^+ current also appears to be reduced, decreasing conduction velocity, and the intercellular coupling channel protein, connexin 40, is reduced in a heterogeneous fashion. The resulting heterogeneous reductions in atrial wavelength make AF more likely to sustain itself and increase the vulnerability of the atria to AF reinduction should the arrhythmia be terminated.

The concept of atrial electrical remodeling due to AF is very important, because it explains why paroxysmal AF tends to become chronic, why longer-lasting AF is more difficult to treat, and why AF recurrence is particularly likely the first few days after electrical cardioversion. However, in order for remodeling to occur, AF needs to be sustained for significant periods of time. Therefore, other mechanisms must be involved in triggering and maintaining AF before remodeling can occur.

STRUCTURAL REMODELING

Autopsy studies of atrial tissues in patients with AF often show extensive atrial fibrotic changes, particularly in association with conditions such as mitral valve disease, congestive heart failure, and senescence. Recent clinical studies suggest that patients with AF have activation of the atrial renin-angiotensin system and related mitogen-activated protein kinases with profibrotic actions. In an animal model of congestive heart failure, sustained AF can be induced fairly readily, with an underlying substrate that involves local conduction abnormalities related to intense atrial interstitial fibrosis. AF in this model sometimes has the appearance of macro-reentry with fibrillatory conduction,[132] consistent with the mechanism illustrated in Figure 15-17*B*. In addition to causing abnormalities in conduction, structural remodeling is often associated with atrial dilatation, favoring AF as illustrated in Figure 15-18*E*. Cellular electrical properties are also altered in heart failure but not in the same way as with atrial tachycardia-induced electrical remodeling.[133] APD is not shortened, and so the wavelength is not reduced; however, the activity of the Na^+, Ca^{2+} exchange current (NCX) is increased. The NCX exchanges Ca^{2+} accumulated in the cell during the AP for extracellular Na^+, with one intracellular Ca^{2+} ion exchanged for three extracellular Na^+ ions. Thus, the NCX carries one net positive charge into the cell with each cycle. When NCX activity is enhanced, delayed after depolarizations and triggered ectopic activity can result. Congestive heart failure,

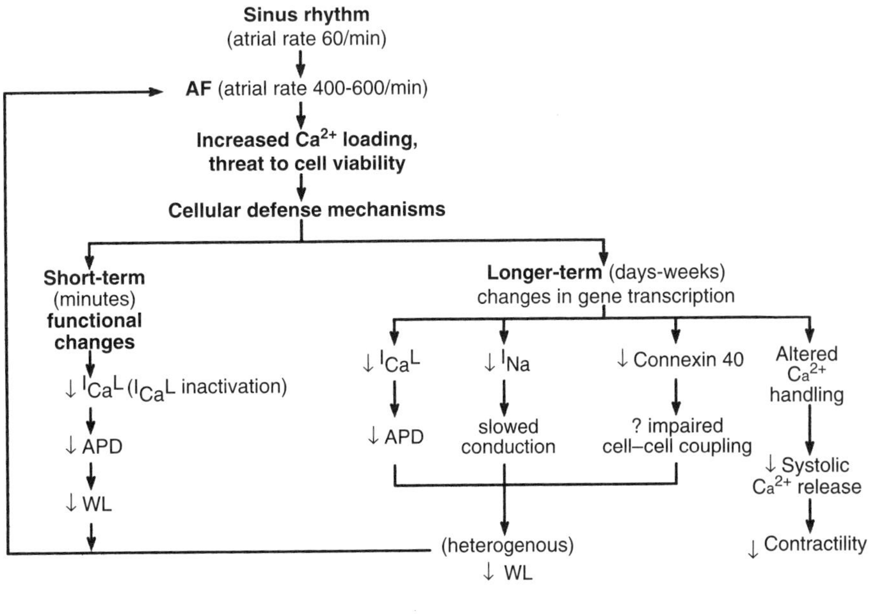

FIGURE 15-19 Mechanisms involved in atrial electrical remodeling induced by atrial fibrillation. Similar mechanisms operate for any form of atrial tachycardia. Atrial tachycardia increases intracellular Ca^{2+} loading, which initiates a series of physiologic changes, some of which (like decreased I_{Ca}) reduce cellular Ca^{2+} entry and prevent potentially lethal Ca^{2+} overload. Protection against Ca^{2+} overload comes at the expense of decreased WL, which promotes AF. AF, atrial fibrillation; WL, wavelength.

therefore, favors AF both by creating a structural substrate for atrial reentry and by producing a functional basis for atrial ectopic activity that can trigger reentry.

SIGNIFICANCE OF ECTOPIC ACTIVITY

It has recently been shown that many patients with AF have enhanced atrial ectopic activity originating in the pulmonary vein region.[17] These ectopic foci can promote AF both by triggering atrial reentry in the presence of a susceptible substrate and by causing rapid atrial tachycardias that cause electrical remodeling, which then creates the reentrant substrate for AF. Less commonly, ectopic activity in patients with AF can originate in other atrial sites, such as the venae cavae and the ligament of Marshall, an anatomic remnant structure close to the great cardiac vein. Irrespective of the site of origin of atrial ectopy, effective ablation of the ectopic site can prevent AF recurrence, indicating its importance in the pathophysiology of the arrhythmia.

DETERMINANTS OF THE VENTRICULAR RESPONSE

In addition to the physiologic factors controlling occurrence and maintenance of the atrial arrhythmia in AF, a very important determinant of the clinical manifestations is the ventricular response rate. There are two largely independent determinants of the ventricular response: (1) the rate and regularity of atrial activity, particularly at the entry to the AV node, and (2) the refractory properties of the AV node itself.

The pattern of atrial input is an important determinant of the ventricular response, but decreases in atrial rate do not always cause a reduced ventricular response. The central, compact portion of the AV node is composed of tissue with a slow, $I_{Ca.L}$-dependent upstroke ("slow-channel tissue"). Conduction through slow-channel tissue is not all-or-none: impulses entering the AV node during its relatively refractory period can penetrate partway and then die out, leaving the AV node partially refractory for the next impulse. Thus, not only does the impulse fail to conduct, but it makes it more difficult for the next impulse reaching the AV node to conduct as well. This "concealed conduction" means that a very rapid, irregular input into the AV node can be associated with a lower ventricular response rate than a slower, more regular atrial activation pattern. Drugs that slow the atrial rate during AF, such as class IC antiarrhythmic agents, can therefore sometimes produce a paradoxical and potentially dangerous increase in the ventricular response rate.

The second important determinant of the ventricular response is the functional state of the AV connection. Normally, it is the refractory properties of the slow-channel tissue in the AV node that determine the ventricular response. Drugs that interfere with conduction through slow channel tissues, like $I_{Ca.L}$-antagonists (diltiazem or verapamil), β-adrenoceptor antagonists (that oppose the conduction-enhancing effect of background β-adrenergic tone), and digitalis (that acts by vagal enhancement), increase the filtering function of the AV node and slow the ventricular response rate. Unlike slow-channel tissue, with a refractory period determined largely by the time required for recovery of $I_{Ca.L}$ following depolarization, the refractory period of "fast-channel" tissues (with I_{Na}-dependent upstrokes) is determined by APD. APD (and consequently the refractory period) of fast-channel tissues decreases markedly at faster rates. Thus, when the atria are connected to the ventricles by an accessory bypass tract of fast-channel tissues (as in the Wolff-Parkinson-White syndrome), the bombardment of the bypass tract at very rapid rates during AF may result in very short refractory periods and dangerously rapid ventricular response rates due to conduction across the accessory pathway during AF.

SYNTHESIS OF PHYSIOLOGIC DETERMINANTS IN ATRIAL FIBRILLATION

Figure 15-20 presents a synthesis of the physiologic determinants of AF described earlier. Ectopic activity

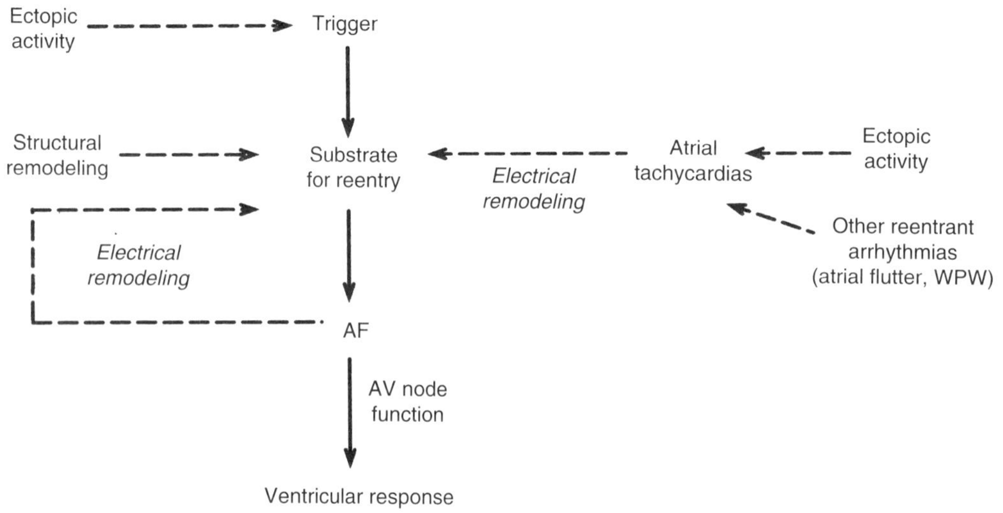

FIGURE 15-20 Synthesis of physiologic determinants of AF and its consequences in terms of the ventricular response. WPW, Wolff-Parkinson-White syndrome.

plays an important role by triggering reentry and AF in the presence of a vulnerable substrate, such as that created by structural remodeling. In addition, continuous atrial ectopic activity can generate atrial tachycardias that cause atrial electrical remodeling which produces a substrate for AF. Thus, ectopic activity can result in both the substrate for AF (atrial remodeling) and the trigger that acts on the substrate to cause the arrhythmia. Other atrial tachyarrhythmias (including AFL, AV node reentry, and AV reentry tachycardias) can also cause tachycardia-dependent remodeling and produce a substrate for AF, accounting (at least in part) for the occurrence of AF in association with these arrhythmias in some patients. If AF is induced by any mechanism, electrical remodeling will ensue, causing heterogeneous wavelength decreases that promote multiple circuit reentry. Thus, electrical remodeling tends to make multiple circuit reentry the final common pathway of AF, irrespective of the initial mechanism of the arrhythmia. As discussed earlier, even though AF may begin as macroreentry with fibrillatory conduction (as in some cases of experimental heart failure–related AF) or as a rapidly discharging focus with fibrillatory conduction, by the time patients get medical attention, they may already have multiple-circuit AF because of remodeling.

Recent discoveries regarding AF mechanisms cast an interesting light on the theories of AF in the early 20th century, as illustrated in Figure 15-17. Recent work suggests that all the original ideas regarding AF mechanisms are probably accurate for some subsets of patients. Furthermore, the subsequent dominance of the multiple circuit reentry concept is also well-founded, since electrical remodeling acts to make multiple circuit reentry the final common pathway of AF in many cases. The dynamic nature of AF mechanisms needs to be considered in order to understand the mechanisms of the arrhythmia and to optimize therapy in any given patient.

TABLE 15-4 Cardiac and Extracardiac Causes and Precipitants of Atrial Tachyarrhythmias

Cardiac	Extracardiac
Hypertension	Thyrotoxicosis
Congestive heart failure	Pheochromocytoma
Coronary artery disease and myocardial infarction	Acute and chronic pulmonary disease (pneumonia, COPD, pulmonary fibrosis)
Valvular heart disease	Pulmonary-vascular disease (pulmonary embolism)
Cardiomyopathy: dilated, hypertrophic	Electrolyte disturbances (hypokalemia)
Myocarditis and pericarditis	Increased sympathetic tone (exercise, adrenergically mediated arrhythmia)
Congenital heart disorders	Increased parasympathetic tone (vagally induced and postprandial arrhythmia)
Wolff-Parkinson-White syndrome	Excessive alcohol consumption ("holiday heart" and long-term use)
Conduction system disorders (sinus node dysfunction, intra-atrial conduction delay)	Caffeine
Cardiac tumors	Smoking
Cardiac surgery for congenital heart disorders, coronary bypass, valvuloplasty, and valve replacement	Recreational drug use
Interventions for treatment or modification of arrhythmias and conduction disorders (permanent ventricular-based pacing, ablation of atrial flutter)	Emotional stress, sleep deprivation

AF, atrial fibrillation; COPD, chronic obstructive pulmonary disease.

Investigations in Atrial Tachyarrhythmias

The initial evaluation of a patient with atrial tachyarrhythmias must first consider the presence and nature of the underlying pathology, which may act as a major clinical cause of the arrhythmia or a precipitant, or both. In a significant proportion of patients, several such factors can be simultaneously involved. Although atrial tachyarrhythmias are more likely to occur in conjunction with cardiovascular disease or cardiovascular intervention, it is not uncommon for them to be provoked or aggravated by various extracardiac factors (Table 15-4).[132] Identification and treatment of underlying pathologies, such as acute illnesses or correctable metabolic disorders, in particular thyrotoxicosis, is integral to the management of patients with atrial tachyarrhythmias. Once the precipitating agent is eliminated, the arrhythmia may terminate spontaneously and usually is unlikely to recur. Next, the duration and pattern of the arrhythmia as paroxysmal, persistent or permanent, the symptomatic status, and clinical impact should be assessed, and the risk of possible complications and the prognosis should be estimated. This assessment is important for the subsequent selection of the most appropriate investigations from the extensive range of possible tests. The assessment allows the determination of the principal management strategies, including the restoration and maintenance of sinus rhythm (the rhythm control strategy), pharmacologic or nonpharmacologic control of the ventricular rate response and lifelong anticoagulation, and the identification of arrhythmias that can be cured by ablation or surgical methods (Table 15-5).

Clinical factors obtained from the history and physical evaluation largely determine: (1) whether the patient should be cardioverted immediately (usually with direct current shock) or later, thereby permitting performance of transesophageal echocardiography (TEE) in order to exclude left atrial thrombosis, particularly when the arrhythmia has lasted more than 24 to 48 hours, and an attempt at pharmacologic cardioversion; (2) the nature and extent of further investigations; and (3) the evaluation of optimal treatment strategies.

Many cases of atrial tachyarrhythmias, particularly AF, are discovered incidentally and are thought to be asymptomatic. Although about one quarter of patients truly may not have symptoms, more often symptoms are not elicited as physicians tend to focus on palpitations as the most important feature of the arrhythmia, but patients are often more aware of mild exertional dyspnea, poor exercise tolerance, and fatigue, especially

TABLE 15-5 Aims of and Tests for the Evaluation of a Patient with Atrial Tachyarrhythmias

Test \ Aim	Rule Out Underlying Disease	Assess the Arrhythmia Pattern	Evaluate Clinical Impact	Decide Whether to Cardiovert Acutely	Decide if Nonpharmacologic Management Can Be First Line	Select Rhythm Versus Rate Control Strategy	Estimate Risk for Proarrhythmia	Estimate Risk for Stroke and Need for AC	Predict the Effectiveness of Treatment
History	+++	+++	+++	++	+	++	++	+++	++
Physical exam	+++	+	+++	+++	+	++	+	+	+
Blood tests	+++	—	++	+	—	—	+	?	—
ECG	+++	++	+	+++	++	+	++	—	+
Holter recording	++	+++	+	+	++	++	++	—	+
Event recorder	?	+++	+	—	++	—	+	—	+
Echocardiogram	+++	—	++	+++	—	++	+	+++	++
X-rays	+++	—	++	++	—	++	—	—	+
Exercise	++	++	++	—	—	+	+	—	+
TEE	+	—	+	+++	—	+	—	+++	+
EP study	++	+++	++	—	+++	++	++	—	++
QoL questionnaire	?	+	+++	—	—	+	—	—	±

+++, very important, almost always should be involved in decision; ++ often should be used, sometimes to answer a specific question; + may be useful for confirmation of existent knowledge or as a source of additional information; ?, not established if the hypothesis of hypercoagulative state is confirmed and reliable markers are established; AC, anticoagulation; EP study, electrophysiological study; QoL, quality of life; TEE, transesophageal echocardiography.

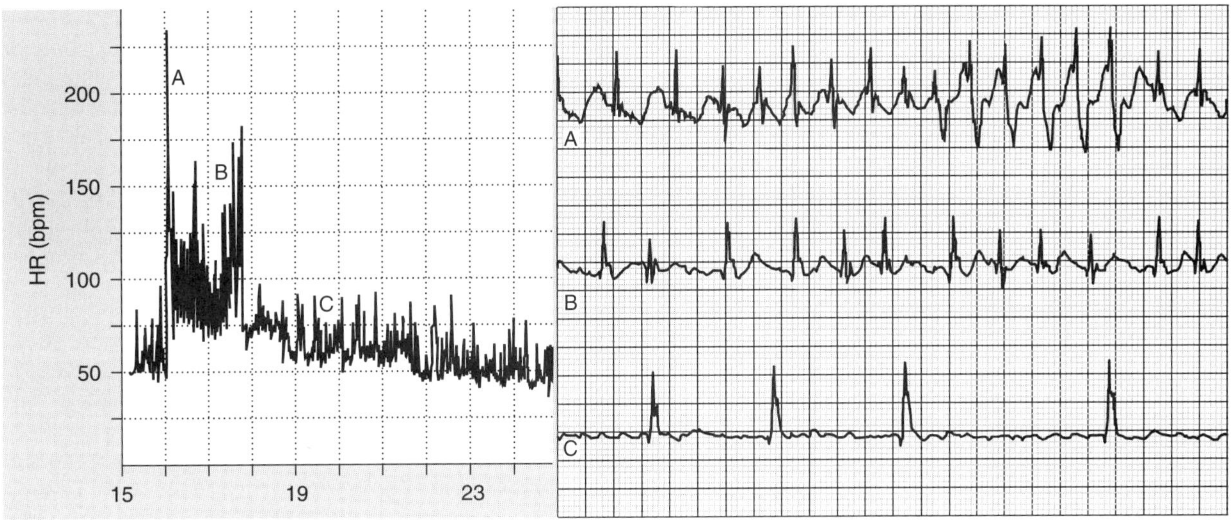

FIGURE 15-21 **A,** Holter monitor tachogram and ECG strips from a patient with atrial fibrillation showing: (**A**) symptomatic episode of the arrhythmia with a rapid ventricular rate response (>200 beats per minute). **B,** Symptomatic episode of the arrhythmia with a slower but irregular ventricular rhythm. **C,** Asymptomatic episode with a slow ventricular rate response. In such "intermittently symptomatic atrial fibrillation," the presence of frequent asymptomatic periods can be mistakenly assumed for sinus rhythm.

if the ventricular rate is slow.[44] The true impact of the disease may be missed, and patients with long-standing unrecognized and untreated arrhythmia are exposed to high risk of thromboembolic complications or tachycardia-induced left ventricular dysfunction, rendering ischemic stroke and heart failure as the first clinical manifestations. However, clinical symptoms cannot be relied upon to differentiate a paroxysmal recurrent form from permanent arrhythmia as patients may complain of episodic palpitations when rapid or highly irregular ventricular rates are present but may not be aware of their arrhythmia when the ventricular rate is slow (Fig. 15-21). Alternatively, pharmacologic therapy can render recurrence of paroxysmal arrhythmia asymptomatic or modify the perception of the arrhythmia by the patient.[50]

Chest x-rays and the ECG are essential in a workup of the patient with suspected tachyarrhythmias. Chest x-rays may help to exclude significant underlying cardiovascular pathology (e.g., cardiac enlargement, aortic dilatation) and to assess the effects of treatment (e.g., rhythm restoration or adequate rate control, reduction in heart size, and improvement of heart failure). In about 20% of patients, uncontrolled atrial tachyarrhythmias may be the sole cause of heart dilatation (tachycardia-induced cardiomyopathy) (Fig. 15-22). The diagnosis of any arrhythmia requires electrocardiographic documentation, the probability of which can be increased by

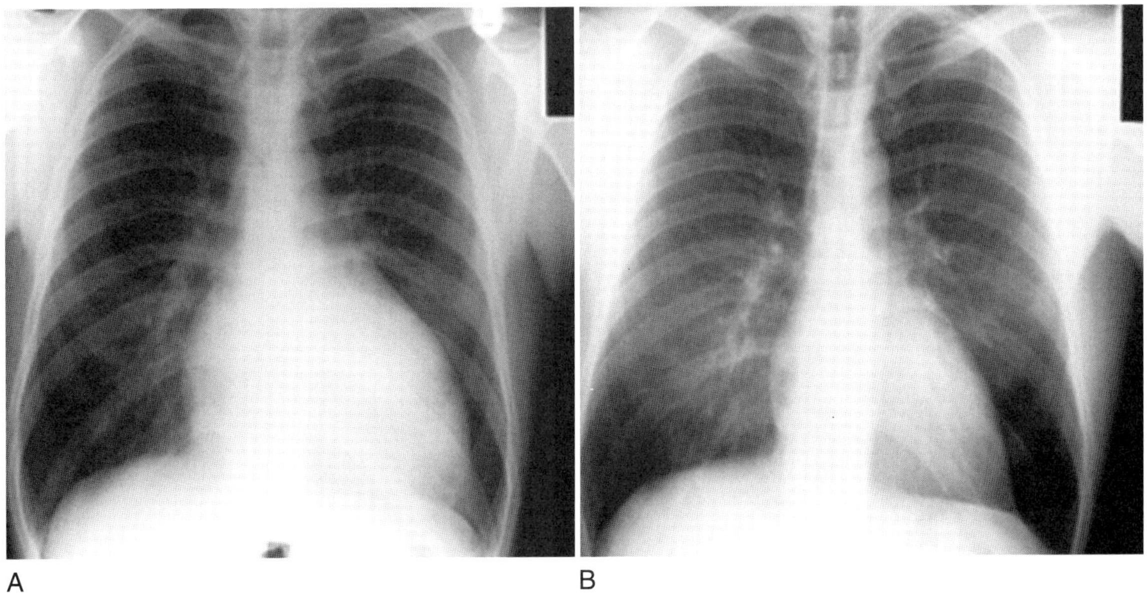

FIGURE 15-22 **A,** Chest x-rays in a patient with suspected tachycardia-induced cardiomyopathy due to persistent atrial fibrillation with rapid ventricular rates. **B,** Same patient 3 months after restoration of sinus rhythm.

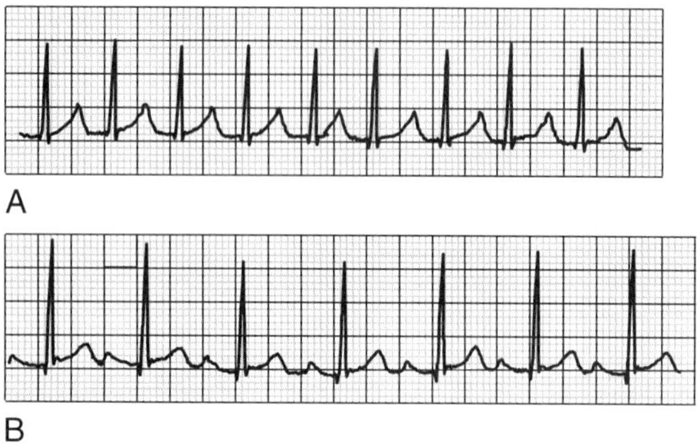

FIGURE 15-23 A, An ECG strip showing atrial tachycardia with ventricular rates of 140 beats per minute (bpm) before treatment. **B,** Atrial tachycardia is still present after the initiation of sotalol, but ventricular rates are significantly slower (100 bpm).

using standard ECG; rhythm strips obtained during the arrhythmia; an ambulatory Holter recording; transtelephonic or telemetric monitoring; and loop event recorders, including implantable monitoring devices.

Proper evaluation of the 12-lead ECG is vital for several reasons. First, the diagnosis must be confirmed, and the type of the arrhythmia should be differentiated, as confusion may arise from other causes of irregular pulse (e.g., heart block or frequent ventricular premature beats). Conversely, a regular, slow pulse can be found in patients with AFL or AT if the ventricular rate is pharmacologically slowed down (Fig. 15-23). AF may coexist with complete heart block where atrial fibrillatory activity is visible as the baseline, but the ventricular rhythm is slow and regular. Second, the presence of preexcitation syndrome on the baseline ECG determines further investigation and management. Third, other abnormalities such as Q waves, low amplitude ventricular complexes, ST-T changes, left ventricular hypertrophy, and axis deviation can be useful for identification of coexistent disease. Finally, monitoring the P–R and Q–T intervals and the QRS complex width is essential to prevent proarrhythmias during antiarrhythmic drug therapy.

Ambulatory 24-hour Holter monitoring is useful for diagnosis of frequent paroxysms of the arrhythmia and for assessment of the effects of rate-slowing agents if rate control was selected as a principal management strategy (Fig. 15-24). In some cases (e.g., "focal" AF), it may provide information pertinent to the principal management of the arrhythmia—documentation of regular atrial tachycardia with a ventricular rate 200 to 250 ms at the onset of AF may be crucial for the selection of nonpharmacologic treatment alternatives (Fig. 15-25). Novel digital ambulatory recorders enable standard 12-lead configurations similar to those of the digital

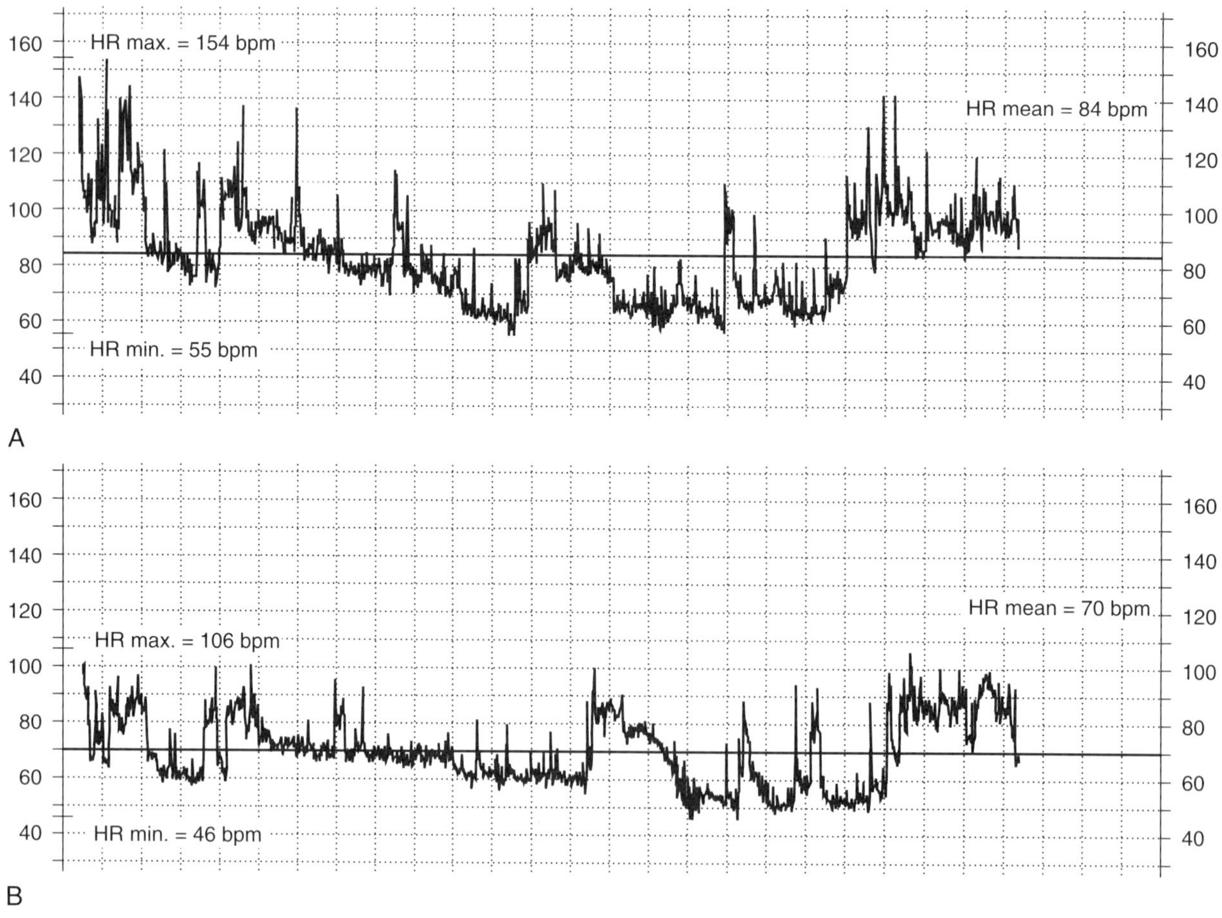

FIGURE 15-24 A, Holter monitor tachograms from a patient with permanent atrial fibrillation before treatment. **B,** During rate control therapy with atrioventricular node blocking agents.

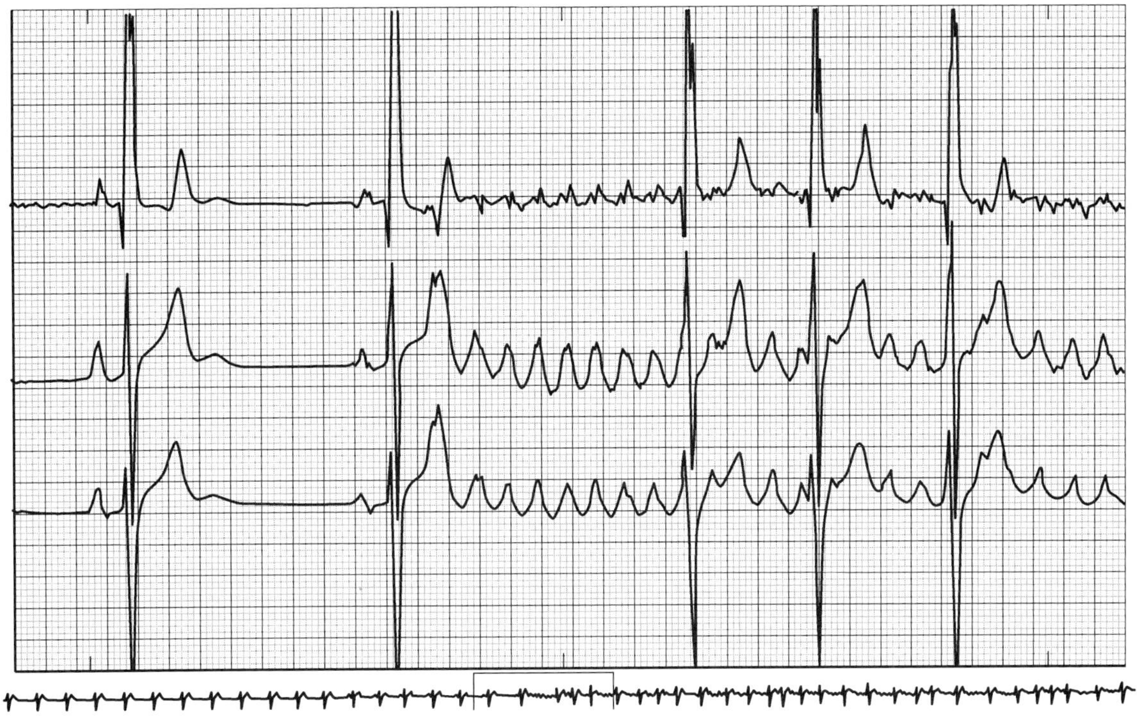

FIGURE 15-25 Holter ECG recording of onset of the arrhythmia in a patient with paroxysmal atrial fibrillation (AF). Note the presence of focal atrial tachycardia with a cycle length of 190 to 200 milliseconds, which initiates AF. As this arrhythmia is likely to originate from the region of the pulmonary veins and may be amenable to radiofrequency catheter ablation, the next step in investigation should be electrophysiological study.

diagnostic-quality 12-lead ECG machines, which may increase the diagnostic accuracy for the detection and localization of arrhythmic events.[116]

When episodes of the arrhythmia are infrequent, a loop event recorder with transtelephonic or telemetric facilities can offer long-term, continuous ECG monitoring and event-specific recording. The implantable loop recorder features both patient-activating and auto-activating working modes for recording the event and is capable of continuous ECG monitoring for up to 18 months. The probability of capturing the arrhythmic event ranges from 65% to 88%.[133] Implantable loop recorders are superior to Holter monitoring systems in the diagnosis of very infrequent arrhythmic attacks and the determination of onset patterns, identification of asymptomatic arrhythmias, assessment of treatment efficacy, and detection of proarrhythmias (Fig. 15-26).

Of note, much data on the prevalence of atrial tachyarrhythmias, transition between different types, modes of onset and termination, underlying rhythms, and potential triggers have come from implantable pacemakers and cardioverter-defibrillators capable of continuous monitoring and accurate detection of arrhythmias. Substantial improvement in the memory function of implantable devices has recently made possible the recognition of a high prevalence of asymptomatic atrial arrhythmias.[49] Analysis of the distribution of potential triggers showed that bradycardia, atrial premature beats producing short-long sequences, and short runs of AT were responsible for triggering the majority of sustained episodes of AF, suggesting that nonpharmacologic treatment options may be available to a wide range of patients with this arrhythmia.[134]

Exercise stress testing can be used for the evaluation of patients with predominantly atrial tachyarrhythmias, for the exclusion of adrenergically mediated or exercise-induced AF in the absence of structural heart disease, and for the assessment of the effectiveness of rate control therapy in patients with permanent AF (Fig. 15-27). It is also generally performed in individuals with coronary heart disease, especially if type IC antiarrhythmic drug therapy is planned, or class IC toxicity (improved AV conduction or impaired intraventricular conduction, or ventricular arrhythmia) is suspected.

Essential blood tests include a full blood count and hemoglobin; serum electrolytes, including serum potassium and magnesium levels; and the assessment of thyroid function. Serum cholesterol is often warranted in

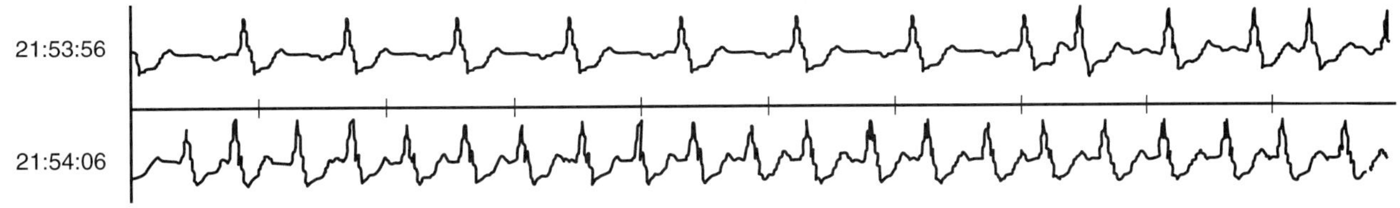

FIGURE 15-26 An ECG strip of the onset of atrial tachycardia recorded by the Reveal device.

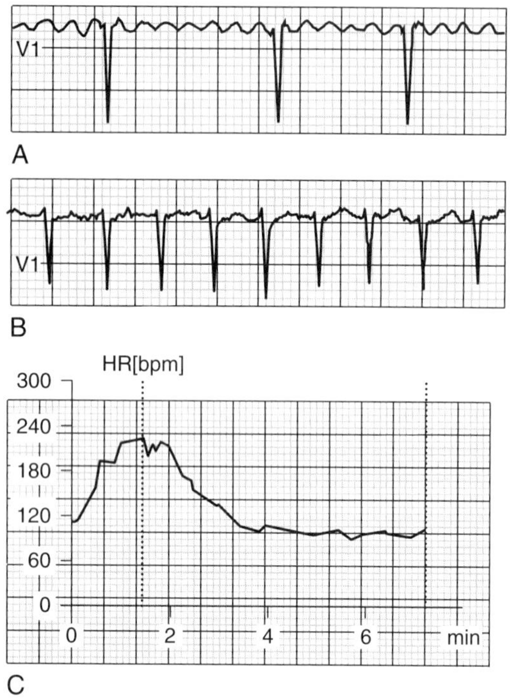

FIGURE 15-27 Inadequate ventricular rate control during exercise in a patient with permanent atrial fibrillation treated with digoxin only. **A,** Baseline ECG. **B,** An ECG after 1½ minutes of exercise stress test. **C,** A tachogram showing a rapid increase in ventricular rates immediately after the beginning of exercise.

patients with risk factors for coronary artery disease. Thyrotoxicosis accounts for 1% to 2% of new cases of AF, but abnormal thyroid function tests can be found in 4% to 12.5% of AF patients.[135,136] Recognition of hyperthyroid state is important because cardioversion and long-term maintenance of sinus rhythm are unlikely as long as the underlying condition persists. Thyroid hormones have a direct positive chronotropic effect and may enhance cardiac sensitivity to catecholamines by increasing β-adrenergic receptor density. In some patients, there are no obvious symptoms of hyperthyroidism, or only subtle signs of an overactive thyroid are present (apathetic hyperthyroidism), and the arrhythmia can be the only clinical marker of the disease. This hypothesis has been supported by data from the Framingham Heart Study, in which a low serum thyroid-stimulating hormone (TSH) level in subjects older than 60 years of age was associated with a 3.1-fold increased risk for AF after 10 years of follow-up.[137] Data from the Canadian Registry of Atrial Fibrillation (CARAF) have shown that asymptomatic presentation of AF was common in patients with a low TSH level (odds ratio, 5.5).[138] Hypothyroidism traditionally has not been associated with risk for AF.

Anemia may contribute to dyspnea or precipitate atrial tachyarrhythmias. The white cell count may suggest the presence of inflammatory mechanisms precipitating the arrhythmia. Measures of the acute phase response, such as the erythrocyte sedimentation rate and C-reactive protein, may give further guidance where inflammatory or malignant processes are suspected. Recent studies suggest that C-reactive protein, which is elevated in patients with AF that is "idiopathic" or associated with structural heart disease, may be a marker of the inflammatory mechanism underlying the arrhythmia.[139,140] Patients with persistent AF had a higher C-reactive protein level (0.34 mg/dL) than those with paroxysmal AF (0.18 mg/dL), and both groups had significantly higher C-reactive protein levels than control subjects (0.096 mg/dL).[140] C-reactive protein may also produce prothrombotic effects by increasing tissue factor expression. Serum liver enzyme levels, in particular γ-glutamyltransferase, may be helpful when alcohol is suspected as a cause of AF. Thyroid and liver function monitoring is mandatory during therapy with amiodarone, and creatinine clearance is essential when adjusting the dose of dofetilide.

Transthoracic echocardiography is essential if active management is contemplated. Valve heart lesions, severe enough to cause atrial tachyarrhythmias, are usually evident on clinical examination, but may not be discovered. Echocardiographic measure of left ventricular hypertrophy is important for guidance of prophylactic antiarrhythmic drug therapy in patients with hypertension and hypertrophic cardiomyopathy. Echocardiography can detect left atrial dilatation and assess the severity of left ventricular dysfunction. Both of these significantly increase the risk of stroke and reduce the likelihood of successful cardioversion. Furthermore, transthoracic echocardiography can be effectively used to assess left atrial mechanical function by measuring the A component of the transmitral pulsed Doppler signal.[141] Although transthoracic echocardiography has proven to be useful for the detection of intraventricular thrombus, the sensitivity of the method for diagnosis of left atrial thrombus is low.

TEE is capable of providing high-quality images of cardiac structure and function and is the most sensitive and specific technique to detect the presence of thrombi in the left atrium and the left atrial appendage, which have been found in 5% to 15% of patients with AF before cardioversion (Fig. 15-28).[142] TEE has proven useful for stratification of patients with AF and flutter with regard to the risk of ischemic stroke, and it has been suggested that TEE should be undertaken before elective cardioversion.

In patients with a suspected history of transient ischemic attacks, computer tomography or magnetic resonance imaging can reveal previous silent cerebral infarcts and influence the decision in favor of lifelong anticoagulation (see Fig. 15-28).

Finally, electrophysiological study is indicated for identification of the principal arrhythmia mechanism when curative catheter ablation is considered for selected patients, including those with typical atrial flutter, focal or reentrant atrial tachycardia, and paroxysmal or persistent AF (Fig. 15-29). Electrophysiological study can identify sinus node dysfunction, define the mechanism of wide QRS complex tachycardia, and exclude preexcitation syndrome as a cause of AF with a rapid ventricular rate response. Electrophysiological study may be required to select patients for pacemaker

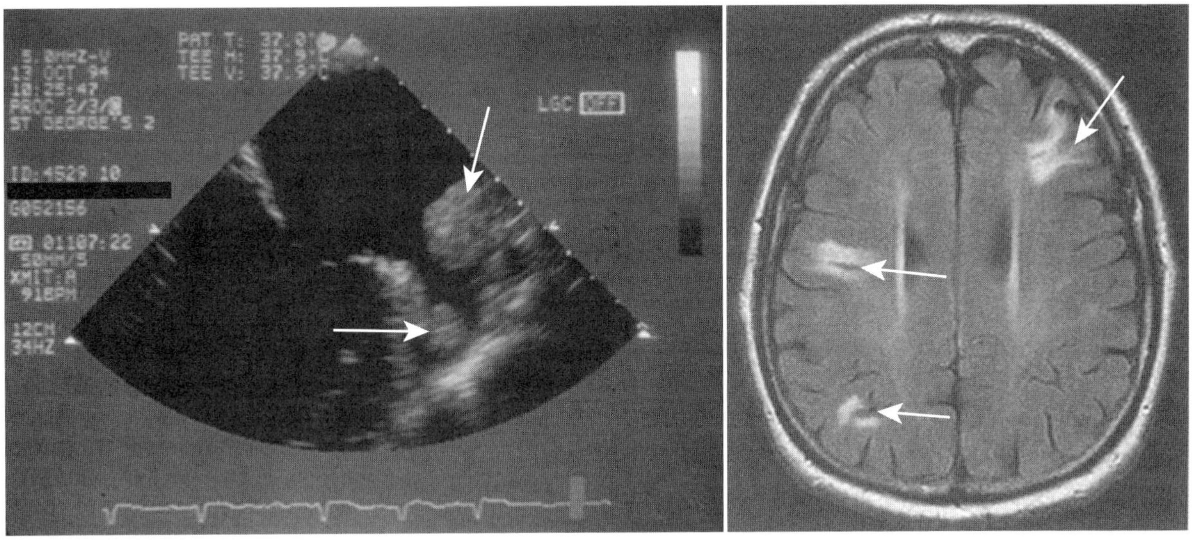

A B

FIGURE 15-28 **A,** Left atrial appendage thrombi detected by transesophageal echocardiography in a patient with atrial fibrillation. Note the presence of a small thrombus within a lumen of the left atrial appendage and a larger one at the orifice *(arrows)*. **B,** Magnetic resonance image of multiple cerebral infarcts *(arrows)*.

or ICD therapy when considering the prevention of atrial tachyarrhythmias.

Clinical Electrophysiology

ATRIAL TACHYCARDIA

Two major types of AT are classified: focal AT and macro-reentry AT. Focal AT is defined as a supraventricular tachycardia originating in the atrial myocardium, including sinus nodal reentrant tachycardia and excluding AV nodal reentrant tachycardia or AV reciprocating tachycardia (Fig. 15-30). Macroreentry AT is defined as AT demonstrated by entrainment criteria and the presence of slow conduction. AT can be diagnosed if the following criteria are present: (1) the atrial activation sequence during tachycardia is different from that which occurs during sinus rhythm or during ventricular pacing; (2) changing P–R interval and R–P interval is related to changing of the tachycardia rate; and (3) AV block occurs without affecting the tachycardia.[34,145-148]

The possible mechanisms of focal AT include abnormal automaticity, triggered activity, and reentry (microreentry). Automatic AT is identified by the following characteristics: (1) episodes of AT with "warm-up" and "cool-down" phenomena; (2) AT can be initiated spontaneously or with isoproterenol; (3) AT cannot be initiated, entrained, or terminated with pacing maneuvers; and (4) AT can be transiently suppressed with overdrive atrial pacing, but it subsequently resumes with a gradual increase in the atrial rate. AT due to triggered activity is diagnosed when the following characteristics are present: (1) AT can be initiated or terminated with atrial pacing or extrastimuli, and its initiation is cycle length dependent; (2) entrainment is not found, but overdrive suppression or termination is noted; and (3) delayed afterdepolarization can be found on the recordings of monophasic AP just before the onset of AT (Fig. 15-31).

Reentrant AT is diagnosed if the following characteristics are present: (1) AT can be reproducibly initiated and terminated with atrial pacing or extrastimuli; and (2) delayed afterdepolarization cannot be found on the recordings of monophasic AP.[34,145,146] The origin of focal AT can be located at any site of either the atrium or thoracic veins with atrial connection. Recent studies have suggested that the predominant origin of focal AT is along the crista terminalis. Other origins of focal AT include the atrial septum, Koch's triangle, the atrial appendages, and the AV (mitral and tricuspid) annulus.[149-151] Furthermore, there are several venous structures connected to both atria, including the superior and inferior vena cava, the coronary sinus, and four pulmonary veins. Atrial myocardial tissue can extend into these venous structures, and they can be the origin of focal atrial tachyarrhythmia.[17,152,153] After the development of three-dimensional mapping systems (Carto and ESI), the true reentry circuit of macroreentry AT has been demonstrated to include normal atrial tissue and scar tissue (with low voltage and slow conduction); furthermore, the circuit can be a single reentry loop or a "figure-of-eight" reentry loop. Most of these AT are associated with atrial myopathy or atrial scar after surgical operation of congenital heart defect. Concealed entrainment with the identical atrial activation sequence and a short post-pacing interval can be obtained when the pacing site is at or close to the exit site of the slow conduction zone.[154]

ATRIAL FLUTTER

Major types of AFL include typical (isthmus-dependent, clockwise, and counterclockwise AFL) and atypical (isthmus-independent) AFL. Cosio et al.[155] first mapped the flutter circuit and demonstrated the slow conduction

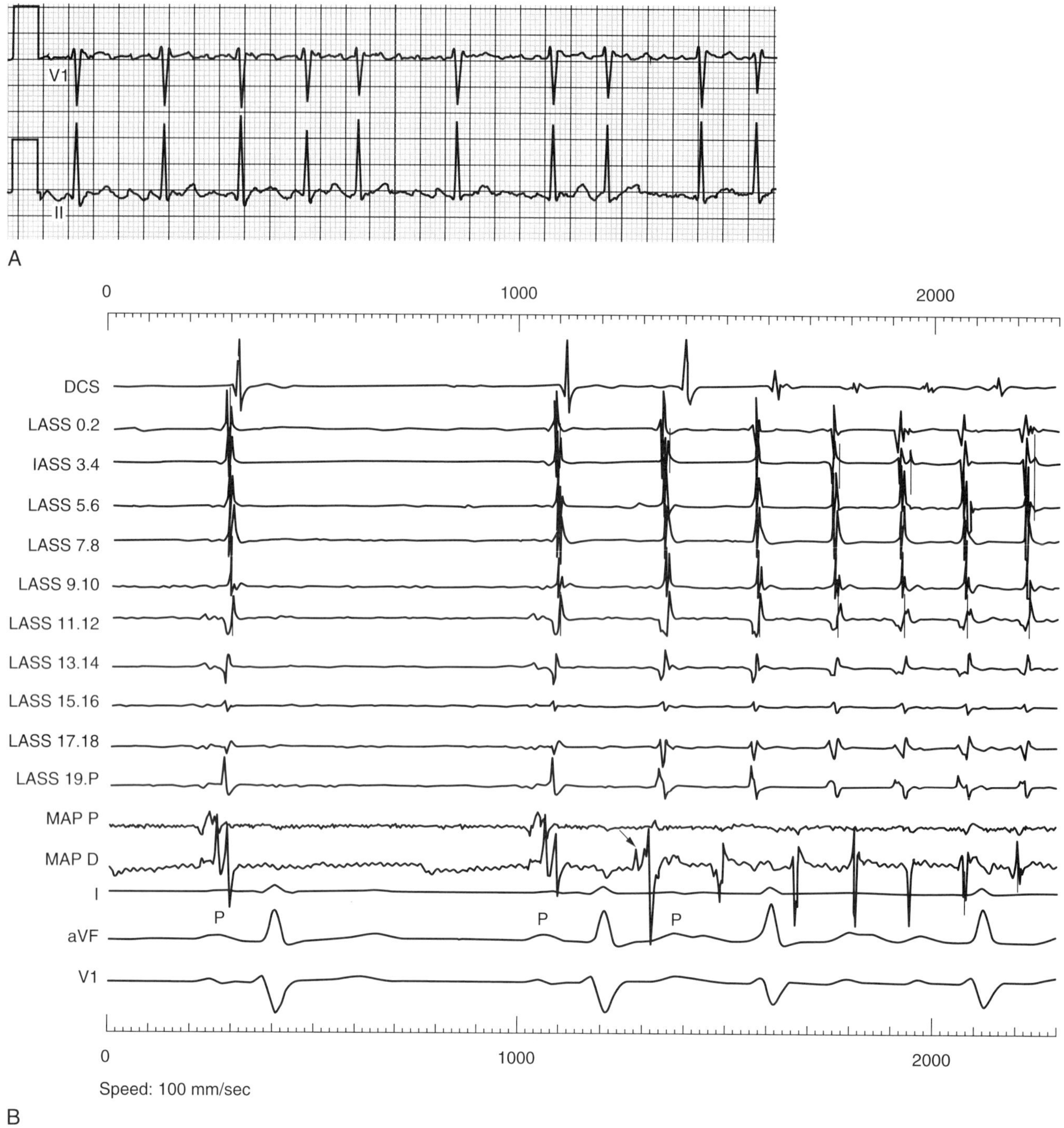

FIGURE 15-29 **A,** Coarse atrial fibrillation with discrete atrial *f* waves in a patient with suspected "focal" atrial fibrillation (AF). **B,** Intracardiac recording during electrophysiological study with attempt at radiofrequency ablation in the same patient. Pulmonary vein tachycardia originating from the left upper pulmonary vein with conduction to the atria that initiates atrial fibrillation. Shown are leads I, avF, V1, proximal and distal electrode pairs of a mapping catheter positioned within the left upper pulmonary vein (MAP P, MAP D), a decapolar Lasso catheter positioned in the left atrium (Lass D,2 ... Lass 19,P), and a distal electrode pair of a coronary sinus catheter (DSC). A premature pulmonary vein depolarization (*arrow*) initiates pulmonary vein tachycardia, which triggers AF.

in low right atrial isthmus. With this slow conduction property in the isthmus, atrial wavefronts may encounter unidirectional block in the isthmus and engage the AFL reentry circuit. Olshansky et al.[156] demonstrated that double potentials recorded in the low right atrium during human AFL represent collision of two different activation wavefronts in a functional center of the AFL reentrant circuit, and therefore may serve as a marker for an area of functional block. Saoudi et al.[157] also proved the slow conduction zone of human AFL. Chen et al.[158] further demonstrated that entrainment pacing from the low right atrium with double potentials during typical AFL revealed manifest entrainment. Using intracardiac echocardiography to place the multipolar

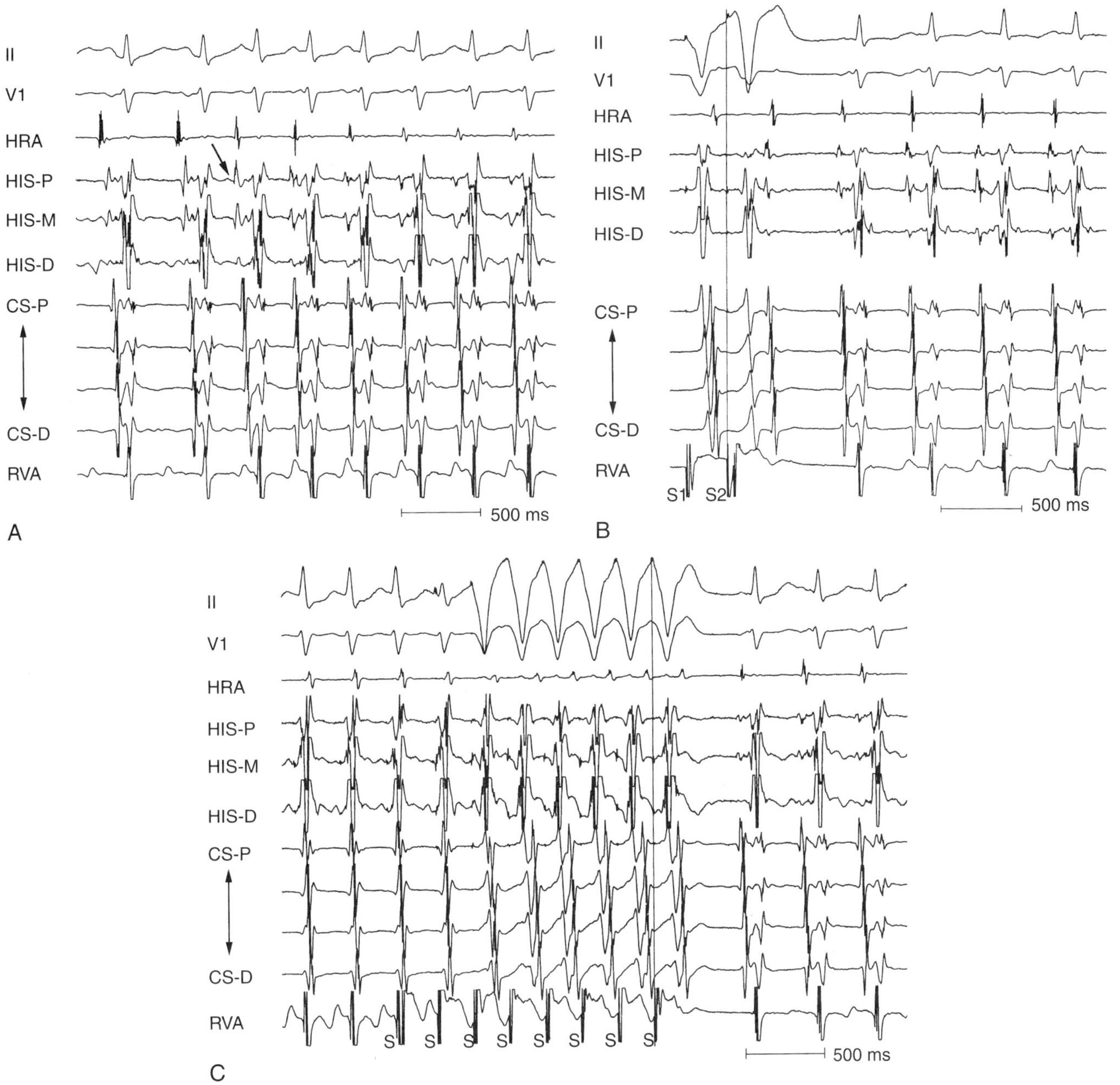

FIGURE 15-30 Diagnosis of atrial tachycardia. **A,** Spontaneous initiation of atrial tachycardia from the His-bundle area (*arrow*). **B,** Intracardiac recordings show atrial tachycardia induced by ventricular stimuli, and the presence of V-A-A-V electrogram sequence is noted after the last ventricular stimulus. **C,** Transition of atrioventricular nodal reentry tachycardia to atrial tachycardia after rapid ventricular pacing. V-A-A-V is also demonstrated in the last ventricular pacing beat. AT, atrial tachycardia; CS, coronary sinus; HIS, His bundle; HRA, high right atrium; P, D, proximal, distal.

catheter, Olgin et al.[159] recorded split potentials along the entire length of the crista terminalis. Tai et al.[160] also demonstrated a line of block along the length of the crista terminalis at a pacing cycle length similar to the AFL cycle length in patients with supraventricular tachycardia, as well as induced AFL (Fig. 15-32). These findings suggest that the crista terminalis is an anatomic structure forming a functional barrier of the typical AFL reentrant circuit; furthermore, poor transverse conduction capacity in the crista terminalis may contribute to the occurrence and maintenance of clinical typical AFL, because it could effectively protect the reentrant circuit from being short-circuited.

Using simultaneous recordings from multiple electrode catheters, application of entrainment criteria, and the three-dimensional mapping systems, several types of atypical AFL have been demonstrated, including AFL in the low right atrial free wall, AFL involving the high right atrium (superior vena cava–atrial septum area), and AFL involving two or four pulmonary vein orifices, mitral annulus isthmus, and the fossa ovalis area.[111,117-119,161,162]

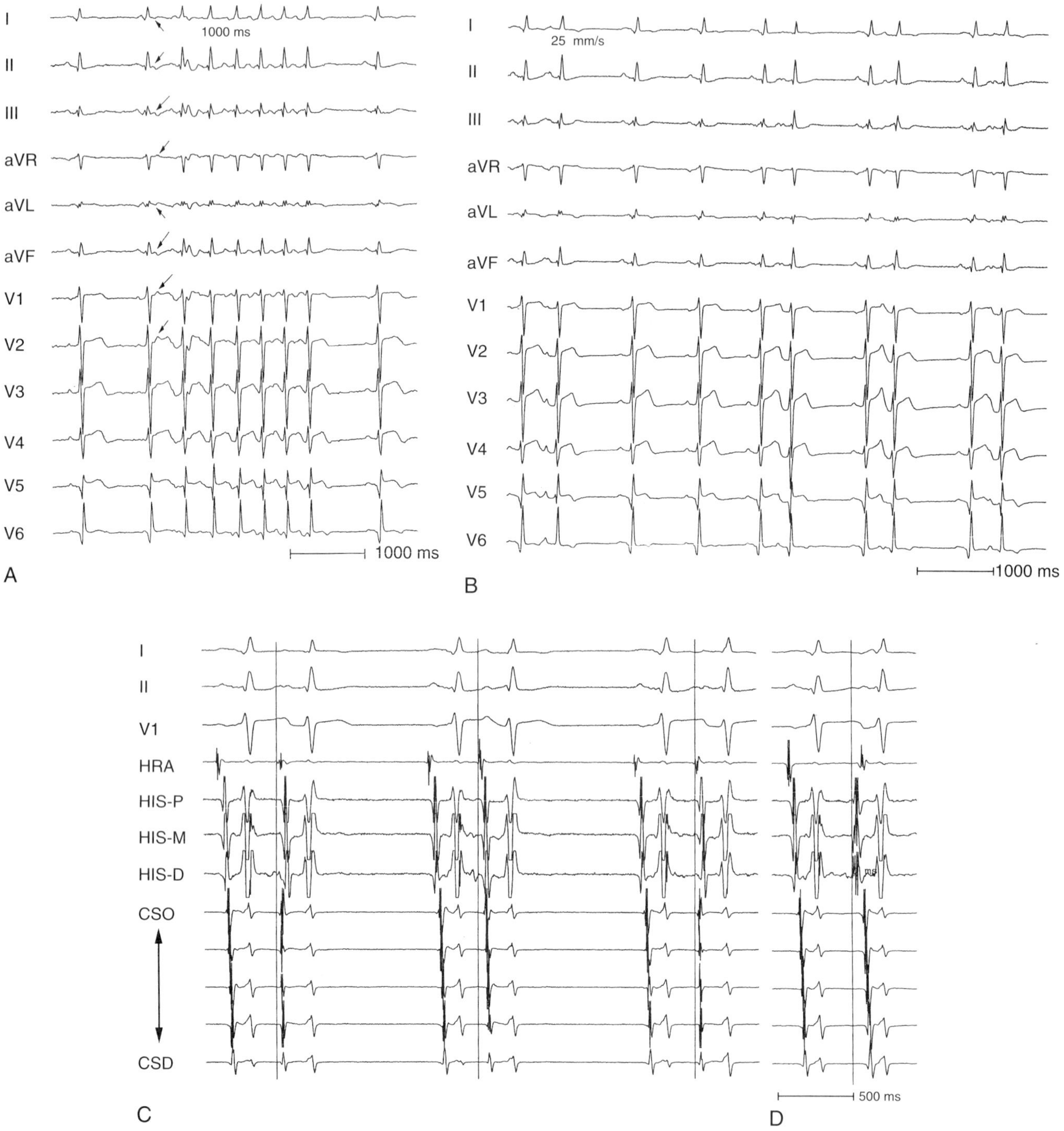

FIGURE 15-31 Atrial tachycardia with multifocal atrial premature beats. **A,** Spontaneous initiation of atrial tachycardia by an atrial premature beat. The arrows indicate P wave morphologies in a 12-lead ECG. **B,** The same patient with frequent atrial premature beats, and P wave morphologies are different for each atrial premature beat. **C** and **D,** The intracardiac activation sequence of different atrial premature beats. The first atrial premature beat (**C**) shows the earliest activation of the high right atrium and almost simultaneous depolarization in the coronary sinus ostium-distal coronary sinus. The second atrial premature beat shows the earliest atrial activation in the high right atrium, and the coronary sinus activation sequence is from the proximal to distal coronary sinus. The third atrial premature beat has a similar activation sequence as the first atrial premature beat. The earliest atrial activation (**D**) is at the His-bundle area. APB, atrial premature beat; AT, atrial tachycardia; CS, coronary sinus; CSD, distal coronary sinus; CSO, coronary sinus ostium; HRA, high right atrium.

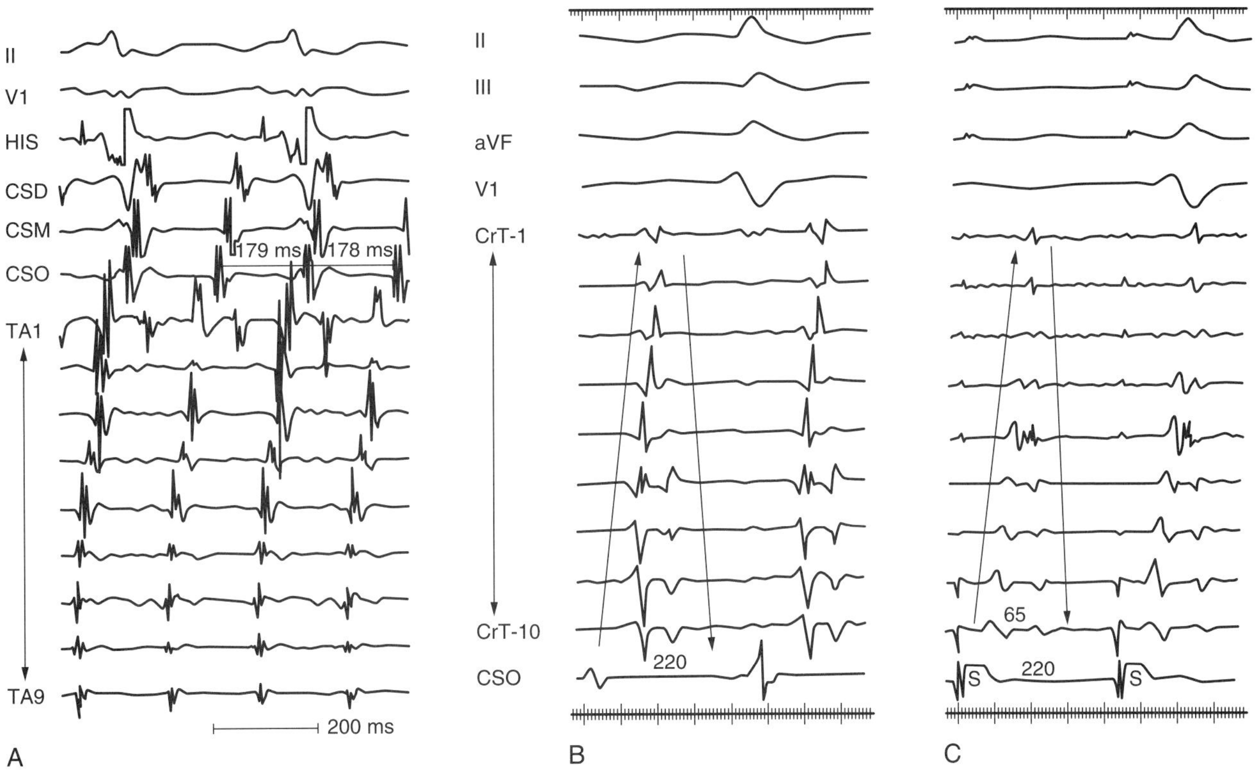

FIGURE 15-32 **A,** Intracardiac recording of the atrial activation sequence during counterclockwise atrial flutter. TA_1 through TA_9 show the recordings along tricuspid annulus. **B,** Double potentials with opposite activation sequences were recorded along the crista terminalis during counterclockwise atrial flutter (the cycle length 220 milliseconds [ms]). **C,** Double potentials with the same activation sequences as those in **panel B** are recorded during pacing from the low posterior right atrium at a cycle length of 220 ms. The maximal interdeflection interval (65 ms) is measured in the recordings from the most proximal electrode dipole located in the lower CrT. AFL, atrial flutter; CrT, crista terminalis. (Reprinted from Tai CT, Chen SA, Chen YJ, et al: Conduction properties of the crista terminalis in patients with typical atrial flutter: Basis for a line of block in the reentrant circuit. J Cardiovasc Electrophysiol 1998;9:811-9.)

ATRIAL FIBRILLATION

Recently, several reports have demonstrated that most paroxysmal AF is initiated by ectopic beats from the thoracic veins or atria and that radiofrequency catheter ablation can effectively cure AF.[18,152,153,163] These ectopic foci include pulmonary veins, superior vena cava, the ligament of Marshall, crista terminalis, coronary sinus, and the atrial wall (Fig. 15-33). This "focal" mechanism of AF is different from the classic "reentry" mechanism of AF. The focal mechanism is important for the initiation and reinitiation of AF, and the reentry mechanism is important to explain the maintenance of AF.[128,164]

The other important concept is atrial electrical remodeling ("AF begets AF").[129] Clinical observations

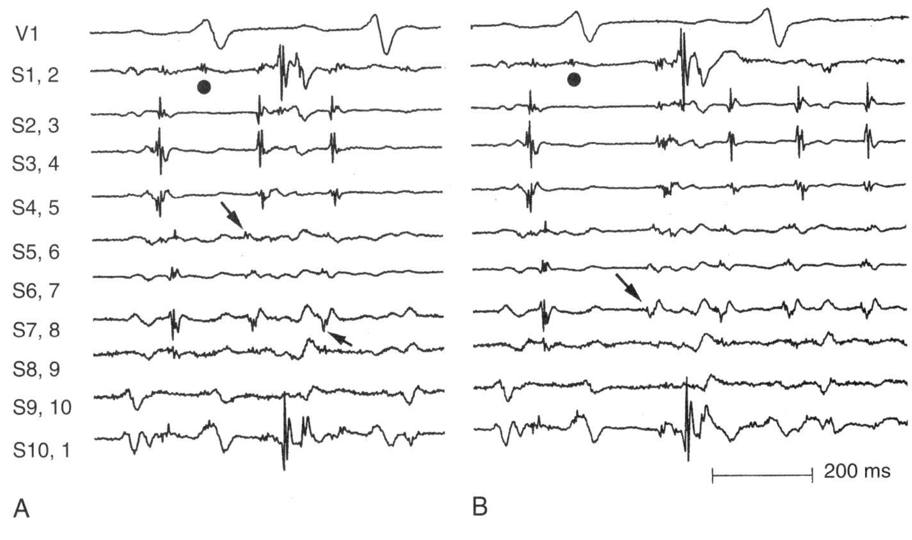

FIGURE 15-33 Circular catheter with multipolar recording along the pulmonary vein ostium shows spontaneous initiation of burst atrial fibrillation from the right superior pulmonary vein. The earliest initiating focus migrates along the ostial area. **A,** The earliest ectopy on S5, 6, then migrates to S7, 8 (*arrows*). **B,** The earliest ectopy on S7, 8 (*arrow*) and the following depolarizations also come from S7,8. The dots on S1,2 represent the late pulmonary vein potential or pulmonary vein ectopy with concealed depolarization. AF, atrial fibrillation; PV, pulmonary vein.

have shown that paroxysmal AF could progress to sustained AF, even in patients without underlying heart disease. In addition, the possibility of conversion and maintenance of sinus rhythm decreases as the duration of AF increases. These findings suggest that AF is a self-perpetuating arrhythmia. Wijffels et al.[129] demonstrated that maintenance of AF by pacing for 2 to 3 weeks led to sustained AF in a chronic goat model. The major electrophysiological changes are shortening of the atrial effective refractory period and loss of its normal rate-dependent characteristic; these changes recover after termination of AF. In humans, a short duration of AF has been demonstrated to cause a significant shortening of the atrial effective refractory period, which recovers within minutes after termination of AF.[165,166] In patients with persistent AF, the atrial refractory period is shorter immediately after conversion to sinus rhythm, but the refractory period returns to normal range within 1 week (Fig. 15-34).[167] The term "electrical remodeling" is used to describe these electrophysiological changes.

Principles of Practice

The fundamental principles of therapy of atrial tachyarrhythmias include: (1) electrical or pharmacologic restoration and maintenance of sinus rhythm; (2) rate control if restoration and maintenance of sinus rhythm is impossible; (3) risk stratification and prevention of thromboembolic complications and stroke; (4) identification and prevention or reduction of risk factors and elimination of precipitating agents; (5) "upstream" therapy of the underlying pathology (e.g., hypertension) and specific atrial pathophysiology (e.g., fibrosis); (6) identification of arrhythmias amenable to nonpharmacologic treatment, including radiofrequency catheter ablation and maze procedure (i.e., typical AFL, focal or macroreentrant AT, "focal" AF); and (7) selection of patients who may benefit from implantable device therapies (antitachycardia and preventative atrial pacing, atrial defibrillators). A summary of therapies currently available for management of atrial tachyarrhythmias is presented in Table 15-6.

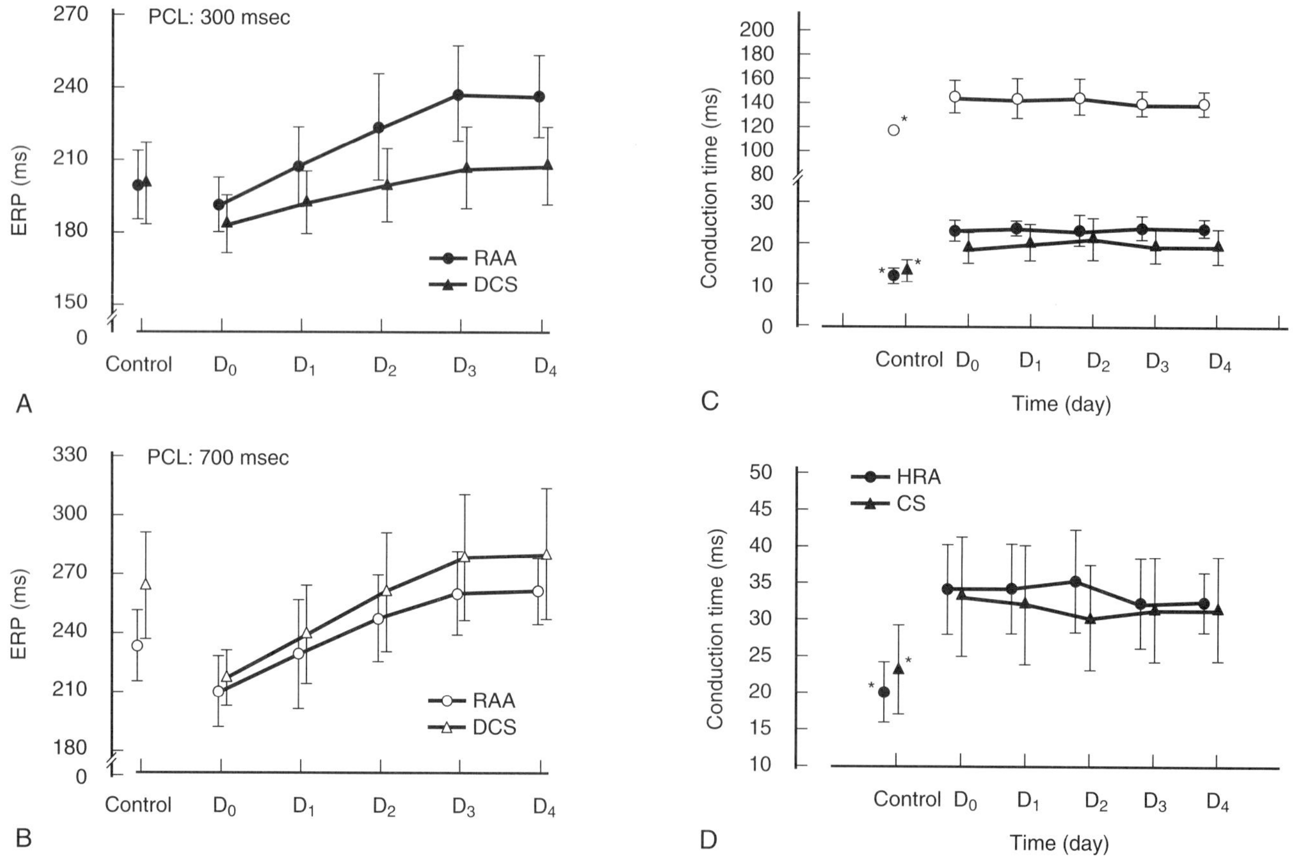

FIGURE 15-34 **A** and **B,** After cardioversion of persistent atrial fibrillation (AF), temporal changes in effective refractory periods of the right atrial appendage and the distal coronary sinus at a pacing cycle length of 300 milliseconds (ms), and those of the right atrial appendage and the distal coronary sinus at a pacing cycle length of 700 ms are shown. The effective refractory periods of both the right atrial appendage and the distal coronary sinus gradually increase and reach a stable level on the third day after cardioversion. **C** and **D,** Right atrial conduction time, left conduction time, and ECG P wave duration in patients converted from persistent AF. These times are longer in patients converted from persistent AF compared to controls. Please note that the data for the control group are on the left side. In addition, there are no significant changes in the conduction properties during a 4-day follow-up postcardioversion. The atrial conduction times were measures at a basic pacing cycle length (**C**); the atrial conduction times were measured at the earliest capture beat (**D**). *, $P < .01$ for comparison between AF and controls. (Reprinted from Yu WC, Lee SH, Tai CT, et al: Reversal of atrial electrical remodeling following cardioversion of long-standing atrial fibrillation in man. Cardiovasc Res 1999;42:470-6.)

TABLE 15-6 Management Strategies in Atrial Tachyarrhythmias

Strategy	Advantages	Disadvantages	Indications
Pharmacologic rhythm control	(1) Symptom relief (2) Normal intracardiac and systemic hemodynamics (3) Prevention or delay of electrophysiological and structural atrial remodeling, atrial enlargement (4) Availability; low cost (5) Relatively wide antiarrhythmic drug choice (6) No intervention; often may be started on an outpatient basis (7) Leaves other, nonpharmacologic options if fails	(1) Low probability of remaining in sinus rhythm in the long term (about 50%-60% at 1-2 yr) (2) Often need for repeat cardioversion (3) No reduction in mortality or thromboembolic events (4) Risk of proarrhythmia and bradycardia (5) Risk of noncardiac adverse effects (6) Need for anticoagulation in high-risk patients if increased likelihood of recurrence or asymptomatic recurrence of the arrhythmia	(1) Recurrent paroxysmal AF (2) Recurrent persistent AF after cardioversion, especially if highly symptomatic (3) Recurrent paroxysmal or persistent AFL if ablation therapy is impossible or ineffective (4) Recurrent paroxysmal or incessant AT if ablation therapy is impossible or ineffective
Pharmacologic rate control	(1) Reduction in symptoms (2) Availability; low cost (3) Easy to start and control (4) Competitive results with pharmacologic rhythm control in terms of mortality and thromboembolic events (4) Leaves other, nonpharmacologic options if fails	(1) Impaired intracardiac and systemic hemodynamics (2) Insufficient rate control with one drug, need for polypharmacy (3) Risk of proarrhythmia and bradycardia (e.g., digoxin) (4) Risk of noncardiac adverse effects (5) Need for lifelong anticoagulation (6) No long-term prevention of atrial remodeling	(1) Permanent AF (2) "Accepted" AF if mildly symptomatic or asymptomatic and in those older than 65 yr (3) Persistent AF in patients with heart failure > NYHA class II (4) "Accepted" AFL if ablation therapy is impossible or ineffective
"Ablate and pace"	(1) Significant symptoms improvement (2) Some improvement in cardiac performance (exercise tolerance, ejection fraction) (3) Improvement in quality of life (4) Mortality does not differ from that with pharmacologic rate control	(1) Renders patients pacemaker dependent (2) Risk of thromboembolic complications; need for anticoagulation (3) Pharmacologic rhythm control may be needed (4) No long-term prevention of atrial remodeling; progression to a permanent form (5) Right ventricular pacing may be deleterious in the long-term (systolic dysfunction)	(1) Highly symptomatic, drug refractory AF (2) Inadequate pharmacologic rate control (3) Contraindications for pharmacologic therapy
AV node modification	Comparable with the "ablate and pace" strategy	(1) High percentage of complete AV block resulting in pacemaker implantation (2) Risk of thromboembolic complications; need for anticoagulation (3) Irregularity of ventricular rate may be accompanied by symptoms (4) Often need for additional pharmacologic rate control (5) Fast AV conduction may restore in the long-term	Highly symptomatic, drug refractory AF
Isthmus-ablation	(1) "Cure" typical AFL with a 95% success rate (2) May reduce the incidence of AF triggered by AFL	(1) AF is not completely abolished (2) The incidence of AF postablation varies from 10% to 74% (3) If AF was converted to AFL by AAD, these should be continued after ablation of AFL	(1) Typical AFL (2) AFL as a result of AAD for AF (3) Incisional reentrant AFL if a vulnerable isthmus is identified
Ablation for focal or macro-reentrant AT	(1) "Cure" AT with a 75%-86% success rate (2) May reduce the incidence of AF triggered by AT	(1) Difficulty with identification of an arrhythmogenic focus or the reentry circuit (2) The rate of recurrence is 8%, but the follow-up is limited (3) The rate of significant complications, including cardiac perforation and AV block, is 1%-2%	(1) Focal AT (2) Macro-reentrant AT after atrial septal defect closure, Fontan and Mustard procedures
Pulmonary vein isolation	Renders AF entirely curable if successful; the greatest success rate is 66%; up to 85% with circumferential pulmonary vein isolation	(1) Difficulty with identification of all arrhythmogenic foci (2) High rate of recurrence; need for repeat procedure (3) Risk of pulmonary vein stenosis or thrombosis (up to 42%; lesser with new techniques, 1%-2%) (4) Risk of thromboembolism (0.5%) (5) Limited to paroxysmal AF (6) Need for the continuation of AAD in 23% of cases	AF triggering by rapid AT from local foci (usually in the pulmonary veins)

Continued

TABLE 15-6 Management Strategies in Atrial Tachyarrhythmias—(*cont'd*)

Strategy	Advantages	Disadvantages	Indications
Intraoperative or catheter-based "maze" procedure	Renders AF entirely curable if successful; the long-term success rate is 80% (32%-93%) and up to 99% with AAD	(1) Need for expanded linear ablation, not limited to one atrium (2) High degree of tissue trauma (3) Incomplete lesions may be arrhythmogenic (4) Increased procedure time (5) Systemic thromboembolic complications (6) Mortality 1%-2%	(1) Highly symptomatic, drug refractory AF (2) Patients undergoing cardiac surgery (coronary by-pass; valve repair or replacement)
Left atrium isolation	"Cure" AF; the success rate is 71%; with transection extended to the right atrium, 85%	Probably same as above	Same as above
Atrial defibrillator	(1) Converting AF or AFL with a synchronized low-energy atrial shock (2) Can be performed on an outpatient basis (3) Shock can be administered promptly or delayed (4) Prevention or delay of electrophysiological and structural atrial remodeling, atrial enlargement, and progression to a permanent form	(1) Considerable discomfort (2) Poor compliance (3) High incidence of early recurrence (4) Need for repeat shock (5) Limited data in patients with structural heart disease	Infrequent, symptomatic, drug refractory AF or AFL not amenable for ablation (probably with minimal structural heart disease)
Atrial or dual-chamber pacemaker	(1) Provide physiologic pacing with normal intracardiac hemodynamics (2) May prevent or reduce progression to a permanent form in the long term (>2 yr), especially in patients with no history of arrhythmias and patients with sick sinus syndrome but not with AV block (3) Painless antitachycardia pacing to terminate organized atrial rhythms (4) A variety of preventative atrial pacing algorithms to modify the substrate and to suppress triggers of AF (5) Potential for pacing from specific sites (Bachmann's bundle, triangle of Koch) or multisite pacing	(1) No reduction in all-cause mortality with DDD(R) vs. VVI(R) pacing modes (2) The efficacy of antitachycardia pacing and preventative pacing has not yet been proven (3) No reduction in arrhythmia burden has been reported (4) Some therapies (e.g., high frequency burst) are available on an inpatient basis only (5) Risk of thromboembolic complications remains; need for lifelong anticoagulation (6) Often need for adjunctive AAD therapy (7) Right ventricular pacing may be deleterious in the long term (9) Technical difficulty with placing additional electrodes (e.g., coronary sinus) (10) Lead complications (e.g., lead dislodgement in 1.5% in dual-site pacing, up to 15% in biatrial pacing)	Conventional bradycardia pacing indications and recurrent arrhythmias, predominantly AF
Dual-chamber ICD	(1) Painless antitachycardia pacing to terminate organized atrial rhythms (2) A variety of preventative atrial pacing algorithms to modify the substrate and to suppress triggers of AF (3) Converting the arrhythmia with a synchronized low-energy atrial shock administered promptly or delayed (4) May prevent degeneration of AF into ventricular fibrillation	(1) The efficacy of antitachycardia pacing and preventative pacing has not yet been proven (2) Some therapies (e.g., high frequency burst) are available on an inpatient basis only (3) Risk of thromboembolic complications remains; need for lifelong anticoagulation (4) Risk of electrical proarrhythmia remains unclear (5) Disadvantages of atrial shock are similar to those with stand-alone defibrillator, but there is a possibility of overdrive pacing to prevent early recurrence (6) Often need for adjunctive AAD therapy to prevent frequent shocks (7) Expensive, not widely available (8) Technical difficulty with placing specific electrodes (e.g., coronary sinus) (9) Potential risk of inappropriate shock due to sensing problems (e.g., far-field oversensing)	Patients with both ventricular and atrial arrhythmias, structural heart disease, impaired left ventricular function, with high risk of proarrhythmia on AAD

AAD, antiarrhythmic drug; AF, atrial fibrillation; AFL, atrial flutter; AT, atrial tachycardia; AV, atrioventricular; ICD, implantable cardioverter-defibrillator.

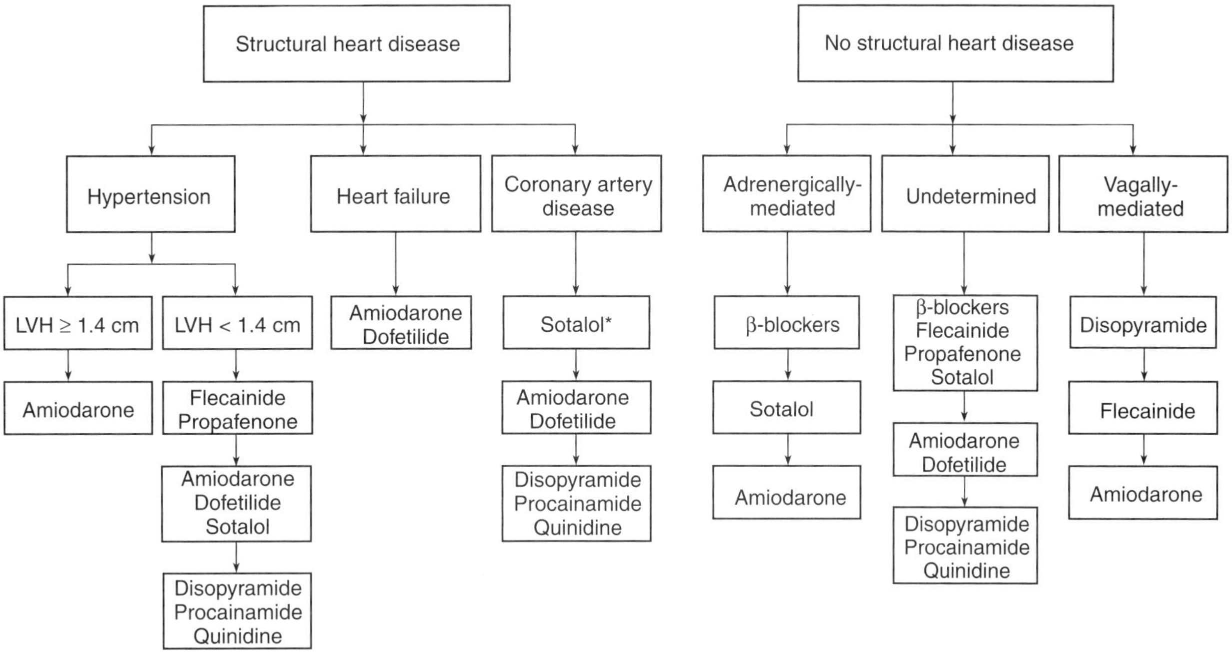

FIGURE 15-35 Selection of the optimal antiarrhythmic agent to prevent atrial tachyarrhythmias. *, no left ventricular dysfunction; LVH, left ventricular hypertrophy. (Modified from Fuster V, Rydén LE, Asinger RV, et al: Task Force Report: ACC/AHA/ESC guidelines for the management of patients with atrial fibrillation. Eur Heart J 2001;22:1852-1923.)

Primary pathology is crucial in decision making for selection of the most appropriate therapeutic strategy in atrial tachyarrhythmias. If the new onset arrhythmia has resulted from a transient cause, it may not require specific treatment. Underlying structural heart disease is essential for the selection of an antiarrhythmic drug if the strategy of restoration and maintenance of sinus rhythm is a therapeutic option (Fig. 15-35).[34] Thus, class IC antiarrhythmic agents or sotalol may be the drug of choice in lone atrial tachyarrhythmias or tachyarrhythmias associated with hypertension without significant left ventricular hypertrophy, disopyramide in vagotonic AF, sotalol in patients with coronary disease, amiodarone or dofetilide in the presence of left ventricular dysfunction and overt congestive heart failure. Recognition of advanced organic heart disease may prompt rate control rather than rhythm control as electrical or pharmacologic cardioversion and subsequent antiarrhythmic drug therapy to maintain sinus rhythm are generally unsuccessful in this setting, and the relapse rate is high. Finally, identification of risk factors for stroke, such as hypertension and heart failure or left ventricular dysfunction, constitutes one of the basic principles of anticoagulation therapy in atrial tachyarrhythmias.[39]

RESTORATION AND MAINTENANCE OF SINUS RHYTHM

Cardioversion can be performed either electively to restore sinus rhythm in patients with persistent atrial tachyarrhythmias or urgently, usually in patients with underlying heart disease, such as acute myocardial infarction or congestive heart failure, in whom AF is associated with rapid progression of symptoms, hypotension, angina pectoris, or worsening left ventricular function (Fig. 15-36). In some patients multiple repeat cardioversions may increase the overall success rate of remaining in sinus rhythm, especially if applied with concomitant antiarrhythmic drug therapy, which can enhance immediate success and suppress early recurrences.

More recently, low-energy transvenous cardioversion has emerged as an alternative option to conventional transthoracic cardioversion with a higher primary success rate, particularly in patients with persistent arrhythmias of long duration in whom external cardioversion failed, patients with severe obstructive pulmonary disease, and overweight individuals.[166] Other indications for transvenous cardioversion may include patients with implanted pacemakers or defibrillators.

Cardioversion can be achieved by means of antiarrhythmic drugs, usually administered intravenously. ACC/AHA/ESC Task Force experts consider it effective if initiated within 7 days after the onset of the arrhythmia, in which case restoration of sinus rhythm can be achieved in about 70% of patients. Although more convenient, pharmacologic cardioversion is less effective than electrical cardioversion and is likely to be associated with the same risk of thromboembolic complications (Table 15-7).[167] A possibility of acute conversion out of hospital using a loading oral dose of antiarrhythmic drug can be considered if the agent has been proven effective and safe during in-hospital experience.

Prophylactic drug therapy to prevent recurrence of the arrhythmia is usually recommended for patients with paroxysmal tachyarrhythmias with frequent paroxysms (one per 3 months), which are associated with significant symptoms or lead to worsening left ventricular function, or both, and for those with a recurrent

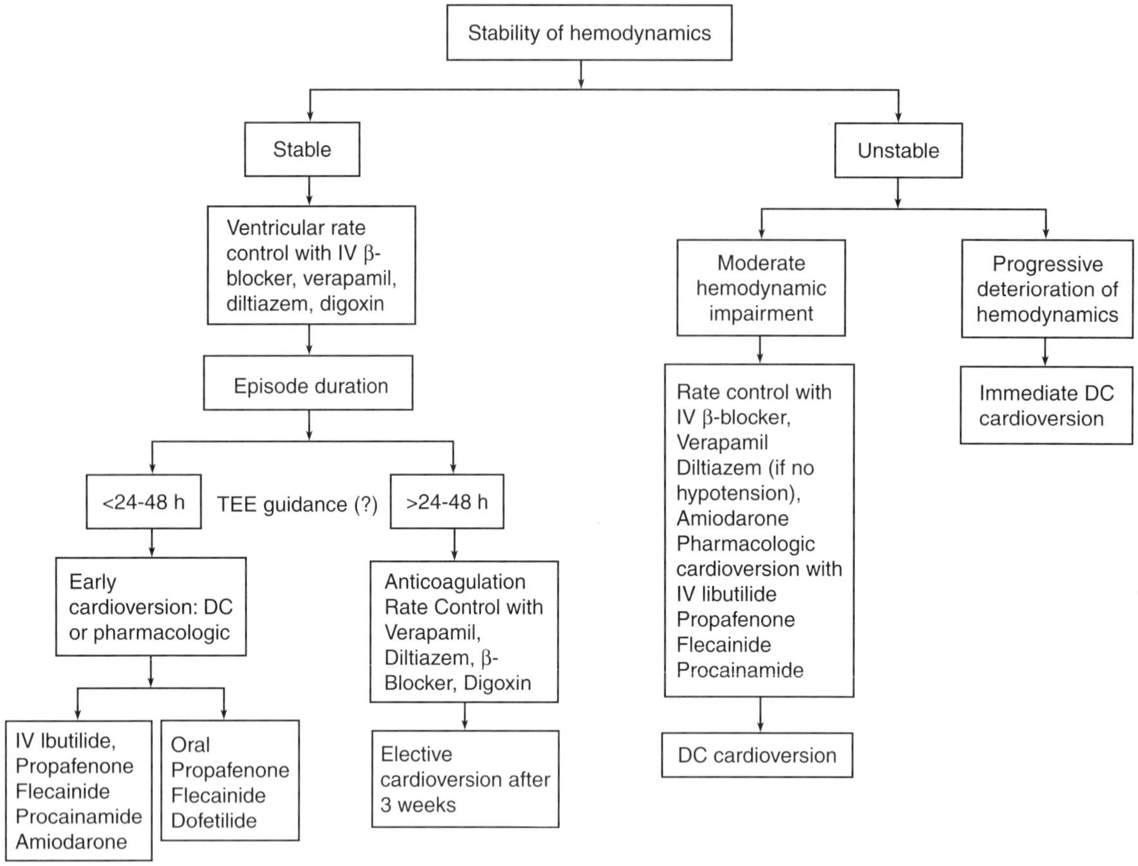

FIGURE 15-36 Cardioversion strategies for atrial tachyarrhythmias. DC, direct current; INR, International normalized ratio.

TABLE 15-7 Advantages and Disadvantages of Pharmacologic versus Electrical Cardioversion of Atrial Tachyarrhythmias

Pharmacologic Cardioversion		**Electrical Cardioversion**	
Advantages	*Disadvantages*	*Advantages*	*Disadvantages*
No need for anesthesia Outpatient initiation* "Pill-in-the pocket" approach† May be repeated immediately if first dose fails to restore sinus rhythm	Delayed effect, especially if oral drug administration Effective for arrhythmias of short duration Risk of proarrhythmia Negative inotropic effect—limited use in patients with advanced structural heart disease Risk of thromboembolism Adverse effects; idiosyncrasy; organ toxicity Transtelephonic monitoring may be required	Immediate effect—preferable use in hemodynamically unstable patients High likelihood of restoration of sinus rhythm in persistent arrhythmias of long duration	Need for general anesthesia Risk of proarrhythmia, although low Immediate/early recurrences Adjunct antiarrhythmic therapy may be required Risk of thromboembolism May cause patient discomfort

*For selected antiarrhythmic agents only.

†After proving feasibility.

persistent form of the arrhythmia when the likelihood of maintenance of sinus rhythm is uncertain, especially if there are risk factors for early recurrences (left atrial enlargement, evidence for depressed atrial mechanical function, left ventricular dysfunction, underlying cardiovascular pathology, long duration of the arrhythmia, advanced age, female gender, and previous cardioversions).[168] Prophylactic drug treatment is usually not recommended after a first episode of the arrhythmia, whether it self-terminates or requires cardioversion. There is no need for antiarrhythmic drug therapy, with exception for a rate-slowing agent, if the cause of the arrhythmia is transient or can be completely eliminated medically (e.g., acute illnesses or thyrotoxicosis).

WHERE TO INITIATE ANTIARRHYTHMIC DRUGS

The issue of the proper site for initiation of antiarrhythmic drug therapy for atrial tachyarrhythmias revolves around considerations of risk and practicality.[169] Antiarrhythmic drugs that prolong repolarization confer an increased risk of proarrhythmia, particularly in the presence of baseline Q–T interval prolongation. The rate of torsades de pointes with class III agents ranges from 0.9% to 3.3%.[170-172] Ibutilide and dofetilide exhibit reverse use-dependence (i.e., a greater potential to increase the effective refractory period at slower heart rates, resulting in proarrhythmic QT prolongation during bradycardia). Amiodarone may be an exception to these observations; although it prolongs the Q–T interval and there have been case reports of drug-induced torsades de pointes, the risk of proarrhythmia appears to be trivial.[173] In contrast to most class III antiarrhythmic drugs, which preferentially prolong the M-cell AP duration of the mid-myocardial layer and exaggerate transmural heterogeneity of repolarization, amiodarone prolongs the AP duration of all ventricular cell subtypes but does so the least in M cells, thereby reducing the transmural dispersion of repolarization. In addition, amiodarone inhibits the slow inward calcium current and suppresses calcium-dependent early afterdepolarizations in Purkinje fibers, which may contribute to the development of torsades de pointes. Class IC agents are known to convert AF into AFL and slow atrial activity to 200 bpm, which may result in 1:1 AV conduction.

Accurate assessment of the efficacy and prompt recognition of adverse effects, such as bradycardia, conduction abnormalities, Q–T interval prolongation, proarrhythmias, and idiosyncrasy, favor in-hospital initiation. Patients at expectedly high risk of developing adverse effects or those in whom sinus node function is unknown should be hospitalized for the initiation of antiarrhythmic drug therapy. For some antiarrhythmic agents (e.g., dofetilide), there is formal mandate for in-hospital initiation. However, in the absence of proarrhythmic concerns and formal labeling, convenience and cost effectiveness favor out-of-hospital initiation (e.g., oral propafenone and flecainide in patients with lone tachyarrhythmias). The same approach is valid for amiodarone, given its long elimination half-life period and low probability of developing torsades de pointes.

TABLE 15-8 Principles of Initiation of Antiarrhythmic Drug Therapy: Inpatient versus Outpatient

Drug	**Underlying Rhythm** *Atrial Fibrillation/Flutter*	*Sinus Rhythm*
Quinidine	Inpatient	Inpatient
Procainamide	Inpatient	Inpatient
Flecainide	Outpatient*	Outpatient
Propafenone	Outpatient*	Outpatient
Sotalol	Inpatient	Outpatient†
Dofetilide	Inpatient	Inpatient
Azimilide‡	Inpatient	Inpatient
Amiodarone	Outpatient†	Outpatient†

*No known sinus node dysfunction.

†Preferably with transtelephonic monitoring.

‡An investigational drug.

Table 15-8 summarizes current recommendations on in-hospital or out-of-hospital initiation of antiarrhythmic drug treatment.[169] As a general rule, antiarrhythmic drugs should be started at a lower dose with upward titration, reassessing the ECG as each dose change is made or concomitant drug therapies are introduced. Transtelephonic monitoring may be used to provide the surveillance of heart rate, P–R and Q–T interval durations, QRS width, and assessment of the efficacy of treatment. Of note, even with in-hospital initiation, antiarrhythmic agents impose risk of developing proarrhythmia later in the course of therapy, which may be facilitated by progression of underlying heart disease, bradycardia, electrolyte abnormalities, drug interactions, and changes in absorption, metabolism, or clearance.

RATE CONTROL

Rate control is pertinent to all atrial tachyarrhythmias and is an essential constituent of management of AF. Rate control as a primary strategy may be appropriate in: (1) patients with a permanent form of the arrhythmia associated with mild symptoms, which can be further improved by slowing heart rate; (2) patients older than 65 years with recurrent atrial tachyarrhythmias (primary "accepted" AF); and (3) patients with persistent tachyarrhythmias who have failed repeat cardioversions and serial prophylactic antiarrhythmic drug therapy and in whom risk/benefit ratio from using specific antiarrhythmic agents is shifted toward increased risk or who are ineligible for nonpharmacologic "curative" therapy (e.g., pulmonary vein ablation) (Table 15-9).[174] Please note that the current guidelines on management of AF have been adjusted in accordance with the results of the Atrial Fibrillation Follow-up Investigation of Rhythm Management (AFFIRM) trial (Fig. 15-37).[175]

The issue of what constitutes adequate ventricular rate control is not universally agreed. Neither is there consensus on what methods need to be employed for the effective assessment of ventricular rates. Little appreciation has been given so far to the fact that adequate rate control encompasses more than the mere prevention of fast ventricular rates. The absence of chronotropic

TABLE 15-9 Preference, Advantages, and Disadvantages of Rate Control for Atrial Fibrillation and Atrial Flutter

Factors Important in the Decision Making	Advantages	Disadvantages
Heart failure >NYHA II Ejection fraction <40% Long arrhythmia duration (>6 mo) AAD and cardioversion refractory (AAD ≥3; CV ≥3) Known proarrhythmic effects of classes I and III AAD Proarrhythmic risk factors Intolerance of classes I and III AAD Left atrial size >5 cm Symptoms secondary to uncontrolled ventricular rate Asymptomatic arrhythmia Sinus node dysfunction No contraindications for lifelong oral anticoagulation	Reduction of symptoms Low proarrhythmic risk Prevention of tachycardia-induced cardiomyopathy Cost-effective	Need for lifelong anticoagulation Impaired hemodynamics; loss of atrial contribution Progression of atrial remodeling Evolution to a permanent form

AAD, antiarrhythmic drugs; CV, cardioversion, NYHA, New York Heart Association.

incompetence and excessive irregularity of rhythm are rarely acknowledged as important constituents of the effective rate control. There have been very few studies specifically addressing these issues, and most of them have been assessing exclusively pharmacologic means of rate control. It is generally assumed that, in order to compensate for loss of atrial contribution, the ventricular rate during AF should probably be about 10% to 20% higher than a corresponding rate during sinus rhythm. Current ACC/AHA/ESC guidelines recommend that the rate is generally considered controlled when the ventricular response ranges between 60 and 80 bpm at rest and between 90 and 115 bpm during moderate exercise.[34]

Rate control therapy in atrial tachyarrhythmias is based mainly on pharmacologic depression of conduction through the AV node. Therapy aimed at rate control requires careful dose titration and, not infrequently, may impose risk of symptomatic bradycardia and need for permanent pacing. Ablation of the AV node and permanent pacemaker implantation may be the treatment of choice in the presence of disturbing symptoms, poorly controlled by rate-slowing agents. Until very recently, there have been no data supported by clinical randomized trials on the most effective rate control strategy. Nonpharmacologic therapies, such as AV node ablation, were systematically underused in major clinical studies attempting to demonstrate the benefits of the rate control strategy.

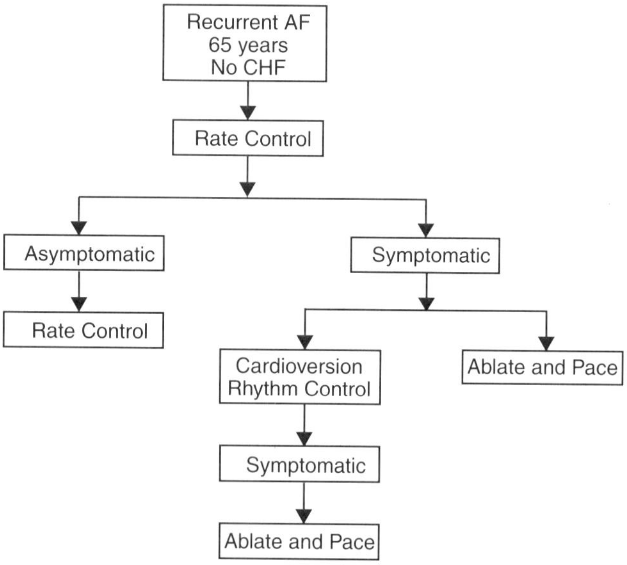

FIGURE 15-37 Revision of the current guidelines on management of atrial fibrillation in accordance with the results of the AFFIRM (Atrial Fibrillation Follow-up Investigation of Rhythm Management) trial. CHF, congestive heart failure.

ANTICOAGULATION

Anticoagulation has now become imperative in a significant proportion of patients with atrial tachyarrhythmias. Risk stratification for stroke is essential for decision making in favor of anticoagulation. A number of risk models have been devised to predict the risk of stroke and the likelihood of benefit from therapy with warfarin or aspirin.[5,34,39,176,177] These include models suggested by: (1) the Atrial Fibrillation Investigators (AFI),[5] supported by the pooled analysis of the five primary prevention trials of warfarin; (2) Stroke Prevention in Atrial Fibrillation (SPAF) Investigators,[176] based on the results of SPAF I–III studies; and (3) the American College of Chest Physicians (ACCP),[39] based on stratification of risk for stroke as high (>7%), intermediate (2.5%), and low (1%). All three models and the recent ACC/AHA/ESC Task Force on AF guidelines agreed on the following risk factors associated with high risks of stroke and systemic embolism (Table 15-10):[176a] (1) age 75 years or older, especially women; (2) previous stroke or transient ischemic attack; (3) valvular heart disease, especially mitral stenosis; (4) hypertension; and (5) congestive heart failure or left ventricular dysfunction. The ACCP Consensus Conference and the ACC/AHA/ESC guidelines also considered diabetes, coronary heart disease, and thyrotoxicosis to be intermediate risk factors. Patients with any one of the major risk factors and particularly those with several major risk factors are at

TABLE 15-10 Risk Stratification for Stroke and Recommendations for Anticoagulation in Atrial Fibrillation

Annual Risk of Stroke	Risk Factors	Recommendations*
Low (1%)	Age younger than 65 years and no major risk factors (prior stroke, TIA, systemic thromboembolism, hypertension, heart failure, ejection fraction <.50)	Aspirin: 227 (132-2500)
Low moderate (1.5%)	Age 65 to 74 years and no major risk factors	Aspirin: 152 (88-1667)†
High moderate (2.5%)	Age 65 to 74 years, no major risk factors but either diabetes mellitus or coronary artery disease	Warfarin: 32 (28-42)
High (6%)	Age younger than 75 years and either hypertension, heart failure, or ejection fraction <.50; or age 75 years or older in the absence of other risk factors	Warfarin: 14 (12-17)
Very high (10%)	Age 75 years or older and either hypertension, heart failure, or ejection fraction <.50; or any age and prior stroke, TIA, or systemic thromboembolism	Warfarin: 8 (7-10)

*Number of patients to treat to prevent one stroke in two years.

†With warfarin, the number to treat is 54 (range, 46 to 69).

TIA, transient ischemic attack.

Adapted from Strauss SE, Majumdar SR, McAlister FA: New evidence for stroke prevention: Scientific review. JAMA 2002;288:1388-95.

high risk for stroke, which is estimated to be 7% to 12% per year, whereas individuals with no risk factors are considered to be at a low risk of 1% per year.

On the basis of this risk stratification, guidelines for antithrombotic therapy in patients with atrial tachyarrhythmias have been developed.[34,39] Anticoagulation with adjusted-dose warfarin (target INR 2.5; range 2.0 to 3.0) is recommended in all high-risk patients without contraindications. Low-risk patients can be treated with an antiplatelet agent, usually aspirin or clopidogrel. Although these stratification schemes identify patients at high and low risk who benefit most and least from lifelong anticoagulation, uncertainties exist regarding this strategy in those with intermediate risk (3% to 5% per year) for whom individual approach balancing risk/benefit of warfarin therapy or aspirin is advocated. Thus, patients with more than one moderate-risk factor are at higher risk of stroke than those with only one risk factor and may, therefore, require anticoagulation with warfarin.

UPSTREAM THERAPY

The "upstream" approach to pharmacologic therapy of arrhythmias is a treatment strategy targeting the underlying disease process that may favor the atrial arrhythmia by disorganized hemodynamics or the development of atrial pathology. There is an increased interest in the beneficial effects on atrial tachyarrhythmias of ACE inhibitors and AT-1 receptor blockers, which may produce direct antiarrhythmic effects additive to improved atrial hemodynamics.[25,178,179] Furthermore, the possibility that AF might be eradicated by primary prevention strategies merits serious consideration. The explosion in the prevalence and incidence of this arrhythmia is undoubtedly related to the pathophysiology of ageing and the comorbid existence of cardiovascular, pulmonary, and other systemic disease. Many of these processes may be slowed or prevented. In this respect, identification and modification of risk factors would make intervention possible early in the course of the disease, when preventative or corrective strategies are most efficient.

NONPHARMACOLOGIC THERAPIES FOR ATRIAL TACHYARRHYTHMIAS

Advances in electrophysiological mapping and radiofrequency catheter ablation techniques have rendered about 95% of cases of typical or reverse typical AFL entirely curable, making isthmus ablation the treatment of choice in most patients with clinically important arrhythmias.[15] The success rate for focal or macro-reentrant AT is slightly lower (75% to 86%).[16] Nonpharmacologic treatment alternatives for the management of AF include: (1) AV node ablation to abolish frequent conduction to the ventricles; (2) catheter ablation of "focal" AF originating from atrial myocardium in the pulmonary veins, venae cavae, crista terminalis, and ligament of Marshall; (3) ablation of an accessory pathway in Wolff-Parkinson-White syndrome; (4) surgical and catheter-based maze procedures aimed at compartmentalization of the fibrillating atria; (5) implantation of a stand-alone atrial defibrillator, dual chamber ICDs, and pacemakers capable of providing preventative and antitachycardia pacing therapies (Fig. 15-38). Radiofrequency ablation of the AV node, followed by implantation of a permanent pacemaker, the "ablate and pace" strategy, is now an established treatment in patients with highly symptomatic, intractable AF when poorly controlled, sustained, rapid ventricular rate response is present and is likely to induce or aggravate myocardial dysfunction.[180] Ablation in or around the pulmonary veins, often in combination with right atrial isthmus ablation, and a modified catheter-based maze procedure can effectively provide curative treatment of AF, but these techniques are limited to a selected patient population with "focal" AF or patients with AF undergoing heart surgery.[17-19]

The high success rate of transvenous cardioversion of AF and AFL using low-energy biphasic atrial shocks delivered between electrodes in the right atrium and the coronary sinus or the left pulmonary artery prompted the development of an implantable atrial defibrillator.[181,182] Such devices, either as a stand-alone atrioverter (Metrix Atrioverter System, InControl/Guidant) or incorporated in an ICD (Jewel AF, Gem III

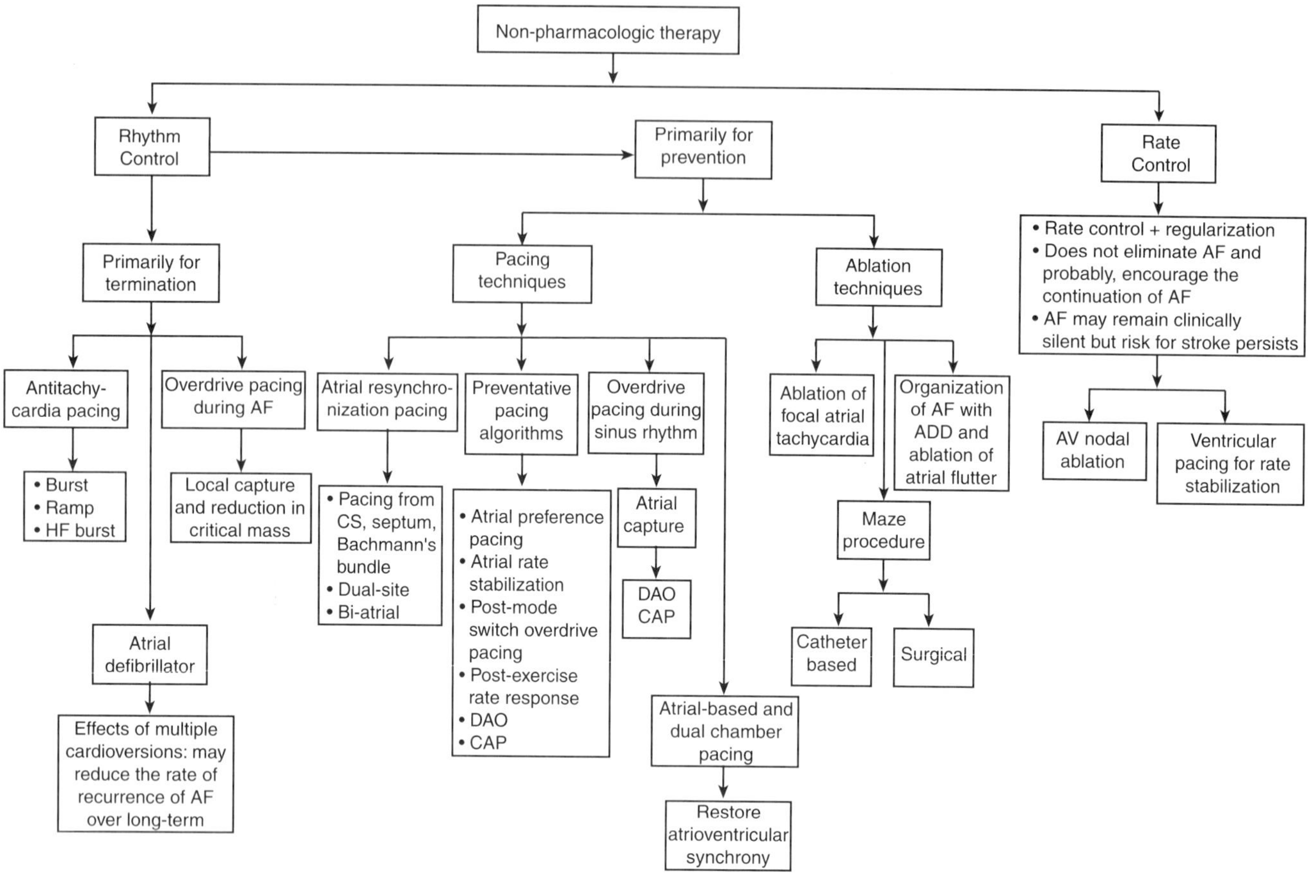

FIGURE 15-38 Nonpharmacologic therapies for atrial fibrillation. AAD, antiarrhythmic drugs; AF, atrial fibrillation; CAP, consistent atrial pacing; CS, coronary sinus; DAO, dynamic atrial overdrive; HF, high frequency.

AT, Medtronic), can restore sinus rhythm expediently in patients with recurrent symptomatic AF and are highly effective for a subgroup of patients failing conventional therapies.

Atrial or dual-chamber physiologic pacing may benefit patients with sinus node dysfunction and intact AV conduction in terms of risk reduction of developing sustained atrial tachyarrhythmias compared with ventricular pacing.[22,23,183] Experimental and clinical evidence has strongly implicated site-dependent nonuniform conduction in the initiation of AF and has suggested that pacing from specific regions with slow conduction, dual-site atrial pacing, or bi-atrial pacing may prevent AF due to improved synchronization of atrial depolarization.[21,184,185] The recognition of potential triggers of AF, such as atrial premature complexes, short-long sequence, and bradycardia, has encouraged the development of novel atrial pacing algorithms designed to prevent the initiation of the arrhythmia on an individual basis.[24,186] The concept of "hybrid therapy" based on the combination of several different therapeutic strategies suggests that antitachycardia pacing therapy, integrated with an atrial defibrillator and preventative atrial pacing modes, may act synergistically to prevent AF.[20]

Evidence-Based Therapy and Management of Atrial Tachyarrhythmias

ACUTE THERAPY

Acute therapy for atrial tachyarrhythmias depends on clinical presentation. Emergent electrical cardioversion is indicated for patients with hemodynamic collapse and progressively deteriorating left ventricular systolic function. If tachyarrhythmia is hemodynamically stable and is of recent onset, pharmacologic cardioversion can be effective. In AFL and known macro-reentrant AT burst overdrive, atrial pacing has proven effective in 82% of cases and is feasible after cardiac surgery, as these patients frequently have epicardial atrial pacing wires, or in patients with implantable pacemakers and dual-chamber ICDs.[66] High-frequency atrial pacing is available in some of the latest models for termination of early onset AF.[24] It must be appreciated that atrial burst overdrive pacing may induce sustained AF, although short periods of AF often precede conversion to sinus rhythm. Rarely, AT can be terminated by vagal maneuvers (e.g., Valsalva's maneuver, carotid sinus massage). Pharmacologic rate control is pertinent to all atrial tachyarrhythmias, particularly if restoration of sinus rhythm is deferred. Anticoagulation is imperative if the arrhythmia persists for more than 24 to 48 hours or if its duration is unknown. This is discussed in detail later.

CARDIOVERSION FOR ATRIAL TACHYARRHYTHMIAS

Spontaneous conversion to sinus rhythm may occur in 66% of patients within 24 hours after onset of the arrhythmia (Fig. 15-39) and in only 17% of those with the arrhythmia of longer duration (odds ratio, 1.8).[187,188] Normal left ventricular function is more common among patients with spontaneous conversion. Advanced age, female gender, structural heart disease, left atrial enlargement, and the use of digoxin decrease the likelihood of spontaneous restoration of sinus rhythm.

Pharmacologic Cardioversion

The evidence base for the efficacy and safety of pharmacologic conversion of atrial tachyarrhythmias has emerged predominantly from studies in AF that included a limited number of patients with AFL. No large studies have been conducted to define the efficacy of pharmacologic cardioversion for AT. Antiarrhythmic drugs proven to be effective for cardioversion of atrial tachyarrhythmias are listed in Table 15-11.

Placebo-controlled, randomized studies of class IC agents show the highest efficacy rate of 60% to 80% for conversion of AF to sinus rhythm.[189-192] Cardioversion rates are 51% to 59% at 3 hours and 72% to 78% at 8 hours after administration of a single loading oral dose of *propafenone* and *flecainide,* respectively, compared with 18% and 39% on placebo.[191] Both oral and IV routes of administration are equally effective, although the IV route is quicker.[192,193] In the third Propafenone in Atrial Fibrillation Italian Trial (PAFIT-3), IV infusion of 2 mg/kg of propafenone resulted in restoration of sinus rhythm in nearly one half the patients within the first hour compared with 14% on placebo.[192] IV flecainide at 2 mg/kg converted 77% of patients within 30 minutes.[193] Class IC agents may cause 1:1 AV conduction with a rapid ventricular response due to slowing the atrial rate, and concomitant administration of AV node blocking agents (β-blockers, verapamil, or diltiazem) is therefore mandatory. Other cardiovascular effects include reversible QRS widening and, rarely, left ventricular decompensation because of the negative inotropic effect.

Class IC drugs usually are ineffective for conversion of AFL. They slow conduction within the reentrant circuit and prolong the flutter cycle length but rarely interrupt the circuit, and there is greater risk of increased (e.g., 2:1 or 1:1) AV conduction. The efficacy rates are reported as low as 13% to 40% with IV flecainide and propafenone.[194,195] There is anecdotal evidence that class IC agents can terminate AT.[196]

The class III agent *ibutilide* has proven effective for conversion of atrial tachyarrhythmias and particularly AFL in randomized, placebo-controlled studies and direct comparisons with procainamide and sotalol (Table 15-12).[197-199] The conversion rate for AFL is up to twice that for AF (63% versus 31%) with ibutilide administered intravenously as a 10-minute infusion of 1 to 2 mg.[197] In direct comparison studies, ibutilide was more effective than procainamide (76% versus 14%) and sotalol (70% versus 19%) for conversion of AFL.[198,199] The success rates for conversion of AF were lower: 51% compared with 21% on procainamide and 44% compared with 11% on sotalol. The rate of successful conversion decreases if the arrhythmia persists for more than 7 days: from 71% to 57% for AFL and from 46% to only 18% for AF. Higher doses of ibutilide administered as two successive infusions of 1 mg are usually required for termination of AF.[197]

Like many of the class III antiarrhythmic drugs, ibutilide may cause significant Q–T interval prolongation and ventricular proarrhythmias. The incidence of sustained polymorphic ventricular tachycardia requiring electrical cardioversion has been reported to be 0.5% to

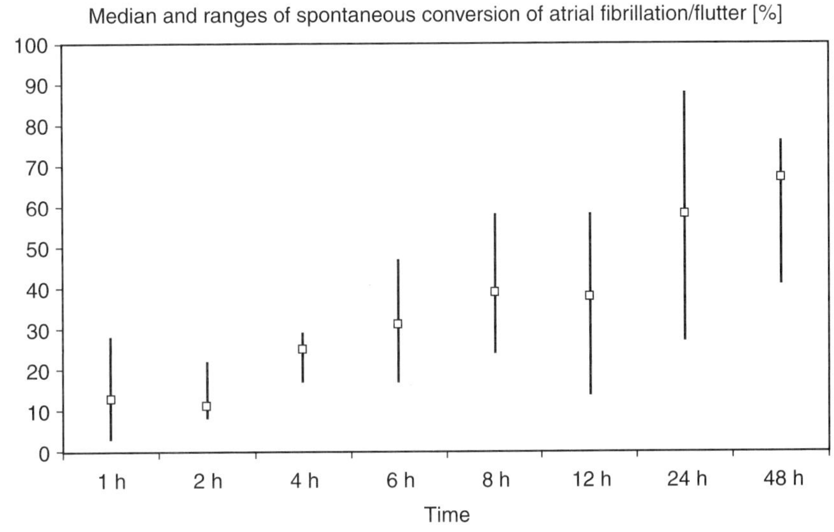

FIGURE 15-39 Incidence of spontaneous conversion to sinus rhythm of recent onset (<7 days) atrial fibrillation/flutter in prospective, randomized, placebo-controlled studies. AF, atrial fibrillation. (Modified from Slavik RS, Tisdale JE, Borzak S: Pharmacologic conversion of atrial fibrillation: A systemic review of available evidence. Prog Cardiovasc Dis 2001;44:121-52.)

TABLE 15-11 Antiarrhythmic Drugs for Pharmacologic Conversion of Atrial Tachyarrhythmias

Drug	Route of Administration	Dose	Class Recommendation for Arrhythmia ≤7 Days	Class Recommendation for Arrhythmia >7 Days	Potential Adverse Effects
Dofetilide	Oral	125-500 mg twice daily*	I	I	QT prolongation; torsades de pointes; contraindicated if creatinine clearance <20 mL/min
Ibutilide	IV	1 mg over 10 min; repeat 1 mg if necessary	I	IIA	QT prolongation; torsades de pointes; hypotension
Flecainide	Oral or IV	Loading dose 200-300 mg or 1.5-3.0 mg/kg over 10-20 min	I	IIB	Rapidly conducted atrial flutter; possible deterioration of ventricular function in the presence of organic heart disease
Propafenone	Oral or IV	Loading dose 450-600 mg or 1.5-2.0 mg/kg over 10-20 min	I	IIB	
Amiodarone	Oral or IV	Inpatient: 1200-1800 mg daily in divided doses until 10 g total; then 200-400 mg daily Outpatient: 600-800 mg daily until 10 g total; then 200-400 mg daily 5-7 mg/kg over 30-60 min intravenously; then 1200-1800 mg daily oral until 10 g total; then 200-400 mg daily	IIA	IIA	Hypotension; bradycardia; QT prolongation; torsades de pointes (rare); gastrointestinal upset; constipation; phlebitis
Quinidine	Oral	750-1500 mg in divided doses over 6-12 hours + rate-slowing drug (verapamil or a β-blocker)	IIB	IIB	QT prolongation; torsades de pointes; QRS widening; rapid atrial flutter; hypotension; gastrointestinal upset
Procainamide†	Intravenous	1000 mg over 30 min (33 mg/min) followed by 2 mg/min infusion	IIB	IIB	QRS widening; torsades de pointes; rapid atrial flutter
Sotalol†	Oral	80 mg initial dose; then 160-320 mg in divided doses	III	III	QT prolongation; torsades de pointes; bradycardia

*Dose depends on creatinine clearance: >60 mL/min—500 mg; 40-60 mL/min—250 mg; 20-40 mL/min—125 mg twice daily.

†Less effective or incompletely studied agents; dose regimens vary in different studies.

1.7%; for nonsustained polymorphic ventricular tachycardia, the incidence is 2.6% to 6.7%.[172,197,199] There are few data to support the use of ibutilide in patients with significant structural heart disease, and there is no evidence of its efficacy in the treatment of AT.

Dofetilide exhibits higher affinity to atrial myocardium and prolongs the effective refractory period twofold more in the atria than in the ventricles, which may explain its greater efficacy in converting atrial tachyarrhythmias to sinus rhythm. In a randomized, placebo-controlled study, dofetilide converted 64% patients with AFL and 24% patients with AF within 3 hours from the start of a 30-minute IV infusion of 8 μg/kg compared with 3.3% on placebo.[200] Based on the results of three large prospective studies, the Danish Investigations of Arrhythmia and Mortality ON Dofetilide (DIAMOND), Symptomatic Atrial Fibrillation Investigative Research on Dofetilide (SAFIRE-D) and European and Australian Multicenter Evaluative Research on Atrial Fibrillation Dofetilide (EMERALD) the oral formulation of dofetilide has attained a class I indication for cardioversion of both recent onset and persistent atrial tachyarrhythmias, including patients with significant structural heart disease (see Table 15-11).[34] In the DIAMOND-AF substudy of 506 patients with AF or AFL at baseline, 75% of whom had symptomatic heart failure and an ejection fraction below 35%, treatment with dofetilide at the maximum dose of 500 μg twice daily was associated with a greater rate of conversion to sinus rhythm (44% versus 14%) and a greater probability of remaining in sinus rhythm at 1 year (79% versus 42%).[201] The dose-efficacy SAFIRE-D study of 325 patients with persistent AF and

TABLE 15-12 Clinical Evidence of Efficacy of Class III Antiarrhythmic Drugs for Conversion or Maintenance, or Both, of Sinus Rhythm in Patients with Atrial Tachyarrhythmias

Study	Drug	Number of Patients	Follow-up	Cardioversion Rate	Remained in Sinus Rhythm	All-cause Mortality	QTc Interval Prolongation	Torsades de Pointes
Ibutilide Repeat Dose Study[197]	Ibutilide 1.5 mg or 2 mg IV	266; 50% with AFL	24 hr	47% on ibutilide vs. 2% on placebo; converted 63% AFL vs. 31% AF	Not assessed	Not applicable	By 106-160 ms	8.3%; 1.7% required DCC
Randomized study of ibutilide vs. procainamide[198]	Ibutilide 1-2 mg IV vs. procainamide 1.2 g IV	127; 33% with AFL	24 hr	58.3% on ibutilide vs. 18.3% on procainamide; AFL 76% vs. 14%, AF 51% vs. 21%	Not assessed	Not applicable	↑ QT requiring withdrawal in 1 patient	0.8%
Ibutilide/Sotalol Comparator Study (ISCS)[199]	Ibutilide 1 mg or 2 mg IV vs. d,l-sotalol 1.5 mg/kg IV	319; 18.5% AFL	72 hr	AFL: 56%-70% on ibutilide vs. 19% on sotalol; AF: 20%-44% on ibutilide vs. 11% on sotalol	Not assessed	Not applicable	Not stated	0.9% on ibutilide 2 mg; 0.5% required DCC; none on sotalol
d,l-Sotalol Atrial Fibrillation/Flutter Study Group[266]	d,l-sotalol 160, 240, 320 mg	253; 20% AFL	1 yr	All in sinus rhythm at randomization	Time to first episode: 106-229 days on sotalol vs. 27 days on placebo	Not applicable	QT ≥520 ms in 7 patients on sotalol	None but strict inclusion criteria
SOCESP[268]	Sotalol 160-320 mg vs. quinidine 600-800 mg	121	6 mo	All cardioverted	93% on sotalol vs. 64% on quinidine in AF ≤72 hr; 33% on sotalol vs. 33% on quinidine in AF >72 hr	Not applicable	Not stated	5% on sotalol vs. 1% on quinidine
PAFAC[269]	Sotalol 320 mg Quinidine 160 mg + verapamil 80 mg bid	1182; 383 sotalol	1 yr; daily transtelephonic ECG monitoring	All cardioverted	Recurrence rates:[§] 49% on sotalol vs. 38% on quinidine + verapamil vs. 78% on placebo	1.6% on sotalol; 1.3% on quinidine + verapamil; 0% on placebo	Not stated	2.3% (all on sotalol)
Bellandi[267]	Sotalol 240 mg Propafenone 450-900 mg	300	1 yr	All cardioverted	73% on sotalol, 3% on quinidine vs. 35% on placebo	Not applicable	QTc >550 ms in 1 patient	4% on sotalol; 2% required DCC
CTAF[270]	Amiodarone 200 mg vs. sotalol 160-320 mg and propafenone 450-600 mg	403; 51% persistent AF	16 mo	All cardioverted	69% vs. 39% on sotalol or propafenone (57% risk reduction)	Not applicable	0.5% vs. 0%	0% vs. 0.5% on placebo

Continued

TABLE 15-12 Clinical Evidence of Efficacy of Class III Antiarrhythmic Drugs for Conversion or Maintenance, or Both, of Sinus Rhythm in Patients with Atrial Tachyarrhythmias—*(cont'd)*

Study	Drug	Number of Patients	Follow-up	Cardioversion Rate	Remained in Sinus Rhythm	All-cause Mortality	QTc Interval Prolongation	Torsades de Pointes
CHF-STAT[208]	Amiodarone 800 mg for 2 wk; 400 mg for 50 wk; then 200 mg	667; 103 AF or AFL	1 yr in the main study; 4-5 yr in the AF substudy	31.3% on amiodarone vs. 7.7% on placebo	31% on amiodarone vs. 8% on placebo	↓ Mortality in patients who converted to SR vs. those in AF	Not stated	Not stated
DDAFF[200]	Dofetilide 8 μg/kg IV	96 (18% flutter)	3 hr	All, 30.3% on dofetilide vs. 3.3% on placebo; 64% AFL vs. 24% AF	Not assessed	Not applicable	In 12% patients	3%
DIAMOND-AF[201]	Dofetilide 250 μg bid	506	1.5 yr	59% on dofetilide vs. 34% on placebo	79% vs. 46%	41% vs. 42%	Usually prolonged QT	3.3% (76% on day 3)
SAFIRE-D[202]	Dofetilide 250, 500, 1000 μg	225	1 yr	6.1%, 9.8%, 19.9% on dofetilide vs. 1.2% on placebo	40%, 37%, 58% on dofetilide vs. 25% on placebo	2.5% vs. 3.5%	4.4%; 3.1% on day 3	0.8% on day 3
EMERALD[203]	Dofetilide 250, 500, 1000 μg	671	Phase 1: 72 hr; Phase 2: 2 yr	6%, 11%, 29% on dofetilide vs. 5% on sotalol at 72 hr	40%, 52%, 66% on dofetilide vs. 21% on placebo at 1 yr	Not applicable	Not stated	3 torsades de pointes; 1 sudden death
ASAP	Azimilide 50, 100, 125 mg	384	180 days	All in sinus rhythm	17%, 38%, 83%*	2.9% vs. 0%	6.9%	0.99% on day 4
Meta-analysis of 4 SVA studies[274]	Azimilide 35, 50, 100, 125 mg	1380; 20%-31% AFL	Time to first recurrence	All in sinus rhythm	Hazard ratio 1.34 for 100 mg; 1.32 for 125 mg† 1.49 for 100 mg; 1.86 for 125 mg‡	Not applicable	Not stated	0.9% at 100 and 125 mg vs. 0% on placebo
ALIVE[275a]	Azimilide 100 mg	3381	1 yr	26.8% vs. 10.8%	New AF: 0.49% on azimilide vs. 1.15% on placebo	11.6% in each group	Not stated	0.3% vs. 0.1% on placebo
DAFNE[273]	Dronedarone 800 mg 1200 mg 1600 mg	199	6 mo	5.8% 8.2% 14.7% vs 3.1% on placebo	35% on dronedarone 800 mg vs. 10% on placebo	Not applicable	6.6%-16.4%	0%

*Indicates a reduction in risk of AF recurrence.

†Indicated hazard ratios for the time to first recurrence of the arrhythmia.

‡In patients with structural heart disease.

§Persistent AF.

AF, atrial fibrillation; AFL, atrial flutter; AT, atrial tachycardia; DCC, direct current cardioversion; ms, milliseconds; SR, sinus rhythm; SVA, supraventricular arrhythmia; TdP, torsades de pointes.

AFL reported cardioversion rates of 6.1%, 9.8%, and 29.9% with 250, 500, and 1000 μg of dofetilide compared with 1.2% of spontaneous conversion in the placebo arm.[202] The EMERALD study tested the efficacy of dofetilide against sotalol and found the former more effective for conversion of persistent AF and AFL at 72 hours after oral administration (29% versus 5%).[203] The incidence of torsades de pointes varies between 0.6% and 3.3%, with more than three quarters of episodes occurring during the first 3 to 4 days of drug ingestion. Dofetilide appears to be safe in patients with myocardial infarction and patients with heart failure; the dosage, however, should be adjusted according to renal function.

The efficacy of *sotalol* for conversion of AF and AFL has not proven to be superior to digoxin and placebo; therefore, it is not recommended for termination of atrial tachyarrhythmias, but it is effective for rate control.[34] The conversion rate was 11% to 13% with sotalol versus 14% on placebo in the double-blinded phase and 30% in the open-labeled phase.[204] Bradycardia and hypotension were the most common adverse effects with an incidence of 6.5% and 3.7%, respectively.[200] Failure of sotalol to terminate atrial tachyarrhythmias may relate to its ability to prolong the atrial effective refractory period at lower atrial rates but not during rapid AF or AFL (reverse use-dependency).

A meta-analysis of 13 randomized controlled studies in 1174 patients has shown that IV *amiodarone* is more effective in converting AF and AFL than placebo with a 44% superiority and it is nearly as effective as class IC antiarrhythmic drugs, but the effect is delayed by 24 hours.[205] At 8 hours, the probability of restoration of sinus rhythm is 65% higher with flecainide or propafenone than with amiodarone. Amiodarone appears to be inferior to IV dofetilide in converting persistent arrhythmias (4% versus 35%).[206] There is anecdotal evidence that using higher than the usual IV dose of amiodarone and a combination of IV and oral routes of administration may enhance the cardioversion rate (class IIA indication).[34] In the largest study of 208 patients with AF, the conversion rate with amiodarone administered intravenously at a dose of 300 mg for 1 hour followed by infusion of 20 mg/kg for 24 hours and then followed by oral treatment for 4 weeks was 80% for AF.[207] However, the conversion rate was also high (40%) in the placebo group. Because of its ability to control the ventricular rate, a low likelihood of torsades de pointes, and the absence of a negative inotropic effect, oral amiodarone may be initiated on an outpatient basis, including patients with significant structural heart disease. In the Congestive Heart Failure Survival Trial of Antiarrhythmic Therapy (CHF-STAT) study, 31% of patients converted to sinus rhythm on amiodarone compared with only 7.7% on placebo.[208] There are no consistent data supporting the use of amiodarone as first-line treatment of AFL and AT, but theoretically it is preferred in patients with poor ventricular function.

Quinidine given in a cumulative dose of up to 1000 to 1200 mg has been shown to cardiovert 60% to 80% of patients with recent onset AF. It was more effective than sotalol and, in some studies, it was as effective as IV amiodarone for conversion of persistent AF (47%).[209-212] The effect usually occurs within 12 hours of treatment. To reduce the risk of 1:1 AV conduction during organization of AF into AFL, quinidine should be administered in conjunction with AV node blocking agents, such as β-blockers or nondihydropyridine calcium antagonists (verapamil and diltiazem). The combination of quinidine and digoxin is not recommended, as it is associated with a lower conversion rate.[212] The poor safety profile of quinidine has led to a decrease in its use for acute cardioversion of atrial tachyarrhythmias. Quinidine poses increased risk of torsades de pointes, which is observed at a rate of 6% and is probably due to an increase in dispersion of ventricular repolarization. Quinidine also widens the QRS complex, causes gastrointestinal side effects in a significant proportion of patients, and is contraindicated in individuals with structural heart disease.

There is evidence from randomized, controlled studies supporting the use of *procainamide* in recent onset AF and AFL (<48 hours) but its efficacy is limited in the arrhythmia of longer duration.[199,213-215] The conversion rate with procainamide administered intravenously at a dose of 1000 mg for 30 minutes and followed by an infusion of 2 mg/min over 1 hour varied from 51% to 70%. Procainamide was superior to placebo (69% versus 38%)[213] and propafenone (69.5% versus 48.7%)[214] but less effective than flecainide (65% versus 92%)[215] and ibutilide (18.3% versus 58.3%).[199]

Adenosine can terminate a significant proportion of AT, including sinus node reentry tachycardia. Adenosine-sensitive AT is usually focal in origin and arises either from the region of the crista terminalis and tricuspid annulus or from diverse atrial sites with a repetitive monomorphic pattern. Adenosine (mean dose 7.3 ± 4.0 mg) terminated AT in 14 (82%) of 17 patients with focal AT but in only 1 of 13 patients with macro-reentrant AT.[215] Focal AT originating from the crista terminalis usually responds to IV verapamil. Tecadenoson (CVT510) is an adenosine derivative with a high specificity to A_1-adenosine receptors and is devoid of the side effects caused by stimulation of A_2-receptors.[217] It is effective in 86.5% of AV reentry tachycardias[218] and may be potentially useful for termination of adenosine-sensitive AT.

Digoxin is not recommended for restoration of sinus rhythm in patients with atrial tachyarrhythmias.[34] In the Digitalis in Acute Atrial Fibrillation (DAAF) study, there was no difference in cardioversion rates at 16 hours between IV digoxin and placebo (51% versus 46%).[219] Furthermore, the drug has been shown to facilitate AF due to its cholinergic effects, which may cause a nonuniform reduction in conduction velocity and effective refractory periods in the atria, and to delay the reversal of remodeling after restoration of sinus rhythm.[220,221]

Electrical Cardioversion

The overall success rate is 90% to 95% with electrical cardioversion for arrhythmias less than 48 hours and

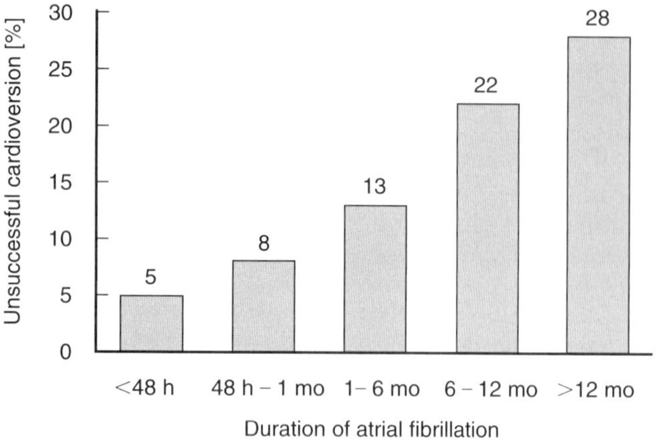

FIGURE 15-40 Prevalence of unsuccessful electrical cardioversion as a function of duration of the arrhythmia. (Modified from Elhendy A, Gentile F, Khanderia BK, et al: Predictors of unsuccessful electrical cardioversion in atrial fibrillation. Am J Cardiol 2002;89;83-6.)

decreases to 72% to 78% if the arrhythmia is present for 1 year (Fig. 15-40).[222] Longer duration of the arrhythmia, greater weight, and higher transthoracic impedance are associated with lower shock success.[222,223] Left atrial size does not predict the outcome of cardioversion, which was successful in 83% of patients with a significantly enlarged left atrium. Advanced age also fails to predict failure to cardiovert. Studies in older patients have demonstrated that the efficacy rate and the incidence of complications of electrical cardioversion was not significantly higher than in younger individuals, suggesting that a patient should not be refused electrical restoration of sinus rhythm merely on the grounds of age.[224]

AFL and AT can be converted with a direct current (DC) shock energy as low as 25 to 50 J, but because a 100 J shock is virtually always successful, it should be considered as the initial shock strength. It should be recognized that DC cardioversion is effective in AT in which a microreentrant circuit or triggered activity is suspected as the underlying mechanism, but it seldom terminates automatic focal AT. In AF of less than 30 days duration sinus rhythm can be restored by a shock of 100 J, but it is recommended that cardioversion should be started with the initial shock energy level of 200 J or greater. In those with AF for more than 180 days, in heavier individuals, or those with chronic obstructive lung disease and pulmonary emphysema, an initial setting of 300 to 360 J is appropriate. Success may occur on the third or subsequent attempt at an intensity that initially proves ineffective. The higher energy levels required for direct current cardioversion of AF compared with AFL and many other organized supraventricular tachyarrhythmias is probably due to the larger amount of atrial tissue that is in a partially refractory state, during which the defibrillation threshold is four times higher than in the resting state.

For successful cardioversion, the current vector must traverse a critical atrial mass, and the anterior-posterior electrode position appears to be the best. The energy requirement tended to be lower and the success rate greater (87% versus 76%) with anterior-posterior (anterior precordium and left subscapular region) compared with anterior-lateral (ventricular apex and right infraclavicular area) alignment of the electrode paddles.[225,226] However, depending on the individual patient, the antero-lateral electrode position may be more effective.

The use of *biphasic* waveform of DC current (Fig. 15-41) appears to achieve a higher rate of conversion to sinus rhythm at a lower energy level than the conventional *monophasic* waveform.[227,228] In contrast to the monophasic waveform, the biphasic waveform is relatively insensitive to changes in transthoracic impedance due to impedance compensation, which ensures a constant current in the first phase. The result is a decrease in defibrillation threshold by approximately one third. First shock efficacy with the 70 J rectilinear biphasic waveform was significantly greater than with the 100 J damped sine monophasic waveform (68% versus 21%), as was the cumulative efficacy after several shocks (94% versus 79%).[227] The following three variables were independently associated with successful cardioversion: use of a biphasic waveform shock, transthoracic impedance, and duration of the arrhythmia. In the randomized, double-blind BiCard study, the success rate was consistently higher for a truncated exponential biphasic shock than for a monophasic shock at the shared energy level of 100 J (60% versus 22%) and 200 J (90% versus 53%) (Fig. 15-42), although this difference disappeared when the duration of the index episode was more than 1 year.[228] The biphasic waveform was also associated with a lower frequency of dermal injury. These observations suggest

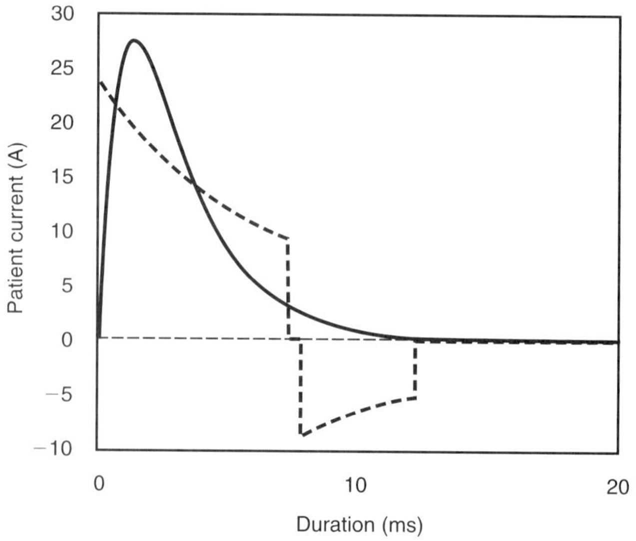

FIGURE 15-41 Superimposed monophasic damped sine waveform (*solid line*) and a biphasic truncated exponential waveform (*dashed line*). Shock energy 150 J, impedance 77 Ω. (Modified from Page RL, Kerber RE, Russell JK, et al, for the BiCard Investigators: Biphasic versus monophasic shock waveform for conversion of atrial fibrillation. J Am Coll Cardiol 2002;39:1956-63.)

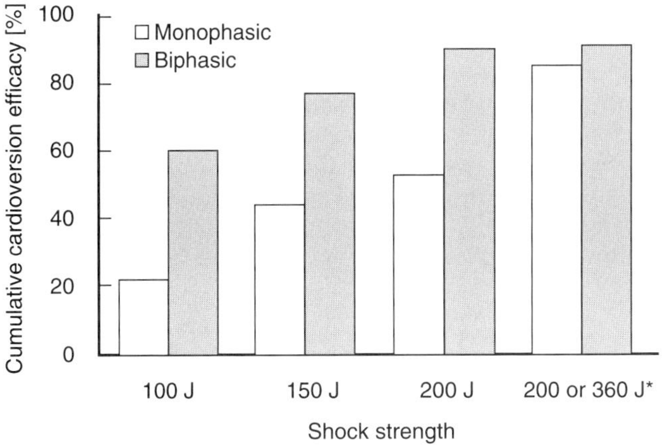

FIGURE 15-42 The efficacy of truncated exponential biphasic shock and monophasic shock at the shared energy level. Patients received up to four shocks, either biphasic or monophasic, as necessary for conversion: 100 J, 150 J, 200 J, and then either 200 J biphasic or 360 J monophasic (*). (Modified from Page RL, Kerber RE, Russell JK, et al, for the BiCard Investigators: Biphasic versus monophasic shock waveform for conversion of atrial fibrillation. J Am Coll Cardiol 2002;39:1956-63.)

that biphasic waveforms can improve the efficacy of transthoracic cardioversion for refractory AF and may reduce the need for transcatheter internal cardioversion procedures.

Internal Cardioversion

Low energy transvenous cardioversion is an alternative option in patients with persistent arrhythmia of longer duration in whom the likelihood of successful external cardioversion is low, patients with severe obstructive pulmonary disease, overweight individuals, and probably in those with implanted pacemakers or defibrillators.[166,229] For internal cardioversion, electrode configuration can be the right atrium–coronary sinus or the right atrium–left pulmonary artery (Fig. 15-43). The energy level usually required for internal cardioversion is in the range of 2 to 6 J, which is approximately 2% to 5% of the energy required for transthoracic cardioversion.

The success rate of internal cardioversion for persistent atrial arrhythmias was reported to be 95% compared with a 79% success rate observed with external cardioversion.[166] The mean energy required for internal cardioversion was 5.0 ± 3.2 J and 313 ± 71 J for external cardioversion. When the shock was delivered between the right atrium and the coronary sinus, the success rate was higher and the energy level was lower compared with the right atrium–left pulmonary artery electrode configuration (84% versus 60%; 2.4 ± 0.9 J versus 4.2 ± 2.0 J).[230] The energy level was significantly higher for conversion of persistent AF compared with paroxysmal AF (3.3 ± 1.3 J versus 1.8 ± 0.7 J). The likelihood of maintenance of sinus rhythm at a 1-year follow-up did not differ between the two methods (48% for internal and 38% for external cardioversion).

Early Recurrence of Arrhythmias

Early recurrence of AF was reported in 10% to 30% of patients, usually within 1 minute following successful cardioversion, resulting in the reduction of the primary success rate of cardioversion.[231-233] The incidence of early recurrence of AF was similar for internal and external methods of cardioversion, with 91% of episodes being initiated by an atrial premature beat with a short

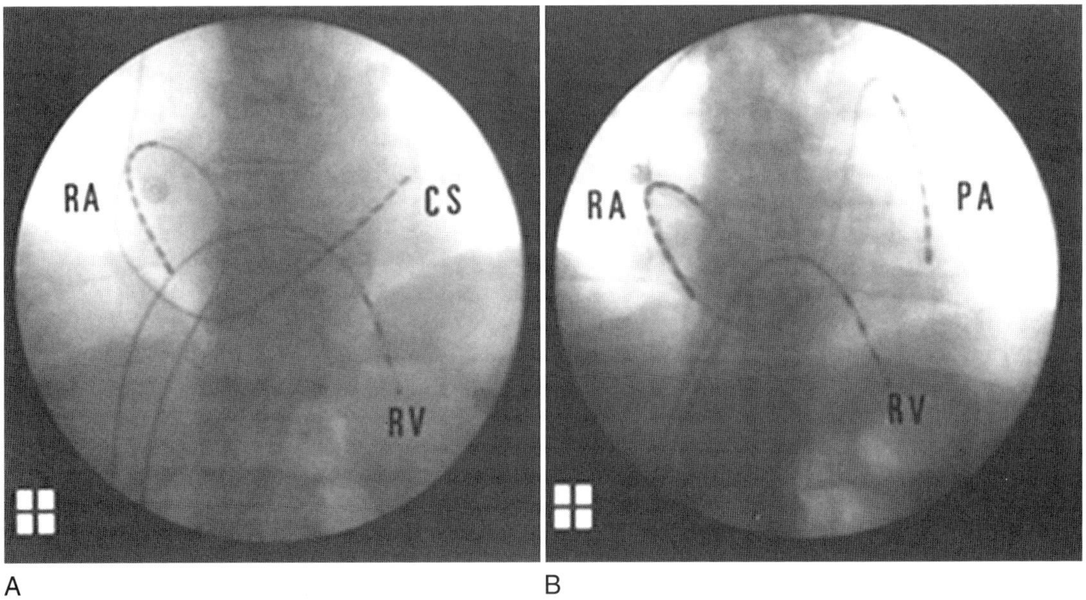

FIGURE 15-43 Electrode configuration for internal cardioversion. **A,** The right atrium (*RA,* cathode)—coronary sinus (*CS,* anode). **B,** The right atrium (*RA,* cathode)—left pulmonary artery (anode). (Reprinted from Lévy S, Ricard P, Guenoun M, et al: Low energy internal cardioversion of spontaneous atrial fibrillation. Circulation 1997;96:253-9.)

coupling interval and 9% occurring after a preceding bradycardia (Fig. 15-44).[233] Early recurrence of AF often required repeat shock, usually at a higher energy level and in combination with antiarrhythmic drugs. There is anecdotal evidence that overdrive atrial pacing after cardioversion may prevent early recurrence of AF by eliminating triggers of the arrhythmia, such as short-long sequences produced by frequent atrial premature beats and bradycardia.[233]

Various antiarrhythmic drugs, including propafenone, flecainide, sotalol, ibutilide, and amiodarone, have recently been shown to prevent early recurrence and to facilitate either external or internal electrical conversion of atrial tachyarrhythmias.[233-237] The synergistic effect with antiarrhythmic drugs may be due to prolongation of the atrial effective refractory period, conversion of AF to a more organized atrial rhythm, and suppression of atrial premature beats immediately after restoration of sinus rhythm. However, pretreatment with antiarrhythmic drugs may favor ventricular proarrhythmias or bradycardia in the presence of sinus node dysfunction.

IV flecainide at a dose of 2 mg/kg when used as adjunctive therapy to internal cardioversion increased the likelihood of restoration of sinus rhythm by 16% and reduced the energy requirement from 4.4 to 3.5 J for conversion of persistent AF and from 1.68 to 0.84 J for paroxysmal AF.[235] Pretreatment with ibutilide before transthoracic cardioversion significantly improved the success rate of cardioversion and lowered the energy requirement by approximately 30%.[236] Furthermore, all patients who failed electrical cardioversion converted to sinus rhythm following the administration of ibutilide and repeat shock. Higher success rates of electrical cardioversion were reported in patients with persistent AF treated with oral amiodarone before cardioversion compared with diltiazem (88% versus 56%).[237]

The synergistic effect of pretreatment with verapamil is likely to be due to the favorable effects on atrial remodeling rather than modification of the electrophysiological properties of the arrhythmia itself. In the Verapamil in Atrial Fibrillation (VerAF) trial, adjunctive treatment with verapamil significantly increased the likelihood of remaining in sinus rhythm during the first week after transthoracic cardioversion (74% versus 39% compared with the control group), but this effect did not relate to the suppression of triggers of the arrhythmia.[238] However, in the Verapamil versus Digoxin Cardioversion Trial (VERDICT), only 47% of patients in the verapamil-treated group and 53% of those in the digoxin-treated group remained free from recurrence of the arrhythmia during the first month after cardioversion.[239] Data regarding the beneficial effects of diltiazem on electrical cardioversion are inconclusive.[240]

Complications of Electrical Cardioversion

Complications associated with electrical cardioversion are mainly related to thromboembolic events, which have been reported to occur at a rate of 1% to 7% in the absence of anticoagulation therapy. In addition, various, usually transient, tachyarrhythmias and bradyarrhythmias such as ventricular and supraventricular premature beats, runs of supraventricular tachycardia, bradycardia, and short periods of sinus arrest may occur.[241,242] Ventricular tachycardia or fibrillation may be precipitated by electrolyte imbalance (hypokalemia, hypomagnesemia) or digitalis intoxication. Transient ST segment elevation and increased plasma levels of cardiac enzymes (primarily creatine kinase), which may occur following electrical cardioversion, are not generally associated with clinically significant myocardial damage. In patients with underlying organic heart disease and significantly impaired left ventricular function, there is a small risk of the development of acute heart failure and pulmonary edema because of a sudden reduction in cardiac output due to postcardioversion relative bradycardia and the absence of a mechanical contribution from the stunned atria. Pulmonary edema of neurogenic origin has also been reported after cardioversion.

Although external cardioversion in patients with implanted pacemakers and defibrillators is believed to be feasible, electricity conducted along an implanted lead may cause local myocardial injury, resulting in an increase of pacing threshold or exit block and loss of capture, which may require reprogramming the device to increase generator output. The pacemaker/defibrillator

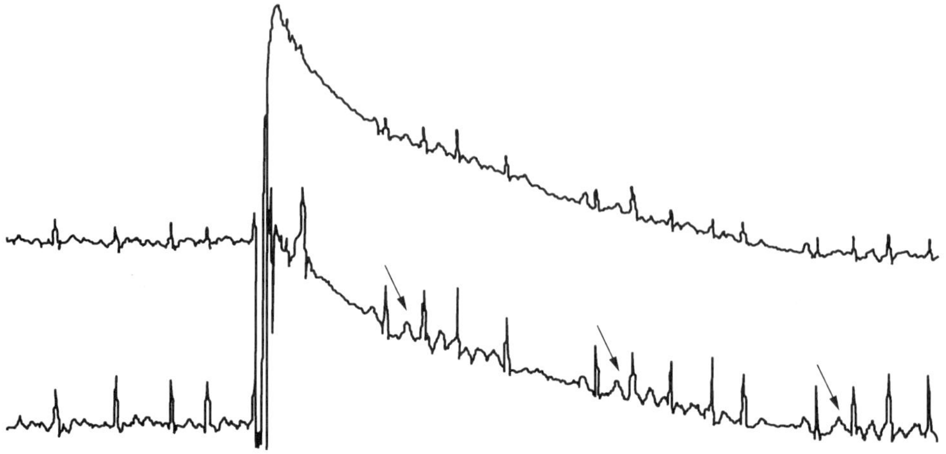

FIGURE 15-44 Early recurrence of atrial fibrillation within seconds after successful electrical cardioversion. Note the presence of atrial premature beats (*arrow*) that initiates atrial fibrillation.

may also be damaged. The risk is lower with an anterior-posterior position of defibrillator electrodes and in pacemakers with bipolar lead systems. The shock vector and the pacemaker lead vectors should be positioned as far from each other as is sensible. The device should be interrogated immediately before and after cardioversion to verify appropriate function. Low-energy internal cardioversion appears not to interfere with function of implanted devices.[243] It was reported that shocks of up to 20 J did not affect pacemaker function.

ACUTE RATE CONTROL

Randomized, controlled studies that included patients with AF and AFL have shown that both IV nondihydropyridine calcium antagonists (*verapamil, diltiazem*) and β-blockers (usually *esmolol*) can accomplish rapid control of the ventricular response rate in the emergency setting (Table 15-13).[34,244-249] The decrease in the ventricular rate (≈ 20% to 30%), time to the maximum effect (20 to 30 minutes), conversion rate (12% to 25%), and the adverse reactions (usually hypotension and bradycardia, although left ventricular dysfunction and high-degree heart block may occur) have been reported to be similar with both classes. β-Blockers are preferable if thyrotoxicosis is suspected as a cause of the arrhythmia. In patients with preexcitation syndrome and AF, AV node blocking agents can improve anterograde conduction in the accessory pathway and produce AF with fast ventricular rates that may degenerate into ventricular fibrillation.

IV *digoxin* is no longer the treatment of choice when rapid rate control is essential because of delayed onset of therapeutic effect (>60 minutes). In a direct comparison with diltiazem, the peak effect of digoxin was not observed until 3 hours, whereas diltiazem significantly decreased the ventricular rate within 5 minutes.[248] However, because of its positive inotropic action, digoxin may be safer to use in patients with poor ventricular function and moderately fast ventricular rates.[249] It may be more difficult to achieve rate control for AFL than AF. Digoxin may convert AFL to AF, in which rate control is easier to accomplish.

There is anecdotal evidence that IV *amiodarone* may be effective in rate control when other AV node blocking agents have no effect on ventricular response or are contraindicated. IV infusion of amiodarone (mean dose 242 ± 137 mg over 1 hour) was associated with a decrease in ventricular rates by 37 bpm and an increase in systolic blood pressure by 24 mm Hg in 38 patients with hemodynamically destabilizing AF and AFL refractory to conventional therapy including electrical cardioversion and IV agents, such as esmolol, diltiazem, digoxin, and procainamide.[250] Eighteen of these patients received IV vasopressors, and seven required intra-aortic balloon pump or left ventricular assist device. *Sotalol,* although failing to terminate atrial tachyarrhythmias, may satisfactorily control the ventricular rate.[198,204]

The evidence base for urgent rate control in AT is scarce. It is generally accepted that β-blockers and calcium antagonists, particularly verapamil, can either terminate AT or produce rate control through AV block, which is often difficult to achieve. *Adenosine* can terminate AT, but the most common response to adenosine is to create AV block and thereby reveal the unaffected AT. The use of adenosine for rate control is limited because of its extremely short half-life. *Tecadenoson* promptly prolongs AV conduction at doses that do not reduce blood pressure or cause bronchospasm. As with adenosine, tecadenoson has an immediate onset of action (<30 seconds) but has a longer half-life (approximately 30 minutes) and may be useful for immediate control of ventricular rates in atrial tachyarrhythmias. *DTI0009* is another adenosine-like agent, but it differs from tecadenoson in that it has a longer

TABLE 15-13 Acute Pharmacologic Rate Control for Atrial Tachyarrhythmias

Drug*	Route of Administration	Dose	Onset	Class Recommendation	Potential Adverse Effects
Digoxin	IV	0.25 mg every 2 hr, max 1.5 mg	2 hr	IIb	Bradycardia; atrioventricular block; atrial arrhythmias; ventricular tachycardia
Diltiazem	IV	0.25 mg/kg over 2 min followed by 5-15 mg/hr infusion	2-7 min	I	Hypotension, bradycardia, heart block, possible deterioration of ventricular function in the presence of organic heart disease
Verapamil	IV	0.075-0.15 mg/kg over 2 min	3-5	I	
Esmolol	IV	0.5 mg/kg over 1 min followed by 0.05-0.2 mg/kg/min infusion	5 min	I	Hypotension, bradycardia, heart block, possible deterioration of ventricular function in the presence of organic heart disease
Metoprolol	IV	2.5-5 mg over 2 min followed by repeat doses if necessary	5 min	I	Hypotension, bradycardia, heart block, possible deterioration of ventricular function in the presence of organic heart disease
Propranolol	IV	0.15 mg/kg	5 min	I	Hypotension, bradycardia, heart block, possible deterioration of ventricular function in the presence of organic heart disease

half-life and appears more suitable for oral application. Initiation of AF is recognized as a potential adverse effect of adenosine (1% to 15%) and probably its derivatives. If the accessory pathway conduction is present it may be blocked or slowed by infusion of sodium channel blocking drugs such as procainamide, propafenone, or flecainide.

PROPHYLACTIC ANTIARRHYTHMIC DRUG THERAPY

The goals of prophylactic antiarrhythmic drug therapy include prevention of recurrence of atrial tachyarrhythmias or modification of paroxysms of the arrhythmia rendering them less symptomatic, less frequent, and less sustained, or self-terminating. If the arrhythmia burden is reduced, there may be a decrease in the progression of left ventricular dysfunction, thromboembolic complications, and mortality. However, the likelihood of remaining free from recurrence of the arrhythmia over the long term is not satisfactory with any of the antiarrhythmic drugs that are presently available (Fig. 15-45).

β-Blockers

β-Blockers are modestly effective in preventing AFL and AF and are mainly used for rate control, usually as adjuvant therapy to digoxin. Exceptions are lone, adrenergically mediated AF[251]; AF associated with coronary bypass surgery[252]; and thyrotoxicosis, in which case β-blockers may be a first-line therapy. There is anecdotal evidence suggesting that β-blockers are more effective than placebo and equally effective as sotalol in preventing AF after electrical cardioversion.[253,254] Fifty-eight percent of patients who received bisoprolol and 59% of those treated with sotalol remained in sinus rhythm during the first year after cardioversion.[254] In the Carvedilol Post-Infarct Survival Control in Left Ventricular Dysfunction (CAPRICORN) study of 1959 patients with myocardial infarction and left ventricular dysfunction, therapy with carvedilol reduced the risk of developing AF or AFL by nearly two thirds, probably contributing to the overall beneficial effect of a β-blocker on prognosis.[255] The available studies pertaining to the long-term prevention of AT are largely observational and often include patients with other paroxysmal supraventricular tachyarrhythmias. There is some evidence that β-blockers (and nondihydropyridine calcium antagonists) may reduce the rate of recurrence in patients with AT.[66]

Propafenone and Flecainide

A number of randomized, controlled studies have addressed the long-term efficacy of class IC antiarrhythmic drugs in atrial tachyarrhythmias, but none reported patients with AFL or AT as a distinct group.[256-263] In the Propafenone Atrial Fibrillation Trial (PAFT) of 102 cardioverted patients, the likelihood of maintenance of sinus rhythm with low-dose *propafenone* (450 mg) was 67% at 6 months compared with 35% with placebo.[257] The Paroxysmal Supraventricular Tachycardia (UK PSVT) study evaluated safety and efficacy of two doses of propafenone: 600 mg and 900 mg daily, and showed that both regimens were effective for the suppression of recurrences of the arrhythmia, but a dose of 900 mg was associated with a less favorable adverse events profile.[258] Consequently, whereas the relative risk of treatment failure on placebo versus low-dose propafenone was 6.0 (95% CI, 1.8 to 20), the corresponding value for placebo versus high-dose propafenone was 1.9 (95% CI, 0.7 to 4.7) because of a 10-fold higher rate of adverse events in the active arm.

The efficacy of a *sustained-release propafenone* formulation that allows twice-daily dosing of 225, 325, and

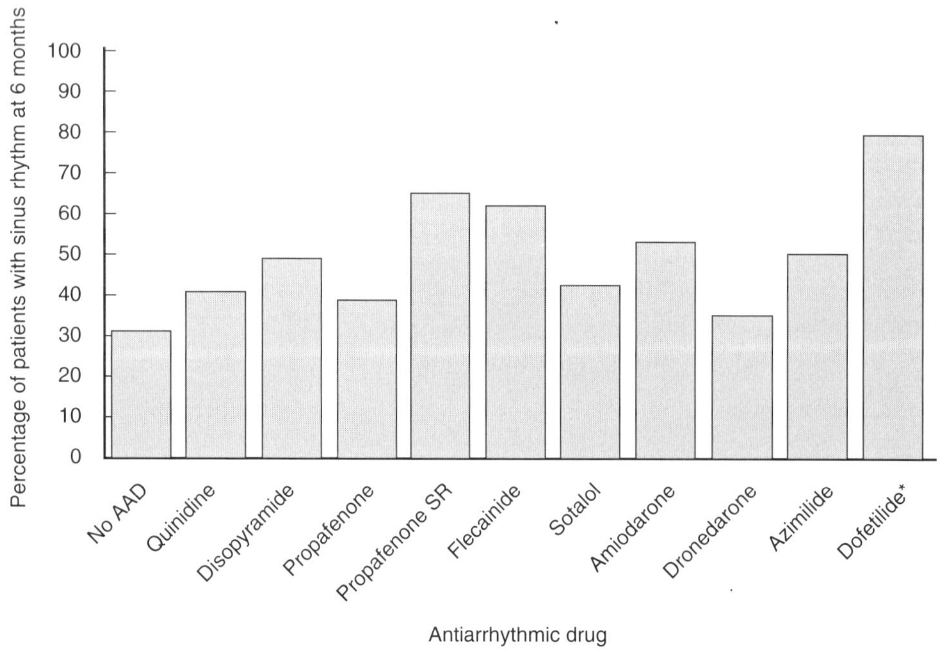

FIGURE 15-45 Likelihood of remaining free from recurrence of atrial fibrillation or atrial flutter with prophylactic antiarrhythmic drug therapy.*, at 1 year. AAD, antiarrhythmic drug.

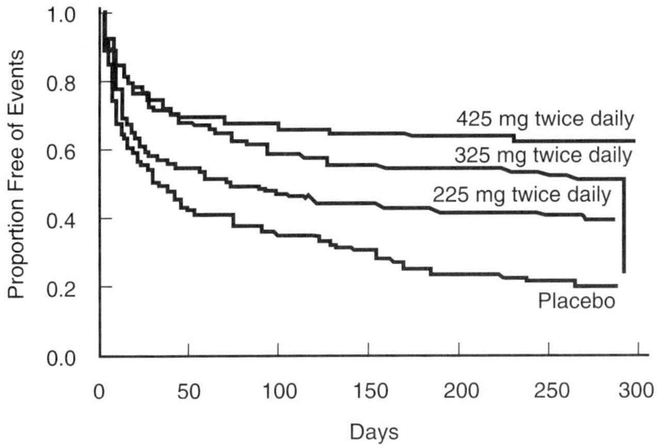

FIGURE 15-46 Time to first symptomatic recurrence of atrial fibrillation or flutter with propafenone SR in the RAFT study. (Reprinted from Pritchett EL, Page RL, Carlson M, et al, for Rythmol Atrial Fibrillation Trial (RAFT) Investigators: Efficacy and safety of sustained-release propafenone (propafenone SR) for patients with atrial fibrillation. Am J Cardiol 2003;92:941-6.)

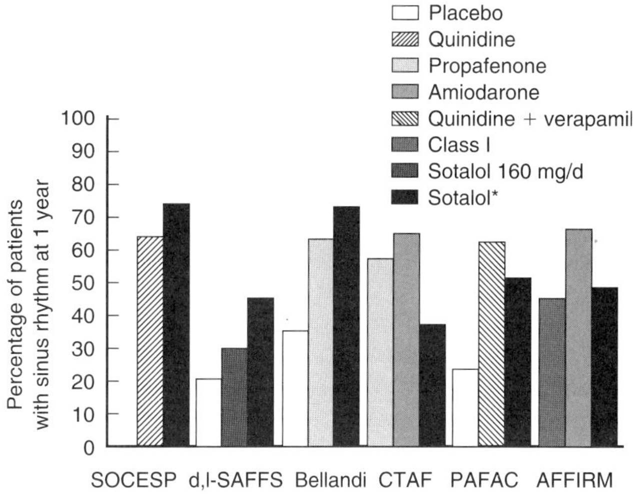

FIGURE 15-47 Randomized, controlled, comparator studies of the efficacy of sotalol in maintenance of sinus rhythm in patients with atrial tachyarrhythmias. *, varying doses of sotalol; 320 mg/d in the d,l-SAFF Study.

425 mg has been studied in the North American Recurrence of Atrial Fibrillation Trial[259] (RAFT) and its European equivalent, ERAFT.[260] Both studies have shown that propafenone SR is superior to placebo in prolonging the time to first symptomatic recurrence of AF or AFL at all doses tested. Sixty-five percent of patients treated with the highest dose of 425 mg twice daily were free of symptomatic AF after 300 days of follow-up compared with only 20% in the placebo arm, although the adverse effect rate was higher than with low-dose regimens (Fig. 15-46).[259] Propafenone SR did not increase the occurrence of AFL with 1:1 or 2:1 AV conduction. The effects of propafenone were consistent in all subgroups, including patients with structural heart disease and long-term duration of the arrhythmia.[260] However, patients with significant underlying heart pathology were minimally represented in both studies.

It is popular belief that *flecainide* is a more potent class IC antiarrhythmic agent than propafenone. Several placebo-controlled and comparator trials, including the Flecainide Multicenter Atrial Fibrillation Study, have consistently reported a 70% to 80% likelihood of continuing in sinus rhythm after 1 year with flecainide with an acceptable risk-benefit relationship.[261-263] In the direct comparison with propafenone, both drugs were equally effective in preventing AF and AFL in patients with little heart disease (77% for flecainide and 75% for propafenone).[263]

The rationale for using class IC antiarrhythmic drugs for long-term management of AT is based mainly on evidence of their efficacy in randomized, controlled trials in AF and is partly supported by observational studies in a limited number of patients with AT.[197,264,265]

Sotalol

A class III antiarrhythmic drug property and the beneficial rate slowing action of *sotalol* have prompted a galaxy of randomized, placebo-controlled and comparator studies, many of which reported a higher efficacy of sotalol than that of placebo, quinidine, and propafenone (see Table 15-12; Fig. 15-47).[266-272] In the d,l-Sotalol Atrial Fibrillation/Flutter dose efficacy study of 253 patients (20% with AFL), sotalol significantly prolonged the time to first symptomatic recurrence of the arrhythmia documented by transtelephonic monitoring from 27 days on placebo to the maximum of 229 days in the active treatment group.[266] The dose of 120 mg twice daily was reported to be both safe and effective in maintaining sinus rhythm compared with a lower and higher dose regimen (160 mg/day and 320 mg/day, respectively). In direct comparisons, sotalol was slightly but significantly superior to propafenone in maintaining sinus rhythm at 1 year (73% versus 63%)[267] and was as effective as quinidine (74% versus 68%).[268] Hypotension and bradycardia were the most common cardiovascular adverse effects of sotalol with an incidence of 6% to 10%. Proarrhythmias associated with Q–T interval prolongation were reported in 1% to 4% of patients, usually within 72 hours after the first dose.[267,269]

However, subsequent studies reported limited efficacy of sotalol compared with amiodarone, dofetilide, and the combination of quinidine and verapamil.[269-271] In the CTAF study, sotalol was significantly inferior to amiodarone for the long-term maintenance of sinus rhythm (37% versus 65%).[270] The PAFAC study demonstrated a 50% recurrence rate with sotalol during 1 year of daily transtelephonic ECG monitoring compared with 38% on the combination of quinidine and verapamil.[269] The efficacy of sotalol was similar to that of class I antiarrhythmic drugs and inferior to that of amiodarone in the AFFIRM substudy (48%, 45%, and 66%, respectively).[271]

Sotalol is generally accepted as an effective antiarrhythmic drug for the long-term treatment of AT;

however, this opinion is not supported by direct evidence as most studies include patients with different supraventricular tachyarrhythmias without specifying the results for each arrhythmia.[272]

Amiodarone and Dronedarone

The potential of *amiodarone* to maintain sinus rhythm in patients with atrial tachyarrhythmias, particularly in association with significant structural heart disease, has been repeatedly shown in observational and prospective, randomized, controlled studies (see Table 15-12). In the CTAF study, therapy with amiodarone reduced the incidence of sustained recurrence of AF by 57% compared with sotalol and propafenone (Fig. 15-48).[270] Data from the CHF-STAT substudy showed that patients who received amiodarone had less recurrence of AF and were twice as less likely to develop new AF compared with placebo.[208] Given its neutral effect on all-cause mortality, amiodarone should be considered a drug of choice for management of atrial tachyarrhythmias in patients with congestive heart failure, hypertrophic cardiomyopathy, and hypertension in the presence of significant left ventricular hypertrophy.[34] In addition to antiarrhythmic effects, the beneficial effects of amiodarone include the ability to control fast ventricular rates, which may be particularly deleterious in patients with advanced heart disease, although its use as a rate-slowing agent is not recommended in the long-term because of multiorgan toxicity. In the CHF-STAT study, amiodarone produced a sustained and significant reduction in the mean and maximal ventricular responses in the ranges of 16% and 20% and 14% and 22%, respectively.

Dronedarone is an investigational agent with multiple electrophysiological effects, in which it is similar to amiodarone but is devoid of iodine substituents and is believed to have a better side effect profile. Dronedarone has been studied in the Dronedarone Atrial FibrillatioN study after Electrical cardioversion (DAFNE) trial, which was designed to determine the most effective dose for maintenance of sinus rhythm and to assess the safety and tolerability of the drug in persistent AF. The primary end point was time to first recurrence of the arrhythmia detected by means of transtelephonic monitoring. Three doses of dronedarone (800, 1200, and 1600 mg daily) were tested in 199 patients who entered the maintenance phase following pharmacologic or electrical cardioversion. An unexpected finding of the study was the absence of consistent effects of higher doses on the time to first AF recurrence, with 800 mg daily being the most effective in prolonging the mean interval to recurrence of the arrhythmia by 55% compared with placebo.[273] Spontaneous conversion to sinus rhythm was dose dependent and occurred in 5.8%, 8.2%, and 14.8% of patients treated with dronedarone compared with 3.1% on placebo. Similarly to amiodarone, dronedarone controlled ventricular rates during recurrence of the arrhythmia. Persistent Q–T interval prolongation of more than 500 ms was noticed with the highest of the three doses. No pulmonary, thyroid, or ocular side effects have been noted. The preliminary results of two larger studies, EURopean trial In atrial fibrillation or flutter patients receiving Dronedarone for the maintenance of Sinus rhythm (EURIDIS) and its American-Australian-African equivalent, ADONIS, showed that dronedarone given 400 mg twice daily was moderately superior to placebo in the prevention of recurrent AF and was also effective in controlling ventricular rates in 1237 patients (preliminary data). EURIDIS and ADONIS were efficacy and safety trials that were not designed to assess mortality and excluded patients with significant left ventricular dysfunction. However, the ANDROMEDA (Antiarrhythmic trial with Dronedarone in Moderate to severe CHF Evaluating Morbidity Decrease) study was stopped prematurely after 627 patients, out of the 1000 planned, were enrolled, because an interim safety analysis revealed a potential excess risk of death in patients on active treatment.

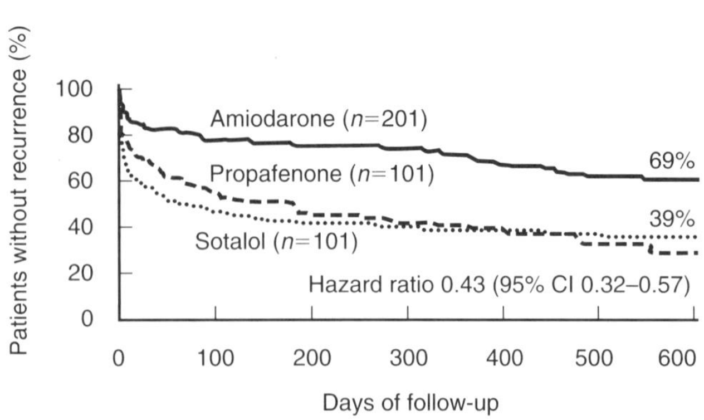

FIGURE 15-48 Probability of remaining free of symptomatic recurrence of atrial fibrillation with amiodarone compared with sotalol and propafenone in the CTAF (Canadian Trial of Atrial Fibrillation) study. (Reprinted from Roy D, Talajic M, Dorian P, et al, for the Canadian Trial of Atrial Fibrillation Investigators: Amiodarone to prevent recurrence of atrial fibrillation. N Engl J Med 2000;342:913-20.)

Dofetilide and Azimilide

Dofetilide is considered a relatively safe and effective class III antiarrhythmic drug for treatment of AF and AFL in patients with significant underlying heart disease (see Table 15-12). In the DIAMOND studies, treatment with dofetilide 500 μg twice daily was associated with a significantly greater probability of remaining in sinus rhythm at 1 year compared with placebo (79% versus 42%) (Fig. 15-49) and a lower incidence of new AF (1.98% versus 6.55%).[171] Consequently, patients in sinus rhythm benefited a 56% reduction in mortality.[201] In the SAFIRE-D study, 58% of patients treated with the maximal dose of dofetilide of 1 mg remained in sinus rhythm after a 1-year follow-up, compared with 25% in the placebo group.[202] The efficacy of dofetilide is nearly twofold higher in the long-term prevention of AFL than AF (73% versus 40%).[202] Dofetilide is associated with substantial risk of torsades de pointes and is contraindicated in patients with low creatinine clearance (<20), hypokalemia, hypomagnesemia, and a prolonged Q–T interval of more than 500 ms at baseline. There are no data regarding the safety and efficacy of dofetilide for management of AT.

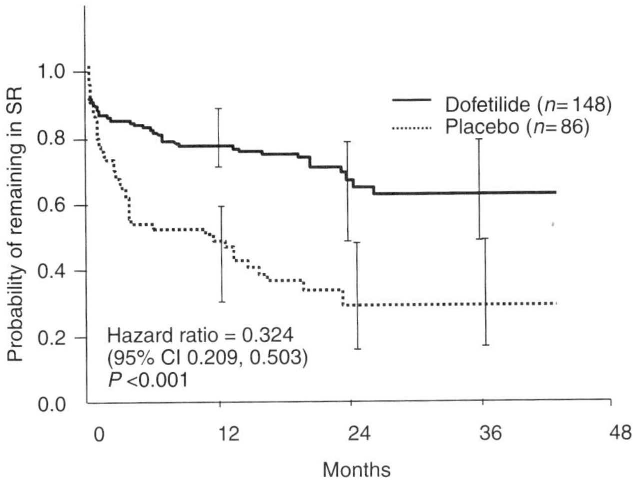

FIGURE 15-49 Probability of remaining in sinus rhythm following spontaneous conversion or electrical cardioversion among patients with atrial fibrillation or flutter in the DIAMOND-CHF study. Vertical bars indicate 95 percent confidence intervals. (Reprinted from Pedersen OD, Bagger H, Keller N, et al, for the Danish Investigations of Arrhythmia and Mortality ON Dofetilide Study Group: Efficacy of dofetilide in the treatment of atrial fibrillation-flutter in patients with reduced left ventricular function: A Danish Investigations of Arrhythmia and Mortality on Dofetilide (DIAMOND) Substudy. Circulation 2001;104:292-6.)

Azimilide, which is now completing the final phase of its clinical development, has proven effective in preventing recurrence of atrial tachyarrhythmias.[274,274a] Meta-analysis of four randomized, controlled dose-efficacy studies in 1380 patients with AF and AFL has shown that each of the two highest doses (100 and 125 mg/day) significantly prolonged the time to symptomatic recurrence of the arrhythmia with the hazard ratio of 1.34 and 1.32, respectively.[274] Patients with a history of coronary artery disease or congestive heart failure had a significantly greater treatment effect from azimilide than those with other underlying cardiovascular pathologies (hazard ratio, 1.49 to 1.86).

Azimilide blocks both I_{Kr} and I_{Ks} currents and is devoid of frequency-dependent effects on APD, suggesting a lower probability of torsades de pointes and an increased efficacy by maintaining AP prolongation at high rates. The ALIVE (AzimiLide post Infarct survival Evaluation) study of more than 3000 high-risk patients with a recent myocardial infarction, impaired ventricular function, and decreased heart rate variability has shown a neutral effect of azimilide 100 mg daily on all-cause mortality, including patients with significantly reduced ejection fraction of less than 20%.[275] The hazard ratio was 1.0, with an 11.6% mortality rate in both the placebo and azimilide arms. Fewer patients who started the trial in sinus rhythm developed AF on azimilide, and there was a clear tendency to higher pharmacologic conversion rates in the azimilide arm than in the placebo arm (26.8% versus 10.8%).[275a] The initiation of therapy with azimilide in-hospital or out-of-hospital did not affect mortality. No dose adjustment was required for renal or hepatic impairment, and the drug was used safely in combination with digoxin and warfarin.

There are now two ongoing studies of azimilide 125 mg, the Supraventricular TachyArrhythmia Reduction (STAR) and the CardiOversion MaintEnance Trial (COMET). The STAR study will look at the time to first recurrence of AF and AF burden, including frequency and severity of symptoms. The COMET study will investigate the conversion rate and the efficacy of azimilide in maintaining sinus rhythm against placebo (the North American study) or against sotalol (the European study) in persistent AF. However, the efficacy of azimilide appears to be moderate and restricted to patients with structural heart disease.

Experimental Antiarrhythmic Agents

Atrial repolarization delaying agents are characterized by highly selective affinity to ion channels involved in the repolarization process in the atrial myocardium, specifically the ultrarapid component of the delayed potassium rectifier current I_{Kur}. RSD 1235 (*Cardiome*) is in the most advanced stage of development.[14] Its mechanism of action involves prolongation of atrial APD with no effect on ventricular refractoriness and, consequently, on the Q–T interval. It also blocks the inward fast sodium current I_{Na}. RSD 1235 slows conduction velocity within the atrium and prolongs its recovery; it is also selective for the fibrillating atria. It is now under investigation for pharmacologic conversions of AF.

In experimental studies, stimulation of atrial 5-HT_4 serotonin receptor subtypes has been shown to produce positive chronotropic effects and to induce AF. The exact mechanism by which 5-HT_4 receptor modulation produces electrophysiological effects is uncertain, but experimental studies have suggested that stimulation of 5-HT_4 receptors increases L-type Ca^{2+} channel current in atrial myocytes via a cAMP-dependent protein kinase, resulting in an increase of intracellular calcium, and, as a consequence, the augmentation of the delayed potassium rectifier current I_{Kr} via a calmodulin-dependent pathway. Conversely, 5-HT_4 receptor blockade may have an antiarrhythmic effect by decreasing intracellular Ca^{2+} concentration and by blocking I_K, thus prolonging atrial refractory period. 5-HT_4 serotonin receptor inhibitor, *piboserod,* which was first developed for the treatment of irritable bowel syndrome, has been suggested as a potential new antiarrhythmic agent, which may block the binding of serotonin in the atrium. Piboserod is not likely to cause excess Q–T interval prolongation and associated ventricular proarrhythmias. In a pig model, piboserod terminated 89% of AF.

Stretch-activated ion channels in the dilated and fibrotic atrium may produce electrophysiological changes favoring reentry and facilitate abnormal automaticity and triggered activity in the atria. There is experimental evidence that the blockade of stretch-activated channels by gadolinium impeded initiation and maintenance of electrically induced AF and suppressed the occurrence of spontaneous AF. Blockade of stretch-activated ion channels may represent a novel antiarrhythmic approach to AF under conditions of elevated atrial pressure or volume. Such agents (e.g., *GsMtx4*) are under development.

Structural remodeling in cardiac gap junctions can result from alterations in connexin transport and/or connexin protein synthesis or degradation under conditions resulting from atrial stretch and prolonged periods of atrial fibrillation. *Antiarrhythmic peptide* (AAP 10) has been shown to modify electrophysiological properties of the myocardium without producing the direct effects on ion channels and APD. AAP 10 increases gap junction conductivity and improves intracellular coupling by activation of protein kinase C via G-protein and enhanced phosphorylation of connexin 43. However, the use of endogenous AAP and synthetic derivatives has been hindered by instability and a very short half-life. Newer gap junction modifiers with prolonged action, such as *GAP-486,* are under investigation.

RATE CONTROL VERSUS RHYTHM CONTROL

Sinus rhythm theoretically offers physiologic rate control, normal atrial activation and contraction, the correct sequence of AV activation, normal hemodynamic and AV valve function, and a regular rhythm. However, sinus rhythm with normal AV conduction may not be an alternative to the arrhythmia since sinus node disease may be the underlying problem, and chronotropic incompetence may well be present. Atrial conduction and mechanical function may be seriously impaired, and atrial contraction may not contribute much to cardiac output. Furthermore, it is not unusual for patients to be relieved of their symptoms, such as palpitations, anxiety, and chest pain, when the arrhythmia is established and becomes permanent. Often the only symptoms that remain are a minor limitation of exercise tolerance and a subtle reduction in the quality of life. Finally, restoration and maintenance of sinus rhythm may incur certain risks, particularly from thromboembolism consequent upon restoration of mechanical activity to previously "stunned" atria, proarrhythmias, and organ toxicity of antiarrhythmic drugs. There is, therefore, a genuine equipoise as to whether it is best to accept the arrhythmia while controlling the ventricular rate and preventing thromboembolic complications with anticoagulant therapy or to restore and maintain sinus rhythm. The issue of rate versus rhythm control especially concerns patients with AF because it is the most common arrhythmia of the elderly, which tends to progress to the permanent form, and cannot be readily cured by nonpharmacologic means.

Randomized Clinical Trials

Several clinical trials have been instigated to specifically address this problem.[175,276-280] All but one trial consistently reported that there was no clear advantage to rhythm control (Table 15-14). Generally, there was a trend toward improved survival and less serious cardiovascular adverse events in association with rate rather than rhythm control. On meta-analysis the reduction in the cerebrovascular accident rate was significantly less with rate control, but this was largely because anticoagulation was often omitted when the patient seemed to be in stable sinus rhythm with an effective antiarrhythmic agent. Furthermore, despite the appearance of freedom from AF, the patients may well have suffered asymptomatic episodes or thromboembolism associated with AF but related to some other factors such as aortic atherosclerotic plaques.

In the Pharmacological Intervention in Atrial Fibrillation (PIAF) study of 252 patients randomized to rate control with diltiazem or rhythm control with amiodarone after electrical or pharmacologic cardioversion, the rhythm control strategy resulted in better exercise performance but did not significantly improve symptoms or quality of life and was associated with increased numbers of hospitalizations for repeat cardioversion and the adverse effects of antiarrhythmic drugs.[276] There was no difference in the occurrence of the combined end point of death, stroke or transient ischemic attack, systemic embolism, and cardiopulmonary resuscitation between the rate control arm and the rhythm control arm (5.54% versus 6.09% per year) in the Strategies of Treatment of Atrial Fibrillation (STAF) pilot study in 200 high-risk patients with persistent AF.[277] However, although there was a similar number of primary end points with rhythm and rate control (9 and 10, respectively), 18 of these occurred while patients were in AF. Furthermore, long-term anticoagulation was not mandatory in this trial, and given the significant proportion of patients with risk factors for stroke and the high incidence of AF recurrence, this might have increased the likelihood of thromboembolism in patients assigned to the rhythm control strategy.

The results of the Rate Control versus Electrical Cardioversion (RACE) study of 522 patients with persistent AF assigned to rate control or the aggressive rhythm control strategy consisting of serial cardioversions and antiarrhythmic drugs, including sotalol, propafenone or flecainide, and amiodarone, has also shown no difference in the primary composite end point of cardiovascular death, hospital admissions for heart failure, thromboembolic events, major bleedings, pacemaker implantation, and adverse effects of antiarrhythmic drug therapy between the two strategies (22.6% versus 17.2%).[278] As in the STAF study, AF was an underlying rhythm in more than two thirds of the RACE patients at the time of the primary end point event. The substudy analysis revealed that patients with hypertension (approximately half the total study population) assigned to rhythm control had a significantly higher event rate compared with those randomized to rate control (30.8% versus 17.3%), whereas this relation was reverted in normotensive participants (12.5% versus 17.1%). Possible explanations may be predisposition of hypertrophied myocardium to adverse effects of antiarrhythmic drug therapy and a higher risk for stroke in the presence of hypertension.

The AFFIRM study of 4060 AF patients aged 65 years or older with at least one risk factor for stroke was the only one designed to assess mortality benefit of different strategies in AF.[175] The mean follow-up was

TABLE 15-14 Clinical Studies of Rhythm Control and Rate Control in Atrial Fibrillation

Study	Number of Patients	AF Duration	Etiology of AF	Follow-up	Number of Patients in SR*	1° End Point (Rhythm Control vs. Rate Control)	2° End Point (Rhythm Control vs. Rate Control)
PIAF[276]	252	7 days-1 yr	HTN 49% CHD 23% VHD 16% CHF 17% Lone 15%	1 yr	56% vs. 10%	Symptomatic improvement 55% vs. 61% (n.s.)	In rhythm control arm ↑ exercise tolerance; hospitalizations 69% vs. 24%; adverse effects of AAD 25% vs. 14%; no difference in QoL
STAF[277]	200	>4 wk, <1 yr	HTN 62.5% CHD 43.5% VHD 13% CHF 12.5% Lone 10.5%	2 yr	40% vs. 12% at 1 yr, 26% vs. 11% at 2 yr; 23% vs. 0% at 3 yr	Composite end point (all-cause death, cardiovascular events, CPR, TE) 5.54% vs. 6.09% (n.s.); 18/19 events occurred during AF	Hospitalizations (54 vs. 26); no difference in QoL, NYHA class, echocardiographic parameters; syncope 0% vs. 0.6%; bleeding 6.8% vs. 4.9%
RACE[278]	522	>24 hr, <1 yr	HTN 55% vs. 43% CHD 26% vs. 29% CHF 7% vs. 11% Lone 21% vs. 21%	3 yr	40% vs. 10%	Composite end point (cardiovascular death, admissions for CHF, TE, bleeding, pacemaker implantation, adverse effects of AAD) 22.6% vs. 17.2% (n.s.); in HTN 30.8% vs. 17.3%	Cardiovascular death 6.7% vs. 7%; admissions for CHF 4.5% vs. 3.5%; TE 7.5% vs. 5.5%; bleeding 3.4% vs. 4.7%; adverse effects of AAD 4.5% vs. 0.8%; pacemaker 3.0% vs. 1.2%
AFFIRM[175]	4060 ≥65 yr old + 1 risk factor for stroke	>6 hr, <6 mo; 1 episode in the past 12 wk	HTN 51% CHD 26%	3.5 yr	60% vs. 38% at 5 yr	All-cause death 27% vs. 26% (n.s.)	Stroke 7.3% vs. 5.7%; hospitalizations 78% vs. 70%; torsades de pointes 0.8% vs. 0.2%; no difference in Qol and functional status
HOT CAFE[279]	205	7 days-2 yr	HTN, CHD, nonsevere VHD, lone	1 yr	75% in the rhythm control arm; not stated in the rate control arm	Not stated	In rhythm control arm: ↑ exercise tolerance and systolic function; hospitalizations 61.5% vs. 4.8%; proarrhythmias 11.4% vs. 0.99%; TE 2.9% vs. 1.0%; bleeding 7.7% vs. 5.0%; death 1.9% vs. 1%

AAD, antiarrhythmic drugs; AF, atrial fibrillation; CHD, coronary heart disease; CHF, congestive heart failure; CPR, cardiopulmonary resuscitation; DCC, direct current cardioversion; HTN, hypertension; MI, myocardial infarction; NYHA, New York Heart Association, QoL, quality of life; SR, sinus rhythm; TE, thromboembolism; VHD, valvular heart disease.

3.5 years, with a maximum of 6 years. There was no difference in the primary end point of all-cause mortality as well as quality of life and functional status between rate and rhythm control (Fig. 15-50). The rhythm control strategy was associated with a slightly higher incidence of stroke (7.3% versus 5.7%), probably due to more frequent discontinuation of oral anticoagulation (despite the presence of at least one risk factor for stroke as a pre-specified inclusion criterion), torsades de pointes (0.8% versus 0.2%), and hospitalizations.

The closely similar primary end point results for the rhythm and rate control strategies were probably due to a general failure to achieve a clear difference with respect to rhythm and rate status in the two arms of the trials. Ideally, the rhythm control arm should have largely comprised patients who were in sinus rhythm, whereas the rate control arm should have comprised mostly patients in AF. This was not usually the case; for example, in the AFFIRM trial, only 60% of patients in the rhythm, control arm were maintained in sinus rhythm,

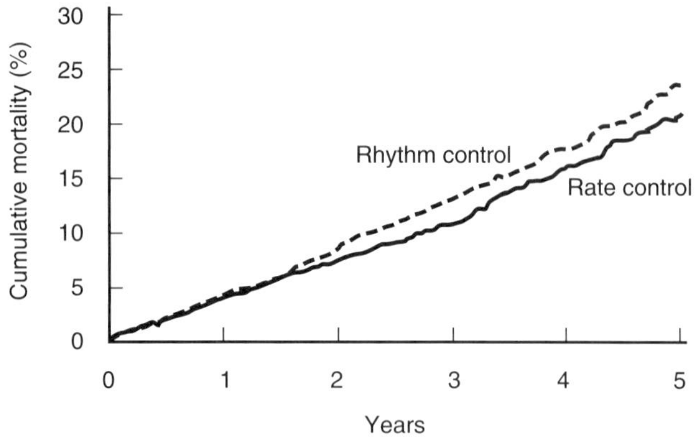

FIGURE 15-50 Kaplan-Meier curves for cumulative all-cause mortality in the rhythm control group and the rate control group in the AFFIRM trial. (Reprinted from The Atrial Fibrillation Follow-up Investigation of Rhythm Management (AFFIRM) Investigators: A comparison of rate control and rhythm control in patients with atrial fibrillation. N Engl J Med 2002;347:1825-33.)

whereas 40% of the rate control arm had reverted spontaneously to sinus rhythm. Moreover, the "on-treatment" analysis of the AFFIRM study has shown that the presence of sinus rhythm is associated with a considerable reduction in the risk of death (hazard ratio .53, 99% confidence intervals .39, .72; $P < .0001$).[280a]

The RACE trial included patients with persistent AF who may have had the arrhythmia as long as 1 year and who may have already undergone serial electrical cardioversions. Despite an aggressive rhythm control strategy, the likelihood of maintenance of sinus rhythm in this selected group of patients is expected to be low, thus favoring the rate control strategy. After 3 years of follow-up, sinus rhythm was present in 40% patients in the rhythm control arm and 10% patients in the rate control arm. Only 40% of patients in the STAF study remained in sinus rhythm at 1 year, and at 2 years this number further decreased to 26%. This gives rise to the most serious criticism of this clutch of recent trials. It is argued that the impression that rate control is marginally better than rhythm control is entirely due to the inadequacy of the rhythm control therapy and that newer, potentially much more effective therapies that are now available may effectively cure AF.

Furthermore, the results of the AFFIRM study, which refer to older patients in whom rate control is generally considered to be preferential, cannot be extrapolated to younger patients who are more likely to be symptomatic and to have impaired quality of life associated with the arrhythmia, even with good rate control. An unexpected finding that rhythm control is beneficial in patients younger than 65 years of age with heart failure may modify the existing guidelines on management of AF. The ongoing AF-CHF study compares rate and rhythm control strategies in this clinical setting and is powered to detect a 25% reduction in mortality in the rhythm control arm compared with the rate control arm in 1450 patients with NYHA class II to IV heart failure and an ejection fraction less than 35%.

What Is Adequate Rate Control?

Most of the studies specifically addressing rate versus rhythm control in atrial tachyarrhythmias have exclusively assessed pharmacologic means of rate control. The agents that may be administered for long-term control of the ventricular rate response in AF are listed in Table 15-15. Some trials, which advocate rate control and anticoagulation as acceptable management for many patients with AF, report an improvement in symptoms, NYHA class, and quality of life when rate control is optimized, but few state the definition of optimum heart rate and how it was monitored (Table 15-16). Furthermore, considering the significant recurrence rate in the rhythm control arm, the rate control protocol for patients who reverted to AF has not been clearly stated. The RACE investigators monitored rate control by means of a resting 12-lead ECG setting the target ventricular rate of less than 100 bpm and

TABLE 15-15 Antiarrhythmic Drugs for Long-Term Rate Control in Atrial Tachyarrhythmias

Drug	Dose	Type of Recommendation	Potential Adverse Effects
Digoxin	Loading dose: 250 µg every 2 hr; up to 1500 µg; maintenance dose 125-250 µg daily	I	Bradycardia; atrioventricular block; atrial arrhythmias; ventricular tachycardia
Diltiazem	120-360 mg daily	I	Hypotension; atrioventricular block; heart failure
Verapamil	120-360 mg daily	I	Hypotension; atrioventricular block; heart failure
Atenolol	50-100 mg daily	I	Hypotension; bradycardia; heart failure; deterioration of chronic pulmonary obstructive or bronchospastic disease
Metoprolol	50-200 mg daily	I	Hypotension; bradycardia; heart failure; deterioration of chronic pulmonary obstructive or bronchospastic disease
Propranolol	80-240 mg daily	I	Hypotension; bradycardia; heart failure; deterioration of chronic pulmonary obstructive or bronchospastic disease
Amiodarone	800 mg daily for 1 wk, then 600 mg daily for 1 wk, then 400 mg daily for 4-6 wk; maintenance dose 200 mg daily	IIB	Bradycardia; QT prolongation; torsades de pointes (rare); photosensitivity; pulmonary toxicity; polyneuropathy; hepatic toxicity; thyroid dysfunction; gastrointestinal upset

TABLE 15-16 Optimal Rate Control in Randomized Studies of Rhythm versus Rate Control in Atrial Fibrillation

Study	Number of Patients*	Age, yr	Follow-up	Rate Control	Method of Control	Optimum Heart Rate	Outcome
PIAF[276]	125	61 ± 9	1 yr	Diltiazem; additional drug at discretion of the treating physician	24-hr Holter	Not stated	Mean VR at baseline 88 beats/min; at 12 mo 81 beats/min Symptomatic improvement in 76 patients; improvement in QoL on 6 scores but no increase in 6-min walk distance
STAF[277]	100	66.2 ± 7.6	2 yrs	β-Blocker Calcium blocker Digoxin (% not stated) AV node ablation 2%	12-lead ECG	Not stated	Mean VR at baseline 81 ± 15 beats/min; no changes at follow-up Improvement in QoL on 5 scores but no changes in symptoms Improvement in NYHA class in 26 patients, worsening NYHA class in 29 patients but no changes in LV function (LV diameter, LV ejection fraction) or the left atrial size
RACE[278]	256	69 ± 9	3 yr	At baseline: β-Blocker 26% Calcium blocker 16% Digoxin 32% β-Blocker + digoxin 15% Calcium-blocker + digoxin 4% AV node ablation 1.17%	12-lead ECG	100 beats/min at rest	Mean VR at baseline 91 ± 21 beats/min; >100 beats/min in 25% patients Mean VR at follow-up 82 ± 16 beats/min No comparisons between the baseline and follow-up except for improvement in QoL on 4 scores have been published
AFFIRM[175]	2027; complete data in 1968	69.8 ± 8.9	3.5 yr	At any time: β-blocker 68.1% Diltiazem 46.1% Verapamil 16.8% Digoxin 70.6%	12-lead ECG 24-hour Holter 6-minute walk	80 beats/min at rest Mean VR 100 beats/min 110 beats/min	Optimum rate control was achieved in 82%; no difference in survival between the groups with achieved and not achieved rate control‡
HOT CAFE[279]	101	61 ± 17	1 yr	β-Blocker 49.5% β-blocker + digoxin 39.6% Digoxin alone 3% Calcium blocker 7.9% AV node ablation 2%	24-hr Holter	Not stated	Improvement in NYHA class but not in exercise tolerance and LV function (fraction shortening) on the echocardiogram
DEHLI[280]	48	39†	1 yr	Diltiazem	Not stated	Not stated	Improvement in NYHA class by ≥1 in 20% patients and in QoL by ≥1 score in 57% patients but no changes in exercise tolerance

*Number of patients assigned to rate control.

†Etiology of atrial fibrillation was rheumatic heart disease.

‡Data available in 680 (34%) of 2,027 patients assigned to rate control.

LV, left ventricle; NYHA, New York Heart Association; QoL, quality of life; VR, ventricular rate.

did not attempt to assess rate control during activity. In the AFFIRM trial, rate control was considered adequate if ventricular rates at rest and during a 6-minute walk did not exceed 80 and 110 bpm, respectively, or if the mean rate during 24-hour Holter monitoring was below 100 bpm. Overall rate control was achieved in 70% of patients given β-blockers as the first drug (with or without digoxin), 54% with calcium channel blockers (with or without digoxin), and 58% with digoxin alone.[282a]

Little is known as to whether different methods employed for the assessment of rate control in AF provide comparable information and whether the combination of two or theoretically more tests may provide better results (Table 15-17). In a small series of 18 patients with permanent AF, a combination of ventricular rate assessment after 5 minutes of rest, a 50-yard walk, and 1 minute of stair stepping exercise was correlated with 24-hour Holter recordings (Fig. 15-51).[281] The ventricular rate at rest during the clinic visit was significantly

TABLE 15-17 Assessment of Rate Control in Atrial Fibrillation

Physical examination (pulse counting during 1 min, apex-radius deficit)
Pulse counting after 6-min walk
12-lead ECG or rhythm strips
24-48 hr Holter monitoring
Heart rate variability (algorithms, specifically designed for atrial fibrillation, are needed)
Exercise stress test
Implantable loop recorders
Transtelephonic and telemetric monitoring
Pacemaker or implantable cardioverter-defibrillator diagnostics
Indirect measures of left ventricular systolic function (e.g., echocardiography)
Quality-of-life questionnaires

less than the mean Holter ventricular rate. The resting heart rate minus 20 bpm correlated well with the minimum heart rate on Holter. The mean of the ventricular heart rate at rest and after a 50-yard walk was similar to the average ventricular rate recorded from a 24-hour Holter monitor. The heart rate after 1 minute of stair stepping was similar to the maximum ventricular rate recorded on 24-hour Holter. However, these findings need further verification.

Nonpharmacologic therapies, such as AV node ablation, were systematically underused in major clinical studies attempting to demonstrate the benefits of the rate control strategy. Only 1% to 2% of patients assigned to rate control in the AFFIRM, RACE, and STAF trials underwent AV node ablation. The recently reported Australian Interventional Randomized Control of Rate in Atrial Fibrillation Trial (AIRCRAFT) study directly compared pharmacologic rate control and AV node ablation, but it focused predominantly on the effects of the two strategies on left ventricular function.[282] It is not surprising that the study showed no benefit from the "ablate and pace" strategy with respect to this end point, as it enrolled patients with relatively good left ventricular function. However, better quality of life and an 18% reduction in symptoms associated with AV node ablation as compared with conservative treatment are encouraging and merit further investigation.

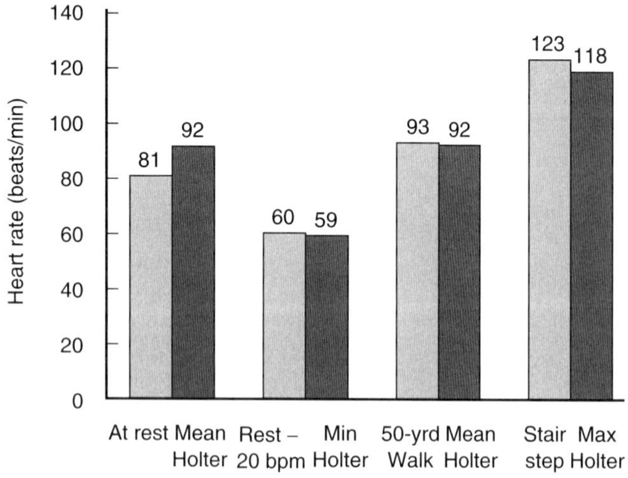

FIGURE 15-51 Correlation among different methods for the assessment of rate control in atrial fibrillation. (Modified from Wasmer K, Oral H, Sticherling C, et al: Assessment of ventricular rate in patients with chronic atrial fibrillation [abstract]. J Am Coll Cardiol 2001;37:93A.)

ANTICOAGULATION THERAPY IN ATRIAL TACHYARRHYTHMIAS

Absence of organized mechanical contraction of the fibrillating atria with a consequent increase in atrial pressure, atrial stretch, and dilatation generate conditions for blood stasis and thrombus formation (Fig. 15-52). Atrial tachyarrhythmias and particularly AF are therefore associated with abnormalities of hemostasis, endothelial function, and platelet activation, adding to the risk of thrombus formation and thromboembolic complications.[283] With TEE, left atrial thrombi can be found in about 14% of patients with AF or AFL, even if they were effectively anticoagulated, and spontaneous echo contrast can be seen in 52% of patients.[284,285]

Anticoagulation has now become imperative in a significant proportion of patients with atrial tachyarrhythmias. A number of large randomized clinical trials have convincingly demonstrated the benefits of therapy with warfarin (Table 15-18).[286-300] Meta-analysis of pooled data from five primary prevention trials and one for secondary prevention of thromboembolic events in patients with nonrheumatic AF has shown a 61% risk reduction in the incidence of stroke with adjusted-dose warfarin compared with placebo and a 36% risk reduction compared with aspirin.[301] The number of patients to treat in order to prevent one stroke has been estimated to be 100 patient-years for those at low risk (≤2% per year), compared with 25 patient-years for those at high risk (≥6% per year).

The efficacy of warfarin for the prevention of ischemic stroke should be balanced against the risk of major hemorrhagic complications, particularly cerebral hemorrhage, which is usually fatal or significantly disabling. The risk of hemorrhagic events is related to the intensity of anticoagulation, patient age, and fluctuations of previously therapeutic INR values due to drug or food interaction.[302] Overall, randomized trials comparing outcomes of therapy with full dose warfarin versus placebo have not shown unacceptably higher rates of major hemorrhagic complications, including intracranial bleeding. However, most of these events occurred in patients older than 75 years of age (see Table 15-18).

The intensity of anticoagulation with INR in the range between 2 and 3 is considered to achieve maximum efficacy in prevention of stroke, whereas INR 1.6 to 2.5 is estimated to be associated with 80% of efficacy achieved with more intense anticoagulation (Fig. 15-53).[303] This strategy has been advocated by some investigators as an alternative for patients older than 75 years of age, as they have both a higher risk of stroke and bleeding compared with younger individuals. However, there is increasing evidence that an INR less than 2.0 confers unacceptably high rates of stroke. Compared with the recommended INR values, an INR of 1.5 to 1.9 was associated with a 3.6-fold increased risk for stroke in patients

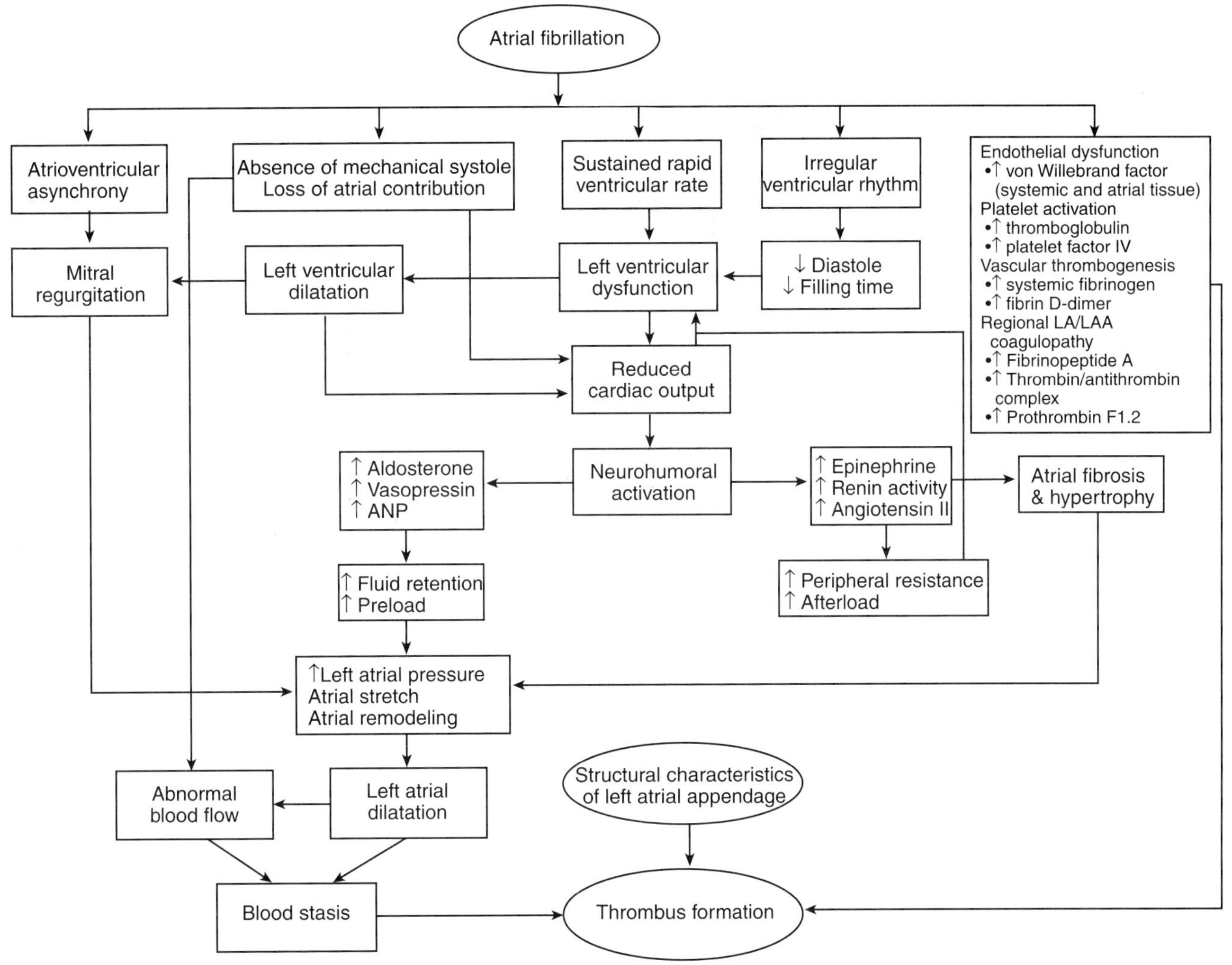

FIGURE 15-52 Pathogenesis of thrombus formation in atrial fibrillation. AF, atrial fibrillation; ANP, atrial natriuretic peptide; LA, left atrium; LAA, left atrial appendage.

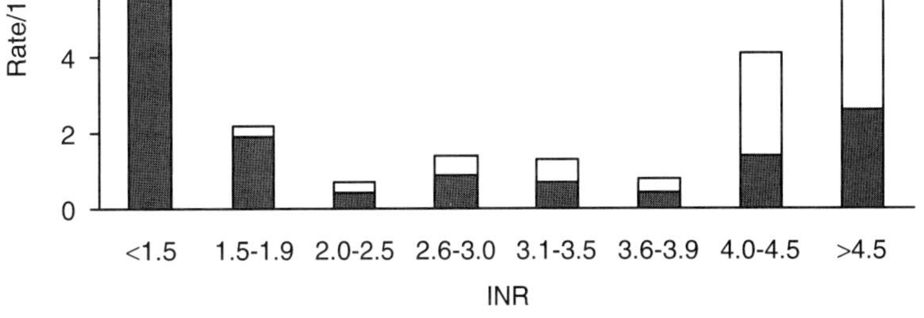

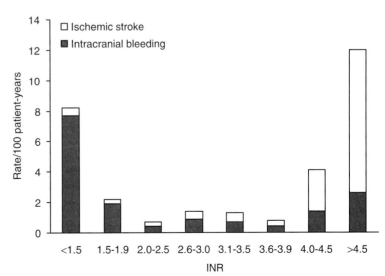

FIGURE 15-53 Effects of intensity of anticoagulation on incidence rates of ischemic stroke and intracranial hemorrhage in a cohort of 13,355 patients with atrial fibrillation. INR, international normalized ratio. (Modified from Hylek EM, Go AS, Chang Y, et al: Effect of intensity of oral anticoagulation on stroke severity and mortality in atrial fibrillation. N Engl J Med 2003;349:1019-26.)

TABLE 15-18 Clinical Studies of Anticoagulation in Atrial Tachyarrhythmias

Study	Sample size	Follow-up (Yr)	Treatment: Warfarin (INR or fixed dose, mg) Aspirin, Control	Risk Reduction/Other Outcome Measures	P	All Major Bleeds (Annual Rate, %)
AFASAK I[286]	1007	1.2	Warfarin 2.8-4.2 Aspirin 75 mg Control	56% Warfarin vs. Control 16% Aspirin vs. Control	<.05 n.s.	Warfarin 0.6% Aspirin 0.30% Control 0%
AFASAK II[287]	677	3.0	Warfarin 2.0-3.0 Aspirin 300 mg Warfarin fixed dose 1.25 mg Warfarin fixed dose 1.25 mg + Aspirin 300 mg	−21% Warfarin full dose vs. Aspirin 13% Warfarin full dose vs. Warfarin fixed dose −6% Warfarin full dose vs. Warfarin fixed dose + Aspirin	n.s. n.s. n.s.	Warfarin full dose 1.7% Aspirin 1.6% Warfarin fixed dose 0.8% Warfarin fixed dose + Aspirin 0.3%
SPAF I[288]	522	1.3	Warfarin 2.0-4.5* Control	67% Warfarin vs. Control	.01	Warfarin 1.5% Control 1.6%
SPAF II[289]	1100	2.7	Warfarin 2.0-4.5* Aspirin 325 mg	≤75 yr: 33% Warfarin vs. Aspirin >75 yr: 27% Warfarin vs. Aspirin	n.s. n.s.	≤75 yr: Warfarin 1.7%, Control 0.9% >75 yr: Warfarin 4.2%, Control 1.6%
SPAF III High risk[290]	1044	1.1	Warfarin 2.0-3.0 Warfarin low dose 1.2-1.5 mg + Aspirin 325 mg	74% Warfarin full dose vs. Warfarin low dose + Aspirin	<.0001	Warfarin 2.1% Warfarin low dose + Aspirin 2.4%
SPAF III Low risk[291]	892	2.0	Aspirin 325 mg	Identification of AF patients with low risk for stroke (0.5%/yr without HTN; 1.4%/yr with HTN vs. approximately 1% in a general population)	—	—
SPINAF[292]	525	1.8	Warfarin 1.4-2.8* Control	79% Warfarin vs. Control	.001	Warfarin 1.3% Control 0.9%
BAATAF[293]	420	2.2	Warfarin 1.5-2.7* Control	86% Warfarin vs. Control	.002	Warfarin 0.4% Control 0.2%
CAFA[294]	383	1.3	Warfarin 2.0-3.0 Control	26% Warfarin vs. Control	n.s.	Warfarin 2.1% Control 0.4%
PATAF[295]	729	2.7	Coumarin 2.5-3.5 Coumarin low dose 1.1-1.6 Aspirin 150 mg	−14% Coumarin full dose vs. Coumarin low dose 19% Coumarin full dose vs. Aspirin 9% Coumarin low dose vs. Aspirin	n.s. n.s n.s.	Coumarin full dose 0.2% Coumarin low dose 0.3% Aspirin 0.3%
EAFT[296]	1007	2.3	Warfarin 2.5-4.0 Aspirin 300 mg Control	47% Warfarin vs. Control 17% Aspirin vs. Control 40% Warfarin vs. Aspirin	.001 n.s. .008	Warfarin 2.6% Aspirin 0.7% Control 0.6%
SIFA[297]	916	1.0	Warfarin 2.0-3.0 Indobufen 400 mg	15% Warfarin vs. Indobufen	n.s.	Warfarin 0.9% Indobufen 0%
ESPS-2 (with AF)[298]	427	1.1	Aspirin 50 mg Control	33% Aspirin vs. Control	n.s.	Aspirin 0.9% Control 0.4%
LASAF pilot[299]	285	1.5	Aspirin 125 mg daily Aspirin 125 mg on alternate days Control	−18% Aspirin 125 mg daily vs. Control 68% Aspirin 125 mg on alternating days	n.s. .05	—
Italian Study[300]	303	1.2	Warfarin 2.0-3.0 Warfarin fixed dose 1.25 mg	Combined event rate: Warfarin full dose vs. Warfarin fixed dose 6.1% vs. 11.1%; Ischemic stroke rate: 0% vs. 3.7%	n.s. .025	Warfarin full dose 2.6% Warfarin fixed dose 1.0%

*Prothrombin time ratio.

AF, atrial fibrillation; INR, international normalization ratio; HTN, hypertension.

younger than 75 years of age and a 2.0-fold increased risk for stroke in patients 75 years of age or older.

Comparisons of the efficacy of aspirin with placebo or with full-dose anticoagulation have shown less impressive results in preventing thromboembolic events. Although it is associated with a lower incidence of major hemorrhagic complications, aspirin has been shown to provide only a modest protection against cardiogenic thromboemboli in the setting of AF, reducing the risk of stroke by 33% for primary prevention and by 11% for secondary prevention.[301] Based on pooled data of five studies, adjusted-dose warfarin compared with aspirin reduced the risk for all strokes by 36% and the risk for ischemic strokes by 46%. Aspirin may benefit patients younger than 65 years of age with no clinical or echocardiographic evidence of cardiovascular disease. The combination of low-dose warfarin (1 to 3 mg/day; INR <1.5) and aspirin 325 mg is ineffective in preventing stroke, whereas the combination of aspirin with a higher dose of warfarin has been shown to accentuate intracranial bleeding, especially in elderly patients. A study exploring the efficacy of a combination of aspirin with clopidogrel, (Atrial Fibrillation Clopidogrel Trial with Irbesartan for prevention of Vascular Events (ACTIVE), is under way. This study will enroll 14,000 patients with AF, and the efficacy of combination therapy over warfarin will be assessed by the first occurrence of cardiovascular death, myocardial infarction, stroke, or a vascular thromboembolic event.

Anticoagulation in Paroxysmal Atrial Fibrillation

Although clinical data suggest that paroxysmal AF may be associated with an increased risk for stroke, and several cohort studies have demonstrated a consistent reduction in the annual stroke rates in patients with paroxysmal AF compared with permanent AF,[304] the risk-benefit ratio for lifelong anticoagulation therapy in a paroxysmal form of the arrhythmia remains uncertain. However, randomized controlled trials of warfarin generally enrolled patients with permanent AF. Thus, the Atrial Fibrillation ASpirin Anticoagulation Kobenhaven[286] (AFASAK) and the Stroke Prevention In Non-rheumatic Atrial Fibrillation[292] (SPINAF) studies exclusively enrolled patients with permanent AF, whereas the proportion of patients with intermittent AF was 7% in the Canadian Atria Fibrillation Anticoagulation (CAFA) study,[294] 16% in the Boston Area AnTicoagulation in Atrial Fibrillation study[293] (BAATAF), and 34% in the SPAF trial.[288] In accordance with recent guidelines, patients with paroxysmal AF should be considered at intermediate risk for stroke in the presence of frequent and prolonged episodes of the arrhythmia and should be considered candidates for anticoagulation based on general recommendations for patients with permanent AF.[39]

Anticoagulation in Atrial Flutter

The potential risk of stroke with AFL has now been established.[40,305,306] This risk is further supported by the fact that AF often coexists with AFL and may recur after radiofrequency catheter ablation of AFL in about one third of patients.

TEE studies have shown that left atrial appendage flow velocity is lower in patients with AFL compared with sinus rhythm but higher than during AF. Whether this accounts for a lower risk of thrombus formation is uncertain. In the Assessment of Cardioversion Using Transesophageal Echocardiography (ACUTE) study, a small proportion of patients with AFL (4.7% of the total study cohort) appeared to have fewer thrombi than those with AF, but this tendency did not reach statistical significance.[307] The results of the Flutter Atriale Società Italiana di Ecografia Cardiovascolare (FLASIEC) study have confirmed the observation that atrial flutter may be associated with atrial thrombi even in a small proportion of patients (2.4%).[308] The same criteria for anticoagulation should, therefore, be applied in patients with AFL and probably with AT as is recommended for AF.

Anticoagulation in Patients with Prosthetic Heart Valves

The strategy of oral anticoagulation targeting INR 3.0 to 4.0 is also pertinent for patients with atrial tachyarrhythmias and rheumatic mitral valve disease and prosthetic heart valves. The results of a study in 1608 patients with mechanical heart valves who were followed up for 6 years showed that the optimal intensity of anticoagulation therapy associated with the lowest incidence of both thromboembolic and hemorrhagic complications was observed with INR values of 2.5 to 4.9.[309] At this level of INR, the rate of all adverse events was 2 per 100 patient-years and rose sharply to 7.5 per 100 patient-years when the INR was 2.0 to 2.4 and to 4.8 per 100 patient-years when the INR increased to 5.0 to 5.5.

Recommendations for Anticoagulation Before Cardioversion

There is evidence for a significantly lower incidence of thromboembolic complications associated with electrical cardioversion for atrial tachyarrhythmias in patients receiving oral anticoagulation compared with those without anticoagulation therapy (0.8% versus 5.3%).[241] The beneficial effects of anticoagulation are probably associated with thrombus organization and adherence to the atrial wall or to complete thrombus resolution after a month of therapy. Therefore, for patients undergoing elective cardioversion, oral anticoagulation with warfarin (target INR, 2.5; range, 2.0 to 3.0) is recommended for 3 weeks before the procedure and should be continued for at least 4 weeks after cardioversion, as this may decrease the likelihood of formation of a new thrombus until the contractility of the left atrium is completely restored during sinus rhythm. Anticoagulation will also prevent thromboembolic complications if the AF recurs soon after cardioversion. In patients presenting with their first AF episode of less than 48 hours duration, the risk of thromboembolism is low. Cardioversion may be performed without delay provided that it is covered with low-molecular-weight heparin. Postcardioversion anticoagulation should be considered if thromboembolic risk factors are present.

Transesophageal Echocardiography

TEE has emerged as the most sensitive and specific imaging technique for detection of left atrial thrombi, also permitting the assessment of left atrial appendage flow. In the ACUTE study, the incidence of detection of atrial thrombi was 13.8%.[159] Several TEE criteria have been associated with thromboembolism: thrombi in the left atrium and left atrial appendage, reduced flow velocity in the left atrial appendage (≤ 20 cm/s), spontaneous echo contrast, and complex atheroma of aorta (mobile, ulcerated, or thick [>4 mm] aortic plaque).[142] The TEE-guided strategy with short-term anticoagulation is a safe and effective alternative in patients for whom early cardioversion is deemed to be clinically beneficial. It is ideal for inpatients with recent onset AF and AFL or individuals at high risk of bleeding complications during prolonged anticoagulation therapy. In the ACUTE study, the rate of embolic events did not differ between patients assigned to TEE-guided cardioversion or the conventional strategy of anticoagulation for 3 weeks before cardioversion, but the incidence of major and minor hemorrhagic complications was significantly lower in the TEE-guided group compared with the conventional therapy (2.9% versus 5.5%).[285]

A strategy of TEE-guided cardioversion in conjunction with the use of low-molecular-weight heparin has been suggested as an alternative for warfarin in low-risk patients.[310] Compared with unfractionated heparin, low-molecular-weight heparin therapy does not involve prolonged IV administration, hospitalization, or laboratory monitoring and has the potential to greatly simplify anticoagulation therapy for atrial tachyarrhythmias, especially pericardioversion.[311] Recent studies have demonstrated that low-molecular-weight heparin can be used safely and effectively in place of unfractionated heparin for acute treatment at the onset of the arrhythmia and during early cardioversion. In patients with AF or AFL, a strategy of immediate administration of dalteparin (100 IU/kg subcutaneously twice daily) continued for 11 days, combined with early TEE and immediate cardioversion in patients with no thrombus, resulted in sinus rhythm in 74% of patients after a median of 7 days. At 1 month, 75% of patients assigned to dalteparin and early cardioversion were in sinus rhythm compared with only 45% of those assigned to the conventional anticoagulation strategy.[312] However, this may reflect the presence of a higher proportion of patients in the conventional arm who had underlying heart disease (45% versus 31%). The ACUTE II pilot study is designed to compare the feasibility and safety of a TEE-guided low-molecular-weight heparin (enoxaparin) approach compared with a TEE-guided unfractionated heparin approach for patients with atrial tachyarrhythmias undergoing immediate cardioversion.[310]

Investigational Drugs and Devices for Prevention of Stroke in Atrial Tachyarrhythmias (Table 15-19)

In the recently completed Stroke Prevention using an ORal Thrombin Inhibitor in atrial Fibrillation (SPORTIF) III trial, treatment with the oral direct thrombin inhibitor *ximelagatran* proved to be at least as effective as tightly controlled warfarin in the prevention of stroke in high risk patients with AF. The primary event (all strokes and systemic embolism) rates by intention-to-treat were 2.3% per year in the warfarin group and 1.6% per year in the ximelagatran group (the absolute difference .66%), reflecting a 29% risk reduction (Fig. 15-54).[313] Rates of disabling or fatal stroke, mortality, and major bleeding did not differ between the groups, but the combined incidence of

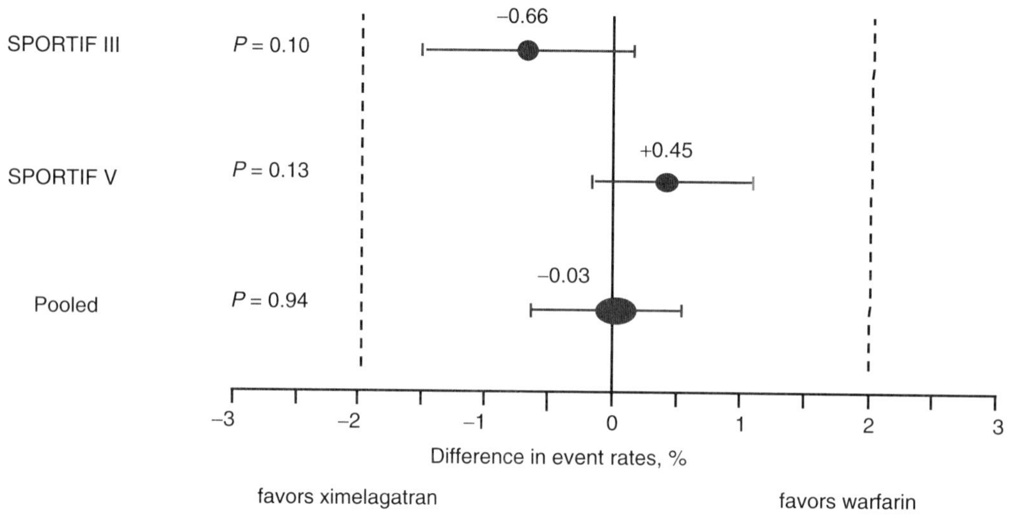

FIGURE 15-54 Noninferiority of ximelagatran versus warfarin for prevention of all strokes and systemic embolism (the primary end point) in the intention-to-treat analysis in the SPORTIF III and SPORTIF V studies and in pooled data from both trials. A prespecified margin for the absolute difference in primary event rates was set at 2% *(dashed lines)*. Note that in each study and in pooled data, the absolute difference in event rates between the warfarin and ximelagatran arm fell within these margins. (Data presented at the American Heart Association Scientific Sessions in 2003.)

TABLE 15-19 Alternative Therapies for Prevention of Stroke in Atrial Fibrillation

Therapy	Advantages and Disadvantages
Clopidogrel	This has shown to be noninferior to warfarin in a small study. There is an ongoing randomized clinical trial of clopidogrel 75 mg daily versus warfarin and clopidogrel + aspirin versus aspirin alone in 14,000 patients (the ACTIVE study).
Ximelagatran	The SPORTIF III and V studies proved noninferiority of ximelagatran 36 mg twice daily versus warfarin, but liver function should be monitored (elevated alanine aminotransferase).
LMWH (enoxaparin, dalteparin)	This has proven effective for short-term use (e.g., before cardioversion) in patients with a high risk of bleeding on warfarin, but no long-term follow-up is available.
Indraparinux	This can be given as a subcutaneous injection once a week; a randomized clinical trial versus warfarin in 5700 patients is under way (the AMADEUS study).
PLAATO	This technique has proven feasible and safe in selected patients, but no long-term follow-up is available. It is unclear whether anticoagulation should be continued.
LAA ligation	This technique has proven feasible in a small number of patients undergoing open-heart surgery, but no long-term follow-up is available. It is unclear whether anticoagulation should be continued.
Maze procedure	This technique has proven feasible in selected patients undergoing open-heart surgery, but no long-term data on the incidence of stroke are available. It is unclear whether anticoagulation should be continued.

ACTIVE, Atrial Fibrillation Clopidogrel Trial with Irbesartan for prevention of Vascular Events; LAA, left atrial appendage; LMWH, low-molecular-weight heparin; PLAATO, Percutaneous Left Atrial Appendage Transcatheter Occlusion; SPORTIF, Stroke Prevention using an ORal Thrombin Inhibitor in atrial Fibrillation.

minor and major bleeding was 14% lower with ximelagatran than with warfarin. In the SPORTIF V trial, which was different from the SPORTIF III study in that it had a double-blind, double-dummy design, the annual primary event rates were 1.6% in patients assigned to ximelagatran and 1.2% in those assigned to warfarin (*preliminary data*). The absolute difference was .45%, which was not statistically significant, thus meeting the criterion of noninferiority of ximelagatran (see Fig. 15-54). Similar to the SPORTIF III results, the rates of death, fatal bleeding, disabling stroke, or hemorrhagic stroke did not differ between the treatment groups, but the combined incidence of major and minor bleeding was significantly less with ximelagatran. In the pooled intention-to-treat analysis of the SPORTIF III and V studies, the net clinical benefit of ximelagatran was calculated by adding the number of primary events, major bleeding episodes, and death. By this analysis, ximelagatran was associated with a 16% relative risk reduction ($P = .038$).

The greatest advantages of ximelagatran are the absence of interactions with other medications, rapid achievement of the therapeutic effect, and no need for coagulation monitoring. Ximelagatran can be given in a fixed dose of up to 60 mg twice daily (36 mg twice daily in the SPORTIF III and V trials) in order to achieve stable anticoagulation. Its active metabolite melagatran can be administered intravenously or subcutaneously. Melagatran inhibits both soluble and fibrin-bound thrombin. In the SPORTIF program, ximelagatran had a favorable adverse effect profile but caused asymptomatic elevations of liver enzymes more than threefold the upper limit in approximately 6% of patients compared with less than 1% of patients on warfarin. Elevations in serum alanine transaminase occurred mainly early in the course of treatment (between 2 and 6 months) and usually returned toward normal over time, even in patients who continued taking the drug. The mechanism responsible for this finding is unknown, but consistency of this abnormality in all trials indicates the need for monitoring liver function at least for several months. Other direct oral thrombin inhibitors with better safety profiles are now at different stages of development.

An alternative strategy to inhibit the coagulation cascade is to intervene at the factor Xa level. A phase III study of the pentasaccharide *idraparinux* for prevention of stroke and systemic embolism (the primary end point), which will enroll approximately 5700 patients with AF, is under way (the AMADEUS study). Indraparinux, which inhibits factor Xa, can be given once a week as a subcutaneous preparation. Oral factor Xa inhibitors are also being developed.

The feasibility and safety of percutaneous left atrial appendage transcatheter occlusion (PLAATO) via transseptal catheterization has been investigated for prevention of thromboembolic complications in patients with contraindications for, or poor tolerance of, lifelong oral anticoagulation. The implant consists of a self-expanding nitinol cage (diameter 15 to 32 mm) covered with an occlusive polymeric membrane that serves both to occlude the orifice of the left atrial appendage and to allow tissue incorporation into the device. The initial experience in 15 patients with AF and contraindications for oral anticoagulation showed that the left atrial appendage was successfully occluded in all patients with no complications related to the device.[314] In experimental studies, histologic examination at 1 month has shown the surface of the implant to be completely smooth and free of mobile thrombi.[315] An extended follow-up of 11 patients showed that the PLAATO implant had no impact on the left atrial structure and function, including the pulmonary veins and the mitral valve.[315a] Left atrial appendage ligation or appendectomy via thoracoscopy or limited sternotomy may also be considered for prevention or reduction of thromboembolism.

UPSTREAM THERAPY

Atrial dilatation and stretch associated with left ventricular dysfunction create the substrate for sustained atrial tachyarrhythmias. As its size increases and its conduction slows, the so-called "critical atrial mass" is achieved.

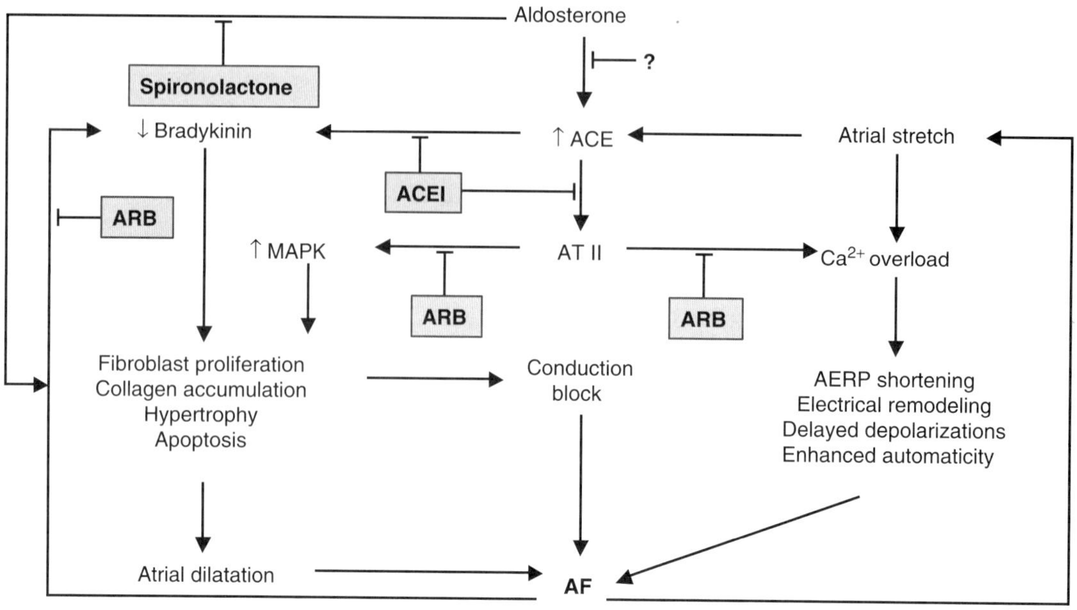

FIGURE 15-55 Angiotensin II in pathogenesis of atrial fibrillation. AERP, atrial effective refractory period; ACEI, angiotensin converting enzyme inhibitors; ARB, angiotensin receptor blockers; MAPK, mitogen-activated proteinkinases. (Modified from Savelieva I, Camm AJ: Atrial fibrillation and heart failure: Natural history and pharmacologic treatment. Europace 2004;6:S5-S19.)

This will accommodate a sufficient number of multiple reentrant wavelets and promote stretch-related abnormal automaticity and triggered activity in the atria to ensure the stability of the fibrillatory process.[316,317] Effective conventional treatment of heart failure resulting in atrial unloading may therefore delay or prevent the development of atrial tachyarrhythmias, particularly AF.

The TRACE study was the first to show the beneficial effects of an ACE inhibitor trandolapril on the frequency of AF in post–myocardial infarction patients with left ventricular dysfunction. In the trandolapril group, significantly fewer patients developed AF during 4 years of follow-up compared with the placebo group (2.8% versus 5.3%), reflecting a 55% risk reduction in the development of the arrhythmia.[61] A retrospective analysis of the selected patient population from the Studies of Left Ventricular Dysfunction (SOLVD) showed that enalapril was associated with a 78% risk reduction of the development of AF, particularly in individuals with mildly impaired ventricular function.[318]

Experimental studies in various animal models and in the human heart have shown increased expression of the angiotensin-converting enzyme and angiotensin II, as well as altered regulation of angiotensin II type 1 and 2 receptors in the fibrillating atria.[178,319-322] The density of angiotensin II receptors is generally greater in the atria than in the ventricles. Activation of the angiotensin II type 1 receptors induces a cascade of phosphorylation processes that activate mitogen-activated protein kinases (MAPK). The three best-defined angiotensin II–dependent MAPK subfamilies include extracellular signal-regulated protein kinases 1 and 2, c-Jun N-terminal kinase, and p38 MAPK. The latter two are also activated by atrial stretch. Activation of MAPK stimulates proliferation of fibroblasts, cellular hypertrophy, and apoptosis (Fig. 15-55).[322a] Angiotensin II–activated phosphoprotein kinase C mediates release of calcium from sarcoplasmic reticulum and phosphorylates L-type calcium channels, which increases calcium influx and potentiates electrical remodeling in the atria.

In experimental tachycardia-induced heart failure and rapid atrial pacing models of AF, both ACE inhibitors and AT-1 receptor blockers have been shown to reduce interstitial fibrosis. This prevents shortening of the atrial effective period and decreases the inducibility of sustained AF.[178,321,322] The beneficial effects on atrial electrical and structural remodeling of angiotensin II inhibition appear to be independent of intra-atrial pressures.

There is increasing evidence that ACE inhibitors and AT-1 receptor blockers can prevent atrial remodeling in patients with atrial tachyarrhythmias (Fig. 15-56).[322a] Pretreatment with an ACE inhibitor was shown to be associated with a higher rate of successful cardioversion in 188 patients with persistent AF and impaired ventricular function.[323] Retrospective data from 8804 patients with hypertension and no history of atrial tachyarrhythmias who participated in the Losartan Intervention For Endpoint reduction in hypertension (LIFE) study, therapy with losartan was associated with a 30% reduction of new onset AF compared with atenolol.[324]

In the first prospective study in 154 patients with persistent AF, pretreatment with an angiotensin II antagonist irbesartan facilitated electrical cardioversion and reduced the incidence of recurrence by 65% (from 44% to 20%) compared with the control group over 1 year of follow-up.[179] However, the study suffered significant limitations. It had an open-label design with no placebo arm, and treatment with irbesartan was additive to amiodarone. Furthermore, there were incidental differences between the irbesartan and nonirbesartan arms.

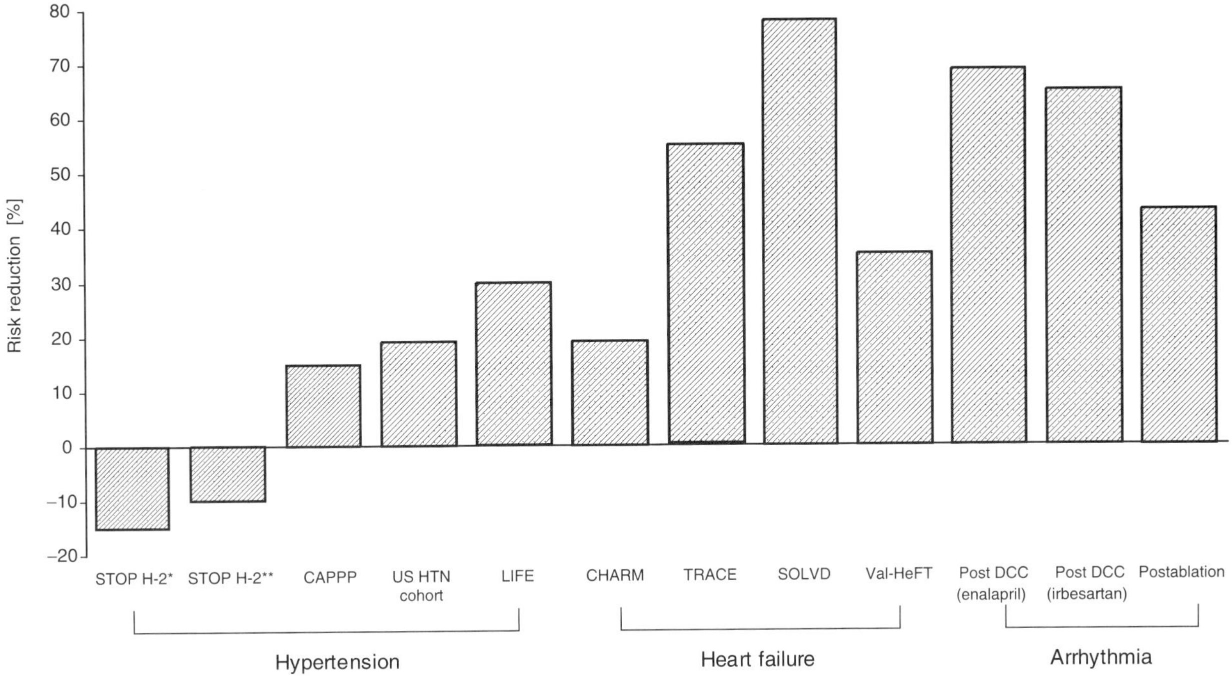

FIGURE 15-56 Prevention of atrial fibrillation with angiotensin-converting enzyme (ACE) inhibitors and angiotensin II type 1 (AT1) receptor blockers in selected randomized clinical trials. CAPPP, STOP H-2, LIFE, and US Hypertension were studies in hypertension. SOLVD, CHARM, and Val-HeFT were studies in heart failure. TRACE included patients with myocardial infarction and left ventricular dysfunction. Three studies shown on the right specifically looked at the effects of ACE inhibitors or AT1 receptor blockers in patients with atrial fibrillation postcardioversion and patients who underwent ablation for atrial flutter. CAPPP, Captopril Prevention Project, CHARM Candesartan in Heart failure assessment of Reduction in Mortality; LIFE, Losartan Intervention For Endpoint reduction in hypertension; SOLVD, Studies Of Left Ventricular Dysfunction; STOP H-2, Swedish Trial in Old Patients with Hypertension-2; TRACE, Trandolapril Cardiac Evaluation; Val-HeFT, Valsartan in Heart Failure Trial; HTN, hypertension. *ACE inhibitors versus conventional antihypertensive drugs; **ACE inhibitors versus calcium antagonists. (Reprinted from Savelieva I, Camm AJ: Atrial fibrillation and heart failure: Natural history and pharmacologic treatment. Europace 2004;6:S5-S19.)

Previous cardioversions were more frequent in patients who were randomized to therapy with amiodarone alone. More patients received β-blockers or calcium antagonists in the irbesartan group, which might also have affected the outcome of cardioversion and the incidence of AF recurrence during the follow-up. However, these observations open the possibility of exploitation of ACE inhibitors and AT-1 receptor blockers to prevent or delay atrial remodeling in patients with AF even in the absence of routine indications, such as heart failure, hypertension, or prevention of ventricular remodeling after acute myocardial infarction. The effects of irbesartan on atrial remodeling and the incidence of AF recurrence are under investigation in the ongoing ACTIVE trial, which is powered to show the superiority of irbesartan over placebo for reducing the risk of recurrent AF and preventing the progression to a permanent form in approximately 10,000 patients.

Furthermore, aldosterone synthesis, stimulated by angiotensin II, can lead to fibroblast proliferation, hypertrophy, and collagen deposition transduced via nuclear-mediated mineralocorticoid receptor pathways. These effects appear to be independent of effects of aldosterone on sodium and fluid retention and can be ameliorated by therapy with aldosterone antagonists.[325,326] There is evidence that restoration of sinus rhythm in patients with AF was associated with a significant reduction in the serum aldosterone and aldosteron/renin index within 48 hours after electrical cardioversion.[326]

Finally, recent advances in our understanding of the mechanisms of AF in the light of time-tested therapies, have led to evolving, promising therapeutic approaches aimed at the protection of the atrial myocardium and the prevention of AF. These include but are not limited to modulating the inflammatory mechanism in the pathogenesis of AF, for example, by low-dose glucocorticoid therapy[326a] and statins,[326b] preventing metabolic changes associated with AF by introducing an experimental partial fatty acid oxidation (pFOX) inhibitor trimetazidine, and exploring the antiarrhythmic membrane-stabilizing properties of omega-3 fatty acids (eicosapentaenoic acid and docosahexaenoic acid).[326c]

NONPHARMACOLOGIC TREATMENT OF ATRIAL TACHYARRHYTHMIAS

Ablation Strategies

Most advances in the nonpharmacologic cure of AFL have come from the understanding of its anatomic and electrophysiological mechanisms.[327] Typical, or isthmus-dependent, AFL involves a macro-reentrant right atrial circuit around the tricuspid annulus. The wavefront circulates down the lateral wall of the right atrium through the eustachian ridge between the tricuspid annulus and the inferior vena cava and up the interatrial septum, giving rise to the most frequent pattern, referred to as *counterclockwise* AFL. Reentry can also occur in the opposite direction (*clockwise or reverse* AFL). Catheter ablation aimed at creating bidirectional conduction block between the tricuspid annulus and the inferior vena cava is the most effective therapy for typical AFL. Advances in both electrophysiological mapping and radiofrequency ablation techniques have resulted in about 95% efficacy of this approach, rendering it the treatment of choice in most patients.[15,328] Catheter ablation is more effective than antiarrhythmic drugs for management of AFL, reducing the recurrence rate from 93% to 5% when used as first-line therapy.[15]

Long-term AF can occur in a proportion of patients, ranging from 10% in the low-risk group to 74% in high-risk individuals.[329] A history of spontaneous AF and left ventricular dysfunction have been shown to be associated with a nearly fourfold increase in risk of AF recurrence. Inducibility of AF during the ablation procedure and severe mitral regurgitation are also independent predictors of postablation AF. However, the incidence of AF is lower after catheter ablation than with pharmacologic therapy (29% versus 60% during 1 year of follow-up).[15] Approximately 15% to 20% of patients with AF treated with propafenone, flecainide, or amiodarone may develop AFL as the dominant arrhythmia, in which case isthmus ablation can facilitate the pharmacologic management of AF.[330]

Non–isthmus-dependent AFL, including AFL originating in the left atrium (pulmonary vein or mitral valve annulus dependent) and "incisional" macro-reentrant AFL, poses substantial difficulty for catheter ablation. The success rates for AFL associated with surgical correction of congenital heart disease were reported between 50% and 88%.[66] Although ablation of left AFL is feasible, the number of patients studied and followed is too limited to draw any conclusions about the efficacy of this approach.[117]

Pooled data from 514 patients who underwent catheter ablation for focal AT, including 18% with left AT and 10% with multifocal AT, showed an initial 86% success rate with an 8% incidence of recurrence after repeat procedure, suggesting that in individuals with incessant or refractory AT, ablation should be considered first-line therapy.[16] Electrophysiological mapping and catheter ablation techniques for nonpharmacologic management of atrial tachyarrhythmias are discussed in Chapters 6 and 7.

"Ablate and Pace" Strategy in Atrial Fibrillation

Radiofrequency ablation of the AV node followed by implantation of a permanent pacemaker has been successfully used for rate control in patients with symptomatic, drug-resistant AF.[180] A meta-analysis of 21 studies of 1181 patients has shown that the "ablate and pace" strategy significantly improved cardiac symptoms, quality of life, and health care resource utilization in patients with highly symptomatic, drug refractory arrhythmia.[331] In the absence of underlying heart disease, survival among patients with AF after ablation of the AV node did not differ from expected survival in the general population.[332]

Physiologic dual-chamber pacing has been suggested to be more beneficial than ventricular-based pacing in patients who have undergone AV node ablation for paroxysmal AF. DDDR pacing with mode switch can provide AV synchrony during sinus rhythm and an adequate increase in the ventricular rate during exercise, prevents atrial tracking during atrial tachyarrhythmias, and delays progression to permanent AF. However, this hypothesis has not been successfully demonstrated in the recent Atrial Pacing Peri-Ablation for Paroxysmal AF (PA^3) study, which showed that DDDR pacing compared with VDD pacing did not prevent the recurrence of paroxysmal AF or delay progression to permanent AF at 1 year of follow-up.[333] Left and right ventricular contractile asynchrony associated with pacing from the right ventricular apex may counteract the beneficial hemodynamic effect of regularization of heart rhythm postablation. The alternative pacing sites (e.g., from the left ventricular free wall, dual-site right ventricular pacing from the apex and the outflow tract, or biventricular pacing) have been subjects of considerable interest in recent trials. In a short-term study of 12 patients who underwent "ablate and pace" therapy (6 with an ejection fraction of ≥ 40% and 6 with dilated cardiomyopathy), both left ventricular pacing and biventricular pacing resulted in significant improvements in left ventricular contractile function and filling compared with conventional right ventricular pacing.[333a] The Post Atrio-Ventricular node ablation Evaluation trial (PAVE) presented at the Annual Scientific Session of the American College of Cardiology (ACC) was the first prospective, randomized study of biventricular pacing in patients treated with AV node ablation. The PAVE study randomized 184 patients in a 2:1 fashion to either biventricular pacing or conventional right ventricular pacing and followed them for 6 months. Patients who were assigned to biventricular pacing benefited from a significant improvement in functional capacity over right ventricular pacing as measured by the 6-minute walk test; peak VO_2; and exercise duration, which was sustained over time and resulted in a better quality of life. The left ventricular ejection fraction was maintained from baseline in the biventricularly paced group but declined from 44.9% to 40.7% in the conventionally paced group.

Modification of the Atrioventricular Node

Although the "ablate and pace" strategy has proven highly effective for management of patients with intractable arrhythmia and poor ventricular rate control or increased risk of drug-induced proarrhythmia, it is a destructive and irreversible procedure that renders patients dependent on a pacemaker and vulnerable to progression of cardiac dysfunction. To avoid the adverse effects of permanent pacing, efforts have been made to control the ventricular rate by radiofrequency modification of the conduction properties of the AV node without inducing complete AV block. The initial attempts to modify the fast pathway of the AV junction by applying radiofrequency energy in the antero-superior area of the intra-atrial septum resulted in a low long-term clinical efficacy of 32% and were associated with the development of heart block in about one third of patients.[334] More recent efforts directed at prolonging nodal refractoriness by ablation of the slow nodal pathway have been successful in 50% to 74% of patients, resulting in a decrease in the mean ventricular rate response by 15% to 35% compared with baseline.[83,335] However, a partial recovery of AV conduction with recurrence of rapid ventricular rates may occur in the longer term. Furthermore, irregular heart rate is associated with compromised hemodynamics, progression of left ventricular dysfunction, and persistence of symptoms. Given the high rate of inadvertent provocation of heart block and incomplete symptom relief, radiofrequency catheter modification of the AV node without pacemaker implantation is only rarely used. Furthermore, AV node ablation has been associated with greater improvement in symptoms and quality of life in the long term compared with the modification procedure.[83]

Pulmonary Vein Ablation and Maze Procedure

Ablation in or around the pulmonary veins to prevent the induction of AF by rapid repetitive atrial ectopic activity emanating from "sleeves" of the atrial myocardium inside the veins has been shown to abolish AF in selected patients.[17,18] In a controlled, nonrandomized study, 589 patients with "focal" AF who underwent this procedure benefited from a 70% reduction in the incidence of AF recurrence and an impressive 54% reduction in mortality compared with their 582 medically treated counterparts.[336] There was a significant decrease in morbidity, particularly from congestive heart failure and ischemic cardiovascular events, and a noticeable improvement in quality of life, suggesting that the indications for pulmonary vein isolation may be extended to a larger number of patients with AF.

The first curative approach to AF was the surgical maze procedure conceived with the idea of modifying the substrate for the arrhythmia by creating lines of conduction block in order to interrupt all possible reentrant circuits responsible for maintenance of AF.[337] Although it was first conducted to cure lone AF, it is presently used mainly in association with mitral valve or coronary bypass surgery, with success rates of 74% to 90% at 2 to 3 years postoperatively and a perioperative mortality less than 1%. A modified maze procedure, using cooled-tip radiofrequency endocardial ablation, in patients with AF and AFL undergoing mitral valve surgery has proven effective in restoring sinus rhythm in 80% of patients, whereas only 27% of those who had mitral valve replacement alone were in sinus rhythm at 1 year.[338] Currently available techniques and prospects of further development and refinement of ablation options and pacing strategies for the prevention or termination, or both, of atrial tachyarrhythmias are discussed in detail in other chapters.

Pacing Strategies for the Prevention or Termination, or Both of Atrial Arrhythmias

Atrial and Dual-Chamber Pacing in Patients with the Conventional Indications for Pacemaker Implantation

A recent detailed review of several retrospective and prospective studies have shown that patients with sinus node dysfunction conferred a substantial benefit from atrial or dual chamber pacing in terms of progression to permanent AF as compared with ventricular pacing.[339] A possible mechanism of the deleterious effect of ventricular pacing is that valvular regurgitation and atrial contraction against closed AV valves during ventricular pacing may result in atrial stretch and atrial enlargement, predisposing to the development of atrial tachyarrhythmias (Fig. 15-57). The beneficial effects of dual-chamber pacing can be multiple, involving both mechanical and electrophysiological factors. Atrial or physiological pacing provides the synchronized filling patterns and ventricular contraction, prevents consistent retrograde conduction, and reduces atrial overload and stretch. Pacing can prevent atrial tachyarrhythmias by eliminating pauses caused by bradycardia and by suppressing atrial ectopic beats.

In a retrospective analysis of nonrandomized trials, atrial-based or physiologic pacing has been shown to be associated with a 62% risk reduction of permanent AF and a 36% risk reduction of death (Fig. 15-58).[339] The results of the small Pacemaker Atrial Tachycardia (PAC-A-TACH) study in patients with tachy-brady syndrome showed an improved survival in physiologically paced

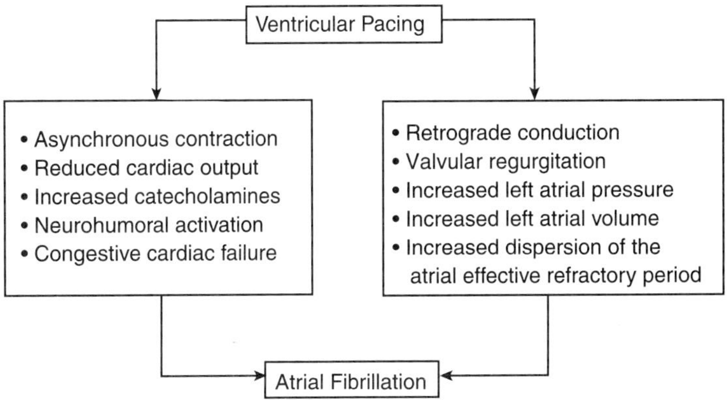

FIGURE 15-57 Pathogenesis of progression of atrial fibrillation during permanent right ventricular pacing.

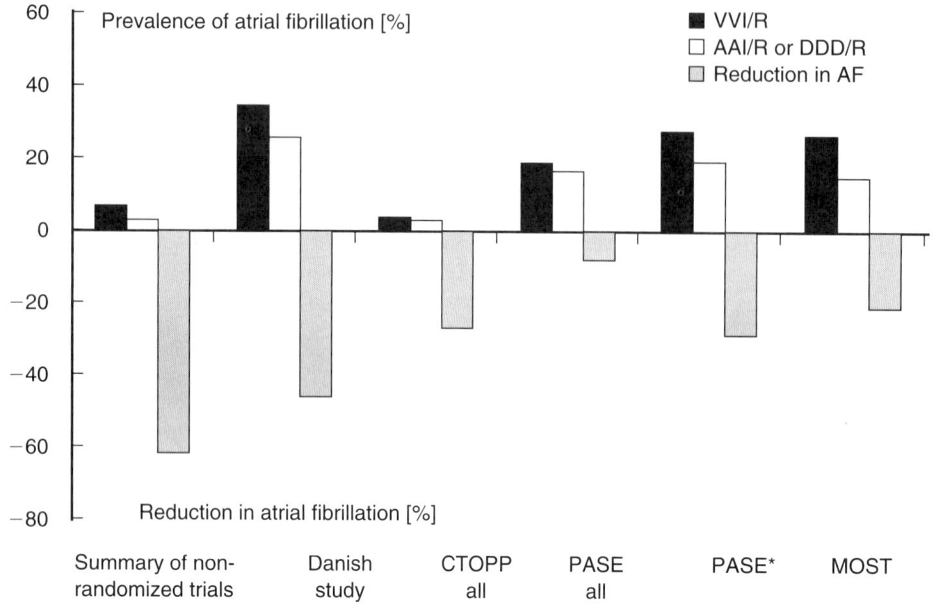

FIGURE 15-58 Prevention of progression of atrial fibrillation (AF) with atrial or dual-chamber pacing compared with ventricular pacing. Note that in the summary of nonrandomized studies, which predominantly included patients with sinus node dysfunction, and the CTOPP (Canadian Trial of Physiologic Pacing) trial, the incidence of AF is expressed as percent per year. *, data for patients with sinus node dysfunction and preserved atrioventricular conduction. (Reprinted from Savelieva I, Camm AJ: The results of pacing trials for the prevention and termination of atrial tachyarrhythmias. Is there any evidence of therapeutic breakthrough? J Interv Card Electrophysiol 2003;8:103-15.)

patients compared with ventricular-paced patients.[340] Two recent large-scale prospective studies, the Canadian Trial Of Physiologic Pacing[341] (CTOPP) and the PAcemaker Selection in the Elderly[342] (PASE) did not show benefit from physiologic pacing on overall mortality (Table 15-20). However, CTOPP reported an 18% reduction in risk of all AF episodes in patients who were randomized to physiologic pacing compared with those in the ventricular pacing group. The subgroup analysis from this study has further shown that physiologic pacing reduced the incidence of developing persistent AF from 3.84% per year to 2.8% per year, reflecting the relative risk reduction of 27.1%.[22] In the PASE study, patients with sinus node dysfunction who were randomized to physiologic pacing had a trend toward a lower incidence of AF compared with those assigned to ventricular

TABLE 15-20 Clinical Trials of Pacing Mode Selection in Patients with the Conventional Indications for Pacemaker Implantation

Study	Number of Patients	Pacing Indications	Randomization	Follow-up	Effects on Mortality/Morbidity	Incidence of AF
Danish Study[183]	225	SND	110 AAI vs. 115 VVI	5.5 yr	All-cause mortality risk reduction 34%	Risk reduction 46% AAI vs. VVI
Pac-A-Tach[340]	198	SND	100 DDDR vs. 98 VVIR	23.7 mo	3.2% vs. 6.8%	48% vs. 43%
PASE[342]	407	SND, AVB	203 DDDR vs. 204 VVIR	1.5 yr	All-cause mortality 17% vs. 16%	All patients: 17% vs. 19%; SND patients: 19% vs. 28%; AVB patients: 16% vs. 11%
CTOPP[22,341]	2568	SND, AVB	1094 DDDR vs. 1474 VVIR	3.1 yr	Cardiovascular death or stroke 4.9% vs. 5.5%	All AF: 5.3% vs. 6.6%; risk reduction 18% Persistent AF: 2.8% vs. 3.84% per yr; risk reduction 27.1%
PA-3[333]	77 AF	Post–AV node ablation	33 DDDR vs. 34 VDD	1 yr	Not evaluated	35% vs. 32%; AF burden 6.93 hr/day DDDR vs. 6.30 hr/day VDD
MOST[23]	2010	SND, 20% AVB	1014 DDDR vs. 996 VVIR	3 yr	All-cause mortality hazard ratio 0.97 (0.80-1.18); cardiovascular mortality 0.93 (0.78-1.13)	21% risk reduction in all patients; 50% risk reduction in patients without AF history
UK PACE (unpublished)	2021 ≥70 yrs	SND, AVB, no established AF	DDD vs VVI	3 yr	No difference	No difference

AF, atrial fibrillation; AVB, atrioventricular block; SND, sinus node dysfunction.

pacing (19% versus 28%, $P = .06$).[342] In both the CTOPP trial[341] and the Danish study[183] of patients with sick sinus syndrome, a significant decrease in the incidence of AF with physiologic or atrial-based pacing was seen only after a prolonged follow-up (2.5 to 5.5 years), suggesting that the biologic effect of atrial or dual-chamber pacing may be delayed.

The results from a recently completed study, MOde Selection Trial (MOST), of 2010 patients with sinus node dysfunction or AV block, or both, who were randomized to DDDR or VVIR pacing and followed up for a mean period of 3 years, have shown that DDDR pacing did not exert any effect on total mortality or cardiovascular death but significantly delayed progression to permanent AF.[23] During physiologic pacing, only 15.2% of patients developed permanent AF compared with 26.7% cases during ventricular pacing, reflecting a 21% risk reduction. The beneficial effect of physiologic pacing on the development of permanent AF was particularly noticeable in patients without a history of atrial arrhythmias before pacemaker implant (risk reduction 50%).

It is unclear whether atrial pacing is superior to dual-chamber pacing in patients with sinus node dysfunction and whether patients with AV block or combined conduction abnormalities would receive benefit from physiologic pacing. The substudy from the MOST trial has revealed a strong association between the percentage of ventricular pacing in the DDDR mode and the incidence of heart failure and AF.[342a] Despite maintenance of AV synchrony, ventricular pacing greater than 40% of the time conferred a 2.6-fold increased risk of hospitalization for heart failure. For each 1% increase in the cumulative percentage of ventricular pacing, the risk of AF increased by 1%. The discrepancy between the relative benefits of atrial pacing compared with dual-chamber pacing is probably due to the adverse effects of abnormal electrical activation of the left ventricle imposed by right ventricular apical pacing, which resembles that in left bundle branch block. The resulting alteration in mechanical activation may lead to impaired hemodynamic performance and mitral regurgitation favoring the development of heart failure and AF. This may explain the discrepancy between the significant benefits of atrial pacing compared with ventricular pacing in the study by Andersen and colleagues[183] and the more modest benefits of dual-chamber pacing in the CTOPP[341] and PASE[342] studies. The ongoing Danish study (DANPACE) will compare the long-term effects of atrial versus dual chamber pacing in nearly 2000 patients with sinus node dysfunction who will be followed up for 5 years.

Multisite Atrial Pacing

The presence in the atria of zones with consistently prolonged activation, such as the coronary sinus ostium and Bachmann's bundle, which result in nonuniform atrial conduction, has been linked to the initiation and perpetuation of AF. It has been suggested that pacing from these sites or others (multisite atrial pacing) may prevent AF due to improved synchronization of atrial depolarization. The Dual-Site Atrial Pacing for Prevention of Atrial Fibrillation (DAPPAF) trial prospectively enrolled 120 patients with documented recurrent symptomatic AF and bradycardia requiring cardiac pacing and randomly assigned them to 6-month periods in three pacing modes: dual-site pacing from the right atrium and coronary sinus, high right atrial pacing (both at a rate of 80 to 90 bpm), and support pacing at 40 bpm.[21] There was no difference in the main outcome measures, including the time to first symptomatic recurrence of AF and quality of life. However, the subgroup analysis showed that some beneficial effects of simultaneous pacing from the coronary sinus and the high right atrium were confined to patients who received antiarrhythmic drug therapy, resulting in a significantly prolonged time to recurrence of AF compared with single-site and support pacing (hazard ratios, 0.46, $P = .004$, and .623, $P = .006$, respectively).

In the New Indication for Preventive Pacing in Atrial Fibrillation (NIPP-AF) study of 22 patients with frequent symptomatic AF episodes and no conventional indications for pacing, dual-site overdrive pacing significantly prolonged the time to first symptomatic recurrence of AF, with a 3.2-fold relative risk of an event in the absence of pacing.[343] It also reduced total AF burden from 45% to 22% but, similarly to the DAPPAF study, provided no effect on quality of life and symptoms. Unfortunately, the study was small and the dual-site pacing concept was combined with an additional feature, such as continuous overdrive atrial pacing, rendering interpretation difficult. In the Pacing In Prevention of Atrial Fibrillation (PIPAF) study of 91 patients with paroxysmal AF, dual-site atrial pacing showed only a trend toward a lower frequency of AF episodes compared with single-site pacing.[344]

Single-site pacing from specific atrial sites has proven effective over the long term in patients with paroxysmal AF. In a multicenter prospective randomized study of 120 patients with class I and II bradycardia indications for pacing and a history of recurrent paroxysmal AF, a significant shortening of the P wave and significantly better rates of survival free from AF have been observed with pacing from the region of Bachmann's bundle compared with conventional pacing from the right atrial appendage during a 1-year follow-up (75% versus 47%).[184] In the Atrial Septal Pacing Efficacy Clinical Trial (ASPECT), 298 patients with bradycardia and paroxysmal AF were randomly assigned to pacing in the region of interatrial septum (in the vicinity of coronary sinus os, Bachmann's bundle, or midseptally) or to high right atrial pacing. The preliminary results of this study have also demonstrated a significant reduction in the total atrial conduction time as determined by a decrease in the P wave duration from 137 ± 16 ms at baseline to 115 ± 15 ms ($P < .001$) and a 39% decrease in the frequency of symptomatic episodes of AF compared with HRA pacing.[345] However, there was no effect on AF burden.

Biatrial pacing has been reported to significantly improve arrhythmia-free survival in the long term: at 9 years of follow-up, 55 (64%) of 86 patients remained in sinus rhythm, including 28 (33%) of patients without documented recurrences.[185]

Overdrive Atrial Pacing

Analysis of the distribution of potential triggers in the Atrial Fibrillation Therapy (AFT) study of 372 patients with paroxysmal AF has shown that 43% of AF episodes were triggered by an atrial premature beat, 22% occurred during preceding bradycardia, and 27% were classified as re-initiation of the arrhythmia which in turn were most often mediated by atrial ectopics.[134] Specific pacing algorithms have been developed to prevent initiation of AF. These algorithms can provide dynamic atrial pacing at a rate which is maintained only slightly higher than constantly monitored intrinsic rate. The Dynamic Atrial Overdrive (DAO) pacing algorithm is based on rate acceleration after a sensed atrial event, overdrive pacing at a plateau level followed by deceleration with rate smoothing until the first sensed atrial event occurs, to ensure that the pacemaker constantly controls the atrial depolarization. In the Atrial Dynamic Overdrive Pacing Trial conducted in America (ADOPT A), the combination of DDDR pacing and the DAO pacing algorithm was found to be superior to DDDR pacing alone for the suppression of symptomatic atrial tachyarrhythmias in 319 patients with sinus dysfunction and paroxysmal or persistent AF.[186] Active DAO was associated with a 65.3% reduction in the burden of organized atrial tachyarrhythmias (AT and AFL) and a trend towards reduction in AF burden by 25%. The mean number of AF episodes, total hospitalizations, adverse events, and mortality did not differ between groups. In the AFT study, the combination of DDD pacing (70 beats/min) with preventative pacing algorithms resulted in a 30.4% reduction in AF burden and a 78.1% decrease in sustained AF episodes.[346]

The important problem with translating the findings of these studies into clinical practice is lack of consistency in the interpretation of the results, which may be due to inconsistencies in the definitions of end points; differences between patient populations; and variable, often mixed pacing protocols used in these studies. The follow-up was usually too short to demonstrate benefits of preventative atrial pacing over time, although the results from the studies with extended follow-up periods, such as the AFT trial, are expected.

In general, randomized clinical trials have shown a neutral or slightly positive effect of specific pacing algorithms in the prevention of AF in addition to physiologic pacing. The results of these trials are not conclusive with respect to the definition of AF populations that would obtain the most advantage from pacing strategies and which of these strategies are the best.[24] An attempt to identify such patients has been undertaken by the PIPAF investigators. They found that patients with left atrial enlargement and patients with relatively preserved atrioventricular conduction conferred the most benefit from dynamic overdrive pacing.[346a] Several ongoing studies have explored the role of preventative atrial pacing in selected patient populations and/or as an additive to other therapies, such as cardiac resynchronization and antiarrhythmic drug therapy. For example, the Management of Atrial fibrillation Suppression in AF-HF COmorbidity Therapy (MASCOT) trial will randomize approximately 500 patients with heart failure and AF to cardiac resynchronization therapy with the AF-Suppression algorithm "on" or "off" and will follow them for at least 2 years. The primary end point is the development of permanent AF. The BREATHE study will assess the effects of atrial pacing in patients with sleep apnea associated with bradycardia and AF.

Combined Preventative Pacing and Antitachycardia Pacing

Electrophysiological studies in AF have shown that discrete atrial complexes separated by isoelectric baseline and characterized by a constant activation sequence over a relatively large area can be seen when AF is present on the surface ECG.[162] The interrogation of implantable devices has demonstrated that 43% of stored atrial electrograms in patients with AF were classified as monomorphic discrete signals separated by an isoelectric line with a cycle length of more than 200 ms.[347] These observations support the concept that transitional, regular or irregular, organized atrial activity observed at the onset of AF and intermittently during AF may be interrupted or modified by antitachycardia pacing. Modern implantable devices feature continuous

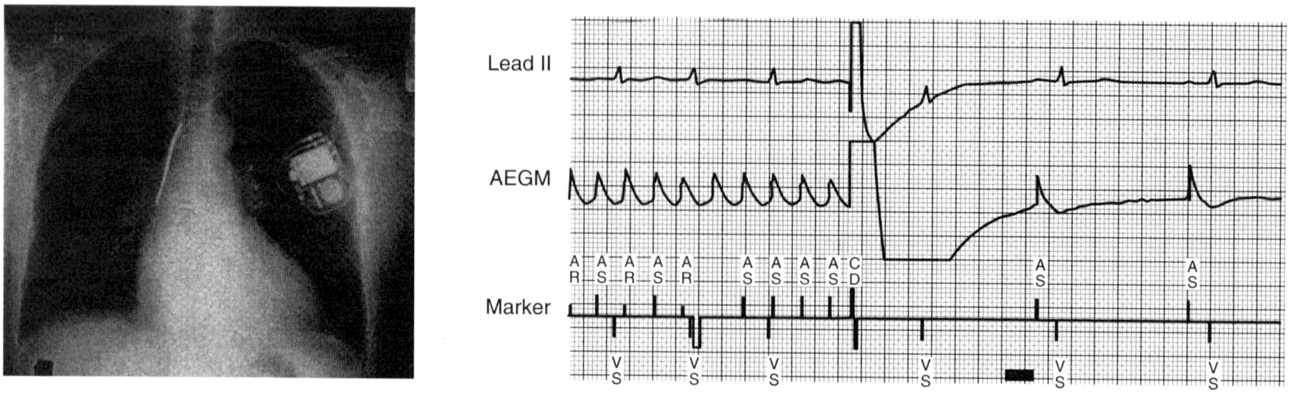

FIGURE 15-59 Internal cardioversion of atrial tachyarrhythmia with a 4 J shock via the implantable dual-chamber cardioverter-defibrillator. AEGM, atrial electrogram.

detection of atrial tachyarrhythmias and various atrial tiered therapies, such as preventative atrial pacing, antitachycardia pacing, and shock (Fig. 15-59).

In the largest study to date of 537 patients implanted with a dual-chamber ICD Jewel AF for ventricular arrhythmias, 74% of whom had concomitant AF or AFL, antitachycardia pacing including burst, ramp, and high frequency pacing terminated 48% of episodes of atrial tachyarrhythmias with no episodes of pacing-induced proarrhythmia.[348] The suppression rates for AT/AFL and AF were 58.5% and 29.8%, respectively. However, in the recently completed Atrial Therapy Efficacy and Safety Trial (ATTEST) of 368 patients with bradycardia and atrial tachyarrhythmias, the 54% success rate of antitachycardia pacing neither translated into reduced AF burden nor to improved quality of life,[352] although better results were reported previously for the combination of antitachycardia pacing, atrial shock, and preventative atrial pacing.[350]

SUMMARY

In conclusion, although theoretically very attractive, atrial pacing for the treatment of atrial tachyarrhythmias remains a debatable indication because the evidence base is too limited to inform definitive management guidelines. The results of randomized clinical trials that have been completed are not conclusive with respect to the beneficial effects of pacing for the prevention of atrial tachyarrhythmias. No specific population that would obtain clear advantage from pacing strategies has yet been defined.[24]

REFERENCES

1. Psaty BM, Manolio TA, Kuller LH, et al: Incidence of and risk factors for atrial fibrillation in older adults. Circulation 1997;96: 2455-61.
2. Granada J, Uribe W, Chyou PH, et al: Incidence and predictors of atrial flutter in the general population. J Am Coll Cardiol 2000;36:2242-6.
3. Poutianen AM, Koistinen MJ, Airaksinen KE, et al: Prevalence and natural course of ectopic atrial tachycardia. Eur Heart J 1999;20:694-700.
4. Dorian P, Jung W, Newman D, et al: The impairment of health-related quality of life in patients with intermittent atrial fibrillation: Implications for the assessment of investigational therapy. J Am Coll Cardiol 2000;36:1303-9.
5. Atrial Fibrillation Investigators: Risk factors for stroke and efficacy of antithrombotic therapy in atrial fibrillation: Analysis of pooled data from five randomized controlled trials. Arch Intern Med 1994;154:1449-57.
6. Gottdiener JS, Arnold AM, Aurigemma GP, et al: Predictors of congestive heart failure in the elderly: The Cardiovascular Health Study. J Am Coll Cardiol 2000;35:1628-37.
7. Dries DL, Exner DV, Gersh BJ, et al: Atrial fibrillation is associated with an increased risk for mortality and heart failure progression in patients with asymptomatic and symptomatic left ventricular systolic dysfunction: A retrospective analysis of the SOLVD trials. J Am Coll Cardiol 1998;32:695-703.
8. Wolf PA, Benjamin EJ, Belanger AJ, et al: Secular trends in the prevalence of atrial fibrillation: The Framingham Study. Am Heart J 1996;131:790-5.
9. Stewart S, MacIntyre K, MacLeod MMC, et al: Trends in hospital activity, morbidity and case fatality related to atrial fibrillation in Scotland, 1986-1996. Eur Heart J 2001;22:693-701.
10. Benjamin EJ, Wolf PA, D'Agostino RB, et al: Impact of atrial fibrillation on risk of death: The Framingham Study. Circulation 1999;98:946-52.
11. Vidaillet H, Granada JF, Chyou PH, et al: A population-based study on mortality among patients with atrial fibrillation or flutter. Am J Med 2002;113:365-70.
12. Task Force of the Working Group on Arrhythmias of the European Society of Cardiology. The Sicilian Gambit: A new approach to the classification of antiarrhythmic drugs based on their action on arrhythmogenic mechanisms. Circulation 1991; 84:1831-51.
13. Members of the Sicilian Gambit. New approaches to antiarrhythmic therapy: Emerging therapeutic applications of the cell biology of cardiac arrhythmias. Eur Heart J 2001;22:2148-63.
14. Savelieva I, Camm AJ: Update on antiarrhythmic treatment of atrial fibrillation. In Tse HF, Lee KLF, Lau CP (eds): Clinical Cardiac Pacing and Electrophysiology. Bologna, Italy, Monduzzi Editore, 2003, pp 923-44.
15. Natale A, Newby KH, Pisano E, et al: Prospective randomized comparison of antiarrhythmic therapy versus first-line radiofrequency ablation in patients with atrial flutter. J Am Coll Cardiol 2000;35:1898-904.
16. Hsieh MH, Chen SA: Catheter ablation of focal AT. In Zipes DP, Haïssaguerre M (eds): Catheter Ablation of Arrhythmias. Armonk, NY, Futura Publishing, 2002, pp 185-204.
17. Haïssaguerre M, Jais P, Shah DC, et al: Spontaneous initiation of atrial fibrillation by ectopic beats originating in the pulmonary veins. N Engl J Med 1998;339:659-66.
18. Pappone C, Rosanio S, Oreto G, et al: Circumferential radiofrequency ablation of pulmonary vein ostia. Circulation 2000;102: 2619-28.
19. Sie HT, Beukema WP, Ramdat Misier AR, et al: The radiofrequency modified maze procedure: A less invasive surgical approach to atrial fibrillation during open-heart surgery. Eur J Cardiothorac Surg 2001;19:443-7.
20. Camm AJ, Savelieva I: Rationale and patient selection for "hybrid" drug and device therapy in atrial and ventricular arrhythmias. J Interv Card Electrophysiol 2003;9:207-14.
21. Saksena S, Prakash A, Ziegler P, et al, for the DAPPAF Investigators: The Dual-site Atrial Pacing for Prevention of Atrial Fibrillation (DAPPAF) trial: Improved suppression of atrial fibrillation with dual-site atrial pacing and antiarrhythmic drug therapy. J Am Coll Cardiol 2001;38:598-9.
22. Skanes AC, Krahn AD, Yee R, et al, for the CTOPP Investigators. Progression to chronic atrial fibrillation after pacing: The Canadian Trial of Physiologic Pacing. J Am Coll Cardiol 2001; 38:167-72.
23. Lamas GA, Lee KL, Sweeney MO, et al, for the Mode Selection Trial in Sinus Node Dysfunction: Ventricular pacing or dual chamber pacing for sinus node dysfunction. N Engl J Med 2002; 346:1854-62.
24. Savelieva I, Camm AJ: The results of pacing trials for the prevention and termination of atrial tachyarrhythmias: Is there any evidence of therapeutic breakthrough? J Interv Card Electrophysiol 2003;8:103-15.
25. Pedersen OD, Bagger H, Køber L, et al, for the TRACE Study Group: Trandolapril reduces the incidence of atrial fibrillation after acute myocardial infarction in patients with left ventricular dysfunction. Circulation 1999;100:376-80.
26. Benjamin EJ, Levy D, Vaziri SM, et al: Independent risk factors for atrial fibrillation in a population-based cohort: The Framingham Heart Study. JAMA 1994;271:840-4.
27. Krahn AD, Manfreda J, Tate RB, et al: The natural history of atrial fibrillation: Incidence, risk factors, and prognosis in the Manitoba Follow-Up Study. Am J Med 1995;98:476-84.
28. Wilhelmsen L, Rosengren A, Lappas G: Atrial fibrillation in the general male population—morbidity and risk factors. J Intern Med 2001;250:382-9.
29. Go AS, Hylek EM, Phillips KA, et al. Prevalence of diagnosed atrial fibrillation in adults. National implications for rhythm management and stroke prevention: The AnTicoagulation and Risk Factors In Atrial Fibrillation (ATRIA) Study. JAMA 2001;285: 2370-5.
30. Furberg CD, Psaty BM, Manolio TA, et al, for the CHS Collaborative Research Group: Prevalence of atrial fibrillation in elderly subjects: The Cardiovascular Health Study. Am J Cardiol 1994;74:236-41.
31. Cleland JGF, Swedberg K, Follath F, et al, for the Study Group on Diagnosis of the Working Group on Heart Failure of the

European Society of Cardiology: The EuroHeart Failure survey programme—a survey on the quality of care among patients with heart failure in Europe. Part 1: Patient characteristics and diagnosis. Eur Heart J 2003;24:442-63.
32. Tsang TS, Gersh BJ, Appleton CP, et al: Left ventricular diastolic dysfunction as a predictor of the first diagnosed nonvalvular atrial fibrillation in 840 elderly men and women. J Am Coll Cardiol 2002;40:1636-44.
33. Saoudi N, Cosio F, Waldo A, et al: A classification of atrial flutter and regular atrial tachycardia according to electrophysiological mechanisms and anatomical bases: A statement from a Joint Expert Group from the Working Group of Arrhythmias of the European Society of Cardiology and the North American Society of Cardiology and the North American Society of Pacing and Electrophysiology. Eur Heart J 2001;22:1162-82.
34. Fuster V, Rydén LE, Asinger RV, et al: Task Force Report: ACC/AHA/ESC guidelines for the management of patients with atrial fibrillation. Eur Heart J 2001;22:1852-923.
35. Lévy S, Maarek M, Coumel P, et al, on behalf of the College of French Cardiologists: Characterization of different subsets of atrial fibrillation in general practice in France: The ALFA Study. Circulation 1999;99:3028-35.
36. Wolf PA, Abbot RD, Kannel WB: Atrial fibrillation as independent risk factor for stroke: The Framingham Study. Stroke 1991;22:983-8.
37. Feinberg WM, Seeger JF, Carmody RF, et al: Epidemiologic features of asymptomatic cerebral infarction in patients with nonvalvular atrial fibrillation. Arch Intern Med 1990;150:2340-4.
38. Ezekowitz MD, James KE, Nazarian SM, et al, for the Veterans Affairs Stroke Prevention in Nonrheumatic Atrial Fibrillation Investigators: Silent cerebral infarction in patients with nonrheumatic atrial fibrillation. Circulation 1995;92:2178-82.
39. Albers GW, Dalen JE, Laupacis A, et al: Antithrombotic therapy in atrial fibrillation. Chest 2001;119:194S-206S.
40. Biblo LA, Yan Z, Quan KJ, et al: Risk for stroke in patients with atrial flutter. Am J Cardiol 2001;87:346-9.
41. Naito M, David D, Michelson EL, et al: The hemodynamic consequences of cardiac arrhythmias: evaluation of the relative roles of abnormal atrioventricular sequencing, irregularity of ventricular rhythm and atrial fibrillation in a canine model. Am Heart J 1983;106:284-91.
42. Clark DM, Plumb VJ, Epstein AE, Kay GN: Hemodynamic effects of an irregular sequence of ventricular cycle lengths during atrial fibrillation. J Am Coll Cardiol 1997;30:1039-45.
43. Van Gelder IC, Crijns HJGM, Blanksma PK, et al: Time course of hemodynamic changes and improvement of exercise tolerance after cardioversion of chronic atrial fibrillation unassociated with cardiac valve disease. Am J Cardiol 1993;72:560-6.
44. Savelieva I, Camm AJ: Clinical relevance of silent atrial fibrillation: Prevalence, prognosis, quality of life, and management. J Interv Card Electrophysiol 2000;4:369-82.
45. Mathew J, Hunsberger S, Fleg J, et al, for the Digitalis Investigation Group: Incidence, predictive factors, and prognostic significance of supraventricular tachyarrhythmias in congestive heart failure. Chest 2000;118:914-22.
46. Benjamin EJ, Wolf PA, D'Agostino RB, et al: Impact of atrial fibrillation on risk of death: The Framingham Study. Circulation 1999;98:946-52.
47. Jouven X, Desnos M, Guerot C, Ducimetiere P: Idiopathic atrial fibrillation as a risk for mortality: The Paris Prospective Study I. Eur Heart J 1999;20:896-9.
48. Maas R, Kuesel T, Patten M: Detection of symptomatic and asymptomatic atrial arrhythmias by continuous transtelephonic self-monitoring [abstract]. Circulation 2000;102:II-801.
49. Defaye P, Dournaux F, Mouton E, for the Aida Multicenter Study Group: Prevalence of supraventricular arrhythmias from the automated analysis of data stored in the DDD pacemakers of 617 patients: The AIDA study. Pacing Clin Electrophysiol 1998; 21:250-5.
50. Fetsch T, Breithardt G, Engberding R, et al: Can we believe in symptoms for detection of atrial fibrillation in clinical routine? The results of the PAFAC study [abstract]. Circulation 2001; 104:II-699.
51. Lin H-J, Wolf PA, Benjamin EJ, et al: Newly diagnosed atrial fibrillation and acute stroke: The Framingham Study. Stroke 1995; 26:1527-30.
52. Cullinane M, Wainwright R, Brown A, et al: Asymptomatic embolization in subjects with atrial fibrillation not taking anticoagulants: A prospective study. Stroke 1998;29:1810-5.
53. Pozzoli M, Cioffi G, Traversi E, et al: Predictors of primary atrial fibrillation and concomitant clinical and hemodynamic changes in patients with chronic heart failure: A prospective study in 344 patients with baseline sinus rhythm. J Am Coll Cardiol 1998;32:197-204.
54. Savelieva I, Camm AJ: Supraventricular and ventricular arrhythmias in heart failure. In Van Veldhuisen DJ, Pitt B (eds): Chronic Heart Failure. Amsterdam, Benecke NI, 2002, pp 111-43.
55. Crijns HJGM, Tjeerdsma G, De Kam PJ, et al: Prognostic value of the presence and development of atrial fibrillation in patients with advanced chronic heart failure. Eur Heart J 2000;21:1238-45.
56. Middlekauff HR, Stevenson WG, Stevenson LW: Prognostic significance of atrial fibrillation in advanced heart failure: a study of 390 patients. Circulation 1991;84:40-8.
57. Carson PE, Johnson GR, Dunkman WB, et al, for the V-HeFT VA Cooperative Studies Group: The influence of atrial fibrillation on prognosis in mild to moderate heart failure: The V-HeFT Studies. Circulation 1993;87:102-10.
58. Crenshaw BS, Ward SR, Granger CB, et al, for the GUSTO-I Trial Investigators: Atrial fibrillation in the setting of acute myocardial infarction: the GUSTO-I experience. J Am Coll Cardiol 1997;30:406-13.
59. Rathore SS, Berger AK, Weinfurt KP, et al: Acute myocardial infarction complicated by atrial fibrillation in the elderly: Prevalence and outcomes. Circulation 2000;101:969-74.
60. Hornestam B, Swedberg K, Snappin S, for the OPTIMAAL Study Group: Significance of atrial fibrillation in patients with acute myocardial infarction and symptomatic or asymptomatic left ventricular dysfunction: Data from the OPTIMAAL Study [abstract]. J Am Coll Cardiol 2002;39:134A.
61. Pedersen OD, Bagger H, Køber L, Torp-Pedersen C: The ocurrence and prognostic significance of atrial fibrillation/-flutter following acute myocardial infarction. TRACE Study group TRAndolapril Cardiac Evaluation. Eur Heart J 1999;20:748-54.
62. Al-Khatib S, Pieper KS, Lee KL, et al: Atrial fibrillation and mortality among patients with acute coronary syndromes without ST-segment elevation: results from the PURSUIT trial. Am J Cardiol 2001;88:76-9.
63. Maron B, Casey S, Poliac LC, et al: Clinical course of hypertrophic cardiomyopathy in a regional United States cohort. JAMA 1999;282:650-5.
64. Olivotto I, Cecchi F, Casey S, et al: Impact of atrial fibrillation on the clinical course of hypertrophic cardiomyopathy. Circulation 2001;104:2517-24.
65. Maron BJ, Olivotto I, Bellone, et al: Clinical profile of stroke in 900 patients with hypertrophic cardiomyopathy. J Am Coll Cardiol 2002;39:301-7.
66. Blomström-Lundvist C, Scheiman MM, Aliot EM, et al: ACC/AHA/ESC guidelines for the management of patients with supraventricular arrhythmias—executive summary. A report of the American College of Cardiology/American Heart Association Task Force on Practice Guidelines and the European Society of Cardiology Committee for Practice Guidelines (Writing Committee to develop guidelines for the management of patients with supraventricular arrhythmias). J Am Coll Cardiol 2003;42:1493-531.
67. Mathew JP, Parks R, Savino JS, et al, for the Multicenter Study of Perioperative Ischaemia Research Group: Atrial fibrillation following coronary artery bypass graft surgery. JAMA 1996;276:300-6.
68. Asher CR, Miller DP, Grimm RA, et al: Analysis of risk factors for development of atrial fibrillation early after cardiac valvular surgery. Am J Cardiol 1998;82:892-5.
69. Van Hare GF, Lesh MD, Ross BA, et al: Mapping and radiofrequency ablation of intra-atrial reentrant tachycardia after the Senning or Mustard procedure for transposition of the great arteries. Am J Cardiol 1996;77:985-91.
69a. Athanasiou T, Aziz O, Mangoush O, et al: Do off-pump techniques reduce the incidence of postoperative atrial fibrillation in elderly patients undergoing coronary artery bypass grafting? Ann Thorac Surg 2004;77:1567-74.
70. Brand FN, Abbott RD, Kannel WB, Wolf PA: Characteristics and prognosis of lone atrial fibrillation: 30-year follow-up in the Framingham Study. JAMA 1985;254:3449-53.

71. Kopecky SL, Gersh BJ, McGoon MD, et al: The natural history of lone atrial fibrillation: A population-based study over three decades. N Engl J Med 1987;317:669-74.
72. Brugada R, Tapscott T, Czernuszewicz GZ, et al: Identification of a genetic locus for familial atrial fibrillation. N Engl J Med 1997; 336:905-11.
73. Ellinor PT, Shin JT, Moore RK, et al: Locus for atrial fibrillation maps to chromosome 6q14-16. Circulation 2003;107:2880-3.
74. Darbar D, Herron KJ, Ballew JD, et al: Familial atrial fibrillation is a genetically heterogeneous disorder. J Am Coll Cardiol 2003; 41:2185-92.
75. Taylor MRG, Fain PR, Sinagra G, et al: Natural history of dilated cardiomyopathy due to lamin A/C gene mutations. J Am Coll Cardiol 2003;41:771-80.
76. Bharati S, Sasaki D, Siedman S, Vidaillet H: Sudden death among patients with a hereditary cardiomyopathy due to a missense mutation in the rod domain of the lamin a/c gene [abstract]. J Am Coll Cardiol 2001:37:174A.
77. Gruver EJ, Fatkin D, Dodds GA, et al: Familial hypertrophic cardiomyopathy and atrial fibrillation caused by Arg663His beta-cardiac myosin heavy chain mutation. Am J Cardiol 1999;83:13H-18H.
78. Lai LP, Su MJ, Yeh HM, et al: Association of the human minK gene 38G allele with atrial fibrillation: Evidence of possible genetic control on the pathogenesis of atrial fibrillation. Am Heart J 2002;144:485-90.
79. Chen YH, Xu SJ, Bendahhou S, et al: KCNQ1 gain-of-function mutation in familial atrial fibrillation. Science 2003; 299:251-4.
80. Kay GN, Bubien RS, Epstein AE, Plumb VJ: Effect of catheter ablation of the atrioventricular junction on quality of life and exercise tolerance. Am J Cardiol 1988;62:741-4.
81. Paquette M, Roy D, Talajic M, et al: Role of gender and personality on quality-of-life impairment in intermittent atrial fibrillation. Am J Cardiol 2000;86:764-8.
82. Fitzpatrick AP, Kourouyan HD, Siu A, et al: Quality of life and outcomes after radiofrequency His-bundle catheter ablation and permanent pacemaker implantation: Impact of treatment in paroxysmal and established atrial fibrillation. Am Heart J 1996;131:499-507.
83. Lee SH, Chen SA, Tai CT, et al: Comparisons of quality of life and cardiac performance after complete atrioventricular junction ablation and atrioventricular junction modification in patients with medically refractory atrial fibrillation. J Am Coll Cardiol 1998;31:637-44.
84. Natale A, Zimerman L, Tomassoni G, et al: Impact on ventricular function and quality of life of transcatheter ablation of the atrioventricular junction in chronic atrial fibrillation with a normal ventricular response. Am J Cardiol 1996;78:1431-3.
85. Savelieva I, Paquette M, Dorian P, et al: Quality of life in patients with silent atrial fibrillation. Heart 2001;85:216-7.
85a. Wolf PA, Mitchell JB, Baker CS, et al: Impact of atrial fibrillation on mortality, stroke, and medical costs. Arch Intern Med 1998; 158:229-34.
85b. Stewart S, Murphy N, Walker A, et al: Cost of an emerging epidemic: An economic analysis of atrial fibrillation in the UK. Heart 2004;90:286-92.
85c. Weerasooriya K, Jaïs P, Le Heuzey JY, et al: Cost analysis of catheter ablation for paroxysmal atrial fibrillation. Pacing Clin Electrophysiol 2003;26:292-94.
86. Bharati S, Lev M: Histologic abnormalities in atrial fibrillation: Histology of normal and diseased atrium. In Falk RH, Podrid PJ (eds): Atrial Fibrillation: Mechanisms and Management. New York, Raven Press, 1992, p 15.
87. Bharati S, Lev M: The pathologic changes in the conduction system beyond the age of ninety. Am Heart J 1992;124:486-96.
88. Bharati S, Lev M: Pathologic changes of the conduction system with aging. Cardiol in the Elderly 1994;2:152.
89. Rosen KM, Rahimtoola SH, Bharati S, Lev M: Bundle branch block with intact atrioventricular conduction. Am J Cardiol 1973;32:782-93.
90. Bharati S, Lev M, Dhingra RC, et al: Electrophysiologic and pathologic correlations in two cases of chronic second degree atrioventricular block with left bundle branch block. Circulation 1975;52:221-9.
91. Lev M, Bharati S: The conduction system in coronary artery disease. In Donosco E, Lipski J (eds): Acute Myocardial Infarction. 1977, pp 1-16.
92. Lev M, Kinare SG, Pick A: The pathogenesis of atrioventricular block in coronary disease. Circulation 1970;42:409-25.
93. Bharati S, Lev M, Dhingra R, et al: Pathologic correlations in three cases of bilateral bundle branch disease with unusual electrophysiologic manifestations in two cases. Am J Cardiol 1976;38:508-18.
94. Kaplan BM, Langendorf R, Lev M, Pick A: Tachycardia-bradycardia syndrome (so-called "sick sinus syndrome"). Am J Cardiol 1973;31:497-508.
95. Bharati S, Lev M: Conduction system in sudden unexpected death a considerable time after repair of atrial septal defect. Chest 1988;94:142-8.
96. Bharati S, Molthan ME, Veasy G, Lev M: Conduction system in two cases of sudden death two years after the Mustard procedure. J Thorac Cardiovasc Surg 1979;77:101-8.
97. Bharati S, Lev M, Denes P, et al: Infiltrative cardiomyopathy with conduction disease and ventricular arrhythmia: Electrophysiologic and pathologic correlations. Am J Cardiol 1980;45:163-72.
98. Bharati S, Scheinman MM, Morady F, et al: Sudden death after catheter-induced atrioventricular junctional ablation in the human. Chest 1985;88:883-9.
99. Bharati S, Rosen KM, Strasberg B, et al: Anatomic substrate for congenital atrioventricular block in middle-aged adults. Pacing Clin Electrophysiol 1982;5:860-9.
100. Amat-Yy-Leon F, Racki AJ, Denes P, et al: Familial atrial dysrhythmia with AV block; intracellular microelectrode, clinical electrophysiologic and morphologic observations. Circulation 1974;50:1097-104.
101. Allessie MA, Boyden PA, Camm AJ, et al: Pathophysiology and prevention of atrial fibrillation. Circulation 2001;103:769-77.
102. Ausma J, Litijens N, Lenders MH, et al: Time course of atrial fibrillation-induced cellular structural remodeling in atrial of the goat. J Mol Cell Cardiol 2001;33:2083-94.
103. Ausma J, Van Der Velden HMW, Lenders MH, et al: Reverse structural and gap-junctional remodeling after prolonged atrial fibrillation in the goat. Circulation 2003;107:2051-8.
104. Elvan A, Zipes DP: Pacing induced chronic atrial fibrillation impairs sinus node function in dogs. Circulation 1996;94:2953-60.
105. Saito T, Waki K, Becker A: Left myocardial extension onto pulmonary veins in humans: Anatomical observations relevant for atrial arrhythmias. J Cardiovasc Electrophysiol 2000;11:888-94.
106. Hassink RJ, Arentz HT, Ruskin J, Keane D: Morphology of atrial myocardium in human pulmonary veins. J Am Coll Cardiol 2003;42:1108-14.
107. Knight BP, Michaud GF, Strickberger SA, Morady F: Electrocardiographic differentiation of atrial flutter from atrial fibrillation by physicians. J Electrocardiol 1999;32:315-19.
108. Saoudi N, Nair M, Abdelazziz A, et al: Electrocardiographic patterns and results of radiofrequency catheter ablation of clockwise type I atrial flutter. J Cardiovasc Electrophysiol 1996;7:931-42.
109. Kalman JM, Olgin JE, Saxon La, et al: Electrocardiographic and electrophysiologic characterization of atypical atrial flutter in man: Use of activation and entrainment mapping and implications for catheter ablation. J Cardiovasc Electrophysiol 1997;8:121-44.
110. Ndrepepa G, Zrenner B, Weyerbrock S, et al: Activation patterns in the left atrium during counterclockwise and clockwise atrial flutter. J Cardiovasc Electrophysiol 2001;12:893-9.
111. Rodriguez LM, Timmermans C, Nabar A, et al: Biatrial activation in isthmus-dependent atrial flutter. Circulation 2001;104: 2545-50.
112. Marine JE, Korley VJ, Obioha-Hgwu O, et al: Different patterns of interatrial conduction in clockwise and counterclockwise atrial flutter. Circulation 2001;104:1153-7.
113. Sippens-Groenewegen A, Lesh MD, Roithinger FX, et al: Body surface mapping of counterclockwise and clockwise typical atrial flutter: A comparative analysis with endocardial activation sequence mapping. J Am Coll Cardiol 2000;35:1276-87.
114. Kolb C, Nurnberger S, Ndrepepa G, et al: Modes of initiation of paroxysmal atrial fibrillation from analysis of spontaneously occurring episodes using a 12-lead Holter monitoring system. Am J Cardiol 2001;88:853-7.

115. Cheng J, Cabeen WR, Scheinman MM: Right atrial flutter due to lower loop reentry. Mechanism and anatomic substrate. Circulation 1999;99:1700-5.
116. Kall JG, Rubenstein DS, Douglas EK, et al: Atypical atrial flutter originating in the right atrial free wall. Circulation 2000;101: 270-9.
117. Jaïs P, Shah DC, Haïssaguerre M, et al: Mapping and ablation of left atrial flutters. Circulation 2000;101:2928-34.
118. Roithinger FX, Karch MR, Steiner PR, et al: Relationship between atrial fibrillation and typical atrial flutter in humans: Activation sequence changes during spontaneous conversion. Circulation 1997;96:3484-91.
119. Lu TM, Tai CT, Hsoej MH, et al: Electrophysiologic characteristics in initiation of paroxysmal atrial fibrillation from a focal area. J Am Coll Cardiol 2001;37:1658-64.
120. Zrenner B, Ndrepepa G, Karch MR, et al: Electrophysiologic characteristics of paroxysmal and chronic atrial fibrillation in human right atrium. J Am Coll Cardiol 2001;38:1143-9.
121. Garrey WE: Auricular fibrillation. Physiol Rev 1924;4:215-50.
122. Lewis T, Feil HS, Stroud WD: Observations upon flutter and fibrillation. II. The nature of auricular flutter. Heart 1920;7: 191-233.
123. Moe GK, Abildskov JA: Atrial fibrillation as a self-sustained arrhythmia independent of focal discharge. Am Heart J 1959; 58:59-70.
124. Allessie MA, Bonke FI, Schopman FJ: Circus movement in rabbit atrial muscle as a mechanism of tachycardia. III. The "leading circle" concept: A new model of circus movement in cardiac tissue without the involvement of an anatomical obstacle. Circ Res 1977;41:9-18.
125. Wang Z, Page P, Nattel S: Mechanism of flecainide's antiarrhythmic action in experimental atrial fibrillation. Circ Res 1992;71:271-87.
126. Rensma PL, Allessie MA, Lammers WJ, et al: Length of excitation wave and susceptibility to reentrant atrial arrhythmias in normal conscious dogs. Circ Res 1988;62:395-410.
127. Wijffels MC, Kirchhof CJ, Dorland R, et al: Atrial fibrillation begets atrial fibrillation: A study in awake chronically instrumented goats. Circulation 1995;92:1954-68.
128. Nattel S: Atrial electrophysiological remodeling caused by rapid atrial activation: Underlying mechanisms and clinical relevance to atrial fibrillation. Cardiovasc Res 1999;42:298-308.
129. Yue L, Feng J, Gaspo R, et al: Ionic remodeling underlying action potential changes in a canine model of atrial fibrillation. Circ Res 1997;81:512-25.
130. Nattel S, Li D, Yue L: Basic mechanisms of atrial fibrillation—very new insights into very old ideas. Annu Rev Physiol 2000;62:51-77.
131. Li D, Melnyk P, Feng J, et al: Effects of experimental heart failure on atrial cellular and ionic electrophysiology. Circulation 2000;101:2631-8.
132. Waktare J, Camm J: Atrial Fibrillation. London, Martin Duniz, 2000.
133. Waktare JEP, Malik M: Holter, loop recorder, and event counter capabilities of implanted devices. Pacing Clin Electrophysiol 1997;20:2658-69.
134. Hoffmann E, Janko S, Hahnewald S, et al: The atrial fibrillation therapy (AFT) trial: Novel information on dominant triggers of paroxysmal atrial fibrillation [abstract]. Circulation 2000;102: II481-II482.
135. Kannel WB, Abbott RD, Savage DD, McNamara PM: Epidemiologic features of chronic atrial fibrillation: The Framingham Study. N Engl J Med 1982;306:1018-22.
136. Ciaccheri M, Cecchi F, Arcangelli C, et al: Occult thyrotoxicosis in patients with chronic and paroxysmal isolated atrial fibrillation. Clin Cardiol 1984;7:413-16.
137. Sawin CT, Geller A, Wolf PA, et al: Low serum thyrotropin concentrations as a risk factor for atrial fibrillation in older persons. N Engl J Med 1994;331:1249-52.
138. Krahn AD, Klein GJ, Kerr CR, et al, for the Canadian Registry of Atrial Fibrillation Investigators: How useful is thyroid function testing in patients with recent-onset atrial fibrillation? Arch Intern Med 1996;156:2221-4.
139. Dernellis J, Panaretou M: C-reactive protein and paroxysmal atrial fibrillation: Evidence of the implication of an inflammatory process in paroxysmal atrial fibrillation. Acta Cardiol 2001; 56:375-80.
140. Chung MK, Martin DO, Sprecher D, et al: C-reactive protein elevation in patients with atrial arrhythmias: Inflammatory mechanisms and persistence of atrial fibrillation. Circulation 2001;104:2886-91.
141. Manning WJ, Silverman DI, Katz SE, et al: Impaired left atrial mechanical function after cardioversion: Relation to the duration of atrial fibrillation. J Am Coll Cardiol 1994;23:1535-40.
142. The Stroke Prevention in Atrial Fibrillation Investigators Committee on Echocardiography. Transesophageal echocardiographic correlates of thromboembolism in high-risk patients with nonvalvular atrial fibrillation. Ann Intern Med 1998; 128:639-47.
143. Chen SA, Chiang CE, Yang CJ, et al: Sustained atrial tachycardia in adult patients. Electrophysiological characteristics, pharmacological response, possible mechanisms, and effects of radiofrequency ablation. Circulation 1994;90:1262-78.
144. Chen SA, Tai CT, Chiang CE, et al: Focal atrial tachycardia: Reanalysis of the clinical and electrophysiologic characteristics and prediction of successful radiofrequency ablation. J Cardiovasc Electrophysiol 1998;9:355-65.
145. Knight BP, Zivin A, Souza J, et al: A technique for the rapid diagnosis of atrial tachycardia in the electrophysiology laboratory. J Am Coll Cardiol 1999;33:775-81.
146. Knight BP, Ebinger M, Oral H, et al: Diagnostic value of tachycardia features and pacing maneuvers during paroxysmal supraventricular tachycardia. J Am Coll Cardiol 2000;36:574-82.
147. Kalman JM, Olgin JE, Karch MR, et al: "Cristal tachycardias": Origin of right atrial tachycardias from the crista terminalis identified by intracardiac echocardiography. J Am Coll Cardiol 1998;31:451-9.
148. Nogami A, Suguta M, Tomita T, et al: Novel form of atrial tachycardia originating at the atrioventricular annulus. Pacing Clin Electrophysiol 1998;21:2691-4.
149. Morton JB, Sanders P, Das A, et al: Focal atrial tachycardia arising from the tricuspid annulus: Electrophysiologic and electrocardiographic characteristics. J Cardiovasc Electrophysiol 2001;12:653-9.
150. Chen SA, Hsieh MH, Tai CT, et al: Initiation of atrial fibrillation by ectopic beats originating from the pulmonary veins: Electrophysiologic characteristics, pharmacologic responses, and effects of radiofrequency catheter ablation. Circulation 1999;100:1879-86.
151. Tsai CF, Tai CT, Hsieh MH, et al: Initiation of atrial fibrillation by ectopic beats originating from the superior vena cava: Electrophysiological characteristics and results of radiofrequency ablation. Circulation 2000;102:67-74.
152. Nakagawa H, Shah N, Matsudaira K, et al: Characterization of reentrant circuit in macroreentrant right atrial tachycardia after surgical repair of congenital heart disease: Isolated channels between scars allow "focal" ablation. Circulation 2001;103:699-709.
153. Cosio FG, Arribas F, Palacios J, et al: Fragmented electrograms and continuous electrical activity in atrial flutter. Am J Cardiol 1986;57:1309-14.
154. Olshansky B, Okumura K, Henthorn EW, et al: Characterization of double potentials in human atrial flutter: Studies during transient entrainment. J Am Coll Cardiol 1990;15:833-41.
155. Saoudi N, Atallah G, Kirkorian G, et al: Catheter ablation of the atrial myocardium in human type I atrial flutter. Circulation 1990;81:762-71.
156. Chen SA, Chiang CE, Wu TJ, et al: Radiofrequency catheter ablation of common atrial flutter: Comparison of electrophysiologically guided focal ablation technique and linear ablation technique. J Am Coll Cardiol 1996;27:860-8.
157. Olgin JE, Kalman JM, Fitzpatrick AP, et al: Role of right atrial endocardial structure as barriers to conduction during human type I atrial flutter: Activation and entrainment mapping guided by intracardiac echocardiography. Circulation 1995;92:1839-48.
158. Tai CT, Chen SA, Chen YJ, et al: Conduction properties of the crista terminalis in patients with typical atrial flutter: Basis for a line of block in the reentrant circuit. J Cardiovasc Electrophysiol 1998;9:811-9.
159. Friedman PA, Luria D, Fenton AM, et al: Global right atrial mapping of human atrial flutter: The presence of posteromedial (sinus venosa region) functional block and double potentials: A study in biplane fluoroscopy and intracardiac echocardiography. Circulation 2000;101:1568-77.

160. Yang Y, Chen J, Bochoeyer A, et al: Atypical right atrial flutter patterns. Circulation 2001;103:3092-8.
161. Hwang C, Karaguezian HS, Chen PS: Idiopathic paroxysmal atrial fibrillation induced by a focal discharge mechanism in the left superior pulmonary vein: Possible roles of the ligament of Marshall. J Cardiovasc Electrophysiol 1999;10:636-48.
162. Konings KTS, Kirchhof CJHJ, Smeets JRLM, et al: High density mapping of electrically induced atrial fibrillation in man. Circulation 1994;89:1665-80.
163. Daoud EG, Knight BP, Weiss R, et al: Effect of verapamil and procainamide on atrial fibrillation-induced electrical remodeling in human. Circulation 1997;5:1542-50.
164. Yu WC, Chen SA, Lee SH, et al: Tachycardia-induced change of atrial refractory period in human: Rate-dependency and effects of antiarrhythmic drugs. Circulation 1998;97:2331-7.
165. Yu WC, Lee SH, Tai CT, et al: Reversal of atrial electrical remodeling following cardioversion of long-standing atrial fibrillation in man. Cardiovasc Res 1999;42:470-6.
166. Lévy S, Ricard P, Guenoun M, et al: Low energy internal cardioversion of spontaneous atrial fibrillation. Circulation 1997; 96:253-9.
167. Van Gelder IC, Tuinenberg AE, Schoonderwoerd BS, et al: Pharmacologic versus direct-current electrical cardioversion of atrial flutter and fibrillation. Am J Cardiol 1999;84: 147R-151R.
168. Van Gelder IC, Crijns HJGM, Van Gilst WH, et al: Prediction of uneventful cardioversion and maintenance of sinus rhythm from direct-current electrical cardioversion of chronic atrial fibrillation and flutter. Am J Cardiol 1991;68:41-6.
169. Reiffel JA: Drug choices in the treatment of atrial fibrillation. Am J Cardiol 2000;85:12D-19D.
170. Pritchett ELC, Page RL, Connolly SJ, et al: Antiarrhythmic effects of azimilide in atrial fibrillation: Efficacy and dose-response. J Am Coll Cardiol 2000;36:794-802.
171. Torp-Pedersen C, Møller M, Bloch-Thomsen PE, et al, for the Danish Investigations of Arrhythmia and Mortality on Dofetilide Study Group: Dofetilide in patients with congestive heart failure and left ventricular dysfunction. N Engl J Med 1999;341:857-65.
172. Kowey PR, Vanderlught JT, Luderer JR: Safety and risk/benefit analysis of ibutilide for acute conversion of atrial fibrillation/flutter. Am J Cardiol 1996;78:46A-52A.
173. Connolly SJ: Evidence-based analysis of amiodarone efficacy and safety. Circulation 1999;100:2025-34.
174. Carlsson J, Neuzner J, Rosenberg YD: Therapy of atrial fibrillation: Rhythm control versus rate control. Pacing Clin Electrophysiol 2000;23:891-903.
175. The Atrial Fibrillation Follow-up Investigation of Rhythm Management (AFFIRM) Investigators. A comparison of rate control and rhythm control in patients with atrial fibrillation. N Engl J Med 2002;347:1825-33.
176. Hart RG, Pearce LA, McBride R, et al, for the Stroke Prevention in Atrial Fibrillation (SPAF) Investigators: Factors associated with ischemic stroke during aspirin therapy in atrial fibrillation: Analysis of 2012 participants in the SPAF I-III clinical trials Stroke Prevention in Atrial Fibrillation (SPAF) Investigators. Stroke 1999;30:1223-9.
176a. Strauss SE, Majumdar SR, McAlister FA: New evidence for stroke prevention: Scientific review. JAMA 2002;288:1388-95.
177. Atrial Fibrillation Investigators. Echocardiographic predictors of stroke in patients with atrial fibrillation: A prospective study of 1,066 patients from three clinical trials. Arch Intern Med 1998;158:1316-20.
178. Li D, Shinagawa K, Pang L, et al: Effects of angiotensin-converting enzyme inhibition on the development of atrial fibrillation substrate in dogs with ventricular tachycardia-induced congestive heart failure. Circulation 2001;104:2608-14.
179. Hernandez-Madrid A, Bueno MG, Rebollo JG, et al: Use of irbesartan to maintain sinus rhythm in patients with long-lasting persistent atrial fibrillation: A prospective and randomized study. Circulation 2002;39:106:331-6.
180. Kay GN, Ellenbogen KA, Giudici M, et al: The Ablate and Pace Trial: A prospective study of catheter ablation of the AV conduction system and permanent pacemaker implantation for treatment of atrial fibrillation. J Interv Cardiac Electrophysiol 1998;2:121-35.
181. Lévy S, Richard P, Lau CP, et al: Multicenter low energy transvenous atrial defibrillation (XAD) trial results in different subsets of atrial fibrillation. J Am Coll Cardiol 1997;27:750-5.
182. Wellens HJJ, Lau CP, Lüderitz B, et al: Atrioverter: An implantable device for the treatment of atrial fibrillation. Circulation 1998;98:1651-6.
183. Andersen HR, Nielsen JC, Thomsen PEB, et al: Long-term follow-up of patients from a randomised trial of atrial versus ventricular pacing for sick sinus syndrome. Lancet 1997;350:1210-6.
184. Bailin SJ, Adler S, Giudici M: Prevention of chronic atrial fibrillation by pacing in the region of Bachmann's bundle: Results of a multicenter randomized trial. J Cardiovasc Electrophysiol 2001;12:912-7.
185. D'Allonnes GR, Pavin D, Leclercq C, et al: Long-term effects of biatrial synchronous pacing to prevent drug-refractory atrial tachyarrhythmia: A nine-year experience. J Cardiovasc Electrophysiol 2000;11:1081-91.
186. Carlson MD, Gold MR, Ip J, et al, for the ADOPT Investigators: A new pacemaker algorithm for the treatment of atrial fibrillation: results of the Atrial Dynamic Overdrive Pacing Trial (ADOPT). J Am Coll Cardiol 2003;20;42:627-33.
187. Danias PG, Caulfield TA, Weigner MJ, et al: Likelihood of spontaneous conversion of atrial fibrillation to sinus rhythm. J Am Coll Cardiol 1998;31:588-92.
188. Slavik RS, Tisdale JE, Borzak S: Pharmacologic conversion of atrial fibrillation: A systematic review of available evidence. Prog Cardiovasc Dis 2001;44:121-52.
189. Boriani G, Biffi M, Capucci A, et al: Oral propafenone to convert recent-onset atrial fibrillation in patients with and without underlying heart disease: A randomized, controlled trial. Ann Intern Med 1997;126:621-5.
190. Azpitarte J, Alvarez M, Baun O, et al: Value of single oral loading dose of propafenone in converting recent-onset atrial fibrillation: Results of a randomized double-blind, controlled study. Eur Heart J 1997;18:1649-54.
191. Capucci A, Boriani G, Botto GL, et al: Conversion of recent-onset atrial fibrillation to sinus rhythm by a single oral loading dose of propafenone or flecainide. Am J Cardiol 1994;74:503-5.
192. Bianconi L, Mennuni M, and PAFIT-3 Investigators: Comparison between propafenone and digoxin administered intravenously to patients with acute atrial fibrillation: Am J Cardiol 1998;82:584-8.
193. Crijns HJ, van Wijk LM, van Gilst WH, et al: Acute conversion of atrial fibrillation to sinus rhythm: Clinical efficacy of flecainide acetate. Comparison of two regimens. Eur Heart J 1988;9:634-8.
194. Crijns HJGH, van Gelder IC, Kingma JH, et al: Atrial flutter can be terminated by a class III antiarrhythmic drug, but not by a class IC drug. Eur Heart J 1995;15:1403-8.
195. Kingma JH, Suttorp MJ: Acute pharmacologic conversion of atrial fibrillation and flutter: The role of flecainide, propafenone, and verapamil. Am J Cardiol 1992;70:56A-60A; discussion 60A-61A.
196. Kuck KH, Kunze KP, Schluter M, Duckeck W: Encainide versus flecainide for chronic atrial and junctional ectopic tachycardia. Am J Cardiol 1988;62:37L-44L.
197. Stambler BS, Wood MA, Ellenbogen KA, et al, and the Ibutilide Repeat Dose Investigators: Efficacy and safety of repeated intravenous doses of ibutilide for rapid conversion of atrial flutter or fibrillation. Circulation 1996;94:1613-21.
198. Volgman AS, Carberry PA, Stambler B, et al: Conversion efficacy and safety of intravenous ibutilide compared with intravenous procainamide in patients with atrial flutter or fibrillation. J Am Coll Cardiol 1998;31:1414-19.
199. Vos MA, Golitsyn SR, Stangl K, et al, for the Ibutilide/Sotalol Comparator Study Group: Superiority of ibutilide (a new class III agent) over d,l-sotalol in converting atrial flutter and atrial fibrillation. Heart 1998;79:568-75.
200. Norgaard BL, Wachtell K, Christensen PD, et al: Efficacy and safety of intravenously administered dofetilide in acute termination of atrial fibrillation and flutter: A multicenter, randomized, double blind, placebo-controlled study. Am Heart J 1999;137:1062-9.
201. Pedersen OD, Bagger H, Keller N, et al, for the Danish Investigations of Arrhythmia and Mortality ON Dofetilide Study Group: Efficacy of dofetilide in the treatment of atrial fibrillation-flutter in patients with reduced left ventricular function: A Danish Investigations of Arrhythmia and Mortality on Dofetilide (DIAMOND) Substudy. Circulation 2001;104:292-6.

202. Singh S, Zoble RG, Yellen L, et al, for the Dofetilide Atrial Fibrillation Investigators: Efficacy and safety of oral dofetilide in converting to and maintaining sinus rhythm in patients with chronic atrial fibrillation or atrial flutter: The Symptomatic Atrial Fibrillation Investigative Research on Dofetilide (SAFIRE-D) Study. Circulation 2000;102:2385-90.
203. Greenbaum RA: European and Australian Multicenter Evaluative Research on Atrial Fibrillation Dofetilide (EMERALD). Presented at 71st Scientific Sessions of the American Heart Association. Circulation 1999;99:2486-91.
204. Sung RJ, Tan HL, Karagounis L, et al, for the Sotalol Multicenter Study Group: Intravenous sotalol for termination of supraventricular tachycardia and atrial fibrillation and flutter: A multicenter, randomized, double-blind, placebo-controlled study. Am Heart J 1995;129:739-48.
205. Chevalier P, Durand-Dubief A, Burri H, et al: Amiodarone versus placebo and classic drugs for cardioversion of recent-onset atrial fibrillation: a meta-analysis. J Am Coll Cardiol 2003;41: 255-62.
206. Vardas PE, Kochiadakis GE, Igoumenidis NE, et al: Amiodarone as a first-choice drug for restoring sinus rhythm in patients with atrial fibrillation: A randomized, controlled study. Chest 2000;117:1538-45.
207. Bianconi L, Castro A, Dinelli M, et al: Comparison of intravenously administered dofetilide versus amiodarone in the acute termination of atrial fibrillation and flutter. A multicentre, randomized, double-blind, placebo-controlled study. Eur Heart J 2000;21:1265-73.
208. Deedwania PC, Singh BN, Ellenbogen K, for the Department of Veterans Affairs CHF-STAT Investigators: Spontaneous conversion and maintenance of sinus rhythm by amiodarone in patients with heart failure and atrial fibrillation: Observations from the Veterans Affairs Congestive Heart Failure Survival Trial of Antiarrhythmic Therapy (CHF-STAT). Circulation 1998;98:2574-9.
209. Hohnloser SH, Van de LA, Baedeker F: Efficacy and proarrhythmic hazards of pharmacologic cardioversion of atrial fibrillation: Prospective comparison of sotalol versus quinidine. J Am Coll Cardiol 1995;26:852-8.
210. Halinen MO, Huttunen M, Paakkinen S, Tarssanen L: Comparison of sotalol with digoxin-quinidine for conversion of acute atrial fibrillation to sinus rhythm (the Sotalol-Digoxin-Quinidine Trial). Am J Cardiol 1995;76:495-8.
211. Kerin NZ, Faitel K, Naini M: The efficacy of intravenous amiodarone for the conversion of chronic atrial fibrillation: Amiodarone vs quinidine for conversion of atrial fibrillation. Arch Intern Med 1996;156:49-53.
212. Innes GD, Vertesi L, Dillon EC, Metcalfe C: Effectiveness of verapamil-quinidine versus digoxin-quinidine in the emergency department treatment of paroxysmal atrial fibrillation. Ann Emerg Med 1997;29:126-34.
213. Kochiadakis GE, Igoumenidis NE, Solomou MC, et al: Conversion of atrial fibrillation to sinus rhythm using acute intravenous procainamide infusion. Cardiovasc Drugs Ther 1998;12:75-81.
214. Mattioli AV, Lucchi GR, Vivoli D, Mattioli G: Propafenone versus procainamide for conversion of atrial fibrillation to sinus rhythm. Clin Cardiol 1998;21:763-6.
215. Madrid AH, Moro C, Marin-Huerta E, et al: Comparison of flecainide and procainamide in cardioversion of atrial fibrillation. Eur Heart J 1993;14:1127-31.
216. Markowitz SM, Stein KM, Mittal S, et al: Differential effects of adenosine on focal and macroreentrant atrial tachycardia. J Cardiovasc Electrophysiol 1999;10:489-502.
217. Lerman BB, Ellenbogen KA, Kadish A, et al: Electrophysiologic effects of a novel selective adenosine A1 agonist (CVT-510) on atrioventricular nodal conduction in humans. J Cardiovasc Pharmacol Ther 2001;6:237-45.
218. Prystowsky EN, Niazi I, Curtis AB, et al: Termination of paroxysmal supraventricular tachycardia by tecadenoson (CVT-510), a novel A1-adenosine receptor agonist. J Am Coll Cardiol 2003; 42:1098-102.
219. The Digitalis in Acute Atrial Fibrillation (DAAF) Trial Group. Results of a randomized, placebo-controlled multicentre trial in 239 patients. Eur Heart J 1997;18:649-54.
220. Sticherling C, Oral H, Horrocks J, et al: Effects of digoxin on acute, atrial fibrillation-induced changes in atrial refractoriness. Circulation 2000;102:2503-8.
221. Tieleman RG, Blaau Y, Van Gelder IC, et al: Digoxin delays recovery from tachycardia-induced electrical remodeling of the atria. Circulation 1999;100:1836-42.
222. Elhendy A, Gentile F, Khanderia BK, et al: Predictors of unsuccessful electrical cardioversion in atrial fibrillation. Am J Cardiol 2002;89:83-6.
223. Kerber RE, Martins JB, Kienzle MG, et al: Energy, current and success in defibrillation and cardioversion: Clinical studies using an automated impedance-based adjustment method. Circulation 1988;77:1038-46.
224. Carlsson J, Tebbe U, Rox J, et al, for the ALKK Study Group: Cardioversion of atrial fibrillation in the elderly. Am J Cardiol 1996;78:1380-4.
225. Botto GL, Politi A, Bonini W, et al: External cardioversion of atrial fibrillation: Role of paddle position on technical efficacy and energy requirements. Heart 1999;82:726-30.
226. Kerber RE, Jensen SR, Grayzel J, et al: Elective cardioversion: Influence of paddle-electrode location and size on success rate and energy requirements. N Engl J Med 1981;305:658-62.
227. Mittal S, Ayati S, Stein KM, et al: Transthoracic cardioversion of atrial fibrillation: Comparison of rectilinear biphasic versus damped sine wave monophasic shocks. Circulation 2000;101: 1282-7.
228. Page RL, Kerber RE, Russell JK, et al, for the BiCard Investigators: Biphasic versus monophasic shock waveform for conversion of atrial fibrillation. J Am Coll Cardiol 2002;39:1956-63.
229. Sopher SM, Murgatroyd FD, Slade AK, et al: Low energy internal cardioversion of atrial fibrillation resistant to transthoracic shocks. Heart 1996;75:635-8.
230. Alt E, Ammer R, Schmitt C, et al: A comparison of treatment of atrial fibrillation with low-energy intracardiac cardioversion and conventional external cardioversion. Eur Heart J 1997;18: 1796-804.
231. Yu WC, Lin YK, Tai CT, et al: Early recurrence of atrial fibrillation after external cardioversion. Pacing Clin Electrophysiol 1999;22:1614-9.
232. Timmermans C, Rodriguez LM, Smeets JLRM, Wellens HJJ: Immediate reinitiation of atrial fibrillation following internal defibrillation. J Cardiovasc Electrophysiol 1998;9:122-8.
233. Tse HF, Lau CP, Ayers GM: Atrial pacing for suppression of early reinitiation of atrial fibrillation after successful internal cardioversion. Eur Heart J 2000;21:1167-76.
234. Bianconi L, Mennuni M, Lukic V, et al: Effects of oral propafenone administration before electrical cardioversion of chronic atrial fibrillation: A placebo-controlled study. J Am Coll Cardiol 1996;28:700-6.
235. Boriani J, Biffi M, Capucci A, et al: Favorable effects of flecainide in transvenous internal cardioversion of atrial fibrillation. J Am Coll Cardiol 1999;33:333-41.
236. Oral H, Souza JJ, Michand GF, et al: Facilitating transthoracic cardioversion of atrial fibrillation with ibutilide pretreatment. N Engl J Med 1999;340:1849-54.
237. Capucci A, Villani GQ, Aschieri D, et al: Oral amiodarone increases the efficacy of direct-current cardioversion in restoration of sinus rhythm in patients with chronic atrial fibrillation. Eur Heart J 2000;21:66-73.
238. Botto GL, Belotti G, Cirò A, et al, on behalf of the VeRAF Study Group: Verapamil around electrical cardioversion of persistent atrial fibrillation: Results from the VeRAF Study [abstract]. J Am Coll Cardiol 2002;39:91A.
239. Van Noord T, Van Gelder IC, Tieleman RG, et al: VERDICT: The Verapamil versus Digoxin Cardioversion Trial: A randomized study on the role of calcium lowering for maintenance of sinus rhythm after cardioversion of persistent atrial fibrillation. J Cardiovasc Electrophysiol 2001;12:766-69.
240. Villani GQ, Piepoli MF, Terraciano C, Capucci A: Effects of diltiazem pretreatment on direct-current cardioversion in patients with persistent atrial fibrillation: a single-blind, randomized, controlled study. Am Heart J 2000;140:437-43.
241. Arnold AZ, Mick MJ, Mazurek RP, et al: Role of prophylactic anticoagulation for direct current cardioversion in patients with atrial fibrillation and flutter. J Am Coll Cardiol 1992;19:851-5.

242. Rabbino MD, Likoff W, Dreifus LS: Complications and limitations of direct current countershock. JAMA 1964;190:417-20.
243. Prakash A, Saksena S, Mathew P, Krol RB: Internal atrial defibrillation: Effect on sinus and atrioventricular function and implanted cardiac pacemakers. Pacing Clin Electrophysiol 1997;20:2434-41.
244. Waxman HL, Myerburg RJ, Appel R, Sung RJ: Verapamil for control of ventricular rate in paroxysmal supraventricular tachycardia and atrial fibrillation or flutter: A double-blind randomized cross-over study. Ann Intern Med 1981;94:1-6.
245. Ellenbogen KA, Dias VC, Plumb VJ, et al: A placebo-controlled trial of continuous intravenous diltiazem infusion for 24-hour rate control during atrial fibrillation and atrial flutter: A multicenter study. J Am Coll Cardiol 1991;185:891-7.
246. Platia EV, Michelson EL, Porterfield JK, Das G: Esmolol versus verapamil in the acute treatment of atrial fibrillation or atrial flutter. Am J Cardiol 1989;63:925-9.
247. Byrd RC, Sung RJ, Marks J, Parmley WW: Safety and efficacy of esmolol (ASL-8052: An ultrashort-acting beta-adrenergic blocking agent) for control of ventricular rate in supraventricular tachycardias. J Am Coll Cardiol 1984;3:394-9.
248. Schreck DM, Rivera AR, Tricarico VJ: Emergency management of atrial fibrillation and flutter: Intravenous diltiazem versus intravenous digoxin. Ann Emerg Med 1997;29:135-40.
249. Pinter A, Dorian P, Paquette M, et al: Left ventricular performance during acute rate control in atrial fibrillation: The importance of heart rate and agent used. J Cardiovasc Pharmacol Ther 2003;8:17-24.
250. Clemo HF, Wood MA, Gilligan DM, Ellenbogen KA: Intravenous amiodarone for acute heart rate control in the critically ill patient with atrial tachyarrhythmias. Am J Cardiol 1998;81:594-8.
251. Coumel P: Autonomic influences in atrial tachyarrhythmias. J Cardiovasc Electrophysiol 1996;7:999-1007.
252. Crystal E, Connolly SJ, Sleik K, et al: Interventions on prevention of postoperative atrial fibrillation in patients undergoing heart surgery: A meta-analysis. Circulation 2002;106:75-80.
253. Kühlkamp V, Schirdewan A, Stangl K, et al: Use of metoprolol CR/XL to maintain sinus rhythm after conversion from persistent atrial fibrillation: A randomized, double-blind, placebo-controlled study. J Am Coll Cardiol 2000;36:139-46.
254. Plewan A, Lehmann G, Ndrepepa G, et al: Maintenance of sinus rhythm after electrical cardioversion of persistent atrial fibrillation: Sotalol vs bisoprolol. Eur Heart J 2001;22:1504-10.
255. McMurray JJ, Dargie HJ, Ford I, et al: Carvedilol reduces supraventricular and ventricular arrhythmias after myocardial infarction: Evidence form the CAPRICORN study [abstract]. Circulation 2001;104:II-700.
256. Pritchett ELC, McCarthy EA, Wilkinson WE: Propafenone treatment of symptomatic paroxysmal supraventricular arrhythmias: A randomized, placebo-controlled, crossover trial in patients tolerating oral therapy. Ann Intern Med 1991;114:539-44.
257. Stroobandt R, Stiels B, Hoebrechts R, on behalf of the Propafenone Atrial Fibrillation Trial Investigators: Propafenone for conversion and prophylaxis of atrial fibrillation. Am J Cardiol 1997;79:418-23.
258. UK Propafenone PSVT Study Group: A randomized, placebo-controlled trial of propafenone in the prophylaxis of paroxysmal supraventricular tachycardia and paroxysmal atrial fibrillation. Circulation 1995;92:2550-7.
259. Pritchett EL, Page RL, Carlson M, et al, for Rythmol Atrial Fibrillation Trial (RAFT) Investigators: Efficacy and safety of sustained-release propafenone (propafenone SR) for patients with atrial fibrillation. Am J Cardiol 2003;92:941-6.
260. Meinertz T, Lip GYH, Lombardi F, et al, on behalf of the ERAFT Investigators: Efficacy and safety of propafenone sustained release in the prophylaxis of symptomatic paroxysmal atrial fibrillation (The European Rythmol/Rytmonorm Atrial Fibrillation Trial [ERAFT] Study). Am J Cardiol 2002;90:1300-6.
261. Pietersen AH, Hellemann H, for the Danish-Norwegian Flecainide Multicenter Atrial Fibrillation Study Group: Usefulness of flecainide for prevention of paroxysmal atrial fibrillation and flutter. Am J Cardiol 1991;67:713-7.
262. Naccarelli GV, Dorian P, Hohnloser S, Coumel P, for the Flecainide Multicenter Atrial Fibrillation Study Group: Prospective comparison of flecainide versus quinidine for the treatment of paroxysmal atrial fibrillation/flutter. Am J Cardiol 1996;77:53A-59A.
263. Chimienti M, Cullen MT Jr, Casadei G: Safety of long-term flecainide and propafenone in the management of patients with symptomatic paroxysmal atrial fibrillation: Report from the Flecainide and Propafenone Italian Study Investigators. Am J Cardiol 1996;77:60A-75A.
264. Creamer JE, Nathan AW, Camm AJ: Successful treatment of atrial tachycardias with flecainide acetate. Br Heart J 1985; 53:164-6.
265. Coumel P, Leclercq JF, Assayag P: European experience with the antiarrhythmic efficacy of propafenone for supraventricular and ventricular arrhythmias. Am J Cardiol 1984;54:60D-66D.
266. Benditt DG, Williams JH, Jin J, et al, for the d,l-Sotalol Atrial Fibrillation/Flutter Study Group: Maintenance of sinus rhythm with oral d,l-sotalol therapy in patients with symptomatic atrial fibrillation and/or atrial flutter. Am J Cardiol 1999;84:270-7.
267. Bellandi F, Simonetti I, Leoncini M, et al: Long-term efficacy and safety of propafenone and sotalol for the maintenance of sinus rhythm after conversion of recurrent symptomatic atrial fibrillation. Am J Cardiol 2001;88:640-5.
268. De Paola AAV, Veloso HH, for the SOCESP Investigators: Efficacy and safety of sotalol versus quinidine for the maintenance of sinus rhythm after conversion of atrial fibrillation. Am J Cardiol 1999;84:1033-7.
269. Fetsch T, Bauer P, Engberding R, et al, for the Prevention of Atrial Fibrillation after Cardioversion Investigators: Prevention of atrial fibrillation after cardioversion: Results of the PAFAC trial. Eur Heart J 2004;25:1385-94.
270. Roy D, Talajic M, Dorian P, et al, for the Canadian Trial of Atrial Fibrillation Investigators: Amiodarone to prevent recurrence of atrial fibrillation. N Engl J Med 2000;342:913-20.
271. The AFFIRM First Antiarrhythmic Drug Substudy Investigators. Maintenance of sinus rhythm in patients with atrial fibrillation: An AFFIRM Substudy of First Antiarrhythmic Drug. J Am Coll Cardiol 2003;42:20-9.
272. Wanless RS, Anderson K, Joy M, Joseph SP: Multicenter comparative study of the efficacy and safety of sotalol in the prophylactic treatment of patients with paroxysmal supraventricular tachyarrhythmias. Am Heart J 1997;133:441-6.
273. Touboul P, Brugada J, Capucci A, et al: Dronedarone for prevention of atrial fibrillation: A dose-ranging study. Eur Heart J 2003;24:1481-7.
274. Connolly SJ, Schnell DJ, Page RL, et al: Dose-response relations of azimilide in the management of symptomatic, recurrent atrial fibrillation. Am J Cardiol 2001;88:974-9.
274a. Page RL, Connolly SJ, Wilkinson WE, et al, and Azimilide Supraventricular Arrhythmia Program (ASAP) Investigators: Antiarrhythmic effects of azimilide in paroxysmal supraventricular tachycardia: Efficacy and dose-response. Am Heart J 2002;143:643-9.
275. Camm AJ, Pratt CM, Schwartz PJ, et al, on behalf of the Azimilide Postinfarct Survival Evaluation (ALIVE) Investigators: Mortality in patients with recent myocardial infarction: A randomized, placebo-controlled trial of azimilide using heart rate variability for risk stratification. Circulation 2004;109: 990-6.
275a. Pratt CM, Singh S, Al-Khalidi HR, et al, on behalf of the ALIVE Investigators: The efficacy of azimilide in the treatment of atrial fibrillation in the presence of left ventricular systolic dysfunction: Results from the Azimilide Postinfarct Survival Evaluation (ALIVE) trial. J Am Coll Cardiol 2004;43:1211-6.
276. Hohnloser SH, Kuck KH, Lilienthal J, for the PIAF Investigators: Rhythm or rate control in atrial fibrillation—Pharmacological Intervention in Atrial Fibrillation (PIAF): A randomised trial. Lancet 2000;356:1789-94.
277. Carlsson J, Miketic S, Windeler J, et al: Randomized trial of rate-control versus rhythm-control in persistent atrial fibrillation: Results from The Strategies of Treatment in Atrial Fibrillation (STAF) Study. J Am Coll Cardiol 2003;41:1690-6.
278. Van Gelder IC, Hagens VE, Bosker HA, et al, for the Rate Control versus Electrical Cardioversion for Persistent Atrial Fibrillation Study Group: A comparison of rate control and rhythm control in patients with recurrent persistent atrial fibrillation. New Engl J Med 2002;347:1834-40.

279. Opolski G, Torbicki A, Kosior D, et al. Rhythm control versus rate control in patients with persistent atrial fibrillation: Results of the HOT CAFE Polish Study. Kardiol Pol 2003;59:1-16.
280. Vora AM, Goyal VS, Naik AM, et al: Maintenance of sinus rhythm by amiodarone is superior to ventricular rate control in rheumatic atrial fibrillation: A blinded placebo controlled study [abstract]. Pacing Clin Electrophysiol 2001;24:546.
280a. The AFFIRM Investigators. Relationships between sinus rhythm, treatment, and survival in the Atrial Fibrillation Follow-up Investigation of Rhythm Management (AFFIRM) study. Circulation 2004;109:1509-13.
281. Wasmer K, Oral H, Sticherling C, et al: Assessment of ventricular rate in patients with chronic atrial fibrillation [abstract]. J Am Coll Cardiol 2001;37:93A.
282. Weerasooriya R, Davis M, Powell A, et al: The Australian Intervention Randomized Control of Rate in Atrial Fibrillation Trial (AIRCRAFT). J Am Coll Cardiol 2003;41:1697-1702.
282a. Olshansky B, Rosenfeld LE, et al, and AFFIRM Investigators: The Atrial Fibrillation Follow-up Investigation of Rhythm Management (AFFIRM) study: Approaches to control rate in atrial fibrillation. J Am Coll Cardiol 2004;43:1201-8.
283. Kamath S, Blann AD, Lip GYH: Platelets and atrial fibrillation. Eur Heart J 2001;22:2233-42.
284. Klein AL, Grimm RA, Murray RD, et al, for the Assessment of Cardioversion Using Transesophageal Echocardiography Investigators: Use of transesophageal echocardiography to guide cardioversion in patients with atrial fibrillation. N Engl J Med 2001;344:1411-20.
285. Seidl K, Rameken M, Drögemüller A, et al: Embolic events in patients with atrial fibrillation and effective anticoagulation: Value of transesophageal echocardiography to guide direct-current cardioversion: Final results of the Ludwigshaften Observational Cardioversion Study. J Am Coll Cardiol 2002;39:1436-42.
286. Petersen P, Boysen G, Godtfredsen L, et al: Placebo-controlled, randomized trial of warfarin and aspirin for prevention of thromboembolic complications in chronic atrial fibrillation: The Copenhagen AFASAK study. Lancet 1989;1:175-9.
287. Gulløv AL, Koefold BG, Petersen P: Bleeding during warfarin and aspirin therapy in patients with atrial fibrillation: The AFASAK-2 study. Arch Intern Med 1999;159:1322-8.
288. Stroke Prevention in Atrial Fibrillation Investigators. Stroke Prevention in Atrial Fibrillation Study: Final results. Circulation 1991;84:527-39.
289. Stroke Prevention in Atrial Fibrillation Investigators. Warfarin versus aspirin for prevention of thromboembolism in atrial fibrillation: Stroke Prevention in Atrial Fibrillation II Study. Lancet 1994;343:687-91.
290. Stroke Prevention in Atrial Fibrillation Investigators. Adjusted-dose warfarin versus low-intensity, fixed-dose warfarin plus aspirin for high-risk patients with atrial fibrillation: Stroke Prevention in Atrial Fibrillation III randomized clinical trial. Lancet 1996;348:633-8.
291. SPAF III Writing Committee for the Stroke Prevention in Atrial Fibrillation Investigators. Patients with non-valvular atrial fibrillation at low risk of stroke during treatment with aspirin: Stroke Prevention in Atrial Fibrillation III Study. JAMA 1998;279:1273-7.
292. Ezekowitz MD, Bridgers SL, James KE, et al: Warfarin in prevention of stroke associated with nonrheumatic atrial fibrillation: Veterans Affairs Stroke Prevention in Nonrheumatic Atrial Fibrillation Investigators. N Engl J Med 1992;327:1406-12.
293. Boston Area Anticoagulation Trial for Atrial Fibrillation Investigators: The effect of low-dose warfarin on the risk of stroke in patients with nonrheumatic atrial fibrillation. N Engl J Med 1990;323:1505-11.
294. Connolly SJ, Laupacis A, Gent M, et al: Canadian Atrial Fibrillation Anticoagulation (CAFA) Study. J Am Coll Cardiol 1991;18:349-55.
295. Hellemons BSP, Langenberg M, Lodder J, et al: Primary prevention of arterial thromboembolism in nonrheumatic atrial fibrillation in primary care: Randomised controlled trial comparing two intensities of coumarin with aspirin. BMJ 1999;319: 958-64.
296. European Atrial Fibrillation Trial Study Group. Optimal oral anticoagulation therapy in patients with nonrheumatic atrial fibrillation and recent cerebral ischemia. N Engl J Med 1995; 333:5-10.
297. Morocutti C, Amabile G, Fattapposta F, et al: Indobufen versus warfarin in the secondary prevention of major vascular events in nonrheumatic atrial fibrillation: SIFA (Studio Italiano Fibrillazione Atriale) Investigators. JAMA 1998;279:1273-7.
298. Diener H, Cunha L, Forbes C, et al: European Stroke Prevention Study 2: Dipyridamole and acetylsalicylic acid in the prevention of stroke. J Neurol Sci 1996;143:1-13.
299. Posada IS, Barriales V: Alternate-day dosing of aspirin in atrial fibrillation: LASAF Pilot Study Group. Am Heart J 1999;138:137-43.
300. Pengo V, Zasso A, Barbero F, et al: Effectiveness of fixed minidose warfarin in the prevention of thromboembolism and vascular death in nonrheumatic atrial fibrillation. Am J Cardiol 1998;82:433-7.
301. Hart RG, Benavente O, McBride R, Pearce LA: Antithrombotic therapy to prevent stroke in patients with atrial fibrillation: A meta-analysis. Ann Intern Med 1999;131:492-501.
302. Hylek EM, Singer DE: Risk factors for intracranial hemorrhage in patients taking warfarin. Ann Intern Med 1994;120:897-902.
303. Hylek EM, Go AS, Chang Y, et al: Effect of intensity of oral anticoagulation on stroke severity and mortality in atrial fibrillation. N Engl J Med 2003;349:1019-26.
304. Hart RG, Pearce LA, Rothbart RM, et al: Stroke with intermittent atrial fibrillation: Incidence and predictors during aspirin therapy. J Am Coll Cardiol 2000;35:183-7.
305. Wood KA, Eisenberg SJ, Kalman JM, et al: Risk of thromboembolism in chronic atrial flutter. Am J Cardiol 1997;79:1043-7.
306. Seidl K, Haver B, Schwick NG, et al: Risk of thromboembolic events in patients with atrial flutter. Am J Cardiol 1998;82:580-4.
307. Murray RD, Tejan-Sie SA, Jasper SE, et al: Atrial mechanical function and eight-week clinical outcomes in patients with atrial flutter undergoing electrical cardioversion: Results from the ACUTE clinical trial [abstract]. J Am Coll Cardiol 2002; 39:376A.
308. Corrado G, Sgalambro A, Mantero A, et al, for the FLASIEC Investigators: Thromboembolic risk in atrial flutter: The FLASIEC (Flutter Atriale Societa Italiana di Ecografia Cardiovascolare) multicentre study. Eur Heart J 2001;22:1042-51.
309. Cannegieter SC, Rosendaal FR, Wintzen AR, et al: Optimal oral anticoagulant therapy in patients with mechanical heart valves. N Engl J Med 1995;333:11-7.
310. Murray RD, Shah A, Jasper SE, et al: Transesophageal echocardiography guided enoxaparin antithrombotic strategy for cardioversion of atrial fibrillation: The ACUTE II pilot study. Am Heart J 2000;139:1-7.
311. Camm AJ: Atrial fibrillation: Is there a role for low-molecular-weight heparin? Clin Cardiol 2001;24:I15-I19.
312. Roijer A, Eskilsson J, Olsson B: Transoesophageal echocardiography-guided cardioversion of atrial fibrillation or flutter: Selection of a low-risk group for immediate cardioversion. Eur Heart J 2000;21:837-47.
313. Olsson SB and Executive Steering Committee on behalf of the SPORTIF III Investigators: Stroke prevention with the oral direct thrombin inhibitor ximelagatran compared with warfarin in patients with non-valvular atrial fibrillation (SPORTIF III): Randomised controlled trial. Lancet 2003;362:1691-8.
314. Sievert H, Lesh M, Trepels T, et al: Percutaneous left atrial appendage transcatheter occlusion to prevent stroke in high risk patients with atrial fibrillation: Early clinical experience. Circulation 2002;105:1887-9.
315. Nakai T, Lesh MD, Gerstenfeld EP, et al: Percutaneous left atrial appendage occlusion (PLAATO) for preventing cardioembolism: First experience in canine model. Circulation 2002;105:2217-22.
315a. Hanna IR, Kolm P, Martin R, et al: Left atrial structure and function after percutaneous left atrial appendage transcatheter occlusion (PLAATO): Six-month echocardiographic follow-up. J Am Coll Cardiol 2004;43:1868-72.
316. Ravelli F, Allessie M: Effects of atrial dilatation on refractory period and vulnerability to atrial fibrillation in the isolated Langendorff-perfused rabbit heart. Circulation 1997;96:1686-95.
317. Satoh T, Zipes DP: Unequal atrial stretch in dogs increased dispersion of refractoriness conductive to developing atrial fibrillation. J Cardiovasc Electrophysiol 1996;7:833-42.
318. Vermes E, Tardif JC, Bourassa MG, et al: Enalapril decreases the incidence of atrial fibrillation in patients with left ventricular dysfunction: Insight from the Studies of Left Ventricular Dysfunction (SOLVD) trials. Circulation 2003;107:2926-31.

319. Goette A, Staack T, Röcken C, et al: Increased expression of extracellular signal-regulated kinase and angiotensin-converting enzyme in human atria during atrial fibrillation. J Am Coll Cardiol 2000;35:1669-77.
320. Goette A, Arndt M, Röcken C, et al: Regulation of angiotensin II receptor subtypes during atrial fibrillation in humans. Circulation 2000;101:2678-81.
321. Nakashima H, Kumagai K, Urara H, et al: Angiotensin II antagonist prevents electrical remodeling in atrial fibrillation. Circulation 2000;101:2612-7.
322. Kumagai K, Nakashima H, Urata H, et al: Effects of angiotensin II type 1 receptor antagonist on electrical and structural remodeling in atrial fibrillation. J Am Coll Cardiol 2003;41:2197-204.
322a. Savelieva I, Camm AJ: Atrial fibrillation and heart failure: Natural history and pharmacological treatment. Europace 2004;6:S5-S15.
323. Van Noord T, Van Gelder IC, Van Den Berg M, et al: Pretreatment with ACE inhibitors enhances cardioversion outcome in patients with persistent atrial fibrillation [abstract]. Circulation 2001;104:II-699.
324. Wachtell K, Lehto M, Hornestam B, et al: Losartan reduces the risk of new onset atrial fibrillation in hypertensive patients with electrocardiographic left ventricular hypertrophy: The LIFE study [abstract]. Eur Heart J 2003;24:504.
325. Pitt B, Zannad F, Remme WJ, et al, for the Randomized Aldactone Evaluation Study Investigators: The effect of spironolactone on morbidity and mortality in patients with severe heart failure. N Engl J Med 1999;341:709-17.
326. Goette A, Hoffmanns P, Enayati W, et al: Effect of successful electrical cardioversion on serum aldosterone in patients with persistent atrial fibrillation. Am J Cardiol 2001;88:906-9.
326a. Dernellis J, Panaretou M: Relationship between C-reactive protein concentrations during glucocorticoid therapy and recurrent atrial fibrillation. Eur Heart J 2004;25:1100-7.
326b. Siu CW, Lau CP, Tse HF: Prevention of atrial fibrillation recurrence by statin therapy in patients with lone atrial fibrillation after successful cardioversion. Am J Cardiol 2003;92:1343-5.
326c. Mozaffarian D, Psaty BM, Rimm EB, et al: Fish intake and risk of incident atrial fibrillation. Circulation 2004;110:368-73.
327. Waldo AL: Pathogenesis of atrial flutter. J Cardiovasc Electrophysiol 1998;9:518-25.
328. Cosio FG, Arribas F, Lopex-Gill M, et al: Atrial flutter mapping and ablation. II. Radiofrequency ablation of atrial flutter circuits. Pacing Clin Electrophysiol 1996;19:965-75.
329. Paydak H, Kall JG, Burke MC, et al: Atrial fibrillation after radiofrequency ablation of type I atrial flutter: Time to onset, determinants, and clinical course. Circulation 1998;98:315-22.
330. Huang DT, Monahan KM, Zimetbaum P, et al: Hybrid pharmacologic and ablative therapy: A novel and effective approach for the management of atrial fibrillation. J Cardiovasc Electrophysiol 1998;9:462-9.
331. Wood MA, Brown-Mahoney C, Kay GN, Ellenbogen KA: Clinical outcomes after ablation and pacing therapy for atrial fibrillation: A meta-analysis. Circulation 2000;101:1138-44.
332. Ozcan C, Jahangir A, Friedman PA, et al: Long-term survival after ablation of the atrioventricular node and implantation of a permanent pacemaker in patients with atrial fibrillation. N Engl J Med 2001;344:1043-51.
333. Gilles AM, Connolly SJ, Lacombe P, et al, for the Atrial Pacing Periablation for Paroxysmal Atrial Fibrillation (PA^3) Study Investigators. Randomized crossover comparison of DDDR versus VDD pacing after atrioventricular junction ablation for prevention of atrial fibrillation. Circulation 2000;102:736-41.
333a. Simantirakis EN, Vardakis KE, Kochiadakis GE, et al: Left ventricular mechanics during right ventricular apical or left ventricular-based pacing in patients with chronic atrial fibrillation after atrioventricular junction ablation. J Am Coll Cardiol 2004;43:1013-8.
334. Duckeck W, Engelstein E, Kuck KH: Radiofrequency current therapy in atrial tachyarrhythmias: Modulation versus ablation of atrioventricular nodal conduction. Pacing Clin Electrophysiol 1993;16:629-36.
335. Della Bella P, Carbucicchio C, Tondo C, et al: Modulation of atrioventricular conduction by ablation of the "slow" atrioventricular node pathway in patients with drug-refractory atrial fibrillation or flutter. J Am Coll Cardiol 1995;25:39-46.
336. Pappone C, Rosanio S, Augello G, et al: Mortality, morbidity, and quality of life after circumferential pulmonary vein ablation for atrial fibrillation: Outcomes from a controlled nonrandomized long-term study. J Am Coll Cardiol 2003;42:185-97.
337. Cox JL, Boineau JP, Schuessler RB, et al: Successful surgical treatment of atrial fibrillation. JAMA 1991;266:1976-80.
338. Deneke T, Khargi K, Grewe PH, et al: Efficacy of an additional MAZE procedure using cooled-tip radiofrequency ablation in patients with chronic atrial fibrillation and mitral valve disease: A randomized, prospective trial. Eur Heart J 2002;23:558-66.
339. Connolly SJ, Kerr C, Gent M, et al: Dual-chamber versus ventricular pacing: Critical appraisal of current data. Circulation 1996;94:578-83.
340. Wharton JM, Sorrentino RA, Campbell P, et al: Effect of pacing modality on atrial tachyarrhythmia recurrence in the tachybradycardia syndrome: Preliminary results of the Pacemaker Atrial Tachycardia (PAC-A-TACH) study [abstract]. Circulation 1998;98:2601.
341. Connolly SJ, Kerr CR, Gent M, et al: Effects of physiologic pacing versus ventricular pacing on the risk of stroke and death due to cardiovascular cause: Canadian Trial of Physiologic Pacing Investigators. N Engl J Med 2000;342:1385-91.
342. Lamas GA, Orav EJ, Stambler BS, et al, for the Pacemaker Selection in the Elderly Investigators: Quality of life and clinical outcomes in elderly patients treated with ventricular pacing as compared with dual-chamber pacing. N Engl J Med 1998;338:1097-104.
342a. Sweeney MO, Hellkamp AS, Ellenbogen KA, et al, for the Mode Selection Trial (MOST) Investigators: Adverse effect of ventricular pacing on heart failure and atrial fibrillation among patients with normal baseline QRS duration in a clinical trial of pacemaker therapy for sinus node dysfunction. Circulation 2003;107:2932-37.
343. Lau CP, Tse HF, Yu CM, et al, for the New Indication for Preventive Pacing in Atrial Fibrillation (NIPP-AF) Investigators: Dual-site atrial pacing for atrial fibrillation in patients without bradycardia. Am J Cardiol 2001;88:371-5.
344. Seidl K, Cazeau S, Gaita F, et al: Dual-site pacing vs mono-site pacing in prevention of atrial fibrillation [abstract]. Pacing Clin Electrophysiol 2002;24:568.
345. Padeletti L, Puerefellner H, Adler S, et al: Atrial septal lead placement and atrial pacing algorithms for prevention of paroxysmal atrial fibrillation [abstract]. Pacing Clin Electrophysiol 2002;24:687.
346. The Hotline Sessions of the 23rd European Congress of Cardiology. Eur Heart J 2001;22:2033-7.
346a. Blanc JJ, De Roy L, Mansourati J, et al: Atrial pacing for prevention of atrial fibrillation: Assessment of simultaneously implemented algorithms. Europace 2004;6:371-9.
347. Israel CW, Ehrlich JR, Grönefeld G, et al: Prevalence, characteristics and clinical implications of regular atrial tachyarrhythmias in patients with atrial fibrillation: Insights from a study using a new implantable devices. J Am Coll Cardiol 2001;38: 355-63.
348. Adler SW, Wolpert C, Warman EN, et al, for the Worldwide Jewel AF Investigators: Efficacy of pacing therapies for treating atrial tachyarrhythmias in patients with ventricular arrhythmias receiving a dual-chamber implantable cardioverter defibrillator. Circulation 2001;104:887-92.
340. Lee M, Weachter R, Pollack S, et al, for the ATTEST Investigators: The effect of atrial pacing therapies on atrial tachyarrhythmia burden and frequency: Results of a randomized trial in patients with bradycardia and atrial tachyarrhythmias. J Am Coll Cardiol 2003;41:1926-32.
350. Friedman PA, Dijkman B, Warman EN, et al, for the Worldwide Jewel AF Investigators: Atrial therapies reduce atrial arrhythmia burden in defibrillator patients. Circulation 2001;104:1023-8.

Chapter 16 Nonsustained Ventricular Tachycardia

A. JOHN CAMM, DEMOSTHENES G. KATRITSIS, WOJCIECH ZAREBA, YONGKEUN CHO, and ARTHUR J. MOSS

Introduction: A. John Camm and Demosthenes G. Katritsis
Epidemiology: Demosthenes G. Katritsis and A. John Camm
Electrocardiography: Demosthenes G. Katritsis and A. John Camm
Investigations: Demosthenes G. Katritsis and A. John Camm
Principles of Practice: Demosthenes G. Katritsis and A. John Camm
Management and Evidence-Based Therapy: Wojciech Zareba, Yongkeun Cho, and Arthur J. Moss

Since the late 1960s, when ambulatory electrocardiographic monitoring became clinically available, nonsustained ventricular tachycardia (NSVT) has been recognized as a usually asymptomatic rhythm disorder detected in various clinical settings. NSVT can be detected in an extremely wide range of conditions, from asymptomatic, apparently healthy, young individuals to patients with significant heart disease and annual mortality rates exceeding 50%.[1] Due to its brevity, NSVT does not produce symptoms in most instances; it derives its clinical importance from the fact that, depending on the underlying pathology, its detection may have important prognostic implications. In several clinical settings NSVT is a marker of increased risk for subsequent sustained tachyarrhythmias and sudden cardiac death.[1] NSVT represents an ominous sign in a patient with reduced left ventricular (LV) function and ejection fraction (EF) below 40%, whereas it may have no prognostic significance in a healthy individual. This makes its clinical interpretation far from easy. The distinction between ranges of normality and deviations that result from disease is not always possible and most of the time remains a matter of conjecture. Rarity of occurrence has often been used to connote disease but could equally well represent an extreme of the shifting spectrum of normality. This could well be true for healthy subjects in whom an NSVT run is an accidental finding not related to physical exercise; however, there is continually accumulating evidence for occult pathology in apparently normal subjects who develop ventricular arrhythmia.[2,3] The main task of the physician is to detect the apparently healthy individuals in whom NSVT represents a sign of occult disease and to risk-stratify patients with known disease who present with this arrhythmia. This, however, is far from an easy task. Today the patient with NSVT represents a clinical challenge in regard to proper management and, as discussed in the following chapters, several crucial questions remain unanswered.

Epidemiology

Reliable epidemiologic data on NSVT are difficult to obtain. Usually, although not invariably, most patients remain asymptomatic, and reproducibility of NSVT recordings is documented in only half of the patients with this arrhythmia. A recent study analyzed the variability of NSVT in a patient population fulfilling the noninvasive criteria adopted by the Multicenter Automatic Defibrillator Implantation Trial (MADIT).[4] The arrhythmia was defined as reproducible when found in at least two recordings of three consecutive Holter ECGs performed in weekly intervals. Reproducible NSVT was identified in only 50% of the patients with NSVT and did not seem to be an independent risk factor for future arrhythmic events.[4] The reported prevalence of NSVT in various clinical conditions is presented in Table 16-1. Most of the available information emerges from older studies based on Holter monitoring. The advent of implantable permanent pacemakers and defibrillators (implantable cardioverter-defibrillators, ICDs) with extensive ECG monitoring capabilities has allowed the accumulation of considerable information from patients with heart disease. Several points, therefore, are being modified.

TABLE 16-1 Prevalence of Nonsustained Ventricular Tachycardia in Different Cardiac Conditions

Condition	Percentage
Apparently healthy individuals	0-3
Acute MI (early phase)	40-70
Acute MI (later than 1 mo)	
Thrombolysed	7
Not thrombolysed	15
IHD (preserved LV function)	5
IHD (LVEF < 40%)	30-90
DCM	40-95
HOCM	25-80
Significant valve disease	25
Hypertension	6
Hypertension and LV hypertrophy	15
Repaired tetralogy of Fallot	17

DCM, dilated cardiomyopathy; HOCM: hypertrophic cardiomyopathy; IHD, ischemic heart disease; LV, left ventricular; MI, myocardial infarction.

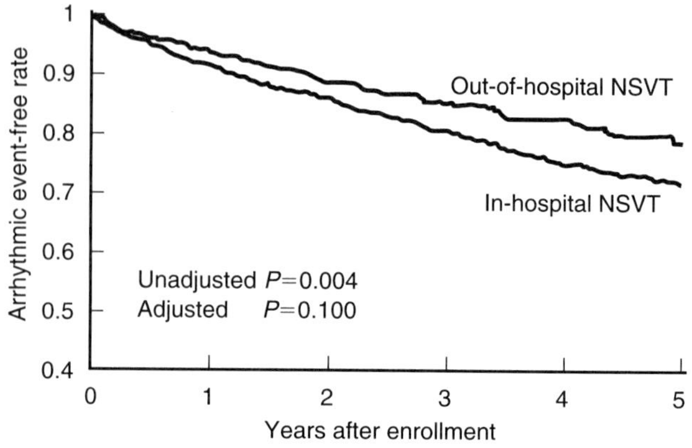

FIGURE 16-1 Kaplan-Meier curves comparing the rates of cardiac arrest and arrhythmic death between untreated patients with in-hospital versus out-of-hospital identified NSVT. Data from the MUSTT cohort. (From Pires LA, Lehmann MH, Buxton AE, et al: Differences in inducibility and prognosis of in-hospital versus out-of-hospital identified nonsustained ventricular tachycardia in patients with coronary artery disease. J Am Coll Cardiol 2001;38:1156-62.)

There is emerging evidence from ICD data that NSVT is a distinct tachyarrhythmia that may cause syncope without causing death in patients with heart disease, and that the incidence of polymorphic NSVT relative to sustained arrhythmia is greater than previously believed.[5] The occurrence of NSVT during the first day of acute myocardial infarction (MI), although frequent (40% to 70% of patients), is not associated with increased risk for subsequent long-term mortality.[6] Following the first 24 hours postinfarction, NSVT is detected in approximately 5% to 10% of the patients, particularly during the first months, and has been considered to have adverse prognostic significance.[7-9] NSVT can be detected in approximately 5% of ischemic patients with preserved LV function, apparently excluding previous MI, but does not appear to indicate an adverse clinical outcome.[10,11] However, NSVT appears now to occur in the majority of ischemic patients with reduced LV function,[12,13] and this is also the case with dilated cardiomyopathy.[14-18] In hypertrophic cardiomyopathy (HCM) 20% to 30% of patients may have NSVT, whereas in patients with a history of cardiac arrest this proportion approaches 80%.[19-22]

Reports from the Analysis of the Multicenter UnSustained Tachycardia Trial (MUSTT)[23] sub-studies have provided valuable information regarding the prevalence as well as the prognostic significance of NSVT in the context of different clinical settings. It seems that it is important not only how often but under what circumstances NSVT occurs. MUSTT data have shown prognostic differences in patients with in-hospital–, as opposed to out-of-hospital–identified NSVT.[24] Overall mortality rates at 2- and 5-year follow-up were 24% and 48% for inpatients and 18% and 38% for outpatients (adjusted $P = .018$) (Fig. 16-1). In patients subjected to surgical coronary revascularization, the occurrence of NSVT within the early (<10 days) postrevascularization period portends a far better outcome than when it occurs later after CABG (10 to 30 days) or in nonpostoperative settings.[25] Indeed, approximate 2-year mortality rates for patients with early postrevascularization NSVT, late postrevascularization NSVT, and nonpostoperative NSVT were 14%, 23%, and 24%, respectively.[25] When, however, sustained VT is inducible in patients with early postoperative NSVT, the outcome is worse as compared to noninducible patients.[26] In the MUSTT racial differences have also been shown to influence outcome of patients with NSVT and reduced LV function.[27]

In patients with valvular disease the incidence of NSVT is considerable (up to 25% in aortic stenosis and in significant mitral regurgitation) and appears to be a marker of underlying LV pathology.[29,30] Up to 65% of patients with surgical repair of tetralogy of Fallot have ventricular ectopy, whereas NSVT occurs in up to 17% of patients either at rest or during exercise.[31,32] In patients with arterial hypertension NSVT is correlated to the degree of cardiac hypertrophy and subendocardial fibrosis.[33,34] Approximately 15% of patients with hypertension and LV hypertrophy present NSVT as opposed to 6% of patients with hypertension alone.[33,35]

Electrocardiography

NSVT represents an electrocardiographic diagnosis based on resting or exercise ECG and Holter monitoring recordings. Two major methodology problems may interfere with the recognition as well as interpretation of this condition. First, a universally accepted, established definition does not exist. Second, morphologic criteria such as those employed in the description of sustained ventricular tachycardia have not been adopted. The nature of the short-lived episodes of arrhythmia may not allow a clear distinction between monomorphic and polymorphic ventricular rhythms. Thus, in most published studies the terms "NSVT episodes" or "complex ventricular ectopy" are used in a general way without attempting to distinguish between regular, monomorphic rhythms and salvos of ventricular depolarizations of variable morphology. This might have important implications in the assessment of the clinical significance of NSVT in various conditions.

DEFINITIONS

Traditionally, the term tachycardia (Greek *tachy* = fast and *cardia* = heart) is used to describe conditions in which the heart rate exceeds the conventional number of 100 beats per minute (bpm) for more than three consecutive beats, either in response to metabolic demand or other stimuli or due to disease.[1] Sustained is defined as tachycardia that lasts more than 30 seconds (unless requiring termination because of hemodynamic collapse), whereas nonsustained tachycardia terminates spontaneously within 30 seconds.[1] Thus, NSVT (Fig. 16-2) has been defined as three (sometimes five) or more consecutive beats arising below the atrioventricular node with an R–R interval of less than 600 milliseconds (ms) (>100 bpm) and lasting less than 30 seconds.[36] This definition is far from universal. For ventricular tachycardia, the cutoff rate is often considered greater than 120 bpm since rhythms between 100 and 120 bpm are used to connote accelerated idioventricular rhythm (i.e., an entity with specific clinical characteristics and significance).[37] As a consequence, NSVT has also been required to exceed the rate of 120 bpm.[38]

The time period of a tachycardia run in order to qualify for NSVT has also been variable. In the Electrophysiologic Study Versus Electrocardiographic Monitoring (ESVEM) study,[39] the time cutoff was 15 seconds as opposed to the conventional number of 30 seconds, which has been used both by the MADIT[40] and MUSTT[23] trials. This implies that NSVT cases according to the MADIT and MUSTT criteria might well have been dealt with as sustained VT episodes in the ESVEM cohort. In other studies the terms *ventricular ectopy* or *NSVT* are often used without strictly defined diagnostic criteria.[41]

ELECTROCARDIOGRAPHIC PATTERNS

Nonsustained Ventricular Tachycardia Associated with "Normal Heart"

In the apparently normal population, nonspecific NSVT episodes without consistent morphology patterns may be recorded either at rest or after exercise and are usually described as *complex ventricular ectopy*.[42,43] When NSVT is documented in the context of a history of established monomorphic VT, it may demonstrate the same morphology as the clinical sustained arrhythmia. Approximately 10% of patients with ventricular tachycardia do not have clinically obvious heart disease to account for the arrhythmia.[28] These tachycardias may be early manifestations of cardiac disease that is otherwise clinically latent or may be due to a primary electrical abnormality of the ventricular muscle.[28] VT presents both in sustained and nonsustained forms. Ventricular tachycardias with left bundle branch block (LBBB) pattern and inferior axis originate in the right ventricular outflow tract (RVOT).[44] LV outflow tract tachycardias or tachycardias originating in the aortic valve cusps produce a right bundle branch block (RBBB) morphology with inferior axis.[3,45] *Fascicular tachycardias* originate within one of the fascicles of the left bundle branch, and usually the posterior fascicle is involved,[46,47] resulting in a tachycardia with RBBB and left axis deviation, but cases with inferior axis (i.e., of anterior fascicular origin) have also been described.[45]

The most systematic attempts to characterize the electrocardiographic pattern of NSVT have been accomplished in reports on patients with the so-called

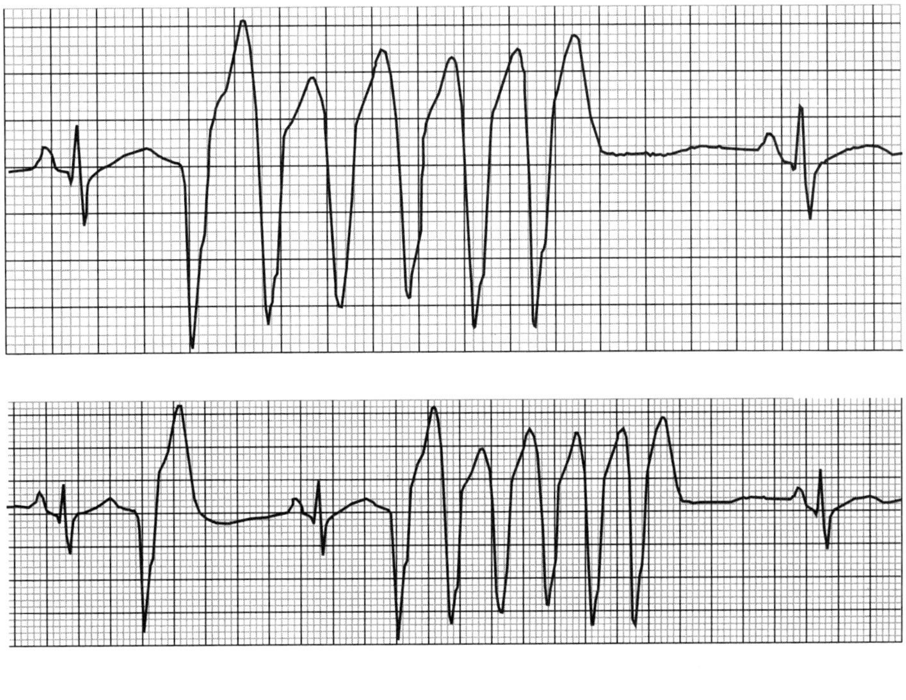

FIGURE 16-2 This trace shows two examples of nonsustained ventricular tachycardia in a patient who had suffered a myocardial infarction 3 months previously. The arrhythmias consist of six wide complex beats at an accelerating rate of up to approximately 200 bpm. The morphology of the complexes is similar but not absolutely identical. These rhythms would be regarded as monomorphic nonsustained ventricular tachycardias.

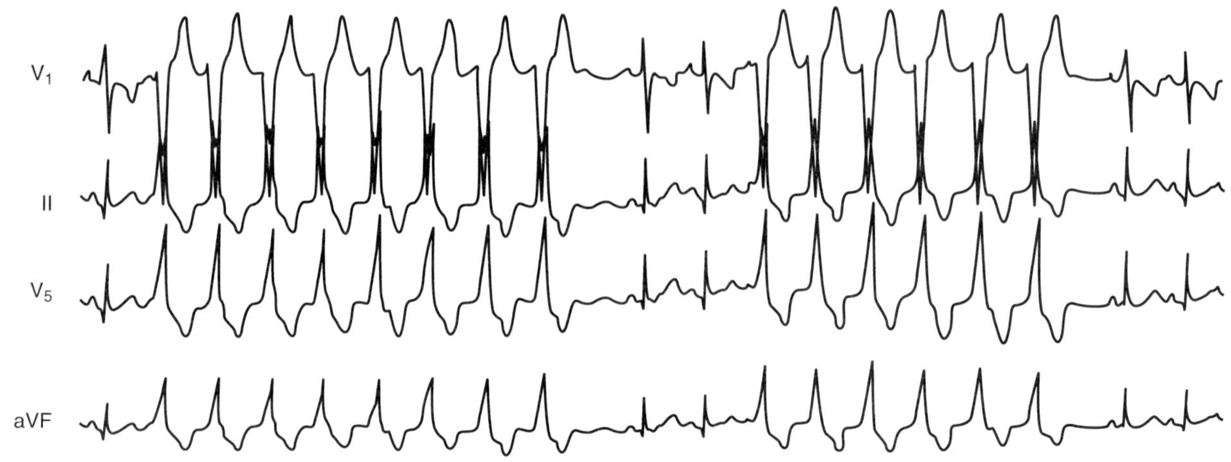

FIGURE 16-3 Repetitive nonsustained episodes of ventricular tachycardia. The LBBB-like QRS complexes suggest that the tachycardia arises from the right ventricle. This patient was asymptomatic, and there was no evidence of any cardiac pathology.

repetitive monomorphic VT[28] (Fig. 16-3). The relatively consistent recording of frequent arrhythmia episodes in these patients has made possible the accumulation of useful information on the morphology of NSVT. Holter studies have revealed that the tachycardia begins with a fusion complex or with an ectopic beat that has the same morphology as the subsequent beats.[48] The salvos of ventricular tachycardia are generally short (3 to 15 beats), and the coupling interval of the first beat is usually long (>400 ms). The coupling intervals of successive beats may gradually become prolonged.[48-50] Many patients mostly have ventricular premature beats with only occasional episodes of tachycardia, whereas others manifest mainly with short runs of ventricular tachycardia where the tachycardia beats far exceed the number of sinus beats. The rate during tachycardia is generally 110 to 150 bpm. The arrhythmia is only present within a critical window of heart rates (upper and lower thresholds).[28] Thus, the tachycardia often occurs during exercise but disappears as the heart rate increases and returns during the recovery period following exercise. Some patients develop sustained episodes (>30 seconds) during the recovery phase, and this behavior differentiates repetitive monomorphic ventricular tachycardia from the *exercise-triggered paroxysmal ventricular tachycardia* first described by Wilson and others[51,52] in patients with apparently normal hearts. It is now known that most cases of repetitive monomorphic VT represent a nonsustained counterpart of RVOT VT. However, repetitive behavior has been documented in various clinical settings, including cardiomyopathy and previous MI,[48] as well as tachycardias originating in the aortic valve cusps.[3]

Ischemic Heart Disease

As discussed earlier in "Epidemiology," NSVT is frequently detected in patients with ischemic heart disease. The QRS morphology during tachycardia may show either RBBB or LBBB pattern or may even be nonspecific (Fig. 16-4). Both RBBB and LBBB patterns can be seen in the same patient when the infarct scar involves the interventricular septum (Fig. 16-5).[53]

Frequent and complex ventricular ectopy as well as NSVT occur more commonly in patients with LV dysfunction. Most patients with polymorphic VT in the absence of QT prolongation have coronary artery disease.[54] At least two groups of patients have been described. In patients with stable coronary artery disease with prior MI but no evidence of acute ischemia, polymorphic VT may occur as an isolated phenomenon.[55] The second group comprises ischemic patients who develop VT/ventricular fibrillation (VF) as a manifestation of acute ischemia.[56] Untreated VF is usually fatal, but spontaneous reversions to sinus rhythm, thus documenting nonsustained VF, can occur (Fig. 16-6). Recent data from patients with an ICD reveal that up to 40% of all VF episodes are nonsustained.[5] During an acute MI, ventricular fibrillation may be spontaneous or even initiated by a ventricular premature beat falling on the vulnerable period (R-on-T), but in patients who have not recently sustained an infarction, the arrhythmia more often results from degeneration of ventricular tachycardia.[57]

Cardiomyopathies and Other Conditions

Ventricular tachycardias (both sustained and nonsustained) in dilated cardiomyopathy may present with multiple morphologies or with an LBBB or RBBB pattern. In approximately one third of the cases of idiopathic dilated cardiomyopathy[58] and probably in a small percentage of ischemic patients, both sustained and nonsustained forms of VT are due to bundle branch reentry. The necessary condition for bundle branch reentry seems to be prolonged conduction in the His-Purkinje system, and this is reflected in the H–V interval that is prolonged during sinus rhythm and prolonged or equal to the baseline sinus rhythm during VT.[59] The circuit involves the right and left bunch bundles with

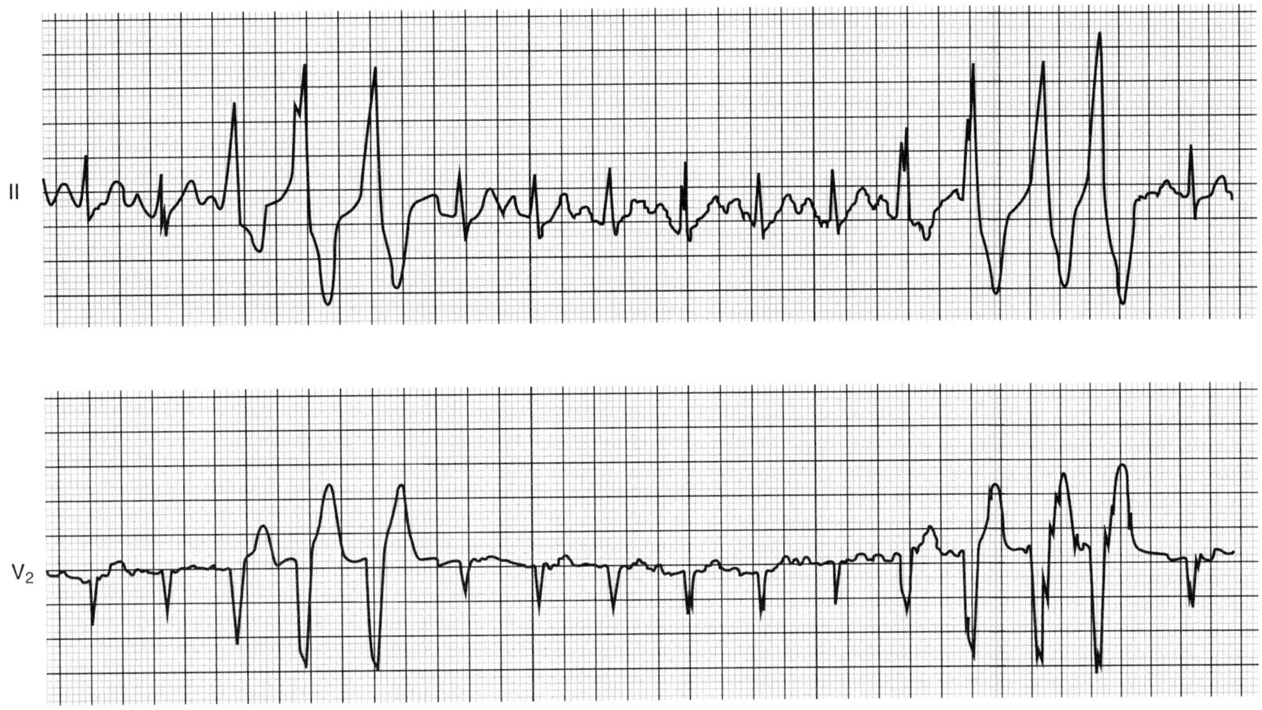

FIGURE 16-4 Two runs of nonsustained ventricular tachycardia from a patient with a previous myocardial infarction. The first beat of tachycardia in the two runs probably represents fusion with sinus beats.

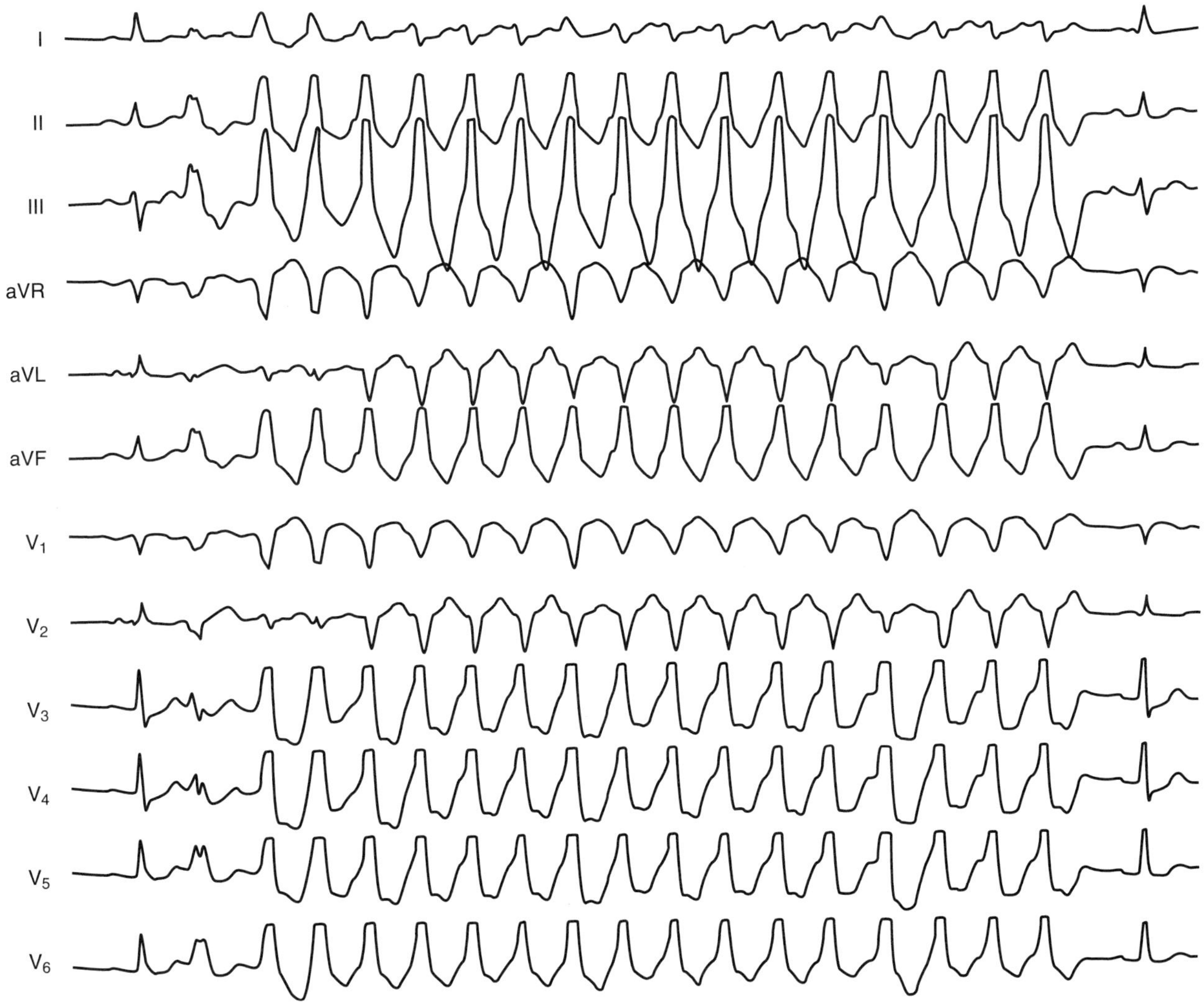

FIGURE 16-5 Nonsustained ventricular tachycardia in a patient with hypertension, LV hypertrophy, and a previous small septal infarct. Early precordial transition and morphology of lead I suggest a septal origin of the tachycardia.

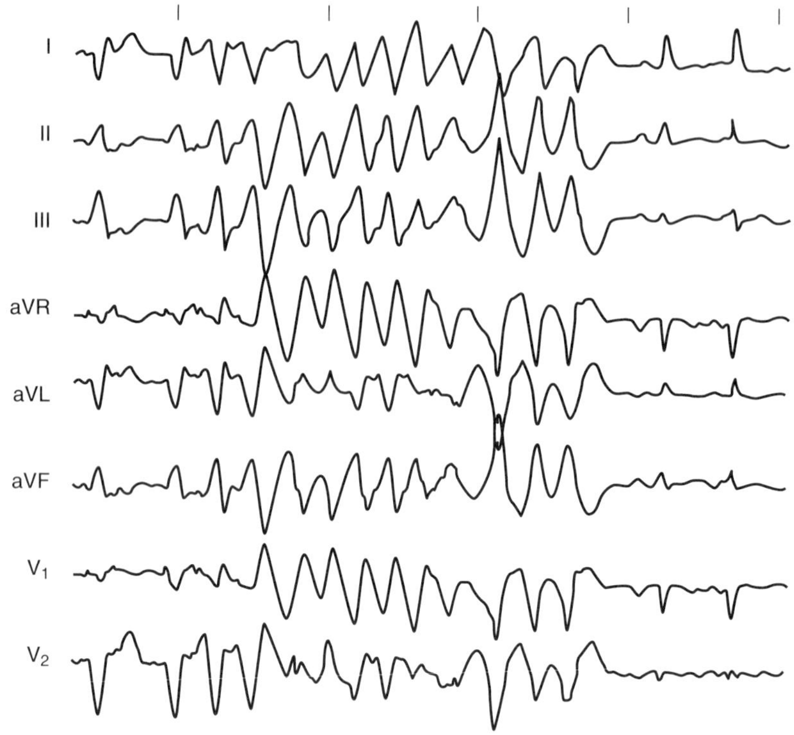

FIGURE 16-6 During electrophysiology testing in a patient with ischemic heart disease and nonsustained ventricular tachycardia (VT) on Holter monitoring, an episode of nonsustained polymorphic VT/ventricular fibrillation is induced.

antegrade conduction occurring most of the time through the RB. As a rule, atrioventricular dissociation is present. These tachycardias are usually unstable, and the 12-lead ECG (when obtainable) may show either LBBB or RBBB pattern depending on the orientation of activation of the bundle branches.[58,59]

There is no characteristic electrocardiographic morphology of NSVT in patients with HCM.[36] Relatively slow, often asymptomatic, nonsustained episodes of monomorphic ventricular tachycardia may be documented on prolonged ambulatory recordings. When nonsustained or sustained VT is induced by programmed stimulation, it is more often polymorphic (60% of patients) than monomorphic (30%).[1]

In other conditions, such as valvular disease[30] and systemic hypertension,[34] ventricular arrhythmias are common but less well characterized and usually represent polymorphic rhythms. Most of the reported patients with mitral valve prolapse demonstrate LBBB morphology during tachycardia, thus raising the possibility that the mitral valve prolapse might be an incidental finding or that the arrhythmia is due to other mechanisms not directly related to the mechanical stress imposed upon the ventricle by the valvular apparatus.[36]

Ventricular tachycardia may occur several years after repair of tetralogy of Fallot. The arrhythmias are probably due to reentry around the ventriculotomy scar, related to the repair of the ventricular septal defect, or to disturbed right ventricular depolarization.[60] No specific patterns have been described in cases of NSVT.

The tachycardias in arrhythmogenic right ventricular dysplasia arise from the right ventricle and typically present with LBBB morphology with a left or even right axis deviation.[61]

Long QT Syndromes

The typical morphology of recorded arrhythmias in the long QT syndromes (congenital or acquired) is that of *torsades de pointes*.[62] The term refers to the electrocardiographic appearance of spikelike QRS complexes that rotate irregularly around the isoelectric line at rates of 200 to 250 bpm (Fig. 16-7). Typically, there is a long (600 to 800 ms) coupling interval of the initial beat of the *torsade*, whereas the last QRS complex of the episode is larger than the normal QRS during sinus rhythm.[62] This is a ventricular tachycardia that is frequently nonsustained and sometimes sustained. The tachycardia, which usually occurs in the setting of bradycardia or long postectopic pauses, is often repetitive and may trigger ventricular fibrillation. It is seen primarily in association with prolongation of the Q–T interval that may be appreciable before the onset of arrhythmias or after a pause. Conventionally, the term *torsades de pointes* is reserved for use with the long QT syndrome only. However, not all patients with the long QT syndrome have polymorphic ventricular tachycardia with a characteristic *torsades de pointes* configuration, and this pattern can be seen in some patients without the long QT syndrome. A variety of *torsades* initiating with a short coupling interval in patients without any evidence of long QT syndrome has also been described.[63]

Investigations

Careful clinical assessment of the patient is the cornerstone of the diagnostic process. Age, general condition, previous medical history, and conditions such as

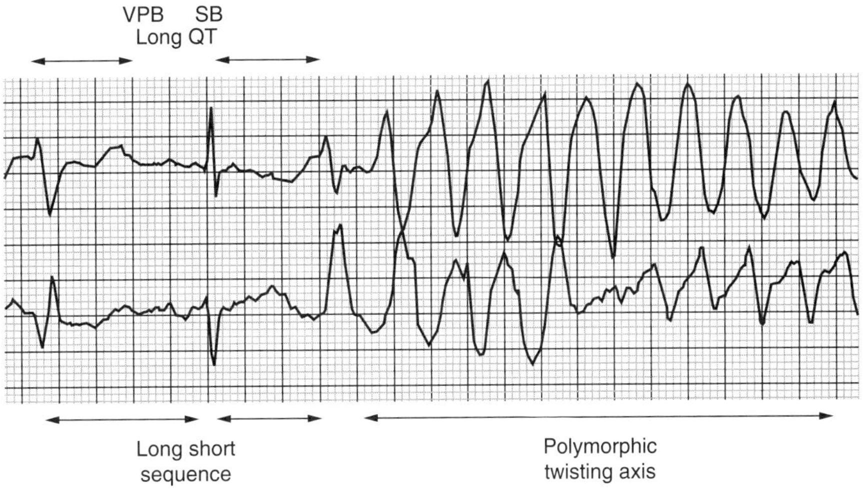

FIGURE 16-7 This is an example of *torsades de pointes*. Note the relatively rapid but variable rate, continually changing QRS morphology, and variable QRS axis. This is a polymorphic ventricular tachycardia, usually diagnosed as *torsades de pointes* when conducted QRS complexes (between episodes of tachycardia) have a prolonged Q–T interval, or the tachycardia is initiated after a preceding pause or, as in this case, a (short-) long-short sequence of QRS complexes. SB, sinus beat; VPB, ventricular premature beat.

electrolyte disturbances, metabolic imbalance, and proarrhythmic effect of drugs should be considered. A standard 12-lead ECG might indicate potential causes of NSVT including signs of previous MI or active ischemia, conduction disturbances, prolonged QT, or other signs of electrical instability suggestive of Brugada syndrome or arrhythmogenic right ventricular dysplasia. In addition, several other clinical parameters and tests can provide specific information about the potential risk of future arrhythmic events in patients who have presented with nonsustained ventricular arrhythmias.

TRANSTHORACIC ECHOCARDIOGRAPHY

Transthoracic echocardiography may detect signs of cardiomyopathy, arrhythmogenic right ventricular dysplasia (ARVD), and other structural abnormalities, as well as impaired LV function.

There has been overwhelming evidence that in patients with heart disease in general, LVEF is the major determinant of cardiac and total mortality.[1] The results of the MADIT and MUSTT trials have also established the importance of LVEF as the most critical prognostic factor in patients with ischemic heart disease and NSVT. LV function assessment, therefore, with echocardiography or radionuclide ventriculography is mandatory for the risk stratification and subsequent management of patients with NSVT.

AMBULATORY ECG MONITORING

Although Holter monitoring is a valuable diagnostic tool in detecting patients with NSVT, its use for the subsequent follow-up and evaluation of treatment is questionable. First, the prognostic significance of the frequency of NSVT runs or other ECG variables such as heart rate and complex ventricular ectopy is unknown. In a recent study on 531 patients with HCM, a relation between the risk of arrhythmic death and the frequency, duration, and rate of NSVT episodes could not be demonstrated.[22] In the Grupo de Estudio de la Sobrevida en la Insuficiencia Cardiaca en Argentina (GESICA) trial, the presence of couplets showed a greater independent prognostic value than NSVT for the incidence of sudden death.[17,64] The Autonomic Tone and Reflexes After Myocardial Infarction (ATRAMI) trial failed to establish a relationship of ventricular ectopy with cardiac mortality in postinfarction patients,[65] whereas results from the smaller Bucindolol Evaluation in Acute myocardial infarction Trial (BEAT) suggest that the number of ventricular beats per hour is significantly associated with mortality, independent of β-blocker therapy.[66] Interestingly, in a recent analysis of 1071 ATRAMI patients mostly subjected to thrombolysis (63%), NSVT (with a prevalence of 13.4%) was found to adversely influence prognosis, independent of reduced LVEF.[150] In another study in which most patients (78%) had revascularization of the infarct-related artery, NSVT early after MI (4 to 16 days) had a low prevalence (9%) and carried a significant but low relative risk for the composite end point of cardiac death, VT, or VF but not for arrhythmic events considered alone.[142] It seems that in the β-blocking era, all common arrhythmia risk variables, including NSVT, have diminished predictive power in identifying postinfarction patients at risk of sudden cardiac death.[67]

Second, the suppression of frequent ventricular ectopy or NSVT runs following β-blockade or amiodarone therapy does not imply a favorable diagnosis. Mortality was increased in the Cardiac Arrhythmia Suppression Trials (CAST and CAST II) despite reduced ectopic activity,[68,69] whereas mortality was not reduced by amiodarone in the Veterans Administration Congestive Heart Failure Survival Trial of Antiarrhythmic Therapy (CHF-STAT) trial despite the elimination of ventricular ectopy.[18] The absence of ventricular premature depolarizations and NSVT in a 24-hour ambulatory ECG monitoring in the GESICA trial was found to indicate a low probability of sudden death.[17,64] However, a correlation between the number and length of NSVT episodes and sudden death was not found.

Tests for Myocardial Ischemia

Functional tests such as treadmill exercise electrocardiography, thallium scintigraphy, and stress echocardiography are required to demonstrate myocardial ischemia in patients with NSVT. Acute myocardial ischemia is an established cause of polymorphic ventricular rhythms.[53,56] The association of monomorphic ventricular tachycardia, a substrate-dependent arrhythmia, with acute ischemia is less well characterized, and the role of ischemia in the induction of monomorphic ventricular tachycardia does not appear to be important.[70-72] However, in experimental studies an interaction between acute ischemia and chronic substrate, such as myocardial scar, has been implicated in the genesis of various ventricular arrhythmias.[73,74]

Apart from demonstration of ischemia, exercise testing may also be helpful in patients with exercise-provocable right ventricular tachycardias. Exertion, as well as isoproterenol infusion, is critical in inducing ventricular tachycardia, which is probably due to catecholamine-related delayed afterdepolarizations.[75] Exercise testing is also needed to further explore a possibility of other causes of arrhythmias such as Wolff-Parkinson-White syndrome, Brugada syndrome, arrhythmogenic right ventricular dysplasia, or catecholaminergic polymorphic VT.[76-79]

Electrophysiology Testing

Electrophysiology testing may be required for establishment of initial diagnosis in patients presenting with nonsustained ventricular rhythms. Indications include the need for differential diagnosis of NSVT from short runs of atrial fibrillation in the context of an accessory pathway or other forms of aberration, drug testing for the diagnosis of the Brugada syndrome, and programmed electrical stimulation for induction of VT.

Induction of sustained arrhythmia by programmed electrical stimulation still retains a predictive power in ischemic patients with impaired LV function. In ischemic patients with NSVT, the induction of sustained ventricular tachycardia was associated with a two- to threefold increased risk of arrhythmia-related death in a previous meta-analysis.[80] In NSVT in the context of reduced LVEF (<40%), inducibility of sustained monomorphic ventricular tachycardia at baseline PES was associated with a 2-year actuarial risk of sudden death or cardiac arrest of 50% compared with a 6% risk in patients without inducible ventricular tachycardia.[81] According to data from the MUSTT trial, in ischemic patients with an LVEF<40% and asymptomatic NSVT, the incidence of inducible sustained ventricular tachycardia approaches 30%.[82]

Noninducible patients may not carry a lower risk for arrhythmic events. Both the MUSTT and MADIT II trials have suggested that noninducible patients at electrophysiological (EP) testing have a similar risk with inducible patients for sudden cardiac death.[23,83] In MUSTT, the inducible group had a 28% total 2-year mortality rate, whereas the noninducible group had a 21% rate. The 21% 2-year total mortality rate in noninducible patients is high enough to consider the need for programmed electrical stimulation at least questionable. Actually, data from MADIT II suggest an even higher rate of appropriate ICD shocks in patients who were noninducible at electrophysiology testing.[12] These findings are consistent with analysis of stored ICD data that have clearly shown there is little association between spontaneous and induced ventricular arrhythmias.[84] Thus, electrophysiology testing cannot select patients with a favorable outcome.

In the ischemic patient with relatively preserved LV function (LVEF > 40%), the role of PES is not established. There has been some evidence that noninducibility of sustained ventricular tachycardia at PES might retain discriminative ability by means of identifying a higher-risk patient cohort.[85,86] In patients treated with amiodarone, noninducibility at PES is associated with a much better prognosis than when ventricular tachycardia can still be induced,[87-90] whereas induction of ventricular fibrillation predicts a high incidence of sudden death.[86] Recently, the MUSTT investigators analyzed the relation of EF and inducible ventricular tachyarrhythmias to mode of death in 1791 patients enrolled in MUSTT who had not already received antiarrhythmic therapy. Total and arrhythmic mortality were higher in patients with an EF less than 30% than to those whose EFs were 30% to 40%. Inducibility of tachyarrhythmia identified patients for whom death is significantly more likely to be arrhythmic. Therefore, this study suggested that the major utility of electrophysiological testing may be restricted to patients having an EF between 30% and 40%.[13]

Similarly, there is currently no established role for PES in patients with NSVT and nonischemic heart disease.

Signal-Averaged Electrocardiography

The clinical value of signal-averaged electrocardiography (SAECG) in patients with NSVT by means of determining prognosis and identifying patients in need for an aggressive antiarrhythmic management is not established. In patients presenting with unexplained syncope, the presence of late potentials is a good predictor of induction of sustained ventricular tachycardia with high sensitivity (80% to 100%) and specificity (80% to 90%),[91-96] and its predictive ability might be higher than that of Holter monitoring.[94,96] However, as happens with postinfarction patients, the positive predictive value of SAECG in this setting is low (40%), whereas its negative predictive value is high, 90%. Thus, a negative SAECG might obviate the need for further investigations when the suspicion of a ventricular arrhythmia is low, but in the case of a high suspicion of ventricular arrhythmia, a negative SAECG is not sufficient evidence for exclusion of nonsustained or sustained ventricular tachycardia as the cause of syncope. In postinfarction patients treated with β-blockers, the positive predictive accuracy of abnormal SAECG for arrhythmic death is low, 13%.[67]

There has been some evidence that in patients with arrhythmogenic right ventricular dysplasia, SAECG can identify those with more extensive disease and a propensity for inducible ventricular tachycardia at PES.[97]

Assessment of Autonomic Tone

Heart rate variability (HRV) has been used as an independent risk factor for cardiac mortality in post-MI patients and has been claimed to be a specific predictor of arrhythmic rather than total cardiac mortality.[98,99] In post-MI patients a depressed baroreflex sensitivity (BRS) has been associated with an increased cardiac mortality and sudden death[100,101] and with a higher predictive power (in values below 3 ms/mm Hg) than LVEF, SAECG, or heart rate variability.[101,102] Analysis of ATRAMI patients has shown that NSVT, heart rate variability, and depressed baroreflex sensitivity were all significantly and independently associated with increased mortality. Depressed baroreflex sensitivity, in particular, identified a subgroup with the same mortality risk as patients with NSVT and reduced LVEF.[150]

Recently, the Multiple Risk Factor Analysis Trial (MRFAT) showed that, in the β-blocking era, the common arrhythmia risk variables, particularly the autonomic and standard ECG markers, have limited predictive power in identifying patients at risk of sudden cardiac death.[67] In a 43 ± 15 months follow-up of 675 patients, sudden cardiac death was weakly predicted only by reduced LVEF (<40%), NSVT, and abnormal SAECG but not by autonomic markers or ECG variables. The positive predictive accuracy of these markers (low LVEF, NSVT, and abnormal SAECG), however, was low—8%, 12%, and 13%, respectively.[67] More data are clearly needed are needed to establish the clinical utility of autonomic markers, particularly in the setting of NSVT.

Other Tests

There is some evidence that in patients with sustained right ventricular outflow tract tachycardia, the detection of an abnormal biopsy may indicate a worse prognosis,[28] but in case ARVD is suspected, MRI is probably the investigation of choice.[78]

Principles of Practice

The physician who takes care of a patient presenting with an episode of NSVT has two tasks. First, it should be established whether underlying occult pathology is responsible for the arrhythmia and, in case of diagnosed heart disease, to risk-stratify the patient in order to propose the appropriate management and therapy. The clinical approach of the patient with NSVT, therefore, should always be considered within the particular clinical context the arrhythmia occurs. In several settings the patient with NSVT represents a clinical challenge in regards to proper management, and several crucial questions remain unanswered. The occurrence of an abrupt ventricular arrhythmia is a multifactorial process involving a continually changing complex substrate of myocardial scarring, ischemia, and adrenergic/genetic factors. We are therefore dealing with a probabilistic event in which each of the currently considered risk factors such as NSVT identifies only a small fraction of the risk process.

APPARENTLY HEALTHY INDIVIDUALS

Initial reports, including longitudinal community-based studies, such as the Busselton[42] and the Framingham,[43] indicated that the presence of complex ventricular ectopy and NSVT increase the risk of subsequently observed heart disease. However, when the presence of occult ischemia or structural heart disease was excluded, NSVT was not found to adversely influence prognosis, with cardiac event rates not exceeding those detected in age-matched populations without the arrhythmia.[103,104] Thus, recording of spontaneous NSVT in apparently healthy individuals does not imply an adverse prognosis provided that occult cardiomyopathy and genetic arrhythmia disorders are excluded, as discussed elsewhere. These conditions, however, may remain latent for several years, and although apparently healthy individuals presenting with NSVT can be reassured about their prognosis, long-term follow-up is advisable.

Exercise-Induced Nonsustained Ventricular Tachycardia

The occurrence of premature ventricular demoralizations during exercise in apparently healthy subjects has not been associated with an increase in cardiovascular mortality and was considered to be a normal response to exertion.[105-108] It appears now that the long-term prognostic implications of exercise-induced ventricular arrhythmias may be adverse. Recently, the Paris Prospective Study,[109] in agreement with earlier data,[110] reported that runs of two or more consecutive ventricular depolarizations during exercise or at recovery may occur in up to 3% of healthy men and indicate an increase in cardiovascular mortality within the next 23 years by a factor of more than 2.5. This increased relative risk persisted even after adjusting for other characteristics and known factors that are predisposed to coronary artery disease. Interestingly, among subjects with a positive exercise test for ischemia, only 3% had ectopy, whereas among subjects with exercise-induced ectopy, only 6% had a positive exercise test for ischemia. Thus, although the precise mechanism of arrhythmia is unknown in this setting, and various forms of occult cardiomyopathy cannot be excluded, it does not appear to be a direct consequence of ischemia. This study argues that exercise-induced nonsustained ventricular arrhythmia may predict coronary artery disease even in the absence of evidence of ischemia in asymptomatic individuals.

Exercise testing may also induce catecholaminergic polymorphic ventricular tachycardia. When recognized, this condition requires aggressive management.[76]

PATIENTS WITH HEART DISEASE

In patients with known heart disease the physician's main task is risk assessment of the patient (i.e., the estimation of the probability that a patient will experience future morbid arrhythmic events). Risk stratification attempts to identify the specific mechanisms of further morbidity in order to predict clinical outcomes and, eventually, propose clinical strategies for their prevention.

In most, but not all, cases of heart disease NSVT does carry an adverse prognostic significance. However, still today, it is not known whether NSVT bears a cause and effect relationship with sustained ventricular tachyarrhythmias or it is merely a surrogate marker of LV dysfunction and electric instability. Even in the case where it does hold prognostic significance, NSVT does not necessarily imply the mechanism of death. There are certain patient groups who carry a high mortality due to progress of their disease. Death in these patients may be arrhythmic, but this does not imply that mere prevention of NSVT will unconditionally prolong life significantly. The cardiac mortality rate at 2 years in the MADIT was still 11% despite the use of defibrillators.[40] Furthermore, reduction of arrhythmic death does not necessarily imply a concomitant reduction in total mortality,[111-113] and even insignificant increase in mortality has been demonstrated despite the reduction of sudden death by amiodarone in acute MI patients.[111]

Ischemic Heart Disease

The main prognostic determinant in ischemic patients presenting with NSVT is the LVEF. In the Multi-center Postinfarction Research Program (MPIP) study, the presence of NSVT was associated with a twofold increase in total and arrhythmic deaths, independent of LVEF.[7,114] In the same analysis, LVEF less than 30% was associated with a fivefold increase in the risk of death, and the combination of low EF and NSVT was associated with the greatest risk of total and arrhythmia-related death, 57% and 34%, respectively. A similar relationship between repetitive forms and mortality, and the joint predictive value of NSVT and low EF was noted in the Multicenter Investigation of Limitation of Infarct Size (MILIS).[115] Recent evidence also suggests that in the patient with ischemic heart disease, NSVT no longer appears to be an independent predictor of death if other factors such as the EF are taken into account. In the GISSI-2 trial, NSVT was a significant predictor of mortality in univariate analysis but not independently in multivariate analysis involving other clinical variables.[9] Similarly, in the ESVEM trial, although univariate analysis suggested that there was an association between the presenting arrhythmia and outcome, multivariate analysis failed to establish the predictive value of the presenting arrhythmia. LVEF was the single most important predictor of arrhythmic death or cardiac arrest in patients with life-threatening arrhythmias who were treated with antiarrhythmic drugs. Over a 6-year follow-up period, 285 of the 486 patients enrolled in the ESVEM trial had an arrhythmia recurrence. Patients with EF greater than 40% had a 5% risk of developing a malignant arrhythmia, whereas for each decrease of 5% in LVEF, the risk of cardiac arrest or arrhythmic death increased by 15%.[39]

It is becoming apparent now that patients with LVEF less than 30% have an adverse prognosis, regardless of the presence of NSVT or not. In ischemic patients ICD implantation may reduce mortality, although this is not the case with dilated cardiomyopathy. Ischemic patients with impaired LV function (EF 30% to 40%) may invariably have NSVT; in this cohort programmed electrical stimulation is useful for risk stratification according to the MUSTT and MADIT trials.[13,23] LV function assessment is the most important investigation in this case, therefore, and methods of improving it (by revascularization or perhaps newer investigative techniques such as autologous bone marrow stem cell transplantation) should be the main therapeutic strategy. In patients with preserved LV function (LVEF > 40%), no hard data exist.

Nonischemic Heart Disease

In conditions other than ischemic heart disease, the independent prognostic significance of NSVT, at least as sudden arrhythmic death is concerned, is even less established. There has been evidence regarding its prognostic significance in young patients with hypertrophic obstructive cardiomyopathy[19] and particularly in the young,[22] but several questions remain in patients with dilated cardiomyopathy[16-18] and other congenital or idiopathic arrhythmogenic conditions.[79,116,117] Clearly, more data are needed for risk stratification of these patients, and ongoing trials are eagerly awaited in this respect.[118,119]

No convincing evidence exists to prove that NSVT is an independent predictor of sudden death in patients with valve disease, repaired congenital abnormalities, or hypertension.[29-35] In theses conditions, therefore, the approach of the patient from the arrhythmia point of view currently remains empirical.

Management and Evidence-Based Therapy

NSVT is an electrocardiographic finding always triggering questions about its etiology, clinical significance, and management.[120,121] The clinical approach to a patient with NSVT depends on several factors including presence or absence of underlying heart disease, associated symptoms, and duration of the episodes. NSVT may occur in patients with structurally normal hearts, but is usually identified in patients with various heart diseases.

Episodes of NSVT usually are asymptomatic, although in patients with compromised LV function they may cause syncope, especially if NSVT is fast and lasting more than few seconds. Although asymptomatic, NSVT may bear significant prognostic meaning, further influencing its management and therapy.

CLINICAL MANAGEMENT OF PATIENTS WITH NSVT

Episodes of NSVT are usually recorded while performing ECG testing for specific clinical reasons. These could include prophylactic ECG, Holter, and exercise testing, as well as ECG recordings ordered due to presence of symptoms (e.g., palpitations, syncope) or presence of underlying disease (e.g., coronary artery disease), or both. Newly diagnosed episodes of NSVT usually require further steps of clinical evaluation, which depend on underlying health conditions.

NSVT identified in otherwise healthy individuals requires clinical evaluation in order to exclude organic heart disease and possible comorbidities. A standard 12-lead ECG might indicate potential causes of NSVT including signs of myocardial hypertrophy, conduction disturbances, prolonged QT, or evidence for myocardial ischemia or necrosis. Structural and functional abnormalities of the heart (evaluated in ECG) may include depressed LV function due to cardiomyopathy, presence of myocardial hypertrophy, or signs of right ventricular dysfunction, to name a few. While considering comorbidities, electrolyte and metabolic imbalance and proarrhythmic effect of drugs should always be considered. In the absence of the abnormalities mentioned previously, exercise testing is warranted to evaluate presence of ischemic heart disease and to further explore a possibility of infrequent causes of arrhythmias including arrhythmogenic right ventricular dysplasia or catecholaminergic polymorphic VT. Frequently, exercise testing needs to be supplemented (or replaced) by stress echocardiography, stress myocardial scintigraphy and radionuclide ventriculography, or coronary angiography in order to rule out or confirm underlying ischemic heart disease.

NSVTs are also found in ECG recordings (Holter, event recorder, implantable recorder) while evaluating patients with syncope. The evaluation process in such cases frequently includes programmed electrical stimulation, which might be useful in further steps of patient management in the presence of ischemic heart disease. Programmed electrical stimulation is generally not useful in subjects with NSVT and other conditions like nonischemic dilated cardiomyopathy and HCM.

The management and evidence-based therapy of patients with NSVT are described in the following sections. Specific underlying conditions are classified as nonischemic and ischemic. Nonischemic conditions with NSVT may include a structurally normal heart, hypertension, mitral valve prolapse and other valvular disorders, HCM, nonischemic dilated cardiomyopathy, long QT syndrome, Brugada syndrome, and other rare genetic disorders. Ischemic disease with NSVT cannot be considered as a uniform condition, and its management depends on the presence or absence of MI and the degree of LV dysfunction.

Nonsustained Ventricular Tachycardia with a Structurally Normal Heart

NSVT is found infrequently (0% to 3%) in apparently healthy asymptomatic subjects with a structurally normal heart and is usually not associated with worse prognosis.[103] NSVT during exercise in otherwise healthy individuals occurs infrequently, mainly in those older than 65 years of age, and it does not adversely alter prognosis in the absence of myocardial ischemia.[105] Subjects with episodes of NSVT before 40 years of age should be evaluated primarily to rule out nonischemic causes of arrhythmia, including HCM, long QT syndrome, idiopathic dilated cardiomyopathy, and arrhythmogenic right ventricular dysplasia. Episodes of polymorphic NSVT may indicate the presence of genetic arrhythmia disorders (long QT syndrome, catecholaminergic polymorphic ventricular tachycardia) or drug-induced repolarization abnormalities. ECG testing, history of sudden death and syncopal episodes in family members, and history of medication being taken may lead to proper diagnosis. NSVT at 40 years of age and older requires that the rare causes mentioned earlier should be ruled, but the main focus is ischemic heart disease.

Sometimes episodes of NSVT in patients without structural heart disease may cause symptoms and require treatment. NSVT originating from the RVOT may occasionally cause syncope, although the risk of death is very low.[122] Subjects with RVOT ventricular tachycardia might have subtle structural and functional abnormalities of RVOT as detected by magnetic resonance imaging.[123] Although changes in autonomic tone have been implicated as underlying these arrhythmias, there is no evidence of abnormal sympathetic innervation in this disorder. The diagnosis of RVOT ventricular tachycardia needs to be differentiated from the diagnosis of arrhythmogenic right ventricular dysplasia/cardiomyopathy, which can be accomplished by a series of noninvasive ECG and imaging studies and, if needed, endomyocardial biopsy. As RVOT ventricular tachycardia does not increase the risk of sudden death, asymptomatic patients do not require treatment. In symptomatic patients, treatment should be directed toward relief of tachycardia-related symptoms with β-blockers or verapamil and adenosine, if needed in an acute setting. If medication is not effective, radiofrequency catheter ablation is recommended—it is successful in more than 80% of cases with a low risk of relapse during follow-up.[124]

Rarely, exercise-induced polymorphic NSVT is observed in some patients with no structural heart disease, and these arrhythmias are associated with syncope and increased risk of sudden cardiac death.[76] β-Blocker therapy is the treatment of choice, although defibrillator implantation might be necessary in addition to β-blockers in some patients with catecholaminergic polymorphic ventricular tachycardia.

Nonsustained Ventricular Tachycardia in Hypertension

Hypertension, especially if accompanied by ECG or echocardiographic signs of LV hypertrophy, is associated with increased incidence of ventricular ectopic beats and NSVT and increased risk of sudden cardiac death.[34] Myocardial fibrosis and increased collagen content seem to contribute to arrhythmogenic conditions in hypertension.[34] The occurrence of NSVT in hypertensive patients requires evaluation of ischemic heart disease. However, the prognostic value of NSVT in patients with lone hypertension (without evidence for concomitant ischemic heart disease) remains unclear. There is no evidence that either reduction of LV hypertrophy or elimination of NSVT improves survival in patients with hypertension.[34] Aggressive treatment of hypertension (including β-blockers) is the therapy of choice in patients with hypertension and NSVT.

Nonsustained Ventricular Tachycardia in Mitral Valve Prolapse and in Valvular Diseases

Initial studies suggested that patients with mitral valve prolapse had an increased incidence of NSVT,[125] and that the presence of NSVT increased the risk of sudden cardiac death in these patients.[126] Kilgfield et al.[127] estimated the annual risk of sudden cardiac death in patients with mitral valve prolapse at 1.9/10,000 patients, lower than the annual risk of sudden cardiac death from all causes in the adult population in the United States. Thus, patients with mitral valve prolapse are not at increased risk of sudden cardiac death. NSVT in these patients is related primarily to coexisting mitral regurgitation and secondarily chronic hemodynamic and myocardial abnormalities, but they do not independently contribute to the risk of death.[127] The majority of patients with mitral valve prolapse and NSVT do not require extensive diagnostic evaluation or treatment. In symptomatic patients (palpitations, syncope) with NSVT, β-blocker is considered the first-line therapy. Other antiarrhythmic agents may be used in cases refractory to β-blocker treatment. Treatment should be directed primarily at improvement of overall quality of life, while considering the potential adverse (potential proarrhythmic) effects of therapy.

NSVT is common in patients with other valvular diseases, particularly aortic stenosis and mitral regurgitation. Presence of NSVT is usually associated with LV hypertrophy or dysfunction, but there is no clear evidence that these arrhythmias indicate increased risk of mortality after valve replacement. Limited data suggest that valve replacement may reduce the risk of sudden death in such patients.

Nonsustained Ventricular Tachycardia in Hypertrophic Cardiomyopathy

The proportion of patients with HCM presenting with NSVT may vary from 20% in unselected subjects to 80% in those with history of syncope or cardiac arrest.[19-21,128] These estimations may not reflect the true incidence of NSVT in HCM since they are based on highly selected referral populations. In addition to a history of syncope, cardiac arrest, or a family history of sudden death, the presence of NSVT in patients with HCM seems to indicate increased risk of sudden death.[19] There is little evidence of benefit from electrophysiology studies in patients with HCM and NSVT. In patients with a history of cardiac arrest, polymorphic VT that deteriorates into VF was induced in only about half.[20,21,128]

The effects of pharmacologic antiarrhythmic therapy in patients with HCM are disappointing.[20,21,128] β-Blockers may decrease the risk of sudden death; however, the risk of death on β-blocker remains significant. Amiodarone was shown to suppress VT induction in some patients while causing conduction abnormalities and facilitating induction of VT in others[20,21,128]; therefore, this drug cannot be considered as therapy of choice in high-risk patients with HCM.

Recent data indicate that defibrillator implantation is the therapy of choice in the primary and secondary prevention of sudden death in patients with HCM.[129] The patients with a history of syncope, a family history of sudden death, or history of NSVT who had received defibrillators as primary prevention had a 5% per year appropriate discharge rate. In patients resuscitated from cardiac arrest (secondary prevention), the rate of appropriate discharges was 11%. More than a third of the patients who had an appropriate discharge of the defibrillator were taking amiodarone, further emphasizing the limited role of antiarrhythmic therapy in these patients. In a recent study on 531 patients with HCM, a relation between arrhythmic mortality and the frequency, duration, and rate of NSVT episodes could not be demonstrated.[22] However, NSVT was associated with a substantial increase in sudden death risk in young patients with hypertrophic obstructive cardiomyopathy.[22] These observations indicate that presence of high-risk factors, including NSVT accompanied by history of syncope or family history of sudden death at young age, may justify defibrillator implantation as primary prevention in patients with HCM.

Nonsustained Ventricular Tachycardia in Nonischemic Dilated Cardiomyopathy

There is a high incidence (40% to 70%) of usually asymptomatic NSVT in patients with nonischemic dilated cardiomyopathy, a patient population with high risk of sudden death.[14,41,130,131] The frequency and complexity of ventricular arrhythmias increases with decreasing function of the left ventricle and with presence of symptoms of congestive heart failure in patients with nonischemic dilated cardiomyopathy. However, cardiac death in patients with nonischemic dilated cardiomyopathy and functional class IV heart failure is more frequently due to bradyarrhythmia or electromechanical dissociation than due to ventricular tachyarrhythmias.[15,16]

There is no consistency in reports investigating the association between NSVT and cardiac mortality in nonischemic dilated cardiomyopathy. Some studies have found that the presence of NSVT is associated with increased cardiac mortality, but there is no consistency in the findings.[14,17,18] In the GESICA trial,[17] the risk of sudden cardiac death was found to be significantly increased in patients with NSVT. In contrast, the CHF-STAT trial investigators did not find increased risk for sudden death in patients with NSVT.[14,18]

The prognostic usefulness of programmed stimulation in patients with nonischemic dilated cardiomyopathy, including those with NSVT, remains controversial, but there is growing evidence that inducibility of ventricular tachyarrhythmias is of limited use.[132-134] Pooled data from few small studies showed that sustained VT is induced in less than 10% of asymptomatic patients with NSVT, and those inducible have higher risk of sudden death than those noninducible. However, recent observations and clinical practice indicate that programmed stimulation has no role in risk stratification in patients with nonischemic dilated cardiomyopathy.[134]

Patients with NSVT in the course of nonischemic dilated cardiomyopathy frequently are treated with amiodarone. Some small studies showed benefit of amiodarone therapy in preventing sudden death. There were two large trials evaluating amiodarone therapy in such patients. In the GESICA trial[17] amiodarone therapy was associated with decreased mortality, whereas no such effect was observed in the CHF-STAT trial.[18] The reasons for different frequency of NSVT in both studies (34% and 79%, respectively) and different effectiveness of amiodarone therapy are not clear and might be related to different characteristics of patient populations (with the GESICA trial enrolling patients with more advanced LV dysfunction and a high incidence of patients with Chagas' disease).

Recently presented results of the Sudden Cardiac Death in Heart Failure Trial (SCD-HeFT)[118] demonstrate that amiodarone is not effective in preventing mortality in nonischemic dilated cardiomyopathy patients (hazard ratio = 1.07 in 813 patients randomized to amiodarone or placebo). The SCD-HeFT data seem to close the controversy regarding the usefulness of amiodarone for primary prevention of sudden death in nonischemic (and also in ischemic) cardiomyopathy patients. There is no benefit from amiodarone as primary prevention management.

Since completion of the Defibrillators in Nonischemic Cardiomyopathy Treatment Evaluation (DEFINITE) and SCD-HeFT studies, patients with nonischemic dilated cardiomyopathy and unexplained syncope was the only group of patients that benefitted from ICD therapy.[135] Unexplained syncope in such patients could be caused by undocumented episodes of NSVT or sustained VT. In a study by Knight et al.,[135] 14 cardiomyopathy patients with unexplained syncope and 19 patients with cardiac arrest received ICDs. During long-term follow-up, seven (50%) of the syncope patients and eight (42%) of the cardiac arrest patients received appropriate ICD therapy for ventricular tachyarrhythmias.

Results from the prematurely stopped AMIOVIRT trial with ICDs compared with amiodarone therapy in patients with nonischemic cardiomyopathy (EF < 35%) and asymptomatic NSVT did not show a difference in all-cause mortality between the groups.[136] However, this study was severely underpowered with just 103 patients randomized to ICD or amiodarone therapy. An additional 75 patients in the trial refused randomization, including 65 on amiodarone and 10 with ICDs. The effect of defibrillator implantation as the primary prevention of cardiac mortality in patients with nonischemic dilated cardiomyopathy was the subject of the recently completed DEFINITE trial.[119] In this study, 458 patients with nonischemic dilated cardiomyopathy with EF less than 36%, symptomatic congestive heart failure, and NSVT or more than 10 premature ventricular contractions per hour were randomized to defibrillator versus no defibrillator therapy (229 patients in each arm). The mean age of the patients was 58 years. They were predominantly males (71%, mean EF = 21%). Regarding congestive heart failure, 22% of the patients were in New York Heart Association (NYHA) class I, 57% in class II, and 21% in class III. NSVT was present in 23% of the patients. During a mean 29-month follow-up, there were 68 deaths, 28 in the ICD arm and 40 in the non-ICD arm (P = .08 from the log-rank test). Two-year mortality was 14.1% in the non-ICD arm and 7.9% in the ICD arm (hazard ratio was .65 with confidence interval: .40-1.06). The hazard ratio for sudden death was .20 with P = .006. Patients with NYHA class III demonstrated significant benefit from ICD therapy with a hazard ratio of .37 (P = .02). The benefit of ICD therapy was observed in addition to optimal pharmacologic therapy (86% of patients were on ACE-inhibitors and 85% on β-blockers). Despite this state-of-the-art therapy, there was a 7% annual mortality in the non-ICD arm, substantially reduced by ICD therapy.

The SCD-HeFT study of 2521 patients with ischemic (52%) and nonischemic (48%) cardiomyopathy (EF < 36% and NYHA class II or III) randomized to amiodarone, ICD therapy, or conventional therapy was recently completed.[118] The trial enrolled 70% of patients with NYHA class II and 30% with class III, with a mean EF of 25%. A history of NSVT was reported in 23% of patients. During a mean 45-month follow-up, there was a 7.2% annual mortality in the placebo group. ICD therapy was associated with a 23% reduction in mortality (hazard ratio = .77; P = .007) in all patients combined. Nonischemic cardiomyopathy patients had a hazard ratio of .73, and ischemic cardiomyopathy patients had a hazard ratio of .79 (no significant difference between the two subgroups).

Results from DEFINITE and SCD-HeFT show that nonischemic cardiomyopathy patients with EF less than 36% and NYHA class II or III have a high 7% annual mortality despite optimal pharmacologic therapy and that the risk of mortality could be substantially reduced by ICD therapy.[118,119] These findings open further indications for ICD therapy in nonischemic cardiomyopathy patients beyond those based on syncope and arrhythmic events.

NSVT in Long QT Syndrome, Brugada Syndrome, and ARVD

Long QT syndrome is a hereditary disorder with prolonged Q–T interval on ECG and high propensity to syncope, cardiac arrest, or sudden death. Usually cardiac events in long QT syndrome patients are due to prolonged polymorphic NSVT or sustained VT (torsades de pointes) that might degenerate into ventricular fibrillation.[137] However, transient, asymptomatic runs of polymorphic NSVT are also common. β-blocker is the therapy of choice in long QT syndrome patients; it is effective in 70% of patients, whereas the remaining 30% of patients continue to have cardiac events despite β-blocker treatment.[138] Those patients with recurrent syncope despite β-blocker therapy and those with cardiac arrest should be considered for ICD therapy, which is successful in preventing death.[139]

Brugada syndrome is another rare genetic disorder associated with high risk of sudden death. Although there are no systematic studies evaluating prognostic significance of NSVT, there is agreement that symptomatic patients (i.e., those with syncope or cardiac arrest presumably due to NSVT or sustained VT) should be treated with ICDs.[140]

Similarly, patients with arrhythmogenic right ventricular dysplasia (or cardiomyopathy) are at high risk of sudden death.[141] ARVD patients may present with asymptomatic NSVT despite only subtle RV abnormalities. The results of antiarrhythmic medications and catheter ablation are disappointing, and ICD seems to be the therapy of choice in patients with documented arrhythmias or those who are symptomatic.[141]

Nonsustained Ventricular Tachycardia in Coronary Artery Disease

From the beginning of risk stratification studies in patients after MI, ventricular arrhythmias, particularly NSVT, were considered as harbingers of cardiac and sudden death. The prevalence of NSVT after MI was reported in 3% to 40% of cases depending on the series of patients, timing of ECG recordings, and therapy evolving over the past 2 decades.[9,114,120] In recently analyzed series, NSVT is present in up to 10% of patients evaluated during hospitalization after the first day of MI.[6,142] The occurrence of NSVT during the first day of acute MI, although frequent (up to 40% of patients), is not associated with increased risk for subsequent long-term mortality. However, the presence of NSVT after the first 13 hours from the onset of MI is associated with significantly increased mortality both in older and newer series of postinfarction patients.[6] In the Multicenter Post-infarction Research Program[114] study involving 820 patients after MI, NSVT was recorded in 11% of patients who had a 21% 3-year incidence of sudden death, significantly higher than patients without NSVT (8%). In multivariate analysis, NSVT was associated with a twofold increase in total mortality and in sudden death. The risk associated with NSVT was much higher in patients with depressed LV function. Frequent ventricular ectopy and NSVT also was found to be associated with adverse outcome in the Gruppo Italiano per lo Studio della Sopravvivenza nell'Infarto miocardico-2 Investigators (GISSI-2) study, conducted in the thrombolytic era.[9]

In contrast to patients with nonischemic heart diseases, postinfarction patients with NSVT may benefit from programmed electrical stimulation to stratify their risk of cardiac events. Denniss et al.[143] showed in a cohort of unselected 403 postinfarction patients undergoing programmed stimulation that patients with inducible sustained VT had a fourfold increase in cardiac death or sustained VT in comparison to noninducible patients. The inducibility and outcome were related to the degree of LV dysfunction.

The combination of impaired ventricular function, spontaneous NSVT, and inducible SMVT identifies a particularly high-risk group of chronic postinfarction patients. In the study of Wilber et al.,[81] 100 patients with spontaneous NSVT, remote MI, and EF less than 40% underwent EP testing. The induction of sustained monomorphic VT was associated with a 43% 2-year actuarial risk of sudden death or cardiac arrest, whereas this risk was 6% in noninducible patients. In a multivariate Cox analysis, the induction of sustained VT was associated with a fourfold increased risk of sudden death or cardiac arrest. The recently completed MUSTT trial also showed that patients with NSVT and inducible ventricular tachycardia had a significantly higher risk of cardiac arrest or death from arrhythmia than those noninducible.[144] The inducible group had 28% whereas noninducible group had 21% 2-year total mortality rate. The 21% 2-year total mortality rate in noninducible patients is high enough to consider the need for programmed electrical stimulation at least questionable. These findings are not surprising in light of observations that there is little association between characteristics of spontaneous and induced episodes of ventricular arrhythmias.[84]

The management of postinfarction patients with NSVT should first include modern therapy for ischemic heart disease including revascularization procedures, the use of β-blockers, statins, and ACE-inhibitors. This therapy is the most effective antiarrhythmic measure in patients with coronary artery disease as demonstrated by the Coronary Artery Bypass Graft (CABG) Patch (CABG-Patch) trial.[145]

Since the presence of ventricular ectopy (including NSVT) was found to be associated with increased mortality, a series of studies were designed to determine whether antiarrhythmic therapy suppressing ventricular arrhythmias reduces mortality.[146] The CAST and CAST II[68,69] trials demonstrated that class IC antiarrhythmic agents (encainide, flecainide, and moricizine) confer increased (not decreased) mortality in postinfarction patients. Subsequently, d-sotalol (class III drug) tested in the Survival With Oral d-Sotalol (SWORD) trial[147] also was associated with increased mortality in comparison to placebo. The European Myocardial Infarct Amiodarone Trial (EMIAT)[113] and the Canadian Amiodarone Myocardial Infarction Arrhythmia Trial (CAMIAT)[148] failed to demonstrate beneficial effect of amiodarone on survival. Therefore, no antiarrhythmic drug is suitable for primary prevention of cardiac death with exception for β-blockers, which were shown in several studies to reduce total and cardiac mortality in postinfarction patients. However, in low- to moderate-risk postinfarction patients with NSVT from the Beta-blocker Heart Attack Trial, there was no reduction in mortality in patients with NSVT.[149]

In the 1990s the implantable defibrillator became a feasible alternative to pharmacologic antiarrhythmic therapy for primary prevention of sudden cardiac death in patients with coronary artery disease and NSVT. Two randomized, controlled trials have provided much-needed evidence regarding treatment of postinfarction patients with NSVT.[23,40] In the MADIT study,[40] 196 patients with prior MI, EF less than or equal to 35%, a documented episode of NSVT, and inducible and nonsuppressible sustained VT were randomized to receive an implanted defibrillator or conventional (primary amiodarone) non-ICD therapy. There were 15 deaths in 95 patients randomized to ICD therapy and 39 deaths in 101 patients randomized to conventional

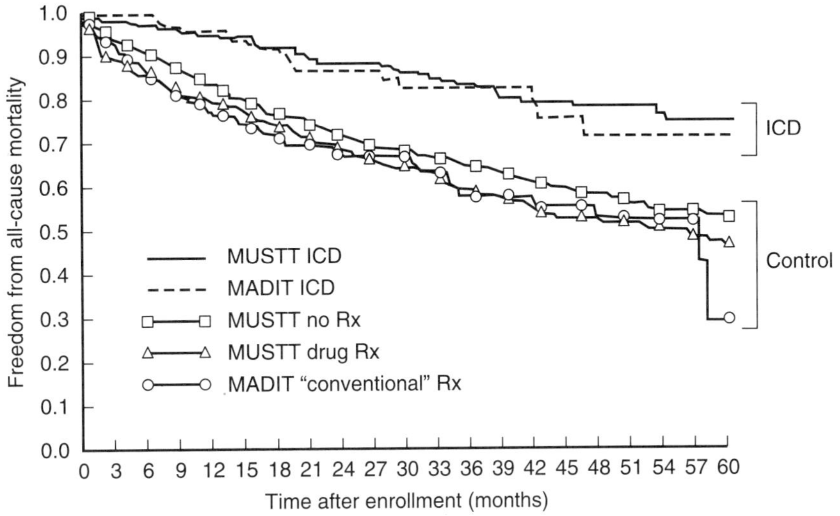

FIGURE 16-8 Kaplan-Meier survival curves for Multicenter UnSustained Tachycardia Trial (MUSTT) and Multicenter Automatic Defibrillator Implantation Trial (MADIT). The *three lower curves* represent conventional therapy in MADIT, and electrophysiological-guided antiarrhythmic drug therapy or no specific antiarrhythmic therapy in MUSTT. The *two upper curves* show survival outcomes for patients treated with ICDs in the two studies. The risk ratio for reduction in mortality for the ICD-treated patients versus the controls was 0.49 ($P < .0001$) for MUSTT and 0.46 ($P < .009$) for MADIT. (From Prystowsky EN, Nisam S: Prophylactic implantable cardioverter defibrillator trials. Am J Cardiol 2000;86:1214-15.)[150]

therapy, yielding a 54% reduction in total mortality associated with ICD during mean 27-month follow-up. This was the first study demonstrating that ICD therapy is beneficial as primary prevention in high-risk postinfarction patients.

Antiarrhythmic therapy guided by invasive EP testing has historically been used to treat patients with NSVT and inducible sustained VT. The MUSTT[23] study employed the concept of EP-guided antiarrhythmic to determine whether this approach can reduce mortality in survivors of MI with EF less than or equal to 40% and NSVT. The main study enrolled 704 patients with coronary artery disease, asymptomatic NSVT, EF less than or equal to 40%, and inducible sustained VT. Patients with induced, sustained VT were randomized to no antiarrhythmic therapy or to EP-guided therapy. The EP-guided therapy included pharmacologic antiarrhythmic therapy and implantable defibrillators as indicated, based on EP study. The risk of cardiac arrest or death from arrhythmia among patients who received treatment with defibrillators in the MUSTT study was significantly lower than that among patients who did not receive ICDs (relative risk, 0.24; 95% confidence interval, 0.13 to 0.45; $P < .001$). Patients assigned to EP study–guided therapy who were treated with only pharmacologic antiarrhythmic therapy had a similar rate of death and cardiac arrest as those in the control group who had not been treated with antiarrhythmic therapy. Figure 16-8 shows the superimposition of cumulative survival curves in patients treated and not treated with ICDs in the MADIT and MUSTT trials.[151] There is a remarkable consistency in the findings of these two studies, further substantiating that ICD indication established by MADIT has been confirmed in a larger study. These studies demonstrate that ICDs used in postinfarction patients with depressed EF and NSVT significantly reduce the risk of SCD and total mortality.

Since the subset of postinfarction patients with depressed EF, documented NSVT, and inducible sustained VT is relatively small, there is a need to identify broader groups of postinfarction patients who may benefit from prophylactic ICD therapy. Two studies, MADIT II and SCD-HeFT, were launched in the late 1990s to address this question. MADIT II[83] hypothesized that postinfarction patients with markedly depressed LV function (EF ≤ 30%), regardless of presence or absence of other risk stratifiers including frequent ventricular ectopy and inducibility of sustained VT, will benefit from ICD therapy. SCD-HeFT[118] was designed to compare the effectiveness of ICD therapy versus amiodarone and versus placebo in preventing total mortality in patients with congestive heart failure (of both ischemic and nonischemic etiology) and EF less than or equal to 35%. The MADIT II trial has been recently completed after enrollment of 1232 postinfarction patients (742 randomized to ICD therapy and 490 to non-ICD arm), demonstrating that ICD therapy was associated with a significant 30% reduction in total mortality during a 20-month follow-up.[83] Recent results of the SCD-HeFT (with 52% of the patients having ischemic cardiomyopathy) confirmed the MADIT II findings indicating substantial benefit from ICD therapy in patients with low EF (<36%) and NYHA classes II and III.

The MADIT II study significantly changes the clinical approach to postinfarction patients with NSVT. In light of MADIT II, postinfarction patients with EF less than or equal to 30% should undergo ICD therapy regardless of the presence or absence of NSVT. Based on MADIT and MUSTT findings, postinfarction patients with NSVT and EF of 31% to 40% should undergo EP testing and, if inducible, should be treated with ICDs. The incidence of NSVT in postinfarction patients with EF greater than 40% is low, and the prognosis in such patients is much better than in those with EF less than or equal to 40%. There are no data regarding optimal therapy in postinfarction patients with NSVT and EF greater than 40%, and the role of EP-guided therapy and ICDs in these patient populations remains to be determined.

SUMMARY

1. NSVT can be recorded in an extremely wide range of conditions, from apparently healthy individuals to patients with significant heart disease.

TABLE 16-2 Management and Evidence-Based Therapy in Patients with Nonsustained Ventricular Tachycardia Depending on Underlying Conditions

NSVT Setting	Management	Evidence-Based Rx
Nonischemic Conditions		
Asymptomatic, no overt heart disease	Evaluate for ischemic heart disease and other organic disorders: ECG; exercise test; ECHO; and other test, if needed	No treatment necessary
Symptomatic with RVOT morphology of NSVT	EP testing Differentiate with ARVD	β-blockers, verapamil, adenosine Rf ablation, if needed
Hypertrophic cardiomyopathy	Evaluate other risk factors: syncope, family history	β-blockers, amiodarone, ICD if other risk factors present
Nonischemic dilated cardiomyopathy	No EP testing needed	
EF <36%		ICD therapy
EF ≥36% and syncope with documented VT/VF or cardiac arrest		ICD therapy
LQTS with recurrent syncope or cardiac arrest	No testing needed	β-blockers, ICD
Brugada syndrome with syncope or cardiac arrest	No testing needed	ICD
ARVD with syncope or cardiac arrest	No testing needed	ICD
Ischemic Conditions		
Acute MI, NSVT < 24 hr	Routine for acute MI	Primary angioplasty, thrombolysis
NSVT > 24 hr until predischarge	Routine for acute MI EP testing if EF < 40%	Optimal CAD therapy* Revascularization
Asymptomatic CAD w/o MI or MI with EF > 40%	No need for specific management	Optimal CAD therapy*
Syncope in CAD w/o MI or MI with EF > 40%	EP testing	Optimal CAD therapy* If EP-inducible: ICD
MI with EF=31%-40%	EP testing	Optimal CAD therapy* If EP-inducible: ICD
MI with EF ≤ 30%	No further testing needed	Optimal CAD therapy* ICD therapy

*Optimal CAD therapy defined as administration of β-blockers, statins, ACE-inhibitors (when needed), and revascularization therapy.

ARVD, arrhythmogenic right ventricular dysplasia; CAD, coronary artery disease; EF, ejection fraction; EP, electrophysiological; ICD, implantable cardioverter-defibrillator; LQTS, long QT syndrome; MI, myocardial infarction; NSVT, nonsustained ventricular tachycardia; RVOT, right ventricular outflow tract.

2. The prognostic significance of NSVT is strongly influenced by the type and severity of underlying heart disease.
3. The current evidence-based therapy in patients with NSVT is summarized in Table 16-2.
4. The ICD therapy remains the mainstay of current antiarrhythmic regimen in high-risk conditions complicated by NSVT.

REFERENCES

1. Camm AJ, Katritsis D: Risk stratification of patients with ventricular arrhythmias. In Zipes DP, Jalife J (eds): Clinical Electrophysiology. From Cell to Bedside, 3rd ed. Philadelphia, WB Saunders, 2000, pp 808-27.
2. Chimenti C, Calabrese F, Thiene G, et al: Inflammatory left ventricular microaneurysms as a cause of apparently idiopathic ventricular tachyarrhythmias. Circulation 2001;104:168-73.
3. Ouyang F, Cappato R, Ernst S, et al: Electroanatomic substrate of idiopathic left ventricular tachycardia: Unidirectional block and macroreentry within the Purkinje network. Circulation 2002; 105:462-9.
4. Senges JC, Becker R, Schreiner KD, et al: Variability of Holter electrocardiographic findings in patients fulfilling the noninvasive MADIT criteria. Multicenter Automatic Defibrillator Implantation Trial. Pacing Clin Electrophysiol 2002;25:183-90.
5. Farmer DM, Swygman CA, Wang PJ, et al: Evidence that nonsustained polymorphic ventricular tachycardia causes syncope (data from implantable cardioverter defibrillators). Am J Cardiol 2003;91:606-9.
6. Cheema AN, Sheu K, Parker M, et al: Nonsustained ventricular tachycardia in the setting of acute myocardial infarction: Tachycardia characteristics and their prognostic implications. Circulation 1998;98:2030-6.
7. Bigger JT, Fleiss JL, Rolnitzky LM: Prevalence, characteristics and significance of ventricular tachycardia detected by 24-hour continuous electrocardiographic recordings in the late hospital phase of acute myocardial infarction. Am J Cardiol 1986;58:1151-60.
8. Burkart F, Pfisterer M, Kiowski W, et al: Effect of antiarrhythmic therapy on mortality in survivors of myocardial infarction with asymptomatic complex ventricular arrhythmias: Basel Antiarrhythmic Study of Infarct Survival (BASIS). J Am Coll Cardiol 1990;16:1711-8.
9. Maggioni AP, Zuanetti G, Franzosi MG, et al, on behalf of GISSI-2 Investigators: Prevalence and prognostic significance of ventricular arrhythmias after acute myocardial infarction in the fibrinolytic era: GISSI-2 results. Circulation 1993;87:312-22.
10. Ruberman W, Weinblatt E, Golberg JD, et al: Ventricular premature beats and mortality after myocardial infarction. N Engl J Med 1977;297:750-7.
11. Califf RM, McKinnis RA, Burks J, et al: Prognostic implications of ventricular arrhythmias during 24 hour ambulatory monitoring in patients undergoing cardiac catheterization for coronary artery disease. Am J Cardiol 1982;50:23-31.
12. Moss AJ: Dead is dead, but can we identify patients at increased risk for sudden cardiac death? J Am Coll Cardiol 2003;42:659-60.
13. Buxton AE, Lee KL, Hafley GE, et al; MUSTT Investigators: Relation of ejection fraction and inducible ventricular tachycardia to mode of death in patients with coronary artery disease: An analysis of patients enrolled in the multicenter unsustained tachycardia trial. Circulation 2002;106:2466-72.

14. Singh SN, Fisher SG, Carson PE, Fletcher RD: Prevalence and significance of nonsustained ventricular tachycardia in patients with premature ventricular contractions and heart failure treated with vasodilator therapy. J Am Coll Cardiol 1998;32:942-7.
15. Sugrue DD, Rodeheffer RJ, Codd MB, et al: The clinical course of idiopathic dilated cardiomyopathy. A population-based study. Ann Intern Med 1992;117:117-23.
16. Tamburro P, Wilber D: Sudden death in idiopathic dilated cardiomyopathy. Am Heart J 1992;124:1035-45.
17. Doval HC, Nul DR, Grancelli HO, et al: Nonsustained ventricular tachycardia in severe heart failure. Independent marker of increased mortality due to sudden death. GESICA-GEMA Investigators. Circulation 1996;94:3198-203.
18. Singh SN, Fletcher RD, Fisher SG, et al: Amiodarone in patients with congestive heart failure and asymptomatic ventricular arrhythmia. Survival Trial of Antiarrhythmic Therapy in Congestive Heart Failure. N Engl J Med 1995;333:77-82.
19. Maron BJ, Savage DD, Wolfson JK, Epstein SE: Prognostic significance of 24 hour ambulatory electrocardiographic monitoring in patients with hypertrophic cardiomyopathy: A prospective study. Am J Cardiol 1981;48:252-7.
20. Maron BJ: Ventricular arrhythmias, sudden death, and prevention in patients with hypertrophic cardiomyopathy. Curr Cardiol Rep 2000;2:522-8.
21. McKenna WJ, Behr ER: Hypertrophic cardiomyopathy: Management, risk stratification, and prevention of sudden death. Heart 2002;87:169-76.
22. Monserrat L, Elliott PM, Gimeno JR, et al: Non-sustained ventricular tachycardia in hypertrophic cardiomyopathy: An independent marker of sudden death risk in young patients. J Am Coll Cardiol 2003;42:873-9.
23. Buxton AE, Lee KL, Fisher JD, et al: A randomized study of the prevention of sudden death in patients with coronary artery disease. Multicenter Unsustained Tachycardia Trial Investigators. N Engl J Med 1999;341:1882-90.
24. Pires LA, Lehmann MH, Buxton AE, et al: The Multicenter Unsustained Tachycardia Trial Investigators. Differences in inducibility and prognosis of in-hospital versus out-of-hospital identified nonsustained ventricular tachycardia in patients with coronary artery disease: Clinical and trial design implications. J Am Coll Cardiol 2001;38:1156-62.
25. Pires LA, Hafley GE, Lee KL, et al: Multicenter Unsustained Tachycardia Trial Investigators: Prognostic significance of nonsustained ventricular tachycardia after coronary bypass surgery in patients with left ventricular dysfunction. J Cardiovasc Electrophysiol 2002;13:757-63.
26. Mittal S, Lomnitz DJ, Mirchandani S, et al: Prognostic significance of nonsustained ventricular tachycardia after revascularization. J Cardiovasc Electrophysiol 2002;13:342-6.
27. Russo AM, Hafley GE, Lee KL, et al; Multicenter UnSustained Tachycardia Trial Investigators: Racial differences in outcome in the Multicenter UnSustained Tachycardia Trial (MUSTT): A comparison of whites versus blacks. Circulation 2003;108:67-72.
28. Katritsis D, Gill JS, Camm AJ: Repetitive monomorphic ventricular tachycardia. In Zipes DP, Jalife J: Clinical Electrophysiology. From Cell to Bedside, 2nd ed. Philadelphia, WB Saunders, 1995, pp 900-6.
29. Klein RC: Ventricular arrhythmias in aortic valve disease: Analysis of 102 patients. Am J Cardiol 1984;53:1079-83.
30. Martinez-Rubio A, Schwammenthal Y, Schwammenthal E, et al: Patients with valvular heart disease presenting with sustained ventricular tachyarrhythmias and syncope. Results of programmed ventricular stimulation and long-term follow-up. Circulation 1997;96:500-8.
31. Garson A, Gillette PC, Gutgesell HP, McNamara DG: Stress-induced ventricular arrhythmias after tetralogy of Fallot repair. Am J Cardiol 1980;46:1006-12.
32. Deanfield JE, McKenna WJ, Hallidie-Smith KA: Detection of late arrhythmias and conduction disturbance after correction of tetralogy of Fallot. Br Heart J 1980;44:248-53.
33. Cameron JS, Myerburg RJ, Wong SS, et al: Electrophysiologic consequences of chronic experimentally induced left ventricular pressure overload. J Am Coll Cardiol 1983;2:481-7.
34. McLenachan JM, Henderson E, Morris KI, Dargie HJ: Ventricular arrhythmias in patients with hypertensive left ventricular hypertrophy. N Engl J Med 1987;317:787-92.
35. Papademetriou V, Price M, Notargiacomo A, et al: Effect of diuretic therapy on ventricular arrhythmias in hypertensive patients with and without left ventricular hypertrophy. Am Heart J 1985; 110:595-9.
36. Buxton AE, Duc J, Berger EE, Torres V: Nonsustained ventricular tachycardia. Cardiol Clin 2000;18:327-36.
37. Grimm W, Marchlinski FE: Accelerated idioventricular rhythm, bidirectional ventricular tachycardia. In Zipes DP, Jalife J (eds): Clinical Electrophysiology. From Cell to Bedside, 3rd ed. Philadelphia, WG Saunders, 2000, pp 673-7.
38. Kinder C, Tamburro P, Kopp D, et al: The clinical significance of nonsustained ventricular tachycardia: Current perspectives. Pacing Clin Electrophysiol 1994;17:637-64.
39. Caruso AC, Marcus FI, Hahn EA, et al: Predictors of arrhythmic death and cardiac arrest in the ESVEM trial. Electrophysiologic Study Versus Electromagnetic Monitoring. Circulation 1997;96: 1888-92.
40. Moss AJ, Hall WJ, Cannom DS, et al: Improved survival with an implanted defibrillator in patients with coronary disease at high risk for ventricular arrhythmia. Multicenter Automatic Defibrillator Implantation Trial Investigators. N Engl J Med 1996;335:1933-40.
41. Teerlink JR, Jalaluddin M, Anderson S, et al: Ambulatory ventricular arrhythmias in patients with heart failure do not specifically predict an increased risk of sudden death. PROMISE (Prospective Randomized Milrinone Survival Evaluation) Investigators. Circulation 2000;101:40-6.
42. Cullen K, Stenhouse NS, Wearne KL, Cumpston GN: Electrocardiograms and 13 year cardiovascular mortality in Busselton study. Br Heart J 1982;47:209-12.
43. Bikkina M, Larson MG, Levy D: Prognostic implications of asymptomatic ventricular arrhythmias: The Framingham Heart Study. Ann Intern Med 1992;117:990-6.
44. Buxton AE, Waxman HE, Marchlinski FE, et al: Right ventricular tachycardia: Clinical and electrophysiologic characteristics. Circulation 1983;68:917-27.
45. Ohe T, Shinomura K, Aihara N, et al: Idiopathic sustained left ventricular tachycardia: Clinical and electrophysiologic characteristics. Circulation 1988;77:560-8.
46. Ward DE, Nathan AW, Camm AJ: Fascicular tachycardia sensitive to calcium antagonists. Eur Heart J 1984;5:896-905.
47. Katritsis D, Ahsan A, Anderson M, et al: Catheter ablation for successful management of left posterior fascicular tachycardia. An approach guided by recording of fascicular potentials. Heart 1996;75:384-8.
48. Buxton AE, Marchlinski FE, Doherty JU, et al: Repetitive monomorphic ventricular tachycardia: Clinical and electrophysiologic characteristics in patients with and patients without organic heart disease. Am J Cardiol 1984;54:997-1002.
49. Rahilly GT, Prystowsky EN, Zipes DP, et al: Clinical and electrophysiologic findings in patients with repetitive monomorphic ventricular tachycardia and otherwise normal electrocardiogram. Am J Cardiol 1982;50:459-68.
50. Zimmermann M, Maisonblanche P, Cauchemez B, et al: Determinants of the spontaneous ectopic activity in repetitive monomorphic idiopathic ventricular tachycardia. J Am Coll Cardiol 1986;7:1219-27.
51. Wilson FN, Wishart SW, Macleod AG, et al: A clinical type of paroxysmal tachycardia of ventricular origin in which paroxysms are induced by exertion. Am Heart J 1932;8:155-62.
52. Wu D, Kou HC, Hung JS: Exercise-triggered paroxysmal ventricular tachycardia. A repetitive rhythmic activity possibly related to afterdepolarization. Ann Intern Med 1981;95:410.
53. Akhtar M: Clinical spectrum of ventricular tachycardia. Circulation 1990;82:1561-73.
54. Sclarovsky S, Stransberg B, Lewin RF, Agmon J: Polymorphous ventricular tachycardia: Clinical features and treatment. Am J Cardiol 1979;44:339-44.
55. Ruskin JN, DiMarco JP, Garan H: Out-of-hospital cardiac arrest: Electrophysiologic observations and selection of long-term antiarrhythmic thearpy. N Engl J Med 1980;303:607-13.
56. Wolfe CL, Nibley C, Bhandari A, et al: Polymorphous ventricular tachycardia associated with acute myocardial infarction. Circulation 1991;84:1543-51.
57. Bayes de Luna A, Coumel P, Leclercq JF: Ambulatory sudden cardiac death: Mechanisms of production of fatal arrhythmia on the basis of data from 157 cases. Am Heart J 1989;117:151-9.

58. Caceres J, Jazayeri M, McKinnie J, et al: Sustained bundle branch reentry as a mechanism of clinical tachycardia. Circulation 1989;79:256-70.
59. Blanck Z, Jazayeri M, Dhala A, et al: Bundle branch reentry: A mechanism of ventricular tachycardia in the absence of myocardial or valvular dysfunction. J Am Coll Cardiol 1993;22: 1718-22.
60. Gillette PC, Yeoman MA, Mullins CE, McNamara DG: Sudden death after repair of tetralogy of Fallot: Electrocardiographic and electrophysiologic abnormalities. Circulation 1977;56:566-71.
61. Marcus FI, Fontaine GH, Guiraudon G, et al: Right ventricular dysplasia: A report of 24 adult cases. Circulation 1982;65:384-98.
62. Jackman W, Clark M, Friday K, et al: Ventricular tachyarrhythmias in the long QT syndromes. In Zipes DP (ed): The Medical Clinics of North America. Philadelphia, WB Saunders, 1984, p 1079.
63. Leenhardt A, Glaser E, Burguera M, et al: Short-coupled variant of torsade de pointes: A new electrocardiographic entity in the spectrum of idiopathic ventricular tachyarrhythmias. Circulation 1993;89:206-15.
64. Doval HC, Nul DR, Grancelli HO, et al: Randomised trial of low-dose amiodarone in severe congestive heart failure. Lancet 1994;344:493-8.
65. La Rovere MT, Bigger JT Jr, Marcus FI, et al: Baroreflex sensitivity and heart-rate variability in prediction of total cardiac mortality after myocardial infarction. ATRAMI (Autonomic Tone and Reflexes After Myocardial Infarction) Investigators. Lancet 1998; 351:478-84.
66. Abildstrom SZ, Jensen BT, Agner E, et al; BEAT Study Group: Heart rate versus heart rate variability in risk prediction after myocardial infarction. J Cardiovasc Electrophysiol 2003;14:168-73.
67. Huikuri HV, Tapanainen JM, Lindgren K, et al: Prediction of sudden cardiac death after myocardial infarction in the beta-blocking era. J Am Coll Cardiol 2003;42:652-8.
68. Cardiac Arrhythmia Suppression Trial (CAST) Investigators. Preliminary report: Effect of encainide and flecainide on mortality in a randomized trial of arrhythmia suppression after myocardial infarction. N Engl J Med 1989;321:406-12.
69. The Cardiac Arrhythmia Suppression Trial-II Investigators: Effect of antiarrhythmic agent moricizine on survival after myocardial infarction. The Cardiac Arrhythmia Suppression Trial II. N Engl J Med 1992;327:227-33.
70. El-Sherif N: Clinical significance of polymorphic ventricular tachycardia induced by programmed stimulation. J Am Coll Cardiol 1993;21:99-101.
71. DeBakker JM, van Cappelle FJ, Janse MJ, et al: Reentry as a cause of ventricular tachycardia in patients with chronic ischemic heart disease: Electrophysiologic and anatomic correlation. Circulation 1988;77:589-606.
72. Mehta D, Curwin J, Gomes JA, Fuster V: Sudden death in coronary artery disease: Acute ischemia versus myocardial substrate. Circulation 1997;96:3215-23.
73. Furukawa T, Moroe K, Mayrovitz HN, et al: Arrhythmogenic effects of graded coronary blood flow reductions superimposed on prior myocardial infarction in dogs. Circulation 1991;84:368-77.
74. Garan H, McComb JM, Ruskin JN: Spontaneous and electrically induced ventricular arrhythmias during acute ischemia superimposed on 2 week old canine myocardial infarction. J Am Coll Cardiol 1988;11:603-11.
75. Palileo EV, Ashley WW, Swiryn S, et al: Exercise provocable right ventricular outflow tract tachycardia. Am Heart J 1982;104:185.
76. Leenhardt A, Lucet V, Denjoy I, et al: Catecholaminergic polymorphic ventricular tachycardia in children. A 7-year follow-up of 21 patients. Circulation 1995;91:1512-9.
77. Vlay SC: Catecholamine-sensitive ventricular tachycardia. Am Heart J 1987;114:455-61.
78. Tandri H, Calkins H, Nasir K, et al: Magnetic resonance imaging findings in patients meeting task force criteria for arrhythmogenic right ventricular dysplasia. J Cardiovasc Electrophysiol 2003;14:476-82.
79. Antzelevitch C, Brugada P, Brugada J, et al: Brugada syndrome: 1992-2002: A historical perspective. J Am Coll Cardiol 2003; 41:1665-71.
80. Kowey PR, Taylor JE, Marinchak RA, Rials SJ: Does programmed stimulation really help in the evaluation of patients with nonsustained ventricular tachycardia? Results of a meta-analysis. Am Heart J 1992;123:481-5.
81. Wilber DJ, Olshansky B, Moran JF, Scanlon PJ: Electrophysiological testing and nonsustained ventricular tachycardia. Use and limitations in patients with coronary artery disease and impaired ventricular function. Circulation 1990;82:350-8.
82. Buxton AE, Lee KL, DiCarlo L, et al: Nonsustained ventricular tachycardia in coronary artery disease: Relation to inducible sustained ventricular tachycardia. Ann Intern Med 1996;125:35-9.
83. Moss AJ, Zareba W, Hall WJ, et al: Prophylactic implantation of a defibrillator in patients with myocardial infarction and reduced ejection fraction. N Engl J Med 2002;346:877-83.
84. Monahan KM, Hadjis T, Hallett N, et al: Relation of induced to spontaneous ventricular tachycardia from analysis of stored far-field implantable defibrillator electrograms. Am J Cardiol 1999;83:349-53.
85. Wellens HJ, Doevendans P, Smeets J, et al: Arrhythmia risk: Elecrophysiological studies and monophasic action potentials. Pacing Clin Electrophysiol 1997;20(II):2560-5.
86. Rodriguez LM, Sternick EB, Smeets JL, et al: Induction of ventricular fibrillation predicts sudden death in patients treated with amiodarone because of ventricular tachyarrhythmias after myocardial infarction. Heart 1996;75:23-8.
87. Borggrefe M, Breithardt G: Predictive value of electrophysiologic testing in the treatment of drug-refractory ventricular arrhythmias with amiodarone. Eur Heart J 1986;7:735-42.
88. Greene HL: The efficacy of amiodarone in the treatment of ventricular tachycardia or ventricular fibrillation. Prog Cardiovasc Dis 1989;31:319-54.
89. Fisher JD, Kim SG, Waspe LE, Johnston DR: Amiodarone: Value of programmed electrical stimulation and Holter monitoring. Pacing Clin Electrophysiol 1986;9:422-35.
90. Horowitz LN, Greenspan AM, Spielman SR, et al: Usefulness of electrophysiologic testing in evaluation of amiodarone therapy for sustained ventricular tachyarrhythmias associated with coronary heart disease. Am J Cardiol 1985;55:367-71.
91. Kuchar DL, Thorburn CW, Sammel NL: Signal-averaged electrocardiogram for evaluation of recurrent syncope. Am J Cardiol 1986;58:949-53.
92. Gang ES, Peter T, Rosenthal ME, et al: Detection of late potentials on the surface electrocardiogram in unexplained syncope. Am J Cardiol 1986;58:1014-20.
93. Lindsay BD, Ambos HD, Schechtman KB, Cain ME: Improved selection of patients for programmed ventricular stimulation by frequency analysis of signal-averaged electrocardiograms. Circulation 1986;73:675-83.
94. Winters SL, Stewart D, Gomes JA: Signal averaging of the surface QRS complex predicts inducibility of ventricular tachycardia in patients with syncope of unknown origin: A prospective study. J Am Coll Cardiol 1987;10:775-81.
95. Nalos PC, Gang ES, Mandel WJ, et al: The signal-averaged electrocardiogram as a screening test for inducibility of sustained ventricular tachycardia in high risk patients: A prospective study. J Am Coll Cardiol 1987;9:539-48.
96. Breithardt G, Schwarzmaier J, Borggrefe M, et al: Prognostic significance of late ventricular potentials after acute myocardial infarction. Eur Heart J 1983;4:487-95.
97. Wichter T, Martinez-Rubio A: Value of the signal-averaged ECG in patients with arrhythmogenic right ventricular disease. Circulation 1992;86:319.
98. Algra A, Tijssen JGP, Roelandt JR, et al: Heart rate variability from 24-hour electrocardiography and the 2-year risk for sudden death. Circulation 1993;88:180-5.
99. Hartikainen JE, Malik M, Staunton A, et al: Distinction between arrhythmic and nonarrhythmic death after acute myocardial infarction based on heart rate variability, signal-averaged electrocardiogram, ventricular arrhythmias and left ventricular ejection fraction. J Am Coll Cardiol 1996;28:296-304.
100. La Rovere MT, Specchia G, Mortara A, Schwartz PJ: Baroreflex sensitivity, clinical correlates, and cardiovascular mortality among patients with a first myocardial infarction. Circulation 1988;78:816-24.
101. Farrell TG, Cripps TR, Malik M, et al: Baroreflex sensitivity and electrophysiological correlates in patients after acute myocardial infarction. Circulation 1991;83:945-52.
102. Farrell TG, Odemuyiwa O, Bashir Y, et al: Prognostic value of baroreflex sensitivity testing after acute myocardial infarction. Br Heart J 1992;67:129-37.

103. Kennedy HL, Whitlock JA, Sprague MK, et al: Long-term follow-up of asymptomatic healthy subjects with frequent and complex ventricular ectopy. N Engl J Med 1985;312:193-7.
104. Montague TJ, McPherson DD, MacKenzie BR, et al: Frequent ventricular ectopic activity without underlying cardiac disease: Analysis of 45 subjects. Am J Cardiol 1983;52:980-4.
105. Fleg JL, Lakatta EG: Prevalence and prognosis of exercise-induced nonsustained ventricular tachycardia in apparently healthy volunteers. Am J Cardiol 1984;54:762-4.
106. Bruce RA, DeRouen TA, Hossack KF: Value of maximal exercise tests in risk assessment of primary coronary heart disease events in healthy men: Five years' experience of the Seattle heart watch study. Am J Cardiol 1980;46:371-8.
107. Drory Y, Pines A, Fisman EZ, Kellermann JJ: Persistence of arrhythmia exercise response in healthy young men. Am J Cardiol 1990;66:1092-4.
108. Nair CK, Aronow WS, Sketch MH, et al: Diagnostic and prognostic significance of exercise-induced premature ventricular complexes in men and women: A four year follow-up. J Am Coll Cardiol 1983;1:1201-6.
109. Jouven X, Zureik M, Desnos M, et al: Long-term outcome in asymptomatic men with exercise-induced premature ventricular depolarizations. N Engl J Med 2000;343:826-33.
110. Udall JA, Ellestad MH: Predictive implications of ventricular premature contractions associated with treadmill stress testing. Circulation 1977;56:985-9.
111. Elizari MW, Martinez JM, Belzitic C, et al: Mortality following early administration of amiodarone in acute myocardial infarction: Results of the GEMICA trial. Circulation 1996;94 (suppl 1):90.
112. Cairns JA, Connolly SJ, Gent M, Roberts R: Post-myocardial infarction mortality in patients with ventricular premature depolarizations. Circulation 1991;84:550-7.
113. Julian DG, Camm AJ, Frangin J, et al: Randomised trial of effect of amiodarone on mortality in patients with left-ventricular dysfunction after recent myocardial infarction: EMIAT. Lancet 1997;349:667-74.
114. The Multicenter Post-infarction Research Group. Risk stratification and survival after myocardial infarction. N Engl J Med 1983;309:331-6.
115. Mukharji J, Rude RE, Poole WK, et al: Risk factors for sudden death after acute myocardial infarction: Two-year follow-up. Am J Cardiol 1984;54:31-6.
116. Katritsis D, Camm AJ: Role of invasive electrophysiology testing in the evaluation and management of idiopathic ventricular fibrillation. Cardiac EP Review 1998;1:452-6.
117. Priori SG, Napolitano C, Gasparini M, et al: Natural history of Brugada syndrome: Insights for risk stratification and management. Circulation 2002;105:1342-7.
118. Sudden Cardiac Death in Heart Failure Trial (SDC-HeFT): Available at http://www.sicr.org; accessed July 26, 2004.
119. Kadish A, Dyer A, Daubert JP, et al: Prophylactic defibrillator implantation in patients with nonischemic dilated cardiomyopathy. N Engl J Med 2004;350:2151-8.
120. Anderson KP, DeCamilla J, Moss AJ: Clinical significance of ventricular tachycardia (3 beats or longer) detected during ambulatory monitoring after myocardial infarction. Circulation 1978;57:890-7.
121. Denes P, Gillis AM, Pawitan Y, et al: Prevalence, characteristics and significance of ventricular premature complexes and ventricular tachycardia detected by 24-hour continuous electrocardiographic recording in the Cardiac Arrhythmia Suppression Trial. CAST Investigators. Am J Cardiol 1991;68: 887-96.
122. Ritchie AH, Kerr CR, Qi A, Yeu-Lai-Wah JA: Nonsustained ventricular tachycardia arising from the right ventricular outflow tract. Am J Cardiol 1989;64:594-8.
123. Carlson MD, White RD, Trohman RG, et al: Right ventricular outflow tract ventricular tachycardia: Detection of previously unrecognized anatomic abnormalities using cine magnetic resonance imaging. J Am Coll Cardiol 1994;24:720-7.
124. O'Connor BK, Case CL, Sokoloski MC, et al: Radiofrequency catheter ablation of right ventricular outflow tachycardia in children and adolescents. J Am Coll Cardiol 1996;27:869-74.
125. Winkle RA, Lopes MG, Fitzgerald JW, et al: Arrhythmias in patients with mitral valve prolapse. Circulation 1975;52:73-81.
126. Duren DR, Becker AE, Dunning AJ: Long-term follow-up of idiopathic mitral valve prolapse in 300 patients: A prospective study. J Am Coll Cardiol 1988;11:42-7.
127. Kligfield P, Levy D, Devereux RB, Savage DD: Arrhythmias and sudden death in mitral valve prolapse. Am Heart J 1987;113: 1298-307.
128. Fananapazir L, Epstein SE: Hemodynamic and electrophysiologic evaluation of patients with hypertrophic cardiomyopathy surviving cardiac arrest. Am J Cardiol 1991;67:280-7.
129. Maron BJ, Shen WK, Link MS, et al: Efficacy of implantable cardioverter-defibrillators for the prevention of sudden death in patients with hypertrophic cardiomyopathy. N Engl J Med 2000;342:365-73.
130. Meinertz T, Hofmann T, Kasper W, et al: Significance of ventricular arrhythmias in idiopathic dilated cardiomyopathy. Am J Cardiol 1984;53:902-7.
131. Olshausen KV, Stienen U, Schwartz F, et al: Long-term prognostic significance of ventricular arrhythmias in idiopathic dilated cardiomyopathy. Am J Cardiol 1988;61:146-51.
132. Grimm W, Glaveris C, Hoffmann J, et al: Arrhythmia risk stratification in idiopathic dilated cardiomyopathy based on echocardiography and 12-lead, signal-averaged, and 24-hour Holter electrocardiography. Am Heart J 2000;140:43-51.
133. Grimm W, Hoffmann J, Menz V, et al: Programmed ventricular stimulation for arrhythmia risk prediction in patients with idiopathic dilated cardiomyopathy and nonsustained ventricular tachycardia. J Am Coll Cardiol 1998;32:739-45.
134. Brilakis ES, Shen WK, Hammill SC, et al: Role of programmed ventricular stimulation and implantable cardioverter defibrillators in patients with idiopathic dilated cardiomyopathy and syncope. Pacing Clin Electrophysiol 2001;24:1623-30.
135. Knight BP, Goyal R, Pelosi F, et al: Outcome of patients with nonischemic dilated cardiomyopathy and unexplained syncope treated with an implantable defibrillator. J Am Coll Cardiol 1999;33:1964-70.
136. Strickberger SA, Hummel JD, Bartlett TG, et al; AMIOVIRT Investigators: Amiodarone versus implantable cardioverter-defibrillator: Randomized trial in patients with nonischemic dilated cardiomyopathy and asymptomatic nonsustained ventricular tachycardia—AMIOVIRT. J Am Coll Cardiol 2003;41: 1707-12.
137. Zareba W, Moss AJ, le Cessie S, et al: Risk of cardiac events in long QT syndrome family members. J Am Coll Cardiol 1995;26:1685-91.
138. Moss AJ, Zareba W, Hall WJ, et al: Effectiveness and limitations of beta-blocker therapy in congenital LQTS. Circulation 2000;101:616-23.
139. Zareba W, Moss AJ, Daubert JP, et al: Implantable cardioverter-defibrillator in high-risk LQTS patients. J Cardiovasc Electrophysiol 2003;14:337-41.
140. Brugada J, Brugada R, Antzelevitch C, et al: Long-term follow-up of individuals with the electrocardiographic pattern of right bundle-branch block and ST-segment elevation in precordial leads V1 to V3. Circulation 2002;105:73-8.
141. Tavernier R, Gevaert S, De Sutter J, et al: Long term results of cardioverter-defibrillator implantation in patients with right ventricular dysplasia and malignant ventricular tachyarrhythmias. Heart 2001;85:53-6.
142. Hohnloser SH, Klingenheben T, Zabel M, et al: Prevalence, characteristics and prognostic value during long-term follow-up of nonsustained ventricular tachycardia after myocardial infarction in the thrombolytic era. J Am Coll Cardiol 1999;33: 1895-902.
143. Denniss AR, Richards DA, Cody DV, et al: Prognostic significance of ventricular tachycardia and fibrillation induced at programmed stimulation and delayed potentials detected on the signal-averaged electrocardiograms of survivors of acute myocardial infarction. Circulation 1986;74:731-45.
144. Buxton AE, Lee KL, DiCarlo L, et al: Electrophysiologic testing to identify patients with coronary artery disease who are at risk for sudden death. Multicenter Unsustained Tachycardia Trial Investigators. N Engl J Med 2000;342:1937-45.
145. Bigger JT Jr: Prophylactic use of implanted cardiac defibrillators in patients at high risk for ventricular arrhythmias after coronary-artery bypass graft surgery. Coronary Artery Bypass Graft (CABG) Patch Trial Investigators. N Engl J Med 1997;337:1569-75.

146. Camm AJ, Yap YG: Clinical trials of antiarrhythmic drugs in postmyocardial infarction and congestive heart failure patients. J Cardiovasc Pharmacol Ther 2001;6:99-106.
147. Waldo AL, Camm AJ, deRuyter H, et al: Effect of d-sotalol on mortality in patients with left ventricular dysfunction after recent and remote myocardial infarction. The SWORD Investigators. Survival With Oral d-Sotalol. Lancet 1996;348:7-12.
148. Cairns JA, Connolly SJ, Roberts R, Gent M: Randomised trial of outcome after myocardial infarction in patients with frequent or repetitive ventricular premature depolarisations: CAMIAT. Canadian Amiodarone Myocardial Infarction Arrhythmia Trial Investigators. Lancet 1997;349:675-82.
149. Viscoli CM, Horwitz RI, Singer BH: Beta-blockers after myocardial infarction: Influence of first-year clinical course on long-term effectiveness. Ann Intern Med 1993;118:99-105.
150. Prystowsky EN, Nisam S: Prophylactic implantable cardioverter defibrillator trials: MUSTT, MADIT, and beyond. Multicenter Unsustained Tachycardia Trial. Multicenter Automatic Defibrillator Implantation Trial. Am J Cardiol 2000;86:1214-5.

Chapter 17 Sustained Ventricular Tachycardia with Heart Disease

PAUL DORIAN, MICHIEL JANSE, SAROJA BHARATI, BRUCE D. LINDSAY, WILLIAM G. STEVENSON, ROBERT J. MYERBURG, and SANJEEV SAKSENA

Introduction: Paul Dorian
Etiology and Pathologic Anatomy: Saroja Bharati
Basic Electrophysiology: Michiel Janse
Clinical Presentation: Sanjeev Saksena
Electrocardiography: Bruce D. Lindsay
Clinical Electrophysiology: William G. Stevenson
Principles of Practice: Paul Dorian
Management: Robert J. Myerburg, Paul Dorian, and Sanjeev Saksena

Sustained ventricular tachycardia (VT) degenerating to ventricular fibrillation (VF) and asystole is the most common cause of premature cardiac death in the western world. It is difficult to definitely establish from epidemiologic studies whether VT or VF more often causes sudden cardiac death. However, it is probable that the identification, immediate treatment, and long-term prevention of sustained VT in patients with preexisting heart disease will have a significant impact on overall cardiac morbidity and mortality.

The definition of sustained VT has undergone evolution. While no formal definition has been adopted for spontaneous sustained VT in contrast to VT induced by programmed stimulation, two versions are in common use. In the Electrophysiologic Study versus Electrocardiographic Monitoring (ESVEM) trial, a 15-second duration was considered sustained VT.[1] Alternatively, many physicians prefer to use 30 seconds duration or evidence of hemodynamic compromise warranting earlier termination, similar to the definition used for induced VT. This chapter provides an overview of the epidemiology, pathology, electrophysiology, and management of sustained VT in patients with structural heart disease. In such patients, VT can be conveniently divided into (1) VT caused by, or related to, a "fixed arrhythmogenic substrate" and (2) VT in whom the substrate is present intermittently during transient abnormalities in cardiac electrophysiological structure and function (e.g., with changes in the neurohormonal milieu).

Etiology and Pathologic Anatomy

In general, the majority of hearts in patients with a previous history of VT reveal varying types of pathologic changes with healthy myocardium interspersed in between the pathologic processes.[2-4] The most common cause of VT is coronary artery disease, either in its acute or chronic phase. During the acute phase, it is believed to occur due to altered physiologic, biochemical, or metabolic states. In the chronic phase, a reentry circuit is created at the junctional areas of the healthy myocardium and scar tissue The infarct zone is surrounded by varying degrees of border zone with surrounding but damaged myocardial fibers. In general, the infarct size is quite large in these patients and may be related to the development of a ventricular aneurysm.[2-4]

VT also occurs frequently in cardiomyopathy. Any disease state that affects the heart in a chronic process can eventually lead to a cardiomyopathy. Thus, one should expect VT to occur in advanced cardiomyopathy during the end stage of heart diseases such as coronary artery disease, myocarditis, valvular disease, hypertensive heart disease, or familial types of cardiomyopathy and many other disease states. This cardiomyopathy can be due to several etiologies; however, the most common type is an idiopathic form. In some patients, "idiopathic" cardiomyopathy may be the end result of a previous viral myocarditis. Pathologically, in most cases of cardiomyopathy, the ventricular myocardial cells show varying types of histologic features such as hypertrophy of cells and varying stages of degenerative changes in the cells and fibrosis, with or without the presence of chronic inflammatory cells. Varying degrees of myocardial fiber damage can provide a substrate for a reentry circuit at the junctional areas with the healthy myocardium.[3]

Myocarditis of any type, either in the acute or chronic phase, may cause VT that may result in sudden death. The heart may be normal at the gross anatomic level, but the microscopic examination may reveal an interstitial type of a myocarditis in the ventricular myocardium including the bundle branches. In many patients, the past clinical history may be essentially unremarkable, or there may be a history of a mild attack of influenza several weeks before death.[2-4]

Fibrotic scars in the ventricular myocardium are often seen in young sudden death victims with an otherwise normal heart. They may be associated with pathologic changes in the peripheral conduction system, such as the branching atrioventricular (AV) bundle and bundle branches. The focal scars in the ventricular myocardium, surrounded by healthy myocardium, may form an anatomic substrate for ventricular arrhythmias, promoting reentry or abnormal automaticity. The etiology of the fibrotic scars in the ventricular myocardium and the beginning of the bundle branches is unknown today. We hypothesize that these may represent an end result of an autoimmune reaction or an allergic state of the individual that may or may not be related to a silent form of a previous myocarditis.[2-4]

UNCOMMON TYPES OF CARDIOMYOPATHY

Less commonly, VT is seen in mitral valve prolapse. There are several pathologic abnormalities in cases of mitral valve prolapse associated with VT.[2-5] A variety of pathologic findings may be seen in these patients, such as a right-sided AV bundle, fibrotic scars in the ventricular septum, degenerative changes in the conduction system, and arteriolosclerosis.

In arrhythmogenic right ventricular dysplasia (ARVD), the anterior wall of the right ventricle can be partially or completely replaced by fibrofatty tissue. There may be some intact myocardial cells scattered within the fatty tissue. Similar findings extend to the ventricular septum, as well as to the left ventricular myocardium. This is often associated with necrosis of cells and mononuclear cell infiltration. Varying amounts of degenerative changes in the myocardium are present in the right and left ventricles. A segmented and looping left-sided AV bundle has been reported. There may also be small vessel disease of the ventricular septum. Thus, in addition to the acquired pathologic changes, there are also congenital abnormalities of the conduction system in the sinoatrial, AV node, and AV bundle.[2-4,6]

In hypertrophic cardiomyopathy the heart is hypertrophied and enlarged. Hypertrophy of the interventricular septum is seen in varying degrees. The AV node may be partly or mostly situated within the central fibrous body and occasionally partly embedded in the tricuspid valve annulus or at the aortic-mitral annulus. The AV nodal artery is usually thickened and narrowed. The sinoatrial and AV nodes are frequently infiltrated with fat. The AV bundle may be on the right side of the ventricular septum with loop formation and fibrosis of the branching bundle. There are focal, fibrotic scars in the ventricular myocardium associated with myocardial fiber disarray and arteriolosclerosis in the summit of the ventricular septum.[2-4,7]

INFILTRATIVE DISEASES OF THE MYOCARDIUM

VT may occur in sarcoidosis and amyloidosis. An infiltrative disease of the myocardium, such as sarcoidosis, may affect not only the branching AV bundle and the bundle branches but also the ventricular myocardium. Likewise, amyloidosis, either in primary or secondary form, may involve the ventricular myocardium including the bundle branches. In 70% of cases of primary amyloidosis, the heart is involved, often with disruption to the entire conduction system. The infiltration of amyloid in the heart eventually results in a restrictive cardiomyopathy, and sudden death is commonly seen.[2-4,8] The heart is affected in approximately 13% to 25% of cases of sarcoidosis. Cardiac involvement is often associated with lymph node and lung involvement. Sarcoidosis has a predilection to affect the posterior wall of the left ventricle with aneurysm formation. Sarcoid granulomas may be present in the conduction system and the surrounding myocardium (Fig. 17-1). In general, infiltrative lesions can cause conduction disturbances and provide a substrate for reentrant or automatic VT.[2-4,8]

PRIMARY ELECTRICAL DISEASES

It is well known that supraventricular arrhythmias occur in preexcitation syndrome. However, ventricular arrhythmias can also occur in this entity. Pathologically, there is the obvious accessory pathway on either the left or right side, associated with cardiomyopathy and fibroelastosis of the left ventricle, along with hypertrophy and degenerative changes of the myocardium.[2-4,9]

Familial QT prolongation is predominantly an autosomal dominant disorder associated with syncopal episodes, VT, and sudden death. Recently, abnormal genes have been identified on several chromosomes in patients with congenital prolonged Q–T interval. Genetic heterogeneity has been documented through linkage studies with loci on several chromosomes. Further, mutations in ion channel genes on chromosomes 3, 7, and others have been identified to be related to other forms of long QT syndrome.[2-4,10] Pathologically, there is marked fatty infiltration in the approaches to the AV node; a lobulated AV bundle with or without loop formation; arteriolosclerosis; and focal fibrosis of the summit of the ventricular septum, seen predominantly on the right side. In addition, fibrotic changes are seen to a varying degree in the AV bundle and bundle branches with chronic inflammatory cells in the ventricular myocardium. *It is important to remember that pathologic findings exist in congenital long QT syndrome with sudden death.*[2-4,10]

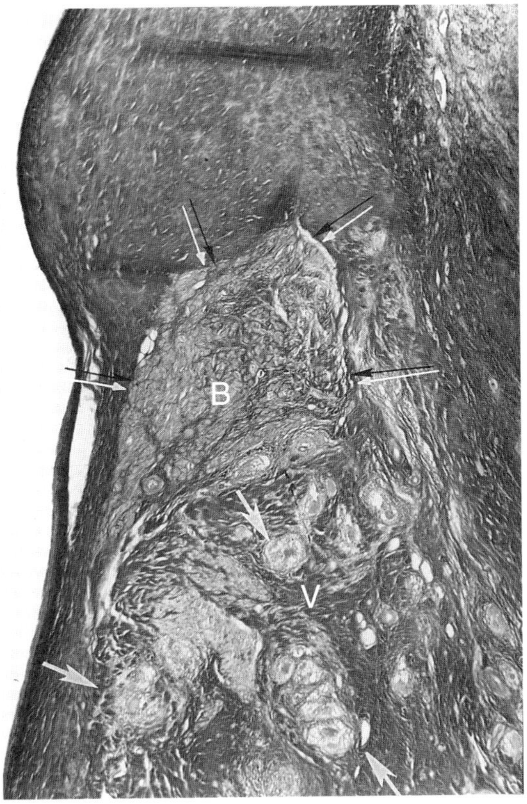

FIGURE 17-1 A 34-year-old male of Italian ancestry died suddenly. He was diagnosed to have right bundle branch block with first-degree and intermittent third-degree atrioventricular (AV) block and recurrent unifocal paroxysmal ventricular tachycardia. During electrophysiological studies with extrastimulus technique, repetitive ventricular tachycardias were induced. Photomicrograph demonstrating sarcoid granulomas of the ventricular myocardium and branching AV bundle. (Weigert-van Gieson stain ×45.) B, branching AV bundle with sarcoid granulomas; V, summit of the ventricular septum with sarcoid granulomas. *Thin white and black arrows* point to sarcoid granulomas in the branching AV bundle and *thick arrows* point to the ventricular septum with granulomas.

SKELETAL MUSCLE DISORDERS

The conduction system is frequently affected in congenital myotonic dystrophy. This progressive, generalized disease is characterized by atrophy of skeletal muscles in a characteristic manner, with associated myotonia. Various types of arrhythmias are known to occur in this disease. Pathologically, degenerative changes in the smooth muscles of the cardiac vessels in the left atrium and the aorta with fatty infiltration in the approaches to the AV node, fibrosis of the bundle branches, summit of the ventricular septum, and varying degenerative changes in the myocardium are present. These could form a substrate for AV block and VT.[3,11]

The Kearns-Sayre syndrome is characterized by progressive external ophthalmoplegia, retinitis pigmentosa, and AV block rather than VT. Degenerative changes that are progressive in nature affect the entire heart. The conduction system is replaced by fibrotic destruction of the bundle branches with fibrofatty replacement. Some muscle cells may reveal hypertrophy, and others become atrophied with perineural and perivascular fibrosis, eventually leading to cardiomyopathy. Although AV block occurs frequently in this disease, electrophysiological studies have demonstrated that the disease affects the entire His-Purkinje system. These findings in the conduction system and ventricular myocardium may form an anatomic basis for VT.[3,12]

TUMORS AND PARASITIC DISEASES OF THE HEART

Any tumor, either primary or secondary in nature, can produce ventricular arrhythmias, as can parasitic infiltration. Hydatid cyst infiltration can result in fibrosis in the perimeter tissues around the cyst. This can permit reentry and may result in VT.[3,13]

ANEURYSM AND DIVERTICULUM OF THE HEART

These may occur in either the right or left ventricle of an otherwise normal heart, and VT may be the first clinical manifestation.[3,13] Pathologic anatomy reveals a large wide-mouthed aneurysm, but its microscopic appearance can vary with the etiology.

POSTOPERATIVE PATIENTS WITH CONGENITAL HEART DISEASE

It is well known that ventricular arrhythmias occur in postoperative patients with congenital heart diseases such as tetralogy of Fallot, aortic stenosis, and many other congenital cardiac anomalies. VT can appear many years following the surgery. Fibrotic scars along with healthy myocardium are seen, for example, at the outflow tract of the right ventricle or in any other area in the heart. This substrate can result in reentry circuits leading to VT.[2-4,14,15] In aortic stenosis, cardiomegaly may occur with myocardial fibrosis.[16]

IATROGENIC DISORDERS

Antiarrhythmic drugs may alter the physiologic, metabolic, and biochemical state of the myocardium or conduction system, or both, and give rise to VT, which may or may not be transient in nature. Likewise, the various resuscitative techniques used and catheter ablation of myocardium may alter the myocardial tissue with formation of fibro-fatty scar and chronic inflammation. This may become a substrate for VT in the future.[2-4,13,17]

FAMILIAL VENTRICULAR TACHYCARDIA

Pathologically, the conduction system and the ventricular myocardium may show degenerative changes with mononuclear cell infiltration and fat to a varying degree. In others, there may be atrophy of the branching part of the AV bundle with almost complete absence of left bundle branch and absence of right bundle branch. These findings suggest a genetic abnormality of the conduction system that leads to degenerative changes, inflammatory phenomena, and susceptibility to ventricular arrhythmias that cause sudden death.[18,19] The anatomic substrate may originate from the myocardial disarray that can be present either in the ventricular septal myocardium or in the conduction system. Left or right ventricular septal hypertrophy may also cause degeneration of the AV node, AV bundle, and the main left bundle branch[3,4,20] (Fig. 17-2).

PREMATURE AGING CHANGES ON THE RIGHT SIDE OF THE VENTRICULAR SEPTUM

Premature aging occurs in the summit of the ventricular septum in some individuals, with degenerative changes of the branching bundle and the right bundle branch. These changes are usually associated with arteriolosclerosis of the summit of the ventricular septum and are often seen in sudden death in teenagers.[3,4,21]

IDIOPATHIC VENTRICULAR TACHYCARDIA DUE TO CONGENITAL ABNORMALITIES OF THE CONDUCTION SYSTEM

VT may occur in many disease states involving the conduction system or the myocardium, or both. Diseased myocardial fibers are usually surrounded, at least in part, by healthy myocardial fibers, thereby creating a substrate for reentry, abnormal automaticity, or slowed conduction. Thus, varying types of congenital abnormalities of the conduction system, such as a right-sided AV bundle; left-sided AV bundle; genetically abnormally formed conduction system; and altered metabolic, biochemical, or physiologic states may result in VT.

Chronic recurrent right VT with QRS morphology of left bundle branch block (LBBB) pattern in a normal heart, normal coronary arteries, and normal cardiac catheterization findings, has been well documented, especially in the young patient. However, an anatomic substrate for this abnormality has only been occasionally reported.[22] In one report, a 13-year-old boy with a history of recurrent VT died suddenly. A right-sided, markedly septated AV bundle was found at autopsy. The AV node formed the AV bundle, and the nodo-bundle junction was ill-defined (Fig. 17-3). The penetrating AV bundle cells were not well defined as typical cells of AV bundle, and the latter remained on the right side and eventually became the right bundle branch. In addition, there were patchy fibrotic scars in the right ventricular myocardium. It has been hypothesized that a right-sided, markedly septated, undifferentiated AV bundle that eventually continues as the right bundle branch may cause recurrent right VT in an otherwise normal heart.[2,3,22]

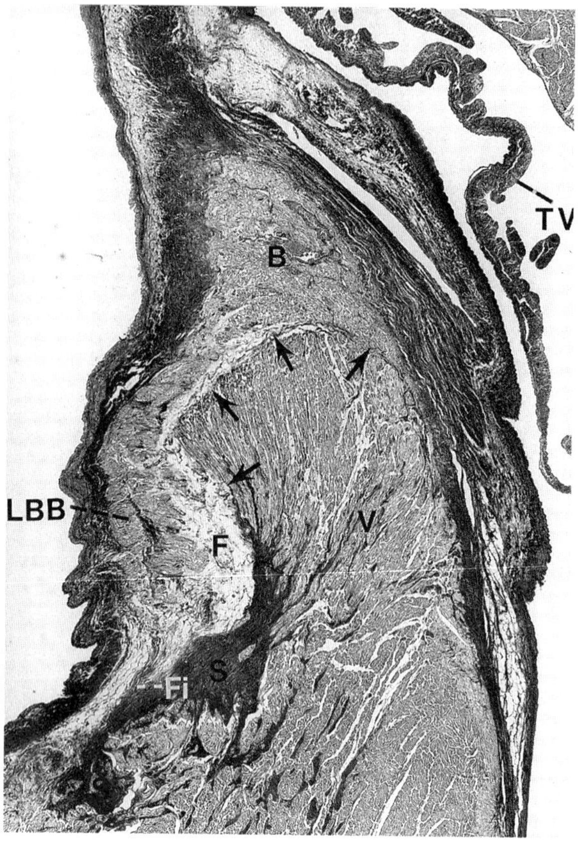

FIGURE 17-2 Nonsustained ventricular tachycardia of familial etiology. A 15½-year-old boy with a history of exercise-related syncope died suddenly while swimming. He had a family history of sudden death involving three consecutive generations, including a brother. The ECG and cardiac catheterization were normal. During electrophysiological studies, with extrastimulus testing, he had polymorphic nonsustained ventricular tachycardia; during stage 5 of the Bruce protocol, he had a run of nonsustained ventricular tachycardia. Photomicrograph of the branching atrioventricular bundle being compressed by the right ventricular septal muscle at the region of the posterior radiation of left bundle branch. (Weigert-van Gieson stain × 22.5.) B, branching bundle; F, fatty metamorphosis; FI, fibrosis and linear change of left bundle branch; LBB, posterior radiation of left bundle branch; S, increased sclerosis on the mid septal area on the left; V, summit of the ventricular septum. *Arrows* point to the pressure of the right ventricular septal hypertrophy on the atrioventricular bundle at the level of the posterior radiation of left bundle branch. (From Brookfield L, Bharati S, Denes P, et al: Familial sudden death, report of a case and review of the literature, Chest 1988;94:989-93.)

Basic Electrophysiology

VT occurs in different disease substrates. The most common and most widely studied experimental models include healed myocardial infarct, myocardial hypertrophy, or myocardial failure. The genesis of VT in each of these substrates is specific mechanisms that provide insight into clinical VT.

ISCHEMIA AND VENTRICULAR TACHYARRHYTHMIAS

The ventricular arrhythmias caused by myocardial ischemia and infarction occur in several distinct phases. A ventricular arrhythmia can be induced by acute ischemia and reperfusion. It usually occurs between 2 and 30 minutes following acute coronary artery occlusion, when the changes caused by ischemia are still reversible. Arrhythmias associated with the development of myocardial infarction can be categorized as delayed

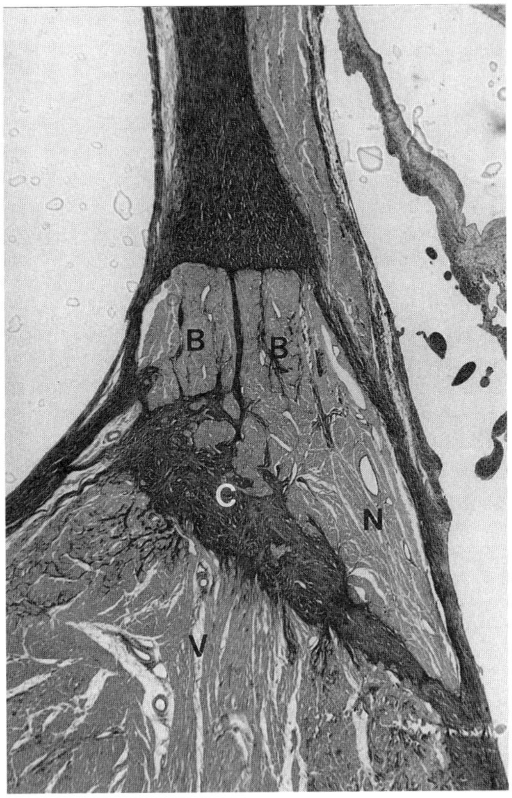

FIGURE 17-3 Idiopathic ventricular tachycardia in an otherwise normal heart. A 13-year-old boy with a history of recurrent ventricular tachycardia died suddenly. Photomicrograph demonstrating a right-sided atrioventricular (AV) node and AV bundle with no sharp line of differentiation between the AV node partly engulfed in the central fibrous body and the markedly septated AV bundle. (Weigert-van Gieson stain × 30.) B, AV bundle; C, central fibrous body; N, AV node; V, ventricular septum. (From Bharati S, Bauernfeind R, Scheinman M, et al: Congenital abnormalities of the conduction system in two patients with tachyarrhythmias. Circulation 1979;59: 593-646.)

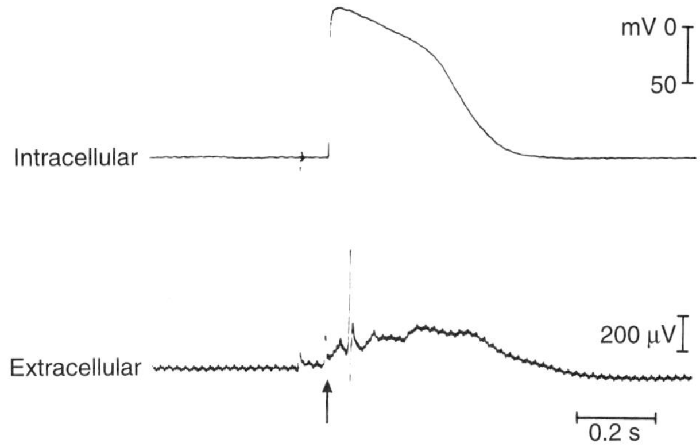

FIGURE 17-4 Transmembrane potential and extracellular electrogram recorded from a resected endocardial preparation from a patient with ventricular tachycardia and a chronic myocardial infarction. The electrogram shows fragmentation. The action potential is close to normal. (From de Bakker JMT, Coronel R, Tasseron S, et al: Ventricular tachycardia in the infarcted, Langendorff-perfused human heart: Role of the arrangement of surviving cardiac fibers. J Am Coll Cardiol 1990;15:1594-1607.)

arrhythmias, usually occurring between 5 and 48 hours, late arrhythmias after days to weeks, and chronic arrhythmias occurring months to years later.[23,24] Delayed arrhythmias such as slow VTs and accelerated idioventricular rhythms rarely degenerate into VF and are due to abnormal automaticity of Purkinje fibers overlying the infarct.

Data obtained in 4- to 5-day-old canine infarcts show that the healing infarct undergoes structural and functional changes. The surviving epicardial cells overlying the infarct have abnormal action potentials with diminished upstrokes with loss of the plateau and shorter action potential duration. The density and kinetics of a number of ion channels are altered, and sodium ion and calcium ion currents are reduced, as are transient outward potassium ion currents and the delayed and inward rectifying potassium ion currents.[25] During this stage, reentrant VT in the so-called *epicardial border zone* (the layer of surviving cells overlying the infarct) can easily be induced by premature stimuli. Both the cellular abnormalities, as well as a redistribution of intercellular gap junctions, play a role in determining the "substrate" for reentry.[26] These VTs may degenerate into VF, especially in the presence of a high sympathetic tone,[24] but this is uncommon.

Over the next few weeks, transmembrane action potentials of the surviving cells gradually return to normal, as most of the ion channels recover, and by 2 months, action potential configuration in both canine and human infarcts is completely normal[27,28] (Fig. 17-4). It is difficult to be certain when the healing phase of myocardial infarction is over and when the fully healed phase begins. It is likely that the electrophysiological substrate for VT gradually develops over several weeks and remains stable from several months to 15 to 20 years.[29]

Most of the electrophysiological data have been obtained from dogs with healing infarcts, and although there are many similarities between canine arrhythmias and those in humans, there are some important differences. In dogs with infarcts, arrhythmias can be induced easily during the first week, but thereafter inducibility decreases. In humans, VTs can only be induced in a small minority after 5 days, and inducibility increases after 3 weeks. In dogs, the reentrant circuit responsible for the tachycardia is located in the epicardial border zone. In patients with healed infarcts, the reentrant circuit is usually located subendocardially, and only about 20% of VTs are due to reentry in the subepicardium.

THE SUBSTRATE FOR REENTRANT TACHYCARDIA IN THE HUMAN HEART WITH A HEALED INFARCT

In the prethrombolytic era, the incidence of sustained monomorphic VT following discharge from hospital in patients surviving a myocardial infarction has been reported to be around 3%, and that of nonsustained VT as 10% to 20%.[30] Although thrombolysis has drastically reduced arrhythmic events early after myocardial infarction, with an incidence of sustained VT lower than 1%,[31] postinfarction arrhythmias have certainly not become irrelevant. With improved therapy in the acute stage of myocardial infarction, more patients survive, but many survivors suffer from left ventricular dysfunction. It is well known that ventricular dysfunction is the most important risk factor for sudden death, as verified in the European Myocardial Infarct Amiodarone Trial (EMIAT). The combination of arrhythmic death and resuscitated cardiac arrest occurred in 8.6% of patients.[32] It is unknown how many of these patients developed sustained VT that degenerated into VF and how many suffered an episode of acute ischemia that induced the lethal arrhythmia. Still, despite intensive therapy, arrhythmic events in postinfarct patients, especially in the presence of ventricular dysfunction, remain an important cause of death.

Reconstruction of the reentrant circuits and the mechanism of slow conduction in the human heart with a healed infarct is based on observations made during mapping-guided surgery in patients with infarct-related VT, as well as studies on isolated, Langendorff-perfused human hearts and isolated papillary muscles from patients with infarcts undergoing cardiac transplantation.[28,33-35] Figure 17-5 shows the activation map of one beat of a VT induced by programmed electrical stimulation in an isolated, Langendorff-perfused heart. The endocardial surface of the left ventricle is schematically depicted with the left ventricle folded out by making a virtual cut along the left anterior descending artery. The black zone indicates the infarcted area. Selected extracellular electrograms, recorded simultaneously from the endocardium by a balloon electrode inserted into the ventricular cavity, are shown as well. The cycle length of this tachycardia was 264 milliseconds (ms). The site of earliest activation during this tachycardia was in a small area near the apex, on the border of the septum and the posterior wall, at the margin of the infarcted region (the area encircled by the 0 ms isochrone). Activation spread from this area to the left in the figure (from *a* to *b*) and continued to the anterior wall (from *b* to *c* to *d* to *e*), to reach the other margin of the infarct after 192 ms (near site *e*). Although it seems that activation at this margin died out, because of the slow positive deflection recorded at site *e*, the presence of small deflections indicated by arrows over the infarcted zone suggests that the spread of activation continued via tracts of surviving muscle within the infarct, through sites *f*, *g*, and *h*, to reach the opposite side of the infarct after 258 ms, reexciting the area of the 0 ms isochrone and completing a large reentrant circuit around the circumference of the ventricle.

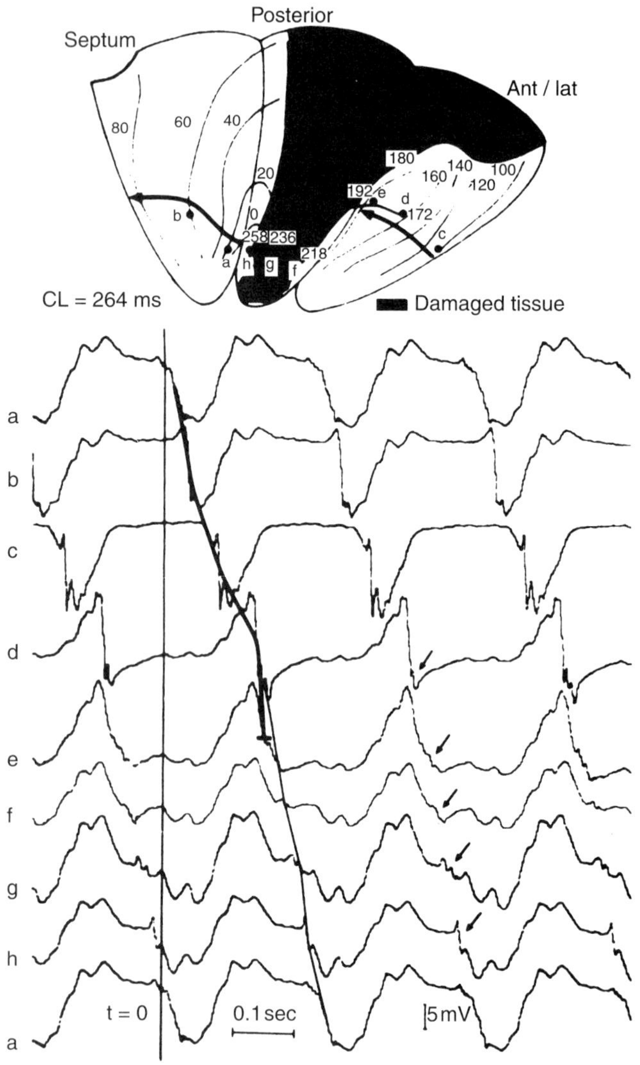

FIGURE 17-5 Endocardial activation pattern of one cycle of sustained ventricular tachycardia induced in a Langendorff-perfused human heart with extensive infero-posterior infarction. Isochrones are in milliseconds (ms) with respect to a time reference ($t = 0$) in the bottom panel. *Thick arrows* on the map indicate main spread of activation. In the bottom panel, endocardial electrograms recorded at sites indicated in the top panel are shown. The *thick line* connects times of activation of subendocardial tissue at the sites *a* to *e*. At site *d*, the main deflection is followed by a second response of small amplitude (*arrow*). At sites *e* to *h* large signals reflect remote activity, but in all signals small deflections are present (*arrows*). The timing of these small deflections is indicated by the *thin line*. (From de Bakker JMT, Coronel R, Tasseron S, et al: Ventricular tachycardia in the infarcted, Langendorff-perfused human heart: Role of the arrangement of surviving cardiac fibers. J Am Coll Cardiol 1990;15:1594-1607.)

Subsequent histologic studies of this region confirmed the presence of surviving myocardial muscle bundles, several millimeters below the endocardial surface that connected both margins of the infarct, completing the reentrant circuit. Figure 17-6 shows superimposed drawings of many histologic sections of this region, where viable muscle is in *black* and connective tissue in *white*. Similar tracts were found in other hearts, embedded within the scar of the healed infarct. Such tracts could be localized in the subendocardium, the subepicardium, and intramurally. Sometimes the myocardial fibers were arranged in a parallel fashion along the long axis of the tract, allowing for relatively rapid transmission of the reentrant impulse from one side of the infarct to the other. However, in other hearts the fibers were oriented transverse to the direction of impulse propagation, and in these cases, transmission

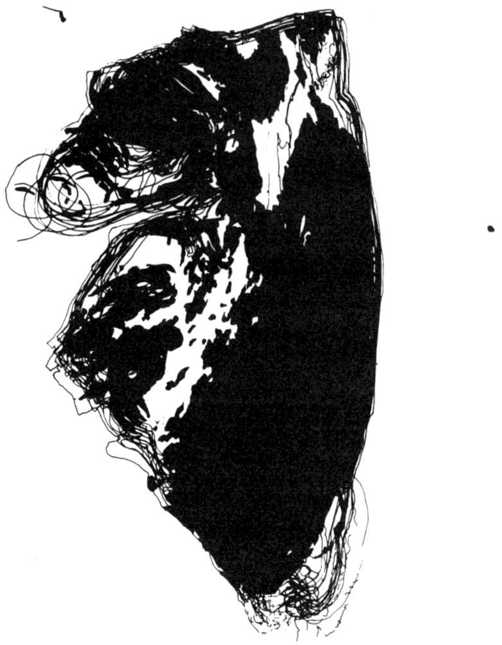

FIGURE 17-6 Superimposed drawings of selected serial sections from the infarcted part of the left ventricle taken from an isolated human heart in which sustained ventricular tachycardia was induced. Viable myocardial tissue is in *black*, scar tissue in *white*. It can be seen that a bundle of surviving myocardium runs through the infarct, connecting the noninfarcted tissue on both sides of the infarct (unpublished data from de Bakker JMT and Tasseron S).

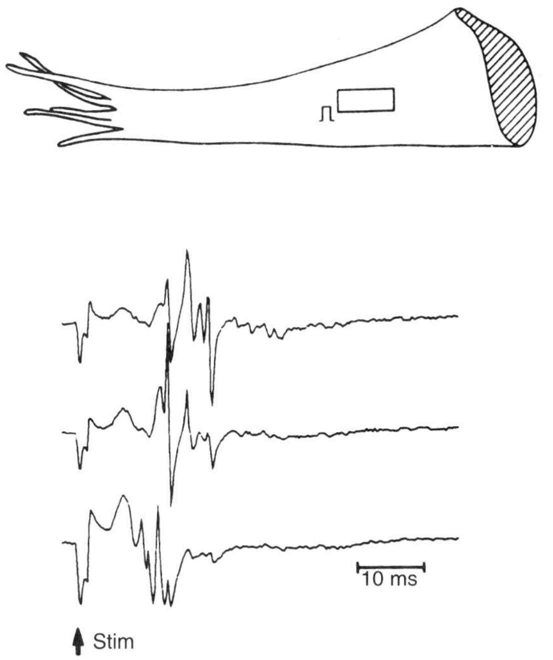

FIGURE 17-7 Selected electrograms recorded at distances of 200 microns in an isolated, superfused infarcted human papillary muscle. The recording area is indicated by the *rectangle*; the site of stimulation by the *square wave*. Note the high degree of fragmentation of the electrograms. (Modified from De Bakker JMT, van Capelle FJL, Janse MJ, et al: Slow conduction in the infarcted human heart: "Zigzag" course of activation. Circulation 1993;88:915-26.)

through the infarct was very slow. Figures 17-7 and 17-8 show the mechanism of such "slow" conduction. Figure 17-7 shows selected electrograms from an infarcted human papillary muscle to illustrate the highly fragmented waveforms, typical for tissue in which connective tissue is intermingled with viable myocardium. In this muscle, the delay between the deflections at the extreme electrode terminals of the multiple electrode (distance 1.4 mm) was 45 ms, which would correspond to an overall conduction velocity on the order of 3 cm/sec. As illustrated in Figure 17-8, this apparent slow conduction was due to "zigzag" conduction in small muscle bundles separated by collagenous septa. Activity in the muscle tracts proceeded both toward the site of stimulation *A*, and away from it. Many tracts were dead-end pathways. The actual pathway of activation is plotted in panel *B*. Although the shortest distance between *A* and *B* was 1.2 mm, the length of the zigzag pathway was 25.2 mm. In 10 papillary muscles, conduction velocity parallel to the fiber orientation was on average 79 cm/sec, which indicates that both the active electrical properties (action potential amplitude and upstroke velocity) and passive electrical properties (longitudinal coupling resistance) were normal. Conduction velocity at the bifurcation points was only 49 cm/sec, indicating that in addition to the pathway length, impedance mismatch at bifurcations also contributed to activation delay.

Thus, in the human heart with a healed infarct, the substrate for reentry is formed by the surviving fibers within the infarct. In addition to a substrate, a trigger is needed to initiate the tachycardia. It is likely that this trigger may occur in the myocardium remote from the infarct. In the presence of progressive left ventricular dysfunction, these myocardial zones can undergo remodeling that can be arrhythmogenic. Furthermore, the properties of the noninfarcted myocardium can determine whether or not VT degenerates into VF. Dispersion of refractoriness in the noninfarcted myocardium is three times larger in postinfarction patients who develop VF than in those in whom the VT remains monomorphic and hemodynamically stable.[36]

ROLE OF CARDIAC REMODELING IN VENTRICULAR TACHYCARDIA

The heart may respond to a variety of abnormal environmental stimuli by altering gene expression or the functional properties of proteins. This ultimately leads to both functional and structural cardiac alterations that may be arrhythmogenic. These processes are referred to as cardiac remodeling. Common to all studies on hypertrophic and failing ventricular myocardium is a prolonged action potential, especially at slow heart rates.[37] This may be considered an adaptive process. In the setting of a prolonged action potential, the intracellular calcium transport is increased, causing enhanced force of myocardial contraction. It can be argued that a prolonged action potential may also protect the heart against reentrant excitation. However, prolonging

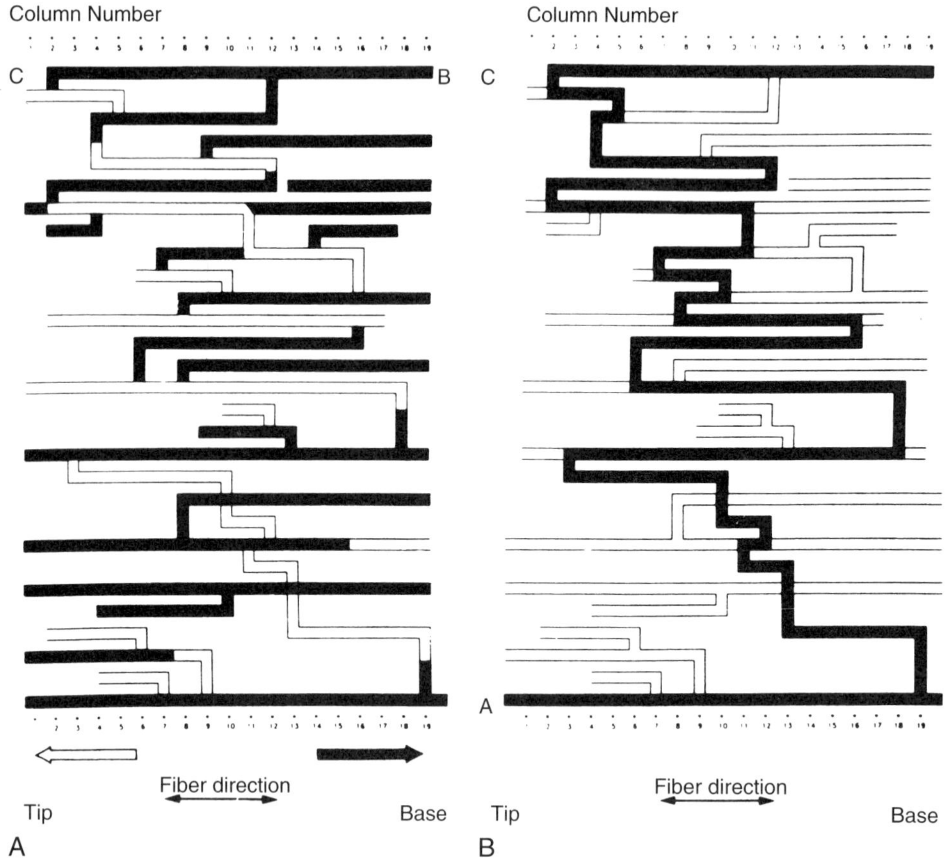

FIGURE 17-8 Map illustrating the spread of activation of the rectangular area shown in the inset of Figure 17-4. In the solid tracts of **panel A,** activation moves away from the site of stimulation (*site A*); in the open tracts, activation moves toward the site of stimulation. **Panel B** shows the tortuous route that activation followed to reach *site B.* (Modified from De Bakker JMT, van Capelle FJL, Janse MJ, et al: Slow conduction in the infarcted human heart: "Zigzag" course of activation. Circulation 1993;88:915-26.)

repolarization may be arrhythmogenic. Drugs, both cardiac and noncardiac, can prolong the action potential. There is a risk in this situation of the development of early afterdepolarizations and torsades de pointes.[38] It is generally accepted that an early afterdepolarization-induced premature beat will only cause reentry in the presence of a marked increase in dispersion of repolarization.[38] In hypertrophic and failing myocardium, increased dispersion in repolarization (or refractoriness) has indeed been reported.[36,39-41] In hearts with a healed infarct, repolarization of normal, lateral border and infarct zone cells is nonuniform, with myocardial cells showing different degrees of action potential disturbance and prolongation. Many of these cells, especially those close to the infarct border, show post–repolarization-refractoriness.[25] There may be several reasons for the increased dispersion in repolarization: unequal distribution of remodeled ion channels[41,42]; decreased expression of gap junctional connexins and an altered distribution of gap junctions,[43,44] with the development of fibrosis; and changes in autonomic innervation.[45,46]

Changes in ionic currents contributing to repolarization are seen. The most consistent finding is a reduction in the transient outward current, It_o.[37,42,43] Whereas this current is very important in determining action potential duration in small mammals, such as the rat, its downregulation probably does not have much effect on action potential duration in hearts of large mammals. It does change the level of the plateau phase of the action potential, and it can affect other currents activated later during the action potential. Both rapid and slow components of the delayed rectifier, Ik_r and Ik_s, are reduced in rabbits with pacing-induced heart failure.[47] The inward rectifier current, I_{K1}, can decrease, remain unchanged, or increase.[37] If indeed I_{K1} is reduced, it will lead to an unstable resting potential, and when it is unregulated, pacemaker activity could result.[48,49] The L-type calcium current is either unchanged or decreased, and it is unlikely that this current contributes to action potential prolongation.[50] There may be a role for the late sodium current, which is increased.[51] For most of these remodeled currents, there are insufficient data on regional differences. However, in the presence of enhanced β-adrenergic activity, transmural dispersion in repolarization increases because of augmentation of residual Ik_s in epicardial and endocardial layers resulting in action potential shortening. But this does not occur in midmural M cells, in which Ik_s is intrinsically weak.[53] A decrease in electrical cell-to-cell coupling, either by a decrease in expression of connexins or by the development of microscopic fibrosis in hypertrophied myocardium, reduces electrotonic current flow and will unmask intrinsic differences in action potential duration. Well-coupled myocardium will attenuate these differences.

Ischemia and infarction result in both afferent and efferent parasympathetic and sympathetic dysfunction in regions apical to the area of infarction. The denervated but otherwise normal myocardium develops adrenergic supersensitivity, and therefore the response to circulating catecholamines is exaggerated.[45]

The nerve-sprouting hypothesis of ventricular arrhythmias and sudden death states that myocardial infarction results in nerve injury followed by sympathetic nerve sprouting and regional, heterogeneous myocardial hyperinnervation, which, together with electrical remodeling, leads to heterogeneous distribution of repolarization and ventricular arrhythmias.[46]

There is general agreement that intracellular calcium handling is compromised in heart failure, but there are contradictory reports on virtually all components involved in calcium homeostasis.[37,53-55] It is generally accepted that ATP-dependent calcium accumulation by the sarcoplasmic reticulum is decreased. In the presence of a prolonged action potential and altered calcium homeostasis, both early and delayed afterdepolarizations may occur.[43,49] In hearts with an infarct, where an anatomic reentrant circuit may be present, premature ventricular depolarizations caused by these afterdepolarizations, or salvos of triggered activity may initiate sustained reentrant tachycardias. In hypertrophied and failing hearts without an infarct, with fibrosis and increased dispersion of repolarization, these triggers may equally initiate reentrant tachycardias. In animal models of heart failure, such as rabbits, combined volume and pressure overload resulted in nonsustained VTs developing in more than 50% of animals, and sudden death was common.[56] However, it is not certain whether sudden death is always due to VT degenerating into VF. In both patients with end-stage heart failure and in rabbit models of heart failure, bradycardia, asystole, and electromechanical dissociation have been documented to cause sudden death.[57-59]

Clinical Presentation

The symptoms associated with sustained VT can range from an asymptomatic patient to cardiovascular collapse resulting in circulatory arrest and unconsciousness. Clinical literature dating to the early period of the 20th century has documented case reports of sustained VT without symptoms or minimal symptoms. Remarkably, some of these episodes may last weeks or even months. These patients may only experience minimal or no palpitations and, when present, can regard this as an insignificant symptom. Figure 17-9 is derived from one of our patients with incessant VT who, in 1981, experienced palpitations with mild dyspnea during a single sustained slow VT episode that lasted 1 week, despite severe left ventricular systolic dysfunction being present. The most common symptoms of sustained VT are palpitations, dyspnea, angina, hypotension, near syncope, or frank syncope. The severity of symptoms is related to several factors, including tachycardia rate, morphology, severity of left ventricular dysfunction, preload, and coexisting diseases such as coronary artery disease. The symptom complex is often defined by the hemodynamic impact of the arrhythmia. In an early study, Saksena and colleagues proved that VT resulted in impaired left ventricular relaxation and subsequently systolic dysfunction and decline in negative and positive dP/dt.[60] Tachycardia rate and preexisting LV dysfunction were important variables, as was preload. Faster VT rates and low preload predisposed to hypotension and syncope. Hamer noted similar associations with syncope during VT.[61] In a study of defibrillator recipients, presyncope or syncope preceding delivery of implantable cardioverter defibrillator (ICD) therapy was determined by faster events in the "ventricular fibrillation" zone.[62] However, the absence of hemodynamic compromise does not exclude the diagnosis of VT. It is an axiom that wide QRS tachycardia in the presence of history or ECG evidence of myocardial infarction is due to VT until proven otherwise by subsequent investigation.

Electrocardiography

A 12-lead ECG during sustained ventricular tachycardia is an important and essential investigational tool for

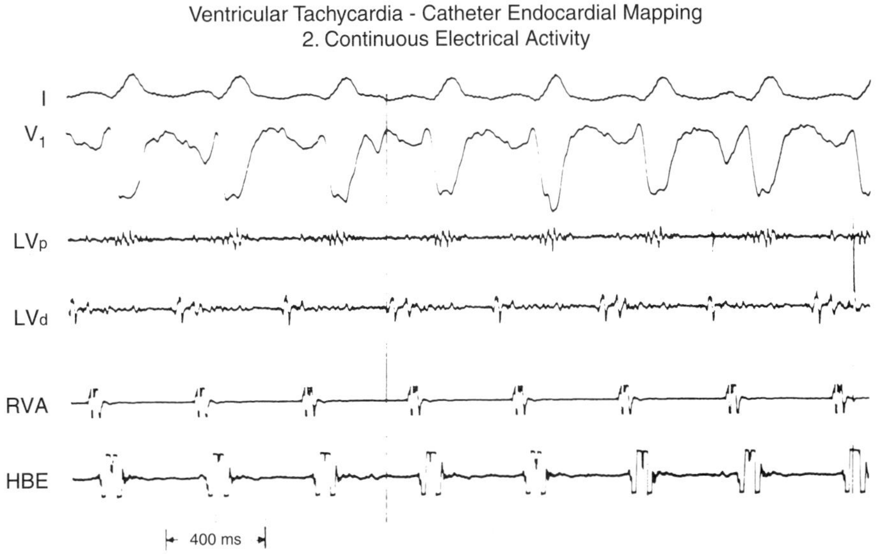

FIGURE 17-9 Incessant ventricular tachycardia (VT) in a patient with coronary artery disease, large left ventricular aneurysm, estimated left ventricular ejection fraction 15%, and congestive heart failure. The patient was in sustained VT for days with minimal symptoms and a slow VT rate. Catheter endocardial mapping was performed during the episode. HB, His-bundle electrogram; RV, right ventricle electrogram.

management of the VT patient. It should be obtained during the tachycardia and compared to a prior recording in sinus rhythm if possible. Careful evaluation can confirm the diagnosis, exclude other tachycardias, and help in establishing its mechanism. The QRS morphology during VT can also assist in determining its site of origin.

ELECTROCARDIOGRAM DIAGNOSIS OF VENTRICULAR TACHYCARDIA

Twelve-lead ECGs are particularly useful to differentiate supraventricular tachycardia (SVT) from VT. It is more difficult to distinguish these arrhythmias when aberrant intraventricular conduction occurs during SVT, but several criteria have been proposed to improve diagnostic accuracy.[63-69] AV dissociation is the most specific ECG criterion for the diagnosis of VT in recordings of wide QRS complex tachyarrhythmias. Certain rare types of SVTs (e.g., AV node reentry with retrograde block, junctional tachycardia using a nodo-ventricular fiber with retrograde atrial block) may mimic this finding. AV dissociation is seen in only 21% of ECG recordings with VT and may be difficult to identify with absolute certainty.[65] When AV dissociation is not apparent, other criteria that depend on whether the QRS resembles left bundle or right bundle branch block (RBBB) must be used.

When the ventricles are structurally normal, the QRS morphology is an excellent marker of the arrhythmia origin. VT with an LBBB configuration in lead V1 can have its origin that is in the right ventricle or the intraventricular septum (either right or left side of the septum) (Fig. 17-10). A QRS frontal plane axis directed inferiorly (dominant R waves in leads II, III, AVF) indicates an origin in the cranial aspect of the heart (e.g., anterior wall of the left ventricle or the right ventricular outflow tract). A QRS frontal plane axis directed superiorly (dominant S waves in leads II, III, and AVF) indicates initial depolarization arising in the inferior wall of the left or right ventricle. Dominant R waves in leads V3 to V4 favor a location of the focus nearer the base of the heart than the apex. Dominant S waves in these leads favor a more apical location. VT with a typical LBBB or RBBB QRS configuration suggests bundle branch reentry as a mechanism. When areas of ventricular scar are present, the QRS morphology is less reliable and is sometimes very misleading.

Kindwall observed several characteristic patterns found in patients with VT complexes resembling LBBB and evaluated four criteria to distinguish VT with LBBB pattern from SVT with aberrant conduction.[66] All patients in the study had a predominantly negative QRS in V1 and a QRS duration greater than 120 ms. Four electrocardiographic criteria were evaluated in this study: (1) R wave in V1 or V2 of greater than 30 ms duration; (2) any Q wave in V6; (3) a duration of greater than 60 ms from the onset of the QRS to the nadir of S wave in V1 or V2; and (4) notching on the downstroke of the S wave in those leads. Table 17-1 summarizes the sensitivity and predictive accuracy of these criteria. None of the criteria alone were very sensitive, but all patients with VT had at least one of these criteria. The specificity remained high (89%) when the combined criteria were employed, and the predictive accuracy was excellent (Fig. 17-11). Left axis deviation was of no value in distinguishing VT from SVT in this study. These criteria are easily measured and provide a very practical approach to the differentiation of VT from SVT.

Brugada[67] prospectively analyzed a series of wide QRS tachycardias that included morphologies resembling

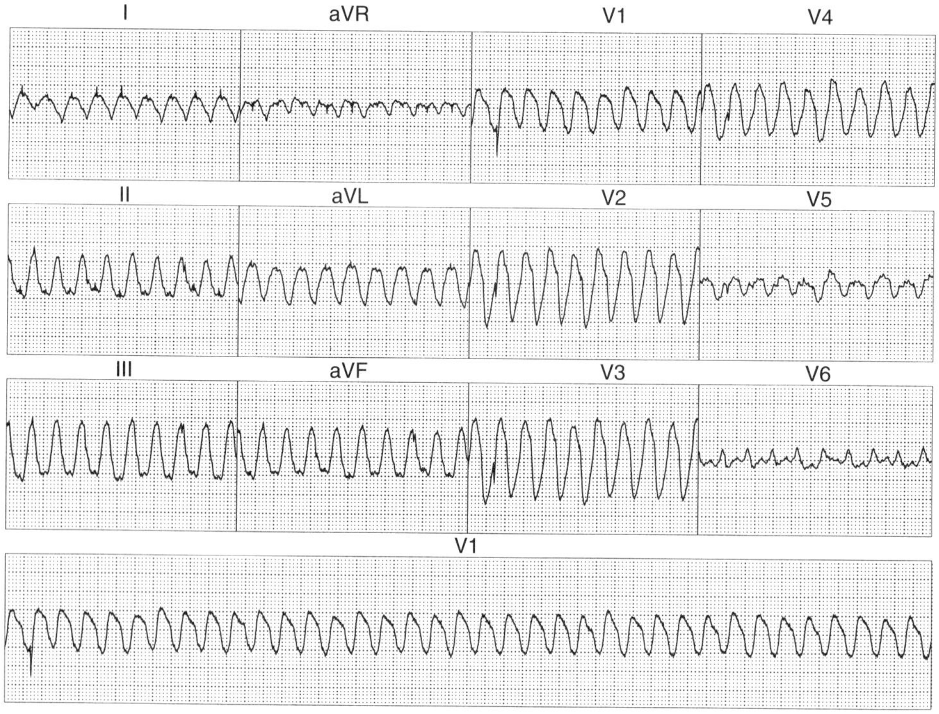

FIGURE 17-10 Twelve-lead ECG of sustained ventricular tachycardia in an elderly man with coronary artery disease and inferoposterior myocardial infarction. The tachycardia cycle length is 280 ms and has a left bundle branch block–like morphology in lead V1 and an inferior axis. Prominent R waves are seen in leads V3 and V4.

TABLE 17-1 Sensitivity and Predictive Accuracy of Electrocardiographic Criteria to Distinguish Ventricular Tachycardia Resembling LBBB from SVT with LBBB Aberration

Criteria*	Sensitivity	Predictive Accuracy
1. R > 30 ms in V1 or V2	36%	100%
2. Any Q in V6	55%	98%
3. >60 ms to S nadir in V1 or V2	63%	98%
4. Notched downstroke S wave	36%	97%
Any of the above present	100%	96%

*Diagnostic criteria for VT with a morphology resembling left bundle branch block.

LBBB, left bundle branch block; ms, milliseconds; SVT, supraventricular tachycardia.

From Kindwall KE, Brown J, Josephson ME: Electrocardiographic criteria for ventricular tachycardia in wide complex left bundle branch block morphology tachycardias. Am J Cardiol 1988;61:1279-83.

both LBBB and RBBB. The criteria used to differentiate VT from SVT included (1) presence of an RS complex in at least one precordial lead; (2) an R–S interval in any precordial lead greater than 100 ms; (3) the detection of AV dissociation; and (4) morphology criteria for VT present in leads V1 to V2 and V6. The morphology criteria used for the diagnosis of VT are summarized in Table 17-2. These criteria were present in V1 to V2 and V6 in 61% of patients with VT. Unfortunately, only one third of VT recordings fulfill these criteria due to discordance in the morphology criteria in different ECG leads. An RS complex was present in at least one precordial lead in all SVT recordings, but only 26% of the VT recordings did not have an RS complex in any precordial lead. None of the SVTs had an R–S interval greater than 100 ms, but this criterion was only met in 52% of VTs. From these observations the authors concluded that an RS complex in all precordial leads or an R–S interval greater than 100 ms in any precordial lead were highly specific for VT.

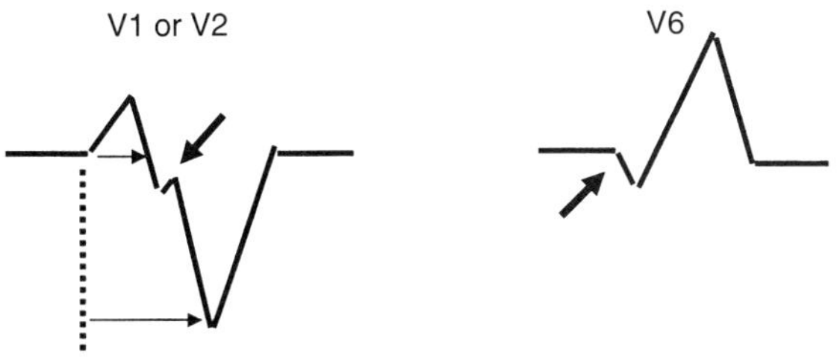

FIGURE 17-11 Four electrocardiographic criteria for ventricular tachycardia with QRS morphology resembling left bundle branch block. (From Kindwall KE, Brown J, Josephson ME: Electrocardiographic criteria for ventricular tachycardia in wide complex left bundle branch block morphology tachycardias: Am J Cardiol 1988;61:1279-83.)

TABLE 17-2 Morphology Criteria for Ventricular Tachycardia

	SN (%)	SP (%)	+PV (%)	−PV (%)
Tachycardia with RBBB-QRS				
Lead V1				
Monophasic R	60	84	78	69
QR or RS	30	98	95	60
Triphasic	82	91	90	83
Lead V6				
R/S ratio < 1	41	94	87	63
QS or QR	29	100	100	60
Monophasic R	1	100	100	52
Triphasic	64	95	93	71
R/S ratio > 1	30	76	58	81
Tachycardia with LBBB-QRS				
Lead V1 or V2 and V6				
Any of following:	100	89	96	—
V1 or V2				
R>30 ms*				
>60 ms to nadir S*				
Notched S				
Q wave in V6				
Lead V6				
QR or QS	17	100	100	52
Monophasic R	100	17	51	100

*Diagnostic criteria for analysis of RS width in precordial leads.

SN, sensitivity; SP, specificity; +PV, positive predictive value; −PV, negative predictive value; RBBB, right bundle branch block.

From Brugada P, Brugada J, Mont L, et al: A new approach to the differential diagnosis of a regular tachycardia with a wide QRS complex. Circulation 1991;83:1649-59.

To address the sensitivity and specific concerns, Brugada[67] employed a stepwise approach to differentiate SVT from VT and prospectively tested this algorithm (Fig. 17-12). The algorithm was both sensitive (99%) and specific (96%). The advantage of the stepwise approach is that it guides the analyst to the correct diagnosis. If an RS complex is not present in any precordial lead, the diagnosis of VT is confirmed without the need for further analysis. If an RS complex is present and the longest R–S interval in any precordial lead exceeds 100 ms, then the diagnosis of VT is confirmed. If the R–S interval is less than 100 ms, then an examination for atrioventricular dissociation must be performed. The presence of this criterion helps diagnose VT. If AV dissociation is not detected, then the diagnosis of VT depends on the presence of morphology criteria in both leads V1 and V6.

ECG LOCALIZATION OF VENTRICULAR TACHYCARDIA ORIGIN

The morphology of monomorphic VT documented by a 12-lead ECG is also useful in predicting the exit site of reentrant VT based on studies in patients being considered for catheter ablation or surgical excision of their arrhythmias. Josephson[68] analyzed the QRS morphology during VT in patients selected for electrophysiological mapping. The VT morphologies can be separated into RBBB and LBBB patterns based on the

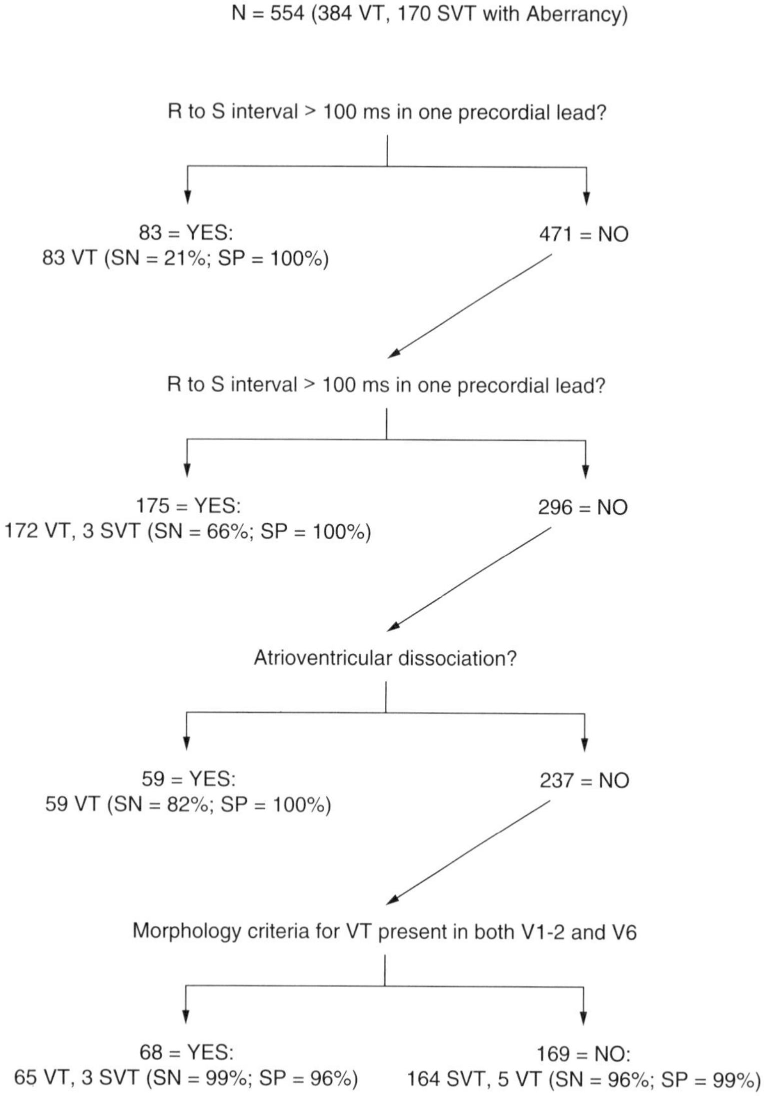

FIGURE 17-12 Algorithm developed by Brugada for diagnosis of ventricular tachycardia. (Modified from Brugada P, Brugada J, Mont L, et al: A new approach to the differential diagnosis of a regular tachycardia with a wide QRS complex. Circulation 1991:83:1649-59.)

QRS morphology in lead V1 (Fig. 17-13). The RBBB pattern includes monophasic, biphasic, or triphasic R waves in V1 or a qR complex in that lead. The LBBB pattern was defined by a QS, rS, or qrS recorded in V1. Sixteen VTs with LBBB pattern were studied in patients with coronary artery disease. The QRS morphology in V1 showed an rS pattern in 10, QS in 4, and qrS in 2. All but one of these arose from sites on or immediately adjacent to the intraventricular septum. The gross electrocardiographic features such as the bundle branch block pattern or QRS axis did not reliably localize the origin of the tachycardia.

Tachycardias were also analyzed according to their site of origin. VT arising from the inferior aspect of the anterior septum was characterized by a superior and leftward axis. In five of the six tachycardias originating from this region, a QS pattern was observed in V1 to V4. The one exception was less apical in origin. None developed substantial R waves in the precordial leads, and all had Q waves in leads I and V6. Four tachycardias arose from the superior aspect of the mid to apical septum. These were characterized by normal or rightward axis and an rS, Qs, or qrS in V1. They all developed prominent R waves in the lateral precordial leads. Five tachycardias arose from the posterobasal region. They all had rS complexes in V1 to V2 and progressive R wave development in the lateral leads. None had a Q wave in leads I or V6. Four of the tachycardias had a superior and leftward axis. In summary, the presence of Q waves in leads I or V6 and a superior and leftward axis was characteristic of the inferior aspect of the anterior septum. Those arising from the superior aspect of the anterior septum had an inferior and rightward axis. VT arising from the posterobasal region exhibited larger R waves in leads I, V2, V3, and V6.

The tachycardias with a right bundle branch pattern all arose from the left ventricle and were grouped according to their origin from anteroseptal, anterolateral, or basal sites. The electrocardiographic patterns recorded from these sites showed substantial overlap and did not reliably distinguish septal from more lateral origins. Anterior and basal origins were distinguished by lead I and the precordial leads. A qR or QS in lead 1 and a Q wave in V1, V2, and V6 suggested an anterior origin.

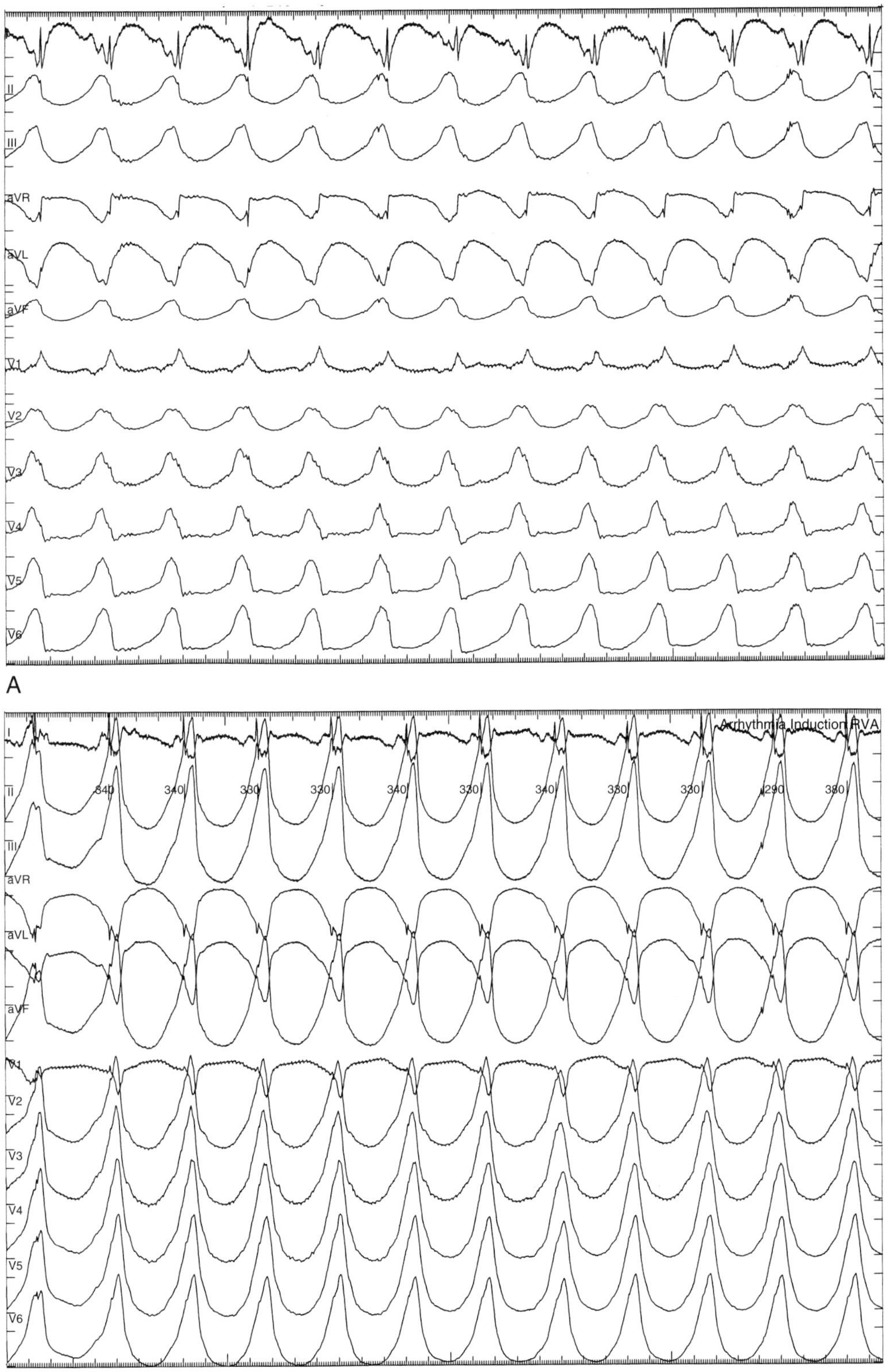

A

B

FIGURE 17-13 Twelve-lead ECGs of two clinical morphologies of ventricular tachycardia in a patient with ischemic cardiomyopathy due to posterobasal left ventricular aneurysm associated with multivessel coronary artery disease. The two hemodynamically stable ventricular tachycardia (VT) morphologies were mapped to different exit points on the margin of the ventricular aneurysm. **A,** Sustained VT with a right bundle branch block–type morphology in lead V1 with an inferior axis and cycle length of 315 milliseconds (ms). Note the positive R wave in ECG leads V1-V6. **B,** Sustained VT with a left bundle branch block–type morphology in lead V1 with an inferior axis and cycle length of 330 ms.

Those arising from basal regions exhibited prominent R waves in V1 to V6.

Kuchar analyzed QRS configurations observed during left ventricular pacing in 22 patients to construct an algorithm to localize the origin of VT, which was then prospectively tested in a second group of 44 episodes of VT.[67] As shown in Figure 17-14, apical versus basal sites were differentiated by QRS configurations in V4 and aVR; anterior versus inferior sites by QRS configuration in leads II, II, and V6; and septal versus lateral sites were distinguished by QRS configuration in leads I, AVL, and V1. Anterior sites were correctly identified in 83%, inferior sites in 84%, septal sites in 90%, and lateral sites in 82%. Apical and basal sites were accurately distinguished in 70%, but the algorithm was not reliable for intermediate sites. Overall, the site of origin was localized in 39%, and in an additional 36% the site of origin was immediately adjacent to the predicted site.

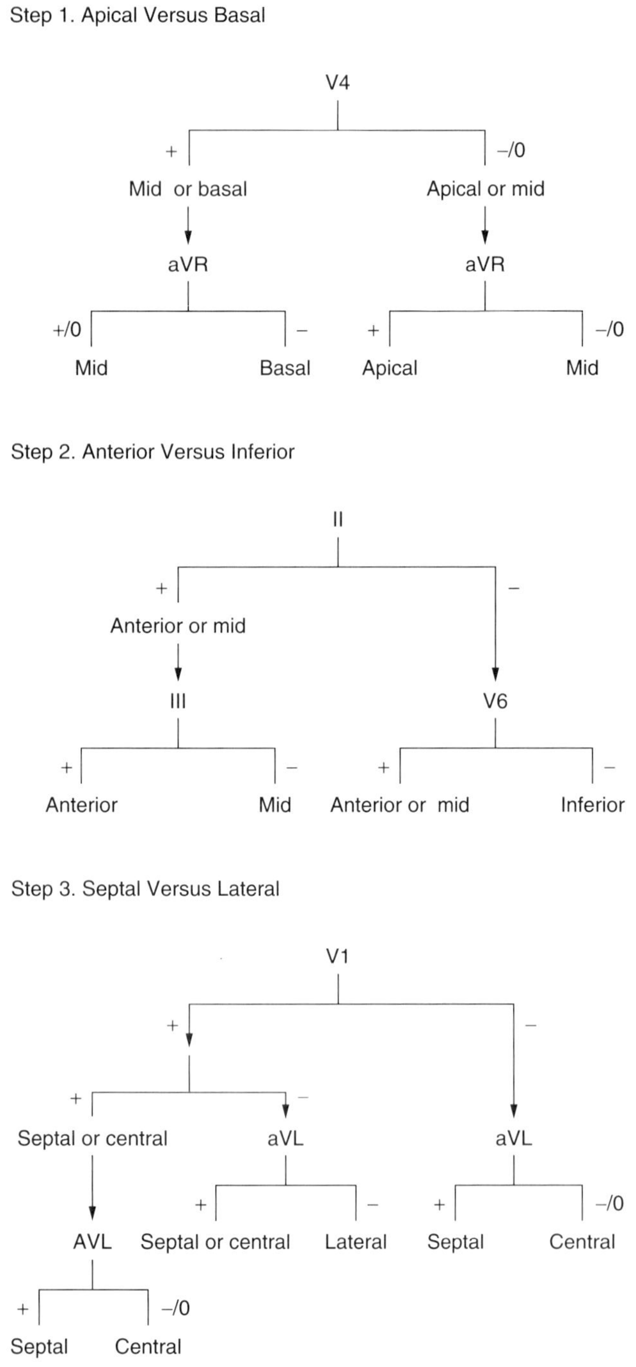

FIGURE 17-14 Algorithm for localization of origin of ventricular tachycardia. (Modified from Kuchar DL, Ruskin JN, Garan H: Electrocardiographic localization of the site of origin of ventricular tachycardia in patients with prior myocardial infarction. J Am Coll of Cardiol 1989;13:893-903.)

Sippensgroenewegen compared 62-lead body surface QRS integral maps and scalar 12-lead ECGs obtained during pace mapping in patients with prior myocardial infarction and VT to determine the accuracy of localizing the origin of VT with these two methods.[70] An evaluation of the localization results obtained during pace mapping was performed to compare 12-lead ECGs with body surface mapping. The site of origin, as determined by endocardial mapping, was compared to the location suggested by pace mapping. With pace mapping used in combination with body surface maps, the site of origin was identified within 2 cm of the origin in 80% of VTs. The remaining 20% were localized to an adjacent (2 to 4 cm) or disparate (>4 cm) site. Results obtained with standard 12-lead ECGs were less accurate. Correct localization (<2 cm) was achieved in only 18%. The localization site was adjacent to the site of origin in 55% and disparate in 27%. The size of the corresponding endocardial area was much smaller with body surface mapping (6.0 ± 4.5 cm) than with the 12-lead ECG (15.1 ± 12.0 cm). These results are concordant with a study by Josephson, which showed that pace mapping with a 12-lead ECG allows localization to a relatively large area of 20 to 25 cm.[2,71]

Further study is needed to determine the utility of body surface mapping for ablation of VT; however, results obtained by Sippensgroenewegen[72] indicate that this approach is relatively accurate. QRS integral maps (62 leads) were compared to activation mapping data that were acquired in 64 episodes of VT. The maps were compared with a previously generated infarct-specific reference database of paced QRS integral maps. Each pattern of the database corresponded to 1 of 18 to 22 segments of the left ventricle. Electrocardiographic localization was compared to intraoperative or catheter endocardial activation mapping. Body surface maps identified the correct segment of origin in 62%, an adjacent segment in 30%, and disparate segments in 8%.

These studies show that the QRS morphology, recorded during VT or ventricular pace-mapping, provides an estimate of the site of origin, but frequently there are major disparities between the origin and the site estimated from 12-lead ECGs. Body surface mapping appears to be more accurate, but is no substitute for the additional information gained during an electrophysiological study, which provides detailed information about specific components of the reentrant circuit that is needed for an ablation procedure.

Clinical Electrophysiology

The majority of ventricular tachycardias associated with heart disease are due to reentrant mechanisms and can usually be initiated and terminated by critically timed ventricular stimuli. This finding allows evaluation in the electrophysiology laboratory under controlled conditions. Induced arrhythmias are classified based on duration and morphology. Arrhythmias that last for 30 seconds or longer or that require earlier termination due to hemodynamic consequences are referred to as *sustained VT*. Sustained VT can be either monomorphic or polymorphic based on QRS morphology.

PROGRAMMED STIMULATION FOR INITIATION OF VENTRICULAR TACHYCARDIA

Most laboratories proceed through a stepwise stimulation protocol initiation of sustained VT. Following ventricular pacing for 8 to 15 beats, a single extrastimulus is used to scan electrical diastole until it encounters refractoriness or reaches a very short coupling interval (<200 ms). A second, third, and, in some cases, fourth stimulus is then added. Two or more different basic pacing rates (drive cycles) preceding premature stimulus delivery are usually employed (e.g., 100 beats per minute [bpm] and 150 bpm). If pacing at the right ventricular apex fails to induce ventricular tachycardia, pacing at a second right ventricular site (e.g., the right ventricular outflow tract) is generally employed. Some laboratories also employ pacing from the left ventricle. The number of stimuli, basic drive cycles, and ventricular pacing sites vary among laboratories. In general, pacing with up to three extrastimuli at two cycle lengths and two sites induces VT in approximately 90% of patients who have had this arrhythmia spontaneously after a myocardial infarction.[73] The addition of rapid burst pacing, left ventricular stimulation, and programmed stimulation during isoproterenol infusion further increases sensitivity.

As the number of extrastimuli increases, the risk of initiating nonspecific, polymorphic VT (Fig. 17-15) or VF increases (see later). Limiting the closest coupling interval to greater than 200 ms reduces the risk of initiating ventricular fibrillation.[73] Although stimulation protocols are relatively sensitive for detecting an inducible VT, the precise number of stimuli and stimuli coupling intervals required often varies over time, such that a change in number of extrastimuli required for initiation of VT is not necessarily a reliable indication that susceptibility to VT has changed.[74,75] VT initiated by isoproterenol infusion or with rapid burst pacing during isoproterenol administration, but not in response to coupled extrastimuli, is felt to be more likely due to abnormal automaticity rather than reentry. Uncommonly, monomorphic VT may be induced by such measures in genetically based cardiac arrest syndromes.

CONFIRMING THE DIAGNOSIS

In contrast to the ECG, where atrial activity may be difficult to detect, direct recordings from the atrium always allow clear delineation of whether AV dissociation is present (Fig. 17-16). When AV dissociation is present, the diagnosis is VT, with the rare exception of junctional ectopic tachycardia with ventriculo-atrial (V-A) block or rare forms of SVT without atrial tissue participation. The former arrhythmia, which occurs most commonly in the early postoperative period after

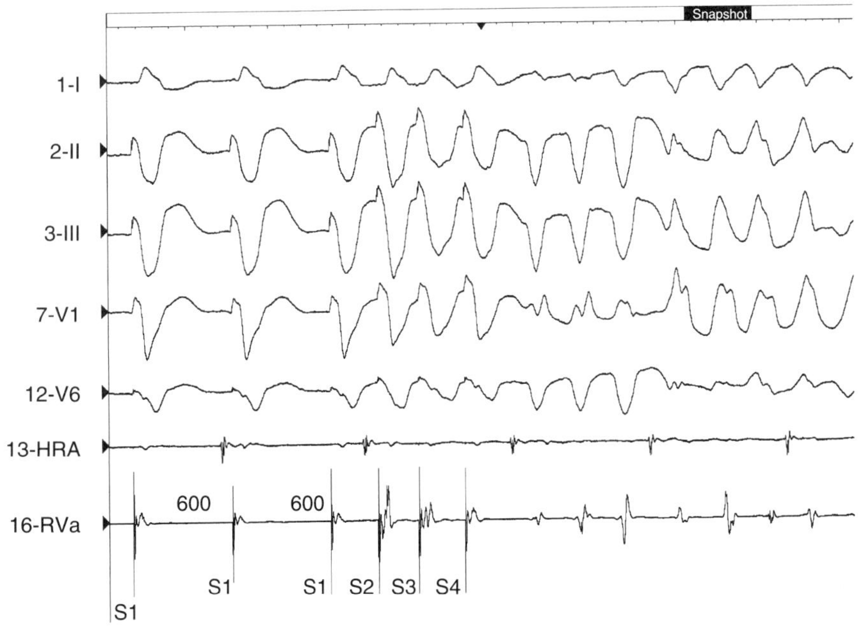

FIGURE 17-15 Initiation of polymorphic ventricular tachycardia (VT) by programmed stimulation from the right ventricular apex is shown. From the top are surface ECG leads I, II, III, V1, and V6 followed by intracardiac recordings from high right atrium (*HRA*) and right ventricular apex (*RVa*). From the left, the last two stimuli (*S1*) of a stimulus train at a cycle length of 600 milliseconds (ms) (100 beats per minute) followed by three consecutive premature stimuli (*S2, S3, S4*) are shown. This initiates polymorphic VT.

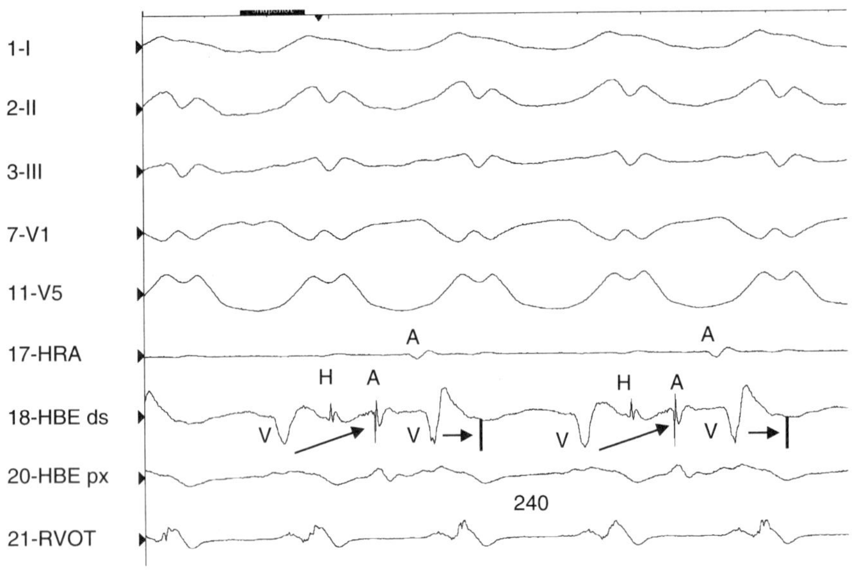

FIGURE 17-16 A tracing from the electrophysiology laboratory during sustained monomorphic ventricular tachycardia (VT) is shown. From the top are surface ECG leads I, II, III, V1, and V5 followed by intracardiac recordings from the high right atrium (*HRA*), and His bundle HBE catheters. VT has a cycle length of 240 milliseconds (250 beats per minute) and a left bundle branch block, inferior axis QRS configuration. During VT 2:1 VA block is present; every other ventricular beat conducts retrogradely into the His bundle (*H*) and then to the atrium (*A*). Thus, the atria and His bundles are not involved in causing the VT.

repair of congenital heart disease in children, is rare in adults. When AV dissociation is present and His bundle depolarization does not precede each QRS complex, the diagnosis of VT is unequivocal (see Fig. 17-16). Diagnostic dilemmas are encountered. If each QRS complex is followed by an atrial depolarization, VT with one-to-one conduction retrogradely through the His-Purkinje system and AV node must be distinguished from antidromic AV reentry using an accessory pathway (see Chapter 14, Paroxysmal Supraventricular Tachycardia and the Preexcitation Syndromes).

Polymorphic Ventricular Tachycardia. Polymorphic VT (see Fig. 17-15) indicates a continually changing ventricular activation sequence. Spontaneous polymorphic VT is most commonly due to myocardial ischemia or torsades de pointes associated with QT prolongation. Torsades de pointes is not reproducibly inducible in the electrophysiology laboratory, and electrophysiological studies are not warranted in the presence of active ischemia. Polymorphic VT also occurs in idiopathic ventricular fibrillation, Brugada syndrome, and hypertrophic as well as other cardiomyopathies.

During ventricular stimulation the significance of initiating polymorphic VT or VF is often difficult to interpret. When sustained, it usually quickly deteriorates to VF. Nonsustained polymorphic VT is a nonspecific response to programmed stimulation that can be observed in patients who do not have a susceptibility to arrhythmias. Sustained polymorphic VT (lasting >30 sec or requiring termination) is initiated less commonly in normals and usually requires aggressive stimulation (3 or more extrastimuli and relatively short stimulus coupling intervals of <200 ms). Initiation of polymorphic VT may be a marker of electrical instability in some situations. It is induced in approximately one third of patients who have been resuscitated from VF associated with depressed ventricular function and coronary artery disease.[76] In patients with prior infarction and inducible polymorphic VT, sustained monomorphic VT became inducible after administration of IV procainamide.[74] Stabilization of a reentry circuit by an antiarrhythmic drug is a possible mechanism, but the clinical relevance of this observation is unclear.

Initiation of polymorphic VT is of possible relevance in patients with suspected Brugada syndrome. Polymorphic VT, often induced with two or fewer extrastimuli, is induced in approximately 80% of patients with Brugada syndrome who have been resuscitated from cardiac arrest.[77] In a patient with other features of the syndrome, inducible polymorphic VT further supports the diagnosis. In patients with hypertrophic cardiomyopathy, sustained VT (usually polymorphic) is inducible in 66% of those who had a prior cardiac arrest, as compared with 23% of patients without a history of cardiac arrest or syncope.[78] The prognostic significance of inducible polymorphic VT in hypertrophic cardiomyopathy and other cardiomyopathies is controversial.[79]

Sustained Monomorphic Ventricular Tachycardia. During sustained monomorphic VT, each QRS complex resembles the preceding and following QRS (see Fig. 17-16). The ventricles are repetitively depolarized in the same sequence. In contrast to polymorphic VT, which can occasionally be induced in the absence of a cardiac abnormality, spontaneous or inducible sustained monomorphic VT indicates the presence of either an abnormal region supporting reentry or a focus of automaticity. Reentry through regions of ventricular scar due to myocardial infarction is the most common cause. Other causes of scar-related VT include arrhythmogenic right ventricular cardiomyopathy, sarcoidosis, scleroderma, Chagas' disease, ventricular incisions after

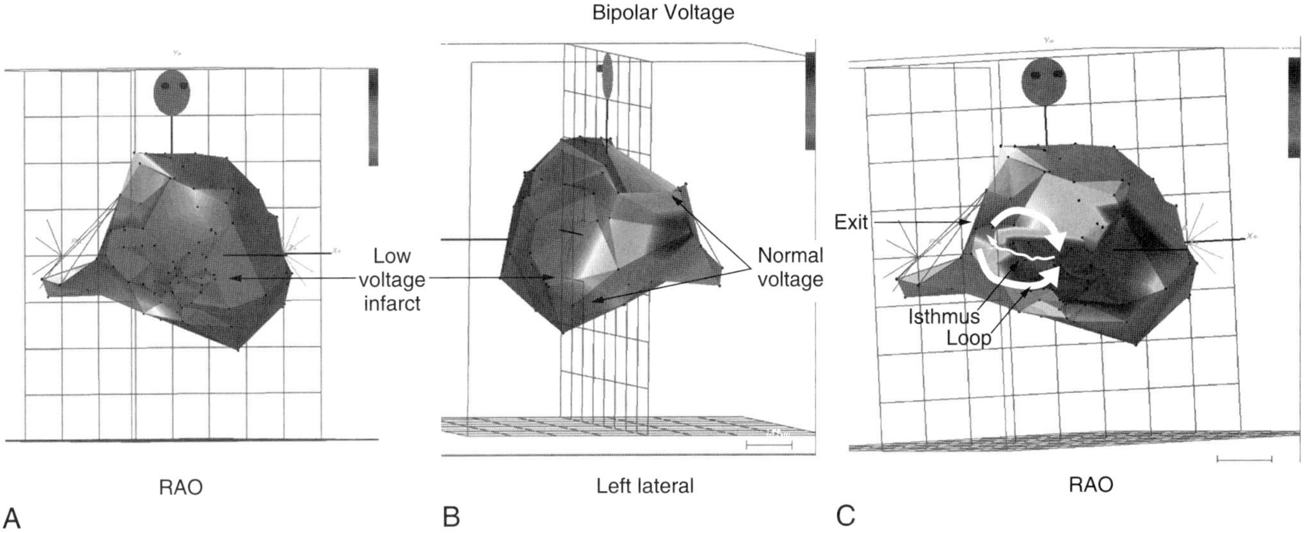

FIGURE 17-17 (See also Color Plate 17-17.) Mapping data from a patient with prior anterior wall infarction and ventricular tachycardia. **A** and **B,** Map of the left ventricle in right anterior oblique (AO) and left anterior oblique (LAO) projections obtained by moving a mapping catheter from point to point. The color-coding indicates electrogram voltage; low voltage areas are *red, yellow, orange*; normal voltage is *green, blue, purple.* A very large area of low voltage, identifying the infarct region, occupies the septum, apex, and anterior wall. **C,** Ventricular activation sequence map obtained during ventricular tachycardia. The color-coding indicates the sequence of activation, and the circuit is indicated by the *white arrows.* The VT circuit isthmus is located on the septum. The reentry wavefront travels from apical to basal along the septum exiting near the base (*red*), then divides into two wavefronts that propagate superiorly and inferiorly and back to the isthmus.

cardiac surgery, or other cardiomyopathic processes.[78] Areas of scarring can often be identified as regions of low electrogram voltage during catheter mapping (Fig. 17-17). Occasionally an incessant, idiopathic VT that is consistent with automaticity will cause tachycardia-induced cardiomyopathy.[80]

BUNDLE BRANCH REENTRY VENTRICULAR TACHYCARDIA

A His bundle deflection is consistently present before each QRS during an uncommon type of VT due to reentry through the bundle branches (Fig. 17-18). The reentry wavefront circulates up the left bundle branch, down the right bundle branch, and then through the intraventricular septum to reenter the left bundle.[81] Ventricular depolarization proceeds from the right bundle, giving rise to a tachycardia that has an LBBB configuration. Rarely, the circuit revolves in the opposite direction (down the left bundle and back up the right bundle) or is confined to the fascicles of the left bundle branch system, giving rise to a tachycardia that has a RBBB configuration. The diagnosis is confirmed by showing that the atria can be dissociated, but that the His bundle depolarization is closely linked to the circuit. Bundle branch reentry should be particularly

FIGURE 17-18 Bundle branch reentry ventricular tachycardia is shown. In the tracing at the left, surface ECG leads I, II, III, V1, and V5 are followed by recordings from the high right atrium and His bundle, respectively. Tachycardia with a cycle length of 295 milliseconds (203 beats per minute) with a left bundle branch block–like configuration in V1 is present. AV dissociation is present as indicated by the slower rate of atrial depolarization (*A*) in the right atrial tracing. Although the atria are dissociated, a His bundle deflection (*H followed by arrow*) precedes each QRS complex. The mechanism is shown in the panel at the right (see text).

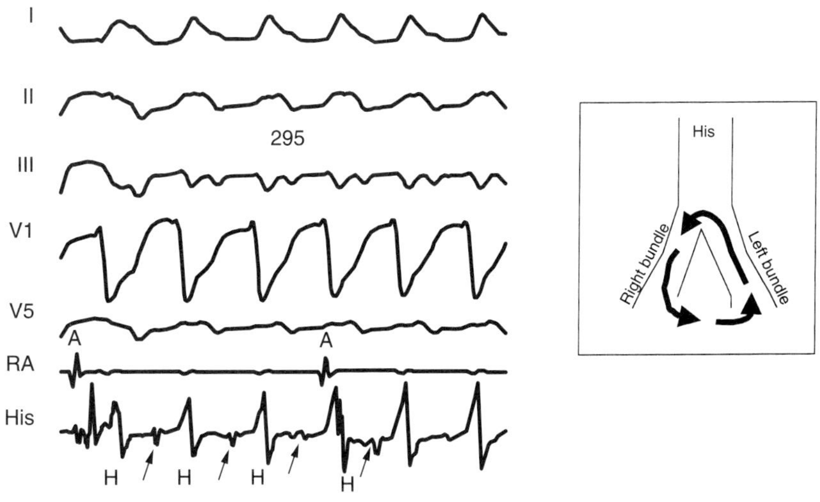

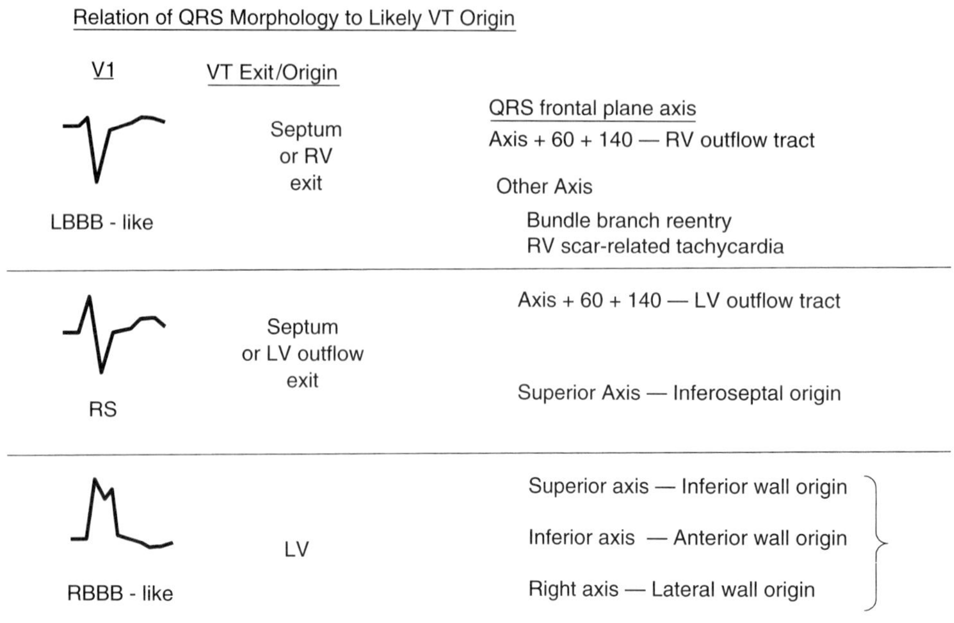

FIGURE 17-19 The QRS morphology suggests the likely origin and often the type of heart disease and ventricular tachycardia. The morphology of V1 provides initial indication of the likely ventricle of origin. Further refinement is based on the frontal plane QRS axis as shown at right. LBBB, left bundle branch block; LV, left ventricle; RBBB, right bundle branch block; RV, right ventricle.

suspected when sustained monomorphic VT occurs in the absence of a large region of ventricular scar or infarction, such as in patients with valvular heart disease,[82] cardiomyopathy, or muscular dystrophy[83] and in association with evidence of His-Purkinje system disease (intraventricular conduction delay) during sinus rhythm.

QRS MORPHOLOGY OF INDUCED VENTRICULAR TACHYCARDIA

The QRS morphology reflects the ventricular activation sequence and often provides an indication of the likely location of the arrhythmogenic region (Fig. 17-19). For automatic or focal VT, activation spreads away from one arrhythmia focus; the QRS morphology is an excellent indicator of the location of the focus. The QRS morphology is less reliable as an indicator of the circuit location for scar-related VT but is still useful.

Scar-related reentry circuits can be modeled as having surviving bundles of myocytes in the region of the scar. The depolarization of these strands is not detected in the surface ECG (see Figs. 17-17 and 17-20). These bundles may form narrow isthmuses in the reentry circuit that are desirable targets for ablation. Recordings from these regions often reveal multiple low-amplitude potentials (Fig. 17-21). During VT the QRS is inscribed after the reentry wavefront emerges from an isthmus at

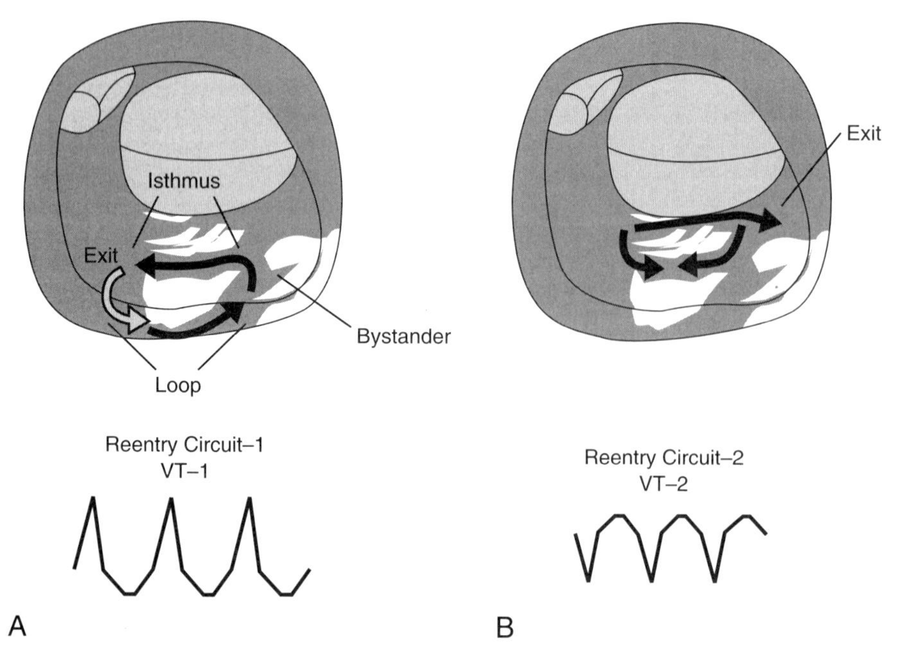

FIGURE 17-20 Schematic of the left ventricle with reentry circuits in an inferior wall region of infarction are shown. **A,** The reentry wavefront propagates through a narrow isthmus and exits at the margin of the infarct to propagate across the ventricles, producing the QRS complex. After exiting, the circulating wavefront propagates along the margin of the infarct and then into the infarct to the proximal portion of the isthmus. The tissue along the infarct border forms a broad loop in the reentry circuit. Bystander areas are adjacent to the infarct, but not in the reentry circuit, often giving rise to abnormal signals that misleadingly appear to be in the circuit based on timing and characteristics. **B,** Illustration of how the same infarct can give rise to a different reentry circuit than shown in **A,** creating multiple morphologies of monomorphic ventricular tachycardia.

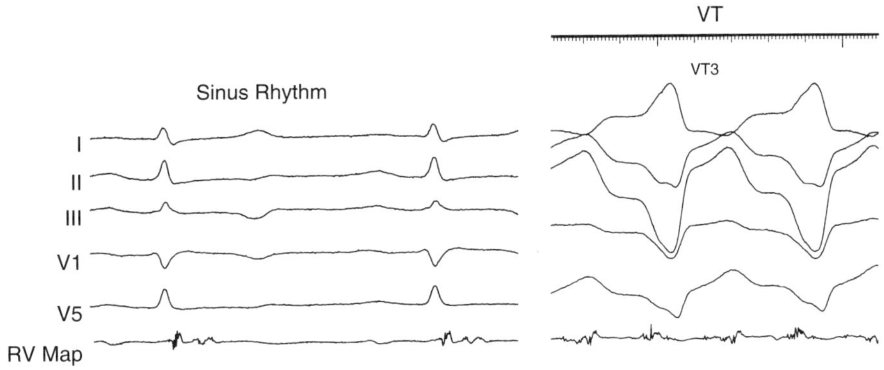

FIGURE 17-21 Recordings from a patient with arrhythmogenic right ventricular cardiomyopathy and ventricular tachycardia. From the top are surface ECG leads *I, II, III, V1,* and *V5* and an intracardiac recording from the mapping catheter positioned at the anterobasal right ventricle near the tricuspid annulus (*RV Map*). During sinus rhythm the right ventricular electrograms recorded from this area are low amplitude and fractionated (multiple deflections), consistent with an area of scar with surviving myocyte bundles. Induced ventricular tachycardia (*right panel*) had a rate of 150 beats per minute. Fractionated potentials are recorded preceding the QRS complex, consistent with an isthmus in the reentry circuit.

its exit and propagates across the ventricles. The QRS morphology indicates the location of the reentry circuit exit (see Fig. 17-20). Patients with scar-related ventricular tachycardias often have multiple morphologies of VT inducible; typically three or more are observed in patients referred for catheter ablation (see Figs. 17-20 and 17-22).[84] The VT that has been observed to occur spontaneously is often referred to as the "clinical VT." The other VTs may arise from the same region of the scar or anatomically separate regions. Some "nonclinical" VTs are subsequently observed to occur spontaneously or are initiated by attempted antitachycardia pacing from an implanted defibrillator in attempts to terminate a "clinical VT."

PACING FOR VENTRICULAR TACHYCARDIA TERMINATION

Monomorphic VT can be terminated reliably by electrical cardioversion, which presumably depolarizes the reentry circuit in advance of the circulating wavefront, which then collides with refractory tissue, extinguishing reentry. Many monomorphic VTs can also be terminated by pacing faster than the tachycardia (overdrive pacing). The stimulated wavefront propagates to the reentry circuit and then splits into wavefronts traveling in the same direction as the reentry circuit wavefronts (orthodromic) and a second wavefront that is traveling in the opposite direction (antidromic). The antidromic wavefront collides with the next reentry wavefront, and both are extinguished. The stimulated orthodromic wavefront continues through the circuit and may reset the circuit so that tachycardia continues (known as resetting or entrainment). If the orthodromic wavefront encounters refractory tissue and is extinguished, the tachycardia terminates. Pacing during tachycardia also has the possibility to accelerate the tachycardia or initiate VF (see Fig. 17-22). Thus, the capability for prompt defibrillation must be present when antitachycardia pacing is employed.

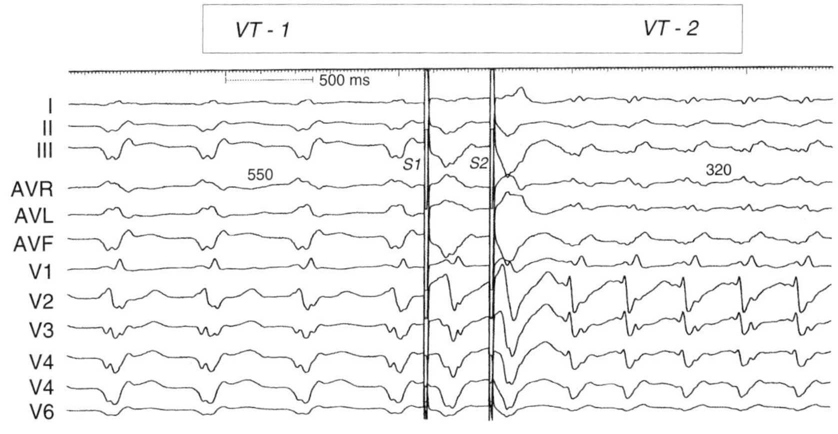

FIGURE 17-22 Twelve-lead ECG recording two morphologies of ventricular tachycardia (VT) in a patient with prior infarction. The initial VT (*VT-1*) has a cycle length of 550 milliseconds (ms). Two premature ventricular stimuli (*S1, S2*) initiate monomorphic *VT-2*, which is substantially faster (cycle length 320 ms) with a different QRS morphology.

CATHETER MAPPING OF VENTRICULAR TACHYCARDIA

When VT is due to increased automaticity, the source of the tachycardia can often be identified by activation sequence mapping (i.e., recording the activation sequence of the ventricle during VT). At the tachycardia focus, ventricular activation preceding the QRS onset is identified. Pacing at this site during sinus rhythm will produce a QRS morphology strikingly similar to that of the VT. Thus, comparing the QRS morphology produced by ventricular pacing at different sites (pace-mapping) can also localize the VT focus.

Activation sequence mapping is more difficult for reentrant VTs. The area of activation prior to the QRS onset usually identifies an area that is near the reentry circuit exit. Low-amplitude signals are often recorded from sites in the circuit that are proximal to the exit. Often, only portions of the circuit are located in the endocardium, because other segments are deep to the endocardium or epicardial in location and cannot be sampled by an endocardial catheter.

Pacing during the tachycardia can be used to help determine the reentry circuit location (entrainment mapping). Pacing trains that capture, but have no effect on, the tachycardia often indicate that the pacing site is not in the tachycardia circuit. Stimulated wavefronts that propagate to the circuit, enter the circuit, and propagate through the circuit will reset the reentry circuit in a characteristic manner. Whether this occurs is dependent on the timing of the stimulus relative to activation in the circuit. The wavefront must reach a portion of the circuit after that area has recovered excitability from the preceding wavefront. A single pacing stimulus may reset the tachycardia, while a train of several stimuli continually resets the tachycardia, a response known as *entrainment*.[84]

Management

PRINCIPLES OF PRACTICE

It is convenient to divide sustained VT in patients with heart disease into "syndromes," which can usually be identified by examining the ECG recorded at the initiation of the tachycardia (usually recorded as a single lead rhythm strip), the 12-lead ECG recorded shortly after treatment of the acute rhythm disturbance, and considering the clinical context in which this arrhythmia arises.

Sustained monomorphic VT occurs in patients with prior left ventricular scarring (most commonly from a prior myocardial infarction); most often arises without a specific predisposing event, often at rest or with modest activity; and generally is not preceded by symptomatic, electrocardiographic, or enzymatic evidence for myocardial ischemia or myocardial infarction.[85] Sustained VT is observed in a similar circadian pattern to that for sudden death and myocardial infarction, with a predilection for the morning hours (at least in patients with implanted defibrillators).[86,87] It tends to occur in "clusters," suggesting that neurohumoral, in particular sympathetic, activation may contribute indirectly to the instantaneous probability of VT occurring in a susceptible patient.[87,88] Occasionally, sustained VT will occur many times within a brief period, a syndrome commonly termed "electrical storm." It is often arbitrarily defined as more than two or three episodes of sustained VT in a 24-hour period. Patients with this syndrome are perceived to have a very poor prognosis, although aggressive therapy, particularly β-blockers and amiodarone, may, in observational studies, allow a prognosis similar to patients who do not develop electrical storm.[88,89] A related syndrome is that of polymorphic VT in patients with severe ventricular scarring, occurring most often in the setting of a generalized cardiomyopathy such as dilated idiopathic cardiomyopathy, or that associated with hypertrophic cardiomyopathy, valvular disease, or hypertensive cardiomyopathy. Although the precipitating events for this type of VT are also not clearly understood, they may be more likely to occur during physical exercise, elevated sympathetic tone, or episodes of worsening heart failure. The management of these patients, in addition to antiarrhythmic therapy, may also be aggressively directed toward improving myocardial function and decreasing sympathetic tone, for example, with afterload reduction, preload reduction, and sympathetic blockade.

A particular pattern of sustained ventricular arrhythmias can be observed in patients with myocardial ischemia, acute myocardial infarction, or severe coronary artery disease. In these patients, VT arises more frequently under conditions of physical or psychological stress; it may be preceded by chest pain or other symptoms of myocardial ischemia; and the ECG before or following treatment of the VT frequently shows sinus tachycardia, ST segment depression, or signs of myocardial infarction. The *electrocardiographic signature* of this arrhythmia (Fig. 17-23) typically involves a relatively short Q–T interval, and the initiating beat of tachycardia being closely coupled to the last normal beat, immediately ushering in the polymorphic VT that may rapidly degenerate to VF. Patients in whom this pattern is observed should be immediately and thoroughly investigated for myocardial ischemia, and treatment directed at ischemia rather than only at the arrhythmia is often sufficient to prevent recurrences.[90,91]

Finally, the syndrome of polymorphic VT associated with prolonged repolarization is discussed in detail in other sections of this book. However, it should be noted that sustained VT in association with delayed repolarization (usually termed "torsades de pointes") may manifest as relatively monomorphic ventricular tachycardia, especially at its outset (Fig. 17-24). The electrocardiographic clues to the etiology as being related to abnormal and delayed repolarization rather than an underlying "arrhythmogenic scar" are also to be found in the initiating sequence of tachycardia. Delayed repolarization–related VT is generally associated with QT prolongation, a "pause-dependent" initiating sequence, and a relatively long coupling interval between the last normal beat and the first beat of tachycardia. This beat typically falls on the late portion of the T wave or T-U complex of a beat with prolonged repolarization.

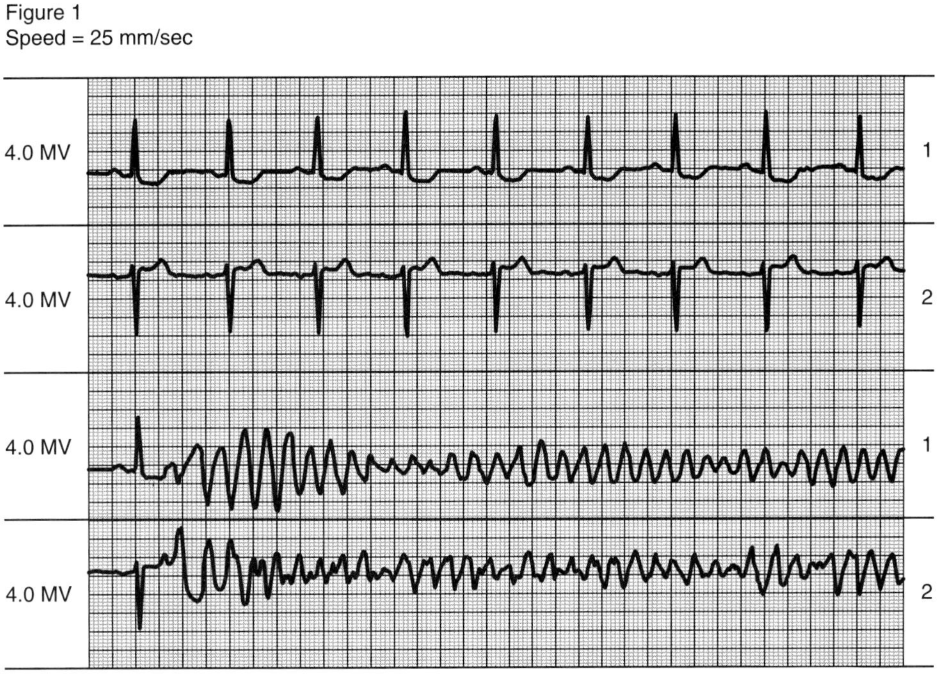

FIGURE 17-23 "R on T" ventricular premature beat inducing ventricular fibrillation from a 70-year-old man with severe three-vessel coronary disease. Note that the Q–T interval is relatively short, and premature ventricular contraction inducing tachycardia has a short coupling interval to the last normal beat.

These arrhythmias should be carefully distinguished from those caused by an "arrhythmogenic scar," because they can be effectively treated with IV magnesium, repletion of potassium as indicated, increasing the heart rate, and treating the underlying disturbance leading to prolonged repolarization, which is often drug therapy that prolongs repolarization or causes bradycardia, or both.

Acute Treatment of Sustained Ventricular Tachycardia

The acute management of VT should first be directed at restoring effective circulation, as outlined in the International Guidelines for Advanced Cardiac Life Support.[92] In brief, all patients with substantial hemodynamic compromise, including diminished level of consciousness, hypotension, and clinical signs suggesting cerebral or vital organ hypoperfusion, heart failure, or signs or symptoms of myocardial ischemia should be treated with urgent synchronized cardioversion. It should be remembered that carefully performed electrical cardioversion is almost completely effective for the treatment of VT, if it can be provided promptly, and is associated with extremely low risk. The perceived need for anesthesia is often seen as a limitation to acute cardioversion but can generally be performed using conscious sedation, not requiring intubation or involving serious anesthetic risk. If drug therapy is chosen either as initial treatment of VT or therapy to prevent recurrences immediately after electrical cardioversion, IV procainamide and IV amiodarone are recommended.[92] Careful monitoring for hypotension and bradycardia, which may be caused by the antiarrhythmic drug therapy, is needed.

There is not a large evidence base of randomized clinical trials to assist the clinician in the treatment of

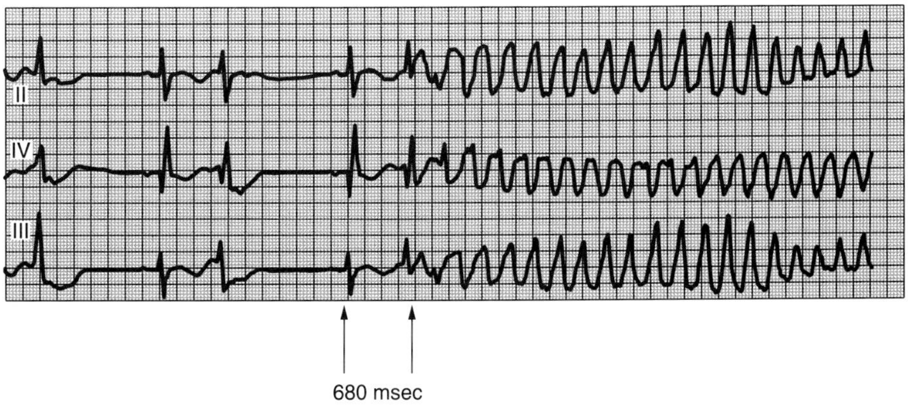

FIGURE 17-24 ECG signature of polymorphic ventricular tachycardia (VT) with delayed repolarization. Note pause-dependent onset of VT with prolonged Q–T interval on the beat initiating tachycardia. From a patient on sotalol and with hypokalemia.

patients immediately following successful conversion from sustained VT to sinus rhythm. The immediate risk of recurrence will vary with the clinical context and be generally relatively low in patients with stable coronary artery disease and left ventricular dysfunction with monomorphic VT. It is higher in patients with acute hemodynamic compromise independent of the tachycardia; ongoing myocardial ischemia; or other metabolic disturbances found in postoperative states, sepsis, pneumonia, or hypoxia. In patients with nonischemic cardiomyopathies, therapy specifically directed at blunting the effects of sympathetic activation on the heart with intravenous β-blockers, or stellate ganglion block, is an often neglected but important aspect of therapy.[89] Limited, controlled clinical trials suggest that IV amiodarone and bretylium are likely effective at preventing recurrent VT.[93] Bretylium, however, is no longer commercially available. In a randomized multicenter trial in patients with "electrical storm" (defined as more than two episodes of sustained VT in a 24-hour period [mean, 4.93], resistant to treatment with IV lidocaine and procainamide), patients were randomized to blinded treatment with IV amiodarone at 1000 mg/day, IV bretylium (2.5 g/day), and low-dose amiodarone (125 mg/day).[93] Although the interpretation of this study is made complicated by the fact that many patients in the low-dose amiodarone group and the bretylium group received open label treatment with unblinded IV amiodarone, the results suggest that IV amiodarone is efficacious at reducing the likelihood of recurrent VT. There was a median of 0 episodes in the first 12 hours after treatment with amiodarone, and 0.48 episodes during the entire study period. The median time to termination of incessant VT was 4.23 hours after the initiation of IV amiodarone therapy. There are no comparable trials of class I drugs or sotalol in electrical storm. Extrapolation from clinical trials of prophylactic therapy suggests that class I drugs (sodium channel blockers) should be avoided in patients with frequently recurring VT or VF.[92]

Although there are no randomized clinical trials of sedation in patients with frequently recurring VT, anecdotal experience suggests that vigorous sedation is an important adjunct in the treatment of this syndrome. As an acute therapy, IV sotalol, IV flecainide, and IV procainamide have all been reported to be effective in the acute termination of sustained VT.[92] Given the negative inotropic and proarrhythmic risks of drugs with sodium channel blocking activity, it seems prudent to use such drugs with great care, if at all, in patients with structural heart disease and sustained VT, especially in the context of coronary artery disease.

Once patients have been stabilized with respect to the acute event, attention can be turned to investigating the underlying cardiac pathology and assessing and optimizing therapy for left ventricular dysfunction and myocardial ischemia.

Impact of Clinical Trials

A recent explosion of epidemiologic insights and information from randomized trials has changed clinical practice in the treatment of VT. For several years, the similarity of patient profiles (stratifying by left ventricular ejection fraction [LVEF], previous myocardial infarction) and risk for sudden cardiac death and sustained VT has been apparent.[85] In 1975 Schaffer et al. reported on 234 patients after successful resuscitation of cardiac arrest. Over a follow-up period of 51 months, 89 episodes (≈38%) of recurrent cardiac arrest or death occurred.[94] A similar rate of recurrence was noted by Myerburg et al. in 1984—the recurrence rate for cardiac arrest was 10% the first year after arrest and 5% per year in each of the following 3 years.[95] In 1993 the CASCADE study[96] reported an improvement in out-of-hospital cardiac arrest survival in patients randomized to amiodarone (versus conventional antiarrhythmic therapy) of 78% at 2 years and 52% at 4 years. Therapeutic options drastically shifted in the 1980s with the emergence of ICDs. Several recent trials have compared the efficacy of antiarrhythmic therapy and ICDs for overall mortality in patients who have sustained a VT event.

In 1995[97] Dutch investigators randomized 60 patients with previous myocardial infarction, cardiac arrest secondary to VT or VF, and inducible ventricular arrhythmia at electrophysiological study to conventional therapy (class IA, IC, and III drugs) or the ICD.[97] The left ventricular EF was approximately 30% in each group. Drug efficacy was assessed by serial drug testing, and nonresponders were given an ICD. An imbalance of coronary revascularization occurred in the conventional therapy group (10%) and the ICD group (26%). Although the numbers of clinical end points (death, prolonged syncope with circulatory arrest, and congestive heart failure requiring transplant) were small, the outcome favored treatment with ICD ($P < .02$).

The Cardiac Arrest Study Hamburg (CASH)[98] began enrollment in 1987 and randomized patients to antiarrhythmic therapy (propafenone, metoprolol, or amiodarone) or ICD. Entry criteria included survivors of cardiac arrest secondary to VT (83%) or VF (16%). All patients underwent electrophysiological testing. The primary end point was total mortality with secondary end points of VT recurrence, need to discontinue medical therapy, recurrence of cardiac arrest, and need for cardiac transplantation. In 1992 the CASH Safety and Monitoring Board discontinued the propafenone randomization arm. This was due to an excess of sudden death, recurrent cardiac arrest, and recurrent VT (all secondary end points) in the patients on class IC drug compared with the ICD arm ($P < .05$), despite no differences in total mortality. The left ventricular ejection averaged for all four groups was more than 40%. The other three treatment arms continued to recruit patients. The final results showed 36.4% mortality in the ICD group and 44.4% in the metoprolol and amiodarone groups combined. A one-sided t-test (which would have amplified any benefit of the ICD) had a P value of .081 as the reported significance value.

Two larger studies, having slightly more heterogeneous populations, also address the issue of ICD versus antiarrhythmic therapy in patients with symptomatic VT. The Canadian Implantable Defibrillator Study (CIDS)[99] randomized 659 patients with documented VF, out-of-hospital cardiac arrest, symptomatic VT, or syncope with inducible VT at electrophysiology study to ICD or amiodarone therapy. Entry criteria in approximately 50% of each group were VF or cardiac arrest. Mean LVEF was 33% in each group. The primary end point was all-cause mortality. An intention-to-treat analysis was used to analyze the data in a one-sided t-test. On treatment analysis showed 94% of those randomized to ICD actually received an ICD, and 85% of patients randomized to amiodarone were still on therapy at their 5-year follow-up. Although β-blocker use was a potential confounder (23% amiodarone group, 53% ICD group), the overall mortality (10.2%/year) in the amiodarone group was not significantly higher than in the ICD group (8.3%, $P = .142$).

The largest study to date has been the Antiarrhythmics versus Implantable Defibrillator (AVID) study,[100] which randomized 1016 patients with resuscitated VF, sustained VT with syncope, and sustained VT with hemodynamic compromise to antiarrhythmic therapy or ICD implantation. The 6035 patients were screened, indicating that 17% were actually randomized. Of the randomized patients, 45% had VF, and 55% had VT at entry. The patients randomized to antiarrhythmic therapy could be further randomized to EP-guided treatment with sotalol or empiric amiodarone at the discretion of the investigator. LVEF was similar in each group (31% and 32%). An on-treatment analysis showed that 98% of the ICD group actually received an ICD. In the antiarrhythmic drug group, 85% received empiric amiodarone, 11% received amiodarone after failing sotalol, and 2.6% were actually treated with sotalol. The study was terminated early in April 1997 due to the recommendation of the Data and Safety Monitoring Board because, unlike the CIDS study, the ICD group showed a significant improvement in total mortality over an average follow-up of 18 months (15.8% ICD group, 24% antiarrhythmic group, $P < .02$).

In summary, the available epidemiologic and clinical trial data evoke several principles applicable to patients with sustained symptomatic VT. The ICD only prevents arrhythmic death. The majority of participants in these trials with low LVEF and structural heart disease have competing modes of death. Antiarrhythmic drugs may have some positive effect on congestive heart failure (amiodarone) or actually worsen arrhythmic death via proarrhythmia (propafenone in the CASH study). Investigator and subject bias is unavoidable in a trial that is not blinded to treatment arm. Furthermore, a heterogeneous study population (as in the AVID and CIDS studies) may not have the same treatment effect across all strata in the group. Domanski et al.[91] reported a differential effect of antiarrhythmic drugs and ICD therapy from the AVID database based on strata of left ventricular EF.[101] In patients with LVEF greater than 35%, the two therapies showed no difference in survival, suggesting that all the treatment effect came from enrolled patients with LVEF less than 35%. Despite the inherent limitations of the published clinical trials, the largest studies favor ICD therapy in patients whose primary mortality risk is from symptomatic ventricular tachyarrhythmia.

In summary, the long-term management of patients with sustained VT and heart disease is very similar to that for patients resuscitated from cardiac arrest. A retrospective analysis of outcome in patients with "tolerated" VT and VT without cardiac arrest in the AVID trial suggests that their outcome is very similar to that in patients with VF or VT with serious cardiac compromise. It seems reasonable to expect that most patients with sustained symptomatic VT or VF, regardless of the severity of symptoms, have a poor prognosis and should be managed as outlined in the section on evidence-based therapy in ventricular fibrillation. However, patients with relatively preserved LV function (EF > 40%) and symptomatic VT without cardiac arrest were not included in the randomized trials of ICD therapy. It is not known if such patients would have improved survival with ICD therapy over that with "best medical therapy" (which should almost certainly include β-blockers and amiodarone). Subanalyses of the CIDS and AVID studies do suggest that a benefit of the ICD over amiodarone, in the medium term, may not be observed in patients with EF greater than 35% and symptomatic VT.[101,102] These latter patients can be reasonably treated with either an ICD or oral amiodarone and a β-blocker.[103] Long-term management of patients with sustained VT without heart disease is detailed in the next section.

Impact of New Device Technology

Concurrent with these important secondary and primary prevention clinical trials have been important technologic advances in device technology.[104] These are discussed in greater detail in Section V, Pharmacologic and Interventional Therapies, of this text. Of greatest importance for patients with sustained VT and left ventricular dysfunction is the advent of biventricular pacing. Biventricular pacing has shown benefits in patients with drug refractory congestive heart failure and coexisting intraventricular and interventricular dyssynchronous wall motion.[105] These patients have been largely identified by the presence of prolonged QRS complexes, though other techniques of wall motion analysis are being increasingly applied. Thus, patients with bundle branch block, particularly with QRS complexes greater than 0.15s and first-degree AV block, which can further compromise ventricular filling, are candidates for this technique. Atrial fibrillation has been an important coexisting arrhythmia in many of these sustained VT patients and causes progressive heart failure and increased mortality. New atrial fibrillation therapies, both preventative and for termination, are now available in ICD devices. Thus, alternate site pacing such as dual-site atrial pacing, novel preventative pacing algorithms for atrial premature beat suppression, antitachycardia pacing for atrial tachycardia termination, and atrial defibrillation therapies are now available. Figure 17-25 shows an ECG from an elderly patient with refractory persistent atrial fibrillation, sustained VT with syncope, and refractory congestive heart failure. A four-chamber ICD device (Medtronic Insync III ICD) with biventricular pacing and dual-site right atrial pacing leads was inserted for management of these clinical syndromes.

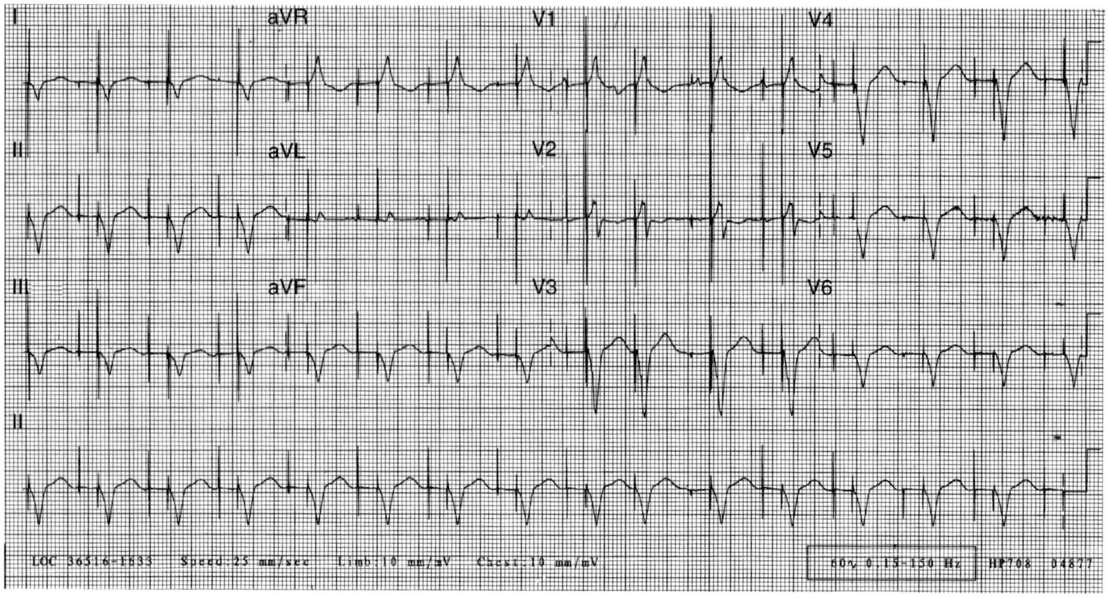

FIGURE 17-25 Twelve-lead ECG of a patient with refractory congestive heart failure, ventricular tachycardia, and refractory atrial fibrillation who had a four-chamber implantable cardioverter defibrillator (ICD) system implanted. Dual-site right atrial and biventricular pacing leads were placed with a Medtronic InSync III ICD. Control of atrial fibrillation and heart failure improved with device therapy.

Data from recent trials such as the COMPANION study (discussed in Chapter 49, Device Technology for Congestive Heart Failure) confirms the morbidity and mortality benefits of biventricular pacing. The availability of new technology and supportive evidence that can reduce the risk of competing (i.e., non-sudden) mortality in the sustained VT and heart disease patient provides an important impetus to the primary usage of ICD devices in patients with left ventricular dysfunction and congestive heart failure.

Impact of Catheter Ablation Techniques

Catheter ablation technologies have been applied for ablation of sustained and hemodynamically stable VT in patients with and without organic heart disease who could undergo catheter mapping.[106,107] Radiofrequency ablation remains the mainstay of this approach. In addition to ECG morphology, catheter mapping with contact catheter and noncontact three-dimensional mapping techniques have facilitated exact localization of the VT substrate. The details of these techniques are reviewed in other chapters in this text. However, these methods now permit rapid mapping and accurate geometric localization of these diseased tissues in a few or even one tachycardia cycle. Thus, increasingly hemodynamically unstable VT episodes can be treated by ablative interventions.[108] Radiofrequency energy delivered by contact electrodes with or without cooled tip technology can now produce moderate-sized lesions in diseased ventricular myocardium. Thus, VT ablation can temporarily suppress, reduce frequency, and even eliminate recurrent sustained VT in selected patients. It can reduce the need for ICD therapies.[109] However, it is not curative, and VT recurrences often occur during long-term follow-up.[110] Thus, VT ablation in patients with organic heart disease is a potential management tool in specific clinical scenarios. It is valuable in drug refractory incessant VT or "electrical storms" unresponsive to medical therapy. In patients with frequent, recurrent ICD shocks despite antiarrhythmic drug therapy, VT ablation will reduce the need for these therapies. Finally, it can be used in a patient who is intolerant to antiarrhythmic drug therapy when frequent VT recurrences need to be addressed. However, VT ablation should be used in conjunction with ICD therapy whenever possible to provide backup rescue therapies for recurrent VT/VF.

Impact on Patient Selection

The selection of the first line of therapy in VT management is now defined by the clinical disease, intercurrent factors, and arrhythmia characteristics. Management of the underlying disease and associated conditions such as heart failure and ischemia is an integral part of patient management. These topics have been discussed earlier in this chapter. Arrhythmia frequency is an important consideration. Frequent or incessant VT requires evaluation for intercurrent precipitating factors such as electrolyte abnormalities, enhanced sympathetic tone, hypoxia, or uncontrolled ischemia. Correction of these factor will often convert this intolerable VT frequency to "occasional event" status. The hemodynamic impact of the arrhythmia also determines the urgency of the response. Hemodynamic collapse requires urgent institution of both pharmacologic and nonpharmacologic therapies. In more stable VT episodes, drug therapy, mapping and ablation, and antitachycardia pacing become competitive options. With reduction of event frequency or in patients with sporadic VT episodes, a longer-term management strategy is appropriate for implementation. Prevention of sudden death with ICD backup is an essential part of the therapeutic prescription in these patients. Antiarrhythmic drugs and catheter ablation can play a supportive role in this phase of management to reduce VT event rates, make it amenable to pacing therapies, and prevent frequent symptoms related to the arrhythmia and ICD therapies.[111]

REFERENCES

1. Mason JW, for the Electrophysiologic Study versus Electrocardiographic Monitoring Investigators: A comparison of electrophysiologic testing with Holter monitoring to predict antiarrhythmic drug efficacy for ventricular tachyarrhythmias. N Engl J Med 1993;329:445.
2. Bharati S: Pathology of the conduction system. In Silver MD, Gotlieb AI, Schoen FJ (ed): Cardiovascular Pathology, 3rd ed. New York, Churchill Livingston, 2001, p 20.
3. Bharati S, Lev M: The pathologic aspects of ventricular tachycardia. In Iwa T, Fontaine G (eds): Cardiac Arrhythmias—Recent Investigation and Management. Amsterdam, Elsevier Science Publishers BV Biomedical Division, 1988, pp 15-27.
4. Bharati S, Lev M: The Cardiac Conduction System in Unexplained Sudden Death. Mt. Kisco, NY, Futura Publishing, 1990, pp 1-416.
5. Bharati S, Granston AS, Liebson PR, et al: The conduction system in mitral valve prolapse syndrome with sudden death. Am Heart J 1981;101:667-70.
6. Bharati S, Feld AW, Bauernfeind R, et al: Hypoplasia of the right ventricular myocardium with ventricular tachycardia. Arch Pathol Lab Med 1983;107:249-53.
7. Bharati S, McAnulty JH, Lev M, Rahimtoola SH: Idiopathic hypertrophic subaortic stenosis with split His bundle potentials: Electrophysiologic and pathologic correlations. Circulation 1980;62:1373-80.
8. Bharati S, Lev M, Denes P, et al: Infiltrative cardiomyopathy with conduction disease and ventricular arrhythmia: Electrophysiologic and pathologic correlations. Am J Cardiol 1980;45:163-73.
9. Bharati S, Strasberg B, Bilitch M, et al: Anatomic substrate for preexcitation in idiopathic myocardial hypertrophy with fibroelastosis of the left ventricle. Am J Cardiol 1981;48:47-58.
10. Bharati S, Driefus L, Bucheleres G, et al: The conduction system in patients with a prolonged QT interval. J Am Coll Cardiol 1985; 6:1110-9.
11. Bharati S, Bump T, Bauernfeind R, Lev M: Dystrophica myotonia. Chest 1984;86:444-50.
12. Gallastegui J, Hariman RJ, Handler B, et al: Cardiac involvement in the Kearns-Sayre syndrome. Am J Cardiol 1987;60:385-8.
13. Bharati S, Lev M: Arrhythmogenic ventricles. Pacing Clin Electrophysiol 1983;6:1035-49.
14. Bharati S, Lev M: Sequelae of atriotomy and ventriculotomy on the endocardium, conduction system and coronary arteries. Am J Cardiol 1982;50:580-6.
15. Bharati S, Lev M: Conduction system in cases of sudden death in congenital heart disease many years after surgical correction. Chest 1986;90:861-8.
16. Bharati S, Lev M: The pathology of congenital heart disease. A personal experience with more than 6300 congenitally malformed hearts. Armonk, NY, Futura Publishing, 1996, 761-800; pp 1445-98.
17. Bharati S, Scheinmann MM, Morady F, et al: Sudden death after catheter-induced atrioventricular junctional ablation. Chest 1985;88:883-9.
18. Gault JH, Cantwell J, Lev M, Braunwald E: Fatal familial cardiac arrhythmias. Am J Cardiol 1972;29:548-53.
19. Husson GS, Blackman MS, Rogers MC, et al: Familial congenital bundle branch system disease. Am J Cardiol 1973;32:365-9.
20. Brookfield L, Bharati S, Denes P, et al: Familial sudden death: Report of a case and review of the literature. Chest 1988;94:989-93.
21. Bharati S, Lev M: Congenital abnormalities of the conduction system in sudden death in young adults. J Am Coll Cardiol 1986; 8:1096-104.
22. Bharati S, Bauernfeind R, Scheinman M, et al: Congenital abnormalities of the conduction system in two patients with tachyarrhythmias. Circulation 1979;59:593-606.
23. Janse MJ, Wit AL: Electrophysiological mechanisms of ventricular arrhythmias resulting from myocardial ischemia and infarction. Physiol Rev 1989;69:1049-169.
24. Wit AL, Janse MJ: The Ventricular Arrhythmias of Ischemia and Infarction. Electrophysiological Mechanisms. Mount Kisco, NY, Futura Publishing, 1993.
25. Pinto JMB, Boyden PA: Electrical remodeling in ischemia and infarction. Cardiovasc Res 1999;42:284-97.
26. Peters NS, Coromilas J, Severs NJ, Wit AL: Disturbed connexin 43 gap junction distribution correlates with the location of reentrant circuits in the epicardial border zone of healing canine infarcts that cause ventricular tachycardia. Circulation 1997;95:988-96.
27. Ursell PC, Gardner PI, Albala A, et al: Structural and electrophysiological changes in the epicardial border zone during infarct healing. Circ Res 1985;56:436-51.
28. De Bakker JMT, Van Capelle FJL, Janse MJ, et al: Reentry as a cause of ventricular tachycardia in patients with chronic ischemic heart disease: Electrophysiologic and anatomic correlation. Circulation 1988;77:589-606.
29. Callans DJ, Josephson ME: Ventricular tachycardia associated with coronary artery disease. In Zipes DP, Jalife J (eds): Cardiac Electrophysiology: From Cell to Bedside, 2nd ed. Philadelphia, WB Saunders, 1995, pp 732-43.
30. Willems AR, Tijssen JGP, van Capelle FJL, et al: Determinants of prognosis in symptomatic ventricular tachycardia or ventricular fibrillation late after myocardial infarction. J Am Coll Cardiol 1990;16:521-30.
31. Hohnloser SH, Franck P, Klingenheben T, et al: Open infarct artery, late potentials and other prognostic factors in patients after acute myocardial infarction in the thrombolytic era. A prospective trial. Circulation 1994;90:1747-56.
32. Julian DG, Camm AJ, Frangin G, et al: Randomized trial of amiodarone on mortality in patients with left-ventricular dysfunction after recent myocardial infarction: EMIAT. Lancet 1997:349:667-74.
33. De Bakker JMT, Coronel R, Tasseron S, et al: Ventricular tachycardia in the infarcted, Langendorff-perfused human heart: Role of the arrangement of surviving cardiac fibers. J Am Coll Cardiol 1990;15:1594-607.
34. De Bakker JMT, van Capelle FJL, Janse MJ, et al: Slow conduction in the infarcted human heart: "Zigzag" course of activation. Circulation 1993;88:915-26.
35. Maglaveras N, de Bakker JMT, van Capelle FJL, et al: Activation delay in healed myocardial infarction: A comparison between model and experiment. Am J Physiol 1995;269: H1441-H1449.
36. Ramdat Misier AR, Opthof T, van Hemel NM, et al: Dispersion of "refractoriness" in noninfarcted myocardium of patients with ventricular tachycardia or ventricular fibrillation after myocardial infarction. Circulation 1995;91;2566-72.
37. Tomaselli GF, Marbán E: Electrophysiological remodeling in hypertrophy and heart failure. Cardiovasc Res 1999;42:270-83.
38. Haverkamp W, Breithardt G, Camm AJ, et al: The potential for QT prolongation and pro-arrhythmia by non-anti-arrhythmic drugs: Clinical and regulatory implications. Cardiovasc Res 2000; 47:219-33.
39. McIntosh MA, Cobbe SM, Smith GL: Heterogeneous changes in action potential and intracellular Ca++ in left ventricular myocyte subtypes from rabbits with heart failure. Cardiovasc Res 2000;45: 397-409.
40. Bryant SM, Shipsey SY, Hart G: Regional differences in electrical and mechanical properties of myocytes from guinea-pig hearts with mild left ventricular hypertrophy. Cardiovasc Res 1997;35:315-23.
41. Keung EC, Aronson RS: Non-uniform electrophysiological properties and electrotonic interaction in hypertrophied myocardium. Circ Res 1981;49:150-8.
42. Qin D, Zhang ZH, Caret EB, et al: Cellular and ionic basis of arrhythmias in postinfarction remodeled ventricular myocardium. Circ Res 1996;79:461-73.
43. Peters NS, Cabo C, Wit AL: Arrhythmogenic mechanisms: Automaticity, triggered activity, and reentry. In Zipes DP, Jalife J (eds): Cardiac Electrophysiology. From Cell to Bedside, 3rd ed. Philadelphia, WB Saunders, 2000, pp 345-56.

44. Saffitz JE, Schuessler RB, Yamada KA: Mechanisms of remodeling of gap junction distribution and the development of anatomic substrates of arrhythmias. Cardiovasc Res 1999;42:309-17.
45. Zipes DP: Influence of myocardial ischemia and infarction on autonomic innervation of heart. Circulation 1990;82;1095-104.
46. Cao J-M, Chen LS, Ken Knight BH, et al: Nerve sprouting and sudden cardiac death. Circ Res 2000;86:816-21.
47. Tsuji Y, Opthof T, Kamiya K, et al: Pacing-induced heart failure causes a reduction of delayed rectifier potassium currents along with a decrease in calcium and transient outward current in rabbit ventricle. Cardiovasc Res 2000;48:300-9.
48. Cerbai E, Pino R, Porciati F, Mugelli A: Characterization of the hyperpolarization-activated current in ventricular myocytes from human failing heart. Circulation 1998;95:568-71.
49. Vermeulen JT, McGuire MA, Opthof T, et al: Triggered activity and automaticity in ventricular trabeculae of failing human and rabbit hearts. Cardiovasc Res 1994;28:1547-54.
50. Wickenden AD, Kaprielien R, Tiassiri Z, et al: Role of action potential prolongation and altered intracellular calcium handling in the pathogenesis of heart failure. Cardiovasc Res 1998;37:312-23.
51. Makielski J, Kamp T, Valdivia CR: Late sodium current is increased in ventricular myocytes from failing human heart compared with non-failing hearts. Jpn J Electrocardiol 2000; 20(suppl 3):99-100.
52. Shimizu W, Antzelevitch C: Cellular basis for the ECG features of the LQT1 form of the long QT syndrome: Effects of beta-adrenergic agonists and sodium channel blockers on transmural dispersion of repolarization and torsade de pointes. Circulation 1998;98:2314-22.
53. Hasenfuss G: Alteration of calcium regulatory proteins in heart failure. Cardiovasc Res 1998;37:279-89.
54. Movsesian M, Schwinger RHG: Calcium sequestration by the sarcoplasmic reticulum in heart failure. Cardiovasc Res 1998;37: 352-9.
55. Phillips RM, Novayam P, Gomez A, et al: Sarcoplasmic reticulum in heart failure: Central players or bystanders? Cardiovasc Res 1998;37:346-51.
56. Opthof T, Coronel R, Rademaker HME, et al: Changes in sinus node function in a rabbit model of heart failure with ventricular arrhythmias and sudden death. Circulation 2000;101:2975-80.
57. Rademaker HME: Arrhythmogenesis During the Development of Heart Failure in Rabbits [thesis]. University of Amsterdam, 1997.
58. Luu M, Stevenson WG, Stevenson LW, et al: Diverse mechanisms of unexpected cardiac arrest in advanced heart failure. Circulation 1989;80:1675-80.
59. Stevenson WG, Stevenson LW, Middlekauf HR, Saxon LA: Sudden death prevention in patients with advanced ventricular dysfunction. Circulation 1993;88:2953-61.
60. Saksena S, Ciccone J, Craelius W, et al: Studies on left ventricular function during sustained ventricular tachycardia. J Am Coll Cardiol 1984;4:501-8.
61. Hamer AW, Rubin SA, Peter T, Mandel WJ: Factors that predict syncope during ventricular tachycardia in patients. Am Heart J 1984;107:997-1005.
62. Krol RB, Saksena S: Clinical Trials of Antiarrhythmic Drugs in Recipients of Implantable Cardioverter-Defibrillators. Interventional Electrophysiology, 2nd ed. Mt. Kisco, NY, Futura 1996, p 371.
63. Sandler A, Marriott HJL: The differential morphology of anomalous ventricular complexes of RBBB type in lead V1.Ventricular ectopy versus aberration. Circulation 1965;31:551-6.
64. Marriott HJL, Sandler IA: Criteria, old and new, for differentiating between ectopic ventricular beats and aberrant ventricular conduction in the presence of atrial fibrillation. Progr Cardiovasc Dis 1966;9:18-28.
65. Wellens JHH, Bar FW, Lie KI: The value of the electrocardiogram in the differential diagnosis of a tachycardia with a widened QRS complex. Am J Med 1978;64:27-33.
66. Kindwall KE, Brown J, Josephson ME: Electrocardiographic criteria for ventricular tachycardia in wide complex left bundle branch block morphology tachycardias. Am J Cardiol 1988;61: 1279-83.
67. Brugada P, Brugada J, Mont L, et al: A new approach to the differential diagnosis of a regular tachycardia with a wide QRS complex. Circulation 1991;83:1649-59.
68. Josephson ME, Horowitz LN, Waxman HL, et al: Sustained ventricular tachycardia: Role of the 12-lead electrocardiogram in localizing the site of origin. Circulation 1981;64:257-71.
69. Kuchar DL, Ruskin JN, Garan H: Electrocardiographic localization of the site of origin of ventricular tachycardia in patients with prior myocardial infarction. J Am Coll of Cardiol 1989;13: 893-903.
70. Sippensgroenewegen A, Spekhorts H, Van Hemel NM, et al: Localization of the site of origin of postinfarction ventricular tachycardia by endocardial pace mapping: Body surface mapping compared with the 12-lead electrocardiogram. Circulation 1993: 88;2290-306.
71. Josephson ME, Waxman HL, Cain ME, et al: Ventricular activation during ventricular endocardial pacing, II: Role of pace-mapping to localize origin of ventricular tachycardia. Am J Cardiol 1982; 50:11-22.
72. Sippensgroenewegen A, Spekhorst H, Van Hemel NM, et al: Value of body surface mapping in localizing the site of origin of ventricular tachycardia in patients with previous myocardial infarction. J Am Coll Cardiol 1994;24:1708-24.
73. Hummel JD, Strickberger SA, Daoud E, et al: Results and efficiency of programmed ventricular stimulation with four extrastimuli compared with one, two, and three extrastimuli. Circulation 1994;90:2827-32.
74. Cooper MJ, Koo CC, Skinner MP, et al: Comparison of immediate versus day to day variability of ventricular tachycardia induction by programmed stimulation. J Am Coll Cardiol 1989;13:1599-607.
75. Wellens HJ, Brugada P, Stevenson WG: Programmed electrical stimulation of the heart in patients with life-threatening ventricular arrhythmias: What is the significance of induced arrhythmias and what is the correct stimulation protocol? Circulation 1985; 72:1-7.
76. Buxton AE, Josephson ME, Marchlinski FE, et al: Polymorphic ventricular tachycardia induced by programmed stimulation: response to procainamide [see comments]. J Am Coll Cardiol 1993;21:90-8.
77. Brugada J, Brugada R, Brugada P: Right bundle-branch block and ST-segment elevation in leads V1 through V3: A marker for sudden death in patients without demonstrable structural heart disease. Circulation 1998;97:457-60.
78. Fananapazir L, Chang AC, Epstein SE, et al: Prognostic determinants in hypertrophic cardiomyopathy. Prospective evaluation of a therapeutic strategy based on clinical, Holter, hemodynamic, and electrophysiological findings. Circulation 1992;86:730-40.
79. Turitto G, Ahuja RK, Caref EB, et al: Risk stratification for arrhythmic events in patients with nonischemic dilated cardiomyopathy and nonsustained ventricular tachycardia: Role of programmed ventricular stimulation and the signal-averaged electrocardiogram. J Am Coll Cardiol 1994;24:1523-8.
80. Delacretaz E, Stevenson WG, Ellison KE, et al: Mapping and radiofrequency catheter ablation of the three types of sustained monomorphic ventricular tachycardia in nonischemic heart disease. J Cardiovasc Electrophys 2000;11:11-7.
81. Blanck Z, Dhala A, Deshpande S, et al: Bundle branch reentrant ventricular tachycardia: Cumulative experience in 48 patients [see comments]. J Cardiovasc Electrophysiol 1993;4:253-62.
82. Narasimhan C, Jazayeri MR, Sra J, et al: Ventricular tachycardia in valvular heart disease: Facilitation of sustained bundle-branch reentry by valve surgery. Circulation 1997;96:4307-13.
83. Merino JL, Carmona JR, Fernandez-Lozano I, et al: Mechanisms of sustained ventricular tachycardia in myotonic dystrophy: Implications for catheter ablation. Circulation 1998;98:541-6.
84. Stevenson WG, Friedman PL, Sager PT, et al: Exploring postinfarction reentrant ventricular tachycardia with entrainment mapping. J Am Coll Cardiol 1997;29:1180-9.
85. Stevenson W, Brugada P, Waldecker B, et al: Clinical, angiographic, and electrophysiologic findings in patients with aborted sudden death as compared with patients with sustained ventricular tachycardia after myocardial infarction. Circulation 1985;71: 1146-52.
86. Nanthakumar K, Newman D, Paquette M, et al: Circadian variation of sustained ventricular tachycardia in patients subject to standard adrenergic blockade. Am Heart J 1997;134:752-7.
87. Greene M, Newman D, Geist M, et al: Is electrical storm in ICD patients the sign of a dying heart? Outcome of patients with clusters of ventricular tachyarrhythmias. Europace 2000;2:263-9.

88. Credner SC, Klingenheben T, Mauss O, et al: Electrical storm in patients with transvenous implantable cardioverter-defibrillators: Incidence, management and prognostic implications. Am Coll Cardiol 1998;32:1909-15.
89. Nademanee K, Taylor R, Bailey WE, et al: Treating electrical storm: Sympathetic blockade versus advanced cardiac life support-guided therapy. Circulation 2000;102:742-7.
90. Holmes DR Jr, Davis K, Gersh BJ, et al: Risk factor profiles of patients with sudden cardiac death and death from other cardiac causes: A report from the Coronary Artery Surgery Study (CASS). J Am Coll Cardiol 1989;13:524-30.
91. Garan H, Ruskin JN, DiMarco JP, et al: Electrophysiologic studies before and after myocardial revascularization in patients with life-threatening ventricular arrhythmias. Am J Cardiol 1983;51: 519-24.
92. Guidelines 2000 for cardiopulmonary resuscitation and emergency cardiovascular care: International consensus on science. Part 6: Advanced cardiac life support; Section 5: Pharmacology I: Agents for arrhythmias. Circulation 2000;102:I112-I1128.
93. Kowey PR, Levine JH, Herre JM, et al: Randomized, double-blind comparison of intravenous amiodarone and bretylium in the treatment of patients with recurrent, hemodynamically destabilizing ventricular tachycardia or fibrillation. The Intravenous Amiodarone Multicenter Investigators Group. Circulation 1995;92:3255-63.
94. Schaffer W, Cobb L: Recurrent ventricular fibrillation and modes of death in survivors of out-of-hospital ventricular fibrillation. N Engl J Med 1975;293:259-62.
95. Myerburg R, Kessler K, Estes D, et al: Long term survival after prehospital cardiac arrest: Analysis of outcome during an 8 year study. Circulation 1984;70:538-46.
96. The CASCADE Investigators: Randomized antiarrhythmic drug therapy in survivors of cardiac arrest (the CASCADE study). Am J Cardiol 1993;72:280-7.
97. Wever EFD, Hauer RNW, van Capelle FJI, et al: Randomized study of implantable defibrillator as first-choice therapy versus conventional strategy in postinfarct sudden death survivors. Circulation 1995;91:2195-203.
98. Kuck KH, Cappato R, Siebels J, Ruppel R: Randomized comparison of antiarrhythmic drug therapy with implantable defibrillators in patients resuscitated from cardiac arrest: The Cardiac Arrest Study Hamburg (CASH). Circulation 2000;102:748-54.
99. Connolly S, Gent M, Roberts R, et al: Canadian implantable defibrillator study (CIDS)—a randomized trial of the implantable cardioverter defibrillator against amiodarone. Circulation 2000;101:1297-302.
100. The AVID investigators: A comparison of antiarrhythmic-drug therapy with implantable defibrillators in patients resuscitated from near-fatal ventricular arrhythmias. N Engl J Med 1997; 337:1576-83.
101. Domanski M, Saksena S, Epstein A, et al: Relative effectiveness of the implantable cardioverter-defibrillator and antiarrhythmic drugs in patients with varying degrees of left ventricular dysfunction who have survived malignant ventricular arrhythmias. J Am Coll Cardiol 1999;34:1090-5.
102. Raj SR, Sheldon RS: The implantable cardioverter-defibrillator: Does everybody need one? Prog Cardiovasc Dis 2001;44:169-94.
103. Amiodarone Trials Meta-Analysis Investigators: Effect of prophylactic amiodarone on mortality after acute myocardial infarction and in congestive heart failure: Meta-analysis of individual data from 6500 patients in randomized trials. Lancet 1997; 350:1417-24.
104. Rao BH, Saksena S: Implantable cardioverter-defibrillators in cardiovascular care: Technologic advances and new indications. Curr Opin Crit Care 2003;9:362-8.
105. Kalinchak DM, Schoenfeld MH: Cardiac resynchronization: A brief synopsis part I: Patient selection and results from clinical trial. J Interv Card Electrophysiol 2003;9:155-61.
106. Delacretaz E, Stevenson WG: Catheter ablation of ventricular tachycardia in patients with coronary heart disease: Part I: Mapping. Pacing Clin Electrophysiol 2001;24:1261-77.
107. Delacretaz E, Stevenson WG: Catheter ablation of ventricular tachycardia in patients with coronary heart disease. Part II: Clinical aspects, limitations, and recent developments. Pacing Clin Electrophysiol 2001;24:1403-11.
108. Sra J, Bhatia A, Dhala A, et al: Electroanatomically guided catheter ablation of ventricular tachycardias causing multiple defibrillator shocks. Pacing Clin Electrophysiol 2001;24:1645-52.
109. Reddy VY, Neuzil P, Taborsky M, Ruskin JN: Short-term results of substrate mapping and radiofrequency ablation of ischemic ventricular tachycardia using a saline-irrigated catheter. J Am Coll Cardiol 2003;41:2228-36.
110. Weinstock J, Wang PJ, Homoud MK, et al: Clinical results with catheter ablation: AV junction, atrial fibrillation and ventricular tachycardia. J Interv Card Electrophysiol 2003;9:275-88.
111. Camm AJ, Savelieva I: Rationale and patient selection for "hybrid" drug and device therapy in atrial and ventricular arrhythmias. J Interv Card Electrophysiol 2003;9:207-14.

Chapter 18 Ventricular Tachycardia and Ventricular Fibrillation without Structural Heart Disease

KELLEY ANDERSON, BRUCE D. LINDSAY, and SANJEEV SAKSENA

Introduction: Sanjeev Saksena
Definition and Classification: Kelley Anderson
Epidemiology: Kelley Anderson and Sanjeev Saksena
Clinical Presentation: Sanjeev Saksena
Electrocardiographic Features: Bruce D. Lindsay
Clinical Electrophysiology: Kelley Anderson and Sanjeev Saksena
Principles of Practice: Kelley Anderson
Management: Kelley Anderson and Sanjeev Saksena

Idiopathic ventricular tachycardia (IVT) and ventricular fibrillation (VF) have been intermittently recognized and reported since the early 1900s. Since Vakil's classic description of ventricular aneurysms and their association with ventricular tachycardia, early literature defined ventricular tachycardia (VT) associated with heart disease.[1-3] The occurrence of paroxysmal VT in young patients, and occasionally with apparently healthy hearts, emerged from time to time.[4,5] The availability of electrocardiographic recordings focused attention on the region of origin of ventricular arrhythmias including ectopic beats as well as VT.[6,7] In a careful review of existing literature of the 20th century, investigators from the highly respected Chicago School of Electrophysiology noted in 1974 that "cardiac stimulatory studies suggest the site of origin of ventricular ectopic beats can be identified by QRS configuration." Ectopic beats were classified according to their bundle branch block–like configuration, with left ventricular origin being ascribed to right bundle branch configurations and right ventricular origin to left bundle branch configurations.[8] The latter were credited with a more favorable outcome. However, occasional reports of deaths in young patients existed, disputing a uniform benign prognosis.

Definition and Classification

In current day parlance, IVT refers to VT of unknown cause that occurs in the absence of significant structural heart disease or transient or reversible arrhythmogenic factors (e.g., electrolyte disorders, myocardial ischemia). In practice, one cannot be strict about excluding patients with evidence of structural heart disease, because advances in diagnostic tests are demonstrating structural changes in some patients with classic IVT syndromes.[9-11] In addition, a patient could have structural disease due to a coexisting cardiac disorder. The term "idiopathic" is sometimes inappropriate given the depth of understanding that now exists for a few of the disorders. For the purposes of this discussion, we assume the syndromes listed in Table 18-1 are distinct IVT syndromes. However, it must be acknowledged that these syndromes consist of heterogeneous subtypes, and some share characteristics with other IVT syndromes so that clear distinctions are not always possible. Moreover, reports of "new" IVTs arise frequently.[12-13] Some of these may be previously unrecognized syndromes, whereas others may be variants of established forms.

Epidemiology

Epidemiologic data on IVT syndromes are scarce due to difficulties in recognition of the syndrome and its demographics. The demographics of the clinical syndrome of IVT derived initial impetus from clinical and later electrophysiological studies. Pietras et al. analyzed 27 patients with chronic recurrent VT classified according to the ECG classification and examined their clinical, ECG, and hemodynamic features.[8] They further tabulated 73 reported cases in then existing literature.

TABLE 18-1 Idiopathic Ventricular Tachyarrhythmia Syndromes

Polymorphic Ventricular Tachycardia/ Ventricular Fibrillation Syndromes
Long QT syndrome
Brugada syndrome
Short-coupled torsades de pointes
Catecholamine-induced polymorphic VT
Other idiopathic polymorphic VT/VF syndromes
Monomorphic VT Syndromes
Right ventricular outflow tract VT*
Left ventricular fascicular tachycardia
Adrenergic monomorphic VT
Other idiopathic monomorphic VT syndromes

*And other adenosine-sensitive VTs.

VT, ventricular tachycardia; VF, ventricular fibrillation.

They noted that the mean age of right-sided VT was lower (mean age 32 years) than left-sided VT (mean age 43 years) with a similar trend in the previously reported data (36 versus 53 years). Females predominated in right-sided VT (52% or 66%, respectively) compared to left-sided VT (33% or 15%, respectively). Organic heart disease was almost invariably present with left-sided VT (100% or 85%, respectively) and absent or uncommon in right-sided VT (25% or 48%, respectively). Mortality was significantly lower in right-sided VT with no deaths in their series and a 12% incidence in the reported literature. Chapman et al. reported electrical and hemodynamic correlates in a series of patients with "idiopathic VT" in 1975.[14] Subsequently, electrophysiological studies in this syndrome have defined the arrhythmia mechanisms, subgroups, novel clinical therapies, and their outcomes.[9,15]

Formal epidemiologic data are lacking, but IVTs are considered rare. It has been estimated, for instance, that Brugada syndrome accounts for 4 to 10 cases of sudden death per 10,000 inhabitants per year in Thailand and Laos, countries with an apparently high incidence. Nevertheless, the problem may be much larger. Autopsy series suggest that in 5% to 15% of cases of sudden death, there is no evidence of heart disease or other likely cause.[16-18] These observations raise the possibility that the IVT syndromes are common but insufficiently recognized. A recent retrospective multicenter study evaluated the profile of pediatric patients with IVT. Mean age at first manifestation of the arrhythmia was 5.4 years with a range of 0.1 to 15 years, with 27% of patients having manifested the disorder in infancy.[19]

Clinical Presentation

The clinical presentation of IVT patients is highly variable. Palpitations in an otherwise healthy young individual are a classic description in this entity. These can culminate in near syncope or syncope, dysthymia, and, on occasion, cardiac arrest. Cardiac arrest can be the primary presentation in some individuals, particularly if a family history of sudden death is present. Yet others will seek medical attention in the asymptomatic state with a family history of sudden death or for an electrocardiographic abnormality such as long QT syndrome or Brugada syndrome. Increasing awareness of the familial forms of the arrhythmia and symptomatic high-density ventricular ectopy or nonsustained VTA can lead to medical evaluation.

Electrocardiographic Features of Idiopathic Ventricular Tachyarrhythmias

The primary forms of IVT include arrhythmias arising from the right or left ventricular outflow tracts and intrafascicular reentry. These three types of IVTs can be distinguished by characteristic ECG patterns that reflect the origin or mechanism of the arrhythmias.

Right Ventricular Outflow Tract. The clinical features and general ECG characteristics of VT arising from the right ventricular outflow tract (RVOT) were described by Buxton[15] in 1983. The ECG exhibits a morphology resembling left bundle branch block with an inferior axis. These arrhythmias are associated with exercise, and in many patients the arrhythmias are nonsustained and repetitive. Others may simply have frequent monomorphic ventricular ectopy. Figure 18-1 was recorded from a patient with repetitive VT arising from the RVOT. Jadonath[20] determined the utility of the 12-lead ECG in localizing the origin of RVOT tachycardia in patients who underwent ablation of their arrhythmias. They divided the RVOT into nine segments and assessed the QRS morphology recorded by pacing at each segment. A QS pattern in lead aVR and monophasic R waves in leads II, II, and aVF were noted in all patients at all nine sites. Pacing at anterior sites produced either a dominant Q wave or qR pattern in lead I, but a monophasic R or Rs complex was never observed. Conversely, a dominant R wave was observed in lead I during pacing from posterior sites. The QRS exhibited a QS pattern in aVL during pacing from anterior sites, but the R wave became progressively larger in aVL with more posterior sites. Early R wave transition in the precordial leads was likely to be observed during pacing from the posterior and superior aspect of the septum. Similar results were reported by Yoshida and colleagues.[21] Coggins[22] observed that VT arising from the septal side of the RVOT had a negative QRS complex in aVL, whereas VT arising from the lateral aspect of the RVOT was associated with a positive QRS in aVL.

Left Ventricular Outflow Tract. Ventricular arrhythmias arising from the left ventricular outflow tract (LVOT) are distinguished from RVOT arrhythmias by differences in the QRS morphology. Earliest ventricular activation may arise in the sinus of Valsalva, below the aortic valve, or on the epicardium. Figure 18-2 was recorded from a patient whose arrhythmia was successfully ablated just below the anterior-lateral aspect of the aortic valve. Callans[23] described four patients with monomorphic VT who were mapped to the LVOT. Two patients, whose arrhythmias originated above the mediosuperior aspect

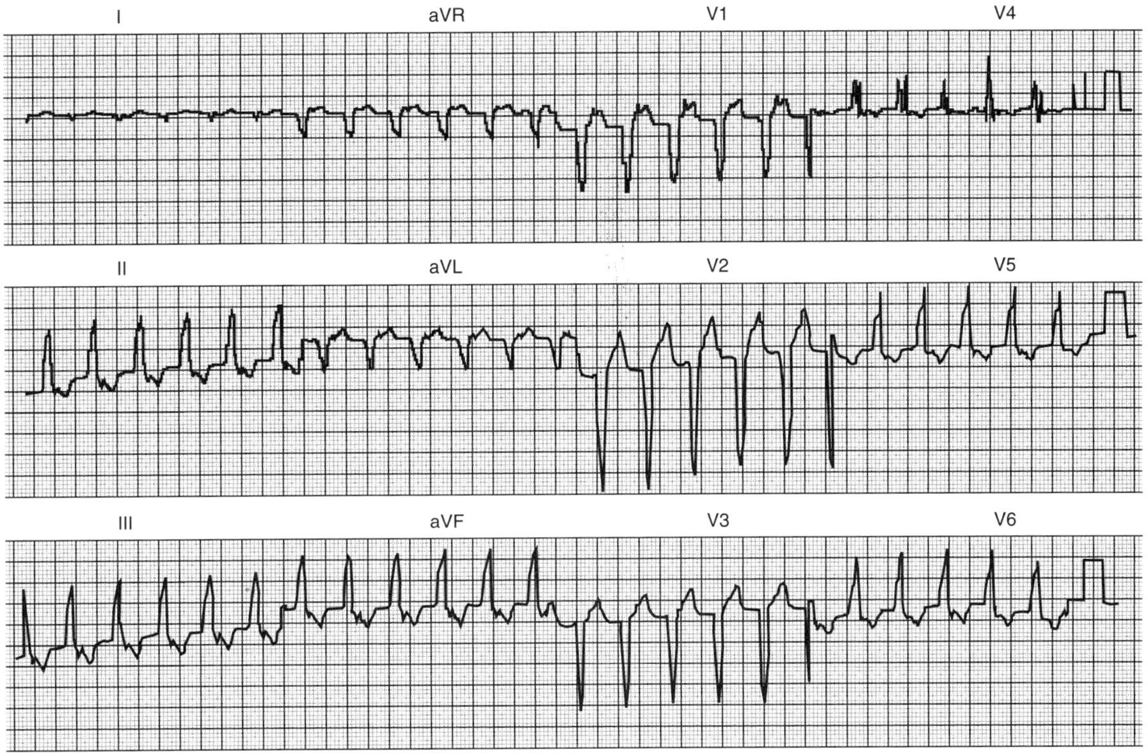

FIGURE 18-1 Twelve-lead ECG of ventricular tachycardia arising from the right ventricular outflow tract. The QRS exhibits a morphology resembling right bundle branch block with an inferior axis.

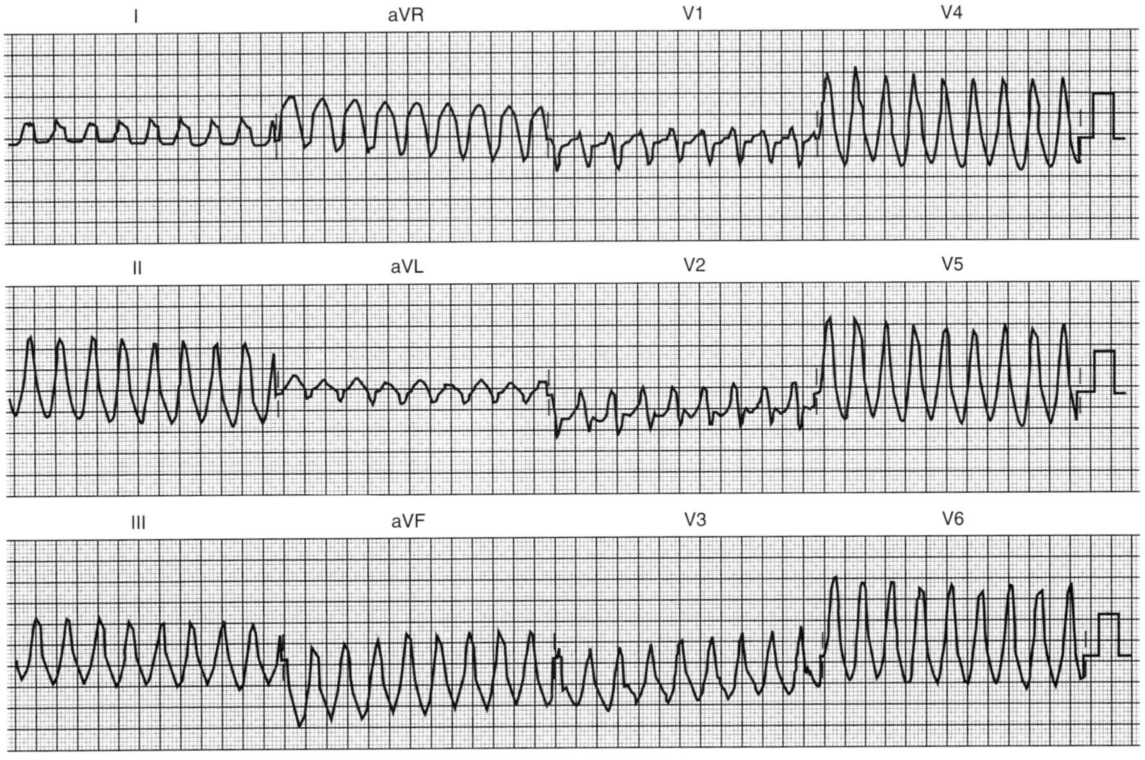

FIGURE 18-2 Twelve-lead ECG of ventricular ectopy that arose from the left ventricular outflow tract. Note the RS configuration in V1 and the prominent R wave in V2. The axis is inferior.

of the mitral annulus, had VT that exhibited a right bundle, inferior-axis VT with a dominant R wave in V1. The ECGs of these patients resembled those recorded from five patients studied by Lamberti[24] in whom intracardiac echocardiography was used to demonstrate that in all patients VT arose close to the left coronary cusp in the area of atrio-mitral continuity. The other two patients described by Callans had VT that arose from the basal aspect of the superior left ventricular septum. ECGs recorded from these patients had a left bundle, inferior-axis QRS morphology with a precordial R wave transition at lead V2. Krebs[25] studied 10 patients with VTs exhibiting a left bundle inferior-axis that were mapped to the LVOT. ECGs recorded from these patients were distinguished from those arising from the right ventricular outflow by an earlier precordial R wave transition (median V3 versus V5), more rightward axes, taller R waves inferiorly, and small R waves in lead V1. Absence of an R wave in V1 and late precordial transition suggested an origin in the RVOT. These results are similar to those reported by Kamakura,[26] who found that if the R/S ratio in V3 was greater than or equal to 1, the origin was likely to be in the LVOT. Kanagaratnam[27] has described a variant of LVOT VT in 12 patients whose arrhythmias arose from the aortic sinus of Valsalva. Two characteristic QRS morphologies were observed. ECGs recorded from all patients had a left bundle inferior-axis, tall monophasic R waves inferiorly, and early precordial lead transition with rS or RS in V1 and Rs pattern in V2 or V3. Those in whom the tachycardia was ablated from the noncoronary sinus were distinguished by a notched R in lead I.

Idiopathic Verapamil Sensitive Left Ventricular Tachycardia. Idiopathic verapamil sensitive VT that has a QRS pattern of right bundle branch block (RBBB) and left axis deviation was described by Lin[28] in 1983. Subsequent studies, which focused on the electrophysiological basis of the arrhythmia and ablation techniques, demonstrated that while the QRS axis is usually leftward and superior, it may be indeterminate or rightward. Electrophysiological data demonstrates that reentry is either intrafascicular or involves abnormal Purkinje tissue with decremental conduction properties is sensitive to verapamil.[29-36] The exact QRS morphology depends on which branches of the fascicles provide the exit point, and unsuccessful attempts to ablate the arrhythmia have shown changes in QRS morphology that suggest the potential for multiple exit points. Figure 18-3 demonstrates idiopathic left VT reentry that was ablated by applying radiofrequency energy at the site of a presystolic "fascicular" potential on the inferior aspect of the septum approximately 2 to 3 cm from the apex.

Clinical Electrophysiology

Remarkable advances have been made in understanding the congenital LQTS and Brugada syndrome. These disorders are described in detail elsewhere in this text, but because they may provide insights into the other IVT syndromes, certain aspects are discussed here. The mechanism of VT (usually torsades de pointes) in patients with LQTS is not known with certainty.

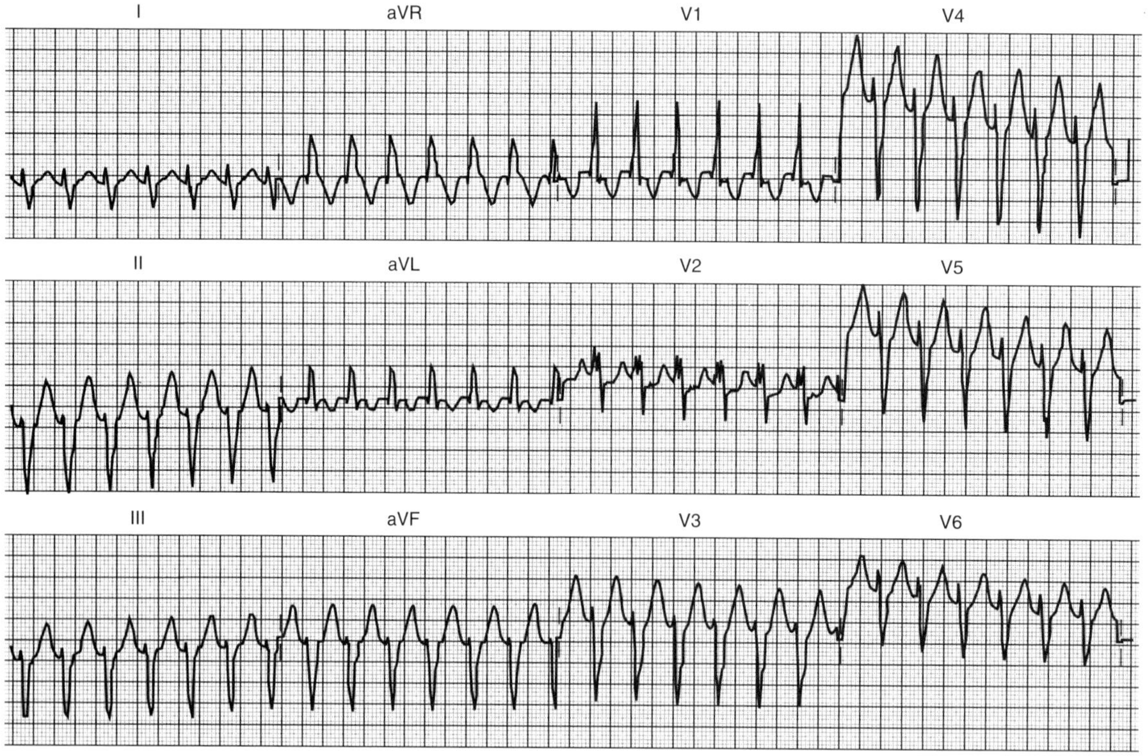

FIGURE 18-3 Twelve-lead ECG of idiopathic VT caused by intrafascicular reentry. The QRS morphology resembles right bundle branch block, and the axis is leftward and superior.

However, one theory is that the mechanism is reentry resulting from increased dispersion of repolarization due to differential prolongation of action potential duration in myocardial cells with particular lengthening in M cells.[37,38] Reentry may be triggered by early afterdepolarizations due to abnormal calcium kinetics resulting from delayed repolarization. This could explain why different mutations producing action potential prolongation by distinct mechanisms result in similar VTs. However, differences exist between the circumstances under which arrhythmias arise. In a model of LQTS, sympathetic activity had different effects in the three subtypes: in LQT1 sympathetic activity was necessary for increased dispersion of repolarization, and torsades de pointes did not develop without it. In LQT2 dispersion of repolarization and tendency for torsades de pointes varied with the level of sympathetic activity, whereas in LQT3, enhanced sympathetic activity reduced dispersion of repolarization and was protective against torsades de pointes.[38]

The mechanism of VT in Brugada syndrome has also been hypothesized to be reentry as a consequence of increased dispersion of repolarization. In this case the increased dispersion of repolarization is thought to result from differential action potential abbreviation with the greatest effect in epicardial cells.[39] Reentry could be triggered by phase 2 reentry. It is known that several mutations may produce the Brugada syndrome phenotype. This may account for the differences in clinical characteristics seen in some nocturnal death disorders encountered in Southeast Asia. Little is known about the short-coupled variant of torsades de pointes[40,41] and the catecholaminergic polymorphic VT syndrome.[42] Based on the appearance of the tachyarrhythmias, one could surmise that reentry due to dispersion of repolarization is a possible mechanism. However, the mechanism by which this occurs and the factors that produce it are unknown.

The utility of clinical electrophysiological studies (EPS) in the IVT syndromes associated with polymorphic VT or VF is limited. First, ventricular tachyarrhythmias cannot be induced in all IVT syndromes, notably LQTS. Second, when induced, it is difficult to confirm that the induced VT is the same as the spontaneous arrhythmia. Third, the induced VTs are difficult to study in a systematic fashion because they are rapid, polymorphic, poorly tolerated, and often require countershock. In contrast to LQTS, VT or VF is often inducible in patients with Brugada syndrome. In addition, a provocative test exists. The characteristic ECG pattern of RBBB and ST elevation may not be present at rest, but it may be provoked by administration of ajmaline, flecainide, pilscainide, or procainamide. VT cannot usually be induced in patients with the short-coupled variant of torsades de pointes. Isoproterenol infusion can be used to provoke VT in patients with the catecholaminergic polymorphic VT syndrome.

The clinical electrophysiology of the monomorphic forms of IVT differ considerably from the forms associated with VF and polymorphic VT. The most common form of monomorphic IVT is RVOT VT. It has been shown to result from cyclic adenosine monophosphate (cAMP)–mediated triggered activity, which is dependent on delayed afterdepolarizations.[9] The underlying pathophysiologic basis for this abnormality remains unclear, particularly the reasons that it arises from localized areas of the heart. A genetic abnormality for a variant form of RVOT VT has been reported,[43] but a familial distribution is not usually observed. EPS has been very useful in the investigation of this syndrome. RVOT VT is often induced by rapid pacing or isoproterenol infusion. In some cases, infusions of aminophylline, atropine, epinephrine, or very high-dose isoproterenol (up to 14 μg/min) may be required.[44] RVOT VT usually presents with a single QRS morphology, typically a left bundle branch block pattern and inferior axis consistent with an RVOT origin. It cannot be entrained. It usually terminates in response to adenosine, verapamil, β-blockers, or vagal maneuvers. Multiple configurations in individual patients have been reported.[45] Mapping of RVOT VT can localize this tachycardia to virtually all regions of the outflow tract. Figure 18-4 is a three-dimensional map of the propagation of a single VT wavefront. Note the site of origin in the outflow tract is focal in nature and well defined by the mapping technique. Noncontact mapping can define the focus in a single cycle, but sequential mapping with magnetic electroanatomic techniques may be employed in stable, sustained VT. Tachycardias arising from the LVOT and epicardial sites have many features in common with RVOT VT, including a response to adenosine,[23,46,47] which suggests a common mechanism and justifies the designation "adenosine-sensitive VTs." Sites of origin can often be predicted by careful examination of the configuration in the 12-lead ECG.[23,25]

Left ventricular fascicular tachycardia is another important monomorphic IVT syndrome. It is less common than RVOT VT, although it is more prevalent in series arising from Asian countries than from Western countries.[45,48,49] This arrhythmia is believed to result from reentry. Although previously thought to result from micro-reentry,[50] recent investigations suggest a macro-reentrant circuit.[51-56] Characteristic features include induction and termination with programmed stimulation, a relatively narrow QRS, the presence of a Purkinje system potential, and entrainment. Left ventricular fascicular tachycardia is more commonly associated with the left posterior fascicle and has an RBBB–left axis configuration. The RBBB–right axis configuration of this VT is associated with the anterior fascicle.[57] Termination with verapamil is usual. Rarely adenosine may terminate this VT if catecholamines were required for its induction.[9] Incessant forms of left ventricular fascicular VT can result in a tachycardia-induced cardiomyopathy.[58]

Adrenergic monomorphic VT can present with a left or right bundle branch QRS pattern. Clinical and electrophysiological characteristics of this VT are poorly defined. This form of VT has electrophysiological properties that would be expected with enhanced automaticity as its basis. It is often initiated by catecholamine infusion and exercise, but it cannot be induced, entrained, or terminated by programmed stimulation.

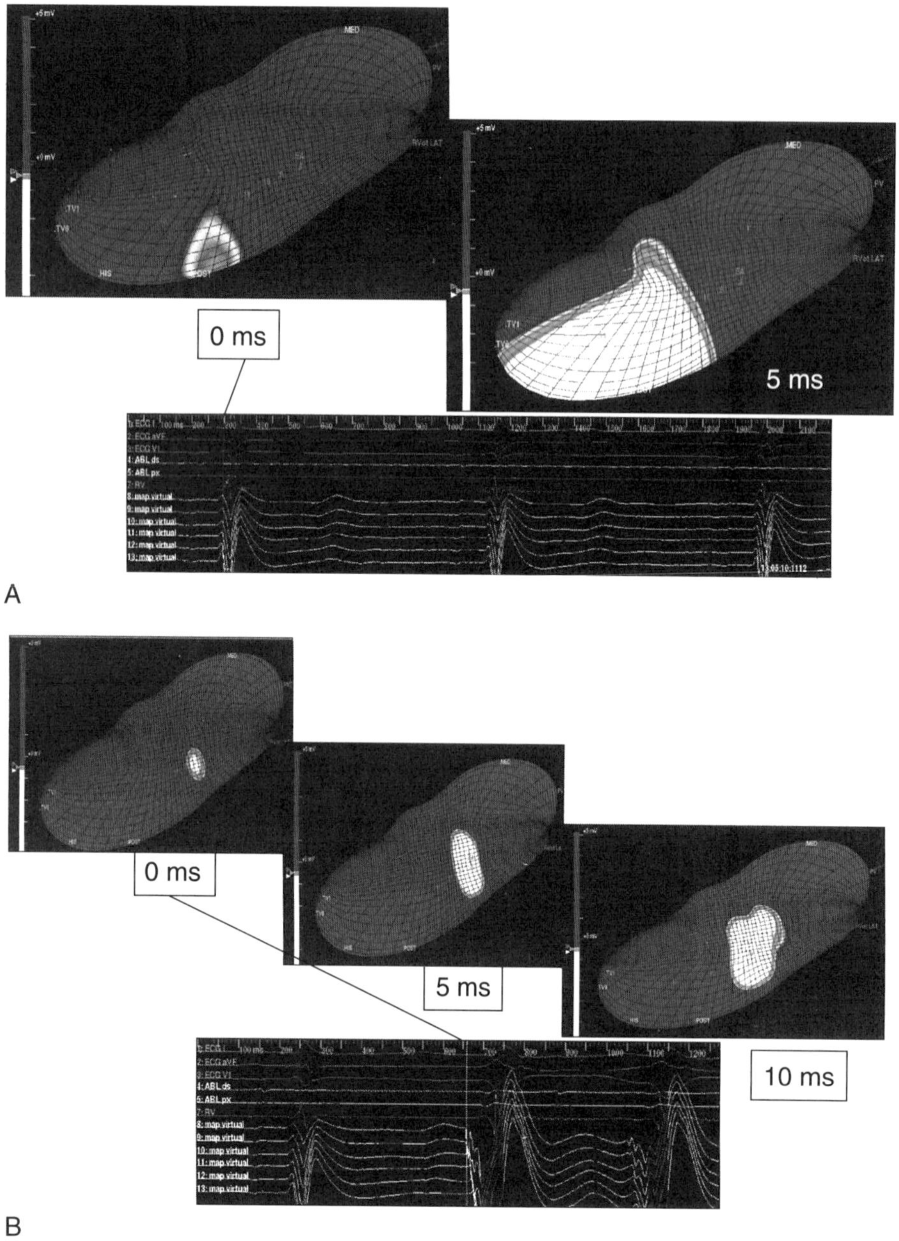

FIGURE 18-4 (See also Color Plate 18-4.) Three-dimensional noncontact mapping of ventricular tachycardia (VT) arising in the right ventricular outflow tract (RVOT). **A,** Sinus rhythm. **B,** VT, arising from RVOT. The noncontact map of the outflow tract is seen as an elliptical oblong chamber with the VT wavefront shown in sequential images. Note the focal origin with centrifugal propagation. RV, right ventricle. Timing is shown in milliseconds after onset. Standard ECG leads, aVF and V, as well as virtual electrograms, are displayed below on three-dimensional map.

Verapamil has no effect on this form of VT,[59] whereas adenosine has been reported to result in transient suppression.[60] It is terminated and suppressed by β-blockers.

Monomorphic IVT with both left and RBBB morphologies is observed in infants and children.[61] These arrhythmias may be incessant, and this can result in a tachycardia-induced cardiomyopathy. More often, they have a good prognosis and may also resolve spontaneously. The extent to which these arrhythmias have similar mechanisms to RVOT VT, left ventricular fascicular VT, and adrenergic monomorphic VT is uncertain. Other forms of monomorphic VT may be encountered in patients without significant structural heart disease in clinical practice. Often these cannot be classified as one of the common forms detailed earlier.

Principles of Practice

The first clinical objective in evaluating a patient who presents with possible IVT is to estimate the immediate risk of arrhythmic death. If the risk is high or uncertain, the patient must be hospitalized and continuously monitored until definitive treatment is established or the risk of death is determined to be low. IVT syndromes are usually considered after common arrhythmic conditions are excluded. However, laboratory studies cannot definitively exclude the presence of an IVT syndrome. Therefore, even when the etiology of VT appears clear, the practitioner must always be cognizant of the possibility that an IVT syndrome may coexist. The goal of the clinical evaluation process is

TABLE 18-2 Disorders that Mimic Nonstructural Ventricular Tachyarrhythmia Syndromes

Extrinsic Disorders
Epilepsy
Substance abuse
Extreme diets
Hyperaldosteronism
Hypercoagulation
Pulmonary embolism
Bradyarrhythmias
Supraventricular tachyarrhythmias
Undetected Structural Heart Disease
Right ventricular dysplasia
Sarcoidosis
Hypertrophic cardiomyopathies
Dilated cardiomyopathies
Myocarditis
Coronary artery spasm
Anomalous coronary artery
Coronary artery disease
Accessory atrioventricular connections

also to identify all potential arrhythmogenic factors. This permits the most precise estimate of sudden death risk and appropriate choice of therapy. Thus, evaluation is a process of both positive identification and exclusion of arrhythmogenic factors. Two classes of conditions associated with VT must be excluded before a presumptive diagnosis of IVT is made (Table 18-2). Extrinsic proarrhythmic factors (e.g., conditions that cause electrolyte disturbances or drugs that prolong repolarization) may be associated with VT in the presence of normal cardiac structure and function. In addition, VT may exist with undetected cardiac disease and can mimic IVT syndromes. A comprehensive evaluation that attempts to exclude all potential conditions that could contribute to arrhythmogenesis requires invasive procedures for definitive diagnosis.[62] This is essential in the patient who presents with cardiac arrest or a hemodynamically compromising VT, but the extent to which it should be applied to patients with less symptomatic presentations or suspected IVT must be judged individually.

EPS can provide information in some cases where the history, ECG, and other tests results are inadequate to guide therapy. EPS have at least three potential indications. First, they are used to screen for various forms of IVT in patients without documented arrhythmias. Various forms of programmed stimulation (extrastimulus testing and rapid pacing) before and during catecholamine infusions are used to initiate VT. Induction of a sustained VT is generally considered to be an abnormal response. However, the induced VT and the hemodynamic response should be consistent with the clinical presentation. Identification of fractionated potentials during sinus rhythm might be taken as evidence of substrate for reentrant arrhythmias. Induction of polymorphic VT or VF is considered a risk factor for spontaneous ventricular arrhythmias in patients with idiopathic VF or Brugada syndrome. However, failure to induce VT or VF does not indicate a low risk. Unfortunately, the sensitivity of EPS for nonspecific screening is low, because many forms of IVT cannot be reliably induced. In addition, if the pretest probability of an IVT is low, as for common symptoms such as syncope, the yield of EPS will be exceedingly low. Moreover, the likelihood of a false-positive response is increased. The advantages of EPS are that the results can be obtained quickly and the results used immediately if the test is positive. It is hoped that new methods of provocation will be developed to expand the utility and enhance the value of EPS. Provocative testing by administration of ajmaline, flecainide, or procainamide has been advocated for patients at increased risk for Brugada syndrome. However, flecainide has also been suggested as a therapy for LQTS type 3. A recent study demonstrated that changes in the ECG suggestive of Brugada syndrome occurred in patients with the LQTS type 3 genotype, raising concerns about the role of flecainide both as a treatment and as a provocative test.[63] Thus, the optimal use of this drug test is not yet defined.

The second function of EPS is to confirm IVT and refine clinical diagnostic possibilities. This is exemplified by the patient who presents with monomorphic IVT. If the arrhythmia can be induced, it can often be characterized by the mode of induction (extrastimulus technique, rapid pacing, requirement for catecholamines), the electrocardiographic pattern (bundle branch pattern and axis), the presence of multiple configurations, the response of tachycardia to pacing maneuvers (entrainment, resetting, termination), and the response to drugs (e.g., adenosine, verapamil, and β-blockers) and vagal maneuvers. EPS may provide critical data needed to distinguish monomorphic IVTs, which are usually associated with a good prognosis, from VT associated with life-threatening disorders. The most common clinical dilemma is to distinguish between RVOT VT and VT due to arrhythmogenic right ventricular dysplasia. VT arising from these two disorders may have similar QRS configuration and response to programmed stimulation, including sensitivity to catecholamines. The most useful differential finding is sensitivity to adenosine, which is rarely observed in other forms of VT.

The third function of EPS in patients with IVT is therapeutic. EPS has been reported to be useful for guiding antiarrhythmic drug therapy in patients with idiopathic VF,[64] although this approach is not widely recommended.[62] Radiofrequency catheter ablation is widely employed in the monomorphic IVT syndromes after mapping of the IVT origin at EPS. Figure 18-5 is a noncontact three-dimensional map of catheter ablation of the RVOT focus. Note that individual lesions are marked at the site of application relative to the focal origin and propagation of the arrhythmia. Catheter ablation may be the therapy of choice for many patients with IVT syndromes.

Genetic testing can provide a more precise diagnosis if a specific mutation is detected. It is limited in that many key genetic abnormalities have not yet been defined. Another major current problem is that this test is not generally available and does not provide results in a reasonable time period. Nevertheless, there is little doubt that genetic testing will enter the clinical realm, because it can improve risk-assessment and

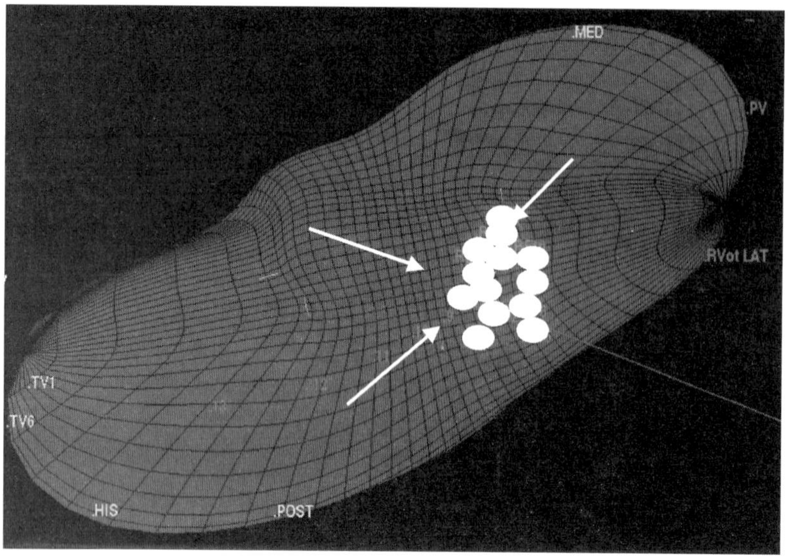

FIGURE 18-5 (See also Color Plate 18-5.) Catheter ablation guided by three-dimensional noncontact mapping in the same patient with ventricular tachycardia (VT) arising from the right ventricular outflow tract shown in Figure 18-4. Ablation lesions (*circular markers*) are shown individually at the focal origin and in the surrounding region. Successful VT ablation was achieved.

treatment of the patient, living relatives, and future generations of the family.

Evidence-Based Therapy

No clinical trials have addressed therapeutic options in patients with IVT syndromes. Retrospective and prospective registries have produced data that are used to guide therapy. The reliability of the results and applicability to future patients are uncertain, because databases are small, and there may be biases created by the referral process. Nevertheless, they provide the best available evidence. The largest registry exists for patients with LQTS. Results from this database support the use of β-blockers as a mainstay of therapy but report a persistently high risk despite β-blocker therapy in patients with a history of cardiac arrest or syncope.[65] A report from an international registry suggests that medical therapy with β-blockers, sodium channel blockers, and amiodarone are ineffective for idiopathic VF.[62] This contradicts a prior study from Belhassen et al.[64] that reported success in EPS-guided treatment with certain sodium channel blockers, especially quinidine.

Management of Ventricular Tachycardia

Prevention of sudden death is the first priority in patients with IVT syndromes. Because affected individuals are often young and have normal cardiovascular function, there are few competing risks, and long-term survival may depend upon arrhythmic death risk and complications resulting from therapy. Unfortunately, precise arrhythmic risk assessment for the IVT syndromes is difficult. First, there is a paucity of data for reliable outcome prediction. Second, there is heterogeneity within the IVT population. Arrhythmic risk may be affected by the particular genotype and by variations among a particular genotype. Penetrance for a given genotype has been found to vary unpredictably.[66] Inadequate diagnosis in families poses another challenge. This is often encountered during the evaluation of living members of a family in which multiple sudden deaths have occurred. Affected living family members are assumed to carry the "malignant" genotype, and prophylactic therapy is often considered. However, it is often not possible to determine the genotype because of an ambiguous phenotype and because genetic testing is not available. In such instances, the individual is either at very high or very low risk. Fortunately, risk assessment should continue to improve as IVT registries collect more data, diagnostic methods improve, and genetic testing becomes a clinical tool.

The potential for sudden death, uncertainty of arrhythmic risk, and limitations of existing therapeutic options complicate the management of IVT syndromes. It is widely assumed that the implantable cardioverter defibrillator (ICD) is an acceptable therapy despite the limited data regarding the risks and benefits in this population of patients. Results from clinical trials support the use of ICDs over amiodarone for reduction of all-cause mortality in survivors of cardiac arrest and patients with certain forms of VT. On the other hand, subgroup analyses from the same trials suggest that ICDs are less effective in patients with a normal left ventricular ejection fraction. However, available clinical trial evidence has been largely obtained from cardiac arrest/VT populations with heart disease and multiple competing mortality risks. Furthermore, there is a relatively short follow-up for IVT populations. These trials may be poor guides to selection of therapy in patients with IVT syndromes.

Patients with IVT are often children or adolescents. ICDs may have a significant advantage in these patients in that they enforce therapeutic compliance. In this group of patients, compliance with medications is often difficult, especially in the absence of symptoms. Devices also offer female patients protection during pregnancy without the need for antiarrhythmic drug therapy. On the other hand, young patients face many ICD generator and lead replacements and a long period of exposure to the complications of ICD therapy. Furthermore, quality of life may be adversely affected by concern about body image, isolation and ridicule from peers, and dependence on medical services.[67] ICD shocks, whether appropriate or inappropriate, may diminish quality of life still further.

Despite reasons for considerable concern regarding effects of ICD therapy on quality of life in the young population and the risks of many years of ICD therapy, no alternatives are likely to provide more reliable protection against arrhythmic death. ICDs are therefore recommended if the risk of arrhythmic death is high. Survivors of cardiac arrest or rapid ventricular tachyarrhythmias are usually assumed to be at high risk. This is borne out by results from the Unexplained Cardiac Arrest Registry of Europe (UCARE), which estimated the rate of recurrence of cardiac arrest or syncope in patients with idiopathic VF to be about 30%.[62]

Decisions regarding the use of ICDs in groups of patients with higher risk-to-benefit ratios are more difficult (e.g., patients who present with syncope, affected asymptomatic family members of cardiac arrest survivors, and asymptomatic members of "high risk" families whose genotype is unknown).

LQTS may be a prototype for IVT syndromes in that it demonstrates how progress improves care but also creates new areas of uncertainty. Only about 70% of LQTS carriers have unequivocal QT prolongation, and approximately 12% have a normal Q–T interval. In addition, sudden death is the first manifestation of the syndrome in at least 10% of affected persons. There are probably several undiscovered genetic defects and sporadic mutations that cause this disorder. Moreover, the types of provocative factors, the risk of sustained VTA, and the response to therapy vary with the specific genetic defect. Failure of medical therapy may result from several sources despite effectiveness of the therapy itself, including incomplete compliance, inadequate drug dose, or exposure to proarrhythmic drugs. Patients who present with VT or syncope, those with symptoms despite β-blockers, and other individuals at high risk are candidates for ICD therapy. Left cervicothoracic ganglionectomy and permanent pacing probably prevent sudden death in some individuals, but failures have been reported. They may be useful adjuncts or alternatives in patients who are intolerant of β-blockers or who cannot comply with drug therapy. Experimental therapies include administration of supplemental potassium chloride for HERG mutations, potassium channel openers for LQT1 and LQT2, and mexiletine or flecainide for LQT subtypes 3, 2, and possibly 1.[37,38]

In contrast to LQTS, there are as yet no widely accepted treatment alternatives to the ICD for preventing arrhythmic death in other IVT syndromes associated with polymorphic VT/VF. In Brugada syndrome, sodium channel blockers are used to provoke characteristic ECG abnormalities and may provoke VT.[68] Other drugs such as β-blockers and amiodarone have not provided adequate protection. Likewise, as noted earlier, the UCARE database results indicate failure of treatment with amiodarone, β-blockers, and sodium channel blockers for patients with idiopathic VF.[62] In contrast, treatment with class I antiarrhythmic drugs, mostly quinidine, guided by EPS was reported to be effective in one small series.[65] It is possible that the distinctive effects of quinidine on the transient outward current were responsible for the excellent protective effect that was observed. However, this approach has not been widely embraced. Verapamil and amiodarone may inhibit VT in the short-coupled variant of torsades de pointes but have not been shown to prevent sudden death.[69] β-Blockers are reportedly effective in patients with catecholaminergic polymorphic VT, but the experience may be too limited to be confident of this approach. There is as yet no well-defined role for surgical intervention (other than left cervicothoracic sympathectomy in LQTS) for the polymorphic VT/VF IVT syndromes.

The risk of arrhythmic death is believed to be low in the monomorphic IVT syndromes such as RVOT VT and its variants. Unless features are present to suggest a malignant potential, such as syncope or very rapid tachycardia, an ICD is not indicated. Instead, the goal of therapy is to relieve symptoms or prevent detrimental effects of incessant tachycardia.[58,61] If symptoms are absent or minimal and tachycardia is infrequent, no treatment may be necessary. Pharmacologic therapy with β-blockers or calcium antagonists alone or in combination is usually safe and often effective. Amiodarone, sotalol, encainide, and flecainide may be effective. Nicorandil, a potassium channel opener, has been reported to terminate and suppress adenosine-sensitive VT.[69] If adequate control is obtained with well-tolerated and presumably safe pharmacologic therapy (e.g., with β-blockers or calcium channel blockers), then this therapy may be preferred in many individuals. If symptoms are intermittent, then drugs could also be used on an "as-needed" basis.

Nonpharmacologic therapy should be considered for patients refractory to or intolerant of drug therapy, those who prefer it to a lifelong need for drugs, and those who wish to become pregnant. Many patients will prefer to discontinue all medications when pregnant, a time when therapy should not be withdrawn because changes in cardiovascular function could exacerbate some forms of VT or cause more severe symptoms. Catheter ablation offers the possibility of permanent correction with a high probability of success (see Fig. 18-5). The technique of mapping and ablation is described in Chapters 42 and 45. In brief, contact catheter mapping or noncontact mapping is used to identify the site of earliest ventricular activation. Presystolic potentials are usually present at this site, and repetitive firing can usually be demonstrated at this location, often after isoproterenol administration.[70] Radiofrequency contact catheter ablation lesions are delivered at this site and in adjoining areas. Occasionally, more than one site may be seen. Efficacy rates can exceed 90% in RVOT VT. Predictors of failure in catheter ablation included multiple morphologies, a δ wave–like beginning of the QRS, and a poor match between the 12-lead ECG during VT and during pacing at the ablation site.[71] Left heart catheterization is not required for most cases of RVOT-VT, which minimizes the risk of complications (2%). Risks may be greater if lesions are delivered in the left ventricle or in epicardial regions. Drug therapy is associated with a low initial risk but greater risk of failure due to ineffectiveness or intolerance and potential for unknown, long-term adverse reactions. Moreover, the cost difference may be minimal due to the long-term need for drugs and the possibility that catheter ablation will be necessary at a later date. In the absence of a controlled comparison of drugs and catheter ablation, each case must be considered individually, and the alternatives discussed thoroughly to determine the most appropriate therapy for the patient.

Device therapy (e.g., pacing inhibition or antitachycardia pacing) could trigger RVOT VT and would not be a logical first-line choice. However, published evidence that confirms or refutes this is lacking.

Idiopathic fascicular VT is also believed to have a benign long-term outlook, although the extent of experience is less than that for RVOT VT. Treatment is indicated for prevention or control of symptoms and for the protection against the potential deleterious effects of incessant tachycardia. Calcium antagonists have been shown to be effective in many patients, as have β-blockers. Drug therapy or catheter ablation are reasonable approaches.

Probably, many varieties of monomorphic IVT do not have characteristics of left ventricular fascicular VT or RVOT VT. The prognosis of such arrhythmias cannot be assumed to be as benign as these VTs. Clinical judgment must be used individually to assess the risk-to-benefit ratio of ICD implantation, catheter ablation, or drug trials in these patients.

SUMMARY

Although not commonly encountered in practice, IVT syndromes are always in the differential diagnosis of patients who present with known or suspected VT. The diagnosis can be obvious if the characteristic ECG findings are present, but these can be occult or absent in many patients. The clinical challenge occurs because there are no tests or algorithms that are capable of excluding the IVT syndromes. It seems likely that a large group of IVT syndromes remain to be discovered. Finally, if IVT syndromes prove to be important contributors to sudden death in the absence of structural heart disease, then substantial effort is justified to discover the mechanisms of the arrhythmias as well as develop tests to identify susceptible individuals.

REFERENCES

1. Vakil RJ: Repetitive paroxysmal ventricular tachycardia with fusion beats. Br Heart J 1957;19:293-5.
2. Thind US, Blakemore WS, Zinsser HF: Ventricular aneurysmectomy for the treatment of recurrent ventricular tachyarrhythmia. Am J Cardiol 1971;27:690-4.
3. Nitter-Hauge S, Storstein O: Surgical treatment of recurrent ventricular tachycardia. Br Heart J 1973;35:1132-5.
4. Rally CR, Walters MB: Paroxysmal ventricular tachycardia without evident heart disease. J Can Med Assoc 1962;86:268-73.
5. Hair TE Jr, Eagen JT, Orgain ES: Paroxysmal ventricular tachycardia in the absence of demonstrable heart disease. Am J Cardiol 1962;9:209-14.
6. Bisteni A, Sodi-Pallares D, Medrano GA, et al: A new approach for the recognition of ventricular premature beats. Am J Cardiol 1960;5:358-69.
7. Rosenbaum MB: Classification of ventricular extrasystoles according to form. J Electrocardiol 1969;2:289-98.
8. Pietras RJ, Mautner R, Denes P, et al: Chronic recurrent right and left ventricular tachycardia: Comparison of clinical hemodynamic and angiographic findings. Am J Cardiol 1977;40:32-7.
9. Lerman BB, Stein KM, Markowitz SM, et al: Recent advances in right ventricular outflow tract tachycardia. Card Electrophysiol Rev 1999;3:210-4.
10. Carlson MD, White RD, Trohman RG, et al: Right ventricular outflow tract ventricular tachycardia: Detection of previously unrecognized anatomic abnormalities using cine magnetic resonance imaging. J Am Coll Cardiol 1994;24:720-7.
11. Markowitz SM, Litvak BL, Ramirez de Arellano EA, et al: Adenosine-sensitive ventricular tachycardia: Right ventricular abnormalities delineated by magnetic resonance imaging. Circulation 1997;96:1192-200.
12. Swan H, Piippo K, Viitasalo M, et al: Arrhythmic disorder mapped to chromosome 1q42-q43 causes malignant polymorphic ventricular tachycardia in structurally normal hearts. J Am Coll Cardiol 1999;34:2035-42.
13. Chiladakis J, Vassilikos V, Maounis T, et al: Unusual features of right and left idiopathic ventricular tachycardia abolished by radiofrequency catheter ablation. Pacing Clin Electrophysiol 1998;21:1831-4.
14. Chapman JH, Schrank JP, Crampton RS: Idiopathic ventricular tachycardia: An intracardiac electrical hemodynamic and angiographic assessment of six patients. Am J Med 1975;59:470-80.
15. Buxton AE, Waxman HL, Marshlinski FE, et al: Right ventricular tachycardia: Clinical and electrophysiologic characteristics. Circulation 1983;68:917-27.
16. Chugh SS, Kelly KL, Titus JL: Sudden cardiac death with apparently normal heart. Circulation 2000;102:649-54.
17. Loire R, Tabib A: Unexpected sudden cardiac death: An evaluation of 1000 autopsies. Arch Mal Coeur Vaiss 1996;89:13-8.
18. Bacci M, Giusti G: Diagnostic possibilities and limitations of the necropsy examination on cadavers exhumed because of sudden death. Pathologica 1996;88:25-8.
19. Pfammatter JP, Paul T: Idiopathic ventricular tachycardia in infancy and childhood: A multicenter study on clinical profile and outcome. Working Group on Dysrhythmias and Electrophysiology of the Association for European Pediatric Cardiology. J Am Coll Cardiol 1999;33:2067-72.
20. Jadonath RL, Schwartzman DS, Preminger MW, et al: Utility of the 12-lead electrocardiogram in localizing the origin of right ventricular outflow tract tachycardia. Am Heart J 1995;130:1107-13.
21. Yoshida Y, Hirai M, Murakami Y, et al: Localization of precise origin of idiopathic ventricular tachycardia from the right ventricular outflow tract by a 12-lead ECG: A study of pace mapping using a multielectrode "basket" catheter. Pacing Clin Electrophysiol 1999;22:1760-8.
22. Coggins DL, Randall JL, Sweeney J, et al: Radiofrequency catheter ablation as a cure for idiopathic tachycardia of both left and right ventricular origin. J Am Coll Cardiol 2994;23:133-41.
23. Callans DJ, Volker M, Schwartzman D, et al: Repetitive monomorphic tachycardia from the left ventricular outflow tract: Electrocardiographic patterns consistent with a left ventricular site of origin. J Am Coll Cardiol 1997;29:1023-7.
24. Lamberti F, Calo L, Pandozi C, et al: Radiofrequency catheter ablation of idiopathic left ventricular outflow tract tachycardia: Utility of intracardiac echocardiography. J Cardiovasc Electrophysiol 2001;12:529-35.
25. Krebs ME, Krause PC, Engelstein ED, et al: Ventricular tachycardias mimicking those arising from the right ventricular outflow tract. J Cardiovasc Electrophysiol 2000;11:45-51.
26. Kamakura S, Shimizu W, Matsuo K, et al: Localization of optimal ablation site of idiopathic ventricular tachycardia from right and left ventricular outflow tract by body surface ECG. Circulation 1998;98:1525-33.
27. Kanagaratnam L, Tomassoni G, Schweiker R, et al: Ventricular tachycardias arising from the aortic sinus of Valsalva: An under-recognized variant of left outflow tract ventricular tachycardia. J Am Coll Cardiol 2001;37:1408-14.

28. Lin FC, Finley CD, Rahimtoola SH, Wu D: Idiopathic paroxysmal ventricular tachycardia with a QRS pattern of right bundle branch block and left axis deviation: A unique clinical entity with specific properties. Am J Cardiol 1983;52:95-100.
29. Ohe T, Shimomura K, Aihara N, et al: Idiopathic sustained left ventricular tachycardia: Clinical and electrophysiologic characteristics. Circulation 1988;77:560-8.
30. Nakagawa H, Beckma KJ, McClelland JH, et al: Radiofrequency catheter ablation of idiopathic left ventricular tachycardia guided by a Purkinje potential. Circulation 1993;88:2607-17.
31. Wen MS, Yeh SJ, Wang CC, et al: Radiofrequency ablation therapy in idiopathic left ventricular tachycardia with no obvious structural heart disease. Circulation 1994;89:1690-6.
32. Chen YJ, Chen SA, Tai CT, et al: Radiofrequency ablation of idiopathic left ventricular tachycardia with changing ECG morphology. Pacing Clin Electrophysiol 1998;21:1668-71.
33. Tsuchiya T, Okumura K, Honda T, et al: Significance of late diastolic potential preceding Purkinje potential in verapamil sensitive idiopathic left ventricular tachycardia. Circulation 1999;99:2408-13.
34. Aiba T, Suyama K, Aihara N, et al: The role of Purkinje and pre-Purkinje potentials in the reentrant circuit of verapamil-sensitive idiopathic LV tachycardia. Pacing Clin Electrophysiol 2001;24:333-44.
35. Nogami A, Naito S, Tada H, et al: Demonstration of diastolic and presystolic Purkinje potentials as critical potentials in a macroreentry circuit of verapamil-sensitive idiopathic left ventricular tachycardia. J Am Coll Cardiol 2000;36:811-23.
36. Miyauchi Y, Kobayashi Y, Ino T, Atarashi H: Identification of the slow conduction zone in idiopathic left ventricular tachycardia. Pacing Clin Electrophysiol 2000;23:481-7.
37. Vincent GM, Timothy K, Zhang L: Congenital long QT syndrome. Card Electrophysiol Rev 1999;1:207-9.
38. Shimizu W, Antzelevitch C: Differential effects of beta-adrenergic agonists and antagonists in LQT1, LQT2 and LQT3 models of the long QT syndrome. J Am Coll Cardiol 2000;35:778-86.
39. Gussak I, Antzelevitch C, Bjerregaard P, et al: The Brugada Syndrome: Clinical, electrophysiologic and genetic aspects. J Am Coll Cardiol 1999;33:5-15.
40. Leenhardt A, Glaser E, Burguera M, et al: Short-coupled variant of torsades de pointes: A new electrocardiographic entity in the spectrum of idiopathic ventricular tachyarrhythmias. Circulation 1994;89:206-15.
41. Von Bernuth G, Bernsau U, Hoffman W, et al: Tachyarrhythmic syncope in children with structurally normal hearts with and without QT prolongation in the electrocardiogram. Eur J Pediatr 1982;138:206-10.
42. Leenhardt A, Lucet V, Denjoy I, et al: Catecholaminergic polymorphic ventricular tachycardia in children: A 7-year follow-up of 21 patients. Circulation 1995;91:1512-9.
43. Lerman BB, Dong B, Stein KM, et al: Right ventricular outflow tract tachycardia due to a somatic cell mutation in G protein subunit$_{12}$. J Clin Invest 1998;101:2862-8.
44. Calkins H: Role of invasive electrophysiologic testing in the evaluation and management of right ventricular outflow tract tachycardias. Card Electrophysiol Rev 2000;4:71-5.
45. Lokhandwala Y, Vora A, Naik A, et al: Dual morphology of idiopathic ventricular tachycardia. J Cardiovasc Electrophysiol 1999;10:1326-34.
46. Yeh S, Wen M, Wang C, et al: Adenosine-sensitive ventricular tachycardia from the anterobasal left ventricle. J Am Coll Cardiol 1997;30:1339-45.
47. Stellbrink C, Diem B, Schauerte P, et al: Transcoronary venous radiofrequency catheter ablation of ventricular tachycardia. J Cardiovasc Electrophysiol 1997;8:916-21.
48. Goy JJ, Tauxe F, Fromer M, et al: Ten-years' follow-up of 20 patients with idiopathic ventricular tachycardia. Pacing Clin Electrophysiol 1990;13:1142-7.
49. Sebastien P, Waynberger M, Beaufils P, et al: Isolated ventricular tachycardia without patent cardiopathy. Arch Mal Coeur Vaiss 1976;69:919-28.
50. Kottkamp H, Chen X, Hindricks G, et al: Radiofrequency catheter ablation of idiopathic left ventricular tachycardia: Further evidence for microreentry as the underlying mechanism. J Cardiovasc Electrophysiol 1994;5:268-73.
51. Han J, Sung RJ: Evaluation and management of idiopathic ventricular tachyarrhythmias—update. Card Electrophysiol Rev 2000;4:61-5.
52. Tsuchiya T, Okumura K, Honda T, et al: Significance of late diastolic potential preceding Purkinje potential in verapamil-sensitive idiopathic left ventricular tachycardia. Circulation 1999;2408-13.
53. Friedman PA, Beinborn DA, Schultz J, Hammil SC: Ablation of noninducible idiopathic left ventricular tachycardia using a noncontact map acquired from a premature complex with tachycardia morphology. Pacing Clin Electrophysiol 2000;23:1311-4.
54. Lai L, Lin J, Hwang J, Huang S: Entrance site of the slow conduction zone of verapamil-sensitive idiopathic left ventricular tachycardia: Evidence supporting macroreentry in the Purkinje system. J Cardiovasc Electrophysiol 1998;9:184-90.
55. Aiba T, Suyama K, Matsuo K, et al: Mid-diastolic potential is related to the reentrant circuit in a patient with verapamil-sensitive idiopathic left ventricular tachycardia. J Cardiovasc Electrophysiol 1998;9:1004-7.
56. Miyauchi Y, Kobayashi Y, Ino T, Atarashi H: Identification of the slow conduction zone in idiopathic left ventricular tachycardia. Pacing Clin Electrophysiol 2000;23:481-7.
57. Nogami A, Naito S, Tada H, et al: Verapamil-sensitive left anterior fascicular ventricular tachycardia: Results of radiofrequency ablation in six patients. J Cardiovasc Electrophysiol 1998;9:1269-78.
58. Toivonen L, Nieminen M: Persistent ventricular tachycardia resulting in left ventricular dilatation treated with verapamil. Int J Cardiol 1986;13:361-5.
59. Sung RJ, Shapiro WA, Shen EN, et al: Effects of verapamil on ventricular tachycardias possibly caused by reentry, automaticity, and triggered activity. J Clin Invest 1983;81:688-99.
60. Lerman BB: Response of nonreentrant catecholamine-mediated ventricular tachycardia to endogenous adenosine and acetylcholine: Evidence for myocardial receptor-mediated effects. Circulation 1993;87:382-90.
61. Pfammatter JP, Paul T: Idiopathic ventricular tachycardia in infancy and childhood: A multicenter study on clinical profile and outcome. J Am Coll Cardiol 1999;33:2067-72.
62. Survivors of out-of-hospital cardiac arrest with apparently normal heart. Need for definition and standardized clinical evaluation. Consensus Statement of the Joint Steering Committees of the Unexplained Cardiac Arrest Registry of Europe and of the Idiopathic Ventricular Fibrillation Registry of the United States. Circulation 1997;95:265-72.
63. Priori SG, Napolitano C, Schwartz PJ, et al: The elusive link between LQT3 and Brugada syndrome—the role of flecainide challenge. Circulation 2000;102:945-7.
64. Belhassen B, Viskin S, Fish R, et al: Effects of electrophysiologic-guided therapy with class IA antiarrhythmic drugs on the long-term outcome of patients with idiopathic ventricular fibrillation with or without the Brugada syndrome. J Cardiovasc Electrophysiol 1999;10:1301-12.
65. Moss AJ, Zareba W, Hall WJ, et al: Effectiveness and limitations of beta-blocker therapy in congenital long-QT syndrome. Circulation 2000;101:616-23.
66. Priori SG, Napolitano C, Schwartz PJ: Low penetrance in the long-QT syndrome—clinical impact. Circulation 1999;99:529-33.
67. Domanski MJ, Saksena S, Epstein AE, et al: Relative effectiveness of the implantable cardioverter-defibrillator and antiarrhythmic drugs in patients with varying degrees of left ventricular dysfunction who have survived malignant ventricular arrhythmias. J Am Coll Cardiol 1999;34:1090-5.
68. Brugada R, Brugada J, Antzelevitch C, et al: Sodium channel blockers identify risk for sudden death in patients with ST-segment elevation and right bundle branch block but structurally normal hearts. Circulation 2000;101:510-15.
69. Kobayashi Y, Miyata A, Tanno K, et al: Effects of nicorandil, a potassium channel opener, on idiopathic ventricular tachycardia. J Am Coll Cardiol 1998;32:1377-83.
70. Lee SH, Tai CT, Chiang CE, et al: Determinants of successful ablation of idiopathic ventricular tachycardias with left bundle branch block morphology from the right ventricular outflow tract. Pacing Clin Electrophysiol 2002;25:1346-51.
71. Rodriguez LM, Smeets JLRM, Timmermans C, Wellens HJJ: Predictors for successful ablation of right- and left-sided idiopathic ventricular tachycardia. Am J Cardiol 1997;79:309-14.

Chapter 19 Ventricular Fibrillation

PAUL DORIAN, SAROJA BHARATI, ROBERT J. MYERBURG, DAVID ROSENBAUM, KARA J. QUAN, MARIAH L. WALKER, BRUCE D. LINDSAY, L. BRENT MITCHELL, WILLIAM G. STEVENSON, and MICHAEL DOMANSKI

Introduction: Paul Dorian
Etiology and Pathology: Saroja Bharati
Epidemiology: Robert J. Myerburg
Basic Electrophysiology: David Rosenbaum, Kara J. Quan, and Mariah L. Walker
Clinical Electrocardiography: Bruce D. Lindsay
Diagnostic Evaluation: L. Brent Mitchell
Electrophysiological Testing: William G. Stevenson
Principles of Practice: Paul Dorian
Evidence-Based Therapy: Paul Dorian and Michael Domanski

Ventricular fibrillation (VF) is the most serious cardiac arrhythmia and has a primary role in mediating sudden cardiac death (SCD). It leads to immediate circulatory arrest with cardiovascular collapse. A variable period may elapse, but cardiac asystole usually supervenes (Fig. 19-1). Spontaneous termination of VF, which is seen in animal experiments, is rare in man. VF is often preceded by an organized, rapid ventricular tachycardia (VT) of variable duration; recordings from implantable devices have now substantiated this. Ischemic injury may trigger VF. If untreated, there is irreversible end-organ damage, including cerebral and myocardial damage after 5 to 7 minutes of VF. Even with optimally performed basic cardiopulmonary resuscitation, mortality is greater than 95% if defibrillation is delayed more than 10 minutes. The rule of survival from out-of-hospital VF is that survival decreases by 10% for each minute before defibrillation. The sine qua non of effective resuscitation from VF is thus prompt defibrillation, delivered as early as possible. Importantly, however, a brief period of cardiopulmonary resuscitation (CPR) *before* defibrillation in out-of-hospital cardiac arrest, especially if the arrest is more than 4 minutes in duration, may increase survival rates.

Etiology

VF occurs in many disease states of the myocardium and the conduction system that can be broadly grouped under two categories: (1) genetic-familial and (2) acquired. Genetically based abnormalities of myocardium or the specialized conduction fibers, or both, may give rise to the clinical manifestations of familial occurrence of ventricular tachycardia and fibrillation. However, the major types of acquired cardiac diseases, such as coronary artery disease, hypertensive heart disease, cardiomyopathy of any etiology with or without heart failure, and other miscellaneous disease states (previously discussed in Chapter 17, Sustained Ventricular Tachycardia with Heart Disease) are the most common causes of VF and sudden death.[1-3] Ventricular tachycardia leading to VF occurs in acute myocarditis of any etiology. However, intractable VT with recurrent cardiac arrest refractory to medications or defibrillator device implantation may be related to chronic myocarditis (Fig. 19-2).[4]

Pathology

The anatomic basis for VF in acquired heart diseases is similar to and has been dealt with in the previous chapter on ventricular tachycardia. In this chapter, however, emphasis is given to the morphologic findings for VF in athletes or healthy youngsters who were seemingly living a "normal," "asymptomatic" life and unfortunately died suddenly. In addition, familial occurrences of sudden death, due to recurrent ventricular tachycardia degenerating to VF is discussed briefly.[1-3,5,6]

SUDDEN DEATH IN ATHLETES OR THE YOUNG AND HEALTHY

The conduction system has been carefully studied in several young victims who were living a "normal life" and died suddenly. Routine autopsies are often unremarkable. In the majority, the heart is hypertrophied and enlarged to a mild, moderate, or marked extent; however, at the gross level, there were no significant

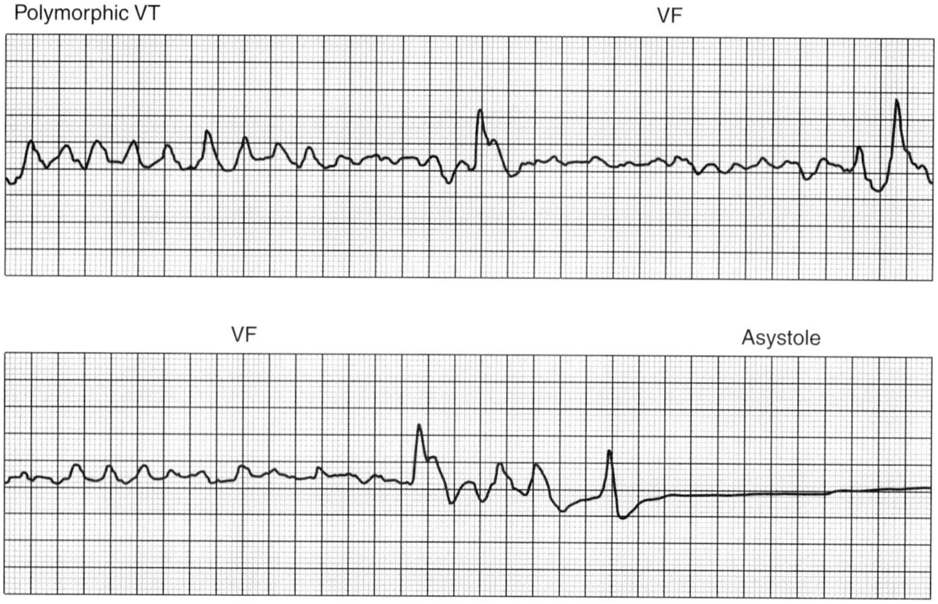

FIGURE 19-1 Evolution of polymorphic ventricular tachycardia into ventricular fibrillation and subsequent development of cardiac asystole.

abnormalities.[3] A serial section examination of the conduction system and the surrounding myocardium reveals varying types of abnormalities, either of a congenital or an acquired nature. Congenital abnormalities can exist in the sinoatrial (SA) node, atrioventricular (AV) node, or the atrioventricular bundle and bundle branches (Fig. 19-3). Abnormal formation of the SA or AV nodes, such as a double SA node or a double AV node or abnormal location of the SA and AV nodes in unexpected areas can be seen pathologically. The AV bundle may be considerably fragmented into several components or abnormally located, or both. In addition, there are acquired pathologic findings in the form of focal myocardial disarray, fat, or fibrosis to a varying degree disrupting or replacing parts of the specialized conduction fibers and the surrounding myocardium (see Fig. 19-3) with arteriolosclerosis (small vessel disease) of the ventricular septum. In general, the findings in the conduction system are accompanied by pathologic findings in the surrounding myocardium. Mononuclear cell infiltration in the approaches to the SA node and the SA node is present.[3]

Despite such pathologic findings in and around the conduction system, these victims were considered to be totally "asymptomatic," and sudden death was the first manifestation of the disorder. Clinically, in almost all, VF was the only common denominator that was observed by the paramedics at the time of resuscitation. It can be hypothesized that varying types of congenital or acquired pathologic findings in and around the conduction system may remain "silent," and these individuals are "asymptomatic" for long periods of time. However, during an altered physiologic or metabolic state, the anatomic or pathologic findings, or both, may trigger an arrhythmia that progresses to ventricular tachycardia, fibrillation, and sudden death.[3]

FAMILIAL SUDDEN DEATH

A genetic tendency for the development of an abnormal conduction system or the surrounding myocardium, or both, as such may also lead to ventricular tachycardia, fibrillation, and sudden death (Fig. 19-4).[5,6] Familial occurrence of ventricular arrhythmias may be related to the many genetic mutations associated with congenital long Q–T interval syndrome or Brugada syndrome. These disorders of ion channel function may also lead to cellular dysfunction and pathologic changes such as fatty infiltration, fibrosis, and disruption of the conduction system and adjoining myocardium.[7,8] These disorders have been individually discussed in other chapters and will be alluded to in subsequent sections of this chapter.

Epidemiology

VENTRICULAR FIBRILLATION AND SUDDEN CARDIAC DEATH

The epidemiology of VF intertwines with the available data on SCD (or cardiac arrest) and its documentation by emergency medical system personnel.[9] In 1998 an epidemiologic prospective study from northeast Italy reported an incidence of cardiac arrest of 95/100,000 persons per year.[10] In this study, VF accounted for 30.2% of the presenting rhythms, asystole in 48.3%, and pulseless electrical activity in 21.5%. A similar study of out-of-hospital cardiac arrest confirmed ventricular tachycardia or VF as the initial rhythm in 59 of 197 patients (30%) of a patient cohort from the Los Angeles area.[11] Although ventricular tachycardia and VF were consistently more likely to be associated with return of spontaneous circulation in these two studies, the time dependency of the initial recorded rhythm is debated. The Ontario Prehospital Advanced Life Support (OPALS) study quoted an annual incidence of 58 out-of-hospital cardiac arrests per 100,000 persons, and in the subgroup of emergency medical services (EMS)–witnessed arrests, ventricular tachycardia and VF accounted for 34.2% of cases.[12] However, a recent analysis from Sweden estimated that patients with an

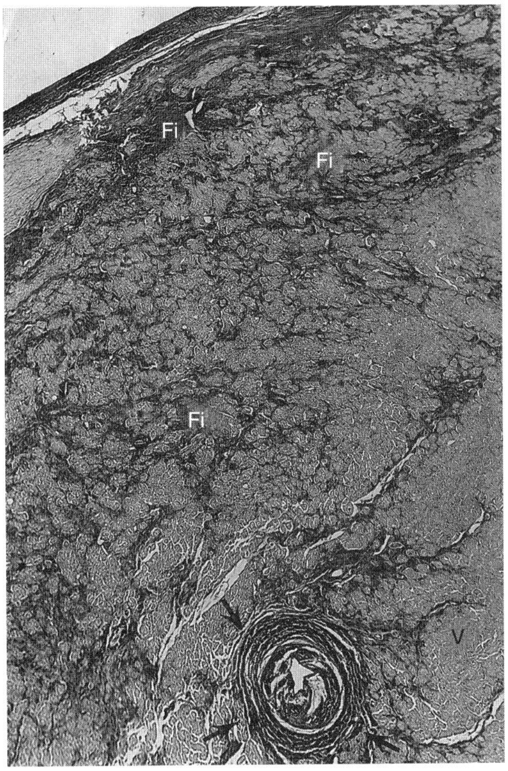

FIGURE 19-2 Chronic myocarditis in a 35-year-old male with polymorphic sustained drug-refractory ventricular tachycardia associated with multiple cardiac arrests, uncontrolled by β-blockers, antiarrhythmic drugs, and defibrillator insertion. The patient underwent heart transplantation. The explanted heart revealed myocarditis of a chronic or a smoldering type in the approaches to the atrioventricular (AV) node, the AV node, and beginning of the right bundle branch. The AV bundle revealed fibro-fatty change and was left sided. There was marked fibrosis of the left bundle branch, chronic epicarditis, arteriolosclerosis of the summit of the ventricular septum, and fibrosis on both sides of the septum. Photomicrograph of ventricular septum showing marked fibrosis and arteriolosclerosis of the septum. (Weigert-van Gieson stain ×45.) Fi, fibrosis in the myocardium; V, ventricular septum. *Arrows* point to arteriolosclerosis. (From Bharati S, Olshansky B, Lev M: Pathological study of an explanted heart due to intractable ventricular fibrillation. J Cardiovasc Electrophysiol 1992;3:437-41.)

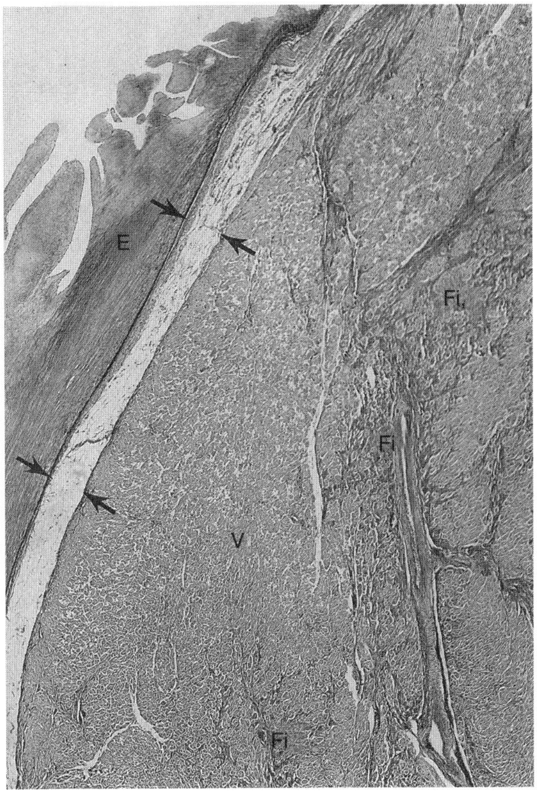

FIGURE 19-3 Conduction system and adjoining myocardial disease in a 25-year-old male who collapsed while playing basketball. He was found to be in ventricular fibrillation and could not be resuscitated. Photomicrograph of the left bundle branch showing falling out of cells. Note the endocardial thickening and numerous focal areas of fibrosis of the ventricular septum. (Weigert-van Gieson stain ×30.) E, endocardium; V, ventricular septum; Fi, fibrosis. *Arrows* point to the loss of left bundle branch fibers. (From Bharati S, Lev M: The Cardiac Conduction System in Unexplained Sudden Death. Mt. Kisco, NY, Futura Publishing, 1990, p 87.)

ECG taken within the first 10 minutes from a witnessed cardiac arrest have an incidence of VF of 50% to 60%. Linear regression further estimated that ECGs taken within the first 4 minutes should have an incidence of 75% to 80%.[13]

VENTRICULAR FIBRILLATION AND POPULATION CONSIDERATIONS

Despite the debate on the true proportion and incidence of VF in prehospital cardiac arrest, event rates range from 250,000 to 500,000 per year in the United States.[14] This absolute number is heavily influenced by population dynamics and the defined subpopulation being discussed. The well-documented "high-risk" subgroups (ischemic heart disease with low left ventricular ejection fraction [LVEF], complex ventricular ectopy, prior hospital admission for congestive heart failure, or previous cardiac arrest) contribute a *minority* of the total number of cardiac arrests, although the subgroup percent incidence per year is the highest (Fig. 19-5).

A screening tool or preventative intervention then would have to be needlessly applied to 999 out of 1000 persons to influence the 1 out of 1000 previously unidentified persons destined for a cardiac arrest. The Framingham study[15] demonstrated a large disparity from the highest to lowest decile (14-fold increase in risk) for SCD measuring the well-known risk factors such as age, family history, gender, tobacco use, hyperlipidemia, and hypertension. Electron beam computed tomography for the detection of coronary artery calcification has been correlated to coronary event risk.[16] A recent publication compared the Framingham risk index and coronary artery calcification in an autopsy series of SCD victims.[17] The two risk assessment methods had modest correlation with each other (63%). Of the cases, 83.5% had a coronary artery calcification score

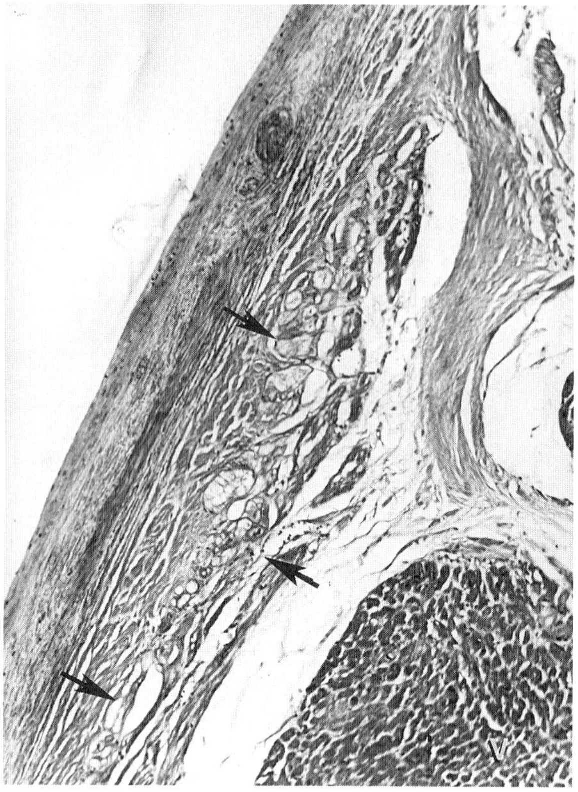

FIGURE 19-4 Cardiac conduction system histology in a 16-year-old girl with a familial history of alternating bidirectional tachycardia died suddenly. Fatal ventricular arrhythmias occurred in three generations, including two siblings and the mother. The conduction system revealed degenerative changes. Photomicrograph demonstrating vacuolar degeneration of left bundle branch in the 16-year-old. (Hematoxylin-eosin stain ×104.) *Arrows* point to vacuolar degeneration of main left bundle branch. (From Gault JH, Cantwell J, Lev M, Braunwald E: Fatal familial cardiac arrhythmias. Am J Cardiol 1972;29:548-53.)

or Framingham risk index above average for age. The remaining 17.5% highlight the need for exploration of new cardiac risk factors (perhaps fibrinogen, homocysteine, infectious agents, or ion channel abnormalities), which could be applied to the population at large.

The "high-risk" subpopulations have been targeted in recent publications for epidemiologic analysis. The Worcester Heart Attack Study described a relatively fixed incidence rate of 4.7% for VF in hospitalized patients with validated acute myocardial infarction (MI) over a 22-year period (1975-1997).[18] The Gruppo Italiano per lo Studio della Sopravvivenza nell'Infarto Miocardico (GISSI-2) database reported an incidence of early-onset VF (<4 hours after onset of acute MI) and late VF (>4 to 48 hours) of 3.1% and 0.6%, respectively. In-hospital prognosis was worse in patients with VF than without VF, but the postdischarge to 6-month mortality was similar for both VF subgroups and controls.[19] The post–cardiac arrest patient has a similar "time-dependence" of risk of recurrent events as was presented by Furukawa et al. (Fig. 19-6).[20]

Low LVEF was a strong predictor of the highest risk of recurrence (11% in the first 6 months). Persistent inducibility at electrophysiological study (EPS) was the better predictor for recurrent events from 6 to 42 months, but the subsequent risk fell to 0.8% over the final 6 months of the study period. In the patients who entered the Antiarrhythmics Versus Implantable Defibrillators (AVID) study with VF,[21] the survival curve shows the same early recurrence (<6 months) in the patients treated with antiarrhythmic drugs as those who did not receive this therapy (Fig. 19-7).

Clinical trials are assessing the "static" risk of cardiac events outside the setting of acute hospitalization.[22] The Multicenter Automatic Defibrillator Implantation Trial (MADIT-II) trial tested implantable cardioverter

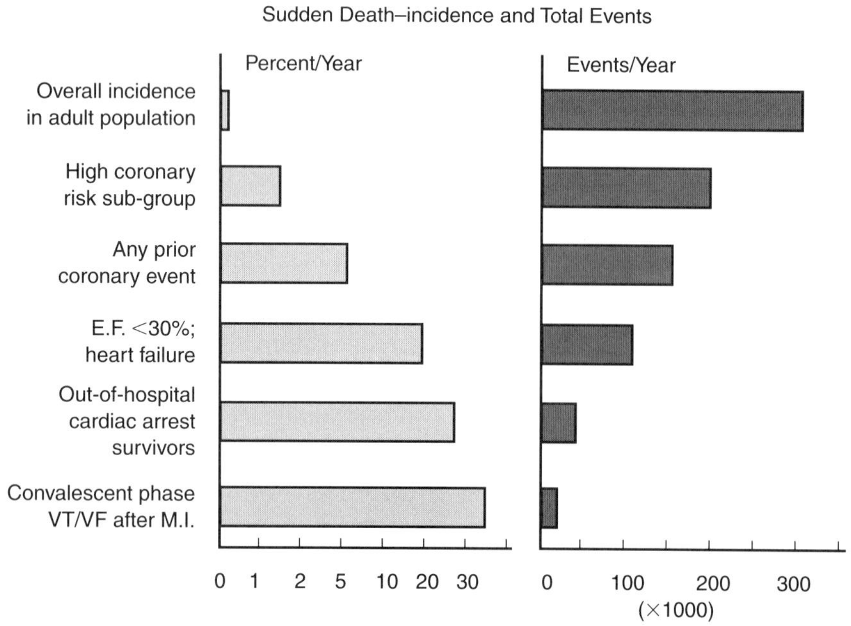

FIGURE 19-5 Risk of sudden cardiac death, with influence of preexisting heart disease on magnitude of risk in middle-aged and older adults. The prevalent etiologies are a function of age (see text). (Modified from Figure 1 in Myerburg RJ, Kessler K, Castellanos A: Sudden cardiac death: Structure, function and time-dependence of risk. Circulation 1992; 85[suppl I]:2-10.)

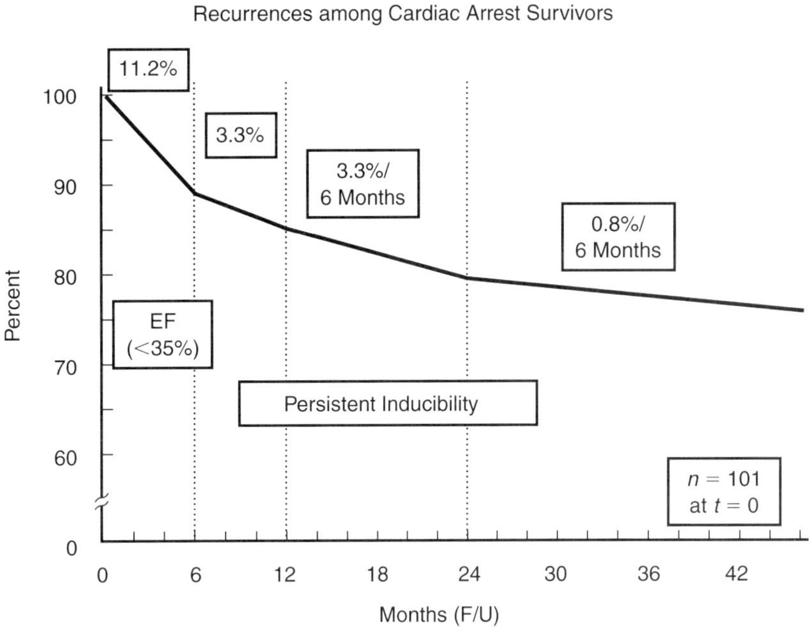

FIGURE 19-6 Risk of recurrent cardiac arrest among survivors of out-of-hospital cardiac arrest. The highest risk of recurrent cardiac arrest is within the first 12 to 24 months. In the earlier part of that period, ejection fraction is the most powerful predictor, whereas persistent inducibility of ventricular tachyarrhythmias during electrophysiological studies becomes an important predictor beyond that period of time. Nonetheless, ejection fraction remains a predictor throughout. (Modified from Figure 1 in Furukawa T, Rozanski JJ, Nogami A, et al: Time-dependent risk of and predictors for cardiac arrest recurrence in survivors of out-of-hospital cardiac arrest with chronic coronary artery disease. Circulation 1989;80:599-608.)

defibrillator (ICD) therapy versus standard medical therapy in patients with ischemic heart disease, LVEF less than 30%, and previous MI. It showed a 31% relative risk reduction with ICD therapy. The Sudden Cardiac Death in Heart Failure Trial (SCD-HeFT) trial compared three arms of therapy in patients with congestive heart failure and LVEF less than 30%—best conventional heart failure therapy, with placebo best conventional heart failure therapy with amiodarone, and best conventional heart failure therapy with ICD.[22] In a preliminary report, there was a 23% relative reduction in all cause mortality with the ICD compared with placebo (from 7.2% to 5.8%/year), and no reduction in mortality with amiodarone compared with placebo. The most recent developments in assessing risk of cardiac arrest have been in the blossoming technology of molecular genetics. The familial long QT syndromes (LQTSs) and their respective genetic bases have been well described elsewhere. An increasingly recognized syndrome of right bundle block, ST elevation, and aborted sudden cardiac death (Brugada syndrome) has also been ascribed to a genetic mutation in the gene encoding the cardiac sodium channel. Its prevalence in a European study by ECG screening was estimated at 0.1% of the general population.[23] If the only effective therapy for the syndrome is ICD implantation, the economic impact on the public health budget is obvious.

The epidemiology of VF and SCD are closely aligned. From a public health perspective, the largest number of cardiac arrest survivors will have the well-described cardiac risk factors from the Framingham study (e.g., hypertension, hyperlipidemia). An intervention for the population as a whole, however, will require treatment of a vast majority to prevent a single cardiac arrest. The "high-risk" subset pre–cardiac arrest (although a minority of the total number) is currently undergoing research investigating prophylactic ICD implantation and the cost effectiveness of such a practice. New screening tools that can be widely applicable, economical, and cost effective are needed for future progress in SCD prevention.

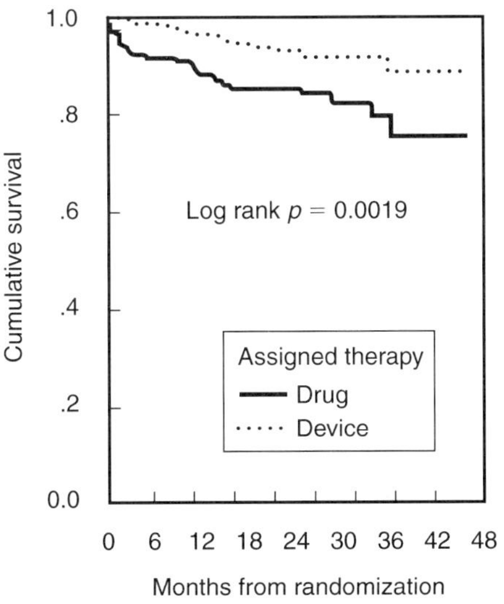

FIGURE 19-7 Survival free of arrhythmic cardiac death in patients with ventricular fibrillation who qualified for the Antiarrhythmics versus Implantable Defibrillators (AVID) study. Nonarrhythmic cardiac and noncardiac deaths are censored. VF, ventricular fibrillation. (From the AVID Investigators: Causes of death in the AVID study. J Am Coll Cardiol 1999;34:1552-9.)

Basic Electrophysiology

During VF the myocardium fails to contract effectively and circulate blood. While certain species such as rats can spontaneously revert from VF to normal sinus rhythm, in humans VF is lethal if not treated promptly.

Thus, analysis of the electrophysiological mechanisms culminating in VF is critical to effective therapy.

There are three major mechanisms by which ventricular arrhythmias can develop. Abnormal impulses may be initiated by (1) "triggered activity" (early or delayed afterdepolarizations); (2) abnormal automaticity, which refers to the inherent ability of various regions of myocardium (other than the sinoatrial node) to act as pacemakers; and (3) reentry. Reentrant excitation is believed to be the primary, and most clinically relevant, mechanism for VF and is therefore the focus of this chapter.

DYNAMICS OF REENTRANT VENTRICULAR FIBRILLATION

VF is generated by self-sustained reentrant circuits that turn around regions of conduction block. In some circumstances, the zone of conduction block can be anatomic in nature, such as the site of a healed infarct or a region of fibrosis. In such cases, the reentrant circuit is anchored to the site of block and results in a stable rotor and monomorphic ventricular tachycardia. When the zone of conduction block is functional in nature, the region of refractoriness is transient both spatially and temporally. Reentrant circuits that form around such zones can therefore meander freely or change in size, providing an important mechanism for VF. However, the nature of reentrant waves and the mechanisms by which they lead to VF are still not entirely clear.

THE NATURE OF FIBRILLATORY WAVEFRONTS

Reentrant excitation is the fundamental mechanism responsible for VF. The precise nature of these wavefronts, their formation, and sustenance are key to our understanding of VF. Various theories exist regarding the nature of reentrant waves. One such theory is the "leading circle" hypothesis, originally put forward by Allessie and coworkers to explain atrial fibrillation (AF).[24] This hypothesis states that a reentrant circuit is the smallest pathway in which the impulse can continue to circulate, and that the core is kept permanently refractory (i.e., there is no excitable gap) (Fig. 19-8). However, Janse[25] questioned whether the core is truly refractory during reentry, since membrane potential is relatively normal in this region and diastolic intervals are relatively long. Furthermore, the presence of an excitable gap allows for wavefronts (spontaneous or induced) to invade the area and terminate or entrain the reentrant circuit, and Allessie et al.[24] were able to terminate tachycardia with premature beats (i.e., antitachycardia pacing). It now appears that in most cases of VF an excitable gap does exist; however, there may still be room for leading circle reentry in a subset of arrhythmic phenomena, particularly regarding AF.[26]

An alternative mechanism for VF is spiral wave reentry. This theory suggests that the wavefront curves, or forms a spiral, because curvature is negatively related to conduction velocity (due to the source-sink relationship). As a result, the wave is highly curved near the core and

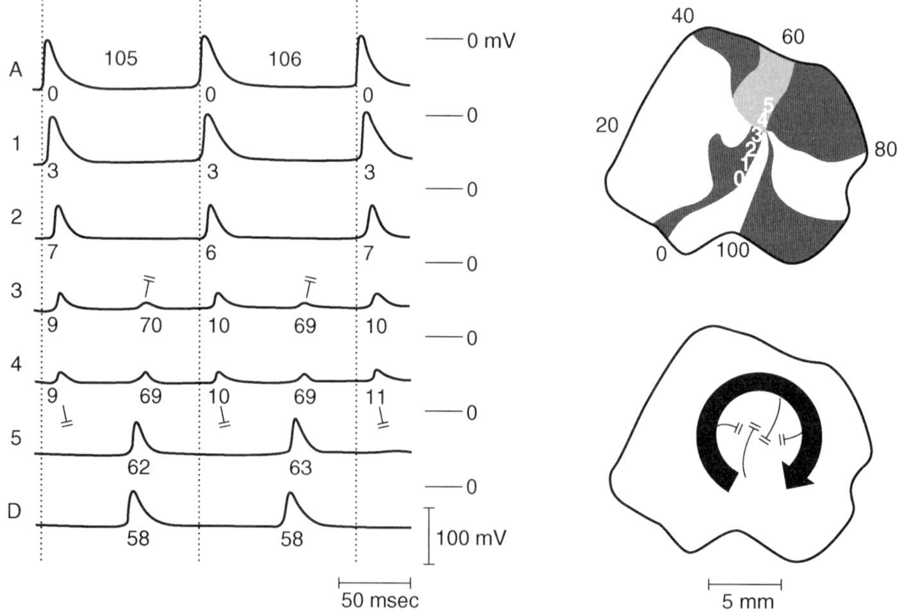

FIGURE 19-8 Leading circle reentry. Activation map **(right)** and action potential recordings **(left)** obtained during steady-state tachycardia in an isolated rabbit atrial preparation. Cells in the central area of the reentrant circuit show double potentials of low amplitude (*tracings 3 and 4*). A schematic activation pattern is shown on the lower right. *Double bars* indicate conduction block; *the black part* is absolutely refractory; *between the head and tail of the reentrant wavefront* there is relatively refractory tissue. Characteristics for leading circle reentry are (1) the core is kept in a permanent state of refractoriness by centripetal wavelets, and (2) the head of the leading circle bites into its own relative refractory tail. (Reproduced from Allessie MA, Bonke FI, Schopman FJ: Circ Res 1977;41:9-18.)

moves slowly, but at the distal end wave speed increases, resulting in a spiral shape. While two-dimensional spiral waves may theoretically exist in surviving two-dimensional layers overlying healed myocardial infarcts, three-dimensional waves (i.e., "scroll waves") are clearly more relevant to cardiac arrhythmias. A scroll wave can be thought of as a stack of spiral waves where the cores line up to form a "filament" at the center. Simple scroll waves, where the filament forms a straight line, have been demonstrated during VF.[27] A scroll wave with a core that is perpendicular to the recording surface will appear as a two-dimensional spiral wave, whereas a filament that is parallel to the recording surface will appear as a plane wave. It has been suggested that during VF, scroll waves are usually oriented parallel to the surface, which may explain why mapping studies sometimes fail to detect reentry during VT or VF. A scroll wave with a ring-shaped filament is called a scroll ring. It is interesting to note that the cross section of such a wave would represent classic "figure-of-eight" reentry. There is only indirect experimental evidence for the existence of scroll rings, likely because of the difficulty of measuring them, as they are never manifest on the surface, only transmurally.[28] The shape of a scroll as it is initiated is strictly dependent on the shape of the inexcitable obstacle encountered by the wavefront; thus, filaments of varying shape can readily develop in myocardium. While scroll rings ultimately collapse upon themselves, simple scroll waves can be stabilized and maintained.[29]

There are many hypotheses regarding the nature of fibrillatory waves that underlie VF. Three of these theories are illustrated in Figure 19-8: the multiple wavelet hypothesis, the meandering spiral wave hypothesis, and the mother-daughter wavelet hypothesis. The "multiple wavelet hypothesis" was put forward in the early 1960s by Moe[30] and describes fibrillation as consisting of multiple, nonstationary wavefronts that are continuously forming, fractionating, and reforming (Fig. 19-9*A*). Spatial dispersion of repolarization was an important condition for initiating multiple wavelet VF. Others have suggested that VF is due to spiral reentrant waves that meander around the myocardium (see Fig. 19-9*B*).[31] According to this theory, the chaotic appearance of the ECG in VF is attributable to meandering of a single spiral wave rotor throughout ventricular myocardium. Jalife[32] suggested that the mechanism of fibrillation may consist of a stable, periodic reentrant wave that gives off "daughter wavelets," which meander and fractionate (see Fig. 19-9*C*). According to this hypothesis, unstable daughter wavelets form due to local gradients of refractoriness and give rise to the fibrillatory patterns of the ECG.

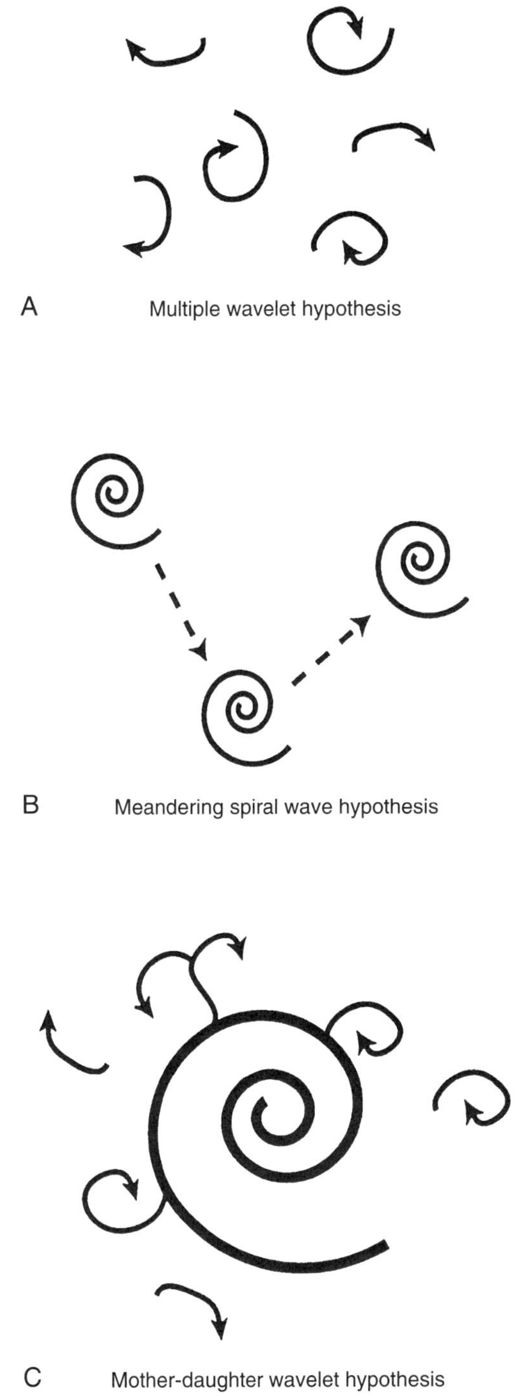

FIGURE 19-9 Schematic illustration of various theories of fibrillatory wavefront behavior. **A,** Multiple wavelet hypothesis. **B,** Meandering spiral wave hypothesis. **C,** Mother-daughter wavelet hypothesis.

SUBSTRATES FOR VENTRICULAR FIBRILLATION

Heterogeneities of Repolarization and Refractoriness

Various anatomic and functional substrates can lead to development of the fibrillatory dynamics described earlier. Early studies of AF[30] suggested that heterogeneity of refractoriness was a necessary substrate for fibrillation in the heart. While it was later shown that VF can occur in hearts without large dispersions of refractoriness (i.e., in normal hearts),[33] or in modeling studies using simulations of electrically homogenous tissue, these studies may not be clinically relevant. It is more difficult to initiate VF in a normal heart, and the vast majority of fibrillatory episodes occur in diseased hearts where gradients of repolarization kinetics (and therefore refractoriness) are almost certainly abnormal. The important role of repolarization in the development of reentrant substrates has recently been emphasized by studies demonstrating the inherent heterogeneities in repolarization properties that exist transmurally.[34] Antzelevitch[35] and others[36] have demonstrated that transmural heterogeneities of repolarization are critical in the development of substrates for reentry under various drug- or disease-induced conditions.

Rosenbaum et al. recently developed a hypothesis that explains how critical heterogeneities of repolarization that provide a substrate for reentry may occur.[37] Beat-by-beat alternation of cardiac action potential duration (APD) has been shown to underlie T-wave alternans, an electrocardiographic indicator that correlates well with risk of SCD.[37] We have demonstrated that when alternans occurs in various regions of myocardium in a "discordant" manner (i.e., some regions are on a long-short-long cycle and others on a short-long-short cycle), steep gradients of repolarization develop and reverse direction on a beat-by-beat basis, which can lead to conduction block, reentry, and VF.[38]

In related work, Laurita et al.[39] demonstrated that premature beats modulate dispersion of repolarization in a manner that has direct effects on vulnerability to VF. As the S_1-S_2 interval is shortened, gradients of repolarization decrease concomitant to a decrease in vulnerability to VF, but then increase with a concomitant increase in VF vulnerability as S_1-S_2 is shortened further (Fig. 19-10). Such biphasic modulation of dispersion relates back to the concept of the "vulnerable period," which arose from critical studies by Moe[30] describing the window of time during the ECG T wave when vulnerability to VF is greatest. Thus, dispersion of repolarization and refractoriness appears to play a major role in the mechanism of reentrant VF.

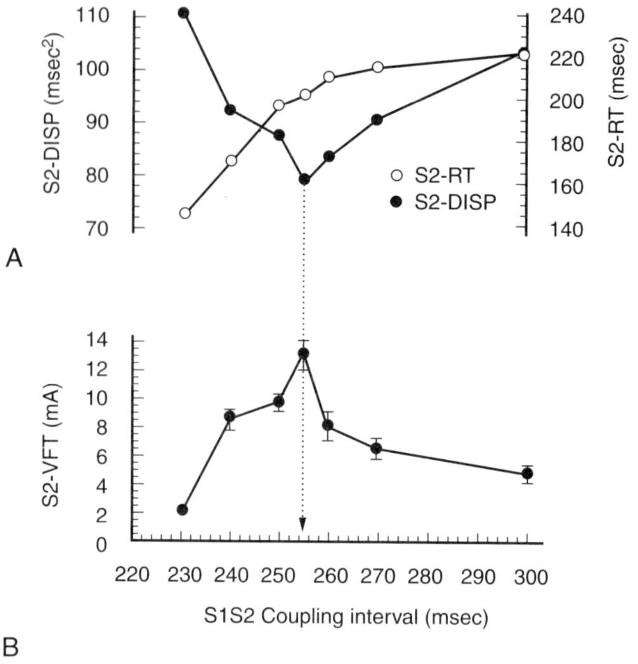

FIGURE 19-10 Modulated dispersion hypothesis and vulnerability to ventricular fibrillation. **A,** Dependence of mean repolarization (*S_2-RT, open circles*) and dispersion of repolarization (*S_2-DISP, filled circles*) on S_1S_2 coupling interval. **B,** Dependence of arrhythmia vulnerability (*S_2-VFT*) on premature coupling interval. Dispersion of repolarization (**A,** *filled circles*) and vulnerability to fibrillation **(B)** were modulated in a similar biphasic fashion, with minimum vulnerability (i.e., maximum S_2-VFT) and minimum dispersion occurring at the same S_1S_2 coupling interval (255 ms, *dashed arrow*). This inherent mechanism may provide protection against premature stimuli delivered at moderate coupling intervals. (From Laurita KR, Girouard SD, Akar FG, et al: Modulated dispersion explains changes in arrhythmia vulnerability during premature stimulation of the heart. Circulation 1998;98:2774-80.)

THE RESTITUTION HYPOTHESIS

Restitution is a property of cardiac myocytes that dictates the APDs of a premature action potential following a period of steady-state rhythm. The APD of the extrasystolic beat is determined not only by the S_1-S_2 interval but also by the APD of the pre–extrasystolic beat. Furthermore, gradients of restitution kinetics are known to exist across the epicardial surface[40] such that cells in different regions of the heart having equal baseline APDs could have different extrasystolic APDs following the same S_1-S_2 interval. In theory, if gradients of restitution kinetics are steep enough, underlying heterogeneities of repolarization would not be necessary to create dispersions of repolarization following a premature beat.[40] However, cardiac tissue is known to exhibit repolarization heterogeneities across epicardial and transmural surfaces, so the role of restitution may be to enhance existing dispersion of repolarization and promote VF.

One possible mechanism to describe the role of restitution in VF is the "restitution hypothesis," which states that when the restitution curve (a plot of APD versus diastolic interval) has a slope >1, VF may be initiated. For example, a wavefront encroaching on the tail end of a previous wave creates a gradient of diastolic interval (DI) between the two (assuming the degree of curvature is not identical), and in regions of tissue where restitution is steep, this will create large dispersions of APD along the second wavefront. Some areas of the wave may have such a short DI that they fail to propagate, and the wavefront fractionates as a result. This was demonstrated by Garfinkel et al.,[41] who also used pharmacologic agents to flatten restitution curve slope and decrease vulnerability to VF. Koller et al.[42] showed that elevating extracellular potassium decreases the portion of the restitution curve where slope is greater than or equal to 1 and converts VF to a periodic rhythm. Other investigators have shown that the biphasic nature of a restitution curve, demonstrated in a number of cardiac tissue types, is also an important determinant of vulnerability to VF.[43] More recently, the restitution hypothesis has been applied to restitution curves for both APD and conduction velocity.[44]

MYOCARDIAL ISCHEMIA

VF is commonly caused by acute myocardial ischemia. Under experimental conditions it is often difficult to provoke VF in normal tissue using premature stimuli; however, in ischemic tissue, premature beats exacerbate dispersion of refractoriness and readily lead to tachyarrhythmias.[45] This is due to the various electrical alterations that occur during ischemia and the consequent formation of various substrates that promote reentry and VF.

At the cellular level, acute ischemia results in depolarization of resting membrane potential, decreased maximum rise rate and amplitude of the action potential, and decreased APD,[46-48] as well as reduced intercellular coupling.[49-51] Depolarization of the membrane is attributed largely to accumulation of extracellular potassium,[52] as is postrepolarization refractoriness.[53] Acidosis can produce a small depolarization of resting membrane potential[54] and also decreases gap junction conductance and cell-to-cell coupling. The net result of the electrical changes that occur during ischemia is the formation of various substrates that can lead to reentry and wavebreak, such as slowed conduction velocity, increased dispersion of refractoriness, and alternans.

HEALED MYOCARDIAL INFARCTS

The process of tissue healing and scar formation following an episode of acute myocardial ischemia involves necrosis of the infarcted region, as well as swelling and hypertrophy of the noninfarcted region as it attempts to compensate for the loss of cardiac muscle.[55] Heterogeneity of repolarization may result from uneven prolongation of APD in hypertrophied post-MI ventricle[56] or changes in expression of gap junction proteins,[57] or both. A healed infarct may also provide an anatomic substrate for arrhythmias. Infarcts vary from simple surviving muscle bundles that form accessory pathways to complex subendocardial sheetlike structures, which are linked to the surrounding myocardium by multiple connecting bundles, creating complex matrices of conductive tissue that may promote multiple reentrant circuits. In particular, the combined effects of slowed conduction and the presence of structural anomalies is significant because, in tissue where excitability is low, a wavefront breaking past an obstacle will curl at the inner ends (where the break was) to create figure-of-eight reentry,[58] whereas in normally excitable medium the wave tends to reform on the other side of the obstacle.

AUTONOMIC MODULATION OF VENTRICULAR FIBRILLATION

It is well established that increases in sympathetic tone increase the risk for SCD, whereas vagal reflexes have the opposite effect.[59-61] Studies in dogs have shown that left cardiac sympathetic denervation, or left stellectomy, increases survival and decreases arrhythmias during acute MI.[62] In contrast, unilateral blockade of the right stellate ganglion by cooling increases the number of arrhythmias following coronary occlusion.[63] Many clinical trials have shown that β-adrenergic receptor blocking drugs have a potential benefit in preventing SCD in MI patients,[64-67] whereas anti–α-adrenergic agents have not been shown to be effective.[68]

It has been suggested that during acute MI, sympathetic nerves that traverse the myocardial wall are damaged, leading to denervation hypersensitivity.[68] This, in turn, leads to spatially inhomogeneous responses to β-adrenergic stimulation across the ventricle, forming potential substrates for reentrant excitation. Further investigation is required to fully elucidate the mechanistic role of β-adrenergic stimulation and arrhythmogenesis.

GENETIC SUBSTRATES FOR VENTRICULAR FIBRILLATION

The Brugada syndrome is a cardiac disease characterized by an elevated ST segment that is unrelated to ischemia, Q–T interval prolongation, electrolyte abnormalities, or structural heart disease.[69] Patients with Brugada syndrome are at increased risk of SCD due to VF. The syndrome is caused by a mutation in the gene encoding the cardiac sodium channel (*SCN5A*)[70] and is common in Southeastern Asia, where it is the second highest cause of death among young males. In Europe the sex-specificity is less distinct, although affected women tend to be less symptomatic than affected males. Brugada patients often die in their sleep, and a possible relationship to sudden infant death syndrome (SIDS) has been suggested.[71]

It is believed that the mechanism of VF in Brugada patients is related to reduced function of the mutant late inward sodium current, forcing the balance of plateau currents in favor of repolarization. This is why any further impairment of sodium channel function (e.g., by sodium blocking drugs) can exacerbate the electrocardiographic and arrhythmogenic phenotype in this disorder.[72] Since epicardial, and not endocardial, cells possess the transient outward current (I_{to}), they are most susceptible to the repolarizing effect of the sodium channel mutation in Brugada syndrome. Marked and selective shortening of epicardial relative to endocardial action potentials produces a transmural voltage gradient during the plateau, which, in turn, is thought to account for the characteristic pattern of ST elevation seen in these patients.[73] Transmural action potential gradients also account for the presumed mechanism of VF (i.e., phase 2 reentry).

In summary, VF is a complex arrhythmia associated with many distinct electrophysiological mechanisms that are no doubt highly dependent on the disease state in question. Although most often associated with acute ischemia or healed MI, VF can occur in the absence of any structural heart disease such as in heritable disorders like Brugada syndrome. Greater understanding of these various mechanisms is required to guide development of novel pharmacologic approaches and targets that can be used ultimately in the treatment and prevention of VF. Until then, we will have to rely on electrical defibrillation, which is highly effective irrespective of the underlying VF mechanism.

Clinical Electrocardiography

The role of electrocardiography has been increasing in patients presenting with VF. ECG features of high-risk patients or transient events that can result in VF are clinically important and detected by ECG. VT may precede VF, and this arrhythmia is discussed in Chapter 17, Sustained Ventricular Tachycardia with Heart Disease.

ELECTROCARDIOGRAPHIC FEATURES OF PATIENTS AT HIGH RISK FOR VENTRICULAR FIBRILLATION

Electrocardiographic recordings obtained at the onset of VF provide insight regarding events that precipitate sudden death. Documentation of these events has been obtained by continuous monitoring in hospital telemetry units and by ambulatory monitoring. The patients involved in these studies had a high incidence of coronary artery disease, and most had frequent or complex ventricular ectopy. The terminal events were associated with sinus arrest, complete heart block, or ventricular asystole in about 10% of patients, while in 90% VF was preceded by VT or ventricular flutter of variable duration.[74-77] ST segment and T wave changes indicative of ischemia related to acute MI or coronary spasm have also been reported to precede terminal ventricular arrhythmias.[74-77] Other studies indicate that bradycardia and electromechanical dissociation are important causes of sudden death in patients with advanced heart failure and nonischemic cardiomyopathy.[78] These observations have significant implications for strategies to reduce mortality in patients at risk for sudden death.

The diagnostic role of the ECG for recognition of patients who have genetic disorders associated with sudden death from ventricular arrhythmias has continued to evolve over the past decade. The electrocardiographic characteristics of LQTS, Brugada syndrome, and arrhythmogenic right ventricular cardiomyopathy have gained widespread attention, which is vital to the prevention of sudden death in young patients with these disorders, particularly if they participate in athletics.

LQTS. The electrocardiographic manifestations of LQTS include QT prolongation, abnormalities in T-wave morphology, increases in QT dispersion, T-wave alternans, and a relative degree of bradycardia in children. (Fig 19-11). The upper limits of normal for the QTc values are 460 to 470 milliseconds (ms) for females and 440 to 460 ms for males.[79-80] Longer QT values may be observed in normal women after puberty.[81] The degree of QT prolongation does not correspond directly with the risk of syncope, but malignant ventricular arrhythmias are more frequent when the QTc exceeds 600 ms.[81] A diagnostic dilemma is that the Q–T interval shows temporal variation in patients with this syndrome, and the QTc may fall within the normal range on a random

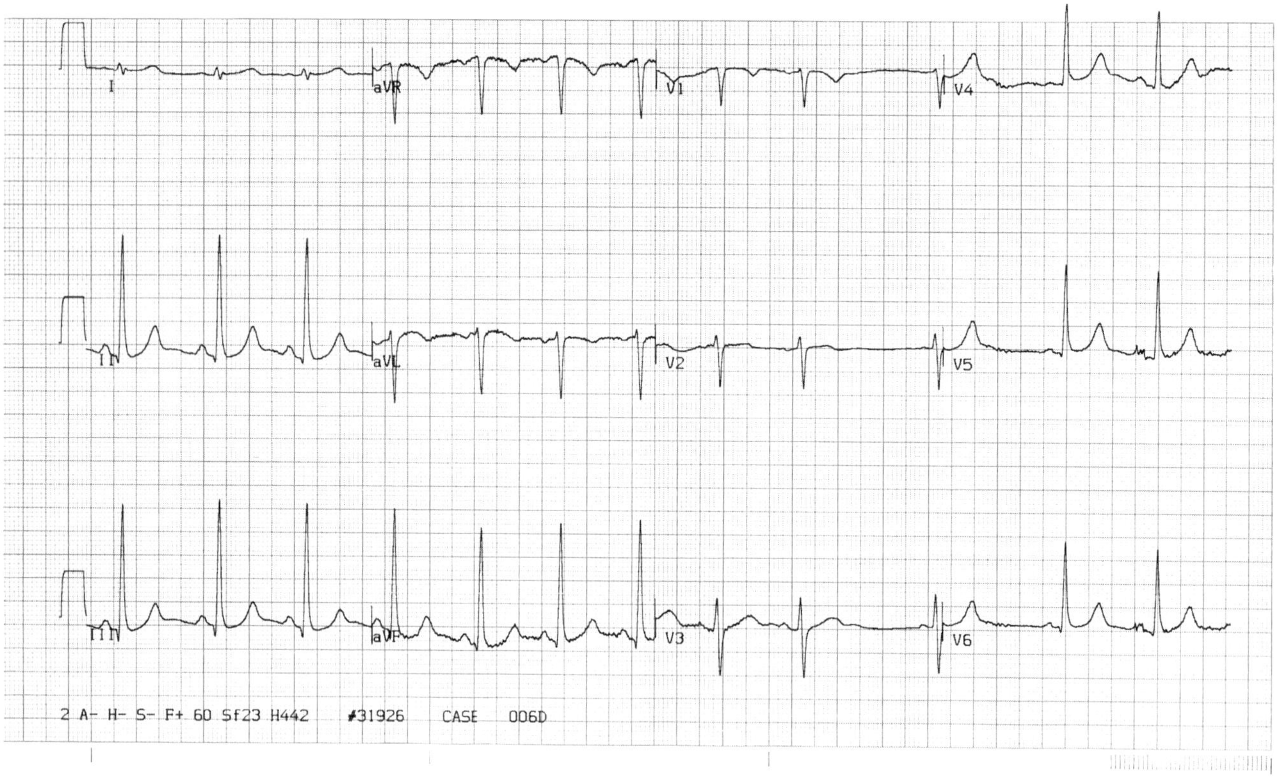

FIGURE 19-11 Twelve-lead ECG of a patient with familial long QT syndrome. Note the prolonged Q–T interval and T wave abnormalities.

recording.[82] Garson[83] reported that 6% of LQTS patients had a normal Q–T interval, and data from the International Registry showed that 10% of family members with a QTc less than 440 ms had a cardiac arrest.[84] Thus, an ECG with a normal QTc does not exclude the diagnosis if there is strong suspicion that a patient has the syndrome, especially if the QTc is on the border of normality. The effect of exercise increases the QTc in patients with LQTS, but this effect is less apparent in LQT3 patients than other genotypes.[85] Approximately 62% of patients with LQTS exhibit T waves that are biphasic or notched, and there is a higher incidence of these abnormalities in patients with cardiac events.[84] The characteristic features are most pronounced in precordial leads V2 to V5. The appearance of notched T waves may be provoked by exercise. The degree of QT dispersion is measured by the difference between the longest and shortest Q–T interval on the 12-lead ECG and is prolonged in patients with LQTS.[87] It is thought to represent increased dispersion of repolarization. Patients who show no change in the degree of QT dispersion when they are treated with β-blockers appear to be at increased risk for cardiac events.[84] T-wave alternans is a beat-to-beat alternation in the amplitude or polarity of the T wave. It appears to be a marker of electrical instability that may precede torsades de pointes.[88] Children with LQTS often have resting heart rates that are lower than normal and may exhibit a blunted chronotropic response to exercise.[81,89] Torsades de pointes, which is the ventricular arrhythmia associated with LQTS, is characterized by the undulating amplitude of the QRS complex, which gives the appearance of twisting about its axis. The onset is frequently associated with pause-dependent ventricular ectopy that falls on the T wave.[90-92]

Brugada Syndrome. The ECG patterns associated with the Brugada syndrome are (1) a terminal R′ in V1 with complete or incomplete right bundle branch block; (2) convex downward (coved) ST segment elevation equal to 0.1 mV in lead V1 or leads V1 and V2; convex upward (saddle-shaped) ST elevation equal to 0.1 mV; (3) J-point elevation followed by downsloping ST segment ending in a negative deflection (triangular shape).[92] Serial ECGs performed on the same patient may show variation from one pattern of ST segment elevation to another, normalization, and progressive development of right bundle branch block. Figure 19-12*A* through *C* shows examples of variable Brugada ECG patterns that occurred during an ajmaline test in the same patient. The prevalence of the Brugada ECG pattern is reported to be 0.07% to 0.7%, and there is a male predominance that is especially marked in Asians.[95-100] The range in prevalence appears to depend on the criteria that are used to make the diagnosis. The saddle-back ST segment elevation is more common. The typical coved pattern was found in 0.1% to 0.26% of community-based populations in Japan and Europe.[98-100] In a Japanese study population that underwent ECGs during health examinations, the prevalence of all types of Brugada ECG patterns was 0.7%.[97] The coved-type ST segment elevation was found in 38% of subjects with the Brugada pattern, and saddle-back type ST segment elevation was seen in 62%. In the same study, the rsR′ pattern in V1 was observed in 41%, and the Rsr′ pattern was recorded in 59%. The prevalence of the coved pattern was 0.26%, and "typical" Brugada ECG pattern with coved ST segment elevation and the rsR′ pattern in V1 was 0.12%. If only male subjects were considered, the criteria for a Brugada ECG pattern was met in 2.14% of the population, but the typical pattern in males was 0.38%.[98] In a European population studied by Hermida,[97] the prevalence of the ST segment elevation was 6.1%; however, only 1 (0.1%) of the 61 subjects who met their criteria for the Brugada pattern had the coved pattern. All of the others had the saddle-back pattern.

The prognostic significance of the Brugada ECG pattern is difficult to assess. Brugada reported an 8% incidence of arrhythmic events in an asymptomatic hospital-based population.[100] The degree and type of ST segment elevation requires further study for risk stratification of asymptomatic individuals in community-based populations. Matsuo[95] evaluated the mortality in patients younger than 50 years old in 1958 who had ECGs recorded during biannual examinations through 1999. A total of 32 patients were identified with the Brugada ECG pattern. Seven of these patients died suddenly or died of an unexplained accident. Although total mortality was not increased in patients with the Brugada ECG pattern, the mortality from unexpected death was significantly higher. No increase in mortality was observed in studies by Miyasaka,[98] Takenaka,[99] or Priori.[101] In the Osaka population, there was one sudden death among the 98 subjects with the Brugada ECG pattern during a mean follow-up of 2.6 years.[98] A 3-year follow-up reported by Atarashi[102] of patients with a Brugada ECG pattern found cardiac event–free rates of 67.6% in symptomatic patients and 93.4% in an asymptomatic group. A coved-typed ST elevation appeared to be related to cardiac events. The higher incidence in hospital-based studies may reflect referral patterns based on a family history of sudden death. Differences in criteria or ECG interpretation affect the diagnosis of this pattern, and the follow-up in most studies is too short to draw definitive conclusions about the long-term prognosis.

Arrhythmogenic Right Ventricular Cardiomyopathy (ARVC). ECG recordings during sinus rhythm in patients with ARVC have several distinctive features (Fig. 19-13).[103] The QRS may be prolonged in the right precordial leads to a greater extent than in leads I or V6. The QRS is often greater than 110 ms in V1 (sensitivity 55%) and a pattern of incomplete right bundle branch block is observed.[104] In 30% of cases the delay in conduction over the right ventricle results in a small potential at the terminal portion of the QRS in V1 that has been termed an *epsilon* (ε) *wave*,[103] which can be amplified with bipolar recordings over the inferior and superior aspects of the sternum. This can be achieved by repositioning the left arm lead over the xiphoid process, positioning the right arm lead over the manubrium sternum, and applying the left leg lead

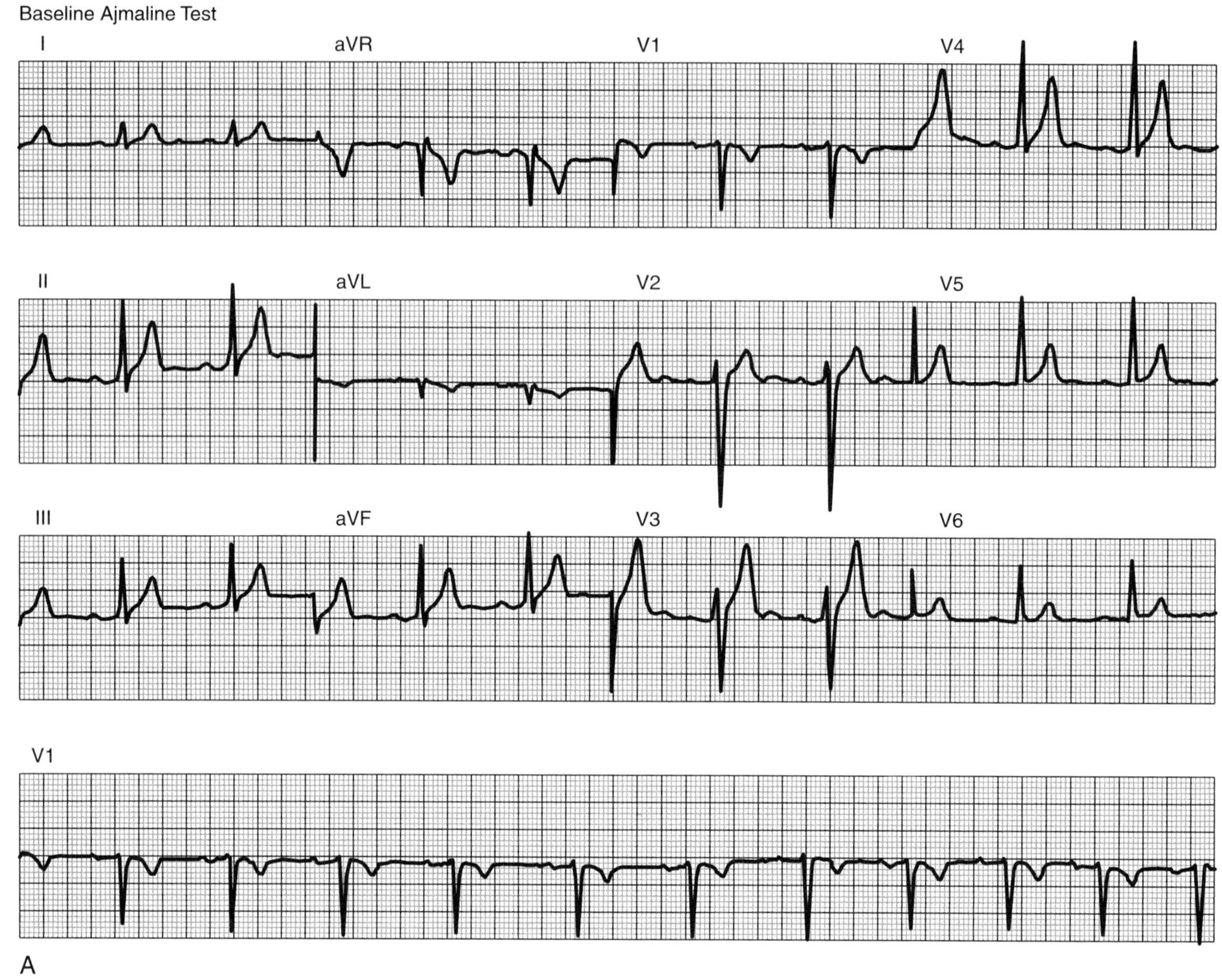

FIGURE 19-12 Serial changes in a Brugada pattern ECG recorded in the same patient during an Ajmaline test. Twelve-lead ECGs are recorded at baseline **(A)**, followed by Ajmaline infusion. At 2 minutes, ST segment elevation that is coving upward is seen in the right precordial leads **(B)**, and this is followed by marked ST segment elevation in V1 to V3 **(C)**.

(Continued)

at the position that is customarily used for V4 or V5.[105] The other major feature of ECGs recorded from patients with ARVC is inversion of the T waves in the precordial leads, which is observed in 42% to 54% of patients.[105,106] Metzger[107] assessed the value of serial 12-lead ECGs in 20 patients to recognize progression of ARVC over a mean of 71 plus or minus 48 months. Abnormalities were detected in 90% of the patients. The most frequent abnormality was T-wave inversion in the precordial leads. No correlation was demonstrated between the ECG and the extent of disease detected by echocardiography. In the 14 patients who had several ECGs recorded over time, there was no clear progression of electrocardiographic abnormalities.

The ventricular arrhythmias associated with ARVC typically show a morphology resembling left bundle branch block in V1 with a variable axis. Nava[106] published a clinical profile and long-term follow-up of 37 families with ARVC that demonstrated a correlation between the severity of echocardiographic findings and the severity of ventricular arrhythmias, which were seen in all patients with the severe form of the disease, 82% with moderate disease, and 23% in those with mild disease. Overall, 60 (45%) of 132 affected living members had ventricular arrhythmias. These included VF in 1, sustained VT in 14, nonsustained VT in 8, ventricular couplets and triplets in 16, and frequent ventricular premature depolarizations in 8. Exercise-induced polymorphic VT was observed in 13 patients with no other documented arrhythmias. Only one patient who was judged to have mild disease had a sustained ventricular arrhythmia. Although the data from this study shows a low incidence of VF, the incidence may be higher. In 19 of the 37 families, the proband died at a young age, and the diagnosis was made at autopsy. One may speculate that some of these subjects had VF. Figure 19-14 shows electrograms recorded from an ICD implanted in a teenage boy with ARVC and frequent nonsustained ventricular arrhythmias. The recording shows the sudden onset and successful termination of VF, which occurred at night while he was asleep. He had no prior history of syncope or sustained ventricular arrhythmias.

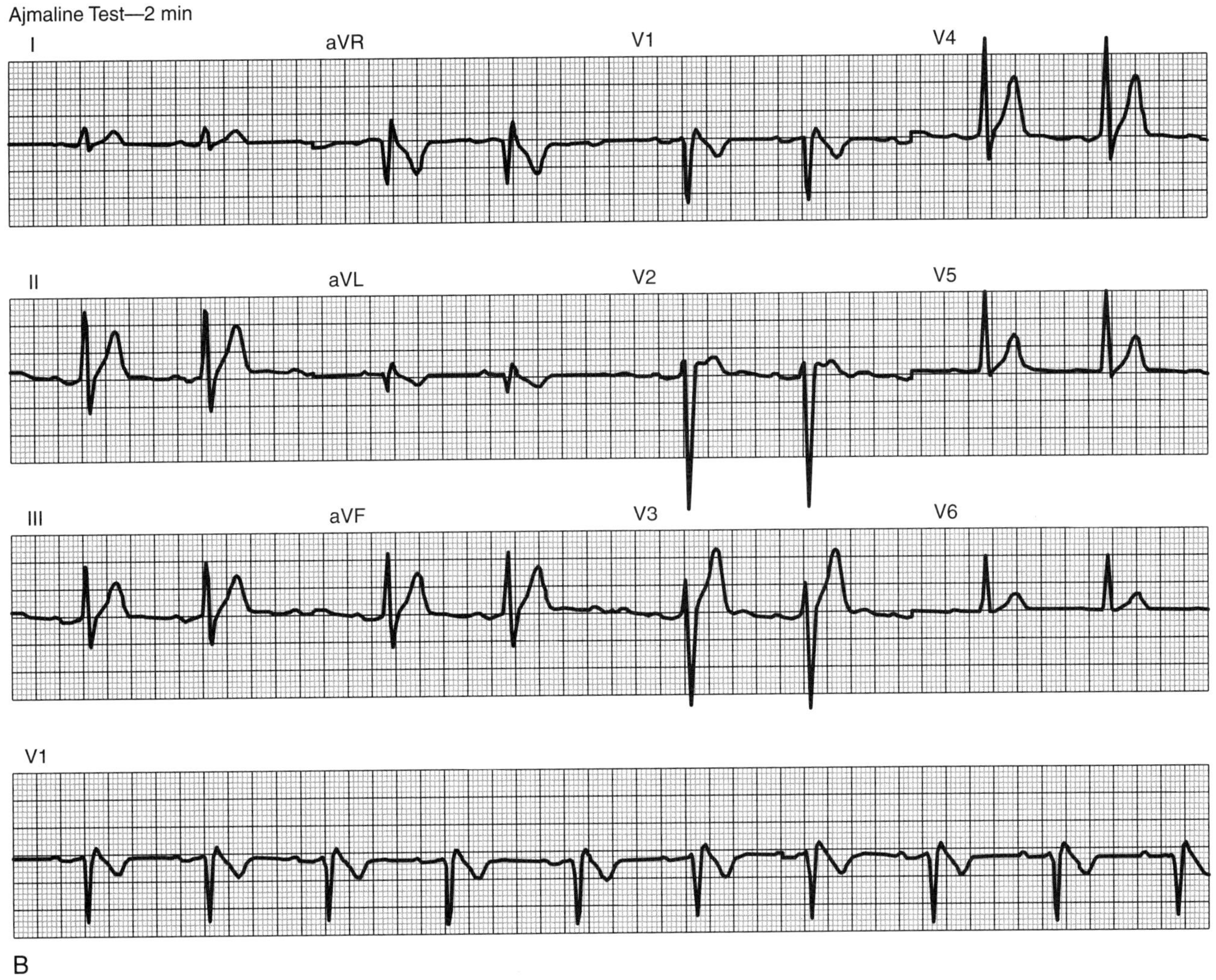

Figure 19-12 cont'd

(Continued)

Diagnostic Evaluation

Investigation of the patient who has been resuscitated from an episode of spontaneous VF is directed toward determination if the episode of VF was due to a transient or reversible cause, identification of the type and extent of underlying structural heart disease, documentation of the mechanism of VF, identification of coexisting disease states that may interact with future antiarrhythmic therapies, and assessment/monitoring of selected antiarrhythmic therapy. In each instance, the importance of a complete history and physical examination is well established and needs no further discussion here.

EVALUATION OF TRANSIENT OR REVERSIBLE CAUSES

VF that occurs secondary to a reversible or transient cause may be adequately treated by correction of the reversible cause or by short-term therapy or close observation while awaiting spontaneous resolution of the transient cause.[108] The causes of VF with these characteristics are usually readily identified with a few focused investigations. Electrolyte abnormalities are identified with serum electrolyte testing performed as soon as possible after resuscitation. The most common electrolyte abnormalities leading to VF are hypokalemia or hypomagnesemia, or both. When considering the temptation to wholly ascribe an episode of VF to hypokalemia or hypomagnesemia, or both, one must recall that the adrenergic discharge state during resuscitated VF results in redistribution of extracellular potassium and magnesium into the intracellular compartment. Accordingly, relative hypokalemia or hypomagnesemia, or both, is very common after VF.[109] Only marked hypokalemia or hypomagnesemia, or both, and confidence that future episodes can be prevented should prompt the belief that the episode of VF was due to a reversible cause.

The early performance of a 12-lead ECG and serum markers of myocardial necrosis (CKMB, Troponin) will permit identification of the patient whose VF has occurred in the acute phase (first 48 hours) of an MI. Evidence that acute MI produces an environment that constitutes only a transient risk of VF is most convincing

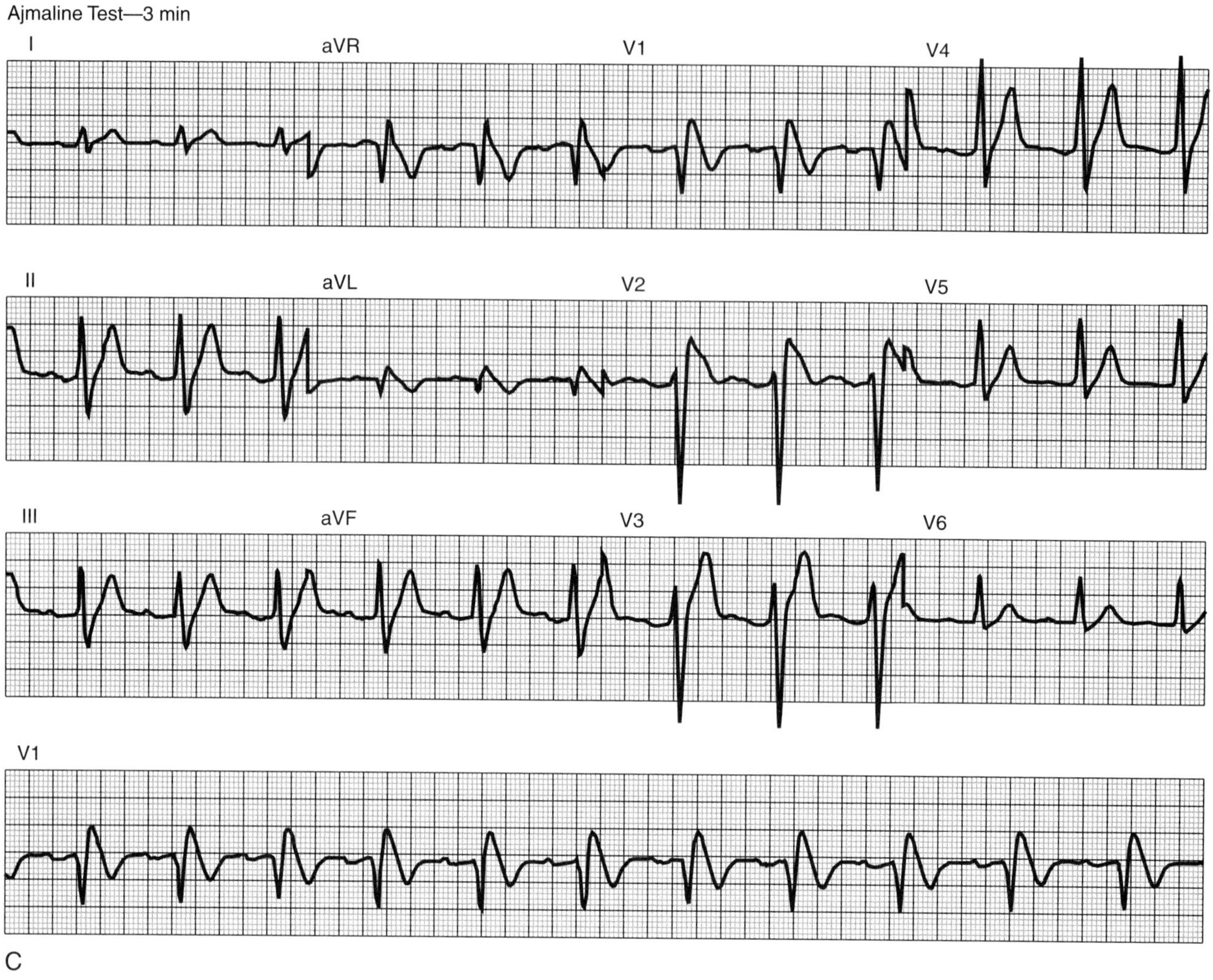

Figure 19-12 cont'd

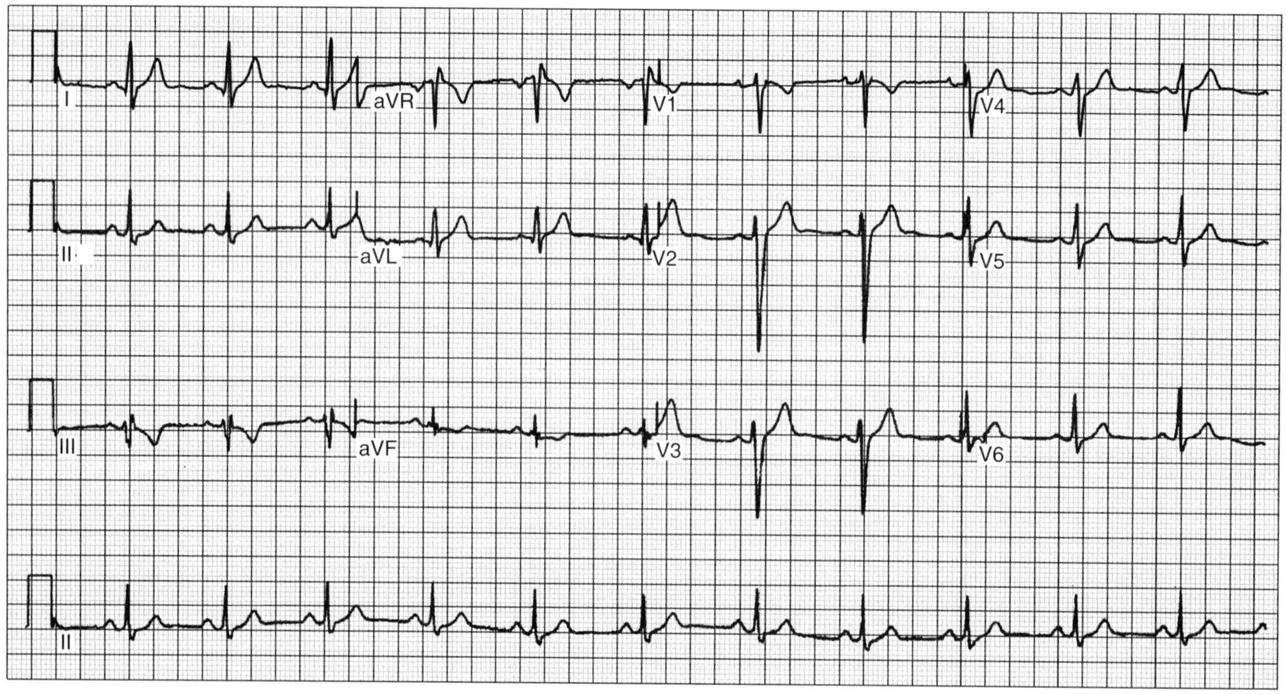

FIGURE 19-13 Twelve-lead ECG recorded from a 19-year-old male with arrhythmogenic right ventricular cardiomyopathy. An ε wave is present in lead V1.

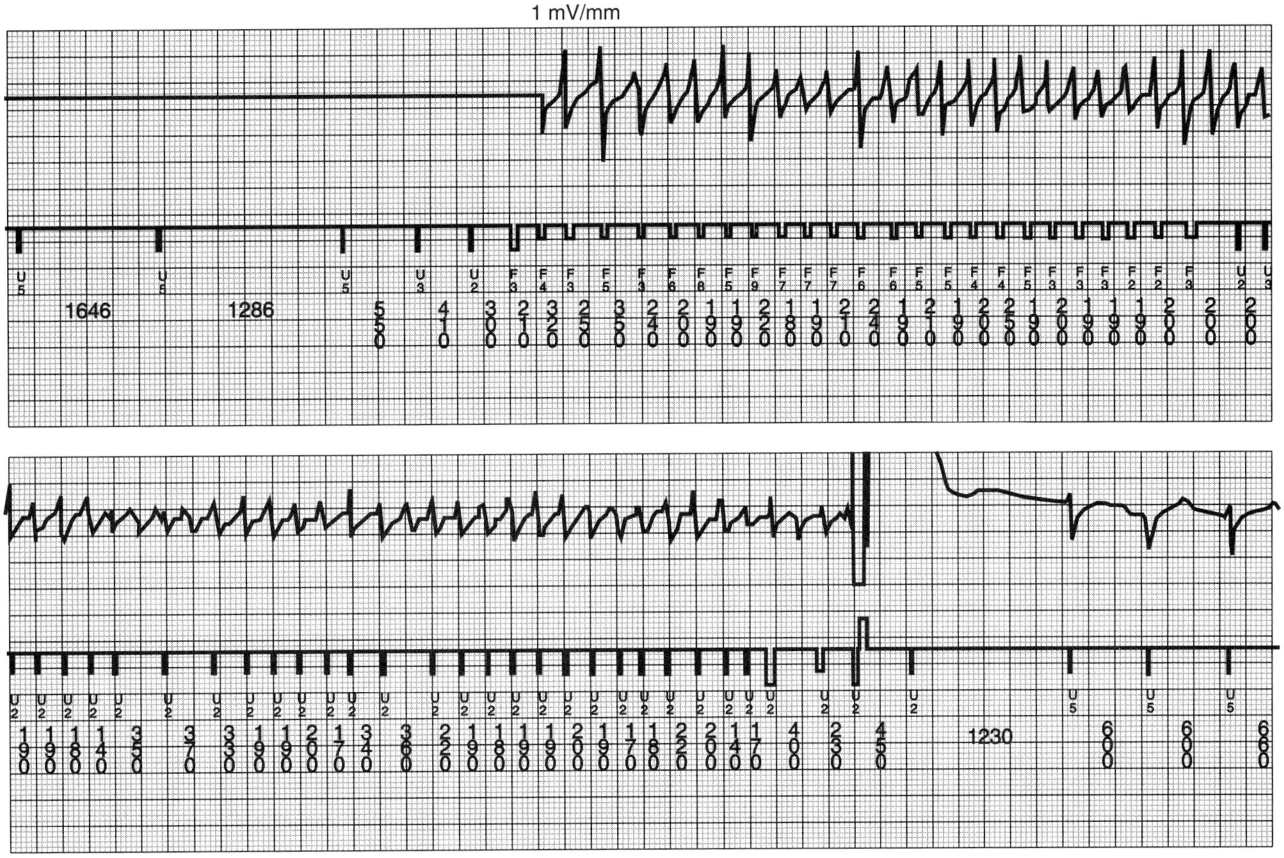

FIGURE 19-14 Electrograms recorded from an implantable cardioverter defibrillator (ICD) in a patient with arrhythmogenic right ventricular cardiomyopathy. The continuous recordings document the sudden onset of ventricular fibrillation, followed by an ICD shock and conversion to sinus rhythm. The upper channel represents the stored intracardiac electrogram. The lower channel marks sensed events and displays the cycle length.

for a Q-wave MI.[110] Nevertheless, the risk of VF may also be transient in the setting of a non–Q-wave MI.[111] However, the practitioner must avoid the temptation to wholly ascribe an episode of VF to an acute MI if the only evidence of acute myocardial necrosis is a marginal elevation in a serum marker of necrosis, as such elevations may be secondary to the VF-induced cardiac arrest rather than represent its cause.

The other common reversible or transient cause of VF is the use of a proarrhythmic drug[112]—a classic "antiarrhythmic" drug; other drugs with electrophysiological effects; and "recreational" drugs, particularly cocaine. Use of these agents is, of course, determined from the history. However, if the practitioner believes that a report of such drug use will not be forthcoming from a high-risk individual, a toxicology screen would be in order. When therapeutic drug use is reported, early determination of a serum concentration of the agent may be important where toxicity related to that agent has a relationship to serum concentrations (i.e., digitalis).

Finally, an episode of VF may be considered to be reversible if it accompanies a state of extreme physiologic derangement that is not expected to be recurrent. Such states include that seen in the immediate postoperative period, with sepsis, and with hemodynamic instability, especially when the treatment of the physiologic derangement required administration of sympathomimetic agents.

IDENTIFICATION OF STRUCTURAL HEART DISEASE

In most of the world, the most common forms of structural heart disease that precipitate VF are atherosclerotic coronary artery disease and either congestive or hypertrophic cardiomyopathy. Accordingly, most patients who have been resuscitated from VF will require an echocardiographic examination, an exercise test (with or without myocardial perfusion imaging), and a cardiac catheterization with coronary angiography.

The echocardiogram serves primarily as an adjunct to the physical examination for identification of myocardial or valvular structural heart disease. It is particularly suited to the documentation of hypertrophic cardiomyopathy and for the detection of other forms of structural heart disease that have escaped clinical detection. Although the echocardiogram is also useful to quantitate left ventricular systolic function, radionuclide ventriculography and contrast ventriculography provide more accurate determinations of this important prognostic variable. Of importance, the echocardiogram is not very sensitive for the detection of early arrhythmogenic right ventricular dysplasia.

Exercise testing of the patient who has been resuscitated from an episode of VF provides information regarding the inducibility of both exercise-related myocardial ischemia and exercise-related arrhythmias.[113] The sensitivity, specificity, and spatial localization of reversible myocardial ischemia can be enhanced by coupling the exercise test with radionuclide or echocardiographic imaging techniques.

Despite the availability of these noninvasive diagnostic procedures, the critical importance of accurate and complete anatomic and functional diagnosis of structural heart disease in a patient who has been resuscitated from VF will usually necessitate cardiac catheterization and coronary angiography. Occasionally, the combination of information available from the history of events surrounding the spontaneous episode of VF (especially when preceded by physical or emotional stress accompanied by angina), evidence of reversible myocardial ischemia on exercise testing (especially when associated with evidence of ventricular electrical instability), and documentation of coronary artery disease (especially when not associated with severe left ventricular dysfunction) will suggest the possibility that the episode of VF was due to reversible myocardial ischemia that can be corrected. Although coronary revascularization in this setting has been reported to prevent further episodes of VF in some patients,[114] the effect is not necessarily predictable. Accordingly, it has become customary to either assume that the revascularization procedure was insufficient for the prevention of VF recurrences and provide chronic antiarrhythmic therapy or to document that VT or VF is not inducible by programmed stimulation performed during a transvenous catheter EPS study after the revascularization procedure—preferably after a preoperative catheter EPS documented the inducibility of sustained VT or VF.

On occasion, specialized procedures are required to identify underlying structural heart disease in selected patients for whom there is a high index of suspicion. These procedures include magnetic resonance imaging for the diagnosis of arrhythmogenic right ventricular dysplasia, catheter endomyocardial biopsy for the diagnosis of myocarditis, and infusion of a class I antiarrhythmic agent for the diagnosis of Brugada syndrome. To date, genetic testing has not yet reached the maturity where it can be recommended for diagnostic purposes in patients with VF. However, the future promise of genetic testing in this regard is high.

DOCUMENTATION OF THE MECHANISM OF VENTRICULAR FIBRILLATION

There is still debate regarding the advisability of offering a baseline transvenous catheter EPS to all patients who have experienced an episode of VF in the absence of a reversible or transient cause.[115] The major potential advantage of performance of electrophysiological testing for patients with VF is the possibility of demonstrating that the VF was actually caused by another arrhythmia that would be treated in another way. Of course, VF may result from the degeneration of other tachyarrhythmias best treated by transcatheter ablation procedures including supraventricular tachyarrhythmias (including AF in the setting of ventricular preexcitation) and certain VTs (including bundle branch reentrant VT). Electrophysiological testing may also help to distinguish patients with a permanent VT substrate (typically those with a myocardial scar who have inducible VT) from those without a permanent VT substrate (typically those without a myocardial scar who do not have inducible monomorphic VT). This distinction may assist the practitioner in the selection of treatment modalities. For example, revascularization of coronary artery disease is rarely, if ever, useful in the former circumstance. Finally, electrophysiological testing may also provide information regarding optimal programming of a subsequently implanted cardioverter defibrillator device. Nevertheless, each of these potential advantages is more likely to be enjoyed by the patient who presents with VT.

IDENTIFICATION OF OTHER DISEASE STATES

Identification of other disease states that interact with antiarrhythmic therapies to be prescribed for the treatment of VF is an important goal of the investigations performed in this patient population. Screening biochemical investigations for renal and hepatic dysfunction are indicated before the prescription of antiarrhythmic drugs that depend upon renal or hepatic metabolism. Similarly, prescription of those therapies whose adverse effect profiles include interaction with other organ systems should be preceded by screening of the integrity of that organ system (i.e., thyroid function testing and pulmonary function testing in preparation for amiodarone therapy). These screening examinations will then be repeated, as necessary, during follow-up and can be referenced to the baseline evaluations to substantiate change.

ASSESSMENT AND MONITORING OF SELECTED ANTIARRHYTHMIC THERAPY

The assessment and monitoring of some forms of antiarrhythmic therapy will require other selected investigations. Although now used infrequently, the selection of antiarrhythmic drug therapy using the approach of suppression of ventricular premature beats will require both baseline, antiarrhythmic–drug-free and drug assessment 24-hour ambulatory electrocardiographic examinations, and exercise tolerance tests.[113,116] Similarly, the selection of antiarrhythmic drug therapy using the approach of suppression of ventricular tachyarrhythmias induced by programmed stimulation will require both baseline, antiarrhythmic–drug-free and drug assessment EPSs.[117,118] Of course, the follow-up of patients with a treated propensity to VF will usually require long-term surveillance—most commonly with repeated 24-hour ambulatory electrocardiographic examinations. However, there is no direct evidence that such surveillance is of value to the patient who presented with VF.

Electrophysiological Testing

The role of electrophysiological testing in the patient presenting with VF depends on the etiology of the arrhythmia. Secondary VF is associated with acute reversible derangement, such as ischemia, electrolyte imbalance, or cardiac trauma. In contrast, primary VF is not associated with any acute precipitant. In secondary VF, the best approach is to treat the specific precipitant that was responsible for VF. The role of electrophysiological testing is often limited in secondary VF. Therefore, the focus of this chapter is on the approach to primary VF in the electrophysiology laboratory. In the setting of primary VF, it is imperative to first define the anatomic substrate. Because healed MI is the most common cause of primary VF, echocardiography, stress testing, and cardiac catheterization are all important diagnostic tools. If myocardial ischemia is demonstrated, revascularization (percutaneous or surgical) can improve long-term survival.[119,120]

INDUCTION OF VENTRICULAR TACHYCARDIA VERSUS VENTRICULAR FIBRILLATION

After defining the underlying heart disease, electrophysiological evaluation is important for risk stratification for primary VF. It is believed that VF is often preceded by VT, which degenerates into VF, so that induction of VT in the electrophysiological laboratory is indicative of the clinical rhythm (Fig. 19-15). Therefore, the principal goal at electrophysiological testing is the induction of sustained monomorphic VT. In the setting of a healed MI, VT can be induced at electrophysiological testing in 20% to 45% of patients. The anatomic and electrophysiological substrate of VT is well described. Areas of healed MI, regions of slow conduction, and inhomogeneities of refractoriness all contribute to the specific responses to programmed stimulation and the induction of VT. As the induction of ventricular tachycardia at electrophysiological testing is already well described (see Chapter 17, Sustained Ventricular Tachycardia with Heart Disease), the focus of this section will be induction of VF.

There is limited correlation between the induction of VF in the electrophysiology laboratory and the incidence of VF as the presenting arrhythmia, perhaps due to transient factors or prior VT in either situation.[121] Among patients with previous cardiac arrest, 20% to 50% do not have VF induced at EPS. VF is inducible in up to 25 percent of cardiac arrest survivors as compared with 3% of patients who presented with monomorphic VT. This suggests that induction of VF in patients who present clinically with VF may be specifically predictive of spontaneous VF episodes, whereas induction of VF in patients who presented with VT may represent

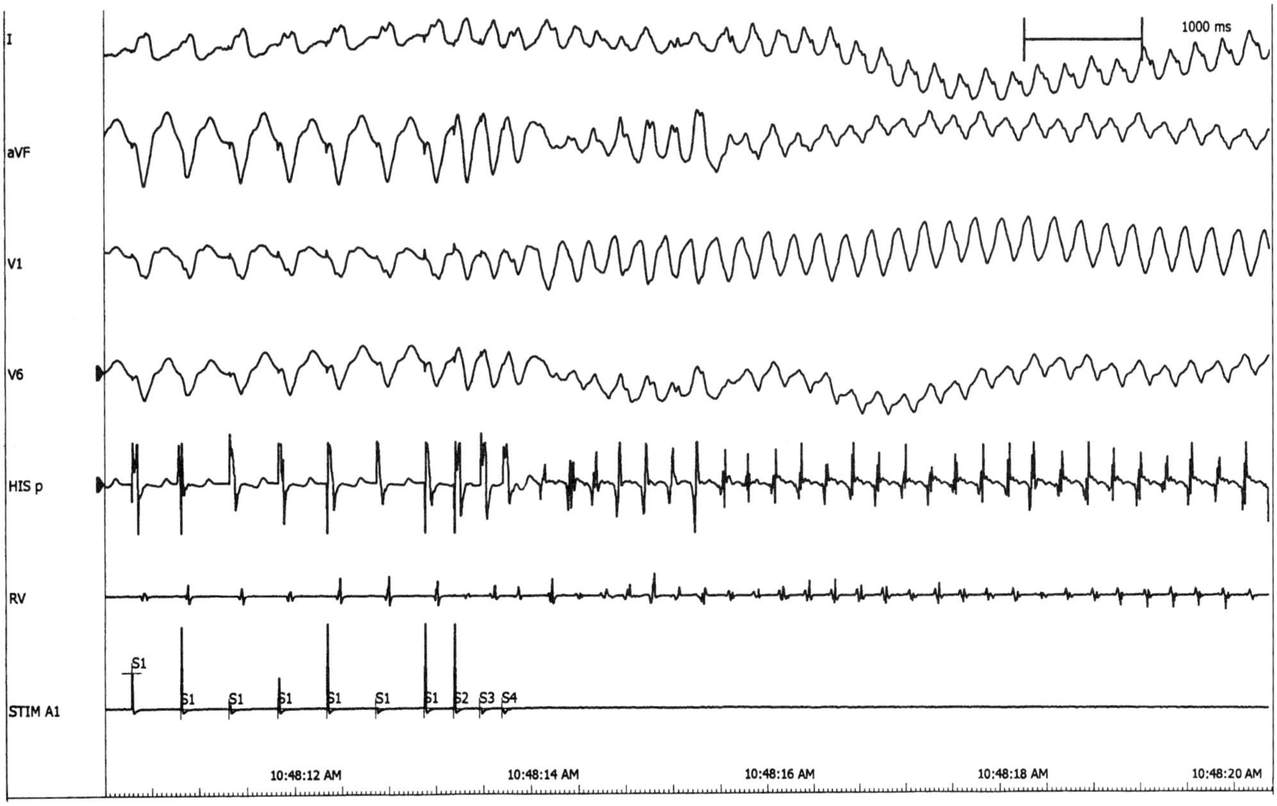

A

FIGURE 19-15 **A,** Induction of rapid monomorphic ventricular tachycardia with triple ventricular extrastimuli in a patient with cardiac arrest.

(Continued)

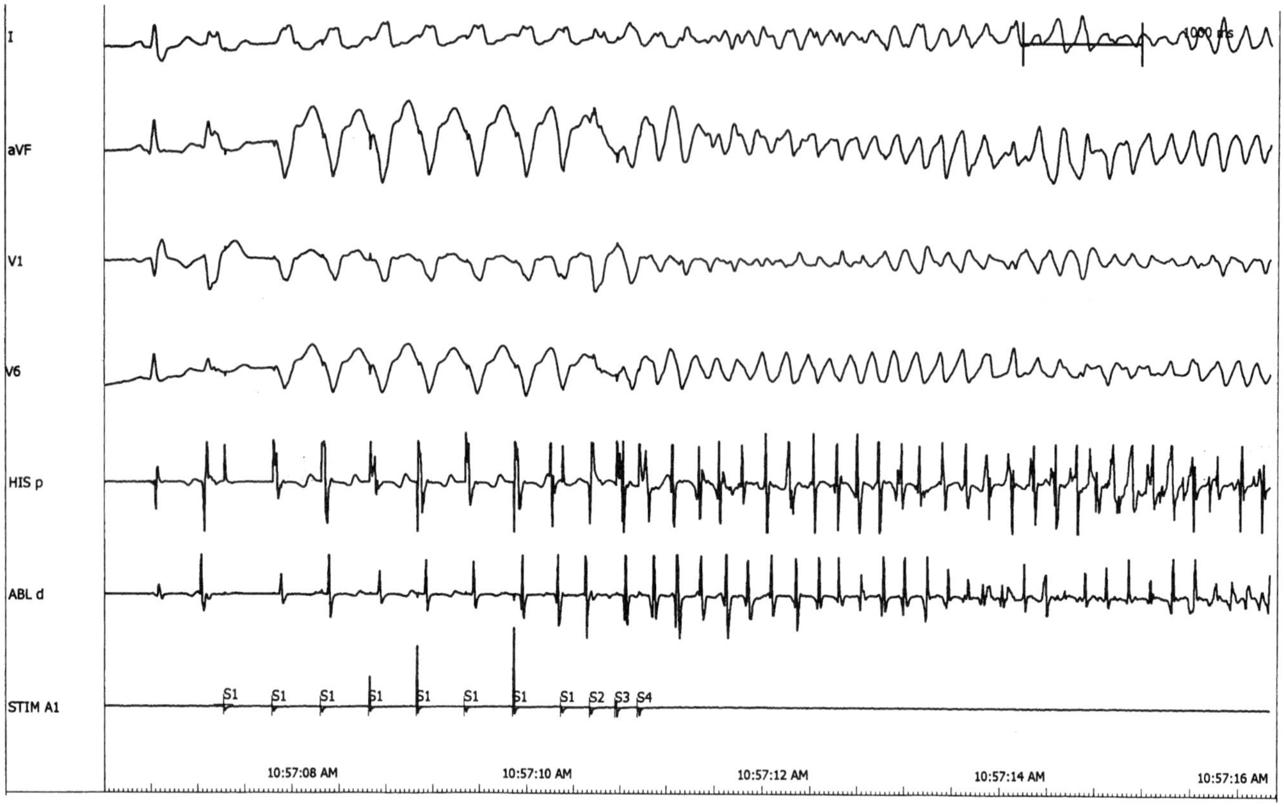

B

FIGURE 19-15 **B,** A subsequent induction attempt with triple extra-stimuli resulted in induction of ventricular fibrillation.

a nonspecific response to programmed electrical stimulation.[122]

In patients with asymptomatic nonsustained VT, coronary artery disease, and ejection fraction of less than 40 percent, VF or polymorphic VT is induced in up to 6% of cases.[117] Clinical variables often fail to distinguish patients with inducible arrhythmias from those without inducible arrhythmias. Ejection fraction is not significantly different in patients with no inducible arrhythmia, inducible nonsustained VT, inducible sustained monomorphic VT, or inducible VF (Fig. 19-16). Of the 6% of patients in whom VF or polymorphic VT was induced, 17% died suddenly during follow-up.[120] The presence of inducible sustained ventricular arrhythmias and the persistence of inducible sustained ventricular arrhythmias on therapy were significant univariate predictors of sudden death. However, only the persistence of inducible sustained arrhythmias on therapy was an independent predictor of SCD (Fig. 19-17).

ELECTROPHYSIOLOGICAL CHARACTERISTICS ASSOCIATED WITH INDUCTION OF VENTRICULAR FIBRILLATION

It is not well understood why VF is induced in some patients in the clinical electrophysiology laboratory and not in others. Because VF can be induced in perfectly normal hearts using either multiple closely coupled premature stimuli, or by shocks applied on the T wave, induction of VF by catheter stimulation is not necessarily associated with a poor prognosis. There are several potential reasons of the nonspecific response of induction of VF at electrophysiological testing, including local graded response in normal muscle or decremental conduction block due to short coupling intervals. This appears to have less to do with dispersion of refractoriness or propagation, and more to do with local graded responses in normal muscle. For this reason, premature coupling intervals are usually not shortened below 180 ms so as to avoid inducing VF that has no diagnostic value.[123] Other causes of nonspecific responses to programmed stimulation and induction of VF include increased conduction latency and prolongation of local activation time in proximity of the stimulus electrode. Induction of VF is preceded by increased latency (Fig. 19-18).[124]

Another hypothesis suggesting why VF is induced with closely coupled premature beats during programmed stimulation is delayed conduction. Conduction slowing by itself causes dispersion of activation time between sites near the stimulating electrode and distant sites. Also, conduction slowing of a premature beat allows distant sites to have a longer coupling interval of the premature beat compared to that from the pacing site. These factors result in dispersion of refractoriness, with the refractory periods following the premature beat being longer at distant sites than at the pacing site.

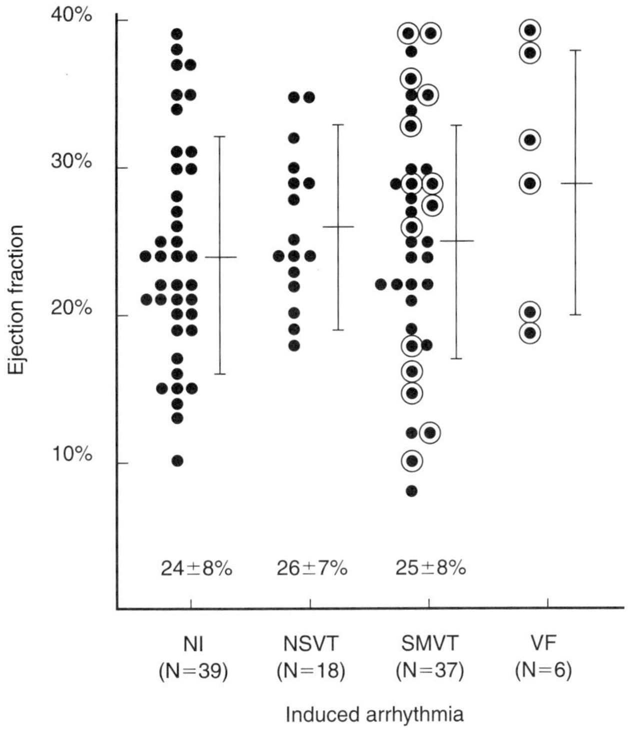

FIGURE 19-16 Scatterplot of left ventricular ejection fraction stratified by baseline-induced ventricular arrhythmia. *Vertical bars* are the mean ± SD of each group. *Circled dots* represent patients in whom inducible arrhythmias were suppressed. NI, noninducible; NSVT, nonsustained ventricular tachycardia; SMVT, sustained monomorphic ventricular tachycardia; VF, ventricular fibrillation). P = not significant among groups. (Reproduced from Wilber DJ, Olshansky B, Moran JF, et al: Electrophysiological testing and nonsustained ventricular tachycardia: Use and limitations in patients with coronary artery disease and impaired ventricular function.)

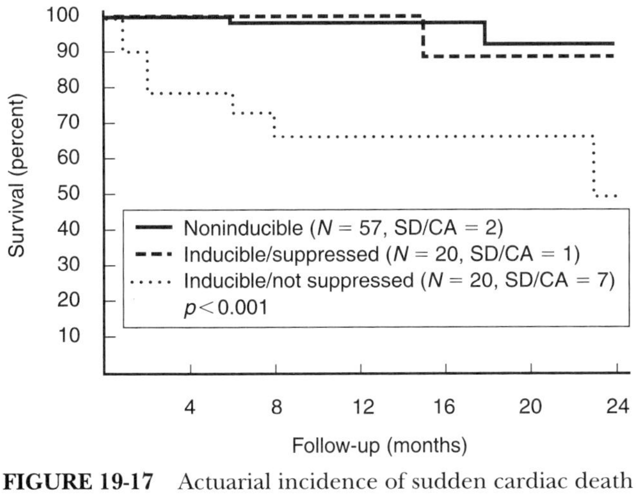

FIGURE 19-17 Actuarial incidence of sudden cardiac death or cardiac arrest in 97 patients, stratified by treatment subgroup. Three patients with cardiac arrest during serial drug testing were not included in analysis. CA, cardiac arrest; SD, sudden cardiac death. (Reproduced from Wilber DJ, Olshansky B, Moran JF, et al: Electrophysiological testing and nonsustained ventricular tachycardia: Use and limitations in patients with coronary artery disease and impaired ventricular function.)

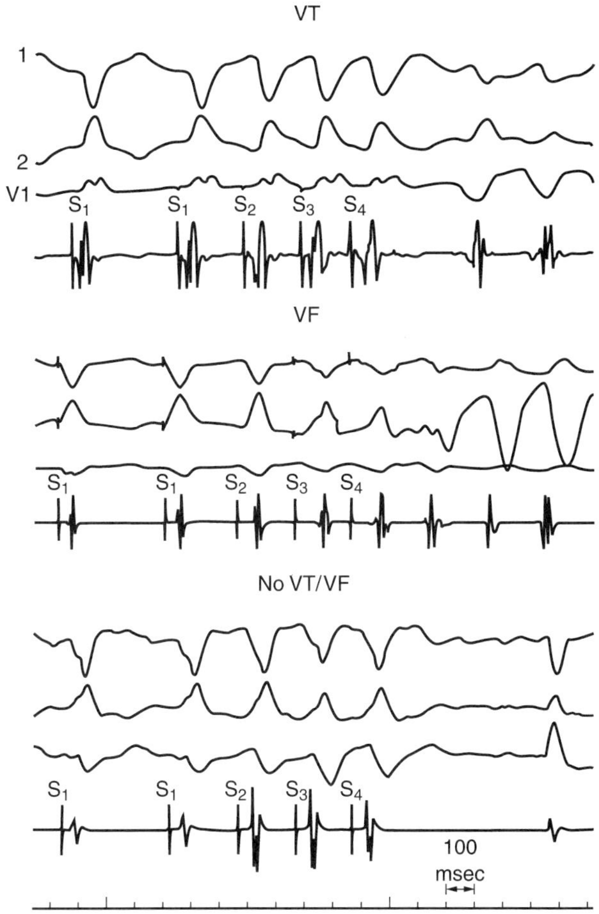

FIGURE 19-18 Tracings of ECG leads I, II, and V1 and ventricular electrograms. *Top panel,* Tracing of the last two ventricular stimuli from the drive (S_1) and the triple extrastimuli that result in the induction of sustained monomorphic ventricular tachycardia. *Middle panel,* Induction of ventricular fibrillation. *Lower panel,* No induction of ventricular tachycardia or ventricular fibrillation. VF, ventricular fibrillation, paper speed 100 ms/cm; VT, ventricular tachycardia. (Reproduced from Avitall B, McKinnie J, Jazayeri M, et al: Induction of ventricular fibrillation versus monomorphic ventricular tachycardia during programmed stimulation: Role of premature beat conduction delay. Circulation 1992;85:1271-8.)

This in turn allows for additional premature beats to induce VF. Therefore, the measurement technique itself alters what we are attempting to measure (Table 19-1).[125] Additionally, there is indirect evidence that the mechanism of VF in the setting of a healed MI may be related to areas of slow conduction and stable reentrant circuits. Occasionally polymorphic VT/VF can transform to monomorphic VT by type I antiarrhythmic medications.[126] These tachycardias have been mapped in the operating room and have been cured by endocardial resection. These findings have not been observed in patients with a normal heart and inducible polymorphic VT/VF. Therefore, use of a type I antiarrhythmic agent may be helpful in differentiating between a nonspecific finding from a significant one in the electrophysiology laboratory.

TABLE 19-1 Normalized Latency and Activation Times for S_1S_2, $S_1S_2S_3$, and $S_1S_2S_3S_4$ (ms) Used to Derive Cumulative Latency and Activation Times

Ventricular Tachycardia	Ventricular Fibrillation	No Ventricular Tachycardia/ Ventricular Fibrillation	Ventricular Tachycardia	Ventricular Fibrillation	No Ventricular Tachycardia/ Ventricular Fibrillation
S_2 4 ± 4*	17 ± 13	1 ± 6*	2 ± 6*	43 ± 27	1 ± 12*
S_2 5 ± 5	10 ± 14	1 ± 6*	6 ± 10*	22 ± 20	1 ± 12†
S_3 4 ± 5*	14 ± 23	5 ± 12*	8 ± 15*	31 ± 32	10 ± 15*
S_2 4 ± 8	8 ± 10	1 ± 6*	4 ± 9†	16 ± 15	1 ± 12†
S_3 12 ± 14	20 ± 20	5 ± 12*	17 ± 17†	37 ± 26	10 ± 15†
S_4 7 ± 18	14 ± 27	7 ± 13	8 ± 22†	33 ± 36	14 ± 17‡

Normalized latency and activation times for S_1S_2, $S_1S_2S_3$, and $S_1S_2S_3S_4$ used to derive cumulative latency and activation times (in ms). Sustained monomorphic VT was initiated with lower latency times than the induction of VF. Total activation times were longer in the inducible VF group with single, double, and triple premature stimuli.

*$P < .01$ versus ventricular fibrillation.

†$P < .001$ versus ventricular fibrillation.

‡$P < .05$ versus ventricular fibrillation.

ms, milliseconds; VF, ventricular fibrillation, paper speed 100 ms/cm; VT, ventricular tachycardia.

Reproduced from Avitall B, et al: Induction of ventricular fibrillation versus monomorphic ventricular tachycardia during programmed stimulation. Role of premature beat conduction delay. Circulation 1992;85:1271-8.

PROGNOSIS AND CLINICAL RELEVANCE

The prognostic value of electrophysiological evaluation has been extensively studied in survivors of cardiac arrest. The reproducibility of inducible VF is strongly predictive of recurrent cardiac arrest. However, the failure to induce arrhythmias with programmed stimulation may not necessarily be associated with a benign prognosis. In a specific subset of patients with coronary artery disease, LVEF less than 40%, asymptomatic nonsustained VT, and no inducible arrhythmias, the relative risk of death was 1.8.[119]

Additionally, there have been several secondary prevention trials supporting the efficacy of defibrillator implantation in out-of-hospital VF survivors irrespective of the outcome of electrophysiological testing. The Cardiac Arrest Study Hamburg (CASH), AVID (discussed earlier), and the Canadian Implant Defibrillator Study (CIDS) support defibrillator placement in cardiac arrest survivors who have no reversible cause of VF.[127,129]

In summary, the usefulness of electrophysiological testing in patients with VF is evolving. If a clearly reversible cause of secondary VF is identified, correction of the underlying problem is needed. With primary VF, it is often recommended that the patient undergo electrophysiological testing. The specificity of induction of VF by programmed stimulation varies with the patient's clinical presentation. The induction of VF may be related to areas of slow conduction/healed MI, local graded responses, decremental conduction block due to short coupling intervals, as well as increased latency and prolongation of activation time. Defibrillator insertion may be indicated despite a negative electrophysiological study in these patients.

Principles of Practice

The main strategic approaches to preventing death from VF include prevention and treatment. *Preventative* therapies of proven benefit include β-blocker therapy in patients with MI or heart failure; ACE inhibition in patients with LV dysfunction; spironolactone therapy in patients with severe heart failure; revascularization (bypass surgery in patients with left main or severe three vessel disease, especially with LV dysfunction); and possibly amiodarone therapy. No other antiarrhythmic drug has been shown, in any controlled study, to reduce the likelihood of VF, sudden death, or overall mortality. *Anticipating* VF (i.e., treating those patients at particularly high risk for this arrhythmia) may include the implantation of an automatic defibrillator; providing family members with a semiautomatic external defibrillator; using "wearable" external automated defibrillators; and from a public health perspective, providing emergency medical services personnel or trained lay rescuers with manually operated semiautomated external defibrillators. These are deployed in locations readily accessible to rescuers in proximity to patients suffering from cardiac arrest. The best and most publicized examples of these "anticipatory" strategies include the provision of semiautomated external defibrillators and trained lay operators in casinos, airports, and on long-haul airplane flights and sophisticated tiered therapy programs with emergency medical services programs. *Therapy* for VF includes both defibrillation and adjunctive therapies, such as measures to support the failing circulation during cardiopulmonary resuscitation, as well as antiarrhythmic drug therapy to enhance the probability of successful resuscitation from VF.

Evidence-Based Therapy

RESCUE OF THE VENTRICULAR FIBRILLATION PATIENT

Optimum resuscitation from VF requires the restoration of a perfusing cardiac rhythm as early as possible, as well as providing the maximum possible cerebral and coronary blood flow during the resuscitation process. Mechanical aids, which improve cardiac output during cardiopulmonary resuscitation, have consistently shown improved outcomes in experimental settings and, in limited human clinical trials, have shown improved early resuscitation and improved survival to hospital admission.[128] Although definitive clinical trials with respect to improving survival to discharge from hospital are lacking, such interventions are likely to be able to improve long-term outcomes in patients with VF, particularly if they are begun relatively early (e.g., at 5 to 10 minutes) following the onset of cardiac arrest from VF.

Mechanical interventions of this type include the active compression–decompression CPR device, interposed abdominal and chest compression devices, the impedance threshold valve, mechanical or pneumatically driven chest compression devices, or devices inserted into the chest cavity to assist in direct cardiac compression.[130] All of these devices increase cardiac output by increasing intrathoracic pressure during the compression phase of CPR or by increasing venous return during the decompression phase, either with a positive abdominal pressure or by causing negative intrathoracic pressure during the decompression phase of CPR.

DRUG THERAPY IN ACUTE MANAGEMENT OF VENTRICULAR FIBRILLATION

The benefit of pharmacologic therapy as an adjunct to defibrillation in out-of-hospital VF is incompletely established. Despite many decades of use and ample laboratory experimental evidence of benefit, there is no clear evidence from controlled clinical trials that either low-dose or high-dose epinephrine is beneficial in the treatment of patients with out-of-hospital VF.[131] Vasopressin, which appears to be superior to epinephrine in experimental models of cardiac arrest in improving survival from experimental VF,[132] appears to be superior to epinephrine in some studies[133] but not in others.[134] However, vasopressin has also not been proven superior to placebo in randomized controlled trials.

Similarly, the evidence base regarding antiarrhythmic therapy is incomplete. The standard therapy used to assist electrical defibrillation has been lidocaine; there is no evidence from controlled clinical trials that lidocaine is superior to placebo or any other agent in improving survival to hospital admission or survival to hospital discharge.[135] In a meta-analysis of prophylactic lidocaine used in the postinfarction period, lidocaine may reduce the incidence of primary VF but appears to increase mortality.[136] Lidocaine likely increases defibrillation thresholds[137,138] and may increase the incidence of asystole. Magnesium is not superior to placebo in in-hospital cardiac arrest due to VF.[139] Bretylium, although studied in ventricular fibrillation, is no longer available.

IV amiodarone for VF resistant to three defibrillation shocks was compared with placebo in a blinded randomized clinical trial by Kudenchuk et al.[140] The 246 patients with shock-resistant, out-of-hospital VF were randomized to amiodarone, 300 mg, by IV bolus versus placebo; the primary study end point was survival to hospital admission, achieved in 44% of amiodarone-treated patients, versus 34% of placebo-treated patients. In a related study, IV amiodarone was compared with IV lidocaine in a blinded randomized study of patients with VF persisting after three shocks, IV epinephrine, and a further defibrillation shock. Survival to hospital admission was achieved in 23% of amiodarone-treated patients versus 12% of lidocaine-treated patients ($P = .009$).[141] Although neither of these studies demonstrated statistically significant improvements in survival to hospital discharge, based on these studies, if any antiarrhythmic drug is to be used in out-of-hospital VF, it seems logical to consider amiodarone as the drug of choice.

It is important to note that resuscitation from VF is a dynamic clinical situation, and many patients will suffer multiple recurrences of VF in the seconds to minutes after initial defibrillation, as well as undergo multiple transitions among VF, asystole, pulseless electrical activity, and a perfusing organized rhythm during the course of a protracted cardiac arrest. The use of adjunctive drug therapy can therefore be considered as aimed to prevent recurrences of VF as much as it is used to assist in defibrillation. In particular, patients with frequent recurrences of VF may benefit from intensive antiadrenergic therapy,[142] IV amiodarone therapy,[143] and occasionally from anti-ischemic therapy, revascularization therapy, or therapy with an intra-aortic balloon pump.

"NONANTIARRHYTHMIC" DRUGS THAT PREVENT SUDDEN CARDIAC DEATH

In the setting of heart failure, there is a cascade of "neurohumoral activation" that initially supports perfusion.[144] Over time, however, the biologically active molecules released in the neurohumoral activation, including the sympathetic nervous system and the renin-angiotensin-aldosterone system, result in progressive left ventricular dysfunction. Pharmacologic antagonism of these two systems has proven to be an effective treatment paradigm.

β-Blockers. Strong evidence for a beneficial effect of β-blocker treatment in patients following an MI was reported in 1981 by the Norwegian Multicenter Study Group.[145] In their study, 1884 patients were randomly assigned to treatment with timolol (10 mg twice a day) or to placebo. Mortality in the control group was 16% compared to 10% in the timolol-treated patients ($P = .0001$), and there was a reduction in SCD from 13.9% to 7.7% ($P = .0001$). In the following year, the Beta-Blocker Heart Attack Trial (BHAT) reported a comparison of propranolol and placebo in post-MI patients.[146] In that study, 3837 patients were randomized to treatment with propranolol or placebo 5 to 10 days following an MI. The trial was stopped 9 months early, after an average follow-up of 25 months because of a highly significant difference in mortality in favor of propranolol (7.2% in the propranolol group versus 9.8% in the placebo group). These trials, along with a series of subsequent studies, were combined in a meta-analysis confirming that β-blockers reduce both total mortality and SCD.[147]

More recently, two studies investigating treatment of heart failure patients with β-blockers have suggested a benefit with respect to reduction of total mortality and prevention of SCD.[148,149] The Cardiac Insufficiency Bisoprolol Study II (CIBIS II) trial randomized 2647 patients with LVEF less than or equal to 0.35 and New York Heart Association (NYHA) functional class III or IV heart failure to the β_1-selective agent bisoprolol or to placebo.[148] All-cause mortality was reduced from 17.3% in the placebo group to 11.8% in the bisoprolol group ($P < .0001$). There were also significantly fewer sudden deaths in the bisoprolol-treated patients compared to placebo-treated patients (3.6% versus 6.3%, $P = .0011$). The Metoprolol Randomized Intervention Trial in Congestive Heart Failure (MERIT-HF) randomized 3991 patients with LVEF less than or equal to 0.40 in NYHA functional class II to IV heart failure to treatment with the β_1 selective agent metoprolol, in its controlled release/extended release form, or to placebo.[149] All-cause mortality was 7.2% per year in the metoprolol group and 11% in the placebo group ($P = .00009$). There was also a reduction in SCD in the metoprolol-treated patients (odds ratio, 0.59; confidence interval, 0.45 to 0.78).

ANGIOTENSIN CONVERTING ENZYME INHIBITORS

The survival advantage conferred by angiotensin converting enzyme (ACE) inhibitors in heart failure patients has been demonstrated in a number of studies.[150-153] However, only in the Tandolapril Cardiac Evaluation Trial (TRACE) did the reduction in SCD reach statistical significance.[153] A recently published meta-analysis has now clearly demonstrated that the ACE inhibitors reduce SCD.[154] In this study 15,104 patients from 15 trials were studied. The relative risk of SCD in the ACE inhibitor–treated patients was 0.80 (confidence interval, 0.70 to 0.92).

ALDOSTERONE ANTAGONISM

The effect on survival of spironolactone was studied in the Randomized Aldactone Evaluation Study (RALES).[155] A total of 822 patients with LVEF less than or equal to .35 on an ACE inhibitor were randomly assigned to treatment with spironolactone or to placebo. After a mean follow-up of 24 months, mortality was 46% in the placebo group and 35% in the spironolactone-treated patients ($P < .001$). The relative risk of SCD in the spironolactone-treated patients was 0.71 (confidence interval, 0.54 to 0.95). Taken together, the data confirm that antagonizing the various components of the neuroendocrine activation that accompanies heart failure results in a reduction in SCD.

ANTITHROMBOTIC AND ANTICOAGULANT THERAPY

Data have accumulated suggesting that ischemia caused by thrombus formation in stenotic coronary arteries may result in SCD. Based on pathologic examinations, greater than 80% of SCD cases in patients with ischemic heart disease may be associated with thrombus formation or plaque fissuring.[156-157] In the Second International Study of Infarct Survival (ISIS-2), aspirin use was associated with a reduction in total mortality and SCD.[158] A multivariate analysis of the 6797 participants in the Studies of Left Ventricular Dysfunction (SOLVD) demonstrated that both antiplatelet and anticoagulant therapy were associated with a reduction in the risk of SCD.[159] This provides a rationale for antithrombotic or antiplatelet therapy, or both, in patients with left ventricular dysfunction.

ANTIARRHYTHMIC DRUGS

The importance of ventricular ectopy as a risk factor for SCD in patients with coronary disease is well established. It was therefore reasonable to hypothesize that suppression of premature ventricular complexes (PVCs) using standard antiarrhythmic agents that block sodium or potassium channels might reduce the risk of SCD.

Class I Agents. The Cardiac Arrhythmia Suppression Trial (CAST)[160] studied patients with a history of MI and frequent PVCs that were suppressible on encainide, flecainide, or moricizine (type IC agents that are sodium channel blockers). Patients more than 90 days post-MI were also required to have an LVEF less than or equal to 0.40. Despite suppression of PVCs, the encainide and flecainide arms of the trial were stopped early because of an increased mortality in patients treated with the antiarrhythmic drugs. The moricizine arm was subsequently stopped, because it became clear that there would not be any benefit and because of concern about an apparent early increase in mortality.[161]

The CAST trial provided four important insights: (1) the mechanism causing PVCs was different from those causing the arrhythmia (presumably a ventricular tachyarrhythmia) that provoked SCD; (2) ventricular ectopy is not an appropriate surrogate end point for SCD in clinical trials; (3) antiarrhythmic drugs could be proarrhythmic, even months after initiation; and (4) a change in the myocardial substrate, presumably ischemia, could make the antiarrhythmic drug proarrhythmic.

New Class III Drugs. In the Survival with Oral D-Sotalol (SWORD) study, the impact on SCD of d-sotalol, a potassium channel blocker without β-blocker properties, was studied.[3,121,162] Patients with a history of MI and LVEF less than 0.40 were randomly assigned to d-sotalol or placebo. The study was terminated because of an excess risk of mortality in the d-sotalol–treated patients (relative risk, 1.65; $P = .006$). Dofetilide was also evaluated in patients with congestive heart failure.[163] In a study of

1518 patients with symptomatic congestive heart failure and severe left ventricular dysfunction randomized to dofetilide or to placebo, there was no difference in mortality, although dofetilide was successful in converting AF to sinus rhythm. Similar results were obtained in post-MI patients without congestive heart failure.[164] These data suggest that, while dofetilide has no role in the prevention of SCD, it may be a useful treatment option in patients with AF who are also at risk for SCD, since it did not increase mortality in these patients. A meta-analysis performed by Teo et al.[165] of antiarrhythmic trials was reported in 1993. Data were drawn from 138 trials and 98,000 patients. The mortality of patients randomized to receive class I agents was significantly higher than that of patients receiving placebo (odds ratio, 1.14; 95% confidence interval, 1.01 to 1.28; $P = .03$).

Given the data presented, it is reasonable to ask whether the reason for the failure of these antiarrhythmic drugs to prevent SCD is caused by intrinsic ineffectiveness of drug action or by the method of guiding selection of the drug. If the problem were the method of guiding selection of potentially effective drugs, it would be expected that, while the drugs might not prevent an arrhythmia, at least they would not be proarrhythmic. However, most antiarrhythmic drugs other than amiodarone were shown to provoke SCD in clinical trials. This suggests that the failure to prevent SCD is based on the molecule rather than the selection paradigm. The results of the Electrophysiologic Study versus Electrocardiographic Monitoring (ESVEM) study provide support for this view.[116] In this study, 486 patients with sustained monomorphic ventricular tachycardia, inducible with programmed electrical stimulation (PES) and greater than or equal to 10 PVCs/hour on Holter monitoring, were randomly assigned to guidance by demonstrating suppression of PES stimulated ventricular tachycardia or Holter recording suppression of ventricular ectopy. While the PES protocol used in ESVEM was limited, the results in ESVEM were no different in either arm, indicating that PES guidance was not better than Holter guidance. In contrast, the Calgary study, using a standard PES protocol, showed superior results with PES-guided therapy in less drug-refractory patients.[117,118]

It can be concluded from the studies presented that the antiarrhythmic drugs, excluding amiodarone, that have been tested may not prevent death from ventricular arrhythmias and, depending upon the drug, may even increase it. As a result, the antiarrhythmics other than amiodarone that have been studied have a limited role in the prevention of SCD and should be considered after other, better treatments have been rejected.

Amiodarone. A number of early studies suggested that amiodarone, unlike other antiarrhythmics, might have a role in suppressing VTs in the post-MI population. The Basel Antiarrhythmic Study of Infarct Survival (BASIS) randomized 312 post-MI patients with complex ventricular ectopy to amiodarone, individualized drug therapy starting with procainamide, or to no antiarrhythmic drug therapy.[166] After 1 year of follow-up, amiodarone-treated patients had a significantly greater survival than patients treated with no antiarrhythmic therapy (95% versus 87%; $P = .048$). The amiodarone-treated patients also had better survival than the individualized therapy patients, but this did not reach statistical significance. Ceremuzynski et al. randomized 613 post-MI patients who were ineligible to receive β-blockers to amiodarone or placebo.[167] There was a statistically significant ($P = .048$) reduction in cardiac mortality in the amiodarone-treated patients, although the reduction in total mortality did not achieve statistical significance. The meta-analysis by Teo et al. discussed earlier showed a significant mortality reduction in amiodarone-treated post-MI patients.[165] Taken together, these studies suggested that amiodarone might reduce arrhythmic death.

This picture was further clarified by the European Myocardial Infarction Amiodarone Trial (EMIAT)[168] and the Canadian Amiodarone Myocardial Infarction Trial (CAMIAT).[173] EMIAT randomized 1486 patients who had an LVEF less than or equal to 0.40, 5 to 21 days after an MI, to treatment with amiodarone or to placebo. There was no difference in the primary end point of total mortality or in cardiac mortality. Arrhythmic death was reduced from 7.0% in the placebo group to 4.0% in the amiodarone-treated patients ($P = .05$). All of the mortality benefit occurred in amiodarone-treated patients, who were also treated with a β-blocker. In CAMIAT 1202 patients with a history of MI and frequent ventricular ectopy (≥10 PVCs/hour or at least one episode of ventricular tachycardia) were randomized to receive amiodarone or placebo. There was a reduction in the end point of resuscitated VF or arrhythmic death from 6.0% in the placebo group to 3.3% in the amiodarone-treated patients ($P = .03$). There was no difference in total mortality. As in EMIAT, only amiodarone-treated patients who were also on a β-blocker appeared to derive benefit. Two recent meta-analyses that provide a quantitative overview of the available randomized trials have been published. The Amiodarone Trials Meta-Analysis Investigators examined 13 randomized trials involving 6553 patients.[170] There were five trials of patients with congestive heart failure and eight trials involving post-MI patients. The mean LVEF in this population was 0.31. They found a 29% reduction in SCD (odds ratio, 0.71; 95% confidence interval, 0.59 to 0.85). There was no difference between the post-MI and CHF populations. Sim et al.[171] studied 15 randomized trials of amiodarone to prevent SCD using the random effects model. Amiodarone was found to significantly reduce the risk of total mortality (odds ratio, 0.77; 95% confidence interval, 0.66 to 0.89; $P < .001$) and SCD (odds ratio, 0.70; 95% confidence interval, 0.58 to 0.85; $P < .001$).

These data suggest that, unlike other antiarrhythmic drugs, amiodarone may reduce arrhythmic death in the post-MI population and that its effect appears to be greater if there is concomitant β-blocker therapy. The fact that the reduction in arrhythmic death is not easily translated into an improvement in total mortality is likely due to the limited protective effect of amiodarone.

However, compared with prior studies, the SCD-HeFT study (detailed above) did not show any mortality benefit from amiodarone versus placebo; the hazard ratio for mortality in 845 amiodarone treated patients followed up to 5 years versus the 847 patients treated with placebo was 1.06 (97.5% confidence interval 0.86, 1.30).

NONPHARMACOLOGIC THERAPY

Revascularization. As discussed earlier, ischemia appears to be a precipitating event in SCD and is frequently caused by plaque disruption and coronary thrombus formation. Data supporting the usefulness of antithrombotic agents (presented earlier) suggest a potential anti-arrhythmic role for revascularization. Interestingly, trial-derived data bearing directly on the issue of revascularization are few. In the Coronary Artery Surgery Study (CASS) registry, revascularization was independently associated with improved survival free of SCD (SCD occurred in 4.9% of patients assigned to medical therapy and 1.6% of patients assigned to surgical therapy).[172] Garan et al. showed that myocardial revascularization can result in elimination of PES-inducible ventricular tachyarrhythmias.[173] Hii et al.[174] found that a patent infarct-related artery was associated with the effective drug suppression of inducible VT at PES. In sum, data supporting a role for ischemia in production of ventricular arrhythmias and studies suggesting a positive treatment effect of eliminating the ischemia lead to the conclusion that elimination of inducible ischemia should be a component of treatment to prevent SCD.

The ICD. Since the first implant of an ICD by Mirowski in 1980,[175] there has been a continuing evolution of these systems. Today, they are generally placed transcutaneously with an implant mortality of less than 1%.[176] Current systems are capable of defibrillation of patients in ventricular fibrillation, tiered therapy of VT (competitive pacing, synchronized cardioversion, defibrillation), and antibradycardia pacing. Increasingly sophisticated electrogram storage and telemetry are useful for rhythm diagnosis.

A number of randomized controlled clinical trials in patients with malignant ventricular arrhythmias including SCD have been performed that provide an increasingly clear picture of the appropriate role of the ICD. The AVID trial was a secondary prevention study that randomized patients with a history of symptomatic VT or VF to ICD implantation or to antiarrhythmic therapy.[125] Randomized patients must have been resuscitated from ventricular fibrillation or have had an episode of ventricular tachycardia that was hemodynamically compromising or that was associated with an LVEF less than 0.40. A total of 1016 patients were randomized to immediate ICD implantation or to antiarrhythmic drug therapy. Although sotalol or amiodarone were permitted in the antiarrhythmic drug arm, only 2.6% of patients randomized to antiarrhythmic drug were discharged on sotalol. Thus, for all practical purposes this was an amiodarone versus ICD trial. The trial was stopped early because of significantly better survival in the ICD-treated patients (75.4% versus 64.1% at 3 years, $P < .02$). Similar patient populations were studied in CIDS[128] and CASH[129] with results that were similar to AVID.

The Multicenter Automatic Defibrillator Implantation Trial (MADIT) was a primary prevention trial that studied 196 post-MI patients with nonsustained ventricular tachycardia and LVEF less than or equal to 0.35 who were inducible into a sustained monomorphic VT at PES not suppressible by procainamide.[177] This population was different from the AVID patients in not having had a spontaneous sustained ventricular arrhythmia but had shown that VT could be induced by PES. These patients were randomly assigned to immediate ICD implantation or to antiarrhythmic drug therapy. At 1 month of follow-up, 74% of the patients in the antiarrhythmic drug therapy group were on amiodarone. Survival was significantly better in the ICD group than in the patients treated with antiarrhythmic drug (mostly amiodarone) (hazard ratio for overall mortality, 0.46; 95% confidence interval, 0.26 to 0.82; $P = .009$).

All of the ICD trials (AVID, MADIT, CIDS, and CASH) presented thus far entered patients known to be at high risk for a fatal arrhythmia, because they had a history of prior VT, either as a presenting problem or at PES. The Coronary Artery Bypass Graft Patch (CABG-Patch) trial examined patients at risk of SCD because of the presence of coronary disease for which they were to undergo CABG, an LVEF less than or equal to 0.35, and an abnormal signal-averaged ECG.[178] In contrast to the patients presented in the other studies, these patients had no history of a sustained VT, either occurring spontaneously or at PES. This study randomly assigned 900 patients to ICD implantation or no implantation at the time of coronary bypass surgery. During an average follow-up of 32 plus or minus 16 months, there was no significant difference in mortality between patients assigned to ICD and those assigned to no ICD. The hazard ratio for total mortality with ICD placement was 1.07 (95% confidence interval, 0.81 to 1.42). In CABG Patch, 71% of the deaths were nonarrhythmic, and this accounts for the absence of benefit from the ICD.[179] It is possible that the revascularization reduced the frequency of VTs by removing the contributing role of inducible ischemia. The results of this trial do not provide support for the use of the signal-averaged ECG in risk stratification.

The Multicenter Unsustained Tachycardia Trial (MUSTT) studied patients with coronary artery disease, LVEF less than or equal to 0.40, and nonsustained VT on Holter recording who had an inducible sustained monomorphic VT or VF.[188] Patients were randomized to standard therapy for coronary disease or to standard therapy in addition to a PES-guided attempt to suppress inducible VT/VF. Patients in whom inducible VT/VF was suppressed or whose VT was hemodynamically tolerated were treated with the drug that suppressed the VT (or rendered it tolerated). Patients who continued to be inducible into a sustained VT at PES that was hemodynamically intolerable had an ICD placed. The primary end point of the study was cardiac arrest or death. A total of 704 patients were randomized. The patients who received the ICD because of nonsuppressibility would be expected to be the highest risk group, yet they had a better survival than either patients not receiving antiarrhythmic therapy or those who were suppressible with PES-guided antiarrhythmic therapy (relative risk, 0.24; 95% confidence interval, 0.13 to 0.45; $P < .001$). There was no difference in death or cardiac arrest between patients who received no antiarrhythmic therapy and those with antiarrhythmic drug suppression of inducibility. Interestingly, unlike the other ICD trials, it is possible to compare ICDs with standard treatment without an antiarrhythmic drug in patients who have had a demonstrated VT (in this case at PES). These data suggest that the ICD is the most effective approach to preventing SCD in these patients. Also, they proved an important additional demonstration of the ineffectiveness of PES-guided antiarrhythmic drug selection. The conclusions drawn from MUSTT suggest that the potentially large group of coronary patients with depressed ejection fraction and nonsustained VT who are inducible into a sustained monomorphic VT or into VF by PES should have an ICD inserted. The Multicenter Automatic Defibrillator Trial II (MADIT II) randomized post-MI patients with LVEF less than or equal to 0.30 to ICD or standard therapy.[181] The primary end point was total mortality. Most patients were in NYHA class I or II. The ICD arm had a 31% relapse risk reduction in total mortality, suggesting an important role for SCD prevention exists in this population.[181] However, most of the benefit was seen in patients with a prolonged QRS complex on the resting ECG. In summary, the accumulated data suggest that the ICD reduces mortality in patients at high risk for a fatal VT and that it is more effective than amiodarone in doing so.

A consideration of these facts leads to the current management approach to patients at risk for SCD:

a) Patients resuscitated from VF or having VT associated with hemodynamic compromise or reduced LVEF (≤.40) should have an ICD inserted. They should also be treated with aspirin, β-blockers, and ACE inhibitors as tolerated, as well as revascularization if inducible ischemia is present.
b) Patients with coronary disease, reduced LVEF (0.31 to 0.40), and nonsustained VT who have a sustained monomorphic VT induced by PES should have also an ICD placed. They should be treated with aspirin, β-blockers, and ACE inhibitors as tolerated, as well as revascularization if inducible ischemia is present. It should be noted that the presence of nonsustained VT may be useful as a screening tool for detecting higher risk for sustained ventricular arrhythmias but is likely superfluous in a patient with a history of sustained arrhythmia, including at PES.
c) Finally, patients with coronary artery disease, prior MI, and severe left ventricular dysfunction (LVEF = 0.30) should be risk stratified (e.g., by QRS width and ≥ 120 ms) or otherwise considered for ICD therapy, or both.

PRIMARY PREVENTION OF SUDDEN DEATH

At this point there is no longer any doubt about the capacity of the ICD to prevent arrhythmic death. This means that any patient destined to have a potentially fatal ventricular arrhythmia (in spite of otherwise appropriate medical therapy) could potentially benefit from having an ICD in place. A central issue for future investigation, then, is the identification of patients at sufficiently high risk of SCD to justify the morbidity and cost of placing a device. Some subpopulations of the patient populations described earlier may be effectively treated with amiodarone. For instance, a posthoc analysis of the AVID database suggests that there is no benefit of ICD over amiodarone for patients with LVEF greater than 0.35.[182] A randomized trial in such patients would be needed to make definitive treatment recommendations.

Certain populations of patients are known to be at risk for SCD but, because of competing risks of death, it is not clear whether the ICD will confer a benefit to the population as a whole.

The Defibrillator in Acute Myocardial Infarction Trial (DINAMIT) study used a risk factor representing autonomic dysfunction as a risk stratifier to allow defibrillation therapy to more effectively focus on the highest-risk patients. A reduction in heart rate variability (HRV) has been shown to portend a worse prognosis in coronary artery disease patients.[183-185] Patients in DINAMIT had low heart rate variability and an ejection fraction of less than or equal to 35%. They were within 40 days of an acute MI. There was no mortality benefit in patients randomized to the ICD ($n = 332$) compared with control patients ($n = 342$), with a hazard ratio for all-cause death of 1.08 (95% confidence interval, 0.76 to 1.55) in the ICD versus control patients, over 30 ± 13 months of follow-up.[186]

In the Defibrillators In Non-Ischemic Cardiomyopathy Treatment Evaluation (DEFINITE) study, 458 patients with nonischemic cardiomyopathy, a left ventricular EF of less than or equal to 35%, and more than 10 ventricular premature complexes per hour or any episode of unsustained VT were randomized to standard medical therapy versus standard therapy plus ICD. The hazard ratio for all-cause mortality after 29 ± 14.4 months of follow-up in the ICD versus control groups was .65 (95% confidence interval, .4 to 1.06; $P = .08$).[187]

The entire discussion to this point has focused on populations known to be at high risk for SCD. Myerberg and others have emphasized the fact that the majority of cardiac arrests occur in the low-risk, but very large, population of patients whose increased risk has not come to clinical recognition.[188] In order to make major inroads into the prevention of SCD, learning how to screen the asymptomatic population inexpensively, but safely and effectively, will be necessary. Identifying high-risk patients in the low-risk, asymptomatic population and identifying members of known high-risk populations who are not at high enough risk to justify ICD placement is the next frontier in SCD prevention.

REFERENCES

1. Bharati S: Pathology of the conduction system. In Silver MD, Gotlieb AI, Schoen FJ (eds): Cardiovascular Pathology. New York, Churchill Livingston, 2001, pp 20:607-28.
2. Bharati S, Lev M: The pathologic aspects of ventricular tachycardia. In Iwa T, Fontaine G (eds): Cardiac Arrhythmias: Recent Investigation and Management. Elsevier Science Publishers BV (Biomedical Division), 1988, pp 2:15-27.
3. Bharati S: Pathology of the conduction system. In Silver MD, Gotlieb AI, Schoen FJ (eds): Cardiovascular Pathology. New York, Churchill Livingston, 2001, pp 20:607-28.
4. Bharati S, Lev M: The pathologic aspects of ventricular tachycardia. In Iwa T, Fontaine G (eds): Cardiac Arrhythmias: Recent Investigation and Management. Elsevier Science Publishers BV (Biomedical Division) 1988;2:15-27.
5. Bharati S, Lev M: The Cardiac Conduction System in Unexplained Sudden Death. Mt. Kisco, NY, Futura Publishing, 1990, pp 1-416.
6. Bharati S, Olshansky B, Lev M: Pathological study of an explanted heart due to intractable ventricular fibrillation. J Cardiovasc Electrophysiol 1992;3:437-41.
7. Gault JH, Cantwell J, Lev M, Braunwald E: Fatal familial cardiac arrhythmias: Histologic observation in the cardiac conduction system. Am J Cardiol 1972;29:548-53.
8. Husson GS, Blackman MS, Rogers MC, et al: Familial congenital bundle branch system disease. Am J Cardiol 1973;32:365-9.
9. Cobb LA, Fahrenbruch CE, Walsh TR, et al: Influence of cardiopulmonary resuscitation prior to defibrillation in patients with out-of-hospital ventricular fibrillation. JAMA 1999;281:1182-8.
10. Kette F, Sbrojavacca R, Rellini G, et al: Epidemiology and survival rate of out-of-hospital cardiac arrest in northeast Italy: The FACS study. Resuscitation 1998;36:153-9.
11. Stratton S, Niemann J: Outcome from out-of-hospital cardiac arrest caused by nonventricular arrhythmias: Contribution of successful resuscitation to overall survivorship supports the current practice of initiating out-of-hospital ACLS. Ann Emerg Med 1998;32:448-53.
12. De Maio V, Stiell I, Wells G, Spaite D for the OPALS study group: Cardiac arrest witnessed by emergency medical services personnel: Descriptive epidemiology, prodromal symptoms, and predictors of survival. Ann Emerg Med 2000;35:138-46.
13. Holmberg M, Holmberg S, Herlitz J: Incidence, duration and survival of ventricular fibrillation in out-of-hospital cardiac arrest patients in Sweden. Resuscitation 2000;44:7-17.
14. American Heart Association: 1999 Heart and Stroke Statistical Update. Dallas, American Heart Association, 1998.
15. Kannel W, Thomas H: Sudden coronary death: The Framingham study. Ann N Y Acad Sci 1982;382:3-21.
16. Secci A, Wong N, Tang W, et al: Electron beam computed tomographic coronary calcium as a predictor of coronary events: Comparison of two protocols. Circulation 1997;96:122-9.
17. Taylor A, Burke A, O'Malley P, et al: A comparison of the Framingham risk index, coronary artery calcification and culprit plaque morphology in sudden cardiac death. Circulation 2000; 101:1243-8.
18. Thompson C, Yarzebski J, Goldberg R, et al: Changes over time in the incidence and case-fatality rates of primary ventricular fibrillation complicating acute myocardial infarction: Perspectives from the Worcester heart attack study. Am Heart J 2000;139:1014-21.
19. Volpi A, Cavalli A, Santoro L, Negri E: Incidence and prognosis of early primary ventricular fibrillation in acute myocardial infarction. Results of the Gruppo Italiano per lo Studio della Sopravvivenza nell'Infarto Miocardico (GISSI-2) Database. Am J Cardiol 1998;82:265-71.
20. Furukawa T, Rozanski J, Nogami A, et al: Time-dependent risk of and predictors for cardiac arrest recurrence in survivors of out-of-hospital cardiac arrest with chronic coronary artery disease. Circulation 1989;80:599-608.
21. The AVID investigators: Causes of death in the Antiarrhythmics versus Implantable Defibrillators trial. J Am Coll Cardiol 1999; 34:1552-9.
22. Klein H, Auricchi A, Reek S, Geller C: New primary prevention trials of sudden cardiac death in patients with left ventricular dysfunction: SCD-HeFT and MADIT-II. Am J Cardiol 1999;83: 91d-97d.
23. Hermida J, Lemoine J, Aoun F, et al: Prevalence of the Brugada syndrome in an apparently healthy population. Am J Cardiol 2000;86:91-4.
24. Allessie MA: Circus movement in rabbit atrial muscle as a mechanism of tachycardia. III. The "leading circle" concept: A new model of circus movement in cardiac tissue without the involvement of an anatomic obstacle. Circ Res 1977;41:9-18.
25. Janse MJ: Functional reentry: Leading circle or spiral wave? J Cardiovasc Electrophysiol 1999;10:621-2.
26. Ravelli F, Disertori M, Cozzi F, et al: Ventricular beats induce variations in cycle length of rapid (type II) atrial flutter in humans: Evidence of leading circle reentry. Circulation 1994;89: 2107-16.
27. Frazier DW, Wolf PD, Wharton JM, et al: Stimulus-induced critical point mechanism for electrical initiation of reentry in normal canine myocardium. J Clin Invest 1988;83:1039-52.
28. Krinsky V: Qualitative theory of reentry. In Zipes D, Jalife J (eds): Cardiac Electrophysiology. From Cell to Bedside, 3rd ed. Philadelphia, WB Saunders, 2000, pp 320-7.
29. Panfilov A, Persov A: Vortex rings in a three-dimensional medium described by reaction-diffusion equations. Dokl Biophys 1984; 274:58-60.
30. Moe GK: On the multiple wavelet hypothesis of atrial fibrillation. Arch Int Pharmacodyn 1962;140:183-8.
31. Perstov AM, Davidenko JM, Salomonsz R, et al: Spiral waves of excitation underlie reentrant activity in isolated cardiac muscle. Circ Res 1993;72:631-50.
32. Jalife J, Berenfeld O, Skanes A, Mandapati R: Mechanisms of atrial fibrillation: Mother rotors or multiple daughter wavelets, or both? J Cardiovasc Electrophysiol 1998;9(suppl):S2-S12.
33. Chen P, Wolf PD, Dixon EG, et al: Mechanism of ventricular vulnerability to single premature stimuli in open-chest dogs. Circ Res 1988;62:1191-209.
34. Antzelevitch C, Sicouri S, Litovsky SH, et al: Heterogeneity within the ventricular wall: Electrophysiology and pharmacology of epicardial, endocardial and M cells. Circ Res 1991;69:1427-49.
35. Antzelevitch C, Nesterenko VV, Yan G-X: Role of M cells in acquired long QT syndrome, U wakes, and torsade de pointes. J Electrocardiol 1996;28(suppl):131-8.
36. Akar FG, Laurita KR, Rosenbaum DS: Cellular basis for dispersion of repolarization underlying reentrant arrhythmias. J Electrocardiol 2000;33:(suppl):23-31.
37. Rosenbaum DS, Jackson LE, Smith JM, et al: Electrical alternans to the genesis of cardiac fibrillation. Circulation 1999;99:1385-94.
38. Pastore JM, Girouard SD, Laurita KR, et al: Mechanism linking T-wave alternans to the genesis of cardiac fibrillation. Circulation 1999;99:1385-94.
39. Laurita KR, Girouard SD, Akar FG, et al: Modulated dispersion explains changes in arrhythmia vulnerability during premature stimulation of the heart. Circulation 1998;98:2774-80.
40. Watanabe MA, Fenton FH, Evans SJ, et al: Mechanisms for discordant alternans. J Cardiovasc Electrophysiol 2001;12:196-206.
41. Garfinkel A, Kim YH, Voroshilovsky O, et al: Preventing ventricular fibrillation by flattening cardiac restitution. Proc Natl Acad Sci USA 2000;97:6061-6.
42. Koller M, Riccio M, Gilmour RJ: Effects of $[K^+]_0$ on electrical restitution and activation dynamics during ventricular fibrillation. Am J Physiol Heart Circ Physiol 2000;279:H2665-H2672.
43. Watanabe M, Otani NF, Gilmour FR Jr: Biphasic restitution of action potential duration and complex dynamics in ventricular myocardium. Circ Res 1995;76:915-21.
44. Dekker L, Coronel R, VanBavel E, et al: Intracellular Ca^{2+} and the delay of ischemia-induced electrical uncoupling in preconditioned rabbit ventricular myocardium. Cardiovasc Res 1999;44:101-12.
45. Han J, Moe GK: Nonuniform recovery of excitability of ventricular muscle. Circ Res 1964;14:44-60.
46. Franz MR, Flaherty JT, Platia EV, et al: Localization of regional myocardial ischemia by recording of monophasic action potentials. Circulation 1984;69:593-604.
47. Mohabir R, Franz MR, Clusin WT: In vivo electrophysiological detection of myocardial ischemia through monophasic action potential recording. Prog Cardiovasc Dis 1991;34:15-28.
48. Dilly SG, Lab MJ: Changes in monophasic action potential duration during the first hour of regional myocardial ischemia in the anaesthetized pig. Cardiovasc Res 1987;21:908-15.

49. Jongsma HJ WR: Gap junctions in cardiovascular disease. Circ Res 2000;86:1193-7.
50. Owens LM, Fralix TA, Murphy E, et al: Correlation of ischemia-induced extracellular and intracellular ion changes to cell-to-cell electrical uncoupling in isolated blood-perfused rabbit hearts. Experimental Working Group. Circulation 1996;94:10-3.
51. Owens L, Murphy E, Fralix TA: Relationship of cellular electrical uncoupling to changes of CA, pH, ATP, and contracture in ischemic rabbit myocardium [abstract]. Circulation 1993; 88:1373.
52. Coronel R, Fiolet JW, Wilms-Schopman FJ, et al: Distribution of extracellular potassium and its relation to electrophysiologic changes during acute myocardial ischemia in the isolated perfused porcine heart. Circulation 1988;77:1125-38.
53. Cascio WE, Yan GX, Kleber AG: Passive electrical properties, mechanical activity, and extracellular potassium in arterially perfused and ischemic rabbit ventricular muscle. Effects of calcium entry blockade or hypocalcemia. Circ Res 1990;66:1461-73.
54. Kagiyama Y, Hill JL, Gettes LS: Interaction of acidosis and increased extracellular potassium on action potential characteristics and conduction in guinea pig ventricular muscle. Circ Res 1982;51:614-23.
55. Anversa P, Olivetti G, Meggs LG, et al: Cardiac anatomy and ventricular loading after myocardial infarction. Circulation 1993; 87(suppl):VII22-VII27.
56. Qin D, Zhang ZH, Caref EB, et al: Cellular and ionic basis of arrhythmias in post-infarction remodeled ventricular myocardium. Circ Res 1996;79:461-73.
57. Peters NS, Green CR, Poole-Wilson PA, Severs NJ: Reduced content of connexin43 gap junctions in ventricular myocardium from hypertrophied and ischemic human hearts. Circulation 1993; 88:864-75.
58. Starobin JM, Zilberter YI, Rusnak EM, Starmer CF: Wavelet formation in excitable cardiac tissue: The role of wavefront obstacle interactions in initiating high-frequency fibrillatory-like arrhythmias. Biophys J 1996;70:581-94.
59. Stramba-Badiale M, Vanoli EDFG, et al: Sympathetic-parasympathetic interaction and accentuated antagonism in conscious dogs. Am J Physiol 1991;260:H335-H340.
60. Vanoli E, De Ferrari G, Stramba-Badiale M, et al: Vagal stimulation and prevention of sudden death in conscious dogs with a healed myocardial infarction. Circ Res 1991;68:1471-81.
61. De Ferrari G, Vanoli E, Stramba-Badiale M, et al: Vagal reflexes and survival during acute myocardial ischemia in conscious dogs with a healed myocardial infarction. Am J Physiol 1991;261: H63-H69.
62. Schwartz PJ, Stone HL: Left stellectomy in the prevention of ventricular fibrillation caused by acute myocardial ischemia in conscious dogs with anterior myocardial infarction. Circulation 1980;62:1256-65.
63. Schwartz PJ, Stone HL, Brown AM: Effects of unilateral stellate ganglion blockade on the arrhythmias associated with coronary occlusion. Am Heart J 1976;92:589-99.
64. GHAT investigators: A randomized trial of propranolol in patients with acute myocardial infarction: I. Mortality results. JAMA 1982; 247:1707-14.
65. Hjalmarsen A: Effects of beta blockade on sudden cardiac death during acute myocardial infarction and the postinfarction period. Am J Cardiol 1997;80:35J-39J.
66. Timolol-induced reduction in mortality and reinfarction in patients surviving acute myocardial infarction. N Engl J Med 1981;304:801-7.
67. Vanoli E, Hull SJ, Foreman R, et al: Alpha-1 adrenergic blockade and sudden cardiac death. J Cardiovasc Electrophysiol 1994; 5:76-89.
68. Kammerling JJ, Green FJ, Watanabe AM, et al: Denervation supersensitivity of refractoriness in noninfarcted areas apical to transmural myocardial infarction. Circulation 1987;76:383-93.
69. Brugada P, Brugada J, Brugada R: The Brugada syndrome. Cardiovasc Drugs Ther 2001;15:15-17.
70. Chen QY, Kirsch GE, Zhang DM, et al: Genetic basis and molecular mechanism for idiopathic ventricular fibrillation. Nature 1998;392:293-6.
71. Priori SG, Napolitano C, Giordano U, et al: Brugada syndrome and sudden cardiac death in children. Lancet 2000;355:808-9.
72. Antzelevitch C: The Brugada syndrome. J Cardiovasc Electrophysiol 1998;9:513-6.
73. Lukas A, Antzelevitch C: Phase 2 reentry as a mechanism of initiation of circus movement reentry in canine epicardium exposed to simulated ischemia. Cardiovasc Res 1996;32:593-603.
74. Pandis IP, Morganroth J: Sudden death in hospitalized patients: Cardiac rhythm disturbances detected by ambulatory electrocardiographic monitoring. J Am Coll Cardiol 1983;2:798-805.
75. Pandis IP, Morganroth J: Holter monitoring and sudden cardiac death. Cardiovasc Rev Rep 1984;5:283-304.
76. Pratt CM, Francis MJ, Luck JC: Observations on sudden cardiac death recorded during ambulatory electrocardiographic monitoring. Circulation 1982;66(suppl):2:26.
77. Savage DD, Dasteli WP, Anderson SJ: Sudden unexpected death during ambulatory electrocardiographic monitoring: The Framingham study. Am J Med 1983;74:148-52.
78. Luu M, Stevenson WG, Stevenson LW, et al: Diverse mechanisms of unexpected sudden cardiac arrest in advanced heart failure. Circulation 1989;80:1675-80.
79. Merri M, Benhorin J, Alberti M, et al: Electrocardiographic quantitation of ventricular repolarization. Circulation 1989;80: 1301-8.
80. Surawicz B, Knoebel SB: Long QT: Good, bad, or indifferent. J Am Coll Cardiol 1984;4:398-413.
81. Priori S, Bloise R, Crotti L: The long QT syndrome. Europace 2001;3:16-27.
82. Schwartz PJ: The long QT syndrome. In Kulbertus HE, Wellens HJJ (eds): Sudden Death. The Hague, M Nijhoff 1980, pp 358-78.
83. Garson A Jr, Dick M II, Fournier A, et al: The long QT syndrome in children: An international study of 287 patients. Circulation 1993;87:11866-72.
84. Moss AJ, Schwartz PJ, Crampton RS, et al: The long QT syndrome: Prospective longitudinal study of 328 families. Circulation 1991;84:1136-44.
85. Schwartz PJ, Priori SG, Locati EH, et al: Long QT syndrome patients with mutations on the SCN5A and HERG genes have differential responses to Na channel blockade and to increases in heart rate. Implications for gene-specific therapy. Circulation 1995;92:3381-6.
86. Malfatto G, Beria G, Sala S, et al: Quantitative analysis of T wave abnormalities and their prognostic implications in the idiopathic long QT syndrome. J Am Coll Cardiol 1994;23:296-301.
87. De Ambroggi L, Negroni MS, Monza E, et al: Dispersion of ventricular repolarization in the long QT syndrome. Am J Cardiol 1991;68:614-20.
88. Priori SG, Napolitano C, Diehl L, Schwartz PJ: Dispersion of the QT interval: A marker of therapeutic efficacy in the idiopathic long QT syndrome. Am Heart J 1975;89:1681-9.
89. Schwartz PJ, Periti M, Maliani A: The long QT syndrome. Am Heart J 1975;378-90.
90. Kay GN, Plumb VJ, Arciniegas JG, et al: Torsade de pointes: The long-short initiating sequence and other clinical features: Observations in 32 patients. Am J Cardiol 1983;806-17.
91. Jackman WM, Friday KJ, Anderson JL, et al: The long QT syndrome: A critical review, new clinical observations, and a unifying hypothesis. Prog Cardiovasc Disease 1986;31:115-72.
92. Guzak I, Antzelevitch C, Bjerregaard P, et al: The Brugada syndrome: Clinical, electrophysiologic and genetic aspects. J Am Coll Cardiol 1999;33:5-15.
93. Atararashi J, Ogawa W, Harumi K, et al: Characteristics of patients with right bundle branch block and ST-segment elevation in right precordial leads. Am J Cardiol 1996;78:581-3.
94. Brugada P, Brugada R, Brugada J: Sudden death in high-risk family members. Brugada syndrome. Am J Cardiol 200;86:40-3.
95. Matsuo K, Akahoshi M, Nakashima E, et al: The prevalence, incidence and prognostic value of the Brugada-type electrocardiogram. A population-based study of incidence and prognostic value of the Brugada-type electrocardiogram. A population-based study of four decades. J Am Coll Cardiol 2001;38:765-70.
96. Monroe MH, Littman L: Two-year case collection of the Brugada syndrome electrocardiogram pattern at a large teaching hospital. Clin Cardiol 2002;23:849-51.
97. Hermida JS, Lemoine JL, Aoun FB, et al: Prevalence of the Brugada syndrome in an apparently healthy population. Am J Cardiol 2000;86:91-4.

98. Myasaka Y, Tsuji H, Yamada K, et al: Prevalence and mortality of the Brugada-type electrocardiogram in one city in Japan. Am J Cardiol 2001;38:771-4.
99. Takenaka S, Hisamatsu K, Nagase S, et al: Relatively benign clinical course in asymptomatic patients with Brugada-type electrocardiogram without family history of sudden death. J Cardiovasc Electrophysiol 2001;12:2-6.
100. Brugada J, Brugada R, Brugada P: Asymptomatic patients with Brugada electrocardiogram: Are they at risk? J Cardiovasc Electrophysiol 2001;12:7-8.
101. Priori SG, Napolitano C, Gasparini M, et al: Clinical and genetic heterogeneity of right bundle branch block and ST-segment elevation syndrome. A prospective evaluation of 52 families. Circulation 2000;102:2509-15.
102. Atarachi H, Ogawa S, Harumi K, et al: Three year follow-up of patients with right bundle branch block and ST segment elevation in the right precordial leads. Japanese registry of Brugada syndrome. Idiopathic ventricular fibrillation investigators. J Am Coll Cardiol 2001;37:1916-20.
103. Macus FI, Fontaine GH, Guiradon G, et al: Right ventricular dysplasia: A report of 24 adult cases. Circulation 1982;65:384-98.
104. Fontaine G, Umemura J, Di Donna P, et al: Duration of the QRS complexes in arrhythmogenic right ventricular dysplasia. A new non-invasive diagnostic marker. Ann Cardiol Angeiol (Paris) 1993;42:399-405.
105. Fontaine G, Fontaliran F, Hebert JL, et al: Arrhythmogenic right ventricular dysplasia. Annu Rev Med 1999;50:17-35.
106. Nava A, Bauce B, Basso C, et al: Clinical profile and long-term follow-up of 37 families with arrhythmogenic right ventricular cardiomyopathy. J Am Coll Cardiol 2000;36:2226-33.
107. Metzger JT, de Chillou C, Cheriex E, et al: Value of the 12-lead electrocardiogram in arrhythmogenic right ventricular dysplasia and absence of correlation with echocardiographic findings. Am J Cardiol 1993;72:964-7.
108. Mitchell LB: Drug therapy of sustained ventricular tachyarrhythmias: Is there still a role? Cardiol Clin 2000;18:357-73.
109. Salerno DM, Katz A, Dunbar DN, et al: Serum electrolytes and catecholamines after cardioversion from ventricular tachycardia and atrial fibrillation. PACE 1993;16:1862-71.
110. Cobb LA, Baum RS, Alvarez H III, et al: Resuscitation from out-of-hospital ventricular fibrillation: 4 years follow-up. Circulation 1975;51(suppl III):III223-III228.
111. Volpi A, Cavalli A, Santoro L, et al: Incidence and prognosis of early primary ventricular fibrillation in acute myocardial infarction—results of the Gruppo Italiano per lo Studio della Sopravvivenza nell'Infarto Miocardico (GISSI-2) database. Am J Cardiol 1998;82:265-71.
112. Kerin NZ, Somberg J: Proarrhythmia: Definition, risk factors, causes, treatment, and controversies. Am Heart J 1994;128:575-85.
113. Podrid PJ, Graboys TB: Exercise stress testing in the management of cardiac rhythm disorders. Med Clin North Am 1984;68:1139-52.
114. Kelly P, Ruskin JN, Vlahakes GJ, et al: Surgical coronary artery revascularization in survivors of pre-hospital cardiac arrest: Its effect on inducible ventricular arrhythmias and long-term survival. J Am Coll Cardiol 1990;15:267-73.
115. Mitchell LB, Gettes LS: Is a baseline electrophysiologic study mandatory for the management of patients with spontaneous, sustained, ventricular tachyarrhythmias? Prog Cardiovasc Dis 1996;38:385-92.
116. Mason J for the Electrophysiologic Study Versus Electrocardiographic Monitoring Investigators: A comparison of electrophysiologic testing with Holter monitoring to predict antiarrhythmic drug efficacy for ventricular tachyarrhythmias. N Engl J Med 1993;329:445-51.
117. Mitchell LB, Duff HJ, Manyari DE, Wyse DG: A randomized clinical trial of the noninvasive and invasive approaches to drug therapy ventricular tachycardia. N Engl J Med 1987;317:1681-7.
118. Mitchell LB, Duff HJ, Miller CE, et al: Drug therapy of ventricular tachycardia: A cost comparison of randomized noninvasive/invasive approaches. Can J Cardiol 1992;8:487-94.
119. Buxton AE, Lee KL, Fisher JD, et al: A randomized study of the prevention of sudden death in patients with coronary artery disease. N Engl J Med 1999;341:1882-90.
120. Wilber DJ, Olshansky B, Moran JF, Scanlong PJ: Electrophysiological testing and nonsustained ventricular tachycardia. Circulation 1990;82:350-8.
121. Swerdlow CD, Bardy GH, McAnulty J, et al: Determinants of induced ventricular arrhythmias in survivor of out-of -hospital VF. Circulation 1987;76:1053-60.
122. Adhar GC, Larson LW, Bardy GH, et al: Sustained ventricular arrhythmias, differences between survivors of cardiac arrest and patients with recurrent sustained ventricular tachycardia. J Am Coll Cardiol 1988;12:159-65.
123. Morady F, DiCarlo LA Jr, Baerman JM, DeBuitleir M: Comparison of coupling intervals that induce clinical and nonclinical forms of ventricular tachycardia during programmed stimulation. Am J Cardiol 1986;57:1269-73.
124. Avitall B, McKinnie J, Jazayeri M, et al: Induction of VF versus monomorphic ventricular tachycardia during programmed stimulation. Circulation 1992;85:1271-8.
125. Avitall B, Levine HJ, Naimi S, et al: Local effects of electrical and mechanical stimulation in the recovery properties of the canine ventricle. Am J Cardiol 1982;50:263-70.
126. Horowitz LN, Greenspan AM, Speilman SR, Josephson ME: Torsades de pointes: Electrophysiologic studies in patients without transient pharmacologic of metabolic abnormalities. Circulation 1981;63:1120.
127. The Antiarrhythmics versus Implantable Defibrillators (AVID) Investigators: A comparison of antiarrhythmic drug therapy with implantable defibrillators in patients resuscitated from near fatal ventricular arrhythmias. N Engl J Med 1997;337:1576-83.
128. Connolly SJ, Gent M, Roberts RS, et al: Canadian Implantation Defibrillator Study (CIDS) Investigators. Circulation 2000; 101:1297-302.
129. Cappato R, Siebels J, Ruppel R, et al. For the Cardiac Arrest Study Hamburg (CASH) Investigators. PACE 2000;23:568.
130. Voelckel WG, Lurie KG, Sweeney M, et al: Effects of active compression-decompression cardiopulmonary resuscitation with the inspiratory threshold valve in a young porcine model of cardiac arrest. Pediatr Res 2002;51:523-7.
131. Gueugniaud PY, Mols P, Goldstein P, et al: A comparison of repeated high doses and repeated standard doses of epinephrine for cardiac arrest outside the hospital. European Epinephrine Study Group. N Engl J Med 1998;339:1595-601.
132. Wenzel V, Lindner KH, Krismer AC, et al: Survival with full neurologic recovery and no cerebral pathology after prolonged cardiopulmonary resuscitation with vasopressin in pigs. J Am Coll Cardiol 2000;35:527-33.
133. Lindner KH, Dirks B, Strohmenger HU, et al: Randomized comparison of epinephrine and vasopressin in patients with out-of-hospital ventricular fibrillation. Lancet 1997;349:535-7.
134. Stiell IG, Hebert PC, Wells GA, et al: Vasopressin versus epinephrine for inhospital cardiac arrest: A randomized controlled trial. Lancet 2001;358:105-9.
135. Weaver WD, Fahrenbruch CE, Johnson DD, et al: Effect of epinephrine and lidocaine therapy on outcome after cardiac arrest due to ventricular fibrillation. Circulation 1990;82:2027-34.
136. Hine LK, Laird N, Hewitt P, Chalmers TC: Meta-analytic evidence against prophylactic use of lidocaine in acute myocardial infarction. Arch Intern Med 1989;149:2694-8.
137. Echt DS, Gremillion ST, Lee JT, et al: Effects of procainamide and lidocaine on defibrillation energy requirements in patients receiving implantable cardioverter defibrillator devices. J Cardiovasc Electrophysiol 1994;5:752-60.
138. Dorian P, Fain ES, Davy JM, Winkle RA: Lidocaine causes a reversible, concentration-dependent increase in defibrillation energy requirements. J Am Coll Cardiol 1986;8:327-32.
139. Thel MC, Armstrong AL, McNulty SE, et al: Randomized trial of magnesium in in-hospital cardiac arrest. Duke Internal Medicine Housestaff. Lancet 1997;350:1272-6.
140. Kudenchuk PJ, Cobb LA, Copass MK, et al: Amiodarone for resuscitation after out-of-hospital cardiac arrest due to ventricular fibrillation. N Engl J Med 1999;341:871-8.
141. Dorian P, Cass D, Schwartz B, et al: Amiodarone as compared with lidocaine for shock-resistant ventricular fibrillation. N Engl J Med 2002;346:884-90.
142. Nademanee K, Taylor R, Bailey WE, et al: Treating electrical storm: Sympathetic blockade versus advanced cardiac life support-guided therapy. Circulation 2000;102:742-7.
143. Kowey PR, Levine JH, Herre JM, et al: Randomized, double-blind comparison of intravenous amiodarone and bretylium in the treatment of patients with recurrent, hemodynamically

destabilizing ventricular tachycardia or fibrillation. The Intravenous Amiodarone Multicenter Investigators Group. Circulation 1995;92:3255-63.
144. Mann D: Mechanisms and models in heart failure. Circulation 1999;100:999-1008.
145. The Norwegian Multicenter Study Group: Timolol-induced reduction in mortality and reinfarction in patients surviving acute myocardial infarction. N Engl J Med 1981;304:801-7.
146. The Beta-Blocker Heart Attack Research Group. A randomized trial of propranolol in patients with acute myocardial infarction, I: Mortality results. JAMA 1982;247:1707-14.
147. Teo K, Yusef S, Furberg C: Effects of prophylactic antiarrhythmic drug therapy in acute myocardial infarction: An overview of results from randomized, controlled trials. JAMA 1993; 270:1589-95.
148. CIBIS-II Investigators and Committees: The Cardiac Insufficiency Bisoprolol Study II (CIBIS II): A randomized trial. Lancet 1999;353:9-13.
149. The MERIT-HF Study Group: Effect of metoprolol CR/XL in chronic heart failure: Metoprolol CR/XL Randomized Intervention Trial in Congestive Heart Failure (MERIT-HF). Lancet 1999;353:2001-7.
150. Pfeffer M, Braunwald E, Moye, et al: Effect of captopril on mortality and morbidity in patients with left ventricular dysfunction after myocardial infarction: Results of the Survival and Ventricular Enlargement Trial. N Engl J Med 1992;327:669-77.
151. The Consensus Trial Study Group: Effects of enalapril on mortality in severe congestive heart failure: Results of the cooperative North Scandinavian Enalapril Survival Study (CONSENSUS). N Engl J Med 1987;316:1429-35.
152. The Studies of Left Ventricular Dysfunction Investigators: Effect of enalapril in survival in patients with reduced left ventricular ejection fraction and congestive heart failure. N Engl J Med 1991;325:293-302.
153. Kober L, Torp-Pederson C, Carlsen J, et al for the Trandolapril Cardiac Evaluation (TRACE) Study Group: A clinical trial of the ACE inhibitor Trandolapril in patients with left ventricular dysfunction after myocardial infarction. N Engl J Med 1995;333: 1670-6.
154. Domanski M, Exner D, Borkowf C, et al: Effect of angiotensin converting enzyme inhibition on sudden cardiac death in patients following a myocardial infarction: A meta-analysis of randomized clinical trials. J Am Coll Card 199;33:598-604.
155. Pitt B, Zannad F, Remme W, et al for the Randomized Aldactone Evaluation Study Investigators: The effect of spironolactone on morbidity and mortality in patients with severe heart failure. N Engl J Med 1999;341:709-17.
156. Davies M: Anatomic features in victims of sudden coronary death: Coronary artery pathology. Circulation 1992;85(suppl I): I19-I24.
157. Farb A, Tang A, Burke A, et al: Sudden coronary death: Frequency of active coronary lesions, inactive coronary lesions and myocardial infarction. Circulation 1995;92:1701-9.
158. ISIS-2 (Second International Study of Infarction Survival) Collaborative Group: Randomized trial of intravenous streptokinase, oral aspirin, both or neither among 17,187 cases of suspected myocardial infarction. Lancet 1988;ii:349-60.
159. Dries D, Domanski M, Waclawiw M, Gersh B: Effect of antithrombotic therapy on risk of sudden coronary death in patients with congestive heart failure. Am J Cardiol 1997;79:909-13.
160. Echt D, Liebson P, Mitchell B, et al: Mortality and morbidity in patients receiving encainide, flecainide, or placebo: The Cardiac Arrhythmia Suppression Trial (CAST I). N Engl J Med 1992;327:227-33.
161. The Cardiac Arrhythmia Suppression Trial II Investigators: Effect of antiarrhythmic agent moricizine on survival after myocardial infarction. N Engl J Med 1992;327:227-33.
162. Waldo A, Camm J, deRuyter H, et al: Survival with oral d-sotalol in patients with left ventricular dysfunction after myocardial infarction: Rationale, design and methods (the SWORD Trial). J Am Coll Cardiol 1995;75:1023-7.
163. Torp-Pedersen C, Moller M, Bloch-Thomsen P, et al: Dofetilide in patients with congestive heart failure and left ventricular dysfunction. N Engl J Med 1999;341:857-65.
164. Kober L, Bloch Thomsen PE, Moller M, et al, Danish Investigations of Arrhythmia and Mortality on Dofetilide (DIAMOND) Study Group: Effect of dofetilide in patients with recent myocardial infarction and left-ventricular dysfunction: A randomised trial. Lancet 2000;356:2052-8.
165. Teo K, Yusef S, Furberg C: Effects of prophylactic antiarrhythmic drug therapy in acute myocardial infarction: An overview of results from randomized controlled trials. JAMA 1993;270:1589-95.
166. Burkart F, Pfisterer M, Kiowski W, et al: Effect of antiarrhythmic therapy on mortality on survivors of myocardial infarction with asymptomatic complex ventricular arrhythmias: Basel Antiarrhythmic Study of Infarct Survival (BASIS). J Am Cardiol 1992;20:1056-62.
167. Ceremuzynski L, Kleczar K, Krzeminska-Pakula M, et al: Effect of amiodarone on mortality after myocardial infarction: A double-blind, placebo-controlled, pilot study. J Am Coll Cardiol 1992; 20:1056-62.
168. Julian D, Camm A, Frangin G, et al for the European Myocardial Infarction Amiodarone Trial Investigators: Randomized trial of effect of amiodarone on mortality in patients with left ventricular dysfunction after recent myocardial infarction: EMIAT. Lancet 1997;349:667-74.
169. Cairns J, Connolly S, Roberts R, Gent M for the Canadian Amiodarone myocardial Infarction Arrhythmia Trial Investigators: Randomized trial of outcome after myocardial infarction in patients with frequent of repetitive premature depolarizations: CAMIAT. Lancet 1997;349:675-82.
170. Amiodarone Trials Meta-Analysis Investigators. Effect of prophylactic amiodarone in mortality after acute myocardial infarction and in congestive heart failure: Meta-analysis of individual data from 6500 patients in randomized trials. Lancet 1997; 350:1417-24.
171. Sim I, McDonald K, Lavori P, Norbutas B: Quantitative overview of randomized trials of amiodarone to prevent sudden cardiac death. Circulation 1997;96:2823-9.
172. Holmes D, Davis K, Gersh B, et al, and participants in the Coronary Artery Surgery Study: Risk factor profiles of patients with sudden death and from other cardiac causes: A report of the Coronary Artery Surgery Study (CASS). J Am Cardiol 1989; 13:524-30.
173. Garan H, Ruskin J, DiMarco J, et al: Electrophysiologic studies before and after myocardial revascularization in patients with life-threatening ventricular arrhythmias. Am J Cardiol 1983; 51:519-24.
174. Hii J, Traboulsi M, Mitchell L, et al: Infarct artery patency predicts outcome of electropharmacological studies in patients with malignant tachyarrhythmias. Circulation 1993;87:764-72.
175. Mirowski M, Reid P, Mower M, et al: Termination of malignant ventricular arrhythmias with an implanted automatic defibrillator in human beings. N Engl J Med 1980;303:322-4.
176. Saksena S: The PCD investigators Group: Clinical outcome of patients with malignant ventricular tachyarrhythmias and a multiprogrammable cardioverter-defibrillator implanted with or without thoracotomy: An international multicenter study. J Am Coll Cardiol 1994;23:1521-30.
177. Moss A, Hall J, Cannom D, et al for the Multicenter Automatic Defibrillator Implantation Trial Investigators: Improved survival with an implanted defibrillator in patients with coronary disease at high risk for ventricular arrhythmia. N Engl J Med 1996; 335:1933-40.
178. Bigger J for the Coronary Artery Bypass Graft (CABG) Patch Trial Investigators: Prophylactic use of implanted cardiac defibrillators in patients at high risk for ventricular arrhythmias after coronary artery bypass graft surgery. N Engl J Med 1997; 337:1569-75.
179. Bigger J, Whang W, Rottman J, et al: Mechanisms of death in the CABG Patch Trial. A randomized trial of implantable cardiac defibrillator prophylaxis in patients at high risk of death after coronary artery bypass graft surgery. Circulation 1999;99:1416-21.
180. Buxton A, Lee K, Fisher J, et al for the Multicenter Unsustained Tachycardia Trial (MUSTT) Investigators: A randomized study of the prevention of sudden death in patients with coronary artery disease. N Engl J Med 1999;341:1882-90.
181. Moss AJ, Zareba W, Hall WJ, et al: Multicenter Automatic Defibrillator Implantation Trial II investigators: Prophylactic implantation of a defibrillator in patients with myocardial infarction and reduced ejection fraction. N Engl J Med 2002;346:877-83. Epub 2002; Mar 19.

182. Domanski M, Saksena S, Epstien A, et al for the AVID Investigators: Relative effectiveness of the implantable cardioverter-defibrillator and antiarrhythmic drugs in patients with varying degrees of left ventricular dysfunction who have survived malignant ventricular arrhythmias. J Am Coll Cardiol 1999;34:1090-5.
183. Hohnloser S, Klingengeben T, Zabel M, Li Y: Heart rate variability used as an arrhythmia risk stratifier after myocardial infarction. Pacing Clin Electrophysiol 1997;20:2594-601.
184. LaRovere M, Bigger J, Marcus F, et al: Baroreflex sensitivity and heart-rate variability in prediction of total cardiac mortality after myocardial infarction. ATRAMI (Autonomic Tone and Reflexes After Myocardial Infarction) Investigator. Lancet 1998;351:478-84.
185. Nolan J, Batin P, Andrews R, et al: Prospective study of heart rate variability and mortality in chronic heart failure: Results in the United Kingdom heart failure evaluation and assessment of risk trial (UK-HEART). Circulation 1998;98:1510-6.
186. Connolly SJ, Hohnloser SH, on behalf of the DINAMIT steering committe and investigators: DINAMIT: Randomized Trial of Prophylactic Implantable Defibrillator Therapy versus Optimal Medical Treatment Early after Myocardial Infarction: The Defibrillator in Acute Myocardial Infarction Trial. Late Breaking Clinical Trial, ACC Annual Scientific Session 2004.
187. Kadish A, Dyer A, Daubert JP, et al: Prophylactic defibrillator implantation in patients with non-ischemic dilated cardiomyopathy. N Engl J Med 2004;350:2151-58.
188. Huikuri HV, Makikallio TH, Raatikainen MJ, et al: Prediction of sudden cardiac death: Appraisal of the studies and methods assessing the risk of sudden arrhythmic deaath. Circulation 2003;108:110-5.

SECTION III

Clinical Syndromes

Chapter 20 Sudden Cardiac Death

VIVEK Y. REDDY and JEREMY N. RUSKIN

In developed countries, the most common cause of natural death stems from cardiovascular disease. This remains true despite an approximately 50% decrease in cardiovascular disease–related mortality in the United States between 1965 and 1990.[1] Of these, *sudden* death from cardiac causes is estimated to account for approximately 50% of the cases.[2,3] In the United States and other developed countries, the majority of sudden cardiac deaths (SCDs) are due to ventricular tachyarrhythmias, most often in the setting of coronary artery disease.[3,4] In less developed countries, the relative paucity of cause-specific mortality data is a major impediment to the estimation of the absolute and relative SCD toll.[5] Some reports suggest that the SCD rate in these countries approximates the incidence of coronary artery disease. However, other studies suggest that that the epidemiology in these developing countries may be more closely linked to non–coronary disease–related causes of heart failure (such as hypertensive heart disease).[6] In either event, it is clear that as life expectancy in developing countries is increasing, the incidence of cardiovascular diseases and SCD is also increasing.

Over the past 20 years, data from various clinical trials has increased our understanding of who is at high risk for SCD. In general, these trials have shown that antiarrhythmic medications are largely ineffective in preventing SCD in those individuals at high risk.[7-13] The only preventative therapy that is highly and consistently efficacious is the implantable cardioverter-defibrillator (ICD). Indeed, the advent of the ICD is among the most important developments in the field of electrophysiology in the past 20 years. However, the efficacy of the ICD has been evaluated only in very high-risk patient subsets that represent a relatively small fraction of the total number of patients who will manifest SCD. That is, the majority of patients at risk for SCD do not fall into the categories of patients that were included in the clinical trials on ICD efficacy.

Definition

SCD refers to unexpected natural death from a cardiovascular cause within a short period of time, generally less than 1 hour from the onset of symptoms.[3-4] By definition, this occurs in individuals who are not afflicted with prior conditions that could be fatal in the short term. Today, SCD is generally defined as death from unexpected cardiovascular collapse, which is usually due to a cardiac arrhythmia, within an hour of the onset of symptoms. The use of 1 hour in this definition is an arbitrary one. However, it is important to distinguish a primary arrhythmic cause of death from, for example, an episode of worsening congestive heart failure that culminates in a life-threatening arrhythmia. Unfortunately, it is frequently difficult to be sure about the exact circumstances surrounding a particular death—a fact that must be remembered when evaluating clinical trials that use SCD as an end point.

Epidemiology of Sudden Cardiac Death

The occurrence of SCD in the United States is generally approximated at 300,000 deaths per year, with estimates ranging from as low as 200,000 to as high as 400,000 cases per year.[3,4,14,15] The figure of 300,000 annual deaths assumes an average of 1 to 2 deaths per 1000 adults older than the age of 35 per year and represents 50% of all heart-related deaths.[15] However, the population substrate of these patients is a changing landscape, as is the array of coronary interventional capabilities that have evolved during the past 25 years.

Current epidemiologic data indicate that in developed countries, structural coronary arterial abnormalities and their consequences account for 80% of fatal arrhythmias.[2-4,15] Dilated and hypertrophic cardiomyopathies account for the next largest group of SCDs. The remainder of the deaths are due to other cardiac disorders such as valvular or congenital heart diseases and primary electrophysiological disorders, including ion "channelopathies." The latter account for a smaller percentage of the population who present with SCD. However, the greatest advances in our understanding of the mechanisms underlying SCD have been made in these "channelopathies."

In developing countries, a major surge in life expectancy is expected to significantly affect both the proportion and absolute mortality rates related to cardiovascular disease and coronary artery disease. For example, life expectancy in India has increased from 41.2 years in 1951-1961 to 61.4 years in 1991-1996—due principally to a decline in childhood deaths secondary to perinatal complications, infectious diseases, and nutritional deficiencies.[5] The experience of urban China is illustrative of how this affects the burden of cardiovascular disease: The number of cardiac deaths has increased from 12.1% in 1957 to 35.8% in 1990.[16] Interestingly, even within these developing countries, there is a diverse profile of cardiovascular "epidemics" in different regions. For example, the incidence of death secondary to coronary heart disease is 42% for Brazilian men younger than the age of 65 years, as compared with 25% in the industrialized countries.[5,17] Conversely, coronary heart disease is less common in Africa, while hypertension-related congestive heart failure is more common.[5,18]

It is important to recognize the inverse relationship between the incidence of SCD and absolute numbers of events in the various epidemiologic categories. While the incidence of SCD in the general adult population is only 0.1% to 0.2% per year, when applied to the entire U.S. population, it accounts for more than 300,000 events per year.[2,3,4,15] Conversely, patients who fall into higher risk profiles represent a smaller percentage of the population. For example, patients who have a history of a myocardial infarction (MI), depressed left ventricular ejection fraction, and inducible ventricular tachycardia (VT) during electrophysiological (EP) testing have a greater than 30% risk of experiencing sudden death. However, the absolute number of patients who fit these demographic criteria is comparatively small in relation to the total number of patients who die suddenly each year. This principle is important to recognize when one considers applying population-based preventive interventions to decrease the incidence of SCD. That is, in order to exert the greatest possible public health impact on SCD, it will be necessary to identify specific risk-stratifying markers applicable to the general population.

The epidemiology of SCD in adolescents and young adults (ages 10 to 30) is distinctly different from the adult population. The incidence of SCD in this population is two orders of magnitude less than in the adult group (1 per 100,000 versus 1 per 1000 individuals annually).[15] Coronary atherosclerosis is an uncommon cause of SCD in this age group. Instead, the most common causes are hypertrophic cardiomyopathy, myocarditis, right ventricular dysplasia, anomalous coronary arteries, Brugada syndrome, long QT syndrome, and *commotio cordis* (blunt, nonpenetrating chest trauma). Since many of these conditions are genetically determined disorders, there appears to be a modest inverse age relationship in this age group, with adolescents having a somewhat higher mortality risk than young adults.[15]

Mechanisms of Sudden Cardiac Death

From a clinical perspective, the causes of SCD can be divided into three broad categories: (1) primary ventricular tachyarrhythmias, (2) secondary ventricular tachyarrhythmias due to supraventricular tachycardias with rapid ventricular conduction over an accessory pathway, and (3) bradyarrhythmias and asystole. The importance of each of these mechanisms in leading to SCD was evaluated in a series of observational studies examining the initial rhythm observed in victims with cardiovascular collapse. These studies revealed that this initial rhythm was dependent on the time elapsed from loss of consciousness to the first ECG recording by paramedics.[14] When the elapsed time was unknown, the initial rhythm identified by the emergency response teams was ventricular fibrillation (VF) in 40% of patients, asystole in 40%, electromechanical dissociation in 20%, and VT in 1%.[19] If the time from collapse was less than 4 minutes, VF was present in 95% of the cases, and asystole in 5%. If the time from collapse was 12 to 15 minutes, VF accounts for 71%, and asystole in 29%.[20] Thus, ventricular tachyarrhythmia is the primary mechanism of SCD, and electromechanical dissociation and asystole appear to manifest secondarily as a result of prolonged VF and hypoxia. Indeed, the prevalence of VF as the initial dysrhythmia portends a better long-term survival (25%) compared with only 6% or 1% survival if the initial rhythm is electromechanical dissociation or asystole, respectively.[21]

Data from Holter recordings have shown that the most common initial arrhythmia is rapid ventricular tachycardia, which degenerates to polymorphic VT and VF.[22,23] This is corroborated by analyses of stored ICD electrogram data from patients with devices implanted for a history of ventricular arrhythmias and aborted SCD.[24,25] However, it is important to note that there is some controversy in this regard. Specifically, it has been suggested that VF is only rarely preceded by VT and may rather initiate as VF itself.[14] This has been suggested because (1) VT is rarely identified by paramedics as the initial rhythm; (2) in 1287 cardiac arrest patients treated by automated external defibrillators, in which the average response time was 2 to 3 minutes, monomorphic VT was detected in only 16 patients (1.2%)[19]; and (3) in a supervised cardiac rehabilitation program in which 25 patients were resuscitated within 30 seconds of collapse, 92% had VF, and only 8% had VT.[26] However, it is impossible to rule out even from this data that VT did precede VF but was simply short-lived.

Secondary VF arrests may also occur due to ventricular preexcitation syndrome and a rapidly conducting accessory pathway with a short refractory period. However, the incidence of SCD due to Wolff-Parkinson-White (WPW) syndrome is low and likely manifests in only approximately 1/1000 patients with the syndrome. Of note, primary bradyarrhythmias such as electromechanical dissociation and asystole are not common causes of SCD, except in patients with severe heart failure.[27] In these patients, VT/VF is a less important cause of SCD than asystole and electromechanical dissociation. Furthermore, patients with SCD due to asystole usually do not survive to the hospitalization.

Risk Factors for Sudden Cardiac Death

Since the majority of cases of SCD result from coronary artery disease, it is not surprising that the risk factors for SCD mirror those of coronary artery disease. Thus, the incidence of SCD increases with age and is more common in men than women.[28] Caucasians tend to have a higher incidence of SCD than other racial groups. The impact of physical activity on SCD is somewhat controversial. Most likely as a result of enhanced platelet aggregatability and adhesiveness, sporadic vigorous exercise can trigger acute MI culminating in SCD.[28] However, habitual exertion is beneficial and even decreases the incidence of SCD during vigorous exertion.[28,29] In canine experiments, regular exercise was able to prevent ischemia-induced VF and death by increasing vagal activity.[30] Indeed, in athletes the incidence of SCD is exceedingly low: 1 per 200,000 to 250,000 annually.

When it does occur, SCD in young athletes is usually the result of hypertrophic cardiomyopathy.[31] One unusual exception is in northern Italy, where congenital arrhythmogenic right ventricular dysplasia (Naxos disease) is the predominant anatomic finding in athletes with SCD.[32] Young athletes who participate in organized sports are also susceptible to *commotio cordis*—blunt, nonpenetrating, and usually innocent-appearing chest blows. The spectrum of *commotio cordis* includes baseball (the most common), softball, hockey, football, soccer, and even extremely unusual circumstances such as attempts to terminate hiccups, and from the head of a pet dog as it greeted a small child.[33] The phenomenon of *commotio cordis* requires "the exquisite confluence of several determinants" including the location of the blow to directly over the heart and precise timing to the vulnerable repolarization phase just before the T wave peak.[34,35,36,37]

In addition to being a risk factor for coronary artery disease, smoking is an important independent risk factor for SCD. In the Framingham study, the annual incidence of SCD increased 2.5 times in people who smoked more than 20 cigarettes per day as compared with nonsmokers. This appears to be mediated by alterations in platelet physiology, catecholamine levels, and by an increased predisposition to atherosclerotic plaque rupture.[38] Emotional stress also appears to be an important trigger for SCD.[39]

One of the most important predictors of risk for SCD is the left ventricular ejection fraction (LVEF). Left ventricular dysfunction is an independent predictor of SCD in both ischemic and nonischemic cardiomyopathy. For example, the Multicenter Automatic Defibrillator Implantation Trial (MADIT) I demonstrated that patients with a history of MI, LVEF less than or equal to 35%, nonsustained VT (NSVT), and inducible VT during EP testing had a greater than 30% annual incidence of SCD.[40] Furthermore, the MADIT II trial recently demonstrated that a history of MI and a low LVEF (<30%) are sufficient to identify a group at high risk for SCD.[41] Similarly, patients with nonischemic cardiomyopathy are also at high risk for SCD after an initial episode of VT or VF.[4] In patients with severely depressed left ventricular function, competing causes of SCD, including electromechanical dissociation and asystole, limit the impact of the ICD in patients with end-stage congestive heart failure.

Pathophysiology of Sudden Cardiac Death

SCD can be viewed as an electrical accident that occurs only when two distinct conditions are present: a vulnerable substrate and an initiating factor. That is, there is a structural abnormality that provides the necessary substrate for a ventricular arrhythmia to manifest. This substrate is modulated by various transient events that may perturb the homeostatic balance and initiate an arrhythmia. Various structural abnormalities exist, the most common of which is the presence of scarred myocardium resulting from a prior MI. Other pathologic states that may result in an anatomic substrate for VT/VF include cardiomyopathy—nonischemic, hypertrophic, and valvular.

Anatomic predispositions to SCD may also occur at an ultrastructural level. Specifically, ion channelopathies that have been identified in several major membrane channels can initiate various electrical disturbances ranging from torsades de pointes in long QT syndrome to idiopathic VF in Brugada syndrome.[3,4] Certain mutations are severe enough that their presence alone is associated with an extremely high risk for SCD. However, other mutations (e.g., recessive long QT mutation in the *HERG* gene) may by themselves be insufficient to cause SCD, but may predispose patients to torsades de pointes in the presence of other factors, such as drugs that prolong the Q–T interval.

The role of ischemia as an initiating factor for these life-threatening arrhythmias has been studied extensively over the past several decades. Ischemia causes immediate, acute, electrophysiological changes at a cellular level that may alter regional conduction velocity and refractoriness. The resulting dispersion of conduction patterns and repolarization establishes the environment for reentrant arrhythmias. Specifically, in animal experiments, there is an arrhythmogenic interval within the first few minutes after coronary occlusion that begins to abate after 30 minutes. The first 30 minutes can be divided broadly: The first 10 minutes are due to direct ischemic injury and the second 20 minutes are related to either reperfusion or to the evolution of injury patterns in the various layers of the myocardium—the epicardium, endocardium, and Purkinje fibers.[4,42,43]

A number of local derangements can occur in the ischemic myocardium, including a dramatic reduction of local tissue pH to less than 6.0, an increase in interstitial potassium levels to greater than 15 mmol/L, increases in intracellular calcium, and other neurohormonal changes. All of these contribute to altered electrophysiological properties including slowed conduction velocity, reduced excitability and prolonged refractoriness, reduced cell-to-cell coupling, and even spontaneous electrical activity.[4,44] In addition to the local micro- and macro-reentrant circuits that may be generated, there also exist regional changes in automaticity and triggered activity due to afterdepolarizations.

In recent years, there have been significant advances in our understanding of the mechanism of VF. It was previously thought that VF simply represented the coexistence of a number of random reentrant wavefronts. However, it has become evident that there is a substantial amount of order to the fibrillating myocardium. In animal models, it has been shown that a single or small number of sources of periodic activity (scroll waves) are responsible for the maintenance of VF and that fibrillatory conduction away from such sources results in multiple short-lived wavelets.[45,46,47] In this model, the high-frequency source is a stable scroll wave rotor that, because of the complexity of the ventricles—scars, papillary muscles, blood vessels, and so on—is evident on the surface ECG as disorganized activity. Whether this represents the primary mechanism of VF in the clinical setting is not fully established but is being actively investigated.

Disease States Leading to Sudden Cardiac Death

Coronary Artery Disease. Nonatherosclerotic coronary arterial diseases include coronary malformations, coronary embolism, inflammatory arteritis, and congenital malformations such as an anomalous origin of a coronary vessel. These typically manifest earlier in life and are relatively uncommon. On the other hand, approximately 80% of patients who experience SCD have coronary atherosclerotic arterial disease as an underlying anatomic substrate. In survivors of SCD, critical flow-limiting coronary stenoses are found in approximately 40% to 86% of patients, depending on the age and sex of the population.[48,49,50] Although less than 50% of the resuscitated patients have evidence of an acute MI, autopsy studies reveal that a recent occlusive coronary thrombus can be found in 20% to 95% of victims of SCD.[14,48,49] In addition, healed infarctions are found in 40% to 75% of hearts of SCD victims at autopsy.[14,48,51]

The extent to which superimposed acute ischemia plays a role in the pathogenesis of SCD is unclear in patients who do not develop enzymatic or electrocardiographic evidence of an acute MI. It is clear that rapid fibrinolysis of a ruptured plaque could occur such that no visible evidence remains at autopsy. Similarly, cholesterol-laden plaque could conceivably rupture and embolize microscopic debris into distal coronary vessels that could lead to microscopic necrosis and ventricular arrhythmias and yet not be visible at autopsy.[52] In general, there is little clinical evidence to support a recent history of worsening angina before SCD or a history of strenuous exercise at the time of SCD.[14,53,54] Furthermore, in those patients who develop VT/VF while being monitored, it is uncommon to note significant ST segment changes.[14,55,56]

Dilated Cardiomyopathy. Idiopathic dilated cardiomyopathy accounts for approximately 10% of cases of SCD in the adult population. Depending on the severity of the myopathy, the annual incidence ranges from 10% to 50%.[57] Bundle branch reentrant VT appears to represent an important cause of ventricular tachyarrhythmias in this population.[58] However, as the myopathy progresses and congestive heart failure worsens, the incidence of VT/VF decreases, and the primary terminal event more frequently becomes electromechanical dissociation or asystole.[59] As with ischemic heart disease, the overall LVEF is an important prognostic factor. In addition, the New York Heart Association functional class and the occurrence of syncope are both important clinical prognostic factors for SCD in patients with dilated cardiomyopathy.[14,56]

Hypertrophic Cardiomyopathy. The annual incidence of mortality in patients with hypertrophic cardiomyopathy is as high as 2% to 3%. A correlation does appear to exist between the degree of ventricular wall hypertrophy and the incidence of SCD. Specifically, a ventricular wall thickness of 30 mm or more is associated with a cumulative 20-year SCD risk of 40%.[60] Other important prognostic factors include a personal or family history of SCD or sustained VT, recurrent syncope, and multiple episodes of NSVT.[61] Recently, it has been demonstrated that certain mutations in the α-tropomyosin as well as in the β-myosin heavy chain gene are associated with SCD.[61,62] Conversely, the apical ("Japanese") variant of hypertrophic cardiomyopathy has been demonstrated to carry a minimal risk for SCD.[63]

Left Ventricular Hypertrophy. When established by ECG or echocardiography, left ventricular hypertrophy is an independent risk factor for SCD.[4,64] Furthermore, NSVT is more common in patients with left ventricular hypertrophy.[4,65] However, the utility of this information for prognostic purposes is not clear.

Long QT Syndrome. Mutations in the *KVLQT1,* human "ether-a-go-go" related gene (*HERG*), cardiac voltage-dependent sodium channel gene (*SCN5A*), *minK* and *MiRP1* genes, respectively, are responsible for the LQT1, LQT2, LQT3, LQT5, and LQT6 variants of the Romano-Ward syndrome—the autosomal dominant disease with no deafness.[66] The less common Jervell-Lange-Nielsen syndrome, which has marked QT prolongation and sensorineural deafness, arises when a child inherits mutant KVLQT1 or mink alleles from both parents. The defect in the sodium channel gene (*SCN5A*) prolongs action potential duration by increasing inward plateau sodium current. The other mutations cause a decrease in net repolarizing current by reducing potassium currents through "loss-of-function" mechanisms. Polymorphic VT associated with a prolonged Q–T interval (torsades de pointes) is thought to be initiated by early afterdepolarizations in the Purkinje system and maintained by transmural reentry in the myocardium.[67] Clinical presentations vary with the specific gene affected and the specific mutation. Some patients with LQTS mutations may not manifest any phenotypic abnormality. This latter situation may be one of high risk for the administration of QT-prolonging drugs or during myocardial ischemia and, in some families, may be associated with high risk for SCD.

Arrhythmogenic Right Ventricular Cardiomyopathy. Arrhythmogenic right ventricular cardiomyopathy (ARVC) is a primary myocardial disease frequently associated with left bundle branch block VT and SCD in young patients.[68] A familial form of this disease was found in northern Italy in association with palmoplantar keratoderma (Naxos disease) and is caused by a defect in the gene for a cellular structural element, plakoglobin.[69] The disease causes progressive fatty replacement of the right ventricular (and sometimes left ventricular) wall. It is best diagnosed by cardiac magnetic resonance imaging to define the presence and degree of fatty infiltration in the myocardium.

Brugada Syndrome. SCD due to primary VF without apparent structural heart disease occurs in Brugada syndrome.[70] The syndrome was first described as "right bundle branch block and persistent ST segment elevation in V1-V3" in patients with structurally normal hearts and a high incidence of VF. This syndrome appears to be responsible for the "sudden and unexpected death syndrome" in Southeast Asian men called lai-tai ("sleep death," Laos), pokkuri ("sudden and unexpected death," Japan), or bangungut ("to rise and moan in sleep," Philippines).[70,71] A striking feature of this syndrome is the remarkably high incidence of recurrent cardiac arrest in patients who manifest one arrhythmic episode.[4]

This syndrome is a primary electrical disease resulting from a mutation in the SCN5A ion channel.[70] The ECG manifestations of the syndrome may completely normalize during routine evaluation. However, the ST segment abnormalities in the anterior precordial leads can be unmasked by administration of sodium-channel blocking agents such as flecainide, Ajmaline or procainamide.[70] The action potential dome in the right ventricular epicardium but not endocardium underlies the J-point and ST segment changes seen. This electrical heterogeneity leads to the development of closely coupled premature ventricular contractions in a phase 2 reentrant mechanism that rapidly precipitates VT/VF.[70]

Wolff-Parkinson-White Syndrome. WPW syndrome can lead to SCD if the accessory pathway is able to conduct rapidly in the presence of an atrial arrhythmia such as atrial fibrillation. While most SCD survivors with WPW have had symptomatic arrhythmias before the event, up to 10% experience SCD as their first manifestation of the disease.[72] Recent studies from the Milan group suggest lower arrhythmic and SCD risk when prophylactic catheter ablation is employed in asymptomatic adults and children with WPW syndrome.[72a,72b] The best predictor for the development of VF is a rapid ventricular response during induced atrial fibrillation—with a shortest R–R interval of less than 250 milliseconds (ms). While this test's sensitivity is virtually 100%, its specificity is low. However, the induction of atrial fibrillation has been suggested to have prognostic value.[72b] The syndrome does not usually follow a familial course, but a mutation in the *PRKAG2* gene has been shown to cause WPW in a family, along with conduction system disease.[73]

Other. Catecholaminergic VT is a rare arrhythmia wherein affected individuals present with syncopal events and with a characteristic pattern of highly reproducible, stress-related, bidirectional VT in the absence of any structural heart disease or prolonged Q–T interval. The responsible mutation has been shown to reside in the cardiac ryanodine receptor gene.[74] Finally, in addition to long QT syndrome and Brugada syndrome, mutations in the *SCN5A* gene have also been shown capable of causing isolated cardiac conduction disturbances.[75]

Testing to Assess the Risk For Sudden Cardiac Death

As previously stated, assessment of overall cardiovascular function is important to gauge the risk of SCD. This includes both New York Heart Association functional class and the overall LVEF. However, as the level of functional impairment increases to severe/class IV heart failure, and the absolute number of sudden deaths increase, the proportion of total deaths due to ventricular tachyarrhythmias decreases. Currently, the LVEF is used in the risk stratification schema of virtually every clinical trial on SCD. In addition, the MADIT II trial used *only* the LVEF (<30%) for risk stratification for inclusion into the protocol. The results of MADIT II indicate that the use of LVEF alone for risk stratification is associated with significant mortality benefit from prophylactic ICD therapy.[41]

The presence of ambient nonsustained ventricular arrhythmias was initially thought to be an important predictor of increased mortality. However, the suppression of ambient ventricular arrhythmias as recorded on ambulatory monitoring devices has not been shown to result in improved survival. Indeed, antiarrhythmic drugs given to specifically suppress premature ventricular contractions (PVCs) were not only ineffective in preventing SCD but actually increased mortality in patients with prior MI.[76] Also, while ambient ventricular activity does correlate with an increased risk of SCD, recent studies have demonstrated that this finding does not provide any additional prognostic information in patients with advance heart failure. The prevalence of ambient ventricular activity (PVCs and NSVT) is higher among patients with severe heart failure than among those with mild-moderate heart failure.[77] However, the proportion of patients who develop SCD due to VT/VF is significantly lower in this severe heart failure group. These data suggest that ambient ventricular arrhythmias may actually reflect the degree of heart failure, rather than providing a specific marker of vulnerability to SCD.[3] Thus, the role of ambient nonsustained ventricular arrhythmias in predicting SCD appears to be limited.

High-resolution signal processing of the surface ECG can be performed to scrutinize both ventricular depolarization and repolarization. In the signal-averaging ECG, the terminal part of the QRS complex is evaluated for microvolt potentials that reflect delayed activation in the scarred myocardial substrate.[3,78] In patients with prior MI, the negative predictive value of this test has been demonstrated to be excellent. However, the usefulness of this test has been limited by its low positive predictive value in many studies.[3,78-80]

Abnormalities of repolarization can be evaluated as either Q–T interval dispersion or T wave alternans. Q–T interval dispersion examines the difference between the maximal and minimal Q–T intervals from various standard ECG leads,[81] while T wave alternans is defined as microvolt changes in the T wave amplitude from beat to beat.[82] In recent studies, Q–T interval dispersion has not been a consistent predictor of SCD risk.[83,84] However, analysis of T wave alternans at heart rates of 105 to 110 bpm in high-risk patients accurately predicts the risk of ventricular arrhythmias.[85-87] Studies are being conducted to more fully define the predictive value of T wave alternans in the larger population of low- to intermediate-risk patients at risk for SCD.

Evidence from both clinical and experimental studies supports a strong role for the autonomic nervous system in the genesis of VT/VF. There is a clear association between increased sympathetic activity or reduced vagal activity, or both, and a greater propensity for VF during myocardial ischemia.[88,89] Specifically, these data support a role for either an increased sensitivity to sympathetic input or abnormal sympathetic function. The two major measures of autonomic function that are currently being tested in clinical studies include heart rate variability (HRV) and baroreflex sensitivity (BRS).[3,14] HRV reflects both sympathetic and parasympathetic effects on the heart and is measured as the standard deviation of R–R intervals in the heart rate over a 24-hour period. BRS is an indicator of the reflex capacity of the autonomic nervous system and is measured using a phenylephrine test. While these tests do not appear to be particularly useful when used alone, they are promising when used in combination with other tests.[3,90-93] Further studies are being performed to gain a better understanding of the usefulness of these measures of autonomic function.

The most common test currently used for the purpose of risk stratification for SCD is invasive EP testing with programmed ventricular stimulation.[94,95] In patients with a history of previous MI, the induction of VT during an electrophysiological study has been shown to predict a high risk for recurrent arrhythmias and SCD. The predictive value of EP testing has been established in a number of clinical situations including patients with a history of cardiac arrest,[96-98] patients with impaired ventricular function and NSVT,[40,99] and patients with impaired ventricular function and syncope.[100]

The predictive value of EP testing is highest in patients with a history of MI, as opposed to those with nonischemic cardiomyopathy.[3] While a positive test predicts increased risk for SCD, a negative study does not exclude risk for SCD, particularly in patients with severe LV dysfunction. In the recently completed MADIT II trial in patients with ischemic cardiomyopathy with LVEF less than 30%, ICD therapy was associated with a significant survival benefit even in the absence of EP testing.[41] Whether this strategy can also be extended to patients with nonischemic cardiomyopathy is unknown but is currently under investigation.

In the past, serial EP testing had been performed to gauge the effectiveness of antiarrhythmic drug therapy. However, this strategy has generally not proven useful and is currently not performed in the vast majority of clinical situations. In addition, EP testing is of limited clinical benefit in patients with normal left ventricular function. In patients with conditions such as left ventricular hypertrophy and hypertrophic cardiomyopathy, the role of EP testing is uncertain, and a negative EP study does not exclude high risk for SCD. During EP testing, VF is the most common ventricular arrhythmia induced in these patient populations and is of uncertain clinical significance.[3,4,14]

Out-of-Hospital Resuscitation

As discussed earlier, the most important prognostic variable in determining the effectiveness of a resuscitation effort is the time from cardiac arrest to initial defibrillation. One of the most effective emergency response systems (EMS) in the world is in Seattle. The Seattle EMS system is organized as a tiered-response system: following a 911 call, the nearest fire truck or ambulance is dispatched with personnel able to perform basic first aid and CPR. These teams carry an automated external defibrillator (AED) and have an initial response time of 2.6 plus or minus 1.7 minutes.[101,102] A more advanced paramedic unit is also dispatched and arrives in 6.0 plus or minus 2.9 minutes. In addition, an extensive training program in the Seattle community has resulted in training 25% to 40% of the population in bystander CPR techniques.[14] It has been shown that even when a first defibrillatory shock is delivered quickly, bystander CPR has an important effect on outcome. For patients who received an appropriate defibrillation shock within 6 minutes of onset of cardiac arrest, the performance of bystander CPR within 3 minutes resulted in 70% survival, as compared with 39% survival when performed after 6 minutes.[21] In Seattle, the implementation of the rapid EMS system and bystander CPR have combined to result in an *in-field* resuscitation rate of 60% for those patients found in VF.[103]

However, only 40% of cardiac arrest victims in the Seattle experience have VF at initial contact, and 26% of these VF victims will be discharged from the hospital neurologically intact.[19] Thus, despite prompt CPR, a good emergency medical system response time, and the use of AEDs, only approximately 10% of all cardiac arrest victims will be satisfactorily discharged from the hospital.[102,103] And this assumes an ideal scenario with what is one of the best EMS in the United States. Many other communities have been less successful in fashioning their emergency medical response systems. For example, New York and Chicago report post–cardiac arrest survival rates of 1.4% and 2%, respectively.[104] The reasons for this disappointing success rate are multifactorial and include population characteristics, traffic congestion, and vertical development of communities and the resulting adverse effect on EMS response times.[15] Indeed, the cumulative U.S. survival rate from cardiac arrest was approximately 1% to 3% in 1991.[15]

In an effort to improve outcomes, there has been increased interest in expanding the availability of AEDs to less experienced operators. For example, in Rochester, Minn, AEDs were placed in police vehicles. Both the paramedics and police responded to EMS calls, and the first team to arrive delivered a shock. In the group of cardiac arrest victims with VF treated by the police, 41% had return of spontaneous circulation, as compared to 28% of those treated by the paramedics. And of these patients, 96% survived to the end of the hospitalization. These success rates correlated directly with the improved response times.[105] Thus, in communities where traffic congestion is a major problem, the placement of AEDs in police vehicles is likely to have a significant benefit on outcomes. For example, in Miami-Dade County, Fla, the mean police response time to a 911 call is 4.9 minutes, as compared with 8.1 minutes for the fire rescue response.[15]

The strategy of broad AED deployment has now extended beyond police vehicles to other public places. For example, the use of AEDs in cases of sudden cardiac arrest in casinos was studied.[106] Casino security officers were instructed in the use of AEDs, and data were collected on 148 consecutive patients. Of this group, 105 had an initial rhythm of VF, while the rest had either electromechanical dissociation (17 victims) or asystole (26 victims). In the VF arrest group, 56 patients (53%) survived to discharge from the hospital. The survival rate was 74% for those who received their first defibrillation no later than 3 minutes after a witnessed collapse and 49% for those who received their first defibrillation after more than 3 minutes. While placement of AEDs in casinos will clearly not have a significant impact on the public health problem of SCD, this example does illustrate that placement of AEDs in densely populated sites can be efficacious even when operated by minimally experienced personnel.

Similarly, the use of AEDs on U.S. airlines was evaluated in a study wherein AEDs were placed aboard selected aircraft on a major airline.[107] The sensitivity and specificity of the device in advising administration of a shock was 100%. All of the patients in whom defibrillation was attempted were successfully defibrillated, and the rate of survival to discharge from the hospital was 40%. Interestingly, only a minority of the patients were found to have "shockable" rhythms. This observation was corroborated in another similar airline observational study in which 78% of the victims were initially found to manifest bradyarrhythmias or asystole.[108] Given the rapid response time to placement of the AED, the reason for this excess of bradyarrhythmic events is not clear. However, it may be related to the lower oxygen saturation that is maintained at cruise altitudes in airliners.[14] There are experimental data demonstrating that low oxygen levels can decrease the time from VF to asystole.[109] The success of these initial experiences has prompted the U.S. government to mandate the placement of AEDs on all major airlines and in airports. Indeed, AEDs are now being placed in many buildings and locations where people congregate. It is hoped that this improved public access to defibrillation will improve the outcome from cardiac arrest in future years.

Preventative Therapy for Sudden Cardiac Death

Because of their importance in the pathogenesis of SCD, treatments that decrease the incidence of coronary artery disease and MI (e.g., lipid-lowering drugs, aspirin, ACE inhibitors, β-blockers) are also important in minimizing the incidence of SCD. For the primary prevention of life-threatening ventricular arrhythmias, two major strategies have been studied: the use of antiarrhythmic drugs and implantable ICDs. Despite the promising results of arrhythmia surgery, the effect of ancillary approaches such as catheter-based ablation procedures on VF and SCD has not been well studied to date. However, as catheter ablation technologies improve, it is likely that attempts will be undertaken to replicate the surgical experience with a catheter approach.

The first experience with pharmacologic prevention of SCD was with β-blocker therapy in patients with prior MIs. While the initial trial designs were not specifically constructed to evaluate SCD, these trials have uniformly demonstrated a clinically important reduction in the incidence of SCD with β-blocker therapy.[4,110,111] The effect is particularly striking in patients with the lowest ejection fractions.[111] However, when the membrane-active antiarrhythmic drugs were used to treat patients with frequent ventricular ectopy, the incidence of SCD actually increased.[7-8] This was true for both the class Ic agents in CAST as well as class III agents such as d-sotalol in SWORD.[7-12] The only membrane-active antiarrhythmic medications approved for the treatment of ventricular arrhythmias that do not carry some increased mortality risk in patients with ischemic heart disease are d,l-sotalol and amiodarone. However, in two major trials examining the routine use of amiodarone in patients post-MI (European Myocardial Infarct Amiodarone Trial [EMIAT] and Canadian Amiodarone Myocardial Infarction Arrhythmia Trial [CAMIAT]), amiodarone did not demonstrate a survival benefit.[112,113] In a higher risk population of patients (Grupo de Estudio de la Sobrevida en la Insuficiencia Cardiaca en Argentina [GESICA]: LVEF <35% and either ischemic or nonischemic cardiomyopathy), empiric amiodarone treatment was shown to favorably affect mortality,[114] although most patients in this study had nonischemic cardiomyopathy. These data were not corroborated by another trial examining a population composed primarily of patients with ischemic cardiomyopathy (Congestive Heart Failure Survival Trial of Antiarrhythmic Therapy [CHF-STAT]: LVEF <40%).[115] This trial found no survival benefit with prophylactic amiodarone in patients with congestive heart failure. Newer antiarrhythmic agents such as dronaderone, which appears to have many of the beneficial antiarrhythmic actions of amiodarone but fewer side effects, may offer some benefit in the future. However, experience to date with pharmacologic prevention of SCD has been largely disappointing, save the use of β-blocking agents.

Other primary prevention trials have focused on the efficacy of ICD therapy in the prevention of SCD. There are two large-scale, randomized clinical trials that have examined the use of ICDs in patients with prior MI, low LVEF, NSVT, and inducible VT at EP testing: MADIT I and Multicenter UnSustained Tachycardia Trial (MUSTT).[40,99] Both trials demonstrated a significant mortality benefit in patients who received ICDs compared with those patients treated with or without antiarrhythmic medications. The results of these two trials have recently been extended by MADIT II, a randomized controlled trial in post-MI patients with LVEF less than 30%. In this prophylactic ICD trial, NSVT was not a required entry criterion, and a significant mortality benefit was nonetheless seen with ICD therapy compared to conventional therapy with β-blockers and ACE inhibitors.[41] Similar to these primary prevention trials, three secondary prevention trials (Antiarrhythmic Versus Implantable Defibrillators [AVID], Canadian Implantable Defibrillator Study [CIDS], and Cardiac Arrest Study Hamburg [CASH]) have been conducted in patients who have survived an episode of sustained VT or VF.[96-98] As with the primary prevention trials, these studies demonstrated a significant survival benefit with ICD therapy compared with antiarrhythmic drug therapy (predominantly amiodarone). The only ICD trial that has not shown a mortality benefit is the Coronary Artery Bypass Graft (CABG)-Patch trial in patients with LV dysfunction undergoing surgical revascularization.[116] This study underscores the importance of risk stratification as well as the potent effect of coronary revascularization in the prevention of SCD.

Despite the efficacy of ICD therapy in very high-risk patient subsets, it is important to realize that these populations represent a select minority of the total number of patients at risk for sudden death. Protection against sudden death by ICD implantation is expensive even in these high-risk populations, and would be substantially more expensive in lower-risk populations. However, substudy analyses of these large trials suggest that much of the benefit of ICD therapy is realized in the sickest patients (i.e., those with the lowest LVEFs).[117,118] Accordingly, efforts are under way to identify and refine risk stratification techniques to better characterize patients at highest risk for SCD, as well as to identify additional patient populations that may also benefit from ICD therapy.

SUMMARY

As we enter the new millennium, SCD continues to be a public health problem of major significance. Our understanding of the mechanisms and triggers for SCD is expanding. It is well established that ICD therapy is more effective than antiarrhythmic drugs in preventing SCD in high-risk patient populations. In order to bring the benefit of ICDs to a wider spectrum of patients, advances will be required in two major areas: (1) better definition of the clinical subsets of patients at highest risk and (2) the availability of less expensive prophylactic ICDs that will make their widespread use economically feasible. While ICDs are likely to dominate the electrophysiology landscape for the foreseeable future, molecular biology will likely play a more prominent role in arrhythmia management in the future. Concurrent with epidemiologic advances, the past 2 decades have witnessed an explosion in our understanding of disease states such as the long Q–T interval and Brugada syndromes at a molecular level. It is possible that an improved understanding of ion channel physiology and pathophysiology may contribute to the development of safer and more specific pharmacologic and gene-based therapies for the prevention of potentially lethal ventricular arrhythmias. Ultimately, the elimination of SCD as an epidemic in the industrialized world will require fundamental advances in vascular biology, which lead to the prevention of atherosclerotic coronary artery disease and its consequences.

REFERENCES

1. Lopez AD: Assessing the burden of mortality from cardiovascular disease. World Health Stat Q 1993;46:91-6.
2. Myerburg RJ, Interian A, Mitrani RM, et al: Frequency of sudden cardiac death and profiles of risk. Am J Cardiol 1997;80:10F-19F.
3. Huikuri HV, Castellanos A, Myerburg RJ: Sudden death due to cardiac arrhythmias. N Engl J Med 2001;345:1473-82.
4. Zipes DP, Wellen HJJ: Sudden cardiac death. Circulation 1998; 98:2334-51.
5. Reddy KS, Yusuf S: Emerging epidemic of cardiovascular disease in developing countries. Circulation 1998;97:596-601.
6. Rotimi O, Ajayi AA, Odesanmi WO: Sudden unexpected death from cardiac causes in Nigerians: A review of 50 autopsied cases. Int J Cardiol 1998;63:111-5.
7. The Cardiac Arrhythmia Suppression Trial (CAST) Investigators. Preliminary Report: Effect of encainide and flecainide on mortality in a randomized trial of arrhythmia suppression after myocardial infarction. N Engl J Med 1989;321:406-12.
8. The Cardiac Arrhythmia Suppression Trial II (CAST II) Investigators: Effect of the antiarrhythmic agent moricizine on survival after myocardial infarction. N Engl J Med 1992;327: 227-33.
9. Waldo AL, Camm AJ, deRuyter H, et al: Effect of d-sotalol on mortality in patients with left ventricular dysfunction after recent and remote myocardial infarction (SWORD). Lancet 1996;348: 7-12.
10. Coplen SE, Antman EM, Berlin JA, et al: Efficacy and safety of quinidine therapy for maintenance of sinus rhythm after cardioversion: A meta-analysis of randomized control trials. Circulation 1990;82:1106-16.
11. Stanton MS, Prystowsky EN, Fineberg NS, et al: Arrhythmogenic effects of antiarrhythmic drugs: A study of 506 patients treated for ventricular tachycardia or fibrillation. J Am Coll Cardiol 1989; 14:209-15.
12. IMPACT Research Group. International mexiletine and placebo antiarrhythmic coronary trial: Report on arrhythmia and other findings. J Am Coll Cardiol 1984;4:1148-63.

13. Teo KK, Yusuf S, Furberg CD: Effects of prophylactic antiarrhythmic drug therapy in acute myocardial infarction. An overview of results from randomized controlled trials (see comments). JAMA 1993;270:1589-95.
14. Poole JE, Bardy GH: Sudden cardiac death. In Zipes DP, Jalife F (eds): Cardiac Electrophysiology: From Cell to Bedside. Philadelphia, WB Saunders, 2000, pp 615-40.
15. Myerburg RJ: Sudden cardiac death: Exploring the limits of our knowledge. J Cardiovasc Electrophysiol 2001;12:369-81.
16. Yao C, Wu Z, Wu J: The changing pattern of cardiovascular diseases in China. World Health Stat Q 1993;46:113-8.
17. Nicholls ER, Peruga A, Restrepo HE: Cardiovascular mortality in the Americans. World Health Stat Q 1993;46:134-50.
18. Walker ARP, Sarell P: Coronary heart disease: Outlook for Africa. J R Soc Med 1997;90:23-7.
19. Weaver WE, Hill D, Fahrenbruch CD, et al: Use of the automatic external defibrillator in the management of out-of-hospital cardiac arrest. N Engl J Med 1988;319:661-6.
20. Hallstrom AP, Eisenberg MS, Bergner L: The persistence of ventricular fibrillation and its implications for evaluating EMS. Emerg Health Serv Q 1983;1:41-7.
21. Weaver WD, Cobb LA, Hallstrom AP, et al: Considerations for improving survival from out-of-hospital cardiac arrest. Ann Emerg Med 1986;15:1181-6.
22. Bayes de Luna A, Coumel P, Leclercq JF: Ambulatory sudden cardiac death: Mechanisms of production of fatal arrhythmia on the basis of data from 157 cases. Am Heart J 1989;117:151-9.
23. Paridis IP, Morganroth J: Sudden death in hospitalized patients: Cardiac rhythm disturbances detected by ambulatory electrocardiographic monitoring. J Am Coll Cardiol 1983;2:798-805.
24. Marchlinski FE, Callans DJ, Gottlieb CD, et al: Benefits and lesions learned from stored electrogram information in implantable defibrillators. J Cardiovasc Electrophysiol 1995;6: 832-51.
25. Hook BG, Callans DG, Kleiman RB, et al: Implantable cardioverter-defibrillator therapy in the absence of significant symptoms: Rhythm diagnosis and management aided by stored electrogram analysis. Circulation 1993;87:1897-1906.
26. Hossaack KF, Hartwig R: Cardiac arrest associated with supervised cardiac rehabilitation. J Cardiac Rehabil 1982;2:402-18.
27. Luu M, Stevenson WG, Stevenson LW, et al: Diverse mechanisms of unexpected cardiac arrest in advanced heart failure. Circulation 1989;80:1675-80.
28. Wang JS, Jen CJ, Kung HC, et al: Different effects of strenuous exercise and moderate exercise on platelet function in men. Circulation 1994;90:2877-85.
29. Albert CM, Mittleman MA, Chae CU, et al: Triggering of sudden death from cardiac causes by vigorous exertion. N Engl J Med 2000;343:1355-61.
30. Schwartz PJ, Zipes DP: Autonomic modulation of cardiac arrhythmias and sudden cardiac death. In Zipes DP, Jalife J (eds): Cardiac Electrophysiology: From Cell to Bedside. Philadelphia, WB Saunders, 2000.
31. Maron BJ, Shirani J, Poliac LC, et al: Sudden death in young competitive athletes: Clinical, demographic, and pathological profiles. JAMA 1996;276:199-204.
32. Corrado D, Basso C, Schiavon M, Thiene G: Screening for hypertrophic cardiomyopathy in young athletes. N Engl J Med 1998; 339:363-9.
33. Maron BJ, Gohman TE, Kyle SB, et al: Clinical profile and spectrum of commotio cordis. JAMA 2002;287:1142-6.
34. Maron BJ, Poliac L, Kaplan JA, Mueller FO: Blunt impact of the chest leading to sudden death from cardiac arrest during sports activities. N Engl J Med 1995;333:337-42.
35. Link MS, Maron BJ, Vanderbrink BA, et al: Impact directly over the cardiac silhouette is necessary to produce ventricular fibrillation in an experimental model of commotio cordis. J Am Coll Cardiol 2001;37;649-54.
36. Link MS, Wang PJ, Pandian NG, et al: An experimental model of sudden death due to low-energy chest-wall impact (commotio cordis). N Engl J Med 1998;338:1805-11.
37. Link MS, Wang PJ, Vanderbrink BA, et al: Selective activation of the KATP channel is a mechanism by which sudden death is produced by low-energy chest-wall impact (commotio cordis). Circulation 1999;100:413-8.
38. Burke AP, Farb BA, Malcom GT, et al: Coronary risk factors and plaque morphology in men with coronary heart disease who died suddenly. N Engl J Med 1997;336:1276-82.
39. Leor J, Poole WK, Kloner RA: Sudden cardiac death triggered by an earthquake. N Engl J Med 1996;334:413-9.
40. Moss AJ, Hall WJ, Cannom DS, et al: Multicenter automatic defibrillator implantation trial investigators. Improved survival with an implanted defibrillator in patients with coronary disease at high risk for ventricular arrhythmia. N Engl J Med 1996;335:1933-40.
41. Moss AJ, Zareba W, Hall WJ, et al: Prophylactic implantation of a defibrillator in patients with myocardial infarction and reduced ejection fraction. N Engl J Med 2002;346:877-83.
42. Kimura S, Bassett AL, Saoudi NC, et al: Cellular electrophysiologic changes and "arrhythmias" during experimental ischemia and reperfusion in isolated cat ventricular myocardium. J Am Coll Cardiol 1986;7:833-42.
43. Kimura S, Bassett AL, Kohya T, et al: Simultaneous recording of action potentials from endocardium and epicardium during ischemia in the isolated cat ventricles. Circulation 1986;64:401-9.
44. Owens LM, Fralix TA, Murphy E, et al: Correlation of ischemia-induced extracellular and intracellular ion changes to cell-to-cell electrical uncoupling in isolated blood-perfused rabbit hearts. Circulation 1996;94:10-3.
45. Chen J, Mandapati R, Berenfel O, et al: High-frequency periodic sources underlie ventricular fibrillation in the isolated rabbit heart. Circ Res 2000;86:86-93.
46. Ohara T, Ohara K, Cao JM, et al: Increased wave break during ventricular fibrillation in the epicardial border zone of hearts with healed myocardial infarction. Circulation 2001;103:1465-72.
47. Zaitsev AV, Berenfeld O, Mironov SF, et al: Distribution of excitation frequencies on the epicardial and endocardial surfaces of fibrillating ventricular wall of the sheep heart. Circ Res 2000; 86:408-17.
48. Davies MJ, Thomas A: Thrombosis and acute coronary artery lesions in sudden cardiac ischemic death. N Engl J Med 1984; 310:1137-40.
49. El Fawal MA, Berg GA, Whealey DJ, et al: Sudden coronary death in Glasgow: Nature and frequency of acute coronary lesions. Br Heart J 1987;57:329-35.
50. Davies M, Bland J, Hangartner J, et al: Factors influencing the presence or absence of acute coronary artery thrombi in sudden ischemic death. Eur Heart J 1989;10:203-8.
51. Reichenbach DD, Moss NS, Meyer E: Pathology of the heart in sudden cardiac death. Am J Cardiol 1977;39:865.
52. Farb A, Tang AL, Burke AP, et al: Sudden coronary death: Frequency of active coronary lesions, inactive coronary lesions, and myocardial infarction. Circulation 1995;92:1701-9.
53. Weaver WD, Cobb LA, Hallstrom AP: Characteristics of survivors of exertion and on exertion-related cardiac arrest: Value of subsequent exercise testing. Am J Cardiol 1982;50:671-6.
54. Cobb LA, Weaver WD: Exercise: A risk for sudden death in patients with coronary artery disease. J Am Coll Cardiol 1986; 7:215-9.
55. Pratt CM, Francis MJ, Luck JC, et al: Analysis of ambulatory electrocardiograms in 15 patients during spontaneous ventricular fibrillation with special reference to preceding arrhythmic events. J Am Coll Cardiol 1983;2:789-97.
56. Gomes JA, Alexopoulos D, Winters SL, et al: The role of silent ischemia, the arrhythmic substrate and the short-long sequence in the genesis of sudden cardiac death. J Am Coll Cardiol 1989; 14:1618-25.
57. Tamburro P, Wilber D: Sudden death in idiopathic dilated cardiomyopathy. Am Heart J 1992;124:1035-45.
58. Blanck Z, Dhala A, Deshpande S, et al: Bundle branch reentrant ventricular tachycardia: Cumulative experience in 48 patients. J Cardiovasc Electrophysiol 1993;4:253-62.
59. Stevenson WG, Stevenson LW, Middlekauff HR, Saxon LA: Sudden death prevention in patients with advanced ventricular dysfunction. Circulation 1993;88:2953-61.
60. Sprito P, Bellone P, Harris KM, et al: Magnitude of left ventricular hypertrophy and risk of sudden death in hypertrophic cardiomyopathy. N Engl J Med 2000;342:1778-85.
61. Spirito P, Seidman CE, McKenna WJ, Maron BJ: The management of hypertrophic cardiomyopathy. N Engl J Med 1997; 336: 775-85.

62. Marian AJ, Robert R: Molecular genetic basis of hypertrophic cardiomyopathy: Genetic markers for sudden cardiac death. J Cardiovasc Electrophysiol 1998;9:88-99.
63. Eriksson MJ, Sonnenberg B, Woo A, et al: Long-term outcome in patients with apical hypertrophic cardiomyopathy. J Am Coll Cardiol 2002;39:638-45.
64. Levy D, Garrison GJ, Savage DD, et al: Prognostic implications of echocardiographically determined left ventricular mass in the Framingham heart study. N Engl J Med 1990;322:1561-6.
65. McLenachan JM, Henderson E, Morris KI, et al: Ventricular arrhythmias in patients with hypertensive left ventricular hypertrophy. N Engl J Med 1987;317:787-92.
66. Chiang E-E, Roden DM: The long QT syndromes: Genetic basis and clinical implications. J Am Coll Cardiol 2000;36:1-12.
67. El-Sherif N, Chinushi M, Caref EB, Restivo M: Electrophysiological mechanism of the characteristic electrocardiographic morphology of torsades de pointes tachyarrhythmias in the long QT syndrome. Detailed analysis of ventricular tridimensional activation pattern. Circulation 1997;96:4392-9.
68. Nava A, Bauce B, Basso C, et al: Clinical profile and long-term follow-up of 37 families with arrhythmogenic right ventricular cardiomyopathy. J Am Coll Cardiol 2000;36:2226-33.
69. McKoy G, Protonotarios N, Crosby A, et al: Identification of a deletion in plakoglobin arrhythmogenic right ventricular cardiomyopathy with palmoplantar keratoderma and wooly hair (Naxos disease). Lancet 2000;355:2119-24.
70. Gussak I, Antzelevitch C, Bjerregaard P, et al: The Brugada syndrome: Clinical, electrophysiologic and genetic aspects. J Am Coll Cardiol 1999;33:5-15.
71. Nademanee K, Veerakul G, Nimmannit S, et al: Arrhythmogenic marker for the sudden unexplained death syndrome in Thai men. Circulation 1997;96:2595-600.
72. Bromberg BI, Lindsay BD, Cain ME, Cox JL: Impact of clinical history and electrophysiologic characterization of accessory pathways on management strategies to reduce sudden death among children with Wolff-Parkinson-White syndrome. J Am Coll Cardiol 1996;27:690-5.
72a. Pappone C, Santinelli V, Manguso F, et al: A randomized study of prophylactic catheter ablation in asymptomatic patients with the Wolff-Parkinson-White syndrome. N Engl J Med 2003;349: 1803-11.
72b. Pappone C, Manguso F, Santinelli R, et al: Radiofrequency ablation in children with asymptomatic Wolff-Parkinson-White syndrome. N Engl J Med 2004;351:1197-1205.
73. Gollob MH, Seger JJ, Gollob TN, et al: Novel PRKAG2 mutation responsible for the genetic syndrome of ventricular preexcitation and conduction system disease with childhood onset and absence of cardiac hypertrophy. Circulation 2001;104:3030-3.
74. Priori SG, Napalitano C, Tiso N, et al: Mutations in the cardiac ryanodine receptor gene (hRyR2) underlie catecholaminergic polymorphic ventricular tachycardia. Circulation 2001;103: 196-200.
75. Tan HL, Bink-Boelkens MTE, Bezzina CR, et al: A sodium-channel mutation causes isolated cardiac conduction disease. Nature 2001;409:1043-4.
76. Echt DS, Liebson PR, Mitchell LB, et al: Mortality and morbidity in patients receiving encainide, flecainide, or placebo: The Cardiac Arrhythmia Suppression Trial. N Engl J Med 1991;324: 781-8.
77. Packer M: Lack of relation between ventricular arrhythmias and sudden death in patients with chronic heart failure. Circulation 1992;85(suppl I):I50-I56.
78. El-Sherif N, Denes P, Katz R, et al: Definition of the best prediction criteria of the time-domain signal-averaged electrocardiogram for serious arrhythmic events in the postinfarction period. J Am Coll Cardiol 1995;25:908-14.
79. Gomes JA, Winter SL, Stewart D, et al: A new noninvasive index to predict sustained ventricular tachycardia and sudden death in the first year after myocardial infarction: Based on signal averaged electrocardiogram, radionuclide ejection fraction and Holter monitoring. J Am Coll Cardiol 1987;10:349-57.
80. McClements BM, Adgey AAJ: Value of signal-averaged electrocardiography, radionuclide ventriculography, Holter monitoring and clinical variables for prediction of arrhythmic events in survivors of acute myocardial infarction in the thrombolytic era. J Am Coll Cardiol 1993;21:1419-27.
81. Day CP, McComb JM, Campbell RW: QT dispersion: An indication of arrhythmia risk in patients with long QT intervals. Br Heart J 1990;63:342-4.
82. Smith JM, Clancy EA, Valeri CR, et al: Electrical alternans and cardiac electrical instability. Circulation 1988;77:110-21.
83. Zabel M, Klingenheben T, Franz MR, Hohnloser SH: Assessment of QT dispersion for prediction of mortality or arrhythmic events after myocardial infarction: Results of a prospective, long-term follow-up study. Circulation 1998;97:2543-50.
84. Glancy JM, Garratt CJ, Woods KL, de Bono DP: QT dispersion and mortality after myocardial infarction. Lancet 1995;345:945-8.
85. Rosenbaum DS, Jackson LE, Smith JM, et al: Electrical alternans and vulnerability to ventricular arrhythmias. N Engl J Med 1994; 330:235-41.
86. Hohnloser SH, Klingenheben T, Li Y-G, et al: T wave alternans as a predictor of recurrent ventricular tachyarrhythmias in ICD recipients: Prospective comparison with conventional risk markers. J Cardiovasc Electrophysiol 1998;9:1258-68.
87. Klingenheben T, Zabel M, D'Agostino L, et al: Predictive value of T-wave alternans for arrhythmic events in patients with congestive heart failure. Lancet 2000;356:651-2.
88. Schwartz PJ: The autonomic nervous system and sudden death. Eur Heart J 1998;78:969-79.
89. Schwartz PJ, La Rovere MT, Vanoli E: Autonomic nervous system and sudden cardiac death: Experimental basis and clinical observations for post-myocardial infarction risk stratification. Circulation 1992;85(suppl I):I77-I91.
90. Bigger JT, Fleiss JL, Steinman RC, et al: Frequency domain measures of heart period variability and mortality after myocardial infarction. Circulation 1992;85:164-71.
91. La Rovere MT, Bigger JT, Marcus FI, et al: Baroreflex sensitivity and heart-rate variability in prediction of total cardiac mortality after myocardial infarction. Lancet 1998;351:478-84.
92. La Rovere MT, Pinna GD, Hohnloser SH, et al: Baroreflex sensitivity and heart rate variability in the identification of patients at risk for life-threatening arrhythmias: Implications for clinical trials. Circulation 2001;103:2072-7.
93. Huikuri HV, Makikallio T, Airaksinen KEJ, et al: Measurement of heart rate variability: A clinical tool or a research toy? J Am Coll Cardiol 1999;34:1878-83.
94. Brugada P, Green M, Abdollah H, Wellens HJJ: Significance of ventricular arrhythmias initiated by programmed ventricular stimulation: The importance of the type of ventricular arrhythmia induced and the number of premature stimuli required. Circulation 1984;69:87-92.
95. Wilber DJ, Garan H, Finkelstein D, et al: Out-of-hospital cardiac arrest: Use of electrophysiological testing in the prediction of long-term outcome. N Engl J Med 1988;318:19-24.
96. The Antiarrhythmics versus Implantable Defibrillators (AVID) Investigators. A comparison of antiarrhythmic-drug therapy with implantable defibrillators in patients resuscitated from near-fatal ventricular arrhythmias. N Engl J Med 1997;337:1576-83.
97. Connolly SJ, Gent M, Roberts RS, et al: Canadian Implantable Defibrillator Study (CIDS): Randomized trial of the implantable cardioverter defibrillator against amiodarone. Circulation 2000;101:1287-1302.
98. Kuck KH, Cappato R, Siebels J, Ruppel R: Randomized comparison of antiarrhythmic drug therapy with implantable defibrillators in patients resuscitated from cardiac arrest: The Cardiac Arrest Study Hamburg (CASH). Circulation 2000;102;748-54.
99. Buxton AE, Lee KL, Fisher JD, et al: A randomized study of the prevention of sudden death in patients with coronary artery disease. N Engl J Med 1999;341:1882-90.
100. Mittal S, Iwai S, Stein KM, et al: Long-term outcome of patients with unexplained syncope treated with an electrophysiologic-guided approach in the implantable cardioverter-defibrillator era. J Am Coll Cardiol 1999;34:1082-9.
101. Hallstrom AP, Cobb LA, Swain M, et al: Predictors of hospital mortality after out-of-hospital cardiopulmonary resuscitation. Crit Care Med 1985;13:927-9.
102. Weaver WD, Copass MK, Hill DL, et al: Cardiac arrest treated with a new automatic external defibrillator by out-of-hospital first responders. Am J Cardiol 1986;57:1017-21.
103. Cobb LA, Weaver WD, Fahrenbruch CE, et al: Community-based interventions for sudden cardiac death: Impact, limitations, and changes. Circulation 1992;85:I98-I102.

104. Becker LB, Han BH, Meyer PM, et al: Racial differences in the incidence of cardiac arrest and subsequent survival. The CPR Chicago Project. N Engl J Med 1993;329:600-6.
105. White RD, Asplin BR, Bugliosi TF, et al: High discharge survival rate after out-of-hospital ventricular fibrillation with rapid defibrillation by police and paramedics. Ann Emerg Med 1996;28:480-5.
106. Valenzuela TD, Roe DJ, Nichol G, et al: Outcomes of rapid defibrillation by security officers after cardiac arrest in casinos. N Engl J Med 2000;343:1206-9.
107. Page RL, Joglar JA, Kowal RC, et al: Use of automated external defibrillators by a U.S. airline. N Engl J Med 2000;343: 1210-16.
108. O'Rourke MF, Donaldson E, Geddes JS: An airline cardiac arrest program. Circulation 1997;96:2849-53.
109. DeBehnke DJ, Hilander SJ, Dobler DW, et al: The hemodynamic and arterial blood gas response to asphyxiation: A canine model of pulseless electrical activity. Resuscitation 1995;30:169-75.
110. Yusuf S, Peto R, Lewis J, et al: Beta blockade during and after myocardial infarction: An overview of the randomized trials. Prog Cardiovasc Dis 1985;27:335-71.
111. Gottlieb SS, McCarter RJ, Vogel RA: Effect of beta-blockade on mortality among high-risk and low-risk patients after myocardial infarction. N Engl J Med 1998;3339:489-97.
112. Julian DG, Camm AJ, Frangin G, et al: Randomized trial of effect of amiodarone on mortality in patients with left ventricular dysfunction after recent myocardial infarction: EMIAT. Lancet 1997;349:667-84.
113. Cairns JA, Connolly SJ, Roberts R, Gent M: Randomized trial of outcome after myocardial in patients with frequent or repetitive ventricular premature depolarizations: CAMIAT. Lancet 1997; 349:675-82.
114. Doval HC, Nul DR, Grancelli HO, et al GESICA-GEMA Investigators: Nonsustained ventricular tachycardia in severe heart failure: Independent marker of increased mortality due to sudden death. Circulation 1996;94:3198-203.
115. Massie BM, Fisher SG, Radford M, et al: Effect of amiodarone on clinical status and left ventricular function in patients with congestive heart failure. Circulation 1996;93:2128-34.
116. Bigger JT: Prophylactic use of implantable cardiac defibrillators in patients at high risk for ventricular arrhythmias after coronary artery bypass graft surgery. N Engl J Med 1997;335: 1569-75.
117. Moss AJ: Implantable cardioverter defibrillator: The sickest patients benefit the most. Circulation 2000;101:1638-40.
118. Domanski MJ, Saksena S, Epstein AE, et al: Relative effectiveness of the implantable cardioverter-defibrillator and antiarrhythmic drugs in patients with varying degrees of left ventricular dysfunction who have survived malignant ventricular arrhythmias. J Am Coll Cardiol 1999;34:1090-5.

Chapter 21 Syncope

HUGH CALKINS

Syncope (from Greek *synkope*, meaning cessation, pause) is a sudden and brief loss of consciousness and postural tone with spontaneous recovery. Syncope is an important clinical problem, as it is common, is costly, impairs quality of life, and may be the first and only warning symptom before an episode of sudden cardiac death (SCD).

Prior studies have estimated that syncope accounts for 3% of emergency room visits and 1% of hospital admissions.[1] Several surveys of young adults have consistently revealed that approximately one third of young persons have experienced syncope. The majority of these are isolated events that never come to medical attention.[2] Syncope is also extremely common among institutionalized elderly persons, who have a 6% annual incidence of syncope.[3] In contrast to the frequency with which syncope occurs in young adults and the very elderly, syncope is remarkably uncommon in middle-aged persons. The Framingham Study performed biennial examinations in free-living adults. The results of this study revealed that at least one syncopal episode occurred in approximately 3% of the population during the 26-year period of follow-up.[4]

The cost of diagnosing and caring for Medicare patients with syncope was recently estimated to be $800 million dollars annually.[5,6] Patients with recurrent syncope also report a markedly reduced quality of life, similar to that experienced by patients with chronic diseases such as rheumatoid arthritis and chronic obstructive pulmonary disease.[7] Some syncopal episodes are benign and self-limited, while others result from serious or life-threatening conditions such as ventricular tachycardia, with its attendant poor prognosis if untreated. Up to one third of patients presenting with syncope from untreated ventricular tachycardia experience SCD within 1 year.[8]

This chapter reviews the most important causes of syncope and provides a diagnostic strategy for the evaluation of this relatively common clinical problem. This chapter also comments briefly on the treatment of patients with syncope. However, a detailed review of treatments for the various causes of syncope is beyond the scope of this chapter. This may be obtained from recent publications by the European Society of Cardiology's Task Force on Syncope.[9,10]

Pathophysiology

Syncope results from a global reversible reduction of blood flow to the reticular activating system of the brain, which is the neuronal network in the brainstem that is responsible for supporting consciousness. In contrast to that of many other organs, the metabolism of the brain is exquisitely dependent on perfusion. As a result, cessation of cerebral blood flow for 10 seconds or longer leads to loss of consciousness. In most cases, the underlying cause for transiently reduced cardiac output results from a variety of circulatory disturbances such as volume depletion, diminished venous return, arrhythmias, cardiac outflow obstruction, or cerebrovascular obstruction.

The elderly are particularly susceptible to syncope as a result of the aging process. Vascular stiffening and diminished cardiac output leads to a 25% fall in cardiac output in the elderly.[11] In addition, neural reflex compensatory mechanisms for maintaining a stable blood pressure are less reliable in the elderly, as is cerebral autoregulation.[11,12]

Differential Diagnosis

Establishing the cause of syncope is a challenging problem for a number of reasons. First, the number of potential causes of syncope is extremely large. Second, the cause of syncope is often multifactorial. And third, it is almost never possible to witness a spontaneous episode of syncope while simultaneously monitoring the heart rate, blood pressure, and electroencephalogram. As a result, clinicians are charged with identifying the "probable cause of syncope."

The causes of syncope are often grouped into two categories: cardiovascular and noncardiovascular. Despite its simplicity, this approach has important limitations. Perhaps the most notable of these limitations is that certain conditions, especially orthostatic hypotension and the neurally mediated syncopal syndromes, are difficult to classify as either cardiovascular or noncardiovascular. In addition, although it is generally accepted that "cardiac" causes of syncope are potentially life threatening, the reflex-mediated syncopal syndromes are benign. Because of these and other limitations with the "cardiovascular–noncardiovascular" approach, we have employed a somewhat different classification scheme (Table 21-1). In this scheme, related etiologic conditions are divided into five main groups: cardiac, metabolic/miscellaneous, neurologic/cerebrovascular, vascular, and syncope of unknown origin. Vascular causes of syncope can be further subdivided into anatomic, orthostatic, and reflex-mediated causes. Likewise, cardiac causes of syncope can be further subdivided into anatomic and arrhythmic causes. Previously, it was well accepted that in up to 50% of patients presenting with syncope, no specific cause will be established after initial evaluation.[8] However, with the advent of tilt table testing as a diagnostic test for vasovagal syncope and availability of other new diagnostic tools such as implantable continuous loop event monitors, the probable cause of syncope can now be identified in approximately 80% of patients.[8,13-19]

TABLE 21-1 Causes of Syncope

Cardiac
Anatomic
Aortic dissection
Aortic stenosis
Atrial myxoma
Cardiac tamponade
Hypertrophic cardiomyopathy
Mitral stenosis
Myocardial ischemia/infarction
Pulmonary embolism
Pulmonary hypertension
Arrhythmias
Bradyarrhythmias
Atrioventricular block
Pacemaker malfunction
Sinus node dysfunction/bradycardia
Tachyarrhythmias
Supraventricular tachycardia
Ventricular tachycardia
Torsades de pointes/long QT syndrome
Neurologic/Cerebrovascular
Arnold-Chiari malformation
Migraine
Seizures (partial complex, temporal lobe)
Transient ischemic attack/vertebral-basilar insufficiency/cerebrovascular accident
Metabolic/Miscellaneous
Metabolic
Hyperventilation (hypocapnia)
Hypoglycemia
Hypoxemia
Drugs/alcohol
Miscellaneous
Psychogenic syncope
Hysterical
Panic disorder
Anxiety disorder
Cerebral syncope
Hemorrhage
Vascular
Anatomic
Subclavian steal
Orthostatic
Drug-induced
Hypovolemia
Primary disorders of autonomic failure
Pure autonomic failure (Bradbury-Eggleston syndrome)
Multiple system atrophy (Shy-Drager syndrome)
Parkinson's disease with autonomic failure
Secondary neurogenic
Postprandial (in the elderly)
Postural orthostatic tachycardia syndrome (POTS)
Reflex mediated
Neurally mediated syncope/vasovagal syncope
Carotid sinus hypersensitivity
Situational (cough, defecation, micturition, swallow)
Glossopharyngeal syncope
Trigeminal neuralgia
Unknown

VASCULAR CAUSES OF SYNCOPE

Vascular causes of syncope, particularly those associated with orthostatic intolerance, are by far the most common. At least one third of all episodes of syncope have a vascular cause.

Orthostatic Hypotension

Orthostatic hypotension, defined as a 20 mm Hg drop in systolic blood pressure or a 10 mm Hg drop in diastolic blood pressure within 3 minutes of standing, results from a defect in the blood pressure control system.[20] Patients with orthostatic hypotension probably have inadequate reflex adaptations to upright posture due to vascular or baroreceptor abnormalities.[21] Most often, the problem is secondary to or aggravated by extrinsic factors. Orthostatic hypotension is particularly common in the elderly.[11,12,22]

Drugs that either cause volume depletion or result in vasodilation (e.g., diuretics, antihypertensive agents) are the most common cause of orthostatic hypotension. Hytrin, which is commonly prescribed for treatment of bladder outlet obstruction, is also a frequent cause of orthostatic hypotension. Elderly patients are particularly susceptible to the hypotensive effects of drugs because of reduced baroreceptor sensitivity, decreased cerebral blood flow, renal sodium wasting, and an impaired thirst mechanism that develops with aging.[2,11,12,21]

Syncope that occurs after meals, particularly in the elderly, may result from a redistribution of blood to the gut. A decline in systolic blood pressure of about 20 mm Hg approximately 1 hour after eating has been reported in up to one third of elderly nursing home residents.[23] While usually asymptomatic, it may result in lightheadedness or syncope. The combination of a large meal and alcohol may be a particularly dangerous combination in susceptible patients, as alcohol has been shown to markedly impair the body's ability to vasoconstrict appropriately.[24] Other secondary factors, such as dehydration, anemia, or other concomitant illnesses, are less frequent but important considerations.

Orthostatic hypotension may also result from primary or secondary autonomic failure.[6,25] Primary autonomic failure is idiopathic, whereas secondary autonomic failure occurs in association with a known biochemical or structural abnormality, or is seen as part of a particular disease or syndrome. Primary autonomic failure consists of three types: (1) pure autonomic failure (Bradbury-Eggleston syndrome), (2) multiple system atrophy (Shy-Drager syndrome), and (3) Parkinson's disease with autonomic failure. Bradbury-Eggleston syndrome is an idiopathic disorder characterized by orthostatic hypotension in conjunction with evidence of more widespread autonomic failure such as disturbances in bowel, bladder, thermoregulation, and sexual function. These patients have reduced supine plasma norepinephrine levels. Shy-Drager syndrome is a sporadic, progressive disorder characterized by autonomic dysfunction, parkinsonism, and ataxia in any combination.

Postural orthostatic tachycardia syndrome (POTS) is a more recently described and milder form of chronic autonomic failure and orthostatic intolerance characterized by the presence of symptoms of orthostatic intolerance, a 28 beat per minute or greater increase in heart rate, and the absence of a significant change in blood pressure within 5 minutes of standing or upright tilt.[27-29] POTS appears to result from a failure of the peripheral vasculature to appropriately vasoconstrict under orthostatic stress. POTS may also be associated with syncope due to neurally mediated hypotension. Recent studies have demonstrated that the symptoms associated with POTS can generally be controlled with medical therapy and that this condition does not progress into a more severe form of autonomic failure.[29]

Reflex-Mediated Syncope

There are a large number of reflex-mediated syncopal syndromes (see Table 21-1). For each of these syndromes, syncope results from activation of a reflex composed of a trigger (the afferent limb) and a response (the efferent limb). The triggers for these forms of syncope may arise from any number of peripheral receptors. For example, swallowing syncope results from afferent neural impulses arising from the upper gastrointestinal tract, micturition syncope results from activation of mechanoreceptors in the bladder, and defecation syncope results from neural inputs from gut wall tension receptors. Perhaps the most widely recognized trigger site is the carotid sinus baroreceptor, which is in large part responsible for carotid sinus syncope. These triggers may also arise from within the central nervous system itself (e.g., syncope associated with anxiety or fear). The reflex-mediated syncopal syndromes are united by the response limb of the reflex, which consists of increased vagal tone and a withdrawal of peripheral sympathetic tone leading to bradycardia, vasodilation, and ultimately presyncope or syncope. The two most common forms of reflex-mediated syncope are neurally mediated (vasodepressor syncope) and carotid sinus syncope.

Neurally mediated syncope is the preferred term to describe a common abnormality of blood pressure regulation characterized by the abrupt onset of hypotension with or without bradycardia syndromes (also known as *neurocardiogenic syncope, vasodepressor syncope, vasovagal syncope,* and *fainting*). Triggers associated with the development of neurally mediated syncope (e.g., sight of blood, pain, prolonged standing, a warm environment, a hot shower, stressful situations) are those that either reduce ventricular filling or increase catecholamine secretion. The proposed mechanism involves a paradoxical reflex that is initiated when ventricular preload is reduced by venous pooling, leading to a reduction in cardiac output and blood pressure. This is sensed by arterial baroreceptors, resulting in increased catecholamine levels. Combined with reduced venous filling, this leads to a vigorously contracting volume-depleted ventricle. The heart itself is involved in this reflex by virtue of the presence of nonmyelinated mechanoreceptors, or C-fibers, found in the atria, the ventricles, and the pulmonary artery.[30-33] It has been proposed that vigorous contraction of a volume-depleted ventricle leads to activation of these receptors in susceptible individuals. These afferent fibers project centrally to the dorsal vagal nucleus of the medulla, leading to a "paradoxical" withdrawal of peripheral sympathetic tone and an increase in vagal tone, which in turn causes vasodilation and bradycardia. The ultimate clinical consequences are syncope or presyncope. Not all neurally mediated syncope results from activation of mechanoreceptors. In humans, it is well known that the sight of blood or extreme emotion can trigger syncope. These observations suggest that higher neural centers can also participate in the pathophysiology of vasovagal syncope. In addition, central mechanisms can contribute to the production of neurally mediated syncope.[32-34]

Carotid sinus syncope results from an abnormal response to stimulation of carotid sinus baroreceptors, located in the internal carotid artery above the bifurcation of the common carotid artery. Three types of abnormal responses have been described: (1) the cardioinhibitory response, characterized by marked bradycardia (>3 second pause), (2) the vasodepressor response, characterized by a 50 mm Hg fall in the systolic blood pressure in the absence of bradycardia, and (3) the mixed response. Although carotid sinus hypersensitivity is commonly detected in patients with syncope, it is important to recognize that pauses longer than 3 seconds are frequently observed in asymptomatic elderly patients.[3] Because of this, the diagnosis of carotid sinus hypersensitivity should be approached cautiously after excluding alternative causes of syncope.

CARDIAC CAUSES OF SYNCOPE

Cardiac causes of syncope are the second most common etiology of syncope, accounting for approximately one fifth of all syncopal episodes. The cardiac causes of syncope can be further subdivided into cardiac arrhythmias (tachyarrhythmias or bradyarrhythmias) and structural cardiovascular/cardiopulmonary disease.

Primary cardiac arrhythmias are probably the second most common cause of syncope, particularly in patients with underlying cardiovascular disease. These arrhythmias include the broad range of tachyarrhythmias and bradyarrhythmias. It is common for tachyarrhythmias to cause syncope at their onset followed by a return of consciousness despite continuation of the arrhythmia.

This results from the assumption of a recumbent position as well as the activation of compensatory mechanisms.

Structural cardiovascular and cardiopulmonary diseases are far less common causes of syncope as compared with cardiac arrhythmias. These causes include conditions such as critical aortic stenosis, atrial myxoma, and pulmonary emboli. Although acute myocardial ischemia and infarction may also cause syncope, the basis of syncope in these situations is usually multifactorial.[36,37]

NEUROLOGIC CAUSES OF SYNCOPE

Neurologic causes of syncope, including migraines, seizures, Arnold-Chiari malformations, and transient ischemic attacks, are surprisingly uncommon causes of syncope, accounting for less than 10% of all cases of syncope. The majority of patients in whom a "neurologic" cause of syncope is established are found, in fact, to have had a seizure rather than true syncope.[8,13]

METABOLIC, MISCELLANEOUS, AND PSYCHIATRIC CAUSES OF SYNCOPE

Syncope due to metabolic causes, including hypoglycemia, hypoxia, and hyperventilation, account for less than 5% of syncopal episodes. The establishment of hypoglycemia as the cause of syncope requires demonstration of hypoglycemia during the syncopal episode. Although the mechanism of hyperventilation-induced syncope has been generally considered to be due to a reduction in cerebral blood flow, a recent study demonstrated that hyperventilation alone was not sufficient to cause syncope, suggesting that hyperventilation-induced syncope may also have a psychological component.[29]

Psychiatric disorders mimicking true syncope are a common cause of spells that remain unexplained even after a thorough evaluation. Psychiatric causes of syncope have been implicated in up to 25% of patients with syncope of unknown origin.[39] The precise cause of these disturbances is unknown, but in most cases they do not result in true syncope. It is important to note, however, that patients who present with these "hysterical" causes of syncope may also have neurally mediated syncope, which further complicates their evaluation and management.

Cerebral syncope is a rare cause of syncope that has only recently been described.[40] This unusual type of syncope occurs during orthostatic stress and is associated with a normal blood pressure. This form of syncope is believed to result from cerebrovascular spasm, which causes cerebral hypoperfusion and loss of consciousness despite a normal blood pressure. Cerebral syncope is difficult to diagnose and may easily be confused with "pseudosyncope" resulting from a psychiatric condition. At The Johns Hopkins Hospital, it is our belief that this diagnosis can only be established by demonstration of syncope in association with a normal hemodynamic response to tilt in conjunction with cerebral hypoperfusion (using transcranial Doppler) and electroencephalogram evidence of cerebral hypoperfusion.[40]

Diagnostic Evaluation

As noted previously, syncope usually occurs sporadically and infrequently. It is therefore extremely difficult to examine or document a patient's heart rhythm and blood pressure at the time of syncope. Because of this, the primary goal in the evaluation of a patient with syncope is to arrive at a presumptive determination of the cause of syncope. The utility of some of the main diagnostic tests used in the evaluation of patients with syncope is outlined as follows.

HISTORY AND PHYSICAL EXAMINATION

The history and physical examination remain the most important components of the evaluation of a patient with syncope.[13,14,41-44] It is generally accepted that more than 50% of syncope diagnoses are established based on the history and physical examination.[13,14] The initial evaluation should begin by determining if the patient actually experienced true syncope. In contrast to syncope, dizziness, presyncope, and vertigo do not result in a loss of consciousness or postural tone. The term "drop attacks" refers to abrupt falls that occur without a loss of consciousness. A cardiac arrest differs from syncope in that the resumption of consciousness does not occur spontaneously but rather requires cardioversion. It is also important to distinguish syncope from seizures (Table 21-2).

When taking a clinical history from a patient with syncope, particular attention should be focused on the following historical features: (1) a history of cardiac disease, (2) a family history of cardiac disease, syncope, or sudden death, (3) identification of medications that may have played a role in syncope, (4) number and chronicity of prior episodes, (5) identification of precipitating factors including body position, and (6) type and duration of prodromal and recovery symptoms. Table 21-2 summarizes the features of the clinical history that are most useful in determining whether syncope resulted from neurally mediated hypotension, an arrhythmia, or a seizure.

It is also important to search for additional precipitating factors that may suggest a specific type of syncope. For example, carotid sinus syncope should be suspected if a patient reports syncope in response to head rotation, and micturition syncope should be considered if syncope occurs in association with urination. Akinetic or petit mal seizures are characterized by the patient's lack of responsiveness in the absence of a loss of postural tone. Temporal lobe seizures last several minutes and are characterized by postictal confusion, changes in the level of consciousness, and autonomic signs such as flushing. Vertebrobasilar insufficiency should be considered as the cause of syncope if syncope occurs in association with other symptoms of brainstem ischemia (e.g., diplopia, tinnitus, dysgeusia, focal weakness or sensory loss, vertigo, dysarthria).

TABLE 21-2 Features of the Clinical History in Patients with Syncope

	Arrhythmias	Neurally Mediated Hypotension	Seizure
Demographics/Clinical Setting	Male > female gender Older age (>54 yr) Fewer episodes (<3) Any setting	Female > male gender Younger age (<55 yr) More episodes (>2) Standing, warm room, emotional upset	Younger age (<45 yr) Any setting
Premonitory Symptoms	Shorter duration (<6 sec) Palpitations less common	Longer duration (>5 sec) Palpitations Blurred vision Nausea Warmth Diaphoresis Lightheadedness	Sudden onset or brief aura (déjà vu, olfactory, gustatory, visual)
Observations During the Event	Blue, not pale Incontinence may occur Brief clonic movements may occur	Pallor Diaphoretic Dilated pupils Slow pulse, low BP Incontinence may occur Brief clonic movements may occur	Blue face, no pallor Frothing at the mouth Prolonged syncope (duration >5 min) Tongue biting Horizontal eye deviation Elevated pulse and BP Incontinence more likely Tonic-clonic movements if grand mal
Residual Symptoms	Residual symptoms uncommon (unless prolonged unconsciousness) Oriented	Residual symptoms common Prolonged fatigue common (>90%) Oriented	Residual symptoms common Aching muscles Disoriented Fatigue Headache Slow recovery

BP, blood pressure.

The physical examination should include defining the patient's level of hydration and orthostatic vital signs, determining if structural heart disease is present, and detecting the presence of significant neurologic abnormalities suggestive of a dysautonomia or a cerebrovascular accident.

LABORATORY TESTS

The routine use of blood tests, such as serum electrolytes, glucose, and hematocrit levels, generally has low diagnostic value. These tests result in a diagnosis in less than 3% of patients.[8,13] Most patients in whom abnormal laboratory test results are found are demonstrated to have a seizure rather than true syncope. Nonetheless, these tests are commonly obtained as part of the evaluation of a patient with syncope.

CAROTID SINUS MASSAGE

Carotid sinus hypersensitivity testing is performed by applying gentle pressure over the carotid pulsation just below the angle of the jaw where the carotid bifurcation is located. Pressure should be applied for 5 seconds. Auscultation to determine if a carotid bruit is present should be performed before carotid sinus massage. Although the risks of a cerebrovascular accident occurring in response to carotid sinus massage are unknown, there is a widely held belief that carotid sinus massage performed in the presence of a carotid bruit may result in an embolic cerebrovascular accident. An abnormal response has been defined as a greater than 3 second pause or a 50 mm Hg fall in systolic blood pressure.

ELECTROCARDIOGRAM

An abnormal ECG is found in up to 50% of patients with syncope, but in most patients it is not diagnostic. It has been estimated that the ECG results in establishment of a diagnosis in approximately 5% of syncope patients.[13,14] Examples of diagnostic ECG findings include Q–T interval prolongation, ventricular preexcitation, evidence of an acute myocardial infarction, and high-grade atrioventricular block. Less specific findings that may suggest potential causes of syncope and can be later confirmed with directed testing include evidence of an old myocardial infarction; bundle branch block; ventricular hypertrophy; T wave inversion in the right precordial leads (suggestive of right ventricular dysplasia); or a right bundle branch block pattern with persistent ST elevation in the right precordial leads, which points to a recently described hereditary arrhythmia syndrome known as "Brugada syndrome." This syndrome is associated with a high incidence of SCD.[45,46] The finding of a normal ECG suggests that a cardiac cause of syncope is unlikely.

The signal-averaged electrocardiogram (SAECG) is a noninvasive technique used for detection of low-amplitude signals in the terminal portion of the QRS complex (late potentials), which suggest the presence of the substrate for malignant ventricular arrhythmias. Available studies suggest that the SAECG may indicate inducibility of ventricular arrhythmias in syncope patients with ischemic heart disease. However, a major limitation of the SAECG is that it serves only as a surrogate for the direct link between arrhythmia and symptoms. Thus, obtaining an SAECG is currently not considered a standard part of the evaluation of a patient with syncope.[47] One notable exception is for patients in whom arrhythmogenic right ventricular dysplasia is suspected. These patients generally present with exercise-induced ventricular tachycardia or syncope, or both.[48,49] The diagnosis is difficult to establish because left ventricular function is normal, and there is not one diagnostic test that can be relied upon to establish or exclude this diagnosis.

AMBULATORY ELECTROCARDIOGRAPHY MONITORING

Ambulatory electrocardiography was introduced by Holter and Gengerilli in 1949.[50] Since that time a great deal of research has demonstrated the value of ECG monitoring devices in the diagnosis and management of cardiac arrhythmias.[41-54] The equipment used to record the ambulatory ECG has evolved with improved recording fidelity, reduction in size and weight, and improved data acquisition and transmission capabilities. The types of devices now available include (1) Holter monitors, which monitor the patient's heart rhythm continuously over a 24- to 48-hour period; and (2) event monitors, which either continuously or intermittently monitor the patient's heart rhythm. When these "event monitors" are activated, they either prospectively record the ECG signals for several minutes or freeze the prior 2 to 5 minutes of ECG signals in memory. These stored ECG recordings can subsequently be transmitted to a monitoring station for analysis and interpretation. Most of these event monitors are provided to patients for approximately 1 month. An exception is the recently developed implantable ECG monitor (Medtronic Reveal, Minneapolis, Minn.). This device is about the size of a lighter and is inserted under the skin. Once inserted, it continuously monitors the ECG signal and stores ECG signals either based on programmed criteria or when activated by the patient, or both.[55]

Holter Monitors

Holter recordings are often performed in an attempt to establish or exclude an arrhythmic cause of syncope. However, because of the infrequent and sporadic nature of syncope, the likelihood of syncope while wearing a Holter recorder in an unselected population of patients with syncope is only approximately 0.1%.[56] Some studies of Holter monitoring have reported symptoms in conjunction with arrhythmias in 4% of patients and symptoms occurring in the absence of an arrhythmia in 17% of patients. Increasing the duration of Holter monitoring to 72 hours did not increase the diagnostic yield of Holter monitoring significantly.[57,58] Although detecting arrhythmias during ambulatory monitoring can suggest a cause of syncope, it is important to recognize that unless these are accompanied by presyncope or syncope, they are likely to be incidental findings and should not be assumed to be causative. Since another episode of syncope is required to establish a diagnosis while wearing a Holter monitor, the clinical situation in which this modality is most likely to be diagnostic is in the occasional patient with very frequent (i.e., daily) episodes of impaired consciousness.

Event Recorders

Continuous loop event recorders are used for long-term monitoring (weeks or months). Event recorders are especially useful for patients with frequent episodes of syncope, particularly once potentially malignant causes of syncope have been excluded.[51-54] In patients with frequently recurring episodes of syncope, arrhythmias were detected at the time of syncope in 8% to 20% of patients, and normal rhythm was recorded at the time of symptoms in 12% to 27% of patients.[53,54] The limitations of these recorders are patient noncompliance, the potential for errors in using the device, and problems with transmission. Another limitation is that repeated use of event monitoring, which requires continuous lead placement, may result in significant skin irritation.

Implantable Monitors

In some patients, syncope occurs only once or twice a year. In these patients, traditional event recorders may not be diagnostic due to the associated problems with prolonged recording periods. Recently, a very small (61 × 19 × 8 mm) implantable continuous loop recorder that is inserted subcutaneously on the chest was developed (Medtronic Reveal, Minneapolis, Minn.). It incorporates two electrodes within its can and has the capability to perform cardiac monitoring for up to 24 months.[55] In a small study its diagnostic yield was found to be higher than that of combined Holter monitoring, tilt testing, and electrophysiological (EP) study.[19]

TILT TABLE TESTING

Tilt table testing has generally been considered the "gold standard" for establishing the diagnosis of neurally mediated syncope, since it provides a means for assessing susceptibility to this condition directly.[59-67] Upright tilt testing is generally performed for 30 to 45 minutes at an angle of approximately 70 degrees. In general, a positive response to tilt table testing is defined as the development of syncope or presyncope in association with hypotension or bradycardia, or both. Positive responses to table testing can be divided into mixed, cardioinhibitory, and vasodepressor categories. In the absence of pharmacologic provocation, tilt table testing at angles of 60 to 70 degrees exhibits a specificity of approximately 90%.[60,67] An American College of Cardiology Expert Consensus Document has proposed indications and methods for tilt testing.[67] Upright tilt table testing is generally agreed upon to be indicated in patients with: (1) recurrent syncope or a single syncopal episode in a high-risk patient who either has no evidence of structural heart disease or in whom other causes of syncope have been excluded; (2) evaluation of patients in whom the causes of syncope have been determined (e.g., asystole), but in whom the presence of neurally mediated syncope on upright tilt would influence treatment; and (3) as part of the evaluation of patients with exercise-related syncope. There is also general agreement that upright tilt testing is not necessary for patients who have experienced only a single syncopal episode that was highly typical for neurally mediated syncope and during which no injury occurred. Tilt table testing is not useful in establishing a diagnosis of situational syncope (e.g., post micturition syncope).[63] In the past, isoproterenol was the provocative agent that was most widely used in conjunction with tilt testing. More recently, nitrates have been administered more commonly.[64-66] This transition to the use of nitrates as provocative agents was fueled in part by the recent lack of availability of isoproterenol. In addition, the administration of a single dose of nitroglycerine sublingually following a 45-minute tilt test significantly simplified the protocol. The use of continuous monitoring of finger systolic arterial pressure may allow a shortened protocol.[68]

ECHOCARDIOGRAMS

Although echocardiograms are commonly used in the evaluation of patients with syncope, little objective evidence exists to support their use in patients with a normal physical examination and a normal ECG.[69-72] Echocardiograms only reveal unsuspected structural heart disease in 5% to 10% of patients, and the discovery of such abnormalities does not necessarily lead to the cause of syncope. Since structural heart disease is a strong predictor of mortality among patients with syncope, an echocardiogram may help to risk stratify the patient by determining whether occult cardiac disease is present when not apparent after the history, physical examination, and electrocardiography. Echocardiograms are most likely to be helpful in the evaluation of patients with a history of structural heart disease, the elderly, and those who present with exercise-related syncope.

STRESS TESTING, CARDIAC CATHETERIZATION

Myocardial ischemia is an unlikely cause of syncope.[13,36,37] Therefore, exercise stress testing and cardiac catheterization are unlikely to establish a diagnosis in patients presenting with syncope unless the clinical suspicion of ischemia is high. Stress testing in syncope patients should be reserved for those in whom syncope or presyncope occurred during or immediately following exertion or in association with ischemic symptoms. It should also be considered in young individuals with recurrent syncope during exertion to rule out anomalous coronary arteries when other causes of syncope have been excluded. It should be noted that even among patients with syncope during exertion, it is highly unlikely that exercise stress testing will trigger another event. Patients suspected of having severe aortic stenosis or obstructive hypertrophic cardiomyopathy should not undergo exercise stress testing as it may precipitate a cardiac arrest.

ELECTROPHYSIOLOGICAL TESTING

In general, electrophysiology (EP) testing in syncope patients is most helpful for identifying potential arrhythmic causes of syncope in individuals with underlying structural heart disease, conduction system disturbances, and the preexcitation syndromes. The indications for EP testing in the evaluation of patients with syncope have recently been established based on an American College of Cardiology/American Heart Association Task Force Report.[73] There is general agreement that EP testing should be performed in patients with suspected structural heart disease and unexplained syncope (class 1 indication) and that EP testing should *not* be performed in patients with a known cause of syncope for whom treatment will not be influenced by the findings of the test (class 3 indication). The role of EP testing in evaluating patients with recurrent, unexplained syncope who do not have structural heart disease and have had a negative tilt test remains controversial.

Approximately one third of patients with syncope of unknown origin referred for EP testing will have a presumptive diagnosis established. Clinical factors that have been identified as predictors of a positive response to EP testing include impaired ventricular function, male sex, prior myocardial infarction, bundle branch block, injury, and nonsustained ventricular tachycardia. The remaining two thirds of patients with a normal response to EP testing are generally considered to be at low risk of SCD. It is important to recognize, however, that EP testing does not always identify the arrhythmic cause of syncope, because transient abnormalities such as those caused by ischemia or fluctuations in autonomic tone may be missed.[73] In addition, recent studies have suggested that EP testing may have less predictive value in patients with markedly impaired ventricular function, particularly those with an idiopathic dilated cardiomyopathy. Several recent studies have reported a high (approximately 50%) incidence of sustained ventricular arrhythmias among patients with an idiopathic dilated cardiomyopathy and syncope, despite a normal response to EP testing.[74-76] These new findings suggest that greater use of implantable defibrillators may be needed in patients with an idiopathic dilated cardiomyopathy who present with syncope.

Sinus Node Function

Identification of sinus node dysfunction as the cause of syncope is an uncommon finding during EP tests (<5%).[77-80] It is important to note that the absence of evidence of sinus node dysfunction during EP testing does not exclude a bradyarrhythmia as the cause of syncope.

Sinus node function is evaluated during EP testing primarily by determining the sinus node recovery time (SNRT). Pacing is usually performed for 30 to 60 seconds at a number of drive cycle lengths between 350 and 600 milliseconds (ms). The SNRT reflects the degree of overdrive suppression of the sinus node and is defined as the interval between the last paced atrial depolarization and the first spontaneous atrial depolarization resulting from activation of the sinus node. A corrected SNRT (CSNRT = SNRT − sinus cycle length) greater than 525 ms is generally considered abnormal.[73] Although the first postpacing cycle length is often normal, subsequent cycle lengths may be prolonged, thus the term "secondary pauses." When identified, these increase sensitivity of SNRT in detecting sinus node dysfunction. Other, less often used measures of sinus node function include the sinoatrial conduction time and the sinus node refractory period.

Carotid Sinus Hypersensitivity

Carotid sinus hypersensitivity (CSH) may be diagnosed by performing carotid sinus massage for approximately 5 seconds. CSH is defined as a sinus pause greater than 3 seconds' duration. However, pauses of 3 to 5 seconds are not infrequently observed in asymptomatic elderly patients. In these patients, CSH remains a presumptive diagnosis.[81] Thus, the diagnosis of CSH should be approached cautiously after excluding alternative causes of syncope. Since patients with isolated CSH may undergo spontaneous remission of symptoms, pacemaker implantation should be reserved for those with recurrent symptomatic episodes. However, among patients who undergo placement of a pacemaker for treatment of syncope associated with carotid sinus hypersensitivity, the recurrence rate of syncope is small. Moreover, in patients older than 50 years of age presenting with nonaccidental falls who are found to have carotid sinus hypersensitivity, pacing reduces the incidence of recurrent falls.[82]

Atrioventricular Block

Atrioventricular (AV) conduction is assessed during EP testing by measuring the conduction time from the AV node to the His bundle (A–H interval) and from the His bundle to the ventricles (H–V interval), as well as determining the response of AV conduction to incremental atrial pacing and atrial premature stimuli. During EP testing, the findings that allow AV block to be established as the probable cause of syncope include an H–V interval greater than 90 ms or infra-Hisian block that is observed spontaneously or during atrial pacing.[73,79,83] The presence of infra-Hisian block has been shown to be predictive of development of subsequent second-degree or third-degree AV block. AV block has been reported to be the probable cause of syncope in approximately 10% to 15% of patients.

Supraventricular Tachycardia

Supraventricular arrhythmias are diagnosed as the probable cause of syncope in less than 5% of patients undergoing EP testing for syncope of unknown origin.[13,83] Although most patients with supraventricular arrhythmias experience symptoms such as palpitations or mild dizziness, syncope is uncommon in these patients, particularly if the heart is normal. Completion of a standard EP test will allow accurate identification of most types of supraventricular arrhythmias that may have caused syncope. The study should be repeated during an isoproterenol infusion to increase the sensitivity of the study, particularly for detecting idiopathic ventricular tachycardia or AV nodal reentrant tachycardia.

Ventricular Tachycardia

Ventricular tachycardia is the most common abnormality uncovered during EP testing in patients with syncope. An EP test is interpreted as positive for ventricular tachycardia when sustained monomorphic ventricular tachycardia is induced. The induction of polymorphic ventricular tachycardia, ventricular fibrillation, and nonsustained ventricular tachycardia have been considered to represent nonspecific responses to EP testing. Among studies that have reported the results of EP testing in evaluating patients with syncope, identification of ventricular tachycardia as the probable cause of syncope is reported in approximately 20% of patients.[13,73,78,79]

NEUROLOGIC TESTS

Syncope as an isolated symptom is rarely due to a neurologic cause. In most large published series, neurologic causes of syncope are established in less than 5% of patients. As a result, widespread use of tests to screen for neurologic conditions rarely are diagnostic. An electroencephalogram provides diagnostic information in less than 2% of patients with syncope,[13,84] the vast majority of whom had symptoms suggestive of a seizure or a history suggestive of a seizure disorder. The diagnostic yield of CT scans is approximately 4%[13] most of which had focal neurologic findings. Transient ischemic attacks involving the carotid or vertebrobasilar arteries rarely result in syncope. To date, no studies have demonstrated the usefulness of carotid Doppler studies in evaluating syncope. In many institutions, CT scans, electroencephalograms, and carotid duplex scans are overused, being obtained in more than 50% of patients with syncope. A diagnosis is almost never uncovered that was not first suspected based on a careful history and neurologic examination. A recent study indicated that 29% of patients with treatment-resistant epilepsy or suspected nonepileptic seizures have an underlying cardiovascular cause of syncope such as neurally mediated hypotension, carotid sinus hypersensitivity, or transient AV block.[85]

Evaluation Strategy For Patients With Syncope

The initial evaluation of a patient with "syncope" involves determining if the patient did, in fact, experience syncope, or whether the patient actually presented with

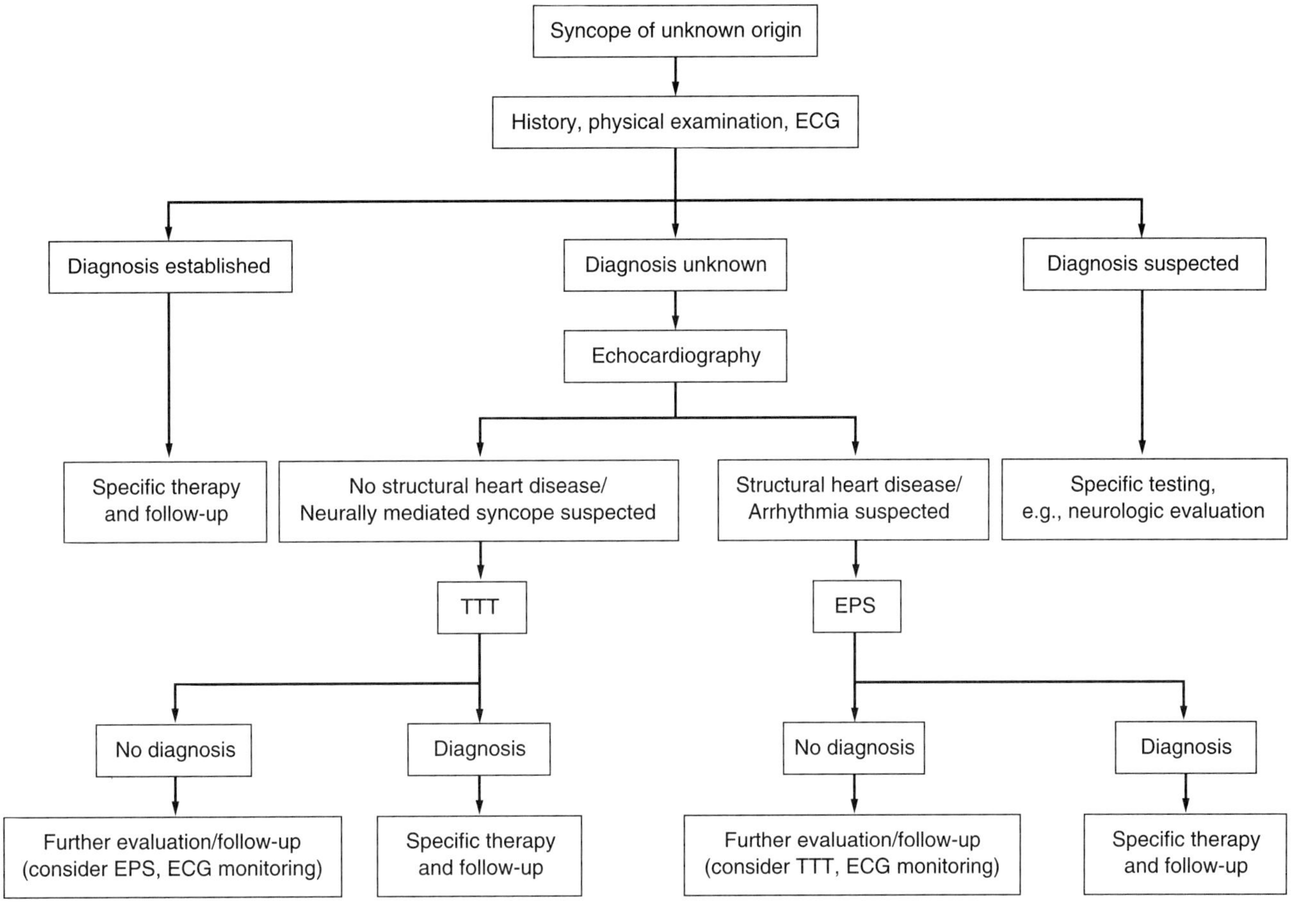

FIGURE 21-1 Clinical management of syncope of unknown origin.

dizziness, presyncope, vertigo, a "drop attack" in which the patient lost postural tone but did not lose consciousness, or a seizure. The next step in the evaluation involves a careful history, physical examination, and 12-lead ECG (Fig. 21-1). Based on this initial evaluation, the probable cause of syncope can be identified in approximately one third of patients. Syncope diagnoses that can often be established at this initial stage include orthostatic hypotension, situational syncope, and often neurally mediated syncope. In many other patients, the probable cause of syncope can be suspected based on this initial evaluation and later confirmed with specific diagnostic testing (i.e., aortic stenosis, right ventricular dysplasia, neurologic causes of syncope). The next step in the evaluation depends on the presence of structural heart disease as well as the physician's clinical suspicion that an arrhythmia may have been the cause of syncope.

The presence of structural heart disease (coronary artery disease, congestive heart failure, congenital heart disease, or valvular heart disease) has emerged as the most important factor for predicting the likelihood of arrhythmias and of death.[8,41] Syncope due to a cardiac cause has been associated with up to a 30% mortality at 1 year.[8] In contrast, syncope due to a noncardiac etiology or in patients without structural heart disease is generally associated with a benign prognosis.

An echocardiogram is often obtained at this point to help determine if structural heart disease is present. In our experience, the echocardiogram is particularly useful if the patient has a clinical history compatible with heart disease, is elderly, or presents with exercise-induced syncope. If the patient has significant structural heart disease or a clinical history suggestive of an arrhythmia, an EP study would be an appropriate next step. On the other hand, if structural heart disease is absent, and the clinical history is not suspicious for an arrhythmia, the evaluation can be continued either with a tilt test, event or implantable ECG monitor, or clinical follow-up depending on the severity and chronicity of the patient's symptoms. Using this approach, a probable cause of syncope can be determined in 75% to 80% of patients.[16,17]

Treatment Considerations

Treatment of a patient with syncope clearly depends on the diagnosis. For example, the appropriate treatment of a patient with syncope due to the Wolff-Parkinson-White syndrome would likely involve catheter ablation, whereas treatment of a patient with syncope due to ventricular tachycardia would likely involve placement of an implantable defibrillator. For other types of syncope, optimal patient management may involve discontinuation of an offending pharmacologic agent. A detailed discussion of the treatments for various causes of syncope has been produced by the Task Force on Syncope of the European Society of Cardiology.[10]

Another issue that must be considered is the need for hospitalization of a patient with syncope. No studies have evaluated the indications for hospital admission among patients with syncope. Generally, hospital admission is indicated when there is concern about an adverse outcome if the evaluation is delayed. In general, patients should be considered for an inpatient evaluation if they have structural heart disease, symptoms suggestive of arrhythmias or ischemia, or abnormal electrocardiographic findings. Patients in whom neurally mediated syncope is suspected as well as patients with no clinical evidence of structural heart disease generally can be evaluated on an outpatient basis.

Physicians also need to address the issue of driving risk. Patients who experience syncope while driving pose a risk both to themselves and to others. Although some would argue that all patients with syncope should never drive again because of the theoretical possibility of developing a recurrence, this is an impractical solution that will be ignored by many patients. Factors that should be considered when making a recommendation for a particular patient include: (1) the potential for recurrent syncope, (2) the presence and duration of warning symptoms, (3) whether syncope occurs while seated or only when standing, (4) how often and in what capacity the patient drives, and (5) any applicable state laws? When considering these issues, physicians should note that acute illnesses including syncope are unlikely to cause a motor vehicle accident. Recently, both the American Heart Association and the Canadian Cardiovascular Society have published guidelines concerning this issue.[86,87] For noncommercial drivers, it is generally recommended that driving be restricted for several months. If the patient remains asymptomatic, driving can be resumed several months later.

SUMMARY

In conclusion, syncope is an important medical problem, because it is common, results in a considerable economic burden, and may be the first and only warning symptom before an episode of SCD. Physicians who care for patients with syncope are charged with establishing the probable cause of syncope in a cost-effective but thorough fashion.

REFERENCES

1. Kapoor W: Evaluation and management of syncope. JAMA 1992;268:2553-2560.
2. Williams RL, Allen PD: Loss of consciousness. Aerospace Med 1962;33:545-551.
3. Lipsitz L, Wei JY, Rowe JW: Syncope in an elderly, 310 institutionalized population: Prevalence, incidence, and associated risk. QJM 1985;55:45.
4. Savage DD, Corwin L, McGee DL, et al: Epidemiologic features of isolated syncope: The Framingham Study. Stroke 1985;16:626-629.
5. Nyman JA, Krahn AD, Gland PC, et al: The costs of recurrent syncope of unknown origin in elderly patients. Pacing Clin Electrophysiol, 1999;22:1386-1394.
6. Calkins H, Byrne M, El-Atassi R, et al: The economic burden of unrecognized vasodepressor syncope. Am J Med 1993;95: 473-479.
7. Linzer M, Pontinen M, Gold, DT, et al: Impairment of psychosocial function in recurrent syncope. J Clin Epidemiology 1991;44:1037-1043.
8. Kapoor WN, Karpf M, Wieand S, et al: A prospective evaluation and follow-up of patients with syncope. N Engl J Med 1983;309: 197-204.
9. Brignole M, Alboni P, Benditt D, et al: Task force on syncope, European Society of Cardiology. Part 1. The initial evaluation of patients with syncope. Europace 2001;3:253-260.
10. Brignole M, Alboni P, Benditt D, et al: Task force on syncope, European Society of Cardiology. Part 2. Diagnostic tests and treatment: Summary of recommendations. Europace 2001;3: 261-268.
11. Scheinberg P, Blackburn I, Saslaw RM: Effects of aging on cerebral circulation and metabolism. Arch Neurol Psychiatry 1953;70:77.
12. Wollner L, McCarthy ST, Soper NDW, et al: Failure of cerebral autoregulation as a cause of brain dysfunction in the elderly. BMJ 1979;1:1117.
13. Kapoor WN: Evaluation and outcome of patients with syncope. Medicine 1990;69:160-175.
14. Linzer M, Yang EH, Estes III M, et al: Diagnosing syncope, part 1: Value of history, physical examination, and electrocardiography. Ann Intern Med 1997;126:989-996.
15. Linzer M, Yang EH, Estes III M, et al: Diagnosing syncope, part 2: Unexplained syncope. Ann Intern Med 1997;127:76-86.
16. Ammirati F, Colivicchi F, Santini M: Diagnosing syncope in clinical practice. Eur Heart J 2000;21:935-940.
17. Sra JS, Anderson AJ, Sheikh SH, et al: Unexplained syncope evaluated by electrophysiologic studies and head-up tilt testing. Ann Intern Med 1991;114:1013-1019.
18. Garcia-Civera R, Ruiz-Granell R, Morell-Cabedo S, et al: Selective use of diagnostic tests in patients with syncope of unknown cause. J Am Coll Cardiol 2003;41:787-790.
19. Krahn AD, Klein GJ, Yee R, Skanes AC: Randomized assessment of syncope trial: Conventional diagnostic testing versus a prolonged monitoring strategy. Circulation 2001;104:46-51.
20. The Consensus Committee of the American Autonomic Society and the American Academy of Neurology: Consensus statement on the definition of orthostatic hypotension, pure autonomic failure, and multiple system atrophy. Neurology 1996;46: 1470-1471.
21. Wieling W, van Lieshout JJ: Maintenance of postural normotension in humans. In Low P (ed): Clinical Autonomic Disorders. Boston, Little Brown, 1993, pp 69-75.
22. Kapoor WN: Syncope in the older person. J Am Geriatr Soc 1994;42:426.
23. Vaitkevicius PV, Esserwein DM, Maynard AK, et al: Frequency and importance of postprandial blood pressure reduction in elderly nursing-home patients. Ann Intern Med 1991;115:865.
24. Narkiewic K, Cooley RL, Somers VK: Alcohol potentiates orthostatic hypotension: Implications for alcohol-related systems. Circulation 2000;101:398-402.
25. Mathias CJ: The classification and nomenclature of autonomic disorders: Ending chaos, resolving conflict and hopefully achieving clarity. Clin Auton Res 1995;5:307-310.
26. Low P, McLeod J: The autonomic neuropathies. In Low P (ed): Clinical Autonomic Disorders. Boston, Little Brown, 1993, pp 395-421.
27. Schondorf R, Low PA: Idiopathic postural orthostatic tachycardia syndrome: An attenuated form of acute pandysautonomia? Neurology 1993;43:132-137.
28. Low P, Opfer-Gehrking T, Textor S, et al: Postural tachycardia syndrome (POTS). Neurology 1995;45:519-525.
29. Sandroni P, Opfer-Gehrking TL, McPhee BR, Low PA: Postural tachycardia syndrome: Clinical features and follow-up study. Mayo Clin Proc 1999;74:1106-1110.
30. Smith ML, Carlson MD, Thames MD: Reflex control of the heart and circulation: Implications for cardiovascular electrophysiology. J Cardiovasc Electrophysiol 1991;2:441-449.

31. Rea R, Thames M: Neural control mechanisms and vasovagal syncope. J Cardiovasc Electrophysiol 1993;4:587-595.
32. Abboud FM: Neurocardiogenic syncope. N Engl J Med 1993; 328:1117-1119.
33. Morillo CA, Eckberg DL, Ellenbogen KA, et al: Vagal and sympathetic mechanisms in patients with orthostatic vasovagal syncope. Circ 1997;96:2509-2513.
34. Hjorth S: Penbutolol as a blocker of central 5HT1A receptor-mediated responses. Eur J Pharmacol 1992;222:121-127.
35. Richardson DA, Beston RS, Shaw FE, Kenny RA: Prevalence of cardioinhibitory carotid sinus hypersensitivity in patients 50 years or over presenting to the accident and emergency department with unexplained or recurrent falls. Pacing Clin Electrophysiol 1997;20:820-823.
36. Pathy MS: Clinical presentation of myocardial infarction in the elderly. Br Heart J 1967;29:190-199.
37. Dixon MS, Thomas P, Sheridon DJ: Syncope is the presentation of unstable angina. Int J Cardiol 1998;9:125-129.
38. Racco F, Sconocchini C, Reginelli R: La sincope in una popolazone generale: Diagnosi ezilogoca e follow-up. Minerva Med 1993;84: 249-261.
39. Kapoor WN, Fortunato M, Hanusa BH, Schulberg HC: Psychiatric illnesses in patients with syncope. Am J Med 1995; 99:505-512.
40. Grubb, BP, Samoil D, Kosinski D, et al: Cerebral syncope: Loss of consciousness associated with cerebral vasoconstriction in the absence of systemic hypotension. Pacing Clin Electrophysiol 1998;21:652-658.
41. Martin TP, Hanusa BH, Kapoor WN: Risk stratification of patients with syncope. Ann Emerg Med 1997;29:459-466.
42. Calkins H, Shyr Y, Frumin H, et al: The value of the clinical history in the differentiation of syncope due to ventricular tachycardia, atrioventricular block, and neurocardiogenic syncope. Am J Med 1995;98:365-373.
43. Hoefnagels WAJ, Padberg GW, Overweg J, et al: Transient loss of consciousness: The value of the history for distinguishing seizure from syncope. J Neurol 1991;238:39-43.
44. Benbadis, SR, Wolgamuth BR, Goren H, et al: Value of tongue biting in the diagnosis of seizures. Arch Intern Med 1995;155: 2346-2349.
45. Brugada J, Brugada P: Further characterization of the syndrome of right bundle branch block, ST segment elevation, and sudden cardiac death. J Cardio Elect 1997;8:325-331.
46. Alings M, Wilde A: Brugada syndrome, clinical data and suggested pathophysiological mechanism. Circulation 1999;99:666-673.
47. Steinberg JS, Prystowsky E, Freedman RA, et al: Use of the signal-averaged electrocardiogram for predicting inducible ventricular tachycardia in patients with unexplained syncope: Relation to clinical variables in a multivariate analysis. J Am Coll Cardiol 1994;23:99-106.
48. Marcus FI, Fontaine G, Guiraudon G, et al: Right ventricular dysplasia: A report of 24 adult cases. Circulation 1982;65: 384-398.
49. McKenna WJ, Thiene G, Nava A, et al: Diagnosis of arrhythmogenic right ventricular dysplasia/cardiomyopathy. Br Heart J 1994;71: 215-218.
50. Holter NJ, Gengerelli JA: Remote recording of physiologic data by radio. Rocky Mt Med J 1949;16:79.
51. Linzer M, Pritchett ELC, Pontinen M, et al: Incremental diagnostic yield of loop electrocardiographic recorders in unexplained syncope. Am J Cardiol 1990;66:214-219.
52. Fogel RI, Evans JJ, Prystowsky EN: Utility and cost of event recorders in the diagnosis of palpitations, presyncope, and syncope. Am J Cardiol 1997;79:207-208.
53. Linzer M, Pritchett ELC, Pontinen M, et al: Incremental diagnostic yield of loop electrocardiographic recorders in unexplained syncope. Am J Cardiol 1990;66:214-219.
54. Brown AP, Dawkins KD, Davies JG: Detection of arrhythmias: Use of a patient-activated ambulatory electrocardiogram device with a solid-state memory loop. Br Heart J 1987;58:251-253.
55. Krahn AD, Klein GJ, Yee R, et al: Use of an extended monitoring strategy in patients with problematic syncope. Circulation 1999; 99:406-410.
56. Gibson TC, Heizman MR: Diagnostic efficacy of 24 hour electrocardiographic monitoring for syncope. Am J Cardiol 1984;53:1013.
57. DiMarco JP, Philbrick JT: Use of ambulatory electrocardiographic (Holter) monitoring. Ann Intern Med 1990;113:53-68.
58. Bass EB, Curtiss EI, Arena VC, et al: The duration of Holter monitoring in patients with syncope: Is 24 hours enough? Arch Intern Med 1990;150:1073-1078.
59. Benditt DG, Ferguson DW, Grubb BP, et al: Tilt table testing for assessing syncope. J Am Coll Cardiol 1996;28:263-275.
60. Natale A, Akhtar M, Jazayeri M, et al: Provocation of hypotension during head-up tilt testing in subjects with no history of syncope or presyncope. Circulation 1995;92:54-58.
61. Ammirati F, Colivicchi F, Biffi A, et al: Head-up tilt testing potentiated with low-dose sublingual isosorbide dinitrate: A simplified time-saving approach for the evaluation of unexplained syncope. Am Heart J 1998;135;671-676.
62. Sutton R, Peterson MEV: The clinical spectrum of neurocardiogenic syncope. J Cardiovasc Electrophysiol 1995;6:569.
63. Sumiyoshi M, Nakata Y, Yoriaki M, et al: Response to head-up tilt testing in patients with situational syncope. Am J Card 1998;82: 1117-1118.
64. Raviele A, Giada F, Brignole M, et al: Comparison of diagnostic accuracy of sublingual nitroglycerin test and low-dose test in patients with unexplained syncope. Am J Cardiol 2000;85: 1194-1198.
65. Del Rosso A, Bartoletti A, Bartoli P, et al: Methodology of head-up tilt testing potentiated with sublingual nitroglycerin syncope. Am J Cardiol 2000;85:1007-1011.
66. Del Rosso A, Bartoli P, Bartoletti A, et al: Shortened head-up tilt testing potentiated with sublingual nitroglycerin in patients with unexplained syncope. Am Heart J 1998;135:564-570.
67. Benditt DG, Ferguson DW, Grubb BP, et al: ACC expert consensus document: Tilt table testing for assessing syncope. J Am Coll Cardiol 1996;28:263-275.
68. Pitzalis M, Massari F, Guida P, et al: Shortened head-up tilting test guided by systolic pressure reductions in neurocardiogenic syncope. Circulation 2002;105:146-148.
69. Ewy GA, Appleton CP, DeMaria AN, et al: ACC/AHA guidelines for the clinical application of echocardiography. Circulation 1990;82:2323-2345.
70. Krumholz HM, Douglas PS, Goldman L, Waksmonski C: Clinical utility of transthoracic two-dimensional and Doppler echocardiography. J Am Coll Cardiol 1994;24:125-131.
71. Panther R, Mahmood S, Gal R: Echocardiography in the diagnostic evaluation of syncope. J Am Soc of Echocardiogr 1998; 11:294-298.
72. Recchia D, Barzilai B: Echocardiography in the evaluation of patients with syncope. J Gen Intern Med 1995;10:649-655.
73. Zipes DP, DiMarco JP, Gillette PC, et al: Guidelines for clinical intracardiac electrophysiological and catheter ablation procedures. J Am Coll Cardiol 1995;26:555-573.
74. Middlekauf HB, Stevenson WG, Saxon LA: Prognosis after syncope: Impact of left ventricular function. Am Heart J 1993; 125:121-127.
75. Middlekauff HR, Stevenson WG, Stevenson LW: Prognostic significance of atrial fibrillation in advanced heart failure. A study of 390 patients. Circulation 1991;84:40.
76. Knight BP, Goyal R, Pelosi F, et al: Outcome of patients with nonischemic dilated cardiomyopathy and unexplained syncope treated with an implantable defibrillator. J Am Coll Cardiol 1999;33:1964-1970.
77. Bachinsky WB, Linzer M, Weld L, Estes III NAM: Usefulness of clinical characteristics in predicting the outcome of electrophysiologic studies in unexplained syncope. Am J Cardiol 1992;69: 1044-1049.
78. Moazez F, Peter T, Simonson J, et al: Syncope of unknown origin: Clinical, noninvasive, and electrophysiologic determinants of arrhythmia induction and symptom recurrence during long-term follow-up. Am Heart J 1991;121:81-88.
79. Dimarco JP, Garan H, Harthorne W, Ruskin JN: Intracardiac electrophysiologic techniques in recurrent syncope of unknown cause. Ann Intern Med 1981;95:542-548.
80. Fujimura O, Yee R, Klein GJ, et al: The diagnostic sensitivity of electrophysiologic testing in patients with syncope caused by transient bradycardia. N Engl J Med 1989;321:1703-1707.
81. Sugrue DD, Gersh BJ, Holmes DR, et al: Symptomatic "isolated" carotid sinus hypersensitivity: Natural history and results of

treatment with anticholinergic drugs or pacemaker. J Am Coll Cardiol 1986;7:158-162.
82. Kenny RA, Richardson DA, Steen N, et al: Carotid sinus syndrome: A modifiable risk factor for nonaccidental falls in older adults (SAFE PACE). J Am Coll Cardiol 2001;38:1491-1496.
83. Leitch JW, Klein GJ, Yee R, et al: Syncope associated with supraventricular tachycardia. Circulation 1992;85:1064-1071.
84. Davis TL, Freemon FR: Electroencephalography should not be routine in the evaluation of syncope in adults. Arch Intern Med 1990;150:2027-2029.
85. Zaidi A, Crampton S, Clough P, et al: Misdiagnosis of epilepsy: Many seizure-like episodes have a cardiovascular cause. J Am Coll Cardiol 2000;36:181-184.
86. Epstein AE, Miles WM, Benditt DG: Personal and public safety issues related to arrhythmias that may affect consciousness: Implications for regulation and physician recommendations. An AHA/NASPE Medical Scientific Statement. Circulation 1996;94:1147-1166.
87. Consensus Conference, Canadian Cardiovascular Society: Assessment of the cardiac patient for fitness to drive. Can J Cardiol 1992;8:406-412.

Chapter 22 Asymptomatic ECG Abnormalities

ALLAN C. SKANES, BRUCE WALKER, PAUL LELORIER, and GEORGE J. KLEIN

Unexpected electrocardiographic abnormalities may frequently appear in patients undergoing Holter monitoring or electrocardiography in the context of "routine screening" or investigation of an unrelated issue. The thoughtful physician will not want to miss an opportunity to prevent a potential bad outcome related to this but also considers that needless attention to a nonsignificant abnormality is also not productive and is potentially harmful. A thoughtful assessment will include such questions as:

1. Is the abnormality a "true bill" or does it merely mimic the abnormality?
2. Is the abnormality associated with potential adverse outcomes?
3. What is the positive predictive value of the abnormality in question for a specific adverse outcome?
4. Is the risk of the abnormality sufficiently compelling to warrant the risk and expense of further investigation?

Asymptomatic Wolff-Parkinson-White Pattern

Illustrative case 1: A 14-year-old boy was seen in the emergency department because of noncardiac chest pain following chest trauma received during a basketball game. The ECG seen in Figure 22-1 was recorded. He was asymptomatic and denied palpitation, symptoms of tachycardia, or syncope. An incidental diagnosis of Wolff-Parkinson-White (WPW) syndrome was made. The most appropriate next step would be:

1. *Reassure the patient and his family with no further follow-up necessary*
2. *Schedule electrophysiological (EP) study*
3. *Schedule echocardiogram*
4. *Advise treadmill exercise test*

The electrocardiographic findings that are now known as ventricular preexcitation (δ wave and short P–R interval) were first described in 1930.[1] Initial interest was in the arrhythmias associated with the ECG pattern, although it was later observed that many individuals with identical ECG patterns remained asymptomatic and had a benign course. Radiofrequency ablation of accessory pathways has become the preferred treatment option for WPW patients, because it is curative with a low associated complication rate. The question remains whether this treatment should be extended to asymptomatic individuals with an electrocardiographic WPW pattern as a prophylactic measure to prevent the small risk of sudden death.

Several studies provide an estimate of the prevalence of the WPW pattern. It is estimated that 0.1% to 0.15% of the population will have the ECG manifestations of WPW, of which approximately 50% have never had symptoms.[2] New cases arise at a rate of approximately 0.004%/year.[2] However, the true incidence is likely higher given that the WPW pattern is frequently intermittent and may be missed on ECG, in addition to individuals who never receive an ECG in their lifetime. The incidence of sudden death among patients with WPW appears to be 0.01% per year by the most generous estimates.[2,3] Interestingly, Munger et al. do not report any sudden deaths in their large (113 patients) retrospective trial. Autopsy studies delving into the issue are generally limited by the difficulty in finding accessory pathways on routine tissue sections, but the great majority of unexpected sudden death in individuals is related to coronary or other unrecognized structural heart disease. As rare as it is, sudden death in an asymptomatic patient with WPW is a tragic and emotionally charged event typically occurring in young, otherwise healthy individuals.

The most common cause of sudden death in patients with WPW is ventricular fibrillation triggered by atrial fibrillation. Atrial fibrillation (AF) is associated with a rapid ventricular response due to at least one accessory pathway with a short refractory period. It follows that only those patients who have accessory pathways capable of mediating a rapid ventricular rate are at risk for developing sudden cardiac death (SCD). It has been found that ventricular fibrillation rarely occurs when the shortest R–R (SRR) interval during induced

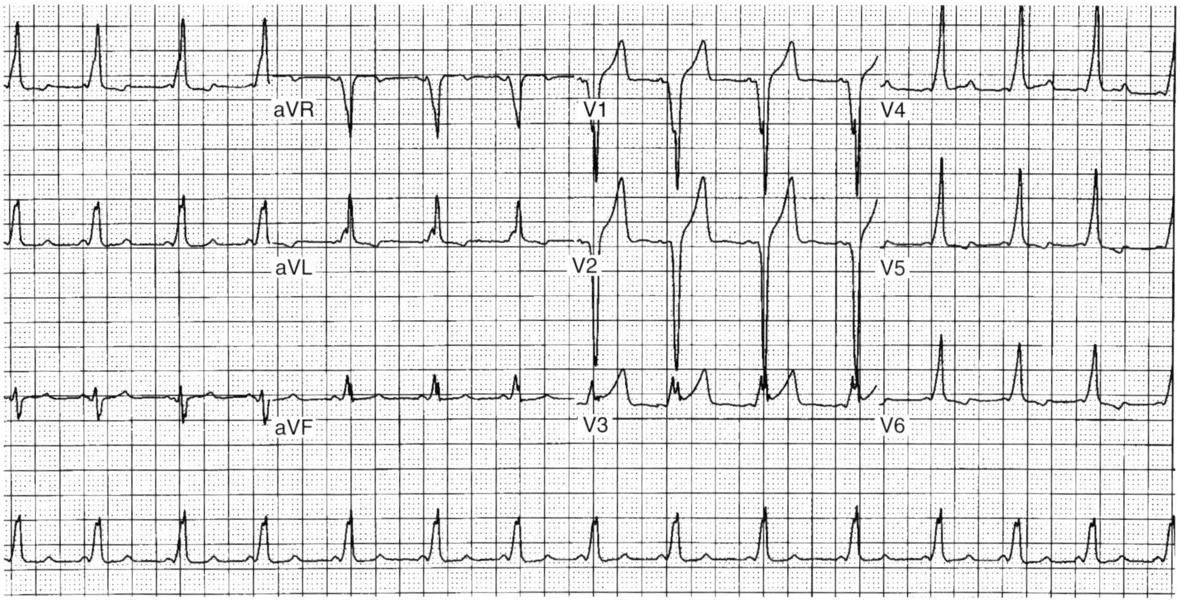

FIGURE 22-1 Wolff-Parkinson-White pattern.

AF is greater than 250 milliseconds (ms).[4] AF is usually preceded by atrioventricular (AV) reentrant tachycardia in these patients, presumably due to the hemodynamic and metabolic consequences of the supraventricular tachycardia. Hence, the inability to induce tachycardia would independently predict a good prognosis. Interestingly, the presence of multiple accessory pathways in a given individual increases the risk of sudden death by an odds ratio of 3.1,[5] presumably because of increased ventricular desynchronization related to multiple preexcitation wavefronts.

Because of the rarity of the outcome in question (<1/1000 patients/year), it is evident that predicting outcomes with any test is problematic. As one would expect from such a low incidence of SCD, the negative predictive value of the EP study is excellent. Klein et al. have broken up the risk of SCD based on the shortest R–R (SRR) interval during AF as follows[6]:

(1) Definite-risk—shortest preexcited R–R interval <220 ms;
(2) Probable risk—shortest preexcited R–R interval <250 ms but >220 ms;
(3) Possible risk—shortest preexcited R–R interval <300 ms but >250 ms; and
(4) Negligible risk—shortest preexcited R–R interval >300 ms.

It must be emphasized that "definite" risk in this stratification is not equivalent to "high" risk. Use of the SRR less than 250 ms, which is widely used in clinical practice, is not as helpful as intuitively believed because 17% of asymptomatic individuals with WPW have SRRs in this range, a far greater proportion than that expected to have an event.[6] Using an SRR less than 250 ms, a sensitivity of 77.8%, a specificity of 48.3% and predictive accuracy of 18.9%[7] can be obtained. Although the presence of other described risk factors for the onset of sudden death (i.e., atrial vulnerability and inducible tachycardia) may theoretically improve the predictive accuracy of SRR less than 250 ms, there are few data to support this approach. Teo et al. report that the combined findings of multiple accessory pathways and SRR less than 250 ms lower the sensitivity (88% to 29%) but increase the specificity (36% to 92%) and the positive predictive value (9% to 22%).[5]

Isoproterenol, which is often used in an attempt to induce tachycardia or atrial fibrillation, or both, will cause an even greater percentage of individuals to have SRR less than 250 ms. Hence, the already poor positive predictive value of SRR less than 250 ms would be expected to worsen.

Despite its limitations, the EP study remains the gold standard for determination of risk of SCD in asymptomatic patients with WPW. Other noninvasive methods for evaluating risk for sudden death have been described: Holter monitoring to determine whether intermittent preexcitation is present (therefore making it extremely unlikely for the SRR of the pathway to be <250 ms); *sudden* loss of preexcitation on an exercise test[8] or with infusion of ajmaline[9] or procainamide[10] (suggesting a high refractory period for the pathway); and transesophageal pacing. With the exception of transesophageal pacing, all methods rely on assessments of accessory pathway conduction in sinus rhythm and are imperfect substitutes for a test that is in itself limited. While transesophageal pacing has proponents,[11] it provides less information than catheter EP study and is arguably of similar risk and discomfort.

How does one, then, approach the asymptomatic patient given that sudden death as a first manifestation of the WPW syndrome is a very rare occurrence? Since serious complications from radiofrequency ablation in experienced centers occur infrequently (≈1%), it becomes difficult to make a categorical recommendation for or against ablation. Clearly, studying all patients with asymptomatic WPW and ablating those in whom an SRR

of less than 250 ms is present will result in morbidity and mortality in patients who may never have developed any symptoms. On the other hand, an approach of *not* studying any asymptomatic individuals (barring those individuals in whom, for occupational reasons, a WPW diagnosis is not acceptable) may result in some preventable deaths. Many physicians may prefer to accept a small procedural risk with a finite risk interval to a longer-term risk of potentially lethal arrhythmias.[10a,10b]

Similarly important is the patient's or parent's ability to accept a small but ongoing risk of sudden death. If not addressed, this may result in anxiety and needless restriction of recreational or occupational activities. Such patients may choose ablation. On the other hand, individuals may be less concerned about the theoretical risk of arrhythmia versus a real procedural risk.

Finally, two other factors may influence the discussion:

1) Some accessory pathways have higher risks associated with their ablation than others: for example, a "septal" accessory pathway in close proximity to the AV node.
2) The incidence of sudden death as a first presentation of WPW syndrome decreases with age: for example, a 60-year-old male is much less likely to develop this outcome than a 10-year-old male.

Since both the risks of treating and not treating are low, it is difficult to give an all-encompassing recommendation regarding ablation of asymptomatic WPW. It is reasonable to recommend ablation if the patient is otherwise prevented from pursuing a professional or important recreational activity. In all other cases, the decision to undergo an EP study and to proceed with ablation should be individualized. Given that several emotional, personal, and social issues enter into the consideration, and that there are minimal risks associated with either approach, the ultimate decision must rest with the patient.

Discussion case 1: As discussed earlier, asymptomatic patients have a small but statistically measurable risk of sudden death. The patient and family need to understand this. Therefore, reassurance alone is not appropriate. The δ wave is prominent and could be easily monitored during a treadmill test. This is the least invasive and most appropriate next step. If loss of preexcitation occurs with exercise, this classifies the accessory pathway as benign and eliminates concerns of sudden death. EP testing is probably premature at this point but may be required if the treadmill test is not prognostically helpful. An echocardiogram is not useful under the circumstances, as most patients with WPW have structurally normal hearts. The best answer is 4.

On the treadmill test, constant preexcitation was seen despite maximal heart rate of 190 beats per minute (bpm). The patient and family wanted further assessment to clarify the ability to play competitive sports. Following discussion of the risks and benefits, the patient underwent EP study. A mid-septal accessory pathway was documented. No tachycardia was inducible. No AF was inducible. Conduction over the accessory pathway failed at a pacing rate of 320 ms, suggesting that rapid conduction during AF would not be possible. Thus, the patient was at low risk for sudden death. The accessory pathway was located very close to the AV node, and no ablation was performed.

Repolarization Abnormalities

Many primary repolarization changes in the 12-lead ECG are benign or normal variants such as early repolarization. Others may result from the presence of structural heart disease (e.g., myocardial infarction, dilated and hypertrophic cardiomyopathy) or the consequence of pharmacologic agents (e.g., antiarrhythmic drugs). The congenital long QT syndrome (LQTS) and Brugada syndrome are uncommon diseases associated with significant and potentially life-threatening ventricular arrhythmias, which are primarily due to abnormal repolarization. The following sections discuss the typical ECG changes and clinical features that accompany these disorders.

Long QT Syndrome

Illustrative case 2: A 36-year-old accountant is referred for assessment after the unexpected death of his 24-year-old brother, who was previously well. The post-mortem was unremarkable, with no obvious cause of death. A 27-year-old sister is alive and well. Their mother died at age 46 as a passenger in a motor vehicle accident and was known to have epilepsy. The ECG shows a very long Q–T interval (QT = 600 ms, QTc = 545 ms, Fig. 22-2A). The accountant is taking no medication and jogs regularly. An ECG done 2 years previously for an insurance physical was reported as normal (Figure 22-2B). The most appropriate course of action at this time would be:

1. *Reassure the patient with no further investigation.*
2. *Schedule treadmill exercise test.*
3. *Schedule EP study.*
4. *Recommend β-blocker therapy.*

The congenital LQTS is a rare disorder (incidence 1:10,000 to 1:15,000) characterized by prolongation of the Q–T interval on the surface ECG, recurrent syncope, and sudden death. There are two major clinical variants: the Romano-Ward syndrome (autosomal dominant inheritance) and the Jervell and Lange-Nielson syndrome (autosomal recessive inheritance), in which patients also have congenital deafness. Over the past 10 years, great advances have been made toward understanding the genetic and cellular basis for LQTS. Six distinct types of LQTS have now been identified (Table 22-1), encompassing more than 170 genetic mutations, such that a genetic diagnosis is now obtainable in 50% to 60% of patients.[12-14] The principal abnormality in LQTS is prolongation of action potential duration caused by a reduction in outward potassium current (LQT1 and LQT2) or, less commonly, a persistent inward sodium current during the plateau phase (LQT3). Action potential prolongation provokes the development of early afterdepolarizations, which may produce a triggered action potential resulting in a premature beat.[15] Such impulses can initiate a polymorphic ventricular tachycardia known as torsades de pointes, which underlies the clinical symptoms of palpitations, syncope, or sudden death due to ventricular fibrillation.

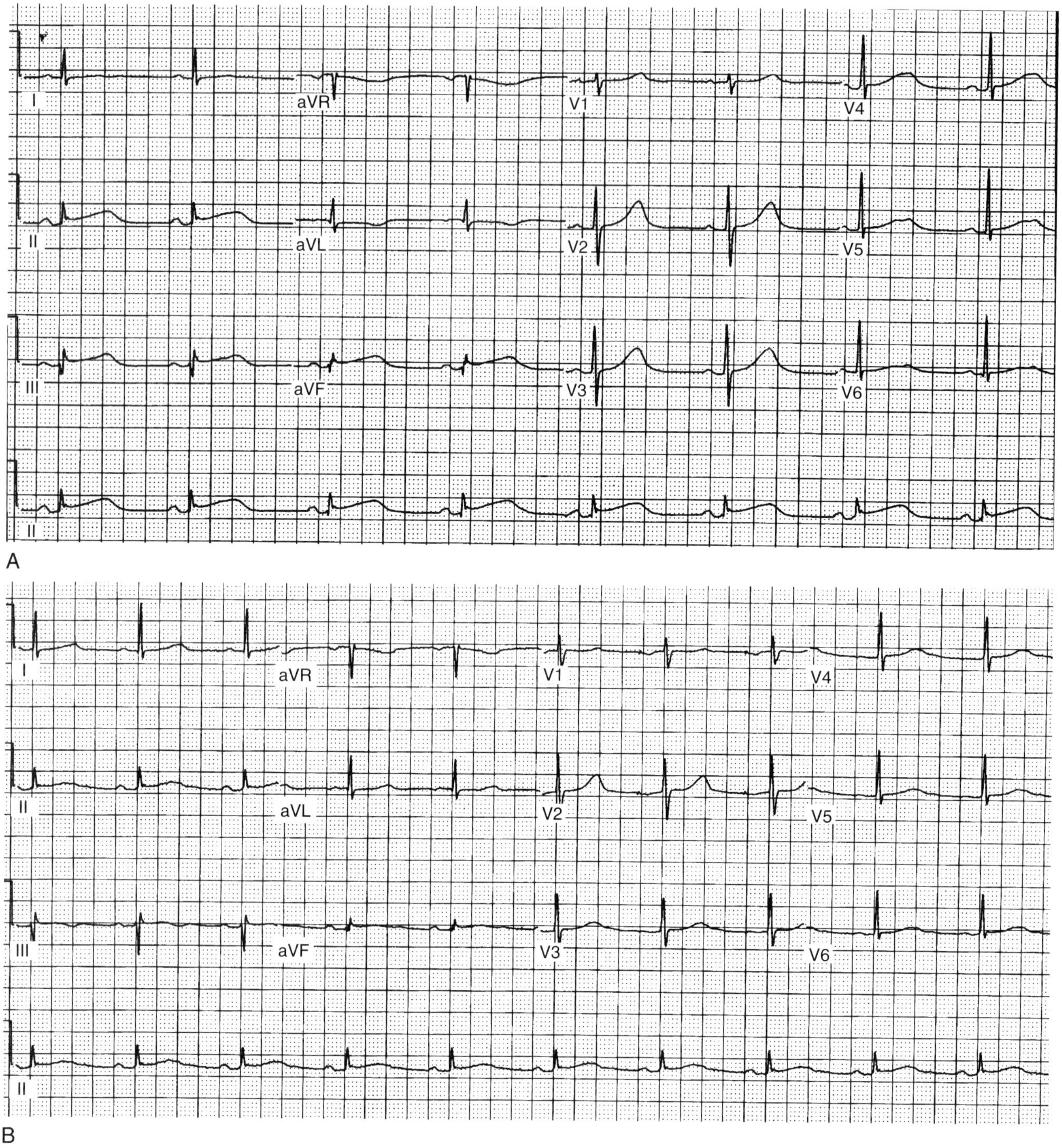

FIGURE 22-2 Long QT syndrome.

TABLE 22-1 Classification of Long QT Syndrome

Type	Locus	Gene	Ion Channel	Mutations, %
LQT1	11p15.5	*KCNQ1 (KVLQT1)*	I_{Ks} (α-subunit)	42
LQT2	7q35-q36	*KCNH2 (HERG)*	I_{Kr} (α-subunit)	45
LQT3	3p21-p23	*SCN5A*	I_{Na} (α-subunit)	8
LQT4	4q25-q27	?	?	
LQT5	21q22.1-p22	*KCNE1 (MinK)*	I_{Ks} (β-subunit)	3
LQT6	21q22.1	*KCNE2 (MiRP1)*	I_{Kr} (β-subunit)	2

Modified from Walker BD, Krahn AD, Klein GJ, et al: Congenital and acquired long QT syndromes. Can J Cardiol 2003;19:76-87.

The classic description of events in LQTS involves exercise- or emotion-induced clinical events. Symptoms often begin in adolescence, though they may begin earlier in LQT1.[14] It is common for patients to be investigated and have a misdiagnosis of seizure disorder. Clinical events in LQTS are often seen in the context of heart rate acceleration precipitated by exercise, emotion, or sudden arousal.[14,16,17] LQT1 patients are particularly prone to events during hyper-adrenergic states, with swimming identified as a specific trigger in such patients.[14,16] Events that occur during rest or sleep are more suggestive of LQT3.[14] The history of arousal to an auditory stimulus (classically, a telephone call at night) is strongly predictive of LQT2.[14,17] Many of these circumstances represent the setting of sudden or marked changes in heart rate that may influence repolarization and subsequent vulnerability to arrhythmias.

The hallmark of this condition is prolongation of the QTc greater than 460 ms, although it has recently been recognized that QTc may be normal in up to one third of genotype-positive patients.[18-20] Since repolarization is affected by factors such as sympathetic outflow, electrolyte balance, and pharmacologic agents, it is not surprising that there may be considerable temporal variation in QTc. In the absence of obvious reversible causes of QT prolongation (Tables 22-2 and 22-3), a diagnosis of LQTS can be made based on the electrocardiographic features and clinical presentation.[21] The mean QTc does not differ between the LQT1, LQT2, and LQT3 types[14] but is significantly longer in Jervell and Lange-Neilson syndrome. In addition to QTc prolongation, other electrocardiographic abnormalities that may be found in LQTS include ST–T wave changes, U waves, T wave alternans, increased QT dispersion, and sinus bradycardia.[15] In particular, characteristic ST–T wave morphologies may allow prediction of LQT type, such that broad-based T waves typify LQT1, low amplitude, notch T waves in LQT2, and long isoelectric ST segment with late-onset T wave in LQT3.[18]

In untreated patients, the annual risk of syncope is 5% and the 10-year mortality rate is estimated at 50%.[18,22]

TABLE 22-2 Secondary Causes of QT Prolongation

Factor	Mechanism
Bradycardia	↑APD ↑APD prolongation with class III agents
Drugs (see Table 22-3)	Mainly I_{Kr} I_{Ks} blockade
Electrolyte disorders (hypokalemia, hypomagnesemia, hypocalcemia)	Hypokalemia ↓I_{Kr} and ↑I_{Kr} sensitivity to pharmacologic blockers
Left ventricular hypertrophy/failure	↓K^+ currents (I_{to}, I_{Kr}, I_{Ks}) Changes to I_{CaL} and intracellular Ca^{2+}
Miscellaneous (e.g., anorexia, cerebrovascular disease, hypothyroidism, ionic contrast media)	
? Reduced repolarization reserve	Long QT syndrome

Modified from Walker BD, Krahn AD, Klein GJ, et al: Congenital and acquired long QT syndromes. Can J Cardiol 2003;19:76-87.

TABLE 22-3 Drugs Associated with QT Prolongation and Torsades de Pointes

Type	Subgroup	Drug
Cardiac	Class Ia	Disopyramide, procainamide, quinidine
	Class Ib	Bretylium
	Class III	Amiodarone, dofetilide, ibutilide, sotalol
	Class IV	Bepridil, lidoflazine, prenylamine
Antimicrobial	Antibiotics	Clarithromycin, doxorubicin, erythromycin, pentamidine, spiramycin, cotrimoxazole
	Antifungals	Fluconazole, itraconazole, ketoconazole
	Antimalarials	Chloroquine, halofantrine
Antihistamines		Astemizole, diphenhydramine, hydroxyzine, terfenadine
Psychiatric drugs	Antidepressants	Amitriptyline, imipramine, fluoxetine, doxepin
	Antipsychotics	Chlorpromazine, droperidol, haloperidol, lithium, pimozide, prochlorperazine, thioridazine, sertindole, trifluoperazine
Other drugs		Amantidine, aminophylline, cisapride, fenoxidil, ketanserin, prednisone, probucol, salbutamol, suxamethonium, terodiline, vincamine, zimelidine

Modified from Walker BD, Krahn AD, Klein GJ, et al: Congenital and acquired long QT syndromes. Can J Cardiol 2003;19:76-87.

It has been estimated that LQTS causes 3000 to 4000 sudden deaths per year in children and young adults in the United States.[23] Survival is dramatically improved by aggressive treatment with β-adrenoreceptor blocking drugs, cardiac pacing, left cervical sympathectomy, and the implantable cardioverter defibrillator (ICD).[24] In such treated patients, mortality can be reduced to 3% to 5% in 5 years.[15]

Discussion case 2: The patient in question clearly has marked QT prolongation in the absence of medication. The family history is disturbing and suggests that his brother and mother (LQTS is frequently misdiagnosed as epilepsy) both had LQTS. Since the family history is probably the best "risk stratifier," the best answer would be number 4. An argument can be made for an ICD in such a patient, although this would not be a universal recommendation at this time. Treadmill testing will not influence the decision to treat, although exercise may be diagnostically useful, especially if arrhythmias are observed. EP testing is generally of no value in the LQTS. This man previously had a normal ECG, and it is not unusual for such patients to have variability of the Q–T interval on different ECGs. Many cases will be less obvious than this, especially when the ECG is borderline. It is useful to keep in mind that the family history is generally a good prognostic guide.

Brugada Syndrome

Illustrative case 3: A 42-year-old man of Southeast Asian descent was referred for assessment of symptoms of atypical chest

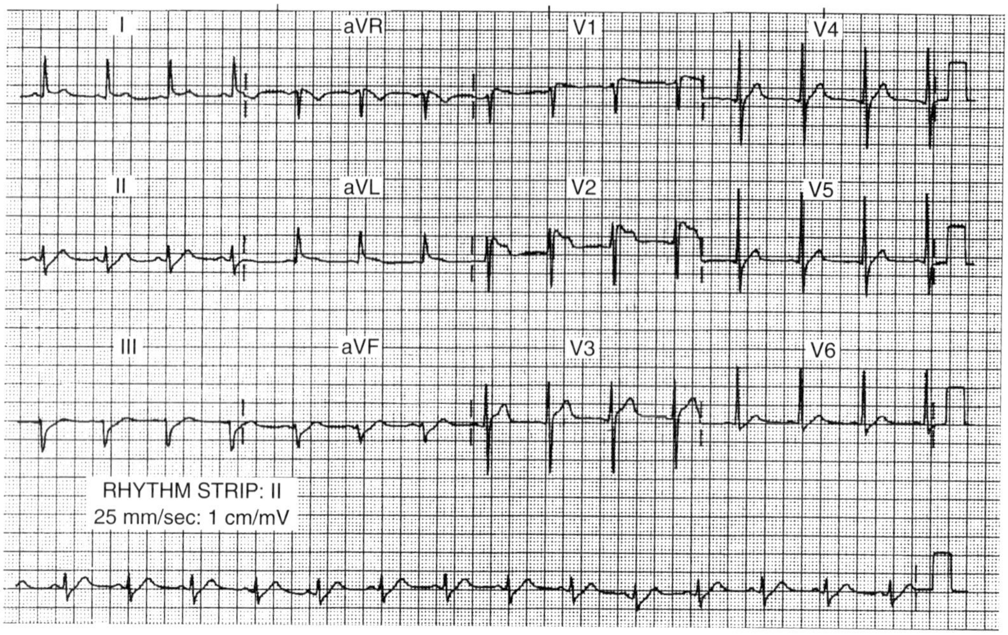

FIGURE 22-3 Brugada syndrome.

pain and ECG abnormalities as shown in Figure 22-3. The patient was otherwise well. His father died suddenly in his mid 40s, with no further details available. The most appropriate next step to clarify the diagnosis of the ECG abnormality would be:

1. *Genetic testing*
2. *Exercise treadmill testing*
3. *Procainamide infusion*
4. *EP testing*

Brugada syndrome is an inherited disorder (autosomal dominant) characterized by syncope and sudden death due to polymorphic ventricular tachycardia and ventricular fibrillation.[25] First described in 1992, it is now a recognized cause of SCD, particularly in southern Europe and Southeast Asia. It is thought to be the underlying cause in up to 50% of patients with "idiopathic" ventricular fibrillation.[26] The primary abnormality is heterogeneous shortening of action potential duration, particularly affecting the right ventricle, which results in phase 2 reentry and polymorphic ventricular tachycardia.[27] DNA sequencing of affected individuals has revealed mutations in the *SCN5A* gene, which encodes the fast inward Na^+ current (I_{Na}).[26] In a canine right ventricular (RV) wedge preparation, it has been shown that the combination of a weaker I_{Na} in the presence of a large transient outward K^+ current (I_{to}) results in dramatic truncation of the epicardial action potential and marked transmural heterogeneity of repolarization.[27]

The ECG may be normal or may demonstrate a characteristic pattern of right bundle branch block and ST segment elevation of greater than 0.1 mV in the precordial leads V1 to V3 (reflecting predominant RV abnormality). The ST elevation has a typical "coved" appearance as distinct from the saddle-shaped appearance of acute pericarditis. The Q–T interval is usually normal, although patients with both Brugada syndrome and LQTS have been described. Importantly, these ECG changes may also be seen in other disease states such as anteroseptal myocardial infarction, acute pericarditis, arrhythmogenic RV dysplasia, and hypothermia; therefore, the disease remains a diagnosis of exclusion.

Pharmacologic provocation may be useful in patients suspected of having Brugada syndrome with normal or nondiagnostic ECGs. Administration of the Na^+ channel blocking agents flecainide, ajmaline, and procainamide reproduces the characteristic ECG changes with a high sensitivity and specificity.[28,29] Other investigations often performed include EP testing in which programmed ventricular stimulation with up to three extrastimuli will often provoke polymorphic ventricular tachycardia or ventricular fibrillation.[30] The signal-averaged ECG will show late potentials in most patients.[31] Exercise stress testing does not usually provoke ventricular arrhythmias, as most events occur at night or during episodes of high vagal tone.[31] Symptomatic patients with Brugada syndrome should be treated with an implantable defibrillator. Asymptomatic patients can undergo risk stratification with the tests mentioned previously in order to guide therapy, with preliminary evidence suggesting that a normal resting ECG and negative EP study indicate a low risk of future events. Clearly, considerable investigation is being performed to determine the best investigations to risk-stratify asymptomatic patients. At present, there is no clear consensus, and such patients should be referred to an experienced center for further evaluation. No antiarrhythmic drug treatment has proved efficacious in preventing ventricular arrhythmias. ICD remains the treatment for those at high risk.

Discussion case 3: The right precordial leads in the ECG demonstrated persistent ST elevation with T wave inversion.

These findings are highly suggestive of Brugada syndrome. In the future genetic testing may prove to be the best diagnostic test. At present, however, genetic testing is not widely available, and not all mutations have been catalogued. Treadmill testing is not useful in this syndrome. Challenge with sodium channel blockade (most commonly IV procainamide in North America) can increase the magnitude of the ST elevation. This is diagnostic of Brugada syndrome and would be most useful. EP testing with pacing maneuvers alone provides little "diagnostic" information. However, as noted later, its utility as a prognostic test is under evaluation. Therefore, number 3 is the best answer.

An echocardiogram was also performed in this patient to eliminate silent previous myocardial infarction with persistent aneurysm as a cause for the ECG abnormalities. The echocardiogram demonstrated no wall motion abnormalities. Challenge with IV procainamide in this case demonstrated further ST elevation and was diagnostic. No arrhythmia was induced during the infusion. An EP study induced nonsustained polymorphic ventricular tachycardia but no sustained arrhythmia. Preliminary data in patients with negative EP studies have suggested a good prognosis, although over a relatively short follow-up.[32,33] Nonetheless, a family history of SCD in his father at or near 40 years of age is suggestive of a malignant form of Brugada syndrome. Thus, an ICD was implanted in this patient.

Asymptomatic Arrhythmias

VENTRICULAR ECTOPY

Illustrative case 4: A 52-year-old man has a Holter monitor performed because of palpitations. No symptoms were noted in the patient's diary, but asymptomatic ectopy similar to that seen on the 12-lead ECG in Figure 22-4A is recorded. His 12-lead ECG without ectopy is also shown in Figure 22-4B. He is referred for further assessment. The most appropriate next step would be:

1. *Reassurance with no follow-up*
2. *Echocardiogram*
3. *Cardiac magnetic resonance imaging (MRI) scan*
4. *Coronary angiography*

Electrocardiography and Holter monitoring frequently record unexpected ventricular ectopy as isolated ventricular premature beats or runs of nonsustained ventricular tachycardia. It is imperative to know in what setting this occurs. Frequent ectopy and nonsustained ventricular tachycardia have been demonstrated to be risk factors for SCD in the setting of all forms of structural heart disease.[34] On the other hand, ectopy in the presence of a structurally normal heart has a benign prognosis and requires no further therapy beyond

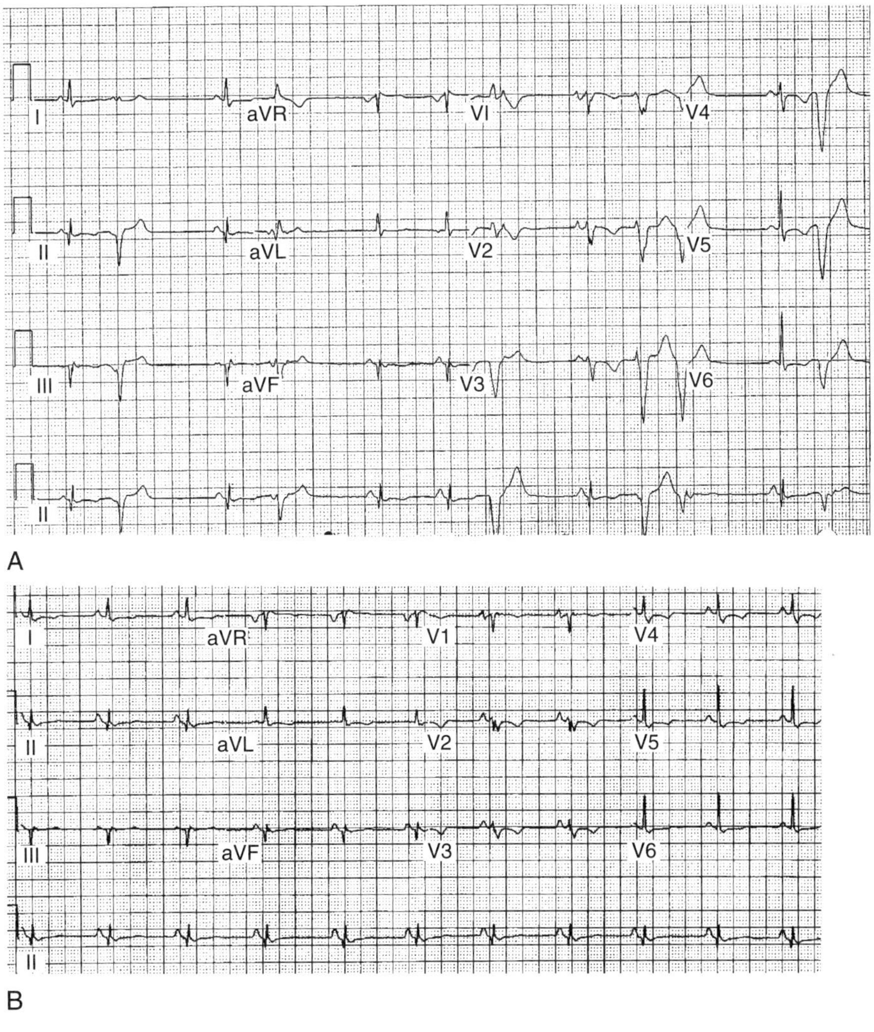

FIGURE 22-4 Ventricular ectopy.

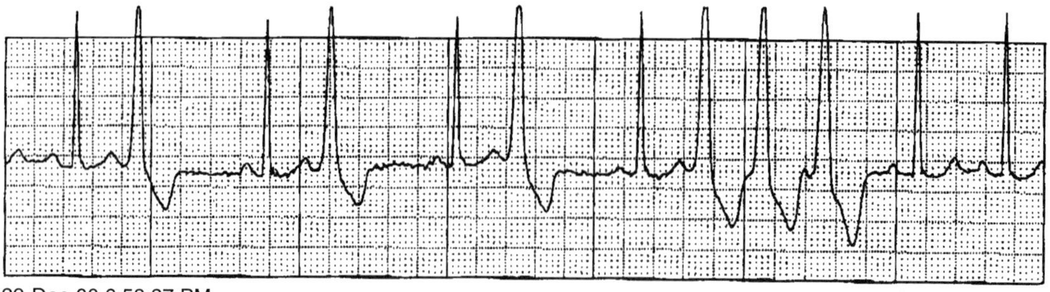

FIGURE 22-5 Ventricular premature beats from the right ventricular outflow tract as recorded in lead II from a Holter monitor recording.

reassurance.[35] Thus, all patients with ventricular ectopy should undergo assessment for the presence of structural heart disease.

Ectopy in the Setting of a Structurally Normal Heart. Forms of benign ventricular tachycardia found in patients with a structurally normal heart have been described to occur in both the left and right ventricles—so-called *idiopathic ventricular tachycardia.* Some forms of idiopathic VT manifest with repetitive bursts of nonsustained monomorphic ectopy, especially that from the RV outflow tract (RVOT).[35] The source of the ectopy produces a characteristic left bundle branch block morphology QRS with large amplitude R waves in the inferior leads with T wave inversions.[35] Figure 22-5 shows ventricular premature beats from the RVOT as recorded in lead II from a Holter monitor recording. Idiopathic forms of left ventricular (LV) VT have also been recognized and can produce asymptomatic ectopy.[35-37] Patients with idiopathic asymptomatic ectopy need no further investigations or therapy.

ARRHYTHMOGENIC RIGHT VENTRICULAR DYSPLASIA (ARVD)

The right ventricle can be the source of serious arrhythmia manifesting as asymptomatic ectopy, and this must be distinguished from idiopathic forms of ectopy, including RVOT tachycardia. Arrhythmogenic RV dysplasia (ARVD) is characterized by palpitations, syncope, and sudden death due to ventricular arrhythmias in the setting of pathologic fatty infiltration and fibrosis of the right ventricle.[38] It is most commonly diagnosed in young males, although cases have been described at all ages and in either sex. Although the pathogenesis is poorly understood, the disease is regarded as a cardiomyopathy that may also affect the left ventricle. Both familial and sporadic forms of ARVD have been described, with the former displaying an autosomal dominant inheritance. Genetic mutations in the short arm of chromosome 3 (3p23) and chromosome 1 (1q42-q43) have recently been described.[39,40] Chromosome 1 mutations involve the cardiac ryanodine receptor (RYR2) gene, which encodes a protein that regulates Ca^{2+} release from the sarcoplasmic reticulum.

The ECG in sinus rhythm may demonstrate QRS prolongation in leads I, V1, V3, and V6, most likely due to delayed RV activation. In the right precordial leads, a distinct wave after the QRS (ε wave) is found in 30% of cases and is thought to represent markedly delayed RV activation.[38] There may also be T wave inversion, RV hypertrophy, and increased P wave duration in the same leads. Late potentials on the signal-averaged ECG are the most sensitive markers of ARVD (85%).[38] Programmed ventricular stimulation at several different RV sites may induce one or more types of sustained monomorphic ventricular tachycardia and, less commonly, ventricular fibrillation. The detection of gross myocardial fatty infiltration and scarring can be made by echocardiography, RV angiography, and nuclear or magnetic resonance imaging. There is, however, considerable controversy over the sensitivity and specificity of such tests. Management may involve antiarrhythmic drug treatment (class I agents, amiodarone, and β-adrenoreceptor blockers), radiofrequency ablation, surgery to disconnect the affected segment, and increasingly frequent use of the implantable defibrillator.

When one returns to the evaluation of a patient with left bundle branch block morphology ventricular ectopy, especially when not from the RVOT, careful historical features must be considered. A malignant family history of ARVD or SCD with diagnostic changes in the right ventricle using imaging technologies (echocardiogram, MRI, or exercise multigated scan) warrants careful investigation with EP testing and consideration of an ICD in those with inducible arrhythmias.

Discussion case 4: The baseline 12-lead ECG has features of ARVD with terminal low amplitude voltage seen best in leads V1 to V3. Accompanying T wave inversions are also seen. The ectopy has a left bundle branch block morphology with left superior axis suggesting a source from the RV apex. It is clearly not from the RVOT and therefore cannot be immediately dismissed as benign. Imaging of the right ventricle is required. Echocardiography is highly variable for imaging the right ventricle and in general is not of sufficient quality to diagnose ARVD. In experienced hands, cardiac MRI is the best imaging modality for ARVD, and diagnostic criteria have been developed.[41,42] Coronary angiography will not provide useful information. Any coronary lesions would have to be interpreted as incidental.

An echocardiogram suggested a dilated hypokinetic right ventricle. A rest and exercise multiple gated acquisition scan confirmed these changes, and an MRI scan demonstrated an enlarged RV with evidence of fibrofatty infiltration consistent with ARVD. There was no family history of SCD. EP testing induced no arrhythmia. β-Blockers were prescribed and the patient was followed closely.

ATRIAL FIBRILLATION

Illustrative case 5: A 59-year-old man presented with a 2-week history of progressive dyspnea. The ECG as seen in Figure 22-6 was recorded. Examination in the emergency department demonstrated findings of congestive heart failure with an elevated jugular venous pulse, a laterally displaced apex, a third heart sound, and inspiratory crackles. His symptoms and findings on examination responded to IV diuretics and an ACE inhibitor. An echocardiogram demonstrated reduced LV function with an estimated ejection fraction of 35%. Appropriate first management would be:

1. *DC cardioversion*
2. *Pharmacologic cardioversion*
3. *Pharmacologic rate control*
4. *AV nodal ablation and pacemaker*

Not infrequently patients are found to be in persistent AF of unknown duration and are unaware that any dysrhythmia is present.[43] A similar presentation can occur in atrial flutter (see Fig. 22-6). On specific questioning patients may describe minimal or even no change in functional capacity. Nonetheless, several issues must be addressed in such patients. These relate to decisions about anticoagulation, occult tachycardia-induced cardiomyopathy, and possible cardioversion.

Rate Control. The syndrome of tachycardia-induced cardiomyopathy has been well described in the setting of rapid atrial fibrillation.[44-46] Importantly, this form of LV dysfunction is usually reversible if rate control can be achieved.[44-46] While the symptoms of AF correlate loosely with the rate, asymptomatic rapid AF can occur, and such patients are at risk for a reduction of LV function based on the rate alone. Assessment of LV function should be made in all such patients, and aggressive rate control should be initiated in patients with a reduction. Repeated Holter monitoring can be used to assess rate control during the patient's usual activities. Ideally, the rate should be below 100 bpm except for periods of exercise. More importantly, repeated assessment of LV function is required. For those in whom repeated pharmacologic attempts to control rate are met with resistance or whose LV function does not improve, DC cardioversion or, alternatively, AV nodal ablation and permanent pacemaker insertion should be considered regardless of the patient's asymptomatic status.

Anticoagulation. Decisions with respect to the use of warfarin or aspirin for the reduction in stroke risk should be based on risk factors for stroke as outlined in the Stroke Prevention in Atrial Fibrillation (SPAF) studies.[47-49] Guidelines in this regard do not consider associated symptoms.[47] It should be noted that patients who have minimal symptoms should stay on warfarin, as it becomes very difficult to determine recurrences of atrial fibrillation.

Cardioversion. There is frequently pressure, especially from patients or referring physicians, to perform DC cardioversion in patients despite their asymptomatic state.

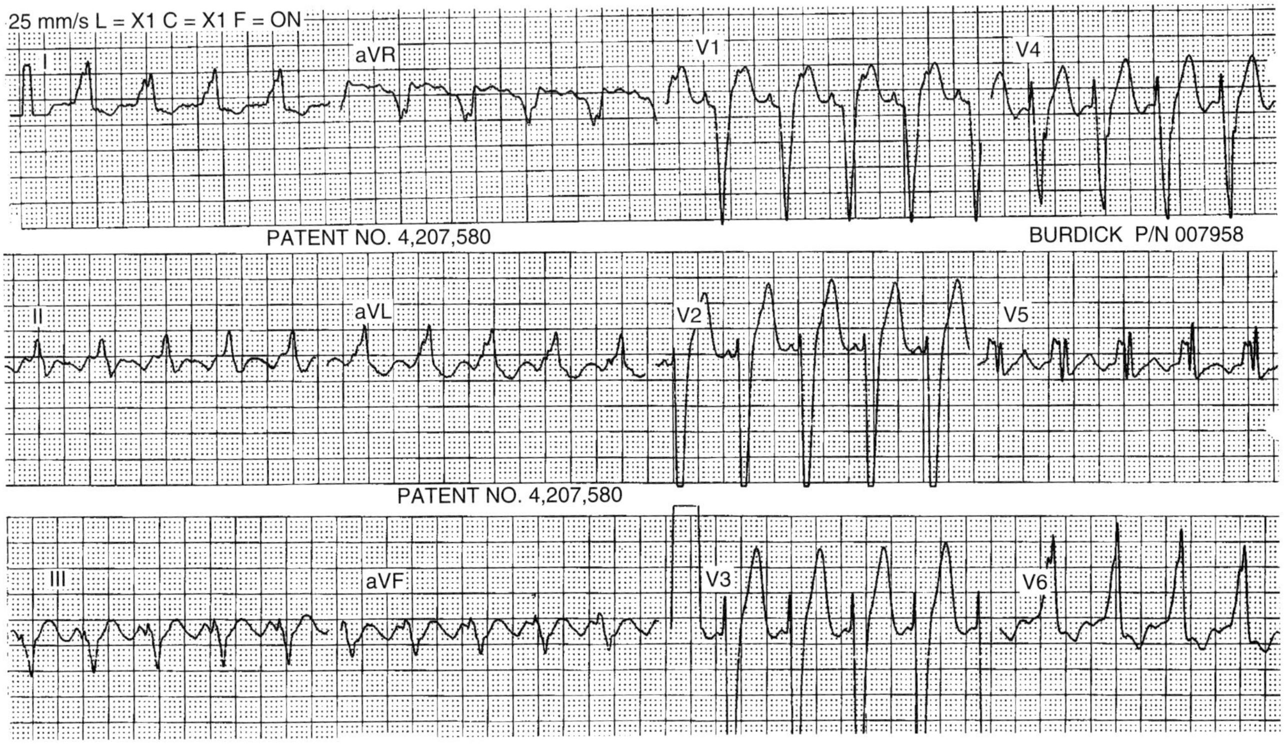

FIGURE 22-6 Atrial flutter.

In patients who are truly asymptomatic, it may be difficult to make a cogent argument in favor of cardioversion. At present, there is no evidence that such patients will have a better prognosis or reduced risk of stroke in sinus rhythm.[50]

Some patients with seemingly asymptomatic AF are unaware that they have had a reduction in functional status until sinus rhythm has been returned. For this reason, consideration is frequently given to perform at least a single cardioversion in minimally symptomatic patients. In general, cardioversion success is known to depend on the duration of AF as well as left atrial dimension. The former is frequently difficult to assess in such patients. Nonetheless, an echocardiographic assessment of left atrial dimension may provide some insight into the likelihood of success of cardioversion and the likelihood of maintenance of sinus rhythm. If patients note no improvement in functional status during sinus rhythm, repeated cardioversion is truly unwarranted.

Asymptomatic Paroxysmal AF (PAF). Electrophysiologists are frequently asked to assess patients who are found to have bursts of rapid, irregular atrial activity on Holter monitoring. Such patients pose a difficult problem, as little information is available in clinical trials regarding the magnitude of thromboembolic risk. Specifically, it is not known how much or little of such AF is necessary to provide an increased risk of stroke. It is likely that patients with asymptomatic bursts of atrial ectopy on Holter monitor were not included or at least underrepresented in the trials investigating anticoagulation. Nonetheless, subgroup analysis from the SPAF studies demonstrate that patients with intermittent AF have a risk of thromboembolic events that is similar to those with chronic AF after correction for risk factors. The same risk factors apply. Based on these data, decisions about anticoagulation must be individualized based on the risk factors for stroke as outlined in the trials.[51]

Incessant Supraventricular Tachycardias. A small group of arrhythmogenic substrates allows for relatively slow but frequently incessant SVTs.[52] Patients who develop these arrhythmias can present with tachycardia-induced cardiomyopathy and insidious symptoms of dyspnea rather than palpitations. Thus, they may present much like those with asymptomatic AF (see earlier). Patients may have a permanent form of junctional reciprocating tachycardia (PJRT) due to a slowly conducting accessory pathway or, alternatively, an incessant atrial tachycardia.[45,54] As noted earlier, the LV dysfunction is usually reversible if it is indeed due to the prolonged periods of tachycardia. In the current era, radiofrequency ablation forms first-line therapy in such patients. Alternatively, medication can be used, but maintenance of sinus rhythm is paramount to improvement of LV function. We recommend that such patients be referred for radiofrequency ablation.

Discussion case 5: The ECG demonstrates typical atrial flutter with 2:1 conduction and left bundle branch block. While it is impossible to know, the LV dysfunction is likely due to tachycardia-induced cardiomyopathy. In the absence of anticoagulation or transesophageal echocardiography to rule out left atrial appendage thrombus, cardioversion is contraindicated. On the other hand, LV dysfunction is known to improve following institution of rate control. As a first measure, AV nodal blocking agents should be initiated. AV nodal ablation with pacemaker insertion can be used to control the ventricular rate in refractory cases, especially in atrial fibrillation. However, typical atrial flutter can be eliminated with standard ablation techniques, and this forms effective therapy without the need for AV nodal ablation.

Digoxin and small doses of atenolol failed to provide adequate ventricular rate control. The patient underwent an EP study, which demonstrated counterclockwise isthmus-dependent atrial flutter. Following successful ablation and resumption of sinus rhythm, LV function returned to normal as documented by echocardiogram 3 months later.

ASYMPTOMATIC ATRIOVENTRICULAR NODAL AND HIS-PURKINJE DISEASE

The discussion of bradyarrhythmias and indications for pacing under such circumstances are discussed elsewhere. Nonetheless, some patients with bradyarrhythmias may present with asymptomatic ECG abnormalities and are therefore discussed briefly here.

Illustrative case 6: An otherwise healthy 54-year-old woman was referred for evaluation of an asymptomatic Holter monitor recording as seen in Figure 22-7. Before referral, a diagnosis of Mobitz II second-degree AV block was made, and the patient was told she may require a permanent pacemaker. Is this correct?

AV Block. Although unusual in the adult population, truly asymptomatic complete heart block occasionally is seen. Older patients with progressive His-Purkinje degeneration as the etiology tend to recount symptoms of exercise intolerance or exertional dyspnea upon directed questioning. The American College of Cardiology/American Heart Association Guidelines for Device Implantation state that patients with asymptomatic third-degree heart block should undergo pacing.[55] Certainly, awake asymptomatic patients with documented asystole greater than 3.0 seconds or escape rate less than 40 have a class I indication for pacing. Asymptomatic third-degree AV block with faster average ventricular rates should also be considered.

Asymptomatic type II second-degree AV block can occur at the level of the AV node or below, within the His-Purkinje system. This is an important distinction, as block at the level of the AV node does not require pacing. Associated bundle branch block or bifascicular block may be helpful to determine the level of block. Invasive EP testing may be required to determine the level of block. The American College of Cardiology/American Heart Association Guidelines for Device Implantation have recognized asymptomatic type II second-degree AV block as a class IIa indication for pacing.[55]

Discussion case 6: Upon close inspection, 2:1 block is seen. It can be difficult to determine the level of block in such cases—at the level of the AV node or in the His-Purkinje system. As noted in the foregoing discussion, the importance is more than academic. In this case, the QRS complexes are narrow throughout. The P–R interval of the last conducted beat before the block is substantially longer than that of conducted beats following the

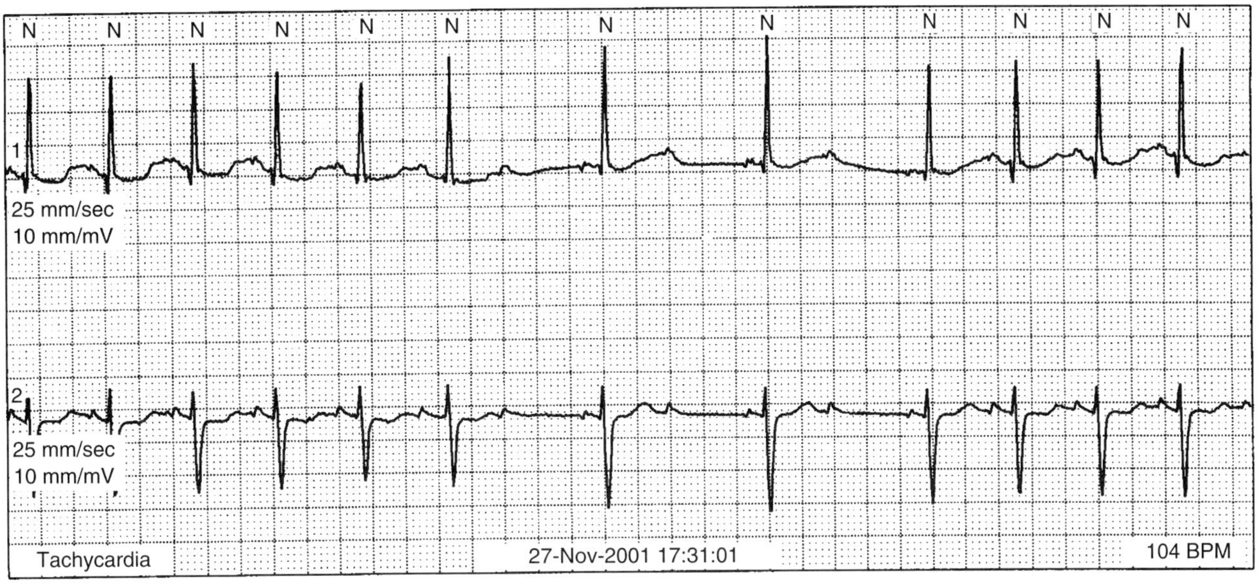

FIGURE 22-7 Atrioventricular nodal and His-Purkinje disease.

dropped beats. This suggests that the 2:1 block is Mobitz I and likely at the level of the AV node. This is very common during sleep in healthy individuals. No pacemaker was inserted. The presence of distal conduction disease (bundle branch block, bifascicular or trifascicular block) suggests that the block is below the level of the AV node (Mobitz II), and pacing may be indicated despite the asymptomatic nature of the abnormality.[55]

Asymptomatic bifascicular and trifascicular disease has been recognized to progress slowly to higher degrees of AV block.[55] Also, no single clinical or laboratory variable is predictive of the progression of AV block. Therefore, asymptomatic bundle branch block alone or bifascicular or trifascicular block are not indications for pacing. Asymptomatic bundle branch block may be the first indication of insidious cardiac disease. Thus, patients should be evaluated noninvasively with consideration given to evaluation of cardiac size and function.

SUMMARY

The abnormalities detected on the ECG in asymptomatic individuals will always remain a challenge. It is important to be certain that the natural history of the asymptomatic condition be considered relative to the emotional distress and potential morbidity of further investigations and therapy. The risk associated with the abnormality must be looked at in this light as it is "difficult to make an asymptomatic individual feel better."

REFERENCES

1. Wolff L, Parkinson J, White P: Bundle branch block with short P–R interval in healthy young people prone to paroxysmal tachycardia. Am Heart J 1930;5:685-704.
2. Munger TM, Packer DL, Hammill SC, et al: A population study of the natural history of Wolff-Parkinson-White syndrome in Olmsted County, Minnesota. Circulation 1993;87:866-873.
3. Klein GJ, Prystowsky EN, Yee R, et al: Asymptomatic Wolff-Parkinson-White. Should we intervene? Circulation 1989;80:1902-1905.
4. Klein GJ, Bashore TM, Sellers TD, et al: Ventricular fibrillation in the Wolff-Parkinson-White syndrome. N Engl J Med 1979;301:1080-1085.
5. Teo WS, Klein GJ, Guiraudon GM, et al: Multiple accessory pathways in the Wolff-Parkinson-White syndrome as a risk factor for ventricular fibrillation. Am J Cardiol 1991;67:889-891.
6. Leitch J, Klein GJ, Yee R, Murdock C: Prognostic value of electrophysiology testing in asymptomatic patients with Wolff-Parkinson-White pattern. Circulation 1990;82:1718-1723.
7. Sharma AD, Yee R, Guiraudon G, Klein GJ: Sensitivity and specificity of invasive and noninvasive testing for risk of sudden death in Wolff-Parkinson-White syndrome. J Am Coll Cardiol 1987;10:373-381.
8. Gaita F, Giustetto C, Riccardi R, et al: Stress and pharmacological test as methods to identify patients with Wolff-Parkinson-White syndrome at risk of sudden death. Am J Cardiol 1989;64:487-490.
9. Eshchar Y, Belhassen B, Laniado S: Comparison of exercise and ajmaline tests with electrophysiologic study in the Wolff-Parkinson-White syndrome. Am J Cardiol 1986;57:782-786.
10. Boahene KA, Klein GJ, Sharma AD, et al: Value of a revised procainamide test in the Wolff-Parkinson-White syndrome. Am J Cardiol 1990;65:195-200.

10a. Pappone C, Manguso F, Santinelli R, et al: Radiofrequency ablation in children with asymptomatic Wolff-Parkinson-White syndrome. N Engl J Med 2004;351:1197-1205.

10b. Wellens HJ: Catheter ablation for cardiac arrhythmias. N Engl J Med 2004;351:1172-1174.

11. Critelli G, Grassi G, Peticone F, et al: Transesophageal pacing for prognostic evaluation of preexcitation syndrome and assessment of protective therapy. Am J Cardiol 1983;51:513-518.
12. Walker BD, Krahn AD, Klein GJ, et al: Congenital and acquired long QT syndromes. Can J Cardiol 2003;19:76-87.
13. Splawski I, Shen J, Timothy KW, et al: Spectrum of mutations in long-QT syndrome genes. KVLQT1, HERG, SCN5A, KCNE1, and KCNE2. Circulation 2000;102:1178-1185.
14. Schwartz PJ, Priori SG, Spazzolini C, et al: Genotype-phenotype correlation in the long-QT syndrome: Gene-specific triggers for life-threatening arrhythmias. Circulation 2001;103:89-95.
15. Roden DM, Lazzara R, Rosen M, et al: Multiple mechanisms in the long-QT syndrome. Circulation 1996;94:1996-2012.
16. Ackerman MJ, Tester DJ, Porter CJ, Edwards WD: Molecular diagnosis of the long-QT syndrome in a woman who died after near-drowning. N Engl J Med 1999;341:1121-1125.
17. Wilde AA, Jongbloed RJ, Doevendans PA, et al: Auditory stimuli as a trigger for arrhythmic events differentiate HERG-related (LQTS2) patients from KVLQT1-related patients (LQTS1). J Am Coll Cardiol 1999;33:327-332.

18. Moss AJ, Schwartz PJ, Crampton RS, et al: The long QT syndrome. Prospective longitudinal study of 328 families. Circulation 1991;84:1136-1144.
19. Locati EH, Zareba W, Moss AJ, et al: Age- and sex-related differences in clinical manifestations in patients with congenital long-QT syndrome: Findings from the International LQTS Registry. Circulation 1998;97:2237-2244.
20. Vincent GM: Role of DNA testing for diagnosis, management and genetic screening in long QT syndrome, hypertrophic cardiomyopathy and Marfan's syndrome. Heart 2001;86:12-14.
21. Schwartz PJ, Moss AM, Vincent GM, Crampton RS: Diagnostic criteria for the long QT syndrome. Circulation 1993;88:782-784.
22. Schwartz PJ: Idiopathic long QT syndrome: Progress and questions. Am Heart J 1985;109:399-411.
23. Vincent GM: The molecular genetics of the long QT syndrome: Genes causing fainting and sudden death. Annu Rev Med 1998; 49:263-274.
24. Priori SG, Barhanin J, Hauer RNW, et al: Genetic and molecular basis of cardiac arrhythmia: impact on clinical management. Circulation 1999;99:518-528.
25. Brugada P, Brugada J: Right bundle branch block, persistent ST elevation and sudden cardiac death: A distinct clinical and electrocardiographic syndrome. J Am Coll Cardiol 1992;20: 1391-1396.
26. Chen Q, Kirsch GE, Zhang D, et al: Genetic basis and molecular mechanism for idiopathic ventricular fibrillation. Nature 1998; 392:293-296.
27. Yan G-X, Antzelevitch C: Cellular basis for the Brugada syndrome and other mechanisms of arrhythmogenesis associated with ST-segment elevation. Circulation 1999;100:1660-1666.
28. Gussak I, Antzelevitch C, Bjerregaard P, et al: The Brugada syndrome: Clinical, electrophysiologic and genetic aspects. J Am Coll Cardiol 1999;33:5-15.
29. Brugada R, Brugada J, Antzelevitch C, et al: Sodium channel blockers identify risk for sudden death in patients with ST-segment elevation and right bundle branch block but structurally normal hearts. Circulation 2000;101:510-515.
30. Brugada J, Brugada R, Brugada P: Right bundle-branch block and ST-segment elevation in leads V1 through V3: A marker for sudden death in patients without demonstrable structural heart disease. Circulation 1998;97:457-460.
31. Ailings M, Wilde A: "Brugada" syndrome. Clinical data and suggested pathophysiological mechanism. Circulation 1999;99:666-673.
32. Brugada P, Geelen P, Brugada R, et al: Prognostic value of electrophysiologic investigations in Brugada syndrome. J Cardiovasc Electrophysiol 2001;12:1004-1007.
33. Priori SG: Foretelling the future in Brugada syndrome: Do we have a crystal ball? J Cardiovasc Electrophysiol 2001;12:1008-1009.
34. Myerburg RJ, Kessler KM, Kimura S, et al: Life-threatening ventricular arrhythmias: The link between epidemiology and pathophysiology. In Zipes DP, Jalife J (eds): Cardiac Electrophysiology: From Cell to Bedside. Philadelphia, WB Saunders, 1995, p 723.
35. Wellens HJJ, Rodriquez L-M, Smeets JL: Ventricular tachycardia in structurally normal hearts. In Zipes DP, Jalife J (eds): Cardiac Electrophysiology: From Cell to Bedside. Philadelphia, WB Saunders, 1995, p 780.
36. Lin FC, Finley CD, Rahimtoola SH, Wu D: Idiopathic paroxysmal ventricular tachycardia with a QRS pattern of right bundle branch block and left axis deviation: A unique clinical entity with specific properties. Am J Cardiol 1983;52:95-100.
37. Belhassen B, Shapira I, Pelleg A, et al: Idiopathic recurrent sustained ventricular tachycardia responsive to verapamil: An ECG-electrophysiologic entity. Am Heart J 1984;108:1034-1037.
38. Fontaine G, Fontaliran F, Lascault G, et al: Arrhythmogenic right ventricular dysplasia. In Zipes DP, Jalife J (eds): Cardiac Electrophysiology: From Cell to Bedside. Philadelphia, WB Saunders, 1995, p 754.
39. Ahmad F, Li D, Karibe A, et al: Localization of a gene responsible for arrhythmogenic right ventricular dysplasia to chromosome 3p23. Circulation 1998;98:2791-2795.
40. Tiso N, Stephan DA, Nava A, et al: Identification of mutations in the cardiac ryanodine receptor gene in families with arrhythmogenic right ventricular cardiomyopathy 2 (ARVD2). Hum Mol Genet 2001;10:189-194.
41. McKenna WJ, Thiene G, Nava A, et al: Diagnosis of arrhythmogenic right ventricular dysplasia/cardiomyopathy. Task Force of the Working Group Myocardial and Pericardial Disease of the European Society of Cardiology and of the Scientific Council on Cardiomyopathies of the International Society and Federation of Cardiology. Br Heart J 1994;71:15-18.
42. Corrado D, Fontaine G, Marcus FI, et al: Arrhythmogenic right ventricular dysplasia/cardiomyopathy. Circulation 2000;101: e101-e106.
43. Humphries KH, Kerr CR, Connolly SJ, et al: New-onset atrial fibrillation: Sex differences in presentation, treatment, and outcome. Circulation 2001;103:2365-2370.
44. Packer D, Bardy G, Worley S, et al: Tachycardia-induced cardiomyopathy: A reversible form of left ventricular dysfunction. Am J Cardiol 1986;57:563-570.
45. Fenelon G, Wijns W, Andries E, Brugada P: Tachycardiomyopathy: Mechanisms and clinical implications. Pacing Clin Electrophysiol 1996;19:95-106.
46. Phillips E, Levine S: Auricular fibrillation without other evidence of heart disease: A cause of reversible heart failure. Am J Med 1949;7:478-489.
47. Laupacis A, Albers G, Dalen J, et al: Antithrombotic therapy in atrial fibrillation. Chest 1998;114(suppl 5):579S-589S.
48. Albers GW, Atwood JE, Hirsh J, et al: Stroke prevention in non-valvular atrial fibrillation. Ann Intern Med 1991;115:727-736.
49. Hart RG, Benavente O, McBride R, Pearce LA: Antithrombotic therapy to prevent stroke in patients with atrial fibrillation: A meta-analysis. Ann Intern Med 1999;131:492-501.
50. The Planning and Steering Committees of the AFFIRM study for the NHLBI AFFIRM investigators: Atrial fibrillation follow-up investigation of rhythm management—the AFFIRM study design. Am J Cardiol 1997;79:1198-1202.
51. The SPAF III Writing Committee for the Stroke Prevention in Atrial Fibrillation Investigators: Patients with nonvalvular atrial fibrillation at low risk of stroke during treatment with aspirin: Stroke Prevention in Atrial Fibrillation III Study. [see comments]. JAMA 1998;279:1273-1277.
52. Gillette PC, Smith RT, Garson A Jr, et al: Chronic supraventricular tachycardia. A curable cause of congestive cardiomyopathy. JAMA 1985;253:391-392.
53. Cruz FE, Cheriex EC, Smeets JL, et al: Reversibility of tachycardia-induced cardiomyopathy after cure of incessant supraventricular tachycardia. J Am Coll Cardiol 1990;16:739-744.
54. Shinbane JS, Wood MA, Jensen DN, et al: Tachycardia-induced cardiomyopathy: A review of animal models and clinical studies. J Am Coll Cardiol 1997;29:709-715.
55. Gregoratos G, Cheitlin MD, Conill A, et al: ACC/AHA guidelines for implantation of cardiac pacemakers and antiarrhythmia devices: A report of the American College of Cardiology/American Heart Association Task Force on Practice Guidelines (Committee on Pacemaker Implantation). J Am Coll Cardiol 1998;31:1175-1209.

Chapter 23 Evaluation and Management of Arrhythmias in Athletes

SAMI FIROOZI, SANJAY SHARMA,
and WILLIAM J. McKENNA

The benefits of regular exercise with respect to cardiovascular health are well established.[1,2] However, judging from media reports, one would expect that athletes are particularly susceptible to cardiac disease, sudden death, and arrhythmias. The news of an athlete dying suddenly or being found to have heart disease has a strong impact on the community, which regards athletes as the healthiest segment of the population and "immune" from cardiovascular diseases.

Arrhythmias are not infrequently documented in athletes and can be a source of morbidity and mortality. Although at times difficult, it is crucial to differentiate benign or physiologic arrhythmias from pathologic arrhythmias, particularly those associated with the risk of sudden cardiac death. The annual risk of sudden death in the athlete is thought to be between 5 and 10 athletes for each million athletes per year.[3] It is vitally important to determine whether or not an athlete should be advised to discontinue the sport at which he or she excels and that may be the key to a college education or the only means of livelihood. Furthermore, arrhythmias may be the presenting feature of an underlying cardiac abnormality (Table 23-1) that may predispose the competitive athlete to sudden cardiac death.

TABLE 23-1 Most Common Causes of Sudden Cardiac Death in Athletes Younger than 35 Years of Age

Hypertrophic cardiomyopathy
Arrhythmogenic right ventricular cardiomyopathy
Coronary artery anomalies
Ion channelopathies (long QT and Brugada syndromes)
Wolff-Parkinson-White syndrome
Marfan's syndrome
Myocarditis
Idiopathic dilated cardiomyopathy
Mitral valve prolapse
Aortic stenosis
Premature coronary artery disease

Intensive prolonged physical training leads to cardiovascular adaptations, which are collectively referred to as the "athletic heart syndrome."[4] Athletic training increases the ability to transport oxygen to the exercising muscles, and this is reflected in maximal oxygen uptake measurements.[5] These high levels of oxygen delivery are achieved by a high cardiac output. As training does not raise the maximal heart rate, this is achieved by a large increase in stroke volume. The ventricles increase in end-diastolic volume with little change in end-systolic volume. The cavity enlargement and increased systolic contraction are associated with an increase in myocardial mass, resulting in some increase in ventricular wall thickness.[6] The ECG of the trained athlete reflects these changes in the heart, typically with high voltages in the left ventricular leads. Furthermore, the high vagal tone induced from prolonged physical training results in a number of bradyarrhythmias including sinus bradycardia at rest and other findings such as sinus arrhythmia, sinoatrial block, a wandering atrial pacemaker, junctional rhythm, and minor degrees of atrioventricular (AV) block.[4,7]

In the following section we provide an update of arrhythmias in the athlete with recommendations based on the 26th Bethesda Conference on Recommendations for Determining Eligibility for Competition in Athletes with Cardiovascular Abnormalities.[8] We also outline the evaluation of an athlete who presents with symptoms suggestive of an arrhythmia and an asymptomatic athlete with an arrhythmogenic substrate.

Bradyarrhythmias

The high vagal tone resulting from prolonged intense training leads to the well-known presence of sinus bradycardia in most elite athletes. Similarly, sinus pauses, sinus arrhythmia, and wandering atrial pacemakers are also recognized and are due to the vagotonic state (Table 23-2).[4,9-11] In contrast, surgical and chemical denervation studies have shown lower intrinsic heart

TABLE 23-2 Frequency of Bradyarrhythmias in Athletes and the General Population

Arrhythmia	General Population, %	Athletes, %
Sinus bradycardia	23	50-85
Sinus arrhythmia	2-40	13-83
Wandering atrial pacemaker	Very rare	7-20
Sinus pauses	6	37-39
First-degree AV HB	0.65	5-35
Second-degree AV HB		
Mobitz I	0.003	0.1-10
Mobitz II	0.003	0.1-8
Third-degree AV HB	0.0002	0.017
Nodal rhythm	0.06	0.03-7

AV HB, atrioventricular heart block.

rates, suggesting an intrinsic cardiac component to the bradycardia.[4,12] Once the vagal tone is withdrawn, as occurs with exertion, the maximal heart rate is unaffected.[9] Further evaluation is not warranted as long as there are no inappropriately slow rates of symptomatic sinus pauses greater than 3 seconds. Longer pauses are uncommon in athletes and, along with inappropriately slow rates, need further evaluation with resting 12-lead and ambulatory ECG monitoring, an exercise test, and an echocardiogram to look for underlying structural heart disease. If the heart rate increases appropriately with exercise and the athlete is asymptomatic, participation in all competitive sports is allowed. In symptomatic athletes with presyncope or syncope, participation in sports should be restricted unless there is definitive treatment of the bradyarrhythmia for at least 3 months.[13]

Vagal hypertonia also accounts for both the first-degree heart block and the Mobitz type I second-degree AV heart block seen in many athletes at rest (see Table 23-2). The degree of block is related to the intensity of training and regresses upon cessation of training.[4,7,10] There is no need to restrict athletic participation as long as the athlete is asymptomatic and the degree of heart block does not worsen with exertion (Table 23-3). Mobitz type II and complete heart block have been reported, and it is likely that they are more common than in the general population (see Table 23-2).[9,14] They generally warrant further evaluation for underlying structural heart disease with echocardiography and an exercise test. A transient or permanent complete heart block, however, does not rule out high levels of physical performance over long periods. In one study on more than 12,000 athletes, 5 had second-degree AV block, and two had complete heart block. All of these athletes showed normalization of AV conduction with exercise and detraining.[15] In general, however, advanced AV block is likely to be associated with underlying structural abnormalities and warrant permanent pacing regardless of the presence or absence of symptoms (see Table 23-3). The exception to this are those athletes with congenital complete heart block who have no underlying structural heart disease, no symptoms, a resting heart rate greater than 40 beats per minute (bpm), and no exercise-induced ventricular arrhythmias. Athletes fitted with a permanent pacemaker are not advised to take part in sports where there is a danger of heavy body collision for obvious reasons (see Table 23-3). Furthermore, in any athlete with bradyarrhythmia, the presence of structural heart disease would independently determine restrictions on participation in sports.[13]

Supraventricular Arrhythmias (Table 23-4)

ATRIAL PREMATURE EXTRASYSTOLES

Atrial premature extrasystoles are common in athletes[4,7] and may be detected by the athletes as extra beats or brief palpitations or picked up on evaluation by physicians. In the absence of evidence from the history or clinical examination suggesting structural heart disease and in the absence of other symptoms other than occasional brief palpitations, evaluation other than with 12-lead ECG is not recommended, and no specific treatment is needed.[13] Athletes with more marked symptoms such as frequent or prolonged palpitations or presyncope could be considered for β-blockers. However, it must be remembered that β-blockers are poorly tolerated by some athletes and are banned in some sports. Athletes with atrial premature extrasystoles can participate in all competitive sports.

TABLE 23-3 Treatment and Restrictions Applied in Athletes with Bradyarrhythmias

Disorder	Symptoms	Diagnosis	Treatment	Competitive Sports
Sinus pause <3 sec	None	ECG, ACM	None	No restrictions
Sinus pause <3 sec	Syncope/LH	ECG, ACM Echo, ET	? PPM	No bodily collision if PPM implanted
Sinus pause >3 sec	None	ECG, ACM Echo, ET	? PPM	No bodily collision if PPM implanted
Sinus pause >3 sec	Syncope/LH	ECG, ACM Echo, ET	PPM	No bodily collision
First-degree HB	None	ECG	None	No restrictions
Wenckebach's HB	None	ECG, ACM	None	No restrictions
Wenckebach's HB	Syncope/LH	ECG, ACM	PPM	No bodily collision
Mobitz II HB/CHB	None	ECG, ACM	PPM	No bodily collision
Mobitz II HB/CHB	Syncope/LH	ECG, ACM	PPM	No bodily collision

ACM, ambulatory cardiac monitor; CHB, complete heart block; ET, exercise test; HB, heart block; LH, lightheadedness; PPM, permanent pacemaker.

TABLE 23-4 Evaluation and Management of Supraventricular Arrhythmias

Condition	Symptoms	ECG	Diagnosis	Treatment	Competition
AVNRT	Palpitations Presyncope Syncope	Normal EPS	ECG monitor	RFA BB, CCA or digoxin	After 3-6 mo if symptomless
WPW	Asymptomatic	Short PR δ Waves	ECG EPS	No therapy If high risk, RFA	Only after EPS excludes high risk
WPW	Palpitations Presyncope Syncope	Short PR δ Waves	ECG EPS	RFA Antiarrhythmics	After 3-6 mo if symptomless
AF	Palpitations	Often normal	ECG monitor	Rate control Anticoagulation Antiarrhythmics	Contact sports should be avoided if heparinized
A/Flutter	Palpitations	Often normal	ECG monitor	Rate control Anticoagulation Antiarrhythmics RFA	Contact sport should be avoided if heparinized
APE	Palpitations	Often normal	ECG monitor	No therapy; BB, if symptoms disabling	Restriction not needed

AF, atrial fibrillation; A/Flutter, atrial flutter; APE, atrial premature extrasystole; AVNRT, atrioventricular node reentry tachycardia; BB, β-blocker; CCA, calcium channel antagonist; EPS, electrophysiological studies; RFA, radiofrequency ablation; WPW, Wolff-Parkinson-White syndrome.

ATRIAL FIBRILLATION AND ATRIAL FLUTTER

Although uncommon, atrial fibrillation or flutter is more prevalent in athletes than in a matched sedentary population.[4,16] Athletes may be at particular risk of vagally mediated atrial fibrillation.[17] Focal atrial tachycardia initiating paroxysmal atrial fibrillation ("focal AF") may also be more common, especially in the younger age groups.[18]

Evaluation of these athletes should include investigation into illicit drug use, thyrotoxicosis, and congenital or acquired structural heart disease and include echocardiography, exercise testing, and ambulatory ECG monitoring.

The therapeutic options are based on the strategy of rhythm control with the aim of maintaining sinus rhythm, or rate control if atrial fibrillation is established or frequent. Although the maintenance of sinus rhythm allows physiologic AV synchrony and avoids the need for anticoagulation, concern remains about the potential proarrhythmic properties of the agents used, particularly in the context of strenuous exercise and athletes.[19] Consequently, in some cases rate control with agents such as digoxin, calcium channel antagonists, or β-blockers is preferred. If anticoagulation is required, athletes should not participate in sports associated with danger of heavy bodily contact. Athletes with atrial fibrillation, with or without structural heart disease, who maintain a ventricular response during exercise comparable to appropriate sinus tachycardia while receiving no therapy or therapy with rate-slowing agents can participate in sports according to the limitations imposed by the structural heart disease.[13] If atrial fibrillation is vagally mediated, atrial pacing could be considered in such athletes. Athletes with focal atrial fibrillation or atrial flutter should be considered for electrophysiological testing with a view to curative radiofrequency ablation of the pulmonary vein AF focus or the flutter circuit.[20] If ablation is successful, an athlete can resume full athletic participation after 3 months.[13]

ATRIOVENTRICULAR NODAL REENTRANT TACHYCARDIA

When an athlete presents with symptomatic supraventricular tachycardia (SVT), invasive electrophysiological evaluation is justified to establish the exact mechanism. In experienced hands, radiofrequency ablation offers success rates exceeding 95% and complication rates of less than 1%. It is generally considered appropriate first-line therapy for AVNRT.[21] This is particularly indicated in athletes with significant symptoms such as presyncope or syncope because of the unpredictable nature of the tachycardia. Radiofrequency ablation offers a potential cure and, if successful with no inducible arrhythmia or recurrences, allows the athlete to return to full competitive sports at 3 to 6 months.[13] The response to drug therapy is less defined and any return to competition should be restricted until the athlete has been adequately treated and has no recurrences for 6 months.[13]

WOLFF-PARKINSON-WHITE SYNDROME

Symptomatic presentation in athletes with Wolff-Parkinson-White syndrome (WPW) is with paroxysmal SVT, syncope with pre-excitation on routine ECG, atrial fibrillation, or rarely cardiac arrest due to ventricular fibrillation. Evaluation should initially include resting ECG and ambulatory electrocardiographic monitoring, exercise testing, and echocardiography to exclude further accompanying cardiac disease.[13] The accessory pathway in WPW, which allows retrograde conduction from the ventricle to the atrium during the tachycardia, may be manifest on the resting ECG or may be concealed.

Whether the accessory pathway is manifest or concealed, in cases of SVT secondary to WPW, catheter ablation is considered the preferred therapy due to the high success and low complication rates. If ablation is successful, with no inducible arrhythmias and no recurrence of SVT in 3 months, athletes can participate in all competitive sports.[13] For athletes treated with drugs, the same guidelines apply as with AV node reentry tachycardia (AVNRT). Further restriction in competitive sports may apply if there is accompanying structural heart disease.[13]

The evaluation and management of asymptomatic athletes with WPW is more controversial, as the risk of sudden death in this group appears to be very low.[22,23] Generally, it is recommended for these individuals to undergo electrophysiological testing to assess whether the accessory pathway is capable of high conduction rates. If conduction rates exceed 240 bpm, radiofrequency ablation of the accessory pathway is considered to eliminate the risk of future potential fatal tachyarrhythmias.[13,24]

Ventricular Arrhythmias

Sudden death in young athletes is uncommon. The mechanism underlying most cases of sudden death in the young is thought to involve ventricular arrhythmias in the setting of an underlying cardiac abnormality (see Table 23-1) that behave as an arrhythmogenic substrate.[25-27] However, there is little direct evidence to support this, as young athletes are rarely electrocardiographically monitored at the time of sudden death. Retrospective studies have, however, shown that the majority of cases of sudden death in young athletes were cardiovascular in nature and were associated with underlying structural heart disease.[28] Hypertrophic cardiomyopathy (HCM) is the most common underlying cause of sudden death in young athletes, accounting for more than 50% of cases (Fig. 23-1). In North America, studies in young athletes have shown anomalous coronary arteries to be the next most common underlying cause.[26] Coronary artery disease is by far the most common cause of sudden death in athletes older than 35 years of age, accounting for more than 80% of cases. These findings are in contrast with the experience in Italy, where arrhythmogenic right ventricular cardiomyopathy (ARVC) is the most common underlying cause of sudden death in young athletes (Fig. 23-2).[25,27] This difference in the prevalence of ARVC and HCM in Europe and North America could be due to the rigorous screening for HCM in athletes in Italy[29] or a result of the underdiagnosis of ARVC in North America. In general it appears that young athletes with HCM, ARVC, or anomalous coronary arteries are at highest risk of ventricular arrhythmias and sudden death.

In most individuals with structural heart disease, exercise can precipitate ventricular arrhythmias and sudden death. In HCM, more than half of arrhythmias and sudden death are associated with exertion.[30] Similar strong associations with exertion are also observed in ARVC,[31] the long QT syndrome (LQTS),[32] idiopathic ventricular fibrillation,[33] and individuals with coronary artery disease.[34]

Two relatively recently described conditions associated with sudden death in young athletes have been identified.

A significant subset of individuals with idiopathic ventricular fibrillation possess Brugada syndrome, which is secondary to a genetic mutation leading to an abnormality of the sodium channel *SCN5A*.[35,36] In this condition, affected individuals have a predisposition for fatal ventricular arrhythmias. The diagnosis is made using the resting ECG, where typically the precordial leads show incomplete right bundle branch block and ST segment elevation.[35]

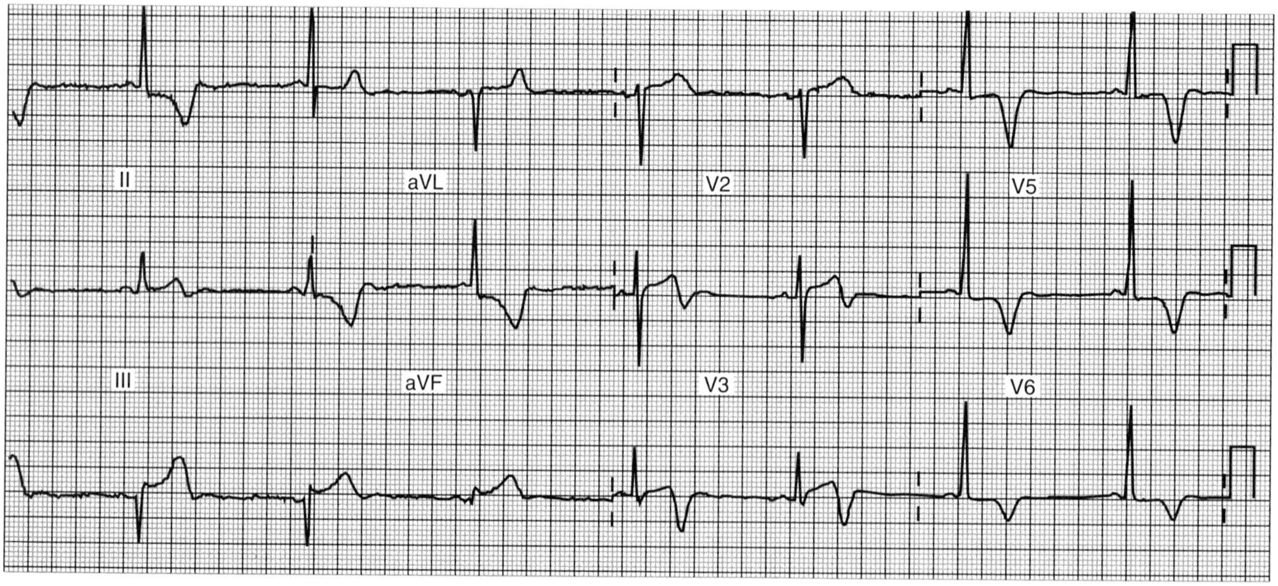

FIGURE 23-1 An ECG of an athlete with hypertrophic cardiomyopathy diagnosed during preparticipation screening.

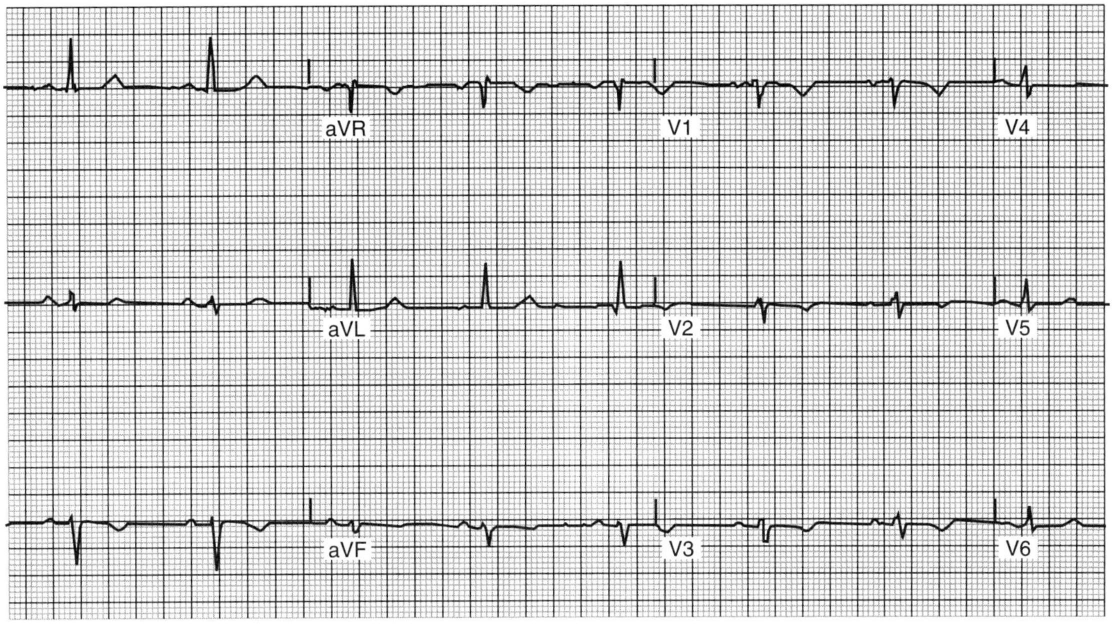

FIGURE 23-2 An ECG of an athlete with arrhythmogenic right ventricular cardiomyopathy showing T wave inversion in the anterior precordial leads.

The other condition is commotio cordis, where sudden death in young athletes is caused by blunt, nonpenetrating trauma to the precordium.[37,38] Most of the reported deaths have been in baseball, but they also occur in hockey, lacrosse, softball, and any sport where direct contact with the chest wall is possible. Sudden death is due to ventricular fibrillation, which is resistant to resuscitation with a survival rate of less than 10%. Studies have identified important variables in the pathogenesis of sudden death due to commotio cordis, including impact timing (impact must occur during a 20 millisecond [ms] window on the upslope of the T wave), speed and location of impact, and hardness of the impact object.[39]

PREMATURE VENTRICULAR EXTRASYSTOLES AND NONSUSTAINED VENTRICULAR TACHYCARDIA

In athletic and sedentary persons, premature ventricular extrasystoles (PVEs) are common, mostly benign, and usually asymptomatic. In the absence of underlying structural heart disease, PVEs do not confer an increased risk of sudden death but may be a sign of underlying cardiac disease that is not readily apparent from clinical evaluation or echocardiography. In cases where PVEs are frequent or multiform, further evaluation using cardiac magnetic resonance imaging (MRI) and cardiac catheterization may be considered to look for structural heart disease such as ARVC and anomalous and atherosclerotic coronary arteries. β-Blockers may be effective in reducing the severity of the symptoms but not necessarily the frequency of the premature extrasystoles. Treatment is generally not needed unless the symptoms are disabling. With regard to competitive sports, athletes with PVEs and no structural heart disease do not require any restriction as long as during exercise the arrhythmias do not worsen and the PVEs do not lead to presyncope or syncope (Table 23-5).[13]

Nonsustained ventricular tachycardia in the absence of structural heart disease also confers no increased risk of sudden death.[40] These athletes are treated in the same manner as athletes with PVEs. If underlying structural heart disease is found in the context of nonsustained ventricular tachycardia, then restriction in competitive sports is based on the underlying cardiac abnormality (see Table 23-5).[13]

SUSTAINED VENTRICULAR TACHYCARDIA AND VENTRICULAR FIBRILLATION

Athletes with sustained ventricular arrhythmia are more difficult to evaluate and treat. Treatment options include antiarrhythmic agents, ablation, and the implantable cardioverter defibrillator (ICD). In general, documented ventricular arrhythmia in an athlete is potentially fatal.[41] The exception to this is an athlete with no structural heart disease and left ventricular tachycardia or right ventricular outflow tract ventricular tachycardia.[42] These cases should be referred for electrophysiological studies with a view to radiofrequency catheter ablation,[43,44] as cure rates reaching 90% are achievable (see Table 23-5). If ablation is successful with no recurrence of the ventricular arrhythmia, the athlete can return to competitive sports within 6 months with no restrictions.[13]

On the other hand, cure of sustained ventricular arrhythmias in athletes with structural heart disease is unlikely. The long-term risk of further ventricular arrhythmias is significant with the potential for sudden death. As a result, these athletes are often treated with prophylactic therapy either in the form of antiarrhythmic agents or the ICD. In patients with life-threatening

TABLE 23-5 Evaluation and Treatment of Ventricular Arrhythmias in Athletes

Arrhythmia	Symptoms	ECG	Treatments	Competitive Sports
VPE	Palpitation	Normal	Reassure	No restrictions β-Blockers SHD assessment
NSVT	Palpitations	Usually normal	SHD assessment If no SHD, reassure If SHD present, further evaluation needed	No restrictions if no SHD If SHD present, restrictions based on type of SHD
VT/VF	Palpitations Presyncope Syncope SCD	Normal or reflecting underlying SHD	RFA, if no SHD If SHD present, ICD or antiarrhythmics	Restricted to low-intensity sports

ICD, implantable cardioverter defibrillator; NSVT, nonsustained ventricular tachycardia; RFA, radiofrequency ablation; SHD, structural heart disease; VF, ventricular fibrillation; VPE, ventricular premature extrasystole; VT, ventricular tachycardia.

arrhythmias, the ICD has been shown to be superior to antiarrhythmic agents.[45-47] Furthermore, antiarrhythmic agents are increasingly feared due to their potential for proarrhythmia.[48] Because of its superiority, athletes with noncurable ventricular arrhythmias should be ideally treated with the ICD.

Regardless of the mode of therapy, a return to competitive sports is usually not recommended for athletes with life-threatening ventricular arrhythmias (see Table 23-5) unless the competitive sport is low intensity, as in golf.[13] Another exception to this is the case of athletes with anomalous coronary arteries that are surgically repaired. These athletes can return to competitive sports after 6 months as long as they remain free of a recurrence of the ventricular arrhythmias.[13] The rationale for avoiding moderate to high exertion includes the probable exacerbation of ventricular arrhythmia with exertion and the concern over the integrity of the ICD and leads.

Evaluation of the Athlete with Symptoms of Arrhythmias

Symptoms of arrhythmia in the athlete are similar to those in the sedentary individual and may range from brief palpitations to syncope and sudden cardiac death. The most important aspects of the evaluation include assessing the severity of the symptoms, the presence of underlying structural heart disease, and the presence of a family history. Generally, the severity of the symptoms is related to the risk of ventricular arrhythmias and sudden death. Thus, palpitations are often benign, while presyncope and syncope are more concerning, and resuscitated sudden death is of utmost worry.

Palpitations are common and may be caused by premature atrial or ventricular extrasystoles, supraventricular or ventricular tachycardias, or anxiety and awareness of the normal heart rhythm. Long-term ECG monitoring with Holter monitors can be useful in those patients who present with frequent or reproducible symptoms. Athletes with intermittent or infrequent symptoms are best evaluated with a continuous loop recording monitor that continuously records 1- to 3-minute segments of a surface ECG. Upon activation of the monitor by the subject, the tape is frozen, and the previous few minutes are recorded.

Syncope without warning or at high levels of exertion is more worrying than gradual onset of syncope or syncope that is orthostatic or situational in nature. Injury as a result of syncope is also more typical of arrhythmic rather than neurocardiogenic syncope (Table 23-6).

Athletes with worrying symptoms should at minimum receive a 12-lead resting ECG and an echocardiogram. Ideally, they should also have an exercise test and ambulatory ECG monitoring.

In athletes older than 35 years of age or in athletes with exertional syncope, an assessment for cardiac ischemia should be made. If the diagnosis of anomalous coronary arteries is suspected, then cardiac catheterization is required. If an athlete is suspected to have ARVC, the investigation of choice would be a cardiac MRI looking for fatty replacement of the right ventricular myocardium.

In cases of resuscitated sudden death, a full cardiac workup including all of the previous assessments in addition to electrophysiological studies is indicated. Consideration should also be given to ajmaline testing to look for accentuation of the ECG changes found in Brugada syndrome.

TABLE 23-6 Clinical Features in Arrhythmic and Nonarrhythmic Syncope

Clinical Feature	Nonarrhythmic Syncope	Arrhythmic Syncope
Prodrome	Presyncope, nausea, warmth	None/brief presyncope
No. of episodes	Often/frequent	Few/once
Situational factors	Upright posture, fear	Exertional
Postsyncopal symptoms	Fatigue	None
Injury	Unusual	Common
Underlying SHD	Unusual	Common

SHD, structural heart disease.

Many cardiac diseases predisposing to arrhythmias and sudden death are genetic in nature; therefore, the family history is very important, as the presence of a familial cardiac abnormality or a family history of sudden death should lead to a detailed cardiac evaluation regardless of symptom status. Some of the genetic conditions include HCM, LQTS, Brugada syndrome, coronary artery disease, and some forms of ARVC and idiopathic dilated cardiomyopathy (DCM).

ELECTROCARDIOGRAM

The ECG in an athlete shows a number of physiologic changes including sinus bradycardia, first-degree or Wenckebach heart block, voltage criteria for left ventricular hypertrophy, and elevation of the J point with minor ST segment elevation.[4,7,49] Electrocardiographic changes seen in cardiac diseases leading to arrhythmias and sudden death are listed in Table 23-7. It is important to note that there is significant overlap between the ECG of the athlete and cardiac disease.[50,51]

INVASIVE ELECTROPHYSIOLOGICAL EVALUATION

The role of invasive electrophysiological study (EPS) in the evaluation of the athlete with ventricular arrhythmia is not absolutely clear. In patients with coronary artery disease, EPS is often performed to ascertain the risk of future arrhythmias and whether the patient merits prophylactic measures. In this setting, ventricular stimulation leading to sustained monomorphic ventricular tachycardia has a high sensitivity and specificity.[52] In patients with ARVC, the sensitivity is lower but still in the region of 70% to 80%.[53] In the case of HCM, idiopathic DCM, and the LQTS, the sensitivity and specificity of EPS is low, limiting its clinical usefulness.[41,54]

TABLE 23-7 Electrocardiographic Changes Observed in a Number of Cardiac Diseases

Cardiac Disease	ECG Abnormalities
Arrhythmogenic right ventricular cardiomyopathy	Anterior T wave inversion Incomplete or complete RBBB ε Waves *Rarely normal*
Hypertrophic cardiomyopathy	Left ventricular hypertrophy Pathological Q waves T wave inversion ST segment depression *Rarely normal*
Idiopathic dilated cardiomyopathy	Intraventricular conduction abnormalities *May be normal*
Long QT syndrome	Prolonged QT
Brugada syndrome	Incomplete or complete RBBB Anterior ST elevation
Anomalous coronary arteries	Usually normal
Coronary artery disease	Usually normal Q waves ST segment changes
Wolff-Parkinson-White syndrome	Short P–R interval δ Waves

RBBB, right bundle branch block.

Therefore, EPS in athletes with coronary artery disease is a useful tool to evaluate symptoms indicating arrhythmia or to prognosticate for future events. However, in athletes with cardiac diseases other than coronary artery disease, EPS does not offer reliable prognostic information to guide treatment.

Management of the Athlete with an Arrhythmogenic Substrate without Documented Ventricular Arrhythmias

Athletes who have underlying cardiac diseases predisposing to ventricular arrhythmia (Table 23-8) and sudden death but are asymptomatic and do not have ventricular arrhythmias are a difficult group to advise and manage. If they have structural heart disease such as HCM, ARVC, or idiopathic DCM, competitive sports and high-intensity physical training are generally disallowed.[13]

HYPERTROPHIC CARDIOMYOPATHY

The most important aspect of the management of the athlete with HCM is the identification of the high-risk athlete by using a risk stratification model looking for markers associated with an increased risk of sudden death.[55] These markers include a family history of sudden death, presentation with syncope or resuscitated cardiac arrest, documented ventricular arrhythmia during ambulatory ECG monitoring, abnormal vascular responses during exercise, and extreme left ventricular hypertrophy (maximal wall thickness ≥30 mm).[56,57] The presence of two or more risk markers indicates an individual at increased risk of sudden death[58] and merits prophylactic therapy either with ICD or amiodarone.[59-61] The ICD has been shown to be superior to amiodarone in this setting and should therefore be the standard therapy.[59-61] The markers used in assessing risk of sudden death are based on experience in specialist centers interested in HCM and not in an athletic population affected with HCM. The application of the risk stratification model to athletes with HCM is open to question but in the absence of further studies remains valid.

Other aspects of the management of the athlete with HCM include treatment of symptoms such as chest pain and dyspnea and screening of first-degree family relatives.[55]

ARRHYTHMOGENIC RIGHT VENTRICULAR CARDIOMYOPATHY

An athlete with ARVC usually presents with palpitations, presyncope, syncope, or cardiac arrest. Some are asymptomatic, however, and need evaluation for the risk of ventricular arrhythmias and sudden death. Unfortunately, the experience with ARVC is more limited in comparison with HCM, and risk stratification is less established. Retrospective data, however, indicate that previous cardiac arrest, a malignant family history, previous syncope, young age at presentation, extensive right ventricular disease with systolic impairment, and

TABLE 23-8 Structural Cardiac Disease in Athletes and Associated Ventricular Arrhythmias

Condition	Symptoms	ECG	VT Morphology	Treatments	Competitive Sports
HCM	Palpitations Syncope Sudden death	LVH Q waves ST–T inv	Right or left bundle	β-Blockers Amiodarone ICD	Low intensity only
ARVC	Palpitations Presyncope Sudden death	T inv RBBB ε wave	Left bundle Inferior axis	β-Blockers ICD RF ablation	Low intensity only
Idiopathic DCM	Palpitations Syncope Sudden death	Often LBBB	Right or left bundle	β-blockers Amiodarone ICD	Low intensity only
Anomalous CAD	Exertional CP Syncope Sudden death	Normal	VF	CABG	No restriction post-CABG
CAD	Chest pain Palpitations Syncope Sudden death	May have ischemic changes	Right or left bundle	Amiodarone ICD	Low intensity only
LQTS	Palpitations Syncope Sudden death	Long QT syndrome	Torsades de pointes	β-Blockers ± PPM ICD	Low intensity only
Idiopathic left ventricular VT	Palpitations Presyncope Syncope	Normal	Right bundle Left axis	RF ablation	No restriction at 3 mo post-RFA
Idiopathic RVOT VT	Palpitations Presyncope Syncope	Normal	Left bundle Inferior axis	RF ablation	No restriction at 3 mo post-RFA

ARVC, arrhythmogenic right ventricular cardiomyopathy; CABG, coronary artery bypass grafting; CAD, coronary artery disease; CP, chest pain; DCM, dilated cardiomyopathy; ICD, implantable cardioverter defibrillator; inv, interval; LBBB, left bundle branch block; LVH, left ventricular hypertrophy; PPM, permanent pacemaker; RBBB, right bundle branch block; RF, radiofrequency; RVOT VT, right ventricular outflow tract VT; VF, ventricular fibrillation; VT, ventricular tachycardia.

left ventricular involvement indicate an adverse prognosis.[53,62,63] In athletes with ARVC the treatment options include β-blockers,[64] radiofrequency ablation,[65] or ICD implantation. If an athlete is suspected to be at significant risk of future ventricular arrhythmias and sudden death, then ICD implantation should be considered.[31,62]

The differential diagnosis of ARVC from right ventricular outflow tract tachycardia in an athlete is important, as the latter is benign and amenable to radiofrequency ablation (Table 23-9).[43,44]

Moderate and high-level physical training and competitive sports are disallowed in athletes with ARVC, particularly in view of the provocative effect of exercise on ventricular arrhythmias.[13]

TABLE 23-9 Differential Diagnosis of Arrhythmogenic Right Ventricular Cardiomyopathy and Right Ventricular Outflow Tract Ventricular Tachycardia

Symptom/ Investigation	RVOT VT	ARVC
Resting ECG	Normal	Abnormal
Syncope	Uncommon	Common
Other symptoms	No symptoms between VT	Often present
Signal-averaged ECG	Normal	Often abnormal
2-D Echocardiogram	Normal	May be abnormal
Cardiac MRI	May show nonspecific changes in RVOT	Fibro-fatty replacement
Prognosis	Good	May be poor
Treatment of VT	RFA, β-blockers	ICD

ICD, implantable cardioverter defibrillator; MRI, magnetic resonance imaging; RFA, radiofrequency ablation; RVOT, right ventricular outflow tract; VT, ventricular tachycardia.

DILATED CARDIOMYOPATHY

The most important factor in the risk of sudden death appears to be the degree of left ventricular impairment.[66] The presence of syncope and nonsustained ventricular tachycardia on Holter monitoring may also be associated with an increased risk.[67] Generally, athletes with DCM who are asymptomatic do not need additional therapy other than with ACE inhibitors and β-blockers, which are established agents in systolic impairment. However, if an athlete with DCM has syncope, then prophylaxis with an ICD should be considered. In athletes with DCM and sustained ventricular arrhythmia, competitive sports and moderate to high-intensity physical training should be avoided.[13]

ION CHANNELOPATHIES

Athletes with LQTS may be at increased risk of ventricular arrhythmia and sudden death. Indicators of high risk include a family history of sudden death, a young age at presentation, previous syncope, or aborted cardiac arrest.[68,69] β-Blockers should be an initial step in the management of any athlete with the LQTS, and if bradycardia or recurrent symptoms occur on β-blockers, then permanent pacing can be used.[70] If an athlete is felt to be at significant risk of sudden death, ICD should be considered.[32]

An athlete found to have Brugada syndrome is advised to have EPS. Prophylaxis with the ICD is recommended if the athlete has had syncope, suffered an aborted cardiac arrest, or has inducible ventricular arrhythmias during EPS.[71]

ANOMALOUS CORONARY ARTERIES

If an athlete with anomalous coronary arteries is asymptomatic, a maximal exercise test should be carried out to seek objective evidence of exertional ischemia. If an athlete is found to have anomalous coronary arteries, surgical revascularization should be considered. If surgery is successful and inducible myocardial ischemia is excluded, a return to competitive sports is allowed after 6 months.[72]

CORONARY ARTERY DISEASE

If an athlete has coronary artery disease, the risk of sudden death is thought to be higher with nonsustained ventricular tachycardia, inducible ventricular tachycardia, and poor ejection fraction. The Multicenter Automatic Defibrillator Implantation Trial study has confirmed improved survival with the ICD in patients with poor ejection fraction and inducible ventricular tachycardia.[73] An athlete with coronary artery disease and ventricular arrhythmias may only participate in low intensity training and sports. If an athlete with coronary artery disease is asymptomatic and has normal systolic function, there is no need for EPS or ICD implantation, and no restriction is necessary.[13]

SUMMARY

Athletes are generally regarded as the healthiest individuals in society, but they are not immune from arrhythmias or cardiac disease. The evaluation and management of athletes with arrhythmia or symptoms suggestive of arrhythmia are uncommon but important issues facing the general physician and the sports cardiologist. The importance of this topic is highlighted by the fact that the diagnosis of significant arrhythmia can disrupt or end an athletic career.

Asymptomatic bradyarrhythmias are common among athletes and are due to the high vagal tone associated with intense physical training. Pathologic bradyarrhythmias are very uncommon and treatable by pacing. Supraventricular tachyarrhythmias in athletes, although usually not fatal, can be symptomatic and are often treatable either using pharmacology or radiofrequency ablation. It is very rare for an athlete to be unable to participate in competitive sports indefinitely due to bradyarrhythmias or supraventricular tachyarrhythmias.

Potentially life-threatening ventricular arrhythmias are uncommon in athletes and are generally associated with underlying cardiac disease. Cure of sustained ventricular arrhythmias is generally only possible in the absence of underlying cardiac disease. The majority of patients with sustained ventricular arrhythmias will therefore not be curable, and their management centers on risk identification and prophylactic therapy. Competitive sports and high-intensity physical training are restricted in the long term for most of these patients.

REFERENCES

1. Paffenbarger RS Jr, Hyde RT, Wing AL, et al: The association of changes in physical activity level and other lifestyle characteristics with mortality among men. N Engl J Med 1993;328:538-545.
2. Sandvik L, Erikssen J, Thaulow E, et al: Physical fitness as a predictor of mortality among healthy, middle-aged Norwegian men. N Engl J Med 1993;328:533-537.
3. Maron BJ: Cardiovascular risks to young persons on the athletic field. Ann Intern Med 1998;129:379-386.
4. Huston TP, Puffer JC, Rodney WM: The athletic heart syndrome. N Engl J Med 1985;313:24-32.
5. Saltin B, Astrand PO: Maximal oxygen uptake in the athlete. J Appl Physiol 1967;23:353.
6. Maron BJ: Structural features of the athletic heart as defined by echocardiography. J Am Coll Cardiol 1986;7:190-203.
7. Oakley CM: The electrocardiogram in the highly trained athlete. Cardiol Clin 1992;10:295-302.
8. 26th Bethesda Conference: Recommendations for determining eligibility for competition in athletes with cardiovascular abnormalities. J Am Coll Cardiol 1996;94.
9. Zehender M, Meinertz T, Keul J, Just H: ECG variants and cardiac arrhythmias in athletes: Clinical relevance and prognostic importance. Am Heart J 1990;119:1378-391.
10. Oakley DG, Oakley CM: Significance of abnormal electrocardiograms in highly trained athletes. Am J Cardiol 1982;50:985-989.
11. Roeske WR, O'Rourke RA, Klein A, et al: Non-invasive evaluation of ventricular hypertrophy in professional athletes. Circulation 1976;53:286-292.
12. Ordway GA, Charles JB, Randall DC, et al: Heart rate adaptation to exercise training in cardiac-denervated dogs. J Appl Physiol 1982;52:1586-1590.
13. Zipes DP, Garson A: Task force 6: Arrhythmias. J Am Coll Cardiol 1996;94:892-899.
14. Cooper JP, Fraser AG, Penny WJ: Reversibility and benign occurrence of complete heart block in athletes. Int J Cardiol 1992;35:118-120.
15. Fenici R, Caselli G, Zeppilli P, Piovano G: High degree A-V block in 17 well-trained endurance athletes. In Lubich T, Venerando A (eds): Sports Cardiology. Bologna, Italy, Aulo Gaggi, 1980, pp 523-537.
16. Furlanello F, Bertoldi A, Dallago M, et al: Atrial fibrillation in elite athletes. J Cardiovasc Electrophysiol 1998;9:S63-S68.
17. Coumel P: Autonomic influences in atrial tachyarrhythmias. J Cardiovasc Electrophysiol 1996;7:999-1007.
18. Haissaguerre M, Jais P, Shah DC, et al: Spontaneous initiation of atrial fibrillation by ectopic beats originating in the pulmonary veins. N Engl J Med 1998;339:659-666.
19. Ganz LI, Antman EM: Anti-arrhythmic drug therapy in the management of atrial fibrillation. J Cardiovasc Electrophysiol 1997;8:1175-1189.
20. Fischer B, Haissaguerre M, Garrigues S, et al: Radiofrequency catheter ablation of common atrial flutter in 80 patients. J Am Coll Cardiol 1995;25:1365-1372.
21. Naccarelli GV, Shih H, Jalal S: Catheter ablation for the treatment of paroxysmal supraventricular tachycardia. J Cardiovasc Electrophysiol 1995;6:951-961.
22. Wellens HJ, Rodriguez LM, Timmermans C, Smeets JL: The asymptomatic patient with the Wolff-Parkinson-White electrocardiogram. Pacing Clin Electrophysiol 1997;20:2082-2086.
23. Munger TM, Packer DL, Hammil SC, et al: A population study of the natural history of Wolff-Parkinson-White syndrome in Olmsted County, Minnesota, 1953-1989. Circulation 1993; 87:866-873.
24. Leitch JW, Klein GJ, Yee R, Murdock C: Prognostic value of electrophysiological testing in asymptomatic patients with Wolff-Parkinson-White pattern. Circulation 1990;82:1718-1723.
25. Maron BJ, Epstein SE, Roberts WC: Causes of sudden death in competitive athletes. J Am Coll Cardiol 1986;7:204-214.
26. Liberthson RR: Sudden death from cardiac causes in children and young adults. N Engl J Med 1996;334:1039-1044.
27. Corrado D, Thiene G, Nava A, et al: Sudden death in young competitive athletes: Clinicopathological correlations in 22 cases. Am J Cardiol 1990;89:588-596.
28. Maron BJ, Shirani J, Poliac LC, et al: Sudden death in young competitive athletes: Clinical, demographic and pathologic profiles. JAMA 1996;276:199-204.
29. Corrado D, Basso C, Schiavon M, Thiene G: Screening for hypertrophic cardiomyopathy in young athletes. N Engl J Med 1998; 339:364-369.
30. Maron BJ, Fananapazir L: Sudden death in hypertrophic cardiomyopathy. Circulation 1992;85[suppl I]:57-63.

31. Link MS, Wang PJ, Haugh CJ, et al: Arrhythmogenic right ventricular dysplasia: Clinical results with implantable cardioverter defibrillators. J Interv Card Electrophysiol 1997;1:41-48.
32. Roden DM, Lazzara R, Rosen M, et al: Multiple mechanisms in the long-QT syndrome. Current knowledge, gaps and future directions. The SADS Foundation Task Force on LQTS. Circulation 1996;94:1996-2012.
33. Viskin S, Belhassen B: Idiopathic ventricular fibrillation. Am Heart J 1990;120:661-671.
34. Thompson PD: The cardiovascular complications of vigorous physical activity. Arch Intern Med 1996;156:2297-2302.
35. Brugada P, Brugada J: Right bundle branch block, persistent ST segment elevation and sudden cardiac death: A distinct clinical and electrocardiographic syndrome. J Am Coll Cardiol 1992; 20:1391-1396.
36. Gussack I, Antzelevitch C, Bjerregaard P, et al: The Brugada syndrome: Clinical, electrophysiologic and genetic aspects. J Am Coll Cardiol 1999;33:5-15.
37. Maron BJ, Poliac LC, Kaplan JA, Mueller FO: Blunt impact to the chest leading to sudden death from cardiac arrest during sports activities. N Engl J Med 1995;333:337-342.
38. Maron BJ, Link MS, Wang PJ, Estes NA III: Clinical profile of commotio cordis: An under-appreciated cause of sudden death in the young during sports and other activities. J Cardiovasc Electrophysiol 1999;10:114-120.
39. Link MS, Wang PJ, Pandian NG, et al: An experimental model of sudden death due to low energy chest wall impact (commotio cordis). N Engl J Med 1998;338:1805-1811.
40. Kinder C, Tamburro P, Kopp D, et al: The clinical significance of non-sustained ventricular tachycardia: Current perspectives. Pacing Clin Electrophysiol 1994;17:637-664.
41. Link MS, Estes NA III: Ventricular arrhythmias. In Estes NA III, Salem DN, Wang PJ (eds): Sudden Cardiac Death in the Athlete. Armonk, NY, Futura Publishing, 1998, pp 253-275.
42. Brooks R, Burgess JH: Idiopathic ventricular tachycardia. Medicine 1988;67:271-294.
43. Klein LS, Miles WM: Ablative therapy for ventricular arrhythmias. Prog Cardiovasc Dis 1995;37:225-242.
44. Manolis AS, Wang PJ, Estes NA III: Radiofrequency catheter ablation for cardiac tachyarrhythmias. Ann Intern Med 1994; 121:452-461.
45. AVID investigators: A comparison of anti-arrhythmic drug therapy with implantable cardioverter defibrillators in patients resuscitated from near-fatal ventricular arrhythmias. N Engl J Med 1997;447:1576-1583.
46. Kuck K-H, Cappato R, Siebels J, Ruppel R: Randomized comparison of anti-arrhythmic drug therapy with implantable cardioverter defibrillator in patients resuscitated from cardiac arrest. Circulation 2000;102:748-754.
47. Connolly SJ, Gent M, Roberts RS, et al: Canadian implantable cardioverter defibrillator study (CIDS): A randomized trial of the implantable cardioverter defibrillator against amiodarone. Circulation 2000;101:1297-1302.
48. Link MS, Homoud M, Foote CB, et al: Anti-arrhythmic drug therapy of ventricular arrhythmias: Current perspectives. J Cardiovasc Electrophysiol 1996;7:653-670.
49. Sharma S, Whyte G, Elliott PM, et al: Electrocardiographic changes in 1000 highly trained junior elite athletes. Br J Sports Med 1999;33:319-324.
50. Pelliccia A, Maron BJ, Culasso F, et al: Clinical significance of abnormal electrocardiographic patterns in trained athletes. Circulation 2000;102:278-284.
51. Foote CB, Michaud G: The athlete's electrocardiogram: Distinguishing normal from abnormal. In Estes NA III, Salem DN, Wang PJ (eds): Sudden Cardiac Death in the Athlete. Armonk, NY, Futura Publishing, 1998, pp 101-114.
52. Brugada P, Green M, Abdollah H, Wellens HJ: Significance of ventricular arrhythmias initiated by programmed ventricular stimulation: The importance of the type of ventricular arrhythmia induced and the number of premature stimuli required. Circulation 1984;69:87-92.
53. Peters S, Reil GH: Risk factors of cardiac arrest in arrhythmogenic right ventricular cardiomyopathy. Eur Heart J 1995;16: 77-80.
54. Anderson KP, Mason JW: Clinical value of cardiac electrophysiological studies. In Zipes DP, Jalife J (eds): Cardiac Electrophysiology. Philadelphia, Saunders, 1995, pp 1133-1150.
55. Spirito P, Seidman CE, McKenna WJ, Maron BJ: The management of hypertrophic cardiomyopathy. N Engl J Med 1997; 336:775-785.
56. Elliott PM, Poloniecki J, Dickie S, et al: Sudden death in hypertrophic cardiomyopathy: Identification of high risk patients. J Am Coll Cardiol 2000;36:2212-2218.
57. Spirito P, Bellone P, Harris KM, et al: Magnitude of left ventricular hypertrophy and risk of sudden death in hypertrophic cardiomyopathy. N Engl J Med 2000;342:1778-1785.
58. Elliott PM, Gimeno Blanes JR, Mahon NG, et al: Relation between severity of left ventricular hypertrophy and prognosis in patients with hypertrophic cardiomyopathy. Lancet 2001;357:420-424.
59. Maron BJ, Shen W-K, Link MS, et al: Efficacy of the implantable cardioverter-defibrillator for the prevention of sudden death in hypertrophic cardiomyopathy. N Engl J Med 2000;342:365-373.
60. McKenna WJ, Oakley CM, Krikler DM, Goodwin JF: Improved survival with amiodarone in patients with hypertrophic cardiomyopathy and ventricular tachycardia. Br Heart J 1985;53:412-416.
61. Cecchi F, Olivotto I, Montereggi A, et al: Prognostic value of non-sustained ventricular tachycardia and the potential role of amiodarone treatment in hypertrophic cardiomyopathy: Assessment in an unselected non-referral based patient population. Heart 1998;79:331-336.
62. Marcus FI, Fontaine G: Arrhythmogenic right ventricular dysplasia/cardiomyopathy: A review. Pacing Clin Electrophysiol 1995; 18:1298-1314.
63. Peters S, Peters H, Thierfelder L: Risk stratification of sudden cardiac death and malignant ventricular arrhythmias in right ventricular dysplasia/cardiomyopathy. Int J Cardiol 1999;71:243-250.
64. Witcher T, Borggrefe M, Haverkamp W, et al: Efficacy of antiarrhythmic drugs in patients with arrhythmogenic right ventricular disease. Circulation 1992;86:29-37.
65. Fontaine G, Tonet J, Gallais Y, et al: Ventricular tachycardia catheter ablation in arrhythmogenic right ventricular dysplasia: A 16-year experience. Curr Cardiol Rep 2000;2:498-506.
66. Borggrefe M, Block M, Breithardt G: Identification and management of the high risk patient with dilated cardiomyopathy. Br Heart J 1994;72(suppl):S42-S45.
67. Middlekauff HR, Stevenson WG, Stevenson LW, Saxon LA: Syncope in advanced heart failure: High risk of sudden death regardless of origin of syncope. J Am Coll Cardiol 1993;21:110-116.
68. Moss AJ: Prolonged QT-interval syndromes. JAMA 1986; 256:2985-2987.
69. Jackman WM, Friday KJ, Anderson JL, et al: The long QT syndromes: A critical review, new clinical observations and a unifying hypothesis. Prog Cardiovasc Dis 1988;31:115-172.
70. Moss AJ, Zareba W, Hall WJ, et al: Effectiveness and limitations of beta-blocker therapy in congenital long-QT syndrome. Circulation 2000;101:616-623.
71. Priori SG: Long QT and Brugada syndromes: From genetics to clinical management. J Cardiovasc Electrophysiol 2000;11: 1174-1178.
72. Graham TP Jr, Bricker JT, James FW, Strong WB: Task force 1: Congenital heart disease. J Am Coll Cardiol 1994;24:867-873.
73. Moss AJ, Hall WJ, Cannom DS, et al: Improved survival with an implanted defibrillator in patients with coronary disease at high risk for ventricular arrhythmia. Multicenter Automatic Defibrillator Implantation Trial Investigators. N Engl J Med 1996; 335:1933-1940.

Chapter 24 Proarrhythmia Syndromes

ANNE M. GILLIS

Proarrhythmia is usually defined as provocation of a new arrhythmia or aggravation of an existing arrhythmia during therapy with a drug at concentrations usually not considered to be toxic.[1-6] However, other therapies or pathophysiologic conditions may also trigger proarrhythmia. The various proarrhythmia syndromes are summarized in Table 24-1 and the drugs and conditions most frequently associated with proarrhythmia are listed in Table 24-2. Most types of proarrhythmia occur in the setting of structural heart disease, but proarrhythmia may also occur in individuals without apparent heart disease.[7] The relatively higher risk of proarrhythmia in the setting of antiarrhythmic drug use compared to other therapies obliges the clinician to weigh the relative risks and benefits of antiarrhythmic drug therapy in each individual patient (Fig. 24-1).

Proarrhythmia Due to Sodium Channel Blocking Drugs

SUDDEN CARDIAC DEATH IN THE POST–MYOCARDIAL INFARCTION POPULATION

The Cardiac Arrhythmia Suppression Trials. The Cardiac Arrhythmia Suppression Trial (CAST) investigators reported that patients with frequent ventricular premature beats following a myocardial infarction (MI) who were treated with flecainide or encainide had a higher mortality than patients treated with placebo (Fig. 24-2).[8] The proarrhythmic risk was not just limited to the early phase of drug therapy but increased over the course of the trial. Subgroup analyses suggested that recurrent myocardial ischemia played an important role in the increased risk of arrhythmic death.[8,9] Total mortality and risk of cardiac arrest were higher in patients who had sustained a non–Q wave MI or in those who had not received thrombolytic therapy.[8,9] The relative risk of death or cardiac arrest was 10.6 in patients sustaining a non–Q wave MI versus 1.45 for those patients sustaining a Q wave MI.[9] This led to the hypothesis that recurrent ischemia in the setting of class IC drug therapy altered the electrophysiological milieu in the infarct region, predisposing to a proarrhythmic response.[3,10] Consistent with this hypothesis, concomitant β-blocker and flecainide or encainide use was not associated with an increased relative mortality risk.[11,12]

CAST-II demonstrated the increased proarrhythmic potential of the class I antiarrhythmic drug moricizine during early dosage titration.[13] Death or nonfatal cardiac arrest was significantly greater during the drug initiation phase in patients treated with moricizine compared to placebo. The relative risk of death or nonfatal cardiac arrest during the initial 2 weeks of moricizine therapy was 5.6. Although survival during long-term follow-up was similar in the moricizine and placebo groups, CAST II was terminated early, because it was deemed unlikely that a survival benefit from moricizine would be observed if the trial were completed.

The CAST trials[8,13] and meta-analyses of class I antiarrhythmic drugs in patients with ischemic heart disease or left ventricular dysfunction,[14,15] or both, indicate that these drugs increase the risk of lethal cardiovascular events. Therefore, such drugs are contraindicated in patients with a history of prior MI. By extension, many clinicians also believe that these drugs are relatively

TABLE 24-1 Proarrhythmia Syndromes

New Sustained Arrhythmia
SODIUM CHANNEL BLOCKING DRUGS
• Ventricular fibrillation
• VT
• Incessant VT
• Atrial flutter with 1:1 AV conduction
ACTION POTENTIAL PROLONGING DRUGS
• Torsades de pointes VT—long QT syndrome
• Ventricular fibrillation
Aggravation of an Existing Arrhythmia
• Conversion of nonsustained arrhythmia to a sustained arrhythmia
• Acceleration of tachycardia
• Aggravation of bradycardia or conduction abnormality
○ Sinus node dysfunction
○ AV block
○ Marked intraventricular conduction delay
○ Increase in pacing thresholds
• Increase in defibrillation thresholds

AV, atrioventricular; VT, ventricular tachycardia.

TABLE 24-2 Drugs and Conditions Associated with Torsades de Pointes Ventricular Tachycardia

Drug Class	Specific Drugs
Antiarrhythmic	
Class IA	quinidine
	procainamide
	disopyramide
Class IC	propafenone
	amiodarone
Class III	d, l-sotalol
	d-sotalol
	dofetilide
	ibutilide
Psychotropic	haloperidol
	phenothiazines
	risperidone
	tricyclic antidepressants
	tetracyclic antidepressants
Diuretics	furosemide
	hydrochlorothiazide
	indapamide
	metolazone
Antimicrobial	erythromycin
	clarithromycin
	trimethoprim-sulfamethoxazole
	pentamidine
Antifungal	ketoconazole
	fluconazole
	itraconazole
Antihistamines	diphenhydramine
	terfenadine
	astemizole
Lipid lowering	probucol*
Cholinergic antagonists	cisapride*
Other Conditions	
Bradycardia	complete heart block or significant bradycardia
Electrolyte abnormalities	hypokalemia
	hypomagnesemia
	hypocalcemia
Nervous system injury	subarachnoid hemorrhage
Starvation	anorexia nervosa, liquid protein diets

*No longer commercially available.

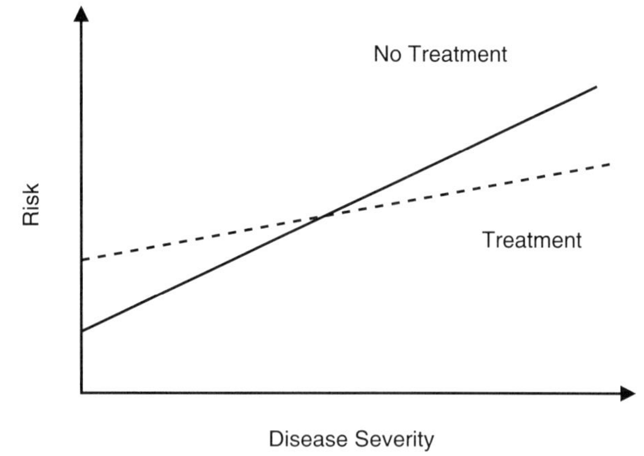

FIGURE 24-1 The risk-benefit of antiarrhythmic drug therapy is based on disease severity. If the risk of a life-threatening arrhythmic event is high, antiarrhythmic drug therapy may reduce that risk substantially compared to the risk of a proarrhythmic event. In contrast, if the risk of a life-threatening arrhythmic event is small, the risk of a proarrhythmic event may exceed the marginal benefits of antiarrhythmic drug therapy.

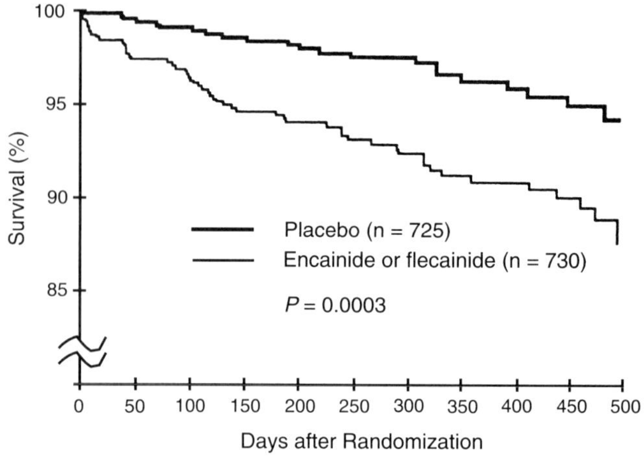

FIGURE 24-2 Survival among 1455 patients randomized to receive encainide or flecainide, or the matching placebo. Patients were more likely to die if receiving encainide or flecainide compared to placebo (P = .0003 by log rank test). (From The Cardiac Arrhythmia Suppression Trial (CAST) Investigators: Preliminary report: Effect of encainide and flecainide on mortality in a randomized trial of arrhythmia suppression after myocardial infarction. N Engl J Med 1989; 3321:406-412.)

contraindicated in any patient with ischemic heart disease or risk factors for ischemic heart disease, as well as in patients with significant left ventricular dysfunction.

Mechanisms of Ventricular Proarrhythmia in Ischemic Heart Disease. Sodium channel blocking agents slow conduction in atrial and ventricular muscle in a use-dependent manner (i.e., greater conduction block occurs at faster heart rates).[16] Some sodium channel blocking agents (encainide and flecainide) produce extensive block at slow physiologic heart rates, which increases the risk of proarrhythmia.[3,4,16-18] Class IC antiarrhythmic drugs, by causing excessive slowing of conduction that exceeds changes in refractoriness in diseased tissue, may stabilize a reentrant circuit that has previously been unable to sustain reentry and thus establish the electrophysiological substrate for sustained ventricular tachycardia (VT) or convert nonsustained VT to sustained monomorphic VT.[3,19-22] This model of reentry assumes that the drug alters only conduction velocity. In reality, many sodium channel blocking drugs alter conduction velocity and refractoriness in a heterogeneous manner that is dependent on the physiologic milieu. Increased dispersion of tissue refractoriness may also promote ventricular reentry.

The proarrhythmic response of class I drugs reported in the CAST trials led to a series of experimental studies of antiarrhythmic drug effects in the setting of prior MI and in the setting of acute ischemia. Several investigators reported that encainide and flecainide enhanced induction of sustained VT in dogs with prior MI without VT inducible at baseline.[20,22-24] These drugs cause preferential conduction slowing in the peri-infarct zone, resulting in unidirectional block or marked slowing of conduction in the direction transverse to fiber orientation that facilitates ventricular reentry. Restivo et al.[25]

reported that unidirectional block occurred in the peri-infarct zone when premature beats were induced, but the impulse encountered refractory tissue and failed to reenter. Flecainide slowed conduction in the infarct zone, permitting the recirculating impulse to reenter and thus allowing VT to be induced. Ranger et al.[20] and Coromilas et al.[21] showed that flecainide caused rate-dependent, localized, transverse conduction block or exaggerated an existing conduction delay, which allowed for the occurrence of spontaneous VT or permitted induction of stable macroreentrant VT.

In normal ventricular myocardium, the effects of flecainide on conduction velocity are independent of fiber orientation. In infarcted tissue, the exaggerated slowing of conduction in the direction transverse to fiber orientation provides the substrate for ventricular reentry.[20,21,25] Infarct size and drug concentration have been reported to be important determinants of proarrhythmia secondary to sodium channel blocking drugs.[3,20] In chronic ischemia, proarrhythmia occurs infrequently at low drug concentrations.

In experimental models of acute ischemia, the presence of a sodium channel blocking agent significantly increases the incidence of spontaneous ventricular fibrillation (VF).[3] Class I drugs including lidocaine,[6,26,27] flecainide,[28,29] and aprindine[30] produce exaggerated effects on conduction velocity and repolarization in ischemic tissue, contributing to the substrate for VF. The class IC drug aprindine administered before coronary artery ligation increases the risk of ischemic VF, whereas aprindine administered following coronary artery ligation does not cause ventricular proarrhythmia. In acute ischemia, substantial drug concentrations accumulate in the ischemic tissue and probably contribute to increased conduction slowing, whereas following coronary artery ligation there is limited drug distribution in ischemic tissue.[31] Ranger et al. evaluated the concentration-dependence of flecainide proarrhythmia in dogs with acute or chronic MI.[20] Proarrhythmia was observed in 79% of dogs with a chronic MI versus 55% of dogs during acute ischemia. However, the concentration-dependence of flecainide proarrhythmia differed substantially between the two groups: proarrhythmia occurred at therapeutic concentrations in the setting of acute MI and was most likely manifest as VF, whereas 20-fold higher concentrations of flecainide were required to induce proarrhythmia in the chronic infarct dog, and proarrhythmia was manifest as induced sustained monomorphic VT.[20]

These experimental studies provide mechanistic insights into the CAST results: Class I antiarrhythmic drugs, when present in sufficient concentrations in the myocardium, increase the risk for VF in patients with coronary artery disease if acute ischemia develops, presumably because of spatially heterogenous effects on conduction velocity and possibly repolarization, which provides the substrate for reentry and VF.

Experimental models also support the concept that β-blockers, in the setting of acute ischemia, reduce the proarrhythmic effects of sodium channel blockers.[32,33] Under normal conditions, isoproterenol reverses flecainide's effect on sodium channel block. Under conditions of membrane depolarization, such as would be expected in ischemia, the block is amplified.[33,34] Thus, enhanced sympathetic tone in the setting of acute ischemia may further modulate the interaction between a class I drug and ischemic tissue to promote proarrhythmia.

Other Cardiac Disease States. There is limited information about the proarrhythmic potential of class I sodium channel blocking drugs in cardiac disease models without ischemic heart disease. However, cardiac hypertrophy and ventricular dysfunction in the setting of volume overload are associated with heterogeneous abnormalities of ventricular conduction and repolarization.[35,36] In the setting of cardiac disease, the resting membrane potential is likely elevated in some cells, and this might predispose them to greater antiarrhythmic drug-induced sodium channel conduction block.[35,37] As well, frequent ventricular premature beats, which may occur in the setting of left ventricular dysfunction, may be associated with greater conduction delays in the presence of class I drugs and thus predispose to ventricular reentry. Some antiarrhythmic drugs, including the class IA drug quinidine and the IC drug propafenone, produce exaggerated slowing of conduction in models of disease states.[38,39] Thus, it is likely that any sodium channel blocking drug might produce exaggerated electrophysiological responses in diseased ventricular tissue and that these responses occur in a heterogeneous manner, thereby predisposing to ventricular proarrhythmia. Therefore, selective sodium channel blocking drugs are relatively contraindicated in all patients with significant left ventricular dysfunction.

INCESSANT VENTRICULAR TACHYCARDIA SECONDARY TO SODIUM CHANNEL BLOCKING DRUGS

Some patients treated with sodium channel blocking drugs, particularly class IC drugs including flecainide, propafenone, or encainide, may develop slow, incessant VT.[17,18,40] This usually occurs in the setting of structural heart disease (e.g., prior MI). Typically, significant slowing of conduction develops; it may manifest as increased lengthening of the QRS interval at rest and more marked prolongation of conduction at increased heart rates.[41] In animal models, conduction slowing in the peri-infarct zone contributes to a macroreentrant circuit.[3,20,21] This arrhythmia usually occurs in the setting of marked conduction slowing, since the circulating wavefront is less likely to encounter refractory tissue and be abolished (Fig. 24-3).

Incessant VT may be hemodynamically tolerated due to the slow rate. However, the arrhythmia may degenerate into a hemodynamically significant VT or VF and be lethal.[40,42] This arrhythmia is usually observed with higher or toxic doses of drug[20] or in the setting of other pathophysiologic conditions that exaggerate the effects of the drug on conduction (e.g., acidosis, hyperkalemia, or concomitant sodium channel blockers including phenytoin). This type of VT may also be observed with tricyclic antidepressant overdose.[43] The anticholinergic effects of the tricyclic antidepressants cause an increase

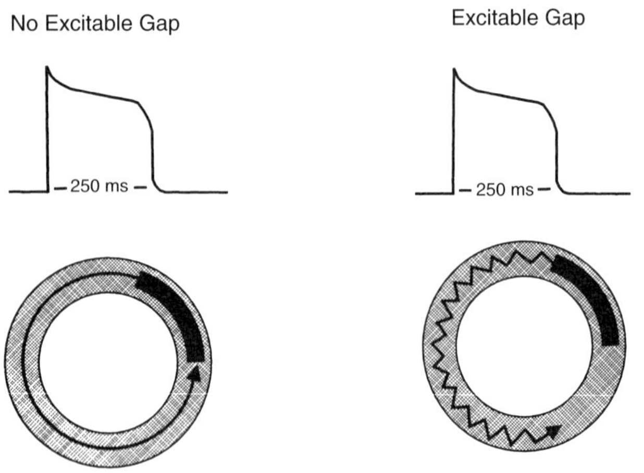

FIGURE 24-3 Mechanism by which class IC antiarrhythmic drugs promote proarrhythmia. The **upper panels** show action potentials with a relative refractory period of 250 milliseconds (ms). The **lower left panel** shows a fixed anatomic obstacle with a path length of 4 cm. The reentrant wavefront circulates around this path. This tissue must recover excitability before the next wavefront arrives in order to permit reentry. If an excitable gap does not exist, reentry cannot occur. If the conduction velocity of the wavefront is 40 cm/sec and if the refractory period of the tissue is 250 ms, then the time to circuit the obstacle is 100 ms, no excitable gap exists, and reentry is not possible. If, however, as shown in the **right lower panel,** a drug significantly reduces conduction velocity in this region to 10 cm/sec without altering the refractory period, then the transit time around the obstacle is now 400 ms, an excitable gap of 150 ms now exits, and reentry is possible.

in sinus rate that results in exaggerated sodium channel block.

Treatment. The antiarrhythmic drug should be discontinued. Electrolyte abnormalities or acidosis should be corrected.[44] Hypertonic saline or sodium bicarbonate may reverse the conduction slowing and terminate the arrhythmia.[44-47] However, sodium may exacerbate heart failure, so caution is required in patients with significant left ventricular dysfunction. Rate slowing using β-blockers may be beneficial, since the magnitude of conduction slowing is less at slower heart rates.[41] Sodium channel blocking drugs should generally be avoided, although theoretically more rapidly inactivating sodium channel blockers may compete with more slowly inactivating sodium channel blockers, thereby reducing the magnitude of sodium channel block.[48]

ATRIAL FLUTTER WITH 1:1 ATRIOVENTRICULAR CONDUCTION

In patients with atrial flutter, sodium channel blocking drugs may significantly slow the atrial flutter cycle length such that 1:1 atrioventricular (AV) conduction develops and the ventricular rate increases (Fig. 24-4).[49-52] This phenomenon has been reported with quinidine and has been attributed to quinidine's vagolytic effects. However, slowing of the atrial flutter cycle length also occurs in association with flecainide and propafenone, particularly when rate controlling antiarrhythmic drugs are not prescribed. Typically, during atrial flutter, the atrial rate is approximately 300 beats per minute (bpm). In most individuals not on AV node blocking drugs, 2:1 AV conduction is present, and the ventricular response is around 150 bpm. Flecainide or propafenone may slow the atrial flutter rate to around 200 to 220 bpm. This may allow the ventricular response to increase from 150 bpm to 200 to 220 bpm with 1:1 AV conduction. An intraventricular conduction delay pattern is often observed due to the marked rate-dependent conduction block associated with these drugs at higher heart rates. In fact, this arrhythmia may be misdiagnosed as VT. This form of proarrhythmia may also occur in patients receiving class I antiarrhythmic drugs for the treatment of atrial fibrillation. Such patients may have intermittent atrial flutter, or the antiarrhythmic drug may alter the electrophysiological substrate in the atria to predispose to atrial flutter.

The diagnosis of atrial flutter with 1:1 AV conduction should be considered in patients with a history of atrial fibrillation/flutter who are treated with class IC drugs. Carotid sinus massage or intravenous adenosine may unmask atrial flutter and establish the diagnosis. Intravenous β-blockers or calcium channel blockers (verapamil or diltiazem) should be administered to control the ventricular rate. Sodium channel blocking drugs should be administered in conjunction with an AV node blocking drug (e.g., a β-blocker or calcium channel blocker) in patients with atrial flutter/fibrillation to prevent the development of this form of proarrhythmia.

Proarrhythmia Secondary to Prolongation of the Action Potential Duration

TORSADES DE POINTES VENTRICULAR TACHYCARDIA

A polymorphic, usually nonsustained, pause-dependent VT may develop in association with drugs or pathophysiologic conditions that excessively prolong the Q–T interval (Fig. 24-5).[3-6,53,54] Although syncope occurring early after the initiation of quinidine therapy was recognized as early as the 1920s, it was not until the advent of continuous electrocardiographic monitoring that "quinidine syncope" was recognized to be due to polymorphic VT.[55] The term *torsades de pointes* was initially used by Dessertenne in 1966 to describe this VT.[56] Torsades de pointes VT may be associated with syncope or sometimes may degenerate into sustained VT or VF, causing sudden cardiac death.

The conditions associated with excessive Q–T interval prolongation and torsades de pointes VT are summarized in Table 24-2. Antiarrhythmic drugs that prolong the ventricular action potential duration (APD), including the class IA drugs quinidine, procainamide, disopyramide; the class IC drug propafenone; and the class III drugs amiodarone, sotalol, dofetilide, and ibutilide have all been reported to cause torsades de pointes VT.[1-3,6,53-71] Associated bradycardia and

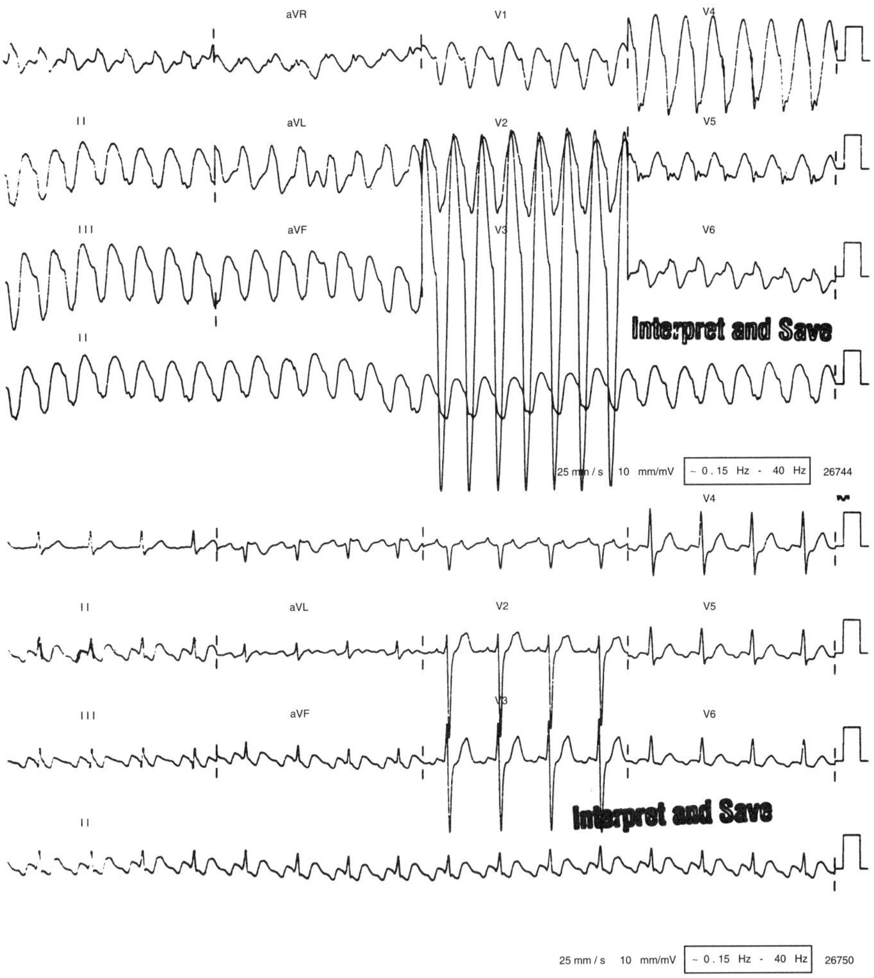

FIGURE 24-4 **Upper panel,** Atrial flutter with 1:1 atrioventricular (AV) conduction during propafenone therapy in a patient without structural heart disease. Propafenone was prescribed for prevention of paroxysmal atrial fibrillation in the absence of AV nodal blocking drugs. Propafenone slowed the atrial flutter cycle length to 320 milliseconds, which results in 1:1 AV conduction with an intraventricular conduction delay abnormality. **Lower panel,** After administration of AV node blocking drugs, flutter with 2:1 AV conduction is apparent, the ventricular rate is now 96 beats per minute, and the intraventricular conduction delay has resolved.

electrolyte abnormalities (hypokalemia or hypomagnesemia, or both) secondary to diuretic use increase the probability of torsades de pointes VT in the setting of class IA or III antiarrhythmic drug use.[72-74] Although amiodarone prolongs the Q–T interval, the incidence of torsades de pointes VT associated with amiodarone use is relatively low.[1,58,69] Diuretics, by virtue of causing profound hypokalemia/hypomagnesemia, may be associated with torsades de pointes VT in the absence of other drug use. The diuretic indapamide, which blocks the slowly activating component of the delayed rectifying current (I_{ks}), may cause excessive Q–T interval prolongation and torsades de pointes VT.[75]

Noncardiovascular drugs associated with torsades de pointes VT include tricyclic antidepressants (imipramine), antipsychotic drugs (haloperidol and phenothiazine),[3,4,53,54,76-79] antihistamines (diphenhydramine, terfenadine, and astemizole),[80-87] macrolide antibiotics (erythromycin, clarithromycin),[88-95] pentamadine,[96,97] antifungal agents,[98] and cisapride.[99-105] Cisapride was recently removed from the market because of the high incidence of significant arrhythmic events. The risk of torsades de pointes VT due to terfenadine or cisapride is due to the accumulation of these drugs in plasma as a consequence of coadministration of a drug (e.g., erythromycin or ketoconazole) that

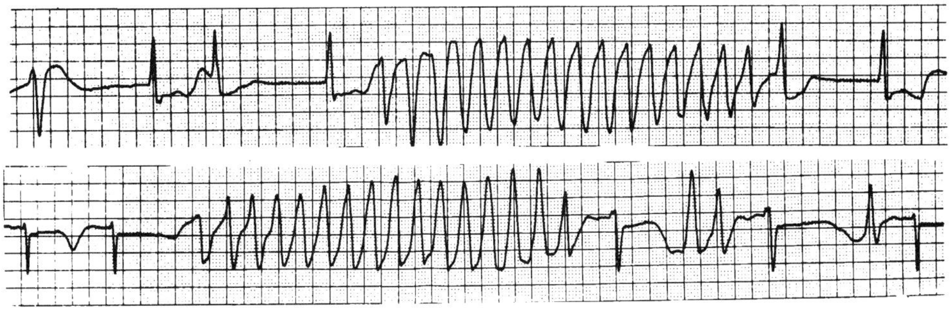

FIGURE 24-5 Torsades de pointes ventricular tachycardia. Note the bradycardia, Q–T interval prolongation, and the late coupled ventricular premature beats, which cause greater prolongation of the Q–T interval and result in repeated runs of spontaneously terminating polymorphic ventricular tachyarrhythmia.

inhibits their biotransformation to noncardioactive metabolites.[53]

Isolated profound bradycardia (e.g., in the setting of complete heart block) may be associated with profound Q–T interval prolongation and torsades de pointes VT, particularly in the setting of drugs or other factors that prolong the Q–T interval. As well, neurologic events such as subarachnoid hemorrhage have been associated with marked repolarization abnormalities, Q–T interval prolongation, and torsades de pointes VT.[106] Adrenergic stimuli in the setting of dobutamine infusion has also been reported to cause torsades de pointes VT.[107-110] Case reports have also implicated anesthetic agents including halothane as a cause for torsades de pointes VT.[111]

Risk Factors for Torsades de Pointes Ventricular Tachycardia. The risk factors for torsades de pointes VT are summarized in Table 24-3.[1,4,68,112,113] Female gender[114-117]; Q–T interval prolongation at baseline; excessive Q–T interval prolongation following initiation of drug therapy; bradycardia; hypokalemia or hypomagnesemia, or both; the presence of left ventricular hypertrophy or left ventricular dysfunction, or both; a prior history of VT; and renal impairment are all associated with an increased risk of developing torsades de pointes VT.[1,3,53,54,118] The presence of multiple risk factors (e.g., female, cardiac hypertrophy, electrolyte abnormalities, and a prior history of VT) dramatically increase the risk of developing torsades de pointes VT. Houltz et al. reported that diuretic use in the setting of normal serum potassium concentrations was a risk factor for torsades de pointes VT.[68] This would be consistent with total body potassium depletion that may not be reflected by serum potassium measures or the direct action potential prolonging effects of some diuretic drugs. These investigators also reported that sequential bilateral bundle branch block, the presence of T wave alternans, and the development of T wave morphologic changes (particularly biphasic T waves in lead V2) were also risk factors for torsades de pointes VT. Interestingly, torsades de pointes VT may frequently occur following conversion of atrial fibrillation to sinus rhythm.[55,119,120]

Women have a twofold to threefold greater risk of developing torsades de pointes VT during treatment with action potential prolonging drugs independent of other risk factors.[114-118] The mechanism of this increased risk is not completely understood. In the general population women have longer Q–T intervals compared to men.[121] Thus, it appears that the process of cardiac repolarization is intrinsically slower in women. Whether this is exaggerated in disease states is currently unknown.

TABLE 24-3 Risk Factors for Torsades de Pointes Ventricular Tachycardia

Female
Baseline Q–T interval prolongation
Excessive QT prolongation on drug
Increased Q–T interval dispersion on drug
Bradycardia/irregular heart rate
Hypokalemia
Hypomagnesemia
Congestive heart failure/left ventricular dysfunction
Ventricular hypertrophy
History of VT/VF
Renal impairment

VF, ventricular fibrillation; VT, ventricular tachycardia.

The risk of torsades de pointes VT is also generally related to drug dose.[53,118] Torsades de pointes VT has been described as an idiosyncratic reaction during quinidine therapy, with syncope secondary to torsades de pointes occurring within hours of initiation of this drug therapy. The development of torsades de pointes VT at low concentrations may reflect quinidine's multiple electrophysiological effects, with predominant I_{Kr} blocking effects occurring at low concentrations and more dominant sodium channel blocking effects occurring at higher drug concentrations.[53] However, the risk of torsades de pointes VT associated with sotalol therapy clearly increases significantly with dose,[118] and this is further exacerbated by other risk factors, including renal failure. The incidence of ventricular proarrhythmia during sotalol initiation at doses up to 320 mg/day was 1.8%; this incidence increased to 4.5% at doses up to 480 mg/day and to 6.8% at doses greater than 640 mg/day. The risk of torsades de pointes VT on other antiarrhythmic drugs including ibutilide and dofetilide is also related to dose.[122]

Some cardiac disease processes such as ventricular hypertrophy or significant left ventricular dysfunction are associated with significant prolongation of the APD.[35,36] These electrophysiological changes are due to changes in repolarizing current densities including decreases in I_{to}, I_{K1}, or I_{Kr}.[36,123] Changes in the densities or balance, or both, of these repolarizing currents may alter the response of specific channel blockers, resulting in exaggerated prolongation of the APD and increased dispersion of ventricular repolarization that provides the substrate for torsades de pointes VT.[124-127]

Mechanisms of Torsades de Pointes Ventricular Tachycardia. A reduction of outward currents or an increase in inward currents, or both, during phase 2 and 3 of the ventricular action potential leads to prolongation of the action potential, which manifests as Q–T interval prolongation on the ECG. The prolongation of the APD occurs in a spatially heterogenous manner due to differences in current densities in specific cell types (Purkinje fibers and endocardial, midmyocardial, or epicardial cells).[128,129] This dispersion of repolarization is further exaggerated in disease states including ventricular hypertrophy.[127] The reduction in net outward currents or increase in inward currents, or both, facilitates the development of early afterdepolarizations that develop in M cells or Purkinje cells due to reactivation of the inward calcium current or activation of the sodium-calcium exchange current.[53,54]

Drugs or conditions that prolong the APD can facilitate intracellular calcium overload, leading to early afterdepolarizations (Fig. 24-6).[128-134] Early afterdepolarizations are more likely to occur in the setting of bradycardia, since the action potential is more prolonged.[130,133]

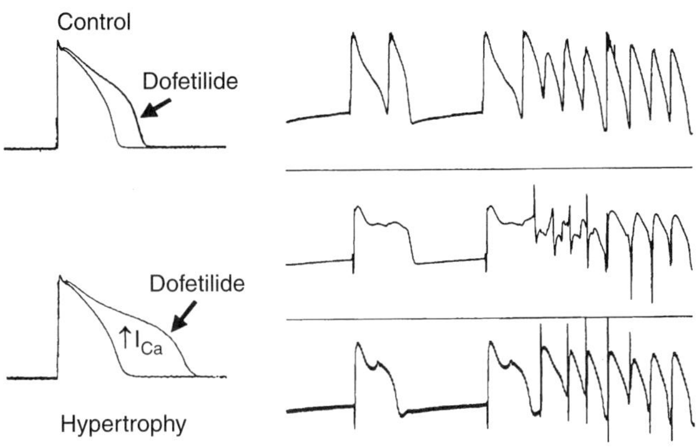

FIGURE 24-6 Mechanism of torsades de pointes ventricular tachyarrhythmia. The action potential duration is prolonged, causing augmentation of the inward calcium current. Intracellular calcium overload initiates early afterdepolarizations, which may generate a premature ventricular beat followed by repetitive premature ventricular beats. Regions of slow conduction or exaggerated dispersion of repolarization may permit ventricular reentry, resulting in a sustained ventricular reentrant tachyarrhythmia.

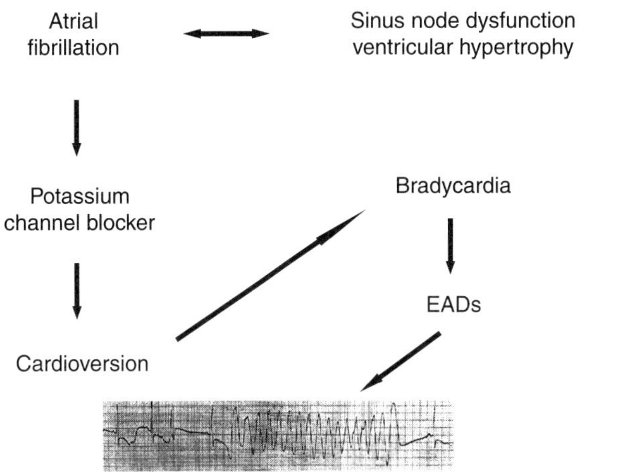

FIGURE 24-7 Conversion from atrial fibrillation to sinus rhythm increases the risk of ventricular proarrhythmia. Resumption of sinus rhythm is often associated with profound bradycardia, which, in the setting of left ventricular hypertrophy or left ventricular dysfunction, is associated with action potential prolongation. Early afterdepolarizations triggering torsades de pointes ventricular tachyarrhythmia are more likely to occur in this setting.

In the setting of marked dispersion of repolarization, these early afterdepolarizations may trigger reentry and initiate torsades de pointes VT.[53,54,128] Marked spatial heterogeneity of the ventricular APD appears to be a prerequisite for torsades de pointes VT.[125-131] Spatial heterogeneity of ventricular repolarization provides the functional substrate for reentry. Functional block may vary on a beat-to-beat basis, contributing to spiral reentrant waves, which explains the polymorphic nature of the arrhythmia.[135]

Multiple ionic mechanisms likely contribute to torsades de pointes VT. Torsades de pointes VT usually occurs with drugs that block I_{Kr}. However, any drug that reduces net repolarizing currents may create a risk for this arrhythmia. Block of multiple repolarizing currents may cause excessive prolongation of ventricular repolarization, predisposing to VT development.[75,127] Studies in patients with congenital long QT syndrome have identified mutations of repolarizing potassium currents or the sodium channel contributing to abnormalities of ventricular repolarization.[136] Indeed, some individuals who develop torsades de pointes may have unrecognized forms of the congenital long QT syndrome.[53] Whether polymorphisms in the genes encoding potassium repolarizing currents confer increased risk for acquired torsades de pointes VT is unknown at present.

Most drugs that cause torsades de pointes VT also cause prolongation of the APD by blocking potassium channels (either the rapidly activating component of the delayed rectifying current I_{Kr} or the slowly activating component of the delayed rectifying current I_{Ks}). However, potassium channel block alone is likely insufficient for the development of torsades de pointes VT. Action potential prolongation secondary to potassium channel block is thought to allow reactivation of calcium channels, allowing inward current to flow and generate the early afterdepolarizations and triggered beats.[6,134,137]

The period immediately following cardioversion to sinus rhythm from atrial fibrillation appears to be a time of great risk for torsades de pointes VT.[119,120] The potential mechanism underlying this event is shown in Figure 24-7. The sustained tachycardia and neurohumoral changes associated with atrial fibrillation may alter some electrophysiological properties including downregulation of repolarizing currents. Restoration of sinus rhythm is often associated with bradycardia and action potential prolongation that could be further exaggerated by downregulation of potassium channels. The action potential prolongation facilitates increased sarcoplasmic calcium, predisposing to triggered activity.

A similar situation may occur after AV node ablation for the management of persistent atrial fibrillation. The early period after ablation is a particularly vulnerable period for sudden cardiac death, which is believed to be due to bradycardia-dependent polymorphic VT.[138,139] Following successful ablation is an abrupt slowing of the heart rate. Sudden slowing of the heart rate and the associated prolongation of the APD following ablation could increase the likelihood of early afterdepolarizations. Consistent with this hypothesis we have observed exaggerated bradycardia-dependent prolongation of the APD in patients with significant left ventricular dysfunction following total AV junction ablation.[140,141]

Electrocardiographic Harbingers of Torsades de Pointes Ventricular Tachycardia. Excessive Q–T interval prolongation and morphologic changes in the T wave usually precede the development of torsades de pointes VT.[68,143.144] Figure 24-8 illustrates the changes in the Q–T interval and T wave morphology during sotalol initiation that precede the development of torsades de pointes VT. Note the development of Q–T interval prolongation and a U wave, which become progressively more prominent over the 48 to 72 hours following

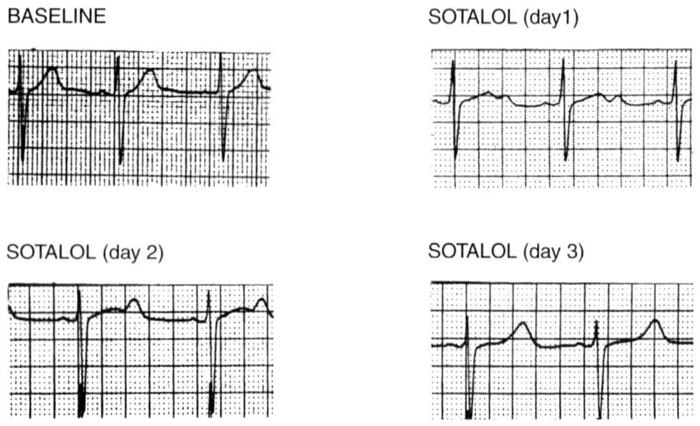

FIGURE 24-8 Serial changes in the Q–T interval following sotalol initiation. Note that the Q–T interval prolongs early following initiation of sotalol **(right upper panel)**. This is followed by the development of more marked U waves, bradycardia, and further Q–T interval prolongation.

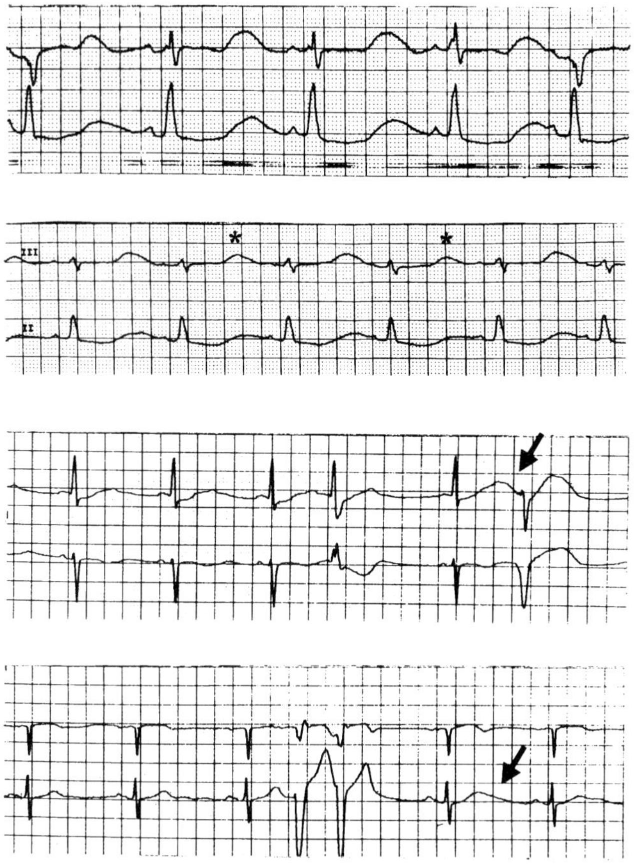

FIGURE 24-9 Electrocardiographic harbingers of torsades de pointes ventricular tachyarrhythmia. Note the marked Q–T interval prolongation, the prolongation of the Q–T interval following a postextrasystolic pause, the T wave alternans in the setting of Q–T interval prolongation, and late coupled premature ventricular beats.

initiation of sotalol. The magnitude of Q–T interval prolongation may be exaggerated by abrupt slowing of the heart rate (e.g., following a compensatory pause). The prominence of the U waves may vary on a beat-to-beat basis and may be modified by abrupt changes in heart rate. Figure 24-9 illustrates and Table 24-4 summarizes some of the electrocardiographic harbingers of torsades de pointes VT. Note the excessive QT prolongation, the development of prominent U waves, T wave alternans,[144,145] and post–extrasystolic pause exaggeration of the QT prolongation, as well as T wave morphologic changes and clustering of repetitive polymorphic ventricular premature beats.

Dispersion of Ventricular Repolarization and Torsades de Pointes Ventricular Tachycardia. Increased Q–T interval dispersion, defined as the difference between the shortest and longest Q–T intervals measured on the 12-lead ECG, may precede the development of torsades de pointes VT. We have shown that class IA antiarrhythmic drugs significantly increased Q–T interval dispersion compared to the drug free state in patients who developed torsades de pointes VT during drug therapy (Fig. 24-10).[58] In contrast, patients who did not develop this proarrhythmia during drug therapy did not manifest a significant increase in Q–T interval dispersion compared to the drug-free state. Amiodarone did not increase Q–T interval dispersion in the patients who developed torsades de pointes VT on class IA drugs.

The low incidence of torsades de pointes VT observed during amiodarone therapy likely reflects its homogenous effects on ventricular repolarization, which may be due in part to the calcium channel and β-adrenergic blocking effects of amiodarone.[58,146,147] Drouin et al. recently evaluated the effects of amiodarone on APD in ventricular myocytes isolated from explanted hearts of transplant recipients, compared to those from control hearts, as well as cells from hearts explanted from patients with heart failure who were not on amiodarone.[146] Amiodarone therapy was associated with less bradycardia-induced prolongation of the action potential of the midmyocardial layer cells (M cells). This resulted in a reduced transmural dispersion of the APD compared to myocytes from patients with heart failure who were not treated with amiodarone and those from normal control hearts (Fig. 24-11). Similar findings have been reported in experimental models.[147,148] Thus, amiodarone appears to produce less heterogeneity of ventricular repolarization, which likely explains the low incidence of torsades de pointes VT occurring in association with amiodarone compared to class I or other class III antiarrhythmic drugs. In keeping with this observation, amiodarone produces a greater decrease in Q–T interval dispersion in patients following MI

TABLE 24-4 Electrocardiographic Harbingers of Torsades de Pointes Ventricular Tachycardia

QTU—prolongation
Increased QT dispersion >60 ms
T wave alternans
Postextrasystolic pause TU changes
Late-coupled polymorphic VPBs
Repetitive polymorphic beats

ms, milliseconds; QTU, QT-U wave interval; TU, T wave-U wave complex; VPB, ventricular premature beats; VT, ventricular tachycardia.

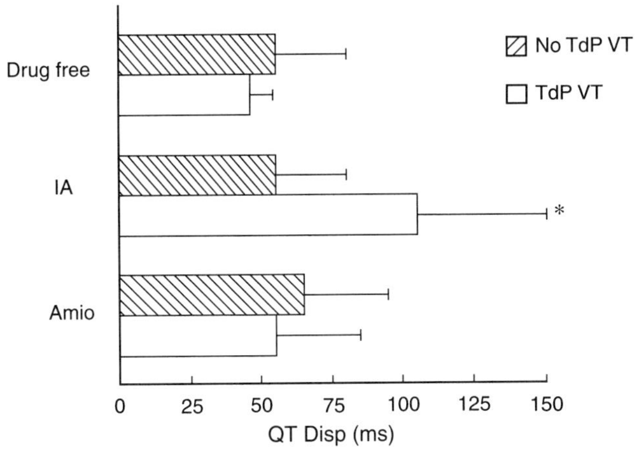

FIGURE 24-10 Precordial Q–T interval dispersion is increased in patients who developed torsades de pointes ventricular tachyarrhythmia during class I antiarrhythmic drug therapy. (From Hii JT, Wyse DG, Gillis AM, et al., Circulation 1993;87:764-772.)

compared to that produced by other class III drugs (sotalol or sematilide).[149,150] We have observed that patients who develop torsades de pointes VT on amiodarone therapy are likely to manifest marked increases in Q–T interval dispersion.[151]

Treatment. The initial treatment for torsades de pointes VT is magnesium sulfate (1 to 2 g IV bolus).[44,152-155] The drug implicated should be discontinued, and other reversible causes should be corrected (e.g., electrolyte abnormalities, bradycardia). Overdrive atrial or ventricular pacing is effective when the arrhythmia recurs despite magnesium administration.[156,157] Atropine has also been reported to eliminate torsades de pointes VT,[158] presumably via the increase in heart rate reversing the QT prolongation. Serum potassium should be maintained in the high normal range, 4.5 to 5 mEq/L. Elevation of the extracellular potassium concentration increases outward potassium currents and reduces

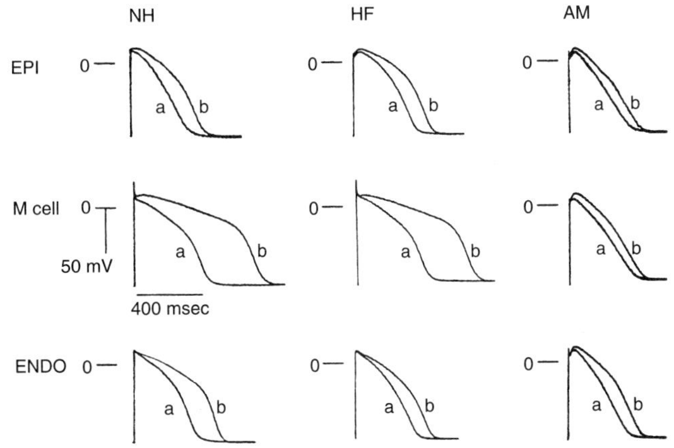

FIGURE 24-11 Effects of amiodarone on heterogeneity of ventricular repolarization. (From Drouin E, Lande G, Charpentier F: Amiodarone reduces transmural heterogeneity of repolarization in the human heart. J Am Coll Cardiol 1998;32:1063-1067.)

the magnitude of drug block of I_{Kr}, thus shortening ventricular repolarization. Modest increases in serum potassium may reverse the repolarization abnormalities associated with some variants of the long QT syndrome or associated with antiarrhythmic drug use and cause Q–T interval shortening as well as reduced Q–T interval dispersion.[159,160] The mechanism by which magnesium prevents torsades de pointes VT is uncertain. Magnesium does not shorten the APD or the Q–T interval and likely exerts its effect by blocking calcium channels.[153-155] Sodium channel blockers have been reported in experimental models to antagonize the effects of class IA and III drugs on APD and to prevent proarrhythmia.[161-164] Whether such interventions have an important clinical role requires further clinical study. Potassium channel activating compounds may also prevent early afterdepolarization and proarrhythmia in animal models.[165] One case report suggested nicorandil, a potassium channel opener, was an effective treatment for torsades de pointes VT.[166]

Prevention. Torsades de pointes VT may be prevented by: (1) avoiding the use of class I/III antiarrhythmic drugs in patients with significant left ventricular dysfunction, (2) reducing the dose or stopping the drug if significant Q–T interval prolongation (>500 milliseconds [ms]) or increased Q–T interval dispersion (>80 ms) is observed during drug therapy, (3) correcting electrolyte abnormalities (hypokalemia/hypomagnesemia) before antiarrhythmic drug initiation, (4) avoiding potassium-wasting diuretics or using potassium-sparing diuretics (amiloride/spironolactone), (5) considering prophylactic magnesium supplements in high-risk patients[167]; (6) preventing significant bradycardia; (7) considering permanent pacing in patients with significant bradycardia in whom class I/III antiarrhythmic drugs are considered important therapies; and (8) monitoring high-risk patients in the hospital during initiation of antiarrhythmic drug therapy.[118,168,169]

Should Antiarrhythmic Drug Therapy Be Initiated in the Hospital? There is controversy over whether antiarrhythmic drug therapy can be initiated safely in the outpatient setting. This issue has been carefully reviewed by Nattel and Thibault.[118] About 50% of episodes of ventricular proarrhythmia occur within 72 hours of initiation of antiarrhythmic drug therapy for atrial fibrillation. They estimated that 1200 patients would have to be monitored for 3 days to prevent one torsades de pointes VT–related death. The Symptomatic Atrial Fibrillation Investigative Research on Dofetilide (SAFIRE-D) investigators reported a low incidence (0.8%) of torsades de pointes VT during the initial 3 days of dofetilide therapy in patients with atrial fibrillation.[170] Nevertheless, the U.S. Food and Drug Administration–approved guidelines for dofetilide initiation require that this drug be initiated in the hospital. Similar recommendations can be found in the product monograph for sotalol approved for the treatment of atrial fibrillation. ECG monitoring during drug initiation should be undertaken for high-risk patients but does not appear to be cost-effective for low-risk patients. Moreover, efforts at prevention of proarrhythmia must be continued throughout the course of antiarrhythmic drug therapy.

INCREASED CARDIAC MORTALITY ASSOCIATED WITH CLASS III DRUGS

The Survival with Oral d-Sotalol (SWORD) investigators reported increased cardiovascular mortality with d-sotalol compared to placebo in patients with left ventricular dysfunction.[171,172] Similar to the CAST study, excess mortality on active drug continued through the course of follow-up. The mechanism or mechanisms of the increased mortality are uncertain. Torsades de pointes VT is certainly one possibility. However, VF in the setting of acute ischemia precipitated by ischemia-mediated exaggeration of antiarrhythmic drug effects (similar to the mechanism proposed in CAST) is also another plausible cause. In contrast, the Danish Investigation of Arrhythmia and Mortality on Dofetilide (DIAMOND) investigators did not observe an increased mortality associated with dofetilide compared to placebo in patients with left ventricular dysfunction.[173] Differences in trial design likely explain these apparently divergent outcomes. In DIAMOND, drug therapy was initiated in the hospital during ECG monitoring, and therapy was stopped if excessive Q–T interval prolongation was observed. In contrast, the SWORD trial design favored drug doses associated with Q–T interval prolongation. Furthermore, SWORD patients were generally healthy and hence less likely to benefit substantially from drug therapy (see Fig. 24-1).

Device—Drug Interactions

Proarrhythmia in Implantable Cardioverter Defibrillators. Antitachycardia pacing therapies for the termination of sustained VT are not always effective and may at times accelerate VT to VF or to a more hemodynamically unstable form of VT.[174-176] The risk of acceleration of VT is relatively small (4%) and is similar for ramp and burst pacing modalities. Syncope in the setting of an implantable defibrillator may be due to ventricular proarrhythmia provoked by antitachycardia pacing therapies and may be avoided by careful programming.[177,178] Sometimes concomitant antiarrhythmic drug therapy by slowing the VT rate may increase the efficacy of antitachycardia pacing therapies and minimize the risk of antitachycardia pacing–induced proarrhythmia. The risk of ventricular arrhythmia associated with implantable atrial defibrillators has been reported to be extremely low when the cardioversion shock is synchronized to the R wave and the shock is delivered at a slow heart rate.[179-182] False-positive detection of atrial fibrillation due to far-field R wave oversensing can lead to inappropriate therapies, which might infrequently initiate atrial fibrillation.[183]

Antiarrhythmic Drug Effects on Defibrillation Thresholds. Some antiarrhythmic drugs, by virtue of their effects on passive and active membrane properties, may alter defibrillation thresholds. Sodium channel blocking drugs have been reported to have no effect on or to increase defibrillation thresholds. These divergent effects appear in part to be dependent on the experimental model or the type of anesthesia used.[184-187] In patients with left ventricular dysfunction, lidocaine has been reported to increase defibrillation thresholds.[188] Drugs that prolong the APD and exert predominantly class III antiarrhythmic drug effects (e.g., procainamide, d-sotalol, azimilide) have been reported to have either no significant or reduced defibrillation energy requirements.[188-191] The effects of amiodarone on defibrillation thresholds are controversial. Some studies have reported no change in defibrillation thresholds during chronic amiodarone therapy, whereas other investigators have reported a significant increase in defibrillation thresholds.[192-194] Patients with marginal defibrillation thresholds should have defibrillation thresholds reassessed if an antiarrhythmic drug known to increase defibrillation threshold is initiated.

Antiarrhythmic Drug Effects and Cardiac Pacemakers. Some antiarrhythmic drugs (e.g., β-blocking agents, calcium channel antagonists) may cause suppression of sinus node automaticity or advanced AV node conduction abnormalities, resulting in profound sinus bradycardia, sinus arrest, or high-grade AV block requiring a permanent pacemaker insertion. These effects may be exaggerated by the concomitant use of class I or III antiarrhythmic drugs prescribed for the management of atrial fibrillation. Some anecdotal reports have suggested that antiarrhythmic drugs may cause an increase in pacing thresholds, particularly in the setting of ischemia, acidosis, or hypoxia.[196,197] Some newer pacemakers have automatic pacing threshold measurement capabilities and the ability for measuring pacing thresholds to maintain an adequate pacing safety margin. In our experience with modern pacing leads, antiarrhythmic drugs do not cause substantial changes in pacing thresholds over the 24-hour period.[198]

Heavy Metals, Toxins, and Proarrhythmia

Organophosphates have been reported to prolong the Q–T interval and cause torsades de pointes VT.[198] Cesium chloride has been used in experimental models to induce polymorphic VT.[199] Cesium chloride has been used in some dietary supplements and alternative therapies for malignancies.[200,201] Polymorphic VT has been reported in a patient with hypokalemia who was taking cesium as a dietary supplement.[200] Barium intoxication has also been reported to cause VT in the setting of hypokalemia.[202] Antimony preparations used in the treatment of parasitic infestations have also been reported to cause torsades de pointes VT under some circumstances.[203,204]

Some chemotherapeutic agents, specifically anthracycline, may cause cardiotoxicity and cardiomyopathy.[205,206] Arrhythmias may arise secondary to the underlying structural heart disease.

Environmental Causes and Proarrhythmia

Environmental disturbances of great magnitude (e.g., earthquake) may trigger an increased incidence of sudden death, presumably secondary to ventricular tachyarrhythmias.[207,208] It has been suggested that such events trigger significant emotional stress, which is associated with marked surges in sympathetic tone that facilitate the development of ventricular tachyarrhythmias.[208-210] Extremes of temperature have also been reported to be associated with an increased incidence of ventricular tachyarrhythmias.[211] It has been postulated that thermal stress is associated with heightened sympathetic tone favoring arrhythmogenesis.

ACKNOWLEDGMENTS

This chapter was supported by the Medical Research Council of Canada Group Grant and the Heart and Stroke Foundation of Alberta. Dr. Gillis is a medical scientist of the Alberta Heritage Foundation for Medical Research.

REFERENCES

1. Hohnloser SH, Singh BN: Proarrhythmia with class III antiarrhythmic drugs: Definition, electrophysiologic mechanisms, incidence, predisposing factors, and clinical implications. J Cardiovasc Electrophysiol 1995;6:920-936.
2. Friedman PL, Stevenson WG: Proarrhythmia. Am J Cardiol 1998;82:50N-58N.
3. Nattel S: Experimental evidence for proarrhythmic mechanisms of antiarrhythmic drugs. Cardiovasc Res 1998;37:567-577.
4. Roden DM: Mechanisms and management of proarrhythmia. Am J Cardiol 1998;82:49I-57I.
5. Skanes AC, Green MS: What have clinical trials taught us about proarrhythmia? Can J Cardiol 1996;12(suppl B):20B-26B.
6. Viskin S: Long QT syndromes and torsade de pointes. Lancet 1999;354:1625-1633.
7. Reiffel JA: Impact of structural heart disease on the selection of class III antiarrhythmics for the prevention of atrial fibrillation and flutter. Am Heart J 1998;135:551-556.
8. The Cardiac Arrhythmia Suppression Trial (CAST) Investigators: Preliminary report: Effect of encainide and flecainide on mortality in a randomized trial of arrhythmia suppression after myocardial infarction. N Engl J Med 1989;321:406-412.
9. Anderson J, Platia E, Hallstrom A, et al: Interaction of baseline characteristics with the hazard of encainide, flecainide and moricizine therapy in patients with myocardial infarction: A possible explanation for increased mortality in the cardiac arrhythmia suppression trial (CAST). Circulation 1994;90:2843-2852.
10. Greenberg HM, Dwyer EM Jr, Hochman JS, et al: Interaction of ischemia and encainide/flecainide treatment: A proposed mechanism for the increased mortality in CAST I. Br Heart J 1995;74:631-635.
11. Kennedy HL, Brooks MM, Barker AH, et al: Beta-blocker therapy in the Cardiac Arrhythmia Suppression Trial. Am J Cardiol 1994;74:674-680.
12. Kennedy HL: Beta-blocker prevention of proarrhythmia and proischemia: Clues from CAST, CAMIAT, and EMIAT. Am J Cardiol 1997;80:1208-1211.
13. The Cardiac Arrhythmia Suppression Trial II (CAST) Investigators: Effect of the antiarrhythmic agent moricizine on survival after myocardial infarction. N Engl J Med 1992;327:227-233.
14. Coplen SE, Antman EM, Berlin JA, et al: Efficacy and safety of quinidine therapy for maintenance of sinus rhythm after cardioversion. A meta-analysis of randomized control trials. Circulation 1990;82:1106-1116.
15. Flaker GC, Blackshear JL, McBride R, et al: Antiarrhythmic drug therapy and cardiac mortality in atrial fibrillation. J Am Coll Cardiol 1992;20:527-532.
16. Nattel S: The molecular and ionic specificity of antiarrhythmic drug actions. J Cardiovasc Electrophysiol 1999;10:272-282.
17. Morganroth J, Horowitz LN: Flecainide: Its proarrhythmic effect and expected changes on the surface electrocardiogram. Am J Cardiol 1984;53:89B-94B.
18. Morganroth J: Risk factors for the development of proarrhythmic events. Am J Cardiol 1987;59:32E-37E.
19. Brugada J, Boersma L, Kirchhof C, et al: Proarrhythmic effects of flecainide. Experimental evidence for increased susceptibility to reentrant arrhythmias. Circulation 1991;84:1808-1818.
20. Ranger S, Nattel S: Determinants and mechanisms of flecainide-induced promotion of ventricular tachycardia in anesthetized dogs. Circulation 1995;92:1300-1311.
21. Coromilas J, Saltman AE, Waldecker B, et al: Electrophysiological effects of flecainide on anisotropic conduction and reentry in infarcted canine hearts. Circulation 1995;91:2245-2263.
22. Duff HJ, Stemler M, Thannhauser T, et al: Proarrhythmia of a class IC drug: Suppression by combination with a drug prolonging repolarization in the dog late after infarction. J Pharmacol Exp Ther 1995;274:508-515.
23. Kidwell GA, Gonzalez MD: Effects of flecainide and d-sotalol on myocardial conduction and refractoriness: Relation to antiarrhythmic and proarrhythmic drug effects. J Cardiovasc Pharmacol 1993;21:621-632.
24. Wallace AA, Stupienski RF III, Kothstein T, et al: Demonstration or proarrhythmic activity with the class IC antiarrhythmic agent encainide in a canine model of previous myocardial infarction. J Cardiovasc Pharmacol 1993;21:397-404.
25. Restivo M, Yin H, Caref EB, et al: Reentrant arrhythmias in the subacute infarction period. The proarrhythmic effect of flecainide acetate on functional reentrant circuits. Circulation 1995;91:1236-1246.
26. Carson DL, Cardinal R, Savard P, et al: Relationship between an arrhythmogenic action of lidocaine and its effects on excitation patterns in acutely ischemic porcine myocardium. J Cardiovasc Pharmacol 1996;8:126-136.
27. Aupetit JF, Timour Q, Loufoua-Moundanga J, et al: Profibrillatory effects of lidocaine in the acutely ischemic porcine heart. J Cardiovasc Pharmacol 1995;25:810-816.
28. Kou WH, Nelson SD, Lynch JJ, et al: Effect of flecainide acetate on prevention of electrical induction of ventricular tachycardia and occurrence of ischemic ventricular fibrillation during the early postmyocardial infarction period: Evaluation in a conscious canine model of sudden death. J Am Coll Cardiol 1987;9:359-365.
29. Lederman SN, Wenger TL, Bolster DE, et al: Effects of flecainide on occlusion and reperfusion arrhythmias in dogs. J Cardiovasc Pharmacol 1989;13:541-546.
30. Elharrar V, Gaum WE, Zipes DP: Effect of drugs on conduction delay and incidence of ventricular arrhythmias induced by acute coronary occlusion in dogs. Am J Cardiol 1977;39:544-549.
31. Nattel S, Pedersen DH, Zipes DP: Alterations in regional myocardial distribution and arrhythmogenic effects of aprindine produced by coronary artery occlusion in the dog. Cardiovasc Res 1981;15:80-85.
32. Myerberg RJ, Kessler KM, Cox MM, et al: Reversal of proarrhythmic effects of flecainide acetate and encainide hydrochloride by propranolol. Circulation 1989;80:1571-1579.
33. Packer DL, Munger TM, Johnson SB, et al: Mechanism of lethal proarrhythmia observed in the cardiac arrhythmia suppression trial: Role of adrenergic modulation of drug binding. PACE 1997;20[part II]:455-467.
34. Cragun KT, Johnson SB, Packer DL: β-Adrenergic augmentation of flecainide-induced conduction slowing in canine Purkinje fibers. Circulation 1997;96:2701-2708.
35. Boyden PA, Jeck CD: Ion channel function in disease. Cardiovasc Res 1995;73:379-385.
36. Tomaselli GF, Marban E: Electrophysiological remodeling in hypertrophy and heart failure. Cardiovasc Res 1999;42:270-278.
37. Task Force of the Working Group on Arrhythmias of the European Society of Cardiology: The Sicilian Gambit. A new approach to the classification of antiarrhythmic drugs based on

their actions on arrhythmogenic mechanisms. Circulation 1991;84:1831-1851.

38. Yuan F, Pinto JM, Li Q, et al: Characteristics of I(K) and its response to quinidine in experimental healed myocardial infarction. J Cardiovasc Electrophysiol 1999;10:844-854.
39. Gillis AM, Lester WM, Keashly R: Propafenone disposition and pharmacodynamics in normal and cardiomyopathic hearts. J Pharmacol Exp Ther 1991;258:722-727.
40. Winkle RA, Mason JW, Griffin JC, et al: Malignant ventricular tachyarrhythmias associated with the use of encainide. Am Heart J 1981;102:857-863.
41. Ranger S, Talijic M, Lemery R, et al: Amplification of flecainide-induced ventricular conduction slowing by exercise. Circulation 1989;80:1571-1579.
42. Oetgen WJ, Tibbits PA, Abt MEO, et al: Clinical and electrophysiologic assessment of oral flecainide acetate for recurrent ventricular tachycardia: Evidence for exacerbation of electrical instability. Am J Cardiol 1983;52:746-750.
43. Nattel S, Keable H, Sasniuk BI: Experimental amitriptyline intoxication: Electrophysiologic manifestations and management. J Cardiovasc Pharmacol 1983;6:83-89.
44. Gillis AM: Intractable ventricular tachycardia: Investigations and treatment. Card Electrophysiol Rev 1999;3:145-148.
45. Bellet S, Hamden G, Somlyo A, et al: A reversal of the cardiotoxic effects of procainamide by molar sodium lactate. Am J Med Sci 1959;237:177-189.
46. Bajaj AK, Woosley RL, Roden DM: Acute electrophysiologic effects of sodium administration in dogs treated with o-desmethyl encainide. Circulation 1989;80:994-1002.
47. Winklemann BR, Leinberger H: Life threatening flecainide toxicity: A pharmacodynamic approach. Ann Intern Med 1987;106:807-814.
48. Clarkson CW, Hondeghem LM: Evidence for a specific receptor site for lidocaine, quinidine, and bupivacaine associated with cardiac sodium channels in guinea pig ventricular myocardium. Circ Res 1985;56:496-506.
49. Crijns HJ, van Gelder IC, Lie KI: Supraventricular tachycardia mimicking ventricular tachycardia during flecainide treatment. Am J Cardiol 1988;62:1303-1306.
50. Feld GK, Chen PS, Nicod P, et al: Possible atrial proarrhythmic effects of class IC antiarrhythmic drugs. Am J Cardiol 1990; 66:378-383.
51. Murdock CJ, Kyles A, Yeung-Lai-Wah JA: Atrial flutter in patients treated for atrial fibrillation with propafenone. Am J Cardiol 1990;66:755-757.
52. Falk RH: Proarrhythmia in patients treated for atrial fibrillation or flutter. Ann Intern Med 1992;117:141-150.
53. Roden DM: Acquired long QT syndromes and the risk of proarrhythmia. J Cardiovasc Electrophysiol 2000;11:938-940.
54. Haverkamp W, Breithardt G, Camm AJ, et al: The potential for QT prolongation and pro-arrhythmia by non-anti-arrhythmic drugs: Clinical and regulatory implications. Report on a Policy Conference of the European Society of Cardiology. Cardiovasc Res 2000;47:219-233.
55. Selzer A, Wray HW: Quinidine syncope—paroxysmal ventricular fibrillation occurring during treatment of chronic atrial arrhythmias. Circulation 1964;30:17-26.
56. Dessertenne F: La tachycardie ventriculaire a deux foyers opposes variables. Arch Mal Coeur Vaiss 1966;59:263-272.
57. Dangman KH, Hoffman BF: In vivo and in vitro antiarrhythmic and arrhythmogenic effects of N-acetyl procainamide. J Pharmacol Exp Ther 1981;217:851-862.
58. Hii JTY, Wyse DG, Gillis AM, et al: Precordial QT interval dispersion as a marker of torsade de pointes: Disparate effects of class la antiarrhythmic drugs and amiodarone. Circulation 1992;86:1376-1382.
59. Hii JTY, Wyse DG, Gillis AM, et al: Propafenone-induced torsade de pointes: Cross reactivity with quinidine. Pacing Clin Electrophysiol 1991;14:1568-1570.
60. Botto GL, Bonini W, Broffoni T, et al: Conversion of recent onset atrial fibrillation with single loading oral dose of propafenone: Is in-hospital admission absolutely necessary? Pacing Clin Electrophysiol 1996;19[part II]:1939-1943.
61. de Paola AAV, Veloso HH for the SOCESP Investigators: Efficacy and safety of sotalol versus quinidine for the maintenance of sinus rhythm after conversation of atrial fibrillation. Am J Cardiol 1999;84:1033-1037.
62. Dancey D, Wulffhart Z, McEwan P: Sotalol-induced torsades de pointes in patients with renal failure. Can J Cardiol 1997;13:55-58.
63. Barbey JT, Sale ME, Woosley RL, et al: Pharmacokinetic, pharmacodynamic, and safety evaluation of an accelerated dose titration regimen of sotalol in healthy middle-aged subjects. Clin Pharmacol Ther 1999;66:91-99.
64. Delacretaz E, Fuhrer J: Fatal torsade de pointes with d, l-sotalol and low potassium. Clin Cardiol 1999;22:423-424.
65. Kuhlkamp V, Mermi J, Mewis C, et al: Efficacy and proarrhythmia with the use of *d,l*-sotalol for sustained ventricular tachyarrhythmias. J Cardiovasc Pharmacol 1997;29:373-381.
66. Pfammatter JP, Paul T, Lehmann C, et al: Efficacy and proarrhythmia of oral sotalol in pediatric patients. J Am Coll Cardiol 1995;26:1002-1007.
67. Norgaard BL, Wachtell K, Christensen PD, et al: Efficacy and safety of intravenously administered dofetilide in acute termination of atrial fibrillation and flutter: A multicenter, randomized, double-blind, placebo-controlled trial. Am Heart J 1999; 137:1062-1069.
68. Houltz B, Darpo B, Swedberg K, et al: Effects of the I_{Kr}-blocker almokalant and predictors of conversion of chronic atrial tachyarrhythmias to sinus rhythm. A prospective study. Cardiovasc Drugs Ther 1999;13:329-338.
69. Middlekauff HR, Stevenson WG, Saxon LA, et al: Amiodarone and torsades de pointes in patients with advanced heart failure. Am J Cardiol 1995;76:499-502.
70. Tomcsanyi J, Merkely B, Tenczer J, et al: Early proarrhythmia during intravenous amiodarone treatment. PACE 1999; 22[part I]:968-970.
71. Chen Y-J, Hsieh M-H, Chiou C-W, et al: Electropharmacologic characteristics of ventricular proarrhythmia induced by ibutilide. T Cardiovasc Pharmacol 1999;34:237-247.
72. Brachmann J, Scherlag BJ, Rosenshtraukh LV, et al: Bradycardia-dependent triggered activity: Relevance to drug-induced multiform ventricular tachycardia. Circulation 1983;68:846-856.
73. Chvilicek JP, Hurlbert BJ, Hill GE: Diuretic-induced hypokalaemia inducing torsades de pointes. Can J Anaesth 1995;42:1137-1139.
74. Eriksson JW, Carlberg B, Hillorn V: Life-threatening ventricular tachycardia due to liquorice-induced hypokalaemia. J Intern Med 1999;245:307-310.
75. Fiset C, Drolet B, Hamelin BA, Turgeon J: Block of IKs by the diuretic agent indapamide modulates cardiac electrophysiological effects of the class III antiarrhythmic drug dl-sotalol. J Pharmacol Exp Ther 1997;283:148-156.
76. Drolet B, Vincent F, Rail J, et al: Thioridazine lengthens repolarization of cardiac ventricular myocytes by blocking the delayed rectifier potassium current. J Pharmacol Exp Ther 1999;288:1261-1268.
77. Jackson T, Ditmanson L, Phibbs B: Torsade de pointes and low-dose oral haloperidol. Arch Intern Med 1997;157:2013-2015.
78. O'Brien JM, Rockwood RP, Suh KI: Haloperidol-induced torsade de pointes. Ann Pharmacother 1999;33:1046-1050.
79. Michalets EL, Smith LK, van Tassel ED: Torsade de pointes resulting from the addition of droperidol to an existing cytochrome P450 drug interaction. Ann Pharmacother 1998;32:761-765.
80. Zreba W, Moss AJ, Rosero SZ, et al: Electrocardiographic findings in patients with diphenhydramine overdose. Am J Cardiol 1997; 80:1168-1173.
81. Vorperian VR, Zhou Z, Mohammad S, et al: Torsade de pointes with an antihistamine metabolite: Potassium channel blockade with desmethylastemizole. J Am Coll Cardiol 1996;28:1556-1561.
82. Khalifa M, Drolet B, Daleau P: Block of potassium currents in guinea pig ventricular myocytes and lengthening of cardiac repolarization in man by the histamine H1 receptor antagonist diphenhydramine. J Pharmacol Exp Ther 1999;288:858-865.
83. Yap YG, Camm AJ: Arrhythmogenic mechanisms of non-sedating antihistamines. Clin Exp Allergy 1999;29(suppl 3):174-181.
84. Feroze H, Suri R, Silverman DI: Torsades de pointes from terfenadine and sotalol given in combination. Pacing Clin Electrophysiol 1996;19:1518-1521.
85. June RA, Nasr I: Torsades de pointes with terfenadine ingestion. Am J Emerg Med 1997;15:542-543.
86. Ikeda S, Oka H, Matunaga K, et al: Astemizole-induced torsades de pointes in a patient with vasospastic angina. Jpn Circ J 1998; 62:225-227.
87. Martin ES, Rogalski K, Black JN: Quinine may trigger torsades de pointes during astemizole therapy. Pacing Clin Electrophysiol 1997;20:2024-2025.

88. Chennareddy SB, Siddique M, Karim MY, et al: Erythromycin-induced polymorphous ventricular tachycardia with normal QT interval. Am Heart J 1996;132:691-694.
89. Antzelevitch C, Zuwo-Qian S, Zi-Qing Z, et al: Cellular and ionic mechanisms underlying erythromycin-induced long QT intervals and torsade de pointes. J Am Coll Cardiol 1996;28:1836-1848.
90. Fazekas T, Krassoi I, Lengyel C, et al: Suppression of erythromycin-induced early afterdepolarizations and torsade de pointes ventricular tachycardia by mexiletine. Pacing Clin Electrophysiol 1998;21:147-150.
91. Drici MD, Ebert SN, Wang WX, et al: Comparison of tegaserod (HTF 919) and its main human metabolite with cisapride and erythromycin on cardiac repolarization in the isolated rabbit heart. Cardiovasc Pharmacol 1999;34:82-88.
92. Granberry MC, Gardner SF: Erythromycin monotherapy associated with torsade de pointes. Ann Pharmacother 1996;30:77-78.
93. Lee KL, Jim MH, Tang SC, Tai YT: QT prolongation and torsades de pointes associated with clarithromycin. Am J Med 1998; 104:395-396.
94. Yap YG, Camm J: Risk of torsades de pointes with non-cardiac drugs. Doctors need to be aware that many drugs can cause QT prolongation. Br Med J 2000;320:1158-1159.
95. Daleau P, Lessard E, Groleau M-F, et al: Erythromycin blocks the rapid component of the delayed rectifier potassium current and lengthens repolarization of guinea pig ventricular myocytes. Circulation 1995;91:3010-3016.
96. Girgis I, Gualbert J, Langan L, et al: A prospective study of the effect of IV pentamidine therapy on ventricular arrhythmias and QTc prolongation in HIV-infected patients. Chest 1997; 112:646-653.
97. Orsuka M, Kanamori H, Sasaki S, et al: Torsades de pointes complicating pentamidine therapy of pneumocystis carinii pneumonia in acute myelogenous leukemia. Intern Med 1997;36: 705-708.
98. Wassmann S, Nickenig G, Bohm M: Long QT syndrome and torsade de pointes in a patient receiving fluconazole. Ann Intern Med 1999;131:797.
99. Carlsson L, Amos GJ, Andersson B, et al: Electrophysiological characterization of the prokinetic agents cisapride and mosapride in vivo and in vitro: Implications for proarrhythmic potential? J Pharmacol Exp Ther 1997;282:220-227.
100. Drolet B, Khalifa M, Daleau P, et al: Block of the rapid component of the delayed rectifier potassium current by the prokinetic agent cisapride underlies drug-related lengthening of the QT interval. Circulation 1998;97:204-210.
101. Piquette RK: Torsade de pointes induced by cisapride/ clarithromycin interaction. Ann Pharmacother 1999;33:22-26.
102. Sekkarie MA: Torsades de pointes in two chronic renal failure patients treated with cisapride and clarithromycin. Am J Kidney Dis 1997;30:437-439.
103. Michalets EL, Williams CR: Drug interactions with cisapride: Clinical implications. Clin Pharmacokinet 2000;39:49-75.
104. Rampe D, Roy M-L, Dennis A, et al: A mechanism for the proarrhythmic effects of cisapride (propulsid): High affinity blockade of the human cardiac potassium channel HERG. FEBS Lett 1997;417:28-32.
105. Thomas AR, Chan L-N, Bauman JL, et al: Prolongation of the QT interval related to cisapride-diltiazem interaction. Pharmacotherapy 1998;18:381-385.
106. Machado C, Baga JJ, Kawasaki R, et al: Torsade de pointes as a complication of subarachnoid hemorrhage. A critical reappraisal. J Electrocardiol 1997;30:31-37.
107. Chao C-L, Chen W-J, Chen M-F, et al: Torsade de pointes in a patient using usual dose of beta agonist therapy. Int J Cardiol 1996;57:295-296.
108. Fujikawa H, Sato Y, Arakawa H, et al: Induction of torsades de pointes by dobutamine infusion in a patient with idiopathic Long QT Syndrome. Intern Med 1998;37:149-152.
109. La Vecchia L, Ometto R, Finocchi G, et al: Torsade de pointes ventricular tachycardia during low dose intermittent dobutamine treatment in a patient with dilated cardiomyopathy and congestive heart failure. Pacing Clin Electrophysiol 1999;22: 397-399.
110. Tisdale JE, Patel R, Webb CR, et al: Electrophysiologic and proarrhythmic effects of intravenous inotropic agents. Progress Cardiovasc Dis 1995;38:167-180.
111. Abe K, Takada K, Yoshiya I: Intraoperative torsade de pointes ventricular tachycardia and ventricular fibrillation during sevoflurane anesthesia. Anesth Analg 1998;86:701-702.
112. Roden DM: Taking the "idio" out of "idiosyncratic": Predicting torsades de pointes. Pacing Clin Electrophysiol 1998;20:1029-1034.
113. Buckingham TA, Bhutto ZR, Telfer EA, et al: Differences in corrected QT intervals at minimal and maximal heart rate may identify patients at risk for torsades de pointes during treatment with antiarrhythmic drugs. J Cardiovasc Electrophysiol 1994;5:408-411.
114. Makker RR, Fromm BS, Steinman RT, et al: Female gender as a risk factor for torsades de pointes associated with cardiovascular drugs. JAMA 1993;270:2590-2597.
115. Kawasaki R, Machado C, Reinoehl J, et al: Increased propensity of women to develop torsades de pointes during complete heart block. J Cardiovasc Electrophysiol 1995;6:1032-1038.
116. Lehmann MH, Hardy S, Archibald D, et al: Sex difference in risk of torsade de pointes with *d, l*-Sotalol. Circulation 1996; 94:2534-2541.
117. Lehmann MH, Timothy KW, Frankovich D, et al: Age-gender influence on the rate-corrected QT interval and the QT-heart rate relation in families with genotypically characterized long QT syndrome. J Am Coll Cardiol 1997;29:93-99.
118. Thibault B, Nattel S: Optimal management with class I and class III antiarrhythmic drugs should be done in the outpatient setting: Protagonist. J Cardiovasc Electrophysiol 1999;10:472-481.
119. Prystowsky EN: Proarrhythmia during drug treatment of supraventricular tachycardia: Paradoxical risk of sinus rhythm for sudden death. Am J Cardiol 1996;78(suppl 8A):35-41.
120. Choy AM, Darbar D, Dell'Orto S, et al: Exaggerated QT prolongation after cardioversion of atrial fibrillation. J Am Coll Cardiol 1999;34:396-401
121. Rautaharj PM, Zhou SH, Wong S, et al: Sex differences in the evolution of the electrocardiographic QT interval with age. Can J Cardiol 1992;8:690-695.
122. VanderLugt JT, Mattioni T, Denker S, et al: Efficacy and safety of ibutilide fumarate for the conversion of atrial arrhythmias after cardiac surgery. Circulation 1999;100:369-375.
123. Gillis AM, Geonzon RA, Mathison HJ, et al: The effects of barium, dofetilide and 4-aminopyridine (4-AP) on ventricular repolarization in normal and hypertrophied rabbit heart. J Pharmacol Exp Ther 1998;289:262-270.
124. Vos MA, Verduyn SC, Gorgels APM, et al: Reproducible induction of early afterdepolarizations and torsade de pointes arrhythmias by d-sotalol and pacing in dogs with chronic atrioventricular block. Circulation 1995;91:864-872.
125. Volders PGA, Sipido KR, Vos MA, et al: Cellular basis of biventricular hypertrophy and arrhythmogenesis in dogs with chronic complete atrioventricular block and acquired torsade de pointes. Circulation 1998;98:1136-1147.
126. Vos MA, de Groot SHM, Verduyn SC, et al: Enhanced susceptibility for acquired torsade de pointes arrhythmias in the dog with chronic, complete AV block is related to cardiac hypertrophy and electrical remodeling. Circulation 1998;98:1125-1135.
127. Gillis AM, Mathison HJ, Kulicz E, et al: Dispersion of ventricular repolarization and ventricular fibrillation in left ventricular hypertrophy: Influence of selective potassium channel blockers. J Pharmacol Exp Ther 2000;292:381-386.
128. Antzelevitch C, Sicouri S: Clinical relevance of cardiac arrhythmias generated by afterdepolarizations. Role of M cells in the generation of U waves, triggered activity and torsade de pointes. J Am Coll Cardiol 1994;23:259-277.
129. Antzelevitch C, Shimizu W, Gan-Xin Y, et al: The M Cell: Its contribution to the ECG and to normal and abnormal electrical function of the heart. J Cardiovasc Electrophysiol 1999;10:1124-1152.
130. El-Sherif N, Bekheit SS, Henkin R: Quinidine-induced long QTU interval and torsade de pointes: Role of bradycardia-dependent early afterdepolarizations. J Am Coll Cardiol 1989; 14:252-257.
131. Davidenko JM, Cohen L, Goodrow R, et al: Quinidine-induced action potential prolongation, early afterdepolarizations and triggered activity in canine Purkinje fibers. Effects of stimulation rate, potassium and magnesium. Circulation 1989;79:674-686.
132. Nattel S, Quantz MA: Pharmacological response of quinidine induced early afterdepolarizations in canine cardiac Purkinje fibers: Insights into underlying ionic mechanisms. Cardiovasc Res 1988;22:808-817.

133. Roden DM, Hoffman BF: Action potential prolongation and induction of abnormal automaticity by low quinidine concentrations in canine Purkinje fibers. Relationship to potassium and cycle length. Circ Res 1985;56:857-867.
134. Carlsson L, Drews L, Duker G: Rhythm abnormalities related to delayed repolarization in vivo: Influence of sarcolemmal Ca^{++} entry and intracellular Ca^{++} overload. J Pharmacol Exp Ther 1996;279:231-239.
135. Asano Y, Davidenko JM, Baxter WT, et al: Optical mapping of drug-induced polymorphic arrhythmias and torsade de pointes in the isolated rabbit heart. J Am Coll Cardiol 1997;29:831-842.
136. Vincent GM, Timothy K, Fox J, et al: The inherited long QT syndrome: From ion channel to bedside. Cardiol Rev 1999; 7:44-55.
137. Volders PG, Vos MA, Szabo B, et al: Progress in the understanding of cardiac early afterdepolarizations and torsades de pointes: Time to revise current concepts. Cardiovasc Res 2000;46: 376-392.
138. Kappos KG, Kranidis AJ, Anthopoulos LP: Torsades de pointes following radiofrequency catheter His ablation. Int J Cardiol 1996;57:177-179.
139. Conti JB, Mills RM Jr, Woodard DA, et al: QT dispersion is a marker for life-threatening ventricular arrhythmias after atrioventricular nodal ablation using radiofrequency energy. Am J Cardiol 1997;79:1412-1414.
140. Dizon J, Blitzer M, Rubin D, et al: Time dependent changes in duration of ventricular repolarization after AV node ablation: Insights into the possible mechanism of postprocedural sudden death. Pacing Clin Electrophysiol 2000;23:1539-1544.
141. Raj SR, Gillis AM, Morck M, et al: Adaptive response of the QT dispersion to sudden rate drop following AV node ablation: Dependence on LV function [abstract]. Can J Cardiol 2000; 16:138F.
142. Haverkamp W, Hordt M, Breithardt G, et al: Torsade de pointes secondary to d, l-sotalol after catheter ablation of incessant atrioventricular reentrant tachycardia: Evidence for a significant contribution to the "cardiac memory." Clin Cardiol 1998; 21:55-58.
143. Trohman RG, Sahu J: Drug-induced torsade de pointes. Circulation 1999;99:E7.
144. Tan HL, Wilde AAM: T wave alternans after sotalol; evidence for increased sensitivity to sotalol after conversion from atrial fibrillation to sinus rhythm. Heart 1998;80:303-306.
145. Shimizu W, Antzelevitch C: Cellular and ionic basis for t-wave alternans under long-QT conditions. Circulation 1999;99: 1499-1507.
146. Drouin E, Charpentier F, Gauthier C, et al: Electrophysiological characteristics of cells spanning the left ventricular wall of human heart: Evidence for the presence of M cells. J Am Coll Cardiol 1995;26:185-192.
147. Sicouri S, Moro S, Litovsky S, et al: Chronic amiodarone reduces transmural dispersion of repolarization in the canine heart. J Cardiovasc Electrophysiol 1997;8:1269-1279.
148. Zabel M, Hohnloser SH, Behrens S, et al: Differential effects of d-sotalol, quinidine, and amiodarone on dispersion of ventricular repolarization in the isolated rabbit heart. J Cardiovasc Electrophysiol 1997;8:1239-1245.
149. Cui G, Sen L, Sager P, Uppal P, Singh BN: Effects of amiodarone, sematilide, and sotalol on QT dispersion. Am J Cardiol 1994;74:896-900.
150. Gillis AM: Effects of antiarrhythmic drugs on QT interval dispersion—relationship to antiarrhythmic action and proarrhythmia. Prog Cardiovasc Dis 2000;42:385-396.
151. Mitchell LB, Gillis AM, Sheldon RS et al: Increased QT interval dispersion in patients with amiodarone-associated torsades de pointes ventricular tachycardia [abstract]. Can J Cardiol 2000;16:227F.
152. Tzivoni D, Banai S, Schuger C, et al: Treatment of torsade de pointes with magnesium sulfate. Circulation 1988;77:392-397.
153. Bailie DS, Inoue H, Kaseda S, et al: Magnesium suppression of early afterdepolarizations and ventricular tachyarrhythmias induced by cesium in dogs. Circulation 1988;77:1395-1402.
154. Verduyn SC, Vos MA, van der Zande J, et al: Role of interventricular dispersion of repolarization in acquired torsade-de-pointes arrhythmias: Reversal by magnesium. Cardiovasc Res 1997; 34:453-463.
155. Po SS, Wang DW, Yang IC-H, et al: Modulation of HERG potassium channels by extracellular magnesium and quinidine. J Cardiovasc Pharmacol 1999;33:181-185.
156. Monraba R, Sala C, Cortes E, et al: Percutaneous overdrive pacing in the out-of-hospital treatment of torsades de pointes. Ann Pharmacother 1999;33:3.
157. Viskin S, Fish R, Roth A, et al: Prevention of torsade de pointes in the congenital long QT syndrome: Use of a pause prevention pacing algorithm. Heart 1998;79:417-419.
158. Tan HL, Wilde AAM, Peters RJ: Suppression of torsades de pointes by atropine. Heart 1998;79:99-100.
159. Yang T, Roden DM: Extracellular potassium modulation of drug block of IKr. Implications for torsade de pointes and reverse use-dependence. Circulation 1996;93:407-411.
160. Tan HL, Alings M, Van Olden RW, et al: Long-term (subacute) potassium treatment in congenital HERG-related long QT syndrome (LQTS2). J Cardiovasc Electrophysiol 1999;10: 229-233.
161. Chezalviel-Guilbert F, Davy J-M, Poirier J-M, et al: Mexiletine antagonizes effects of sotalol on QT interval duration and its proarrhythmic effects in a canine model of torsade de pointes. J Am Coll Cardiol 1995;26:787-792.
162. Hallman K, Carlsson L: Prevention of class III-induced proarrhythmias by flecainide in an animal model of the acquired long QT syndrome. Pharmacol Toxicol 1995;77:250-254.
163. Sicouri S, Antzelevitch D, Heilmann C, et al: Effects of sodium channel block with mexiletine to reverse action potential prolongation in in vitro models of the long QT syndrome. J Cardiovasc Electrophysiol 1997;8:1280-1290.
164. Assimes TL, Malcolm I: Torsade de pointes with sotalol overdose treated successfully with lidocaine. Can J Cardiol 1998;14: 753-756.
165. Brosch SF, Studenik C, Heistracher P: Abolition of drug-induced early afterdepolarizations by potassium channel activators in guinea-pig Purkinje fibers. Clin Exp Pharmacol Physiol 1998; 25:225-230.
166. Watanabe O, Okumura T, Takeda H, et al: Nicorandil, a potassium channel opener, abolished torsades de pointes in a patient with complete atrioventricular block. Pacing Clin Electrophysiol 1999;22:686-688.
167. White CM, Xie J, Chow MSS, et al: Prophylactic magnesium to decrease the arrhythmogenic potential of class III antiarrhythmic agents in a rabbit model. Pharmacotherapy 1999;19:635-640.
168. Maisel WH, Kuntz KM, Reimold SC, et al: Risk of initiating antiarrhythmic drug therapy for atrial fibrillation in patients admitted to a university hospital. Ann Intern Med 1997;127: 281-284.
169. Pinski SL, Helguera ME: Antiarrhythmic drug initiation in patients with atrial fibrillation. Prog Cardiovasc Dis 1999; 42:75-90.
170. Singh S, Zoble RG, Yellen L, et al: Efficacy and safety of oral dofetilide in converting to and maintaining sinus rhythm in patients with chronic atrial fibrillation or atrial flutter: The symptomatic atrial fibrillation investigative research on dofetilide (SAFIRE-D) study. Circulation 2000;102:2385-2390.
171. Waldo AL, Camm AJ, de Ruyter H, et al: Effect of d-sotalol on mortality in patients with left ventricular dysfunction after recent and remote myocardial infarction. Lancet 1996;348:7-12.
172. Pratt CM, Camm AJ, Cooper W, et al: Mortality in the survival with oral d-sotalol (SWORD) trial: Why did patients die? Am J Cardiol 1998;81:869-876.
173. Torp-Pedersen C, Moller M, Bloch-Thomsen PE, et al: Dofetilide in patients with congestive heart failure and left ventricular dysfunction. Danish Investigations of Arrhythmia and Mortality on Dofetilide Study Group. N Engl J Med 1999; 341:857-865.
174. Gillis AM, Leitch JW, Sheldon RS, et al: A prospective randomized comparison of autodecremental pacing to burst pacing in device therapy for chronic ventricular tachycardia secondary to coronary artery disease. Am J Cardiol 1993;72:1146-1151.
175. Gillis AM: The current status of the implantable cardioverter defibrillator. Annu Rev Med 1996;47:85-93.
176. Dougherty AH: Interactions between antiarrhythmic drugs and implantable cardioverter-defibrillators. Curr Opin Cardiol 1996;11:2-8.

177. Olatidoye AG, Verroneau J, Kluger J: Mechanisms of syncope in implantable cardioverter-defibrillator recipients who receive device therapies. Am J Cardiol 1998;82:1372-1376.
178. Vlay LC, Vlay SC: Pacing induced ventricular fibrillation in internal cardioverter defibrillator patients: A new form of proarrhythmia. Pacing Clin Electrophysiol 1997;20:132-133.
179. Simons GR, Newby KH, Kearney MM, et al: Safety of transvenous low energy cardioversion of atrial fibrillation in patients with a history of ventricular tachycardia: Effects of rate and repolarization time on proarrhythmic risk. Pacing Clin Electrophysiol 1998;21:430-437.
180. Levy S, Ricard P, Gueunoun M, et al: Low-energy cardioversion of spontaneous atrial fibrillation. Immediate and long-term results. Circulation 1997;96:253-259.
181. Ayers GM, Alferness CA, Ilina M, et al: Ventricular proarrhythmic effects of ventricular cycle length and shock strength in a sheep model of transvenous atrial defibrillation. Circulation 1994;89:413-422.
182. Tse HF, Lau CP, Sra JS, et al: Atrial fibrillation detection and R-wave synchronization by Metrix implantable atrial defibrillator: Implications for long-term efficacy and safety. The Metrix Investigators. Circulation 1999;99:1446-1451.
183. Wolpert C, Jung W, Scholl C, et al: Electrical proarrhythmia: Induction of inappropriate atrial therapies due to far-field R wave oversensing in a new dual chamber defibrillator. J Cardiovasc Electrophysiol 1998;9:859-863.
184. Natale A, Jones DL, Kim YH, et al: Effects of lidocaine on defibrillation threshold in the pig: evidence of anesthesia related increase. Pacing Clin Electrophysiol 1991;14:1239-1244.
185. Natale A, Jones DL, Kleinstiver PW, et al: Effects of flecainide on defibrillation threshold in pigs. J Cardiovasc Pharmacol 1993;21:573-577.
186. Stevens SK, Haffajee CI, Naccarelli GV, et al: Effects of oral propafenone on defibrillation and pacing thresholds in patients receiving implantable cardioverter-defibrillators. Propafenone Defibrillation Threshold Investigators. J Am Coll Cardiol 1996; 28:418-422.
187. Fain ES, Dorian P, Davy JM, et al: Effects of encainide and its metabolites on energy requirements for defibrillation. Circulation 1986;73:1334-1341.
188. Echt DS, Gremillion ST, Lee JT, et al: Effects of procainamide and lidocaine on defibrillation energy requirements in patients receiving implantable cardioverter defibrillator devices. J Cardiovasc Electrophysiol 1994;5:752-760.
189. Dorian P, Newman D, Sheahan R, et al: d-Sotalol decreases defibrillation energy requirements in humans: A novel indication for drug therapy. J Cardiovasc Electrophysiol 1996;7:952-961.
190. Qi XQ, Newman D, Dorian P: Azimilide decreases defibrillation voltage requirements and increases spatial organization during ventricular fibrillation. J Interv Card Electrophysiol 1999;3:61-67.
191. Dorian P, Newman D, Sheahan R, et al: d-Sotalol decreases defibrillation energy requirements in humans: A novel indication for drug therapy. J Cardiovasc Electrophysiol 1996;7:952-961.
192. Huang SK, Tan de Guzman WL, Chenarides JF, et al: Effects of long-term amiodarone therapy on the defibrillation threshold and the rate of shocks of the implantable cardioverter-defibrillator. Am Heart J 1991;122:720-727.
193. Jung W, Manz M, Pizzulli L, et al: Effects of chronic amiodarone therapy on defibrillation threshold. Am J Cardiol 1992;70: 1023-1027.
194. Dorian P: Amiodarone and defibrillation thresholds: A clinical conundrum. J Cardiovasc Electrophysiol 2000;11:741-743.
195. Danilovic D, Ohm OJ: Pacing threshold trends and variability in modern tined leads assessed using high resolution automatic measurements: Conversion of pulse width into voltage thresholds. Pacing Clin Electrophysiol 1999;22:567-587.
196. Stokes KB, Kay GN: Artificial electrical cardiac stimulation. In Ellenbogen KA, Kay GN, Wilkoff BL (eds): Clinical Cardiac Pacing and Defibrillation, 2nd ed. Philadelphia, WB Saunders, 2000, pp 17-52.
197. Gillis AM, Lundstrom R, Tonder L, et al: Circadian variation of ventricular pacing thresholds [abstract]. Can J Cardiol 2000;16:191F.
198. Kiss Z, Fazekas T: Organophosphates and torsade de pointes ventricular tachycardia. J R Soc Med 1983;76:984-985.
199. Eckardt L, Haverkamp W, Borggrefe M, Breithardt G: Experimental models of torsade de pointes. Cardiovasc Res 1998;39:178-193.
200. Saliba W, Erdogan O, Niebauer M: Polymorphic ventricular tachycardia in a woman taking cesium chloride. Pacing Clin Electrophysiol 2001;24:515-517.
201. Olshansky B, Shivkumar K. Patient—heal thyself? Electrophysiology meets alternative medicine. Pacing Clin Electrophysiol 2001;24:403-405.
202. Jan IS, Jong YS, Lo HM: [Barium intoxication: A case report] J Formos Med Assoc 1991;90:908-910.
203. Sundar S, Sinha PR, Agrawal NK, et al: A cluster of cases of severe cardiotoxicity among kala-azar patients treated with a high-osmolarity lot of sodium antimony gluconate. Am J Trop Med Hyg 1998;59:139-143.
204. Ortega-Carnicer J, Alcazar R, De la Torre M, Benezet J: Pentavalent antimonial-induced torsade de pointes. J Electrocardiol 1997; 30:143-145.
205. Minotti G, Cairo G, Monti E: Role of iron in anthracycline cardiotoxicity: New tunes for an old song? FASEB J 1999; 13:199-212.
206. Hershko C, Pinson A, Link G: Prevention of anthracycline cardiotoxicity by iron chelation. Acta Haematol 1996;95:87-92.
207. Leor J, Poole WK, Kloner RA: Sudden cardiac death triggered by an earthquake. N Engl J Med 1996;334:413-419.
208. Saksena S: Calamities and cardiac arrhythmias: The medical and scientific impact of epidemics of suffering. J Int Cardiovasc Electrophysiol 2002;6:7-8.
209. Muller JE, Verrier RL: Triggering of sudden death—lessons from an earthquake. N Engl J Med 1996;334:460-461.
210. Lampert R, Jain D, Burg MM, et al: Destabilizing effects of mental stress on ventricular arrhythmias in patients with implantable cardioverter-defibrillators. Circulation 2000;101:158-164.
211. Fries RP, Heisel AG, Jung JK, Schieffer HJ: Circannual variation of malignant ventricular tachyarrhythmias in patients with implantable cardioverter-defibrillators and either coronary artery disease or idiopathic dilated cardiomyopathy. Am J Cardiol 1997;79:1194-1197.

Chapter 25 Arrhythmias During Pregnancy

JAMIE BETH CONTI and ANNE B. CURTIS

In general, the approach to the evaluation of arrhythmias in pregnant patients is similar to that in patients who are not pregnant. However, pregnancy carries with it several unique considerations that must be kept in mind when deciding upon appropriate treatment. These factors include the presence of the fetus; the characteristic hemodynamic changes seen in pregnant women; the effect of antiarrhythmic therapy on labor, delivery, and lactation; and the effects of these drugs on the unborn child. In addition, the physician must consider the stage of pregnancy, as the risk of teratogenicity is higher during the first trimester of pregnancy. The first 8 weeks after fertilization is the period of greatest risk; after 8 weeks, organogenesis is complete, and any risk to the fetus is substantially reduced. Teratogenic abnormalities may occur depending upon which drug the fetus is exposed to, the duration of exposure, and genetic predisposition.[1] To further complicate management, the absorption, distribution, and excretion of drugs is dramatically altered during pregnancy, making monitoring of drug levels difficult.

Fortunately, life-threatening rhythm disorders in women of childbearing age are relatively rare. Sustained symptomatic bradyarrhythmias in this population are also exceedingly rare, and most appear long before the patient reaches childbearing age.[2] For this reason, most adult cardiologists have not had extensive experience in treating this patient population. For some, the knowledge that therapy may affect an unborn child is intimidating. This need not be the case if the following principles of evaluation and management are followed. Most patients with arrhythmias during pregnancy can be managed with an excellent outcome.

Physiologic Changes During Pregnancy

Total body water increases by approximately 5 to 8 L in normal pregnancy, more so in patients with clinical edema.[3] Increased blood volume results in an increase in cardiac output of approximately 40%, significantly increasing the demands on the maternal heart. The increase in cardiac output begins during the first trimester of pregnancy, probably around the 5th to 10th week. The rise in cardiac output early in pregnancy is disproportionately greater than the increase in heart rate, most likely secondary to augmentation in stroke volume (Table 25-1).[4] Cardiac output continues to increase until midpregnancy and then stabilizes, with a possible small decline in the last week. Peripheral vascular resistance is reduced throughout pregnancy, resulting in a decrease in blood pressure in early pregnancy that reaches a minimum in midpregnancy and then returns to baseline levels at term.

Physiologic changes during pregnancy make absorption, distribution, and elimination of drugs rather unpredictable (Fig. 25-1).[5] Gastrointestinal absorption of drugs, and thus bioavailability, is altered during pregnancy in an unpredictable way because of changes in gastric motility and secretion. In some patients, gastric motility is decreased, leading either to an increase in absorption (increased transit time through the small bowel), a decrease in absorption (destruction of drug), or no change.[6] Changes in gastric pH may alter the rate and degree of absorption of specific drugs.

The increased blood volume and resultant increased volume of distribution may lower the concentration of drugs in the central compartment and increase the elimination half-life. At the same time, decreases in plasma protein concentration result in less protein binding of drug.[4,7,8] The increase in cardiac output causes an increase in renal blood flow, and thus may dramatically affect metabolism of drugs that are excreted renally. Renal blood flow in midpregnancy is 60% to 80% greater than in the nonpregnant state. The glomerular filtration rate reaches a peak of 50% higher than the nonpregnant level, a change that is sustained until the end of pregnancy.[4] In addition, increased levels of progesterone may increase hepatic clearance of drugs, leading to a reduction in drug concentration.[5]

The combined effects of the these changes make careful assessment of clinical drug effect essential, because the measured serum drug concentration may be misleading due to altered protein binding. Although total measured drug concentration may be lower

TABLE 25-1 Cardiocirculatory Changes During Normal Pregnancy

Parameter	Changes* at Various Times (wk)					
	5	12	20	24	32	38
Heart rate	↑	↑↑↑	↑↑↑	↑↑↑	↑↑↑↑	↑↑↑↑
Systolic blood pressure	↔	↓	↓	↔	↑	↑↑
Diastolic blood pressure	↔	↓	↓↓	↓	↔	↑↑
Stroke volume	↑	↑↑↑↑↑	↑↑↑↑↑↑	↑↑↑↑↑↑	↑↑↑↑↑	↑↑↑↑↑
Cardiac output	↑↑	↑↑↑↑↑↑	↑↑↑↑↑↑↑	↑↑↑↑↑↑↑	↑↑↑↑↑↑↑	↑↑↑↑↑↑↑
Systemic vascular resistance	↓↓	↓↓↓↓↓	↓↓↓↓↓↓	↓↓↓↓↓↓	↓↓↓↓↓↓	↓↓↓↓↓
Left ventricular ejection fraction	↑	↑↑	↑↑	↑↑	↑	↑

*↑, ≤5%; ↑↑, 6%-10%; ↑↑↑, 11%-15%; ↑↑↑↑, 16%-20%; ↑↑↑↑↑, 21%-30%; ↑↑↑↑↑↑, >30%; ↑↑↑↑↑↑↑, >40%.

Modified from Robson et al: Am J Physiol 1989;256:H1060-H1065.

From Elkayam U, Gleicher N: Hemodynamics and cardiac function during normal pregnancy and the puerperium. In Elkayam U, Gleicher N (eds): Cardiac Problems in Pregnancy, 3rd ed. New York, Wiley Liss, 1988, p 4.

secondary to decreased protein binding, the active free fraction may remain unchanged. Thus, although the drug level may appear to be low, therapeutic efficacy may have been achieved, making increases in drug dosage unnecessary and perhaps dangerous.

Principles of Evaluation and Management

First, correct diagnosis and arrhythmia documentation are vitally important as *"there is no place for empirical treatment in pregnant patients."*[9] Patients who complain of palpitations, but in whom no arrhythmia can be documented using noninvasive testing, have a low likelihood for having a life-threatening arrhythmia, and no further evaluation is warranted.[1] Some patients can be evaluated with a standard ECG. For other patients, Holter monitors or event monitors may be used to correlate symptoms with arrhythmias. It is critical to establish a relationship between symptoms and documented arrhythmia before the initiation of medical therapy.

Second, the need for treatment must be clear. The severity of the symptoms allows one to judge whether the risks of therapy outweigh the benefits. The only

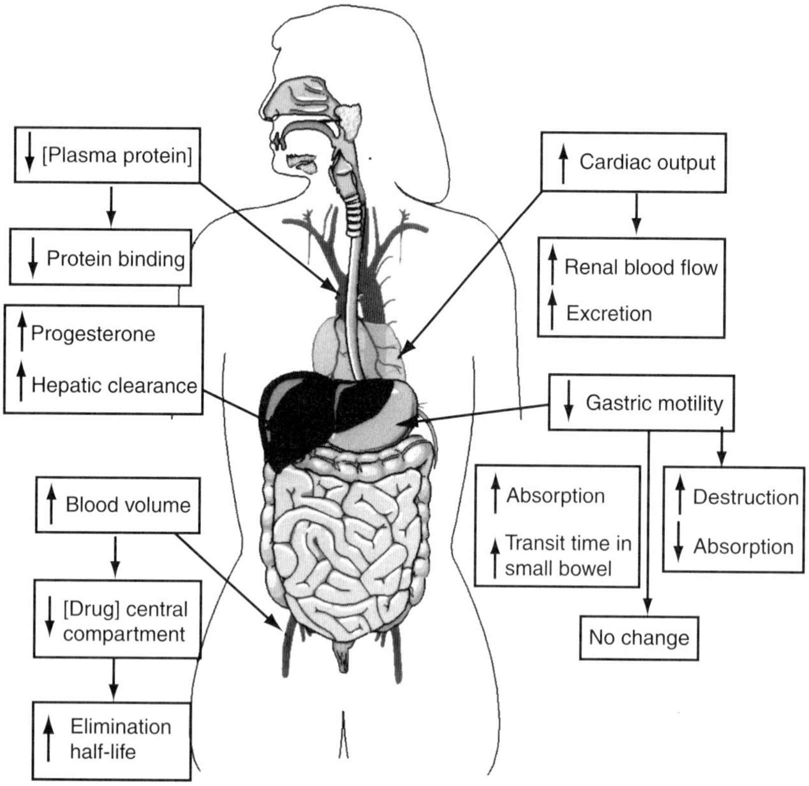

FIGURE 25-1 Schematic drawing representing the physiologic changes during pregnancy affecting absorption, distribution, and elimination of drugs.

difference in pregnancy is that the physician must consider the risk-benefit ratio for both the mother and the fetus. An arrhythmia that is hemodynamically compromising to the mother constitutes a major concern because of inadequate maternal as well as placental blood flow.

Third, use as few drugs at the lowest doses as are effective. Although this is true in the treatment of all patients, it is particularly important in pregnant patients. This approach exposes them to the least number of potential toxins possible. Pharmacologic therapy should be reserved for patients with hemodynamically compromising arrhythmias, whereas bothersome symptoms in a patient with a normal heart should be treated with reassurance if feasible.

Fourth, choose a drug that has been used frequently in pregnant patients. The medications that have been clinically available for many years have the most safety data. The majority of antiarrhythmic drugs are classified as U.S. Food and Drug Administration (FDA) "category C," which means that risk cannot be ruled out as there are either animal studies suggesting risk but no human studies or no controlled studies in either humans or animals (Table 25-2).[10] The categories have been criticized as misleading, because they convey the impression that there is graded risk as one crosses categories, and that there is similar risk among drugs in the same category. This is not accurate, as there is a wide range of severity of adverse effects within classes, and often no distinction between teratogenic and other toxic effects. Thus, if given a choice of using a category C drug that is new, and one that has a long, safe history of use, the drug with the history of safe use is recommended.

TABLE 25-2 U.S. Food and Drug Administration Use-in-Pregnancy Ratings

Category	Interpretation
A	Controlled studies show no risk. Adequate, well-controlled studies in pregnant women have failed to demonstrate a risk to the fetus in any trimester of pregnancy.
B	No evidence of risk in humans. Adequate, well-controlled studies in pregnant women have not shown increased risk of fetal abnormalities despite adverse findings in animals, or, in the absence of adequate human studies, animal studies show no fetal risk. The chance of fetal harm is remote but remains a possibility.
C	Risk cannot be ruled out. Adequate, well-controlled human studies are lacking, and animal studies have shown a risk to the fetus or are lacking as well. There is a chance of fetal harm if the drug is administered during pregnancy, but the potential benefits may outweigh the potential risk.
D	Positive evidence of risk. Studies in humans or investigational or postmarketing data have demonstrated fetal risk. Nevertheless, potential benefits from the use of the drug may outweigh the potential risk. For example, the drug may be acceptable if needed in a life-threatening situation or serious disease for which safer drugs cannot be used or are ineffective.
X	Contraindicated in pregnancy. Studies in animals or humans or investigational or postmarketing reports have demonstrated positive evidence of fetal abnormalities or risk, which clearly outweighs any possible benefit to the patient.

From the Physicians' Desk Reference: PDR, 55th ed. Oradell, NJ, Medical Economics, 2001, p 344.

HISTORY

The history is particularly important in the evaluation of the pregnant patient. Establishment of the intensity of symptoms is crucial when deciding when and whom to treat with drug therapy. The history should include detailed questioning of time, length and degree of symptoms, as well as aggravating and alleviating factors. Symptoms indicating hemodynamic instability, such as dizziness, syncope, and near syncope imply impaired fetal circulation and possible compromise as well. Initial assessment should also focus on symptoms indicating concomitant cardiovascular disease, as arrhythmias during pregnancy may be the first manifestation of underlying heart disease. Other precipitants of arrhythmias, such as thyrotoxicosis, should be considered as well.

Increased ectopic activity is commonly seen in both the atrium and ventricle throughout pregnancy.[11] Generally, this ectopy is benign and well tolerated, although life-threatening arrhythmias have been described. However, women with previously diagnosed tachyarrhythmias are at an increased risk of recurrence of their arrhythmia, possibly triggered by the increasing ectopy. In addition, patients with no previous history may present for the first time with a symptomatic arrhythmia during this period.[12,13]

PHYSICAL EXAMINATION

Physical examination of the pregnant patient is essentially identical to that of a nonpregnant patient from a cardiovascular standpoint. In addition, normal findings for pregnancy include a widely split first heart sound with expiratory splitting of the second heart sound late in pregnancy, a soft systolic flow murmur that is thought to represent increased flow across the pulmonary artery and valve, and mild peripheral edema.[14] Findings that would be of specific interest in a patient with an arrhythmia would include the murmur of mitral stenosis (atrial fibrillation), signs consistent with congestive heart failure (ventricular tachycardia), and murmurs consistent with congenital heart disease (reentrant tachycardias associated with repair in childhood).

DIAGNOSTIC STUDIES

Electrocardiography

The ECG during pregnancy often shows a shift in the QRS axis to the left in the frontal plane with a small Q wave and inverted T wave in lead III (Figs. 25-2*A* and *B*). These are normal changes in the ECG and can

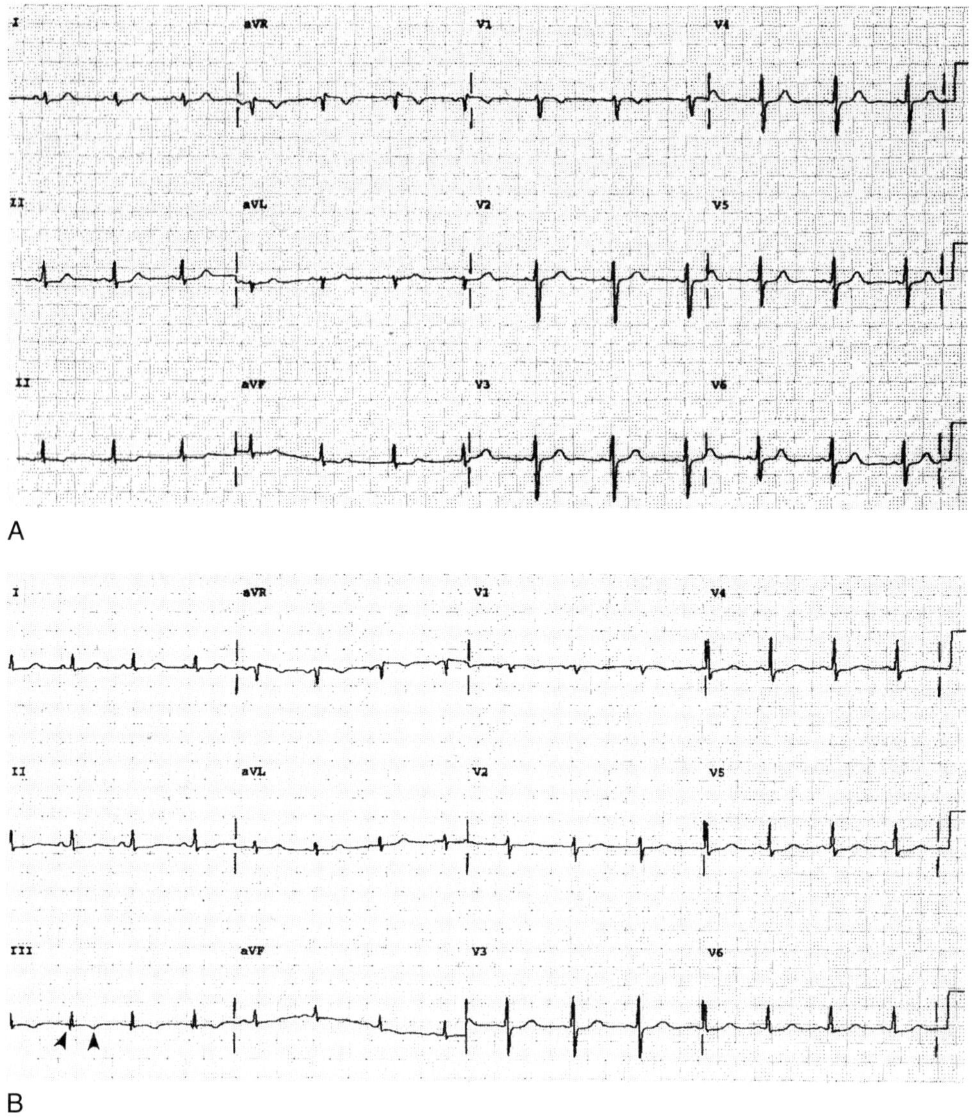

FIGURE 25-2 **A,** Baseline ECG in 23-year-old patient before pregnancy. The axis is slightly rightward, with normal T waves seen in lead III. **B,** ECG in the same patient obtained when she was 7 months pregnant. The axis has shifted leftward. A small Q wave and an inverted T wave are seen in lead III (*arrows*).

be related to a gradual shift in the position of the heart in the thorax.[15]

A routine ECG in some patients with palpitations may be sufficient to reveal premature atrial or ventricular beats as the cause. Likewise, sustained tachyarrhythmias may be captured, allowing a diagnosis to be made in many cases. Narrow, regular complex tachyarrhythmias with abrupt onset and termination are most likely due to atrioventricular (AV) nodal reentrant tachycardia or AV reentrant tachycardia. Ventricular preexcitation may indicate that supraventricular tachycardia (SVT) is due to AV reentrant tachycardia. Atrial fibrillation, although unusual during pregnancy, is readily recognized in most cases. Patients with ventricular tachyarrhythmias, on the other hand, may shown a prolonged Q–T interval during sinus rhythm, or short runs of ventricular tachycardia from the right ventricular outflow tract on the ECG that may prompt more specific evaluation and management (Fig. 25-3).

Echocardiography

An echocardiogram is of obvious value in ruling out a potential congenital or acquired cardiac defect and should be a standard part of the arrhythmia workup. In a patient with a suspected arrhythmia, echocardiography is useful because of its noninvasive nature. Establishment of normal ventricular function is important for several reasons. First, the choice of treatment is unrestricted in a patient with normal wall motion and function. Second, the possibility of cardiomyopathy is excluded as the underlying etiology of symptoms.

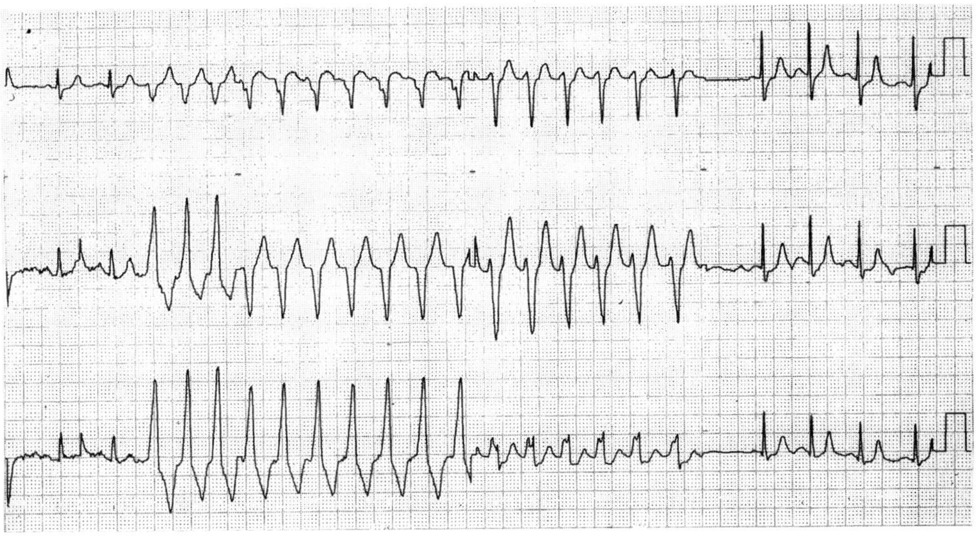

FIGURE 25-3 Nonsustained ventricular tachycardia seen on a routine ECG in a patient who is 8 months pregnant. The tachycardia has a left bundle branch block morphology with a rightward axis, consistent with right ventricular outflow tract tachycardia.

Exercise Treadmill Testing

Exercise treadmill testing is rarely indicated in the evaluation of the pregnant patient with an arrhythmia. However, if documentation of an arrhythmia has been otherwise unsuccessful, exercise testing may be useful in patients with exercise-induced arrhythmias. Other unusual circumstances in which exercise testing may be helpful include patients with suspected ischemia as the substrate for their arrhythmia or for assessment of chronotropic incompetence. Moderate exercise testing during pregnancy is not contraindicated, although the patient should be carefully monitored for symptoms and stopped for any sign of hemodynamic compromise.

Cardiac Catheterization

There are infrequent case reports of left heart catheterization performed during pregnancy.[16] Although rarely indicated for the evaluation of an arrhythmia, measurement of intracardiac pressures or knowledge of the coronary anatomy may be crucial to the management of certain patients. An example would be the patient with aborted sudden cardiac death during pregnancy and a suspected coronary anomaly. In addition, as maternal age increases, the prevalence of ischemia during pregnancy may rise.

Ambulatory Monitoring

In patients with palpitations of unknown etiology, Holter monitoring is one of the most useful tests to establish the etiology of symptoms. However, many arrhythmias occur sporadically and are unsuited to documentation with a 24-hour Holter monitor. For patients with less frequent symptoms, event monitors provide a noninvasive approach to documenting arrhythmias. Unless the presenting problem is syncope with no prodromal symptoms, a hand-held postevent recorder will usually be adequate to correlate symptoms with an arrhythmia. Some patients with troublesome symptoms who have not had success recording their rhythm during symptoms may need to use a continuous loop monitor, which can save rhythm information retroactively for a short period of time when patients have very transient symptoms. If no arrhythmia can be documented using such noninvasive means, it is unlikely that a life-threatening arrhythmia is present.

Electrophysiological Study

Although electrophysiological studies have become routine in the diagnosis and treatment of arrhythmias, the use of fluoroscopy for positioning catheters for both diagnostic and catheter ablation procedures significantly limits this option during pregnancy.[1,9] Radiation exposure during the first and second trimester of pregnancy has been linked to congenital malformations and mental retardation.[17] In addition, radiation exposure in utero increases the risk of childhood malignancies,[18,19] particularly leukemia.[20,21] If absolutely necessary, an electrophysiological study can be performed using echocardiographic guidance for catheter placement.[22] If this technique is used, the subcostal approach is most effective.[23]

Interestingly, radiofrequency catheter ablation of an accessory pathway during pregnancy has been reported, with a safe outcome for mother and fetus.[24] In this case report, the patient was 31 years old and 20 weeks pregnant, with a recurrent, hemodynamically unstable supraventricular arrhythmia that was refractory to medical management. During the procedure, the patient's abdomen was draped with a lead apron of 5-mm width. A single 7 Fr quadripolar catheter was used for mapping and delivery of radiofrequency energy. Ablation was accomplished with 70 seconds of fluoroscopy time.

The authors concluded that the procedure is safe in pregnant patients if proper precautions are taken to shield the fetus from the effects of radiation. This conclusion should be interpreted with caution as there was no long-term follow-up of the child.

Obviously, rapid identification of the appropriate ablation site is of paramount importance in this situation, yet it is not always easy to accomplish quickly. Electromagnetic mapping has been reported to result in much lower fluoroscopy times in patients undergoing radiofrequency ablation procedures[25] and thus should be considered in the rare circumstance in which the procedure could not wait until after delivery. Because of the small, but finite, increased risk of childhood malignancy, routine use of fluoroscopy, even with appropriate shielding, cannot be recommended.

Insertable Loop Recorder

There are no reports of these monitors being implanted in pregnant women. However, because the monitor is implanted without the use of fluoroscopy, there is no known contraindication to its use. General considerations on the use of anesthesia would apply.

Tilt Table Testing

Tilt table testing of pregnant patients with syncope is well documented in the literature and appears to be safe.[26] Because of the characteristic hemodynamic changes of pregnancy, vasodepressor syncope can be seen in the postpartum period as well. Grubb et al.[27] described 12 women who developed episodic hypotension resulting in syncope in the immediate postpartum period. All 12 had hypotension during baseline tilt table testing. Neurocardiogenic syncope resolves in the majority of these patients as their blood volume and peripheral vascular resistance return to baseline values.

ANTIARRHYTHMIC DRUGS IN THE MANAGEMENT OF ARRHYTHMIAS

The FDA classifies most antiarrhythmic drugs available for use today as category C during pregnancy (see Table 25-2). This categorization signifies that "... risk cannot be ruled out. Adequate, well-controlled human studies are lacking, and animal studies have shown a risk to the fetus or are lacking as well. There is a chance of fetal harm if the drug is administered during pregnancy; but the potential benefits may outweigh the potential risk."[28] Although all antiarrhythmic medications have potential adverse effects in both the mother and the fetus, it is unlikely that comprehensive clinical trials will ever be performed, for obvious ethical reasons. Therefore, pharmacologic therapy should be reserved for patients who demonstrate significant symptoms or for those patients in whom the arrhythmia may be life threatening. Table 25-3 lists the available antiarrhythmic medications and their compatibility for use in pregnancy and during lactation.

TABLE 25-3 Antiarrhythmic Drugs During Pregnancy

Antiarrhythmic Drug	Vaughn Williams Classification	FDA Category	Safety During Lactation
Disopyramide	IA	C	S
Procainamide	IA	C	S
Quinidine	IA	C	S
Lidocaine	IB	B	S
Mexiletine	IB	C	S
Flecainide	IC	C	S
Moricizine	IC	B	?
Propafenone	IC	C	?
Amiodarone	III	D	NS
Azimilide	III	?	?
Dofetilide	III	?	?
Ibutilide	III	C	?
Sotalol	III	B	S
Adenosine	—	C	?
Verapamil	IV	C	S
Diltiazem	IV	C	S

A, B, C, and D, U.S. Food and Drug Administration classification of drugs for use in pregnancy; S, generally regarded as safe, and maternal medication usually compatible with breast-feeding; NS, generally regarded as unsafe and contraindicated or requires cessation of breast-feeding; —, not cardioselective.

Atrioventricular Nodal Blockers

Adenosine

Although there are only limited data, it appears that adenosine has no direct effect on the fetus when fetal monitoring is performed during bolus IV administration.[29,30] Adenosine is a purine nucleoside present in all human cells that characteristically depresses AV nodal conduction and sinus node automaticity. These electrophysiological effects have been successfully exploited for the termination of SVTs involving the AV node as part of a reentrant circuit.[31] Because of its rapid onset and short duration of action, adenosine appears to be a safe drug for use during pregnancy,[32] although it remains an FDA category C drug.

Verapamil

Verapamil is a calcium channel blocker with a long history of use in the management of SVT. It crosses the placenta and has been reported to affect fetal cardiovascular activity.[33] It is rapidly absorbed but has high first-pass metabolism, with only a small proportion excreted unchanged in the urine. Ninety percent of the drug is plasma protein bound. There are no reports of congenital defects in association with verapamil. Although efficacious in the chronic management of SVT, the longer onset of action when compared to adenosine makes this drug a second-line choice in the acute termination of SVT. Generic verapamil as well as all the long-acting formulations are FDA category C.

Diltiazem

Because diltiazem is a newer drug than verapamil, there is less history of use in pregnancy. Nevertheless, it has been used in the treatment of premature labor without report of congenital abnormalities.[34] Diltiazem and its long-acting formulations are FDA category C.

Digoxin

Cardiac glycosides can be used in the management of supraventricular arrhythmias. They have a long history of use in pregnant patients, although they remain classified FDA category C. Elimination is predominantly renal. Digoxin crosses the placenta readily, with fetal plasma concentrations similar to maternal values within 30 minutes.[35] These agents have also been administered maternally to manage fetal tachyarrhythmias.[36,37] Digoxin is not teratogenic.

β-Blockers

β-Blockers are class II antiarrhythmic agents that have been used extensively in pregnancy. Adverse outcomes with their use have been reported, predominantly fetal bradycardia, hypotonia, apnea, and hypoglycemia,[1,6] but no studies or case reports to date implicate any β-blocker in fetal malformations. The incidence of adverse outcomes is low,[6] although there are some data to suggest that the cardioselective agents should be used preferentially, because they may interfere less with B_2-mediated peripheral vasodilation and uterine relaxation.[38,39]

Acebutolol and pindolol have recently been reclassified by the FDA as category B drugs (Table 25-4). This classification designates that there is no evidence of risk in humans. Specifically, adequate, well-controlled studies in pregnant women have not shown an increased risk of fetal abnormalities despite adverse findings in animals, or, in the absence of adequate human studies, animal studies show no fetal risk. The chance of fetal harm is remote but remains a possibility. In contrast, atenolol has recently been reclassified as a category D drug, meaning that there is positive evidence of risk. Thus, atenolol should not be used in pregnant patients, given that there are multiple alternative β-blockers available. Of the β-blockers available, acebutolol, being both FDA category B and cardioselective, and propranolol, with its long history of safe use in pregnancy, would be better choices for β-blocker therapy in pregnancy.

TABLE 25-4 β-Blocker Therapy in Pregnant Patients

β-Blocker	FDA Category	Safety During Lactation	Cardioselectivity
Acebutolol	B	S	+
Atenolol	D	S	+
Bisoprolol	C	Unknown	+
Esmolol	C	Unknown	+
Inderal	C	S	–
Labetalol	C	S	–
Lopressor	C	S	+
Metoprolol	C	S	+
Nadolol	C	S	–
Pindolol	B	unknown	–
Propranolol	C	S	–
Timolol	C	S	–

A, B, C, and D, U.S. Food and Drug Administration classification of drugs for use in pregnancy; S, generally regarded as safe, and maternal medication usually compatible with breast-feeding; +, cardioselective, –, not cardioselective.

Class IA Antiarrhythmic Drugs

Quinidine

Approximately 70% to 80% of an oral dose of quinidine is absorbed from the gastrointestinal tract, with 80% of the drug bound to plasma proteins. Elimination is predominantly hepatic. Quinidine readily crosses the placenta, a property that has been successfully exploited to terminate fetal arrhythmias.[40] Of the IA agents, quinidine has the longest safety record during pregnancy. Although isolated cases of adverse effects have been reported, such as fetal thrombocytopenia and eighth nerve toxicity, the drug is considered relatively safe.[10] Given its long history of safe use in pregnant patients, quinidine is probably the drug of choice if one elects to use a class IA agent.

Procainamide

Procainamide is 75% to 90% absorbed in the intestines, with 15% of the drug bound to plasma proteins. Elimination is predominantly renal. Although data are limited, procainamide appears to cross the placenta readily, and thus it is also used to treat fetal tachyarrhythmias.[41] There are no reports of adverse fetal outcomes with the use of procainamide.

Disopyramide

Experience with the use of disopyramide in pregnancy is small. It is a class IA antiarrhythmic agent with 60% to 83% absorption in the gastrointestinal tract and variable protein binding.[6] Elimination is predominantly renal. Although it is not teratogenic, it can cause uterine contractions. This latter effect makes disopyramide less desirable to use during pregnancy than the other available antiarrhythmic drugs.

Class IC Antiarrhythmic Drugs

Both flecainide and propafenone cross the placenta rather easily. Fetal levels of flecainide have been reported to be as high as 86% of maternal levels.[42] Although flecainide is classified as a category C drug, there are many reports of its safe use during pregnancy with no adverse outcome.[43] Neither of the class IC antiarrhythmic drugs has been reported to be teratogenic. Although use of these medications in pregnant patients should be approached cautiously, flecainide is a reasonable choice in patients with normal hearts who require treatment with an antiarrhythmic drug.

Class III Antiarrhythmic Drugs

Sotalol

Sotalol has relatively simple pharmacokinetics. It is nearly completely absorbed after oral administration, while approximately 80% to 90% is excreted unchanged in the urine. Sotalol does cross the placenta.[44] In patients with renal insufficiency, the clearance of sotalol is reduced and the elimination half-life prolonged. The only class I or class III antiarrhythmic agent classified by the FDA as category B, its use in pregnant patients has been reported with no adverse outcome.

TABLE 25-5 Adverse Fetal Outcome with Amiodarone Use

Retardation of psychomotor development
Bradycardia
Prematurity
Prolonged Q–T interval
Hypothyroidism
Congenital malformations
Fetal death

Amiodarone

Amiodarone is a highly lipophilic compound that accumulates in the liver, adipose tissue, and cell membranes. It accumulates in placental tissue as well, although it achieves fetal concentrations of only 9% to 14% of the maternal serum concentration.[45] It is 35% to 65% bioavailable, and 98% protein bound. Clearly an effective antiarrhythmic drug in the general population, the safety profile of amiodarone is of particular concern in pregnant patients. Limited information is available, but it appears that it may have serious adverse effects on the fetus, the most dangerous being neonatal hypothyroidism. Whereas amiodarone-induced hypothyroidism is reversible in adults, the newborn infant may have irreversible brain damage and seriously compromised respiration caused by a large goiter. Other possible effects are listed in Table 25-5. In addition, congenital abnormalities have been reported in neonates who were exposed to amiodarone.[46,47] Amiodarone is classified FDA category D. Given these considerations, amiodarone use in pregnancy should be reserved for those patients with drug refractory, symptomatic, or potentially lethal tachyarrhythmias. If amiodarone is used, fetal thyroid function should be monitored in utero to assess for the early development of hypothyroidism.

Ibutilide

Ibutilide is available only for IV use for the conversion of atrial fibrillation or flutter to normal sinus rhythm. No clinical trials have examined the safety of its use in pregnant women. However, laboratory animals have demonstrated a teratogenic effect from its infusion.[48] In view of the limited data and the possibility of a teratogenic effect, the use of ibutilide in pregnant patients cannot be recommended.

Dofetilide

Dofetilide is a newly released class III agent. At present, there are no data to support its use in human pregnancy. Animal data suggest that fetal malformations and embryonic death can occur secondary to drug-induced bradycardia and embryonic hypoxia/ischemia.[49]

MANAGEMENT OF SPECIFIC ARRHYTHMIAS

Palpitations

The majority of pregnant patients have structurally normal hearts. Symptomatic palpitations may be evaluated noninvasively with Holter monitoring or event monitors. Premature atrial, ventricular, and junctional contractions are best treated with reassurance. In patients with frequent arrhythmias during pregnancy, the premature beats may substantially resolve in the postpartum period.[11] If medical therapy is absolutely necessary for patient comfort, a β-blocker is the drug of choice. If more aggressive management is necessary, therapy should be directed at the mechanism of the arrhythmia.

Supraventricular Tachycardia

AV reentrant tachycardia (AV reciprocating tachycardia, orthodromic reentrant tachycardia) and AV nodal reentrant tachycardia are the most common sustained arrhythmias seen in pregnancy.[50] The physiologic stress of pregnancy increases the risk of presenting with new onset SVT or experiencing an exacerbation of a previously documented SVT. Tawam et al.[12] reported this exacerbation of SVT during pregnancy in a retrospective study that included 60 patients, 38 with SVT and 22 controls. All patients developed SVT after the age of 16 years. Thirteen of 38 (22%) patients had the onset of paroxysmal SVT during pregnancy, most often during the first pregnancy (eight patients). Four patients had the onset of paroxysmal SVT during the first trimester of pregnancy, six during the second trimester, and three during the third trimester. The authors found that the relative risk of onset of symptoms of paroxysmal SVT during pregnancy compared to the nonpregnant state was 5.1 ($P<.001$). There was no significant difference in the risk of onset of tachycardia during pregnancy when comparing the different types of paroxysmal SVT. Eleven additional patients had exacerbations of preexisting paroxysmal SVT during pregnancy. Other investigators[13] have confirmed this exacerbation of preexisting paroxysmal SVT during pregnancy. Risk is approximately equally distributed throughout pregnancy, with the risk similar regardless of the mechanism of SVT.[9]

Several mechanisms have been postulated to explain the increased incidence of arrhythmias during pregnancy. These include hemodynamic, autonomic, hormonal, and emotional changes occurring in the pregnant patient.[10] Although no single mechanism has been definitively cited, increased stress, intravascular volume shifts, elevated estrogen levels, and physiologic increases in heart rate have all been implicated.

Other SVTs are less often problems during pregnancy. Sinus tachycardia is a normal response to pregnancy, so that it would be rare to find "inappropriate" sinus tachycardia that would require specific treatment. Atrial tachycardia, atrial flutter, and atrial fibrillation would be encountered most often in patients with structural heart disease, either repaired congenital heart disease or valvular heart disease.

Figure 25-4 outlines a possible treatment algorithm for the acute management of SVT. The mainstay of initial therapy is adenosine. Alternatively, β-blockers or calcium channel blockers can be used for acute termination of SVT and are recommended in the recently published Adult Cardiac Life Support (ACLS) guidelines as second-line therapy.[51] In addition, digoxin,

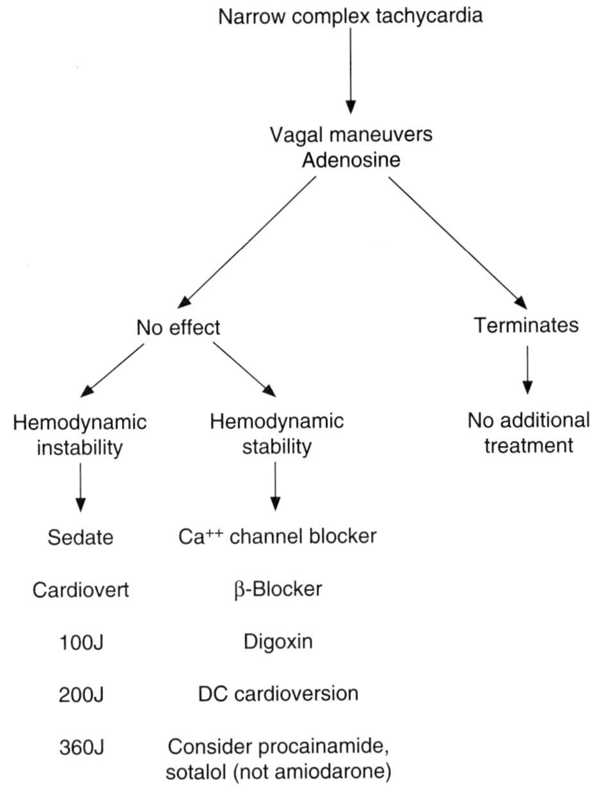

FIGURE 25-4 Proposed algorithm for the management of supraventricular tachycardia in pregnant patients. The algorithm includes the recent changes in advanced cardiac life support recommendations but excludes the use of amiodarone. DC, direct current.

procainamide, and sotalol are available alternatives. One deviation from the guidelines that should be noted is the use of amiodarone in the SVT algorithm. In pregnant patients, even a short infusion of IV amiodarone should be avoided if possible. Electrical cardioversion may be necessary for refractory SVT, and it is always indicated for hemodynamic instability.

Chronic management of these arrhythmias is similar to that in the nonpregnant patient. If overt preexcitation is not present on the surface ECG, any of the AV nodal blockers can be used. The use of membrane-sensitive antiarrhythmic drugs is reserved for patients with clearly defined, symptomatic arrhythmias. In pregnant patients with structurally normal hearts, β-blockers are a good choice for initial therapy. Other antiarrhythmic drugs that could be used include flecainide, quinidine, procainamide, and sotalol, the latter particularly in patients with structural heart disease.

Wolff-Parkinson-White Syndrome

As for AV nodal reentrant tachycardia, an increased incidence of arrhythmias has been reported in pregnant women with Wolff-Parkinson-White syndrome.[52] For patients with Wolff-Parkinson-White syndrome and AV reentrant tachycardia, β-blockers are the best choice for initial therapy. If additional therapy is necessary, flecainide is a good choice because of its common depressant effects on conduction over accessory pathways (Figs. 25-5*A* and *B*).[53,54]

Verapamil and digoxin are best avoided in this situation because of the possibility of enhancing conduction over accessory pathways. In particular, if preexcited atrial fibrillation is the clinical arrhythmia, neither the use of calcium channel blockers nor digoxin is advisable. β-blockers in conjunction with a class IA or IC antiarrhythmic drug would be the best course of therapy.

Atrial Flutter and Atrial Fibrillation

Although common in the general population, atrial flutter and fibrillation are rare in women of childbearing age with no underlying heart disease. Diltiazem, verapamil, β-blockers, and digoxin are all acceptable agents to use in achieving rate control. If spontaneous conversion to sinus rhythm does not occur, early cardioversion (within 48 hours of the onset of symptoms) should be performed to avoid the need for anticoagulation. During cardioversion and immediately after cardioversion, the fetus should be monitored for signs of fetal distress. Cardioversion with up to 300 J has been demonstrated to be safe during pregnancy.[55]

An alternative to electrical cardioversion is the use of oral flecainide (300 mg) or propafenone (600 mg) to promote conversion to sinus rhythm.[56] The use of ibutilide and amiodarone is not recommended during pregnancy, as explained earlier. If the arrhythmia is recurrent, or does not easily convert to sinus rhythm, it is almost always due to underlying rheumatic heart disease. In patients in whom cardioversion is impossible, the therapeutic goal should be adequate rate control to prevent hemodynamic compromise of placental blood supply (ideally <90 beats per minute [bpm] at rest and <140 bpm with exercise).[9]

For maintenance of sinus rhythm, the same drugs that are acceptable for the treatment of SVT are used for atrial fibrillation and flutter. Quinidine and procainamide have a long history of use in the obstetric community and have proven to be safe in this population. Both flecainide and sotalol, although relatively new, appear to be safe as well. Both amiodarone and dofetilide have potential adverse effects on the fetus and should not be used for the treatment of supraventricular arrhythmias during pregnancy.

Anticoagulation

Anticoagulation should be instituted in the presence of chronic atrial fibrillation in patients with risk factors (diabetes, hypertension, congestive heart failure, previous stroke, or rheumatic heart disease) and should be maintained throughout the pregnancy. Importantly, warfarin therapy is labeled FDA category X, meaning that it is contraindicated in pregnancy. The drug passes the placental barrier and may lead to spontaneous abortion, fetal hemorrhage, mental retardation, and birth malformations.[57] High-dose subcutaneous heparin has been used as a substitute, particularly in the first trimester of pregnancy, as its large molecular weight prevents placental transfer. It should be discontinued

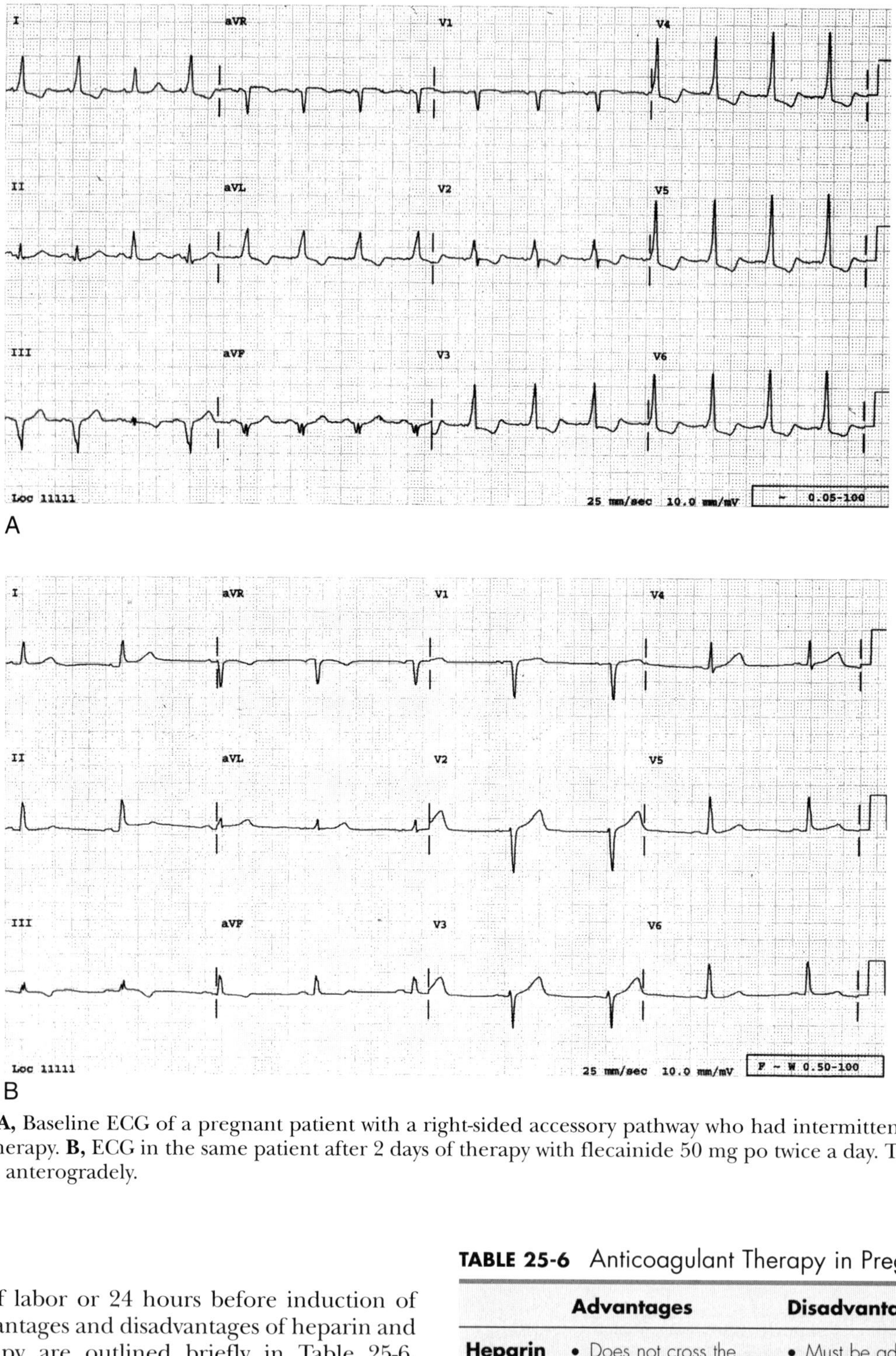

FIGURE 25-5 **A,** Baseline ECG of a pregnant patient with a right-sided accessory pathway who had intermittent atrial fibrillation requiring therapy. **B,** ECG in the same patient after 2 days of therapy with flecainide 50 mg po twice a day. The pathway no longer conducts anterogradely.

at the onset of labor or 24 hours before induction of labor. The advantages and disadvantages of heparin and warfarin therapy are outlined briefly in Table 25-6. Given that the risk of thromboembolism in chronic atrial flutter is lower than for atrial fibrillation, and anticoagulation is more difficult in pregnancy, treatment with warfarin or heparin is not usually recommended during pregnancy in patients with atrial flutter.

Enoxaparin, a relatively new low-molecular-weight heparin, is classified FDA category B in pregnant patients. Currently, it does not have an indication for atrial fibrillation but could be considered for use in the patient in whom heparin is contraindicated.

TABLE 25-6 Anticoagulant Therapy in Pregnancy

	Advantages	Disadvantages
Heparin	• Does not cross the placenta • Easily and rapidly reversed • Short half-life	• Must be administered parenterally • Bleeding risk to mother • Maternal osteopenia • Maternal thrombocytopenia • Maternal valve thrombosis • Systemic infection
Warfarin	• Most effective for preventing thromboembolic events	• Crosses the placenta • Cannot be easily or rapidly reversed • Adverse effects on fetus • Teratogenicity, first trimester • Increased hemorrhage during labor and delivery

Ventricular Tachycardia

In 1921 Mackenzie reported that nonsustained ventricular arrhythmias were present in approximately 50% of pregnant women.[58] The recognized causes of recurrent/paroxysmal ventricular tachycardia during pregnancy include arrhythmogenic right ventricular dysplasia, long QT syndrome (LQTS), hypertrophic cardiomyopathy, peripartum cardiomyopathy, and coronary artery disease (rarely). Patients without organic heart disease who experience paroxysmal nonsustained ventricular tachycardia during pregnancy are at low risk for subsequent morbidity and mortality. Withholding drug therapy in this subset of patients, particularly if they are only mildly symptomatic, is therefore reasonable.[59]

There are multiple reports of sustained ventricular tachycardia during pregnancy.[60] Fortunately, very few women during their childbearing years have coronary artery disease, with an even smaller number having reentrant ventricular tachycardia. The vast majority of these patients have normal hearts and no history of ventricular tachycardia before pregnancy. There are data to suggest that some of these patients have idiopathic ventricular tachycardia that is catecholamine sensitive and responsive to β-blocker therapy.[60] This tachycardia typically has a left bundle branch block and inferior axis morphology and arises from the right ventricular outflow tract (Fig. 25-6). Marchlinski et al. reported that right ventricular outflow tachycardia may be triggered by states of hormonal flux.[61] In their study population, two women had the onset of their arrhythmia during pregnancy. Symptoms were predominant during the first trimester in one patient and during the second trimester in the other. Both patients noted complete resolution of their symptoms after delivery. Marchlinski et al. suggested that there may be sex-specific triggers for this arrhythmia, the pregnant state being one of them. They hypothesized that an increase in sensitivity to catecholamines may contribute to the increase in ventricular ectopic activity. If therapy is required for catecholamine-sensitive ventricular tachycardia, β-blockers are the drugs of choice, specifically cardioselective β-blockers.

Another commonly observed ventricular tachycardia in patients with structurally normal hearts is fascicular tachycardia. This tachycardia, which often arises from the left posterior fascicle, has a typical right bundle branch block appearance with a superior axis and may be responsive to calcium channel blockers. This form of ventricular tachycardia has not been reported in pregnant patients.

In the presence of hemodynamically unstable ventricular tachycardia, immediate direct current cardioversion is appropriate. If the ventricular tachycardia is hemodynamically stable, IV antiarrhythmic drugs may be used. Although ACLS guidelines have recently been revised,[62] lidocaine is still the drug of choice in pregnant patients if antiarrhythmic drug use is considered. Few fetal side effects have been reported with the use of lidocaine, even in early pregnancy and despite significant placental drug transfer.[63]

Chronic antiarrhythmic therapy may be unavoidable in the face of sustained or recurrent ventricular tachycardia. Most of the drugs currently used in the treatment of ventricular tachycardia remain category C for use in pregnancy.[63] Exceptions include atenolol and amiodarone, which are FDA category D, and sotalol and lidocaine, which are classified as FDA category B. Nevertheless, many antiarrhythmic drugs have been used in pregnancy and have been found to be relatively safe at normal therapeutic doses. For example, quinidine has long been used without serious fetal side effects. Review of the literature also demonstrates that procainamide and flecainide have been used in pregnancy with good effect and no adverse fetal outcome,[43,64]

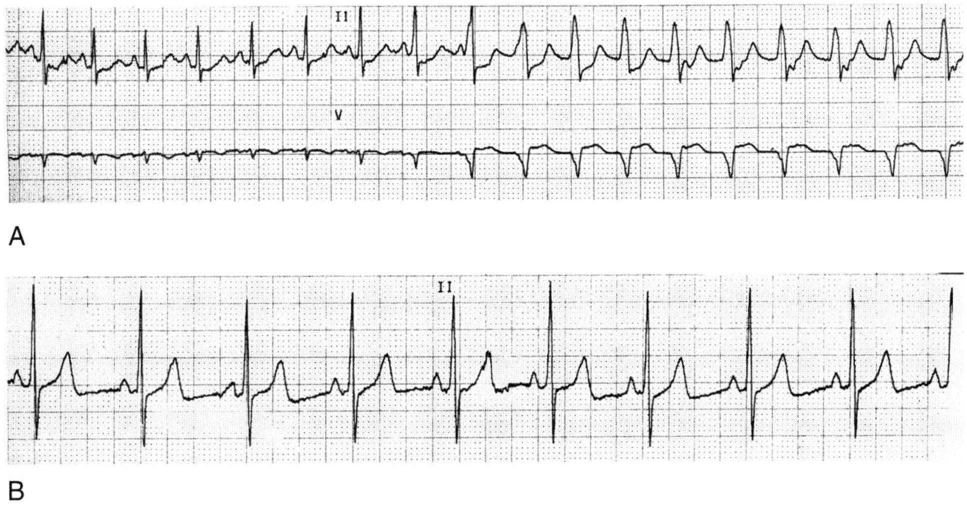

FIGURE 25-6 **A,** Documentation on telemetry of a right ventricular outflow tract tachycardia that is intermittent. In this patient, the tachycardia caused near syncope, thus requiring therapy. **B,** Telemetry tracing of the same patient, 2 days after the initiation of β-blocker therapy. The tachycardia has ceased.

although flecainide is contraindicated in structurally abnormal hearts. Because sotalol is now classified as an FDA category B antiarrhythmic drug, it may be a good choice in women with normal renal function. A number of sources report its safe use during pregnancy.[65]

Our own preference for medical therapy after β-blockers would be flecainide in patients with normal hearts and sotalol in patients with structural heart disease. Inpatient monitoring is advisable for the initiation of antiarrhythmic therapy for ventricular arrhythmias, particularly for drugs such as sotalol that have the potential for QT prolongation and torsades de pointes. Amiodarone should be avoided in all but the most refractory cases because of the significant risk of adverse fetal effects.

The development of smaller implantable defibrillators and prepectoral implantation allows consideration of this mode of therapy in patients with life-threatening ventricular arrhythmias. In light of the Antiarrhythmics versus Implantable Defibrillator (AVID) Trial results,[66] implantable cardioverter defibrillators (ICDs) may be appropriate in women whose ventricular tachycardia is secondary to a cardiomyopathy, either ischemic or nonischemic.[67] If this therapy is chosen, lead placement should be accomplished with echocardiographic guidance to prevent radiation exposure to the fetus. In selected cases, fluoroscopic guidance with appropriate shielding of the fetus may allow more efficient implantation of the device.

Cardiac Arrest

Fortunately, cardiac arrest is a rare occurrence in women of childbearing age, occurring in approximately 1/30,000 deliveries.[68] The acute management, particularly in the late stages of pregnancy, involves several special considerations. Resuscitation with the patient in the traditional supine position, as with standard cardiopulmonary resuscitation (CPR), may be less successful in pregnant women, particularly in late-term pregnancy. In these patients, there may be significant aortocaval obstruction, resulting in reduced venous return and forward blood flow with chest compression.[69,70] Successful resuscitation may be aided by performing CPR with the patient tilted on her side using either a wedge, overturned chairs, or another rescuer's knees for support (Fig. 25-7).[62,71] Chest compressions are generally performed higher on the chest than usual to accommodate the shift of pelvic and abdominal contents towards the head.

Aside from the change in position for CPR, standard ACLS guidelines for medications, intubation, and defibrillation apply, with one other exception. The most recent ACLS guidelines list amiodarone as the drug of choice in cases of refractory ventricular fibrillation.[62] Given the adverse fetal effects with amiodarone, pregnant patients should receive lidocaine or procainamide first if defibrillation is unsuccessful. However, if the fetus is not yet viable, and the death of the mother is imminent because of refractory ventricular fibrillation, amiodarone would certainly be worth a try.

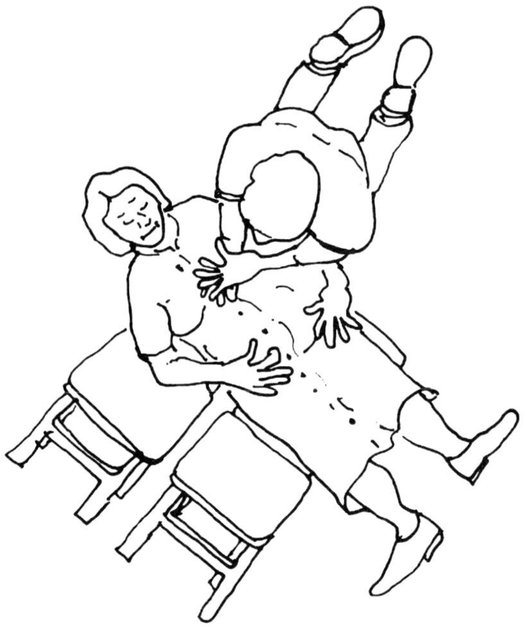

FIGURE 25-7 Schematic drawing of resuscitation of a pregnant patient after cardiac arrest. Overturned chairs are used to displace the uterus leftward, thus allowing increased venous return through the inferior vena cava.

It is important to recognize that the etiology of cardiac arrest in pregnancy may be different than in the general population. Potential causes are listed in Table 25-7.

Until the fetus becomes viable, at approximately 25 weeks, CPR should be performed as in the nonpregnant patient. After 25 weeks, caesarean section to save the fetus should be seriously considered within 5 minutes of the arrest. It may facilitate the successful resuscitation of the mother and rescue the fetus as well.[9] Neonatal and obstetric personnel should be involved early in the resuscitation.

If defibrillation is performed, there is literature that documents success with no adverse effect to mother or fetus with energies as high as 300 J.[55] The risk of inducing a fetal arrhythmia with cardioversion is small, primarily because only a fraction of the total energy reaches the fetal heart.[72] Nevertheless, the fetal ECG should be monitored during the procedure.

Patients with a Previously Implanted Implantable Cardioverter Defibrillator

Rarely, patients who are pregnant have a previously implanted ICD. The outcome of these women has been

TABLE 25-7 Etiology of Cardiac Arrest in Pregnancy

Amniotic fluid embolism	Hemorrhage
Pulmonary embolism	Magnesium sulfate toxicity
Eclampsia	Trauma
Peripartum cardiomyopathy	Epidural anesthesia
Aortic dissection	

evaluated in a multicenter retrospective analysis of 44 women.[73] In this study, 42 of the women had abdominally implanted generators, and two had a prepectoral device. Thirty had epicardial lead sensing systems, and 14 had been implanted transvenously. Thirty-six (82%) experienced no complications during pregnancy, and eight (18%) had a medical or device-related complication. ICD-related problems included tenderness at the ICD pocket scar (two patients), generator migration (one), and pericarditis secondary to the epicardial patches (one). Thirty-nine babies were born healthy, one was stillborn, two were small for gestational age, and one had transient hypoglycemia; one woman had a therapeutic abortion unrelated to the ICD. During pregnancy, 33 women received no ICD therapy, 8 had one shock, one had five discharges, one had 11 shocks, and one had five shocks. The authors concluded that the presence of an ICD should not prevent a woman from becoming pregnant unless she has underlying structural cardiac disease that is considered a contraindication. They added that pregnancy does not increase the risk of major ICD-related complications or result in a high number of ICD discharges.

There are no published data on the implantation of an ICD during pregnancy. In the event that implantation of an ICD became medically necessary, implantation of lead systems has been described using echocardiographic guidance.[23,74]

Long QT Syndrome

Management of patients with the hereditary form of the LQTS was recently reviewed by Rashba et al. for the LQTS Investigators.[75] Because the optimal management of this patient population during pregnancy and the postpartum period was essentially unknown, these investigators evaluated the effect of pregnancy on the incidence of cardiac events in women with LQTS. The study was retrospective and included 422 women, 111 affected patients and 311 first-degree relatives. Cardiac events were defined as the combined incidence of LQTS-related death, aborted cardiac arrest, and syncope. The investigators compared 40-week intervals of the prepartum, pregnancy, and postpartum time periods. They found that the pregnant state was not significantly associated with an increase in risk for cardiac events. However, the postpartum interval was associated with an increase in risk in affected patients compared to the other two time periods studied (odds ratio 40.8; 95% confidence interval, 3.1 to 5.4; $P = .01$). They suggested that management of these patients should include β-blocker therapy throughout the pregnant and postpartum periods. Successful pregnancy has been reported in a patient with LQTS after a left stellate ganglionectomy.[76]

Other investigators have published case reports on the management of LQTS patients should they develop an arrhythmia while pregnant.[76–78] Syncope is a particularly alarming presentation and should be interpreted as heralding possible sudden death. Many of these patients will require continuous hospitalization until after delivery, when definitive therapy such as ICD implantation can be undertaken. Telemetric monitoring during labor and delivery is essential. Prolongation of the QTc interval during labor has been described,[79] suggesting that the physical or emotional stress, or both, during labor might cause prolonged QTc intervals in women with LQTS.

Arrhythmogenic Right Ventricular Dysplasia

Reports of pregnancy in patients with arrhythmogenic right ventricular dysplasia (Uhl's anomaly) are rare but available. One such report described a 26-year-old woman with severe enlargement and hypokinesis of the right ventricle, tricuspid regurgitation, and mild pulmonic insufficiency.[80] She was advised to avoid pregnancy, but became pregnant 3 months later. She was delivered of a healthy baby girl at 37 weeks' gestation by cesarean section. Her pregnancy was complicated only by mild peripheral edema. No arrhythmias were noted during labor and delivery, despite a history of recurrent syncope.

Another case report described a young woman with a moderate to severe form of arrhythmogenic right ventricular dysplasia who was receiving flecainide for ventricular arrhythmias.[81] During the pregnancy, no significant ventricular arrhythmias were noted. A normal baby was delivered at term by cesarean section.

Hypertrophic Cardiomyopathy

Hypertrophic cardiomyopathy during pregnancy is fairly well described in the literature.[82-84] The physiologic changes that are most devastating for patients with hypertrophic cardiomyopathy, particularly those with an obstructive component, are peripheral vasodilation and a reduction in preload. In contrast, pregnancy characteristically causes an increase in intravascular volume and a decrease in systemic vascular resistance. Thus, the vast majority of patients with this cardiomyopathy do quite well during pregnancy. However, there are isolated case reports in the literature describing maternal and fetal death from ventricular arrhythmias associated with hypertrophic cardiomyopathy.[83] Although rare, ventricular arrhythmias in this patient population are particularly dangerous precisely because of the physiologic changes of pregnancy. Aortocaval compression near the end of pregnancy or large blood losses during delivery can decrease preload dramatically and impede successful resuscitation.

BRADYARRHYTHMIAS AND PREGNANCY

Vasodepressor Syncope

Vasovagal syncope is one of the most common causes of symptomatic bradycardia in women of childbearing age and may be responsible for more than 20% of unexplained syncope in the general population. Nevertheless, it is rare during pregnancy. Evaluation of the patient with suspected vasovagal syncope may include tilt table testing, with therapy then tailored to the results obtained. Many patients can simply be managed with education as to the warning signs of imminent syncope and measures to take to abort the episode, in addition to staying well hydrated. These patients are best treated with β-blockade as initial therapy, if treatment is necessary.

Congenital Complete Heart Block

Although most cases of congenital heart block are diagnosed during childhood, some are discovered incidentally during pregnancy. In general, these patients are asymptomatic, and no acute intervention is required during pregnancy. For symptomatic patients in the first and second trimesters, there are no definite guidelines, but permanent pacemaker implantation is probably indicated, using electrocardiographic or echocardiographic guidance for lead placement.[23,74] In symptomatic women who present at or near term, temporary pacing with induction of labor as soon as possible is the procedure of choice.[85]

Labor and Delivery

During labor, both cardiac output and blood pressure increase. After delivery, cardiac output initially increases, but it begins to decrease within the first hour to reach baseline levels 2 weeks postpartum. The sudden onset of supraventricular and ventricular tachycardia during labor and delivery has been reported.[86] Management of these arrhythmias depends on the etiology of the arrhythmia. For example, if the patient develops an SVT that uses the AV node as part of the reentrant circuit, adenosine would be the drug of choice. Recurrent, rapid SVT that could potentially compromise the fetus would best be managed by cesarean section rather than administration of multiple drugs to control the arrhythmia.

Patients with underlying heart disease should have continuous electrocardiographic monitoring during labor and delivery, even if no previous arrhythmia has been documented. For example, in one patient with mitral insufficiency due to rheumatic disease, a Holter monitor worn during labor and delivery demonstrated frequent runs of ventricular tachycardia. Routine bedside examination of the pulse did not unmask the problem.[86]

Complete heart block occurring during labor and delivery is an uncommon occurrence, although it has been described.[86] In one published case, the patient had underlying rheumatic heart disease, with mitral stenosis complicated by atrial fibrillation. Treatment for the arrhythmia was 0.5 mg of digoxin per day. The authors speculated that digoxin toxicity was the etiology of the complete heart block, although no levels were drawn to document this supposition. Treatment for sudden complete heart block during labor and delivery is limited to those patients who become hemodynamically unstable, and would consist of insertion of a temporary pacemaker under echocardiographic guidance.

Lactation and Breast-Feeding

Antiarrhythmic drug use while breast-feeding has similar considerations as in the use of these drugs during pregnancy (see Table 25-4). Many antiarrhythmic drugs and AV nodal blockers are excreted in breast milk, and thus continued use while breast-feeding must be considered on a case by case basis. Both warfarin and heparin are safe for use in nursing mothers.

When to Hospitalize the Patient

The indications for hospitalization of the pregnant patient are similar to those for nonpregnant individuals. Patients with a cardiac arrest, hemodynamically unstable supraventricular arrhythmias, or sustained ventricular tachycardia should be hospitalized for evaluation and therapy. In most cases, pregnant patients should be hospitalized for the initiation of antiarrhythmic therapy, with the possible exception being the class 1C antiarrhythmic drugs for supraventricular arrhythmias. Patients do not need to be hospitalized for evaluation of palpitations or for hemodynamically stable supraventricular arrhythmias that are self-terminating or that terminate with vagal maneuvers.

How to Manage the Patient After the Pregnancy

All arrhythmic problems that were encountered during pregnancy should be readdressed early after delivery in order to decide upon a treatment plan. Although many patients' symptoms will subside when the pregnancy is over, the substrate for the arrhythmia may remain. For example, in patients with reentrant SVT, radiofrequency ablation should be considered before the next pregnancy, the patient with congenital complete heart block should be considered for pacemaker therapy, and so forth. Even in those patients who do not plan another pregnancy, attempts should be made to address the arrhythmia once the child has been safely delivered.

SUMMARY

Arrhythmias during pregnancy can be safely managed both pharmacologically and nonpharmacologically with rare exceptions. Correct diagnosis of the arrhythmia is vital. Many of the currently available antiarrhythmic drugs are safe for both mother and fetus. If the symptoms are severe enough and therapy is warranted, choose drugs with the most historical use in pregnancy, use the least number of medications possible, and use the lowest dose of drug that is effective. Cardioversion has been proved safe during pregnancy and can be used as an early treatment option in arrhythmia management.

REFERENCES

1. Rotmensch HH, Elkayam U, Frishman W: Antiarrhythmic drug therapy during pregnancy. Ann Intern Med 1983;98:487.
2. Michaelsson M, Engle MA: Congenital complete heart block: An international study of the natural history. Cardiovasc Clin 1972;4:85.
3. Krauer B, Krauer F: Drug kinetics in pregnancy. Clin Pharmacokinet 1977;2:167.
4. Dean M, Stock B, Patterson RJ, et al: Serum protein binding of drugs during and after pregnancy in humans. Clin Pharmacol Ther 1980;28:253.
5. Cox JL, Gardner MJ: Treatment of cardiac arrhythmias during pregnancy. Prog Cardiovasc Dis 1993;36:137.
6. Mitani GM, Steinberg I, Lien EJ, et al: The pharmacokinetics of antiarrhythmic agents in pregnancy and lactation. Clin Pharmacokinet 1987;12:253.
7. Chen SS, Perucca E, Lee JN, et al: Serum protein binding and free concentration of phenytoin and phenobarbitone in pregnancy. Br J Clin Pharmacol 1982;13:547.
8. Herngren L, Ehrnebo M, Boreus LO: Drug binding to plasma proteins during human pregnancy and in the perinatal period. Studies on cloxacillin and alprenolol. Dev Pharmacol Ther 1983;6:110.
9. Anderson MH: Rhythm disorders. In Oakley C (ed): Heart Disease in Pregnancy. London, BMJ Publishing Group, 1997, p 248.
10. Joglar JA, Page RL: Treatment of cardiac arrhythmias during pregnancy: Safety considerations. Drug Saf 1999;20:85.
11. Shotan A, Ostrzega E, Mehra A, et al: Incidence of arrhythmias in normal pregnancy and relation to palpitations, dizziness, and syncope. Am J Cardiol 1997;79:1061.
12. Tawam M, Levine J, Mendelson M, et al: Effect of pregnancy on paroxysmal supraventricular tachycardia. Am J Cardiol 1993; 72:838.
13. Lee SH, Chen SA, Wu TJ, et al: Effects of pregnancy on first onset and symptoms of paroxysmal supraventricular tachycardia. Am J Cardiol 1995;76:675.
14. Hunter S, Robson S: Adaptation of the cardiovascular system to pregnancy. In Oakley C (ed): Heart Disease in Pregnancy. London, BMJ Publishing Group, 1997, p 5.
15. Nihoyannopoulos P: Cardiovascular examination in pregnancy and the approach to diagnosis of cardiac disorder. In Oakley C (ed): Heart Disease in Pregnancy. London, BMJ Publishing Group, 1997, p 52.
16. Shalev Y, Ben-Hur H, Hagay Z, et al: Successful delivery following myocardial ischemia during the second trimester of pregnancy. Clin Cardiol 1993;16:754.
17. Miller RW: Intrauterine radiation exposures and mental retardation. Health Phys 1988;55:295.
18. Stewart A, Webb J, Giles D: Malignant disease in childhood and diagnostic irradiation in utero. Lancet 1956;2:447.
19. Doll R, Wakeford R: Risk of childhood cancer from fetal irradiation. Br J Radiol 1997;70:130.
20. Brent RL: The effect of embryonic and fetal exposure to x-ray, microwaves, and ultrasound: Counseling the pregnant and nonpregnant patient about these risks. Semin Oncol 1989;16:347.
21. Brent R, Meistrich M, Paul M: Ionizing and nonionizing radiations. In Paul M (ed): Occupational and Environmental Reproductive Hazards: A Guide for Clinicians. Baltimore, Williams & Wilkins, 1993, p 165.
22. Lee MS, Evans SJ, Blumberg S, et al: Echocardiographically guided electrophysiologic testing in pregnancy. J Am Soc Echocardiogr 1994;7:182.
23. Drinkovic N: Subcostal echocardiography to determine right ventricular pacing catheter position and control advancement of electrode catheters in intracardiac electrophysiologic studies. Am J Cardiol 1981;47:1260.
24. Dominguez A, Iturralde P, Hermosillo AG, et al: Successful radiofrequency ablation of an accessory pathway during pregnancy. Pacing Clin Electrophysiol 1999;22:131.
25. Marchlinski F, Callans D, Gottlieb C, et al: Magnetic electroanatomical mapping for ablation of focal atrial tachycardias. Pacing Clin Electrophysiol 1998;21:1621.
26. Grubb BP: Neurocardiogenic syncope. In Grubb BP, Olshansky B (eds): Syncope: Mechanisms and Management. Armonk, NY, Futura Publishing Company, 1998, p 73.
27. Grubb BP, Kosinski D, Samoil D, et al: Postpartum syncope. Pacing Clin Electrophysiol 1995;18:1028.
28. Physicians' Desk Reference: PDR, 54th ed. Oradell, NJ, Medical Economics, 2000, p 344.
29. Afridi I, Moise KJ Jr, Rokey R: Termination of supraventricular tachycardia with intravenous adenosine in a pregnant woman with Wolff-Parkinson-White syndrome. Obstet Gynecol 1992;80:481.
30. Mason BA, Ricci-Goodman J, Koos BJ: Adenosine in the treatment of maternal paroxysmal supraventricular tachycardia. Obstet Gynecol 1992;80:478.
31. DiMarco JP, Sellers TD, Lerman BB, et al: Diagnostic and therapeutic use of adenosine in patients with supraventricular tachyarrhythmias. J Am Coll Cardiol 1985;6:417.
32. Elkayam U, Goodwin M: Safety and efficacy of intravenous adenosine therapy for supraventricular tachycardia during pregnancy—results of a national survey. J Am Coll Cardiol 1994;23(suppl A):91A.
33. Klein V, Repke JT: Supraventricular tachycardia in pregnancy: Cardioversion with verapamil. Obstet Gynecol 1984;63 (suppl 3):16S.
34. El-Sayed YY, Holbrook RH Jr, Gibson R, et al: Diltiazem for maintenance tocolysis of preterm labor: Comparison to nifedipine in a randomized trial. J Matern Fetal Med 1998;7:217.
35. Saarikoski S: Placental transfer and fetal uptake of ^{3}H-digoxin in humans. Br J Obstet Gynaecol 1976;83:879.
36. Heaton FC, Vaughan R: Intrauterine supraventricular tachycardia: Cardioversion with maternal digoxin. Obstet Gynecol 1982;60:749.
37. Wladimiroff JW, Stewart PA: Treatment of fetal cardiac arrhythmias. Br J Hosp Med 1985;34:134.
38. Page RL: Treatment of arrhythmias during pregnancy. Am Heart J 1995:130;871.
39. Pruyn SC, Phelan JP, Buchanan GC: Long-term propranolol therapy in pregnancy: Maternal and fetal outcome. Am J Obstet Gynecol 1979;135:485.
40. Hill LM, Malkasian GD Jr: The use of quinidine sulfate throughout pregnancy. Obstet Gynecol 1979;54:366.
41. Dumesic DA, Silverman NH, Tobias S, et al: Transplacental cardioversion of fetal supraventricular tachycardia with procainamide. N Engl J Med 1982;307:1128.
42. Kofinas AD, Simon NV, Sagel H, et al: Treatment of fetal supraventricular tachycardia with flecainide acetate after digoxin failure. Am J Obstet Gynecol 1991;165:630.
43. Ahmed K, Issawi I, Peddireddy R: Use of flecainide for refractory atrial tachycardia of pregnancy. Am J Crit Care 1996;5:306.
44. MacNeil DJ: The side effect profile of class III antiarrhythmic drugs: Focus on d,l-sotalol. Am J Cardiol 1997;80:90G.
45. Chow T, Galvin J, McGovern B: Antiarrhythmic drug therapy in pregnancy and lactation. Am J Cardiol 1998;82:58I.
46. Magee LA, Downar E, Sermer M, et al: Pregnancy outcome after gestational exposure to amiodarone in Canada. Am J Obstet Gynecol 1995;172:1307.
47. Ovadia M, Brito M, Hoyer GL, et al: Human experience with amiodarone in the embryonic period. Am J Cardiol 1994;73:316.
48. Ellsworth AJ, Witt DM, Dugdale DC, Oliver LM: Mosby's Complete Drug Reference, 7th ed. St. Louis, Mosby-Yearbook, 1997.
49. Webster WS, Brown-Woodman PD, Snow MD, et al: Teratogenic potential of almokalant, dofetilide, and d-sotalol: Drugs with potassium channel blocking activity. Teratology 1996;53:168.
50. Wilbur SL, Marchlinski FE: Adenosine as an antiarrhythmic agent. Am J Cardiol 1997;79:30.
51. The American Heart Association in collaboration with the International Liaison Committee on Resuscitation: Guidelines 2000 for cardiopulmonary resuscitation and emergency cardiovascular care. Part 6: Advanced cardiovascular life support: 7D: The tachycardia algorithms. Circulation 2000;102(suppl 8):I158.
52. Widerhorn J, Widerhorn AL, Rahimtoola SH, et al: WPW syndrome during pregnancy: Increased incidence of supraventricular arrhythmias. Am Heart J 1992;123:796.
53. Goldberger J, Helmy I, Katzung B, et al: Use-dependent properties of flecainide acetate in accessory atrioventricular pathways. Am J Cardiol 1994;73:43.
54. Manolis AS, Estes NA III: Reversal of electrophysiologic effects of flecainide on the accessory pathway by isoproterenol in the Wolff-Parkinson-White syndrome. Am J Cardiol 1989;64:194.
55. Curry JJ, Quintana FJ: Myocardial infarction with ventricular fibrillation during pregnancy treated by direct current defibrillation with fetal survival. Chest 1970;58:82.

56. Reiffel JA: Drug choices in the treatment of atrial fibrillation. Am J Cardiol 2000;85(10 suppl 1):12.
57. Shaul WL, Hall JG: Multiple congenital anomalies associated with oral anticoagulants. Am J Obstet Gynecol 1977;127:191.
58. Mackenzie J: Heart Disease and Pregnancy. London, Henry Frowde & Hodder & Stoughton, 1921, pp 92-116.
59. Chandra NC, Gates EA, Thamer M: Conservative treatment of paroxysmal ventricular tachycardia during pregnancy. Clin Cardiol 1991;14:347.
60. Brodsky M, Doria R, Allen B, et al: New-onset ventricular tachycardia during pregnancy. Am Heart J 1992;123: 933.
61. Marchlinski FE, Deely MP, Zado ES: Sex-specific triggers for right ventricular outflow tract tachycardia. Am Heart J 2000;139:1009.
62. American Heart Association and the International Liaison Committee on Resuscitation: Guidelines 2000 for cardiopulmonary resuscitation and emergency cardiovascular care. Part 8: Advanced challenges in resuscitation. Section 3: Special challenges in ECC. 3F: Cardiac arrest associated with pregnancy. Circulation 2000;102(suppl 8):I247.
63. Briggs GG. Freeman RK, Yaffe SJ: Drugs in Pregnancy and Lactation, 5th ed. Baltimore, Williams & Wilkins, 1998.
64. Fagih B, Sami M: Safety of antiarrhythmics during pregnancy: Case report and review of the literature. Can J Cardiol 1999;15:113.
65. O'Hare MF, Murnaghan GA, Russell CJ, et al: Sotalol as a hypotensive agent in pregnancy. Br J Obstet Gynaecol 1980;87:814.
66. The Antiarrhythmics versus Implantable Defibrillators (AVID) Investigators: A comparison of antiarrhythmic-drug therapy with implantable defibrillators in patients resuscitated from near-fatal ventricular arrhythmias. N Engl J Med 1997;337:1576.
67. The AVID Investigators: Causes of death in the Antiarrhythmics versus Implantable Defibrillators (AVID) Trial. J Am Coll Cardiol 1999;34:1552.
68. Kloeck W, Cummins RO, Chamberlain D, et al: Special resuscitation situations: An advisory statement from the International Liaison Committee on Resuscitation. Circulation 1997;95:2196.
69. Kerr MG, Scott DB, Samuel E: Studies of the inferior vena cava in late pregnancy. BMJ 1964;1:532.
70. Lee RV, Rodgers BD, White LM, et al: Cardiopulmonary resuscitation of pregnant women. Am J Med 1986;81:311.
71. Goodwin AP, Pearce AJ: The human wedge. A manoeuvre to relieve aortocaval compression during resuscitation in late pregnancy. Anaesthesia 1992;47:433.
72. DeSilva RA, Graboys TB, Podrid PJ, et al: Cardioversion and defibrillation. Am Heart J 1980;100:881.
73. Natale A, Davidson T, Geiger MJ, et al: Implantable cardioverter-defibrillators and pregnancy: A safe combination? Circulation 1997;96:2808.
74. Jordaens LJ, Vandenbogaerde JF, Van de Bruaene P, et al: Transesophageal echocardiography for insertion of a physiological pacemaker in early pregnancy. Pacing Clin Electrophysiol 1990;13:955.
75. Rashba EJ, Zareba W, Moss AJ, et al for the LQTS Investigators: Influence of pregnancy on the risk for cardiac events in patients with hereditary long QT syndrome. Circulation 1998; 97:451.
76. Bruner JP, Barry MJ, Elliott JP: Pregnancy in a patient with idiopathic long QT syndrome. Am J Obstet Gynecol 1984;149:690.
77. McCurdy CM Jr, Rutherford SE, Coddington CC: Syncope and sudden arrhythmic death complicating pregnancy. A case report of Romano-Ward syndrome. J Reprod Med 1993;38:233.
78. Ganta R, Roberts C, Elwood RJ, et al: Epidural anesthesia for cesarean section in a patient with Romano-Ward syndrome. Anesth Analg 1995;81:425.
79. Minakami H, Nakayam T, Ohno T, et al: Effect of vaginal delivery on the QTc interval in a patient with the long QT (Romano-Ward) syndrome. J Obstet Gynaecol Res 1999;25:251.
80. Koenig C, Katz M, Gertsch M, et al: Pregnancy and delivery in a patient with Uhl anomaly. Obstet Gynecol 1991;78;932.
81. Villanova C, Muriago M, Nava F: Arrhythmogenic right ventricular dysplasia: Pregnancy under flecainide treatment. G Ital Gardiol 1998;28:691.
82. Kolibash AJ, Ruiz DE, Lewis RP: Idiopathic hypertrophic subaortic stenosis in pregnancy. Ann Intern Med 1975;82:791.
83. Shah DM, Sunderji SG: Hypertrophic cardiomyopathy and pregnancy: Report of a maternal mortality and review of literature. Obstet Gynecol Surv 1985;40:444.
84. Maron BJ, Bonow RO, Cannon RO III, et al: Hypertrophic cardiomyopathy. Interrelations of clinical manifestations, pathophysiology, and therapy. N Engl J Med 1987;316:844.
85. Dalvi BV, Chaudhuri A, Kulkarni HL, et al: Therapeutic guidelines for congenital complete heart block presenting in pregnancy. Obstet Gynecol 1992;79:802.
86. Spritzer RC, Seldon M, Mattes LM, et al: Serious arrhythmias during labor and delivery in women with heart disease. JAMA 1970;211:1005.

Chapter 26

Evaluation and Management of Arrhythmias in a Pediatric Population

VICTORIA L. VETTER and LARRY A. RHODES

Appropriate evaluation and management of arrhythmias in pediatric patients requires an understanding of the presentation, clinical correlations, and unique responses of children to the variety of potential treatments. The primary focus of this review of arrhythmias in the pediatric population is on arrhythmias that are seen more frequently or behave differently in children than adults. In those in which there are shared characteristics, such as presentation and management strategies, we refer you to the excellent chapters in this text dealing with each specific rhythm disturbance.

Supraventricular Tachycardia

PRESENTATION AND EVALUATION

The presentation of the pediatric patient with supraventricular tachycardia (SVT) varies from complaints of palpitations to signs of marked cardiac failure. This variability is in part related to the patient's age at presentation and his or her ability to communicate symptoms, as well as the type of SVT. Symptoms in those old enough to voice complaints include palpitations ("heart beeping"), chest pain, tiredness, dizziness, and syncope. Parents of infants with SVT occasionally voice concerns over changes in feeding habits or irritability as symptoms.

The evaluation of pediatric patients includes a 12- or 15-lead ECG as a baseline screen for Wolff-Parkinson-White (WPW) syndrome or other abnormalities such as an ectopic atrial rhythm. Twenty-four hour ambulatory (Holter) monitoring is useful if the patient is having frequent symptoms (i.e., on a daily basis) or for screening for occult arrhythmias. It is also used to assess heart rate ranges and variability. Transtelephonic event monitors are often helpful in those with sporadic episodes of palpitations and offer the advantage of allowing the patient to record symptoms as they occur. Exercise tolerance testing is useful in those with palpitations associated with activities. More invasive testing such as transesophageal or intracardiac electrophysiology studies (EPSs) are used to induce the arrhythmia and determine the particular mechanism.

SVTs are arrhythmias that originate above the ventricles and involve the atria, atrioventricular (AV) node, or accessory bypass tracts. SVTs represent the most common tachycardias seen in infants and children, with an incidence of 1 in 250 to 1000.[1] SVTs are classified by identifying the components of the heart that are required to maintain the tachycardia. These include primary atrial tachycardias (those in which the AV node or the ventricles are not required), AV reciprocating tachycardias (those that require both the atrium and the ventricle), and those that do not require the atria as part of the circuit.

MECHANISMS

Three primary mechanisms are involved in SVT: enhanced automaticity, triggered automaticity, and reentry. Automaticity is the ability of cardiac myocytes to spontaneously depolarize and lead to myocardial contractions. The sinus and AV nodes are the primary sites of automaticity in the heart. Enhanced automaticity occurs when myocytes outside the sinus or the AV node depolarize spontaneously. Single contractions lead to atrial or ventricular ectopic beats or SVT and ventricular tachycardias (VTs) when repetitive.[2]

Tachycardias arising secondary to enhanced automaticity have characteristics that are distinct from those of reentry. They are very catecholamine sensitive, with warm-up and cool-down phases. This leads to variable rates during the tachycardia. These tachycardias are not inducible with programmed stimulation nor are they terminated with overdrive pacing or with direct current cardioversion. Automatic tachycardias can arise from all areas of the heart, with those from the atria referred to as *ectopic atrial*, those from the AV junction as *junctional ectopic*, and those from the ventricles as *automatic VT*.

Triggered automaticity results from spontaneous myocardial contractions that occur secondary to oscillations during repolarization reaching threshold and leading to a depolarization.[3,4] These oscillations are referred to as *afterdepolarizations*. Arrhythmias arising from triggered automaticity have characteristics shared by both enhanced automaticity and reentry. They are very catecholamine sensitive and have warm-up and cool-down phases with a wide variation in heart rate, similar to other forms of automaticity. They can be induced and terminated with pacing maneuvers and respond to direct current cardioversion, as do reentrant arrhythmias. Triggered automaticity is thought to play a role in the arrhythmias seen in digoxin toxicity.[5]

Reentry occurs when a wavefront of electrical activation travels through tissue for a distance and then reenters the original tissue and propagates through the circuit again. Reentrant SVT is the most common type of tachycardia seen in pediatrics. In order to have a reentrant circuit, there must be at least two pathways with distinct conduction properties and refractory periods. The sequence of events in a reentrant circuit is as follows: a stimulus encounters two distinct pathways with one pathway refractory to conduction (the pathway with the longer refractory period) and one ready for conduction (the pathway with the shorter refractory period). The impulse is conducted along the pathway with the shorter refractory period having enough conduction delay to allow the first pathway to recover and to conduct the impulse back to the original site of entry into the circuit. The impulse then reenters the original pathway and transverses the circuit again.

Reentry can occur in the sinus node, atrial tissue, the AV node, ventricular tissue, or between the atrium and the ventricle. Those reentry circuits between the atrium and the ventricle involve specialized conduction tissue referred to as accessory pathways. In AV nodal reentrant tachycardia (AVNRT), the "accessory pathway" is in the AV node. Moe and colleagues initially described the evidence of dual AV node physiology in 1956.[6] Reentry tachycardias have characteristics that are distinct from those seen in automatic tachycardias. They often have a sudden onset and termination, with patients frequently feeling that they are "switched" on and off. The tachycardia is very regular, with little variation in rate. Reentry tachycardias often respond to vagal maneuvers by slowing or terminating the tachycardia. These tachycardias are easily induced and terminated with pacing protocols ranging from single premature stimuli to burst pacing or by direct current cardioversion.

Primary Atrial Tachycardia

Ectopic Atrial Tachycardia

Ectopic atrial tachycardia (EAT), or automatic atrial tachycardia, is an arrhythmia arising from either atrium with inappropriately fast atrial rates. This tachycardia represents approximately 10% of the SVT seen in the overall population.[7] The heart rates seen in EAT vary based on the patient's age and catecholamine state during the tachycardia. The heart rates in automatic atrial tachycardias will be inappropriately fast for the patient's level of activity and are generally in the 130 to 210 beat per minute (bpm) range. In general, the rates are not as fast as those seen with reentrant tachycardias. The pulse rate may not be reflective of the atrial rate because of variable atrial conduction through the AV node.

Tachycardias arising from foci of increased automaticity can be found in all areas of the atria. There appear to be areas in which the automatic foci are more likely to be located. In the right atrium, they are frequently found in the right atrial appendage, and in the left atrium they are often mapped to the orifices of the pulmonary veins.[8-10]

Presentation. EAT is not seen frequently in young infants, but the overall incidence of automatic atrial tachycardia reported by Gillette et al.[11] was 18% of all SVT seen in children. EATs usually have a "warm-up" phase in which the heart rate will gradually increase to its maximum rate. During termination, it will "cool down" with a gradual slowing of the rate and become indistinguishable from that of sinus rhythm. This gradual increase in heart rate and termination, which is markedly different from the sudden onset and termination seen in reentrant SVT, makes it difficult for the patient to recognize the tachycardia. Failure to perceive the arrhythmia and its incessant nature can lead to significant depression of myocardial function, which results in a tachycardia-mediated cardiomyopathy. Many patients with EAT present with signs of congestive heart failure[12] with decreased left ventricular (LV) contractility, AV valve regurgitation, and atrial dilation. If the tachycardia is not treated aggressively, the myocardial function can continue to decline, resulting in an irreversible cardiomyopathy.

Evaluation. ECGs in patients with an EAT generally show a P wave axis distinct from sinus rhythm. When the focus arises from the left atrium, the P wave is negative in lead I; those with the focus in the low right atrium show a negative P wave axis in lead aVF with a positive P wave in lead I. Occasionally, the focus is in an area close to the sinus node or in the high right atrium with the P wave axis similar to sinus tachycardia (0 to 90 degrees in the frontal plan). This can lead to a delay in diagnosis and institution of therapy, when the rhythm is thought to be sinus tachycardia. EAT, as well as other primary atrial tachycardias, can show variable degrees of AV conduction on ECG and ambulatory monitoring, which is often helpful in establishing the diagnosis. Figure 26-1 represents an ECG of a patient with EAT.

Ambulatory (Holter) monitoring is very helpful in determining the frequency of EAT, as well as in establishing the diagnosis. A careful review of the patient's diary of activities along with the corresponding heart rates allows one to determine if rates are inappropriately elevated. The overall average heart rate over 24 hours provides an indication of the amount of time the patient is in SVT. Special attention should be given to the heart rate during sleep. These noninvasive evaluations may be the only way to assess the patient with EAT. Exercise testing is frequently not useful in bringing out EAT, because as the sinus heart rate

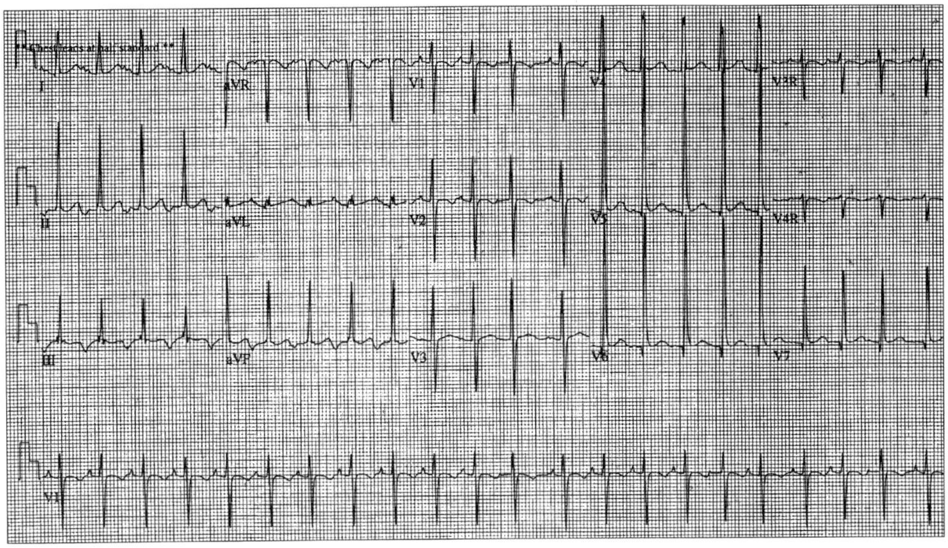

FIGURE 26-1 ECG of ectopic atrial tachycardia in a 10-year-old. Note atrial rhythm originating from a low right atrial focus.

increases, the automatic focus is suppressed. In most cases, EAT cannot be induced in the cardiac catheterization laboratory using conventional pacing protocols but may be induced by rapid atrial pacing of the heart (if the mechanism is triggered automaticity) or isoproterenol infusion.

A patient with EAT should have a very complete echocardiographic evaluation. This evaluation should rule out congenital heart disease, especially the types that could lead to an abnormal sinus node (and therefore abnormal P wave axis) such as heterotaxy syndrome. A careful assessment of ventricular function, as well as the presence of AV valve regurgitation, should be performed. This evaluation should be done as a baseline study, before medical intervention, in order to document progression of the disease or the possible deterioration of ventricular function secondary to the negative inotropic properties of the medical management.

Treatment. The medical management of EAT may be problematic, with complete control of the tachycardia difficult to attain. Two strategies are used in the treatment of EAT. One strategy is an attempt to slow the ventricular rate by slowing conduction through the AV node. The primary medication used in this effort is digoxin, which slows the conduction through the AV node by enhancement of vagal activity. Digoxin is a positive inotropic agent and helps to improve ventricular performance. Calcium channel blockers can increase AV block but are negative inotropes and should be used with caution.

Medications that have the potential to slow the tachycardia include β-blockers and class I and class III antiarrhythmics. β-Blockers have the potential to slow the tachycardia by blocking the effect of catecholamines on the ectopic focus. Class IA antiarrhythmic medications such as procainamide decrease automaticity, prolong refractoriness, and slow conduction velocity.[13,14] Class IC medications such as flecainide and propafenone have shown some success in the management of EAT.[15,16] Amiodarone and sotalol (class III) have been used with modest success by slowing conduction throughout the myocardium as well as slowing AV conduction.[17,18,19] All of these medications have the potential to decrease myocardial performance and must be used with caution in patients with decreased LV function.

Because of the difficulty in the medical management of EAT and the sequelae of uncontrolled tachycardia, there is increasing enthusiasm for nonpharmacologic methods of treatment. These procedures include surgical ablation and cryoablation procedures, which have reasonable success rates but require an open chest procedure and, frequently, cardiac bypass.[20,21] Radiofrequency catheter ablation (RFA) has become the treatment of choice for patients with poorly controlled EAT.[10,22,23]

Certain technical aspects may make RFA of EAT challenging. It is difficult to induce EAT. If the arrhythmia does not occur spontaneously, it cannot be mapped and ablated in the electrophysiology laboratory. EAT is quite catecholamine sensitive. It is not uncommon for the tachycardia to "go to sleep" when the patient does, making sedation an issue in the laboratory.

Multifocal Atrial Tachycardia or Chaotic Atrial Tachycardia

Multifocal atrial tachycardia (MAT), or chaotic atrial tachycardia (CAT), is a primary atrial tachycardia arising from multiple areas of enhanced automaticity in the atria. By definition, the tachycardia must have at least three distinct P wave morphologies to be considered a MAT. These tachycardias are similar to other primary atrial tachycardias in that there may be variable conduction from the atrium to the ventricle. Multifocal atrial tachycardia is occasionally confused with atrial fibrillation in that there are multiple P wave morphologies and variable P–R and R–R intervals. The most common presentation in the newborn is with a large percentage resolving spontaneously over time.[24]

Presentation and Management. MAT is similar to EAT in that patients frequently present with signs and symptoms of congestive failure. These are often very difficult to treat. The management strategies are similar to those used in the treatment of EAT (i.e., slowing AV conduction or suppressing the tachycardia focus). In our experience, it often takes a combination of three or more medications to control the rate. This aggressive treatment strategy will often slow the underlying rhythm to the point that the patient may require the use of a pacemaker. Because of the multiple foci, RFA has not been as useful. There are difficulties in mapping the ectopic foci in that the site of activation is constantly changing.

Yeager et al. reported a 17% incidence of sudden death in patients with MAT.[25] Two of these deaths were thought to be bradycardia mediated. The bradycardia that is encountered may be related to the aggressive medical therapy required to control the rapid heart rates.

Atrial Flutter

Atrial flutter is a reentry circuit confined to the atria. Although atrial flutter is seen fairly commonly in adults with structurally normal hearts, the majority of atrial flutter seen in the pediatric population occurs in patients with congenital heart disease. In "classic" atrial flutter there are typical ECG findings that include negative P waves in the inferior leads (II, III, and AVF), with atrial rates of 250 to 450 bpm. This type of atrial flutter usually involves the isthmus between the inferior vena cava and the tricuspid valve as part of its circuit through the atrium.

Neonates present in atrial flutter in utero, at the time of birth or shortly thereafter. In most cases, the patient has a structurally normal heart, although congenital heart disease may be present. Atrial flutter represents 15% to 50% of all fetal SVT,[26-28] often resulting in fetal hydrops (a form of congestive heart failure). Atrial flutter seen in the newborn is generally converted to sinus rhythm by medication, transesophageal pacing, or direct current cardioversion. In the majority of these patients, reccurrence is rare.

Patients with congenital heart disease may develop atrial flutter before or following surgical interventions. Congenital defects associated with development of atrial flutter in the preoperative patient include those with marked right atrial enlargement. These include lesions with a single ventricle physiology (e.g., tricuspid atresia, pulmonary atresia, double inlet single left ventricle [right ventricle], double outlet right ventricle [left ventricle], mitral atresia, aortic atresia), tricuspid stenosis, tricuspid regurgitation, and pulmonary hypertension. Patients with structural heart disease do not have the classic saw-toothed ECG findings and have variable atrial rates between 150 and 450 bpm. The flutter circuits in repaired or palliated congenital heart disease travel between scars created by suture lines made during surgical procedures.

A number of types of congenital heart disease require atrial surgery. These range from procedures as simple as atrial septal repair to more complex procedures including the Mustard or Senning palliation of transposition of the great vessels and the Fontan procedure for palliation of single ventricle physiology.[29-32] These patients have multiple atrial flutter circuits. The conduction may change its course in the atrium, causing the rhythm to resemble a combination of atrial flutter and atrial fibrillation. These different "flutter" characteristics have led many experts to refer to this as *intra-atrial reentry tachycardia* rather than atrial flutter.

Presentation. The presentation of patients with atrial flutter is variable, with symptoms ranging from episodes of syncope and aborted sudden death to no symptoms at all. Symptoms are related to the ventricular rate during the tachycardia and the underlying health of the myocardium. The heart rate is secondary to the AV conduction of the atrial impulses to the ventricles. In patients with brisk AV conduction, the atrial flutter can be conducted in a 1:1 or 2:1 ratio depending on the flutter rate. This rapid AV conduction can lead to dizziness, syncope, or sudden death, especially in patients with poorly functioning ventricles and little myocardial reserve. If there is variable AV conduction, the patient will occasionally complain of palpitations.

Patients may present with decreased exercise tolerance or the complaint of feeling unwell. This is most likely secondary to the loss of AV synchrony with the loss of the atrial contraction. Patients with single ventricle physiology appear to be more dependent on the atrial contraction to maintain adequate cardiac output than patients with two functioning ventricles.

Many patients with structural heart disease have some component of AV node disease or are on medications such as digoxin that slow AV conduction. Atrial flutter rates in congenital heart disease are often slower than those seen in classical atrial flutter and result in ventricular rates not elevated to the point that the patient notices the elevated rate from the atrial flutter. This lack of recognition of the tachycardia is very worrisome in that there is a significant risk for the formation of atrial thrombi, especially in those patients who have sluggish blood flow in their atria secondary to marginal cardiac output.

Treatment. Before attempting to convert atrial flutter, one should rule out the presence of atrial thrombi to prevent emboli with conversion of the rhythm. The most effective means to evaluate intracardiac thrombi is a transesophageal echocardiogram. Acute treatment of atrial flutter includes the use of IV medications, overdrive pacing, or direct current cardioversion. Medications that have been used include those from class A such as procainamide, class C (propafenone), and class III (amiodarone, sotalol). Before using these medications, one should block rapid AV nodal conduction by using a medication such as digoxin to slow AV nodal conduction.

Chronic treatment of patients with congenital heart disease and atrial flutter is difficult, leading to the use of multiple medications (alone and in combination), pacemakers, RFA, and surgical interventions. The first-line medication used in the chronic treatment of atrial flutter is digoxin, which works by slowing the conduction through the AV node, preventing rapid ventricular rates, or preventing the initiation of atrial flutter. Frequently, digoxin alone will not control atrial flutter in these patients. Other medications such as class A agents (procainamide, quinidine, and disopyramide) have been used. These medications have potential deleterious side effects. Each has the propensity to prolong the Q–T interval and can be proarrhythmic with the development of torsades de pointes. Disopyramide is a potent negative inotrope, which may be detrimental to the patient with marginal ventricular function. Class C medications such as flecainide and propafenone have been used, but they have negative inotropic properties. Class III medications (amiodarone and sotalol) have been used with varying degrees of success in the management of patients with chronic atrial flutter. These medications have the benefit of slowing conduction through the AV node. Sotalol, with β-blocking properties, can be a significant negative inotropic agent. As the use of any of these medications can lead to the development of significant bradycardia, patients should be monitored closely for sinus bradycardia, sinus arrest, or default or junctional escape rhythms. Pacemakers should be placed in those patients who develop symptomatic bradycardia.

Antitachycardia pacing has also been used in the treatment of chronic atrial flutter in patients with congenital heart disease.[33,34] These devices have been successful in the termination of atrial flutter in patients with congenital heart disease. There are disadvantages to using these devices. The currently manufactured antitachycardia pacing devices require the use of bipolar pacing, requiring the patient to have two atrial epicardial wires placed on the atrium. There have also been reports of atrial flutter being converted to atrial fibrillation with rapid ventricular conduction.[33] Newer antitachycardia devices have the ability to offer backup defibrillation if there is degeneration of the rhythm to atrial or ventricular fibrillation (VF).

In most patients with complex congenital heart disease, including those who have the Fontan procedure or those with right-to-left intracardiac shunts, transvenous pacing may be contraindicated because of the potential for embolic phenomenon.

RFA has been used in the treatment of patients with atrial flutter. The circuits for atrial flutter in patients with congenital heart disease are distinctly different from those seen in classical atrial flutter. There are multiple reentrant circuits in these patients. These circuits transverse between scar lines created from the surgical procedures and natural barriers of conduction such as the crista terminalis and the orifices of the superior vena cava and inferior vena cava. In congenital heart patients, because the atrial tissue is thick, it is possible that energy delivered with conventional radiofrequency generators may not be sufficient to create a lesion that will disrupt the circuit. These unique problems lead to an overall success rate that is less than that seen with conventional atrial flutter. The acute success rate is somewhere between 70% and 80%, with recurrence rates as high as 50%.[35]

Atrial Fibrillation

Atrial fibrillation is a primary atrial tachycardia that involves a number of micro reentry circuits in the atrial tissue. Atrial fibrillation is thought to arise primarily from the left atrium. It is seen much more commonly in adults with acquired heart disease than in the pediatric population. It is a very common arrhythmia following rheumatic mitral valve disease and in patients with poor LV function. The classic finding of atrial fibrillation is a patient with an atrial rate of 350 to 400 bpm with a very irregular ventricular rate, secondary to the variable conduction through the AV node.

There is a population of adolescents with structurally normal hearts who present with idiopathic atrial fibrillation. Most of these patients will not experience recurrences once the rhythm is converted, but some will have recurrences and require anticoagulation in order to protect them from the development of thromboembolic phenomenon and antiarrhythmics to control the arrhythmia.

Junctional Ectopic Tachycardia

Junctional ectopic tachycardia (JET) is an automatic tachycardia that arises from the AV junction. There appear to be two distinct types of JET seen in childhood. The first is a familial form that occurs in early infancy and may be associated with congenital heart defects in up to 50% of patients.[36] The second type is seen in the early postoperative period following repair of congenital heart disease.[37] In both forms, the tachycardia appears to be secondary to enhanced automaticity. In those patients who present with the familial type of JET, the heart rates will range from 180 to 240 bpm. The ECG findings in JET show a tachycardia with a ventricular rate that is faster than the atrial rate with a narrow QRS complex similar to that seen in the patient's normal sinus rhythm. The findings of JET are shown in Figure 26-2. Rarely, patients with JET develop rate-related aberrancy, and some postoperative patients have a bundle branch block that will lead to a wide complex tachycardia. If the QRS is wide, the diagnosis of VT must be considered. If there is a wide complex tachycardia, VT must be ruled out by comparison to the QRS in normal sinus rhythm or by pacing the atrium faster than the ventricular rate to demonstrate wide complex conduction through the His-Purkinje system.

Patients with JET can present with signs and symptoms of congestive heart failure secondary to the persistently elevated heart rate. Sudden death has been reported in this patient population.[38]

Junctional ectopic tachycardia in the postoperative period is often very transient, lasting anywhere from 24 to 72 hours.[37,39,40] Although this tachyarrhythmia is transient, it can be fatal. The combination of a depressed cardiac function following cardiac surgery; an accelerated rhythm, with rates as high as 250 bpm; and a loss of AV synchrony can lead to significant and sometimes irreversible deterioration in the patient's hemodynamic status. JET has been seen following cardiac surgery for all cardiac defects but appears to be more prevalent in those patients having repair of defects that include a VSD, young patient age, and transient AV block.[41]

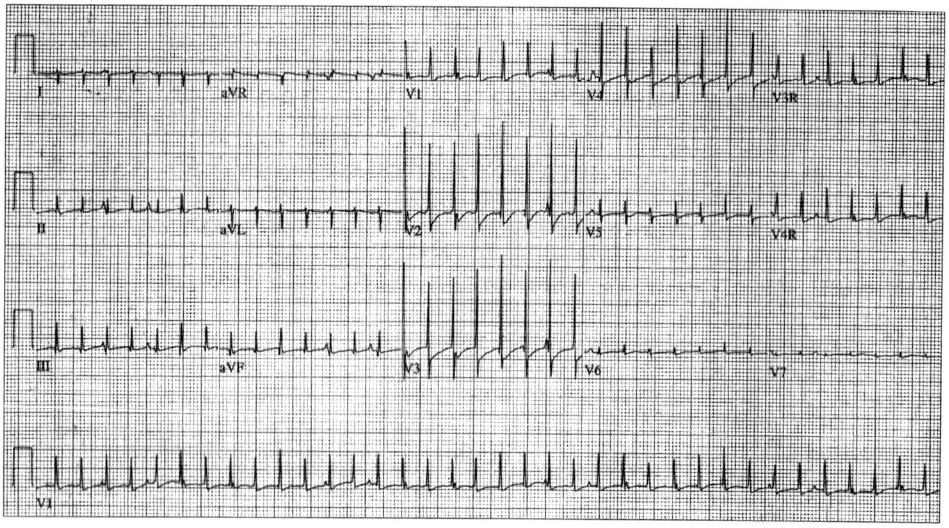

FIGURE 26-2 ECG of junctional ectopic tachycardia. Note narrow complex tachycardia with atrioventricular dissociation. The ventricular rate is 240 bpm, and the atrial rate is 120 to 140 bpm.

Treatment

Treatment strategies for the familial form of JET include using digoxin to slow the rhythm and provide inotropic support. Digoxin alone may not be sufficient to manage this arrhythmia, and the addition of a class IA, class IC, or class III antiarrhythmic agent is required. It has been our experience that a combination of these medications is required to control the tachycardia. The amount of antiarrhythmic medication needed to control the rate to prevent decompensation of cardiac function may suppress the sinus node sufficiently to require a pacemaker. Patients with JET have been treated with RFA.[42,43] Because of the proximity of the AV node and His-bundle to the area of enhanced automaticity responsible for JET, RFA in this setting carries a high risk of causing complete heart block. It is possible that over time the junctional rate will slow to a point that the patient may be weaned from the chronic medications.

Numerous treatment strategies have been developed in an attempt to treat JET in the postoperative patient. These have included the use of medications such as digoxin and IV procainamide. There have been proponents of the use of this combination of medications along with surface cooling.[39,41] Paired ventricular pacing to decrease the ventricular rate has been used.[44] IV amiodarone[45] has been effective in the treatment of these patients. In using IV antiarrhythmics, one must be aware of the potential negative inotropic effect of these medicines on the postoperative patient's heart. If the patient can be stabilized for a period of 48 to 72 hours, the rhythm disturbance will often spontaneously revert to a normal sinus rhythm.

Atrioventricular Reciprocating Tachycardias

Presentation and Treatment

There is some variability in the presentation of patients with AV reciprocating tachycardia in childhood. This may relate more to the age of the patient than to the specific type of tachycardia. For this reason, the presentation and treatment of pediatric patients with AV nodal reentry, concealed bypass tract, and WPW reentrant tachycardia are combined in the following overview, with emphasis on the different age-related presentations and treatment strategies.

Patients can present with tachycardia as early as during fetal development. The diagnosis of fetal SVT may be made by auscultation of the baby's heart rate during a maternal examination and is confirmed by fetal ultrasonography. The combination of M-mode and two-dimensional echocardiography can determine the tachycardia rate and possibly the mechanism of the tachycardia. These patients may present with hydrops fetalis, a form of congestive heart failure. The fetus with hydrops is considered to have a medical emergency and must be treated aggressively with attempts to terminate the tachycardia by treatment of the mother. Medications used include digoxin; β-blockers; calcium channel blockers; and classes I and III antiarrhythmics, either alone or in combination. Variations in transplacental transport of medications complicate treatment of the fetus. This difficulty in transplacental transport to the fetus is compounded by the fetal hydrops, further decreasing absorption. Another problem is the development of side effects in the mother, who must be monitored closely for proarrhythmic effects such as those seen with the classes I and III agents. If the fetus doesn't respond to medications in a timely manner and hydrops persists, premature delivery may be required.

Neonates and infants with tachycardia are often very sick, presenting with congestive heart failure. Tachycardia may be present for 48 to 72 hours before the patient receives medical care. This is secondary to the infant's inability to communicate and the family's lack of awareness that the child could have a significant medical condition. Parents may note that the infant was acting somewhat different than normal, more irritable, or not eating well. This is often interpreted as colic or some other "normal" childhood problem. At presentation, these babies often are acidotic from decreased cardiac output and may need aggressive resuscitation including artificial ventilation. These patients require rapid termination of the tachycardia. IV adenosine is effective in the acute termination of SVT in this population, but IV access is often difficult in a 3 to 4 kg baby in congestive heart failure. Transesophageal overdrive pacing has proven very helpful. Once the rhythm is restored to normal and the cardiac function has begun to recover, IV access becomes easier, and IV medications can be employed.

Digoxin is a first-line medication used in the treatment of SVT in infants with decreased myocardial performance. It is contraindicated in patients with WPW syndrome; we do not discharge patients with WPW syndrome on digoxin. Other medications used acutely to treat SVT include IV β-blockers such as esmolol, IV procainamide, and IV amiodarone. These medications must be used with caution because of their negative inotropic effects, which can lead to worsening cardiac function. The use of IV calcium channel blockers is contraindicated in infants as there have been reports of sudden death.[46,47]

The long-term treatment of infants with AV reciprocating tachycardia includes the use of oral preparations of the medications in Table 26-1. During initiation of therapy with oral antiarrhythmics, we monitor the patient in the hospital for at least five half-lives of the medication, looking for side effects and educating the family to administer the medicine and to recognize the tachycardia. When starting therapy with oral β-blockers in young children, one must monitor the patient for hypoglycemia and educate the parents in recognizing hypoglycemia symptoms. We ask the family to notify medical personnel if there are periods of decreased oral intake or periods of vomiting or diarrhea in patients on β-blockers. It is not our practice to discharge patients with heart rate monitors. Families can assess their children for recurrent SVT without the need for continuous monitoring. Most infants are treated with oral medications for 10 to 12 months, with the dosage adjusted based on their weight. At 10 months to 1 year of age, if there have been no recurrences of the SVT, the patient will be weaned from the medication unless WPW syndrome is still present on the ECG. Approximately one third of all patients who develop SVT in the first 3 months of life will outgrow it by 1 year of age.[48-51]

Presentation in Older Children

Older children often present with a complaint of palpitations or may complain of dizziness. It is not uncommon to hear a 3- or 4-year-old complain that their heart is "beeping" fast. These patients are generally not in SVT long enough to develop congestive heart failure, as seen in infants, because they can communicate to their caregivers that they are experiencing something abnormal. With the exception of those with WPW syndrome with rapid conduction down their accessory pathway during an atrial tachycardia, patients rarely present with syncope during SVT.

Treatment

The acute and chronic medical management of this age group is similar to that outlined earlier in infants. The exception is that RFA becomes a therapeutic option around 5 years of age. Ablation can and has been performed in younger children, but the apparent overall consensus is that it should be reserved for patients 4 years of age or younger with medically refractory arrhythmias.

AV Nodal Reentry

AV nodal reentrant tachycardia is a reentrant tachycardia in which the reentry circuit is the region of the AV node. There are two separate pathways in which there

TABLE 26-1 Chronic Antiarrhythmic Agents for Supraventricular and Ventricular Tachycardia

Arrhythmia	Agent	Dose (Oral)	Level
SVT	Digoxin	Dose is age dependent Give in 3 doses (½ TDD, ¼ TDD, ¼ TDD) Preterm infant: 10-20 μg/kg TDD Term newborn–adolescent: 30-40 μg/kg TDD oral to maximal TDD of 1-1.5 mg (IV ¾ po) Oral maintenance: 10 μg/kg/day q12h	1-2.5 ng/mL
	Verapamil	2-8 mg/kg/day tid	100-300 ng/mL
VT	Phenytoin	Loading dose: 10-20 mg/kg/day q12h × 2 d Maintenance: 5-10 mg/kg/day q12h	10-20 μg/mL
	Mexiletine	5-15 mg/kg/day q8h	0.5-2.0 μg/mL
SVT or VT	Propranolol	0.5-2 mg/kg/dose q6h	50-100 μg/L
	Nadolol	0.25 mg/kg/dose q12h	0.03-.13 μg/mL
	Atenolol	0.5-1 mg/kg/day qd	
	Procainamide	20-100 mg/kg/day q4-6h	PA: 4-10 mg/L;NAPA = 4-8 mg/L
	Quinidine	20-60 mg/kg/day q6-8h	2-5 mg/L
	Disopyramide	5-15 mg/kg/day q6h	2-4 μg/mL
	Flecainide	50-200 mg/m²/day or 3-6 mg/kg/day q12h	0.2-1.0 mg/L
	Amiodarone	Loading dose:10-20 mg/kg/day q12h × 7 days Maintenance: 5-10 mg/kg/dose q d	RT_3 <90 ng/dL
	Sotalol	2-8 mg/kg/day q12h	

IV, intravenous; NAPA, N-acetylprocainamide; PA, procainamide; RT_3, reverse T_3; SVT, supraventricular tachycardia; TDD, total digitalizing dose; VT, ventricular tachycardia.

is conduction from the atrium to the ventricle in the AV nodal region. One of these pathways has a long refractory period and fast conduction, while the second pathway has a shorter refractory period with slow conduction. AV nodal reentrant tachycardia occurs more commonly in young adults than younger children. It is seen most frequently in the pediatric population in the adolescent years. It is rarely seen in patients during the neonatal period. The classic finding on an ECG is a tachycardia in which there is very short R to P interval on the ECG. This is secondary to the fact that the retrograde limb of the pathway conducts very rapidly from the ventricle to the atrium.

Concealed Bypass Tachycardia

Concealed bypass tachycardias are tachycardias in which there is unidirectional conduction in an accessory pathway. In these pathways, the conduction is only from the ventricle to the atrium (retrograde), with no antegrade conduction. There is no indication of the accessory pathway on a resting ECG. These accessory pathways can occur anywhere along the AV groove, either on the right or left side of the heart or in the septal region. These tachycardias are seen throughout all phases of childhood. Patients can present with concealed bypass tract tachycardias in the neonatal period, as well as throughout adolescence.

Wolff-Parkinson-White Syndrome

Presentation. WPW syndrome in children is very similar to that in adults. This syndrome involves an accessory pathway between the atrium and the ventricle that has bidirectional conduction properties. The reported incidence of WPW syndrome ranges from 1 to 4 per 1000 live births.[52] Of all patients who present with WPW syndrome in childhood, one fifth to one third will have associated cardiac abnormalities. The most common congenital lesions include Ebstein's anomaly of the tricuspid valves and L-transposition of the great arteries.[51] Patients with WPW syndrome present anywhere from fetal life through adolescence. The patients who present during fetal life and early infancy have reentrant SVT. Toddlers and adolescents also present with reentrant SVT but can occasionally present with atrial fibrillation with rapid conduction down the accessory pathway. Approximately 10% of children will present with an antidromic tachycardia, with approximately one half of these patients having multiple accessory pathways.[53] A number of asymptomatic patients are noted to have the pattern of WPW syndrome noted on an ECG obtained for other reasons. Occasionally, an infant presents with a narrow complex tachycardia with no evidence of WPW syndrome on a baseline ECG, but preexcitation becomes obvious when the patient is treated with a medication such as digoxin that slows AV nodal conduction.

Treatment. The treatment strategies for patients with WPW syndrome vary based on the patient's age and symptoms at time of presentation. The mainstay of treatment in infants and young children with WPW has been the use of antiarrhythmic medications. We refrain from the use of digoxin or calcium channel blockers in patients with WPW because there may be enhancement of conduction down the accessory pathway, along with block of the conduction down the AV node. This is thought to lead to an increased risk of rapid conduction of atrial fibrillation or premature atrial contractions through the accessory pathway. We recommend use of β-blockers as a first-line therapy unless there is a contraindication such as severe reactive airway disease. When β-blocker therapy is contraindicated or fails to control the tachycardia, we use other medications such as flecainide, amiodarone, or sotalol. Occasionally, a neonate with WPW syndrome presents with significant signs of cardiac decompensation with cardiogenic shock. In those circumstances, we use digoxin while the patient is still in the hospital to improve ventricular function, as well as suppress the tachycardia. We always convert the patient from digoxin to another antiarrhythmic medication before discharge to home.

In the young adolescent or older child who presents with WPW syndrome, we offer RFA as a first-line therapeutic option. It has become our practice in children starting at 8 years of age with known WPW syndrome to assess the conduction or refractory period of the accessory pathway through their accessory pathway by the use of exercise testing, esophageal pacing, or intracardiac EPS. Many patients, when given the option, decide to undergo an intracardiac EPS with the thought that an RFA procedure can be performed and the WPW addressed definitively.

A special consideration in the pediatric population is the patient who presents with asymptomatic WPW syndrome. These patients present secondary to having an ECG obtained for some other indication such as chest pain or for a school physical. We recommend that patients older than 8 years of age have the evaluation mentioned earlier to assess whether they have rapid conduction down their accessory pathways. Generally, we take these patients to the cardiac catheterization laboratory and measure their minimum cycle length of preexcitation both with atrial pacing and during atrial fibrillation. We use a minimum cycle length of preexcitation of less than 220 milliseconds (ms) during atrial fibrillation as a marker that the patient may have a significant risk of sudden death. This is based on studies performed by Bromberg et al[54] and Paul et al,[55] which demonstrate that these values are helpful though not fully predictive of cardiac arrest and syncope.

Ventricular Arrhythmias

Ventricular arrhythmias include premature ventricular contractions (PVCs), couplets, nonsustained VT (NSVT), sustained VT, and VF. PVCs may be seen in 15% of normal newborns, one third of normal adolescents, and two thirds of adolescents and adults with repaired congenital heart disease.[56] PVCs may occur without identifiable cause in children and are often benign. They may be associated with acute conditions or more chronic conditions (Tables 26-2 and 26-3). There is a marked difference in prognosis between PVCs in children with normal and abnormal hearts, so investigation

TABLE 26-2 Causes of Acute Ventricular Tachycardia

Drugs/Toxins	**Myocardial Ischemia**
General Anesthetics	Abnormal coronaries/infarction
Antiarrhythmics	Kawasaki disease
Caffeine	**Hyperlipidemia**
Nicotine	**Infectious**
Sympathomimetics/catecholamine infusions	Myocarditis
Psychotropic agents: tricyclic	Pericarditis
Antidepressants/phenothiazines	Rheumatic fever
Cocaine	**Idiopathic**
Digoxin toxicity	
Metabolic	
Hypoxia	
Acidosis	
Hypoglycemia	
Hypocalcemia	
Trauma	
Blunt: cardiac contusion	
Thoracic surgery	
Cardiac catheters	

of children with PVCs for associated conditions should be undertaken.

Evaluation

An echocardiogram should be obtained to look at associated factors and conditions including structural abnormalities, such as hypertrophic cardiomyopathy and abnormalities in cardiac function that might accompany a myocarditis or a dilated cardiomyopathy. Rarely, cardiac tumors such as rhabdomyomas are identified. The evaluation should include a standard ECG on which the Q–Tc interval is carefully hand measured. A 24-hour ambulatory monitor will determine the amount and complexity of the ectopy. In the presence of a normal heart, less than 20% ectopy usually does not interfere with cardiac function and can be followed. More than 30% ectopy may result in ventricular dysfunction over time. It may be that patients with this degree of ectopy have underlying cardiac disease that is not initially diagnosed. Magnetic resonance imaging may be indicated if right ventricular dysplasia (RVD) is suspected. In patients with more frequent PVCs, an exercise stress test should be performed. Generally, suppression of PVCs during exercise is a positive finding, while an increase in ventricular arrhythmia with exercise is not. Prolongation of the corrected Q–T interval, especially in the recovery phase, may be seen in patients with long QT syndrome (LQTS). EPSs would rarely be indicated unless symptoms suggest more complex arrhythmias or the PVCs are associated with conditions that might predispose the patient to VT or VF.

Prognosis

Long-term follow-up suggests that PVCs and VT disappear over time in 37% to 65% of patients with normal hearts.[57] Sudden death is rare in normal children with PVCs but has been reported. In children with abnormal hearts, PVCs may be precursors of more serious arrhythmias, especially if they are complex-multiform, coupled, or associated with VT.

TABLE 26-3 Etiology of Chronic Ventricular Tachycardia

Congenital Heart Disease	Ebstein's anomaly Tetralogy of Fallot, absent PV leaflets Aortic valve disease, AI/AS Mitral valve prolapse Hypertrophic cardiomyopathy/IHSS Coronary artery anomalies Eisenmenger's syndrome, pulmonary hypertension
Postoperative CHD	Tetralogy of Fallot, DORV Ventricular septal defects AV canal defects Aortic valve disease, stenosis and insufficiency Single ventricle complexes s/p Fontan repair d-TGA s/p intra-atrial repair
Acquired Heart Disease	Rheumatic heart disease Lyme disease Myocarditis Kawasaki disease
Cardiomyopathies	Hypertrophic RV dysplasia Postviral Connective tissue disease: SLE Marfan's syndrome Muscular dystrophy, Friedrich's ataxia
Tumors and Infiltrates	Rhabdomyoma Hemosiderosis: thalassemia, sickle cell disease Oncocytic cardiomyopathy Leukemia
Idiopathic/Structurally Normal Heart	RV outflow tract VT LV septal VT/fascicular tachycardia
Primary Arrhythmias	LQTS Congenital complete heart block Familial VT
Other	Myocardial ischemia/infarction

AI, aortic insufficiency; AS, aortic stenosis; AV, atrioventricular; CHD, congenital heart disease; DORV, double outlet left ventricle; d-TGA, d-transposition of the great arteries; IHSS, idiopathic hypertrophic subaortic stenosis; LQTS, long QT syndrome; LV, left ventricular; PV, pulmonary valve; RV, right ventricular; SLE, systemic lupus erythematosus; VT, ventricular tachycardia.

Clinical Signs and Symptoms

Children with PVCs are frequently asymptomatic and unaware of their arrhythmias, especially if they are under 5 years of age. Older children may complain of a skipped or hard beat or a fluttering in their chest. Some children perceive PVCs as painful.

Treatment

PVCs do not need intervention unless they are frequent enough to interfere with the cardiac output, are closely coupled, or frequently fall on the T wave in a patient judged to be vulnerable to such occurrences (LQTS). PVCs associated with heart disease such as myocarditis, cardiomyopathy, or congenital heart disease may require further investigation and treatment, especially if they are frequent or occur in runs resulting in hemodynamic instability. Treatment of PVCs is addressed, along with treatment for VT.

Ventricular Tachycardia

As with PVCs, VT occurs in both acute and chronic situations (see Tables 26-2 and 26-3). Ventricular arrhythmias are less common than supraventricular arrhythmias in children but appear to be occurring more frequently in recent years or are being recognized to a greater extent. There is an increase in VT in patients after congenital heart surgery, as survival after complex surgery has increased.[58] Improved methods of surveillance and diagnosis of arrhythmias have allowed recognition of various etiologies of VT in children, with the most common being the congenital LQTS, hypertrophic and dilated cardiomyopathies, RVD, myocarditis, abnormal foci or circuits in structurally normal hearts, or idiopathic etiologies.

ELECTROCARDIOGRAPHIC MANIFESTATIONS

The electrocardiographic diagnosis of VT is made most easily in the presence of a wide QRS tachycardia with AV dissociation. Many children have ventriculoatrial (VA) conduction with relatively rapid 1:1 retrograde VA conduction, and AV dissociation may not occur. VT must be differentiated from all other forms of wide QRS tachycardias, including SVT with bundle branch aberrancy; antidromic SVT using an accessory AV connection; SVT using a nodoventricular, nodofascicular, or atriofascicular connection; or atrial flutter with aberrant conduction. Although other mechanisms of wide QRS tachycardia have been described, until proved otherwise, a wide QRS tachycardia in a child must be considered to be VT. It must be remembered that the normal QRS duration in infants and young children is 40 to 80 ms, so a wide QRS in an infant might only be 80 ms. The rates of VT in pediatrics vary from 120 to 300 bpm. The T wave vector is divergent from the QRS vector, but opposite polarity will not occur in every lead. Left bundle branch block (LBBB) is the most common morphology, but right bundle branch block (RBBB) or alternating RBBB and LBBB may occur. The presence of PVCs during sinus rhythm with the same configuration as VT is a suggestive sign. AV dissociation is suggestive of VT, but 1:1 VA conduction is common, especially in young children. Fusion beats are commonly noted at the onset or termination of the VT. VT may be sustained (>30 consecutive complexes) or nonsustained (3 to 30 consecutive complexes).

Further differentiation is made according to the morphology, with VT being described as monomorphic or polymorphic. Two types of polymorphic VT have been described, torsades de pointes and bidirectional VT (Fig. 26-3). Torsades de pointes is associated with LQTS and receives its name from its twisting, undulating nature. Bidirectional VT has been associated with digoxin toxicity, familial hyperkalemic paralysis, or catecholamine sensitivity. The ECG shows beat-to-beat variation in the QRS axis.[59]

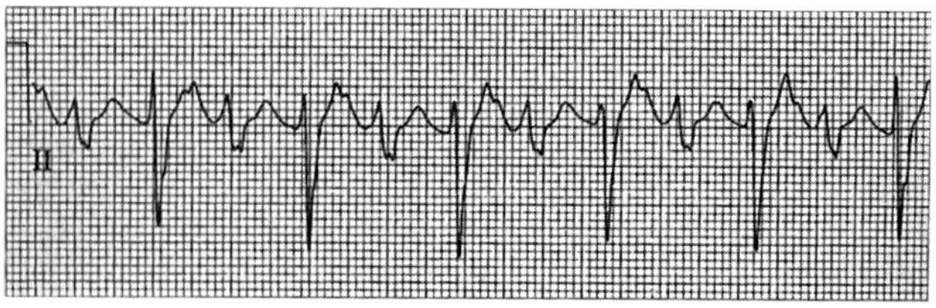

FIGURE 26-3 ECG of bidirectional ventricular tachycardia. Note the two distinctly different wide QRS morphologies.

Etiology of Ventricular Tachycardia

Causes of acute VT not associated with congenital heart defects are shown in Table 26-2. These most commonly include metabolic and electrolyte abnormalities; infectious processes such as myocarditis, which may cause LV microaneurysms[60]; HIV infections[61]; blunt cardiac trauma including commotio cordis[62-64]; coronary ischemia, especially in association with Kawasaki disease; and drugs such as caffeine, inhalation anesthetics, and recreational drugs including amphetamines and cocaine.

The causes of chronic or recurrent VT include congenital heart disease, both preoperatively and postoperatively acquired heart diseases; metabolic disorders including disorders of fatty acid metabolism[65]; neuromuscular disorders such as Duchenne's muscular dystrophy, cardiomyopathies including ARVD (ARVC); hypertrophic cardiomyopathy (HCM); tumors and infiltrates; VT associated with structurally normal hearts originating in both the right ventricle and left ventricle; VT associated with LQTS; and other primary electrophysiological abnormalities such as Brugada syndrome, familial polymorphic VT, and catecholaminergic VT. A detailed list is shown in Table 26-3. It is not uncommon to be unable to identify a specific cause in children.[66-69] Persistence in evaluations often identifies pathology in patients initially not thought to have an identifiable etiology.[70]

MECHANISM OF VENTRICULAR TACHYCARDIA

VT has been reported to result from reentry, triggered automaticity, and abnormal automaticity. The EPS is helpful in differentiating these mechanisms. The mechanisms of VT in children include reentry in 60% and abnormal automaticity in 40%.[71] Reentry is most often the mechanism in postoperative congenital heart patients, related to reentry circuits that develop around suture lines and ventriculotomy scars.

CLINICAL CORRELATIONS

Presentation of patients with VT varies and depends to a large extent on the underlying etiology and clinical status with regard to underlying myocardial function and structure. In one study of patients with VT with structurally normal hearts, presentation was most common in infancy (48%), with 58% being younger than 6 months.[66] Associated findings were heart failure in 30%, hemodynamic compromise or collapse in 23%, and in utero diagnosis in 18%. Thirty percent had incidental diagnosis. No specific etiology was found in 50%, with cardiomyopathy or myocarditis (20%) being the most common etiology identified.

CLINICAL SIGNS AND SYMPTOMS

The type and degree of symptoms appear to be rate related, with symptoms most common in patients with rates greater than 150 bpm. Except for those patients with underlying cardiac disease, patients with VT have symptoms similar to those with SVT with the degree of symptom relating more to the rate than the mechanism of tachycardia. Symptoms include dyspnea, shortness of breath, chest or abdominal pain, palpitations, dizziness, syncope, and cardiac arrest or sudden death. Older children may exhibit exercise intolerance or easy fatigability. Infants may feed poorly and be irritable or lethargic. Patients with VT and heart disease usually have symptoms, while only one third with normal hearts and VT have symptoms. The type of symptom relates to both the tachycardia rate (rare at <150 bpm) and the underlying state of the myocardium. Sudden death occurs most commonly in the presence of an abnormal heart[72] but has been reported in patients with normal hearts.[67,73] Children younger than 5 years of age or those in incessant tachycardia may not have a perception of a fast heart rate or be able to accurately express what they are feeling. Signs include palpitations; sensation of a rapid heart rate, tachypnea, or hypotension with accompanying pallor and diaphoresis; and signs of congestive heart failure. Although VT usually has a sudden onset, it may occur during exercise and be difficult to perceive. It may gradually warm up or increase in rate.

SPECIFIC ASSOCIATED CONDITIONS

Accelerated Ventricular Rhythm

An accelerated ventricular rhythm is a rhythm originating from the ventricle with all the characteristics of VT but with a rate that is only slightly more rapid than the underlying sinus rhythm, usually less than 120 bpm. It is often seen in children with normal hearts. This arrhythmia is not uncommon in the neonate and is self-limited, resolving from 2 weeks to 3 months after birth. These early ventricular arrhythmias probably relate to developmental factors associated with the autonomic nervous system. In older children, these arrhythmias may relate to unidentified viral infections with myocarditis that affects only the conduction system. This arrhythmia is seen around puberty and probably relates to autonomic and hormonally mediated factors. In addition, accelerated ventricular rhythms have been reported in association with metabolic abnormalities, medication, ARVD, and myocardial infarction.[74] In pediatric patients, this arrhythmia is generally thought to be benign, even in the occasional patient who has congenital heart disease.[74,75]

Evaluation and Treatment

Evaluation should include ECG, 24-hour ambulatory monitoring, and exercise stress testing in those older than 5 years of age who can cooperate. These patients generally require no therapy but should be followed, because an occasional patient will have acceleration of their VT to a much higher rate and develop symptoms. Treatment of this arrhythmia and restriction of activity is not required in the majority of patients, especially those with normal hearts.

Right Ventricular Dysplasia

An unusual cause of VT known as RVD or arrhythmogenic RVD (ARVD) was first described in 1978.[76] The VT is an LBBB pattern in most instances. A pediatric series reported RVD in 30% of its patients with VT and an apparently normal heart,[77] although it is much less common in most other pediatric series. RVD is a familial form of right ventricular (RV) cardiomyopathy associated with sudden death. It has an autosomal dominant genetic pattern with variable penetrance and variable expression. The pathologic lesion involves massive replacement of the RV wall by fibrous or fatty tissue, or both. There may be focal myocardial changes including necrosis, degeneration, or hypertrophy and chronic inflammatory infiltrates. The process is most severe in the subepicardium and progresses toward the subendocardium in its later stages. Although many cases are familial, sporadic cases have been reported. RVD has not been commonly reported in young patients, but this condition should be suspected in previously healthy children or adolescents who present with VT.

Evaluation

Patients suspected of this condition should have an ECG, echocardiogram, and MRI. Because of the localized nature of this condition, echocardiography may not be diagnostic. MRI or cine MRI may be more helpful by demonstrating thinning of the RV myocardium replaced by fatty tissue or showing localized areas of hypokinesis in the infundibulum, free wall, or RV apex, accompanied by RV dilation and decreased contractility. LV free wall and septal involvement in this process has been noted.[78] Other potential diagnostic modalities include exercise testing, contrast ventriculography, signal-averaged ECG, and single photon emission computed tomography (SPECT) analysis.[79,80-85] Research on genetic identification is ongoing. Children in affected families should be evaluated by obtaining ECGs, 24-hour ambulatory monitors, echocardiograms, and MRIs.[70,86]

Treatment and Follow-Up

Variable medical therapies including β-blockers and sotalol have been suggested for patients with frequent, symptomatic, or potentially threatening ventricular arrhythmias. Automatic implantable cardioverter defibrillators (ICDs) have been used[87] and can be life saving in these patients.[88] Extensive surgical procedures, including a complete electrical disconnection of the RV free wall, have been reported.[89] The prognosis and clinical course reported in these patients have been quite variable. Continued surveillance with periodic Holter monitoring and exercise stress testing are important in the follow-up of this patient group as the incidence of serious arrhythmias increases with age.

Long QT Syndrome

The congenital LQTS is an inherited condition characterized by syncope, seizures, and sudden death, associated in most individuals with a prolongation of the Q–T interval on the ECG.[90] An example of the ECG in LQTS is shown in Figure 26-4. In addition to the prolongation of the QTc, these patients often have bizarre or notched T wave morphology with prominent U waves or T wave alternans. They develop life-threatening VT, known as *torsades de pointes*, or VF. This syndrome includes the Jervell and Lange-Nielson syndrome, described in 1957,[91] associated with congenital deafness and thought to demonstrate an autosomal recessive inheritance and the Romano-Ward syndrome described in 1963 and 1964 demonstrating autosomal dominant inheritance, without hearing deficit.[92,93]

In 1993 statistics from a group of 287 children were compiled from a number of medical centers.[94] The initial presentation was cardiac arrest (9%), syncope (26%), seizures (10%), presyncope, or palpitations (6%). Eighty-eight percent had exercise-related symptoms. Thirty-nine percent were identified because of family history or the identification of other family members with the syndrome. Thirty-nine percent of the patients were asymptomatic at presentation. Of these, 4% experienced sudden death compared to 8% overall. The strongest predictors of sudden death were QTc greater than 0.60 and noncompliance with taking recommended medication.

Bradycardia is commonly seen in these patients, and some may develop or present with second-degree AV block. This is more common in neonates who may have second- or third-degree AV block[95] but it may be seen in older children, especially with exercise.

One series reported sudden death occurring in 73% without treatment,[96] and others have reported sudden death in 21% of symptomatic patients in the first year after presenting with syncope.[97]

Diagnosis and Evaluation

The diagnosis of this syndrome is made from a variety of criteria. Schwartz et al. provided criteria and suggested a scale for identifying these patients.[98] All criteria involve measurement of the Q–T interval and a careful history for syncope, seizures, and arrhythmias in the patients and their families. Commonly, a complete history may reveal a family history of sudden death in young relatives or history of syncope or seizures associated with exercise or emotional stress.

Additional studies such as 24-hour monitoring and exercise stress testing may provide helpful information in the form of significantly prolonged Q–T intervals, especially during recovery from exercise or the occurrence of polymorphic ventricular arrhythmias during or after exercise. The use of provocative tests such as isoproterenol or epinephrine infusions remains controversial.[99,100]

A high level of suspicion is needed to diagnose LQTS in these patients. Any patient who presents with syncope during or immediately after exercise or VT, especially of the polymorphic or torsades de pointes type, or in association with physical or emotional stress should have an ECG with corrected Q–T intervals determined. Evaluation of a resting ECG may not be sufficient as more than 12% of gene carriers may have normal ECGs.[101] A more worrisome study by Priori showed an even lower penetrance of the gene in Italian families with probands initially thought to have sporadic occurrence of LQTS. Genetic studies revealed multiple family members who were genetic carriers but with normal ECGs.[102]

Recent genetic studies have identified a number of abnormal genes that encode for proteins that modulate potassium or sodium ion channels, causing the LQTS by altering cardiac repolarization and increasing the risk for ventricular arrhythmias. These genes include *KVLQTI, HERG, SCN5A, minK,* and *MiRp1* (Table 26-4).[103-106] In 1991 Keating et al. linked a DNA marker on the short arm of chromosome 11 (11p15.5 near the Harvey RAS-1 locus) in a group of families with the LQTS (LQT1).[107] Genetic heterogeneity was found to be present as families with LQTS that were not linked to chromosome 11p15.5 were identified.[108,109] In 1994 additional linkages were found to chromosome

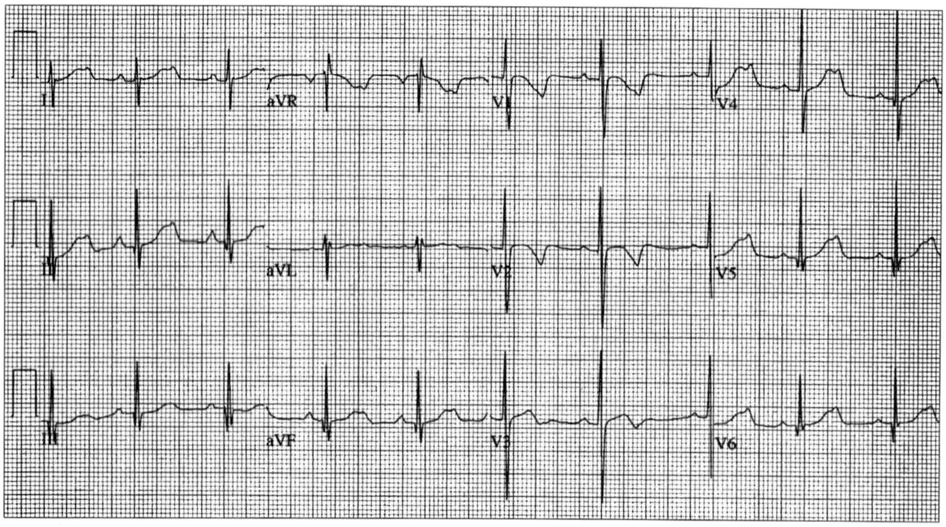

FIGURE 26-4 ECG of patient with long QT syndrome. The QTc interval measures 490 milliseconds. Note the long, notched T waves.

TABLE 26-4 Long QT Syndrome

Type	Inheritance	Gene (alternate name)	Protein
LQT1	AD	*KvLQT1 (KCNQ1)*	$I_{Ks}K^+$ channel α subunit
LQT2	AD	*HERG (KCNH2)*	$I_{Kr}K^+$ channel α subunit
LQT3	AD	*SCN5A*	$I_{Na}K^+$ channel α subunit
LQT4	AD	*LQT4*	Unknown
LQT5	AD	*MinK (KCNE1)*	$I_{Ks}K^+$ channel β subunit
LQT6	AD	*MiRP1 (KCNE2)*	$I_{Kr}K^+$ channel β subunit
Lange Jervell-Nielson (LQTS with deafness)	AR	*KVLQT1*	$I_{Ks}K^+$ channel β subunit

AD, autosomal dominant; AR, autosomal recessive.

7 (7q35-36) (LQT2) and chromosome 3 (4q21-24) (LQT3).[104] A fourth locus (LQT4) has been identified at 4q25-27 in a single large kindred in Japan, but no specific gene has been found.[106] Genes associated with LQT5 (minK/KCNE1) and LQT6 (MiRP1/KCNE2) have been identified.[110,111]

Not all families with known LQTS have shown linkage to these known loci, suggesting the presence of additional genes yet to be discovered. In addition to these genes that affect ionic channels altering the repolarization phase of the cardiac action potential and resulting in the development of ventricular arrhythmias, an imbalance or oversensitivity of the myocardium to sympathetic stimulation appears to play a role in the development of ventricular arrhythmias. The trigger for arrhythmia in the LQTS is believed to be spontaneous secondary depolarizations that arise during or just following the prolonged plateau phase of action potentials, early afterdepolarizations. Increased sympathetic tone may increase early afterdepolarizations with these spontaneous repolarizations triggering a sustained arrhythmia.

Specific Genetic Defects

KvLQT1 (LQT1) and *MinK* (LQT5)

The *KvLQT1 (KCNQ1)* gene is located on chromosome 11p5.5 and encodes the voltage-gated potassium channel subunits.[103,112] *MinK (KCNE1)* encodes a much smaller potassium channel subunit. *MinK* subunits assemble with *KvLQT1* subunits to form cardiac I_{ks} channels.[110,113,114] Abnormalities of either or both of these genes inhibit channel function and prolong repolarization by affecting I_{ks} by a greater than 50% reduction in channel function, a dominant-negative effect. The molecular mechanism resulting in reduced *KvLQT1* function occurs from synthesis of abnormal subunits that do not assemble with normal subunits, with a reduction in the number of functional channels and a loss of function. Other mutations cause synthesis of *KvLQT1* subunits with structural abnormalities causing loss of function. Current evidence suggests that greater than 40% of affected LQTS families have *KvLQT1* mutation.[115] Approximately 5% of mutations to date have involved *minK*.[115,116] It has been reported that a homozygous mutation of *KvLQT1* or *minK* causes the Lange-Jervell-Nielsen syndrome.[117,118] Both *KvLQT1* and *minK* are expressed in the inner ear. Homozygous mutations of *KvLQT1* or *minK* have no functional I_{ks} channels. This leads to inadequate endolymph production and deterioration of the organ of Corti with neural deafness.[119]

HERG (LQT2) and *MIRP1* (LQT6)

In 1994 Jiang et al. showed linkage of a cohort of families with LQTS to chromosome 7 (LQT2).[104] This locus, 7q 35-36 was determined to be the Human Ether -a-go-go-Related Gene.[105] *HERG* mutations represent 45% of the total number of LQTS mutations found to date. This gene encodes for the subunits that form the cardiac potassium channel delayed rectifier I_{Kr} channel, the second of two channels responsible for termination of the plateau phase of the action potential.[120,121] These mutations result in decreased outward potassium current preventing termination of the plateau phase of the action potential. *HERG* subunits assemble with the *MiRP1* (*minK*-related protein1) also known as *KCNE2* located on chromosome 21 close to *minK*, to form cardiac I_{kr} channels.[111] *MiRP1* mutations represent 2% of identified LQTS mutations.

Multiple drugs are known to prolong the Q–T interval and potentially induce arrhythmias. The HERG channel is the K^+ channel most commonly blocked. The structure of *HERG* channels appears be predisposed to blockage by multiple drugs.[122] Drugs that block I_{kr} are associated with pause-dependent torsades de pointes.

SCN5A (LQT3)

Jiang et al.[104] reported an additional group of LQTS families with linkage to chromosome 3 (3p21-24), which was shown to be the human cardiac sodium channel gene that encodes the subunit of sodium channel responsible for initiating cardiac action potentials.[123,124] Gain of function mutations, especially in the inactivation gate between domains III and IV of the sodium channel, have been reported with abnormal gain of function resulting in continued inward sodium current prolonging the action potential and predisposing to ventricular arrhythmias.[125,126]

Clinical Associations of Genetic Findings

T wave changes have been associated with specific genetic mutations[127] as shown in Figure 26-5. Overlap exists, limiting the specificity of this finding. The influence of genotype on clinical course is being elucidated.[128] The frequency of cardiac events is higher among subjects with LQT1 (63%) or LQT2 (46%) than among LQT3 patients (18%). The likelihood of dying during a cardiac event was higher among LQT3 patients (20%) than among LQT1 or LQT2 (4%) patients. Cardiac events in LQT1 patients occur frequently during exercise (62%), especially swimming.[129,130] Only 3% occurred during sleep. LQT2 and LQT3 patients were less likely

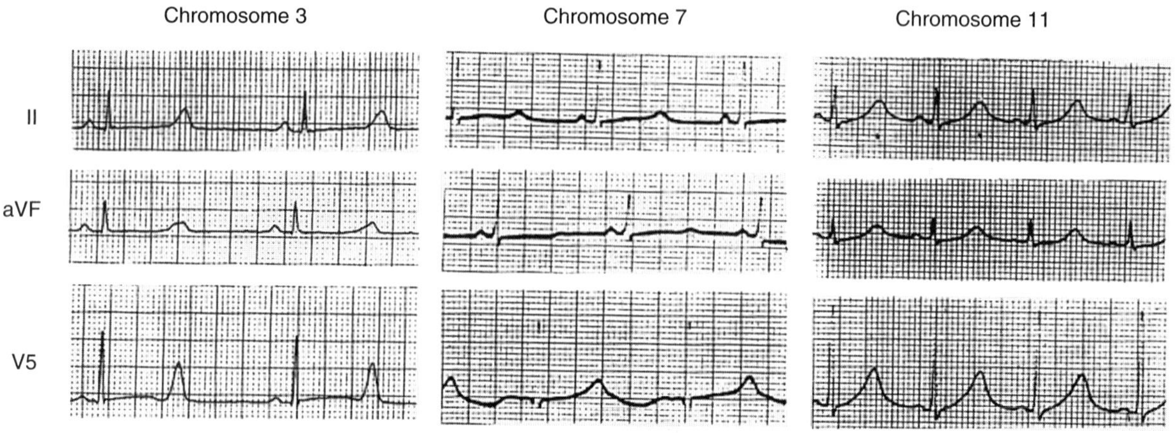

FIGURE 26-5 Electrocardiographic recordings from leads II, aVF, and V5 in three patients with long QT syndrome, linked to chromosomes 3, 7, and 11. The patient with chromosome 3 linkage has late-onset T waves with a Q–Tc interval of 570 milliseconds (ms). The patient with chromosome 7 linkage has low-amplitude T waves and a Q–Tc interval of 583 ms. The patient with chromosome 11 linkage has early-onset broad T waves with a Q–Tc interval of 573 ms. (From Moss AJ, Zareba W, Benhorin J, et al: ECG-T wave patterns in genetically distinct forms of the hereditary long QT syndrome. Circulation 1995;92:2931.)

to have events during exercise (13%) and more likely to have events during rest/sleep (29% and 39%). The percentage of patients who were free of recurrence with β-blocker therapy was higher and the death rate was lower among LQT1 patients (81% and 4%, respectively) than among LQT2 (59% and 4%) and LQT3 patients (50% and 17%).[129] LQT3 patients have more cardiac events at rest or during sleep, while LQT2 patients experience more events during exercise or stress. LQT2 events are more likely to be stimulated by loud noises.[131]

Vincent's study in 1992 in the *New England Journal of Medicine* indicated that some patients may be genetic carriers for this syndrome without significant prolongation of the Q–T interval.[267,101] Some noncarriers (15%) have abnormal prolongation of the Q–T interval above 0.44. A QTc of greater than .47 seconds had a 100% positive predictive value in gene carriers. Above 0.47, there were no false-positives in this study. Only 6% of gene carriers had Q–Tc intervals of less than 0.44 seconds.[101] These genetic tests may make it possible to identify more specifically many patients with LQTS. Priori identified families with only a 25% penetrance where 33% of family members considered to be unaffected on clinical grounds were gene carriers.[102]

Another interesting association that has come from the International Registry relates to the association of asthma and LQTS. The occurrence of asthma in LQTS patients increases with QTc duration. Asthma comorbidity in LQTS patients is associated with an increased risk of cardiac events. This risk is diminished after initiation of β-blocker therapy.[132]

Sudden Infant Death Syndrome

Schwartz reported on ECGs on 34,442 Italian babies on days 3 or 4 of life with 24 subsequently having sudden infant death syndrome (SIDS) and 12 having prolongation of the Q–T interval greater than 440 ms.[133] Additional reports have confirmed the *SCNA5* and *HERG* mutations in a few babies who experienced SIDS.[133-135]

Evaluation and Diagnosis

Diagnosis is made by a careful history, both of the individual's episodes and a complete family history looking for sudden unexplained death, syncopal episodes in family members, unusual seizure disorders, or hearing deficits.

All patients suspected of LQTS should have a standard ECG with careful QTc measurement, 24-hour ambulatory monitoring, and exercise stress testing if age appropriate. We do not routinely use other provocative testing such as isoproterenol or epinephrine infusions, reserving it instead for isolated cases with a high index of suspicion but negative testing otherwise.

Careful evaluation of the ECG and 24-hour ambulatory monitor is essential. The Q–Tc interval is measured by hand using Bazett's formula.[136] The longest Q–T interval in any lead is divided by the square root of the preceding R–R interval. The measurement should be made manually and calculated, as the computerized values are frequently incorrect. We use a value greater than 0.45 in any lead as abnormal on the resting or exercise ECG. Although a number of methods have been proposed for calculating the QTc in the presence of sinus arrhythmia,[137] we make every attempt to record an ECG not in sinus arrhythmia. On the Holter monitor, as different filters are used, we consider a value greater than 0.50 as abnormal. In addition to Q–T interval prolongation, the Holter may be helpful in illustrating T wave abnormalities, R on T phenomenon, short runs of nonsustained VT, sustained VT, or torsades de pointes.

Provocative testing should include exercise stress testing in children able to exercise. Exercise will generally obliterate sinus arrhythmia, and a strip can be obtained in which a reasonable QTc calculation can be made. The recovery period with heart rates around 120 to 130 bpm seems to demonstrate the greatest degree of QTc prolongation in many patients. Exercise may uncover abnormal T waves, polymorphic PVCs, or VT. If suspicion for LQTS is high and other testing has not been definitive, isoproterenol or epinephrine infusion may help to identify these patients. T wave abnormalities,[138] in addition to QTc prolongation or the development of ventricular arrhythmias during these provocative tests, may occur. We use a value of greater than 0.50 as abnormal on isoproterenol testing.

Efforts to identify patients at high risk for syncope and sudden death continue. High-risk electrocardiographic markers have included QTc greater than 0.60, T wave alternans, and QTc dispersion.[139,140] Dispersion of the Q–T interval has been correlated with high risk in LQTS patients.[138,139,141] QT dispersion, which indicates heterogeneity of repolarization, could predispose to the development of torsades de pointes. In Priori's study, patients not responding to β-blocker had a significantly higher dispersion of repolarization than responders.[139] In Shah's study, LQTS patients at high risk for developing critical ventricular arrhythmias had a QT or JT dispersion greater than 55 ms.[141] Little information is available on microvolt T wave alternans in LQTS patients, although visible T wave alternans is known to be a high risk factor.

Treatment

Emergent treatment of these patients includes lidocaine and cardioversion. Magnesium may be used to treat torsades de pointes.[142-144] Intravenous propranolol and phenytoin have also been successfully used in these patients. The class I agents, which are known to prolong the Q–T interval in normal patients, should be avoided in these LQTS patients. In fact, a number of drugs that may produce this form of VT are shown in Table 26-5. This is felt to be related to QTc prolongation with associated bradycardia or ventricular arrhythmias, or both. Temporary pacing and removal of the offending agent are effective therapies.

TABLE 26-5 Pharmacologic Agents for Acute Treatment of Supraventricular Tachycardia

Agent	Initial Treatment (IV)
Adenosine	50-100 μg/kg, increase by 50 μg/kg increments q 2 min to 400 μg/kg or 12 mg maximal dose
Amiodarone	IV: 5 mg/kg over 1 hour, followed by 5-10 μg/kg/min infusion
Digoxin	Dose is age dependent Give in 3 doses (½ TDD, ¼ TDD, ¼ TDD) Preterm infant: 10-20 μg/kg TDD Term newborn-adolescent: 30-40 μg/kg TDD oral to maximal TDD of 1-1.5 mg (IV 3/4 po) Oral maintenance: 10 μg/kg/day q12h
Esmolol	IV Load: 200-500 μg/kg/min over 2-4 min. May increase in 50-100 μg/kg/min increments (maximum dose = 1000 μg/kg/min) Maintenance infusion: 50-200 μg/kg/min
Phenylephrine	100 μg/kg bolus Infusion: 10 μg/kg/min
Procainamide	5 mg/kg over 5-10 min or 10-15 mg/kg over 30-45 min Infusion: 20-100 μg/kg/min
Propranolol	0.05-0.1 mg/kg over 5 min q6h
Verapamil	0.05 mg/kg-0.30 mg/kg over 3-5 min Maximal dose = 10 mg

TDD, total digitalizing dose.

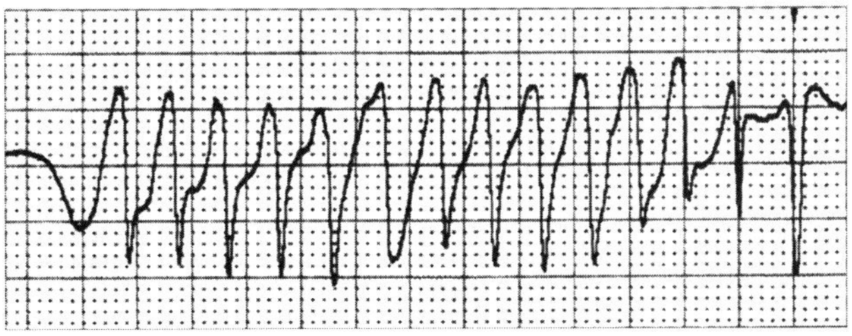

FIGURE 26-6 ECG rhythm strip of torsades de pointes from a patient with long QT syndrome. This ventricular rhythm is rapid, irregular, and polymorphic.

Sudden death is secondary to the ventricular arrhythmias (torsades de pointes) of the type shown in Figure 26-6, which frequently degenerate to VF.

The standard long-term treatment in this condition is the use of β-blockers. Those most commonly used are propranolol and nadolol. Some have suggested long-acting propranolol or atenolol. A concern about once-daily dosing relates to the lowest levels being present in early morning hours, a particularly high-risk time for some patients. Therefore, we would suggest twice daily dosing at least in this group of patients. One study did suggest a less than favorable result with atenolol.[145] In those patients in whom compliance is a factor, once-daily β-blocker would be preferred to none. The dose of β-blocker required is variable and is usually greater per kilogram in younger patients. Most teenagers only require 10 to 20 mg of nadolol twice daily. We titrate the appropriate dose by the heart rate response to maximal exercise testing, aiming for a blunted maximal heart rate response on therapy of 150 to 160 bpm. Treatment with β-blockers can lower the mortality to less than 4%.[146] Other series indicate that β-blockers are not protective in every segment of this population.[147]

Our patients are followed yearly or twice yearly with exercise stress tests and Holter monitoring to look for adequacy of treatment or the development of significant ventricular arrhythmias, or both.

Patients who do not respond to β-blockers may be treated with mexiletine, phenytoin, or pacing. Rarely, other antiarrhythmics may be used, but those known to prolong the Q–T interval should be avoided. Potassium therapy may be helpful. Left stellate ganglionectomy is a controversial treatment with variable success.[148-151]

Permanent pacing has been shown to be an effective adjunctive treatment in these patients, especially those with severe bradycardia either from the syndrome itself or from the β-blocker therapy.[152,153] The rate of the pacing should be at least 10% to 20% higher than the sinus rate and in severe cases should control the rhythm as much of the time as possible. Pauses should be avoided. Using this treatment, episodes of torsades de pointes may be reduced or eliminated.

In patients known to have had a cardiac arrest or frequent or significant syncope associated with ventricular arrhythmias, we recommend the implantation of an automatic internal cardioverter/defibrillator device. These devices can recognize VT or VF according to programmed criteria and provide a series of shocks to convert the patient to sinus rhythm. Some can provide backup pacing but at present are not appropriate for continuous higher rate pacing, as the battery will be depleted prematurely. Their size led to limited use in smaller children, but improved technology now allows even small children to benefit from this technology. This is not a therapy to be undertaken lightly at this time, as follow-up and possible false discharges can significantly affect a child's life and lifestyle. Groh reported on 35 patients with LQTS who had ICDs and were followed for a mean of 31 months.[154] The major indication was aborted sudden death. Sixty percent of patients had at least one appropriate discharge in the follow-up period. Two patients had multiple discharges and required additional therapies. No patients died. These results were similar to those reported earlier by Silka.[155]

A greater understanding of the molecular mechanisms of LQTS has prompted studies to identify more specific or gene-directed therapies.[156] The sodium channel blocker mexiletine has been used.[157] The Q–T interval was noted to decrease in LQT3 patients with mexiletine treatment. Sato et al. had reported using a potassium channel opener, nicorandil, in a patient with LQTS.[158] Trials are under way with potassium supplementation and spironolactone.

It is generally recommended that competitive athletics be avoided by patients with LQTS. In those with documented LQTS and symptoms or arrhythmias, we would certainly agree. However, as more "carriers" or asymptomatic patients are being discovered, who have only a prolonged Q–T interval and no family history of sudden death or ventricular arrhythmias, individual exercise and sports participation recommendations may be made. The most important aspect of the care of these patients is continued surveillance. This is true for young family members who appear to have normal Q–T intervals on initial evaluation. We have seen the Q–T interval change with age and would recommend periodic ECGs and appropriate 24-hour and exercise electrocardiography in children and adolescent members of LQTS families whose initial evaluation is negative unless genetic testing has definitively ruled out LQTS.

It is recommended that patients with LQTS avoid caffeine, adrenergic stimulants such as epinephrine, and over-the-counter stimulants such as decongestants. Medications that prolong the Q–T interval should be avoided.

Brugada Syndrome

SCNA5A (Brugada/Familial Ventricular Fibrillation)

The association of an ECG pattern of RBBB and ST segment elevation in ECG leads V1 to V3 with sudden death was reported in 1992 and has been labeled *Brugada syndrome*.[159] Patients die during sleep, presumed secondary to VF. The syndrome is inherited with an autosomal dominant mode. A mutation of *SCNA5* causes loss of function with slowing of conduction velocity.[160] Patients with this condition have normal Q–T intervals, ST segment elevation, RV conduction delay, and a propensity for sudden death. Sodium channel blocking agents such as procainamide or flecainide have been used to unmask this condition.[161,162] Placement of the precordial leads several intercostal spaces higher than usual may unmask the condition as well. The syndrome occurs most commonly in males and Asians. Patients with unexplained syncope or aborted sudden death or a family history of these occurrences should have this condition considered. Antiarrhythmics have not decreased the incidence of sudden death in these patients and implanted defibrillators are recommended.[163]

Catecholamine-Induced Ventricular Tachycardia

Catecholamine-induced VT is a genetic disorder associated with stress-induced, bidirectional VT that may degenerate into VF and result in sudden death. This condition usually occurs in childhood, adolescence, or young adulthood. Reports of mutation in the ryanodine receptor gene *RyR1* mapped to 1q42-q43 have been found in families with catecholamine-induced VT.[164] Abnormalities of this gene would result in an abnormality of intracellular calcium handling leading to calcium overload and tachyarrhythmias.

Evaluation

Evaluation should include ECG, Holter, and exercise stress testing with periodic surveillance testing in affected individuals.

Treatment

Treatment includes β-blocker medication and ICD implant in patients who have had significant symptomatic episodes.

Myocarditis

Patients with myocarditis present another special problem. Of patients with ventricular ectopy, 14% to 50% have been shown to have histologic evidence of myocarditis.[165,166] The most common causes of viral myocarditis include Coxsackie A and B and adenovirus.

A large number of these patients present with ventricular arrhythmias, usually single PVCs or nonsustained VT and only mild or no impairment of ventricular function.

Evaluation

Resting ECGs may show ST-T wave changes or low voltage in addition to PVCs or VT. Twenty-four-hour monitoring will indicate the extent and complexity of the arrhythmia. In a hospitalized patient, telemetry monitoring may pick up runs of sustained or nonsustained VT or periods of AV block.

Treatment

Steroid therapy has been beneficial in some patients.[167] The use of IV immune globulin may improve recovery of LV function and improve survival in the first year after presentation in these patients.[168]

In instances of simple ventricular arrhythmias, no treatment may be needed. Patients with more complex or frequent arrhythmias may need treatment. We have used β-blockade or mexiletine successfully in these patients. Some patients have slow rates from sinus node dysfunction or AV block and develop rapid ventricular arrhythmias as the heart rate slows. Temporary pacing may be necessary in these patients and may result in control of the ventricular arrhythmia without the need for pharmacologic agents.

When ventricular arrhythmias in these patients are potentially threatening or impair ventricular function, immediate treatment may be needed. Often these patients with diminished myocardial function require inotropic support to maintain cardiac output. Although each patient has unique sensitivities, if possible, a supportive agent with the least arrhythmogenic characteristics should be chosen. For example, dobutamine is usually less arrhythmogenic than dopamine, which is less arrhythmogenic than epinephrine or isoproterenol. Ventricular arrhythmias may occur in patients with myocarditis and associated complete heart block and slow escape rhythms. In these instances, an increase in the heart rate with a temporary transvenous pacemaker may be all that is needed to control the ventricular arrhythmia. In general, pressor agents should not be used just to increase the heart rate because of the arrhythmogenic potential in this subset of patients.

Postoperative Ventricular Tachycardia

With few exceptions, patients who undergo intracardiac surgery risk the development of postoperative arrhythmias and conduction defects.[169] A history of previous surgery on the ventricle or previous elevation of LV pressure before surgery is seen in most patients with VT after congenital heart surgery. Most of the instances of sudden death in postoperative congenital patients are seen after repair of aortic stenosis, coarctation, transposition of the great arteries, or tetralogy of Fallot.[170] Postoperative VT is more likely to occur in response to the effects of long-standing RV hypertension, healing of ventriculotomy scars under pressure, postoperative volume overload of the right ventricle, and residual LV pressure overload. Any surgery with repair of a ventricular septal defect, AV canal defects, aortic stenosis, idiopathic hypertrophic subaortic stenosis, Ebstein's anomaly, coronary artery anomalies, single ventricle complexes after Fontan repair, and d-transposition of the great arteries (d-TGA) may develop VT.[171] In addition, patients after intra-atrial repair of complete transposition of the great arteries and the Fontan repair are now presenting with ventricular arrhythmias. Recent studies have shown an incidence of perioperative arrhythmias in patients undergoing Ross repairs for aortic stenosis of 29%.[172] A 20-year follow-up of patients after switch repair of d-TGA showed sudden death in 2.7% and ventricular arrhythmias in 2.8%. Coronary artery abnormalities were identified in these patients.[173]

Ventricular Tachycardia In Postoperative Tetralogy of Fallot

Postoperatively, the most common congenital lesion associated with VT is tetralogy of Fallot.[174,175] Among these susceptible patients, 10% to 15% have VT postoperatively. Sudden death occurs in 5% to 10%.[176] Ventricular arrhythmias are believed to be responsible for sudden death in patients with these associated conditions.

Risk Factors and Clinical Correlations

In tetralogy of Fallot, risk factors associated with the development of VT and sudden death include older age at repair, a longer postoperative period, RV systolic pressure greater than 60 mm Hg at rest, RV end-diastolic pressure greater than 10 mm Hg at rest, depressed RV systolic function, and moderate to severe pulmonary or tricuspid regurgitation.[174,177-179] Recent studies have suggested an association between abnormal signal-averaged electrograms with late potentials and the development of VT.[180] A wide QRS duration of greater than 180 ms has been associated with VT.[179] This has been correlated with severe pulmonary insufficiency leading to RV dilation. QRS duration and degree of pulmonary regurgitation seem to be the greatest risk factors for VT and sudden death.[181] Pulmonary valve replacement does not shorten QRS duration, although the QRS duration does not further increase. Valve replacement decreases the incidence of episodes of VT and atrial flutter.[182] Others have correlated dilated right ventricles (RVEDV), QT dispersion (QTd), and increased QRS duration with VT.[183] It is thought that the ventriculotomy, myocardial resection, and subsequent scarring provide the electrophysiological substrate of slow conduction and block that predisposes the patient to develop reentrant arrhythmias. Pathologic studies in patients with tetralogy of Fallot who have died suddenly have revealed extensive fibrosis of the RV myocardium at the ventriculotomy site, RV outflow tract (RVOT), and septum.[184] Serious ventricular arrhythmias occasionally occur despite good hemodynamic results, although sudden death occurs most commonly in VT associated with poor hemodynamics. Patients repaired earlier in life seem to have a lower incidence of VT.[185-187] It has been suggested that early repairs may decrease the incidence of VT in these patients.[188]

Evaluation

All postoperative patients, especially those noted earlier at highest risk, should have periodic follow-up (usually yearly) with standard ECGs. Holter monitoring should be performed every 2 to 3 years in those without known arrhythmias and every year in those in whom arrhythmias have been identified. Those being treated for arrhythmias may need more frequent monitoring. Patients with arrhythmias on ambulatory monitoring who are old enough to exercise should perform an exercise stress test. Those with complex arrhythmias (nonsustained VT, polymorphic PVCs, or polymorphic VT) or monomorphic VT should undergo further testing. This would include an EPS as described in the following section.

Electrophysiological Studies and Radiofrequency Catheter Ablation in Postoperative Patients

EPSs have been used to evaluate the propensity of these patients to develop VT, evaluate the efficacy of specific pharmacologic therapies, and locate the site of origin of the arrhythmia in patients who are candidates for ablative therapy.[189-191]

The predictive value of EPS in this population is not clear, but EPSs may be helpful in determining the need for implantation of an automatic cardioverter-defibrillator. A negative study does not guarantee that VT/VF or sudden death will not occur.[192] A multicenter retrospective trial did not show a correlation between sudden death and inducible VT, but a variety of protocols were used.[193] The usefulness of EPS in risk stratification of this population awaits the appropriate controlled study.

Fifteen percent of patients have inducible VT after repair of tetralogy of Fallot. The site of origin of the VT has been localized to the RVOT in most instances and to the ventricular septum in others.[190,194] The VT in these patients can be reproducibly initiated and terminated at EPS and has been presumed to be reentrant.[176] Factors associated with inducibility include RV systolic or volume overload and a longer period from the time of surgery.[195] Continuous electrical activity as illustrated in Figure 26-7 has been noted in these patients in the RVOT.[190] Fragmented, prolonged, low amplitude electrograms from the right ventricle have been reported and occur more commonly in patients with VTs as shown in Figure 26-8.

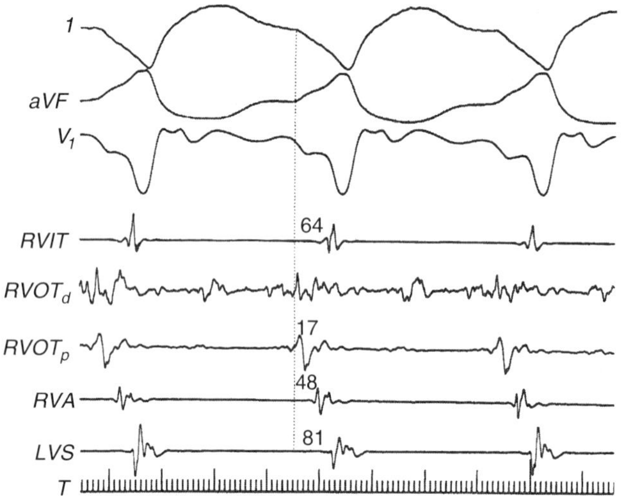

FIGURE 26-7 Endocardial catheter mapping during ventricular tachycardia (VT) in a patient with VT following repair of tetralogy of Fallot. ECG leads *1, AVF,* and *V1* are shown with electrograms recorded at the right ventricular inflow tract (*RVIT*), distal right ventricular outflow tract (*$RVOT_d$*), proximal right ventricular outflow tract (*$RVOT_p$*), right ventricular apex (*RVA*), and left ventricular mid septum (*LVS*) with 10-millisecond time lines (*T*). The *dotted line* indicates the onset of the QRS complex, and the numbers indicate the activation times at these sites. Continuous electrical activity was present in the distal outflow tract electrogram. The earliest discrete electrogram occurred in the proximal right ventricular outflow tract, which was the site of origin of the tachycardia. The left ventricle was activated after the right ventricle. (From Horowitz LN, Vetter VL, Harken AH, Josephson ME: Electrophysiologic characteristics of sustained ventricular tachycardia occurring after repair of tetralogy of Fallot. Am J Cardiol 1980;46:450.)

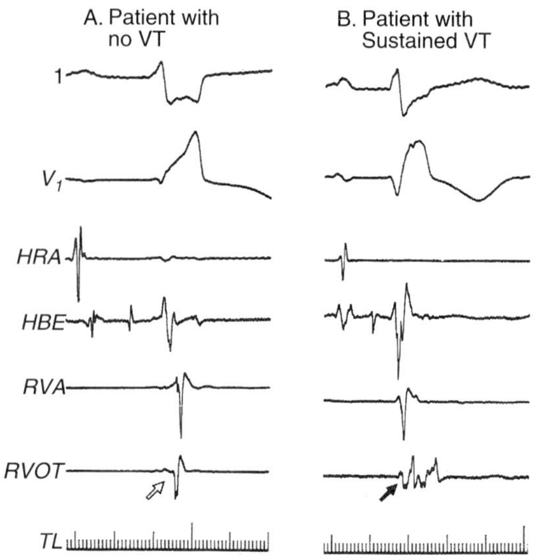

FIGURE 26-8 Electrograms recorded during normal sinus rhythm after repair of tetralogy of Fallot in a patient without ventricular tachycardia (VT) (*A*) and in a patient with ventricular tachycardia (*B*). In both panels, electrocardiographic leads *I* and *V_1* are shown with electrograms recorded in the high right atrium (*HRA*), His bundle area (*HBE*), right ventricular apex (*RVA*), and right ventricular outflow tract (*RVOT*) with 10 millisecond time lines (*TL*). In **panel A,** recorded in a patient with no clinical evidence of VT and in whom VT could not be induced in the laboratory, the RVOT electrogram is relatively narrow and smooth in contour (*open arrow*). In contrast **(panel B),** in a patient who had recurrent VT and in whom the tachycardia could not be induced in the clinical electrophysiology laboratory, the electrogram in the RVOT is fragmented and prolonged, exceeding 100 milliseconds in duration (*solid arrow*). Such an electrogram is typical of patients in whom VT occurs after repair of tetralogy of Fallot. (From Vetter VL, Horowitz LN: Electrophysiologic residua and sequelae of surgery for congenital heart defects. Am J Cardiol 1982;50:594.)

Management and Treatment of Patients with Postoperative Ventricular Tachycardia

Because of the high incidence of ventricular arrhythmias in postoperative patients, their precise role in the occurrence of sudden death is unclear. It is known after tetralogy repair that exercise stress testing or ambulatory monitoring will uncover a 25% to 70% incidence of ventricular arrhythmias. In 1985 Garson suggested that treatment of greater than 10 PVCs/hour on ambulatory monitoring had decreased the incidence of sudden death in their population.[196] On the other hand, Sullivan and others have reported no increase in the incidence of sudden death by not treating similar patients.[197] The final story awaits a large controlled study of these patients that is yet to be performed. The presence of frequent or complex ventricular ectopy probably identifies a high-risk group, but at present our ability to further identify those patients at highest risk is limited. It appears that patients with QRS duration above 180 ms and severe pulmonary regurgitation represent high-risk patients. The best time for valve replacement in children has not been determined.

Although there is no absolute agreement as to the indications for treatment, we have adopted a policy of treating patients with clinical episodes of VT or with symptoms and inducible VT. We generally treat patients with complex ventricular arrhythmias and abnormal hemodynamics or patients with significant symptoms and abnormal hemodynamics, or both. We use the EPS to determine the efficacy of specific drug regimens or the need for an ICD in patients with ineffective drug therapy or hemodynamic deterioration.

Treatment can include a combination of pharmacologic agents, radiofrequency ablation, surgical repair, and ablation or implantation of a pacemaker or cardioverter defibrillator.

The most commonly used drug regimens include β-blockers, mexiletine, or class IA or IC agents. In the 1970s phenytoin was found to be an effective drug in this population, but mexiletine is more commonly used at present. Amiodarone has been found to be an effective drug in some refractory patients. Recently, surgical or catheter ablation has been used effectively in these patients. ICDs are used in patients who have had syncopal events or aborted sudden death or those in whom unstable VT or VF is induced in the electrophysiology laboratory.

Radiofrequency Ablation in Postoperative Patients with Ventricular Tachycardia

Radiofrequency ablation may be an effective treatment in selected postoperative patients. The abnormal anatomy and hemodynamic instability in some of these patients complicate the procedure, but newer noncontact mapping techniques including electroanatomic mapping (CARTO) may help to identify the reentrant circuits, areas of scar, and anatomic barriers more precisely. There are reports of successful ablations in small groups of patients.[198-200,201]

Ventricular Tachycardia and Tumors

There is a known association of VT and cardiac rhabdomyomas. An incessant form of VT has been reported in infants and young children secondary to myocardial hematomas. These patients can be treated successfully with surgical ablation.[202] We have found that aggressive medical management including combination drug therapy may be used to control this type of VT. These tumors and rhabdomyomas may regress over time. A more diffuse infiltrative type of disease known as *histiocytoid or oncocytic cardiomyopathy of infancy* is known to be manifest as incessant VT in infancy and is usually fatal.[203]

Any patient with a diagnosis of tuberous sclerosis, which is known to be associated with cardiac rhabdomyoma, should have an ECG and echocardiogram. If tumors are noted, 24-hour ambulatory monitoring should be performed.

Ventricular Tachycardia and Mitral Valve Prolapse

In adults the incidence of sudden death associated with mitral valve prolapse (MVP) is 1.4%.[204] There are reports of sudden death in children and adolescents, especially in athletes with MVP.[205] The sudden death is proposed to be secondary to ventricular arrhythmias.[206] A recent study found MVP in 25% of patients with idiopathic VT. On follow-up, no patient developed sudden death.[207]

Twenty-four hour monitoring has revealed PVCs or more complex ventricular arrhythmias in as many as 46% of children with MVP.[208] Twenty-three percent of these arrhythmias were considered life threatening. Exercise stress testing revealed serious ventricular arrhythmias in 20% of patients.

β-blockers have been the drugs of choice for control of these arrhythmias.

Left Ventricular Aneurysms

Aneurysms or diverticula of the left ventricle have been reported in children, especially newborns, and may be associated with ventricular arrhythmias. Treatment is indicated by the clinical condition of the patient. Many of these diverticula will regress over time.[209]

Ventricular Tachycardia and Cardiomyopathies

Hypertrophic Cardiomyopathy

There is a high incidence of sudden death in children with hypertrophic obstructive cardiomyopathy (HCM), also known as *idiopathic hypertrophic subaortic stenosis (IHSS)*, which is presumed secondary to ventricular arrhythmias.[210-214] Hypertrophic cardiomyopathy is the most common cause of sudden death in young competitive athletes.[211,215]

Clinical Correlations. A family history of sudden death ventricular arrhythmias including nonsustained VT on 24-hour ambulatory monitors, previous syncope, extensive generalized hypertrophy, or ventricular systolic or diastolic dysfunction may identify a high-risk population.[214] Asymptomatic VT on Holter monitoring may or may not be predictive.[216] Abnormal blood pressure response to exercise may predict sudden death. Exercise stress testing may uncover ST segment depression or an abnormal systolic blood pressure response, including a small difference between peak and resting systolic blood pressure. Predictors of sudden cardiac death include higher LVOT pressure gradient at rest and failure to increase systolic blood pressure during exercise testing.[217] Two events that predict subsequent sudden death include a combination of inducible sustained VT at EPS and a history of cardiac arrest or syncope.[218] More than one third of patients who experience syncope VT/VF or resuscitated VF will die within 7 years from sudden cardiac death or progressive heart failure.[219] Unfortunately, many high-risk patients will not have inducible VT at EPS but may still be at high risk. It has been suggested that fractionation of electrograms with programmed stimulation may be a marker for high-risk patients.[220] Recent studies suggest that myocardial bridging of the left anterior descending coronary arteries may cause myocardial ischemia in children with HCM.[221] There remains a divergence of opinion whether bridging or compression of branches from LV hypertrophy is the primary problem.[222]

Hypertrophic cardiomyopathy is a genetic disease caused by mutations in contractile sarcomeric proteins. It is hoped that recent genetic identification of hypertrophic cardiomyopathy will increase understanding and identify high-risk patients.[223] The first gene identified for HCM is the β-myosin heavy chain MHC located on the long arm of chromosome 14.[224] The MHC gene (*MyHC*) is responsible for 35% of the families with HCM.[225] Multiple other genes that code the proteins of the sarcomere have been found to be responsible for HCM. These include the gene of cardiac T-troponin (*cTnT*),[226] the gene of alpha-tropomyosin, and the gene of binding protein C (*MyBP-C*).[227] Specific mutations are associated with a higher incidence of sudden death.[228,229]

Evaluation. All family members and identified patients should be followed with periodic ECGs and echocardiograms. In identified patients, yearly Holter monitoring should be performed. Exercise stress testing in those who are of an appropriate age should be performed. EPSs should be performed in those who have had syncope, documented complex arrhythmias, or symptoms suggesting ventricular arrhythmias.

Treatment. Treatment includes antiarrhythmic therapy,[230-232] pacemakers,[233,234] myotomy/myectomy,[235] and implantable defibrillators.[155,236] Maron reported a retrospective multicenter study of efficacy of ICDs in preventing sudden death in 128 patients with HCM who were judged to be at higher risk for sudden death and had implantable defibrillators. Twenty-three percent had appropriate shocks or antitachycardia pacing or 7% inappropriate shocks per year. In those in whom secondary prevention was provided after cardiac arrest or VT, 11% were appropriately activated. The interval between implantation and the first appropriate discharge was substantially prolonged, 4 to 9 years in six patients. The defibrillators were highly effective in terminating life-threatening ventricular arrhythmias.[210]

Dilated Cardiomyopathy

The most common etiology of dilated cardiomyopathy (DCM) in children is idiopathic,[237] but other causes include antecedent myocarditis, familial or genetic associations, and immune regulatory abnormalities.[238-241] The prognosis for children with DCM is poor, and reported mortality has been estimated at 24% to 50%. In a study of 62 children and adolescents (followed for 3.9 ± 4.5 years), 50% died, 16% recovered, 27% had residual decreased LV function, and 7% underwent orthotopic heart transplantation.[242] In another follow-up study of 41 children followed for a median of 2.5 years, 30% died, 32% recovered, and 38% survived with decreased LV function.[243] Mortality rates at follow-up were 24% at 1 year and 29% at 5 years. Poorer prognosis was associated with the clinical severity at presentation, lower mean shortening fraction, and severe arrhythmias.[243] Little information is available on risk stratification in children with dilated cardiomyopathy.

Clinical Correlations

In adults, a variety of high-risk factors have been suggested including presence of ventricular arrhythmia, LV end-diastolic diameter of greater than 70 mm, and nonsustained VT on Holter monitoring. Low ejection fraction of less than 30% and nonsustained VT are high-risk factors as well.[244]

Treatment

Treatment includes antiarrhythmic therapy, biventricular pacing, and implantable defibrillators. Ten percent of patients will require heart transplant.

Biventricular pacing has been reported recently as a successful treatment modality in adults with symptomatic drug-refractory heart failure secondary to dilated cardiomyopathy and associated interventricular conduction delay.[245-247] Several adult electrophysiologists are recommending a combination of biventricular pacing or implantable defibrillators, or both.[236,248] The utility of biventricular pacing in pediatric patients with a dilated cardiomyopathy, interventricular conduction delays, ventricular asynchrony, and congestive heart failure has not been studied but may provide a new treatment in these patients.

Several large multicenter trials of biventricular pacing are ongoing, including trials to evaluate the effects of biventricular pacing on ventricular arrhythmias.[249-251] Although early results in adults are encouraging, many questions remain.

Idiopathic Ventricular Tachycardia

VT in the absence of underlying heart or known genetic disease is unusual in childhood. Of that population, 27% present in infancy with mean age of presentation at 5 years.[252] Severe heart failure was uncommon but did occur in 12%, with 36% having some evidence of LV dysfunction. Resolution can be expected in two thirds of patients over time. The most favorable prognosis is in patients with RV VT and in infants.

Ventricular Tachycardia and Structurally Normal Hearts

Two sites have been associated with VT and structurally normal hearts. The first is located in the RVOT, usually in the anterior portion, most commonly anteroseptal, but also in the anterolateral and anterior regions.[253] The morphology of the VT on ECG is LBBB, most commonly with an inferior axis but also with a superior or rightward axis. RBBB pattern with superior or rightward axis is also seen. Clinically, both sustained and nonsustained VT are seen. It has been suggested that RVOT tachycardias may resolve over time. On the other hand, Drago reported biopsies positive for acute myocarditis, ARVD fatty infiltration, and other histologic abnormalities in 67% of patients without obvious heart disease and RVOT tachycardia.[254]

The tachycardia may be induced as nonsustained or sustained VT by programmed electrical stimulation isoproterenol in 40% to 80% of selected patients. Pace mapping can help identify the specific site of the VT in the RVOT.

The second form of VT in patients without heart disease originates from the left ventricle, most commonly in the septum in the mid to inferior position. The morphology of the VT is generally RBBB with left axis deviation. It can be induced in a similar fashion to the RVOT VT. These VTs are often sensitive to verapamil. VT from the basal aspect of the superior LV septum can appear as repetitive monomorphic VT, revealing a dominant R wave pattern in V1, an inferior axis, and a precordial R wave transition at or before lead V2.[255] These VTs may be responsive to verapamil or adenosine, suggesting a triggered mechanism.[256] Some left ventricular VT originate in the LVOT with an LBBB pattern.

Evaluation

Holter monitoring and exercise stress testing should be used to evaluate patients with these arrhythmias. EPSs may be necessary to identify the site of the tachycardia if RFA is being considered.

Treatment

Antiarrhythmic therapy may be appropriate in many of these patients, especially those who are not yet adolescents. Some of the RVOT tachycardias will resolve, and conservative therapy may be appropriate in those with slower VT that is easily controlled with medication. RVOT VT may respond to β-blockade or mexiletine. LV VT may respond to β-blockade or verapamil.

Both forms of VT are amenable to treatment with RFA.[59,255,257-259] Those with LBBB pattern and inferior axis and early precordial transition may be ablated from the left or noncoronary aortic sinus of Valsalva.[260] For LV septal VTs, ablation at the pre-Purkinje potential recording site during VT seems to be most successful.[261]

RV VT may be ablated using standard entrainment, pace mapping, or noncontact mapping, including electroanatomic mapping techniques. A higher success rate has been seen in the RV tachycardias.

Acute Treatment of Ventricular Tachycardia

VT should be treated emergently unless the rate is slow and the patient is clinically stable. If an extracardiac cause such as an electrolyte abnormality or acidosis has been identified, the underlying abnormality should be corrected. The correction usually results in conversion of the ventricular arrhythmia to sinus rhythm. In patients with cardiac compromise, IV lidocaine at 1 mg/kg should be given immediately. If the lidocaine is effective, a continuous infusion of lidocaine at 10 to 50 μg/kg should be started to maintain an adequate level of lidocaine. Lidocaine levels should be monitored carefully to prevent toxicity.

Synchronized cardioversion at 2 to 5 watt-seconds/kg should be performed if the lidocaine does not result in immediate conversion or if an IV site is not available.

Procainamide has been used to treat acute episodes of VT. Because of the associated negative inotropic effects, the patients must be monitored very carefully during the infusion. More recently, amiodarone has been shown to be effective when given intravenously for VT. Other drugs used in acute therapy are shown in Table 26-6.

TABLE 26-6 Acute Treatment of Ventricular Tachycardia

Initial Treatment	Dosage	Level
Lidocaine	1-2 mg/kg IV bolus every 5-15 min IV infusion: 20-50 μg/kg/min	1.5-5.0 mg/L
Cardioversion	1-5 watt-sec/kg Double if ineffective	
Secondary Treatment	**IV Dosage**	**Level**
Amiodarone	5 mg/kg over 1 hour, followed by 5-10 μg/kg/min infusion	
Bretylium	5 mg/kg bolus every 15 min Infusion: 5-10 mg/kg over 10 min q6h	
Magnesium	0.25 mEq/kg over 1 min, followed by 1 mEq/kg over 5 hours to achieve Mg^{++} level of 3-4 mg/dl	
Phenytoin	3-5 mg/kg over 5 min, not to exceed 1 mg/kg/min	10-20 μg/mL
Procainamide	5 mg/kg over 5-10 min or 15 mg/kg over 30-45 min Infusion: 20-100 μg/kg/min	NAPA—4-10 mg/L PA—4-8 mg/L
Propranolol	0.05-0.1 mg/kg over 5 min q6h	

NAPA, n-acetylprocainamide; PA, procainamide.

Rapid ventricular pacing may be used to convert the rhythm to sinus if pharmacologic therapy fails or is contraindicated.

Long-Term Treatment

Once the arrhythmia has been converted, choice of an appropriate long-term regimen is essential to maintaining stability of the patient. The drugs most commonly used are shown in Table 26-1. If lidocaine has been successful, it may be maintained until adequate levels of a chronic regimen have been reached or the acute causative agent is no longer present. When switching to mexiletine, the lidocaine should be gradually weaned, because the mexiletine is loaded to prevent combined toxicity of these two drugs as the side effects are similar. Propranolol or other β-blockers are especially effective in patients whose arrhythmia is sensitive to adrenergic stimuli.[262] The class I agents and amiodarone are effective in more refractory cases.[263,264] A combination of amiodarone and propranolol has been effective in infants and children.[265]

As sudden death occurs in up to 30% of patients with VT and congenital heart defects, these patients should be placed on a long-term regimen. In patients who are immediately postoperative, this type of arrhythmia is poorly tolerated but generally responds to lidocaine and correction of other underlying hemodynamic and metabolic abnormalities. Studies have shown that patients with early postoperative VT are likely to develop this arrhythmia in the late postoperative period, so long-term therapy is generally recommended. The postoperative patient who presents months to years after surgery also requires long-term therapy. A thorough investigation is needed to rule out underlying hemodynamic abnormalities, as VT is tolerated less well in this group of patients. Mexiletine has been shown to be an effective long-term drug in patients after tetralogy of Fallot repair, as is the β-blocker class of drugs.[266] Class I agents may be effective in refractory cases, as may amiodarone. The EPS may be a helpful guide to medical therapy.[189,190] In patients refractory to drug therapy, the EPS can determine the site of origin of the tachycardia and direct treatment by surgical ablation. Patients with life-threatening episodes or those refractory to medical or ablative therapy may require implantation of a cardioverter-defibrillator.

Electrophysiological Study of Ventricular Tachycardia

The specific indications for performance of EPSs in children with VT are included in the Table 26-7. Intracardiac recordings during VT show the absence of His bundle deflections consistently preceding ventricular depolarizations. Atrial capture at rates more rapid than the tachycardia normalizes the QRS complex. One of the most significant differences between adults and children is in the different mechanisms responsible for VT.[71] In adults, more than 90% have inducible VT. In children, only 30% to 60% have inducible or reentrant VT, while the rest have triggered or automatic VT. Inducible VT in children is more commonly associated with structural heart disease.

Radiofrequency Ablation of Ventricular Tachycardia

In the pediatric population, the categories of VT that are amenable to RFA include those in structurally normal hearts that originate primarily in the RVOT or in the left ventricle, especially in the mid to inferior septum. The technique of ablation involves identifying an activation site during VT that is earlier than any surface ventricular electrogram. Newer electroanatomic mapping or other forms of noncontact mapping may be used to identify the earliest activation site. At this site, pace mapping should be performed to identify an ECG that has the same morphology with pacing as the clinical VT.[253,267,268]

The other category of patients who may have VT that is amenable to RFA are those who have postoperative VT, particularly in the RVOT, commonly postoperative patients with tetralogy of Fallot.[190] Scarring from the right ventriculotomy or infundibular resection allows a zone of slow conduction producing the substrate for reentry. Successful RFA of VT in a tetralogy patient was reported in 1993.[269] Successful sites in these patients include those with earliest activation during VT, areas of slow conduction with low-amplitude fractionated electrograms and those sites that produce a pace map identical to the clinical or inducible VT.[199,269,270] A more recent report has noted success in 15 of 16 patients.[271] In the Pediatric Radiofrequency Ablation Registry, 54 patients with VT and heart disease have been ablated with 59% success.[272]

TABLE 26-7 Electrophysiological Study of Ventricular Tachycardia

Indications for Electrophysiological Study (EPS)
1. Documented ventricular tachycardia (VT) >150 bpm or wide QRS tachycardia, except in association with acute metabolic or electrolyte abnormalities, myocarditis, or long QT syndrome; value of EPS in cardiomyopathies has not been determined but may be helpful in selected cases
2. Nonsustained VT or complex premature ventricular depolarizations in a patient with an abnormal heart
3. Suspected VT in the presence of syncope or cardiac arrest of unknown etiology
4. Symptoms suggestive of VT in vulnerable patient with an abnormal heart (e.g., postoperative tetralogy of Fallot)
5. Follow-up of patient with inducible, documented VT to test efficacy of long-term medication
6. For radiofrequency ablation in selected patients

Prognosis of Ventricular Tachycardia

The prognosis of VT depends on the underlying condition. Reviews of VT in children have reported a high incidence of death of 10% to 47%.[58] The higher incidence occurs in those patients with underlying structural or postoperative heart disease and poor hemodynamic results. Sudden death has been reported in patients with normal hearts at an incidence of 6% to 8%.[171]

In the absence of structural heart disease, especially in the neonate or young child, spontaneous resolution of VT may occur. The outlook for infants and children with VT is excellent if the VT can be controlled with treatment.[66]

Sinus Node Dysfunction

Sinus node dysfunction, although not common, is seen in the pediatric patient. Patients can present with significant bradycardia with long pauses secondary to sinus arrest, with chronotropic incompetence leading to exercise intolerance or with intermittent periods of tachycardia and bradycardia (sick sinus syndrome). Etiologies of sinus node dysfunction include congenital defects of the sinus node, traumatic damage to the sinus node or its blood supply, inflammatory processes, or idiopathic causes. There are also instances in which the patient may present with signs and symptoms consistent with sinus node dysfunction wherein the sinus node is normal but there is abnormal autonomic input to the sinus node.

PRESENTATION

The pediatric patient with sinus node dysfunction may present with symptoms of fatigue and decreased exercise tolerance, dizziness, syncope or palpitations.

Some patients are completely asymptomatic, and the abnormality is noted on screening for other cardiac abnormalities. Occasionally, infants and toddlers present with a history consistent with breath holding spells that progress to frank syncope. These episodes are often brought on by a noxious stimulus such as pain, scolding, or the child not getting his or her way. The patient begins to cry and then proceeds to syncope. In evaluating these children, we have found that these events are often secondary to periods of prolonged sinus arrest lasting as long as 15 to 20 seconds. It appears that these periods of asystole are likely secondary to increased vagal tone and not related to abnormalities in the sinus node.

EVALUATION

The work-up of a patient with suspected sinus node dysfunction should include an ECG to determine if the underlying rhythm is sinus. If the rhythm is atrially derived with a P wave axis not between 0 and 90 degrees, further evaluation to rule out congenital heart diseases should be performed. Patients should also undergo 24-hour ambulatory monitoring documenting low and high rates to evaluate heart rate variability and look for significant periods of bradycardia or long pauses and for default to secondary pacers such as ectopic atrial, junctional, or ventricular rhythms. If the patient is old enough, an exercise tolerance test (generally 5 years of age or older) should be performed to evaluate chronotropic competence. Occasionally, patients undergo intracardiac electrophysiological evaluation to access sinus node function. In our laboratory, we evaluate sinus node function using sinus node recovery times both at a baseline state as well as while on isoproterenol, with or without atropine. Sinus node recovery times are helpful in determining if there is a primary defect in the sinus node versus an autonomic etiology for the bradycardia. It is our impression that these electrophysiological parameters should be used in combination with other testing modalities such as exercise stress testing and 24-hour ambulatory monitoring.

TREATMENT

The mainstay of therapy for patients with significant sinus node dysfunction is the use of pacemakers. Attempts have been made to use pharmacologic agents such as caffeine, β-sympathomimetics (e.g., theophylline), and oral vagolytic agents such as glycopyrrolate or atropine. These interventions generally have proved to be ineffective or only partially effective in treating significant bradycardia, with the potential side effect of being proarrhythmic in a population that may be at an increased risk of tachyarrhythmias. We have had some positive results with the use of atropine given orally in patients with hypervagotonia, without significant side effects.

With the primary therapeutic option being a pacemaker, the criteria for treatment should be well defined. We use the following as criteria for pacing in sinus node dysfunction: symptoms of dizziness or syncope that appear to be related to the bradycardia and not some other trigger such as vasodepressor syncope or hypervagotonia, bradycardia leading to tachycardia that requires treatment, or significant changes in exercise tolerance.

Although pacemakers can and have been used with great success in pediatrics, there are special issues that arise unique to this population.[273] The first and possibly most significant is patient size. It is possible to place a single transvenous pacing lead in someone as small as 10 kg.[274] In patients smaller than 5 kg, it is generally prudent to place the lead epicardially, which requires a sternotomy or thoracotomy. These procedures are much more invasive and result in considerably more discomfort than the use of transvenous pacing systems. The second issue is that many patients with congenital heart disease have anatomy that precludes the use of transvenous pacing such as residual right to left shunting or connections that do not allow the entry of the ventricle from veins (Fontan procedure).[275] These patients also require the use of epicardial pacing. The final consideration is that children are much more active than adults and are more likely to develop lead fractures and other technical difficulties such as lead displacement.

Atrioventricular Block

Abnormalities of AV conduction are uncommon in children but do occur. The etiologies include congenital abnormalities, inflammatory processes, or trauma.

FIRST- AND SECOND-DEGREE ATRIOVENTRICULAR BLOCKS

First- and second-degree heart block may be congenital or acquired. Acquired etiologies include traumatic damage associated with surgery for congenital heart defects and inflammatory processes such as rheumatic heart disease or Lyme carditis. EPSs in pediatric patients have generally localized the delay to the AV node.[276] Delay in His-Purkinje system with prolongation of the H–V interval may be more significant, indicating a predisposition to the development of complete heart block.[277,278] Patients with first- and second-degree AV blocks are generally asymptomatic unless the ventricular rate is significantly decreased. The low heart rates are especially significant in patients with compromised myocardial function, where the cardiac output may be insufficient to meet the patient's needs.

Evaluation

Patients with first- and second-degree AV block should have 24-hour ambulatory monitoring to determine the longest pause present and the lowest rate. The tracing should also be evaluated for periods of higher-grade AV block. Exercise stress testing should be used to determine the maximal heart rate and to look for potential ventricular arrhythmias.

Treatment

If a higher rate is needed, pharmacologic agents such as atropine may be helpful, especially if the block is in the AV node and partially mediated by vagal influences. Isoproterenol may increase the heart rate by increasing the rate of the escape pacemaker. As mentioned earlier, oral medications such as caffeine, theophylline, and antivagolytics are often not very useful in changing the heart rate or AV conduction. Acutely, temporary transcutaneous or transvenous pacing may be necessary. For persistent symptomatic or high-grade AV block, permanent pacing may be needed. In some patients with second-degree AV block, progression to complete AV block may occur.[277,278]

COMPLETE HEART BLOCK

Complete heart block is the most common cause of significant bradycardia in children. The ventricular rate is usually 40 to 80 bpm, depending on the patient's age. The QRS morphology and heart rate are related to the location of the escape pacemaker. The higher the origin of the pacemaker, the faster the ventricular rate and more narrow the QRS complex. Wider QRS escape complexes usually originate from the His bundle or below. The usual rate in the neonate is 60 to 80 bpm.

Complete heart block may be either congenital or acquired and occurs in one per 20,000 live births. The same causes of acquired heart block that cause second-degree AV block discussed earlier can cause complete heart block. In a series of 599 patients with congenital complete heart block, followed for more than 10 years, there was 92% survival in the patients without associated structural heart disease.[279] The greatest risk of mortality was during the first weeks of life, with half of the deaths occurring during the first year. Highest risk was indicated by associated cardiac anomalies, cardiomegaly, ventricular bradycardia less than 55 bpm, and atrial tachycardia greater than 140 bpm in infants.

There has been a strong association noted with connective tissue disease in the mother and congenital complete heart block. The prevalence has been reported as high as 85% in mothers of affected infants.[280,281] Only one half of these affected mothers are symptomatic, whereas the other half, although asymptomatic, are serologically positive. A mother with systemic lupus erythematosus has a 1:20 risk of having a child with complete heart block if she is anti-Ro positive.[282] Maternal immunoglobulin G antibodies to soluble tissue ribonucleoprotein antigens, found in the cytoplasm or nucleus of human cells (anti-Ro:SS-A and anti-La:SS-B), cross the placenta after the 12th to 16th week of gestation. This results in an inflammatory response in the fetal heart, particularly the conduction system, leading to destruction of the AV node.[283-285]

Buyon has demonstrated increased risk when the maternal antibodies target specific portions of the ribonuclear complex, particularly the 48-kd La (SS-B), the 52-kd Ro (SS-A), and the 60-kd Ro (SS-A). The highest risk was reactivity to both the 48-kd La (SS-B) and the 52-kd Ro (SS-A) components.[286] Because of this immunologic factor, dexamethasone and plasmapheresis have been recommended as an effective treatment in the mother of an affected fetus.[287,288] Although the heart block has not resolved with these treatments, the clinical course of the fetus has improved with decreased pleuropericardial effusions and other signs of fetal hydrops.

Presentation

Some cases will be diagnosed in utero because of low fetal heart rates, which usually develop after the 17th week of gestation. Fetal echocardiography may show fetal hydrops in 15% to 61%, with the higher incidence being seen in association with structural heart disease, especially AV valve regurgitation. The survival with hydrops is less than 10% unless the fetus can be delivered immediately. Even with early delivery, these patients are often very ill and difficult to manage and require an aggressive team approach using the skills of neonatologists, cardiologists, and cardiothoracic surgeons. With this aggressive management, the outlook is relatively good if the fetal heart rate is greater than 55 bpm and if no structural heart disease is present.

Structural heart disease is present in one third to one fourth of infants with congenital heart block. The associated congenital heart defects most commonly include those with l-transposition of the great arteries or the heterotaxy syndromes. With associated heart disease, mortality in the first year was 29%.[279] Complex postoperative congenital heart lesions or those with an unusual course of the AV conduction system may develop postoperative block.[289] After ventricular septal defect repair, the presence of RBBB and left anterior hemiblock is 7% to 17% with a 1% incidence of complete heart block.[290] Complete heart block has been reported to occur as late as 14 years after surgical repair.[291] Complete heart block occurs more commonly after repair of AV canal defects, probably because of the unusual course of the conduction system in these lesions, and may be seen in up to 7% of patients.[292] Corrected transposition of the great arteries is associated with AV conduction disturbances ranging from first-degree to complete heart block in 30% to 60% of patients.

These conduction disturbances may be present at birth, develop insidiously, or occur during or after the surgical correction of associated defects. The AV bundle and conduction system have been found to cross the pulmonary outflow tract and descend along the anterior rim of the ventricular septal defect or along the right margin of the foramen ovale between the main and outflow chambers in a single ventricle with an outflow chamber. Careful attention to these facts and intraoperative mapping have decreased the incidence of this form of postoperative block.

Evaluation

When low fetal heart rates are noted, fetal echocardiography should be performed. Signs of cardiac decompensation include evidence of fetal hydrops with pleural or pericardial effusions or ascites. Cardiac function may be decreased, and tricuspid regurgitation is often present.

In a study of patients with congenital heart block who were followed for more than 30 years, only 10% did not require a pacemaker. There was a 20% incidence of sudden death. Mitral regurgitation and prolonged QTc interval were poor prognostic findings. Pacemakers were recommended even in asymptomatic adults.[293]

Treatment

Although many infants with congenital complete heart block are asymptomatic at birth, others show findings typical of congestive heart failure in an infant and require treatment. A subset may develop severe congestive heart failure and cardiovascular collapse. These distressed infants may require intubation and ventilation, treatment of acidosis, and catecholamine support of heart rate and blood pressure.

In emergencies, immediate transthoracic pacing can be accomplished. The transcutaneous pacemaker may be effective for short-term emergency situations, but should be replaced with another pacing method as soon as possible. Placement of a temporary transvenous pacemaker, either through the umbilical vein or femoral vein under direct fluoroscopic observation, is preferred. Although infants with rates lower than 50 bpm or slightly higher rates and associated congenital heart defects or cardiomyopathies may require pacing, this decision should not be made on the basis of rate alone. Older children with congenital heart block may develop evidence of exercise intolerance, easy fatigability, or syncope. Not all patients with congenital complete heart block need a pacemaker, and many are not required until an older age.[293]

Pacemakers may be placed because of associated ventricular arrhythmias, either during sleep or exercise, easy fatigability or exercise intolerance, syncope, or presyncope.[294-296] Other indicators for pacing that have been associated with sudden death include severe bradycardia, ventricular ectopy especially with exercise, prolonged pauses, increased QRS width, prolongation of the QTc interval,[297] or a junctional recovery time of greater than 3 seconds.[298] The acquired heart block associated with inflammatory disease such as viral myocarditis, Lyme disease, or rheumatic fever may be transient and only require temporary pacing.

With postoperative heart block, temporary pacing is usually performed for 7 to 10 days. Permanent pacing should be performed if sinus rhythm does not return because of the high incidence of sudden death in this group of postoperative patients if they are not paced.[299] As the return of sinus rhythm does not ensure that heart block will not return at a later time, close follow-up is indicated.

Abnormalities in intraventricular conduction may be congenital or acquired. Congenital RBBB may be inherited[300] or found in association with Ebstein's anomaly of the tricuspid valve.[301] Children with Kearns-Sayre syndrome are likely to develop RBBB that progresses to complete heart block[302]; prophylactic placement of a pacemaker is recommended in this group of patients to prevent sudden death. Postoperative RBBB is common after repair of many cardiac anomalies[303] and is either central or peripheral.[304] Over the years, there has been a great deal of interest in the coexistence of left anterior hemiblock and RBBB in postoperative patients, especially after repair of tetralogy of Fallot.[305,306] Some of these patients are susceptible to the development of complete heart block. At present, close follow-up is recommended.

Treatment of Bradycardia

Acute medical therapy consists of atropine (0.02 to 0.04 mg/kg IV) or isoproterenol (0.01 to 2.0 μg/kg per minute). Temporary atrial pacing may be performed by the transcutaneous, esophageal, or intracardiac routes. Long-term medical therapy is rarely indicated, and persistent symptomatic bradycardia should be treated by permanent pacing.

Implantable Pacemakers

The use of permanent pacemakers in children was first reported in the early 1960s and predated many recent advances in lead and generator technology.[307-310] These advances, including programmability and miniaturization, have significantly increased pacemaker use in the pediatric population.[311-314]

Pacemakers are generally implanted in children who meet one of three criteria: congenital complete heart block,[294] postoperative, or acquired heart block.[274,303,315-318] Indications for pacemaker implantation are shown in Table 26-8. These recommendations may be revised over time. For example, asymptomatic patients with congenital complete heart block and awake rates of less than 50 bpm and evidence of escape pacemaker instability should be a class 1 indication due to their high incidence of sudden death.[293,307] Other possible new indications may include LQTS with bradycardia or uncontrolled arrhythmias and congenital AV block with prolonged Q–T intervals.

Once the decision is made to place a pacemaker, the lead route and type of pacing system must be selected. Epicardial pacing leads, those placed on the outside of the heart via a thoracotomy or sternotomy, were the first type used in children. These systems often had high pacing thresholds. This, in combination with the fact that children require higher pacing rates, resulted in early generator battery depletion.[273] The development of transvenous leads led to the ability to place pacemakers without a thoracotomy. These leads have much lower pacing thresholds and have improved the longevity of the pacemaker.[319,320] Esperer recently reported that the rate of lead-related complications was 35% with no significant difference between the epicardially and endocardially paced groups. Epicardially paced children had a higher rate of exit block, whereas endocardial leads were more likely to dislocate.[321] The newer steroid-eluting leads available in both systems should significantly decrease problems with exit block due to less scarring and decreased pacing thresholds.

TABLE 26-8 Indications for Permanent Pacing in Children and Adolescents

Class I
Advanced second- or third-degree atrioventricular (AV) block associated with symptomatic bradycardia, ventricular dysfunction, or low cardiac output. *(Level of evidence: C)*
Sinus node dysfunction with correlation of symptoms during age-inappropriate bradycardia. The definition of bradycardia varies with the patient's age and expected heart rate. *(Level of evidence: B)*
Postoperative advanced second- or third-degree AV block that is not expected to resolve or persists at least 7 days after cardiac surgery. *(Levels of evidence: B, C)*
Congenital third-degree AV block with a wide QRS escape rhythm, complex ventricular ectopy, or ventricular dysfunction. *(Level of evidence: B)*
Congenital third-degree AV block in the infant with a ventricular rate <50 to 55 bpm or with congenital heart disease and a ventricular rate <70 bpm. *(Levels of evidence: B, C)*
Sustained pause-dependent VT, with or without prolonged QT, in which the efficacy of pacing is thoroughly documented. *(Level of evidence: B)*
Class IIa
Bradycardia-tachycardia syndrome with the need for long-term antiarrhythmic treatment other than digitalis. *(Level of evidence: C)*
Congenital third-degree AV block beyond the first year of life with an average heart rate <50 bpm or abrupt pauses in ventricular rate that are two or three times the basic cycle length or associated with symptoms due to chronotropic incompetence. *(Level of evidence: B)*
Long QT syndrome with 2:1 AV or third-degree AV block. *(Level of evidence: B)*
Asymptomatic sinus bradycardia in the child with complex congenital heart disease with resting heart rate <40 bpm or pauses in ventricular rate >3 sec. *(Level of evidence: C)*
Class IIb
Transient postoperative third-degree AV block that reverts to sinus rhythm with residual bifascicular block. *(Level of evidence: C)*
Congenital third-degree AV block in the asymptomatic infant, child, adolescent, or young adult with an acceptable rate, narrow QRS complex, and normal ventricular function. *(Level of evidence: B)*
Asymptomatic sinus bradycardia in the adolescent with congenital heart disease with resting heart rate <40 bpm or pauses in ventricular rate >3 sec. *(Level of evidence: C)*
Class III
Transient postoperative AV block with return of normal AV conduction. *(Level of evidence: B)*
Asymptomatic postoperative bifascicular block with or without first-degree AV block. *(Level of evidence: C)*
Asymptomatic type I second-degree AV block. *(Level of evidence: C)*
Asymptomatic sinus bradycardia in the adolescent with longest R–R interval <3 sec and minimum heart rate >40 bpm. *(Level of evidence: C)*

From Gregoratos G, Abrams J, Epstein AE, et al: AAC/AHA/NASPE 2002 guideline update for implantation of cardiac pacemakers and antiarrhythmia devices: summary article: A report of the American College of Cardiology/American Heart Association Task Force on Practice Guidelines (ACC/AHA/NASPE Committee to Update the 1998 Pacemaker Guidelines). Circulation 2002;106:2145-61.

Epicardial atrial or ventricular pacing systems can be implanted through a subxiphoid incision with a small sternotomy or small left thoracotomy. Although technically possible, we do not advise intracardiac leads in very small children. Thrombosis, or narrowing of the venous system, occurs in a substantial number of these patients.[322-324] Pediatric patients, unlike adults, may require pacemakers for 60 or more years. Early obstruction of the venous system is undesirable and will lead to complications when the pacing system or lead requires replacement. Thrombus formation may develop with transvenous leads and become a potential source of emboli. This is a serious potential risk in patients with a right-to-left shunt, which may lead to cerebral vascular accidents. The risk of thrombus formation and emboli is compounded in patients with sluggish circulation such as those who have had Fontan procedures. We do not recommend the use of transvenous leads in either of these populations.

Both single- and dual-chamber pacing systems are used in pediatrics. Dual-chamber pacemakers provide AV synchrony and may improve cardiac output.[324] This may be especially significant in patients with single ventricle physiology and in those with poor ventricular function. A variety of programmable pacemaker functions are available. The most common modes used in pediatrics include single-chamber programming in the VVI or VVIR or AAI or AAIR modes. Dual-chamber pacing is generally programmed in the DDD, DDI, DDIR modes, or DDDR with or without mode switch. Patients with congenital or acquired complete heart block with normal sinus node function are often paced in a VDD mode, which can be performed with two pacing leads or with a single transvenous lead that has the ability to sense the atrium and pace the ventricle. The advantage of the single lead is that there is less cross-sectional area of the vein occupied, leading to a decrease in the risk of thrombus formation or vascular narrowing.

Many of the newer pacemakers provide the ability to perform temporary rapid pacing or noninvasive programmed electrophysiological testing, or both, which has proven helpful in the termination of certain SVT. Atrial antitachycardia pacemakers have been used in the treatment of patients with atrial flutter and some other SVTs.[325] There is a risk of acceleration of the atrial tachycardia or degeneration to a more serious rhythm. These pacemakers should be used with caution in these types of arrhythmias.[326]

ICDs have been used in pediatric patients. In 1993 Silka reported on 177 patients younger than 20 years of age who had an ICD with follow-up in 125 patients.[155] The indications for use included aborted sudden cardiac death (76%), drug refractory VT (10%), and syncope with inducible VT (10%). The most common associated cardiovascular diseases were hypertrophic and dilated cardiomyopathy in 54%, primary electrical disease, LQTS (26%), and congenital heart defects (18%).[154] Sixty-eight percent of patients experienced an appropriate discharge during follow-up. There were nine deaths. The use of transvenous nonthoracotomy leads has made implantation less invasive, but epicardial patches are still necessary in smaller children due to the size of the transvenous lead system.[327]

PACEMAKER FOLLOW-UP

Pacemakers in children, as in adults, should be followed in a systematic and organized fashion. Follow-up visits aid in the detection of pacemaker malfunction or battery depletion and allow one to determine the most optimal settings to provide pacemaker benefit and longevity. At follow-up visits, pacemaker thresholds, sensitivity, and graphic analysis of rate ranges over time need to be evaluated. Transtelephonic pacemaker checks are used in children but are not sufficient as the only follow-up. Careful surveillance for wire fracture, impedance changes, and threshold changes require on-site visits. In addition, Holter monitoring, exercise stress tests, and echocardiograms obtained at different programmed settings are used to assess proper function and determine the most appropriate settings.

REFERENCES

1. Ludomirsky A, Garson A Jr: Supraventricular tachycardia. In Garson A Jr, Gillette PC (eds): Pediatric Arrhythmias: Electrophysiology and Pacing. Philadelphia, WB Saunders, 1990, pp 380-426.
2. Akhtar M, Tchou PJ, Jazayeri M: Mechanism of clinical tachycardias. Am J Cardiol 1988;61:9A-19A.
3. Damiano BP, Rosen M: Effects of pacing on triggered activity induced by early after depolarization. Circulation 1984; 69:1025.
4. Gilmour RF, Heger JJ, Prystowsky EN, et al: Cellular electrophysiologic abnormalities of diseased human ventricular myocardium. Am J Cardiol 1983;51:137-44.
5. Ferrier GR: Digitalis arrhythmias: Role of oscillatory afterpotentials. Prog Cardiovasc Dis 1977;19:459.
6. Moe GK, Preston JB, Burlington H: Physiologic evidence for a dual AV transmission system. Circulation 1956;4:357.
7. Gillette PC: The mechanisms of supraventricular tachycardia in children. Circulation 1976;54:133-9.
8. Mehta AV, Sanchez GR, Sacks EJ, et al: Ectopic automatic atrial tachycardia in children: Clinical characteristics, management and follow-up. J Am Coll Cardiol 1988;11:379-85.
9. Walsh EP, Saul JP, Hulse JE, et al: Transcatheter ablation of ectopic atrial tachycardia in young patients using radiofrequency current. Circulation 1992;86:1138.
10. Silka MJ, Gillette PC, Garson A Jr, Zinner A: Transvenous catheter ablation of a right atrial automatic ectopic tahcycardia. J Am Coll Cardiol 1997;5:999-1001.
11. Gillette PC, Garson A Jr: Electrophysiologic and pharmacologic characteristics of automatic ectopic atrial tachycardia. Circulation 1977;56:571.
12. Gillette PC, Smith RT, Garson A Jr, et al: Chronic supraventricular tachycardia: A curable cause of congestive cardiomyopathy. JAMA 1985;253:391-2.
13. Wellens HJJ, Durer DR, Liem KL, et al: Effects of digitalis in patients with paroxysmal atrioventricular nodal tachycardia. Circulation 1975;52:779.
14. Rosen MR, Merker C, Gelband H: Effects of procainamide on the electrophysiologic properties of the canine ventricular conduction system. J Pharmacol Exp Ther 1973;185:438.
15. Perry JC, McQuinn R, Smith RT: Flecainide acetate for resistant arrhythmias in the young: Efficacy and pharmacokinetics. J Am Coll Cardiol 1989;14:185.
16. Zeigler V, Gillette PC, Ross BA, et al: Flecainide for supraventricular and ventricular arrhythmias in children and young adults. Am J Cardiol 1989;14:185-191.
17. Coumel P, Fidelle J: Amiodarone in the treatment of cardiac arrhythmias in children: One hundred thirty-five cases. Am Heart J 1992;100:1063-9.
18. Bernuth GV, Engelhardt W, Kramer HH, et al: Atrial automatic tachycardia in infancy and childhood. Eur Heart J 1992;13:1410-5.
19. Colloridi V, Perri C, Ventriglia F, Critelli G: Oral sotalol in pediatric atrial ectopic tachycardia. Am Heart J 1992;123:254-6.
20. Gillette PC, Garson AJ, Hesslein PS, et al: Successful surgical treatment of atrial, junctional and ventricular tachycardia unassociated with accessory connections in infants and children. Am Heart J 1981;102:984.
21. Ott DA, Gillette PC, Garson A Jr, et al: Surgical management of refractory supraventricular tachycardia in infants and children. J Am Coll Cardiol 1985;5:124-9.
22. Walsh EP, Saul JP, Hulse JE, et al: Transcatheter ablation of ectopic atrial tachycardia in young patients using radiofrequency current. Circulation 1992;86:1138-46.
23. Margolis PO, Roman CA, Moulton KP, et al: Radiofrequency catheter ablation of left and right ectopic atrial tachycardia [abstract]. Circulation 1991;82:718.
24. Liberthson RR, Colan SD: Multifocal or chaotic atrial rhythm. Pediatr Cardiol 1982;2:179-84.
25. Yeager SB, Hougen TJ, Levy AM: Sudden death in infants with chaotic atrial rhythm. Am J Dis Child 1984;138:689-92.
26. Dunnigan A, Benson DW Jr, Benditt DG: Atrial flutter in infancy: Diagnosis, clinical features, and treatment. Pediatrics 1985; 75:725.
27. Mendelsohn A, Macdonald DII, Serwer GA: Natural history of isolated atrial flutter in infancy. J Pediatr 1991;119:386-91.
28. Soyeur DJ: Atrial flutter in the human fetus: Diagnosis, hemodynamic consequences, and therapy. J Cardiovasc Electrophysiol 1996;7:989-98.
29. Vetter VL, Tanner CS, Horowitz LN: Inducible atrial flutter after the Mustard repair of complete transposition of the great arteries. Am J Cardiol 1988;61:428.
30. Garson A Jr, Bink-Boelkens M, Hesslein PS: Atrial flutter in the young: A collaborative study of 380 cases. J Am Coll Cardiol 1985;6:871.
31. Gewillig M, Wyse RK, de Leval MR, Deanfield JE: Early and late arrhythmias after the Fontan operation: Predisposing factors and clinical consequences. Br Heart J 1992;67:72-9.
32. Balaji S, Gewillig M, Bull C, et al: Arrhythmias after the Fontan procedure: Comparison of total cavopulmonary connection and atriopulmonary connection. Circulation 1991;84:III-162-III-167.
33. Rhodes LA, Walsh EP, Gamble WJ, et al: Benefits and potential risks of atrial antitachycardia pacing after repair of congenital heart disease. Pacing Clin Electrophysiol 1995;18:1005-16.
34. Fisher JD, Johnston DR, Kim SG, et al: Implantable pacers for tachycardia termination: Stimulation techniques and long-term efficacy. Pacing Clin Electrophysiol 1986;9:1325.
35. Triedman JK, Saul JP, Weindling SN, Walsh EP: Radiofrequency ablation of intra-atrial reentrant tachycardia after surgical palliation of congenital heart disease. Circulation 1995;91:707-14.
36. Garson A Jr, Gillette PC: Junctional ectopic tachycardia in children: Electrocardiography, electrophysiology and pharmacologic response. Am J Cardiol 1979;44:298.
37. Case CL, Gillette PC: Automatic atrial and junctional tachycardias in the pediatric patient: Strategies for diagnosis and management. Pacing Clin Electrophysiol 1993;16:1323-35.
38. Villain E, Vetter VL, Garcia JM, et al: Evolving concepts in the management of junctional ectopic tachycardia: A multicenter study. Circulation 1990;81:1544.
39. Balaji S, Sullivan I, Deanfield JE, James I: Moderate hypothermia in the management of resistant automatic tachycardias in children. Br Heart J 1991;66:224.
40. Till JA, Rowland E: Atrial pacing as an adjunct to the management of post-surgical His bundle tachycardia. Br Heart J 1991; 66:225-9.
41. Walsh EP, Saul JP, Sholler GF, et al: Evaluation of a staged treatment protocol for rapid automatic junctional tachycardia after

operation for congenital heart disease. J Am Coll Cardiol 1997;29:1046-55.
42. Van Hare GF, Velvis H, Langberg JJ: Successful transcatheter ablation of congenital junctional ectopic tachycardia in a ten-month-old infant using radiofrequency energy. Pacing Clin Electrophysiol 1990;13:730.
43. Rychik J, Marchlinski F, Sweeten TL, et al: Transcatheter radiofrequency ablation of congenital junctional ectopic tachycardia in a neonate. Pediatr Cardiol 1996;17:220-2.
44. Waldo AF, Krongrad E, Kupersmith J, et al: Ventricular paired pacing to control rapid ventricular heart rate following open heart surgery. Observations on ectopic automaticity. Report of a case in a four-month-old patient. Circulation 1976;53:176-80.
45. Raja P, Hawker RE, Chaikitpinyo A, et al: Amiodarone management of junctional ectopic tachycardia after cardiac surgery in children. Br Heart J 1994;72:265.
46. Epstein MC, Kiel EA, Victoria BE: Cardiac decompensation following verapamil therapy in infants with supraventricular tachycardia. Pediatrics 1985;75:737.
47. Abinader E, Borochowitz Z, Berger A: A hemodynamic complication of verapamil therapy in a neonate. Helv Paediatr Acta 1981;36:451.
48. Perry JC, Garson A Jr: Supraventricular tachycardia due to Wolff-Parkinson-White syndrome in children: Early disappearance and late recurrence. J Am Coll Cardiol 1990;16:1215-20.
49. Wolff GS, Han J, Gurran J: Wolff-Parkinson-White syndrome in the neonate. Am J Cardiol 1978;41:559.
50. Nadas AD, Daeschner CW, Roth A, Blumenthal SL: Paroxysmal tachycardia in infants and children. Pediatrics 1952;9:167-81.
51. Mantakas ME, McCue CM, Miller WW: Natural history of Wolff-Parkinson-White syndrome discovered in infancy. Am J Cardiol 1978;41:1097.
52. Chung KY, Walsh TJ, Massie E: Wolff-Parkinson-White syndrome. Am Heart J 1965;69:116.
53. Benson DW Jr, Smith WM, Dunnigan A, et al: Mechanisms of regular, wide QRS tachycardia in infants and children. Am J Cardiol 1982;49:1778.
54. Bromberg BI, Lindsay BD, Cain ME, Cox JL: Impact of clinical history and electrophysiologic characterization of accessory pathways on management strategies to reduce sudden death among children with Wolff-Parkinson-White syndrome. J Am Coll Cardiol 1996;27:690-5.
55. Paul T, Guccione P, Garson A Jr: Relation of syncope in young patients with Wolff-Parkinson-White syndrome to rapid ventricular response during atrial fibrillation. Am J Cardiol 1990;65: 318-21.
56. Alexander ME, Berul CI: Ventricular arrhythmias: When to worry. Pediatr Cardiol 2000;21:532-41.
57. Tsuji A, Nagashima M, Hasegawa S, et al: Long-term follow-up of idiopathic ventricular arrhythmias in otherwise normal children. Jpn Circ J 1995;59:654-62.
58. Vetter VL: Ventricular arrhythmias in pediatric patients with and without congenital heart disease. In Horowitz LN (ed): Current Management of Arrhythmias. Philadelphia, BC Decker, 1990, pp 208-20.
59. Benson DW Jr, Gallagher JJ, Sterba R, et al: Catecholamine induced double tachycardia: Case report in a child. Pacing Clin Electrophysiol 1980;3:96-103.
60. Chimenti C, Calabrese F, Thiene G, et al: Inflammatory left ventricular microaneurysms as a cause of apparently idiopathic ventricular tachyarrhythmias. Circulation 2001;104:168-73.
61. Saidi AS, Moodie DS, Garson AJ, et al: Electrocardiography and 24-hour electrocardiographic ambulatory recording (Holter monitor) studies in children infected with human immunodeficiency virus type 1. The Pediatric Pulmonary and Cardiac Complications of Vertically Transmitted HIV-1 Infection Study Group. Pediatr Cardiol 2000;21:189-96.
62. Perron AD, Brady WJ, Erling BF: Commodio cordis: An underappreciated cause of sudden cardiac death in young patients: Assessment and management in the ED. Am J Emerg Med 2001; 19:406-9.
63. Kohl P, Nesbitt AD, Cooper PJ, Lei M: Sudden cardiac death by commotio cordis: Role of mechano-electric feedback. Cardiovasc Res 2001;50:280-9.
64. Link MS, Wang PJ, Maron BJ, Estes NA: What is commotio cordis? Cardiol Rev 1999;7:265-9.
65. Bonnet D, Martin D, Pascale De Lonlay, et al: Arrhythmias and conduction defects as presenting symptoms of fatty acid oxidation disorders in children. Circulation 1999;100:2248-53.
66. Davis AM, Gow RM, McCrindle BW, Hamilton RM: Clinical spectrum, therapeutic management, and follow-up of ventricular tachycardia in infants and young children. Am Heart J 1996;131:186-91.
67. Fulton DR, Kyung JC, Burton ST: Ventricular tachycardia in children without heart disease. Am J Cardiol 1985;55:1328.
68. Rocchini AP, Chun PO, Dick M: Ventricular tachycardia in children. Am J Cardiol 1981;47:1091-7.
69. Noh CI, Gillette PC, Case CL, Zeigler VL: Clinical and electrophysiological characteristics of ventricular tachycardia in children with normal hearts. Am Heart J 1990;120:1326-33.
70. Corrado D, Basso C, Thiene G: Sudden cardiac death in young people with apparently normal heart. Cardiovasc Res 2001;50:399-408.
71. Vetter VL, Josephson ME, Horowitz LN: Idiopathic recurrent sustained ventricular tachycardia in children and adolescents. Am J Cardiol 1981;47:315.
72. Garson A Jr, Smith RT, Moak JP, et al: Ventricular arrhythmias and sudden death in children. J Am Coll Cardiol 1985;5:130B.
73. Deal BJ, Scott MM, Scagliotti D, et al: Ventricular tachycardia in a young population without overt heart disease. Circulation 1986; 73:1111.
74. MacLellan-Tolbert SG, Porter CJ: Accelerated idioventricular rhythm: A benign arrhythmia in childhood. Pediatrics 1995; 96:122-5.
75. Nakagawa M, Yoshihara T, Matsumura A, et al: Accelerated idioventricular rhythm in three newborn infants with congenital heart disease. Chest 1993;104:322-3.
76. Frank R, Fontaine G, Vedel J, et al: Electrocardiology of 4 cases of right ventricular dysplasia inducing arrhythmia. Arch Mal Coeur Vaiss 1978;71:963-72.
77. Dungan WT, Garson A Jr, Gillette PC: Arrhythmogenic right ventricular dysplasia: A cause of ventricular tachycardia in children with apparently normal hearts. Am Heart J 1981;102:745-50.
78. Lobo FV, Silver MD, Butany J, Heggtveit HA: Left ventricular involvement in right ventricular dysplasia/cardiomyopathy. Can J Cardiol 1999;15:1239-47.
79. Casset-Senon D, Babuty D, Alison D, et al: Delayed contraction area responsible for sustained ventricular tachycardia in an arrhythmogenic right ventricular cardiomyopathy: Demonstration by Fourier analysis of SPECT equilibrium radionuclide angiography. J Nucl Cardiol 2000;7:539-42.
80. Sekiguchi K, Miya Y, Kaneko Y, et al: Evaluation of signal-averaged electrocardiography for clinical diagnosis in arrhythmogenic right ventricular dysplasia. Jpn Heart J 2001;42:287-94.
81. Turrini P, Angelini A, Thiene G, et al: Late potentials and ventricular arrhythmias in arrhythmogenic right ventricular cardiomyopathy. Am J Cardiol 1999;83:1214-9.
82. Toyofuku M, Takaki H, Sunagawa K, et al: Exercise-induced ST elevation in patients with arrhythmogenic right ventricular dysplasia. J Electrocardiol 1999;32:1-5.
83. Groh WJ, Silka MJ, Oliver RP, et al: Use of implantable cardioverter-defibrillators in the congenital long QT syndrome. Am J Cardiol 1996;78:703-6.
84. Fontaine G, Fontaliran F, Hebert JL, et al: Arrhythmogenic right ventricular dysplasia. Annu Rev Med 1999;50:17-35.
85. Casset-Senon D, Philippe L, Babuty D, et al: Diagnosis of arrhythmogenic right ventricular cardiomyopathy by Fourier analysis of gated blood pool single-photon emission tomography. Am J Cardiol 1998;82:1399-404.
86. Li D, Ahmad F, Gardner MJ, et al: The locus of a novel gene responsible for arrhythmogenic right-ventricular dysplasia characterized by early onset and high penetrance maps to chromosome 10p12-p14. Am J Hum Genet 2000;66:148-56.
87. Link MS, Wang PJ, Haugh CJ, et al: Arrhythmogenic right ventricular dysplasia: Clinical results with implantable cardioverter defibrillators. J Interv Card Electrophysiol 1997;1:41-8.
88. Tavernier R, Gevaert S, De Sutter J, et al: Long term results of cardioverter-defibrillator implantation in patients with right ventricular dysplasia and malignant ventricular tachyarrhythmias. Heart 2001;85:53-6.
89. Guiradon GM, Klein GJ, Gulamhusein SS, et al: Total disconnection of the right ventricular free wall: Surgical treatment of right ventricular tachycardia associated with right ventricular dysplasia. Circulation 1983;67:463-7.

90. Schwartz PJ. Idiopathic long QT syndrome: Progress and questions. Am Heart J 1985;109:399.
91. Jervell A, Lange-Nielsen F: Congenital deaf-mutism, functional heart diseases, with prolongation of the QT interval and sudden death. Am Heart J 1957;54:59-68.
92. Romano C, Gemme G, Pongiglione R: Aritmie cardiache rare delle¢eta pediatrica. Clin Pediatr 1963;45:565.
93. Ward O: A new familial cardiac syndrome in children. J Irish Med Assoc 1964;54:103.
94. Garson A Jr, Macdonald DII, Fournier A, et al: The long QT syndrome in children: An international study of 287 patients. Circulation 1993;87:1866-72.
95. Scott WA, Macdonald DII: Two:one atrioventricular block in infants with congenital long Q-T syndrome. Am J Cardiol 1987; 60:1409-10.
96. Moss AJ, Schwartz PJ, Crompton RS, et al: The long QT syndrome: A prospective international study. Circulation 1985;71:17.
97. Moss AJ, Robinson JL: Long QT syndrome. Heart Dis Stroke 1992;1:309-14.
98. Schwartz PJ, Moss AJ, Vincent GM, Crampton RS: Diagnostic criteria for the long QT syndrome: An update. Circulation 1993; 88:782-4.
99. Vetter VL, Berul CI, Sweeten TL: Response of corrected QT intervals to exercise, pacing and isoproterenol. Cardiol Young 1993;3(suppl 1):63.
100. Callans DJ, Schwartzman D, Marchlinski FE: Is an isoproterenol induced QT abnormality specific to patients with the long QT syndrome? [abstract] Circulation 94[suppl I], I-432. 1996.
101. Vincent GM, Timothy KW, Leppert M, Keating M: The spectrum of symptoms and QT intervals in carriers of the gene for the long QT syndrome. N Engl J Med 1992;327:846-52.
102. Priori SG, Napolitano C, Schwartz PJ: Low penetrance in the long-QT syndrome: Clinical impact. Circulation 1999;99:529-33.
103. Keating M, Atkinson D, Dunn C, et al: Linkage of a cardiac arrhythmia, the long QT syndrome, and the Harvey ras-1 gene. Science 1991;252:704-6.
104. Jiang C, Atkinson D, Towbin JA, et al: Two long QT syndrome loci map to chromosomes 3 and 7 with evidence for further heterogeneity. Nat Genet 1994;8:141-7.
105. Curran ME, Splawski I, Timothy KW, et al: A molecular basis for cardiac arrhythmia: HERG mutations cause long QT syndrome. Cell 1995;80:795-803.
106. Schott J, Charpentier F, Peltier S, et al: Mapping of a gene for long QT syndrome to chromosome 4q25-27. Am J Hum Genet 1995;57:1114-22.
107. Keating M, Dunn C, Atkinson D, et al: Consistent linkage of the long-QT syndrome to the Harvey ras-1 locus on chromosome 11. Am J Hum Genet 1991;49:1335-9.
108. Towbin JA, Li H, Taggart RT, et al: Evidence of genetic heterogeneity in Romano-Ward long QT syndrome. Analysis of 23 families. Circulation 1994;90:2635-44.
109. Vincent GM: Heterogeneity in the inherited long QT syndrome. J Cardiovasc Electrophysiol 1995;6:137-46.
110. Barhanin J, Lesage F, Guillemare E, et al: K(V)LQT1 and IsK (minK) proteins associate to form the I(Ks) cardiac potassium current. Nature 1996;384:78-80.
111. Abbott GW, Sesti F, Splawski I, et al: MiRP1 forms IKr potassium channels with HERG and is associated with cardiac arrhythmia. Cell 1999;97:175-87.
112. Wang Q, Curran ME, Splawski I, et al: Positional cloning of a novel potassium channel gene: KVLQT1 mutations cause cardiac arrhythmias. Nat Genet 1996;12:17-23.
113. Sanguinetti MC, Curran ME, Zou A, et al: Coassembly of K(V)LQT1 and minK (IsK) proteins to form cardiac I(Ks) potassium channel. Nature 1996;384:80-3.
114. Sanguinetti MC, Jurkiewicz NK: Two components of cardiac delayed rectifier K^+ current. J Gen Physiol 1990;96:195-215.
115. Keating MT, Sanguinetti MC: Molecular and cellular mechanisms of cardiac arrhythmias. Cell 2001;104:569-80.
116. Splawski I, Shen J, Timothy KW, et al: Spectrum of mutations in long-QT syndrome genes KVLQT1, HERG, SCN5A, KCNE1, and KCNE2. Circulation 2000;102:1178-85.
117. Schulze-Bahr E, Wang Q, Wedekind H, et al: KCNE1 mutations cause Jervell and Lange-Nielsen syndrome. Nat Genet 1997; 17:267-8.
118. Splawski I, Tristani-Firouzi M, Lehmann MH, et al: Mutations in the hminK gene cause long QT syndrome and suppress IKs function. Nat Genet 1997;17:338-40.
119. Vetter DE, Mann JR, Wangemann P, et al: Inner ear defects induced by null mutation of the isk gene. Neuron 1996;17: 1251-64.
120. Sanguinetti MC, Jiang C, Curran ME, Keating MT: A mechanistic link between an inherited and an acquired cardiac arrhythmia: HERG encodes the IKr potassium channel. Cell 1995;81: 299-307.
121. Trudeau MC, Warmke JW, Ganetzky B, Robertson GA: HERG, a human inward rectifier in the voltage-gated potassium channel family. Science 1995;269:92-5.
122. Mitcheson JS, Chen J, Lin M, et al: A structural basis for drug-induced long QT syndrome. Proc Natl Acad Sci U S A 2000;97: 12329-33.
123. Wang Q, Shen J, Splawski I, et al: SCN5A mutations associated with an inherited cardiac arrhythmia, long QT syndrome. Cell 1995;80:805-11.
124. Gellens ME, George AL Jr, Chen LQ, et al: Primary structure and functional expression of the human cardiac tetrodotoxin-insensitive voltage-dependent sodium channel. Proc Natl Acad Sci U S A 1992;89:554-8.
125. Splawski I, Shen J, Timothy K, et al: Spectrum of mutations in Long-QT syndrome genes. Circulation 2000;102:1185.
126. Bennett PB, Yazawa K, Makita N, et al: Molecular mechanism for an inherited cardiac arrhythmia. Nature 1995;376:683-5.
127. Moss AJ, Zareba W, Benhorin J, et al: ECG-T wave patterns in genetically distinct forms of the hereditary long QT syndrome. Circulation 1995;92:2929-34.
128. Zareba W, Moss AJ, Schwartz PJ, et al: Influence of genotype on the clinical course of the long-QT syndrome. International Long-QT Syndrome Registry Research Group. N Engl J Med 1998;339:960-5.
129. Schwartz PJ, Priori SG, Spazzolini C, et al: Genotype-phenotype correlation in the long-QT syndrome: Gene-specific triggers for life-threatening arrhythmias. Circulation 2001;103:89-95.
130. Ackerman MJ, Tester DJ, Porter CJ: Swimming, a gene-specific arrhythmogenic trigger for inherited long QT syndrome. Mayo Clin Proc 1999;74:1088-94.
131. Wilde AA, Jongbloed RJ, Doevendans PA, et al: Auditory stimuli as a trigger for arrhythmic events differentiate HERG- related (LQTS2) patients from KVLQT1-related patients (LQTS1). J Am Coll Cardiol 1999;33:327-32.
132. Rosero SZ, Zareba W, Moss AJ, et al: Asthma and the risk of cardiac events in the long QT syndrome. Long QT Syndrome Investigative Group. Am J Cardiol 1999;84:1406-11.
133. Schwartz PJ, Priori SG, Dumaine R, et al: A molecular link between the sudden infant death syndrome and the long-QT syndrome. N Engl J Med 2000;27:262-7.
134. Schwartz PJ, Priori SG, Bloise R, et al: Molecular diagnosis in a child with sudden infant death syndrome. Lancet 2001; 358:1343.
135. Schwartz PJ, Stramba-Badiale M, Segantini A, et al: Prolongation of the QT interval and the sudden infant death syndrome. N Engl J Med 1998;338:1709-14.
136. Bazett HC: An analysis of the time-relations of electrocardiograms. Heart 1920;7:353-70.
137. Garson A Jr: How to measure the QT interval—what is normal? Am J Cardiol 1993;72:14B-16B.
138. Malfatto G, Beria G, Sala S, et al: Quantitative analysis of T wave abnormalities and their prognostic implications in the idiopathic long QT syndrome. J Am Coll Cardiol 1994;23:296-301.
139. Priori SG, Napolitano C, Diehl L, et al: Dispersion of QT interval. A marker of therapeutic efficacy in the idiopathic long QT syndrome. Circulation 1994;89:1681-8.
140. Moss AJ: Measurement of the QT interval and the risk associated with QT_c interval prolongation: A review. Am J Cardiol 1993;72: 23B-25B.
141. Shah MJ, Wieand TS, Rhodes LA, et al: QT and JT dispersion in children with long QT syndrome. J Cardiovasc Electrophysiol 1997;8:642-8.
142. Banai S, Tzivoni D: Drug therapy for torsade de pointes. J Cardiovasc Electrophysiol 1993;4:206-10.
143. Kothari SS, Krishnaswami S: Magnesium sulfate therapy in torsades de pointes. Indian Heart J 1988;40:210-1.

144. Crawford MH, Karliner JS, O'Rouke RA, et al: Prolonged QT interval syndrome: Successful treatment with combined ventricular pacing and propranolol. Chest 1975;68:369.
145. Trippel DL, Gillette PC: Atenolol in children with ventricular arrhythmias. Am Heart J 1990;119:1312-6.
146. Moss AJ, Robinson J: Clinical features of the idiopathic long QT syndrome. Circulation 1992;85(suppl):I140-I144.
147. Moss AJ, Zareba W, Hall WJ, et al: Effectiveness and limitations of beta-blocker therapy in congenital long-QT syndrome. Circulation 2000;101:616-23.
148. Locati EH, Schwartz PJ, Moss AJ, et al: Long-term survival after left cervico-thoracic sympathectomy in high risk long Q-T syndrome patients with refractory ventricular arrhythmias. J Am Coll Cardiol 1986;7:235A.
149. Packer DL, Coltori F, Smith MS, et al: Sudden death after left stellectomy in the long Q-T syndrome. Am J Cardiol 1984;54:1365-6.
150. Ouriel K, Moss AJ: Long QT syndrome: An indication for cervicothoracic sympathectomy. Cardiovasc Surg 1995;3:475-8.
151. Schwartz PJ, Locati EH, Moss AJ, et al: Left cardiac sympathetic denervation in the therapy of congenital long QT syndrome. A worldwide report. Circulation 1991;84:503-11.
152. Eldar M, Griffin JC, Abbott JA: Permanent cardiac pacing in patients with the long QT syndrome. J Am Coll Cardiol 1987;10:600.
153. Moss AJ, Liu JE, Gottlieb S, et al: Efficacy of permanent pacing in the management of high-risk patients with long QT syndrome. Circulation 1991;84:1524-9.
154. Groh WJ, Silka MJ, Oliver RP, et al: Use of implantable cardioverter-defibrillators in the congenital long QT syndrome. Am J Cardiol 1996;78:703-6.
155. Silka MJ, Kron J, Dunnigan A, Macdonald DII: Sudden cardiac death and the use of implantable cardioverter-defibrillators in pediatric patients. Circulation 1993;87:800-7.
156. Roden DM, Lazzara R, Rosen M, et al: Multiple mechanisms in the long-Qt syndrome: Current knowledge, gaps, and future directions. Circulation 1996;94:1996-2012.
157. Schwartz PJ, Priori SG, Locati EH, et al: Long QT syndrome patients with mutations of the SCN5A and HERG genes have differential responses to Na^+ channel blockade and to increases in heart rate. Circulation 1995;92:3381-6.
158. Sato T, Hata Y, Yamamoto M, et al: Early afterdepolarizations abolished by potassium channel opener in a patient with idiopathic long QT syndrome. J Cardiovasc Electrophysiol 1995;6:279-82.
159. Brugada P, Brugada J: Right bundle branch block, persistent ST segment elevation and sudden cardiac death: A distinct clinical and electrocardiographic syndrome. A multicenter report. J Am Coll Cardiol 1992;20:1391-6.
160. Chen Q, Kirsch GE, Zhang D, et al: Genetic basis and molecular mechanism for idiopathic ventricular fibrillation. Nature 1998;392:293-6.
161. Brugada R: Use of intravenous antiarrhythmics to identify concealed Brugada syndrome. Curr Control Trials Cardiovasc Med 2000;1:45-47.
162. Naccarelli GV, Antzelevitch C: The Brugada syndrome: Clinical, genetic, cellular, and molecular abnormalities. Am J Med 2001;110:573-81.
163. Kakishita M, Kurita T, Matsuo K, et al: Mode of onset of ventricular fibrillation in patients with Brugada syndrome detected by implantable cardioverter defibrillator therapy. J Am Coll Cardiol 2000;36:1646-53.
164. Priori SG, Napolitano C, Tiso N, et al: Mutations in the cardiac ryanodine receptor gene (hRyR2) underlie catecholaminergic polymorphic ventricular tachycardia. Circulation 2001;103:196-200.
165. Balaji S, Wiles HB, Sens MA, Gillette PC: Immunosuppressive treatment for myocarditis and borderline myocarditis in children with ventricular ectopic rhythm. Br Heart J 1994;72:354-9.
166. Wiles HB, Gillette PC, Harley RA, Upshur JK: Cardiomyopathy and myocarditis in children with ventricular ectopic rhythm. J Am Coll Cardiol 1992;20:359-62.
167. Ino T, Okubo M, Akimoto K, et al: Corticosteroid therapy for ventricular tachycardia in children with silent lymphocytic myocarditis. J Pediatr 1995;126:304-8.
168. Drucker NA, Colan SD, Lewis AB, et al: Globulin treatment of acute myocarditis in the pediatric population. Circulation 1994;89:252-7.
169. Vetter VL, Horowitz LN: Electrophysiologic residua and sequelae of surgery for congenital heart defects. Am J Cardiol 1982;50:588.
170. Silka MJ, Hardy BG, Menashe VD, Morris CD: A population-based prospective evaluation of risk of sudden cardiac death after operation for common congenital heart defects. J Am Coll Cardiol 1998;32:245-51.
171. Vetter VL: Ventricular arrhythmias in patients with congenital heart disease. In Greenspon AJ, Waxman HL (eds): Contemporary Management of Ventricular Arrhythmias. Philadelphia, FA Davis, 1991, pp 255-73.
172. Bockoven JR, Wernovsky G, Vetter VL, et al: Perioperative conduction and rhythm disturbances after the Ross procedure in young patients. Ann Thorac Surg 1998;66:1383-8.
173. von Bernuth G: Twenty-five years after the first arterial switch procedure: Mid-term results. Thorac Cardiovasc Surg 2000;48:228-32.
174. Garson A Jr, Nihill MR, McNamara DG, et al: Status of the adult and adolescent after repair of tetralogy of Fallot. Circulation 1979;59:1232.
175. Gillette PC, Yeoman MA, Mullins CE, et al: Sudden death after repair of tetralogy of Fallot. Circulation 1977;56:566.
176. Vetter VL: Postoperative arrhythmias after surgery for congenital heart defects. In Zipes DP (ed): Cardiology in Review. Philadelphia, Williams & Wilkins, 1994, pp 83-97.
177. Deanfield JE, McKenna WJ, Presbitero P, et al: Ventricular arrhythmia in unrepaired tetralogy of Fallot: Relation to age, timing of repair and hemodynamic status. Br Heart J 1984;52:77.
178. Vaksmann G, Fournier A, Davignon A, et al: Frequency and prognosis of arrhythmias after operative "correction" of tetralogy of Fallot. Am J Cardiol 1990;66:346-9.
179. Gatzoulis MA, Till JA, Somerville J, Redington AN: Mechanoelectrical interaction in tetralogy of Fallot. QRS prolongation relates to right ventricular size and predicts malignant ventricular arrhythmias and sudden death. Circulation 1995;92:231-7.
180. Zimmermann M, Friedl B, Adamec R, et al: Ventricular late potentials and induced ventricular arrhythmias after surgical repair of tetralogy of Fallot. Am J Cardiol 1991;67:873-8.
181. Gatzoulis MA, Balaji S, Webber SA, et al: Risk factors in arrhythmia and sudden cardiac death late after repair of tetralogy of Fallot: A multicenter study. Lancet 2000;356:975-81.
182. Therrien J, Siu SC, Harris L, et al: Impact of pulmonary valve replacement on arrhythmia propensity later after repair of tetralogy of Fallot. Circulation 2001;103:2489-94.
183. Daliento L, Rizzoli G, Menti L, et al: Accuracy of electrocardiographic and echocardiographic indices in predicting life threatening ventricular arrhythmias in patients operated for tetralogy of Fallot. Heart 1999;81:650-5.
184. Deanfield JE, Ho S, Anderson RH, et al: Late sudden death after repair of tetralogy of Fallot: A clinicopathologic study. Circulation 1983;67:626-31.
185. Alexiou C, Mahmoud H, Al-Khaddour A, et al: Outcome after repair of tetralogy of Fallot in the first year of life. Ann Thorac Surg 2001;71:494-500.
186. Walsh EP, Rockenmacher S, Keane JF, et al: Late results in patients with tetralogy of Fallot repaired during infancy. Circulation 1998;77:1062-7.
187. Joffe H, Georgakopoulos D, Celermajer DS, et al: Late ventricular arrhythmia is rare after early repair of tetralogy of Fallot. J Am Coll Cardiol 1994;23:1146-50.
188. Walsh EP, Rockenmacher S, Keane JF, et al: Late results in patients with tetralogy of Fallot repaired during infancy. Circulation 1988;77:1062-7.
189. Garson A Jr, Porter CJ, Gillette PC, et al: Induction of ventricular tachycardia during electrophysiologic study after repair of tetralogy of Fallot. J Am Coll Cardiol 1983;1:1493.
190. Horowitz LN, Vetter VL, Harken AH, et al: Electrophysiologic characteristics of sustained ventricular tachycardia occurring after repair of tetralogy of Fallot. Am J Cardiol 1980;46:446.
191. Deal BJ, Scagliotti D, Miller SM, et al: Electrophysiologic drug testing in symptomatic ventricular arrhythmias after repair of tetralogy of Fallot. Am J Cardiol 1987;59:1380-5.

192. Lucron H, Marcon F, Bosser G, et al: Induction of sustained ventricular tachycardia after surgical repair of tetralogy of Fallot. Am J Cardiol 1999;83:1369-73.
193. Chandar JS, Wolff GS, Garson A Jr, et al: Ventricular arrhythmias in postoperative tetralogy of Fallot. Am J Cardiol 1990;65:655-71.
194. Kugler JD, Pinsky WW, Cheatham JP, et al: Sustained ventricular tachycardia after repair of tetralogy of Fallot: New electrophysiologic findings. Am J Cardiol 1983;51:1137-43.
195. Marie PY, Marçon F, Brunotte F, et al: Right ventricular overload and induced sustained ventricular tachycardia in operatively "repaired" tetralogy of Fallot. Am J Cardiol 1992;69:785-9.
196. Garson A Jr, Randall DC, Gillette PC, et al: Prevention of sudden death after repair of tetralogy of Fallot: Treatment of ventricular arrhythmias. J Am Coll Cardiol 1985;6:221-7.
197. Sullivan I, Presbitero P, Gooch VM, et al: Is ventricular arrhythmias in repaired tetralogy of Fallot an effect of operation ora consequence of the course of the disease? Br Heart J 1987;58:40-4.
198. Stevenson WG, Delacretaz E, Friedman PL, Ellison KE: Identification and ablation of macroreentrant ventricular tachycardia with the CARTO electroanatomical mapping system. Pacing Clin Electrophysiol 1998;21:1448-56.
199. Biblo LA, Carlson MD: Transcatheter radiofrequency ablation of ventricular tachycardia following surgical correction of tetralogy of Fallot. Pacing Clin Electrophysiol 1994;17:1556-60.
200. Burton ME, Leon AR: Radiofrequency catheter ablation of right ventricular outflow tract tachycardia late after complete repair of tetralogy of Fallot using the pace mapping technique. Pacing Clin Electrophysiol 1993;16:2319-25.
201. Oda H, Aizawa Y, Murata M, et al: A successful electrical ablation of recurrent sustained ventricular tachycardia in a postoperative case of tetralogy of Fallot. Jpn Heart J 1986;27:421-8.
202. Garson A Jr, Gillette PC, Titus JL, et al: Surgical treatment of ventricular tachycardia in infants. N Engl J Med 1984;310:1443.
203. Malhotra V, Ferrans VJ, Virmani R: Infantile histiocytoid cardiomyopathy: Three cases and literature review. Am Heart J 1994;128:1009-21.
204. Jeresaty RM: Sudden death in the mitral valve prolapse click syndrome. Am J Cardiol 1976;37:317.
205. Maron BJ, Roberts WC, McAllister HA, et al: Sudden death in young athletes. Circulation 1980;62:218.
206. Swartz MH, Teichholz LE, Donoso E: Mitral valve prolapse. A review of associated arrhythmias. Am J Med 1977;62:377.
207. La Vecchia L, Ometto R, Centofante P, et al: Arrhythmic profile, ventricular function, and histomorphometric findings in patients with idiopathic ventricular tachycardia and mitral valve prolapse: Clinical and prognostic evaluation. Clin Cardiol 1998;21:731-5.
208. Kavey RW, Sandheimer HM, Blackman MS: Detection of dysrhythmia in pediatric patients with mitral valve prolapse. Circulation 1980;62:582.
209. Papagiannis J, Van Praagh R, Schwint O, et al: Congenital left ventricular aneurysm: Clinical, imaging, pathologic, and surgical findings in seven new cases. Am Heart J 2001;141:491-9.
210. Maron BJ, Shen WK, Link MS, et al: Efficacy of implantable cardioverter-defibrillators for the prevention of sudden death in patients with hypertrophic cardiomyopathy. N Engl J Med 2000; 342:365-73.
211. Maron BJ, Roberts WC, Epstein SE: Sudden death in hypertrophic cardiomyopathy: A profile of 78 patients. Circulation 1982;65:1388.
212. McKenna WJ, Deanfield JE, Faruqui A, et al: Prognosis in hypertrophic cardiomyopathy: Role of age and clinical, electrocardiographic and hemodynamic features. Am J Cardiol 1981; 47:532.
213. Maron BJ, Savage DD, Wolfson JK, et al: Prognostic significance of 24 hour ambulatory electrocardiographic monitoring in patients with hypertrophic cardiomyopathy: A prospective study. Am J Cardiol 1981;48:252.
214. McKenna WJ, England D, Doi YL, et al: Arrhythmias in hypertrophic cardiomyopathy: Influence on prognosis. Br Heart J 1981;46:168.
215. Maron BJ, Fananapazir L: Sudden cardiac death in hypertrophic cardiomyopathy. Circulation 1992;85:I-57-I-63.
216. McKenna WJ, Franklin RCG, Nihoyannopoulos P, et al: Arrhythmia and prognosis in infants, children and adolescents with hypertrophic cardiomyopathy. J Am Coll Cardiol 1988; 11:147-53.
217. Yetman AT, Hamilton RM, Benson LN, McCrindle BW: Long-term outcome and prognostic determinants in children with hypertrophic cardiomyopathy. J Am Coll Cardiol 1998;32:1943-50.
218. Fananapazir L, Chang AC, Epstein SE, McAreavey D: Prognostic determinants in hypertrophic cardiomyopathy: Prospective evaluation of a therapeutic strategy based on clinical, Holter, hemodynamic, and electrophysiological findings. Circulation 1992;86:730-40.
219. Elliott PM, Sharma S, Varnava A, et al: Survival after cardiac arrest or sustained ventricular tachycardia in patients with hypertrophic cardiomyopathy. J Am Coll Cardiol 1999;33:1596-601.
220. Saumarez RC: Electrophysiological investigation of patients with hypertrophic cardiomyopathy: Evidence that slowed intraventricular conduction is associated with an increased risk of sudden death. Br Heart J 1994;72(suppl):S19-S23.
221. Yetman AT, McCrindle BW, MacDonald C, et al: Myocardial bridging in children with hypertrophic cardiomyopathy—a risk factor for sudden death. N Engl J Med 1998;339:1201-9.
222. Mohiddin SA, Begley D, Shih J, Fananapazir L: Myocardial bridging does not predict sudden death in children with hypertrophic cardiomyopathy but is associated with more severe cardiac disease. J Am Coll Cardiol 2000;36:2270-8.
223. Marian AJ: Sudden cardiac death in patients with hypertrophic cardiomyopathy: From bench to bedside with an emphasis on genetic markers. Clin Cardiol 1995;18:189-98.
224. Geisterfer-Lowrance AA, Kass S, Tanigawa G, et al: A molecular basis for familial hypertrophic cardiomyopathy: A beta cardiac myosin heavy chain gene missense mutation. Cell 1990;62: 999-1006.
225. Maron BJ, Moller JH, Seidman CE, et al: Impact of laboratory molecular diagnosis on contemporary diagnostic criteria for genetically transmitted cardiovascular diseases: Hypertrophic cardiomyopathy, long-QT syndrome, and Marfan syndrome. Circulation 1998;98:1460-71.
226. Thierfelder L, Watkins H, MacRae C, et al: Alpha-tropomyosin and cardiac troponin T mutations cause familial hypertrophic cardiomyopathy: A disease of the sarcomere. Cell 1994;77:701-12.
227. Marian AJ, Roberts R: The molecular genetic basis for hypertrophic cardiomyopathy. J Mol Cell Cardiol 2001;33:655-70.
228. Epstein ND, Cohn GM, Cyran F, Fananapazir L: Differences in clinical expression of hypertrophic cardiomyopathy associated with two distinct mutations in the b-myosin heavy chain gene: A 908 Leu-Val mutation and a 403 Arg-Gln mutation. Circulation 1992;86:345-52.
229. Watkins H, Rosenzweig A, Hwang D, et al: Characteristics and prognostic implications of myosin missense mutations in familial hypertrophic cardiomyopathy. N Engl J Med 1992;326:1108-14.
230. McKenna WJ, Oakley CM, Krikler DM, Goodwin JF: Improved survival with amiodarone in patients with hypertrophic cardiomyopathy and ventricular tachycardia. Br Heart J 1985;53: 412-6.
231. Pelliccia F, Cianfrocca C, Romeo F, Reale A: Hypertrophic cardiomyopathy: Long-term effects of propranolol versus verapamil in preventing sudden death in "low-risk" patients. Cardiovasc Drugs Ther 1990;4:1515-8.
232. Rosing DR, Condit JR, Maron BJ, et al: Verapamil therapy: A new approach to the pharmacologic treatment of hypertrophic cardiomyopathy: III. Effects of long-term administration. Am J Cardiol 1981;48:545-53.
233. Fananapazir L, Cannon ROI, Tripodi D, Panza JA: Impact of dual-chamber permanent pacing in patients with obstructive hypertrophic cardiomyopathy with symptoms refractory to verapamil and beta-adrenergic blocker therapy. Circulation 1992; 85:2148-61.
234. Jeanrenaud X, Goy JJ, Kappenberger L: Effects of dual-chamber pacing in hypertrophic obstructive cardiomyopathy. Lancet 1992;339:1318-23.
235. Cohn LH, Trehan H, Collins JJ Jr: Long-term follow-up of patients undergoing myotomy/myectomy for obstructive hypertrophic cardiomyopathy. Am J Cardiol 1992;70:657-60.
236. Gaita F, Bocchiardo M, Porciani MC, et al: Should stimulation therapy for congestive heart failure be combined with defibrillation backup? Am J Cardiol 2000;86(9 suppl 1):K165-K168.

237. Felker MG, Hu W, Hare JM, et al: The spectrum of dilated cardiomyopathy: The Johns Hopkins experience with 1,278 patients. Medicine 1999;78:270-83.
238. Elliot P: Diagnosis and management of dilated cardiomyopathy. Heart 2000;84:106-12.
239. Bione S, D'Adamo P, Maestrini E, et al: A novel X-linked gene, G4.5, is responsible for Barth syndrome. Nat Genet 1996; 12:385-9.
240. Keeling PJ, McKenna WJ: Clinical genetics of dilated cardiomyopathy. Herz 1994;19:91-6.
241. Ortiz-Lopez R, Li H, Su J, et al: Evidence for a dystrophin missense mutation as a cause of X-linked dilated cardiomyopathy. Circulation 1997;95:2434-40.
242. Arola A, Jokinen E, Ruuskanen O, et al: Epidemiology of idiopathic cardiomyopathies in children and adolescents. A nationwide study in Finland. Am J Epidemiol 1997;146:385-93.
243. Nogueira G, Pinto FF, Paixao A, Kaku S: Idiopathic dilated cardiomyopathy in children: Clinical profile and prognostic determinants. Revista Portuguesa de Cardiologia 2000;19: 191-200.
244. Grimm W, Glaveris C, Hoffmann J, et al: Arrhythmia risk stratification in idiopathic dilated cardiomyopathy based on echocardiography and 12-lead, signal-averaged, and 24-hour Holter electrocardiography. Am Heart J 2000;140:43-51.
245. Cazeau S, Ritter P, Lazarus A, et al: Multisite pacing for end-stage heart failure: Early experience. Pacing Clin Electrophysiol 1996; 19:1748-57.
246. Blanc JJ, Etienne Y, Gilard M, et al: Evaluation of different ventricular pacing sites in patients with severe heart failure. Circulation 1997;96:3273-7.
247. Nelson GS, Curry CW, Wyman BT, et al: Predictors of systolic augmentation from left ventricular preexcitation in patients with dilated cardiomyopathy and intraventricular conduction delay. Circulation 2000;101:2703-9.
248. Lozano I, Bocchiardo M, Achtelik M, et al: Impact of biventricular pacing on mortality in a randomized crossover study of patients with heart failure and ventricular arrhythmias. Pacing Clin Electrophysiol 2000;23:1711-2.
249. Gras D, Cazeau S, Mabo P, et al: Long-term benefit of cardiac resynchronization in heart failure patients: The 12 month results of the InSync™ trial [abstract]. J Am Coll Cardiol. 2000; 35(suppl A):230.
250. Auricchio A, Stellbrink C, Sack S, et al: The pacing therapies for congestive heart failure (PATH-CHF) study: Rationale, design, and endpoints of a prospective randomized multicenter study. Am J Cardiol 1999;83:130D-135D.
251. Saxon LA, Boehmer JP, Hummel J, et al: Biventricular pacing in patients with congestive heart failure: Two prospective randomized trials. Am J Cardiol 1999;83:120D-123D.
252. Pfammatter JP, Paul T: Idiopathic ventricular tachycardia in infancy and childhood: A multicenter study on clinical profile and outcome. Working Group on Dysrhythmias and Electrophysiology of the Association for European Pediatric Cardiology. J Am Coll Cardiol 1999;33:2067-72.
253. Klein LS, Shih H, Hackett FK, et al: Radiofrequency catheter ablation of ventricular tachycardia in patients without structural heart disease. Circulation 1992;85:1666-74.
254. Drago F, Mazza A, Gagliardi MG, et al: Tachycardias in children originating in the right ventricular outflow tract: Lack of clinical features predicting the presence and severity of the histopathological substrate. Cardiol Young 1999;9:273-9.
255. Callans DJ, Menz V, Schwartzman D, et al: Repetitive monomorphic tachycardia from the left ventricular outflow tract: Electrocardiographic patterns consistent with a left ventricular site of origin. J Am Coll Cardiol 1997;29:1023-7.
256. Lauer MR, Liem LB, Young C, et al: Cellular and clinical electrophysiology of verapamil-sensitive ventricular tachycardias. J Cardiovasc Electrophysiol 1992;3:500-14.
257. Klein LS, Zipes DP, Miles WM: Ablation of ventricular tachycardia in patients without coronary artery disease and ventricular tachycardia due to bundle branch reentry. In Huang SKS (ed): Radiofrequency Catheter Ablation of Cardiac Arrhythmias. Basic Concepts and Clinical Applications. Armonk, NY, Futura, 1994, pp 479-90.
258. Wen M, Yeh S, Wang C, et al: Radiofrequency ablation therapy in idiopathic left ventricular tachycardia with no obvious structural heart disease. Circulation 1994;89:1690-6.
259. Coggins DL, Lee RJ, Sweeney J, et al: Radiofrequency catheter ablation as a cure for idiopathic tachycardia of both left and right ventricular origin. J Am Coll Cardiol 1994;23:1333-41.
260. Kanagaratnam L, Tomassoni G, Schweikert R, et al: Ventricular tachycardias arising from the aortic sinus of Valsalva: An under-recognized variant of left outflow tract ventricular tachycardia. J Am Coll Cardiol 2001;37:1408-14.
261. Aiba T, Suyama K, Aihara N, et al: The role of Purkinje and pre-Purkinje potentials in the reentrant circuit of verapamil-sensitive idiopathic LV tachycardia. Pacing Clin Electrophysiol 2001;24:333-44.
262. Kornbluth A, Frishman WH, Ackerman M: Beta-Adrenergic blockade in children. Cardiol Clin 1987;5:629.
263. Garson A Jr, Gillette PC, McVoy P: Amiodarone treatment of critical arrhythmias in children and young adults. J Am Coll Cardiol 1984;4:749.
264. Moak JR, Smith RT, Garson AJ: Newer antiarrhythmic drugs in children. Am Heart J 1986;113:179.
265. Drago F, Mazza A, Guccione P, et al: Amiodarone used alone or in combination with propranolol: A very effective therapy for tachyarrhythmias in infants and children. Pediatr Cardiol 1998;19:445-9.
266. Malcolm ID, Stubington D, Gibbons JE: Mexiletine for ventricular arrhythmia after repair of tetralogy of Fallot. CMAJ 1980; 123:530.
267. Morady F, Kadish AH, DiCarlo L, et al: Long-term results of catheter ablation of idiopathic right ventricular tachycardia. Circulation 1990;82:2093-9.
268. Gursoy S, Brugada J, Souza O, et al: Radiofrequency ablation of symptomatic benign ventricular arrhythmias. Pacing Clin Electrophysiol 1992;15:738-41.
269. Burton ME, Leon AR: Radiofrequency catheter ablation of right ventricular outflow tract tachycardia late after complete repair of tetralogy of Fallot using the pace mapping technique. Pacing Clin Electrophysiol 1993;16:2319-25.
270. Goldner BG, Cooper R, Blau W, Cohen TJ: Radiofrequency catheter ablation as a primary therapy for treatment of ventricular tachycardia in a patient after repair of tetralogy of Fallot. Pacing Clin Electrophysiol 1994;17:1441-6.
271. Gonska B, Cao K, Raab J, et al: Radiofrequency catheter ablation of right ventricular tachycardia late after repair of congenital heart defects. Circulation 1996;94:1902-8.
272. Van Hare G, Kugler J: Pediatric Radiofrequency Ablation Registry. 2002. Personal communication.
273. Williams WG, Hesslein PS, Kormos R. Exit block in children with pacemakers. Clin Prog Electrophysiol Pacing 1986;4: 478-89.
274. Smith RT Jr: Pacemakers for children. In Gillette PC, Garson A Jr (eds): Pediatric Arrhythmias: Electrophysiology and Pacing. Philadelphia, WB Saunders, 1990, pp 532-58.
275. Cohen MI, Vetter VL, Wernovsky G, et al: Epicardial pacemaker implantation and follow-up in patients with a single ventricle after the Fontan operation. J Thorac Cardiovasc Surg 2001; 121:804-11.
276. Gillette PC, Reitman MJ, Gutgesell HP, et al: Intracardiac electrocardiography in children and young adults. Am Heart J 1975;89:36.
277. Kelly DT, Brodsky SJ, Krovetz LJ: Mobitz type II atrioventricular block in children. Pediatrics 1971;79:972.
278. Young D, Eisenberg R, Fish B, et al: Wenckebach atrioventricular block (Mobitz type 1) in children and adolescents. Am J Cardiol 1977;40:393.
279. Michaëlsson M, Engle MA: Congenital complete heart block: An international study of the natural history. In Engle MA (ed): Pediatric Cardiology. Philadelphia, FA Davis, 1972, pp 85-101.
280. Chameides L, Truex RC, Vetter VL, et al: Association of maternal systemic lupus erythematosus with congenital complete heart block. N Engl J Med 1977;297:1204.
281. McCue CM, Mantakas ME, Tingelstad JB, Ruddy S: Congenital heart block in newborns of mothers with connective tissue disease. Circulation 1977;56:82-90.

282. Ramsey-Goldman R, Hom D, Deng J, et al: Anti-SS-A antibodies and fetal outcome in maternal systemic lupus erythematosus. Arthritis Rheum 1986;29:1269-73.
283. Scott JS, Maddison PJ, Taylor PV, et al: Connective-tissue disease, antibodies to ribonucleoprotein, and congenital heart block. N Engl J Med 1983;309:209-12.
284. Taylor PV, Scott JS, Gerlis LM, et al: Maternal antibodies against fetal cardiac antigens in congenital complete heart block. N Engl J Med 1986;315:667-72.
285. Litsey SE, Noonan JA, O'Connor WN, et al: Maternal connective tissue disease and congenital heart block. Demonstration of immunoglobulin in cardiac tissue. N Engl J Med 1985;312:98-100.
286. Buyon JP, Winchester RJ, Slade SG, et al: Identification of mothers at risk for congenital heart block and other neonatal lupus syndromes in their children. Comparison of enzyme-linked immunosorbent assay and immunoblot for measurement of anti-SS-A/Ro and anti-SS-B/La antibodies. Arthritis Rheum 1993;36:1263-73.
287. Buyon JP, Winchester R: Congenital complete heart block. A human model of passively acquired autoimmune injury. Arthritis Rheum 1990;33:609-14.
288. Buyon JP, Swersky SH, Fox HE, et al: Intrauterine therapy for presumptive fetal myocarditis with acquired heart block due to system lupus erythematosus. Arthritis Rheum 1987;30:44-9.
289. Fryda RJ, Kaplan S, Helmsworth JA: Postoperative complete heart block in children. Br Heart J 1971;22:456.
290. Kulbertus HE, Coyne JJ, Hallidie-Smith KA: Conduction disturbances before and after surgical closure of ventricular septal defect. Am Heart J 1969;77:123-30.
291. Moss AJ, Klyman J, Emmanouilides JC: Late onset of complete heart block. Am J Cardiol 1972;30:884-7.
292. Levy MJ, Cuello L, Tuna N, et al: Atrioventricularis communis. Am J Cardiol 1964;14:587-98.
293. Michaëlsson M, Jonzon A, Riesenfeld T: Isolated congenital complete atrioventricular block in adult life. Circulation 1995;92:442-9.
294. Karpawich PP, Gillette PC, Garson A Jr, et al: Congenital complete atrioventricular block: Clinical and electrophysiologic predictors of need for pacemaker insertion. Am J Cardiol 1981; 48:1098-102.
295. Dewey RC, Capeles MA, Levy AM: Use of ambulatory electrocardiographic monitoring to identify high-risk patients with congenital complete heart block. N Engl J Med 1987; 316:835.
296. Levy AM, Camon AJ, Keane JF: Multiple arrhythmias detected during nocturnal monitoring in patients with congenital complete heart block. Circulation 1977;55:247.
297. Sholler GF, Walsh EP: Congenital complete heart block in patients without anatomic cardiac defects. Am Heart J 1989; 118:1193-8.
298. Benson DW Jr, Spach MS, Edwards SB, et al: Heart block in children. Evaluation of subsidiary ventricular pacemaker recovery times and ECG tape recordings. Pediatr Cardiol 1982; 2:39-45.
299. Lillehei CW, Sellers RD, Bonnabeau RC, et al: Chronic postsurgical complete heart block. J Thorac Cardiovasc Surg 1963;46:436-56.
300. Esscher E, Hardell L, Michaëlsson M: Familial, isolated, complete right bundle branch block. Br Heart J 1976;37:745-7.
301. Giuliani ER, Fuster V, Brandenburg RO, Mair DD: Ebstein's anomaly: The clinical features and natural history of Ebstein's anomaly of the tricuspid valve. Mayo Clin Proc 1979;54:163-73.
302. Gallastegui J, Hariman RF, Handler B, et al: Cardiac involvement in the Kearns-Sayre syndrome. Am J Cardiol 1987;60:385-8.
303. Krongrad E: Prognosis for patients with congenital heart disease and postoperative intraventricular conduction defects. Circulation 1978;57:867-70.
304. Horowitz LN, Alexander JA, Edmonds LH: Postoperative right bundle branch block: Identification of three levels of block. Circulation 1980;60:319-28.
305. Friedli B, Bolens M, Taktak M: Conduction disturbances after correction of tetralogy of Fallot: Are electrophysiologic studies of prognostic value? J Am Coll Cardiol 1988;11:162-5.
306. Quattlebaum TG, Varghese PJ, Neill CA, Donahoo JS: Sudden death among postoperative patients with tetralogy of Fallot. Circulation 1976;54:289-93.
307. Kugler JD, Danford DA: Pacemakers in children: An update. Am Heart J 1989;17:665-79.
308. Taber RE, Estoye LR, Green ER, Gahagan T: Treatment of congenital and acquired heart block with an implantable pacemaker. Circulation 1964;29:182-5.
309. Kangos JJ, Griffiths SP, Blumenthal S: Congenital complete heart block: A classification and experience with 18 patients. Am J Cardiol 1967;20:632-8.
310. Glenn WL, de Leuchtenberg N, van Heeckeren DW, et al: Heart block in children: Treatment with a radiofrequency pacemaker. J Thorac Cardiovasc Surg 1969;58:361-73.
311. Furman S, Young D: Cardiac pacing in children and adolescents. Am J Cardiol 1977;39:550-8.
312. Young D: Permanent pacemaker implantation in children: Current status and future considerations. Pacing Clin Electrophysiol 1981;4:61-7.
313. Beder SD, Hanisch DG, Cohen MH, et al: Cardiac pacing in children: A 15-year experience. Am Heart J 1985;109:152-6.
314. Goldman BS, Williams WG, Hill T, et al: Permanent cardiac pacing after open heart surgery: Congenital heart disease. Pacing Clin Electrophysiol 1985;8:732-9.
315. Daicoff GR, Aslami A, Tobias JA, et al: Management of postoperative complete heart block in infants and children. Chest 1974;66:639-41.
316. Driscoll DJ, Gillette PC, Hallman GL, et al: Management of surgical complete AV block in children. Am J Cardiol 1979; 43:1175-80.
317. Hofschire PJ, Nicoloff DM, Moller JH: Postoperative complete heart block in 64 children treated with and without cardiac pacing. Am J Cardiol 1977;39:559-62.
318. Flinn CJ, Wolff GS, Dick M II, et al: Cardiac rhythm after the Mustard operation for complete transposition of the great arteries. Circulation 1984;310:1635-8.
319. Gillette PC, Shannon C, Blair H, et al: Transvenous pacing in pediatric patients. Am Heart J 1983;105:843-7.
320. Epstein ML, Knauf DG, Alexander JA: Long-term follow-up of transvenous cardiac pacing in children. Am J Cardiol 1986;57:889-90.
321. Esperer HD, Singer H, Riede FT, et al: Permanent epicardial and transvenous single- and dual-chamber cardiac pacing in children. Thorac Cardiovasc Surg 1993;41:21-7.
322. Ward DE, Jones S, Shinebourne EA: Long-term transvenous pacing in children weighing less than 10 kg. Circulation 1985; 72(suppl 3):340.
323. Walsh C, McAlister H, Andrews C, et al: Pacemaker implantation in children: A 21-year experience. Pacing Clin Electrophysiol 1988;11(suppl 2):1940-4.
324. Karpawich PP, Perry BL, Farooki ZQ, et al: Pacing in children and young adults with nonsurgical atrioventricular block: Comparison of single-rate ventricular and dual-chamber modes. Am Heart J 1987;113:316-21.
325. Gillette PC, Ziegler VL, Case CL, et al: Atrial antitachycardia pacing in children and young adults. Am Heart J 1997;122:844-9.
326. Rhodes LA, Saul JP, Gamble WJ, et al: Benefits and potential risk of atrial antitachycardia pacing after repair of congenital heart disease. J Am Coll Cardiol 1993;21:107A.
327. Kron J, Silka MJ, Ohm OJ, et al: Preliminary experience with nonthoracotomy implantable cardioverter defibrillators in young patients. Pacing Clin Electrophysiol 1994;17:26-30.

Chapter 27 Genetics and Cardiac Arrhythmias

JEFFREY A. TOWBIN, MATTEO VATTA, HUA LI, and NEIL E. BOWLES

Cardiac arrhythmias are major causes of morbidity and mortality, including sudden cardiac death. Sudden cardiac death in the United States occurs with a reported incidence of greater than 300,000 persons per year.[1] Although coronary heart disease is a major cause of death, other etiologies contribute to this problem. In many of these non–ischemia-related cases, autopsies are unrevealing. Interest in identifying the underlying cause of the death in these instances has been focused on cases of unexpected arrhythmogenic death, which is estimated to represent 5% of all sudden deaths. In cases in which no structural heart disease can be identified, the long QT syndrome (LQTS),[1] ventricular preexcitation (Wolff-Parkinson-White syndrome),[2] and idiopathic ventricular fibrillation (IVF) or Brugada syndrome[3] (characterized by ST-segment elevation in the right precordial leads with or without right bundle branch block [RBBB]) are most commonly considered as likely causes. Another important disease in which arrhythmias are thought to play a central role is sudden infant death syndrome (SIDS),[4] a disorder with no structural abnormalities.

Arrhythmogenic right ventricular dysplasia (ARVD) is also a significant cause of sudden death[5] and is considered to be a primary electrical disease despite being associated with fibrosis and fatty infiltration of the right ventricle. The arrhythmias associated with ARVD also occur in other disorders in which structurally normal myocardium is seen, such as catecholaminergic ventricular tachycardia (VT).[6] The purpose of this chapter is to describe the current understanding of the clinical and molecular genetic aspects of inherited diseases in which arrhythmias are prominent features.

The Long QT Syndromes

CLINICAL DESCRIPTION

LQTS's are inherited or acquired disorders of repolarization identified by the electrocardiographic (ECG) abnormalities of prolongation of the Q–T interval corrected for heart rate (QTc), usually above 460 to 480 milliseconds (ms); relative bradycardia; T wave abnormalities (Fig. 27-1); and episodic ventricular tachyarrhythmias,[7] particularly *torsades de pointes* (Fig. 27-2). The inherited form of LQTS is transmitted as an autosomal dominant or autosomal recessive trait. Acquired LQTS may be seen as a complication of various drug therapies or electrolyte abnormalities. Whether the abnormality is genetic or acquired, the clinical presentation is similar.[1,7] The initial presentation of LQTS is heterogeneous and most commonly includes syncope, which in many instances is triggered by emotional stress, exercise, or auditory phenomena. Other presenting features include seizures or palpitations. Some individuals have sudden death as their first symptom, while other cases are diagnosed by surface ECG as a family screening evaluation due to family history of LQTS or sudden death.

CLINICAL GENETICS

Two differently inherited forms of familial LQTS have been reported. The Romano-Ward syndrome is the most common of the inherited forms of LQTS and

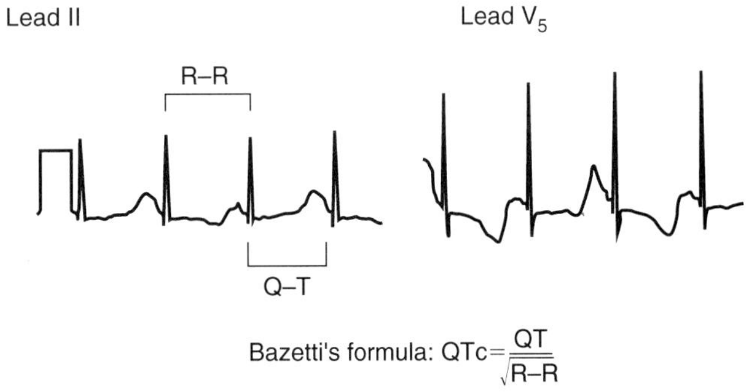

FIGURE 27-1 ECG leads II and V_5 demonstrating prolonged Q–T intervals and T wave alternans. Bazett's formula, used to calculate the Q–T interval corrected for heart rate (QTc), is noted.

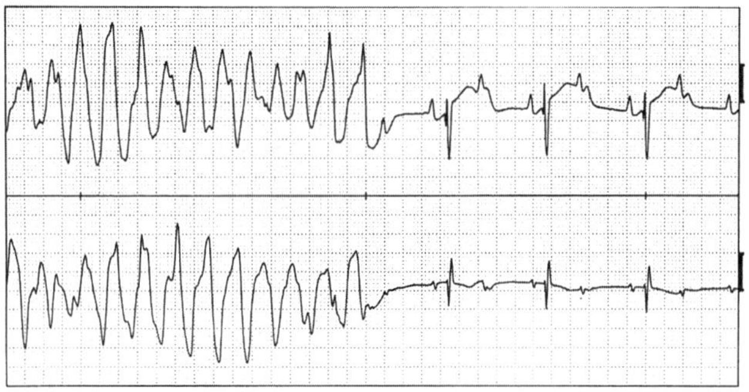

FIGURE 27-2 Polymorphic ventricular tachycardia in a patient with long QT syndrome.

appears to be transmitted as an autosomal dominant trait.[8,9] In this disorder, the disease gene is transmitted to 50% of the offspring of an affected individual. However, low penetrance has been described, and therefore gene carriers may, in fact, have no clinical features of disease.[10] Individuals with Romano-Ward syndrome have the pure syndrome of prolonged Q–T interval on ECG with the associated symptom complex of syncope, sudden death, and in some patients, seizures.[11,12] Occasionally, other noncardiac abnormalities such as diabetes mellitus,[13,14] asthma,[15] or syndactyly[16] may also be associated with QT prolongation. LQTS may also be involved in some cases of SIDS,[5,17,18] which, in some cases, appear in several family members.

The Jervell and Lange-Nielsen syndrome (JLNS) is a relatively uncommon inherited form of LQTS. Classically, this disease has been described as having apparent autosomal recessive transmission.[19-21] These patients have the identical clinical presentation as those with Romano-Ward syndrome but also have associated sensorineural deafness. Clinically, patients with JLNS usually have longer Q–T intervals as compared to individuals with Romano-Ward syndrome and also have a more malignant course. Priori and colleagues have reported autosomal recessive cases of Romano-Ward syndrome[22] as well, thus changing one of the sine qua nons of JLNS.

GENE IDENTIFICATION IN ROMANO-WARD SYNDROME

KVLQT1 or *KCNQ1*: The *LQT1* Gene

The first of the genes mapped for LQTS, termed *LQT1*, required 5 years from the time that mapping to chromosome 11p15.5 was first reported[23] to gene cloning. This gene, originally named *KVLQT1* but more recently called *KCNQ1* (Table 27-1), is a novel potassium channel gene that consists of 16 exons, spans approximately

TABLE 27-1 Genetics and Cardiac Arrhythmias

Disease	Rhythm Abnormality	Inheritance	Chromosome Location	Gene
Ventricular Arrhythmias				
Romano-Ward syndrome	TdP, VF	AD	11p15.5, 7q35, 3p21, 4q25, 21q22*	*KVLQT1* (11p15.5); *HERG* (7q35); *SCN5A* (3p21); *minK* (21q22); *MiRP1* (21q22)
Jervell and Lange-Nielsen syndrome	TdP, VF	AD/AR†	11p15.5, 21q22	*KVLQT1* (11p15.5); *minK* (21q22)
Brugada syndrome	VT, VF	AD	3p21	*SCN5A*
Sudden infant death syndrome	VT/VF	AD	3p21	*SCN5A*
Familial VT	VT	AD	?	?
Familial bidirectional VT	VT	AD	1q42	*RYR2*
Familial polymorphic VT	VT	AD	1q42*	*RYR2*
Arrhythmogenic RV dysplasia	VT	AD	1q42	*RYR2*
Naxos disease	VT	AR	17q21, 6p24	Plakoglobin (17q21); Desmoplakin (6p24)
Supraventricular Arrhythmias				
Familial atrial fibrillation	AF	AD	10q22	?
Familial total atrial standstill	SND, AF	AD	?	?
Familial absence of sinus rhythm	SND, AF	AD	?	?
Wolff-Parkinson-White syndrome	AVRT, AF, VF	AD	7q3	*AMPK*
Familial PJRT	AVRT	AD	?	?
Conduction Abnormalities				
Familial AV block	AVB, AF, SND, VT, SD	AD	19q13	?
Isolated AV block	AVB, AF, SND, VT, SD	AD	3p21	*SCN5A*
Lev-Lenègre's syndrome	AVB, AF, SND, VT, SD	AD	3p21	*SCN5A*
Familial bundle branch block	RBBB	?	?	?

AD, autosomal dominant; AF, atrial fibrillation; AR, autosomal recessive; AVB, atrioventricular block; AVRT, atrioventricular reciprocating tachycardia; PJRT, permanent form of junctional reciprocating tachycardia; RBBB, right bundle branch block; SD, sudden death; SND, sinus node dysfunction; TdP, torsades de pointes; VF, ventricular fibrillation; VT, ventricular tachycardia.

*At least one other unknown.

†Jervell and Lange-Nielsen syndrome: autosomal dominant rhythm abnormality and autosomal recessive sensorineural deafness.

400 kilobases (kb), and is widely expressed in human tissues including heart, inner ear, kidney, lung, placenta, and pancreas but not in skeletal muscle, liver, or brain.[24] Although most of the mutations are "private" (i.e., only seen in one family), there is at least one frequently mutated region (called a "hot spot") of *KVLQT1*.[25-27] This gene is the most commonly mutated gene in LQTS.

Analysis of the predicted amino acid sequence of *KVLQT1* suggests that it encodes a potassium channel α-subunit with a conserved potassium-selective pore-signature sequence flanked by six membrane-spanning segments similar to shaker-type channels (Fig. 27-3).[24,27-29] A putative voltage sensor is found in the fourth membrane-spanning domain (S4), and the selective pore loop is between the fifth and sixth membrane-spanning domains (S5,S6). Biophysical characterization of the *KVLQT1* protein confirmed that *KVLQT1* is a voltage-gated potassium channel protein subunit, which requires coassembly with a β-subunit called *minK* to function properly.[28,29] Expression of either *KVLQT1* or *minK* alone results in either inefficient or no current development. When *minK* and *KVLQT1* are coexpressed in either mammalian cell lines or *Xenopus* oocytes, however, the slowly activating potassium current [I_{Ks}] is developed in cardiac myocytes.[28,29] Combination of normal and mutant *KVLQT1* subunits forms abnormal I_{Ks} channels, and these mutations are believed to act through a dominant-negative mechanism (the mutant form of *KVLQT1* interferes with the function of the normal wild type form through a "poison pill"–type mechanism) or a loss-of-function mechanism (only the mutant form loses activity).[30]

Since *KVLQT1* and *minK* form a unit, mutations in *minK* could also be expected to cause LQTS, and this fact was subsequently demonstrated (discussed later in "*minK* or *KCNE1*: The *LQT5* Gene").[31]

HERG or *KCNH2*: The *LQT2* Gene

The *LQT2* gene was initially mapped to chromosome 7q35-36 by Jiang et al,[27,32] and subsequently, candidate gene screening identified the disease-causing gene *HERG* (*h*uman *e*ther-a-go-go-*r*elated *g*ene), a cardiac potassium channel gene to be the *LQT2* gene (see Table 27-1). *HERG* was originally cloned from a brain cDNA library[33] and found to be expressed in neural-crest–derived neurons,[34] microglia,[35] a wide variety of tumor cell lines,[36] and the heart.[37] LQTS-associated mutations were identified

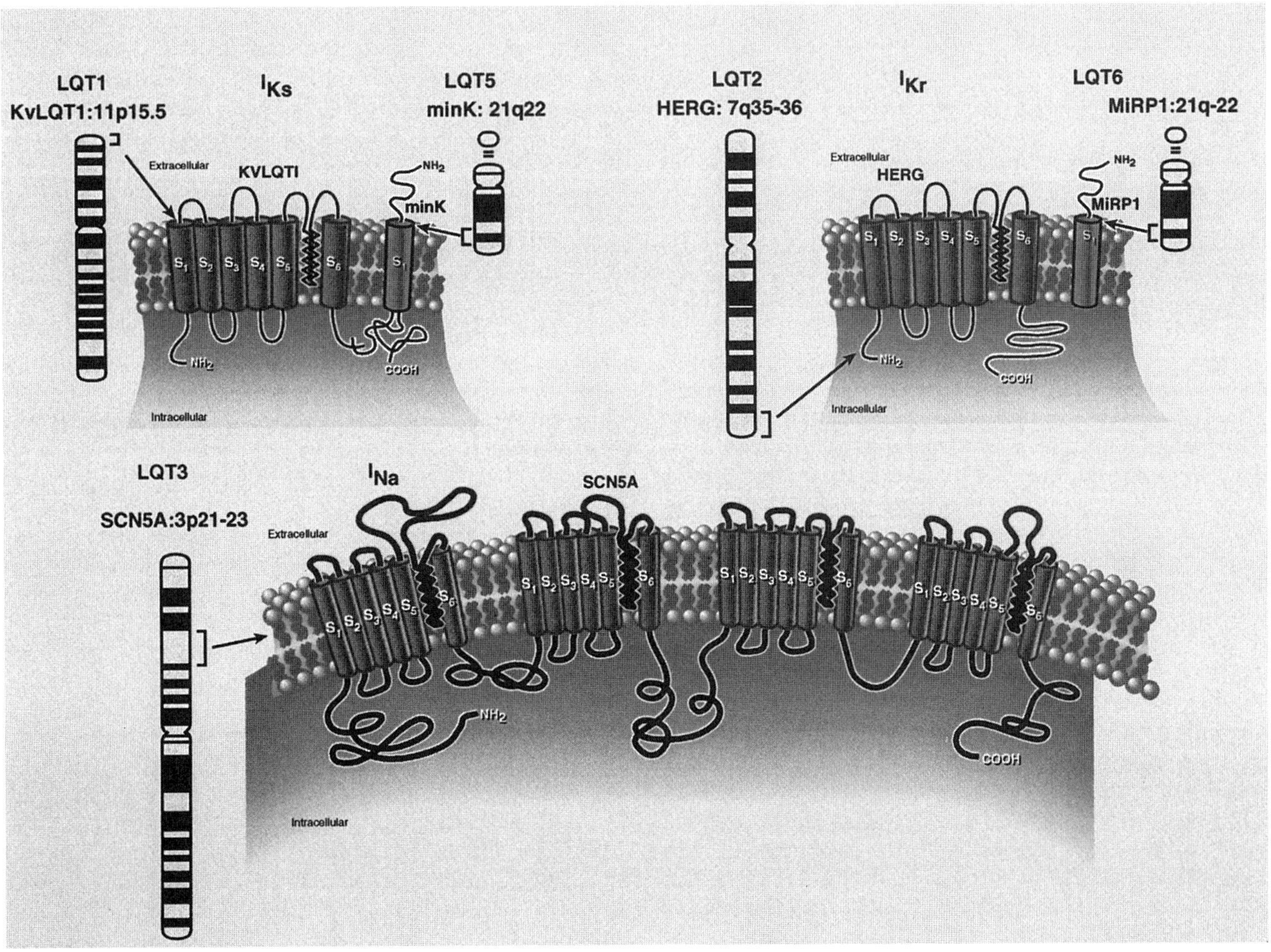

FIGURE 27-3 (See also Color Plate 27-3.) Genetics of ventricular arrhythmias. The genetic loci and known genes identified for long QT syndrome are shown along with the ion channel protein structure. Note that the potassium channel α-subunits *KvLQT1* and *HERG* require association with β-subunits (*minK* and *KVLQT1*=IKs; *MiRP1* and *HERG*=IKr for normal function).

in *HERG* throughout the gene, including missense mutations, intragenic deletions, stop codons, and splicing mutations.[27,37,38] Currently, this gene is thought to be the second most common gene mutated in LQTS (second to *KVLQT1*). As with *KVLQT1*, "private" mutations that are scattered throughout the entire gene without clustering preferentially are seen.

HERG consists of 16 exons and spans 55 kb of genomic sequence.[37] The predicted topology of *HERG* (see Fig. 27-3) is similar to *KVLQT1*. Unlike *KVLQT1*, *HERG* has extensive intracellular amino-and-carboxyl termini, with a region in the carboxyl-terminal domain having sequence similarity to nucleotide binding domains (NBDs).

Electrophysiological and biophysical characterization of expressed *HERG* in *Xenopus* oocytes established that *HERG* encodes the rapidly activating delayed rectifier potassium current I_{Kr}.[39-41] Electrophysiological studies of LQTS-associated mutations showed that they act through either a loss-of-function or a dominant-negative mechanism.[41,42] In addition, protein trafficking abnormalities have been shown to occur.[43,44] This channel has been shown to coassemble with β-subunits for normal function, similar to that seen in I_{Ks}. McDonald et al[45] initially suggested that the complexing of *HERG* with *minK* is needed to regulate the I_{Kr} potassium current. Bianchi et al provided confirmatory evidence that *minK* is involved in regulation of both I_{Ks} and I_{Kr}.[46] Abbott et al[47] identified *MiRP1* as a β-subunit for *HERG* (discussed later in "*MiRP1* or *KCNE2:* The *LQT6* Gene").

SCN5A: The *LQT3* Gene

The positional candidate gene approach was also used to establish that the gene responsible for chromosome 3–linked LQTS (*LQT3*) is the cardiac sodium channel gene *SCN5A* (see Table 27-1).[48,49] *SCN5A* is highly expressed in human myocardium and brain but not in skeletal muscle, liver, or uterus.[50-52] It consists of 28 exons that span 80 kb and encodes a protein of 2016 amino acids with a putative structure that consists of four homologous domains (DI-DIV), each of which contains six membrane-spanning segments (S1-S6) similar to the structure of the potassium channel α-subunits (see Fig. 27-3).[27,39] Linkage studies with *LQT3* families and *SCN5A* initially demonstrated linkage to the *LQT3* locus on chromosome 3p21-24,[50,51] and multiple mutations were subsequently identified. Biophysical analysis of the initial three mutations were expressed in *Xenopus* oocytes, and it was found that all mutations generated a late phase of inactivation-resistant, mexiletine- and tetrodotoxin-sensitive whole-cell currents through multiple mechanisms.[53,54] Two of the three mutations showed dispersed reopening after the initial transient, but the other mutation showed both dispersed reopening and long-lasting bursts.[54] These results suggested that *SCN5A* mutations act through a gain-of-function mechanism (the mutant channel functions normally, but with altered properties such as delayed inactivation) and that the mechanism of chromosome 3–linked LQTS is persistent noninactivated sodium current in the plateau phase of the action potential. Later, An et al[55] showed that not all mutations in *SCN5A* are associated with persistent current and demonstrated that *SCN5A* interacted with β-subunits.

minK or *KCNE1*: The *LQT5* Gene

minK (*IsK or KCNE1*) was initially localized to chromosome 21 (21q22.1) and found to consist of three exons that span approximately 40 kb (see Table 27-1). It encodes a short protein consisting of 130 amino acids and has only one transmembrane-spanning segment with small extracellular and intercellular regions (see Fig. 27-3).[30,31,56] When expressed in *Xenopus* oocytes, it produces potassium current that closely resembles the slowly activating delayed-rectifier potassium current I_{Ks} in cardiac cells.[56,57] The fact that the *minK* clone was only expressed in *Xenopus* oocytes and not in mammalian cell lines raised the question whether *minK* is a human channel protein. With the cloning of *KVLQT1* and coexpression of *KVLQT1* and *minK* in both mammalian cell lines and *Xenopus* oocytes, it became clear that *KVLQT1* interacts with *minK* to form the cardiac slowly activating delayed-rectifier I_{Ks} current[28,29]; *minK* alone cannot form a functional channel but induces the I_{Ks} current by interacting with endogenous *KVLQT1* protein in *Xenopus* oocytes and mammalian cells. Bianchi et al showed that mutant *minK* results in abnormalities of I_{Ks}, I_{Kr}, and protein trafficking abnormalities.[46] McDonald et al[45] showed that *minK* also complexes with *HERG* to regulate the I_{Kr} potassium current. Splawski et al[31] demonstrated that *minK* mutations cause *LQT5* when they identified mutations in two families with LQTS. In both cases, missense mutations (S74L, D76N) were identified; they reduced I_{Ks} by shifting the voltage dependence of activation and accelerating channel deactivation. This was supported by the fact that a mouse model of *minK*-defective LQTS was also created.[58] The functional consequences of these mutations include delayed cardiac repolarization and, hence, an increased risk of arrhythmias.

MiRP1 or *KCNE2*: The *LQT6* Gene

MiRP1, the *minK*-related peptide 1, or *KCNE2* (see Table 27-1), is a novel potassium channel gene recently cloned and characterized by Abbott and colleagues.[47] This small integral membrane subunit protein assembles with *HERG* (*LQT2*) to alter its function, enabling full development of the I_{Kr} current (see Fig. 27-3). *MiRP1* is a 123 amino acid channel protein with a single predicted transmembrane segment similar to that described for *minK*.[56] Chromosomal localization studies mapped this *KCNE2* gene to chromosome 21q22.1, within 79kb of *KCNE1* (*minK*) and arrayed in opposite orientation.[47] The open reading frames of these two genes share 34% identity, and both are contained in a single exon, suggesting that they are related through gene duplication and divergent evolution.

Three missense mutations associated with LQTS and ventricular fibrillation (VF) were identified in *KCNE2* by Abbott et al,[47] and biophysical analysis demonstrated that these mutants form channels that open slowly and close rapidly, thus diminishing potassium currents. In one case, the missense mutation, a C to G transversion at nucleotide 25, which produced a glutamine (Q) to glutamic acid (E) substitution at codon 9 (Q9E) in the putative extracellular domain of *MiRP1*, led to the development of *torsades de pointes* and VF after intravenous clarithromycin infusion (i.e., drug-induced).

Therefore, like *minK*, this channel protein acts as a β-subunit but, by itself, leads to ventricular arrhythmia risk when mutated. These similar channel proteins (i.e., *minK* and *MiRP1*) suggest that a family of channels that regulates ion channel α-subunits exists. The specific role of this subunit and its stoichiometry remains unclear and is currently being hotly debated.

GENETICS AND PHYSIOLOGY OF AUTOSOMAL RECESSIVE LONG QT SYNDROME (JERVELL AND LANGE-NIELSEN SYNDROME)

Neyroud et al[59] reported the first molecular abnormality in patients with JLNS when they reported on two families in which three children were affected by JLNS and in whom a novel homozygous deletion-insertion mutation of *KVLQT1* in three patients was found. A deletion of 7 bp and an insertion of 8 bp at the same location led to premature termination at the C-terminal end of the *KVLQT1* channel. At the same time, Splawski et al[60] identified a homozygous insertion of a single nucleotide that caused a frameshift in the coding sequence after the second putative transmembrane domain (S2) of *KVLQT1*. Together, these data strongly suggested that at least one form of JLNS is caused by homozygous mutations in *KVLQT1* (see Table 27-1). This has been confirmed by others.[27,30,61,62]

As a general rule, heterozygous mutations in *KVLQT1* cause Romano-Ward syndrome (LQTS only), while homozygous (or compound heterozygous) mutations in *KVLQT1* cause JLNS (LQTS and deafness). The hypothetical explanation suggests that although heterozygous *KVLQT1* mutations act by a dominant-negative mechanism, some functional *KVLQT1* potassium channels still exist in the stria vascularis of the inner ear. Therefore, congenital deafness is averted in patients with heterozygous *KVLQT1* mutations. For patients with homozygous *KVLQT1* mutations, no functional *KVLQT1* potassium channels can be formed. It has been shown by in situ hybridization that *KVLQT1* is expressed in the inner ear,[60] suggesting that homozygous *KVLQT1* mutations can cause the dysfunction of potassium secretion in the inner ear and lead to deafness. However, it should be noted that incomplete penetrance exists, and not all heterozygous or homozygous mutations follow this rule.[11,22]

As with Romano-Ward syndrome, if *KVLQT1* mutations can cause the phenotype, it could be expected that *minK* mutations could also be causative of the phenotype (JLNS). Schulze-Bahr et al,[63] in fact, showed that mutations in *minK* result in JLNS syndrome as well, and this was confirmed subsequently (see Table 27-1).[60] Hence, abnormal I_{Ks} current, whether it occurs due to homozygous or compound heterozygous mutations in *KVLQT1* or *mink*, results in LQTS and deafness.

Genotype-Phenotype Correlations in Long QT Syndrome

CLINICAL FEATURES

To a significant extent, the clinical features of LQTS depend on the gene mutated as well as the intragenic position of the mutation and its effect on the channel protein. Several studies have clarified the specific associations including clinical severity of probands and their parents and siblings, as well as modifying influences on severity.

Kimbrough et al[64] recently reported on the study of 211 probands with LQTS and classified the severity in the probands, affected parents, and siblings. Importantly, they showed that the severity of the disease in the proband did not correlate with the clinical severity seen in first-degree relatives, specifically their parents and siblings. In fact, variable intrafamily penetrance was noted, consistent with other genetic and environmental factors playing a role in modulating and modifying clinical manifestations in members of the same family. Several stratifiers were identified as important.

The length of the QTc in affected parents and siblings was shown to be associated with significant risk of LQTS-related cardiac events. They also confirmed that genotype, age, and sex influence the course of disease in affected family members. For instance, male probands were found to have their first cardiac event at younger mean age than female probands (13 years versus 19 years), but female probands had a higher frequency of cardiac arrest or LQTS-related death by 40 years of age. For affected parents, they found that female sex and QTc length were risk factors for events, while QTc duration was the only risk factor for siblings. Affected mothers of LQTS probands displayed an ongoing cardiac event risk well after the birth of the proband, but affected fathers did not display this ongoing risk.

These findings complement the findings previously described by Zareba et al.[65] In this study, the authors provided evidence of clinical outcome, age of onset, and frequency of events based on genotype. Patients with mutations in *LQT1* had the earliest onset of events and the highest frequency of events followed by mutations in *LQT2*. The risk of sudden death in these two groups was relatively low for any event. Mutations in *LQT3* resulted in a paucity of syncopal events, but events commonly resulted in sudden death. In addition, mutations in *LQT3* resulted in the longest QTc duration. Mutations in *LQT1* and *LQT2* appeared to be associated with stress-induced symptoms,[66] with *LQT1* associated with exercise and swimming[67,68] and *LQT2* associated with auditory triggers.[68-71] *LQT3* appeared to be associated with sleep-associated symptoms and events.

ELECTROCARDIOGRAPHIC AND BIOPHYSICAL FEATURES

In 1995 Moss and colleagues[72] provided the first evidence that mutations in different genes cause differing ECG features. Specifically, the authors focused on the different types of T waves seen in patients with *LQT1* versus *LQT2* versus *LQT3*. ECGs of *LQT1* patients were shown to display broad-based T waves, *LQT2* patients had low-amplitude T waves, and those with *LQT3* mutations had distinctive T waves with late onset. More recently, Zhang et al[73] showed that there are actually four different ST-T wave patterns. Using these definitions, they were able to identify 88% of *LQT1* and *LQT2* patients accurately by surface ECG and 65% of *LQT3* carriers. Prospectively, these authors correctly predicted genotype in 100% of patients.

Further insight into ECG findings and genotype were reported by Lupoglazoff et al[74] using Holter monitor analysis. Analysis of 133 *LQT1* patients and 57 *LQT2* carriers, as well as 100 control individuals, led the authors to conclude that T wave morphology was normal in most *LQT1* patients (92%) and normal controls (96%), but the vast majority of *LQT2* patients had abnormal T waves (19% normal, 81% abnormal). In the largest percentage of *LQT2* patients, T wave notching was identified in which the T wave protuberance is above the horizontal, while another subset had a bulge at or below the horizontal. In the former case, young age, missense *LQT2* mutations, and mutations in the core domain of *HERG* predicted morphology, while potential diagnostic clues gained by the latter morphology included amino-terminal or carboxy-terminal mutations or frameshifts in *HERG*.

ANIMAL MODELS OF LONG QT SYNDROME

Using an arterially perfused canine left ventricular wedge preparation developed pharmacologically, induced animal models of *LQT1, LQT2,* and *LQT3* have been created.[75,76] Using chromanol 293B, a specific I_{Ks} blocker, a model that mimics *LQT1* was produced.[75] In this model, I_{Ks} deficiency alone was not enough to induce torsades de pointes, but addition of β-adrenergic influence (i.e., isoproterenol) predisposed the myocardium to torsade by increasing transmural dispersion of repolarization. Addition of β-blocker or mexiletine reduced the ability to induce torsades de pointes, suggesting that these medications might improve patient outcomes.

Models for *LQT2* and *LQT3* were created by using D-sotalol (*LQT2*) or ATX-II (*LQT3*) in this wedge preparation.[76] Both drugs preferentially prolong M cell action potential duration, with ATX-II also causing a sharp rise in transmural dispersion. Mexiletine therapy abbreviated the Q–T interval prolongation in both models and reduced dispersion. Spontaneous torsades de pointes was suppressed, and the vulnerable window during which torsades de pointes induction occurs was also reduced in both models. These models support the current understanding of the different subtypes of LQTS and provide an explanation for potential therapies.

Therapeutic Options in Long QT Syndrome

Currently, the standard therapeutic approach in LQTS is the initiation of β-blockers at the time of diagnosis.[7] Recently, Moss et al[77] demonstrated significant reduction in cardiac events using β-blockers. However, syncope, aborted cardiac arrest, and sudden death do continue to occur. In cases in which β-blockers cannot be used, such as in patients with asthma, other medications have been tried, such as mexiletine.[78] When medical therapy has failed, left sympathectomy or implantation of an automatic cardioverter defibrillator has been used.[7]

Genetic-based therapy has also been described. Schwartz et al[78] showed that sodium channel blocking agents (i.e., mexiletine) shorten the QTc in patients with *LQT3,* while exogenous potassium supplementation[79] or potassium channel openers[80] have been shown to be potentially useful in patients with potassium channel defects. However, long-term potassium therapy with associated potassium-sparing agents has been unable to keep the serum potassium above 4 mmol/L due to renal potassium homeostasis. This suggests that long-term potassium therapy may not be useful. In addition, no definitive evidence that these approaches (i.e., sodium channel blockers, exogenous potassium, or potassium channel openers) improve survival has been published.

Andersen Syndrome (LQT7)

CLINICAL ASPECTS

Andersen and colleagues (1971)[190] identified a complex phenotype including ventricular arrhythmias, potassium-sensitive periodic paralysis, and dysmorphic features. The dysmorphisms included hypertelorism, broad nasal root, defects of the soft and hard palate, as well as short stature. More recently, skeletal abnormalities have broadened the phenotype (Andelfinger et al).[191] The associated cardiac abnormalities include QTc prolongation, ventricular tachycardia, ventricular fibrillation, and atrial arrhythmias. Torsades de pointes and bidirectional VT have been seen. In addition, repolarization abnormalities affecting late repolarization and resembling giant U waves are common. Sudden death has not been reported as a major risk in this disorder.

GENETIC ASPECTS

Andersen syndrome was originally mapped to chromosome 17q23-q24.2 by Plaster et al[192] using genome-wide linkage analysis. The critical region within this locus was narrowed, and candidate gene mutation screening identified mutations in *KCNJ2,* which encodes an inward rectifier potassium channel called Kir2.1 (Tristani-Firouzi et al).[193] This channel is highly expressed in the heart and plays a role in phase 3 repolarization and in the resting membrane potential. Multiple gene mutations have been identified to date with relatively high penetrance noted. Functional studies have demonstrated reduction or suppression of Ik_1, by a haploinsufficiency or dominant negative effect. This gene may play a role in developmental signalling pathways as well, which is believed to cause dysmorphisms.

Brugada Syndrome

CLINICAL ASPECTS OF BRUGADA SYNDROME

The first identification of the electrocardiographic pattern of RBBB with ST-elevation in leads V1 to V3 was reported by Osher and Wolff.[81] Shortly thereafter, Edeiken[82] identified persistent ST-elevation without RBBB in 10 asymptomatic males, and Levine et al[83] described ST-elevation in the right chest leads and conduction block in the right ventricle in patients with severe hyperkalemia. The first association of this ECG pattern with sudden death was described by Martini et al[84] and later by Aihara et al.[85] This association was further confirmed in 1991 by Pedro and Josep Brugada,[86] who described four patients with sudden death and aborted sudden death with ECGs demonstrating RBBB and persistent ST-elevation in leads V1 to V3 (Fig. 27-4). In 1992 these authors characterized what they believed to be a distinct clinical and electrocardiographic syndrome.[3]

The finding of ST-elevation in the right chest leads has been observed in various clinical and experimental settings and is not unique to or diagnostic of Brugada syndrome by itself.[87] Situations in which these ECG findings occur include electrolyte or metabolic disorders, pulmonary or inflammatory diseases, and abnormalities of the central or peripheral nervous system.

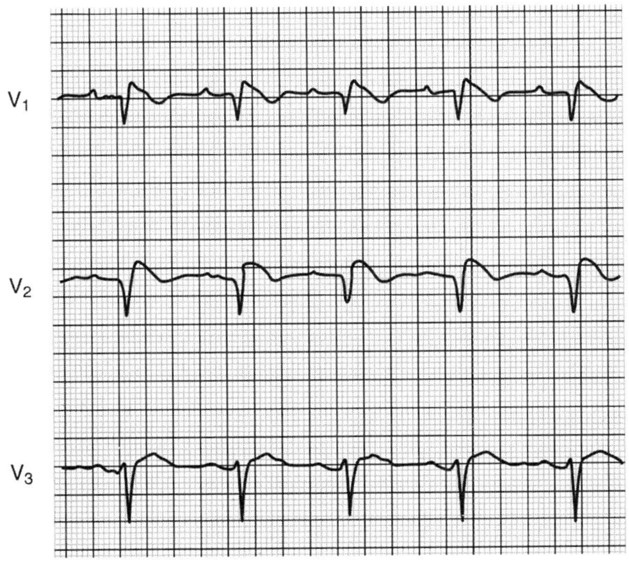

FIGURE 27-4 Electrocardiographic features of Brugada syndrome. Note the ST-segment elevation (cove-type) in leads V_1 and V_2.

In the absence of these abnormalities, the term *idiopathic ST-elevation* is often used and may identify Brugada syndrome patients.

The ECG findings and associated sudden and unexpected death have been reported as common problems in Japan and Southeast Asia, where it most commonly affects men during sleep.[88] This disorder, known as sudden and unexpected death syndrome (SUDS) or sudden unexpected nocturnal death syndrome (SUNDS), has many names in Southeast Asia: bangungut (to rise and moan in sleep) in the Philippines; non-laitai (sleep-death) in Laos; lai-tai (died during sleep) in Thailand; and in Japan, pokkuri (sudden and unexpectedly ceased phenomena). General characteristics of SUNDS include young, healthy males in whom death occurs suddenly with a groan, usually during sleep late at night. No precipitating factors are identified, and autopsy findings are generally negative.[89] Life-threatening ventricular tachyarrhythmias as a primary cause of SUNDS have been demonstrated, with VF occurring in most cases.[90]

The risk of sudden death associated with Brugada syndrome and SUNDS reported for European and Southeast Asian individuals has been reported to be extremely high; approximately 75% of patients reported by Brugada et al[3,86,91] survived cardiac arrest. In addition, symptomatic and asymptomatic patients have been considered to be at equal risk. Priori et al[92] have disputed this claim, however. In a study of 60 patients with Brugada syndrome, asymptomatic patients had no episodes or events. The importance of this difference is its impact on therapeutic decision making, as currently all patients receive ICD therapy. Should the data of Priori et al[92] hold up, selective use of ICDs would be appropriate. If selective use of ICDs were to beconsidered, other diagnostic tests for risk stratification would be necessary.

Kakishita et al[93] studied a high-risk group of patients with 37% having spontaneous episodes of VF. As the majority of patients had ICD placement, the authors were able to show that 65% of episodes were preceded by premature ventricular complexes (PVCs), which were essentially identical to the initiating PVCs of VF in morphology. In fact, the PVCs initiating all VF episodes arose from terminal portions of the T wave, and pause-dependent arrhythmias were rare. This suggests that vigilant evaluation by Holter monitoring could identify at-risk patients. In addition, the authors suggested that the use of radiofrequency ablation targeting the initiating PVCs could be helpful in reducing risk and reducing the need for ICD placement.

CLINICAL GENETICS OF BRUGADA SYNDROME

Most of the families thus far identified with Brugada syndrome have apparent autosomal dominant inheritance.[94-96] In these families, approximately 50% of offspring of affected patients develop the disease. Although the number of families reported has been small, it is likely that this is due to under-recognition as well as premature and unexpected death.[87,97,98]

MOLECULAR GENETICS OF BRUGADA SYNDROME

In 1998 our laboratory reported the findings on six families and several sporadic cases of Brugada syndrome.[96] The families were initially studied by linkage analysis using markers to the known ARVD loci and LQT loci, and linkage was excluded. Candidate gene screening using the mutation analysis approach of single strand conformation polymorphism (SSCP) analysis and DNA sequencing was performed, and *SCN5A* was chosen for study based on the suggestions of Antzelevitch.[87,95,97-99] In three families, mutations in *SCN5A* were identified (see Table 27-1)[96] including: (1) a missense mutation (C-to-T base substitution) causing a substitution of a highly conserved threonine by methionine at codon 1620 (T1620M) in the extracellular loop between transmembrane segments S3 and S4 of domain IV (DIVS3-DIVS4), an area important for coupling of channel activation to fast inactivation; (2) a two-nucleotide insertion (AA), which disrupts the splice-donor sequence of intron 7 of *SCN5A*; and (3) a single nucleotide deletion (A) at codon 1397, which results in an in-frame stop codon that eliminates DIIIS6, DIVS1-DIVS6, and the carboxy-terminus of *SCN5A* (Fig. 27-5). Since the initial report, multiple confirming mutations have been identified. Mutations have also been found in *SCN5A* in children with sudden cardiac death.[100]

Biophysical analysis of the mutants in *Xenopus* oocytes demonstrated a reduction in the number of functional sodium channels in both the splicing mutation and one-nucleotide deletion mutation, which should promote development of reentrant arrhythmias. In the missense mutation, sodium channels recover from inactivation more rapidly than normal. Subsequent experiments conducted in modified HEK cells revealed that at physiological temperatures (37°C), reactivation of the T1620M mutant channel was actually slower, while inactivation of the channel was importantly accelerated. These alterations leave the transient outward current unopposed and thus able to effect an all-or-none repolarization of the action potential at the end of phase 1.[101] Failure of the sodium channel to express, as with the insertion and deletion mutations, results in similar electrophysiological changes. Reduction of the sodium channel I_{Na} current causes heterogeneous loss of the action potential dome in the right ventricular epicardium, leading to a marked dispersion of depolarization and refractoriness, an ideal substrate for development of reentrant arrhythmias. Phase 2 reentry produced by the same substrate is believed to provide the premature beat necessary for initiation of the VT and VF responsible for symptoms in these patients. Interestingly, however, Kambouris et al[102] identified a mutation in essentially the same region of *SCN5A* as the T1620M mutation (R1623H), but the clinical and biophysical features of this mutation were found to be consistent with *LQT3* and not Brugada syndrome.[102] More recently, mutations in *SCN5A* in which both Brugada syndrome and LQT3 features were seen in the same patient has been described.[103] Hence, there clearly remains a gap in our understanding of these entities.

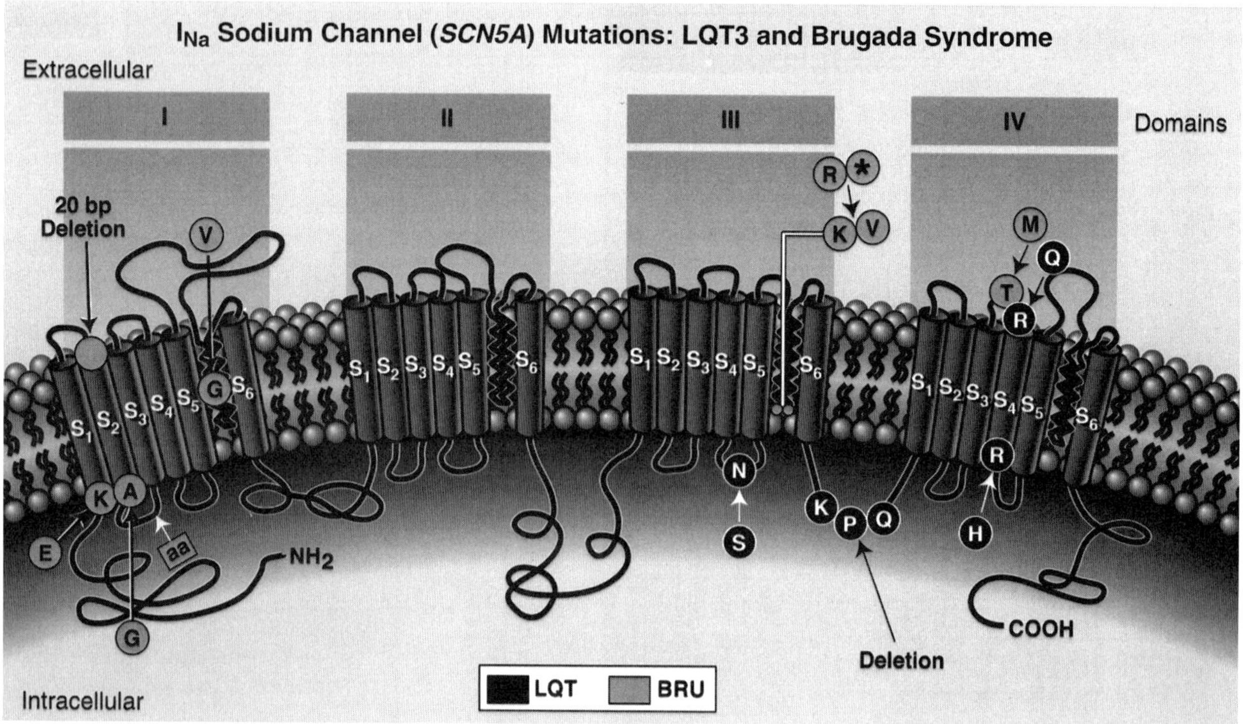

FIGURE 27-5 The molecular architecture of the cardiac I_{Na} sodium channel encoded by *SCN5A* and the mutations causing *LQT3* (*black*) and Brugada syndrome (*gray*).

Risk Stratification in Brugada Syndrome

Most symptomatic or at-risk patients with Brugada syndrome manifest an ECG with a "coved-type" ST-segment elevation with or without provocation using sodium channel blocking agents such as ajmaline or flecainide (Fig. 27-6).[104] The other form of ST-segment elevation, the so-called "saddle-type," is not associated with definitive Brugada syndrome unless it transitions into a "coved-type" by provocation or independently. Brugada et al[86,91,104] have suggested, however, that the risk of sudden death is not different between symptomatic patients and asymptomatic patients, including those with concealed forms of disease.

Various other risk stratifiers have also been identified.[105-108] Assessment of noninvasive markers by Ikeda et al[106] demonstrated that late potentials noted using signal-averaged ECGs were present in 24/33 (73% of) patients with a history of syncope or aborted sudden death. Using multivariate logistic regression, the authors showed that the presence of late potentials were significantly correlated with the occurrence of life-threatening events in patients with Brugada syndrome. The evaluation of these same patients with microvolt T wave alternans and corrected Q–T-interval dispersion failed to correlate with outcome. These findings were supported by others as well.[108]

Finally, spontaneous episodes of VF in patients with Brugada syndrome were shown to be triggered by PVCs with specific morphologies. Kakishita et al[93] suggested that the use of ICD therapy not only could be lifesaving but also could record the specific triggering events. They suggested that this knowledge could define risk and potentially lead to either ablative therapy or the ability to stratify risk of sudden death.

Hence, the identification of "coved-type" ST-segment elevation on surface ECG, the identification of late potentials on signal-averaged ECG, and the finding of triggering PVCs could provide insight into those patients with Brugada syndrome at high risk. Addition of family history could allow for further improvements of risk stratification.

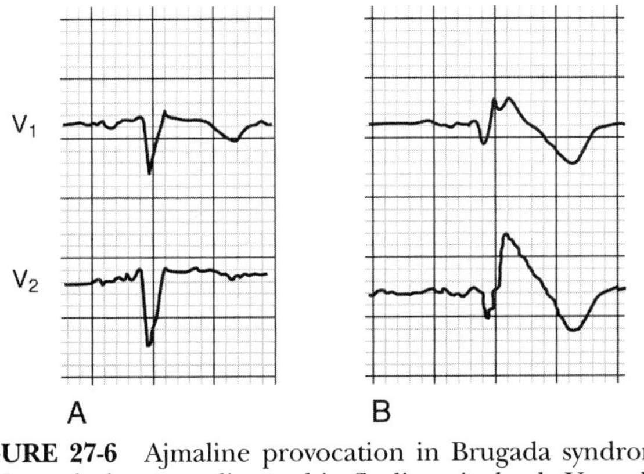

FIGURE 27-6 Ajmaline provocation in Brugada syndrome. **A,** Normal electrocardiographic findings in leads V_1 and V_2 in an individual with asymptomatic concealed Brugada syndrome and an *SCN5A* mutation. **B,** The response to ajmaline, identifying ST-segment elevations in leads V_1 and V_2.

Cardiac Conduction Disease

Syncope and sudden death may also occur due to bradycardia. The most common form of life-threatening bradycardias include disorders in which complete atrioventricular block occurs.[109] These disorders require pacemaker implantation for continued well-being. Two major forms of conduction system disease in which no congenital heart disease is associated include isolated forms of conduction disease[109,110] associated with dilated cardiomyopathy.[111]

Progressive cardiac conduction defect, also known as Lev-Lenègre's disease, is one of the most common cardiac conduction disorders.[109,112] This disorder is characterized by progressive alteration of conduction through the His-Purkinje system with development of RBBB or left bundle branch block with widening of the QRS complexes. Ultimately, complete atrioventricular block occurs, resulting in syncope and sudden death. Lev-Lenègre's disease represents the most common cause of pacemaker implantation worldwide, accounting for 0.15 implants per 1000 population yearly in developed countries. This disorder has been considered to be a primary degenerative disease, an exaggerated aging process with sclerosis of the conduction system, or an acquired disease. The first gene identified for Lev-Lenègre's disease was reported in 1999 by Schott et al.[112] They identified a missense mutation and deletion mutation, respectively, in *SCN5A* (see Table 27-1), the cardiac sodium channel gene, in two families with autosomal dominant inheritance (Fig. 27-7). Although the authors suggested that the biophysical abnormality was channel loss of function, no electrophysiological analysis was provided. As *SCN5A* also causes *LQT3*,[50] Brugada syndrome,[61,103,113] and SIDS (see Fig. 27-7),[114] all diseases in which ventricular tachyarrhythmias result in syncope and sudden death,[110] the association of conduction disturbance with *SCN5A* mutations was initially surprising. However, it is now known that conduction disturbance occurs in these disorders as well.

A similar disorder, known as isolated cardiac conduction disease, also results in complete atrioventricular block, syncope, and sudden death. This disorder has been considered to be genetically inherited (autosomal dominant trait) and not acquired. Brink et al and de Meeus et al independently mapped a gene to chromosome 19q13.3 in families with isolated conduction disturbance in 1995,[115,116] but the gene has remained elusive (see Table 27-1). Recently, however, Tan and colleagues[110] identified a mutation in *SCN5A* in this disorder (see Fig. 27-7) and also presented biophysical analysis (see Table 27-1). This mutation, a G to T transversion in exon 12 of *SCN5A*, resulted in a change from glycine to cysteine at position 514 (G514C) encoding an amino acid within the DI-DII intercellular linker of the cardiac sodium channel.

Biophysical characterization of the mutant channel demonstrated abnormaiities in voltage-dependent

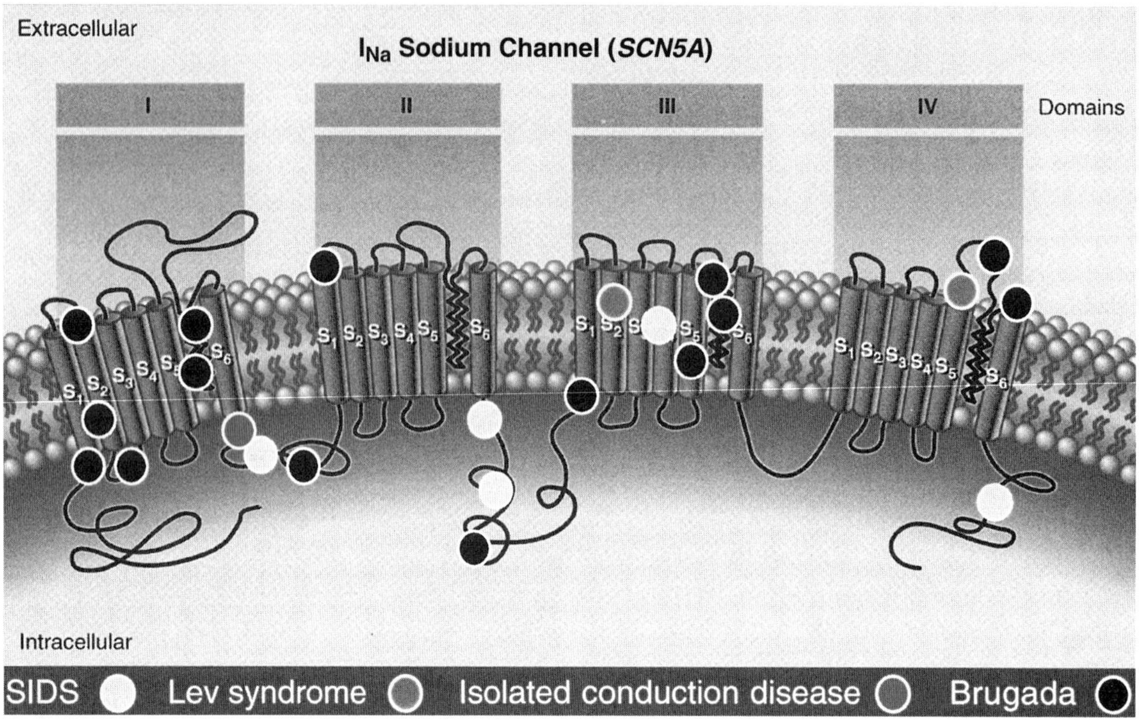

FIGURE 27-7 (See also Color Plate 27-7.) Cardiac sodium channel (*SCN5A*) gene mutations associated with cardiac arrhythmias and conduction system diseases. Cardiac arrhythmias: SIDS, sudden infant death syndrome (*yellow*); Brugada syndrome (*black*); conduction system disease: Lev syndrome (*green*); isolated conduction disease (*red*). Note that mutations for all the disorders are scattered throughout the channel protein domains. They are found within the transmembrane portions and pore regions of the channel, as well as within intracellular and extracellular regions of the protein.

gating behavior. The sodium current (I_{Na}) was found to decay more rapidly than the wild-type channel. In the mutant, open-state inactivation was hastened, while closed state inactivation was reduced and destabilized. Computational analysis predicted that the gating defects selectively slowed myocardial conduction without provoking the rapid cardiac arrhythmias seen in LQTS and Brugada syndrome. When comparing Brugada syndrome, *LQT3*, and conduction disease biophysics, the following findings are notable. In Brugada syndrome, *SCN5A* mutations cause reduction in I_{Na}, hastening epicardial repolarization and causing the development of VT and VF. In contradistinction, *LQT3* mutations in *SCN5A* result in excessive I_{Na}, delaying repolarization and torsades de pointes VT. Importantly, the G514C mutation evokes gating shifts reminiscent of both *LQT3* and Brugada syndrome, including an activation gating shift responsible for reducing I_{Na} and destabilization of inactivation that causes an increase in I_{Na}. Tan et al[110] showed that these voltage-dependent gating abnormalities may be partially corrected by dexamethasone, consistent with the known salutary effects of glucocorticoids on the clinical phenotype. It is also worth noting again that some patients with *LQT3* and Brugada syndrome have been reported to have conduction disturbances.

Finally, patients with conduction disease and dilated cardiomyopathy present a conundrum of what comes first, conduction abnormalities leading to cardiomyopathy or vice versa.[111] Clinically, these patients tend to develop variable degrees of atrioventricular block in their teen years or 20s with progression of this block over another 1 or 2 decades before developing the signs and symptoms of heart failure consistent with the cardiomyopathic phenotype. To date, only the gene *lamin A/C* located on chromosome 1q21 has been confirmed to cause this disease.[117,118] *Lamins A* and *C* are members of the intermediate filament multigene family, which are encoded by a single gene. *Lamins A* and *C* polymerize to form part of the nuclear lamina, a structural filamentous network on the nucleoplasmic side of the inner nuclear membrane. The specific cause of conduction disease and myocardial dysfunction are not currently known but could be due to progressive degeneration of cardiac tissue analogous to that described in Lev-Lenègre's disease.

Sudden Infant Death Syndrome

Sudden infant death syndrome (SIDS) is defined as the sudden death of an infant younger than 1 year of age that remains unexplained after performance of a complete autopsy, review of clinical and family history, and examination of the death scene. Although the incidence of SIDS has been dramatically reduced from 1.6 per 1000 live births to 0.64 per 1000 live births as reported in 1998 in the United States, it is still one of the most common causes of death among children between 1 month and 6 months of age. Death usually occurs during sleep.[4,119]

The potential causes of sudden death in infants are many, including cardiac disorders, respiratory abnormalities, gastrointestinal diseases, metabolic disorders, traumatic injury, brain abnormality, or child abuse. One of the most referenced etiologic speculations was that described by Schwartz in 1976.[120] He proposed that a developmental abnormality in cardiac sympathetic innervation predisposed some infants to lethal cardiac arrhythmias. Specifically, an imbalance in the sympathetic nervous system was speculated to result in prolongation of the Q–T interval on the ECG and in potentially lethal ventricular arrhythmias.[17,18] In 1998 Schwartz et al[4] published data collected from 1976-1994 in which ECGs were recorded on the third or fourth day of life in 34,442 Italian newborns. These babies were followed for 1 year; during that period, 34 children died. Evaluation of these 34 children demonstrated that 24 died of SIDS. These 24 SIDS victims were found to have longer QTc measurements than controls or other infants dying of other causes. In 12 of these 24 cases, the QTc was clearly prolonged, and the authors suggested that QTc prolongation during the first week of life is associated with SIDS.[4,121]

Although this suggestion linking SIDS and LQTS was roundly criticized,[122-128] the authors were subsequently able to identify a mutation in *SCN5A* (see Table 27-1) in one patient with aborted SIDS.[114] In addition, Priori et al[100] reported identification of an *SCN5A* mutation in an infant with Brugada syndrome. More recently, Ackerman et al[129] reported a molecular epidemiology study of 95 cases of SIDS in which myocardium obtained at autopsy was screened for ion channel gene mutations. In 4 of 93 cases, mutations in *SCN5A* were identified postmortem, and the authors suggested that 4.3% of this SIDS cohort was due to mutations in this known arrhythmia-causing gene. Hence, it appears that ion channel mutations, particularly *SCN5A*, result in SIDS in some infants (see Fig. 27-7). Biophysical analysis identified a sodium current characterized by slower delay, and a twofold to threefold increase in late sodium current similar to that seen in LQTS. The fact that these children die during sleep is consistent with the features seen for this channel when mutations result in LQTS (*LQT3*). *SCN5A* mutations in SIDS have been further confirmed recently.[130] It is likely that other ion channel gene abnormalities will be found in infants with SIDS, and that there is wide etiologic heterogeneity.

Arrhythmogenic Right Ventricular Dysplasia/Cardiomyopathy

Arrhythmogenic right ventricular dysplasia/cardiomyopathy (ARVD/C) is characterized by fatty infiltration of the right ventricle, fibrosis, and ultimately thinning of the wall with chamber dilatation (Fig. 27-8).[5] It is the most common cause of sudden cardiac death in the young in Italy[131] and is said to account for about 17% of sudden death in the young in the Unites States.[132] Rampazzo et al[133] mapped this disease in two families, one to 1q42-q43 and the other on chromosome 14q23-q24.[134] A third locus was mapped to 14q12.[135] A large Greek family with arrhythmogenic right ventricular dysplasia and Naxos disease was recently mapped to 17q21.[136] Two loci responsible for ARVD/C in North America were subsequently mapped at 3p23[137] and the other at 10p12.[138]

ARVD/C is a devastating disease, since the first symptom is often sudden death. Electrocardiographic abnormalities include inverted T waves in the right precordial leads, late potentials, and right ventricular arrhythmias with left bundle branch block. In many cases, the ECG looks similar to that seen in Brugada syndrome, with ST elevation[139] in V1 to V3. The issue of sudden death is compounded by the great difficulty in making the diagnosis of ARVD/C even when occurring in a family with the disease history. Since the disease affects only the right ventricle, it is difficult to detect by most diagnostic modalities. There is no diagnostic definitive standard at present. The right ventricular biopsy

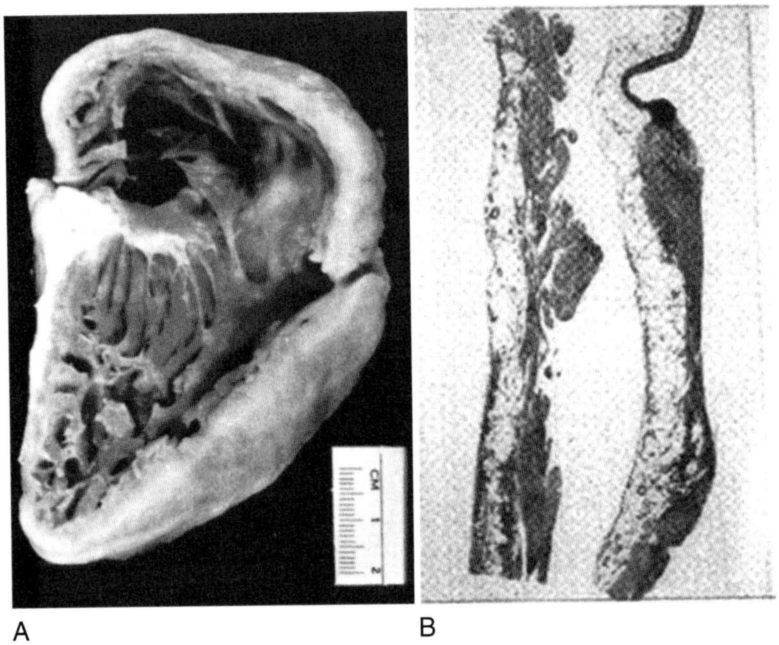

FIGURE 27-8 (See also Color Plate 27-8.) Arrhythmogenic right ventricular dysplasia/cardiomyopathy (ARVD/C). **A,** Gross anatomy with thinned right ventricle and fatty infiltration. **B,** Histology identifying fibrofatty infiltration of right ventricle.

may be definitive when positive but often gives a false-negative diagnosis, since the disease initiates in the epicardium and spreads to the endocardium of the right ventricular free wall, making it inaccessible to biopsy. A consensus diagnostic criteria that includes right ventricular biopsy, MRI, echocardiography, and electrocardiography was developed.[140]

The genetic basis of ARVD/C has started to unravel recently. The first gene causing ARVD/C was identified by Tiso et al[141] for the chromosome 1q42-1q43-linked ARVD2 locus in 2001 (Fig. 27-9). This gene (see Table 27-1), the cardiac ryanodine receptor gene (*RYR2*), a 105 exon gene that encodes the 565 kd monomer of a tetrameric structure interacting with four FK-506 binding proteins called FKBP12.6, is fundamental for intracellular calcium homeostasis and for excitation-contraction coupling. This large protein physically links to the dihydropyridine (DHP) receptor of the t-tubule, where the DHP receptor protein, a voltage-dependent calcium channel, is activated by plasma membrane depolarization and induces a calcium influx.[142,143] The *RYR2* protein, activated by calcium, induces release of calcium from the sarcoplasmic reticulum into the cytosol. Hence, mutations in *RYR2* would be expected to cause calcium homeostasis imbalance and result in abnormalities in rhythm as well as excitation-contraction coupling and myocardial dysfunction.[143] This causative gene, therefore, is in many ways similar to the mutant genes responsible for the ventricular arrhythmias of LQTS, Brugada syndrome, and SIDS, in which ion channel mutations cause the clinical phenotype. In those instances, potassium channel and sodium channel dysfunction occurs, while in ARVD2, intracellular calcium channel dysfunction plays a central role.

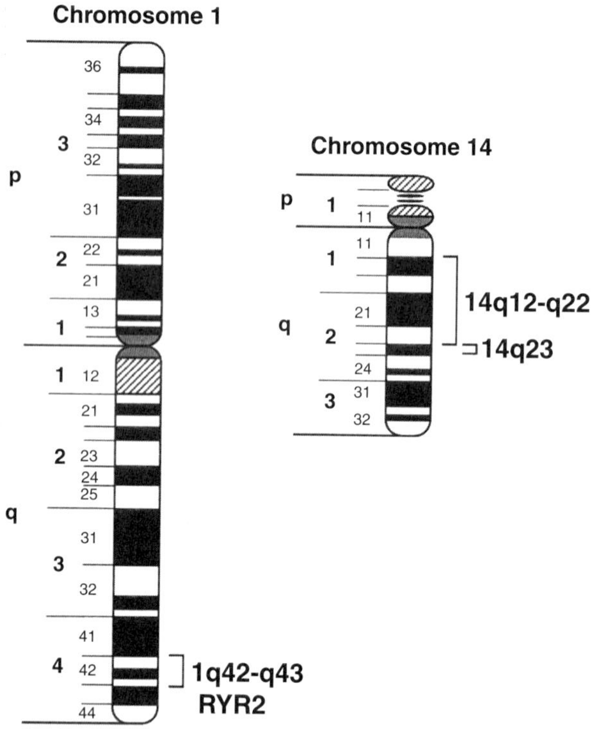

FIGURE 27-9 Chromosomal positions of *ARVD1* (chromosome 14q23), *ARVD2* (chromosome 1q42-q43), and *ARVD3* (chromosome 14q12-q22) with identification of the *ARVD2* gene, ryanodine receptor (*RYR2*).

Two other genes associated with arrhythmogenic cardiomyopathy have also been described. The first of these, plakoglobin, was shown to cause the chromosome 17q21-linked autosomal recessive disorder called Naxos disease (see Table 27-1).[144] This disorder is characterized by ARVD/C in association with abnormalities of skin (palmoplantar keratoderma) and hair (woolly hair) and therefore is not exactly the same as isolated ARVD/C, being a more complex phenotype.

Plakoglobin is a cell adhesion protein thought to be important in providing functional integrity to the cell. This protein is found in many tissues, including the cytoplasmic plaque of cardiac junctions and the dermal-epidermal junctions of the epidermis, and it has a potential signaling role in the formation of desmosomal junctions. It is believed that plakoglobin serves as a linker molecule between the inner and outer portions of the desmosomal plaque by binding tightly to the cytoplasmic domains of cadherins. The mutations identified, a homozygous 2 bp deletion, resulted in a frameshift and premature termination of the protein.[144] Support for this gene being causative of Naxos disease comes from a mouse model with null mutations, which exhibit the heart and skin abnormalities seen in affected patients. The mutated protein is thought to cause disruption of myocyte integrity, leading to cell death and fibrofatty replacement, with secondary arrhythmias occurring due to the abnormal myocardial substrate.

The last gene identified, desmoplakin,[145] is another desmosomal protein with similarities to plakoglobin (see Table 27-1). Homozygous mutations in this gene resulted in a Naxos-like disorder, although the cardiac features occurred in the left ventricle instead of the right ventricle. The affected protein is an important protein in cell adhesion and appears to function similarly to that described for plakoglobin. While mutation in this gene and in plakoglobin can easily be speculated to cause the myocardial abnormalities, it remains unclear why differences in ventricular chamber specificity occurs and how the ventricular tachyarrhythmias develop.

Brugada Syndrome and Arrhythmogenic Right Ventricular Dysplasia

Controversy exists concerning the possible association of Brugada syndrome and ARVD, with some investigators arguing that these are the same disorder or at least one is a forme-fruste of the other.[139,146-150] However, the classic echocardiographic, angiographic, and magnetic resonance imaging findings of ARVD are not seen in Brugada syndrome patients. In addition, Brugada syndrome patients typically are without the histopathologic findings of ARVD. Further, the morphology of VT/VF differs.[91,139] Finally, the genes identified to date differ.[61,141,144,145]

Polymorphic Ventricular Tachycardia

Familial polymorphic VT, an autosomal dominant disorder characterized by episodes of bidirectional (Fig. 27-10) and polymorphic VT, typically in relation to adrenergic stimulation or physical exercise, was first described by Coumel et al in 1978.[151] This disorder occurs in childhood and adolescence most commonly, presenting with syncope and sudden death.[6,152] Mortality rates of 30% to 50% by age 20 to 30 years of age have been reported, suggesting this to be a highly malignant disorder. Autopsy data demonstrate this disorder to have no associated structural cardiac abnormalities.

Mutations in the cardiac ryanodine receptor (*RYR2*), the same gene responsible for ARVD2 (see Fig. 27-9), were recently independently identified by Laitinen et al[153] and Priori et al[154] in multiple families linked to chromosome 1q42 (see Table 27-1).[155,156] Interestingly, ARVD2 typically is considered to be the one form of ARVD/C in which catecholaminergic input is important in the development of symptoms. Why patients with familial polymorphic VT have no associated structural cardiac abnormalities and patients with ARVD/C have classic fibrofatty replacement in the right ventricle is not clear at this time.

Mutations in another member of the ryanodine receptor gene family, *RYR1*, which is expressed in skeletal muscle, result in malignant hyperthermia and central core disease.[157] The mutations in this gene appear to cluster in three regions of the gene, regions similar to the mutations found in *RYR2* in the cases of VT reported, suggesting these to be functionally critical regions.

Wolff-Parkinson-White Syndrome

This disorder is the second most common cause of paroxysmal supraventricular tachycardia (SVT), with a prevalence of 1.5 to 3.1 per 1000 individuals.[2] In some parts of the world, such as China, Wolff-Parkinson-White syndrome (WPW) is even more common, being responsible for up to 70% of cases of SVT.[158] Tachycardia presents typically in a bimodal fashion, with onset common in infancy as well as during the teen years. Symptoms most commonly include syncope, presyncope, shortness of breath, palpitations, and sudden death.[109]

WPW has long been described to be due to accessory pathways derived from muscle fibers providing direct continuity between atrial and ventricular myocardium.[2,159] These accessory pathways may be identified by the peculiar ECG findings seen in WPW, including short P–R interval, widened QRS complexes, and the classic δ wave in which an abnormal initial QRS vector is notable (Fig. 27-11).[2,159-161] In a significant percentage of patients, conduction abnormalities including high-grade sinoatrial or atrioventricular block occur,[162] necessitating pacemaker implantation. In most patients with WPW and SVT, radiofrequency ablation of the accessory pathways is curative.[163]

Some cases of WPW are associated with other primary disorders, such as hypertrophic cardiomyopathy[162,164] or left ventricular noncompaction cardiomyopathy,[165] or the congenital cardiac disorder Ebstein's anomaly. Whether the underlying cause of WPW is similar in these cases compared to pure cases of WPW has been discussed for many years, but no definitive answers have been provided.

The first gene in patients with WPW was recently identified by Gollob et al[166] and Blair et al[167] independently in familial forms of WPW. In both reports, autosomal dominant inheritance was reported. Interestingly, this gene, which maps to chromosome 7q34-7q36,[164] was found to cause WPW and hypertrophic cardiomyopathy in a significant percentage of patients in both reports (see Table 27-1). The gene, the γ2 subunit of adenosine monophosphate–activated protein kinase, consists of 569 amino acids, is 63 kd in size, and functions as a metabolic sensor in cells, responding to cellular energy demands by regulating ATP production and utilization.[168-172] Confusion exists as to whether this is a primary hypertrophic cardiomyopathy–causing gene or WPW gene, particularly since the initial mapping of this locus was in patients with hypertrophic cardiomyopathy and associated WPW. Clearly, this is not the only gene responsible for WPW, and the functional and physiologic abnormalities responsible for the resultant WPW are not yet obvious.

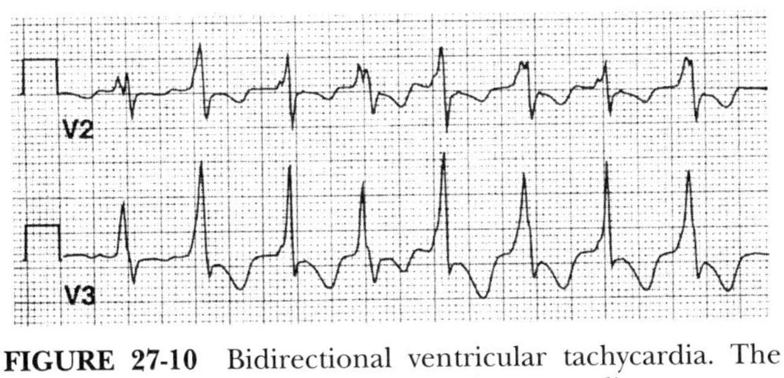

FIGURE 27-10 Bidirectional ventricular tachycardia. The genetic cause, mutations within the ryanodine receptor (*RYR2*) gene, is the same gene responsible for the chromosome 1–linked Arrhythmogenic right ventricular dysplasia cardiomyopathy.

Final Common Pathway Hypothesis

Clearly, long QT syndrome is a disease of the ion channel. Similarly, patients with Brugada syndrome, catecholaminergic VT, and conduction abnormalities have primary ion channelopathies. Patients with other cardiac disorders, such as familial dilated cardiomyopathy (FDCM) and familial hypertrophic cardiomyopathy (FHCM), have been shown to have mutations in genes encoding a consistent family of proteins as well. In FDCM, cytoskeletal protein-encoding genes and sarcomeric proteins have been speculated to be causative

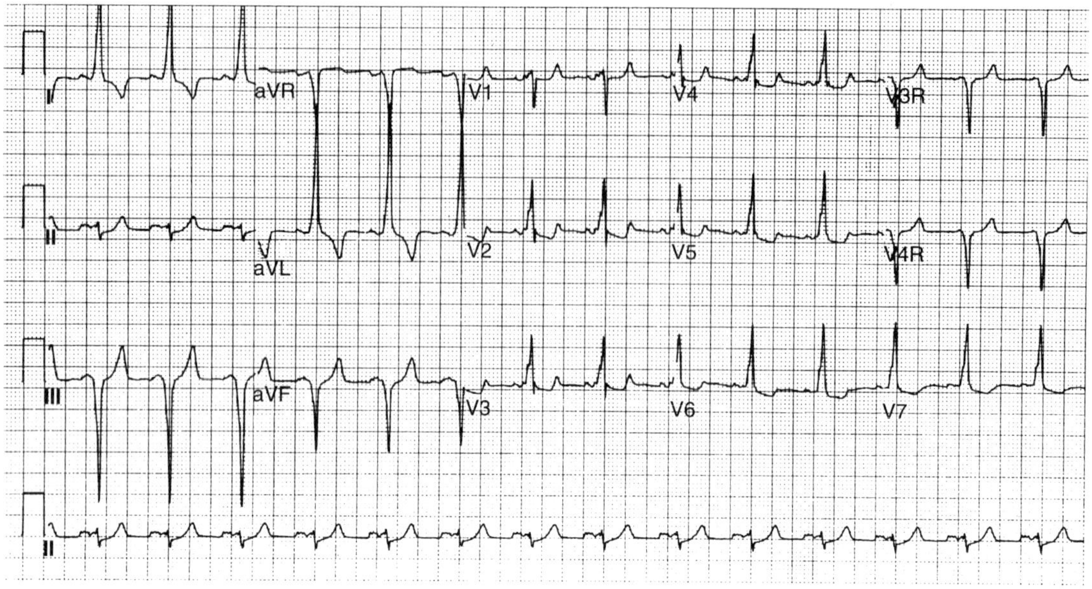

FIGURE 27-11 ECG showing the short P–R interval and δ wave associated with Wolff-Parkinson-White syndrome.

(i.e., cytoskeletal/sarcomyopathy).[173] In addition, FHCM has been shown to be a disease of the sarcomere. Hence, the final common pathways of these disorders include the sarcomere and cytoskeletal proteins, similar to the ion channelopathies in arrhythmias and conduction disease. Intermediate disorders such as ARVD/ARVC appear to connect the primary electrical and primary muscle disorders mechanistically. In addition, it appears that cascade pathways are involved directly in some cases (i.e., mitochondrial abnormalities in HCM and DCM), while secondary influences are likely to result in the wide clinical spectrum seen in patients with similar mutations. In HCM, mitochondrial and metabolic influences are probably important. Additionally, molecular interactions with such molecules as calcineurin, sex hormones, and growth factors are probably involved in development of clinical signs, symptoms, and age of presentation. In the future, these factors are expected to be uncovered, allowing for development of new therapeutic strategies.

Relevance

The relevance of the hypothesis is its ability to classify disease entities on a molecular and mechanistic basis. This reclassification of disorders on the basis of molecular abnormalities such as "ion channelopathies," "sarcomyopathies," or "cytoskeletopathies" could lead to more focused approaches to gene discovery and future therapeutic interventions. For instance, on the basis of the understanding of the molecular aspects of long QT syndrome, we considered the possibility that all ventricular arrhythmias are the result of ion channel abnormalities. On the basis of the hypothesis, we studied the possibility that the cardiac sodium channel gene (*SCN5A*) was mutated in patients with the idiopathic VF disorder called Brugada syndrome, identifying mutations in three separate, unrelated families. Use of this hypothesis for disorders such as inherited DCM is likely to more narrowly focus efforts at gene identification. In the near future when the human genome project is completed, this will allow for investigators to more rapidly identify disease-responsible genes. Once the genes are known, and the mechanisms causing the clinical phenotype and natural history are known, improved pharmacologic therapies based on the actual disease mechanism can be produced and used. At that time, the impact of molecular genetics on clinical practice and patient care will become fully evident.

New Clinical Directions from Molecular Studies

New information is rapidly accumulating in the understanding of the genetic basis of several cardiac arrhythmias. The impact of this information on the clinician is now becoming apparent with further characterization of these defects in relation to the different clinical syndromes.[175-189] In familial bidirectional (polymorphic) VT, different ryanodine receptor mutations have been detected and were absent in unaffected family members.[176] In screening of family members and probands, genotyping may define subgroups at differing risk of clinical events and earlier diagnosis of asymptomatic carriers.[177,178] These mutations alter the regulation of the channel, resulting in increased sarcoplasmic reticulum calcium leak during sympathetic stimulation, which has been associated with a reduced binding protein affinity explaining the exercise relationships of the clinical arrhythmia.[176,179]

There are therapeutic implications of these findings as well. Potential implication of apoptosis in the molecular pathogenesis of arrhythmogenic right ventricular cardiomyopathy has led to consideration of a class of drugs that could permit a new direction in therapy.[181] The inheritability of these disorders may also be defined by genotyping. Mutations in desmoplakin may produce different clinical phenotypes and different modes of inheritance.[183,184] Differing complex arrhythmia phenotypes may be related to variation in the genetic defect at the same locus. New mutations in the *SCN5A* gene and the gene that encodes the gamma 2 regulatory subunit of adenosine monophosphate–activated protein kinase have been associated with atrioventricular block alone or with ventricular preexcitation, respectively.[185,188] Genetic defects affecting the inward rectifying potassium current can also affect cardiac and skeletal muscle, resulting in a distinct arrhythmia–muscular dystrophy phenotype and enhancing our understanding of its pathogenesis.[189]

SUMMARY

Ventricular arrhythmias appear to result from ion channel abnormalities. Whether this is necessarily a primary abnormality or can occur secondary is becoming better understood as the genes responsible for ARVD/C are identified. Therapeutic options are likely to be expanded once this knowledge has matured. Similarly, conduction system abnormalities have been shown to occur secondary to mutations in the ion channel gene *SCN5A*, as well as a result of mutations in the intermediate filament protein lamin A/C. The recent finding of Malhotra et al[174] that *SCN5A*, the α-subunit of the cardiac sodium channel, interacts with accessory subunits β1 and β2, which act as a junction protein interactor, could explain how these apparently disparate genes (*SCN5A*, *lamin A/C*) result in similar clinical findings. The unraveling of these questions will lead to improved ability to develop rational therapies in the future.

REFERENCES

1. Priori SG, Barhanin J, Hauer RNW, et al: Genetic and molecular basis of cardiac arrhythmias: Impact on clinical management (Parts I and II). Circulation 1999;99:518-528.
2. Al-Khatib SM, Pritchett EL: Clinical features of Wolff-Parkinson-White syndrome. Am Heart J 1999;138:403-413.
3. Brugada P, Brugada J: Right bundle-branch block, persistent ST segment elevation and sudden cardiac death: A distinct clinical and electrocardiographic syndrome. A multicenter report. J Am Coll Cardiol 1992;20:1391-1396.
4. Schwartz PJ, Stramba-Badiale M, Segantini A, et al: Prolongation of the QT interval and the sudden infant death syndrome. N Engl J Med 1998;338:1709-1714.
5. Thiene G, Basso C, Danieli G, et al: Arrhythmogenic right ventricular cardiomyopathy. Trends Cardiovasc Med 1997;7:84-90.
6. Fisher JD, Krikler D, Hallidie-Smith KA: Familial polymorphic ventricular arrhythmias: A quarter century of successful medical treatment based on serial exercise-pharmacologic testing. J Am Coll Cardiol 1999;34:2015-2022.
7. Schwartz PJ, Locati EH, Napolitano C, Priori SG: The long QT syndrome. In Zipes DP, Jalife J (eds): Cardiac Electrophysiology: From Cell to Bedside. Philadelphia, WB Saunders, 1996, pp. 788-811.
8. Romano C, Gemme G, Pongiglione R: Antmie cardiache rare in eta pediatrica. Clin Pediatr 1963;45:656-683.
9. Ward OC: A new familial cardiac syndrome in children. J Ir Med Assoc 1964;54:103-106.
10. Priori SG, Napolitano C, Schwartz PJ: Low penetrance in the long-QT syndrome. Clinical impact. Circulation 1999;99:529-533.
11. Singer PA, Crampton RS, Bass NH: Familial Q-T prolongation syndrome: Convulsive seizures and paroxysmal ventricular fibrillation. Arch Neurol 1974;31:64-66.
12. Ratshin RA, Hunt D, Russell RO Jr, Rackley CE: QT-interval prolongation, paroxysmal ventricular arrhythmias, and convulsive syncope. Ann Intern Med 1971;75:19-24.
13. Bellavere F, Ferri M, Guarini L, et al: Prolonged QT period in diabetic autonomic neuropathy: A possible role in sudden cardiac death. Br Heart J 1988;59:379-383.
14. Ewing DJ, Boland O, Neilson JMM, et al: Autonomic neuropathy, QT interval lengthening, and unexpected deaths in male diabetic patients. Diabetologia 1991;34:182-185.
15. Weintraub RG, Gow RM, Wilkinson JL: The congenital long QT syndromes in children. J Am Coll Cardiol 1990;16:674-680.
16. Marks ML, Trippel DL, Keating MT: Long QT syndrome associated with syndactyly identified in females. Am J Cardiol 1995;10: 744-745.
17. Schwartz PJ, Segantini A: Cardiac innervation, neonatal electrocardiography and SIDS. A key for a novel preventive strategy? Ann N Y Acad Sci 1988;533:210-220.
18. Schwartz PJ, Stramba-Badiale M, Segantini A, et al: Prolongation of the QT interval and the sudden infant death syndrome. N Engl J Med 1998;338:1709-1714.
19. Jervell A, Lange-Nielsen F: Congenital deaf-mutism, function heart disease with prolongation of the Q-T interval and sudden death. Am Heart J 1957;54:59-68.
20. Jervell A: Surdocardiac and related syndromes in children. Adv Intern Med 1971;17:425-438.
21. James TN: Congenital deafness and cardiac arrhythmias. Am J Cardiol 1967;19:627-643.
22. Priori SG, Schwartz PJ, Napolitano C, et al: A recessive variant of the Romano-Ward long-QT syndrome. Circulation 1998;97: 2420-2425.
23. Keating MT, Atkinson D, Dunn C, et al: Linkage of a cardiac arrhythmia, the long QT syndrome, and the Harvey ras-1 gene. Science 1991;252:704-706.
24. Wang Q, Curran ME, Splawski I, et al: Positional cloning of a novel potassium channel gene: *KVLQT1* mutations cause cardiac arrhythmias. Nature Genet 1996;12:17-23.
25. Li H, Chen Q, Moss AJ, et al: New mutations in the *KVLQT1* potassium channel that cause long QT syndrome. Circulation 1998;97:1264-1269.
26. Choube C, Neyroud N, Guicheney P, et al: Properties of *KVLQTI* K^+ channel mutations in Romano-Ward and Jervell and Lange-Nielsen inherited cardiac arrhythmias. EMBO J 1997;16:5472-5479.
27. Chiang CE, Roden DM: The long QT syndromes: Genetic basis and clinical implications. J Am Coll Cardiol 2000;36:1-12.
28. Barhanin J, Lesage F, Guillemare E, et al: *KVLQT1* and IsK (*minK*) proteins associate to form the I_{Ks} cardiac potassium current. Nature 1996;384:78-80.
29. Sanguinetti MC, Curran ME, Zou A, et al: Coassembly of *KvLQT1* and *minK* (IsK) proteins to form cardiac I_{Ks} potassium channel. Nature 1996;384:80-83.
30. Wollnick B, Schreeder BC, Kubish C, et al: Pathophysiological mechanisms of dominant and recessive *KVLQTI* K^+ channel mutations found in inherited cardiac arrhythmias. Hum Mol Genet 1997;6:1943-1949.
31. Splawski I, Tristani-Firouzi M, Lehmann MH, et al: Mutations in the *minK* gene cause long QT syndrome and suppress I_{Ks} function. Nat Genet 1997;17:338-340.
32. Jiang C, Atkinson D, Towbin JA, et al: Two long QT syndrome loci map to chromosome 3 and 7 with evidence for further heterogeneity. Nat Genet 1994;8:141-147.
33. Warmke JE, Ganetzky B: A family of potassium channel genes related to *eag* in *Drosophila* and mammals. Proc Natl Acad Sci U S A 1994;91:3438-3442.
34. Arcangeli A, Rosati B, Cherubini A, et al: *HERG*- and IRK-like inward rectifier currents are sequentially expressed during neuronal crest cells and their derivatives. Eur J Neurosci 1997; 9:2596-2604.
35. Pennefather PS, Zhou W, Decoursey TE: Idiosyncratic gating of *HERG*-like K^+ channels in microglia. J Gen Physiol 1998;111: 795-805.
36. Bianchi L, Wible B, Arcangeli A, et al: *HERG* encodes a K^+ current highly conserved in tumors of different histogenesis: A selective advantage for cancer cells? Cancer Res 1998;58:815-822.

37. Curran ME, Splawski I, Timothy KW, et al: A molecular basis for cardiac arrhythmia: *HERG* mutations cause long QT syndrome. Cell 1995;80:795-803.
38. Schulze-Bahr E, Haverkamp W, Funke H: The long-QT syndrome. N Engl J Med 1995;333:1783-1784.
39. Sanguinetti MC, Jiang C, Curran ME, Keating MT: A mechanistic link between an inherited and an acquired cardiac arrhythmia: *HERG* encodes the I_{Kr} potassium channel. Cell 1995;81:299-307.
40. Trudeau MC, Warmke J, Ganetzky B, Robertson G: *HERG*, a human inward rectifier in the voltage-gated potassium channel family. Science 1995;269:92-95.
41. Sanguinetti MC, Curran ME, Spector PS, Keating MT: Spectrum of *HERG* K^+ channel dysfunction in an inherited cardiac arrhythmia. Proc Natl Acad Sci U S A 1996;93:2208-2212.
42. Roden DM, Balser JR: A plethora of mechanisms in the HERG-related long QT syndrome genetics meets electrophysiology. Cardiovasc Res 1999;44:242-246.
43. Furutani M, Trudeau MC, Hagiwara N, et al: Novel mechanism associated with an inherited cardiac arrhythmia. Defective protein trafficking by the mutant *HERG* (G601S) potassium channel. Circulation 1999;99:2290-2294.
44. Zhou Z, Gong Q, Epstein ML, January CT: *HERG* channel dysfunction in human long QT syndrome. J Biol Chem 1998;263: 21061-21066.
45. McDonald TV, Yu Z, Ming Z, et al: A *minK-HERG* complex regulates the cardiac potassium current I_{Kr}. Nature 1997;388:289-292.
46. Bianchi L, Shen Z, Dennis AT, et al: Cellular dysfunction of *LQT5-minK* mutants: Abnormalities of I_{Ks}, I_{Kr} and trafficking in long QT syndrome. Hum Mol Genet 1999;8:1499-1507.
47. Abbott GW, Sesti F, Splawski I, et al: *MiRP1* forms I_{Kr} potassium channels with *HERG* and is associated with cardiac arrhythmia. Cell 1999;97:175-187.
48. Gellens M, George AL, Chen L, et al: Primary structure and functional expression of the human cardiac tetrodotoxin-insensitive voltage-dependent sodium channel. Proc Natl Acad Sci U S A 1992;89:54-58.
49. George AL, Varkony TA, Drakin HA, et al: Assignment of the human heart tetrodotoxin-resistant voltage-gated Na channel-subunit gene (SCN5A) to band 3p21. Cytogenet Cell Genet 1995; 68:67-70.
50. Wang Q, Shen J, Splawski I, et al: *SCN5A* mutations associated with an inherited cardiac arrhythmia, long QT syndrome. Cell 1995;80:805-811.
51. Wang Q, Shen J, Li Z, et al: Cardiac sodium channel mutations in patients with long QT syndrome, an inherited cardiac arrhythmia. Hum Mol Genet 1995;4:1603-1607.
52. Hartmann HA, Colom LV, Sutherland ML, Noebels JL: Selective localization of cardiac *SCN5A* sodium channels in limbic regions of rat brain. Nat Neurosci 1999;2:593-595.
53. Bennett PB, Yazawa K, Makita N, George AL Jr: Molecular mechanism for an inherited cardiac arrhythmia. Nature 1995;376:683-685.
54. Dumaine R, Wang Q, Keating MT, et al: Multiple mechanisms of sodium channel-linked long QT syndrome. Circ Res 1996;78: 916-924.
55. An RH, Wang XL, Kerem B, et al: Novel *LQT-3* mutation affects Na^+ channel activity through interactions between alpha- and beta 1-subunits. Circ Res 1998;83:141-146.
56. Honore E, Attali B, Heurteaux C, et al: Cloning, expression, pharmacology and regulation of a delayed rectifier K^+ channel in mouse heart. EMBO J 1991;10:2805-2811.
57. Arena JP, Kass RS: Block of heart potassium channels by clofilium and its tertiary analogs: Relationship between drug structure and type of channel blocked. Mol Pharmacol 1988;34:60-66.
58. Vetter DE, Mann JR, Wangemann P, et al: Inner ear defects induced by null mutation of the IsK gene. Neuron 1996;17: 1251-1264.
59. Neyroud N, Tesson F, Denjoy I, et al: A novel mutation in the potassium channel gene *KVLQT1* causes the Jervell and Lange-Nielsen cardioauditory syndrome. Nat Genet 1997;15:186-189.
60. Splawski I, Timothy KW, Vincent GM, et al: Brief report: Molecular basis of the long-QT syndrome associated with deafness. N Engl J Med 1997;336:1562-1567.
61. Chen Q, Zhang D, Gingell RL, et al: Homozygous deletion in *KVLQT1* associated with Jervell and Lange-Nielsen syndrome. Circulation 1999;99:1344-1347.
62. Tyson J, Tranebjaerg L, Bellman S, et al: *IsK* and *KVLQT1*: Mutation in either of the two subunits of the slow component of the delayed rectifier potassium channel can cause Jervell and Lange-Nielsen syndrome. Hum Molec Genet 1997;12: 2179-2185.
63. Schulze-Bahr E, Wang Q, Wedekind H, et al: *KCNE1* mutations cause Jervell and Lange-Nielsen syndrome. Nat Genet 1997; 17:267-268.
64. Kimbrough J, Moss AJ, Zareba W, et al: Clinical implications for affected parents and siblings of probands with long-QT syndrome. Circulation 2001;104:557-562.
65. Zareba W, Moss AJ, Schwartz PJ, et al: Influence of the genotype on the clinical course of the long-QT syndrome. N Engl J Med 1998;339:960-965.
66. Tanabe Y, Inagaki M, Kurita T, et al: Sympathetic stimulation produces a greater increase in both transmural and spatial dispersion of repolarization in LQT1 than LQT2 forms of congenital long QT syndrome. J Am Coll Cardiol 2001;37:911-919.
67. Ackermann MJ, Tester DJ, Porter CJ: Swimming, a gene-specific arrhythmogenic trigger for inherited long QT syndrome. Mayo Clin Proc 1999;74:1088-1094.
68. Moss AJ, Robinson JL, Gessman L, et al: Comparison of clinical and genetic variables of cardiac events associated with loud noise versus swimming among subjects with the long QT syndrome. Am J Cardiol 1999;84:876-933.
69. Wilde AAM, Jongbloed RJE, Doevendans PA, et al: Auditory stimuli as a trigger for arrhythmic events differentiate HERG-related (LQTS2) patients from KVLQT1-related patients (LQTS1). J Am Coll Cardiol 1999;33:327-332.
70. Wilde AAM, Roden DM: Predicting the long-QT genotype from clinical data: From sense to science. Circulation 2000;102: 2796-2798.
71. Ali RH, Zareba W, Moss AJ, et al: Clinical and genetic variables associated with acute arousal and nonarousal-related cardiac events among subjects with the long QT syndrome. Am J Cardiol 2000;85:457-461.
72. Moss AJ, Zareba W, Benhorin J, et al: ECG T-wave patterns in genetically distinct forms of the hereditary long-QT syndrome. Circulation 1995;92:2929-2934.
73. Zhang L, Timothy KW, Vincent GM, et al: Spectrum of ST-T-wave patterns and repolarization parameters in congenital long-QT syndrome: ECG findings identify genotypes. Circulation 2000; 102:2849-2855.
74. Lupoglazoff JM, Denjoy I, Berthet M, et al: Notched T waves on Holter recordings enhance detection of patients with LQT2 *(HERG)* mutations. Circulation 2001;103:1095-1101.
75. Shimizu W, Antzelevitch C: Sodium channel block with Mexiletine is effective in reducing dispersion of repolarization and preventing torsade de pointes in LQT2 and LQT3 models of the long-QT syndrome. Circulation 1997;96:2038-2047.
76. Shimizu W, Antzelevitch C: Differential effects of beta-adrenergic agonists and antagonists in LQT1, LQT2, and LQT3 models of the long QT syndrome. J Am Coll Cardiol 2000;35:778-786.
77. Moss AJ, Zareba W, Hall WJ, et al: Effectiveness and limitations of beta-blocker therapy in congenital long-QT syndrome. Circulation 2000;101:616-623.
78. Schwartz PJ, Priori SG, Locati EH, et al: Long-QT syndrome patients with mutations of the SCN5A and HERG genes have differential responses to Na^+ channel blockade and to increases in heart rate: Implications for gene-specific therapy. Circulation 1995;92:3381-3386.
79. Compton SJ, Lux RL, Ramsey MR, et al: Genetically defined therapy of inherited long-QT syndrome: Correction of abnormal repolarization by potassium. Circulation 1996;94:1018-1022.
80. Shimizu W, Kurita T, Matsuo K, et al: Improvement of repolarization abnormalities by K^+ channel opener in the long QT syndrome. Circulation 1998;97;1581-1588.
81. Osher HL, Wolff L: Electrocardiographic pattern simulating acute myocardial injury. Am J Med Sci 1953;226:541-545.
82. Edeiken J: Elevation of RS-T segment, apparent or real in right precordial leads as probable normal variant. Am Heart J 1954; 48:331-339.
83. Levine HD, Wanzer SH, Merrill JP: Dialyzable currents of injury in potassium intoxication resembling acute myocardial infarction or pericarditis. Circulation 1956;13:29-36.

84. Martini B, Nava A, Thiene G, et al: Ventricular fibrillation without apparent heart disease. Description of six cases. Am Heart J 1989;118:1203-1209.
85. Aihara N, Ohe T, Kamakura S: Clinical and electrophysiologic characteristics of idiopathic ventricular fibrillation. Shinzo 1990;22:80-86.
86. Brugada P, Brugada J: A distinct clinical and electrocardiographic syndrome: Right bundle-branch block, persistent ST segment elevation with normal QT interval and sudden cardiac death. Pacing Clin Electrophysiol 1991;14:746.
87. Antzelevitch C, Brugada P, Brugada J: The Brugada Syndrome. Armonk, NY, Futura Publishing, 1999.
88. Nademanee K, Veerakul G, Nimmannit S, et al: Arrhythmogenic marker for the sudden unexplained death syndrome in Thai men. Circulation 1997;96:2595-2600.
89. Gotoh K: A histopathological study on the conduction system of the so-called Pokkuri disease (sudden unexpected cardiac death of unknown origin in Japan). Jpn Circ J 1976;40:753-768.
90. Hayashi M, Murata M, Satoh M, et al: Sudden nocturnal death in young males from ventricular flutter. Jpn Heart J 1985;26:585-591.
91. Brugada J, Brugada P: Further characterization of the syndrome of right bundle branch block, ST segment elevation, and sudden death. J Cardiovasc Electrophysiol 1997;8:325-331.
92. Priori SG, Napolitano C, Gasparini M, et al: Clinical and genetic heterogeneity of right bundle branch block and ST-segment elevation syndrome: A prospective evaluation of 52 families. Circulation 2000;102:2509-2515.
93. Kakishita M, Kurita T, Matsuo K, et al: Mode of onset of ventricular fibrillation in patients with Brugada syndrome detected by implantable cardioverter defibrillator therapy. J Am Coll Cardiol 2000;36:1647-1653.
94. Kobayashi T, Shintani U, Yamamoto T, et al: Familial occurrence of electrocardiographic abnormalities of the Brugada-type. Intern Med 1996;35:637-640.
95. Gussak I, Antzelevitch C, Bjerregaard P, et al: The Brugada syndrome: Clinical, electrophysiological, and genetic considerations. J Am Coll Cardiol 1999;33:5-15.
96. Chen Q, Kirsch GE, Zhang D, et al: Genetic basis and molecular mechanism for idiopathic ventricular fibrillation. Nature 1998; 392:293-296.
97. Antzelevitch C: The Brugada Syndrome. J Cardiovasc Electrophys 1998;9:513-516.
98. Antzelevitch C: The Brugada syndrome: Diagnostic criteria and cellular mechanisms. Eur Heart J 2001;22:356-363.
99. Antzelevitch C: Ion channels and ventricular arrhythmias: Cellular and ionic mechanisms underlying the Brugada syndrome. Curr Opin Cardiol 1999;14:274-279.
100. Priori SG, Napolitano C, Giordano U, et al: Brugada syndrome and sudden cardiac death in children. Lancet 2000;355:808-809.
101. Dumaine R, Towbin JA, Brugada P, et al: Ionic mechanisms responsible for the electrocardiographic phenotype of the Brugada syndrome are temperature dependent. Circ Res 1999; 85:803-809.
102. Kambouris NG, Nuss HB, Johns DC, et al: Phenotypic characterization of a novel long-QT syndrome mutation (R1623Q) in the cardiac sodium channel. Circulation 1998;97:640-644.
103. Bezzina C, Veldkamp MW, van Den Berg MP, et al: A single Na^+ channel mutation causing both long-QT and Brugada syndromes. Circ Res 1999;85:1206-1213.
104. Brugada R, Brugada J, Antzelevitch C, et al: Sodium channel blockers identify risk for sudden death in patients with ST segment elevation and right bundle branch block but structurally normal heart. Circulation 2000;1:510-515.
105. RuDusky BM: Right bundle branch block, persistent ST-segment elevation, and sudden death. Am J Cardiol 1998;82:407-408.
106. Ikeda T, Sakurada H, Sakabe K, et al: Assessment of noninvasive markers in identifying patients at risk in the Brugada syndrome: Insight into risk stratification. J Am Coll Cardiol 2001;37: 1628-1634.
107. Remme CA, Wever EFD, Wilde AAM, et al: Diagnosis and long-term follow-up of Brugada syndrome in patients with idiopathic ventricular fibrillation. Eur Heart J 2001;22:400-409.
108. Gussak I, Bjerregaard P, Hammill SC: Clinical diagnosis and risk stratification in patients with Brugada syndrome. J Am Coll Cardiol 2001;37:1635-1638.
109. Zipes DP, Wellens HJJ: Clinical cardiology: New frontiers. Sudden cardiac death. Circulation 1998;98:2334-2351.
110. Tan HL, Bink-Boelkens MTE, Bezzina CR, et al: A sodium-channel mutation causes isolated cardiac conduction disease. Nature 2001;409:1043-1047.
111. Kass S, MacRae C, Graber HL, et al: A gene defect that causes conduction system disease and dilated cardiomyopathy maps to chromosome 1p1-1q1. Nat Genet 1994;7:546-551.
112. Schott JJ, Alshinawi C, Kyndt F, et al: Cardiac conduction defects associate with mutations in SCN5A. Nat Genet 1999;23:20-21.
113. Deschenes I, Baroudi G, Berthet M, et al: Electrophysiological characterization of SCN5A mutations causing long QT (E1784K) and Brugada (R1512W and R1432G) syndromes. Cardiovasc Res 2000;46:55-65.
114. Schwartz PJ, Priori SG, Dumaine R, et al: A molecular link between the sudden infant death syndrome and the long QT syndrome. N Engl J Med 2000;343:262-267.
115. Brink PA, Ferreira A, Moolman JC, et al: Gene for progressive familial heart block type I maps to chromosome 19q13. Circulation 1995;91:1633-1640.
116. de Meeus A, Stephan E, Debrus S, et al: An isolated cardiac conduction disease maps to chromosome 19q. Circ Res 1995;77: 735-740.
117. Fatkin D, MacRae C, Sasaki T, et al: Missense mutations in the rod domain of the lamin A/C gene as causes of dilated cardiomyopathy and conduction system disease. N Engl J Med 1999;341: 1715-1724.
118. Brodsky GL, Muntoni F, Miocic S, et al: Lamin A/C gene mutation associated with dilated cardiomyopathy with variable skeletal muscle involvement. Circulation 2000;101:473-476.
119. Towbin JA, Ackerman MJ: Editorial. Cardiac sodium channel gene mutations and SIDS. Circulation 2001;104:1092-1093.
120. Schwartz PJ: Cardiac sympathetic innervation and the sudden infant death syndrome. A possible pathogenetic link. Am J Med 1976;60:167-172.
121. Towbin JA, Friedman RA: Prolongation of the QT interval and the sudden infant death syndrome. N Engl J Med 1998;338: 1760-1761.
122. Lucey JF: Comments on a sudden infant death article in another journal. Pediatrics 1999;103:812.
123. Martin RJ, Miller MJ, Redline S: Screening for SIDS: A neonatal perspective. Pediatrics 1999;103:812-813.
124. Guntheroth WG, Spiers PS: Prolongation of the QT interval and the sudden infant death syndrome. Pediatrics 1999;103: 813-814.
125. Hodgman JE: Prolonged QTc as a risk factor for SIDS. Pediatrics 1999;103:814-815.
126. Hoffman JIE, Lister G: The implications of a relationship between prolonged QT interval and the sudden infant death syndrome. Pediatrics 1999;103:815-817.
127. Shannon DC: Method of analyzing QT interval can't support conclusions. Pediatrics 1999;103:819.
128. Southall DP: Examine data in Schwartz article with extreme care. Pediatrics 1999;103:819-820.
129. Ackerman MJ, Siu B, Sturner WQ, et al: Postmortem molecular analysis of SCN5A defects in sudden infant death syndrome. JAMA 2001; 2264-2269.
130. Wedekind H, Smits JPP, Schulze-Bahr E, et al: De novo mutation in the SCN5A gene associated with early onset of sudden infant death. Circulation 2001;104:1158-1164.
131. Thiene G, Nava A, Corrado D, et al: Right ventricular cardiomyopathy and sudden death in young people. N Engl J Med 1988; 318:129-133.
132. Shen WK, Edwards WD, Hammill SC: Right ventricular dysplasia: A need for precise pathological definition for interpretation of sudden death. J Am Coll Cardiol 1994;23:34.
133. Rampazzo A, Nava A, Erne P, et al: A new locus for arrhythmogenic right ventricular cardiomyopathy (ARVD2) maps to chromosome 1q42-q43. Hum Mol Genet 1995;4:2151-2154.
134. Rampazzo A, Nava A, Danieli GA, et al: The gene for arrhythmogenic right ventricular cardiomyopathy maps to chromosome 14q23-q24. Hum Mol Genet 1994;3:959-962.
135. Severini GM, Krajinovic M, Pinamonti B, et al: A new locus for arrhythmogenic right ventricular dysplasia on the long arm of chromosome 14. Genomics 1996;31:193-200.

136. Coonar AS, Protonotarios N, Tsatsopoulou A, et al: Gene for arrhythmogenic right ventricular cardiomyopathy with diffuse nonepidermolytic palmoplantar keratoderma and woolly hair (Naxos disease) maps to 17q21. Circulation 1998;97:2049-2058.
137. Ahmad F, Li D, Karibe A, et al: Localization of a gene responsible for arrhythmogenic right ventricular dysplasia to chromosome 3p23. Circulation 1998;98:2791-2795.
138. Li D, Ahmad F, Gardner MJ, et al: The locus of a novel gene responsible for arrhythmogenic right ventricular dysplasia characterized by early onset and high penetrance maps to chromosome 10p12-p14. Am J Hum Genet 2000;66:148-156.
139. Corrado D, Nava A, Buja G, et al: Familial cardiomyopathy underlies syndrome of right bundle branch block, ST segment elevation and sudden death. J Am Coll Cardiol 1996;27:443-448.
140. McKenna WJ, Thiere G, Nava AA, et al: Diagnosis of arrhythmogenic right ventricular dysplasia/cardio-myopathy. Br Heart J 1994;71:215-218.
141. Tiso N, Stephan DA, Nava A, et al: Identification of mutations in cardiac ryanodine gene in families affected with arrhythmogenic right ventricular cardiomyopathy type 2 (ARVD2). Hum Mol Genet 2001;10:189-194.
142. Stokes DL, Wagenknecht T: Calcium transport across the sarcoplasmic reticulum—structure and function of $Ca2^{+}$-ATPase and the ryanodine receptor. Eur J Biochem 2000;267:5274-5279.
143. Missiaen L, Robberecht W, Van Den Bosch L, et al: Abnormal intracellular $Ca2^{+}$ homeostasis and disease. Cell Calcium 2000;28:1-21.
144. McKoy G, Protonotarios N, Crosby A, et al: Identification of a deletion in plakoglobin in arrhythmogenic right ventricular cardiomyopathy with palmoplantar keratoderma and woolly hair (Naxos disease). Lancet 2000;355:2119-2124.
145. Norgett EE, Hatsell SJ, Carvajal-Huerta L, et al: Recessive mutation in desmoplakin disrupts desmoplakin-intermediate filament interactions and causes dilated cardiomyopathy, woolly hair and keratoderma. Hum Mol Genet 2000;9:2761-2766.
146. Naccarella F: Malignant ventricular arrhythmias in patients with a right bundle-branch block and persistent ST segment elevation in V1-V3: A probable arrhythmogenic cardiomyopathy of the right ventricle [editorial comment]. G Ital Cardiol 1993; 23:1219-1222.
147. Fontaine G: Familial cardiomyopathy associated with right bundle branch block, ST segment elevation and sudden death [letter]. JACC 1996;28:540.
148. Scheinman MM: Is Brugada syndrome a distinct clinical entity? J Cardiovasc Electrophysiol 1997;8:332-336.
149. Ohe T: Idiopathic ventricular fibrillation of the Brugada type—an atypical form of arrhythmogenic right ventricular cardiomyopathy [editorial]. Intern Med 1996;35:595.
150. Fontaine G, Piot O, Sohal P, et al: Right precordial leads and sudden death. Relation with arrhythmogenic right ventricular dysplasia. Arch Mal Coeur Vaiss 1996;89:1323-1329.
151. Coumel P, Fidelle J, Lucet V, et al: Catecholaminergic-induced severe ventricular arrhythmias with Adams-Stokes syndrome in children: Report of four cases. Br Heart J 1978;40(suppl):28-37.
152. Leenhardt A, Lucet V, Denjoy I, et al: Catecholaminergic polymorphic ventricular tachycardia in children: A 7-year follow-up of 21 patients. Circulation 1995;91:1512-1519.
153. Laitinen PJ, Brown KM, Piippo K, et al: Mutations of the cardiac ryanodine receptor (RyR2) gene in familial polymorphic ventricular tachycardia. Circulation 2001;103:485-490.
154. Priori SG, Napolitano C, Tiso N, et al: Mutations in the cardiac ryanodine receptor gene (hRyR2) underlie catecholaminergic polymorphic ventricular tachycardia. Circulation 2001;103: 196-200.
155. Swan H, Piippo K, Viitasalo M, et al: Arrhythmic disorder mapped to chromosome 1q42-q43 causes malignant polymorphic ventricular tachycardia in structurally normal hearts. J Am Coll Cardiol 1999;34:2035-2042.
156. Bauce B, Nava A, Rampazzo A, et al: Familial effort polymorphic ventricular arrhythmias in arrhythmogenic right ventricular cardiomyopathy map to chromosome 1q42-43. Am J Cardiol 2000;85:573-579.
157. McCarthy TV, Quane KA, Lynch PJ: Ryanodine receptor mutations in malignant hyperthermia and central core disease. Hum Mutat 2000;15:410-417.
158. Wan Q, Wu N, Fan W, et al: Clinical manifestations and prevalence of different types of supraventricular tachycardia among Chinese. Chin Med J (Engl) 1992;105:284-288.
159. Gollob MH, Bharati S, Swerdlow CD: Accessory atrioventricular node with properties of a typical accessory pathway: Anatomic-electrophysiologic correlation. J Cardiovasc Electrophysiol 2000;11:922-926.
160. Packard JM, Graettinger JS, Graybiel A: Analysis of the electrocardiograms obtained from 1000 young healthy aviators: Ten year follow-up. Circulation 1954;10:384-400.
161. Hejtmancik MR, Hermann GR: The electrocardiographic syndrome of short P-R interval and broad QRS complexes: A clinical study of 80 cases. Am Heart J 1957;54:708-721.
162. Khair GZ, Soni JS, Bamrah VS: Syncope in hypertrophic cardiomyopathy. II. Coexistence of atrioventricular block and Wolff-Parkinson-White syndrome. Am Heart J 1985;110:1083-1086.
163. Jackman WM, Wang X, Friday KJ, et al: Catheter ablation of accessory atrioventricular pathways (Wolff-Parkinson-White syndrome) by radiofrequency current. N Engl J Med 1991;324:1605-1611.
164. MacRae CA, Ghaisas N, Kass S, et al: Familial hypertrophic cardiomyopathy with Wolff-Parkinson-White syndrome maps to a locus on chromosome 7q3. J Clin Invest 1995;96:1216-1220.
165. Ichida F, Hamamichi Y, Miyawaki T, et al: Clinical features of isolated noncompaction of the ventricular myocardium: Long-term clinical course, hemodynamic properties, and genetic background. J Am Coll Cardiol 1999;34:233-240.
166. Gollob MH, Green MS, Tang A S-L, et al: Identification of a gene responsible for familial Wolff-Parkinson-White syndrome. N Engl J Med 2001;344:1823-1831.
167. Blair E, Redwood C, Ashratian H, et al: Mutations in the $\gamma 2$ subunit of AMP-activated protein kinase cause familial hypertrophic cardiomyopathy: Evidence for the central role of energy compromise in disease pathogenesis. Hum Mol Genet 2001; 10:1215-1220.
168. Lang T, Yu L, Tu Q, et al: Molecular cloning, genomic organization, and mapping of *PRKAG2*, a heart abundant γ_2 subunit of 5′-AMP-activated protein kinase, to human chromosome 7q36. Genomics 2000;70:258-263.
169. Hardie DG, Carling D: The AMP-activated protein kinase-fuel gauge of the mammalian cell? Eur J Biochem 1997;246:259-73.
170. Kemp BE, Mitchelhill KI, Stapleton D, et al: Dealing with energy demand: The AMP-activated protein kinase. Trends Biochem Sci 1999;24:22-25.
171. Winder WW, Holmes BF, Rubink DS, et al: Activation of AMP-activated protein kinase increases mitochondrial enzymes in skeletal muscle. J Appl Physiol 2000;88:2219-2226.
172. Cheung PC, Salt IP, Davies SP, et al: Characterization of AMP-activated protein kinase gamma-subunit isoforms and their role in AMP binding. Biochem J 2000;346:659-669.
173. Bowles NE, Bowles KR, Towbin JA: The "Final Common Pathway" hypothesis and inherited cardiovascular disease: The role of cytoskeletal proteins in dilated cardiomyopathy. Herz 2000;25:168-175.
174. Malhotra JD, Chen C, Rivolta I, et al: Characterization of sodium channel α- and β-subunits in rat and mouse cardiac myocytes. Circulation 2001;103:1301-1310.
175. Laitinen PJ, Brown KM, Piippo K, et al: Mutations of the cardiac ryanodine receptor (RyR2) gene in familial polymorphic ventricular tachycardia. Circulation 2001;103:485-490.
176. Marks AR, Priori S, Memmi M, et al: Involvement of the cardiac ryanodine receptor/calcium release channel in catecholaminergic polymorphic ventricular tachycardia. J Cell Physiol 2002; 190:1-6.
177. Priori SG, Napolitano C, Memmi, et al: Clinical and molecular characterization of patients with catecholaminergic polymorphic ventricular tachycardia. Circulation 2002;106:69-74.
178. Bauce B, Rampazzo A, Basso C, et al: Screening for ryanodine receptor type 2 mutations in families with effort induced polymorphic ventricular arrhythmias and sudden death: Early diagnosis of asymptomatic carriers. J Am Coll Cardiol 2002; 40:341-349.
179. Wehrens XH, Lehnart SE, Huang F, et al: FKBP12.6 deficiency and defective calcium release channel (ryanodine receptor) function linked to exercise induced sudden cardiac death. Cell 2003;113:829-840.

180. Tiso N, Stephan DA, Nava A, et al: Identifications of mutations in the cardiac ryanodine receptor gene in families affected with arrhythmogenic right ventricular cardiomyopathy type 2. Hum Mol Genet 2001;10:189-194.
181. Danieli GA, Rampazzo A: Genetics of arrhythmogenic right ventricular cardiomyopathy. Curr Opin Cardiol 2002;17:218-221.
182. Protonarios N, Tsatsopoulou A, Anastasakis A, et al: Genotype-phenotype assessment in autosomal recessive arrhythmogenic right ventricular cardiomyopathy (Naxos disease) caused by deletion in plakoglobin. J Am Coll Cardiol 2001;38:1477-1484.
183. Rampazzo A, Nava A, Malacrida S, et al: Mutation in human desmoplakin binding to plakoglobin causes a dominant form arrhythmogenic right ventricular cardiomyopathy. Am J Hum Genet 2002;71:1200-1206.
184. Alcalai R, Metzger S, Rosenheck S, et al: A recessive mutation in desmoplakin causes arrhythmogenic right ventricular dysplasia, skin disorder and wooly hair. J Am Coll Cardiol 2003;42:319-327.
185. Wang DW, Viswanathan PC, Balser JR, et al: Clinical, genetic and biophysical characterization of SCN5A mutations associated with atrioventricular block. Circulation 2002;105:341-346.
186. Arbustini E, Pilooto A, Repetto A, et al: Autosomal dominant dilated cardiomyopathy with atrioventricular block: A lamin A/C defect related disease. J Am Coll Cardiol 2002;39:981-990.
187. Gollob MH, Green MS, Tang AS, et al: Identification of a gene responsible for familial Wolff-Parkinson-White syndrome. N Engl J Med 2001;344:1823-1831.
188. Gollob MH, Green MS, Tang AS, Roberts R: PRKAG2 cardiac syndrome: Familial ventricular preexcitation, conduction system disease and cardiac hypertrophy. Curr Opin Cardiol 2002;17: 229-234.
189. Ai T, Fujiwara Y, Tsuji K, et al: Novel KCNJ2 mutation in familial periodic paralysis with ventricular dysrhythmia. Circulation 2002;105:2592-2594.
190. Andersen ED, Krasilnikoff PA, Overvad H: Intermittent muscular weakness, extrasystoles, and multiple developmental anomalies: A new syndrome? Acta Paediat Scand 1971;60:559-564.
191. Andelfinger G, Tapper AR, Welch RC, et al: KCNJ2 mutation results in Andersen syndrome with sex-specific cardiac and skeletal muscle phenotypes. Am J Hum Genet 2002;71:663-668.
192. Plaster NM, Tawil R, Tristani-Firouzi M, et al: Mutations in Kir2.1 cause the developmental and episodic electrical phenotypes of Andersen's syndrome. Cell 2001;105:511-519.
193. Tristani-Firouzi M, Jensen JL, Donaldson MR, et al: Functional and clinical characterization of KCNJ2 mutations associated with LQT7 (Andersen syndrome). J Clin Invest 2002;222: 381-388.

SECTION IV

Disease States Associated with Cardiac Arrhythmias

Chapter 28 Arrhythmias in Coronary Artery Disease

TIM BETTS and ALAN KADISH

Other than hypertensive heart disease, coronary artery disease (CAD) is the most common cause of structural heart disease in the United States. Because most patients who experience life-threatening arrhythmias have underlying structural heart disease, it is not surprising that the majority of patients presenting with sustained ventricular arrhythmias have underlying CAD. For example, a recent autopsy series that reviewed 270 consecutive patients presenting with cardiac arrest after age 20 demonstrated that 256 of them had underlying structural heart disease and of the 180 in whom there were specific pathologic findings, 65% had coronary disease.[1] Patients with coronary disease may also have less severe forms of arrhythmias including bradycardia and supraventricular and ventricular arrhythmias.

Arrhythmias Associated with Acute Ischemia and Myocardial Infarction

Acute myocardial ischemia usually results from threatened or actual coronary occlusion with a subsequent imbalance between myocardial oxygen supply and demand. The term *acute coronary syndrome* (ACS) refers to any constellation of clinical symptoms that are compatible with acute myocardial ischemia and that encompass myocardial infarction (MI) (ST-segment elevation and depression, Q wave and non-Q wave) as well as unstable angina.[2] Acute coronary syndromes may result in arrhythmias during acute coronary occlusion, reperfusion, myocardial infarct evolution, or the healing phase after infarction. Changes in heart rate, the size of the ischemic area, the presence or absence of collateral vessels, and the presence of concurrent structural heart disease all contribute to arrhythmia formation by influencing the interaction between ischemia-induced changes in ions, metabolites, ion channels, gap junctions, and cellular and tissue architecture.

Mechanisms

Much of our understanding of the cellular mechanisms of arrhythmogenesis in the setting of acute ischemia comes from experiments performed in animal models. The very nature of the condition and its associated mortality have prevented similar studies in humans.

Cellular Changes Following Coronary Artery Occlusion

Acute ischemia following coronary artery occlusion results in local tissue hypoxia, a fall in pH, and a rise in extracellular potassium. The normal cardiac resting membrane potential decreases from −80 mV to around −50 mV.[3,4] The action potential amplitude falls and there is a decrease in maximal upstroke velocity (dV/dt max).[5] Within the first 2 minutes the fall in resting membrane potential results in an increase in conduction velocity.[6] As the action potential upstroke velocity falls over the next 10 minutes, conduction velocities decrease by up to 50%.[7] Importantly, the effects of ischemia on the electrophysiological properties of myocardial cells are inhomogeneous. Action potential duration and upstroke velocity are more reduced in subepicardial cells than in subendocardial cells.[8] Within the central zone of ischemia, refractory periods are prolonged.[9] In the nearby border zone between ischemic and non-ischemic tissue the refractory period may become shortened.

Changes in the degree of cell-to-cell coupling and tissue architecture also cause slowing and then failure of electrical propagation. The extracellular compartment shrinks and extracellular resistance increases.[10] Gap junction disruption results in cellular uncoupling.

Inhomogeneities in changes in intracellular and extracellular resistance are particularly pronounced in the border zone around the ischemic area. The autonomic nervous system also plays an important part in arrhythmia development during MI.[11] Increased sympathetic stimulation and accumulation of catecholamines are now known to be arrhythmogenic.

Stages of Ventricular Arrhythmogenesis Following Coronary Artery Occlusion

There are two distinct phases of arrhythmogenesis (Table 28-1) that occur during the initial 30 minutes of ischemia.[7] The first phase, resulting in type 1a arrhythmias, occurs from 2 to 10 minutes following coronary artery occlusion (peak 5 to 6 minutes). The sustained ventricular arrhythmias that occur in this phase are due to reentry within ischemic myocardium resulting from the inhomogeneity of refractory periods in normal and ischemic tissue. Mapping studies reveal the presence of low amplitude fractionated electrograms.[6,7]

The second phase, called type 1b arrhythmias, occurs from 10 to 30 minutes following coronary artery occlusion (peak 15 to 20 minutes). The precise mechanism of type 1b ventricular arrhythmias is unclear. By this stage the inhomogeneities in subepicardial refractoriness and conduction have improved to near normal values.[7] Despite this fact, arrhythmias remain common. Because of the important role of catecholamines in arrhythmogenesis, it has been postulated that abnormal automaticity is the arrhythmia mechanism.[12] Studies in canine hearts during the first 30 minutes following coronary artery occlusion suggest that up to 60% of ventricular tachycardias (VTs) are focal in origin, arising from Purkinje fibers.[13] Generally, during these two phases no structural damage occurs, and on reperfusion, ischemic cells survive and generally recover function. However, toward the end of phase 1b, changes in internal axial resistance of cardiac tissue are first noted, indicating the onset of irreversible cellular and gap junction damage.[14]

The subacute or delayed phase occurs between 6 and 72 hours following coronary artery occlusion (peak 12 to 24 hours).[15] It coincides with the onset of cell death; reperfusion at this stage does not reduce the amount of cell damage. Although substantial myocardial cell death occurs in the infarcted region, subendocardial Purkinje fibers survive with altered electrophysiological properties predisposing to arrhythmia generation.[16] A reduced resting membrane potential and spontaneous membrane depolarizations lead to abnormal automaticity. Delayed afterdepolarizations resulting in triggered activity have also been demonstrated.[17] In addition, inhomogeneity of conduction and refractoriness at the interface of dead and still-viable myocardium may lead to reentrant arrhythmias.

Reperfusion Arrhythmias

The advent of reperfusion therapies (either with thrombolytic agents or mechanical intervention) has led to the recognition of reperfusion arrhythmias. Reperfusion arrhythmias are more common after short ischemic episodes than after long ischemic periods.[18] Although reperfusion arrhythmias in the setting of acute MI are relatively rare, they remain an important cause of sudden death. In the canine model, reperfusion arrhythmias occur in two stages. Immediately following restoration of perfusion after coronary artery occlusion, ventricular fibrillation (VF) may occur due to multiple wavelet reentry. This occurs as a result of a rapid but inhomogeneous return of action potentials to previously unexcitable cells within the ischemic zone and a shortening of refractory periods in the border zone brought about by the washout of K^+ and metabolites from the extracellular space.[19] In addition, premature depolarizations may be induced by triggered activity. Although overall electrical function can return to normal at this stage, gap junction injury may persist with a corresponding inhomogeneous delay in conduction properties.

Accelerated idioventricular rhythms are commonly seen following reperfusion in the canine model. This arrhythmia may be due to the increased adrenergic stimulation of Purkinje fibers near the ischemic region causing enhanced automaticity or triggered activity.[20] As accumulation of catecholamines is required, these arrhythmias typically occur after 20 to 30 minutes of occlusion. Compared with to the canine model, the incidence of early reperfusion arrhythmias in the human population is significantly lower. This probably reflects

TABLE 28-1 Phases of Arrhythmogenesis

	Stage 1a	Stage 1b	Reperfusion	Subacute	Chronic
Timing	2-10 minutes	10-30 minutes	up to 6-12 hours	6-72 hours	Long-term
Arrhythmia	VT = VF	VT > VF	VF > VT, AIVR	VT > VF, AIVR	VT > VF
Mechanism	Reentry, triggered activity	Abnormal automaticity	Abnormal automaticity, triggered activity, and reentry	Abnormal automaticity, triggered activity, and reentry	Reentry
Substrate	Local hypoxia, acidosis, and increased [K^+]. Increased sympathetic tone. Increase in extracellular resistance, cell-to-cell uncoupling.		Endothelial damage, catecholamine accumulation. (Electrolyte shifts with washout)	Onset of cell death	Chronic scar ± aneurysm. Acute-on-chronic ischemia

the longer occlusion times and less rapid or incomplete reperfusion typically seen in patients presenting with acute MI.

CLINICAL CHARACTERISTICS OF VENTRICULAR ARRHYTHMIAS IN ACUTE CORONARY SYNDROME

Ventricular arrhythmias are present in 64.1% of patients following infarction.[21] More than 10 PVCs/hour may be seen in 19.7% and nonsustained VT (NSVT) in 6.8% of patients. Sustained VT or VF occurs in 10.2% of admissions, with an incidence of 1.9% within the first 24 hours.[22,23] Older age, systemic hypertension, previous MI, Killip class, anterior infarct, and depressed ejection fraction are associated with a higher risk of sustained VT and VF.[22] Based on data from the TIMI Phase II trial, VT and fibrillation do not act as markers for reperfusion after thrombolytic therapy because they are associated with occlusion, not patency, of the infarct-related artery.[23] Ventricular arrhythmias are more common in patients with signs of extensive left ventricular damage. However, early mortality is increased in those patients who develop VT and fibrillation, even in the absence of congestive heart failure and hypotension. Ventricular arrhythmias may also occur in the setting of unstable angina, both during episodes of pain, and when patients are pain free.[24]

Premature Ventricular Contractions

Premature ventricular contractions (PVCs) are seen in the majority of cases of acute MI and do not appear to adversely affect short-term prognosis.[25] In the human heart, R-on-T PVCs are rarely observed, accounting for only 1.8% of PVCs during the first 24 hours of admission.[26] They occur more frequently during, rather than after, thrombolysis and at a higher rate during thrombolysis in nonreperfused than in perfused patients (based on ECG criteria). Although they rarely serve as triggers of severe ventricular tachyarrhythmias, in a small series of patients, 5/7 VF and 3/5 VT episodes started with R-on-T PVCs. [27] In the canine heart, R-on-T PVCs during the first 12 minutes of ischemia (phase 1a) accounted for only 8% of PVCs, and initiated only 4% of spontaneous episodes of VT and fibrillation.[28] In contrast, between 12 and 30 minutes (phase 1b) R-on-T PVCs accounted for 24% of PVCs and were responsible for the initiation of 34% of spontaneous episodes of VT and fibrillation. Thus, underlying electrophysiological derangements appear to be of primary importance in determining both the frequency and relative malignancy of R-on-T PVCs during acute myocardial ischemia.

Accelerated Idioventricular Rhythm

AIVR is commonly witnessed in the first 12 hours after admission for AMI. Although more common in patients with successful reperfusion therapy, it is not a specific marker, with 63% of patients with occluded arteries still demonstrating the arrhythmia.[29] The presence of AIVR does not affect prognosis.

Nonsustained Ventricular Tachycardia

The presence of NSVT identifies patients at risk of in-hospital cardiac arrest. NSVT that occurs within the first two to three hours does not carry an adverse prognosis, whereas NSVT that occurs beyond several hours after admission does, particularly in patients with prior MI (Fig. 28-1). NSVT occurring 24 hours after AMI carries a worse prognosis than NSVT occurring within the first 24 hours following AMI (see Fig. 28-1). NSVT in the setting of AMI occurs in 1% to 7%, and possibly in as many as 75% of patients (Fig. 28-2).[25] This is contrary to the commonly held belief that arrhythmias occurring within the first 48 hours post-MI do not carry an adverse long-term prognosis. NSVT in the setting of healing MI (7 to 10 days post-MI) is also associated with a poorer prognosis.

Ventricular Tachycardia, Polymorphic Ventricular Tachycardia, and Ventricular Fibrillation

The incidence of documented sustained monomorphic VT within the first 48 hours of AMI is 0.3% to 1.9%. It may indicate extensive myocardial damage and serve as an independent predictor of mortality. Polymorphic VT occurs in 0.3% to 2% of patients and may be a marker of ongoing ischemia and, therefore, can often be effectively managed by anti-ischemic interventions. It is more often seen in patients who also develop VF.[30] In a case series of 11 patients with polymorphic VT, none had sinus bradycardia, but 3 of 11 had a sinus pause preceding the onset.[31] None had prolonged Q–T interval, hypokalemia or abnormal serum magnesium or calcium. Nine of eleven had signs of recurrent ischemia immediately before arrhythmia onset.

VF occurs in up to 3% of acute infarctions, with a peak incidence within the first 4 hours.[32] In data taken

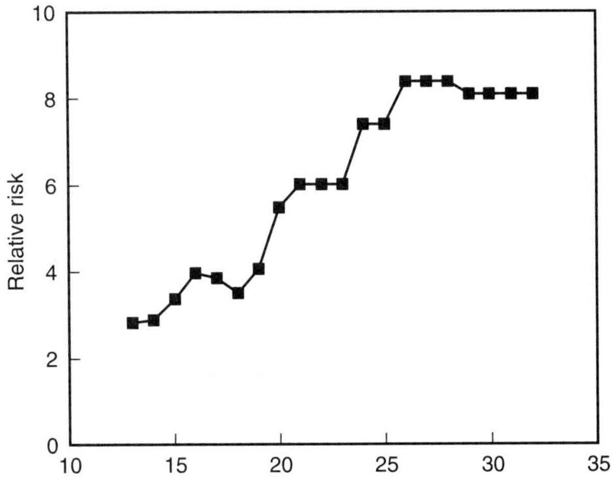

FIGURE 28-1 Plot of relative risk of nonsustained VT using various time cutoffs from presentation. All time cutoffs <13 hours were significant. (From Cheema A, Sheu K, Parker M, et al: Nonsustained ventricular tachycardia in the setting of acute myocardial infarction: Tachycardia characteristics and their prognostic implications. Circulation 1998;98(19): 2030-2036.)

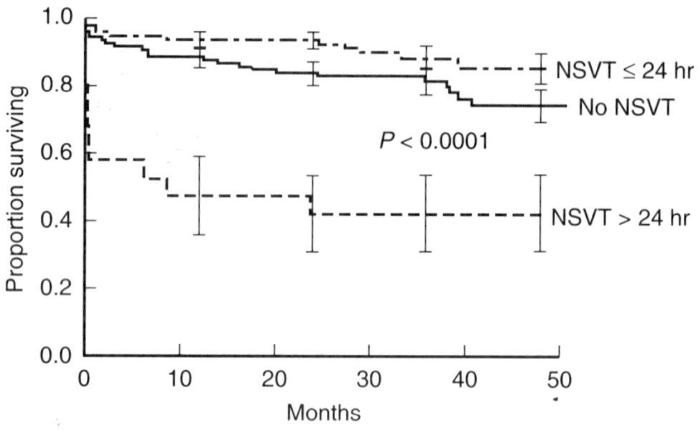

FIGURE 28-2 Kaplan-Meier survival curves for control and case patients stratified by time to occurrence of nonsustained VT less than or greater than 24 hours from presentation (from time of admission). Patients with nonsustained VT greater than 24 hours after presentation had poorer survival rates ($P<.0001$) than the other two groups. (From Cheema A, Sheu K, Parker M, et al: Nonsustained ventricular tachycardia in the setting of acute myocardial infarction: Tachycardia characteristics and their prognostic implications. Circulation 1998;98(19):2030-2036.)

from GISSI-2 trial, VF within the first 48 hours was associated with an increased in-hospital mortality.[33] VF occurring in the hyperacute phase of infarction was not associated with increased post-discharge mortality, possibly because high-risk patients die during their initial hospital stay. VF was recurrent in 11% to 15% of patients. In the first 4 hours of admission, VF was more likely to occur in the setting of hypokalemia, low blood pressure, larger infarct size, current smoking, and a younger age. VF was more common in inferoposterior infarcts, possibly due to greater autonomic upset. Initial bradycardia was also associated with early fibrillatory risk, fitting with the observation that vagal overactivity may precede VF. VF at all stages of infarct evolution is more common in patients with larger infarcts as determined by serial cardiac enzyme measurements.[34]

The temporal relationship between arrhythmia occurrence and whether it manifests as VT or fibrillation may reflect the underlying mechanisms. In the first 24 hours, VF and VT occur in almost equal frequencies; however VF is typically seen within the first 2.5 hours, and VT after the first 2.5 hours following the initiation of thrombolysis.[27] Between day 2 and 14, VF is rare, occurring in 0.5% of patients, whereas VT may be documented in 2.5%. Angiographic studies in a small number of patients suggest that VF tends to occur in patients with occluded arteries, whereas VT occurs in patients with at least a degree of reperfusion.[35] This finding is supported by a comparison of patients who received thrombolytics or placebo in the early thrombolytic trials. The likelihood of developing VF in the early hours after thrombolysis (i.e., before or at the onset of possible reperfusion) is similar in patients receiving thrombolytics or placebo.[36] However, throughout the hospital course, the risk of VF is greater in patients receiving placebo, whereas the risk of VT is higher in patients receiving thrombolysis (and, thus are more likely to have achieved at least a degree of reperfusion).

Patients with both VT and VF have worse outcomes than those with VT or VF. Among patients who survive hospitalization, the 1-year mortality rate is significantly higher in those who experienced VT alone than in those with both VT and VF.[22]

SUPRAVENTRICULAR ARRHYTHMIAS IN THE SETTING OF ACUTE ISCHEMIA

Holter monitoring reveals that 15% of patients recovering from MI have supraventicular tachycardia (ranging from single APBs to sustained atrial fibrillation) during their hospital stay.[37] The prevalence of SVT increases during the first month after MI. The increase is most pronounced in patients with residual myocardial ischemia. The advent of thrombolytic therapy and introduction of larger clinical trials, which often included an angiographic substudy, has provided much information regarding the incidence and consequences of atrial fibrillation in the acute MI setting. In a study of 152 patients, treatment with streptokinase reduced AF occurrence from 16% to 3%, although life-threatening arrhythmias occurred with equal frequency in the two groups.[38] Overall, the incidence of AF in AMI in the modern era is 7.8% to 28%.[39-44]

Mechanism of Atrial Fibrillation

The pathophysiology of AF that occurs in the course of acute MI has many components. Inflammation (pericarditis), changes in hemodynamics (atrial stretch and dilation), and atrial ischemia (site of culprit lesion and coexistent disease, and changes at a cellular level in response to ischemia) may all play roles.[45-48] Following significant ventricular damage, end-diastolic volume and pressure rise, causing an increase in atrial pressure and wall tension. This situation predisposes to atrial fibrillation, and also explains the close relationship between heart failure and atrial fibrillation in the setting of MI. There is also a relationship between the culprit coronary artery lesion and occurrence of AF that suggests that atrial ischemia promotes the development of atrial fibrillation. In an angiographic study, AF that occurred during inferior MI was more likely to occur in the setting of an occluded proximal left circumflex artery with or without right CAD if it was combined with impaired perfusion of the AV nodal artery.[49] Patients with distal left circumflex artery occlusion, or a proximal circumflex artery occlusion but good AV nodal blood supply, did not get AF. AF did not occur in patients with right coronary artery occlusions if the circumflex artery was unobstructed. In a series of 266 patients, all 12 who developed atrial arrhythmias had inferior infarction. In the vast majority, the sinus node artery was distal to the site of right coronary occlusion, suggesting that sinus node ischemia may also play a role.[50] Evidence of atrial infarction in the 12-lead ECG (manifest as PR segment displacement) may also predict the onset of AF during AMI.[51]

Risk factors include increasing age,[41,43,44,52] the presence of congestive heart failure,[39,41,43,44,52] 3-vessel coronary disease,[43] RCA occlusion,[39] female gender,[52] anterior Q-wave MI, previous MI,[41] and previous coronary artery bypass graft (CABG).[41]

Consequences of Atrial Fibrillation During Acute Myocardial Infarction

Development of atrial fibrillation results in the loss of atrial contraction. Atrial contraction is an important component of ventricular filling, particularly in failing hearts. In the ischemic canine heart, induced AF caused a reduction in cardiac output, a fall in mean aortic pressure, and a fall in mean myocardial blood flow.[53] This may precipitate a vicious downward spiral, with AF exacerbating heart failure, which, in turn, promotes atrial fibrillation. Both will increase the ischemic burden and the likelihood of ventricular arrhythmias.

In patients who suffer an acute MI, hospital mortality is significantly higher in patients with atrial fibrillation than in those without it (Fig. 28-3).[39,41,42,44,54] Atrial fibrillation occurs in patients with signs of heart failure and larger infarctions. The signs of heart failure usually appear before the onset of AF. Although it was initially believed that the degree of heart failure, not the presence of AF, accounted for the worst prognosis, it is now recognized that in patients with heart failure, the occurrence of AF causes a further decline in hemodynamics. In more recent, large scale trials the negative impact of atrial fibrillation has been shown to be independent of other variables.[43,52] However, it is possible that the increase in in-hospital mortality is restricted to those patients with new-onset AF (after admission) rather than preexisting AF (see Fig. 28-3).[43] Not only the occurrence of, but also the prognosis of, patients with paroxysmal AF is dependent on the severity of hemodynamic disturbance imposed by infarction.

Chest pain and ST segment depression are extremely common findings in patients presenting to the emergency department with AF and have limited power to predict MI. In contrast, ECG evidence of ST segment elevation or depression >2 mm appears to be a reliable discriminator of which patients are at risk for MI. Patients without significant ST segment changes are at very low risk for MI and may not require investigation for acute ischemic events or hospital admission if they are clinically stable.[55]

BRADYARRHYTHMIAS

High degree atrioventricular block is seen in a significant proportion of patients presenting with acute inferior MI. Studies from the prethrombolytic era record figures of 17% to 22% of patients with acute inferior wall infarction suffering transient or permanent advanced AV block (second degree or third degree).[56-59] Approximately 11% of patients with acute inferior wall infarction suffered complete heart block (CHB).[44] The incidence of advanced AV block has probably decreased in the thrombolytic era, in one series falling from 5.6% to 3.7% of all AMI patients.[60] In the 1990s there was still a considerable range of reported incidence, from 7.3% of inferior wall MI patients having second or third degree block, to 13% of inferior MI patients having complete heart block.[61,62]

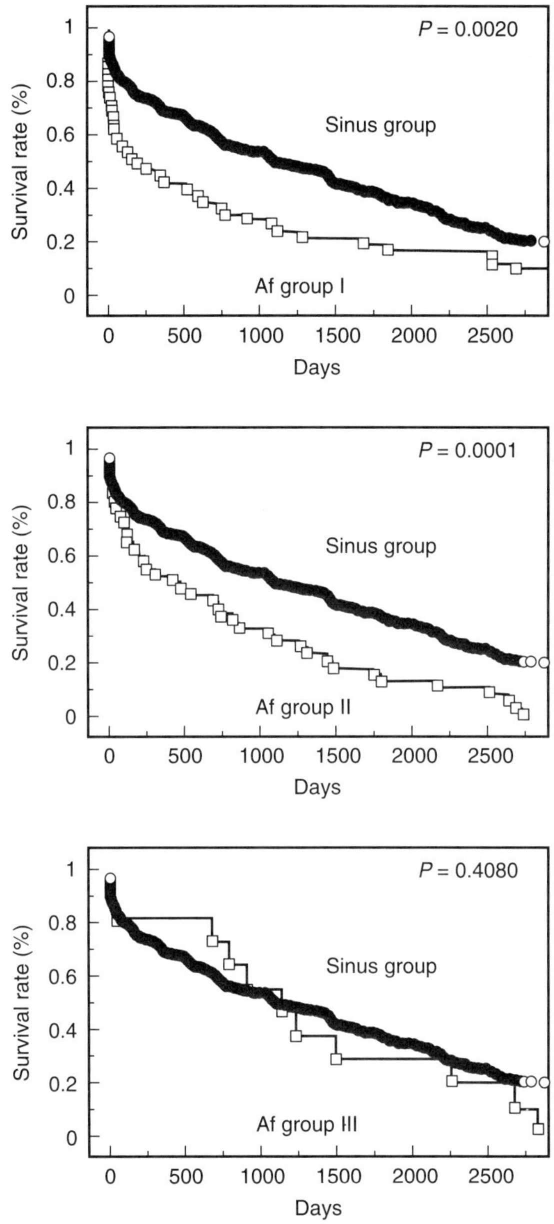

FIGURE 28-3 Kaplan-Meier analysis of cumulative survival rates for patients with myocardial infarction during 8-year follow-up. Survival rates are stratified by presence or absence of atrial fibrillation (AF). These patients were divided into AF group 1 (45 patients who developed AF within 24 hours of onset of AMI; 5 ± 2 hours, range 1 to 15), AF group 2 (41 patients who developed AF 24 hours after onset of AMI; 6 ± 4 days, range 2 to 14), and AF group 3 (14 who developed AF before the onset of AMI). The remaining 939 patients were classified as sinus patients. (From Sakata K, Kurihara H, Iwamori K, et al: Clinical and prognostic significance of atrial fibrillation in acute myocardial infarction. Am J Cardiol 1997; 80:1522-1527.)

All studies examining patients with heart block after infarction have found an association with a greater degree of myocardial damage, whether measured by cardiac enzymes, echocardiography, or nuclear scintography.[56,44,57,59,62,63] As it has long been recognized that heart block is most prevalent in patients with inferior wall MI (occurring in 7.3% of inferior versus 3.0% of anterior AMI patients)[64] the majority of studies have confined themselves to this population of patients. Within this group, right ventricular involvement also appears to be associated with the development of advanced AV block.[57] A higher incidence has also been reported in women and patients older than 70 years of age[4] and in patients with hyperkalemia.[65]

Patients with inferior MI and coexisting left anterior descending coronary artery obstruction have a sixfold chance of developing heart block in the acute phase of infarction than do patients with inferior infarction without such obstruction.[58] The site of left anterior descending artery occlusion is usually proximal to the origin of the first septal perforator. These findings suggest that the proximal AV conduction system has a dual arterial blood supply from both the right and left anterior descending coronary arteries, and may explain the transient behavior of heart block and lack of necrosis of the AV node seen in many inferior MI patients. Interestingly, following early thrombolytic therapy (<6 hours from onset of symptoms), patency rates of the infarct artery are similar in patients with and without CHB.[62] A histopathologic study of hearts with posteroinferior MI has shown a lack of correlation between AV block and infarction of the specialized conducting system. Instead, the presence of AV block during the clinical course of MI appears to be strongly related to atrial infarction in the region of the inputs to the AV node.[66]

In the setting of inferior infarction, patients with complete heart block have more episodes of VF or tachycardia, sustained hypotension, pulmonary edema, pericarditis, and atrial fibrillation than patients without heart block.[44,59,61,62] They also have a higher in-hospital mortality.[44,59,61-63] In one study, the excess mortality was limited to those patients who also had right ventricular infarction.[67] Although the increased in-hospital mortality may just be a reflection of a larger infarct, it remains an independent predictor of death. On the other hand, in those patients with inferior MI who survive to hospital discharge, the presence of heart block has no effect on long-term mortality.[59,61,62]

There are differences between those patients who develop heart block early and those who develop it late in the course of their AMI. Different studies, however, reveal conflicting data. Sclarovsky and associates report that patients who developed *early* advanced block, defined as occurring while still having hyperacute changes of AMI on the ECG, had CHB that was of short duration, unresponsive to atropine, and often required pacemaker therapy.[68] Symptoms of syncope, heart failure, and cardiogenic shock were frequently present. Patients with late block typically had second degree heart block of longer duration, a positive response to atropine, and rarely required pacemaker therapy. The mortality rate in the early group was high (23%) compared with the late group (7%). In contrast, the study by Feigl and coworkers divided patients with inferior MI who developed second or third degree block into early and late using a 6-hour cut-off time limit from admission.[69] Early patients all had transient AV block that appeared suddenly, had disappeared by 24 hours, and displayed a positive response to atropine. In the late group, heart block was often preceded by first degree block, lasted longer, had a relatively fast ventricular escape rhythm, and had little response to atropine. A third study, dividing patients by whether AV block appeared before or after 24 hours from admission, found no significant difference in hospital mortality.[70]

The mechanisms responsible for atrioventricular (AV) block during acute inferior MI would, therefore, appear to be multiple and related to the time course. Along with acute necrosis of the perinodal atrial myocardium or specialized conduction tissue, increased parasympathetic tone is a factor that is usually postulated; however, persistence of AV block after atropine administration is frequently observed. It has been demonstrated that endogenously released adenosine in oxygen-deprived myocardium can cause AV block.[71] Thus, not surprisingly, it has been reported that aminophylline may be successful in restoring sinus rhythm in atropine-resistant patients with inferior infarction.[72-74]

Arrhythmias in Chronic Coronary Artery Disease

Patients with chronic CAD may develop ventricular arrhythmias during episodes of acute myocardial ischemia due to either increases in oxygen demand or decreases in myocardial blood flow. In addition, prior MI may provide a nidus for the development of tachyarrhythmias in the setting of chronic CAD. It is often challenging for the clinician to determine the extent to which chronic infarction or acute ischemia contribute to a particular arrhythmic event. However, a determination as to whether ischemia, chronic scarring, or a combination is responsible for arrhythmia in patients with CAD can help direct therapy.[75]

Patients with chronic CAD may present with PVCs and non-sustained VT, that are either asymptomatic or associated with mild palpitations.[76] However, more serious forms of arrhythmia include nonsustained polymorphic VT, nonsustained or sustained monomorphic VT, and VF (Fig. 28-4).[77]

Arrhythmia Mechanisms

The mechanisms of ventricular arrhythmias in patients with chronic coronary disease are diverse. The contribution of ischemia to arrhythmogenesis is discussed earlier. The development of myocardial fibrosis can lead to VT through a number of mechanisms. In both experimental models and human tissue, activation in healed infarction takes a zigzag course in which fibrous septa separate bundles of muscle and in which gap junction number is decreased to produce slow conduction creating the substrate for reentry.[78-80,81] This phenomenon is associated with the recording of fractionated electrograms.[82,83] The poor coupling among surviving myocardial cells may also allow the development of unidirectional block. Whether action potential characteristics are also

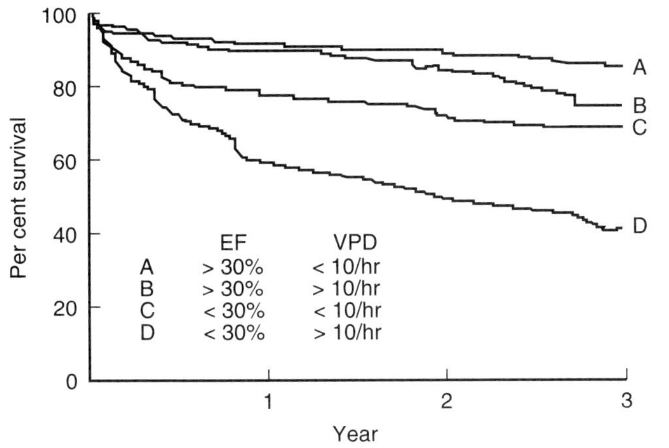

FIGURE 28-4 Influences of ejection fraction and premature ventricular contraction (PVC) frequency on mortality in patients after myocardial infarction. (From Bigger J, Fleis J, Kleiger R, et al: The relationships among ventricular arrhythmias, left ventricular dysfunction, and mortality in the two years after myocardial infarction. Circulation 1984;69:250-258.)

depressed in chronic infarction and whether remodeling in response to infarction leads to changes in the expression of ion channel proteins (other than connexins) is still controversial.

The mechanism of beats initiating ventricular tachyarrhythmias in patients with CAD remains unclear. There are experimental models in which triggered activity, abnormal automaticity, or reentry due to functional block in sinus rhythm all have occurred in healed MI.[84] One clinical study of monomorphic VT has demonstrated that the QRS configuration of the beat that initiates sustained VT was similar to that occurring in sustained tachycardia. This suggests the possibility that the initiating beat is indeed reentrant. One careful intraoperative three-dimensional mapping study of nonsustained VT has been performed. Ten patients with nonsustained VT in the setting of CAD were studied and in one half of these a reentrant circuit was identified and in the other half, a focal origin to the tachycardia appeared to be present.[85] However, because of the difficulty in mapping isolated premature beats, even with advanced mapping techniques and because of the inferential nature of the evidence required to determine the mechanism of isolated premature beats, the mechanism of isolated PVCs, short episodes of nonsustained VT, or the beats that initiate VT or VF has not been established with certainty in humans.

More is known about the mechanism of sustained VT. Mapping studies in the electrophysiology laboratory, operating room, and inferential studies using pacing techniques suggest that most cases of myocardial sustained VT in patients or experimental animals with healed MI are due to reentry.[4,82,86,87] Both fixed and functional block may contribute to sustained VT in patients with coronary disease.[88] In support of the potential importance of functional reentry is the observation that pace mapping in sinus rhythm may produce a QRS configuration that is different from that during VT even when pacing is performed at the site of origin of the tachycardia. However, one study has suggested that whether or not QRS configuration during pacing is similar to that of VT depends not on whether functional block is present but on where in a fixed reentrant circuit pacing is performed.[89] Experimental studies on the contribution fixed or functional block also yield conflicting results depending on the time after MI and on the animal model selected.[15,90] Although the overwhelming weight of evidence suggests that VT in patients with healed MI is reentrant in nature, whether fixed or functional reentry is more common will require further study.

A number of experimental studies have suggested that VF at least shortly after its origin is due to reentry.[91-94] The mechanism of initiating beats is unclear but VF in most experimental models appears to be maintained by reentry. Some epicardial mapping studies in the operating room have suggested similar results, but further study of human VF will be required to further elucidate this phenomenon.

Clinical Syndromes of Ventricular Arrhythmias in Chronic Coronary Artery Disease

PVCs and Nonsustained Ventricular Tachycardia

Brief episodes of ventricular arrhythmia (PVCs or episodes of nonsustained VT lasting 15 beats or less) in patients with chronic CAD may be important for two reasons: (1) they may indicate an adverse prognosis or (2) they may cause intolerable symptoms.

PVCs are common in patients with healed MI.[95] The prevalence of PVCs varies in different studies and is weakly related to the extent of left ventricular dysfunction. The GISSI-2 study, in which all patients received thrombolysis of some type, examined over 16,000 patients undergoing Holter monitoring a mean of 17 days after MI.[96] Over 64% of patients had some PVCs. Twenty percent had more than 10 PVCs per hour and 6.8% had nonsustained VT. In this and other studies, frequent PVCs (>10 premature beats/hour) and complex ventricular arrhythmias increase the risk of sudden cardiac death.[96-99] In the GISSI-2 study, Holter monitor recordings from 8676 patients were analyzed. By multivariate analysis, the presence of more than 10 PVCs per hour 1 week after infarction was associated with an odds ratio of 1.62 for total mortality and an odds ratio of 2.22 for sudden mortality. Interestingly, in this study, nonsustained VT (defined as 3 beats to 30 seconds of VT) was associated with an increased mortality during follow-up by univariate but not multivariate analysis. On the other hand, only a small fraction of postinfarction patients (<10%) with tachyarrhyythmias during Holter monitoring die suddenly (low positive predictive value). However, the European Infarct Study Group[100] showed that fewer than 1% of patients in whom Holter were normal, died suddenly during the first year after MI. Thus, the absence of arrhythmia in the healing phase of infarction predicts a good prognosis.

The anatomic and electrophysiological characteristics of MI evolve in the first several months after infarction.[101] Thus, findings on the prognostic significance of spontaneous arrhythmia shortly after infarction may not apply to healed infarction. Data on patients with PVCs or nonsustained VT after the subacute phase of MI are based on studies of smaller numbers of patients and are inconsistent.[76] Nonetheless, several recent studies have used nonsustained VT in association with left ventricular dysfunction and inducible sustained VT at electrophysiological testing to risk stratify patients with chronic CAD.[102,103]

The symptomatology associated with PVCs and nonsustained VT may vary dramatically from patient to patient. In some patients, every PVC noted by Holter monitor is recorded in a diary. In other cases, patients may have ten to twenty thousand PVCs in a 24-hour period without symptoms. The factors associated with sensation of premature beats have not been well studied. However, there are no data to suggested that the extent of cardiac awareness has any prognostic significance. Because of these observations, symptoms are an imperfect guide to the treatment of PVCs in nonsustained VT. In general, the primary goal of treatment of PVCs and nonsustained VT is suppression of symptoms. Unless patients are experiencing palpitations, dizziness, or heart failure (the latter two being uncommon but real symptoms of frequent PVCs) there is no indication for specific antiarrhythmic therapy for the suppression of PVCs. In patients whose symptoms are significant enough to warrant therapy, a trial of β-blockers or calcium blockers is an appropriate first approach. Therapy for PVCs is discussed in a later section.

Monomorphic Ventricular Tachycardia

The clinical presentation of monomorphic VT is also variable. Some patients, especially those with large MIs, may have stable monomorphic VT at slow rates that is hemodynamically reasonably well tolerated. This is particularly true if patients are treated with antiarrhythmic drugs. In other patients, sustained VT is associated with presyncope, syncope, or cardiac arrest.[104] A variety of factors may affect the hemodynamic tolerance of sustained VT. These include the rate of the tachycardia, atrial synchrony, and left ventricular function.[105] The site of origin of VT may also alter hemodynamic tolerance of VT but extensive data in this regard are lacking. The variability in hemodynamic consequences of sustained monomorphic VT from patient to patient and perhaps in the same patient, raises a management challenge in dealing with sustained VT.

There are conflicting data on the prognostic significance of patients with sustained hemodynamically well-tolerated VT.[106] Some studies have suggested that these patients have a better long-term prognosis than those who present with hemodynamically unstable VT.[107] However, other studies have suggested that even patients who present with hemodynamically well-tolerated tachycardia may also be at high risk for subsequent cardiac arrest. The differences and results of these studies may relate to differences in the patient population studied. However, it is most likely that while presentation with sustained hemodynamically well-tolerated VT does indicate a substantial increase in the risk of sudden cardiac death, this risk may not be as high as in patients who present with a cardiac arrest. In the recent AVID substudy, patients who presented with sustained well-tolerated VT had at least as poor a prognosis as patients presenting with cardiac arrest.[108] This subanalysis of a large, prospective study strongly suggests that patients who present with stable VT are also at risk for life threatening tachyarrhythmias.

Cardiac Arrest and Ventricular Fibrillation

Despite a decline in the incidence of cardiovascular disease, 300,000 sudden deaths still occur annually in the United States.[109] The precise arrhythmia initiating sudden death in patients with chronic CAD (and in patients with other types of structural heart disease) is not completely known. Data from the Seattle Heart Watch Project and from several Holter monitoring studies from the late 1970s and 1980s demonstrated that VT and/or VF is responsible for 40% to 50% of cardiac arrests and that Q-wave infarction is only present in 20% of these.[110,111] It has been postulated that many of the 40% of patients with cardiac arrests who present with asystole actually represent a terminal rhythm following prior VT/VF.[112,113] Holter monitoring data support this contention.[114] One study by Luu and colleagues in patients with advanced heart failure demonstrated that bradycardia and/or electromechanical dissociation may be responsible for 50% or more of cardiac arrests but it is likely that these data do not apply to the majority of patients suffering cardiac arrest.[115] Thus, most patients with chronic CAD who suffer cardiac arrest have VT or VF as the mechanism of death. Even in the absence of infarction, ischemia may be a frequent contributing factor.[116]

Holter monitoring data have suggested that VT almost always preceded the onset of VF during cardiac arrest.[114,117-120] Kempf and coworkers studied 27 patients who were found to have a cardiac arrest on Holter monitoring. Twenty-one had CAD. Of the 27 patients, 20 developed VF that was always preceded by VT. However, the VT was often polymorphic and often only 3 beats in length. Two smaller series of ten patients in total showed that eight had VF. Although monomorphic VT is often observed in patients successfully resuscitated from cardiac arrest, it appears that it is uncommon as a cause of sudden death. More recent observations have also suggested that in patients presenting with life threatening arrhythmias that VT is often polymorphic or that VF may occur *de novo*. A change in the epidemiology of MI may be a contributing factor. Before the onset of mechanical or pharmacologic reperfusion, patients experiencing acute coronary occlusions often experienced *completed infarction*. This resulted in large, dense areas of scar and the development of left ventricular aneurysm that can form the substrate for slow, sustained monomorphic VT. When patients presenting with monomorphic VT are compared with those presenting with VF, they more often have lower ejection fractions and ventricular aneurysms. The development of reperfusion therapy has led to a change in the pathophysiologic substrate of ventricular tachyarrhythmias in patients with CAD. Patients now more often have mottled infarctions without the development of ventricular aneurysm. Thus, polymorphic VT or VF opposed to monomorphic VT may be even more common as initiating arrhythmia in cardiac arrest than a decade or two ago.

Polymorphic Ventricular Tachycardia

Polymorphic VT is defined as VT in which QRS configuration varies from beat to beat but in which a clearly defined QRS complex (as opposed to ventricular flutter or fibrillation) can be detected. Polymorphic VT is often associated with a congenital or acquired long QT syndrome, which is rarely caused by CAD.[121] However, isolated case reports have described long QT syndrome and associated polymorphic VT in patients with coronary disease. Acute myocardial ischemia can also classically cause polymorphic VT. However, some patients with healed MI may present with nonsustained polymorphic VT or episodes of polymorphic VT degenerating to VF.[122] There are no clear data to suggest that an early description of the morphology of the ventricular arrhythmias alters prognosis. Dr. Lown has developed a classification of ventricular arrhythmias that suggests that polymorphic VT may be a higher grade than uniform premature beats or uniform nonsustained VT and portend a worse prognosis.[123] However, there are few data that suggest that the morphology of episodes of nonsustained VT alters prognosis or symptoms in patients with coronary disease and healed MI. Clinical symptoms, indications for treatment (or for withholding treatment), and the prognostic value of nonsustained episodes of polymorphic VT are likely the same as those of monomorphic VT in patients with healed CAD.

EXERCISE-INDUCED ARRHYTHMIAS

Exercise induced arrhythmias represent a potentially life-threatening problem in patients with CAD. While physical training in general decreases total mortality from heart disease, the relative risk of sudden death during exercise is increased.[124] The increases in myocardial oxygen demand and sympathetic stimulation that occur during exercise can provoke arrhythmias even if exercise in general decreases the extent of CAD. Less severe forms of arrhythmia may also occur with exercise. Data on the prognostic significance of PVCs or nonsustained VT during exercise are controversial.[125] Elhendy studied 171 patients who underwent stress testing. Three patients had nonsustained VT and 43 patients had complex ventricular arrhythmias. The presence of myocardial ischemia in dyskinetic regions was associated with the occurrence of arrhythmia. However, this finding has not been reproduced in all studies. The weight of evidence suggests that PVCs and nonsustained VT that occur during exercise may be multifactorial. In some cases, these arrhythmias may be due to myocardial ischemia, whereas others may be due to the effects of scarring and CAD. The extent to which ventricular arrhythmias indicate myocardial ischemia in patients with CAD or to the degree to which the presence of these arrhythmias affects prognosis is not completely clear at the present time.

Treatment

Although the treatment of serious ventricular arrhythmias is discussed elsewhere in this text, a few principles regarding the treatment (or lack thereof) of less serious arrhythmias in patients with chronic CAD may be useful.

There are two potential indications for treatment of patients with premature beats and nonsustained ventricular tachycardia: (1) improvement of symptoms and (2) prolongation of life. The overwhelming majority of patients with isolated premature beats or nonsustained VT are asymptomatic or have mild symptoms that are not clinically or hemodynamically significant. Holter monitoring studies showing that up to 10% of elderly patients even in the absence of structural heart disease may have ventricular premature beats and that ventricular premature beats are extremely common after MI, confirm that most patients with ventricular ectopy do not need treatment. However, a subset of patients with PVCs or nonsustained VT have highly symptomatic palpitations or impairment in left ventricular function due to extremely frequent ventricular ecotopy.[126,127] In these patients, suppression of ventricular premature beats to control symptoms may be appropriate. Although the therapy of arrhythmias is discussed in detail in a later chapter, β-blockers should be the therapy of first choice for the suppression of symptoms related to PVCs or nonsustained VT in patients with CAD. Studies cited above suggested that spontaneous ventricular arrhythmias may represent an independent risk factor for the prediction of sudden death in patients who have coronary disease and healed MI. Based on these observations, a hypothesis was developed that suppression PVCs would result in a decrease in mortality. The cardiac arrhythmia suppression trial (CAST) was designed to test this hypothesis.[128] Unfortunately, this study showed that encainide and flecainide increased mortality and that other drugs had, at best, a neutral effect. A meta-analysis of antiarrhythmic drug therapy in patients with coronary disease revealed that all drugs available in 1993, except for amiodarone, increased mortality in patients with CAD.[129,130] At that time, only preliminary data were available using amiodarone therapy. Since then, several studies have been performed examining the ability of amiodarone to prolong life in patients with CAD. Two large studies were performed after MI: the EMIAT and CAMIAT studies.[131,132] Although both showed some trends, especially when secondary endpoints were examined, neither study showed that amiodarone improved survival when given to patients after MI with left ventricular dysfunction and/or spontaneous ectopy. Three recent meta-analyses have been performed on the use of amiodarone therapy for the prevention of sudden death. They show that amiodarone appears to decrease mortality by approximately 10% when administered prophylactically.[133]

Dofetilide, the potassium channel blocking drug has also been studied extensively in patients with prior MI and heart failure. Although prognosis is not improved, dofetilide does not cause increased mortality in patients with CAD.[134] However, dofetilide may have adverse effects in patients with baseline prolonged Q–T intervals.[135]

The available data do not support the routine use of antiarrhythmic drugs for the prevention of sudden death in patients with coronary disease and spontaneous arrhythmias. However, if drug therapy is required to suppress symptoms in patients with CAD, β-blockers, dofetilide, or amiodarone are all drugs that have been shown to have the beneficial or neutral effect on survival.

Atrial Fibrillation and Coronary Artery Disease

Coronary artery disease is commonly cited as one of the principle causes of atrial fibrillation. Although the role of acute ischemia in the development of atrial fibrillation is undisputed (see earlier), the role of chronic CAD is much more controversial. A review of the Coronary Artery Surgery Study (CASS) data found only a 0.6% prevalence of sustained AF in over 18,000 patients undergoing CABG.[136] The presence of AF was positively associated with older age, male gender, mitral regurgitation, and functional impairment due to congestive heart failure. The number of diseased coronary arteries was inversely related to the presence of AF. These findings are consistent with data from the Framingham Heart Study, which also reported that older age, mitral valve disease, and congestive heart failure are the predominant risk factors for the development of atrial fibrillation. Importantly, when these factors are taken into account, the presence of CAD has no significant impact.[137] In a randomly selected group of 9000 Swedish adults between the ages of 32 and 64 years, 0.28% were found to have AF. Although the presence of heart failure was higher in the AF population, the incidence of CAD was equal in the AF group and in an age- and sex-matched control group who were in sinus rhythm.[138] Thus, it would appear that CHF, which is one of the potential consequences of CAD, predisposes to atrial fibrillation, but CAD itself does not.

Advanced Atrioventricular Block and Coronary Artery Disease

Chronic CAD is one of the causes of AV block, although it is a much less common cause than idiopathic fibrosis of the conduction system. AV block in coronary disease is usually related to extensive infarction and necrosis of the distal conduction system rather than ischemia. In 30 patients aged 45 to 65 years with complete heart block, who were referred for pacing and without symptoms of coronary disease, coronary angiography disclosed the presence of severe CAD in 13 patients (43%). Myocardial revascularization was undertaken in six patients but did not result in any sustained improvement in atrioventricular conduction.[139] In contrast, there are many case reports in the literature of patients with paroxysmal or exercise-induced heart block that have complete resolution of their symptoms after angioplasty of a lesion in the right coronary artery, suggesting that ischemia of the AV node may have a role.[140-143] Most patients with AV block and coronary disease require pacemaker therapy.

REFERENCES

1. Chugh SS, Kelly KL, Titus JL: Sudden cardiac death with apparently normal heart. Circulation 2000;102:649-654.
2. Braunwald E, Antman EM, Beasley JW, et al: ACC/AHA guidelines for the management of patients with unstable angina and non-ST-segment elevation myocardial infarction. A report of the American College of Cardiology/American Heart Association Task Force on Practice Guidelines (Committee on the Management of Patients with Unstable Angina). J Am Coll Cardiol 2000;36:970-1062.
3. Kleber A: Resting membrane potential, extracellular potassium activity and intracellular sodium activity during acute global ischemia in isolated perfused guinea pig hearts. Circulation Research 1983;52:442-450.
4. Downar E, Janse MJ, Durrer D: The effect of acute coronary artery occlusion on subepicardial transmembrane potentials in the intact porcine heart. Circulation 1977;56:217-224.
5. Kleber A, Janse MJ, van Capelle FJ: Mechanism and time course of ST and TQ segment changes during acute regional myocardial ischemia in the pig heart determined by extracellular and intracellular recordings. Circ Res 1978;42:603-613.
6. Janse MJ, Wit AL: Electrophysiological mechanism of ventricular arrhythmias resulting from myocardial ischemia and infarction. Physiol Rev 1989;69:1049-1169.
7. Kaplinsky E, Ogawa S, Balke CW, et al: Two periods of early ventricular arrhythmia in the canine acute myocardial infarction model. Circulation 1979;60:397-403.
8. Gilmour RF, Zipes DP: Different electrophysiologic responses of canine myocardium and epicardium to combined hyperkalemia, hypoxia, and acidosis. Circ Res 1980;1980:814.
9. Lazzara R, el Sherif N, Hope RR: Ventricular arrhythmias and electrophysiological consequences of myocardial ischemia and infarction. Circ Res 1978;42:740-749.
10. Kleber A, Riegger CB, Janse MJ: Electrical uncoupling and increase in extracellular resistance after induction of ischemia in isolated, arterially perfused rabbit papillary muscle. Circ Res 1987;61:271-279.
11. Tsien R: Adrenaline-like effects on intracellular iontophoresis of cyclic AMP in cardiac Purkinje fibres. Nature-New Biol 1973;245:120-122.
12. Penny WJ: The deleterious effects of myocardial catecholamines on cellular electrophysiology and arrhythmias during ischemia and reperfusion. Eur Heart J 1984;5:960-973.
13. Arnar DO, Bullinga JR, Martins JB: Role of the Purkinje system in spontaneous ventricular tachycardia during acute ischemia in a canine model. Circulation 1997;96:2421-2429.
14. Spear JF, Kleval RS, Moore EN: The role of myocardial anisotropy in arrhythmogenesis associated with myocardial ischemia and infarction. J Cardiovasc Electrophysiol 1992;3:579-588.
15. Wit AL, Janse MJ: Experimental models of ventricular tachycardia and fibrillation caused by ischemia and infarction. Circulation 1992;85:I32-I42.
16. Friedman PL, Stewart JR, Fenoglio J: Survival of subendocardial Purkinje fibers after extensive myocardial infarction in dogs. Circ Res 1973;33:597-611.
17. el Sherif N, Gough WB, Zeiler RH: Triggered ventricular rhythms in 1-day old myocardial infarction in the dog. Circ Res 1983;52:566-579.
18. Birnbaum Y, Leor J, Kloner RA: Pathobiology and clinical impact of reperfusion injury. J Thromb Thrombol 1997;4:185-195.
19. Ehlert FA, Goldberger JJ: Cellular and pathophysiological mechanisms of ventricular arrhythmias in acute ischemia and infarction. Pacing Clin Electrophysiol 1997;20:966-975.
20. Kaplinsky E, Ogawa S, Michelson EL: Instantaneous and delayed ventricular arrhythmias after reperfusion of acutely ischemic myocardium: Evidence for multiple mechanisms. Circulation 1981;63:333-340.
21. Maggioni AP, Zuanetti G, Franzosi MG, et al: Prevalence and prognostic significance of ventricular arrhythmias after acute myocardial infarction in the fibrinolytic era. GISSI-2 results. Circulation 1993;87:312-322.
22. Newby KH, Thompson T, Stebbins A, et al: Sustained ventricular arrhythmias in patients receiving thrombolytic therapy: Incidence and outcomes. The GUSTO Investigators. Circulation 1998;98:2567-2573.

23. Berger PB, Ruocco NA, Ryan TJ, et al: Incidence and significance of ventricular tachycardia and fibrillation in the absence of hypotension or heart failure in acute myocardial infarction treated with recombinant tissue-type plasminogen activator: Results from the Thrombolysis in Myocardial Infarction (TIMI) Phase II trial. J Am Coll Cardiol 1993;22:1773-1779.
24. Lopes MG, Harrison DC, Schroeder JS: Ventricular arrhythmias during unstable angina pectoris. Arch Int Med 1975;135: 1548-1553.
25. Cheema A, Sheu K, Parker M, et al: Nonsustained ventricular tachycardia in the setting of acute myocardial infarction: Tachycardia characteristics and their prognostic implications. Circulation 1998;98(19):2030-2036.
26. Chiladakis JA, Karapanos G, Davlouros P, et al: Significance of R-on-T phenomenon in early ventricular tachyarrhythmia susceptibility after acute myocardial infarction in the thrombolytic era. Am J Cardiol 2000;85:289-293.
27. Heidbuchel H, Tack J, Vanneste L, et al: Significance of arrhythmias during the first 24 hours of acute myocardial infarction treated with alteplase and effect of early administration of a beta-blocker or a bradycardiac agent on their incidence. Circulation 1994;89:1051-1059.
28. Naito M, Michelson EL, Kaplinsky E, et al: Role of early cycle ventricular extrasystoles in initiation of ventricular tachycardia and fibrillation: Evaluation of the R on T phenomenon during acute ischemia in a canine model. Am J Cardiol 1982;49:317-322.
29. Miller FC, Krucoff MW, Satler LF, et al: Ventricular arrhythmias during reperfusion. Am Heart J 1986;112:928-932.
30. Bluzhas J, Lukshiene D, Shlapikiene B, et al: Relation between ventricular arrhythmia and sudden cardiac death in patients with acute myocardial infarction: The predictors of ventricular fibrillation. J Am Coll Cardiol 1986;8:69A-72A.
31. Wolfe CL, Nibley C, Bhandari A, et al: Polymorphous ventricular tachycardia associated with acute myocardial infarction. Circulation 1991;84:1543-1551.
32. Volpi A, Cavalli A, Turato R, et al: Incidence and short-term prognosis of late sustained ventricular tachycardia after myocardial infarction: Results of the Gruppo Italiano per lo Studio della Sopravvivenza nell'Infarto Miocardico (GISSI-3) Database. Am Heart J 2001;142:87-92.
33. Volpi A, Cavalli A, Santoro L, et al: Incidence and prognosis of early primary ventricular fibrillation in acute myocardial infarction—results of the Gruppo Italiano per lo Studio della Sopravvivenza nell'Infarto Miocardico (GISSI-2) database. Am J Cardiol 1998;82:265-271.
34. Herlitz J, Hjalmarson A, Swedberg K, et al: Relationship between infarct size and incidence of severe ventricular arrhythmias in a double-blind trial with metoprolol in acute myocardial infarction. Int J Cardiol 1984;6:47-60.
35. Chen S, Tai C, Yu W, et al: Right atrial focal atrial fibrillation: electrophysiologic characteristics and radiofrequency catheter ablation. J Cardiovasc Electrophysiol 1999;10:328-335.
36. Solomon SD, Ridker PM, Antman EM: Ventricular arrhythmias in trials of thrombolytic therapy for acute myocardial infarction. A meta-analysis. Circulation 1993;88:2575-2581.
37. Jespersen CM, Vaage-Nilsen M, Hansen JF: The significance of myocardial ischaemia and verapamil treatment on the prevalence of supraventricular tachyarrhythmias in patients recovering from acute myocardial infarction. Danish Study Group on Verapamil in Myocardial Infarction. Eur Heart J 1992;13:1427-1430.
38. Nielsen FE, Sorensen HT, Christensen JH, et al: Reduced occurrence of atrial fibrillation in acute myocardial infarction treated with streptokinase. Eur Heart J 1991;12:1081-1083.
39. Sakata K, Kurihara H, Iwamori K, et al: Clinical and prognostic significance of atrial fibrillation in acute myocardial infarction. Am J Cardiol 1997;80:1522-1527.
40. Pizzetti F, Turazza F, Franzosi M, et al: Incidence and prognostic significance of atrial fibrillation in acute myocardial infarction: the GISSI-3 data. Heart 2001;86:527-532.
41. Rathore SS, Berger AK, Weinfurt KP, et al: Acute myocardial infarction complicated by atrial fibrillation in the elderly: Prevalence and outcomes. Circulation 2000;101:969-974.
42. Pedersen O, Bagger H, Kober L, et al: Trandolapril reduces the incidence of atrial fibrillation after acute myocardial infarction in patients with left ventricular dysfunction. Circulation 1999; 100:376-380.
43. Crenshaw BS, Ward SR, Granger CB, et al: Atrial fibrillation in the setting of acute myocardial infarction: The GUSTO-I experience. Global Utilization of Streptokinase and TPA for Occluded Coronary Arteries. J Am Coll Cardiol 1997;30:406-413.
44. Behar S, Zahavi Z, Goldbourt U, et al: Long-term prognosis of patients with paroxysmal atrial fibrillation complicating acute myocardial infarction. SPRINT Study Group. Eur Heart J 1992; 13:45-50.
45. Nagahama Y, Sugiura T, Takehana K, et al: The role of infarction-associated pericarditis on the occurrence of atrial fibrillation. [see comments]. Eur Heart J 1998;19:287-292.
46. Widimsky P, Gregor P: Recent atrial fibrillation in acute myocardial infarction: A sign of pericarditis 1993;35:230-232.
47. Sugiura T, Iwasaka T, Takahashi N, et al: Factors associated with atrial fibrillation in Q wave anterior myocardial infarction. Am Heart J 1991;121:1409-1412.
48. Solti F, Kekesi V, Juhasz-Nagy A: The effect of atrial dilatation on reperfusion arrhythmias: Development of supraventricular tachycardias on reperfusion with atrial stretching. Acta Medica Hungarica 1992;49:159-170.
49. Hod H, Lew AS, Keltai M, et al: Early atrial fibrillation during evolving myocardial infarction: A consequence of impaired left atrial perfusion. Circulation 1987;75:146-150.
50. Kyriakidis M, Barbetseas J, Antonopoulos A, et al: Early atrial arrhythmias in acute myocardial infarction: Role of the sinus node artery. Chest 1992;101:944-947.
51. Nielsen FE, Andersen HH, Gram-Hansen P, et al: The relationship between ECG signs of atrial infarction and the development of supraventricular arrhythmias in patients with acute myocardial infarction. Am Heart J 1992;123:69-72.
52. Pedersen OD, Bagger H, Kober L, et al: The occurrence and prognostic significance of atrial fibrillation/flutter following acute myocardial infarction. TRACE Study group. TRAndolapril Cardiac Evaluation. Eur Heart J 1999;20:748-754.
53. Friedman HS, Kottmeier S, Melnicker L, et al: Effects of atrial fibrillation on myocardial blood flow in the ischemic heart of the dog. J Am Coll Cardiol 1984;4:729-734.
54. Wong CK, White HD, Wilcox RG, et al: New atrial fibrillation after acute myocardial infarction independently predicts death: The GUSTO-III experience. Am Heart J 2000;140:878-885.
55. Zimetbaum PJ, Josephson ME, McDonald MJ, et al: Incidence and predictors of myocardial infarction among patients with atrial fibrillation. J Am Coll Cardiol 2000;36:1223-1227.
56. Tans AC, Lie KI, Durrer D: Clinical setting and prognostic significance of high degree atrioventricular block in acute inferior myocardial infarction: A study of 144 patients. Am Heart J 1980; 99:4-8.
57. Strasberg B, Pinchas A, Arditti A, et al: Left and right ventricular function in inferior acute myocardial infarction and significance of advanced atrioventricular block. Am J Cardiol 1984;54:985-987.
58. Bassan R, Maia IG, Bozza A, et al: Atrioventricular block in acute inferior wall myocardial infarction: Harbinger of associated obstruction of the left anterior descending coronary artery. J Am Coll Cardiol 1986;8:773-778.
59. Dubois C, Pierard LA, Smeets JP, et al: Long-term prognostic significance of atrioventricular block in inferior acute myocardial infarction. Eur Heart J 1989;10:816-820.
60. Harpaz D, Behar S, Gottlieb S, et al: Complete atrioventricular block complicating acute myocardial infarction in the thrombolytic era. SPRINT Study Group and the Israeli Thrombolytic Survey Group. Secondary Prevention Reinfarction Israeli Nifedipine Trial. J Am Coll Cardiol 1999;34:1721-1728.
61. Rathore SS, Gersh BJ, Berger PB, et al: Acute myocardial infarction complicated by heart block in the elderly: Prevalence and outcomes. Am Heart J 2001;141:47-54.
62. Clemmensen P, Bates ER, Califf RM, et al: Complete atrioventricular block complicating inferior wall acute myocardial infarction treated with reperfusion therapy. TAMI Study Group. Am J Cardiol 1991;67:225-230.
63. Kaul U, Hari Haran V, Malhotra A, et al: Significance of advanced atrioventricular block in acute inferior myocardial infarction—a study based on ventricular function and Holter monitoring. Int J Cardiol 1986;11:187-193.
64. Rathore SS, Gersh BJ, Berger PB, et al: Acute myocardial infarction complicated by heart block in the elderly: Prevalence and outcomes. Am Heart J 2001;141:47-54.

65. Sugiura T, Iwasaka T, Takahashi N, et al: Factors associated with late onset of advanced atrioventricular block in acute Q wave inferior infarction. Am Heart J 1990;119:1008-1013.
66. Bilbao FJ, Zabalza IE, Vilanova JR, et al: Atrioventricular block in posterior acute myocardial infarction: A clinicopathologic correlation. Circulation 1987;75:733-736.
67. Mavric Z, Zaputovic L, Matana A, et al: Prognostic significance of complete atrioventricular block in patients with acute inferior myocardial infarction with and without right ventricular involvement. Am Heart J 1990;119:823-828.
68. Sclarovsky S, Strasberg B, Hirshberg A, et al: Advanced early and late atrioventricular block in acute inferior wall myocardial infarction. Am Heart J 1984;108:19-24.
69. Feigl D, Ashkenazy J, Kishon Y: Early and late atrioventricular block in acute inferior myocardial infarction. J Am Coll Cardiol 1984;4:35-38.
70. Altun A, Ozkan B, Gurcagan A, et al: Early and late advanced atrioventricular block in acute inferior myocardial infarction. Coron Artery Dis 1998;9:1-4.
71. Belardinelli L, Mattos EC, Berne RM: Evidence for adenosine mediation of atrioventricular block in the ischemic canine myocardium. J Clin Invest 1981;68:195-205.
72. Altun A, Kirdar C, Ozbay G: Effect of aminophylline in patients with atropine-resistant late advanced atrioventricular block during acute inferior myocardial infarction. Clin Cardiol 1998;21:759-762.
73. Goodfellow J, Walker PR: Reversal of atropine-resistant atrioventricular block with intravenous aminophylline in the early phase of inferior wall acute myocardial infarction following treatment with streptokinase. Eur Heart J 1995;16:862-865.
74. Bertolet BD, McMurtrie EB, Hill JA, et al: Theophylline for the treatment of atrioventricular block after myocardial infarction. [see comments]. Ann Int Med 1995;123:509-511.
75. Mehta D, Curwin J, Gomes JA, et al: Sudden death in coronary artery disease: Acute ischemia versus myocardial substrate. Circulation 1997;96:3215-3223.
76. Denes P, Gillis AM, Pawitan Y, et al: Prevalence, characteristics and significance of ventricular premature complexes and ventricular tachycardia detected by 24-hour continuous electrocardiographic recording in the Cardiac Arrhythmia Suppression Trial. CAST Investigators. Am J Cardiol 1991;68:887-896.
77. Kannel W, Abbott R, Savage D, et al: Epidemiologic features of chronic atrial fibrillation: The Framingham study. N Engl J Med 1982;306:1018-1022.
78. Gardner PI, Ursell PC, Fenoglio JJ Jr, et al: Electrophysiologic and anatomic basis for fractionated electrograms recorded from healed myocardial infarcts. Circulation 1985;72:596-611.
79. de Bakker JM, van Capelle FJ, Janse MJ, et al: Slow conduction in the infarcted human heart: 'Zigzag' course of activation. Circulation 1993;88:915-926.
80. Kawara T, Derksen R, de Groot JR, et al: Activation delay after premature stimulation in chronically diseased human myocardium relates to the architecture of interstitial fibrosis. Circulation 2001;104:3069-3075.
81. Kanno S, Saffitz JE: The role of myocardial gap junctions in electrical conduction and arrhythmogenesis. Cardiovasc Pathol 2001;10:169-177.
82. Gardener P, Ursell P, Fenoglio JJ, et al: Electrophysiologic and anatomic basis for fractionated electrograms recorded from healed myocardial infarcts. Circulation 1985;72:596-611.
83. Ursell PC, Gardner PI, Albala A, et al: Structural and electrophysiological changes in the epicardial border zone of canine myocardial infarcts during infarct healing. Circ Res 1985;56:436-451.
84. Janse MJ, Wit AL: Electrophysiological mechanisms of ventricular arrhythmias resulting from myocardial ischemia and infarction. Physiol Rev 1989;69:1049-1169.
85. Pogwizd SM: Focal mechanisms underlying ventricular tachycardia during prolonged ischemic cardiomyopathy. Circulation 1994;90:1441-1458.
86. Mickleborough L, Harris L, Downar E, et al: A new intraoperative approach for endocardial mapping of ventricular tachycardia. J Thorac Cardiovasc Surg 1988;95:271-280.
87. Josephson ME, Zimetbaum P, Huang D, et al: Pathophysiologic substrate for sustained ventricular tachycardia in coronary artery disease. Japanese Circ J 1997;61:459-466.
88. Callans DJ, Zardini M, Gottlieb CD, et al: The variable contribution of functional and anatomic barriers in human ventricular tachycardia: An analysis with resetting from two sites. J Am Coll Cardiol 1996;27:1106-1111.
89. Bogun F, Bahu M, Knight BP, et al: Response to pacing at sites of isolated diastolic potentials during ventricular tachycardia in patients with previous myocardial infarction. J Am Coll Cardiol 1997;30:505-513.
90. Kanaan N, Robinson N, Roth SI, et al: Ventricular tachycardia in healing canine myocardial infarction: Evidence for multiple reentrant mechanisms. Pacing Clin Electrophysiol 1997; 20:245-260.
91. Gray R, Jalife J, Pafilov A, et al: Mechanisms of cardiac fibrillation. Science 1995;270:1222-1223.
92. Gray R, Pertsov A, Arkady M, et al: Spatial and temporal organization during cardiac fibrillation. Nature 1998;392:75-78.
93. Damle R, Kanaan N, Robinson N, et al: Spatial and temporal linking of epicardial activation directions during ventricular fibrillation in dogs: Evidence for underlying organization. Circulation 1992;86:1547-1558.
94. Lee J, Kamjoo K, Hough D, et al: Reentrant wave fronts in Wiggers' Stage II ventricular fibrillation: Characteristics and mechanisms of termination and spontaneous regeneration. Circulation 1996;78:600-675.
95. Statters DJ, Malik M, Redwood S, et al: Use of ventricular premature complexes for risk stratification after acute myocardial infarction in the thrombolytic era. Am J Cardiol 1996;77:133-138.
96. Maggioni A, Zuanetti G, Franzosi M, et al: Prevalence and prognostic significance of ventricular arrhythmias after acute myocardial infarction in the fibrinolytic era. GISSI-2 results. Circulation 1993;87:312-322.
97. Bigger J, Fleis J, Kleiger R, et al: The relationships among ventricular arrhythmias, left ventricular dysfunction, and mortality in the two years after myocardial infarction. Circulation 1984; 69:250-258.
98. Hofmann A, Buehler F, Burckhardt D: High grade ventricular ectopic activity and 5-year survival in patients with chronic heart disease and in healthy subjects. Cardiology 1983;70(Suppl. 1):82-87.
99. Hohnloser SH, Klingenheben T, Zabel M, et al: Prevalence, characteristics and prognostic value during long-term follow-up of nonsustained ventricular tachycardia after myocardial infarction in the thrombolytic era. J Am Coll Cardiol 1999;33: 1895-1902.
100. Andresen D, Bethge K, Boissel J, et al: Importance of quantitative analyses of ventricular arrhythmias for predicting the prognosis in low-risk postmyocardial infarction patients. Eur Heart J 1990; 11:529-536.
101. Kadish A: Polymorphic ventricular tachycardia. Card Electrophysiol Rev 1999;3:140-142.
102. Buxton AE, Lee KL, DiCarlo L, et al: Electrophysiologic testing to identify patients with coronary artery disease who are at risk for sudden death. Multicenter Unsustained Tachycardia Trial Investigators. N Engl J Med 2000;342:1937-1945.
103. Moss J, Hall W, Cannom D, et al: Improved survival with an implanted defibrillator in patients with coronary disease at high risk for ventricular arrhythmia. N Engl J Med 1996;335:1933.
104. Kolettis TM, Kyriakides ZS, Popov T, et al: Importance of the site of ventricular tachycardia origin on left ventricular hemodynamics in humans. Pacing Clin Electrophysiol 1999;22:871-879.
105. Holmes J, Kubo S, Cody R, et al: Arrhythmias in ischemic and nonischemic dilated cardiomyopathy: Prediction of mortality by ambulatory electrocardiography. Am J Cardiol 1985;55:146-151.
106. Cherry J, Green M: Survival in patients presenting with sustained monomorphic ventricular tachycardia. PACE 1987;10:617.
107. Sarter BH, Finkle JK, Gerszten RE, et al: What is the risk of sudden cardiac death in patients presenting with hemodynamically stable sustained ventricular tachycardia after myocardial infarction? J Am Coll Cardiol 1996;28:122-129.
108. Raitt M, Renfroe E, Epstein A, et al: "Stable" Ventricular Tachycardia Is Not a Benign Rhythm: Insight from the Antiarrhythmics versus Implantable Defibrillation (AVID) Registry. Circulation (online) 2001;103:244-252.
109. Myerburg RJ, Interian A, Mitrani RM, et al: Frequency of sudden cardiac death and profiles of risk. [see comments]. Am J Cardiol 1997;80:10F-19F.

110. Eisenberg M, Bergner L, Hallstrom A: Paramedic programs and out-of-hospital cardiac arrest: I. Factors associated with successful resuscitation. Am J Pub Health 1979;69:30-38.
111. Pratt C, Greenway P, Schoenfeld M, et al: Exploration of the precision of classifying sudden cardiac death: Implications for the interpretation of clinical trials. Circulation 1996;93(3): 519-524.
112. Peckova M, Fahrenbruch CE, Cobb LA, et al: Circadian variations in the occurrence of cardiac arrests: Initial and repeat episodes. [see comments]. Circulation 1998;98:31-39.
113. De Maio VJ, Stiell IG, Wells GA, et al: Cardiac arrest witnessed by emergency medical services personnel: Descriptive epidemiology, prodromal symptoms, and predictors of survival. OPALS study group. Ann Emerg Med 2000;35:138-146.
114. Kempf FC, Josephson ME: Cardiac arrest recorded on ambulatory electrocardiograms. Am J Cardiol 1984;53:1577-1582.
115. Luu M, Stevenson W, Stevenson L, et al: Diverse mechanisms of unexpected cardiac arrest in advanced heart failure. Circulation 1989;80(6):1675-1680.
116. Huikuri HV, Castellanos A, Myerburg RJ: Sudden death due to cardiac arrhythmias. N Engl J Med 2001;345:1473-1482.
117. Nikolic G, Bishop RL, Singh JB: Sudden death recorded during Holter monitoring. Circulation 1982;66:218-225.
118. Panidis I, Morganroth J: Sudden death in hospitalized patients: Cardiac rhythm disturbances detected by ambulatory electrocardiographic monitoring. J Am Coll Cardiol 1983;2:798-805.
119. Pratt C, Francis M, Luck J, et al: Analysis of ambulatory electrocardiograms in 15 patients during spontaneous ventricular fibrillation with special reference to preceding arrhythmic events. J Am Coll Cardiol 1983;2:789-797.
120. Schmidinger H, Weber H: Sudden death during ambulatory Holter monitoring. Int J Cardiol 1987;16:169-176.
121. Kadish A, Quigg R, Schaechter A, et al: DEFibrillators In Non-Ischemic Cardiomyopathy Treatment Evaluation (DEFINITE). Personal communication, 1999.
122. Passman R, Kadish A: Polymorphic ventricular tachycardia, long QT syndrome, and torsades de pointes. Med Clin North Am 2001;82:321-341.
123. Lown B, Graboys TB: Abolishing advanced grades of ventricular premature beats protects patients with recurring malignant ventricular arrhythmias. Int J Cardiol 1983;4:345-350.
124. Albert CM, Mittleman MA, Chae CU, et al: Triggering of sudden death from cardiac causes by vigorous exertion. [see comments]. N Engl J Med 2000;343:1355-1361.
125. Elhendy A, Sozzi FB, van Domburg RT, et al: Relation among exercise-induced ventricular arrhythmias, myocardial ischemia, and viability late after acute myocardial infarction. Am J Cardiol 2000;86:723-729.
126. CIBIS Investigators and Committees: A randomized trial of beta-blockade in heart failure: The Cardiac Insufficiency Bisoprolol Study (CIBIS). Circulation 1994;90:1765-1773.
127. Viscoli CM, Horwitz RI, Singer BH: Beta-blockers after myocardial infarction: Influence of first-year clinical course on long-term effectiveness. Ann Int Med 1993;118:99-105.
128. The Cardiac Arrhythmia Suppression Trial (CAST) Investigators. Preliminary Report: Effect of encainide and flecainide on mortality in a randomized trial of arrhythmia suppression after myocardial infarction. N Engl J Med 1989;321:406-412.
129. Teo K, Yusuf S, Furberg C: Effects of prophylactic antiarrhythmic drug therapy in acute myocardial infarction. An overview of results from randomized controlled trials. JAMA 1993;270 (13).
130. Waldo AL, Camm AJ, deRuyter H, et al: Effect of d-Sotalol on mortality in patients with left ventricular dysfunction after recent and remote myocardial infarction. The SWORD Investigators. Survival With Oral d-Sotalol. [see comments]. [erratum appears in Lancet 1996 Aug 10;348(9024):416]. Lancet 1996;348:7-12.
131. Boutitie F, Boissel J, Connolly S, et al: Amiodarone interaction with beta-blockers: Analysis of the merged EMIAT (European Myocardial Infarct Amiodarone Trial) and CAMIAT (Canadian Amiodarone Myocardial Infarction Trial) databases. Circulation 1999;99 (17):2268-2275.
132. Cairns J, Connolly S, Roberts R, et al: Randomised trial of outcome after myocardial infarction in patients with frequent or repetitive ventricular premature depolarisations: CAMIAT. Lancet 1997;349:675-682.
133. Anonymous: Effect of prophylactic amiodarone on mortality after acute myocardial infarction and in congestive heart failure: Meta-analysis of individual data from 6500 patients in randomised trials. Amiodarone Trials Meta-Analysis Investigators. [see comments]. Lancet 1997;350:1417-1424.
134. Pedersen OD, Bagger H, Keller N, et al: Efficacy of dofetilide in the treatment of atrial fibrillation-flutter in patients with reduced left ventricular function: A Danish investigation of arrhythmia and mortality on dofetilide (diamond) substudy. Circulation 2001;104:292-296.
135. Brendorp B, Elming H, Jun L, et al: Qtc interval as a guide to select those patients with congestive heart failure and reduced left ventricular systolic function who will benefit from antiarrhythmic treatment with dofetilide. Circulation 2001;103: 1422-1427.
136. Cameron A, Schwartz MJ, Kronmal RA, et al: Prevalence and significance of atrial fibrillation in coronary artery disease (CASS Registry). Am J Cardiol 1988;61:714-717.
137. Kannel WB, Abbott RD, Savage DD, et al: Coronary heart disease and atrial fibrillation: The Framingham Study. Am Heart J 1983;106:389-396.
138. Onundarson PT, Thorgeirsson G, Jonmundsson E, et al: Chronic atrial fibrillation—epidemiologic features and 14 year follow-up: A case control study. Eur Heart J 1987;8:521-527.
139. Ginks W, Sutton R, Siddons H, et al: Unsuspected coronary artery disease as cause of chronic atrioventricular block in middle age. Br Heart J 1980;44:699-702.
140. Moreyra AE, Horvitz L, Presant SB, et al: Resolution of complete heart block after right coronary artery angioplasty. Am Heart J 1988;115:179-181.
141. Coplan NL, Morales MC, Romanello P, et al: Exercise-related atrioventricular block: Influence of myocardial ischemia. Chest 1991;100:1728-1730.
142. Deaner A, Fluck D, Timmis AD: Exertional atrioventricular block presenting with recurrent syncope: Successful treatment by coronary angioplasty. Heart 1996;75:640-641.
143. Kovac JD, Murgatroyd FD, Skehan JD: Recurrent syncope due to complete atrioventricular block, a rare presenting symptom of otherwise silent coronary artery disease: Successful treatment by PTCA. Cath Cardiovasc Diag 1997;42:216-218.

Chapter 29 Arrhythmias Associated with Hypertrophic Cardiomyopathy

BARRY J. MARON, IACOPO OLIVOTTO,
and FRANCO CECCHI

Hypertrophic cardiomyopathy (HCM) is a genetic cardiac disease with particularly heterogeneous clinical presentation and diverse natural history.[1-4] Sudden and unexpected death has been recognized as a prominent and devastating consequence of HCM since the initial description of this disease more than 40 years ago.[5] Occurrence of ventricular tachyarrhythmias has always represented a focus of concern in this disease, primarily because of the perceived causal relationship between such arrhythmias and sudden death.[6-21] In addition, it is now obvious that supraventricular tachyarrhythmias (particularly atrial fibrillation) commonly occur in HCM and can represent important determinants of clinical course.[18,20,21] This chapter focuses on recent observations linking tachyarrhythmias with sudden death in HCM and the potential role of preventive interventions.

Ventricular Arrhythmias and Sudden Death

HISTORICAL CONTEXT

Since Teare's original pathologic report of this disease,[5] recognition that a small but important subgroup of patients with HCM is at increased risk for sudden cardiac death has for many years generated considerable interest in the role of arrhythmias and the process of risk stratification,[1,3,6-14] as well as a continuing debate regarding the most appropriate measures for effective prevention of these unexpected catastrophes.[15] Over the years, many investigators have emphasized that sudden death in HCM, presumably due to arrhythmias, occurs not uncommonly in young asymptomatic patients[5-8,12,16,22,23] with annual mortality rates reportedly as high as 4% to 6% in tertiary center referral populations disproportionately composed of high-risk patients.[6,18]

Ventricular arrhythmias are a devastating feature of HCM due to the associated risk for sudden cardiac death (Fig. 29-1). On routine ambulatory (Holter) monitoring, up to 95% of HCM patients demonstrate ventricular arrhythmias, which are often substantial or complex, including premature ventricular depolarizations or ventricular couplets (each in about 50%) and nonsustained ventricular tachycardia (in 20% to 33%).[1,3,4,11,12,17,18,24] These ventricular arrhythmias appear disproportionate to the occurrence of sudden cardiac death in HCM patient populations.[24] Nevertheless, short bursts of nonsustained ventricular tachycardia (usually 3 to 6 beats) have been identified as markers for sudden death risk by two studies from tertiary HCM centers in the early 1980s, focusing attention on ventricular tachyarrhythmias identified primarily on ambulatory (Holter) monitoring as premonitory to major cardiac events,[11,12,17] thereby triggering an era of antiarrhythmic drug treatment. This review represents a contemporary accounting of the nature and significance of arrhythmias occurring in

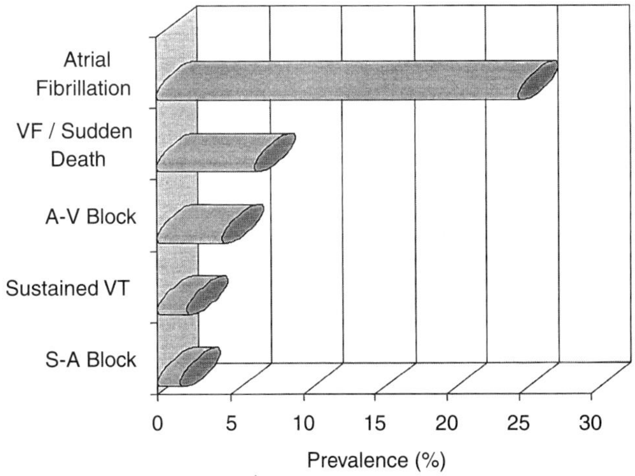

FIGURE 29-1 Estimated prevalence of arrhythmias and conduction abnormalities occurring in an HCM population, assembled from previously published reports. HCM, hypertrophic cardiomyopathy.

patients with HCM, as well as considerations for management.

NONSUSTAINED VENTRICULAR TACHYCARDIA

Ventricular arrhythmias in patients with HCM are largely clinically occult and documented by 24- or 48-hour ambulatory (Holter) electrocardiography. Indeed, nonsustained ventricular tachycardia evident on Holter ECG as short bursts commonly occurs in 20% to 25% of HCM patients.[9,11,12,17,24] Early studies in adult HCM patients linked short runs of asymptomatic ventricular tachycardia with an enhanced risk for sudden cardiac death, focusing interest in the ambulatory ECG for stratifying risk of sudden death; in HCM. For example, ventricular tachycardia conferred about an eightfold increased risk for sudden death; i.e., 8% per year compared with less than 1% per year in the absence of ventricular tachycardia, demonstrating high negative and low positive predictive value (96% and 26%, respectively).[11,12,17] The cited low positive predictive value substantiates the significant heterogeneity of risk within that subset of HCM patients with nonsustained ventricular tachycardia.

Despite its limitations, the finding of ventricular tachycardia on Holter monitoring nevertheless can be regarded as a useful and practical noninvasive screening test for the risk of sudden cardiac death in adult HCM patients. The presence of ventricular tachycardia potentially places the patient in a high-risk group for which further risk evaluation is justified (including serial Holter recordings to judge the overall arrhythmia profile). In those patients in whom nonsustained ventricular tachycardia on Holter is a risk factor, it may be regarded as a primary arrhythmogenic substrate and trigger mechanism for sudden cardiac death. Conversely, the absence of ventricular tachycardia on Holter monitoring in an adult patient with an otherwise low-risk clinical profile (e.g., no history of syncope or familial sudden death and absence of extreme hypertrophy) permits prognostic reassurance.

Some investigators have previously advocated treating nonsustained runs of ventricular tachycardia with various antiarrhythmic agents, including low-dose amiodarone.[11,12,17,25,26] However, because of concerns about proarrhythmia, it is no longer customary to recommend pharmacologic treatment for all infrequent or sporadic bursts of nonsustained ventricular tachycardia.[3,9,26] Nevertheless, it is perhaps reasonable to assume that frequent and multiple (>5 bursts) or prolonged (>10 beats) episodes of nonsustained ventricular tachycardia on serial Holter recordings, suggest high-risk status and may be considered a potential justification for prophylactic treatment with amiodarone or an implantable defibrillator.

POTENTIAL MECHANISMS OF SUDDEN DEATH

The determinants of sudden cardiac death in HCM remain incompletely defined. Although the pathogenesis of sudden death is likely a complex and multifactorial process, catastrophic events ultimately involve ventricular tachyarrhythmias that are regarded as primary.[9,11,12,15,17,24,27,28] Identifying those arrhythmias linked to sudden and unexpected collapse in HCM has proved difficult, given the paucity of ECG recordings during clinical events.[27,29]

More recently, arrhythmia sequences have been documented with stored electrocardiographic recordings in patients with implantable cardioverter-defibrillators (ICD) experiencing appropriate device interventions (Fig. 29-2).[15,28] These observations offer a unique window to understanding the mechanisms responsible for sudden death in HCM. A multicenter ICD trial in high-risk HCM patients showed ventricular tachycardia or fibrillation triggered appropriate device activations in each case, supporting the long-standing hypothesis that primary ventricular tachyarrhythmias (ventricular fibrillation and/or tachycardia) are most commonly responsible for unexpected catastrophes in this disease.[1,15] It was not possible to conclusively exclude bradycardia-mediated events in that analysis because of the back-up pacing capability operative in many of the devices, and it is possible that other more diverse arrhythmia mechanisms may ultimately prove to be involved.[27-29] Clinically evident and sustained monomorphic ventricular tachycardia is a rare spontaneous arrhythmia in HCM,[1,3] and has largely been reported in association with the rare occurrence of left ventricular apical aneurysm.[30-33] Complete heart block and other conduction abnormalities, such as accelerated atrioventricular conduction due to accessory pathways, are acknowledged but are particularly rare causes of syncope or sudden death in HCM.[18,34,35]

In HCM, ventricular tachyarrhythmias probably emanate primarily from a substrate of electrical instability and distorted electrophysiological transmission created by the disorganized left ventricular myocardial architecture,[5,36-38] or from bursts of myocardial ischemia (probably due to structurally abnormal, narrowed intramural arterioles) leading to myocyte necrosis and repair in the form of replacement fibrosis (Fig. 29-3).[38] This myocardial substrate may be vulnerable to a variety of triggers, either intrinsically—related to the HCM disease process, such as abrupt increase in outflow obstruction—or to extrinsic environmental factors such as intense physical exertion[1,3]; also, a variable component of individual patient susceptibility undoubtedly plays a role in determining which HCM patients experience clinical events at particular moments in their lives.

PREVENTION OF SUDDEN DEATH

Pharmacologic Treatment

Historically, the management of high-risk HCM patients had been limited to prophylactic pharmacologic treatment with β-blockers, verapamil, and antiarrhythmic agents such as procainamide and quinidine, and more recently with amiodarone.[4,25,26] However, there are only limited data in HCM supporting the efficacy of prophylactic treatment for sudden death with drugs.[7,11,15,17,25,26]

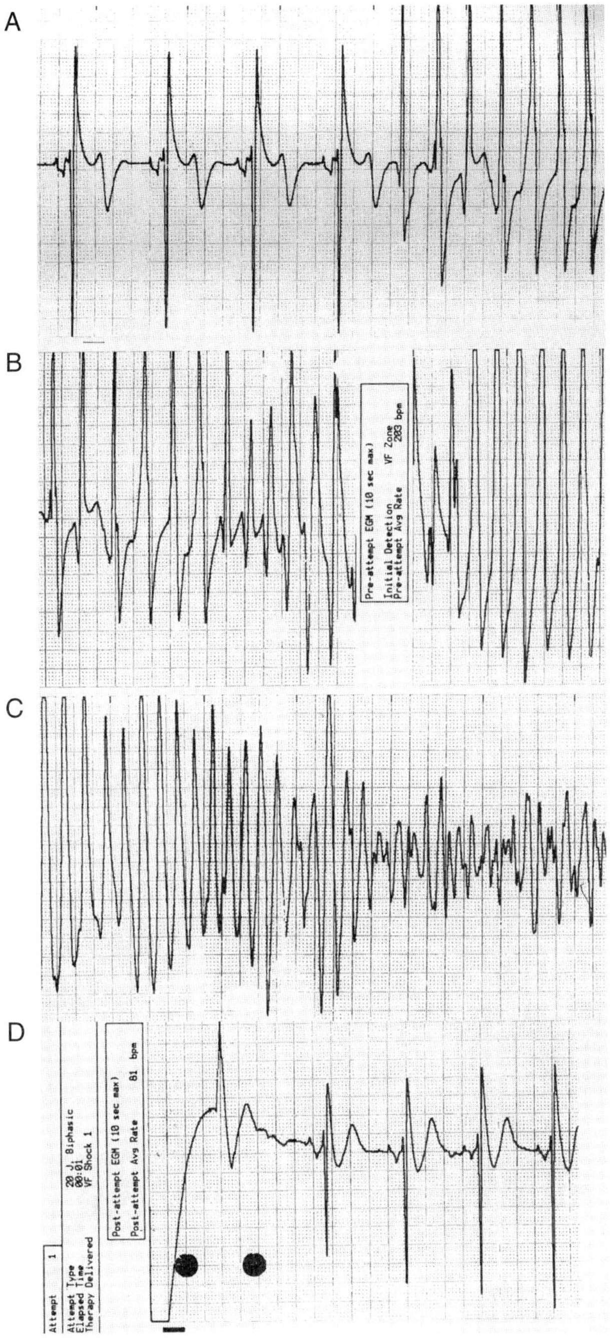

FIGURE 29-2 Primary prevention of sudden cardiac death in HCM. Stored ventricular electrogram from asymptomatic 35-year-old man who received an ICD prophylactically because of a family history of HCM-related sudden death and marked ventricular septal hypertrophy (i.e., wall thickness 31 mm). Intracardiac electrogram was triggered 4 years and 8 months after the defibrillator implant (at 1:20 AM during sleep). Continuous recording at 25 mm/sec, shown in four contiguous panels with the tracing recorded left-to-right in each segment. **A,** Begins with 4 beats of sinus rhythm and, thereafter, ventricular tachycardia begins abruptly (at 200 beats/min); **B,** Device senses ventricular tachycardia and charges; **C,** Ventricular tachycardia deteriorates into ventricular fibrillation; **D,** Defibrillator discharges appropriately (20-J shock) during ventricular fibrillation, and restores sinus rhythm immediately. HCM, hypertrophic cardiomyopathy; ICD, implantable cardioverter defibrillator. (From Maron BJ, et al[15]; reproduced with permission of the Massachusetts Medical Society.)

For example, no controlled studies address the effects of β-blockers or verapamil on sudden death. Type IA antiarrhythmic agents (such as procainamide and quinidine) have been abandoned as prophylactic treatment of HCM patients with isolated or infrequent nonsustained ventricular tachycardia on Holter monitoring, due to the known proarrhythmic effects of these drugs.[12,25,26]

It is possible that amiodarone may reduce the risk for sudden death in HCM.[25] However, the sole report on this drug in HCM,[25] proposing protective drug effects against sudden death in symptomatic or mildly symptomatic patients with nonsustained ventricular tachycardia was 15 years ago and used a retrospective and nonrandomized study design with historic controls. The efficacy of amiodarone in preventing sudden death in HCM has been called into question by an observation as part of an implantable cardioverter-defibrillator (ICD) study,[15] that about 50% of HCM patients experiencing an appropriate device discharge for ventricular tachycardia/fibrillation had also been taking amiodarone at that time. The frequent adverse consequences associated with the chronic administration of amiodarone severely limits its application to sudden death prevention for young patients with HCM who harbor characteristically long periods of risk.

Based on the paucity of efficacy data, and the evolving perceptions regarding the nonspecificity of infrequent short bursts of ventricular tachycardia in HCM, pharmacologic treatment for ventricular tachyarrhythmias has been largely abandoned. Indeed, the prevention of sudden death in HCM has remained a major management challenge for clinicians for many years.

Implantable Defibrillator

Since its introduction by Michel Mirowski[39] more than 20 years ago, the ICD has achieved widespread acceptance as a preventive treatment for sudden death, by virtue of indisputable evidence of efficacy in terminating life-threatening ventricular tachyarrhythmias and prolonging life, principally in high-risk patients with ischemic heart disease.[40,41] In such patients, the superiority of the ICD to antiarrhythmic drug treatment has recently been documented in prospective, randomized trials.[40,41] Of particular importance in this regard has been the evolution of the ICD from a thoracotomy-based procedure with epicardial leads to a transvenous endocardial electrode system with pectoral implantation of the pulse generator, which has greatly facilitated its clinical employment, particularly for primary prevention strategies. However, despite the widespread and increasing use of the ICD in subsets of patients with coronary artery disease, there had been relatively little systematic application of the device to less common genetic conditions, which also constitute risks for sudden death, such as long QT syndrome, Brugada syndrome, arrhythmogenic right ventricular dysplasia, and HCM.[15,42-44]

Defibrillator Trials in HCM

The efficacy of the ICD in sensing and automatically terminating potentially lethal tachyarrhythmias was

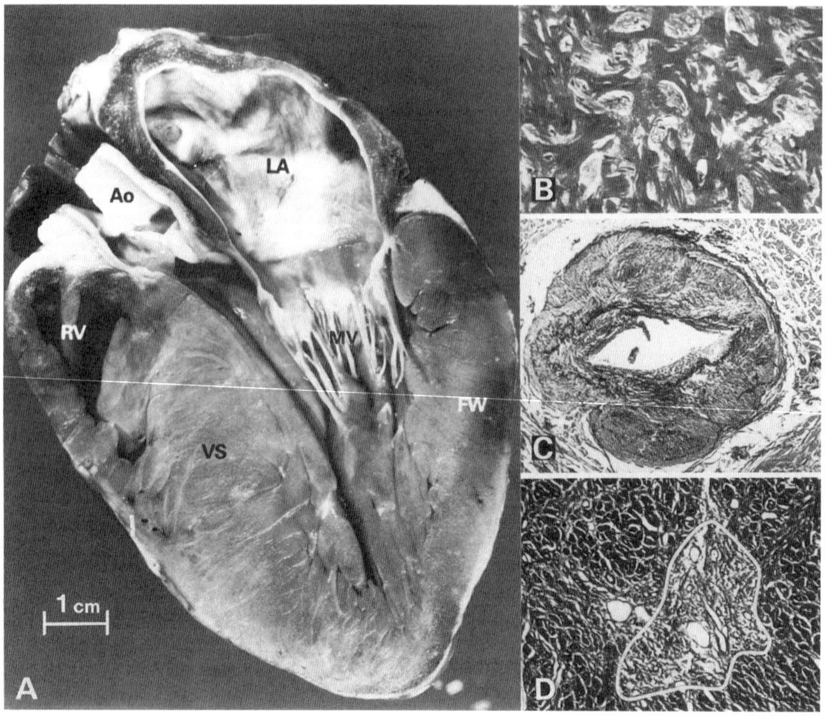

FIGURE 29-3 Morphologic components of the disease process in HCM, which is the most common cause of sudden death in the young. **A,** Gross heart specimen sectioned in a cross-sectional plane similar to that of the echocardiographic (parasternal) long axis; left ventricular wall thickening shows an asymmetrical pattern and is confined primarily to the ventricular septum (VS), which bulges prominently into the left ventricular outflow tract. Left ventricular cavity appears reduced in size. FW, left ventricular free wall. **B, C,** and **D,** Histologic features characteristic of left ventricular myocardium and representative of the arrhythmogenic substrate in HCM. **B,** Markedly disordered architecture with adjacent hypertrophied cardiac muscle cells arranged at perpendicular and oblique angles; **C,** intramural coronary artery with thickened wall, due primarily to medial hypertrophy and an apparently narrowed lumen; **D,** replacement fibrosis (after cell death) in an area of ventricular myocardium adjacent to an abnormal intramural coronary artery. Ao, aorta; LA, left atrium; RV, right ventricle. (Reproduced with permission from Maron BJ: Hypertrophic cardiomyopathy. Lancet 1997;350:127-133.)

recently investigated in a large group of HCM patients judged to be at high risk for sudden death as part of a retrospective, multicenter study in the United States and Italy (Fig. 29-4).[15] The study group of 128 HCM patients, all of whom had ICDs implanted for sudden death prevention, were followed for an average period of 3 years. Appropriate device discharges (either defibrillation shocks or antitachycardia pacing), triggered by ventricular tachyarrhythmias, occurred in almost 25% of patients, with an average discharge rate of 7% per year. About 60% of those patients who received defibrillator therapy experienced multiple appropriate interventions. Of note, the demonstrated efficacy of the ICD in HCM occurred despite the substantially increased heart mass characteristic of this disease.[16,22,23,45,46] Data from a European registry limited largely to ICD therapy following resuscitated cardiac arrest,[47] and from preliminary reports of high-risk pediatric patients,[43] describe similar beneficial clinical experience with prophylactic ICD therapy.

At the time of appropriate defibrillator interventions more than one half the patients were taking amiodarone or other antiarrhythmic drugs. This ancillary observation substantiates the superiority of the ICD in

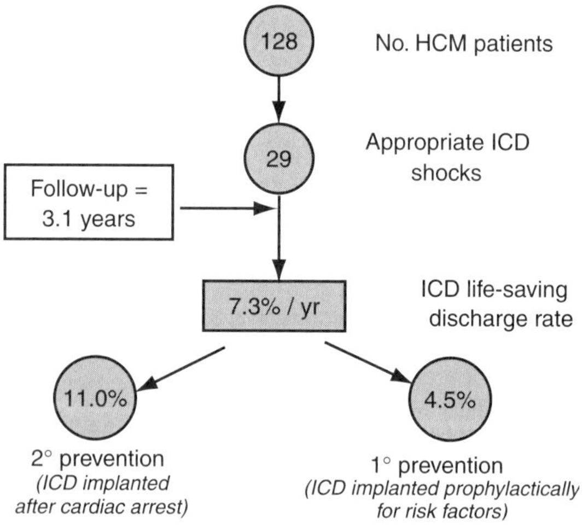

FIGURE 29-4 Flow-diagram showing outcome of 128 high-risk HCM patients with implantable defibrillators for primary prevention (with ≥1 risk factor[s]) or secondary prevention (following cardiac arrest or sustained ventricular tachycardia). Two patients *(not represented here)* died of HCM with refractory end-stage heart failure and systolic dysfunction (despite the ICD). HCM, hypertrophic cardiomyopathy; ICD, implantable cardioverter defibrillator.

preventing sudden death, as well as disputing earlier, exaggerated claims that amiodarone is absolutely protective against sudden death in HCM patients.[25]

Secondary Prevention

Not unexpectedly, lifesaving defibrillator interventions were most frequent in those patients implanted specifically for secondary prevention—i.e., following fortuitous resuscitation from cardiac arrest (with documented ventricular fibrillation), or after an episode of spontaneous and sustained ventricular tachycardia; more than 40% of such patients received defibrillator therapy for secondary prevention during the relatively short 3-year follow-up period. Such frequent recurrences of potentially lethal ventricular tachyarrhythmias following cardiac arrest are consistent with a previously reported experience involving similar HCM patients, but in the pre-ICD era.[48]

Primary Prevention

Of particular note, those patients receiving ICDs solely for primary prevention also showed a substantial appropriate device intervention rate of about 5% per year. These primary prevention strategies represented prophylactic implantation due to perceived high-risk status based on a clinical profile with one or more identifiable risk factors for sudden death: family history of HCM-related sudden death in a close relative; syncope, if exertional or repetitive, and in young patients; multiple-repetitive or prolonged nonsustained ventricular tachycardia on repeated ambulatory (Holter) ECGs; extreme left ventricular hypertrophy (maximal wall thickness, ≥30 mm)[16,22,23]; and, possibly hypotensive blood pressure response to exercise. The presence, magnitude, or absence of outflow obstruction has not proved to be a sole risk factor specifically for sudden cardiac death in HCM.[49]

THE RISK PERIOD IN HYPERTROPHIC CARDIOMYOPATHY

Crucial to understanding the role of the ICD within the broad HCM disease spectrum is an appreciation of certain demographic distinctions from ICD therapy in coronary artery disease. The latter patients are of relatively advanced age at the time of implant (average, about 65 years), and often with advanced disease as a consequence of earlier myocardial infarction. In sharp contrast, HCM is characterized by an extended period of risk for sudden death that is predominant in patients younger than 30 years of age, but importantly includes those in midlife and even beyond[8,49]; indeed, no particular age itself appears to confer immunity to sudden death. Therefore, HCM represents a much different clinical circumstance compared with coronary artery disease, in which at-risk patients are often young and with few or no symptoms before collapse. In HCM, the mean age at implant is about only 40 years (almost 25% <30 years old), and the age at the time of first appropriate device intervention is also 40 years.[15]

Although annual appropriate intervention rates for HCM patients are lower than those reported in coronary artery disease,[40,41] they nevertheless are significant, given that the ICD experience in HCM must be considered in the context of a much younger patient population usually free of significant congestive heart failure (with preserved systolic function) who are exposed to long periods of potential risk—and protected by the ICD—could survive many decades with normal or near-normal life expectancies.

By extrapolating the reported primary prevention discharge rate for HCM,[15] it can be estimated that within 10 years about 50% of the defibrillators prophylactically implanted in young patients will intervene and abort a sudden death event. Indeed, the 5% annual discharge rate achieved in this subset of high-risk patients implanted with an ICD represents a figure remarkably similar to that reported for sudden death in the selected and at-risk HCM patient cohorts at tertiary referral centers.[6,12,18,50] It should be emphasized that prophylactic ICD employment, as described in HCM, represents a particularly pure form of primary prevention, with device implantations performed in advance of any major cardiovascular event.

Of particular note, the time interval between implant and first appropriate ICD intervention may be quite variable, with particularly long time delays of up to 4 to 9 years for the first lifesaving intervention not uncommon (Fig. 29-5).[15] This observation underlines the unpredictable timing of sudden death in HCM in which the ICD may remain dormant for substantial time periods before ultimately intervening appropriately. Also, the decision to prophylactically implant an ICD in an HCM patient is often based on the precise time at which risk stratification is undertaken and high-risk status is recognized. For example, a patient identified as high risk at age 20 (and implanted with a device prophylactically at that time) will still be young, and at an increased risk by age 35, even if the ICD has not been triggered appropriately during that

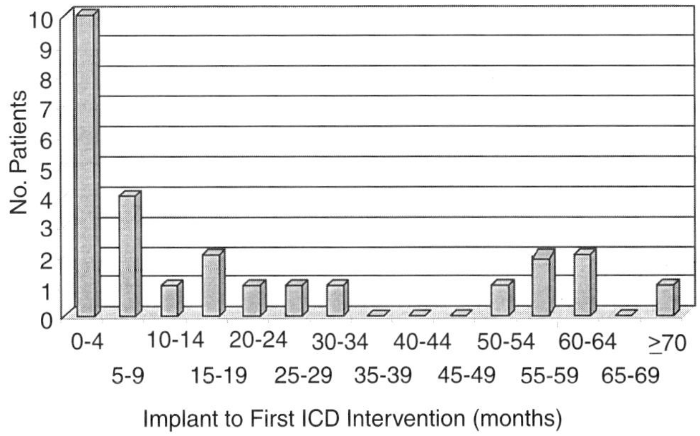

FIGURE 29-5 Interval between implantation of the defibrillator and the first appropriate discharge in 29 HCM patients. Note the substantial proportion of patients with initial defibrillation shock occurring 4 to 9 years after implant. HCM, hypertrophic cardiomyopathy. (From Maron BJ, et al[15]; reproduced with permission of the Massachusetts Medical Society.)

15-year period. Once the decision to implant an ICD in a high-risk HCM patient is made, it is likely to represent a lifelong preventive measure, and should be considered in this context.

STRATEGIC LIMITATIONS

Risk Stratification

Although there is now little reason to doubt the efficacy of the ICD in preventing sudden death in uncommon genetic diseases such as HCM, several important issues regarding prophylactic treatment remain incompletely resolved. For example, the question of precisely which clinical markers most definitively identify high risk and which patients within the broad HCM disease spectrum should receive implantable devices for primary prevention is constrained somewhat by certain imperfections in available risk stratification profiles—which ultimately emanate from the relatively low prevalence and heterogeneous clinical expression of HCM.[1-4,8,45,46,49,51]

Given the absence of risk stratification data that can be applied to all possible at-risk clinical circumstances in individual HCM patients, the ultimate decision regarding ICD implantation for each and every case may, on occasion, be challenging and unavoidably reside with the best clinical judgment of the treating cardiologist. For example, a partially unresolved risk factor is that of a family history of HCM-related sudden death as a justification for a primary prevention ICD implant. Should *only* one sudden death in a close relative of a surviving affected individual be sufficient to trigger the decision to implant an ICD, or should two or more such deaths be required to initiate this treatment strategy?[1,3,52] Should *all* affected members in a large HCM family be offered an ICD because of a familial occurrence of one sudden death? Data governing such focused (but critical) questions are sparse, and definitive answers are not presently available. Consequently, individual physician judgment and patient motivation for an ICD are often important factors in resolving such dilemmas in clinical decision making. It is reasonable to conclude, that the potency and sophistication of the ICD to effectively abort sudden cardiac death exceeds the power of available risk stratification profiles to reliably discriminate all appropriate candidates with HCM for the ICD.

Further clinical trials with much larger numbers of patients will be required to define with greater precision those individual HCM patients among the broad disease spectrum who should be targeted for (and would benefit most from) prophylactic ICD therapy. Such investigations will necessarily be retrospective and nonrandomized because of the particularly long potential risk period that is characteristic of young high-risk HCM patients, as well as the obvious ethical considerations.[8,15]

Inducible Ventricular Arrhythmia

The role of electrophysiological testing with programmed ventricular stimulation, and the significance of inducible ventricular tachyarrhythmias to identify the substrate and mechanisms for sudden death and routinely and independently target those HCM patients at increased risk[53] is a strategy that has been largely abandoned.[1,3,54,55] Limitations to this technique include the infrequency with which monomorphic ventricular tachycardia is inducible in HCM (only about 10% of patients)[53] and the fact that the electrical response of the HCM substrate appears to be highly dependent on the precise stimulation protocol used.[1] For example, aggressive electrophysiological testing using three premature extrastimuli can be expected to frequently trigger sustained polymorphic ventricular tachycardia or ventricular fibrillation in a substantial proportion of patients; these inducible arrhythmias are largely regarded as nonspecific in other more common cardiac conditions including coronary artery disease.[1,3,54,55] The precise clinical significance that should be attached to ventricular arrhythmias induced with two extrastimuli is unresolved in HCM. Therefore, given the inherent risks and inconvenience associated with programmed ventricular stimulation, the considerable uncertainty surrounding the significance of induced arrhythmias, and the fact that most high-risk patients can be identified using noninvasive clinical markers independent of programmed stimulation, routine use of such laboratory testing to replicate clinical arrhythmias would appear to have little practical value in assessing risk and predicting outcome in HCM.[1]

Complications and Other Considerations

It is also important to recognize the potential complications of ICD therapy that may affect implant decisions, including inappropriate and spurious device discharges (reported frequency = 25% in HCM patients),[15] fractured or disrupted leads, and infection. Other important considerations that may influence clinical decision making include the substantial cost of the ICD (particularly over the long time periods required for primary prevention in young patients), as well as varying physician and patient attitudes toward ICDs and access to such devices, within different countries and cultures.[56] Of course, all these issues must be weighed against the ultimate potential benefits of the ICD for individual high-risk patients.

Atrial Fibrillation

Paroxysmal supraventricular tachyarrhythmias with or without accessory atrioventricular pathways occur commonly in about 30% to 50% of adult HCM patients on ambulatory (Holter) ECG,[21,24] with atrial fibrillation (AF) the most common of these arrhythmias (see Fig. 29-1). Many of these arrhythmias may be brief and clinically occult. However, AF may be expressed clinically and of substantial prognostic importance to a considerable proportion of HCM patients by virtue of its potential for acute hemodynamic decompensation and also for producing heart failure and embolic stroke over the long term.[20,21,49,57-66]

AF is the most common and perhaps the most important sustained arrhythmia encountered in HCM.[21,49,65-67] Studies addressing the prognosis of AF in HCM patients are limited and in many instances have reached conflicting conclusions. Early observational studies (confined largely to patients with outflow obstruction) emphasized the association of AF onset with severe acute clinical deterioration.[57] However, one study reported no difference in mortality among patients with AF compared to an HCM control group in sinus rhythm.[61] In that investigation, potentially deleterious consequences of AF may have been obscured by the low survival rate of control patients.

PREVALENCE AND DEMOGRAPHICS

Based on reports from a number of centers, AF commonly appears in HCM with a prevalence of about 20% to 25%, and an incidence of 2% of new cases annually (Table 29-1).[21] Patients with HCM appear to have an overall four- to sixfold greater likelihood of developing AF when compared with the general population.[68-71] Average age for AF onset in HCM is 55 (±15 years); AF increases in frequency progressively with age and is predominant in patients older than 60 years of age, but is not uncommon in patients ≤50 years. However, AF appears to be infrequent in HCM patients younger than age 25 (Fig. 29-6).

PREDISPOSING FACTORS

Atrial fibrillation has proved to have some measure of predictability as a complication of HCM, and is related to advanced age, earlier congestive symptoms, and increased left atrial size at the time of diagnosis.[57,61,63,65-67] The most powerful of these predictors for AF is left atrial size. Modest enlargement of the left atrium (i.e., in the range of about 40 to 45 mm) is common in HCM, usually associated with sinus rhythm, and probably the consequence of impaired diastolic function due to the thickened and poorly compliant ventricular chambers.[72] The determinants of marked and progressive left atrial enlargement, which ultimately predisposes to AF in some HCM patients, remain unresolved (i.e., those factors responsible for the left atrium to dilate in the range of 50 to 60 mm). Neither the degree of mitral regurgitation nor the presence of outflow obstruction reliably predicts the development of AF in HCM. Moderate-to-severe mitral regurgitation occurs in only a minority of HCM patients with AF (i.e., about 15%), and the proportion of patients with outflow obstruction is similar among patients with or without AF.[21]

TABLE 29-1 Prevalence of Atrial Fibrillation (AF) in HCM Populations

Study	No. Patients	Prevalence of AF (%)	Follow-up (yrs)
Glancy et al[57]	167	10	3
Savage et al[19]	100	12	n/a
McKenna et al[17]	86	14	2.6
Cecchi et al[66]	202	28	10.1
Maron et al[49]	277	18	8.1
Olivotto et al[21]	480	22	9.1

From Olivotto I, Maron BJ, Cecchi F. Clinical significance of atrial fibrillation in hypertrophic cardiomyopathy. In Current Cardiology Reports (Current Science), Philadelphia, 2001;3:141-146.

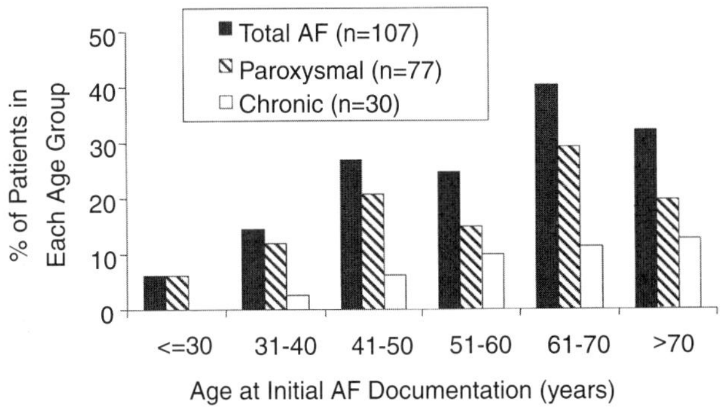

FIGURE 29-6 Age at development of AF in HCM. Bars express the proportion of patients in each age group with paroxysmal or chronic AF. Patients evolving from paroxysmal to chronic AF are considered paroxysmal (i.e., as the initial manifestation of the arrhythmia). AF, Atrial fibrillation; HCM, hypertrophic cardiomyopathy. (From Olivotto I, et al[21] with permission of the American Heart Association.)

It is also possible that specific HCM-causing mutations may increase the predisposition to AF, possibly by creating an intrinsic atrial myopathy associated with prolonged and fragmented atrial conduction.[64] Such hypotheses could also explain the development of AF in the absence of left atrial enlargement, a scenario observed in a minority of HCM patients (about 10%).[21]

It is not possible to reliably identify those HCM patients in sinus rhythm who are ultimately predisposed to develop AF in the near future—or at least not to a degree sufficient to justify prophylactic intervention. Nevertheless, based on available data, the combination of three noninvasive parameters (left atrial enlargement, P-wave prolongation on signal-averaged ECG, and supraventricular tachyarrhythmias on Holter ECG) may allow identification of a patient subset with a higher risk of developing AF.[65]

ATRIAL FIBRILLATION AND HYPERTROPHIC CARDIOMYOPATHY-RELATED MORTALITY AND MORBIDITY

The development of AF is a key determinant of HCM-related mortality and limiting symptoms and may represent a turning-point dominating the clinical course of some patients, and decisively influencing long-term outcome (Fig. 29-7). Patients with AF demonstrate a fourfold increase in the risk for HCM-related death compared to matched HCM control patients in sinus rhythm, independent of age and symptomatic state (odds ratio = 3.7),[21] but reflecting a significant increase only in heart failure and stroke-related mortality.[67] AF is significantly associated with an increased risk for severe functional disability due to congestive symptoms (odds ratio, 2.8).[21]

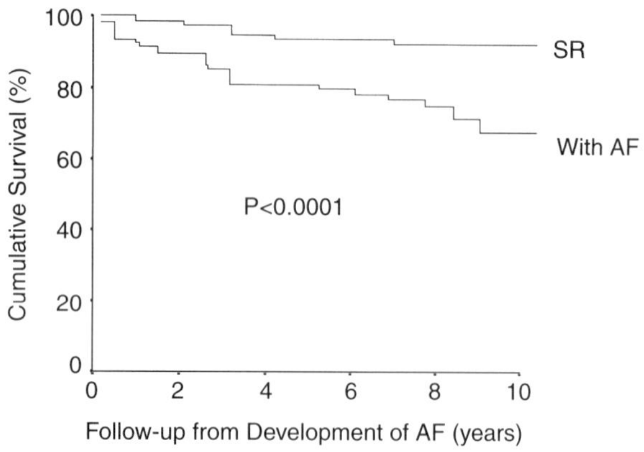

FIGURE 29-7 Impact of AF on overall HCM-related mortality. Cumulative survival of patients with HCM and AF compared to a matched group of HCM patients in sinus rhythm (SR). (From Olivotto I, Cecchi F, Casey SA, et al: Impact of atrial fibrillation on the clinical course of hypertrophic cardiomyopathy. Circulation 2001;104:2517-2524.)

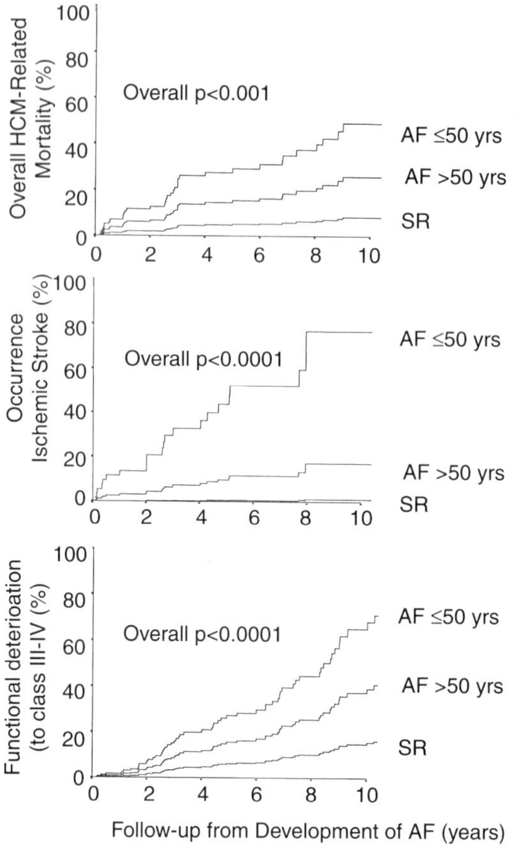

FIGURE 29-8 Relation of early (≤50 years) versus late (>50 years) development of AF to overall HCM-related mortality **(top panel),** stroke **(middle panel)** and progression of NYHA class III/VI **(bottom panel),** compared to HCM patients remaining in sinus rhythm (SR). NYHA, New York Heart Association. (From Olivotto I, Cecchi F, Casey SA, et al: Impact of atrial fibrillation on the clinical course of hypertrophic cardiomyopathy. Circulation 2001;104:2517-2524.)

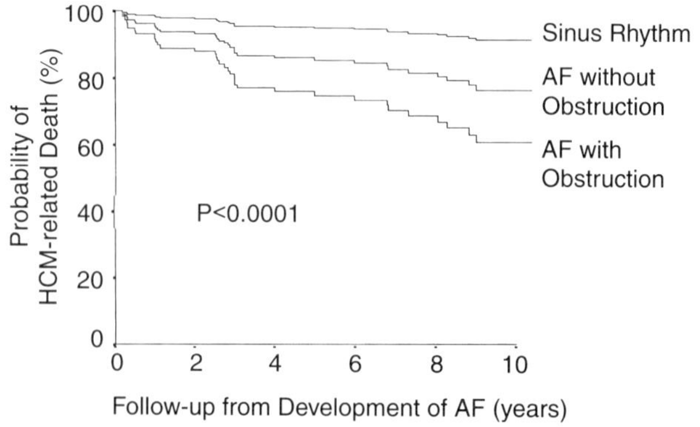

FIGURE 29-9 Combined impact of AF and basal outflow obstruction (gradient ≥30 mm Hg) to overall HCM-related mortality in patients with HCM and AF, as compared to a control group of HCM patients in sinus rhythm. AF, Atrial fibrillation; HCM, hypertrophic cardiomyopathy. (From Olivotto I, Cecchi F, Casey SA, et al: Impact of atrial fibrillation on the clinical course of hypertrophic cardiomyopathy. Circulation 2001;104:2517-2524.)

HCM-related mortality, complications, and clinical deterioration are not uniformly distributed among a cohort of AF patients, but rather appear to preferentially affect certain patient subgroups, such as those with early development of the arrhythmia (before age 50) (Fig. 29-8).[21] Although it is uncertain why more severe consequences of AF are associated with its development earlier in life, it is possible that AF represents a marker of generally more aggressive disease,[21,29] particularly when in combination with basal outflow obstruction (Fig. 29-9).[21,57] In this regard, the dependence of patients with obstructive HCM on left atrial contraction for adequate left ventricular filling[72] may exceed that of nonobstructive patients, and thus, they may be more prone to long-term deterioration following development of AF.

Of particular note, there does not appear to be a clear linkage between clinically evident AF and the occurrence of sudden unexpected death. Although a causal link between AF and potentially lethal ventricular tachyarrhythmias has been suggested in individual patient reports,[14,29,48] population-based data do not support AF as a consistent trigger for sudden death in an HCM patient population. Stafford and coworkers[58] described the case of an adolescent boy who survived cardiac arrest with documented ventricular fibrillation. During a subsequent electrophysiological study, AF with rapid ventricular response was induced, eliciting evidence of myocardial ischemia on ECG, and then rapidly degenerated into ventricular fibrillation. A potential association between AF and sudden death is also suggested by the observation that 20% of those HCM patients who survived a documented cardiac arrest had demonstrated supraventricular arrhythmias (including AF) at the onset of symptoms preceding their collapse.[48]

The pathophysiologic mechanisms by which recurrent and chronic AF may affect long-term clinical outcomes of HCM patients are undoubtedly complex (Fig. 29-10) and incompletely understood, and require

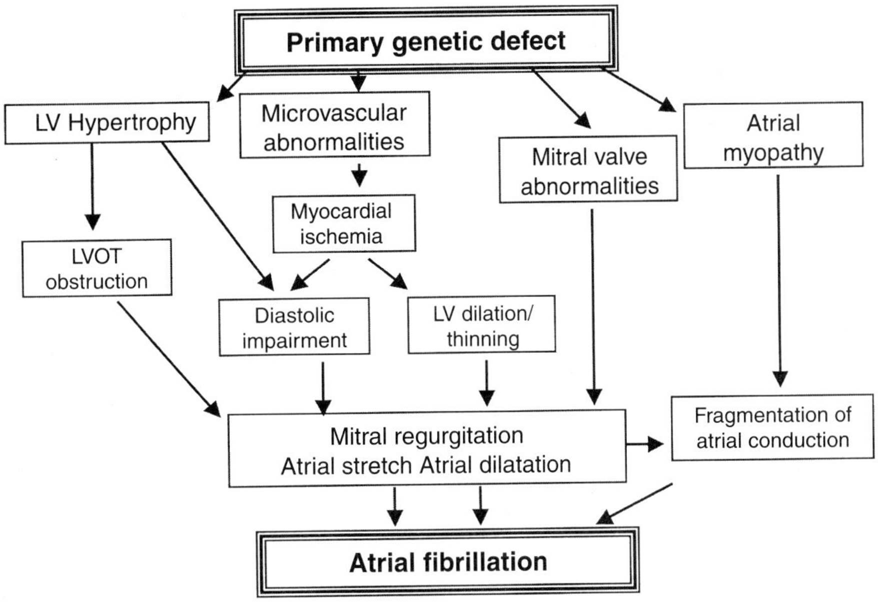

FIGURE 29-10 Arrhythmogenic substrates and potential triggers for AF in HCM patients. Proposed hypothetical sequence of events evolving from the primary gene mutation to the development of AF via multiple pathophysiologic processes known to occur in HCM. LV, Left ventricular; LVOT, left ventricular outflow tract. (From Olivotto I, Maron BJ, Cecchi F: Clinical significance of atrial fibrillation in hypertrophic cardiomyopathy. In Current Cardiology Reports (Current Science), Philadelphia, 2001;3:141-146.)

further investigation. However, a disproportionate increase in mean ventricular rate and reduced cardiac output on effort, recurrent myocardial ischemia, as well as the well-documented risk for thromboembolic complications associated with AF are likely to play important roles.

As a group, HCM patients with AF show distinctive left ventricular morphology with only modest and usually localized hypertrophy—i.e., 18 to 22 mm wall thickness in the anterior basal septum unassociated with basal outflow obstruction.[63] Chronic AF is uncommonly observed in the presence of extreme left ventricular hypertrophy.[16]

CLINICAL VARIABILITY OF ATRIAL FIBRILLATION

Characteristically, the impact of AF on long-term prognosis in individual HCM patients shows substantial heterogeneity (Fig. 29-11) and should not be regarded as invariably unfavorable.[1,21,61,63] Although AF is strongly associated with HCM-related mortality and clinical deterioration, about one third of these patients tolerate AF (free of stroke or severe symptoms) and may experience a generally uneventful course. The likelihood of benign outcome is significantly higher in patients with exclusively paroxysmal AF, as compared to those patients progressing to chronic AF.[21] Therefore, therapeutic efforts aimed at preventing or delaying the transition from paroxysmal to chronic AF carry the potential to improve patient outcome.

Given the available data, it is not possible to define in precise terms the importance of ventricular rate control in HCM patients with chronic AF, although excess mortality in patients with chronic AF might be expected in the presence of persistently elevated ventricular rates.

The question of whether the clinical impact of AF can be explained primarily by its hemodynamic impact on left ventricular filling,[72-75] myocardial ischemia,[76,77] or to a more aggressive underlying cardiomyopathy process, also remains unresolved. The substantial clinical diversity associated with AF in HCM suggests, however, that multiple variables probably contribute differently in many patients to promote the final clinical outcome after the development of AF.[21,49,57,58,66]

ATRIAL FIBRILLATION AND RISK OF STROKE

HCM patients with AF show as much as an eightfold increase in the risk of ischemic stroke (of probable cardioembolic origin), compared to HCM patients in sinus rhythm.[21,59,60,62,67] The risk of stroke is similar in patients with chronic or paroxysmal AF and does not show a direct relation to the number of clinically documented AF episodes.[21] Therefore, patients with a single paroxysm of AF appear to have a risk for stroke similar to patients with multiple episodes. Such observations suggest that the threshold for initiation of anticoagulant treatment in HCM patients should be low[21] and agree with the findings in a wide range of clinical studies in patients with AF and heart diseases other than HCM.[68-71] Although warfarin treatment for primary prevention of stroke should be strongly considered (regardless of the number of documented AF episodes) such clinical decisions need to be tailored in the individual patient after considering the risk for hemorrhagic complications, lifestyle modification, and expectations for compliance.[68] Long-term administration of low-dose amiodarone (200 mg/day) may represent the most efficacious strategy for maintaining sinus rhythm in HCM patients following the development of paroxysmal AF, based largely on prospective data in patients with other cardiac conditions.[68,78]

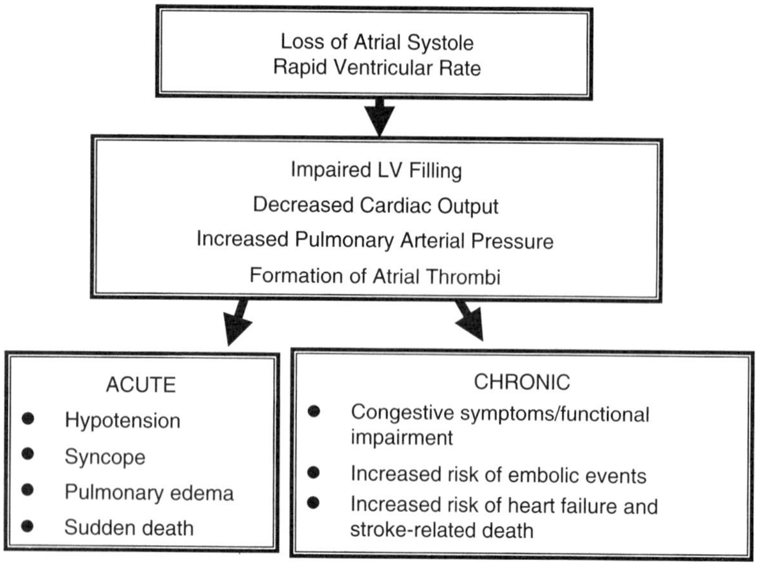

FIGURE 29-11 Acute and chronic pathophysiologic consequences of atrial fibrillation in HCM patients. HCM, hypertrophic cardiomyopathy; LV, left ventricular. (From Olivotto I, Maron BJ, Cecchi F: Clinical significance of atrial fibrillation in hypertrophic cardiomyopathy. In Current Cardiology Reports (Current Science), Philadelphia, 2001;3: 141-146.)

ATRIAL FIBRILLATION AND ACUTE DETERIORATION

On occasion, patients with HCM may experience acute and severe hemodynamic and clinical deterioration with the abrupt and unexpected onset of AF, which may involve heart failure, pulmonary edema, and marked functional impairment, as well as syncope or acute myocardial ischemia (see Fig. 29-11).[21,57,79] This dramatic change in clinical condition is attributable to the sudden loss of atrial systole (and its contribution to ventricular filling) and a rapid increase in ventricular rate, which necessarily accompany AF onset, and may trigger a fall in cardiac output when superimposed on diastolic dysfunction (see Fig. 29-11).[21,29,57,58,72,74,75]

Treatment for the acute onset of AF in HCM patients does not differ substantially from the standard strategies most commonly employed in AF patients with other cardiac conditions or in the absence of structural heart disease.[68-71] In the presence of acute clinical deterioration, urgent DC cardioversion is mandatory and may be lifesaving. In stable patients with AF lasting less than 48 hours, pharmacologic cardioversion may be attempted using class IC agents (propafenone or flecainide) or class III agents (amiodarone, sotalol, or dofetilide),[80] although efficacy data for specific drugs in such HCM patients are still lacking. In patients with AF of 48 hours or more (or when of unknown duration) the goal of urgent treatment is to rapidly achieve control of the ventricular rate using verapamil or β-blockers (for AV node blockade); restoration of sinus rhythm by DC cardioversion (often associated with amiodarone administration) should be postponed in such patients until optimal anticoagulation has been achieved with warfarin for at least 3 weeks. Chronic treatment with β-blockers or verapamil, and amiodarone, may be warranted even after restoration of sinus rhythm to prevent AF recurrences.

SUMMARY

Primary ventricular tachyarrhythmias arising from an unstable myocardial substrate appear to be the basis of sudden and unexpected death in HCM. Prevention of sudden death is now an achievable aspiration for many patients with HCM, due to application of the implantable defibrillator to this relatively uncommon genetic disease. It is evident that the ICD is efficacious and lifesaving in HCM, with an important role established in high-risk patients for secondary, as well as primary prevention of sudden cardiac death. However, the ICD should not be regarded as a treatment strategy for all (or even most) patients with HCM, and should be confined to the high-risk subset in accordance with current risk stratification profiles.

Atrial fibrillation is the most common sustained arrhythmia encountered in an HCM patient population, and is due to complex arrhythmogenic pathophysiology (see Fig. 29-10). Atrial fibrillation is more common in older patients, but is more adverse when onset is in younger patients. Although the long-term consequences of AF are not uniformly unfavorable, the powerful independent association of paroxysmal or chronic AF with HCM-related mortality, stroke, and severe functional disability (particularly in the presence of outflow obstruction) underlines the necessity for aggressive therapeutic strategies.

REFERENCES

1. Maron BJ: Hypertrophic cardiomyopathy: A systematic review. JAMA 2002;287:1308-1320.
2. Wigle ED, Rakowski H, Kimball BP, Williams WG: Hypertrophic cardiomyopathy: Clinical spectrum and treatment. Circulation 1995;92:1680-1692.
3. Spirito P, Seidman CE, McKenna WJ, Maron BJ: The management of hypertrophic cardiomyopathy. N Engl J Med 1997;336: 775-785.

4. Maron BJ, Bonow RO, Cannon RO, et al: Hypertrophic cardiomyopathy: Interrelation of clinical manifestations, pathophysiology, and therapy. N Engl J Med 1987;316:780-789; 844-852.
5. Teare D: Asymmetrical hypertrophy of the heart in young patients. Br Heart J 1958;20:1-8.
6. McKenna WJ, Deanfield JE: Hypertrophic cardiomyopathy: An important cause of sudden death. Arch Dis Child 1984;59: 971-975.
7. Maron BJ, Roberts WC, Epstein SE: Sudden death in hypertrophic cardiomyopathy: Profile of 78 patients. Circulation 1982;65:1388-1394.
8. Maron BJ, Olivotto I, Spirito P, et al: Epidemiology of hypertrophic cardiomyopathy-related death: Revisited in a large non-referral based patient population. Circulation 2000;102:858-864.
9. Spirito P, Rapezzi C, Autore C, et al: Prognosis of asymptomatic patients with hypertrophic cardiomyopathy and nonsustained ventricular tachycardia. Circulation 1994;90:2743-2747.
10. Olivotto I, Maron BJ, Montereggi A, et al: Prognostic value of systemic blood pressure response during exercise in a community-based patient population with hypertrophic cardiomyopathy. J Am Coll Cardiol 1999;33:2044-2051.
11. Maron BJ, Savage DD, Wolfson JK, Epstein SE: Prognostic significance of 24-hour ambulatory electrocardiographic monitoring in patients with hypertrophic cardiomyopathy: A prospective study. Am J Cardiol 1981;48:252-257.
12. McKenna WJ, Camm AJ: Sudden death in hypertrophic cardiomyopathy: Assessment of patients at high risk. Circulation 1989;80:1489-1492.
13. Saumarez RC, Chojnowska L, Derksen R, et al: Sudden death in noncoronary heart disease is associated with delayed paced ventricular activation. Circulation 2003;107:2595-2600.
14. Maron BJ, Cecchi F, McKenna WJ: Risk factors and stratification for sudden cardiac death in patients with hypertrophic cardiomyopathy. Br Heart J 1994;72(Supplement):S-13-S-18.
15. Maron BJ, Shen W-K, Link MS, et al: Efficacy of implantable cardioverter-defibrillators for the prevention of sudden death in patients with hypertrophic cardiomyopathy. N Engl J Med 2000;342:365-373.
16. Spirito P, Bellone P, Harris KM, et al: Magnitude of left ventricular hypertrophy predicts the risk of sudden death in hypertrophic cardiomyopathy. N Engl J Med 2000;324:1778-1785.
17. McKenna WJ, England D, Doi YL, et al: Arrhythmia in hypertrophic cardiomyopathy: I—influence on prognosis. Br Heart J 1981;46:168-172.
18. Stewart JT, McKenna WJ. Arrhythmias in hypertrophic cardiomyopathy. J Cardiovasc Electrophysiol 1991;2:516-524.
19. Savage DD, Seides SF, Maron BJ, et al: Prevalence of arrhythmias during 24-hour electrocardiographic monitoring and exercise testing in patients with obstructive and nonobstructive hypertrophic cardiomyopathy. Circulation 1979;59:866-875.
20. Olivotto I, Maron BJ, Cecchi F. Clinical significance of atrial fibrillation in hypertrophic cardiomyopathy. In Current Cardiology Reports (Current Science), Philadelphia, 2001;3:141-146.
21. Olivotto I, Cecchi F, Casey SA, et al: Impact of atrial fibrillation on the clinical course of hypertrophic cardiomyopathy. Circulation 2001;104:2517-2524.
22. Elliott PM, Gimeno JR, Mahon NG, et al: Relation between severity of left-ventricular hypertrophy and prognosis in patients with hypertrophic cardiomyopathy. Lancet 2001;357:420-424.
23. Elliott PM, Poloniecki J, Dickie S, et al: Sudden death in hypertrophic cardiomyopathy: Identification of high risk patients. J Am Coll Cardiol 2000;36:2212-2218.
24. Adabag AS, Casey SA, Maron BJ: Sudden death in hypertrophic cardiomyopathy: Patterns and prognostic significance of tachyarrhythmias on ambulatory Holter ECG. Circulation 2002; 106(Suppl II):II-710.
25. McKenna WJ, Oakley CM, Krikler DM, Goodwin JF: Improved survival with amiodarone in patients with hypertrophic cardiomyopathy and ventricular tachycardia. Br Heart J 1985;53: 412-416.
26. Cecchi F, Olivotto I, Montereggi A, et al: Prognostic value of non-sustained ventricular tachycardia and the potential role of amiodarone treatment in hypertrophic cardiomyopathy: Assessment in an unselected non-referral based patient population. Heart 1998;79:331-336.
27. Nicod P, Polikar R, Peterson KL. Hypertrophic cardiomyopathy and sudden death. N Engl J Med 1988;318:1255-1256.
28. Elliott PM, Sharma S, Varnava A, et al: Survival after cardiac arrest in patients with hypertrophic cardiomyopathy. J Am Coll Cardiol 1999;33:1596-1601.
29. Stafford WJ, Trohman RG, Bilsker M, et al: Cardiac arrest in an adolescent with atrial fibrillation and hypertrophic cardiomyopathy. J Am Coll Cardiol 1986;7:701-704.
30. Alfonso F, Frenneaux M, McKenna WJ: Clinical sustained monomorphic ventricular tachycardia in hypertrophic cardiomyopathy: Association with left ventricular aneurysm. Br Heart J 1989;61:178-181.
31. Maron BJ, Hauser RG, Roberts WC: Hypertrophic cardiomyopathy with left ventricular apical diverticulum. Am J Cardiol 1996;77:117-119.
32. Ando H, Imaizumi T, Urabe , et al: Apical segmental dysfunction in hypertrophic cardiomyopathy: Subgroup with unique clinical features. J Am Coll Cardiol 1990;16:1579-1588.
33. Fighali S, Krajcer Z, Edelman S, Leachman RD: Progression of hypertrophic cardiomyopathy into a hypokinetic left ventricle: Higher incidence of patients with midventricular obstruction. J Am Coll Cardiol 1987;9:288-294.
34. Krikler DM, Davies MJ, Rowland E, et al: Sudden death in hypertrophic cardiomyopathy: Associated accessory atrioventricular pathways. Br Heart J 1980;43:245-251.
35. Maron BJ, Connor TM, Roberts WC: Hypertrophic cardiomyopathy and complete heart block in infancy. Am Heart J 1981;101:857-860.
36. Maron BJ, Roberts WC: Quantitative analysis of cardiac muscle cell disorganization in the ventricular septum of patients with hypertrophic cardiomyopathy. Circulation 1979;59:689-706.
37. Maron BJ, Anan TJ, Roberts WC: Quantitative analysis of the distribution of cardiac muscle cell disorganization in the left ventricular wall of patients with hypertrophic cardiomyopathy. Circulation 1981;63: 882-894.
38. Maron BJ, Wolfson JK, Epstein SE, Roberts WC: Intramural ("small vessel") coronary artery disease in hypertrophic cardiomyopathy. J Am Coll Cardiol 1986;8:545-557.
39. Mirowski M, Reid PR, Mower MM, et al: Termination of malignant ventricular arrhythmias with an implanted automatic defibrillator in human beings. N Engl J Med 1980;303:322-324.
40. The Antiarrhythmics Versus Implantable Defibrillators (AVID) Investigators: A comparison of antiarrhythmic-drug therapy with implantable defibrillators in patients resuscitated from near-fatal ventricular arrhythmias. N Engl J Med 1997;337:1576-1583.
41. Moss AJ, Hall WJ, Cannom DS, et al: Improved survival with an implanted defibrillator in patients with coronary disease at high risk for ventricular arrhythmia. N Engl J Med 1996;335:1933-1940.
42. Moss AJ, Daubert JP: Internal ventricular defibrillation. N Engl J Med 2000;342:398.
43. Silka MJ, Kron J, Dunnigan A, et al: Sudden cardiac death and the use of implantable cardioverter-defibrillators in pediatric patients. Circulation. 1993;87:800-807.
44. Link MS, Wang PJ, Haugh CJ, et al: Arrhythmogenic right ventricular dysplasia: Clinical results with implantable cardioverter defibrillators. J Inter Cardiovasc Elect 1997;1:41-48.
45. Klues HG, Schiffers A, Maron BJ: Phenotypic spectrum and patterns of left ventricular hypertrophy in hypertrophic cardiomyopathy: Morphologic observations and significance as assessed by two-dimensional echocardiography in 600 patients. J Am Coll Cardiol 1995;26:1699-1708.
46. Maron BJ, Gottdiener JS, Epstein SE: Patterns and significance of distribution of left ventricular hypertrophy in hypertrophic cardiomyopathy: A wide angle, two-dimensional echocardiographic study of 125 patients. Am J Cardiol 1981;48:418-428.
47. Borggrefe MM, McKenna WJ: Survival after cardiac arrest in patients with hypertrophic cardiomyopathy: Role of the implantable cardioverter-defibrillator (abstract). Circulation 1999;100 (Suppl I):I-76.
48. Cecchi F, Maron BJ, Epstein SE: Long-term outcome of patients with hypertrophic cardiomyopathy successfully resuscitated after cardiac arrest. J Am Coll Cardiol 1989;13:1283-1288.
49. Maron BJ, Casey SA, Poliac LC, et al: Clinical consequences of hypertrophic cardiomyopathy in an unselected regional United States cohort. JAMA 1999;281;650-655.

50. Maron BJ, Spirito P: Impact of patient selection biases on the perception of hypertrophic cardiomyopathy and its natural history. Am J Cardiol 1993;72:970-972.
51. Maron BJ, Gardin JM, Flack JM, et al: Assessment of the prevalence of hypertrophic cardiomyopathy in a general population of young adults: Echocardiographic analysis of 4111 subjects in the CARDIA Study. Circulation 1995; 92:785-789.
52. Maron BJ, Lipson LC, Roberts WC, et al: "Malignant" hypertrophic cardiomyopathy: Identification of a subgroup of families with unusually frequent premature deaths. Am J Cardiol 1978;41:1133-1140.
53. Fananapazir L, Chang AC, Epstein SE, McAreavey D: Prognostic determinants in hypertrophic cardiomyopathy: Prospective evaluation of a therapeutic strategy based on clinical, Holter, hemodynamic and electrophysiologic findings. Circulation 1992;86:730-740.
54. Geibel A, Brugada P, Zehender M, et al: Value of programmed electrical stimulation using standardized ventricular stimulation protocol in hypertrophic cardiomyopathy. Am J Cardiol 1987; 60:738-739.
55. Kuck K-H, Kunze K-P, Schluter M, et al: Programmed electrical stimulation in hypertrophic cardiomyopathy: Results in patients with and without cardiac arrest or syncope. Eur Heart J 1988;9:177-185.
56. Camm AJ, Nisam S. The utilization of the implantable defibrillator—a European enigma. Eur Heart J 2000;21: 1998-2004.
57. Glancy DL, O'Brien KP, Gold HK, et al: Atrial fibrillation in patients with idiopathic hypertrophic subaortic stenosis. Br Heart J 1970;32:652-659.
58. Losi MA, Betocchi S, Aversa M, et al: Determinants of atrial fibrillation development in patients with hypertrophic cardiomyopathy. Am J Cardiol 2004;94:895-900.
59. Furlan AJ, Craciun AR, Raju NR, et al: Cerebrovascular complications associated with idiopathic hypertrophic subaortic stenosis. Stroke 1984;15:282-284.
60. Higashikawa M, Nakamuri Y, Yoshida M, et al: Incidence of ischemic strokes in hypertrophic cardiomyopathy is markedly increased if complicated by atrial fibrillation. Jpn Circ J 1997;61:673-681.
61. Robinson KC, Frenneaux MP, Stockins B, et al: Atrial fibrillation in hypertrophic cardiomyopathy: A longitudinal study. J Am Coll Cardiol 1990;15:1279-1285.
62. Greenspan AM: Hypertrophic cardiomyopathy and atrial fibrillation: A change of perspective (editorial). J Am Coll Cardiol 1990;15:1286-1287.
63. Spirito P, Lakatos E, Maron BJ: Degree of left ventricular hypertrophy in patients with hypertrophic cardiomyopathy and chronic atrial fibrillation. Am J Cardiol 1992;69:1217-1222.
64. Gruver EJ, Fatkin D, Dodds GA, et al: Familial hypertrophic cardiomyopathy and atrial fibrillation caused by Arg663His beta-cardiac myosin heavy chain mutation. Am J Cardiol 1999;83: 13H-18H.
65. Cecchi F, Montereggi A, Olivotto I, et al: Risk for atrial fibrillation in patients with hypertrophic cardiomyopathy assessed by signal-averaged P-wave. Heart 1997;78:44-49.
66. Cecchi F, Olivotto I, Montereggi A, et al: Hypertrophic cardiomyopathy in Tuscany: Clinical course and outcome in an unselected population. J Am Coll Cardiol 1995;26:1529-1536.
67. Maron BJ, Olivotto I, Bellone P, et al: Clinical profile of stroke in 900 patients with hypertrophic cardiomyopathy. J Am Coll Cardiol, 2002;39:301-307.
68. Levy S, Breithardt G, Campbell RW, et al: Atrial fibrillation: Current knowledge and recommendations for management. Working Group on Arrhythmias of the European Society of Cardiology. Eur Heart J 1998;19:1294-1320.
69. Kannel WB, Abbot RD, Savage DD, et al: Epidemiologic features of chronic atrial fibrillation: The Framingham Study. N Engl J Med 1982;306:1018-1022.
70. Benjamin EJ, Wolf PA, D'Agostino RB, et al: Impact of atrial fibrillation on the risk of death: The Framingham Heart Study. Circulation 1998;98:946-952.
71. Cameron A, Schwartz MJ, Kronmal RA, Kosinski AS: Prevalence and significance of atrial fibrillation in coronary artery disease (CASS Registry). Am J Cardiol 1988;61:714-717.
72. Bonow RO, Frederick TM, Bacharach SL, et al: Atrial systole and left ventricular filling in patients with hypertrophic cardiomyopathy: Effect of verapamil. Am J Cardiol 1983;51:1386-1391.
73. Wigle ED, Rakowski H, Kimball BP, et al: Hypertrophic cardiomyopathy: Clinical spectrum and treatment. Circulation. 1995;92:1680-1692.
74. Maron BJ, Spirito P, Green KJ, et al: Noninvasive assessment of left ventricular diastolic function by pulsed Doppler echocardiography in patients with hypertrophic cardiomyopathy. J Am Coll Cardiol 1987;10:733-742.
75. Losi MA, Betocchi S, Grimaldi M: Heterogeneity of left ventricular filling dynamics in hypertrophic cardiomyopathy. Am J Cardiol 1994;73:987-990.
76. Cannon RO, Rosing DR, Maron BJ, et al: Myocardial ischemia in hypertrophic cardiomyopathy: Contribution of inadequate vasodilator reserve and elevated left ventricular filling pressures. Circulation 1985;71:284-243.
77. O'Gara PT, Bonow RO, Maron BJ, et al: Myocardial perfusion abnormalities in patients with hypertrophic cardiomyopathy: Assessment with thallium-201 emission computed tomography. Circulation 1987;76:1214-1223.
78. Roy D, Talajic M, Dorian P, et al: Amiodarone to prevent recurrence of atrial fibrillation. Canadian Trial of Atrial Fibrillation Investigators. N Engl J Med 2000;342:913-920.
79. Brembilla-Perrot B, Terrier de la Chaise A, Beurrier D: Paroxysmal atrial fibrillation: Main cause of syncope in hypertrophic cardiomyopathy. Arch Mal Coeur Vaiss 1993;86:1573-1576.
80. Reiffel JA: Drug choices in the treatment of atrial fibrillation. Am J Cardiol 2000;85:12D-19D.

Chapter 30 Evaluation and Management of Arrhythmias in Dilated Cardiomyopathy and Congestive Heart Failure

STEFAN H. HOHNLOSER

Dilated cardiomyopathy (DCM) is a disease that is characterized by enlargement of the cardiac chambers and impaired systolic function of the left or of both ventricles. DCM is a major source of morbidity and mortality from cardiovascular disease and is one of the prime indications for heart transplantation. The presumed incidence of DCM is approximately 5 to 8 per 100,000 population per year with a likely recent increase in its prevalence.[1] The precise prevalence may be even higher because mild or asymptomatic cases may be underreported in the literature.[2] In many patients, the cause of DCM cannot be defined with certainty. There are a large variety of specific diseases of heart muscle that can produce clinical manifestation of DCM. Accordingly, this condition is likely to represent a final common pathway of many diseases resulting in myocardial damage.[1]

Mortality and Causes of Death in Dilated Cardiomyopathy

Mortality in patients with DCM ranges between 10% and 50% annually, mainly determined by the underlying causes[3] and the severity of the disease. Recent years have seen an improvement in the prognosis of patients afflicted with DCM. Redfield and associates have demonstrated that secular trend and referral bias affect the apparent natural history of DCM.[4] In that study, survival in referral patients with this disease was significantly better than previously described.[4] In 1992, Tamburro and Wilber compiled 14 studies including a total of 1432 patients and demonstrated an average mortality rate over 4 years of 42%.[5] Of particular importance, 28% of all fatalities were classified in that analysis as sudden deaths (Table 30-1).[5] In recent years, several randomized clinical trials evaluating potential survival benefits of device therapy in DCM patients have demonstrated that all-cause mortality rates have declined dramatically.[17-19] In fact, annual mortality rates of as low as 6% to 7% have been reported. The reason for this marked improvement in prognosis is for the most part due to adherence to optimal medical therapy, comprising the administration of ACE-inhibitors, β-blockers, and aldosterone antagonists. Despite this significant decline in all-cause mortality, DCM continues to represent the substrate for approximately 10% of all sudden cardiac deaths in the adult population.[20] Likely mechanisms of sudden death in DCM include rapid ventricular tachycardia (VT), ventricular fibrillation (VF), or primary bradyarrhythmias or electromechanical dissociation. The specific pathophysiologic mechanisms of underlying sudden death are dependent on disease severity. For example, Luu and coworkers reported the mechanisms of 21 episodes of unexpected cardiac arrest in 20 patients hospitalized for cardiac transplantation evaluation and management of advanced heart failure, including 7 with DCM.[21] The rhythm at the time of arrest was severe bradycardia or electromechanical dissociation in 13 (62%) patients. Only 38% of arrests in that particular patient population were caused by VT or VF. Of particular note, cardiac arrest was due to a bradyarrhythmia or electromechanical dissociation in every patient with DCM in that series. Precipitating factors included coronary artery embolus or thrombosis, electrolyte disarrangement, pulmonary emboli, and others. Thus, the proportion of tachyarrhythmic sudden death appears to decline in advanced stages of congestive heart failure, whereas death due to VT or VF is responsible for the majority of sudden case fatalities in patients with New York Heart Association (NYHA) II and III functional classes.

Pathophysiology of Arrhythmias in Dilated Cardiomyopathy

The pathogenesis of ventricular arrhythmias in DCM is incompletely understood. Table 30-2 lists potential

TABLE 30-1 Mortality in Dilated Cardiomyopathy

Study	Ref.	Year	# Pts (% male)	% Mortality at 1-8 yr (% sudden deaths)			
				1 yr	*2 yr*	*3 yr*	*8 yr*
Fuster et al	(6)	1981	104 (62)	32	48	51	70
Huang SK et al	(7)	1983			11 (50)		
Franciosa et al	(8)	1983	87 (100)	22	47	70 (42)	
Schwarz et al	(9)	1984	68 (81)			34 (35)	
Neri et al	(10)	1986	65 (80)			29 (21)	
Diaz et al	(11)	1987	169 (78)	28	43	50	63 (9)
Stevenson et al	(12)	1987	28 (64)	54	76 (82)		
Hofmann et al	(13)	1988	101 (93)			49 (51)	
Romeo et al	(14)	1989	104 (75)			34 (40)	
Komajda et al	(15)	1990	201 (81)	1	5	10	30 (31)
Sugrue et al	(16)	1992	40 (64)	5	8	10	41 (33)
Baensch et al*	(17)	2002	104 (83)	3.7 (0)†			
Strickberger et al*	(18)	2003	103 (70)	10†		13†	
Kadish et al*	(19)	2004	458 (71)	6.2†	14.1*		

*Randomized controlled clinical trial.

†Standard therapy group.

contributing factors. Autopsy studies have revealed extensive subendocardial scarring in the left ventricle in one third of patients and multiple patchy areas of replacement fibrosis in more than one half of the patients.[22] Such areas contribute to disparities in conduction and refractoriness, thereby providing the substrate for reentrant arrhythmias. This has been recently confirmed in an elegant study in explanted human hearts from patients who underwent transplantation for end-stage DCM.[23] These investigators performed high-resolution epicardial mapping and demonstrated that increased fibrosis provides sites for conduction block, leading to the continuous generation of reentry.

TABLE 30-2 Arrhythmogenic Factors in Dilated Cardiomyopathy

Structural Abnormalities
Fibrosis and scarring
Increased dispersion of conduction and repolarization
Altered response to antiarrhythmic drugs
Cellular Abnormalities
Action potential prolongation
Altered density of repolarization currents (i.e., I_{to}, I_{K1})
Altered gap junction structure and distribution
Altered response to antiarrhythmic drugs
Hemodynamic Factors
Altered refractoriness during changes in loading
Stretch-induced afterdepolarizations
Altered Cardiac Autonomic Tone
Sympathetic activation
Reduction in vagal tone
Increased disparity of repolarization
Altered response to antiarrhythmic drugs
Electrolyte Imbalances
Polypharmacotherapy

Occlusion of small intramyocardial arteries by thrombosis or emboli has also been demonstrated and may be associated with acute ischemia and chronic scarring.[24] Alterations in ventricular geometry and mechanical function may also contribute to reentrant arrhythmias through variable wall tension and stretch-dependent shortening of ventricular refractoriness.[25] Action potential prolongation in the absence of other significant electrophysiological changes is a hallmark of failing ventricular myocardium.[26,27] It is associated with a functional downregulation of the Ca^{++}-independent transient outward potassium current (I_{to}) and inward rectifier potassium current (I_{K1}).[26,28] The molecular basis of this I_{to} downregulation in human heart failure has recently been reported by Kääb and associates.[29] Heart failure–associated alterations of gap junction spatial distribution and decreased connexin[43] in regions of normal gap junction density have also been reported.[30] Changes in cardiac autonomic tone, particularly surges in sympathetic activity, are associated with heart failure. There is both experimental and clinical evidence that sympathetic activation is closely linked to the occurrence of spontaneous ventricular arrhythmias.[31,32] Sympathetic activation may promote reentrant arrhythmias by enhancing regional differences in conduction and refractoriness and by facilitation of early and delayed afterdepolarizations. Finally, many patients with congestive heart failure are subjected to polypharmacotherapy, which may result in electrolyte disorders, most notably in hypokalemia and hypomagnesemia. Both may further contribute to the genesis of ventricular arrhythmias. Of particular note, proarrhythmic side effects of antiarrhythmic drugs have been demonstrated to occur much more frequently in patients with reduced left ventricular function.[33]

The most characteristic form of VT in DCM is bundle branch reentry tachycardia (BBRT). This particular form of reentry produces VT via a macro-reentrant circuit involving the His-Purkinje system, usually with antegrade conduction over the right bundle branch and retrograde conduction over the left bundle. In patients with DCM, this particular form of VT may account for up to 25% of all VTs.[34] BBRT is usually a very rapid VT associated with syncope in the majority of cases. Degeneration to VF can occur. The importance of recognizing this particular arrhythmia is due to the fact that there is a cure for this VT by means of catheter ablation of the right bundle branch (see later).

Risk Stratification for Arrhythmogenic Death in Dilated Cardiomyopathy

The assessment of risk of sudden death compared with the risk of any death is probably one of the greatest challenges facing the clinician who is taking care of a patient with DCM. With the relatively crude risk stratification methods available at present, it is very difficult to determine the life expectancy of an individual patient afflicted with this common disease.

CLINICAL RISK FACTORS

Medical history and the physical examination offer few hints as to which patients are at increased mortality risk and particularly at increased risk for arrhythmogenic death. This may be due to the subjective nature of such data. For instance, the functional status as assessed by the NYHA class has been found in some studies to be predictive of mortality,[35] whereas other investigators reported no correlation.[13] The only clinical sign that has been demonstrated to have predictive power is a history of syncope. Brambilla-Perrot and associates prospectively followed 103 DCM patients over a mean period of 24 months.[36] Ten patients had developed VT or had sudden death; of note, 7 of these 10 patients had had a history of syncope. These observations have been subsequently confirmed by other investigators.[37,38]

Left Ventricular Function and Functional Status

The degree of left ventricular dysfunction is, by far, the most important predictor of mortality in patients with DCM and congestive heart failure. Various indices of left ventricular function including pulmonary artery wedge pressure, left- and right-sided pressures, cardiac index, and right atrial pressure have been correlated with outcome in clinical studies. Most often, left ventricular ejection fraction (LVEF) has been used to assess the predictive power of hemodynamic function. For instance, Hofmann and colleagues showed that 84% of DCM patients with an LVEF of less than 35% died over an average follow-up time of 39 months compared with 46% of patients with better preserved LV-function.[13] Unfortunately, measures of LV function failed to correlate with the specific cause of subsequent mortality; i.e., sudden versus nonsudden death. Only one study found a correlation of depressed right ventricular function and subsequent risk of sudden death.[14]

Electrocardiography

Several abnormalities diagnosed from the surface ECG have been described as risk factors for mortality in patients with DCM. First- or second-degree AV block has been reported as an independent risk factor both for sudden and nonsudden cardiac deaths.[39] Similarly, intraventricular conduction defects have been found to be associated with increased mortality by some investigators[40] but not by others.[13,14] The finding of atrial fibrillation on the standard ECG seems to be associated with poor outcome. For instance, Hofmann and coworkers reported that the presence of atrial fibrillation significantly increased the risk for both: death from congestive heart failure and from arrhythmias.[13] There are, however, conflicting results stemming from various studies for all of these ECG-derived risk factors. Accordingly, the standard ECG is not a particularly helpful tool for risk stratification in DCM.

Spontaneous Ventricular Arrhythmias

Spontaneous ventricular arrhythmias, including episodes of nonsustained VT, are found in most patients with DCM. This high prevalence makes their prognostic significance difficult to assess and generally results in a low specificity of these arrhythmias. Several investigators have used detailed analyses of the 24-hour Holter ECG in an attempt to improve specificity. For instance, Meinertz and coworkers quantified ventricular ectopic activity in 74 patients with DCM and found that 93% of patients had ventricular arrhythmias, 35% had more than 1000 ventricular premature beats per 24 hours, and 49% had at least one episode of nonsustained VT.[41] There was a significant inverse correlation between the number of extrasystoles and LVEF. Patients who subsequently suffered from sudden death had a significantly higher number of episodes of nonsustained VT, ventricular pairs, and total extrasystoles. Particularly, the presence of nonsustained VT on Holter monitoring has subsequently been reevaluated in a plethora of studies (Table 30-3). The analysis of these investigations, however, demonstrates the heterogeneity of findings and the diverging results with respect to the predictive power of nonsustained VT episodes for subsequent sudden death (see Table 30-3). On careful reassessment of studies evaluating the predictive power of spontaneous arrhythmias for subsequent arrhythmic death, it seems that spontaneous ventricular arrhythmias predominantly reflect the extent of the underlying structural changes, thereby implicating both tachyarrhythmias and pump failure as causes of subsequent death.

Signal-Averaged Electrocardiography

In patients with coronary artery disease, detection of late potentials in the signal-averaged ECG (SAECG) is considered to be a marker of slow conduction and has been found to be linked to the occurrence of sustained VT.

TABLE 30-3 Nonsustained VT as a Predictor of Sudden Death

Study	Ref	Year	Pts (n)	Incidence of NsVT (%)	Follow-up (months)	No. of Arrhythmic Events (%)	Association with SD
Huang et al	(7)	1983	35	21 (60)	34 ± 17	2 (6)	no
Von Olshausen et al	(42)	1984	60	25 (42)	12 ± 5	3 (5)	no
Unverferth et al	(40)	1984	69	11 (16)	12	22 (32)	yes
Hofmann et al	(13)	1988	110	49 (45)	39 ± 21	25 (23)	yes
Von Olshausen et al	(43)	1988	73	31 (42)	≥36	14 (19)	no
Romeo et al	(14)	1989	104	19 (18)	46 ± 42	14 (14)	yes
Brembilla-Perrot et al	(36)	1991	103	56 (54)	24 ± 8	10 (10)	no
De Maria et al	(44)	1992	218	67 (27)	29 ± 16	12 (6)	no
Kadish et al	(45)	1993	43	43 (100)	20 ± 14	7 (16)	no
Schoeller et al	(39)	1993	85	27 (32)	49 ± 37	13 (15)	yes
Turitto et al	(46)	1994	80	80 (100)	22 ± 26	9 (11)	no
Baker et al	(47)	2000	380	125 (33)	57 ± 40	n/a	yes
Fauchier et al	(48)	2000	176	n/a	47 ± 39	22 (13)	no

NsVT, nonsustained ventricular tachycardia; SD, sudden death.

A number of studies have similarly evaluated the value of this technique to risk-stratify patients with DCM. Generally, a lower incidence of late potentials has been described compared with postinfarction patients. In some studies in DCM patients, SAECG has been found to have predictive value for the development of sustained ventricular arrhythmias and sudden death, but careful analysis of these studies reveals a wide range of sensitivities and specificities.[49-52] For instance, Fauchier and associates recently reported results from a study comprising 131 DCM patients of whom 43 suffered from arrhythmic events or sudden death during a mean follow-up of 54±41 months.[48] Late potentials were associated with all-cause cardiac death (relative risk 3.3, 95% C.I. 1.5 to 7.5) and with arrhythmic events (relative risk 7.2, 95% C.I. 2.6 to 19.4). Of note, however, this study also included patients with a previous history of VT or VF. Yi and colleagues recently introduced wavelet decomposition of the SAECG as a more sophisticated type of analysis for risk stratification in DCM.[53] Although they reported this methodology to be superior over conventional SAECG analysis, they considered it unlikely that wavelet decomposition of the SAECG would allow prospective identification of DCM patients at high risk for subsequent arrhythmic events. Accordingly, the predictive value of the SAECG in DCM remains controversial at best.

Measures of Cardiac Autonomic Tone

Heart rate variability (HRV) and baroreflex sensitivity have proved to be useful tools to study cardiac autonomic tone.[31] Reduction of one or both parameters in survivors of myocardial infarction has been demonstrated to be associated with increased risk of death, particularly arrhythmogenic death. In patients with DCM, only sparse data exist regarding the predictive power of these two noninvasive tests. For instance, Hoffmann and coworkers measured HRV in 71 patients with DCM, 10 of whom developed VT or VF or sudden death during follow-up.[54] Analysis of time-domain parameters indicated lower values in patients compared with those without arrhythmic events. Another study confirmed that, in general, HRV is reduced in DCM and is related to disease severity.[55] However, depressed HRV was related only to death from progressive heart failure. Fauchier and associates reported that reduced HRV was an independent marker for both nonsudden and sudden death in a cohort of 131 DCM patients.[48] Actually, on multivariate analysis, a history of previous VT or VF, depressed HRV, and late potentials were the only independent risk predictors in that study. BRS has not been systematically examined as a risk stratifier in patients with DCM. One study showed that BRS is weakly associated with HRV in DCM, indicating that both methods may reflect different aspects of cardiac autonomic control.[56] Given the established role of autonomic measures in patients after myocardial infarction, more studies in patients with DCM are necessary and appear worthwhile.

T Wave Alternans

Analysis of microvolt level T wave alternans (MTWA) from the surface ECG has been introduced as a new approach to evaluate the risk of ventricular arrhythmias.[57] MTWA involves detecting alterations in T wave morphology that occur on an every-other-beat basis. MTWA is thought to reflect the occurrence of localized action potential alternans, which creates dispersion of recovery and, in turn, promotes the development of reentrant arrhythmias. Recently, Adachi and colleagues examined the clinical significance of MTWA in 58 patients with DCM using bicycle exercise testing.[58] In an attempt to compare the results of MTWA assessment with other risk stratifiers, they also determined the functional class of their patients, recorded an SAECG, determined QT dispersion from the surface ECG, and noted the grade of ventricular arrhythmia from Holter recordings. Of the 58 patients,[23] tested positive for MTWA. Univariate analysis revealed that the percentage of patients with ventricular tachycardia (sustained and nonsustained) was significantly higher in patients with MTWA as compared with MTWA-negative patients (61% versus 8%; $P < .001$). Unfortunately, this study is limited by the fact that no prospective follow-up was performed.

Yap and coworkers studied different repolarization parameters in patients with DCM.[59] MTWA, QT dispersion, and principal component analysis ratio were analyzed in 34 individuals. MTWA-positive patients had larger left ventricular diastolic and systolic diameters. No association was found, however, between the three repolarization measurements, suggesting that these studies examined different components of ventricular repolarization.[56]

More recently, Kitamura and colleagues studied 104 patients with DCM and observed that during a follow-up period of 21±14 months, 12 arrhythmic events occurred. On multivariate analysis, a positive MTWA finding at an onset heart rate of less than 101 bpm and LVEF were independent predictors of these events.[60] In a similar study from our institution, 137 DCM patients were subjected to a battery of noninvasive risk-stratification methods. Over the average observation period of 14±6 months, only the result of MTWA assessment proved to independently predict ventricular tachyarrhythmic events.[61] The relative risk for an arrhythmic event was 3.44 in patients who tested MTWA positive. In contrast to these three studies, Grimm and associates reported that in their series of DCM patients, assessment of MTWA did not yield predictive power for future arrhythmic events.[62] One major discrepancy between the Grimm study and the other three studies is that β-blockers were withheld in the first but not in the other studies before assessment of MTWA. It may be speculated that in the Grimm study[62] the number of MTWA-positive patients would have been substantially reduced if all patients who took β-blockers during follow-up had been tested while on β-blockers; by the same token, the event rate may have been substantially higher among patients who still tested MTWA positive despite β-blockers. In summary, therefore, analysis of MTWA by noninvasive means holds promise for being an important risk stratifier in patients with stable DCM. Prospective, randomized, interventional trials are currently under way to prove this hypothesis.

Electrophysiological Testing

Programmed ventricular stimulation is of proven value for risk stratification in patients with coronary artery disease, previous myocardial infarction, reduced LVEF, and asymptomatic nonsustained VT on Holter monitoring.[63,64] In DCM, however, induction of ventricular tachyarrhythmias in the electrophysiological laboratory does not reliably predict future arrhythmic events or sudden death. Most notably, this is true even for those patients with DCM who have survived an episode of cardiac arrest.[36,65] Sustained monomorphic VT is only rarely induced in DCM.[36,66] The induction of polymorphic VT or VF is more often observed in these patients, but these arrhythmias are considered nonspecific and have no prognostic power. The low sensitivity, specificity, and positive predictive value of programmed ventricular stimulation for risk assessment in DCM are detailed in Table 30-4. Specifically, patients with DCM and a history of previous syncope deserve a comment in this context. Knight and colleagues demonstrated in 14 patients with DCM with a history of previous syncope and no inducible arrhythmias a high incidence of subsequent spontaneous tachyarrhythmic events.[37] Mittal and coworkers performed programmed ventricular stimulation in 21 such patients.[38] Eight patients were inducible for monomorphic VT, whereas 13 were not. During follow-up, 8 patients had VT or VF and 5 died. The results of electrophysiological testing were not helpful in identifying those patients with VT or VF or sudden death.

Management

ANTIADRENERGIC THERAPY

The rationale for the use of β-blockers in nonischemic DCM is to counteract the deleterious effects of increased sympathetic activation in patients with congestive heart failure. Waagstein and associates were the first to propose the use of β-blockers in this patient population.[71] Following these initial observations, several small studies showed improvement in symptoms in patients with nonischemic cardiomyopathy (CMP)

TABLE 30-4 Predicted Value of Electrophysiological Testing in DCM

			Inducible Arrhythmias				Arrhythmic Events During Follow-up		
Study	**Year**	**Pts (n)**	*NsVT (%)*	*SmVT (%)*	*PmVT/VF (%)*	**(Months)**	*Ind. SmVT SmVT (%)*	*Ind. PmVT (%)*	*Noninducible (%)*
Meinertz (66)	1985	42	35	0 (0)	1 (2)	16 ± 7	0 (0)	0 (0)	2 (5)
Stamato (67)	1986	15	93	0 (0)	0 (0)	19 ± 4	0 (0)	0 (0)	2 (13)
Poll (65)	1986	20	100	2 (10)	4 (20)	17 ± 14	0 (0)	2 (50)	5 (36)
Das (68)	1986	24	n/a	1 (4)	4 (17)	12 ± 6	1 (100)	1 (25)	2 (11)
Gossinger (69)	1990	32	100	4 (13)	0 (0)	21	1 (25)	0 (0)	2 (7)
Brembilla-Perrot (36)	1991	92	46	3 (3)	5 (5)	24 ± 8	2 (67)	2 (40)	3 (4)
Kadish (45)	1993	43	100	6 (14)	n/a	20 ± 14	1 (17)	n/a	6 (16)
Turitto (46)	1994	80	100	10 (13)	7 (9)	22 ± 26	3 (30)	0 (0)	6 (10)
Grimm (70)	1997	34	100	3 (9)	10 (29)	24 ± 13	1 (33)	3 (30)	5 (24)

Ind, inducible; NsVT, nonsustained ventricular tachycardia; PmVT, polymorphic ventricular tachycardia; SmVT, sustained monomorphic ventricular tachycardia.

TABLE 30-5 Deaths in the Large β-Blocker Mortality Analyses

	USCP		CIBIS II		MERIT-HF	
	Placebo (n=398)	*Carvedilol (n=696)*	*Placebo (n=1320)*	*Bisoprolol (n=1327)*	*Placebo (n=200)*	*Metoprolol (n=1990)*
All	31	22	228	156	217	145
Cardiovascular (%)	31 (100)	20 (91)	161 (71)	119 (76)	203 (94)	128 (88)
Sudden (%)	15 (48)	12 (55)	83 (36)	48 (31)	132 (61)	79 (54)
Pump failure (%)	13 (42)	5 (23)	47 (12)	36 (23)	58 (27)	30 (21)

HF, heart failure.

treated with this medication. Analysis of pooled data stemming from 9 studies with 144 patients showed improved symptomatic status in 66% of patients, no change in symptoms in 19%, and worsened symptoms in the remaining 15%.[72] Recently, three large-scale prospective randomized placebo-controlled trials have been conducted in patients with congestive heart failure due to either ischemic or nonischemic CMP in order to evaluate the effects of β-blockers added to the optimal standard treatment of heart failure on all-cause mortality (Table 30-5).[73-75] All three trials showed major reductions in mortality and morbidity in patients receiving β-blockers when compared with control patients. In all trials, a varying proportion of patients with nonischemic DCM was included. Actually, the largest proportion of DCM patients (35%) was enrolled in the MERIT-HF trial.[75] The benefits derived from β-blockers were similar in patients with DCM compared with those with ischemic CMP. Of particular note, in CIBIS-II and in MERIT-HF decreases in sudden death were 44% and 41%, respectively, compared with a reduction in heart failure death of 26% and 49%, respectively.[74,75] These results confirm results of earlier experimental and clinical works demonstrating the antifibrillatory effects of antiadrenergic therapy. Accordingly, initiated at a low dose and titrated slowly, β-blockers are well tolerated and are considered a mainstay of contemporary medical therapy of patients with DCM and congestive heart failure.

ANTIARRHYTHMIC DRUGS

During the past 10 to 15 years, randomized trials have investigated the ability of several antiarrhythmic drugs to reduce sudden cardiac death in patients deemed to be at high risk.[76-79] Apart from β-blockers, however, no other agent has been conclusively demonstrated reduced mortality. Actually, there have been clear increases in mortality with some classes I and III antiarrhythmic drugs.[77-79] Although these trials do not all pertain specifically to patients with DCM, antiarrhythmic pharmacotherapy has been subjected to a reassessment in this patient population. It is now well appreciated that cardiac side effects, specifically proarrhythmic effects of these compounds, are much more frequently encountered in patients with depressed left ventricular function compared with patients with well-preserved LVEF.[80] Furthermore, most class I antiarrhythmic drugs can exacerbate left ventricular dysfunction. Accordingly, antiarrhythmic drug development has focused on new class III substances that prolong ventricular repolarization and refractoriness as their sole or predominant mode of action. In the clinical setting, amiodarone is probably the most commonly used drug, both for treatment of symptomatic supraventricular and ventricular arrhythmias, as well as for primary prevention of sudden death in patients with congestive heart failure.

In patients with significant ventricular dysfunction and congestive heart failure, several randomized trials have evaluated the effects of amiodarone on mortality. In the first prospective study enrolling exclusively patients with DCM, Neri and associates described a significant decrease in spontaneous ventricular arrhythmias and sudden death.[81] Subsequently, several randomized trials evaluated the beneficial effects of amiodarone in patients with congestive heart failure.[82-85] The proportion of patients with nonischemic CMP varied between 28.6% and 63%.[86] The two largest studies were the GESICA[84] and the CHF-STAT trials[82]; in the first, amiodarone reduced total mortality 28% (95% C.I. 4% to 45%) with death from congestive heart failure and sudden mortality reduced by 23% and 27%, respectively.[83] In a further analysis of these data, it was shown that these beneficial effects of amiodarone were particularly evident in patients with elevated resting heart rates equal to or greater than 90 bpm.[87]

In contrast, the CHF-STAT trial failed to demonstrate a reduction in all-cause or sudden death mortality in the entire study cohort despite a trend toward reduced mortality among patients with nonischemic CMP.[82] In a substudy, the effects of amiodarone on left ventricular function were carefully evaluated by means of repeated LVEF determinations.[88] Compared with placebo, amiodarone resulted in a significant improvement in left ventricular dysfunction at each time point during follow-up.

In a recent meta-analysis comprising 13 prospective randomized placebo-controlled studies (8 postinfarction and 5 congestive heart failure trials) with a total of 6553 patients, amiodarone was associated with a 13% reduction in overall mortality compared with placebo.[86] Arrhythmia or sudden death was reduced by 29%. Discontinuation of amiodarone, mostly due to side effects, varied between 4.6% and 40.2% in individual studies.[82]

Probably the most convincing evidence regarding the use of amiodarone for the primary prevention of sudden death stems from the recently completed SCD-HeFT trial.[89] In that trial, patients with either ischemic or nonischemic CMP, LVEF of less than 36%, and a history of congestive heart failure were randomized to receive placebo, amiodarone, or an implantable cardioverter defibrillator (ICD). Importantly, drug therapy was administered in a double-blind manner. Almost 50% of enrolled patients suffered from nonischemic DCM. Over an observation period of 6 years, there was no difference in all-cause mortality in patients assigned to receive amiodarone therapy compared with those on placebo. In contrast, there was a significant benefit of ICD therapy (see later). In summary, therefore, amiodarone seems to be of no benefit when used for the primary prevention of sudden death in patients with congestive heart failure due to DCM. The lack of significant negative inotropic effects and its low proarrhythmic potential make this substance a suitable antiarrhythmic drug for the therapy of symptomatic supraventricular (particularly atrial fibrillation) or nonsustained symptomatic ventricular arrhythmias in this patient population.

CATHETER ABLATION

As already indicated, bundle branch reentry tachycardia (BBRT) is most frequently found in patients with DCM. BBRT is a form of VT in which the right and left bundle branches, the ventricular septum, and the distal His bundle constitute a macro-reentrant circuit. VT is generally rapid, and syncope or cardiac arrest is the usual clinical presentation. If BBRT is diagnosed according to established criteria,[90,91] radiofrequency catheter ablation is the appropriate first-line therapy. Radiofrequency current is applied to the right bundle when a right bundle potential is recorded by the distal bipole of the ablation catheter. This procedure is effective and safe, although the published experience with catheter ablation of BBRT is limited.[92-94] Following ablation, some patients will require dual chamber pacing, particularly those with a preexisting wide left bundle branch block.

Recently, Delacretaz and colleagues reported their experience in 26 consecutive patients with nonischemic CMP referred for management of recurrent VT.[95] These investigators described three distinct mechanisms of VT: BBRT (19% of patients), VT based on reentrant mechanisms (62%), and VT due to focal automaticity (27%). Success of radiofrequency current application was 100% for patients with BBRT, 86% in patients with focal VT, and 53% in patients with VT due to reentrant mechanisms. In this series, the overall success rate of catheter ablation was 77% (20 of 26 patients). In these patients, clinical VT recurrences were abolished during an average follow-up time of 15±12 months.[95]

Another study reported results of catheter ablation in eight DCM patients with nine different VTs.[96] In four of these patients, VT was incessant. VT mapping was performed using endocardial mapping during sinus rhythm, activation and entrainment mapping during VT, and pace mapping. After radiofrequency current application all incessant VTs were rendered noninducible, but only two of five chronic recurrent VTs were rendered noninducible. During follow-up, however, complete prevention of VT could be achieved in only two of the eight patients.

IMPLANTABLE CARDIOVERTER DEFIBRILLATOR THERAPY FOR SECONDARY PREVENTION OF SUDDEN DEATH

Over the past 2 decades, the ICD has become an important treatment modality for secondary prevention of cardiac arrest and sudden death. Three large randomized trials of ICD therapy versus medical treatment for prevention of death in survivors of VF or sustained VT have been reported.[97-99] AVID was the largest of these three trials, but it was stopped early because of the observation of a greater-than-expected benefit of the ICD. Accordingly, this trial has the shortest follow-up of all. The percentage of DCM patients enrolled was approximately 9.5% in CIDS and 12% in CASH, respectively. A subanalysis of AVID reported mortality figures for 162 patients with nonischemic CMP—not all of whom had DCM.[100] Of these patients, 77 were randomized to drug therapy and 75 received an ICD. All-cause mortality was reduced among patients with an ICD: 9% versus 12% at 1 year, 11% versus 18% at 3 years, and 14% versus 38% at 3 years. A meta-analysis of all three large secondary ICD prevention trials was conducted based on the individual data of all patients who were entered into a master database.[101] This analysis comprised a total of 1866 patients. This meta-analysis showed that results from individual trials were consistent with each other. Specifically, there was a significant reduction in death from any cause with the ICD with a summary hazard ratio of 0.72 (95% C.I. 0.60 to 0.87; $P = .0006$). This 28% reduction in the relative risk of death with the ICD was almost entirely due to a 50% reduction in arrhythmic death (hazard ratio 0.50 [95% C.I. 0.37 to 0.67; $P < .0001$]). The subgroup analysis of patients with nonischemic CMP (225 patients) revealed a similar benefit for ICD patients compared with the larger group of patients with ischemic heart disease (1493 patients).[101] There was, however, an important interaction between LVEF and ICD benefit. Whereas patients with better-preserved left ventricular function obtained only little or no benefit from device therapy, moderate or severe left ventricular dysfunction was associated with a significant positive treatment effect of the ICD. Because of the retrospective nature of this analysis, however, the authors emphasize that these results should be interpreted cautiously and need confirmation by a prospective study.

Primary Prevention of Sudden Death in Dilated Cardiomyopathy

The high incidence of sudden arrhythmogenic death in DCM has prompted the design of several prospective randomized trials that aim to compare the benefits of the ICD with those of the best medical therapy in patients with DCM but with no history of sustained ventricular tachyarrhythmias. The CAT trial (Cardiomyopathy Trial) in Germany looked for patients with newly diagnosed DCM (<9 months) and an LVEF of less than or equal to 35%. A total of 104 patients were enrolled in the pilot phase of this study, randomly assigned to receive an ICD (50 patients) or to serve as controls (54 patients) and were followed for a mean of 5.5 ± 2.2 years.[17] At the end of the observation period, there was no significant difference in terms of all-cause mortality between both groups. In a similar study, Strickberger and colleagues randomized 103 patients with DCM to receive either amiodarone or the ICD.[18] This trial was stopped when the prospective stopping rule for futility was reached. The percentage of patients surviving at 1 year (90% versus 96%) and at 3 years (88% versus 87%) were not statistically significantly different from each other. The largest primary prevention trial in DCM patients is the DEFINITE (Defibrillators in Nonischemic Cardiomyopathy Treatment Evaluation) trial.[19] This multicenter study enrolled patients with DCM, an LVEF of less than 35%, a history of symptomatic heart failure, and spontaneous arrhythmia (>10 premature ventricular contractions per hour or nonsustained VT). In this trial, 458 patients were randomly assigned to receive the ICD (229 patients) or best medical therapy (229 patients). Patients were followed for 29 ± 14.4 months, during which time 68 patients died: 28 in the ICD group, 40 in the control group. The hazard ratio for all-cause mortality was 0.65 (95% C.I. 0.40 to 1.06; $P = 0.08$).[19] Sudden death mortality was significantly reduced in the ICD group with a hazard ratio of 0.20 (95% C.I. .06 to .71; $P = .006$). Importantly, the average duration of heart failure was 2.8 years, which is in contrast to the much shorter duration in the CAT trial.

The aforementioned SCD-HeFT trial, which is by far the largest primary prevention ICD study, enrolled 2521 patients with symptomatic congestive heart failure, ischemic or nonischemic cardiomyopathy, and an ejection fraction of less than or equal to 35%.[89] One half of the patients were, in fact, suffering from nonischemic DCM. Patients were randomized to receive either placebo or amiodarone or an ICD on top of the best medical therapy. At the end of the observation period, there was no survival benefit in patients allocated to amiodarone when compared with placebo. However, in ICD recipients there was a significant 23% relative risk reduction for all-cause mortality compared with placebo (hazard ratio .77, 95% C.I. .62 to .96; $P = .007$). In summary, therefore, there is now good evidence that the ICD, when used for primary prevention purposes, prolongs life in patients with DCM. However, the absolute mortality reduction achieved by the ICD in SCD-HeFT was 6.9% over 5 years (corresponding to an annual rate of approximately 1.5%). In order to improve the cost-to-benefit ratio, there is the continued need for better identification of DCM patients at particularly high risk for sudden death who are likely to derive the greatest benefit from device therapy.

An important subgroup of DCM patients in which the efficacy of the ICD for primary prevention of sudden death has been examined comprises patients with DCM and unexplained syncope. As mentioned earlier, these patients have a high risk of dying suddenly. Knight and coworkers implanted an ICD in 14 patients with DCM, unexplained syncope, and a negative electrophysiological study and followed the patients for a mean of 24 ± 13 months.[37] Outcome in this group was compared with 19 DCM patients who received their ICD for secondary prevention of sudden death. In the group of patients with syncope, 7 of 14 (50%) received appropriate ICD shocks during follow-up compared with 8 of 19 patients with previous cardiac arrest ($P = .1$). The mean time from ICD implantation to first appropriate device therapy was actually shorter in the syncope patients (10 ± 14 months compared with 48 ± 47 months, $P = .06$). Most notably, recurrent syncope was always associated with ventricular tachyarrhythmias. These findings were subsequently confirmed by a similar study in 21 patients with DCM and unexplained syncope undergoing electrophysiological testing.[38] These authors reported inducibility of monomorphic VT in 40% of their patients. More importantly, however, there was a high incidence of VT or VF in these patients, which was often fatal in the absence of an ICD. Accordingly, primary prevention of sudden cardiac death by ICD implantation appears to be appropriate in patients with DCM and unexplained syncope in the absence of previously documented ventricular tachyarrhythmias.

Pacemaker Therapy

Besides established standard indications for permanent pacing, recent interest in pacing in patients with DCM has focused on a purely hemodynamic indication for pacing. The clinical results of standard dual-chamber pacing for hemodynamic reasons, however, have been mixed—mainly attributable to the small populations studied, marked heterogeneity of patients included, highly variable study designs, and short follow-up periods.[102-106] Hochleitner and colleagues were the first to advocate pacing in DCM for hemodynamic improvement.[102,103] They demonstrated improvements in NYHA class, LVEF, and exercise tolerance. The mechanisms responsible for the clinical benefits are multifactorial. Among others, they may be related to elimination of the presystolic component of mitral and tricuspid regurgitation, providing increased ventricular filling and thus, stroke volume; preservation of AV synchrony; and elevation of diastolic and systolic blood pressures—thereby allowing increases in dosage of angiotensin-converting enzyme inhibitors and β-blockers.[34,104] However, other studies did not find beneficial effects of dual-chamber pacing for improvement of left ventricular function in patients with congestive heart failure.[105,106] No firm indications for dual-chamber permanent pacing for hemodynamic improvement have emerged so far.

More recently, resynchronization therapy by means of biventricular pacing has been proposed as a means to treat patients with DCM and severe congestive heart failure. A large proportion of DCM patients have intraventricular conduction disturbances resulting in a prolonged QRS duration and are thus suitable candidates for resynchronization therapy. Cardiac resynchronization therapy was shown in several small uncontrolled studies to improve exercise capacity, functional class, and quality of life. Bradley and coworkers performed a meta-analysis of the first four clinical trials that randomized patients to cardiac resynchronization therapy or control.[102] Their analysis comprised of 1634 patients and showed that resynchronization therapy reduced mortality from progressive heart failure in patients with symptomatic left ventricular dysfunction.

More recently, the COMPANION study was published in which 1520 patients with NYHA classes III and IV heart failure and a QRS of greater than 120 ms were randomized (in a 1:2:2 fashion) to optimal medical therapy alone or in addition to cardiac resynchronization therapy with or without defibrillator backup.[108] The primary composite endpoint of that study was the time to death from or hospitalization for heart failure. The secondary study endpoint was all-cause mortality. The results showed a significant reduction in the time to the primary endpoint for both methods of resynchronization therapy. All-cause mortality, however, was only significantly reduced in those patients who received an ICD along with resynchronization therapy. Of note, between 41% and 45% of patients enrolled in this trial suffered from nonischemic cardiomyopathy. Among patients with nonischemic cardiomyopathy, pacemaker-defibrillator therapy was associated with a significantly lower risk of death from any cause as compared with pharmacologic therapy alone (hazard ratio .50, 95% C.I. .29 to .88; P = .015). The hazard ratio for patients with ischemic cardiomyopathy was .73 (95% C.I. .52 to 1.04; P = .082). These results suggest that patients with nonischemic DCM as the underlying cause of heart failure may be particularly suited for cardiac resynchronization therapy.

ACKNOWLEDGMENT

The author expresses his gratitude to Sabine Zach-Lampson for her expert editorial assistance.

REFERENCES

1. Wynne J, Braunwald E: The cardiomyopathies and myocarditides. In Braunwald E (ed): Heart Disease. Philadelphia, WB Saunders, 1997, pp 1404-1463.
2. Richardson P, McKenna W, Bristow M, et al: Report of the 1995 World Health Organization/International Society and Federation of Cardiology Task Force on the Definition and Classification of Cardiomyopathies. Circulation 1996;93:841-842.
3. Felker GM, Thompson RE, Hare JM, et al: Underlying causes and long-term survival in patients with initially unexplained cardiomyopathy. N Engl J Med 2000;342:1077-1084.
4. Redfield MM, Gersh BJ, Bailey KR, et al: Natural history of idiopathic dilated cardiomyopathy: Effect of referral bias and secular trend. J Am Coll Cardiol 1993;22:1921-1926.
5. Tamburro P, Wilber D: Sudden death in idiopathic dilated cardiomyopathy. Am Heart J 1992;124:1035-1045.
6. Fuster V, Gersh BL, Giuliani ER, et al: The natural history of idiopathic dilated cardiomyopathy. Am J Cardiol 1981;47:525-531.
7. Huang SK, Messer JV, Denes P: Significance of ventricular tachycardia in idiopathic dilated cardiomyopathy: Observations in 35 patients. Am J Cardiol 1983;51:507-512.
8. Franciosa JA, Wilen M, Ziesche S, Cohn JN: Survival in men with severe chronic left ventricular failure due to either coronary artery disease or idiopathic dilated cardiomyopathy. Am J Cardiol 1983;51:831-836.
9. Schwartz F, Mall G, Zebe H, et al: Determinants of survival in patients with congestive cardiomyopathy: Quantitative morphologic findings and left ventricular hemodynamics. Circulation 1984;70:923-928.
10. Neri R, Mestroni L, Salvi A, Camerini F: Arrhythmias in dilated cardiomyopathy. Postgrad Med J 1986;62:593-597.
11. Diaz R, Obasohan A, Oakley C: Prediction of outcome in dilated cardiomyopathy. Br Heart J 1987;58:393-399.
12. Stevenson LW, Fowler MB, Schroeder JS, et al: Poor survival of patients with idiopathic cardiomyopathy considered too well for transplantation. Am J Med 1987;83:871-876.
13. Hofmann T, Meinertz T, Kasper W, et al: Mode of death in idiopathic dilated cardiomyopathy: A multivariate analysis of prognostic determinants. Am Heart J 1988;116:1455-1463.
14. Romeo F, Pelliccia F, Cianfrocca C, et al: Predictors of sudden death in idiopathic dilated cardiomyopathy. Am J Cardiol 1989;63:138-140.
15. Komajda M, Jais JP, Reeves F, et al: Factors predicting mortality in idiopathic dilated cardiomyopathy. Eur Heart J 1990; 11(9):824-831.
16. Sugrue DD, Rodeheffer RJ, Codd MB, et al: The clinical course of idiopathic dilated cardiomyopathy: A population-based study. Ann Intern Med 1992;117:117-123.
17. Baensch D, Antz M, Boczor S, et al: Primary prevention of sudden cardiac death in idiopathic dilated cardiomyopathy. The Cardiomyopathy Trial (CAT), Circulation 2002;105:1453-1458.
18. Strickberger SA, Hummel JD, Bartlett TG, et al: Amiodarone versus implantable cardioverter defibrillator: Randomized trial in patients with nonischemic dilated cardiomyopathy and asymptomatic nonsustained ventricular tachycardia — AMIOVIRT. J Am Coll Cardiol 2003;41:1707-1712.
19. Kadish A, Dyer A, Daubert JP, et al: Prophylactic defibrillator implantation in patients with nonischemic dilated cardiomyopathy. N Engl J Med 2004;350:2151-2158.
20. Zipes DP, Wellens HJJ: Sudden cardiac death. Circulation 1998;98:2334-2351.
21. Luu M, Stevenson WG, Stevenson LW, et al: Diverse mechanisms of unexpected cardiac arrest in advanced heart failure. Circulation 1989;80:1675-1680.
22. Roberts WC, Siegel RJ, McManus BM: Idiopathic dilated cardiomyopathy: Analysis of 152 necropsy patients. Am J Cardiol 1987;60:1340-1355.
23. Wu TJ, Ong JJ, Hwang C, et al: Characteristics of wave fronts during ventricular fibrillation in human hearts with dilated cardiomyopathy: Role of increased fibrosis in the generation of reentry. J Am Coll Cardiol 1998;32:187-196.
24. Schwartz CJ, Gerity RG: Anatomical pathology of sudden unexplained cardiac death. Circulation 1975;51:18-29.
25. Franz MR, Burkoff D, Yue DT, Sagawa K: Mechanically induced action potential changes and arrhythmia in insolated in situ canine hearts. Cardiovasc Res 1989;23:213-223.
26. Beuckelmann DJ, Näbauer M, Erdmann E: Alterations of K+ currents in isolated human ventricular myocytes from patients with terminal heart failure. Circ Res 1993;73:379-385.
27. KääB S, Nuss HB, Chiamvimonvat N, et al: Ionic mechanisms of action potential prolongation in ventricular myocytes from dogs with pacing-induced heart failure. Circ Res 1996;78:262-273.
28. Näbauer M, Beuckelmann DJ, Erdmann E: Characteristics of transient outward current in human ventricular myocytes from patients with terminal heart failure. Circ Res 1993;73:386-394.
29. KääB S, Dixon J, Duc J, et al: Molecular basis of transient outward potassium current downregulation in human heart failure. Circulation 1998;98:1383-1393.
30. Sevres NJ: Gap junction alterations in the failing heart. Eur Heart J 1994;15:D53-D57.
31. Schwartz PJ, La Rovere MT, Vanoli E: Autonomic nervous system and sudden cardiac death. Circulation 1992:85 (Suppl. I. :I-77-I-91).
32. Meredith IT, Esler MD, Jemmimgs GL, Broughton A: Evidence of a selective increase in cardiac sympathetic activity in patients with sustained ventricular arrhythmias. N Engl J Med 1991; 325:618-624.
33. Hohnloser SH, Singh BN: Proarrhythmia with class III antiarrhythmic drugs: Definition, electrophysiologic mechanisms, incidence, predisposing factors, and clinical implication. J Cardiovasc Electrophysiol 1995;6:920-936.

34. Galvin JM, Ruskin JN: Ventricular tachycardia in patients with dilated cardiomyopathy. In Zipes DP, Jalife J (eds): Cardiac Electrophysiology—From Cell to Bedside. Philadelphia, WB Saunders, 2000, pp 537-546.
35. Wilson JR, Schwartz S, Sutton MS, et al: Prognosis in severe heart failure: Relation to hemodynamic measurements and ventricular ectopic activity. J Am Coll Cardiol 1983;2:403-410.
36. Brembilla-Perrot B, Donetti J, Terrier de la Chaise A, et al: Diagnostic value of ventricular stimulation in patients with idiopathic dilated cardiomyopathy. Am Heart J 1991;121: 1124-1131.
37. Knight BP, Goyal R, Pelosi F, et al: Outcome of patients with non-ischemic dilated cardiomyopathy and unexplained syncope treated with an implantable defibrillator. J Am Coll Cardiol 1999;33:1964-1970.
38. Mittal S, Stephenson K, Stein KM, et al: Long-term outcome of patients with a non-ischemic cardiomyopathy and unexplained syncope. PACE 2000;23:560 (abstract).
39. Schoeller R, Andresen D, Büttner P, et al: First- or second-degree atrioventricular block as a risk factor in idiopathic dilated cardiomyopathy. Am J Cardiol 1993;71:720-726.
40. Unverfehrt DV, Magorien RD, Moeschberger ML, et al: Factors influencing the one-year mortality of dilated cardiomyopathy. Am J Cardiol 1984;54:147-152.
41. Meinertz T, Hofmann T, Kasper W, et al: Significance of ventricular arrhythmias in idiopathic dilated cardiomyopathy. Am J Cardiol 1984;53:902-907.
42. Von Olshausen K, Schäfer A, Mehmel HC, et al: Ventricular arrhythmias in idiopathic dilated cardiomyopathy. Br Heart J 1984;51:195-201.
43. Von Olshausen K, Stienen U, Schwarz F, et al: Long-term prognostic significance of ventricular arrhythmias in idiopathic dilated cardiomyopathy. Am J Cardiol 1988;61:146-151.
44. DeMaria R, Gavazzi A, Caroli A, et al: Ventricular arrhythmias in dilated cardiomyopathy as an independent prognostic hallmark. Am J Cardiol 1992;69:1451-1457.
45. Kadish A, Schmaltz S, Calkins H, Morady F: Management of non-sustained ventricular tachycardia guided by electrophysiological testing. PACE 1993;16:1037-1050.
46. Turitto G, Ahuja RK, Caref EB, El-Sherif N: Risk stratification for arrhythmic events in patients with nonischemic dilated cardiomyopathy and nonsustained ventricular tachycardia: Role of programmed ventricular stimulation and the signal-averaged electrocardiogram. J Am Coll Cardiol 1994;24:1523-1528.
47. Baker R, Grossman M, Chough S, et al: High risk subset predicted by ambulatory Holter monitoring in patients with non-ischemic cardiomyopathy. (Abstr.) PACE 2000;23:267.
48. Fauchier L, Babuty D, Cosnay P, et al: Long-term prognostic value of time domain analysis of signal-averaged electrocardiography in idiopathic dilated cardiomyopathy. Am J Cardiol 2000; 85:618-623.
49. Poll DS, Marchlinski FE, Falcone RA, et al: Abnormal signal-averaged ECG in nonischemic congestive cardiomyopathy: Relationship to sustained ventricular tachyarrhythmias. Circulation 1986;72:1308-1313.
50. Middlekauf HR, Stevenson WG, Woo MA, et al: Comparison of frequency of late potentials in idiopathic dilated cardiomyopathy and ischemic cardiomyopathy with advanced congestive heart failure and their usefulness in predicting sudden death. Am J Cardiol 1990;66:1113-1117.
51. Winters SL, Goldman DS, Banas JS: Prognostic impact of late potentials in nonischemic dilated cardiomyopathy. Circulation 1993;87:1405-1407.
52. Mancini DM, Wong KLO, Simson MB: Prognostic value of an abnormal signal-averaged electrocardiogram in patients with nonischemic congestive cardiomyopathy. Circulation 1993; 87:1083-1092.
53. Yi G, Hnatkova H, Mahon NG, et al: Predictive value of wavelet decomposition of the signal-averaged electrocardiogram in idiopathic dilated cardiomyopathy. Eur Heart J 2000;21(12):1015-1022.
54. Hoffmann J, Grimm W, Menz V, et al: Heart rate variability and major arrhythmic events in patients with idiopathic dilated cardiomyopathy. PACE 1996;19:1841-1844.
55. Yi G, Goldman JH, Keeling PJ, et al: Heart rate variability in idiopathic dilated cardiomyopathy: Relation to disease severity and prognosis. Heart 1997;77:108-114.
56. Hoffmann J, Grimm W, Menz V, et al: Heart rate variability and baroreflex sensitivity in idiopathic dilated cardiomyopathy. Heart 2000;83:531-538.
57. Rosenbaum D, Albrecht P, Cohen RJ: Predicting sudden cardiac death from T wave alternans of the surface electrocardiogram: Promise and pitfalls. J Cardiovasc Electrophysiol 1996;7: 1095-1111.
58. Adachi K, Ohnishi Y, Shima T, et al: Determinant of microvolt-level T-wave alternans in patients with dilated cardiomyopathy. J Am Coll Cardiol 1999;34:374-380.
59. Yap YG, Aytemir K, Mohan N, et al: Values of T wave alternans, QT dispersion, and principal component analysis ratio in patients with dilated cardiomyopathy. PACE 1999;22(suppl.II):883.
60. Kitamura H, Ohnishi Y, Okajima K, et al: Onset heart rate of microvolt-level T-wave alternans provides clinical and prognostic value in nonischemic dilated cardiomyopathy. J Am Coll Cardiol 2002;39:295-300.
61. Hohnloser SH, Klingenheben T, Bloomfield D, et al: Usefulness of microvolt T-wave alternans for prediction of ventricular tachyarrhythmic events in patients with dilated cardiomyopathy: Results from a prospective observational study. J Am Coll Cardiol 2003; 41:2220-2224.
62. Grimm W, Christ M, Bach J, et al: Noninvasive arrhythmia risk stratification in idiopathic dilated cardiomyopathy. Circulation 2003;108:2883-2891.
63. Moss AJ, Hall WJ, Cannom DS, et al: For the Multicenter Automatic Defibrillator Implantation Trial Investigators. Improved survival with an implanted defibrillator in patients with coronary disease at high risk for ventricular arrhythmia. N Engl J Med 1996;335:1933-1940.
64. Buxton AE, Lee KL, Fischer JD, et al: For the Multicenter Unsustained Tachycardia Trial Investigators: A randomized study of the prevention of sudden death in patients with coronary artery disease. N Engl J Med 1999;341:1882-1890.
65. Poll DS, Marchlinski FE, Buxton AE, Josephson ME: Usefulness of programmed stimulation in idiopathic dilated cardiomyopathy. Am J Cardiol 1986;58:992-997.
66. Meinertz T, Treese N, Kasper W, et al: Determinants of prognosis in idiopathic dilated cardiomyopathy as determined by programmed electrical stimulation. Am J Cardiol 1985;56: 337-341.
67. Stamato NJ, O'Connel JB, Murdock DK, et al: The response of patients with complex ventricular arrhythmias secondary to dilated cardiomyopathy to programmed electrical stimulation. Am Heart J 1986;112:505-508.
68. Das SK, Morady F, Dicarlo L, et al: Prognostic usefulness of programmed ventricular stimulation in idiopathic dilated cardiomyopathy without symptomatic ventricular arrhythmias. Am J Cardiol 1986;58:998-1000.
69. Gossinger HD, Jung M, Wagner L, et al: Prognostic role of inducible ventricular tachycardia in patients with dilated cardiomyopathy and asymptomatic nonsustained ventricular tachycardia. Int J Cardiol 1990;29:215-220.
70. Grimm W, Hoffmann J, Menz V, et al: Programmed ventricular stimulation for arrhythmia risk prediction in patients with idiopathic dilated cardiomyopathy and nonsustained ventricular tachycardia. J Am Coll Cardiol 1997;32:739-745.
71. Waagstein F, Hjalmarson A, Varnauskas E, Wallentin F: Effect of chronic beta adrenergic receptor blockade in congestive cardiomyopathy. Br Heart J 1975;37:1022-1036.
72. Hjalmarson A, Waagstein F: New therapeutic strategies in chronic heart failure: Challenge of long-term beta-blockade. Eur Heart J 1991;12(suppl F):63-69.
73. Packer M, Bristow MR, Cohn JN, et al: The effect of carvedilol on morbidity and mortality in patients with chronic heart failure. N Engl J Med 1996;334:1349-1355.
74. CIBIS-II Investigators and Committees. The cardiac insufficiency bisoprolol study II (CIBIS-II): A randomized trial. Lancet 1999; 353:9-13.
75. MERIT-HF study group. Effect of metoprolol CR/XL in chronic heart failure: Metoprolol CR/XL randomized intervention trial in congestive heart failure (MERIT-HF). Lancet 1999;353:2001-2007.
76. Teo KK, Yusuf S, Furberg CD: Effects of prophylactic antiarrhythmic drug therapy in acute myocardial infarction: An overview of results of randomised controlled clinical trials. JAMA 1993; 270:1589-1595.

77. The cardiac arrhythmia suppression trial (CAST) Investigators. Effect of encainide and flecainide on mortality in infarction. N Engl J Med 1989:321:406-412.
78. The cardiac arrhythmia suppression trial II Investigators: Effect of antiarrhythmic agent moricizine on survival after myocardial infarction. N Engl J Med 1992;327:227-233.
79. Waldo AL, Camm AJ, de Ruyter H, et al: Effect of d-sotalol on mortality in patients with left ventricular dysfunction after recent and remote myocardial infarction. Lancet 1996:348:7-12.
80. Hohnloser SH, Klingenheben T, Singh BN: Amiodarone-associated proarrhythmic effects: A review with special reference to torsades de pointes tachycardia. Ann Int Med 1994;121:529-535.
81. Neri R, Mestroni L, Salvi A, et al: Ventricular arrhythmias in dilated cardiomyopathy: Efficacy of amiodarone. Am Heart J 1987;113: 707-715.
82. Singh SN, Fletcher RD, Fisher SG, et al: For the survival trial in congestive heart failure study. Amiodarone in patients with congestive heart failure and asymptomatic ventricular arrhythmia. N Engl J Med 1995;333:77-82.
83. Doval HC, Nul DR, Grancelli HO, et al: For grupo de estudio de la sobrevida en la insufficiencia cardiaca en argentina (GESICA). Randomised trial of low-dose amiodarone in severe congestive heart failure. Lancet 1994;344:493-498.
84. Garguichevich JJ, Ramos JL, Gambarte A, et al: For the Argentine pilot study of sudden death and amiodarone investigators: Effect of amiodarone therapy on mortality in patients with left ventricular dysfunction and asymptomatic complex ventricular arrhythmias: Argentine pilot study of sudden death and amiodarone (EPAMSA). Am Heart J 1995;130:494-500.
85. Nicklas JM, McKenna WJ, Stewart RA, et al: Prospective, double-blind, placebo-controlled trial of low-dose amiodarone in patients with severe heart failure and asymptomatic frequent ventricular ectopy. Am Heart J 1991;122:1016-1021.
86. Amiodarone Trials Meta-analysis Investigators. Effect of prophylactic amiodarone on mortality after acute myocardial infarction and in congestive heart failure: Meta-analysis of individual data from 6500 patients in randomised trials. Lancet 1997;350: 1417-1424.
87. Nul DR, Doval HC, Grancelli HO, et al: Heart rate is a marker of amiodarone mortality reduction in severe heart failure. JACC 1997;29:1199-1205.
88. Massie BM, Fisher SG, Deedwania PC, et al: Effect of amiodarone on clinical status and left ventricular function in patients with congestive heart failure. Circ 1996;93:2128-2134.
89. Bardy GH, Lee KL, Mark DB and the SCD-HeFT pilot investigators. The sudden cardiac death in heart failure trial: Pilot study. (Abstr.) PACE 1997;20:1148.
90. Caceres J, Jazayeri M, Mckinnie J, et al: Sustained bundle branch reentry as a mechanism of clinical tachycardia. Circ 1989; 79:256-270.
91. Simons GR, Klein GJ, Natale A: Ventricular tachycardia: Pathophysiology and radiofrequency catheter ablation. PACE 1997; 20:534-551.
92. Blanck ZA, Dhala A, Deshpande S, et al: Bundle branch reentrant ventricular tachycardia: Cumulative experience in 48 patients. J Cardiovasc Electrophysiol 1993;4:253-262.
93. Blanck ZA, Jazayeri M, Dhala A, et al: Bundle branch reentry: A mechanism of ventricular tachycardia in the absence of myocardial of valvular dysfunction. JACC 1993;22:1718-1722.
94. Cohen TJ, Chien WW, Lurie KG, et al: Radiofrequency catheter ablation for treatment of bundle branch reentrant ventricular tachycardia: Results and long-term follow-up. JACC 1991;18: 1767-1773.
95. Delacretaz E, Stevenson WG, Ellison KE, et al: Mapping and radiofrequency catheter ablation of the three types of sustained monomorphic ventricular tachycardia on nonischemic heart disease. J Cardiovasc Electrophysiol 2000;11:11-17.
96. Kottkamp H, Hindricks G, Chen X, et al: Radiofrequency catheter ablation of sustained ventricular tachycardia in idiopathic dilated cardiomyopathy. Circ 1995;92:1159-1168.
97. The Antiarrhythmics Versus Implantable Defibrillators (AVID) Investigators. A comparison of antiarrhythmic-drug therapy with implantable defibrillators in patients resuscitated from near-fatal ventricular arrhythmias. N Engl J Med 1997;337: 1576-1583.
98. Connolly SJ, Gent M, Roberts RS, et al: Canadian Implantable Defibrillator Study (CIDS): A randomized trial of the implantable cardioverter defibrillator against amiodarone. Circulation 2000;101:1297-1302.
99. Kuck KH, Cappato R, Siebels J, Rüppel R: For the CASH Investigators. A randomized comparison of antiarrhythmic drug therapy with implantable defibrillators resuscitated from cardiac arrest: The Cardiac Arrest Study Hamburg (CASH). Circulation 2000;102:748-754.
100. Steinberg JS, Ehlert FA, Cannon DS, et al: Dilated cardiomyopathy vs. coronary artery disease in patients with VT/VF: Differences in presentation and outcome in the antiarrhythmic versus implantable defibrillators (AVID) registry. Circulation 1997; 96:I-715 (abstract).
101. Connolly AJ, Hallstrom AP, Cappato R, et al: Meta-analysis of the implantable cardioverter defibrillator secondary prevention trials. Eur Heart J 2000;21:2071-2078.
102. Hochleitner M, Hortnagl H, Hg C, et al: Usefulness of physiologic dual-chamber pacing in drug-resistant idiopathic dilated cardiomyopathy. Am J Cardiol 1990;66:198-202.
103. Hochleitner M, Hortnagl H, Friedrich L, Gschnitzer F: Long-term efficacy of physiologic dual-chamber pacing in the treatment of end-stage idiopathic dilated cardiomyopathy. Am J Cardiol 1992;70:1320-1325.
104. Nishimura RA, Hayes DL, Holmes DR, Tajik AJ: Mechanism of hemodynamic improvement by dual-chamber pacing for severe left ventricular dysfunction: An acute Doppler and catheterization hemodynamic study. J Am Coll Cardiol 1995; 25:281-288.
105. Linde C, Gadler F, Edner M, et al: Results of atrioventricular synchronous pacing with optimized delay in patients with severe congestive heart failure. Am J Cardiol 1995;75:919-923.
106. Gold MR, Feliciano Z, Gottlieb SS, et al: Dual-chamber pacing with a short atrioventricular delay in congestive heart failure: A randomized study. J Am Coll Cardiol 1995;26:967-973.
107. Blanc JJ, Etienne Y, Gilard M, et al: Evaluation of different ventricular pacing sites in patients with severe heart failure: Results of an acute hemodynamic study. Circulation 1997; 96:3273-3277.
108. Bristow MR, Saxon LA, Boehmer J, et al: Cardiac resynchronization therapy with or without an implantable defibrillator in advanced chronic heart failure. N Engl J Med 2004;350: 2140-2150.

Chapter 31 Arrhythmogenic Right Ventricular Cardiomyopathy

PAUL TOUBOUL and MARJANEH FATEMI

The concept of myocardial disease confined to the right ventricle first emerged in 1952 after Uhl[1] reported on an 8-month-old infant deceased from congestive heart failure and in whom pathologic examination demonstrated "almost total absence of myocardium of the right ventricle." This report evoked the parchment heart already described by Osler in 1905.[2] The relationship between right ventricular disease and malignant tachyarrhythmias was subsequently emphasized from isolated cases. In 1977, Fontaine and colleagues[3] truly defined a new clinical entity called arrhythmogenic right ventricular dysplasia, an event that triggered intense research and contributed to the revisitation of right ventricle diseases. The initial report described six patients with sustained, drug-resistant ventricular tachycardias and "apparently normal hearts." Dilatation of the right ventricle and wall motion abnormalities could be demonstrated in all cases. In the three patients who had surgery, there was an unusual fatty appearance of the right ventricular free wall. Later on, right ventricular dysplasia was also held responsible for unexpected sudden death in young adults.[4] Thus, the potential for life-threatening events in this setting was firmly established. However, the cause of the disease remained a matter of debate. Genetic defects appeared to play a role, but the mechanisms underlying the loss of myocardium are unclear and may be multifactorial—justifying the preferred use of cardiomyopathy (rather than dysplasia) to designate the disease. The importance of the contribution of the Padova Medical School to evolving concepts has to be stressed with special reference to the pathologic studies conducted by Thiene and coworkers.[5] A better characterization of the substrates has ensued, playing a significant role in our current understanding of arrhythmogenic right ventricular cardiomyopathy and opening new avenues of research.

Definition and Epidemiology

The term "arrhythmogenic right ventricular cardiomyopathy" (ARVC) suggests a myocardial process confined within the right ventricle and predisposing to cardiac arrhythmias. In common practice, such a denomination applies to clinical situations combining ventricular arrhythmias and the notion of right ventricular involvement as sole possible causes. Because right ventricular abnormalities may be latent, this is to be differentiated from so-called idiopathic rhythm disorders. Due to the current limitations of the diagnostic tools and the lack of pathologic findings, it is likely that a significant number of cardiomyopathy cases go unrecognized. Conversely, borderline cases may lead to false-positive diagnoses. As a rule, except for pathologic reports, the physician deals with probable diagnoses carrying variable margins of certainty. Moreover, some forms of the disease may not be arrhythmogenic, despite a clear-cut pattern of right ventricular cardiomyopathy. Assuming that the potential for arrhythmias is present but latent, these patients can be maintained in the same entity as those with clinical rhythm abnormalities. Coexisting damage to the left ventricle does not preclude the diagnosis of right ventricle cardiomyopathy if the right-sided abnormalities are suggestive, predominant, and further supported by relevant clinical markers. The incidence of this association is likely to increase in the late stages of the disease.[6] Finally, even the histopathologic criteria are a matter of debate. Replacement of the myocardium by fatty and fibrous tissue is characteristic, but the limit between normal and abnormal infiltration is unclear (see later). This provides bases for the discrepancies between the recognition of ARVC as a true entity and the current lack of well-accepted definition. Further research is needed to improve our knowledge and better define the profile of this condition. Due to these uncertainties, the incidence and prevalence of ARVC are difficult to assess and its actual epidemiologic status is unknown. The pathologic reports are only indicative of the role of the disease as a cause of death.[4,5,6] In this view, the results of a prospective Italian study of sudden death[5] in the young are of interest. This investigation was conducted in the Veneto region. Up to 20% of deaths in young people and athletes were attributed to previously undiagnosed right ventricular cardiomyopathy, which emphasizes the significant impact of the disease on juvenile mortality. In a more general context, ARVC was reportedly found in 5% of autopsies for unexpected sudden cardiac death.

Pathologic Findings

The pathologic diagnosis of ARVC is based on the loss of myocardium and the presence of gross and/or histologic evidence of regional or diffuse transmural fibrofatty infiltration[4,5,6] affecting predominantly the right ventricular free wall, without valve, coronary, and pericardial disease (Figs. 31-1 through 31-4). The regions of the right ventricle most frequently involved are the right inflow, the apex, and the infundibulum, known as the triangle of dysplasia.[4] Basso and associates[6] defined two different patterns: the fibrofatty variety characterized by loss of myocardium replaced by both fibrous and fatty tissue, and the fatty variety in which myocardial cells are replaced almost exclusively by adipose tissue located intramyocardially and epicardially.

In the fibrofatty form (see Figs. 31-3 and 31-4), loss of myocardial cells due to necrosis and apoptosis[7] and their replacement by fibrofatty tissue result in right ventricular wall atrophy. Wall thinning may account for aneurysmal dilatation in 50% of autopsy cases.[8] Aneurysms typically affect the inferior wall beneath the posterior leaflet of the tricuspid valve, as well as the apical and infundibular portions of the right ventricle. Due to these unique characteristics, right ventricular aneurysms can be considered pathognomonic. The pathologic process is extended to the left ventricle (see Figs. 31-1 and 31-5) in almost half of the cases, but the interventricular septum is often spared.[8] Fibrofatty tissue development and myocardial fiber disorganization are likely to provide an appropriate substrate for slow conduction and reentrant ventricular arrhythmias.

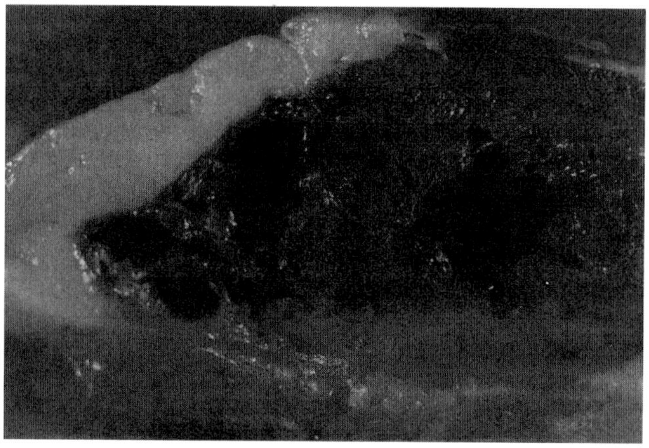

FIGURE 31-1 Cross section of the heart from a patient with arrhythmogenic right ventricular cardiomyopathy (ARVC) who died suddenly. Note the massive and almost circumferential fatty infiltration of the right ventricle, with extension to the left ventricle.

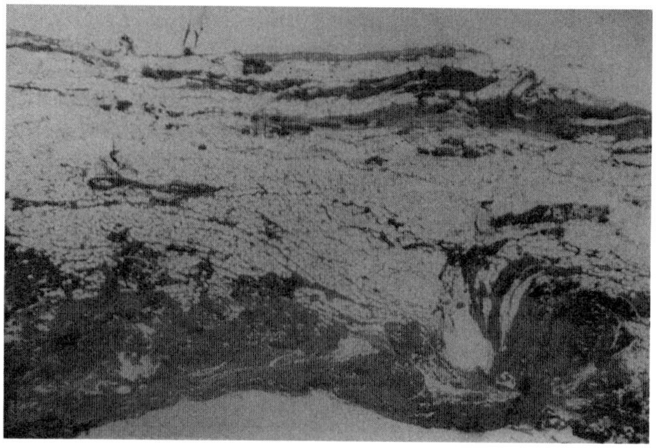

FIGURE 31-3 The fibrofatty form of ARVC: the microscopic view of the right ventricle shows fatty infiltration of the epicardium adjacent to subendocardial fibrosis associated with myocyte loss. There are also areas of epicardial fibrosis. Magnification: 1×. Stain: hematoxylin-eosin-safranin.

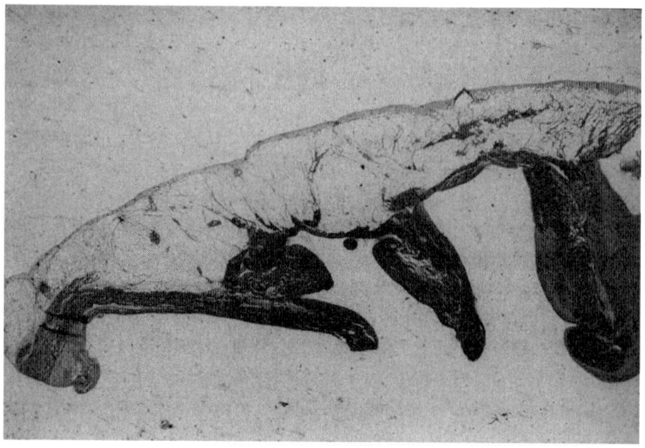

FIGURE 31-2 The fatty form of ARVC: panoramic histologic view of the right ventricular free wall showing transmural fatty replacement. Note the epicardial coronary vessels embedded in the adipose infiltration. Stain: hematoxylin-eosin-safranin.

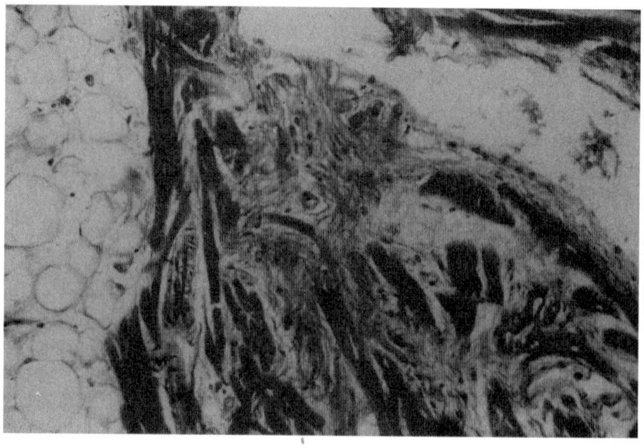

FIGURE 31-4 The fibrofatty form of ARVC: adipose infiltration adjacent to islands of residual myocytes interspersed with interstitial fibrosis. Magnification: 10×. Stain: hematoxylin-eosin-safranin.

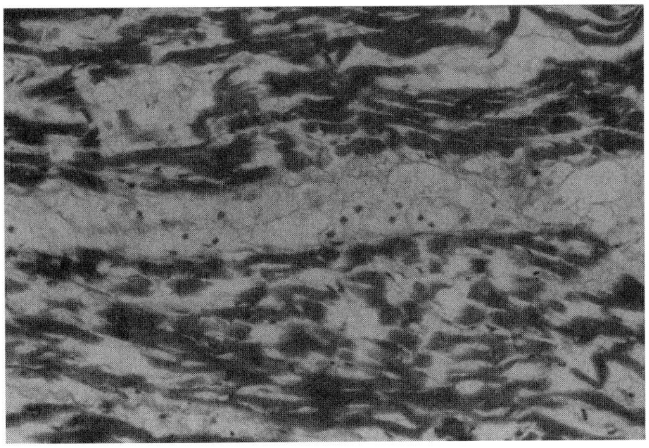

FIGURE 31-5 ARVC extended to the left ventricle: myocytolysis associated with interstitial edema and lymphocytic infiltrates. Magnification: 10×. Stain: hematoxylin-eosin-safranin.

Lymphocytic myocarditis is a frequent finding in this pattern (Fig. 31-6) and may precede fibrosis.[9]

The fatty variety (see Fig. 31-2) is characterized by little inflammation, and absence of ventricular wall thinning. On the contrary, myocardial thickness is normal or even increased. Therefore, the incidence of aneurysms is low. Whether this form of ARVC carries an increased risk of sudden death remains a conflicting issue. However, compared with the fibrofatty variety, no differences were noted in some reports regarding the occurrence of ventricular arrhythmias and mode of death.[9]

The mechanisms underlying these morphologic changes are still debated. Originally, an analogy with Uhl's disease was postulated, leading to the belief that the myocardial aspect corresponded to a congenital structural defect. Hence, the term *dysplasia* was proposed implying maldevelopment as a cause of the pathologic abnormality. This term was exclusively used for years and contributed to the recognition of the new entity.[10] However, unlike Uhl's anomaly, where the

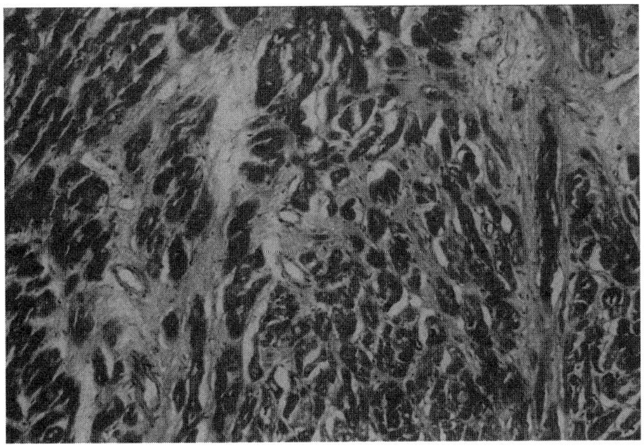

FIGURE 31-6 ARVC extended to the left ventricle: interstitial fibrosis infiltrating the myocardium mimicking a healed myocarditis. Magnification: 10×. Stain: hematoxylin-eosin-safranin.

ventricular wall is paper thin, in ARVC, the distance between the epicardium and the endocardium is normal or slightly decreased. Given the progressive nature of the disease and the presence of foci of inflammation, degeneration, and necrosis, the term *cardiomyopathy* was preferred to dysplasia.[11]

Two theories have been proposed: (1) The degenerative theory, where the loss of myocardium is considered to result from progressive myocyte death due to some metabolic or ultrastructural defect of genetic cause. This theory is in accordance with the familial occurrence of the disease implying genetic mutation with autosomal dominant transmission and variable penetrance and expression.[12-15] (2) The inflammatory theory, where the fibrofatty replacement is viewed as a healing process in the setting of chronic myocarditis. The disappearance of the right ventricular myocardium would be the consequence of an inflammatory injury followed by fibrofatty repair.[16-18] An infectious (viral) and/or immune myocardial reaction is involved in the pathogenesis of the disease. This hypothesis is supported by the presence of coxsackie virus genome in the myocardium of some patients with ARVC.[19] This is not in contrast with a familial occurrence, because genetic predisposition to viral infection eliciting immune reaction cannot be excluded. It is noteworthy that some experimental myocarditis is exclusively limited to the right ventricle.[20] Genetic factors may also play roles in the site of cardiac involvement.

Pathologic features associated with ARVC can occasionally be misleading. In the case of localized involvement of the heart, the myocardial changes may elude gross examination, unless careful attention is paid to the right ventricle. Ultimately, the final diagnosis will result from systematic histopathologic studies. The notion of fatty infiltration is a matter of debate. It is well-known that fatty infiltration of the right ventricle occurs in more than 50% of normal hearts in the elderly, especially in the anteroapical region. Fatty deposits are predominantly located in the subepicardium. There is controversy regarding how much fat should be considered as abnormal. Angelini and colleagues[21] claimed that fatty tissue exceeding 3.21% was highly suspect for ARVC in right ventricular endomyocardial biopsies, whereas, Burke and coworkers[9] reported up to 15% fat in the anterior apex of the right ventricle from normal hearts. According to Loire and Tabib,[22] fatty tissue that is extended beyond the coronary vessels in the epicardium is highly abnormal. Due to these uncertainties, it is likely that the disease is largely overlooked by routine autopsy.

Presentation

Patients with ARVC have a wide spectrum of clinical presentations varying from no symptoms to significant arrhythmic events including frequent ventricular ectopies, episodes of ventricular tachycardias, or sudden death. Atrioventricular block,[23,24] supraventricular arrhythmias,[25] and heart failure[29] are other possible signs of the disease.

Ventricular tachyarrhythmias associated with ARVC probably result from electrical instability and the widespread disruption of the myocardium by fibrofatty infiltration. This causes heterogeneous electrical conduction and is an ideal substrate for reentrant ventricular tachyarrhythmias. The propensity for occurrence of ventricular arrhythmias on physical exertion[26] is probably related to both hemodynamic and neurohumoral factors. Physical exercise results in an increase of right ventricular afterload and cavity enlargement, which may trigger ventricular arrhythmias by stretching the right ventricular myocytes.[27] Progression of the disease from epicardium to endocardium may also account for functional and/or structural sympathetic denervation, decreased catecholamine reuptake, and enhanced sensitivity to catecholamines and arrhythmogenicity during sympathetic stimulation.[28] Heart failure[29] due to extensive myocardial disease has been observed and coexisting ventricular tachyarrhythmias may further aggravate the clinical course.

VENTRICULAR ECTOPIES AND/OR SUSTAINED OR NONSUSTAINED VENTRICULAR TACHYCARDIA

Ventricular arrhythmias associated with ARVC are usually of left bundle branch block morphology[4,5] indicating the right ventricular origin (Fig. 31-7). The QRS axis is inferior or normal when the ectopic beats originate from the right ventricular outflow tract, and superior when arising from the apex or the diaphragmatic wall of the right ventricle.[30] The diagnosis of ARVC should be considered in patients with apparently normal hearts and so-called *benign idiopathic ventricular arrhythmias.*[30] ARVC has been described as the underlying cause of ventricular tachycardia in children who are devoid of a clear cardiac abnormality.[31] A family history of ARVC or sudden death in a close relative may be a clue to the diagnosis.

Initiation of episodes of sustained ventricular tachycardias usually coincides with increased sympathetic tone as demonstrated by the increased heart rate and the shortening of the coupling intervals of the first cycles before the tachycardia initiation. Leclercq and associates[32] showed that a stronger sympathetic stimulation was needed to produce sustained ventricular tachycardias than to elicit couplets or nonsustained runs of ventricular tachycardia. The tolerance of ventricular arrhythmias is variable. Palpitations and chest discomfort are common features. In cases of sustained tachycardia, faintness, dizziness, the inability to sustain physical activity, and hypotension can be noticed. Occurrence of syncope may be indicative of hemodynamic deterioration due to high heart rates. Very rapid ventricular tachycardias occurring during strenuous physical activity may degenerate into ventricular fibrillation and, therefore, precede sudden death in young athletes with undiagnosed ARVC.

SUDDEN DEATH DUE TO VENTRICULAR FIBRILLATION

Sudden death was reported to occur in 5%[33] of patients with ARVC with an annual rate of 2% to 3%.[34] Basso and coworkers,[35] based on their experience of juvenile sudden death in the Veneto region, found that ARVC accounted for 12.5% of total cases, being the second most frequent cardiovascular cause of sudden death in the young (≤35 years), after coronary atherosclerosis, and the most common cause of sudden death among athletes. Conversely, sudden death due to ARVC seems to be anecdotal in American series.[36-37]

Unlike other causes of sudden death, such as coronary artery disease (congenital anomalies or atherosclerotic disease), where prodroma in affected patients are seldom noticed, sudden death due to ARVC is rarely the first manifestation of the disease and earlier symptoms may have been present but overlooked because of their *benign* nature and the lack of awareness of the disease. Often, victims of sudden death were found retrospectively to have a family history of sudden death or warning signs, such as palpitations or syncope on exertion, ECG-documented ventricular arrhythmias, and ECG abnormalities (inverted T waves in right

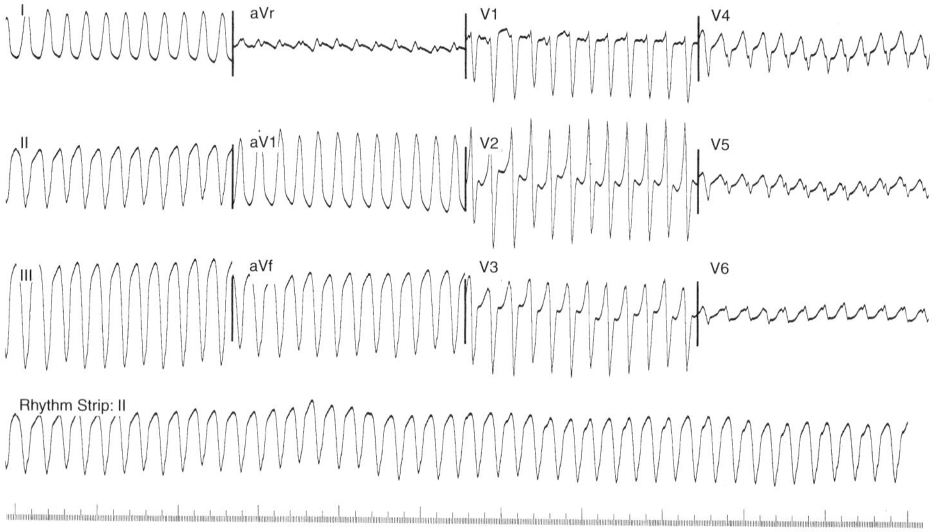

FIGURE 31-7 Twelve-lead ECG during ventricular tachycardia from a patient with ARVC and characterized by left bundle branch block morphology.

precordial leads) that could have led to the diagnosis of ARVC.

HEART FAILURE

Heart failure is a rare manifestation of ARVC. Preferential involvement of the right ventricle is a striking feature of ARVC and represents one of the most important diagnostic criteria. In most cases, patients present with symptoms of isolated right heart failure, engorgement of the systemic veins, and absence of pulmonary hypertension.[29] Cases of ARVC with biventricular involvement mimicking dilated cardiomyopathy[29,38] have also been described. Pinamonti and colleagues[39] reported mild left ventricular involvement in 14 of 39 patients during initial evaluation (the lowest left ventricular ejection fraction being 45%). After approximately 4 years, left ventricular dysfunction progressed in some patients and the lowest left ventricular ejection fraction was 24%. In a series of 121 ARVC patients reported by Peters and associates,[29] heart failure occurred in 16 patients: 13 patients with isolated right heart failure and three patients with biventricular failure within a time course of 4 to 8 years after developing complete right bundle branch block.

LATENT FORMS

In some asymptomatic patients, presence of the disease can be documented by noninvasive tests, indicating that the arrhythmogenic substrate is present but silent. These patients may have ECG signs of ARVC and late potentials detected by signal averaging (see next page). The value of positive tests in screening asymptomatic family members of patients with a history of sudden death needs to be further investigated.[94]

Diagnostic Tools

PHYSICAL EXAMINATION

Although physical examination in patients with ARVC is usually unremarkable, a split first or second heart sound, a systolic murmur and/or a third or fourth heart sound occasionally may be present.[4]

CHEST RADIOGRAPH

In the majority of patients, the chest radiograph is usually normal. However, a moderate cardiac enlargement without pulmonary vascular redistribution may be present in some patients. In young adults practicing sports, this cardiomegaly may be wrongly attributed to training. The cardiothoracic index is less than 0.6 in most cases.[10]

ELECTROCARDIOGRAM IN SINUS RHYTHM

The following electrocardiographic abnormalities have been reported in patients with suspected ARVC (Figs. 31-8 and 31-9):

1. Prolonged QRS duration = 110 ms in lead V1 has a sensitivity of 55% and a specificity of 100%. Generally, the QRS duration is more prolonged in lead V1 as compared with leads I and V6.[40] Complete or incomplete right bundle branch blocks are also common findings and may result from parietal block rather than disease of the right bundle branch.[41]
2. In 30% of patients, an abnormal deflection due to delayed right ventricular activation (epsilon wave or postexcitation wave)[3] is present at the end of the QRS complex. This feature can be easily overlooked.
3. Low voltage QRS amplitude (defined as QRS amplitude <5 mm in leads I, II, III, or <10 mm in precordial leads) may indicate a widespread cardiomyopathic process.[42]
4. T wave inversion in right precordial leads is frequently encountered (about 60% of patients with T wave abnormalities in leads V1 through V3 and up to 80% in leads V1 and V2). The significance of ST-segment elevation in the right precordial leads for the diagnosis of ARVC is unsettled,[43] except for dynamic changes recorded during exercise (see later).

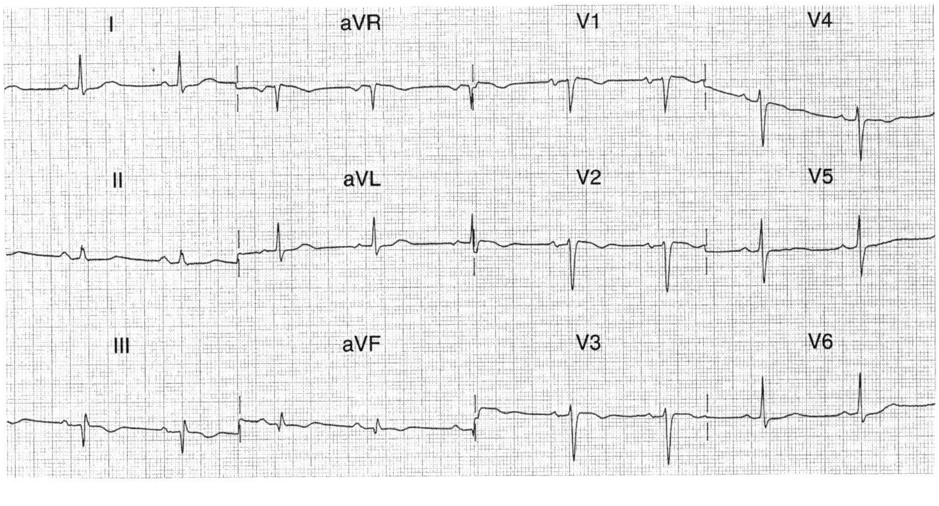

FIGURE 31-8 Twelve-lead ECG obtained during sinus rhythm from a patient diagnosed with ARVC. Note T wave inversion in leads V1 to V4 and the epsilon wave at the end of the QRS complex, in lead V1.

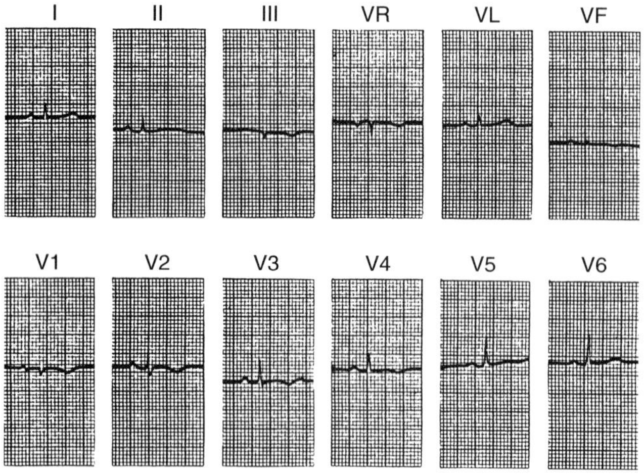

FIGURE 31-9 Twelve-lead ECG obtained during sinus rhythm from another patient with extensive ARVC. Beside the T wave inversion in leads V1 to V4, note the low voltage QRS complexes.

5. Increased QT dispersion was noted in patients with ARVC compared with control subjects. However, the degree of dispersion was not related to the severity of symptoms, nor was it influenced by antiarrhythmic treatment with sotalol.[44]

In a follow-up study of 20 patients with symptomatic ARVC,[42] electrocardiographic abnormalities were present in 90% of patients. However, no correlation was found between abnormalities on the initial 12-lead electrocardiogram and the echocardiographic extent and location of the right ventricular involvement. Serial electrocardiographic recordings over a 71-month follow-up period did not provide information regarding anatomic progression of the disease. Jaoude and coworkers[45] reported normal ECGs in up to 40% of their patients referred for arrhythmic events due to ARVC. During a 9.5-year follow-up, ECG changes (negative T waves, new left axis deviation, QRS enlargement, right atrial hypertrophy, and atrial fibrillation) were observed in 56% of patients and were correlated to the length of follow-up after the initial symptom.

EXERCISE STRESS TEST

The exercise stress test is helpful in exacerbating ventricular arrhythmias in about one half of patients with ARVC, but lack of worsening of ventricular arrhythmias during exercise does not exclude this condition.[46]

Local or diffuse wall motion abnormalities in patients with ARVC may induce ST-segment elevation in response to exercise. Toyofuku and colleagues[47] observed exercise-induced ST-segment elevations of more than 0.1 mV, in 65% of their patients with ARVC. This finding was more frequently observed in right precordial leads. Because exercise-induced ST-segment elevation is a rare phenomenon in normal subjects and noninvasive imaging techniques, such as echocardiography and radionuclide angiography, cannot always detect the abnormal morphology of the right ventricle, this observation is useful in helping diagnose patients with ARVC.

SIGNAL-AVERAGED ECG

The fibrofatty replacement characteristic of ARVC interrupts the electrical continuity of myocardial fibers, which accounts for conduction delay. These electrophysiological abnormalities produce delayed ventricular potentials responsible for the appearance of either epsilon waves on the surface electrocardiogram or late potentials recorded by signal-averaged ECG.

Mehta and coworkers[48] found abnormal signal-averaged ECGs using time domain analysis, in 90% of patients with ARVC and ventricular tachycardia of right ventricular origin. There was a strong correlation between all signal-averaged ECG parameters and right ventricular cavity dimensions.

Nava and associates[49] determined the sensitivity, specificity, and the predictive value of abnormal signal-averaged ECG in 138 patients with different degrees of ARVC (minor, moderate, and extensive) compared with 146 healthy subjects. The signal-averaged ECG had an overall sensitivity of 57%, a specificity of 95%, and a positive predictive value of 92%. However, a closer correlation was noted with the extent of disease rather than with the presence of ventricular arrhythmias: late potentials were present in 94.4% of patients with the extensive form of the disease, in 77.7% of patients with the moderate form, and only in 31.8% of patients with the minor form. A greater proportion of patients with ventricular fibrillation or sustained ventricular tachycardia had abnormal signal-averaged ECGs (71% and 72%, respectively), and among them 94.4% of these had ventricular tachycardia with a superior axis.

Similarly, Turini and colleagues[50] found that low right ventricular ejection fraction was the most powerful predictor of late potentials. Both a right ventricular ejection fraction of less than or equal to 0.5 and the root-mean-square voltage of the terminal 40 ms at 25 Hz, were predictive factors for the occurrence of sustained ventricular arrhythmias.

The incidence of late potentials in family members of patients diagnosed with ARVC has also been studied. Hermida and coworkers[51] found a higher incidence of late potentials in asymptomatic family members of patients with ARVC compared with healthy subjects (16% versus 3%). The clinical significance of positive signal-averaged ECG for late potentials in apparently healthy relatives of patients with ARVC remains to be investigated by long-term follow-up studies.

ELECTROPHYSIOLOGICAL TESTING

The ability to induce and terminate ventricular tachycardias related to ARVC by programmed ventricular stimulation supports a reentrant mechanism (Fig. 31-10). Electrophysiological testing aims to provide evidence for an arrhythmogenic substrate, to define the electrophysiological characteristics of the arrhythmia, and, in selected cases, to guide therapy. However, the value of programmed electrical stimulation to predict occurrence of spontaneous ventricular arrhythmias is quite low. Pacing should be performed at more than one right ventricular site and in some cases from the left

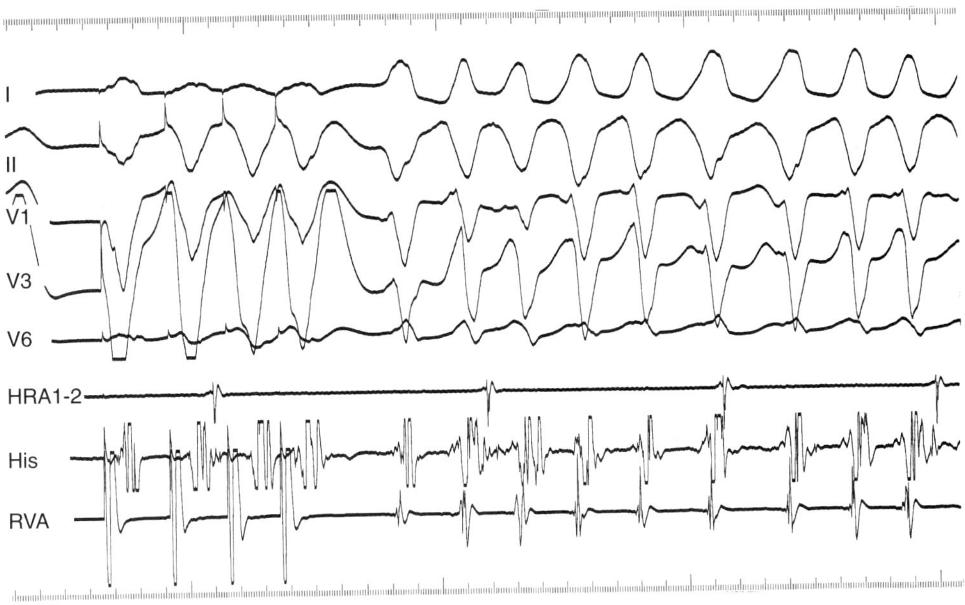

FIGURE 31-10 Initiation of ventricular tachycardia of left bundle branch block morphology in a patient with ARVC, by programmed ventricular stimulation at 600 ms cycle length with three extrastimuli. Surface leads I, II, V1, V3, and V6 are shown along with intracardiac electrograms from the high right atrium (HRA), the His bundle (HBE), and the right ventricular apex (RVA). Note the atrioventricular dissociation during tachycardia.

ventricle, using different basic cycle lengths, one to three extrastimuli, and burst pacing, in order to improve the sensitivity of the test. In patients with mild forms of ARVC, DiBiase and coworkers[52] found that programmed stimulation induced ventricular tachycardia mainly in the subgroup with spontaneous sustained ventricular tachycardia. Only 13% of patients with repetitive ventricular ectopies had inducible sustained tachyarrhythmias. Occasionally, induction of sustained ventricular tachcyardia may require administration of isoproterenol.[53] Similarly, Peters and associates[54] reported inducible ventricular tachycardia in 90% of patients with spontaneous sustained tachycardia episodes. In their study, sustained ventricular tachycardia was noninducible in all patients with nonsustained attacks and ventricular ectopies and in 30% of patients with previous cardiac arrest—even after isoproterenol infusion. It is noteworthy that ventricular fibrillation is sometimes initiated by programmed stimulation, but the prognostic value of this response is unclear.[55]

Electrophysiological testing can shed some insight into the arrhythmogenic substrate. The vast majority of ventricular tachycardias in patients with ARVC occur at sites with abnormal delayed, fractionated, and low voltage electrograms during sinus rhythm. These diastolic potentials are tiny deflections recorded after the QRS complex and exhibit rate-dependent properties.[56] Such abnormal electrograms may be widespread in the right ventricle, and their predictive value to determine the site of origin of ventricular tachycardia is poor.

Pace-mapping techniques[57] suggest that ventricular tachycardia in ARVC shares many of the characteristics seen in postmyocardial infarction patients. Electrical stimulation at the site of diastolic potentials may result in a long time interval between the electrical stimulus and the ventricular response due to delayed emergence of the impulse from the slow conduction area. During tachycardia, entrainment may ensue which further helps to delineate the ablation target. Finally, programmed electrical stimulation has also been shown to induce atrial tachyarrhythmias in patients with ARVC.[58] The significance of this phenomenon is unknown.

ECHOCARDIOGRAPHIC FINDINGS

Echocardiographic studies may show localized abnormalities of the right heart cavities, increased thickness of the moderator band, and trabeculations of the right ventricular apex. There may be obvious irregular dilatation of the outflow tract with an increased right ventricle/left ventricle ratio. However, in minor forms, the abnormality is difficult to detect. The structural changes in most cases of ARVC are moderate.[59] They will be recognized only if systematically sought, especially by measuring diameters at several strategic points of the right ventricle. These measurements should then be compared with normal values, according to a protocol by Foale and colleagues.[60]

Morphologic abnormalities found on two-dimensional echocardiography include dilatation of the right ventricle (Fig. 31-11A), thin ventricular wall with akinetic areas, the presence of aneurysms during diastole (see Fig. 31-11B), and dyskinetic areas in the inferobasal region, particularly specific when observed below the tricuspid valve. Parasternal views with the patient lying on the right lateral should be performed systematically when ARVC is suspected. With the subcostal approach, it is possible to see the free wall of the right ventricle both in two-dimensional mode and on time motion echocardiography. Rotating the probe 90 degrees enables one to visualize the long axis of the right ventricle, which is useful for studying the infundibulum and pulmonary artery. Particular attention must be paid to tricuspid valve prolapse and regurgitation in the severe forms of the disease, as well as mitral valve prolapse, which has been recently recognized as a condition frequently associated with ARVC.[61] Contrast echocardiography may help to evaluate right ventricular

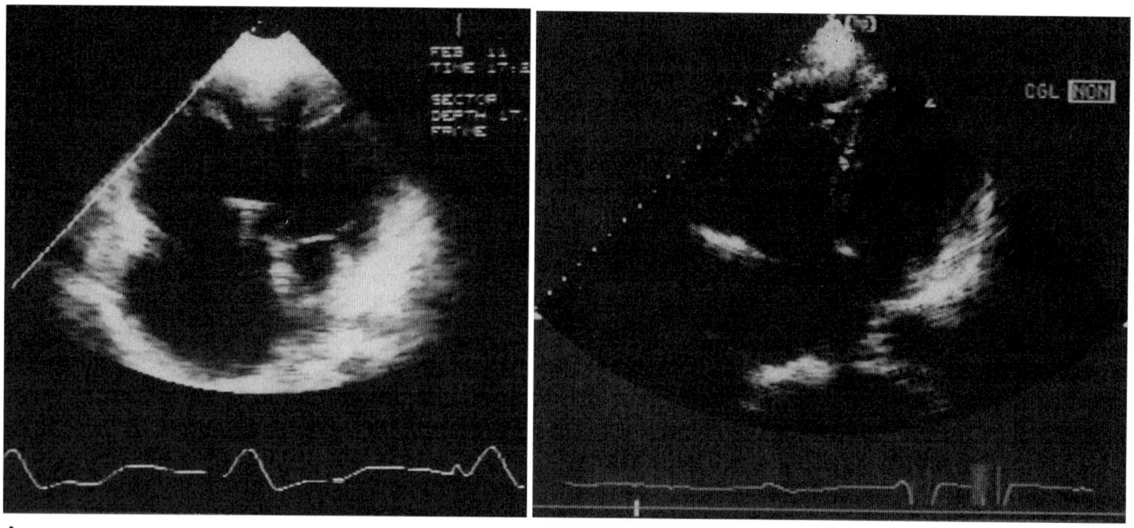

A B

FIGURE 31-11 **A,** Modified apical four-chamber echocardiogram obtained from a patient with extensive ARVC, showing severely dilated right cavities and a right-to-left ventricular ratio greater than one. **B,** Apical four-chamber echocardiogram from a different patient demonstrating a moderately enlarged right ventricle with a dyskinetic area at its apex.

regional or global function and may better outline the right ventricle to allow measurement of the right ventricular volume. Transesophageal echocardiography may be more sensitive than a transthoracic approach in detecting wall motion abnormalities. Three-dimensional echocardiography, particularly combined with the transesophageal approach, is being investigated to enhance the diagnostic accuracy of this technique.[62]

ENDOMYOCARDIAL BIOPSY

In most cases, the confirmation of diagnosis by endomyocardial biopsy is not required. This technique may be useful to demonstrate the typical, highly diagnostic, histologic abnormalities of ARVC, provided the biopsy is directed to the affected areas. However, the endomyocardial biopsy provides samples from the subendocardium, and the subendocardial appearance does not necessarily reflect the transmural pattern of disease. The presence of specimens showing only fibrosis is not surprising. In the fibrofatty variant of ARVC, fibrous tissue tends to be prevalently distributed within the subendocardial area, whereas fatty tissue predominates in the mid- or subepicardial layers. Because ARVC is a focal disease, the biopsy may not be a sensitive tool for the degree of fibrofatty infiltration. It is also an invasive procedure, with potential risk of perforation and cardiac tamponade, considering the thin and atrophic right ventricular walls of these patients. Therefore, this procedure should be restricted to very controversial cases and directed to the affected areas as demonstrated by imaging techniques.

RIGHT CONTRAST VENTRICULOGRAPHY

Cineangiographic investigation is currently used to evaluate the morphologic and functional status of cardiac cavities and parietal walls. Cardiomegaly, deep fissures, margin changes, localized akinetic or dyskinetic bulges, outpouchings, dilatation of the right ventricular outflow tract, and slow dye evacuation are the most frequently reported angiocardiographic parameters associated with the fibrofatty replacement in the right ventricle of patients diagnosed with ARVC (Fig. 31-12).[63] Daliento and associates[64] found that transversally arranged hypertrophic trabeculae separated by deep fissures were associated with the highest probability of ARVC. Posterior subtricuspid and infundibular wall bulgings were the other independent diagnostic variables for ARVC. In their study, coexistence of these signs was associated with 87.5% sensitivity and 96% specificity.

RADIONUCLIDE ANGIOGRAPHY AND COMPUTED TOMOGRAPHY

Radionuclide angiography is another useful diagnostic tool that may overcome limitations of contrast ventriculography. Because of the particular morphology and trabeculations of the right ventricle, the presence of frequent premature beats during angiography and the difficulty in obtaining a technically adequate angiogram, the analysis of right ventricular segmental wall motion is complex. Furthermore, the reproducibility of contrast ventriculography is not optimal.[65] Radionuclide angiography permits mathematic fitting of the pixel ventricular curve by means of Fourier analysis. This analysis has been demonstrated to be of value in detection of right and left wall motion abnormalities and aneurysms.[66,67] These abnormalities reflect on the distribution histogram that depicts a typical double peak morphology or late regional phase distribution, as opposed to the one-peak homogeneous phase distribution histogram observed in normal subjects. In a

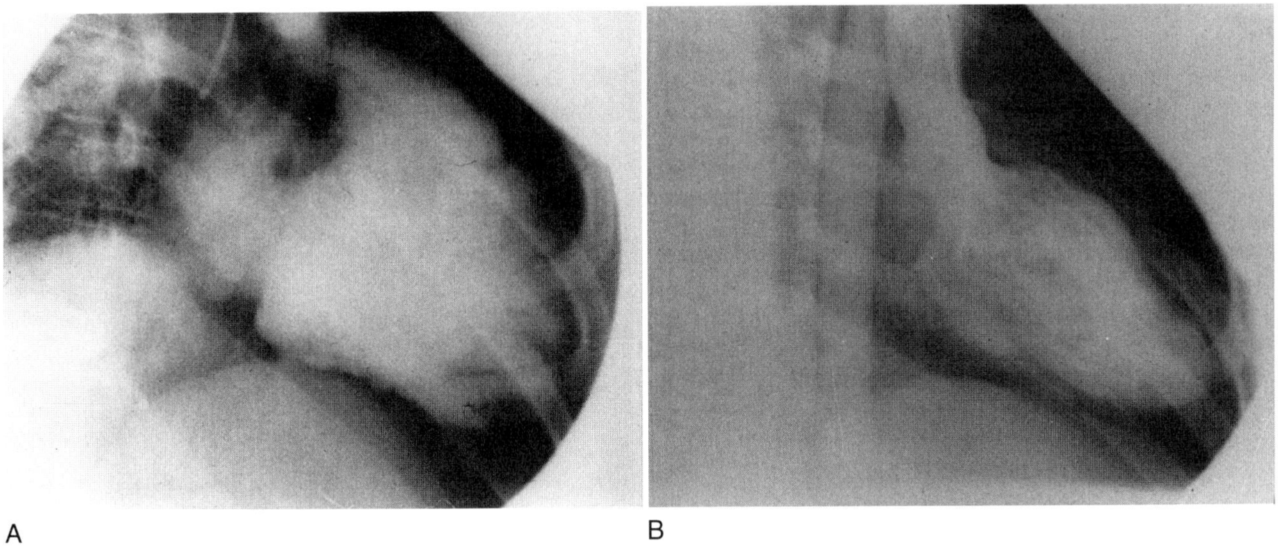

FIGURE 31-12 **A,** Right contrast ventriculography performed from the right atrium (right anterior oblique view). Note deep fissures and the marked right ventricular dilatation compared with the left ventricle (**B**).

prospective study on patients with ventricular tachycardia originating from the right ventricle, Le Guludec and colleagues[68] compared contrast and radionuclide angiography for the diagnosis of ARVC. The sensitivity of the latter technique was 94.3%, specificity 90%, and positive and negative predictive values were 96% and 85.7%, respectively. Agreement between the two techniques for the location of wall motion abnormalities was 60% for the apex, 76% for the outflow tract, 82% for the inferior wall, and 74% for the free wall. The reproducibility of the technique (96.2%) is also of interest in making it possible to follow the progression of right ventricular wall motion abnormalities. Reliable simultaneous analysis of the left ventricle is important to detect cases of ARVC with left ventricular involvement.

However, given the complex morphology of the right ventricle and the difficulty of tricuspid annulus delineation on planar study, a tomographic method appears to be more helpful. Gated blood pool single-photon emission tomography (GBP-SPECT) has been demonstrated to be more accurate than planar studies for right volume determination but also for detection of left ventricular wall motion abnormalities. Casset-Senon and coworkers[69] found the following abnormalities on GBP-SPECT, in 18 patients with ARVC: significantly decreased right ventricular ejection fraction, right ventricle dilatation, nonsynchronized contraction of the ventricles, increased right ventricular contraction dispersion, presence of right ventricular wall motion disorders and/or phase delays (Fig. 31-13), and occasional left ventricular abnormalities. They proposed a scoring system with these criteria to diagnose ARVC. In addition, multiharmonic and factor analyses can be used to enhance the diagnostic capabilities (third harmonic Fourier analysis for detecting outflow tract and inferior wall motion abnormalities, and factor analysis for the right ventricular apex).[70]

^{123}I-META-IODOBENZYLGUANIDINE SCINTIGRAPHY

In patients with ARVC, regional abnormalities of sympathetic innervation are frequent and can be demonstrated by ^{123}I-MIBG scintigraphy. Wichter and coworkers[28] observed a regional reduction in ^{123}I-MIBG uptake in 40 of 48 patients with ARVC that they studied with this technique, whereas the tracer distribution was found to be homogeneous in all control subjects. Sympathetic denervation appears to be the underlying mechanism of reduced ^{123}I-MIBG uptake. Therefore, in patients with ARVC, the noninvasive detection of localized sympathetic denervation by ^{123}I-MIBG imaging may have implications for the early diagnosis of the disease.

MAGNETIC RESONANCE IMAGING

Magnetic resonance imaging (MRI) is an emerging noninvasive technique providing cardiac images with a high spatial resolution that enables an accurate morphofunctional assessment of the right ventricle[71] and detection of features characteristic of ARVC, such as trabecular disarray, bulges, and aneurysms. The spin-echo technique is used to identify areas of fatty replacement (hyperintensity).[72] Therefore, MRI is the only noninvasive tool to provide information on tissue composition. The identification of abnormal areas may be helpful to guide the cardiac electrophysiologist for the mapping of ventricular arrhythmias. Cine-MRI using gradient echo sequences improves temporal resolution sufficiently to identify end systole and end diastole, thereby enabling measurements of global and regional right and left ventricular function.[72] However, the sensitivity of this technique is too low and its potential in diagnosing concealed forms of ARVC remains controversial.

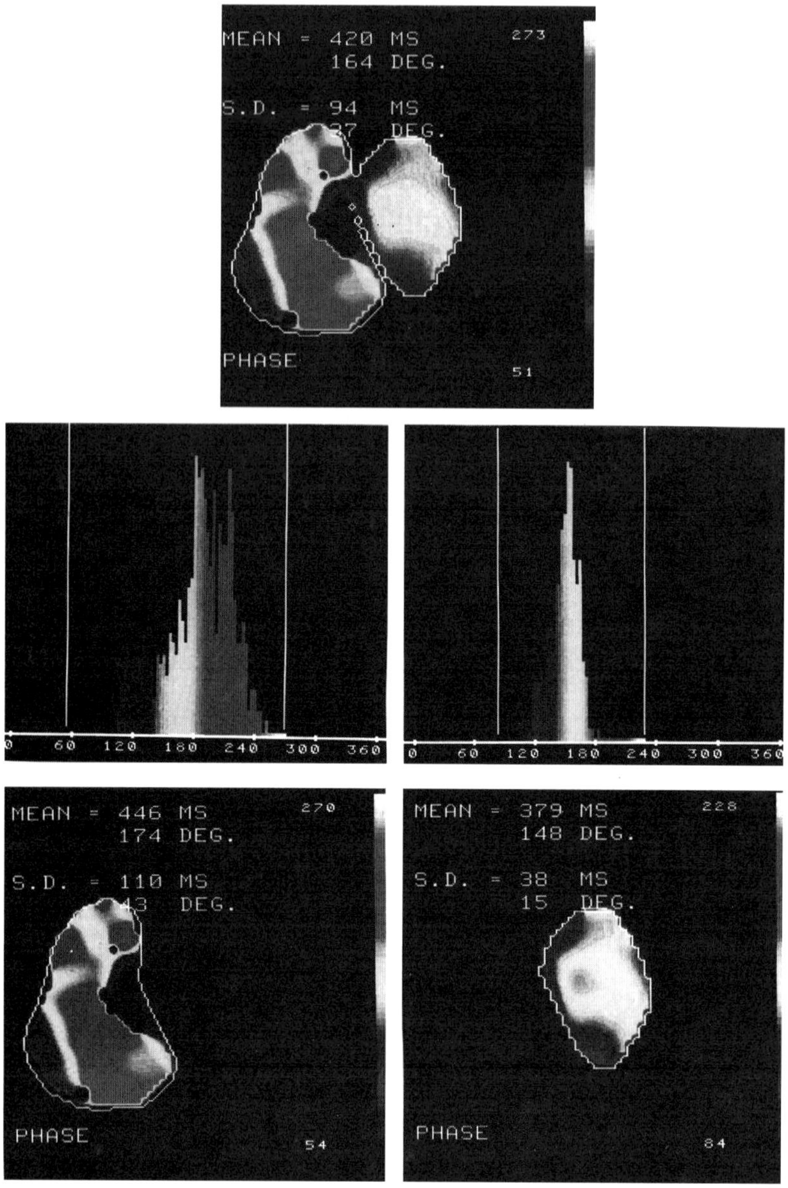

FIGURE 31-13 (See also Color Plate 31-13.) Gated blood pool tomographic data (short axis) obtained from a patient diagnosed with ARVC with no left ventricular involvement. **Top:** Phase analysis of the right and left ventricles showing significant contraction delay in the right ventricle relative to the left ventricle. **Middle:** Phase histograms of the right ventricle **(left)** and the left ventricle **(right):** the histogram corresponding to the left ventricle shows a homogeneous distribution of contraction phases, whereas, the right ventricular histogram is much more dispersed, due to heterogeneity of contraction. **Bottom:** Phase dispersion more pronounced in the right ventricle **(left)** as compared with the left ventricle **(right),** expressed as mean phase and phase standard deviation.

GENETICS

In 30% to 50% of patients, a family history of ARVC is present.[12-15] The most common form of transmission is autosomal dominant with incomplete penetrance and variable expression, which accounts for the appearance of minor forms of the disease. An autosomal recessive pattern has also been reported. Although the gene has not been identified, linkage analysis has associated the disease with 8109 (ARVC1-8) on chromosomes 1, 2, 3, 6, and 14[73-76,95,100] for the dominant form and on chromosomes 17 and 14 for the recessive variant of the disease[77,99]. One recessive form (Naxos disease) is characteristically associated with epidermal abnormalities, such as palmoplantar keratosis and woolly hair. It has been associated with a mutation of the gene that codes for plakoglobin a protein involved in cytoskeletal structure.[96] In Naxos disease, signs of the disease are more severe and penetrance in family members is 90%. More recently, a mutation of desmoplakin, a protein that interacts with plakoglobin, has been described in a family with ARVC.[95] This suggests that at least some ARVC is due to genetic disease of the cytoskeleton. Some families are not linked to these aforementioned loci, which suggests further genetic heterogeneity.

In addition, ARVC2, has been associated with mutations of the human ryanodine receptor, responsible for the release of calcium ions from the sarcoplasmic reticulum into the cytosol.[97] This remains controversial due to the association of mutations of this gene with nonstructural cardiac disease (catecholaminergic polymorphic ventricular tachycardia).[98]

Clinical Approach to ARVC

The incidence of ARVC was found to be significantly different in Europe and North America. The higher incidence of the disease in northern Italy compared with North America could be due either to clustering of the disease in some geographic areas or, more likely, to the fact that this entity is underdiagnosed both pathologically and clinically.

Despite tremendous technologic improvement of different diagnostic methods, diagnosis of ARVC remains a clinical challenge at its early stage in patients with minimal right ventricular abnormalities at echocardiographic or angiographic examination. Although endomyocardial biopsy has the potential to provide in vivo demonstration of fibrofatty infiltration of the heart, samples are usually taken, for safety reasons, from the septum, an area rarely involved by the disease. Signal-averaged ECG may detect late potentials in the vast majority of patients with widespread disease, but is negative in a significant number of focal forms. MRI appears to be a promising method to evaluate right ventricular wall composition and wall motion abnormalities, as well as right ventricular volume and shape. However, normal range limits for right ventricular volumes and regional function are debated. Furthermore, the relevant studies are of limited size and there is considerable heterogeneity of findings. Currently, the sensitivity and specificity of MRI still need to be defined.

Because of these difficulties, a set of diagnostic criteria for ARVC was proposed by the task force of the working group on myocardial and pericardial diseases of the European Society of Cardiology and the Scientific Council of Cardiomyopathies of the International Society and Federation of Cardiology.[78] "Major" and "minor" criteria were distinguished (Table 31-1). According to this consensus, the diagnosis of ARVC can be made if there is a combination of two major criteria, or one major plus two minor criteria, or four minor criteria. A clinical validation of the proposed criteria on a large series of patients is not available, but this stepwise approach should enhance the diagnostic capabilities and help standardize ARVC identification.

DIFFERENTIAL DIAGNOSIS

Differentiation between ARVC and the so-called right ventricular *outflow tract tachycardias*, a usually benign and nonfamilial condition, is a practical problem. Patients present with paroxysms of tachyarrhythmias with left bundle branch block morphology. Long-term outcome is favorable with drug therapy. Radiofrequency catheter ablation can be successfully performed for the treatment of refractory forms. Whether the right ventricular outflow tract tachycardia is a variant of ARVC is still debated.[79]

Another matter of discussion is *Uhl's disease*.[46] In this condition, aplasia of the right ventricular wall consists of apposition of endocardium and epicardium without intervening muscle fibers. This anomaly is considered to be congenital due to a dysontogenetic process. There is no fat or inflammatory reaction in the myocardium. Unlike ARVC, the major clinical manifestation

TABLE 31-1 Criteria for the Diagnosis of ARVC*

I Global and/or Regional Dysfunction and Structural Alteration	**IV Depolarization/Conduction Abnormalities**
MAJOR	MAJOR
Severe dilatation and reduction of right ventricular ejection fraction with no (or only mild) LV impairment.	Epsilon waves or localized prolongation (>110 ms) of the QRS complex in right precordial leads (V1-V3)
Localized right ventricular aneurysms (akinetic or dyskinetic areas with diastolic bulging)	MINOR
Severe segmental dilatation of the right ventricle	Late potentials (signal-averaged ECG)
MINOR	**V Arrhythmias**
Mild global right ventricular dilatation and/or ejection fraction reduction with normal left ventricle	MINOR
II Tissue Characterization of Walls	Left bundle branch block type ventricular tachycardia (sustained and nonsustained) (ECG, Holter, exercise testing)
MAJOR	Frequent ventricular extrasystoles (more than 1000/24 hr) (Holter)
Fibrofatty replacement of myocardium on endomyocardial biopsy	**VI Family History**
III Repolarization Abnormalities	MAJOR
MINOR	Familial disease confirmed at necropsy or surgery
Inverted T waves in right precordial leads (V2 and V3) (people older than 12 yrs; in the absence of right bundle branch block)	MINOR
	Family history of premature sudden death (<35 yrs) due to suspected right ventricular dysplasia.
	Family history (clinical diagnosis based on present criteria)

From McKenna WJ, Thiene G, Nava A, et al: Diagnosis of arrhythmogenic right ventricular dysplasia/cardiomyopathy. Br Heart J 1994;71:215.

of this disease is right heart failure that is associated with a high mortality rate. Whereas ARVC occurs predominantly in males, there is no such gender difference in Uhl's disease. Although the pathologic features of Uhl's disease are unique, such an entity remains affiliated to right ventricular myocardial diseases.

ARVC should also be distinguished from *Brugada syndrome,*[80,81] which is another clinical entity characterized by increased risk of syncope or cardiac arrest due to ventricular fibrillation, associated with right bundle branch block and ST-segment elevation in right precordial leads without any evidence of structural heart disease. However, the ST-segment elevation is inconsistently present and can be provoked pharmacologically with class I antiarrhythmic drugs (ajmaline, procainamide, and flecainide). Occasionally, in patients diagnosed with Brugada syndrome, in whom routine diagnostic methods showed no abnormalities, MRI could suggest fatty replacement in the right ventricular infundibulum consistent with ARVC. Mutations affecting the cardiac sodium channel gene (*SCN-5A*) have been found in three small families and individual patients. Patients are mostly male and have a first arrhythmic event around the fourth decade of life. The recurrence rate of new arrhythmic events is as high as 40%. Pharmacologic treatment does not seem to protect effectively against relapses, and currently, implantation of a cardioverter defibrillator appears to be the only effective therapy to prevent sudden death.

Natural History and Prognosis of the Disease

There are many problems regarding the natural history of ARVC. The true onset of the disease is unknown because a concealed phase probably precedes the symptomatic phase. The long-term follow-up studies are relatively few and may have selection bias because the majority include patients with severe arrhythmias.

Although the long-term prognosis of ARVC appears to be more favorable than that for patients with ventricular tachyarrhythmias in the setting of ischemic heart disease or dilated cardiomyopathy, the mortality rate due to sudden arrhythmic death and heart failure is significant. Blomström-Lundqvist and colleagues[82] reported a 20% mortality rate after 8.8 years of follow-up in patients on empirical antiarrhythmic treatment. Marcus and coworkers[4] reported a 17% mortality rate during long-term follow-up. Although patients with more concealed forms of the disease appear to carry a lower risk of serious arrhythmic event and cardiac arrest, sudden death has also been reported in this group of patients.[30]

Sudden death is an essential factor influencing the prognosis of the disease. Therefore, identification of risk factors is a major step toward prevention. This issue remains controversial. Some studies failed to identify any predictor of sudden death in affected patients.[83] However, the risk of sudden death may be increased in patients with syncope and/or family history of sudden death. Bettini and associates[84] stressed the role of sustained or polymorphic ventricular arrhythmias and that of left ventricular abnormalities. In a retrospective study on a cohort of 121 patients, Peters and colleagues[34] found that right ventricular dilatation and additional left ventricular abnormalities demonstrated by echocardiography and/or angiography, increased QRS duration (≥110 ms), precordial T wave inversion beyond V3, QT dispersion (≥50 ms), and JT dispersion (≥30 ms) on the ECG were strong predictors of arrhythmic events.

In another report, Peters and Reil[54] could relate risk of sudden death to definite enlargement of the right ventricle, reduced global right ventricular ejection fraction (<40%), hypo- or akinesia of three or more right ventricular segments, end-diastolic/end-systolic outpouchings in more than two segments, and inducible sustained ventricular arrhythmia during programmed ventricular stimulation. This subgroup of patients represented a typical form of the disease with progressive worsening of right ventricular function and increasing electrical imbalance. Strenuous exercise and sports activities remain, however, the most important precipitating factors.

Long-term follow-up data from clinical studies seem to indicate that ARVC is a progressive disease. The right ventricle may become more diffusely involved with time leading to right-sided heart failure. Left ventricular impairment appears at all functional and morphologic stages of the disease, even in cases with only slight or moderate right ventricular dysfunction. Heart failure tends to worsen progressively, and is more frequent with increasing age. In a series of patients reported by Pinamonti and coworkers,[24] the presence of heart failure was found to be associated with an increased mortality rate (≈50%) compared with patients with no symptoms or with arrhythmias (15%). In a series of patients with ARVC reported by Peters and colleagues,[29] heart failure (mostly right-sided) appeared in 11% of patients over a follow-up period of 4 to 8 years.

Therapy

As previously mentioned, the outcome of right ventricular cardiomyopathy, although often favorable, remains affected by the long-term risk of sudden death and the development of progressive heart failure. Thus, the notion of persisting risk must be kept in mind when selecting a therapeutic option. However, because of the rarity of the disease, the series reported in the literature are limited and the data that may form the basis of an appropriate management algorithm are insufficient. The clinical markers that predict outcome are debated. Similarly, selection of therapy may have rational grounds, but commonly lacks scientific evidence. The treatment of patients with arrhythmogenic right ventricular cardiomyopathy is, therefore, empirical. Various strategies can ensue based on the individual experience of different centers. Like other aspects of the disease, therapy remains uncertain, which leaves open the prospect for future evaluation studies.

PHARMACOLOGIC THERAPY

Selection of a therapeutic strategy is dependent on the clinical presentation. An apparently benign presentation is that of a patient with frequent or complex ventricular ectopic activities combined with signs suggestive of right ventricular cardiomyopathy. This can occasionally be seen in the setting of familial forms of the disease. The subsequent arrhythmic risk is difficult to evaluate. The possibility of underlying cardiomyopathy justifies further exploration of the significance of the clinical arrhythmia. The presence of a potentially malignant arrhythmogenic substrate is supposedly related to the inducibility of sustained ventricular tachycardia following exercise stress testing or electrophysiological study. A positive response to testing supports the use of prophylactic antiarrhythmic drug therapy despite the lack of definite evidence. In this view, an agent such as sotalol, which combines class III and β-blocking properties, might be preferred.[85] However, testing to initiate therapy may not be performed. An alternative consists of restricting the prescription of drugs to only symptomatic patients. Conversely, systematic β-blocking therapy has been recommended prophylactically,[86] although the risk-benefit ratio of this measure is unknown. In every case, sports activities must be limited.

Most frequently, the clinician deals with patients with episodes of sustained ventricular tachyarrhythmia. The hemodynamic tolerance of the attacks may be used as a guide to select therapy. Following a first well-tolerated episode of tachycardia, drug therapy is recommended. Because the arrhythmia is commonly dependent on increased adrenergic tone, β-blockers should be considered. In the case of recurrent or poorly tolerated tachycardia, electrophysiological testing is recommended to select the antiarrhythmic agent, suppression of ventricular arrhythmia inducibility being the end point. The value of a guided approach in this setting is, however, unproven. Class Ic drugs or sotalol is better evaluated by the electrophysiological techniques than is amiodarone. Combinations of antiarrhythmics have been studied by some investigators.[85,86] Irrespective of the antiarrhythmic used, the long-term drug efficacy is unknown. However, the incidence of ventricular tachycardia recurrences appears to be low. In a prospective study on 81 patients with ARVC, Wichter and coworkers[85] evaluated the short- and long-term efficacy of antiarrhythmic drugs. In patients with inducible sustained ventricular tachycardia during programmed ventricular stimulation, sotalol suppressed the arrhythmia in 68% of patients, whereas classes Ia and Ib drugs were effective in only 6%, class Ic drugs in 12%, and amiodarone in 15% of patients. Drug combination did not offer any additional benefit. In this group, the nonfatal arrhythmia recurrence rate was 10% during a follow-up of 34 months. In patients with no electrically inducible ventricular arrhythmia, in whom Holter monitoring and exercise testing were used to assess drug action, sotalol was again the most effective. Within 14 months, 12% of patients in this group had nonfatal arrhythmia recurrences. Besides the pharmacologic approach, other therapeutic options are available.

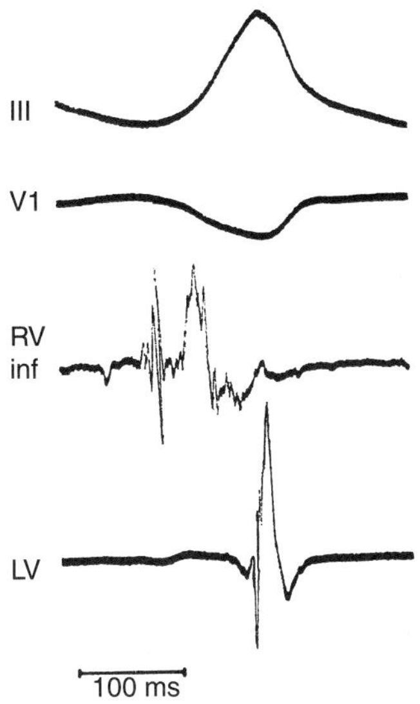

FIGURE 31-14 One beat of ventricular tachycardia shown in surface leads III and V1. Note the early ventricular activation of the right ventricular infundibulum (RV inf) that shows a fractionated presystolic potential preceding the onset of the QRS complex by 50 ms. This site was found to be the site of origin of the ventricular tachycardia. Note also the late activation of the left ventricle (LV).

RADIOFREQUENCY CATHETER ABLATION

Radiofrequency catheter ablation has been proposed as an alternative therapy in patients with drug-refractory and hemodynamically stable sustained ventricular tachycardia.[87] Mapping and entrainment techniques (Figs. 31-14 and 31-15) can be used to characterize reentrant circuits and to guide ablation.[57,88] Acute suppression of tachycardia by current delivery at sites of diastolic activity is feasible. Isolation of critical areas

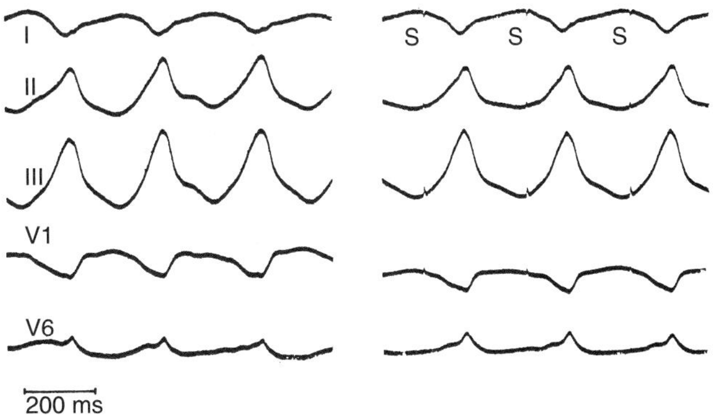

FIGURE 31-15 **Left panel:** Ventricular tachycardia of left bundle branch block configuration with a right inferior axis, induced following programmed stimulation. Surface leads I, II, III, V1, and V6 are shown. **Right panel:** Pace-mapping performed during sinus rhythm, from the right ventricular infundibulum, where abnormal electrograms were recorded, reproduced a QRS morphology identical to the ventricular tachycardia.

by radiofrequency ablation has also been reported.[89] However, relevant series are limited and a definite knowledge of long-term results is lacking. Ablation should be viewed as adjunctive treatment. Furthermore, potential hazards may lead to avoidance of ablation in those patients with severe right ventricular dysplasia.

SURGICAL DISCONNECTION OF THE RIGHT VENTRICLE

Before catheter ablation, surgical procedures had been designed to treat drug-refractory patients with ARVC. Total or partial disarticulation of the right ventricular free wall was first described by Guiraudon and associates[90] and aimed at isolating the abnormal myocardial zones from the rest of the heart. The rationale for this method was to prevent transmission of abnormal rhythms to the left ventricle. The results in terms of long-lasting suppression of ventricular tachycardia are unknown. Furthermore, the progression of the disease is likely to counteract the early benefit of surgery. This surgical approach has, therefore, been abandoned.

IMPLANTABLE CARDIOVERTER DEFIBRILLATOR

For patients with near-fatal arrhythmias and those who have survived cardiac arrest, the most effective safeguard against sudden cardiac death admittedly is the implantable cardioverter defibrillator (ICD), even though there is no definitive evidence supporting this guideline. In the recurrent forms of sustained ventricular tachyarrhythmia, the availability of ICDs has radically changed the approach to refractory patients. In this case, the rationale of device therapy is based on the suppressibility of tachycardias by rapid pacing, with shock delivery used only as a last resort. In a series of 82 patients with ARVC reported by Breithardt and colleagues,[91] an ICD was implanted in 18 patients. Indications for ICD implantations consisted of drug-refractory malignant ventricular arrhythmias, previous cardiac arrest without inducible ventricular tachycardia or fibrillation, and failure of catheter ablation. During a follow-up of 17 months, 50% of patients received appropriate electrical therapy.

Although the perioperative complication rate is low, pacing thresholds may be higher and R wave amplitudes lower in patients with ARVC, thus requiring testing at two or more ventricular sites.[92] On the other hand, Link and associates[92] found no significant differences in defibrillation thresholds in ARVC patients compared with a cohort of control patients.

Thus, in addition to defibrillation, the devices can eliminate most tachycardias, thereby preventing life-threatening events and repeat hospitalizations. Moreover, combined drug therapy may be helpful for decreasing the frequency and rate of ventricular arrhythmia episodes. Currently, the actual role of defibrillation therapy in arrhythmogenic right ventricular cardiomyopathy needs further elucidation.

TREATMENT OF ARVC ASSOCIATED WITH RIGHT OR BIVENTRICULAR HEART FAILURE

When heart failure is present due to severe right ventricular or biventricular systolic dysfunction, treatment is that of any patient with cardiac decompensation including diuretics, angiotensin-converting enzyme inhibitors, digitalis, and possibly anticoagulant therapy. The occurrence of ventricular arrhythmia episodes may cause further aggravation, rendering the control of the disease increasingly difficult. If long-lasting stabilization is possible, patients are occasionally candidates for heart transplantation.

Prevention

ARVC is a rare entity, which is associated with a lethal risk. Any preventive approach to the disease is impaired by a number of problems. The clinical features are imprecisely defined, rendering the epidemiologic data questionable. The size of the pathologic population is unknown. Definite assessment of sudden death risk is, accordingly, impossible. However, among patients with known ARVC treated with drugs, the frequency of arrhythmic deaths can reportedly reach 2% to 3% per year. This provides a target for prevention measures. In accordance with the conventional terminology, prevention applied to patients who have suffered from malignant arrhythmias or resuscitated cardiac arrest is named *secondary*. Currently any proof of the efficacy of antiarrhythmic drugs against the risk of sudden cardiac death is lacking. Implantable defibrillators may provide the most appropriate tool in this setting. Primary prevention concerns affected subjects who have not yet experienced life-threatening events. Those patients who are at risk need to be identified. In fact, no indisputable markers have been defined thus far. However, a history of familial sudden death, recurrent syncope, or left ventricular involvement may be of significance. The sudden death risk due to ARVC is enhanced by intense sports activities. The subset of young competitive athletes is, therefore, of special interest. The disease, in this context, may actually be suspected on the basis of family history, arrhythmic prodromal symptoms, and electrocardiographic features. With effective screening, including increased awareness of the disease, preventing the related sudden death risk in young athletes appears to be a reasonable expectation.

International Registry

Given the uncertainties inherent to the prevalence and the natural history of ARVC, as well as limitations of current techniques to diagnose the disease or to evaluate the efficacy of therapeutic approaches, an international registry has been established by the Study Group on ARVD/C of the Working Group on Myocardial and Pericardial Diseases and Arrhythmias of the European Society of Cardiology and the Scientific Council on

Cardiomyopathies of the World Health Federation, that commenced officially on January 1, 2001.[93] The objective of the ARVD/C International Registry is to follow-up for more than 10 years a large patient population, with the following aims: to prospectively validate criteria for the clinical diagnosis of ARVC; to assess the natural course of the disease in different clinical subgroups, including asymptomatic family members; to identify high risk groups for sudden death and poor prognosis; and to improve clinical management by evaluating long-term efficacy of empirical or test-guided antiarrhythmic agents or nonpharmacologic therapies. This project may also facilitate pathologic, molecular, and genetic research on the causes of the disease. Furthermore, availability of an international database should enhance awareness of this condition among cardiologists.

ACKNOWLEDGMENTS

Anatomopathologic data were provided by Dr. A. Tabib, Division of Pathology; GPB-SPECT images were prepared by Dr. L. Bontemps, Division of Nuclear Medicine; Echocardiographic images were selected by Dr. M. Perinetti, Division of Echocardiography, Hôpital Cardiovasculaire et Pneumologique Louis Pradel, Lyon-France.

REFERENCES

1. Uhl HSM: A previously undescribed congenital malformation of the heart: Almost total absence of the myocardium of the right ventricle: Bull Johns Hopkins Hospital 1952;91:197.
2. Osler WLM: The principles and practice of medicine, 6th ed. New York, D' Appleton, 1905, p 820.
3. Fontaine G, Guiraudon G, Frank R, et al: Stimulation studies and epicardial mapping in ventricular tachycardia: Study of mechanisms and selection for surgery. In HE Kulbertus (ed): Reentrant Arrhythmias: Mechanisms and Treatment. Lancaster, Pa, MTP Publishers, 1977, p 334.
4. Marcus FI, Fontaine G, Guiraudon G, et al: Right ventricular dysplasia : A report of 24 adult cases. Circulation 1982;65:384-399.
5. Thiene G, Nava A, Corrado D, et al: Right ventricular cardiomyopathy and sudden death in young people. N Engl J Med 1988; 318:129-133.
6. Basso C, Thiene G, Corrado D, et al: Arrhythmogenic right ventricular cardiomyopathy: Dysplasia, dystrophy, myocarditis? Circulation 1996;94:983.
7. Mallat Z, Tedgui A, Fontaliran F, et al: Evidence of apoptosis in arrhythmogenic right ventricular dysplasia. N Engl J Med 1996; 335:1190.
8. Basso C, Thiene G, Nava A, Dalla Volta S: Arrhthmogenic right ventricular cardiomyopathy: A survey of the investigations at the University of Padua. Clin Cardiol 1997;20:333.
9. Burke AP, Farb A, Tashko G, Virmani R: Arrhythmogenic right ventricular cardiomyopathy and fatty replacement of the right ventricular myocardium: Are they different diseases? Circulation 1998;97:1571.
10. Fontaine G, Guiraudon G, Frank R, et al: Dysplasie ventriculaire droite arythmogène et maladie de Uhl. Arch Mal Cœur 1982; 4:361.
11. Richardson P, McKenna W, Bristow M, et al: Report of the 1995 World Health Organization/International Society and Federation of Cardiology Task Force on the definition and classification of cardiomyopathies. Circulation 1996;93:841.
12. Rakover C, Rossi L, Fontaine G, et al: Familial arrhythmogenic right ventricular disease. Am J Cardiol 1986;58:377.
13. Nava A, Scognamiglio R, Thiene G, et al: A polymorphic form of familial arrhythmogenic right ventricular dysplasia. Am J Cardiol 1987;59:1405.
14. Nava A, Thiene G, Canciani B, et al: Familial occurrence of right ventricular dysplasia: A study involving nine families. J Am Coll Cardiol 1988;12:1222.
15. Fontaine G, Fontaliran F, Lascault G, et al: Dysplasie transmise et dysplasie acquise. Arch Mal Cœur 1990;83:915.
16. Thiene G, Corrado D, Nava A, et al: Right ventricular cardiomyopathy: Is there evidence of an inflammatory etiology? Eur Heart J 1991;12(Supplement D):22.
17. Sabel KG, Blomström-Lundqvist C, Olsson SB, Enestrom S: Arrhythmogenic right ventricular dysplasia in brother and sister: Is it related to myocarditis? Pediatr Cardiol 1990;11:113.
18. Hofmann R, Trappe HJ, Klein H, Kemmitz J: Chronic (or healed) myocarditis mimicking right ventricular dysplasia. Eur Heart J 1993;14:717.
19. Grumbach IM, Heim A, Vonhof S, et al: Coxsackievirus genome in myocardium of patients with arrhythmogenic right ventricular dysplasia/cardiomyopathy. Cardiology 1998;89:241.
20. Matsumori A, Kawai C: Coxsackievirus B3 perimyocarditis in BALB/c mice: Experimental model of chronic perimyocarditis in the right ventricle. J Pathol 1980;131:97.
21. Angelini A, Thiene G, Boffa G, et al: Endomyocardial biopsy in right ventricular cardiomyopathy. Int J Cardiol 1993;40:273.
22. Loire R, Tabib A: Arrhythmogenic right ventricular dysplasia and Uhl disease. Anatomic study of 100 cases after sudden death. Ann Pathol 1998;18:165.
23. Akazawa H, Ikeda U, Minezaki K, et al: Right ventricular dysplasia with complete atrioventricular block: Necessity and limitation of left ventricular epicardial pacing. Clin Cardiol 1998;21:604.
24. Pinamonti B, Lenarda A, Sinagra G, et al: Long-term evolution of right ventricular dysplasia-cardiomyopathy. Am Heart J 1995; 129:412.
25. Tonet J, Castro Miranda R, Iwa T, et al: Frequency of supraventricular tachyarrhythmias in arrhythmogenic right ventricular dysplasia. Am J Cardiol 1991;67:1153.
26. Corrado D, Thiene G: Sudden death in children and adolescents without apparent heart disease. N Trends Arrhythmias 1991; 7:209.
27. Sarubbi B, Ducceschi V, Santangelo L, Iacono A: Arrhythmias in patients with mechanical ventricular dysfunction and myocardial stretch: Role of mechano-electric feedback. Can J Cardiol 1998; 14:245.
28. Wichter T, Hindricks G, Lerch H, et al: Regional myocardial dysinnervation in arrhythmogenic right ventricular cardiomyopathy: An analysis using ^{123}I-MIBG scintigraphy. Circulation 1994;89:667.
29. Peters S, Peters H, Thierfekder L: Heart failure in arrhythmogenic right ventricular dysplasia-cardiomyopathy. Int J Cardiol 1999;71:251.
30. Nava A, Thiene G, Canciani B, et al: Clinical profile of concealed form of arrhythmogenic right ventricular cardiomyopathy with apparently idiopathic ventricular arrhythmias. Int J Cardiol 1992; 35:195.
31. Scognamiglio R, Fasoli G, Nava A, et al: Contribution of cross-sectional echocardiography to the diagnosis of right ventricular dysplasia at the asymptomatic stage. Eur Heart J 1989;10:538.
32. Leclercq JF, Potenza S, Maison-Blanche P, et al: Determinants of spontaneous occurrence of sustained monomorphic ventricular tachycardia in right ventricular dysplasia. J Am Coll Cardiol 1996; 28:720.
33. Kullo IJ, Edwards WD, Seward JB: Right ventricular dysplasia: The Mayo Clinic experience. Mayo Clin Proc 1995;70:541.
34. Peters S, Peters H, Thierfelder L: Risk stratification of sudden cardiac death and malignant ventricular arrhythmias in right ventricular dysplasia-cardiomyopathy. Int J Cardiol 1999; 71:243.
35. Basso C, Corrado D, Thiene G: Cardiovascular causes of sudden death in young individuals including athletes. Cardiol Rev 1999; 7:127.
36. Burke AP, Farb A, Virmani R, et al: Sports-related and non–sports-related sudden cardiac death in young adults. Am Heart J 1991;121:568.
37. Maron BJ, Shirani J, Poliac LC, et al: Sudden death in young competitive athletes: Clinical, demographic and pathological profile. JAMA 1996;276:199.

38. Nemec J, Edwards BS, Osborn MJ, Edwards WD: Arrhythmogenic right ventricular dysplasia masquerading as dilated cardiomyopathy. Am J Cardiol 1999;84:237.
39. Pinamonte B, Sinagra G, Salvi A, et al: Left ventricular involvement in right ventricular dysplasia. Am Heart J 1992;123:711.
40. Fontaine G, Umemura J, Di Donna P, et al: La durée des complexes QRS dans la dysplasie ventriculaire droite arythmogène. Un nouveau marqueur diagnostic non invasif. Ann Cardiol Angeiol 1993;42:399.
41. Fontaine G, Frank R, Guiraudon G, et al: Signification des troubles de conduction intraventriculaires observés dans la dysplasie ventriculaire droite arythmogène. Arch Mal Cœur 1984;77:872.
42. Metzger JT, de Chillou C, Cheiex E, et al: Value of the 12-lead electrocardiogram in arrhythmogenic right ventricular dysplasia and absence of correlation with echocardiographic findings. Am J Cardiol 1993;72:964.
43. Fontaine G, Piot O, Sohal P, et al: Dérivations en précordiales droites et mort subite. Relation avec la dysplasie ventriculaire droite arythmogène. Arch Mal Cœur 1996;89:1323.
44. Benn M, Hansen PS, Pedersen AK: QT dispersion in patients with arrhythmogenic right ventricular dysplasia. Eur Heart J 1999; 20:764.
45. Jaoude SA, Leclercq JF, Coumel P: Progressive ECG changes in arrhythmogenic right ventricular disease. Eur Heart J 1996; 17:1717.
46. Marcus FI, Fontaine G: Arrhythmogenic right ventricular dysplasia/cardiomyopathy: A review. Pacing Clin Electrophysiol 1995;18:1298.
47. Toyofuku M, Takaki H, Sunagawa K, et al: Exercise-induced ST elevation in patients with arrhythmogenic right ventricular dysplasia. J Electrocardiol 1999;32:1.
48. Mehta D, Goldman M, David O, Gomes JA: Value of quantitative measurement of signal-averaged electrocardiographic variable in arrhythmogenic right ventricular dysplasia: Correlation with echocardiographic right ventricular cavity dimensions. J Am Coll Cardiol 1996;28:713.
49. Nava A, Folino F, Bauce B, et al: Signal-averaged electrocardiogram in patients with arrhythmogenic right ventricular cardiomyopathy and ventricular arrhythmias. Eur Heart J 2000; 21:58.
50. Turini P, Angelini A, Thiene G, et al: Late potentials and ventricular arrhythmias in arrhythmogenic right ventricular cardiomyopathy. Am J Cardiol 1999;83:1214.
51. Hermida JS, Minassian A, Jarry G, et al: Familial incidence of late ventricular potentials and electrocardiographic abnormalities in arrhythmogenic right ventricular dysplasia. Am J Cardiol 1997; 79:1375.
52. Di Biase M, Favale V, Massari G, et al: Programmed stimulation in patients with minor forms of right ventricular dysplasia. Eur Heart J 1989;10(Supplement D):49.
53. Haissaguerre M, Le Metayer P, D'Ivernois C, et al: Distinctive response of arrhythmogenic right ventricular disease to high dose isoproterenol. Pacing Clin Electrophysiol 1990;13(Pt II):2119.
54. Peters S, Reil H: Risk factors of cardiac arrest in arrhythmogenic right ventricular dysplasia. Eur Heart J 1995;16:77.
55. Panidis IP, Greenspan AM, Mintz GS, Ross J Jr: Inducible ventricular fibrillation in arrhythmogenic right ventricular dysplasia. Am Heart J 1985;110:1067.
56. Rosenfeld LE, Batsford WP: Intraventricular Wenckebach conduction and localized reentry in a case of right ventricular dysplasia with recurrent ventricular tachycardia. J Am Coll Cardiol 1983;2:585.
57. Ellison KE, Friedman PL, Ganz LI, Stevenson WG: Entrainment mapping and radiofrequency ablation of ventricular tachycardia in right ventricular dysplasia. J Am Coll Cardiol 1998;32:724.
58. Brembilla-Perrot B, Jacquemein L, Houplon P, et al: Increased atrial vulnerability in arrhythmogenic right ventricular disease. Am Heart J 1998;135:748.
59. Fontaine G, Fontaliran F, Hébert JL, et al: Arrhythmogenic right ventricular dysplasia. Ann Rev Med 1999;50:17.
60. Foale RA, Nihoyannopoulos P, McKenna WJ, et al: Echocardiocardiographic measurement of the normal adult right ventricle. Br Heart J 1986;56:33.
61. Fontaine G, Fontaliran F, Frank R: Arrhythmogenic right ventricular cardiomyopathies. Clinical forms and main differential diagnoses (editorial). Circulation 1998;97:1532.
62. De Piccoli B, Rigo F, Caprioglio F, et al: The usefulness of transesophageal echocardiography in the diagnosis of arrhythmogenic right ventricular cardiomyopathy. G Ital Cardiol 1993; 23:247.
63. Dobrinski G, Werdiere C, Fontaine G, et al: Diagnostic angiographique des dysplasies ventriculaires droites. Arch Mal Cœur 1985;78:544.
64. Daliento L, Rizzoli G, Thiene G, et al: Diagnostic accuracy of right ventriculography in arrhythmogenic right ventricular cardiomyopathy. Am J Cardiol 1990;66:741.
65. Bloström-Lundqvist C, Selin K, Jonsson R, et al: Cardioangiographic findings in patients with arrhythmogenic right ventricular dysplasia. Br Heart J 1988;59:556.
66. Pavel D, Byrom E, Law W, et al: Detection and quantification of regional wall motion abnormalities using phase analysis of equilibrium gated cardiac studies. Clin Nucl Med 1983;8:315.
67. Bourguignon MH, Sebag C, Le Guludec D, et al: Arrhythmogenic right ventricular dysplasia demonstrated by phase mapping of gated equilibrium radioventriculography. Am Heart J 1986; 111:997.
68. Le Guludec D, Slama MS, Frank R, et al: Evaluation of radionuclide angiography in diagnosis of arrhythmogenic right ventricular cardiomyopathy. J Am Coll Cardiol 1995;26:1476.
69. Casset-Senon D, Philippe L, Babuty D, et al: Diagnosis of arrhythmogenic right ventricular cardiomyopathy by Fourier analysis of gated blood pool single-photon emission tomography. Am J Cardiol 1998;82:1399.
70. Daou D, Lebtahi R, Petegnief Y, et al: Angioscintigraphy (RNA) in arrhythmogenic right ventricular cardiomyopathy: one (H1) and three (H3) harmonics Fourier phase analysis versus factor analysis. Eur J Nucl Med 1997;24:971.
71. Markiewicz W, Sechtem U, Higgins CB: Evaluation of the right ventricle by magnetic resonance imaging. Am Heart J 1987;113:8.
72. Molinari F, Sardanelli F, Gaita F, et al: Right ventricular dysplasia as a generalized cardiomyopathy? Findings on magnetic resonance imaging. Eur Heart J 1995;16:1619.
73. Rampazzo A, Nava A, Danieli GA, et al: The gene for arrhythmogenic right ventricular cardiomyoapthy maps to chromosome 14q23-q24. Hum Mol Genet 1994;3:959.
74. Rampazzo A, Nava A, Erne P, et al: A new locus for arrhythmogenic right ventricular cardiomyopathy maps to chromosome 1q42-q43. Hum Mol Genet 1995;4:2151.
75. Rampazzo A, Nava A, Miorin M, et al: A new locus for arrhythmogenic right ventricular cardiomyopathy (ARVD4) maps to chromosome 2q32. Genomics 1997;45:259.
76. Ahmad F, Li D, Karibe A, et al: Localization of a gene responsible for arrhythmogenic right ventricular dysplasia to chromosome 3p23. Circulation 1998;98:2791.
77. Coonar AS, Protonotarius N, Tsatsopoulou A, et al: Gene for arrhythmogenic right ventricular cardiomyopathy with diffuse nonepidermolytic palmoplantar keratoderma and woolly hair (Naxos disease) maps to chromosome 17q21. Circulation 1998;97:2049.
78. McKenna WJ, Thiene G, Nava A, et al: Diagnosis of arrhythmogenic right ventricular dysplasia/cardiomyopathy. Br Heart J 1994;71:215.
79. Carlson MD, White RD, Throhman RG, et al: Right ventricular outflow tract ventricular tachycardia: Detection of previously unrecognised anatomic abnormalities using cine magnetic resonance imaging. J Am Coll Cardiol 1994;24:720.
80. Brugada J, Brugada P: Right bundle branch block, persistent ST segment elevation and sudden cardiac death: A distinct clinical and electrocardiographic syndrome—A multicenter report. J Am Coll Cardiol 1992;20:1391.
81. Alings M, Wilde A: "Brugada" syndrome. Clinical data and suggested pathophysiological mechanism. Circulation 1999;99:666.
82. Blomström-Lundqvist C, Sabel KG, Olsson SB: A long-term follow-up of 15 patients with arrhythmogenic right ventricular dysplasia. Br Heart J 1987;58:477.
83. Leclercq JF, Denjoy I, Maison-Blanche P, et al: La recherche de potentiels tardifs chez les sujets ayant des troubles du rythme ventriculaires sur cœur apparemment sain. L'information cardiologogique 1990;14:23.
84. Bettini R, Furlanello F, Vergara G, et al: Arrhythmologic study of 50 patients with arrhythmogenic disease of the right ventricle: Prognostic implications. G Ital Cardiol 1989;19:567.

85. Wichter T, Borggrefet M, Haverkamp W, et al: Efficacy of antiarrhythmic drugs in patients with arrhythmogenic right ventricular disease. Results in patients with inducible and noninducible ventricular tachycardia. Circulation 1992;86:29.
86. Fontaine G, Zenati O, Tonet J, et al: The treatment of ventricular arrhythmias. In Nava A, Rossi L, Thiene G (eds): Arrhythmogenic Right Ventricular Cardiomyopathy-Dysplasia:315. Amsterdam-Netherlands, Elsevier, 1997.
87. Feld GK: Expanding indications for radiofrequency catheter ablation: Ventricular tachycardia in association with right ventricular dysplasia. J Am Coll Cardiol 1998;32:729.
88. Harada T, Aonuma K, Yamauchi Y, et al: Catheter ablation of ventricular tachycardia in patients with right ventricular dysplasia: Identification of target sites by entrainment mapping techniques. Pacing Clin Electrophysiol 1998;21(Pt II):2547.
89. Fukushima K, Emori T, Morita H, et al: Ablation of ventricular tachycardia by isolating the critical site in a patient with arrhythmogenic right ventricular cardiomyopathy. J Cardiovasc Electrophysiol 2000;11:102.
90. Guiraudon GM, Klein GJ, Gulamhusein S, et al: Total disconnection of the right ventricular free wall: Surgical treatment of right ventricular tachycardia associated with right ventricular dysplasia. Circulation 1983;67:463.
91. Breithardt G, Wichter T, Haverkamp W, et al: Implantable cardioverter defibrillator therapy in patients with arrhythmogenic right ventricular cardiomyopathy, long QT syndrome or no structural heart disease. Am Heart J 1994;127:1151.
92. Link MS, Wang PJ, Haugh CJ, et al: Arrhythmogenic right ventricular dysplasia: Clinical results with implantable cardioverter defibrillators. J Interv Card Electrophysiol 1997;1:41.
93. Corrado D, Fontaine G, Marcus FI, et al: Arrhythmogenic right ventricular dysplasia/cardiomyopathy: Need for an international registry. Circulation 2000;101:e101.
94. Hamid MS, Norman M, Quraishi A, et al: Prospective evaluation of relatives for familial arrhythmogenic right ventricular cardiomyopathy/dysplasia reveals a need to broaden diagnostic criteria. JACC 2002;40(8):1445-1450.
95. McKoy G, Protonotarios N, Crosby, et al: Identification of a deletionin plakoglobin in arrhythmogenic right ventricular cardiomyopathy with palmo-plantar karatodermia and woolly hair (Naxos disease). Lancet 2000;355:2119-2124.
96. Rampazzo A, Nava A, Malacrida S, et al: Mutation in human desmoplakin domain binding to plakoglobin causes a dominant form of arrhythmogenic right ventricular cardiomyopathy. Am J Hum Genet 2002;71:1200-1206.
97. Tiso N, Stephan DA, Bagattin A, et al: Identification of mutations in the cardiac ryanodine receptor gene in families affected with ARVC type 2. Hum Mol Genet 2001;10:189-194.
98. Priori SG, Napolitano C, Tiso N, et al: Mutations in the cardiac ryanodine receptor gene (hRg R2) XXX Catecholaminergic polymorphic ventricular tachycardia. Circulation 2001;103(2): 196-200
99. Frances R, Rodriguez Benitez AM, Cohen DR: Arrhythmogenic right ventricular dysplasia and anterior wall cataract. Am J Med Genet 1997;73:125-126.
100. Severini GM, Krajinovic M, Pinamonti B, et al: A new locus for arrhythmogenic right ventricular dysplasia on the long arm of chromosome 14. Genomics 1996;31:193-200.
101. Li D, Ahmad F, Gardner MJ, et al: The locus of a novel gene responsible for arrhythmogenic right ventricular dysplasia characterised by early onset and high penetrance maps to chromosome 10p12-p14. Am J Hum Genet 2000;66:148-156.
102. Melberg A, Oldfors A, Blomström-Lundqvist C, et al: Autosomal dominant myofibrillar myopathy with ARVC linked to chromosome 10q. Am Neurol 1999;46:684-692.

Chapter 32 Postoperative Arrhythmias After Cardiac Surgery

DAVID B. BHARUCHA and PETER R. KOWEY

Arrhythmias occur frequently following cardiac surgery. This chapter discusses postoperative atrial and ventricular tachyarrhythmias as well as bradyarrhythmias. The incidence, prognosis, potential mechanisms of pathogenesis, and management of these arrhythmias are described. Postoperative atrial arrhythmias are associated with increased morbidity, length of hospitalization, and medical costs, whereas prophylactic therapy is frequently given in an attempt to avoid these problems and be cost effective. Whether these goals are met is discussed in the chapter. Treatments with pharmacologic and nonpharmacologic modalities are considered. Management strategies for atrial arrhythmias that nonetheless occur postoperatively are discussed. This chapter also discusses the incidence, risk factors, and potential etiologies of ventricular arrhythmias and bradyarrhythmias that occur following cardiac surgery, risk stratification techniques, and treatment for these issues.

Atrial Arrhythmias

INCIDENCE AND PREDICTORS

Atrial arrhythmias occur frequently after most types of cardiac surgery, with a prevalence as high as 40% following coronary artery bypass grafting (CABG).[1-4] Atrial fibrillation (AF) and flutter are the most common arrhythmias; atrial tachycardias, including multifocal atrial tachycardia are also observed. Clinical variables that convey higher risk for the development of postoperative AF are described in Table 32-1.[5-14]

Passman and colleagues[11] have proposed a series of nomograms to assess the degree of risk for postoperative atrial fibrillation based on multiple preoperative clinical and ECG variables. The investigators performed a chart review of 229 consecutive patients who underwent CABG and found that independent predictors for postoperative AF included advanced age; left main or proximal right coronary artery stenoses; history of AF or heart failure; or preoperative ECG findings of a P–R interval greater than or equal to 185 ms or a P wave duration in lead V1 greater than or equal to 110 ms. Univariate analysis indicated that, in this cohort, a history of chronic obstructive pulmonary disease (COPD), left main or proximal right coronary artery stenoses, or frequent premature atrial contractions or left atrial abnormality on ECG were not significant predictors for postoperative AF. Initial reports indicated that a minimally invasive approach to CABG (versus a conventional sternotomy) does not lessen the incidence of postoperative AF, when corrected for disease severity.[15,16] However, more recent investigations indicate that the incidence of postoperative AF is lessened with minimally invasive techniques for CABG[17,18] and valvular surgery.[19,20]

Coronary artery bypass grafting without the use of cardiopulmonary bypass (CPB) has been associated with a decreased incidence of postprocedural AF in situations of minimally invasive techniques,[21] reoperative

TABLE 32-1 Predictors of the Development of Postcardiac Surgical Atrial Arrhythmias

Advanced age
Valvular heart disease (particularly mitral valvular disease and stenosis)
Increased left atrial size
Sinus ECG showing: PR ≥185 ms, p wave duration in lead V1 ≥110 ms, atrial premature depolarizations, or left atrial abnormality
Roentgenographic cardiomegaly
History of congestive heart failure
Previous atrial arrhythmias
Previous cardiac surgery
Long bypass times
Method of cardioprotection, hypothermia, and cardiac venting via right superior pulmonary vein during bypass
Right coronary artery grafting
Left main coronary stenosis grafting
Postoperative inotropic support and endotracheal intubation time
Elevated postoperative adrenergic tone
Postoperative chest infection
Absence of β-blocker treatment (or withdrawal of previous treatment)
Chronic obstructive pulmonary disease
Chronic renal failure
Electrolyte disturbances: hypokalemia, hypomagnesemia
Pericarditis
Length of hospital stay

("redo") single-vessel revascularization,[22] and octogenarian patients.[23] Ascione and colleagues[10] investigated, in a prospective, randomized trial, whether the use of cardiopulmonary bypass and cardioplegic arrest influenced the incidence of postoperative AF. Two hundred patients were randomized to the use of CPB (with normothermic CPB versus cardioplegic arrest CPB) and off-pump, "beating heart" surgery. In this study, a risk factor for postoperative AF was observed to be the use of CPB with cardioplegic arrest, not CPB in general. Other risk factors for postoperative AF described by this study include postoperative inotropic support, intubation time, chest infection, and length of hospital stay. In summary, similar to patients with postoperative AF observed after conventional CABG, patients with AF that occurs following minimally invasive techniques have a higher in-hospital morbidity rate, length of stay, and mortality rate compared with patients without AF; however, AF following cardiac surgery performed minimally invasively or without CPB or cardioplegia can occur with a lower frequency.

Risk stratification for postoperative atrial arrhythmias can be performed using clinical characteristics (see Table 32-1) or by laboratory methods. One example of a stratification method is based on P wave duration, as calculated directly from the surface electrocardiogram[24] or from signal-averaged data.[25] An investigation by Buxton and Josephson[24] assessed P wave duration from standard electrocardiographic leads in 99 cardiac surgical patients and found that the mean total P wave duration of patients who developed AF or atrial flutter was significantly longer than patients who remained in sinus rhythm (mean total P wave duration of 160 ms and 126 ms [$P < .001$], respectively). A significantly prolonged P wave duration was observed to be a sensitive (83%) but not specific (43%) predictor of postoperative atrial arrhythmias. Prolonged P wave duration in patients who develop postoperative atrial arrhythmias may be a reflection of underlying preoperative atrial disease.

PROGNOSIS

Postoperative AF usually arises 1 to 5 days following surgery with a peak incidence on day 2 and usually has a self-limited course.[26-28] More than 90% of patients with AF following cardiac surgery who have no history of atrial arrhythmias are in sinus rhythm 6 to 8 weeks following their operation.[29] Rubin and associates[30] followed postcardiac surgical patients for an average of 26 months and observed no differences in cardiovascular or cerebrovascular morbidity or mortality between patients with postoperative AF versus patients without AF. However, other observations have been made showing an increased rate of early and late postoperative stroke in association with AF[14,21,31,32] (see later under section on postarrhythmia therapy). Postoperative atrial arrhythmias are usually considered to increase morbidity, ICU length of stay, length of hospitalization, and medical costs. Almassi and coworkers, in a series of 3855 cardiac surgical patients, found that postoperative AF was associated with a longer ICU stay (3.6 versus 2 days for patients without AF, $P < .001$), an increased rate of ICU readmission (13% versus 4%), a greater incidence of perioperative myocardial infarction (7.4% versus 3.4%), more persistent congestive heart failure (4.6% versus 1.4%), and a higher rate of reintubation (10.6% versus 2.5%).[14] In an investigation by Abreu and colleagues there was an increased length of stay of 4.9 days due to postcardiac surgical AF, which was calculated to augment medical costs by $10,000.[33] It should be noted that, in slight distinction, a study by Kim and associates observed a lesser lengthening in hospital stay (1 to 1.5 days in this study) attributed to postoperative AF.[34] Villereal and colleagues[35] found the occurrence of AF after CABG identifies a subset of patients who are at an increased risk for both early and long-term cardiovascular events.

ETIOLOGY

There are many perioperative factors that have been described in the pathogenesis of postoperative AF (Table 32-2) but there are no definitive data. The pathophysiology of postoperative AF is probably related to preexisting age-related degenerative cardiac changes in many patients, coupled with perioperative abnormalities in several electrophysiological parameters, such as dispersion of atrial refractoriness, atrial conduction velocity, and atrial transmembrane potential. Postoperative pericarditis is generally felt to be an etiology or at least a potentiating factor for atrial fibrillation. Perioperative hypokalemia has been shown to be associated with atrial arrhythmias with an odds ratio of 1.7 (even after adjusting for confounding factors) in a multicenter trial that followed over 2400 patients through cardiac surgery.[36] Potential mechanisms whereby hypokalemia might alter atrial electrophysiology include increased phase 3 depolarization, increased automaticity, and decreased conduction velocity.

PROPHYLACTIC THERAPY WITH PHARMACOLOGIC AGENTS

Given the high incidence of postoperative AF, it is strongly recommended to consider prophylactic treatment, especially in the presence of the risk factors described in Table 32-1. The use of β-blockers, in the presence or absence of digitalis, has been demonstrated to decrease AF from 40% for CABG patients and 60% for valvular surgery patients to 20% and 30%,

TABLE 32-2 Potential Factors Implicated in the Pathogenesis of Postcardiac Surgical Atrial Arrhythmias

Pericarditis
Atrial injury from surgical handling or from cannulation
Acute atrial enlargement from pressure or volume overload
Inadequate cardioprotection during bypass
Atrial infarction or ischemia
Hyperadrenergic state
Pulmonary issues (e.g., infection, atelectasis, pleural effusion, hypoxia)

respectively.[2,27,28] The effect of β-blockade to reduce postoperative AF, both alone or with digoxin, has been demonstrated in multiple meta-analyses.[37,38] Whereas it might be assumed that a preventive strategy of β-blocker administration would save both medical resources and decrease length of hospital stay, such a benefit has not yet been demonstrated. Specifically, in the β-blocker length of stay study (BLOSS) trial,[12] 1000 patients undergoing cardiac surgery were randomized in a double-blinded fashion to receive either metoprolol (possibly in an up-titrated dose) or placebo. In this trial, patients treated with β-blockers had a decreased incidence of atrial arrhythmias but this did not translate into a decreased length of stay. Although investigators have found an association between postoperative AF and cardiovascular and other morbidities, broadly accepted therapies that reduce the incidence of AF have not yet been demonstrated to have long-term benefit or cost effectiveness.[35]

Digitalis given pre- or postoperatively has been shown only to be *possibly* helpful and not to the same extent or with the same reliability as β-blockers.[37] Postoperative verapamil given to patients in sinus rhythm has been observed to slow the rate of AF if it occurs, but not to alter the prevalence.[39] Other antiarrhythmic agents, such as procainamide,[40] have been studied in a prophylactic role but have been associated with varying benefits in different reports. There are no comprehensive data available regarding propafenone or flecainide.

dl-Sotalol, has β-blocker and class III activities and may have a role in the prophylactic treatment of postoperative AF. Preliminary data[41,42] indicate that oral sotalol may reduce the incidence of AF following cardiac surgery. Sotalol may be more effective than metoprolol,[43] speaking to a potential incremental benefit of the activity of sotalol class III. Conversely, a relatively large study of 429 consecutive patients by Suttorp and associates[44] demonstrated no dramatic difference in the benefit of low- versus high-dose sotalol when compared with low- versus high-dose β-blockers, suggesting that the potential benefit of dl-sotalol arises from its β-blocker effect.

The potential prophylactic role of sotalol was further examined in a recent prospective, randomized, double-blind, placebo-controlled study of 85 postcardiac surgical patients by Gomes and coworkers.[45] In this group of patients, a significant reduction in postoperative AF was observed with sotalol treatment when compared with either placebo or β-blocker treatment. Also, there was no detected increase in ventricular arrhythmias, suggesting that the membrane effect of sotalol class III, when given in this study's setting was not a liability. However, this study excluded patients with heart failure or marked left ventricular dysfunction (characteristics that might predict a high risk for postoperative AF or for ventricular proarrhythmia), suggesting that although sotalol may be useful, it cannot be broadly applied.[46]

Amiodarone—both oral and parenteral formulations—have been evaluated for prophylaxis against perioperative atrial arrhythmias. Daoud[47] and colleagues, in a placebo-controlled trial, assessed the potential benefit of preoperatively administered amiodarone in 124 patients undergoing cardiac surgery. Patients who received amiodarone, which had been initiated at least 7 days preoperatively, had a lower incidence of AF (25%) when compared with patients who had received placebo (53%). Amiodarone administration was also associated with a decreased length of hospitalization and resultant decreased hospital costs. Although these data suggest a possible benefit of outpatient preoperative medication with oral amiodarone to decrease the incidence of AF, the observation of a high incidence of atrial arrhythmias in the control group, despite a high rate of use of β-blockers—an approach shown to prevent at least 50% of AF in almost every trial in which it has been studied[2,37]—suggests that these observations may not be broadly applicable. Distinct disadvantages of preoperative amiodarone treatment exist: the need to identify patients well in advance of their procedure, potential bradyarrhythmic hazards (especially in an outpatient setting and in the elderly), and, although rare, a risk of perioperative pulmonary toxicity.[48] The latter risk may be due to a potentiated risk from amiodarone of cardiopulmonary bypass-associated adult respiratory distress syndrome, which has a poor prognosis.

The Amiodarone Reduction in Coronary Heart (ARCH) trial investigated, in a placebo-controlled, double-blinded study of 300 patients, whether postoperative administration of intravenous (IV) amiodarone reduced the incidence of AF.[49] Results showed a significant decrease in the incidence of AF in patients given amiodarone (35%) compared with those given placebo (47%), without significant risk from the active agent. However, the size of the benefit did not result in reduced length of hospital stay in this study. The relatively modest benefit of IV amiodarone in this report[49] would probably have been even smaller had a greater number of patients received β-blockers.

In any clinical situation, potential benefits have to be balanced against risks. Antiarrhythmic medications have possible proarrhythmic effects, particularly classes I and III agents in the context of structural heart disease, myocardial ischemia, and metabolic dysfunction. Caution is urged, especially with these agents and circumstances.[50]

Beta-blockers have been shown to be the most effective prophylactic agent[2,37] and carry a lower risk relative to other antiarrhythmic agents. Based on these data, we believe that β-blocker prophylaxis should be widely applied. Use of other drugs for prophylaxis needs to be further investigated.

PROPHYLACTIC THERAPY WITH PACING

The potential role of nonpharmacologic therapy for the prevention of postoperative AF has been examined in several studies. Single- and multiple-site atrial pacing has been shown to be helpful in some cases of nonperioperative paroxysmal AF.[51] Investigations into the potential role of single- and multiple-site atrial pacing for the prevention of postoperative AF have yielded results showing varying benefits.

Initial reports indicated that atrial pacing might not be beneficial for postoperative AF.[52] An investigation of

86 post-CABG patients[52] found that atrial pacing via single-site atrial epicardial wires, at a rate of at least 80 beats/minute and always above the intrinsic sinus rate ("overdrive pacing"), was not associated with a different incidence of postoperative atrial arrhythmias when compared with the absence of pacing. A recent study of 100 post-CABG patients, randomized to no atrial pacing versus atrial pacing at equal to or greater than 10 beats/minute above the resting heart rate, indicated that atrial pacing significantly increased atrial ectopy and did not attenuate the rate of atrial fibrillation occurrence.[53] The potential role of bi-atrial overdrive pacing in the prevention of postoperative AF has also been investigated in several studies. One prospective, randomized trial by Kurz and colleagues[54] examined the effect of bi-atrial pacing in a group of postcardiac surgical patients, assessing the incidence of AF and the possible proarrhythmic effects of pacing. Unfortunately, after only 21 of the planned 200 patients were randomized, the study was terminated because this study's pacing protocol was observed to promote AF, a possible consequence of undersensing of atrial signals by the epicardial pacing system leading to asynchronous atrial pacing. An investigation by Gerstenfeld and coworkers[55] studied 61 post-CABG patients who were randomized to right atrial pacing, left and right atrial pacing, or no pacing, using epicardial wires. There was no significant difference in the incidence of AF between groups, although there was a trend toward less atrial arrhythmia in paced patients also receiving a β-blocker.

Some studies, however, have shown a potential benefit from atrial pacing in the prevention of postoperative AF. Greenberg and associates[56] performed an investigation of 154 patients following CABG or CABG plus aortic valve replacement. Patients were randomized to no pacing, right atrial pacing, left atrial pacing, or bi-atrial pacing for 72 hours postoperatively, and efforts were made to administer β-blocker medications. Any pacing modality reduced the incidence of AF from 37.5% to 17% and the length of hospitalization from 7.8 to 6.1 days. In this study, multivariate analysis indicated that the most effective sites of pacing were right atrial followed by left atrial followed by bi-atrial. Of note, patients in this study did not have significant left ventricular dysfunction (average ejection fraction 53±10%). In contrast, an investigation by Blommaert and colleagues[57] examined the course of 96 postoperative patients who had a wide range of left ventricular function. Patients were randomized to no pacing versus 24 hours of atrial pacing using a dynamic overdrive algorithm.[57] Attention was paid to the use of β-blocker medication. Pacing was associated with a lower incidence of AF (10%) versus an absence of pacing (27%). Multivariate analyses showed that the beneficial effect of atrial pacing was observed particularly in patients with preserved left ventricular function and in older patients.

Findings regarding the potential benefit of bi-atrial pacing in the prevention of postoperative AF have been varied. In contrast to the findings of Greenberg and coworkers,[56] Fan and colleagues[58] observed a greater benefit with bi-atrial versus single-site atrial pacing. Fan and associates studied 132 postoperative patients without a history of AF and randomized them to no pacing, bi-atrial pacing, left atrial pacing, or right atrial pacing. After overdrive atrial pacing for 5 days, the incidence of AF was 41.9, 12.5, 36.4, and 33.3%, respectively. Reductions in rates of postoperative AF translated into reductions in hospital lengths of stay in this study. Also, patients who remained in sinus rhythm had significant reductions in P-wave duration and variability in P-wave duration following pacing therapy.

One component of the potential benefit of postoperative pacing in the prevention of AF might be its allowance for more thorough β-blocker dosing. Clearly, additional studies are necessary to investigate the potential role of overdrive pacing in the prevention of postoperative AF.

POSTARRHYTHMIA THERAPY

Rate Control Treatment for Postoperative Atrial Fibrillation

Given the self-limited course of postoperative AF for the vast majority of patients with no history of preoperative atrial arrhythmias,[29] treatment to control the ventricular response rate in postoperative AF is a useful strategy. Rate control therapy with β-blockade should be the first-line choice, with the relative benefit partly due to treatment of the hyperadrenergic postoperative state and prevention of the well demonstrated phenomenon of β-blocker withdrawal. Rapid administration of IV digoxin is occasionally mentioned as being helpful in restoring sinus rhythm, although the data are not supportive.[59] AV-nodal blocking agents, such as calcium channel blockers and digoxin, have roles in the control of the ventricular rate in AF, but are not more effective than β-blockers; calcium channel blockers or digoxin may be useful when β-blockers cannot be given (for instance, in situations of bronchospasm). A randomized, double-blinded investigation by Tisdale and associates,[60] comparing parenteral diltiazem and digoxin in post-CABG patients with AF, indicated that this calcium channel blocker results in rate control of AF more rapidly than does digoxin, but that after 12 and 24 hours, there is no significant difference in effect, nor is there a difference in length of hospital stay.

Administration of IV beta-blocking agents with short half-lives (e.g., esmolol) can be particularly useful if there is potential for bronchospasm, bradyarrhythmias, or hypotension, as the IV administration can be discontinued with a resultant rapid disappearance of the pharmacologic effect.

An investigation by Clemo and colleagues[61] assessed the potential benefits of IV amiodarone in critically ill patients with atrial arrhythmias with rapid ventricular response rates, some of whom were postoperative following cardiac surgery. The data were retrospectively obtained from 38 patients with atrial arrhythmias in an intensive care unit setting who had AF with resultant hemodynamic destabilization, despite previous use of conventional AV nodal blocking agents for rate control. Observations showed that IV amiodarone administration was associated with improved rate control, peripheral blood pressure, cardiac filling pressures, and cardiac output. There was, however, no significantly increased rate of spontaneous reversion to sinus rhythm after IV amiodarone treatment. In summary, this investigation showed that IV amiodarone has a beneficial role in ventricular rate slowing in AF in critically ill patients, likely to include groups of postcardiac surgical patients, particularly when previous AV nodal blocking drugs have not been fully effective. Although sole treatment with IV amiodarone cannot be relied on for conversion from AF,[62] it is likely to be quite helpful in maintaining sinus rhythm and it (or some antiarrhythmic agent) should be considered before electrical cardioversion.

Electrical Cardioversion for Postoperative Atrial Fibrillation

Conversion from well tolerated postoperative AF is generally not actively pursued, both because of the high recurrence rate and the self-limited course. For patients with symptoms, however, therapies are similar to those employed in nonpostoperative circumstances except that there should be a greater emphasis on postconversion pharmacologic therapy because causative factors inevitably persist to cause a recurrence. If conversion is necessary, atrial defibrillation or, if atrial flutter or tachycardia is present, pace termination can be employed.[63] In a case of atrial fibrillation that is difficult to convert by usual external techniques, consideration should be given to (1) internal defibrillation using transvenous coils if available; (2) a "double defibrillator" technique in which two pairs of orthogonally placed transthoracic, external patch electrodes are discharged simultaneously[64,65]; or (3) pretreatment with IV ibutilide.[66] Information is emerging on low-energy atrial defibrillation via operatively implanted temporary epicardial coils in animal models[67] and in clinical studies[68]; this may ultimately be an effective strategy for high-risk patients (see Table 32-1) who are unable to tolerate pharmacologic therapy or as an adjunct to prophylactic therapy.

Pharmacologic Cardioversion for Postoperative Atrial Fibrillation

Pharmacologic measures for the conversion of AF should be considered, especially if the patient's respiratory status makes anesthesia for an electrical conversion potentially hazardous or some other contraindication to general anesthesia exists. Medications that have been shown to be potentially useful include newer class III agents (such as ibutilide) and investigational agents such as tedisamil and tercetilide. A recent study by the Ibutilide Investigators[69] compared the use of increasing doses of IV ibutilide with placebo in the treatment of postoperative atrial arrhythmias. Ibutilide was significantly more effective than placebo in a dose-responsive fashion. This intravenous type III agent was also observed to be more efficacious in atrial flutter than in AF, as is often the case with class III agents. Ibutilide carries a risk of ventricular proarrhythmia (which occurs most often as sustained or nonsustained torsades de pointes) in about 2% to 4% of patients; this occurs particularly in situations of bradyarrhythmia, hypokalemia, hypomagnesemia,[69] and female gender. However, when this medication is used carefully (with attention to the above risk factors and with telemetry observation during and following its administration), proarrhythmic risk can be mitigated. Indeed, the Ibutilide Investigators group[69] observed that ibutilide-treated patients with higher heart rates had a lower incidence of ventricular arrhythmias than did rate-controlled patients, a likely consequence of bradyarrhythmia-induced torsades de pointes in the latter group.

Anticoagulation for Postoperative Atrial Fibrillation

Although one might expect a relatively low risk for thromboembolic events in association with a limited course of postoperative AF, several studies have shown, with both prospective case series and case-controlled retrospective analyses, an increased rate of post-CABG stroke in association with postoperative AF,[14,31,32] even after correction for comorbid risk factors. Given the potentially devastating consequences of a thromboembolic event, anticoagulation with Coumadin (warfarin) should be considered for postoperative AF, particularly for patients with mitral valvular disease or prosthesis, left atrial enlargement, marked left ventricular dysfunction, previous thromboembolic events, and age older than 65 years. Because of the risk of bleeding in postoperative patients, anticoagulation must be performed carefully and IV heparin is often not employed. If cardioversion is performed for postoperative AF, conventional recommendations for anticoagulation should be followed. A summary of therapeutic measures for postcardiac surgical atrial arrhythmias, including treatments for prophylaxis and treatments for postoperatively occurring AF, are summarized in Table 32-3.

TABLE 32-3 Summary of Therapeutic Measures for Postcardiac Surgical Atrial Arrhythmias

PROPHYLAXIS (pre- and/or postoperative administration)
Proved benefit: β-blockers
Possible benefit: sotalol, amiodarone, digitalis, procainamide
To date, no benefit proved: propafenone, flecainide, atrial pacing
Ineffective: calcium channel blockers
TREATMENT (for postoperative atrial arrhythmia)
Ventricular rate control with β-blocker, calcium channel blocker, digitalis, or amiodarone
Anticoagulation
Electrical or pharmacologic conversion if marked symptoms or hemodynamic problems, preferably after drug loading; administration of the agent that is used for maintenance of sinus rhythm following conversion will probably be needed for at least 1-2 months
—*example of medication for pharmacologic conversion: ibutilide*
—*examples of medications for maintenance of sinus rhythm following conversion: amiodarone, sotalol*

Ventricular Arrhythmias

INCIDENCE AND PROGNOSIS

Sustained and unsustained ventricular arrhythmias—both monomorphic ventricular tachycardia (VT) and ventricular fibrillation (VF)—are observed following cardiac surgery. The reported incidence of *de novo* sustained and nonsustained ventricular arrhythmias is 0.7% to 3%[70-74] and 36%,[75] respectively. Much earlier observations show an incidence of postoperative sustained ventricular arrhythmias of up to 6%,[76] consistent with the known development of cardiac surgical techniques and postoperative care since the 1960s. Nonetheless the occurrence of ventricular arrhythmias predicts significant mortality. For example, Tam and colleagues[71] observed *de novo*, sustained ventricular arrhythmias in only 16 of 2364 patients in the first week following cardiac surgery (most notably in patients with marked left ventricular dysfunction preoperatively). However, of those patients with ventricular arrhythmias, 75% had recurrences of sustained VT or VF, and there was a 19% mortality rate. Kron and coworkers,[72] in studying 1251 postoperative patients, observed an in-hospital mortality rate of 44% in patients with unprecedented and sustained ventricular arrhythmias. Recurrences of VT beyond the immediate postoperative period have been observed in 40% of patients with early postoperative VT who are also inducible to VT at electrophysiological study (EPS).[77]

ETIOLOGIES

Etiologies of postoperative ventricular arrhythmias include structural heart disease, such as previous infarction, fibrosis and dilation, and hypertrophy. Potential acute precipitants of postoperative ventricular arrhythmias are also described in Table 32-4.[70-74,78-80] Several electrolyte and metabolic abnormalities are associated with postoperative ventricular arrhythmias, most notably hypokalemia[81] (but not low potassium levels intracellularly).[81]

DIAGNOSTIC ISSUES

In the postcardiac surgical patient with a wide-complex tachycardia, the diagnosis of VT is most often favored because of the high frequency of structural heart disease. In situations where surface ECG recordings are ambiguous, a diagnosis of VT versus supraventricular tachycardia or rapidly conducted AF with aberrancy can often be made with the use of epicardial atrial wire recordings.[82,83]

TABLE 32-4 Potential Acute Precipitants of Postcardiac Surgical Ventricular Arrhythmias

Cardiac ischemia
Acute myocardial infarction
Reperfusion injury
Electrolyte or metabolic abnormalities
Acidemia
Extreme hemodynamic instability
Hypoxia
Pharmacologic toxicity (particularly with drugs prolonging the Q-T interval)
High levels of catecholamines (endogenous or administered sympathomimetics)
Asynchronous ventricular pacing

SUSTAINED MONOMORPHIC VENTRICULAR TACHYCARDIA

A study by Steinberg and associates[70] prospectively enrolled consecutive patients undergoing CABG and found that 3.1% of patients had at least one episode of sustained VT at a mean of 4.1 days following the operation; these patients had a 25% in-hospital mortality rate. Predictors of postoperative VT included previous myocardial infarction, marked heart failure, significant left ventricular dysfunction (defined as an ejection fraction <.40), and revascularized myocardium previously supplied by a noncollateralized native vessel. When the first three of these predictive factors were present, the incidence of VT increased to 30%.

VENTRICULAR FIBRILLATION OR POLYMORPHIC VENTRICULAR TACHYCARDIA

There is almost always an underlying arrhythmogenic substrate for monomorphic VT (e.g., myocardial scarring), whereas polymorphic VT is more likely related to transient perioperative abnormalities.[73,74,84] Kron and colleagues observed that acute cardiac ischemia following heart surgery was more highly associated with primary VF than VT.[72] The perioperative factors that may contribute to polymorphic ventricular arrhythmias are described earlier and in Table 32-4. There are no clear data speaking to whether polymorphic VT or VF events that occur immediately after cardiopulmonary bypass, in the absence of additional risk factors for ventricular arrhythmia, predict an increased risk for ventricular arrhythmic events in the long term.

ACUTE MANAGEMENT

There have been no comprehensive or controlled studies of the acute treatment for postoperative VT. However, commonly accepted and consensus treatments include ACLS-based algorithms[85,86] including synchronized DC cardioversion and pace-termination (facilitated by the presence of epicardial pacing wires). Prevention of additional episodes can be aided by attention to abnormalities, such as ischemia, fluid overload, electrolyte disturbances, and polypharmacy with potentially proarrhythmic cardiac or other[87] agents. Effective acute treatment can also be found with antiarrhythmic agents, such as parenteral amiodarone,[88,89] β-blockers, sotalol,[90] magnesium sulfate (specifically for torsades de pointes),[91,92] and procainamide.[40] Recurrent VT can also be prevented or treated with the use of overdrive ventricular pacing or burst ventricular pacing, respectively.[93] Mechanical supportive measures, such as intra-aortic balloon counter-pulsation,[94-96] intra-thoracic ventricular-assist devices,[97] or external cardiopulmonary support[98] have been used. There is a case report[99] of simultaneous intra-aortic balloon and external cardiopulmonary therapy allowing for high-dose β-blocker infusion in a patient with refractory postoperative VT, although this technique is not widely applied.

POSTOPERATIVE RISK STRATIFICATION AND TREATMENT

Longer-term treatment for patients who have manifested monomorphic VT postoperatively may involve risk stratification with an assessment of left ventricular function and/or EPS.[74,77] Post-hospital discharge recurrences of VT have been noted in 40% of patients with postoperative VT who were also inducible to VT at postoperative EPS[77] suggesting a value to long-term treatment (e.g., with an implantable cardioverter-defibrillator) although data showing a direct benefit in mortality or other measures are lacking. Of note, a discordance between inducibility of VT at EPS by programmed electrical stimulation (PES) performed via epicardial pacing at one cardiac site versus endocardial pacing at two sites has been observed. Sheppard and colleagues[100] performed EPS on 26 postcardiac surgical patients using both operatively implanted epicardial pacing wires and endocardial electrophysiological catheters. Despite similar epicardially and endocardially measured effective and functional refractory periods, concordant results of inducibility to VT between the techniques were obtained in only 70% of patients; 40% of patients inducible with endocardial PES from two sites had been uninducible via epicardial PES at one site, suggesting that single-site PES from epicardial wires may not fully assess postoperative risk for VT.

The presence of late potentials on postoperative signal-averaged ECG has been reported as an independent predictor for VT early after CABG,[101] but there are no comprehensive data on longer-term VT occurrences or outcomes in patients with an abnormal signal-averaged ECG. The modality of exercise testing is also employed, although there is no direct evidence showing its utility in predicting risk of postoperative VT. The timing of postoperative risk stratification is important, with some investigators suggesting a wait of at least 1 week to allow for both healing and equilibration of substrate to occur.[75,102]

Lifelong treatment with antiarrhythmic medication and/or implantable cardioverter-defibrillator placement is likely useful for patients who are found to be at high risk for ventricular arrhythmia, such as those who have manifested postoperative ventricular arrhythmias and who also have significant left ventricular dysfunction and/or inducible monomorphic ventricular arrhythmias at EPS. The significant rate of subsequent clinical ventricular arrhythmic events in patients with postoperative VT who are also inducible to VT at postoperative EPS[77] suggests a value to long-term treatment, although direct data showing a benefit in mortality or other measures for patients with VT that emerges postoperatively are lacking.

Bradyarrhythmias

INCIDENCE AND PROGNOSIS

Cardiac conduction abnormalities and sinus bradyarrhythmias are reported in 17% to 34% of postoperative patients, depending on a multitude of factors[103,104] (Table 32-5), however, there is a significant rate of reversibility. Persistent bradyarrhythmias requiring permanent pacing occur in 0.8% to 4% of postcardiac surgical patients.[105-110] Risk for the need for permanent pacemaker implantation following cardiac surgery is elevated following valvular surgery, with preexisting conduction system disease, and in several other situations, as described in Table 32-5.[105-108,111-114] Bradyarrhythmias necessitating permanent pacing have been associated with longer postoperative courses in the intensive care unit and in the hospital,[110] as well as with inevitably increased postoperative economic costs, although there are no direct data on cost.

It is likely that recovery from postoperative bradyarrhythmias occurs in a significant proportion of patients. An investigation by Glikson and colleagues[111] of long-term pacemaker dependency in a group of 120 patients who had received a permanent pacemaker following cardiac surgery indicated that 41% of patients eventually became pacemaker nondependent. Patients with postoperative complete heart block as the indication for pacemaker implantation were more likely to remain pacemaker dependent in this study. In contrast, in a smaller study of 93 consecutive postcardiac surgical patients by Baerman and Morady,[115] three of three patients who underwent pacemaker implantation because of third-degree heart block were no longer in heart block after 2 months. The issue of eventual use of a permanent pacemaker implanted postcardiac surgery was further explored in a study by Gordon and coworkers.[108] This investigation, which involved a prospective data collection for 10,421 consecutive patients who had cardiac operations included a logistic regression analysis of independent and multivariate predictors of permanent pacing following cardiac surgery (with predictors including the risk factors listed

TABLE 32-5 Risk Factors for the Need for Permanent Pacemaker Implantation Following Cardiac Surgery

Preexisting left bundle branch block or other signs of AV conduction disease
Valvular surgery (procedures in descending order of risk: multiple valvular, tricuspid, aortic, mitral)
CABG with concomitant valvular surgery
LV aneurysmectomy or arrhythmia ablative surgery
Repeat cardiac operation
Advanced age
Type of cardioplegia solution and if hypothermia was used
Prolonged cardiopulmonary bypass and cross-clamp duration
Number of bypass grafts (perhaps related to operative times)
Presence of active endocarditis
Female gender

AV, Atrioventricular; CABG, coronary artery bypass grafting; LV, left ventricle.

in Table 32-5). The study also found that their predictive model correlated highly with eventual pacemaker use, suggesting that the presence of a greater number of risk factors for pacemaker implantation following cardiac surgery predicts a higher rate of actual pacemaker use.

RISK STRATIFICATION AND MANAGEMENT STRATEGIES

The decision to implant a permanent pacemaker in a postcardiac surgical patient with high-grade AV block and a poor escape rhythm who is otherwise ready for hospital discharge is relatively straightforward. However, there are no clear data on the optimal length of time to wait in more equivocal situations. Some authors[111] have recommended for situations of complete AV block that a decision regarding permanent pacemaker implantation occur no later than the sixth and ninth postoperative days for wide-complex and narrow-complex escape rhythms, respectively. In a study by Bonatti and associates[109] of pediatric patients with CHB following operation for congenital heart disease (with most cases involving the closure of a ventricular septal defect), the researchers recommend permanent pacing after at least a 2-week observation period with temporary pacing in that population.

Any decision regarding timing of implantation of a permanent pacemaker will be impacted by the stability of the temporary pacing system. Epicardial electrode temporary pacing systems frequently display increasing pacing thresholds with time, and endocardial pacing catheters can have similar problems plus infective and vascular complications. A prospective study of 30 postcardiac surgical patients by Kosmas and colleagues[116] found that effective atrial, ventricular, and dual-chamber pacing via epicardial wires could not be performed by the fifth postoperative day in 39%, 38%, and 61% of patients, respectively.

In cases of resolved or resolving heart block, the decision regarding whether there will be persistent risk of bradyarrhythmias and whether permanent pacemaker implantation should be pursued can be influenced by several factors. First, consideration should be given to whether there is a need for concomitant antiarrhythmic medication (which would be expected to exacerbate even borderline bradyarrhythmias). Second, consideration should be made to performing an EPS (an abbreviated but effective form of which can be performed at the bedside with atrial pacing via operatively placed epicardial wires) to help detect significant atrioventricular conduction and/or sinus node. Electrophysiological study results suggesting that permanent pacing is needed are similar to generally applied criteria, including an H–V interval greater than 75 ms, infra-Hisian block occurring spontaneously or at an atrial paced cycle length of less than or equal to 400 ms (300 ms in a pediatric population),[109] or a corrected sinus node recovery time equal to or greater than 550 ms. Third, an exercise study can be performed in order to assess chronotropic competence and AV conduction at elevated heart rates.

ACKNOWLEDGMENT

The authors would like to thank Ms. Rose Marie Wells for her expert assistance with the preparation of this chapter.

REFERENCES

1. Daudon P, Corcos T, Gardjbakah I, et al: Prevention of atrial fibrillation or flutter by acebutolol after coronary bypass grafting. Am J Cardiol 1986;58:933-936.
2. Andrews TC, Reimold SC, Berlin JA, Antman EM: Prevention of supraventricular arrhythmias after coronary artery bypass surgery. Circulation 1991;84(Suppl III):III-236-III-243.
3. Asher CR, Miller DP, Grimm RA, et al: Analysis of risk factors for development of atrial fibrillation early after cardiac valvular surgery. Am J Cardiol 1998;82:892-895.
4. Creswell LL, Schuessler RB, Rosenbloom M, Cox JL: Hazards of postoperative atrial arrhythmias. Ann Thorac Surg 1993;56:539-549.
5. Dixon FE, Genton E, Vacek JL, et al: Factors predisposing to supraventricular tachyarrhythmias after coronary artery bypass grafting. J Thorac Cardiovasc Surg 1990;100:338-342.
6. Leitch JW, Thomson D, Baird DK, Harris PJ: The importance of age as a predictor of atrial fibrillation and flutter after coronary artery bypass grafting. J Thorac Cardiovasc Surg 1990;100:338-342.
7. Fuller JA, Adams GC, Buxton B: Atrial fibrillation after coronary artery bypass grafting: Is it a disorder of the elderly? J Thorac Cardiovasc Surg 1989;97:821-825.
8. Roffman JA, Feldman A: Digoxin and propranolol in the prophylaxis of supraventricular tachydysrhythmias after coronary artery bypass surgery. Ann Thorac Surg 1981;31:496-501.
9. Aranki SF, Shaw DP, Adams DH, et al: Predictors of atrial fibrillation after coronary artery surgery. Current trends and impact on hospital resources. Circulation 1996;94:390-397.
10. Ascione R, Caputo M, Calori G, et al: Predictors of atrial fibrillation after conventional and beating heart coronary surgery: A prospective, randomized study. Circulation 2000;102:1530.
11. Passman RS, Beshai JF, Parker M, et al: A prediction rule for post-CABG atrial fibrillation. PACE 2001;24(4).
12. Connolly SJ: The beta-blocker length of stay study (BLOSS) trial: A randomized, placebo controlled trial of beta-blocker treatment for reduction of hospitalization after cardiac surgery. J Am Coll Cardiol 2000;36:313.
13. Hattori Y, Yang Z, Sugimura S, et al: Terminal warm blood cardioplegia improves the recovery of myocardial electrical activity. A retrospective and comparative study. Jpn J Thorac Cardiovasc Surg 2000;48:1-8.
14. Almassi GH, Schowalter T, Nicolosi AC, et al: Atrial fibrillation after cardiac surgery: A major morbid event?. Ann Surg 1997;226:501-513.
15. Tamis JE, Vloka ME, Malhorta S, et al: Atrial fibrillation is common after minimally invasive direct coronary artery bypass surgery. (abstract) J Am Coll Cardiol 1998;31:116A.
16. Cohn WE, Sirois CA, Johnson RG: Atrial fibrillation after minimally invasive cardiac artery bypass grafting: A retrospective, matched study. J Thorac Cardiovasc Surg 1999;117:298.
17. Abreu JE, Reilly J, Salzano RP, et al: Comparison of frequencies of atrial fibrillation after coronary artery bypass grafting with and without the use of cardiopulmonary bypass. Am J Cardiol 1999;83:775.
18. Chauhan VS, Gill I, Woodend KA, Tang AS: Lower incidence of atrial fibrillation after minimally invasive direct coronary artery bypass surgery (MID-CAB) than bypass surgery (CABG) (abstract). Circulation 1997;96:I-263.
19. Asher CR, DiMengo JM, Arheart KL, et al: Atrial fibrillation early postoperatively following minimally invasive cardiac valvular surgery. Am J Cardiol 1999;84:744.
20. Cohn LH, Adams DH, Couper GS, et al: Minimally invasive cardiac valve surgery improves patient satisfaction while reducing costs of cardiac valve replacement and repair. Ann Surg 1997;226:421-428.
21. Stamou SC, Dangas G, Hill PC, et al: Atrial fibrillation after beating heart surgery. Am J Cardiol 2000;86:64.
22. Stamou SC, Pfister AJ, Dangas G, et al: Beating heart versus conventional single-vessel reoperative coronary artery bypass. Ann Thorac Surg 2000;69:1383-1387.

23. Stamou SC, Dangas G, Dullum MK, et al: Beating heart surgery in octogenarians: Perioperative outcome and comparison with younger age groups. Ann Thorac Surg 2000;69:1140-1145.
24. Buxton AE, Josephson ME: The role of P wave duration as a predictor of post-operative atrial arrhythmias. Chest 1981;80:68-73.
25. Oshima H, Usi A, Murakami F, et al: Value of the regional p-wave recorded by signal-averaged ECG on the atrium for predicting atrial fibrillation after cardiac surgery. (abstract) Circulation 1996;94:I-69.
26. Landymore RW, Howell F: Recurrent atrial arrhythmias following treatment for post-operative atrial fibrillation after coronary bypass operations. Eur J Cardiothorac Surg 1991;5:436-439.
27. Matangi MF, Neutze JM, Graham IC, et al: Arrhythmia prophylaxis after aorta-coronary bypass: The effect of minidose propranolol. J Thorac Cardiovasc Surg 1985;89:439-443.
28. Stephenson LW, MacVaugh H, Tomasello DN, et al: Propranolol for prevention of post-operative cardiac arrhythmias: A randomized study. Ann Thorac Surg 1980;29:113-115.
29. Stebbins D, Igidbashian L, Goldman S, et al: Clinical outcome of a patient who developed atrial fibrillation after coronary artery bypass surgery. (Abstract) PACE 1995;18:811.
30. Rubin DA, Nieminski KE, Reed GE, Herman MV: Predictors, prevention, and long-term prognosis of atrial fibrillation after coronary artery bypass graft operations. J Thorac Cardiovasc Surg 1987;94:331-335.
31. Reed GL, Singer DE, Picard EH, et al: Stroke following coronary-artery bypass surgery: A case-control estimate of the risk from carotid bruits. N Engl J Med 1988;319:246-250.
32. Taylor GJ, Malik SA, Colliver JA, et al: Usefulness of atrial fibrillation as a predictor of stroke after isolated coronary artery bypass grafting. Am J Cardiol 1987;60:905-907.
33. Abreu JE, Reilly J, Salzano RP, et al: Comparison of frequencies of atrial fibrillation after coronary artery bypass grafting with and without the use of cardiopulmonary bypass. Am J Cardiol 1999; 83:775.
34. Kim MH, Deeb GM, Morady F, et al: Effect of postoperative atrial fibrillation on length of stay after cardiac surgery (the postoperative atrial fibrillation in cardiac surgery study). Am J Cardiol 2001;87:881.
35. Villareal RP, Lee V-V, Kar B, et al: Post-operative atrial fibrillation is a predictor of early and late mortality. PACE 2001;24.
36. Wahr JA, Parks R, Boisvert D, et al: Preoperative serum potassium levels and perioperative outcomes in cardiac surgical patients. JAMA 1999;281:2203-2210.
37. Kowey PR, Taylor JE, Rials SJ, Marinchak RA: Meta-analysis of the effectiveness of prophylactic drug therapy in preventing supraventricular arrhythmia early after coronary artery bypass grafting. Am J Cardiol 1992;69:963-965.
38. Andrews TC, Reimold SC, Berlin JA, et al: Prevention of supraventricular arrhythmias after coronary artery bypass surgery: A meta-analysis of randomized control trials. Circulation 1991;84 Suppl III:236-244.
39. Davison R, Hertz R, Kaplan K, et al: Prophylaxis of supraventricular tachyarrhythmia after coronary bypass surgery with oral verapamil: A randomized, double-blinded trial. Ann Thorac Surg 1985;39:336.
40. Gold MR, O'Gara PT, Buckley MJ, DeSanctis RW: Efficacy and safety of procainamide in preventing arrhythmias after coronary artery bypass surgery. Am J Cardiol 1996;78:975-979.
41. Nystrom U, Edvardsson N, Berggren H, et al: Oral sotalol reduces the incidence of atrial fibrillation after coronary artery bypass surgery. Thorac Cardiovasc Surg 1993;41:34-47.
42. Suttorp MJ, Kingma JH, Peels HO, et al: Effectiveness of sotalol in preventing supraventricular tachyarrhythmias shortly after coronary artery bypass grafting. Am J Cardiol 1991;68:1163-1169.
43. Parikka H, Toivonen L, Heikkila L, et al: Comparison of sotalol and metoprolol in the prevention of atrial fibrillation after coronary artery bypass surgery. J Cardiovasc Pharmacol 1998;31:67-73.
44. Suttorp MJ, Kingma JH, Gin TJ, et al: Efficacy and safety of low- and high-dose sotalol versus propranolol in the prevention of supraventricular tachyarrhythmias early after coronary artery bypass operations. J Thorac Cardiovasc Surg 1990;100:921-926.
45. Gomes JA, Ip J, Santoni-Rugiu F, et al: Oral d,l sotalol reduces the incidence of post-operative atrial fibrillation in coronary artery bypass surgery patients: A randomized, double-blind, placebo-controlled study. J Am Coll Cardiol 1999;34:334-339.
46. Podrid PJ: Prevention of post-operative atrial fibrillation: What is the best approach? J Am Coll Cardiol 1999;34:340-342.
47. Daoud EG, Strickberger A, Man KC, et al: Preoperative amiodarone as prophylaxis against atrial fibrillation after heart surgery. N Engl J Med 1997;337:1785-1791.
48. Greenspon AJ, Kidwell GA, Hurley W, Mannion J: Amiodarone related post-operative adult respiratory distress syndrome. Circulation 1991;84 (Suppl III):III-407-III-415.
49. Guarnieri T, Nolan S, Gottlieb SO, et al: Intravenous amiodarone for the prevention of atrial fibrillation after open heart surgery: The ARCH (Amiodarone Reduces Coronary Artery Disease Bypass Grafting Hospitalization) trial. J Am Coll Cardiol 1999;34:343-347.
50. Humphries JO: Unexpected instant death following successful coronary artery bypass graft surgery (and other clinical settings): Atrial fibrillation, quinidine, procainamide, et cetera, and instant death. Clin Cardiol 1998;21(10):711-718.
51. Defaut P, Saksena S, Krol RB: Long-term outcome of patients with drug-refractory atrial flutter and fibrillation after single- and dual-site right atrial pacing for arrhythmia prevention. J Am Coll Cardiol 1998;32:1900-1908.
52. Schweikert RA, Grady TA, Gupta N, et al: Atrial pacing in the prevention of atrial fibrillation after cardiac surgery: Results of the second post-operative pacing study (POPS-2). (abstract) J Am Coll Cardiol 1998;31:117A.
53. Chung MK, Augostini RS, Asher CR, et al: Ineffectiveness and potential proarrhythmia of atrial pacing for atrial fibrillation prevention after coronary artery bypass grafting. Ann Thorac Surg 2000;69:1057-1063.
54. Kurz DJ, Naegeli B, Kunz M, et al: Epicardial biatrial synchronous pacing for prevention of atrial fibrillation after cardiac surgery. Pacing Clin Electrophysiol 1999;22:721-726.
55. Gerstenfeld EP, Hill MR, French SN, et al: Evaluation of right atrial and biatrial temporary pacing for the prevention of atrial fibrillation after coronary artery bypass surgery. J Am Coll Cardiol 1999;33:1981-1988.
56. Greenberg MD, Katz NM, Iuliano S, et al: Atrial pacing for the prevention of atrial fibrillation after cardiovascular surgery. J Am Coll Cardiol 2000;35:1416-1422.
57. Blommaert D, Gonzalez M, Mucumbitsi J, et al: Effective prevention of atrial fibrillation by continuous atrial overdrive pacing after coronary artery bypass surgery. J Am Coll Cardiol 2000;35: 1411-1415.
58. Fan K, Lee KL, Chiu CSW, et al: Effects of biatrial pacing in prevention of postoperative atrial fibrillation after coronary artery bypass surgery. Circulation 2000;102:755.
59. Weiner P, Bessan MM, Jacchovsky J, et al: Clinical course of acute atrial fibrillation treated with digitalization. Am Heart J 1983; 105:223-227.
60. Tisdale JE, Padhi ID, Goldberg AD, et al: A randomized, double-blind comparison of intravenous diltiazem and digoxin for atrial fibrillation after coronary artery bypass surgery. Am Heart J 1998 May;135(5 Pt 1):739-747.
61. Clemo HF, Wood MA, Gilligan DM, Ellenbogen KA: Intravenous amiodarone for acute heart rate control in the critically ill patient with atrial tachyarrhythmias. Am J Cardiol 1998;81:594-598.
62. Galve E, RiusT, Ballester R, et al: Intravenous amiodarone in treatment of recent-onset atrial fibrillation: Results of a randomized, controlled study. J Am Coll Cardiol 1996;27:1079-1082.
63. Waldo AL, MacLeen WA, Cooper TB, et al: Use of temporarily placed epicardial atrial wire electrodes for the diagnosis and treatment of arrhythmias following open heart surgery. J Thorac Cardiovasc Surg 1978;76:500-505.
64. Saliba WI, Juratli N, Chung MK, et al: Higher energy synchronized external direct current cardioversion with 720 joules for atrial fibrillation refractory to standard cardioversion. (abstract) J Am Coll Cardiol 1999;34:2031-2034.
65. Bjerregaard P, El-Shafei A, Janosik DL, et al: Double external direct-current shocks for refractory atrial fibrillation. Am J Cardiol 1999;83:972-974.
66. Oral H, Souza JJ, Michaud GF, et al: Facilitating transthoracic cardioversion of atrial fibrillation with ibutilide pretreatment. N Engl J Med 1999;340:1849-1854.
67. Mehmanesh H, Bauernschmitt R, Lange R, Hagl S: Adjustable atrial and ventricular temporary electrode for low-energy termination of tachyarrhythmias early after cardiac surgery. Pacing Clin Electrophysiol 1999;22:1802-1807.

68. Liebold A, Wahba A, Birnbaum: Low-energy cardioversion with epicardial wire electrodes: New treatment of atrial fibrillation after open heart surgery. Circulation 1998;98:883-886.
69. VanderLugt JT, Mattioni T, Denker S, et al: Efficacy and safety of ibutilide fumarate in patients with atrial flutter or fibrillation following cardiac surgery. Circulation 1999;100:369-375.
70. Steinberg JS, Gaur A, Sciacca R, Tan E: New-onset sustained ventricular tachycardia after cardiac surgery. Circulation 1999 Feb 23;99(7):903-908.
71. Tam SK, Miller JM, Edmunds LH: Unexpected, sustained ventricular tachyarrhythmia after cardiac operations. J Thorac Cardiovasc Surg 1991;102:883-889.
72. Kron IL, DiMarco JP, Harman PK, et al: Unanticipated postoperative ventricular tachyarrhythmias. Ann Thorac Surg 1984;38: 317-322.
73. Pires LA; Wagshal AB; Lancey R; Huang SK: Arrhythmias and conduction disturbances after coronary artery bypass graft surgery: Epidemiology, management, and prognosis. Am Heart J 1995;129:799-808.
74. Topol EJ, Lerman BB, Baughman KL, et al: De novo refractory ventricular tachyarrhythmias after coronary revascularization. Am J Cardiol 1986;57:57-59.
75. Carlson MD, Biblo LA, Waldo AL: Post open heart surgery ventricular arrhythmias. Cardiovasc Clin 1992;22:241-253.
76. Favaloro RG, Effler DB, Groves LK, et al: Direct myocardial revascularization with saphenous vein autograft: Clinical experience in 100 cases. Dis Chest 1969;56:279.
77. Costeas XF, Schoenfeld MH: Usefulness of electrophysiologic studies for new-onset sustained ventricular tachyarrhythmias shortly after coronary artery bypass grafting. Am J Cardiol 1993; 72:1291-1294.
78. Ducceschi V, D'Andrea A, Liccardo B, et al: Perioperative correlates of malignant ventricular tachyarrhythmias complicating coronary surgery. Heart Vessels 1999;14:90-95.
79. Ducceschi V, D'Andrea A, Liccardo B, et al: Ventricular tachyarrhythmias following coronary surgery: Predisposing factors. Int J Cardiol 2000;31;73:43-48.
80. Preisman S, Cheng DC: Life-threatening ventricular dysrhythmias with inadvertent asynchronous temporary pacing after cardiac surgery. Anesthesiology 1999;91:880-883.
81. Johnson RG, Shafique T, Sirois C, et al: Potassium concentrations and ventricular ectopy: A prospective, observational study in post-cardiac surgery patients. Crit Care Med 1999;27:2430-2434.
82. Malcolm ID, Cherry DA, Morin JE: The use of temporary atrial electrodes to improve diagnostic capabilities with Holter monitoring after cardiac surgery. Ann Thorac Surg 1986;41:103-105.
83. Waldo AL, Henthorn RW, Epstein AE, et al: Diagnosis and treatment of arrhythmias during and following open heart surgery. Med Clin North Am 1984;68:1153-1170.
84. Azar RR, Berns E, Seecharran B, et al: De novo monomorphic and polymorphic ventricular tachycardia following coronary bypass grafting. Am J Cardiol 1997;80(1):76-78.
85. Advanced Cardiovascular Life Support, Part 6. Circulation 2000;102(I):I-86-I-171.
86. ACLS Subcommittee: Textbook of Advanced Cardiac Life Support. American Heart Association, 1994.
87. Perrault LP, Denault AY, Carrier M, et al: Torsades de pointes secondary to intravenous haloperidol after coronary bypass grafting surgery. Can J Anaesth 2000;47:251-254.
88. Kowey PR, Levine JH, Herre JM, et al: A randomized, double-blind comparison of intravenous amiodarone and bretylium in the treatment of patients with refractory, hemodynamically destabilizing ventricular tachycardia or fibrillation. Circulation 1995;92:3255-3263.
89. Perry JC, Knilans TK, Marlow D, et al: Intravenous amiodarone for life-threatening tachyarrhythmias in children and young adults. J Am Coll Cardiol 1993 Jul;22(1):95-98.
90. Evrard P, Gonzalez M, Jamart J, et al: Prophylaxis of supraventricular and ventricular arrhythmias after coronary artery bypass grafting with low-dose sotalol. Ann Thorac Surg 2000;70:151.
91. Speziale G, Ruvolo G, Fattouch K, et al: Arrhythmia prophylaxis after coronary artery bypass grafting: Regimens of magnesium sulfate administration. Thorac Cardiovasc Surg 2000;48:22-26.
92. Thompson LD, Cohen AJ, Bellasis RM: Polymorphous ventricular tachycardia following cardiopulmonary bypass. J Natl Med Assoc 1996;88:49-51.
93. Fisher JD, Mehra R, Furman S: Termination of ventricular tachycardia with bursts of rapid ventricular pacing. Am J Cardiol 1978;41:94-102.
94. Kaplan JA, Craver JM, Jones EL, et al: The role of the intra-aortic balloon in cardiac anesthesia and surgery? Am Heart J 1979; 98:580-586.
95. Craver JM, Kaplan JA, Jones EL, et al: What role should the intra-aortic balloon have in cardiac surgery? Ann Surg 1979; 189:769-776.
96. Fotopoulos GD, Mason MJ, Walker S, et al: Stabilisation of medically refractory ventricular arrhythmia by intra-aortic balloon counterpulsation. Heart 1999;82:96-100.
97. Swartz MT, Lowdermilk GA, McBride LR: Refractory ventricular tachycardia as an indication for ventricular assist device support. J Thorac Cardiovasc Surg 1999;118:1119-1120.
98. Hamid BA, Azuma SS, Hong RA, Lau JM: Persistent postoperative ventricular tachycardia treatment by using external cardiopulmonary support. Hawaii Med J 1994;53:10-11.
99. Kurose M, Okamoto K, Sato T, et al: Successful treatment of life-threatening ventricular tachycardia with high-dose propranolol under extracorporeal life support and intraaortic balloon pumping. Jpn Circ J 1993;57:1106-1110.
100. Sheppard RC, Nydegger CC, Kutalek SP, Hessen SE: Comparison of epicardial and endocardial programmed stimulation in patients at risk for ventricular arrhythmias after cardiac surgery. Pacing Clin Electrophysiol 1993;16:1822-1832.
101. Elami A, Merin G, Flugelman MY, Adar L, et al: Usefulness of late potentials on the immediate postoperative signal-averaged electrocardiogram in predicting ventricular tachyarrhythmias early after isolated coronary artery bypass grafting. Am J Cardiol 1994;1;74:33-37.
102. Bhandari AK, Au PK, Rose JS, et al: Decline in inducibility of sustained ventricular tachycardia from two to twenty weeks after acute myocardial infarction. Am J Cardiol 1987;59:284-290.
103. Wexelman W, Lichstein E, Cunningham JN, et al: Etiology and clinical significance of new fascicular conduction defects following coronary bypass surgery. Am Heart J 1986;111:923-927.
104. Chu A, Califf RM, Priyor DB, et al: Prognostic effect of bundle branch block related to coronary artery bypass grafting. Am J Cardiol 1987;59:798-803.
105. Emlein G, Huang SK, Pires LA, et al: Prolonged bradyarrhythmias after isolated coronary artery bypass graft surgery. Am Heart J 1993;126:1084-1090.
106. Baerman JM, Kirsh MM, de Buitleir M, et al: Natural history and determinants of conduction defects following coronary artery bypass surgery. Ann Thorac Surg 1987;44:150-153.
107. Gundry SR, Sequeira A, Coughlin TR, McLaughlin JS: Postoperative conduction disturbances: A comparison of blood and crystalloid cardioplegia. Ann Thorac Surg 1989;47:384-390.
108. Gordon RS, Ivanov J, Cohen G, Ralph-Edwards AL: Permanent cardiac pacing after a cardiac operation: Predicting the use of permanent pacemakers. Ann Thorac Surg 1998;66:1698-1704.
109. Bonatti V, Agnetti A, Squarcia U: Early and late postoperative complete heart block in pediatric patients submitted to open-heart surgery for congenital heart disease. Pediatr Med Chir 1998;20:181-186.
110. Del Rizzo DF, Nishimura S, Lau C, et al: Cardiac pacing following surgery for acquired heart disease. J Card Surg 1996;11:332-340.
111. Glikson M, Dearani JA, Hyberger LK, et al: Indications, effectiveness, and long-term dependency in permanent pacing after cardiac surgery. Am J Cardiol 1997;15;80:1309-1313.
112. O'Connell JB, Wallis D, Johnson Sa, et al: Transient bundle branch block following use of hypothermic cardioplegia in coronary artery bypass surgery: High incidence without perioperative myocardial infarction. Am Heart J 1982;103:85-91.
113. Ellis RJ, Mavroudis C, Gardner C, et al: Relationship between atrioventricular arrhythmias and the concentration of K+ ion in cardioplegia solution. J Thorac Cardiovasc Surg 1980;80:517-526.
114. Moore SL, Wilkoff BL: Rhythm disturbances after cardiac surgery. Semin Thorac Cardiovasc Surg 1991;3:24-28.
115. Baerman JM, Kirsh MM, de Buitleir M, et al: Natural history and determinants of conduction defects following coronary artery bypass surgery. Ann Thorac Surg 1987;44:150-153.
116. Kosmas CE, Ryder RG, Poon MJ, et al: Time-limited efficacy of pacing electrodes following open heart surgery. Indian Heart J 1996;48:681-684.

Chapter 33 Arrhythmias and Electrolyte Disorders

DIONYSSIOS A. ROBOTIS, GIOIA TURITTO, and NABIL EL-SHERIF*

Cardiac arrhythmias are an expression of the same fundamental electrophysiological principles that underlie the normal electrical behavior of the heart. The electrical activity of the heart depends on transmembrane ionic gradients and the time- and voltage-dependent alterations of their conductance. The resting membrane potential (V_m) is calculated by the Goldman constant field equation[1]

$$V_m = \frac{RT}{zF} \ln \frac{P_K a_{Ko} + P_{Na} a_{Nao} + P_{CL} a_{Cli}}{P_K a_{Ki} + P_{Na} a_{Nai} + P_{CL} a_{Clo}},$$

which incorporates the permeability (P) and activity (a) of all ionic species that contribute to it.

Electrolyte abnormalities may generate or facilitate clinical arrhythmias, even in the setting of normal cardiac tissue. Furthermore, electrolyte aberrations are more likely to interact with abnormal myocardial tissue to generate their own cadre of cardiac arrhythmias. Antiarrhythmics exert their actions by modulating the conduction of ions across specific membrane channels; hence, abnormal ionic gradients across the membrane will augment or mitigate their antiarrhythmic effects and potentiate or alleviate their proarrhythmic sequelae.

POTASSIUM

Potassium is the most abundant intracellular cation and the most important determinant of the resting membrane potential (RMP). The electrophysiological effects of potassium depend not only on its extracellular concentration, but also on the direction (hypokalemia versus hyperkalemia) and rate of change. Hoffman and Suckling[2] have noted that the effect of potassium on the RMP is modulated by the simultaneous calcium concentration. The interrelationship is such that an elevated calcium level decreases the depolarizing effect of an elevated potassium level, and low calcium levels diminish the depolarization produced by hypokalemia. When extracellular potassium levels are higher than normal, the cell membrane behaves as a potassium electrode, as described by the Nernst equation:

$$V_m = -61.5 \log [K^+]_i/[K^+]_o.$$

At levels of less than ~3 mM, the transmembrane potential (Vm) is less than that predicted by the Nernst equation,[3] because hypokalemia reduces membrane permeability to potassium (P_k).

Indeed, potassium currents are modulated by the potassium gradient itself and other electrolytes as well (Table 33-1). The conductance of the inward rectifier current (i_{K1}) is proportional to the square root of the extracellular K concentration $[K^+]_o$.[5,6] The dependence of the activation of the delayed rectifier current (I_{Kr}) on the extracellular potassium concentration $[K^+]_o$ helps explain why the action potential duration (APD) is shorter at higher $[K^+]_o$ and longer at low $[K^+]_o$ concentrations[7] (see Table 33-1). As important as the time factor may be on the electrophysiological impact of different potassium levels, it is equally important to note that rapid fluctuations in extracellular potassium levels do occur, especially through transcellular shifts (Table 33-2). Insulin, β-adrenergic agonists, aldosterone, and changes in blood pH may independently affect serum potassium levels.[8]

Hypokalemia

Hypokalemia is the most common electrolyte abnormality encountered in clinical practice. Potassium values of less than 3.6 mmol/L are seen in over 20% of hospitalized patients.[9] As many as 10% to 40% of patients on thiazide diuretics[10] and almost 50% of patients resuscitated from out-of-hospital ventricular fibrillation[11] have low potassium levels. Hypokalemia results from decreased potassium intake, transcellular shift, and, most commonly, increased renal or extrarenal losses (Table 33-3).

*Supported in part by VA MERIT and REAP grants to Nabil El-Sherif.

TABLE 33-1 Modulation of Potassium Currents by Electrolyte Concentration

I_{K1}, Inward rectifier	Its conductance is proportional to the square root of $[K^+]_o$ The instantaneous inward rectification on depolarization is due to the Mg^{2+} block at physiologic $[Mg^{2+}]_i$[4]
I_{Kr}, Delayed rectifier	Low $[K^+]_o$ decreases the delayed rectifier current (I_{Kr})
I_{to}, Transient outward	One type is voltage activated and modulated by neurotransmitters The other type is activated by intracellular Ca
$IK_{(Ca)}$	Opens in the presence of high levels of intracellular calcium
$IK_{(Na)}$	Opens in the presence of high levels of intracellular sodium

TABLE 33-3 Causes of Hypokalemia

Decreased intake	
Potassium shift into the cell (see Table 33-2)	
Renal potassium loss	
Increased mineralocorticoid effects	**Increased flow to distal nephron**
Primary or secondary aldosteronism	Diuretics
Ectopic ACTH-producing tumor or	Salt-losing nephropathy
Cushing's syndrome	**Hypomagnesemia**
Bartter's syndrome	Nonresorbable anion
Licorice	Carbenicillin, penicillin
Renovascular or malignant hypertension	**Renal tubular acidosis (type I or II)**
Congenital abnormality of steroid metabolism	**Congenital defect of distal nephron**
Renin-producing tumor	Liddle's syndrome
Extrarenal potassium loss	
Vomiting, diarrhea, laxative abuse	
Villous adenoma, Zollinger-Ellison syndrome	

ELECTROPHYSIOLOGICAL EFFECTS OF HYPOKALEMIA

Hypokalemia leads to a higher (more negative) RMP and, at least during electrical diastole, a decrease in membrane excitability as a result of widening of the RMP and the threshold potential (TP) difference. Low extracellular potassium decreases the delayed rectifier current (I_{Kr}), resulting in an increase in the APD and a delay in repolarization. It has been suggested that extracellular K^+ ions are required to open the delayed rectifier channel.[7]

Most importantly, hypokalemia alters the configuration of the action potential (AP), with the duration of phase 2 first increasing and subsequently decreasing, whereas the slope of phase 3 decelerates. The latter effect leads to an AP with a long "tail," resulting in an increase in the relative refractory period (RRP) and a decrease in the difference of the resting membrane potential from the threshold potential during the terminal phase of the AP. Thus, cardiac tissue demonstrates increased excitability with associated ectopy for a considerable portion of the AP. Conduction slows because depolarization begins in incompletely repolarized fibers. Furthermore, hypokalemia prolongs the plateau in the Purkinje fibers but shortens it in the ventricular fibers.[12] The AP tail of the conducting system prolongs more than that of the ventricles, increasing the dispersion of repolarization. Hypokalemia increases diastolic depolarization in Purkinje fibers, thereby increasing automaticity.

In summary, the electrophysiological effects of hypokalemia are: (1) a decrease in conduction velocity; (2) shortening of the effective refractory period (ERP); (3) prolongation of the relative refractory period; (4) increased automaticity; and (5) early afterdepolarizations (EADs) (Table 33-4).

TABLE 33-2 Factors That Affect the Transcellular Shift of Potassium

From Inside to Outside	From Outside to Inside
Acidosis	Alkalosis
α-adrenergic receptor stimulation	β_2-adrenergic receptor stimulation
Digitalis	Insulin
Solvent drag	

ELECTROCARDIOGRAPHIC MANIFESTATIONS OF HYPOKALEMIA

The electrocardiographic manifestations of hypokalemia[13] can be conceptualized as those due to its effects on repolarization and those emanating from its effects on conduction (Table 33-5).

The ECG changes resulting from its effects on repolarization include: decreased amplitude and broadening of the T waves; prominent U waves; ST segment depression; and T and U wave fusion, which is seen in severe hypokalemia (Fig. 33-1). When the U wave exceeds the T wave in amplitude, the serum potassium is below 3.0. Electrocardiographic changes due to conduction abnormalities are seen in more advanced stages of hypokalemia and include: increase in QRS duration without a concomitant change in the QRS configuration; atrioventricular block; cardiac arrest; increase in P wave amplitude and duration; and slight prolongation of the P–R interval.

ARRHYTHMOGENIC POTENTIAL AND CLINICAL IMPLICATIONS OF HYPOKALEMIA

Hypokalemia-induced hyperexcitability is clinically manifested by an increase in supraventricular and ventricular ectopy. In the Framingham Offspring Study, potassium and magnesium levels were inversely related

TABLE 33-4 Electrophysiological Effects of Hypokalemia

Decrease in conduction velocity
Shortening of the effective refractory period
Prolongation of the relative refractory period
Increased automaticity
Early afterdepolarizations (EADs)

TABLE 33-5 ECG Manifestations of Hypokalemia

Repolarization Changes
Decreased amplitude and broadening of the T waves
Prominent U waves
ST segment depression
T and U waves fusion (in severe hypokalemia)
Conduction Abnormalities
Increase in QRS duration
Atrioventricular block
Cardiac arrest
Increase in P wave amplitude and duration
Slight prolongation of the P–R interval

to the occurrence of complex or frequent ventricular premature complexes (VPCs) after adjustment for covariates.[15]

Hypokalemia facilitates reentry by slowing conduction during the prolonged RRP and by causing an increase in the dispersion of refractoriness. Its suppressant effect on the Na-K pump leads to intracellular Ca^{2+} overload and this facilitates the development of delayed afterdepolarizations (DADs) via a transient inward current (I_{ti}). Hypokalemia enhances the propensity for ventricular fibrillation in the normal, as well as the ischemic, canine heart.[17] An association between hypokalemia and ventricular fibrillation in patients with acute myocardial infarction has been well established.[18-20]

Dofetilide and quinidine were found to exert increased block of I_{kr} in the setting of low $[K^+]_o$ providing a mechanism that explains the link between hypokalemia and torsades de pointes[7] (Figs. 33-2 and 33-3).

Hyperkalemia

Less common than hypokalemia, hyperkalemia is seen mainly in the setting of compromised renal function, particularly in association with the administration of a variety of nephrotoxic medications. Hyperkalemia may result from either decreased excretion or a shift of potassium from within the cell (Table 33-6).

ELECTROPHYSIOLOGICAL EFFECTS OF HYPERKALEMIA

The disproportional effects of varying levels of hyperkalemia on the RMP and the TP explain the initial increase in excitability and conduction velocity followed by their decrease as the potassium level increases further (Table 33-7). Mild-to-moderate levels of hyperkalemia decrease the RMP (less negative) more than the TP, thereby diminishing the difference between the two and increasing excitability. The decrease in the slope of the upstroke of the AP ($\delta V/\delta T$), one of the major determinants of conduction velocity, is counterbalanced by a decrease in the RMP to TP difference, resulting in an ultimate increase in conduction velocity. Severe hyperkalemia is associated with an increase in the difference between the RMP and the TP, leading to a decrease in excitability. Further decrement in the AP upstroke ($\delta V/\delta T$) overwhelms the positive effect of the TP decrease on the conduction velocity, resulting in a definitive decrease of the latter.

Hyperkalemia is associated with increased membrane permeability to potassium, a consequence of an increase of the inward going rectifier [i_{k1}] and the delayed rectifier current (I_{Kr}).[5-7] This accelerates the rate of repolarization and shortens the AP duration. Hyperkalemia preferentially shortens the plateau of the Purkinje fibers, thereby decreasing the dispersion of repolarization in the ventricle.[12] Furthermore, it slows the diastolic depolarization of the Purkinje fibers.

The effects of hyperkalemia depend on the tissue involved, with the atrial myocardium being the most sensitive, the ventricular myocardium less sensitive, and the specialized tissue (sinoatrial node and His bundle) the least sensitive. In other words, the depression of excitability and conduction in the atrium occur at lower extracellular potassium levels than in other types of myocardial tissue, as exemplified by the

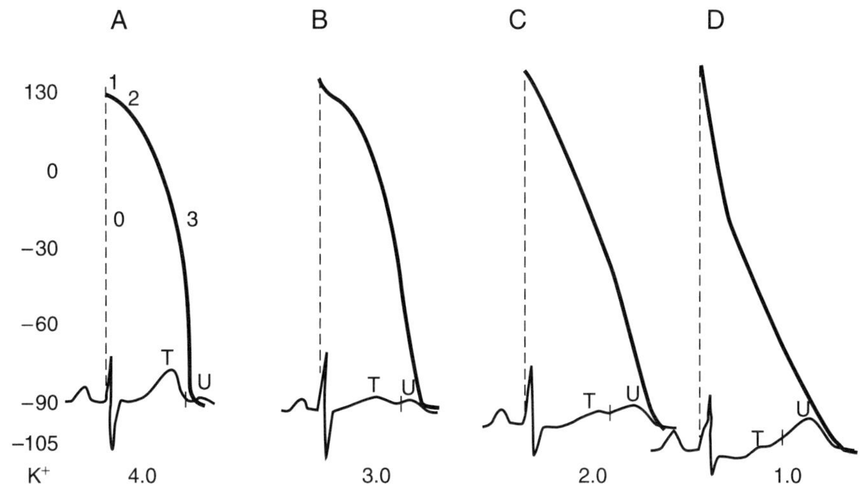

FIGURE 33-1 Diagram of the ventricular action potential superimposed on the ECG at different extracellular potassium concentrations (4.0, 3.0, 2.0, mEq/L). The numbers on the left designate the transmembrane potential in millivolts. (From Surawicz B: Relation between electrocardiogram and electrolytes. Am Heart J 1967;73:814-834.)

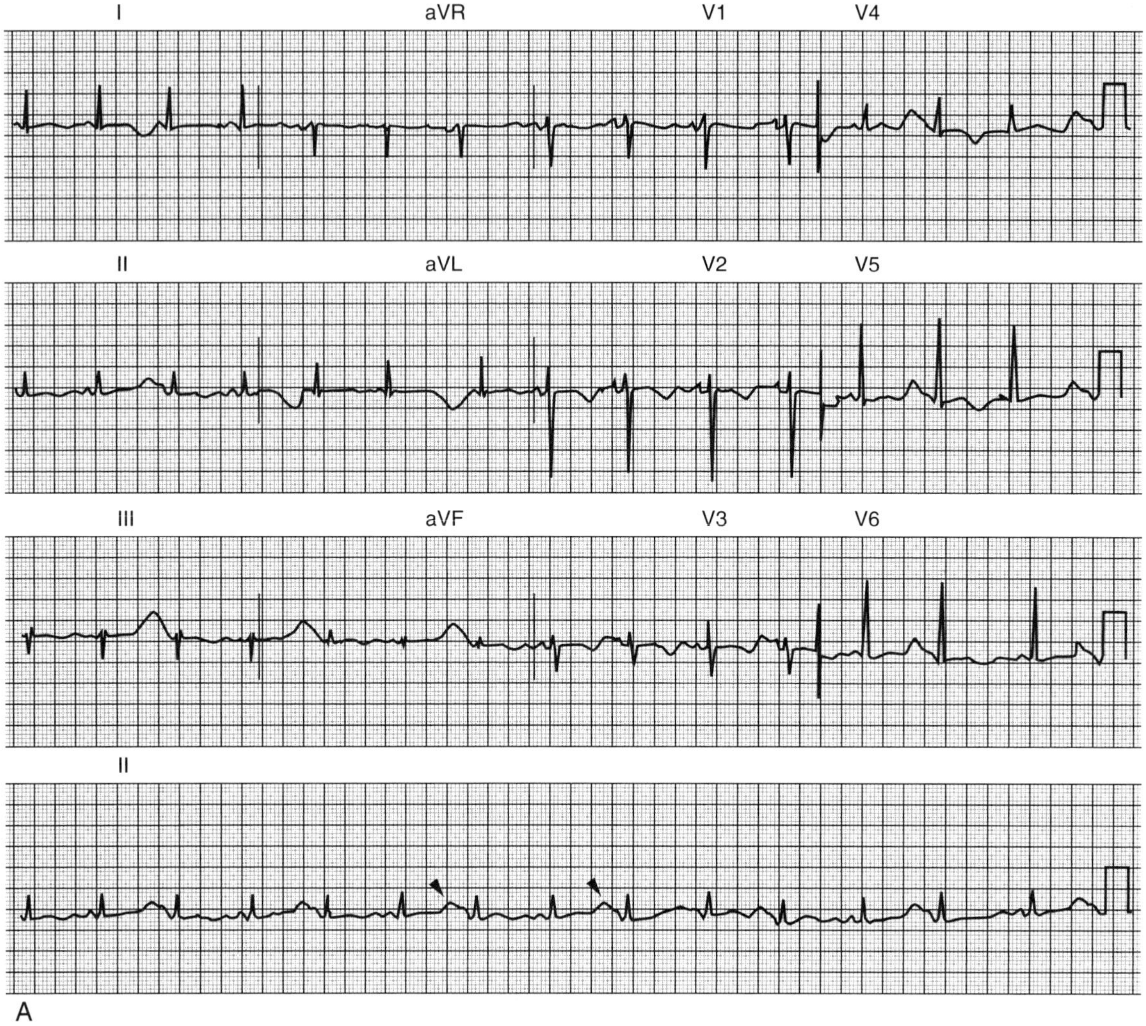

FIGURE 33-2 **A,** Twelve-lead ECG from a patient with hypokalemia and hypomagnesemia showing marked QTU prolongation and QTU alternans (*marked by arrowheads*).

occasional pacemaker case of atrial noncapture and ventricular capture.[22] The sympathetic nervous system seems to contribute to the sinus node resistance to hyperkalemia.[23]

Many investigators have observed regional differences in repolarization time in the nonischemic myocardium in response to hyperkalemia. Sutton and colleagues have recorded monophasic action potentials (MAPs) from the endocardium and epicardium in open-chest dogs during graded intravenous infusion of potassium to a plasma level of 9 mM. Their results suggest that the regional differences in repolarization times are mainly a result of local changes in activation times rather than a direct effect on the APD.[24]

An interesting phenomenon, the Zwaardemaker-Libbrecht effect, is the result of a change from a low to high extracellular potassium level and manifests itself by a transient arrest of pacemaker cells, abbreviation of APD, and hyperpolarization.[25-27] This phenomenon underscores the fact that the rate of intravenous administration of potassium is more important—from a proarrhythmic standpoint—than the absolute amount of potassium administered and the final level of extracellular potassium.

ELECTROCARDIOGRAPHIC MANIFESTATIONS OF HYPERKALEMIA

The electrocardiogram is not a sensitive indicator of hyperkalemia; 50% of patients with potassium levels greater than 6.5 mEq/L will not manifest any electrocardiographic changes (Table 33-8).[14] The ECG changes due to mild potassium elevations (K = 5.5 – 7.0 mEq) include tall, peaked, narrow-based T waves and fascicular blocks (LAFB and LPFB). Moderate hyperkalemia (K = 7.5 – 10.0 mEq) is associated with first-degree AV block and diminished P wave amplitude. As the potassium level increases further, the P wave disappears, and sinus arrest, as well as ST segment depression, may develop. Atypical bundle branch blocks (LBBB and RBBB), intraventricular conduction delays (IVCD), ventricular tachycardia (see Fig. 33-3), ventricular

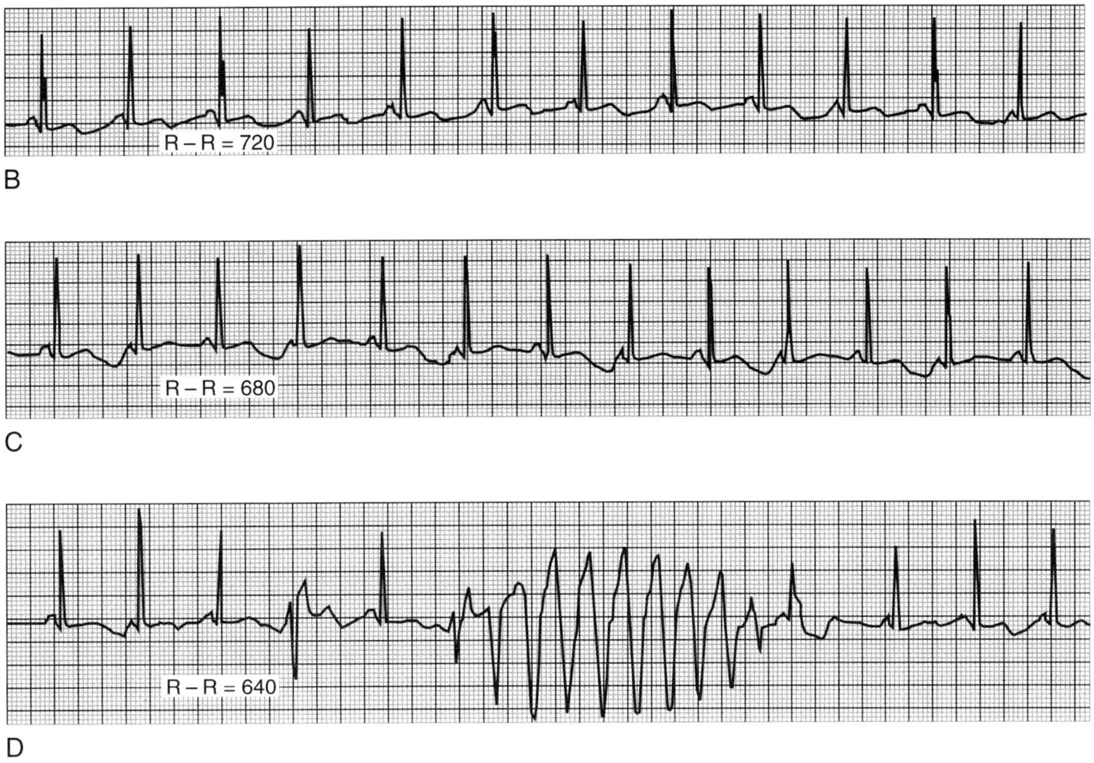

FIGURE 33-2 (cont'd) **B-D,** Representative rhythm strips from the same patient as in **A,** showing tachycardia-dependent QTU alternans and torsades de pointes.[16]

fibrillation, and idioventricular rhythm are more commonly seen in cases of severe hyperkalemia ($K^+ >$ 10.0 mEq). The bundle branch blocks associated with hyperkalemia are atypical in the sense that they involve the initial and terminal forces of the QRS complex. Shifts in the QRS axis indicate disproportionate conduction delays in the left bundle fascicles. These manifestations of intraventricular conduction delay correlate with a prolonged intracardiac H–V interval.[28] As hyperkalemia progresses, depolarization merges with repolarization, expressed in the ECG with QT shortening and apparent ST segment elevation simulating acute injury. The latter disappears with hemodialysis—*dialyzable current* of injury.[29] The parallel changes in the AP and ECG can be appreciated better by reviewing Figure 33-4.[14]

Arrhythmogenic Potential and Clinical Implications of Hyperkalemia

POTASSIUM AND MYOCARDIAL ISCHEMIA

In the early phases of an ischemic insult, the cardiac membrane becomes increasingly permeable to potassium. After coronary ligation in dogs and pigs,[30,31] the rise in extracellular potassium concentration results in currents of injury, refractory period shortening, conduction slowing, and ventricular fibrillation. During ischemia, APD shortening is more pronounced and the conduction velocity is slower in failing than in control myocardium. Extracellular potassium $[K^+]_o$ reaches higher values during acute ischemia in failing versus normal myocardium. Increased spatial dispersion in electrophysiological parameters and $[K^+]_o$ over the ischemic border in failing hearts may explain the higher propensity for reentrant arrhythmias during acute regional ischemia.[32]

POTASSIUM AND CALCIUM ABNORMALITIES

The combination of hyperkalemia and hypocalcemia has a cumulative effect on the atrioventricular and intraventricular conduction delay and facilitates the development of ventricular fibrillation. Hypercalcemia, through its membrane-stabilizing effect, counteracts the effects of hyperkalemia on AV and intraventricular conduction and averts the development of ventricular fibrillation. This protective effect of calcium is immediate and its intravenous administration should be the first therapeutic measure in cases of hyperkalemia.

POTASSIUM, DIGITALIS, AND QUINIDINE

Hyperkalemia inhibits glycoside binding to (Na^+, K^+) ATPase, decreases the inotropic effect of digitalis, and suppresses digitalis-induced ectopic rhythms. Alternatively, hypokalemia increases glycoside binding to (Na^+,K^+) ATPase, decreases the rate of digoxin elimination, and potentiates the toxic effects of digitalis. Finally, hyperkalemia and hypokalemia augment quinidine toxicity.

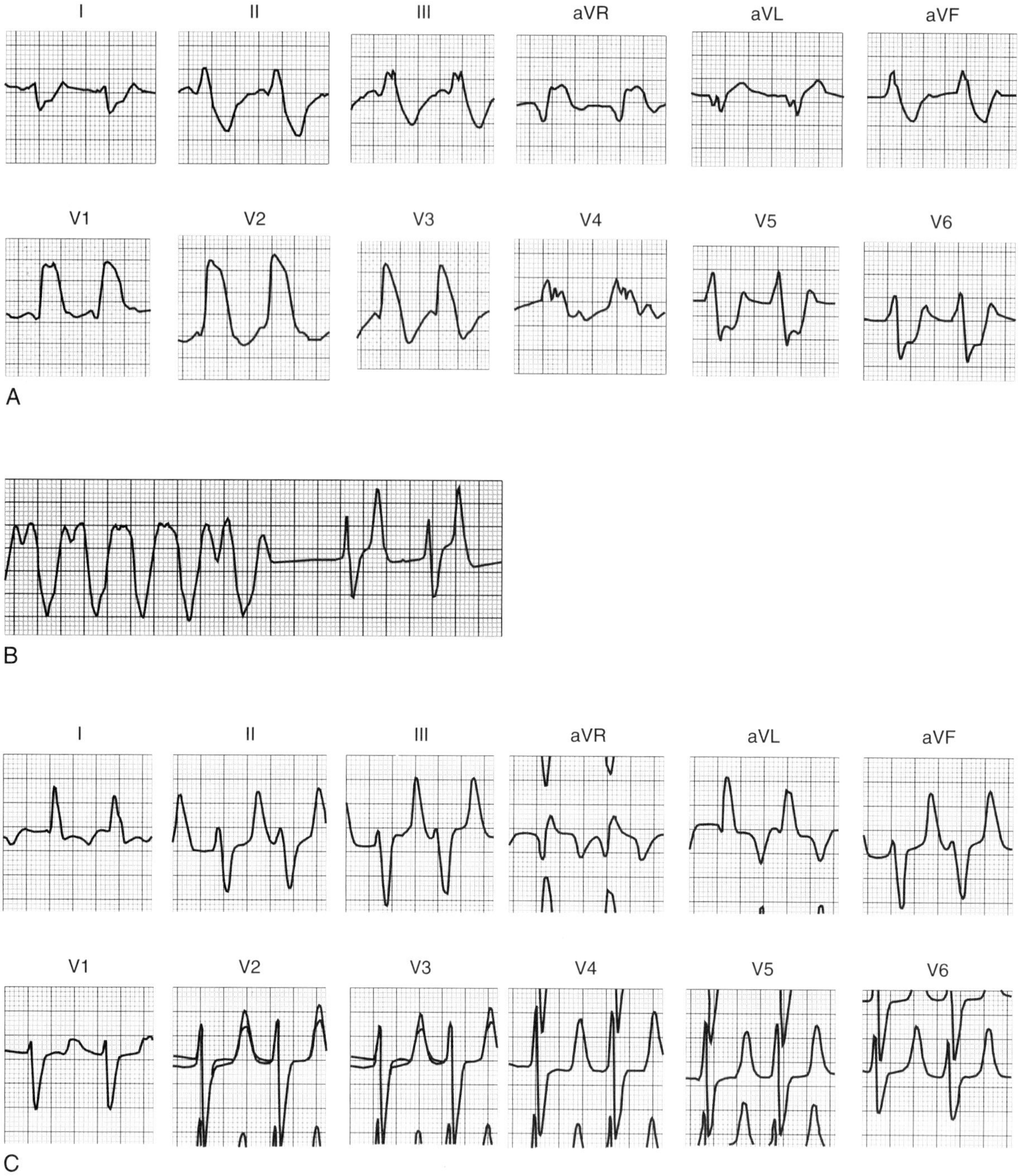

FIGURE 33-3 **A, B,** and **C** are tracings from a 39-year-old patient with end stage renal disease on dialysis who presented with weakness and palpitations. Chemistry showed a serum potassium level of 10 mEq/L. **A,** Atypical bundle branch block; the QRS is wide and is merging with the T wave. **B,** Ventricular tachycardia that failed to respond to adenosine and procainamide given by paramedics. **C,** Resolution of above ECG changes with dialysis and normalization of serum potassium level.

CALCIUM

Isolated abnormalities of calcium concentration produce clinically significant electrophysiological effects only when they are extreme in either direction. As expected by reviewing the Goldman equation, extracellular calcium concentrations in the physiologic range have no appreciable effect on the resting membrane potential. Hypocalcemia and hypercalcemia have opposing effects on the APD and ERP by affecting intracellular calcium concentration and modulating potassium currents.[2]

Hypocalcemia

Hypocalcemia is most frequently seen in the setting of chronic renal insufficiency and is usually associated with

TABLE 33-6 Causes of Hyperkalemia

Decreased Excretion
Renal failure
Renal secretory defects
Hyporeninemic hypoaldosteronism
Heparin
Drugs (ACE inhibitors, spironolactone, triamterene, NSAIDs, trimethoprim)
Shift of Potassium from Within the Cell (see Table 33-2)
Massive release of intracellular potassium
Hypertonicity
Insulin deficiency
Hyperkalemic periodic paralysis

other electrolyte abnormalities. Generally, hypocalcemia may result from decreased intake or absorption or an increase in calcium loss (Table 33-9).

ELECTROPHYSIOLOGICAL EFFECTS OF HYPOCALCEMIA

Low extracellular calcium decreases the slow inward current and intracellular calcium concentration during the AP plateau. The latter decreases outward current, possibly via $I_{k(Ca)}$, prolonging phase 2 of the AP, the total APD, and the duration of the ERP. As a consequence of low intracellular calcium, contractility decreases. Moreover, hypocalcemia slightly decreases the rate of diastolic depolarization in the Purkinje fibers and increases excitability through a direct interaction with the sarcolemma.

ELECTROCARDIOGRAPHIC MANIFESTATIONS OF HYPOCALCEMIA

The ECG changes of hypocalcemia involve a prolongation of the ST segment and QTc interval and T wave alterations, including upright, low, flat, or sharply inverted T waves in leads with an upright QRS complex (Fig. 33-5).

TABLE 33-7 Effect of $[K^+]_o$ on RMP, TP, and V_{max} in Ventricular Muscle Fibers of Guinea Pigs

$[K^+]_o$ (mM)	RMP (mV)	TP (mV)	RMP—TP (mV)	V_{max} (V/s)
2.0	99.4	72.7	26.7	236
5.4	83.0	65.1	17.9	219
10.0	65.8	51.4	14.4	178
11.5	62.8	45.6	17.2	154
13.0	60.7	41.9	18.8	103
16.2	55.4	34.7	20.7	45

RMP, Resting membrane potential; TP, threshold potential; $[K^+]_o$, extracellular potassium concentration.

From Kishida H, Surawicz B, Fu LT: Effects of K^+ and K^+-induced polarization on $(dV/dt)_{max}$, threshold potential, and membrane input resistance in guinea pig and cat ventricular myocardium. Circ Res 1979;44:800-814.

TABLE 33-8 ECG Manifestations of Hyperkalemia

Mild Hyperkalemia (K = 5.5 - 7.5 mEq)
Tall, peaked, narrow-based T waves
Fascicular blocks (LAFB, LPFB)
Moderate Hyperkalemia (K = 7.5 - 10.0 mEq)
First-degree AV block
Decreased P wave amplitude followed by disappearance of the P waves and sinus arrest
ST segment depression
Severe Hyperkalemia (K > 10.0 mEq)
Atypical bundle branch block (LBBB, RBBB), IVCD
Ventricular tachycardia, ventricular fibrillation, idioventricular rhythm

LAFB, Left anterior fascicular block; LPFB, left posterior fascicular block; LBBB, left bundle branch block; RBBB, right bundle branch block; IVCD, intraventricular conduction delay.

Hypercalcemia

Hypercalcemia occurs most commonly in the setting of hyperparathyroidism or as a consequence of several malignancies. A plethora of other causes account for the remaining clinically encountered cases (Table 33-10).

ELECTROPHYSIOLOGICAL EFFECTS AND ELECTROCARDIOGRAPHIC MANIFESTATIONS OF HYPERCALCEMIA

High extracellular calcium levels shorten the AP plateau, the total APD, and, consequently, the duration of the ERP. A study of the effect of hypercalcemia on the guinea pig ventricular AP suggested that a decrease in the inward Na^+/Ca^{++} exchange current might be largely responsible for the shortening of the AP.[33] Elevated extracellular Ca^{2+} concentration has a stabilizing effect on the membrane, increasing the extent of depolarization needed to initiate an AP. In addition, hypercalcemia has a positive inotropic effect, decreases excitability,[34] and slightly increases the rate of diastolic depolarization in the Purkinje fibers. The electrocardiographic changes as a result of hypercalcemia are limited to shortening or elimination of the ST segment and decreased QTc interval.

MAGNESIUM

Magnesium is the second most abundant intracellular cation after potassium. The significance of magnesium disorders has been debated because of difficulties in accurate measurement and their frequent association with other electrolyte abnormalities.[35,36] It is an important cofactor in several enzymatic reactions contributing to normal cardiovascular physiology. Magnesium deficiency is common, but its electrophysiological sequelae have evaded even the closest scrutiny. Magnesium therapy in pharmacologic doses has been beneficial in treating torsades de pointes. Magnesium toxicity rarely occurs except in patients with renal dysfunction.

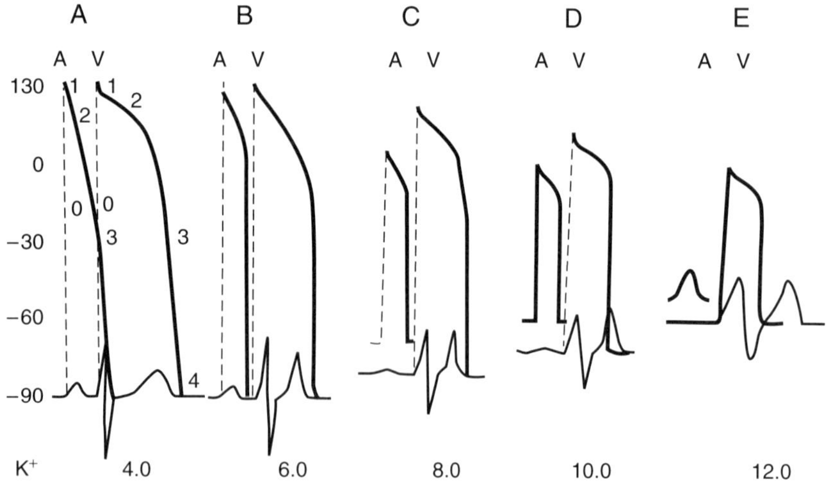

FIGURE 33-4 Diagram of an atrial and ventricular action potential superimposed on the ECG for extracellular potassium at 4.0, 6.0, 8.0, 10.0, and 12.0 mEq/L. The numbers on the left designate the transmembrane potential in millivolts, and the numbers at the bottom designate extracellular potassium concentration (mEq/L) (From Surawicz B: Relation between electrocardiogram and electrolytes. Am Heart J 1967;73:814-834.)

ELECTROPHYSIOLOGICAL EFFECTS AND ELECTROCARDIOGRAPHIC MANIFESTATIONS OF HYPOMAGNESEMIA AND HYPERMAGNESEMIA

In the presence of extremely low extracellular calcium concentrations, magnesium exerts an effect on the current or currents that modulate the duration of the ventricular AP plateau. Hoffman and Suckling[2] found that in the presence of normal calcium concentrations, magnesium deficiency had little effect on the AP of the canine papillary muscle. However, when the calcium concentration was lowered to 1/10 of normal, complete omission of magnesium in the superfusate prolonged the AP plateau, which was already increased in duration by low calcium, from a normal value of 100 to 150 ms to 1000 ms or more.

Magnesium blocks the calcium channel, shifts the steady-state inactivation curve of the fast sodium channel in the hyperpolarizing direction, modifies the effect of hyperkalemia, and exerts modulating effects on several potassium currents. Hypermagnesemia depresses AV and intraventricular conduction. DiCarlo and coworkers observed the following electrocardiographic effects of intravenous administration of magnesium in patients with normal baseline serum magnesium and other electrolyte levels:[37] (1) prolongation of sinus node recovery time (SNRT) and corrected sinus node recovery time (CSNRT); (2) prolongation of the AV nodal functional, relative, and effective refractory periods; (3) a small increase in QRS duration during ventricular pacing at cycle lengths of 250 and 500 ms; and (4) a significant increase in the atrial-His interval and the atrial paced cycle length causing AV node Wenckebach conduction.

Kulick and associates,[38] in studying healthier hearts, noted the following ECG effects of intravenous magnesium administration: (1) significant prolongation of the P–R interval from 145 to 155 ms after magnesium infusion; (2) prolongation of the atrial-His interval; (3) prolongation of the sinoatrial conduction time (SACT); (4) prolongation of the AV nodal effective refractory periods; and (5) no significant increase in SNRT in the atrial paced cycle length causing AV node Wenckebach conduction or in QRS duration. Hypermagnesemia and hypomagnesemia do not produce specific ECG changes.

TABLE 33-9 Causes of Hypocalcemia

Decreased Intake or Absorption
Malabsorption
Decreased absorptive area (small bowel bypass, short bowel)
Vitamin D deficit (decreased absorption, decreased 25-hydroxy-D or 1,25-dihydroxy-D production)
Increased Loss
Alcoholism
Chronic renal insufficiency
Diuretic therapy (furosemide, bumetanide)
Endocrine Disease
Hypoparathyroidism or pseudohypoparathyroidism
Medullary carcinoma of the thyroid (calcitonin secretion)
Familial hypocalcemia
Physiologic Causes
Associated with decreased serum albumin (normal calcium ion concentration)
Decreased end-organ response to vitamin D
Sepsis; hyperphosphatemia; induced by aminoglycoside antibiotics, plicamycin, loop diuretics, foscarnet

Magnesium and Torsades de Pointes

The administration of intravenous magnesium sulfate to patients with prolonged Q–T interval and torsades de pointes, whether the initial magnesium level is normal or low, may suppress ventricular tachycardia. Takanaka and colleagues studied the effects of magnesium and lidocaine on the APD and on barium-induced EADs in canine Purkinje fibers. Their data suggest that hypomagnesemia may be arrhythmogenic when combined with hypokalemia and bradycardia, and magnesium administration may suppress triggered activity, mainly

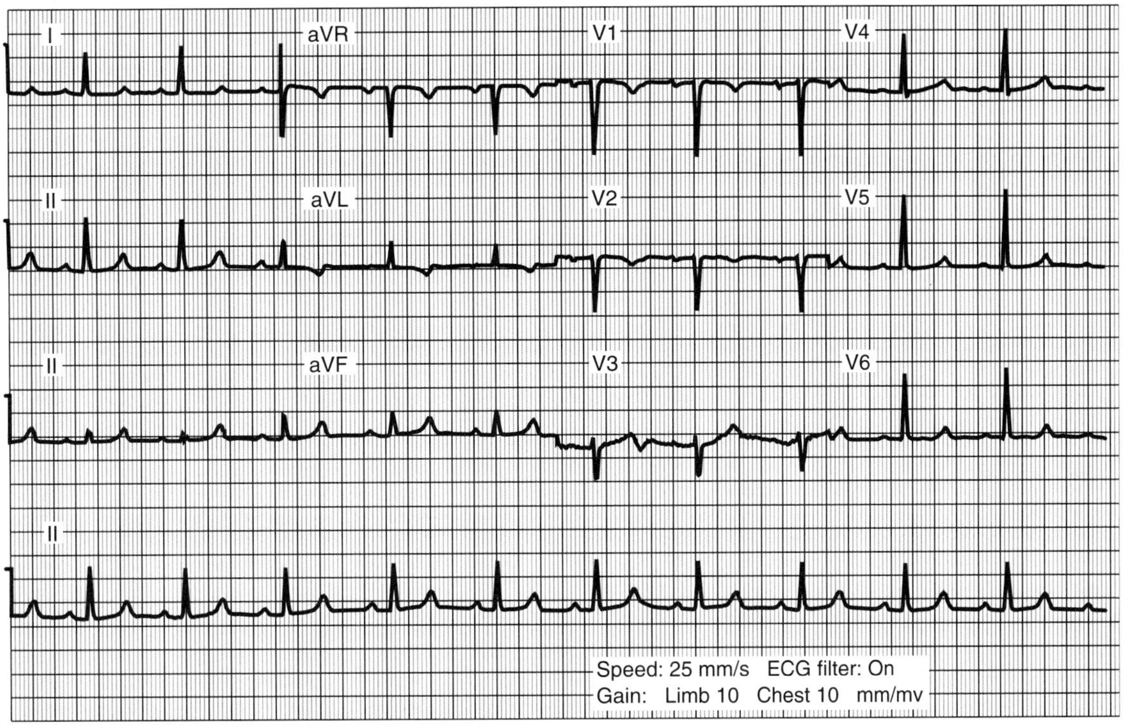

FIGURE 33-5 This 12-lead ECG is from a 24-year-old black female patient with sickle cell disease, end stage renal disease, and hyperparathyroidism. The patient's serum calcium level was 7 mg/dL and the potassium level was 6.5 mEq/L. This ECG demonstrates a prolongation of the ST segment and QTc interval (leads V5 and V6).

by directly preventing the development of triggered APs.[39] In conclusion, magnesium sulfate is a very effective and safe treatment for torsades de pointes (TdP).[40]

Magnesium and Heart Failure

In a sample of ambulatory patients with heart failure, magnesium depletion in serum and tissue did not appear to occur more commonly in patients with serious ventricular arrhythmias than in patients without serious ventricular arrhythmias.[41] In patients with moderate to severe heart failure, serum magnesium does not appear to be an independent risk factor for either sudden cardiac death or all-cause mortality.[42] Hypomagnesemia was found to be associated with an increase in frequency of ventricular couplets, but it did not lead to a higher incidence of clinical events.[42]

TABLE 33-10 Causes of Hypercalcemia

Increased Intake or Absorption	Miscellaneous Causes
Vitamin D or vitamin A excess	Thiazides Sarcoidosis Paget's disease of the bone Hypophosphatasia Familial hypocalciuric hypercalcemia Iatrogenic
Endocrine Disorders	
Primary hyperparathyroidism or secondary hyperparathyroidism (renal insufficiency, malabsorption) Acromegaly Adrenal insufficiency	
Neoplastic Diseases	
Tumors producing PTH-related protein (ovary, kidney, lung) Multiple myeloma (osteoclast-activating factor)	

Magnesium and Myocardial Infarction

Magnesium administration has been found to have a positive effect on the consequences of myocardial infarction in experimental models. Its effect in the clinical setting of myocardial infarction has been controversial. In the LIMIT-2 study,[43-45] magnesium administration was noted to have a positive effect on mortality rate, a finding that has not been reproduced in the ISIS-4 trial.[43,46-48]

Relationship Between Potassium, Magnesium, and Cardiac Arrhythmias

No specific electrophysiological effects or arrhythmias have been linked to isolated magnesium deficiency. Nonetheless, magnesium may influence the incidence of cardiac arrhythmias through a direct effect, by modulating the effects of potassium, or through its action as a calcium channel blocker. Magnesium deficiency is thought to interfere with the normal functioning of membrane ATPase, and thus the pumping of sodium out of the cell and potassium into the cell. This affects the transmembrane equilibrium of potassium, which may result in changes in the resting membrane potential, changes in potassium conductance across the cell membrane, and disturbances in the repolarization phase.[49]

SODIUM

Sodium is the most abundant extracellular cation, and the sodium current determines the phase 0 and amplitude of the AP. Its conductance increases precipitously with the initiation of the AP, allowing its transmembrane gradient to determine the first phase of the AP, and, consequently, its ultimate configuration. Hence, hypernatremia increases and hyponatremia decreases phase 0 of the AP by altering the transmembrane sodium gradient. The upstroke of the AP is determined by the sodium gradient and the transmembrane potential. Therefore, hypernatremia, by increasing the sodium gradient, negates many of the effects of hyperkalemia, which decreases the transmembrane potential. By increasing the amplitude of the AP, high sodium levels prolong the APD. In vitro, an increased Na concentration restores the normal configuration of AP that is altered by previous treatment with sodium channel blockers. Despite the frequency of sodium abnormalities, particularly hyponatremia, its electrophysiological effects are rarely of clinical significance.

LITHIUM

Although lithium is not a naturally occurring electrolyte, it is frequently encountered clinically in the treatment of manic-depressive disorders and, as such, its potential adverse effects on the sinoatrial node should not be overlooked. Lithium is associated with sinoatrial node dysfunction (sinus bradycardia, sinoatrial arrest, or exit block, either type I or type II) and reversible T wave changes.[50-52]

REFERENCES

1. Goldman DE: Potential, impedance and rectification in membranes. J Gen Physiol 27:37-60.
2. Hoffman BF, Suckling EE: Effect of several cations on transmembrane potentials of cardiac muscle. Am J Physiol 186:317-324.
3. Sheu SS, Korth M, Lathrop DA, Fozzard HA: Intra- and extracellular K^+ and Na^+ activities and resting membrane potential in sheep cardiac Purkinje strands. Circ Res 1980;47:692-700.
4. Ishihara K, Mitsuiye T, Noma A, Takano M: The Mg2+ block and intrinsic gating underlying inward rectification of the K+ current in guinea-pig cardiac myocytes. J Physiol (Lond) 1989;419:297-320.
5. Sakmann B, Trube G: Conductance properties of single inwardly rectifying potassium channels in ventricular cells from guinea pig heart. J Physiol 1984;347:641-657.
6. Pelzer D, Trautwein W: Currents through ionic channels in multicellular cardiac tissue and single heart cells. Experientia 1987; 43:1153-1162.
7. Yang T, Roden DM. Extracellular potassium modulation of drug block of I_{kr}. Circulation 1996;93:407-411.
8. Brown MJ, Brown DC, Murphy MB: Hypokalemia from $beta_2$-stimulation by circulating epinephrine. N Engl J Med 1983; 309:1414-1419.
9. Paice BJ, Paterson KR, Onyanga-Omara F, et al: Record linkage study of hypokalemia in hospitalized patients. Postgrad Med J 1986;62:187-191.
10. Schulman M, Narins RG: Hypokalemia and cardiovascular disease. Am J Cardiol 1990;65:4E-9E.
11. Thompson RG, Gobb LA: Hypokalemia after resuscitation out-of-hospital ventricular fibrillation. JAMA 1982;248:2860-2863.
12. Gettes L, Surawicz B: Effects of low and high concentrations of potassium on the simultaneously recorded Purkinje and ventricular action potentials of the perfused pig moderator band. Circ Res 1968;23:717-729.
13. Surawicz B, Braun HA, Crum WB, et al: Quantitative analysis of the electrocardiographic pattern of hypopotassemia. Circulation 1957;16:750-763.
14. Surawicz B: Relation between electrocardiogram and electrolytes. Am Heart J 1967;73:814-834.
15. Tsuji H, Venditti FJ Jr, Evans JC, et al: The associations of levels of serum potassium and magnesium with ventricular premature complexes (the Framingham Heart Study). Am J Cardiol (United States) 1994;74:232-235.
16. Habbab MA, El-Sherif N: TU Alternans, long QTU, and torsades de pointes. PACE 1992;15:916-931.
17. Hohnloser SH, Verrier RL, Lown B, Raeder EA: Effect of hypokalemia on susceptibility to ventricular fibrillation in the normal and ischemic canine heart. Am Heart J 1986;112:32-35.
18. Nordrehaug JE, Johannessen K, Von der Lippe G: Serum potassium concentration as a risk factor of ventricular arrhythmias early in acute myocardial infarction. Circulation 1985;71:645-649.
19. Cooper WD, Kuan P, Reuben SR, Vanderburg MJ: Cardiac arrhythmias following acute myocardial infraction: Association with serum potassium level and prior diuretic therapy. Eur Heart J 1984;5:464-469.
20. Nordrehaug JE, Von der Lippe G: Hypokalemia and ventricular fibrillation in acute myocardial infarction. Br Heart J 1983;50:525-529.
21. Kishida H, Surawicz B, Fu LT: Effects of K^+ and K^+-induced polarization on $(dV/dt)_{max}$, threshold potential, and membrane input resistance in guinea pig and cat ventricular myocardium. Circ Res 1979;44:800-814.
22. Barold S, Falkoff MD, Ong LS, Heinle RA: Hyperkalemia-induced failure of atrial capture during dual-chamber cardiac pacing. J Am Coll Cardiol 1987;10:467-469.
23. Vassalle M, Greineder JK, Stuckey JH: Role of the sympathetic nervous system in the sinus node resistance to high potassium. Circ Res 1973;32:348-355.
24. Sutton PM, Taggart P, Spear DW, et al: Monophasic action potential recordings in response to graded hyperkalemia in dogs. Am J Physiol 1989;256(4 Pt2):H956-H961.
25. Zwaardemaker H: On physiological radioactivity. J Physiol 53:273-289.
26. Libbrecht W: Le paradoxe cardiaque. Arch Int Physiol 16: 448-452.
27. Ito S, Surawicz B: Transient, "paradoxical" effects of increasing extracellular K^+ concentration on transmembrane potential in canine cardiac Purkinje fibers. Circ Res 1977;41:799-807.
28. Ettinger PO, Regan TJ, Oldewurtel HA: Hyperkalemia, cardiac conduction and the electrocardiogram: A review. Am Heart J 1974;88:360-371.
29. Levine HD, Wanzer SH, Merrill JP: Dialyzable currents of injury in potassium intoxication resembling acute myocardial infarction or pericarditis. Circulation 1956;13:29-36.
30. Hill JL, Gettes LS: Effect of acute coronary artery occlusion on local myocardial extracellular K^+ activity in swine. Circulation 1980;61:768-778.
31. Coronel R, Fiolet JW, Wilms-Schopman FJ, et al: Distribution of extracellular potassium and its relation to electrophysiologic changes during acute myocardial ischemia in the isolated perfused porcine heart. Circulation 1988;77:1125-1138.
32. Vermeulen JT, Tan HL, Rademaker H, et al: Electrophysiologic and extracellular ionic changes during acute ischemia in failing and normal rabbit myocardium. J Mol Cell Cardiol 1996;28:123-131.
33. Leitch SP, Brown HF: Effect of raised extracellular calcium on characteristics of the guinea pig ventricular action potential. J Mol Cell Cardiol 1996;28:541-551.
34. Weidmann S: Effects of calcium ions and local anesthetics on electrical properties of Purkinje fibres. 1955;129:568-582.
35. Keller PK, Aronson RS: The role of magnesium in cardiac arrhythmias. Prog Cardiovasc Dis 1990;32:433-448.

36. Surawicz B, Lepeschkin E, Herrlich HC: Low and high magnesium concentrations at various calcium levels: Effect on the monophasic action potential, electrocardiogram and contractility of isolated rabbit hearts. Circ Res 9:811-818.
37. DiCarlo LA Jr, Morady F, DeBuitleir M, et al: Effects of magnesium sulfate on cardiac conduction and refractoriness in humans. J Am Coll Cardiol 1986;7:1356-1362.
38. Kulick DL, Hong R, Ryzen E: Electrophysiologic effects of intravenous magnesium in patients with normal conduction systems and no clinical evidence of significant cardiac disease. Am Heart J 1988;115:367-373.
39. Takanaka C, Ogunyankin KO, Sarma JS, Singh BN: Antiarrhythmic and arrhythmogenic actions of varying levels of extracellular magnesium: Possible cellular basis for the differences in the efficacy of magnesium and lidocaine in torsades de pointes. J Cardiovasc Pharmacol Ther 1997;2:125-134.
40. Tzivoni D, Banai S, Schuger C, et al: Treatment of torsade de pointes with magnesium sulfate. Circulation 1988;77:392-397.
41. Ralston MA, Murname MR, Unverferth DV, Leier CV: Serum and tissue magnesium concentrations in patients with heart failure and serious ventricular arrhythmias. Ann Int Med 1990;113:841-846.
42. Eichhorn EJ, Tandon PK, DiBianco R, et al: Clinical and prognostic significance of serum magnesium concentration in patients with severe chronic congestive heart failure: The PROMISE study. J Am Coll Cardiol 1993;21:634-640.
43. Baxter GF, Sumeray MS, Walker JM: Infarct size and magnesium: Insights into LIMIT-2 and ISIS-4 from experimental studies [see comments]. Lancet 1996;348:1424-1426.
44. Woods KL, Fletcher S: Long-term outcome after intravenous magnesium sulphate in suspected acute myocardial infarction: The second Leicester Intravenous Magnesium Intervention Trial (LIMIT-2) [see comments]. Lancet 1994;343:816-819.
45. Woods KL, Fletcher S, Roffe C, Haider Y: Intravenous magnesium sulphate in suspected acute myocardial infarction: Results of the second Leicester Intravenous Magnesium Intervention Trial (LIMIT-2) [see comments]. Lancet 1992;339:1553-1558.
46. ISIS-4: A randomised factorial trial assessing early oral captopril, oral mononitrate, and intravenous magnesium sulphate in 58,050 patients with suspected acute myocardial infarction. ISIS-4 (Fourth International Study of Infarct Survival) Collaborative Group [see comments]. Lancet 1995;345:669-685.
47. Fourth International Study of Infarct Survival: Protocol for a large simple study of the effects of oral mononitrate, of oral captopril, and of intravenous magnesium. ISIS-4 collaborative group. Am J Cardiol 1991;68:87D-100D.
48. Flather M, Pipilis A, Collins R, et al: Randomized controlled trial of oral captopril, of oral isosorbide mononitrate and of intravenous magnesium sulphate started early in acute myocardial infarction: Safety and haemodynamic effects. ISIS-4 (Fourth International Study of Infarct Survival) Pilot Study Investigators [see comments]. Eur Heart J 1994;15:608-619.
49. Dyckner T, Wester PO: Relation between potassium, magnesium and cardiac arrhythmias. Acta Med Scand [Suppl] 1981;647:163-169.
50. Tilkian AG, Schroeder JS, Kao J, Hultgren H: Effect of lithium on cardiovascular performance: Report on extended ambulatory monitoring and exercise testing before and during lithium therapy. Am J Cardiol 1976;38:701-708.
51. Wellens HJ, Cats VM, Duren DR: Symptomatic sinus node abnormalities following lithium carbonate therapy. Am J Med 1975;59:285-287.
52. Carmeliet EE: Influence of lithium ions on transmembrane potential and cation content of cardiac cells. J Gen Physiol 1964;47:501-530.

Chapter 34 The Long QT Syndrome

PETER J. SCHWARTZ

The congenital long QT syndrome (LQTS) is an arrhythmogenic disease that is relatively uncommon, but certainly not rare, and very important for the cardiologist interested in cardiac arrhythmias and for the clinical electrophysiologist. Since 1975[1] it has included under the unifying designation "long QT syndrome" two hereditary variants. One is associated with deafness[2] and one is not[3,4]; they are referred to as the Jervell and Lange-Nielsen syndrome (J-LN) and as the Romano-Ward syndrome (R-W), respectively.

There are several reasons for the current widespread interest in LQTS. One is represented by the dramatic manifestations of the disease, namely syncopal episodes, which often result in cardiac arrest and sudden death and usually occur in conditions of either physical or emotional stress in otherwise healthy young individuals, mostly children and teenagers. Another reason is that, although LQTS is a disease with a very high mortality rate among untreated patients, very effective therapies are available; this makes unacceptable and inexcusable the existence of symptomatic and undiagnosed or misdiagnosed patients. Finally, the identification of some of the genes responsible for LQTS, and moreover the realization that so far they all encode ion channels, has provided a new stimulus for clinical cardiologists and basic scientists. The impressive correlation between specific mutations and critical alterations in the ionic control of ventricular repolarization makes this syndrome a unique paradigm that allows correlation of genotype and phenotype, thus providing a direct bridge between molecular biology and clinical cardiology in the area of sudden cardiac death.

Molecular Biology of LQTS

GENE IDENTIFICATION AND EXPRESSION STUDIES

The first report indicating the existence of genetic linkage for LQTS on chromosome 11 appeared in 1991.[5] This was soon followed by evidence of linkage, also on chromosomes 7, 3, and 4.[6,7] The identification of four of the responsible genes took place between 1995 and 1999. It was then that the genes for LQT1 (on chromosome 11), for LQT2 (on chromosome 7), for LQT3 (on chromosome 3), and for LQT5 and LQT6 (both on chromosome 21) were identified.[8-11] The gene for LQT4 on chromosome 4 has been identified,[7] but there are growing questions related to the fact that this appears to be a disease different from LQTS.

KvLQT1 and *minK*

The gene for LQT1 is *KvLQT1*, and the gene for LQT5 is *minK*. LQT1 and LQT5 may share common clinical features because the two gene products coassemble to form the ion channel conducting the I_{Ks} current.[12,13]

Mutations in *KvLQT1* represent the most common genetic variant of LQTS; however, its exact prevalence has not yet been determined. Mutations in *minK* are rare: in 140 LQTS genotyped LQTS patients, only 3 *minK* mutations were identified, thereby suggesting a prevalence of less than 3% for LQT5.[14]

Mutations in the *KvLQT1*, gene have been identified in families from all over the world. The reported mutations are mainly single residue substitutions leading to an amino acid change; however, insertions, deletions, and splice-donor errors have also been identified. Expression studies have shown that mutations may result in nonfunctional proteins that do not coassemble with the wild type protein. In this case, the loss of function equals a 50% reduction because the wild type allele is fully functional. However, other mutations result in poorly functional proteins that interfere in a dominant negative fashion with the wild type protein and produce a loss of function greater than 50%. Thus, the more severe molecular defects, producing a nonfunctional protein, may result in a less severe electrophysiological defect.

Mutations in *KvLQT1*[10] and *minK*[12,13] are also responsible for the J-LN syndrome,[15] when present as homozygous or compound heterozygous defects.[16] A likely explanation for the clinical symptom of deafness involves the fact that *KvLQT1* is expressed in the stria vascularis of mouse inner ear and probably plays a pivotal role in normal hearing through maintenance of the endolymph homeostasis.[15]

An interesting twist of these molecular findings is that, contrary to previous concepts and independent of being symptomatic, both parents of every J-LN patient are, by definition, affected by the R-W syndrome. In other words, the cardiac phenotype (QT prolongation) of the J-LN syndrome is inherited as a dominant trait, whereas deafness is inherited as a recessive trait. To add further complexity, not all homozygous mutations in *KvLQT1* result in deafness.[17]

HERG and *MiRP1*

The gene for LQT2 is *HERG*, the α-subunit of a potassium channel that carries the I_{Kr} current. This outward current plays a major role in the rapid repolarization of phase 3 of the action potential. The gene for LQT6, a rare variant, is *MiRP1*.[11] Apparently, *HERG* and *MiRP1* coassemble to form the ion channel conducting the I_{Kr} current. The uncertainties about the role of *MiRP1* depend on the difficulties in detecting it in cardiac tissue.

The mutations identified in *HERG*[9,18-22] are mainly single residue substitutions leading to changes in highly conserved amino acids. Deletions, frameshifts, and splice-donor errors have also been reported.[9] Point mutations have been identified in all four transmembrane segments, and expression studies have shown that functionally relevant reductions in I_{Kr} current are caused by minimal amino acid changes.

The expression studies that have compared different mutations[23,24] indicate that the magnitude of the effect on repolarization varies according to the specific mutation. In a counterintuitive manner there seems to be no correlation between the severity of the clinical manifestations and the spectrum of *HERG* or *KvLQT1* dysfunction as assessed in vitro.[25]

SCN5A

The gene for LQT3 is *SCN5A*, the cardiac sodium channel. At variance with mutations on *HERG*, *KvLQT1*, or *minK*, which reduce or abolish channel function, *SCN5A* mutations in LQTS produce a gain of function.[23,24] The mutations identified result in a small, sustained inward current likely to disrupt the normal balance between inward and outward currents during the plateau phase, and hence, prolong cardiac action potential. Of potential clinical relevance, this sustained inward current is sensitive to mexiletine.[24,26] Mutations in the cardiac sodium channel not resulting in a gain of function cause another arrhythmogenic disease, the Brugada syndrome,[27] and can also cause progressive familial AV block.[28,29]

Clinical Presentation

The large increase in the number of patients diagnosed as affected has significantly modified the perception of the natural clinical history of the disease. Since the early 1970s, when the first consistent series of patients were reported,[1] the typical clinical presentation of LQTS was considered to be the occurrence of syncope or cardiac arrest, precipitated by emotional or physical stress, in a young individual with a prolonged Q–T interval on the surface electrocardiogram. If these symptomatic patients were left untreated, the syncopal episodes would recur and eventually prove fatal in the majority of cases. This concept was largely due to the fact that the patients initially diagnosed were those most severely affected. It has become progressively evident that this traditional picture represents an oversimplification, and it now appears that, in addition to the severe manifestations described earlier, there are also numerous patients with a very benign course. Unfortunately, the occasional occurrence of sudden death cannot yet be predicted. It is likely that modifier genes, possibly associated with the release of autonomic mediators, contribute to the highly different severities of the clinical manifestations of LQTS.[30]

When family screening is performed, prolongation of the Q–T interval often can be detected and a history of fainting episodes or of sudden unexpected deaths in early age are often recorded within the family. However, there is a significant number (approximately 30%) of sporadic cases; i.e., patients with syncope and a prolonged Q–T interval but without clinical evidence of familial involvement.[31]

The clinical history of repeated episodes of loss of consciousness under emotional or physical stress is so typical and unique that it is almost unmistakable, provided that the physician is aware of LQTS. However, the clinical presentation is not always so clear, and sometimes the diagnosis may be uncertain. There are two cardinal manifestations of LQTS: syncopal episodes and electrocardiographic abnormalities.

SYNCOPAL EPISODES

The syncopal episodes are due to torsades de pointes (TdP) often degenerating into ventricular fibrillation. The characteristics of the onset of TdP have been well analyzed, particularly for the acquired forms.[32-35] It should be noted that in LQTS, TdP or ventricular fibrillation can initiate without changes in heart rate and without specific sequences, such as short-long-short interval, even though a pause does often precede the onset of TdP.[36]

The syncopal episodes are characteristically associated with sudden increases in sympathetic activity, such as during strong emotions (particularly fright, but also anger) or physical activity (notably swimming).[37] This type of trigger is most frequent for LQT1 patients.[38] Sudden awakening is an almost specific trigger for some patients. A higher incidence of syncope in association with menses has been noted,[37] and also in the postpartum period.[39] The relatively frequent occurrence of convulsions has too frequently led to a diagnosis of epilepsy, which represents the most common misdiagnosis. Families and sporadic cases have been reported in which cardiac arrests occur almost exclusively either at rest or, more frequently, during sleep.[40,41] These events at rest are rare in LQT1, whereas they are much more frequent in LQT2 and LQT3 patients.[38]

ELECTROCARDIOGRAPHIC ABNORMALITIES

The bizarre electrocardiogram of many LQTS patients should be easily recognized (Fig. 34-1). Clearly, there is

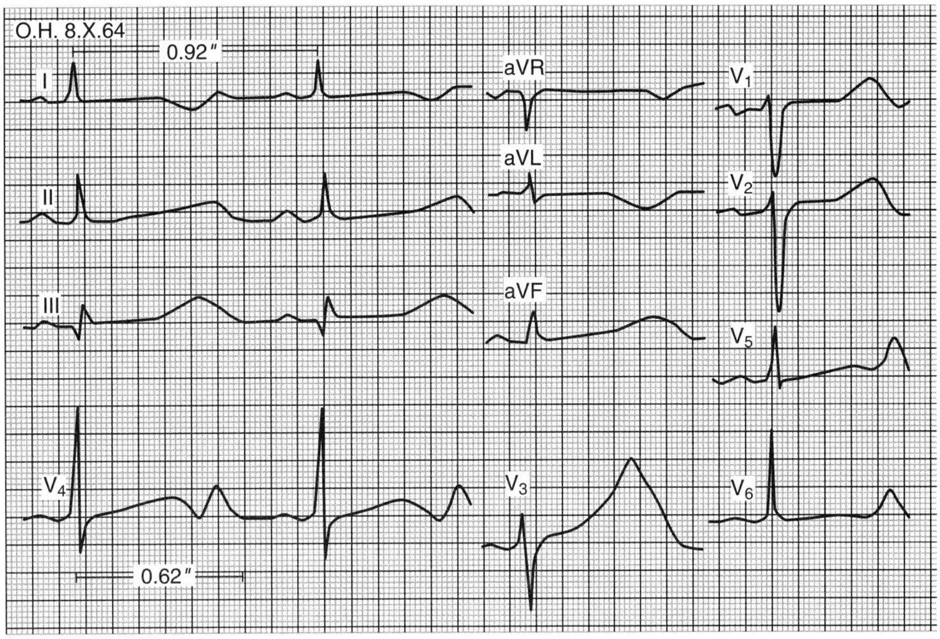

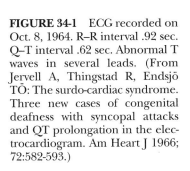

FIGURE 34-1 ECG recorded on Oct. 8, 1964. R–R interval .92 sec. Q–T interval .62 sec. Abnormal T waves in several leads. (From Jervell A, Thingstad R, Endsjö TÖ: The surdo-cardiac syndrome. Three new cases of congenital deafness with syncopal attacks and QT prolongation in the electrocardiogram. Am Heart J 1966; 72:582-593.)

much more than a mere prolongation of ventricular repolarization. The T wave has several morphologic patterns easily recognizable on the basis of clinical experience; they are difficult to quantify but very useful for diagnosis. Moreover, gene-specific ECG patterns appear to exist,[42,43] as will be discussed later. LQT1 patients tend to have smooth, broad-based T waves, whereas LQT2 patients frequently have low amplitude and notched T waves; LQT3 patients have a more distinctive pattern characterized by a late onset of the T wave. However, these morphologic differences are not always so well defined; there is some degree of overlap, and in some families extreme heterogeneity of T wave morphology may be observed (Fig. 34-2).

QT Duration

The Bazett's correction for heart rate, despite 70 years of relentless criticism, continues to be a useful clinical tool.[44] Traditionally, QTc values in excess of 440 ms were considered prolonged; however, values up to 460 ms may still be normal among females.[45] The longer QT values in women[45] become evident only after puberty but are absent at birth.[46] The extent of QT prolongation is variable and is not strictly correlated with the likelihood of syncopal episodes, even though the occurrence of malignant arrhythmias is more frequent in patients with very marked prolongations (QTc in excess of 550 to 600 ms).[47,48]

The initial and understandable concept that QT prolongation was the essential cornerstone of LQTS has been challenged. On the basis of theoretical considerations, it was proposed in 1980[49] that some patients might be affected by LQTS and still have a normal Q–T interval on the surface electrocardiogram. The validity of this unorthodox concept, supported early on by anecdotal observations of family members of LQTS patients who had a normal Q–T interval but who nonetheless developed syncopal episodes, has now been fully validated by the existence of gene carriers with a normal Q–T interval. In 1989, data from the International Registry for LQTS indicated that 6% of the LQTS family members with a QTc less than 440 ms had syncope or cardiac arrest.[50] Similarly, the report by Garson and colleagues[48] on 287 LQTS patients indicated that 6% of them had a normal QTc. Vincent and colleagues[51] demonstrated in three families that some gene carriers had a QTc less than 440 on the first recorded ECG. More recently, while studying nine genotyped families with an apparently sporadic proband (all other family members had a normal Q–T interval), we found that in five of these families several family members were "silent" gene carriers; penetrance was as low as 17%.[52] Finally, our recent analysis of 401 patients, all symptomatic and directly genotyped, revealed that 8% of them have a QTc less than 450 ms. Thus, variable expressivity and low penetrance is present in LQTS as hypothesized 20 years ago.[49]

This concept has important practical and medicolegal implications because, e.g., it no longer allows a cardiologist to state that a sibling of an affected patient with a normal QTc *is definitely not affected by LQTS.*

T Wave Morphology

In LQTS, not only the duration of repolarization is altered but also its morphology. In its most typical presentation, the T wave may be biphasic or notched (Fig. 34-3), suggesting regional differences in the time course of ventricular repolarization. These abnormalities are particularly evident in the precordial leads and contribute to the diagnosis of LQTS; they often are more immediately striking than the sheer prolongation of the Q–T interval.

Compared to healthy individuals of the same age and sex, the LQTS patients have biphasic or notched T waves more frequently (62% versus 15%, $P < .001$).[53,54]

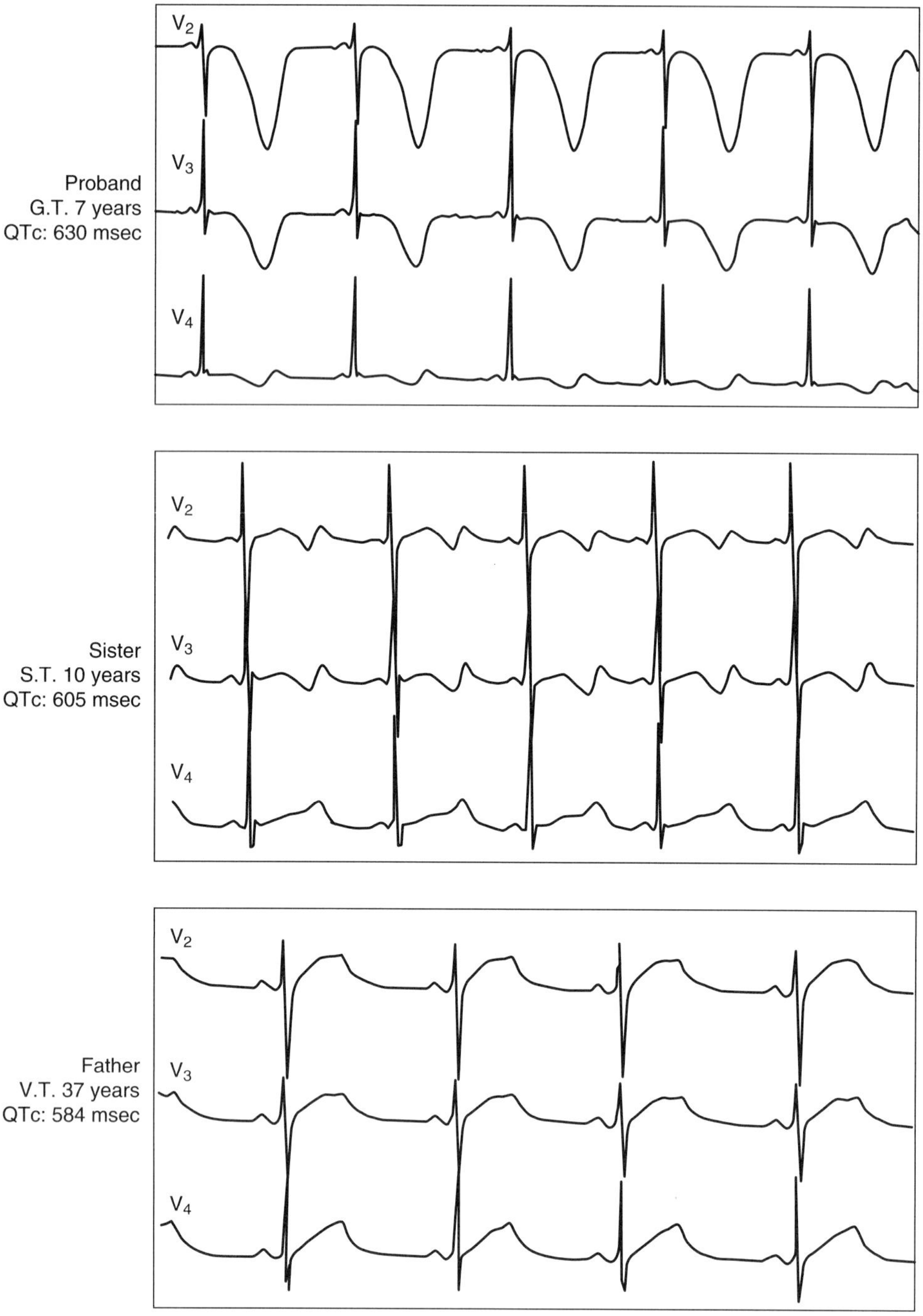

FIGURE 34-2 Different T wave morphologies in affected members of the same family. The proband had a documented cardiac arrest as first manifestation of LQTS. His sister is still asymptomatic, whereas his father has had two syncopal episodes. (From Schwartz PJ, Priori SG, Napolitano C: The long QT syndrome. In Zipes DP, Jalife J (eds): Cardiac Electrophysiology. From Cell to Bedside, 3rd ed. Philadelphia, WB Saunders, 2000, pp 597-615.)

When these patterns are present in control subjects, they are usually limited to leads V2 and V3, whereas among LQTS patients, they are usually visible from lead V2 to V5 with a predominant prevalence in leads V3 and V4. The presence of these repolarization abnormalities is more frequent in LQTS patients with cardiac events (81% versus 19%, $P < .005$). Of diagnostic importance, the appearance of notched T waves in the recovery phase of exercise is markedly more frequent (85% versus 3%, $P < .0001$) among patients than among healthy controls.

T Wave Alternans

Beat-to-beat alternation of the T wave, in polarity or amplitude, may be present at rest for brief moments but most commonly appears during emotional or physical stresses (Fig. 34-4) and may precede TdP. It is a marker of major electrical instability and identifies patients at particularly high risk.

In 1975, we proposed that T wave alternans represents the second characteristic electrocardiographic feature of LQTS.[55] The transient nature of T wave alternans

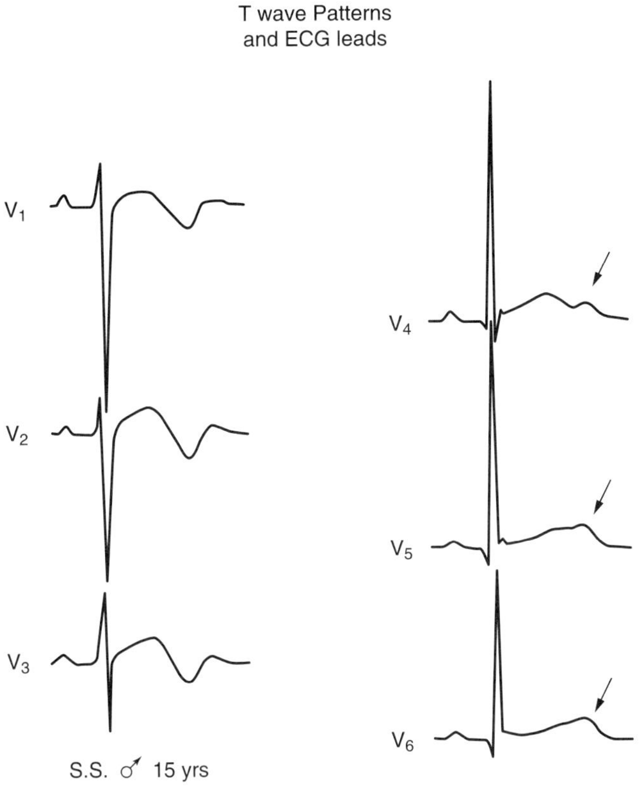

FIGURE 34-3 Example of different T wave morphologies across the precordial leads in the same LQTS patient. In leads V2-V4, the T wave is distinctly biphasic, whereas in V5 and V6 a clear notch is visible, as indicated by the *arrow*, which produces a TU complex. (From Malfatto G, Beria G, Sala S, et al: Quantitative analysis of T wave abnormalities and their prognostic implications in the idiopathic long QT syndrome. J Am Coll Cardiol 1994;23:296-301.)

limits the possibility of its observation, and its absence even in a 24-hour Holter recording does not exclude its occurrence under different circumstances in the same patient. T wave alternans may appear with a diversity of morphologies (see Figs. 34-4 and 34-5). This is a rather gross phenomenon that should not go unnoticed, when present.

Sinus Pauses

Several LQTS patients have sudden pauses in sinus rhythm exceeding 1.2 seconds that are not related to sinus arrhythmia[49] and that may contribute to the initiation of arrhythmias in LQTS patients. In several patients, these pauses are usually followed by the appearance of a notch on the T wave, and it is mostly from these notches that repetitive ventricular beats take off (Fig. 34-6).[56] Of note, these notches can be abolished by LCSD.[56] Often the pauses precede the onset of TdP.

Heart Rate

In 1975 Schwartz and associates[1] called attention to the presence of a lower-than-normal heart rate in most patients, a phenomenon particularly striking in children. This is a somewhat ubiquitous finding.[57] During exercise, most LQTS patients reach a heart rate level lower than that achieved by healthy controls matched by age and sex.[58]

Whether such an abnormality represents a genetically controlled developmental alteration in cardiac innervation or, more likely, a not yet defined interaction between the ion channel abnormalities produced by the various mutations and adrenergic receptors in the sinus node, remains speculative.

Molecular Diagnosis in LQTS

MOLECULAR GENETICS AND RISK STRATIFICATION

Risk stratification may become an important contribution of molecular genetics to patient management. In LQTS, the large heterogeneity of mutations within each disease-related gene has so far prevented the ability to extrapolate prognostic information and to define the risk of sudden death. A possible exception exists: It has been suggested that mutations in the C-terminal part of the LQTS genes may produce less severe forms of the disease compared with mutations occurring in the pore region.[59] In contrast, risk stratification appears possible in a gene-specific manner.

Data from the International Registry on 246 gene carriers show that LQT1 and LQT2 gene carriers have more cardiac events than LQT3 gene carriers. However, as LQT3 patients have the same death rate as LQT1 and LQT2 patients, lethality appears to be higher in LQT3.[60] In another set of data based on more than 800 symptomatic patients of known genotype, we found that the J-LN patients are those with the earliest occurrence of cardiac events followed by LQT1, LQT2, and LQT3.[38,61] By ages 5 and 10, 85% of J-LN and 61% of LQT1, respectively, have had their first syncope. In contrast, by age 10 this occurred in only 25% of LQT2 patients. From the Pavia database[62] on 647 genotyped patients, it appears that a large portion of LQT1 patients may remain asymptomatic through life, partly because in this group the percentage of "silent gene carriers" is unusually high, almost 40%. These data have allowed us to identify subgroups at higher risk, for instance, all patients with a QTc > 500 ms and LQT2 females.

DIAGNOSIS OF ASYMPTOMATIC PATIENTS

A major impact of molecular diagnosis is the identification of asymptomatic gene carriers, particularly when they have a normal Q–T interval. The percentage of LQTS asymptomatic gene carriers who will develop symptoms if left untreated is still unknown. The management of these individuals poses special problems, as discussed in the section on therapy. Even if one decides to leave gene carriers untreated, knowledge of their status will be useful to favor lifestyle changes, avoiding stressful conditions, and strenuous exercise. The information of being a gene carrier is also valuable for parental planning.

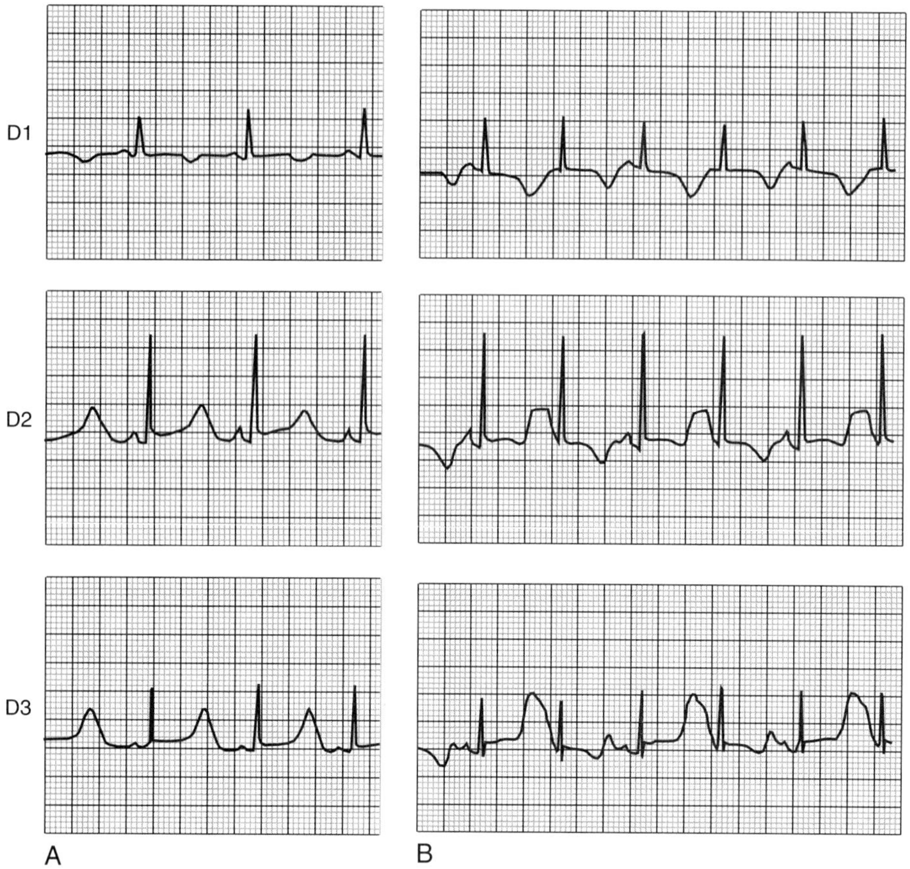

FIGURE 34-4 **A**, Control conditions, QTc .61. **B**, During fright. Tracings in **B** are simultaneous. Alternation of the T wave, in amplitude (D1) and in polarity (D2 and D3), is evident. (From Schwartz PJ, Malliani A: Electrical alternation of the T-wave: Clinical and experimental evidence of its relationship with the sympathetic nervous system and with the long QT syndrome. Am Heart J 1975;89:45-50.)

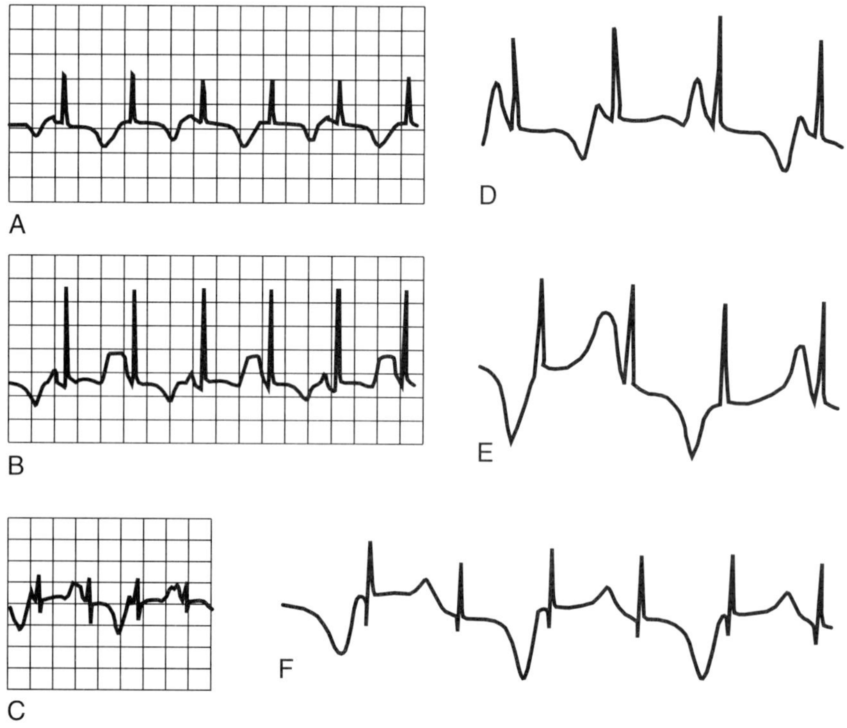

FIGURE 34-5 Examples of T wave alternans in LQTS patients. **A** and **B**, 9-year-old girl: alternation of T wave occurred during an unintentionally induced episode of fear. **C**, 3-year-old boy. **D** and **E**, 14-year-old girl 1 minute (**D**) and 3 minutes (**E**) after induced fright. **F**, 7-year-old girl: T wave alternans occurred during exercise. (**A-C**, Modified from Schwartz PJ, Malliani A: Electrical alternation of the T-wave: Clinical and experimental evidence of its relationship with the sympathetic nervous system and with the long QT syndrome. Am Heart J 1975;89:45-50; **D** and **E**, Modified from Fraser GR, et al: Q J Med, 1964;33:361-385; **F**, Modified from Jervell A, Adv Intern Med 1971;17: 425-438.)

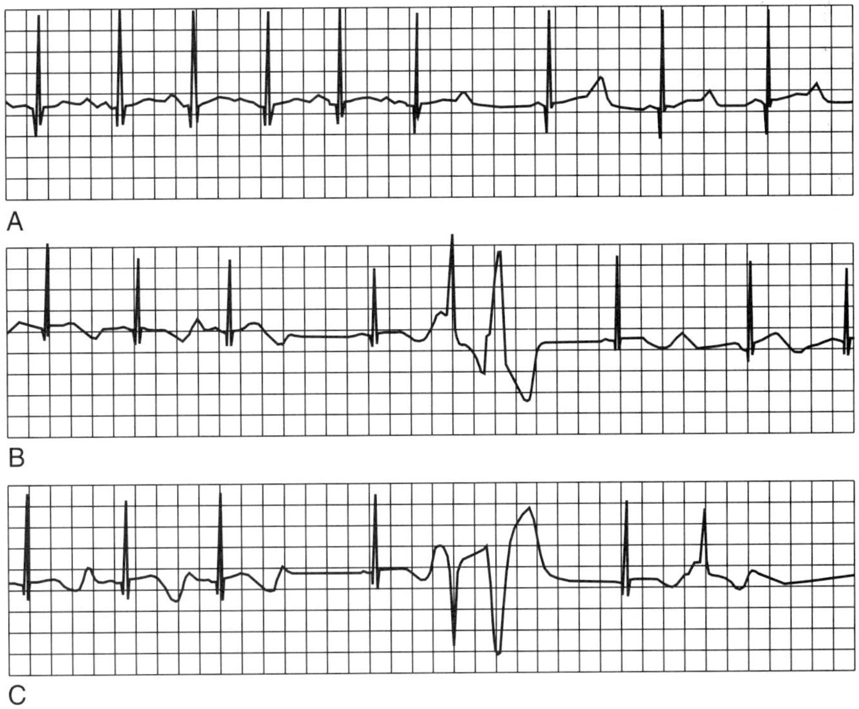

FIGURE 34-6 These strips are taken from a Holter recording obtained in the absence of therapy. They show the changes in the T wave morphology induced by sinus pauses (**A**), and the presence of ventricular arrhythmias coincident with the end of the T wave (**B** and **C**). (From Malfatto G, Rosen MR, Foresti A, Schwartz PJ: Idiopathic long QT syndrome exacerbated by β-adrenergic blockade and responsive to left cardiac sympathetic denervation: Implications regarding electrophysiologic substrate and adrenergic modulation. J Cardiovasc Electrophysiol 1992;3:295-305.)

GENOTYPE-PHENOTYPE CORRELATION AND THERAPEUTIC IMPLICATIONS

There has been a flourishing of genotype-phenotype correlation studies and some have been better developed or are of special clinical interest.

Gene-Specific ECG patterns. In 1995, and based on only six families, Moss and colleagues proposed that the three major LQTS genes were producing different electrocardiographic phenotypes.[42] This led to a large-scale study involving almost 300 gene carriers, which confirmed the existence of typical ST-T wave patterns in the majority of genotyped LQTS patients, thereby allowing correct identification of individual patients in approximately 80% of cases.[43] The enthusiasm for these percentages is tempered by the reminder that 90% of genotyped LQTS families belong to one of two subgroups (LQT1 or LQT2); thus, even by blind picking one would be right 50% of the time. This study also showed that typical patterns in family-grouped ECGs are most effective in correctly identifying the genotype of families with a known genotype.

The ECG morphology may be useful in suggesting to look first for mutations on a certain gene, thus saving time for the molecular diagnosis. However, ECG morphology does not represent a valid surrogate for actual genotyping, and it should not be used to make a molecular diagnosis.

Gene-Specific QT-Shortening with Na^+ Channel Blockers. Within 1 month from the identification in LQTS patients of mutations in *SCN5A*, the cardiac Na^+ channel gene, we attempted to shorten the Q–T interval in LQT3 patients by using the Na^+ channel blocker mexiletine.[26,63] The rationale for this approach was strengthened by the evidence that most *SCN5A* mutations increase a late Na^+ inward current,[23,24] that this delayed inactivation can be antagonized by mexiletine,[24] and that in the first experimental model for LQT2 and for LQT3, mexiletine effectively counteracts the action potential prolongation produced by interventions mimicking the effects of mutations on *SCN5A*.[64] Our preliminary results indicate that mexiletine shortens the Q–T interval in all subgroups but with significant (and gene-specific) differences. The shortening in LQT1 and LQT2 patients averages 30 ms, and it has no clinical significance, whereas in LQT3 patients it averages 90 ms, which is of potential clinical significance (Fig. 34-7).

On this basis we have cautiously begun to add mexiletine to β-blockers in the management of LQT3 patients. QT shortening, albeit encouraging, could represent just a cosmetic effect and does not imply at all a protective effect. Mexiletine should not be used as sole therapy in LQTS patients; at this time, we can draw no conclusions from the preliminary data available, but we are already aware of its failures to afford protection to some patients, even in combination with β-blockers.

It has been suggested that another Na^+ channel blocker, flecainide, might be even more effective than mexiletine in shortening the Q–T interval of LQT3 patients.[65,66] Because flecainide is also used as a somewhat specific tool to unmask the ECG phenotype of the Brugada syndrome, another disease involving the *SCN5A* gene, we explored its presumed specificity in 13 LQT3 patients.[67] We found that flecainide shortened QTc by a mean of 31 ms in 12 of 13 patients but that also produced a Brugada-like ST segment elevation in leads V1 through V3 (≥2 mm) in 6 of the 13 patients. This finding raises concerns and calls for caution before considering initiation of long-term treatment with flecainide in LQT3 patients.

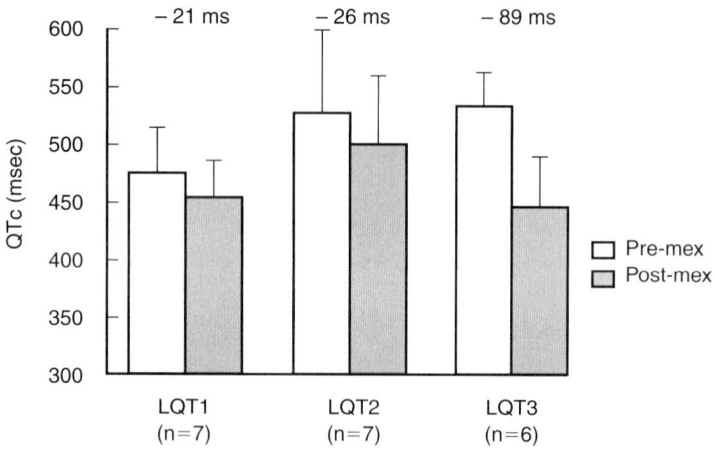

FIGURE 34-7 QTc values in control condition and during acute oral drug testing with mexiletine in LQT3 (linked to chromosome 3, n = 8), in LQT2 (linked to chromosome 7, n = 7), and in LQT1 (linked to chromosome 11, n = 13) patients. Data are expressed as mean ± SD. Mex, mexiletine. (Modified from Schwartz PJ, Priori SG, Locati EH, et al: Long QT syndrome patients with mutations on the *SCN5A* and *HERG* genes have differential responses to Na+ channel blockade and to increases in heart rate. Implications for gene-specific therapy. Circulation 1995;92:3381-3386.)

Gene-Specific Changes in QT Adaptation to Heart Rate Changes. The Q–T interval adapts to changes in cycle length by shortening when heart rate increases. We compared to control individuals the responses of patients belonging to the three main subgroups. Several LQT1 patients shortened their QT much less than normal—as expected given their reduced I_{Ks} current—but there is a significant overlap with normal individuals. Heart rate increases surprisingly produce a somewhat marked shortening of the Q–T interval among LQT3 patients; in agreement with our experimental observations in isolated myocytes,[64] this effect is much less evident among LQT2 patients (Fig. 34-8).

We examined the same question from a different perspective and determined the night-day change in QTc in the three genotypes.[68] While LQT1 patients had no obvious change, among LQT2 patients there was a trend toward a nighttime increase, which became clear and significant among LQT3 patients (Fig. 34-9). A reasonable inference from these two observations is that LQT3 patients might be at lesser risk of syncope during physical exercise, when the progressive heart rate increase may allow appropriate QT shortening, but could be at higher risk at nighttime when their heart rate decreases. For the same reason they might also be less effectively protected by β-blockers that could produce an excessively slow heart rate and could prevent an adequate heart rate increase during exercise. The next section provides validation for these inferences.

Gene-Specific Triggers for Life-Threatening Arrhythmias. When we first suggested that a gene-specific difference appeared to exist in relation to the conditions (triggers) associated with cardiac events in LQTS,[26] we also indicated the need to validate this intriguing hypothesis in a large population of genotyped and symptomatic patients. Thanks to a large cooperative effort, we did

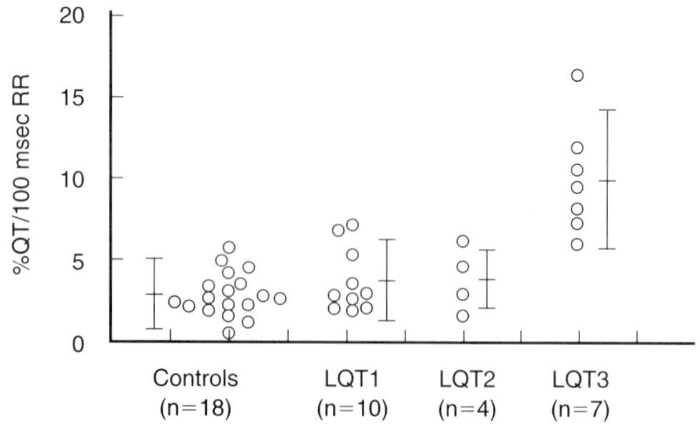

FIGURE 34-8 Individual and mean values (±SD) of the response to heart rate increase in controls (n = 18) and in LQT3 (linked to chromosome 3, n = 7), LQT2 (linked to chromosome 7, n = 4), and in LQT1 (linked to chromosome 11, n = 10) patients. Data are expressed as percent QT shortening for each 100 ms reduction in R–R interval. (Modified from Schwartz PJ, Priori SG, Locati EH, et al: Long QT syndrome patients with mutations on the *SCN5A* and *HERG* genes have differential responses to Na+ channel blockade and to increases in heart rate. Implications for gene-specific therapy. Circulation 1995;92:3381-3386.)

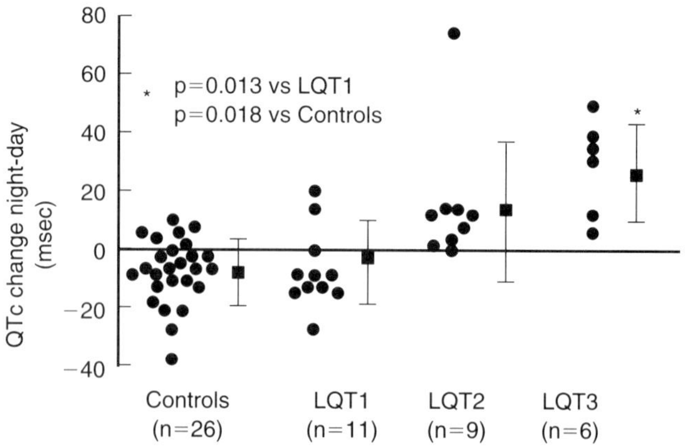

FIGURE 34-9 Difference between QTc during sleep and QTc during wakefulness in healthy controls, LQT1, LQT2, and LQT3 patients. Individual values (*open circles*) and mean ±1 standard error of QTc changes in each group are shown. (From Stramba-Badiale M, Priori SG, Napolitano C, et al: Gene-specific differences in the circadian variation of ventricular repolarization in the long QT syndrome: A key to sudden death during sleep? Ital Heart J 2000;1:323-328.)

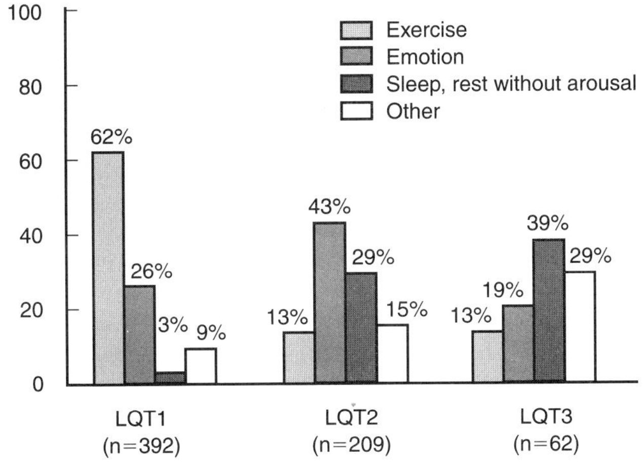

FIGURE 34-10 Triggers for cardiac events according to the three genotypes. Numbers in parenthesis are triggers, not patients. (From Schwartz PJ, Priori SG, Spazzolini C, et al: Genotype-phenotype correlation in the long QT syndrome. Gene-specific triggers for life-threatening arrhythmias. Circulation 2001;103:89-95.)

gather adequate information on 670 genotyped and symptomatic LQTS patients and met this objective.[38]

We identified three main factors associated with cardiac events (syncope, cardiac arrest, or sudden death): exercise, emotions, and events occurring during either sleep or at rest without arousal. Striking differences became evident among the three subgroups. LQT1 patients had most of their cardiac events during exercise (62%) and very few (3%) during rest or sleep. As shown in Figure 34-10, these percentages were almost reversed among LQT2 and LQT3 patients, who were less likely to have cardiac events during exercise (both 13%) and more likely to have them during rest/sleep (29% and 39%). These differences are further accentuated when the analysis is limited to lethal events (Fig. 34-11). The relative similarities between LQT2 and LQT3 probably reflect the fact that both groups have a preserved I_{Ks} current, and this facilitates QT adaptation during heart rate increases. Highly specific triggers also emerged. Swimming is a cause of cardiac events almost solely among LQT1 patients (99%), whereas auditory stimuli are very often (80%) responsible for events among LQT2 patients.

The fact that homozygous mutations on *KvLQT1* or on *minK* are responsible for the J-LN has prompted us to examine whether or not the same triggering factors play an important role for LQT1 and J-LN patients.[61] Our data on 136 such patients indicate that emotion and exercise together account for most of the events and that only 3% of them occur at sleep/rest. This figure is remarkably similar to that of 3% observed in 371 LQT1 patients. This finding represents a further support for the use of β-blockers in LQT1 and J-LN patients.

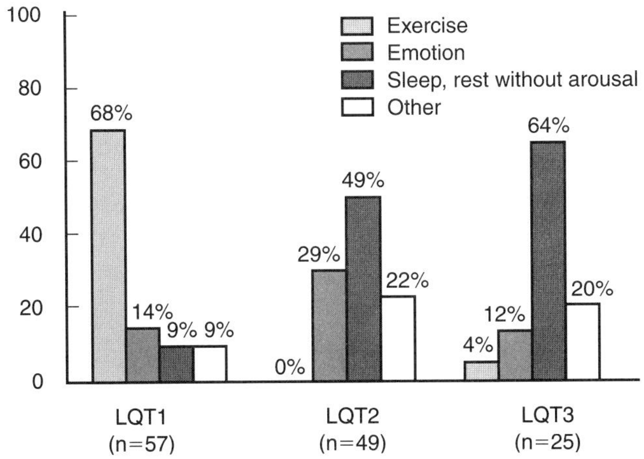

FIGURE 34-11 Lethal cardiac events according to the three classified triggers in the three genotypes. (From Schwartz PJ, Priori SG, Spazzolini C, et al: Genotype-phenotype correlation in the long QT syndrome. Gene-specific triggers for life-threatening arrhythmias. Circulation 2001;103:89-95.)

INCOMPLETE PENETRANCE AND VARIABLE EXPRESSIVITY

This issue has already been mentioned in reference to QT prolongation. The 1980-1985 hypothesis[49,69] that some patients with LQTS could present a normal Q–T interval duration has been definitely demonstrated by the molecular evidence, in several families with only one individual clinically affected, of several gene-carriers with a normal Q–T interval.[52] In these families the penetrance of LQTS ranged between 17% and 45%, suggesting that, at least in some families, for each patient identified with clinical methods there are two to four family members who are currently inappropriately reassured of not being affected by LQTS (Fig. 34-12).

PROARRHYTHMIA AS A MANIFESTATION OF A "FORME FRUSTE" OF LQTS

The existence of incomplete penetrance of LQTS mutations suggests that individuals who carry silent mutations may not present the phenotype of LQTS and yet they may be highly susceptible to ventricular arrhythmias under specific circumstances. Indeed, some of the patients who develop drug-induced TdP might be carriers of such a silent mutation, and we have already provided evidence for this possibility.[70] A similar observation has been reported by Schulze-Bahr and colleagues.[71] These observations confirm the hypothesis proposed by Moss and Schwartz in 1982 that a significant number of drug-induced TdP might be due to a "forme fruste" of LQTS.[72]

Mortality

Mortality is high among the untreated symptomatic patients. In the first review of 203 cases of LQTS,[1] mortality was approximately 5% per year, during an estimated average follow-up of about 5 years. This represented an early survey of a large group of high-risk patients (high proportion of congenital deafness and symptomatic patients, *mostly not treated*) and was limited by the absence

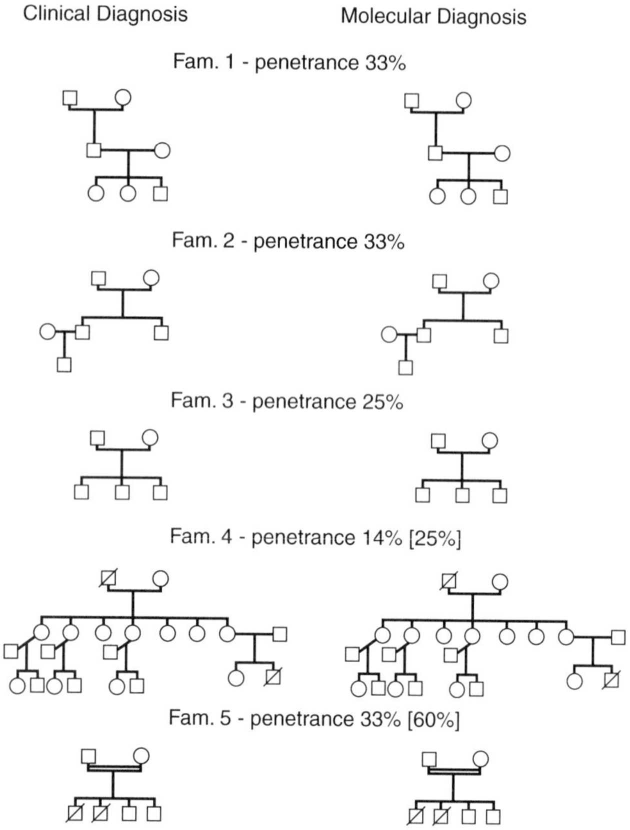

FIGURE 34-12 Family trees of five families in whom clinical evaluation (*left side*) led to identification of proband as the only affected member within each family. The results of the molecular screening are depicted on the *right side*. *Solid symbols* represent the affected individuals identified with the two approaches. The *hatched symbols* represent premature (<35 years) sudden death. Family 5 has been described in detail elsewhere.[31] Two values of penetrance are shown for families 4 and 5: value *within brackets* includes premature sudden deaths. (From Priori SG, Napolitano C, Schwartz PJ: Low penetrance in the long QT syndrome. Clinical impact. Circulation 1999;99:529-533.)

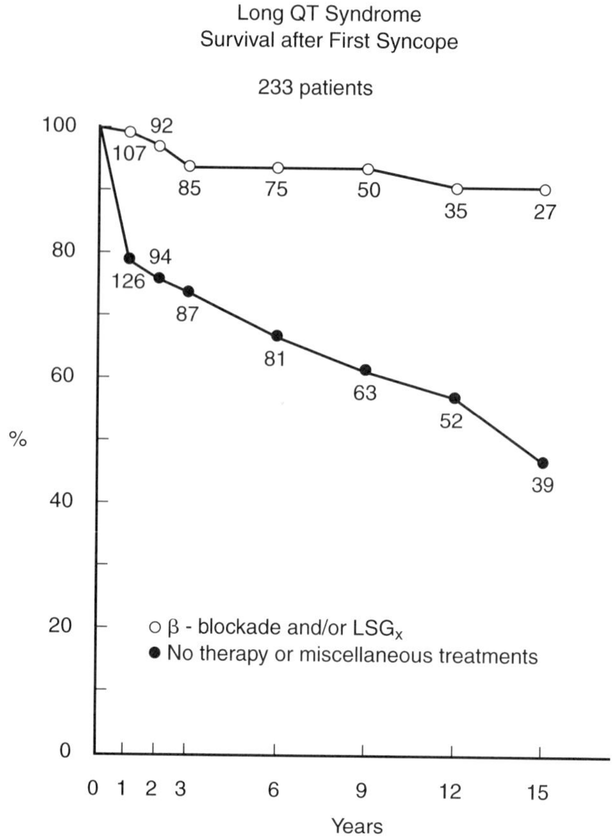

FIGURE 34-13 Effect of therapy on the survival, after the first syncopal episode, of 233 patients affected by the congenital LQTS. The protective effect of β-adrenergic blockade and of left stellectomy (LSGx) is evident. The mortality 3 years after the first syncope is 6% in the group treated with antiadrenergic interventions and 26% in the group treated differently or not treated. Fifteen years after the first syncope, the respective mortality rates are 9% and 53%. (From Schwartz PJ: Idiopathic long QT syndrome: Progress and questions. Am Heart J 1985;109:399-411.)

of precise information on the time of exposure; i.e., the time elapsed from the first syncopal episode.

This limitation was overcome in a subsequent analysis based on 126 patients for whom the exact date of the first syncope was known and who were either untreated or received nonantiadrenergic therapy.[73] Figure 34-13 shows that within 1 year after the first syncope more than 20% of these patients died suddenly, and mortality exceeded 50% at 15 years. The average age at first syncope was 14 years; thus, this is largely a population of otherwise healthy children and teenagers that have a very poor prognosis, *if left without proper treatment.* These data are unique because they will remain our only information about the natural history of this disease among symptomatic patients. Most of these patients came to medical attention in the late 1960s or early 1970s, when the use of β-blockers was not yet widely established. Today, this type of information can no longer be obtained; moreover, a fairly large number of low-risk patients are diagnosed, and this provides a different mortality picture.

Diagnosis

Given the characteristic features of LQTS, the typical cases present no diagnostic difficulty for the physicians aware of the disease. However, borderline cases are more complex and require the evaluation of multiple variables besides clinical history and surface electrocardiogram. To overcome these difficulties, diagnostic criteria were first proposed in 1985[69] and were subsequently updated in 1993.[74]

The new diagnostic criteria are listed in Table 34-1, with relative points assigned to various electrocardiographic, clinical, and familial findings. The score ranges from a minimum value of 0 and a maximum value of 9 points. Based on our experience, we have arbitrarily divided the point score into three probability categories: less than or equal to 1 point = low probability of LQTS; greater than 1 to 3.5 points = intermediate probability of LQTS; and greater than or equal to 4 points = high probability of LQTS. As QTc overcorrects at fast heart rates, additional diagnostic caution is necessary when

TABLE 34-1 1993 LQTS Diagnostic Criteria

Electrocardiographic Findings*		Points
A. QTc†	>480 ms½	3
	460-470 ms½	2
	450 (male) ms½	1
B. Torsade de pointes‡		2
C. T wave alternans		1
D. Notched T wave in 3 leads		1
E. Low heart rate for age§		0.5
Clinical History		
A. Syncope‡	With stress	2
	Without stress	1
B. Congenital deafness		0.5
Family History‖		
A. Family members with definite LQTS¶.		1
B. Unexplained sudden cardiac death below age 30 among immediate family members.		0.5

*In the absence of medications or disorders known to affect these electrocardiographic features.

†QTc calculated by Bazett's formula, where QTc = QT/√RR.

‡Mutually exclusive.

§Resting heart rate below the 2nd percentile for age.

‖The same family member cannot be counted in A and B.

¶Definite LQTS is defined by an LQTS score >4.

SCORING

<1 point = low probability of LQTS.

2 to 3 points = intermediate probability of LQTS.

>4 points = high probability of LQTS.

From Schwartz PJ, Moss AJ, Vincent GM, Crampton RS: Diagnostic criteria for the long QT syndrome: An update. Circulation 1993;88:782-784.

dealing with tachycardic patients or infants with fast heart rates.

When a patient receives a score of 2 to 3 points, based on the Q–T interval measurement, serial ECG records should be obtained because the QTc value in LQTS patients may vary from time to time. In this group with an intermediate probability of LQTS, the presence of the more recently described abnormalities, such as excessive QT dispersion, abnormal body surface mapping, appearance of notched T waves during recovery from exercise, or echocardiographic abnormalities[75,76] may help the physician in the diagnostic decisions. It should be remembered that symptoms often appear in the first few years of life, may resemble epileptic convulsions,[37] and could be mistaken for hysteric reactions[37]; e.g., following a stressful moment at school or a parental punishment.

Any set of diagnostic criteria involving a quantitative score has an unavoidable arbitrary component. Nonetheless, these new diagnostic criteria provide a uniform diagnostic standard. The unusual combination of an often lethal disease for which effective therapies exist and of a rather elementary diagnosis makes inexcusable the existence of undiagnosed and, therefore, untreated patients.

Relation to Sudden Infant Death Syndrome

The Sudden Infant Death Syndrome (SIDS) remains the leading cause of sudden death during the first year of life in the Western world.[77,78] Despite a large number of theories mostly focused on abnormalities in the control of respiratory or cardiac function,[78] the causes of SIDS remain unknown. In 1976, I proposed that an undefined number of SIDS victims might die because of an arrhythmic death favored by a prolongation of the Q–T interval.[79] The mechanism proposed originally involved a developmental imbalance in cardiac sympathetic innervation.

To test this hypothesis we determined prospectively the Q–T interval during the first week of life in 34,442 newborns and followed them for the possible occurrence of SIDS. The study lasted 19 years, and 1-year follow-up data were available for 33,034 infants.[80] There were 34 deaths, of which 24 were due to SIDS. The infants who died of SIDS had a longer QTc (measured blindly to the outcome) than the survivors and the victims from other causes. Moreover, 12 of the 24 SIDS victims but none of the other dead infants had a prolonged QTc (defined a priori as exceeding 440 ms). To avoid the potential pitfalls of the Bazett's correction with fast heart rates, we also determined the absolute Q–T intervals for similar cardiac cycle lengths and found that 12 of the 24 SIDS victims had a QT value exceeding the 97.5 percentile for the study group as a whole. The odds ratio for SIDS for infants with a prolonged QTc was 41, reaching 47 for male infants.

The unavoidable conclusion of the study was that a QT prolongation in the first week of life represents a major risk factor for SIDS. Besides significant implications specific to SIDS, such as the issue of considering routine neonatal ECG screening and the potential value of prophylactic β-blockade for the infants with a prolonged QT, the study also raised the question about the cause(s) of QT prolongation in infancy. Possible mechanisms could involve a developmental[79] or genetic[81] alteration in cardiac sympathetic innervation, or a genetic abnormality of ion channel genes. The parents of the SIDS victims all had a normal Q–T interval, thus ruling out a traditional LQTS family history.

Two recent cases shed new light on this issue. The first was a typical case of "near-miss" for SIDS: a 2-month-old infant found apneic and pulseless by the parents who rushed him to a nearby hospital where an ECG showed the presence of ventricular fibrillation and, after defibrillation, a markedly prolonged Q–T interval.[82] The diagnosis of LQTS was followed by molecular screening, which revealed the presence of a *de novo* mutation in *SCN5A* in the infant but not in the parents (paternity confirmed). If this infant had died, a certainty with ventricular fibrillation, the normal QT of both parents would have prevented even the suspicion of LQTS. The second was a 4-month-old SIDS victim who died without an ECG. Postmortem molecular diagnosis revealed the presence of another *de novo* mutation, this time on *KvLQT1*; the same mutation is present in one of the LQTS families followed by our center in Pavia.[83]

These two cases represent *proof of concept* for the hypothesis[79] that a number of SIDS cases may actually be cases of LQTS. This implies that these infants could very probably be saved if they could be identified and treated early on. This could be accomplished by performing an ECG in newborns before the end of the first month of life. This molecular evidence for LQTS obtained postmortem in "SIDS victims" also carries significant medicolegal implications.[83]

Ongoing molecular studies on large numbers of SIDS victims suggest that the prevalence of apparent SIDS deaths actually due to LQTS is likely to be in the range of 10% to 15%. The extremely important practical aspect lies in the fact that these infants could be identified by an ECG performed before the end of the first month of life and thus saved by appropriate preventive therapy.[84]

Finally, there is growing evidence linking stillbirths to LQTS.[85]

Therapy

The trigger for most episodes of life-threatening arrhythmias of LQTS is represented by a sudden increase in sympathetic activity. Indeed, antiadrenergic therapies provide the greatest degree of protection. However, some patients have syncopal episodes while being asleep or at rest and, in some cases, the arrhythmias are pause dependent.

ANTIADRENERGIC INTERVENTIONS

The most significant information on therapy still comes from the 1985 study discussed in the section on mortality.[73] That analysis included 233 symptomatic patients with detailed clinical information on the time of first syncope and with an adequate follow-up. Figure 34-13 shows the dramatic change in survival produced by pharmacologic or surgical antiadrenergic therapy when compared to any other therapy or no treatment.[69] The mortality at 15 years after the first syncope was 9% in the group treated by antiadrenergic therapy (β-blockers and/or sympathectomy), and more than 53% in the group not treated or treated by miscellaneous therapies not including β-blockers. These data do not distinguish between the efficacy of β-blockade and that of left cervical sympathetic denervation (LCSD). Nonetheless, they conclusively demonstrated that pharmacologic and/or surgical antiadrenergic therapy radically modifies prognosis for symptomatic patients with LQTS.

β-Adrenergic Blockade

β-Adrenergic blocking agents still represent the first choice therapy in symptomatic LQTS patients, unless specific contraindications are present. Propranolol is still the most widely used drug, at a full daily dosage of 3 mg/kg, which sometimes is increased to 4 mg/kg. The main advantages of propranolol are its lipophilicity, which allows it to cross the blood-brain barrier, and its well-known tolerability for chronic therapy. Its main disadvantages are the need for multiple daily administrations and the contraindications for patients with asthma and diabetes. We use nadolol with increasing frequency because its longer half-life allows twice daily administration at a daily dose slightly lower than that of propranolol. This increases compliance, particularly when dealing with teenagers who may easily forget to take their medication in the afternoon.

There is, however, a minority of LQTS patients who do not tolerate β-blockers because of excessive bradycardia or in whom a very low heart rate raises concerns for the possibility of favoring EAD-induced TdP. This group is likely to include most LQT3 patients and others as well. For these patients the association between β-blockers and cardiac pacing becomes appropriate and necessary.

The response to β-blockers is not uniform among genetic subgroups (Table 34-2). Thus, β-blockers are extremely effective for LQT1 patients.[38] The impairment in the I_{Ks} current makes these patients particularly sensitive to catecholamines[86] and quite responsive to β-blockade. They may seldom need more than antiadrenergic therapy. Despite a higher recurrence rate, LQT2 patients also respond well to β-blockers, whereas the LQT3 patients appear insufficiently protected.

Left Cardiac Sympathetic Denervation

Following a small incision in the left subclavicular region, LCSD is performed by an extrapleural approach, which makes thoracotomy unnecessary. The average time for the complete operation is 35 to 40 minutes.[87] Adequate LCSD requires removal of the first three to four thoracic ganglia. Left stellectomy, which removes only the left stellate ganglion, provides an insufficient denervation. On the other hand, there is no need to ablate the cephalic portion of the left stellate ganglion. In this way, the Horner's syndrome can almost always be avoided.[87] In approximately 30% of patients there is a very modest (1 to 2 mm) ptosis,

TABLE 34-2 Genotype and β-Blocker Therapy

Genotype	No recurrence n (%)	Recurrences n (%)	CA/SCD n (% of all pts; % of pts with recurrences)
LQT1 (n = 162)	131(81)	31(19)	7(4;23)
LQT2 (n = 91)	54(59)	37(41)	4(4;11)
LQT3 (n = 18)	9(50)	9(50)	3(17;33)

CA/SCD, Cardiac arrest/Sudden cardiac death; pts, patients.

From Schwartz PJ, Priori SG, Spazzolini C, et al: Genotype-phenotype correlation in the long QT syndrome. Gene-specific triggers for life-threatening arrhythmias. Circulation 2001;103:89-95.

which can be noted only by close examination but fully escapes notice in normal social interactions.

Following the first analysis in 1991,[87] we have recently reviewed the existing data worldwide.[88]

We identified 147 LQTS patients who underwent LCSD. This was a group of patients at an especially high risk—99% of them were symptomatic and 48% had a cardiac arrest, their average QTc was extremely prolonged (543 ± 65 ms), and 75% of those treated with β-blockers continued to have cardiac events. The mean follow-up after surgery was 8 years. After LCSD, 46% became asymptomatic. Syncope occurred in 31%, aborted cardiac arrest in 16%, and sudden death in 7%. The mean yearly number of cardiac events decreased by 91% ($P < .001$).

As today patients who have survived a cardiac arrest have an indication for ICD implant, the most practically important data are those concerning the patients with syncope only before LCSD. The 5-year mortality in this group was 3%. LCSD shortened QTc duration by an average of 39 ms, thus showing that its antifibrillatory action results not only by interference with the trigger but also by a favorable modification of the substrate ("electrical remodeling"). The patients whose QTc 6 months after surgery is below 500 ms appear to be at extremely low risks for subsequent events, whereas if their QTc remains above 500 ms they remain at risk and an ICD should be considered (Fig. 34-14). Finally, in patients previously implanted with an ICD and receiving multiple discharges, LCSD reduced the count of shocks by 95% ($P = .02$) from a median number of 25 to 0 per patient.

These data show that LCSD is associated with a significant and long-term reduction in the incidence of aborted cardiac arrest and syncope, even though it is not entirely effective in preventing sudden death. LCSD should be considered in all LQTS patients who experience syncope despite β-blockers and in those who have arrhythmia storms and shocks with an ICD. Before deciding on the next therapeutic step in patients with syncope despite β-blockers, the patients and/or their families should have the right to be informed about the long-term benefits and limitations of both ICDs and LCSD.

It is unfortunate that, despite its relative simplicity, LCSD is not performed in many high-risk patients just because the physicians or the thoracic surgeons in their hospitals are not familiar with the procedure. The consequence is that all too often the choice goes to what appears the easiest approach, usually an ICD, even when this choice is not the best for the patient and their quality of life.

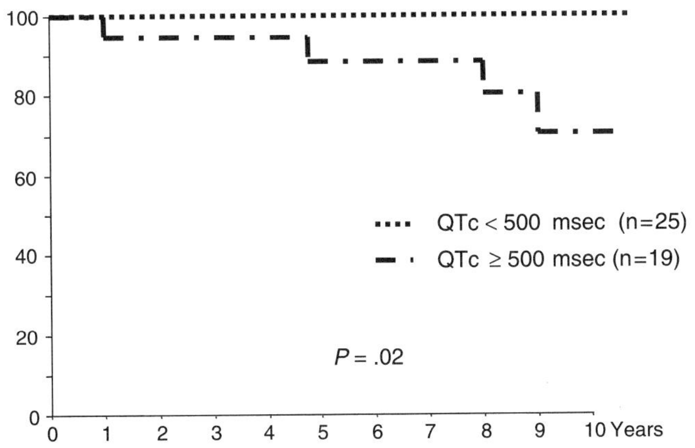

FIGURE 34-14 Kaplan-Meier curves of event-free survival and survival according to post-LCSD QTc in patients with only syncope or aborted cardiac arrest pre-LCSD. (Modified from Schwartz PJ, Priori SG, Cerrone M, et al: Left cardiac sympathetic denervation in the management of high-risk patients affected by the long QT syndrome. Circulation 2004;109: 1826-1833.)

CARDIAC PACING

It has been reported that cardiac pacing in LQTS has a beneficial effect.[89,90] Unfortunately, the fact that pacemaker implantation has been almost always associated with the institution of or with an increase in dosage of β-blocking therapy or with LCSD, makes it difficult to establish the independent therapeutic effect of cardiac pacing. The data by Garson and associates[48] do not support the use of pacing when antiadrenergic therapy fails. Cardiac pacing is clearly indicated in LQTS patients with AV block and whenever there is evidence of pause-dependent malignant arrhythmias. The fact that the onset of TdP is very often preceded by a pause[36] adds to the rationale of using a pacemaker as an adjunct to the therapy of some LQTS patients.

Pacemakers should never be used as a sole therapy for LQTS. They should be regarded as an adjuvant to β-blocking therapy to prevent the occurrence of excessive bradycardia in selected patients. However, in these patients one should also consider LCSD because this selective denervation does not reduce heart rate. The role of pacemakers is further superseded by the availability of ICDs with pacing capabilities, which obviously provide more complete protection.

IMPLANTABLE CARDIOVERTER-DEFIBRILLATORS

There has been a large increase, often not fully justified, in the number of automatic implantable cardioverter defibrillators (ICDs) used in LQTS patients. The strongest rationale for the use of an ICD is when a patient has survived a cardiac arrest or continues to have syncope despite full-dose β-blockade and LCSD.

It should be remembered that the ICD does not prevent the occurrence of malignant arrhythmias and that TdP is frequently self-terminating in LQTS, as indicated by the high incidence of syncopal episodes with spontaneous recovery in LQTS patients. The recurrence of numerous ICD discharges during the same day has led to suicidal attempts in teenagers. Examples of questionable use of ICD in LQTS patients have been recently presented and discussed.[91] The massive release of catecholamines, triggered by pain and fear, that follows an ICD discharge in a conscious patient leads to further arrhythmias and to further discharges that produce a dramatic vicious circle. To overcome these limitations we have very recently designed a new algorithm for an ICD that is specifically tailored for the young LQTS patients, and that is currently under clinical evaluation. We have allowed, prior to the discharge, sufficient time for the possibility of a spontaneous return to sinus rhythm after the onset of the arrhythmia and have instituted automatic and specific measures, such as a period of relatively rapid pacing, to prevent the reinitiation of life-threatening arrhythmias favored by the release of catecholamines due to the ICD shock.

Unless a cardiac arrest has occurred, special caution is necessary before choosing an ICD for a child with LQTS. Such a decision is seldom justified before a proper trial with combined antiadrenergic therapy, i.e., full-dose β-blockade and LCSD, which prevents sudden death in 96% to 97% of symptomatic patients.[69,73,88] However, the nature of the disease is such that cardiac arrest may recur with lethal consequences. Thus, there are concerns for the life of the patient and medicolegal considerations. The risk-benefit ratio of an ICD should be clearly explained to the patient or to the parents, including the need for several battery replacements, which constitutes a significant problem with young patients. Both in the United States and in Europe, registries for LQTS patients implanted with ICDs are collected and provide novel information about the current use, and sometimes misuse, of the ICDs in this special population.[92,93] Our current policy is to implant an ICD whenever syncope recurs despite β-blockade and LCSD and whenever cardiac arrest (requiring resuscitation) has occurred, whether or not the patient has been on β-blocking therapy.

TREATMENT OF ASYMPTOMATIC LQTS PATIENTS

For many years, one of the most difficult aspects of the management of LQTS was the choice between treatment and no treatment for those asymptomatic individuals who had a prolonged Q–T interval and who were siblings or close relatives of fully symptomatic LQTS patients. Data from more than 200 families followed in Pavia indicates that sudden death may be the first manifestation of the disease in 12% of the patients. Thus, we now recommend initiation of therapy with β-blockers in all patients diagnosed as affected by LQTS. This will undoubtedly lead to the unnecessary treatment of many patients who will remain asymptomatic, but it should importantly prevent a significant number of sudden deaths occurring during the first arrhythmic episode. A reasonable exception may be constituted by male patients older than 20 years of age because they are unlikely to become symptomatic.

We always instruct family members in cardiopulmonary resuscitation with particular emphasis on the importance of the thump-version. Finally, we provide a list of cardiac and noncardiac drugs that prolong the Q–T interval and that should always be avoided by the patient affected by LQTS.

OVERVIEW ON THERAPY

For nongenotyped individuals, the most important recent information is an unexpectedly high incidence of sudden death at first episode (12%) and by the relatively high recurrence rate after cardiac arrest, often despite β-blocker therapy. The two direct considerations of appropriate management are the need to initiate therapy with β-blockers in every patient diagnosed with LQTS and to implant an ICD—for safety reasons—in any patient who, independent of being on or off therapy, has had a cardiac arrest.

LQT1 patients should avoid situations involving strenuous exercise, competitive sports, and physical stress in general. They appear to be particularly responsive to β-blockers and should be treated with full-dose β-blockade. This should apply to the J-LN patients as well, given the implications of having mutations affecting the same current; however, this group has very severe presentations and shows a higher recurrence rate despite β-blocker therapy. Thus, an ICD may be the best choice for this high-risk group.

LQT2 patients should avoid, as much as possible within the constraints of modern life, the use of alarm clocks[94,95] and receiving phone calls while they sleep.

A potential note of concern comes from the observation that most LQT3 patients have a tendency toward marked QT prolongation at long cardiac cycles[96] because it occurs during bradycardia or sleeping periods. This observation is in agreement with the increased risk that LQT3 patients have when they are at rest or asleep[38] and raises the disturbing possibility that they not only may not benefit from β-blockers, but they may actually be at higher risk during β-blocker treatment because of the long cardiac cycles. This has led our group to treat these patients with LCSD because this procedure, besides being quite effective in all LQTS patients,[87] exerts its powerful antiadrenergic and antifibrillatory effect without reducing heart rate.[97] If β-blockers are used in association with a pacemaker, then the potentially adverse effects of the reduction in heart rate are avoided. For those LQT3 patients with mutations causing a delayed inward Na^+ current, mexiletine may be useful in shortening the Q–T interval.[26,96]

When dealing with an adult, genotyped, asymptomatic patient, consideration about initiating therapy may still be an issue. After 20 years of age, the probability of becoming symptomatic is minimal if the patient has the J-LN syndrome and is modest if the patient is a male with LQT1. This is not true with the other genotypes.

A new complication is gene carriers with a normal Q–T interval. Should they be treated or be left alone? Cardiac events, mostly syncopal episodes, are known to occur in some of these individuals. To date, we do not have sufficient data to make sound decisions. Our current choice, after performance of repeated ECGs and Holter recordings to exclude even transient repolarization abnormalities such to suggest arrhythmic risk, is not to initiate therapy in gene carriers with a consistently normal Q–T interval. It seems fair to conclude that it is becoming possible to look at the arrhythmic risk for LQTS patients in a novel way. The information originating from the genotype-phenotype studies provides new insights for a better-tailored approach to therapy for the individual patient affected by LQTS.

REFERENCES

1. Schwartz PJ, Periti M, Malliani A: The long Q-T syndrome. Am Heart J 1975;89:378-390.
2. Jervell A, Lange-Nielsen F: Congenital deaf-mutism, functional heart disease with prolongation of the Q-T interval, and sudden death. Am Heart J 1957;54:59-68.
3. Romano C, Gemme G, Pongiglione R: Aritmie cardiache rare dell'età pediatrica. Clin Pediat 1963;45:656-683.
4. Ward OC: A new familial cardiac syndrome in children. J Irish Med Assoc 1964;54:103-106.
5. Keating M, Atkinson D, Dunn C, et al: Linkage of a cardiac arrhythmia, the long QT syndrome, and the Harvey ras-1 gene. Science 1991;252:704-706.
6. Jiang C, Atkinson D, Towbin JA, et al: Two long QT syndrome loci map to chromosomes 3 and 7 with evidence for further heterogeneity. Nat Genet 1994;8:141-147.
7. Mohler PJ, Schott JJ, Gramolini AO, et al: Ankyrin-B mutation causes type 4 long-QT cardiac arrhythmia and sudden cardiac death. Nature 2003;421:634-639.
8. Wang Q, Shen J, Splawski I, et al: *SCN5A* mutations associated with an inherited cardiac arrhythmia, long QT syndrome. Cell 1995;80:805-811.
9. Curran ME, Splawski I, Timothy KW, et al: A molecular basis for cardiac arrhythmia: *HERG* mutations cause long QT syndrome. Cell 1995;80:795-803.
10. Wang Q, Curran ME, Splawski I, et al: Positional cloning of a novel potassium channel gene: *KvLQT1* mutations cause cardiac arrhythmias. Nat Genet 1996;12:17-23.
11. Abbott GW, Sesti F, Splawski I, et al: *MiRP1* forms I_{Kr} potassium channels with *HERG* and is associated with cardiac arrhythmia. Cell 1999;97:175-187.
12. Barhanin J, Lesage F, Guillemare E, et al: KvLQT1 and IsK (minK) proteins associate to form the I_{Ks} cardiac potassium current. Nature 1996;384:78-80.
13. Sanguinetti MC, Curran ME, Zou A, et al: Coassembly of KvLQT1 and minK (IsK) proteins to form cardiac I_{Ks} potassium channel. Nature 1996;384:80-83.
14. Priori SG, Napolitano C, Ronchetti E, et al: Characterization of the prevalence of *minK* polymorphism and mutations in 140 long QT syndrome families. PACE 1998;21:799.
15. Neyroud N, Tesson F, Denjoy I, et al: A novel mutation in the potassium channel gene *KvLQT1* causes the Jervell and Lange-Nielsen cardioauditory syndrome. Nat Genet 1997;15:186-189.
16. DeJager T, Corbett CH, Badenhorst JCB, Brink PA: Evidence of a long QT gene with varying phenotypic expression in South African families. Mol Genet 1996;33:567-573.
17. Priori SG, Schwartz PJ, Napolitano C, et al: A recessive variant of the Romano-Ward long QT syndrome? Circulation 1998;97: 2420-2425.
18. Sanguinetti C, Curran ME, Spector PS, Keating MT: Spectrum of *HERG* K^+-channel dysfunction in an inherited cardiac arrhythmia. Proc Natl Acad Sci USA 1996;93:2208-2212.
19. Schulze-Bahr E, Haverkamp W, Weibusch H, et al: Frequency and phenotype of *HERG* mutations in congenital long QT syndrome (LQTS). Circulation 1996;94(Abstr suppl):I-719.
20. Priori SG, Schwartz PJ, Napolitano C, et al: Molecular analysis of *HERG*-gene in forty-eight unrelated long QT syndrome patients: Genotype/phenotype correlation in two families with novel mutations. J Am Coll Cardiol 1997;29[Suppl A]:184A.
21. Napolitano C, Priori SG, Schwartz PJ, et al: Identification of a mutational hot spot in *HERG*-related long QT syndrome (LQT2): Phenotypic implications. Circulation 1997;96[Abstr Suppl]:212.
22. Benson DW, MacRae CA, Vesley MR, et al: Missense mutation in the pore region of *HERG* causes familial long QT syndrome. Circulation 1996;93:1791-1795.
23. Bennett PB, Yazawa K, Makita N, George AL Jr: Molecular mechanism for an inherited cardiac arrhythmia. Nature 1995;376: 683-685.
24. Dumaine R, Wang Q, Keating MT, et al: Multiple mechanisms of Na^+ channel-linked long-QT syndrome. Circ Res 1996;78: 916-924.
25. Priori SG, Napolitano C, Brown AM, et al: The loss of function induced by *HERG* and *KvLQT1* mutations does not correlate with the clinical severity in the long QT syndrome. Circulation 1998;98[Suppl I]:I-457.
26. Schwartz PJ, Priori SG, Locati EH, et al: Long QT syndrome patients with mutations on the *SCN5A* and *HERG* genes have differential responses to Na^+ channel blockade and to increases in heart rate. Implications for gene-specific therapy. Circulation 1995;92:3381-3386.
27. Brugada J, Brugada R, Brugada P: Right bundle-branch block and ST-segment elevation in leads V_1 through V_3. A marker for sudden death in patients without demonstrable structural heart disease. Circulation 1998;97:457-460.
28. Brink PA, Ferreira A, Moolman JC, et al: Gene for progressive familial heart block type I maps to chromosome 19q13. Circulation 1995;91:1633-1640.
29. Schott JJ, Alshinawi C, Kyndt F, et al: Cardiac conduction defects associate with mutations in *SCN5A*. Nat Genet 1999;23:20-21.
30. Schwartz PJ: Another role for the sympathetic nervous system in the long QT syndrome? J Cardiovasc Electrophysiol 2001;12: 500-502.
31. Roden DM, Lazzara R, Rosen MR, et al: The SADS Foundation Task Force on LQTS: Multiple mechanisms in the long-QT syndrome. Current knowledge, gaps, and future directions. Circulation 1996;94:1996-2012.
32. Jackman WM, Friday KJ, Anderson JL, et al: The long QT syndromes: A critical review, new clinical observations and a unifying hypothesis. Prog Cardiovasc Dis 1988;31:115-172.
33. Cranefield PF, Aronson RS (eds): Cardiac Arrhythmias: The Role of Triggered Activity and Other Mechanisms. Mount Kisco, NY, Futura, 1988, pp 553-579.
34. Gilmour RF Jr, Riccio ML, Locati EH, et al: Time and rate-dependent alterations of the QT interval precede the onset of torsades de pointes in patients with acquired QT prolongation. J Am Coll Cardiol 1997;30:209-217.
35. El-Sherif N, Chinushi M, Caref EB, Restivo M: Electrophysiological mechanism of the characteristic electrocardiographic morphology of torsades de pointes tachyarrhythmias in the long-QT syndrome. Detailed analysis of ventricular tridimensional activation patterns. Circulation 1997;96:4392-4399.
36. Viskin S, Alla SR, Barron HV, et al: The mode of onset of torsades de pointes in the congenital long QT syndrome. J Am Coll Cardiol 1996;28:1262-1268.
37. Schwartz PJ, Zaza A, Locati E, Moss AJ: Stress and sudden death. The case of the long QT syndrome. Circulation 1991;83[Suppl II]: II71-II80.
38. Schwartz PJ, Priori SG, Spazzolini C, et al: Genotype-phenotype correlation in the long QT syndrome: Gene-specific triggers for life-threatening arrhythmias. Circulation 2001;103:89-95.
39. Rashba EJ, Zareba W, Moss AJ, et al: The LQTS Investigators. Influence of pregnancy on the risk for cardiac events in patients with hereditary long QT syndrome. Circulation 1998;97:451-456.
40. Kappenberger LJ, Gloor HO, Steinbrunn W: A new observation on long QT-syndrome. [abstr] Circulation 1984;70:II-251.
41. Viersma JW, May JF, de Jongste MJL, et al: Long-QT syndrome and sudden death during sleep in one family. [abstr] Eur Heart J 1988;9[Suppl 1]:45.
42. Moss AJ, Zareba W, Benhorin J, et al: ECG T-wave patterns in genetically distinct forms of the hereditary long QT syndrome. Circulation 1995;92:2929-2934.
43. Zhang L, Timothy KW, Vincent GM, et al: Spectrum of ST-T-wave patterns and repolarization parameters in congenital long QT syndrome. ECG findings identify genotypes. Circulation 2000; 102:2849-2855.
44. Butrous GS, Schwartz PJ (eds): Clinical Aspects of Ventricular Repolarization. London, Farrand Press, 1989, pp 498.
45. Merri M, Benhorin J, Alberti M, et al: Electrocardiographic quantitation of ventricular repolarization. Circulation 1989;80: 1301-1308.
46. Stramba-Badiale M, Spagnolo D, Bosi G, Schwartz PJ: The MISNES Investigators. Are gender differences in QTc present at birth? Am J Cardiol 1995;75:1277-1278.
47. Moss AJ, Schwartz PJ, Crampton RS, et al: The long QT syndrome: Prospective longitudinal study of 328 families. Circulation 1991;84:1136-1144.
48. Garson A Jr, Dick M II, Fournier A, et al: The long QT syndrome in children. An international study of 287 patients. Circulation 1993;87:1866-1872.
49. Schwartz PJ: The long QT syndrome. In Kulbertus HE, Wellens HJJ (eds): Sudden Death. M Nijhoff, The Hague, 1980, pp 358-378.

50. Schwartz PJ, Moss AJ, Locati E, et al: The long QT syndrome international prospective registry. J Am Coll Cardiol 1989; 13[Suppl A]:20A.
51. Vincent GM, Timothy KW, Leppert M, Keating M: The spectrum of symptoms and QT intervals in carriers of the gene for the long QT syndrome. N Engl J Med 1992;327:846-852.
52. Priori SG, Napolitano C, Schwartz PJ: Low penetrance in the long QT syndrome. Clinical impact. Circulation 1999;99:529-533.
53. Malfatto G, Beria G, Sala S, et al: Quantitative analysis of T wave abnormalities and their prognostic implications in the idiopathic long QT syndrome. J Am Coll Cardiol 1994;23:296-301.
54. Lehmann MH, Suzuki F, Fromm BS, et al: T wave "humps" as a potential electrocardiographic marker of the long QT syndrome. J Am Coll Cardiol 1994;24:746-754.
55. Schwartz PJ, Malliani A: Electrical alternation of the T-wave: Clinical and experimental evidence of its relationship with the sympathetic nervous system and with the long Q-T syndrome. Am Heart J 1975;89:45-50.
56. Malfatto G, Rosen MR, Foresti A, Schwartz PJ: Idiopathic long QT syndrome exacerbated by β-adrenergic blockade and responsive to left cardiac sympathetic denervation: Implications regarding electrophysiologic substrate and adrenergic modulation. J Cardiovasc Electrophysiol 1992;3:295-305.
57. Vincent GM: The heart rate of Romano-Ward syndrome patients. Am Heart J 1986;112:61-64.
58. Locati E, Pancaldi A, Pala M, Schwartz PJ: Exercise-induced electrocardiographic changes in patients with the long QT syndrome. Circulation 1988;78(Suppl II):42.
59. Berthet M, Denjoy I, Donger C, et al: C-terminal *HERG* mutations: The role of hypokalemia and a *KCNQ1*-associated mutation in cardiac event occurrence. Circulation 1999;99:1464-1470.
60. Zareba W, Moss AJ, Schwartz PJ, et al: The International Long-QT Syndrome Registry Research Group. Influence of the genotype on the clinical course of the long QT syndrome. N Engl J Med 1998;339:960-965.
61. Cerrone M, Schwartz PJ, Priori SG, et al: Natural history and genetic aspects of the Jervell and Lange-Nielsen syndrome. [Abstr] Circulation 2001;104 (Suppl II):II-462.
62. Priori SG, Schwartz PJ, Napolitano C, et al: Risk stratification in the long-QT syndrome. N Engl J Med 2003;348:1866-1874.
63. Schwartz PJ, Locati EH, Priori SG, Cantù F: Can Na^+ channel blockers normalize the prolonged QT interval of long QT syndrome patients linked to chromosome 3p21-24? Circulation 1995;92(Abstr Suppl):I-275.
64. Priori SG, Napolitano C, Cantù F, et al: Differential response to Na^+ channel blockade, β–adrenergic stimulation, and rapid pacing in a cellular model mimicking the *SCN5A* and *HERG* defects present in the long QT syndrome. Circ Res 1996;78:1009-1015.
65. Benhorin J, Taub R, Goldmit M, et al: Effects of flecainide in patients with new *SCN5A* mutation: Mutation-specific therapy for long-QT syndrome? Circulation 2000;101:1698-1706.
66. Windle JR, Geletka RC, Moss AJ, et al: Normalization of ventricular repolarization with flecainide in long QT syndrome patients with *SCN5A*:ΔKPQ mutation. Ann Noninvasive Electrocardiol 2001;6:153-158.
67. Priori SG, Napolitano C, Schwartz PJ, et al: The elusive link between LQT3 and Brugada syndrome. The role of flecainide challenge. Circulation 2000;102:945-947.
68. Stramba-Badiale M, Priori SG, Napolitano C, et al: Gene-specific differences in the circadian variation of ventricular repolarization in the long QT syndrome: A key to sudden death during sleep? Ital Heart J 2000;1:323-328.
69. Schwartz PJ: Idiopathic long QT syndrome: Progress and questions. Am Heart J 1985;109:399-411.
70. Napolitano C, Schwartz PJ, Brown AM, et al: Evidence for a cardiac ion channel mutation underlying drug-induced QT prolongation and life-threatening arrhythmias. J Cardiovasc Electrophysiol 2000;11:691-696.
71. Schulze-Bahr E, Haverkamp W, Hordt M, et al: Do mutations in cardiac ion channel genes predispose to drug-induced (acquired) long QT syndrome? Circulation 1997;96[Abstr Suppl]:I-211.
72. Moss AJ, Schwartz PJ: Delayed repolarization (QT or QTU prolongation) and malignant ventricular arrhythmias. Mod Conc Cardiovasc Med 1982;51:85-90.
73. Schwartz PJ, Locati E: The idiopathic long QT syndrome. Pathogenetic mechanisms and therapy. Eur Heart J 1985; 6[Suppl. D]:103-114.
74. Schwartz PJ, Moss AJ, Vincent GM, Crampton RS: Diagnostic criteria for the long QT syndrome: An update. Circulation 1993; 88;782-784.
75. Nador F, Beria G, De Ferrari GM, et al: Unsuspected echocardiographic abnormality in the long QT syndrome: Diagnostic, prognostic, and pathogenetic implications. Circulation 1991;84: 1530-1542.
76. De Ferrari GM, Nador F, Beria G, et al: Effect of calcium channel block on the wall motion abnormality of the idiopathic long QT syndrome. Circulation 1994;89:2126-2132.
77. Schwartz PJ: The quest for the mechanism of the sudden infant death syndrome. Doubts and progress. Circulation 1987;75: 677-683.
78. Schwartz PJ, Southall DP, Valdes-Dapena M: The sudden infant death syndrome: Cardio-respiratory mechanisms and interventions. Ann NY Acad Sci 1988;533:474.
79. Schwartz PJ: Cardiac sympathetic innervation and the sudden infant death syndrome. A possible pathogenetic link. Am J Med 1976;60:167-172.
80. Schwartz PJ, Stramba-Badiale M, Segantini A, et al: Prolongation of the QT interval and the sudden infant death syndrome. N Engl J Med 1998;338:1709-1714.
81. Schwartz PJ: QT prolongation, sudden death, and sympathetic imbalance: The pendulum swings. J Cardiovasc Electrophysiol 2001;12:1074-1077.
82. Schwartz PJ, Priori SG, Dumaine R, et al: A molecular link between the sudden infant death syndrome and the long QT syndrome. N Engl J Med 2000;343:262-267.
83. Schwartz PJ, Priori SG, Bloise R, et al: Molecular diagnosis in victims of sudden infant death syndrome. Lancet 2001;358: 1342-1343.
84. Schwartz PJ, Garson A Jr, Paul T, et al: Guidelines for the interpretation of the neonatal electrocardiogram. Eur Heart J 2002;23:1329-1344.
85. Schwartz PJ: Stillbirths, sudden infant deaths, and long QT syndrome: Puzzle or mosaic, the pieces of the jigsaw are being fitted together. Circulation 2004;109:2930-2932.
86. Shimizu W, Antzelevitch C: Cellular basis for the ECG features of the LQT1 form of the long QT syndrome. Effects of β-adrenergic agonists and antagonists and sodium channels blockers on transmural dispersion of repolarization and torsades de pointes. Circulation 1998;98:2314-2322.
87. Schwartz PJ, Locati EH, Moss AJ, et al: Left cardiac sympathetic denervation in the therapy of congenital long QT syndrome: A worldwide report. Circulation 1991;84:503-511.
88. Schwartz PJ, Priori SG, Cerrone M, et al: Left cardiac sympathetic denervation in the management of high-risk patients affected by the long QT syndrome. Circulation 2004;109:1826-1833.
89. Eldar M, Griffin JC, Abbott JA, et al: Permanent cardiac pacing in patients with the long QT syndrome. J Am Coll Cardiol 1987;10:600-607.
90. Moss AJ, Liu JE, Gottlieb S, et al: Efficacy of permanent pacing in the management of high-risk patients with long QT syndrome. Circulation 1991;84:1524-1529.
91. Schwartz PJ: The long QT syndrome. In Camm AJ (ed): Clinical Approaches to Tachyarrhythmias. Armonk, NY, Futura, 1997.
92. Zareba W, Moss AJ, Hall JW, et al: Long-term follow-up of high-risk long QT syndrome patients with and without implantable cardioverter-defibrillators. Circulation 2000;102(Suppl. II):II-675.
93. Crotti L, Spazzolini C, De Ferrari GM, et al: Is the implantable defibrillator appropriately used in the long QT syndrome? Data from the European Registry. Health Rhythm 2004; 1(Suppl):582.
94. Wilde AA, Jongbloed RJ, Doevendans PA, et al: Auditory stimuli as a trigger for arrhythmic events differentiate *HERG*-related (LQTS2) patients from *KVLQT1*-related patients (LQTS1). J Am Coll Cardiol 1999;33:327-332.
95. Wellens HJ, Vermeulen A, Durrer D: Ventricular fibrillation occurring on arousal from sleep by auditory stimuli. Circulation 1972;46:661-665.
96. Schwartz PJ, Priori SG, Napolitano C: The long QT syndrome. In Zipes DP, Jalife J (eds): Cardiac Electrophysiology. From Cell to Bedside, 3rd ed. Philadelphia, WB Saunders, 2000, pp 597-615.
97. Schwartz PJ: The rationale and the role of left stellectomy for the prevention of malignant arrhythmias. Ann NY Acad Sci 1984; 427:199-221.

Chapter 35 Arrhythmias Associated with Congenital Heart Disease

GEORGE F. VAN HARE

In patients with congenital heart disease, the management of cardiac arrhythmias poses many challenges for the clinician. The occurrence of arrhythmias that are difficult to adequately control may be very disappointing and distressing to the patient, as such patients often have excellent hemodynamic surgery results but continue to require intensive medical attention due to their arrhythmias.

The mechanisms of arrhythmias in patients with congenital heart disease may be quite varied, with different arrhythmias seen in preoperative versus postoperative patients. The principles of management, however, share many features between the two states. This chapter will address the substrates for arrhythmia in both the preoperative and postoperative states, and then discusses the various treatment modalities available.

Arrhythmia Substrates in Patients with Congenital Heart Disease

ARRHYTHMIA SUBSTRATES IN UNREPAIRED PATIENTS WITH CONGENITAL HEART DISEASE

Accessory Pathways

Congenital heart disease is relatively common in the general population (1% or 2% of births) and accessory pathways are also somewhat common (1.6 to 3 of 1000 live births). Therefore, one would expect to observe a certain incidence of Wolff-Parkinson-White (WPW) syndrome and of supraventricular tachycardia (SVT) on the basis of these incidences alone. Indeed, in the Pediatric Radiofrequency Ablation Registry, ablations for accessory pathways have been reported in patients with most types of defects.[1] However, it is well known that certain types of congenital cardiac disease are more commonly associated with WPW syndrome. The most prominent of these defects is Ebstein's anomaly of the tricuspid valve.[2,3] In such patients, there is a prevalence of WPW of approximately 9%, based on a review of records from Children's Hospital in Boston (J.P. Saul, personal communication). Other defects reported to demonstrate an increased association with WPW include L-transposition of the great vessels (ventricular inversion and congenitally corrected transposition)[4-6] and hypertrophic cardiomyopathy.[7,8] In patients with Ebstein's anomaly, accessory pathways are often multiple and are generally right-sided, although occasionally a posteroseptal pathway location is present. In L-transposition, there is an increased incidence of Ebstein's anomaly of the left-sided (systemic) atrioventricular (AV) valve, which is morphologically always a tricuspid valve and is related to a left-sided, morphologic right ventricle. In such patients, it is thought that the increased incidence of WPW is explained by the coexistence of Ebstein's anomaly.[4-6] Finally, WPW has perhaps been over-diagnosed in patients with hypertrophic cardiomyopathy due to the common QRS abnormality existing in such patients that may resemble preexcitation. However, SVT mediated by accessory pathways does occur in such patients, and in them, supraventricular tachycardia may be very poorly tolerated due to coexisting hemodynamic abnormality.[7]

Other Substrates for Tachycardia

Other arrhythmias are occasionally seen in patients with congenital heart disease who have not undergone surgical repair. Atrial flutter, when seen, usually is a complication of atrial dilation on hemodynamic grounds. For example, patients with Ebstein's anomaly or other causes of severe tricuspid regurgitation may have atrial flutter, which in turn is most likely due to right atrial dilation.[3] Patients with mitral valve disease, and in particular mitral stenosis, are at risk for atrial fibrillation, based on their left atrial dilation. Patients with left ventricular failure and those with pulmonary hypertension and/or suprasystemic right ventricular pressure may also have increased atrial pressures, with resulting atrial arrhythmias, and may also have ventricular ectopy, ventricular tachycardia, or ventricular fibrillation.

Atrioventricular Block

Certain groups of patients are predisposed to the development of complete atrioventricular block, independent of attempted surgical repair. First, patients who have L-transposition, also known as ventricular inversion or congenitally corrected transposition of the great vessels, are at risk for the development of complete AV block spontaneously throughout their lives. The yearly incidence has been estimated at approximately 2% per year.[9] This is thought to be due to the abnormalities of the conduction system, in which there is malalignment between the atrial and the ventricular septum, and the normal compact AV node cannot make contact during embryologic development with the distal conducting system. A more anterior AV node forms instead and is thought to be more fragile.

Second, a syndrome of familial AV block associated with various septal defects has recently been described. Heterozygous mutations in NKX2.5, a homeobox transcription factor, lead to the spontaneous development of complete AV block, as well as associated cardiac defects, of which atrial septal defects (ASDs) are the most common.[10] Ventricular septal defects (VSDs) may also be seen, particularly those associated with tetralogy of Fallot.

ARRHYTHMIA SUBSTRATES FOLLOWING SURGERY FOR CONGENITAL HEART DISEASE

When a child is born with congenital heart disease and requires surgical repair of the heart defect, most of the focus of the surgeon, cardiologist, and parents is on obtaining as good a hemodynamic result as possible. For many defects, such as ASDs, or VSDs, atrioventricular canal defects, and tetralogy of Fallot, surgical results are excellent and one can expect to have a perfect or nearly perfect hemodynamic result. With more complex defects, such as truncus arteriosus and transposition of the great arteries, surgery can also be expected to produce a normal or nearly normal hemodynamic situation, with an excellent long-term prognosis for the child. In even the most complex defects, including the single ventricle lesions, good results can also be obtained, prolonging life into adulthood and beyond. It is in this setting that one must consider the impact of late postoperative arrhythmias, both on the child who has undergone repair, as well as on the parents of that child. These arrhythmias can be annoying, debilitating, or even life threatening.

Because late postoperative arrhythmias contribute significantly to morbidity and mortality, it is logical that every effort is made in patients with such arrhythmias to treat them and prevent their recurrence. Such treatments have included medical therapy with antiarrhythmic agents, implantation of anti-tachycardia pacemakers, catheter ablation, and surgical ablation. Unfortunately, there are serious limitations to the effectiveness, applicability, and safety—both of antiarrhythmic drug therapy and of anti-tachycardia pacing. Not surprisingly, ablative techniques, which potentially offer a curative treatment, are favored, but such techniques have not been as successful as initially hoped.

Principal Patient Groups

Postoperative tachyarrhythmias tend to fall into four main groups, based on type of defect, type of tachycardia, and current success rates of radiofrequency (RF) ablation attempts. The first group, referred to as *simple atriotomy-based atrial flutter,* includes patients who have had simple cardiac repairs, such as ASDs, VSDs, tetralogy of Fallot, atrioventricular canal defects, and related defects. The second group, termed *intra-atrial reentry following atrial repair of transposition,* includes patients who have had either the Mustard or the Senning procedure. The third group, termed *intra-atrial reentry following the Fontan procedure,* includes patients who have undergone the various forms of the Fontan procedure. Finally, a fourth group, termed *ventricular tachycardia following tetralogy repair,* includes tetralogy of Fallot patients, as well as those with related lesions, such as certain types of double outlet right ventricle. For each group, the anatomic details that support the tachyarrhythmia and that are important in ablation will be discussed.

Anatomic and Developmental Considerations

As atrial flutter and nearly all types of postoperative atrial tachycardia seem to involve only the anatomic right atrium, the exact anatomy of the right atrium becomes important in understanding these arrhythmias. Therefore, a detailed knowledge of right atrial anatomy is essential for effective mapping and ablation of intra-atrial reentrant tachycardia (IART) and atrial flutter.

During cardiac embryologic development, the right atrium is thought to derive from three sources.[11] The primitive right atrium forms adjacent to the tricuspid annulus and gives rise to the heavily trabeculated right atrial free wall and right atrial appendage. The sinus venosus is incorporated into the right atrium and provides the origin for the smooth-walled portion of the right atrium (sinus venarum) that exists between the cava posterior to the primitive right atrial structures. Finally, septation of the primitive common atrium is accomplished by the formation of the atrial septum from septum primum and septum secundum. The ostium secundum is a foramen that forms in the septum primum, which is subsequently closed by the septum secundum, which forms a flap over this ostium to create the foramen ovale. The foramen ovale provides a route for right atrial blood to cross to the left atrium in the fetal circulation. All along the junction between the primitive right atrium and the sinus venosus portion of the right atrium is the crista terminalis ("terminal crest"), which appears as a ridge along the inner surface of the right atrium. The crista terminalis runs from superior to inferior along the lateral wall of the right atrium. At its superior edge, near the superior vena cava-right atrial junction, is the sinus node pacemaker complex. As it arches inferiorly toward the inferior vena cava, it gives rise to the eustachian valve ridge (EVR), which appears as more of a flap than a ridge. The EVR is a remnant of the primitive right sinoatrial valve, guarding the ostium between the sinus venosus and the primitive right atrium. The EVR runs anterior to the inferior vena caval orifice and posterior to the posterior portion of the tricuspid valve annulus. As such, in the fetal circulation, the EVR acts to direct inferior vena caval (IVC) flow away from the tricuspid annulus and toward the foramen ovale. As the EVR arches toward the inferior atrial septum, it passes just superior to the ostium of the coronary sinus. It joins with the valve of the coronary sinus to form the tendon of Todaro, which inserts on the atrial septum near the His bundle. With the coronary sinus ostium and the tricuspid annulus, the tendon of Todaro forms the triangle of Koch, and at the apex of this triangle, the compact atrioventricular node is found.

Adult-Type Atrial Flutter

Classic atrial flutter is characterized by atrial rates of approximately 300 beats/minute with typical and very characteristic saw-tooth flutter waves visible on the surface electrocardiogram. This suggests the presence of nearly continuous atrial activity, due to the relative lack of a long atrial isoelectric interval. In typical atrial flutter, the flutter waves are prominent and are negative in leads II, III, and AVF, suggesting inferior to superior atrial activation. Although initially thought to represent reentry around the caval veins or around the tricuspid valve annulus, the work of multiple investigators has clearly established the actual circuit. Impulses emerge from an isthmus of atrial tissue between the inferior vena cava and tricuspid annulus to spread up the atrial septum, activating the atrium at the site where the His bundle is recorded, and then down the right atrial free wall to enter the isthmus again.[12] This counterclockwise activation has recently been categorized as *typical* atrial flutter, whereas atrial flutter that uses the same circuit but in the clockwise order of activation has been categorized as *reverse typical* atrial flutter.[13] In addition, further details have been provided recently, using techniques of entrainment pacing, which depend on the demonstration of equivalence of the postpacing interval (PPI) during entrainment and the tachycardia cycle length (TCL) to establish that any given site is in the circuit (PPI = TCL). These studies have demonstrated the importance of the crista terminalis and EVR as sites of conduction block during atrial flutter.[14-16] Conduction block is suggested by the demonstration that there are sites along the ridge where double potentials can be recorded.[15-18] The importance of such areas of conduction block is strengthened by entrainment pacing, demonstrating that atrial myocardium on one or another side of the line of conduction block is part of the circuit. The critical nature of these lines of block is proved by RF ablation lesions that are designed to bridge from one line of block to another, with resultant abolition of the atrial flutter.[19-21] These criteria have been satisfied with both typical and reverse typical atrial flutter, and the features of this arrhythmia circuit now seem well characterized.

As previously described, the wave of activation leaves the region of the tricuspid valve-inferior vena cava isthmus to climb the interatrial septum and enter the heavily trabeculated right atrial free wall in the region of the superior vena cava. It then spreads down the right atrial free wall, with the crista terminalis behind and the tricuspid annulus in front, turning counterclockwise around the tricuspid annulus when viewed from below (left anterior oblique view fluoroscopically). As the wave of activation turns posterior, it enters a "funnel," as described by Nakagawa and colleagues,[16] created because the distance between the crista and the tricuspid annulus becomes progressively shorter. The wave is funneled to the isthmus between the inferior vena cava and tricuspid valve annulus, now with the annulus anterior and inferior, and the EVR (the extension of the crista terminalis) posterior and superior. It is important to note that at this site, the EVR bisects the isthmus between the IVC and tricuspid valve, and that it is the EVR, not the IVC, that provides the critical site of conduction block. As it enters the interatrial septum, the wave again spreads in a superior fashion along the septum for the next circuit. Atrial flutter can be effectively attacked using RF ablation either at the septal site of EVR insertion, by lesions that bridge from the tricuspid valve to the EVR,[16] or posterior, from the tricuspid annulus down to the IVC.[19]

Simple Atriotomy-Based Atrial Flutter

When performing surgery to repair a simple secundum ASD, the surgeon typically places a long incision in the right atrial free wall, which is oblique and runs from the right atrial appendage laterally down toward, but not to, the tricuspid annulus or IVC. Care is typically taken to avoid the sinus node, and this concern results in the crista terminalis not being incised. This incision gives adequate exposure for repair of ASDs, and is also used for the atrial approach to repair VSDs, either alone or as part of tetralogy of Fallot. The ASD itself is commonly closed using sutures, but for large defects a patch may be employed.

This surgical approach clearly creates a long line of permanent conduction block that is entirely in the trabeculated right atrium, anterior to the crista terminalis. This anatomy potentially creates a tunnel of atrial tissue between the crista and the atriotomy, and another between the atriotomy and the tricuspid annulus. Such tunnels can easily be imagined as the required protected zones of conduction, mediating IART. Numerous patients have now been reported who exhibit "incisional" IART in which the atriotomy seems to act as the critical barrier.[22] In such patients, RF application from the atriotomy to either the IVC, tricuspid annulus, or SVC has been successful in terminating tachycardia and preventing reinduction.

Because the atrial structures that support typical atrial flutter are also present, and because such patients often have the other risk factors for the development of flutter (atrial dilation, fibrosis, etc.), postoperative patients may also have typical atrial flutter. Furthermore, as has become apparent in patients with otherwise structurally normal hearts, such IART and flutter circuits can run in either direction (counterclockwise or clockwise). Indeed, several recent series have reported a more common occurrence of typical or reverse typical atrial flutter than true incisional flutter in such patients.[23,24] In patients who have undergone patch closure of a secundum ASD, the patch itself has been reported to be a possible site of conduction block, mediating tachycardia,[19] although this is much less common. The potential variability in circuits and rotation creates the possibility for several distinct P-wave morphologies and atrial tachycardia cycle lengths.

Patients may, of course, have several reentrant circuits. One commonly observes typical atrial flutter, and after successful ablation, a second IART with a longer or shorter tachycardia cycle length and noninvolvement of the flutter isthmus, or of any structure posterior to the crista terminalis. The slower cycle length of such atrial tachycardias may be due to conduction that is confined to the heavily trabeculated atrial free wall (in which atrial conduction may be slower), longer circuits due to dilation, or may reflect slow conduction due to fibrosis.

Intra-atrial Reentry Following Atrial Repair of Transposition

The Senning and Mustard procedures are similar operations meant to address the hemodynamic abnormality in transposition, by directing systemic return to the left ventricle and pulmonary artery, and directing pulmonary venous return to the right ventricle and aorta.[25,26] Although very successful, these operations are rarely performed, in part because of the success of the arterial switch procedure, and in part because of the high incidence of sinus node dysfunction, atrial arrhythmias, and increased risk of sudden death. However, there has recently been interest in the so-called *double switch* procedure as a strategy for managing patients with L-transposition (congenitally corrected transposition). A Senning atrial baffle is constructed in this procedure.[27] This surgical substrate may not, in fact, disappear. In the Mustard procedure, after a long atriotomy anterior to the crista terminalis and resection of the atrial septum, a baffle is constructed and sewn into place around each caval vein, through the isthmus between the IVC and tricuspid annulus, and to the posterior wall of the left atrium, so that caval flow is direct to the mitral annulus.[26] Pulmonary venous flow travels around the baffle and finds the tricuspid annulus. It is important to note that the baffle, where it is sewn into place along the tricuspid annulus, has the same function as the EVR in fetal life, which is to prevent IVC flow from reaching the tricuspid valve. Furthermore, surgical technique is directed at avoiding injury to the sinus node, so the crista terminalis is not disturbed. Finally, various approaches are used to avoid atrioventricular block, and often these lead to the coronary sinus being incorporated into the pulmonary rather than the systemic venous atrium.[28] These details leave the entire right atriotomy, as well as the isthmus of atrial tissue between the EVR and tricuspid annulus, in the new pulmonary venous atrium.[29,30] The one exception is the situation in which the coronary sinus drainage is the systemic venous atrium, in which a catheter can reach the flutter isthmus from the IVC.[29]

In most respects, the Senning procedure is similar electrophysiologically to the Mustard procedure. The Senning procedure was designed to use mostly atrial tissue versus artificial material to construct the baffle.[25] In order to accomplish this, two atrial incisions are made. The first is in the right atrium, longitudinal, parallel, and anterior to the crista terminalis. The second, in the left atrium, is parallel to the first and between the right pulmonary veins and interatrial septum. A U-shaped incision in the atrial septum is made, just above the coronary sinus, leaving the flutter isthmus intact. This flap of atrial septum is sewn to the back of the left atrium, to the left of the left pulmonary veins. The flap of right atrial free wall is sewn into place near or at the site of the EVR, preventing IVC flow from crossing the tricuspid valve. The left atrial incision is closed by sewing to the other edge of the right atrial incision. As in the Mustard procedure, both the flutter isthmus and the right atriotomy are part of the new pulmonary venous atrium.

Typically, and predictably based on the foregoing, one commonly finds two types of IART in such patients. First, typical or reverse typical atrial flutter is usually present, using the usual anatomic structures as barriers to support reentry. Second, true "incisional" IART is often found, and is confined to the anatomic trabeculated right atrium, with the wave of activation passing between the lower limit of the atriotomy and the tricuspid annulus.

Intra-atrial Reentry Following the Fontan Procedure

The Fontan procedure has changed many times since its development as a palliative procedure for patients without two functional ventricles, as a way of relieving ventricular volume overload and of normalizing arterial saturations.[31] Initially, it was thought that the right atrium could be used as an effective pumping chamber, provided that pulmonary artery pressures were low (atriopulmonary connection). Largely as a result of an extremely high incidence of atrial arrhythmias after such procedures, as well as concerns about hydraulic energy loss in the system and pulmonary venous obstruction, this approach has been abandoned in favor of approaches that bypass the heart entirely (total cavopulmonary connection via the lateral tunnel or via an external conduit).[32-34] Within each of the two categories, many modifications exist. Despite the approach of total cavopulmonary connection, atrial arrhythmias continue to be observed, and surgical details are critical in planning mapping and ablation procedures in such patients.

With the various forms of atriopulmonary connection, there is a long atriotomy placed. In patients who had a conduit from the right atrium to the pulmonary artery, and those in whom the right atrial appendage was connected directly to the pulmonary artery or right ventricular outflow tract, this atriotomy was anterior to the crista terminalis. Often in such patients, patch augmentation of the right atrium was performed, using a piece of pericardium or other material incorporated into the closure. Invariably, closure of a large ASD was also necessary. As in the simpler situation of ASD repair (see earlier) both typical atrial flutter, as well as incisional reentry around the anterior atriotomy, are possible and have been observed. In excellent multisite mapping studies using basket catheters, slow conduction up the lateral wall has been demonstrated by Triedman and colleagues,[35] and this configuration fits the concept of conduction in a long isthmus bounded by the atriotomy and the crista terminalis. Reentry around the ASD patch is also possible. Finally, patch closure of the tricuspid annulus has been occasionally performed in patients with single ventricle without tricuspid atresia, potentially creating areas of slow atrial conduction on the other side of the suture line.

Ventricular Tachycardia Following Tetralogy Repair

Ventricular tachycardia continues to be a difficult management problem in patients who have previously undergone surgical repair of tetralogy of Fallot and other related congenital heart defects. The actual etiology of sudden death in patients following tetralogy repair is still somewhat uncertain. However, because of the frequent occurrence of premature ventricular contractions (PVCs), nonsustained and sustained ventricular tachycardia in patients who have undergone complete repair of tetralogy of Fallot , and related defects, such as double outlet right ventricle,[36-42] ventricular tachycardia has been implicated in the etiology of sudden death in this patient group. It is known that postoperative tetralogy of Fallot is the single most common condition seen in children between the ages of 1 and 16 years who have experienced sudden death.[43]

The majority of available information concerning patients with ventricular tachycardia and congenital heart disease pertains to tetralogy of Fallot, as compared with other forms of congenital heart disease. Ventricular arrhythmias do occur, but are much rarer in patients with other lesions.[44] For the purposes of management, tetralogy of Fallot can be viewed as an archetype for other lesions when patients with other lesions present with ventricular arrhythmias in the setting of ventriculotomy and/or right ventricular dysfunction.

Great controversy still exists regarding the role of various risk factors for the occurrence of ventricular arrhythmias and sudden death, the exact relationship between ventricular arrhythmias and sudden death, the role of electrophysiology study and other procedures for risk stratification, and, ultimately, the appropriate management of postoperative patients with ventricular tachycardia.

Patients with tetralogy of Fallot prior to repair have a VSD with (usually severe) right ventricular outflow tract obstruction, leading to chronic cyanosis. The placement of a systemic-to-pulmonary artery shunt as a palliative procedure adds the element of potential left ventricular volume overload. Correction of the defect involves patch closure of the VSD with relief of right ventricular obstruction. In nearly all patients, this requires resection of a large amount of right ventricular muscle, and early in the experience, this was not done via an atriotomy with retraction of the tricuspid valve, but instead required a ventriculotomy. Finally, in tetralogy of Fallot, the pulmonary annulus is typically smaller than normal. This has been approached by the placement of a transannular patch, which leads to chronic pulmonic insufficiency. Pulmonic insufficiency may be very severe if it is associated with downstream obstruction related to significant pulmonary arterial stenosis. It has been hypothesized that ventricular arrhythmias are due to the effects of years of chronic cyanosis, followed by the placement of a ventriculotomy, with elevation of right ventricular pressures due to inadequate relief of obstruction, and severe pulmonic regurgitation with right ventricular dysfunction and enlargement.[36,45-47] Such factors as wall stress and chronic cyanosis, coupled with the passage of time, may lead to myocardial fibrosis and result in the substrate for reentrant ventricular arrhythmias. This hypothesis is supported by histologic examination of the hearts of patients with tetralogy of Fallot who had died suddenly. These studies have shown extensive fibrosis.[48] The hypothesis is also supported by the observation of fractionated electrograms and late potentials recorded from the right ventricle at electrophysiology study, suggesting the presence of slow conduction.[49,50] Although there is a 5% incidence of coronary artery abnormalities in tetralogy of Fallot, putting the left anterior descending coronary artery or other large branches at risk at the time of complete repair, such potential damage has never been implicated in the etiology of ventricular arrhythmias or of sudden death.

Careful electrophysiology studies in patients with ventricular tachycardia following tetralogy surgery have supported the concept that the mechanism of ventricular tachycardia is macro-reentry, which involves the right ventricular outflow tract, either at the site of anterior right ventriculotomy or at the site of a VSD patch. Transient entrainment has been documented, with constant fusion at the paced cycle length and progressive fusion at decreasing cycle lengths. Furthermore, the evaluation of post-pacing intervals strongly suggests that sites in the right ventricular outflow tract are part of a macro-reentrant circuit.[51,52]

Early reports noted the frequent occurrence of PVCs in patients who had previously undergone tetralogy of Fallot repair. Gillette and coworkers identified PVCs on routine electrocardiograms in 18% of patients.[36] With exercise testing, the incidence may increase—to around 20% in one study.[41] With Holter monitoring, the incidence of ventricular ectopy is reported to be as high as 48%.[53] In about one half of these patients, ventricular ectopy is complex, defined as multiform beats, couplets, or ventricular tachycardia. In the great majority of patients, this ventricular ectopy is entirely asymptomatic.

Many investigators have tried to correlate the incidence of ventricular ectopy with various factors, including age at presentation, age at time of repair, and various hemodynamic features. Four factors seem to be the most important: (1) age at initial repair; (2) time since repair; (3) presence of residual right ventricular obstruction; and (4) presence of significant pulmonic insufficiency. Older age at time of repair, especially beyond 10 years of age, was associated with nearly a 100% incidence of ventricular arrhythmias, regardless of follow-up interval, in Chandar and colleagues' multicenter study.[53] In the same study, time since repair also predicted the occurrence of ventricular ectopy, which occurred in 4 of 4 patients followed for more than 16 years, despite repair in infancy. Walsh and colleagues, however, showed that in a group of patients repaired at less than 18 months of age, ventricular ectopy was rare on electrocardiogram (1%) but more common on Holter (31%) after an average of 5 years of follow-up.[54]

Garson and associates, in a study of 488 patients with repaired tetralogy of Fallot, showed that the incidence of ventricular arrhythmias was closely related to right ventricular hemodynamics.[55] The incidence of ventricular arrhythmias was significantly higher in those with a right ventricular systolic pressure greater than 60 mm Hg, and in those with a right ventricular end-diastolic pressure greater than 8 mm Hg, suggesting that residual right ventricular outflow tract obstruction, as well as pulmonic insufficiency, negatively influences outcome. They also found a relationship to age at surgery, but this was not as important as the follow-up interval. Zahka and coworkers, in a prospective study of 59 patients with tetralogy of Fallot repaired prior to 11 years of age, found that the degree of pulmonary regurgitation was, by far, the most important predictor of the frequency and severity of spontaneously occurring ventricular arrhythmias.[45] Although the degree of residual right ventricular outflow tract obstruction was not a predictor in this study, significant residual obstruction was rare in their study group.

Spontaneously occurring sustained ventricular tachycardia is, in fact, fairly uncommon among patients with repaired tetralogy, despite the high incidence of ventricular ectopy. The best data in this regard comes from a report by Harrison and associates of patients with repaired tetralogy of Fallot attending an adult congenital heart disease clinic.[56] Eighteen of 210 patients (8.6%) had either documented sustained ventricular tachycardia, syncope or near-syncope, with palpitations and inducible sustained monomorphic ventricular tachycardia at electrophysiology study. Ventricular tachycardia was closely related to right ventricular hemodynamics, and, in particular, right ventricular outflow tract aneurysms and pulmonic insufficiency. This finding is consistent with the earlier report by Zahka and coworkers[45] that emphasized the importance of pulmonic insufficiency as a risk factor for ventricular ectopy.

The occasional but persistent observation of sudden, unexpected death in this group of patients with repaired congenital heart disease, along with the high incidence of spontaneously occurring ventricular arrhythmias, both simple and complex, has lead to the hypothesis that sudden death in such patients is due to ventricular tachycardia. During the 1970s and 1980s, it was the standard practice to perform electrophysiology studies in a large proportion of patients who had undergone repair of tetralogy of Fallot or related congenital defects. Antiarrhythmic drug therapy was often prescribed based on the results of such studies. This approach has, for the most part, been abandoned due to the lack of strong evidence supporting the proposition that sudden death can be prevented with this approach, as well as worries about the proarrhythmic effect of the antiarrhythmic medications chosen for treatment. In a large multicenter review, Chandar and colleagues reported the experience with 359 postoperative tetralogy of Fallot patients who underwent invasive electrophysiology testing.[53] Ventricular tachycardia could be induced in 17% of patients, but not in any patient who was asymptomatic and had a normal 24-hour electrocardiogram. Although late sudden death occurred in five patients, none of these patients had inducible ventricular tachycardia at electrophysiology study.

Recent data suggest that one can assess the risk of ventricular tachycardia from the surface electrocardiogram QRS duration. Gatzoulis and associates found that in a group of 48 well-studied postoperative tetralogy patients, those with a QRS duration of greater than 180 ms had a greatly increased risk of spontaneous ventricular tachycardia and/or sudden death.[46] Similarly, Balaji and coworkers showed that a QRS duration of greater than 180 ms predicts the finding of inducible sustained monomorphic ventricular tachycardia at electrophysiology study.[57] The cause of this relationship is uncertain.

There is some evidence that the risk of sudden death may increase at late follow-up. In a careful study of 490 survivors of tetralogy of Fallot repair at a single center, Nollert and colleagues constructed actuarial survival curves out to 36 years following surgery.[58] In this study, the yearly actuarial mortality rate during the first 25 years was 0.24% per year, but mortality increased dramatically after 25 years to 0.94% per year. Most mortality was due to sudden death. The mortality risk was also related to date of repair (highest before 1970), the degree of preoperative polycythemia (highest with hematocrit >48) and the use of a right ventricular outflow tract patch (highest with a patch). The last factor is most likely related to the presence of pulmonic insufficiency, as suggested earlier.

Surgically-Induced Atrioventricular Block

The risk of complete AV block as a result of surgery is dependent on the type of repair that is attempted. At highest risk are patients who undergo closure of atrioventricular septal (canal or endocardial cushion) defects and those having closure of VSDs involving the perimembranous region of the ventricular septum.[44,59,60] In such defects, the distal conducting system is in a location that is difficult to avoid in the process of surgical closure of the defect. The repair of subaortic stenosis is often complicated by AV block, due to the need, in many cases, to resect muscle from the left side of the ventricular septum. Muscular VSDs and supracristal (doubly committed subarterial) VSDs are distant from the conducting system and, therefore, are not highly associated with AV block. Likewise, the repair of ASDs is only rarely complicated by postoperative AV block.[44]

Postoperative Sinus Node Dysfunction

Any operation that involves surgery on the right atrium may lead to postoperative sinus node dysfunction; this problem is often not evident early after surgery, but may take years to become manifest. The mechanism of sinus node damage is not well known, but most likely results either from direct injury to the sinus node pacemaker complex, or to interference with the blood supply to this structure, or both. The repair of simple ASDs is only rarely followed by sinus node dysfunction, whereas the risk is higher with the repair of sinus venosus ASDs. Sinus node dysfunction is most common following the Senning or Mustard procedure for transposition, the hemi-Fontan procedure, and the Fontan procedure.[44]

Treatment of Arrhythmias in Patients with Congenital Heart Disease

ANTIARRHYTHMIC DRUG THERAPY

Decision Regarding Treatment

Sustained tachycardia may be more poorly tolerated in this patient population due to poor systemic ventricular function. In addition, the use of antiarrhythmic agents is problematic in this patient population, due both to the increased prevalence of sinus node dysfunction after atrial surgery, such as the Mustard, Senning, or Fontan procedures,[61] and to the potential for proarrhythmia, which is exacerbated by systemic ventricular dysfunction. These issues all need to be evaluated in the choice of antiarrhythmic agents in this population.

The treatment decision in patients with ventricular ectopy and/or ventricular tachycardia following correction of congenital heart disease may be directed at one or both of two goals: the prevention of sudden death, and the suppression or termination of symptomatic episodes of ventricular tachycardia. Treatment may involve antiarrhythmic medication, RF catheter ablation, surgical cryoablation, or implantation of a tiered-therapy device providing antitachycardia pacing and cardioversion/defibrillation.

It is difficult to know which patients deserve electrophysiology study and/or treatment. When one considers that there are no prospective studies demonstrating that sudden death can be prevented in any group of postoperative patients by treatment of any type, it does not seem that one should recommend routine electrophysiology study and treatment of asymptomatic patients with ventricular ectopy or other arrhythmias. Although it seems likely that certain subgroups exist in whom carefully chosen antiarrhythmic therapy may exert a beneficial effect in lowering mortality, it is also quite possible that there are subgroups for whom the proarrhythmic potential of antiarrhythmic medications more than makes up for any beneficial effect, a result similar to that seen in the Cardiac Arrhythmia Suppression (CAST) trial.[62,63] Until such prospective, controlled trials are performed, therapy with drugs, such as flecainide, propafenone, and quinidine, cannot be recommended for most asymptomatic patients. An argument can be made for selecting certain patients for prophylactic treatment with β-blockers, as Garson and associates[64] have suggested. Examples of such patients might include those with significant pulmonic regurgitation and/or residual right ventricular obstruction and complex ventricular ectopy. One can also argue that such patients should be considered for cardiac surgery to correct hemodynamic abnormalities, such as residual obstruction, especially branch pulmonary artery stenosis, and should also be considered for placement of a homograft valve to eliminate pulmonic insufficiency.

If the goal of therapy is to prevent recurrence of clinically documented ventricular tachycardia, then one might consider proceeding in a fashion similar to that with other forms of ventricular tachycardia. Electrophysiology study for induction of ventricular tachycardia with subsequent drug testing is reasonable, bearing in mind that there very well may be some proarrhythmic risk to the use of certain medications, especially quinidine, procainamide, flecainide, and propafenone. One might also use exercise testing as a means of inducing ventricular tachycardia in certain patients. The choice of a pharmacologic agent for long-term treatment is then based both on the effect of suppression of the arrhythmia by the drug, and on its proarrhythmic potential. In the former consideration, investigators have reported some success with most class I agents, and especially quinidine, procainamide, propafenone, and flecainide, as well as with β-blockers. The latter consideration, of proarrhythmia, is difficult to judge. There was a high incidence of documented proarrhythmia with flecainide and encainide in patients with structural cardiac disease, as reported by Fish and associates in a pediatric series.[65] Furthermore, cardiac arrest and deaths occurred predominantly among patients with underlying heart disease. Sotalol is also known to be proarrhythmic, particularly in patients with significant ventricular dysfunction.[66] D-sotalol has been found to increase mortality in post-infarction patients with ventricular dysfunction (the SWORD trial).[67] These findings suggest that sotalol and the class IC agents should be used with extreme caution in repaired tetralogy of Fallot patients with ventricular dysfunction. This, of course, leaves amiodarone, which has the advantage of very little reported proarrhythmia, even in patients with significant ventricular dysfunction. Concern over long-term side effects, such as pulmonary fibrosis, ocular abnormalities, thyroid dysfunction, transaminase elevations, and significant bradycardia, have been raised in children and young adults,[68-70] and the use of this agent in a young person means that there is a likelihood that the medication will be needed for several decades at least. These concerns naturally lead to consideration of nonpharmacologic therapy.

Digoxin

Digoxin has a long history of use in children, and in particular in patients with atrial flutter and intra-atrial reentry. Although it exerts a direct effect on the cardiac cell membrane, most of its antiarrhythmic properties are a result of its indirect actions mediated through the autonomic nervous system.[71] The direct and indirect effects of digoxin on the atrioventricular node prolong the refractory period and slow conduction. In general, digoxin is not an effective drug for the acute conversion of atrial flutter. In the collaborative study reported by Garson and coworkers[72] digoxin used alone was successful in preventing recurrences of atrial flutter in only 44% of patients. Of the conventional agents, the most effective treatment for preventing recurrences (53%) was the combination of digoxin with a type IA agent, such as quinidine or procainamide. However, the use of digoxin may be beneficial in the control of ventricular rate during atrial flutter.

Class I Antiarrhythmic Medications

When using the sodium channel blocking agents to manage arrhythmias in patients with congenital heart disease, one needs to carefully balance the likelihood of success against the possibility of proarrhythmia. The class IA agents (quinidine, procainamide, and disopyramide), as well as the class IC agents (flecainide and propafenone) are useful in atrial flutter and ventricular arrhythmias, as well as in controlling SVT mediated by concealed or manifest accessory pathways. Because of vagolytic properties, class IA medications, particularly disopyramide, need to be given in combination with digoxin preparations or other AV node blocking agents to lessen the likelihood of 1:1 atrioventricular conduction if atrial flutter should occur. Class IB agents (lidocaine, tocainide, mexiletine, phenytoin, and moricizine) however, are not considered particularly useful in atrial arrhythmias, although moricizine has been used effectively in some studies. These agents are primarily used in ventricular arrhythmias. They are all effective in suppressing PVCs, and for many years, phenytoin has been a favorite antiarrhythmic medication for suppressing PVCs in patients following tetralogy of Fallot repair.[55] Whether suppression of PVCs is an important goal in the population is open to speculation. There is no clear evidence that ventricular tachycardia and sudden death can be prevented by the use of these agents. Furthermore, classes 1A and 1C agents share a propensity to proarrhythmia, particularly in patients with ventricular dysfunction, and so should be used with caution.[65]

Class II Antiarrhythmic Medications

The β-adrenergic blocking agents; such as propranolol and atenolol, may be useful for rate control in patients with chronic atrial flutter, and have the advantage of not being proarrhythmic. For other forms of SVT, the β-blockers are often effective in preventing recurrences. Patients with concomitant sinus node dysfunction may require permanent pacing if β-adrenergic blocking agents are used. There is some evidence that the use of propranolol is safe and effective in the management of ventricular arrhythmias following tetralogy of Fallot repair.[64]

Class III Antiarrhythmic Medications

Class III agents are thought to exert their actions primarily by prolonging action potential duration and refractoriness without significantly affecting conduction. The agents in this category are amiodarone and sotalol. Garson and colleagues[73] had an opportunity to use amiodarone in a group of 39 patients with congenital heart disease, critical tachyarrhythmias, and arrhythmias not responsive to conventional agents. Sixteen of the 39 patients had recurrent atrial flutter, and of these 16, 15 had complete elimination of the flutter. Sotalol has been used more recently with good results.[74,75] It combines class III action with β-adrenergic blocking activity, and so may not be tolerated by patients with poor ventricular dysfunction. Both these agents also may exacerbate sinus node dysfunction.

Class IV Antiarrhythmic Medications

Class IV drugs (e.g., verapamil) have been used almost exclusively as an acute intervention given intravenously. Conversion of atrial flutter to normal sinus rhythm occurs rarely after the intravenous use of verapamil. However, what does occur is a lowering of the ventricular rate secondary to delayed conduction through the atrioventricular node or conversion from atrial flutter to atrial fibrillation with a reduction in conducted impulses. Verapamil may be an effective agent for a patient with atrial flutter and 1:1 atrioventricular conduction whose condition is deteriorating rapidly. However, one must be extremely cautious when using this medication in younger children,[76] and verapamil should never be given to those younger than the age of 6 months. Another option for rapid control of ventricular rate is the continuous infusion of diltiazem, another class IV agent.

Results of Antiarrhythmic Drug Therapy

Unfortunately, medical therapy has been disappointing in the management of patients with postoperative arrhythmias. In general, one observes a fairly low success rate with a variety of antiarrhythmic agents for the prevention of recurrent atrial flutter or intra-atrial reentry tachycardia. Garson and associates showed that even if an antiarrhythmic agent was found to be successful in suppressing episodes of atrial flutter, there was a significant incidence of sudden death, suggesting that life-threatening proarrhythmia may be a serious potential problem in postoperative patients.[72] Finally, although Garson and colleagues presented retrospective evidence that the suppression of ventricular ectopy was associated with a lower incidence of sudden death among tetralogy of Fallot patients,[55] no prospective trials have been performed to support this concept.

ABLATION OF ACCESSORY PATHWAYS IN UNREPAIRED PATIENTS WITH CONGENITAL HEART DISEASE

In considering the technical factors that lead to success in the ablation of accessory pathways in patients with associated congenital heart disease, we may consider those relating to venous and arterial access, those relating to visceral situs, those relating to the location of the atrioventricular conducting tissue, and those relating to the specific cardiac defect. In general, however, one may state that no two patients are identical, neither in the cardiac anatomy confronting the electrophysiologist, nor in the approach that will be necessary to successfully ablate the pathway. The electrophysiologist needs to combine a detailed knowledge of the patient's exact intracardiac anatomy with the ability to be creative and persistent in the ablation attempt.

Venous and Arterial Access

Patients who have had multiple prior procedures, or who have had long stays in the intensive care unit with indwelling lines, may have limited venous access due to iliofemoral thrombosis. When bilateral, this problem may prevent a normal approach to the right atrium. Patients who have undergone the bidirectional Glenn procedure will have no direct access to the right atrium from the superior vena cava due to the direct connection of the superior vena cava to the pulmonary artery. In both situations, one may consider approaching the right atrium from the other cava, for example, from the superior vena cava when there is bilateral iliofemoral thrombosis. Reports of the use of a transhepatic approach for diagnostic and interventional catheterization[77,78] suggest that this route might also be efficacious for catheter ablation. Left-sided accessory pathways may, of course, be ablated by a retrograde approach in larger patients.[79]

Situs

The existence of situs abnormalities can render a catheter ablation procedure potentially confusing, due to the nonstandard location of veins, arteries and the heart itself, but these problems are not insurmountable if one possesses a knowledge of congenital cardiac pathology. Such abnormalities are, unfortunately, not limited to simple mirror-image arrangements. Standard fluoroscopy planes for normal anatomy may make little sense in the setting of situs abnormalities. When needed, transthoracic or transesophageal echocardiography may be used to confirm catheter tip locations.[80-82] Finally, patients with heterotaxy syndromes may have interruption of the inferior vena cava with azygous continuation. In such patients, a catheter passed from the femoral vein will traverse the azygous system to join the superior vena cava and enter the atrium from above. In such patients, one may choose to introduce an ablation catheter from the internal jugular or subclavian vein to facilitate catheter manipulation.

Atrioventricular Conduction Structures

The ability to successfully ablate septal accessory pathways without the complication of atrioventricular block in patients with normal intracardiac anatomy depends in large part on the detailed knowledge of the location of the compact atrioventricular node and bundle of His in relation to other structures in the heart. In several forms of congenital heart disease, the AV conducting tissue is located differently, and this anatomy must be kept in mind by the electrophysiologist.[83] For example, patients with complete AV canal as well as those with ostium primum ASD have AV conducting tissue displaced posteriorly toward the coronary sinus.[84] Ablation of posteroseptal accessory pathways would, therefore, likely carry a higher risk of complete AV block. In tetralogy of Fallot, the conducting system is at risk as well, being at the margin of the VSD.[85] Patients with L-transposition have AV conducting tissue located more anteriorly, and it is thought to be more fragile and prone to accidental damage during catheterization.[86] The anatomy for other rarer forms of congenital heart disease has also been defined.[83]

Defect-Specific Factors

Factors that are characteristic of particular defects are clearly quite numerous, and cannot be listed here in entirety. However, one can point out that in some situations, the specific congenital heart defect may make the approach to ablation more straightforward. For example, in the presence of an ostium primum or secundum ASD, the left atrium is quite easily entered and mapped. Similarly, the presence of a left superior vena cava (LSVC) to coronary sinus connection renders the coronary sinus very large and easy to enter, although catheter contact may not be as good as in the normally-sized coronary sinus, and the anterolateral left AV groove may be difficult to map due to the tendency of the catheter to enter the LSVC from the coronary sinus. Conversely, abnormalities of particular structures may dictate a different catheter course than is usual for a particular operator. For example, a patient with aortic stenosis or regurgitation should have ablation of a left-sided accessory pathway by an antegrade (transseptal) route, to avoid crossing the abnormal valve.

The situation with Ebstein's anomaly can be quite challenging. The presence of significant tricuspid regurgitation may make stable catheter position difficult on the right AV groove. Similarly, the downward displacement of the tricuspid valve leaflets (principally the septal and posterior leaflets) makes stable positioning on the AV groove difficult. However, the single most difficult factor in such patients is the difficulty in achieving an adequate temperature at the catheter tip despite maximal voltage, most likely due to the capacious right atrium and atrialized right ventricle. This, combined with the propensity of Ebstein's patients to have multiple accessory pathways and a tendency to atrial fibrillation, often make such procedures long, grueling, and ultimately unsuccessful. One must be prepared to try a variety of catheter approaches, both from the IVC as well as from the internal jugular or right subclavian vein, and one might consider the use of long venous sheaths to allow for better catheter stability.[87] For the fluoroscopic identification of the right AV groove, and for precise mapping of signals at the AV groove, some investigators have used a 2 French custom mapping wire introduced directly into the right coronary artery and advanced around the AV groove.[88-90] The use of temperature monitoring and/or temperature control is mandatory in patients with Ebstein's anomaly to allow one to differentiate between two scenarios: lack of success due to incorrect catheter position versus inadequate temperature.[91]

In patients with L-transposition, the malalignment of the atrial and the ventricular septum creates a complex anatomy when dealing with septal accessory pathways. Typically, the ventricular septum is in the sagittal plane, whereas the atrial septum is more normally positioned, except close to the AV groove. The coronary sinus may be difficult to enter. One must remember the propensity to AV block in such patients (mentioned earlier). The presence of significant left-sided AV valve regurgitation with left-sided Ebstein's anomaly may dictate an antegrade approach to the ablation of left-sided accessory pathways because positioning under the left-sided tricuspid leaflets might not allow close enough contact with the true AV groove. One might also choose this approach out of a desire to avoid creating further regurgitation with catheter ablation attempts. Transseptal puncture can certainly be accomplished, but the angle of attack may not be standard due to the abnormal septal orientation. When in doubt, pulmonary angiography to define the left atrium on levophase, and/or intraoperative echocardiography,[81,82] may be used.

Patients with more complex anatomy in whom the presence of dual atrioventricular conducting systems, which may mediate AV reciprocating tachycardia, have been documented by Saul and associates. These have been patients with atrioventricular discordance with or without atrial situs inversus.[92] The second, or accessory, AV node and distal conducting system resemble Mahaim-type atriofascicular accessory pathways seen in patients with otherwise normal cardiac anatomy, and can be mapped and ablated with similar techniques.

RADIOFREQUENCY ABLATION OF ATRIAL ARRHYTHMIAS IN POSTOPERATIVE PATIENTS WITH CONGENITAL HEART DISEASE

With the advent of catheter procedures using RF energy to eliminate the substrate for conditions, such as WPW syndrome and atrioventricular node reentry,[93-96] ablation of atrial flutter and intra-atrial reentrant tachycardia became possible. Using the surgical experience as a guide, as well as techniques for demonstrating concealed entrainment, Feld and coworkers were able to demonstrate initial success with type 1 ("typical") atrial flutter,[20] by placing RF lesions at the isthmus between the inferior vena cava and tricuspid valve annulus. Radiofrequency lesions placed in these regions were often successful in terminating atrial flutter, and long-term success has been accomplished. Subsequently, other centers have also demonstrated that high success rates are possible in adults with type 1 flutter, with an acceptable incidence of recurrence.[16,97]

These concepts have further been extended to patients with atypical atrial flutter following extensive atrial surgery.[22,90,98-100] In such patients, it is thought that the substrate for atrial flutter consists of natural anatomic obstacles to impulse propagation, such as the inferior vena cava, coronary sinus, tricuspid annulus, and so on, as well as surgically created obstacles, such as atriotomy sites, intra-atrial baffles, and conduits. Transient entrainment is used to demonstrate the reentrant mechanism, and sites of slow conduction are sought where concealed entrainment (entrainment without visible fusion on the surface ECG) can be demonstrated. Lesions at these sites are placed in an attempt to bridge the zone of slow conduction and terminate tachycardia. Encouraging initial results have been reported, but there is a significant incidence of recurrence after an initially successful procedure.

Many such patients are found, by careful mapping, to have atrial flutter, which involves the typical isthmus between the tricuspid annulus and inferior vena cava, despite having had an atriotomy. This is, in fact, quite common in patients with simple atrial surgery, as described earlier. In such patients, when typical or reverse typical atrial flutter is documented and clearly involves this isthmus, the ablation procedure may proceed by standard methods for the ablation of typical atrial flutter, and documentation of bidirectional block in the isthmus is a goal of ablation. When such an approach is taken, the results of ablation may be very good.[23]

In patients with the Mustard or Senning procedure for transposition who also have atrial flutter, most of the critical structures that support atrial reentry are in the new pulmonary venous atrium. One can speculate that the presence of a suture line at the site of the EVR in such patients might cause fibrosis and conduction delay, perhaps dramatically increasing the likelihood that atrial reentry will occur. In any case, typical and reverse typical atrial flutters are also quite common in this patient population. The approach for ablation is not straightforward when the target is in the pulmonary venous atrium and, often, the arrhythmia must be approached either via a leak or separation in the baffle, or by a retrograde transaortic approach.[29] In addition, patients have been studied who exhibited a sudden shift from one tachycardia, involving the flutter isthmus to a second tachycardia not involving this isthmus but instead involving the atriotomy, resulting from successful RF ablation at the flutter isthmus. Such a phenomenon may be an indication of true "figure of eight" reentry, with ablation of one, but not both, limbs of the figure of eight.

Catheter ablation in the atriopulmonary connection type of Fontan has been quite disappointing, in contrast to the experience with ASD repair and with Mustard and Senning procedure patients.[22,29,98,101,102] It is uncertain why this is the case, but multiple tachycardia circuits and a high incidence of recurrence after initial success have been observed. It may be that with high atrial pressures, the resulting thickening of the atrial wall due to atrial hypertrophy prevents the development of full transmural lesions. Alternatively, sluggish blood flow may not allow adequate tip cooling, limiting energy delivery and resulting in ineffective lesions.

More recent innovations involving total cavopulmonary connection by the lateral tunnel technique are clearly associated with better hemodynamics and lower atrial pressures. Unfortunately, IART is still frequently observed in such patients. In order to exclude the atrium, the SVC is connected directly to the pulmonary artery, and a tunnel is created which directs IVC flow to the underside of the pulmonary artery. The baffle that accomplishes this is similar to that used in the Mustard or Senning procedure, with a line of sutures going through the region of the EVR and with the baffle directing IVC flow away from the tricuspid annulus. The long atriotomy used to construct the lateral tunnel is closed and this suture line is in the new pulmonary venous atrium. This anatomy creates the potential for reentry in the usual flutter circuit, as well as incisional IART involving the right atriotomy, which has been elegantly demonstrated in an animal model by Rodefeld and colleagues.[103] There is not yet sufficient experience with RF ablation in this particular anatomic substrate to comment on the effectiveness, but one would expect similar results to those reported for the Senning and Mustard procedures.

RADIOFREQUENCY ABLATION OF VENTRICULAR ARRHYTHMIAS IN POSTOPERATIVE PATIENTS WITH CONGENITAL HEART DISEASE

If ventricular tachycardia is easily inducible and well tolerated hemodynamically, one may consider RF ablation. Because most evidence supports the concept of macro-reentry as the mechanism of such well-tolerated ventricular tachycardia, the use of entrainment pacing and mapping techniques is indicated. Investigators have reported successful procedures using RF energy.[104-109] Successful sites have included the area between the pulmonic annulus and outflow tract patch,[106] the isthmus of ventricular tissue between an outflow tract patch and the tricuspid annulus,[109] and the region of the VSD patch.[104]

Although well-tolerated ventricular tachycardia can be mapped in the electrophysiology laboratory, many patients have ventricular dysfunction and/or rapid ventricular tachycardia rates, and will not tolerate this. Several investigators have reported intraoperative mapping and ablation.[104,110-113] In particular, Downer and colleagues have used intraoperative mapping of the right ventricular outflow tract in the beating heart employing an endocardial electrode balloon and a simultaneous epicardial electrode sock array.[112] Ablation was carried out by cryotherapy lesions during normothermic cardiopulmonary bypass with the heart beating, or during anoxic arrest, with good success in three patients.

DEVICE THERAPY

Pacemaker Therapy

Patients with the sick sinus syndrome who require an antiarrhythmic agent other than digoxin to prevent recurrences of arrhythmia are at risk for very slow heart rates. The Joint American College of Cardiology/American Heart Association Task Force on Assessment of Cardiovascular Procedures recommended that these patients have implantation of a pacemaker.[114] Furthermore, many of these patients will have symptoms due to slow heart rates, such as exercise intolerance or syncope, and should also have pacemaker implantation. The loss of AV synchrony may not be well tolerated in patients with borderline ventricular function, and pacing is sometimes recommended for patients with few symptoms as a means of optimizing hemodynamic function.

All patients who have persistent atrioventricular block as a result of cardiac surgery should have implantation of permanent pacemakers. However, spontaneous resolution of AV block in the immediate postoperative period is often observed and, therefore, a reasonable period of postoperative observation is recommended prior to making the decision concerning permanent pacing; this observation period should be at least 5 to 6 days.

One type of implantable pacemaker is the antitachycardia device, which can be used to effect paced conversion of atrial flutter to sinus rhythm or an atrial paced rhythm. The techniques for conversion are programmable and may include underdrive to overdrive pacing, and programmed extrastimuli to scanning methods.[115] Antitachycardia pacing is generally chosen if an arrhythmia is refractory to medication, the patient is intolerant of medication, or the attacks are frequent and of long duration. An additional advantage is that many patients have the tachycardia-bradycardia syndrome, and after overdrive pacing of atrial flutter, the pacemaker is on standby to begin pacing if the patient's spontaneous rate is not adequate. Gillette and coworkers have reported their extensive experience with antitachycardia pacing in patients with congenital heart disease and atrial flutter,[115,116] using the Intertach II device (Intermedics). Although many patients with this type of pacemaker still required visits to the hospital for cardioversion, these visits were clearly much less frequent. Of some concern is the observation by Rhodes and associates of a patient whose antitachycardia pacemaker converted an episode of atrial flutter to atrial fibrillation with a rapid ventricular response, resulting in sudden death.[117] Such pacemakers must be used with concomitant administration of effective AV node blocking agents, such as propranolol, diltiazem, or verapamil. Another approach is to program the device to be manually activated, so that the patient can be under medical observation when attempts at pace conversion occur.

Although the Intertach II device is no longer available, a new generation of devices is becoming available that provides dual chamber pacing and antitachycardia pacing (AT-500, Medtronic). In addition, the latest generation of implantable cardioverter-defibrillators has the capability to perform atrial antitachycardia pacing, as well as tiered therapy in the ventricle. Finally, implantable defibrillators that provide cardioversion shocks to the atrium are in development and may eventually be useful in the population of patients with repaired congenital heart disease.

Implantable Cardioverter-Defibrillators in Tetralogy

With the rapid changes recently in lead and generator technology, implantable cardioverter-defibrillators (ICDs) have become a more viable option for the treatment of patients with ventricular tachycardia following repair of congenital heart disease. Patients with repaired congenital heart disease made up 18% of patients with ICDs in the multicenter review by Silka and colleagues of 125 pediatric patients.[118] Most reports, to date, have dealt mainly with devices attached to epicardial patches, but recent reports have included patients with transvenous systems.[119] Sudden death continues to be a serious concern in the postoperative patient, and in particular, in patients following tetralogy of Fallot repair who are at least 25 years beyond surgical repair,[58] as well as in patients with the Mustard and Senning procedures.[72] The role of ICDs in the management of patients with congenital heart disease is undergoing rapid evolution, as implantable units become smaller and more multifunctional. Now that transvenous lead technology is well developed, implantation of ICDs in patients with prior sternotomies becomes less problematic.[120] However, congenital heart disease poses many challenges to the placement of such leads. Clearly, patients with persistent intracardiac shunting are not candidates for transvenous defibrillator leads, due to the risk of thromboembolism. The use of active-fixation leads may well be necessary, due to distortions of ventricular anatomy, or the need to place such leads in a smooth-walled, morphologic left ventricle. In the past, the frequent occurrence of atrial arrhythmias in this patient population limited the usefulness of ICDs because of the tendency for atrial tachyarrhythmias with brisk atrioventricular conduction to be detected within the rate criteria for ICD therapy, leading to inappropriate shocks. Now that so-called dual chamber ICDs are becoming available,[121,122] with atrial sensing via an atrial lead, more patients with congenital heart disease will undoubtedly receive implants. The indications for such implants should be similar to those used in the adult population; namely aborted sudden death. However, large multicenter studies may very well identify subpopulations of postoperative congenital heart disease patients who are at high enough risk of sudden death to justify prophylactic ICD implantation.

SUMMARY

It is clear that the field of both catheter-based and surgically-based cardiac ablation is evolving rapidly. In most cases, definitive treatment by either type of ablation should be preferable to long-term antiarrhythmic therapy, especially when one considers the numerous potential side effects of antiarrhythmic medication in this patient population. The results of catheter ablation, unfortunately, are not as good as for ablation of more routine arrhythmias in the population of patients with otherwise normal hearts. Although this may partly be due to a lack of understanding of the exact macro-reentrant circuits that exist in each patient, it is more likely that the ability to make long linear and transmural lesions is limited. Future progress in catheter and lesion formation technology, as well as further experience with surgical ablation, may allow more such patients to benefit from the advantages of definitive cure.

The field of catheter ablation in patients with arrhythmias following repair of congenital heart disease is ripe for innovation. Success in this endeavor will require the development of energy sources, which allow for the formation of deep, transmural lesions, as well as lesions that are linear and that can be designed to bridge the gap between anatomic and surgically created obstacles to cardiac conduction. Success will also require an extensive understanding of both the preexisting cardiac anatomy as well as the details of the surgical procedures that have been performed, coupled with mapping systems that allow for detailed reconstruction of the conduction patterns that exist in these patients. Fortunately, the recent development of exciting new technology, both in new mapping systems and new energy sources, promises to accelerate progress in this field over the next several years.

REFERENCES

1. Kugler JD, Danford DA, Deal BJ, et al: Radiofrequency catheter ablation for tachyarrhythmias in children and adolescents. N Engl J Med 1994;330:1481-1487.
2. Lev M, Givson S, Miller RA: Ebstein's disease with Wolff-Parkinson-White syndrome. Am Heart J 1955;49:724-741.
3. Scheibler GL, Adams P, Anderson RC, et al: Clinical study of twenty three cases of Ebstein's anomaly of the tricuspid valve. Circulation 1959;19:165-187.
4. Benson D Jr, Gallagher JJ, Oldham HN, et al: Corrected transposition with severe intracardiac deformities with Wolff-Parkinson-White syndrome in a child: Electrophysiologic investigation and surgical correction. Circulation 1980;61:1256-1261.
5. Bokeriia LA, Revishvili A, Makhmudov MM: Syndrome of ventricular preexcitation and corrected transposition of the great vessels with insufficiency of the arterial atrioventricular valve of the Ebstein anomaly type. Kardiologiia 1984;24:94-95.
6. Keller N, Soorensen MR: Corrected transposition of the great arteries with a left-sided Ebstein-like anomaly and WPW-syndrome: A case diagnosed by two-dimensional echocardiography. Ugeskrift for Laeger 1981;143:1971-1972.
7. Perosio AM, Suarez LD, Bunster AM, et al: Pre-excitation syndrome and hypertrophic cardiomyopathy. J Electrocardiol 1983;16:29-40.
8. MacRae CA, Ghaisas N, Kass S, et al: Familial hypertrophic cardiomyopathy with Wolff-Parkinson-White syndrome maps to a locus on chromosome 7q3. J Clin Invest 1995;96:1216-1220.
9. Huhta JC, Maloney JD, Ritter DG, et al: Complete atrioventricular block in patients with atrioventricular discordance. Circulation 1983;67:1374-1377.
10. Benson DW, Silberbach GM, Kavanaugh-McHugh A, et al: Mutations in the cardiac transcription factor NKX2.5 affect diverse cardiac developmental pathways. J Clin Invest. 1999;104:1567-1573.
11. Moore KL: The Developing Human. Clinically Oriented Embryology. Philadelphia, WB Saunders, 1974.
12. Cosio FC: Endocardial mapping of atrial flutter. In Touboul P, Waldo AL (eds): Atrial Arrhythmias. St. Louis, Mosby Year Book, 1990, pp 229-240.
13. Saoudi N, Cosio F, Waldo A, et al: A new classification of atrial tachycardias based on electrophysiologic mechanisms. Eur Heart J 2000, in press.
14. Olgin JE, Kalman JM, Fitzpatrick AP, Lesh MD: Role of right atrial endocardial structures as barriers to conduction during human type I atrial flutter: Activation and entrainment mapping guided by intracardiac echocardiography. Circulation 1995;92: 1839-1848.

15. Kalman JM, Olgin JE, Saxon LA, et al: Activation and entrainment mapping defines the tricuspid annulus as the anterior barrier in typical atrial flutter. Circulation 1996;93:398-406.
16. Nakagawa H, Lazzara R, Khastgir T, et al: The role of the tricuspid annulus and the eustachian valve/ridge on atrial flutter: Relevance to catheter ablation of the septal isthmus and a new technique for rapid identification of ablation success. Circulation 1996;93:407-424.
17. Feld GK, Shahandeh-Rad F: Mechanism of double potentials recorded during sustained atrial flutter in the canine right atrial crush-injury model. Circulation 1992;86:628-641.
18. Olshansky B, Okumura K, Henthorn RW, Waldo AL: Characterization of double potentials in human atrial flutter: Studies during transient entrainment. J Am Coll Cardiol 1990; 15:833-841.
19. Lesh MD, Van Hare GF, Epstein LM, et al: Radiofrequency catheter ablation of atrial arrhythmias: Results and mechanisms. Circulation 1994;89:1074-1089.
20. Feld GK, Fleck P, Chen PS, et al: Radiofrequency catheter ablation for the treatment of human type 1 atrial flutter: Identification of a critical zone in the reentrant circuit by endocardial mapping techniques. Circulation 1992;86:1233-1240.
21. Klein G, Guiraudon G, Sharma A, Milstein S: Demonstration of macroreentry and feasibility of operative therapy in the common type of atrial flutter. Am J Cardiol 1986;57:587-591.
22. Kalman JM, Van Hare GF, Olgin JE, et al: Ablation of "incisional" reentrant atrial tachycardia complicating surgery for congenital heart disease: Use of entrainment to define a critical isthmus of slow conduction. Circulation 1996;93:502-512.
23. Chan DP, Van Hare GF, Mackall JA, et al: Importance of atrial flutter isthmus in postoperative intra-atrial reentrant tachycardia. Circulation 2000;102:1283-1289.
24. Collins KK, Love BA, Walsh EP, et al: Location of acutely successful radiofrequency catheter ablation of intraatrial reentrant tachycardia in patients with congenital heart disease. Am J Cardiol 2000;86:969-974.
25. Senning A: Surgical correction of transposition of the great vessels. Surgery 1959;45:966.
26. Mustard WT, Keith JD, Trusler GA, et al: The surgical management of transposition of the great vessels. J Thorac Cardiovasc Surg 1964;48:953.
27. Karl TR, Weintraub RG, Brizard CP, et al: Senning plus arterial switch operation for discordant (congenitally corrected) transposition. Ann Thorac Surg 1997;64:495-502.
28. Ebert PA, Gay WA, Engle MA: Correction of transposition of the great arteries: Relationship of the coronary sinus and postoperative arrhythmias. Ann Surg 1974;180:433-438.
29. Van Hare GF, Lesh MD, Ross BA, et al: Mapping and radiofrequency ablation of intraatrial reentrant tachycardia after the Senning or Mustard procedure for transposition of the great arteries. Am J Cardiol 1996;77:985-991.
30. Kanter RJ, Papagiannis J, Carboni MP, et al: Radiofrequency catheter ablation of supraventricular tachycardia substrates after Mustard and Senning operations for d-transposition of the great arteries. J Am Coll Cardiol 2000;35:428-441.
31. Fontan F, Baudet E: Surgical repair of tricuspid atresia. Thorax 1971;26:240.
32. McElhinney DB, Petrossian E, Reddy VM, Hanley FL: Extracardiac conduit Fontan procedure without cardiopulmonary bypass. Ann Thorac Surg 1998;66:1826-1828.
33. van Son JA, Reddy M, Hanley FL: Extracardiac modification of the Fontan operation without use of prosthetic material. J Thorac Cardiovasc Surg 1995;110:1766-1768.
34. Jonas RA, Castaneda AR: Modified Fontan procedure: Atrial baffle and systemic venous to pulmonary artery anastomotic techniques. J Cardiac Surg 1988;3:91.
35. Triedman JK, Jenkins KJ, Saul JP, et al: Right atrial mapping in humans using a multielectrode basket catheter. PACE. 1995; 18:800 [Abstract].
36. Gillette PC, Yeoman MA, Mullins CE, McNamara DG: Sudden death after repair of tetralogy of Fallot: Electrocardiographic and electrophysiologic abnormalities. Circulation 1977;56:566-571.
37. James FW, Kaplan S, Chou TC: Unexpected cardiac arrest in patients after surgical correction of tetralogy of Fallot. Circulation 1975;52:691-695.
38. Quattlebaum TG, Varghese J, Neill CA, Donahoo JS: Sudden death among postoperative patients with tetralogy of Fallot: A follow-up study of 243 patients for an average of twelve years. Circulation 1976;54:289-293.
39. Marin-Garcia J, Moller JH: Sudden death after operative repair of tetralogy of Fallot. Br Heart J 1977;39:1380-1385.
40. Deanfield JE, McKenna WJ, Hallidie-Smith KA: Detection of late arrhythmia and conduction disturbance after correction of tetralogy of Fallot. Br Heart J 1980;44:248-253.
41. Garson A Jr, Gillette PC, Gutgesell HP, McNamara DG: Stress-induced ventricular arrhythmia after repair of tetralogy of Fallot. Am J Cardiol 1980;46:1006-1012.
42. Shen WK, Holmes DR Jr, Porter CJ, et al: Sudden death after repair of double-outlet right ventricle. Circulation 1990;81:128-136.
43. Garson A Jr, McNamara DG: Sudden death in a pediatric cardiology population, 1958 to 1983: Relation to prior arrhythmias. J Am Coll Cardiol 1985;5:134B-137B.
44. Vetter VL, Horowitz LN: Electrophysiologic residua and sequelae of surgery for congenital heart defects. Am J Cardiol 1982;50: 588-604.
45. Zahka KG, Horneffer PJ, Rowe SA, et al: Long-term valvular function after total repair of tetralogy of Fallot: Relation to ventricular arrhythmias. Circulation 1988;78:III14-19.
46. Gatzoulis MA, Till JA, Somerville J, Redington AN: Mechanoelectrical interaction in tetralogy of Fallot: QRS prolongation relates to right ventricular size and predicts malignant ventricular arrhythmias and sudden death. Circulation 1995; 92:231-237.
47. Gatzoulis MA, Till JA, Redington AN: Depolarization-repolarization inhomogeneity after repair of tetralogy of Fallot: The substrate for malignant ventricular tachycardia? Circulation. 1997;95: 401-404.
48. Deanfield JE, Ho SY, Anderson RH, et al: Late sudden death after repair of tetralogy of Fallot: A clinicopathologic study. Circulation 1983;67:626-631.
49. Deanfield J, McKenna W, Rowland E: Local abnormalities of right ventricular depolarization after repair of tetralogy of Fallot: A basis for ventricular arrhythmia. Am J Cardiol 1985;55:522-525.
50. Zimmermann M, Friedli B, Adamec R, Oberhansli I: Ventricular late potentials and induced ventricular arrhythmias after surgical repair of tetralogy of Fallot. Am J Cardiol 1991;67:873-878.
51. Kremers MS, Wells PJ, Black WH, Solodyna MA: Entrainment of ventricular tachycardia in postoperative tetralogy of Fallot. Pacing Clin Electrophysiol 1988;11:1310-1314.
52. Aizawa Y, Kitazawa H, Washizuka T, et al: Conductive properties of the reentrant pathway of ventricular tachycardia during entrainment from outside and within the zone of slow conduction. Pacing Clin Electrophysiol 1995;18:663-672.
53. Chandar JS, Wolff GS, Garson A Jr, et al: Ventricular arrhythmias in postoperative tetralogy of Fallot. Am J Cardiol 1990;65:655-661.
54. Walsh EP, Rockenmacher S, Keane JF, et al: Late results in patients with tetralogy of Fallot repaired during infancy. Circulation 1988;77:1062-1067.
55. Garson A Jr, Randall DC, Gillette PC, et al: Prevention of sudden death after repair of tetralogy of Fallot: Treatment of ventricular arrhythmias. J Am Coll Cardiol 1985;6:221-227.
56. Harrison DA, Harris L, Siu SC, et al: Sustained ventricular tachycardia in adult patients late after repair of tetralogy of Fallot. J Am Coll Cardiol 1997;30:1368-1373.
57. Balaji S, Lau YR, Case CL, Gillette PC: QRS prolongation is associated with inducible ventricular tachycardia after repair of tetralogy of Fallot. Am J Cardiol 1997;80:160-163.
58. Nollert G, Fischlein T, Bouterwek S, et al: Long-term survival in patients with repair of tetralogy of Fallot: 36-year follow-up of 490 survivors of the first year after surgical repair. J Am Coll Cardiol 1997;30:1374-1383.
59. Goldman BS, Williams WG, Hill T, et al: Permanent cardiac pacing after open heart surgery: Congenital heart disease. Pacing Clin Electrophysiol 1985;8:732-739.
60. Bonatti V, Agnetti A, Squarcia U: Early and late postoperative complete heart block in pediatric patients submitted to open-heart surgery for congenital heart disease. Pediatr Med Chir 1998;20:181-186.
61. Garson A Jr: Medicolegal problems in the management of cardiac arrhythmias in children. Pediatrics 1987;79:84-88.

62. Preliminary report: Effect of encainide and flecainide on mortality in a randomized trial of arrhythmia suppression after myocardial infarction. The Cardiac Arrhythmia Suppression Trial (CAST) Investigators. N Engl J Med 1989;321:406-412.
63. Effect of the antiarrhythmic agent moricizine on survival after myocardial infarction. The Cardiac Arrhythmia Suppression Trial II Investigators. N Engl J Med 1992;327:227-233.
64. Garson A Jr, Gillette PC, McNamara DG: Propranolol: The preferred palliation for tetralogy of Fallot. Am J Cardiol 1981;47: 1098-1104.
65. Fish FA, Gillette PC, Benson DW Jr: Proarrhythmia, cardiac arrest and death in young patients receiving encainide and flecainide. The Pediatric Electrophysiology Group. J Am Coll Cardiol 1991; 18:356-365.
66. Hohnloser SH: Proarrhythmia with class III antiarrhythmic drugs: Types, risks, and management. Am J Cardiol 1997;80:82G-89G.
67. Waldo AL, Camm AJ, deRuyter H, et al: Effect of d-sotalol on mortality in patients with left ventricular dysfunction after recent and remote myocardial infarction. The SWORD Investigators. Survival with Oral d-Sotalol. Lancet 1996;348:7-12.
68. Daniels CJ, Schutte DA, Hammond S, Franklin WH: Acute pulmonary toxicity in an infant from intravenous amiodarone. Am J Cardiol 1997;80:1113-1116.
69. Celiker A, Kocak G, Lenk MK, et al: Short- and intermediate-term efficacy of amiodarone in infants and children with cardiac arrhythmia. Turk J Pediatr 1997;39:219-225.
70. Villain E: Amiodarone as treatment for atrial tachycardias after surgery. Pacing Clin Electrophysiol 1997;20:2130-2132.
71. Simpson RJ Jr, Foster JR, Woelfel AK, Gettes LS: Management of atrial fibrillation and flutter: A reappraisal of digitalis therapy. Postgrad Med 1986;79:241-253.
72. Garson A Jr, Bink BM, Hesslein PS, et al: Atrial flutter in the young: A collaborative study of 380 cases. J Am Coll Cardiol 1985;6: 871-878.
73. Garson A Jr, Gillette PC, McVey P, et al: Amiodarone treatment of critical arrhythmias in children and young adults. J Am Coll Cardiol 1984;4:749-755.
74. Tipple M, Sandor G: Efficacy and safety of oral sotalol in early infancy. Pacing Clin Electrophysiol 1991;14:2062-2065.
75. Maragnes P, Tipple M, Fournier A: Effectiveness of oral sotalol for treatment of pediatric arrhythmias. Am J Cardiol 1992;69: 751-754.
76. Abinader E, Borochowitz Z, Berger A: A hemodynamic complication of verapamil therapy in a neonate. Helv Paediatr Acta 1981;36:451-455.
77. Shim D, Lloyd TR, Cho KJ, et al: Transhepatic cardiac catheterization in children: Evaluation of efficacy and safety. Circulation 1995;92:1526-1530.
78. Sommer RJ, Golinko RJ, Mitty HA: Initial experience with percutaneous transhepatic cardiac catheterization in infants and children. Am J Cardiol 1995;75:1289-1291.
79. Lesh MD, Van Hare GF, Scheinman MM, et al: Comparison of the retrograde and transseptal methods for ablation of left free wall accessory pathways. J Am Coll Cardiol 1993;22:542-549.
80. Lai WW, al-Khatib Y, Klitzner TS, et al: Biplanar transesophageal echocardiographic direction of radiofrequency catheter ablation in children and adolescents with the Wolff-Parkinson-White syndrome. Am J Cardiol 1993;71:872-874.
81. Drant SE, Klitzner TS, Shannon KM, et al: Guidance of radiofrequency catheter ablation by transesophageal echocardiography in children with palliated single ventricle. Am J Cardiol 1995; 76:1311-1312.
82. Tucker KJ, Curtis AB, Murphy J, et al: Transesophageal echocardiographic guidance of transseptal left heart catheterization during radiofrequency ablation of left-sided accessory pathways in humans. Pacing Clin Electrophysiol 1996;19:272-281.
83. Davies MJ, Anderson RH: Conduction system in congenital heart disease. In Davies MJ, Anderson RH (eds): The Conduction System of the Heart. London, Butterworths, 1983, pp 95-166.
84. Thiene G, Wenink AC, Frescura C, et al: Surgical anatomy and pathology of the conduction tissues in atrioventricular defects. J Thorac Cardiovasc Surg 1981;82:928-937.
85. Dickinson DF, Wilkinson JL, Smith A, et al: Variations in the morphology of the ventricular septal defect and disposition of the atrioventricular conduction tissues in tetralogy of Fallot. Thorac Cardiovasc Surg 1982;30:243-249.
86. Anderson RH, Becker AE, Arnold R, Wilkinson JL: The conducting tissues in congenitally corrected transposition. Circulation 1974;50:911-923.
87. Saul JP, Hulse JE, De W, Weber AT, et al: Catheter ablation of accessory atrioventricular pathways in young patients: Use of long vascular sheaths, the transseptal approach and a retrograde left posterior parallel approach. J Am Coll Cardiol 1993;21:571-583.
88. Swartz JF, Cohen AI, Fletcher RD, et al: Right coronary epicardial mapping improves accessory pathway catheter ablation success. Circulation 1989;80 Suppl II:II-430 [Abstract].
89. Lesh MD, Van Hare GF, Chien WW, Scheinman MM: Mapping in the right coronary artery as an aid to radiofrequency ablation of right-sided accessory pathways. PACE 1991;14:671 [Abstract].
90. Van Hare GF, Lesh MD, Stanger P: Radiofrequency catheter ablation of supraventricular arrhythmias in patients with congenital heart disease: Results and technical considerations. J Am Coll Cardiol 1993;22:883-890.
91. Calkins H, Prystowsky E, Carlson M, et al: Temperature monitoring during radiofrequency catheter ablation procedures using closed loop control. Atakr Multicenter Investigators Group. Circulation 1994;90:1279-1286.
92. Saul JP, Walsh EP, Triedman JK: Mechanisms and therapy of complex arrhythmias in pediatric patients. J Cardiovasc Electrophysiol 1995;6:1129-1148.
93. Van Hare GF, Lesh MD, Scheinman MM, Langberg JJ: Percutaneous radiofrequency catheter ablation for supraventricular arrhythmias in children. J Am Coll Cardiol 1991;17: 1613-1620.
94. Dick MD, O'Connor BK, Serwer GA, et al: Use of radiofrequency current to ablate accessory connections in children. Circulation 1991;84:2318-2324.
95. Saul JP, Walsh EP, Langberg JJ, et al: Radiofrequency ablation of accessory atrioventricular pathways: Early results in children with refractory SVT. Circulation 1990;82, Supplement III:III-222 [Abstract].
96. Walsh EP, Saul JP: Transcatheter ablation for pediatric tachyarrhythmias using radiofrequency electrical energy. Pediatr Ann 1991;20:386, 388-392.
97. Okumura K, Henthorn RW, Epstein AE, et al: Further observations on transient entrainment: Importance of pacing site and properties of the components of the reentry circuit. Circulation 1985;72:1293-1307.
98. Treidman JK, Saul JP, Weindling SN, Walsh EP: Radiofrequency ablation of intra-atrial reentrant tachycardia after surgical palliation of congenital heart disease. Circulation 1995;91:707-714.
99. Triedman JK, Jenkins KJ, Colan SD, et al: Intra-atrial reentrant tachycardia after palliation of congenital heart disease: Characterization of multiple macroreentrant circuits using fluoroscopically based three-dimensional endocardial mapping. J Cardiovasc Electrophysiol 1997;8:259-270.
100. Baker BM, Lindsay BD, Bromberg BI, et al: Catheter ablation of clinical intraatrial reentrant tachycardias resulting from previous atrial surgery: Localizing and transecting the critical isthmus. J Am Coll Cardiol 1996;28:411-417.
101. Balaji S, Johnson TB, Sade RM, et al: Management of atrial flutter after the Fontan procedure. J Am Coll Cardiol 1994;23: 1209-1215.
102. Case CL, Gillette PC, Douglas DE, Liebermann RA: Radiofrequency catheter ablation of atrial flutter in a patient with postoperative congenital heart disease. Am Heart J 1993; 126:715-716.
103. Rodefeld MD, Bromberg BI, Schuessler RB, et al: Atrial flutter after lateral tunnel construction in the modified Fontan operation: A canine model. J Thorac Cardiovasc Surg 1996;111:514-526.
104. Ressia L, Graffigna A, Salerno-Uriarte JA, Vigano M: The complex origin of ventricular tachycardia after the total correction of tetralogy of Fallot. Giornale Italiano di Cardiologia 1993; 23:905-910.
105. Burton ME, Leon AR: Radiofrequency catheter ablation of right ventricular outflow tract tachycardia late after complete repair of tetralogy of Fallot using the pace mapping technique. Pacing Clin Electrophysiol 1993;16:2319-2325.
106. Biblo LA, Carlson MD: Transcatheter radiofrequency ablation of ventricular tachycardia following surgical correction of tetralogy of Fallot. Pacing Clin Electrophysiol 1994;17:1556-1560.

107. Goldner BG, Cooper R, Blau W, Cohen TJ: Radiofrequency catheter ablation as a primary therapy for treatment of ventricular tachycardia in a patient after repair of tetralogy of Fallot. Pacing Clin Electrophysiol 1994;17:1441-1446.
108. Gonska BD, Cao K, Raab J, et al: Radiofrequency catheter ablation of right ventricular tachycardia late after repair of congenital heart defects. Circulation 1996;94:1902-1908.
109. Horton RP, Canby RC, Kessler DJ, et al: Ablation of ventricular tachycardia associated with tetralogy of Fallot: Demonstration of bidirectional block. J Cardiovasc Electrophysiol 1997;8:432-435.
110. Frank G, Schmid C, Baumgart D, et al: Surgical therapy of life-threatening tachycardic cardiac arrhythmias in children. Monatsschrift Kinderheilkunde 1989;137:269-274.
111. Lawrie GM, Pacifico A, Kaushik R: Results of direct surgical ablation of ventricular tachycardia not due to ischemic heart disease. Ann Surg 1989;209:716-727.
112. Downar E, Harris L, Kimber S, et al: Ventricular tachycardia after surgical repair of tetralogy of Fallot: Results of intraoperative mapping studies. J Am Coll Cardiol 1992;20:648-655.
113. Misaki T, Tsubota M, Tanaka M, et al: Surgical treatment of ventricular tachycardia after radical correction of tetralogy of Fallot. Nippon Kyobu Geka Gakkai Zasshi Journal of the Japanese Association for Thoracic Surgery 1990;38:130-134.
114. Frye RL, Collins JJ, DeSanctis RW, et al: Guidelines for permanent cardiac pacemaker implantation, May 1984. A report of the Joint American College of Cardiology/American Heart Association Task Force on Assessment of Cardiovascular Procedures (Subcommittee on Pacemaker Implantation). Circulation 1984;70:331A-339A.
115. Gillette PC, Zeigler VL, Case CL, et al: Atrial antitachycardia pacing in children and young adults. Am Heart J 1991;122: 844-849.
116. Gillette PC: Antitachycardia pacing. Pacing Clin Electrophysiol 1997;20:2121-2124.
117. Rhodes LA, Walsh EP, Gamble WJ, et al: Benefits and potential risks of atrial antitachycardia pacing after repair of congenital heart disease. Pacing Clin Electrophysiol 1995;18:1005-1016.
118. Silka MJ, Kron J, Dunnigan A, Dick MD: Sudden cardiac death and the use of implantable cardioverter-defibrillators in pediatric patients. The Pediatric Electrophysiology Society. Circulation 1993;87:800-807.
119. Wilson WR, Greer GE, Grubb BP: Implantable cardioverter-defibrillators in children: A single-institutional experience. Ann Thorac Surg 1998;65:775-778.
120. Silka MJ: Implantable cardioverter-defibrillators in children. A perspective on current and future uses. J Electrocardiol 1996;29 Suppl:223-225.
121. Li HG, Thakur RK, Yee R, Klein GJ: Ventriculoatrial conduction in patients with implantable cardioverter defibrillators: Implications for tachycardia discrimination by dual chamber sensing. Pacing Clin Electrophysiol 1994;17:2304-2306.
122. Lavergne T, Daubert JC, Chauvin M, et al: Preliminary clinical experience with the first dual chamber pacemaker defibrillator. Pacing Clin Electrophysiol 1997;20:182-188.

Chapter 36 The Brugada Syndrome

PEDRO BRUGADA, RAMON BRUGADA, and
JOSEP BRUGADA

Sudden cardiac death is a major cause of mortality in the Western world, with an approximate incidence of 1 per 1000 people per year. Up to 25% of all natural deaths are sudden. The most common cause of sudden death is an acute ischemic event, but sudden death may also be caused by a ventricular arrhythmia without the need for ischemia. Patients suffering from (near) sudden death usually have some form of structural heart disease. However, in 10% to 20% of cases, no structural heart disease is found at autopsy even after extensive medical investigation of the survivors. In 1992, we reported a new syndrome causing sudden death in individuals with structurally normal hearts.[1] This syndrome is responsible for approximately 50% of sudden deaths in individuals without structural heart disease. It is the leading cause of natural death in males younger than 50 years in South Asia, with annual incidences of up to one death per 1000 inhabitants in countries such as Laos. This syndrome is hereditary and known in the medical literature as Brugada syndrome.

The syndrome is characterized by: (1) an electrocardiogram resembling a right bundle branch block with a peculiar type of ST segment elevation in the right precordial leads V1 through V3 (Fig. 36-1); (2) polymorphic ventricular arrhythmias that cause syncope when self-terminating, and sudden death when long lasting and not terminated by cardiopulmonary resuscitation; (3) a structurally normal heart as assessed by invasive and noninvasive investigations and cardiac biopsies; (4) a familial occurrence in about one half of patients with an autosomal mode of inheritance in at least one half of the familial cases; and (5) mutations in the gene *SCN5A* that encodes for the human cardiac sodium channel, and the genes encoding for some potassium channels (unpublished observations). Other channels may also be involved indicating that the disease is heterogeneous. Electrophysiologically, Brugada syndrome is the mirror image of the LQT3 variant of the long QT syndrome, an allelic disorder to Brugada syndrome that also affects the gene *SCN5A*. Additional features of the Brugada syndrome include: (1) a prolonged P–R interval on the electrocardiogram caused by a prolongation of the H–V interval as shown by electrophysiological investigations; (2) a terminal negative T wave in leads V1 through V3; (3) a great variability of the electrocardiogram over time depending on the autonomic balance, administration of antiarrhythmic and other drugs affecting channel function, body temperature, and other factors not clear at present; and (4) monomorphic, instead of polymorphic, ventricular tachycardia, particularly while on antiarrhythmic drugs.

Recent Advances in the Diagnosis of Brugada Syndrome

The diagnosis of Brugada syndrome is obvious when the electrocardiogram looks like the one in Figure 36-1.

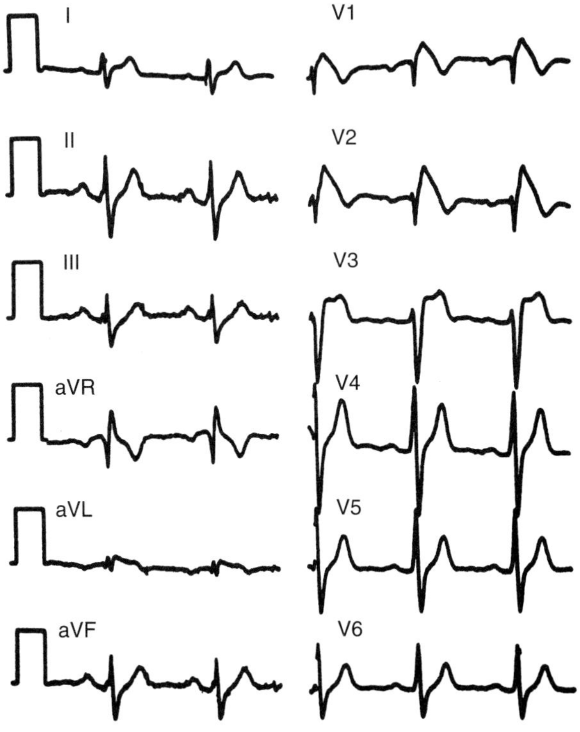

FIGURE 36-1 Typical electrocardiogram of Brugada syndrome. Please note the prolongation of the P–R interval, the ST segment elevation in leads V1 through V3 with a terminal negative T wave.

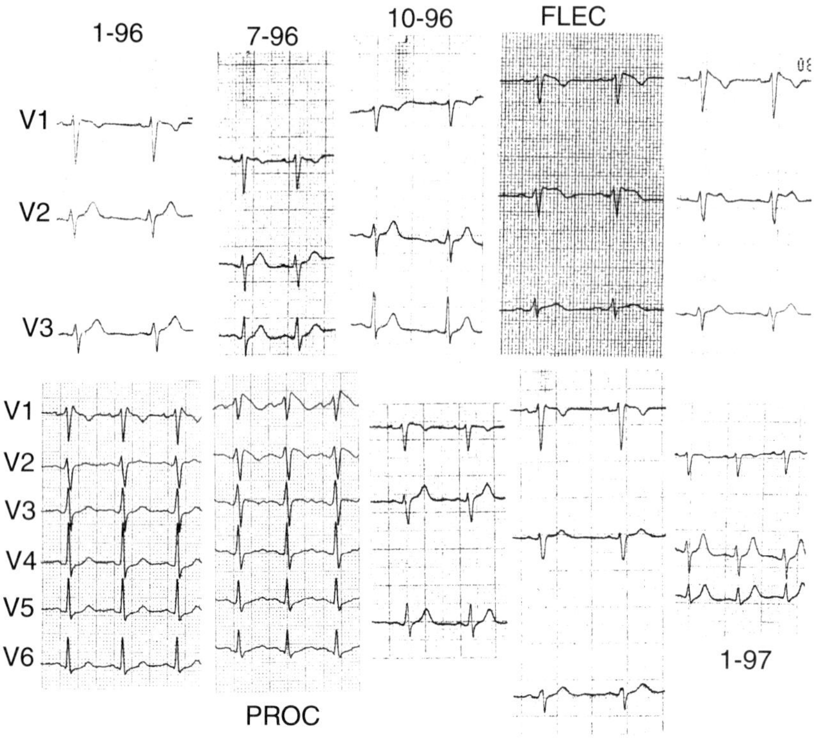

FIGURE 36-2 Spontaneous and drug-induced variations of the electrocardiogram in a patient with Brugada syndrome. Please note the "saddle type" ST segment elevation in 7 of 96 as compared with the "coved type" ST segment elevation in other moments (Flec and Proc).

However, in recent publications it has become clear that the electrocardiogram may change from day to day under various influences. The J wave elevation may be the most prominent feature giving a "saddle-type" ST segment elevation (Fig. 36-2) instead of the coved type shown in Figure 36-1.[2-7] A large variety of drugs influence the electrocardiographic aspect, including antiarrhythmic drugs, tricyclic antidepressants, all blockers of sodium and potassium currents, adrenergic and vagal stimulation, and exercise.[8-19] Even body temperature has effects on the electrocardiographic aspects (unpublished observations). Administration of class I drugs may result in the spontaneous occurrence of ventricular tachycardia or ventricular fibrillation.[20] However, controlled administration of these drugs is of extreme value for the diagnosis of Brugada syndrome in the concealed forms of the disease (Fig. 36-3). Variants have been described with ST segment elevation in the inferior leads, bifid T waves, alternans of the T wave, or only J wave elevation. Monomorphic, instead of polymorphic ventricular, tachycardia has been described.[20-27]

It is important to exclude other causes of ST segment elevation before making the diagnosis of Brugada syndrome. These other causes include ischemia, mediastinal tumors, hypothermia, and right ventricular disease.[28-46] Some investigators have suggested that Brugada syndrome is a possible variant of right ventricular dysplasia; however, new findings have shown that Brugada syndrome and right ventricular dysplasia are genetically unrelated (see later). Most publications suggesting right ventricular dysplasia as the cause of Brugada syndrome have reported only nonspecific findings.

The diagnosis of Brugada syndrome strongly depends on the degree of suspicion by the physician. Patients with *syncope of unknown cause* should be challenged with an intravenous class I drug to exclude Brugada syndrome. Ajmaline, flecainide, or procainamide can be used for that purpose. The doses are given in Table 36-1. Unfortunately, only procainamide is available in the United States. Procainamide seems to be the weakest drug of the three to uncover the electrocardiographic abnormalities. No drugs are approved by the FDA to test for Brugada syndrome. Patients with *ventricular fibrillation of unknown cause and a normal heart* (idiopathic ventricular fibrillation) should also be challenged with a class I drug to unmask Brugada syndrome. Many cases of idiopathic ventricular fibrillation are not familial, but Brugada syndrome is frequently familial. Failure to make the diagnosis may result in failure to test relatives of the patient. That may also result in severe legal problems if another family member develops syncope or (aborted) sudden death. First degree relatives of patients with Brugada syndrome have to be tested for the disease. They should undergo a pharmacologic challenge (see Table 36-1) and blood should be drawn for genetic testing. Positive individuals should undergo programmed ventricular stimulation of the heart and, if inducible, should receive an implantable defibrillator—independent of symptoms. To the best of our knowledge asymptomatic noninducible carriers of the disease need no therapy but should be carefully followed. These thoughts may change when new follow-up data become available.

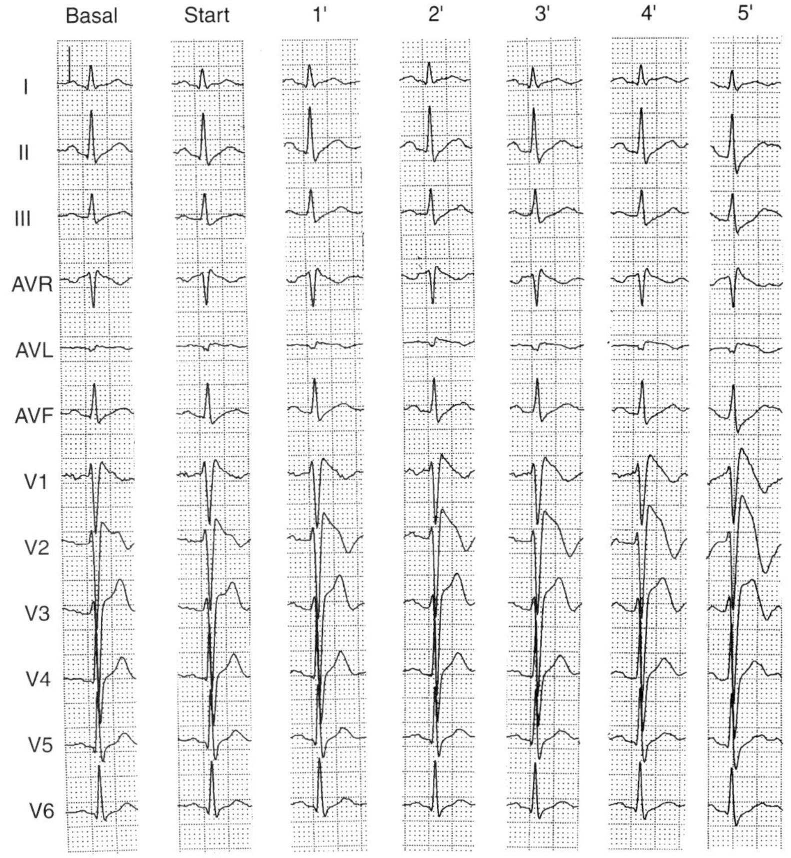

FIGURE 36-3 Effects of the acute intravenous administration of 50 mg ajmaline in a patient with the concealed form of Brugada syndrome. Please note the acute worsening of ST segment elevation in leads V1 through V3.

Advances in the Understanding of the Etiology and Molecular Electrophysiology of Brugada Syndrome

The Brugada syndrome is a genetically determined disease. A genetic cause was suspected from the very first publication.[1] A family history of sudden death was present in several of the first eight reported patients, two were siblings with the same manifestations. The genetic nature of the disease was proved just 6 years later when we reported that the syndrome is caused by mutations in the gene *SCN5A* encoding for the human cardiac sodium channel.[47] Many reports have since confirmed that patients with Brugada syndrome suffer from a channelopathy.[48-49] Affected loci are different from the ones affected in familial forms of right ventricular dysplasia,[50] chronic forms of heart failure, nonspecific conduction disturbances, or congestive cardiomyopathy.[51-52] Some genetically determined conduction disturbances are allelic to Brugada syndrome, also affecting the gene *SCN5A*.[53] Also LQT3, a variant of the long QT syndrome, affects the same gene but has mirror-image electrophysiological effects, with gain of function versus loss of function as in Brugada syndrome.[54] Other crossover syndromes associated with mutations of *SCN5A* exhibit features of LQT3, Brugada syndrome, and conduction abnormalities.[85,86] The apparent paradox has been explained by subtle alterations of ion channel gating behavior in a heterogeneous myocardium.[85,87]

The electrophysiological mechanisms of Brugada syndrome have become understood thanks to the discovery of the genetic defect and the expression of these defects in *Xenopus* oocytes and thanks to the experimental studies by the group of Antzelevitch. In summary, Brugada syndrome causes a loss of function of the cardiac sodium channel. The loss of function is temperature dependent, worsening at higher temperatures and, thus, helping to explain cases of Brugada syndrome with ventricular fibrillation during febrile episodes. The loss of function of the sodium channel leaves the transient outward potassium current, I_{to}, unopposed, particularly in cells of the right ventricular epicardium where it is prominent. The unopposed I_{to} shortens markedly the duration of the epicardial action potential, creating heterogeneity of action potential duration between the epicardium and endocardium. This sets the stage for phase 2 reentry and ventricular fibrillation (for further details and reviews

TABLE 36-1 Doses of the Antiarrhythmic Drugs Used to Unmask Brugada Syndrome

Ajmaline	0.7 mg/kg body weight IV in 5 min
Flecainide	2 mg/kg body weight IV in 10 min
Procainamide	10 mg/kg body weight IV in 10 min

see references 55 through 71). The pathophysiologic mechanisms explain the effects (or lack thereof) of antiarrhythmic and other drugs in Brugada syndrome. The same holds true for recently discovered potassium channel mutations (unpublished observations).

Other Aspects Related to Mechanisms of Brugada Syndrome

As for the long QT syndrome, it has been suggested that autonomic disturbances may play a role in Brugada syndrome.[72] There exists, indeed, a circadian pattern of development of ventricular fibrillation in Brugada syndrome, with most episodes occurring at rest (as in the LQT3 variant of the long QT syndrome).[73] However, the disease itself is clearly explained by the genetic defects and the role of the autonomic nervous system is, as in the long QT syndrome, simply one of a modulator.

It is important to realize that Brugada syndrome is not related to other forms of polymorphic ventricular arrhythmias that have been recently reported in individuals with normal hearts.[74-75] Unfortunately, in these new reports, no pharmacologic challenge was done to exclude Brugada syndrome.

Advances in the Understanding of the Prevalence and Distribution of the Disease

A review of the literature shows that the number of patients recognized to suffer from Brugada syndrome grows exponentially. In the year 2000, more than 2000 publications on Brugada syndrome or related to it appeared in peer-reviewed journals (for a list visit *www.brugada.org*). Patients have been reported from everywhere in the world with the exception of Africa, although patients do exist and were reported to our registry. From all this literature, two articles require special attention. Viskin and coworkers recently showed that a "Brugada ECG is a definite marker of sudden death"[77] as already was shown by Nademanee and colleagues in a study from South Asia. Hermida and associates[78] found a frightening incidence of 1 per 1000 electrocardiograms with diagnostic Brugada syndrome without pharmacologic challenge in an asymptomatic and healthy French population. They discussed the possible medical, sociologic, and even economic consequences of the prevalence of the disease and its recognition in asymptomatic individuals.

Relation with Sudden Infant Death Syndrome (SIDS) and Sudden Unexpected Death Syndrome (SUDS)

For years, the causes of SIDS have remained controversial, particularly because the circumstances of death have been difficult to study and the methods to study the causes were inappropriate. In our first publication, three of the eight reported patients were children with aborted sudden death.[1] That immediately established a link between Brugada syndrome and SIDS, which has been recently confirmed by other investigators.[79] That makes clear that an electrocardiogram should be a routine examination of each newborn to search for Brugada syndrome and long QT syndrome. We have to recognize that the sensitivity and specificity of the electrocardiographic findings are unknown.

SUDS was first described among South Asian refugees. Nademanee and colleagues[80] showed that SUDS is Brugada syndrome. While the genetic defects in South Asian patients are being intensively investigated, preliminary data show the same mutations as in European families with Brugada syndrome. These patients suffer frequently from sudden death at night, whereas in Europe, patients with Brugada syndrome suffer from death during a variety of circumstances. Also, in South Asia, affected individuals are almost exclusively male, whereas in the rest of the world a ratio of eight males to one female exists.

Further Advances in Treatment

Unfortunately, and in spite of the major advances in the diagnosis and understanding of the pathophysiology of Brugada syndrome, a definite breakthrough in curative treatment (genetic manipulation) has not been reached. Thus, the implantable cardioverter defibrillator remains the only therapy to prevent sudden death. We have shown that β-blockers and amiodarone do not prevent recurrences of ventricular arrhythmias in these patients. Although flecainide is good for LQT3,[81] it is bad for Brugada patients, as are propafenone, disopyramide, and other class I drugs—including pilsicainide. Despite promising reports showing the value of quinidine in idiopathic ventricular fibrillation, our personal experience with Brugada syndrome has been that quinidine is ineffective in preventing either recurrences or even electrical storms (unpublished observations in three patients). Electrical storms remain a problem in Brugada syndrome patients due to the psychological, physical, and even social effects of multiple defibrillations. The only therapy that has shown some effect during the electrical storm is an infusion of isoproterenol (personal observations). Bretylium tosylate seemed effective in one patient but failed to prevent recurrences shortly thereafter. All class I antiarrhythmic drugs and intravenous amiodarone worsen electrical storms (unpublished data). Repolarization changes in Brugada syndrome may trigger inappropriate device therapy because of oversensing of the T wave caused by the automatic gain control of the device. Alternative therapies do not exist at present, although it may be possible to localize the site of origin and ablate the initiating ventricular premature beat thereby preventing the polymorphic arrhythmias.

Treatment of *symptomatic individuals* with an implantable defibrillator is generally accepted, but the treatment of *asymptomatic carriers* of the disease remains a major problem. It has been suggested that asymptomatic carriers without any evidence of ECG changes spontaneously present can be left untreated.[88] Certainly among asymptomatic carriers a spontaneously abnormal ECG carries a poorer prognosis.[89] A family history of sudden death is important to make decisions but only when sudden death is related to the disease. That is not always obvious because many sudden deaths in families with Brugada syndrome are of an unclear cause or not related to the disease.[82] That is certainly also the case in other diseases, such as the long QT syndrome or hypertrophic cardiomyopathy. There is growing evidence that classifying a patient as symptomatic or asymptomatic is just a matter of the timing of the diagnosis. If the patient is seen before an event, we call him *asymptomatic*, but if seen after an event (syncope or aborted sudden death) we call him *symptomatic*. Analysis of electrocardiograms in symptomatic individuals recorded before the symptoms start shows that they already had the disease before the symptoms. This is quite a logical observation in a genetically determined disease. The clustering of first onset of symptoms around 40 years of age (as in many other genetically determined diseases) is really worrying. We are, at present, confronted with the second and third generations of carriers of the disease. The asymptomatic carriers are, on average, 15 years younger than the symptomatic carriers. Thus, we may need a 15-year follow-up interval to show that the young asymptomatic carriers will become symptomatic. Meanwhile, we recommend electrophysiological investigation of asymptomatic carriers. If a sustained ventricular arrhythmia is induced, an implantable defibrillator should be recommended because we do not know better and the patient should not pay for our ignorance. This is not always an easy decision, particularly in children. In a recent publication[83] we have reported the long-term follow-ups of 334 individuals with the syndrome. Of them, 190 were asymptomatic. Sixteen sudden deaths (or aborted sudden deaths) occurred, an incidence of 8% at a mean of 27 months' follow-up (standard deviation 29 months). The predictors of sudden death in asymptomatic individuals were: a spontaneous abnormal ECG, (all events occurred in this group with an incidence of 14%), inducibility during programmed ventricular stimulation, and a prolonged H–V interval.

SUMMARY

Sudden cardiac death has epidemic proportions. When it affects individuals at the onset of their most productive period in life, the consequences are devastating at all levels. That is why intensive research should continue to find a definitive cure for this genetically determined disease. The pessimistic attitudes of some investigators[84] should not be taken into account. The necessary funds should be allocated to support research to fight sudden cardiac death, because one day we will have a cure for Brugada syndrome, a monogenic disease, and later on also for the most complex, multigenetic determined forms of cardiac disease and sudden death.

ACKNOWLEDGMENTS

This work was supported in part and equally by: The Ramon Brugada Senior Foundation (Belgium, Spain, USA), the Mapfre Medicine Foundation (Spain), and the Cardiovascular Research and Teaching Institute Aalst (Belgium).

REFERENCES

1. Brugada P, Brugada J: Right bundle branch block, persistent ST segment elevation and sudden cardiac death: A distinct clinical and electrocardiographic syndrome. J Am Coll Cardiol 1992;20: 1391-1396.
2. Bjerregaard P, Gussak I, Kotar SL, et al: Recurrent syncope in a patient with prominent J wave. Am Heart J 1994;127: 1426-1430.
3. Brugada P, Geelen P: Some electrocardiographic patterns predicting sudden cardiac death that every doctor should recognize. Acta Cardiol 1997;52:473-484.
4. Bjerregaard P, Gussak I, Antzelevitch C: The enigmatic manifestation of Brugada syndrome. J Cardiovasc Electrophysiol 1998;9: 109-110.
5. Matduo K, Shimizu W, Kurita T, et al: Dynamic changes of 12-lead electrocardiograms in a patient with Brugada syndrome. J Cardiovasc Electrophysiol 1998;9:508-512.
6. Prieto-Solis JA, Martin-Duran A: Multiples cambios en la morfologia del segmento ST en un paciente con sindrome de Brugada. Rev Esp Cardiol 2000;53:136-138.
7. Brugada P: Brugada syndrome: An electrocardiographic diagnosis not to be missed. Heart 2000;84:1-3.
8. Miyazaki T, Mitamura H, Miyoshi S, et al: Autonomic and antiarrhythmic modulation of ST segment elevation in patients with Brugada syndrome. J Am Coll Cardiol 1996;27: 1061-1070.
9. Nakazato Y, Shimada K, Nakazato K, et al: Abnormal ST elevation during oral pilsicainide treatment. Ther Research 1997;18: 357-362.
10. Nakazato Y, Nakata Y, Yasuda M, et al: Safety and efficacy of oral flecainide acetate in patients with cardiac arrhythmias. Jpn Heart J 1997;38:379-385.
11. Kasanuki H, Ohnishi S, Ohtuka M, et al: Idiopathic ventricular fibrillation induced with vagal activity in patients without obvious heart disease. Circulation 1997;95:2277-2285.
12. Goethals P, Debruyne P, Saffarian M: Drug-induced Brugada syndrome. Acta Cardiol 1998;53:157-160.
13. Nakamura W, Segawa K, Ito H, et al: Class Ic antiarrhythmic drugs, flecainide and pilsicainide, produce ST segment elevation simulating inferior myocardial ischemia. J Cardiovasc Electrophysiol 1998;9:855-858.
14. Krishnan S, Josephson M: ST segment elevation induced by class Ic antiarrhythmic agents: Underlying electrophysiologic mechanisms and insights into drug-induced proarrhythmia. J Cardiovasc Electrophysiol 1998;9:1167-1172.
15. Eckardt L, Kirchhof P, Johna R, et al: Transient local changes in right ventricular monophasic action potentials due to ajmaline in a patient with Brugada syndrome. J Cardiovasc Electrophysiol 1999;10:1010-1015.
16. Fujiki A, Usui M, Nagasawa H, et al: ST segment elevation in the right precordial leads induced with class Ic antiarrhythmic drugs: Insight into the mechanism of Brugada syndrome. J Cardiovasc Electrophysiol 1999;10:214-218.
17. Roden D, Wilde A: Drug-induced J point elevation: A marker for genetic risk of sudden death or ECG curiosity? J Cardiovasc Electrophysiol 1999;10:219-223.
18. Brugada R, Brugada J, Antzelevitch A, et al: Sodium channel blockers identify risk for sudden death in patients with ST segment elevation and right bundle branch block but structurally normal hearts. Circulation 2000;101:510-515.
19. Matana A, Goldner V, Stanic K, et al: Unmasking effect of propafenone on the concealed form of the Brugada phenomenon. PACE 2000;23:416-418.
20. Pinar Bermudez E, Garcia-Alberola A, Martinez Sanchez J, et al: Spontaneous sustained monomorphic ventricular tachycardia after administration of ajmaline in a patient with Brugada syndrome. PACE 2000;23:407-409.

21. Miyanuma H, Sakurai M, Odaka H: Two cases of idiopathic ventricular fibrillation with interesting electrocardiographic findings. Kokyu to Junkan 1993;41:287-291.
22. Brembilla-Perrot B, Beurrier D, Jacquemin L, et al: Incomplete bundle branch block and ST-segment elevation: Syndrome associated with sustained monomorphic ventricular tachycardia in patients with apparently normal heart. Clin Cardiol 1997;20: 407-410.
23. Brugada J, Brugada R, Brugada P: Right bundle branch block and ST segment elevation in leads V1-V3: A marker for sudden death in patients with no demonstrable structural heart disease. Circulation 1998;97:457-460.
24. Gussak I, Antzelevitch C, Bjerregaard P, et al: The Brugada syndrome: Clinical, electrophysiologic and genetic aspects. J Am Coll Cardiol 1999;33:5-15.
25. Brugada J, Brugada P, Brugada R: The syndrome of right bundle branch block, ST segment elevation in V1 to V3 and sudden death—the Brugada syndrome. Europace 1999;1:156-166.
26. Grace AA: Brugada syndrome. The Lancet 1999;354:445-446.
27. Brugada P, Brugada J, Brugada R: The Brugada syndrome. Cardiac Electrophysiology Review 1999;3:42-44.
28. Itoh E, Suzuki K, Tanabe Y: A case of vasospastic angina presenting Brugada-type ECG abnormalities. Jpn Circ J 1999; 63:493-495.
29. Tarin N, Farre J, Rubio JM, et al: Brugada-like electrocardiographic pattern in a patient with a mediastinal tumor. PACE 1999;22:1264-1266.
30. Tomaszewski W: Changements electrocardiographiques observes chez un homme mort de froid. Arch Mal Coeur 1938;31:525-528.
31. Anguera I, Valls V: Giant J waves in hypothermia. Circulation 2000;101:1627-1628.
32. Nacarella F: Malignant ventricular arrhythmias in patients with a right bundle branch block and persistent ST segment elevation in V1-V3: A probable arrhythmogenic cardiomyopathy of the right ventricle. G Ital Cardiol 1993;23:1219-1222.
33. Corrado D, Nava A, Buja G, et al: Familial cardiomyopathy underlies syndrome of right bundle branch block, ST segment elevation and sudden death. J Am Coll Cardiol 1996;27:443-448.
34. Brugada J, Brugada P: Further characterisation of the syndrome of right bundle branch block, ST segment elevation and sudden cardiac death. J Cardiovasc Electrophysiol 1997;8:325-331.
35. Brugada P, Brugada J, Brugada R. "Brugada syndrome": A structural cardiomyopathy or a functional electrical disease? In Raviele A (ed): Cardiac Arrhythmias. Springer-Verlag 1997.
36. Fontaine G, Aouate P, Fontaliran F: Arrhythmogenic right ventricular dysplasia, torsades de pointes and sudden death: New concepts. Ann Cardiol Angeiol 1997;46:531-538.
37. Lorga Filho A, Brugada P: Right bundle branch block, ST segment elevation from V1 to V3 and sudden death: What we know about this peculiar and clinical electrocardiographic syndrome. Arq Bras Cardiol 1997;68:205-208.
38. Lorga Filho A, Prino J, Brugada J, Brugada P: Right bundle branch block, the elevation of the ST segment in V1 to V3 and sudden death: The diagnostic and therapeutic approach. Rev Port Cardiol 1997;16:443-447.
39. Tada HT, Aihara N, Ohe T, et al: Arrhythmogenic right ventricular cardiomyopathy underlies syndrome of right bundle branch block, ST-segment elevation, and sudden death. Am J Cardiol 1998;81:519-522.
40. Zipes D, Wellens HJJ: Sudden cardiac death. Circulation 1998;98:2334-2351.
41. Eriksson P, Hansson P, Eriksson H, Dellborg M: Bundle-branch block in a general male population. Circulation 1998;98:2494-2500.
42. Corrado D, Buja G, Baso C, et al: What is the Brugada syndrome? Cardiol Rev 1999;7:191-195.
43. Izumi T, Ajiki K, Nozaki A, et al: Right ventricular cardiomyopathy showing right bundle branch block and right precordial ST segment elevation. Intern Med 2000;39:28-33.
44. Brugada J, Brugada P, Brugada R: El sindrome de Brugada y las cardiopatias derechas como causa de muerte subita. Rev Esp Cardiol 2000;53:272-285.
45. Yoshioka N, Tsuchihashi K, Yuda S, et al: Electrocardiographic and echocardiographic abnormalities in patients with arrhythmogenic right ventricular cardiomyopathy and their pedigrees. Am J Cardiol 2000;85:885-889.
46. Izumi T, Ajiki K, Nozaki A, et al: Right ventricular cardiomyopathy showing right bundle branch block and right precordial ST segment elevation. Intern Med 2000;39:28-33.
47. Chen Q, Kirsch GE, Zhang D, et al: Genetic basis and molecular mechanisms for idiopathic ventricular fibrillation. Nature 1998; 392:293-296.
48. Bezzina C, Veldkamp MW, van den Berg MP, et al: A single Na+ channel mutation causing both long-QT and Brugada syndrome. Circ Res 1999;85:1206-1213.
49. Rook MB, Alshinawi CB, Groenewegen WA, et al: Human SCN5A gene mutations alter cardiac sodium channel kinetics and are associated with the Brugada syndrome. Cardiovasc Res 1999; 44:507-517.
50. Coonar A, Protonotarios N, Tsatsopoulou A, et al: Gene for arrhythmogenic right ventricular cardiomyopathy with diffuse nonepidermolytic palmoplantar keratoderma and woolly hair (Naxos disease) maps to 17q21. Circulation 1998;97:2049-2058.
51. Stephan E, Chedid R, Loiselet J, Bouvagnet P: Génétique clinique et moléculaire d' un bloc de branche familial lié au chromosome 19. Arch Mal Coeur 1998;91:1465-1474.
52. Barinaga M: Tracking down mutations that can stop the heart. Science 1998;281:32-34.
53. Probst V, Hoorntje TM, Hulsbeek M, et al: Cardiac conduction defects associate with mutations in SCN5A. Nat Genet 1999; 23:20-21.
54. Priori S, Barhanin J, Hauer R, et al: Genetic and molecular basis of cardiac arrhythmias: Impact on clinical management. Parts I and II. Circulation 1999;99:518-528.
55. Litovsky SH, Antzelevitch C: Rate dependence of action potential duration and refractoriness in canine ventricular endocardium differs from that of epicardium: Role of the transient outward current. J Am Coll Cardiol 1989;14:1053-1066.
56. Litovsky SH, Antzelevitch C: Differences in the electrophysiologic response of canine ventricular subendocardium and subepicardium to acetylcholine and isoproterenol: A direct effect of acetylcholine in ventricular myocardium. Circ Res 1990;67:615-627.
57. Yan G-X, Antzelevitch C: Cellular basis for the electrocardiographic J wave. Circulation 1996;93:372-379.
58. Krishnan SC, Antzelevitch C: Flecainide-induced arrhythmia in canine ventricular epicardium: Phase 2 re-entry? Circulation 1992;87:562-572.
59. Matsuo K, Shimizu W, Kurita T, et al: Increased dispersion of repolarisation time determined by monophasic action potentials in two patients with familial idiopathic ventricular fibrillation. J Cardiovasc Electrophysiol 1998;9:74-83.
60. Antzelevitch A: The Brugada syndrome. J Cardiovasc Electrophysiol 1998;9:513-516.
61. Okazaki O, Yamauchi Y, Kashida M, et al: Possible mechanism of ECG features in patients with idiopathic ventricular fibrillation studied by heart model and computer simulation. J Electrocardiol 1998;30:98-104.
62. Yan GX, Antzelevitch C: Cellular basis for the Brugada syndrome and other mechanisms of arrhythmogenesis associated with ST segment elevation. Circulation 1999;100:1660-1666.
63. Antzelevitch C, Brugada P, Brugada J, et al: The Brugada syndrome. Futura, Armonk, NY, 1999.
64. Dumaine R, Towbin J, Brugada P, et al: Ionic mechanisms responsible for the electrocardiographic phenotype of the Brugada syndrome are temperature dependent. Circ Res 1999; 85:803-809.
65. Balser JR: Sodium "channelopathies" and sudden death: Must you be so sensitive? Circ Res 1999;85:872-874.
66. Antzelevitch C, Yan GX, Shimizu W: Transmural dispersion of repolarization and arrhythmogenicity: The Brugada syndrome versus the long QT syndrome. J Electrocardiol 1999;32:158-165.
67. Makita N, Shirai N, Wang DW, et al: Cardiac Na+ channel dysfunction in Brugada syndrome is aggravated by B1 subunit. Circulation 2000;101:54-60.
68. Baroudi G, Carbonneau E, Pouliot V, Chahine M: SCN5A mutation (T1620M) causing Brugada syndrome exhibits different phenotypes when expressed in Xenopus oocytes and mammalian cells. FEBS Lett 2000;467:12-16.
69. Deschenes I, Batoudi G, Berthet M, et al: Electrophysiological characterization of SCN5A mutations causing long QT (E1784K) and Brugada (R1432G) syndromes. Cardiovasc Res 2000;46:55-65.

70. Shimizu W, Matsuo K, Takagi M: Body surface distribution and response to drugs of ST segment elevation in Brugada syndrome: Clinical implication of eighty-seven-lead body surface potential mapping and its application to twelve-lead electrocardiograms. J Cardiovasc Electrophysiol 2000;11:396-404.
71. Gonzalez Rebollo JM, Hernandez Madrid A, Garcia A, et al: Recurrent ventricular fibrillation during a febrile illness in a patient with the Brugada syndrome. Rev Esp Cardiol 2000;53: 755-757.
72. Sao N, Akasaka K, Kawashima E, et al: A case of idiopathic ventricular fibrillation with possible mechanism of autonomic dysfunction. Jpn J Electrocardiol 1994;14:206-217.
73. Matsuo K, Kurita T, Inagaki M, et al: The circadian pattern of the development of ventricular fibrillation in patients with Brugada syndrome. Eur Heart J 1999;20:465-470.
74. Leenhardt L, Glaser E, Burguera M, et al: Short-coupled variant of torsades de pointes: A new electrocardiographic entity in the spectrum of idiopathic ventricular arrhythmias. Circulation 1994; 89:206-215.
75. Fisher JD, Krikler D, Hallidie-Smith KA: Familial polymorphic ventricular arrhythmias. J Am Coll Cardiol 1999;34:2015-2022.
76. Swan H, Piippo K, Viitsalo M, et al: Arrhythmic disorder mapped to chromosome 1q42-q43 causes malignant polymorphic ventricular tachycardia in structurally normal hearts. J Am Coll Cardiol 1999;34:2035-2042.
77. Viskin S, Fish R, Eldar M, et al: Prevalence of the Brugada sign in idiopathic ventricular fibrillation and healthy controls. Heart 2000;84:31-36.
78. Hermida J, Lemoine J, Aoun FB, et al: Prevalence of the Brugada syndrome in an apparently healthy population. Am J Cardiol 2000;86:91-94.
79. Priori S, Napolitano C, Giordano U, et al: Brugada syndrome and sudden cardiac death in children. The Lancet 2000;355:808-809.
80. Nademanee KK, Veerakul G, Nimmannit S, et al: Arrhythmogenic marker for the sudden unexplained death syndrome in Thai men. Circulation 1997;96:2595-2600.
81. Benhorin J, Taub R, Goldmit M, et al: Effects of flecainide in patients with new SCN5A mutation: Mutation-specific therapy for long QT syndrome? Circulation 2000;101:1698-1706.
82. Brugada P, Brugada R, Brugada J: Sudden death in patients and relatives with the syndrome of right bundle branch block, ST segment elevation in the precordial leads V1 to V3 and sudden death. Eur Heart J 2000;21:321-326.
83. Brugada J, Brugada R, Antzelevitch C, et al:. Long-term follow-up of individuals with the electrocardiographic pattern of right bundle-branch block and ST segment elevation in precordial leads V1 to V3. Circulation 2002;105:73-78.
84. Holtzman NA, Marteau TM: Will genetics revolutionize medicine? N Engl J Med 2000;343:141-144.
85. Grant AO, Carloni MP, Neplioleva V, et al: Long QT Syndrome, Brugada syndrome and conduction system disease are linked to a single sodium channel mutation. JCI 2002;110(8): 1201-1209.
86. Baroudi G, Chahine M: Biophysical phenotypes of SCN5A mutations causing long QT and Brugada syndromes. EBS Letters 2000;487(2):224-228.
87. Clancy CE, Rudy Y. Na(+) channel mutation that causes both Brugada and long QT syndrome phenotypes: A simulation study of mechanism. Circulation 2002;105(10):1208-1213.
88. Priori SG, Napolitano C, Gasparini M, et al: Natural history of Brugada syndrome: Insights for risk stratification and management. Circulation 2002;105(11):1342-1347.
89. Brugada J, Brugada R, Antzdevitch C, et al: Long-term follow-up of individuals with the electrocardiographic pattern of right bundle branch block and ST segment elevation in precordial leads V1 to V3. Circulation 2002;105(1):73-78.

SECTION V

Pharmacologic and Interventional Therapies

Chapter 37 Antiarrhythmic Drugs

BRAMAH N. SINGH

There has been a considerable change in recent years in the role of antiarrhythmic drugs and a number of factors have played critical roles in the ongoing reorientation of therapy in disorders of cardiac rhythm. In the case of supraventricular tachyarrhythmias, a precise understanding of the mechanisms of these arrhythmias has led to electrode catheter ablation with cure in most patients. On the one hand, there is now less emphasis on chronic suppression except for symptom relief. On the other hand, available drugs given intravenously may induce conversion in nearly all patients with reentrant supraventricular tachyarrhythmias but not to the same extent in atrial flutter (AFL) and atrial fibrillation (AF), which may often require electrical conversion.

In the case of life-threatening ventricular arrhythmias, there is now an increasing focus on the use of implantable devices to prevent sudden death, and drugs are used increasingly as adjunctive therapy to prevent symptoms from shocks. Unlike the case of supraventricular tachyarrhythmias, conversion of ventricular tachycardia (VT) and ventricular fibrillation (VF) is largely effected by electrical energy because of the left-threatening nature of the arrrhythmias—pharmacologic conversion by intravenous drugs being used largely in asymptomatic patients with stable hemodynamics. There is an increasing focus on prophylactic therapy in the case of patients at high risk for sudden arrhythmic deaths, a setting in which pharmacologic therapy may still be pre-eminent.

Reorientation of drug therapy has also been necessary in the wake of the knowledge that certain antiarrhythmic agents, while suppressing supraventricular and ventricular arrhythmias, may increase mortality by their associated proarrhythmic effects. This was demonstrated in the Cardiac Arrhythmia Suppression Trial.[1] The findings have important clinical implications with respect to the choice of antiarrhythmic agents in the acute, the chronic, and the prophylactic control of arrhythmias. The issue is critical in the choice of intravenous therapy of ventricular arrhythmias not only in hospitalized patients but also in those developing out-of-hospital VT or VF, especially in the choice of agents as an integral part of the current Advanced Cardiovascular Life Support (ACLS) guidelines.[2]

CLASSIFICATION OF ANTIARRHYTHMIC DRUGS REVISITED

Interest in the mechanisms of action of pharmacologic agents has provided the basis for their classification since the 1970s.[3-9] The original intent was to classify *mechanisms of action of antiarrhythmic drugs and not the drugs themselves.* Such an approach permitted grouping of drugs by their dominant action with the inevitable realization that many antiarrhythmic drugs in the clinic exerted single actions (e.g., β-blockers) or a varying spectrum of actions (sotalol and amiodarone) with their defined overall clinical effects both beneficial as well as deleterious.

The attempts at classification of the manner in which antiarrhythmic agents might work beneficially in the control of disorders of rhythm followed closely in the wake of a series of experimental studies of a number of structurally disparate compounds having certain discrete electrophysiological actions.[3-5,8,9] These compounds were local anesthetics or sodium channel blockers, β-adrenergic receptor blocking drugs, drugs that were also found to have antifibrillatory actions in the context of their property to selectively prolong cardiac repolarization, and those that were found to selectively block calcium channels at the membrane level. The premise of the initial classification was that each of the mechanisms described exerted, albeit in differing ways, an antifibrillatory effect in a standardized experimental model of VF.[5] The evolution of the steps in the classification of antiarrhythmic mechanisms of drugs that are now widely used stems from the original framework,

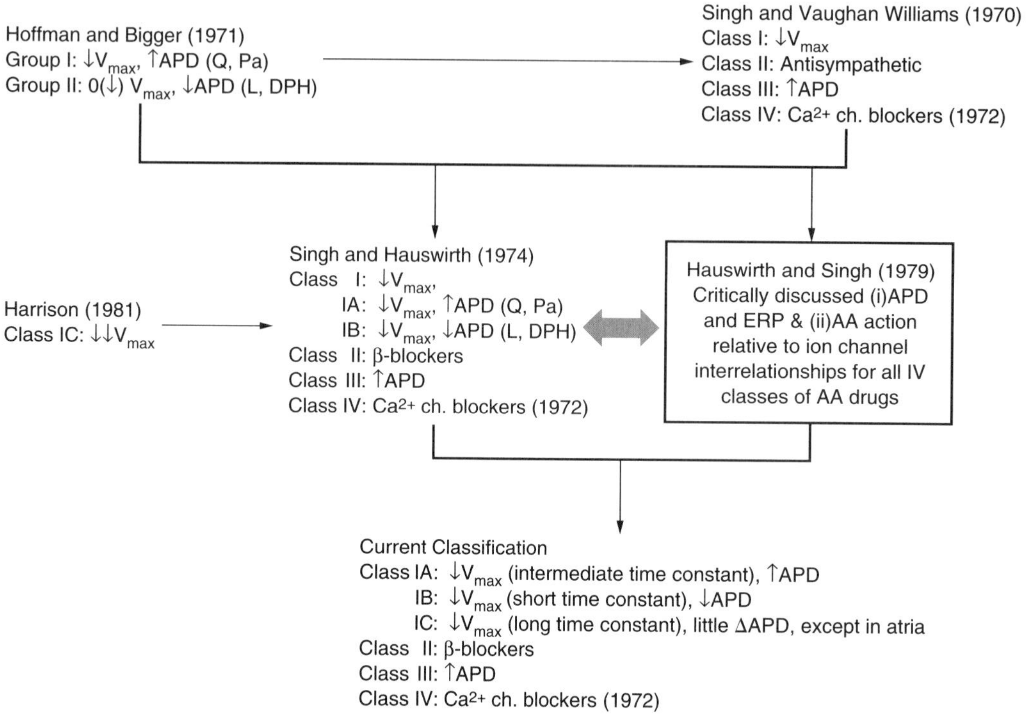

FIGURE 37-1 Evolution of the current classification of antiarrhythmic drugs. V_{max}, maximal rate of change in the velocity of depolarization; APD, action potential duration; Q, quinidine; P, procainamide; L, lidocaine; DPH, diphenylhydantoin; ERP, effective refractory period; AA, antiarrhythmic; ↑, increase; ↓, decrease. (Modified from Nattel S, Singh BN: Evolution, mechanisms, and classification of antiarrhythmic drugs: Focus on class III actions. Am J Cardiol 1999; 84:11-19.)

which has[3-5] subsequently been modified slightly,[8] the most recent being by Nattel and Singh.[10] The final step completed in 1979[12] is shown in Figure 37-1. The need for the classification of antiarrhythmic drug mechanisms was conceived almost simultaneously and independently on both sides of the Atlantic. It will be noted that the Hoffman and Bigger[11] attempt at classifying antiarrhythmic drugs placed compounds such as quinidine and procainamide, which had been known to slow conduction as well as to prolong repolarization in one category (type 1), and lidocaine and diphenylhydantoin (DPH), which also inhibited conduction but *shortened repolarization* into another category (type 2). Because propranolol in high doses exhibited potent conduction blocking properties, it was included among type 1 antiarrhythmic compounds.

It was suggested that type 1 compounds terminated or prevented arrhythmias by converting unidirectional block into a bidirectional one in an arrhythmia circuit, whereas the type 2 agents acted by eliminating unidirectional block by *increasing* conduction velocity based on their findings in the experimental laboratory. Subsequently, as an integral part of a series of studies on which the conventional antiarrhythmic classification has been based,[9] it was found that this observation might have occurred as a consequence of the use of K^+ concentration of 2.7 mEq/L in the perfusion media, because the retesting of the effects of lidocaine and DPH in media containing physiologic levels of potassium ions, clearly established that both the compounds had their predominant actions as inhibitors of sodium channel-mediated conduction. It also became evident that in clinically significant concentrations, propranolol exerted no local anesthetic actions.[8]

In contrast, in the comprehensive classification scheme suggested by Singh,[5] and Singh and Vaughan Williams,[3,7] all drugs that exerted local anesthetic actions in the nerve as well as in the myocardial membrane (including lidocaine, quinidine, procainamide, disopyramide, and DPH among numerous others) were thought to act via class I actions, and propranolol and other β-receptor blocking compounds that exhibited sympatholytic actions were characterized as acting via class II antiarrhythmic actions.

When it was found that compounds, such as sotalol (a β-blocker that prolonged repolarization) and amiodarone, which prolonged repolarization, exerted unequivocal antifibrillatory actions not accounted for by class I or class II actions, the possibility of a class III action was suggested.[3] Subsequently, on the basis of the electropharmacologic effects of the drug verapamil, a fourth class of antiarrhythmic actions was added by Singh and Vaughan Williams in 1972.[7] Hauswirth and Singh[12] subdivided class I compounds into those that suppressed sodium-channels in all cardiac fast-channel tissues as typified by quinidine and procainamide (class IA agents) and those that inhibited sodium channels only in diseased or depolarized tissues (class IB: lidocaine and diphenylhydantoin). Another difference between these two subclasses was that in the case of the 1A agents, repolarization was also prolonged, whereas in the case of 1B agents, repolarization was abbreviated. Harrison[13] completed the subclassification by assigning the potent class I agents, flecainide and encainide, into a group designated as class IC. Experimental studies by Campbell[14] validated the separation of subclasses of sodium-channel blockers on the biophysical basis; he found that class IA agents had Na-channel blocking kinetics between those of class IB agents (very fast) and IC compounds (very slow).

This classification of antiarrhythmic drugs is still widely used in clinical practice. It has been the basis for the development of newer antiarrhythmic compounds. It was recognized early that a direct extrapolation of the experimental data to the clinical setting in this regard might not be readily possible. However, it was considered that for advances in understanding drug action, it was desirable to explore the effects of various classes of antiarrhythmic drugs in myocardial cells and membranes in terms of ion channels and related parameters on the one hand[12] and relate them to their clinical effects on the other hand.[8] This approach was first suggested in detail by Hauswirth and Singh in 1979,[12] and it re-emerged subsequently in the form of the Sicilian Gambit in 1991,[15] incorporating an increasing understanding of membrane currents with the introduction of the patch clamp technique.

The conventional classification has been of fundamental importance for the synthesis and initial characterization of new antiarrhythmic compounds (as has been the case of pure class III agents) as well as in terms of a choice of an agent for the management of a particular arrhythmia. The classification does appear to have direct relevance in terms of impact on mortality in patients with significant cardiac disease. For example, in patients with cardiac disease, class I agents may increase mortality via the development of proarrhythmic reactions, especially in patients with coronary artery disease. As a consequence, the role of sodium-channel blockers is declining—being restricted to alleviating arrhythmia symptoms in patients without heart disease. In contrast, class II agents (β-blockers) uniformly prolong survival in numerous subsets of patients by their multiplicity of actions that include antifibrillatory effects in patients with a varying spectrum of severity of heart disease.

Amiodarone and sotalol—two unique compounds—the dominant electrophysiological property of which is the prolongation of repolarization, formed the basis for the class III action.[3,4,7,16-18] Amiodarone and sotalol have provided the background for the synthesis and characterization of simpler compounds (such as dofetilide and azimilide) as well as the impetus to develop other agents with similar properties while having safer electropharmacologic profiles. The main properties of the major agents are discussed in this chapter relative to their roles in the control of supraventricular and ventricular tachyarrhythmias. Certain electrophysiological properties of antiarrhythmic compounds are of much importance in the clinical area. They will be emphasized at the outset.

HEART RATE DEPENDENCY OF ACTION OF ANTIARRHYTHMIC AGENTS

The electrophysiological property that also appears to be of major clinical interest is the rate-dependent effect of the compounds on the action potential duration and refractoriness.[19] Rate dependency refers to a different effect at varying heart rates (less effect at higher rates). The differences in this parameter among the various agents may be of therapeutic relevance. This issue is of particular relevance in the case of their role in the control of AF. The drugs that exert the classic reverse-rate dependency (dofetilide, quinidine, and dl-sotalol) appear to exhibit a similar ceiling of efficacy for maintaining stability of sinus rhythm and a similar propensity for inducing torsades de pointes (TdP). Their effect on prolonging the action potential duration (APD) and the effective refractory period (ERP) in the atrial muscle decline as the stimulation frequency increases. All such compounds are powerful I_{Kr} blockers and they are likely to be more effective in terminating AFL than AF when acutely administered. They are moderately effective in preventing recurrences of AF and AFL in paroxysmal and persistent forms of these arrhythmias.

In contrast, compounds such as azimilide and possibly ambasilide which may also inhibit the slow component of the delayed rectifier current (I_{Ks}) exhibit a different pattern of rate dependency of action with respect to the APD and associated refractoriness. Under the action of the drugs, there is a parallel increase in the APD and the ERP as shown in Figure 37-2. Whether this is clinically significant may require a direct comparison of these compounds with those of other so-called class III compounds such as dofetilide. In the case of azimilide, the available data suggest a low incidence of TdP, a neutral effect on mortality[20] in high risk postmyocardial infarction patients, and at least moderate efficacy in maintaining stability of sinus rhythm in patients with AF.[21] Large pivotal placebo-controlled trials with respect to azimilide are in progress.

Of particular interest, the action of amiodarone and that of dronedarone[22] with respect to rate-dependent effects are similar, being associated with a parallel shift (i.e., ERP increases in a parallel fashion over a wide range of frequencies), as noted in the case of azimilide and ambasilide.[10] In the case of amiodarone, there is now expanding evidence for the effectiveness of the drug for maintaining stability of sinus rhythm. Its potency (see later) appears the highest among all class III antiarrhythmic compounds and it is associated with the lowest incidence of TdP in the context of the longest Q–T interval that amiodarone may produce during the course of chronic treatment. The precise reason for this combination of drug effects remains unclear but the observations are clearly of theoretical importance for the purposes of developing future compounds. Perhaps of much importance also is the observation that in the case of flecainide (presumably also propafenone) is the nature of their effects on the ERP and APD as a function in atrial tissue. Wang and coworkers[23] found that in a variety of mammalian atria, flecainide (and presumably propafenone acts in a similar manner) had the property of prolonging the APD and ERP as a function of rate—for example, greater effects were seen as the stimulation rates were increased (Fig. 37-3). The available data suggest that the rate-related effects of antiarrhythmic drugs may vary in differing

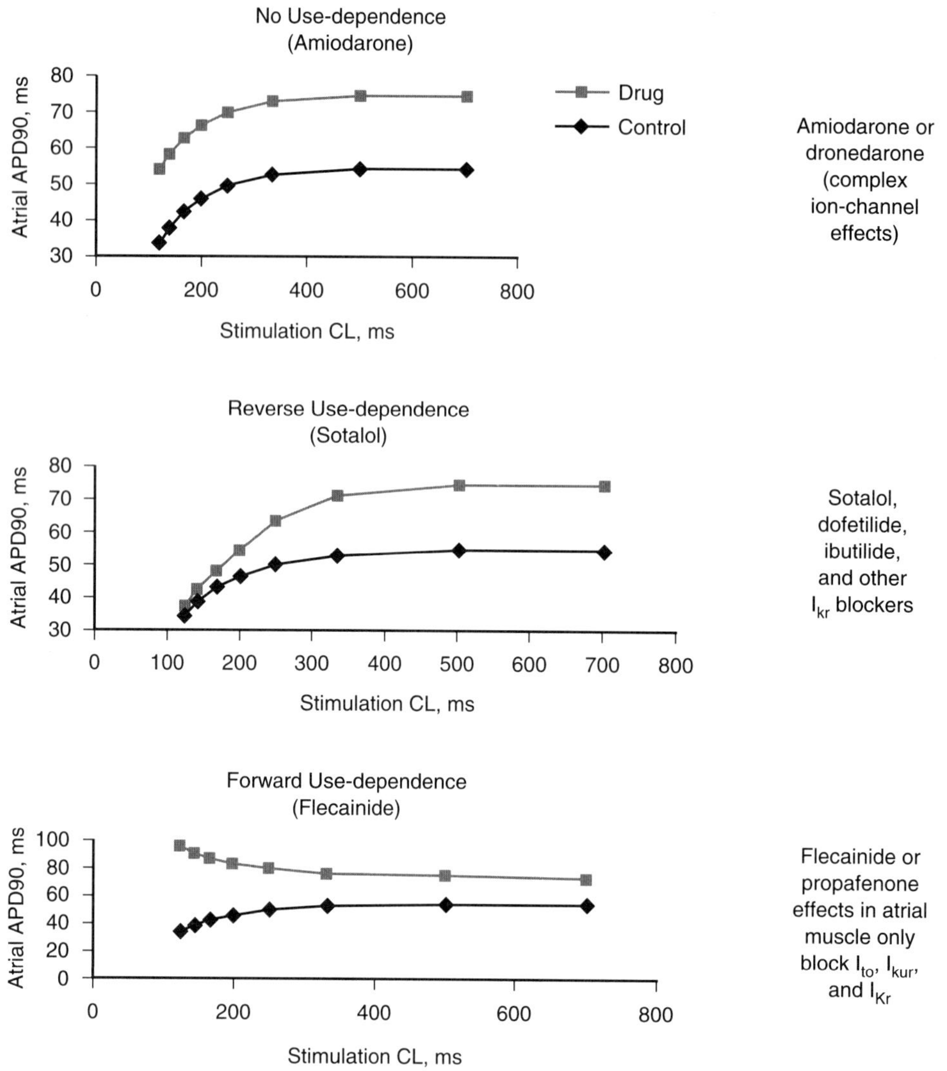

FIGURE 37-2 Influence of changes in stimulation frequency in time course of myocardial atrial repolarization induced by various antiarrhythmic drugs. Three patterns are illustrated: no use dependence, reverse rate, and forward use-dependence. In the first, as illustrated by amiodarone, there is no effect on action potential duration or refractory period as the cardiac frequency increases; in the cases of sotalol and other I_{kr} blockers, the action potential duration (APD) and effective refractory period (ERP) shortens as the cardiac frequency increases and there is forward use-dependence as has been demonstrated in the case of flecainide in the atria. These differences may be of much clinical significance, as well as for the development of atrial-specific antifibrillatory agents. (From Sarma JSM, Singh BN, 2001, unpublished observations.)

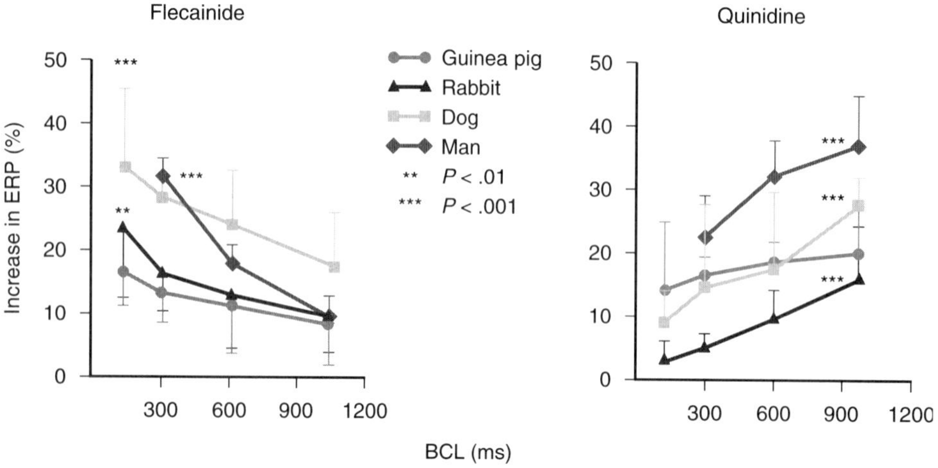

FIGURE 37-3 Rate-dependent effects of flecainide and quinidine on effective refractory period in mammalian atria. The data contrast the forward use-dependence of flecainide with the reverse use-dependence of quinidine. See Figure 37-2 and the text for further details and significance. (From Wang Z, Pelletier LC, Talajic M, Nattel S: Effects of flecainide and quinidine on human atrial potentials: Role of rate dependence and comparison with guinea pig, rabbit and dog tissues. Circulation 1990;82:274-283.)

cardiac tissues, possibly a reflection of the differential action of drugs on ion channels that they may inhibit to varying extents.

Class I Antiarrhythmic Compounds

As a class, this group of agents share one common and dominant property: they slow conduction in those myocardial tissues in which conduction velocity is controlled by the fast sodium channels with variable, often inconsistent effects on the refractory period in ventricular tissues. The proarrhythmic effects may also vary in relation to the additional electrophysiological properties certain class I agents might have—as is the case with prolongation of repolarization in quinidine, procainamide, or disopyramide (class IA drugs), which may contribute to the prolongation of the refractory period but with the propensity to induce TdP.

Conversely, lidocaine and its oral congeners, mexiletine and tocainide, and diphenylhydantoin (class IB agents), in actuality *shorten* repolarization and hence refractoriness, in addition to their property of slowing conduction. The actions of class I agents are usually more intense in diseased tissues that are partially depolarized, such as in myocardial ischemia. Here, their proarrhythmic effects in settings of ischemia or LV dysfunction are often much greater and may prove fatal—as in the case of flecainide, encainide, and propafenone (class IC compounds), which are contraindicated in patients with significant ventricular dysfunction. As a class of antiarrhythmic drugs, the use of sodium-channel blockers is declining because none of the agents has the potential to increase survival by controlling cardiac arrhythmias. If they are administered to patients with potentially serious cardiac disease, mortality may be increased.[24] Their main utility is in the conversion of AF to sinus rhythm and in maintaining stability of sinus rhythm in patients in whom the arrhythmia occurs with normal or near normal ventricular function. Relevant aspects of individual agents are presented subsequently.

Quinidine

Quinidine has dual electrophysiological properties of prolonging repolarization and slowing conduction by blocking inward sodium current, thereby slowing conduction and by blocking a variety of outward potassium currents; the effects of the drug on myocardial ion channels are compared with those of other antiarrhythmic drugs in Table 37-1. The drug is available in injectable and oral forms, but the injectable formulation is now rarely used. Quinidine may increase heart rate and facilitate AV nodal conduction by its vagolytic actions and increase ventricular response in AF. The major electrophysiological effect of clinical utility relative to the drug's antiarrhythmic actions is the prolongation of the action potential duration and the effective refractory period (a class III action); the effects on both are attenuated as the heart rate increases. The drug has been used in patients with impaired ventricular function but it may depress myocardial contractility. The elimination half-life of quinidine is 8 to 9 hours, its metabolism being largely by hepatic hydroxylation. The usual oral dose of the drug is 1.2 to 1.6 gm/day in 8 to 12 hourly divided doses relative to the preparation in use.

Most common side effects of quinidine include diarrhea, nausea, and vomiting, which occur in one third of all patients. Quinidine can also cause cinchonism (headache, dizziness, and tinnitus) and quinidine syncope, a syndrome characterized by lightheadedness and fainting. The major shortcomings of quinidine are the severe diarrhea that the drug may frequently produce and in 2% to 5% the rate of TdP that develops in association with the prolonged Q–T interval. Uncommonly, it might be fatal. In a detailed meta-analysis of the trials in AF involving quinidine, an increase in mortality has been reported.[25] For these reasons, when considered in conjunction with the limited ceiling of effectiveness of the drug both in ventricular as well as supraventricular arrhythmias, the clinical utility of quinidine is limited. Justification for its continued use is marginal.

Procainamide

This drug acts largely by prolonging the APD and refractoriness in atrial and ventricular tissues with little or no effect on nodal tissues; it does affect conduction with a modest anticholinergic effect. It has no anti-adrenergic actions. Its major metabolite is *N*-acetyl-procainamide (NAPA), which contributes to the overall

TABLE 37-1 Effects of I_{Kr} Inhibition on Effectiveness in Maintaining Sinus Rhythm in AF Relative to the Development of Torsades de Pointes and Total Mortality with and without Blocking Other Ionic Currents

Antiarrhythmic Agent	Ionic Currents or Receptor Blocked	Effectiveness in AF (1yr)	Torsades de Pointes	Mortality
Amiodarone	I_{Kr}, I_{Ks}, I_{to}, I_{Na}, beta-receptor	60%-70%	<0.5%	Neutral or lower
dl-Sotalol	I_{Kr}, beta-receptor	<50%	>2%-3%	Neutral
d-Sotalol	I_{Kr}	20%-30%	>2%-3%	Increased
Dofetilide	I_{Kr}	50%-60%	>2%-3%	Neutral
Azimilide	I_{Kr}, I_{Ks}, I_{ca}	50%	?1%-2%	Neutral (ALIVE)
Quinidine	I_{Kr}, I_{Kl}, I_{kur}, I_{to}	50%	3%-5% or higher	1%-3% (Meta-analysis)

electrophysiological component of the parent compound. The elimination (renal) half-life of procainamide is less than 4 hours, although slow-release formulations (6-hourly dosing) have been in use. The electrophysiological properties of the drug resemble those of quinidine.

Available as an oral form (1 gm loading and 500 mg 3 hourly or equivalent of slow-release formulation 6 hourly) and injectable (100 mg bolus up to 25 mg/min to 1gm in the first hour, then 2 to 2 to 6 mg/min), procainamide has been used in the past for treatment of supraventricular and ventricular arrhythmias. Although procainamide produces a lower incidence of TdP than quinidine, its clinical utility has declined substantively in recent years. Its role in the acute conversion of AFL and AF has been superseded by that of intravenous class III agents, especially ibutilide in most, if not all, clinical settings. The intravenous drug, however, is still useful in the conversion of monomorphic VT in a patient who is not severely hypotensive and in the conversion of AF, complicating the Wolff-Parkinson-White syndrome. The oral formulation of the drug exhibits a significant hypotensive effect by virtue of its vasodilator and negative inotropic actions; however it cannot be used for prolonged periods of time because hematologic adverse reactions and, in particular, systemic lupus-like syndromes can complicate its prolonged use. Thus, the oral form of the drug is now obsolete.

Disopyramide

As in the case of quinidine and procainamide, disopyramide as a class IA antiarrhythmic agent has limited clinical use. Its electropharmacologic profile is similar to that of quinidine but it has a more potent anticholinergic action and it produces a lesser degree of gastrointestinal disturbance. The elimination half-life of the drug is about 8 hours but slow-release formulations have been introduced. The effects of the drug in arrhythmia control have not been extensively studied except for the suppression of premature ventricular contractions (PVCs). Disopyramide has been used largely in the oral form at a dose of 100 to 200 mg 6-hourly but it is now rarely used for the control of ventricular tachyarrhythmias. Like quinidine, it does prolong the QT interval and may produce a variable incidence of TdP; of particular importance, in patients with impaired ventricular function and in heart failure, it is likely to severely aggravate cardiac decompensation by its potent negative inotropic actions. However, such a property may be of therapeutic value in the syndrome of idiopathic hypertrophic cardiomyopathy (HCM), a setting in which the drug produces a greater negative inotropic effect than do β-blockers with which disopyramide can be combined. Although not approved in the United States, disopyramide may have a significant antifibrillatory effect in AF in two settings where its role in controlling this arrhythmia may be unique—in HCM patients who develop AF and in patients whose AF is entirely due to excess vagal tone.[26] In the latter setting, the drug's anticholinergic actions may be a specific therapy, which might be strikingly effective. It should be emphasized, however, that the drug's anticholinergic actions may lead to serious adverse reactions in the form of urinary retention, worsening of glaucoma or myasthenia gravis, and severe constipation.

LIDOCAINE, MEXILETINE, AND TOCAINIDE

The properties and the roles of intravenous lidocaine and its two orally-active congeners, mexiletine, and tocainide, as class IB compounds, will be discussed briefly because their roles have declined considerably in recent years and are likely to become merely of historic interest. The data continue to accrue regarding the adverse effects on mortality, despite the suppression of ventricular arrhythmias that class IB agents produce.[27,28,30] Lidocaine, a class I antiarrhythmic, is available as an injection. Lidocaine action is characterized by a fast, on-off block of sodium ion channels in both the activated and inactivated states. It also shortens the duration of the APD (hence the Q–T interval), and refractoriness, and has been shown in the experimental setting to *elevate* the ventricular defibrillation threshold. Thus, the drug and its oral congeners (also available as intravenous formulations) may act largely by slowing conduction rather than prolonging refractoriness. However, their actions in terminating and preventing arrhythmias may also stem from the conversion of unidirectional block to a bidirectional block in the arrhythmia re-entrant circuit although the proof for this in the clinical setting is lacking. Lidocaine is metabolized rapidly in the liver, and it is administered generally as an initial bolus of 100-200 mg, followed by 2 to 4 mg/min for varying durations of time (usually 24 to 36 hours) aiming at serum levels of 1.4 to 5 μg/mL.

Lidocaine gained prominence as a prophylactic agent in the early days of the inception of coronary care units and subsequently as the first-line agent in the control of VT/VF in all acute care units, including the emergency department, and in the resuscitation of those developing out-of-hospital cardiac arrest. The use of lidocaine and, indeed, other antiarrhythmic drugs in these settings was not based on data from controlled clinical trials. It is of interest to note that in a direct blinded comparison, intravenous lidocaine converted 18% of VT to sinus rhythm compared with 69% conversion with intravenous sotalol.[29] The Cardiac Arrhythmias Suppression Trial[1] with the drugs encainide, flecainide, and moricizine indicated a dichotomy between arrhythmia suppression and mortality. There is much data now to suggest that intravenous lidocaine and mexiletine are similarly ineffective in the setting of postmyocardial infarction,[27,28,30] and raise the issue that they might also be similar in the control of destabilizing VT/VF as well as in the survivors of out-of-hospital cardiac arrest. It is not surprising that in line with the data, the most recent AHA/ACC/ACLS Guidelines have relegated the use of lidocaine as being "indeterminate."[2] This issue will be discussed in depth later in this chapter when the effects of intravenous lidocaine and intravenous amiodarone will be discussed relative to the broader implications of the choice antiarrhythmic agents in the control of life-threatening ventricular arrhythmias in various clinical settings.

CLASS 1C AGENTS: FLECAINIDE AND PROPAFENONE

This class of antiarrhythmic agents developed in the wake of the belief that the suppression of ambient arrhythmias, whether PVCs, sustained or nonsustained VT, by antiarrhythmic drugs should result in prevention of arrhythmic death and in the prolongation of survival. The fact that such a hypothesis was not vindicated has had a number of therapeutic consequences (1) the fact that near-complete or complete suppression of PVCs in the postinfarct patient is associated with an increased mortality in patients with cardiac disease indicate that this class of drugs cannot be used with impunity even for control of symptoms due to arrhythmias; (2) increased observed mortality induced by class IC agents may be a property of class I agents in general, and is also found in the case of lidocaine; and (3) the data do not exclude the possibility for using such agents for controlling arrhythmia symptoms or for restoring and maintaining sinus rhythm in AF or AFL *in patients without structural heart disease.* This is especially so in the case of flecainide and propafenone. Moricizine is often included in the category of class IC agents but its properties are difficult to classify and it may not have advantages over flecainide or propafenone. These agents slow conduction velocity profoundly but have little, if any, effect on refractoriness in ventricular tissues and, theoretically, eliminate reentry by slowing conduction to a point where the impulse is extinguished and cannot propagate further. This may be the basis of their effectiveness in markedly reducing ventricular ectopy. Their actions in atrial tissues differ markedly compared with those in the ventricle. *They increase the ERP in the atria as a function of increases in heart rate,*[23] an effect that is likely to be the basis of their effectiveness in restoring and maintaining sinus rhythm. This is their major clinical utility as anti-arrhythmic drugs.

Flecainide

Flecainide is a powerful blocker of sodium channels in virtually all cardiac tissues but it does not significantly inhibit the pacemaker current or the calcium channels. It has minimal effects on K^+ channels but without a major effect on repolarization, except in the atria in which repolarization is prolonged and the effective period is lengthened. The drug has a significant negative inotropic effect especially at higher doses. The plasma half-life of the drug is 13 to 19 hours; two thirds of it is metabolized in the liver. The usual dose of flecainide is 100 to 300 mg daily, given twice daily. The intravenous formulation is available but is not used routinely. One clinical utility of flecainide is in the control of intractable symptoms due to PVCs in patients with no significant cardiac disease. Perhaps, its greatest value is in the restoration and prevention of recurrences of AF but only in patients without structural disease in view of its serious proarrhythmic reactions in patients with cardiac disease.[31] In the prophylactic control of AF or flutter, the drug should be combined with an AV nodal blocking drug (β-blockers or calcium calcium channel blockers) to avoid the development facilitated conduction across the AV node as the atrial rate slows under the action of the drug. In recent years, single oral doses (200 to 300 mg) of the drug have been successfully used for the acute termination of paroxysms of AF ("pill in the pocket approach").

Propafenone

This drug is also a powerful class IC agent with a electrophysiological activity profile similar to that of flecainide, while having a mild degree of β-blocking action that does not appear to be clinically relevant.[32] Its actions in the atria may also be similar to those of flecainide and the drug does not prolong repolarization in ventricular tissues. The drug is rapidly absorbed with the bioavailability of 50%, elimination half-life being 2 to 10 hours in normal subjects and 12 to 32 hours in poor metabolizers. Orally, propafenone is administered as 150 to 300 mg three times daily and a longer-acting formulation has been introduced. The drug can also be administered intravenously, and single oral doses (300 mg or 600 mg) for the acute conversion of paroxysms of atrial AF. As in the case of flecainide, propafenone should not be used in patients with significant cardiac disease and its two main indications are for the suppression of resistant symptomatic PVCs and for the restoration and maintenance of sinus rhythm in the case of AF in patients with structurally normal hearts. Its efficacy rivals that of flecainide, although direct comparisons have not been made. As in the case of flecainide, propafenone should be combined with an AV nodal blocking drug to reduce the possibility of accelerated conduction with the slowing of atrial rate induced by the drug.

β-ADRENERGIC BLOCKERS AS ANTIARRHYTHMIC AND ANTIFIBRILLATORY COMPOUNDS

β-Blockers as a class of drugs exert distinctive antiarrhythmic and antifibrillatory effects with the properties of not only consistently alleviating symptoms but equally consistently prolonging survival in a wide subset of patients. The electrophysiological and antifibrillatory effects are most striking in the clinical context of most intense sympathetic stimulation. They prevent the development of VF in a variety of experimental animal models. The beneficial effects in this setting could not be accounted for by any known electrophysiological mechanisms. However, in the clinic, the antiarrhythmic potential of β-blockade was greatly overshadowed by its anti-ischemic actions for which this class of drugs was synthesized. Thus, the recognition of the fact that blunting the effects of catecholamines on the heart might be a potent antifibrillatory mechanism was slow in coming. That increased activity of the sympathetic nervous system might induce cardiovascular morbidity and mortality through a variety of mechanisms has been known for many years. For example, given the appropriate pathologic substrate, increased sympathetic activity may be associated with sudden arrhythmic death.[33-35] This is now known to be particularly striking in the case of patients developing myocardial infarction with or without heart failure.

The intrinsic effects of β-blockers may be modified to varying extents by the associated pharmacologic properties that individual agents may have. For example, in the case of propranolol at high concentrations, there is significant inhibition of the Na channel. There are agents (e.g., acebutolol, atenolol, bisoprolol, carvedilol, and metoprolol) which are relatively cardioselective for blocking β_1-adrenoceptors, and others that are nonselective (propranolol, nadolol, timolol, and sotalol) with respect to β_1- and β_2-adrenoceptors. However, there is little evidence that these varying degrees of selectivity of action significantly alter the antiarrhythmic and antifibrillatory actions of β-blockers in patients. On the other hand, there is little doubt that the presence of marked agonist actions (e.g., in pindolol) is associated with often significant *increases* rather than decreases in heart rate that largely offset the antiarrhythmic and antifibrillatory actions of a given agent.[36]

Electropharmacologic Properties and Antifibrillatory Actions

The precise effects of β-antagonists on ionic currents in differing myocardial cells in terms of depolarization and repolarization are difficult to characterize.[37] It should be recognized that for a given agent within the broad class of these compounds, the effects may stem from their intrinsic properties of blocking β-receptors as well as from their associated pharmacologic actions. The dominant effect, however, is from β-receptor blockade, which has minimal effects on calcium channels ($I_{Ca\text{-}L}$) or various potassium channels (I_K). It is known that sympathomimetic amines exert a stimulant effect on the pacemaker current (I_f), which is markedly inhibited by β-blockers. Indeed, the blocking effect of the pacemaker current by β-blockers is the most readily defined pharmacodynamic property of this class of drugs and one that correlates well with the beneficial effects on mortality in patients with cardiac disease. The effects of β-blockade on the nodal tissues are, therefore, significant with the slowing of the heart rate by effects in the sinoatrial (SA) node and by prolonging refractoriness in the AV node which may be of clinical utility in terminating supraventricular tachycardias and slowing the ventricular response in the setting of AFL and AF.

The electrophysiological effects in other tissues are variable in terms of changes in conduction and refractoriness. The effects are minimal or modest in atrial and ventricular muscle; thus, acutely administered they do not consistently convert AFL and AF or ventricular tachycardia to sinus rhythm, although the latter has not been widely studied. Similarly, they have little effect on Purkinje fibers or the accessory bypass tracts. The acute effects of β-blockade on repolarization in isolated tissues and in the intact heart are variable and minor increases or decreases in the Q–T interval after chronic continuous drug administration have been reported.

Antiarrhythmic Actions of β-Blockers

It is inherently likely that the major basis for the salutary effects of β-blockers in cardiac arrhythmias stems from their property for counteracting the arrhythmogenic effects of catecholamines. However, their exact mechanism may differ in various disorders of rhythm in differing clinical settings. As a class, β-blockers exert a modest effect in suppressing ventricular and supraventricular arrhythmias (PVCs, PACs, and nonsustained VT) documented on Holter recordings. However, they increase VF threshold and reduce dispersion of repolarization especially in the ischemic myocardium. On the one hand, they have very little effect in preventing inducibility of ventricular tachycardia and ventricular fibrillation (VT/VF) in patients with sustained symptomatic VT/VF. On the other hand, they are the most potent antiarrhythmic and antifibrillatory agents with consistent effects on sudden death and all-cause mortality without the propensity for the development of discernible proarrhythmic reactions.

Perhaps, the antiarrhythmic and antifibrillatory actions of β-blockers are best characterized in terms of their property of attenuating the deleterious effects of excess catecholamines. The electrophysiological consequences of sympathetic hyperactivity have been extensively documented in numerous experimental[33] and clinical studies.[34-35] In the experimental setting, they have included (1) shortening of the ventricular action potential duration, and hence the refractory period; (2) augmenting ventricular conduction; (3) increasing ventricular automaticity; (4) reducing vagal tone; (5) decreasing VF threshold; and (6) the reversal or attenuation of the effects of antiarrhythmic drugs being administered in the expectation of preventing arrhythmic deaths. Conversely, it is known that the depletion of the adrenergic transmitters to the heart increases VF threshold and in experimental models in which VF could be induced reproducibly, the arrhythmia is preventable by sympathetic blockade.[33] Aspects of antifibrillatory effects of β-adrenergic blocking drugs are shown in Figure 37-4.[36] In fact, the appreciation of such an antifibrillatory effect was the basis for classifying it as a class II antiarrhythmic action.[3-8]

β-Blockade: Impact on Sudden Death and Total Mortality in Survivors of Myocardial Infarction and in Patients with Heart Failure

β-Adrenergic blocking drugs are effective in reducing mortality in many subsets of patients with manifest arrhythmias and in those at high risk of dying from arrhythmic deaths.[37-42] For example, they reduce death rates in patients with congenital long Q–T interval syndrome,[39,40] in survivors of cardiac arrest, and in selected cases of VT,[41-43] although the data supporting these conclusions have not always been from controlled clinical trials. However, they are in line with the compelling data from randomized, placebo-controlled β-blocking trials in which there have been consistent and significant decreases in mortality, especially in survivors of acute myocardial infarction and in those with congestive cardiac failure.[44-51] These trials have usually been of adequate sample size and of acceptable protocol design including features that include double-blind comparison in placebo-controlled studies. They have

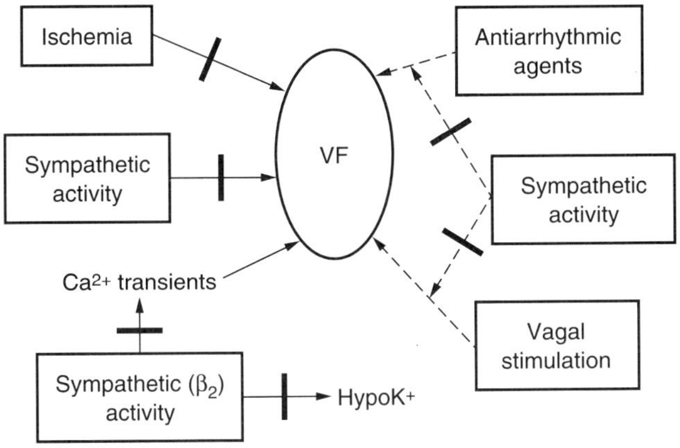

FIGURE 37-4 Model of possible antifibrillatory effects of β-blockade. Myocardial ischemia and increased adrenergic tone favor the development of ventricular fibrillation (VF) shown by *solid arrows*. Such effects are attenuated by the anti-ischemic and adrenergic antagonism of β-blockers indicated by *crossed lines*. β_2-Agonists also favor the development of VF by promoting calcium transients or hypokalemia. These effects are blocked by nonselective β-blockers. Vagal stimulation and some antiarrhythmic agents act to prevent the development of VF (*dotted lines*) but these actions are reversed by sympathomimetic stimulation. β-Blockade is likely to prevent this reversal and would again be antifibrillatory. (From Reiter MJ: Antiarrhythmic impact of anti-ischemic, antifailure and other cardiovascular strategies. Cardiac Electrophysiol Rev 2000;194-205.)

shown that these agents as a class not only effect a beneficial change in total mortality but also on sudden, presumably arrhythmic death (Table 37-3). The overall clinical utility and versatility of this class of antiadrenergic compounds are well documented as indicated in Table 37-4.[36]

β-Blockers as a class consistently reduce the incidence of sudden death and total mortality in the survivors of acute myocardial infarction and in patients with cardiac failure of ischemic, as well as of nonischemic, origin. In the case of patients with myocardial infarction, beneficial effect on sudden death and total mortality has been documented at the time of diagnosis out of the hospital, during hospital stay, and following discharge from the hospital. In the latter case, sudden death and total mortality were reduced 18% to 39% during the first year. It is presumed that the benefit in survival is from the prevention of VF. It is noteworthy, that mortality with β-blockers is reduced over the first 10 days when the drugs are given in an initial intravenous dose and followed by oral therapy, as well as when they are given in conjunction with thrombolytic therapy during the early stages of acute myocardial infarction. Two other features of the response to β-blockers in this setting should be emphasized. First, unlike the trials such as CAST,[1] β-blocker trials have not been arrhythmia suppression trials, but the degree of benefit on survival correlated linearly with the degree of heart rate reduction, both in the case of myocardial infarction as well as with cardiac failure. Second, in two postmyocardial infarction trials,[52,53] there is evidence that total mortality reduction was greater when amiodarone was administered in combination with β-blockers versus when it was used alone in the drug treatment limb.[54]

As summarized recently,[5] β-blockers, given in controlled and graduated dose regimens, have been shown to variably but significantly reduce mortality in subsets of patients with advanced congestive cardiac failure.[55-61] The outcomes of three recent trials with three different β-blockers[56,57,60] should be emphasized (see Table 37-2).

In the trial involving the use of metoprolol in 3991 patients with largely class II-IV (NYHA) heart failure (LVEF < 40%), randomized to long-acting metoprolol (n = 1990) in graduated doses of up to 200 mg/day or placebo (n = 2001), the primary end point was all-cause mortality, analyzed by intention to treat. The mean follow-up was 1 year. During the period of follow-up, there were 147 deaths (7.2%) on metoprolol versus 217 deaths (11.0%) on placebo (P < .00009). There was a statistically significant reduction in sudden death on metoprolol when compared with placebo.[56]

Similar data have been reported for the drug bisoprolol in 2647 patients in a double-blind, multicenter study[57] randomized to the β-blocker (n = 1327) or placebo (n = 1320). The patients were all in classes III and IV with left ventricular ejection fraction of 35% or lower and all were taking diuretics and ACE inhibitors as in the case of the MERIT-HF trial. All-cause mortality was 11.8% (156 deaths) versus 17.3% (17.3%). This difference was significant (P < .0001). The drug was effective in reducing sudden death significantly. There were significantly fewer cardiovascular deaths on bisoprolol, fewer were admitted to the hospital, and the difference for the combined endpoints was also significant. The numbers of permanent treatment withdrawals were similar for bisoprolol

TABLE 37-2 Results of Some End-point Trials of β-Blockers

Trial/Drug	N	β-Blocker	Total Mortality Reduction	Sudden Death Reduction
CUMULATIVE (Post-MI)	>50,000	Various (45)	23% (P < 0.01)	23% (P < 0.001)
BHAT-CHF	710	Propranolol (55)	27% (P < 0.05)	28% (P < 0.05)
MC Carvedilol Trials	1,094	Carvedilol (59)	65% (P < 0.001)	55% (P <= 0.001)
CIBIS-II	2,647	Bisoprolol (57)	34% (P < 0.001)	44% (P = 0.0011)
MERIT-HF	3,991	Metoprolol (56)	34% (P = 0.0002)	34% (P = 0.0002)
COPERNICUS	2,289	Carvedilol (58)	35% (P = 0.0014)	41% (P = 0.0001)
BEST	2,708	Bucindolol (60)	8% (NS)	10% (NS)

Post-MI, Post-myocardial infarction; NS, not significant; NR, not reported.

TABLE 37-3 Principal Uses of β-Blockers to Reduce Life-Threatening Ventricular Arrhythmias and in the Control of Postoperative Atrial Fibrillation Following Cardiac Surgery

Indications	Comments
Cardiac arrest survivors	No controlled clinical trials but uncontrolled data compelling.
Adjunctive therapy to ICDs	Controlled data still to be obtained, but support for VT/VF strong from large uncontrolled database
Congenital long Q–T interval syndrome	Compelling but uncontrolled data used a single drug modality, now being replaced with ICDs with which they often need to be combined.
Post-MI survivors	Most consistent antiarrhythmic agents increase survival in this setting, reduce total mortality, sudden death, and reinfarction rates. Overwhelmingly positive data from controlled clinical trials.
Congestive cardiac failure	Reduce total mortality and sudden death in CHF of ischemic and nonischemic origins: role well established.
Prevention of atrial fibrillation	Prophylactic administration reduces the incidence of AF in postoperative cardiac patient undergoing cardiac surgery

AF, Atrial fibrillation; ICDs, implantable cardioverter defibrillators; MI, myocardial infarction; VT/VF, ventricular tachycardia/ventricular fibrillation.

and placebo. In the case of carvedilol,[60] there have been a number of relatively smaller individual trials but the cumulative data have been sufficiently compelling for making it the first β-blocker to be approved for use in congestive heart failure in the United States. The composite data from all the trials of β-blockers performed to date support the routine use of this mode of prophylactic therapy in all patients in whom β-blockers are not contraindicated.

Drugs Acting by Prolonging Repolarization

Of the currently available antiarrhythmic compounds, β-blockers, sotalol, amiodarone, ibutilide, and dofetilide are likely to remain the prototypes of the drugs of the future. Amiodarone[62-65] and sotalol[66,67] are of the greatest interest; they prolong repolarization and refractoriness in atria and ventricles while blocking sympathetic stimulation, albeit by differing mechanisms. Their propensities to decrease heart rate and to modulate the atrioventricular node function and thereby slow ventricular response in AF and AFL have additional value for the control of many arrhythmias. Thus, they have the greatest clinical utility for controlling ventricular, as well as supraventricular, arrhythmias. Their respective roles are further bolstered by controlled trials[68,69] showing superiority over class I agents in patients with VT/VF. The perception that, although being powerful sympathetic antagonists, sotalol and amiodarone acted dominantly by prolonging the APD, led to the search for pure compounds devoid of other associated properties having simpler side effect profiles.[70-73] Intravenous amiodarone has recently been studied extensively in clinical trials in destabilizing VT/VF in the hospital and in shock-resistant VF following out-of-hospital cardiac arrest.[2,74-78]

There are now data suggesting that compounds that induce an isolated or "pure" prolongation of the action potential duration accompanied by a proportional lengthening of myocardial refractoriness may be antifibrillatory.[37] Under certain circumstances, they have advantages over complex compounds such as sotalol or amiodarone. This has been the case with dofetilide, ibutilide, and azimilide.[79] The properties of tedisamil, which is also under development, are somewhat more complex.[37] Such compounds may be the result of the block of a single repolarizing ionic current (e.g., dofetilide) or multiple currents (ibutilide, tedisamil, or azimilide). Thus, for the present, it would appear that two subclasses of such agents might be appropriate to develop—those that are either safer and more effective chemical derivatives of sotalol and amiodarone or those that are relatively pure prolongers of repolarization and refractoriness. All such agents may need to produce a homogeneous lengthening of repolarization by selectively inhibiting multiple or single ionic channels in the myocardium with a minimal proclivity to cause TdP.

TABLE 37-4 Actions of Intravenous Amiodarone Compared with Chronic Amiodarone

Actions	Intravenous Amiodarone	Chronic Amiodarone
Repolarization (QT interval) prolongation (atria and ventricles)	±	++++
Conduction velocity (atria and ventricles reduced)	++	++ (as a function of rate)
Sinus rate reduced	+	+++
AV nodal conduction slowed	+	++
AV nodal refractoriness increased	++	+++
Atrial refractoriness increased	+	+++
Ventricular refractoriness increased	±	+++
Noncompetitive α- and β-blocking activities	+	++

AMIODARONE AS A VERSATILE AND COMPLEX CLASS III AGENT

Amiodarone was synthesized as an antianginal drug in 1962, but attention was not drawn to the compound for its unique properties until 1970 when Singh and Vaughan Williams[4] showed the properties of the drug to prolong cardiac repolarization, while having antifibrillatory properties together with the potential for slowing heart rate due to nonspecific antiadrenergic actions. Since these early studies, it has become increasingly evident that amiodarone is an exceedingly complex antiarrhythmic compound (see Figure 37-5).

The electrophysiological effects of amiodarone administered acutely and after a period of chronic drug therapy differ markedly (see Table 37-4), both associated with distinct and potent antiarrhythmic actions.[36] The electropharmacologic effects of the compound are multifaceted.[80] The most striking property long term is the ability of the drug to increase the APD in atrial and ventricular tissues following chronic drug administration. There are little or no effects in Purkinje fibers[81] and M cells,[82] a phenomenon that leads to the reduction in QT dispersion.[83] The drug is also unique in that its effect on repolarization is not influenced by heart rate[84,85]; in isolated myocardial fibers, it eliminates the tendency for the development of early afterdepolarization (EAD) produced in Purkinje fibers and in M cells. The ionic channels that amiodarone blocks after chronic drug administration are not fully defined but it has been shown to block I_{to} and I_K as well as I_{Na} and I_{ca}.[64] It abolishes EADS induced in isolated cardiac tissue.[84] In humans, despite producing marked slowing of the heart rate and considerable increases in the Q–T/QTc interval, the drug has a very low propensity for producing TdP.[84] Amiodarone rarely exhibits proarrhythmic actions typical of class I agents while being a fairly potent sodium-channel blocker, especially at high heart rates. The salient pharmacodynamic features of amiodarone are summarized in Figure 37-5.

The clinical pharmacologic effects are complex with an elimination half-life of 30 to 60 days (longer for the main metabolite, desethylamiodarone), bioavailablity of about 40%, with a large volume of distribution, over 98% protein bound. Very little of the drug is excreted by the kidney and elimination is largely by hepatic metabolism, the biotransformation occurring by desethylation via human cytochrome (CYP) 34A. Neither the levels of the parent compound nor those of the metabolite levels accurately predict therapeutic effects of the drug. There are numerous drug–drug interactions involving amiodarone, perhaps the most clinically significant being with warfarin, digoxin, and verapamil. The drug is best not combined with other compounds that also prolong cardiac repolarization. In clinical practice, a variable loading regimen is desirable; in the past a dose as high as 2 gm/day had been used but such regimens are now rarely used. In practice, the aim is to use the lowest dose that produces the defined therapeutic effect.

Amiodarone has a broad-spectrum antiarrhythmic effect with a varied side effect profile, generally dose and duration dependent. Included in Table 37-5 are the main approved and potential clinical indications. Intravenous amiodarone is effective in the control of hemodynamically significant and refractory VT/VF (to lidocaine and procainamide) with a potency at least as high or higher than that of intravenous bretylium.[74,75] The drug exerts a powerful suppressant effect on PVCs, nonsustained ventricular tachycardia (NSVT), and provides control in 60% to 80% of recurrent VT/VF when conventional drugs have failed[86] during continuous oral therapy. In only a small number of patients does the drug prevent inducibility of VT/VF, there being little or no systematic relationship between the prevention of inducibility of VT/VF and the long-term clinical outcome.[87]

The properties of amiodarone during chronic administration permit predictable control of recurrent paroxysmal supraventricular tachycardias, slowing of the ventricular response in AF and AFL, and in maintaining stability of sinus rhythm after their chemical or electrical conversion.[88] Amiodarone is the most potent drug for maintaining stability of sinus rhythm in patients converted from AF. The drug is currently being compared with placebo and sotalol in a number of major multicenter studies.

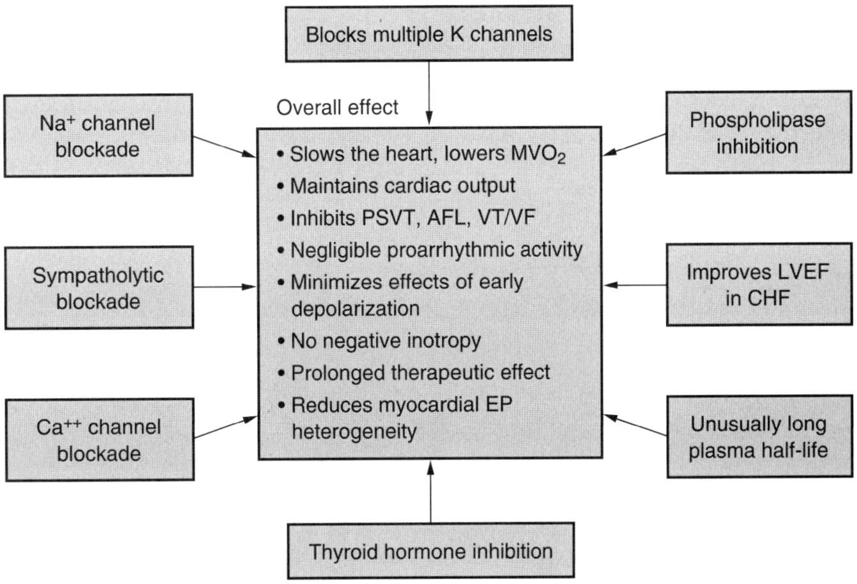

FIGURE 37-5 Multiplicity and complexity of amiodarone action. Note that the drug exhibits all four actions cited in the conventional classification of antiarrhythmic actions but not the deleterious effects of the same actions occurring independently in other compounds. The drug does not have the proarrhythmic actions of class I agents, or minimal or negligible rate of TdP, despite lengthening the QT markedly and slowing the heart significantly; it improves ventricular function and is effective in almost the entire spectrum of atrioventricular and ventricular arrhythmias.

TABLE 37-5 Major Uses of Amiodarone in the Control of Ventricular and Supraventricular Arrhythmias

Indications	Comments
Cardiac arrest survivors	Superior to class I agents (guided therapy); tolerated in VT/VF and in all levels of LVEF and CHF; low incidence of torsades de pointes. Increasingly perceived as the most potent antiarrhythmic agent, especially in the setting of low LVEF but being increasingly replaced with ICDs for primary therapy in many subsets of patients with VT/VF. May increase VT defibrillation threshold; clinical significance unclear.
Adjunctive therapy to ICDs	Controlled data still to be obtained but support for VT/VF is strong from large uncontrolled database; preferred agent in patients with significantly impaired LVEF and CHF.
Conversion of VT to SR by IV regimen	Precise efficacy not known
Intravenous amiodarone in destabilizing refractory VT/VF	Superior to lidocaine and procainamide; efficacy equal to that of bretylium but with less hypotension; increases survival to hospitalization in patients with out-of-hospital cardiac arrest (ARREST and ALIVE trials). First-line antiarrhythmic agent in cardiac arrest resuscitation.
Post-MI survivors	Numerous uncontrolled positive mortality studies; two recent placebo-controlled studies—EMIAT and CAMIAT—significant reduction in arrhythmic deaths, trend toward total mortality reduction in one of the studies, augmented effect if combined with β-blockers. Meta-analysis of all data consistent with reduction in total mortality and in sudden death while on drug.
Congestive cardiac failure	Two controlled trials, one blinded placebo-controlled (CHF STAT) and one blinded (GESICA) indicate a spectrum of effects from reduction in total mortality (GESICA) and neutral (CHF STAT) with a trend in selective favorable effects in nonischemic cardiomyopathy. Amiodarone increased LVEF.
Atrial fibrillation (AF):	
Acute conversion	Variable efficacy but systematic placebo-controlled study against a comparator agent not available.
Maintenance of SR	Up to 60% or greater at 12 months after restoration SR; maintains ventricular on relapse to atrial fibrillation on oral drug (controlled trials in progress)
Effect on AF prevention in postoperative cardiac surgery	Preoperative administration of oral drug reduces incidence of postoperative AF; IV drug given postoperatively may shorten hospital stay by reducing AF.

AF, Atrial fibrillation; CHF, congestive heart failure; ICD, implantable cardioverter-defibrillator; IV, intravenous; LVEF, left ventricular ejection fraction; MI, myocardial infarction; SR, sinus rhythm; VT/VF, ventricular tachycardia and ventricular fibrillation.

Amiodarone is effective in converting AF to sinus rhythm both in valve disease[88] as well as in patients with nonrheumatic AF.[89] With respect to AF occurring after cardiac surgery, the drug has been studied in three different trials.[90-92] Hohnloser and colleagues[90] studied amiodarone in a randomized, double-blind, placebo-controlled study in 77 patients after coronary artery bypass surgery. They gave a 300 mg bolus of intravenous amiodarone followed by 1200 mg daily every 24 hours for 2 days and 900 mg daily for the next 2 days. At the end of the study period, the overall incidence of AF in the placebo group ($n = 38$) was 21%; in the amiodarone group ($n = 39$) it was 5% ($P < .05$).

More recently, Guarnieri and his associates[91] have provided further supportive data from a larger double-blind study in which they randomized 300 patients to placebo ($n = 142$) and to intravenous amiodarone ($n = 158$); 2 gm of amiodarone were given over 2 days (1 gm/day) through a central venous line starting shortly after the completion of open heart surgery. In the placebo group, 67 of 142 patients developed AF (47%) compared with 56 of 158 (35%) in the amiodarone limb ($P < .039$). The time to the onset of AF in the amiodarone limb was significantly longer than in the placebo (2.9 versus 2.2 days; $P < .006$).

Daoud and colleagues[92] recently reported the first of the two studies on the effects of orally administered amiodarone in preventing postoperative AF in patients undergoing heart surgery. Their study was performed in 124 patients randomized to placebo ($n = 60$) and to amiodarone ($n = 64$) for a minimum of 7 days preoperatively (600 mg/day), followed by 200 mg/day until discharge from the hospital. Postoperative AF developed in 32 of the 60 patients (53%) in the placebo group and in 16 of the 64 patients (25%) in the group given amiodarone. Patients in the amiodarone group were hospitalized for significantly fewer days compared with those on placebo.

The unique pharmacodynamic profile of amiodarone holds much interest as the prototype of complex compounds that might be necessary to develop for antifibrillatory actions in the atria and the ventricles[93-95] provided the drug's side effect profile can be improved. Clearly, along with sotalol, amiodarone has now emerged as one of the two leading antiarrhythmic drugs currently in use for the control of life-threatening ventricular tachyarrhythmias. However, their therapeutic roles in this setting need to be placed in perspective relative to the expanding indications in the use of implantable cardioverter defibrillators (ICDs) in preventing arrhythmic deaths in patients with significant structural heart disease.[95]

Intravenous Amiodarone

Intravenous amiodarone has now emerged as the most powerful antiarrhythmic drug in the treatment of destabilizing VT/VF in hospitalized patients and as an antiarrhythmic agent for the resuscitation of patients developing cardiac arrest out of the hospital.[74-77] It has also been shown to convert recent onset AF to sinus rhythm, although it is not used widely for this indication. However, the intravenous drug is used prophylactically for the prevention of AF and AFL in patients who have undergone cardiac surgery.[90,91] The major utility of intravenous amiodarone now is in the acute control of destabilizing VT/VF in hospitalized patients and in the resuscitation and continued control of pulseless VT/VF in patients surviving out-of-hospital cardiac arrest.[74-77]

A number of studies were performed in the early 1990s in destabilizing VT/VF, including those from electrical storm in patients resistant to lidocaine or procainamide. In particular, two amiodarone dose-response studies (125 mg, 500 mg, and 1000 mg or 2000 mg per 24 hours) with the drug, and one comparative trial with bretylium (2500 mg/24 hours), provided the initial data that intravenous amiodarone may exert a superior effect in controlling VT/VF (including electrical storm) in patients who were refractory to lidocaine or procainamide. In a direct comparison with intravenous bretylium in 305 patients in the randomized double-blind study, amiodarone showed a strong trend for a greater efficacy in reducing VT/VF events, but the difference did not reach statistical significance. However, the data provided the rationale for the exploration of the role of intravenous amiodarone in this setting. Two blinded randomized studies conducted in patients with persistent or recurrent pulseless VT/VF in patients developing out-of-hospital cardiac arrest have now been reported.[76,77] In ARREST (*A*miodarone in out-of-hospital *R*esuscitation of *RE*fractory *S*ustained Ventricular *T*achyarrhythmias), 504 patients incurring out-of-hospital cardiac arrest with recurrent or persistent pulseless VT/VF were enrolled in a prospective randomized double-blind study in which the effects of intravenous amiodarone (300 mg) were compared with those of placebo against the background of ACLS regimen, which included lidocaine in both limbs of the trial. The data revealed that amiodarone-treated patients were more likely to survive to reach hospital compared with those in the placebo group (44% versus. 34%, $P = .03$). The difference was larger among the 221 patients in whom the arrest was witnessed ($P = .008$), possibly due to the shorter interval from the occurrence of the arrest and the administration of drug therapy.

ARREST did not directly compare the efficacy of amiodarone to that of lidocaine because the latter was an integral part of *both* treatment limbs. However, superiority of amiodarone demonstrated in ARREST was considered compelling enough to be acknowledged in the "Guidelines 2000 for Cardiopulmonary Resuscitation and Emergency Cardiovascular Care."[2] The evidence ("fair to good evidence") was considered to be in favor of amiodarone as a class IIB indication; the drug was considered "acceptable, safe and useful," and the evidence with respect to lidocaine was deemed "indeterminate" because the "research quantity/quality fell short of supporting a final class decision."

The ALIVE trial was undertaken by Dorian and associates[77] to compare the effects of lidocaine and amiodarone in shock-resistant patients with VF with out-of-hospital cardiac arrest. In ALIVE, 347 patients were enrolled if they were resistant to three shocks, intravenous epinephrine, and further shock or if they had recurrent VF after initial successful defibrillation. The randomization was in a double-blinded manner so that the drugs were given either as intravenous amiodarone plus lidocaine placebo versus lidocaine plus amiodarone placebo. The primary end point was the percentage of patients who survived to reach the hospital. On amiodarone, 22.8% of 188 patients reached the hospital, compared with 12% of 167 patients on lidocaine. The difference was highly statistically significant ($P = .009$; odds ratio 2.17; 95% confidence interval, 1.21 to 3.83). The comparative data from the ALIVE trial are summarized in Figure 37-6.

The cumulative clinical investigative data on intravenous lidocaine and intravenous amiodarone in the setting of cardiac resuscitation and in the control of ventricular tachyarrhythmias in the hospital and out of the hospital when interpreted in light of the outcomes of the ARREST and ALIVE trials, permit a number of conclusions. The most compelling is the issue that intravenous lidocaine is inferior to intravenous amiodarone in increasing the numbers of cardiac arrest survivors reaching hospital admission. Lidocaine does not appear to have a role in the control of life-threatening ventricular tachyarrhythmias and its continued use in this setting is no longer justifiable.[30] The current data support the premise that intravenous amiodarone should now supersede intravenous lidocaine as the first-line antiarrhythmic drug in the resuscitation of patients with shock-resistant VT/VF.[30,77]

Sotalol as an Antiarrhythmic Drug

The overall action of sotalol is dominated by its dual propensity for competitively blocking β-receptors without

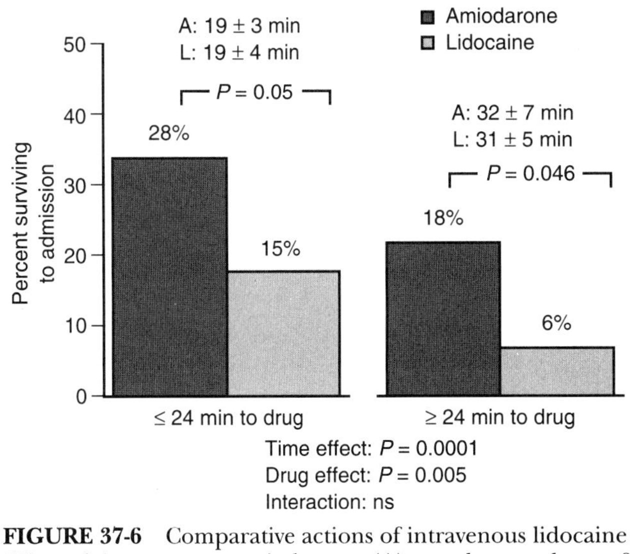

FIGURE 37-6 Comparative actions of intravenous lidocaine (L) and intravenous amiodarone (A) on the numbers of patients with shock-resistant persistent or recurrent VT/VF after out-of-hospital cardiac arrest on reaching the hospital. Note that at both lower and upper median (24 minutes) times from dispatch to study drug administration, the numbers of patients reaching the hospital are significantly greater in the case of intravenous amiodarone compared with intravenous lidocaine. (From Dorian P, Cass D, Schwartz, et al: Amiodarone as compared with lidocaine for shock-resistant ventricular fibrillation. N Engl J Med 2002;346:884-890.)

the predilection for β_1 or β_2 receptors and by the property for prolonging the myocardial APD. Sotalol is a racemate of d- and l-isomers, both of which have equal class III activity, only the l-isomer having significant β-adrenoceptor-blocking activity.[96,97]

Electrophysiological Actions

Sotalol is a nonselective β-adrenoceptor-blocking drug with little or no sodium channel-blocking actions without intrinsic sympathomimetic activity. In 1970, Singh and Vaughan Williams[3] found that sotalol markedly prolonged APD in isolated atrial and ventricular multicellular preparations. This was classified as a class III antiarrhythmic mechanism. Sotalol slows the sinus node frequency by depressing phase 4 depolarization. Sotalol increases APD primarily by blocking the time-dependent outward delayed rectifier K-current (I_k). The effect of increased frequency of stimulation leading to decreases in the APD has been termed reverse rate dependency.[19] The drug prolongs the Q–T and QTc intervals without significant effects on QRS or P–R intervals,[98] with no effect on His-Purkinje (H–V interval), or ventricular (Q–R interval) conduction velocity.

Pharmacokinetics of Sotalol and Optimal Dosing

Because the drug is fully bioavailable, is not metabolized and not bound to plasma proteins, fluctuations in serum concentration are small, the plasma half-life long (10 to 15 hours), and plasma levels are linearly related to dose. The usual starting dose is 80 mg bid, rarely 40 mg bid), with stepwise increases (120 mg bid, 160 mg bid, rarely higher) for optimal effect. Because sotalol is largely excreted by the kidneys in an unchanged form, the plasma concentrations of the drug vary linearly with creatinine clearance. Dose adjustment is necessary in proportion to the degree of renal dysfunction present. For a creatinine clearance greater than 60 mL/min, 12 hour dosing intervals are appropriate. The interval may be increased to 12 to 24 hours when the creatinine clearance is 30 to 69 mL/min and 36 to 48 hours for patients with the clearance of 10 to 30 mL/min. However, it may be clinically prudent not to use the drug when creatinine clearance is less than 60 mL/min. Emerging data suggest that a close monitoring of the Q–T interval relative to the dose may be effective in preventing the development of TdP. Numerous pharmacodynamic interactions with numerous cardioactive compounds are likely. This likelihood is especially so in the case of other antiadrenergic drugs (verapamil, diltiazem, and amiodarone) and other QT prolonging agents (class III agents, including amiodarone, tricyclic antidepressants, and phenothiazines). Coadministration with erythromycin and other QT prolonging drugs should also be avoided. Intravenously (not approved by the FDA) it is given in a dosage of 0.2 to 1.5 mg/kg of dl-sotalol over 2 to 3 minutes under electrocardiographic and hemodynamic control. Although in the past doses of up to 960 mg/d had been used, especially in investigative protocols, it is now clear that higher doses are rarely necessary and may be associated with a greater incidence of TdP, heart failure (New York Heart Association classes III and IV), or severe bradycardia.[66-67] Although in rare patients, 80 mg/d may be effective, the range of total daily doses of 240, 360, and 480 mg administered in two equally divided doses allows flexibility of optimal dosing in most patients. In a recent study in AF, 240 mg (as 120 mg bid) was found to be the optimal daily dosage in terms of efficacy and safety.

Control of Cardiac Arrhythmias with Sotalol

The main clinical indications for the use of sotalol in supraventricular and in ventricular arrhythmias are shown in Table 37-6. The spectrum of sotalol effects in arrhythmias is wider than that of conventional β-blockers[67] on the one hand and wider than that for the pure class III agents on the other hand. Sotalol has a modest suppressant effect on PVCs; it is generally not used for this indication. It exerts a variable effect in suppressing inducible VT/VF but this approach is no longer used in

TABLE 37-6 Major Uses of Sotalol in the Control of Ventricular and Supraventricular Arrhythmias

Indications	Comments
Cardiac arrest survivors	Superior to class I agents (guided therapy); VT/VF may be comparable with amiodarone (no significant direct comparison but incidence of TdP higher. Now being increasingly replaced with ICDs for primary therapy in many subsets of patients
Adjunctive therapy to ICDs	Controlled data still to be obtained but support for VT/VF is strong from large uncontrolled database
Conversion of VT to SR by IV regimen	Superior to lidocaine; use not approved in the United States.
Post-MI survivors	Positive in one study with an 18% reduction in total mortality but not statistically significant.
Congestive cardiac failure	No controlled studies performed
Atrial fibrillation:	
Acute conversion	Variable efficacy but systematic placebo-controlled study or against a comparator agent not available.
Maintenance of SR	Up to 50% at 12 months after restoration of SR; maintains ventricular rate on relapse to atrial fibrillation. Results of placebo-controlled trials imminent.
Prevention of postoperative sequelae	Up to 50% reduction when drug given orally before period of cardiac surgery and after cardiac surgery

ICD, Implantable cardioverter-defibrillator; IV, intravenous; MI, myocardial infarction; SR, sinus rhythm; TdP, torsades de pointes; VT/VF, ventricular tachycardia and ventricular fibrillation. The role of sotalol in supraventricular arrhythmias, other than atrial fibrillation, is not discussed in this review.

the control of VT/VF. There are two main uses of the drug in the control of cardiac arrhythmias: (1) as an adjunctive therapy with the ICD for reducing the number of shocks, and for (2) the maintenance of sinus rhythm in patients with AF restored to sinus rhythm either chemically or by DC cardioversion. Both indications have been substantiated in controlled clinical trials.[99-101]

Experience has indicated that concomitant antiarrhythmic therapy with an ICD is an integral part of therapy of recurrent VT/VF.[101] Sotalol may slow the VT rate and aid termination by antitachycardia pacing; it may decrease the defibrillation threshold and facilitate termination at lower energy levels and reduce the number of shocks by suppressing episodes of supraventricular (including AF) and ventricular tachyarrhythmias. The overall effects of sotalol (as is the case with any other pharmacologic agent now being introduced) in supraventricular arrhythmias should be viewed in relation to the changes that have occurred in the electrode catheter ablation of paroxysmal supraventricular tachycardias (PSVTs) with or without accessory tracts in the heart and many cases of AFL.[98] Because these newer approaches carry an extremely high frequency of success with prospects for cure, prophylactic drug therapy of PSVTs is likely to be used much less often than previously in areas of the world where ablative techniques are readily available.

The predictive accuracy of the acute response for the long-term effects of the drug remains unclear, however. The antifibrillatory effects of sotalol as a result of its class III action[65-67] are likely to make it more effective in maintaining sinus rhythm in patients with AF and AFL after cardioversion. There is a well-defined electrophysiological rationale for the potential utility of sotalol for the acute conversion of AFL and AF to sinus rhythm and for maintaining stability of sinus rhythm during long-term drug prophylaxis.[66,67] Conversion rates have been variable, but a slowing of the ventricular rates has been consistent as might be expected for a β-blocker. The conversion rate may be up to 40% to 50% of selected patients, which compares favorably with intravenous procainamide and possibly with intravenously administered newer class III agents (ibutilide or dofetilide) or class IC agents (e.g., flecainide). The maintenance of sinus rhythm after conversion is potentially of greater practical importance. Sotalol is as effective as propafenone or quinidine in maintaining sinus rhythm in patients with AF[102-104] but appears to be less effective than amiodarone[93,94] in direct comparisons.

Of particular interest are two placebo-controlled trials that deal with stability of sinus rhythm under the action of dl-sotalol. Benditt and colleagues[99] conducted a double-blinded placebo-controlled trial in which they compared the efficacy, safety, and dose-response relationship of three doses of dl-sotalol (80, 120, and 160 mg twice daily) for the maintenance of sinus rhythm in 253 patients converted to sinus rhythm. Compared with the placebo (27 days) the time to recurrence was significantly prolonged by two doses of the drug—226 days on 120-mg dose (P = .001) and 175 days on 160-mg dose (P = .012). There were no deaths or cases of TdP sustained VT or VF. A number of controlled studies have indicated the efficacy of sotalol in preventing the occurrence of AF and flutter developing in patients under cardiac surgery.[105,106] Although there were no cases of TdP, it is known that from studies in patients receiving the drug for other arrhythmias, the incidence of TdP varies with dose[107] and can be higher than 5% at the highest doses of the drug sometimes used.

Bretylium Tosylate

Developed as an antihypertensive agent, bretylium gained prominence for many decades as an intravenous antifibrillatory agent for the control of destabilizing VT/VF, especially in the resuscitation of patients incurring out-of-hospital cardiac arrest. Its introduction into therapeutics was based largely on the basis of studies in animal models and comparative studies with lidocaine in the acute control of VT/VF.[108,109] No superiority over lidocaine has been demonstrated. Bretylium has recently been compared with intravenous amiodarone in a multicenter study of patients with VT-VF refractory to or intolerant of lidocaine and procainamide. Bretylium showed equivalent efficacy, measured as time to recurrence and survival, but caused more side effects (e.g., hypotension) than amiodarone.[75] The drug is no longer included in the list of antiarrhythmic drugs in the latest ACLS Guidelines in the control of destabilizing VT/VF developing in or out of the hospital and is no longer available.[2] Thus, along with lidocaine, as an antiarrhythmic drug, it is likely to become only of historic interest.[30]

"Pure" Class III Antiarrhythmic Drugs

Compounds such as dofetilide,[71] ibutilide,[72] and azimilide[73] among others are examples of pure class III agents. Isolated or pure prolongation of the APD accompanied by a proportional lengthening of myocardial refractoriness is now accepted as an antifibrillatory.[16,17,18] Under certain circumstances, drugs with such effects may have advantages over complex compounds such as sotalol or amiodarone.

IBUTILIDE, A PROTOTYPE OF INTRAVENOUS CLASS III AGENTS

Ibutilide was the first of the "pure" class III agents approved by the FDA for the acute termination of AFL and fibrillation.[110-117] It does this by prolonging the atrial action potential and refractoriness.[72,110]

Electropharmacology and Pharmacokinetics

The class III action of the drug stems from its property of blocking the rapid component of the delayed rectifier current (I_{Kr}); at much lower concentrations the drug also activates a slow inward sodium current which is not blocked by I_{Kr} blockers.[72,110] In isolated atrial and ventricular muscle preparations, ibutilide increases the APD and the ERP, effects on both of which are attenuated at increasing rates. In man, the drug produces concentration-related increases in the Q–T and the QTc intervals.[111,112] Ibutilide, like most other class III compounds, is likely to produce antifibrillatory as well as profibrillatory actions.[72]

Ibutilide exerts little or no effect on sinus rates or AV nodal conduction in healthy volunteers and patients with a minimal influence on the P–R or the QRS intervals[114]; nor does the drug lower blood pressure or worsen heart failure because it does not exert a negative inotropic effect in the atria or the ventricles.[72] It exhibits no significant interaction with the autonomic nervous system. The property that is critical for its therapeutic utility is the acute prolongation of the atrial ERP leading to rapid termination of AF and AFL,[112] and the lowering of the energy for atrial defibrillation when these arrhythmias are not amenable solely to electrical conversion.[72]

Ibutilide is only available as an intravenous formulation as the drug is rapidly metabolized in the liver as a first-pass phenomenon. For this reason, the oral formulation is unlikely to be effective. The elimination half-life is variable (2 to 12 hours, mean 6 hours). The drug is extensively metabolized and excreted largely by the kidney but intravenous dosing does not require dose adjustment relative to changes in renal and hepatic function. Coadministration of ibutilide with digoxin, calcium channel blockers, or β-blockers exerts effects in the pharmacokinetics, safety, or the efficacy of the drug for the conversion of AF or flutter. Similarly, patient age, gender, or the nature of the arrhythmia may have an effect on the drug's pharmacokinetics.[72]

Ibutilide in Conversion of Atrial Fibrillation and Flutter

Effectiveness of ibutilide in converting AF and AFL to sinus rhythm when given intravenously has been established in two pivotal controlled studies.[111,112] For example, intravenous ibutilide was studied in 200 patients who were hemodynamically stable, free of heart failure and angina but who had AF and AFL with a duration of between 3 hours and 90 days. They were randomized in equal numbers to placebo and four doses of the drug (0.005, 0.010, 0.015, and 0.025 mg/kg given over 10 minutes). The conversion rate on placebo was 3%. In those for the four ascending doses of ibutilide the corresponding rates of conversions were 12%, 33%, 45%, and 46%, respectively.[110] The overall rate of conversion in the case of AF was 29% and 38% for AFL. The mean time to conversion was 19 minutes, conversion being accompanied by the prolongation of the Q–T and QTc intervals on the ECG; six patients in the study developed polymorphic ventricular tachycardia (3.6%).

The second study[112] involved 266 patients who had AF or AFL, with a duration of 3 hours to 45 days. They were randomized to be given up to two 10-minute infusions of ibutilide (1.0 mg and 0.5 mg, or 1 mg and 1 mg) or placebo. In most patients, there was a history of cardiac disease with an enlarged left atrium. The conversion rate on placebo was 2%, and 47% on ibutilide (63% in AFL and 31% in AF). The mean time to conversion was 27 minutes with an incidence of TdP of 8.3% (12.5% in AFL, 6.2% in AF), 1.7% being sustained requiring DC cardioversion. These experiences with ibutilide have been confirmed in subsequent clinical experience.

The major issues regarding the use of pure class III agents in the acute conversion of AF and AFL have been discussed critically by Singh[116] and Roden.[117] The conversion rates exceed 30% in the case of AF and about 50% in the case of AFL of recent onset, respectively. The associated rate of TdP developing during the conversion has been as low as 2% to 3% in the case of AF and as high as 8% to 12% in AFL. The data provide the basis for the use of intravenous ibutilide in the acute conversion of AF and AFL.[118] To date, this is the only intravenous compound that has been formally approved in recent years by the FDA for the standardized chemical conversion of AF and AFL and meaningful comparisons with other drug regimens, oral or intravenous, have not been established. It is of much interest that the conversion of AF and AFL by intravenous ibutilide in patients chronically treated with amiodarone has recently been reported by Glatter and associates.[119] In 70 such patients, the conversion of AF by ibutilide was 54% in AFL and 39% in AF, there being one instance of nonsustained TdP (1.4%) in the entire study. The data suggest that, despite the fact that the combination of amiodarone and ibutilide produced a further increase in the Q–T interval, amiodarone appeared to decrease the incidence of TdP usually associated with the IV ibutilide.

Facilitation of Transthoracic Cardioversion with Ibutilide Pretreatment

Pure class III agents predictably lower the threshold for defibrillation in the atria and ventricles. Recently, Oral and colleagues[114] reported that patients with AF who have failed to be cardioverted electrically may be successfully cardioverted if they are pretreated with intravenous ibutilide. They randomized 100 such patients who had AF for a mean duration of 117 days, to undergo transthoracic cardioversion with or without pretreatment with 1 mg of ibutilide. The protocol for conversion was a step-up process involving sequential shocks of 50, 100, 200, 300, and 360 joules. If the transthoracic shock failed to cardiovert in the absence of ibutilide, the drug was given and the shock protocol was repeated. Cardioversion to sinus rhythm in the group who had not been given ibutilide was 36 of 50 (76%) patients but all 50 patients pretreated with ibutilide could be cardioverted (100%; $P < .001$). It was of interest that all 14 patients who did not cardiovert with the initial DC cardioversion could be cardioverted following ibutilide pretreatment. The results reported by Oral and coworkers[114] are striking and need to be confirmed, as the approach will undoubtedly have major implications for clinical practice especially in conjunction with biphasic cardioversion shocks.

Adverse Effects

The safety profile of intravenous ibutilide on the basis of experience with the drug in 586 patients involved in

Phase II and III clinical trials has been reported.[120] Noncardiac side effects were indistinguishable from those of placebo, as were hypotension, conduction block, or bradycardia. The most significant adverse reaction was the overall incidence of 4.3% in the case of TdP, 1.7% requiring cardioversion. In almost all patients the onset of the arrhythmia was within 40 minutes of the commencement of the drug infusion.

Drug Dosage and Administration

There is now considerable experience with intravenous ibutilide for the conversion of recent onset AF. A number of precautions increase the safety margin of the drug's use in this setting. Clearly, patients at high risk for the development of proarrhythmia include those with prolonged baseline QTc (>440 ms), severe bradycardia, low levels of serum K and Mg (unless correctable), and the background use of other QT-prolonging drugs (amiodarone may be an exception), or in patients with a previous history of TdP induced by other antiarrhythmic drugs. The recommended dose of the drug is 1 mg which is administered over a 10-minute period in patients weighing more than 60 kg. Generally, the same dose is repeated if the arrhythmia termination does not occur at the end of the first infusion. Patients are generally monitored for a period of 4 hours following drug administration and QT-prolonging drugs should not be commenced until after this period of time.

The Use of Dofetilide in Maintaining Stability of Sinus Rhythm in Atrial Fibrillation

Dofetilide (Tikosyn) is the prototype of pure or simple class III antiarrhythmic agent. Its oral formulation was recently approved by the FDA for the maintenance of sinus rhythm in patients with paroxysmal and persistent atrial AF and AFL. Like ibutilide, the drug is effective in terminating recent onset AF and flutter,[121] although this indication has not yet been approved for routine clinical use.

Pharmacodynamics and Pharmacokinetics

The drug is a highly selective agent that delays repolarization in the atria, ventricles, and Purkinje fibers by I_{Kr} blockade. Such an I_{kr} block is the sole identifiable action of the drug in cardiac muscle, although it is likely that M cell action potential duration may be prolonged by reduction of I_{Ks}. The compound exerts no effects on sodium and calcium channels, and as such, has minimal effects on conduction velocity. It has no significant effect on the antagonizing autonomic transmitters.[70,122] Thus, it does not alter the PR interval or the QRS duration of the surface ECG.[30] Its sole measurable electrophysiological action is the lengthening of cardiac repolarization as reflected in the Q–T/QTc interval, an effect that correlates directly with the lengthening of the refractory period. The drug has no effect on myocardial contractility or on systemic hemodynamics. It does not depress systemic blood pressure. Therefore, the drug can be used in the control of AF in patients with heart failure. The drug is excreted largely via the kidneys and dose adjustment is necessary in patients with impaired renal function.

The impact of the drug on mortality in the postmyocardial infarct patient has been found to be neutral in patients with left ventricular dysfunction with or without cardiac failure.[71] However, these studies have shown that the safety of the compound is critically dependent on the use of appropriate doses of the drug relative to renal function, therapy initiation in the hospital, and subsequent monitoring of the patient by following the changes in the Q–T/QTc intervals with adjustment of drug dose.

Effectiveness of Dofetilide in Atrial Fibrillation and Atrial Flutter

Two blinded placebo-controlled pivotal studies have provided the pivotal data on the effectiveness of the compound. The first trial, European and Australian Multicenter Evaluative Research on Atrial Fibrillation of Dofetilide (or EMERALD), the effects of three doses of dofetilide (25, 250, and 500 μg bid) were compared with placebo over a 12-month period in 534 patients with AF/AFL durations of between 1 week and 2 years since onset.[123] The conversion rate to sinus rhythm at the highest dose was 29% compared with 1% in placebo (P = .001). Those not converting on the drug, were restored to sinus rhythm by electrical conversion; in all, 427 patients who did convert by either means, were followed for arrhythmia recurrence. At the highest dose, 66% remained in sinus rhythm compared with 26% on placebo (P = .001) at the end of the first year. The median time to relapse of AF/AFL at the two higher doses of the drug was greater than 365 days compared with 34 days for placebo.

In the second study (Symptomatic Atrial Fibrillation and Randomized Evaluation of Dofetilide or SAFIRE-D) involving 325 patients, three doses of the drug (as in EMERALD) were compared with placebo for the conversion and maintenance of sinus rhythm in patients with chronic AF/AFL of durations between 2 weeks to 6 months. Thirty percent of patients converted on the 500 μg bid dose, 70% of such conversions occurring during the first 24 hours. At 48 hours patients not converting on drug alone were cardioverted. Those not converting were excluded from the trial. The 250 patients who achieved sinus rhythm were followed for stability of sinus rhythm. The response showed a dose dependence, but only at the highest dose (500 μg bid) was the overall effect significantly different from that of placebo. The probability of the patients on 500 μg bid remaining in sinus rhythm at the end of 12 months was 58% in the case of dofetilide compared with 25% on placebo (P= 0.001). The median time for the patients to relapse into AF/AFL was over 365 days on the active drug compared with 27 days on placebo.

The data from the EMERALD and SAFIRE-D studies[123,124] indicate, therefore, that dofetilide in a defined group of patients with AF/AFL is a useful anfibrillatory agent for restoring and maintaining sinus rhythm. Supportive data have also been reported from the Danish

Trial in Acute Myocardial Infarction on Dofetilide (DIAMOND).[71] The major aspect of this study focused on mortality rate in two groups of patients after myocardial infarction—those with ventricular dysfunction and those with overt congestive heart failure. There were 1518 patients in the study and patients were randomized to dofetilide or placebo. After a median follow-up of 18 months, there was no impact on mortality. In the patients ($n = 506$) who had AF or developed AF during the course of the study, dofetilide was effective in converting AF to sinus rhythm (12% versus 2% on placebo); once the sinus rhythm was restored in these groups either chemically or electrically, the 1-year maintenance of sinus rhythm was 79% in the group taking dofetilide versus 42% on placebo. These data are consistent with the effects of the drug given intravenously to patients with AF/AFL; in 91 patients (75 with AF and 16 with AFL), Falk and colleagues[121] found a conversion rate of 31% in AF patients given 8 μg/kg of dofetilide intravenously in a double-blinded study; the conversion rate was 12.5% on 4 μg/kg, there being no conversions on placebo. In AFL, the conversion rate was 54%. However, the major significance of dofetilide in clinical therapeutics is likely to be the maintenance of sinus rhythm in patients with paroxysmal or persistent AF as indicated by the outcomes of placebo-controlled clinical trials discussed above. Indication for its use in the prophylactic therapy of AF is likely to be the major niche for dofetilide in patients with or without heart failure.[125]

Adverse Reactions and Contraindications

The pattern of adverse reactions noted in the placebo-controlled studies has been 5% to 10%, including headache, chest pain, dizziness, respiratory infection, dyspnea, and nausea. Quantitatively these were indistinguishable from those in the placebo limb of the clinical trials. The most significant side effect attributable to dofetilide has been the occurrence of TdP. In the two pivotal trials (EMERALD and SAFIRE-D) on AF/AFL, 9 of the 11 cases of TdP were symptomatic and 8 required intervention, but there were no deaths attributable to the arrhythmia. In the DIAMOND trials involving 518 patients, in the drug limb involving 762 postinfarction patients with ventricular dysfunction or heart failure, 29 of the 32 patients with TdP were symptomatic, 23 requiring intervention for termination. In this group there were two deaths, but there were no deaths attributable to TdP in the DIAMOND-AF subgroup.

The concomitant use of certain drugs during dofetilide therapy is contraindicated, especially those that may substantially increase the plasma concentrations of dofetilide. The prominent among these are verapamil, ketoconazole, cimetidine, trimethoprim/sulfamethoxazole, prochloperazine, and magesterol. Dofetilide is also contraindicated in patients with severe renal impairment and in those with acquired or congenital long QT syndrome. A previous history of TdP either on dofetilide or any other QT-prolonging compounds is also a contraindication to the use of the drug for the treatment of AF/AFL.

Dosing Recommendations and Enhancing Safety

Because there is a reasonably linear relationship between the plasma concentrations, drug dose, and the Q–T/QTc intervals and all of them are determinants for the development of TdP, it is critical to determine the appropriate initiating and steady-state dosing regimens for the safe use of dofetilide. As mentioned earlier, the drug dose is adjusted relative to the renal function. For patients with the creatinine clearance greater than 60 mL/min, the recommended starting dose is 500 μg bid, for clearances between 40 and 60 mL/min, the starting dose is reduced to 250 μg bid, and it is further reduced to 125 μg bid for clearance between 20 and 40 mL/min. A baseline Q–T or QTc (determined when sinus rhythm is present) interval exceeding 440 ms (or >500 ms in cases of intraventricular conduction defect) or a creatinine clearance of less than 20 mL/min are contraindications for the use of dofetilide. The bulk of the cases of TdP have been noted in patients during the initiation of therapy which, of necessity, should be in a monitored setting in the hospital for a period of 3 days or for five doses of the drug. A reduction in the dose of the drug is also recommended if the Q–T/QTc intervals increase by more than 15% or prolong beyond 500 ms. These precautions have been shown to markedly reduce the incidence of TdP during the use of dofetilide in patients with AF/AFL treated for maintaining stability of sinus rhythm.

CALCIUM CHANNEL BLOCKERS, ADENOSINE, AND DIGOXIN

The electropharmacologic properties and the clinical utility of structurally disparate compounds will be considered together because the bulk of their actions relative to their antiarrhythmic effects involves the inhibition of the atrioventricular (AV) node to varying extents. Transient complete block of anterograde conduction at the AV node by these agents has been used for termination of re-entrant supraventricular tachycardias either for therapeutic purposes or for differentiating narrow-QRS tachycardia on the surface electrocardiogram.[38] The blocking actions of verapamil, diltiazem, and adenosine, induce a prompt and predictable conversion of PSVT in 80% to 100% of cases of the arrhythmia. The conversion rate is effected by intravenous therapy with 3 to 5 mg (children) to 10 to 15 mg (in adults) with verapamil, 17 to 25 mg of diltiazem, and 6 to 12 mg of adenosine. Adenosine is now the most frequently used compound in this setting because of near complete efficacy and ultrashort elimination half-life accounting for the nature of the drug's transient side effects. However, there are clinical settings in which the use of the drug may not be appropriate. Adenosine is preferred in patients with depressed ventricular function and if they have recently received β-blockers and in neonates. Alternatively, for termination of PSVT, verapamil may be preferable in patients on drugs known to interfere with the actions of adenosine or its metabolism or in patients with bronchospasm. In patients in whom the diagnosis of PSVT is suspected but not certain, it might be preferable to use verapamil or diltiazem because it will not produce sustained hypotension.

Two calcium channel blockers (diltiazem and verapamil) act by blocking the L-type calcium channel and, to a lesser degree, by nonspecific antiadrenergic actions. The latter action is significant because it offsets the reflex increase in heart rate due to the peripheral vasodilator actions of these agents. In the case of adenosine, the AV block is effected by the inhibiting purinergic receptors; in the case of digoxin, the effect stems largely from the augmentation of vagal actions by the cardiac glycoside. The modulation of the AV nodal refractoriness by the calcium channel blockers and digoxin, given intravenously or orally, results in slowing of the ventricular response in AF and flutter. The reduction of the ventricular response is the basis of termination of recent onset AF, an effect that does not stem from their intrinsic antifibrillatory actions in atrial muscle.

Calcium Channel Blockers as Antiarrhythmic Drugs

The agents, verapamil and diltiazem, have no significant electrophysiological effects on atrial, ventricular, or His-Purkinje fiber refractoriness or conduction. However, they may shorten the APD in the atria and uncommonly induce AF.[38] They slow the phase 4 depolarization in the SA and AV nodes with slowing of conduction mediated by the block of L-type calcium channels. Their major effect is in the AV node in which they reduce conduction and prolong the effective and functional refractory periods in the anterograde, as well as retrograde, directions. The major depressant effects of verapamil and diltiazem on the AV node are also used in three other specific settings (1) prevention of the recurrences of episodes of PSVT. Here they can be combined with digoxin or β-blockers, but this use of the drugs is declining with the increasing preference for cure with radio-frequency ablation of the arrhythmia; (2) rapid acute control of ventricular response in the case of AFL and fibrillation; and (3) chronic modulation of the ventricular rate in these arrhythmias when rate control is deemed to be the preferred approach in treatment. In this context, calcium channel blockers can be combined with varying doses of digoxin and/or β-blockers.[126] There are limited data suggesting that verapamil may be of value in the acute control of multifocal atrial tachycardia but its efficacy in the chronic prophylaxis of the arrhythmia remains uncertain.[38] There are recent data that suggest the interesting possibility that the use of calcium-lowering drugs, such as verapamil given during AF, may reduce the recurrence of AF after electrical cardioversion.[127,128]

Because their electrophysiological effects are minimal in the ventricular muscle, they are unlikely to be potent antiarrhythmic agents in most types of ventricular tachyarrhythmias.[128,129] For the same reasons, these compounds do not appear to induce proarrhythmic reactions and have not been shown to adversely affect mortality, although this may possibly occur by virtue of the negative inotropic actions in patients with advanced levels of heart failure. The role of calcium channel blocking drugs in the treatment of ventricular arrhythmias is limited—as might be expected from the nature of their actions in ventricular muscle. They are poor suppressants of premature ventricular premature beats and nonsustained or sustained VT. It is possible that in some subsets of patients with ischemic heart disease, calcium channel blockers may prevent VT/VF by the anti-ischemic actions, but such a possibility has not be tested in relevant clinical models.

In addition to their use in supraventricular tachyarrhythmias, including for rate control in AFL and AF,[126] there are at least two relatively uncommon forms of ventricular tachycardia that respond to calcium channel blockers.[128,129] Such arrhythmias occur in the context of what appears to be a structurally normal heart. The first is the syndrome of the left ventricular septal VT. Such a VT occurs largely in males and electrocardiographically it has a right bundle branch block pattern, left axis deviation; it can be induced by rapid atrial pacing or by programmed electrical stimulation. It can be terminated by intravenous verapamil but not by adenosine. The arrhythmia can be controlled by oral calcium channel blockers especially verapamil, but the primary mode of treatment is by radio-frequency ablation. The other idiopathic ventricular tachycardia that may respond to calcium channel blockers (also to β-blockers) is the right ventricular outflow tract tachycardia (RVOT). It has the left bundle branch block pattern on the ECG with a vertical axis, and occurs more frequently in females. The arrhythmia is not readily induced by programmed electrical stimulation but can be induced by exercise or by isoproterenol infusion. The arrhythmia is terminated predictably and promptly to intravenous verapamil or adenosine. Again, the primary mode of therapy is catheter ablation but it is also controlled by β-blockade or calcium channel blockers and may respond to β-blockers.

Verapamil

The electrophysiological properties of this compound formed the basis for the so-called class IV antiarrhythmic actions.[7] When verapamil is administered intravenously (5 to 20 mg over 2 minutes), the peak effects on the AV node occur in 10 to 15 minutes, the effects lasting for 6 hours. After oral administration, the drug acts in hours, with a peak effect occurring at 3 hours, with an elimination half-life of 3 to 7 hours, but the effects last much longer as a function of duration of drug administration. For sustained oral therapy for modulating the ventricular response, the usual dose range is 80 to 120 mg 3 times daily or 4 times daily. Alternatively, single oral doses of the long-acting preparations (240 to 480 mg/daily) may be used. The major cardiovascular adverse effect of the drug relates to excessive depressant action on the AV node. Drug–drug interactions are with digoxin and amiodarone.

Diltiazem

The electrophysiological properties of diltiazem are similar to those of verapamil with possibly a similar degree of negative inotropic actions. Its antiarrhythmic actions have not been as widely explored, having been limited to supraventricular arrhythmias and with its use focused largely on the acute and chronic control of the ventricular response in patients with AFL and fibrillation. After oral administration, the drug is more than 90% absorbed but with a bioavailability of 45%, onset of action is 15 to 30 minutes, peak action is 1 to 2 hours, and elimination half-life is 4 to 7 hours. Only 35% of the drug is eliminated by the kidneys, the remainder by the gastrointestinal tract. For the rapid control of the ventricular response, a bolus dose of 20 mg IV is given for 2 minutes, with a repeat bolus for 15 minutes if required, followed by a 5 to 15 mg/hour infusion for prolonged effect. Oral therapy now is usually a sustained-release diltiazem once daily (90, 120, 180, 240, or 300 mg). The major therapeutic utility of diltiazem in the control of arrhythmias is the chronic control of ventricular response either as a single agent, or in combination with digoxin or a β-blocker. The side effect profile of diltiazem is similar to that with verapamil, but the drug does not have major drug interactions with digoxin, quinidine, or amiodarone.

Adenosine

The electrophysiological actions of adenosine are mediated via a receptor-effector mechanism that includes the A_1 receptor and a guanine nucleotide-binding G protein. The primary direct actions of adenosine are the activation of an outward potassium current (I_{Kado}) present in the atria and the SA and AV nodes but in the ventricle. In the AV node, the drug depresses the nodal action potential and in the sinus node, there is slowing of the sinus rate with shifts of the pacemaker and hyperpolarization. As mentioned, the major therapeutic effect of adenosine is the consistent and predictable termination of all forms of PSVT in which the antegrade limb of the arrhythmia is in the AV node. It should be emphasized that adenosine does shorten the action potential duration in the atria making them susceptible to the initiation of AF, which in the case of patients with the WPW syndrome, may be potentially dangerous. Adenosine does not have a major role in the control of arrhythmia generated in the ventricles. The role of the drug in the acute termination and the diagnosis of supraventricular tachyarrhythmias is now well established.[130,131]

Digoxin

Classifying digoxin as an antiarrhythmic agent has always been controversial. The major effects of the cardiac glycoside are mediated by the drug's central and peripheral actions to increase vagal activity.[132] Such an action has two electrophysiological effects: in the atria, there is shortening of the atrial refractory period (conducive to the development of AF) and on the atrioventricular node the vagal effect leads to delay in conduction and an increase in the effective refractory period, which slows the ventricular response in AF and flutter, a slowing that may lead to conversion of these arrhythmias to sinus rhythm. Such a conversion does not appear to be due to direct antifibrillatory actions of the drug. This may also be the mechanism of conversion of recent onset AF and flutter by β-blockers, as well calcium channel blockers. The effects of digoxin on the myocardium stems from the drug's propensity to inhibit sodium-potassium adenosine triphosphatase with an increase in the intracellular concentration of calcium by the modulation of the calcium channels and the inhibition of sodium-calcium exchange. This may be the basis for the drug's known, albeit weak, positive inotropic effects in the ventricular myocardium. The effect of digoxin on the ventricular myocardium is in therapeutic concentrations (0.8 to 2.0 ng/mL) but at higher doses it may produce electrocardiographic changes, and in toxic doses it may induce premature atrial and ventricular arrhythmias, such as atrial tachycardia with block, premature atrial and ventricular contractions, and bidrectional ventricular tachycardia. However, in a large clinical trial of heart failure patients, it had a significant adverse effect on total mortality.

The elimination half-life of digoxin is 1.5 days, and its excretion is largely renal. Antiarrhythmic drugs, such as amiodarone, quinidine, propafenone, and verapamil affect its pharmacokinetics. Barring the use of digoxin in the treatment of heart failure, digoxin is used largely for the control of ventricular response in patients with AF, especially in combination with β-blockers and calcium-channel blockers; the effects of such combinations are additive and may possibly be synergistic.

Newer Antiarrhythmic Drugs Under Development

Currently, there is much interest in the possibility that so-called class III agents, which might exert their actions on the myocardial membrane by simultaneously blocking multiple ion channels, could have a more favorable electrophysiological profile in terms of their proarrhythmic potential. Two such agents currently under development might have such a potential. The first is azimilide.

Azimilide

The structure of this compound does not include the methane sulfonamide group present in sotalol, dofetilide, or ibutilide. Although not a benzofuran, azimilide, in some respects, resembles amiodarone, which, like azimilide, maintains a class III effect at high stimulation frequencies.[133-136] Azimilide is likely to exhibit a lower incidence of TdP than the other pure class III agents (d-sotalol, dofetilide, and ibutilide). The terminal half-life of the drug is 4 to 5 days and it can be administered once daily. It is eliminated by hepatic metabolism.[137] It takes approximately 2 weeks for the drug to achieve steady-state if no loading regimen is administered. There are no clinically significant pharmacokinetic interactions between azimilide and warfarin or digoxin.

The compound prolongs the myocardial action potential duration by predominantly blocking the slow component of the delayed rectifier current (I_{Ks}) with presumably somewhat smaller effect on the rapid component (I_{Kr}). Such a property may not be associated with reverse use and rate dependency of action on repolarization. Thus, azimilide is likely to be less "torsadogenic" compared with other specific I_{Kr} blockers. The effect of the drug on the M cells in the mid-myocardial region of ventricular tissue is not fully defined, but in other tissues the drug prolongs the cardiac action potential and refractoriness. Like other pure class III agents, the drug does not slow conduction across the AV node. In isolated human atrial and ventricular myocytes, azimilide produces a concentration-dependent inhibition of both the I_{Ks} and I_{kr}. In intact animal models, azimilide has been shown to suppress both atrial[37] and ventricular arrhythmias and has the potential to prevent sudden death following coronary artery occlusion. Of particular interest, the drug has been found to be unusually and consistently effective in terminating AF and AFL in various canine experimental models.

The available clinical data on the hemodynamic and electrophysiological effects of azimilide are predictable on the basis of its known electropharmacologic properties. In healthy volunteers, oral azimilide in doses of up to 200 mg/day was well tolerated and produced a maximal increase in the QT between 24% and 28%, although individual values ranged from 4% to 42%. The QT increases were dose dependent without significant increases in the P–R or QRS intervals or in heart rate or blood pressure, suggesting that the drug does not have sodium- or calcium-channel blocking actions or significant influence on the sympathetic or parasympathetic nervous systems.

In three regimens (50 mg, 100 mg, and 125 mg) in patients, as in healthy volunteers, azimilide has been shown to produce consistent prolongation of Q–T and QTc intervals in a dose-dependent manner in the absence of clinically significant effects on heart rate, P–R, and QRS intervals.

As indicated previously,[63] in the currently changing therapeutic landscape of arrhythmia control, azimilide may have a particular value in the acute conversion of AFL and fibrillation with a potential for maintaining of sinus rhythm after pharmacologic or electrical conversion of these arrhythmias. The overall efficacy of the drug is likely to rival that of dofetilide (see Table 37-3), and as in the case of other pure class III agents, it is likely to be of value in the reduction of the number of shocks in patients with ICDs for ventricular tachycardia and fibrillation. On theoretical grounds, the drug is of much interest as the first example of a class III agent that blocks both the I_{ks} and I_{Kr}. Thus, its effects in patients with recent myocardial infarction in a controlled study to improve survival by reducing the risk for sudden arrhythmic deaths will be of much interest and significance.

It should be emphasized that as in the case of other pure class III agents, the greatest utility of the drug is likely to be in the prophylactic therapy of AF. There have been a number of studies in which the effect of 35 to 125 mg/day of azimilide on the time to first recurrence of a symptomatic atrial arrhythmia has been determined. Analyses of combined data for 100 mg doses and data for the 125 mg dose have shown statistically significant differences from placebo. There has been a dose-dependent response, placebo patients having an 83% greater recurrence rate versus patients treated with 125 mg of azimilide. To date, the most effective dose of azimilide in preventing recurrences of AF 125 mg/day. For example, Pritchett and colleagues[138] found that in symptomatic patients, the mean recurrence time for AF was 17 days for placebo, 22 days for 50 mg dose, 41days for 100 mg dose (all nonsignificant), but 130 days when the dose was 125 mg/day ($P = .002$).

The risk of mortality was similar between azimilide (0.9% [9 of 1004 patients]) and placebo (0.7% [4 of 569 patients]) in completed SVA placebo-controlled studies. Based on the adverse events (AE) reported in completed SVA studies, once-daily doses of 100 or 125 mg of azimilide are safe and generally well-tolerated in patients with AF, AFL and/or PSVT. The most frequently reported AEs included headache and asthenia; both occurred at rates similar to placebo. Other significant AEs included TdP, neutropenia, and mild increases in liver enzymes. The incidence of TdP and other ventricular arrhythmic events in patients treated with azimilide were low and consistent with class III antiarrhythmic agents. TdP was reported in less than 1% of azimilide patients. Risk factors for TdP included female gender, use of diuretics, and bradycardia.

Dronedarone

This compound is the noniodinated derivative of amiodarone, a compound that was created to reduce the side effect profile of amiodarone without the loss of its complex electrophysiological and pharmacologic profile.[22,139-140] Besides the deletion of the iodine in the benzene ring, a methane sulfonamide group has been included in the benzofuran ring, and the ethyl groups in the side chain have been replaced by butyls in dronedarone. Such structural changes have led to the shortening of the elimination half-life to 20 to 30 hours, no effect on thyroid hormone metabolism, and the propensity to block M_2 receptors so that, despite a significant noncompetitive antiadrenergic action, there is a somewhat lower heart rate-reducing effect compared with amiodarone. Unlike amiodarone, which causes constipation at supratherapeutic doses, dronedarone produces diarrhea. Importantly, in the experimental setting, as well as during the initial clinical studies, pulmonary toxicity of the form noted with amiodarone has not been reported with dronedarone.

In other respects experimental and preliminary clinical studies have shown that the electropharmacologic effects of dronedarone and amiodarone are similar.[138] For example, in acute electrophysiological studies, in vitro dronedarone shortened the APD in cardiac muscle but, like amiodarone, induced significant prolongation of the APD and the ERP in the atrial and ventricular myocardium, as well as in the AV node, with the depression of the phase 4 depolarization in the sinoatrial node following 1 week of drug administration. The effects of dronedarone on the Purkinje fibers, as well as in the M cells, have been reported to be similar to those in the case of amiodarone, suggesting that the drug is likely to exhibit a negligible incidence of *torsades de pointes,* although the clinical experience with the drug has not been extensive.

As in the case of amiodarone and other so-called class III compounds, two properties of dronedarone, Q–T/QTc intervals and heart rate are of much importance in the role of the drug as an antifibrillatory agent for the treatment of AF and flutter on one hand and the ventricular tachycardia and fibrillation on the other. Beat-by-beat analysis over 24-hour Holter recordings has established that dronedarone increases heart rate Q–T/QTc intervals as a function of dose (1200 to 3200 mg/daily), the magnitude of changes being somewhat lower than those on steady state amiodarone administration and the effects being discernible at higher drug doses.[141] The clinical effects of the drug are under study in several large studies in AF and on mortality in high risk patients with cardiac disease.

REFERENCES

1. The Cardiac Arrhythmia Suppression Trial II Investigators: The Cardiac Arrhythmia Suppression Trial (CAST) Investigators. Preliminary Report: Effect of encainide and flecainide on mortality in a randomized trial of arrhythmia suppression after myocardial infarction. N Engl J Med 1989;321:227-233; 406-412.
2. Guidelines 2000 for Cardiopulmonary Resuscitation and Emergency Cardiovascular Care: An international consensus on science. 6. Advanced cardiovascular life support. 5. Pharmacology 1: Agents for arrhythmias. *Circulation* 2000;102:Suppl 1: 1-112-128.
3. Singh BN, Vaughan Williams EM: A third class of anti-arrhythmic action: Effects on atrial and ventricular intracellular potentials, and other pharmacological actions on cardiac muscle, of MJ 1999 and AH 3474. Br J Pharmacol 1970;39:675-687.
4. Singh BN, Vaughan Williams: The effect of amiodarone, a new antianginal drug, cardiac muscle. Brit J Pharmacol 1970;39;657-667.
5. Singh BN: Pharmacological Actions of Certain Cardiac Drugs and Hormones: Focus on Antiarrhythmic Mechanisms. D. Phil. Thesis, 1971; Hertford College & University of Oxford. UK. Published also by Futura, Mt. Kisco, NY, 1991, pp 1-98.
6. Vaughan Williams EM: Classification of antiarrhythmic drugs. E. Sandoe, E Flenstedt-Johnson, Olesen KH (eds): Symposium on Cardiac Arrhythmias. AB Astra, Sodertalje, Sweden, 1970, pp 440-469.
7. Singh BN, Vaughan Williams EM: A fourth class of anti-dysrhythmic action? Effect of verapamil on ouabain toxicity, on atrial and ventricular intracellular potentials, and on other features of cardiac function. Cardiovasc Res 1972;6:109-119.
8. Singh BN, Hauswirth O: Comparative mechanisms of action of antiarrhythmic drugs. Am Heart J 1974;87:367-382.
9. Singh BN, Williams EM: Effect of altering potassium concentration on the action of lidocaine and diphenylhydantoin on rabbit atrial and ventricular muscle. Circ Res 1971;29:286-295.
10. Nattel S, Singh BN: Evolution, mechanisms, and classification of antiarrhythmic drugs: Focus on class III actions. Am J Cardiol 1999;84:11-19.
11. Hoffman BF, Bigger JT Jr. Antiarrhythmic drugs. In DiPalma JR (ed): Drill's Pharmacology in Medicine, 4th ed. New York, McGraw-Hill, 1971, pp 824-852.
12. Hauswirth O, Singh BN: Ionic mechanisms in heart muscle in relationship to the genesis and the pharmacological control of cardiac arrhythmias. Pharmacol Rev 1979;30:5-63.
13. Harrison DC: Is there a rational basis for the modified classification of antiarrhythmic drugs? In Morganroth J, Moore EN (eds): Cardiac Arrhythmias: New Therapeutic Drugs and Devices. Boston, Martinus Nijhoff, 1985, pp 36-48.
14. Campbell TJ: Kinetics of onset of rate-dependent effects of class I antiarrhythmic drugs are important in determining their effects on refractoriness in guinea-pig ventricle, and provide a theoretical basis for their subclassification. Cardiovasc Res 1983;17:344-352.
15. The Task Force of the Working Group on Arrhythmias of the European Society of Cardiology. The Sicilian Gambit: A new approach to the classification of antiarrhythmic drugs based on their actions on arrhythmogenic mechanisms. Circulation 1991:84;1831-1851.
16. Singh BN: The coming of age of class III antiarrhythmic principle: Retrospective and future trends. Am J Cardiol 1996;78 (Suppl 4a):17-27.
17. Lazzara R: From first class to third class: Recent upheaval in antiarrhythmic therapy—lessons from clinical trials. Am J Cardiol 1996;78:28-33.
18. Singh BN: Expanding indications for the use of class III antiarrhythmic agents in patients at high risk for sudden death. J Cardiovasc Electrophysiol 1995;6:887-900.
19. Hondeghem LM, Snyders DJ: Class III anti-arrhythmic agents have a lot of potential but a long way to go: Reduced effectiveness and dangers of reverse use dependence. Circulation 1990;81:686.
20. Camm AJ, Karam R, Pratt CM: The azimilide post-infarct survival evaluation (ALIVE) trial. Am Cardiol 1998;81:35D-39 D.
21. Pritchett E, Page R, Connolly S, Marcello S: Azimilide treatment of atrial fibrillation. Circulation 1998;98(17):1:633.
22. Sun W, Sarma JSM, Singh BN: Chronic and acute effects of dronedarone in the action potential of rabbit atrial muscle preparations: Comparison with amiodarone. J Cardiovasc Pharmacol 2002;39:677-684.
23. Wang Z, Pelletier LC, Talajic M, Nattel S: Effects of flecainide and quinidine on human atrial potentials: Role of rate dependence and comparison with guinea pig, rabbit and dog tissues. Circulation 1990;82:274-283.
24. Teo KK, Yusuf S, Furberg CD: Effects of prophylactic antiarrhythmic drug therapy in acute myocardial infarction: An overview of results from randomized controlled trials. JAMA 1993;270: 1589-1595.
25. Coplen SE, Antman EM, Berlin JA: Efficacy and safety of quinidine therapy for maintenance of sinus rhythm after cardioversion: A meta-analysis of randomized control trials. Circulation 1990;82:1106-1116.
26. Kirk MM, Gold MR.: Disopyramide phosphate. Card Electrophysiol Rev 2000;4:269-271.
27. MacMahon S, Collins R, Perot R, et al: Effects of prophylactic lidocaine in suspected acute myocardial infarction: An overview of results from randomized, controlled trials. JAMA 1988; 260:1910-1916.
28. Impact Research Group: International mexiletine and placebo antiarrhythmic coronary trial: Report on arrhythmias and other findings. J Am Coll Cardiol 1984;4:1148-1156.
29. Ho DS, Zecchin RP, Richards DA, et al: Double-blind trial of lignocaine versus sotalol for acute termination of spontaneous sustained ventricular tachycardia. Lancet 1994;344:18-23.
30. Singh BN: Routine prophylactic lidocaine administration in acute myocardial infarction: An idea whose time is all but gone? Circulation 1992;86:1033-1035.
31. Ruffy R: Flecainide—2000 Update. Card Electrophysiol Rev 2000;4:277-279.
32. Valderrabano M, Singh BN: Electrophysiologic and antiarrhythmic effects of propafenone: Focus on atrial fibrillation. J Cardiovasc Pharmacol Ther 1999;4:183-198.
33. Schwartz PJ, La Rovere MT, Vanoli E: Autonomic nervous system and sudden cardiac death: Experimental basis and clinical observations for post-myocardial infarction risk stratification. Circulation 1992;85(1 Suppl):I77-187.
34. Meredith IT, Broughton A, Jennings GL, Esler MD: Evidence of a selective increase in cardiac sympathetic activity in patients with sustained ventricular arrhythmias. N Engl J Med 1991; 325:618-624.
35. Reiter MJ, Reiffel JA: Importance of beta-blockade in the therapy of serious ventricular arrhythmias. Am J Cardiol 1998;82: 91-105.
36. Reiter MJ: Antiarrhythmic impact of anti-ischemic, antifailure and other cardiovascular strategies. Cardiac Electrophysiol Rev 2000; 194-205.
37. Singh BN: Current antiarrhythmic drugs: An overview of mechanisms of action and potential clinical utility. J Cardiovasc Electrophysiol 1999;10:283-301.

38. Singh BN, Sarma JSM: Beta-blockers and calcium-channel blockers as antiarrhythmic drugs. In DP Zipes and Jalife J (eds): Cardiac Electrophysiology. From Cell to the Bedside. Philadelphia, WB Saunders, 2000, pp 903-921.
39. Schwartz PJ: The idiopathic long QT interval syndrome: Progress and questions. Am Heart J 1985;109:399-405.
40. Tan HL, Hou CJ, Lauer MR, Sung RJ: Electrophysiologic mechanisms of the long QT interval syndromes and torsades de pointes. Ann Int Med 1995;122(9):701-714.
41. Wiesfeld AC, Crijns HJ, Tuininga YS, Lie KI: Beta adrenergic blockade in the treatment of sustained ventricular tachycardia or ventricular fibrillation. Pacing Clin Electrophysiol 1996;19: 1026-1035.
42. Steinbeck G, Andersen D, Bach P, et al: A comparison of electrophysiologically guided antiarrhythmic drug therapy with betablocker therapy to patients with symptomatic sustained tachyarrhythmias. N Engl J Med 1992;327:987-993.
43. Hallstrom AP, Cobb LA, Yu BH, et al: An antiarrhythmic drug experience in 941 patients resuscitated from an initial cardiac arrest between 1970 and 1985. Am J Cardiol 1991;68:1025-1031
44. Yusuf S, Wittes J, Friedman L: Overview of results of randomized clinical trials in heart disease. I. Treatments following myocardial infarction. JAMA 1988;260:2088-2093.
45. Norwegian Multicenter Study Group: Timolol-induced reduction in mortality and reinfarction in patients surviving acute myocardial infarction. N Engl J Med 1981;304:801-807.
46. Beta-blocker Heart Attack Trial Research Group: A randomized trial of propranolol in patients with acute myocardial infarction. I. Mortality results. JAMA 1982;247:1707-1714.
49. Kendall MJ, Lynch KP, Hjalmarson A, Kjekshus J: Beta-blockers and sudden cardiac death. Ann Int Med 1995;123:358-367.
50. Hjalmarson A: Effects of beta blockade on sudden cardiac death during acute myocardial infarction and the postinfarction period. Am J Cardiol 1997;80(9B):35J-39J.
51. Gottlieb SS, McCarter RJ, Vogel RA: Effect of beta-blockade on mortality among high-risk and low-risk patients after myocardial infarction. N Engl J Med 1998;339:489-495.
52. Julian DG, Camm AJ, Franglin G, et al: For the European Myocardial Infarct Amiodarone Trial Investigators. Randomized trial of effect of amiodarone on mortality in patients with left-ventricular dysfunction after recent myocardial infarction: EMIAT. Lancet 1997;349:667-673.
53. Cairns JA, Connolly SJ, Roberts RJ, Gent M: For the Canadian Amiodarone Myocardial Infarction Arrhythmia Trial Investigators. Randomized trial of outcome after myocardial infarction in patients with frequent or repetitive ventricular premature depolarizations: CAMIAT. Lancet 1997;394:675-682.
54. Boutitie F, Boissel JP, Connolly SJ, et al., and the EMIAT and CAMIAT Investigators. Amiodarone interaction with β-blockers: Analysis of the merged EMIAT and CAMIAT databases. Circulation 1999;99:2268-2275.
55. Teerlink JR, Bassie BM: Beta-adrenergic blocker mortality trials in congestive heart failure. Am J Cardiol 1999;8:94R-102R.
56. Chadda K, Goldstein S, Byington R, Curb JD: Effect of propranolol after acute myocardial infarction in patients with congestive heart failure. Circulation1986;73:503-513.
57. MERIT-HEFT Study Group: Effect of metoprolol CR/XL in chronic heart failure: Metoprolol CR/XL randomized intervention trial in congestive heart Failure (MERIT-HF). Lancet 1999;353;2001-2007.
58. CIBIS II Investigators and Committees: The cardiac insufficiency bisoprolol study CIBIS II. Lancet 1999;353:9-13.
59. Packer M, Bristow Mr, Cohn JN, et al: For the US Carvedilol Heart Failure Study group: The effect of carvedilol on morbidity and mortality in patients with chronic heart failure. N Engl J Med 1996;334:1349-1355-1360.
60. Packer M, Coats, Fowler, et al: The Carvedilol, Prospective Randomized Cumulative Survival Study Group. N Engl J Med 2001;344:1651-1658.
61. The Best Steering Committee Design of the Beta Blocker Evaluation Trial (BEST). Am J Cardiol 1995;75:75I:122-132.
62. Singh BN: The coming of age of the class III antiarrhythmic principle: Retrospective and future trends. Am J Cardiol 1996: 78(Suppl 4A):17-27.
63. Singh BN: Antiarrhythmic drugs: A re-orientation in light of recent developments in the control of disorders of rhythm. Am J Cardiol 1998;81(6A):3D-13D.
64. Kodama I, Kamiya K, Toyama J: Cellular electropharmacology of amiodarone. Cardiovasc Res 1997;35:13-29.
65. Singh BN: Expanding indications for the use of class III antiarrhythmic agents in patients at high risk for sudden death. J Cardiovasc Electrophysiol 1995;6:887-900.
66. Singh BN: Antiarrhythmic action of dl-sotalol in ventricular and supraventricular arrhythmias. J Cardiovasc Pharmacol 1992;2:590.
67. Singh BN: Sotalol: Current status and expanding indications. J Cardiovasc Pharmacol Ther 1999;4:59-65.
68. The CASCADE Investigators. The Cascade Study-Randomized Anti-Arrhythmic Drug Therapy in Survivors of Cardiac Arrest in Seattle. Am J Cardiol 1993;72:280-287.
69. Mason JW, and the ERVEM Investigators: A randomized comparison of electrophysiologic study to electrocardiographic monitoring for prediction of antiarrhythmic drug efficacy in patients with ventricular tachyarrhythmias. N Engl J Med 1993;329: 445-451.
70. Singh BN, Ahmed R: Class III antiarrhythmic drugs. Curr Opin Cardiol 1994;9:12-22.
71. Torp-Pedersen C, Moller M, Bloch-Thomsen PE, et al: For the Danish Investigations of Arrhythmia and Mortality on Dofetilide Study group. Dofetilide in patients with congestive heart failure and left ventricular dysfunction. N Engl J Med 1999;341:857-865.
72. Murray KT: Ibutilide. Circulation 1998;97:493-497.
73. Nattel S, Liu L, St George D: Effects of the novel antiarrhythmic agent azimilide on experimental atrial fibrillation and atrial electrophysiologic properties. Cardiovasc Res 1998;37:627-635.
74. Scheinman MM, Levine JH, Cannom DS, et al: Intravenous Amiodarone Multicenter Investigative Group: Dose ranging study of intravenous amiodarone in patients with life-threatening ventricular tachyarrhythmias. Circulation 1995;92:326.
75. Kowey PR, Levine JH, Herre JM, et al: Intravenous Amiodarone Multicenter Investigative Group. Randomized, double-blind comparison of intravenous amiodarone and bretylium in the treatment of patients with recurrent hemodynamically destabilizing ventricular tachycardia or fibrillation. Circulation 1995;92:3255-3263.
76. Kudenchuk PJ, Cobb LA, Copass MK, et al: Amiodarone for resuscitation after out-of-hospital cardiac arrest due to ventricular fibrillation. N Engl J Med 1999;341:871.
77. Dorian P, Cass D, Schwartz, et al: Amiodarone as compared with lidocaine for shock-resistant ventricular fibrillation. N Engl J Med 2002;346:884-890.
78. Singh BN: Initial antiarrhythmic drug therapy during resuscitation from sudden cardiac death: A time for a fundamental change in strategy? J Cardiovasc Pharmacol Ther 2000;5:3-9.
79. Singh BN: What niche will newer class III antiarrhythmic drugs occupy? Curr Cardiol Rep 2001;3:314-323.
80. Singh BN: Antiarrhythmic actions of amiodarone: A profile of a paradoxical agent. Am J Cardiol 1996;78:41-53.
81. Papp JG, Nemeth M, Krassoi I, et al: Differential electrophysiologic effects of chronically administered amiodarone on canine Purkinje fibers versus ventricular muscle. J Pharmacol Exp Ther 1996;1:187-196.
82. Sicouri S, Moro S, Litovsky S, et al: Chronic amiodarone reduces transmural dispersion of repolarization in the canine heart. J Cardiovasc Electrophysiol 1997;8:1269-1279.
83. Cui G, Sen L, Sager PT, et al: Effects of amiodarone, sematilide and sotalol on QT dispersion. Am J Cardiol 1995;75:465-469.
84 Hohnloser SH, Singh BN: Proarrhythmia with class III antiarrhythmic drugs: Definition, electrophysiologic mechanisms, incidence, predisposing factors, and clinical implications. J Cardiovasc Electrophysiol 1995;6:920-936.
85. Sager PT, Uppal P, Follmer CT, et al: The frequency-dependent electrophysiologic effects of amiodarone in humans. Circ 1993; 88:1063-1068.
86. Connolly SJ: Evidence based-analysis of amiodarone efficacy and safety. Circulation 1999;100:2025-2034.
87. Nasir N, Swarna US, Boahene KA, et al: Therapy of sustained ventricular arrhythmias with amiodarone: Prediction of efficacy with serial electrophysiologic studies. J Cardiovasc Pharmacol Ther 1996:1:123-133.
88. Skoulargis J, Rothlisberger C, Skudicky D, et al: Effectiveness of amiodarone and electrical cardioversion for chronic rheumatic atrial fibrillation after mitral valve surgery. Am J Cardiol 1993; 72:423-427.

89. Chun SH, Sager PT, Stevenson WG, et al: Long-term efficacy of amiodarone for the maintenance of sinus rhythm in patients with refractory atrial fibrillation or flutter. Am J Cardiol 1995; 76:47-50.
90. Hohnloser SH, Meinertz T, Dummbacher T, et al: Electrocardiographic and antiarrhythmic effects of intravenous amiodarone: Results of prospective, placebo-controlled study. Am Heart J 1991;121:89-90.
91. Guarnieri T: Intravenous amiodarone reduces CABG hospitalization: The ARCH Trial. Presentation at the NASPE Annual Scientific Meeting, San Diego, May 1998.
92. Daoud EG, Strickenberger AS, Man C, et al: Preoperative amiodarone as prophylaxis against atrial fibrillation after heart surgery. N Engl J Med 1997;337:1785-1791.
93. Roy D, Talajic M, Dorian P, et al: For the Canadian Trial of Atrial Fibrillation Investigators. Amiodarone to prevent recurrence of atrial fibrillation. N Engl J Med 2000;342:913-918.
94. Kochiadakis GE, Igoumenidis NE, Marketou ME, et al: Low-dose amiodarone versus sotalol for suppression of recurrent symptomatic atrial fibrillation. Am J Cardiol 1998;81:995-1005.
95. Singh BN: A Symposium: Approaches to Controlling Cardiac Arrhythmias: Focus on Amiodarone—the last 15 Years. Am J Cardiol 1999;84(Supp 9A);1R-174R.
96. Kato R, Yabek L, Ikeda N, et al: Electrophysiologic effects of dextro- and levo-isomers of sotalol in isolated cardiac muscle and their in vivo pharmacokinetics. J Am Coll Cardiol 1986;7:116-126.
97. Antonaccio MJ, Gomoll AW: Pharmacology, pharmacodynamics and pharmacokinetics of sotalol. Am J Cardiol 1990;65:12A-20A.
98. Nademanee K, Feld G, Hendrickson JA, et al: Electrophysiologic and antiarrhythmic effects of sotalol in patients with life-threatening ventricular tachyarrhythmias. Circulation 1985; 72:555-564.
99. Benditt DG, Williams JH, Jin J, et al: For the dl-Sotalol Atrial Fibrillation/Flutter Study Group. Maintenance of sinus rhythm with oral dl-sotalol therapy in patients with symptomatic atrial fibrillation and flutter: A dose-response study. Am J Cardiol 1999;84:270-277.
100. Pacifico A, Hohnloser S, Williams JH, et al: For the dl-sotalol implantable cardioverter-defibrillator study group. N Engl J Med 1999;340:1855-1862.
101. Movsowitz C, Marchlinski FE: Interactions between implantable cardioverter-defibrillators and class III antiarrhythmic drugs. Am J Cardiol 1998;82:411.
102. Singh BN, Deedwania P, Nademanee K, et al: Sotalol: A review of its pharmacodynamic and pharmacokinetic properties, and therapeutic use. Drugs 1987;34:311-330.
103. Reimold SC, Cantillon CO, Priedman PL, et al: Propafenone versus sotalol for suppression of recurrent symptomatic atrial fibrillation. Am J Cardiol 1993;71:558.
104. Juul-Moller S, Edvardsson N, Ahlberg NR: Sotalol versus quinidine for the maintenance of sinus rhythm after direct current conversion of atrial fibrillation. Circulation 1990;82:1932.
105. Suttorp MJ, Kignma JH, Peels HOJ: Effectiveness of sotalol in preventing supraventricular tachyarrhythmias shortly after coronary artery bypass grafting. Am J Cardiol 1991;68:1163.
106. Singh BN, Lopez B, Sarma JSM: Significance and prevention of atrial fibrillation occurring after surgery: A time for fundamental change in strategy? J Cardiovasc Pharmacol Therapeut 1998; 3:259.
107. Kehoe R, Zheutlin T, Dunnington C, et al: Safety and efficacy of sotalol in patients with drug refractory sustained ventricular tachyarrhythmias. Am J Cardiol 1990;65:58A.
108. Haynes RE, Chinn RL, Copass MK, et al: Comparison of bretylium tosylate and lidocaine in management of out-of-hospital ventricular fibrillation: A randomized clinical trial. Am J Cardiol 1981; 48:353.
109. Olson DW, Thompson BM, Darin JC, et al: A randomized comparison study of bretylium tosylate and lidocaine in resuscitation of patients from out-of-hospital ventricular fibrillation in a paramedic system. Ann Emerg Med 1984;13:807-811.
110. Lee KS, Gibson JK: Unique ionic mechanism of action of ibutilide on freshly isolated heart cells. Circulation 1995;92:2755-2757.
111. Ellenbogen KA, Sambler BS, Wood MA, et al: Efficacy of intravenous ibutilide for rapid termination of atrial fibrillation and flutter: A dose-response study. J Am Coll Cardiol 1996; 28:130-136.
112. Stambler BS, Wood MA, Ellenbogen KA, et al: Efficacy of safety of repeated intravenous doses of ibutilide for rapid conversion of atrial flutter or fibrillation. Circulation 1996;94:1613-1621.
113. Ellenbogen KA, Clemo HF, Stambler BS, et al: Efficacy of ibutilide termination of atrial fibrillation and flutter. Am J Cardiol 1996;78:42-45.
114. Oral H, Souza HJ, Michaud GF, et al: Facilitating transthoracic cardioversion of atrial fibrillation with ibutilide treatment. N Engl J Med 1999;340:1849-1854.
115. Nacarelli GV, Lee KS, Gibson JK, et al: Electrophysiology and pharmacology of ibutilide. Am J Cardiol 1996;78:12-16.
116. Singh BN: Acute conversion of atrial flutter and fibrillation: Direct current cardioversion versus intravenously administered pure class III agents. J Amer Coll Cardiol 1997;29:391-393.
117. Roden DM: Ibutilide and the treatment of atrial arrhythmias. A new drug—almost unheralded—is now available to US physicians. Circulation 1996;94:1499-1502.
118. Stambler BS: Update on intravenous ibutilide. Cardiac Electrophysiol Rev 2000;4:243-247.
119. Glatter KA, Chatterjee K, Huang S, et al: Is it safe to use intravenous ibutilide in patients receiving chronic amiodarone therapy? Circulation 198;98:4417.
120. Kowey PR, Vanderlugt JT, Luderer JR: Safety and risk/benefit analysis of ibutilide for acute conversion of atrial flutter and fibrillation. Am J Cardiol 1996;78:46-52.
121. Falk RH, Pollak A, Singh SN, Friedrich T. Intravenous dofetilide, a class III antiarrhythmic agent, for the termination of sustained atrial fibrillation or flutter. J Am Coll Cardiol 1997;29:385-390.
122. Carmeliet E: Voltage- and time-dependent block of the delayed rectifier K^+ current in cardiac myocytes by dofetilide. J Pharmacol Ther 1992;262:809-815.
123. Data on file in Pfizer database, New York, New York, 2001.
124. Singh SN, Zoble RG, Yellen L, et al: For the Dofetilide Atrial Fibrillation Investigators. Efficacy and safety of oral dofetilide in converting to and maintaining sinus rhythm in patients with chronic atrial fibrillation or flutter. The Symptomatic Atrial Fibrillation Investigative Research on Dofetilide (SAFIRE-D) study. Circulation 2000;102:2383-2390.
125. Doshi S, Singh BN: Pure class III antiarrhythmic drugs: Focus on dofetilide. J Cardiovasc Pharmacol Ther 2000;5:237-247.
126. Farshi R, Kistner D, Sarma JS, et al: Ventricular rate control in chronic atrial fibrillation during daily and programmed exercise: A crossover open-label study of five drug regimens. J Am Coll Cardiol 1999;33:304-310.
127. Tieleman RG, Van Gelder IC, Crijn HJ, et al: Early recurrences of atrial fibrillation after electrical conversion: A result of fibrillation-induced electrical remodeling of the atria? J Am Coll Cardiol 1998;31:167-173.
128. Wyse DG: Calcium channel blockers. Cardiac Electrophysiol Rev 2000;4:308-311.
129. Lee SH, Chen SA, Tai CT, et al: Electropharmacologic characteristics and radiofreqency catheter ablation of sustained ventricular tachycardia in patients without structural heart disease. Cardiology 1996;87:33-41.
130. Wilbur SL, Marchlinski FE: Adenosine as an antiarrhythmic agent. Am J Cardiol 1997;(12A):30-37.
131. Conti JB, Belardinelli L,C, Curtis AB: Usefulness of adenosine in the diagnosis of tachyarrhythmias. Am J Cardiol 1995; 75: 952-955.
132. Smith TW: Digitalis: Mechanisms of action and clinical use. N Engl J Med 1988;318:358-365.
133. Fermini B, Jurkiewicz NK, Jow B: Use dependent effect of the class III antiarrhythmic agent NE-10064 (azimilide) on cardiac repolarization block or delayed rectifier potassium and L-type calcium currents. J Cardiovasc Pharmacol 1995;26:259-267.
134. Salata JJ, Brooks RR: Pharmacology of azimilide dihydrochloride (NE-10064), a class III antiarrhythmic agent. Cardiovas Drug Rev 1997;15:137-156.
135. Black SC, Butterfield JL, Lucchesi BR: Protection against programmed electrical stimulation-induced ventricular tachycardia and sudden cardiac death by NE-10064, a class III antiarrhythmic drug. J Cardiovasc Pharmacol 1993;22:810-818.
136. Restivo M, Hegazy M, El-Hamamy M: Antiarrhythmic efficacy of azimilide dihydrochloride on functional circus movement atrial flutter in the canine right atrial enlargement model. PACE. 1996;19:664.

137. Corey AE, Al-Khalidi H, Brezovic C, et al: Azimilide pharmacokinetics and pharmacodynamics upon multiple oral dosing. Clin Pharmacol Ther 1997;61:205-212.
138. Pritchett E, Page P, Connelly S, et al: Azimilide treatment in atrial fibrillation. Circulation 1999;98(Suppl)1633(Abstract).
139. Sun W, Sarma JSM, Singh BN: Electrophysiologic effects of dronedarone (SR33589), on non-diodinate benzfuran derivative, in the rabbit heart: Comparison with amiodarone. Circulation 1999;100:2276-2283.
140. Guillemare E, Martion A, Nisato D, et al: Acute effects of dronedarone and amiodarone on IK1, I Kr , and IKs in guinea pig ventricular myocytes. Fund Clin Pharmacol 1999;13: 389-395.
141. Singh BN, Sarma JSM: Mechnaims of action of antiarrhythmic drugs relative to the origin and perpetuation of cardiac arrhythmias. J Cardiovasc Pharmacol Ther 2001;6:69-87.

Chapter 38 Pacing Technology and Its Indications: Advances in Threshold Management, Automatic Mode Switching, and Sensors

CHU-PAK LAU

Since the first endocardial pacing lead implantation in 1958, pacemaker therapy has undergone remarkable technologic advances. For example, the number of circuitry components has increased from a mere two to three transistors in early pacemakers to nearly 1 million components with RAM size up to 124,000 bytes.[1] This increased sophistication has led to pacemaker features that the average pacemaker implanter may not have the time either to understand or to program appropriately. In addition, threshold and sensor assessment may take up to 40% of time in an average follow-up (Fig. 38-1).[2]

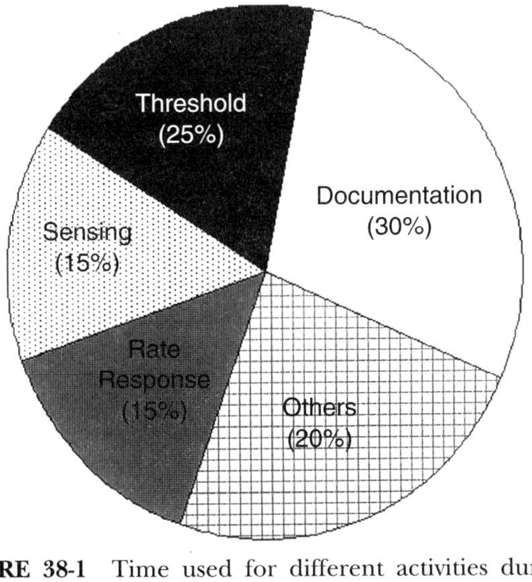

FIGURE 38-1 Time used for different activities during a routine pacemaker follow-up. (From Marshall M, Butts L, Flaim G, et al: Predictors of time requirements for pacemaker clinic evaluation [abstract]. Pacing Clin Electrophysiol 1995;18:952.)

Thus, there is a need for automatic optimization of many pacing parameters. This chapter reviews the current state of art in three important modern pacemaker functions: capture management, automatic mode switching (AMS) and implantable sensors. Particular attention is given to their indications and automaticity in programming.

Capture Management

The primary function of a pacemaker is to pace effectively at an efficient energy output. This depends on the pacing threshold, which varies significantly between individuals, and within an individual over time. The latter may occur because of the spontaneous threshold rise after implantation, the occurrence of gross or microdislodgment, diurnal changes, and the changes introduced by drugs and myocardial ischemia.[3-4] Thus, the ability to track threshold automatically will maximize patient safety, minimize battery drain for pacing, and, importantly, simplify programming. Table 38-1 shows a list of why automatic capture management is required.

TABLE 38-1 Potential Benefits of Capture Management

Increase in battery drain (e.g., sensors, electrogram monitoring, and multisite pacing)
Increase in battery longevity
Two-third of patients will be alive at the time of battery replacement
Pacing for populations such as those with AF and after atrioventricular nodal ablation
Reduction in battery size
Physiologic/medical variation in threshold
Reduction in time for pacemaker programming

An increase in demand for battery energy can result from some sensors. Whereas the piezoelectric sensor is energy inexpensive, the impedance sensor, such as is used to monitor minute ventilation, requires significant current consumption. More energy is required for device monitoring purpose, particularly for electrogram storage, which is becoming important for patients with atrial fibrillation (AF). With the use of multisite pacing (atrial pacing [Ap] for AF and ventricular pacing [Vp] for heart failure), minimizing pacing energy becomes critical. The longer survival of patients means that at least two thirds of those who receive a pacemaker will live to have a replacement within the usual battery life of 7 years. Atrioventricular (AV) nodal ablation followed by permanent pacing provides symptomatic relief and enhancement of quality of life. This group of patients is younger, and a longer battery life is advantageous. All these changes occur simultaneously with an overall effort by manufacturers to reduce the size of devices.

From a clinical standpoint, variation in threshold may lead to an inadequate safety margin of stimulation. Such changes may result from the usual rise of threshold after implantation, from ischemia, and from antiarrhythmic medications. Finally, threshold measurement remains time consuming, and if an alternative and safe method is available, the burden of programming can be reduced.

TYPES OF CAPTURE MANAGEMENT

Several manufacturers have introduced algorithms for detecting ventricular and/or atrial thresholds. The detection of an evoked response is based on either evoked response or impedance. The threshold data are used either on a beat-by-beat basis to ensure a paced response or intermittently to adjust output parameters.

St. Jude/Pacesetter Autocapture

After a ventricular pacing stimulus, the autocapture algorithm opens an evoked response (ER) detection window for 47.5 ms after a 15 ms blanking period. Detection of an ER is used to diagnose capture. In the event that an ER is not detected, a high energy back-up pulse of 4.5V is discharged. If two consecutive back-up pulses are delivered, the algorithm starts a stimulation threshold search by increasing the output to effect two consecutive captures (Fig. 38-2). In single chamber devices (Microny and Regency SR), a margin of 0.3V is added. In addition, to avoid pacing at high output due to diurnal fluctuation in threshold, the device automatically performs a threshold search once every 8 hours. Again a safety margin of 0.3V is added to the detected threshold. In dual chamber devices, the A–V interval is shortened to 50 ms (Ap) or 25 ms (As) to ensure overdrive of intrinsic ventricular rhythm. In the Affinity DR, automatic decrements and increments of output during threshold search are 0.25 and 0.125V, respectively (see Fig. 38-2).

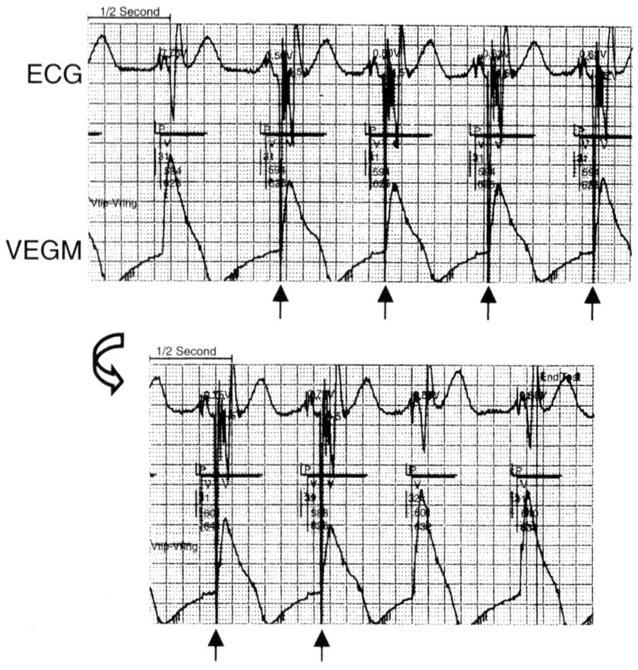

FIGURE 38-2 Autocapture threshold determination in a patient implanted with Affinity. Successful ventricular capture was documented in the first beat. A reduction of ventricular output to 0.5 V resulted in failed capture (second beat), and a back-up pulse (*arrow*) was emitted. An upward search then begins at increments of 0.125 V, and the search ends when two consecutive capture beats are achieved at 0.88 V. VEGM: Ventricular electrogram (*tip to ring*).

Efficacy

Factors that affect Autocapture detection of ER are listed in Table 38-2. A large electrode polarization artifact relative to size of ER can affect ER detection. This can be reduced with the use of low polarization electrodes (made possible by increasing the microscopic electrode-tissue interface area).[5] An alternative is to use a biphasic waveform that comprises a fast precharge followed by a negative postcharge to minimize polarization effect.[6] In one study,[7] the effect of a modified fast prepulse on Autocapture was tested in 45 patients with leads from two manufacturers (Medtronic 4024 Cap Sure, and Pacesetter 1450 K/T and 1470 T leads). Whereas the ER was independent of the type of pacing pulse, the polarization artifact was significantly less during the modified pulse compared with the conventional pacing pulse, resulting in an improved efficacy of the

TABLE 38-2 Factors Affecting Capture Detection

Electrode polarization
Fusion beats (false negative →↑ output)
Ventricular capture + intrinsic beat
Pseudofusion beats (false positive →↓ output)
Pacing spike (and failure of capture) + intrinsic beat
Algorithm related: Unipolar pacing, bipolar sensing
Adequate ER (>2.5 mV)
Other applications: atrial, epicardial, and left ventricle

Autocapture algorithm (94% versus 71% successful ER detection).

An adequate ER amplitude of greater than 2.5 mV is recommended before activation of the autocapture algorithm, and this was present in 93% of 60 patients in one study.[8] Neither the clinical data nor the conventional electrical parameters were effective in predicting the size of the ER signal. Body posture and exercise had relatively little effect on the ER.[9] Atrial and epicardial applications remain investigational. The most important influence on the Autocapture is due to the presence of intrinsic rhythm, which can confuse ER detection.

In a multicenter study,[10] 113 patients received the Pacesetter Microny SR+ and were followed up for 1 year. Evoked response was satisfactory for Autocapture in 102 of 113 patients. Evoked response was stable over time, but correlated poorly with the R wave at the time of implantation. Acute and chronic pacing thresholds measured at the clinic using VARIO significantly correlated with that derived from Autocapture, although the Autocapture threshold was higher (0.11 ± 0.22 V) owing to the way in which threshold was derived. During Holter recordings, there was no failure of ventricular capture, and back-up pulses were used in 1.1% of all paced beats. Most of these were due to fusion or pseudofusion beats (87%), undersensing of either R wave or ER (4.6%), and truly due to loss of capture in only 7%. Although these did not affect pacing performance, the need for back-up pulses may negate the energy saving by the Autocapture itself. Similar positive results from the Autocapture algorithm in medium term for safety and efficacy have been published.[11-12]

Projected increases in battery longevity due to Autocapture have been reported.[13-14] Compared with the factory-set pacemaker setting of 5 V, Autocapture reduced the energy drain in the Microny SR+ (with 0.35 Ah) and increased device longevity by 53%. For the Regency SR+ with a larger battery (0.79 Ah), the increase in device longevity was even more significant (245%). Alternatively, when the conventional output was reduced to 2.5 V, the benefit of Autocapture on battery life was much less impressive.[13-14]

Clinical Implications

The main benefit of any automatic capture management algorithm is patient safety, ensuring effective capture during threshold changes. Programming of threshold can be simplified as the Autocapture threshold was significantly correlated with bedside threshold assessment. The energy saving would be more important in patients with chronic high thresholds. Conversely, fusion/pseudofusion beats appear to be the main limitation, not only in limiting battery energy reduction, but also they may lead to erroneous threshold determination.[15]

Biotronik Capture Control

The Logos pacemakers measure the ER signals from several successful capture beats, to generate a reference curve, against which failure of capture is compared.[16-17] There are no back-up pacing pulses, but persistent loss of capture results in increase of pulse output in 2 V steps. After a programmable period of time, the output is reduced to the programmed value. This algorithm ensures patient safety through beat-by-beat capture verification.

MEDTRONIC VENTRICULAR CAPTURE MANAGEMENT

The Kappa 700 pacemakers incorporate a threshold assessment based on ER: The Pacing Threshold Search (ambulatory) and Capture Management Threshold Test (bedside). During the procedure, the threshold at the Rheobase is determined at 1 ms by amplitude decrement until loss of capture, and then by amplitude increment until capture is confirmed. The Chronaxie is then determined by doubling the programmed amplitude, and decreasing the pulse width (followed by increasing amplitude to capture). A recommended pacing setting is then determined (Fig. 38-3). The physician can use the ambulatory threshold data to automatically adjust the threshold (adaptive), or to use for monitoring only, or the algorithm can be turned off. A minimal adapted output needs to be programmed. The ventricular capture management can be activated once every 15 minutes for 42 days, and is not a beat-by-beat threshold tracking algorithm.

Automatic Mode Switching

Because the ventricular response of a DDD pacemaker is dependent on the atrial rate, rapid ventricular pacing can occur in a DDD pacemaker during episodes of atrial tachycardias (AT), especially during atrial fibrillation (AF) (Fig. 38-4). This is managed in contemporary pacemakers using an algorithm known as automatic mode switching (AMS). There are several reasons why patients with dual chamber pacemakers will develop AT:

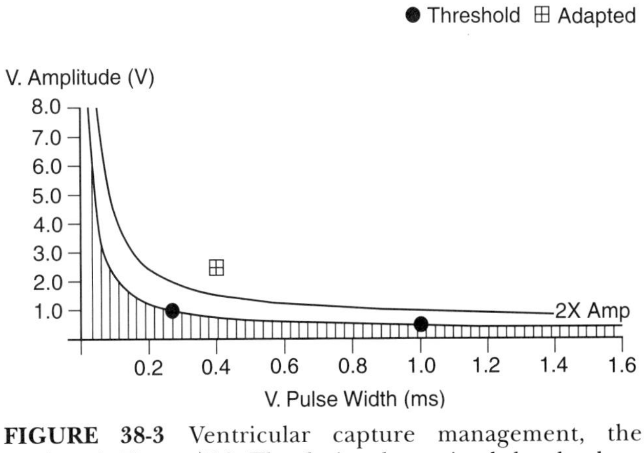

FIGURE 38-3 Ventricular capture management, the Medtronic Kappa 700. The device determined the rheobase (at 1 ms) and chronaxie, and recommends a safety margin of twice the amplitude threshold.

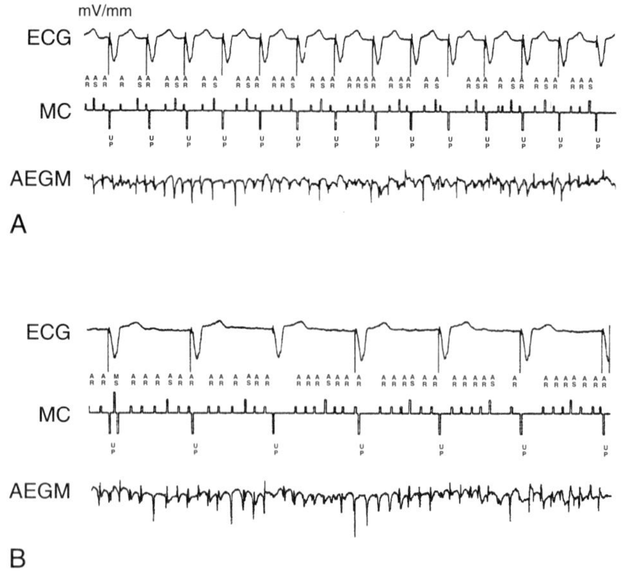

FIGURE 38-4 **A,** Rapid ventricular response in a patient with complete atrioventricular (AV) block with a Medtronic Kappa 400 pacemaker who developed an episode of atrial fibrillation (AF) when the AMS function was turned off. AF was detected by the atrial channel and ventricular pacing occurred at a UR of 112 beats/minute. **B,** Activation of AMS function in the pacemaker resulted in conversion to the DDIR mode with a ventricular rate at 60 beats/minute. AEGM, Atrial electrogram; AMS, automatic mode switching; ECG, surface electrocardiogram; MC, marker channel. (Reproduced with permission from Lau CP, Leung SK, Tse HF, Barold SS: Automatic mode switching of implantable pacemakers. I: Principles of instrumentation, clinical and hemodynamic considerations. Pacing Clin Electrophysiol 2002;25:967-983.)

(1) Nearly one half of patients receiving pacemakers have sinus node disease, and a substantial proportion of these patients have the bradycardia-tachycardia syndrome; (2) As many as 30% of patients with complete AV block either have coexistent bradycardia-tachycardia syndrome, or will develop this problem with time[18]; (3) A dual chamber pacemaker is often used in patients after AV nodal ablation for refractory AT; and (4) The incidence of AF increases markedly with age.[19]

Several registries and controlled trials have generated data on the incidence of AF. From 1988 to 1990, 12.9% of Medicare pacemaker recipients who received dual chamber pacemakers had underlying paroxysmal AF.[20] After implantation of a dual chamber pacemaker there is an overall 2% to 3% per year risk of developing AF. In patients with sinus node disease, the risk of developing paroxysmal or persistent AF is increased to 8% per year.[21-24]

Conventional or specially designed pacemakers can convert automatically to another pacing mode under a variety of circumstances.[25-32] The term *automatic mode switching* is now used to define an automatic function whereby a device is designed to switch temporarily to a nonatrial tracking destination mode during an AT, and to switch spontaneously back to the original mode on resumption of sinus rhythm.

COMPONENTS OF AN AUTOMATIC MODE SWITCHING ALGORITHM

The components of AMS function include (1) AT detection; (2) pacemaker response during AMS; and (3) resynchronization to sinus rhythm or atrial paced rhythm at AT termination.

Atrial Tachycardia Detection

There are four main ways for a device to detect AT (Table 38-3). (1) Most devices use a "rate cut-off" criterion in which a sensed atrial rate exceeding a programmable value will result in AMS. Some systems are designed to avoid mode switch during atrial ectopic beats or short runs of AT. For example, four short cycles out of seven consecutive beats are required before AMS occurs in the Medtronic Kappa 700. Interval number summation is used in the incremental and decremental counter of the Meta DDDR for short cycles and long cycles, respectively. (2) Some devices use a "mean atrial rate" or matched atrial rate based on a moving value related to the duration of the prevailing sensed atrial cycle as a criterion to move toward AMS. AMS will occur when the matched atrial interval has shortened to a predetermined duration. This algorithm is used in the Medtronic Thera DR, Kappa 400 (GEM DR ICD), and the St. Jude Trilogy DR+/Affinity family. Because the process is gradual, the rapidity of AMS will depend not only on the AT detection rate or interval, but also on the preexisting sinus rate. It is easier for the matched atrial interval to reach the tachycardia detection interval when AT occurs in the setting of a higher resting sinus rate than from a sinus bradycardia. This is because the matching atrial interval starts from a shorter baseline duration on its gradual way to reach the tachycardia detection interval. (3) Sensors can be used to determine the physiologic rate (e.g., Diamond, and SmarTracking in Marathon). To take into account the fluctuation in sinus rate, a physiologic heart rate range based on the sensor-indicated rate is used to define sinus rhythm, and rates beyond the upper end of physiologic range will activate AMS. (4) Complex algorithms are either a combination of algorithms or using additional criteria (often from implantable ICDs) to distinguish between AF and other rhythms. For example, a PR logic and a rate criterion are instrumented in the AT500 (Medtronic, Inc.) to detect AF and AT.[33]

Destination Mode

Either the VVI(R) or DDI(R) mode is used. There is no study on the relative merits of VVI(R) versus DDI(R) mode. Obviously, during AMS there is no AV synchrony and the DDI(R) is functionally equivalent to the VVI(R) mode. Theoretically, the DDI(R) mode may avoid AV dissociation when a sinus pause occurs at AT termination if the AMS algorithm has not yet resynchronized to sinus rhythm. The VVI mode during AMS has been described as VDI because the maintenance of atrial sensing (As) controls the perpetuation or termination of AMS but this designation does not strictly conform to

TABLE 38-3 Classification of Different Methods of AT Detection in Current AMS Algorithms

Criterion	Examples	Indications for Mode Switching
Rate cut-off	Pulsar/Vigor/Meridian/Discovery	Incremental/decremental counter
	Inos/Logos	Ratio of short/total cycles (e.g., 4 of 7 consecutive cycles)
	Kappa 400/700	Ratio of short/total cycles (e.g., 4 of 7 consecutive cycles)
	Marathon DR	Consecutive short cycles
	Meta DDDR (Model 1254/1256)	Incremental/decremental counter
Running average rate	Thera DR	Matched atrial interval computed from prevailing atrial rate
	Trilogy DR+ Affinity	Filtered atrial interval
Sensor-based physiologic rate	Clarity/Diamond	Single beat outside a physiologic rate band (15 or 30 bpm)
	Marathon DR	SmarTracking rate range (accelerometer sensor)
	Meta DR (Model 1250)	Sensor controlled PVARP
	Living 1/Living 1 plus	Sensor indicated rate to define tachycardia detection
Complex	Marathon DR	SmarTracking and rate cut-off
	AT 500	Rate cut-off and PR relationship

Models and manufacturers

Affinity DR	Model 5330/1 (St. Jude Medical Pacesetter)
AT 500 DDDRP	Model AT500/500C (Medtronic Inc., MN, USA)
Clarity DDDR	Model 860/2/5 (Vitatron BV, Dieren, The Netherlands)
Chorum DR	Model 7034/7134 (ELA Medical, Rougement, France)
Diamond	Model 800/801/820/840 (Vitatron BV)
Discovery	Model 1273/4/5 (Guidant CPI, St. Paul, MN, USA)
Inos	Biotronik GmbH & Co. (Berlin, Germany)
Kappa	Model 400 and 700 (Medtronic Inc., MN, USA)
Living 1	Sorin Biomedica (Saluggia, Italy)
Living 1 Plus	Sorin Biomedica
Marathon DR	Model 294-09 (Intermedics Inc, Freeport, TX, USA)
Meridian DR	Model 1276 (Guidant CPI)
Meta DDDR	Models 1250, 1254, 1256 (Telectronics Pacing System, Englewood, CO, USA)
Pulsar Max DR	Model 1270 (Guidant CPI)
Talent DR	Model 223 (ELA Medical)
Thera DR	Model 7960/1/2 (Medtronic Inc.)
Trilogy DR+	Model 2364 (St. Jude Medical Pacesetter)
Vigor DR	Model 1230/2/5 (Guidant CPI)

AMS, Automatic mode switching; AT, Atrial tachycardia.

the standard pacemaker code. However, when AF is undersensed during AMS, atrial pacing in DDI(R) destination mode may paradoxically perpetuate paroxysmal AF. Apart from passive handling of AT in terms of the AMS algorithm, some devices deliver an active pacing intervention on the detection of frequent atrial ectopic beats that are thought to herald the onset of AF, or an active pacing to terminate AT.

Resynchronization

Some AMS algorithms use the same onset criteria to resynchronize after AT termination, whereas others use slower criteria of resynchronization to avoid intermittent AMS during short runs of AT, for example, Thera DR.

THE IDEAL AUTOMATIC MODE SWITCHING ALGORITHM

An ideal AMS algorithm (Table 38-4) should have an appropriate *onset speed* after AT begins. Prolonged rapid ventricular pacing due to a slow algorithm, or oscillation between mode switching and tracking during short-lasting AT in a fast algorithm will result in undesirable ventricular rate fluctuation and/or AV dissociation. It is clear that speed of response and *rate stability* are two competing parameters. The atrial and ventricular responses during AMS should result in a pacing rate appropriate to the pathophysiologic state of the patient. In general, this rate is sensor determined. At the termination of AT, and to avoid AV dissociation during the process, the algorithm should *resynchronize* to sinus rhythm at the earliest opportunity. Many algorithms incorporate a rate fallback mechanism to ensure smooth rate control during mode transitions. With ideal sensing and programming, these characteristics are dependent entirely on the AMS algorithm.

TABLE 38-4 Characteristics of an Ideal AMS Algorithm

Characteristics	Remarks
Onset	Rapid to avoid high-rate ventricular pacing without causing frequent mode oscillations during unsustained AT
Response	Avoidance of excessive rate fluctuation
	Avoidance of inappropriate atrial pacing
Resynchronization	Restore AV synchrony to sinus rhythm at the earliest opportunity
Sensitivity	Ability to sense AT of varying rate and signal sizes
	Ability to sense atrial flutter
Specificity	Avoidance of switching during VA crosstalk, sinus tachycardia, and extraneous electrical noises

AMS, Automatic mode switching; AV, atrioventricular; AT, atrial tachycardia.

TABLE 38-5 Factors Affecting AT Detection in Dual Chamber Pacemakers with AMS Algorithms

	Under-Detection (Sensitivity)	Over-Detection (Specificity)
Arrhythmia-related	Atrial flutter	Sinus tachycardia
	AF with small or widely varying signal amplitudes	—
	Slow atrial tachycardia (actual or drug-induced)	—
Pacemaker-related		
Lead configurations	—	Unipolar sensing (far-field sensing and myopotentials) Low atrial lead positions Lead in the coronary sinus Dual site atrial or bi-atrial sensing
Atrial sensitivity	Insufficient atrial sensitivity to sense AF	
Atrial blanking	Reduced sensed AT rate	Inadequate blanking in A–V interval and after ventricular pacing
A–V interval	Reduced sensed AT rate (some devices)	—
PVARP	Reduced sensed AT rate (some devices)	PVARP mediated AMS: AMS can occur during sinus tachycardia or ectopy
VA crosstalk	—	Increased sensed atrial rate

AF, Atrial fibrillation; AMS, automatic mode switching; AT, atrial tachycardia; AV, atrioventricular.

In clinical practice, however, arrhythmia-related and sensing-related issues affect significantly the *sensitivity* and *specificity* of the AMS response. Sensitivity of an AMS algorithm refers to its ability to detect AT (i.e., avoid false-negative response), whereas specificity refers to the absence of AMS during sinus rhythm (i.e., avoid false-positive response) (Table 38-5). For AMS algorithms, the greater the sensitivity, the lower the specificity and vice versa.

Automatic Mode Switching Sensitivity

As most AMS algorithms detect AT by a rate cut-off criterion, a slow atrial rate (e.g., atrial rate slowing after antiarrhythmic medications) may fall below the tachycardia detection rate, and AMS will not occur. Conversion of AF to atrial flutter is a special situation in which alternate flutter waves coincide with the PVAB, and the effectively detected atrial rate falls below the tachycardia detection rate and prevents AMS (Fig. 38-5).

In contemporary devices, the PVARP is opened (completely or on a conditional high rate) to enhance AMS. In other words, sensing occurs in the second part of the PVARP beyond the PVAB. The latter is designed to prevent far-field R-wave sensing. Atrial undersensing may occur because of an inappropriately long PVAB. Undersensing can often be avoided by appropriate adjustment of the PVAB provided the atrial channel exhibits no far-field sensing. The widely varying amplitude of atrial electrogram in AF, both temporally in a patient and between patients, can result in AMS failure when the atrial sensitivity is programmed incorrectly. During electrophysiological study,[34] acutely recorded atrial electrograms during AF and sinus rhythm showed similar mean amplitude, but the variability in amplitude was substantially wider, and the minimum amplitude considerably smaller in AF compared with sinus rhythm (minimum atrial electrogram: 1.4 ± 1.1 and. 2.0 ± 0.8 mV, respectively). A high-programmed atrial sensitivity may cause atrial sensing of far-field signals or noise, whereas a low atrial sensitivity can lead to undersensing during AF (Fig. 38-6).[35] Optimal programming of atrial sensitivity for AMS requires three times the safety margin compared with two times for sinus P-wave sensing (Fig. 38-7).[35]

Far-field sensing of the tail end of the QRS complex by the atrial channel is the commonest cause of a false-positive AMS response. Such far-field sensing of the QRS complex (almost always from a paced beat) causes VA crosstalk opposite in direction to the well-known form of AV crosstalk. Several investigators have studied the incidence of far-field R wave as recorded by an atrial lead or a single lead VDD system.[36-38] In general, unipolar atrial sensing, paced QRS complex, longer dipole lengths (30 versus 17.8 mm), septal and low right atrial implants may predispose to far-field R wave sensing. In one study,[38] at an atrial sensitivity of 0.1 mV, 30/30 bipolar leads had a far-field R wave sensed. The median far-field QRS complex sensing threshold was 0.3 mV, and occurred at 67 to 202 ms following the ventricular pacing stimuli. These have implications on the highest

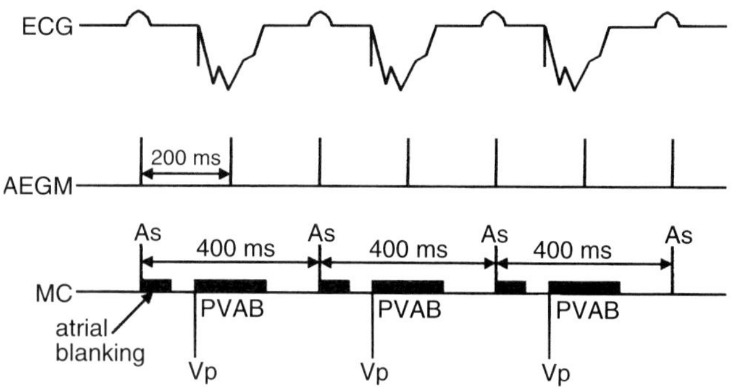

FIGURE 38-5 Sensing of atrial flutter (2:1) by a DDD pacemaker. The atrial flutter cycle length is shorter than the sum of the A–V interval plus the PVAB. The blanking periods are shown in *black*. Every alternate flutter wave falls in the PVAB, resulting in 2:1 sensing. AEGM, Atrial electrogram; ECG, surface electrocardiogram; MC, marker channel; PVAB, postventricular atrial blanking period. (Reproduced with permission from Lau CP, Leung SK, Tse HF, Barold SS: Automatic mode switching of implantable pacemakers. I: Principles of instrumentation, clinical and hemodynamic considerations. Pacing Clin Electrophysiol 2002;25:967-983.)

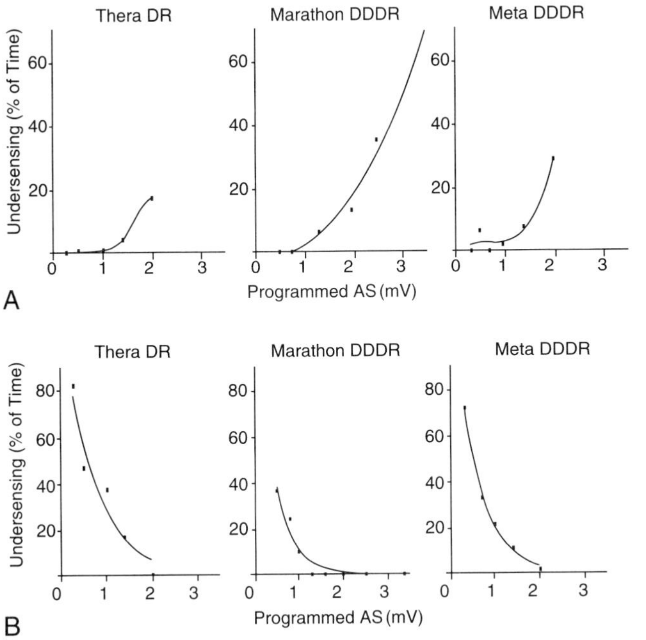

FIGURE 38-6 Effect of programmed atrial sensitivity in three mode switching pacemakers during persistent atrial fibrillation (AF). **A,** Significant undersensing of AF began to occur when the sensitivity was above 1 mV. **B,** Minimum oversensing of noise occurred when sensitivity level was above 2 mV. (Reproduced with permission from Leung SK, Lau CP, Lam CT, et al: Programmed atrial sensitivity: A critical determinant in atrial fibrillation detection and optimal automatic mode switching. Pacing Clin Electrophysiol 1998;21:2214-2219.)

atrial sensitivity and the duration of blanking period needed. In addition to atrial sensitivity, VA crosstalk occurs either when the PVAB is too short or the QRS is too long. Any circumstance that prolongs the QRS complex (e.g., flecainide, amiodarone, or hyperkalemia)

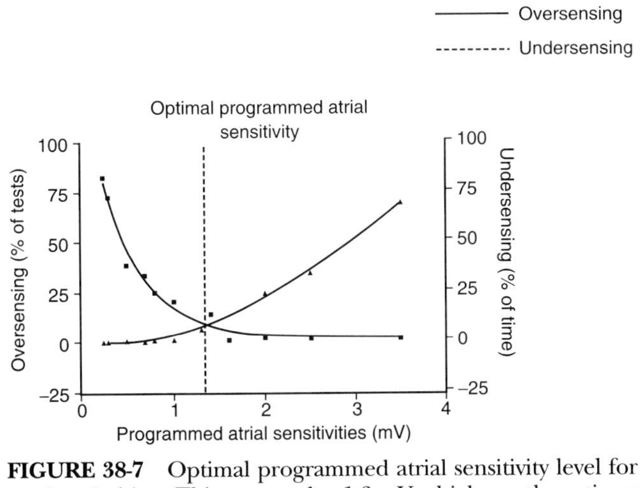

FIGURE 38-7 Optimal programmed atrial sensitivity level for mode switching. This occurred at 1.3 mV, which was three times the safety margin compared with the sinus P wave. (Reproduced with permission from Leung SK, Lau CP, Lam CT, et al: Programmed atrial sensitivity: A critical determinant in atrial fibrillation detection and optimal automatic mode switching. Pacing Clin Electrophysiol 1998;21:2214-2219.)

favors such VA crosstalk. Without PVAB programmability, false AMS from such far-field R wave sensing can often be corrected by *decreasing* the atrial sensitivity.

Less commonly, oversensing of atrial signals can occur within the A–V interval. In such a case, the atrial blanking period connected to the initial part of the A–V interval (postatrial blanking period) must terminate before the A–V interval has timed out. Thus, double sensing of the P wave (near-field), especially due to a large after potential following Ap or sensing of the early part of the spontaneous QRS complex (far field) can occur within the A–V interval. Far-field sensing of the spontaneous QRS complex during the A–V interval is probably less common than sensing the terminal part of the QRS complex beyond the PVAB. The atrial channel can sense the early part of the spontaneous QRS only if it is detected before the ventricular channel senses it as a near-field signal.

In some devices with algorithms based on the matched atrial rate or interval, AMS can occur following a series of short-long cycles, where the short cycles occur within the A–V interval (As-Ar or Ar-As), despite the fact that the long cycles exceed the duration of the tachycardia detection interval. Problems related to atrial sensing within the A–V interval could be eliminated by extending the atrial blanking period to encompass the entire A–V interval—either as a factory-set feature or by means of a special programmable option. Signal detection during the A–V interval beyond the atrial blanking period is designed to optimize the detection of AF. Improved AMS algorithms should now make sensing within the A–V interval unnecessary and obviate atrial oversensing.

A low-lying atrial lead or one in the coronary sinus may pick up both atrial and ventricular signals. The recent use of dual site right atrial (with a posteriorly situated lead near the coronary sinus) and biatrial pacing necessitates special algorithms or optimal lead positioning to avoid far-field R-wave sensing or even double sensing of the P wave. The latter may occur when the atrial conduction time between the two atrial sites is longer than the atrial blanking period (Fig. 38-8). The sensitivity and specificity of AMS can now be validated by electrograms and data storage of current implanted devices.

AUTOMATIC MODE SWITCHING DIAGNOSTICS

Automatic mode switching diagnostics provide an assessment of the frequency and pattern of AT episodes. These data may be useful for consideration of pacemaker mode reprogramming, such as from a dual chamber mode to the VVIR mode when a patient develops permanent AF, and to assess the need for adjunctive antiarrhythmic drug therapy and anticoagulation in patients with a large number and/or long duration of AF episodes. Indeed, recent studies[39-40] suggest that asymptomatic episodes of AF occur with an incidence at least 12 times more frequently than the symptomatic episodes in patients with implantable rhythm control devices. Although these episodes may not be symptomatically relevant, their impact on thromboembolism

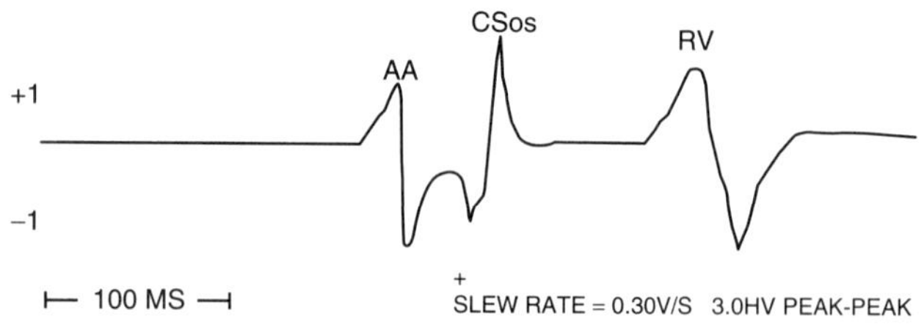

FIGURE 38-8 An atrial electrogram recorded from a pacing system analyzer from a wide atrial bipole (low right atrial as anode, high right atrial as cathode) as used in dual site right atrial pacing. Note the wide separation of the atrial electrogram and the far-field R wave. AA, Right atrial appendage electrogram; CS_{OS}, atrial electrogram just below CS_{OS}; RV, far-field right ventricular electrogram. (Reproduced with permission from Lau CP, Leung SK, Tse HF, Barold SS: Automatic mode switching of implantable pacemakers. I: Principles of instrumentation, clinical and hemodynamic considerations. Pacing Clin Electrophysiol 2002;25:967-983.)

and heart failure is probably similar to symptomatic episodes and may require similarly aggressive treatment. Indeed, AMS was used 66% of the time in patients with a known history of AT/AF, and in 55% of the time in patients without this history.[40] However, the critical issue is the specificity of recorded episodes characterized as AT/AF, which is now better defined through recording of atrial electrograms.

Event Counters

In general, event counters record the number of intrinsic and pacemaker-mediated events that occur during an event recording period. These counters are either triggered by the onset of a high atrial rate or AMS (Fig. 38-9). Some current devices also allow patients to trigger the event counter, using an external magnet, to record data for a preset number of beats. This feature is useful for documentation and assessment of the pattern of symptomatic AF episodes.[41] Either the AMS counter or the atrial high rate episode monitor can be used for detection and assessment of AT/AF episodes.

The AMS counter records the actual number of mode switches that occur. Prior studies on the Medtronic Thera DR pacemaker have demonstrated that 12% to 40% of patients with mode-switching episodes were not attributed to AT. As described above, the majority of inappropriate mode switching was due to far-field R wave, or near-field A wave sensing of atrial paced beats. Furthermore, the mode switch count recorded is also affected by the speed of response, sensitivity of algorithm, and speed of resynchronization to sinus rhythm or atrial paced rhythm.[42-50]

Theoretically, the atrial high rate episode monitor should be independent of the mode switch algorithm and should be more accurate than the AMS counter. For example, intermittent atrial undersensing can mimic frequent short episodes of paroxysmal AF by registering repeated AMS (Fig. 38-10). Seidl and colleagues[44] suggested that optimal programming of the atrial high rate episode monitor in the Medtronic Thera DR pacemaker could reliably detect AT with high sensitivity and specificity. However, false-negative detection during short episodes of AF and false-positive detection due to far-field R-wave oversensing were still observed. A specific pattern of oscillations in the atrial rate profile consistent with atrial oversensing has been described (Fig. 38-11). With additional criteria to exclude oversensing, off-line analysis of the recorded signals can significantly reduce false-positive detection to 2.9%.[45]

Histograms

Histograms give the pacing operation of the device (e.g., As-Vs, As-Vp, Ap-Vs, Ap-Vp, and ventricular

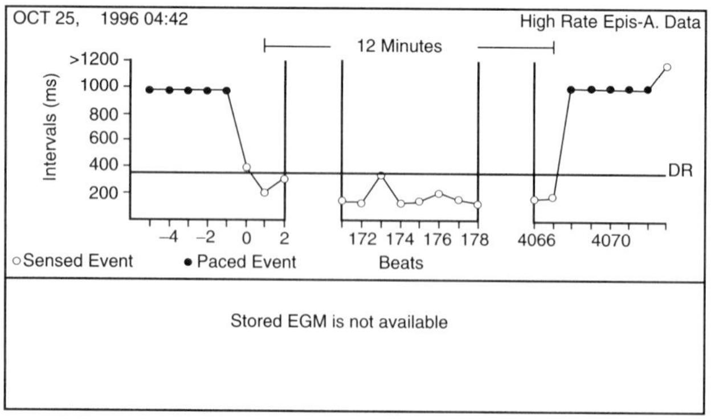

FIGURE 38-9 A recording of the atrial cycle length at the onset, during and at termination of an episode of atrial fibrillation (AF) in a patient with paroxysmal AF with a Medtronic Thera DR pacemaker. An episode of high atrial rate beyond the atrial detection rate (DR) of 350 ms that lasted for more than 4000 cycles was recorded. (Reproduced with permission from Lau CP, Leung SK, Tse HF, Barold SS: Automatic mode switching of implantable pacemakers. I: Principles of instrumentation, clinical and hemodynamic considerations. Pacing Clin Electrophysiol 2002;25:967-983.)

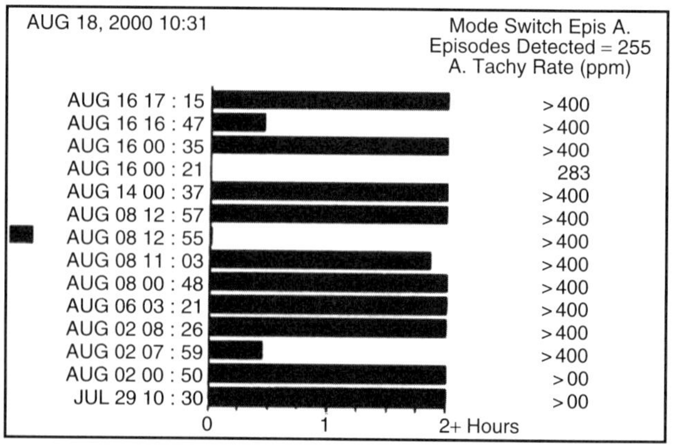

FIGURE 38-10 Mode switch episode-log in the same patient who developed persistent atrial fibrillation (AF) 4 years later. Note the very close spacing between each recorded episode, which gave an impression of very frequent paroxysms of AF. However, this was probably due to mode switching in and out of persistent AF due to AF undersensing.

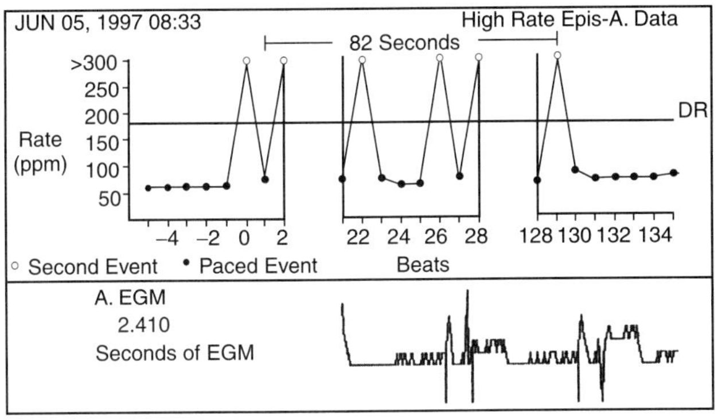

FIGURE 38-11 Double atrial sensing in a dual site pacing device leading to rate fluctuation prior to atrial mode sensing (AMS). The short–long sequence induced rate oscillation, and when the mean atrial rate was reached, sustained AMS was induced. Atrial electrogram clearly showed double P wave sensing. (Reproduced with permission from Lau CP, Leung SK, Tse HF, Barold SS: Automatic mode switching of implantable pacemakers. I: Principles of instrumentation, clinical and hemodynamic considerations. Pacing Clin Electrophysiol 2002;25:967-983.)

ectopy) and the details of the AMS operations and the number of AF episodes (Fig. 38-12). The rate histogram during the mode switch episodes may be useful for identification of inappropriate mode switch. When mode switching episodes occur in the atrial rate of 175 to 250 beats/minute, it may represent double counting because of oversensing either far-field or near-field events. Episodes where the atrial rate is greater than 400 beats/minute are likely to represent true AF,[46] if lead fracture and myopotential sensing can be excluded.

Stored Atrial Electrogram

A number of devices have telemetered atrial electrograms that allow on-line assessment of pacemaker operations. These are very useful to assess atrial sensing issues when adjusting the parameters of AMS. The incorporation of the stored atrial electrogram data is very useful in confirming the etiology of the recorded arrhythmia (see Figs. 38-11 and 38-13). Atrial stored electrograms can increase the accuracy of the event counters and identify the type of atrial arrhythmias.[47-50] They may provide important insight on the onset and termination of the arrhythmias (see Fig. 38-13). In a recent study,[44] when atrial electrograms were available for confirmation, it was found that as many as 62.7% of mode switch episodes were erroneously executed. In patients implanted with a single-lead VDD system,[50] only 35% of 235 episodes of suspected AF were confirmed to be AF, whereas the other episodes were diagnosed to be atrial undersensing. At present, most devices provide only limited duration of stored atrial electrograms due to the limitation of pacemaker memory capability.

In Guidant pacemakers, the number of events and the duration of the two channel (atrial and ventricular) electrograms are programmable, being limited to a cumulative maximum of 40 seconds (2 electrograms of 20-seconds' duration or up to 20 electrograms of 2-seconds' duration). Some other devices allow the recording of a single channel compound electrogram consisting of superimposed atrial and ventricular components in a single channel (AV or summed electrogram). This combined arrangement may make it difficult to differentiate atrial flutter from far-field R-wave sensing by the atrial channel. Electrogram recordings can be

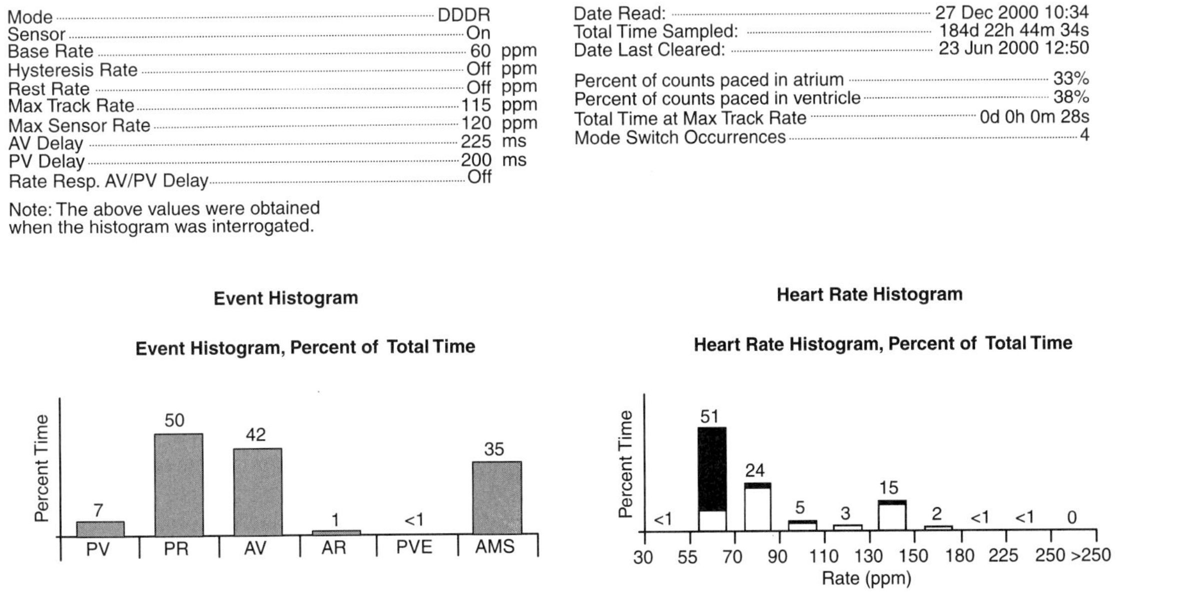

FIGURE 38-12 Pacemaker operation histogram from Affinity DR (St. Jude Medical). Four episodes of mode switches were recorded. In addition, the histograms showed that the patient was in atrial mode switching for 35% of the time, giving an idea of the amount of atrial fibrillation burden during the recording period.

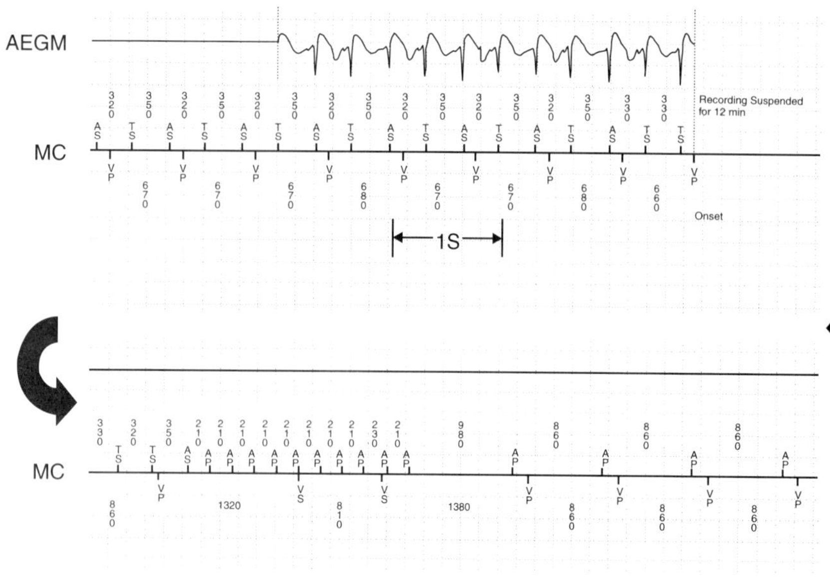

FIGURE 38-13 Atrial electrograms recording at the onset of AT (Model AT500, Medtronic, Inc.), validating an appropriate AT detection. The AT lasted for about 12 minutes (recording suspended) and terminated spontaneously. AEGM, Atrial electrogram; As, atrial sense; Ap, atrial pace, AT, atrial tachycardia; MC, marker channel; Vp, ventricular pace; Vs, ventricular sense; Ts, tachycardia sense.

programmed to be triggered by AT, by other arrhythmias, and by an externally applied magnet during symptomatic episodes. The stored electrograms are able to demonstrate the actual signals that triggered the recording whether or not there was an arrhythmia. Electrogram storage in many pacemakers now begins at the time of the trigger. The standard for pacemakers should be the storage of electrograms with annotations (these do not consume much memory) at the start of the arrhythmia or from a predetermined time before the trigger. Such pacemakers are now becoming available.

Atrial Fibrillation Burden

Several clinical studies[51-52] have attempted to use AMS diagnostics to evaluate the effects of atrial pacing on the total burden of AF (both symptomatic and asymptomatic episodes). This provides a powerful tool to assess the effectiveness of a therapy to treat AF (Fig. 38-14). However, the accuracy of the measurement of AF burden by using the AMS diagnostics is limited by the following factors: (1) the specificity and sensitivity of the AMS diagnostics as described; (2) the availability of the atrial stored electrogram to confirm AF; and (3) the storage capacity of the pacemaker data logs, because the event counters in most pacemakers can easily be saturated.

CLINICAL BENEFITS

The hemodynamic benefits of maintaining a regular ventricular rate have been reported.[53-57] There are anecdotal reports of the symptomatic benefit of AMS, and improvement of tachycardia-related symptoms by avoiding a rapid-paced ventricular rate at the onset of AT. A randomized, crossover, prospective study on the addition of AMS in patients after AV nodal ablation and DDDR pacing has been reported.[58] In this study, 48 patients were randomly assigned to DDDR pacing with and without activation of AMS. It was found that the VVIR mode alone was the least well tolerated, whereas the DDDR with mode switch was the most acceptable (Fig. 38-15). Patient-perceived well-being was superior with AMS activated than with AMS inactivated and early crossover was observed in 3% and 19% of patients, respectively. This study documented a short term symptomatic benefit of AMS over conventional DDDR mode in a population with high AT incidence.

Whether different AMS algorithms may have an impact on symptoms was the subject of another study.[59] Three different AMS onset criteria (mean atrial rate, "4-of-7" and "1-of-1") of an AMS algorithm were injected into an implanted DDDR pacemaker (Thera DR models 7142, 7952, and 7960i; Medtronic, Inc.) in a group of patients with frequent AT episodes. The faster AMS criteria, 4-of-7 or 1-of-1 were better tolerated than the slow onset criteria using mean atrial rate (Fig. 38-16). Conversely, the fast-responding algorithms lead to shorter but more frequent episodes of AF sensed, suggesting that the device may be switching in and out of AMS due to the lack of stability. This study demonstrated that different types of AMS algorithms might also have an impact on patients' symptoms. An inappropriate AMS algorithm can have a deleterious effect on patient symptoms. For example, in the first version of the DDDR Meta, frequent mode switching can result in AV dissociation, which may cause more symptoms than AF itself.[60] In patients with a high sinus rate, undesirable mode switching will occur during sinus rhythm. The extended ventricular escape interval due to mode switching will prevent ventricular pacing,[61] as the device will function in a DDI/VVI mode. In the presence of a prolonged P–R interval, this may lead to a very long A–V interval and a pacemaker syndrome.[62]

Two recent studies[63-64] on the stability of the sinus node after AV nodal ablation questioned the long-term clinical benefit of AMS. In one study,[63] after AV nodal ablation and implantation of a DDDR device, patients were symptomatically better and achieved better quality of life and NYHA class compared with a control group of patients with continuation of medical therapy.

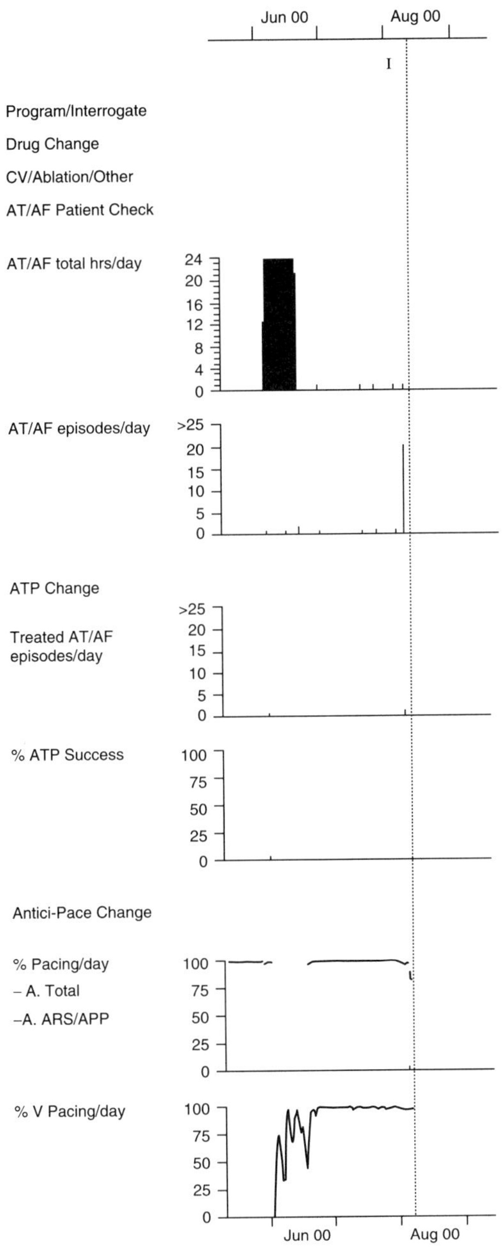

FIGURE 38-14 Atrial fibrillation diagnostics from a DDDR pacemaker with AT/AF therapy (AT500, Medtronic Inc.). A period of persistent AF (June, 2000) was documented by the total AF duration of nearly 24 hr per day. The corresponding atrial pacing was minimal in this pacemaker-dependent patient. AF, Atrial fibrillation; AT, atrial tachycardia.

The long-term maintenance of sinus rhythm was poor in the ablated and paced group. As many as 24% developed chronic AF by 12 months, necessitating reprogramming DDDR pacemakers to the VVIR mode. In another study,[64] 12 of 37 patients developed chronic AF at 6 weeks after AV nodal ablation versus 0 of 19 in those who continued medical treatment. The cause was most likely due to the withdrawal of antiarrhythmic medications after AV nodal ablation, but also possibly due to an atrial proarrhythmic effect of the ablation and pacing procedure. However, should a dual chamber device be implanted after AV nodal ablation, AMS remains an integral part of device therapy to avoid the occurrence of rapid ventricular pacing during AF.

ILLUSTRATIVE TYPES OF AUTOMATIC MODE SWITCHING

This discussion focuses mainly on AT detection by algorithms developed by several major pacemaker companies to illustrate their complexity and evaluation. Clinical results on these algorithms are presented when available.

Telectronics Automatic Mode Switching Function

The Telectronics algorithms give a historical perspective of AMS development. The first algorithm (Model 1250, Meta DDDR) used the postventricular atrial refractory (PVARP) to monitor AT, and an atrial event sensed in the PVARP resulted in a mode switch from DDDR to VVIR. The PVARP duration was further controlled by the minute ventilation sensor, giving a sensor-determined rate band for defining a pathologic rate.[32] Although this algorithm gave an immediate AMS response at the onset of AT, it suffered from frequent undesirable mode changes in the presence of sinus tachycardia and frequent atrial premature beats. In one report on 24 patients,[60] Holter monitoring showed that patients spent up to 50% of the time in effective VVIR pacing due to low specificity of the AMS algorithm. A long A–V interval and PVARP contributed to undesirable AMS in sinus rhythm. Although appropriate adjustment of baseline PVARP and sensor setting could reduce false AMS by automatically hastening PVARP shortening during sinus tachycardia,[65] this algorithm was superseded in subsequent models by an algorithm using a rate cut-off criterion.[66]

In the Model 1254 Meta DDDR, an atrial rate monitor continuously measures atrial rates above and below the AT detection rate. The atrial monitor counts every sensed atrial interval less than AT detection interval increments and every interval greater than AT detection interval decrements. When the number of counts exceeds a programmable number (5 or 11), AMS will occur. The destination mode is VVIR. After AT termination, resynchronization to sinus rhythm occurs after three sensed atrial events with intervals greater than AT detection interval or an atrial pause greater than 1 second.

Thera and Kappa 400

The Medtronic devices differ significantly between the earlier Thera and Kappa 400 series and the more recent Kappa 700 family. In the Thera and Kappa 400 family of pacemakers, the device uses an artificial atrial rate (the so-called *mean atrial rate* calculated by the device), which bears a constantly changing relationship to the true or sensed atrial rate. This artificial atrial interval must eventually match the tachycardia detection interval for AMS to occur. Thus, the mean atrial rate (interval) is best described in terms of a matching atrial rate (interval). When AMS is programmed on,

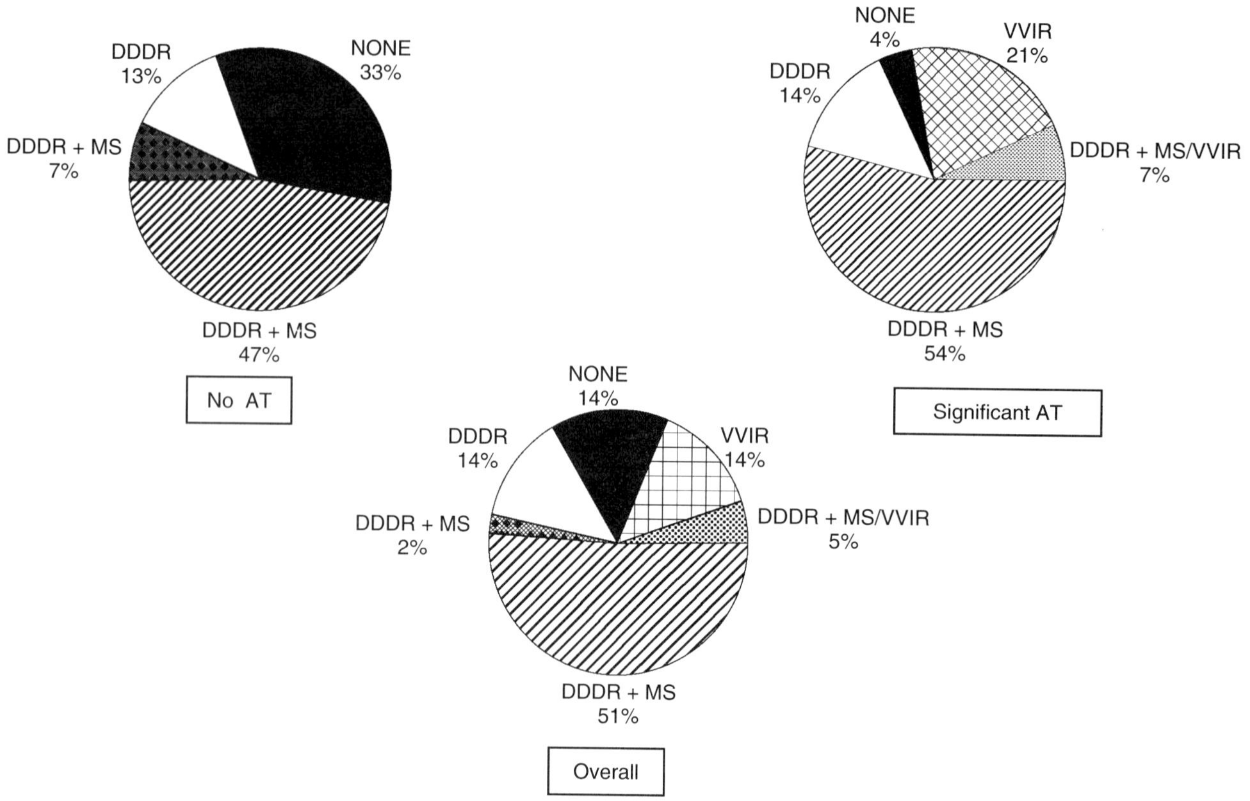

FIGURE 38-15 Preferred pacing modes in patients with and without atrial tachycardia (AT) during a randomized mode study. DDDR with atrial mode sensing activated (DDDR + MS) was the most preferred mode, independent of the presence of AT before implantation. (Reproduced with permission from Kamalvand K, Tam K, Kotsakis K, et al: Is mode switching beneficial? A randomized study in patients with paroxysmal atrial tachyarrhythmias. J Am Coll Cardiol 1997;30:496-504.)

the pacemaker continuously monitors the interval between two consecutive atrial events except for As-Ap and Ar–Ap intervals. The matching atrial interval (MAI) shortens by 24 ms when it is equal or longer than the atrial cycle. When the MAI is shorter than the atrial cycle, it increases by 8 ms (Fig. 38-17). This process continues until the MAI has shortened to the value of the AT detection interval. At this point AMS occurs. Thus, the MAI dithers around the rate of the tachycardia and not the AT detection rate. If the atrial rate (sinus rhythm or AT) is stable the dither is always in a 3:1 ratio. When the MAI becomes equal to the P–P interval, it shortens by 24 ms. The subsequent 3 MAIs will be shorter than the P–P interval and each will increase by 8 ms ($8 \times 3 = 24$). The third 8 ms increment then causes the MAI to equal the P–P interval and this triggers a 24 ms subtraction in the MAI and the process repeats itself. The algorithm is designed to weather intermittent undersensing and the AMS algorithm is resistant to sudden changes in atrial rate. Several detected long

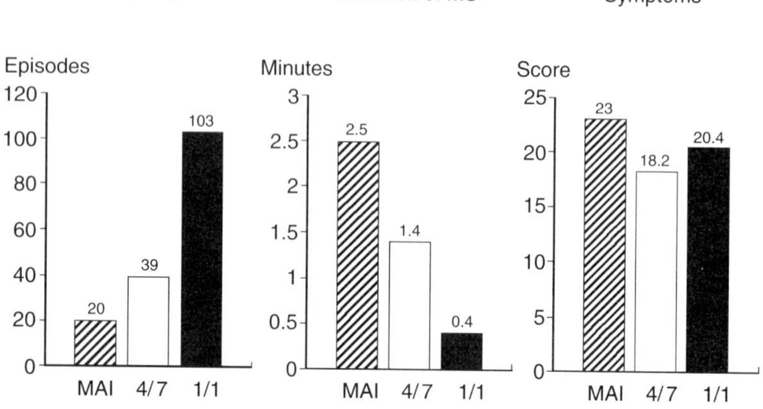

FIGURE 38-16 Effect of different mode onset criteria on mode switch frequency and duration and symptoms. See text for description. MA1, 4/7, 1/1, AMS algorithm based on the mean atrial rate, four out of seven high atrial rate events or one atrial ectopy only. AMS, Automatic mode switching; MAI, matching atrial interval. (Reproduced from Marshall HJ, Kay GN, Hess M, et al: Mode switching in dual chamber pacemakers: Effect of onset criteria on arrhythmia-related symptoms. Europace 1999;1:49-54.)

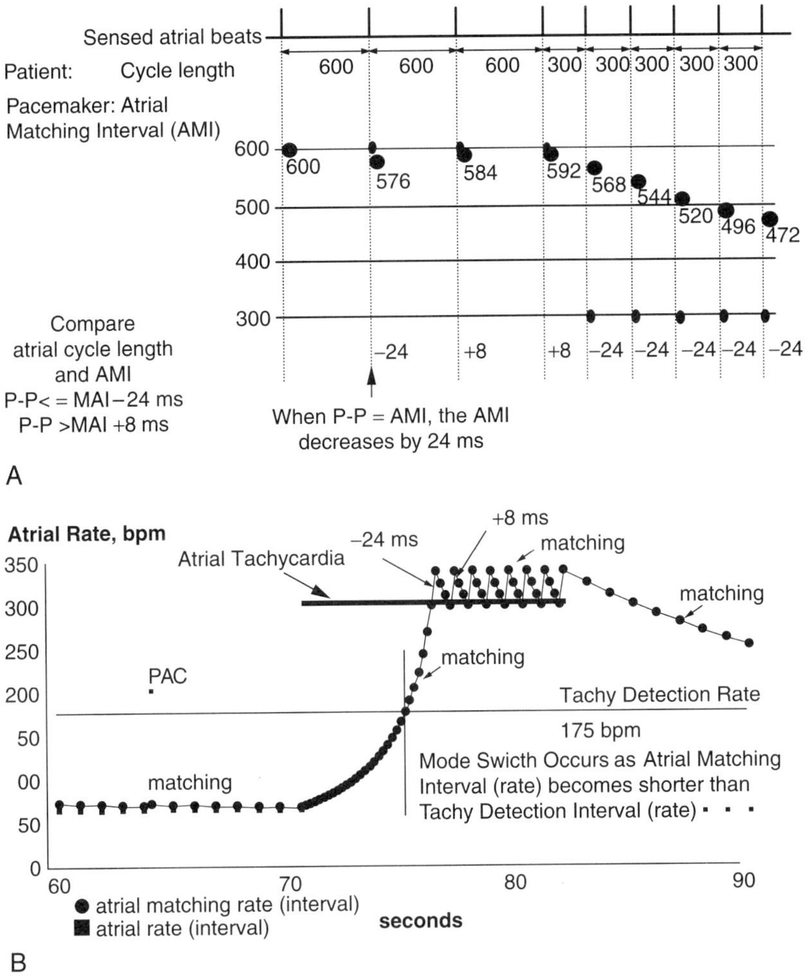

FIGURE 38-17 **A,** Diagrammatic representation of the matching atrial rate of Medtronic Thera pacemaker during supraventricular tachycardia. The *large solid circles* represent the value of the matching atrial interval (MAI). The *smaller oval circles* represent the instantaneous atrial interval. On each interval, the MAI is compared with the instantaneous interval, resulting in a change. First beat: The MAI is the same as the atrial interval, resulting in a decrease of 24 ms in the MAI. Second beat: The MAI is shorter (corresponding rate faster) than the atrial interval, resulting in an increase of 8 ms in the MAI. Fourth beat: The 300 ms atrial interval is shorter (corresponding rate faster) than the MAI, resulting in a decrease of 24 ms in the MAI. Fifth to eight beats:—same. The MAI decreases by 24 ms with each cycle. **B,** Should the postventricular blanking period (PVAB) cause failure to sense an occasional atrial signal, the sensed atrial interval would double to 600 ms. In that case, an 8 ms addition to the MAI would occur but the 24 ms. Subtractions on shorter 300 ms atrial intervals guarantee that the MAI continues to shorten to meet the spontaneous tachycardia interval. Mode switching occurs when the MAI reaches the tachycardia detection interval. (Reproduced with permission from Lau CP, Leung SK, Tse HF, Barold SS: Automatic mode switching of implantable pacemakers. II: Clinical performance of current algorithms and their programming. Pacing Clin Electrophysiol 2001;25:1094-1113.)

As–As intervals secondary to undersensing can be compensated with just one short As–As interval.

The time to AMS depends on the following parameters: (1) starting atrial rate—a faster rate with a correspondingly shorter MAI requires less ground to cover to reach the AT detection rate; (2) the AT detection rate. AMS will take longer to occur with a faster AT detection rate. Exit from AMS at the termination of AT depends on the reverse process with the MAI gradually lengthening with each cycle until resynchronization occurs.

When AMS of the Thera device is programmed, the obligatory AV algorithm shortens the As–Vp interval according to the sensed atrial rate to a minimum of 30 ms to enhance sensing of AT, especially atrial flutter. A–V interval abbreviation is controlled by the MAI algorithm so that shortening of the As–Vp interval is gradual. However, shortening of the As–Vp interval may occur during brief episodes of AT that shorten the MAI without allowing it to reach the level for AMS. This should be considered when As–Vp interval changes on the ECG cannot be explained on the basis of the pacing rate alone.

In the Medtronic Thera and Kappa 400, blanking following Ap or As varies from 50 to 100 ms and is a function of the sensed signal amplitude. Atrial sensing within the A–V interval was designed to provide an opportunity to detect additional atrial events for updating the MAI. However, such a window may cause double sensing of P waves (and possible detection of polarization after atrial pacing) and sensing of far-field R waves preceding near-field sensing of the R waves by the ventricular channel.

Long-Short Sequences

A long-short sequence is characterized by an atrial interval that is longer and then shorter than the AT detection interval (Fig. 38-18). The algorithm of the Thera and Kappa 400 devices, therefore, subtracts 24 ms from the MAI with shorter atrial cycles and adds 8 ms to the MAI with longer atrial cycles. The system is, therefore, biased toward tachycardia detection. This process eventually allows the MAI to reach the AT detection interval and AMS becomes established. Although this algorithm is helpful in detecting an irregular tachycardia with detection gaps (atrial beats undetected in the blanking period or one associated with intermittent undersensing), it may also induce AMS in circumstances without AT, as with frequent atrial ectopy.

Kappa 700

The Kappa 700 series of dual chamber pacemakers employs a different AMS algorithm based on a “4 of 7”

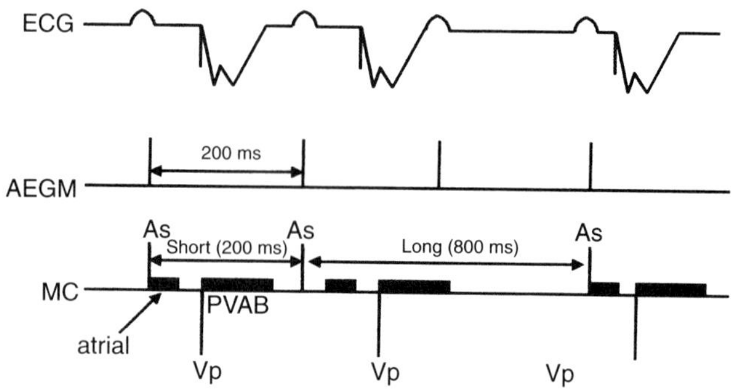

FIGURE 38-18 In the Medtronic Thera and Kappa 400 devices, atrial mode sensing can occur due to long–short–long–short sequences. In this example; 3:2 Wenckebach UR response results in pacemaker detection of alternating long and short intervals. The third atrial signal is not detected because it falls in the PVAB. The matched atrial interval decreases by 24 ms, then increases by 8 ms, then decreases by 24 ms, then increases by 8 ms, until it reaches the tachycardia detection interval, even though the long cycle exceeds the tachycardia detection interval. AEGM, Atrial electrogram; PVAB, postventricular atrial blanking period. (Reproduced with permission from Lau CP, Leung SK, Tse HF, Barold SS: Automatic mode switching of implantable pacemakers. II: Clinical performance of current algorithms and their programming. Pacing Clin Electrophysiol 2001;25:1094-1113.)

protocol (a rate cut-off criterion). This algorithm requires that at least four of the previous seven sensed atrial beats be above the AT detection rate for AMS to the DDI(R)/VVI(R) modes. (Intervals that begin with As or Ar and end with Ap are not counted as part of the seven intervals). There is a programmable time delay (from 0 to 60 seconds) from the detection of AT to the start of AMS to confirm that the AT is sustained. AMS will occur if the pacemaker reconfirms that the AT detection criteria are still met after this time window. AMS will terminate if either seven out of seven sensed atrial events are slower than the AT detection rate or if five consecutive atrial events are paced. If the sinus rate after conversion is different from the sensor-indicated rate, the pacing rate will be gradually adjusted to the sinus rate by 39 ms per beat. The "4 of 7" algorithm provides a faster AMS response compared with the system in the Thera and Kappa 400.

The A–V interval in the Kappa 700 is adjusted by the MAI as in the Thera and Kappa 400. The pacemaker calculates a desired 2:1 AV block point (30 beats faster than the matched atrial rate) but never less than 100 beats/minute and never more than 35 beats/minute above the upper rate (UR) (subject to Wenckebach's operation above UR). The PVARP is then adjusted so that the TARP provides the desired 2:1 block point. If the sensed AV interval is too long and PVARP would need to be very short, the PVARP is shortened to a programmed minimum value and then the sensed AV interval is shortened to maintain the desired 2:1 block response. Every 34 beats, a new TARP is calculated equal to the interval corresponding to the matched atrial rate + 30 beats/minute. In this case PVARP enhances sensing of AT by diminishing the total blanking periods (in A–V interval and PVARP) per cycle. A long A–V interval at rest allows optimal hemodynamics, which is dynamically shortened to reduce blanking during AT. In addition, a long PVARP at rest provides control of the ventricular response at the onset of an arrhythmia.

Blanked Flutter Search

Undersensing of atrial flutter is a common cause of AMS failure. Atrial flutter may be concealed to the pacemaker by synchronization of alternate flutter waves with the postventricular atrial blanking period (PVAB). An automatic algorithm to search for atrial flutter has been incorporated in the Kappa 700 series. This algorithm is based on modification of synchronization for one cycle to bring out atrial signals (flutter waves) masked by blanking of the pacemaker. The pacemaker continuously reviews atrial intervals in comparison with the A–V interval and the PVAB. When the P–P interval is less than 2 × (AV interval + PVAB) and less than 2 × (tachycardia detection interval), it is possible that an atrial rate is being sensed on a 2:1 basis. If this condition persists for eight beats, the PVARP is extended to 400 ms to reveal and detect the true atrial interval (Fig. 38-19).

Exit from Automatic Mode Switching

In the Medtronic Thera and the Kappa 400 and 700 devices any A–V interval ending in Ap except for Ap-Ap is not counted. The Ap–Ar interval is not counted in the Kappa 700, only in the situation described in Table 38-6. In the Thera and Kappa 400, the algorithm for detection of tachycardia termination is controlled by the MAI. Resynchronization will occur when the MAI is greater than the upper rate interval (URI). There is a separate, parallel termination method called "5 consecutive atrial paces." If five Ap events occur in a row, with no intervening As or Ar events, AMS ends immediately and the MAI is reinitialized to the lower rate (LR).

Medtronic AT 500 Pacemaker

The AT500 uses the PR Logic algorithm found in the Jewel AF and Gem series of defibrillators to control mode switching operation. The device evaluates two criteria: first, the median of the most recent 12 atrial intervals must be faster than the programmed AT detection zone (the zone the device can be programmed to treat with antitachycardia pacing). If the P-P median is less than this programmed parameter, the device evaluates the "AT/AF Evidence Counter." The Evidence Counter increases when there are two or more intrinsic atrial events between ventricular events and there is no evidence of far-field R-wave sensing. The latter is checked with an additional algorithm. When the counter reaches 3, and the P-P median is less than the programmed AT detection interval, the device mode switches to the DDIR mode. During DDIR mode, the ventricular rate is constantly regulated. Following each intrinsic ventricular event, the escape interval is set to the cycle length of that event + 70 ms. So, if a ventricular-sensed event occurs at 600 ms, the next escape interval

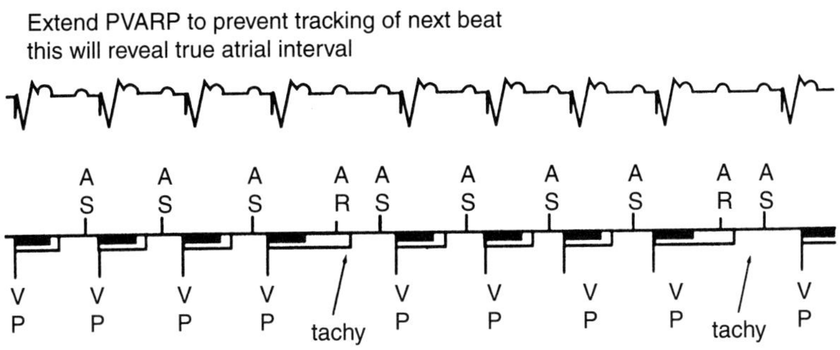

FIGURE 38-19 Automatic mode switching of a Medtronic Kappa 700 pacemaker secondary to a blanked atrial flutter search. Whenever eight atrial intervals are less than twice the total atrial blanking interval (AV * + PVAB), the pacemaker extends the PVARP. This uncovers Ar, which is then followed by As at an Ar–As interval one half of the previous As–As intervals. The device makes the diagnosis of atrial flutter on the basis of the true atrial interval and activates mode switching to the DDIR mode. The end of the PVARP is shown on the marker channel as unlabeled negative tick. Atrial depolarization not marked on the marker channel falls within the PVAB. Ar, Atrial rate; As, atrial sensing; AV, atrioventricular; PVARP, postventricular atrial blanking. (Courtesy of Medtronic, Inc., Minneapolis, MN)

TABLE 38-6 Major Classes of Sensors Used in Rate-Responsive Pacing*

		Examples	
Methods	**Physiologic Parameters**	*MODELS*	*MANUFACTURERS*
Vibration sensing	Body movement	Activitrax, Legend, Thera, DX2, Kappa	Medtronic, Inc.
		Sensolog, Sensorhythm, Trilogy	Pacesetter
		Relay, Dash, Marathon	Intermedics
		Excel	CPI
		Ergos	Biotronik
		Swing	Sorin
Impedance sensing	Respiratory rate	Biorate	Biotec
	Minute ventilation	Chorus RM	ELA Medical
		Legend plus, DX2, Kappa	Medtronic, Inc.
	Stroke volume, preejection period, right ventricular ejection time	Precept	CPI
	Ventricular inotropic parameter	Diplos, Inos	Biotronik
Ventricular evoked response	Evoked Q–T interval	TX, Quintech, Rhythmx	Vitatron
	Evoked R-wave area ("gradient")	Prism CL	Telectronics
Special sensors on pacing electrode	*Physical Parameters*		
	Central venous temperature	Kelvin 500	Cook Pacemakers
		Nova MR	Intermedics
		Thermos	Biotronik
	dP/dt	Deltatrax, Model 2503	Medtronic, Inc.
	Right atrial pressure		
	Pulmonary arterial pressure		
	Peak endocardial acceleration	Best of living	Sorin
	Chemical Parameters		
	PH		
	Mixed venous oxygen saturation	OxyElite	Medtronic Inc.
	Catecholamine levels		Siemens

*Classified according to method of technical realization.

Manufacturers:
- Biotec, S.P.A., Bologna, Italy
- Biotronik, GmbH & Co., Berlin, Germany
- Cardiac Pacemakers Inc., St. Paul, MN
- Cook Pacemakers Corporation, Leechburg, PA
- ELA Medical, Rougemont, France
- Intermedics Inc., Angleton, TX
- Medtronic Inc., Minneapolis, MN
- Siemens Pacesetter Ltd., Solna, Sweden
- Sorin Biomedica, Saluggia, Italy
- Telectronics Pacing Systems, Englewood, CO
- Vitatron Medical B.V., Dieren, The Netherlands

will be 670 ms, then 740 ms, then 810 ms, until the sensor rate or LR is reached.

Clinical Results

In a study on 26 patients with the Thera DR (Model 7940 and 7960 i),[67] appropriate AMS occurred as confirmed by stored atrial electrogram (five patients) and 12-lead ECG (nine patients). However, a more recent report[68] showed that false mode switching occurred in 4 of 30 Holter recordings with documented sinus rhythm due to far-field R-wave sensing. Of the 125 episodes of AT, 93 (74%) were appropriately detected, whereas 32 (36%) short-lasting episodes were not. The sensitivity and specificity of the Thera AMS algorithm were improved in the new algorithm in the Kappa 700 series, with an increase in both the speed of onset of AMS and the speed of resynchronization to sinus rhythm (see later).

Pacesetter/St. Jude AMS Algorithms

The Trilogy and Affinity families used the "running average" principle to define the AT detection rate.[69-71] This averaging rate (the "filtered atrial rate") interval shortens by 38 ms (Trilogy DR+) or 39 ms (Affinity DR) whenever a sensed atrial cycle is shorter than the preceding atrial cycle. When the sensed atrial cycle is longer than the preceding cycle, the filtered atrial rate interval will lengthen by 25 ms for Trilogy DR+ and 16 ms for Affinity DR. When the filtered atrial rate exceeds the AT detection rate, AMS occurs and the device functions in the DDI(R) mode. When AT terminates, the filtered atrial rate interval lengthens by a nonprogrammable interval (25 and 16 ms for the Trilogy DR+ and Affinity DR, respectively), and atrial tracking occurs when the filtered atrial rate interval is longer than the atrial driven URI or sensor-controlled URI, whichever is shorter. Thus, the system is biased toward more rapid AT detection, but slower in resynchronizing to sinus rhythm. In the Trilogy pacemaker the microprocessor begins to sense only in the unblanked part of the PVARP when there is a rapid sensed atrial rate that initiates Wenckebach's UR behavior. This can be a single cycle with an early P wave resulting in prolongation of the AV interval initiated by atrial sensing to conform to UR behavior. At that point the microprocessor *wakes up* and begins to look in the unblanked PVARP for additional P waves. If the rhythm jumps abruptly from 1:1 to 2:1 atrial tracking and never traverses through even a brief period of Wenckebach behavior, the microprocessor remains quiescent with respect to the AMS algorithm. In the Affinity device, the unblanked PVARP is always receptive to promote high atrial rate detection. PVAB can be programmed in all devices with AMS from a nominal of 100 ms. This allows the selection of the shortest atrial blanking period to maximize AT detection and to limit VA crosstalk.

The Frontier biventricular pacemaker presently under clinical investigation was designed with a special timing cycle to prevent VA crosstalk within the A–V interval. This interval is called the *preventricular atrial blanking period,* though it is a timing cycle and not really a true blanking period. The design is aimed at canceling the counting of far-field R wave detected by the atrial channel. The preventricular blanking is initiated whenever a P-wave or sensed atrial signal (such as a far-field R wave) is detected either inside or outside the unblanked refractory period of the atrial channel. If a ventricular depolarization is detected by the ventricular channel within the preventricular atrial blanking period, the P-wave or atrial signal that initiated the preventricular atrial blanking will be invalidated for counting purposes. In other words, it will not trigger an atrial output pulse and it will not be included in the calculation of the filtered atrial rate. The preventricular atrial blanking is programmable between 0 and 62 ms, and is cancelled as soon as Vs is sensed by the ventricular channel as a near-field event.

Intermedics Automatic Mode Switching Algorithms

The Intermedics devices offer two methods of handling AT, a true AMS that switches mode when consecutive atrial cycles fall below a programmable AT detection interval, and an upper rate switching mechanism based on sensor activity.

Automatic Mode Switching Function

The AMS function (Fig. 38-20) is available in DDD(R), DDDR + Hysteresis, and VDD(R) modes. ATs are detected based on a "rate and run" criterion (see Table 38-3). The pacemaker monitors intrinsic atrial events that continue consecutively for the programmed number of cycles at or above a specified AT detection rate (AMS count). AMS to the VVIR mode will occur when the sensed atrial rate exceeds this AT detection rate and counts.[72] The AT detection rate is programmable from 90 to 300 beats/minute at 10-beat intervals and the AMS count from 1 to 7 counts in steps of 1. The PVAB measures 100 ms and an atrial minirefractory period, which is in reality a blanking period after As is set automatically to 70 ms or one half of the AT detection interval (whichever is less) after the sensed events, to prevent the device from sensing an atrial event twice. When the pacemaker has detected one sensed atrial cycle longer than the AT detection interval, or when a paced atrial event has occurred, it will return to the original dual-chamber mode.

Automatic Rate Switching

The SmarTracking algorithm uses an automatic rate switching function (Fig. 38-21) that limits the ventricular rate during AT.[72] This function is an improved version of the "conditional ventricular tracking limit" introduced earlier by the same manufacturer.[73] The SmarTracking rate is programmable at variable rates above the sensor-indicated rate, and thus adjusts according to physical activity. This rate defines not only the rate of an AT above which rate switching occurs, but also represents the upper ventricular tracking rate when an AT occurs. Automatic rate switching involves

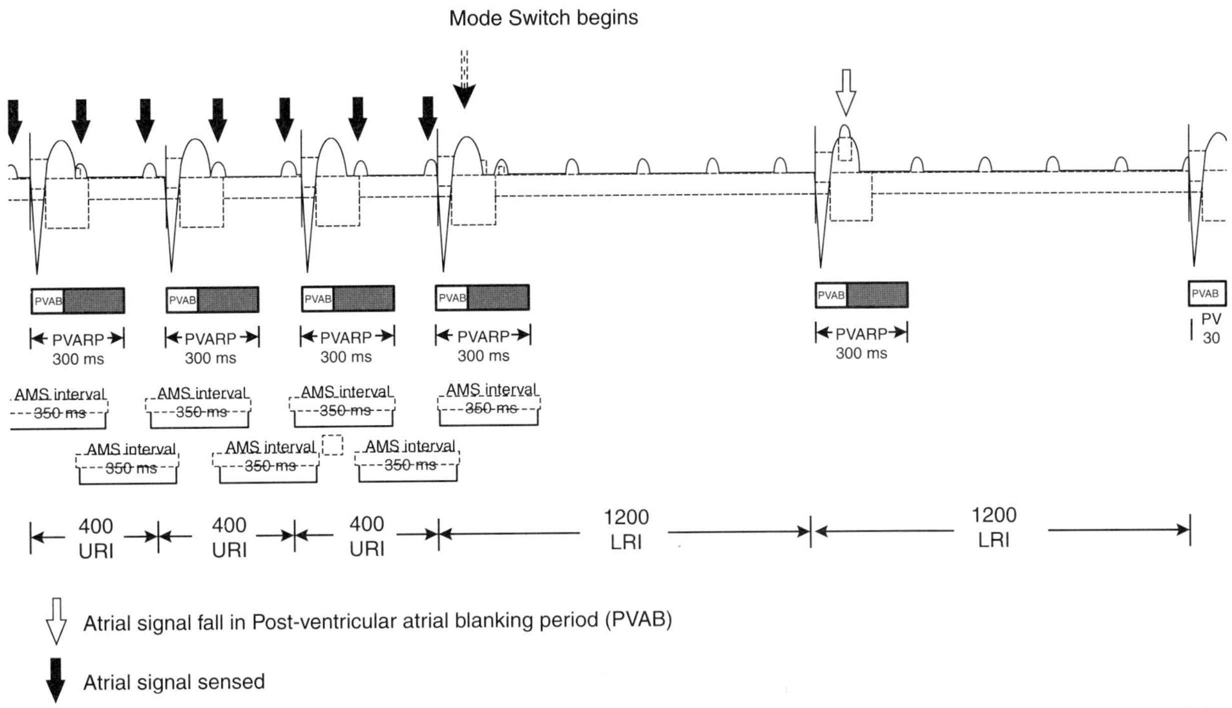

FIGURE 38-20 The AMS algorithm in Marathon DDDR pacemaker. The AMS rate is programmable from 90 to 300 beats/minute and the AMS counts from one to seven counts. In this example, AT at a cycle length of 300 ms (*second arrow from the left*) starts after a normally sensed beat. The cycle length is shorter than the programmed AT detection rate interval of 300 ms, AMS occurs and the pacemaker switches to the lower rate limit (LRL) after a total of seven atrial events with a short cycle. A short period of ventricular tracking at an upper rate limit (URL) occurred before AMS became stabilized. AMS, Automatic mode switching; AT, atrial tachycardia. (Reproduced with permission from Leung SK, Lau CP, Lam CT, et al: Is automatic mode switching effective for atrial arrhythmias occurring at different rates? A study on efficacy of automatic mode and rate switching to simulated atrial arrhythmias by chest wall stimulation. Pacing Clin Electrophysiol 2000;23:823-831.)

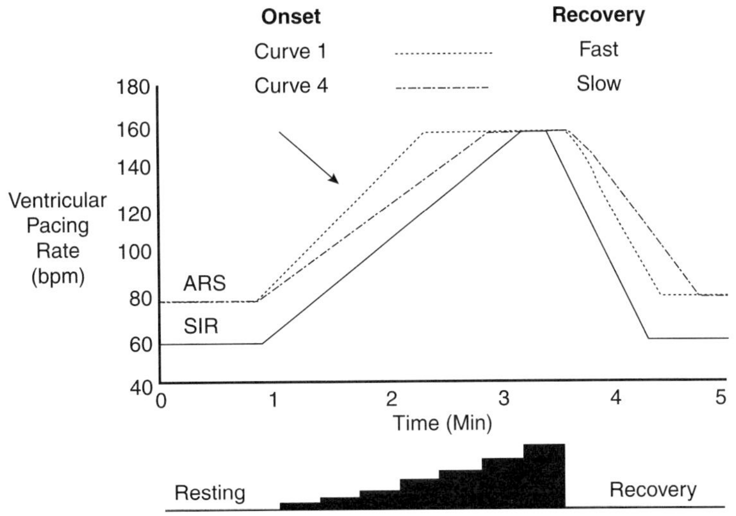

FIGURE 38-21 Schematic representation of the dynamic automatic rate switching algorithm (ARS) in the marathon DDDR. At rest, the ARS rate must be at least 20 beats/minute above the lower rate limit, and an atrial event that exceeds the ARS rate will be tracked only at this rate. During exercise, the ARS rate will be increased according to the programmable ARS response curve (1–4, 1 = most conservative, 4 = most aggressive), which is always kept above and related to the sensor-indicated rate (SIR). When the patient reaches maximal exercise, the ARS rate and the SIR will be equal to the programmed upper rate (UR) of the device (150 beats/minute in this example). Atrial rate occurring above the ARS rate at any point will be considered unphysiologic and will be tracked along at that rate. At the end of exercise, the ARS recovery rate becomes operative and stays above the sensor-indicated rate depending on the programmable ARS. (Reproduced with permission from Leung SK, Lau CP, Lam CT, et al: Is automatic mode switching effective for atrial arrhythmias occurring at different rates? A study on efficacy of automatic mode and rate switching to simulated atrial arrhythmias by chest wall stimulation. Pacing Clin Electrophysiol 2000;23:823-831.)

a change in the UR without a mode change, and is, therefore, a form of fallback rather than AMS.

SmarTracking prevents tracking of unphysiologic rapid atrial rates unless the sensor detects that the patient is active. This feature complements the associated AMS algorithm, and may cause less variation and oscillation of the ventricular rate due to intermittent atrial undersensing during AF.[72]

Clinical Results

The clinical results of the conditional ventricular tracking limit of the automatic rate switch algorithm have been reported.[73] The algorithm responded rapidly to the onset of AT. However, during sinus tachycardia or after exercise when the sensor is not active, Wenckebach's UR behavior will occur as the algorithm operates when the sensor is inactive. This is minimized by the use of the SmarTracking algorithm in which the sensor adjusts the AT detection rate, which is set at a programmed value above the sensor-indicated rate. As the ventricular rate after switching (fall-back rate) is necessarily above the LR, there is concern that the patient may have high rate pacing even after appropriate rate switching. For this reason, rate switching is now used to supplement the AMS function in the Marathon.[72]

ELA MEDICAL ALGORITHMS

AMS function in the Chorum and Talent DR pacemakers combines an initial UR switch at the onset of AT, followed by AMS based on the detection of a sustained atrial rate above a preset AT detection rate.[74] The AT detection interval is termed the *diagnosis of atrial rhythm acceleration period.* This varies from 62.5% (for sinus rate ≤ 80 beats/minute) to 75% of the preceding P–P interval (for sinus rate > 80 beats/minute). At the onset of AT, only events outside the AT detection rate will be sensed and ventricular pacing will be triggered. This, in effect, switches the URI to the AT detection interval (termed *temporary mode switch* in the device). When 28/32 or 36/64 consecutive beats above AT detection rate are detected, AMS will be initiated (termed *permanent mode switch* in the device). A further refinement during AMS allows the pacer to function in the DDI(R) mode for spontaneous ventricular rate (≤100 beats/minute), and VVI(R) mode when this rate is greater than 100 beats/minute to avoid atrial competitive pacing.

Resynchronization to sinus rhythm occurs if 24 consecutive atrial cycles are below 110 beats/minute. This counter will be reset if premature beats are sensed within this confirmation period until the 24 atrial/ventricular cycles below 110 beats/minute are satisfied. The sensitivity and specificity of the AMS function of the Chorum and Talent were recently reported in 194 patients.[75] All patients underwent 24 hour ECG recording with simultaneous activation of the event recording counters. Based on the documented AT on the Holter monitors in 23 patients, AMS exhibited a sensitivity of 96%. In the remaining patients without AT on Holter monitoring, AMS occurred in six patients: VA crosstalk due to sensing of far-field ventricular signal occurred in five patients, and sinus tachycardia in one. This gives a specificity of 96%. Adjustments of the atrial sensitivity or the PVARP were reported to be effective in controlling these false-positive AMS.

Guidant CPI Algorithm

AMS function in the Pulsar Max, Vigor, Meridian, and Discovery DR pacemakers employs a rate cut-off and counter-based algorithm. Atrial events above the AT-detection rate (UR) increments the detection counter, whereas events below the AT detection rate decrements the counter. AT is detected when the counter reaches 8. During this period, atrial tracking results in Wenckebach's or 2:1 block. Once AT is confirmed, pacing is allowed at the UR for a programmable duration from 0 to 2048 ventricular cycles, during which the atrial rate is monitored before AMS is implemented. This is designed to avoid AMS from nonsustained AT. Following this, "fall-back" occurs over a programmable time between 1 to 5 minutes (nominal 30 seconds), from the AV Wenckebach rate to the LR in the DDD mode or sensor-indicated rate in VVI(R) or DDI(R) mode. When AT terminates, the AT detection counter will decrement with each atrial event below the AT detection rate. Resynchronization to sinus rhythm occurs when the counter drops from 8 to 0. This device is relatively slow to activate AMS following the onset of AT, but allows rate smoothing to be effected to minimize the beat-to-beat change in pacing rate prior to AMS, as well as on AMS termination before resynchronization to sinus tracking.

In the Pulsar Max pacemaker system, there is an optional second tier AT management (atrial flutter response) on top of the previously mentioned AT response. This employs an instantaneous fall-back for rates above 230 beats/minute through the use of an atrial flutter window. As soon as an atrial event is detected inside the PVARP, the device will start an atrial flutter response window of 260 ms. If a second event is seen inside the atrial flutter window, the event will be classified as high-rate AT and will trigger another atrial flutter response window. Atrial pacing and atrial tracking are not allowed when the atrial flutter window is in operation, which is further extended as long as the sensed atrial rate is above 230 beats/minute. This causes the pacemaker to function effectively in the VVI(R) mode. This *retriggerable* flutter window (effectively an extended atrial refractory period) will terminate if one atrial flutter window times out without any atrial-sensed event. Apart from AMS algorithm, some CPI devices (e.g., Discovery II) also retain a rate stabilization algorithm that limits the beat-to-beat change in ventricular cycle length. Based on the prevailing ventricle cycle length, an early atrial event will be tracked only at a cycle length, which is a percentage (commonly 12%) of the current cycle length. The same is operative when the atrial rate suddenly decreases, such as at the termination of an AT, to prevent sudden rate drop. This should be differentiated from the AMS algorithm because rate stabilization is operative at the baseline tracking (or the sensor-indicated) rate, and does not require the atrial rate to exceed the UR.

Vitatron Algorithm

The Vitatron Diamond and Clarity DDDR pacemakers have a different AMS algorithm that employs a physiologic band to define normal versus pathologic atrial rate.[76-77] A physiologic atrial rate is calculated from the running average of the actual sensed or paced atrial beats and the rate change in this moving average is limited to 2 beats/minute (Fig. 38-22). The physiologic band is defined by an upper boundary equal to the physiologic rate plus 15 beats/minute (minimum value of 100 beats/minute) and the lower boundary by the physiologic rate minus 15 beats/minute (or the sensor-indicated rate if it is higher). As the physiologic rate during sensor driven pacing will be determined by the sensor, it follows that AT detection is sensor-based when the sensor is active. If an atrial event occurs above the upper boundary of the physiologic band, AMS to DDIR mode will immediately occur. The ventricular escape rate is the sensor-indicated rate or lower boundary of the physiologic band, whichever is higher. The pacemaker may also provide an additional atrial pacing stimulus following a premature atrial contraction to prevent irregular ventricular rate. The mode switch feature can be programmed as either "auto" or "fixed." If the automatic mode is programmed, the AT detection rate varies with the upper boundary of the physiologic band and allows AMS to occur below the UR. When programmed to the fixed mode, the AT detection rate is equal to the UR, and the mode switch is allowed only when the atrial rate exceeds the UR. The Diamond pacemaker also has a fallback mechanism to minimize changes in ventricular cycle length when AMS occurs. Resynchronization to sinus rhythm occurs when the atrial rate falls within the physiologic band for two consecutive beats.

Clinical Results

The improved sensitivity of the auto AMS function compared with the fixed AMS function of the Diamond pacemaker was recently reported in 12 patients.[77] All patients had drug refractory paroxysmal AF who underwent radiofrequency ablation of the AV mode and received a Diamond DDDR pacemaker. During 4 minutes of spontaneous or induced AT, both at rest and during exercise, auto AMS resulted in significantly slower atrial tracking ventricular rate compared with the fixed AMS, both at rest (4 + 4% versus 35 + 14%) and during exercise (2 + 3% versus 24 + 11%). During a follow-up of 5 months, there were no symptoms reported among the eight patients who had spontaneous episodes of AF when the auto AMS was activated. In another study,[78] 26 patients with paroxysmal AF or atrial flutter with AV nodal ablation or high-grade AV block were observed for a period of 3 to 12 months. All patients underwent 24-hour postimplant continuous ECG monitoring, and received Holter monitoring and exercise testing after discharge. Eight of 26 patients developed AF, and all exhibited reliable sensing of AF and did not show fast AF tracking at the onset or during AF or atrial flutter. None of them complained of palpitations. In two other patients, programming to DDIR mode was required due to intermittent tracking of persistent AT, which resulted in asymptomatic, irregular ventricular-paced rhythms below 100 beats/minute.

Biotronik Algorithms

The Biotronik Actros and Kairos family DDDR pacemakers use a *retriggerable* atrial refractory period algorithm to provide protection against AT.[79] This algorithm is simple and responds rapidly to the onset and termination of AT. If the pacemaker detects a P wave in the TARP, an A–V interval is not initiated and the atrial

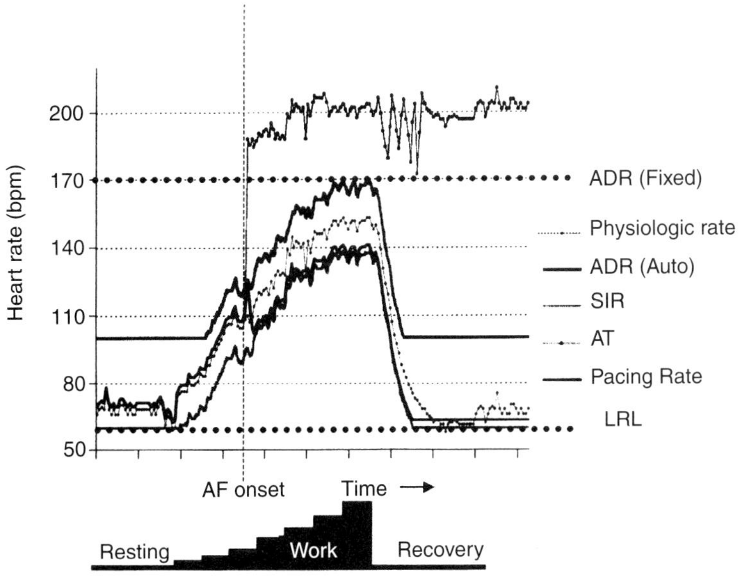

FIGURE 38-22 AMS algorithm for the Diamond pacemaker (Vitatron Medical B.V.). The pacemaker continuously calculates a moving average of the sensed or paced atrial rate—the physiologic rate. The AMS algorithm can be programmed to the Auto or Fixed mode. In the Auto mode, the AT detection "rate (ADR) is the upper boundary of the physiologic band (physiologic rate + 15 beats/minute). In the fixed AMS mode, the ADR is the UR. Mode switch to DDIR mode at the sensor-indicated rate occurs when atrial rate exceeds the ADR. (UR = 170 beats/minute, LR = 60 beats/minute). AMS, Automatic mode switching; SIR, sensor-indicated rate; UR, upper rate.

refractory period is extended by an amount equal to the TARP. If further atrial events are sensed within the new TARP (outside the atrial blanking period), the TARP is further extended, causing the pacemaker to function in the DVI(R) mode with ventricular-based lower rate timing. Instantaneous resynchronization occurs when an atrial event occurs outside the TARP, or when the LR has been reached and an atrial paced event is initiated. Although this algorithm functions effectively in a different mode (DVIR), it provides a sensitive and fast-reacting response to onset and termination of AT. However, it has a low specificity and may result in frequent switching to DVIR pacing during noise and atrial ectopy. In addition, competitive (asynchronous) atrial pacing occurs during tachycardia, and it may paradoxically induce AF should AF terminate spontaneously.

In the Biotronik Inos and Logos pacemakers, a statistical "x out of y" criterion is used for detection of AT. AT is detected if a predefined number of sensed atrial intervals (five for Inos and programmable between two to seven in Logos) out of eight sensed atrial intervals are shorter than the AT-detection interval. During AMS, Inos devices switch to the VVI mode and Logos devices to DDI mode, and the ventricular rate decreases beat by beat gradually to the LR at 2 beats/minute. Resynchronization to sinus rhythm occurs when eight consecutive sensed atrial events are below the AT detection rate or are paced. This "x out of y" criterion provides a higher specificity compared with the retriggerable refractory period.

Sorin Algorithm

The Sorin Living 1 and Living 1 Plus DDDR pacemakers measure the "peak endocardial acceleration" with an implantable acceleration sensor at the tip of the right ventricular pacing lead.[80-81] In addition, they also have a gravitational type of activity sensor. AT is defined as atrial events at a rate exceeding the UR. In the DDD mode, on detection of AT, the pacemaker will mode switch to the VVI mode at the UR for 1000 cycles before it falls back gradually to a nonprogrammable ventricular pacing rate of 77 bpm.

In the DDDR mode using the activity sensor, the device will temporarily mode switch to the VVI mode at the UR for 250 cycles. Should the sinus rate fall below UR during this period, AMS will not occur. This is designed to avoid AMS during nonexercise-related sinus tachycardia. If the atrial rate remains above UR for more than 250 cycles, the sensor-indicated interval is then checked. If there is a discrepancy between it and the sensed atrial rate, the pacing rate will fall back to the sensor-indicated rate.

When the peak endocardial acceleration sensor DDDR mode is activated, the sensed atrial rate above the UR will be compared with the sensor-indicated rate. When the atrial rate is above the sensor-indicated rate, the pacemaker will immediately mode switch to VVIR mode and the pacing rate will fall back to the sensor-indicated rate. In both activity and peak endocardial acceleration DDDR modes, atrial tracking resumes when three sensed atrial cycles are at rates lower than the UR.

Comparative Evaluation of Automatic Mode Switching

Implementation of AMS differs widely among devices depending on their methods of AT detection. It is uncertain if all AMS algorithms can provide equal protection during AT. The clinical benefit of AMS may be limited by the sensitivity and specificity of AMS algorithms. The episodic nature of AT hampers comparative studies of different algorithms in a clinical setting.

A more scientific method involves recording the atrial electrogram during sinus rhythm and AT and *replaying* it into different in-vitro pacemakers to compare and evaluate various AMS algorithms. We have studied 260 recordings of atrial electrograms during AT, AF, and sinus rhythm in patients during electrophysiological studies, and replayed the data into the atrial port of three dual-chamber pacemakers with different AMS algorithms (Thera DR, Marathon DDDR, and Meta DDDR) (Fig. 38-23).[82] The rate cut-off algorithm was significantly faster to achieve AMS (latency 2.5 seconds for Marathon DR, 15 seconds for Meta DDDR, and 26 seconds for Thera DR, $P < .0001$). When AF terminated, the Thera DR resynchronized after 143 seconds versus 3.4 seconds for Marathon DR, and 5.9 seconds for Meta DDDR, resulting in long periods of AV dissociation for the Thera DR. However, during sustained AF, the slower algorithm resulted in the most stable and regular ventricular rhythm, whereas the fast responding ones showed intermittent oscillation during AMS and DDDR mode (cycle length variations: 44 ± 2 seconds for Thera DR versus 346 ± 109 seconds for Marathon DR versus 672 ± 84 seconds for Meta DDDR, $P < .05$). Hence, the rapid-responding algorithms exhibit rate instability after AMS, whereas slow-responding algorithms activate only AMS after a relatively long delay. After AF or AT termination, the slow-responding algorithms are also slow to exit from AMS and carry the risk of AV dissociation for the time it takes to restore resynchronization after AT termination (Fig. 38-24).

A comparative evaluation of AMS behavior of different Medtronic algorithms *replaying* 52 episodes of recorded atrial rhythms has been reported.[83] Models tested included Thera DR and Gem DR (model 7271), which used a "running average" algorithm, and Kappa 700 and Jewel AF (Model 7250), which used the rate cut-off algorithm (4 out of 7). The sensitivity of mode switching is higher for the rate cut-off than the mean running average algorithm (94% and 85% for Kappa 700 and Thera, respectively). This is associated with a prolonged time to the onset of AMS and resynchronization with sinus rhythm. The rapidity to AMS can be shortened in the running atrial rate algorithm if there is prevailing sinus tachycardia.

As AT may occur at rates above and below the tachycardia detection rate due to undersensing or the rate-slowing effect of antiarrhythmic medications, a varying sensor-controlled tachycardia detection rate may have

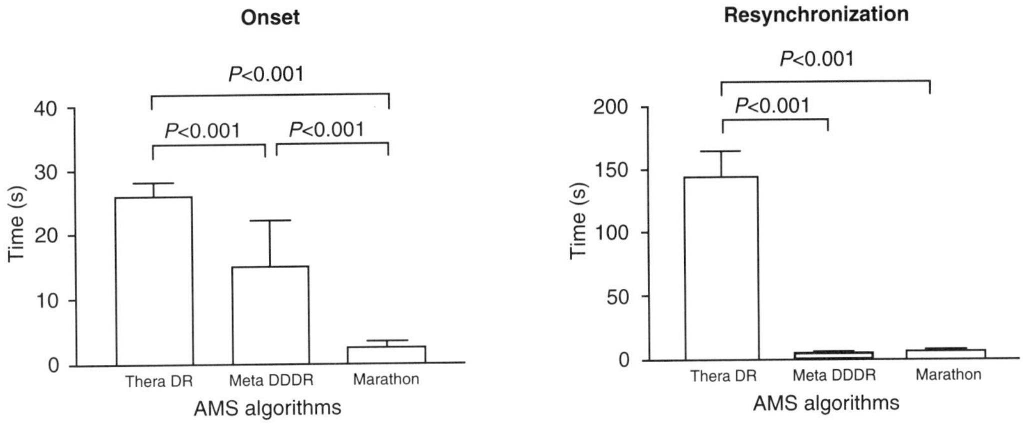

FIGURE 38-23 Comparison of onset latency of AMS response to AF and time for resynchronization to sinus rhythm for the three different AMS algorithms in Thera DR, Meta DDDR, and Marathon DR. An almost immediate response was observed in Marathon DR using a simple rate cut off—"rate and run criteria," and a latency of nearly 30 seconds in Thera DR with the running average algorithm. The AMS response in Meta DDDR, which also uses the rate cut-off criteria, is slower than the Marathon, as its rate counter can be decremented if the sensed atrial late falls below the AT. Similarly, at the termination of AF, Thera DR showed a long delay before resynchronization occurred, whereas the Meta DDDR and Marathon showed an immediate resynchronization at the return of sinus rhythm. AMS, Automatic mode switching; AT, atrial tachycardia. (Reproduced with permission from Leung SK, Lau CP, Lam CT, et al: A comparative study on the behavior of three different automatic mode switching dual chamber pacemakers to intracardiac recordings of clinical atrial fibrillation. Pacing Clin Electrophysiol 2000;23:2086-2096.)

benefit by allowing a lower tachycardia detection rate to prevail at rest and a faster one with exercise to enhance tracking of sinus tachycardia. Another way to allow for slower AT is to add another protective algorithm for the slower rate, such as a rate switching algorithm, as in the SmarTracking algorithm of the Marathon DR and in the temporary mode switch of the Chorum and Talent DR. Using external chest wall stimulation to simulate AT, we tested the efficacy of combined AMS and automatic rate switching algorithms in 33 patients with a Marathon DDDR pacemaker (Intermedics, Inc.).[72] During the study, the pacemaker was programmed

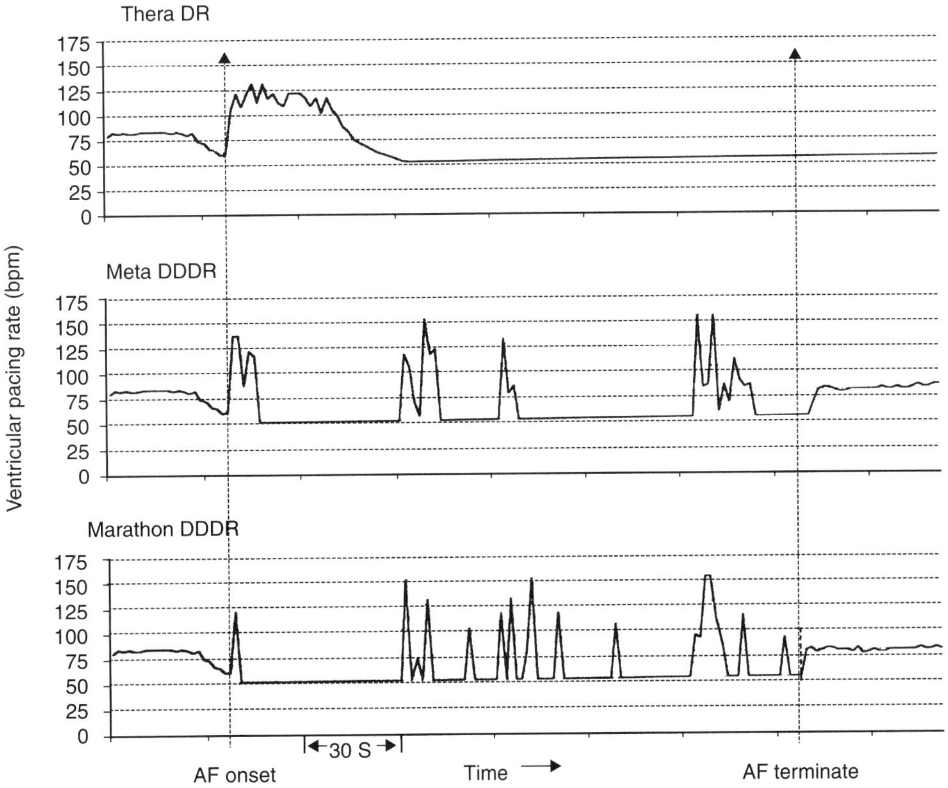

FIGURE 38-24 A typical example of ventricular pacing rate profiles for the three different AMS algorithms in Figure 38-23 in response to sinus rhythm and AF recorded from one patient. The running average algorithm of the Thera DR is low sensitivity but more specific. It has a long latency but is less susceptible to mode switch oscillation. Sensitive algorithms (Meta and Marathon) using the rate cut-off with a rapid AMS onset and termination has low specificity and AMS oscillation. The *arrows* indicate the onset and termination of AF. AF, Atrial fibrillation; AMS, automatic mode switching. (Reproduced with permission from Leung SK, Lau CP, Lam CT, et al: A comparative study on the behavior of three different automatic mode switching dual chamber pacemakers to intracardiac recordings of clinical atrial fibrillation. Pacing Clin Electrophysiol 2000;23:2086-2096.)

randomly to DDDR mode with AMS alone, automatic rate switching alone and their combination. The study showed the addition of automatic rate switching to the AMS function improves ventricular rate control to simulated AT at a rate slower than the AT detection rate for mode switch. Automatic rate switching can also provide back-up rate control during intermittent failure of the AMS response.[72]

FOLLOW-UP AND TROUBLESHOOTING

A detailed evaluation of atrial arrhythmias both preimplantation and postimplantation (based on pacemaker diagnostics and ambulatory recordings) is essential to select the type of AMS algorithms and the programmed settings. It is essential during implantation to maximize the amplitude of the atrial electrogram in sinus rhythm and minimize the far-field R-wave signals by appropriate atrial lead placement. A bipolar atrial lead should be implanted in a patient expected to use AMS frequently. It may also be preferable to use a closely-spaced atrial bipole to minimize far-field R wave sensing.

Because of the highly variable AF amplitudes, a routine setting of nominal or 2 times the atrial sensitivity margin may not be adequate for optimal AT or AF detection. If an episode of AF becomes available, sensitivity adjustment can be guided by the measured amplitude of the atrial electrogram. Leung and associates[35] have suggested that the optimal atrial sensitivity setting for AMS appears to be at three times versus two times the detected amplitude of the atrial electrogram in sinus rhythm.

Incorrect programming of these parameters (or suboptimal pacemaker design) may create a number of problems. These include loss of specificity, such as sensing sinus tachycardia on exercise as pathologic tachycardia, withholding AMS because of inappropriate selection of atrial sensitivity and/or atrial blanking periods and related timing cycles.

AMS failure due to atrial events occurring within the atrial blanking periods was reported in a series of seven patients with Meta DDDR (Model 1254).[60] All patients experienced palpitations. In six of seven patients, AMS failure was due to atrial flutter, with alternate flutter waves falling within the atrial blanking periods (values: 120 ms after atrial sensed/paced event and 150 ms for PVAB). In these patients, relatively slow atrial flutter was related to drugs used to control AF. By either shortening the A–V interval, or prolonging the A–V interval and PVARP so that flutter waves could be sensed outside the atrial blanking interval, normal AMS function was restored. In one of six patients, the cause of AMS failure was due to a low atrial electrogram amplitude during AF, and it was corrected by reprogramming atrial sensitivity.

Palma and colleagues[84] systematically evaluated the effect of varying atrial sensitivity, A–V interval, and detection criteria for AMS in 18 patients. Pacemaker models studied included Meta DDDR (Model 1254, Medtronic, Inc.), Thera DR (Model 7940, Medtronic, Inc.) and Relay (Model 293-03 Intermedics, Inc.). Although an atrial sensitivity of 1 mV allowed correct sensing of sinus rhythm in 13 of 14 patients (93%), only 43% of them did so during AF. A–V intervals between 120 and 200 ms were found to be most effective to detect AF, whereas AF sensing was reduced when a longer A–V interval was used. The use of stringent criteria for AF detection might interfere with AMS onset. This study points out the importance to adjust the conventional pacing parameters to optimize AMS function.

Overall experience with AMS has been satisfactory so far. Barring economic considerations, all pacemakers should possess an AMS function as one of its programmable options. AMS allows the benefits of AV synchrony to be extended to a population with existing or threatened AF. These devices are indicated in all patients with the brady-tachy syndrome. They should also be considered in patients with sinus node disease without paroxysmal AF, obstructive hypertrophic cardiomyopathy, or any condition that predisposes patients to paroxysmal AF. Indeed, like the rate-adaptive function (R), it could be argued that all patients should receive a device with AMS capability because one cannot predict which patients will eventually develop AF (or atrial chronotropic incompetence in the case of the rate-adaptive function). Many algorithms have been used by different manufacturers, and they do not behave similarly. Optimal care of the pacemaker patient requires a thorough knowledge of his or her arrhythmia history, atrial electrogram amplitude (in sinus rhythm and AT), basic timing cycles required for all AMS algorithms, and the characteristics of the various available algorithms.

Implantable Sensors

In the presence of abnormal cardiac automaticity and conduction, physiologic pacing aims to maintain the heart rate and restore the sequence of cardiac activation. It is assumed that sensors should mimic the behavior of the healthy sinus node response to exercise and nonexercise needs. The atrial electrogram can be used for rate control when sinoatrial function is adequate. However, a high proportion of pacemaker recipients have either established or progressive abnormal sinoatrial function occurring either at rest or during exercise. Such chronotropic incompetence commonly occurs due to medications or ischemia, or in association with sick sinus syndrome. In addition, the atrium may be unreliable for sensing or pacing in some patients (such as during atrial fibrillation or paroxysmal atrial tachycardias); hence, the use of sensors as an alternative means to simulate sinus node responsiveness is required. These problems prompted the development of nonatrial implantable sensors for cardiac pacing. In addition, the role of sensors has also been expanded to include functions other than rate augmentation, such as the monitoring of cardiac hemodynamics during heart failure and ventricular arrhythmias. Although a number of implantable sensors to detect exercise have been proposed or instrumented in pacing devices, there is, to date, no single sensor that can simulate the ideal sinus rhythm behavior.

SENSOR CLASSIFICATION

Sensors derived from a common technical principle share similar hardware requirement and sensor stability, and similar drawbacks and limitations.[85] Thus, a practical classification is to classify sensors according to the technical methods that are used to measure the sensed parameter (see Table 38-6). During isotonic exercise, body movements (especially those produced by heel strike during walking) result in changes in acceleration forces that act on the pacemaker. Sensors that are capable of measuring the acceleration or vibration forces in the pulse generator are broadly referred to as activity sensors. The sensing of *body vibrations* is, therefore, a simple way to indicate the onset of exercise. Technically, detection of body movement can be achieved using a piezoelectric crystal, an accelerometer, a tilt switch, or an inductive sensor. Each of these devices transduces motion of the sensor either directly into voltage or indirectly into measurable changes in the electrical resistance of the crystal. Although a tertiary sensor, activity is the most widely used control parameter in rate-adaptive pacing because of its ease of implementation and its compatibility with standard unipolar and bipolar pacing leads. Minimal or no energy expenditure is required for such a system, and because the sensor does not need to have contact with body fluid, it is usually stable over time. Other advantages are that all the hardware is within the pacemaker case, it does not depend on the electrode arrangement, and can be used with standard unipolar and bipolar pacing electrodes. This makes it ideal for combining with other sensors and for pacemaker upgrading. It is used often as a back-up sensor when a new sensor is being investigated. However, as a group, because body movement has only a loose relationship to workload, the sensor-indicated rate has low proportionality, and physical activities that do not involve body movement will not be detected by these sensors.

Impedance is a measure of all factors that oppose the flow of electric current and is derived by measuring resistance to an injected subthreshold electric current across a tissue. The impedance principle has been used extensively for measuring respiratory parameters[86] and relative stroke volume[87] in situations involving invasive monitoring. The elegant simplicity of impedance has enabled it to be used with implantable pacing leads, including both standard pacing leads and specialized multielectrode catheters. The pulse generator casing has been used as one electrode for the measurement of impedance in most of these pacing systems. Impedance can be used to detect relative changes in either ventilatory mechanics or right ventricular mechanical function or their combination. Relative motions between electrodes for impedance sensing also lead to changes in impedance measured, and this is inversely related to the number of electrodes used to measure impedance. In rate-adaptive pacemakers, motion artifacts are usually the result of arm movements that cause the pulse generator to move within the prepectoral pocket,[88] thereby changing the relative electrode separation between the pacemaker and the intracardiac electrodes. Because arm movement accompanies normal walking, these artifacts in the impedance signal occur with both walking and upper limb exercises. Similarly, exogenous electrical interference, such as diathermy, may lead to erroneous sensing.

The *intracardiac ventricular electrogram* resulting from a suprathreshold pacing stimulus has been used to provide several parameters that can guide rate modulation. The area under the curve inscribed by the depolarization phase of the paced ventricular electrogram (the intracardiac R wave) has been termed the *ventricular depolarization gradient* or *paced depolarization integral* (PDI).[89] In addition to depolarization, the total duration of depolarization and repolarization can be estimated by the interval from the pacing stimulus to the intracardiac T wave (the Q–T or stimulus–T interval). Both of these parameters are sensitive to changes in heart rate and circulating catecholamines and can be derived from the paced intracardiac electrogram with conventional pacing electrodes. Because a large polarization effect occurs after a pacing stimulus, a modified waveform of the output pulse that compensates for afterpotentials is needed to eliminate this effect so that these parameters can be accurately measured.

The last group of sensors are those that are incorporated into the pacing lead. These dedicated sensors allow chemical compositions of the blood stream or intracardiac hemodynamics to be measured, and may result in a more *physiologic* sensor system. However, the long term stability of these sensors is questionable. They are also energy expensive. Examples of these specialized leads include thermistors (used to measure blood temperature), piezoelectric crystals (used to measure right ventricular pressure), optical sensors (used to measure SvO_2), and accelerometers at the tip of pacing leads. Some of these sensors measure highly physiologic parameters. These include the measurement of pH, catecholamines, oxygen (SvO_2) and right ventricular pressure.[90-91]

These sensors have only a limited application in the field of rate-adaptive pacing because their long-term stability in the blood environment is questionable, but they may open the possibility for ambulatory monitoring of intracardiac environment using an implanted device; some important developments have occurred in this field. Included in this group are sensors that detect right ventricular dP/dt as an index of contractility,[92-94] and an accelerator sensor at the tip of a pacing lead to monitor the peak endocardial acceleration, which has the possibility for optimizing the AV interval.[80,81,95-98]

CHARACTERISTICS OF AN IDEAL RATE-ADAPTIVE PACING SYSTEM

The normal human sinus node increases the rate of its spontaneous depolarization during exercise in a manner that is linearly related to VO_2. Because this response undoubtedly has evolutionary advantages, the goal of rate-adaptive pacemakers that modulate pacing rate by artificial sensors has been to simulate the chronotropic characteristics of the sinus node (Table 38-7). It should be plainly stated, however, that it is uncertain whether the

TABLE 38-7 Characteristics of an Ideal Sensor for Rate-Responsive Pacing

Considerations	Examples and Remarks
Sensor Considerations	
Proportionality	Oxygen saturation sensing has good proportionality
Speed of response	Activity sensing has the best speed of response
Sensitivity	QT sensing can detect nonexercise-related changes such as anxiety reaction
Specificity	Activity sensing is affected by environmental vibration
	Respiratory sensing is affected by voluntary hyperventilation
Technical Considerations	
Stability	Stability of early pH sensor was a problem
Size	Large size or requirement for additional electrodes may be a problem
	Energy consumption must not unduly harm pacemaker longevity
Biocompatibility	Important for sensor in direct contact with the blood stream
Ease of programming	Difficult programming in early QT sensing pacemakers

sinus node provides the ideal rate response in patients who require permanent pacemakers. Nevertheless, until there is evidence indicating otherwise, rate-adaptive pacemakers will strive to reproduce this physiologic standard. Keeping these uncertainties in mind, the ideal rate-adaptive pacing system should provide pacing rates that are *proportional* to the level of metabolic demand. One of the best indicators of sensor proportionality is the correlation between the sensor-indicated pacing rate and the level of oxygen consumption during exercise (Fig. 38-25). In general, parameters such as minute ventilation and the paced Q–T interval are proportional sensors. Some sensors using specialized pacing leads are also highly proportional. For example, a properly functioning SvO_2 sensor will result in a rate closely related to oxygen consumption during exercise.

In addition, an appropriate speed of response of the pacing rate to the onset of and recovery from exercise is an essential feature of a rate-adaptive pacing system. The change in pacing rate should occur with kinetics (or *speed of response*) of a sensor that is similar to that of the sinus node. An anticipatory response of the heart rate occurs in many individuals prior to exercise. With both supine and upright isotonic exercise, heart rate and cardiac output increase within 10 seconds of the onset of exercise.[99-100] Both cardiac output and sinus rate increase exponentially, with a half-time that ranges from 10 to 45 seconds, the rate of rise being proportional to the intensity of work.[99] At the termination of upright exercise there is a delay of approximately 5 to 10 seconds before cardiac output starts to decrease, followed by an exponential fall with a half-time of 25 to 60 seconds. If the rate decay is faster than is physiologically appropriate, adverse hemodynamic consequences may occur in the presence of a substantial drop in heart rate. In one study in which pacing rate was reduced

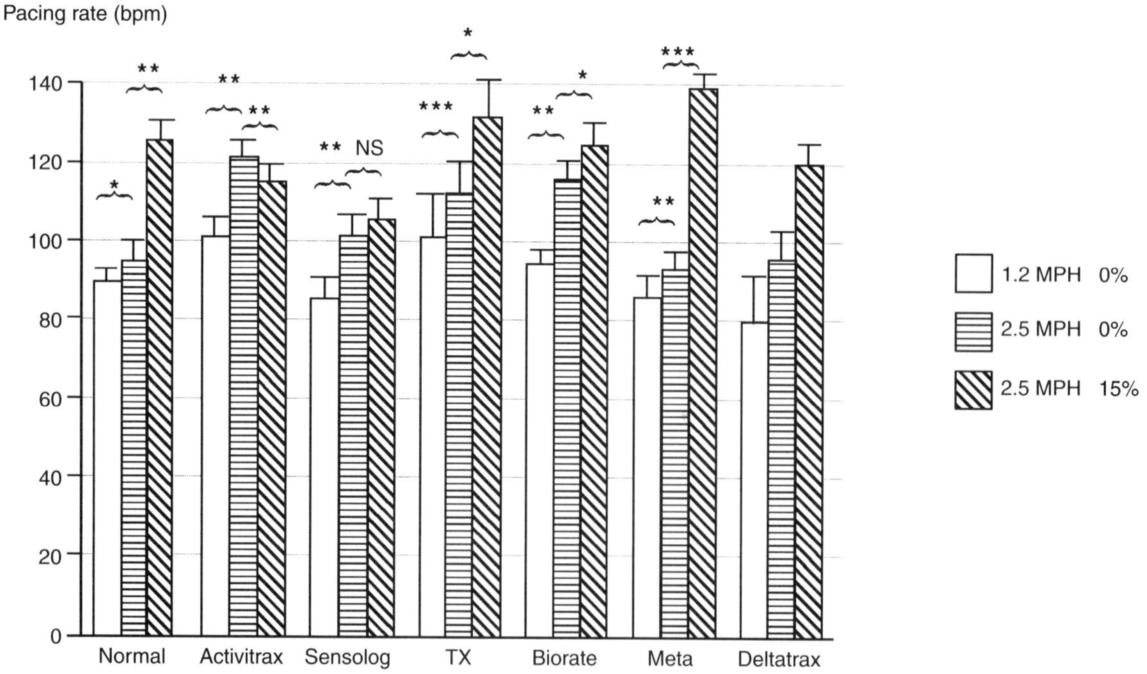

FIGURE 38-25 Brief activities are used to evaluate the proportionality of rate response of some common sensors. Maximal heart rate was derived from a 3-minute walking test done at different speeds (1.2 and 2.5 mph) and on different slopes (0% and 15%). There was no significant change in pacing rate when patients with the Sensology pacemaker ascended an incline, whereas the pacing rate decreased significantly in patients with the Activitrax during the same activity. NS, Not significant; *, $P < .05$; **, $P < .01$; ***, $P < .001$; Tx, QT sensing pacemaker. (Reproduced with permission from Lau CP, Butrous GS, Ward DE, et al: Comparison of exercise performance of six rate-adaptive right ventricular cardiac pacemakers. Am J Cardiol 1989;63:833-838.)

either abruptly or gradually after identical exercise, it was shown that an appropriately modulated rate recovery was associated with a higher cardiac output, lower sinus rate, and faster lactate clearance than with an unphysiologic rate recovery pattern. Appropriate adjustment of rate recovery curve is important to enhance recovery from exercise.

The exercise responses of six different types of rate-adaptive pacemakers (with sensors for activity, Q–T interval, respiratory rate, minute ventilation, and right ventricular dP/dt) were compared with normal sinus rate in one study (Fig. 38-26).[103] The results of this study demonstrated that the activity-sensing pacemakers best simulated the normal speed of rate response at the start of exercise. The rate response of activity sensors is usually immediate (no delay time), and the time needed to attain half of the maximal change in rate occurred within 45 seconds from the onset of exercise. The maximal change in pacing rate is reached within 2 minutes of beginning an ordinary activity such as walking. The respiratory rate and the right ventricular dP/dt sensors had a longer delay time. The slowest sensor to respond to exercise was an early version of the QT-sensing pacemaker, which required up to 1 minute to initiate a rate response, and the maximal change in pacing rate was attained only in the recovery period following a short duration of exercise.

Exercise is but one of the many physiologic requirements for variation in heart rate. Emotions, such as anxiety, may trigger a substantial change in heart rate. The sinus rate is higher when an individual moves from the supine to the upright posture, with a fall in the cardiac output. Isometric exercise also results in an increase in cardiac output and heart rate in most individuals. The changes of heart rate that occur during various physiologic maneuvers (e.g., Valsalva's maneuver), baroreceptor reflexes, and anxiety reactions may also be potentially important. An appropriate compensatory heart rate response is especially important in pathophysiologic conditions such as anemia, acute blood loss, or other causes of hypovolemia. The artificial sensor should be *sensitive* enough to detect the exercise and nonexercise needs for changes in heart rate and yet be *specific* enough not to be affected by unrelated signals arising from both the internal and external environment. Although the ideal sensor should provide these functional characteristics, it must also be technically feasible to implement with a reliability that is acceptable with modern implantable pacemakers (see Table 38-7).

The overall chronotropic response obtained with a rate-adaptive pacing system is dependent on three major factors: (1) the intrinsic properties of the rate-control parameter; (2) the algorithm used to relate changes in the sensed parameter into changes in pacing rate; and (3) the way in which the pacing system is programmed. Each of these factors can dramatically change the rate modulation actually observed with a rate-adaptive pacing system. In addition, an appropriate rate-control algorithm or clever programming can overcome the intrinsic limitations of many sensors. In contrast, inappropriate programming of a pacing system can distort the chronotropic response of an otherwise ideal sensor so that a poor clinical result is observed. Thus, clinical input is required to achieve optimal results with any rate-adaptive pacemaker.

SENSOR COMBINATION

Important issues concerning sensors have emerged in the last few years. First, the only sensors used clinically are those that can be used with a standard lead.

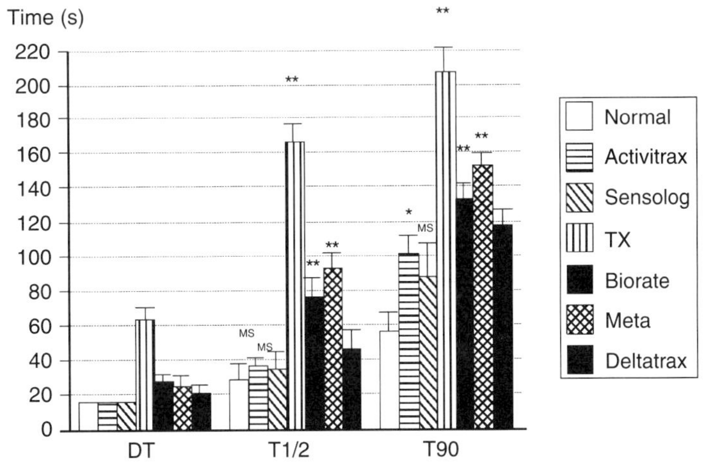

FIGURE 38-26 Speed of rate response of different pacemakers during walking at a nominal speed (2.5 mph at 0% gradient). The normal sinus rate responds almost immediately, one half of the change being achieved in less than 30 seconds and most of the change within 1 minute. This speed of response was most closely simulated by the activity-sensing pacemakers. Significant differences were derived by comparing the response times of each pacemaker (T½ and T90) with those of the normal sinus response. Ninety percent of the maximal rate for this exercise was reached within the exercise period in all patients, except those with the QT-sensing pacemaker (TX), who achieved this pacing rate only in the recovery phase. DT, delay time; T½ and T90, times needed to reach 50% and 90% of maximal heart rate. (Reproduced with permission from Lau CP, Butrous GS, Ward DE, et al: Comparison of exercise performance of six rate-adaptive right ventricular cardiac pacemakers. Am J Cardiol 1989;63:833-838.)

The instability of special-lead sensors prevents their use for rate adaptation. The surviving sensors include activity, minute ventilation, and QT sensors (used only in combinations). Second, rate adaptation using combinations of clinical sensors is feasible and is superior to rate adaptation with a single sensor. Third, automatic programming of sensors is feasible and effective, making the complexity of multisensor pacing programming an insignificant issue.

TABLE 38-8 Relative Advantages of Clinical Sensors

	Speed	Proportionality	Specificity	Sensitivity
Activity	H	L	L	L
Minute Ventilation	M	H	M	L
QT	L	M	H	M

H, high; L, low; M, moderate.

Justification for Sensor Combination

There has been significant improvement in instrumentation, and rate adaptive algorithms have been incorporated in the "clinical sensors" to address the issues of speed, proportionality, specificity, and sensitivity of sensor response. For example, piezoelectric crystal for activity sensing, especially using a "peak counting" algorithm is limited by inability of the sensor to differentiate between different levels of workload and susceptibility to external vibration.[101] However, because body movement has no direct relationship to metabolic requirement, this sensor remains inadequate to detect isometric exercise and nonexercise needs. Minute ventilation, as measured by impedance, delivers appropriate rate adaptive therapy proportional to workload.[101-104] Criticism has centered on the slower initial heart rate response to exercise as compared with activity sensors, with a 30-second delay when compared with the response of the sinus node. It is also potentially influenced by conditions that may not be directly relevant to cardiac output, such as phonation and voluntary respiration.[105] The new, faster algorithm with rate augmentation factor and programmable speed of response has improved this slow response in the early stages of exercise.[106] This, however, leads to a more rapid recovery with a significantly shortened recovery time to resting heart rate. The main limitation of the QT sensor is the relatively slow speed of onset of rate response[107] and the susceptibility of the Q–T interval to drugs and ischemia. With the use of curvilinear rate-responsive curves using a higher slope at the onset of exercise, the lag in the onset of rate response is reduced[108] but the speed of rate response may still be slow during brief periods of exercise.[109] Thus, the new generation of clinical sensors remains imperfect, even when only speed and proportionality of rate response are considered. In addition, apart from the QT sensor, which reacts to emotional changes,[110] none of the other clinical sensors is sensitive to nonexercise needs, such as changes induced by postural, postprandial vagal maneuvers, fever, and circadian variations.

The limitations of currently available sensors mean that none of them is suitable for every patient under all circumstances. Despite the fact that the response of a sensor can be significantly enhanced by fine tuning the characteristics of the sensor and the algorithms used to translate sensor output into modulation of pacing rate, the "clinical" sensors are mainly limited because a fast responding sensor is not proportional, whereas a proportional sensor is relatively slow (Table 38-8). In addition, an activity sensor is relatively insensitive to nonexercise stress, nonspecific, and can be affected by external interference. Because the sinus node behavior is the result of multiple stimulatory or inhibitory reflexes, in replacing the failing sinus node with a pacing rate simulating the physiologic heart rate, a combination of sensors may theoretically best replicate the response of the sinus node during a variety of physiologic needs. Therefore, further technical improvements will not be limited to refinement or development of single-sensor systems but must inevitably lead to a combination of complementary systems.

Dual sensor pacemaker technology aims to create a sensor system that best simulates the sinus node response in normal individuals by combining the strong points and eliminating the weak points of each sensor in combination. The sensor combination aims at improving the speed of rate response, proportionality to workload, sensitivity to physiologic changes induced by exercise and nonexercise requirements, and specificity in rate adaptation (Fig. 38-27). A more specific sensor to exercise can be used to prevent false-positive rate acceleration by a more sensitive, yet relatively nonspecific sensor. In the absence of the specific sensor indicating exercise, the response of the other sensor can either be nullified or restrained to a limited level or duration. Multisensor pacing may also offer the possibility of selecting an alternative sensor, should one sensor fail or become inappropriate for an individual patient. In addition, an appropriate rate recovery can shorten repayment of oxygen debt and promote lactate clearance. Therefore, it will be necessary to incorporate sensors that have more proportionality for this part of exercise. The potential for combining sensors for purposes other than rate regulation during exercise is another strong incentive for the development of multisensor pacemakers.

Two basic methods for combining sensors to control chronotropic response during exercise have been used. The types of sensor combination can be "faster win" or "blending." In the faster win form, the inputs from two sensors are compared and the faster rate is chosen. A differential combination (blending) combines the input of two sensors, either at a fixed ratio or at a different maximal magnitude. For example, a fast-responding sensor (such as activity) can be used to modulate pacing rate at the onset of exercise, and a second sensor (such as minute ventilation) may modulate the pacing rate during more prolonged exercise. The pacing rate may increase to an interim or intermediate value when the faster sensor detects the onset of exercise, whereas a more proportional rate increase will occur when the slower, more proportional sensor

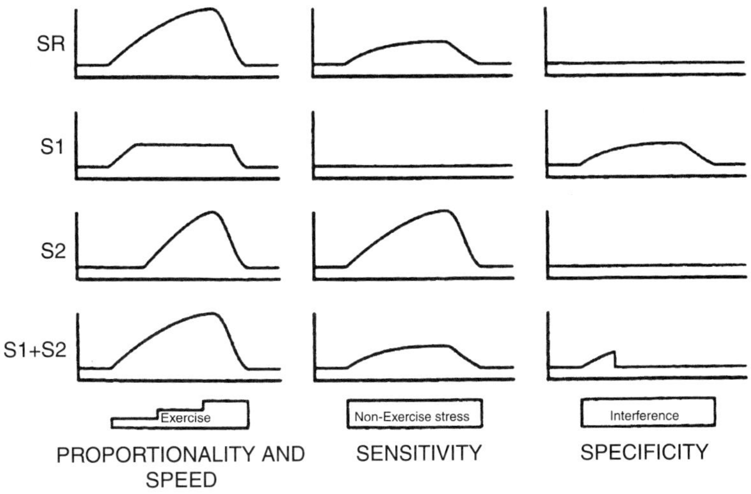

FIGURE 38-27 Different algorithms for sensor combinations needed to achieve better (1) proportionality and speed of response; (2) sensitivity; and (3) specificity. The graphs (*top to bottom*) depict the responses of the sinus node (SR), sensor 1 (S1), sensor 2 (S2), and the combined rate profile of S1 and S2. SR shows ideal proportionality, speed of rate response, and freedom from interference. S1 is a rapidly responding sensor, although it is neither proportional nor sensitive and is susceptible to interference. S2 is a proportional and sensitive sensor, although it has a slow response. It is also specific to exercise. Note the improved ability of the combined sensor approach in simulating the sinus rate under different conditions. (From Lau CP: Rate-Adaptive Cardiac Pacing: Single- and Dual-Chamber. Mt. Kisco, NY, Futura, 1994.)

"catches up." A variation of this approach is to calculate the output of each sensor in a relative proportion so that the ultimate rate profile is a blend of both.

The pacing rate can be controlled by two sensors that have different sensitivities to exercise and nonexercise physiologic stresses so that the system can respond to both exertional and emotional needs. It is conceivable that a separate rate-adaptive slope (or different upper and lower rates) can be programmed for modulation of rate in response to exertional and emotional needs. The algorithm can be designed to weigh the output of both sensors to diagnose a nonexercise physiologic stress and provide a different pattern of rate adaptation.

The response of one sensor can also be checked against the output from another sensor to improve the specificity of the chronotropic response. A more specific sensor may be used to cross-check with a nonspecific sensor to avoid false rate acceleration. In this instrumentation, the rate adaptation of a less specific sensor is allowed only for a restricted magnitude and duration of rate adaptation. In the absence of the other sensor(s) indicating exercise, a diagnosis of false-positive rate acceleration with the first sensor is made, and the pacing rate will return to the baseline so that prolonged high-rate pacing will be avoided. Such sensor cross-checking can be reciprocal between the two sensors so that either sensor may limit the chronotropic response that results from the other. In practice, cross-checking is usually applied to limit only the less specific of the sensors by a more specific one.

Dual Sensor Rate Adaptive Pacemakers

QT and Activity

In the Topaz and Diamond pacemakers (Models 515 and 800, Vitatron Medical BV, Dieren, The Netherlands), a piezoelectric sensor is used for activity sensing. The algorithms of sensor combinations are both blending and cross-checking. Activity and QT input can be programmed at different contributions: Activity less than QT, activity equal to QT, and activity greater than QT, represents ratios at 30:70, 50:50, and 70:30, respectively. To avoid false rate acceleration of activity sensor, the pacemaker allows only activity rate response for a short duration of time unless confirmed by changes in QT sensor (sensor cross-checking).

The blending of the QT and activity sensor shows a quick rate response at exercise onset, and a more proportional rate response during the latter part of the exercise and recovery period.[111] The fast-rate adaptive response during the first stage of exercise is due primarily to the activity sensing with a high correlation between the activity sensor counts and the mean pacing rate ($r = 0.94$), whereas the QT sensor predominates during the latter stages of exercise with a low correlation between activity sensor counts to pacing rate ($r = 0.14$).[111] A multicenter study[112] of 79 patients with Topaz showed that exercise in the dual-sensor mode produced a more gradual rate response than exercise in the activity mode alone. Rate profile during treadmill exercise testing resulting from dual sensor pacing was improved over single-sensor pacing.[112-113] In one study, simultaneous recording and comparison of combined sensor and sinus rates during daily life activities and standardized exercise testing were performed in twelve patients, and an improved correlation between the dual sensor-indicated rate and the sinus rate during daily life activities and treadmill exercise compared with individual sensors was observed (Fig. 38-28).[114] Furthermore, an inappropriate high rate response from the activity sensor due to external vibrations could be limited by sensor blending and cross-checking.[115-116] However, with a sensitive activity sensor setting, activity counts may be registered at rest when the QT sensor is inactive. This may result in the cross-checking of the activity sensor by the QT sensor at rest, and when exercise begins, this cross-checking can delay the speed of dual sensor rate response.[116]

Minute Ventilation and Activity

A piezoelectric activity sensor has been combined with a minute ventilation sensor to improve the initial response time, while allowing a proportional rate response to a higher workload. In the Legend-Plus VVIR

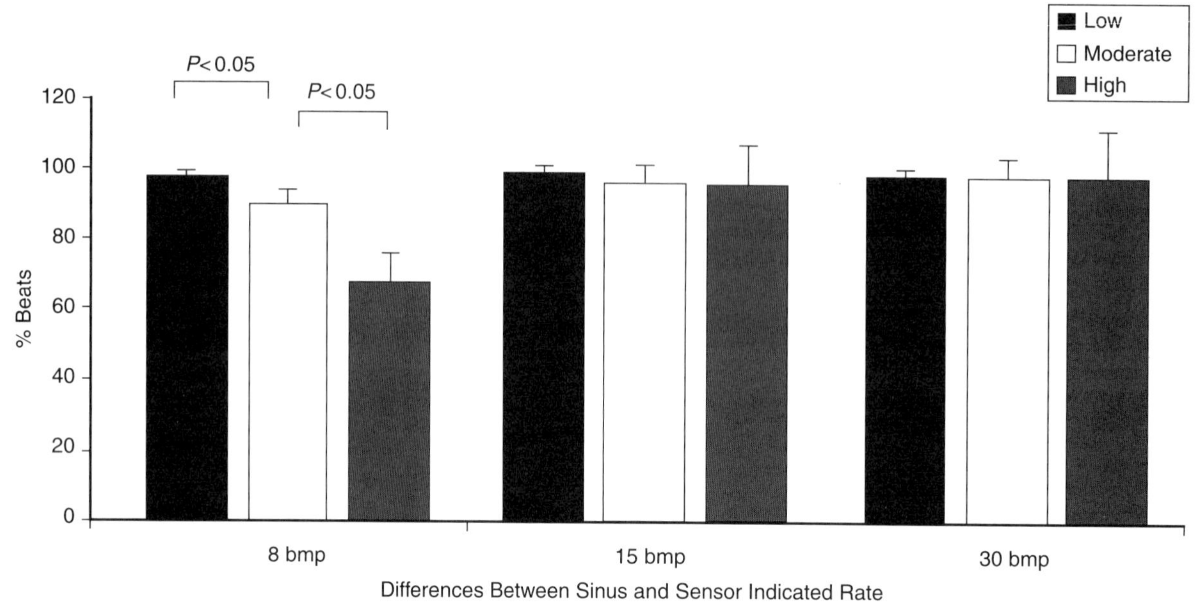

FIGURE 38-28 Sinus-sensor differences from 12 patients during daily activities, classified according to magnitude of differences (8, 15, and 30 beats/minute) and levels of activities, low (L), moderate (M), and heavy (H). At low levels of exercise, most of the sensor-indicated rates were within 8 beats/minute of the sinus rate, although the sinus-sensor differences increased at higher exercise workloads. The difference between the sensor-indicated rate and the sinus rate was almost always within 15 beats/minute at all rate ranges. (Reproduced with permission from Lau CP, Leung SK, Guerola M, et al: Comparison of continuously recorded sensor and sinus rates during daily life activities and standardized exercise testing: Efficacy of automatically optimized rate adaptive dual sensor pacing to simulate sinus rhythm. Pacing Clin Electrophysiol 1996;19:1672-1677.)

pacemaker (Model 8446, Medtronic, Inc.) an additive sensor combination algorithm is used. A sensor-indicated rate for each sensor is calculated, and the faster one determines the pacemaker sensor-driven pacing rate. The rate response settings in each system are determined by using a 3-minute walking test. Impedance values and the activity counts during the exercise are recorded, and the recommended activity and minute ventilation rate-response settings are calculated and displayed on the programmer.

A more sophisticated sensor-combination algorithm is used in the DX2 and Kappa KR (Medtronic, Inc.) pacemaker. The pacing rate in this investigational model is determined by automatic and nonprogrammable "blending" of activity and minute ventilation contribution on rate, using daily activities as a guide. Both sensors in the DX2 and Kappa KR contribute to the sensor-indicated rate between the lower and an interim rate limit, the so called *activities of daily living* (ADL) rate.[117] The influence of the activity sensor diminishes and shifts toward the minute ventilation sensor as the integrated sensor-indicated rate increases toward the ADL rate, thereafter rate adaptation is driven completely by the minute ventilation sensor. The activity sensor will also cross-check against high MV sensor-driven pacing rates. Sensor cross-checking against the minute ventilation sensor is effected if the activity sensor counts are low; minute ventilation-driven rate will then be limited in magnitude and duration. This minimizes the influence of nonphysiologic minute ventilation signals associated with upper body motion, such as with arm movement, or during hyperventilation.[117]

During treadmill exercise and stair climbing, the combined sensor mode in the Legend-Plus was shown to be more proportional to low and high workload activities compared with the activity-VVIR sensor mode, but are similar to the minute ventilation-VVIR sensor.[118] A faster speed of rate response with a shorter delay time, time to 50% rate response and time to 90% of rate response in the DX2 pacemaker was reported compared with the minute ventilation sensor during submaximal exercise and ADL, although the rate kinetics between the activity sensor and combined minute ventilation and activity sensor are similar.[117] The average maximal sensor rates were significantly higher for the dual sensor mode compared with either activity and minute ventilation modes during daily activities (Fig. 38-29).

AUTOMATICITY

An appropriate rate-control algorithm or careful programming can overcome some intrinsic limitations of many sensors. In contrast, inappropriate programming of a pacing system can distort the chronotropic response of an otherwise ideal sensor. On the physician side, programming of a rate adaptive sensor is often time consuming. This may involve multiple exercise tests, especially if a more objective sensor adjustment using exercise parameters is used. The use of a dual sensor may more than double this effort. There is no simple standard by which a programming setting for one sensor can be adjusted, and often the sensor is programmed to achieve a rough output based on the physician's assessment of the patient's overall physical

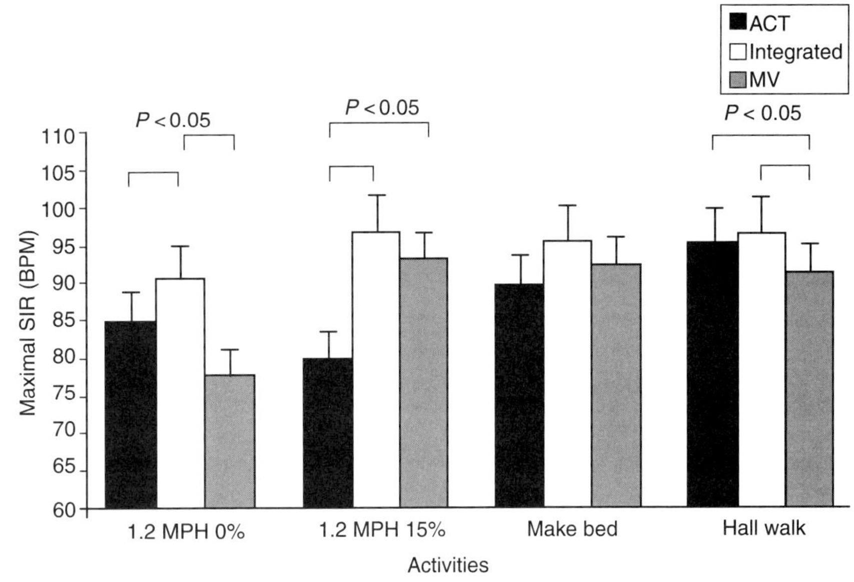

FIGURE 38-29 Comparison of the average maximal sensor-indicated rate (SIR) during submaximal exercise and activities of daily living. The MV + ACT mode gave a better average maximal sensor rate during the submaximal exercise and hall walk compared with the MV and ACT sensor modes. (Reproduced with permission from Leung SK, Lau CP, Tang MO, et al: New integrated sensor pacemaker: Comparison of rate responses between an integrated minute ventilation and activity sensor and single sensor modes during exercise and daily activities and nonphysiological interference. Pacing Clin Electrophysiol 1996;19: 1664-1671.)

state and activity level. On the patient's side, apart from the inconvenience of multiple programming, the actual requirement by the patient may change over time and in disease states, at a time during which the physician may not be available. Thus, the ability of a self-adjusting sensor is not only a clinician's convenience but a clinical need. Automaticity can be achieved using a closed loop sensor (theoretical only), semiautomatic adjustment, or autoprogramming.

Autoprogrammability

There are currently two methods of automatic programming for open loop sensors: one using sensor matching at upper and lower rate limits, the other using a *target rate distribution* approach.

An automatic slope adaptation mechanism has been incorporated in the new version of the QT-sensing pacemaker (Rythmx, Vitatron Medical B.V.), and combined QT and activity sensing pacemakers (Topaz VVIR and Diamond DDDR, Vitatron Medical B.V.). This algorithm involves two different rate-adaptive slopes, one slope designed for low-exercise workloads and the other for higher workloads. The QT and activity slopes at the upper and lower rate limits are automatically adjusted by a daily learning process. The pacemaker monitors the dynamics of the Q–T interval and activity counts and continuously updates its maximum and minimum sensor values.[119] The self-learning process adjusts the rate-response slope at the upper rate and lower rate limits, respectively (Fig. 38-30). Each time the pacemaker reaches the upper rate limit, it continues to monitor the Q–T interval and the activity counts. Further shortening of the Q–T interval or increase in activity counts while pacing at the upper rate limit will indicate that the upper rate limit has been reached too soon and the rate response slope at the upper rate is automatically decreased by one step, which will give a more gradual approach to the upper rate limit. Conversely, if the upper rate limit has not been reached for more than 8 days, this slope is automatically increased by one step. Similarly, the Q–T interval is measured regularly at the lower rate interval, with the patient presumably at rest. If the Q–T interval continues to lengthen at rest, this suggests the lower rate limit is reached too soon, and the slope for the lower rate limit is decreased.

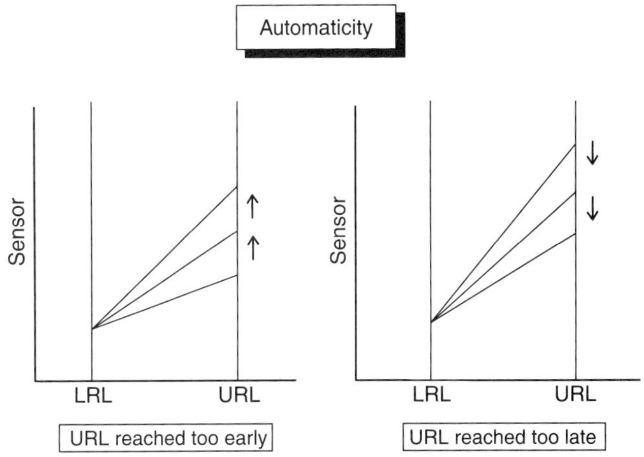

FIGURE 38-30 Automatic adjustment of the QT rate response slope aims to pace at the lower rate limit when the Q–T interval indicates no physical or mental stress, and to pace at the upper rate limit when the Q–T interval indicates maximal workload. At the lower rate limit (LRL), the QT/rate relationship is assessed once every night, and the slope is adjusted automatically one step in the direction of change. During maximal exercise at the upper rate limit (URL), if further QT shortening occurs, the slope-declining factor will be advanced one step further so that the URL will be attained later in a repeat exercise **(left panel)**. Conversely, if the upper rate is not attained in 8 days, the slope-declining factor will be reduced by one step **(right panel)**.

The changes of the QT and activity slopes from factory settings have been reported.[119] Most of the changes occurred within the first 2 weeks and stabilized within a 6-week time frame. Starting at a lower rate of 60 or 70 beats/minute, the combined sensor reached the programmed maximal rate of 110 or 120 beats/minute at 2 to 5 weeks (mean = 19 days).

The clinical efficacy of self-learning automatic programming has been shown to result in a pacing rate close to sinus rate for daily activities (see Fig. 38-28).[114] This comparative study on the efficacy of self-learning versus manual programming was conducted in 12 patients with complete heart block and normal sinus node function who received a combined activity and QT DDDR pacemaker (Model 800, Diamond, Vitatron BV) system. Patients underwent treadmill exercise and 12 ADL in the VDD mode. Sensor-indicated rates during these activities as derived from automatic programming and from manual optimization using a submaximal treadmill exercise were compared.

A similar automatic rate adaptation algorithm matching sensor output at upper and lower pacing rate limits was also being used by the minute ventilation sensing and the combined accelerometer and minute ventilation sensing DDDR pacemaker (Chorum 7234 and Talent DR 213, ELA Medical, Montrouge, France). The pacemaker constantly monitored the minute ventilation signal at the upper and lower rates. The slope of the minute ventilation rate response decreased by one step every eight pacing cycles; the minute ventilation signals continued to rise when the pacemaker paced at the upper rate and increased by 3% per day when the upper pacing rate was not reached in 24 hours. This automatic algorithm could stabilize the rate-response slope in the first month.[120] The pacing response, by combining the acceleration sensor with the automatically calibrated minute ventilation sensor, was shown to give a good correlation to sinus rate during exercise.

The main limitation of this self-learning approach is that the algorithm assumes every patient would exercise up to the programmed upper rate during their daily activities. In practice, however, for those patients who perform maximal exertion infrequently or who have been confined to bed, the automatic optimization could potentially over adjust the rate-response slope at the upper rate range, resulting in an inappropriately fast pacing rate when the patient resumes usual daily activities.

Automatic Rate Optimization by Target Rate Distribution

In normal individuals, a characteristic rate distribution occurs over 24 hours, depending on their sex, age, and activity and fitness levels.[121] Heart rate profiles had been recorded by 48 hours ambulatory ECG monitoring in groups of normal subjects during regular daily activities and during seven upper and lower extremity exercises. It was found that the daily heart rate behavior was mainly submaximal but with transient heart rate increases that exceeded 55% of the heart rate reserve.[121] The less fit subjects had a greater increase in heart rate and used greater heart rate reserve during ADL than did the average and fit subjects. Patients whose age was greater than 65 years also had a longer duration of heart rate increase to greater than 110 beats/minute during normal activities. Based on these data, a *nominal* target rate profile resembling the normal rate profile for the population with different physical fitness—as determined by age, sex, and different activity levels—was derived and used to adjust the desired distribution of pacing rate for an individual patient, using the following two approaches.

Automatic Slope Optimization by Target Distribution of Heart Rate Reserve

In this instrumentation, it is assumed that the heart rate of most individuals exceeds the 23rd percentile of the heart rate reserve only 1% of the time. The Trilogy DR+ (Pacesetter, Inc.) activity sensing pacemaker maintains a 7-day histogram of the sensor level and automatically adjusts the sensor slope once every 7 days such that 99% of the sensor activity is within the initial 23% of the heart rate reserve (Fig. 38-31). The maximal adjustment is limited to two slopes and the slope changes are not made if the patient is inactive (as defined by absence of sensor activities). In addition, offsets from −1 to +3 are available for fine titration of sensor response in individual patients. Positive offsets increase, whereas negative offsets decrease the functional slope.

In 93 patients implanted with the Pacesetter Trilogy DR+ pacemaker, the sensor-indicated rate during a brisk walk after automatic slope optimization was compared with a desired sensor rate selected by the clinician. The automatic slope optimization provided the desired sensor rate in 82.1% of the evaluation and in 75.6% of patients.[122] The rate modulation provided by the automatic slope optimization was appropriate in 83.1% of evaluation and 76.3% of patients during follow-up. Despite the use of automatic slope optimization, approximately one half of the patients required further programming of the slope offset to titrate the sensor rate response. Some of this further programming was needed because of the relatively conservative upper rate programming. In this study, the desired sensor rate and the appropriateness of the rate modulation were decided subjectively at the discretion of the clinician rather than objectively evaluated.

Rate Profile Optimization by Target Rate Histogram

Based on the *nominal* rate histogram, another device was introduced to automatically adjust the sensor setting to match this target rate histogram (Model 7970 DX2 and Kappa DR 403, Medtronic, Inc.). The DX2 and Kappa DR are combined activity and minute-ventilation sensing DDDR pacemakers. In addition to sensor autoprogramming, the device initiates implant detection through the detection of lead impedance and lead polarity recognition.[123] After confirming implantation by measuring stable lead impedance over 6 hours, the

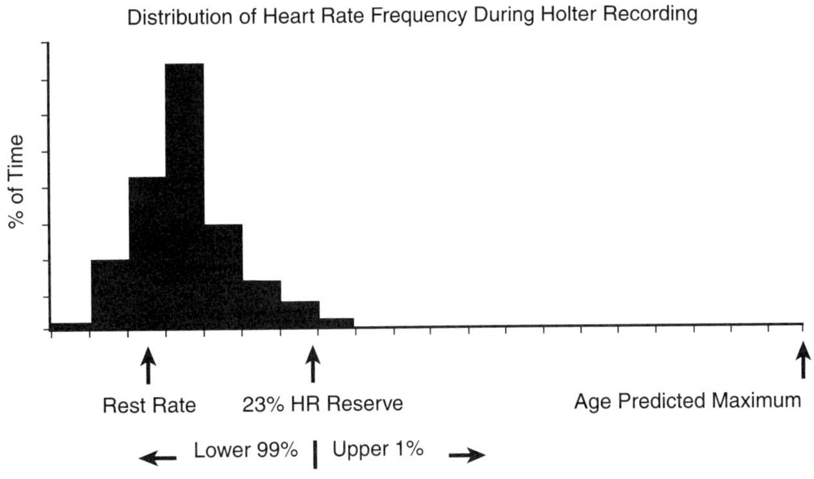

FIGURE 38-31 An illustration of the distribution of heart rate frequency. The automatic rate adaptive algorithm of the Triology DR+ was developed based on this algorithm. The sensor is automatically adjusted so that a rate response to occupy 23% of the heart rate (HR) reserve is attained.

devices were automatically programmed to DDDR mode, and the baseline minute ventilation was automatically measured. An optimal heart rate profile based on the patient's activity level and frequency of exercise is programmed as the "target rate histogram" (Fig. 38-32). This template was used to adjust the submaximal rate response during both daily activities and more vigorous exercise. Once each day, the pacemaker evaluates the percent of time spent pacing in both the submaximal and maximal rate ranges by comparing the sensor rate profile with the target rate histogram. From this comparison, the pacemaker automatically controls how rapidly the sensor-indicated rate increases and decreases in these ranges.

The reliability of implant management in providing automatic detection of lead polarity and sensor initiation has been reported.[124] The efficacy of automatic optimization of the rate response was compared with manual programming in a prospective study by measuring the rate kinetics during ADL, including maximal and submaximal treadmill exercise in seven patients who received this device. The rate changes derived by automatic programming and by manual adjustment were compared. After automatic rate profile optimization, the pacing rate during a hall walk increased from 78 ± 3 at predischarge to 90 ± 5 beats/minute at 2 weeks and 98 ± 3 beats/minute at the 3-month follow-up (Fig. 38-33). Pacing rate during maximal treadmill exercise increased from 89 ± 6 beats/minute to 115 ± 5 at 3 months after implantation with a significant increase in exercise duration from 7.2±1.0 minute to 9.6 ± 2.0 minute. The accuracy of automatic programming versus manual programming was reassessed at 1 month, and the average maximal pacing rates attained and the speed of rate response for the dual sensor after automatic and manual optimization did not differ significantly during the maximal exercise, submaximal exercise, and ADL between the two methods of programming.

The main limitation of the automatic rate optimization by target rate distribution approach is that a nominal population standard for an individual patient still has to be programmed, and a wider patient population is necessary to ascertain the safety of this approach. An inherent risk of using a histogram as the target for sensor setting is the uncertainty as to whether rate response had occurred at the appropriate time because only the rate distribution rather than the rate profile is

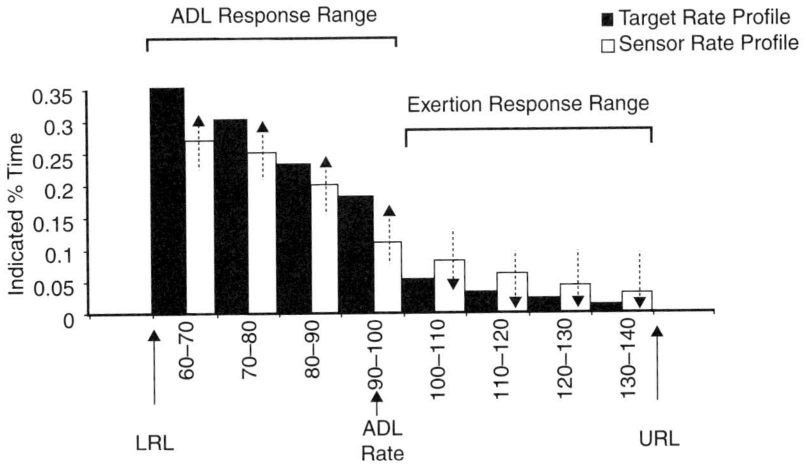

FIGURE 38-32 An illustration of the sensor rate profile matched against a target rate profile. The ADL ranges are moderate pacing rates between the lower rate (LRL) and the ADL rate. The exertion rates are rates from the ADL rate to the upper sensor rate (URL). By comparing the sensor rate profile against the target rate profile once each day, the pacemaker automatically controls how rapidly the sensor-indicated rate increases and decreases in these ranges. (Reproduced with permission from Leung SK, Lau CP, Tang MO, et al: An integrated dual sensor system automatically optimized by target rate histogram. Pacing Clin Electrophysiol 1998;21:1559-1566.)

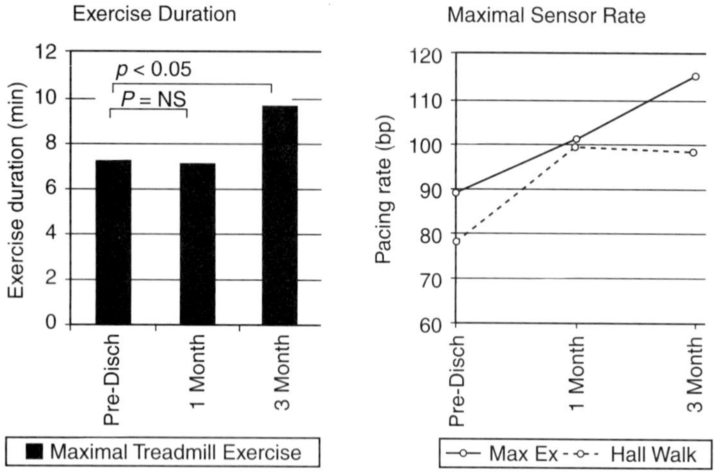

FIGURE 38-33 **Left panel:** Changes in the exercise duration during maximal exercise with the implant duration. There is an increase in exercise duration at 3 months compared with the predischarged exercise test with automatic rate optimization. The exercise duration at 1 month did not differ significantly from the predischarged exercise. **Right Panel:** Changes in sensor rate response to hall walk and maximal treadmill exercise with the duration of implant. Whereas the daily activity range was optimized by 1 month, continuous adaptation of the sensor to achieve a higher rate response occurred up to 3 months of implantation. (Reproduced with permission from Leung SK, Lau CP, Tang MO, et al: An integrated dual sensor system automatically optimized by target rate histogram. Pacing Clin Electrophysiol 1998;21:1559-1566.)

used as the template. In addition, the onset and recovery patterns, which are important characteristics of the sensor rate response characteristics, are not addressed by the histogram approach.

Clinical Outcome

The ultimate goals of the pacemaker therapy is to improve symptoms and quality of life and these criteria have been used to compare pacing modes. In terms of symptomatology, VVIR pacing is superior to VVI pacing. However, the overall contribution of improved control of symptoms to enhanced quality of life is probably small in the usual pacemaker recipient in whom quality of life is already close to that of age-matched normal individuals.[125] There is still no comparative study of different sensors in regard to their effects on symptomatology or quality of life. A patient-randomized, double-blinded crossover study was done in ten patients using the combined activity and minute ventilation dual sensor VVIR pacemaker for high grade AV block and chronic or persistent paroxysmal atrial fibrillation. These patients performed 2 weeks of out-of-hospital activity in the activity only, minute ventilation only, and dual sensor VVIR and VVI modes.[305] Patients were assessed on their perceived general well-being using the visual analog scale scores, Specific Activities Scale functional status questionnaire, and objective improvement by standardized daily activity protocols and graded treadmill testing. Subjective perception of exercise capacity and functional status was significantly reduced in the VVI mode, compared with the VVIR modes. However, there is no clear advantage of dual sensor VVIR pacing over activity sensor pacing. Four of the 10 patients preferred the activity VVIR mode, 3 preferred the dual sensor mode, and 3 had no preference. Three patients found dual sensor VVIR least acceptable, three patients found minute ventilation least acceptable, and one patient found both dual sensor and minute ventilation sensor pacing unacceptable. There was no significant difference in the objective performance between the 3 VVIR modes. These not unexpected results, suggest that there are no major differences between sensor and their combination in gross clinical terms. However, the overall numbers of patients being studied is small and does not have sufficient statistical power to unveil less than major differences, which may be important for the long-term effects of a pacing mode. In addition, the difficulty of multiple comparisons and the order of pacing modes studied are limitations. Lukl and colleagues[126] also assessed the quality of life with regard to cardiovascular symptoms, physical activity, psychosocial and emotional functioning, and self-perceived health during DDD and dual sensor VVIR pacing. Significant improvement during DDD pacing was demonstrated in all subgroups of patients (sick sinus syndrome, chronotropically competent and incompetent patients, and patients with high-degree AV block). The overall result shows that DDD pacing offers better quality of life than does dual sensor VVIR pacing. Thus, a dual sensor VVIR pacing cannot compensate for the lack of AV synchrony. Because most rate-adaptive pacemakers have an activity sensor, it is of interest if the addition of a minute ventilation sensor contributes to better clinical performance. A preliminary study[127] suggests that, although the maximal exercise capacity may not be affected by the combined sensor, better rate adaptation profiles may enhance quality of life.

REFERENCES

1. Ohm OJ, Danilovic D: Improvements in pacemaker energy consumption and functional capability: Four decades of progress. Pacing Clin Electrophysiol 1997;20:2-9.
2. Marshall M, Butts L, Flaim G, et al: Predictors of time requirements for pacemaker clinic evaluation [abstract]. Pacing Clin Electrophysiol 1995;18:952.
3. Preston TA, Fletcher RD, Lucchesi BR, Judge RD: Changes in myocardial threshold. Physiologic and pharmacologic factors in patients with implanted pacemakers. Am Heart J 1967;74:235-242.
4. Dohrmann ML, Goldschlager NF: Myocardial stimulation threshold in patients with cardiac pacemakers: Effect of physiologic variables, pharmacological agents, and lead electrodes. Cardiol Clin 1985;3:527-537.
5. Schaldach M, Hubmann M, Weikl A, et al: Sputter-deposited TiN electrode coatings for superior sensing and pacing performance. Pacing Clin Electrophysiol 1990;13:1891-1895.

6. Curtis AB, Vance F, Miller K: Automatic reduction of stimulus polarization artifact for accurate evaluation of ventricular evoked responses. Pacing Clin Electrophysiol 1991;14:529-537.
7. Provenier F, Germonpre E, De Wagter X: Improved differentiation of the ventricular evoked response from polarization by modification of the pacemaker impulse. Pacing Clin Electrophysiol 2000;23:2073-2077.
8. Schuchert A, Ventura R, Meinertz T: Automatic threshold tracking activation without the intraoperative evaluation of the evoked response amplitude. Pacing Clin Electrophysiol 2000;23:321-324.
9. Schuchert A, Ventura R, Meinertz T: Effects of body position and exercise on evoked response signal for automatic threshold activation. Pacing Clin Electrophysiol 1999;22:1476-1480.
10. Clarke M, Liu B, Schuller H, Binner L, et al: Automatic adjustment of pacemaker stimulation output correlated with continuously monitored capture thresholds: A multicenter study. Pacing Clin Electrophysiol 1998;21:1567-1575.
11. Sermasi S, Marconi M, Libero L, et al: Italian experience with autocapture in conjunction with a membrane lead. Pacing Clin Electrophysiol 1996;19:1799-1804.
12. Lau C, Cameron DA, Nishimura SC, et al: A cardiac evoked response algorithm providing threshold tracking: A North American multicenter study. Pacing Clin Electrophysiol 2000; 23:953-959.
13. Boriani G, Biffi M, Branzi A, et al: Benefits in projected pacemaker longevity and in pacing related costs conferred by automatic threshold tracking. Pacing Clin Electrophysiol 2000;23(II): 1783-1787.
14. Simeon L, Duru F, Fluri M, et al: The impact of automatic threshold tracking on pulse generator longevity in patients with different chronic stimulation thresholds. Pacing Clin Electrophysiol 2000; 23(II):1788-1791.
15. Duru F, Bausersfeld U, Schuller H, Candinas R: Threshold tracking pacing based on beat by beat evoked response detection: Clinical benefits and potential problems. JICE 2000;4:511-522.
16. Guilleman D, Bussillet H, Scanu P, et al: Output adjustment with the DDD pacemaker with automatic capture-control algorithm. Prog in Biomed Res 1999;4:291-294.
17. Novak M, Kamaryt P, Haeuser T, Mach P: Simplifying pacemaker follow-up using automatic threshold determination in ventricle. Prog Biomed Res 1999;4:287-290.
18. Gross JN, Moser S, Benedek ZM, et al: DDD pacing mode survival in patients with a dual-chamber pacemaker. J Am Coll Cardiol 1992;19:1536-1541.
19. Benjamin EJ, Levy D, Vaziri SM, et al: Independent risk factors for atrial fibrillation in a population-based cohort. The Framingham Heart Study. JAMA 1994;271:840-844.
20. Lamas GA, Pashos CL, Normand SL, et al: Permanent pacemaker selection and subsequent survival in elderly Medicare pacemaker recipients. Circulation 1995;91:1063-1069.
21. Connolly SJ, Kerr C, Gent M, et al: Dual-chamber versus ventricular pacing. Critical appraisal of current data. Circulation 1996; 94:578-583.
22. Andersen HR, Nielsen JC, Thomsen PE, et al: Long-term follow-up of patients from a randomised trial of atrial versus ventricular pacing for sick-sinus syndrome . Lancet 1997;350:1210-1216.
23. Lamas GA, Orav EJ, Stambler BS, et al: Quality of life and clinical outcomes in elderly patients treated with ventricular pacing as compared with dual-chamber pacing. Pacemaker Selection in the Elderly Investigators. N Engl J Med 1998;338:1097-1104.
24. Connolly ST, Kerr CR, Gent M, et al: Effects of physiologic pacing versus ventricular pacing on the risk of stroke and death to cardiovascular causes. N Engl J Med 2000;342:1385-1391.
25. Van Wyhe G, Sra J, Rovang K, et al: Maintenance of atrioventricular sequence after His-bundle ablation for paroxysmal supraventricular rhythm disorders: A unique use of the fallback mode in dual chamber pacemakers. Pacing Clin Electrophysiol 1991;14:410-414.
26. Mayumi H, Uchida T, Shinozaki K, et al: Use of a dual chamber pacemaker with a novel fallback algorithm as an effective treatment for sick sinus syndrome associated with transient supraventricular tachyarrhythmia. Pacing Clin Electrophysiol 1993;16:992-1000.
27. Barold SS: Automatic mode switching during antibradycardia pacing in patients without supraventricular tachycardia. In Barold SS, Mugica J (eds): New Perspectives in Cardiac Pacing, 3rd ed. Mt. Kisco, NY, Futura, 1993, pp 455-481.
28. Barold SS, Mond HG: Optimal antibradycardia pacing in patients with paroxysmal supraventricular tachyarrhythmias: Role of fallback and automatic mode switching mechanisms. In Barold SS, Mugica J (eds): New Perspectives in Cardiac Pacing, 3rd ed. Mt. Kisco, NY, Futura, 1993, pp 483-518.
29. Mond HG, Barold SS: Dual chamber, rate adaptive pacing in patients with paroxysmal supraventricular tachyarrhythmias: Protective measures for rate control. Pacing Clin Electrophysiol 1993;16:2168-2185.
30. Lau CP, Leung SK, Tse HF, Barold SS: Automatic mode switching of implantable pacemakers. I: Principles of instrumentation, clinical and hemodynamic considerations. Pacing Clin Electrophysiol 2002;25:967-83.
31. Lau CP, Leung SK, Tse HF, Barold SS: Automatic mode switching of implantable pacemakers. II: Clinical performance of current algorithms and their programming. Pacing Clin Electrophysiol 2001;25:1094-1113.
32. Lau CP, Tai YT, Fong PC, et al: Atrial arrhythmia management with sensor controlled atrial refractory period and automatic mode switching in patients with minute ventilation sensing dual chamber rate adaptive pacemakers. Pacing Clin Electrophysiol 1992;15:1504-1514.
33. Medtronic AT500 Technical Manual. Medtronic, Minn., USA.
34. Kerr CR, Mason MA: Amplitudes of atrial electrical activity during sinus rhythm and during atrial flutter fibrillation. Pacing Clin Electrophysiol 1985;8:348-355.
35. Leung SK, Lau CP, Lam CT, et al: Programmed atrial sensitivity: A critical determinant in atrial fibrillation detection and optimal automatic mode switching. Pacing Clin Electrophysiol 1998; 21:2214-2219.
36. Frohlig G, Helwani Z, Kusch O, et al: Bipolar ventricular far-field signals in the atrium. Pacing Clin Electrophysiol 1999;22(11): 1604-1613.
37. Theres H, Sun W, Combs W, et al: P wave and far-field R wave detection in pacemaker patient atrial electrograms. Pacing Clin Electrophysiol 2000;23:434-440.
38. Brandt J, Worzewski W: Far-field QRS complex sensing: Prevalence and timing with bipolar atrial leads. Pacing Clin Electrophysiol 2000;23(3):315-320.
39. Saveliera I, Camm AJ: Clinical relevance of silent atrial fibrillation: Prevalence, prognosis, quality of life, and management. JICE 2000;4:369-382.
40. Love CJ, Wilkoff BL, Hoggs KS et al: Incidence of mode switch in a general pacemaker population [abstract]. Pacing Clin Electrophysiol 1997;20:1137.
41. Machado C, Sullivan C, Johnson D, et al: Pacemaker patient triggered event recording: Accuracy and utility for pacemaker follow-up clinic [abstract]. Pacing Clin Electrophysiol 1996;19:720.
42. Ricci R, Puglisi A, Azzolini P, et al: Reliability of a new algorithm for automatic mode switching from DDDR to DDIR pacing mode in sinus node disease patients with chronotropic incompetence and recurrent paroxysmal atrial fibrillation. Pacing Clin Electrophysiol 1996;19:1719-1723.
43. Ovsyscher IE, Ketz A, Bondy C: Initial experience with a new algorithm for automatic mode switching from DDDR to DDIR mode. Pacing Clin Electrophysiol 1994;17:1908-1912.
44. Seidl K, Meisel E, Van Agt E, et al: Is the high rate episode diagnostic feature reliable in detecting paroxysmal episodes of atrial tachyarrhythmias? Pacing Clin Electrophysiol 1998;21: 694-700.
45. Fitts SM, Hill MRS, Mehra R, et al: High rate atrial tachyarrhythmia detections in implantable pulse generators: Low incidence of false-positive detections. Pacing Clin Electrophysiol 2000;23: 1080-1086.
46. Ellenbogen KA, Wood MA, Mond HG, et al: Clinical applications of mode-switching for dual-chamber pacemakers. In Singer I, Barold SS, Camm AJ (eds): Nonpharmacological Therapy of Arrhythmias for the 21st Century. The State of the Art. Armonk, NY, Futura, 1998, p 819.
47. Mabo P, Daubert C, Limousin M, et al: Atrial electrogram storage: A new tool for atrial arrhythmia diagnosis in pacemaker patients [abstract]. Pacing Clin Electrophysiol 1996;19:721.
48. Petersen B, Huikuri H, Benzer W, et al: Specificity of pacemaker diagnostics verified with stored E-grams [abstract]. Europace 2000;1:D214.

49. Huikuri H: Effect of stored electrograms on management of the paced patient. Am J Cardiol 2000;86[9 Supp 1]:K101-K103.
50. Israel CW, Gascon D, Nowak B, et al: Diagnostic value of stored electrograms in single-lead VDD systems. Pacing Clin Electrophysiol 2000;23:1801-1803.
51. Gillis AM, Wyse DG, Connolly ST, et al: Atrial pacing periablation for prevention of paroxysmal atrial fibrillation. Circulation 1999; 99:2553-2558.
52. Lau CP, Tse HF, Yu CM, et al: Dual-site right atrial pacing in paroxysmal atrial fibrillation without bradycardia [NIPP-AF study]. Am J Cardiol 2001;88:371-375.
53. Lau CP, Leung WH, Wong CK, et al: Haemodynamics of induced atrial fibrillation: A comparative assessment with sinus rhythm, atrial and ventricular pacing. Eur Heart J 1990;11:219-224.
54. Clark DM, Plumb VJ, Epstein AE, et al: Hemodynamic effects of an irregular sequence of ventricular cycle lengths during atrial fibrillation. J Am Coll Cardiol 1997;30:1039-1045.
55. Wittkampf FH, de Jongste MJ, Lie HI, et al: Effect of right ventricular pacing on ventricular rhythm during atrial fibrillation. J Am Coll Cardiol 1988;11:539-545.
56. Lau CP, Jiang ZY, Tang MO: Efficacy of ventricular rate stabilization by right ventricular pacing during atrial fibrillation. Pacing Clin Electrophysiol 1998;21:542-548.
57. Brunner-La Rocca HP, Rickli H, Weilenmann D, et al: Importance of ventricular rate after mode switching during low intensity exercise as assessed by clinical symptoms and ventilatory gas exchange. Pacing Clin Electrophysiol 2000;23:32-39.
58. Kamalvand K, Tam K, Kotsakis K, et al: Is mode switching beneficial? A randomized study in patients with paroxysmal atrial tachyarrhythmias. J Am Coll Cardiol 1997;30:496-504.
59. Marshall HJ, Kay GN, Hess M, et al: Mode switching in dual chamber pacemakers: Effect of onset criteria on arrhythmia-related symptoms. Europace 1999;1:49-54.
60. Pitney MR, May CD, Davis MJ: Undesirable mode switching with a dual chamber rate responsive pacemaker. Pacing Clin Electrophysiol 1993;16:729-737.
61. Leung SK, Lau CP, Leung WH, et al: Apparent extension of the atrioventricular interval due to sensor-based algorithm against supraventricular tachyarrhythmias. Pacing Clin Electrophysiol 1994;17(3):321-330.
62. Jais P, Barold S, Shah DC, et al: Pacemaker syndrome induced by the mode switching algorithm of a DDDR pacemaker. Pacing Clin Electrophysiol 1999;22(4):682-685.
63. Brignole M, Menozzi C, Gianfranchi L, et al: Assessment of atrioventricular junction ablation and VVIR pacemaker versus pharmacological treatment in patients with heart failure and chronic atrial fibrillation: A randomized, controlled study. Circulation 1998;98:953-960.
64. Marshall HJ, Harris ZI, Griffith MJ, et al: Prospective randomized study of ablation and pacing versus medical therapy for paroxysmal atrial fibrillation: Effects of pacing mode and mode-switch algorithm. Circulation 1999;99:1587-1592.
65. Lau CP: Sensor and pacemaker mediated tachycardias [editorial]. Pacing Clin Electrophysiol 1991;14:495-498.
66. Provenier F, Jordaens L, Verstraeten T, et al: The "automatic mode switch" function in successive generations of minute ventilation sensing dual chamber rate responsive pacemakers. Pacing Clin Electrophysiol 1994;17:1913-1919.
67. Ricci R, Puglisi A, Azzolini P, et al: Reliability of a new algorithm for automatic mode switching from DDDR to DDIR pacing mode in sinus node disease patients with chronotropic incompetence and recurrent paroxysmal atrial fibrillation. Pacing Clin Electrophysiol 1996;19:1719-1723.
68. Seidl K, Meisel E, Van Agt E, et al: Is the high rate episode diagnostic feature reliable in detecting paroxysmal episodes of atrial tachyarrhythmias? Pacing Clin Electrophysiol 1998;21:694-700.
69. Levine PA, Bornzin GA, Barlow J, et al: A new automode switch algorithm for supraventricular tachycardias. Pacing Clin Electrophysiol 1994;17:1895-1899.
70. Pacesetter Inc. Trilogy DR+ Model 2364 Technical Manual. Sylmar, Calif.
71. Pacesetter AB Affinity DR Model 5330 L/R User's Manual. Vedelesta, Sweden.
72. Leung SK, Lau CP, Lam CT, et al: Is automatic mode switching effective for atrial arrhythmias occurring at different rates? A study on efficacy of automatic mode and rate switching to simulated atrial arrhythmias by chest wall stimulation. Pacing Clin Electrophysiol 2000;23:823-831.
73. Lau CP, Tai YT, Fong PC, et al: Clinical experience with an activity sensing DDDR pacemaker using an accelerometer sensor. Pacing Clin Electrophysiol 1992;15:334-343.
74. Gencel L, Geroux L, Clementy J, et al: Ventricular protection against atrial arrhythmias in DDD pacing based on a statistical approach: Clinical results. Pacing Clin Electrophysiol 1996;19: 1729-1734.
75. Geroux L, Limousin M, Cazeau S: Clinical performance of a new mode switch function based on a statistical analysis of the atrial rhythm. Herzchr Elektrophys 1999;10[Suppl 1]:15-21.
76. Begemann MJ, Thijssen WA, Haaksma J: The influence of test window width on atrial rhythm classification in dual chamber pacemakers. Pacing Clin Electrophysiol 1992;15:2158-2163.
77. den Dulk K, Dijkman B, Pieterse M, et al: Initial experience with mode switching in a dual sensor, dual chamber pacemaker in patients with paroxysmal atrial tachyarrhythmias. Pacing Clin Electrophysiol 1994;17:1900-1907.
78. Brignole M, Gianfranchi L, Menozzi C, et al: A new pacemaker for paroxysmal atrial fibrillation treated with radiofrequency ablation of the AV junction. Pacing Clin Electrophysiol 1994; 17:1889-1894.
79. Jayaprakash S, Sparks PB, Kalman JM, et al: Dual demand pacing using retriggerable refractory periods for ventricular rate control during paroxysmal supraventricular tachyarrhythmias in patients with dual chamber pacemakers. Pacing Clin Electrophysiol 2000; 23:1156-1163.
80. Rickards AF, Bombardinin, T, Corbucci G, et al: An implantable intracardiac accelerometer for monitoring myocardial contractility. The Multicenter PEA Study Group. Pacing Clin Electrophysiol 1996 Dec;19[12 Pt 1]:2066-2071.
81. Leung SK, Lau CP, Lam C, et al: Automatic optimization of resting and exercise atrioventricular interval using a peak endocardial acceleration sensor: Validation with Doppler echocardiography and direct cardiac output measurements. Pacing Clin Electrophysiol 2000 Nov;23[11 Pt 2]:1762-1766.
82. Leung SK, Lau CP, Lam CT, et al: A comparative study on the behavior of three different automatic mode switching dual chamber pacemakers to intracardiac recordings of clinical atrial fibrillation. Pacing Clin Electrophysiol 2000;23:2086-2096.
83. Wood MA, Ellenbogen KA, Dinsmoor D, et al: Influence of autothreshold sensing and sinus rate on mode switching algorithm behavior. Pacing Clin Electrophysiol 2000;23:1473-1478.
84. Palma EC, Kedarnath V, Vankawalla V, et al: Effect of varying atrial sensitivity, AV interval, and detection algorithm on automatic mode switching. Pacing Clin Electrophysiol 1996;19:1734-1739.
85. Lau CP: The range of sensors and algorithms used in rate adaptive cardiac pacing. Pacing Clin Electrophysiol 1992;15:1177-1211.
86. Pacela AF: Impedance pneumography—a survey of instrumentation techniques. Med Biol Eng 1966;4:1-15.
87. Rushmer RF, Crystal DK, Wagner C, et al: Intracardiac impedance plethysmography. Am J Physiol 1953;174:171-174.
88. Lau CP, Ritche D, Butrous GS, et al: Rate modulation by arm movements of the respiratory dependent rate responsive pacemaker. Pacing Clin Electrophysiol 1988;11:744-752.
89. Callaghan F, Vollmann W, Livingston A, et al: The ventricular depolarization gradient: effects of exercise, pacing rate, epinephrine, and intrinsic heart rate control on the right ventricular evoked response. Pacing Clin Electrophysiol 1989;12: 1115-1130.
90. Stangl K, Wirtzfeld A, Heinze R, et al: First clinical experience with an oxygen saturation controlled pacemaker in man. Pacing Clin Electrophysiol 1988;11:1882-1887.
91. Lau CP, Tai YT, Leung WH, et al: Rate adaptive cardiac pacing using right ventricular venous oxygen saturation: Quantification of chronotropic behavior during daily activities and maximal exercise. Pacing Clin Electrophysiol 1994;17:2236-2246.
92. Bennett T, Sharma A, Sutton R, et al: Development of a rate adaptive pacemaker based on the maximum rate-of-rise of right ventricular pressure (RV dP/dtmax). Pacing Clin Electrophysiol 1992;15:219-234.
93. Ovsyshcher I, Guetta V, Bondy C, et al: First derivative of right ventricular pressure, dP/dt, as a sensor for a rate adaptive VVI pacemaker: Initial experience. Pacing Clin Electrophysiol 1992;15:211-218.

94. Kay GN, Philippon F, Bubien RS, et al: Rate modulated pacing based on right ventricular dP/dt: Quantitative analysis of chronotropic response. Pacing Clin Electrophysiol 1994;17: 1344-1354.
95. Bongiorni MG, Soldati E, Arena G, et al: Is local myocardial contractility related to endocardial acceleration signals detected by a transvenous pacing lead? Pacing Clin Electrophysiol 1996; 19:1682-1688.
96. Clementy J: Dual chamber rate responsive pacing system driven by contractility: Final assessment after 1-year follow-up. The European PEA Clinical Investigation Group. Pacing Clin Electrophysiol 1998;21:2192-2197.
97. Langenfeld H, Krein A, Kirstein M, et al: Peak endocardial acceleration-based clinical testing of the "BEST" DDDR pacemaker. European PEA Clinical Investigation Group. Pacing Clin Electrophysiol 1998;21:2187-2191.
98. Binner L: One year follow-up of a new DDDR pacemaker based on contractility: A multicenter European study on Peak Endocaridal Acceleration (PEA) [Abstract]. Pacing Clin Electrophysiol 2000;23:1762-1766.
99. Loeppky JA, Greene ER, Hoekenga DE, et al: Beat-by-beat stroke volume assessment by pulsed Doppler in upright and supine exercise. J Appl Physiol 1981;50:1173-1182.
100. Miyamoto Y, Tamura T, Takahashi T, et al: Transient changes in ventilation and cardiac output at the start and end of exercise. Jpn J Physiol 1981;31:153-168.
101. Lau CP, Mehta D, Toff WD, et al: Limitations of rate response of an activity-sensing rate-responsive pacemaker to different forms of activity. Pacing Clin Electrophysiol 1988;11:141-150.
102. Lau CP, Antoniou A, Ward DE, et al: Initial clinical experience with a minute ventilation sensing rate modulated pacemaker: Improvements in exercise capacity and symptomatology. Pacing Clin Electrophysiol 1988;11:1815-1822.
103. Lau CP, Butrous GS, Ward DE, et al: Comparison of exercise performance of six rate-adaptive right ventricular cardiac pacemakers. Am J Cardiol 1989;63:833-838.
104. Mond H, Strathmore N, Kertes P, et al: Rate responsive pacing using a minute ventilation sensor. Pacing Clin Electrophysiol 1988;11:1866-1874.
105. Lau CP, Antoniou A, Ward DE, et al: Reliability of minute ventilation as a parameter for rate responsive pacing. Pacing Clin Electrophysiol 1989;12:321-330.
106. Slade AK, Pee S, Jones S, et al: New algorithms to increase the initial rate response in a minute volume rate adaptive pacemaker. Pacing Clin Electrophysiol 1994;17:1960-1965.
107. Mehta D, Lau CP, Ward DE, et al: Comparative evaluation of chronotropic responses of QT sensing and activity sensing rate responsive pacemakers. Pacing Clin Electrophysiol 1988;11: 1405-1412.
108. Baig MW, Wilson J, Boute W, et al: Improved pattern of rate responsiveness with dynamic slope setting for the QT sensing pacemaker. Pacing Clin Electrophysiol 1989;12:311-320.
109. Bellamy CM, Roberts DH, Hughes S, et al: Comparative evaluation of rate modulation in new generation evoked QT and activity sensing pacemakers. Pacing Clin Electrophysiol 1992; 15:993-999.
110. Hedman A, Nordlander R, Pehrsson SK: Changes in Q-T and Q-T intervals at rest and during exercise with different modes of cardiac pacing. Pacing Clin Electrophysiol 1985;8:825-831.
111. Provenier F, van-Acker R, Backers J, et al: Clinical observations with a dual sensor rate adaptive single chamber pacemaker. Pacing Clin Electrophysiol 1992;15:1821-1825.
112. Connolly DT: Initial experience with a new single chamber, dual sensor rate responsive pacemaker. The Topaz Study Group. Pacing Clin Electrophysiol 1993;16:1833-1841.
113. Leung SK, Lau CP, Wu CW, et al: Quantitative comparison of rate response and oxygen uptake kinetics between different sensor modes in multisensor rate adaptive pacing. Pacing Clin Electrophysiol 1994;17:1920-1927.
114. Lau CP, Leung SK, Guerola M, et al: Comparison of continuously recorded sensor and sinus rates during daily life activities and standardized exercise testing: Efficacy of automatically optimized rate adaptive dual sensor pacing to simulate sinus rhythm. Pacing Clin Electrophysiol 1996;19:1672-1677.
115. Cowell R, Morris TJ, Paul V, et al: Are we being driven to two sensors? Clinical benefits of sensor cross-checking. Pacing Clin Electrophysiol 1993;16:1441-1444.
116. Lau CP, Leung SK, Lee IS: Delayed exercise rate response kinetics due to sensor cross-checking in a dual sensor rate adaptive pacing system: The importance of individual sensor programming. Pacing Clin Electrophysiol 1996;19:1021-1025.
117. Leung SK, Lau CP, Tang MO, et al: New integrated sensor pacemaker: Comparison of rate responses between an integrated minute ventilation and activity sensor and single sensor modes during exercise and daily activities and nonphysiological interference. Pacing Clin Electrophysiol 1996;19:1664-1671.
118. Alt E, Combs W, Fotuhi P, et al: Initial clinical experience with a new dual sensor SSIR pacemaker controlled by body activity and minute ventilation. Pacing Clin Electrophysiol 1995;18:1487-1495.
119. van-Krieken FM, Perrins JP, Sigmund M: Clinical results of automatic slope adaptation in a dual sensor VVIR pacemaker. Pacing Clin Electrophysiol 1992;15:1815-1820.
120. Ritter P, Anselme F, Bonnet JL, et al: Clinical evaluation of an automatic slope calibration function in a minute ventilation controlled DDDR pacemaker [abstract]. Pacing Clin Electrophysiol 1997;20:1173.
121. Mianulli M, Birchfield D, Yakimow K, et al: The relationship between fitness level and daily heart rate behavior in normal adults—Implication for rate-adaptive pacing [abstract]. Pacing Clin Electrophysiol 1995;18:870.
122. Gentzler RD, Lucas EH: Automatic sensor adjustment in a rate modulated pacemaker. North American Trilogy DR+ Phase I Clinical Investigators. Pacing Clin Electrophysiol 1996;19: 1809-1812.
123. Lau CP, Pietersen A, Ohm OJ: Automatic implant detection for initiating lead polarity programming and rate adaptive sensors: Multicentre study [abstract]. Pacing Clin Electrophysiol 1996;19:592.
124. Leung SK, Lau CP, Tang MO, et al: An integrated dual sensor system automatically optimized by target rate histogram. Pacing Clin Electrophysiol 1998;21:1559-1566.
125. Lau CP, Rushby J, Leigh JM, et al: Symptomatology and quality of life in patients with rate-responsive pacemakers: A double-blind, randomized, crossover study. Clin Cardiol 1989;12:505-512.
126. Lukl J, Doupal V, Heinc P: Quality-of-life during DDD and dual sensor VVIR pacing. Pacing Clin Electrophysiol 1994;17:1844-1848.
127. Leung SK, Lau CP, Lam C, et al: Does dual sensor rate adaptive pacing improve quality of life? [abstract]. Europace 2000;1:D195.

Chapter 39 Pacemaker Insertion, Revision, Extraction, and Follow-Up

MARK H. SCHOENFELD

It may well be hubris to presume to undertake a single chapter addressing such diverse pacemaker themes as implantation, revision, extraction, and follow-up. Having elected, nonetheless, to proceed with this task, I have chosen to present a personal perspective on the subject(s). The goal is to be readable for clinicians and students in the field, without being unduly encyclopedic.

Implantation

As with most things in life, preparation is everything; alternatively stated, one must "plan the work, then work the plan." In determining to embark on a pacemaker implantation, there is the tacit assumption that the implanter has had the appropriate degree of training,[1-3] has maintained proficiency in this area,[3] and understands the specific indications for a particular procedure as delineated in the ACC/AHA/NASPE guidelines for device implantation.[4] What is clear is that maintenance of a minimum procedural volume is essential and indeed the number of implants performed by an individual is inversely related to the development of complications at the time of implantation.[5] In addition, preparation entails having anticipated all the resources and personnel essential to perform the operation, including a sterile operating arena (whether operating room or catheterization laboratory),[6,7] standard operating room instruments, pacemaker leads and generators, corresponding pacer programmers, pacer systems analyzers, anesthesia/sedation, fluoroscopy, and emergency equipment (e.g., pericardiocentesis sets). The presence of an experienced nurse or technician who can make the necessary threshold determinations is also indispensable; the role of the "pacemaker manufacturer's representative" in this regard remains highly controversial[8] and continues to raise both quality assurance issues and ethical concerns. Under discussion as well are appropriate lengths of stay, particularly with increasing pressures imposed by managed care; although ambulatory implant procedures have been employed by some, their widespread applicability and safety remain subjects for debate.[9-11]

Deciding on an operating strategy is crucial. This is especially the case in patients in whom a complicated case may be envisioned (patients with preexisting pacer or defibrillator systems with pacer-dependence, with congenital cardiovascular abnormalities—such as persistent[12] superior vena cava syndrome with previous demonstration of subclavian vein thrombosis) or in whom individualization of therapy dictates something "special" (e.g., contralateral placement in a mastectomy patient, placement of the device as a function of the patient's left- or right-handedness, or cosmetic concerns necessitating a submammary implant).

The use of antibiotic prophylaxis before implant has long been debated. A recent meta-analysis, however, suggests that it may indeed be very important in minimizing device-related infections.[13] Generally speaking, an anti-staphylococcal antibiotic is administered "on-call" to the procedure with continuation of said antibiotic for 24 hours following the implant.[14-16] Also to consider preoperatively is the site of peripheral intravenous placement—this should be ipsilateral to the anticipated site of implant so as to allow for dye injection and radiographic imaging of central venous anatomy in cases where percutaneous subclavian or cephalic vein access is difficult. The anticoagulated patient poses a particular challenge both pre- and perioperatively:[17] warfarin is generally held to achieve an International Normalized Ratio (INR) of less than 2, cephalic cutdown and strict hemostasis are encouraged, and warfarin is resumed postoperatively, preferably *without* the adjunct of intravenous heparin (with which significant pocket hematomas have been appreciated).[18]

The choice (and number) of leads must also be considered. Passive fixation leads are typically associated with less in the way of acute threshold rises but active-fixation leads may be selected when potential dislodgment is a significant concern. This might be anticipated, for example, in cases of smooth-walled dilated right ventricles, amputated atrial appendages in patients having undergone earlier open-heart surgery, or right ventricular outflow tract lead positioning. In addition, active fixation leads may be more easily removed in the unlikely (and unhappy) event that extraction is a future consideration. Pacing configuration (unipolar versus bipolar) is also an important issue to contemplate. On the one hand, unipolar leads are typically smaller in diameter and easier to introduce; in many cases, two unipolar leads may be introduced primarily through a single venotomy or even through a single peel-away introducer. However, unipolar systems have many potential disadvantages including myopectoral stimulation, myopotential inhibition, sensing of unipolar spikes by simultaneously implanted defibrillators, and so on. On the other hand, bipolar lead systems have had a number of associated advisories related to insulation degradation and subclavian crush syndrome. Compatibility of the selected lead(s) with the designated pulse generator must be assured.[19-21]

Epicardial or subxiphoid placement of pacer systems are typically reserved for those individuals who cannot undergo effective pacing via the transvenous route (e.g., a patient who has a mechanical tricuspid valve replacement) or who are simultaneously undergoing thoracotomy for other reasons. Limited surgical approaches using a transatrial implantation technique have also been described.[22]

Transvenous pacing is usually the preferred route, either with use of the cephalic vein[23] or subclavian vein[24]; it is rare to employ the external or internal jugular veins.[25] Placing a tourniquet around the arm and identifying a reasonably sized brachial/antecubital vein is a good predictor that the patient will have a usable cephalic vein more proximally. After meticulous prepping and draping and the application of local anesthetic, a circumlinear incision is made and dissection follows down to the deltopectoral fat pad, beneath which courses the cephalic vein (Fig. 39-1). Many implanters choose to obtain venous access by entering the subclavian vein percutaneously *before* making such an incision, believing that the anatomy of the clavicle and first rib are better appreciated "from the outside." There is nothing, however, that precludes using the subclavian approach from *within* the wound if the cephalic vein is not encountered. Indeed, there are distinct advantages to looking for the cephalic vein first: (1) it avoids the potential risk of pneumothorax and/or subclavian artery puncture associated with attempts at subclavian vein puncture, (2) it may avoid trauma to the lead incurred in the subclavian crush syndrome (see later) or associated with the peel-away introducer technique, and (3) it provides yet another avenue of access not available to those implanters who are accustomed only to the subclavian puncture technique, and (4) in cases of subsequent revision, it allows employment of an unused subclavian vein. Ligatures are applied proximally and distally to the site of the venotomy (Fig. 39-2). It is particularly important to tie off the distal ligature (usually 3-0 silk) before introducing the lead to prevent significant back bleeding should a small cephalic vein avulse with manipulation. Vein lifts that are typically supplied with the pacemaker electrodes can be abrasive—my preference is to use a curved iris forceps. The lead should be visually inspected before introducing it to ensure that there are no defects from the outset before manipulation.

It is often the case that a venotomy through the cephalic vein will accommodate both atrial and ventricular leads. Occasionally, only one (or neither) may be introduced primarily, either because the vein is too small or tortuous. In such cases, consideration may be made of introducing a guidewire through the cephalic vein and then using a peel-away introducer technique through the cephalic vein to introduce one or both leads (Ong-Barold technique) (Figs. 39-3 and 39-4).[26]

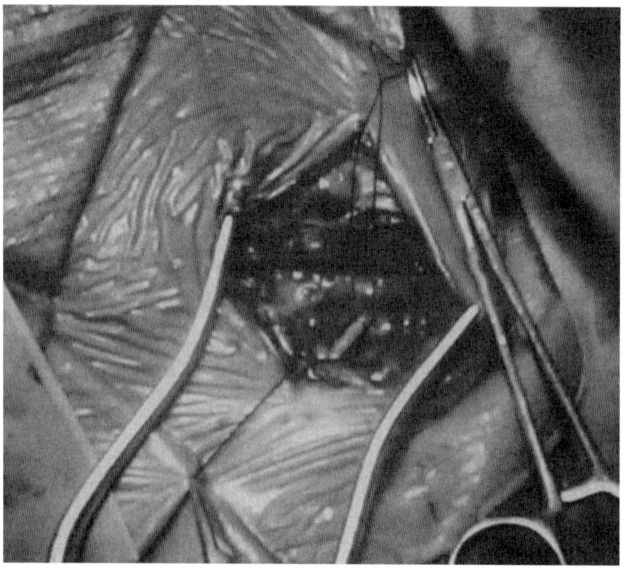

FIGURE 39-1 Isolation of cephalic vein.

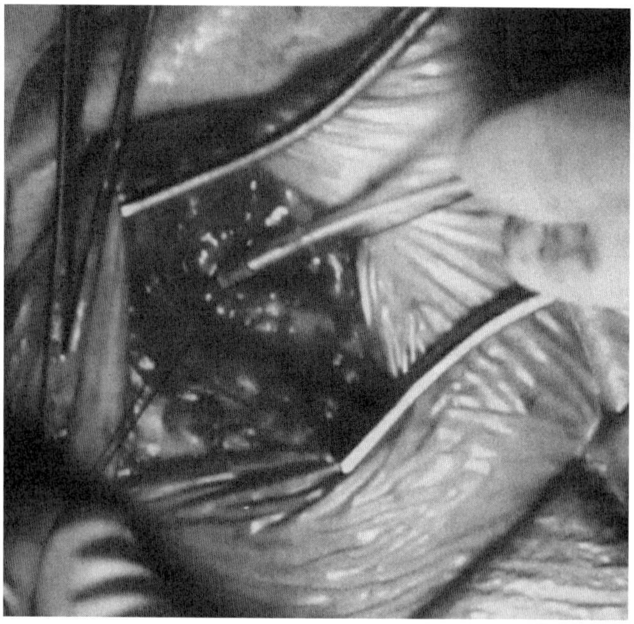

FIGURE 39-2 Insertion of lead through venotomy in cephalic vein.

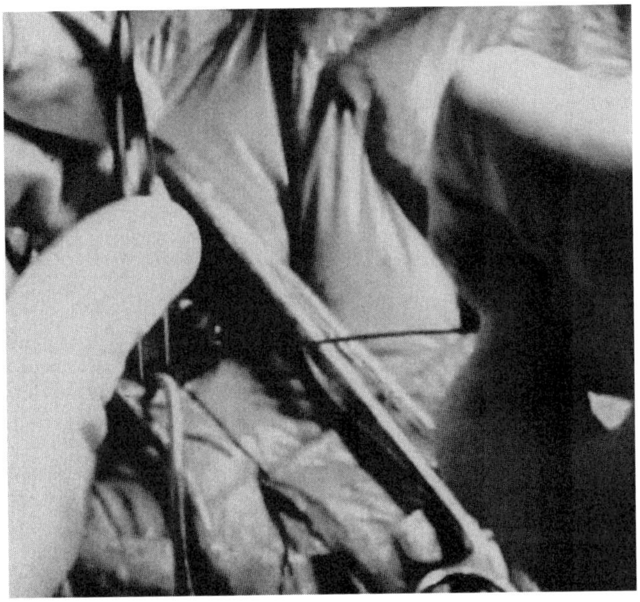

FIGURE 39-3 Insertion of guidewire through cephalic vein; first lead already inserted through venotomy.

Again, great care should be used to advance the introducer (use of a Gerard forceps to raise the flap of the venotomy may be helpful here) to avoid avulsing the cephalic vein.

Percutaneous subclavian vein access has been enabled largely by the peel-away-introducer technique,[27,28] which allows access to be achieved through the Seldinger technique, and removal of the sheath subsequently from the pacemaker lead. A variety of techniques has been reported. The subclavian window approach entails puncture near the apex of the angle formed by the first rib and clavicle, aiming medially and in the cephalic direction.[29] The medial aspect of this approach increases the success rate and reduces the risk of pneumothorax and vascular injury compared with a more lateral entry because the vein is a larger target and the apex of the lung is more laterally situated. The tighter binding between the first rib and clavicle may, however, result in subclavian crush phenomenon with insulation failure particularly in bipolar coaxial polyurethane leads.[30,31] The "safe introducer technique" as described by Byrd[32] relates to a safety zone between the first rib and clavicle, extending laterally from the sternum in an arc. More lateral approaches may avoid soft tissue entrapment (subclavius muscle, costocoracoid ligament, and costoclavicular ligament) and may, therefore, extend lead longevity.[33] Cannulation of the extrathoracic portion of the subclavian vein (the axillary vein) may also be considered, with puncture anteriorly to the first rib, maneuvering posteriorly and medially but remaining lateral to the juncture of the first rib and clavicle; in the more lateral locations, care must be taken to avoid puncture of the subclavian artery.[34,35] Radiography has been advocated during introduction of needles via whichever technique to confirm point of entry and subsequent trajectory (Fig. 39-5), and, in some cases, dye injection with venography may facilitate venipuncture(Fig. 39-6),[36-38] particularly in cases where access has been difficult and the question of subclavian vein thrombosis has been raised. A recent study found this to be particularly effective in guiding venipuncture using the axillary vein approach.[39] In patients who have bona fide superior vena cava and/or innominate vein obstruction, stent dilation has been reported as one option by which to achieve access without going to the contralateral chest.[40]

Once access to the subclavian vein has been achieved, an incision must be made to allow for development of the pocket if the puncture was percutaneous and not from within the wound. The peel-away-introducer technique then allows for passage of the introducer over the guidewire and subsequent passage of the pacemaker lead through the introducer. If the leads are small

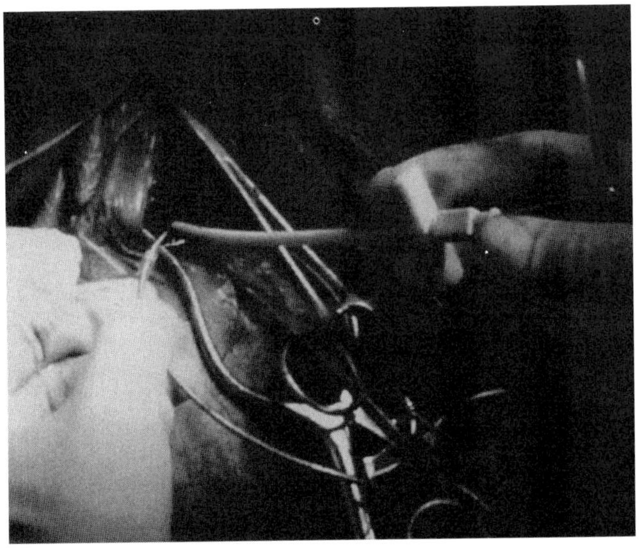

FIGURE 39-4 Insertion of peel-away introducer over guidewire through cephalic vein; first lead already inserted through venotomy.

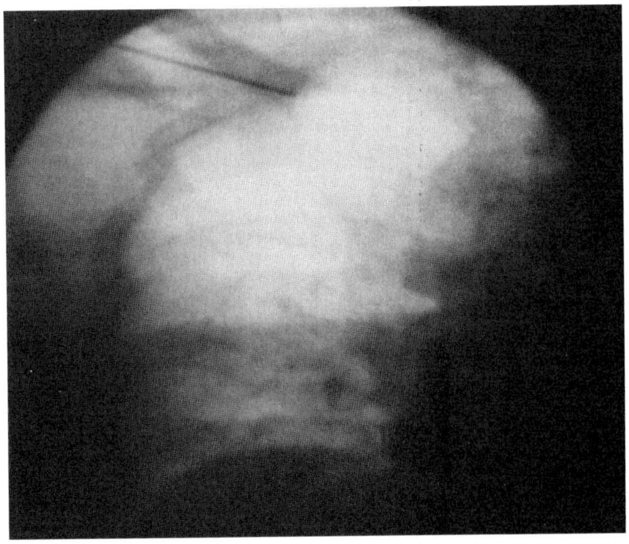

FIGURE 39-5 Fluoroscopic visualization of introducer needle for obtaining subclavian vein access.

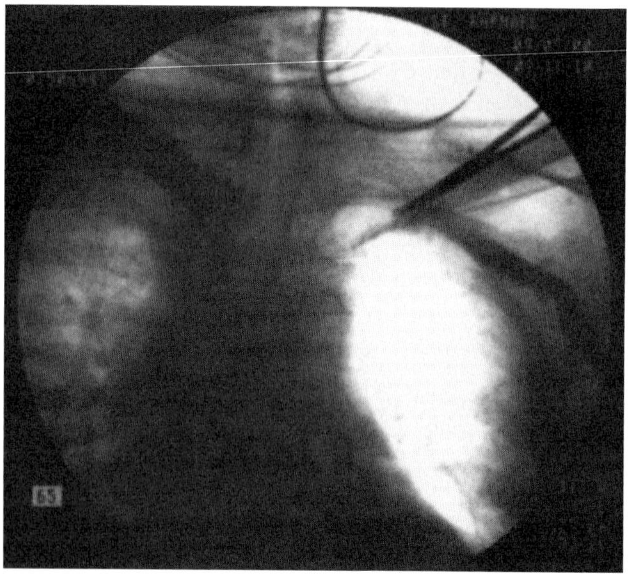

FIGURE 39-6 Peripheral dye injection for venography to guide needle for subclavian vein access.

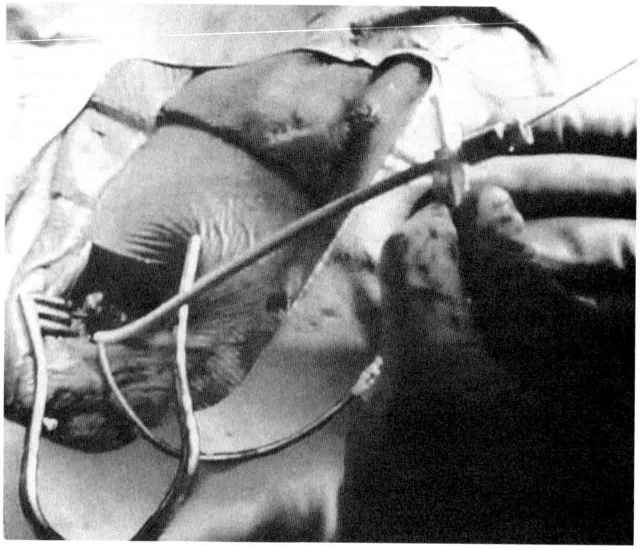

FIGURE 39-7 Peel-away introducer technique with retained wire for subclavian access.

(especially if unipolar), both may be guided through a single introducer. More commonly, the guidewire is retained after passage of the first lead to allow for passage of a second introducer and pacemaker lead, obviating the need for a second venipuncture (Fig. 39-7)[41]; alternatively, two guidewires may be passed through a single percutaneous introducer with subsequent sliding of the introducer over each guidewire separately to provide independent access for each of two pacing leads.[42] Occasionally there are problems with a recently introduced lead (e.g., a need to switch from a passive fixation to active fixation lead) with loss of previously established venous access. In such cases, one may use a blade to slit the insulation of the lead to be sacrificed, wedge a guidewire through this insulation, and advance the lead with guidewire as a unit so that the guidewire enters the vascular space. Holding the guidewire in place, the lead to be sacrificed may be advanced slightly so as to disengage it from the guidewire—in this manner the guidewire is retained within the vascular space, the old lead may be removed, and a new lead may be placed through a peel-away introducer placed over the guidewire (Fig. 39-8).

The technique required for lead manipulation may vary as a function of the lead employed and the cardiac chamber to be accessed. For right ventricular apex positioning, one approach is to prolapse the lead across the tricuspid valve (Fig. 39-9). The other commonly employed approach is to put a curve on the stylet and guide the lead to the right ventricular outflow tract, with subsequent withdrawal of the lead until it drops

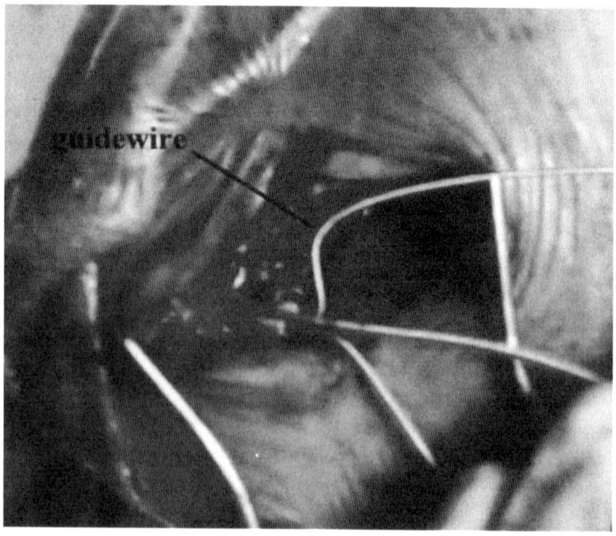

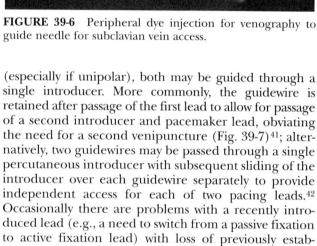

FIGURE 39-8 Insertion of guidewire under insulation of lead to be sacrificed, to re-establish venous access.

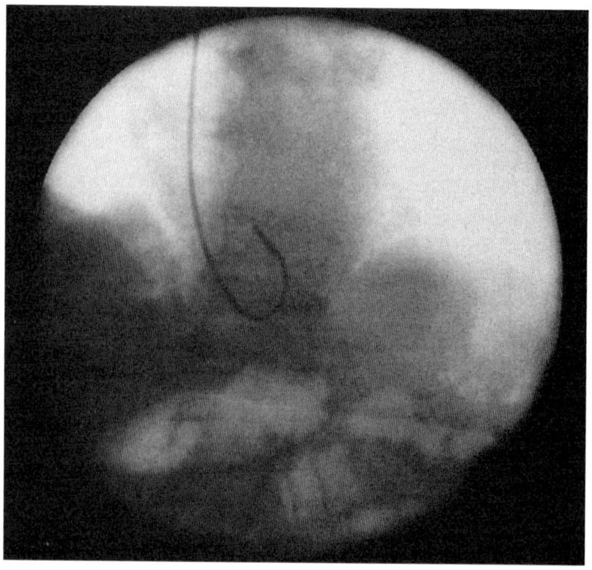

FIGURE 39-9 Prolapse of ventricular lead across tricuspid valve.

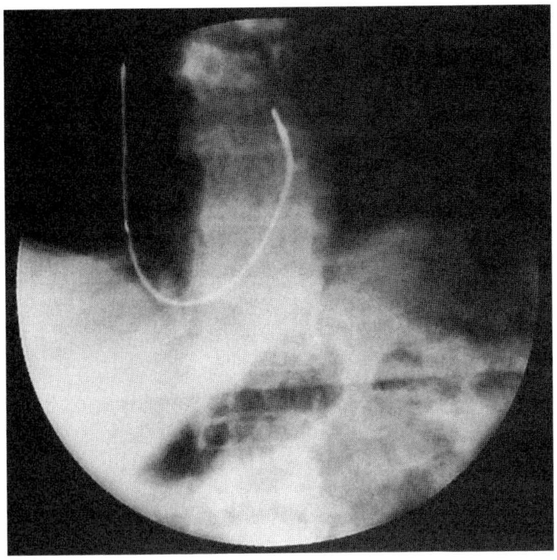

FIGURE 39-10 Passage of ventricular lead into right ventricular outflow tract, with plan to pull back and allow lead to "drop" into right ventricular apex (subtraction image).

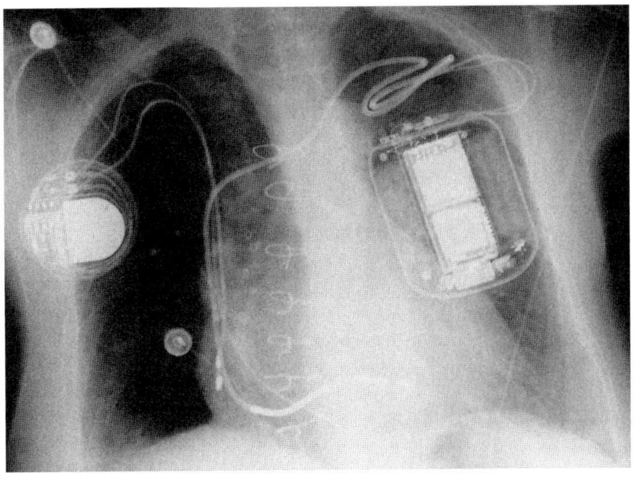

FIGURE 39-11 Patient with right-sided pacer and left-sided defibrillator, with active-fixation ICD lead positioned in right ventricular outflow tract to minimize interaction with right ventricular apical pacing lead. ICD, Implantable cardioverter defibrillator.

into the right ventricular apex position (Fig. 39-10); this approach ensures that passage has been achieved into the right side of the heart and reduces the inadvertent placement of the lead into the coronary sinus. With any lead manipulation, great care must be undertaken to avoid dislodgment of other leads already present, particularly if these have been freshly placed. In addition, placement of the leads may result in induction of ventricular arrhythmias and or trauma to the atrioventricular node or right bundle (of great concern in patients with preexistent left bundle branch block), dictating the need for back-up equipment for external pacing and defibrillation. Placement of the ventricular lead in patients with persistent left superior vena cava may be particularly challenging: a loop-de-loop technique is employed whereby the lead enters the dilated coronary sinus from the left superior vena cava and is banked off the right atrial wall to re-enter the right ventricular apex. An active fixation lead is recommended for stabilization of position in this case (see Fig. 39-34).[43] Occasionally patients may require placement of the lead in the right ventricular outflow tract; for example, to minimize electrical interaction with a previously implanted right ventricular apex defibrillator lead—an active fixation bipolar lead should be employed (Fig. 39-11). There has been some discussion as to the optimal position for right ventricular pacing to promote cardiac hemodynamics (e.g., apical versus outflow tract) but to date there is no definitive answer.[44-46] There is growing interest in biventricular pacing as a method to treat congestive heart failure.[47-49] Placement of a second ventricular lead to enable left ventricular pacing has been accomplished epicardially but also through transvenous placement into the coronary sinus[50] or transseptally.[51]

Manipulation of the atrial lead primarily entails using a straight stylet to pass the lead through the vein into the right atrium, followed by replacement with a curved stylet. Great care must be exercised, as with manipulation of any stylet, to avoid getting heme on the stylet lest it occlude the lumen of the lead and thereby prevent the passage of other stylets that might be required. In the case of a preformed (J-shape) passive fixation atrial lead, engagement of the atrial appendage is the goal and may be appreciated fluoroscopically by a "dog-wag" appearance to the engaged lead as it moves with atrial contraction. For active fixation leads, it is important to test the screw mechanism before passage. Active fixation atrial leads are either preformed J-shape or require the implanter to form a J by passage of a J-shaped stylet; in either case, it is useful to watch the screw mechanism extend under a magnified fluoroscopic image and then torque the body of the lead clockwise slightly for further stabilization. It is also important to test fluoroscopically for stability of the lead fixation with gentle traction, particularly as the J-shaped stylet is being withdrawn from the atrial lead.

As to special positioning of the atrial leads, there has been preliminary work addressing the potential benefit of preventing atrial fibrillation by pacing at Bachmann's bundle and/or interatrial septum pacing.[52,53] Perhaps even more promising has been the observation that dual site right atrial pacing and bi-atrial synchronous pacing may have incremental benefit compared with single site right and left atrial pacing for the prevention of atrial fibrillation. Typically, dual site right atrial pacing consists of pacing simultaneously from leads positioned in the high right atrium and coronary sinus ostium; the proposed mechanism responsible for antifibrillatory effects is a reduction of activation times in the atrium, especially in areas of conduction delay.[54-56]

A question that frequently arises is how best to place atrial leads in a patient who is in atrial fibrillation or flutter at the time of implant. One approach is to acutely cardiovert the patient to sinus mechanism, provided there is no significant risk vis-à-vis thromboembolic phenomena (as assessed by recency of the dysrhythmia, transesophageal echocardiography, previous anticoagulation, etc.). Another is to blindly position the lead and use an active-fixation lead, in particular, to allow for a variety of test positions as assessed by atrial signals and electrogram mapping.[57] Cardioversion could be undertaken more leisurely thereafter in the latter case, or may occur spontaneously in the case of the paroxysmal fibrillator.

In some situations, the implanter may opt for a single-lead dual chamber system, particularly for patients with heart block but normal sinus node function and without atrial dysrhythmias.[58,59] Current models allow for bipolar atrial sensing with ventricular pacing and sensing (unipolar or bipolar), although leads that allow for atrial pacing as well are under active investigation. The advantage of this approach is that atrioventricular synchrony may be achieved with a single lead. It is important to size a patient with regard to cardiac configuration so as to choose a lead with appropriate spacing between the atrial bi-pole and ventricular electrodes. It is ideal to have the atrial electrodes contacting the atrial wall (as opposed to sitting in the blood pool), but also allow enough slack to accommodate changes in atrial electrode positioning that may arise with respiration or change in body position.

Following placement of the lead, use of the pacer system analyzer is crucial in evaluating sensing thresholds, pacing thresholds, lead impedance, and for the possibility of diaphragmatic stimulation at high output—either through atrial leads (phrenic nerve stimulation) or ventricular leads (through a thin-walled right ventricle).

Perhaps the most important step in a pacemaker implant is lead anchoring. An anchoring sleeve minimizes trauma by the suture (typically an 0-silk) to the underlying lead insulation and/or conductor (Fig. 39-12). The right balance must be achieved between anchoring the sleeve too securely versus not securely enough; it is easier to bury the sleeves with a purse-string suture with leads implanted via the cephalic route than via the subclavian route. In addition, repeat fluoroscopy should be undertaken once anchoring has been performed to make sure that lead position and/or redundancy remains stable, both at baseline and with deep breathing. Enough slack should be provided to accommodate tall individuals, particularly with large respiratory excursions so as to prevent undue tension or retraction of the lead. The leads are then connected to the connector block of the generator; it is critical to ensure that the set screw sites are adequately tightened (including both poles in the case of bipolar leads) by gentle tugging on the lead. Application of sterile adhesive over the set screw site may minimize current drain through this locale.[60]

Creation of the generator pocket, either with Metzenbaum scissors/forceps or blunt finger dissection is followed by placement of the generator within. Any redundant or excess lead should be placed *under* the can so that future generator replacements will not be compromised by inadvertent slicing of the lead(s). In some cases, placement of the generator and lead within a Dacron pouch ("Parsonnet pouch") (Fig. 39-13) may allow for more ideal stabilization and anchoring of the system in the prepectoral fat and thereby minimize generator migration and extrusion or so-called Twiddler's syndrome.[61,62] In some patients, consideration of a submammary generator placement may be made for cosmetic reasons; the lead electrodes may be tunneled from the infraclavicular site of entry to the inframammary incision using a long needle, guidewire, and introducer/dilator technique analogous to the retained guidewire technique described earlier (see Fig. 39-35).[63,64]

Potential complications associated with a pacemaker implant are legion, but thankfully are uncommon and have been well described elsewhere.[65,66] There is clearly

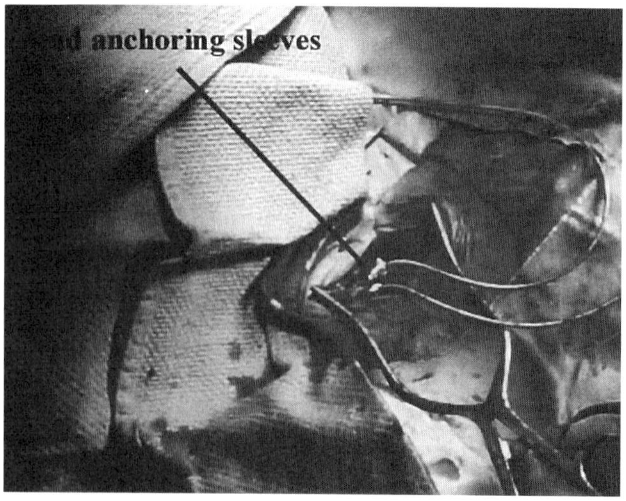

FIGURE 39-12 Use of anchoring sleeves for fixation of pacer leads to underlying fascia.

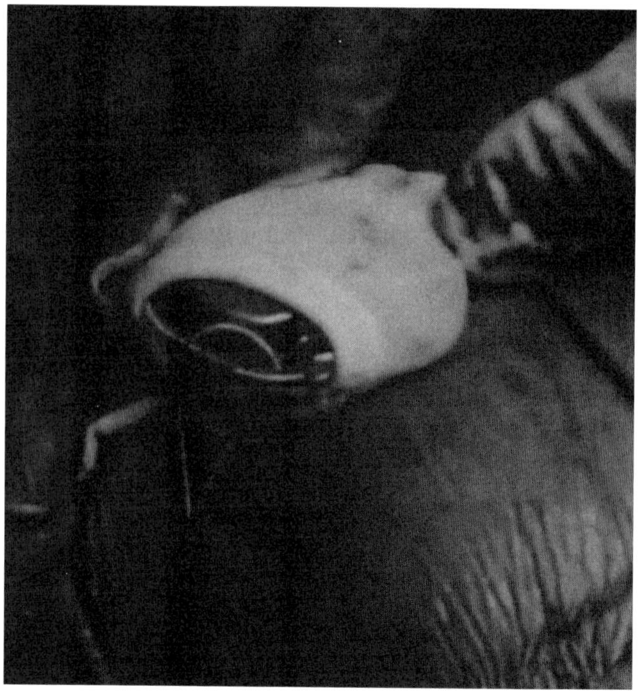

FIGURE 39-13 Use of Parsonnet pouch.

an inverse relationship between operator experience and complication rate, as well as a direct relationship associated with the use of subclavian puncture.[5] Preparedness for any eventuality, as discussed previously, remains the primary goal. Although problems with bleeding are usually readily apparent, as is the case for complications due to arrhythmia induction, the two most worrisome issues likely to be encountered are those of sudden hypoxemia and/or hypotension. The former raises the possibility of pneumothorax, oversedation, and air embolism (when the introducer technique is employed), whereas the latter may reflect a variety of issues, including ongoing bleeding (suspected or otherwise), hypovolemia, medication, left ventricular dysfunction, pneumothorax, or cardiac tamponade. Some of these complications may lead to a vicious spiral (downward); it is, therefore, incumbent on the implanter to keep close scrutiny on the vital signs, oximetry, and rhythm and remain in constant communication with those other individuals in the operating arena who are performing such monitoring.

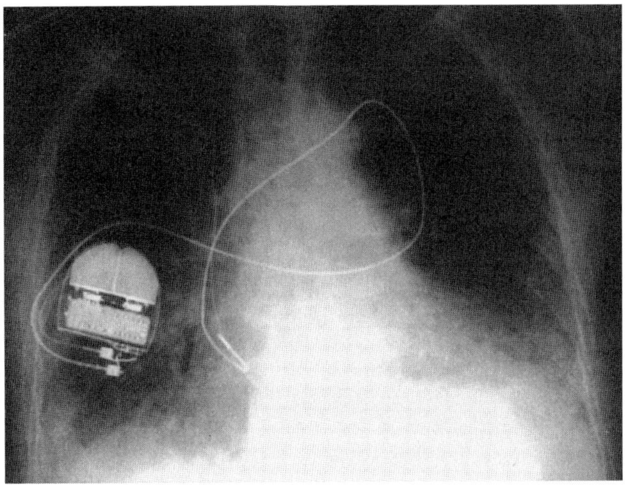

FIGURE 39-14 Right-sided pacer generator and ventricular lead, with atrial lead placed via left side and tunneled over to contralateral generator site.

Pacemaker Upgrades, Revisions, and Generator Replacements

On occasion, it may be necessary to upgrade an existing pacemaker system. This is most commonly encountered in patients with single chamber ventricular systems who experience pacemaker syndrome,[67-69] overt or subclinical,[69] thereby warranting revision to a dual chamber system. A good bit of operator experience is required for this; initial reports of upgrade procedures were associated with a remarkably high complication rate, which one can hope has lessened over time.[70,71] Obtaining previous operative notes is extremely useful to ascertain the vein previously employed. If the original route was subclavian, was it because there was no identifiable cephalic vein or was it the implanter's preference? If the latter, there may well be an accessible cephalic vein to accommodate the new atrial lead, requiring a cutdown technique as previously described. If there is no cephalic vein, the subclavian vein must be used. If there is any doubt as to the viability of the vein vis-à-vis potential thrombosis, dye injection of the ipsilateral brachial or axillary vein is helpful. Otherwise, percutaneous subclavian vein entry through the skin or from within the wound is undertaken, with great care to avoid needling or cutting the insulation of the previously implanted lead. Radiographic visualization of the searching needle trajectory is particularly useful in this regard.

If there is subclavian vein thrombosis or, alternatively, the infraclavicular space is too tight to accommodate a second lead (even the smallest unipolar available), then access may be attempted via the contralateral chest with tunneling of the new lead under the skin to the original pacer site (Fig. 39-14).[72] The alternative would be to abandon the original site altogether and place a new dual chamber system via the contralateral chest. If the new lead is to be implanted and tunneled back, a smaller incision can be used; once again, anchoring is paramount. For tunneling, a Kelly clamp can be advanced bluntly in the subcutaneous tissue from the receiving (original) site to the satellite (contralateral) site. The free terminal end of the newly placed atrial lead may be placed through a Penrose drain with a gentle tie applied around the drain just distal to the lead connector. The free end of the Penrose drain is then grasped and pulled through the tunnel, back to the original implantation site; the tie is released and the drain is then removed. Alternative approaches include use of a guidewire and peel-away introducer technique (introduced from the original site to the satellite location) with passage of the terminal end of the new lead back through the sheath to the original site; a chest tube may also serve as the tunneling conduit. Great care must be taken to strive for a tunnel that is deep (as close as possible to the overlying muscle) and to avoid traumatizing the lead during the tunneling process.[73]

Less commonly, the contralateral site may be used for the second (new) lead, as well as for creation of a new pacemaker pocket with placement of the new generator. Under these circumstances, the original lead is then tunneled under the skin, using the same procedure just described, to the new (satellite) location. Depending on the available remaining length of the original lead, a lead extender may be required to traverse the distance to the new site.

Not infrequently, existing lead(s) may go bad, either because of insulation breach or conductor fracture. Placement of a new lead is then required, either through the original site or via the contralateral chest. The same techniques apply as discussed with an upgrade from single chamber to dual chamber systems. The implanter may decide to abandon the original site altogether and place a new system via the contralateral chest, either leaving the original system intact, removing the old generator (with capping and anchoring of the old lead[s]), or removal of all the previous hardware (generator and leads). Provided there is no evidence of infection and/or erosion, old leads to be abandoned may be retained; i.e., without mandating extraction (see Fig. 39-31).[74-77]

Generator replacement is a commonly undertaken procedure that is often minimized but, in fact, requires advance thought. This is particularly the case in the patient with pacer dependence. It is important to establish whether there is any escape rhythm by slowly lowering the programmed rate. If no spontaneous rhythm emerges, consideration must be made as to how to maintain perfusion when the old generator is disconnected from the chronic leads. Temporary transvenous pacing or noninvasive external (Zoll) pacing are alternative solutions to quickly changing from the old to new generators (i.e., hoping that the implanter has nimble hands!). For unipolar systems, as soon as the generator is lifted out of the pocket, capture will be lost; to facilitate conversion to a new generator, a "money-clip" may be externally attached to the generator with its associated alligator clip connected to the skin retractor at the pocket site. The new generator must be compatible, either primarily or through adapters, with the existing leads[78] and knowledge of the previous system is, therefore, critical—both with regard to lead manufacturer, nature of the terminal pin, polarity, and so on. Visual inspection of the leads and determination of thresholds and impedances—both at baseline as well as with gentle traction on the leads—are all important maneuvers that must be undertaken to make sure that lead replacement is not required (in addition to the generator replacement). Rarely, repair of a conductor fracture may be undertaken using splicing techniques and, occasionally, a terminal pin modification may be made with splicing techniques if an otherwise-needed adapter is unavailable. Not uncommonly, an anchoring sleeve may be applied with sterile silicone adhesive to repair an insulation breach in the lead. One note of caution is warranted and applies to generator changeovers: an alarmingly higher incidence of infections is noted compared with primary implants because insufficient attention is paid to meticulous aseptic technique during these more ambulatory procedures.

Management of Pocket Hematoma, Erosion, Infection, and Pacer Extraction

In the perioperative time frame, it is not uncommon to encounter pacer pocket hematomas, particularly if the procedure has been associated with venous back bleeding at the lead insertion site, if arterial bleeding has been encountered, or if a tear has been made outside the fascial plane during creation of the pocket;[79] the use of anticoagulation or aspirin will also predispose to problems in this regard. Arterial bleeding will result in the rapid formation of a sizable and enlarging hematoma and requires immediate exploration. For less dramatic hematomas, cautious observation is warranted to watch for undue tension on the overlying pocket wall and subsequent tissue necrosis. Needle aspiration of the pocket may result in some decompression but is generally to be avoided because it will not remove clots and may result in the introduction of infection.

Rarely, the decision may be made to remove a generator without replacing it. For example, there are certain patients who elect to have the device removed because of ongoing discomfort related to the implant that does not satisfactorily dissipate with time (so-called *pacemakerodynia*). It is always wise to review the original indications for pacer implantation to address the advisability of removing a device without replacement, so as not to second-guess the past.[80,81]

What is more difficult is deciding whether to remove one or more extant leads, let alone actually performing the corresponding procedure. Unequivocally, device-associated infection is best treated by removal of all associated hardware.[82,83] To do less; i.e., clean the site and pacemaker and aim for reimplantation at the same or distant site, has been reported but is not recommended.[84,85] Once the skin barrier has been broken, colonization occurs and infection is likely; if the erosion or infection occurred because of previous indolent infection at the time of implant or by secondary seeding, then clearly infection is present at the outset.

Extraction of leads may be undertaken transvenously or through a more extensive open-heart surgical approach (i.e., sternotomy or thoracotomy).[86-91] An intermediate procedure called the "transatrial approach" has been described by Byrd[87]; it is reserved for patients whose leads are not accessible or removable through the inferior or superior vena cava. During this procedure the right atrium is exposed by removing the third or fourth right costal cartilage, the pericardium is opened, and a pursestring suture is placed in the right atrium; a biopsy instrument is then inserted through the pursestring and the lead is pulled out of the atrium.

Some clinicians have drawn the distinction between extraction and simple lead removal: in the former the leads are more difficult to take out of the patient. A North American Society of Pacing and Electrophysiology policy conference has proposed that "extraction" apply to the removal of any transvenous lead that has been implanted in excess of 1 year or that requires the use of tools beyond standard stylets or simple traction.[92] Indeed, the procedure can be quite challenging and potentially dangerous, even in experienced hands. In one large database, the risk of incomplete or failed extraction increased with implant duration, less experienced physicians, ventricular leads, noninfected patients and younger patients. Major complications were reported for 1.4% of patients (less than 1% at centers with more than 300 cases), with complication risk associated with increased number of leads removed, female gender, and less experienced physicians.[93,94] This includes death, nonfatal hemopericardium or tamponade, hemothorax, arteriovenous fistula, need for transfusion, pulmonary embolism, and stroke. Unequivocally, there is a learning curve associated with extraction techniques.[95] The procedure is rarely emergent. As is the case with pacer implantation, the physician embarking on extraction must be prepared in advance to address any emergent complication, with particular emphasis on defibrillation, pericardiocentesis, and/or cardiac surgical back-up. Large-bore intravenous access and arterial line monitoring are essential. Temporary pacing leads should be inserted in all patients who are pacer dependent.

Three transvenous approaches have been described for lead removal: mechanical, laser, and electrosurgical.[89] Mechanical approaches entail traction on the lead; removal by simple pulling was more easily achieved decades ago with leads that were silicone rubber non-tined or short-tined. Where applied, direct traction by pulling on the lead (preferably with a standard lead stylet inserted within) is undertaken with just enough force to feel the tugging of the heart without inducing chest pain, mechanical embarrassment with hypotension/tamponade, or ventricular arrhythmias. Such force may be applied for minutes, although in years past, various weights or elastic bands were used to allow constant traction for a period of days.[96,97] Excessive force may result in stretching or tearing the lead, or inducing damage either in the vein or at the tissue-electrode interface.

Improved fixation techniques used to minimize lead dislodgment have ironically led to increased difficulty with lead extraction when required.[93] As a result, countertraction and counterpressure techniques and devices have evolved to facilitate extraction. The use of locking stylets has made it far easier to apply direct traction.[98,99] The superior lead extraction approach entails opening the pocket and using lead clippers to remove the terminal connector from the pacer lead to be extracted. The insulators and outer conductor coil (in the case of coaxial leads) are trimmed back to expose the inner conductor coil, and the opening of this coil is then broadened with a coil expander. Sizing of the inner conductor coil is then undertaken, typically with a set of variably sized gauge pins, to select the largest locking stylet that will fully enter the inner conductor coil. The locking stylet is then introduced and advanced to the lead tip. In some stylets, counterclockwise rotation may be required (Cook Vascular, Leechburg, PA), whereas in others (Wilkoff stylet, Cook Vascular, Leechburg, PA) removal of a latch pin and pushing the locking cannula forward is required to activate the locking mechanism. In yet another model (Spectranetics, Colorado Springs, CO), a lead locking device stylet has a fine, wire mesh stretched over its entire length; the mesh is released once the stylet is advanced and by bunching up holds the lead along its entire length (Fig. 39-15). Tugging on the stylet to ensure the adequacy of locking is followed by ligating the end of the lead insulation with 2.0 suture and tying the suture ends to the locking stylet loop handle. This handle is flattened and placed through a preselected sheath set (telescoping inner and outer sheaths).

Standard sheaths are made of a plastic, such as Teflon (E.I. DuPont, Wilmington, DE) or stainless steel. The latter are used typically when initially entering the central venous circulation and are used to cut through dense fibrosis near the subclavian site before exchange with the more flexible plastic sheaths that are used to negotiate through the various curves encountered along the venous path. Continuous and moderate tension is applied on the locking stylet as first the metal set and then plastic sheaths are advanced to disrupt fibrotic deposits (the outer sheath has a sharp cutting edge that is advanced over its inner sheath). Fluoroscopic guidance is essential, recognizing that misdirected dilators may result in serious vascular injury, particularly at curvatures in the vein where the more flexible plastic sheaths are required (Fig. 39-16). If too much tension is applied to the lead/locking stylet combination, the risks attendant to conventional external traction may arise: tearing the lead, avulsing the vein or heart wall, or disengaging the locking stylet from the lead. The larger sheath is advanced over the inner sheath, always keeping the smaller sheath on the leading edge in a telescoping forward movement. A snow plow effect[89] may arise with scar tissue that is peeled away from the venous wall and pushed in front of the sheaths, thereby preventing further advancement of the dilators.

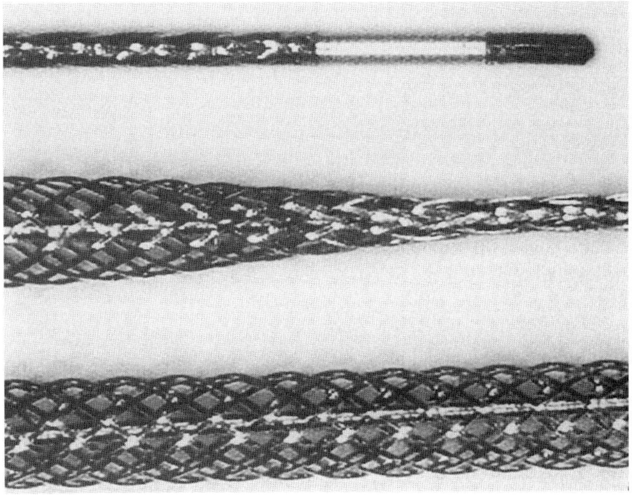

FIGURE 39-15 Special mesh locking stylet that expands along entire length of lead to be extracted.

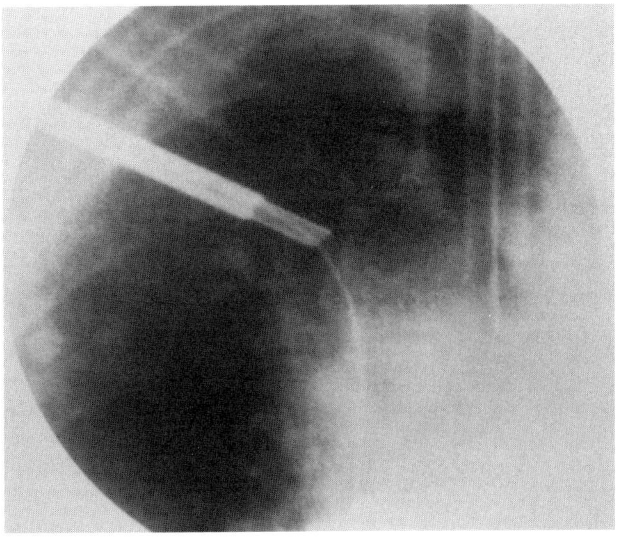

FIGURE 39-16 Use of telescoping (inner and outer) extraction sheaths over lead to be removed.

When the sheath set approaches the lead tip near the tissue-electrode interface, the outer sheath is removed and reversed so that its blunt end faces the myocardium and countertraction is then applied against the myocardial surface to release the lead tip. In some cases after a chronic (noninfected) lead is removed, the retained sheath may be used to place a guidewire, then a peel away dilating sheath to guide the placement of a new pacemaker lead; occasionally balloon venoplasty may be required as an adjunct.[100]

Powered sheaths have been developed to reduce the need for traction and counterpressure.[101-102] The laser sheath is used to deliver excimer laser energy fiberoptically to the distal end of the sheath to dissolve the encapsulating fibrotic tissue in circumferential fashion (Fig. 39-17). The laser has a very short depth of penetration, affecting only tissue that is in direct contact with the end of the sheath. This releases the lead from its tissue attachments, thereby facilitating advancement of the sheath over the lead. More recently, an electrosurgical dissection sheath is under active investigation using radiofrequency energy with a more directed sheath tip.

Snares may be employed to apply indirect traction in extracting lead(s) using a femoral approach. The snare device is used to encircle the free tip of the pacing lead or a loop of the lead with traction then applied from below (Figs. 39-18 and 39-19). The approach may be challenging because it is often difficult to ensnare a lead, but may be less injurious because the coring out of surrounding fibrotic tissue is not required using this approach; nonetheless, the operator must be aware of the associated risks of femoral vein injury and thromboembolism, as well as the same potential complications associated with traction of the lead from above. The technique is indicated if the lead to be extracted is not accessible from the venous entry site, if the lead has been cut or fractured, or if there has been retraction of the lead into the central venous circulation and/or intracardiac space. We have found the technique to be particularly well suited to removing atrial leads under the Accufix advisory associated with fracture of the inner retention J wire, particularly if this wire is protruding.[103-104] If the lead is still anchored in the pectoral region, it must first be freed. Thereafter, a sheath with a hemostatic valve is placed via the femoral vein to the inferior vena cava level just below the atrium and through this the grasping tool is deployed. Tools available for this include a deflecting tip guidewire and Dotter helical basket retriever, pigtail catheter, or Amplatz catheter; special countertraction sheaths are also available.[105-107]

What remains clear at this writing is that lead extraction is a procedure that is not undertaken lightly, may be hazardous, and requires specialized training and careful deliberation and preparation. Indeed, in some series (e.g., Accufix leads), the risk of extraction over time has exceeded that associated with retention of

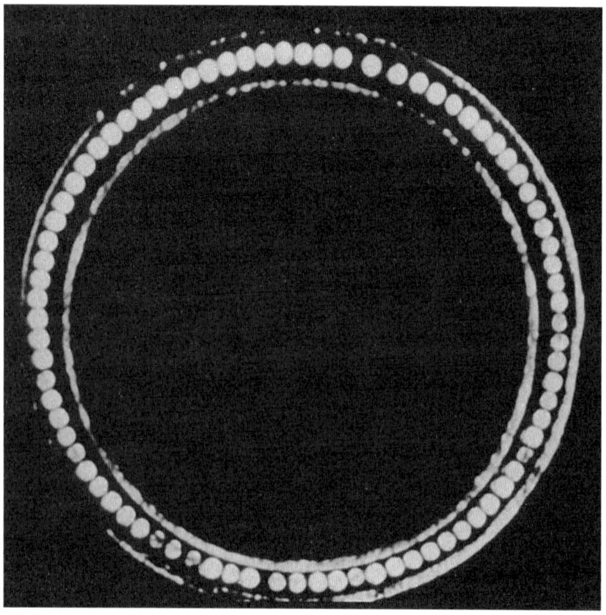

FIGURE 39-17 Cross-section of excimer laser sheath.

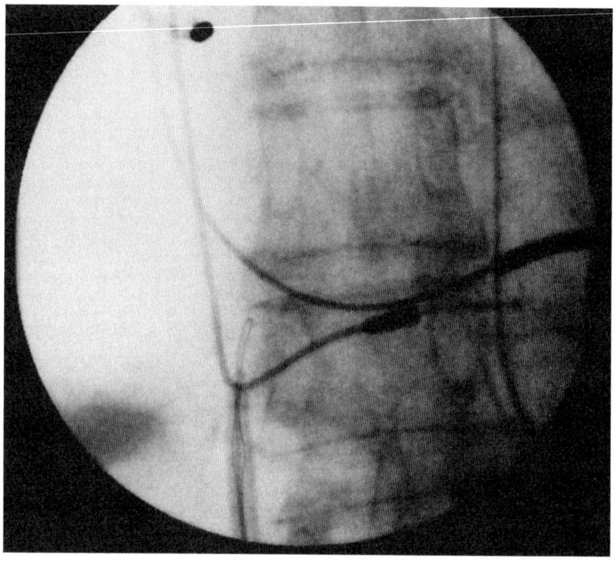

FIGURE 39-18 Fluoroscopy of right ventricular apical lead being extracted from femoral approach/snare technique in patient with dual chamber pacing system and separate defibrillator lead in right ventricular outflow tract.

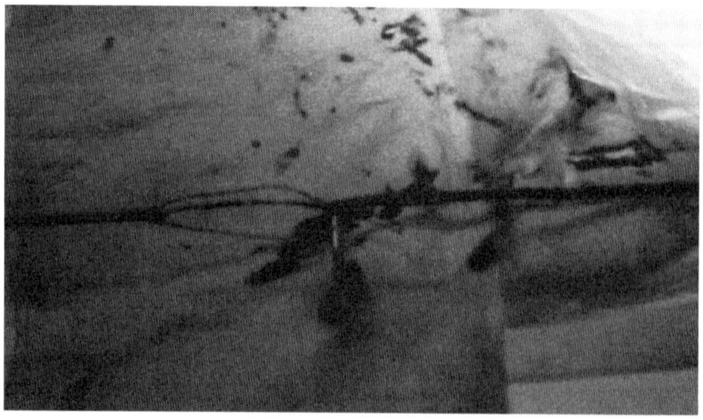

FIGURE 39-19 Right ventricular pacing lead after removal via snare technique (same as Fig. 39-18).

the lead.[108] In some cases (e.g., infected pacer systems), extraction is essentially unavoidable, but there is certainly a precedent for considering abandonment without extraction of noninfected leads.

Follow-Up of the Pacemaker Patient

In the realm of follow-up, there is all too often a tendency to pay attention to the device and its function without adequately acknowledging the condition of the patient. A multitude of factors has conspired to diminish the importance of interpersonal interaction: less available time, less physician reimbursement, the widespread use of transtelephonic monitoring, and increasing reliance on pacemaker automaticity.[109] It remains incumbent on the physician to elicit symptoms from the patient to determine whether the pacemaker is meeting the needs for which it was originally indicated *and*, perhaps less commonly, to ensure that the system is not *contributing* to new problems. Follow-up is intended to document existing pacer system malfunction, if any, as well as to *anticipate* potential problems before they result in patient compromise. The nuances of follow-up have received insufficient emphasis at a time when these devices have, ironically, become increasingly complex.[110-114] It remains unclear as to what constitutes a sufficient core knowledge base for *current* pacemaker follow-up and who is qualified to perform such tasks.[1-3]

The outpatient clinic should allow for history taking and physical examination and be fully equipped to allow for assessment of device function. Equipment required includes electrocardiography, radiography and fluoroscopy, transtelephonic monitoring, ambulatory ECG recording capabilities, and physician's manuals and programmers corresponding to *every pacer model being followed*. A routine pacer follow-up schedule should be devised to allow for the recognition of *potential* problems well before they become *actual* problems. Record-keeping is paramount and should include demographics such as name, age, address, phone number, indications for pacer therapy, pacer generator and lead models/serial numbers, operative notes, follow-up histories and physical examinations, chest radiograph reports, ECG recordings with and without magnets, interrogated data, threshold data, real-time telemetry data, and printouts of programmed parameter settings. The ability to store records in a computer has allowed for databasing, which may prove useful in evaluating systematic issues with device performance, whether model or manufacturer specific.

Transtelephonic monitoring (TTM) of free-running and magnet rates has facilitated patient care,[115-117] particularly for patients who live remote from the pacer clinic but, as mentioned, cannot fully supplant the role of direct physician–patient contact. Sensing and capture can be addressed through this modality as well as an evaluation of ongoing generator viability (Figs. 39-20 through 39-24). In addition, correlation of symptoms with real-time transtelephonic transmissions has been used to identify potential pacer system malfunction or to detect otherwise unsuspected arrhythmias corresponding to a problem such as palpitations or dizziness.

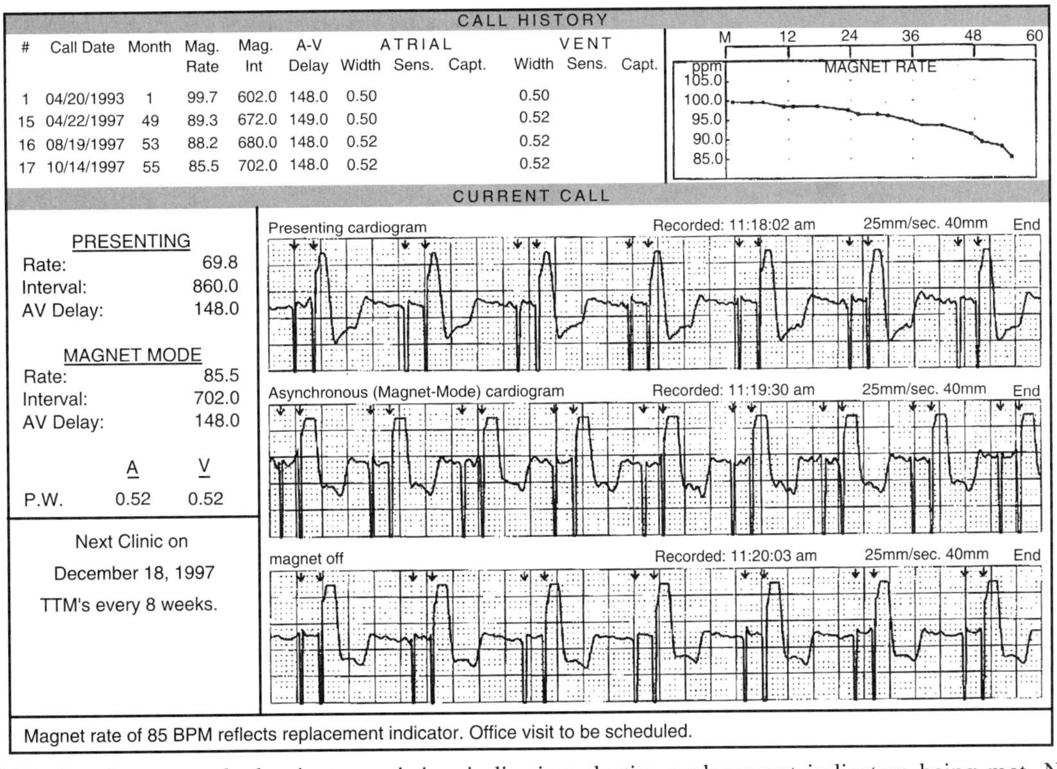

FIGURE 39-20 Routine transtelephonic transmission indicating elective replacement indicators being met. Note gradual decline in magnet rate on graph.

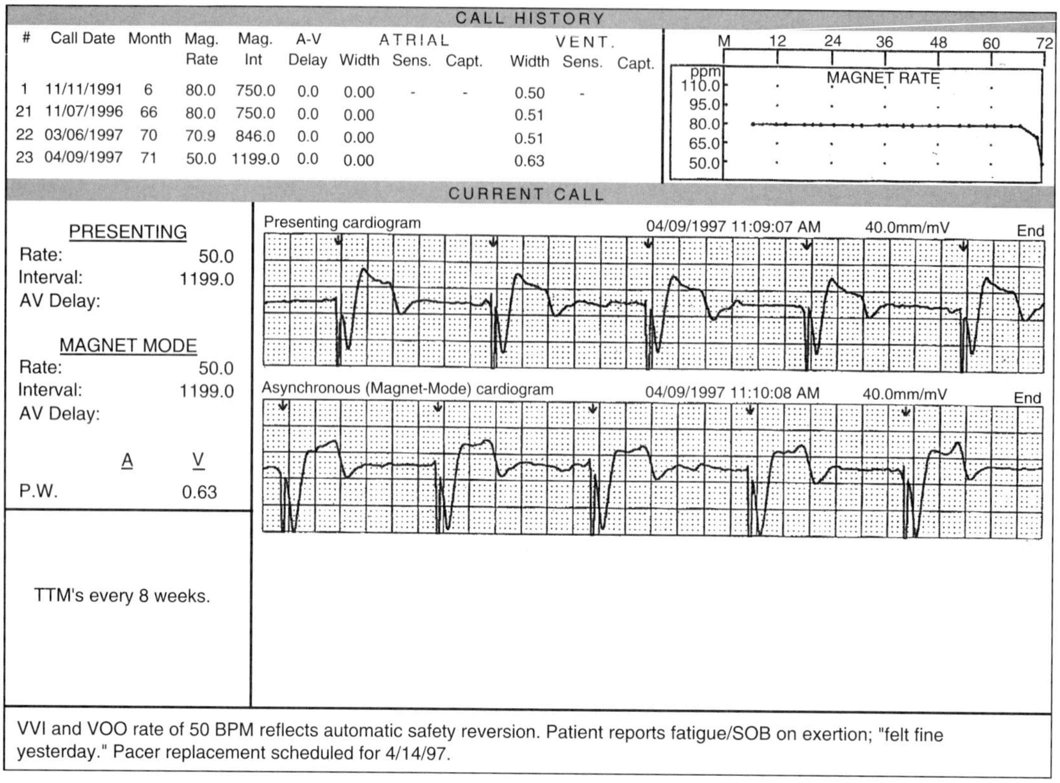

FIGURE 39-21 Transtelephonic transmission precipitated by symptoms of fatigue and dyspnea indicate reversion to VVI pacing at 50 BPM, responsible for symptoms and indicating elective replacement indicators.

The transmission entails connecting electrodes placed on wrists, ankles, or fingertips to a transmitter over which the telephone mouthpiece is positioned; the adequacy of the transmission may be influenced by insufficient voltage (depending on the lead used), 60-Hz interference, and noise from telephone or motion/tremor artifact. Transtelephonic transmissions and office records may be stored in computer format and allow for graphic representation of changes in magnet rate over time (see Figs. 39-20, 39-21, and 39-23A).[118] Specific Medicare guidelines exist for both pacer clinic follow-up and TTM scheduling:[119] the frequency of clinic visits is "the decision of the patient's physician, taking into account, among other things, the medical condition of the patient. However, Medicare carriers can develop monitoring guidelines that will prove useful in screening claims." The guidelines remain subject to debate, with some suggesting that less frequent TTM follow-ups may suffice.[120-121]

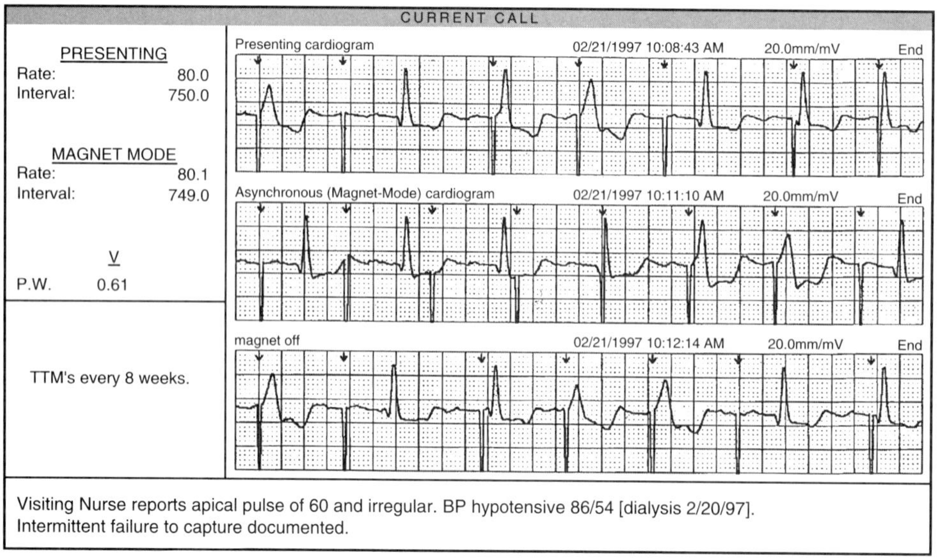

FIGURE 39-22 Tanstelephonic transmission for irregular pulse and hypotension documents intermittent loss of capture by ventricular lead.

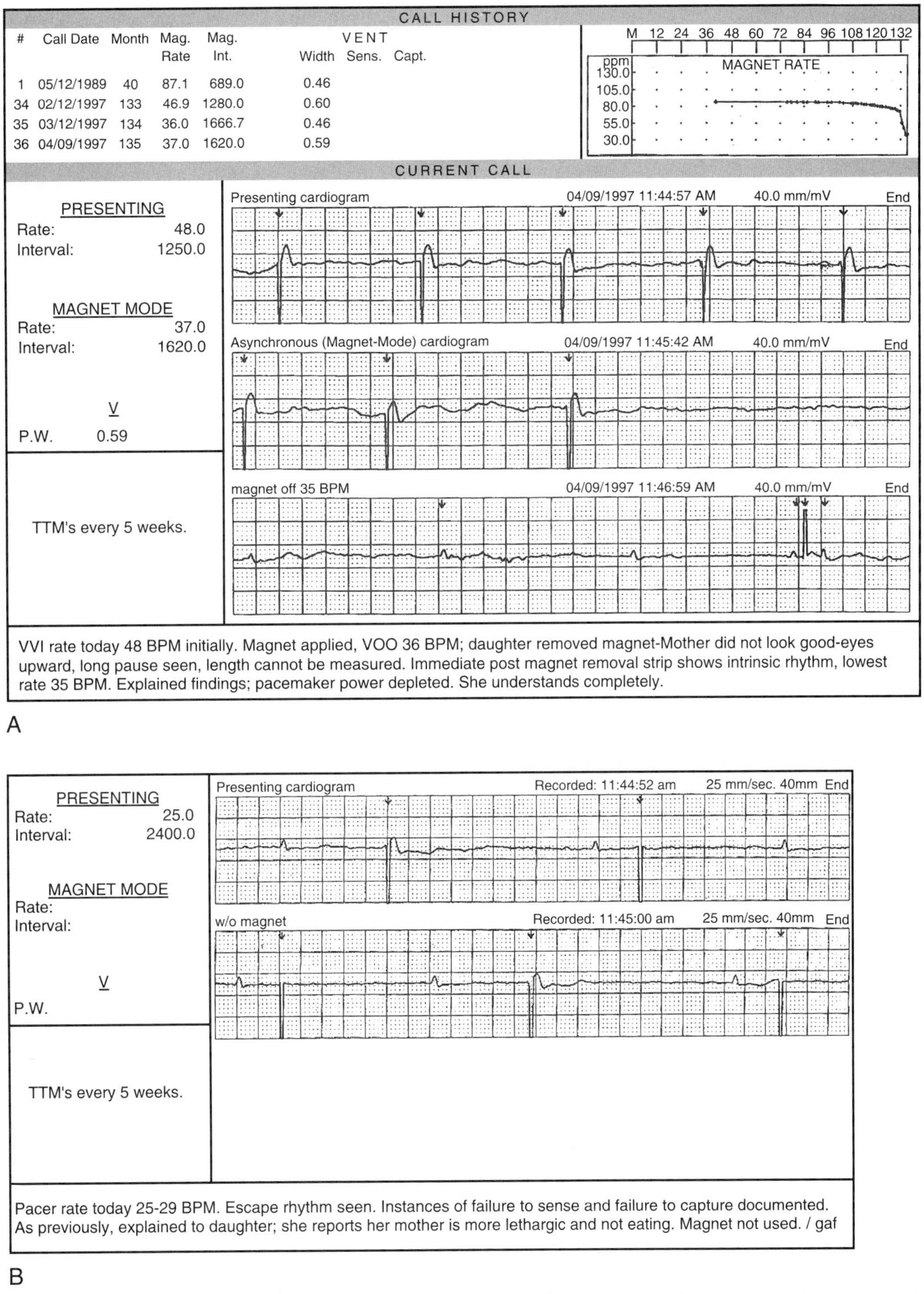

FIGURE 39-23 **A,** Transtelephonic transmission for elderly patient with failure to thrive documenting progressive decline of magnet rate and free-running rate—patient was refusing generator replacement. **B,** Same patient as in **A,** with further decline and continued refusal of generator replacement—daughter requested transmission.

The Patient History

The concept of physician as *sleuth* applies to the role of the pacemaker specialist—in many cases, present or impending difficulties with a pacer system may go unnoticed by the patient, and careful questioning, examination, and device interrogation is required to unearth otherwise unsuspected difficulties. Some patients may presume that the way they are feeling is normal and not volunteer any symptoms; until a reprogramming is undertaken they may not appreciate how unwell they truly felt. Although the pacer may be addressing the

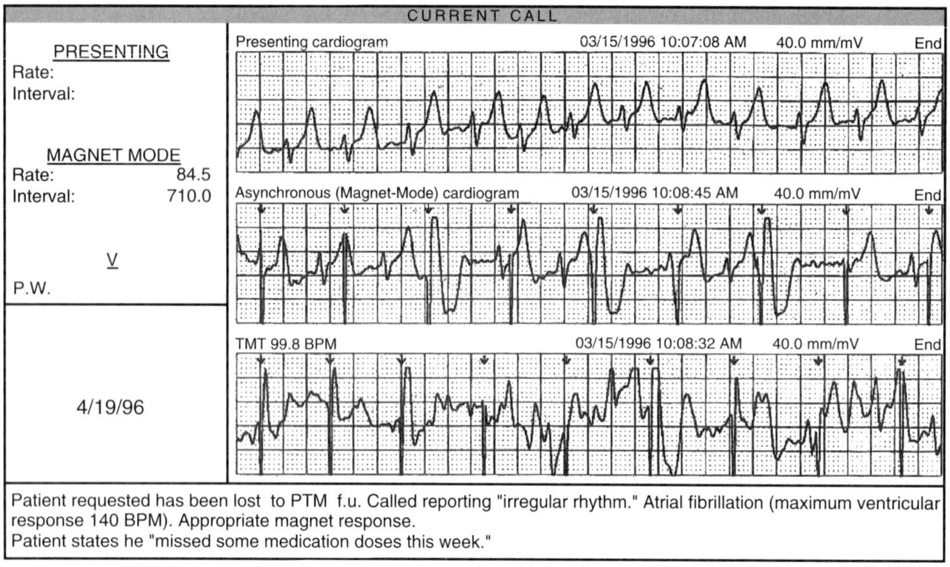

FIGURE 39-24 Patient with symptomatic irregular rhythm with transtelephonic transmission documenting atrial fibrillation.

problem of bradycardia, it may not be addressing some of the other conditions faced by the patient and indeed may, at times, exacerbate such conditions. Such "patient-pacer mismatch" is exemplified by inappropriately low upper rate limits for younger patients or excessively sensitive rate-adaptive parameters in older patients with angina.

The cardinal symptoms of dizziness, presyncope, and syncope are important to address for the pacer patient, though other concerns such as easy fatigability, angina, and dyspnea may be less obvious.[122] These latter symptoms may dictate the need for reprogramming. For example, angina may be precipitated by a pacer system if the upper rate limits are inappropriately high or unduly low, precluding adequate time for diastolic coronary perfusion, and patients with rate-limited cardiac outputs may suffer from congestive heart failure if their backup rates are set too low (Fig. 39-25). Very importantly, a pacer system may actually make the patient feel worse despite apparently normal electrical function because of the so-called *pacemaker syndrome.*[123-126] This most commonly reflects the loss of optimal atrioventricular synchrony in patients with single chamber ventricular pacing systems. The pathophysiology of this phenomenon includes systemic hypotension, atrioventricular (AV) valvular regurgitation, reduced cardiac output, unpleasant cannon A waves in the neck resulting from atrial contraction against a closed AV valve, and pulmonary and hepatic congestion (Fig. 39-26). Such impairment may be particularly pronounced in patients whose cardiac output is critically dependent on "atrial kick" because of underlying diastolic heart conditions such as hypertrophic cardiomyopathy or aortic stenosis.[127] However, even patients with structurally normal hearts may suffer from pacemaker syndrome if they manifest ventriculoatrial (VA) retrograde conduction with *ventricular* pacing or if they have coexistent AV nodal disease that results in Wenckebach block with single chamber *atrial* pacing (often requiring reprogramming to slower atrial paced rates or upgrade to dual chamber pacing systems).

Other symptoms reflect more mechanical concerns related to the pacing system. These include diaphragmatic stimulation or hiccoughs (from phrenic nerve stimulation by an atrial lead, from pacing across a thin right ventricular wall by a ventricular lead, or from direct

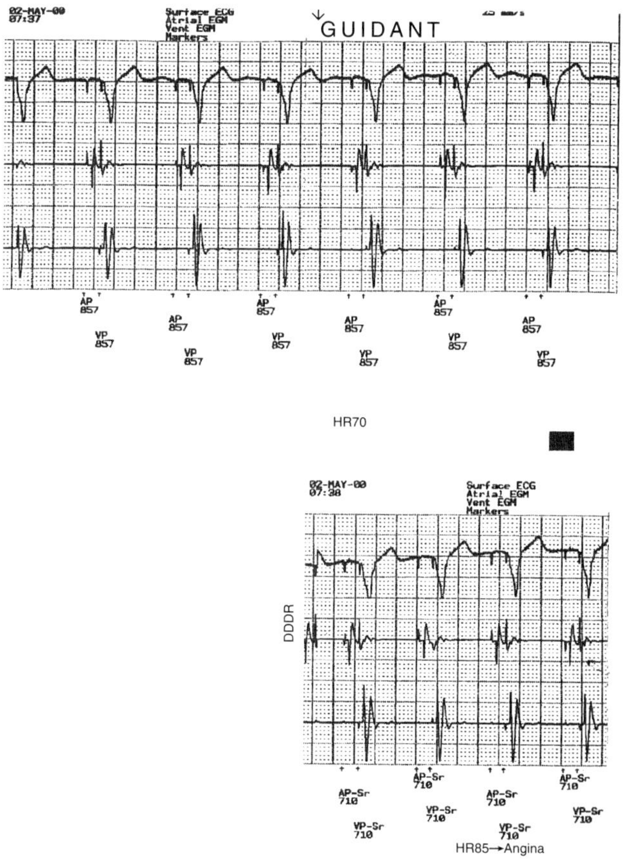

FIGURE 39-25 Patient with chronotropic incompetence without symptoms when DDD pacing at 70 BPM, develops angina when rate-adaptive mode enabled with DDDR at 85 BPM.

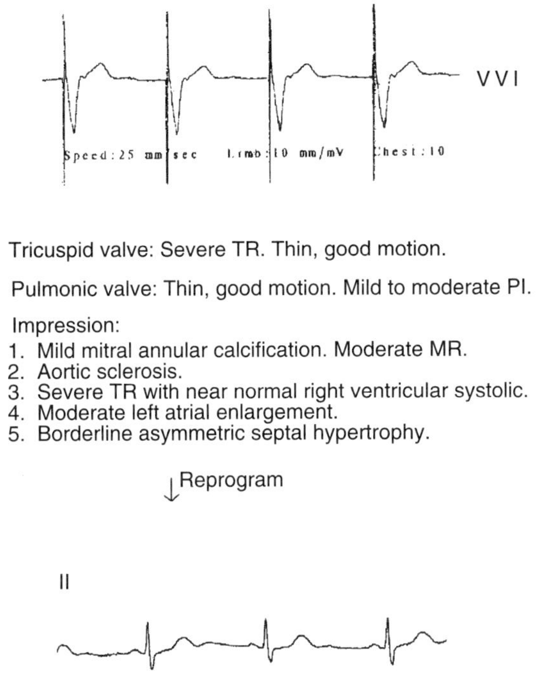
VVI
Speed: 25 mm/sec
Limb: 10 mm/mV
Chest: 10
Tricuspid valve: Severe TR. Thin, good motion.
Pulmonic valve: Thin, good motion. Mild to moderate PI.
Impression:
1. Mild mitral annular calcification. Moderate MR.
2. Aortic sclerosis.
3. Severe TR with near normal right ventricular systolic.
4. Moderate left atrial enlargement.
5. Borderline asymmetric septal hypertrophy.
Reprogram
II
Impression:
1. Mild mitral annular calcification. Moderate MR.
2. Mild to moderate TR.
3. Mild PI.
4. Increased right ventricular systolic pressure.
5. Aortic sclerosis.
6. Since previous test, decreased TR.

FIGURE 39-26 Patient referred for ventricular lead extraction because of severe tricuspid regurgitation thought to be secondary to lead traversal of tricuspid valve; note retrograde ventriculoatrial conduction with pacing. Tricuspid regurgitation significantly diminished when pacer was reprogrammed to lower rate, allowing for normal atrioventricular activation.

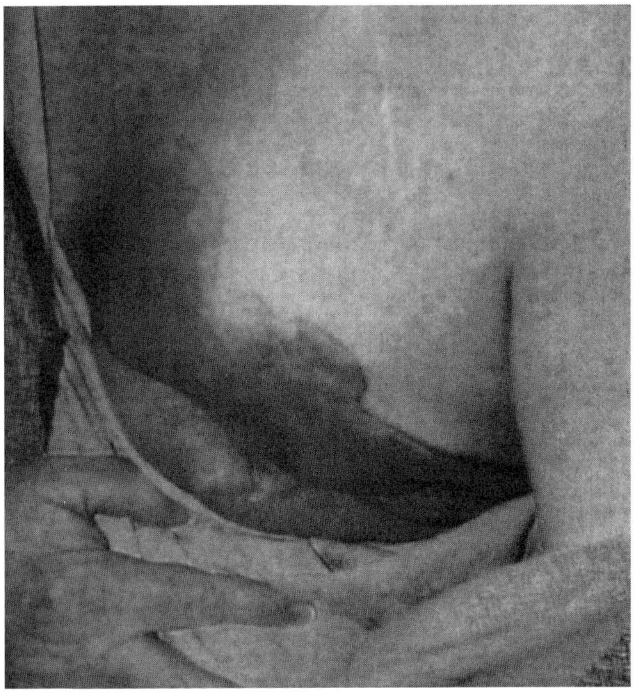

FIGURE 39-27 Patient with breast cancer and prosthetic breast expander implant who developed postoperative cellulitis, just inferior to her separate pectoral implantable cardioverter defibrillator (ICD).

stimulation by a perforated ventricular lead), myopectoral stimulation (most commonly with unipolar pacing systems), and wound-related symptoms such as swelling, discharge, erosion, or pain.

Physical Examination

The status of the incision and pacer pocket must be addressed, especially in patients with potential impairment of wound healing due to steroid dependence or diabetes. Anticoagulation and/or aspirin therapy may also affect the healing process with resultant pocket hematomas.[128] The wound should be examined for warmth, tenderness, erythema, hematoma, or evidence of incipient or overt erosion. In addition, adjacent processes that may affect the wound should be identified, such as skin or breast malignancies or skin infections (Figs. 39-27 and 39-28). Superficial proliferation of veins near the pacer site may suggest occlusion of the central venous access site. Caudal migration or superficiality of the pacing leads and/or anchoring sleeves are commonly observed but do not usually require any intervention. The approach to pocket hematoma and system erosion has been addressed earlier.

Diaphragmatic stimulation may be evident on physical examination, and may usually be addressed by reducing the ventricular output (in the setting of a thin-walled right ventricle), or by reducing atrial output (in cases of phrenic nerve stimulation); occasionally

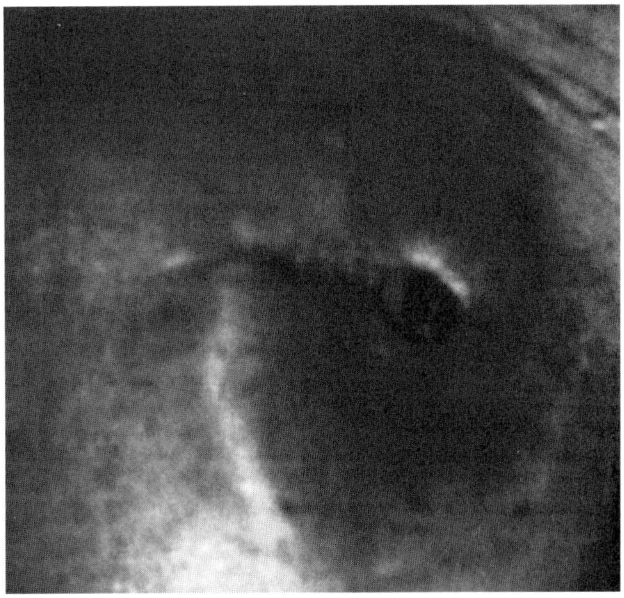

FIGURE 39-28 Erosion of pacer generator after generator replacement that was complicated by tense hematoma from anticoagulation used to treat atrial fibrillation.

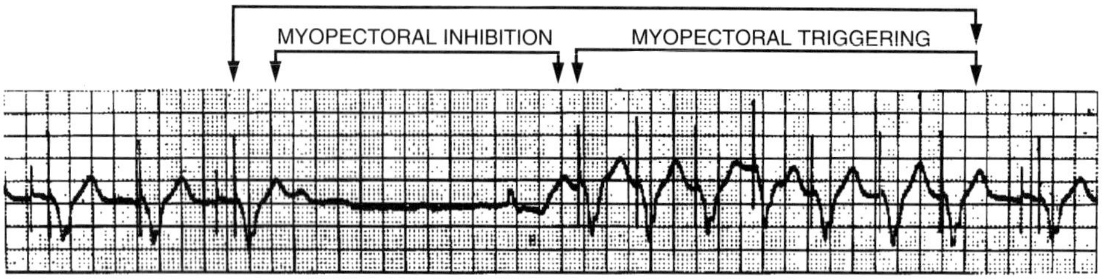

FIGURE 39-29 Isometric arm exercise inducing both myopectoral inhibition and myopectoral triggering in the same patient with a unipolar dual chamber pacing system.

invasive repositioning may be required, especially if the lead has displaced or perforated. Myopectoral stimulation coincident with the paced rhythm is most commonly seen in unipolar systems. In some cases, the generator has been inadvertently implanted with the uncoated side down; less commonly the generator may have flipped or been "twiddled" upside down.[129-130] In other cases, there may be an actual lead fracture resulting in direct stimulation of muscle by the conductor. In most situations, however, the problem reflects excessive stimulation of adjacent muscle by the unipolar (anode) can and may be addressed by reprogramming to a lower output.

Other important aspects of the physical examination are the vital signs, particularly pulse, to confirm pacing rates and blood pressure under various pacing modes (e.g., DDD versus VVI), as well as clues to the presence or absence of pacemaker syndrome. As discussed previously, this would include evidence of cannon A waves in the neck, pulmonary congestion, hepatomegaly, or AV valvular regurgitation. Paradoxical splitting of the second heart sound would be consistent with right ventricular pacing; rubs should be listened for to rule out potential cardiac perforation. It is unusual to detect tricuspid regurgitation merely due to traversal of the ventricular lead across the tricuspid valve. Indeed, recently I was asked to see a patient with severe tricuspid regurgitation on echocardiography as to whether her right ventricular lead should be extracted (see Fig. 39-26); the problem was corrected totally by reprogramming her VVI pacer to a lower rate to eliminate retrograde VA conduction. Occasionally, musical systolic murmurs can be heard, possibly reflecting movement of a pacing lead within the right ventricular cavity.[131,132] The generator/insertion site should be listened to for any evidence of bruits. Carotid sinus massage may be used during the physical examination to slow sinus or AV nodal conduction and elicit/assess pacemaker activity in patients who are otherwise not pacer dependent. The arm ipsilateral to the pacer site should be examined for swelling suggestive of central venous thrombosis; this may be treated with arm elevation but, in more severe cases, may require short- or long-term anticoagulation therapy.[133] Pulmonary thromboembolism from cephalic or subclavian vein thrombosis is unusual.

One of the most important and under-appreciated components of the physical examination is direct manipulation of the system (Figs. 39-29 and 39-30), as a check both on the integrity of the leads and their insertion into the generator. Insulation or conductor fractures may go otherwise unsuspected unless or until traction is applied to the can (see Fig. 39-30). Loose set-screw connections may similarly be revealed. Patients should be examined with deep breathing and coughing to confirm adequacy of pacing during these maneuvers to ensure that the lead position is stable and with adequate slack. In general, such maneuvers should be undertaken under electrocardiographic monitoring, and occasionally under fluoroscopic guidance (in cases where lead fracture or inadequate slack is suspected). The possibility of myopotential inhibition in unipolar single chamber systems or myopotential triggering in

MEASURED DATA	
Pacer Rate	59.7 ppm
Ventricular:	
Pulse Amplitude	3.8 Volts
Pulse Current	9.8 mAmperes
Pulse Energy	13 μJoules
Pulse Charge	4 μCoulombs
Lead Impedance	388 Ohms
Atrial:	
Pulse Amplitude	3.9 Volts
Pulse Current	6.0 mAmperes
Pulse Energy	8 μJoules
Pulse Charge	2 μCoulombs
Lead Impedance	646 Ohms
Battery Data: [W.G. 8077 - NOM. 1.8 AHR]	
Voltage	2.80 Volts
Current	17 μAmperes
Impedance	<1 KOhms

MEASURED DATA With traction on generator	
Pacer Rate	59.7 ppm
Ventricular:	
Pulse Amplitude	3.2 Volts
Pulse Current	40.7 mAmperes
Pulse Energy	25 μJoules
Pulse Charge	11 μCoulombs
Lead Impedance	<250 Ohms
Atrial:	
Pulse Amplitude	3.9 Volts
Pulse Current	6.1 mAmperes
Pulse Energy	9 μJoules
Pulse Charge	2 μCoulombs
Lead Impedance	648 Ohms
Battery Data: [W.G. 8077 - NOM. 1.8 AHR]	
Voltage	2.79 Volts
Current	24 μAmperes
Impedance	<1 KOhms

FIGURE 39-30 Traction on generator unmasks a lead impedance problem (drop to less than 250 ohms) responsible for a patient's device malfunction.

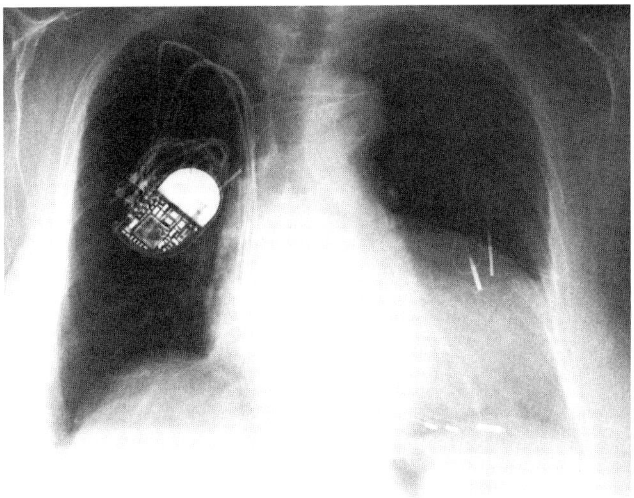

FIGURE 39-31 Chest radiograph of patient with three pacing leads inserted via the right chest and two via the left chest; three of the leads have been "abandoned."

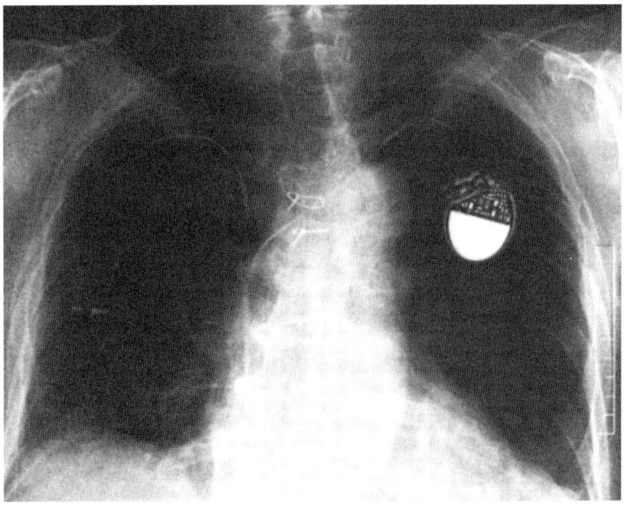

FIGURE 39-32 Chest radiograph of patient with abandoned lead in the right chest and fracture of two leads inserted via left subclavian, just under left clavicle.

dual chamber unipolar systems should be examined during various isometric pectoral muscle maneuvers such as weight lifting, abduction, pulling, or pushing (see Fig. 39-29). Tapping of the generator in rate-adaptive systems sensitive to activity may confirm the ability of the pacer to increase its rate appropriately with sensed deformation of the generator can.

Radiography

The chest radiograph remains indispensable to the follow-up of the pacemaker patient (Figs. 39-31 through 39-35).[134] Aside from screening for problems related to the pacer system, concomitant disease in other areas may occasionally be "picked up," as in the case of a lung malignancy. The ideal timetable/frequency for out-patient radiographic examinations remains unclear. The chest radiograph serves as a check on the stability of the pacer leads from the time of implant and confirms both the site of venous entry (e.g., external jugular versus cephalic or subclavian) and the site of lead tip anchoring. Both intentional (in the case of biventricular pacing) as well as inadvertent passage of a lead through the coronary sinus may be identified by a posteriorly directed lead in the lateral projection; the presence of a persistent left superior vena cava with subsequent entry of the lead into the coronary sinus may be similarly appreciated (see Fig. 39-34). Intentional active fixation of the lead into the right ventricular outflow tract has been increasingly used, particularly in an effort to position pacing leads away from preexistent leads in the right ventricular apex (e.g., in patients with both ICDs and pacers) and the radiograph is useful to confirm adequate placement as well as appropriate advancement of the active fixation screw-tip (see Fig. 39-33). Placement of the generator may also be appreciated, including unorthodox locations such as submammary or abdominal (see Fig. 39-35). The lead connection into the set-screw site of the generator connector block may also be examined radiographically.

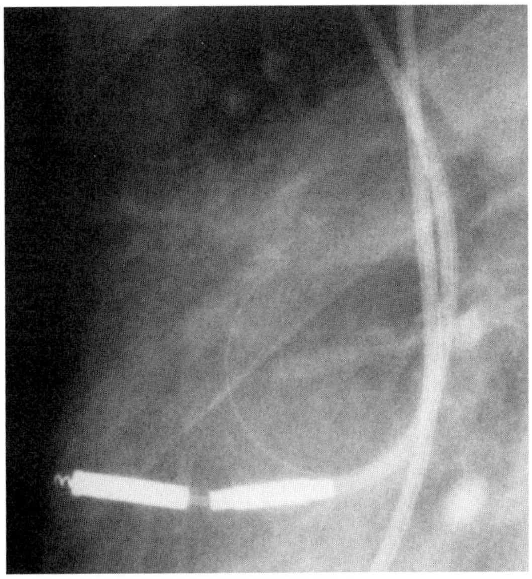

FIGURE 39-33 Protrusion of inner retention wire in Accufix active fixation atrial lead; note advancement of helical screw allowing for active fixation.

In addition to positioning, the chest radiograph may be useful in identifying the model of the generator; if unknown, different systems have various characteristic shapes or battery configurations and some models have model-specific codes that may be visible on the film. Multiple leads (especially if abandoned) may otherwise go unsuspected without radiography (see Fig. 39-31). Polarity (and as already mentioned, fixation) of the leads can also be evaluated, especially important in unsuspected situations where bipolar leads have been "down-programmed" to the unipolar configuration. Occasionally fractures of lead conductors may be seen, particularly at areas of acute angulation of the lead (see Fig. 39-32); sometimes real-time fluoroscopy with the

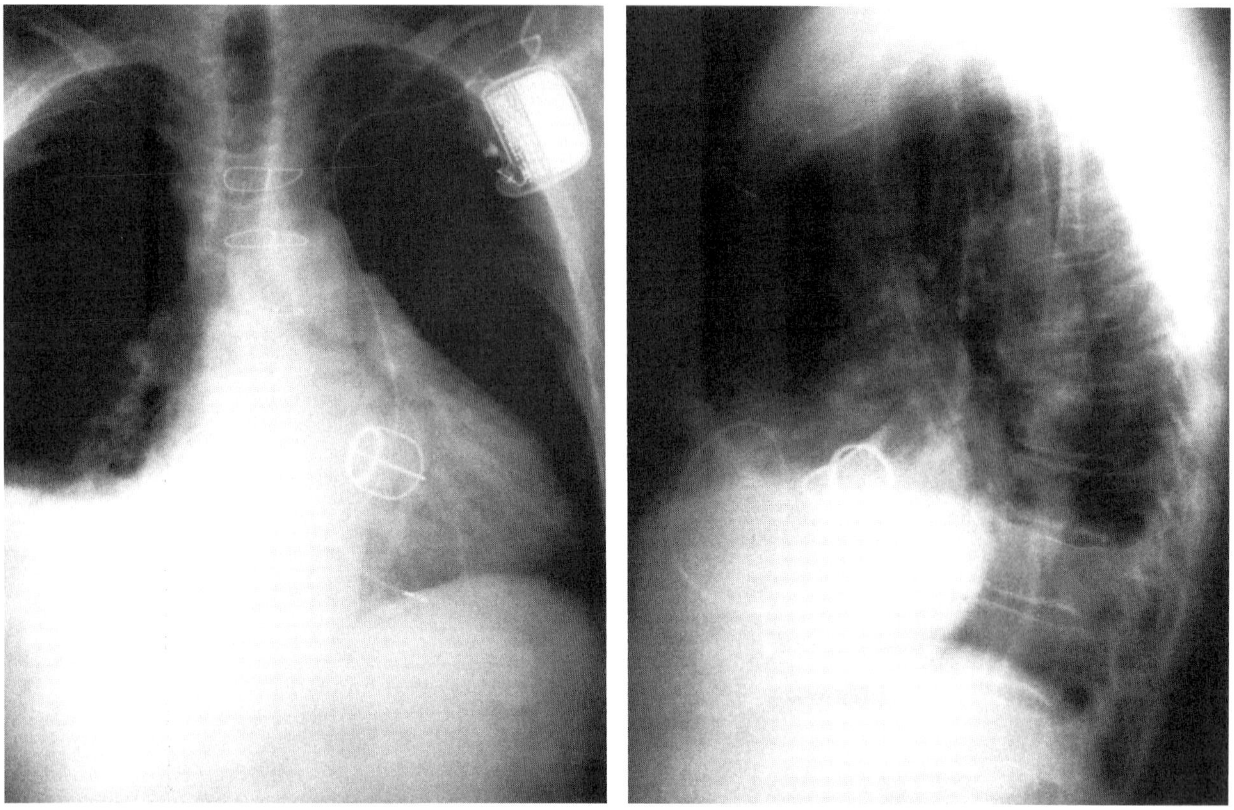

FIGURE 39-34 Active fixation ventricular pacing lead inserted via persistent left superior vena cava, through coronary sinus, and via loop-de-loop into the right ventricular apex in patient with Starr-Edwards valve prosthesis.

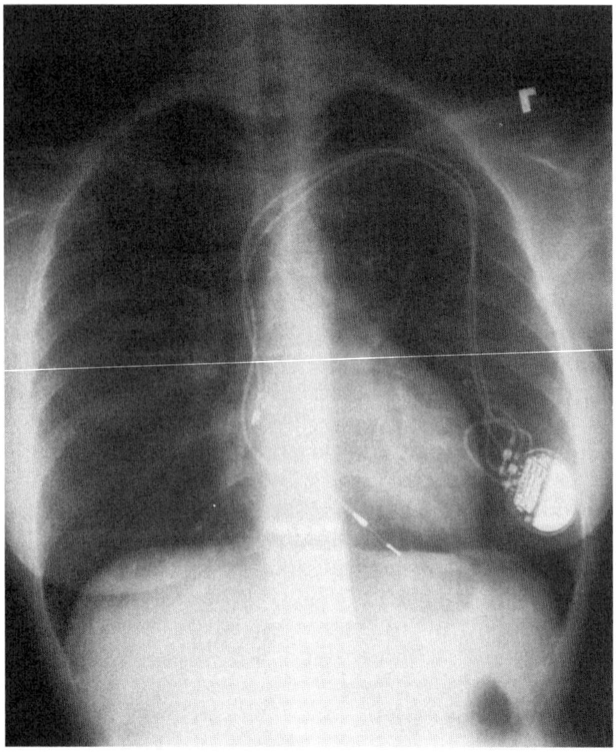

FIGURE 39-35 Atrial and ventricular leads inserted via left subclavian approach, tunnelled to a left submammary location for generator placement.

application of traction to the lead may be necessary to identify this problem. Fluoroscopy has been increasingly important as well to demonstrate fractures of an inner retention wire associated with an advisory in certain active fixation J-shaped atrial leads (Telectronics Accufix) (see Fig. 39-33).[103,104]

ELECTROCARDIOGRAPHY AND THE USE OF MAGNETS

The twelve-lead electrocardiogram is essential in the follow-up of the paced patient.[135] It provides information on the ability of the system to both sense, and thereby inhibit or trigger pacing, as well as to pace/capture. More than this, however, it offers clues as to lead positioning. For example, a right bundle branch block configuration to the paced complex may suggest a left ventricular location, either via an epicardial , transseptal, coronary sinus route or because of cardiac perforation (Fig. 39-36). Occasionally right bundle branch block morphology in the early precordial leads may reflect acceptable right ventricular pacing from the apical septum. An inferior axis to the paced complex may indicate a right ventricular outflow tract position.

Analysis of pacemaker spikes with regard to axis, amplitude, and timing may provide clues as to programmed polarity and potential problems with lead integrity.[136] For example, spike attenuation with prolongation of spike-to-spike intervals by an exact multiple of the automatic interval may indicate a problem with partial

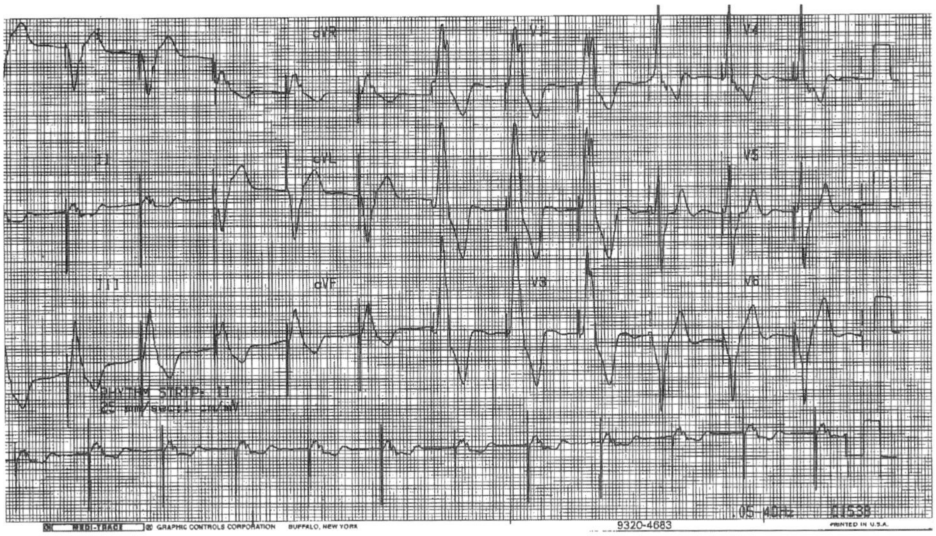

FIGURE 39-36 12-lead electrocardiogram of epicardial pacing system indicating right bundle branch block pacing configuration.

electrode conductor fracture, and increased spike amplitude in bipolar systems may indicate an insulation breach.

Magnet application enables asynchronous pacing and, coupled with electrocardiography, is extremely helpful in evaluating the pacer system. Though theoretically possible, it is uncommon to provoke significant ventricular arrhythmias with the use of magnets. It must be remembered that patients who have a separate implantable defibrillator may, depending on the model, experience implantable cardioverter defibrillator (ICD) inactivation if exposed to magnets and, therefore, great caution should be exercised in patients with dual devices. Magnet application confirms the ability of asynchronous pacing to capture the cardiac chamber at the programmed output settings, provided the pacer spike falls outside the refractory period of the chamber. In addition, various devices have predetermined magnet responses that aid in the identification of the pacer model and provide information as to the estimated longevity of the system: the magnet rate is an indicator of end-of-life or elective replacement that is both model and manufacturer specific.

Beyond this, magnet application may be useful diagnostically to determine pacing mode (e.g., single versus dual chamber) in a patient whose native rhythm otherwise inhibits pacer activity. If a pacer system is experiencing problems due to oversensing (of T waves, polarization voltage, or far-field signals), magnet application with resultant asynchronous pacing may correct the problem, thereby confirming the diagnosis of malsensing. Magnets may be used diagnostically to confirm a propensity for pacemaker-mediated tachycardias by momentarily creating AV dyssynchrony and, as well, be used therapeutically to terminate pacemaker-mediated tachycardia resulting from tracking of atrial fibrillation or retrogradely conducted atrial activity.[137] Rarely, magnet application may be used to terminate slow ventricular tachycardia in a pacemaker patient by underdrive pacing. Not uncommonly, continuous magnet application may be used clinically to ensure pacing in a pacer-dependent patient during situations where electromagnetic interference is anticipated, for example in a patient receiving electrocautery during surgery.

Determination of Pacemaker Dependency

Pacemaker dependency is an important condition that requires periodic reappraisal. It is a condition that may vary as a function of changes in the patient's underlying anatomic or electrophysiological milieu, autonomic state, or alterations in concomitant medications that may facilitate or depress electrical conduction (Fig. 39-37). Strictly speaking, it is defined by the development of significant bradycardia or ventricular asystole when pacing is stopped. The physician should not presume that the patient is pacer dependent merely because he or she is always paced at the programmed lower rate. Furthermore, when assessing the underlying rhythm, pacing rates should be *gradually slowed* to allow for the emergence of an escape rhythm. Abrupt termination of pacing is more likely to result in ventricular asystole and mask the ability of the patient to manifest an

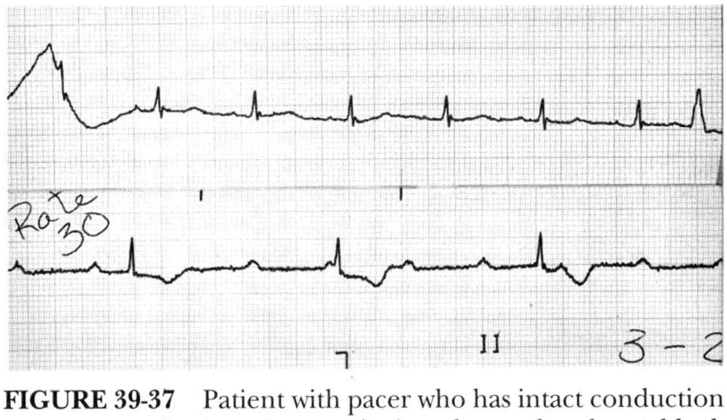

FIGURE 39-37 Patient with pacer who has intact conduction at one point in time **(top tracing)** and complete heart block when pacer is inhibited at a separate clinic visit **(lower tracing)**.

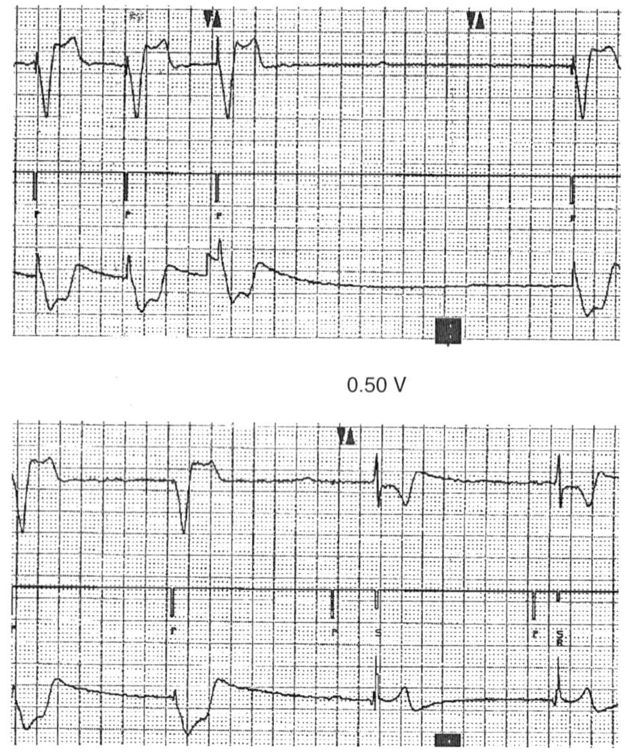

FIGURE 39-38 Ventricular asystole resulting from abrupt cessation of pacing **(upper panel)**; emergence of escape rhythm in same patient **(lower panel)** by starting at slower pacing rate and then reducing output to subthreshold levels.

escape rhythm (Fig. 39-38). If programming is not available, inhibition of permanent pacing may be attempted by chest wall stimulation through skin electrodes (one overlying the pacer generator) that are connected to a temporary pacing box via alligator clips. The sensed electromagnetic signal may inhibit the pacemaker and unmask the patient's native rhythm (Fig. 39-39).

Use of the Programmer: Telemetry of Programmed Settings, Measured Data, Histograms, Electrograms, and Marker Channels

Before any active intervention vis-à-vis reprogramming is contemplated, it is best to interrogate the pacer device and print out its current programmed settings. This allows for a better understanding of a pacer patient's presenting symptoms and provides clues as to how modifications in the program might be made.

Guidant				PULSAR
				20-JUL-00 16:48
Institution	ARRYTHMIA CENTER OF CONNECTICUT			
Patient	ANN		CPI Programmer	002980
Model 1270	Serial	613669	2890 Software	3.27

Patient Data File

PATIENT NAME	ANN	
PULSE GENERATOR DATA		
Model	1270	
Serial	613669	
PG Implant Date	04 JUL 2000	
Last Interrogation Date	11- JUL- 00	
INDICATIONS		
Atrial	N. R.	
AV Nodal	N. R.	
Ventricular	N. R.	
LEAD DATA	ATRIUM	VENTRICLE
Lead Implant Date	JUL 2000	JUL 2000
Lead Manufacturer	GUIDANT CPI	GUIDANT CPI
Lead Model	4461	4451
Connector Type	IS-1 BIPOLAR	IS-1 BIPOLAR
Lead Serial Number	204604	208514
Impedance at PG Implant	360	550 Ω
Threshold at PG Implant	0.50	<0.50 V
Wave Amplitude at PG Implant	2	12 mV
Slew Rate at PG Implant	N.R.	N.R. V/s
PHYSICIAN		
Physician	CONBOY	
Hospital	GREENWICH	
Phone		
NOTES		

End of Report

FIGURE 39-40 Interrogated data from pacemaker indicating patient demographics, lead/generator model, serial numbers, and implant data.

It should not be presumed that the patient's last office program is his current program—hospitalizations and/or encounters with other physicians may result in modifications unknown to the pacer physician and often forgotten by the patient (so-called *phantom reprogramming*).[110,138-141] With some of the older devices with limited telemetry capabilities, telemetry of the program did not always reflect what the device was doing, emphasizing the need to confirm the programmed settings by electrocardiographic assessment. This was commonly encountered, for example, in cases where the device was reset by environmental noise. "Administrative" data may also be interrogated, including patient demographics, implant threshold values, and nominal programmed values for certain models (Fig. 39-40).

Programmers are typically manufacturer specific; unfortunately, universal programming has yet to be universally accepted.[139-140] In some cases, a specific ROM module is required that is model specific in order to interrogate the device. The newer programmers typically have the capabilities of automatically determining and selecting the correct model number if at least the manufacturer is known and providing summary screens of the most essential features of programmed settings

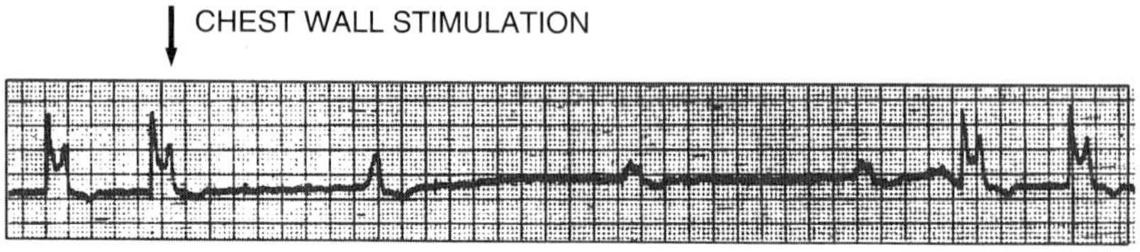

FIGURE 39-39 Use of external chest wall stimulation to inhibit pacer and demonstrate underlying escape rhythm.

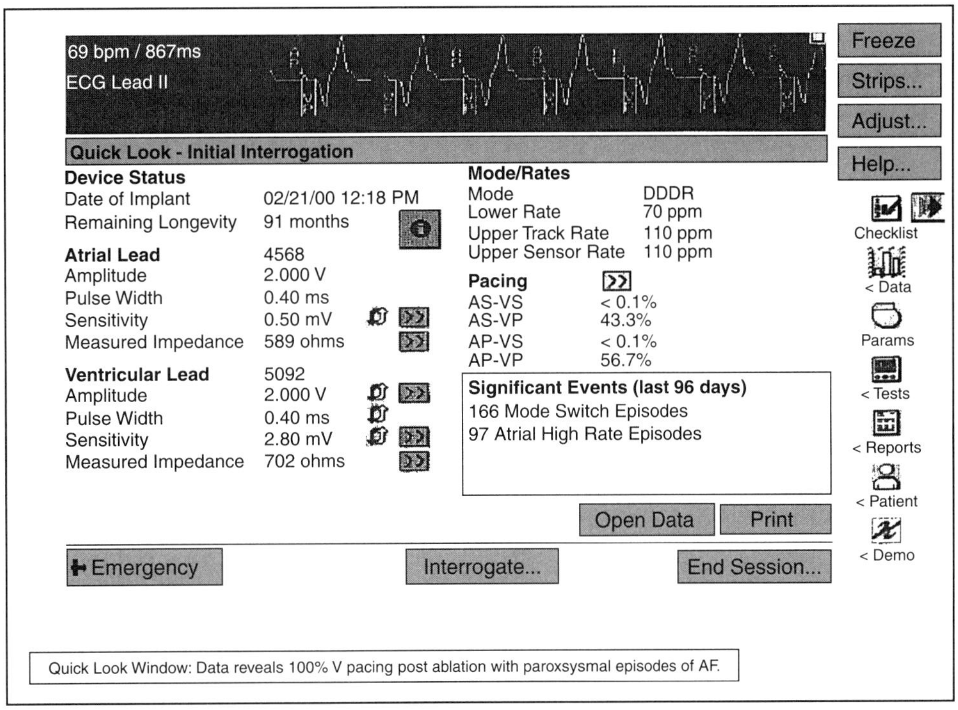

FIGURE 39-41 "Quick Look" interrogation from one manufacturer (Medtronic Inc.) highlighting salient programmed settings and measured data. (Courtesy of Medtronic, Inc.)

and telemetered measurements (Fig. 39-41). Otherwise, the pacer physician must depend on identification by the patient's pacer identification card, or on familiarity with certain generator models by the way they *feel* (size and shape on physical inspection) or *appear* (radiographically—some models have identifying codes, whereas others have characteristic shapes and battery configurations).

Measured data available through telemetry include battery voltage, current and impedance, as well as lead impedance. These measures convey important information with regard to generator longevity and lead integrity.[142-147] Battery voltage progressively declines over time, whereas battery impedance is inversely related to battery voltage (Fig. 39-42). A low lead impedance suggests a breach in insulation (Fig. 39-43); this phenomenon has been particularly problematic in certain bipolar polyurethane leads, especially when implanted using the subclavian approach (so-called *subclavian crush syndrome*).[30,31] An elevated lead impedance revealed by telemetry suggests either a conductor fracture or loose connection at the set-screw site which may otherwise be inapparent radiographically. Some of the newest devices have the capability to telemeter trends in lead impedance over time, which may otherwise be unappreciated because problems with lead impedance may be intermittent (Figs. 39-44 through 39-46).[144] In devices without such trend analyses, multiple determinations of impedance should be undertaken when pacer malfunction is suspected.

MEDTRONIC 9790 PROGRAMMER 9891A320 5/26/00 6:42
Copyright (c) Medtronic, Inc. 1993
---------------BATTERY/LEAD REPORT--------------- Page 1 of 1

Pacemaker Model : Thera DR 7940 Serial Number :

Battery/Lead Values : Collected : 5/26/00 6:42

Battery Status: Replace Pacer	
Estimated Time to Replacement (Average)	Replace Pacer
Battery Voltage	2.74 V
Battery Current	10.5 uA
Battery Impedance	5288 Ohms

Lead Status:	Atrial	Ventricular
Pulse Duration		0.49 ms
Pulse Amplitude		2.78 V
Output Energy		5.4 uJ
Lead Current		4.3 mA
Lead Impedance		598 Ohms
Pacing Configuration		Bipolar

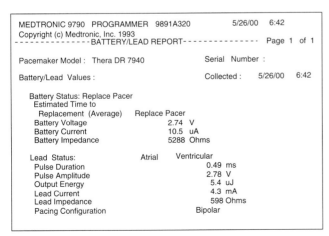

FIGURE 39-42 Interrogation revealing reduced battery voltage, elevated battery impedance and need to "replace pacer." (Courtesy of Medtronic, Inc.)

MEASURED DATA

Pacer Rate	70.5 ppm
Ventricular:	
Pulse Amplitude	2.7 Volts
Pulse Current	12.2 mAmperes
Pulse Energy	10 µJoules
Pulse Charge	4 µCoulombs
Lead Impedance	<250 Ohms
Atrial:	
Pulse Amplitude	5.3 Volts
Pulse Current	9.2 mAmperes
Pulse Energy	25 µJoules
Pulse Charge	5 µCoulombs
Lead Impedance	582 Ohms
Battery Data: (W.G. 8077 - NOM. 1.8 AHR)	
Voltage	2.73 Volts
Current	41 mAmperes
Impedance	<1 KOhms

FIGURE 39-43 Interrogation revealing reduced ventricular lead impedance indicative of insulation failure.

Daily Measurement - Data Table

Date	Atrial Amplitude (mV)	Atrial Impedance (Ω)	Ventricular Amplitude (mV)	Ventricular Impedance (Ω)
26-JUN-00	2.4	SENSED	PACED	590
25-JUN-00	2.7	SENSED	>9.0	610
24-JUN-00	1.7	SENSED	>9.0	640
23-JUN-00	1.3	SENSED	>9.0	610
22-JUN-00	1.8	SENSED	PACED	640
21-JUN-00	1.8	490	PACED	640
20-JUN-00	2.1	SENSED	PACED	590
20-JUN-00	2.0	490	>9.0	610
13-JUN-00	2.0	540	>9.0	650
06-JUN-00	1.8	SENSED	>9.0	610
30-MAY-00	1.8	SENSED	>9.0	620
23-MAY-00	1.4	500	>9.0	600
16-MAY-00	1.8	490	>9.0	610
09-MAY-00	2.0	SENSED	>9.0	610
02-MAY-00	2.3	SENSED	>9.0	640
25-APR-00	2.0	SENSED	PACED	640
18-APR-00	2.3	500	>9.0	640
11-APR-00	2.0	530	>9.0	650
04-APR-00	2.0	540	>9.0	620
28-MAR-00	2.1	540	>9.0	620
21-MAR-00	1.6	540	>9.0	640
14-MAR-00	>3.5	530	>9.0	660
07-MAR-00	1.8	550	>9.0	650
29-FEB-00	2.3	520	>9.0	640
22-FEB-00	1.8	500	PACED	640
15-FEB-00	1.7	490	PACED	610
08-FEB-00	1.8	500	PACED	640
01-FEB-00	1.6	500	PACED	650
25-JAN-00	1.7	490	>9.0	640
18-JAN-00	1.8	490	PACED	650
11-JAN-00	1.8	490	>9.0	620
04-JAN-00	2.1	460	>9.0	640
28-DEC-99	2.1	540	PACED	620
21-DEC-99	1.8	540	>9.0	640
14-DEC-99	1.6	520	PACED	610
07-DEC-99	2.1	550	>9.0	660
30-NOV-99	2.6	530	>9.0	640
23-NOV-99	2.1	540	>9.0	660
16-NOV-99	2.4	530	PACED	650
09-NOV-99	1.6	500	PACED	620
02-NOV-99	2.7	SENSED	>9.0	650
26-OCT-99	2.6	SENSED	>9.0	660
19-OCT-99	3.0	520	>9.0	680
12-OCT-99	2.3	SENSED	PACED	660

FIGURE 39-44 Interrogation revealing daily amplitude and impedance assessments from one manufacturer's device (Guidant, Inc.).

Event counters and histograms are also available through telemetry and provide an understanding of how long the patient spends at particular rates and/or rhythms (paced or otherwise); in essence, the generator serves as a mini-Holter or event recorder. Events are subcategorized into event histograms/event counts (evaluates frequency of pacing since last visit) (Fig. 39-47); sensor-indicated rate histograms (demonstrate how often different rates would occur during rate-adaptive pacing at a particular activity-sensing threshold) (Fig. 39-48); and event records (determine when a specific event occurred and at what rate, such as a tachycardia, automatic mode switch event, or search hysteresis episode) (Fig. 39-49). On the basis of such data retrieval, reprogramming may be undertaken to alter the degree of pacing, modulate rate-responsiveness of the pacer, or address arrhythmias detected by the device.[145] In some pacing systems, symptomatic events may be retrieved and evaluated after the patient triggers storage of an episode via a magnet or other special telemetric equipment.

Occasionally, the ability to telemeter real-time intracardiac electrograms may be helpful.[148-149] The size of the electrogram provides a rough idea as to the sensing capabilities of the lead and may also demonstrate far-field sensing. They may also be useful in evaluating retrograde VA conduction and assisting in rhythm identification, as in the case of supraventricular tachycardias (Figs. 39-50 and 39-51), and may also detect myopotential or other unusual sensing phenomena. Some devices now allow for the telemetry of stored electrograms as well (Figs. 39-52 and 39-53) that may correspond with a patient's symptomatic event. Intrinsic signal amplitudes with actual P or R wave millivolt determinations can be telemetered in some models (see Fig. 39-44).

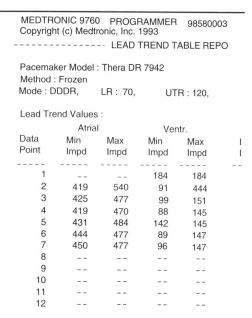

MEDTRONIC 9760 PROGRAMMER 98580003
Copyright (c) Medtronic, Inc. 1993
--------------- LEAD TREND TABLE REPO

Pacemaker Model : Thera DR 7942
Method : Frozen
Mode : DDDR, LR : 70, UTR : 120,

Lead Trend Values :

Data Point	Atrial Min Impd	Atrial Max Impd	Ventr. Min Impd	Ventr. Max Impd
1	--	--	184	184
2	419	540	91	444
3	425	477	99	151
4	419	470	88	145
5	431	484	142	145
6	444	477	89	147
7	450	477	96	147
8	--	--	--	--
9	--	--	--	--
10	--	--	--	--
11	--	--	--	--
12	--	--	--	--

MEDTRONIC 9760 PROGRAMMER 98580003 11/14/94 14:35
Copyright (c) Medtronic, Inc. 1993
------------- LEAD TREND GRAPHICS REPORT -------------- Page 1

Pacemaker Model : Thera DR 9742 Serial Number :

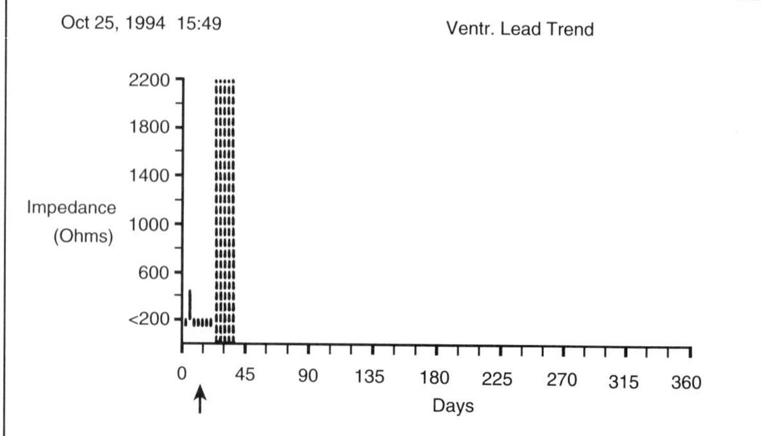

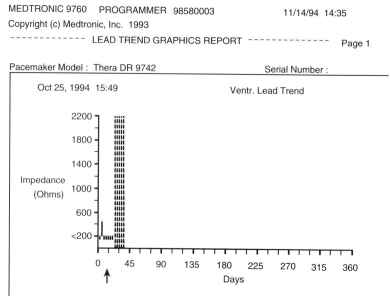

FIGURE 39-45 Impedance trend and graphics report from another manufacturer's device (Medtronic, Inc.) indicating a drop in ventricular lead impedance. (Courtesy of Medtronic, Inc.)

FIGURE 39-46 Decline in atrial lead impedance detected in trend analysis. (Courtesy of Medtronic, Inc.)

The two dashed vertical lines indicate a temporary interruption in data collection caused by application of the programming head (or a magnet) over the pacemaker.

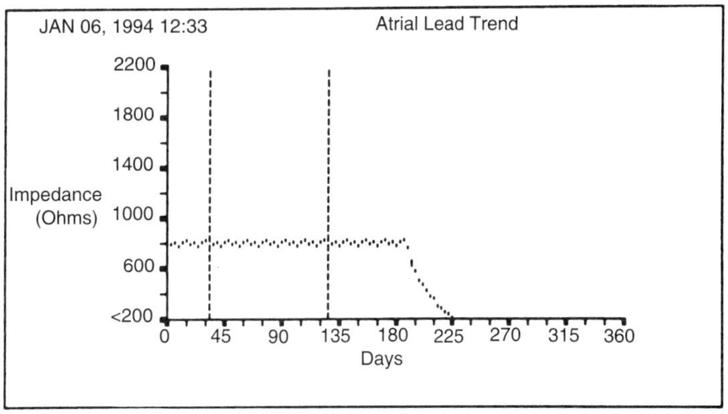

Lead Trend Report Example

Marker channels may be particularly useful in denoting when a particular channel, either atrial or ventricular, is sensing or pacing, whether appropriately or not. Thus, extraneous signals such as polarization voltage (Fig. 39-54), may be falsely sensed and other signals may be ignored because they fall into a refractory period—marker channels may help to sort out this type of pacer behavior. The designation of a paced event on the marker channel does not, however, indicate that the chamber has necessarily been successfully depolarized.

USE OF THE PROGRAMMER TO ASSESS SENSING AND PACING THRESHOLDS

Determination of pacing thresholds (Figs. 39-55 and 39-56) is important because a reduction in output of the device, provided an adequate safety margin is assured, may significantly affect the longevity of the generator. In addition, output programmability will allow for correction of noncapture in patients with elevated pacing thresholds, thereby avoiding surgical revision. Chronic thresholds are typically achieved by 3 months postimplant, allowing for the calculation of strength-duration curves for pacing (see Fig. 39-56). Some systems allow for threshold testing by decrementing either pulse width or amplitude, and some systems actually will compute a strength-duration curve with demonstration of what constitutes a satisfactory safety margin (see Fig. 39-56). Generally speaking, at short pulse durations, small changes in pulse width will produce large differences in voltage and current thresholds. Shorter pulse widths result in increased stimulation efficiency and lower impedance values. Most current devices allow for programmable voltage, and pulse duration thresholds can be determined at half amplitude (2.5 volts), thereafter programming to that pulse duration at 5 volts to provide an adequate safety margin. Another approach, increasingly employed because of low thresholds achieved with steroid-eluting leads, has been to

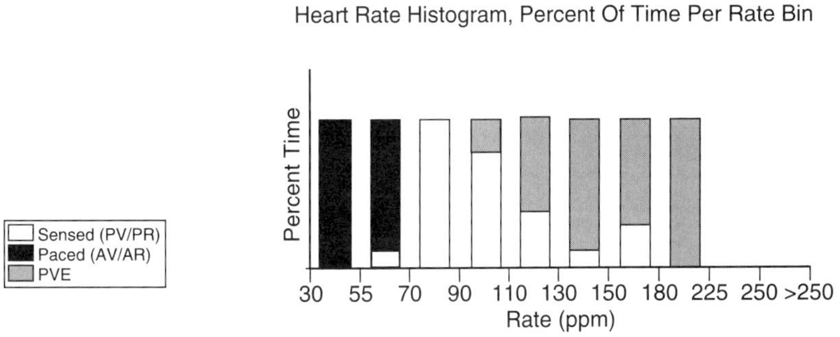

Event Counts

Rate (ppm)	PV	PR	AV	AR	PVE
30 - 54	0	0	158	1,411	0
55 - 69	59,513	158,516	1,067,609	437,575	0
70 - 89	116,499	572,079	956	1,809	1,217
90 - 109	1,984	28,241	108	0	9,307
110 - 129	156	2,581	0	0	4,495
130 - 149	0	271	0	0	2,115
150 - 179	0	20	0	0	50
180 - 224	0	0	0	0	2
225 - 249	0	0	0	0	0
> 250	0	0	0	0	0
Total:	178,152	761,708	1,068,831	440,795	17,186

Total Event Count: 2,466,672

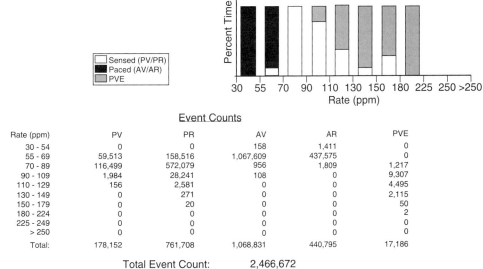

FIGURE 39-47 Heart rate histogram from one manufacturer (Pacesetter, Inc.) indicating percentage of time spent in sensed/paced rhythm.

St. Jude Medical
Cardiac Rhythm Management Division
©1983-1999. St. Jude Medical, Inc.

Page 3a
Affinity® SR Model: 5130 Serial: 22457
3500 Serial: 7517 (3303 - 1.02)

Sensor Indicated Rate Histogram

Mode	VVIR	
Sensor	On	
Base Rate	70	ppm
Max Sensor Rate	100	ppm
Threshold	Auto (+0.0)	
Measured Average Sensor	2.4	
Slope	Auto (+0)	
Measured Auto Slope	4	
Reaction Time	Fast	
Recovery Time	Medium	

Date Read:	Jul 12 1999 2:10 pm
Total Time Sampled:	31d 18h 67m 16s
Date Last Cleared:	Jun 10 1999 7:12 pm

Note: The above values were obtained when the histogram was interrogated.

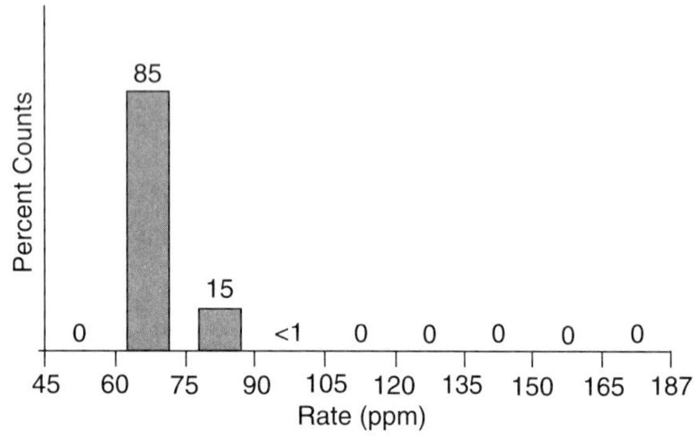

Bin Number	Range (ppm)	Time	Sample Counts
1	45 - < 60	0d 0h 0m 0s	0
2	60 - < 75	27d 3h 27m 10s	1,172,615
3	75 - < 90	4d 15h 21m 58s	200,459
4	90 - < 105	0d 0h 8m 8s	244
5	105 - < 120	0d 0h 0m 0s	0
6	120 - < 135	0d 0h 0m 0s	0
7	135 - < 150	0d 0h 0m 0s	0
8	150 - < 165	0d 0h 0m 0s	0
9	165 - 187	0d 0h 0m 0s	0
		Total:	1,373,318

Jul 12 1999 2:10 pm

FIGURE 39-48 Sensor indicated rate histogram interrogated from Pacesetter/St. Jude Medical, Inc. device.

St. Jude Medical
Cardiac Rhythm Management Division
©1983-1999. St. Jude Medical, Inc.

Page 3
Affinity® DR Model: 5330 Serial: 2609
3510 Serial: 01913 (3303 - 1.02)

Auto Mode Switch Histogram

Mode	DDDR	
Sensor	On	
Base Rate	70	ppm
Max Track Rate	115	ppm
Max Sensor Rate	115	ppm
A. Sensitivity	0.5	mV
Auto Mode Switch	DDIR	
Atrial Tachycardia Detection Rate	150	ppm
Post Vent. Atrial Blanking (PVAB)	165	ms

Date Read:	Nov 23 1999 15:33
Mode Switch Occurrences:	510
Date Last Cleared:	Jul 1 1999 12:34

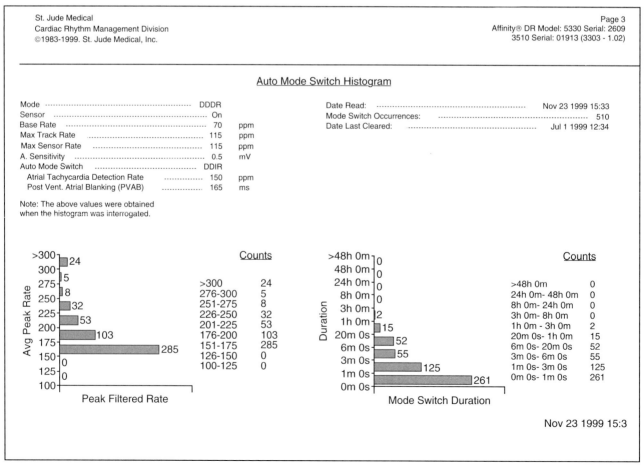

Note: The above values were obtained when the histogram was interrogated.

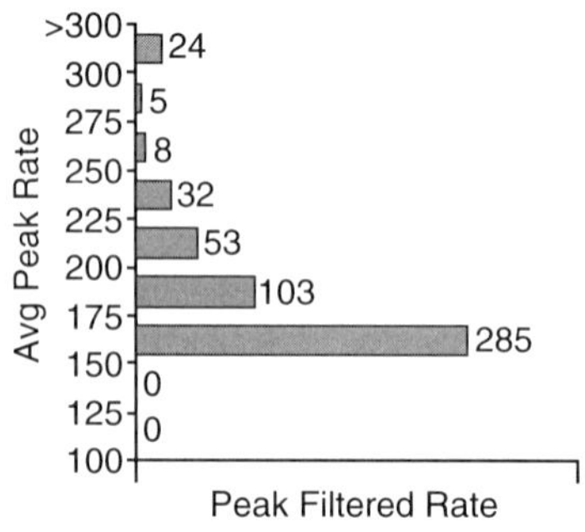

	Counts
>300	24
276-300	5
251-275	8
226-250	32
201-225	53
176-200	103
151-175	285
126-150	0
100-125	0

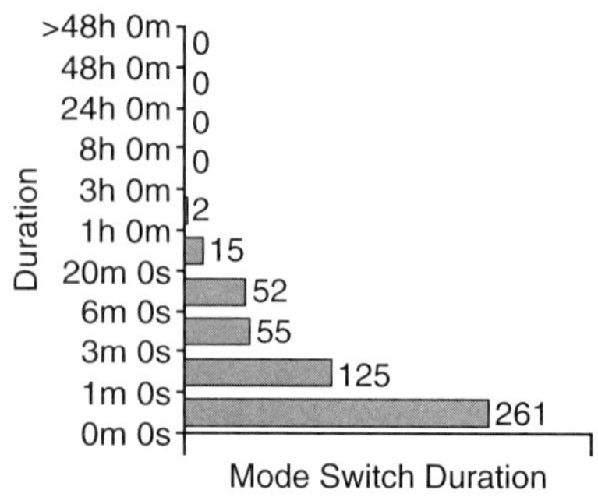

	Counts
>48h 0m	0
24h 0m- 48h 0m	0
8h 0m- 24h 0m	0
3h 0m- 8h 0m	0
1h 0m - 3h 0m	2
20m 0s- 1h 0m	15
6m 0s- 20m 0s	52
3m 0s- 6m 0s	55
1m 0s- 3m 0s	125
0m 0s- 1m 0s	261

Nov 23 1999 15:3

FIGURE 39-49 Histogram of auto mode switch episodes interrogated from St. Jude Medical, Inc. device.

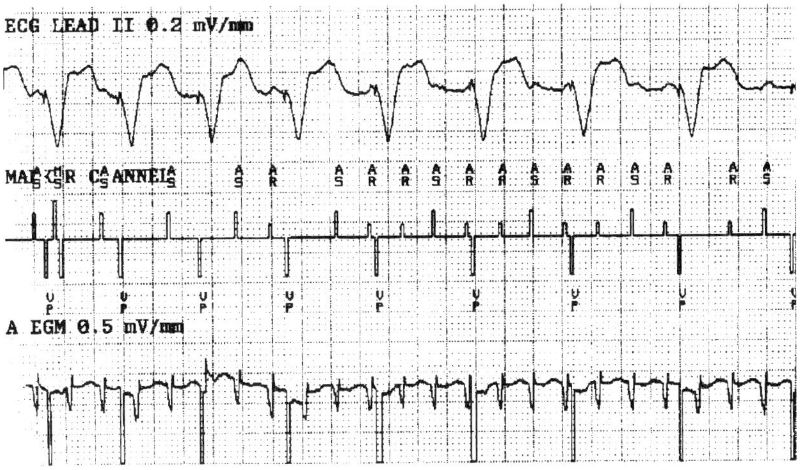

FIGURE 39-50 Intracardiac electrograms and marker channels confirming the diagnosis of atrial flutter.

reduce voltage while extending the pulse width somewhat if the pulse duration threshold is very low (e.g., less than 0.1 ms) at a given output voltage. Generally speaking, an output voltage of less than 2.5 volts does not significantly improve longevity but should be used if diaphragmatic or myopectoral stimulation are observed at higher outputs. Alternatively, situations associated with elevated pulse duration thresholds may require increasing the output voltage to allow for safe pacing because the higher pulse widths approaching rheobase are neither effective nor energy efficient. Very importantly, the pacer physician should note that certain drugs, most notably the antiarrhythmic agents flecainide and amiodarone, elevate the pacing thresholds, necessitating reprogramming of the system.

To determine capture thresholds, the pacing rate must be, at least temporarily, greater than the native rate. Atrial capture may be difficult to assess because of artifact following the atrial spike. In patients with intact AV conduction, atrial capture may be verified by observing the corresponding (conducted) QRS complex. In patients with AV block, however, it may be necessary to pace the atrium more rapidly to dissociate overlap with the patient's QRS complexes; this may result in an even higher-degree block with longer periods of ventricular asystole and may further constrain the ability to determine the atrial pacing threshold. In some patients, loss of atrial capture in dual chamber systems may become apparent because ventricular capture with preceding atrial noncapture allows for retrograde conduction to the atrium, which is detected on the surface ECG or through marker channels/intracardiac electrograms (see Fig. 39-55).

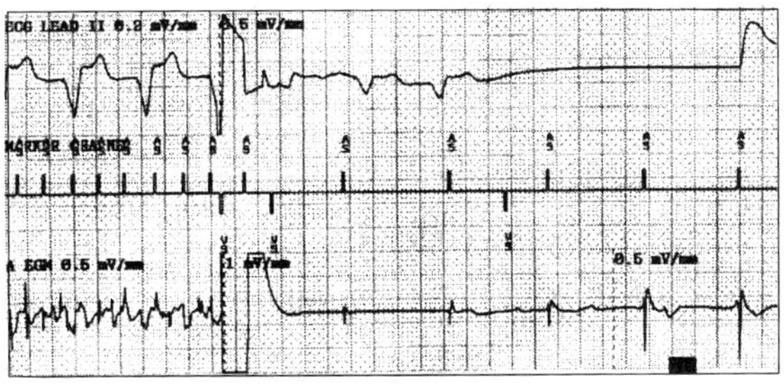

FIGURE 39-51 Intracardiac electrogram taken during cardioversion from atrial fibrillation to sinus rhythm.

Perhaps most exciting has been the recent elaboration of devices that allow for frequent and automatic determination of pacing thresholds (Figs. 39-57 and 39-58).[150] In one such system (see Fig. 39-58), the presence of capture is confirmed on a beat-to-beat basis by monitoring the evoked response associated with the ventricular output. The detection of this evoked response (ER) (cardiac depolarization produced from a pacemaker stimulus) requires the use of a bipolar low polarization pacing lead, allowing the pacing system to discriminate between the ER signal and polarization artifact. A back-up safety pulse is delivered if loss of capture has occurred. If such loss is detected on two consecutive primary output pulses, the device automatically increases the ventricular pulse amplitude to regain capture. When capture is confirmed for two consecutive beats, a threshold search (automatic decrement in pulse amplitude in 0.25 volt steps until two consecutive loss-of-capture events occur) is then initiated to determine the new capture threshold. Based on this threshold, the device automatically upregulates output by 0.25 volt as a working margin. Such automatic stimulation threshold searches are conducted after two consecutive losses of capture or every 8 hours.

Sensing thresholds may be determined by programming the inhibited mode to a rate slower than the native rate; sensitivity is then progressively decreased on the progammer until pacer output is no longer appropriately inhibited (Fig. 39-59). The triggered mode may be used as an alternative approach: a pacer spike is triggered whenever the native beat is sensed, and the sensitivity is progressively decreased until the sensed event no longer triggers pacer output (in essence, the spike serves as a marker channel for a sensed depolarization) (Fig. 39-60). Adjustments of sensitivity to higher values (i.e., less sensitive) may be required in patients subject to oversensing (e.g., T-wave sensing or myopotential inhibition); rarely, the triggered mode may be required to prevent oversensing of extraneous signals, especially in the pacer-dependent patient. Increased sensitivity (lower values) may be required in patients with undersensing

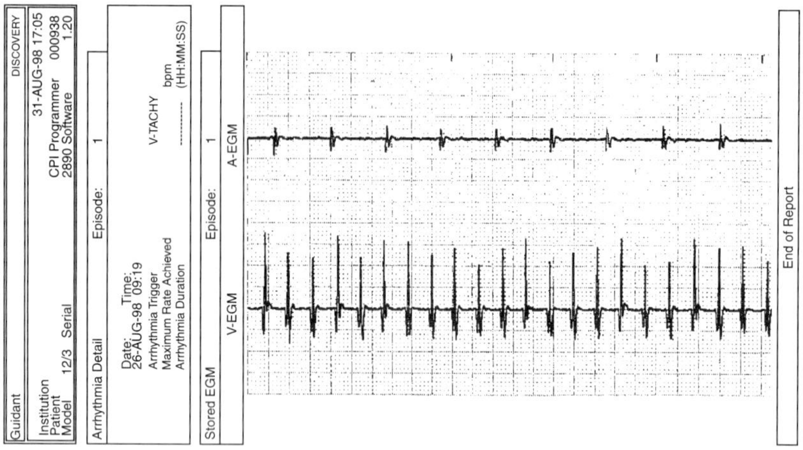

FIGURE 39-52 Stored electrograms from Guidant, Inc. device demonstrating ventricular tachycardia with atrioventricular dissociation.

(commonly seen with atrial leads where smaller electrograms are obtained).

A new and potentially exciting development has been the production of devices allowing for automatic sensitivity measurement and adjustment (Fig. 39-61).[151] Intrinsic depolarization waves may vary in amplitude because of lead maturation, exercise, variations between sinus and atrial fibrillation rhythms, changes in medication, exercise, and lead microdislodgment. Changes in position and respiration may affect the atrial signal, particularly in single lead dual chamber systems (VDD). Automatic adjustments of the sensitivity of the sense amplifier based on a target sensing margin may allow for optimal management of patients subject to such variations in amplitude (especially patients with frequent paroxysms of atrial fibrillation, with a consequent reduction in atrial signal sensing, can be enhanced during atrial fibrillation to promote automatic mode switching while avoiding oversensing during sinus mechanism) (see Fig. 39-61) and may also prove of benefit to minimize extraneous noise sensing in patients prone to electromagnetic interference or far-field sensing.

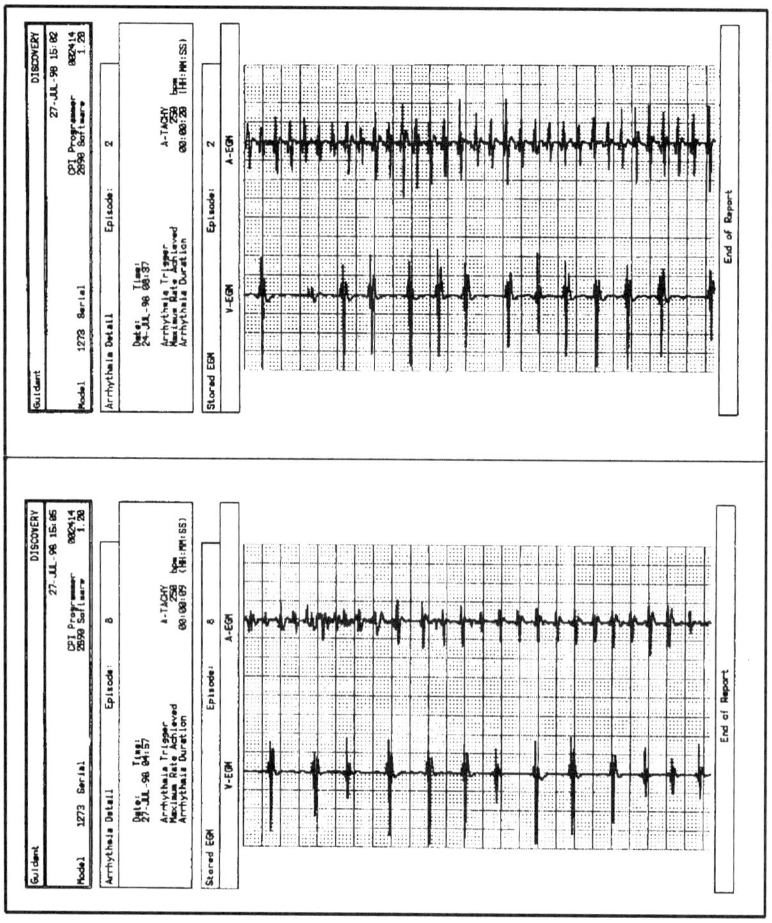

FIGURE 39-53 Stored electrograms from Guidant, Inc. device demonstrating atrial fibrillation with variable ventricular response.

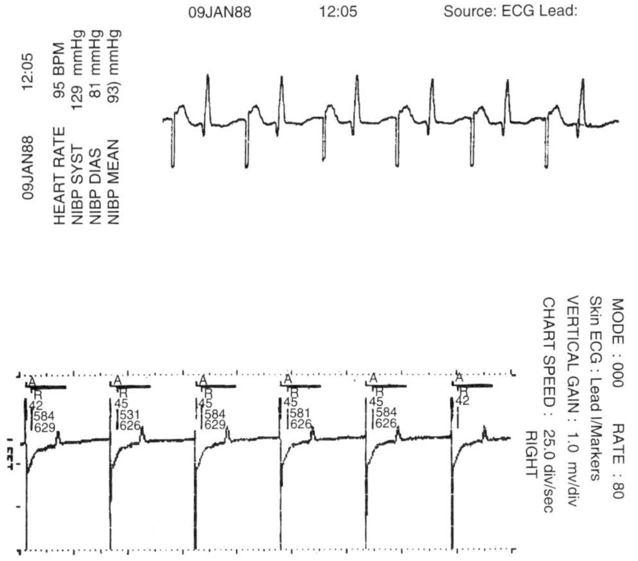

FIGURE 39-54 Marker channels demonstrating crosstalk of atrial lead by ventricular channel in dual chamber pacing system. Pacer is programmed to AV paced at 80 BPM but ventricular channel missenses atrial polarization voltage as ventricular activity (R) indicates sensed ventricular activity coinciding with atrial spike, causing resetting of VA timer and more rapid atrial pacing at 95 BPM. VA, Ventriculoatrial.

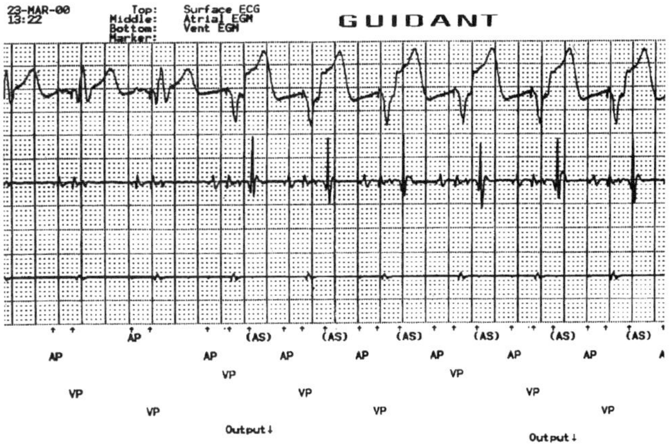

FIGURE 39-55 Atrial pacing threshold test during DDD pacing, with loss of capture on atrial lead on third atrial paced complex, resulting in fully paced ventricular complex and retrograde activation of atrium as confirmed by atrial electrogram analysis *(middle set of recordings).*

Reprogramming of Pacing Mode, Polarity, Atrioventricular Interval, and Refractory Periods

A variety of clinical situations may dictate the need for a change in the pacing mode. As noted previously, problems with severe oversensing in a pacer-dependent patient may require changing to the triggered mode to avoid undue inhibition of pacer output. Patients with dual chamber systems who tolerate ventricular pacing and have uncorrectable atrial undersensing may be reprogrammed to the ventricular demand mode; if atrial sensing is acceptable but pacing thresholds are excessive, the VDD mode may be selected to preserve atrioventricular synchrony. Dual chamber systems in patients with frequent paroxysms of atrial fibrillation may warrant a change to the ventricular demand mode or DDI mode to prevent upper rate tracking, although newer programs allow for automatic mode switching (see later).

Changing the polarity of the pacer has become increasingly programmable in the newer generators, provided the implanted lead is bipolar to begin with (Figs. 39-62 and 39-63). Unipolar pacing provides for a larger pacing spike, often facilitating electrocardiographic interpretation. Alternatively, unipolar pacing can promote myopectoral stimulation, prompting a change to bipolar pacing. Problems with far-field or myopotential inhibition or triggering may be addressed by conversion from unipolar sensing to bipolar sensing.

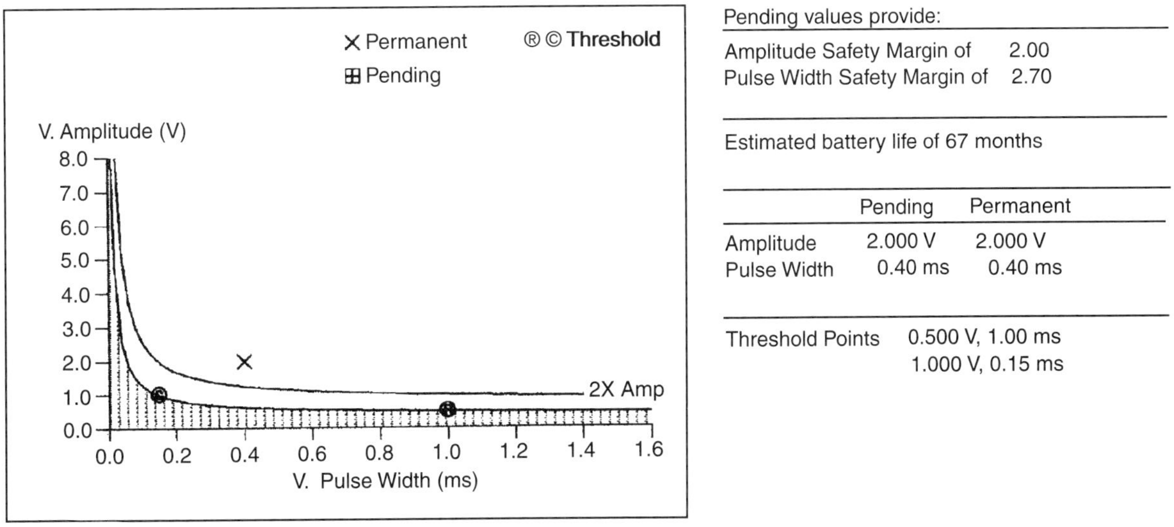

FIGURE 39-56 Strength duration curve for ventricular pacing thresholds in a Medtronic, Inc. pacing system, calculated by the corresponding pacer programmer. (Courtesy of Medtronic, Inc.)

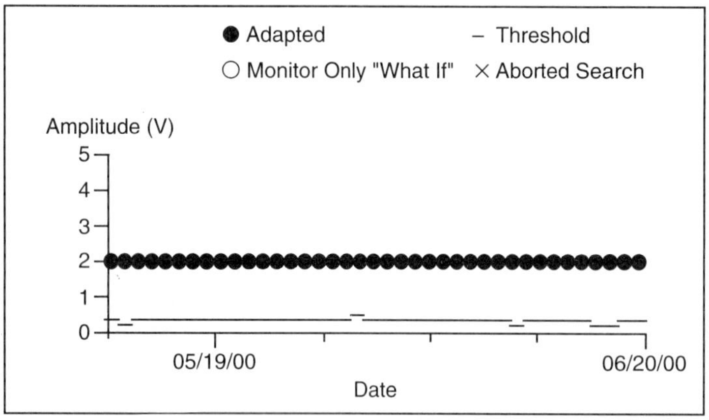

Initial Interrogation

Capture Management	Adaptive
V. Amplitude	2.000 V
V. Pulse Width	0.40 ms
Amplitude Margin	1.5x
Pulse Width Margin	1.5x
Min. Adapted Amplitude	2.000 V
Min. Adapted Pulse Width	0.40 ms
Acute Phase Completed	04/23/00

Data Collected

Every Day at Rest
Measured at 1 ms Pulse Width.
Strip available for last search.

Date/Time	Amplitude Threshold (V)	Pulse Width Threshold (ms)	Results (V, ms)
03/16/00 7.39 PM	0.25	0.27	5, 0.64 – Adapted
03/17/00 7.09 PM	0.25	0.34	5, 0.64 – Adapted
03/18/00 8.39 PM	0.25	0.27	5, 0.64 – Adapted
03/19/00 8.09 PM	0.25	0.27	5, 0.64 – Adapted
03/20/00 8.09 PM	0.25	0.27	5, 0.64 – Adapted

FIGURE 39-57 Automatic threshold determinations by a Medtronic, Inc. pacing system, with graph indicating trend from last month, and table *(below)* indicating daily measurements for preceding 3 months (entire table not reproduced). (Courtesy of Medtronic, Inc.)

In patients suffering from the subclavian crush syndrome, with reduction in lead impedance because of polyurethane insulation degradation, a reprogramming to unipolar pacing may augment the impedance to more acceptable values and prevent loss of capture and undersensing (see Fig. 39-62). Some devices allow for automatic reprogramming from bipolar to unipolar mode when a significant drop in lead impedance is detected.[152] Ultimately, however, lead revision is usually required, with the reprogramming being a transitional solution.

Atrioventricular intervals may be programmed for optimization of hemodynamics in certain patients; echocardiography may be useful to establish the ideal interval for a given patient.[153] There is potential benefit of allowing intrinsic AV conduction and a normal ventricular depolarization sequence to occur in patients with intact AV conduction (i.e., programming to a longer AV interval)[154-156]; this must be counterbalanced by the potential worsening of hemodynamics due to excessively long native AV intervals.[157] Some devices now employ a search mechanism that periodically extends the programmed sensed or paced AV delay to look for a return of optimal intact AV nodal conduction so as to minimize ventricular pacing.[158-159] Also programmable is the differentiation between a sensed versus a paced AV interval. Sensed AV delays are usually shorter than paced AV delays because of the latency between the pacing stimulus and subsequent atrial depolarization in the case of paced AV delays.[160] Another factor to consider is potential interatrial conduction delays in patients with atrial disease; although the AV interval reflects conduction between the right atrium and ventricle, delayed conduction to the left atrium will affect left atrial depolarization and left ventricular filling with particularly important consequences in patients with hypertrophic cardiomyopathy.[161] Another automatic feature available in many devices is a rate-modulated AV delay resulting in shortening of the AV delay when faster rates are achieved to maintain atrial transport and improved exercise tolerance at faster heart rates; newer models allow for programmability of the degree of progressive shortening of the AV or PV delays.[162-163]

In single chamber systems, refractory periods are used to prevent sensing of polarization voltage, T wave sensing for VVI systems, or far-field ventricular sensing in AAI systems. Dual chamber systems employ a PVARP (post ventricular atrial refractory period) when the atrial channel is refractory after a paced or sensed ventricular event so as to prevent inappropriate atrial tracking (e.g., of retrogradely conducted atrial activity) leading to pacemaker-mediated tachycardias; some systems also use a post-PVC PVARP extension. Another refractory

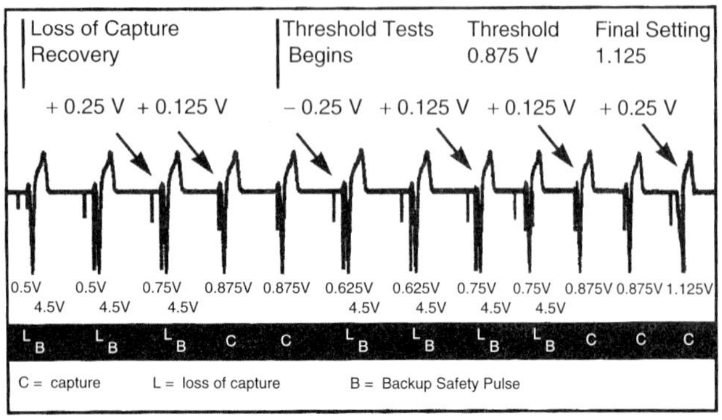

FIGURE 39-58 Depiction of St. Jude Medical, Inc. evoked response algorithm for determination of loss of capture recovery and pacing threshold.

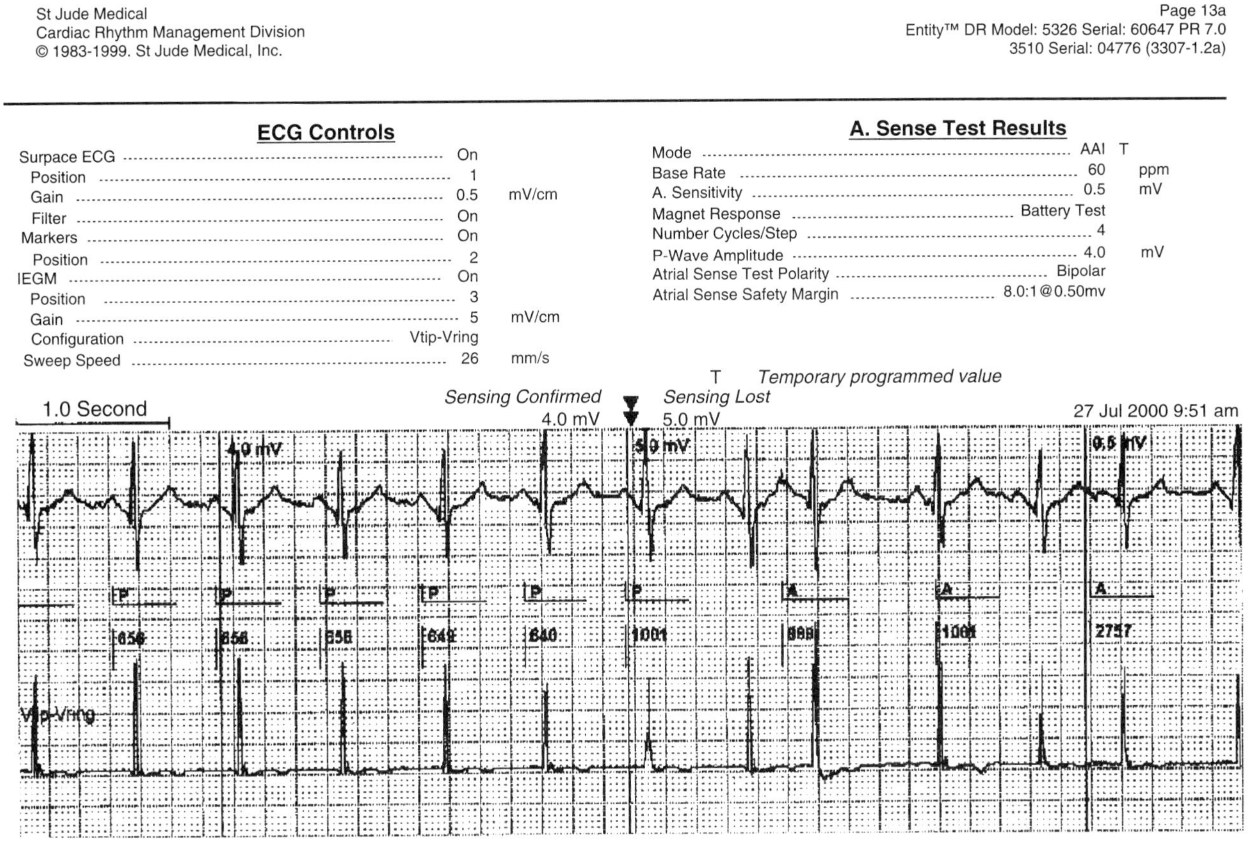
St Jude Medical
Cardiac Rhythm Management Division
© 1983-1999. St Jude Medical, Inc.
Page 13a
Entity™ DR Model: 5326 Serial: 60647 PR 7.0
3510 Serial: 04776 (3307-1.2a)
ECG Controls
Surpace ECG On
Position 1
Gain 0.5 mV/cm
Filter On
Markers On
Position 2
IEGM On
Position 3
Gain 5 mV/cm
Configuration Vtip-Vring
Sweep Speed 26 mm/s
A. Sense Test Results
Mode AAI T
Base Rate 60 ppm
A. Sensitivity 0.5 mV
Magnet Response Battery Test
Number Cycles/Step 4
P-Wave Amplitude 4.0 mV
Atrial Sense Test Polarity Bipolar
Atrial Sense Safety Margin 8.0:1@0.50mv
T Temporary programmed value
1.0 Second
Sensing Confirmed 4.0 mV
Sensing Lost 5.0 mV
27 Jul 2000 9:51 am
4.0 mV
5.0 mV
0.5 mV
P
A
656
656
656
648
640
1001
2757
Vtip-Vring

FIGURE 39-59 Sensing threshold test (St. Jude Medical, Inc.) with failure to demonstrate a "P" (i.e., sensed P wave) when sensing is temporarily programmed to a reduced value of 5.0 mV.

period is the atrial blanking period when the ventricular channel is blinded to the atrial spike to reduce the likelihood of crosstalk between the two channels. The incorporation of automatic mode switch algorithms into pacers has led to the converse blanking period, namely a postventricular atrial blanking (PVAB) period following the ventricular spike. This PVAB is the first portion of the PVARP and is adjusted to prevent far field R wave detection by the atrial channel and yet allow for detection of rapid atrial rates to guide the mode-switching algorithm. Programmability of the refractory periods exists, to varying degrees, in modern day pacemaker systems.

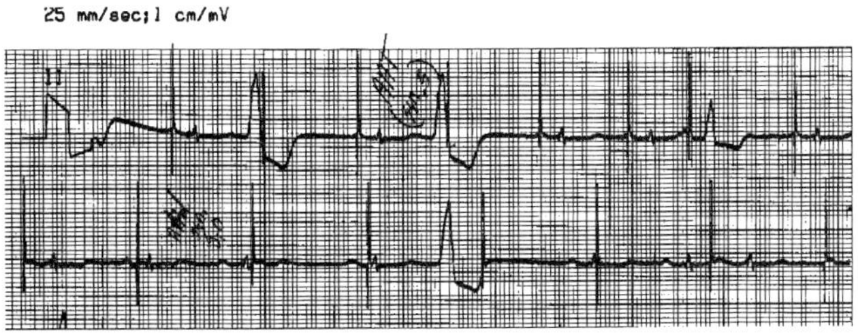
25 mm/sec; 1 cm/mV

FIGURE 39-60 Atrial sensing threshold determination by programming to triggered mode (AAT). At maximal sensitivity (0.5, **upper panel**), an atrial spike is triggered with every P wave, whereas at reduced sensitivity (5.0 mV) atrial spikes no longer coincide with P waves.

Special Considerations in the Dual-Chamber System: Assessing Crosstalk, Retrograde Ventriculoatrial Conduction, and Pacemaker-Mediated Tachycardia

The propensity for crosstalk[164] (sensing of electrical signals generated in one cardiac chamber by the sense amplifier in the other cardiac chamber) and for pacemaker-mediated tachycardias (PMTs) is especially important in the patient with a dual chamber pacer system. To explore for these potential phenomena, the ventricular channel should be programmed to its greatest sensitivity and the atrial output should be maximized. The base rate should exceed the patient's native rate to allow for atrial pacing and the AV interval should be programmed short enough to promote ventricular pacing; the ventricular blanking period should also be set to the shortest value where programmable. Under such settings, crosstalk inhibition of ventricular output because of ventricular sensing of atrial output may be appreciated as a potential problem (Fig. 39-64); if it does not emerge, then it is unlikely to surface as an important clinical problem at the patient's normal programmed settings. This test should be conducted with caution in the pacer-dependent patient insofar as transient ventricular asystole may be observed. The approach

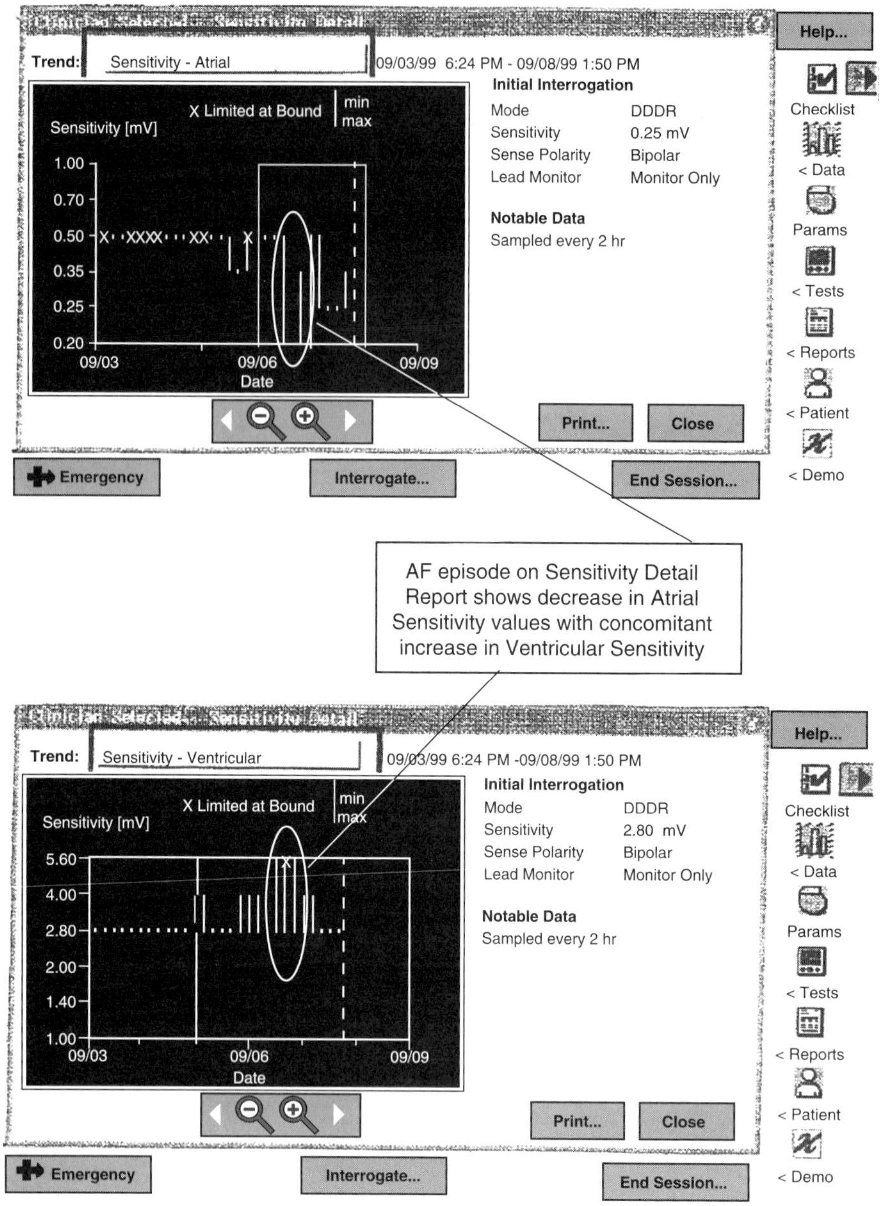

FIGURE 39-61 Sensitivity detail from Medtronic, Inc. device with sensing assurance capabilities, demonstrating drop in atrial sensitivity (and corresponding increase in ventricular sensitivity) during episode of atrial fibrillation. (Courtesy of Medtronic, Inc.)

to a patient with clinically demonstrated crosstalk inhibition, for example, on routine Holter or ECG recordings, is to reprogram a reduction in atrial output, reduction in ventricular sensing, lengthening of the ventricular blanking period, changing from DDD to VDD mode (to eliminate atrial pacing) or changing from unipolar to bipolar pacing/sensing.

Retrograde VA conduction may be sensed on the atrial channel and thereafter trigger a pacemaker-mediated tachycardia (Fig. 39-65).The propensity for this conduction should be assessed by programming to VVI pacing or AV pacing at subthreshold atrial output and should be examined at different pacing rates because the phenomenon may be somewhat dynamic.[165] Retrograde atrial activity may be appreciated in the inferior leads of a 12-lead ECG recording or may be confirmed by looking at marker channels or at atrial electrograms during AV pacing at subthreshold atrial outputs.

PMTs can be generated in the follow-up clinic by inducing transient AV dyssynchrony to promote retrograde VA conduction. If a sufficiently short PVARP is used, PMT may be observed in response to a spontaneous PVC, followed by momentary magnet application and removal, or by programming atrial output to subthreshold values. In the patient with a demonstrated potential for PMT, the ability to terminate with magnet or PMT-terminating pacer algorithms should be explored. Very importantly, the upper tracking rate should be limited, the postventricular atrial refractory period should be lengthened beyond the VA interval, or nontriggered modes, such as DDI, should be considered. PMTs due to tracking of rapid atrial fibrillation or flutter may be addressed by reprogramming to VVI mode until cardioversion has been achieved, limiting the upper tracking rate while maintaining DDD mode, attempting to terminate atrial flutter by temporarily

MEDTRONIC 9790 PROGRAMMER 9891A320 6/08/00 10:58
Copyright (c) Medtronic, Inc. 1993
---------------- BATTERY/LEAD REPORT ---------------- Page 1 of 1

Pacemaker Model: Thera DR 7962i Serial Number: PDD100515

Battery/Lead Values: Collected: 6/08/00 10:58

Battery Status: OK

Estimated Time to Replacement(Average)	28 months(Past History)
Battery Voltage	2.75 V
Battery Current	32.9 uA
Battery Impedance	842 Ohms

Lead Status:	Atrial	Ventricular
Pulse Duration	0.40	0.40 ms
Pulse Amplitude	8.18	2.67 V
Output Energy	27.5	3.6 uJ
Lead Current	> 20.0	3.6 mA
Lead Impedance	< 100	702 Ohms
Pacing Configuration	Bipolar	Unipolan

A

MEDTRONIC 9790 PROGRAMMER 9891A320 6/08/00 10:59
Copyright (c) Medtronic, Inc. 1993
---------------- BATTERY/LEAD REPORT ---------------- Page 1 of 1

Pacemaker Model: Thera DR 7962i Serial Number: PDD100515

Battery/Lead Values: Collected: 6/08/00 10:59

Battery Status: OK

Estimated Time to Replacement(Average)	43 months(Past History)
Battery Voltage	2.77 V
Battery Current	20.1 uA
Battery Impedance	86 Ohms

Lead Status:	Atrial	Ventricular
Pulse Duration	0.40	0.40 ms
Pulse Amplitude	3.88	2.68 V
Output Energy	16.5	4.0 uJ
Lead Current	12.1	4.0 mA
Lead Impedance	282	638 Ohms
Pacing Configuration	Unipolar	Unipolar

B

FIGURE 39-62 **A,** Interrogation of device demonstrates markedly reduced atrial lead impedance (less than 100 ohms) in bipolar configuration. **B,** Impedance increases in same lead when reprogrammed to unipolar configuration, a temporizing measure. (Courtesy of Medtronic, Inc.)

294 - 03 SN 001777 JAN 26 '93 5:12 PM
RELAY TELEMETRY DATA

PACING RATE	70 PPM
PACING INTERVAL	857 MSEC
CELL VOLTAGE	2.72 VOLTS
CELL IMPEDANCE	2.67 KOHMS
CELL CURRENT	17.3 UA

	ATRIAL (Uni)	VENTRICULAR (B1)
SENSITIVITY	1.0	5.0 MV
LEAD IMPEDANCE	635	1815 OHMS
PULSE AMPLITUDE	3.99	4.03 VOLTS
PULSE WIDTH	0.35	0.35 MSEC
OUTPUT CURRENT	6.1	2.2 MA
ENERGY DELIVERED	8.1	3.8 UJ
CHARGE DELIVERED	2.16	0.77 UC

294-03 SN 001777 JAN 26 '93 5:14 PM
RELAY TELEMETRY DATA

PACING RATE	70 PPM
PACING INTERVAL	857 MSEC
CELL VOLTAGE	2.70 VOLTS
CELL IMPEDANCE	2.98 KOHMS
CELL CURRENT	20.3 UA

	ATRIAL (Uni)	VENTRICULAR (Uni)
SENSITIVITY	1.0	5.0 MV
LEAD IMPEDANCE	652	508 OHMS
PULSE AMPLITUDE	3.97	3.94 VOLTS
PULSE WIDTH	0.35	0.35 MSEC
OUTPUT CURRENT	5.9	7.5 MA
ENERGY DELIVERED	7.8	9.6 UJ
CHARGE DELIVERED	2.09	2.64 UC

RETURN

FIGURE 39-63 Elevated ventricular impedance of 1815 ohms in ventricular lead programmed to bipolar configuration, significantly reduced to 508 ohms by temporarily programming to unipolar configuration until definitive lead revision is undertaken for pacer malfunction.

burst pacing from the atrial channel (where available), or activating the device's automatic mode switch function (if available).

Reprogramming of Other Advanced and Automatic Functions: Sleep Mode, Rate Responsiveness, Automatic Mode Switching, Search Hysteresis/Rate Drop Response

Sleep mode is an algorithm currently available in many devices that will allow for a decrease in the lower rate of the pacer during sleeping compared with waking hours and thereby minimize unnecessary and potentially adverse nocturnal pacing (Fig. 39-66).[166-167] Some models allow for programming of the specific time at which the base rate slows and increases (theoretically of concern if the patient alters his sleeping schedule, changes time zones,[168] and so on, whereas others actually differentiate

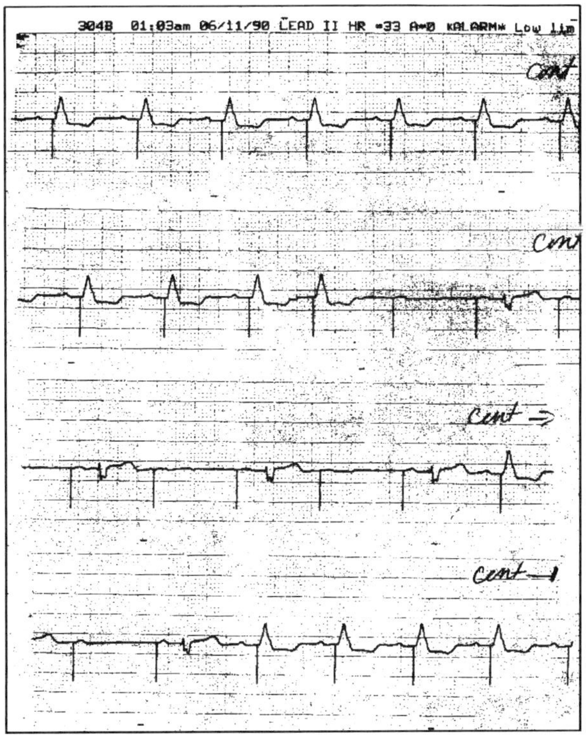

FIGURE 39-64 Crosstalk inhibition of ventricular output by oversensing of atrial unipolar spike by ventricular channel, resulting in periods of ventricular asystole in patient with complete heart block.

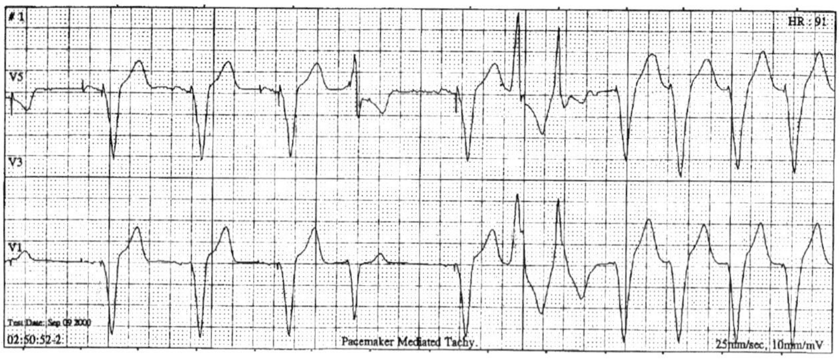

FIGURE 39-65 Pacemaker mediated tachycardia induced by ventricular ectopy that temporarily dissociates subsequent atrial and ventricular activities, with next ventricular paced complex producing retrograde conduction to atrium and engendering pacemaker-mediated tachycardia (PMT).

rest/sleep periods from activity periods using the sensor activity level and adjust the base rate accordingly.[169]

Rate-adaptive pacemakers have been used increasingly to optimize pacing rates in response to one or, in some models, dual sensors, particularly important in patients suffering from chronotropic incompetence.[170] It is beyond the scope of this chapter to discuss the nuances of the various sensors. Different sensors will respond differently to the same stimuli, leading some to prefer the use of dual sensors to better simulate a healthy sinus node response.[171] The lower rate should be set to allow a patient's native rhythm to predominate if at all possible to promote generator longevity. The upper rate is guided by the age-predicted maximal heart rate, adjusted downward in the patient with exertion-related symptoms such as angina or if there is comorbid orthopedic disease.

The threshold should be chosen to be least sensitive at rest but prompt in its response to activity; this may be influenced by underlying cardiac conditions (e.g., the desire to avoid excessive rate-responsiveness in the patient with angina) or coexistent noncardiac conditions (e.g., higher threshold setting required in the tremulous patient). Essentially, this parameter distinguishes false activity (noise) from true activity. The sensor slope determines the incremental heart rate response to a change in sensor activation; it should result in a fairly linear change in rate from lowest to highest rate over the course of the patient's activity range. Thus, the patient who can do more activities should be programmed to a smaller slope, whereas patients with the greatest limitations should be given a steep sensor slope.[172] Acceleration times may also be programmed (slower in patients who have angina or who are elderly and faster in the pediatric patient) as well as deceleration times. Exercise testing, Holter monitoring, and use of event histograms and counters have all been used to assess the adequacy of programmed rate-adaptive parameters.[173-174]

Automatic mode switching has been incorporated into newer pacemakers to preempt tracking of pathologic atrial tachycardias in dual chamber systems; on the detection of an atrial dysrhythmia the system reverts to ventricular demand pacing and resumes atrial tracking when sinus mechanism returns (Fig. 39-67).[175] The frequency of such mode switches can be assessed by event counters as already discussed, potentially providing

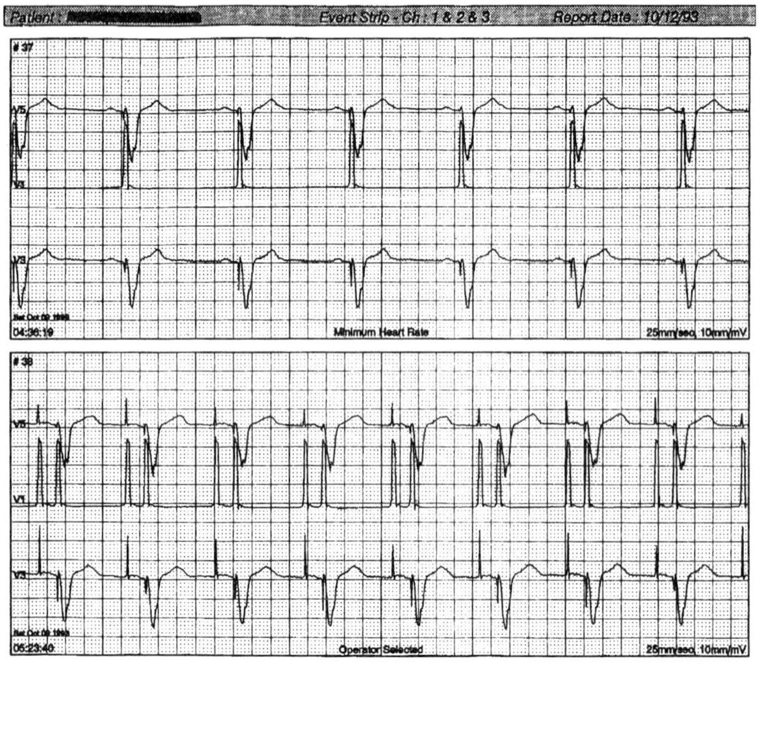

FIGURE 39-66 Patient with sleep mode response/circadian rhythm algorithm—**upper panel** is with patient sleeping and pacer is set to lower atrial paced rate, allowing for endogenous atrial rhythm to emerge; pacer has been programmed to achieve a faster lower rate after 5 A.M., resulting in atrioventricular (AV) paced rhythm seen in **lower panel**.

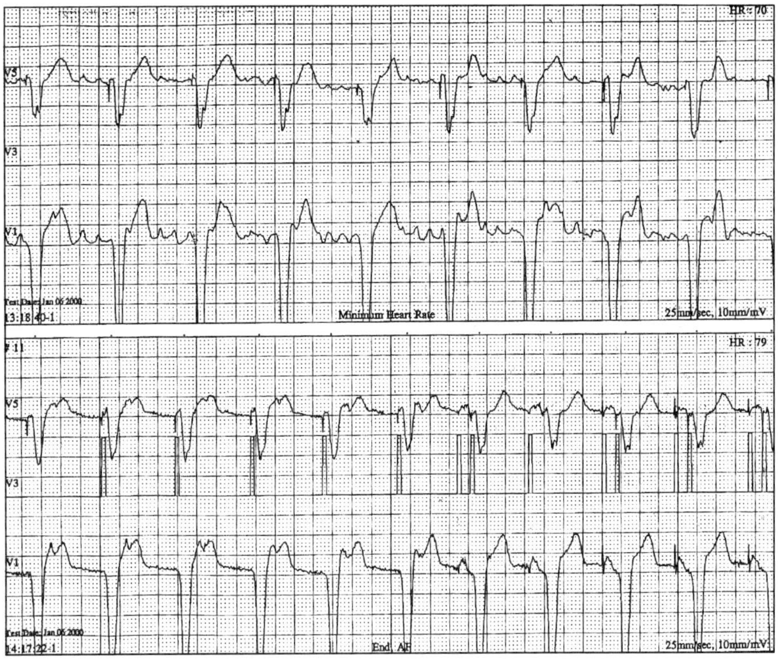

FIGURE 39-67 Demonstration of automatic mode switch—patient in atrial fibrillation with VVI pacing in **upper panel,** with transition back to atrioventricular (AV) pacing when sinus mechanism returns.

information on the atrial rate prompting mode switch, the duration of such episodes, and the time spent in tracking versus nontracking modes.

Search hysteresis/rate drop response functions have been applied increasingly in the treatment of patients with severe cardioinhibitory responses associated with vasovagal syncope.[176-179] Programming allows for the initiation of dual chamber pacing at an elevated rate when a rate drop is detected through a programmable detection window or when a programmable number of consecutive lower rate paces is detected. The patient's rate is allowed to dip to a low value before pacing at a much faster pacing rate, with scanning automatically undertaken to cease pacing when the patient's intrinsic rhythm returns to normal (Fig. 39-68).

It should be noted that programming of certain values or parameters may preempt the enabling of others: essentially, some pacing parameters may be mutually exclusive. Thus, in the case of rate drop response, in one model the drop rate cannot be less than the programmed lower rate. In the same model, activation of the rate drop response makes automatic mode switch unavailable and, furthermore, rate adaptive AV, sensor-varied PVARP, and automatic PVARP are also disabled. In an earlier series of the same manufacturer, rate drop response could be activated only in the DDD mode but not in the DDDR mode.

Adjuncts for Follow-Up: Ambulatory ECG Recording, Event Monitors, and Treadmill Testing

As previously noted, pacers have increasingly taken on the role of mini-Holters, with the abilities to store and retrieve events and even electrograms (even those that are patient designated). Nonetheless, not all systems have refined these capabilities; as such, the ambulatory ECG recording remains an indispensable tool in evaluating a possible correlation of a patient's symptoms with a pacer or rhythm abnormality, but even more importantly may reveal a previously unsuspected system malfunction (Figs. 39-69 through 39-71). Such phenomena as myopotential sensing and atrial undersensing may exist and yet be asymptomatic (see Figs. 39-69 and 39-71).[180-181] Holter monitors may give a sense of pacer-dependence at the programmed settings and help in the assessment of rate adaptiveness of pacers. Concomitant arrhythmias that may complicate pacer management can, on occasion, be revealed. Event recorders with memory capabilities may be used to correlate a patient's symptoms with a potential arrhythmia or pacer malfunction; real-time transtelephonic monitoring offers an alternative approach, provided the event lasts long enough to be transmitted (see Fig. 39-24).

Exercise testing may be used to assess a patient's chronotropic competence (or lack thereof) and aid in the prescription of a rate-adaptive pacer program with regard to slopes and sensing thresholds. Upper rate behavior with activity may also be appreciated on the treadmill, and may indicate the need to reprogram (e.g., to a higher rate in a younger individual). Rarely, exercise-induced arrhythmias may be demonstrated.

Elective Replacement Indicators

The end of life of a pacer is associated with gross malfunction or loss of function and is to be distinguished from the elective replacement indicator (ERI), which indicates that end of life is imminent, though with an advanced window of time, typically weeks to months. Real-time telemetry of increased battery impedance

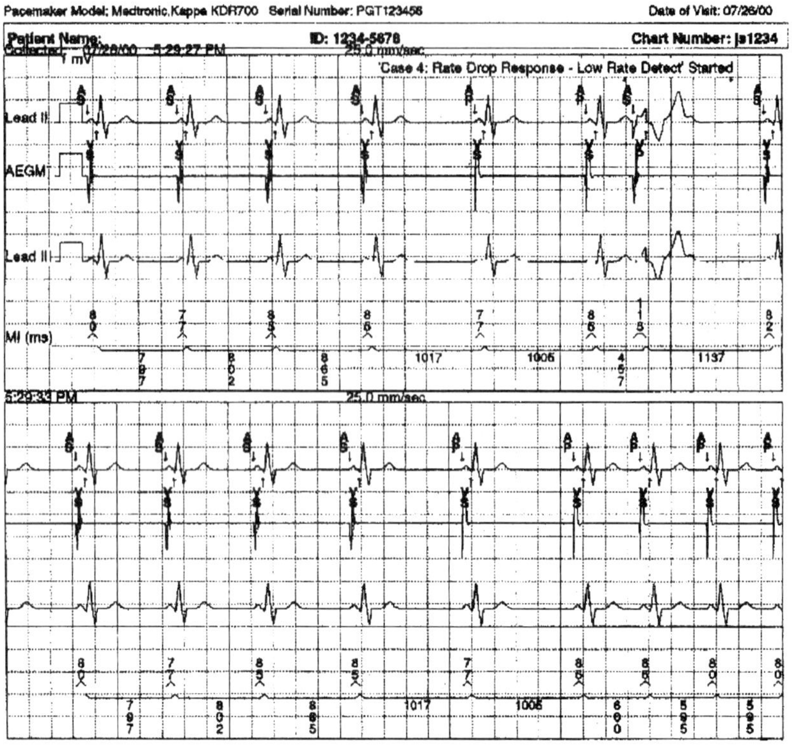

FIGURE 39-68 Demonstration of rate drop response with detection of drop in rate triggering more rapid atrial pacing in a patient with recurrent vasovagal syncope (Medtronic, Inc. device). (Courtesy of Medtronic, Inc.)

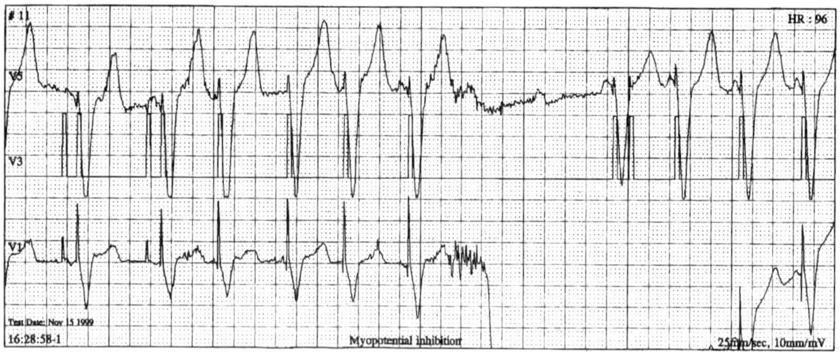

FIGURE 39-69 Myopotential inhibition of ventricular output resulting in transient ventricular asystole in patient with complete heart block, detected on Holter monitoring.

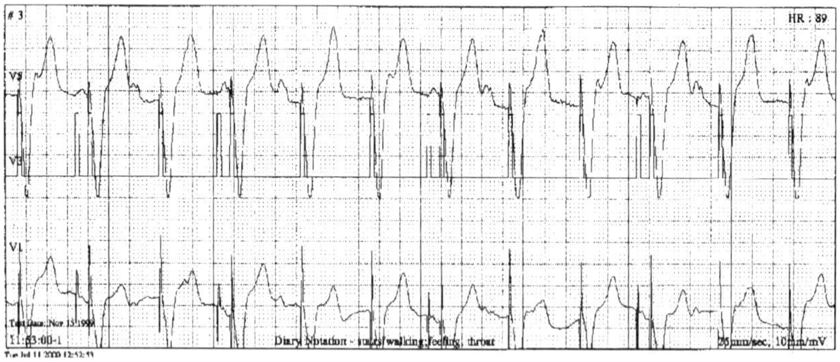

FIGURE 39-70 Holter monitoring demonstrating atrial undersensing, resulting in dissociation of atrial and ventricular activity responsible for patient's symptoms of "funny feelings in throat."

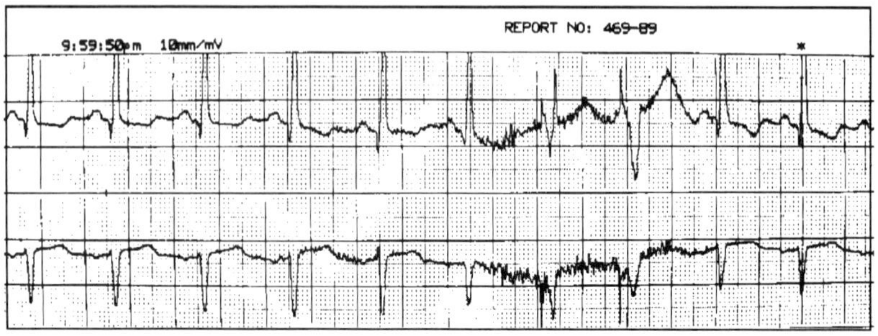

FIGURE 39-71 Myopotential triggering of ventricular paced activity, detected on Holter monitoring.

may dictate the need for replacement. More commonly, however, various models and manufacturers employ different indicators (which may be found by referring to the corresponding physician's manual), including gradual or stepwise declines in free-running or magnet rates, rate drops to preset values, or changes in pacing modes (e.g., DDD to VVI—often the patient will experience symptoms suggestive of pacemaker syndrome when this type of ERI is tripped). It is incumbent on the pacer physician to familiarize himself with the ERI endpoints associated with the devices he is following, as well as the mechanisms governing ERI.[182-183]

Lead/Generator Advisories and Recalls

Advisories and recalls are inescapable aspects of a pacer practice. Although various manufacturers have their own performance evaluations, under-reporting of device malfunction and delayed notification are frequently encountered. The lack, to date, of a national device/lead database (as originally proposed by the North American Society of Pacing and Electrophysiology) places even greater responsibility on the part of the pacer physician to keep his own databases and track device performance.[184,185] Such responsibilities have been codified in a NASPE consensus conference[186] and include: confirmation of all unimplanted versus implanted devices; identification of all patients being followed; notification of centers or physicians to whom patient care may have been transferred; notification of the patient or designate by registered mail and telephone contact; communication to the patient of the magnitude of the problem, natural history, risks of intervention and nonintervention and plans for investigation and management; ensuring patient follow-up; and provision of follow-up to the manufacturer and task forces associated with a device recall. If a patient is pacer-dependent, closer follow-up is mandated, either through the clinic or transtelephonically. There may be a lower threshold for invasively intervening to revise a *potentially* affected pacer system in such a patient, although the risks of intervention and the statistical likelihood of a future device failure must be carefully weighed when making this decision.

Special Considerations: Cardioversion, EMI, MRI, Radiotherapy, and Lithotripsy

Both electrocautery and cardioversion/defibrillation can affect pacer function, either transiently or permanently.[187-190] This includes resetting of the generator to a noise reversion mode (Fig. 39-72) or changes in sensing or pacing thresholds, either due to transmission of current down the lead to the tissue-electrode interface or because of direct damage to the generator. The recommendations for cardioversion are to use anteroposterior paddles with the least amount of energy possible. Recommendations for the use of electrocautery are to use it sparingly, with the ground plate as close

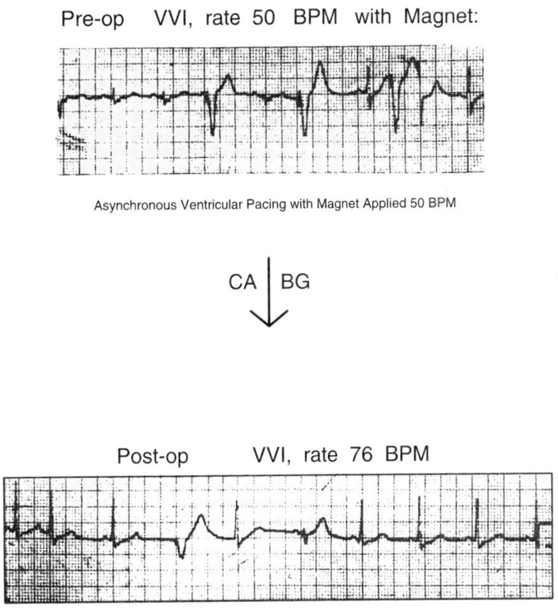

FIGURE 39-72 Patient with Intermedics, Inc. pacer set to VVI rate 50 BPM (magnet response also 50 BPM), reset to VVI 76 BPM because of electrocautery encountered during bypass surgery.

as possible to the operating site and far from the generator. *Bipolar* cautery should be used and, in the pacer-dependent patient, consideration of preoperative programming to the asynchronous or triggered modes should be considered to avoid problems with oversensing.

Electromagnetic interference in the medical environment has increasingly become an issue from other sources as well. Radiofrequency current has been used with increasing application for the ablation of cardiac arrhythmias. Pacer inhibition (even in the asynchronous mode), device resetting, undersensing, asynchronous pacing, and induction of rapid pacing have all been observed[191]; in one trial, no significant impact on pacing thresholds or impedances was observed.[192] Once again, the ground plate should be distanced from the pulse generator; back-up temporary pacing should be considered; and the pacer system should be reassessed postablation to make certain that device resetting has not occurred.

Radiation therapy for lung or breast cancer ipsilateral to the generator site requires extensive discussion with the radiation therapist. Shielding of the generator is often possible, but some patients (Fig. 39-73) may require repositioning of the device. Generators may be damaged randomly, in cumulative fashion, and unpredictably, with consequent loss of output, device reprogramming, or runaway pacing.[193-197] Transcutaneous electrical nerve stimulation infrequently is an issue with device function.[198-199]

Magnetic resonance imaging[200-205] should be avoided in the paced patient, because the magnetic fields may result in physical movement of the internal components of the generator, especially the reed switch; despite this, no reported component damage has been

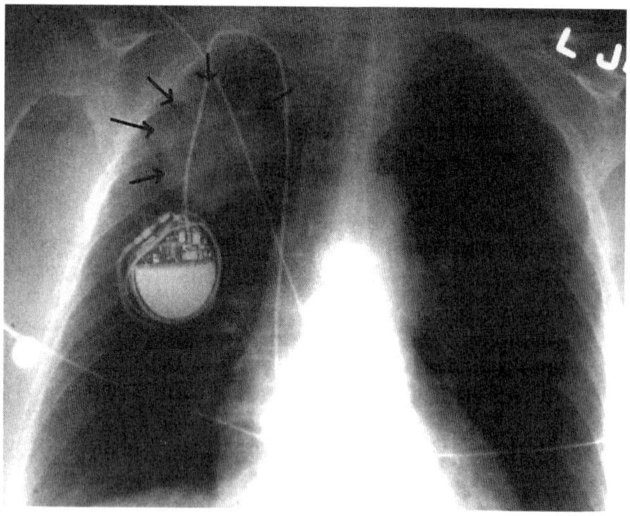

FIGURE 39-73 Chest radiograph of pacer generator close to radiation portal for patient with lung cancer.

reported when tested. Most commonly, asynchronous pacing is reported and/or rapid pacing rates, as well as generator resetting. When MRI is unavoidable, the generator should be tested in vitro to predict its in vivo response, the device should be programmed to magnet "off" if available, and the asynchronous mode should be selected for the pacer-dependent patient.

Cellular telephones and electronic article surveillance systems have become increasingly recognized sources of EMI in the nonmedical environment. Where reported, cellular telephone interference may be observed in up to 20% of tests, with associated symptoms overall in 7.2% of patients when the telephone is placed directly over the generator.[206] It is not felt, however, to be a clinically important phenomenon when the telephone is held in the normal position (i.e., to the ear, particularly the contralateral ear). Regarding electronic article surveillance systems, triggering of device activity (especially discharge of an ICD device) may be observed rarely. Certain dual chamber pacer systems revert to VOO when exposed to magnets, including magnetic security systems, causing a type of pacemaker syndrome in the patient intolerant of ventricular pacing. The best advice to offer to a device patient is not to linger when encountering an electronic article surveillance system.[207-208]

Lithotripsy may result in oversensing or irregular sensing, especially in the case of rate-adaptive systems using piezoelectrode crystals in the generator.[209-210] Oversensing of shock waves may result in rapid pacing rates but the crystals may, as well, be susceptible to shatter from the waves. Whenever possible, lithotripsy should be avoided in the case of abdominal implants and the rate-adaptive mode should be inactivated. The single chamber VVI mode should be selected to prevent triggering of the shock-wave output by the atrial output pulse, which may delay delivery of the shock wave and produce oversensing and inhibition on the ventricular channel in DDD systems. Distancing of the pulse generator from the focal point of the lithotripsy is also recommended, as well as synchronization of the lithotriptor to the QRS complex or pacer spike to minimize inhibition or triggering by the discharge spike (may otherwise occur at other times outside the ventricular refractory period). Once again, back-up temporary pacing should be considered in the pacer-dependent patient and the device should be reevaluated post procedure.

The Future of Pacemaker Follow-Up

There has been an increasing dependence on device automaticity in the follow-up of patients, as indicated earlier. This refers to the algorithmic regulation of pacer function based on patient conditions and pacer system conditions without the need for clinician input. Examples have included algorithms for rate-adaptive pacing, automatic mode switching, and automatic determination of sensing and pacing thresholds. To be anticipated is the automatic capability of a pacing system to gather increasing diagnostics, recognize problems and correct them, both reporting to the physician and indicating when the physician must take some other corrective action. Also to be anticipated are parallel enhancements of programmer technology, with increasingly automatic function and the ability to update pacers to the latest software and track the performance of a particular pacing system.

Automaticity cannot supplant direct physician-patient contact: a history must still be taken as to the presence or amelioration of symptoms, wound healing must be surveyed, and radiographic assessment remains paramount. It may, however, extend the time between follow-ups and undoubtedly will reduce the required time for a particular visit.[109] We must be cautious, however, lest we place undue reliance on automatic technology and insufficient reliance on ourselves—the primacy of the physician in the follow-up of the patient with a pacemaker must be maintained.

ACKNOWLEDGMENTS

The author would like to thank Mike Dabbraccio, David Kalinchak, and Ronald Quain for their excellent technical and photographic assistance.

REFERENCES

1. Schoenfeld MH: Quality assurance in cardiac electrophysiology and pacing: A brief synopsis. Pacing Clin Electrophysiol 1994;17:267-269.
2. Schoenfeld MH: Quality assurance in pacing and electrophysiology. Card Electrophysiol Rev 1998;2:94-96.
3. Hayes DL, Naccarelli GV, Furman S, et al: Report of the NASPE policy conference: Training requirements for permanent pacemaker selection, implantation, and follow-up. Pacing Clin Electrophysiol 1994;17:6-12.
4. Gregoratos G, Cheitliln MD, Conill A, et al: ACC/AHA guidelines for implantation of cardiac pacemakers and antiarrhythmia devices. A report of the American College of Cardiology/American Heart Association task force on practice guidelines (Committee on Pacemaker Implantation) J Am Coll Cardiol 1998;31;1175-1209.

5. Parsonnet V, Bernstein AD, Lindsay B: Pacemaker-implantation complication rates: An analysis of some contributing factors. J Am Coll Cardiol 1989;13:917-921.
6. Yamamura KH, Kloosterman EM, Alba J, et al: Analysis of charges and complications of permanent pacemaker implantation in the cardiac catheterization laboratory versus the operating room. Pacing Clin Electrophysiol 1999;22:1820-1824.
7. Parsonnet V, Furman S, Smyth NP, Bilitch M: Optimal resources for implantable cardiac pacemakers: Pacemaker Study Group. Circulation 1983;68:226A-244A.
8. Bernstein AD, Parsonnet V: Survey of cardiac pacing and defibrillation in the United States in 1993. Am J Cardiol 1996;78: 1887-1896.
9. Hayes DL, Vlietstra RE, Trusty JM, et al: A shorter hospital stay after cardiac pacemaker implantation. Mayo Clin Proc 1988; 63:236-240.
10. Belott PH: Outpatient pacemaker procedures. Int J Cardiol 1987;17:169-176.
11. Irwin ME, Gulamhusein SS, Senaratne MP, St Clair WR: Outcomes of an ambulatory cardiac pacing program: Indications, risks, benefits, and outcomes. Pacing Clin Electrophysiol 1994; 17:2027-2031.
12. Spotnitz HM, Ott GY, BiggerT Jr, et al: Methods of implantable cardioverter-defibrillator-pacemaker insertion to avoid interactions. Ann Thorac Surg 1992;53:253-257.
13. DaCosta A, Kirkorian G, Cucherat M, et al: Antibiotic prophylaxis for permanent pacemaker implantation: A meta-analysis. Circulation 1998;97:1796-1801.
14. Mounsey JP, Griffith MJ, Tynan M, et al: Antibiotic prophylaxis in permanent pacemaker implantation: A prospective randomised trial. Br Heart J 1994;72:339-343.
15. Classen DC, Evans RS, Pestotnik SL, et al: The timing of prophylactic administration of antibiotics and the risk of surgical-wound infection. N Engl J Med 1992;326:281-286.
16. Fu EY, Shepard RK: Permanent pacemaker infections. Card Electrophysiol Rev 1999;3:39-41.
17. Goldstein DJ, Losquadro W, Spotnitz HM: Outpatient pacemaker procedures in orally anticoagulated patients. Pacing Clin Electrophysiol 1998;21:1730-1734.
18. Michaud GF, Pelosi F Jr, Noble MD, et al: A randomized trial comparing heparin initiation 6h or 24h after pacemaker or defibrillator implantation. J Am Coll Cardiol 2000;35:1915-1918.
19. Costeas XF, Schoenfeld MH:Undersensing as a consequence of lead incompatibility: Case report and a plea for universality. Pacing Clin Electrophysiol 1991;14:1681-1683.
20. Van Gelder BM, Bracke FA, El Gamal MI: Adapter failure as a cause of pacemaker malfunction. Pacing Clin Electrophysiol 1993;16:1961-1965.
21. Sweesy MW, Forney RC, Erickson S, Batey RL: Performance of a 3.2 mm to 6 mm adaptor. Pacing Clin Electrophysiol 1994; 17:138-140.
22. Byrd CL, Schwartz SJ: Transatrial implantation of transvenous pacing leads as an alternative to implantation of epicardial leads. Pacing Clin Electrophysiol 1990;13:1856-1859.
23. Furman S: Venous cutdown for pacemaker implantation. Ann Thorac Surg 1986;41:438-439.
24. Miller FA Jr, Holmes DR Jr, Gersh BJ, Maloney JD: Permanent transvenous pacemaker implantation via the subclavian vein. Mayo Clin Proc 1980;55:309-314.
25. Said SA, Bucx JJ, Stassen CM: Failure of subclavian venipuncture: The internal jugular vein as a useful alternative. Int J Cardiol 1992;35:275-278.
26. Ong LS, Barold SS, Lederman M, et al: Cephalic vein guide wire technique for implantation of permanent pacemakers. Am Heart J 1987;114:753-756.
27. Parsonnet V, Roelke M: The cephalic vein cutdown versus subclavian puncture for pacemaker/ICD lead implantation. Pacing Clin Electrophysiol 1999;22:695-697.
28. Littleford PO, Parsonnet V, Spector SD: Method for rapid and atraumatic insertion of permanent endocardial electrodes through the subclavian vein. Am J Cardiol 1979;43:980-982.
29. Belott PH, Reynolds DW: Permanent pacemaker and implantable cardioverter-defibrillator implantation. In Ellenbogen KA, Kay GN, Wilkoff BL: Clinical Cardiac Pacing and Defibrillation, 2nd ed. Philadelphia, WB Saunders, 2000, pp 573-644.
30. Fyke FE 3rd: Simultaneous insulation deterioration associated with side by side subclavian placement of two polyurethane leads. Pacing Clin Electrophysiol 1988;11:1571.
31. Antonelli D, Rosenfeld T, Freedberg NA, et al: Insulation lead failure: Is it a matter of insulation coating, venous approach or both? Pacing Clin Electrophysiol 1998;21:418-421.
32. Byrd CL: Safe introducer technique for pacemaker lead implantation. Pacing Clin Electrophysiol 1992;15:262.
33. Magney JE, Staplin DH, Flynn DM, Hunter DW: A new approach to percutaneous subclavian venipuncture to avoid lead fracture or central venous catheter occlusion. Pacing Clin Electrophysiol 1993;16:2133-2142.
34. Byrd CL: Clinical experience with the extrathoracic introducer insertion technique. Pacing Clin Electrophysiol 1993;16:1781.
35. Gardini A, Benedini G: Blind extrathoracic subclavian venipuncture for pacemaker implant: A 3-year experience in 250 patients. Pacing Clin Electrophysiol 1998;21:2304-2308.
36. Lamas GA, Fish RD, Braunwald NS: Flouroscopic technique of subclavian venous puncture for permanent pacing: A safer and easier approach. Pacing Clin Electrophysiol 1988;11: 1398-1401.
37. Spencer WH 3rd, Zhu DW, Kirkpatrick C, et al: Subclavian venogram as a guide to lead implantation. Pacing Clin Electrophysiol 1998;21:499-502.
38. Higano ST, Hayes DL, Spittell PC: Facilitation of the subclavian-introducer technique with contrast venography. Pacing Clin Electrophysiol 1990;13:681-684.
39. Ramza BM, Rosenthal L, Hui R, et al: Safety and effectiveness of placement of pacemaker and defibrillator leads in the axillary vein guided by contrast venography. Am J Cardiol 1997;80:892-896.
40. Ing FF, Mullins CE, Grifka RG, et al: Stent dilation of superior vena cava and innominate vein obstructions permits transvenous pacing lead implantation. Pacing Clin Electrophysiol 1998;21: 1517-1530.
41. Belott PH: A variation on the introducer technique for unlimited access to the subclavian vein. Pacing Clin Electrophysiol 1981; 4:43-48.
42. Terada Y, Mitsui T: Placement of two pacing leads with one venipuncture in dual-chamber pacemaker implantation. Ann Thorac Surg 1997;64:563.
43. Zerbe F, Bornakowski J, Sarnowski W: Pacemaker electrode implantation in patients with persistent left superior vena cava. Br Heart J 1992;67:65-66.
44. Barin ES, Jones SM, Ward DE, et al: The right ventricular outflow tract as an alternative permanent pacing site: Long-term follow-up. Pacing Clin Electrophysiol 1991;14:3-6.
45. Giudici MC, Thornburg GA, Buck DL, et al: Comparison of right ventricular outflow tract and apical lead permanent pacing on cardiac output. Am J Cardiol 1997;79:209-212.
46. Victor F, Leclercq C, Mabo P, et al: Optimal right ventricular pacing site in chronically implanted patients: A prospective randomized crossover comparison of apical and outflow tract pacing. J Am Coll Cardiol 1999;33:311-316.
47. Alonso C, Leclercq C, Victor F, et al: Electrocardiographic predictive factors of long-term clinical improvement with multisite biventricular pacing in advanced heart failure. Am J Cardiol 1999;84:1417-1421.
48. Auricchio A, Stellbrink C, Sack S, et al: The pacing therapies for congestive heart failure (PATH-CHF) study: Rationale, design, and endpoints of a prospective randomized multicenter study. Am J Cardiol 1999;83:130D-135D.
49. Gras D, Mabo P, Tang T, et al: Multisite pacing as a supplemental treatment of congestive heart failure: Preliminary results of the Medtronic Inc. InSync Study. Pacing Clin Electrophysiol 1998;21:2249-5225.
50. Daubert JC, Ritter P, LeBreton H, et al: Permanent left ventricular pacing with transvenous leads inserted into the coronary veins. Pacing Clin Electrophysiol 1998;21:239-245.
51. Leclercq F, Hager FX, Macia JC, et al: Left ventricular lead insertion using a modified transseptal catheterization technique: A totally endocardial approach for permanent biventricular pacing in end-stage heart failure. Pacing Clin Electrophysiol 1999;22:1570-1575.
52. Bailin SJ, Adler S, Giudici M: Prevention of chronic atrial fibrillation by pacing in the region of Bachmann's bundle: Results of

a multicenter randomized trial. J Cardiovasc Electrophysiol 2001; 12:912-917.
53. Padeletti L, Porciani MC, Michelucci A, et al: Interatrial septum pacing: A new approach to prevent recurrent atrial fibrillation. J Int Card Electrophysiol 1999;3:35-43.
54. Saksena S, Delfaut P, Prakash A, et al: Multisite electrode pacing for prevention of atrial fibrillation. J Cardiovasc Electrophysiol 1998;9:S155-162.
55. Saksena S, Prakush A, Hill M, et al: Prevention of atrial fibrillation with dual-site right atrial pacing. J Am Coll Cardiol 1996; 28:687.
56. Yu WC, Chen SA, Tai CT, et al: Effects of different atrial pacing modes on atrial electrophysiology: Implicating the mechanism of biatrial pacing in prevention of atrial fibrillation. Circulation 1997;96:2992-2996.
57. Kindermann M, Frohlig G, Berg M, et al: Atrial lead implantation during atrial flutter or fibrillation? Pacing Clin Electrophysiol 1998;21:1531-1538.
58. Longo E, Catrini V: Experience and implantation techniques with a new single-pass lead VDD pacing system. Pacing Clin Electrophysiol 1990;13:927-936.
59. Tse HF, Lau CP: The current status of single lead dual chamber sensing and pacing. J Int Card Electrophysiol 1998;2:255-267.
60. Harthorne JW, Dicola V: Infrequent cause of erratic pacemaker performance. Pacing Clin Electrophysiol 1983;6:655-656.
61. Parsonnet V, Bernstein AD, Neglia D, Omar A: The usefulness of a stretch-polyester pouch to encase implanted pacemakers and defibrillators. Pacing Clin Electrophysiol 1994;17:2274-2278.
62. Saliba BC, Ghantous AE, Schoenfeld MH, Marieb MA: Twiddler's syndrome with transvenous defibrillators in the pectoral region. Pacing Clin Electrophysiol 1999;22:1419-1421.
63. Belott PH, Bucko D: Inframammary pulse generator placement for maximizing cosmetic effect. Pacing Clin Electrophysiol 1983;6:1241-1244.
64. Roelke M, Jackson G, Harthorne JW: Submammary pacemaker implantation: A unique tunneling technique. Pacing Clin Electrophysiol 1994;17:1793-1796.
65. Aggarwal RK, Connelly DT, Ray SG, et al: Early complications of permanent pacemaker implantation: No difference between dual and single chamber systems. Br Heart J 1995;73:571-575.
66. Spittel PC, Hayes DL: Venous complications after insertion of a transvenous pacemaker. Mayo Clin Proc 1992;67:258-265.
67. Ausubel K, Furman S: The pacemaker syndrome. Ann Intern Med 1985;103:420-429.
68. Kenny RS, Sutton R: Pacemaker syndrome. Br Med J 1986; 293:902-903.
69. Sulke N, Dritsas A, Bostock J, et al: "Subclinical" pacemaker syndrome: A randomised study of symptom free patients with ventricular demand (VVI) pacemakers upgraded to dual chamber devices. Br Heart J 1992;67:57-64.
70. Gribbin GM, McComb JM, Bexton RS: Ventricular pacemaker upgrade: Experience, complications, and recommendations (letter). Heart 1999;80:420-421.
71. Hildick-Smith D, Lowe M, Newell SA, et al: Ventricular pacemaker upgrade: Experience, complications, and recommendations. Heart 1998;79:383-387.
72. Belott PH: Use of the contralateral subclavian vein for placement of atrial electrodes in chronically VVI paced patients. Pacing Clin Electrophysiol 1983;6:781-783.
73. Belott PH, Reynolds DW: Permanent pacemaker and implantable cardioverter-defibrillator implantation. In Ellenbogen KA, Kay GN, Wilkoff BL (eds): Clinical Cardiac Pacing and Defibrillation, 2nd ed. Philadelphia, WB Saunders, 2000, pp 573-644.
74. Furman S, Behrens M, Andrews C, Klementowicz P: Retained pacemaker leads. J Thorac Cardiovasc Surg 1987;94:770-772.
75. Parry G, Goudevenos J, Jameson S, et al: Complications associated with retained pacemaker leads. Pacing Clin Electrophysiol 1991;14: 1251-1257.
76. DeCock CC, Vinkers M, van Campe LCMC, et al: Long-term outcome of patients with multiple (≥3) noninfected transvenous leads: A clinical and echocardiographic study. Pacing Clin Electrophysiol 2000;23:423-426.
77. Levine PA: Should lead explantation be the practice standard when a lead needs to be replaced? Pacing Clin Electrophysiol 2000;23:421-422.
78. Kutalek SP, Kantharia BK, Maquilan JM: Approach to generator change. In Ellenbogen KA, Kay GN, Wilkoff BL (eds): Clinical Cardiac Pacing and Defibrillation, 2nd ed. Philadelphia, WB Saunders, 2000, pp 645-668.
79. Byrd CL: Management of implant complications. In Ellenbogen KA, Kay GN, Wilkoff BL (eds): Clinical Cardiac Pacing and Defibrillation, 2nd ed. Philadelphia, WB Saunders, 2000, pp 669-694.
80. Iskos D, Lurie KG, Sakaguchi S, Benditt DG: Termination of implantable pacemaker therapy: Experience in five patients. Ann Intern Med 1997;126:787-790.
81. Schoenfeld MH: Deciding against defibrillator replacement: Second-guessing the past? Pacing Clin Electrophysiol 2000;23:2019-2021.
82. Spratt KA, Blumberg EA, Wood CA, et al: Infections of implantable cardioverter defibrillators: Approach to management. Clin Infect Dis 1993;17:679-685.
83. Chua JD, Wilkoff BL, Lee I, et al: Diagnosis and management of infections involving implantable electrophysiologic cardiac devices. Ann Intern Med 2000;133:604-608.
84. Taylor RL, Cohen DJ, Widman LE, et al: Infection of an implantable cardioverter-defibrillator: Management without removal of the device in selected cases. Pacing Clin Electrophysiol 1990;13: 1352-1355.
85. Griffith MJ, Mounsey JP, Bexton RS, Holden MP: Mechanical, but not infective, pacemaker erosion may be successfully managed by re-implantation of pacemakers. Br Heart J 1994;71:202-205.
86. Frame R, Brodman R, Furman S, et al: Surgical removal of infected transvenous pacemaker leads. Pacing Clin Electrophysiol 1993;16:2243-2248.
87. Byrd CL, Schwartz SJ, Sivina M, et al: Technique for the surgical extraction of permanent pacing leads and electrodes. J Thorac Cardiovas Surg 1985;89:142-144.
88. Saliba WI, Erdogan O, Wilkoff BL: Extraction and revision. Card Electrophysiol Rev 1999;3:34-35.
89. Byrd CL, Wilkoff BL: Techniques and devices for extraction of pacemaker and implantable cardioverter-defibrillator leads. In Ellenbogen KA, Kay GN, Wilkoff BL(eds): Clinical Cardiac Pacing and Defibrillation, 2nd ed. Philadelphia, WB Saunders, 2000, pp 695-709.
90. Love CJ: Current concepts in extraction of transvenous pacing and ICD leads. Cardiol Clin 2000;18:193-217.
91. Byrd CL, Schwartz SJ, Hedin N: Lead extraction: Indications and techniques. Cardiol Clin 1992;10:735-748.
92. Love CJ, Wilkoff BL, Byrd CL, et al: Recommendations for extraction of chronically implanted transvenous pacing and defibrillator leads: Indications, facilities, training. North American Society of Pacing and Electrophysiology Lead Extraction Conference Faculty. Pacing Clin Electrophysiol 2000;23:544-551.
93. Madigan NP, Curtis JJ, Sanfelippo JF, Murphy TJ: Difficulty of extraction of chronically implanted tined ventricular endocardial leads. J Am Coll Cardiol 1984;3:724-731.
94. Byrd CL, Wilkoff BL, Love CJ, et al: Intravascular extraction of problematic or infected permanent pacemaker leads: 1994-1996. U.S. Extraction Database, MED Institute. Pacing Clin Electrophysiol 1999;22:1348-1357.
95. Bracke FA, Meijer A, Van Gelder B: Learning curve characteristics of pacing lead extraction with a laser sheath. Pacing Clin Electrophysiol 1998;21:2309-2313.
96. Kaizuka H, Kawamura T, Kasagi Y, et al: An experience of infected catheter removal with continuous traction method and silicon rubber wearing. Kyobu Geka 1983;36:309-311.
97. Bilgutay AM, Jensen MK, Schmidt WR, et al: Incarceration of transvenous pacemaker electrode: Removal by traction. Am Heart J 1969;77:377-379.
98. Byrd CL, Schwartz SJ, Hedin NB, et al: Intravascular lead extraction using locking stylets, sheaths, and other techniques. Pacing Clin Electrophysiol 1990;13:1871-1872.
99. Fearnot NE, Smith HJ, Goode LB, et al: Intravascular lead extraction using locking stylets, sheaths, and other techniques. Pacing Clin Electrophysiol 1990;13:1864-1870.
100. Pace JN, Maquilan M, Hessen SE, et al: Extraction and replacement of permanent pacemaker leads through occluded vessels: Use of extraction sheaths as conduits—balloon venoplasty as an adjunct. J Interv Card Electrophysiol 1997;1:271-279.

101. Wilkoff BL, Byrd CL, Love CJ, et al: Pacemaker lead extraction with the laser sheath: Results of the pacing lead extraction with the excimer sheath (PLEXES) trial. J Am Coll Cardiol 1999; 33:1671-1676.
102. Epstein LM, Byrd CL, Wilkoff BL, et al: Initial experience with larger laser sheaths for the removal of transvenous pacemaker and implantable defibrillator leads. Circulation 1999;100:516-525.
103. Saliba BC, John R, Venditti F, Schoenfeld MH: Predictors of fracture in the Accufix-TM atrial "J" lead. Am J Cardiol 1997; 80:229-231.
104. Ardesia RJ, Saliba BC, Schoenfeld MH: Evolution of fracture in the atrial "J" lead retention wire. Pacing Clin Electrophysiol 1995;18:2227-2228.
105. Taliercio CP, Vlietstra RE, Hayes DL: Pigtail catheter for extraction of pacemaker lead (letter). J Am Coll Cardiol 1985; 5:1020.
106. Espinosa RE, Hayes DL, Vlietstra RE, et al: The Dotter retriever and pigtail catheter: Efficacy in extraction of chronic transvenous pacemaker leads. Pacing Clin Electrophysiol 1993;16:2337-2342.
107. Ramsdale DR, Arumugam N, Pidgeon JW: Removal of fractured pacemaker electrode tip using Dotter basket. Pacing Clin Electrophysiol 1985;8:759-760.
108. Kay GN, Brinker JA, Kawanishi DT, et al: Risks of spontaneous injury and extraction of an active fixation pacemaker lead: Report of the Accufix Multicenter Clinical Study and Worldwide Registry. Circulation 1999;100:2344-2352.
109. Schoenfeld MH, Markowitz HT: Device follow-up in the age of automaticity. Pacing Clin Electrophysiol 2000;23:803-806.
110. Schoenfeld MH: Follow-up of the paced patient. In Ellenbogen KA, Kay GN, Wilkoff BL (eds): Clinical Cardiac Pacing and Defibrillation, 2nd ed. Philadelphia, WB Saunders, 2000, pp 895-929.
111. Griffin JC, Schuenemeyer TD, Hess KR, et al: Pacemaker follow-up: Its role in the detection and correction of pacemaker system malfunction. Pacing Clin Electrophysiol 1986;9:387-391.
112. Furman S: Pacemaker follow-up. In Furman S, Hayes DL, Holmes DR (eds): A Practice of Cardiac Pacing, 3rd ed. Mount Kisco, NY, Futura, 1993, pp 571-603.
113. Schoenfeld MH: Follow-up of the pacemaker patient. In Ellenbogen KA (ed): Cardiac Pacing, 2nd ed. Cambridge, Mass, Blackwell Scientific, 1996, pp 456-500.
114. Levine PA: Proceedings of the policy conference of the North American Society of Pacing and Electrophysiology on programmability and pacemaker follow-up programs. Clin Prog Pacing Elecrophysiol 1984;2:145-191.
115. Gessman LJ, Vielbig RE, Waspe LE, et al: Accuracy and clinical utility of transtelephonic pacemaker follow-up. Pacing Clin Electrophysiol 1995;18:1032-1036.
116. Dreifus L, Pennock R, Feldman M: Experience with 3835 pacemakers utilizing transtelephonic surveillance. Am J Cardiol 1975;35:133.
117. Platt S, Furman S, Gross JN, et al: Transtelephone monitoring for pacemaker follow-up 1981-1994. Pacing Clin Electrophysiol 1996;19:2089-2098.
118. MacGregor CD, Covvey HD, Noble EJ, et al: Computer-assisted reporting system for the follow-up of patients with cardiac pacemakers. Pacing Clin Electrophysiol 1980;3:5688.
119. Medicare Cardiac pacemaker evaluation services effective Oct. 1, 1984, Coverage Issues Manual HCFA Publication No.6, section 50-1.
120. Hayes DL, Hyberger LK, Lloyd MA: Should the Medicare transtelephonic pacemaker follow-up schedule be altered? Pacing Clin Electrophysiol 1997;20:1153.
121. Vallario L, Leman R, Gillette P, Kratz J: Pacemaker follow-up and adequacy of medical guidelines. Am Heart J 1988;116:11.
122. Hoffman A, Jost M, Pfisterer, et al: Persisting symptoms despite permanent pacing: Incidence, causes, and follow-up. Chest 1984;85:207-210.
123. Ausubel K, Furman S: The pacemaker syndrome. Ann Intern Med 1985;103:420-429.
124. Kenny RS, Sutton R: Pacemaker syndrome. Br Med J 1986; 293:902-903.
125. Patel NR, Sulke N: Pacemaker syndrome. Card Electrophysiol Rev 1999;3:50-52.
126. Ellenbogen KA, Thames MD, Mohanty PK: New insights into pacemaker syndrome gained from hemodynamic, humoral and vascular responses during ventriculoatrial pacing. Am J Cardiol 1990;65:53-59.
127. Cohen SI, Frank HA: Preservation of active atrial transport: An important clinical consideration in cardiac pacing. Chest 1982;81:51.
128. Byrd CL, Schwartz SJ, Gonzalez M, et al: Pacemaker clinic evaluations: Key to early identification of surgical problems. Pacing Clin Electrophysiol 1986;9:1259-1264.
129. Bayliss CE, Beanlands DA, Baird RJ: The Pacemaker-twiddler's syndrome: A new complication of implantable transvenous pacemakers. Can Med Assoc J 1968;99:371.
130. Saliba BC, Ghantous AE, Schoenfeld MH, Marieb MA: Twiddler's syndrome with transvenous defibrillators in the pectoral region. Pacing Clin Electrophysiol 1999;22:1419-1421.
131. Misra KP, Korn M, Ghahramani AR, et al: Auscultatory findings in patients with cardiac pacemakers. Ann Intern Med 1971; 74:245.
132. Cheng TO, Ertem G, Vera A: Heart sounds in patients with cardiac pacemakers. Chest 19972;62:66.
133. Mazzetti H, Dussaut A, Tentori C, et al: Superior vena cava occlusion and/or syndrome related to pacemaker leads. Am Heart J 1993;125:831.
134. Grier D, Cook PG, Hartnell GG: Chest radiographs after permanent pacing: Are they really necessary? Clin Radiol 1990;42:244-249.
135. Barold SS, Falkoff MD, Ong LS, Heinle RA: Normal and abnormal patterns of ventricular depolarization during cardiac pacing. In Barold SS (ed): Modern Cardiac Pacing. Mount Kisco, NY, Futura, 1985, pp 545-569.
136. Barold SS, Falkoff MD, Ong LS, Heinle RA: The abnormal pacemaker stimulus. In Barold SS (ed): Modern Cardiac Pacing. Mount Kisco, NY, Futura, 1985, pp 571-586.
137. Van Gelder LM, El Gamal MIH: Magnet application—A cause of persistent arrhythmias in physiological pacemakers: Report of 2 cases. Pacing Clin Electrophysiol 1982;5:710.
138. Fieldman A, Dobrow R: Phantom pacemaker programming. Pacing Clin Electrophysiol 1978;1:166.
139. Schoenfeld MH: A primer on pacemaker programmers. Pacing Clin Electrophysiol 1993;16:2044-2052.
140. Schoenfeld MH, Padula LE, Harthorne JW: Pacemaker programmers. In Ellenbogen KA, Kay GN, Wilkoff B (eds): Clinical Cardiac Pacing. WB Saunders, Boston, 1995, pp 127-138.
141. Schoenfeld MH: Pacemaker programmers: An updated synopsis. Card Electrophysiol Rev 1999;3:20-23.
142. Sanders R, Martin R, Fruman H, et al: Data storage and retrieval by implantable pacemakers for diagnostic purposes. Pacing Clin Electrophysiol 1984;7:1228.
143. Haddad L, Padula LE, Moreau M, Schoenfeld MH: Troubleshooting implantable cardioverter defibrillator system malfunctions: The role of impedance measurements. Pacing Clin Electrophysiol 1994;17:1456-1461.
144. Waktare JE, Malik M: Holter, loop recorder, and event counter capabilities of implanted devices. Pacing Clin Electrophysiol 1997;20:2658-2669.
145. Hayes DL, Higano ST, Eisenger G: Utility of rate histograms in programming and follow-up of a DDDR pacemaker. Mayo Clin Proc 1989;64:495.
146. Levine PA: The complementary role of electrogram, event marker, and measured data telemetry in the assessment of pacing system function. J Electrophysiol 1987;1:404.
147. Ben-Zur UM, Platt SB, Gross JN, et al: Direct and telemetered lead impedance. Pacing Clin Electrophysiol 1994;17:2004-2007.
148. Luceri RM, Castellanos A, Thurer RJ: Telemetry of intracardiac electrograms: Applications in spontaneous and induced arrhythmias. J Electrophysiol 1987;1:417.
149. Levine PA, Sholder J, Jundan JL: Clnical benefits of telemetered electrograms in the assessment of DDD function. Pacing Clin Electrophysiol 1984;7:1170.
150. Clarke M, Lieu B, Schuller H, et al: Automatic adjustment of pacemaker stimulation output correlated with continuously monitored capture thresholds: A multicenter study of the Microny SR + Model 2425T pacemaker. Pacing Clin Electrophysiol 1998; 21:1567-1575.

151. Castro A, Liebold A, Vincente J, et al: Evaluation of autosensing as an automatic means of maintaining 2:1 sensing safety margin in an implanted pacemaker. Pacing Clin Electrophysiol 1996;19:1708-1713.
152. Mauser JF, Huang SKS, Risser T, et al: A unique pulse generator safety feature for bipolar lead fracture. Pacing Clin Electrophysiol 1993;16:1368-7132.
153. Pearson AC, Janosik DL, Redd RR, et al: Doppler echocardiographic assessment of the effect of varying atrioventricular delay and pacemaker mode on left ventricular filling. Am Heart J 1988;115;611-621.
154. Rosenqvistr M, Isaac K, Botvinick EH, et al: Relative importance of activation sequence compared to atrioventricular synchrony in left ventricular function. Am J Cardiol 1991;67:148-156.
155. LeClerq C, Gras D, Le Helloco A, et al: Hemodynamic importance of preserving the normal sequence of ventricular activation in permanent cardiac pacing. Am Heart J 1995;129:1133-1141.
156. Jutzy RV, Feenstra L, Pai R, et al: Comparison of intrinsic versus paced ventricular function. Pacing Clin Electrophysiol 1992;15:1919-1922.
157. Pieterse MG, den Dulk K, van Gelder BM, et al: Programming a long paced atrioventricular interval may be risky in DDDR pacing. Pacing Clin Electrophysiol 1994;17:252-257.
158. Stierle U, Kruger D, Vincent AM, et al: An optimized AV delay algorithm for patients with intermittent atrioventricular conduction. Pacing Clin Electrophysiol 1998;21:1035-1043.
159. Mayumi H, Kohno H, Yasui H, et al: Use of automatic mode change between DDD and AAI to facilitate native atrioventricular conduction in patients with sick sinus syndrome or transient atrioventricular block. Pacing Clin Electrophysiol 1996;19: 1740-1747.
160. Linde C: The clinical utility of positive and negative AV/PV hysteresis. In Santini M: Progress in Clinical Pacing, 1996. Armonk, NY, Futura Media Services, 1997, pp 339-345.
161. Nishimura RA, Trusty JM, Hayes DL, et al: Dual chamber pacing for hypertrophic cardiomyopathy: A randomized, double-blind, crossover trial. J Am Coll Cardiol 1997;29:435-441.
162. Sulke AN, Chambers JB, Sowton E: The effect of atrio-ventricular delay programming in patients with DDDR pacemakers. Eur Heart J 1992;13:464-472.
163. Ritter P, Daubert C, Mabo P, et al: Hemodynamic benefit of rate adapted AV delay in dual chamber pacing. Eur Heart J 1989; 10:637-646.
164. Batey FL, Calabria DA, Sweesy MW, et al: Crosstalk and blanking periods in a dual chamber pacemaker. Clin Prog Pacing Electrophysiol 1985;3:314-318.
165. Klementowicz P, Ausubel K, Furman S: The dynamic nature of ventriculoatrial conduction. Pacing Clin Electrophysiol 1986;9:1050-1054.
166. Chew PH, Bush DE, Engel BT, et al: Overnight heart rate and cardiac function in patients with dual chamber pacemakers. Pacing Clin Electrophysiol 1996;19:822-828.
167. Lee MT, Baker R: Circadian rhythm variation in rate adaptive pacing systems. Pacing Clin Electrophysiol 1990;13:1797-1801.
168. Cohen TJ: Circadian rhythm: A programmable feature in some pacemakers with some limitations. J Invas Cardiol 1998; 10:409.
169. Bornzin GA, Arambula ER, Florio J, et al: Adjusting heart rate during sleep using activity variance. Pacing Clin Electrophysiol 1994;17:1933-1938.
170. Furman S: Rate modulated pacing. Circulation 1990;82: 1081-1094.
171. Benditt DG, Mianulli M, Lurie K, et al: Multiple-sensor systems for physiologic cardiac pacing. Ann Intern Med 1994;121; 960-968.
172. Wilkoff BL, Firstenberg MS: Cardiac chronotropic responsiveness. In Ellenbogen KA, Kay GN, Wilkoff BL (eds): Clinical Cardiac Pacing and Defibrillation, 2nd ed. Philadelphia, WB Saunders, 2000, pp 508-532.
173. Kay G: Quantitation of chronotropic response: Comparison of methods for rate-modulating permanent pacemakers. Am Coll Cardiol 1992;20:1533-1541.
174. Ahern T, Nydegger C, McCormick DJ, et al: Incidence and timing of activity parameter changes in activity responsive pacing systems. Pacing Clin Electrophysiol 1992;15:762-770.
175. Kamalvand K, Tan K, Kotsakis A, et al: Is mode switching beneficial? A randomized study in patients with paroxysmal atrial tachyarrhythmias. J Am Coll Cardiol 1997;30:496-504.
176. Connnolly SJ, Sheldon R, Roberts RS, et al: The North American Vasovagal Pacemaker Study (VPS): A randomized trial of permanent cardiac pacing for the prevention of vasovagal syncope. J Am Coll Cardiol 1999;33:16.
177. Benditt DG, Sutton R, Gammage MD, et al: Clinical experience with Thera DR rate-drop response pacing algorithm in carotid sinus syndrome and vasovagal syncope. Pacing Clin Electrophysiol 1997;20:832.
178. Petersen MEV, Chamberlain-Webber R, Fitzpatrick AP, et al: Permanent pacing for cardioinhibitory malignant vasovagal syndrome. Br Heart J 1994;71:274.
179. Sheldon RS, Koshman ML, Wilson W, et al: Effect of dual chamber pacing with automatic rate-drop sensing on recurrent neurally-mediated syncope, Am J Cardiol 1998;81:158.
180. Famularo MA, Kennedy HL: Ambulatory electrocardiography in the assessment of pacemaker function. Am Heart J 1982; 104:1086-1094.
181. Gaita F, Asteggian R, Bocchiardo M, et al: Holter monitoring and provocative maneuvers in the assessment of unipolar demand pacemaker myopotential inhibition. Am Heart J 1984;107:925.
182. Barold SS, Schoenfeld MH: Pacemaker elective replacement indicators: Latched or unlatched? Pacing Clin Electrophysiol 1989;12:990-995.
183. Barold SS, Schoenfeld MH, Falkoff MD, et al: Elective replacement indicators of simple and complex pacemakers. In Barold SS, Mugica J (eds): New Perspectives in Cardiac Pacing, 2nd ed. Mt Kisco, NY, Futura, 1991, p. 493-526.
184. Schoenfeld MH: Recommendations for implementation of a North American multicenter arrhythmia device/lead database. Pacing Clin Electrophysiol 1992;15:1633-1635.
185. Blitzer ML, Marieb MA, Schoenfeld MH: Inability to communicate with ICDs: An underreported failure mode. Pacing Clin Electrophysiol, 2001;24:13-15.
186. Goldman BS, Newman D, Fraser J, et al: Management of intracardiac device recalls: A consensus conference. Pacing Clin Electrophysiol 1996;19:7-17.
187. Furman S: External defibrillation in implanted cardiac pacemakers. Pacing Clin Electrophysiol 1981;4:485.
188. Barold SS, Ong LS, Scovil J, et al: Reprogramming of implanted pacemaker following external defibrillation. Pacing Clin Electrophysiol 1978;1:514.
189. Yee R, Jones DL, Klein GJ: Pacing threshold changes after transvenous catheter shock. Am J Cardiol 1984;53:503.
190. Lamas GA, Antman EM, Gold JP: Pacemaker backup mode reversion and injury during cardiac surgery. Ann Thorac Surg 1986;41:155.
191. Chin MC, Rosenqvist M, Lee MA, et al: The effect of radiofrequency catheter ablation on permanent pacemakers: An experimental study. Pacing Clin Electrophysiol 1990;13:23.
192. Ellenbogen KA, Wood MA, Stambler BS, et al: Acute effects of radiofrequency ablation of atrial arrhythmias on implanted permanent pacing systems. Pacing Clin Electrophysiol 1996;19:1287-1295.
193. Adamac R, Haefliger JM, Killisch JP, et al: Damaging effect of therapeutic radiation on programmable pacemakers. Pacing Clin Electrophysiol 1982;5:146.
194. Calfee RF: Therapeutic radiation in pacemakers. Pacing Clin Electrophysiol 1982;5:160.
195. Lee RW, Huang SK, Mechling E, et al: Runaway atrial sequential pacemaker after radiation therapy. Am J Med 1986;81:833.
196. Shehata WM, Daoud GL, Meyer RL: Radiotherapy for patients with cardiac pacemakers: Possible risks. Pacing Clin Electrophysiol 1986;9:919.
197. Venselaar JLM, VanKerkoerle HLMJ, Vet AJTM: Radiation damage to pacemakers from radiotherapy. Pacing Clin Electrophysiol 1987;10:538.
198. Rasmussen MJ, Hayes DL, Fiestsra RE, et al: Can transcutaneous electrical nerve stimulation be safely used in patients with permanent cardiac pacemakers? Mayo Clin Proc 1988;63:433.
199. Philbin D, Marieb MA, Aithal K, Schoenfeld MH: Inappropriate therapy delivered by an ICD as a result of sensed potentials from

a transcutaneous electronic nerve stimulation unit. Pacing Clin Electrophysiol 1998;21:2010-2011.
200. Fetter J, Aram G, Holmes DR Jr, et al: The effects of nuclear magnetic resonance imagers on external and implantable pulse generators. Pacing Clin Electrophysiol 1984;7:720.
201. Holmes DR, Hayes DL, Gray J, et al: The effects of magnetic resonance imaging on implantable pulse generators. Pacing Clin Electrophysiol 1986;9:360-370.
202. Hayes DL, Holmes DR Jr, Gray JE: Effect of 1.5 Tesla nuclear magnetic resonance imaging scanner on implanted permanent pacemakers. J Am Coll Cardiol 1987;10:782.
203. Erlebacher JA, Cahill PT, Pannizzo F, Knowles JR: Effect of magnetic resonance imaging on DDD pacemakers. Am J Cardiol 1986;57:437.
204. Iberer F, Justich E, Stenzl W, et al: Behavior of the Activityrax pacemaker during nuclear magnetic resonance investigation. First International Symposium on Rate-Responsive Pacing, Munich. Pacing Clin Electrophysiol 1987;10:1215.
205. Iberer F, Justich E, Tscheliessnigg KH, Wasler A: Nuclear magnetic resonance imaging in pacemaker patients. In Atlee JL, Gombotz H, Tschelissnigg KH (eds): Perioperative Management of Pacemaker Patients. Berlin, Springer-Verlag, 1992, pp 86-90.
206. Hayes DL, Wang PJ, Reynolds DW, et al: Interference with cardiac pacemakers by cellular telephones. N Engl J Med 1997;336: 1473-1479.
207. Groh WJ, Boschee SA, Engelstein ED, et al: Interactions between electronic article surveillance systems and implantable cardioverter-defibrillators. Circulation 1999;100:387-392.
208. McIvor ME, Reddinger J, Floden E, Sheppard RC: Study of pacemaker and implantable cardioverter-defibrillator triggering by electronic article surveillance devices (SPICED TEAS). Pacing Clin Electrophysiol 1998;221:1847-1861.
209. Cooper D, Wilkoff B, Masterson M, et al: Effects of extracorporeal shock-wave lithotripsy on cardiac pacemakers and its safety in patients with implanted cardiac pacemakers. Pacing Clin Electrophysiol 1988;11:1607-1616.
210. Langberg J, Arber J, Thuroff JW, Griffin JC: The effects of extracorporeal shock-wave lithotripsy on pacemaker function. Pacing Clin Electrophysiol 1987;10:1142.

Chapter 40 Implantable Cardioverter Defibrillators: Technology, Indications, Implantation Techniques, and Follow-Up

NANDINI MADAN, SANJEEV SAKSENA, and
MARK PREMINGER

Implantable cardioverter defibrillator (ICD) devices were originally developed for and are still most widely used for secondary prevention of sudden cardiac death (SCD). They were initially used for the treatment of patients with malignant ventricular arrhythmias and in survivors of SCD. More recently, this secondary prevention application is being expanded to include primary prevention in high-risk subgroups. Sudden cardiac death, which is usually secondary to a sustained malignant ventricular tachyarrhythmia, remains a major public health challenge worldwide. It claims over 300,000 lives a year in the United States alone.[1] The implantable cardioverter-defibrillator has now emerged as the primary therapeutic option for this condition.[2,3]

Refinements in ICD lead and generator technology have led to a progressive expansion of indications for their use. Recent clinical reports have demonstrated the benefits of ICD devices in patients with syncope due to ventricular tachyarrhythmias, high-risk patients with coronary artery disease, including those with nonsustained ventricular tachycardias and/or impaired left ventricular function, and examined its value in the treatment of atrial fibrillation. In addition, ICD systems have undergone a rapid technological evolution to improve functionality, and decrease the morbidity and mortality associated with implantation. However, for optimal device performance, follow-up requires intimate knowledge of these devices and their operations, as well as the clinical electrophysiology of the arrhythmia being managed. Recent practice guidelines have defined indications and follow-up needs for these systems.[4] This chapter will summarize current technology, indications for device implantation, insertion techniques, and principles for follow-up of ICD systems.

Implantable Cardioverter Defibrillator Technology: Evolution and Status

ICD device technology has evolved since the 1980s from a nonprogrammable generator requiring epicardial lead implantation by thoracotomy or at cardiac surgery to a multifunctional therapeutic and monitoring device that employs anatomically distinct transvenously implanted lead systems for atrial and ventricular pacing, defibrillation, and monitoring. Developments in this technology have encompassed generator systems as well as lead refinements. The original ICD generators delivered shock therapy alone, and later added pacing for bradycardia prevention and tachycardia termination. Subsequent generations have included capabilities for atrial pacing, rate responsive pacing and, most recently, atrial anti-tachycardia pacing and defibrillation. Complex new algorithms for the detection and discrimination of supraventricular and ventricular tachyarrhythmias, as well as those for prevention of arrhythmias, are now available. In the same period, diagnostic and monitoring capabilities of the device have expanded. Devices can now evaluate their own component functions and permit automated testing functions for system performance, record spontaneous arrhythmic events, deliver interventional therapies, and store information relative to patient or arrhythmia status. ICD generator longevity has now improved with most devices rated for 4 or more years depending on the current drain for activated features. In this regard, the monitoring functions constitute a significant current drain. With the multitude of features and multifunctional nature of the device, it has become necessary for optimal ICD function to define the pivotal features clearly.

ICD technology can be classified in two categories: that related to ventricular defibrillation and that related to combined atrial and ventricular defibrillation. Specific component features of devices will be discussed for each category.

ICDS FOR VENTRICULAR DEFIBRILLATION

Developments of the original ICD device have now improved its functionality .The transition from an epicardial lead system to wholly nonthoracotomy placement occurred in 1987, with the inclusion of a left extrathoracic patch lead that permitted a larger electrode surface area combined with a new left-to-right shock current vector for successful endocardial defibrillation.[5] In the original version, a three-electrode configuration permitted two simultaneous shock vectors, right-to-right and right-to-left, for bidirectional current delivery to improve defibrillation thresholds (DFTs). Refinements to both generator and lead systems in the last decade simplified implementation of endocardial defibrillation with wholly transvenous implantation. This included use of a biphasic shock waveform as the standard defibrillation shock waveform, which helped to reduce defibrillation energy, and reduced the need for multiple electrodes for optimal energy transfer to the myocardium.[6] New transvenous lead configurations have permitted transvenous defibrillation within the maximal leading edge shock voltage (750 to 800 volts) available in most ICD generators, in over 98% of all patients.[7] The device can now serve as an integral component of the shock energy delivery system—becoming an active pectoral lead replacing left thoracic patch electrode insertion in the original systems.[7] ICD generator size has consistently been reduced with device volumes now well under 40 cc (Fig. 40-1A). Although sub-30 cc devices are in contemplation, the trade-off between generator size and the surface area needed for optimal DFTs is now becoming an issue.

SINGLE CHAMBER VENTRICULAR IMPLANTABLE CARDIOVERTER DEFIBRILLATOR TECHNOLOGY

In its simplest iteration, the ICD device has capabilities for detection of lethal ventricular tachyarrhythmias and permits ventricular pacing and monitoring of ventricular rhythms alone. This requires insertion of a single, combined ventricular pacing and defibrillation lead via cephalic or subclavian vein placement with a pectoral generator site—the generator can being an active defibrillation shock electrode (see Fig. 40-1B). The defibrillation lead includes a large surface area electrode in the right ventricle, integrated or true bipolar sensing, and an active or passive fixation mechanism.[8] Because the shock current is transmitted between the right ventricular electrode and generator, a right-to-left shock vector reduces defibrillation energy. Left-sided placement of the generator is thus preferred whenever feasible, although right-sided placement is possible. In studies on human DFTs with different lead configurations, the mean energy for defibrillation varies with electrode configuration and location. In addition, electrode surface area is a critical determinant of current and voltage requirements for defibrillation by virtue of its effects on the electrode-myocardial interface and shock energy vector. Thus, DFTs vary in inverse proportion to electrode surface area with defined boundaries for such effects in either direction. Addition of a third superior venacaval or right atrial electrode to permit bidirectional shock vectors may reduce DFTs by 10% to 20% in some studies and larger increments in individual patients (see Fig. 40-1C). Similarly, reversal of shock polarity may not reduce DFTs in population studies but can alter thresholds in specific patients. Both these maneuvers are of value in patients with high DFTs. Several endocardial defibrillation leads routinely include two defibrillation electrodes for right ventricular and superior venacaval location, resulting in a bidirectional shock configuration. Reduction in generator size for cosmetic acceptability and implant simplicity may require attention to optimal generator location for successful and low DFTs. In our early studies on left-sided electrode location for defibrillation, the lowest DFTs were achieved with left axillary locations in the anterior to midaxillary line in the vicinity of the fourth intercostal space, with slightly higher values in the left infraclavicular location in the second intercostal space.[9] This latter location is most commonly used for generator placement today. Highest thresholds are obtained with an apical placement in the left midclavicular line at the fourth or fifth intercostal spaces. Thus, the thoracic site of placement or relocation of a generator can have an impact on DFTs that should be considered by an implanting physician.

Ventricular rhythm detection in these systems is based on ventricular electrogram rate, regularity, morphology, and patterns of electrogram interval changes. Initial detection in all devices is based on absolute ventricular rate, with most devices allowing up to three zones for discrimination between different tachyarrhythmias. In general, a minimum of two-zone programming is usually performed for discrimination of a monomorphic ventricular tachycardia from ventricular fibrillation. When empiric programming is sought, we prefer to establish three zones for "slow" ventricular tachycardia, "fast" monomorphic ventricular tachyarrhythmias, and ventricular fibrillation as illustrated in Figure 40-2.[10] Such discrimination is useful for both clinical and therapeutic purposes. The slowest zone is usually associated with nonsyncopal rhythms and these are often responsive to anti-tachycardia pacing. The second zone has symptomatic rhythms but early cardioversion with a lower energy shock with rapid charge times may abort syncope. Very rapid anti-tachycardia pacing can also be considered in this zone. Finally, ventricular fibrillation is usually syncopal in many patients, and requires a highly effective shock for immediate termination based on threshold determination. While nominal detection rate values are provided by manufacturers, it is important to individualize these values based on the patient's sinus rhythm mechanism, its chronotropic competence, exercise response and activity level of the patient, and the presence of coexisting

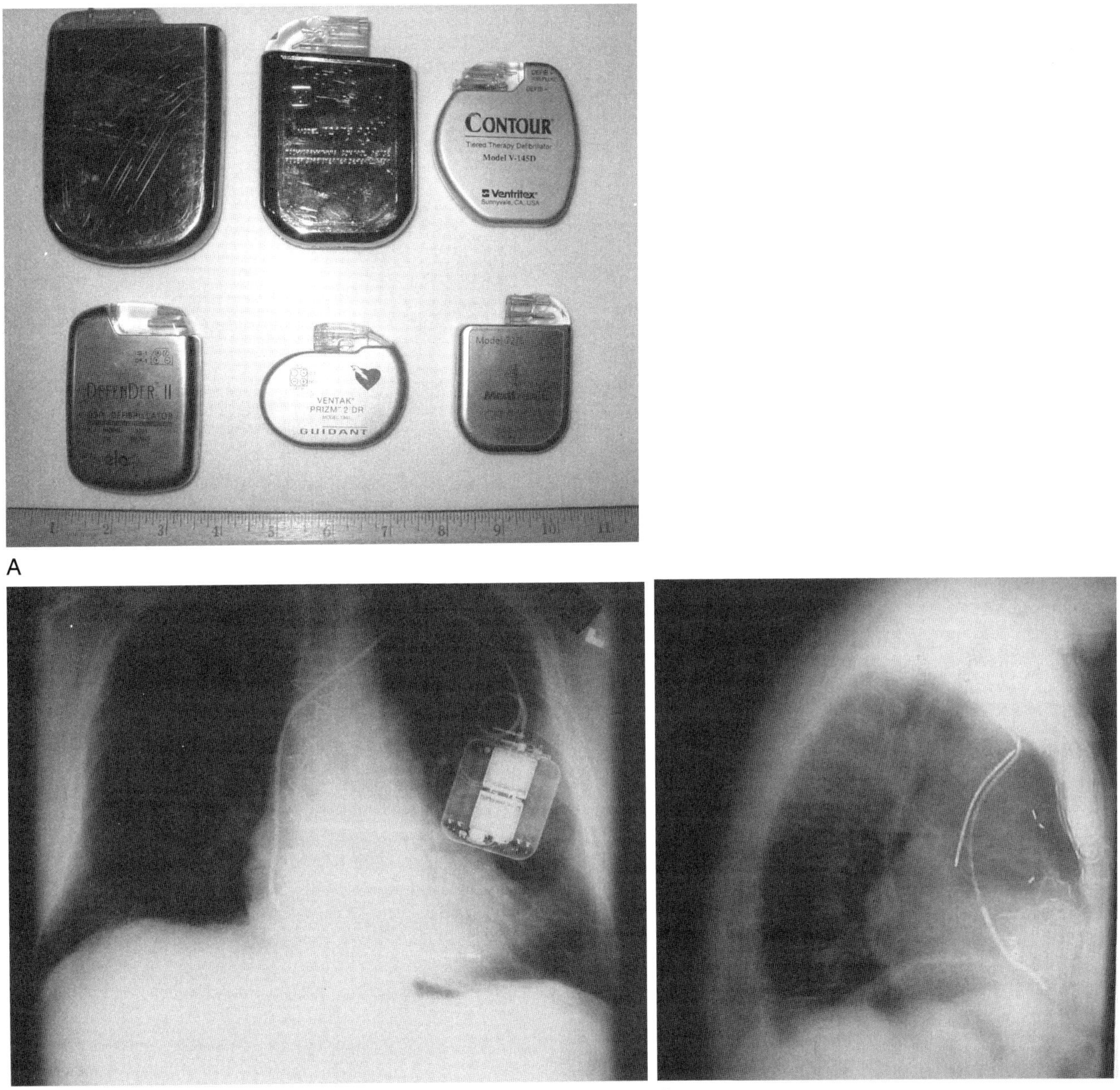

FIGURE 40-1 **A**, Decreasing ICD generator size with progressive technologic advances. The first pacemaker-defibrillator (Telectronics model 4201—*far left upper row*) was succeeded by full-feature pacemaker-defibrillators with monophasic shocks (Medtronic PCD—*middle upper row*) and subsequent biphasic devices. The first dual chamber ICD (Ela Defender—*far left lower row*) was succeeded by down-sized versions and the atrioventricular defibrillator (Medtronic Gem 3 AT). **B**, Chest radiograph (posteroanterior view) of a single-chamber ventricular ICD with single right ventricular pacing and defibrillation lead. **C,** Chest radiograph (lateral view) of single chamber ventricular ICD with additional defibrillation lead in the superior vena cava. ICD, Implantable cardioverter defibrillator.

supraventricular tachyarrhythmias and their ventricular rates. It is generally preferable to ensure a 15 to 20 beats/minute difference between the maximal sinus or supraventricular arrhythmia rate and the initial threshold rate for ventricular tachycardia detection, provided the latter arrhythmia rate falls above this value by at least 10 to 15 beats/minute. If this level of discrimination is not possible based on rate alone, and overlapping rates are present between supraventricular and ventricular tachycardias, other algorithms are available in most devices to assist in identification of the arrhythmia.

All devices offer sudden onset criteria, which identify an abrupt cycle length change in ventricular cycle, to discriminate a pathologic tachycardia from sinus tachycardia. In addition, electrogram morphology in supraventricular and ventricular tachyarrhythmias may be matched to the sinus rhythm template. In the absence

Configuration

Configuration	Defib with Tach A & Tach B
Arrhythmia Sensing	Dual Chamber

Detection Criteria

Fib Detection	270 ms / 222 bpm
Tach B Detection	330 ms / 182 bpm for 12 intervals
Tach A Detection	375 ms / 160 bpm for 12 intervals
SVT Upper Limit	Same as Fib
Post Fib/Tach B Detection	Same as Tach B
MTD	30 sec (Tach Therapy)
MTF	Same as Tach A for 30 sec

SVT Criteria

V < A Rate Branch	
VT Diagnosis Criteria	If Any
Morphology	On (60 %, 5 of 8)
Interval Stability	On (80 ms), (60 ms), 12 intervals
V = A Rate Branch	
VT Diagnosis Criteria	If Any
Morphology	On (60 %, 5 of 8)
Sudden Onset	On (100 ms)
Template	NOT PRESENT

Tachyarrhythmia Therapy

Fib/MTF:	[1] Defib	20.0 J (625 V)
	[2] Defib	33.0 J (801 V)
	[3] Defib x 4	33.0 J (801 V)
Tach B:	[1] CVRT	5.0 J (311 V)
	[2] CVRT	20.0 J (625 V)
	[3] CVRT	33.0 J (801 V)
	[4] CVRT x 2	33.0 J (801 V)
Tach A:	[1] ATP	
	[2] CVRT	10.0 J (443 V)
	[3] CVRT	33.0 J (801 V)
	[4] CVRT x 2	33.0 J (801 V)

Tach A ATP

Output	7.5 V, 1.0 ms
BCL	81 %
Min BCL	200 ms
No. Bursts	3 bursts
Stimuli	8 stimuli
Scanning	12 ms
Ramp	Off

Shock Waveform

Biphasic, Fixed Tilt
RV (+) to SVC/Can (-)
Defib: 65 % / 65 %
CVRT: Same as Defib

Stored EGM

EGM #1	A Sense/Pace, ± 4.0 mV
EGM #2	V Sense/Pace, ± 13.4 mV
Events	Fib, MTD, Tach B, Tach A
Settings	Detection, 16 sec Pre, 1 min Max

Mode

Mode	DDD
Sensor	Passive

Basic Timing

Base Rate	60 ppm / 1000 ms
Rest Rate	Off
Max Track Rate	110 ppm / 545 ms
Hysteresis Rate	Off
AV/PV Delay	250 ms / 225 ms
Rate Responsive AV/PV Delay	Off

Stimulation & Refractory

A. Output	5.0 V, 0.5 ms
V. Output	5.0 V, 0.5 ms
Pace Refractories	PVARP 280 ms/V. 250 ms
Rate Responsive Refractory (V.)	Off

Post-Shock Pacing

Post-Shock Mode	DDD
Post-Shock Base Rate	70 ppm / 857 ms
Post-Shock Pause	1 sec
Post-Shock A. Output	5.0 V, 0.5 ms
Post-Shock V. Output	7.5 V, 1.9 ms
Post-Shock Duration	30 sec

Sensor

Max Sensor Rate	110 ppm / 545 ms
Threshold	Auto (+0.0)
Measured Average Sensor	N/A
Reaction Time	Slow
Recovery Time	Very Slow
Slope	Auto (+0)
Measured Auto Slope:	8

Special Functions

Special EGM Events	AMS Entry/Exit, PMT Termination
Capacitor Maintenance Charge Interval	3 months (800 V)

Extended Parameters

Auto Mode Switch	DDI
Mode Switch Detection Rate	180 bpm / 333 ms
PMT Options	A Pace on PMT
PMT Detection Rate	90 bpm
PVC Options	A Pace on PVC
Ventricular Noise Reversion Mode	Pacer Off
Ventricular Safety Standby	On

Real-Time Measurements

Unloaded Battery Voltage: 3.15 V
Pacing Lead Impedance: A. 440 Ω / V. 350 Ω
Signal Amplitudes: ≥ 3.0 mV P-waves / 8.9 mV R-waves

Real-Time Device Status

Morphology On/Passive; Active Template Not Present

FIGURE 40-2 Three zone tachyarrhythmia programming for discrimination and treatment of slow and fast ventricular tachycardias and ventricular fibrillation. Dual chamber pacing is also programmed.

of intraventricular conduction abnormalities, supraventricular rhythms can often be identified by similar ventricular electrogram morphology.

Less valuable, but often used, is the duration of the ventricular electrogram as a discriminating factor. Ventricular electrogram duration may increase in ventricular tachyarrhythmias due to slower intraventricular conduction without the use of the specialized conduction system (Fig. 40-3). However, few systematic data validate this belief and "narrow" complex ventricular tachycardias are surprisingly common. Most devices consider sustained high-rate events as ventricular in origin for patient safety and trigger a therapeutic strategy. In many patients, ventricular tachyarrhythmia rates can be variable, particularly with changing autonomic tone and anti-arrhythmic drug therapy. Presentation with a monomorphic tachycardia rarely predicts subsequent freedom from ventricular fibrillation. In fact, in clinical studies the mortality and event rates in patients with hemodynamically "stable" ventricular tachycardia were comparable with cardiac arrest victims.[11]

Tachyarrhythmia and bradyarrhythmia therapies are available in these devices. Ventricular demand pacing with or without rate response is present. Ventricular pacing

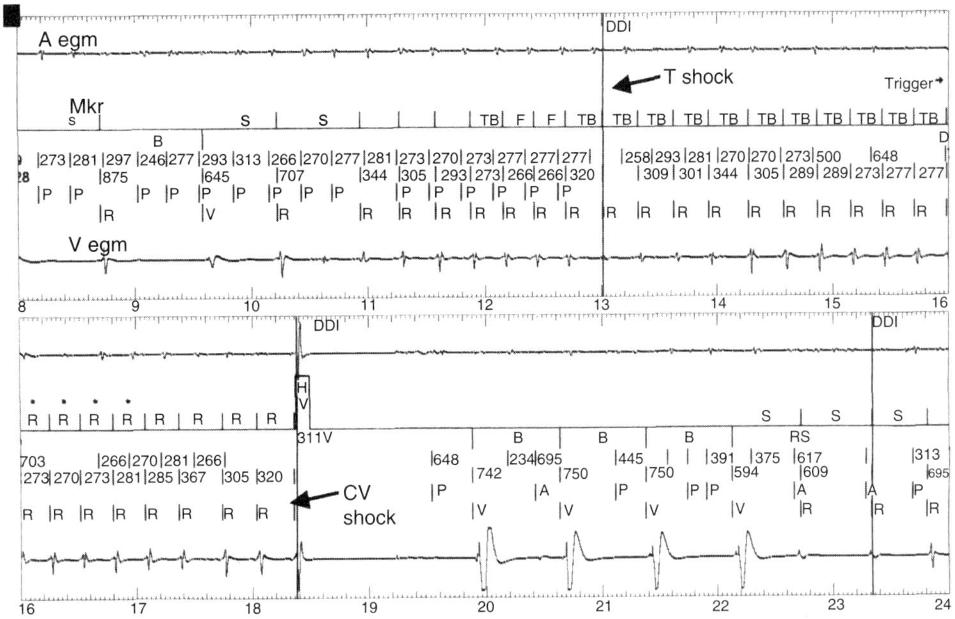

FIGURE 40-3 Duration of ventricular tachycardia with a T wave shock. Note the ventricular electrogram duration is not prolonged. Termination by a high voltage 311 volt shock results in wider V electrograms. *Top trace*: Intracardiac atrial electrogram (A egm). Marker (Mkr) channel P–P and R–R intervals. *Bottom trace*: Intracardiac ventricular electrogram (V egm); CV, cardioversion.

output can be altered for specific situations, such as anti-tachycardia pacing or post-shock bradycardia pacing. Tachyarrhythmia therapies include anti-tachycardia pacing for termination of monomorphic ventricular tachycardia, and low- and high-energy programmable shock therapies for cardioversion of ventricular tachycardia and defibrillation of ventricular fibrillation.[12] Anti-tachycardia pacing is most effective in termination of monomorphic ventricular tachycardia, especially with rates below 180 beats/minute, although individual episodes with rates above this may respond.[12] Several types of anti-tachycardia pacing are available in these devices. It is not uncommon for several attempts or different algorithms to be deployed during such efforts at tachycardia termination in an individual patient. Allowances for these repetitive efforts must include the hemodynamic stability of the patient during anti-tachycardia pacing programs. These algorithms include burst pacing, which may be rate adaptive or fixed rate and low energy cardioversion shocks (Fig. 40-4). In rate-adaptive algorithms, the pacing rate is determined by the tachycardia rate and is generally a percentage of that rate. Commonly used adaptive algorithms use adaptive rates, which can vary from 95% to 70% of the tachycardia rate, for a given number of pacing stimuli.[13] Most commonly used adaptive rates for efficacy vary between 75% and 90% with 3 to 20 delivered pacing stimuli. Ramp pacing modes vary the intervals during a burst to an accelerating (positive) or decelerating (negative) ramp. Other algorithms use scanning

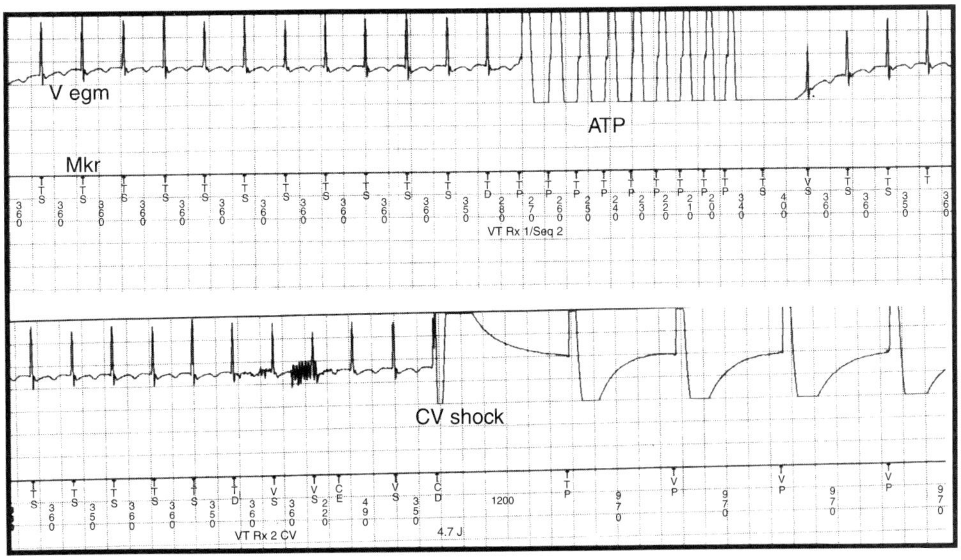

FIGURE 40-4 Failure of termination of ventricular tachycardia with anti-tachycardia pacing (ATP) and subsequent cardioversion by a low energy shock (CV shock). V egm, ventricular electrogram; Mkr, marker channel with intervals in ms; TS, tachycardia sense.

programmed extrastimuli, usually based on electrophysiological testing or ramp or burst pacing with terminal programmed scanning extrastimuli. In general, burst or ramp pacing, or variations thereof, provide the greatest likelihood of efficacy, especially if extensive electrophysiological testing is not employed for pace-termination windows and algorithms. Recent trends toward empirical ICD programming have intensified their usage. However, these algorithms merit significant device testing to ensure efficacy. We prefer to establish a greater than 80% likelihood of efficacy for pace termination before final programming of this algorithm. At a minimum, ten episodes of induced tachycardia should be tested in the laboratory for pace-termination and higher standards have been previously advocated. For cardioversion of fast ventricular tachycardia, studies have shown that 5-joule biphasic waveform shocks achieve approximately 80% success. Should these be ineffective, a high-energy biphasic shock is used for cardioversion.

Initial defibrillation energy programming is based on the DFT. This threshold is not a fixed value and is a probabilistic value in an individual patient at a specific moment. The defibrillation efficacy curve has been well demonstrated in human and animal studies to be sigmoid in nature with the threshold being at the superior but still rising end of the curve. The principles and mechanisms of defibrillation have been discussed in Chapter 4, Fundamental Concepts and Advances in Defibrillation. Thus, the objective of reliable and reproducible defibrillation in a patient requires objective determination of defibrillation efficacy during testing and use of initial shock energies that are highly likely to be successful. This mandates use of energies at or close to this threshold. Repeated efficacy of the lowest successful energy is needed to place the shock near the threshold, with three successive or repeated successful terminations needed to place the shock energy at this level of efficacy. Another approach is to statistically predict the likelihood of success with a maximal energy shock and limit testing. Thus, two successes at 10 joules or less or a single success at 5 joules is highly likely to predict success with a 30-joule shock. In unstable patients in whom testing is to be limited by clinical imperatives, knowledge of defibrillation protocols is invaluable in defining an optimal testing protocol. Step-down DFT testing is used in our laboratory commencing at the mean value for a particular electrode configuration to minimize the number of induced ventricular fibrillation episodes. This is most often 10 joules for left-sided pectoral devices and 15 joules for right-sided implants (Fig. 40-5). Induced ventricular fibrillation is obtained using the noninvasive induction sequences available in the device. The initial shock energy or voltage is tested and based on outcome when subsequent inductions are performed. Should the initial shock be unsuccessful, rescue high energy device-based or external defibrillation should be employed. After hemodynamic stability is restored and 3 to 5 minutes elapse, further testing is performed in step-down decrements if the initial shock was successful, and step-up increments are used if it failed for

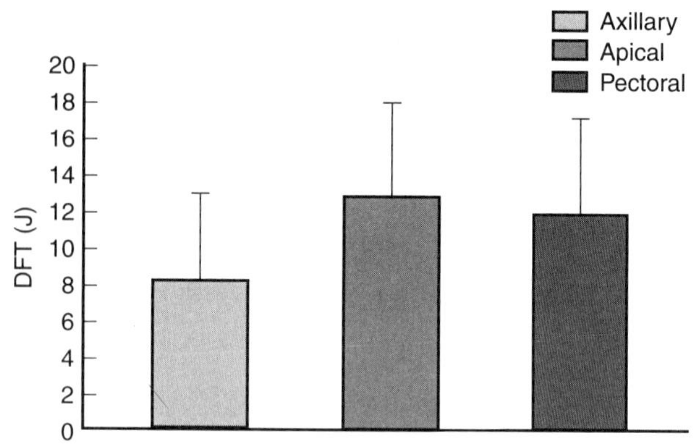

FIGURE 40-5 Defibrillation thresholds (DFTs) with differing thoracic electrode locations. (Adapted from Saksena et al: Circulation 1993;88:2655).

the initial shock efficacy. Other techniques employ a binary protocol with an up-down approach for minimizing shocks and arrhythmia episodes. Once the lowest successful shock energy or voltage is identified, we recommend two additional attempts to confirm reproducible efficacy. Three successful terminations at this level or three out of four such attempts place the shock at or above 90% of the DFT.

When patient stability issues intrude and abort extensive testing, reproducibility of a single energy level is sought. In either instance, a 10-joule safety margin is programmed above the successful shock energy to ensure efficacy. The second and subsequent defibrillation shocks can, and should, be programmed to maximal shock energy for the potential rise in defibrillation energy that is observed with prolonged ventricular fibrillation events. This usually supervenes when such events exceed 30 seconds in clinical studies. Reversal of successive shock polarity may improve efficacy in some situations, although the optimal shock polarity should be determined at implant testing. With an increasing trend to limit implant testing, knowledge of defibrillation energy requirements may compromise the ability to safely introduce new drugs or maintain confidence in the device's ability to defibrillate when intercurrent clinical scenarios, such as heart failure or ischemia, potentially raise DFTs.

Monitoring and testing functions in these devices have improved ease of use and device performance and have provided for more automated device-based testing with expanded capabilities, which include continuous arrhythmia detection and noninvasive electrophysiological testing. Many of these features are present in current pacemakers and have been simply transposed to the ICD. These include battery status, lead impedance measurement on a frequent and even daily basis, pacing thresholds, and extent of pacing. Diurnal pacing rates, rate response, and hysteresis rates are present in many devices. Additional features specific to defibrillators include the capacitor charge time to maximal shock delivery. This was once a key manual test performed at follow-up but now it can be automatically initiated by the device at prespecified intervals of time. Charge times in excess of 12 seconds merit close attention and

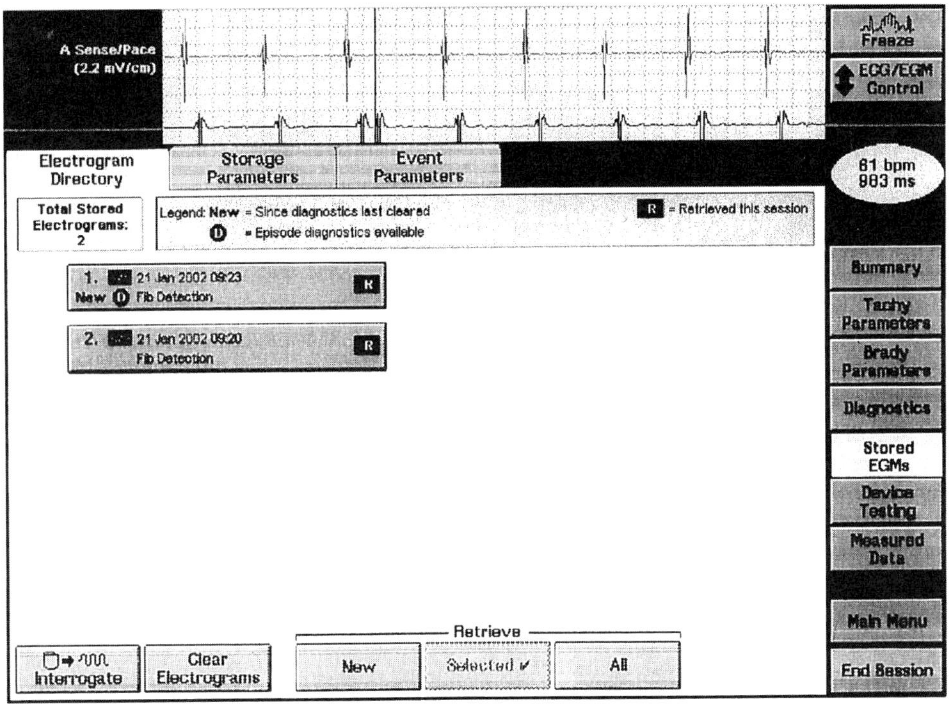

FIGURE 40-6 Arrhythmia event log in device memory of an implantable dual chamber cardioverter defibrillator. The *right panel* is a categorization of programmable device functions. Telemetered atrial and ventricular electrograms are seen in the two traces (*top*).

frequent follow-up, and suggest impending end-of-battery life. We routinely recommend battery replacement if charge times exceed 15 seconds because this prolonged interval permits longer arrhythmic episode continuation before intervention, and can approach intervals in which defibrillation energy requirements can rise, substantially increasing the likelihood of failure. Tachyarrhythmic events are logged with a time and date stamp and their duration recorded (Fig. 40-6). In most devices, recorded ventricular electrograms are available for a segment of the event. Recording capabilities for such digitized graphics require substantial device memory and this has been rapidly expanding with over 120 seconds available in some newer devices. These monitoring capabilities also contribute to substantial current drain on the battery and, if used indiscriminately, can limit longevity. During follow-up evaluations, the patient's arrhythmia history since the last visit or even since implant is readily available.

Noninvasive electrophysiological testing is now available for device-based testing for tachycardia induction and termination. During such testing, noninvasive communication with the device is established by a hand-held wand and driven by a software-based external programmer The present day ICD device can be temporarily programmed to a ventricular stimulation mode for programmed ventricular stimulation for ventricular tachycardia induction. Ventricular fibrillation induction is achieved using shocks delivered in the vulnerable period on the T wave during ventricular pacing, 50-Hz pacing bursts, or standard high rate ventricular pacing bursts (Fig. 40-7). This permits testing of the preprogrammed tachyarrhythmia detection algorithm and selected therapy for appropriate function of the device.

DUAL AND TRIPLE CHAMBER VENTRICULAR IMPLANTABLE CARDIOVERTER DEFIBRILLATOR TECHNOLOGY

Conventional dual chamber pacing has been available in ICD devices for more than 10 years and has gradually expanded in programmability. These devices require the insertion of an additional bipolar atrial pacing lead in the right atrium, akin to a conventional pacemaker (Fig. 40-8A). Device header size is minimally enhanced with an additional port for this lead. Electrode spacing on this lead has been reduced to avoid far-field R wave sensing and lead placement in a high and lateral right atrial location has been advised for the same concern. Inappropriate atrial lead placement can seriously impair tachyarrhythmia functions of this device.

The ability to perform atrial pacing has provided atrioventricular synchronous or simply atrial demand pacing, which may be particularly valuable in patients with impaired ventricular function. It has also allowed improved arrhythmia detection and new algorithms for

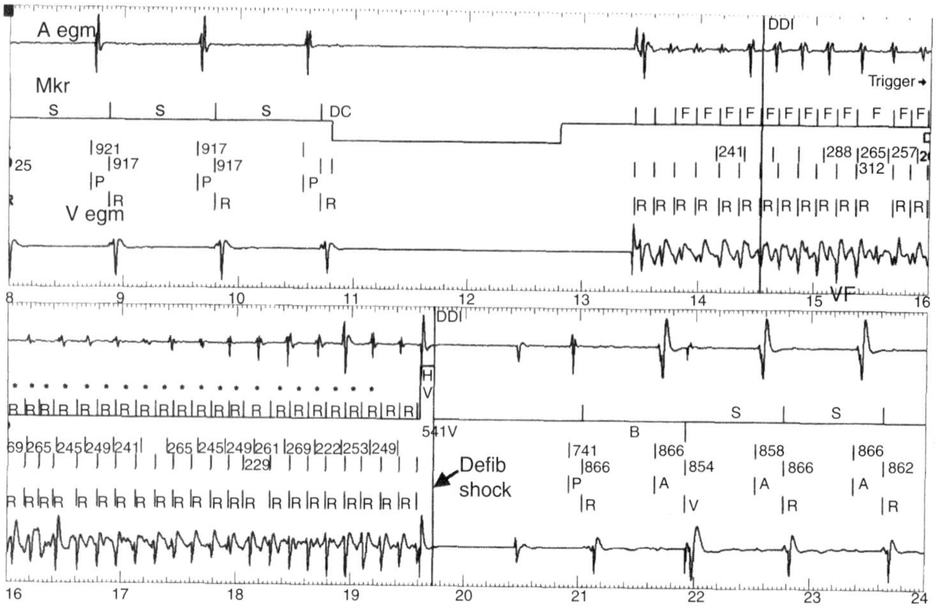

FIGURE 40-7 Induction of ventricular fibrillation (VF) with high frequency direct current (DC) and termination by high energy defibrillation shock (Defib shock) with device-based testing. Electrograms of device-based DFT testing demonstrate induction of VF and termination by shock. *Top trace*: Intracardiac atrial electrogram (A egm). Marker (Mkr) channel P–P and R–R intervals. *Bottom trace*: Intracardiac ventricular electrogram (V Egm). DFT, Defibrillation threshold.

discrimination based on atrial and ventricular electrograms. Pacing capabilities now approach conventional dual chamber pacemakers. Rate response is usually present but the range of upper rate programmability is usually more limited than in dual chamber pacemakers. This is usually due to potential for conflict with tachycardia detection, although models that are more recent have overcome such software conflicts. Rate response is usually based on an activity sensor alone and multisensor devices are still awaited. Future sensors could include ischemia, oxygen, or pressure hemodynamic sensors.

Triple chamber pacing is now available with biventricular pacing ICDs (see Fig. 40-8B), which are discussed fully in Chapter 49 with indications and implant techniques. These ICDs are devised to treat drug refractory congestive heart failure populations with ventricular dyssynchrony.

Tachycardia discrimination in these devices now employs the traditional rate-based detection algorithms in both chambers overlaid with pattern analysis of atrial and ventricular electrogram relationships (see Fig. 40-7). Rate detection alone can easily differentiate rapid and unrelated atrial activity, such as that seen in atrial fibrillation or ectopic atrial tachycardia with varying block from ventricular tachyarrhythmias. Difficulties remain when AV relations are fixed and have identical rates.[14] In such situations, ventricular tachycardia with 1:1 retrograde atrial conduction is hard to differentiate from an atrial tachyarrhythmia with antegrade 1:1 conduction. This could include atrial tachycardia or atrial flutter for 1:1 conduction. Rules for discrimination have been established in device logic and are highly individualized for each device. For example, the initial PR logic algorithm in the Medtronic Gem device series employed a rule for the timing of the atrial event in the ventricular cycle, with the midpoint discriminating an antegrade and retrograde atrial activation. This rule was subsequently discarded with the recognition that AV conduction intervals can vary and prolong substantially in very rapid atrial arrhythmias, reducing algorithm efficiency. In contrast, the PARAD algorithms in the ELA devices examine the onset relationship of the atrium and ventricle and note the chamber showing initial acceleration, and use this as an important determinant of tachycardia origin. Whereas atrial triggers can initiate a ventricular tachyarrhythmia in an occasional patient, this algorithm has proved to be quite robust in clinical practice. Finally, the most recent devices provide retrieval of both atrial and ventricular electrograms based on operator-selected parameters for arrhythmic episodes. These provide important discriminatory information with respect to device detection and therapy. More importantly, new device intervention-induced tachyarrhythmias are now more commonly detected, as are successive distinct arrhythmias in an individual patient (Fig. 40-9).

Current indications for dual chamber ICD insertion in patients with ventricular arrhythmias include standard indications for dual chamber pacing, such as sinus node dysfunction with conduction system disease or when AV nodal blocking drugs are concomitantly administered, in advanced AV block when physiologic pacing is desired, or in patients with left ventricular dysfunction.[4] They are particularly desirable in patients with coexisting atrial and ventricular tachyarrhythmias for discrimination of the two rhythms.

DUAL CHAMBER ATRIOVENTRICULAR IMPLANTABLE CARDIOVERTER DEFIBRILLATOR TECHNOLOGY

This device technology is the latest arrival on the ICD technology scene. The first generation atrial defibrillation device has been replaced by a commercially available dual chamber atrioventricular defibrillator as illustrated in Figure 40-9.[15,16] The technology permits

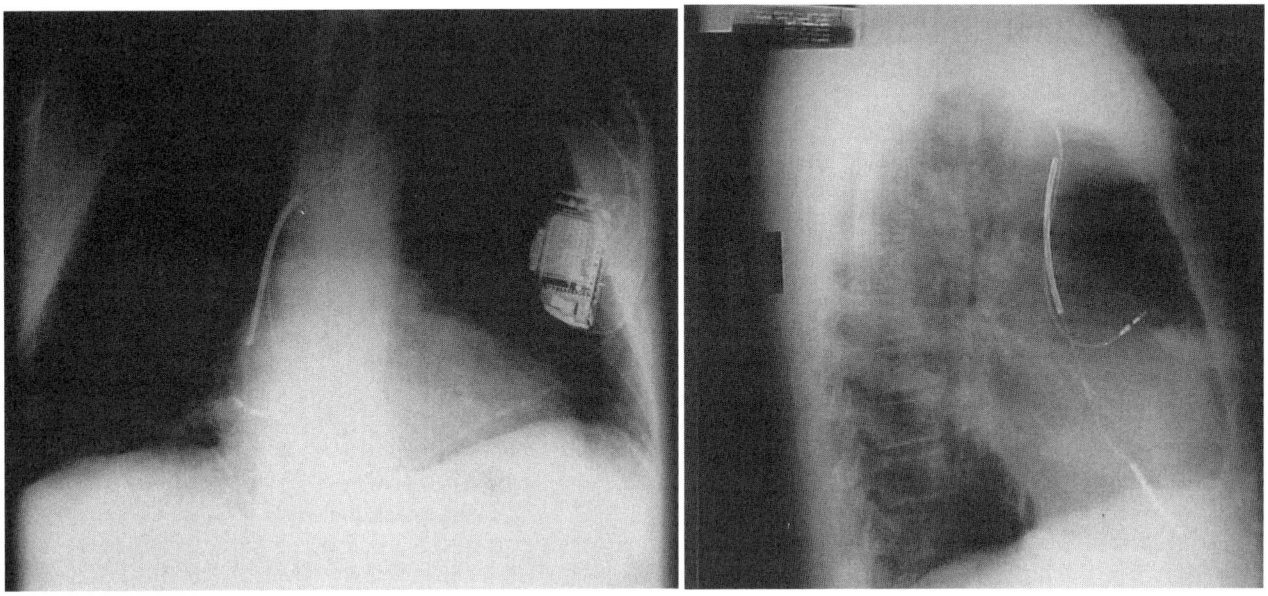

A

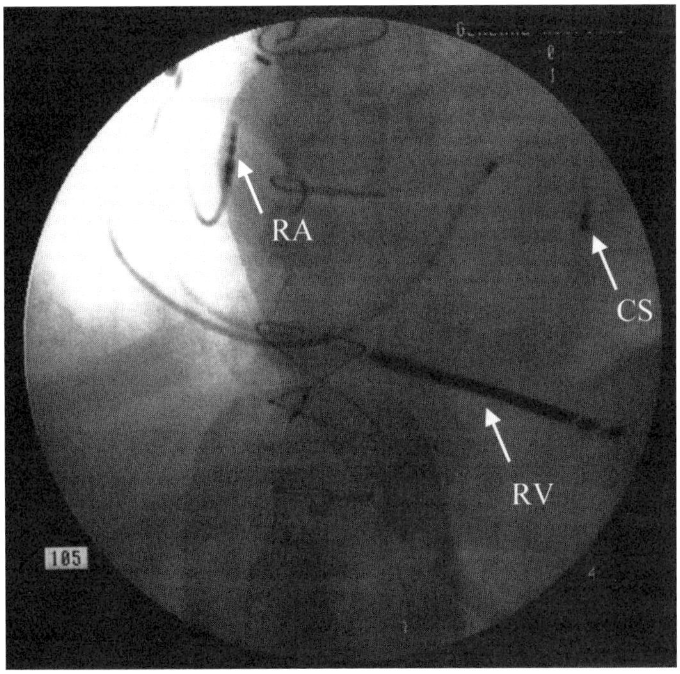

B

FIGURE 40-8 Different types of dual chamber ICD systems. **A,** Posteroanterior and lateral chest radiographs of dual chamber ICD with right atrial pacing and sensing lead and right ventricular pacing and defibrillation lead. **B,** Fluoroscopic view of biventricular ICD demonstrating pacing lead in coronary sinus (CS) in addition to right atrial (RA) and right ventricular (RV) pacing and/or defibrillation leads. ICD, Implantable cardioverter defibrillator.

individualized atrial and ventricular anti-tachycardia therapies to be delivered, extending the device population beyond discrimination of coexisting atrial and ventricular tachyarrhythmias in the same patient.[16,17] Initial studies have been conducted in patients with atrial fibrillation who may or may not have coexisting lethal ventricular tachyarrhythmias.[16,17] Because the population pool of atrial fibrillation is very large and atrial cardioversion is a commonly used procedure, the future of this technology in a hybrid therapy format is quite bright. Stand alone atrial defibrillation devices, while evaluated in prototype form, have great limitations in clinical capability, application, and safety and have been largely discarded.[15] Dual chamber defibrillators are approved for use in patients with drug-refractory and symptomatic atrial fibrillation and in patients with coexisting symptomatic atrial and ventricular tachyarrhythmias.

The atrial channel now permits classification and zone-based therapy of atrial tachyarrhythmias in a manner quite akin to the single chamber ventricular ICD. It requires insertion of an additional atrial defibrillation electrode. This electrode can be mounted on separate and distinct pacing and defibrillation leads placed

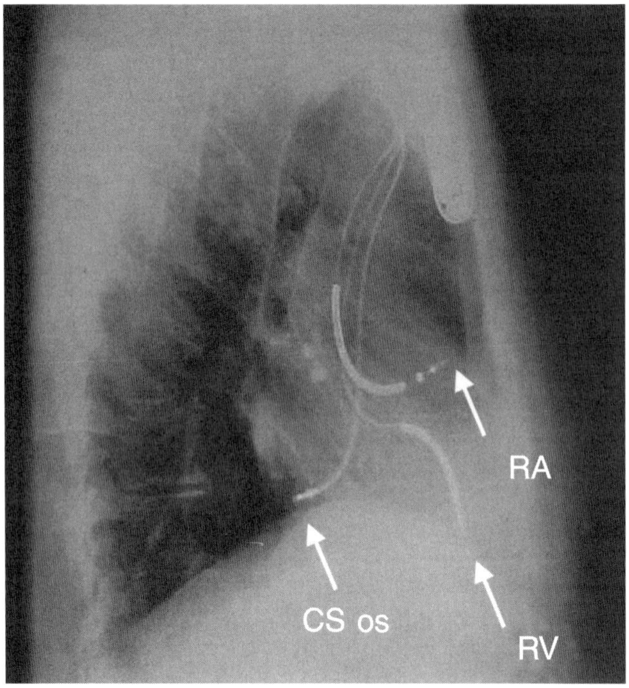

FIGURE 40-9 Chest radiograph, lateral view of dual chamber atrioventricular ICD with dual atrial leads. Note the additional atrial pacing lead at the coronary sinus ostium (CS os) to right atrial (RA) and right ventricular (RV) pacing/defibrillation leads. ICD, Implantable cardioverter defibrillator.

in the right atrium and extending into the superior vena cava for pacing, detection, and shock delivery. Alternatively, the superior vena cava electrode on a conventional ventricular defibrillation lead can provide the atrial defibrillation electrode and a conventional pacing lead used in the atrium. The use of coronary sinus defibrillation leads was widely advocated initially due to modestly lower DFTs.[18,19] However, the reduction in thresholds had few clinical advantages, except to manage the patient with high atrial DFTs. Energy thresholds were still too high for patient tolerance for repeated use. Many clinical disadvantages with the existing technology have limited use of this approach. These include difficult lead insertion with prolonged procedure times, sensing, and pacing issues from the coronary sinus, which can require highly specific and sophisticated algorithms to avoid cross-talk with the ventricular channel and pose significant risks in lead extraction. These issues become particularly important when the atrial fibrillation population has coexisting ventricular tachyarrhythmias.

Atrial tachyarrhythmia detection is based on a two-zone rate stratification structure, with a monomorphic tachycardia zone for atrial tachycardias and flutter categorization and a fibrillation detection zone. Cycle lengths in these two zones are selected based on spontaneous or induced arrhythmias and therapies are individualized for each zone. Anti-tachycardia pacing is available in the tachycardia zone using burst, ramp, or 50 Hz trains. The efficacy of the former two has been proved in prior experiences with anti-tachycardia atrial pacemakers and in laboratory testing.[20] Burst and ramp pacing has been effective in both intraatrial reentrant tachycardia and common atrial flutter with a lower efficacy rate in nonisthmus-dependent atypical atrial flutter. Fifty hertz trains for 1 to 4 seconds have been demonstrated to be effective—the latter arrhythmia with repetitive application, with efficacy rates of up to 60%.[21] Back-up shock therapy is used if pacing therapies are ineffective with programmable shocks from 0.1 to 27 joules. In the AF zone, 50 Hz pacing and shock therapy alone are available. Efficacy of 50 Hz trains in this zone is based on the categorization of very rapid monomorphic arrhythmias in the zone (Fig. 40-10, panel A). It can also accelerate these rhythms into AF (see Fig. 40-10, panel B). Established AF rarely responds to this modality. Atrial defibrillation shocks have similar principles of efficacy to ventricular defibrillation. A sigmoid defibrillation efficacy curve exists and thresholds vary widely with lead configuration. Inclusion of a coronary sinus electrode with a right atrial to coronary sinus vector provides lower thresholds than right-sided shocks using right atrial to right ventricle, or superior vena cava to right ventricle, or configurations with a can or left pectoral electrode.[19]

In clinical studies, reliable atrial defibrillation has been obtained with shock energies of up to 27 joules with the Medtronic Jewel AF device. Newer iterations of these devices include pacing prevention algorithms, such as continuous atrial pacing for AF prevention. Enhanced monitoring capabilities of these devices include atrial and ventricular arrhythmia detection and categorization and electrogram detection in either chamber as programmed. Noninvasive electrophysiological stimulation is available in both the atrium and the ventricle. This permits testing of detection and therapies for both atrial and ventricular tachyarrhythmias. Finally, a hand-held patient activator is available for delivery of shock therapy on demand by the patient or physician. This device can activate shock therapy as programmed by the physician prescription. Thus, symptoms and duration of AF as determined by the user can be used in deciding cardioversion shock delivery.

Indications for Implantable Cardioverter Defibrillator Therapy

Indications for ICD therapy in 2004 now include patients at risk for recurrent and symptomatic ventricular or atrial tachyarrhythmias. Indications in ventricular tachyarrhythmias can be considered in the two general categories of either secondary or primary prevention of sudden cardiac death and sustained ventricular tachycardia (Table 40-1). For atrial tachyarrhythmias, these indications are largely in the area of secondary prevention and are currently in evolution. Updated guidelines for the use of the ICD in ventricular tachyarrhythmias were published in 1998 by the expert panel from the American College of Cardiology/American Heart Association Joint Task Force.[4] These guidelines, which

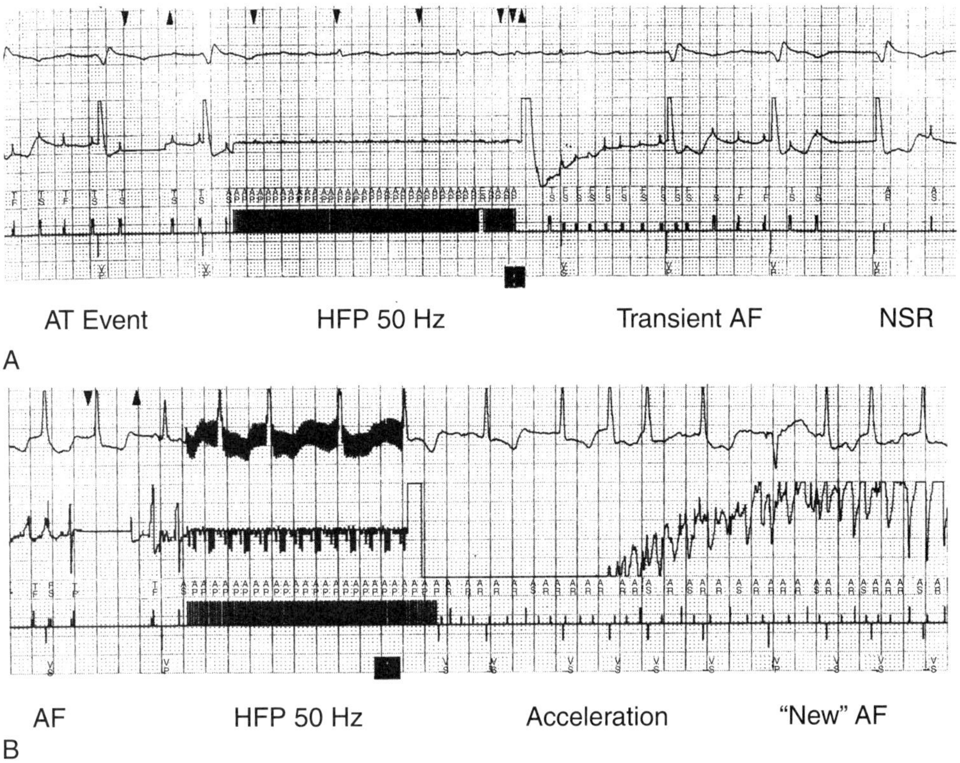

FIGURE 40-10 **A,** Termination of atypical atrial flutter (AT) by high frequency burst (HFP) at 50 Hz. This converts flutter to transient atrial fibrillation (AP) that terminated spontaneously. **B,** High frequency pacing accelerating atypical flutter in the AF zone to atrial fibrillation (AF).

represent the combined opinions of several experts, divide the indications into three major classes:

Class 1: ICD therapy is indicated based on expert consensus or evidence.
Class 2: ICD therapy is a therapeutic option but consensus does not exist.
Class 3: Clinical situations in which the use of ICD therapy is generally not justified.

At the time of publication of these guidelines, atrial tachyarrhythmia applications were considered investigational, and subsequently, commercial release of these devices has occurred with progressive clinical application.

Table 40-1 summarizes ACC/AHA practice guideline indications for ICD therapy. Important elements of these guidelines will be discussed in this chapter. However, it is pertinent to note that these guidelines, when compared with the earlier ones issued in 1991, reflect the monumental changes in our understanding of mechanisms and risks of sudden death in high risk coronary populations, nonischemic heart diseases, and primary electrical disease of the heart.[22] Thus, use of ICD therapy has now been expanded to large groups of patients with left ventricular dysfunction, nonischemic cardiomyopathy, and inherited cardiac disorders with a propensity to suffer from malignant ventricular arrhythmias and sudden death. New clinical trials continue to expand the primary prevention applications for sudden cardiac death. Finally, recent developments have expanded the use of these devices to patients with refractory atrial fibrillation. A number of investigational applications are currently under study and will be briefly mentioned.

IMPLANTABLE DEFIBRILLATORS FOR THE SECONDARY PREVENTION OF VENTRICULAR ARRHYTHMIAS

Sustained Symptomatic Ventricular Tachycardia and Survivors of Cardiac Arrest

The first established indication for the use of the ICD was in patients who have had spontaneous sustained and symptomatic ventricular tachycardia, ventricular fibrillation, or who were survivors of a cardiac arrest. In these patients, ICD devices have terminated sustained ventricular tachyarrhythmias, either by anti-tachycardia pacing or by shock therapy. Various studies have documented pace termination of VT in 89% to 91% of all episodes,[12,23] with the residual events being converted by shock therapy. Programmed ICD therapy has also successfully converted VF in more than 98% of episodes.[12,23] Failure to induce VT or VF at electrophysiological testing occurs in up to 40% of all SCD survivors. [24] However, VT/VF and/or SCD recurs in 8% to 50% of these patients and over one half of them die at the time of recurrence.[24-26]

In a large body of cumulative experience, the sudden death rate reported with device therapy has been in the range of 1% to 2% (Fig. 40-11) per year with a cumulative incidence of less than 10% at 5 years,[12,23,27] with a significant projected survival benefit compared with untreated populations.[27,28] Thus, data from large multicenter

TABLE 40-1 Indications for Implantable Cardioverter Defibrillator Therapy

Class 1

a. Cardiac arrest due to VF or VT not due to a transient or reversible cause
b. Spontaneous sustained VT in association with structural heart disease
c. Recurrent syncope of undetermined origin in a patient with hemodynamically significant or sustained VT or VF induced at electrophysiological study when drug therapy is ineffective, not tolerated, or not preferred
d. Nonsustained VT in patients with coronary disease, prior myocardial infarction, and left ventricular dysfunction who have inducible VF or sustained VT at electrophysiological study that is not suppressible by a class 1 antiarrhythmic drug

Class 2A

Patients with left ventricular ejection fraction of less than or equal to 30% at least 1 month post myocardial infarction and 3 months after coronary revascularization surgery.

Class 2B

a. Implantation of the ICD in patients with cardiac arrest presumed to be due to VF, or documented VF when electrophysiological testing is precluded by other medical conditions
b. Prophylactic implantation in a patient awaiting cardiac transplantation with symptoms attributed to sustained ventricular tachyarrhythmias
c. Patients with long QT syndrome and other familial or inherited conditions believed to be at high risk for life-threatening ventricular tachyarrhythmias
d. Nonsustained VT in patients with coronary artery disease, prior myocardial infarction, left ventricular dysfunction, and who have inducible sustained VT or VF at electrophysiological study
e. Recurrent syncope of undetermined origin in the presence of ventricular dysfunction and inducible ventricular arrhythmias at electrophysiological study
f. Recurrent syncope of unexplained origin or family history of sudden death in association with typical or atypical right bundle branch block and ST segment elevations (the Brugada syndrome)
g. Syncope in patients with advanced structural heart disease in whom thorough invasive and noninvasive investigations have failed to define a cause

Class 3

a. Recurrent syncope of undetermined cause in a patient without inducible ventricular tachyarrhythmias or heart disease
b. Incessant VT or VF
c. VF or VT due to arrhythmias amenable to surgical or catheter ablation—e.g., atrial arrhythmias associated with the Wolff-Parkinson-White syndrome, right ventricular outflow tract VT, idiopathic left ventricular VT, fascicular VT, or VT due to bundle branch reentry tachycardia or in tetralogy of Fallot
d. Ventricular tachyarrhythmias due to a transient or reversible disorder—e.g., acute myocardial infarction or electrolyte imbalance
e. Patients with significant psychiatric illnesses that may be aggravated by device implant or preclude systematic follow-up
f. Patients with other terminal illnesses with projected life expectancies of ≤6 months
g. Patients with coronary artery disease with LV dysfunction and prolonged QRS duration in the absence of spontaneous or inducible VT undergoing bypass surgery
h. NYHA class IV CHF patients who are not candidates for cardiac transplantation

ICD, Implantable cardioverter defibrillator; VF, ventricular fibrillation; VT, ventricular tachycardia.

Adapted from Gregoratos G, Cheitlin C, Conill A, et al: ACC/AHA guidelines for the implantation of cardiac pacemakers and antiarrhythmia devices: A report of the American College of Cardiology/American Heart Association (Committee on Pacemaker Implantation) J Am Coll Cardiol 1998;31:1175 and Circulation 2002;106:2145-2161.

randomized trials comparing ICD therapy with various drugs for the secondary prevention of SCD in patients with VT or VF consistently indicate that device therapy is superior to guided or empiric medical therapy for these patients. In all three trials addressing this subject, the Antiarrhythmics versus Implantable Defibrillator (AVID) trial, Cardiac Arrest Study of Hamburg (CASH), and the Canadian Implantable Defibrillator Study (CIDS), the total mortality showed an average 30% relative risk reduction in the ICD arm of the study.[29-31] Individual patient groups have often varied between trials. AVID excluded patients with sustained but minimally symptomatic or hemodynamically well-tolerated VT and syncope with induced sustained VT. It required presence of hemodynamic instability or left ventricular dysfunction with VT. In contrast, CIDS included many of these subgroups. Subsequent analyses of the AVID registry data, which tracked excluded subgroups, showed that patients with hemodynamically well-tolerated VT had a comparable mortality without device therapy with other groups that were included in the study.

In the current guidelines, all these subgroups are considered indications for device therapy. These and other large clinical trials have shown that the implant risk with current ICD systems is less than 0.5%.[12,23,29] Table 40-2 provides the complications observed in the ICD arm in the AVID study.[32] The major clinical complication observed has been inappropriate device therapy, occurring typically for atrial fibrillation with rapid ventricular response. This occurs in about 5% to 11% of all patients, and most often in combination with appropriate device activations in the same patients in these studies. Careful device follow-up and refinements in technology have further minimized device-related complications. Thus, given the low perioperative mortality, limited morbidity, and the overwhelming evidence for survival benefit, the latest guidelines classify cardiac arrest or sustained VT/VF as class 1 indications for ICD therapy.

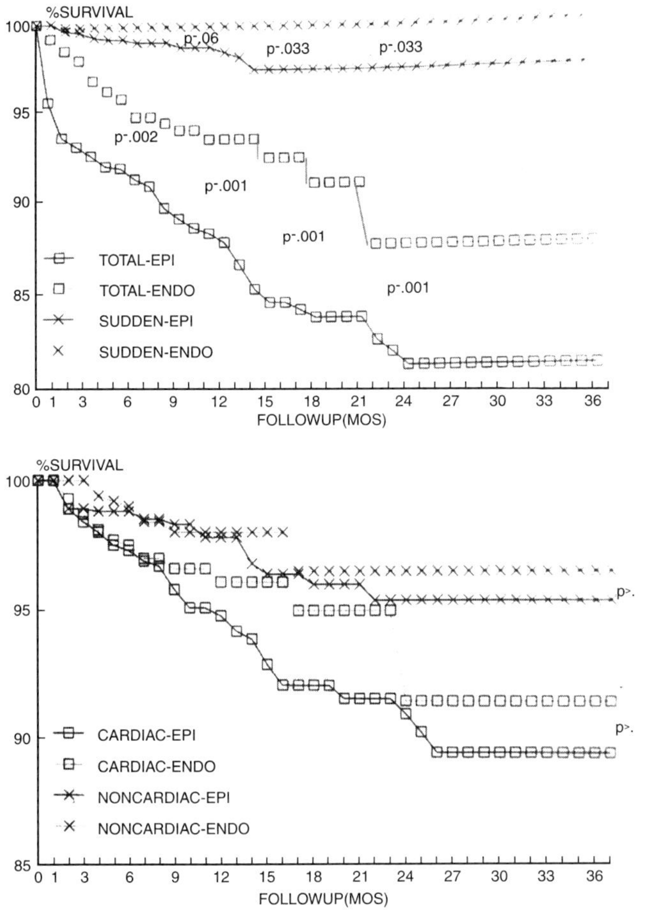

FIGURE 40-11 Survival in the Medtronic PCD trial for ICDs with Epicardial (EPI) and nonthoractomy (ENDO) lead systems. Sudden death survival **(top panel)** and total survival **(bottom panel)** are improved with nonthoracotomy lead systems. ICD, Implantable cardioverter defibrillator.[12]

Syncope with Inducible Sustained Ventricular Tachycardia

In patients with recurrent syncope, electrophysiological evaluation may reveal inducible sustained ventricular tachycardia which may be the mechanism of syncope. In the CIDS study, patients were included with this presentation and derived similar benefits from ICD therapy as did other groups.[30]

IMPLANTABLE DEFIBRILLATORS FOR THE PRIMARY PREVENTION OF VENTRICULAR ARRHYTHMIAS

Several patient subgroups have been or are being actively evaluated for primary prevention of malignant ventricular tachyarrhythmias. These include several indications now listed in classes 1 and 2 of the ACC/AHA guidelines. Coronary artery disease is the most common underlying condition in ICD recipients. Because survival rates after cardiac arrest are dismal, varying from 1% to 25%, with many large cities reporting less than 10% survival rates, there has been a sustained thrust to identify high-risk patients before cardiac arrest

TABLE 40-2 Frequency of the ICD Complications in the First 539 Implantations in the AVID Trial

Type of Complication	No. of Patients (%)
Lead fracture	15 (2.8)
Infection	14 (2.8)
Bleeding/hematoma	8 (1.5)
Lead dislodgment	8 (1.5)
Pneumothorax with chest tube	6 (1.1)
Thrombosis	2 (0.4)
Cardiac perforation	2 (0.4)
Generator migration/erosion	3 (0.6)
Generator failure*	4 (0.7)
TOTAL	62 (11.5)

*An additional six patients had their generators recalled.

Modified from Kron J, Herre J, Renfroe EG, et al: Lead- and device-related complications in the antiarrhythmias versus implantable defibrillator trial. Am Heart J 2001;141:92-98.

for prophylactic device insertion or "primary prevention" of sudden death. In patients with coronary artery disease, risk stratification is performed using clinical and electrophysiological markers to identify those who would benefit from ICD implantation. Other noncoronary populations or high-risk clinical syndromes have been included based on recent clinical trial data or expert consensus. More categories are under study and will be briefly mentioned in the following discussion.

Nonsustained Ventricular Tachycardia with Coronary Artery Disease and Left Ventricular Dysfunction

This patient group is the first primary prevention category adopted as a consequence of the Multicenter Automatic Implantable Defibrillator Trial-1 study.[33] This subgroup has long been recognized to have a high propensity for sudden death and inducible sustained ventricular tachycardia. In early studies, Wilber and coworkers recognized that electrophysiological provocation was capable of stratifying risk of sudden death in this population with the induction of sustained ventricular tachycardia.[34] The MADIT-1 study hypothesized that ICD therapy would improve survival in this high-risk population. It demonstrated a 54% reduction in relative risk of death in these patients compared with conventional drug therapy, such as amiodarone. The study was prematurely terminated and the indication has now been widely adopted. It has been estimated that approximately 3% to 7% of survivors of acute myocardial infarction in different series will eventually be stratified into this subgroup. Implant experiences have confirmed the MADIT data that virtually 50% of these patients experience a sustained ventricular tachyarrhythmia after prophylactic ICD implant within 18 months, confirming the original hypothesis.[35] Cost-effectiveness analyses have been very favorable with an estimated $27,000 per life year saved with transvenous implantation. Recent trends with declining device costs and shorter hospitalizations in these patients are likely to further improve the cost effectiveness of this therapy.

Familial Syndromes with High Risk of Sudden Death

The guidelines recognized several important, but small, patient groups that have familial or acquired diseases that predispose them to sudden death and malignant ventricular tachycardias. In most of these categories, small clinical series or pilot data and expert consensus led to adoption of the indication. This includes high-risk patients with congenital long QT or the Brugada syndromes, hypertrophic cardiomyopathy, and arrhythmogenic right ventricular dysplasia. A family history of sudden death is a key element in selection of the ICD as a primary prevention therapy. Patients with prior symptomatic sustained ventricular tachycardia or cardiac arrest are considered to have declared themselves and would be included in secondary prevention.

Patients Awaiting Cardiac Transplantation for Drug-Refractory Heart Failure

Clinical data from cardiac transplant centers have long documented an inordinately high risk of sudden death in individuals awaiting cardiac transplantation after candidacy has been established by screening. In these individuals, pilot data have shown appropriate ICD prevention of malignant ventricular tachycardias. In view of this small subgroup unlikely to be suitable for clinical trial, expert consensus has supported application of the ICD in these patients.

Coronary Artery Disease with Left Ventricular Dysfunction

In the secondary prevention studies, the role of left ventricular dysfunction as an important determinant of benefit of defibrillator therapy in patient survival was highlighted.[36-38] Survival of ICD recipients is strongly influenced by left ventricular function. Patients with left ventricular ejection fractions of less than 30% have inferior survival rates at 3 years of follow-up when compared with those with better left ventricular function. However, both populations appear to derive a significant survival benefit from ICD implantation. In an early analysis, we demonstrated that patients with left ventricular ejection fractions of less than 36%, derived the benefit of defibrillator therapy over drug therapy.[39] Similar data were noted in the CIDS and MADIT 1 studies. The recently completed MADIT 2 study evaluated the hypothesis that patients with coronary disease and myocardial infarction who had an LV ejection fraction of less than 31% would have improved survival with defibrillator therapy compared with conventional heart failure therapy.[40] The trial demonstrated a 31% reduction in relative risk, which was estimated to decline from a projected 19% 2-year mortality rate to an actual 12% mortality rate with defibrillator insertion. However, important subgroup analyses have identified the subgroup deriving most benefit. These appear to be patients with prolonged QRS duration and left ventricular dysfunction.

Heart Failure Populations

The Sudden Cardiac Death in Heart Failure Trial (SCD-HeFT) compared survival after ICD insertion or empiric amiodarone with placebo in patients with ischemic and nonischemic NYHA classes II and III heart failure and left ventricular ejection fraction below 36%.[41] The overall mortality in the placebo arm (7.2% per year) was not altered by amiodarone therapy but was greatly reduced by ICD insertion ($P < .007$, hazard ratio, .77). This effect was more obvious in class II patients (hazard ratio, .54) left ventricular ejection fraction below 31%, and prolonged QRS duration (hazard ratio, .67). The Comparison of Medical Therapy, Pacing and Defibrillator Trial (COMPANION)[42] demonstrated improved survival in NYHA classes III and IV heart failure patients with a prolonged QRS, left ventricular ejection fraction below 36%, and a diameter greater than 61 mm with a biventricular ICD compared with pacing or conventional heart failure drug therapy.[42] There was a 43% reduction in mortality with biventricular ICD insertion as prophylaxis for sudden death. Practice guidelines incorporating these trial results are awaited.

Specific Disease States and Implantable Cardioverter Defibrillator Therapy

FAMILIAL SYNDROMES ASSOCIATED WITH SUDDEN CARDIAC DEATH

The long QT syndromes represent a spectrum of electrophysiological disorders characterized by the propensity for the development of malignant ventricular arrhythmias, especially torsades de pointes.[43,44] Younger patients also manifest a more malignant form of the prolonged QT syndrome.[44] Because this is a primary electrical disorder, usually with no evidence of structural heart disease or left ventricular dysfunction, effective control of malignant arrhythmias can ensure an excellent long-term prognosis. Most patients can be effectively treated with β-blockers, permanent pacing, or left cervicothoracic sympathectomy.[43] Device therapy is recommended for selected patients in whom recurrent syncope, sustained ventricular arrhythmias, or sudden cardiac death occur despite the drug therapy.[4] However, the use of the ICD as primary therapy should be considered in a subgroup of patients, such as those with a familial history of sudden death, or when compliance or intolerance for drugs is a concern.[45] In addition, as genotypic and molecular analyses permit increasingly accurate risk stratification of these patients, ICD therapy may be considered in the primary prevention of sudden death in selected individuals even in the absence of a sentinel event.

PRIMARY ELECTRICAL DISEASE

Primary electrical diseases, often-prolonged QT syndrome or idiopathic VF are frequent indications for ICD implantation in younger patients.[4,46,47] It has been estimated that 10% of young patients, despite extensive evaluation, do not reveal an etiology for VF.[46] Electrophysiological testing in these patients with idiopathic VF usually reveals polymorphic VT or VF, which is often suppressible with class 1A drugs.[48] However, the long-term efficacy of drug therapy remains unknown. Given the guarded prognosis, even with supposedly effective drug therapy (annual rate of sudden cardiac death estimated to be as high as 11%), the limited clinical data appear to support the use of ICDs in such patients.[48,49]

HYPERTROPHIC CARDIOMYOPATHY

The hypertrophic cardiomyopathies are a diverse group of disorders that have as a common feature primary hypertrophy of the left ventricle. Their prevalence has recently been estimated to be as high as 1 in 500, making this group of diseases the most common genetically transmitted cardiovascular diagnosis.[50] Sudden death, which may be the first manifestation of the disease in an asymptomatic individual, has been reported annually in up to 4% of these patients.[51,52] Malignant ventricular tachyarrhythmias have been described as a mechanism for sudden death in adults with this condition.[52] Risk factors for sudden death include syncope, a very young age at presentation, extreme degrees of ventricular hypertrophy, a strong family history of sudden death from cardiac causes, and nonsustained ventricular tachycardia.[51,52] Studies of patients resuscitated from cardiac arrest indicate that many patients will experience another event. In contrast to other cardiomyopathies, electrophysiological testing may be of prognostic significance, as in some studies inducible sustained ventricular arrhythmias appear to be associated with cardiac arrest and syncope.[53] Pharmacologic therapy, in the form of β-blockers or calcium channel antagonists, has frequently been used. Although outflow tract gradients may diminish with medical therapy, it has, at best, marginal efficacy in preventing sudden death. The empirical use of amiodarone has been recently reported to be associated with improved survival.[54] However risk stratification and the prediction of drug efficacy remains difficult and controversial. High-risk patients with hypertrophic cardiomyopathy and sudden death survivors should be considered for ICD therapy instead of or in conjunction with drug therapy.[4] The limited efficacy of medical therapy combined with compliance issues associated with long-term drug administration in young patients makes device therapy an attractive option. Maron and associates published a multicenter retrospective study of the efficacy of ICD therapy in 128 HCM patients thought to be at high risk for sudden death.[52] Thirty-four percent of these patients had ICDs implanted for secondary prevention following cardiac arrest, whereas 66% had devices implanted for primary prevention because of the presence of risk factors listed earlier. During a mean follow-up of 3 years, appropriate therapy was delivered in 23% of patients. Two patients died suddenly despite ICD therapy. On a cautionary note, there was a significant rate of complications associated with ICD therapy, with inappropriate therapy being delivered to 25% of patients. The results of a large, randomized comparative experience with empiric amiodarone are awaited.

ICD therapy is indicated in HCM patients for secondary prevention after sustained ventricular tachycardia or fibrillation.[4] Primary prevention is considered a class 2B indication for ICD implantation. The recently published results of ICD therapy in HCM should lead to greater use of device therapy in high-risk patients.[52] Future directions of research include better characterization of the specific myosin gene mutations associated with shorter life expectancy, which may target candidates for primary prevention with early use of amiodarone and/or defibrillator therapy.[55]

CORONARY ARTERY DISEASE

Patients with coronary artery disease comprise the majority of device patients in most reports.[12,23,29-31] Device implantation is widely accepted as improving the outcomes of these patients. Patients with reduced left ventricular function may experience greater benefit with ICD therapy than with drug therapy.[38-40] In order to limit patient risk during defibrillation efficacy testing it is important to preclude the presence of active ischemia before proceeding with device implantation. Furthermore, optimal anti-ischemic treatment further enhances patient quality of life as well as survival. Ventricular function should be assessed before device implantation, although depressed function is not a contraindication to device therapy. Defibrillation threshold testing should be minimized in patients with elevated pulmonary capillary wedge pressures or severely compromised cardiac output to prevent deterioration of functional status.[56]

DILATED CARDIOMYOPATHY

Dilated cardiomyopathy is associated with a high mortality within 2 years of diagnosis with a minority of patients surviving 5 years. Approximately one half of these deaths are sudden and unexpected.[57] The combination of poor left ventricular function and frequent episodes of nonsustained VT in these patients is associated with an increased risk of sudden death.[58] Unlike in ischemic heart disease, the value of electrophysiological studies is limited.[59] The efficacy of drug therapy is low in the presence of impaired left ventricular function and is difficult to predict based on invasive or noninvasive testing. ICD implantation may be preferred in the management of symptomatic VT/VF patients with this condition. ICD therapy can also be used as the bridge to orthotopic heart transplantation in many of these patients. The recent Cardiomyopathy Trial (CAT) was undertaken as a pilot study to evaluate the potential benefit of prophylactic ICD implantation on all-cause mortality in patients with DCM. The trial was terminated early because no appreciable benefit of ICD implantation was observed. However, the trial group was small, the event rate was extremely low, and the follow-up was abbreviated.[60] In contrast, in the AVID study, these patients derived similar benefits to the coronary disease patients and had similar event rates.

The optimal management of idiopathic dilated cardiomyopathy, especially the role of ICD in the primary prevention of sudden death, has been defined by large recently completed trials (DEFINITE, SCD-HeFT). These trials enrolled DCM patients with moderate left ventricular dysfunction (ejection fraction ≤ 35%) with or without ventricular arrhythmias. While the DEFINITE trial showed no overall mortality benefit with the ICD, a large subgroup (NYHA class III patients) showed a favorable trend to survival enforcement with the ICD (P = .06).[61] In the SCD-HeFT trial, this population showed survival benefit with the ICD irrespective of QRS duration (unpublished data). These data support the prophylactic use.

SYNCOPE WITH INDUCIBLE SUSTAINED VENTRICULAR TACHYCARDIA

Patients with syncope of undetermined etiology in whom clinically relevant VT/VF is induced at electrophysiological study may be candidates for ICD therapy. In these patients, the induced arrhythmia is presumed to be the cause for syncope.[62,63] Follow-up studies of these patients have established that their cardiovascular mortality rate averages 20% annually with a large proportion of sudden, presumably arrhythmic, deaths.[63] The first line of therapy is usually pharmacologic, however, in some patients, antiarrhythmic treatment is limited by inefficacy, intolerance, or noncompliance. In this patient subset, ICD therapy is often applied with comparable results with those obtained in patients with sustained VT.[30] Currently, ICD therapy is a class 1 indication for patients with syncope of unknown etiology and inducible VT/VF.[4]

Implantation of Implantable Cardioverter Defibrillator Device Systems

The first generation of ICD systems used epicardial leads placed surgically via a transthoracic approach. In 1988, the first nonthoracotomy implantation involving transvenous leads and a submuscular patch was described.[5] This permitted a right-to-left shock vector with successful defibrillation using monophasic shock waveforms. Further technologic advances in the last decade resulted in the use of biphasic shock waveforms in conjunction with endocardial defibrillation leads.[6] This further reduced the energy requirements for defibrillation allowing for smaller size of pulse generators. The reduction in generator size coupled with the replacement of patch leads with active generator can electrodes, permitted pectoral rather than abdominal implantation of ICD systems. The use of the epicardial approach is now limited to very few patients, such as those with complex congenital heart lesions that do not permit transvenous lead placement, or patients who lack vascular access required for lead placement. Even in patients undergoing routine cardiac surgery for any other indication, a postoperative transvenous implantation is preferred over intraoperative epicardial device placement. This is a result of the superior lead performance and durability associated with transvenous leads. The following section will provide a detailed description of the transvenous implantation technique. Epicardial placement techniques will be briefly summarized but the reader is referred elsewhere for a more detailed description of that procedure.[64]

IMPLANTATION FACILITY

Implantation of internal defibrillators requires a dedicated team consisting of an implanting physician (either surgeon or electrophysiologist), a fully trained electrophysiologist (if not the implanting physician), surgical nursing support, and technical support staff for ICD implant and testing. At the present time, defibrillator system implants are performed either in the operating room, or in properly equipped electrophysiology or catheterization suites.[65,66] Limited data suggest that there are no significant differences in complication rates between procedures performed in the operating room environment and the electrophysiology laboratory.[66] Regardless of the site, the implanting location should be equipped with general anesthetic capabilities, appropriate air filtering, surgical scrub areas, surgical sterilization and lighting, and high-quality fluoroscopy. Electrocardiographic and hemodynamic equipment permitting arterial pressure monitoring and intracardiac signal recording should be available in the suite as necessary. ICD implants should be performed in hospitals with electrophysiology programs and rapid access to cardiac surgical services in order to respond to the potential complications of the procedure.

PREOPERATIVE ASSESSMENT

The preceding sections have described in detail the advances in ICD technology that have taken place since the 1980s. An important part of the preoperative assessment is to ascertain which system is appropriate for a particular patient based on status as well as future needs. This is based on the recognition that each new generation of devices is associated with increased complexity of implant and follow-up procedures as well as accelerated battery drainage. Thus, it is important to establish that the device and the patient are the best possible fit.

An integral part of the preoperative assessment is the evaluation of the patient for the presence of any chest deformities, thickness of subcutaneous tissue, presence of any skin lesions in an anticipated implant region, presence of any physical or laboratory signs of active infection, and body and heart sizes as well as general body habitus. All of these factors may influence the technique of system insertion, including the selection of the pulse generator (based on size and its energy output), and lead system (size and length).

Patient preparation commences before entering the operating suite with the removal of cutaneous hair, followed by cleaning of the proposed insertion site. Before the procedure, adhesive cutaneous pacing and defibrillation electrodes are placed to enable electrocardiographic monitoring and external defibrillation and an arterial line is usually placed for hemodynamic monitoring. A prophylactic dose of an appropriate broad-spectrum antibiotic is administered intravenously.

IMPLANTATION TECHNIQUE

Implantation of any ICD system involves insertion and positioning of the leads followed by testing of their pacing and sensing functions. The generator pocket is then created and the leads connected to the generator. This is followed by DFT testing as outlined earlier and final modifications to the programming. The wound is then closed. We will discuss in detail the implantation procedure for a single chamber ICD. The modifications of the implantation technique required for the more advanced devices will be discussed later.

Single Chamber Ventricular Implantable Cardioverter Defibrillator Implantation

ICD implants are generally performed in the left prepectoral area to establish a left-to-right defibrillation shock vector. The generator is placed in a prepectoral pocket and the leads are inserted transvenously from the cephalic and/or subclavian veins. Using standard sterile technique, a prepectoral incision is performed suitable for cephalic or subclavian access. (Fig. 40-12A). The initial incision is usually below the clavicle. The right pectoral region may be used in select left-handed individuals or if local abnormalities, infection, or venous access preclude using the standard technique. Using standard sterile technique, a prepectoral incision is performed suitable for cephalic or subclavian access. The incision is usually performed below the clavicle from the midclavicular line to the deltopectoral groove and is deepened until the prepectoral fascia is identified. Transvenous lead insertion is then performed either via the cephalic or subclavian routes as described subsequently.

The use of cephalic vein access for transvenous lead insertion is preferred to subclavian vein puncture because the latter can be associated acutely with a small risk of pneumothorax and arterial complications and in the long term *subclavian crush* injury to the leads.[67] Subclavian crush occurs when leads inserted via the subclavian vein are entrapped within the costoclavicular complex or under the subclavius muscle. This phenomenon, which is almost never seen with the use of the cephalic vein, is more common with large diameter defibrillating electrodes than with pacemaker leads.[67]

Dissection in the deltopectoral groove is used to isolate the cephalic vein. The vein is controlled proximally and distally to the site of venous entry with ligatures. The vein is opened with a small incision. The pacing and defibrillator electrodes are then introduced and advanced via the subclavian vein into the right side of the heart into the pulmonary artery under fluoroscopic guidance (see Fig. 40-12B). The lead is withdrawn and fixed in the right ventricular apex. If the cephalic vein or venous valves do not allow lead insertion, a guidewire can be passed into the right atrium. A split-sheath introducer is advanced over the wire permitting lead insertion. Ong and Barold described a modified cephalic vein guidewire technique for the introduction of one or more electrodes.[68] In this technique, insertion of the guidewire into the cephalic vein is followed by insertion

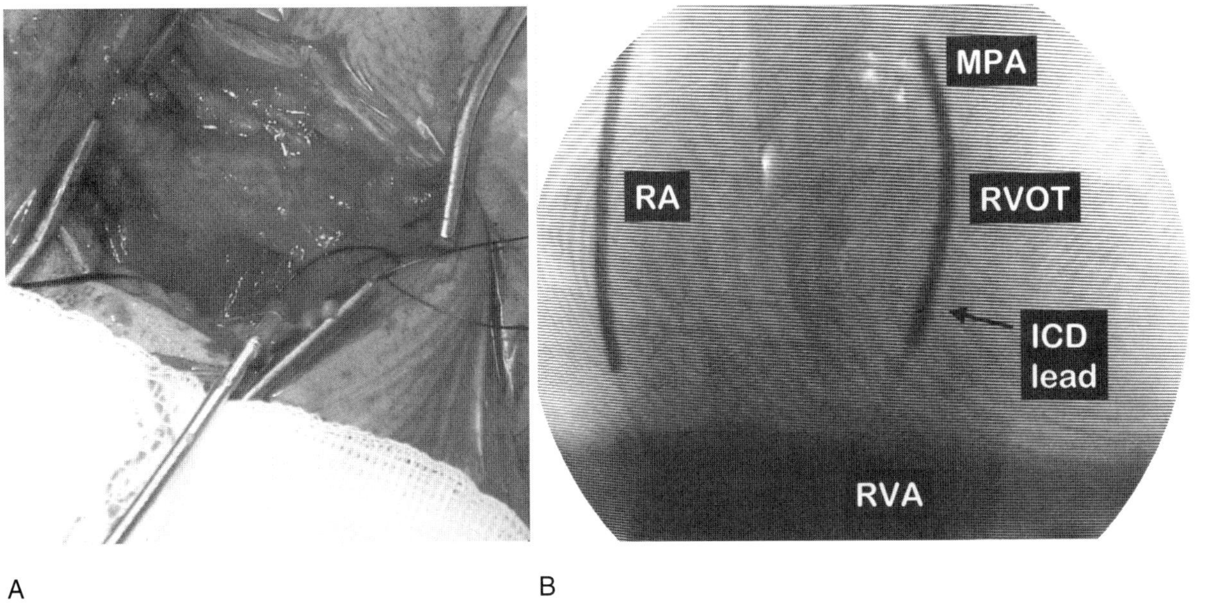

FIGURE 40-12 **A**, Cephalic vein dissection for lead insertion. Note the vein in the deltopectoral groove has been opened with lead introduction. **B**, Fluoroscopic defibrillation lead positioning in the pulmonary artery before withdrawal to right ventricle. ICD, Implantable cardioverter defibrillator; MPA, main pulmonary artery; RA, right atrium; RVA, right vertebral artery; RVOT, right ventricular outflow tract.

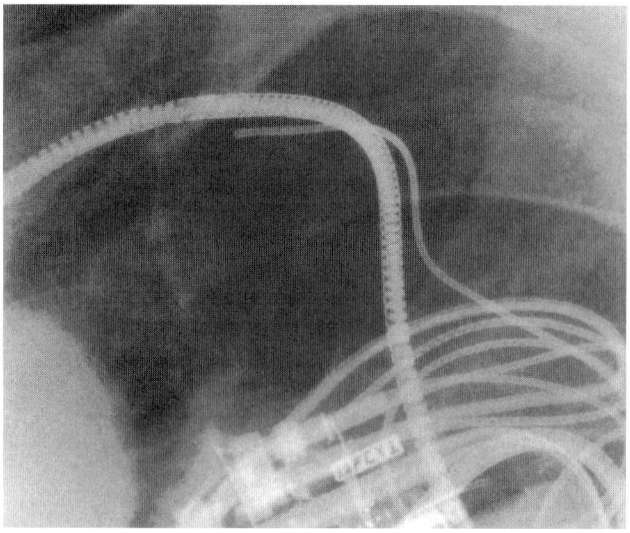

FIGURE 40-13 Subclavian crush fracture of defibrillation electrode in superior vena cava. The fractured lead remnant is seen below the clavicle.

of the introducer and invagination into extrathoracic segment of the subclavian vein with sacrifice of the cephalic vein.

If a subclavian puncture is required for the placement of the lead, it should be performed as far lateral as possible to minimize the risks of subclavian crush syndrome (Fig. 40-13).[69] Before vein puncture, the patient is placed in a Trendelenburg position, distending the vein, facilitating puncture, and avoiding air embolism. Fluoroscopy with or without contrast venography using peripheral injection of 20 to 30 cc of contrast can assist in localizing the vein.[70] In rare instances, when both cephalic and subclavian venous accesses cannot be obtained, the internal jugular, external jugular, or the axillary vein can be used. These alternative approaches require dissection that is more extensive and lead tunneling to reach the pectoral pocket.

Pacing and defibrillator leads are of larger diameters and greater stiffness than pacemaker leads. Great care must be taken to avoid damage to the leads as well as the surrounding vascular structures during lead insertion. A curved stylet is inserted into the lead and the lead is carefully advanced across the tricuspid valve into the right ventricular outflow tract and pulmonary artery. Retracting the stylet to soften the lead tip can facilitate crossing of the tricuspid valve. The curved stylet is replaced with a straight stylet and the lead slowly withdrawn until it drops down toward the right ventricular apex. It is then gently advanced to the right ventricular apex preferably with the stylet slightly withdrawn from the tip. The lead position chosen should place the defibrillation coil electrode in close proximity to the inferior right ventricular myocardium. After optimal lead placement has been achieved, active fixation leads are anchored at the apex. The lead is then evaluated for stability with deep inspiration and after coughing. Pacing thresholds and diaphragmatic stimulation with pacing are assessed.

In patients with high DFTs, the insertion of additional leads in an attempt to increase the active area of the defibrillation electrodes and improve current distribution may be necessary. Commonly employed leads are an additional superior vena caval coil electrode positioned at the atrial junction, a coronary sinus defibrillation lead, or a subcutaneous patch or array. The subcutaneous patch is usually positioned below the axilla, in the second to fourth intercostal space posterior to the midaxillary line. This requires a separate incision, formation of a subcutaneous pocket, and tunneling of the lead connector to the pectoral pocket. If the generator header cannot accept three or more leads, the additional lead is connected to the generator via a Y connector, which is placed in the prepectoral pocket. After demonstrating satisfactory lead positioning, and electrogram stability, as well as acceptable pacing and sensing parameters, the lead is fixed at the venous entry site. Silk sutures are used to anchor the sleeve to the lead and muscular fascia surrounding the lead.

A subcutaneous prepectoral pocket is used for generator location. This is usually on the medial aspect anterior to the plane of the pectoral fascia. The generator is placed inferior to the clavicle to prevent restriction of shoulder motion and medial to avoid interference with arm motion. The pocket should be appropriate for the size of the device. A tight pocket can lead to generator erosion, whereas an oversized pocket may permit migration, seroma formation, and "Twiddler syndrome." The leads are connected to the generator header. Tissue blood or other fluids should not be interposed between the leads and the header ports. It is important to verify that the leads are inserted into the correct ports and securely connected. The pocket is irrigated with antibiotic solution and the generator is placed in the pocket. Excess lead length is looped under the device to avoid lead injury during future operations at this site. Pacing, sensing, and DFT testing is now performed. After demonstration of satisfactory device function, the pocket is closed and sterile adhesive strips can be applied to ensure incision apposition.

Defibrillation threshold testing, which has been described previously, is performed under deep sedation (e.g., propofol) or sometimes general anesthesia. The implanting physician must demonstrate that ventricular fibrillation can appropriately be detected by the device and reproducibly terminated using the lead configuration and energy capabilities of the device. A safety margin of 10 joules above DFT is usually programmed in the first shock therapy. There are many maneuvers, which may be employed to lower the threshold. These include reversing the shock polarity and changing pulse duration or electrode configuration. If these programming options fail to provide an adequate safety margin, it may be necessary to add another lead to the system or to reposition the ventricular and/or superior venacaval electrodes. Finally, a pulse generator with a higher energy output may be used to obtain satisfactory defibrillation.

Based on the results of the DFT testing, final programming of the device is then performed and the pocket is closed. Postoperatively, the patient's arm is placed in a sling and an intravenous antibiotic is administered. A chest radiograph is obtained immediately after the implantation to define lead location and to exclude complications, such as pneumothorax. Posterior–anterior and lateral radiographs should be obtained before discharge for future reference. Patients are usually discharged within 24 hours of implantation. Predischarge ICD testing is recommended for final programming of ventricular tachycardia and fibrillation detection and therapy. In addition, patients may be allowed to experience a device shock or pacing therapies to facilitate psychological adjustment to device function. Figure 40-14 illustrates a typical checklist of postoperative orders for patients following ICD implantation at our institution.

Dual Chamber Ventricular Implantable Cardioverter Defibrillator Insertion

These devices incorporate conventional dual chamber pacing capabilities along with defibrillation. The initial steps in device placement are similar to those with a single ventricular lead. However, these devices require insertion of an additional atrial lead, which is similar to the leads used in conventional pacemakers. The atrial lead should be placed following the insertion of the ventricular lead and, to avoid cross-talk, should be placed in the high right atrium or superior aspect of the atrial appendage. Pacing and sensing thresholds are tested as in conventional pacemakers. During DFT testing, cross-talk between atrial stimuli and ventricular detection circuits should be assessed.

Dual Chamber Atrioventricular Defibrillation Devices

These are the newest generation of devices that incorporate atrial pacing and defibrillation algorithms in patients with combined atrial fibrillation and ventricular tachycardia and/or fibrillation. The atrial defibrillation shock is delivered via an atrial and ventricular defibrillation electrode. The atrial defibrillation coil is mounted on a combined atrial pacing and defibrillation lead (see Fig. 40-8) or incorporated in the ventricular defibrillation lead with a separate atrial pacing lead being inserted in the patient. The generator size is slightly larger than the standard single or dual chamber ventricular defibrillator.

Insertion techniques are similar to those of dual chamber ICD leads. The atrial defibrillator coil is usually located in the superior vena cava and the high right atrium. It is important to maximize atrial sensitivity during device testing to accurately detect atrial fibrillation. This is best accomplished by reducing atrial sensitivity to 0.3 mV.

POSTOPERATIVE PACEMAKER OR ICD IMPLANT ORDERS	
ONLY CHECKED ITEMS WILL BE IMPLEMENTED	
PROCEDURE: DATE:	
PHYSICAN: ALLERGIES:	
☐ **DIET**:	
☐ **ACTIVITY**:	
☐ **WOUND CARE**:	
☐ **ABDOMINAL IMPLANT**: Maintain supine position × _______ hrs.	
☐ **PECTORAL IMPLANT**: Apply arm sling to affected side × _______ days.	
☐ Initial surgical dressing to be removed by physician	
☐ Cleanse surgical site daily with normal saline. Apply Bacitracin ointment and dry sterile dressing	
☐ **LABORATORY AND DIAGNOSTIC STUDIES**:	
☐ Stat portable chest x-ray	
☐ Stat ECG	
☐ Chest x-ray PA and Lateral in AM	
☐ Other:	
☐ **MEDICATIONS**:	
☐ Ancef 1gm IV q8hrs × 3 doses followed by Keflex 500mg PO TID × _______ days.	
☐ Other:	
IF PATIENT IS PENICILLIN ALLERGIC:	
☐ Vancomycin 500mg IV q12hrs × 2 doses followed by Erythromycin 500mg PO TID × _____ days.	
☐ Other:	
☐ Percocet 1 tablet (5/325mg) PO q4hrs PRN for moderate pain	
☐ Tylenol 2 tablets PO q4hrs PRN for mild pain	
☐ Anticoagulation:	
☐ Other:	

FIGURE 40-14 Postoperative orders for implantable cardioverter defibrillator (ICD) management after system insertion.

Postoperative Management

After device insertion, device behavior and limitations on specific physical activity should be reviewed with the patient. Recent guidelines recommend restrictions on driving for a minimum of 3 months and preferably 6 months to determine pattern of recurrent VT/VF events.[71] Device interactions with electromagnetic sources, environmental issues, and antibiotic prophylaxis for device infections should also be discussed. ICD recipients should be encouraged to carry proper device identification at all times. Patients receiving these devices can experience transient or sustained behavioral disturbances including depression and anxiety. Education and psychological support before, during, and after ICD insertion are highly desirable and can improve the patient's quality of life.[72,73]

Follow-Up Techniques for Implantable Defibrillator Patients and Device Systems

Follow-Up Program

All patients with ICD devices require periodic and meticulous follow-ups to ensure patient safety and optimal device performance because the consequences of device failure are potentially catastrophic.[4] The goals of ICD follow-up include monitoring of device system function, optimizing performance for maximal clinical effectiveness and system longevity, minimizing complications, anticipating replacement of system components, ensuring timely intervention for clinical problems, patient tracking, education and support, and maintenance of records relative to the ICD system. The need for device surveillance and management should be discussed with patients before ICD insertion. Compliance with device follow-up is an important element in establishing appropriate candidates for device therapy and obtaining the best long-term result. ICD follow-up is best achieved in an organized follow-up program at outpatient clinics analogous to pacemaker follow-up.[4,74] More recently, Internet-based data management approaches have become available to simplify device data management.

Institutions performing implantation of these devices should maintain these facilities for inpatient and outpatient use. Such facilities should obtain and maintain implant and follow-up support monitoring equipment for all devices used at that facility. The facility should be staffed or supported by a fully trained clinical cardiac electrophysiologist who may work in conjunction with trained associated professionals.[4,75] Ideally, access to these services should be available on both a regularly scheduled and emergent 24-hour basis. The implanting and/or follow-up facility should be able to locate and track patients who have received ICD devices or have entered the follow-up program.

The follow-up of an ICD patient must be individualized in conjunction with the clinical status of the patient. In general, device programming is initiated at implantation and should be reviewed at predischarge or subsequent postoperative electrophysiological testing. Routine follow-up includes a wound check shortly after ICD implantation. The risks of lead dislodgment are highest within the first few days following implantation. The risks of early lead displacement and changes in DFT may warrant noninvasive ICD testing with VT, and VF induction, in the electrophysiology laboratory soon after implantation. Patients experiencing device activation with or without therapy delivery should be evaluated shortly after the event until a regular, acceptable pattern of patient symptomatology during—and tolerance for—such events is established. Furthermore, device behavior must be deemed reliable, safe, and effective. Devices should be followed periodically with the exact frequency depending on the device model and clinical status of the patient. Transtelephonic follow-up should always be supplemented with clinic visits at a minimum of 3-month intervals for patient and device evaluation.[4] More frequent evaluation is needed when elective replacement indicator values are being reached. Manufacturer's guidelines for device follow-up vary with individual models and should be reviewed.

ELEMENTS OF AN IMPLANTABLE CARDIOVERTER DEFIBRILLATOR FOLLOW-UP

ICD follow-up involves routine surveillance visits in asymptomatic or minimally symptomatic patients and troubleshooting in patients with suspected device malfunction, symptoms, or deterioration of clinical status. Routine outpatient ICD follow-up should follow a logical sequence aimed at assessing the unit's functions and lead integrity, as well as evaluation of any problems that may have arisen in the interim period. The interval history should be reviewed for symptoms suggestive of arrhythmias and for other illnesses, such as onset or progression of heart failure. The latter may have resulted in alteration of ventricular function or in the institution of a medical regimen, which may affect device therapy. Changes in antiarrhythmic therapy may have important consequences, because these agents can change defibrillator thresholds and also can slow tachycardia cycle lengths. The lower cut-off limits for tachycardia detection may need to be reprogrammed to ensure appropriate device therapy. Antiarrhythmic drugs may also cause significant sinus bradycardia and chronotropic incompetence that can lead to pacemaker syndrome in patients with a single chamber ICD.

Clinical examination should include careful evaluation of the ICD generator site for signs of infection or device erosion. These complications are infrequent. However, early detection can lead to appropriate intervention with a reduction in morbidity and mortality (Fig. 40-15). An electrocardiogram is performed at all follow-up visits to determine the baseline rhythm. It is also reviewed for potential toxic effects of concomitant antiarrhythmic therapy and to confirm appropriate pacing and sensing functions of the device. An overpenetrated chest radiograph should periodically be performed in these patients. Transvenous lead fractures can be seen but are often subtle and difficult to detect. Complications associated with epicardial systems, such as patch fractures and "crinkling," can often be detected radiologically. These complications affect defibrillation efficacy and result in failure of therapy.

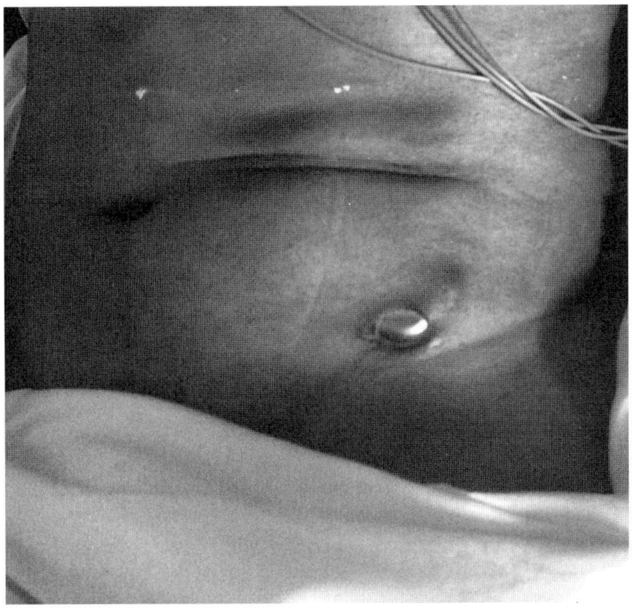

FIGURE 40-15 Postoperative wound infection in an abdominal device pocket resulting in device erosion.

Evaluation of Implantable Cardioverter Defibrillator Function

The device is interrogated to evaluate tachycardia and bradycardia functions as well as patient and device monitoring data. Particular attention should be given to reviewing sensing parameters, programmed defibrillation and pacing therapies, device activations, and event logs. Technical elements requiring review include battery status, lead system parameters, and end-of-life markers. Electrogram amplitudes (P, QRS, and T waves) should remain stable. However, these can change with the development of bundle branch block (Fig. 40-16), and some patients demonstrate a small but progressive diminution in electrogram amplitude over time, resulting in sensing dysfunction. It is also important to assess the pacing lead impedances. The current generation of devices, use sub-threshold stimuli to measure high-voltage lead impedance and, thus, the integrity of the shocking lead circuit. A fall in R wave amplitude coupled with a significant fall in pacing lead impedance (e.g., more than 200 ohms) suggests the possibility of lead insulation failure.

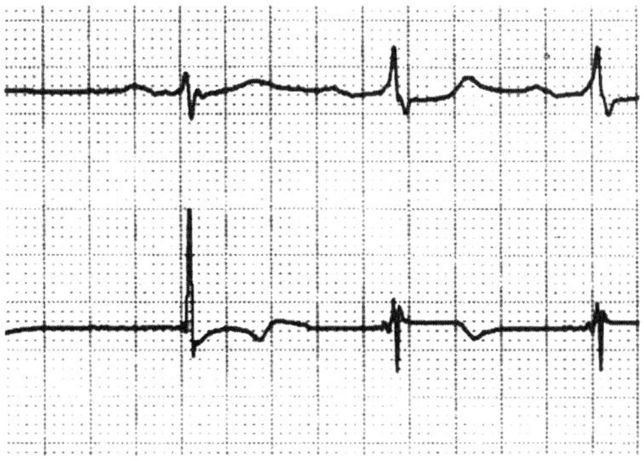

FIGURE 40-16 Changes in intracardiac ventricular electrogram amplitudes with the development of bundle branch block. *Top trace*, Surface ECG. *Bottom trace*, Telemetered intracardiac ventricular electrogram.

It is often necessary to reprogram the initially selected parameters either in the outpatient clinic or by electrophysiological testing. Reprogramming therapies in the outpatient setting should be performed with great caution. Changes in tachycardia detection rates can only be performed with the knowledge of the cycle lengths both of the clinical tachycardia as well as those induced at previous electrophysiological testing. It is important to recognize that clinical tachycardia, especially in patients receiving antiarrhythmic medications tends to be slower than that seen on baseline electrophysiological testing. Thus, the lowest VT detection rate should be set at least 50 ms below the slowest detected VT cycle length to ensure an adequate safety margin. It is also inadvisable to empirically program the VT detection rate upward in response to inappropriate therapy for supraventricular tachycardia. Finally, it is imperative to incorporate a meticulous, easily accessible record-keeping system documenting ICD history, programming data, and characteristics of tachycardia into any successful ICD follow-up program.

Evaluation of Implantable Cardioverter Defibrillator Therapy

Approximately 50% to 75% of patients will receive a shock within the first 2 years following ICD insertion. The occurrence of a single shock does not require evaluation. However, patients should be recalled for their first ICD discharge if they remain symptomatic following the shock, if they receive multiple shocks, or if they experience severe anxiety following the shock. When evaluating shock therapy, it is important to obtain a history of the clinical events surrounding the shock and then interrogate the device. The advent of stored electrograms has improved the accuracy of diagnosis of the rhythm that resulted in the delivery of therapy.[77,78] Dual chamber ICDs can provide both atrial and ventricular electrical activity during the event (see Fig. 40-3). A combination of diagnostic information, either including the onset interval, the stability and rate of the tachycardia, individual R–R intervals singly or tabulated in a graphic form, and the morphology of the stored ventricular electrograms allow for an accurate diagnosis in most instances. Evaluation of "rate sensing" or "shock surface" electrograms can be effective in providing an accurate reflection of the rhythm leading to the delivery of therapy. Each of these modalities when used alone can provide 90% diagnostic accuracy; however, when the device stores both atrial and ventricular electrograms the diagnostic accuracy is very high.[16,17]

TROUBLESHOOTING (Table 40-3)

Inappropriate Therapy

In some instances, ICD therapy can be delivered in response to arrhythmias or detected electrical signals

TABLE 40-3 Troubleshooting ICD Function—Inappropriate Therapy

Cause	Analysis
Asymptomatic tachyarrhythmia SVT, slow VT	Analysis of event EGM including atrial channel if available (R–R interval and morphology) and outcome of shock
Sinus tachycardia	Analysis of event EGM with particular respect to event onset and morphology Exercise test if indicated
T and P wave oversensing	Analysis of marker channel at different heart rates. Re-evaluate leads for fractures and/or insulation failure and R wave signal stability
Oversensing pacemaker spikes/double counting QRS	Analysis of marker channel during a 100% paced event at maximal pacemaker output and analysis of ECG along with intracardiac EGM marker channel
Electromagnetic interference or myopotentials	History combined with analysis of event EGM for high frequency signals Recreate activities that initiate shocks

EGM, electrogram; SVT, supraventricular tachycardia; VT, ventricular tachycardia.

unrelated to ventricular arrhythmias. Studies have documented that as many as 40% of patients may be asymptomatic before an appropriate ICD discharge.[79,80] However, up to 30% of all ICD shocks may be inappropriate in select patient populations.[80] In general, single shocks, whether preceded by symptoms or not, are most often due to the appropriate detection and treatment of a ventricular tachyarrhythmia. Conversely, multiple ICD discharges often result from the detection of other arrhythmias or signals that are inaccurately classified as a ventricular tachyarrhythmia. Thus, ICD therapy is often inappropriately delivered for sinus tachycardia or other supraventricular tachycardias with rapid AV conduction. Stored ICD diagnostic data including electrograms usually provide adequate information to enable accurate detection of these rhythms. It may be necessary, at times, to perform exercise testing to ensure that the maximal sinus rate during exercise is below the programmed detection rate to avoid inappropriate shocks.

Therapies delivered in the absence of documented arrhythmias are most often related to oversensing of noise leading to inappropriate detection by the device (Fig. 40-17). Disruption in the silicone or polyurethane insulation surrounding the lead can result in inappropriate detection of myopotential artifacts. These can result from pressure damage from an overlying device or lead or by an acute angle made by the lead as it leaves the header, or by inadvertent laceration at the time of initial implantation. Insulation leaks will often be accompanied by falls in the measured electrogram amplitudes and lead impedance. Inappropriate sensing of diaphragmatic muscle activity can initiate ICD therapy occurring, for example, while straining as during a bowel movement or during a coughing paroxysm (Fig. 40-18). Stored EGMs from the sensing lead will generally show the noise artifacts, and marker channels can confirm the origin of these signals. However, if the device lacks electrogram storage capabilities it may become necessary to record real time electrograms while manipulating the pocket and have the patient simulate activities of daily living that trigger device activation.

Environmental noise is another, although relatively infrequent, cause of inappropriate device activation. Electronic surveillance systems can sometimes trigger inappropriate shocks, although inhibition of demand pacing is a more common problem.[81,82] In the operating suite, the use of electrocautery can also cause inappropriate shocks. "Phantom shocks" are therapies perceived by the patient in the absence of actual device activation. This phenomenon has been reported both by patients anticipating their first shock as well as by those who have recently received multiple shocks. Interrogation of the device to confirm the absence of device activation followed by reassurance and patient education about device function usually result in disappearance of these symptoms and the associated anxiety.

Failure of Implantable Cardioverter Defibrillator Therapy

Failure of ICD therapy may be related to failure to deliver therapy or actual failure to generate electrical therapy (Table 40-4). Device interrogation in cases of

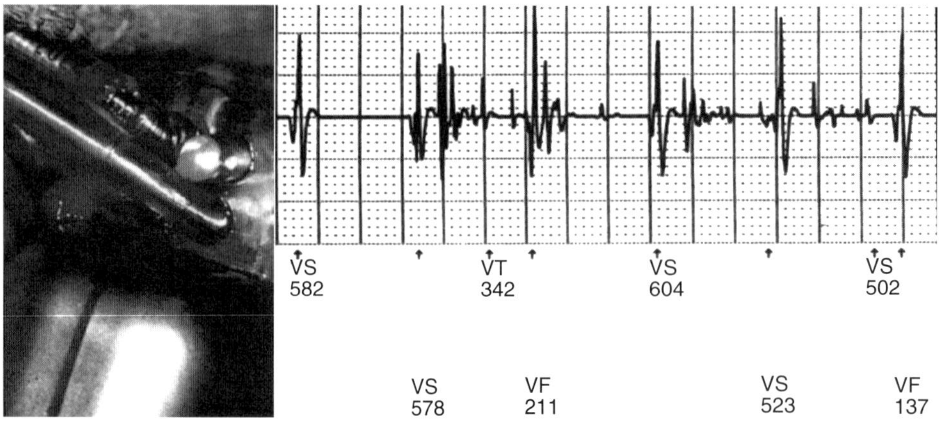

FIGURE 40-17 Oversensing of myopotentials **(right panel)** on ventricular sensing channel due to disruption of lead insulation in an ICD system **(left panel)**. VF, Sensed event in VF zone; VS, ventricular sensed event; VT, sensed event in VT zone. ICD, Implantable cardioverter defibrillator.

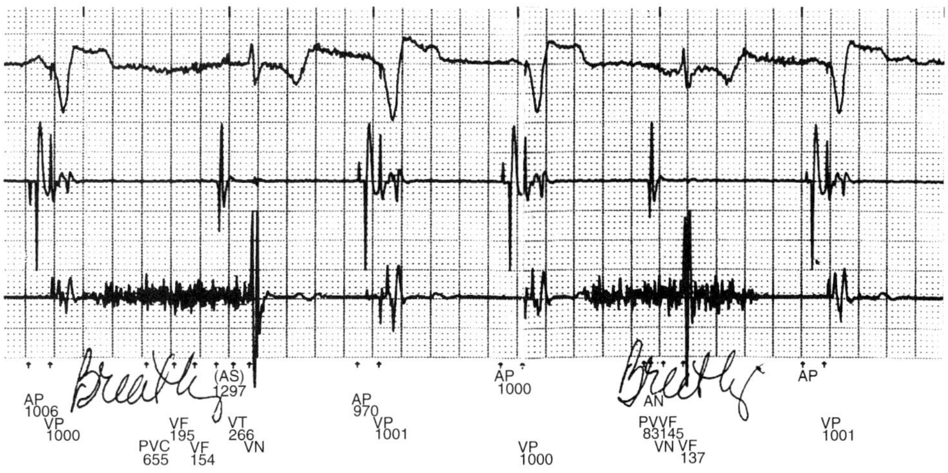

FIGURE 40-18 Oversensing of noise on the ventricular sensing channel during breathing from myopotentials. Note that the counters record electrical signals with markedly shortened intervals (below 100 ms) and in the ventricular fibrillation zone (*top tracing*). Surface ECG (*middle tracings*) show atrial and ventricular electrograms. *Bottom* annotations show atrial and ventricular sensed events.

failure to deliver therapy should confirm the proper programming of detection parameters following implantation and exclude accidental device deactivation during interval interventions. At each surveillance visit, it is imperative to evaluate the sensing parameters, because sensing failure is not usually documented in stored electrogram data. Sudden loss of sensing during episodes may herald the problem. Lead fracture may manifest as failure to pace, rise in impedance, and changing pacing artifact (Fig. 40-19). If device interrogation during sinus rhythm fails to pinpoint the source of the problem and radiographic lead integrity is confirmed, it may be necessary to induce ventricular fibrillation and repeat testing in some patients. Pacing thresholds may rise over time, resulting in exit block and failure of anti-tachycardia pacing. Drug therapy may reduce ventricular tachycardia rate, resulting in detection failure and failure to deliver appropriate therapy.[83] In addition, drugs may raise DFTs resulting in failed defibrillator shocks, particularly if the initial defibrillation safety margin was low.[84,85] Defibrillation thresholds can also be raised by myocardial ischemia and acute heart failure. Mechanical system failure may lead to failure to deliver therapy or in ineffective therapy. Migration, folding, or wrinkling of epicardial or subcutaneous patches can significantly increase the DFTs, as can migration or dislodgment of transvenous defibrillating electrodes. Treatment may require lead replacement with removal of the fractured leads and the possible addition of other leads and/or patch electrodes. It is important to recognize that lead malfunction may be intermittent and may not be detected unless extensive testing is undertaken. Generator failure, although rare, is not unknown. Component failures, especially in capacitors can lead to failed shock delivery. Immediate reoperation and generator replacement are indicated in such cases.

TABLE 40-4 Major Causes of Failure to Deliver Appropriate Therapy

Device inactivated or battery depleted
Failure to sense ventricular EGM
—low amplitude signal with arrhythmia or following a failed high energy discharge
—lead or generator malfunction
Inappropriate programming
—inappropriately high rate for detection of ventricular tachyarrhythmias
—failure to satisfy multiple criteria before delivering therapy
—ineffective anti-tachycardia pacing algorithms (change in threshold) or inappropriately low programmed shocks for VT
—failure to recognize drug- or disease-induced rise in DFT
Mechanical failure
ICD component failure resulting in failure to deliver appropriate therapy

DFT, defibrillation threshold; EGM, electrogram; ICD, implantable cardioverter defibrillator; VT, ventricular tachycardia.

DEVICE SYSTEM INFECTION

Infections of ICD systems are devastating and potentially fatal complications. Cellulitis at the generator pocket incision, if detected early, can be treated effectively using aggressive and prolonged antibiotic therapy. Once there is evidence of significant pocket or lead infection and systemic dissemination, explantation of the system is mandatory. Generator pocket aspiration should be avoided unless strictly necessary. The likelihood of introducing infection into a sterile hematoma exceeds the diagnostic yield obtained by performing this procedure in cases with diagnostic uncertainty. Other techniques, such as gallium scanning, echocardiography, and repeated blood cultures can assist in the diagnosis. The detailed methodology of device and lead explantation procedures are beyond the scope of this chapter and have previously been discussed in Chapter 39, Pacemaker Insertion, Revision, Extraction, and Follow-Up. However, larger caliber pacing and defibrillation leads are more difficult to remove by mechanical methods. Specialized techniques, such as laser lead extraction may be required. After system explantation, the pocket should be allowed to heal fully, if feasible. In some instances, a new system can be inserted on the contralateral side after a few days or weeks if there is evidence of resolution of the active septic process. Long-term antibiotic therapy is necessary to obtain proper

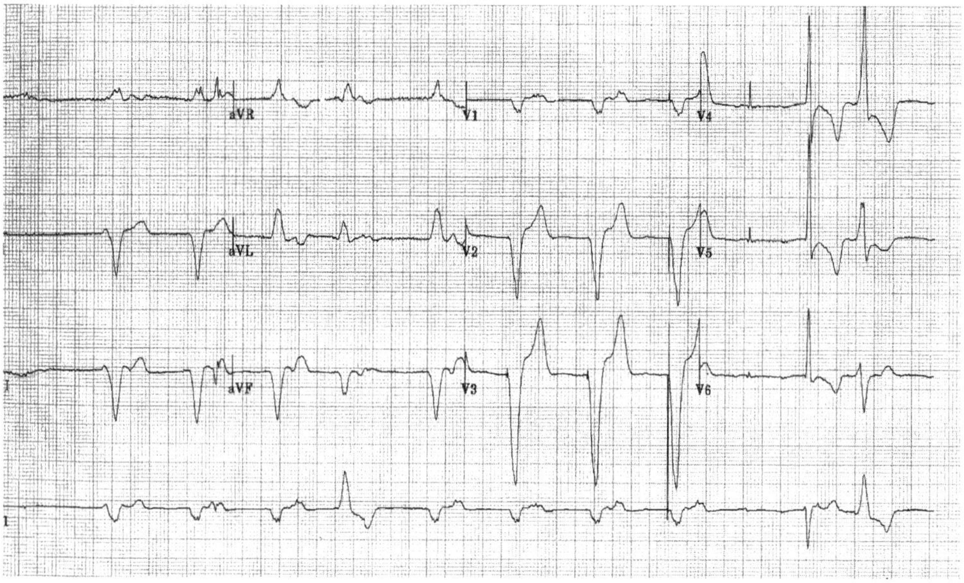

FIGURE 40-19 Twelve-lead electrocardiogram showing intermittent pacing failure with varying pacing amplitudes, and a rise in lead impedance identified a fractured pace/sense lead in this patient.

healing of the previously explanted site and to prevent seeding of the new system location.

Future Directions

ICD therapy will be used increasingly for the primary prevention of sudden death in high-risk populations based on improved methods of risk stratification (e.g., genotype analysis in genetic diseases and heart rate variability analysis). A goal for the next generation of ICDs will be to provide intervention before arrhythmia onset to prevent the discomfort of cardioversion or defibrillation. This may be achieved through pacing therapy to avoid long-short coupling, intermittent antiarrhythmic therapy, or biventricular pacing to improve hemodynamics. This should be coupled with a decrease in device cost, making device therapy more cost effective in a wide variety of clinical conditions.

REFERENCES

1. Cobb LA, Baum RS, Alvarez H 3rd, et al: Resuscitation from out-of-hospital ventricular fibrillation: 4 years follow-up. Circulation 1975;52:223.
2. Lehmann MH, Steinman RT, Schuger CD, et al: The automatic implantable cardioverter defibrillator as the treatment modality of choice for survivors of cardiac arrest unrelated to acute myocardial infarction. Am J Cardiol 1998;62:803.
3. Myerberg RH, Castellanos A: Evolution, evaluation and efficacy of the implantable cardioverter defibrillator technology. Circulation 1992;86:691.
4. Gregoratos G, Cheitlin C, Conill A, et al: ACC/AHA guidelines for the implantation of cardiac pacemakers and antiarrhythmia devices. A report on the American College of Cardiology/American Heart Association (Committee on Pacemaker Implantation). J Am Coll Cardiol 1998;31:1175.
5. Saksena S, Parsonnet V: Implantation of an implantable cardioverter/defibrillator without thoracotomy using a triple electrode system. J Am Med Assoc 1988;259:69-72.
6. Saksena S, An H: Clinical efficacy of dual electrode systems for endocardial cardioversion of ventricular tachycardia: A prospective randomized crossover trial Am Heart J 1990;119:15-22.
7. Munsif AN, Saksena S, DeGroot P, et al: Low-energy endocardial defibrillation using dual, triple, and quadruple electrode systems. Am J Cardiol 1997;79:1632-1639.
8. Saksena S, An H, Mehra R, DeGroot P, et al: Prospective comparison of biphasic and monophasic shocks for implantable cardioverter-defibrillators using endocardial leads. Am J Cardiol 1992;70:304–310.
9. Saksena S, Chandran P, Shah Y, et al: Comparative efficacy of transvenous cardioversion and pacing in patients with sustained ventricular tachycardia: A prospective, randomized, crossover study. Circulation 1985;72:153–160.
10. Neglia JJ, Krol RB, Giogorberedze I, et al: Evaluation of a programming algorithm for the third tachycardia zone in a fourth generation implantable cardioverter defibrillator 1997;1:49-54.
11. Raitt MH, Renfroe EG, Epstein AE, et al: The AVID Investigators. Ventricular tachycardia is not a benign rhythm: Insights from the Antiarrhythmics Versus Implantable Defibrillators (AVID) Registry. Circulation 2001;103:244-252.
12. Saksena S: For the PCD Investigator Group: Clinical outcome of patients with malignant ventricular tachyarrhythmias and a multiprogrammable cardioverter-defibrillator implanted with or without thoracotomy: An international multicenter study. J Am Coll Cardiol 1994;23:1521-1530.
13. Lindsay BD, Saksena S, Rothbart ST, et al: Prospective evaluation of a sequential pacing and high energy bidirectional shock algorithm for transvenous cardioversion in patients with ventricular tachycardia. Circulation 1987;76:601-609.
14. Hook BG, Callans DJ, Kleinmann RB, et al: Implantable cardioverter defibrillator; implantable cardioverter defibrillator therapy in the absence of significant symptoms: Rhythm diagnosis and management aided by stored electrograms analysis. Circulation 1993;87:1897-1906.
15. Timmermans C, Tavenier R, and the Worldwide Metrix Investigators: Ambulatory use of the Metrix automatic implantable atrial defibrillator to treat episodes of atrial fibrillation. PACE 1998;21:811.
16. Swerdlow C, Schls W, Dijkmann B, et al: The Worldwide Jewel AF investigators: Detection of atrial fibrillation and flutter by a dual-chamber implantable cardioverter-defibrillator Circulation 2000;101:878.
17. Adler SW, Wolpert C, Warmann E, et al: The Worldwide Jewel AF Investigators: Atrial therapies reduce atrial arrhythmia burden in defibrillator patients. Circulation 2001;104:1023-1028.
18. Cooper R, Smith W, Ideker RE: Internal cardioversion of atrial fibrillation: Marked reduction in defibrillation threshold with dual current pathways. Circulation 1997;96:2693-2700.

19. Saksena S, Prakash A, Mongeon, et al: Clinical efficacy and safety of atrial defibrillation using biphasic shocks and current nonthoracotomy endocardial lead configurations. Am J Cardiol 1995; 76:913-921.
20. Boccadamo R: Antitachycardia pacemakers for supraventricular tachycardias. In Luderitz B, Saksena S (eds): Interventional Electrophysiology. Mount Kisco, NY, Futura, 1991, pp. 213-224.
21. Giorgberidze I, Saksena S, Mongeon L, et al: Effects of high-frequency atrial pacing in atypical atrial flutter and atrial fibrillation. J Intervent Card Electrophysiol 1997;1:111-123.
22. Dreifus L: Guidelines for implantation of cardiac pacemakers and antiarrhythmia devices: Report of the American College of Cardiology/American Heart Association Task Force on the assessment of cardiovascular procedures (Committee on Pacemaker Implantation). J Am Coll Cardiol 1991;18:1-13.
23. Zipes DP, Roberts D: The PCD investigators: Results of the international study of the implantable pacemaker-cardioverter-defibrillator: A comparison of epicardial and endocardial lead systems. Circulation 1995;92:59.
24. Elder M, Sauve MJ, Scheinmann MM: Electrophysiologic testing and follow-up in patients with aborted sudden-death. J Am Coll Cardiol 1987;10:291.
25. Swerdlow CD, Winkle RA, Mason JW: Determinants of survival of patients with ventricular tachyarrhythmias. N Engl J Med 1983; 308:1436.
26. The CASCADE Investigators: Cardiac arrest in Seattle: Conventional versus amiodarone drug evaluation. Am J Cardiol 1991;67:578.
27. Saksena S, Breithardt GB, Dorian P, et al: Nonpharmacologic therapy for malignant ventricular arrhythmias: Implantable defibrillator trials. Prog Cardiovasc Dis 1996;38:429.
28. Nisam S, Kay SA, Mower MM, et al: AICD automatic cardioverter-defibrillator clinical update: 14 years experience in over 34,000 patients. PACE 1995;18:142.
29. The Antiarrhythmics Versus Implantable Defibrillators (AVID) investigators: A comparison of antiarrhythmic-drug therapy with implantable defibrillators in patients resuscitated from near-fatal ventricular arrhythmias. N Engl J Med 1997;337:1576.
30. Connolly SJ, Gent M, Roberts R, et al: The CIDS Investigators Canadian Implantable Defibrillator Study (CIDS): A randomized trial of implantable defibrillator against amiodarone. Circulation 2000;101:1297.
31. Siebels Luderitz B, Saksena S: Proceeding of a symposium. Electrical device therapy for cardiac arrhythmias: New concepts, problems, and alternatives. Suppl Am Heart J 1994;127:969.
32. Kron J, Herre J, Renfroe EG, et al: The AVID Investigators: Lead and device-related complications in the antiarrhythmic versus implantable defibrillators Trial. Am Heart J 2001;141:92.
33. Moss AJ, Hall WJ, Cannom DS, et al: Improved survival with an implanted defibrillator in patients with coronary disease at high risk for ventricular arrhythmia. Multicenter Automatic Defibrillator Implantation Trial investigators. N Engl J Med 1996;335:15.
34. Wilber DJ, Garan H, Ruskin JN: Electrophysiologic testing in survivors of cardiac arrest. Circulation 1987;75(4 P2):III146-53.
35. Mushlin AI, Hall JW, Zwanger J, et al: The cost-effectiveness of automatic implantable cardioverter defibrillators: Results from MADIT. Circulation 1998;97:2129-2135.
36. Tchou PJ, Kadri N, Anderson J, et al: Automatic implantable cardioverter defibrillators and survival of patients with left ventricular dysfunction and malignant ventricular arrhythmias. Ann Intern Med 1988;109:529.
37. Axtell K, Tchou P, Akhtar M: Survival in patients with depressed left ventricular function treated by implantable cardioverter-defibrillator. PACE 1991;14:291.
38. Mehta D, Saksena S, Krol RB: Survival of implantable cardioverter-defibrillator recipients: Role of left ventricular function and its relationship to device use. Am Heart J 1992;124:1608.
39. Domanski MJ, Saksena S, Epstein AE, et al: The AVID Investigators: Relative effectiveness of the implantable cardioverter-defibrillator and antiarrhythmic drugs in patients with varying degrees of left ventricular dysfunction who have survived malignant ventricular arrhythmias. J Am Coll Cardiol 1999: 34:1090-1095.
40. Moss AJ, Zareba W, Hall WJ, et al: Prophylactic implantation of a defibrillator in patients with myocardial infarction and reduced ejection fraction. N Engl J Med 2002;346:877-833.
41. Bardy G, Lee K, Mark D, et al: The Sudden Cardiac Death in Heart Failure Trial. Presented at the Late Breaking Clinical Trials Session, American College of Cardiology Annual Scientific Session, March 8, 2004, New Orleans, LA.
42. Bristow MR, Saxon LA, Boehmer J, et al: Cardiac resynchronization therapy with or without an implantable defibrillator in advanced heart failure. N Engl J Med 2004;350:2140-2150.
43. Garson A Jr, Dick M 2nd, Fournier A, et al: The long QT syndrome in children: An international study of 287 patients. Circulation 1993;87:1866.
44. Weintraub RG, Gow RM, Wilkinson JL: The congenital long QT syndrome in childhood. J Am Coll Cardiol 1990;16:674.
45. Kron J, Silka MJ, Ohm OJ, et al: Preliminary experience with nonthoracotomy ICDs in young patients. The Medtronic Transvene investigators. PACE 1994;17:26.
46. Benson DW, Benditt DG, Anderson R, et al: Cardiac arrest in young, ostensibly healthy patients: Clinical, hemodynamic, and electrophysiologic findings. Am J Cardiol 1983;52:65.
47. Wever EFD, Hauer RNW, Oomen A, et al: Unfavorable outcome in patients with primary electrical disease who survived an episode of ventricular fibrillation. Circulation 1993;88:1021.
48. Viskin S, Belhassen B: Idiopathic ventricular fibrillation. Am Heart J 1990;120:662.
49. Wellens HJJ, Lemery R, Smeets JL, et al: Sudden arrhythmic death without overt heart disease. Circulation 1992;85 [suppl]:92.
50. Maron B, Gardin J, Flack J, et al: Prevalence of hypertrophic cardiomyopathy in a general population of young adults: Echocardiographic analysis of 4111 subjects in the CARDIA study. Circulation 1995;92:785.
51. Maron B, Roberts WC, Epstein SE: Sudden unexpected death in hypertrophic cardiomyopathy: A profile of 78 patients. Circulation 1982;67:1388.
52. Maron B, Shen WK, Link M, et al: Efficacy of implantable-cardioverter defibrillator for the prevention of sudden death in patients with hypertrophic cardiomyopathy. N Engl J Med 2000;342:365.
53. Fananapazir L, Epstein SE: Hemodynamic and electrophysiologic evaluation of patients with hypertrophic cardiomyopathy surviving cardiac arrest. Am J Cardiol 1991;13:1283.
54. McKenna WJ, Franklin RCG, Nihoyannopoulos P, et al: Arrhythmias and prognosis in infants, children and adolescents with hypertrophic cardiomyopathy. J Am Coll Cardiol 1988; 11:147.
55. Watkins H, Rosenzweig A, Hwang DS, et al: Characteristics and prognostic implications of myosin missense mutations in familial hypertrophic cardiomyopathy. N Engl J Med 1992;326:1108.
56. Steinbeck G, Dorwarth U, Mattke S, et al: Hemodynamic deterioration during ICD implant: Predictors of high-risk patients. Am Heart J 1994;127 [suppl]:1064.
57. Fuster V, Gersh BJ, Guiliani ER, et al: The natural history of idiopathic dilated cardiomyopathy. Am J Cardiol 1981;47:525.
58. Follansbee WP, Michelson EL, Morganroth J: Nonsustained ventricular tachycardia in ambulatory patients: Characteristics and association with sudden cardiac death. Ann Intern Med 1980;92:741.
59. Milner PG, DiMarco JP, Lerman BB: Electrophysiological evaluation of sustained ventricular tachyarrhythmias in idiopathic dilated cardiomyopathy. PACE 1988;11:562.
60. Dietmar B, Antz M, Boczor B, et al: Prophylactic use of cardioverter-defibrillator in patients with implantable-cardioverter defibrillators in patients with idiopathic dilated cardiomyopathy: The Cardiomyopathy Trial (CAT). J Am Coll Cardiol 2001; 38:598.
61. Kadish A, Dyer A, Daubert JP, et al: Prophylactic defibrillator implantation in patients with nonischemic dilated cardiomyopathy. N Engl J Med 2004;350:2151-2158.
62. DiMarco JP, Garan H, Harthorne JW, et al: Intracardiac electrophysiologic techniques in recurrent syncope of unknown cause. Ann Intern Med 1981;95:542.
63. Akhtar M, Shenasa M, Denker J, et al. Role of cardiac electrophysiologic studies in patients with unexplained recurrent syncope. PACE 1983;6:192.
64. Watkins L, Mirowski M, Mower MM, et al: Implantation of the automatic defibrillator: The subxiphoid approach. Ann Thorac Surg 1982;34:515.

65. Trappe H, Pfitzner P, Heintze J, et al: Cardioverter-defibrillator implantation in the catheterization laboratory: Initial experience in 46 patients. Am Heart J 1995;129:259.
66. Strickberger SA, Niebauer M, Ching Man K, et al: Comparison of implantation of nonthoracotomy defibrillators in the operating room versus electrophysiologic laboratory. Am Heart J 1995;75:25.
67. Magney JE, Flynn DM, Parsons JA, et al: Anatomical mechanisms explaining damage to pacemaker leads, defibrillator leads, and failure of central venous catheters adjacent to the sternoclavicular joint. PACE 1993;16:445.
68. Ong LS, Barold SS, Lederman M, et al: Cephalic vein guidewire technique for implantation of permanent pacemakers. Am Heart J 1987;114(4 Part 1):753.
69. Magney JE, Staplin DH, Flynn DM, et al: A new approach to percutaneous subclavian venipuncture to avoid lead fracture or central venous catheter occlusion. PACE 1993;16:2133.
70. Higano ST, Hayes DL, Spittell PC: Facilitation of the subclavian-introducer technique with contrast venography. PACE 1990; 13:681.
71. Epstein AE, Miles WM, Benditt DM, et al: Personal and public safety issues related to arrhythmias that may affect consciousness: Implications for regulation and physician recommendations. Circulation 1996;94:1147-1156.
72. Vlay SC, Olson LC, Fricchione GL, et al: Anxiety and anger in patients with ventricular tachyarrhythmias: Responses after automatic internal cardioverter-defibrillator implantation. PACE 1989;12:366.
73. Luderitz B, Jung W, Deister A, et al: Patient acceptance of the implantable-cardioverter defibrillator in ventricular tachyarrhythmias. PACE 1993;16:1815.
74. Winters WL, Achord JL, Boone AW, et al: American College of Cardiology/American Heart Association Clinical Competence Statement on Invasive Electrophysiology Studies, Catheter Ablation and Cardioversion. J Am Coll Cardiol 2000;36;1725.
75. Faust MM, Fraser J, Schurig LS, et al: Educational guidelines for the clinically associated professional in cardiac pacing and electrophysiology. PACE 1990;17:6.
76. Fetter JG, Stanton MS, Benditt DG, et al: Transtelephonic monitoring and transmission of stored arrhythmia detection and therapy data from an implantable cardioverter defibrillator. PACE 1995;18:1531.
77. Marchlinski FE, Callans DJ, Gottlieb CD, et al: Benefits and lessons learned from stored electrogram information in implantable defibrillators. J Cardiovasc Electrophysiol 1995;6:832-835.
78. Callans DJ, Hook BJ, Marchlinski FE: Use of bipolar recordings from patch-patch and rate sensing leads to distinguish rhythms in patients with implantable cardioverter defibrillators. PACE; 1991;14:1917-1922.
79. Grimm W, Flores BF, Marchilinski FE: Symptoms and electrocardiographically documented rhythm preceding spontaneous shocks in patients with implantable-cardioverter defibrillator. Am J Cardiol 1993;71:1415-1418.
80. Grimm W, Flores BF, Marchilinski FE: Electrocardiographically demonstrated unnecessary spontaneous shocks in 241 patients with implantable cardioverter defibrillators. PACE 1992;15: 1667-1673.
81. Mathew P, Lewis C, Neglia J, et al: Interactions between electronic surveillance systems and implantable defibrillators: Insights from a fourth generation ICD. Pacing Clin Electrophysiol 1997;29: 2857-2859.
82. McIvor ME, Reddington J, Floden E, et al: Study of Pacemaker and Implantable Cardioverter Defibrillator Triggering by Electronic Article Surveillance Devices (SPICED TEAS). Pacing Clin Electrophysiol 1998;21:1847-1861.
83. Gottlieb CD, Horowitz LN: Potential interactions between antiarrhythmic medications and the automatic implantable cardioverter defibrillator. PACE 1991;898-904.
84. Haberman RJ, Veltri EP, Mower MM: The effect of amiodarone on defibrillation threshold. J Electrophysiol 1988;5:415-419.
85. Jung W, Manz M, Pizulli L, et al: Effects of long-term amiodarone on defibrillation threshold. Am J Cardiol 1992;70:1023.
86. Kron J, Herre J, Renfroe EG, et al: Lead- and device-related complications in the antiarrhythmias versus implantable defibrillator trial. Am Heart J 2001;141:92-98.

Chapter 41 Devices for the Management of Atrial Fibrillation

WERNER JUNG, BERNDT LÜDERITZ, HYGRIV B. RAO, and SANJEEV SAKSENA

Atrial fibrillation (AF) is the most common sustained arrhythmia in clinical practice and occurs most often in association with cardiovascular disease. Atrial fibrillation incidence increases with age and affects 2% of the population older than 65 years and 4% of the population older than 80 years of age, with a median age of the AF population being about 75 years.[1,2] In patients with AF, there is a near doubling of cardiovascular mortality in men and a 50% increase in women.[3] In addition to symptoms of palpitations, patients with AF have an increased risk of stroke and may also develop decreased exercise tolerance and left ventricular dysfunction. AF is a frequent and costly health care problem representing the most common arrhythmia resulting in hospital admissions accounting for more than one third of all admissions for arrhythmias.

The importance of antithrombotic therapy is now undisputed, but the management of the arrhythmia itself remains controversial. Although prospective studies as RACE,[4] STAF,[5] and AFFIRM[6] have not shown improved morbidity or mortality rates with rhythm control, effective rhythm control was not often achieved in these studies due to the use of an antiarrhythmic drug-based strategy. The inability to maintain rhythm control is due to the limited efficacy of antiarrhythmic drugs, which has been repeatedly documented in clinical trials. In the STAF study, only 23% of the patients actually achieved freedom from AF on amiodarone therapy.[5] Recently, an increasing number of nonpharmacologic options have become available and can supplement or even attempt to replace drug therapy in selected patients. Currently, available nonpharmacologic strategies revolve around implantable device therapy and ablative approaches. Implantable devices that have been used to manage AF include cardiac pacemakers as well as atrial or atrioventricular defibrillators.

Pathophysiologic Basis for Device Therapy in Atrial Fibrillation

Significant new knowledge has emerged regarding potential mechanisms of AF in different patient populations, and the impact of therapies on these mechanisms. Human AF is dependent on the interaction between triggers and a vulnerable substrate for facilitating the process of functional, anatomic, or spiral wave reentry. These triggers consist of automatic or autonomically triggered atrial premature depolarizations (APDs) or organized automatic or reentrant atrial tachycardias. Triggers may emanate from the right or left atrium, particularly in the presence of structural heart disease.[7] Enhanced sympathetic tone increases the frequency of triggers, and conditions of vagotonic bradycardia can shorten atrial effective refractory periods and increase dispersion of atrial refractoriness. Perpetuation of AF is associated with structural and electrical remodeling of the substrate.[8] These changes include areas of intra-atrial or interatrial conduction delay, shortening and dispersion of the atrial refractoriness, and a loss of rate adaptation. Atrial fibrosis is common, increases with age, and progresses to heart failure. This substrate provides the milieu for conduction delay, reentry, and/or vortex shedding of daughter wavelets. In consequence, organized monomorphic atrial tachycardias (Fig. 41-1) have been demonstrated during the onset and maintenance of AF.[9,10]

The rationale for pacing therapy can now be refined on the basis of this new knowledge. Atrial-based pacing can prevent AF by several mechanisms. Thus, we can examine its effect on triggers, initiating or onset arrhythmias, mechanisms that maintain AF and their contributions in different patient populations. Trigger density is a critical element in the high event rate for asymptomatic and symptomatic AF events seen in refractory patients with AF. Although antiarrhythmic drugs have been commonly used to suppress atrial ectopic beats, overdrive atrial pacing and novel atrial pacing algorithms can reduce ectopy (Fig. 41-2). Overdrive pacing can also suppress triggers, such as automatic atrial tachycardias or APDs.[11,12] Alternate site pacing can reduce atrial conduction delay and shorten the activation time. Atrial-based pacing can also affect the electrophysiological properties of the substrate by preexciting regions critical to reentry, such as sites of conduction delay.[13,14] Reduced interatrial and intra-atrial

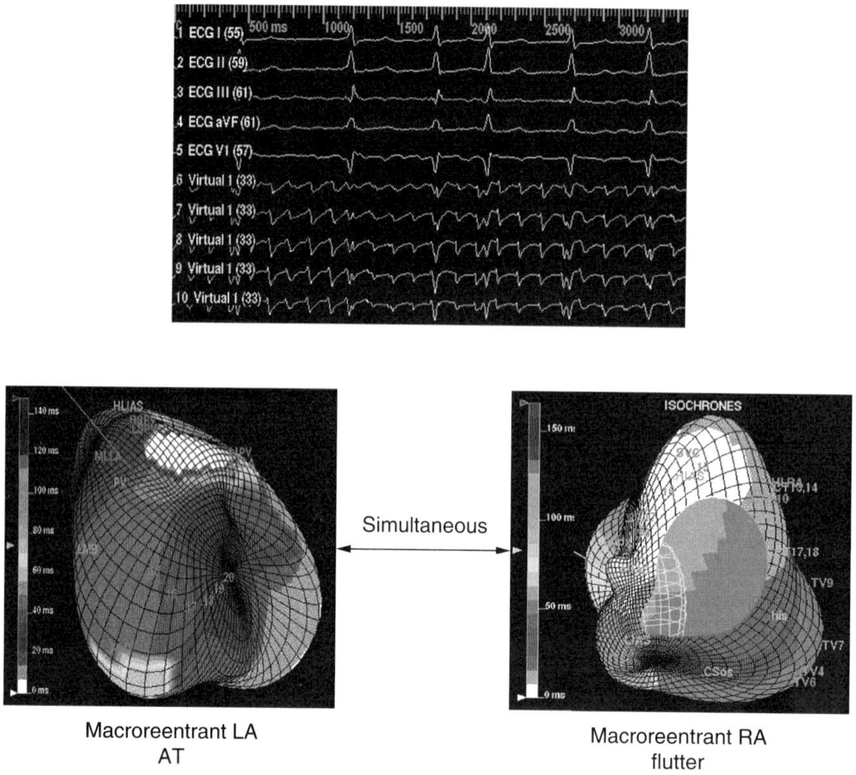

FIGURE 41-1 (See also Color Plate 41-1.) Three-dimensional map showing focal left atrial tachycardia and macro-reentrant right atrial flutter coexisting simultaneously in a patient with permanent atrial fibrillation. The **upper panel** shows surface ECG leads and virtual electrograms at numbered sites in the atrial maps. The color coding shows activation times at different right and left atrial regions during the cycle.

conduction delay and dispersion of refractoriness have been shown with septal and multisite atrial pacing (Fig. 41-3A). With dual site RA or biatrial pacing, regional atrial activation is resynchronized. Recently, Hesselson and Wharton demonstrated acute suppression of pulmonary vein foci with overdrive right atrial pacing, but suppression of AF occurred only when dual site atrial pacing was employed.[14] Dual site right atrial pacing and Bachmann's bundle pacing reduce or markedly truncate the window of AF induction by atrial premature depolarizations.[13,15] Dual site RA pacing has also been shown to improve atrial mechanics by increasing atrial filling velocities and preventing the LV and LA dilatation seen with RV apical pacing (see Fig. 41-3B).[16]

Recent data from the Atrial Fibrillation Therapy (AFT) trial clearly indicates that sinus bradycardia preceded onset of AF in 22% of patients.[17] This deviation in sinus rate can be modest, greater than 5 beats per minute or more and transient (5 minutes or more) but can create the opportunity for triggering AF. Demand atrial pacing can prevent this event. Finally, common right atrial isthmus-dependent atrial flutter is an important

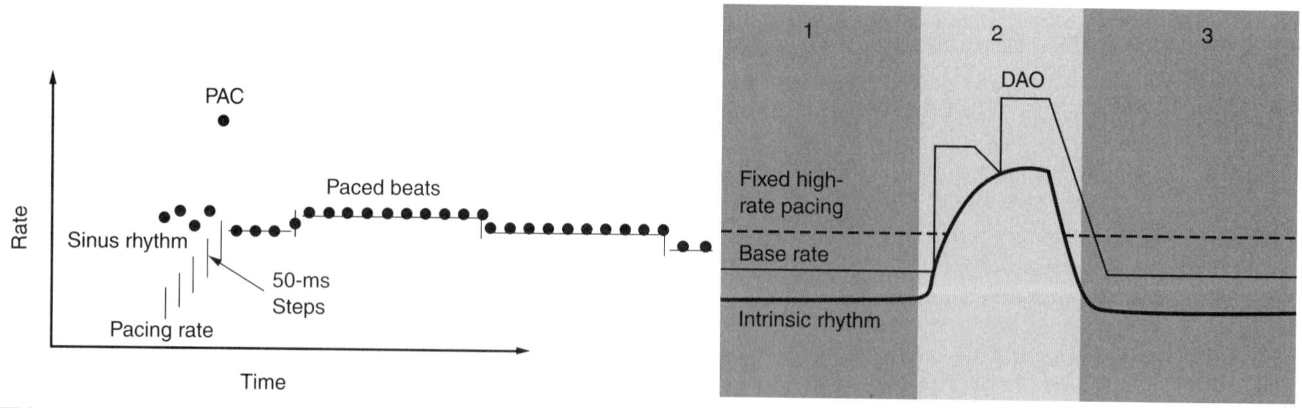

FIGURE 41-2 Two trigger suppression pacing algorithms available in pacemakers with AF therapies. The **left panel** shows the function of the atrial rate stabilization algorithm (Medtronic Inc, Minneapolis, MN) demonstrating an increase in atrial pacing rate in 50-ms increments after a premature atrial complex (PAC) to overdrive suppress subsequent premature beats. The **right panel** shows the dynamic atrial overdrive algorithm (St. Jude Medical, St. Paul, MN), which demonstrates the dynamic increase in overdrive rate as related to underlying changes in sinus cycle length. The algorithm periodically steps down until the sinus rate is encountered and can step up again as needed. The **two panels** on the next page show the response to multiple premature beats with a Vitatron Selection pacemaker (St. Paul, MN) algorithm.

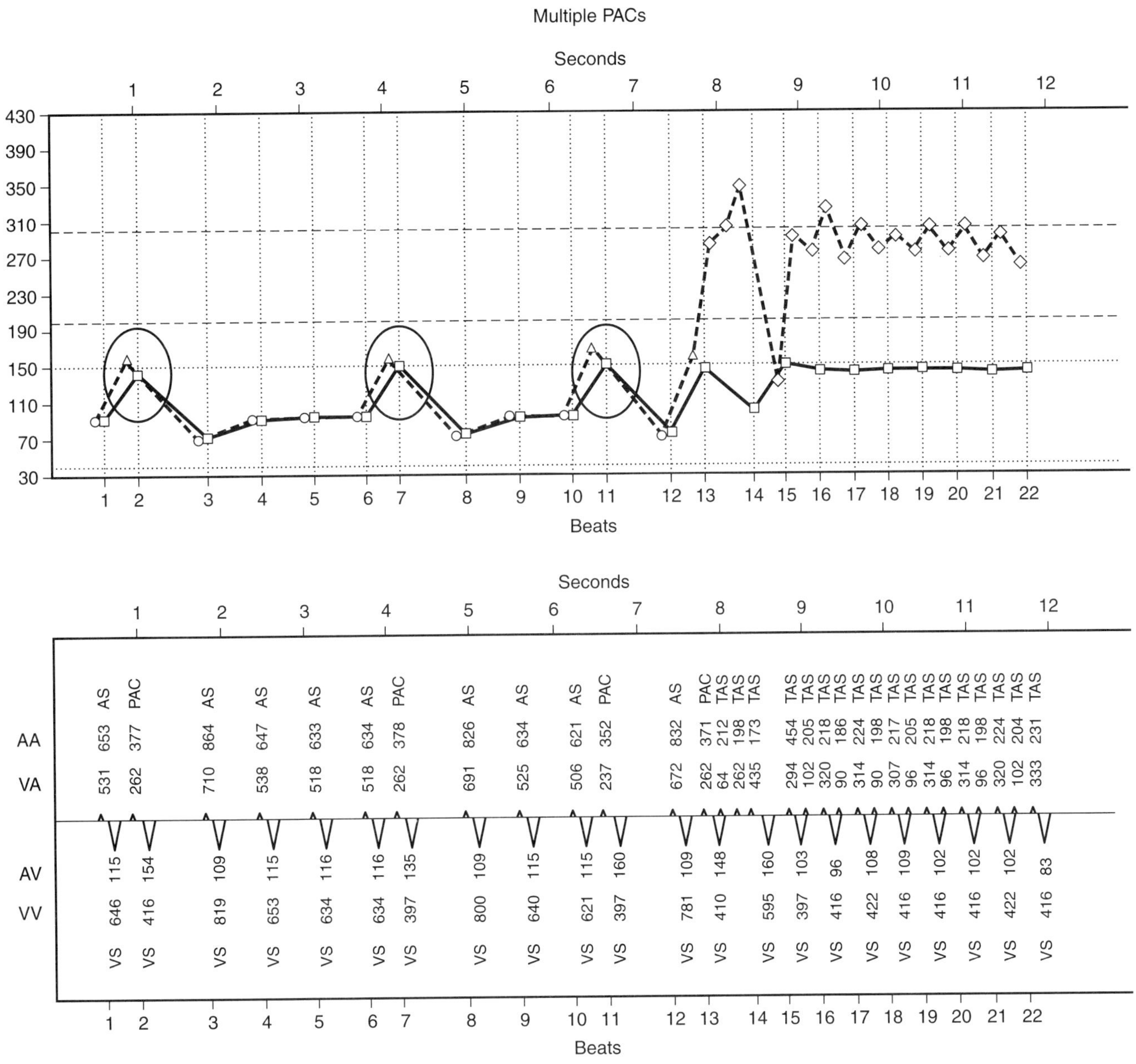

FIGURE 41-2 cont'd

and hitherto under-recognized trigger for AF. Treatment of this rhythm with anti-tachycardia pacing can be effective in its termination and AF prevention.

The second part of the arrhythmogenic chain is the onset or transitional tachycardia in AF. This tachycardia can be transient or prolonged in duration and can simulate AF on the electrocardiogram. Mapping has demonstrated such rhythms in both spontaneous and induced AF.[7,18] Three-dimensional mapping of individual atria has characterized these rhythms as macro-reentrant atrial tachycardias (see Fig. 41-1), common atrial flutter, and atypical atrial flutter.[18,19] Multisite pacing can reduce or eliminate the initiation of these arrhythmias by altering substrate properties (Fig. 41-4). Anti-tachycardia pacing can be used for termination of these rhythms and prevent progression to sustained AF (Fig. 41-5).[20]

Evolution to sustained AF requires electrophysiological mechanisms that result in perpetuation of the arrhythmia. More than one reentrant tachycardia may coexist in this phase, or a single tachycardia may persist, and both models can demonstrate fibrillatory conduction.[21] Critical requirements for such persistence require structural and/or electrophysiological remodeling of the AF substrate. Atrial conduction delays manifest as prolonged or abnormal P waves or prolonged atrial conduction intervals and are a prime predictor of AF recurrence. Resynchronization therapy using multisite atrial pacing can reduce this delay in multiple atrial regions. Dual site right atrial pacing therapy may have novel effects that lead to reverse remodeling of the atrial substrate by improving atrial transport and reducing atrial stretch.

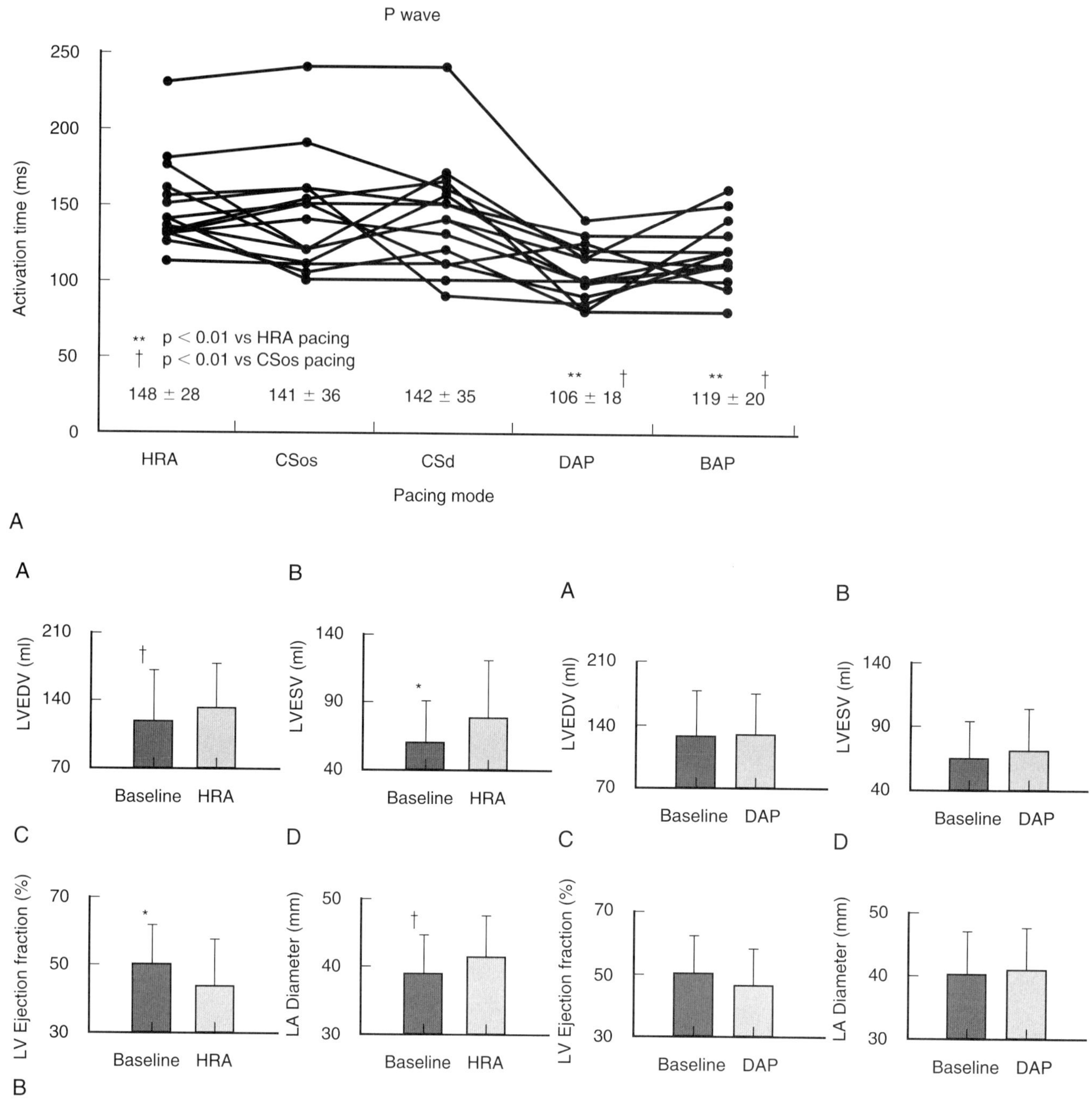

FIGURE 41-3 **A,** P wave duration during different atrial pacing modes in patients with atrial fibrillation. Total P wave duration is shown on the Y axis. HRA, high right atrium; CSos, coronary sinus ostium; CSd, distal coronary sinus; DAP, dual site right atrial pacing; BAP, biatrial pacing. **B, Left panel:** Changes in atrial and ventricular function during single site high right atrial pacing (HRA) compared with unpaced baseline data. Data are mean ± SD; $*P < .05$ versus. baseline, † $P < .10$ versus baseline. **Right panel:** Changes in atrial and ventricular function during support pacing compared with unpaced baseline data. LA Diameter, left atrial diameter; LVEDV, Left ventricular end-diastolic volume; LVESV, left ventricular end-systolic volume; LV, left ventricular; *P,* nonsignificant for all comparisons. (**A,** from Prakash A, Delfaut P, Krol RB, Saksena S: Regional right and left atrial activation patterns during single and dual site atrial pacing in patients with atrial fibrillation. Am J Cardiol 1998;82:1197-1204.)

Termination therapies, such as anti-tachycardia pacing, can terminate common and non-isthmus dependent atrial flutter by burst, ramp, or combination rapid pacing sequences. This can interrupt the chain of AF development or its maintenance. In our laboratory, 50-Hertz pacing (Fig. 41-6) has been shown to terminate atypical nonisthmus-dependent atrial flutter but not sustained AF.[22] Atrial defibrillation shocks produce conduction delay and block (Fig. 41-7). Although extensive mapping data are unavailable for internal atrial defibrillation, mapping of external shocks demonstrates that even ineffective shocks modify the underlying atrial activation patterns in AF, resulting in transient or persistent slowing.[23] Successive shocks may be effective when the first one modifies the sustained AF mechanisms resulting in vulnerability to the next

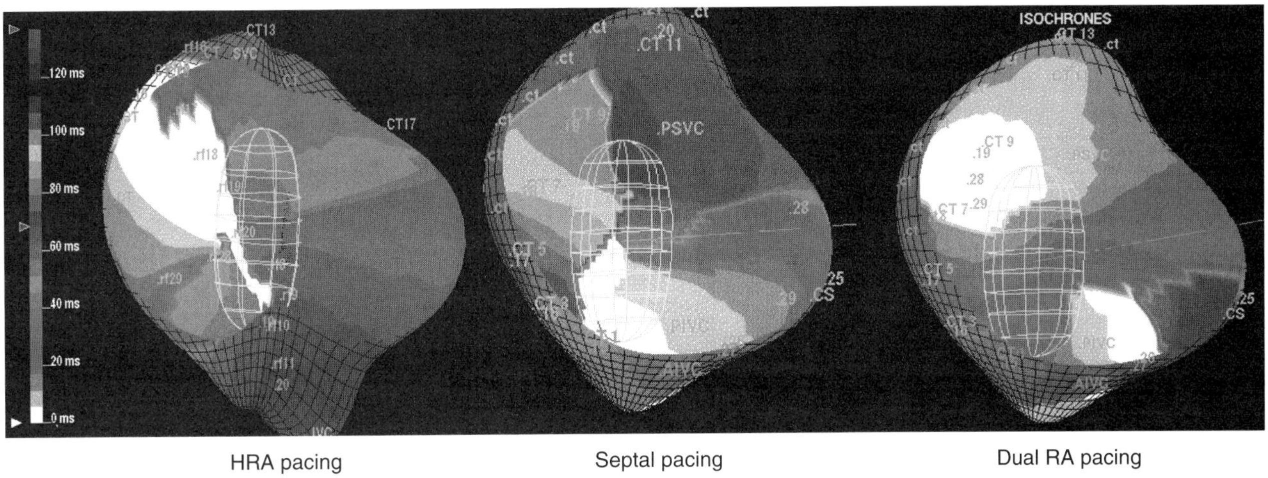

FIGURE 41-4 (See also Color Plate 41-4.) Right atrial three-dimensional activation maps during high right atrial, septal, and dual site right atrial pacing. Note the shorter activation times and two simultaneous activation wavefronts in the dual site pacing mode that abbreviate and synchronize atrial activation.

shock or even spontaneous termination mediated by an unstable circuit or concealed penetration by an atrial premature beat (Fig. 41-8).

Initial studies on catheter atrial defibrillation utilized two small surface area defibrillation electrodes and monophasic shocks. Defibrillation could be accomplished with shock energies of less than 1 joule but was poorly reproducible.[24] Ventricular fibrillation was occasionally induced (2.4%) when shocks were delivered greater than 115 ms after QRS onset. Kumagai and coworkers reported efficacy rates of 47% at energies <0.5% and 74% at 1 joule.[25] Biphasic shocks are associated with lower thresholds than monophasic shocks.[26] Experimental studies have demonstrated that the right atrium to coronary sinus defibrillation vector is associated with the lowest defibrillation thresholds.[27] Using this optimized waveform with individual phase durations of 3 ms and the optimal shock vector, E50 has been estimated at 1.3 joules. In the sheep model, biphasic shocks with a right atrium-left thoracic patch vector had a 47% efficacy at 1 joule.[28] Epicardial atrial defibrillation thresholds have been studied and have shown extremely low energy needs

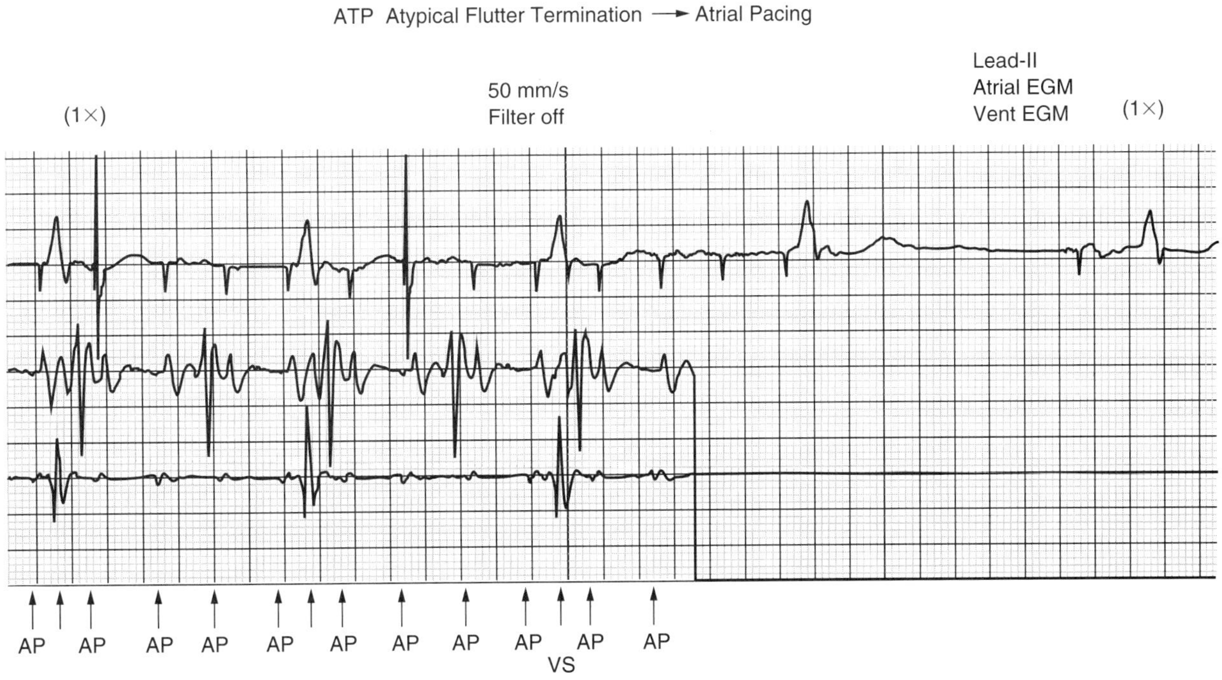

FIGURE 41-5 Rapid atrial pacing (ATP) terminates atypical atrial flutter in this patient with permanent atrial fibrillation (see Figure 41-1) implanted with a Guidant pacemaker and later with a Guidant Vitality AVT dual defibrillator with dual site atrial pacing. The patient reverts to an atrial paced rhythm.

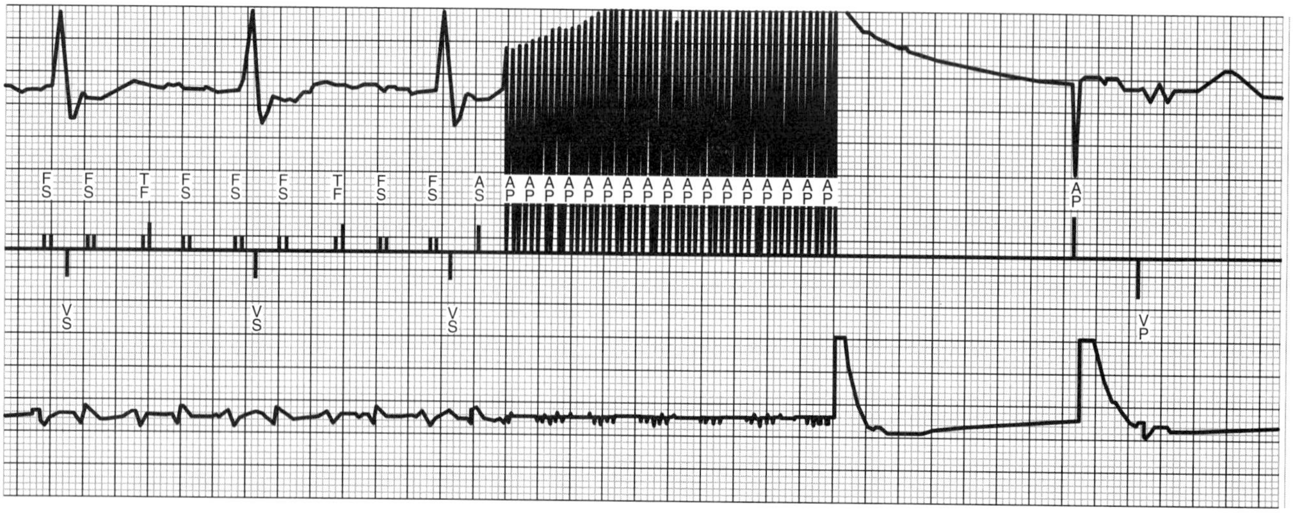

FIGURE 41-6 Termination of atypical atrial flutter in the autodiscrimination zone of a Medtronic Jewel AF dual defibrillator using a 50-Hertz high frequency pacing burst. Note the regularity of the rhythm despite surface ECG suggesting atrial fibrillation. FS, Fibrillation zone marker; TF, tachycardia zone marker.

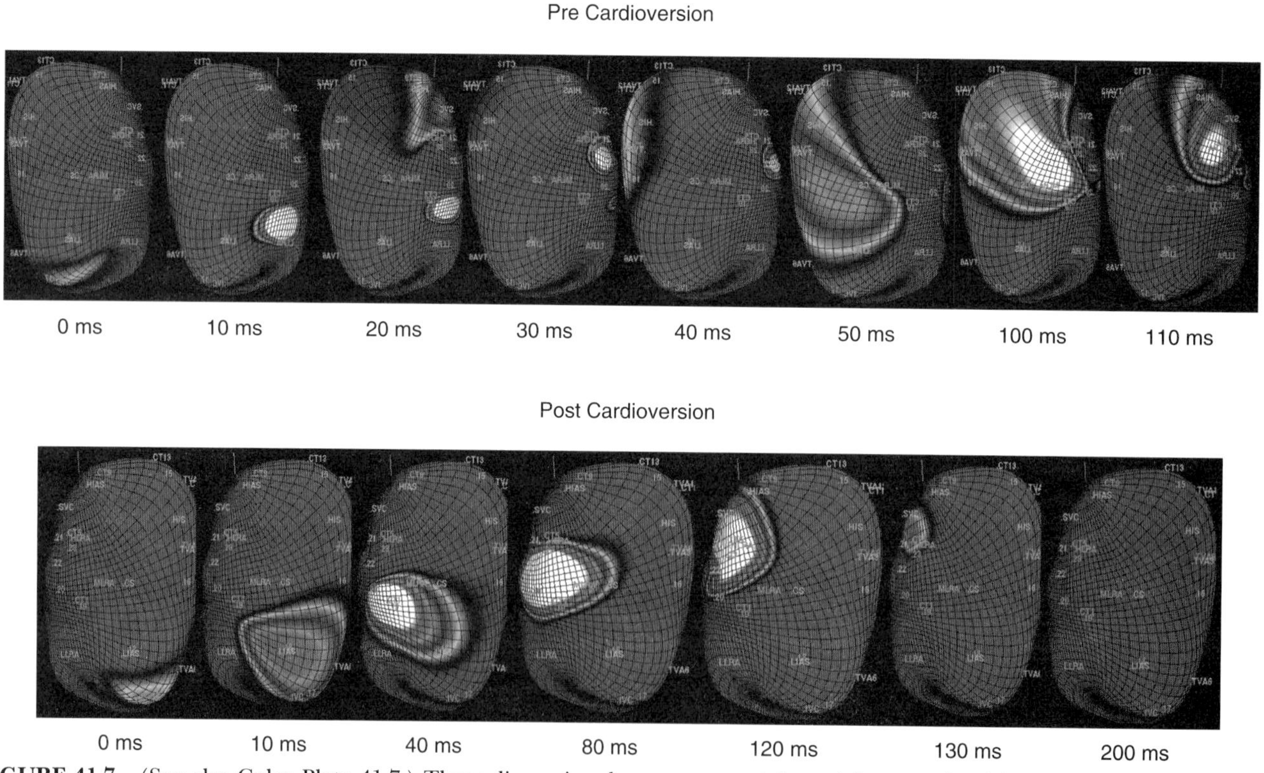

FIGURE 41-7 (See also Color Plate 41-7.) Three-dimensional non-contact right atrial maps of atrial activation patterns before and after delivery of a cardioversion shock. Before cardioversion more than one wavefront is present and after the shock delivery, only a single macro-reentrant wavefront is left. Although classified as a failed cardioversion, delivery of the shock fundamentally altered the tachyarrhythmias. Individual frames of the activation wavefront and their timing in ms relative to shock delivery are shown.

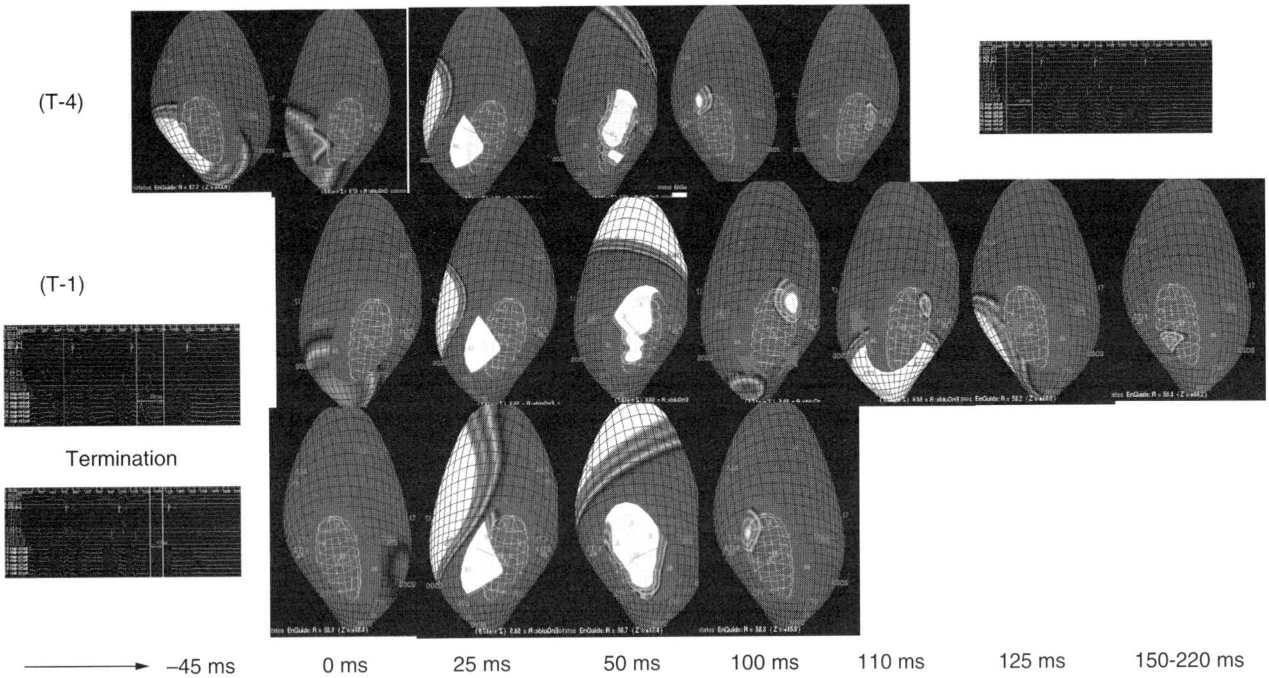

FIGURE 41-8 (See also Color Plate 41-8.) Three-dimensional noncontact right atrial maps of atrial activation patterns before spontaneous termination of an AF event. Select individual frames from the last four cycles before termination are shown and their timing relative to each other in ms are presented. Before termination, only a single wavefront is present in the fourth cycle before cessation. In the cycle immediately before termination, a concealed atrial premature beat invades the excitable gap in the single circuit and gives rise to antidromic and orthodromic wavefronts. The former collides and terminates the original tachycardia and the latter circulates for one cycle and spontaneously stops propagating, resulting in atrial fibrillation episode termination.

with an E 42% in the 0.3 joule range.[29] In contrast to clinical data, these experimental studies suggested that atrial defibrillation could be achieved at values below the pain perception threshold for shock therapy.

Atrial Pacing for the Primary Prevention of Atrial Fibrillation

SINGLE SITE ATRIAL PACING

Atrial pacing performed from the high right atrium can reduce the incidence of persistent or permanent AF in patients receiving permanent pacemakers for the treatment of bradycardias.[30-32] The Danish Trial of Physiologic Pacing in sick sinus syndrome was the first prospective trial comparing atrial-based pacing in the AAI mode with an active control arm using ventricular demand (i.e., VVI pacing). Despite initial equivalence, long-term follow-up (mean 5.5 years) demonstrated a reduced incidence of AF with atrial pacing.[30] These findings suggest a slow reverse atrial remodeling effect with atrial pacing. More recent prospective, randomized trials have corroborated this finding. This result has been corroborated in two larger controlled studies, the MOST[31] and CTOPP[32] trials, although the former compared the dual chamber physiologic pacing (i.e., DDDR mode) with rate responsive ventricular demand (i.e., VVIR pacing). It is particularly effective in patients with sick sinus syndrome alone (Fig. 41-9), reducing the relative risk of AF development by 50%.[31] These observations prompted consideration of the hypothesis that atrial pacing alone would be effective in patients presenting with symptomatic AF with or without bradycardias. Primary prevention trials using high right atrial pacing are summarized in Table 41-1. They uniformly demonstrate a relative risk reduction in the propensity to develop AF in patients with sinus node dysfunction during long-term follow-up. Interestingly, a greater risk reduction was seen with single site atrial pacing than with dual chamber pacing modes. In contrast, patients with atrioventricular block do not show this reduction in AF incidence with dual chamber pacing. This could be due to potentially lower use of atrial pacing in this population or disease localized to the atrioventricular junction.

MULTISITE ATRIAL PACING

In patients who have undergone open-heart surgery, right atrial overdrive pacing has been ineffective in AF prevention. In this population, temporary simultaneous bi-atrial pacing has been demonstrated to reduce the occurrence of AF compared with placebo or right or left atrial triggered/inhibited pacing.[33,34] These data suggest that dual site pacing may be effective in primary prevention in certain populations and could enhance efficacy. An ongoing trial, the PASTA trial, is reexamining this issue comparing single site high right atrial pacing, high and low septal pacing, and dual site right atrial pacing in patients with sick sinus syndrome.[35]

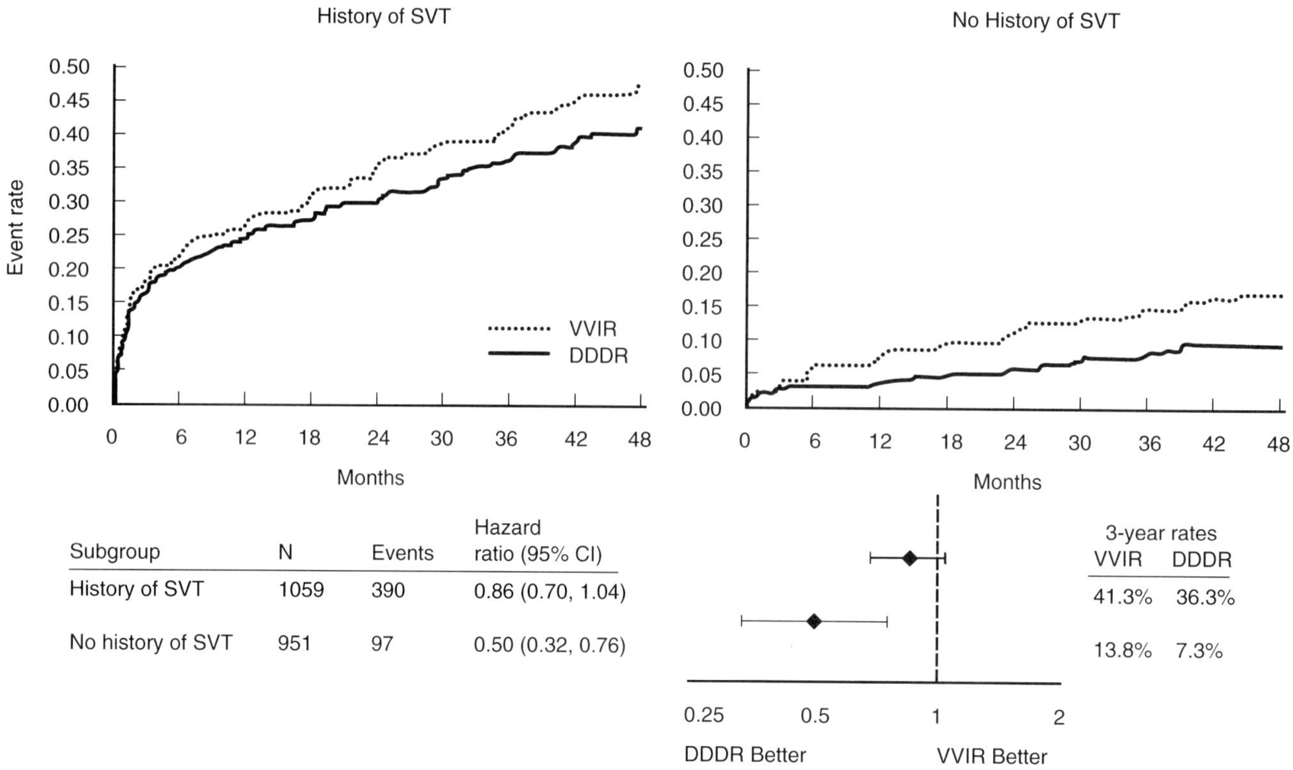

FIGURE 41-9 Primary prevention of permanent atrial fibrillation (AF) by atrial-based pacing in the DDDR mode in the MOST trial. Note that the relative risk of progression to permanent AF is reduced by 50% in patients with sick sinus syndrome that do not have a prior history of supraventricular tachyarrhythmias (SVTs).

Pacing for the Secondary Prevention of AF

Clinical investigation of atrial pacing techniques for management of AF in symptomatic AF populations has been performed in several high profile prospective clinical trials.[36-41] Analysis of the clinical benefit of atrial pacing is complicated by limited knowledge of the natural history of AF in different AF populations and by a lack of standardized end points for quantifying clinical benefit. Many studies lack a control group without atrial pacing therapy to judge efficacy. Approaches to date have included standard high right atrial or dual

TABLE 41-1 Trials of Physiologic Pacing

Trials	Number of Patients	Trial Design and Patient Population	Outcome
Primary Prevention Trials			
Danish Trial of Physiological Pacing (1988-96)	225	AAI vs. VVI in SND patients	34% relative risk reduction in all-cause mortality and 46% relative risk reduction for the development of AF in 5.5 years
Canadian Trial of Physiological Pacing (CTOPP 1997-2000)	2568	DDDR vs. VVIR in SND and AVB patients	18% reduction in annual rate of AF 27% reduction in chronic AF
MODE Selection Trial (MOST 1998-2002)	2010	DDDR vs. VVIR in SND and AVB patients	21% relative risk reduction for patients to develop AF in 3 years
United Kingdom Pacing and Cardiovascular Events (UK PACE 1998-2003)	2021	Single vs. dual chamber pacing in elderly patients with AVB	No differences in mortality and AF incidence at 3 years
STOP-AF	350	DDD(R)/AAI(R) vs. VVI (R) or VDD(R) with AVD = 300 ms	Primary end point — established AF
DANPACE	1900	DDD(R) vs. AAI(R)	Primary end point — quality of life
Secondary Prevention Trials			
Atrial Pacing Periablation for Paroxysmal AF (PA³ 1999-2000)	77	DDDR vs. VDD in patients with AV nodal ablation	43% developed AF in 1 year with no difference from control arm
PRevention by OVEdrive Study (PROVE 2000)	78	Overdrive pacing resulted in 84% prevalence of pacing	No decrease in total arrhythmic episodes; 34% decrease in mode switches; 48% reduction in duration of arrhythmia episodes

AF, atrial fibrillation; AVB, atrioventricular conduction block; AVD, atrioventricular delay; ms, milliseconds; SND, sinus node dysfunction.

TABLE 41-2 Trials with Novel Atrial Pacing Algorithms—Secondary Prevention Trials

Trials	Number of Patients	Pacing Algorithms and Patient Population	Outcome
Atrial Therapy Efficacy and Safety Trial (ATTEST 2003)	368	DDDRP with atrial preventive and anti-tachycardia pacing algorithms ON vs. OFF	No decrease in AT/AF episodes. (1.3 vs.1.2/month)
Adopt A (2001)	400	DDDR vs. DDDR with DAO; resulted in 92.9% Pacing.	Initial 26% reduction in AF burden and 65% reduction in organized atrial arrhythmias.
Italian study (2001)	61	Use of CAP algorithm to increase pacing percentage from 77%-96%	No decrease in AF (78% vs. 75%)
Atrial Fibrillation Trial (AFT 2001)	372	Preventive Atrial Pacing algorithm with DDDR.	30% Reduction in mean AF burden; 68% reduction in AF episodes by using pacing algorithm
AT 500 Verification Study Investigators (2001)	325	Preventive pacing and atrial ATP algorithms in AT 500 Device	Percentage atrial pacing increased from 62 to 97, terminating 53% of AT episodes. AF episodes and time in AF not different

AF, atrial fibrillation; AT, atrial tachycardia; CAP, continuous atrial pacing; DAO, dynamic atrial overdrive; DAP, dual site atrial pacing; HR, hazard ratio; HRA, high right atrial; mo, months; RAA, right atrial appendage.

chamber pacing alone or with rate response (see Table 41-1), right atrial pacing with novel pacing algorithms (Table 41-2), or alternate site (high or low septal pacing) and dual site atrial pacing (Table 41-3).

STANDARD RIGHT ATRIAL PACING

The efficacy of standard high right atrial pacing alone for prevention of symptomatic paroxysmal AF has been evaluated in clinical studies and remains currently unproven. High right atrial pacing alone for prevention of symptomatic drug-refractory paroxysmal AF patients without bradycardias who are awaiting AV junctional ablation was evaluated in a prospective, randomized, crossover study by Gillis and coworkers. They observed no prolongation in the time to recurrent AF compared with placebo.[36] In patients with refractory AF as the sole arrhythmia, only the Jewel AF device trial showed that high right atrial pacing appeared to reduce frequency but not AF burden initially, but more detailed analysis failed to confirm long-term benefit.[37] PROVE, a randomized crossover trial, evaluated a similar device, the Talent DR 213 pacemaker, combining atrial overdrive pacing with an automatic rest rate function.[38]

TABLE 41-3 Alternate and Multisite Pacing Trials—Secondary Prevention Trials

Trials	Number of Patients	Pacing Site and Patient Population	Outcome
Dutch Trial (2000)	26	Single site vs. dual site with antiarrhythmic drugs	Less need for cardioversion for AF >24 hrs with dual site pacing
Bailin (2001)	120	Bachmann's bundle vs. RAA pacing; 67.5% SND	Survival free from permanent AF increased to 75% from 47% with septal pacing 25% progressed to permanent AF in 1 yr
New Indication for Preventive Pacing in AF (NIPPAF 2001)	22	DAP with CAP. Algorithm vs. HRA pacing or no pacing	Reduction of AF events by 1.5 times
Pacing In Prevention of AF (PIPAF 2002)	91	DAP vs. DAP and DAO algorithm vs. HRA pacing	No significant differences. Trend toward lesser AF in dual site group
Dual Site Atrial Pacing for Prevention of AF (DAPPAF 2002)	120	DAP vs. HRA vs. Support pacing as a part of hybrid therapy	Improved mode adherence (5.8 vs. 4.7 vs.3.3 mo) Improved AF free survival (HR: 0.715 vs. 0.835 vs. 0.835) Longer time to recurrence (1.77 vs. 0.62 vs. 0.44 mo)
Atrial Septal Pacing Efficacy Clinical Trial (ASPECT 2002)	298	Septal vs. nonseptal pacing	No decrease in overall AF episodes and no effect on AF burden
Primary Prevention Trial			
Pacing of Atria in Sick Sinus Syndrome Trial. (PASTA)		High right atrial vs. septal vs. CSOs vs. dual site right atrial pacing	Results awaited

AF, atrial fibrillation; DAO, dynamic atrial overdrive; DAP, dual site atrial pacing; HR, hazard ratio; HRA, high right atrial; mo, months; RAA, right atrial appendage.

Overdrive pacing resulted in an improved atrial pacing percentage (84%) but did not reduce the total number of episodes, although there appeared to be a reduction in AF episode duration.

Another approach using high right atrial pacing in combination with other antiarrhythmic therapies, such as drugs, can also be considered but has not been formally studied in prospective studies. In an early experience from our group, antiarrhythmic drug therapy combined with high right atrial or septal pacing prolonged arrhythmia-free intervals, but no long-term data were available on rhythm control.[39]

NOVEL PACING ALGORITHMS

In an effort to improve efficacy, several new directions have evolved for single site atrial pacing. These include several new algorithms for ensuring overdrive atrial pacing. Trials using the novel algorithms are summarized in Table 41-2. The AT 500 pacemaker (Medtronic Inc., Minneapolis, MN) is a DDDRP device with two preventive pacing algorithms: atrial preference pacing and atrial rate stabilization (Fig. 41-10). In addition to having anti-tachycardia pacing capabilities, this device can be used for AF prevention using standard atrial pacing leads. Atrial preference pacing changes the base pacing rate in response to atrial premature beats, with a programmable increment. Atrial rate stabilization intercedes after premature beats altering the post-ectopic escape interval by reducing it markedly and then slowly easing down to the base-pacing rate. Anti-tachycardia pacing, as illustrated in Figure 41-5, can terminate common and nonisthmus-dependent atrial flutter by burst, ramp, or combination rapid pacing sequences.

In a nonrandomized study, the AT500 Verification Study evaluated 325 patients for the efficacy of atrial preventive and anti-tachycardia pacing, device safety, and reliability of atrial tachyarrhythmia detection.[40] Although preventive pacing algorithms were found to increase the median percentage of atrial pacing from 62% to 97%, the frequency and duration of AF episodes were unchanged. Fifty-three percent of atrial tachycardia episodes were terminated with anti-tachycardia pacing; there was an 88% complication-free survival rate at 3 months and 97% reliable detection of atrial tachyarrhythmia episodes. This study documented a role for anti-tachycardia pacing in AF device therapy. The ATTEST study[41] randomized patients with the AT500 pacemaker after implantation to preventive pacing plus anti-tachycardia pacing activation or standard high right atrial pacing in the DDDR mode. Anti-tachycardia

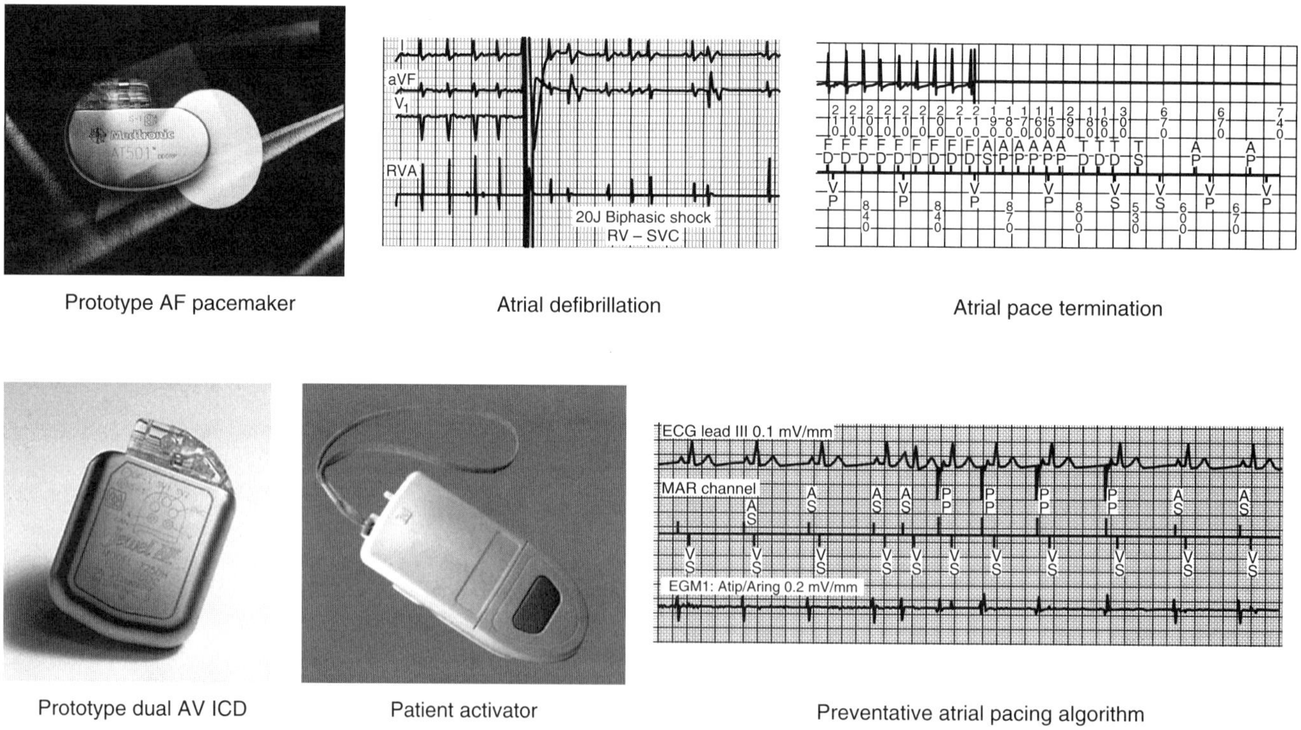

FIGURE 41-10 First generation technology for cardiac pacemakers and dual chamber atrioventricular defibrillators with atrial fibrillation (AF) detection and therapy features. The **left panel** shows illustrations of the implantable pulse generator (**top,** Medtronic AT 500 series pacemaker; **bottom,** Medtronic Jewel AF defibrillator). The center panel (**top**) shows the preventive pacing algorithm present in both devices (atrial rate stabilization, see Figure 41-2 for details) operating in a patient with premature atrial complexes. The **center panel (bottom)** shows a prototype patient activator for the defibrillator device that can be used for AF detection and treatment under patient or physician guidance. The **right panel (top)** shows burst atrial anti-tachycardia pacing for atrial tachyarrhythmia termination present in both devices. The **right panel (bottom)** shows delivery of an atrial defibrillation shock in a patient with spontaneous AF with successful AF termination and sinus rhythm (SR), but frequent atrial premature beats after shock delivery lead to early recurrence of AF (ERAF).

pacing in this study also terminated 53% of episodes, and the positive predictive value for atrial tachycardia detection was 99%. Although quality of life improved in both groups, there were no significant differences in frequency or burden of atrial tachyarrhythmia episodes. In a multicenter trial using continuous atrial pacing algorithms, initial results showed 84% prevalence of atrial pacing. There was a 79% reduction in atrial premature beats with no effect on symptomatic AF events, but a mean of 48% shortening of episode duration and a slight improvement in quality of life.[42] The ADOPT-A study evaluated a new pacing algorithm, dynamic atrial overdrive (St. Jude Medical, St. Paul, MN) in patients with AF with bradycardias. It showed an approximate 25% decrease in AF burden and some modest improvement in quality of life.[43] However, these effects do not appear to be clinically adequate methods for long-term rhythm control.

ALTERNATE SITE ATRIAL PACING

Alternate site and multisite atrial pacing methods, such as dual site right atrial pacing and bi-atrial synchronous pacing have been evaluated for AF and atrial flutter prevention (Table 41-3). Right atrial pacing at a single septal location was performed in two prospective randomized studies. Bailin and colleagues[44] randomized 120 patients with paroxysmal AF and bradycardias to high septal pacing or right atrial appendage pacing. Patients with high septal pacing had improved freedom from permanent AF at 1 year compared with standard right atrial pacing (75% versus 47%), but there was no observed decrease in symptomatic AF event frequency. Despite improvement, a significant proportion (25%) of patients progressed to permanent AF. A control arm without pacing therapy was absent in this trial. The Atrial Septal Pacing Efficacy Clinical Trial (ASPECT) randomized patients to low septal or standard right atrial pacing and employed preventive pacing algorithms using the Medtronic AT500 pacemaker. In this study, low septal pacing was not associated with a reduction in AF frequency or burden.[45]

DUAL SITE RIGHT ATRIAL PACING

For dual site right atrial pacing, an additional atrial pacing lead is inserted just outside the coronary sinus ostium for stability and left atrial stimulation (Fig. 41-11A). During simultaneous pacing of the high right atrium and coronary sinus ostium, the ECG shows a biphasic P wave in the inferior leads with abbreviation of P wave duration. (see Fig. 41-11B). Several small, randomized trials were initially reported in addition to single center pilot experiences. In our pilot experience, in patients with drug refractory symptomatic AF and bradycardias, trends to benefit with dual site right atrial pacing appeared in 3-month crossover interim analyses but significant benefit of dual site over high right atrial or septal pacing was obvious only after 1 year.[39] In a short-term randomized, 12-week comparative study of symptomatic AF patients without bradycardia, Lau and coworkers reported an increase in mean time to first AF recurrence from 15 to 50 days in patients during dual site pacing when compared with no pacing.[46] These singlecenter experiences have been followed by short-term randomized, multicenter studies. SYNBIAPACE, a randomized crossover study, compared synchronous bi-atrial, high right atrial, and demand pacing during 3-month treatment arms. Patients were included if they had a standard indication for pacing, two or more episodes of AF in 3 months, and a P wave duration of at least 120 milliseconds. There was a nonsignificant increase in AF-free interval with bi-atrial pacing alone compared with high-right atrial pacing.[47]

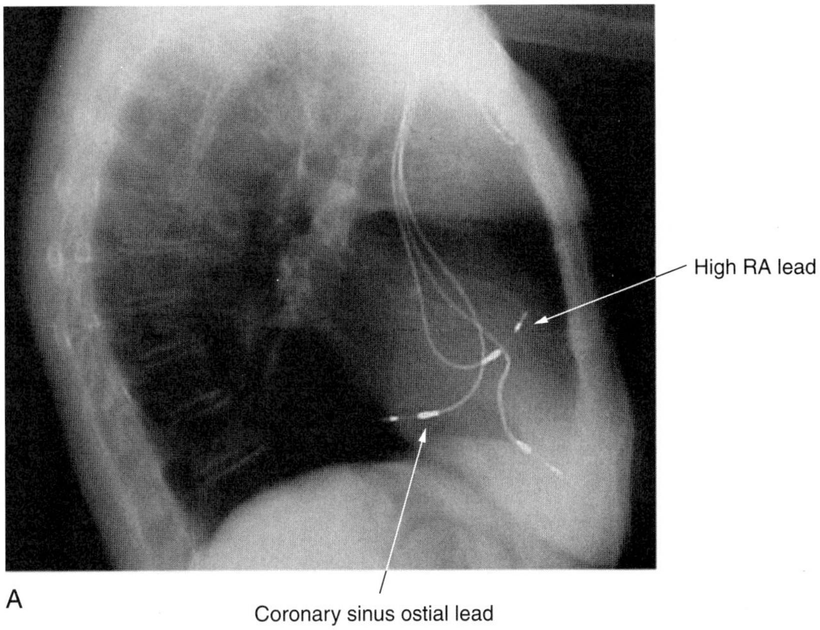

FIGURE 41-11 **A,** Radiograph of the chest in a patient with an implanted dual site right atrial (RA) pacing system. Note the dual atrial leads in the high right atrial appendage and outside the ostium of the coronary sinus for atrial resynchronization therapy.

(Figure continues on next page)

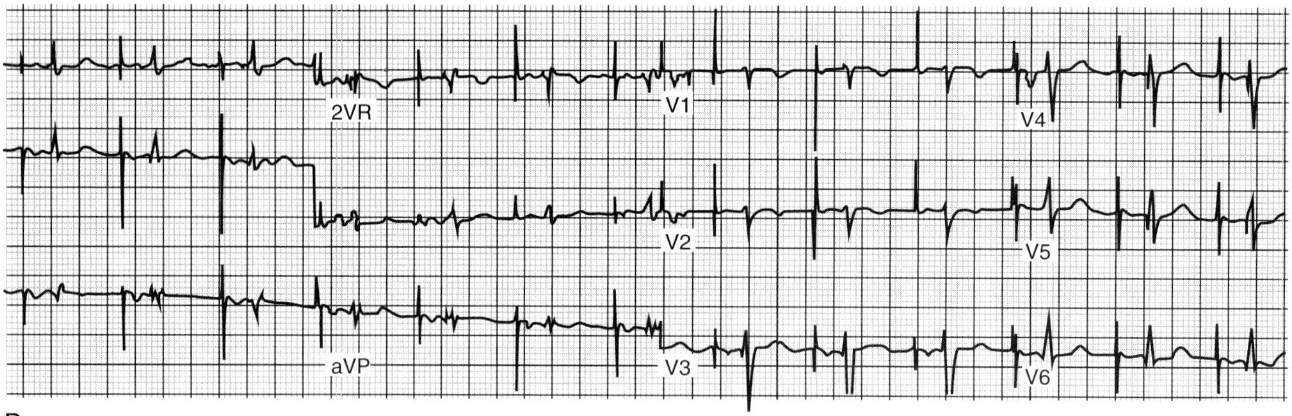

B

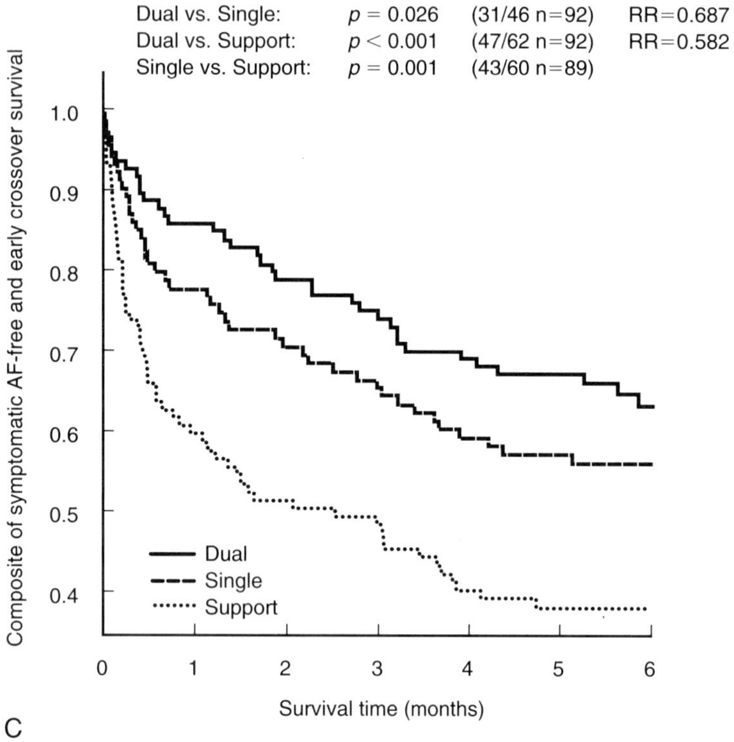

C

FIGURE 41-11 **B,** Electrocardiogram showing the typical inversion and biphasic nature of the atrial paced P wave in the inferior leads in a patient with dual site right atrial pacing leads. **C,** Composite endpoint of tolerance for atrial pacing therapy and freedom from recurrent atrial fibrillation in the dual site pacing for prevention of atrial fibrillation trial (DAPPAF). Note the superiority of dual site RA pacing over single site height RA pacing and support pacing.

The Dual Site Atrial Pacing for Prevention of Atrial Fibrillation Trial (DAPPAF) was a longer term multicenter crossover study with 6-month treatment arms comparing dual site right atrial, high-right atrial, and support pacing.[48] It enrolled patients with frequent, symptomatic, and drug-refractory AF with bradyarrhythmias requiring cardiac pacemaker insertion. After dual site right atrial pacing system implant, optimization of drug and pacing therapies was performed. The three modes of pacing were then randomly selected for 6-month periods. Patient tolerance and adherence to the pacing mode were superior in dual RA pacing when compared with support ($P < .001$) and high RA pacing ($P = .006$) (see Fig. 41-11C). Freedom from any symptomatic AF recurrence tended to be greater with dual RA (hazard ratio 0.72, $P = .07$) but not with high RA pacing ($P = .19$) compared with support pacing. Combined symptomatic and asymptomatic AF frequency in patients was significantly reduced during dual RA pacing as compared with high RA pacing ($P < .01$). However, in antiarrhythmic drug-treated patients, dual RA pacing increased symptomatic AF-free survival compared with support pacing ($P = .011$), and high RA pacing (hazard ratio 0.67, $P = .06$). In drug-treated patients with less than one AF event per week, dual RA pacing significantly improved AF suppression compared with support pacing (hazard ratio 0.46, $P = .004$) and high RA pacing (hazard ratio 0.62, $P = .006$). Lead dislodgment was uncommon (1.7%) with coronary sinus and high RA lead stability being comparable. Thus, DAPPAF showed improved adherence to pacing and rhythm control in the dual site mode, especially when combined with antiarrhythmic drugs. This study supports the use of a hybrid approach to rhythm management. In another prospective crossover trial with 6-month treatment periods performed in

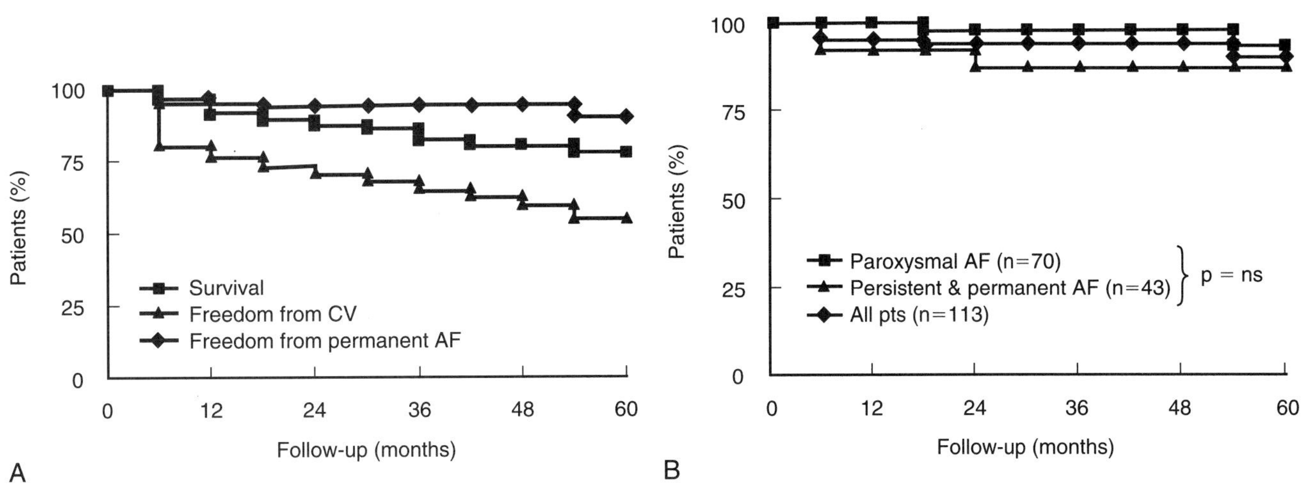

FIGURE 41-12 **A,** Long-term rhythm control, survival from all-cause mortality, and need for cardioversion therapy in study population by actuarial analysis. X-axis, percent of patients; Y-axis, follow-up in months. **B,** Comparison of long-term rhythm control in atrial fibrillation (AF) subpopulations with paroxysmal AF and persistent or permanent AF. The total population data are shown as a reference. (From Madan N, Saksena S: Long-term rhythm control of drug refractory atrial fibrillation with "hybrid therapy" incorporating dual site right atrial pacing, antiarrhythmic drugs and right atrial ablation. Am J Cardiol 2004;100: 569-575.)

patients with recurrent symptomatic AF without structural heart disease, Ramdat Misier and coworkers have shown a significant increase in time to recurrent AF and interventions for symptomatic AF recurrence.[49]

Longer term experience in secondary prevention of drug-refractory AF with dual site RA pacing is now available (Fig. 41-12). This experience has been acquired in patients with recurrent, symptomatic, and frequent drug-refractory AF, with or without concomitant bradycardias. In our long-term experience, now involving over 125 patients with follow-ups ranging to 10 years, the overall patient survival is 80% at 5 years. Rhythm control as documented by device data logs was achieved in more than 90% of patients at 3 years or more of follow-up. The overall stroke incidence was 1.0% per year.[50] Similar efficacy rates can be achieved in paroxysmal, persistent, and permanent AF (see Fig. 41-12B). However, occasional cardioversion and linear right atrial ablation was needed in approximately one half of the patients with persistent and permanent AF for rhythm control. A nonrandomized parallel cohort experience from Europe in patients with bradycardias who required pacing and paroxysmal AF, supports long-term efficacy with dual site right atrial pacing.[51] Of 83 patients, 30 had dual site right atrial pacing systems and 53 had single site high right pacemakers implanted. Patients with dual site systems had longer duration of AF (8.1 versus 3.8 years for high RA systems, $P < .001$) and more failed drug trials (2.4 versus 1.6 for high RA systems, $P < .05$). During a mean follow-up of 18 months, symptomatic paroxysmal AF recurred in nine patients after dual RA pacing when compared with 24 patients after high RA pacing ($P = .03$). Permanent AF supervened in only one patient after dual RA pacing and in 12 patients after high RA pacing ($P < .05$). Bi-atrial triggered pacing in patients with intra-atrial conduction delay and recurrent atrial flutter and fibrillation performed by Revault d'Allonnes and colleagues, resulted in a 64% incidence of rhythm control at a mean follow-up of 33 months.[52] A majority of these patients were on antiarrhythmic drug therapy.

The safety of dual site right atrial pacing has been established by these studies. Lead dislodgment rates are well within estimates for any type of atrial pacing, and long-term dislodgment concerns have been obviated by the dual right atrial lead technique. The remaining complications have been largely similar to those in any pacemaker implant procedure. In contrast, bi-atrial pacing is associated with higher lead dislodgment rates (up to 15%) and greater potential for ventricular oversensing.

AF Termination Therapies

ANTI-TACHYCARDIA PACING

Anti-tachycardia pacing (ATP) in AF termination has been evaluated in randomized and nonrandomized studies. Several observational studies, including prospective multicenter studies, demonstrated the successful termination of atrial tachyarrhythmias in the atrial flutter zone with anti-tachycardia pacing and modest success with high frequency burst pacing in the AF zone. However, prospective randomized trials have not shown that pacing termination strategies reduce or eliminate symptomatic AF or total AF burden. The ATTEST study prospectively randomized ATP therapy in bradycardia patients with symptomatic atrial tachyarrhythmias implanted with the Medtronic AT500 device (see Table 41-2). As mentioned previously, the positive-predictive accuracy for atrial arrhythmia detection was 99.9% and ATP efficacy was 54%. However, prevention combined with ATP therapies failed to reduce atrial tachyarrhythmia burden and AF frequency, despite an increase in the relative amount of atrial pacing.

It would be reasonable to conclude that atrial antitachycardia pacing is effective in select patients and often in a subgroup of atrial tachyarrhythmia events. This therapy is a valuable adjunctive component in devices when used in a hybrid therapy algorithm.

ATRIAL DEFIBRILLATION

Clinical Efficacy

Catheter-based internal cardioversion of atrial fibrillation was first reported by Jain and coworkers in 1969 and early studies were performed by Mirowski and colleagues.[53] High energy shocks delivered via a right atrial catheter electrode and a thoracic patch electrode were reported by Levy, who noted a 90% success rate for atrial defibrillation with 200 to 300 joules damped sinusoidal waveform shocks.[54] Epicardial and wholly endocardial electrodes greatly reduced energy requirements for atrial cardioversion (see Fig. 41-10). For right atrial to coronary sinus shock vectors the mean defibrillation threshold was 4.7 joules for monophasic shocks and 2.5 joules for biphasic shocks in acute feasibility studies using induced AF.[55] Initial prospective randomized controlled studies of different shock vectors in AF patients, however, showed that defibrillation thresholds routinely exceeded 10 joules for right-sided vectors, and right atrium to pulmonary artery configurations averaged 8.6 joules.[56] This was well above the pain threshold and patients required sedation for shock tolerance. Lower atrial defibrillation thresholds were seen in the absence of heart disease, higher left ventricular ejection fraction, and lower left ventricular dimensions.

The safety and tolerance of atrial defibrillation shocks were extensively debated in the early years. Early studies performed in patients without structural heart disease and right atrium to coronary sinus shock vectors showed virtually no proarrhythmic risk. Subsequently, low and intermediate energy shocks delivered in an R wave synchronous mode using a right ventricle to right atrium or right atrium to thoracic cutaneous electrode induced nonsustained ventricular tachycardia. A rare episode of sustained ventricular tachycardia has been reported. In AF populations with structural heart disease, ventricular tachyarrhythmias have been observed for the first time after AF device therapy had been instituted. This has been used as an argument to have ventricular defibrillation capability available when atrial defibrillation is performed. This has also led to the use of dual chamber atrioventricular defibrillators in preference to atrial defibrillators in this population. Bradyarrhythmias are more common after atrial defibrillation shocks. We have observed a 28% incidence of sinus node or atrioventricular conduction delays.[57] Thus, demand ventricular pacing should be available when atrial defibrillation shocks are employed.

Shock Tolerance

Tolerance for intracardiac shocks was examined initially when cardioversion for ventricular tachyarrhythmias was evaluated.[58,59] In 1985, our studies showed that shocks of more than 1 joule were poorly tolerated by 88% of patients.[60] Nathan reported that intra-atrial shocks of even less than 1 joule were often poorly tolerated by awake patients.[61] In our prospective study, 20% of patients reported pain at 1 joule, 40% at 2 joules, and the majority experienced pain by 3 joules.[59] Subsequently, a variety of efforts have been made to improve tolerance to these shocks. Although shock waveform changes have had modest beneficial effects, it has been demonstrated that repeated shock delivery enhances pain perception and shock intolerance. Thus, a single, high-energy successful shock is better tolerated than several low-energy shocks of lesser efficacy. Thus, current therapeutic strategies are designed for successful first shock or nocturnal shock delivery strategies to improve patient acceptance of this therapy and, more importantly, to consider patient-activated shocks for outpatient management of recurrent AF.

THE IMPLANTABLE ATRIAL DEFIBRILLATOR

Device Technology

The development of implantable device technology was achieved in the late 1990s and a prototype device was used in pilot studies. The prototype devices, Metrix models 3000 and its successor 3020, differed principally in their maximum energy outputs, 3 joules for the Metrix 3000 and 6 joules for the Metrix 3020.[62-64] Both delivered a biphasic truncated exponential waveform of 3/3 ms and 6/6 ms duration, respectively, which accounts for the increased energy output of the model 3020. The device, with a weight of 79 grams and a volume of 53 cc, was implanted in the pectoral region like a conventional pacemaker. Graded shock therapy of up to eight shocks (two at each level) for each episode of AF programmable in 20 volt increments up to 300 volts were permitted. Atrial defibrillation was accomplished by a shock delivered between electrodes in the right atrium and the coronary sinus. One defibrillation lead was fixed in the right atrium. The second defibrillation lead was placed in the coronary sinus and had a spiral configuration for retention in the coronary sinus. A separate bipolar right ventricular lead is used for R wave synchronization and demand pacing. The Metrix defibrillator induced AF by using R wave synchronous shocks and atrial defibrillation threshold testing. Defibrillation therapy could be programmed into one of the following five operating modes: fully automatic, patient activated, monitor mode, bradycardia pacing only, and off. The device could also store intracardiac electrograms for up to 2 minutes from the most recent six AF episodes (Fig. 41-13).

The device employed extensive signal processing for AF detection and R wave synchronization.[65] Two detection algorithms are run in series. The first, "quiet interval analysis algorithm" discriminates between a sinus beat and another atrial rhythm in the 8-second electrogram segment. The second algorithm, the "baseline crossing test," is invoked to detect AF (see Fig. 41-13). This latter algorithm examined electrical activity during

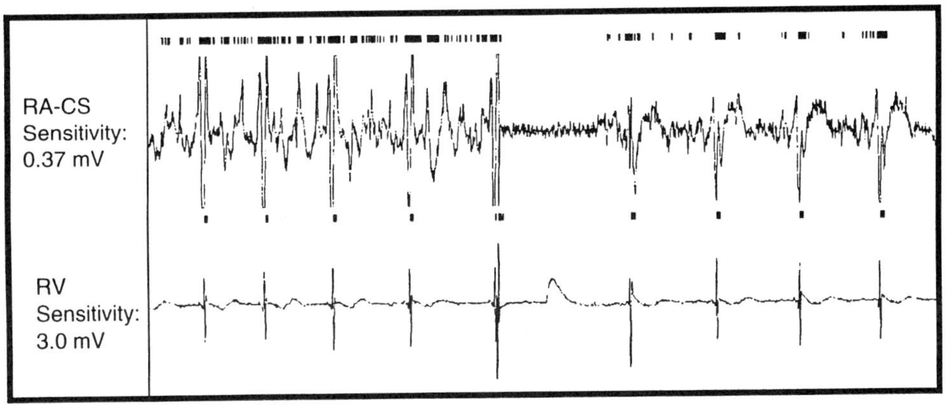

FIGURE 41-13 Stored electrogram of a successful conversion of atrial fibrillation with the automatic implantable atrial defibrillator. Sinus rhythm is restored by application of an internal automatic shock. *Upper trace:* Recordings of the intracardiac electrograms with their corresponding markers between a right atrial (RA) and a coronary sinus (CS) electrode using a sensitivity of 0.37 mV. *Lower trace:* Recordings of the right ventricular bipolar lead and marker signals with a sensitivity of 3.0 mV.

parts of the cardiac cycle that are quiescent during sinus rhythm and most organized atrial arrhythmias, but not during AF where atrial activity is random and unrelated to the cardiac cycle. The first algorithm, the quiet interval algorithm, is very sensitive for detection of AF and highly specific for sinus rhythm. The second algorithm, the baseline crossing algorithm, is highly specific for AF. The result was a highly sensitive and extremely specific detection algorithm for AF. The Metrix device used a dual channel synchronization algorithm to achieve high specificity at the expense of sensitivity. Thus, the algorithm was designed to ensure that all shocks will be delivered only to correctly synchronized R waves to avoid risk of proarrhythmia. Before synchronization was attempted, the two electrograms are evaluated simultaneously in real time for integrity and data quality.

Patient Selection, Follow-up, and Outcomes

In the pilot clinical study, 51 patients from 19 centers in 9 different countries were enrolled. Patients had to meet specific inclusion criteria: (1) Prior episodes of symptomatic AF that had spontaneously terminated or been converted to normal sinus rhythm with intervals of recurrence between 1 week and 3 months; (2) Previous treatment with at least one class I or class III antiarrhythmic agent that proved ineffective or was not tolerated because of side effects. Postoperative evaluation, including AF detection and R wave synchronization tests, was performed at predischarge, 1, 3, and 6 months and thereafter at 6-month intervals until completion of the study. Atrial shock energy was programmed either at the maximal output of the device or at least well above the atrial defibrillation threshold obtained at the time of implant.

During this initial study, a total of 3719 shocks were delivered for AF induction, atrial defibrillation, and testing or termination of spontaneous AF episodes.[66] Of these 3719 shocks, 3049 were delivered during testing and 670 were delivered for treatment of spontaneous episodes of AF. All shocks for spontaneous episodes were given during physician observation. There were no reported cases of induction of ventricular arrhythmias or inaccurately synchronized shocks during the study. A larger and longer term experience confirmed these initial observations. Correct synchronization was observed for all of the marked R waves and the accuracy of synchronization was 100% during both sinus rhythm and AF or atrial flutter. Analysis of the AF detection algorithm performance revealed 100% specificity for the recognition of sinus rhythm and 92.3% sensitivity for the detection of AF. From the same data, the positive-predictive value of the AF detection algorithm was 100% and the negative-predictive value was 92.6%.

One important aspect of the device was its ability to permit prolonged AF monitoring. In the early experience, one or more valid monitoring periods were available from 46 of 51 patients. A total of 1161 valid episodes of AF was obtained during a mean follow-up period of 260 ± 144 days (average recurrence rate: 3.9 ± 5.0 episodes per patient-month). The median duration of the 190 treated episodes falling within a valid monitoring period was 17.6 hours and 3 hours for the 971 nontreated episodes. Of the 190 treated episodes, 28% were equal to 8 hours in length as compared with 78% of the nontreated episodes. Five patients with a total of 29 episodes documented in the device memory did not seek therapy treatment for any of these episodes. These findings demonstrated for the first time the pattern and density of symptomatic and asymptomatic AF events in the symptomatic AF population.[67] These patterns have now been confirmed in larger experiences and with other devices.

Forty-one of the 51 patients had spontaneous episodes of AF that were treated with device therapy (average 5.6 episode per patient with a range of 1 to 26 episodes). A total of 670 shocks were delivered for the treatment of 227 episodes, with an average of three shocks per episode. Following shock delivery, 95.6% of the

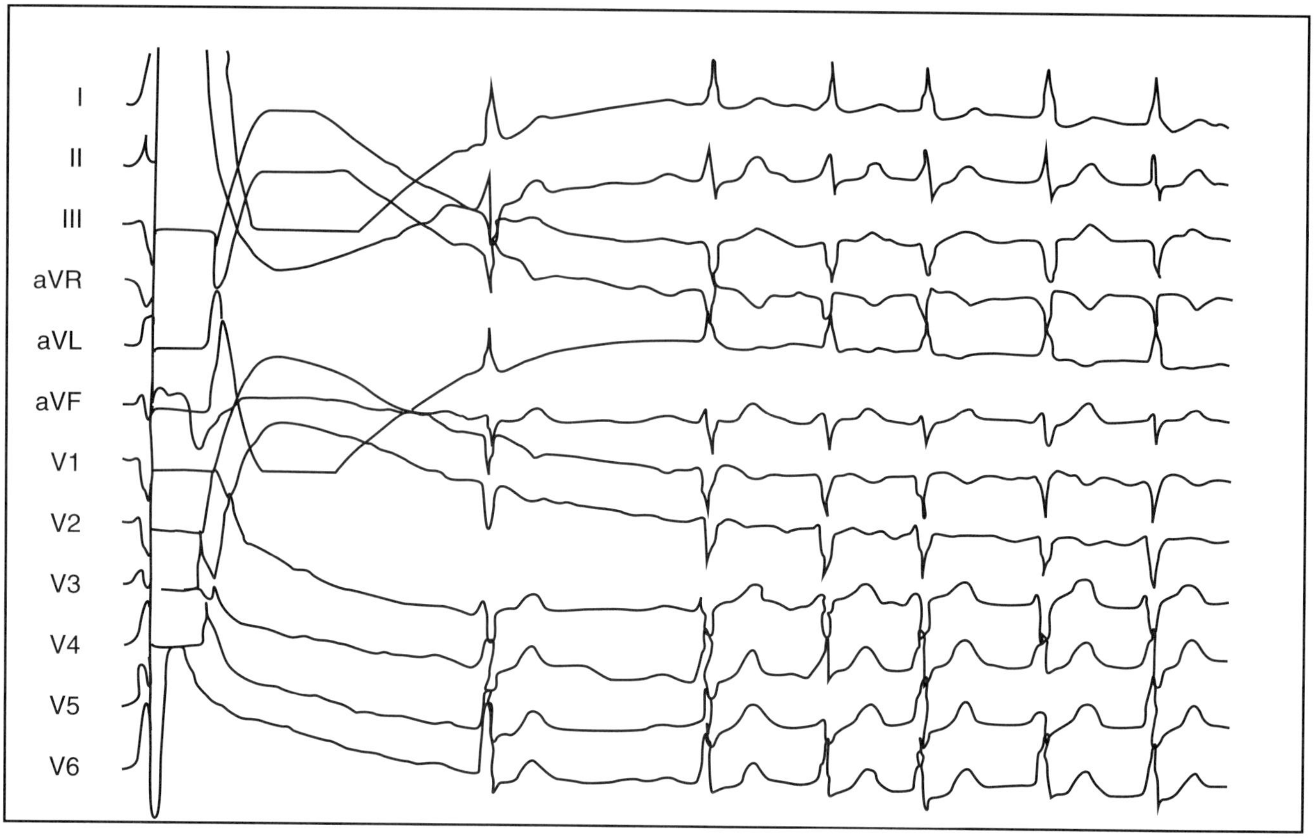

FIGURE 41-14 Early recurrence of atrial fibrillation after shock.

episodes were terminated. Recurrence of AF within 4 minutes following successful defibrillation was observed in 27% of AF events and occurred in 51% of patients (Fig. 41-14). Considering only those episodes where sustained sinus rhythm was restored, the clinical efficacy of the defibrillation therapy for these spontaneous episodes was 86.3%. By October 1998, more than 200 Metrix systems had been implanted worldwide. Safety and efficacy data for the first 186 implants confirmed the initial experience. Most Metrix patients had highly symptomatic drug-refractory AF, usually without cardiovascular disease. The conversion to sinus rhythm and the clinical success rate are similar during long-term experience with an electrical success rate of 93% and a clinical efficacy rate of 84%.

The atrial defibrillator also provided a programmable, patient-controlled mode. As of August 1998, 57 patients had their Metrix devices programmed to this mode. This option enabled these patients to deliver therapy when and where it was appropriate and convenient for them. Two hundred and seventy-six spontaneous episodes of AF were treated in this manner. Most patients chose to deliver shocks with no analgesia or sedation. Most AF episodes treated outside the hospital terminated with a single shock, with a mean number of 1.7 shocks per episode. No inappropriate therapy was delivered in this mode. Patients administering therapy outside the hospital successfully terminated 81% of spontaneous AF episodes versus 84% in the hospital.

Although the initial atrial cardioverter permitted shocks of up to 6 joules and ventricular pacing, higher energy requirements in many patients or when other lead systems were used, coupled with the risk of ventricular proarrhythmia without ventricular defibrillation and pain with atrial defibrillation, limited wider adoption. Although ventricular proarrhythmia resulting from atrial defibrillation shocks was rare in animal and human studies, it had been documented in both experimental and clinical studies in diseased hearts. Thus, the initial clinical experience with atrial defibrillators was encouraging and suggested that effective and safe atrial defibrillation was feasible.

DUAL CHAMBER ATRIOVENTRICULAR DEFIBRILLATOR

Device Technology

The first generation atrial defibrillation device was succeeded by a commercially available dual chamber atrioventricular defibrillator.[68] The prototype device was the Jewel AF (Medtronic Inc. Minneapolis, MN), with a weight of 93 grams and a volume of 55 mL (Figs. 41-10 and 41-15). This device requires a two- or three-pacing/defibrillation lead system, including atrial pacing/sensing electrodes, ventricular pacing/sensing electrodes, atrial or superior venacaval high voltage electrode(s), and a ventricular high voltage electrode. An additional lead may be placed in the coronary sinus, if needed, to lower atrial defibrillation thresholds. As opposed to the Metrix device, the can of the Jewel AF device is an active electrode. This device, shown in Figure 41-10, permits atrial and ventricular antitachycardia pacing, cardioversion, and defibrillation. The most important new features of the Medtronic Jewel AF system and its successor the Gem 3 AT or the Guidant Vitality AVT devices include dual chamber pacing,

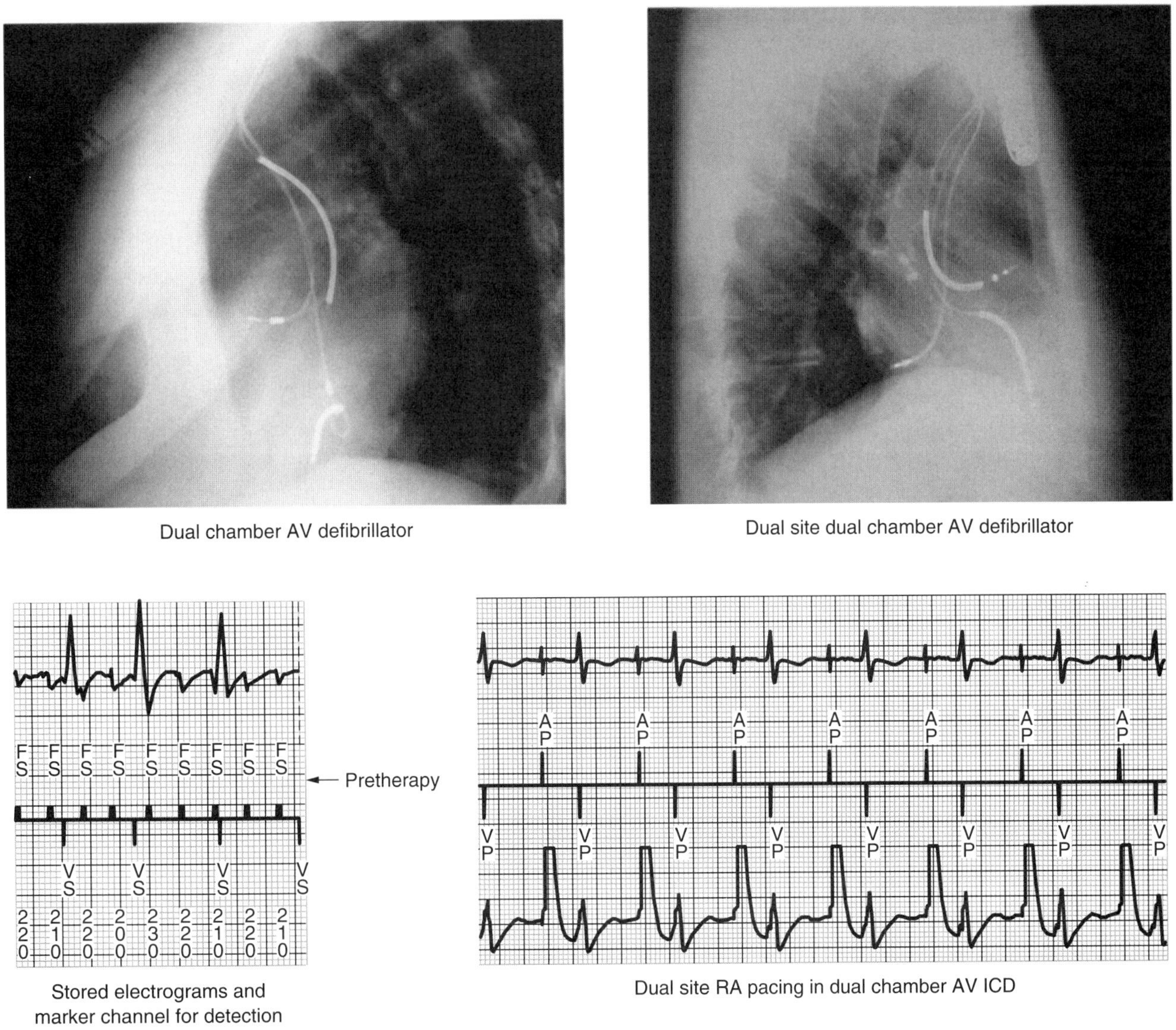

FIGURE 41-15 **Upper panel:** Lateral radiographs of atrioventricular (AV) defibrillator in situ. **Left:** Atrial and ventricular pacing and defibrillation leads are seen in situ. **Right:** A third coronary sinus ostial lead is present to achieve dual-site pacing with an atrioventricular implantable cardioverter defibrillator (ICD). **Lower panel:** Stored electrograms and marker channel in atrioventricular ICD **(left)**. The *top trace* shows atrial flutter with variable AV conduction as seen by device-detected intracardiac electrograms. The *middle trace* is a marker channel with upward deflections marking atrial potential detections and downward deflections marking ventricular potential detections. The *bottom trace* is a numeric record of atrial electrogram cycle lengths. *Right:* telemetered electrograms (bottom channel) marker channel (as in *left tracing*) and surface electrocardiogram (top channel) in dual site dual chamber AV defibrillator. Dual site atrial pacing is in progress in the overdrive mode.

a new dual chamber detection algorithm for rejection of supraventricular tachycardias, detection and painless treatment modalities of atrial arrhythmias, prevention strategies for atrial arrhythmias, and automatic or patient-activated atrial defibrillation with complete ventricular back-up defibrillation. The detection algorithm uses P:R pattern recognition and timing rules for rhythm classification and has been shown to provide improved dual chamber detection.[68-71] Figure 41-16 presents the hierarchical structure of one of the dual chamber detection algorithms. Atrial and ventricular AV pattern analysis and cycle lengths are analyzed. The ventricular tachyarrhythmia detection conditions are examined first. Specificity-enhancing rules make use of AV pattern recognition to determine if a high ventricular rate is due to AF, atrial tachycardia, sinus tachycardia or some other 1:1 supraventricular tachycardia. If the criteria for either ventricular tachycardia or ventricular fibrillation are satisfied and no specificity-enhancing rule is satisfied, ventricular tachyarrhythmia therapy is initiated. This therapy is also initiated if rules for dual tachycardia detection are satisfied (e.g., ventricular tachyarrhythmia arising during the presence of a supraventricular tachyarrhythmia). If a ventricular tachyarrhythmia or dual tachycardia are not detected, the supraventricular tachyarrhythmia detection conditions are then evaluated. These rules are designed to detect and discriminate between AF and other atrial tachycardias without 1:1 conditions, using atrial cycle length thresholds and P:R pattern information. The median atrial cycle length is used to define three detection zones: the AF detection zone, the atrial tachycardia zone, and the autodiscrimination zone formed by the overlap of the two zones. When the cycle length for a detected rhythm is in the autodiscrimination zone, the rhythm is classified as AF if the atrial cycle lengths are irregular, and atrial tachycardia if the cycle lengths are regular. The criterion for atrial cycle length regularity is evaluated on each ventricular event. P:R pattern information is incorporated into the algorithm by an evidence counter that uses a pattern recognition algorithm that is specific for supraventricular tachycardia without 1:1 atrioventricular (AV) conduction. This counter has up-down counting properties and is incremented on each ventricular event where the AV pattern shows evidence of a supraventricular tachyarrhythmia. If there is no evidence, the evidence counter is decremented. The counter has two stages of operation. The first stage is preliminary detection, which defines the start of an AT/AF episode, when the evidence counter reaches a predefined threshold. Once the start of an episode is defined, the episode

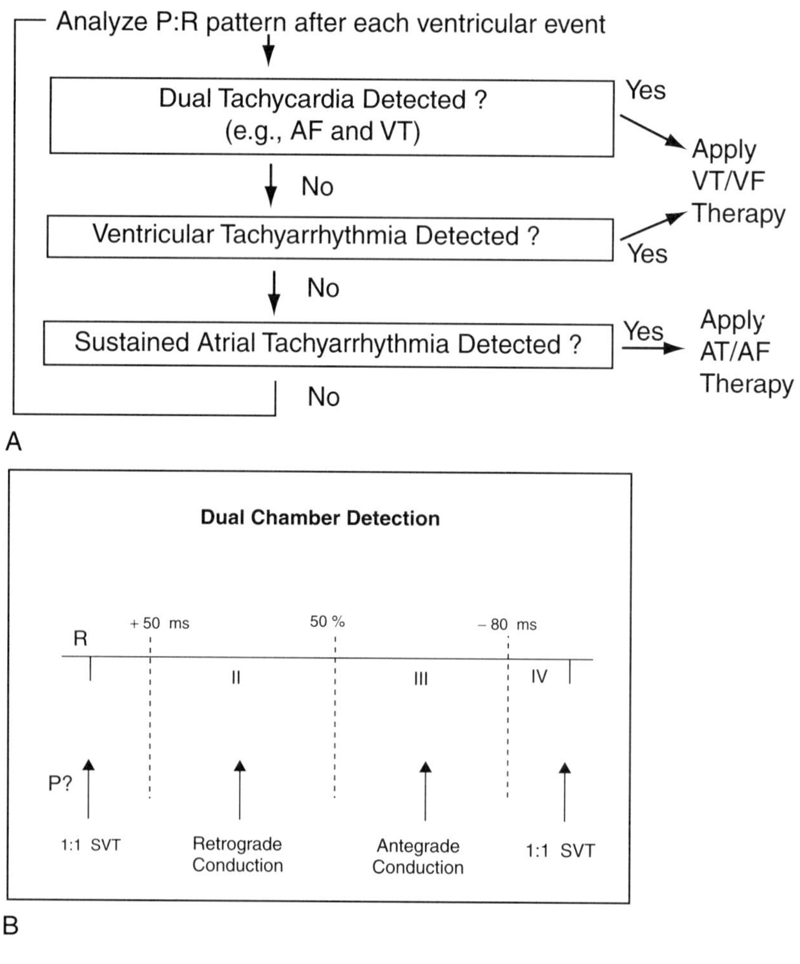

FIGURE 41-16 **A,** Hireachy of dual chamber detection in the Jewel AF. The device is constantly monitoring the rhythm in the atrium as well as in the ventricle. When a dual tachycardia or a ventricular tachyarrhythmia is detected, the device will deliver a ventricular therapy. If not, the device looks for possible atrial arrhythmia and it may deliver an atrial therapy. AT = atrial tachycardia, AF = atrial fibrillation, VT = ventricular tachycardia, VF = ventricular fibrillation. **B,** Definition of time intervals for interval codes. The detection algorithm looks for the position of the P wave relative to the R wave. The R–R interval is divided into four time intervals, I-IV. Atrial senses are expected to occur in intervals I and IV for junctional rhythms. Interval II (normal sinus rhythm, sinus tachycardia) is the normal antegrade conduction interval which extends from the midpoint of the R–R interval to the beginning of interval IV. Atrial events resulting from retrograde conduction of a ventricular event are expected to be sensed in interval II. So called "couple codes" based on the interval codes of the current and previous R–R intervals are used to describe rhythm syntaxes. Sequences of those couple codes are used to classify the rhythm. SVT, supraventricular tachycardia.

duration timer begins incrementing and the evidence counter switches to the sustained detection stage. During this stage, the evidence counter employs criteria that are less strict, allowing for some atrial undersensing.

Patient Selection and Clinical Outcomes

Initial clinical studies were conducted in patients with lethal ventricular tachyarrhythmias who may or may not have coexisting AF. In the initial studies, 211 patients with a history of ventricular tachyarrhythmias only (n = 53) or both a history of ventricular tachyarrhythmias and AF (n = 158) were enrolled in a worldwide study from 37 centers.[71] For inclusion in the latter group, patients had to have ECG-documented spontaneous atrial tachyarrhythmias occurring within the year before implant and a history of such arrhythmias.

Overall, 90% of the patients enrolled in the study received a two-lead defibrillation system, and less than 8% received a three-lead defibrillation system. The mean atrial defibrillation threshold in a representative subgroup of 42 patients was 6.1 ± 4.3 joules. The positive-predictive value of the detection algorithm was 93% for AF and AT and 88% for ventricular tachyarrhythmias. During the course of the study, 21 patients experienced a total of 165 spontaneous, appropriately treated AT episodes where the last therapy to treat the episode was either anti-tachycardia pacing (99 episodes) or high frequency burst (i.e., 50-Hertz pacing) (66 episodes). The success rate for spontaneous atrial tachycardia episodes as defined by the device was 86% with the use of antitachycardia pacing. An additional 35% of episodes were terminated by 50-Hertz pacing. Only 11 patients had 22 episodes of AF as classified by the device, of which 20 were successfully terminated by an atrial shock therapy, approaching an efficacy rate of 91%. During 410 R wave synchronized atrial shocks delivered for both spontaneous and induced AF episodes, no ventricular proarrhythmia was seen.[72,73]

Multicenter trials have evaluated the use of the dual chamber AV defibrillator in patients with recurrent atrial tachyarrhythmias who had other indications for a ventricular implantable cardioverter defibrillator (ICD) insertion. Device therapy resulted in a significant reduction in atrial tachyarrhythmia burden.[74,75] One of these other trials also recently demonstrated an improvement in quality of life scores.[75] Moreover, these improvements were not attenuated by shock therapy. However, no control arm existed in this trial. More recent data suggest a reduction in AF hospitalizations with these devices.[76] Thus, the future of this technology in a hybrid therapy format remains cautiously promising. Dual chamber atrioventricular defibrillators are currently approved for use in patients with drug-refractory, symptomatic atrial fibrillation, and coexisting ventricular tachyarrhythmias, or in patients with refractory atrial fibrillation. The outcomes in various trials using AV defibrillators are summarized in Table 41-4.

SPECIFIC DEVICE FEATURES

Device Monitoring of AF

An important feature of new devices for AF therapy is their ability to provide extensive monitoring capabilities for AF and atrial tachyarrhythmia detection. Figure 41-17 shows data logs from implantable pacemakers or dual chamber AV defibrillators that can provide the clinical practitioner with extensive time stamped and quantitative information on AF events. The practitioner can program the arrhythmia detection criteria. Duration, frequency, and total time elapsed in the arrhythmia (referred to as the arrhythmia burden) can be monitored between clinic visits or after interventions. This feature allows quantitation of asymptomatic and

TABLE 41-4 Studies on Dual Atrioventricular Defibrillators

Trial	Number of Patients	Follow-up period (Mos.)	Patient Population	LV EF%	Efficacy of Pacing for AT/AF (%)	Efficacy of Defibrillation for AT/AF (%)	Lead Dislodgment (%)	Survival (%)
Atrioverter-Metrix Investigators (1999)	179	9	Recurrent drug refractory AF	58±11	—	86.3	4	—
JEWEL VT/VF/AF Trial World Wide Investigators (2001)	293	7.9±4.7	Qualifying ventricular arrhythmia with two documented AT episodes in the last year	38±17	45.4 AT = 59 AF = 30	74	2.4	91
JEWEL AF only Trial (2001)	144	12	Drug refractory AF with no VT/VF	51±18	58.4 AT = 49 AF = 23	86.7	6.8	97.6
Italian study (2002)	112	11±9	68% with IHD and 55% with prior AT	40±11	37.9 AT = 70.9 AF = 24.3	76	1.8	95.5
GEM III AT World Wide Investigators (2002)	151	2.6±1.3	—	38±16	35 AT = 40 AF = 26	Used only for two induced episodes	Not reported	—

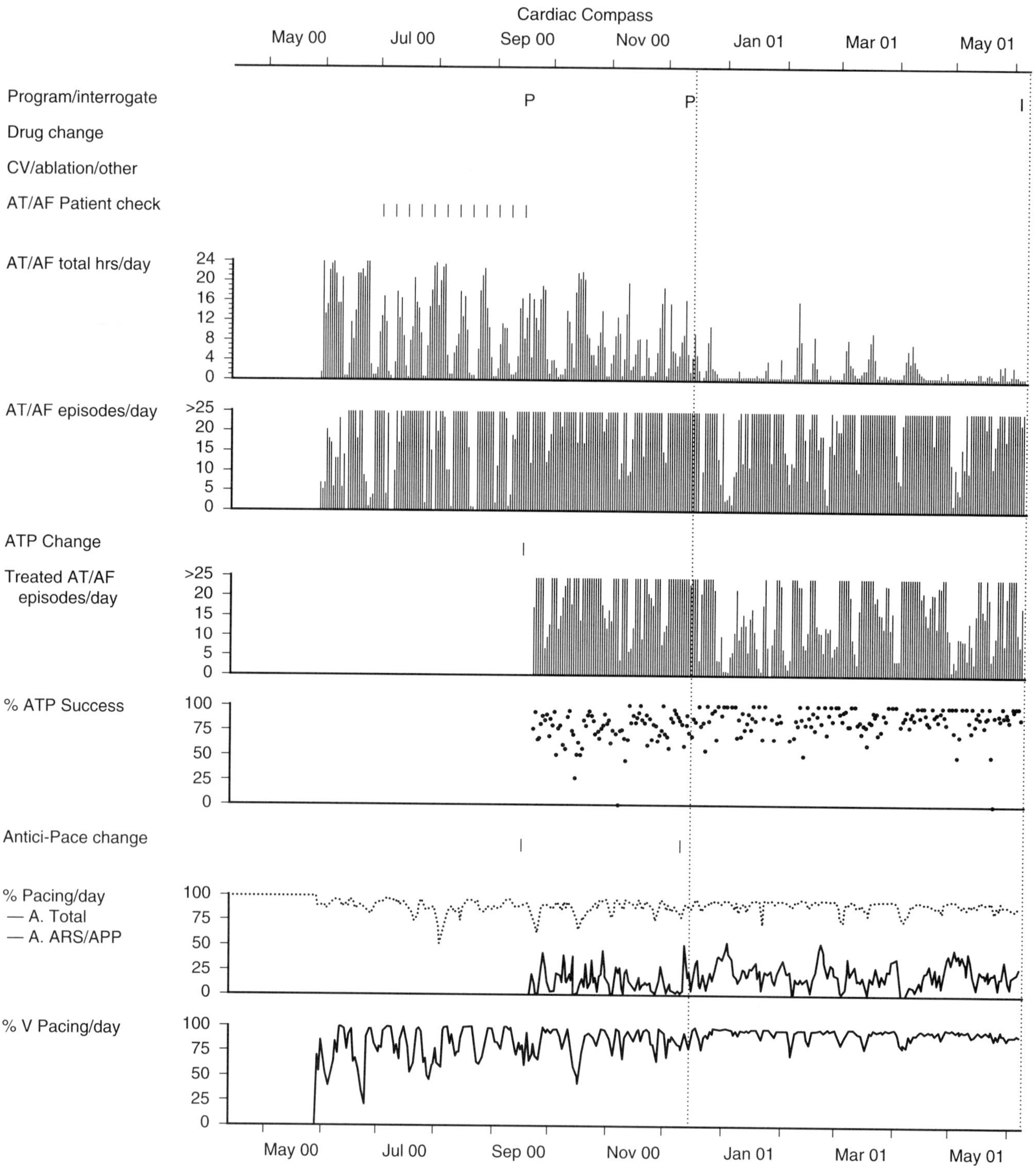

FIGURE 41-17 Implantable device data log report stored in device memory for quantitation of AF burden (hrs/day) frequency (episodes/day) anti-tachycardia therapy outcomes, percent success, percent pacing in atrium and ventricle. These reports make management of asymptomatic AF possible and provide important insights into AF event behavior in an individual patient.

symptomatic AF events, and permits accurate assessment of rhythm control after therapy.

Patient Activator

This handheld device has evolved from a simple activator that could initiate therapies as shown in Figure 41-10, to a device that allows patients to know their cardiac rhythm status and activate therapies at home or proceed to a medical facility for further therapy. This allows for increased patient control of their arrhythmia management and permits outpatient therapy in long-term management. It is expected that this feature will reduce the hospitalizations or emergency department visits now that frequently punctuate the long-term management of AF patients.

HYBRID THERAPY

Devices have been increasingly used in the "hybrid therapy" strategy for the management of atrial fibrillation (Table 41-5). Although the most frequent combination therapy has been with antiarrhythmic drugs, atrial ablation techniques have also been combined with device and drug therapy.[50,78] The former strategy has been detailed in the clinical evaluation trials for these devices as well as the observational studies for dual site right atrial pacing. Catheter-based linear ablation with drug therapy and dual site/high right atrial pacing has also been examined in several reports. There is a decrease in progression to permanent AF, restoration of rhythm control, and resolution of persistent and permanent AF in patients who underwent right atrial maze and atrial flutter ablation procedures in conjunction with device therapy (see Table 41-5). These studies suggest that the use of a hybrid approach may be effective for rhythm management. For a more detailed treatment of this subject, the reader is referred elsewhere.[19,24]

Role of Implantable Device Therapy in Atrial Fibrillation

The selection of device therapy and the preferred technology should depend on the type of AF, clinical characteristics as atrial or ventricular systolic function, the presence of structural heart disease, age, and the presence of coexisting bradycardias. Devices may be implanted empirically in patients with bradycardias. Based on current data, dual site pacing systems are preferable for secondary prevention of AF. Electrophysiological studies in individual patients with refractory AF can characterize the triggers and substrate and identify monomorphic tachycardias responsible for AF initiation and maintenance. Mapping guided ablation or anatomically based linear ablation for atrial compartmentalization or isolation of triggers or substrate can be performed prior to device implant. AF populations can now be divided into specific categories for purposes of device therapy:

Primary Prevention of Atrial Fibrillation

In patients with symptomatic bradyarrhythmias due to sinus node dysfunction, the relative merits of high right atrial pacing have been demonstrated. In addition, bi-atrial pacing can be beneficial in AF prevention after coronary bypass surgery.

Secondary Prevention of Atrial Fibrillation

In patients with refractory symptomatic paroxysmal AF, single site atrial pacing from the high right atrial or septum has not provided clinically relevant benefit. Hybrid device therapy using dual site right atrial pacing in combination with antiarrhythmic drug therapy has shown more efficacy in maintenance of sinus rhythm in this population.

In drug-refractory persistent and permanent AF patients, another hybrid approach using a combination of antiarrhythmic medications, catheter-based right or left atrial linear ablation, and dual site right atrial pacing using pacemakers or atrioventricular defibrillators can restore or improve rhythm control. In patients with symptomatic drug-refractory AF and coexisting ventricular arrhythmias, dual chamber atrioventricular defibrillators have been applied with the ability to manage both arrhythmias—usually with the use of adjunctive antiarrhythmic pharmacologic therapy.

In summary, device therapy now offers new opportunities for restoration of rhythm control in patients with drug-refractory AF. These often require combination of devices with preexisting drug therapy or catheter ablation methods.

TABLE 41-5 Hybrid Therapy Strategy Trials Incorporating Device Therapy

Series	Device Technique	Pts Population	Mean	Follow-up (mos)	Rhythm Control	Complications %
Metrix	Atrial ICD	186	Persist AF	9	84%	6 (device related)
Jewel AF	A-V ICD	537	VT/VT+AF	11±8	9	?
Madan	Dual RA pacing	113	PAF Persist/Perm AF	30±23	90% at 5 yrs	Early = 3.5% Late = 2.7%
D'Allones	Biatrial pacing	64	PAF	33	64%	?
Prakash	DAP+Isth ABL	40	PAF/AFL	26±14	75% at 3 yrs	
Filipecki	AP/ICD+RA Maze	25	Persist/Perm AF	17±10	75% at 18 mos	

Future Directions

Due to its extensive electrical therapy options and long-term AF monitoring capability, the future of device technology in a hybrid therapy format for AF management is quite promising. Dual chamber AV defibrillators are approved for use in patients with drug-refractory and symptomatic AF and in patients with coexisting symptomatic atrial and ventricular tachyarrhythmias. Newer iterations include prevention algorithms, such as continuous atrial pacing for AF prevention. Enhanced monitoring capabilities include patient notification of atrial and ventricular arrhythmia detection. Device-based testing is available and a hand-held patient activator permits monitoring of AF and delivery of shock therapy on demand by the patient or physician. These options offer the opportunities for reducing AF hospitalizations and safe outpatient management of recurrent AF. Devices may offer important new directions for improvement in AF management.

REFERENCES

1. Wolf PA, Abbott RD, Kannel WB: Atrial fibrillation as an independent risk factor for stroke: The Framingham study. Stroke 1991;22:983-988.
2. Bialy D, Lehmann MH, Schumacher DN, et al: Hospitalization for arrhythmias in the United States: Importance of atrial fibrillation. [Abstract] J Am Coll Cardiol 1992;19:41A.
3. Benjamin EJ, Wolf PA, D'Agostino RB, et al: Impact of atrial fibrillation on the risk of death: The Framingham Heart Study. Circulation 1998 Sep 8;98:946-952.
4. Van Gelder IC, Hagens VE, Bosker HA, et al: Rate Control versus Electrical Cardioversion for Persistent Atrial Fibrillation Study Group: A comparison of rate control and rhythm control in patients with recurrent persistent atrial fibrillation. N Engl J Med 2002;347:1834-1840.
5. Carlsson J, Miketic S, Windeler J, et al: STAF Investigators: Randomized referenced trial of rate-control versus rhythm-control in persistent atrial fibrillation: The Strategies of Treatment of Atrial Fibrillation (STAF) study. J Am Coll Cardiol 2003;41:1690-1696.
6. Wyse DG, Waldo AL, DiMarco JP, et al: Atrial Fibrillation Follow-up Investigation of Rhythm Management (AFFIRM) Investigators: A comparison of rate control and rhythm control in patients with atrial fibrillation. N Engl J Med 2002;347:1825-1833.
7. Saksena S, Prakash A, Krol RB, Shankar A: Regional endocardial mapping of spontaneous and induced atrial fibrillation in patients with heart disease and refractory atrial fibrillation. Am J Cardiol 1999; 84:880-889.
8. Shinagawa K, Shi YF, Tardif JC, et al: Dynamic nature of atrial fibrillation substrate during development and reversal of heart failure in dogs. Circulation 2002;105:2672-2678.
9. Shinagawa K, Li D, Leung TK, Nattel S: Consequences of atrial tachycardia-induced remodeling depend on the preexisting atrial substrate. Circulation 2002;105:251-257.
10. Saksena S, Skadsberg N, Rao BH: Biatrial and three-dimensional mapping of spontaneous atrial arrhythmias in patients with refractory atrial fibrillation. J Cardiovasc Electrophysiol 2005 (in press).
11. Murgatroyd FD, Nitzsche R, Slade AK, et al: A new pacing algorithm for overdrive suppression of atrial fibrillation. Chorus Multicentre Study Group. Pacing Clin Electrophysiol 1994;1966-1973.
12. Mehra R, Hill MRS: Prevention of atrial fibrillation/flutter by pacing techniques. In Saksena S, Luderitz B (eds): Interventional Elecctrophysiology: A Textbook, 2nd ed. Armonk, NY, Futura, 1996, pp 521-540.
13. Prakash A, Delfaut P, Krol RB, Saksena S: Regional right and left atrial activation patterns during single and dual-site atrial pacing in patients with atrial fibrillation. Am J Cardiol 1998;82: 1197-1204.
14. Hesselson A, Wharton JM: Dual site pacing suppression of focal atrial fibrillation. [abstract]. Pacing and Clin Electrophysiol 2001; 24:554.
15. Prakash A, Saksena S, Hill M, et al: Acute effects of dual-site right atrial pacing in patients with spontaneous and inducible atrial flutter and fibrillation. J Am Coll Cardiol 1997;29:1007-1014.
16. Prakash A, Saksena S, Ziegler P, et al: Dual site atrial pacing for prevention of atrial fibrillation (DAPPAF) trial: Echocardiographic evaluation of atrial and ventricular function during a randomized trial of support, high right atrial and dual site right atrial pacing. [abstract]. Pacing Clin Electrophysiol 2001;24:553.
17. Camm AJ, Hoffmann E, Janko S, et al: The hotline sessions of the 23rd European Congress of Cardiology. Eur Heart J 2001;22: 2033-2037.
18. Saksena S, Giorberidge I, Mehra R, et al: Electrophysiology and endocardial mapping of induced atrial fibrillation in patients with spontaneous atrial fibrillation. Am J Cardiol 1999;83: 187-193.
19. Saksena S, Madan N: Hybrid therapy for atrial fibrillation: Algorithms and outcome. J Int Card Electrophysiol 2003;9: 235-247.
20. Boccadamo R, Toscano S: Prevention and interruption of supraventricular tachycardia with antitachycardia pacing. In Luderitz B, Saksena S: Interventional Electrophysiology. Futura, Mt. Kisco, NY, 1991, p 213.
21. Filipecki A, Saksena S, Prakash A, Philip G: Right and left atrial initiation of atrial fibrillation: Three dimensional atrial activation analysis. Eur Heart J 2002;2:1315.
22. Giorgberidze I, Saksena S, Mongeon L, et al: Effects of high frequency atrial pacing in atypical atrial flutter and atrial fibrillation. J Interv Card Electrophysiol 1997;1:111-123.
23. Prakash A, Saksena S, Krol R, Philip G: Right and left atrial activation during external direct-current cardioversion shocks delivered for termination of atrial fibrillation in humans. Am J Cardiol 2001:87:1080-1088.
24. Dunbar DN, Tobler HG, Fetter J, et al: Intracavitary electrode catheter cardioversion of atrial tachyarrhythmias in the dog. J Am Coll Cardiol 1986;7:1015.
25. Kumagai K, Yamanouchi T, Tashiro N, et al: Low energy synchronous transcatheter cardioversion of atrial flutter/fibrillation in the dog. J Am Coll Cardiol 1990;16:97.
26. Keane D, Boyd E, Anderson D, et al: Comparision of biphasic and monophasic waveforms in epicardial atrial defibrillation. J Am Coll Cardiol 1994;24:171.
27. Cooper RAS, Alferness CA, Smith WM, Ideker RE: Internal cardioversion of atrial fibrillation in sheep. Circulation 1993; 87:1673.
28. Powell AC, Garan H, McGovern BA, et al: Low energy conversion of atrial fibrillation in sheep. J Am Coll Cardiol 1992;20:707.
29. Ortiz J, Sokoloski MC, Niwano S, et al: Successful atrial defibrillation using temporary pericardial electrodes. J Am Coll Cardiol 1994;23:125A.
30. Andersen HR, Nielsen JC, Thomsen PEB, et al: Long-term follow-up of patients from a randomized trial of atrial versus ventricular pacing for sick sinus syndrome. Lancet 1997;350:1210-1216.
31. Lamas GA, Lee K, Sweeney MO, et al: Ventricular pacing or dual-chamber pacing for sinus-node dysfunction. The Mode Selection Trial (MOST) in sinus node dysfunction. N Engl J Med 2002;346:1854-1862.
32. Connolly SJ, Kerr CR, Gent M, et al: Effects of physiologic pacing versus ventricular pacing on the risk of stroke and death due to cardiovascular causes. Canadian Trial of Physiologic Pacing Investigators. N Engl J Med 2000;342:1385-1391.
33. Daoud EG, Dabir R, Archambeau M, et al: Randomized, double blind trial of simultaneous right and left atrial epicardial pacing for prevention of post-open heart surgery atrial fibrillation. Circulation 2000;102:761-765.
34. Fan K, Lee KL, Chiu CS, et al: Effects of biatrial pacing in prevention of postoperative atrial fibrillation after coronary artery bypass surgery. Circulation 2000;102:755-760.
35. Spitzer SG, Gazarek S, Wacker P: Pacing the atria in sick sinus syndrome trial: Preventive strategies for atrial fibrillation. PASTA trial group. Pacing Clin Electrophysiol 2003;26:268-271.

36. Gillis AM, Connolly SJ, Lacombe P, et al: Randomized crossover comparison of DDDR versus VDD pacing after atrioventricular junction ablation for prevention of atrial fibrillation: The atrial pacing peri-ablation for paroxysmal atrial fibrillation (PA3) study investigators. Circulation 2000;102:736-742.
37. Saksena S, Sulke N, Manda V, et al: Worldwide Jewel AF Investigators: Reduction in frequency of atrial tachyarrhythmia episodes using novel prevention algorithms of an atrial pacemaker defibrillator. PACE 2000;23:581.
38. Funck RC, Adamec R, Lurje L, et al: On behalf of the PROVE study group. Atrial overdriving is beneficial in patients with atrial arrhythmias: First results of the PROVE study. Pacing Clin Electrophysiol 2000;23:1891-1893.
39. Delfaut P, Saksena S, Prakash P, Kroll R: Long-term outcome of patients with drug-refractory atrial flutter and fibrillation after single and dual-site right atrial pacing for arrhythmia prevention. Am Coll Cardiol 1998;32:1900-1908.
40. Israel CW, Hugl B, Unterberg C, et al: AT500 Verification Study Investigators. Pace-termination and pacing for prevention of atrial tachyarrhythmias: Results from a multicenter study with implantable device for atrial therapy. J Cardiovasc Electrophysiol 2001;12:1121-1128.
41. Lee MA, Weachter R, Pollak S, et al: The effect of atrial pacing therapies on atrial tachyarrhythmia burden and frequency: Results of a randomized trial in patients with bradycardia and atrial tachyarrhythmias (ATTEST). J Am Coll Cardiol 2003;41: 1926-1932.
42. Ricci R, Santini M, Puglisi A, et al: Impact of consistent atrial pacing algorithm on premature atrial complexes number and paroxysmal atrial fibrillation recurrences in brady-tachy syndrome: A randomized prospective cross-over study. J Interv Card Electrophysiol 2001;5:33-44.
43. Carlson MD, Gold MR, Ip J, et al: For the ADOPT-A investigators. Dynamic atrial overdrive pacing decreases symptomatic atrial arrhythmia burden in patients with sinus node dysfunction. Circulation 2001;23:383.
44. Bailin SJ, Adler S, Guidici M: Prevention of chronic atrial fibrillation by pacing in the region of Bachmann bundle: Results from a multicenter randomized trial. J Cardiovasc Electrophysiol 2001;12:912-917.
45. Padeletti L, Pieragnoli P, Ciapetti C, et al: Randomized crossover comparison of right atrial appendage pacing versus interatrial septum pacing for prevention of paroxysmal atrial fibrillation in patients with sinus bradycardia. Am Heart J 2001;142: 1047-1055.
46. Lau CP, Tse HF, Yu CM, et al: For the New Indication for Preventive Pacing in Atrial Fibrillation (NIPP-AF) Investigators: Dual-site atrial pacing for atrial fibrillation in patients without bradycardia. Am J Cardiol 2001;88:371-375.
47. Mabo P, Paul V, Jung W, et al: Biatrial synchronous pacing for atrial arrhythmia prevention: The SYMBIAPACE study. Eur Heart J 1999;20:4.
48. Saksena S, Prakash A, Ziegler P, et al: For the DAPPAF investigators. The Dual-site Atrial Pacing for Prevention of Atrial Fibrillation (DAPPAF) trial: Improved suppression of recurrent atrial fibrillation with dual site atrial pacing and antiarrhythmic drug therapy. J Am Coll Cardiol 2002;40:1140-1150.
49. Ramdat Misier AR, Beukema WP, Oude Luttikhuis HA, et al: Multisite atrial pacing: An option for atrial fibrillation prevention? Preliminary results of the Dutch dual-site right atrial pacing for prevention of atrial fibrillation study. Am J Cardiol 2000; 86:K20-24.
50. Madan N, Saksena S: Long-term rhythm control of drug-refractory atrial fibrillation with "hybrid therapy" incorporating dual site right atrial pacing, antiarrhythmic drugs and right atrial ablation. Am J Cardiol 2004;100:569-575.
51. Leclercq JF, De Sisti A, Fiorello P, et al: Is dual site better than single site pacing in the prevention of atrial fibrillation. PACE 2000;23:2101-2107.
52. D'Allonnes GR, Pavin D, Leclercq C, et al: Long term effects of biatrial synchronous pacing to prevent drug-refractory atrial tachyarrhythmia: A nine-year experience. J Cardiovasc Electrophysiol 2000;11:1081-1091.
53. Mirowski M, Mower MM, Langer AA: Low energy catheter cardioversion of atrial tachyarrhythmias. Clin Res 1974;22:290.
54. Levy S, Lacombe P, Cointe R, Bru P: High energy transcatheter cardioversion of chronic atria fibrillation. J Am Coll Cardiol 1988;11:151-157.
55. Johnson EE, Yarger MD, Wharton JM: Monophasic and biphasic waveform for low energy internal cardioversion of atrial fibrillation in humans. Circulation 1993;88(suppl 1):I-592.
56. Saksena S, Mongeon L, Krol R, et al: Clinical efficacy and safety of atrial defibrillation using current non-thoracotomy endocardial lead configurations: A prospective randomized study. J Am Coll Cardiol 1994;23:125A.
57. Prakash A, Saksena S, Krol RB, Philip G: Right and left atrial activation during external direct-current cardioversion shocks delivered for termination of atrial fibrillation in humans. Am J Cardiol 2001;87:1080-1088.
58. Murgatroyd FD, Slade AK, Sopher SM, et al: Efficacy and tolerability of transvenous low energy cardioversion of paroxysmal atrial fibrillation in humans. J Am Coll Cardiol 1995:1347-1353.
59. Levy S, Ricard P, Lau CP, et al: Multicenter low energy transvenous atrial defibrillation (XAD) trial results in different subsets of atrial fibrillation. J Am Coll Cardiol 1997;29:750-755.
60. Saksena S, Chandran P, Shah Y, et al: Comparative efficacy of transvenous cardioversion and pacing in patients with sustained ventricular tachycardia: A prospective randomized crossover study. Circulation 1985;72:153.
61. Nathan AW, Bexter RS, Spurrell RA, Camm AJ: Internal transvenous low energy cardioversion for treatment of cardiac arrhythmias. Br Heart J 1984;52:377.
62. Lau CP, Tse HF, Lok NS, et al: Initial clinical experience with an implantable human atrial defibrillator. PACE 1977:20:220-225.
63. Jung W, Wolpert C, Esmailzadeh B, et al: Specific considerations with the automatic implantable atrial defibrillator. J Cardiovasc Electrophysiol 1998;9:S193-S201.
64. Jung W, Wolpert C, Esmailzadeh B, et al: Clinical experience with implantable atrial and combined atrioventricular defibrillators. J Interv Card Electrophysiol 2000:4:185-195.
65. Sra JS, Maglio C, Dhala A, et al: Feasibility of atrial fibrillation detection and use of a preceding synchronization interval as a criterion for shock delivery in humans with atrial fibrillation. J Am Coll Cardiol 1996;28:1532-1538.
66. Wellens HJJ, Lau CP, Lüderitz B, et al: For the Metrix Investigators: The atrioverter, an implantable device for treatment of atrial fibrillation. Circulation 1998;98:1651-1656.
67. Timmermans C, Levy S, Ayers GM, et al: For the Metrix Investigators: Spontaneous episodes of atrial fibrillation after implantation of the Metrix Atrioverter: Observations on treated and nontreated episodes. J Am Coll Cardiol 2000;35:1428-1433.
68. Jung W, Lüderitz B: Implantation of a new arrhythmia management system in patients with supraventricular and ventricular tachyarrhythmias. Lancet 1997;349:853-854.
69. Swerdlow CD, Schöls W, Dijkman B, et al: For the World Wide Jewel AF Investigators: Detection of atrial fibrillation and flutter by a dual-chamber implantable cardioverter-defibrillator. Circulation 2000;101:878-885.
70. Ricci R, Pignalberi C, Disertori M, et al: Efficacy of a dual chamber defibrillator with atrial antitachycardia functions in treating spontaneous atrial tachyarrhythmias in patients with life threatening ventricular tachyarrhythmias. Eur Heart J 2002; 1471-1479.
71. Schoels W, Swerdlow CD, Jung W, et al: Worldwide clinical experience with a new dual chamber implantable cardioverter-defibrillator system. J Cardiovasc Electrophysiol 2001;12:521-528.
72. Wharton M, Santini M: For the Worldwide Jewel AF Investigators. Treatment of spontaneous atrial tachyarrhythmias with the Medtronic 7250 Jewel AF: Worldwide Clinical Experience. Circulation 1998;98:I-190.
73. Jung W, Wolpert C, Tenzer D, et al: Incidence of spontaneous atrial tachyarrhythmias in patients treated with a combined atrioventricular defibrillator. Circulation 1998;98:I-190.
74. Adler SW, Wolpert C, Warman EN, et al: Efficacy of pacing therapies for treating atrial tachyarrhythmias in patients with ventricular arrhythmias receiving a dual-chamber implantable cardioverter defibrillator. Circulation 2001;104:887-892.
75. Friedman PA, Dijkman B, Warman EN, et al: Atrial therapies reduce atrial arrhythmia burden in defibrillator patients. Circulation 2001;104:1023-1028.

76. Ricci R, Quesada A, Carlo P, et al: Dual defibrillator improves quality of life and dercreases hospitilizations in patients with drug refractory atrial fibrillation. J Interv Card Electrophysiol 2004;10:85-92.
77. Prakash A, Saksena S, Krol RB, et al: Catheter ablation of inducible atrial flutter in combination with atrial pacing and antiarrhythmic drugs (hybrid therapy) improves rhythm control in patients with refractory atrial fibrillation. J Interv Card Electrophysiol 2002;6:165-174.
78. Filipecki A, Saksena S, Prakash A, Philip G: Improved rhythm control with overdrive atrial pacing and right linear right atrial ablation in patients with persistent and permanent atrial fibrillation. Am J Cardiol 2002;6:165-172.

Chapter 42 Catheter Mapping Techniques

VIAS MARKIDES, ANTHONY W.C. CHOW,
RICHARD J. SCHILLING, and D. WYN DAVIES

The technique of mapping cardiac arrhythmias has evolved from preoperative catheter mapping followed by detailed analysis of the arrhythmia substrate at open-heart surgery, to one that achieves accurate delineation of an arrhythmia as an integral part of a therapeutic percutaneous ablation procedure. Although the purpose of mapping is to characterize and localize the arrhythmogenic substrate, this is achieved by a variety of different methods. These techniques differ in their complexity, and usually mirror the complexity of the target arrhythmia. Additionally, for many arrhythmias, such as typical atrial flutter, ablation is increasingly performed on anatomic as well as electrophysiological data. Table 42-1 summarizes the potential role of the various possible mapping techniques in different cardiac arrhythmias.

Conventional Contact Catheter Mapping

This is the most widely applied mapping method because it predominantly employs the information obtained from the catheter electrodes that are placed for the diagnostic portion of an electrophysiology procedure. To convert these data into an understanding of the arrhythmia requires operator experience and coordination with imaging for localization. Assuming these are available, this approach is adequate for arrhythmias arising in normal hearts when the substrate tends to be simple. In fact, when the preprocedural ECG provides enough information, as in some cases of Wolff-Parkinson-White syndrome, it may be possible to achieve adequate mapping with a solitary ablation catheter.

VASCULAR ACCESS

Access to the right heart is usually obtained through the femoral and/or the subclavian veins, the latter being particularly useful for catheterization of the coronary sinus. Retrograde catheterization through a femoral artery and across the aortic valve is the usual method to access the left ventricle and occasionally the left atrium. Both left-sided chambers can also be reached across the interatrial septum, either via a patent foramen ovale or, most often, by a trans-septal puncture.

FLUOROSCOPIC VIEWS

Although fluoroscopic systems generally provide two-dimensional information regarding catheter position, using multiple views and knowing how catheter manipulation will affect its position in three dimensions, an experienced operator can form a mental three-dimensional map of the cardiac chamber(s) where catheters are positioned. The frontal projection is particularly useful for advancing catheters from the access site to the heart and allows good visualization of the right ventricular outflow tract. In the right anterior oblique view, the mitral, tricuspid, and aortic valves are seen side on and can thus be crossed with relative ease. This view can also be used for positioning the His bundle catheter, which is seen in profile at the superior margin of the tricuspid ring. In the left anterior oblique projection, the mitral and tricuspid valve rings are seen face on next to each other, with the septum in the middle, the lateral tricuspid annulus anteriorly forming the right heart border, and the lateral mitral valve annulus posteriorly forming the left heart border. This view is ideal for mapping the mitral or tricuspid valve annuli, the septal, lateral, and basal parts of the left ventricle, and for intubating the coronary sinus.

CONVENTIONAL CATHETERS

Fixed curve quadripolar catheters are most commonly used for diagnostic electrophysiological studies. Deflectable diagnostic catheters are easier to manipulate, but significantly more expensive. The spacing between the electrode poles typically varies between 2 and 10 mm and can be equal or variable. Deflectable ablation catheters with different curves are available that reach and remain stable at various targets within cardiac chambers, such as the "Halo" catheters for recording around the tricuspid annulus, and the "Lasso" catheter developed for circumferential mapping of pulmonary veins (PVs). Catheter manipulation needs to be

TABLE 42-1 Potential Roles of Various Mapping Techniques

Arrhythmia	Contact Mapping	Baskets	Noncontact Mapping	Electroanatomic	LocaLisa	RPM	Comments
Focal Atrial Tachycardia	Sequential point-by-point mapping to find earliest site of activation relative to the P wave or reference atrial electrogram	Simultaneous mapping of atrium Electrogram timing and color maps analyzed	Global activation using a single beat of tachycardia Isopotential and isochronal maps and virtual electrograms analyzed to determine focus of origin	Sequential point-by-point acquisition during stable tachycardia (relative to reference catheter (e.g., CS) Isochronal maps	No data available	No data available	The noncontact system permits mapping of nonsustained and hemodynamically unstable arrhythmias
Macro-Reentrant Atrial Tachycardia	Point-by-point mapping to define the endocardial exit site Entrainment mapping to define the circuit and identify the narrowest part of the circuit	Simultaneous mapping of atrium Electrogram timing and color maps analyzed	Global isopotential map defines the endocardial exit Circuit visualized from map	Sequential point-by-point acquisition during stable tachycardia (relative to reference catheter (e.g., CS) Isochronal maps	Standard catheter mapping techniques to identify circuit Circuit borders marked and linear lesions created guided by the system	No data available	Multiple arrhythmias and unstable circuits make contact mapping and entrainment challenging
Atrial Flutter	Anatomic approach with confirmatory entrainment studies Analysis of double potentials, local electrograms and pacing maneuvers with multielectrode catheters to identify gaps in the ablation line	Simultaneous mapping of atrium and cavotricuspid isthmus Pacing to confirm isthmus block Color maps	Isopotential maps and virtual electrogram analysis may permit spot welding of gaps capable of conduction and high-density isthmus recording with a limited number of catheters	Tricuspid annulus (TA) and inferior vena cava (IVC) marked. Linear RF lesions between TA and IVC, guided and marked using system	No data available	Tricuspid annulus (TA) and inferior vena cave (IVC) marked.Linear RF lesions between TA and IVC, marked using system	Contact mapping and traditional approaches highly successful Noncontact mapping useful when multiple circuits possible and with difficulty in identifying gaps
Atrial Fibrillation	**Paroxysmal/Focal Atrial Fibrillation:** Multielectrode sequential mapping in the PVs of spontaneously occurring premature beats **Endocardial Maze:** Anatomy-based linear ablative lesions	**Paroxysmal/Focal Atrial Fibrillation:** Deployment inside a PV allows localization of focus and of PV-LA connections for PV isolation **Endocardial Maze:** Deployment in RA/LA may help guide formation of linear lesions	**Paroxysmal/Focal Atrial Fibrillation:** Global activation map of left atrium of a single spontaneous initiation of atrial fibrillation used to identify the PV sector of origin **Endocardial Maze:** Used to identify gaps and elucidate mechanism of post maze atrial flutters	**Paroxysmal/Focal Atrial Fibrillation:** Infrequent ectopy nonsustained nature of PV and atrial triggers makes mapping with this technique difficult. Used to guide creation of circumferential lesions in LA around PV ostia **Endocardial Maze:** Used to guide creation of linear lesions in LA and RA and circumferential LA lesions around PV ostia	No data available	No data available	Paucity of spontaneous initiators make noncontact mapping with the ability to discern origin with a single beat desirable; no contact-irritation ectopy Noncontact mapping cannot define the exact site of origin from within a PV, limited to ostial venous segments

							Role of mapping PV "spikes" in the absence of ectopy not well defined; utility of noncontact mapping for "spike" identification has not been systematically assessed
AVNRT/AVRT	Contact mapping of earliest activation, pathway potentials and pacing maneuvers to identify potential ablation sites for AVRT Fluoroscopic and slow pathway potential guided ablation for AVNRT	No data available	Retrograde slow pathway can be identified using noncontact mapping in AVNRT. Very little experience with noncontact mapping for AVRT or AVNRT due to success of conventional methods	Tricuspid or mitral annulus marked using system. Earliest ventricular activation around annulus during atrial pacing or antidromic tachycardia or earliest atrial activation during ventricular pacing or orthodromic tachycardia identified	**AVNRT:** Position of CS ostium and proximal His bundle marked. System used for catheter navigation	No data available	Contact mapping techniques highly successful; consequently there is little or no role for advanced mapping techniques
Ventricular tachycardia	Point-by-point mapping during tachycardia –reentrant VT: identify diastolic potentials; confirm position in diastolic pathway using entrainment mapping –focal/micro-reentrant VT: identify earliest site	Conventional mapping during tachycardia but with simultaneous signal recording from multiple splines and ability to pace/entrain from these Color maps	Global activation map of LV or RV Identify isthmus in reentrant VT Identify earliest site in focal/micro-reentrant VT	Voltage maps to help identify areas of scarring Sequential point-by-point acquisition during stable tachycardia –identify isthmus in reentrant VT –identify earliest site in focal/micro-reentrant VT Linear lesions at critical isthmus	No data available	Standard catheter mapping techniques to identify source or reentrant circuit System used for catheter localization and navigation	Noncontact mapping may be used to identify the diastolic pathway even in hemodynamically unstable and nonsustained VTs. Ablation may then be performed in sinus rhythm

Abbreviations: AVNRT, atrioventricular nodal reentry tachycardia; AVRT, atrioventricular reentry tachycardia; CS, coronary sinus; LA, left atrium; PV, pulmonary vein; RA, right atrium; RF, radiofrequency; RPM, realtime position management.

performed without force so that trauma and perforation of cardiac chambers, venous structures, and the arterial intima can be avoided.

RECORDING SYSTEMS

A number of commercially available recording systems accept inputs from surface ECG and intracardiac channels. Signals can be amplified and filtered over a wide and variable frequency range (from under 1 Hz to 200 kHz), digitized, displayed, and stored.

ELECTROGRAM RECORDINGS

The operator may choose to display unipolar or bipolar signals from a contact catheter. In the case of unipolar signals, the recording is taken between the relevant pole of the catheter as the positive input and an extracardiac reference as the negative one. Sensing of extraneous electrical *noise,* such as skeletal muscle activity and alternating current is reduced if the extracardiac reference electrode is intravascular rather than extracorporeal; the signal may be further improved if the reference electrode is placed on the intravascular portion of the catheter shaft.[1] A positive deflection is produced by a wavefront approaching a unipolar electrode, followed by a negative deflection as the wavefront passes away from the electrode. In the case of the bipolar signals, the recording is made between two closely spaced poles of a multipolar catheter, both of which are either in contact with, or close to, the endocardium. As such, bipolar recordings reflect the differences between the two unipolar electrograms that would have been recorded individually from the two poles.

Unipolar recordings have significant advantages, as well as disadvantages, over bipolar recordings. Not only do they allow more precise spatial localization, but examination of the morphology of unipolar signals can also provide valuable information. Unipolar electrograms recorded at the source of focal tachycardias have a QS morphology, whereas at distant sites, an RS morphology is seen. For this to be valid there needs to be adequate contact with the endocardium, and minimal high-pass filtering. When the tissue surrounding the unipolar electrode is diseased as, for example, in the case of ischemic ventricular tachycardia (VT), larger amplitude far-field signals from surrounding tissue may obscure local lower amplitude signals, thus rendering unipolar recordings difficult to interpret without high-pass filtering.

CATHETER MAPPING

Mapping involves recording of electrical signals from electrodes usually positioned at multiple sites in the heart. These are then analyzed to determine the activation timing of electrodes at key sites and to create a mental map of the activation sequence of the whole heart or relevant parts of it. When an arrhythmia is anatomically well defined, such as typical atrial flutter or atrioventricular reentry in the Wolff-Parkinson-White syndrome, where ventricular pre-excitation suggests the accessory pathway's location (see later), much of the procedure depends on reaching anatomic targets rather than electrophysiological ones. This will be discussed in more detail under each individual arrhythmia.

Conventional Mapping Methods

Conventional contact catheter mapping uses one or more of three main methods:

1. Activation sequence mapping
2. Pace mapping
3. Entrainment mapping

Activation sequence mapping, compares the timing of electrograms recorded from a roving catheter during tachycardia with the timing of a reference signal, which could be a signal from the surface ECG or a catheter positioned in a relevant cardiac chamber and in a stable position throughout mapping. The roving catheter is moved and the timing of electrograms continuously compared with the reference signal in an attempt to identify the earliest possible signal or a progression of activation around a well-defined macro-reentrant circuit, such as typical atrial flutter. In the case of focal tachycardias, unipolar signals recorded at the focus have a QS morphology, as previously discussed, as long as minimal high-pass filtering is used. Activation sequence mapping is ideal for focal arrhythmias arising in structurally normal hearts, such as focal atrial tachycardia, focally initiated atrial fibrillation, and right ventricular outflow tract tachycardia. Although activation sequence mapping is also useful for reentrant circuits arising in the presence of structural heart disease, such as ischemic VT where, for example, electrograms at the exit site of the diastolic pathway precede QRS onset; due to the fact that the reentrant circuit is active throughout the cardiac cycle, one cannot rely solely on activation sequence mapping with contact catheters to fully define the circuit. Potentials with a diastolic (pre-QRS) timing are found (but not exclusively) within the diastolic pathway (Fig. 42-1). Such signals are also found in *blind alleys* that are bystanders and not parts of the reentrant circuit. Especially in such patients, but also in those with tachycardias arising in normal hearts, it is useful for such activation data to be complemented with pace mapping techniques.

Pace mapping is especially useful in identifying the source of focal tachycardias in structurally normal hearts. The principle of the technique is that the cardiac activation sequence generated by the arrhythmia will be reproduced by pacing at its focal origin and at a similar cycle length. In the case of an atrial tachycardia, the tachycardia P wave morphology on the surface ECG will be reproduced by such pacing. This can be difficult to confirm, however, so atrial activation usually needs to be recorded from multiple endocardial sites to aid confirmation of a good "pace match." In the case of VT arising within a normal heart, comparison of surface QRS morphologies is easier so that pacing at the source of the

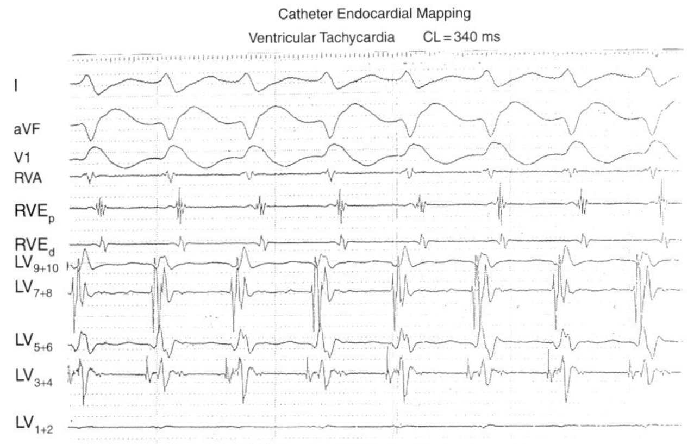

Figure 42-1 Catheter endocardial activation map during sustained ventricular tachycardia. Surface ECG leads 1, aVF, and V1 are shown along with right ventricular (RV) and left ventricular (LV) bipolar endocardial electrograms. LV electrograms show presystolic electrical activity on LV electrodes 3 and 4 at the margin of infarct myocardium. LV 1 and 2 electrodes are in an electrically silent region (infarct-aneurysm), whereas right ventricular activation occurs later in the QRS complex. RVA, right ventricular apex; RVE, right ventricular mapping catheter; p, proximal; d, distal.

arrhythmia can readily be seen to exactly match the QRS morphology generated by tachycardia. When there is structural heart disease, tachycardia is more likely to have a reentry mechanism and be dependent on areas of block to conduction, which are variable, and may not be present during sinus rhythm—so-called *functional block*.[2] Thus, cardiac activation resulting from pace mapping during sinus rhythm may not encounter the same areas of conduction block as those present during tachycardia; it will, therefore, be different from the activation sequence of tachycardia, generating different P or QRS waves, even if the pacing site is on the circuit.

Entrainment mapping is used to map tachycardias dependent on macro-reentrant circuits that, by definition, have an excitable gap. The ability to entrain a tachycardia by pacing confirms a reentrant mechanism for the tachycardia; four criteria for entrainment have been defined. Since the presence of an excitable gap is a prerequisite for reentry, it is possible to introduce stimuli from sites within or outside the reentrant circuit that capture excitable myocardium. If pacing is performed within the reentrant circuit, two wavefronts are produced, one of which travels antidromically and collides with a returning orthodromic wavefront, whereas the other travels orthodromically within the circuit *resetting* the tachycardia. If pacing is successfully performed outside the circuit, the wavefront propagates through the intervening myocardium, reaches the circuit, and propagates in orthodromic and antidromic directions, again resetting the tachycardia. If a train of stimuli is applied at a rate just below the tachycardia cycle length such that all stimuli fall within the excitable gap of the tachycardia, it is possible to continuously reset, or *entrain,* the tachycardia.[3]

Waldo and coworkers first described transient entrainment in 1977 in patients with atrial flutter.[4] The same group subsequently described their observations in a number of other reentrant arrhythmias, including reentrant atrial tachycardia,[5] atrioventricular reentry,[6] and reentrant VT.[3,7-10] Entrainment has also been demonstrated in patients with atrioventricular nodal reentry tachycardia.[7,11] Observations during pacing proximal to the site of slow conduction in a reentrant circuit led Waldo's group to propose first a set of three criteria,[12] and subsequently a fourth criterion for transient entrainment,[8] as follows:

1. The presence of constant fusion beats on the ECG while pacing during tachycardia, at a constant rate that is faster than the tachycardia and that fails to interrupt it, except for the last paced beat which is entrained but not fused;
2. The demonstration of progressive fusion while pacing during tachycardia at two rates that are faster than the tachycardia but do not terminate it;
3. Interruption of a tachycardia during pacing at a rate faster than that of the tachycardia is associated with localized conduction block to a site for one beat, followed by activation of that site by the next pacing impulse from a different direction and with a shorter conduction time;
4. While pacing during tachycardia from a constant site at two rates, both of which are faster than the tachycardia and do not interrupt it, a change in electrogram morphology at, and conduction to, an electrogram recording site is observed.

It should be emphasized that transient entrainment may be present without any of these criteria being present. The site from which pacing is performed is particularly important in this respect: although entrainment may occur while pacing from sites either proximal or distal to the area of slow conduction of a reentrant circuit, the entrainment criteria are demonstrable only while pacing from sites proximal to the area of slow conduction—pacing distal to this area is associated with block of the antidromic wavefront in the region of slow conduction.

Because entrainment can occur during pacing within or outside the reentrant circuit, the presence of entrainment is of no value in localizing the circuit, but merely confirms a reentrant mechanism. Therefore, other criteria, such as the QRS morphology during entrainment and post-pacing interval after entrainment have to be considered in order to establish whether or not the pacing site is within the circuit. For this, pacing is typically performed at cycle lengths just shorter (10 to 50 ms) than the tachycardia cycle length. To confirm that pacing has occurred within the tachycardia circuit, the result is examined for the following criteria:

1. The activation sequence of the chambers paced should be identical to that seen during tachycardia so that, for example in VT, the paced and tachycardia QRS complexes will be identical (concealed

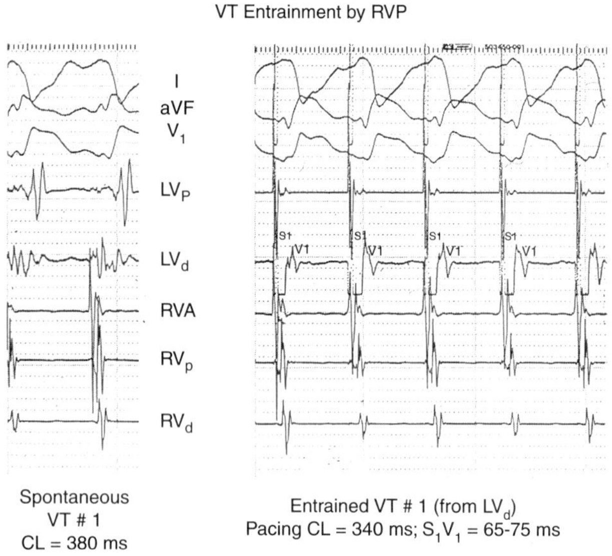

Figure 42-2 Spontaneous ventricular tachycardia (*left panel*) and entrained ventricular tachycardia (*right panel*) show identical electrocardiographic morphology. While only 3 perpendicular leads of the 12 leads are displayed, the morphology was identical in all twelve leads. See text for details. *Abbreviations:* RVA, right ventricular apex; RVE, right ventricular mapping catheter; p, proximal; d, distal; RVP, rapid ventricular pacing.

entrainment or entrainment with concealed fusion) (Fig. 42-2);

2. The interval between the pacing stimulus and a fixed reference point (e.g., onset of surface QRS in VT) will be identical to the interval present during tachycardia between the electrogram recorded at the pacing site and the same reference point;
3. The return cycle (that generated by the last pacing stimulus) and which is ideally measured from the last pacing stimulus to the time of subsequent activation at the pacing site should be equal (within 30 ms) to the tachycardia cycle length. This is because it reflects the time the paced wavefront takes to travel through the circuit and return to the pacing site without being interrupted by another pacing stimulus. Measurement of this phenomenon exactly at the pacing site can be made difficult by post-pacing polarization effects. Thus, acceptable surrogates such as the "N + 1 difference" can be used.[12] With the electrogram and QRS complex inscribed during or immediately after the pacing stimulus assigned as N, and subsequent electrograms and QRS complexes described as N + 1, N + 2, and so on, the "N + 1 difference" is measured as the difference between the intervals from *stimulus* to QRS_{N+1} and *electrogram*$_{N+1}$ to QRS_{N+2}.

The latitude in allowing these intervals (2 and 3) not to be exactly the same during pacing from the circuit and during tachycardia reflects the fact that there may be decremental conduction within the circuit in response to the shorter pacing cycle length. It is to minimize these effects that such pacing is performed at a cycle length just shorter than the tachycardia cycle length.

Conventional Mapping for Specific Arrhythmias

MAPPING OF ATRIOVENTRICULAR NODAL REENTRY TACHYCARDIA (AVNRT)

Mapping and ablation of AVNRT are usually performed in one combined procedure. Typically, quadripolar electrode catheters are positioned in the high right atrium, the area of the His bundle, and the right ventricular apex. A fourth multipolar catheter positioned in the coronary sinus helps identify alternative or additional arrhythmogenic substrates during the diagnostic phase of the procedure and helps mark the position of the coronary sinus ostium to aid slow pathway ablation (see Clinical Electrophysiology, Chapter 14).

With the diagnostic catheters in situ, basic intervals including the A–H interval are measured in sinus rhythm. Programmed stimulation is then performed from the right ventricular apex and high right atrium. This usually involves an eight-beat drive train at two cycle lengths and an atrial or ventricular extrastimulus that is decremented by 10 to 20 ms with every drive train. Sensed atrial and ventricular extrastimuli during sinus rhythm are also helpful. Finally, decremental atrial and ventricular pacing may be performed. Regardless of the methods used, one may see evidence of functional longitudinal dissociation of AV node physiology with at least a 50-ms increase in A–H interval after a 10 to 20 ms shortening in the A1–A2 interval. Because evidence of dual AV node physiology is also frequently encountered in normal individuals without AVNRT, the presence of dual AV node physiology in isolation cannot be used in the diagnosis of AVNRT. Conversely, dual AV node physiology may not be demonstrable if the two "fast and slow" pathways have overlapping conduction and refractory properties. In patients with AVNRT, fast pathway block is followed by tachycardia initiation in up to 70% of patients. If single echo beats are seen but AVNRT does not initiate, administration of isoproterenol may result in sustained arrhythmia with further programmed stimulation. Radiofrequency ablation is generally not performed if AVNRT cannot be induced, even in patients with documented supraventricular tachycardia and dual AV node physiology, in part because of lack of an endpoint if tachycardia is not inducible. In patients with ongoing symptoms, the tachycardia often proves inducible during a subsequent procedure.

When typical AVNRT is induced there is synchronous atrial and ventricular activation with long A–H and short H–A' intervals (see Fig. 14-14). These reflect anterograde conduction via the slow pathway, retrograde conduction occurring via the fast pathway and simultaneously with anterograde His-Purkinje conduction causing ventricular activation. When there is a bundle branch block, the synchronicity of atrial and ventricular activation is affected with right or left ventricular activation being delayed by ipsilateral bundle branch block. Atypical AVNRT is less commonly seen; rarely, both forms are seen in the same patient. In atypical AVNRT, the fast pathway conducts anterogradely and the slow pathway retrogradely (see Figs. 14-15 and 14-16). Thus, atrial and ventricular activation are separated as fast pathway conduction is followed by rapid His-Purkinje conduction, whereas the atria are eventually depolarized after retrograde conduction via the slow pathway. This is, therefore, a form of "long RP' tachycardia."

Introduction of ventricular extrastimuli synchronous with or slightly preceding His depolarization during typical AVNRT will not affect the timing of the next atrial depolarization because the His bundle and AV node will be refractory and there is no accessory pathway by which to advance atrial activation. If the tachycardia is atypical AVNRT, such ventricular extrastimuli may either advance or even delay the next atrial depolarization depending on the amount of decremental conduction encountered in the slow pathway.

Catheter modification of the slow pathway is the method of choice for the ablation of the AVNRT reentrant circuit. This may be performed primarily based on electrograms, fluoroscopic position, or a combination of the two. Successful ablation with a predominantly anatomic approach has been reported with the ablation catheter positioned in the posteromedial tricuspid annulus, with applications starting near the floor of the coronary sinus os (Fig. 42-3) and, if necessary, from there performing further energy applications progressively nearer, but as far as possible from, the compact AV node.[13] Another anatomic approach involves positioning the ablation catheter so that in an LAO projection it lies in a 6-o'clock position on the inferior tricuspid annulus with an atrial electrogram smaller than the ventricular one and delivering energy along this line. If the tachycardia remains inducible after energy applications in this position, further applications are made as necessary, gradually progressing toward a 4-o'clock position.[14]

An alternative approach based predominantly on electrograms is preferred by other electrophysiologists. Having crossed the tricuspid valve in the RAO projection and with the ablation catheter in the right ventricle, the catheter is deflected, and slowly withdrawn with clockwise torque so that it lies along the inferior annulus near the os of the coronary sinus. This area is then mapped to identify target sites with atrial electrograms that are small relative to the ventricular components and show slow pathway potentials. Slow pathway potentials are seen as single or multiple, *usually* high-frequency deflections that follow or fuse with the local atrial electrogram, which is typically of lower frequency (Fig 42-4). Such potentials are usually recorded at the posterior or midseptum near the tricuspid annulus, usually anterior to, but occasionally posterior to or even within the coronary sinus os.[15,16]

A combined approach, based on electrograms as well as anatomic information is now the preferred approach for many centers, including our own. With this approach, the mapping catheter is positioned in the inferoposterior part of Koch's triangle on the tricuspid annulus (TA), such that atrial electrograms are small in amplitude, long in duration, later than the atrial electrogram recorded by the His bundle catheter,[17] and ideally

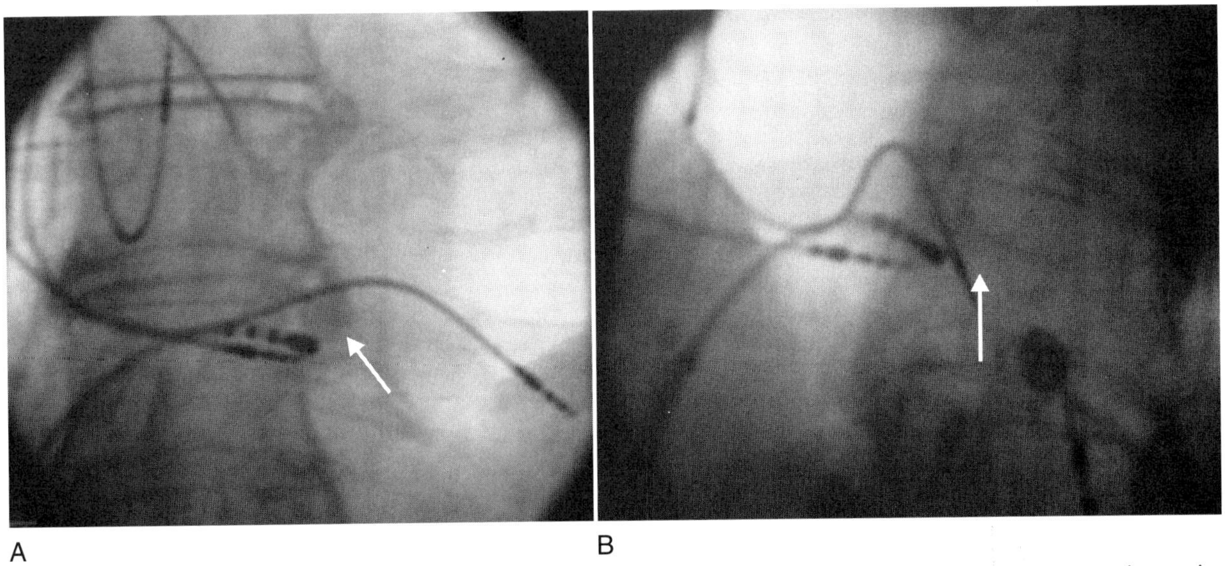

A B

Figure 42-3 Anatomically located ablation electrode catheter *(arrow)* as seen fluoroscopically in the anteroposterior projection **(A)**, and left anterior oblique view **(B)**. Note the location in proximity to the ostium of the coronary sinus at the posteromedial corner of the Triangle of Koch. The pacing lead is fixed outside the ostium of the coronary sinus.

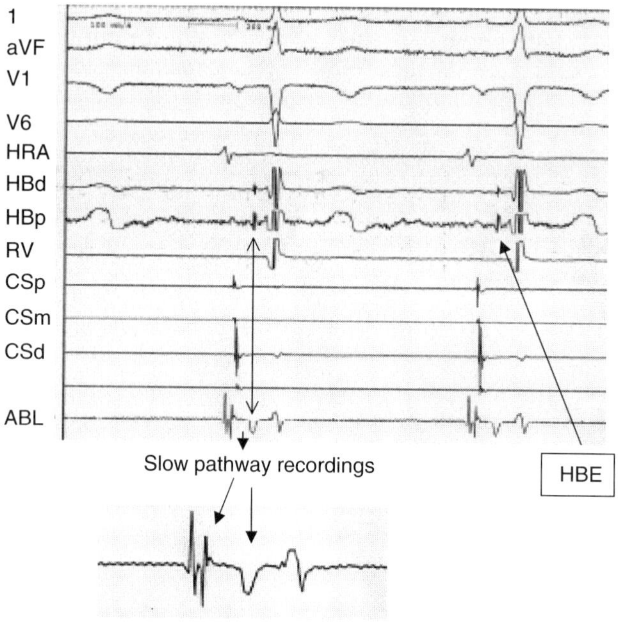

Figure 42-4 Mapping for slow pathway potentials using the ablation (ABL) catheter during sinus rhythm in a patient with recurrent paroxysmal supraventricular tachycardia due to atrioventricular nodal reentry. Surface ECG leads and intracardiac electrogram recordings are shown. Note the high frequency and low frequency electrogram recordings distinct from atrial potentials, easily differentiated from the His bundle electrogram in timing and morphology and located at a distance from the His bundle. The inset below shows the recordings in magnified detail. HB, His bundle; RV, right ventricle; CS, coronary sinus; p, proximal; d, distal; m, mid.

followed by a slow pathway potential. The slow pathway potential is eliminated after energy delivery (Fig. 42-5). If there is no response to ablation in this region, the more superior midseptal TA is mapped and treated in a similar fashion.[18]

Junctional rhythm during energy delivery is associated with successful ablation. In one study, junctional activity was always seen during energy delivery at all successful sites, but also in 65% of unsuccessful sites. In the same series, junctional activity was longer at effective sites than at ineffective sites.[19] Thus, although lack of junctional activity during energy delivery almost invariably suggests an ineffective application, the converse is not necessarily true. Midseptal ablation sites appear to be associated with a higher and earlier occurrence of junctional rhythm and higher success rates than more posterior sites.[20] The cycle length of junctional activity during slow pathway ablation (around 500 ms) tends to be longer than that observed with fast pathway ablation (around 400 ms) and has lower cycle length dispersion than either fast pathway or total AV node ablation.[21] Fast junctional tachycardia with cycle lengths below 350 ms suggests proximity to the compact AV node and is associated with the development of conduction block.[22] VA block during energy delivery and junctional activity is associated with the development of atrioventricular block and, therefore, requires immediate discontinuation of energy application.

WOLFF-PARKINSON-WHITE SYNDROME AND CONCEALED ACCESSORY PATHWAYS

ECG Algorithms

There are now several algorithms that, based on the morphology of a preexcited surface ECG, can help predict the area of insertion of accessory pathways and, hence, the likely target area for ablation (see Chapter 14). Some algorithms are based entirely on the polarity of the QRS complex,[23-25] whereas others use a combination of QRS and polarity and delta wave vector data.[26-28] In a prospective evaluation of one of the most recently published algorithms,[26] the algorithm could localize the accessory pathway to one of ten sites around the tricuspid or mitral valve annuli, the coronary sinus or its tributaries, with overall sensitivity and specificity rates of 90% and 99%, respectively. Diker and coworkers recently demonstrated that localization of posteroseptal pathways on the right or left side of the septum was possible based on the morphology of the QRS, which was always negative in lead V1 for right-sided and biphasic or positive in V1 for left-sided pathways,

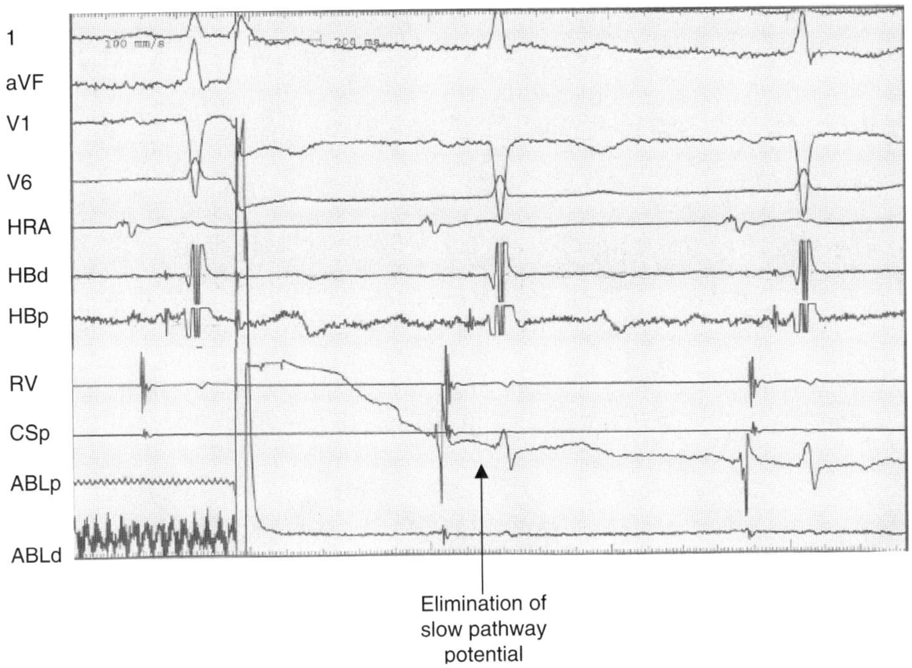

Figure 42-5 Elimination of the slow pathway potentials after radiofrequency energy is applied at the location shown in Figure 42-4. This was accompanied by the disappearance of dual atrioventricular nodal conduction physiology. HB, His bundle; RV, right ventricle; CS, coronary sinus; p, proximal; d, distal; m, mid.

respectively.[29] These algorithms are useful in that they can help identify the likely target area for ablation and thus allow planning for the procedure and guide mapping. It is important to remember that the degree of preexcitation is influenced by the heart rate so that such analysis should ideally be done when pre-excitation is maximized by high rate atrial pacing.

Catheter Mapping

Quadripolar catheters are typically positioned in the high right atrium, right ventricular apex, and His bundle area. A multipolar catheter is inserted into the coronary sinus and initially positioned so that its proximal pole is located at the os. If the accessory pathway is left sided, the coronary sinus catheter may have to be advanced during the procedure so that electrode poles lie on both medial and lateral sides of the accessory pathway, thus enabling it to *bracket* the earliest atrial or ventricular activation. Cappato and colleagues have reported their experience with 2F multielectrode catheters that could be advanced into the coronary sinus and its tributaries, including the great cardiac vein. They found that these catheters could help localize accessory pathways and were particularly useful for anterior pathways or difficult cases.[30]

Programmed stimulation is performed sequentially from the right ventricle and atrium to assess retrograde and antegrade conduction, localize the accessory pathway, and induce tachycardia. Further mapping is then performed with a deflectable catheter. If the accessory pathway is felt to be right-sided or if there is uncertainty regarding its position from surface ECG and preliminary mapping data, then the right atrioventricular ring is mapped. If the pathway is felt to be left-sided, mapping of the left atrioventricular ring is performed (after heparinization) by retrograde catheterization, via a patent foramen ovale or, if necessary, after transeptal puncture. In patients with preexcitation, mapping may be performed during sinus rhythm, atrial pacing, orthodromic or antidromic tachycardia, or ventricular pacing (Fig. 42-6A). Because of the electrophysiological properties of concealed accessory pathways, mapping of such pathways may be performed only during orthodromic tachycardia or ventricular pacing.

If pathway activation is antegrade during mapping, the site showing the shortest atrioventricular conduction time or earliest ventricular activation around the

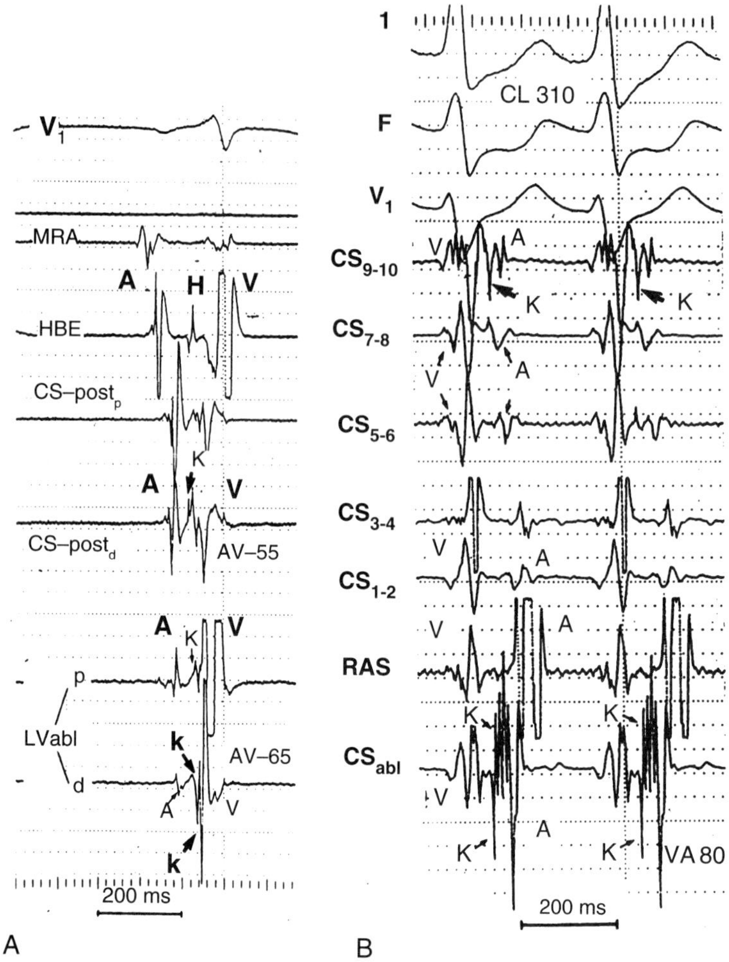

Figure 42-6 Mapping of accessory pathway locations using retrograde and antegrade activation patterns. **A,** Antegrade accessory pathway activation during sinus rhythm with overt preexcitation. Surface ECG leads and intracardiac electrogram recordings are shown. The accessory pathway potential is marked as the Kent (k) potential bracketed by atrial (A) and ventricular (V) electrograms. **B,** Retrograde conduction over the accessory pathway during orthodromic atrioventricular reentrant tachycardia. Surface ECG leads and intracardiac electrogram recordings are shown. The accessory pathway potential is marked as the Kent (k) potential bracketed by ventricular (V) and atrial (A) electrograms. HB, His bundle; RV, right ventricle; CS, coronary sinus; p, proximal; d, distal; m, mid.

right or left atrioventricular ring is sought with the mapping catheter (see Fig. 42-6B). In the case of left free wall pathways, the coronary sinus bipole showing earliest ventricular activation can be used to guide the mapping catheter to the region of interest. It is thus important to position the coronary sinus catheter such that there are electrode pairs bracketing the accessory pathway on either side. If pathway activation is retrograde during mapping (during ventricular pacing or orthodromic AVRT), then the same principles are used to identify the region of shortest ventriculoatrial conduction time or earliest atrial activation. It is important to note that because left-sided accessory pathways often angle from left to right as they traverse the AV groove, with the atrial connection closer to the coronary sinus os than the ventricular connection,[31] and because accessory pathways may have multiple atrial or ventricular attachments, the ablation site may need to be on a different radial point adjacent to the AV ring depending on whether the atrial (ventriculoatrial activation mapping) or ventricular (atrioventricular activation mapping) attachments of the pathway are mapped and targeted.

In the case of right free wall accessory pathways, although access to the left heart is not required, the procedure may actually be more difficult due to a number of factors including the lack, on the right side, of a coronary sinus equivalent that provides anatomic and electrophysiological guidance and difficulties in obtaining a stable position on the tricuspid annulus. Moreover, right free-wall accessory pathways are more often multiple and are occasionally associated with structural abnormalities such as Ebstein's anomaly. In the case of manifest accessory pathways, initial guidance to the likely region of the accessory pathway is obtained by analysis of the maximally preexcited ECG (see Fig. 42-2). The tricuspid ring is subsequently mapped with a deflectable quadripolar catheter, either during atrial pacing with maximum preexcitation in the case of manifest accessory pathways, or during ventricular pacing, as described earlier. Although most pathways may be mapped and ablated on the atrial side of the tricuspid annulus, an alternative approach, whereby the mapping catheter is advanced into the right ventricle and then deflected and looped back on itself so that it lies under the tricuspid valve apparatus, may be preferable in some cases.

Septal accessory pathways may be classified as anteroseptal, midseptal, or posteroseptal. Anteroseptal accessory pathways are almost invariably right sided and, although midseptal pathways (inferior to the His bundle catheter but anterior to the coronary sinus os) may be right or left sided, they are more commonly right sided. The retrograde atrial activation sequence during ventricular pacing can be helpful in distinguishing between anteroseptal and midseptal pathways. Retrograde atrial activation is often recorded simultaneously in the

His bundle catheter and a bipole at the coronary sinus ostium in the case of midseptal pathways, whereas it is earlier in the His bundle catheter signals in the case of anteroseptal pathways.[32]

True posteroseptal pathways may connect the right atrium to the left ventricle in the area of the coronary sinus os, whereas in adjacent areas, right or left paraseptal pathways connect the right atrium and ventricle or left atrium and ventricle, respectively. Accessory atrioventricular connections may be found within the coronary venous system, either within normal structures such as the coronary sinus or middle cardiac vein, or in abnormal structures, such as a coronary sinus diverticulum and may fall into the true posteroseptal or left paraseptal categories. Earliest atrial activation during tachycardia may be at the coronary sinus os either during AVRT due to a posteroseptal accessory pathway, or during atypical AVNRT. Introduction of ventricular extrastimuli synchronous with, or slightly earlier than the His spike during tachycardia, when the His bundle is refractory, may be useful at differentiating between the two. In the case of AVRT due to a posteroseptal accessory pathway, atrial activation is significantly advanced by the extrastimuli, whereas in the case of AVNRT atrial activation may not be advanced and may even be delayed as described. When mapping the mid- and posteroseptal regions it is customary to start below the position of the His catheter in the anteroseptal area and proceed toward and below the coronary sinus os. The proximal coronary sinus and ostium of the middle cardiac vein are subsequently mapped, looking for the earliest possible ventricular signal during atrial pacing, or earliest atrial signal during ventricular pacing (Fig. 42-7).

Signal averaging of intracardiac electrograms recorded during sinus rhythm from potential ablation sites over a number of beats aids in assessing catheter stability, and also increases the probability of seeing the pathway potentials of manifest accessory pathways; it would aid in mapping except for the time required to perform it.[33]

ATRIAL FLUTTER

Although atrial flutter was first described by Jolly and Ritchie in 1911,[34] and only a few years later Lewis and coworkers suggested that it was probably due to reentry within the right atrium,[35] it has taken decades for the circuit of common atrial flutter to be fully described in humans.

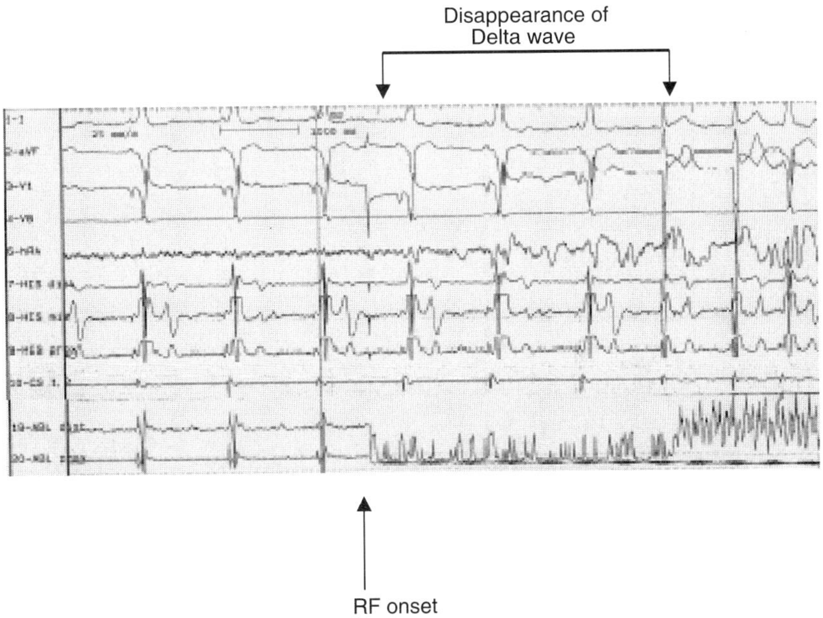

Figure 42-7 Radiofrequency ablation of accessory pathway during sinus rhythm. Surface ECG leads and intracardiac electrogram recordings are shown. The ventricular (V) and atrial (A) electrograms are virtually fused in the ablation recordings with an intervening accessory pathway potential. HB, His bundle; RV, right ventricle; CS, coronary sinus; p, proximal; d, distal; m, mid.

Typical (Type 1) Atrial Flutter

It is now known that the reentrant circuit is formed partly by anatomic barriers and partly by areas of functional block. The anterior boundary is formed by the tricuspid valve. The posterior boundary, traditionally thought to be formed by the crista terminalis in the posterolateral right atrium, may actually be an area of functional block in the sinus venosa region of the posteromedial right atrium.[36] A critical isthmus[37] is formed by the tricuspid annulus, inferior vena cava, the eustachian ridge, and the ostium of the coronary sinus.[38,39] The superior turning point of the circuit, which is often near the superior vena cava, has been described in humans more recently, using three-dimensional mapping.[40,41] Septal activation is caudocranial in common flutter, when the circuit rotates in a counterclockwise direction, and craniocaudal in uncommon atrial flutter when the direction of rotation is clockwise.

MAPPING OF ATRIAL FLUTTER

Mapping of atrial flutter is classically performed with a deflectable multielectrode catheter positioned circumferentially around the tricuspid annulus, a mapping/ablation catheter and a catheter positioned in the coronary sinus. In common type 1 atrial flutter, the Halo activation pattern demonstrates counterclockwise activation around the tricuspid annulus (Fig. 42-8), and clockwise activation in uncommon flutter. In both, concealed entrainment during pacing from the cavotricuspid isthmus should be demonstrable, confirming it as a critical area in the circuit that can be targeted for ablation. Concealed entrainment requires acceleration of the tachycardia cycle length to the pacing cycle length without a change in the pattern of endocardial activation, the morphology of the atrial electrograms, or surface ECG F waves. Concealed entrainment from the cavotricuspid isthmus is seen with a long stimulus to

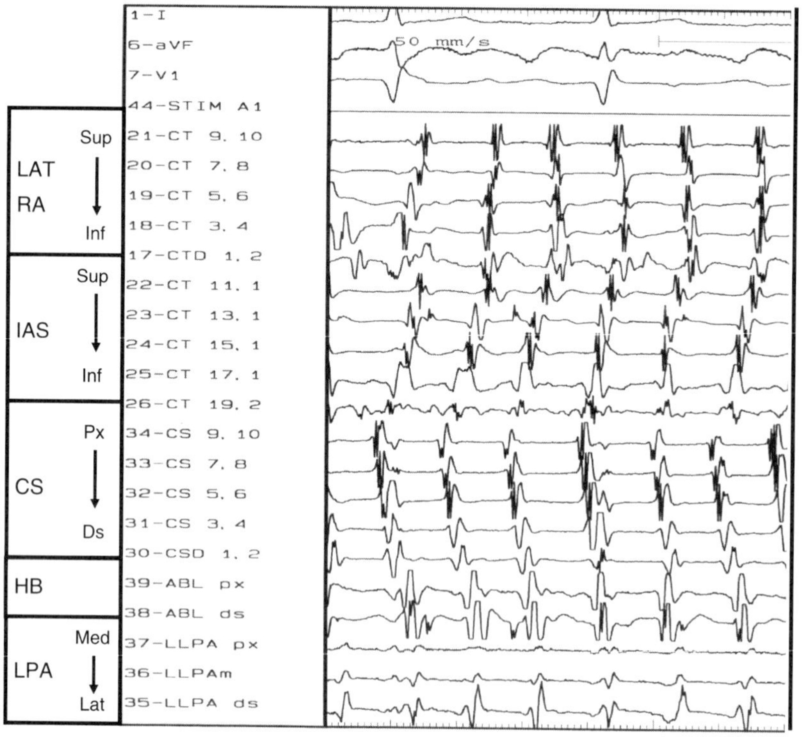

Figure 42-8 Catheter endocardial mapping of common type 1 atrial flutter. Counterclockwise circulation of the flutter wavefront in the right atrium is seen with proximal to distal activation of the coronary sinus (CS) and medial to lateral activation of the superior left atrium via the left pulmonary artery (LPA). LAT, lateral; IAS, interatrial septum; CT, crista terminalis; LLPA, left lower pulmonary artery; HB, His bundle; RV, right ventricle; CS, coronary sinus; px, proximal; ds, distal; m, mid; inf, inferior; sup, superior.

F wave interval.[42] Double potentials indicating anatomic or functional block, are often recorded along the eustachian ridge during atrial flutter.[38]

MAPPING TO GUIDE RADIOFREQUENCY ABLATION

Three catheters are required at this stage: A quadripolar catheter in the coronary sinus, a catheter around the tricuspid annulus and mapping/ablation catheter. Mapping of the isthmus is helped by recording unipolar electrograms from the mapping/ablation catheter against a reference electrode in the inferior vena cava. Ablation is started at the cavotricuspid isthmus where an annular electrogram is recorded with large ventricular and small atrial components. Serial and contiguous energy applications are made along a line, between the ventricular aspect of the tricuspid annulus and the inferior vena cava. Ablation can be performed during atrial flutter or during pacing from the coronary sinus os or low lateral right atrium. If ablation is performed during atrial flutter, tachycardia termination may not necessarily be accompanied by the development of bidirectional conduction block within the cavotricuspid isthmus. Initially, during pacing from the coronary sinus, the activation pattern around the tricuspid annulus is that of fusion of two wavefronts traveling in both clockwise and counterclockwise directions from the coronary sinus os, with latest activation being seen at the lateral tricuspid annulus. Once conduction block across the isthmus has been created, while pacing from the coronary sinus os, the pattern of activation around the tricuspid annulus shows exclusively counterclockwise activation, with the distal pole of the circumferential mapping catheter showing the latest activation. Conversely, when pacing from the lateral cavotricuspid isthmus or low lateral right atrium in the presence of isthmus conduction block, the pattern of tricuspid activation shows progressively later clockwise activation with the proximal coronary sinus being activated last. When, after ablation, conduction block across the isthmus remains, the ablation line itself is mapped. It is here that the unipolar electrograms are helpful. The catheter is placed along the ablation line and the electrograms recorded are examined. In the presence of conduction block, the catheter will record double potentials, with components contributed from either side of the line. If the line of conduction block between the tricuspid annulus and the inferior vena cava were complete, then the spacing of the double potentials recorded from each of the four poles of the catheter would be equal. When the line is incomplete, an clcctrode close to the gap in the line of block will record a shorter interval between the two components of the double potential than one further from the gap. This interval increases as the distance from the break in the line increases. Ablation to the site recording the double potential with the closest spacing (or even continuous activity between the two components) may complete the line of conduction block (Fig. 42-9).

Further testing and ablation are thus carried out during pacing from the coronary sinus os or low lateral right atrium until the pattern of activation around the tricuspid annulus demonstrates bidirectional isthmus block.

Atypical Atrial Flutter

In atypical atrial flutter circuits, reentry occurs around other anatomic obstacles, such as scars created in the right or left atria by previous cardiac surgery. These include the site of cannulation of the inferior right atrium for cardiopulmonary bypass, atriotomy scars, and patches. Although the techniques of atrial flutter mapping described earlier including activation, entrainment, and linear mapping principles may provide valuable information that helps define the reentrant circuit, it is usually helpful to resort to advanced mapping techniques described later in this chapter.

ATRIAL TACHYCARDIAS AND FOCALLY INITIATED ATRIAL FIBRILLATION

Right Atrial Tachycardias

Monomorphic paroxysmal atrial tachycardias (PATs) arise from the right atrium, or the left atrium and PVs with approximately equal frequency.[43] The upper crista terminalis is the most common source of right-sided PAT.[44] A right atrial source of the arrhythmia may be determined during activation sequence mapping. In the case of cristal tachycardias, atrial electrograms in the high, but occasionally the mid- or low, posterolateral right atrium, are seen to lead the electrograms recorded from the coronary sinus and right pulmonary artery.

Focal Left Atrial Tachycardias and Focally Initiated Atrial Fibrillation

The PVs are now known to be the predominant source of left-sided paroxysmal atrial tachycardias,[43] and focally initiated atrial fibrillation.[45,46] Mapping of focally initiated atrial fibrillation can currently be performed during sinus rhythm with ectopy or during atrial tachycardia, but it is not possible during atrial fibrillation. Recent studies have shown that internal cardioversion can allow mapping and successful ablation of atrial fibrillation, not only in patients with a short paroxysm of atrial fibrillation, but also in a substantial proportion of patients with chronic atrial fibrillation.[46] The principle is to identify and ablate sources of atrial ectopy and/or tachycardia on the assumption that such activity initiates AF at other times.[47]

Quadripolar catheters are usually advanced into the high right atrium (HRA) and His bundle area and a decapolar catheter into the coronary sinus (CS). The approximate site of origin of ectopic beats can often be inferred from the P-wave vector. More accurate algorithms for predicting the likely source of ectopy before transseptal puncture have also been developed. Lee and colleagues recently designed and prospectively evaluated such an algorithm, based on the relative activation sequences of atrial electrograms recorded in

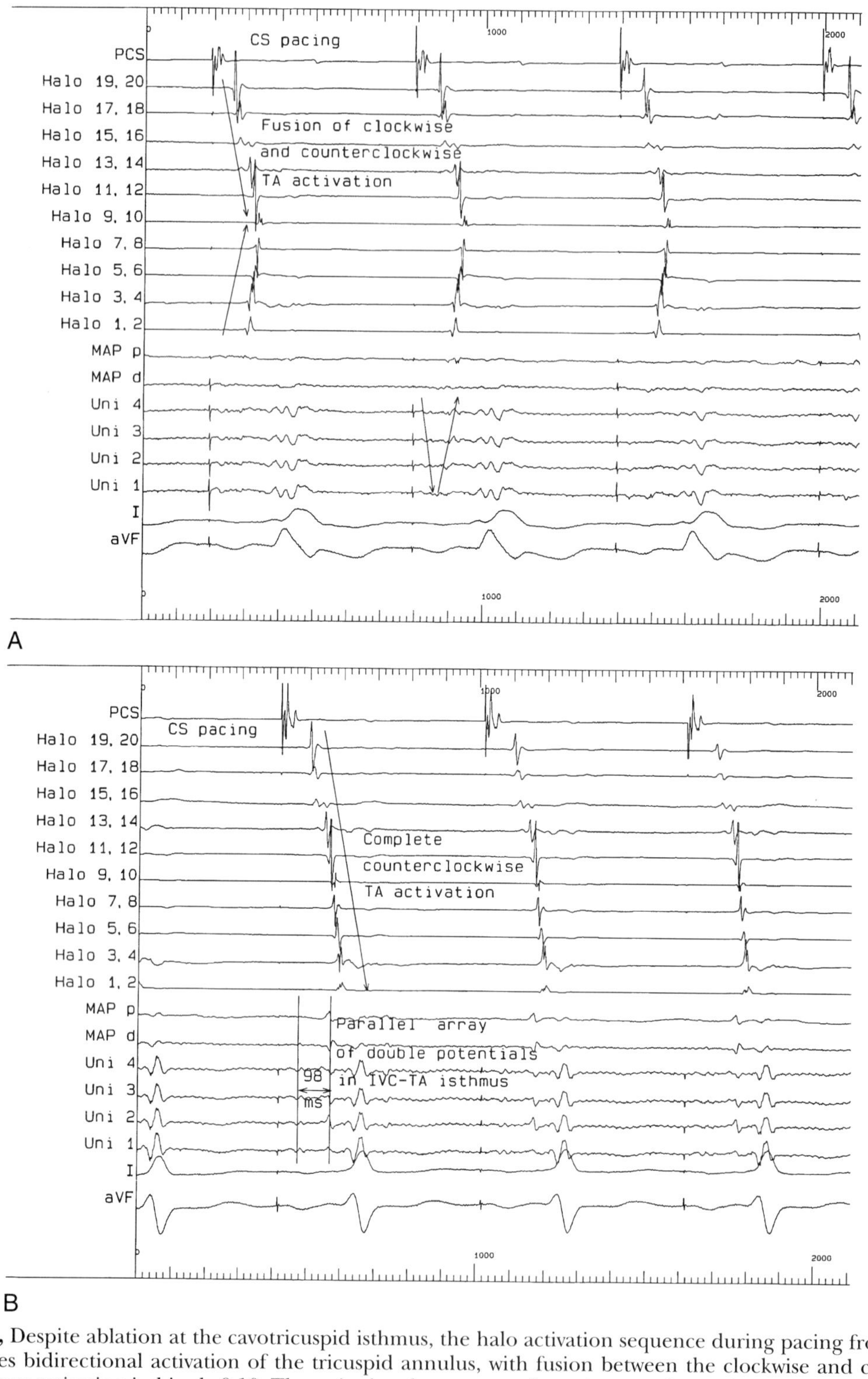

FIGURE 42-9 **A,** Despite ablation at the cavotricuspid isthmus, the halo activation sequence during pacing from the coronary sinus demonstrates bidirectional activation of the tricuspid annulus, with fusion between the clockwise and counterclockwise wavefronts and latest activation in bipole 9,10. The unipolar electrograms from the mapping catheter show double atrial potentials, with the closest spacing in unipole 1, suggesting that unipole 1 was closest to the gap in the ablation line and that unipole 4 (with the widest spacing between the double potentials) was furthest from the gap. **B,** After ablation at the site of unipole 1 in panel **A,** conduction block is achieved at the cavotricuspid isthmus, and the halo activation sequence during pacing from the coronary sinus demonstrates only counterclockwise activation of the tricuspid annulus, with latest activation in the distal bipole. A parallel array of widely split unipolar electrograms is now recorded from the mapping catheter, which can only be found when there is either a complete line of conduction block or there is a very slowly conducting broad wavefront of activation. The latter explanation is extremely unlikely given the change between **A** and **B** produced by ablation at unipole 1. Map d and Map p, bipolar electrograms from distal and proximal bipoles of the mapping catheter, respectively; PCS, proximal coronary sinus; Uni 1-4, unipolar electrograms from the 4 poles of the mapping catheter.

the HRA, His bundle, and proximal and distal CS.[48] They found that if the interval between HRA and His bundle atrial electrograms lengthened during ectopy as compared with sinus rhythm, the focus was in the superior vena cava or crista terminalis in all cases. Conversely, if ectopy was associated with shortening of this interval, the focus was left sided. For left-sided foci, if the CS activation sequence was proximal to distal, then the focus was always in a right PV, whereas if the activation sequence was distal to proximal, the focus was in a left PV in more than 90% of cases (Fig. 42-10). If the earliest point of activation during ectopy is thought to be on the left side, mapping of the left atrium and PVs is performed, either via a patent foramen ovale, or after transseptal puncture.

During sinus rhythm, electrogram recordings taken inside PVs often show a low-frequency potential reflecting adjacent atrial activation, followed by a high-frequency component reflecting local activation of PV muscular sleeves. Although PV potentials are seen more frequently in arrhythmogenic than nonarrhythmogenic veins, the presence of a PV potential in sinus rhythm by no means implies that the vein is arrhythmogenic, nor does lack of a PV potential in sinus rhythm guarantee that the vein will not be arrhythmogenic—many left-sided PVs require pacing from the distal CS to reveal the PV potential.[49]

During ectopy from an arrhythmogenic PV, the above activation sequence is reversed only when the mapping catheter is within the arrhythmogenic vein, when the PV potential occurs prior to the atrial potential (Fig. 42-11). The earliest PV potential signal during ectopy is seen at the site of the arrhythmogenic focus inside the relevant PV trunk. This can be seen to propagate toward the PV ostium, where the signal is recorded progressively later. PV electrograms recorded from successful ablation sites during ectopy often show marked prematurity compared with the high right atrial (HRA) catheter or the surface ECG P wave. Endocardial activation times of more than 100 ms earlier than HRA are almost invariably seen at points of successful ablation of arrhythmogenic veins, whereas activation times of less than 30 ms earlier than HRA are hardly ever recorded at successful sites. An endocardial activation time 75 ms earlier than the HRA signal has a sensitivity and specificity for identifying arrhythmogenic PVs of approximately 80%.[50]

Use of multiple catheters facilitates mapping, especially in patients with infrequent ectopy (Fig. 42-12). If two mapping catheters are used, these are initially positioned, one in a right and the other in a left PV. The catheter that records the later activation during ectopy is then moved to the unmapped vein on the other side of the left atrium to map the next ectopic beat. The earlier vein activating during ectopy at that stage is targeted for ablation.[51] Pulmonary venous mapping with up to four catheters, positioned in each of the PVs, has been reported.[52] Once the arrhythmogenic vein has been identified, multiple (usually two) catheters may be positioned in different parts of the same vein to facilitate further mapping and ablation.

Lack of sufficient ectopic activity to allow accurate identification of all arrhythmogenic PVs is perhaps the most common and important difficulty encountered during mapping of focal left-sided tachycardias and focally initiated AF. A number of maneuvers can be performed in order to provoke ectopy. These include deep breathing, carotid sinus massage, the Valsalva maneuver, slow- and high-rate pacing, DC cardioversion, and drug administration. The Valsalva maneuver, pacing, and DC cardioversion appear to be moderately effective. Intravenously administered isoproterenol appears to be the single most effective maneuver at provoking ectopy.[53] However, in our experience, provoked ectopic beats, and those induced by isoproterenol in particular, have a poor specificity, unless associated with initiation of atrial fibrillation. In patients with infrequent ectopy, use of noncontact mapping (see later) can greatly facilitate identification of arrhythmogenic PVs.[54]

All of the above mapping maneuvers were used to facilitate the identification of "culprit" PVs. With the recognition of both the difficulties in identification of the true culprit vein and that multiple veins can initiate AF in one patient, attention turned to the approach of achieving electrical isolation of all PVs. To aid this goal, circumferential pulmonary venous mapping catheters were developed (Lasso, Biosense Webster, Inc., Diamond Bar, CA). These greatly facilitate the identification of ostial sectors of the PVs that contain the muscular sleeves electrically connecting the foci to the left atrium.[55] This has important advantages that, by allowing ostial PV ablation, include a lower risk of PV stenosis and the ability to perform mapping to guide ablation in sinus rhythm or coronary sinus pacing, even in the absence of ectopy. The technique involves recording the PV potentials around its circumference near its os. The ostial sector showing earliest PV potentials is then targeted for ablation. Elimination of PV potentials, or their electrical disconnection from the left atrium can then be used as endpoints for successful ablation, irrespective of the presence or absence of AF.[49]

MAPPING OF VENTRICULAR TACHYCARDIA

Ischemic heart disease is the most common cause of VT. Because of this close association, as well as the risks associated with induction of hemodynamically unstable VT or ventricular fibrillation in patients with severe coronary disease, patients with ischemic heart disease and suspected VT should undergo coronary angiography followed, if necessary, by percutaneous or surgical revascularization prior to mapping and ablation of VT.

Mapping of VTs remains challenging because of a combination of factors that include the frequent presence of multiple VT morphologies, the hemodynamic instability of the tachycardia, especially in patients with impaired ventricular function and, in the case of ischemic heart disease-related VT, the frequent existence of functional as well as anatomic areas of block that make mapping during sinus rhythm unreliable—although regions of low amplitude potentials, and fractionated electrograms indicate sites most likely to be critical to generating VT.

Such tachycardias are usually reentrant and thus readily inducible and pace-terminable by programmed

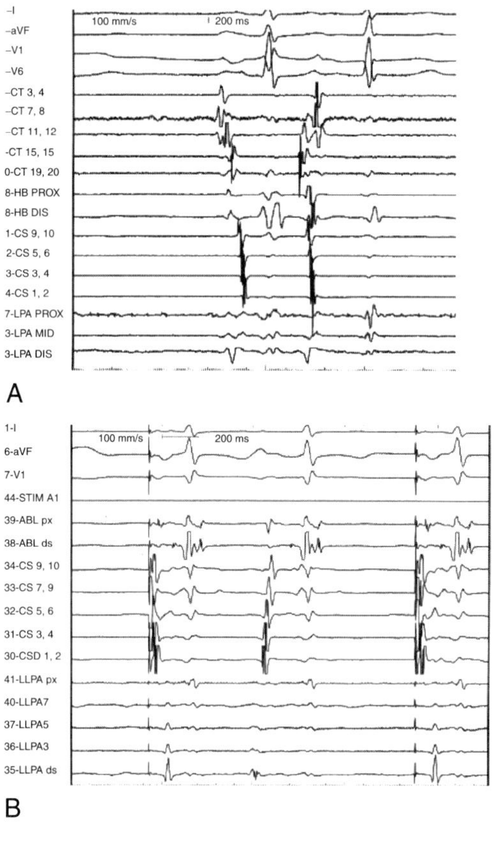

Figure 42-10 Emergence of disparate pulmonary vein foci in a patient with persistent atrial fibrillation and no structural heart disease at two studies. **A,** Atrial premature beaat showing "double" potential in recording from high interatrial septum (CT 11, 12) and proximal (CS 9, 10) to distal (CS 1, 2) activation of the coronary sinus. **B,** Atrial premature beat during atrial paced rhythm showing distal (CS 1, 2) to proximal (CS 9, 10) coronary sinus activation. Abbreviations and catheter locations are listed in Figure 42-8.

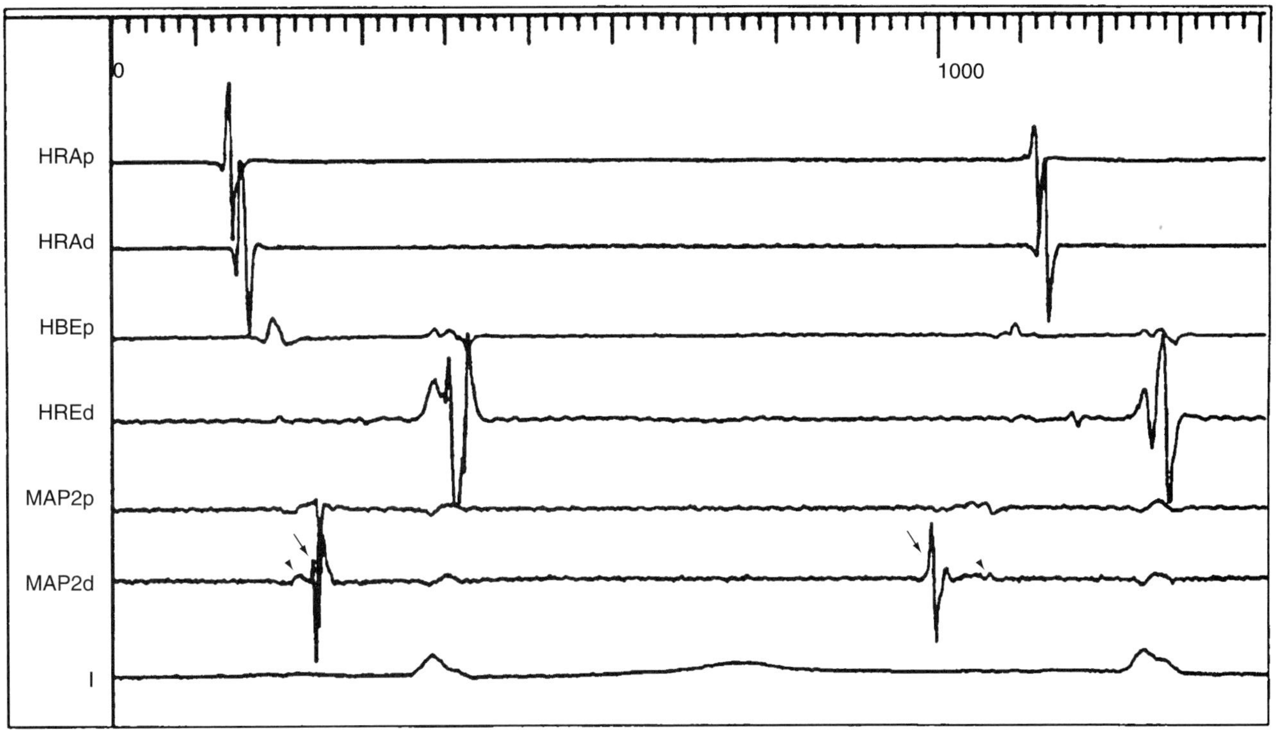

FIGURE 42-11 Pulmonary vein potentials. During sinus rhythm (*left*), pulmonary vein potentials (*arrows*) follow the far-field low frequency atrial potentials (*arrowheads*). During ectopy arising from the pulmonary vein where the mapping catheter is placed (*right*), the high-frequency local muscular sleeve potential (*arrow*) is seen to lead the far-field atrial signal (*arrowhead*).

stimulation. They can also be entrained, usually with a delay between the stimulus and QRS. Bundle branch reentrant VTs usually have a typical left or right bundle branch block pattern with onset of ventricular depolarization preceded by a fascicular or His potential, as well as other diagnostic features described subsequently. Tachycardias due to triggered activity or increased automaticity may not be inducible by programmed stimulation and cannot be entrained.[56]

Idiopathic Ventricular Tachycardia

The term idiopathic VT refers to tachycardias that arise from ventricles without apparent structural abnormalities. The group consists of several distinct entities, including the most common form, which originates from the right ventricular outflow tract (RVOT) and accounts for up to 80% of cases of idiopathic VT. RVOT tachycardias may be further subdivided into repetitive monomorphic VT and paroxysmal sustained VT tachycardias. It is important to distinguish these idiopathic tachycardias from right-sided VTs associated with structural abnormalities of the right ventricle, such as the presence of fatty tissue, wall thinning, dyskinetic segments, or saccular aneurysms, consistent with the diagnosis of arrhythmogenic right ventricular dysplasia. These structural changes are best seen on magnetic resonance images of the heart. Their locations sometimes correlate with sites of successful ablation.[57,58] Other idiopathic VTs include idiopathic left VT, which often arises from the region of the left posterior fascicle, and an automatic form that may originate from either ventricle.

Repetitive monomorphic VT is the most common form of idiopathic VT and is characterized by frequent ventricular ectopy and salvos of nonsustained VT interspersed with sinus rhythm. The tachycardia is often seen in young or middle-aged patients and most often originates from the right ventricular outflow tract. The surface ECG thus shows left bundle branch morphology and an inferiorly directed axis. Although it often occurs at rest, the arrhythmogenic mechanism appears to involve cyclic AMP-mediated triggered activity.[59] Programmed stimulation in these patients is variably successful at inducing tachycardia and, when possible, induction is greatly facilitated by isoproterenol administration.[59,60] In patients with *paroxysmal sustained VT,* the tachycardia is commonly induced by exercise or stress. Programmed stimulation in these patients often initiates the tachycardia, but this again may be possible only after isoproterenol administration. The tachycardia may terminate in response to administration of adenosine, verapamil, or β-adrenoceptor antagonists and also appears to involve cAMP-mediated triggered activity as an arrhythmogenic mechanism (see Chapter 18).[61,62]

Catheter mapping of RVOT tachycardias usually requires a quadripolar catheter placed in the right ventricular apex that is used for programmed stimulation and as a reference catheter. Two deflectable mapping/ablation catheters are then sequentially repositioned in the right ventricular outflow tract to identify the earliest point of activation of this region during tachycardia

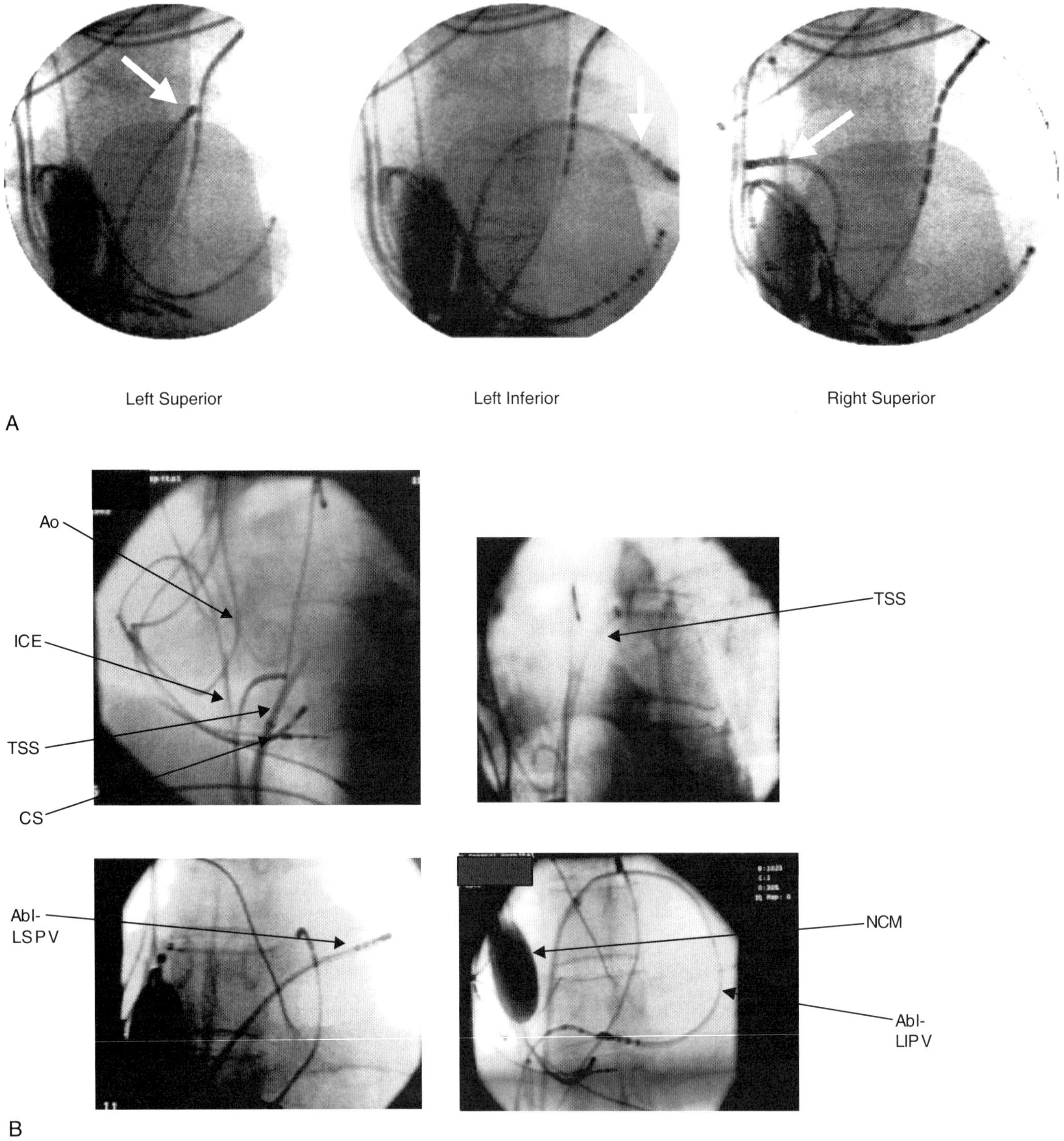

Figure 42-12 **A,** Fluoroscopic images of catheter positioning in individual pulmonary veins shown by arrows in the anterior-posterior projection. **B,** Fluoroscopic image sequence during catheter and three-dimensional balloon mapping, transseptal sheath placement (TSS), ablation (Abl) in the left atrium guided by fluoroscopy, and intracardiac echocardiography (ICE). Ao, aorta; CS; coronary sinus; LSPV, left superior pulmonary vein; RIPV, right inferior pulmonary vein; TSS, transseptal sheath; NCM, noncontact balloon electrode.

or morphologically identical ventricular ectopy. Once the earliest site is identified, pace mapping from this site is performed and, if the pace map is identical in configuration to the tachycardia, radiofrequency energy is delivered.[63,64] The right ventricular inflow, near the His bundle, appears to be the second most common source of idiopathic VT.[65] Careful mapping of this region needs to be performed in order to avoid the development of atrioventricular block with radiofrequency ablation. If the tachycardia does appear to arise from the right ventricular inflow, the His spike is identified and the mapping catheter is advanced to a more superior and ventricular position, such that neither His nor right bundle branch potentials can be recorded. Activation sequence and pace mapping are then performed prior to radiofrequency ablation.

Idiopathic left VT is characterized by symptomatic sustained tachycardia, usually in relatively young patients with no evidence of structural disease (see Chapter 18). The ECG characteristically shows right bundle branch block morphology, and left axis deviation. The origin of the tachycardia is usually at the inferior midseptum or apex of the left ventricle, near the left posterior hemibundle. Fascicular tachycardia can usually be initiated by programmed ventricular stimulation[66] and occasionally by atrial extrastimuli, but may require adrenergic stimulation. It can be terminated by high-rate ventricular pacing. Activation sequence and pace mapping during tachycardia are used to identify the origin of the tachycardia. Concealed entrainment can also be demonstrable from this region, in keeping with its reentrant mechanism.[67] Finally, late diastolic potentials, representing either entry potentials into an area of slow conduction that forms a critical part of the circuit or fascicular potentials, appear to be useful markers of successful ablation sites.[68] Two less common types of idiopathic left VT are also recognized. The first is an adenosine- and verapamil-sensitive tachycardia that may originate from deep within the interventricular septum, exits from the left side of the septum, and is thought to be due to cAMP-mediated triggered activity. The other tachycardia responds to β-adrenoceptor antagonists and transiently to adenosine, but is not terminated by verapamil. It is neither initiated nor terminated by programmed stimulation and is believed to be due to an automatic mechanism.[61] Activation sequence and pace mapping are used to localize the source and guide radiofrequency ablation.[69,70]

Ventricular Tachycardia Secondary to Bundle Branch Reentry

A macro-reentrant circuit involving the His bundle and both its branches is the mechanism underlying this VT,[71] which often occurs in association with idiopathic and other cardiomyopathies,[72,73] mitral and aortic valve disease—both before and after surgery[74]—and only occasionally in patients with structurally normal hearts.[75,76] During sinus rhythm, most patients have complete or partial left bundle branch pattern on the surface ECG and a prolonged H–V interval.[77] During tachycardia, which is sustained, monomorphic, and often associated with hemodynamic instability or collapse, the surface ECG shows a typical left or right bundle branch block pattern. There is atrioventricular dissociation and the onset of ventricular depolarization is preceded by His bundle and right or left bundle branch potentials. Whenever there is a change in the cycle length of tachycardia, the His–His cycle length variation exactly matches the change in the V–V intervals of the associated complexes.[78] If the tachycardia meets these criteria, then mapping of the right bundle branch territory, with a mapping/ablation catheter positioned across the tricuspid valve with its distal electrode tip near the right bundle branch, as far away from the His bundle as possible, is performed with a view to ablation of the right bundle branch (see Fig. 45-7).[77] Successful mapping and ablation of the left bundle branch for treatment of bundle branch reentry has also been reported.[79]

Ventricular Tachycardia Complicating Ischemic Heart Disease

Conventional techniques can be used to map hemodynamically stable VTs. Hemodynamically unstable tachycardias usually require the implantation of an automatic defibrillator (ICD) although in selected cases, VT ablation may prove necessary to reduce the frequency of ICD discharges.[80,81]

Because of the complexity of the reentrant circuit(s) of VT in ischemic heart disease and the lack of reliable diagnostic mapping criteria for identifying the circuit with any one technique, multiple mapping techniques are often required to identify critical parts of the circuit that can subsequently be targeted for ablation. Moreover, full delineation of the reentrant circuit may not be possible during endocardial mapping because parts of, or occasionally the complete circuit, may have an intramural or subepicardial location.

In addition to 12-lead surface ECG, a quadripolar catheter is usually placed in the right ventricular apex as a timing reference and for pacing. Two additional mapping/ablation catheters are then positioned in the left ventricle, one via the retrograde approach through the aortic valve and, in our center, one via the left atrium after transseptal puncture. Electrophysiological testing is performed to confirm that VT can be reliably and reproducibly induced, using well described protocols.[82,83] Systemic and pulmonary arterial blood pressures are invasively monitored throughout the procedure. Elevation of pulmonary artery pressures during VT is the most sensitive marker of hemodynamic deterioration and indicates the necessity to terminate VT and allow recovery during a period of sinus rhythm before VT is reinduced and mapped again.

ACTIVATION SEQUENCE MAPPING

The reentrant circuit during VT complicating healed myocardial infarction consists of a central, common diastolic pathway between entry and exit sites that forms a critical part of the reentrant circuit. The circuit may be completed either by loops that travel within tissues damaged by the infarct (inner loops) or around them through normal myocardium (outer loops). The QRS results from activation of the ventricles after the reentrant wavefront exits the diastolic pathway. The onset of the QRS complex is, thus, usually used as a reference point during activation sequence mapping (see Fig. 42-1).

Complete mapping of the reentrant circuit with conventional techniques in human infarct-related VT is often difficult because it is usually not possible to identify complete VT circuits and because of hemodynamic intolerance of the arrhythmia. The target for ablation is the diastolic pathway. Based on entrainment mapping data, *diastolic potentials* have been noted at approximately 50% of reentry central common pathway (proximal, central, or exit) sites, but less than 10% of inner and outer loop sites (see Fig. 45-3). Typically, they are of low amplitude, short duration, but of high frequency content. The presence of diastolic potentials at sites of radiofrequency energy application increases the likelihood of VT termination. The demonstration of diastolic potentials during activation mapping is only a preliminary stage in mapping of VT in patients with coronary heart disease. In these patients, the cornerstone of mapping is the need to demonstrate concealed entrainment as described earlier (see Fig. 45-4). As an example of its importance, diastolic potentials are also present in 45% of adjacent bystander sites (blind-ending branches of the diastolic pathway). Entrainment mapping is a useful means to differentiate between such bystanders and sites that are critical to the circuit.[84] Pacing from within the diastolic pathway during VT will generate QRS complexes of identical morphology to those seen in VT and the interval between the pacing stimulus will be identical to that between the diastolic potential recorded by that catheter during VT and the onset of the VT QRS. When pacing from within a bystander pathway, the VT QRS morphology may be reproduced, but the interval between the stimulus and QRS onset will be longer than the interval between the recorded diastolic potential and VT QRS because activation has to traverse the additional length of at least part of the bystander pathway.

PACE MAPPING

QRS Morphology of Pacing Stimuli Introduced During Sinus Rhythm

Although the QRS morphology resulting from pacing stimuli introduced during sinus rhythm may be a useful guide to the site of focal tachycardias or reentrant tachycardias completely dependent on areas of anatomic block, both animal and human studies have shown that many postinfarct VT circuits consist of areas of both functional and anatomic block.[85,86] As a result, pace mapping during sinus rhythm, when functional lines of block may be incomplete or not present will produce a different pattern of ventricular activation compared with that seen during tachycardia. Thus, the QRS morphology will be different from that seen in VT and such pacing may not be able to accurately identify the source of a tachycardia.[87]

QRS Morphology During Entrainment

In contrast to pacing stimuli introduced during sinus rhythm as described previously, pace mapping during VT can provide useful information regarding the site of the circuit. Pacing from sites remote from the circuit produces a wavefront that collides with that exiting the reentrant circuit, resulting in a QRS morphology that represents fusion between the two wavefronts. As long as within the excitable gap, the degree to which the QRS morphology is dictated by the pacing stimulus increases with progressively faster pacing rates.

At sites within the reentrant circuit or near its exit site, pacing can entrain the tachycardia, which appears to be accelerated to the pacing rate, without a change in QRS morphology.[87,88] The stimulated antidromic wavefront collides with the orthodromic wavefront very near the pacing site, often within the circuit and as it does not depolarize any significant portion of the ventricular myocardium, it is not detectable on the surface ECG. Similarly, the stimulated orthodromic wavefront travels within the circuit and emerges at its exit site before activating the bulk of ventricular myocardium in the same sequence as the tachycardia. Wavefront fusion is thus concealed, resulting in *entrainment with concealed fusion,* or *concealed entrainment.*[10,89]

STIMULUS TO QRS INTERVAL DURING CONCEALED ENTRAINMENT

During entrainment with pacing performed within the reentrant circuit, the stimulus to QRS (S-QRS) interval reflects the conduction time from the pacing site to the exit site of the reentrant circuit and, hence, provides some information regarding the likely distance between these two sites. Moreover, the S-QRS interval during pacing with concealed entrainment is approximately equal to the electrogram to QRS (E-QRS) interval during tachycardia. When pacing from bystander sites results in concealed entrainment, the S-QRS interval recorder from the same site usually exceeds the E-QRS interval. During pacing from sites that are not within the circuit, the E-QRS interval does not reflect conduction time from the site of the catheter to the circuit exit site and entrainment is not concealed.

POSTPACING INTERVAL AFTER ENTRAINMENT

The interval between the last paced stimulus during entrainment and the next presystolic electrogram recorded at the pacing site (postpacing interval (PPI), or return cycle) reflects the sum of the time required for propagation of the paced stimulus to the reentrant circuit, the tachycardia cycle length (TCL) and the time required for the wavefront exiting the circuit to return to the pacing site. In postinfarct VT, the PPI is, thus, within 30 ms of the tachycardia cycle length when pacing is performed within the circuit, but increases progressively with increasing distance between the pacing site and the circuit.[90] The N + 1 difference, which is a modification of the PPI-TCL interval, may facilitate entrainment mapping by reducing postpacing polarization effects.[91]

A Combined Approach

Although when used individually for mapping postinfarct VT, the mapping techniques described earlier cannot reliably guide the operator to sites of successful ablation; the same techniques can become very powerful tools, however, in guiding the operator to successful ablation sites when used in combination. A recent study has demonstrated that, in patients with coronary artery disease and VT, the presence of three criteria at potential ablation sites reliably predicted success with ablation. The three criteria were an exact QRS match during entrainment, a return cycle within 10 ms of the VT cycle length, and the presence of presystolic potentials with E-QRS interval less than or equal to 10 ms of the S-QRS interval. When all three criteria were met, VT could be terminated with a single radiofrequency lesion in all patients, whereas the lesion was almost invariably unsuccessful when all three criteria were not met.[88]

Anatomic Approach

A predominantly anatomically based approach for conventional mapping of hemodynamically unstable VT in patients with coronary disease has recently been described. The procedure, which was moderately successful in a small group of patients, involved defining areas of scarring by means of left ventricular angiography and the presence of low-frequency, fragmented electrograms. The approximate exit site of the tachycardia circuit was then identified by introducing double extrastimuli at the tachycardia cycle length during sinus rhythm and trying to obtain the best possible pace match. Linear lesions were then created with serial energy applications, which aimed to connect the presumed exit regions to areas of scar or anatomic structures, such as the mitral valve annulus.[92] This method extends the application of the approach described by Wilber and colleagues in patients with VT complicating healed inferior myocardial infarction, whereby ablation in the region between the infarction scar and the mitral annulus successfully terminated the tachycardia.[93]

Use of Radiofrequency Ablation to Guide Mapping

As the temperature rises during delivery of radiofrequency energy, impairment of tissue function is observed before further heating results in a permanent lesion. Because radiofrequency lesions are generally less than 1 cm in diameter,[94,95] termination of tachycardia during energy delivery suggests that the ablation site forms part of a narrow isthmus that may be a critical part of the circuit.

EPICARDIAL MAPPING

Percutaneous Epicardial Mapping

A technique that allows access to the pericardial space for mapping of epicardial VT circuits has been described for patients with Chagas' disease[96] and, more recently, for patients with VT related to previous inferior myocardial infarction.[97] The technique has not only been used to guide endocardial ablation of the tachycardia, but also to successfully ablate tachycardias epicardially. Epicardial identification of the circuit is based on techniques described above for endocardial mapping.

Mapping Through Cardiac Veins

Epicardial mapping can also be performed through the coronary sinus tributaries. Following coronary sinus occlusion venography, multipolar microelectrode catheters (Cardima, Inc., Fremont, CA) may be advanced into coronary veins via a guiding catheter positioned in the proximal coronary sinus. Using standard mapping techniques, including activation sequence mapping and pace mapping techniques, the location of epicardial electrodes can help guide roving endocardial catheters to sites of successful ablation.[98]

New Mapping Technologies

Conventional catheter mapping techniques rely on electrical information gathered from a small number of sources (the electrodes of contact catheters) and anatomic information gathered from two-dimensional fluoroscopic images, in which the endocardial contours cannot actually be visualized. New mapping technologies generally provide electrical information from a much larger number of sources, be it simultaneously or sequentially, thus markedly increasing mapping resolution. Additionally, many systems can localize catheters or electrodes in three-dimensional space. They are thus capable of reconstructing three-dimensional endocardial activation maps as well as accurately localizing and guiding mapping catheters to sites that are suitable for ablation.

These new mapping technologies can be broadly subdivided into methods that combine electrophysiological data with anatomic information, which include CARTO (Biosense-Webster, Diamond Bar, CA), Realtime Position Management (RPM) (Cardiac Pathways-Boston Scientific EP MedSystems, Sunnyvale, CA) and LocaLisa (Medtronic-Cardiorhythm, Minneapolis, MN) and methods that provide continuous and complete logs of all electrophysiological events within a cardiac chamber, including basket maps (Cardiac Pathways, EP Technologies, Sunnyvale, CA) and noncontact mapping (EnSite 3000, Endocardial Solutions Inc, St. Paul, MN). Non-contact mapping also combines anatomic and electrophysiological information.

The first group of methods catalogs a catheter's position in space by means of sensing changes in its position within a magnetic field (CARTO), assessing the relative position of catheters with ultrasonic transducers (RPM), or by sensing impedance changes between the catheter and reference points (LocaLisa) (Fig 42-12). In the cases of CARTO and RPM, this information is allied to the electrograms at each catheter site. In the case of CARTO, this enables the generation of activation maps, whereas with RPM, the electrogram recorded at every catheter position can be recalled for later analysis and comparison. With LocaLisa, important catheter locations are recorded as color-coded dots on a three-dimensional grid requiring imagination of the operator. Where maps are created, or electrograms recovered, these are rhythm dependent and those systems require specific catheters, whereas the LocaLisa data can be obtained with any catheter and the data applied to any rhythm. With all of these systems, a picture of the rhythm has to be built up from sequentially acquired data points, the number of which, therefore, governs the definition of the process.

In contrast, the second group of methods acquires global data so that the rhythm can be characterized from only one beat. The definition of the maps created by the basket systems depends on the number of splines on the basket, the number of electrodes on each spline, and the percentage of both that achieve endocardial contact. The noncontact system is based around a mid-cavity sensor (a special multielectrode array [MEA]) that detects far-field endocardial activity. This is then subjected to inverse-solution mathematics that transform the far-field information into computed equivalents of over 3300 contact "near-field" points on a "virtual" endocardium. The definition of the map is thus influenced by the size of the chamber and the distance between the MEA and the endocardium.

All of the new mapping methods have successfully created maps of all four cardiac chambers, guiding successful ablation in each. Ultimately, a combination of their different strengths will further aid the mapping and successful ablation of the most challenging arrhythmias.

CARTO

The CARTO system consists of an external magnetic field emitter, and a catheter with a miniature passive magnetic field sensor at its tip connected to a processing unit. The magnetic field emitter (locator pad), which includes three coils, is located under the operating table and generates ultra low magnetic fields that code the space around the patient's chest with temporal and spatial characteristics. The catheter is a multipolar 7 F deflectable catheter (Cordis-Webster, Miami Lakes, FL) with the magnetic sensor in addition to a thermocouple near its tip. The magnetic technology used by the system allows accurate determination of the location (x, y, and z) and orientation (roll, pitch, and yaw) of the catheter, and simultaneously records intracardiac local electrograms from its tip.[99] The system thus serves as an endocardial mapping system that allows the creation of three-dimensional electroanatomic maps of the cardiac chambers and can help navigate the roving catheter, reducing the need for fluoroscopy, and guiding it to sites suitable for ablation.

Mapping is performed with a reference catheter usually placed in a stable position in the coronary sinus or right ventricular apex and a second, roving catheter placed in the chamber to be mapped. The location of the roving catheter is gated to a reliable point in the cardiac cycle, such as the R wave on the surface ECG, and recorded relative to the location of the fixed catheter at that time. The system is thus able to compensate both for cardiac and patient movement. The mapping catheter is dragged over the endocardium with simultaneous recording of local electrograms and their respective positions in space. Catheter stability may be assessed by examining the gated position of the catheter during consecutive cardiac cycles. At each point, the local activation times are calculated as the interval between a fixed point on a surface ECG or intracardiac electrogram and the maximal negative dV/dT of the unipolar electrogram recorded from the mapping catheter. This information is then color-coded by the system, with red representing the earliest and purple the latest points of activation and superimposed on the reconstructed three-dimentional geometry of the chamber (see Fig. 45-5).

The system has been validated[99] and used to facilitate mapping of normal and paced rhythms and a variety of arrhythmias in all cardiac chambers, both in animal models and in humans. It has proved a useful tool for mapping VT or areas of scarring and guiding the formation of linear lesions in the left and right ventricles.[100,101] Facilitation of mapping and ablation of focal atrial tachycardias,[102] as well as atrial reentrant circuits related to scarring late after repair of congenital heart disease, have also been reported.[103] In patients with atrial fibrillation, the system has been used not only to guide formation of linear lesions in the right and left atria,[104] but also to create circumferential lesions around the PV ostia and, thus, electrically disconnect the PVs from the left atrium.[105] Finally, electroanatomic mapping has been shown to be effective at reducing fluoroscopy times and facilitating ablation of typical atrial flutter.[106] Software refinements allow activation maps to be generated[107] and maps to be created according to electrogram amplitudes, thereby enabling the identification of scarring.[108-111]

LOCALISA (Figure 42-13)

The LocaLisa system is a nonfluoroscopic catheter positioning system that allows localization of a catheter in three-dimensional space by measuring the voltage drop that occurs between the chest wall and the catheter in three orthogonal planes. Three low current (1 mA) fields, each with a characteristic frequency approximately 30 kHz, are applied at right angles to each other through pairs of skin electrodes. The mixture of 30-kHz signals recorded from each electrode is digitally separated to measure the amplitude of the individual frequency components. Electrode position is determined by dividing each of the three amplitudes by the corresponding electrical field strength and is averaged over 1 to 2 seconds. This reduces cyclic cardiac, but not respiratory, variations.

Although LocaLisa is only a catheter positioning system, it has the significant advantage that no special catheters or arrays need to be used. It has been validated[112,113] and used to create linear lesions to treat incision-related atrial reentry tachycardias.[114] Additionally, the system has recently been shown to be helpful in reducing patient and operator exposure to radiation during mapping.[115]

REALTIME POSITION MANAGEMENT

The Realtime Position Management system is, as its name suggests, a system that allows localization of catheters and endocardial points in three-dimensional

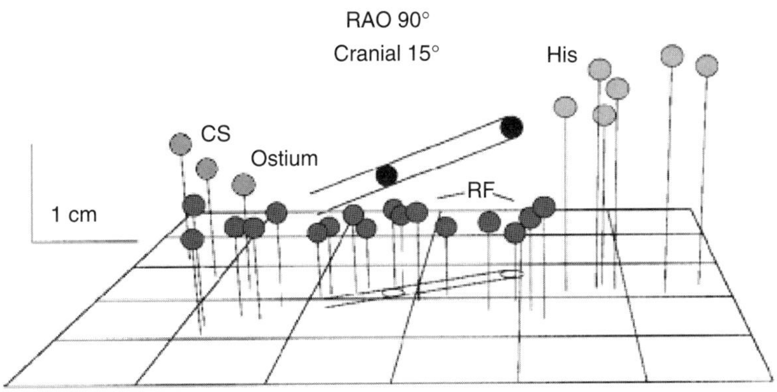

Figure 42-13 LocaLisa navigation system showing spatial location of anatomic landmarks, His bundle (His) and the coronary sinus (CS) relative to lesions delivered during radiofrequency ablation of the cavotricuspid isthmus. The catheter locations are shown. (Courtesy Medtronic Electrophysiology Division, Minneapolis, MN).

space. It is based on ultrasound technology, measuring the time required for conduction of ultrasound from reference points to the mapping catheter. With the conduction velocity of ultrasound in blood as a constant, the system can calculate the distance as time multiplied by velocity. Two 6 F fixed curve catheters are used for reference, one normally positioned in the right ventricular apex and the other in the right atrial appendage or coronary sinus. The reference catheters are equipped with four ultrasound transducers, whereas a deflectable 7 F cooled catheter used for mapping and ablation has three transducers. The catheters are connected to a position management and mapping system, which consists of an ultrasound transmitter and receiver unit and an acquisition module, both of which are connected to a dedicated computer. The system can simultaneously process ultrasound and electrogram data from seven position catheters, in addition to input from surface ECG and two pressure transducers, and it displays electrograms and catheter position information on a computer monitor. Ablation sites and other landmarks can be marked on the display, helping mapping and the creation of linear lesions (Fig. 42-14). The system has been successfully used to aid mapping and ablation of atrial flutter, VT, and accessory pathways.[116]

NONCONTACT MAPPING

The noncontact mapping system uses a multielectrode array, which is positioned in the chamber being mapped, a proprietary recorder and amplifier system amplifier and a Silicon Graphics workstation. The multielectrode array consists of a 9 F catheter with a 7.6 mL ellipsoid balloon. The balloon is surrounded by a braid of 64 electrically insulated wires, each with one small laser-etched break in insulation, allowing them to function as unipolar electrodes. Far-field electrographic data from the array are fed into the amplifier system, sampled at 1.2 kHz, and filtered. The amplifier has 16 inputs for electrograms from contact catheters and 12 inputs for surface ECG. A ring electrode on the proximal shaft of the array catheter in the inferior vena cava is used as a reference for unipolar electrogram recordings. Because the far-field electrograms detected by the array are of low amplitude and frequency, the potentials are enhanced and resolved mathematically. An inverse solution to Laplace's equation predicts how a signal detected at a remote point will have appeared at the source, and a boundary element method applies the inverse solution to resolve a matrix of such signals from a source at a known boundary, namely that between blood and the endocardium.[117] A low-current *locator* signal at 5.68 kHz, passed between a conventional mapping catheter and alternately between ring electrodes situated proximal and distal to the array on the noncontact catheter, is used to locate the mapping catheter in space. The locator signal angles as determined by the system are used to position the source. By moving the mapping catheter to trace the contour of the endocardium, the system records the position in three-dimensional space of a number of points on the endocardial surface, which are used to reconstruct a model of the endocardium of any cardiac chamber. The geometry matrix defines the relationship between the location of the 64 electrodes on the array and 3360 points on the endocardium where the reconstruction is computed, thus allowing the construction of high-resolution endocardial isopotential and isochronal maps. Using the same locator signal, the system can also guide the mapping catheter without the need for fluoroscopy to points on the virtual endocardium that may be suitable targets for ablation.

The noncontact system has been validated in vitro,[118] in animal experiments[118,119] and in humans.[120,121] In contrast to the CARTO system, LocaLisa, and RPM, the noncontact system can analyze the patterns of endocardial activation even from a single beat of tachycardia. Information can be displayed as reconstructed virtual electrograms, series of isopotential maps, or isochronal maps. The system has been used to help map macroreentrant VT complicating ischemic heart disease, where it has proved invaluable in identifying and guiding ablation to the region of the diastolic pathway,[122] as well as other VTs (Fig. 42-15).[123] With the array deployed in the left atrium, the noncontact system can identify the arrhythmogenic PVs and other origins of focally initiated atrial fibrillation.[54,124] Mapping of atrial

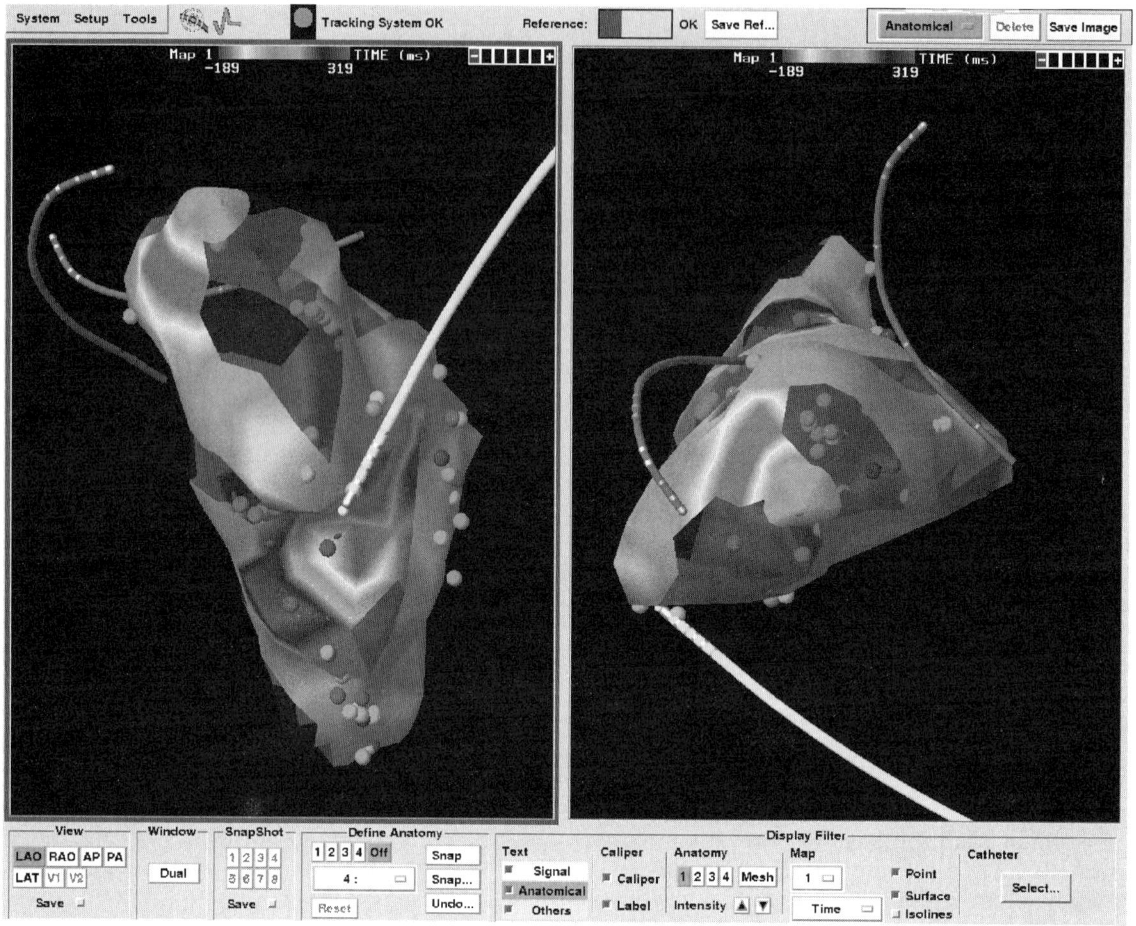

Figure 42-14 (See also Color Plate 42-14.) Realtime positioning management (RPM) system showing three-dimensional spatial navigation and catheter position.

fibrillation and atrial flutter in the right atrium has also been reported using the system.[125-127] Noncontact mapping has been used to guide mapping and ablation of tachycardias due to intra-atrial reentry after Fontan surgery.[128] Recent applications have highlighted its value in refinement of contact catheter mapping and novel information gleaned from combined electroanatomic mapping of a variety of tachyarrhythmias in vivo.[129-139] Difficult arrhythmia sites for ablation (e.g., at the sinus node) may be more precisely localized allowing for new procedural strategies and results may show reduced complications, such as sinus node dysfunction.[129] Differentiation of mechanisms of tachycardias can become easier if the arrhythmia circuit can be directly visualized and effects of interventions can be determined.[130]

Upper loop reentry as a mechanism of atrial tachycardia and its ablation can now be facilitated. Mapping times are now well under 30 minutes for many complex atrial arrhythmias at some centers.[131,132] Ablation of right ventricular outflow tract tachycardia is occasionally difficult. Noncontact mapping assists in differentiating epicardial and endocardial sites.[133] New physiologic data on normal activation in the human heart and its relationship to anatomy in vivo as well as the effects of new interventions affecting this activation pattern in real time are now becoming possible.[134,135] New hybrid mapping techniques are now possible that permit, for the first time, beat-to-beat analysis of hitherto unmappable arrhythmias, such as fibrillation. Saksena and colleagues have combined contact catheter bi-atrial mapping with three-dimensional noncontact mapping and demonstrated the feasibility and provided information on organization of atrial fibrillation.[136-139] The mechanisms of synergy in different interventional therapies can be demonstrated by this technique, opening new avenues in the development of targeted hybrid therapies.[140]

BASKET CATHETERS

Although multielectrode expandable basket catheters have been available for a number of years, they have been used clinically with increasing frequency relatively recently. The most commonly used catheter (Cardiac Pathways, EP Technologies, Sunnyvale, CA) has 64 electrodes on eight highly flexible splines and is capable of acquiring electrophysiological data from multiple sites simultaneously. In addition to the basket catheter, the mapping system consists of an acquisition module connected to a dedicated computer and is capable of simultaneously processing 32 bipolar electrograms from the

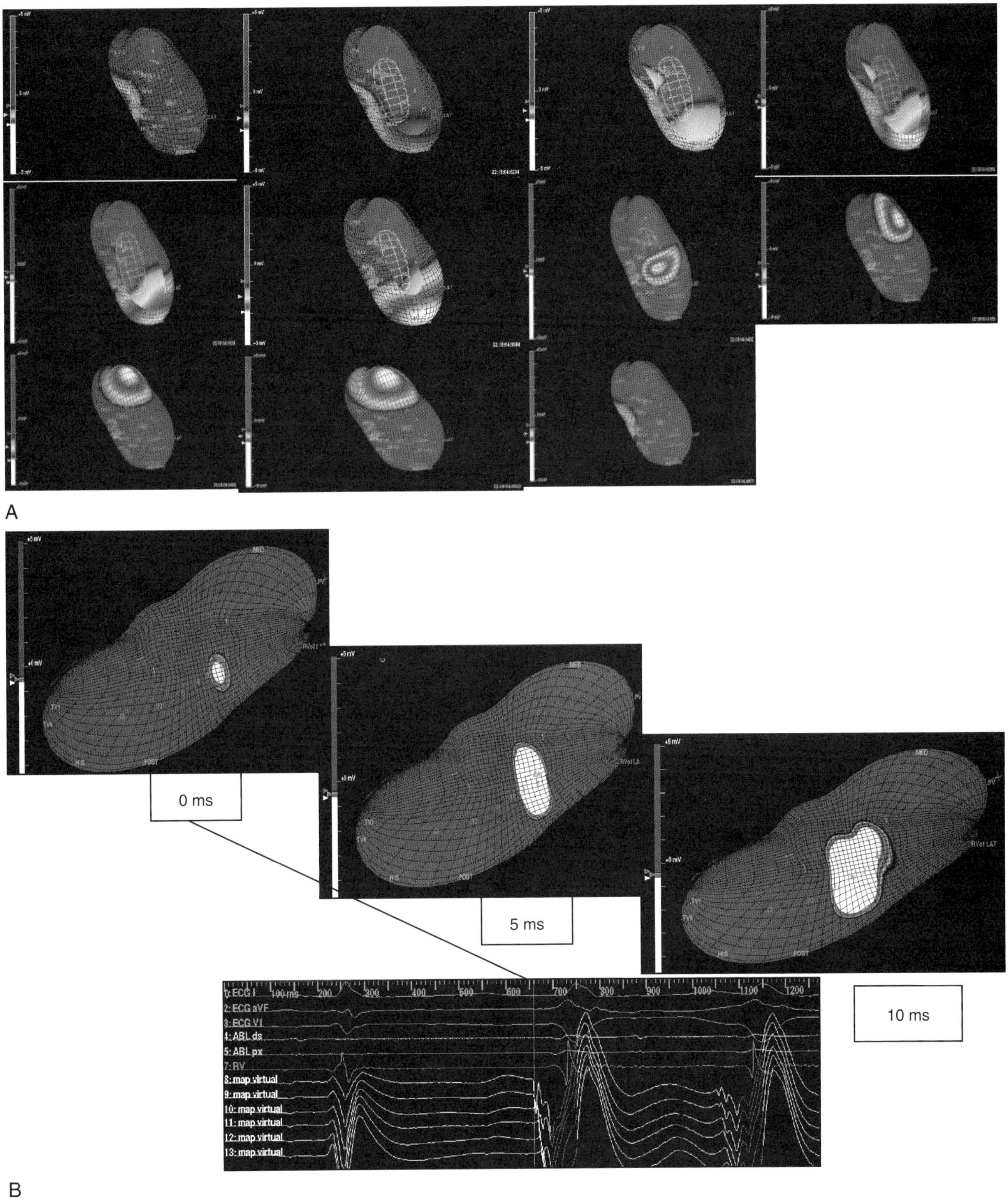

Figure 42-15 (See also Color Plate 42-15.) **A,** Sequence of three-dimensional mapping images during ischemic ventricular tachycardia cycle. The wavefront progression is shown on the left ventricular endocardial surface showing the head to tail relationship seen in a reentrant circuit. Slower and more rapid conduction and spatial locations of regions involved in the tachycardia circuit can be visualized. **B,** Sequence of three-dimensional mapping images during idiopathic ventricular tachycardia. Focal origin is demonstrated, and wavefront progression in centrifugal fashion on the right ventricular endocardial surface can be visualized. The lower panel shows surface ECG leads, intracardiac, and virtual electrograms seen during the mapping process.

basket catheter, as well as electrogram signals from other contact catheters and surface ECG. Electrograms and color-coded activation maps are reconstructed online and are displayed on a computer monitor. The system has been successfully used to aid mapping of human VT,[141] atrial tachycardia, and flutter[142] and even inside PVs to map and ablate atrial fibrillation-initiating foci.[143]

SUMMARY

The limitations of catheter ablation have been reduced by the development of complex computerized mapping systems that have helped in the successful treatment of complex arrhythmias, such as scar-related reentry and VT in association with structural heart disease. They have also increased our understanding of some arrhythmias. However, although they may better define the targets for ablation with reduced fluoroscopy and procedure times, use of these systems, especially in confirming that the chosen target is, indeed, an appropriate site for attempted ablation, benefits from an understanding of conventional principles of cardiac electrophysiology and mapping.

REFERENCES

1. Farre J, Rubio JM, Navarro F, et al: Current role and future perspectives for radiofrequency catheter ablation of postmyocardial infarction ventricular tachycardia. Am J Cardiol 1996;78:76-88.
2. Stevenson WG, Sager PT, Natterson PD, et al: Relation of pace mapping QRS configuration and conduction delay to ventricular tachycardia reentry circuits in human infarct scars. J Am Coll Cardiol 1995;26:481-488.
3. Waldo AL, Henthorn RW, Plumb VJ, et al: Demonstration of the mechanism of transient entrainment and interruption of ventricular tachycardia with rapid atrial pacing. J Am Coll Cardiol 1984;3:422-430.
4. Waldo AL, MacLean WA, Karp RB, et al: Entrainment and interruption of atrial flutter with atrial pacing: studies in man following open heart surgery. Circulation 1977;56:737-745.
5. Henthorn RW, Plumb VJ, Arciniegas JG, et al: Entrainment of "ectopic atrial tachycardia:" evidence for re-entry. Am J Cardiol 1982;49:920-926.
6. Waldo AL, Plumb VJ, Arciniegas JG, et al: Transient entrainment and interruption of the atrioventricular bypass pathway type of paroxysmal atrial tachycardia: A model for understanding and identifying reentrant arrhythmias. Circulation 1983;67:73-83.
7. Brugada P, Waldo AL, Wellens HJ: Transient entrainment and interruption of atrioventricular node tachycardia. J Am Coll Cardiol 1987;9:769-775.
8. Henthorn RW, Okumura K, Olshansky B, et al: A fourth criterion for transient entrainment: The electrogram equivalent of progressive fusion. Circulation 1988;77:1003-1012.
9. Okumura K, Olshansky B, Henthorn RW, et al: Demonstration of the presence of slow conduction during sustained ventricular tachycardia in man: Use of transient entrainment of the tachycardia. Circulation 1987;75:369-378.
10. Waldo AL, Henthorn RW: Use of transient entrainment during ventricular tachycardia to localize a critical area in the reentry circuit for ablation. Pacing Clin Electrophysiol 1989;12:231-244.
11. Portillo B, Mejias J, Leon-Portillo N, et al: Entrainment of atrioventricular nodal reentrant tachycardias during overdrive pacing from high right atrium and coronary sinus: With special reference to atrioventricular dissociation and 2:1 retrograde block during tachycardias. Am J Cardiol 1984;53:1570-1576.
12. Okumura K, Henthorn RW, Epstein AE, et al: Further observations on transient entrainment: Importance of pacing site and properties of the components of the reentry circuit. Circulation 1985;72:1293-1307.
13. Jazayeri MR, Hempe SL, Sra JS, et al: Selective transcatheter ablation of the fast and slow pathways using radiofrequency energy in patients with atrioventricular nodal reentrant tachycardia [see comments]. Circulation 1992;85:1318-1328.
14. Wathen M, Natale A, Wolfe K, et al: An anatomically guided approach to atrioventricular node slow pathway ablation. Am J Cardiol 1992;70:886-889.
15. Haissaguerre M, Gaita F, Fischer B, et al: Elimination of atrioventricular nodal reentrant tachycardia using discrete slow potentials to guide application of radiofrequency energy [see comments]. Circulation 1992;85:2162-2175.
16. Jackman WM, Beckman KJ, McClelland JH, et al: Treatment of supraventricular tachycardia due to atrioventricular nodal reentry, by radiofrequency catheter ablation of slow-pathway conduction. N Engl J Med 1992;327:313-318.
17. Hintringer F, Hartikainen J, Davies DW, et al: Prediction of atrioventricular block during radiofrequency ablation of the slow pathway of the atrioventricular node. Circulation 1995;92: 3490-3496.
18. Willems S, Shenasa H, Kottkamp H, et al: Temperature-controlled slow pathway ablation for treatment of atrioventricular nodal reentrant tachycardia using a combined anatomical and electrogram guided strategy. Eur Heart J 1996;17:1092-1102.
19. Jentzer JH, Goyal R, Williamson BD, et al: Analysis of junctional ectopy during radiofrequency ablation of the slow pathway in patients with atrioventricular nodal reentrant tachycardia. Circulation 1994;90:2820-2826.
20. Poret P, Leclercq C, Gras D, et al: Junctional rhythm during slow pathway radiofrequency ablation in patients with atrioventricular nodal reentrant tachycardia: beat-to-beat analysis and its prognostic value in relation to electrophysiologic and anatomic parameters. J Cardiovasc Electrophysiol 2000;11:405-412.
21. Schumacher B, Tebbenjohanns J, Pfeiffer D, et al: Junctional arrhythmias in radiofrequency modification of the atrioventricular node. Z Kardiol 1995;84:977-985.
22. Lipscomb KJ, Zaidi AM, Fitzpatrick AP, et al: Slow pathway modification for atrioventricular node re-entrant tachycardia: Fast junctional tachycardia predicts adverse prognosis. Heart 2001;85:44-47.
23. Iturralde P, Araya-Gomez V, Colin L, et al: A new ECG algorithm for the localization of accessory pathways using only the polarity of the QRS complex. J Electrocardiol 1996;29:289-299.
24. d'Avila A, Brugada J, Skeberis V, et al: A fast and reliable algorithm to localize accessory pathways based on the polarity of the QRS complex on the surface ECG during sinus rhythm. Pacing Clin Electrophysiol 1995;18:1615-1627.
25. Xie B, Heald SC, Bashir Y, et al: Localization of accessory pathways from the 12-lead electrocardiogram using a new algorithm. Am J Cardiol 1994;74:161-165.
26. Arruda MS, McClelland JH, Wang X, et al: Development and validation of an ECG algorithm for identifying accessory pathway ablation site in Wolff-Parkinson-White syndrome. J Cardiovasc Electrophysiol 1998;9:2-12.
27. Chiang CE, Chen SA, Teo WS, et al: An accurate stepwise electrocardiographic algorithm for localization of accessory pathways in patients with Wolff-Parkinson-White syndrome from a comprehensive analysis of delta waves and R/S ratio during sinus rhythm. Am J Cardiol 1995;76:40-46.
28. Fitzpatrick AP, Gonzales RP, Lesh MD, et al: New algorithm for the localization of accessory atrioventricular connections using a baseline electrocardiogram [published erratum appears in J Am Coll Cardiol 1994 Apr;23(5):1272]. J Am Coll Cardiol 1994;23: 107-116.
29. Diker E, Ozdemir M, Tezcan UK, et al: QRS polarity on 12-lead surface ECG: A criterion for the differentiation of right and left posteroseptal accessory atrioventricular pathways. Cardiology 1997;88:328-332.
30. Cappato R, Schluter M, Weiss C, et al: Mapping of the coronary sinus and great cardiac vein using a 2-French electrode catheter and a right femoral approach. J Cardiovasc Electrophysiol 1997; 8:371-376.
31. Damle RS, Choe W, Kanaan NM, et al: Atrial and accessory pathway activation direction in patients with orthodromic supraventricular tachycardia: Insights from vector mapping. J Am Coll Cardiol 1994;23:684-692.

32. Miles WM, Yee R, Klein GJ, et al: The preexcitation index: An aid in determining the mechanism of supraventricular tachycardia and localizing accessory pathways. Circulation 1986;74:493-500.
33. Berger RD, Nsah E, Calkins H: Signal-averaged intracardiac electrograms: A new method to detect Kent potentials. J Cardiovasc Electrophysiol 1997;8:155-160.
34. Jolly WA, Ritchie WJ: Auricular flutter and fibrillation. Heart 1911;2:177-221.
35. Lewis T, Drury AN, Iliescu CC: A demonstration of circus movement in clinical flutter of the auricles. Heart 1921;8:341-359.
36. Friedman PA, Luria D, Fenton AM, et al: Global right atrial mapping of human atrial flutter: The presence of posteromedial (sinus venosa region) functional block and double potentials: A study in biplane fluoroscopy and intracardiac echocardiography. Circulation 2000;101:1568-1577.
37. Feld GK, Mollerus M, Birgersdotter-Green U, et al: Conduction velocity in the tricuspid valve—inferior vena cava isthmus is slower in patients with type I atrial flutter compared to those without a history of atrial flutter. J Cardiovasc Electrophysiol 1997;8: 1338-1348.
38. Nakagawa H, Lazzara R, Khastgir T, et al: Role of the tricuspid annulus and the eustachian valve/ridge on atrial flutter: Relevance to catheter ablation of the septal isthmus and a new technique for rapid identification of ablation success [see comments]. Circulation 1996;94:407-424.
39. Olgin JE, Kalman JM, Saxon LA, et al: Mechanism of initiation of atrial flutter in humans: Site of unidirectional block and direction of rotation. J Am Coll Cardiol 1997;29:376-384.
40. Schilling RJ, Peters NS, Goldberger J, et al: Characterization of the anatomy and conduction velocities of the human right atrial flutter circuit determined by noncontact mapping. J Am Coll Cardiol 2001;38:385-393.
41. Shah DC, Jais P, Haissaguerre M, et al: Three-dimensional mapping of the common atrial flutter circuit in the right atrium. Circulation 1997;96:3904-3912.
42. Feld GK, Fleck RP, Chen PS, et al: Radiofrequency catheter ablation for the treatment of human type 1 atrial flutter: Identification of a critical zone in the reentrant circuit by endocardial mapping techniques. Circulation 1992;86:1233-1240.
43. Natale A, Breeding L, Tomassoni G, et al: Ablation of right and left ectopic atrial tachycardias using a three-dimensional nonfluoroscopic mapping system. Am J Cardiol 1998;82:989-992.
44. Kalman JM, Olgin JE, Karch MR, et al: "Cristal tachycardias:" origin of right atrial tachycardias from the crista terminalis identified by intracardiac echocardiography. J Am Coll Cardiol 1998; 31:451-459.
45. Haïssaguerre M, Jaïs P, Shah DC, et al: Spontaneous initiation of atrial fibrillation by ectopic beats originating in the pulmonary veins. N Engl J Med 1998;339:659-666.
46. Lau CP, Tse HF, Ayers GM: Defibrillation-guided radiofrequency ablation of atrial fibrillation secondary to an atrial focus. J Am Coll Cardiol 1999;33:1217-1226.
47. Haïssaguerre M, Jaïs P, Shah DC, et al: Catheter ablation of chronic atrial fibrillation targeting the reinitiating triggers. J Cardiovasc Electrophysiol 2000;11:2-10.
48. Lee SH, Tai CT, Lin WS, et al: Predicting the arrhythmogenic foci of atrial fibrillation before atrial transseptal procedure: Implication for catheter ablation. J Cardiovasc Electrophysiol 2000;11:750-757.
49. Haïssaguerre M, Jaïs P, Shah DC, et al: Electrophysiological end point for catheter ablation of atrial fibrillation initiated from multiple pulmonary venous foci. Circulation 2000;101:1409-1417.
50. Tse HF, Lau CP, Kou W, et al: Comparison of endocardial activation times at effective and ineffective ablation sites within the pulmonary veins. J Cardiovasc Electrophysiol 2000;11:155-159.
51. Hsieh MH, Chen SA, Tai CT, et al: Double multielectrode mapping catheters facilitate radiofrequency catheter ablation of focal atrial fibrillation originating from pulmonary veins. J Cardiovasc Electrophysiol 1999;10:136-144.
52. Kumagai K, Gondo N, Matsumoto N, et al: New technique of simultaneous catheter mapping of four pulmonary veins for catheter ablation in focal atrial fibrillation. Circulation 1999;100[18 Suppl I], I-66.2-11 [Abstract].
53. Shah DC, Haissaguerre M, Jaïs P, et al: Variability of provocative manoeuvers for inducing pulmonary vein ectopy. Eur Heart J 1999;20[S]219 [abstract].
54. Markides V, Schilling RJ, Chow AWC, et al: Non-contact mapping of the human left atrium to guide ablation of focal atrial fibrillation. Circulation 2000;102[18 Suppl. II], II-575 [abstract].
55. Haïssaguerre M, Shah DC, Jais P, et al: Electrophysiological breakthroughs from the left atrium to the pulmonary veins. Circulation 2000;102:2463-2465.
56. Delacretaz E, Stevenson WG, Ellison KE, et al: Mapping and radiofrequency catheter ablation of the three types of sustained monomorphic ventricular tachycardia in nonischemic heart disease [see comments]. J Cardiovasc Electrophysiol 2000;11:11-17.
57. Carlson MD, White RD, Trohman RG, et al: Right ventricular outflow tract ventricular tachycardia: Detection of previously unrecognized anatomic abnormalities using cine magnetic resonance imaging. J Am Coll Cardiol 1994;24:720-727.
58. Globits S, Kreiner G, Frank H, et al: Significance of morphological abnormalities detected by MRI in patients undergoing successful ablation of right ventricular outflow tract tachycardia. Circulation 1997;96:2633-2640.
59. Lerman BB, Stein K, Engelstein ED, et al: Mechanism of repetitive monomorphic ventricular tachycardia. Circulation 1995;92: 421-429.
60. Rahilly GT, Prystowsky EN, Zipes DP, et al: Clinical and electrophysiologic findings in patients with repetitive monomorphic ventricular tachycardia and otherwise normal electrocardiogram. Am J Cardiol 1982;50:459-468.
61. Lerman BB, Stein KM, Markowitz SM: Adenosine-sensitive ventricular tachycardia: A conceptual approach. J Cardiovasc Electrophysiol 1996;7:559-569.
62. Lerman BB, Belardinelli L, West GA, et al: Adenosine-sensitive ventricular tachycardia: Evidence suggesting cyclic AMP-mediated triggered activity. Circulation 1986;74:270-280.
63. Callans DJ, Menz V, Schwartzman D, et al: Repetitive monomorphic tachycardia from the left ventricular outflow tract: Electrocardiographic patterns consistent with a left ventricular site of origin. J Am Coll Cardiol 1997;29:1023-1027.
64. Movsowitz C, Schwartzman D, Callans DJ, et al: Idiopathic right ventricular outflow tract tachycardia: Narrowing the anatomic location for successful ablation. Am Heart J 1996;131:930-936.
65. Klein LS, Shih HT, Hackett FK, et al: Radiofrequency catheter ablation of ventricular tachycardia in patients without structural heart disease. Circulation 1992;85:1666-1674.
66. Ohe T, Shimomura K, Aihara N, et al: Idiopathic sustained left ventricular tachycardia: Clinical and electrophysiologic characteristics [published erratum appears in Circulation 1988 Aug;78(2):A5]. Circulation 1988;77:560-568.
67. Okumura K, Matsuyama K, Miyagi H, et al: Entrainment of idiopathic ventricular tachycardia of left ventricular origin with evidence for reentry with an area of slow conduction and effect of verapamil. Am J Cardiol 1988;62:727-732.
68. Tsuchiya T, Okumura K, Honda T, et al: Significance of late diastolic potential preceding Purkinje potential in verapamil-sensitive idiopathic left ventricular tachycardia. Circulation 1999; 99:2408-2413.
69. Lerman BB, Stein KM, Markowitz SM: Mechanisms of idiopathic left ventricular tachycardia. J Cardiovasc Electrophysiol 1997;8: 571-583.
70. Sato M, Sakurai M, Yotsukura A, et al: Diastolic potentials in verapamil-sensitive ventricular tachycardia: True potentials or bystanders of the reentry circuits? Am Heart J 1999;138:560-566.
71. Akhtar M, Damato AN, Batsford WP, et al: Demonstration of reentry within the His-Purkinje system in man. Circulation 1974;50: 1150-1162.
72. Touboul P, Kirkorian G, Atallah G, et al: Bundle branch reentry: A possible mechanism of ventricular tachycardia. Circulation 1983;67:674-680.
73. Blanck Z, Dhala A, Deshpande S, et al: Bundle branch reentrant ventricular tachycardia: Cumulative experience in 48 patients. J Cardiovasc Electrophysiol 1993;4:253-262.
74. Narasimhan C, Jazayeri MR, Sra J, et al: Ventricular tachycardia in valvular heart disease: Facilitation of sustained bundle-branch reentry by valve surgery. Circulation 1997;96:4307-4313.
75. Blanck Z, Jazayeri M, Dhala A, et al: Bundle branch reentry: A mechanism of ventricular tachycardia in the absence of myocardial or valvular dysfunction. J Am Coll Cardiol 1993;22: 1718-1722.

76. Wang PJ, Friedman PL: "Clockwise" and "counterclockwise" bundle branch reentry as a mechanism for sustained ventricular tachycardia masquerading as supraventricular tachycardia. Pacing Clin Electrophysiol 1989;12:1426-1432.
77. Tchou P, Jazayeri M, Denker S, et al: Transcatheter electrical ablation of right bundle branch: A method of treating macroreentrant ventricular tachycardia attributed to bundle branch reentry. Circulation 1988;78:246-257.
78. Caceres J, Jazayeri M, McKinnie J, et al: Sustained bundle branch reentry as a mechanism of clinical tachycardia. Circulation 1989;79:256-270.
79. Blanck Z, Deshpande S, Jazayeri MR, et al: Catheter ablation of the left bundle branch for the treatment of sustained bundle branch reentrant ventricular tachycardia. J Cardiovasc Electrophysiol 1995;6:40-43.
80. Stevenson WG, Friedman PL, Sweeney MO: Catheter ablation as an adjunct to ICD therapy. Circulation 1997;96:1378-1380.
81. Strickberger SA, Man KC, Daoud EG, et al: A prospective evaluation of catheter ablation of ventricular tachycardia as adjuvant therapy in patients with coronary artery disease and an implantable cardioverter-defibrillator. Circulation 1997;96:1525-1531.
82. Wellens HJ, Schuilenburg RM, Durrer D: Electrical stimulation of the heart in patients with ventricular tachycardia. Circulation 1972;46:216-226.
83. The ESVEM trial. Electrophysiologic Study Versus Electrocardiographic Monitoring for selection of antiarrhythmic therapy of ventricular tachyarrhythmias. The ESVEM Investigators. Circulation 1989;79:1354-1360.
84. Kocovic DZ, Harada T, Friedman PL, et al: Characteristics of electrograms recorded at reentry circuit sites and bystanders during ventricular tachycardia after myocardial infarction. J Am Coll Cardiol 1999;34:381-388.
85. Callans DJ, Zardini M, Gottlieb CD, et al: The variable contribution of functional and anatomic barriers in human ventricular tachycardia: An analysis with resetting from two sites. J Am Coll Cardiol 1996;27:1106-1111.
86. Ciaccio EJ, Scheinman MM, Fridman V, et al: Dynamic changes in electrogram morphology at functional lines of block in reentrant circuits during ventricular tachycardia in the infarcted canine heart: A new method to localize reentrant circuits from electrogram features using adaptive template matching. J Cardiovasc Electrophysiol 1999;10:194-213.
87. Morady F, Kadish A, Rosenheck S, et al: Concealed entrainment as a guide for catheter ablation of ventricular tachycardia in patients with prior myocardial infarction. J Am Coll Cardiol 1991;17:678-689.
88. El Shalakany A, Hadjis T, Papageorgiou P, et al: Entrainment/mapping criteria for the prediction of termination of ventricular tachycardia by single radiofrequency lesion in patients with coronary artery disease. Circulation 1999;99:2283-2289.
89. Stevenson WG, Friedman PL, Sager PT, et al: Exploring postinfarction reentrant ventricular tachycardia with entrainment mapping. J Am Coll Cardiol 1997;29:1180-1189.
90. Stevenson WG, Khan H, Sager P, et al: Identification of reentry circuit sites during catheter mapping and radiofrequency ablation of ventricular tachycardia late after myocardial infarction Circulation 1993;88:1647-1670.
91. Soejima K, Stevenson WG, Maisel WH, et al: The N + 1 difference: A new measure for entrainment mapping. J Am Coll Cardiol 2001;37:1386-1394.
92. Furniss S, Anil-Kumar R, Bourke JP, et al: Radiofrequency ablation of hemodynamically unstable ventricular tachycardia after myocardial infarction. Heart 2000;84:648-652.
93. Wilber DJ, Kopp DE, Glascock DN, et al: Catheter ablation of the mitral isthmus for ventricular tachycardia associated with inferior infarction. Circulation 1995;92:3481-3489.
94. Bartlett TG, Mitchell R, Friedman PL, et al: Histologic evolution of radiofrequency lesions in an old human myocardial infarct causing ventricular tachycardia. J Cardiovasc Electrophysiol 1995;6:625-629.
95. Nath S, DiMarco JP, Haines DE: Basic aspects of radiofrequency catheter ablation. J Cardiovasc Electrophysiol 1994;5:863-876.
96. Sosa E, Scanavacca M, d'Avila A, et al: Endocardial and epicardial ablation guided by nonsurgical transthoracic epicardial mapping to treat recurrent ventricular tachycardia. J Cardiovasc Electrophysiol 1998;9:229-239.
97. Sosa E, Scanavacca M, d'Avila A, et al: Nonsurgical transthoracic epicardial catheter ablation to treat recurrent ventricular tachycardia occurring late after myocardial infarction. J Am Coll Cardiol 2000;35:1442-1449.
98. de Paola AA, Melo WD, Tavora MZ, et al: Angiographic and electrophysiological substrates for ventricular tachycardia mapping through the coronary veins. Heart 1998;79:59-63.
99. Gepstein L, Hayam G, Ben Haim SA: A novel method for nonfluoroscopic catheter-based electroanatomical mapping of the heart: In vitro and in vivo accuracy results. Circulation 1997;95:1611-1622.
100. Stevenson WG, Delacretaz E, Friedman PL, et al: Identification and ablation of macroreentrant ventricular tachycardia with the CARTO electroanatomical mapping system. Pacing Clin Electrophysiol 1998;21:1448-1456.
101. Marchlinski FE, Callans DJ, Gottlieb CD, et al: Linear ablation lesions for control of unmappable ventricular tachycardia in patients with ischemic and nonischemic cardiomyopathy. Circulation 2000;101:1288-1296.
102. Marchlinski F, Callans D, Gottlieb C, et al: Magnetic electroanatomical mapping for ablation of focal atrial tachycardias. Pacing Clin Electrophysiol 1998;21:1621-1635.
103. Sokoloski MC, Pennington JC III, Winton GJ, et al: Use of multisite electroanatomic mapping to facilitate ablation of intra-atrial reentry following the Mustard procedure. J Cardiovasc Electrophysiol 2000;11:927-930.
104. Pappone C, Oreto G, Lamberti F, et al: Catheter ablation of paroxysmal atrial fibrillation using a 3D mapping system. Circulation 1999;100:1203-1208.
105. Pappone C, Rosanio S, Oreto G, et al: Circumferential radiofrequency ablation of pulmonary vein ostia: A new anatomic approach for curing atrial fibrillation. Circulation 2000;102:2619-2628.
106. Kottkamp H, Hugl B, Krauss B, et al: Electromagnetic versus fluoroscopic mapping of the inferior isthmus for ablation of typical atrial flutter: A prospective randomized study. Circulation 2000;102:2082-2086.
107. Sra J, Bhatia A, Dhala A, et al: Electroanatomic mapping to identify breakthrough sites in recurrent typical human flutter. Pacing Clin Electrophysiol 2000;23:1479-1492.
108. Marchlinski FE, Callans DJ, Gottlieb CD, et al: Linear ablation lesions for control of unmappable ventricular tachycardia in patients with ischemic and nonischemic cardiomyopathy. Circulation 2000;101:1288-1296.
109. De Groot NM, Kuijper AF, Blom NA, et al: Three-dimensional distribution of bipolar atrial electrogram voltages in patients with congenital heart disease. Pacing Clin Electrophysiol 2001;24:1334-1342.
110. Nakagawa H, Shah N, Matsudaira K, et al: Characterization of reentrant circuit in macroreentrant right atrial tachycardia after surgical repair of congenital heart disease: Isolated channels between scars allow "focal" ablation. Circulation 2001;103:699-709.
111. Willems S, Weiss C, Ventura R, et al: Catheter ablation of atrial flutter guided by electroanatomic mapping (CARTO): A randomized comparison to the conventional approach. J Cardiovasc Electrophysiol 2000;11:1223-1230.
112. Wittkampf FH, Wever EF, Derksen R, et al: Accuracy of the LocaLisa system in catheter ablation procedures. J Electrocardiol 1999;32 Suppl:7-12.
113. Wittkampf FH, Wever EF, Derksen R, et al: LocaLisa: New technique for real-time 3-dimensional localization of regular intracardiac electrodes. Circulation 1999;99:1312-1317.
114. Kammeraad J, van Driel V, Ramanna H, et al: Catheter ablation of incisional atrial tachycardia using a novel mapping system: LocaLisa. Eur Heart J 21[Abstr. Suppl.], 368, 2000.
115. Wittkampf FH, Wever EF, Vos K, et al: Reduction of radiation exposure in the cardiac electrophysiology laboratory. Pacing Clin Electrophysiol 2000;23:1638-1644.
116. de Groot N, Bootsma M, van der Velde ET, et al: Three-dimensional catheter positioning during radiofrequency ablation in patients: First application of a real-time position management system. J Cardiovasc Electrophysiol 2000;11:1183-1192.
117. Khoury DS, Taccardi B, Lux RL, et al: Reconstruction of endocardial potentials and activation sequences from intracavitary probe measurements: Localization of pacing sites and effects of myocardial structure. Circulation 1995;91:845-863.

118. Gornick CC, Adler SW, Pederson B, et al: Validation of a new noncontact catheter system for electroanatomic mapping of left ventricular endocardium. Circulation 1999;99:829-835.
119. Kadish A, Hauck J, Pederson B, et al: Mapping of atrial activation with a noncontact, multielectrode catheter in dogs. Circulation 1999;99:1906-1913.
120. Schilling RJ, Peters NS, Davies DW: Simultaneous endocardial mapping in the human left ventricle using a noncontact catheter: Comparison of contact and reconstructed electrograms during sinus rhythm. Circulation 1998;98:887-898.
121. Asirvatham S, Packer DL: Validation of non-contact mapping to localize the site of simulated pulmonary vein ectopic foci. Circulation 2000;102, II-441.
122. Schilling RJ, Peters NS, Davies DW: Feasibility of a noncontact catheter for endocardial mapping of human ventricular tachycardia. Circulation 1999;99:2543-2552.
123. Betts TR, Roberts PR, Allen SA, et al: Radiofrequency ablation of idiopathic left ventricular tachycardia at the site of earliest activation as determined by noncontact mapping. J Cardiovasc Electrophysiol 2000;11:1094-1101.
124. Schneider MA, Ndrepepa G, Zrenner B, et al: Noncontact mapping-guided catheter ablation of atrial fibrillation associated with left atrial ectopy. J Cardiovasc Electrophysiol 2000;11:475-479.
125. Schilling RJ, Kadish AH, Peters NS, et al: Endocardial mapping of atrial fibrillation in the human right atrium using a non-contact catheter. Eur Heart J 2000;21:550-564.
126. Schilling RJ, Peters NS, Goldberger J, et al: Characterization of the anatomy and conduction velocities of the human right atrial flutter circuit determined by noncontact mapping. J Am Coll Cardiol 2001;38:385-393.
127. Schumacher B, Jung W, Lewalter T, et al: Verification of linear lesions using a noncontact multielectrode array catheter versus conventional contact mapping techniques. J Cardiovasc Electrophysiol 1999;10:791-798.
128. Betts TR, Roberts PR, Allen SA, et al: Electrophysiological mapping and ablation of intra-atrial reentry tachycardia after Fontan surgery with the use of a noncontact mapping system. Circulation 2000;102:419-425.
129. Marrouche NF, Beheiry S, Tomassoni G, et al: Three-dimensional nonfluoroscopic mapping and ablation of inappropriate sinus tachycardia: Procedural strategies and long-term outcome. J Am Coll Cardiol 2002;39:1046-1054.
130. Iwai S, Markowitz SM, Stein KM, et al: Response to adenosine differentiates focal from macroreentrant atrial tachycardia: Validation using three-dimensional electroanatomic mapping. Circulation 2002;106:2793-2799.
131. Tai CT, Huang JL, Lin YK, et al: Noncontact three-dimensional mapping and ablation of upper loop re-entry originating in the right atrium. J Am Coll Cardiol 2002;40:746-753.
132. Hoffmann E, Reithmann C, Nimmermann P, et al: Clinical experience with electroanatomic mapping of ectopic atrial tachycardia. Pacing Clin Electrophysiol 2002;25:49-56.
133. Friedman PA, Asirvatham SJ, Grice S, et al: Non contact mapping to guide ablation of right ventricular outflow tract tachycardia. J Am Coll Cardiol 2002:39:1808-1812.
134. Betts TR, Ho SY, Sanchez-Quintana D, et al: Three-dimensional mapping of right atrial activation during sinus rhythm and its relationship to endocardial architecture. J Cardiovasc Electrophysiol 2002;1152-1159.
135. Garrigue S, Reuter S, Efimov IR, et al: Optical mapping technique applied to biventricular pacing: Potential mechanisms of ventricular arrhythmias occurrence. Pacing Clin Electrophysiol 2003;(1 Pt 2):197-205.
136. Filipecki A, Saksena S, Prakash A, Krol R, Philip G: 3-D Noncontact and multicatheter contact maps of right atrial activation. Europace 2001;Suppl. B,105 [Abstract].
136. Saksena S: Electrophysiologic study in patients with atrial fibrillation: An idea whose time has come yet again. (Editorial) J Int Card Electrophysiol 1999;3:101-107.
137. Saksena S, Prakash A, Krol RB, Shankar A: Regional endocardial mapping of spontaneous and induced atrial fibrillation in patients with heart disease and refractory atrial fibrillation. Am J Cardiol 1999;84:880-889.
138. Saksena S, Shankar A, Prakash A, Krol RB: Catheter mapping of spontaneous and induced atrial fibrillation in man. J Int Card Electrophysiol 2000;4(Suppl):21-28.
139. Saksena S, Filipecki A, Prakash A, Philip G: Organized and distinctive right atrial activation patterns are seen with 3D mapping depending on right or left atrial initiation of atrial fibrillation in patients with heart disease. Circulation 2002;19:2019 [Abstract].
140. Filipecki A, Saksena S, Prakash A, Philip G: Three-dimensional noncontact mapping demonstrates synergistic electrophysiologic effects of multisite atrial pacing and linear atrial ablation in patients with refractory atrial fibrillation. J Am Coll Cardiol 2003;41:1017 [Abstract].
141. Schalij MJ, van Rugge FP, Siezenga M, et al: Endocardial activation mapping of ventricular tachycardia in patients: First application of a 32-site bipolar mapping electrode catheter. Circulation 1998;98:2168-2179.
142. Zrenner B, Ndrepepa G, Schneider M, et al: Computer-assisted animation of atrial tachyarrhythmias recorded with a 64-electrode basket catheter. J Am Coll Cardiol 1999;34:2051-2060.
143. Michael MJ, Haines DE, DiMarco JP, et al: Elimination of focal atrial fibrillation with a single radiofrequency ablation: Use of a basket catheter in a pulmonary vein for computerized activation sequence mapping. J Cardiovasc Electrophysiol 2000;11:1159-1164.

Chapter 43 Ablation Technology

DAVID E. HAINES

Catheter ablation is a therapeutic modality that has enjoyed wide acceptance among patients and physicians for the treatment of a variety of arrhythmias. Ablation therapy is based on the concept that all arrhythmias arise from a focal source, or depend on electrical conduction through a critical isthmus of tissue. If the focal source or isthmus can be mapped, then damaged or destroyed, then the arrhythmia should no longer be initiated or able to propagate. It is paramount that the anatomy and physiology of the targeted arrhythmia are fully delineated by the operator before performing any ablation. A small, localized lesion that incorporates the critical anatomic substrate but results in minimal collateral damage is ideal. Other factors that should improve ablation outcomes include excellent performance of the catheter or other delivery platform to optimize positioning of the ablating instrument in close contiguity to the arrhythmia source. Presently, success rates for catheter ablation of common arrhythmias, such as paroxysmal supraventricular tachycardia exceed 95%. However, challenges in the ablation of complex arrhythmias, such as reentrant atrial fibrillation and ventricular tachycardia (VT) still remain. In order to meet these challenges, improve success rates, and decrease complications and procedure times, technologies for catheter ablation continue to improve. The following section will present data about the biophysics and pathophysiology of radiofrequency (RF) catheter ablation lesion formation, and discuss alternative technologies that are under investigation.

Biophysics of Radiofrequency Catheter Ablation

Radiofrequency energy is a form of electromagnetic electrical energy in a relatively low frequency range (Table 43-1). The energy is transmitted through electrical conducting media (transmission lines, electrode catheters, and tissue), and is dissipated as heat in the transmission circuit in regions with the highest electrical impedance. Tissue has higher impedance than the electrical wires, thus, most of the RF energy is dissipated in the tissue, and relatively little is lost in the transmission line. The electrical contacts in the tissue are generally the tip of the ablation electrode at the targeted site, and a dispersive electrode on the patient's skin. Because RF current is alternating, there is no fixed anode or cathode. Although any low-frequency electrical current would result in a similar electrical current distribution between the catheter tip and dispersive electrode, frequencies of at least 300 kHz are selected to minimize stimulation of the electrically responsive tissues, such as muscle and nerve cells, in order to minimize arrhythmias and pain.

As RF electrical current passes through resistive tissue, the energy is dissipated as heat. The amount of tissue heat is proportional to the square of the current density, and accordingly diminishes in proportion to the distance from the electrode to the fourth power.[1] Thus, the magnitude of direct volume heating drops off precipitously. Most of the ablative lesion is formed instead from heat conduction from the rim of direct volume heating. The edge of the pathologic lesion forms at the 50° C isotherm.[2] The temperature gradient can be shifted to deeper tissue levels and yield a larger lesion by increasing the depth of volume heating. This can be accomplished by increasing power, increasing the temperature at the electrode-tissue contact point, or increasing the size of the electrode.[3] The power delivery is limited, however, by excessive heating at the electrode-tissue interface. When the temperature reaches 100° C, boiling and coagulum form at the interface. This is electrically insulating and results in a sudden rise in electrical impedance.[4] With longer electrodes, there is an uneven distribution of current density along the electrode with highest current and greatest heating at the electrode edges.[5] For that reason, a single tip temperature sensor can underestimate the actual peak temperature by as much as 20° C, and, therefore, the maximal selected temperature should not exceed 80° C. Average RF lesions grow in a monoexponential function with a half time of about 9 seconds.[6] After termination of energy delivery, lesion growth continues for several more seconds due to conductive heating of deeper tissue planes by superficial hotter ones.[7] This is an important phenomenon, because if an adverse response to ablation is observed (e.g., heart block during atrioventricular [AV] nodal modification), tissue destruction will continue, despite prompt cessation of energy delivery.

TABLE 43-1 Factors Affecting Radiofrequency Catheter Ablation

	Relationship to Lesion Depth	Mechanism	Comments
Power amplitude	Linear increase	Greater depth of volume heating	Dependent on impedance and convective cooling
Peak temperature	Linear increase	Greater depth of volume heating	Coagulum formation and fall off of energy delivery if 100° C is exceeded.
Electrode size	Linear increase	Higher power delivery without excessive power density and overheating at surface. Larger source of volume heating	Long, thin geometries may have uneven power distribution and uneven tissue heating
System impedance	Inversely proportional	Higher impedance results in lower power density	Rapid impedance rise due to coagulum formation causes cessation of meaningful tissue heating as the power and current fall off precipitously.
Contact pressure	Variable increase of lesion size with increased pressure	Better energy coupling with tissue with increased contact pressure	If power can be increased to counter the effects of convective cooling by poor contact, lesion size may be larger
Duration of energy delivery	Monoexponential increase	Half time of lesion growth is 8-10 sec	Longer leads to larger, but steady state is reached in 1-2 mins. Further heating beyond that time does not increase lesion size further.
Convective cooling	Decrease	At fixed power, more cooling dissipates ablative energy and a smaller lesion is created	Protective effect of convective cooling on coronary arteries prevents vascular injury during most ablations.
Convective cooling with increased power	Variable increase	Cooling dissipates ablative energy at the surface and allows higher power delivery without impedance rise. A larger lesion is created	Depending on the magnitude of power delivery and power loss due to convective cooling, lesions may be larger, the same size, or even smaller.

The major phenomenon opposing tissue heating is convective cooling by the circulating blood pool. If the RF source is not closely coupled with the tissue, then the energy will be dissipated into the blood rather than into the tissue, and tissue heating will be limited. If energy delivery is not power limited, the phenomenon of convective cooling can paradoxically result in increased lesion size.[8] Cooling of the electrode surface allows delivery of much higher power to the electrode without exceeding 100° C and causing electrical impedance to rise. This phenomenon occurs spontaneously, as well in the setting of sliding catheter contact, or when the catheter is positioned at a site with high-volume blood flow. However, if RF energy delivery is power limited, cooling the electrode tip will just further dissipate energy and result in a smaller lesion.

Pathophysiology of Radiofrequency Energy Lesion Formation

The predominant mechanism of tissue injury appears to be thermal. The effects of hyperthermia on the heart have effects on several levels:

Tissue Effects of Hyperthermia

Appearance of myocardial tissue acutely after RF catheter ablation shows a central section of pallor and volume loss due to tissue desiccation. Surrounding this zone is a region of hemorrhagic tissue that shows evidence of coagulation necrosis, nuclear pyknosis, and basophilic stippling on histologic examination. After several hours, the lesion shows infiltration of the lesion border zone by inflammatory cells.[9] The secondary inflammatory response with either healing of the ablation border zone or progressive damage to this region may account for the clinical observations of recovery of conduction after an initially successful ablation procedure,[10] or a "delayed cure" after an initially unsuccessful procedure.[11] Ultimately, after 8 weeks, the lesion becomes uniformly fibrotic with further volume loss.[12] The lesion edges are well demarcated from normal tissue, thus chronic hyperthermic lesions are not arrhythmogenic.

Hyperthermic exposure has marked effects on microvascular perfusion of myocardium. Within the acute pathologic lesion, tissue blood flow is absent. Beyond the border of the acute pathologic lesion, there is evidence of diminished microvascular blood flow that extends as far as 6 mm from the lesion edge. Ultrastuctural evaluation of the coronary microvascular endothelium shows evidence of basement membrane and plasma membrane dissolution and red blood cell stasis.[13] Thus, vascular injury accounts for the edema that may be observed acutely after RF lesion formation. Part of the formation of the final pathologic lesion may be due to an ischemic mechanism.

Cellular Effects of Hyperthermia

A variety of ultrastructural changes are observed in eukaryotic cells in response to hyperthermia, and are both time and temperature dependent. In particular, structural proteins, metabolic proteins, and lipid membranes seem to be the most thermally sensitive. In erythrocytes, loss of the typical bi-concave morphology occurs reproducibly at 50° C due to thermal inactivation of the cytoskeletal protein, spectrin.[14] Creatine kinase (thermally inactivated at 62 °C[15]) and other metabolic proteins, such as adenylate cyclase, undergo conformational changes in response to heating that render them inactive, and may contribute to cell death. The structure with the most thermal sensitivity appears to be the sarcolemmal membrane. Increased membrane fluidity is observed in eukaryotic cells at 46° to 50° C.[16] Myocytes exposed to brief episodes of hyperthermia begin to depolarize at 45° C and show irreversible contracture above 50° C.[17] Similarly, conduction velocity begins to drop at 45° C, and conduction block occurs at approximately 50° C.[18] This depolarization appears to be mediated by nonspecific cellular entry of cations, particularly calcium. The influx of calcium into the cell can be partially buffered by the sarcoplasmic reticulum and other intracellular reservoirs in the intermediate hyperthermic range (45° to 50° C) and allow for reversibility of the cellular dysfunction. However, these mechanisms are overwhelmed at higher temperatures and cells die.[19]

Advanced Ablation Technologies

The safety and efficacy profile of catheter ablation has been shaped by the skill of the operators, but also to a great extent the characteristics of RF energy in this application (Table 43-2). As described above, RF power delivery through a catheter electrode is limited by the propensity for coagulum and char formation. Once the peak temperature has reached 100° C, electrically insulating coagulated tissue and blood proteins adhere to the electrode surface and diminish the active electrode surface area. As the local power density emanating from the remaining available electrode surface area increases, the coagulation increases. Rapidly, the entire electrode surface becomes encased in nonconducting coagulum, electrical impedance at the electrode-tissue interface rises rapidly, and overall tissue power density rapidly falls. This results in termination of effective tissue heating.[4] Higher power delivery and higher target temperatures only increase the coagulum and char formation, but cannot increase lesion size. Thus, lesion size and depth of conventional RF catheter ablation are limited by their biophysical properties.

Operators have come to appreciate RF catheter ablation as a highly effective modality for the elimination of arrhythmogenic substrates that are relatively superficial to the endocardial surface. However, deeper sites of arrhythmia origin as are found in some patients with reentrant VT, atrial flutter, and accessory pathways located in the posteroseptal space may require larger ablative lesions to effectively eliminate all arrhythmogenic myocardium. To achieve this goal, investigators have examined an array of ablation modalities. Hyperthermic ablation can be achieved with a wide variety of electromagnetic radiation, varying from low frequency RF current, to intermediate frequency (microwave), to high frequency (laser) and very high frequency (β and γ radiation). In addition, conventional RF energy delivery can be modified to improve its depth and distribution of volume heating.

LARGE ABLATION ELECTRODES

As described above, the size of an RF ablative lesion is proportional to the electrode size. With this in mind, investigators have tested a variety of large ablation

TABLE 43-2 Advanced Ablation Technologies

	Mode of Ablation	Advantages	Disadvantages
Large RF ablation electrodes	Greater electrode-tissue contact area. Higher power. More volume heating and larger lesions.	Conventional power supply. Temperature monitoring useful.	Uneven heating along long electrodes. Side contact difficult. High power required.
Cooled tip RF electrodes	Surface cooling with electrode tip irrigation or perfusion allows higher power amplitude. More volume heating and larger lesions.	Conventional power supply. Can produce dramatic increase in lesion size.	Poor controllability. Temperature monitoring of little value. Risk of subendocardial superheating and "pop" lesions.
Phased RF	Phase shift of RF wave between contiguous bipoles leads to blend of unipolar and bipolar energy delivery.	Fairly uniform RF field along a long RF catheter.	Catheter-tissue contact remains major limiting factor in successful lesion formation.
Microwave	Dielectric heating of water in tissue from radiated electromagnetic energy. More volume heating and larger lesions.	Should create large and deep lesions.	New power supply. Difficult to design/build antenna catheters. Heating of transmission line.
Laser	Optical heating from fiberoptic transmission line.	Large amount of ablative energy can be transmitted through small, flexible fiberoptic.	Not omnidirectional. Narrow endocardial contact point with scatter in tissue produces "flask-shaped" lesion. Control difficult. Conventional generators expensive.
Ultrasound	Mechanical heating from sound waves transmitted from piezoelectric crystal. Large depth of volume heating and larger lesions.	Preferential heating of tissue versus blood. Excellent volume heating adjustable by selected ultrasound frequency.	May be power limited for distant targets. Near field microcavitation may limit power transmission.
Cryothermy	Ice crystal formation disrupts cells.	Reversible effects with moderate cooling allows "ice mapping." Painless. Ablation tip sticks to targeted tissue.	Long ablation times. Past systems have been cumbersome. Late recovery of conduction.

RF, Radiofrequency.

electrodes to enhance lesion size and procedure efficacy. The initial studies described the value of a conventional 2 mm-tip electrode to a 4 mm-tip electrode for standard ablation, and reported that there was a higher efficacy and a lower rate of sudden impedance rise.[20] Subsequently, investigators have employed very large-tip electrode catheters (8 to 10 mm) to attempt to improve procedure efficacy and decrease procedure time. A randomized comparison of catheter ablation of atrial flutter with conventional 4-mm tip electrodes versus 8-mm tip electrodes was performed. Fewer RF energy deliveries and shorter procedure times were required to achieve successful conduction block in the tricuspid-inferior vena cava isthmus when the large tip electrode was used.[21] Linear atrial ablation for the curative therapy of atrial fibrillation has been attempted with long coil electrodes with a side-contact orientation.[22] Although these electrodes can effectively create transmural continuous linear atrial lesions, close electrode-tissue contact along the entire length of the electrode is necessary to take advantage of this geometry. A novel clamping bipolar RF electrode system has been introduced for intraoperative linear atrial ablation. This system employs two long, thin electrodes in a parallel orientation mounted on the jaws of a spring-loaded clamp. As RF energy is delivered across the tissue interposed between the clamp jaws, the tissue desiccates, and the regional electrical conductance falls. The RF current shifts to the remaining unablated tissue along the line. The RF energy continues for 5 to 15 seconds until the ablation along the entire line is complete, at which time the electrical conductance along the line falls precipitously and the power delivery is terminated.[23] This system assures 100% continuity and transmurality of linear lesion formation, but it is only available for an open surgical approach at this time.

Balloon catheters with a RF power source have been employed to act as large virtual RF electrodes. One system employed a semi-permeable membrane for the balloon material, and filled the balloon with highly conductive medium (9% saline). Despite high current densities close to the electrode within the balloon, most of the heating occurred on the balloon surface where there was the transition from a low-impedance to a high-impedance medium. Thus, a large ablation electrode could be created on a catheter that passes through a conventional introducer sheath.[24] Other investigators have used RF thermal balloon ablation for the ablation of pulmonary veins. In one study, a 13.56-MHz energy source was employed. This creates capacitive heating along the points where the balloon contacts the endocardium. It was successful in 83% of the experimental pulmonary vein ablations after three ablations per vein. No complications were observed with this system.[25]

COOLED TIP RADIOFREQUENCY CATHETER ABLATION

As described earlier, convective cooling decreases the efficiency of RF energy coupling with the tissue because a portion of that energy is dissipated in the circulating blood pool rather than heating tissue. However, this phenomenon may be used to the advantage of the operator because surface cooling permits application of higher power without boiling, char, and coagulum formation at the catheter tip. The peak tissue temperature is found below the endocardial surface in these cases, and the subsequent volume of resistive heating may be much larger. Thus, cooling of the catheter tip can result in significant increases in lesion depth and diameter.[26] Two approaches have been pursued to achieve this in the clinical setting: tip perfusion with saline that is continuously flushed through tiny holes in the catheter tip; and tip cooling with closed circulation of saline through the catheter tip with return out the back of the catheter. The irrigated tip catheter was compared with the standard 4 mm-tip catheter in a randomized crossover trial of atrial flutter ablation in 50 patients. Crossover after 21 unsuccessful energy applications was required in 15% of patients assigned to conventional catheters, but none assigned to irrigated tip catheters. The irrigated system shortened procedure time and fluoroscopy exposure.[27] Another report described the successful ablation with irrigated-tip catheters in 18 patients with accessory pathways in whom ablation with conventional systems had failed. Success was achieved in 17 of these 18 patients.[28] A similar trial reported the use of the irrigated-tip catheter for the ablation of resistant VT. Success was achieved in five of eight patients (63%) of whom none had clinical recurrence over 6.5 ± 4.0 months of follow up.[29] In a large series of VT ablation with a perfused ablation tip system, all VTs were eliminated in 41% of the 146 patients enrolled, and the clinical, mappable VTs were eliminated in 75%. There was no comparison group. This study had a 9% rate of major adverse events, but it is not clear that any of them were directly related to the perfused tip catheter technology.[30] Complications that have been anecdotally linked with perfused-tip catheters and their associated increased depth of volume heating and occasional subendocardial steam pops include coronary arterial injury and perforation with tamponade. However, if used in carefully selected patients, the perfused tip-catheters certainly should improve overall ablation outcomes.

PHASED RADIOFREQUENCY CATHETER ABLATION

Radiofrequency electrical current is a variety of high frequency alternating current. When RF current is delivered in a unipolar fashion from multiple electrodes, a voltage gradient between the ablation electrodes and the indifferent electrode (ground pad) exists throughout energy delivery, but no electrical current passes from one ablation electrode to another. If, however, the alternating current is out of phase between contiguous electrodes, then current not only passes from the ablation electrode to the indifferent electrode, but it also passes between contiguous electrodes. Thus, if multi-electrode ablation is performed, a blend of unipolar and bipolar energy delivery can be achieved, and lesion contiguity among ablation electrodes may be enhanced. However, because similar biophysical constraints exist between unipolar and multipolar biphasic energy delivery, it would not be anticipated that this approach would yield deeper ablative lesions.

The phased RF technology has been applied to linear ablation attempts for the cure of atrial fibrillation. Preliminary studies have employed a system of phased RF ablation with multiple 3-mm ring electrodes on catheters positioned tangential to the atrial walls. In preliminary human trials, acute success was achieved in 14 of 15 patients with right atrial ablation. Atrial fibrillation recurred during short-term follow-up on 12 patients.[31] Subsequent phased RF ablation in the right and left atrium (including lesions encircling the pulmonary veins) has achieved arrhythmia cure in 33% of patients, with arrhythmia improvement in an additional 56%. Thus, phased RF energy delivery may be a useful technology for continuous and contiguous linear lesion formation, but its ultimate role in this arena still needs to be determined.

MICROWAVE CATHETER ABLATION

Microwave energy is a variety of electromagnetic energy that can accomplish ablation with tissue heating. The FCC-approved frequencies for medical microwave are 915 MHz and 2450 MHz. Unlike RF energy, microwave energy heats tissue with dielectric heating. Dielectric heating is accomplished by inducing oscillatory motion of charged molecules (water) with the microwave energy, changing electromagnetic energy into kinetic energy (heat). Because the mechanism of dielectric heating is different from electrical (resistive) heating with RF ablation, the promise of deeper tissue heating exists. Specifically, microwave energy is radiated into tissue from an antenna, rather than conducted into tissue with a standard electrode. While the power density of any energy point source will decrease with the square of the distance from that source, the dissipation of that electromagnetic energy into the tissue as heat is dependent on the properties of the radiation. Higher frequencies yield less penetration, or skin depth. Lower frequencies (such as 915 MHz) would be predicted to produce deep tissue heating without the risk of overheating at the antenna–tissue interface and coagulum formation. A study of microwave catheter ablation in vitro evaluated both frequencies and the thermodynamics of ablation with this energy source. The duration of energy delivery required to achieve maximal lesion size was longer because of increased volume heating and slower rates of deep tissue heating from a large volume heating source.[2]

The many technical constraints of microwave transmission to a small catheter-tip antenna have limited the value of this technology. High frequency electromagnetic radiation rapidly dissipates along conduction lines (within the catheter body). The curving course of a microwave catheter in the body alters the optimal "tuning" of the power source to the antenna. Without optimal tuning, much of the power that should be transmitted from the antenna at the catheter tip can instead be reflected back to the power supply. Reflected power can produce constructive interference and further heating of the catheter body. Therefore, the majority of power may be lost in the transmission line, and the power that reaches the tip antenna may be dissipated within the antenna as heat. The efficiency of microwave energy delivery to the tissue with intracardiac catheter delivery system appears to be low.

Initial studies with microwave catheter ablation systems have been promising, but the difficulties in adapting the antenna and energy transmission to a reliable catheter platform have impeded their transition to human trials. One study used an 8.5 F catheter system and temperature feedback power control to create experimental ablations in 11 dogs. Relatively large lesions (mean depth 8.8 ± 4.2 mm, 44% transmural) were achieved with mean powers of 9.3 ± 4.4 W.[32] Hand-held microwave probes avoid many of the flexibility, size, and transmission line problems that are found with catheter platforms. Successful lesion sets were completed within 13 ± 5 minutes in 90 patients undergoing intraoperative microwave linear atrial ablation. Atrial fibrillation cure rates were 74% and 67% at 6 and 12 months, respectively.[33,34]

LASER CATHETER ABLATION

Laser energy is another mode of ablative energy delivery to the heart. Laser is a form of concentrated light energy, and also ablates tissue by heating. Laser energy heats tissue optically, in the same way that sunshine warms our environment. As is the case with microwave energy, the depth of penetration of light energy before it is dissipated as heat is dependent on its wavelength. Short wavelength light, such as is produced by an argon laser (488 or 514 nm) has a depth of penetration in tissue of about 2 mm. In contrast, longer wavelength laser sources, such as neodymium-YAG (1064 nm) can penetrate about 5 mm into tissue. An advantage of laser catheter ablation is its ability to deliver high power through a flexible, small diameter fiberoptic. Disadvantages include the difficulty in control of lesion formation, the geometry of laser lesions, and the entry cost of laser generators. Depending on the quality of catheter-tissue contact and the frequency of laser energy selected, varying amounts of energy will be dissipated in the circulating blood pool. With variable efficiency of energy coupling with the tissue and no effective methods of assessing peak tissue temperature, control of laser catheter ablation is challenging. With conventional laser, there is relatively little scatter of the light energy at the endocardial surface. Also, the site of maximal convective cooling of tissue by circulating blood flow occurs at the endocardial surface. These characteristics typically yield a flask-shaped lesion with a narrow endocardial neck and a broad intramural base. Diffuser crystals at the catheter tip can create more superficial beam spread at the cost of heating deeper tissue planes.[35] Side-fire laser catheters have been employed to create linear atrial ablation,[36] but variability of catheter-electrode contact will affect energy coupling to the tissues with this technology similar to other linear ablation systems. The high cost of laser power generators has been a major impediment to laser catheter ablation in the past, but the ready availability of diode lasers now may lead to a renewed interest in this technology.

Clinical experience with laser ablation has been limited. Saksena and colleagues employed argon laser intraoperatively along with limited surgical resection to successfully ablate 38 VTs in 20 patients.[37] Epicardial laser photocoagulation with Nd-YAG was employed to ablate VT in a small series of patients, thus avoiding the need for ventriculotomy.[38] A novel Nd-YAG laser catheter system has been tested in vivo with some success. Reversible AV block could be achieved with brief energy delivery (9.7 ± 1.1 seconds) and persistent block with longer durations of ablation (28.6 ± 7.9 seconds) in experimental animals.[39] More recently, laser energy has been employed to create a circumferential lesion in pulmonary veins using a centering balloon catheter and a diffuser at the terminus of the optical transmission fibers. The investigators were able to successfully create transmural continuous circumferential lesions in vivo in canine pulmonary veins with 30 to 50 W of laser power administered for 60 to 90 seconds of energy delivery.[40] This may become a useful modality for the treatment of atrial fibrillation of focal origin.

ULTRASOUND CATHETER ABLATION

Ultrasound energy is created by passing alternating current through a piezoelectric crystal and generation of oscillating sound waves. Thus, ultrasound energy is mechanical energy, not electromagnetic energy. Some ultrasound energy passes through tissue, some is reflected, and some is dissipated as heat. The power levels used in diagnostic ultrasound are too low to produce meaningful tissue heating. However, if high power is employed, ultrasound energy becomes an effective mode of tissue heating and ablation. Ultrasound energy is well suited for myocardial ablation. It has a 10:1 tissue to blood energy absorption ratio, so the energy is preferentially dissipated in the targeted tissues, not in the circulating blood pool. The depth of tissue penetration can be adjusted by changing the ultrasound frequency (lower frequency produces deeper penetration and deeper heating). Theoretically, therapeutic ultrasound could be coupled with ultrasound imaging to assist in catheter placement and assessment of lesion formation. The predominant challenge of ultrasound catheter ablation lies in its engineering. It is technically challenging to transmit enough ultrasound power to cause tissue heating using a crystal that is small enough to fit on the tip of a catheter. Also, microcavitation in the near field can cause ultrasound waves to scatter and prevent transmission of energy to deeper tissue layers.

One clinically tested ultrasound catheter ablation system employs a specially-designed cylindrical piezoelectric crystal with a central resonating chamber to accomplish circumferential ablation of pulmonary veins in patients with atrial fibrillation of focal origin. The ultrasonic source was positioned on the tip of a balloon catheter and positioned at the origin of pulmonary veins. The balloon was inflated with saline and the ultrasound energy was transmitted around the perimeter of the vein. Fifteen patients with refractory atrial fibrillation underwent ablation of three or four pulmonary veins. There were 14.7 ± 12.6 (range 3 to 39) energy deliveries per patient, the procedure time was 224 ± 89 minutes, and the fluoroscopy time was 62 ± 39 minutes. The acute procedural success rate with ultrasound alone was 93%. In follow-up, 10 of 15 patients (66%) remained free of recurrent arrhythmias.[41] An analysis of procedure failures identified inadequate heating due to possible inadequate power delivery or convective cooling from pulmonary vein blood flow, and mismatch between pulmonary vein anatomy and balloon morphology as was found with large veins, funnel-shaped veins, and veins with early branching. Improved systems with better centering of the ultrasound energy source, larger occlusive balloons, and higher power ultrasound power delivery should improve the efficacy of ultrasound balloon catheter ablation. Controllability of ultrasound heating and determination of optimal power titration to avoid complications (e.g., pulmonary vein stenosis) remain issues, however. Ultrasound systems for ablation of non-pulmonary vein arrhythmogenic substrates are in development, but are not available for clinical use.

CRYOTHERMIC CATHETER ABLATION

Cryothermy has been a technology commonly employed in surgical ablation over the past decades. It is only recently that catheter-based systems have developed to the point that adequate joule cooling can be accomplished so that lesions can be created. Most cryothermic catheters are designed with a lumen for introduction of refrigerant into the tip of the catheter. An evaporation chamber is located at the catheter tip. The phase change of the refrigerant removes heat from the tissue in contact with the catheter tip, and a second lumen under vacuum removes the refrigerant. During freezing, the catheter tip adheres to the contiguous tissue and an ice ball gradually forms. Tissue injury occurs both during the freezing and rewarming phases of the ablation. The intracellular ice matrix produces cytoskeletal disruption. The extracellular ice matrix leads to sarcolemmal membrane disruption. The kinetics of tissue freezing are slow, with continued lesion growth over 10 minutes of application time. Cryothermic ablation has the advantages over other forms of catheter ablation that one can create reversible lesions during mapping and thereby confirm optimal catheter placement before creating irreversible injury. The lesions are homogeneous pathologically and, therefore, are not arrhythmogenic. The ablations are painless. The disadvantages of this technology are the longer ablation times required for success and possibly a higher rate of arrhythmia recurrence.

Preliminary clinical studies have employed cryothermic catheter ablation for creation of complete AV block, and for the treatment of AV nodal reentrant tachycardia. The first study of 12 patients achieved acute procedural success in 83% and overall success in 66%.[42] In contrast, AV junctional ablation with conventional RF ablation systems achieves chronic success in 95% of patients. In another study of 18 patients with AV nodal reentrant tachycardia, successful ablation was achieved in 94%, and there were no arrhythmia recurrences. Importantly, inadvertent fast pathway ablation was averted in two patients in whom transient fast pathway block was observed during cooling, with recovery of normal function on rewarming.[43] Great interest presently exists with cryothermic ablation of pulmonary veins in patients with atrial fibrillation of focal origin. It is thought that pulmonary vein conduction block may be achieved effectively with no risk of pulmonary vein stenosis. Clinical studies in this arena are pending.

SUMMARY

The goal of catheter ablation is to create controlled destruction of an adequate volume to tissue to capture a critical arrhythmogenic substrate that is responsible for initiation or propagation of an arrhythmia. Radiofrequency catheter ablation has a well-established history of high efficacy and an excellent safety profile, and is appropriate for the ablation of most arrhythmias. For selected arrhythmia, larger and/or deeper lesion formation may be desirable. However, all technologies that increase lesion size are limited with regard to controllability, and may be associated with increased procedural risk. New technologies may be useful for the ablation of new or specialized applications, such as pulmonary vein ablation, atrial flutter, or VT. As our understanding of the anatomy and physiology of arrhythmias evolves and the ablation technologies improve, indications for curative catheter ablative therapy will continue to expand. Future procedures will be faster, more effective, and carry lower risks of procedure related complications.

REFERENCES

1. Haines DE, Watson DD: Tissue heating during RF catheter ablation: A thermodynamic model and observations in isolated perfused and superfused canine right ventricular free wall. PACE 1989; 12:962-976.
2. Whayne JG, Nath S, Haines DE: Microwave catheter ablation of myocardium in vitro: Assessment of the characteristics of tissue heating and injury. Circulation 1994;89:2390-2395.
3. Haines DE, Watson DD, Verow AF: Electrode radius predicts lesion radius during radiofrequency energy heating: Validation of a proposed thermodynamic model. Circ Res 1990;67:124-129.
4. Haines DE, Verow AF: Observations on electrode-tissue interface temperature and effect on electrical impedance during radiofrequency ablation of ventricular myocardium. Circulation 1990; 82:1034-1038.
5. McRury ID, Mitchell MA, Panescu D, Haines DE: Non-uniform heating during radiofrequency ablation with long electrodes: Monitoring the edge effect. Circulation 1997;96:4057-4064.
6. Haines DE: Determinants of lesion size during radiofrequency catheter ablation: The role of electrode-tissue contact pressure and duration of energy delivery. J Cardiovasc Electrophys 1991; 2:509-515.
7. Wittkampf FH, Nakagawa H, Yamanashi WS, et al: Thermal latency in radiofrequency ablation. Circulation 1996;93:1083-1086.
8. Petersen HH, Chen X, Pietersen A, et al: Lesion dimensions during temperature-controlled radiofrequency catheter ablation of left ventricular porcine myocardium: Impact of ablation site, electrode size, and convective cooling. Circulation 1999;99: 319-325.
9. Huang SK, Graham AR, Wharton K: Radiofrequency catheter ablation of the left and right ventricles: Anatomic and electrophysiologic observations. PACE 1988;11:449-459.
10. Langberg JJ, Calkins H, Kim YN, et al: Recurrence of conduction in accessory atrioventricular connections after initially successful radiofrequency catheter ablation. J Am Coll Cardiol 1992; 19:1588-1592.
11. DeLacey WA, Nath S, Haines DE, et al: Adenosine and verapamil-sensitive ventricular tachycardia originating from the left ventricle: Radiofrequency catheter ablation. PACE 1992;15: 2240-2244.
12. Huang SK, Bharati S, Lev M, et al: Electrophysiologic and histologic observations of chronic atrioventricular block induced by closed-chest catheter desiccation with radiofrequency energy. PACE 1987;10:805-816.
13. Nath S, Whayne JG, Kaul S, et al:. Effects of radiofrequency catheter ablation on regional myocardial blood flow: Possible mechanism for late electrophysiological outcome. Circulation 1994;89:2667-2672.
14. Coakley WT, Deeley JOT: Effects of ionic strength, serum protein and surface charge on membrane movements and vesicle production in heated erythrocytes. Biochimica et Biophysica Acta 1980;602:355-375.
15. Haines DE, Whayne JG, Walker J, et al: The effect of radiofrequency catheter ablation on myocardial creatine kinase activity. J Cardiovasc Electrophysiol 1995;6:79-88.
16. Grzelinska E, Bartosz G, Leyko W, Chapman IV: Effect of hyperthermia and ionizing radiation on the erythrocyte membrane. Int J Radiat Biol 1982;42:45-55.
17. Nath S, Lynch C III, Whayne JG, Haines DE: Cellular electrophysiologic effects of hyperthermia on isolated guinea pig papillary muscle: Implications for catheter ablation. Circulation 1993;88:1826-1833.
18. Simmers TA, De Bakker JM, Wittkampf FH, Hauer RN: Effects of heating on impulse propagation in superfused canine myocardium. J Am Coll Cardiol 1995;25:457-464.
19. Everett TH, Nath S, Lynch C, et al: The role of calcium in acute hyperthermic myocardial injury. J Cardiovasc Electrophys 2001;12:563-569.
20. Jackman WM, Wang XZ, Friday KJ, et al: Catheter ablation of atrioventricular junction using radiofrequency current in 17 patients: Comparison of standard and large-tip catheter electrodes. Circulation 1991;83:1562-1576.
21. Tsai CF, Tai CT, Yu WC, et al: Is 8-mm more effective than 4-mm tip electrode catheter for ablation of typical atrial flutter? Circulation 1999;100:768-771.
22. Mitchell MA, McRury ID, Haines DE: Linear atrial ablations in a canine model of chronic atrial fibrillation: Morphologic and electrophysiologic observations. Circulation 1998;97:1176-1185.
23. Haines DE, Gabbard J, Wolf R, Damiano RJ: Rapid, continuous and transmural linear ablation with a novel bipolar radiofrequency ablation system. Pacing Clin Electrophys 2001; 24:II-608.
24. Everett TH, Mangrum JM, Haines DE: A novel balloon electrode catheter for left ventricular radiofrequency catheter ablation. Circulation 1998;98:I-435.
25. Tanaka K, Satake S, Saito S, et al: A new radiofrequency thermal balloon catheter for pulmonary vein isolation. J Am Coll Cardiol 2001;38:2079-2086.
26. Nakagawa H, Yamanashi WS, Pitha JV, et al: Comparison of in vivo tissue temperature profile and lesion geometry for radiofrequency ablation with a saline-irrigated electrode versus temperature control in a canine thigh muscle preparation. Circulation 1995;91:2264-2273.
27. Jais P, Shah DC, Haissaguerre M, et al: Prospective randomized comparison of irrigated-tip versus conventional-tip catheters for ablation of common flutter. Circulation 2000;101:772-776.
28. Yamane T, Jais P, Shah DC, et al: Efficacy and safety of an irrigated-tip catheter for the ablation of accessory pathways resistant to conventional radiofrequency ablation. Circulation 2000;102:2565-2568.
29. Nabar A, Rodriguez LM, Timmermans C, Wellens HJ: Use of a saline-irrigated tip catheter for ablation of ventricular tachycardia resistant to conventional radiofrequency ablation: Early experience. J Cardiovasc Electrophys 2001;12:153-161.
30. Calkins H, Epstein A, Packer D, et al: Catheter ablation of ventricular tachycardia in patients with structural heart disease using cooled radiofrequency energy: Results of a prospective multicenter study. J Am Coll Cardiol 2000;35:1905-1914.

31. Calkins H, Hall J, Ellenbogen K, et al: A new system for catheter ablation of atrial fibrillation. Am J Cardiol 1999;83:227D-236D.
32. VanderBrink BA, Gilbride C, Aronovitz MJ, et al: Safety and efficacy of a steerable temperature monitoring microwave catheter system for ventricular myocardial ablation. J Cardiovasc Electrophysiol 2000;11:305-310.
33. Knaut M, Spitzer SG, Karolyi L, et al:. Intraoperative microwave ablation for curative treatment of atrial fibrillation in open heart surgery—the MICRO-STAF and MICRO-PASS pilot trial. Thorac Cardiovasc Surg 1999;47:379-384.
34. Knaut M, Tugtekin SM, Spitzer SG, et al: Curative treatment of chronic atrial fibrillation in patients with simultaneous cardiosurgical disease with intraoperative microwave ablation. J Am Coll Cardiol 2001; 37:109A.
35. Ware DL, Boor P, Yang C, et al: Slow intramural heating with diffused laser light: A unique method for deep myocardial coagulation. Circulation 1999;99:1630-1636.
36. Keane D, Ruskin JN: Linear atrial ablation with a diode laser and fiberoptic catheter. Circulation 1999;100:59-60.
37. Saksena S, Gielchinsky I, Tullo NG: Argon laser ablation of malignant ventricular tachycardia associated with coronary artery disease. Am J Cardiol 1989;64:1298-1304.
38. Pfeiffer D, Moosdorf R, Svenson RH, et al: Epicardial neodymium:YAG laser photocoagulation of ventricular tachycardia without ventriculotomy in patients after myocardial infarction. Circulation 1996;94:3221-3225.
39. Weber H, Enders S, Keiditisch E: Percutaneous Nd:YAG laser coagulation of ventricular myocardium in dogs using a special electrode laser catheter. PACE 1989;12:899-910.
40. Fried NM, Tsitlik A, Rent KC, et al: Laser ablation of the pulmonary veins by using a fiberoptic balloon catheter: Implications for treatment of paroxysmal atrial fibrillation. Las Surg Med 2001;28(3):197-203.
41. Natale A, Pisano E, Shewchik J, et al: First human experience with pulmonary vein isolation using a through-the-balloon circumferential ultrasound ablation system for recurrent atrial fibrillation. Circulation 2000;102:1879-1882.
42. Dubuc M, Khairy P, Rodriguez-Santiago A, et al: Catheter cryoablation of the atrioventricular node in patients with atrial fibrillation: A novel technology for ablation of cardiac arrhythmias. J Cardiovasc Electrophysiol 2001;12:439-444.
43. Skanes AC, Dubuc M, Klein GJ, et al: Cryothermal ablation of the slow pathway for the elimination of atrioventricular nodal reentrant tachycardia. Circulation 2000;102:2856-2860.

Chapter 44 Curative Catheter Ablation for Supraventricular Tachycardia: Techniques and Indications

DIPEN SHAH, MICHEL HAISSAGUERRE, PIERRE JAIS, and MELEZE HOCINI

The major part of the credit for the current popularity of radiofrequency (RF) catheter ablation must rightly go to its contribution to the management of supraventricular tachycardias. Many of the original concepts on which the present day procedure is based can be traced to surgical interventions. However, by integrating an interventional arm into a hitherto solely diagnostic armamentarium, the electrophysiological laboratory was uniquely able to not only ablate (as could the surgeons) but also evaluate the results and reablate if necessary.

In this context this chapter attempts to review the basic principles of performing curative catheter ablation for supraventricular tachycardias. The arrhythmias considered below include: arrhythmias involving accessory atrioventricular (AV) connections, AV nodal reentry tachycardias, atrial tachycardias including typical flutter and other macro-reentrant right and left atrial tachycardias, focal atrial tachycardias, and atrial fibrillation.

Accessory Atrioventricular Connections

The anatomic substrate of accessory AV connections is myocardium bridging the AV annuli, which in normal individuals are fibrous and electrically insulating.[1] By inserting into ordinary myocardium and bypassing the normal septally inserting and insulated His-Purkinje system, the sequence of initial ventricular septal depolarization is altered by conduction through such connections. The relatively slow spread of activation through ordinary myocardium contrasts with the coordinated septal endocardial breakthrough of Purkinje ramifications and results in the delta wave in the surface ECG. Its slow upstroke is due to activation of the working myocardium and, therefore, resembles the QRS of ventricular tachycardia. In addition to providing an additional route for impulse conduction between the atria and ventricles, nearly all accessory connections exhibit conduction properties that are different from those of the AV node. Decremental conduction is not ordinarily seen—that is, with increasing frequency or shortening coupling intervals, the conduction interval across the pathway does not significantly increase. This is reflected in the short R–P interval narrow QRS tachycardia, which is characteristic of orthodromic AV reentrant tachycardia.

VENTRICULAR PREEXCITATION

Preexcitation in the electrophysiology (EP) laboratory involves demonstrating that during sinus rhythm or atrial pacing, ventricular myocardium is activated by a pathway other than the His-Purkinje system. The normal H–V interval is made up of the time required for activation to proceed from the His bundle recording site down the bundle branches to the distal ramifications of the Purkinje fibers before exiting to depolarize working myocardium. Thus, preexcitation is inferred if the H–V interval is abnormally short, but the H–V interval may be normal if the preexcited myocardium is too small (consequently generating feeble voltage) to be evident on the surface ECG. In this situation, an increased frequency of supraventricular impulses results in more ventricular myocardium being preexcited through the accessory connection because of the nondecremental nature of conduction through such pathways. Incremental atrial pacing should, therefore, be a routine preliminary evaluation which will shorten the H–V interval and typically increase preexcitation, thus allowing the optimal surface ECG evaluation of the localization of the pathway insertion.

It follows that in the case of accessory connections with long anterograde conduction times or short conduction times through the AV node, preexcitation may be difficult to discern on the surface ECG in sinus rhythm. As indicated earlier, however, preexcitation should be detectable on intracardiac recordings. If no preexcitation is detectable even on intracardiac recordings, the pathway may not be capable of antegrade conduction.

An accessory pathway with a long antegrade conduction time can be responsible for an isoelectric interval separating the end of the P wave from the onset of ventricular activation: the so called P–delta interval which may persist even during atrial pacing or atrial tachycardia. Theoretically, the same surface ECG morphology can also result from an electrical connection between the node and ventricular myocardium bypassing the His-Purkinje system with the isoelectric interval resulting from conduction delay within the AV node.

Retrograde Conduction

The other indicator of an accessory AV connection is the presence of conduction from the ventricles to the atria over a pathway other than the normal AV conduction axis. Normal retrograde activation activates the atria from the septal region and typically exhibits a decrement of at least 20 ms with faster stimulation, while free-wall activation (so called eccentric activation), which is nondecremental, suggests ventriculoatrial (VA) conduction over an accessory AV connection. To distinguish between septally situated accessory pathways and normal routes of VA conduction, other dynamic maneuvers are required. During sinus rhythm, ventricular extrastimuli resulting in atrial activation preceding retrograde His bundle activation indicates an accessory connection.[2] On moving the ventricular pacing site from the apex toward the septum, if the stimulus to atrial activation time decreases instead of increases, this also suggests an accessory connection. This results from moving away from the apex which increases the conduction time to the normal AV conduction system in the form of the distal Purkinje myocardial interface, whereas the conduction time to the annular insertion of an accessory pathway actually decreases.[3] Similarly, high output due to capture of the right bundle/His bundle compared with lower output ventricular myocardial capture at the same site can show changes in atrial activation sequence in retrograde His-to-atrial activation time as well as in stimulus-to-atrial activation time, suggesting the presence of more than one retrograde pathway of VA conduction.[4] An unchanged atrial activation sequence coupled with a constant H–A interval and prolongation of the St–A interval resulting from loss of His–RB capture indicate the presence of normal VA conduction alone. Conversely, the absence of changes in any of the intervals and sequence indicates the sole presence of accessory pathway retrograde conduction. Such inferences should be drawn in the knowledge that, if the accessory pathway is remote from the pacing site or is captured only with a long conduction time or if conduction through the AV node is very rapid, then conduction through the accessory pathway may be completely masked. In practice, left free-wall pathways remote from a right ventricular pacing site may fulfill these conditions.

During a tachycardia, evidence of conduction through an accessory AV connection can be obtained by delivering late ventricular extrastimuli coincident with or 10 ms before activation of the His bundle—thus being certain of encountering absolute refractoriness within the His bundle (and even if the extrastimulus is earlier, the lack of advancement of the ventricular and/or His bundle electrogram confirms this). If this ventricular extrastimulus advances atrial activation or delays it, or terminates the tachycardia without conduction to the atria, this indicates the presence of conduction through an accessory connection.[5] The latter finding also indicates participation of the accessory pathway in the tachycardia as does advancement of the succeeding ventricular or His bundle electrogram.

Having established the presence of an accessory connection, it is necessary to assess its pathophysiologic role and arrhythmogenic potential. This evaluation is based on parameters reflecting the antegrade refractory period of the pathway, and the inducibility of arrhythmias in which it can be shown to participate. The indications for curative ablation of these pathways, therefore, depend chiefly on their proven threat—preexcited AF degenerating to VF—or potential threat—indicated by R–R intervals shorter than 200 to 250 ms during atrial fibrillation[6] or the presence of clinical tachycardias.

Successful ablation of an accessory AV connection is naturally based on localizing it, and the surface ECG is a vital starting point. Although many algorithms have been described, those using the delta wave vector are more difficult to use than the mean QRS vector during full preexcitation. ECG pattern recognition and iterative refinement resulting from the precise localization inherent to endocardial mapping and eventual successful ablation, allow the development of a strategy specific to the presumed location.

Left Free-Wall Accessory Pathways

Left free-wall pathways can be found typically in the arc extending from about 12 o'clock to about 7 o'clock on the mitral annulus viewed in the left anterior oblique view. The anterior and septal aspect of the mitral annulus is the region of aortomitral fibrous continuity which means that accessory pathways in this area do not normally occur. These pathways have been considered to be the most straightforward of catheter ablations, primarily because these pathways are far away from the AV conduction axis and their location is easily pinpointed by a coronary sinus catheter. However, the proximity of the mitral valve and the circumflex coronary artery and the possibility of a true mid-myocardial pathway all represent technical challenges.

In our laboratory, a retrograde arterial approach is preferred and the transseptal approach is reserved in case of failure. Transseptal access may be used as the first line approach in case of aortic or arterial abnormalities, such as prosthetic valves, aortic stenosis, or severe aortic/femoral/iliac atherosclerosis. In small children as well, the transseptal approach may be preferred in order to avoid injury to the aortic valve, which is delicate compared with the smallest ablation catheters available.

Passing the aortic valve represents an important step of the retrograde arterial approach. When the catheter is brought down the root of the aortic valve, it meets the resistance of the aortic valve and a slight flexing of the catheter facilitates the formation of a loop with continued gentle pressure. Before the catheter tip bends 180°, the loop generally crosses the aortic valve into the LV cavity. If not, this may be facilitated by gentle torquing. It is imperative to avoid catheter tip entry into a coronary artery and wise to withdraw the catheter gently in case of any doubt. A position within the left coronary system may be confused with a left atrial position and complications have been reported. Conversely, if the catheter tip is not flexed as it is brought down, this can permit unrecognized entry into the right coronary artery ostium, particularly if it has a downward takeoff. Occasionally, catheter entry into the left ventricle may be accompanied by mechanical trauma to the AV conduction axis resulting in local block, which becomes evident on ablating the accessory pathway responsible for preexcitation. Fortunately, spontaneous recovery of normal conduction is the most common outcome. Once the catheter has crossed the level of the aortic valve, it should be straightened before being advanced gently toward the LV free wall—this is the natural direction the catheter tends to take because of the curvature of the ascending aorta and the left ventricular outflow tract. Progressive and gentle flexion of the catheter tip as it touches the free wall will bring the catheter tip near or at the level of the mitral annulus and, generally, under the mitral valve leaflet. The recording of a significant atrial electrogram signals the achievement of an appropriate level and mapping can now begin.

Typically, this retrograde route is well known for catheter stability but conversely mapping the annulus freely may not be possible without relaxing the catheter bend and withdrawing it from under the leaflet. Clockwise rotation positions the catheter more laterally and anteriorly, whereas counterclockwise rotation brings it around more medially. In order to map the annulus, the catheter must be advanced and repositioned under the mitral leaflet. It is also obvious that the size of the catheter curve is important; a large curve does not allow the catheter tip to reach the annulus level, whereas a small curve means that the catheter "floats" without achieving a stable position. A more atrial position (where the catheter probably sits on the atrial side of the mitral leaflet) can be achieved by torquing the catheter counterclockwise so that it slips medially onto the atrial side—probably facilitated by the oblique orientation of the mitral annulus with the posterior commissure being lower than the anterior. This also means that the further anterior the accessory pathway, the more difficult it is to reach the annulus when approached retrogradely. Therefore, this is one situation that may call for a larger curve or a transseptal approach. When on top of the mitral leaflet—as indicated by a significantly larger atrial electrogram—the catheter can be much more freely rotated in order to map the annulus, but typically, the catheter is less stable than when positioned under the leaflet. Ectopy, not uncommon during RF delivery in this position, risks dislodging the catheter rather easily.

The transseptal route probably offers greater ease of mapping the left AV annulus and permits the direct use of bipolar electrogram polarity reversal to bracket its atrial insertion. Long sheaths ameliorate catheter stability and permit precise mapping. This route permits easier access to pathways which are anteriorly and laterally located, although septal and posteriorly located pathways are more difficult to access and large-curve catheters and sheaths may be needed. A catheter position within the left inferior pulmonary vein (PV) should be distinguished—an absence of catheter mobility with cardiac contraction should raise suspicion and allow confirmation in the left anterior oblique view. In our laboratory, a pattern indicative of a left free-wall pathway permits a two catheter approach including one femoral vein and another retrograde arterial access. A pattern suggestive of an anteroseptal pathway may call for a subclavian or jugular vein access, a right free-wall pattern the need for a long sheath, and posteroseptal pathways the possible need for coronary sinus angiography.

While a multi-catheter approach can cover most of both AV annuli and provide corroboration of localization by the achievement of bracketing, successful and rapid ablation can be achieved even in apparently complex cases using fewer catheters—typically two or three. In the case of evident preexcitation, a single ablation catheter may be successfully used followed by an adenosine test; however, although in some cases VA conduction may be evident on the surface ECG, the assessment of retrograde VA conduction usually requires an additional intracardiac catheter.

When the best unipolar and bipolar endocardial electrograms are not good enough, an epicardial or intramyocardial pathway insertion may have to be evaluated or considered. It is important to realize that the accessory pathway itself may be endocardially situated or close to the endocardium, whereas the ventricular electrogram itself at this site can be late—not only because the insertion may be far from the endocardium, but also in case of an endocardial insertion of an oblique pathway. A simultaneous comparison of endocardial with epicardial recordings obtained from within the coronary sinus is useful. Ablation within the coronary sinus may be necessary,[7] and conventional RF delivery here is performed at low power, allowing lower success rates but also possibly fewer complications.

Localization based on antegrade conduction involves sampling the annulus of interest for the shortest local A–V intervals (Fig. 44-1) and the earliest V (local ventricular)–delta intervals. Of interest is the fact that V-delta times at sites of successful ablation vary depending on pathway localization—probably relating to the amount of initially preexcited myocardium. Some mid- or posteroseptal pathways exhibit long AV times at successful sites suggesting slow conduction through the accessory pathway.

Both bipolar and unipolar electrograms should be used for mapping; the former because of their higher signal-to-noise ratio and the latter because of their simple morphologic pattern recognition-based analysis.[8] The interpretation of bipolar electrograms hinges on the distinction between atrial and ventricular electrograms—and while the two can be distinguished using late coupled ventricular and atrial extrastimuli, these maneuvers can at times be difficult to perform or

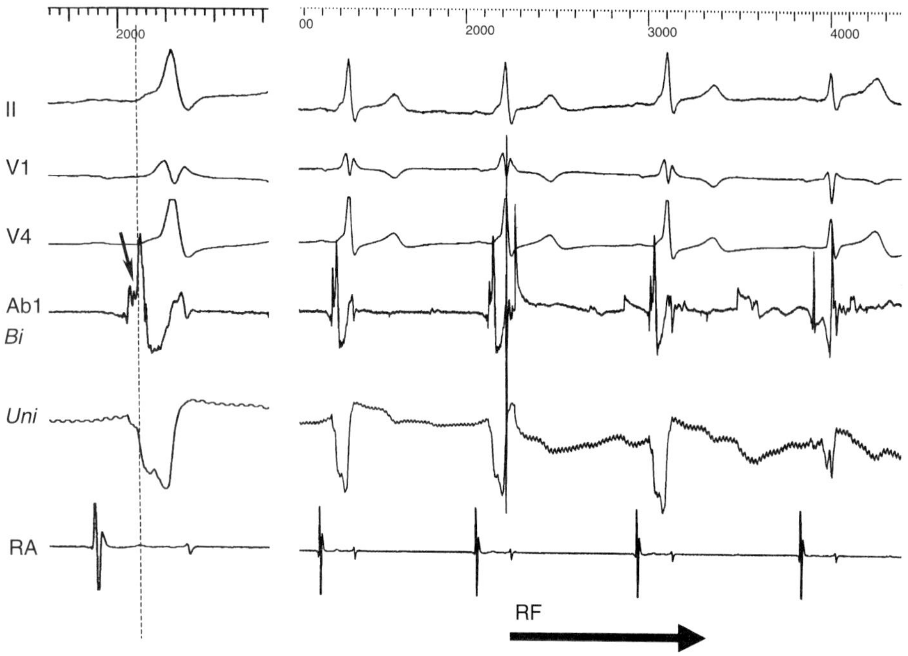

FIGURE 44-1 The site of successful ablation for a preexcited left lateral accessory AV connection is shown **(left)** and prompt elimination of preexcitation during RF delivery is shown **(right)**. Intrinsic deflection of the ventricular electrogram (*dashed line*) precedes the delta wave onset by less than 10 ms. Note the QS morphology in the unipolar electrograms. The *arrow* indicates a presumptive accessory pathway potential interposed between the atrial and ventricular deflections, identified respectively by comparison after ablation (atrial electrogram) and correlation with the unipolar (*Uni*) electrogram (ventricular electrogram).

analyze and may even induce arrhythmias. The contribution of the proximal ring electrode to bipolar electrograms from the distal bipole can be misleading. Atrial electrograms can be clearly distinguished from ventricular electrograms using unfiltered unipolar electrograms from the distal electrode, which also resolves the issue of the contribution of the proximal ring electrode to bipolar electrograms. Unipolar electrograms are particularly useful for localizing the site of ventricular insertion during sinus rhythm, pacing, or even ongoing atrial fibrillation and also in patients with Ebstein's anomaly who exhibit low amplitude fractionated bipolar electrograms on the tricuspid annulus (TV), based on a steep QS morphology (the absence of an initial R wave) (see Fig. 44-1).

It is important to remember that unipolar electrograms are to be used with wide band filters—0.05 to 500 Hz: because the low frequency content contributes importantly to the generation of RS, or QS patterns, whereas higher frequencies are important to the recording of the intrinsic deflection. In order to optimize the signal-to-noise ratio, instead of the Wilson's central terminal, a remote cutaneous or inferior vena cava (IVC) electrode may be useful as a ground—allowing common mode rejection of contaminating 50/60 Hz line noise. Notch filters should also be used with caution, if at all. The ability to distinguish atrial and ventricular electrograms means that the remaining electrograms if any represent accessory pathway potentials (see Fig. 44-1)[9]; and an intervening bipolar electrogram component that does not coincide with unipolar atrial or ventricular intrinsic deflections, can be considered a presumptive accessory pathway potential. Certainly, the best validation should be prompt abolition of accessory pathway conduction by RF ablation at this site (assuming appropriate power delivery and contact), thereby effectively consigning accessory pathway potential validation to retrospective analysis.

Sequential sampling of the right atrial free wall at the annulus level and that on the His bundle at the septum not only provide approximate localization but also provide an indicator of the margin of safety vis à vis the normal conduction axis. Note that in order to correctly assess the timing of ablation catheter electrograms, one must use the surface ECG lead showing the maximum preexcitation. In general, high amplification allows the recognition of low amplitude potentials—a situation typical of endocardially recorded accessory pathway potentials—and if the higher amplitudes are left unclipped, they provide a single glance assessment of timing relative to the onset of the delta wave by their overlap with the surface ECG.

For patients without preexcitation, the target of choice is the shortest V–A interval during orthodromic AV reentrant tachycardia (Fig. 44-2), because this effectively rules out fusion of activation through the normal AV axis with that through the accessory connection. If no tachycardia is inducible and retrograde conduction through the normal AV axis can be excluded or distinguished, earliest atrial activation during ventricular pacing is a reasonable target. Notwithstanding the above generalizations, individual strategies need to be applied to specific pathway locations.

Other Pathway Locations

Right free-wall pathways are defined by a location within the arc of the TV extending from about 12 o'clock to approximately 6 o'clock as viewed from the left anterior oblique 45° view. Certain anatomic distinctions characterize this location and the strategy for ablation of

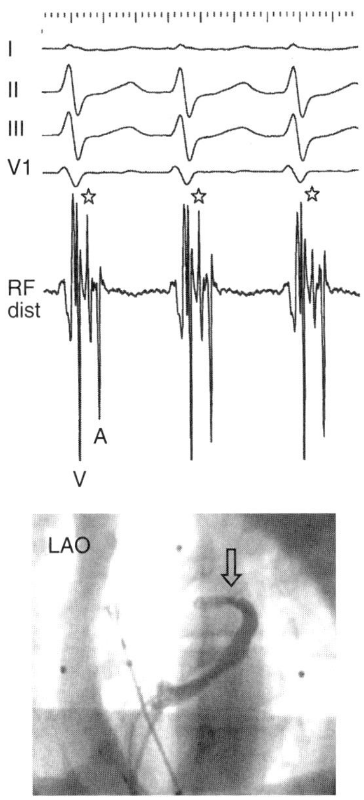

FIGURE 44-2 An epicardial concealed left-sided accessory AV connection successfully ablated within the distal coronary sinus at about the 2-o'clock position in the LAO view—coronary sinus angiogram with catheter position (*arrow*) shown below. The **top panel** shows the recording at the ablation site within the coronary sinus (RF dist): the *star* marks the AP potential, which was found to be dissociated from atrial and ventricular activity in sinus rhythm after successful ablation. LAO, Left anterior oblique.

pathways in this region. The TV has a much more vertical orientation than the mitral annulus, the right ventricle (RV) is much thinner than the left ventricle, and there is no counterpart of the coronary sinus. The access to the right AV annulus is much more direct than on the left side, but a position underneath the tricuspid valve leaf is more difficult to achieve, particularly from femoral access. Additionally right atrial free-wall contraction tends to dislodge the catheter tip. Long sheaths with a curve adapted to the annulus, and/or large curve catheters do facilitate ablation by improving stability—in practice a relatively stiff and torquable sheath is much more useful than a thin-walled sheath that does not respond to torque and kinks easily. Clues to the level of the annulus are useful, including the radiolucent shadow of annular fat, particularly when annular electrograms are fractionated and low voltage as in Ebstein's anomaly. Unipolar electrograms can be of help, and even recordings from within the right coronary artery have been used—although the latter runs the risk of significant complications. Unlike the usual QS morphology at successful ablation sites of other pathways, the ventricular electrogram has a two-stepped deflection; the first indicating local right ventricular activation whereas the second represents far-field activation, probably originating from the left ventricle and interventricular septum.

The site of successful ablation for right free-wall pathways shows ventricular electrograms with timings about 20 to 30 ms earlier than for left free-wall pathways[10]; because this atrial location is activated within the first half of the sinus P wave (so that right ventricular preexcitation actually begins within the P wave) and because preexcitation of the thin-walled right ventricle does not become evident on the surface ECG as early on as it does on the left side with a greater ventricular mass. Local ventricular activation of −10 to −20 ms (preceding the QRS) is usually not good enough for pathways in this location. Another characteristic of ablation at this location is the lower temperature recorded from the catheter during RF delivery—related to catheter contact, electrode orientation, and stability. This results in high power being delivered during temperature-controlled RF delivery. Mechanical block may also be produced during catheter manipulation. With nearly inevitable recovery and no target being available in the interim to direct the ablation, an eventual recurrence is more likely in spite of so called "security" RF applications at anatomically appropriate locations.

Septally situated pathways have been divided into anteroseptal, midseptal, and posteroseptally located pathways. The former are commonly defined by a location typically on the TV annulus (because of aortomitral fibrous continuity on the left side) situated between the 12-o'clock position and the His bundle region—with the provision that if a His potential is recorded at the site, it should be less than 0.1 mv, otherwise this defines the pathway as para-Hisian. Posteroseptal pathways are defined by a location between the coronary sinus to about the 6-o'clock position on either AV annulus. Midseptal pathways are more difficult to define in terms of location and are broadly considered to be between His and coronary sinus locations—excluding para-Hisian pathways. Except for anteroseptal pathways, these so-called septal pathways have in common the possibility of being more or less accessible from either annulus and, proximity to the AV node or bundle of His. In addition, an epicardial location is more common for posteroseptal pathways and, therefore, coronary sinus and middle cardiac vein mapping are frequently necessary.

Anteroseptal pathways as defined earlier can frequently be accessed and ablated with advantage from the superior vena cava (SVC) approach, with active catheter flexion bringing the catheter tip in apposition with the annulus, whereas from the femoral approach, relaxing catheter tip flexion is required to achieve contact and unless the catheter has bidirectional steering, this is a passive movement providing much less stable contact. It may also be easier to achieve a position under the tricuspid valve by making a loop in the RV using an approach from above.

Midseptal pathways, in general, are better tackled from below and the main concern here is to avoid damage to the AV node and normal conduction axis. Here, as for para-Hisian pathways, proximity to the His bundle allows an estimation of this risk; although for midseptal pathways, proximity to the compact AV node is difficult to estimate in the absence of a suitable marker. The occurrence of junctional rhythm is a clear warning that should prompt a cessation of RF delivery; and a narrow QRS complex without a preceding P wave should not be mistaken for loss of preexcitation. Using conventional RF, the strategy for pathways estimated to be in close proximity to the AV node or His bundle should center around careful mapping for the best electrograms and delivering low RF power at sites thought to be furthest from the conduction axis. At prospective ablation sites, it is useful to verify the presence and/or amplitude of a His bundle deflection concealed by preexcitation by using programmed stimulation to induce antegrade pathway block or sustained orthodromic AV reentrant tachycardia. RF power may be increased cautiously in steps of 5 watts. Stopping RF delivery immediately becomes necessary in case of junctional rhythm or if there is no prompt loss of preexcitation in spite of adequate power (about 20 to 25 watts, although it is difficult to define a specific value). Low-power thermal mapping may be one solution to distinguish between otherwise similar sites.[11] Recently, a catheter-based cryoablation technique has become available, which by allowing catheter tip adherence to the endocardium and reversible cryomapping (at temperatures of −30° C) allows verification of the efficacy of the eventual definitive lesion (which requires temperatures of at least −70° C) and, therefore, reduces the risk to normal AV conduction.

The higher likelihood of an epicardial course or insertion of pathways in posteroseptal locations and the anatomic boundaries of the posterior pyramidal space mean that frequently a choice has to be made between right- and left-sided endocardial sites and sites within the proximal coronary sinus or the middle cardiac vein. This may be suspected by the observation of a steep QS complex in lead II or an rS complex in V5/V6 during preexcitation.[12] If endocardial sites on the right and the left side are either not good enough or are unsuccessful, mapping within the coronary sinus and middle cardiac vein is performed, guided by a coronary sinus angiogram. Although an angiogram performed distal to an occluding balloon placed within the proximal coronary sinus allows the best documentation of its anatomy and branches, adequate visualization of the proximal coronary sinus and the middle cardiac vein can be achieved from the femoral approach using an Amplatz catheter that is ordinarily used for selective coronary angiography (see Fig. 44-2). Successful ablation sites are frequently clustered in proximity to venous anomalies, such as aneurysms or diverticula, and such sites exhibit small atrial electrograms because they are removed from the AV annulus. If accessing the middle cardiac veins proves difficult from the femoral approach, a superior approach providing a relatively straight and vertical catheter course may be successful. Ablation within the coronary sinus or cardiac veins frequently has to contend with the ineffective combination of low delivered powers and high electrode temperatures as a consequence of low blood cooling of the electrode. In addition, middle cardiac vein sites are typically close to the posterior descending and posterior left ventricular branches of the distal right coronary artery. Ineffective low power delivery can be overcome by using an irrigated tip catheter allowing power to be titrated up to a limit of 30 watts in order to avoid occlusion or narrowing of nearby coronary arteries (within 2 to 3 mm of the site of ablation).

SPECIFIC SITUATIONS

Specific problems encountered during an ablation procedure include having to cope with sustained atrial fibrillation. There may be no alternatives to electrical cardioversion, particularly for the ablation of concealed accessory pathways—classes I and III drugs may alter accessory pathway properties to the extent of eliminating conduction through them. If antegrade conduction through the pathway is present, mapping for the earliest ventricular activation during the widest QRS (indicating maximum preexcitation) can be successful, particularly if guided by unipolar electrograms. Verification of bidirectional conduction block (assessment of VA conduction), however, is not possible during atrial fibrillation.

Multiple pathways are not common but may be encountered, particularly in association with Ebstein's anomaly. Changing patterns of preexcitation and or V–A intervals and sequences are important clues. The same principles of mapping and ablation described earlier are effective.

"Mahaim" (atrioventricular or atriofascicular) pathways and the substrate responsible for permanent junctional reciprocating tachycardias (PJRTs) are both thought to be particular variants of accessory pathways with long conduction times—antegrade times in case of atriofascicular or AV Mahaim and retrograde in case of PJRT. In addition, these two accessory pathway variants share another characteristic—that of one way conduction only. The few available histologic studies suggest that the anatomic substrate of PJRTs is a long and tortuous muscular fascicle, whereas in the case of the "Mahaim" fiber, an accessory node-like structure is thought to exist at its atrial origin.[13] The latter are probably most effectively ablated by targeting potentials generated by the pathway at the level of the annulus; they resemble His bundle potentials but continue to precede ventricular activation even during preexcitation. PJRTs are ablated by targeting the earliest atrial activation during tachycardia and as with every ablation within that posteroseptal area, care must be taken to ensure a reasonable distance from the normal AV conduction axis. Apart from the usual localized insertion (typically less than 2 to 3 mm), occasionally an accessory pathway with a large[14] or multiply branching insertion may be encountered. Multiple coalescent lesions are required; each of which modifies local electrogram parameters—in contrast to ineffective lesions.

Another uncommon variant is the appendage to ventricular connection characterized by an insertion bridging the appendage tip to the ventricle—resulting in an insertion characteristically removed from the annulus.[15] Careful mapping aided if necessary by three-dimensional mapping can clarify the exact nature of this connection. Similarly, there is the unusual variant of surgically acquired preexcitation—encountered rarely after right atrial appendage anastomosis to the right ventricular outflow tract (performed as a palliative procedure for tricuspid atresia). In the appropriate surgical context and with the preexcited QRS resembling an RV outflow tract tachycardia, careful mapping will allow successful ablation.

Sometimes additional arrhythmia substrate may coexist—AVNRT or an atrial tachycardia. The electrophysiological manuevers described above can assist in deciding whether the accessory pathway participates in the tachycardia, but in practice, elimination of the accessory pathway substrate typically unmasks the AVNRT or atrial tachycardia, which can then be ablated in standard fashion.

Atrioventricular Nodal Reentrant Tachycardias

Atrioventricular nodal reentrant tachycardia (AVNRT) is considered to be the most common form of paroxysmal supraventricular tachycardia (except atrial fibrillation) in adults. It is generally accepted that AVNRT is the result of a reentry circuit in the AV junctional region, despite debate about anatomic delimitations. Functional heterogeneity of AV junctional tissues—primarily with respect to conduction velocity and refractory periods—permits the sustenance of an excitable gap reentry circuit. Because of anatomic factors and the lack of distinct electrophysiological markers of activation, it is difficult to delineate the circuit/circuits entirely. Nevertheless, available evidence suggests that the perinodal atrium, the compact AV node, and possibly a part of the proximal His bundle are involved. Although no significant anatomic differences have been found in patients with AVNRT, evidence is now available to indicate multiple posteriorly situated pathways or approaches to the AV node.[16]

At least three different types of AVNRTs characterized by different circuits have been described: (1) slow antegrade/fast retrograde, (2) fast antegrade/slow retrograde and (3) slow antegrade/slow retrograde. In addition to differences in conduction velocity and refractory periods, the fast and slow pathways manifest relatively disparate anterior and posterior retrograde exit sites. Each of the above tachycardia types has antegrade conduction properties and retrograde atrial activation sites, which indicate the antegrade and retrograde limbs. Most laboratories today establish baseline evidence of antegrade "slow" pathway conduction in the form of a long A–H interval (>200 ms) with or without a discontinuity (50 ms increment in A–H interval for a 10 ms decrement in coupling interval). Additionally an H–A interval ranging from 25 to 80 ms during tachycardia is indicative of a typical slow/fast AVNRT. Variants of AVNRT (fast/slow or slow/slow) need to be differentiated from AV reentrant and atrial tachycardias by the response to ventricular extrastimuli introduced during the AV nodal refractory period, by the atrial activation sequence, and by tachycardia behavior during AV block.

Earliest retrograde activation at the anterosuperior TV during typical AVNRT localizes the fast pathway exit site and analogously retrograde activation near the posteromedial TV during fast/slow AVNRT localizes the slow pathway exit site. Techniques of AV nodal modification have, therefore, targeted these sites to produce selective fast or slow pathway ablation.

The fast pathway exit can be approached by slow withdrawal of the catheter a few millimeters from the His bundle position, while concurrently applying clockwise torque to maintain good contact. A monitoring catheter kept in the His bundle recording position is a convenient reference. Because there are no accepted electrogram markers of fast pathway activation, indirect parameters such as an A:V electrogram amplitude ratio greater than 1 and His deflection of less than 0.05 mv are used to assure relative separation from the His bundle. RF energy applied for a short period at such sites results in PR prolongation and elimination or marked attenuation of retrograde ventriculoatrial conduction. Frequently a junctional tachycardia is noted; this may require atrial pacing to allow monitoring of AV conduction because retrograde conduction cannot provide such assessment. P–R interval prolongation by greater than 50% or AV block make prompt discontinuation of RF energy delivery (which should be titrated in steps of 5 watts) mandatory.

Evaluation following fast pathway ablation typically reveals abolition or marked attenuation of VA conduction accompanied by an increase in the A–H interval and elimination of dual AV nodal physiology. VA conduction is eliminated in more than one third of patients, whereas the VA block cycle length is increased in the remainder. Similarly, the A–H interval is prolonged markedly (<50%) but without significant change in the H–V interval, AV nodal effective refractory period (ERP) or anterograde Wenckebach cycle length. Patients who lose VA conduction without elimination of dual AV node physiology or change in A–H intervals, may have undergone truly selective retrograde fast pathway ablation.

Although ablation is successful in more than 90% of patients, complete heart block occurs in up to 21%.[17] In case of transient conduction block, inpatient telemetric monitoring may be advisable for 1 to 2 days following ablation to watch for delayed complete heart block. The posterior exit of the slow pathway has similarly been targeted by selective ablation and, probably because of the greater distance from the compact AV node, the incidence of complete heart block is consistently lower. Slow pathway ablation has, therefore, become the therapeutic procedure of choice.

Two approaches have been used: the anatomic and the electrophysiological. The anatomic approach uses fluoroscopic landmarks to guide the positioning of the ablation catheter with the target area being the junction of the mid- and posterior thirds of the medial interatrial septum at the level of the TV or in the vicinity of the ostium of the coronary sinus. Assessment of the slow pathway and attempted reinduction is performed after RF delivery for 60 to 90 seconds. If unsuccessful, subsequent RF energy is delivered nearer the AV node;

however, the most posterior successful ablation site is the safest.

The electrophysiological approach uses mapping of the earliest atrial activation during retrograde slow pathway conduction (during fast/slow AVNRT or ventricular pacing) and is limited by the difficulty in inducing consistent retrograde slow pathway conduction adequate for precise mapping. This is possible in only about 10% of patients. Therefore, more commonly, characteristic electrograms representative of slow pathway conduction have been used to guide RF energy application. Two distinct types of slow pathway potentials have been described. One is a sharp spike-like potential (Asp) preceded by a lower frequency, lower amplitude potential (A) during sinus rhythm.[18] Asp usually follows A by 10 to 40 ms and such double potentials are recorded in the vicinity of the CS os near the TV. Their behavior is suggestive of disparate origins. During retrograde conduction over the slow pathway (fast/slow AVNRT, reverse ventricular echo beats or ventricular pacing with retrograde slow pathway conduction—associated with retrograde fast pathway block) the sequence of these double potentials is reversed; that is, Asp now precedes A. RF energy applied to the latest Asp potential close to the TV successfully eliminated AVNRT in 99% of patients (with only one AV block). Experimental data has shown that similar double potentials are produced by asynchronous activation of muscle bands flanking the mouth of the coronary sinus.[19] These cells have histologic characteristics of atrial cells and intracellular recordings demonstrate a rapid upstroke (phase 0) without decremental behavior during stimulation.

The other type of slow pathway potentials are characteristically low amplitude, low frequency signals concealed within or following the atrial electrogram occupying some or all of the A–V interval in sinus rhythm.[20] With high amplification recordings they are easily found by withdrawing the catheter posteriorly from the His bundle position. In the posterior septum, they are usually hump shaped, whereas more anteriorly they are rapid, narrower, often biphasic, and with a superimposed His bundle deflection. They are most typically recorded at the junction of the anterior two thirds and the posterior one third of the area between the His bundle and the CS os. Characteristically, and most importantly, these slow potentials during incremental atrial pacing separate from atrial electrograms, prolong in duration, and decline in amplitude. They fractionate and disappear at rapid pacing rates so that they are not discernible during tachycardia. Animal studies indicate that low frequency signals coincide with activation of cells around the TV possessing AV node-like properties. During reverse ventricular echoes these cells are activated before earliest atrial activation during retrograde slow pathway conduction but fail to be activated during antegrade conduction over the fast pathway. In light of their wide recording area they may represent dead-end pathway activation. Activation of the more posterior part of the slow atrionodal approaches is possibly represented by the Asp potential and that this (transitional) tissue anteriorly (beyond the CS os) gives rise to low-frequency slow potentials.

Following successful ablation of the slow pathway, an increase in the antegrade AV block cycle length and the AV nodal ERP is usually noted without a change in baseline A–H intervals or retrograde conduction. The maximum A–H interval during incremental atrial pacing is characteristically curtailed. However, in about 50% of patients, there is evidence of residual slow pathway conduction in the form of persistent antegrade AV nodal duality and/or single AV nodal echoes, although AVNRT remains typically noninducible even with an intravenous infusion of isoproterenol.

RF ablation guided by either approach offers essentially equivalent results although fewer applications may be required when ablation is guided by low frequency potentials. Complete AV block during slow pathway ablation is definitely uncommon but may be related to an abnormally posteriorly situated fast pathway. To avoid AV block, a careful search should be performed for the most posterior site with typical slow potentials—without His deflections, of course, but also avoiding sites that exhibit slow potentials, which persist and coincide with the end of the A–H interval during rapid pacing. Junctional ectopy is elicited at 70% to 90% of effective sites and it is important to monitor VA conduction during this rhythm. VA block—even intermittent—and faster rates are useful markers of impending AV block. In case of doubt it may be wise to stop RF delivery to check the PR interval during sinus rhythm. Atrial pacing may be helpful although the flip side is that it may mask junctional rhythm and prevent monitoring of VA conduction, which appears to be an earlier sign of encroachment on normal conduction. In the event of AV block, early recovery (within 2 to 3 minutes) indicates a good prognosis.

Recurrence after successful ablation is uncommon—about 2%—and may be lower if complete elimination of the slow pathway is achieved. However, tachycardia noninducibility—in spite of isoproterenol infusion—is an adequate endpoint. The lower incidence of AV block (about 1%) makes slow pathway ablation the technique of choice, although fast pathway ablation may be considered if the slow pathway approach is ineffective. Cautious ablation of the slow pathway is usually effective, even in patients with prolonged P–R intervals at baseline, probably because of persisting so called *intermediate* pathways that permit AV conduction.

Symptomatic patients wishing to be free of drug therapy or intolerant to standard drug therapy may be offered this intervention. The procedure should probably not be performed for initial or infrequent episodes of AVNRT because of the small, but definite, risk of AV block.

Atrial Flutter

The stereotypic ECG pattern of negative sawtooth flutter waves in leads II and III and aVF at a rate of between 200 to 350 beats per minute characterizes typical atrial flutter, and the absence of a diastolic isoelectric baseline distinguishes it from other supraventricular tachycardias. Typical flutter includes both the counterclockwise, as well as the clockwise, form of cavotricuspid isthmus-dependent flutter. Beyond this characterization, macro-reentrant atrial tachycardias have also been considered to be forms of atrial flutter, although termed *atypical* to distinguish them from the foregoing and will be considered separately.

Typical Atrial Flutter

The realization that the anatomically limited and easily accessible cavotricuspid isthmus is critical for the maintenance of typical atrial flutter, coupled with the understanding that stable isthmus conduction block is necessary, has led to the widespread applicability of catheter ablation to eliminate typical atrial flutter.

Present day sophisticated mapping techniques have merely served to confirm data derived originally by surface and oesophageal ECG that the macro-reentrant circuit of typical atrial flutter is confined to the right atrium, resulting in counterclockwise activation when viewed in the left anterior oblique (LAO) perspective with the tricuspid valve *en face*. In analogy with Mines' ring models, the tricuspid valve could be considered the outer and the posterior intercaval right atrium–crista terminalis complex the inner boundary of a ring of reentrant activation.[21]

The surface ECG morphology of counterclockwise typical flutter is remarkably consistent allowing assumptions about the circuit being entirely within the right atrium and obligatory activation through the cavotricuspid isthmus. However, in contrast, the surface ECG morphology of clockwise flutter is more variable and difficult to distinguish from nonisthmus-dependent flutters. As a result, intracardiac activation mapping and entrainment mapping are often necessary as confirmation. Importantly, because both the clockwise and the counterclockwise forms share the same obligatory cavotricuspid isthmus, a common strategy of ablation is clearly applicable and has been shown to be effective.

Methodology

The aim of catheter ablation for typical atrial flutter is to create complete and stable bidirectional cavotricuspid isthmus block, because terminating or interrupting flutter is not enough, and recurrences are likely if isthmus conduction persists. The procedure itself can be subdivided into RF delivery and lesion creation; identification and *filling-in* of residual conducting gaps; and assessment of isthmus conduction.

RF DELIVERY AND LESION CREATION

In our laboratory, we use an irrigated 4-mm tip electrode for temperature-controlled sequential point-by-point RF application targeted at the isthmus between the IVC and the TV. RF delivery can be started at the RV annulus where a large ventricular electrogram is recorded. To achieve a complete block, a linear and contiguous series of point applications needs to be created. This may be facilitated by fluoroscopic monitoring or by nonfluoroscopic methods (e.g., using the three-dimensional mapping systems) of monitoring catheter positioning with or without the use of long introducer sheaths for superior stability. In addition to fluoroscopic monitoring, during counterclockwise flutter, RF is delivered point-by-point from the TV annulus on electrograms within the isthmus region coinciding with the center of the surface ECG flutter wave plateau all the way from the TV annulus to the IVC edge.[22] This ensures a lesion perpendicular to the advancing wavefront and catheter displacement to either side can be instantly and nonfluoroscopically recognized by the altered timing of the site electrogram—for example, in lateral displacement, the site electrogram now coincides with the beginning of the surface ECG plateau (and with the end of the plateau in the case of medial displacement). Recognition of changes in position is facilitated by the naturally lower conduction velocities in this region during flutter. During low lateral right atrial pacing in sinus rhythm, sequential RF is similarly delivered at about the 6-o'clock position in LAO in the isthmus region with electrograms, with a constant stimulus electrogram time from the TV annulus to the IVC edge.[23]

It is important to recognize, however, that the mere fact of having delivered RF energy at a given point does not ensure a transmural lesion; the efficacy of RF varies according to contact, local blood flow, delivered power, and myocardial thickness. During unidirectional activation in atrial myocardium (e.g., in the isthmus during typical flutter or during pacing from the low lateral right atrium or the ostium of the coronary sinus), a local transmural RF lesion of significant size (probably slightly smaller than the distal bipole size) can be recognized by double potentials separated by an isoelectric interval; the second potential being produced by activation detouring around the lesion.[24]

RF power and/or target temperature may need to be manipulated to achieve transmural lesions at each delivery site. Conventional temperature-controlled RF delivery is subject to the vagaries in local convective cooling and topography, which limit the mean power delivered either by achievement of the target temperature and/or by coagulum formation and impedance rise. Irrigating the ablation electrode substantially reduces the latter and permits the delivery of desired (and relatively higher) mean RF power, irrespective of variations in convective cooling. This results in consistent electrogram changes: splitting into double potentials. The use of irrigated tip catheters with a relatively limited power ceiling (40-45 watts) has been shown to be clinically effective and safe, both as a rescue strategy for resistant cases as well as a first line strategy for typical atrial flutter.[25,26] In the latter situation, a significant reduction in procedure and fluoroscopy durations is achieved. Larger lesions with each application may also shorten the procedure: to this end 8-mm electrodes and irrigated tip catheters may be similarly useful.

IDENTIFICATION AND ABLATION OF RESIDUAL GAPS

Because of variations in isthmus anatomy and the inability to create consistent continuous transmural lesions, isthmus conduction frequently persists, despite apparently sufficient ablation. Locating and ablating residual gaps in the ablation line is, therefore, necessary. During typical atrial flutter, such residual gaps can be identified by local electrograms with a single or a fractionated potential centered on or spanning the isoelectric interval of adjacent double potentials. This has allowed parsimonious ablation of flutter recurring after previous ablation. The same approach has been used during pacing from either side of the isthmus, targeting single or fractionated potentials adjacent to double potentials and centered on their isoelectric intervals (Fig. 44-3); the aim being to establish a continuous corridor of double potentials with isoelectric intervals across the full width of the isthmus.[23]

ASSESSMENT OF ISTHMUS CONDUCTION

As indicated earlier, termination of flutter during RF delivery is not a sufficient endpoint because in more than 50% of cases, it does not mean stable isthmus block.[27] Transient block or conduction slowing within the isthmus (or ectopics) may be enough to terminate flutter without completely eliminating or even affecting the substrate. During pacing from one side of the ablation lesion, a long time to activation on the opposite side and an activation sequence within the right atrium demonstrating a 180° change in direction of activation on the other side have been used to diagnose isthmus block. This can be documented both sequentially by using rove mapping and simultaneously with a duodecapolar electrode catheter during low lateral atrial or CS pacing. More recently, local electrogram-based criteria have been effectively used as the most sensitive markers of block in the cavotricuspid isthmus.[23]

The progressive refinement of procedural endpoints has allowed the effective elimination of atrial flutter by catheter ablation. Beginning with a high recurrence rate when flutter termination and noninducibility were considered sufficient endpoints, the demonstration of isthmus block and further refinement with the routine use of local electrogram-based criteria—mapping double potentials supplemented with differential pacing—has led to a reduction in recurrence rates to less than 5%.

The criteria for demonstrating the presence of complete isthmus conduction block may be classified into primary and secondary; the primary criteria divided into right atrial and local electrogram criteria, whereas the secondary criteria include surface ECG changes as well as changes in the activation sequence of the coronary sinus (left atrial criteria).[22] Modification of the sequence of lateral or septal right atrial activation is necessary but not enough to indicate isthmus block. Local criteria more directly evaluate the linear lesion with greater sensitivity and specificity. While the sensitivity of local electrogram criteria derives from being assessed right on the linear lesion, their specificity relates to relative independence from catheter positioning as well as from posterior intercaval conduction. The presence of such conduction may shorten the timing to the second component of double potentials but cannot produce false positive gap electrograms on the ablation line. The extent of reduction in timing of the second component depends chiefly on the level of posterior intercaval conduction. More than one technique can be used to demonstrate the criteria of isthmus block ranging from multielectrode, multicatheter techniques to basket catheters to electroanatomic sequential mapping as well as noncontact mapping. Essentially, all these different technologies present the same activation data.

Not withstanding the different criteria, the choice of pacing site is instrumental in maximizing the sensitivity and specificity for complete isthmus block. The chosen site must be as close as possible to the line of block to maximize the probability and sensitivity for detecting

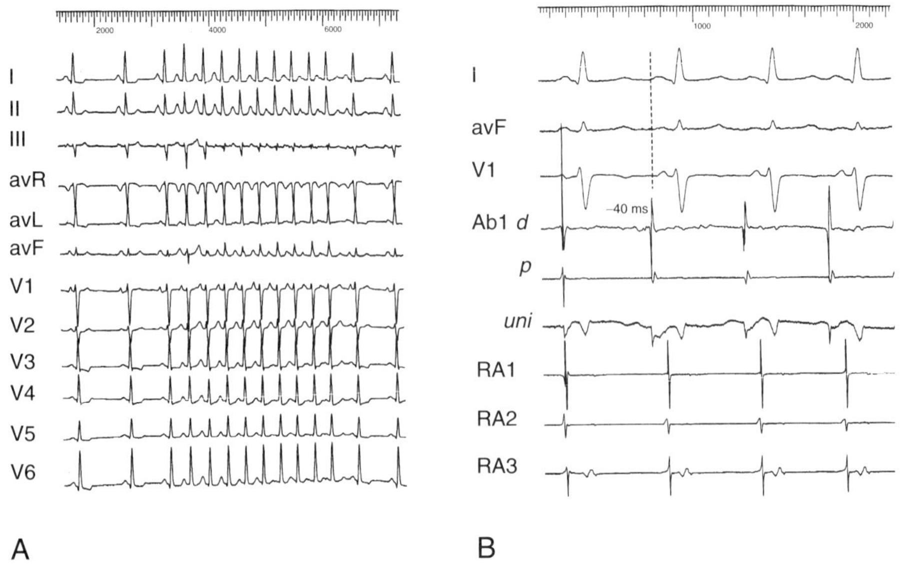

FIGURE 44-3 A non-reentrant left atrial tachycardia that exhibited characteristic warm-up and stop-start behavior (**A,** 12-lead ECG). Successful ablation site (**B**), which was at the junction of the right superior and the right inferior pulmonary veins. The first is a sinus beat; the remaining are during atrial tachycardia. Note the characteristic distal (*d*) to proximal (*p*) activation sequence in the two bipoles of the ablation catheter (Abl) along with a sharp "QS"-type morphology of the unipolar atrial electrogram (*uni*).

slow conduction through the line and avoid its being concealed or masked by the wavefront going around the lesion with a relatively shorter conduction time. If the pacing site is positioned optimally close to the line, the choice of a low lateral RA pacing site versus a coronary sinus ostial pacing site is probably immaterial. With a coronary sinus catheter, however, the most proximal stable pacing site is within the coronary sinus and tends to be at a significant distance from the line—this can be easily assessed by the stimulus to first potential time (st-DP1) on the line. An st-DP1 of 30 ms or less is optimal.[28] Alternatively, the change in the surface ECG P wave produced by isthmus block is maximal when pacing from the low lateral right atrium—and is particularly sensitive to conduction recovery because of the location of the coronary sinus input to the left atrium which is just adjacent to, but on the opposite side of, the lesion only when pacing from the low lateral right atrium. Left atrium activation from the CS input acts as a surface ECG amplifier of P wave change—both for conduction block/delay and particularly for conduction recovery (see later).

Differential pacing involves assessing the response of local onsite electrograms to shifting the site of activation origin (pacing): by advancing or withdrawing it from its original position near the ablation line.[28] If activation on both sides of the line (indicated by local double potentials) is linked directly by a conducting gap, withdrawing the pacing site will increase the activation time to both sides and by approximately the same magnitude. Conversely, if there is no conducting gap across the line, withdrawing the pacing site will certainly increase the activation time upstream of the line, but will either shorten the activation time on the other side of the line or leave it unchanged, but will not increase it.

The demonstration of functional linking by changing pacing sites depends on the relative conduction times to both flanks of the ablation line, and, therefore, may be affected by the selection of the pacing position, relative conduction velocities, and length of the activation detour, as well as by intervening areas of slow conduction or block, which affect only one of the two pacing positions. The pacing catheter should, therefore, be positioned as close as possible to the lesion line and the magnitude of displacement of the pacing position limited (15 mm) so that the stimulus to the first potential time is about 40 ms during distal pacing and 60 ms during proximal pacing. To detect very slow conduction through the isthmus (e.g., 0.05 ms or less) both pacing sites may need to be even closer to the ablation line—that is, with shorter stimulus to first potential times.

Very slow conduction through the isthmus cannot be absolutely ruled out by any technique or criteria and, although we did not find any instances of false-positive diagnoses of persisting conduction, this is theoretically possible in the presence of a conduction delay affecting only activation from the second pacing position.

Differential pacing is a single site assessment technique that is best used as a complement to local electrogram assessment to provide an onsite evaluation of each double- or triple-fractionated potential without having to move the recording ablation catheter from the recording site or perform supplemental mapping. This is an advantage when a gap electrogram is validated to represent persistent conduction through the ablation lesion instead of bystander slow conduction and it permits prompt ablation, whereas recognizing conduction block in spite of triple or fractionated potentials prevents unnecessary ablation. It is also useful for confirming the achievement of block, particularly when limited mapping is performed or to rule out slow conduction when right atrial criteria are used to evaluate conduction.

The standard for complete (and stable) isthmus conduction block is, however, ultimately the absence of recurrence of typical atrial flutter—which can only be evaluated at a distance from the ablation procedure. The stability of the achieved conduction block significantly affects the recurrence rate. The probability of conduction recovery across a composite lesion can be estimated by the number of constituent lesions (mean of 6 to 10 point lesions) times their individual probability of recovery. If the latter is estimated to be 2% based on data from Wolff-Parkinson-White ablation (so called *point* ablation), this works out to 12% to 20%. Recent data have indicated rather high rates of conduction recovery after the achievement of complete block (as also after the termination of flutter by RF delivery in the cavotricuspid isthmus). This mandates monitoring of the stability of isthmus conduction block after ablation: the exponential reduction in the incidence of recovery with time suggests an empirical cutoff for the duration of the monitoring period. Keeping in mind that 97% of conduction recovery was found to occur within 15 minutes, this duration of monitoring is compatible with the present low recurrence rates observed in our laboratory.[27] An extended period of monitoring for recovery followed by reablation might reduce recurrence rates even further.

The procedure as described above is very well tolerated; a few patients require IV sedation and or analgesics for pain relief during RF delivery. Few side effects have been reported; and mostly minor ones relate to femoral venous catheterization. One exception is the small, but significant, risk of AV block when ablating in the so called *septal isthmus*, falling to zero as a more lateral target is selected.

Because the present day ablation procedure is well tolerated, indications have expanded. Although nonpharmacologic therapy was limited earlier to refractory and incapacitated patients with hemodynamic consequences, presently RF catheter ablation is being offered to more and more patients with symptomatic and, at least, single drug refractory atrial flutter. It may now be legitimately considered as alternative first-line therapy for all those with symptomatic sustained typical atrial flutter.

Atypical Atrial Flutter

Because the most common form of atrial reentry is cavotricuspid isthmus-dependent atrial flutter, it has been termed *typical atrial flutter*. Other macro-reentrant atrial tachycardias have been considered to be forms of atrial flutter because, like typical flutter, electrical activation can be recorded at all instances throughout the cycle (i.e., without a diastolic pause characteristic of nonreentrant arrhythmias), although the qualification of *atypical* has been used to distinguish from cavotricuspid isthmus-dependent flutter, irrespective of whether it is right or left atrial in origin.

DIAGNOSTIC CRITERIA

The diagnosis of atypical flutter is based on excluding cavotricuspid isthmus dependence. Essentially, activation through the cavotricuspid isthmus must be shown not to be essential for maintenance of the reentry circuit. This may be indicated by the demonstration of bidirectional activation of the isthmus during tachycardia resulting in collision/fusion (within the isthmus) by activation wavefronts from opposing directions—the low lateral right atrium and the coronary sinus ostium. The recording of double potentials, separated by an isoelectric and constant interpotential interval through the full extent of the isthmus during tachycardia reflects the presence of isthmus block. The demonstration of stable isthmus conduction block during sinus rhythm is strong evidence of exclusion, as is the demonstration by entrainment of the isthmus being out of the circuit. Entrainment mapping at non-right atrial locations—within the coronary sinus or in the left atrium—serves to elucidate their participation, whereas documentation of activation spanning the full reentry cycle length in the right atrium provides support for the right atrium being the locus of the reentry circuit.

Documentation of less than 60% of the cycle length in the right atrium and evidence of intermittent dissociation of the right atrial free wall from the tachycardia also support the diagnosis of left atrial reentry.[29] Passive activation of the right atrium by septally originating wavefronts colliding on the right atrial free wall is also characteristic. However, conduction block in the cavotricuspid isthmus can mask the lower septally originating wavefront.

BARRIERS IN THE ATRIA

Some form of fixed or functional central barrier is a prerequisite for reentry. The most common internal circuit barrier in the right atrium is the posterior intercaval crista terminalis complex around which the circuit of the most frequent right atrial macro-reentrant tachycardia—typical atrial flutter—is formed. The annulus of the tricuspid valve constitutes the external barrier that constrains the activation wavefront of this arrhythmia. Other naturally occurring fixed barriers (independent of the precise form of activation and present also in sinus rhythm) in the right atrium include the inferior and the superior vena cava; of note, only the latter possesses a sleeve of electrically active myocardium continuous with the right atrium. Acquired barriers in this chamber include surgical incisions or patches as well as mute regions devoid of electrical activity.[29,30] The etiology of these mute regions is uncertain in the absence of surgical interventions or coronary artery disease. As opposed to these fixed barriers, the roles of functional activation inhomogeneities in the right atrium are unclear, although the crista terminalis region has been shown to permit conduction across it during sinus rhythm (as well as during certain forms of reentry). Each of the above barriers or inhomogeneities—whether functional or fixed—can potentially support a reentry circuit around it, provided an appropriate trigger is present and the conditions of sufficient conduction times (wavelength) around it are met.

In the left atrium, there are no data available about zones or preferential areas of slow conduction and/or block; however, the left atrium certainly lacks the anatomic equivalent of a crista terminalis and, perhaps, as a result, left atrial macro-reentry is less common. Left atrial flutter as it is commonly termed has been documented to occur in predominantly two situations: (1) as a sequel to ablation in the left atrium by RF delivery, and (2) in subjects with left-sided structural heart disease.[31] Nearly 75% of a cohort of patients undergoing left and right atrial linear ablation for paroxysmal atrial fibrillation subsequently developed left atrial flutter; with endocardial activation mapping indicating that the macro-reentrant circuit was dependant on gaps in incomplete or recovered linear lesions. Of note, although the morphology varied, there were instances when the surface ECG was similar to typical counterclockwise atrial flutter. Characteristically, most patients had more than one morphology (hence circuit) of reentry, and this was concordant with the frequent demonstration of multiple loops with entrainment mapping being used to make the choice of the dominant circuit. Anatomic obstacles, such as the PVs or the mitral valve, buttressed centrally or laterally by the ablation lesion, formed the core of the circuit. In patients who had not undergone linear ablation, macro-reentrant circuits were anchored around relatively large, electrically silent areas in addition to anatomic obstacles, such as the PVs and the mitral valve. Although the exact nature of the etiopathogenesis of these so called *mute zones* remains to be determined, infarction or a myocarditic inflammatory scar have been evoked. Coronary angiography has, however, failed to show an appropriately located coronary occlusion (or lesion) and in the absence of other evidence of myocarditis, other, perhaps hemodynamic (related to elevated pressures) or even genetic reasons may need to be considered. Irrespective of whether the reentrant circuits are a sequel to linear ablation or not, they share the attributes of multiple morphologies and multiple complete and incomplete loops.

Clinical clues that may be of value include a history of a surgical procedure involving a (right) atriotomy (other than cannulation atriotomies), a successful ablation of the cavotricuspid isthmus with demonstrated complete conduction block, or an ECG tracing obviously different from typical flutter. Clockwise cavotricuspid isthmus-dependent flutter, as well as more rarely, counterclockwise isthmus-dependent flutter with an atypical ECG tracing (probably because of altered left atrial activation)—so called *pseudoatypical* flutter—need to be excluded.

Left atrial macro-reentrant tachycardias should be suspected in the setting of an atypical flutter in a patient with structural heart disease affecting the left side of the circulation—particularly if the ECG exhibits low voltages, particularly in the limb leads. V1 commonly shows a dominantly or completely positive deflection reflecting passive activation of the right atrium. So-called *pseudotypical* flutter is a left atrial flutter resembling typical flutter in morphology, although this can be suspected if coronary sinus activation is from distal to proximal—opposite that during typical flutter.

MAPPING STRATEGY

In the EP lab, the mapping strategy depends on the presence or inducibility of arrhythmia and its stability. A stable and sustained arrhythmia allows the sequential acquisition of data—both anatomic and electrical—so as to enable the reconstruction of a three-dimensional activation map. Entrainment mapping, which is a sequential data acquisition technique, is also useful in this situation. Conversely, an arrhythmia that is unstable in terms of either activation sequence or conduction times, or both, does not lend itself to sequential analysis; accordingly, data acquired simultaneously from multiple sites (by multielectrode catheters) must be used to determine the chamber of interest. Some form of three-dimensional electroanatomic mapping can be pursued after cardioversion in sinus rhythm to locate fixed barriers. In patients without inducible arrhythmia, the same strategy may also need to be followed.

The aim of mapping is to determine the complete reentrant circuit that may be defined as the spatially shortest route of unidirectional activation returning to the site of earliest activation and encompassing the complete cycle length of the tachycardia in terms of activation timing. Double- or multiple-loop reentry is defined by the presence of more than one activation front that fulfills the above conditions.[30] Activation fronts that do not fulfill these conditions are bystander fronts and not critical to the arrhythmia. However, incomplete mapping may lead to confusion as to the bystander status of a given activation front or loop, with an incomplete loop being mistaken for a complete one. High-density mapping or entrainment mapping can clarify this situation by documenting wavefront collision and long postpacing intervals, respectively. Multiple, fixed barriers produce multiple isthmuses and because wavefront collision may be present during one arrhythmia only to resolve completely and allow active participation during another circuit, recurrence may result from another arrhythmia or reentry circuit transformation that uses another of the multiple isthmuses.

The behavior of the tachycardia can also provide some clues as to its nature. A single loop tachycardia with a fixed barrier as its core typically remains stable and unchanged during catheter manipulation and may even be difficult to pace terminate, although mechanical *bump* termination (without extrasystoles) rendering the arrhythmia noninducible suggests a restricted and relatively fragile isthmus. A change in ECG morphology without change in cycle length may be due to a transformation of a multiloop tachycardia by interruption of one loop, or a change in bystander activation sufficient to be visible on the surface ECG (e.g., by the change in coronary sinus and left atrial activation observed during incomplete isthmus ablation for typical counterclockwise flutter) or activation of the same circuit in the opposite direction. The observation of significant change in activation within the circuit distinguishes circuit transformation or antidromic activation around the same circuit from bystander activation. Similarly, variations in cycle length can suggest variations in activation pathways resulting from circuit transformation or simply changes in conduction time, the latter usually manifests as cycle length alternans. The absence of ECG changes accompanying changes in activation sequences may occur because of an insufficient change in electromotive force—either because of distance from recording electrodes or because of insufficient electrically active tissue.

The assessment of critical isthmi during sustained stable reentry is relatively easy: guided by activation mapping in case of single loop reentry and supplemented by entrainment mapping when more than one loop is documented by activation mapping; however, evaluation during sinus rhythm is less clear. Barriers potentially capable of serving as central obstacles for reentry can generally be recognized fairly readily. Natural orifices are obvious but electrically inactive barriers may not be. Significantly large (in two dimensions) areas devoid of electrical activity can be easily recognized as electrical scars provided catheter contact is verified, but narrow lines of block (thinner than the recording field of clinically used bipoles) may easily be missed unless the conduction delay across them is maximized by the appropriate choice of optimal pacing sites. In order to avoid overlooking any such scar, it may be advisable to perform mapping during more than one form of activation (e.g., during both proximal coronary sinus and low lateral RA pacing).

CHARACTERISTICS OF ARRHYTHMOGENIC BARRIERS IN THE RIGHT ATRIUM

Once a scar or fixed barrier is localized, its potential for supporting reentry is important in determining whether the isthmuses formed around it need ablation. In a study of 22 consecutive patients with atrial tachycardia after surgical closure of an atrial septal defect, three-dimensional electroanatomic mapping of the right atrium was performed during stable sustained tachycardia and in sinus rhythm to study the properties of electrical scars.

The characteristics of the free-wall line of block resulting from the surgical atriotomy played a significant role in determining the kind of arrhythmia circuit that developed.[32] Specifically, a right atrial free-wall periatriotomy reentry circuit was favored if the scar was relatively long, resulting in a restricted isthmus bounded inferiorly by the IVC, and if the scar was vertical or oblique and relatively anteriorly placed. Nearly all these patients also have peritricuspid reentry. Peritricuspid reentry alone is observed in the absence of any electrophysiological evidence of a right free-wall atriotomy or if this is too small or posterior. If the right atrial free-wall atriotomy is long enough to extend to the IVC inferiorly, thus eliminating the inferior isthmus, periatriotomy reentry cannot occur. Whether the near universal presence of peritricuspid reentry in this cohort of patients is due to diffuse substrate alterations or results from the presence of a (posteriorly placed) atriotomy buttressing the often functional block zone of the crista terminalis is not clear.

The above data suggest that even if sustained reentrant arrhythmias are not inducible in the EP laboratory, in the case of a right atrial free-wall line of block, both isthmuses (inferior end of scar to IVC and the cavotricuspid isthmus) should be ablated with the endpoint of complete and stable block. Alteration of the right atrial substrate (other than the atriotomy) probably plays a lesser role in determining the specific kind of arrhythmia that develops.

Patients with more pronounced hemodynamic loads, such as after corrective or palliative surgery (e.g., Fontan procedure), have also been documented to have the substrate for reentry in the form of small *channels* or isthmuses in the altered milieu of a low-voltage area of the right atrial free wall.[33] It may be difficult to distinguish multiple small channels within this low voltage area without three-dimensional mapping, and even there, high signal to noise ratios are necessary.

The role of reentry confined to or within the interatrial septum is relatively small, at least within the type of patient population described earlier. The remaining minority of atypical right atrial reentrant tachycardias include so called *pericrista reentry,* as well as small and functional reentry circuits in various parts of the right atrium. It is likely that block within the cavotricuspid isthmus facilitates the occurrence of this form of reentry. Mapping during reentry is useful to document the circuit and RF delivery laterally, at the site of conduction across the crista region of block or more medially between the ostium of the coronary sinus.

Transient variations of the typical flutter circuit have been described in the form of upper and lower loop reentry, which are characterized, respectively, by circuits around part of the crista and the SVC or the IVC.[34] Upper loop reentry, therefore, is analogous to pericrista reentry described above. Ablation requires elimination of critical conducting gaps across the crista terminalis. During double wave reentry, a large excitable gap permits two wavefronts to circulate the same circuit—thus cavotricuspid isthmus ablation is curative.

ABLATION PROCEDURE

A standard cure-all lesion is difficult to envision in the face of multiple permutations. Complete endocardial mapping is, therefore, necessary to determine the reentrant circuit and, in order to plan and achieve interruption by ablation, an accurate rendition of the anatomy is equally crucial. Although it may be possible to terminate some circuits based on double potential mapping combined with entrainment mapping, this is generally limited to the subset of right atrial free-wall macro-reentries where the position and extent of the atriotomy scar are well standardized as well as relatively widely appreciated.[30] Additionally, while such empirical ablation may be successful in achieving interruption and termination of the tachycardia, it is difficult to evaluate the achievement of complete conduction block. The situation is even less favorable for the left atrium where multicatheter access is typically limited—for all practical purposes by a transseptal puncture or through a patent foramen ovale.

Some form of three-dimensional reconstruction of the anatomy, combined with activation data are, therefore, useful and multiple versions using different forms of localizing technology and are now commercially available. These systems allow the chamber of interest to be mapped to provide a good idea of the anatomy and of the circuit. The most convenient to access anatomically bounded portion critical to the circuit (all across its width) is chosen for ablation. Entrainment mapping may be helpful to choose between the various loops, whereas high density mapping can reveal that a given segment of the circuit is, in fact, functionally narrower than the anatomy would suggest by demonstrating the presence of lateral boundaries in the form of zones of block.[33]

While it is customary to consider the narrowest segment of the circuit as the target of choice, in practice, there are other issues to be considered. The choice of ablation sites should be among those segments of the reentry circuit that offer the most convenient and safest opportunities for creating conduction block. Among other factors, isthmus size, anticipated catheter stability, and possibility of collateral damage to bystander structures (e.g., such as the phrenic nerve) or to the sinus or AV node should all be assessed in order to make the correct choice. Similarly, the anticipated difficulty of creating a complete conduction block across the chosen site needs to be taken into account—areas of catheter instability due to mechanical contraction (e.g., the right atrial free wall in the region of the tricuspid valve annulus) or regions of thick tissue (particularly in patients with congenital heart disease before or after surgical repair) rendering the achievement of this endpoint that much more difficult.

The width of the targeted isthmus affects both the duration of the procedure and the success rate. It is commonly estimated by anatomic as well as electrophysiological landmarks. Three-dimensional mapping can be of considerable help, although the accuracy depends both on the mapped resolution (number of points sampled) as well as the anatomic resolution of the system reconstruction. Electrophysiological guidance based on local electrograms can be particularly useful. Double potentials that can be traced in a convergent configuration (with a progressively decreasing interpotential interval) and culminate in a fractionated electrogram continuum, indicate one end of a line of block and activation through the resulting isthmus or around a pivot point at the end of the line of block, respectively. Long-duration fractionated electrograms suggest a protected corridor of slow conduction, with single electrograms suggesting a wider and relatively larger ablation target. Electrophysiological signals are more reliable than anatomic guidance for determining the target site because an electrophysiologically defined isthmus may be smaller than an anatomically defined one.

Radiofrequency energy can be delivered sequentially point-by-point to span the targeted segment or by dragging during continuous energy administration. Lesion contiguity and continuity are dependent on the coalescence of multiple transmural lesions, and as for typical flutter, this is best assured by documenting the breakdown of the target electrogram (at each site) into double potentials, and continuing RF delivery at this point for about 30 to 40 seconds more to ensure a stable lesion. Electrogram breakdown into double potentials is dependent on the direction of activation relative to the ablation lesion and is maximized by activation orthogonal to the lesions. An irrigated tip catheter is preferable in order to deliver a cooling infusion of saline and effectively and reproducibly permit the delivery of higher power necessary for consistent transmural lesions.

ASSESSMENT OF OUTCOME

During the delivery of RF lesions, the tachycardia may terminate, or its cycle length may increase—transiently or permanently. Both phenomena indicate that the delivered lesions have affected the reentry circuit and should be followed by continuation of RF or extension of the lesion to ensure the achievement of complete conduction block.

Although inability to reinduce the original arrhythmia is clearly one parameter of successful treatment, this may reflect conduction delay, or variations in autonomic parameters and, of course, is inapplicable if the original arrhythmia is either noninducible from the beginning or was inadvertently terminated mechanically. Therefore, complete stable conduction block within the reentry path is the most objective endpoint for ablation of these reentrant arrhythmias[35]—as well as for accessory AV connections, AV nodal reentry, or typical atrial flutter. In our experience, this has clearly correlated with lower rates of recurrence in the population with incisional reentrant tachycardias.

As shown for typical atrial flutter, an inversion of the activation sequence downstream of ablation and/or local electrogram changes maximized by the choice of an optimal pacing site (one within about 30 ms of conduction time of the ablation lesion) are necessary. In the case of a typical vertically oriented right atrial free-wall atriotomy, a multielectrode halo-type catheter can document activation sequences anterior and posterior to the scar, so that when pacing close to the contiguous ablated isthmus, the absence of a wavefront penetrating the scar and isthmus is used to indicate conduction block. The sensitivity of assessing conduction block is dependent on the distance between the recording and ablation sites and, therefore, local electrogram changes recorded from the site of ablation—wide double potentials separated by isoelectric intervals that respond to a withdrawal of pacing site by a shortening of the activation time to the second or the latest onsite potential (differential pacing)—provide greater accuracy in assessment of conduction through the ablated segment.

Because available multielectrode catheters do not adapt well to mapping parts of the atria other than around the TV, local electrogram criteria or three-dimensional mapping are particularly useful for the purpose of documenting conduction block in other parts of both atria. To maximize the accuracy of three-dimensional mapping, a greater sampling density is desirable close to the ablation site, in addition to the previously indicated necessity for a pacing site close enough to produce a short activation time to the first potential on the ablated segment. Downstream activation advancing toward the ablation segment and any contiguous fixed line of block all along their length is considered to indicate conduction block, although differential pacing may be necessary to rule out very slow conduction. To ensure the stability of block, reverification is advisable after a waiting period; and because careful three-dimensional mapping may be time consuming, an initial assessment by local electrogram criteria followed about 15 minutes later by mapping is an effective and time-saving strategy.

It is clear that other than isthmus-dependent flutter, a multiplicity of reentrant circuits may be encountered in the right and left atria which require ablation to be individually tailored to the specific activation pattern. To achieve high success rates, and avoid recurrence, multiple isthmuses may require ablation coupled with the documentation of stable conduction block through each of them. In situations where this may be difficult to achieve, an empirical linear lesion—maze-like solution, including multiple lesions to block all anatomic isthmuses may be necessary. In spite of this, at present, success rates for ablation of typical atrial flutter remain significantly higher than for atypical flutter; and keeping in mind its stereotyped circuit, this is perhaps not so difficult to understand.

Symptomatic and drug refractory atypical flutters are usually considered for ablation, although the threshold is usually lower in the presence of tachycardiomyopathy or postoperative congenital heart disease where effective cardiac output is dependent on AV synchrony at normal rates. However, the complexity of mapping and ablation in such cases means that the experience and success rates of individual centers should be taken into account.

Non-Reentrant Atrial Tachycardias

This arrhythmia subset must be distinguished from reentrant atrial tachycardias because the approach to their successful ablation is clearly different. The most clear difference is the demonstration of a radial pattern of activation during these tachycardias.[36] The atria, unlike the ventricles, can be considered to be two-dimensional for practical purposes and, therefore, the absence of a significant part of the cycle length indicates a diastolic pause characteristic of abnormal impulse generation resulting from triggered activity or abnormal automaticity. This presupposes an adequate and complete exploration of the endocardium, including the great thoracic veins; the coronary sinus cannot be considered an adequate surrogate for the left atrium. Mapping reveals radial atrial activation confined to electrical systole—that is, coinciding with the surface ECG P wave—originating from the center of this pattern of radial activation. The sequence of activation of the two bipoles of the rove mapping and ablation catheter can be used to advantage: a distal to proximal activation persisting with catheter advancement indicates a vector originating from that direction (Fig. 44-4), whereas a proximal to distal activation sequence suggests that the wavefront originates from the direction of the proximal bipole. An iterative mapping sequence documenting the transition from the former to the latter activation direction is characteristic of the source of radial activation and requires returning to the initial catheter position.

The inference of a non-reentrant mechanism is further strengthened by a pattern of arrhythmia bursts with cycle length irregularity and/or warm-up behavior, unlike a stable reentrant mechanism (see Fig. 44-4).[37] Identical activation sequences for the first and subsequent beats of the tachycardia further support a non-reentrant mechanism. If the arrhythmia is sustained, demonstration of entrainment is supportive of reentry. The most specific evidence of reentry is the demonstration of unidirectional block at initiation with resulting unidirectional circuitous activation. If the reentry circuit is small, the activation of the rest of the atria being radial cannot be distinguished from a non-reentrant mechanism. Clearly, recognition of reentry depends on the mapping resolution. If the reentrant circuit is, in fact, confined to an unmapped area—for example, in the left atrium without access to it or as recently reported within the SVC—the arrhythmia mechanism may indeed be difficult to recognize.[38] In the experimental laboratory, termination of the tachyarrhythmia by sectioning a critical part of the circuit is very specific evidence of reentry but clinically, the difficulty in consistently achieving a completely transmural incision-like lesion by RF delivery reduces the value of this criterion.

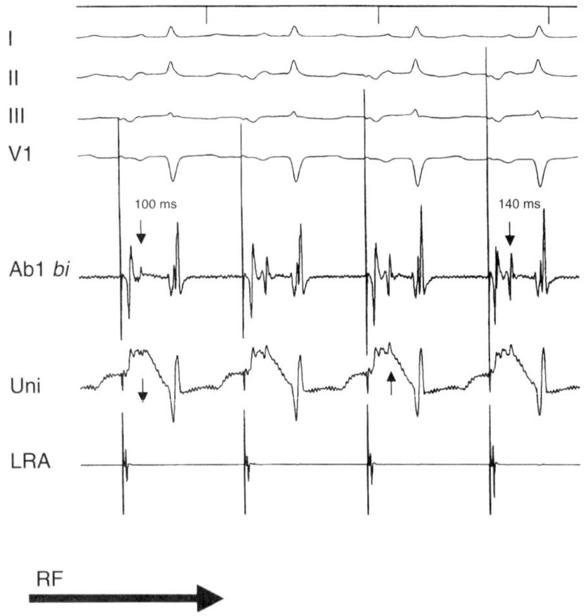

FIGURE 44-4 Ablation of the final gap in the cavotricuspid isthmus ablation line completes the block in a patient undergoing ablation for typical atrial flutter. A fractionated continuous electrogram is recorded from the distal bipole of the ablation catheter (Abl *bi*), which correlates with a multideflection negative unipolar electrogram (Uni) and RF delivery at this site splits the electrograms into two components: the second with a prolonged timing of 140 ms and a completely positive deflection in the unipolar electrogram, together indicating the achievement of complete isthmus block.

The surface ECG with a clearly visible P wave in all 12 leads is very valuable for localization. A cleary evident isoelectric baseline separating individual P waves is compatible with non-reentrant mechanisms; however, some very rapid non-reentrant tachycardias may not exhibit such a pattern, rendering their ECG analysis that much more difficult. P wave polarities in leads V1, I, and aVL are important in assigning the atrium of origin:[39] a late positive, dominantly positive, or completely positive P wave in V1 indicates left atrial origin (see Fig. 44-4).[40] Septally originating tachycardias exhibit narrower P waves and the frontal plane axis is helpful by indicating a superior or inferior origin. Localization can be further refined with the help of pace mapping—using both the surface ECG as well as patterns of intracardiac activation.

Multicatheter activation mapping allows a quicker assessment of approximate localization, although the selection of an appropriate ablation site still requires careful mapping. In the presence of sufficient arrhythmia, sequential mapping with reference to an appropriate intracardiac electrogram can be as effective as multicatheter coverage. The optimal ablation site is one with the earliest bipolar and unipolar activation, and the question of just what timing is early enough can be answered by looking for a timing preceding the surface ECG onset of the P wave, and a QS morphology on the unipolar electrogram that does not precede bipolar activation at that site—indicating that there is no nearby earlier activation (see Fig. 44-4).

Sites on or close to the interatrial septum, as well as the posterior mitral annulus, may need to be mapped from more than one side in order to choose the better site—the former from the right and the left atrium and the latter from endocardium as well as epicardium (coronary sinus). In most instances, focal ablation suffices. The corollary of targeting the earliest activation during arrhythmia is that arrhythmia elimination is the only endpoint. A specific situation is, however, that of a venous tachycardia, particularly a pulmonary venous tachycardia that can even trigger and maintain atrial fibrillation. An expeditious solution is to disconnect the vein from the left atrium (right atrium in case of the SVC) making sure that the disconnection is proximal to the source of the arrhythmia. Thus, the endpoint of demonstrating disconnection is effective even in the absence of arrhythmia.[38]

The role of complex mapping systems is unclear. Those systems relying on sequential data acquisition such as the Biosense system, require sufficient reproducible arrhythmia, whereas simultaneous data acquiring techniques, such as multielectrode basket arrays or the Ensite noncontact system, have the advantage if fewer arrhythmias are available for analysis. The resolution of the latter system is, however, still not good enough to allow selection of the ablation site in the absence of ambient arrhythmia.

Atrial Fibrillation

Atrial fibrillation has become the target of curative ablation procedures based on the assumption that surgical incisions with a knife can be duplicated by catheter ablation. Because surgical lesions are created under direct visual control, it is not surprising to note the difficulty encountered in reproducing surgical results by catheter-based techniques.

The essential lessons learned from initial attempts at catheter-based linear ablation were that creating continuous linear lesions to duplicate surgical atriotomies was difficult; that right atrial lesions alone were safe but ineffective; and that left atrial lesions improved success rates although at significant morbidity and even mortality costs. Most importantly, these studies showed the feasibility of cure by catheter-based techniques in patients with paroxysmal and persistent atrial fibrillation.[41,42]

When shortened paroxysms of atrial fibrillation were studied at close quarters with mapping catheters in the left atrium, stereotyped initiations were traced in nearly all patients to sleeves of atrial myocardium encasing the ostia of one or more PVs. Isolated or multiple discharges culminating in paroxysms of atrial fibrillation were observed.[43] The exact mechanism of such activity is uncertain—hence the term *discharges.* However, activation mapping is suggestive of triggered automaticity followed by apparently random PV activation believed to be related to wandering rotors or fleeting reentry. An intervening or transitional phase of reentry within or around the PV of origin is a likely possibility, particularly because short refractory periods in the PVs coexisting with long conduction times (at typical short coupling intervals) to the left atrium provide a milieu conducive to reentry.

Trigger Burden and Mapping

Irrespective of the exact arrhythmia mechanism(s) involved, in selected patients ablation of the initiating source within a single PV suffices to eliminate short, coupled atrial extrasystoles and bouts of atrial tachycardia, as well as atrial fibrillation. In these patients, a very fast and irregular atrial tachycardia can masquerade as atrial fibrillation on the surface ECG—with an undulating baseline without discernible P waves because of fusion and/or fibrillatory conduction of ectopic P waves—*focal atrial fibrillation.* Ablation at the source terminates ongoing arrhythmia, thereby testifying to its role in maintaining the arrhythmia as well.[44]

Most patients with longer paroxysms of atrial fibrillation (lasting hours to days) however exhibit involvement of two or more PVs. The frequently elusive behavior of trigger discharges and the requirement for stereotypic (and nonsustained) arrhythmias from a single source to distinguish bystander sites/branches within the territory of a given vein trunk render it difficult to identify each of the multiple focal sources. Therefore, a practical solution is to systematically ablate and disconnect the myocardial sleeve of all the PVs during sinus rhythm.[45]

Electrical Disconnection of Pulmonary Vein Musculature

Anatomic considerations dictate that the more proximal the level of ablation, the greater the extent of disconnected myocardium, but this requires more extensive ablation probably because of increasing diameter and myocardial coverage proximally. Electrophysiologically definable sites of preferential inputs to the veins, however, enable disconnection to be achieved without circumferential ablation in the majority of patients. In effect, these preferential inputs demarcate the electrophysiological equivalent of the left atrial–PV junction and can be considered the electrophysiologically defined ostium.

To localize these inputs, venous activation can best be appreciated with some form of circumferential mapping—as opposed to longitudinal mapping typified by a multielectrode catheter placed along the length of the vein. A thin, (low profile and nontraumatic) preshaped circumferential multielectrode catheter allows continuous assessment of activation around the full circumference of the vein, in addition to providing a fluoroscopically visible marker for the vein with a verifiable three-dimensional orientation. One or two bipole(s) or electrode(s) typically show earliest activation with later and sequential spread to the rest of the venous circumference; ablation proximal to this site delays activation in this sector with the antipodal sector (usually) now becoming the earliest activated. Ablation of this secondarily manifested input eliminates all evidence of activation distally, indicating the presence of interconnections beyond the site of ablation. Two sectors of early activation separated by one of later activation suggest two distinct breakthroughs; conversely, multiple near-simultaneous and contiguously activated sectors suggest multiple or coalescent (broad inputs). Because of the cul-de-sac nature of electrical activation in the PV, the disappearance of all distal (circumferentially recordable) potentials (entrance block) is a clear and unarguable indicator of conduction block.

Distinguishing target PV potentials from far-field atrial potentials is important to avoid unncecessary ablation, which could result in vein stenosis or avoidable collateral damage (e.g., to the lung or the phrenic nerve). In the right-sided PVs, both right as well as left, atrial potentials can be identified (by mapping both sides of the interatrial septum), whereas in the left PVs, the nearby appendage is the most common origin of non-PV potentials: distal coronary sinus pacing by anticipating left atrial appendage activation can distinguish the two (Fig. 44-5). Demonstration of entrance block into the vein from the larger current source of the left atrium probably reflects complete bidirectional block (see Fig. 44-5). The demonstration of PV to left atrium block during sinus rhythm may be hampered by the occurrence of threshold effects when pacing from the PVs—at low outputs, PV capture may or may not occur (depending on the proximity to PV muscle—difficult to ascertain in case of a discrete fascicle), whereas at higher outputs, direct (electrotonic) capture of the adjacent left atrial appendage and sometimes posterior left atrium can occur—even in the complete absence of PV potentials.[46]

Circumferential mapping allows electrophysiologically guided disconnection of the four PVs to be accomplished successfully and rapidly in nearly 100% of veins. A strategy of PV disconnection without proof of arrhythmogenicity can be justified only by a low risk of side effects: notably PV stenosis. Limiting RF power, minimizing the circumferential extent of ablation, and targeting the most proximal segment (usually the largest diameter) are all important in limiting the frequency of this difficult-to-treat complication in 1% to 2% of ablated PVs. Nonocclusive stenosis limited to a single PV (typically draining about one half of one lung) usually has no clinical consequences.

Following disconnection, provocative testing is performed with isoprenaline infusion and rapid atrial pacing. Spontaneous arrhythmia at this juncture is necessarily of nonpulmonary venous origin. Such non–PV-initiating (and maintaining) focal sources are preferentially found in the posterior left atrium surrounding the ostia of the PVs. Less commonly, they may be found extra left atrially, but with decreasing frequency as a function of distance from the PV ostia. They may rarely be located epicardially within the coronary sinus, or within a persistent left superior vena cava draining into the coronary sinus (in fact, an abnormal adult remnant of a usually vestigial left common cardinal vein which in the normal adult is the vein or ligament of Marshall) or traced to the septum or, in the right side of the heart, the (right) superior vena cava, or to the vicinity of the crista terminalis.[47]

Supplemental ostial ablation aimed at late or fractionated potentials can reduce these recurrences; for some others, opportune mapping during periods of nonsustained (thus mappable) arrhythmia can allow successful ablation. When these non-PV sources are multiple- or trigger-sustained atrial fibrillation, they are difficult to map during the few opportunities offered by early or immediate reinitiation following cardioversion and different options may have to be tried. Better mapping techniques may allow localization and elimination by wide or even anatomic ablation of these non-PV sources. Adjuvant linear ablation—a necessary component for ablation of chronic atrial fibrillation—may also terminate or reduce the duration of paroxysms and facilitate mapping, although at the cost of possible left atrial proarrhythmia (in the case of incomplete lines) and, perhaps, a variable effect on left atrial contractile function. Previously ineffective antiarrhythmic drug therapy can eliminate residual atrial fibrillation in nearly one half of these patients.

Pulmonary vein isolation is easily demonstrable and routinely possible with a low risk of PV stenosis and the

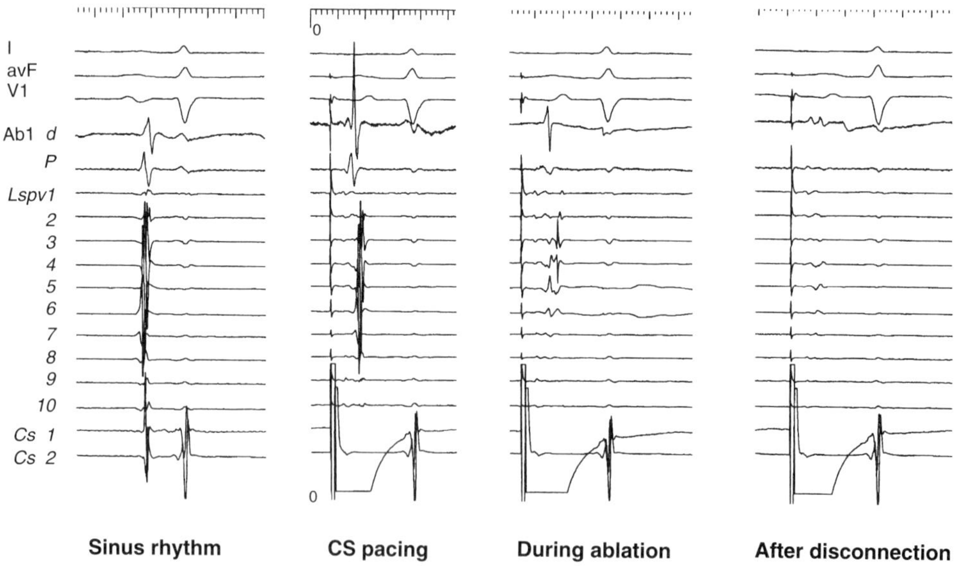

FIGURE 44-5 One example of ostial disconnection of the left superior pulmonary vein from the left atrium. Surface ECG recordings accompanied by recordings from the distal (*d*) and proximal (*P*) bipoles of the ablation catheter, 10 bipoles from a decapolar circumferential catheter placed within the vein (*Lspv 1* through *10*) as well as two bipoles from the coronary sinus (*Cs 1* and *Cs 2*). Sinus rhythm electrograms show single potentials, which split into two components (during CS pacing) with a low-voltage initial far-field component and a second sharp PV potential with earliest activation at bipoles 6 and 7. Ablation first delays and then eliminates all evidence of PV activation indicating complete isolation. PV, Pulmonary vein.

outcome compares favorably with the risk of persistent AF, antiarrhythmic, and anticoagulant drug therapy. In patients with unsuccessful results after PV isolation, the burden of nonpulmonary focal sources must be neutralized, with or without supplementary linear ablation, to increase cure rates.

INDICATIONS

With increasing knowledge of the electrophysiological mechanisms of atrial fibrillation, this curative technique is being offered to more and more patients. However, at the present time, given the complexity of the procedure and the caveats of operating on the left side of the circulation, this procedure should be restricted to symptomatic patients who are resistant to antiarrhythmic drug therapy. Typically one or preferably two of the most effective drugs available at present (classes IC and III) should have been tried. Better results have been obtained in most studies in patients with paroxysmal or self-terminating atrial fibrillation as opposed to those with persistent atrial fibrillation. Patients with long-standing atrial fibrillation and greatly dilated left atria, who may not be good candidates even for a surgical maze procedure or its equivalent, certainly stand little chance for a successful outcome.

REFERENCES

1. Gallagher JJ, Pritchett ELC, Sealy WC, et al: The preexcitation syndromes. Prog Cardiovasc Dis 1978;20:285-327.
2. Benditt DG, Benson DW Jr, Dunnigan A, et al: Role of extrastimulus site and tachycardia cycle length in inducibility of atrial preexcitation by premature ventricular stimulation during reciprocating tachycardia. Am J Cardiol 1987; 60:811-819.
3. Martinez-Alday JD, Almendral J, Arenal A, et al: Identification of concealed posteroseptal Kent pathways by comparison of ventriculoatrial intervals from apical and posterobasal right ventricular sites. Circulation 1994;89:1060-1067.
4. Hirao K, Otomo K, Wang X, et al: Para-Hisian pacing: A new method for differentiating retrograde conduction over an accessory AV pathway from conduction over the AV node. Circulation 1996;94:1027-1035.
5. Sellers TD Jr, Gallagher JJ, Cope GD, et al: Retrograde atrial preexcitation following premature ventricular beats during reciprocating tachycardia in the Wolff-Parkinson-White syndrome. Eur J Cardiol 1976;4:283-294.
6. Klein GJ, Bashore TM, Sellers TD, et al: Ventricular fibrillation in the Wolff-Parkinson-White syndrome. N Engl J Med 1979;301: 1080-1085.
7. Haissaguerre M, Gaita F, Fischer B, et al: Radiofrequency catheter ablation of left lateral accessory pathways via the coronary sinus. Circulation 1992;86:1464-1468.
8. Haissaguerre M, Dartigues JF, Warin JF, et al: Electrogram patterns predictive of successful catheter ablation of accessory pathways: Value of unipolar recording mode. Circulation 1991;84:188-202.
9. Jackman WM, Friday KJ, Yeung-Lai-Wah JA, et al: New catheter technique for recording left free wall accessory atrioventricular pathway activation: Identification of pathway fiber orientation. Circulation 1988;78:598-611.
10. Haissaguerre M, Gaita F, Marcus FI, Clementy J: Radiofrequency catheter ablation of accessory pathways : A contemporary review. J Cardiovasc Electrophysiol 1994;5:532-552.
11. Choi KJ, Shah DC, Jais P, et al: Case report: Successful ablation of Hisian ectopy identified by a reversed His bundle activation sequence. J Int Card Electrophysiol 2002;6:183-186.
12. Takahashi A, Shah DC, Jais P, et al: Specific electrocardiographic features of manifest coronary vein posteroseptal accessory pathways. J Cardiovasc Electrophysiol 1998;9:1015-1025.
13. Tchou P, Lehmann MH, Jazayeri M, Akhtar M: Atriofascicular connection or a nodoventricular Mahaim fiber? Electrophysiological elucidation of the pathway and associated reentrant circuit. Circulation 1988,77:837-848.
14. Gaita F, Haissaguerre M, Scaglione M, et al: Catheter ablation in a patient with a congenital giant right atrial diverticulum presenting as Wolff-Parkinson-White syndrome. Pacing Clin Electrophysiol 1999;22:382-385.
15. Goya M, Takahashi A, Nakagawa H, Iesaka Y: A case of catheter ablation of accessory atrioventricular connection between the right atrial appendage and right ventricle guided by a three dimensional electroanatomic mapping system. J Cardiovasc Electrophysiol 1999;10:1112-1118.
16. Wu J, Wu J, Olgin J, et al: Mechanisms underlying the reentrant circuit of atrioventricular nodal reentrant tachycardia in isolated canine atrioventricular nodal preparations using optical mapping. Circ Res 2001;88:1189-1195.
17. Jazayeri MR, Hempe SL, Sra JS, et al: Selective transcatheter ablation of the fast and slow pathways using radiofrequency energy in patients with atrioventricular nodal reentrant tachycardia. Circulation 1992;85:1318-1328.
18. Jackman WM, Beckman KJ, McClelland JH, et al: Treatment of supraventricular tachycardia due to atrioventricular nodal reentry, by radiofrequency catheter ablation of slow pathway conduction. N Engl J Med 1992;327:313-318.
19. McGuire MA, de Bakker JM, Vermeulen JT, et al: Origin and significance of double potentials near the atrioventricular node: Correlation of extracellular potentials, intracellular potentials, and histology. Circulation 1994;89:2351-2360.
20. Haissaguerre M, Gaita F, Fischer B, et al: Elimination of atrioventricular nodal reentrant tachycardia using discrete slow potentials to guide application of radiofrequency energy. Circulation 1992;85(6):655-656.
21. Shah DC, Jais P, Haissaguerre M, et al: Three dimensional mapping of the common atrial flutter circuit in the right atrium. Circulation 1997;96:3904-3912.
22. Shah DC, Haissaguerre M, Jais P, et al: Atrial flutter: Contemporary electrophysiology and catheter ablation. Pacing Clin Electrophysiol 1999;22(2);344-359.
23. Shah DC, Takahashi A, Jais P, et al: Local electrogram based criteria of cavotricuspid isthmus block. J Cardiovasc Electrophysiol 1999;10:662-669.
24. Shah DC, Haissaguerre M, Jais P, et al:. High density mapping of activation through an incomplete isthmus ablation line. Circulation 1999;99(2):211-215.
25. Jais P, Haissaguerre M, Shah DC, et al: Successful irrigated tip catheter ablation of atrial flutter resistant to conventional radiofrequency ablation. Circulation 1988;98:835-838.
26. Jais P, Shah DC, Haissaguerre M, et al: Prospective randomized comparison of irrigated tip versus conventional tip catheters for ablation of common flutter. Circulation 2000;101:772-776.
27. Shah DC, Takahashi A, Jais P, et al: Tracking dynamic conduction recovery across the cavotricuspid isthmus. J Am Coll Cardiol 2000;35:1478-1484.
28. Shah D, Haissaguerre M, Takahashi A, et al: Differential pacing for distinguishing block from persistent conduction through an ablation line. Circulation 2000;102:1517-1522.
29. Jais P, Shah DC, Haissaguerre M, et al: Mapping and ablation of left atrial flutters. Circulation 2000;101:2928-2934.
30. Shah D, Jais P, Takahashi A, et al: Dual loop intra-atrial reentry in humans. Circulation 2000;101:631-639.
31. Jais P, Shah DC, Haissaguerre M, et al: Efficacy and safety of septal and left atrial linear ablation for atrial fibrillation. Am J Cardiol 1999;84:139R-146R.
32. Shah DC, Jais P, Hocini M, et al:. Catheter ablation of atypical right atrial flutter. In Zipes DP, Haissaguerre M: Catheter Ablation of Arrhythmias, 2nd ed. Armonk, NY, Futura, 2001, pp 153-168.
33. Nakagawa H, Shah N, Matsudaira K, et al: Characterization of reentrant circuit in macroreentrant right atrial tachycardia after surgical repair of congenital heart disease: Isolated channels between scars allow "focal" ablation. Circulation 2001;103: 699-709.
34. Yang Y, Cheng J, Bochoeyer A, et al: Atypical right atrial flutter patterns. Circulation 2001;103:3092-3098.

35. Ouyang F, Ernst S, Vogtmann T, et al: Characterization of reentrant circuits in left atrial macroreentrant tachycardia: Critical isthmus block can prevent atrial tachycardia recurrence. Circulation 2002;105:1934-1942.
36. Goldreyer BN, Gallagher JJ, Damato AN: The electrophysiologic demonstration of atrial ectopic tachycardia in man. Am Heart J 1973, 85:205-215.
37. Wyndham CRC, Arnsdorf MR, Levitsky S, et al: Successful surgical excision of focal paroxysmal atrial tachycardia: Observations in vivo and in vitro. Circulation 1980;62:1365-1372.
38. Shah DC, Haissaguerre M, Jais P, Clementy J: High resolution mapping of tachycardia originating from the superior vena cava: Evidence of electrical heterogeneity, slow conduction, and possible circus movement reentry. J Cardiovasc Electrophysiol 2002; 13:388-392.
39. Tang CW, Scheinman MM, Van Hare GF, et al: Use of P wave configuration during atrial tachycardia to predict site of origin. J Am Coll Cardiol 1995:26:1315-1324.
40. Yamane T, Shah DC, Peng JT, et al: Morphological characteristics of P waves during selective pulmonary vein pacing. J Am Coll Cardiol 2001;31:1505-1510.
41. Haissaguerre M, Jais P, Shah DC, et al: Right and left atrial radiofrequency catheter therapy of paroxysmal atrial fibrillation. J Cardiovasc Electrophysiol 1996;7:1132-1144.
42. Swartz JF, Pellersels G, Silvers J, et al: A catheter based curative approach to atrial fibrillation in humans [abstract]. Circulation 1994;90(pt ii):I-335.
43. Haissaguerre M, Jais P, Shah DC, Takahashi A, et al: Spontaneous initiation of atrial fibrillation by ectopic beats originating in the pulmonary veins. N Engl J Med 1998;339:659-666.
44. Jais P, Haissaguerre M, Shah DC, et al: A focal source of atrial fibrillation treated by discrete radiofrequency ablation. Circulation 1997;95:572-576.
45. Haissaguerre M, Shah DC, Jais P, et al: Electrophysiological breakthroughs from the left atrium to the pulmonary veins. Circulation 2000;102:2463-2465.
46. Shah DC, Haissaguerre M, Jais P, et al: Left atrial appendage activity masquerading as pulmonary vein potentials. Circulation 2002;105:2821-2825.
47. Shah DC, Haissaguerre M, Jais P: Current perspectives on curative catheter ablation of atrial fibrillation. Heart 2002;87:6-8.

Chapter 45 Ventricular Tachycardia

MARCIE BERGER, JASBIR SRA, RYAN COOLEY, ZALMEN BLANCK, SANJAY DESHPANDE, ANWER DHALA, and MASOOD AKHTAR

Radiofrequency ablation of ventricular tachycardia (VT) has been primarily reserved for palliation and cure of monomorphic and hemodynamically well-tolerated arrhythmias. For purposes of ablation, ventricular tachycardia may be categorized as having a mechanism that either depends on a relatively small, localized region or a more diffuse region of myocardium (Table 45-1). Mapping and ablation strategies, as well as ablation outcomes, vary significantly according to the tachycardia mechanism. Ventricular tachycardia arising from a small focus within the heart may have a variety of underlying cellular mechanisms, including automaticity, triggered activity, or micro-reentry. In contrast, tachycardias that incorporate larger regions of myocardium as part of their critical substrate use relatively broad reentrant circuits for arrhythmia induction and propagation. This latter mechanism applies to the majority of ventricular tachycardias associated with myocardial scarring, and to bundle branch reentry ventricular tachycardia.

TABLE 45-1 Mechanisms of Ventricular Tachycardia

VT Substrate Within a Small Region	Probable Mechanism
Idiopathic RVOT tachycardia	(cAMP)-mediated triggered activity
Idiopathic LV tachycardia	Intrafascicular reentry
Repetitive monomorphic VT	(cAMP)-mediated triggered activity
Abnormal automaticity	Enhanced automaticity
Scar-related VT	Micro-reentry
VT Substrate Within a Broad Region	
Reentry Involving HPS	Macro-reentry
Bundle branch reentry	
Interfascicular reentry	
Scar-Related VT	
Prior myocardial infarction	Reentry, (automaticity)
ARVD	Reentry
Chagas' disease, sarcoidosis	Reentry
Repaired CHD	Reentry

ARVD, arrhythmogenic right ventricular dysplasia; CHD, congenital heart disease; HPS, His-Purkinje system; LV, left ventricular; RVOT, right ventricular outflow tract; VT, ventricular tachycardia.

As the dominant mechanism of clinical ventricular tachycardia, reentry represents the most extensively studied model of VT.

Theoretical Basis for Reentrant Circuits Associated with Myocardial Scarring

The majority of ventricular tachycardias associated with myocardial scarring likely arise from reentrant circuits localized to the interface between normal and abnormal myocardium. Initial in vivo evidence for this mechanism came from epicardial recordings in dogs three to seven days following acute infarct. Using this model, El-Sherif and colleagues recorded electrical activity over infarcted tissue bridging the diastolic interval between consecutive ventricular tachycardia beats, providing compelling evidence for reentry.[1] The clinical relevance of reentry was subsequently confirmed by intraoperative mapping studies of human ventricular tachycardia, using high-density endocardial electrode arrays.[2]

THEORETICAL MODEL OF REENTRY

Stevenson and coworkers have proposed a theoretical model of reentry that clarifies the components of ventricular tachycardia reentrant circuits and provides a useful framework to analyze the appropriateness of targets for ventricular tachycardia ablation.[3] According to this model, reentrant ventricular tachycardia circuits contain a region of slow conduction, representing activation of a relatively small mass of myocardial tissue, critical to the initiation and propagation of ventricular tachycardia. Mapping studies in canine models and in humans suggest that these slowly conducting components of reentrant circuits are often localized to narrow isthmi, isolated from adjacent myocardium by arcs of anatomic or functional block.[4,5] During ventricular tachycardia, impulse propagation through these regions occurs in diastole. Scar tissue commonly provides the anatomic boundaries for these isthmi of slow conduction in the post-infarct setting, however, certain cardiac structures may also contribute. In fact, Wilber and associates demonstrated that in four of twelve patients with ventricular tachycardia and earlier inferior wall infarction, the isthmus defined by the mitral valve annulus and the infarct represented a region of slow conduction critical to the reentrant circuit.[6] Because these isthmi of slow conduction are narrow in diameter, bounded by natural arcs of conduction block, and essential to reentrant circuits, they represent attractive targets for ablation.

When the excitation wavefront emerges from the region of slow conduction to depolarize the remainder of the myocardium, at a site designated by Stevenson's model as the "exit" point, the QRS complex is inscribed on the surface ECG. Subsequently, the reentrant wavefront may propagate through normal tissue along the border of the scar (outer loop), or through the infarct region (inner loop), eventually returning to the region of slow conduction. Stevenson terms excitable tissue within the infarct that does not form an essential component of the reentrant circuit "bystander" tissue.

Reentrant circuits may be complex. Downar and colleagues observed that strands of surviving myocardial fibers within infarct zones, providing the region of slow conduction required for reentry, may use alternate entry and exit points during tachycardia.[7,8] The multiplicity of potential reentrant pathways that results may produce spontaneous changes in tachycardia cycle length and in ventricular tachycardia morphology. Moreover, the lack of a single defined exit point suggests that exit sites represent poor radiofrequency ablation targets.

ANISOTROPIC CONDUCTION

Essential to Stevenson's model is the requirement for a region of slow conduction, permitting reentry to occur within the size constraints of the ventricle. It is increasingly clear that the nonuniform anisotropic conduction properties of diseased myocardium may provide the irregular, slow propagation of action potentials that satisfy this requirement.[9] Anisotropic conduction refers to the preferential longitudinal conduction that is observed in adult cardiac myocytes.[10-12] In mature ventricular muscle cells, activation wavefronts may propagate across intercellular junctions both longitudinally and transversely. Because of the elongated configurations of ventricular myocytes, however, wavefronts propagating transversely encounter comparatively more intercellular junctions and, thus, travel more slowly than wavefronts moving an equal distance longitudinally. Therefore, normal myocardial conduction is described as uniformly anisotropic, with advancing wavefronts that are smooth in all directions, but slower transversely than longitudinally, resulting in teardrop-shaped isochrones (Fig. 45-1A). Anisotropic conduction may, under certain conditions, become nonuniform, predisposing to reentrant arrhythmias. For instance, with the deposition of connective tissue associated with advancing age or myocardial infarction, primarily arranged in parallel to the longitudinal axis of the cardiac fibers, lateral impulse propagation encounters greater resistance and becomes extremely slow and irregular (see Fig. 45-1B).[10,13]

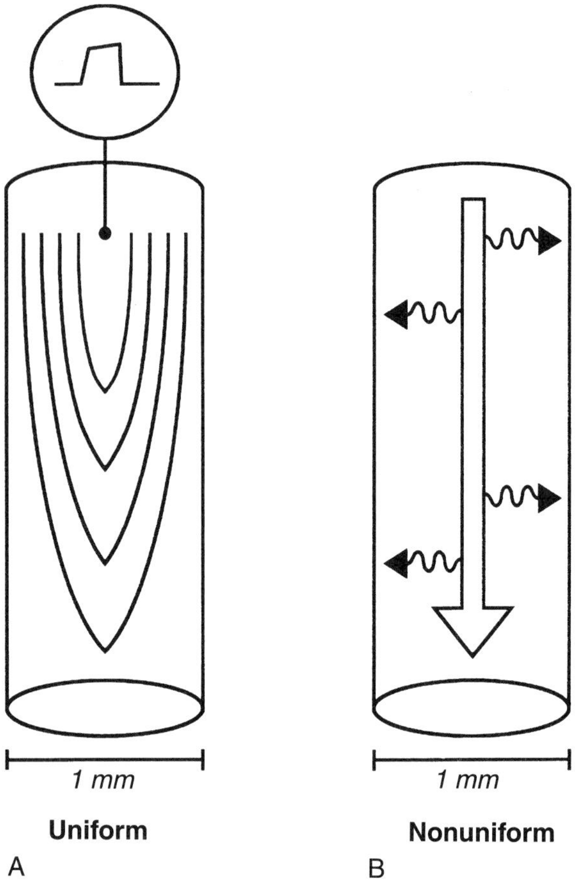

FIGURE 45-1 Anisotropic conduction in cardiac muscle. Schematic depicts propagation characteristics for ventricular myocyte bundles with uniform (**A**) and nonuniform (**B**) anisotropic properties. Uniform anisotropic spread produces advancing wavefronts, depicted by isochrones in **A**, that spread smoothly in all directions, albeit more slowly transversely than longitudinally. The slow irregular spread of transverse excitation characteristic of nonuniform anisotropic conduction is depicted by the sawtooth curves in **B**.

This remodeling of cellular interconnections and electrical uncoupling, producing regions of slow conduction, may play a central role in establishing conditions for reentry.

Mapping Techniques

There are several available methods for localizing ventricular tachycardia. Mapping strategies may be complimentary, but should be tailored according to the patient's clinical hemodynamic presentation and putative tachycardia mechanism. Electrocardiographic documentation of spontaneous ventricular tachycardia greatly facilitates the mapping process.

PACE-MAPPING

Pacing from the tip of the mapping catheter during sinus rhythm permits a comparison between paced QRS complexes and the QRS complexes during clinical ventricular tachycardia, referred to as *pace-mapping*.[14-16] The pacing rate during mapping is performed at a rate similar to that of the clinical tachycardia because local myocardial conduction velocities and refractory periods may change with the frequency of activation, altering the surface ECG morphology. The ECGs during pacing and tachycardia are then compared. Optimal pace maps are those with the closest match between QRS morphologies, including both R/S ratio and fine notching, in each of the 12 leads. When a relatively small region of myocardium is required for initiation and propagation of ventricular tachycardia, as is the case for idiopathic right ventricular outflow tract tachycardia, pacing the tachycardia origin produces a QRS morphology identical to that of the clinical arrhythmia, and is helpful in guiding ablation.[16]

ACTIVATION-MAPPING

Arrhythmias depending on a small region of myocardium for tachycardia initiation and maintenance may also be mapped during tachycardia by locating the earliest local intracardiac bipolar electrograms, which ordinarily precede the onset of the QRS complex on the surface ECG by at least 15 ms.[17-19] Alternately, if unfiltered unipolar electrograms are recorded, the electrograms near the origin of tachycardia have a characteristic qS configuration. Recent case reports indicate that CARTO (Biosense, Tirat Ha-Carmel, Israel), a computerized nonfluoroscopic mapping system, may dramatically facilitate the laborious process of detailed activation sequence mapping. The CARTO system is a cardiac mapping technique, consisting of an ablation catheter with a built-in location sensor, an external magnetic field emitter, and a processing unit that permits acquisition of both catheter-tip electrograms and spatial coordinates. These data may then be used to construct a three-dimensional electro-anatomic model of the heart, localizing the region activated earliest during tachycardia (Fig. 45-2).[20-24]

PURKINJE POTENTIALS

Verapamil-sensitive, idiopathic left ventricular tachycardia is thought to represent a micro-reentrant tachycardia originating, most commonly, from the Purkinje fiber network of the left posterior fascicle.[26-28] Consonant with this theory, some investigators have successfully mapped potentials generated by the Purkinje fibers, which precede earliest ventricular activation during tachycardia, to select ablation target sites.[18,29,30] Nakagawa and coworkers in their study of eight patients with idiopathic left ventricular tachycardia identified earliest Purkinje potentials as more powerful predictors of

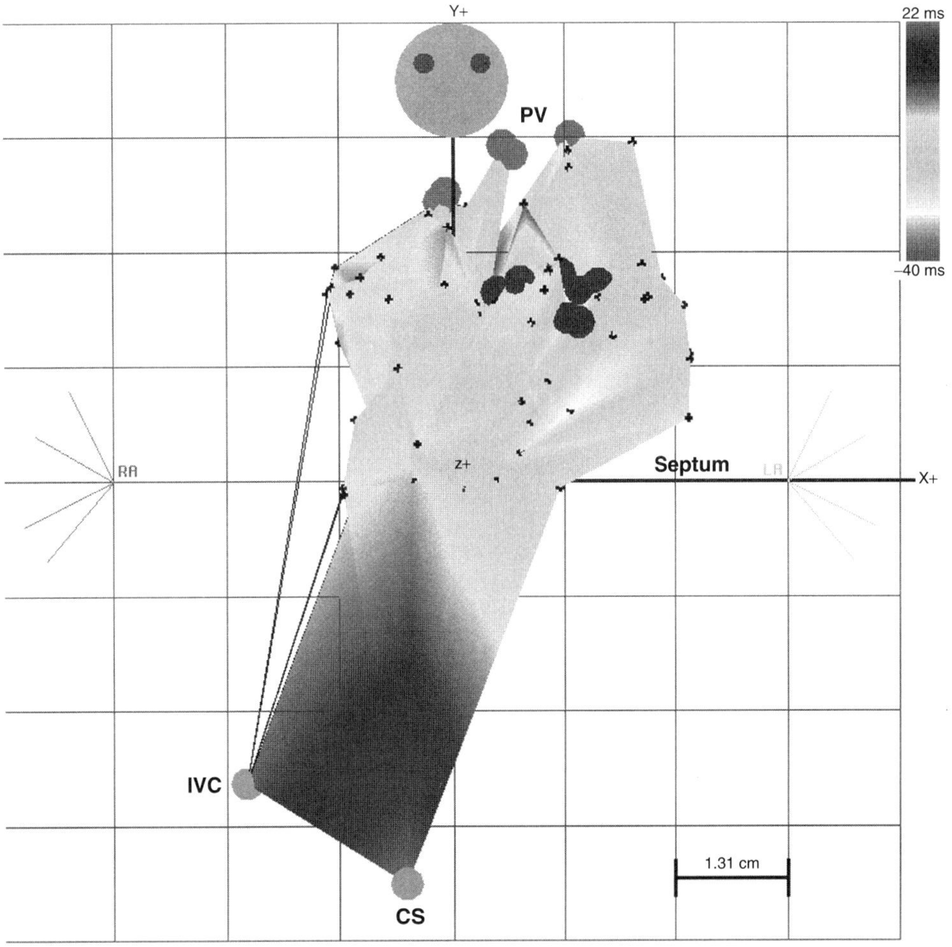

FIGURE 45-2 (See also Color Plate 45-2.) Activation mapping of idiopathic right ventricular outflow tract (RVOT) tachycardia. The CARTO electroanatomic map of the lower right atrium and RVOT is shown during tachycardia. Red indicates areas with earliest endocardial activation and orange, yellow, green, blue, and purple indicate progressively delayed activation sequence. Shown are locations of inferior vena cava (IVC), coronary sinus (CS), and the pulmonic valve (PV). (From Varanasi S, Dhala A, Blanck Z, et al: Electroanatomic mapping for radiofrequency ablation of cardiac arrhythmias. J CV Electrophys 1999;10:538-544.)

successful ablation site than the timing of ventricular activation, or pace-mapping.[30]

MAPPING SCAR-RELATED REENTRANT VENTRICULAR TACHYCARDIA CIRCUITS

Pace-mapping and activation-mapping, useful techniques for guiding ablation of tachycardias originating from relatively small foci, have proved less useful for identifying optimal ablation sites in scar-related ventricular tachycardia. These techniques may fail to localize critical regions of slow conduction, important ablation targets in these reentrant circuits. In one series of 18 patients with postmyocardial infarction ventricular tachycardia, pace maps resembled clinical tachycardia in only 9% of sites where radiofrequency current terminated tachycardia.[15] The poor predictive value of pace maps in post-infarction reentrant ventricular tachycardia may be explained by either the relatively large dimensions of these circuits, or alternately, by the presence of functional conduction block during tachycardia that is absent during sinus rhythm. Stevenson and colleagues observed that slow local conduction at the site of pace-mapping, evidenced by a stimulus to QRS interval of greater than 40 ms, was observed at many critical reentrant circuit sites.[15] This finding, however, does not specifically define an essential component of the reentrant circuit. Analogous to pace-mapping, activation-mapping localizes the general region of scar-related ventricular tachycardia, but does not reliably identify appropriate ablation targets.[31] Given these limitations, a number of strategies have been developed for identifying critical regions of slow conduction within postmyocardial infarction ventricular tachycardia reentrant circuits. Although the predictive value of individual criteria for localizing suitable ablation targets remains modest, the use of complementary mapping strategies may be effectively used to guide ablation.

LATE POTENTIALS AND FRACTIONATED LOCAL ELECTROGRAMS DURING SINUS RHYTHM

As impulses propagate slowly through islands of viable myocardium within infarcted regions during sinus rhythm, low amplitude, fractionated endocardial potentials result, consistent with the underlying asynchronous activation of myofiber bundles.[32] This slow, fractionated local depolarization may even persist following the QRS complex to produce the late potentials that are sometimes recorded endocardially.[33] Stevenson and associates demonstrated that pacing in regions with fractionated sinus rhythm electrograms commonly results in stimulus to QRS (S–QRS) intervals of greater than 40 ms, evidence that these electrograms are associated with slow conduction, which may permit reentrant ventricular tachycardia.[32] In their study of 24 patients with reentrant ventricular tachycardia, Harada and coworkers recorded late potentials in 71% of sites classified by entrainment-mapping as appropriate targets for ablation. The incidence of late potentials at bystander sites approached 33%, limiting their usefulness in predicting successful sites for ablation.[33]

ISOLATED-DIASTOLIC POTENTIALS

Slow conduction through an isthmus bounded by lines of functional or anatomic block during reentrant ventricular tachycardia may produce low amplitude, isolated-diastolic endocardial potentials (IDPs) preceding the QRS onset by as much as hundreds of milliseconds (Fig. 45-3).[34-36] Although "bystander" regions may generate IDPs, potentials that cannot be dissociated from ventricular tachycardia despite repeated induction of arrhythmia may signify essential components of the reentrant circuit and important ablation targets. Fitzgerald and coworkers isolated such potentials in seven of 14 post-infarct ventricular tachycardias. In each case, ablation at the site of these IDPs was successful.[38] Subsequently, Bogun and associates observed uniform concordance between sites of IDPs that could not be dissociated from post-infarct ventricular tachycardia and sites of concealed entrainment, providing additional evidence that isolated diastolic potentials originate in critical regions of slow conduction.[35] Interestingly, discrete diastolic potentials recorded prior to Purkinje potentials during idiopathic left ventricular tachycardia appear to also represent important markers for successful radiofrequency ablation of this micro-reentrant arrhythmia.[39,40]

ENTRAINMENT WITH CONCEALED FUSION

Entrainment-mapping depends on the presence of an "excitable gap" in reentrant circuits.[41] During tachycardia, a time interval exists during which the myocardium within a discrete portion of the reentrant circuit has recovered from the preceding activation wavefront, but has not yet been depolarized by the next orthodromic wavefront. When a train of pacing stimuli is applied to a reentrant circuit during tachycardia at a rate exceeding the tachycardia cycle length, such that each stimulus falls within this "excitable gap," the orthodromic wavefronts initiated by pacing propagate through the circuit, whereas the antidromic wavefronts collide with returning orthodromic impulses. The resulting acceleration of QRS complexes to the pacing rate during tachycardia with paced QRS complex morphology identical to ventricular tachycardia morphology constitutes entrainment with concealed fusion.

To verify concealed fusion, it is essential to pace at several cycle lengths during entrainment-mapping.[3] Pacing at slower rates outside the reentrant circuit may produce minimal QRS fusion, and masquerade as concealed fusion, as wavefronts propagating away from the pacing site and wavefronts emerging from the tachycardia circuit collide close to the pacing site. Pacing at faster rates in these regions, however, should produce progressive fusion as the point of wavefront collision moves away from the pacing site.

Entrainment with concealed fusion alone does not specifically localize critical regions of slow conduction within reentrant circuits. Theoretically, pacing adjacent bystander sites may produce entrainment with concealed fusion (Fig. 45-4). Supporting this concept, Bogun and colleagues observed that in 14 patients with

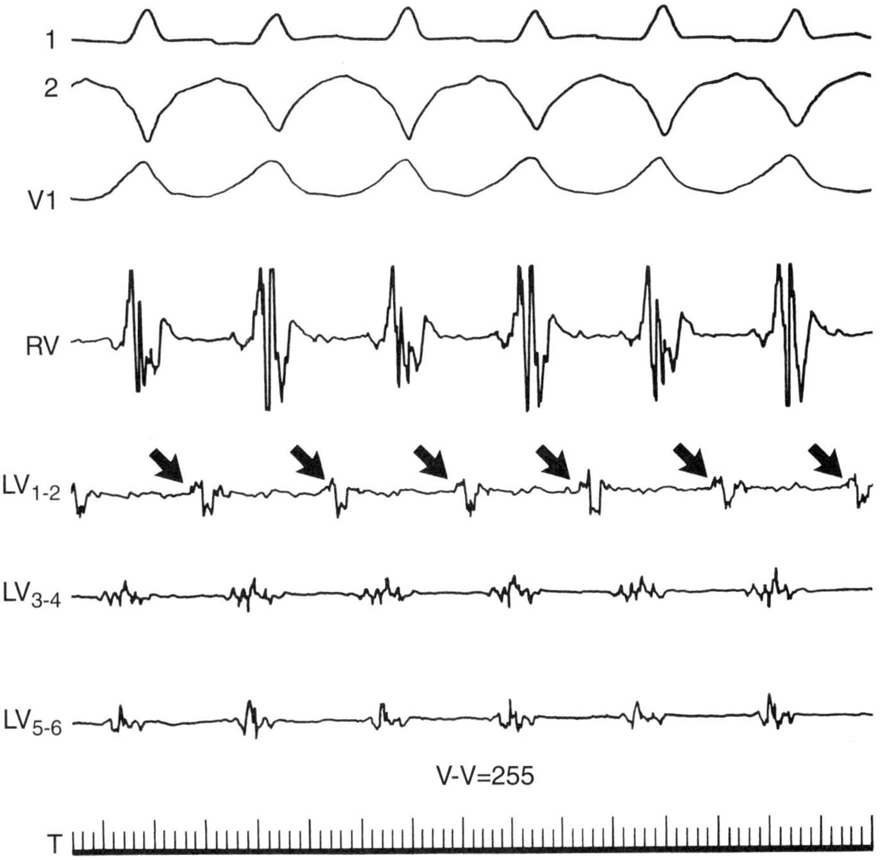

FIGURE 45-3 Endocardial catheter mapping of ventricular tachycardia. Displayed from *top to bottom* are ECG leads 1, 2, and V1, intracardiac recordings from the right ventricle (RV), left ventricular ablating catheter (LV 1-2, LV 3-4, LV 5-6), and time lines (T). Isolated diastolic potentials (*arrows*) are recorded in the ablating catheter (LV 1-2). Delivery of radiofrequency current at this site terminated the ventricular tachycardia. (From Blanck Z, Akhtar M: Therapy of ventricular tachycardia in patients with nonischemic cardiomyopathies. J Cardiovasc Electrophysiol 1996;7:671-683.)

post-infarct monomorphic ventricular tachycardia, the positive predictive value of concealed entrainment for localizing successful radiofrequency ablation targets was only 54%.[42] Therefore, additional criteria have been developed to enhance specificity.

POST–PACING INTERVAL

After demonstration of concealed entrainment, the further demonstration of a post–pacing interval equal to the tachycardia cycle length suggests that a pacing site represents an essential portion of the reentrant circuit.[34] After pacing within a reentrant circuit, the stimulated orthodromic wavefront makes one revolution through the circuit before returning to the pacing site where a local electrogram is recorded. The time from the last captured pacing stimulus in a pacing train, to the next local electrogram recorded at the pacing site, the post–pacing interval, should equal the tachycardia cycle length. When a site outside the reentrant circuit is paced, the paced wavefront must propagate into the circuit, around the circuit, and back to the pacing site. This would result in a post–pacing interval exceeding the tachycardia cycle length (see Fig. 45-4).

The approximation of tachycardia cycle length by the post–pacing interval depends on the maintenance of a consistent reentrant pathway and stable conduction velocities during tachycardia and pacing. Slowing of conduction or the development of functional block will prolong the post–pacing interval, irrespective of the pacing site. In canine models of ventricular tachycardia, these confounding factors have been demonstrated to occur during rapid pacing.[43,44] Therefore, it is critical to use the slowest stimulus trains, usually 20 to 40 ms shorter than the ventricular tachycardia cycle length for analysis of the post–pacing interval.[3]

Errors in the measurement of the post–pacing interval may also arise from stimulation and recording techniques.[45] Unless unipolar stimulation during entrainment is used, the point from which excitation spreads remains uncertain. Although the same distal electrode used for pacing should ideally be used to record the unipolar electrogram for measurement of the post–pacing interval, the unipolar stimulation artifact often obscures this signal, necessitating substitution of unipolar or bipolar signals recorded from more proximal poles. When closely spaced electrodes are used for stimulation and recording, minimal error is introduced to the measurement.[46] If the recording and stimulation electrodes are further apart, particularly in the presence of depressed myocardial conduction velocity, substantial error may result. Finally, the predictive accuracy of the post–pacing interval depends heavily on the ability to differentiate near- from far-field recordings.

Stevenson and coworkers reported that a post–pacing interval within 30 ms of the ventricular tachycardia cycle length was strongly associated with successful

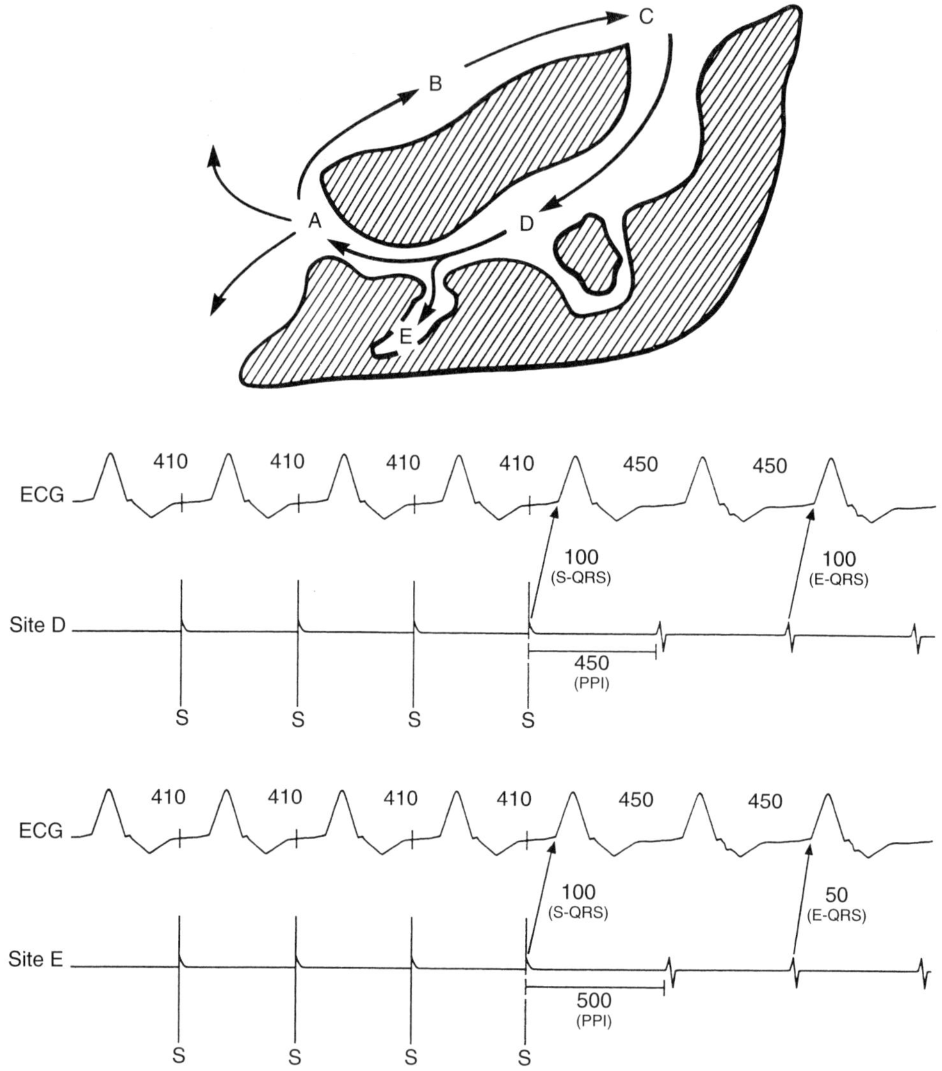

FIGURE 45-4 Schematic contrasting entrainment with concealed fusion at a site within a theoretical reentrant circuit (site D), and at a bystander site (site E). The circuit contains a protected region bounded by inexcitable tissue from site C to sites D and A. Tachycardia wavefronts exit the protected region at site A. A stimulus at site D or E produces an orthodromic wavefront that propagates toward site A, where it exits the protected isthmus to activate the remainder of the myocardium and propagate through the circuit from site A to sites B and C, resetting the tachycardia. Whether the stimulated wavefront originates at site D or site E, the impulse exits the circuit at site A, the tachycardia exit site. Therefore, the QRS–intervals during pacing and tachycardia are identical, as demonstrated in the tracings below the schematic. The post–pacing interval and S–QRS intervals may be used to differentiate sites within the tachycardia circuit and bystander regions. Following entrainment at site D, the stimulated wavefront propagates from the pacing site to sites A, B, and C, returning to site D resulting in a post–pacing interval (PPI) equal to the tachycardia cycle length. After stimulating site E, in contrast, the orthodromic wavefront must travel into the tachycardia circuit, traverse the circuit, then return to site E before a local electrogram is recorded at the tip of the pacing catheter. The PPI would, therefore, exceed the tachycardia cycle length. The interval between the stimulus and the onset of the QRS (S–QRS) represents the conduction time from the pacing site to the exit site (site A). When stimulating within the circuit at site D, the time from the local electrogram recorded during tachycardia at the pacing site to the QRS onset (E–QRS) approximates the S–QRS interval. Pacing bystander site E produces an S–QRS interval that exceeds the E–QRS interval.

radiofrequency ablation-mediated termination of tachycardia.[34] This finding was not confirmed by Bogun and associates who found that the prevalence of post–pacing interval within 30 ms of the ventricular tachycardia cycle length did not differ substantially between effective and ineffective ablation target sites.[47] Although these discrepant findings may be due to important methodologic differences, the value of the post–pacing interval remains controversial.

S–QRS INTERVAL

During entrainment with concealed fusion, the interval between the pacing stimulus and the QRS onset (S–QRS interval) represents the conduction time from pacing site to exit site for a reentrant circuit. When pacing within a circuit, the time from the local electrogram recorded at the pacing site to the QRS onset during tachycardia is ordinarily within 20 ms of the

S–QRS interval. In contrast, pacing a bystander site does not produce an S–QRS interval that approximates the local electrogram to QRS (E–QRS) interval during tachycardia (see Fig. 45-4).[3,34]

Recently, El-Shalakany and colleagues evaluated the accuracy of entrainment-mapping criteria for predicting termination of ventricular tachycardia by a single radiofrequency lesion. In 15 consecutive patients with coronary artery disease, they attempted radiofrequency ablation of 20 monomorphic ventricular tachycardias. They looked for the following at each potential ablation site: (1) exact QRS match during entrainment, (2) post-pacing interval approximating ventricular tachycardia cycle length, and (3) E–QRS interval equal to S–QRS interval but less than 70% of ventricular tachycardia cycle length. At 19 of 19 sites meeting all three criteria, tachycardia terminated with a single radiofrequency application. Ablation failed to terminate ventricular tachycardia, in contrast, at 24 of 25 sites not meeting all three criteria.[48] Bogun and coworkers prospectively evaluated predictors of effective ablation in the presence of concealed entrainment. Studying 14 patients with coronary artery disease and hemodynamically stable monomorphic ventricular tachycardia, they found that concealed entrainment alone carried a positive predictive value of 54% for successful ablation. This increased to 72% when the S–QRS/ventricular tachycardia cycle length ratio was less than 70% to 82% when the S–QRS interval matched the E–QRS interval, and to 89% when mid-diastolic potentials were associated with the ventricular tachycardia.[42]

MAPPING HEMODYNAMICALLY UNSTABLE VENTRICULAR TACHYCARDIA

Conventional mapping techniques described up to this point are suited primarily for mapping hemodynamically stable arrhythmias, although they represent fewer than 10% of clinically relevant ventricular tachycardias.[14] As such, hemodynamic instability associated with tachycardia has traditionally been considered a relative contraindication to radiofrequency ablation. One recent small series that applied conventional fluoroscopic-guided mapping techniques to poorly tolerated ventricular tachycardias reported suboptimal results. The authors used fractionated sinus rhythm electrograms and pace-mapping to identify ablation targets, but were ultimately successful in abolishing tachycardia recurrence in only three of five patients.[49]

With these limitations in mind, the CARTO nonfluoroscopic cardiac mapping system has recently been used to delineate margins of myocardial scarring, and to precisely tag areas of interest, facilitating the localization of ablation target sites in patients with hemodynamically unstable ventricular tachycardia. The mapping catheter is introduced into the left ventricle using either a transseptal or retrograde transaortic approach. During sinus rhythm, the catheter tip is dragged over the endocardium, acquiring both tip locations as well as local electrograms. In this manner, a color-coded voltage map is created, and areas of scar tissue are defined. Infarcted myocardium may be distinguished from healthy myocardium by a reduction in electrical voltage (bipolar voltage ≤1.0 mV).[50] Pace-mapping around the perimeter of the scar tissue may then be performed, tagging the anatomic sites with pace maps most closely approximating the target ventricular tachycardia morphology. Next, a series of radiofrequency lesions is delivered at these sites, obliquely transecting the edges of the scar. (Fig. 45-5) Creating linear lesions in this manner may interrupt a critical portion of the reentrant circuit. Each radiofrequency lesion is precisely tagged on the electroanatomic model, facilitating return to this location should the catheter move, and promoting the creation of continuous lesions. Marchlinski and associates recently applied this approach to 16 patients with drug-refractory, monomorphic ventricular tachycardia not amenable to conventional mapping based on associated hemodynamic instability or inability to sustain tachycardia during mapping. Between eight and 87 radiofrequency lesions were applied per patient, with the creation of between one and nine linear lesions. During a median follow-up of 8 months, 12 patients (75%) have remained free of ventricular tachycardia.[51]

A novel noncontact computerized mapping system has also been successfully applied to the mapping and ablation of hemodynamically unstable ventricular tachycardia. The mapping system (Endocardial Solutions, Inc., St. Paul, MN) uses a computerized electrophysiology recording system along with a 64-electrode, noncontact balloon catheter, to create a three-dimensional model of the left ventricular endocardium, identify the spatial coordinates of a conventional intracardiac ablation catheter tip within the model, and compute 3360 virtual endocardial electrograms.[52,53] A color-coded dynamic isopotential map derived from the virtual electrograms, representing the electrical potential of each virtual electrogram throughout the cardiac cycle, is displayed. The leading edge of myocardial depolarization is identified as a negative potential. After defining the spatial geometry of the left ventricle during sinus rhythm, the system requires only a single beat of ventricular tachycardia to create an isopotential map, facilitating the mapping of hemodynamically unstable arrhythmias. Using the color-coded displays and virtual electrograms, which may be selected for viewing from any portion of the three-dimensional endocardial reconstruction, it is possible to identify isolated diastolic potentials, sites of presystolic endocardial activation during ventricular tachycardia, and reentrant circuit exit sites (Fig. 45-6).[54,55]

EPICARDIAL-MAPPING AND ABLATION

Conventional endocardial mapping strategies are poorly suited to localize and ablate epicardial ventricular tachycardia. Clearly, critical epicardial reentrant circuits may be present in monomorphic ventricular tachycardia associated with Chagas' disease, dilated cardiomyopathy, and inferior wall infarcts.[56,57] Select idiopathic left ventricular tachycardias also appear to have an epicardial origin.[58] Techniques permitting epicardial mapping and ablation are, therefore, receiving interest. Stellbrink and coworkers successfully ablated

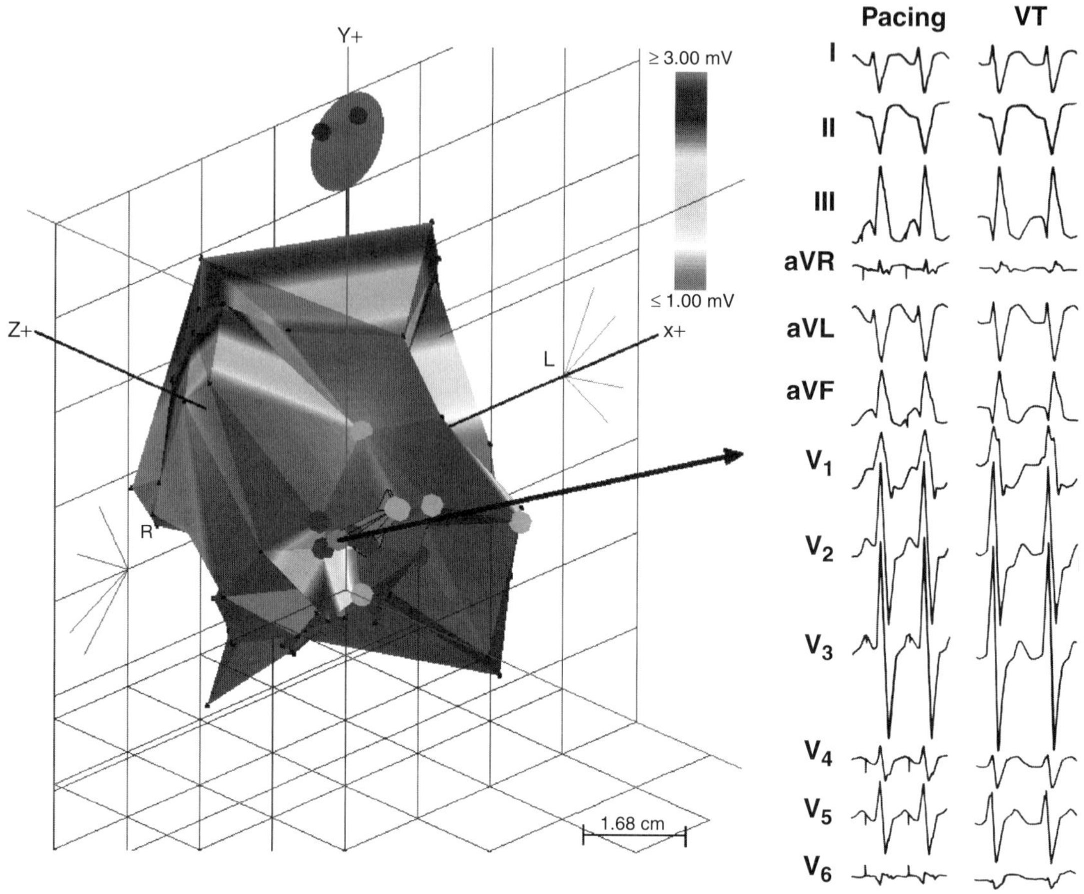

FIGURE 45-5 (See also Color Plate 45-5.) Voltage map of the LV from patient with VT and coronary disease. Purple represents normal endocardium (amplitude ≥3.0 mV); red, dense scar (amplitude ≤1.0 mV); and ranges between purple and red, border zone. Electrograms of pace map and VT are shown to the right of the voltage map, identifying appropriate site for linear ablation at the edge of the dense scar. Linear lesions are identified by adjacent red circles.

an epicardial adenosine-sensitive ventricular tachycardia arising from the lateral wall of the left ventricle using a transcoronary venous approach.[59] The coronary sinus, however, provides access to only a limited epicardial region. The transpericardial approach described by Sosa and associates, in contrast, allows direct access to the entire left ventricular epicardium. After introducing a catheter into the pericardial space, activation-mapping, concealed entrainment, and mid-diastolic potentials may be used to define epicardial circuits.[56,57,60,61] Alternately, the CARTO system may be suitable for electroanatomic mapping of the epicardium.[62] Using the transpericardial approach, Sosa and colleagues successfully performed mapping and ablation of epicardial reentrant circuits in six patients with ventricular tachycardia associated with Chagas' disease, and seven patients with ventricular tachycardia related to an earlier inferior wall infarct.[56,57] Among the 53 patients in whom these investigators performed epicardial mapping between 1995 and 1999, four sustained accidental right ventricular perforation.[57] Although epicardial ablation carries the potential for damage to the coronary arteries, coronary complications have not been reported.

Applications and Outcomes of VT Ablation

Monomorphic ventricular tachycardia may occur in the setting of a spectrum of cardiac diseases, including coronary artery disease, cardiomyopathy, valvular disease, arrhythmogenic right ventricular dysplasia, congenital heart disease, or even in structurally normal hearts. Clinical presentations range from slow, minimally symptomatic ventricular tachycardia to sudden cardiac death. The role of ablation and therapeutic outcomes in this heterogeneous group vary according to the underlying cardiac disorder of the patient.

IDIOPATHIC VENTRICULAR TACHYCARDIA

By definition, idiopathic ventricular tachycardia occurs in individuals lacking structural heart disease. Most commonly, these tachycardias originate from the right ventricular outflow tract or the posteroseptal aspect of the left ventricle. The experience with radiofrequency ablation of these ventricular tachycardias suggests that they may be successfully ablated with

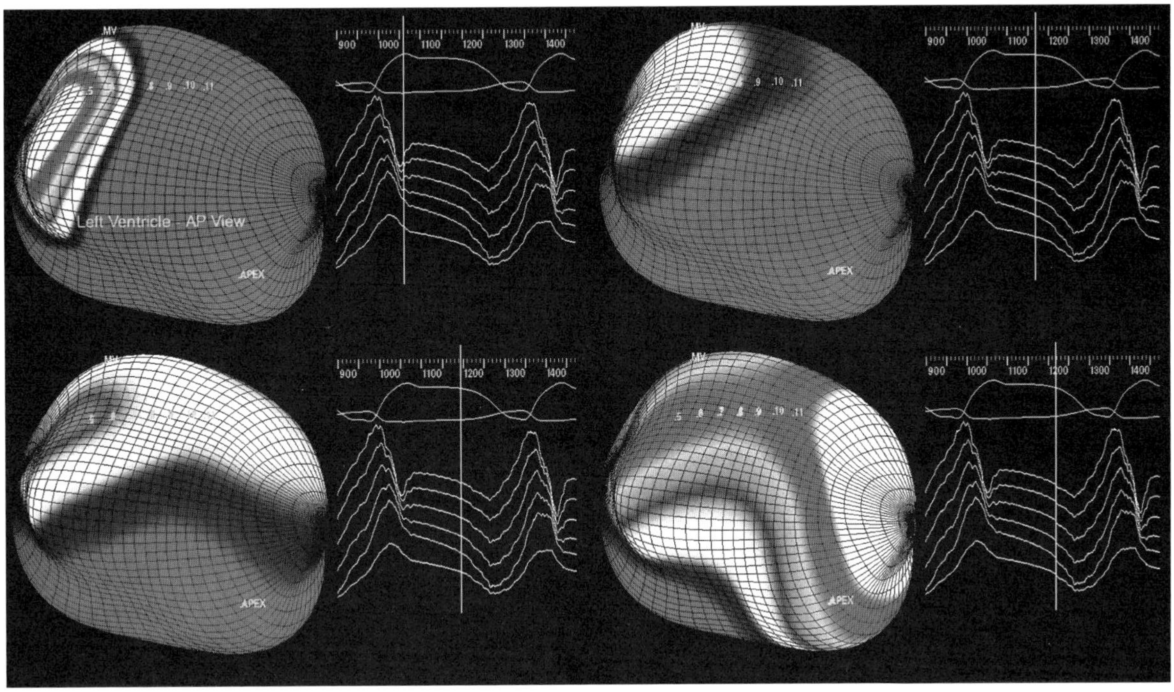

FIGURE 45-6 (See also Color Plate 45-6.) Surface electrograms, virtual electrograms, and instantaneous isopotential maps at four time points, depicted by vertical lines through the electrograms, trace the path of a VT reentrant circuit during diastole and at the onset of the QRS complex. The VT shown in this example had a cycle length of 370 ms and was associated with marked hypotension. The **top panel on the left** shows endocardial activation 190 ms preceding the onset of the QRS complex during VT. The **top panel on the right** and the **bottom left panel** depict the slow spread of presystolic endocardial activation. The **bottom right panel** shows the VT exit site, corresponding to the onset of the QRS interval on the surface ECG. Linear radiofrequency lesions were delivered to transect the region of slow diastolic activation, as guided by the locator signal. This VT was no longer inducible and did not recur spontaneously following ablation.

a high degree of success, and limited procedural morbidity (Table 45-2).

Idiopathic right ventricular tachycardia, originating from the right ventricular outflow tract (RVOT) free wall or septum, carries a relatively benign prognosis. Associated symptoms range from palpitations to recurrent syncope, but sudden death is uncommon. Prognostically, it is essential to distinguish this tachycardia from arrhythmogenic right ventricular dysplasia (ARVD), characterized by fibro-lipomatous myocardial replacement and progressive right ventricular failure.[66] RVOT tachycardia is typically sensitive to calcium channel blockers and may be treated pharmacologically. When disabling symptoms persist, despite optimal medical therapy, or the initial presentation includes syncope, curative catheter ablation may be indicated. In reported series,

TABLE 45-2 Ablation of Idiopathic Ventricular Tachycardia

Reference	Method	No. of Patients	VT Origin	Acute Success (% of Patients)	Risk for VT Recurrence (%)	Duration of Follow-Up (Months)	Complication
63	DC/RF	8	LV	100	13	17	None
16	RF	8	LV	100	25	10.1	1 AI,2 RBBB
—	—	20	RV	85	6	9.9	1 perforation/death
64	DC/RF	7	RV	100	0	16	None
17	RF	31	LV	85	9	28	None
—	—	30	RV	84	19	28	None
18	RF	35	RV	83	14	30	None
—	—	13	LV	92	0	36	None
30	RF	8	LV	100	13	10.5	1 MR
19	RF	—	LV	91	13	41	None
—	—	—	RV	91	10	41	None
39	RF	5	LV	80	0	10	None
65	RF	20	RV	100	5	7	None

AI, aortic insufficiency; DC, direct current; LV, left ventricle; MR, mitral regurgitation; RBBB, right bundle branch block; RF, radiofrequency; RV, right ventricle; VT, ventricular tachycardia.

acute success of ablation ranges from 83% to 100%, with recurrence rates of 0% to 19%. Reasons for failure include inaccurate diagnosis (actual left ventricular septal origin), insufficient localization of endocardial focus, and intramural or epicardial focus. Although infrequent, wall perforation and tamponade may complicate ablation of the RVOT—particularly the free wall.

Idiopathic left ventricular tachycardia is thought to originate from the Purkinje network of the left posterior or anterior fascicle. Accordingly, the earliest endocardial activation during ventricular tachycardia commonly localizes to the posterior or anterior left ventricular septum.[39,63,67] Symptoms at presentation again include palpitations, presyncope, and syncope. These tachycardias are sensitive to verapamil, but like RVOT tachycardia, may also be treated with ablation. Given the favorable prognosis associated with idiopathic left ventricular tachycardia, ablation is generally reserved for patients with severe symptoms, persistent symptoms despite pharmacologic therapy, or drug intolerance. Acute success rates ranging from 80% to 100% may be expected, with recurrence rates of 0% to 25%. Significant complications are uncommon but include thromboembolic events, as well as damage to the valvular apparatus.

POSTMYOCARDIAL INFARCTION VENTRICULAR TACHYCARDIA

For patients with recurrent infarct-related ventricular tachycardia refractory to medical therapy, radiofrequency ablation represents an important therapeutic alternative. Traditionally, hemodynamically stable and reproducibly inducible monomorphic ventricular tachycardia was required for mapping. This necessarily limited the number of patients who were candidates for this modality. More recently, novel mapping techniques and substrate-based approaches have expanded the application of radiofrequency ablation to patients with hemodynamically unstable post-infarct ventricular tachycardia.[51,55]

Patients with coronary disease may have several inducible ventricular tachycardia morphologies at the time of electrophysiological study. Occasionally, ablation in one region abolishes multiple ventricular tachycardia morphologies.[14,68] These tachycardias are thought to derive from either wandering exit sites within a single reentrant circuit or multiple circuits sharing a common critical isthmus. More often, distinct reentrant loops produce each ventricular tachycardia morphology and must be individually targeted for ablation.

One approach to ablation targets only clinically documented ventricular tachycardias. There are practical difficulties associated with this strategy, including the often-limited documentation of the clinical arrhythmia morphology. Also, *nonclinical* ventricular tachycardias induced at electrophysiological study have been subsequently documented in the nonlaboratory setting. Therefore, several groups advocate ablation of all mappable, induced ventricular tachycardias.[68,69] Using this approach, Stevenson and coworkers reported acute success in 71% of their patients, with 31% of patients requiring multiple ablation sessions.[68] Despite this approach, their arrhythmia recurrence rate remained 33% over three years.

Clinical investigations examining ablation efficacy in highly selected postinfarct patients with predominantly stable, monomorphic ventricular tachycardias have reported initial success rates ranging from 71% to 93% (Table 45-3). Notably, the number of tachycardias targeted per patient, and thus the definition of primary success, varies significantly between studies. Despite these encouraging numbers, 18% to 45% of patients subsequently experienced ventricular tachycardia or sudden cardiac death. In some cases, these events were due to ventricular tachycardia morphologies not

TABLE 45-3 Ablation of Postmyocardial Infarction Ventricular Tachycardia

Reference	No. of Patients	Survival (% of Patients)	Primary Success (% of Patients)	Risk for VT Recurrence (%)	EF (%)	Complications	Duration of Follow-Up (Months)
14	15	100	73	18	27	None	9.1
68	52	80	71	33	33	1 Death, 1 TIA, 2 Respiratory insufficiencies 1 Pseudoaneurysm	18
70	21	90	76	31	22	1 Heart block	11.8
69	35	82	86	33	24	3 Hematomas 4 Heart blocks 1 Mesenteric ischemia	14.2
71	136	—	72	16	36	2 Deaths 1 CVA 7 Pericardial effusions 5 Transient AVBs 1 Femoral artery closure	24
72	21	95	81	45	31	1 Death	13.2
73	15	80	93	26	26	None	15
74	66	—	71	—	28	2 Tamponades 1 Respiratory insufficiency 2 CVAs/TIAs	—

AVB, atrial ventricular block; CVA, cerebral vascular accident; TIA, transient ischemic attack.

previously targeted by ablation. Although persistent inducibility of *nonclinical* tachycardia following successful ablation predicts future arrhythmias,[69,71] Stevenson and associates have reported recurrences in patients rendered noninducible following ablation, including one sudden cardiac death.[68]

Thus, radiofrequency ablation remains a palliative, rather than curative, modality in the treatment of postinfarct ventricular tachycardia. As such, decreasing the frequency of ICD discharges represents a legitimate and cost-effective goal of radiofrequency ablation.[75] In 21 patients with ICDs undergoing ventricular tachycardia ablation, Strickberger and associates reported that the frequency of ICD therapies decreased from a mean of 60 times per month down to 0.1 times per month. Not surprisingly, this was accompanied by marked improvements in quality-of-life scores.[70]

Theoretically, the application of multiple radiofrequency lesions during ventricular tachycardia ablation in patients with severely depressed left ventricular function may produce significant additional myocardial damage. In 52 patients with a mean left ventricular ejection fraction of 33% undergoing ventricular tachycardia ablation, Stevenson and colleagues reported a 10% incidence of death from congestive heart failure during follow-up.[68] Although this level of mortality due to congestive heart failure is not excessive for the population studied, the impact of ablation on ventricular function remains unresolved. Following radiofrequency ablation of postinfarct ventricular tachycardia, Morady and coworkers and Kim and colleagues found only minimal to modest increases in serum CK-MB levels.[14,72] Minimizing the number of radiofrequency energy applications and confining ablation to regions of myocardium with abnormal electrograms may limit the volume of iatrogenic myocardial injury. Other potential complications of ventricular tachycardia ablation include cardioembolic events, tamponade, heart block, and death (see Table 45-3).

Clearly, the efficacy of ablation is lower for scar-related ventricular tachycardia than idiopathic ventricular tachycardia. Mapping large reentrant circuits is inherently more difficult than localizing arrhythmias originating from relatively small regions, accounting for some of the disparity. Other barriers to postinfarct ventricular tachycardia ablation include difficulty penetrating scar tissue to create effective lesions, the presence of wide zones of critical conducting tissue, intramural or epicardial reentrant circuits, and insufficient localization of the target site. Compared with radiofrequency ablation, patients with prior infarcts surviving surgical subendocardial resection and cryoablation for ventricular tachycardia have a low recurrence risk, 8% to 11% over one to four years.[41] The greater efficacy of surgery for arrhythmia control points to the small volume of tissue affected by endocardial ablation as a major reason for treatment failures.[76,77]

The optimal management of patients following successful radiofrequency ablation of clinically manifest postinfarct ventricular tachycardia remains poorly defined. At the time of ablation, patients are frequently receiving antiarrhythmic drugs that are failing to suppress the *clinical* tachycardia. Withdrawal of these agents following ablation may be problematic because the drugs may be suppressing other ventricular tachycardia morphologies. Additionally, even in patients rendered completely noninducible at electrophysiological study following successful ablation, some centers recommend ICD implantation because arrhythmia recurrences may be life threatening.

BUNDLE BRANCH REENTRY VENTRICULAR TACHYCARDIA

Bundle branch reentry tachycardia is a form of ventricular tachycardia characterized by a well-defined reentrant circuit encompassing the right and left bundle branches, the interventricular septum, and the His bundle.[78] Patients commonly present with syncope, sustained palpitations, or sudden cardiac death. Bundle branch reentry has been reported as the mechanism of sustained monomorphic ventricular tachycardia in approximately 6% of patients, predominantly affecting those with idiopathic dilated cardiomyopathy, coronary disease, and valvular disease.[79,80] Although bundle branch reentry tachycardia periodically occurs in patients without myocardial or valvular dysfunction, conduction abnormalities in the His-Purkinje system, producing His–Ventricular (H–V) interval prolongation, are universally observed.[80-82]

As for other reentrant circuits, induction of bundle branch reentry tachycardia requires antegrade block in one limb or bundle, with critical delay in the other. The QRS configuration during tachycardia depends on the bundle branch that is used for antegrade propagation. In most cases, impulse propagation proceeds antegradely down the right bundle branch, with subsequent transseptal activation of the left ventricle, and retrograde impulse conduction up the left bundle branch to activate the distal His bundle and reinitiate the reentrant circuit. The surface ECG morphology corresponding to this activation sequence would be a left bundle branch ventricular tachycardia. In a smaller number of cases, the reverse order of activation would lead to a right bundle branch morphology.[80] A related macroreentrant ventricular tachycardia, interfascicular tachycardia, uses the left bundle branch fascicles as the components of the reentrant circuit, activating the right bundle branch as a bystander.[83,84] Again, conduction delays in the His-Purkinje system are prerequisite to the development of sustained tachycardia.

Bundle branch reentry ventricular tachycardia, with its anatomically well-defined circuit, is easily eliminated using radiofrequency ablation. In most patients, catheter ablation of the right bundle eliminates ventricular tachycardia, and is the preferred mode of therapy.[80,82] To perform right bundle branch ablation, the catheter is positioned at the anterior interventricular septum to record a right bundle potential distally. An atrial potential should not be present in the distal bipole recording, and the His-right bundle potential interval should be at least 20 ms. Radiofrequency ablation in this location should produce right bundle branch block, and terminate bundle branch reentry ventricular tachycardia (Fig. 45-7).

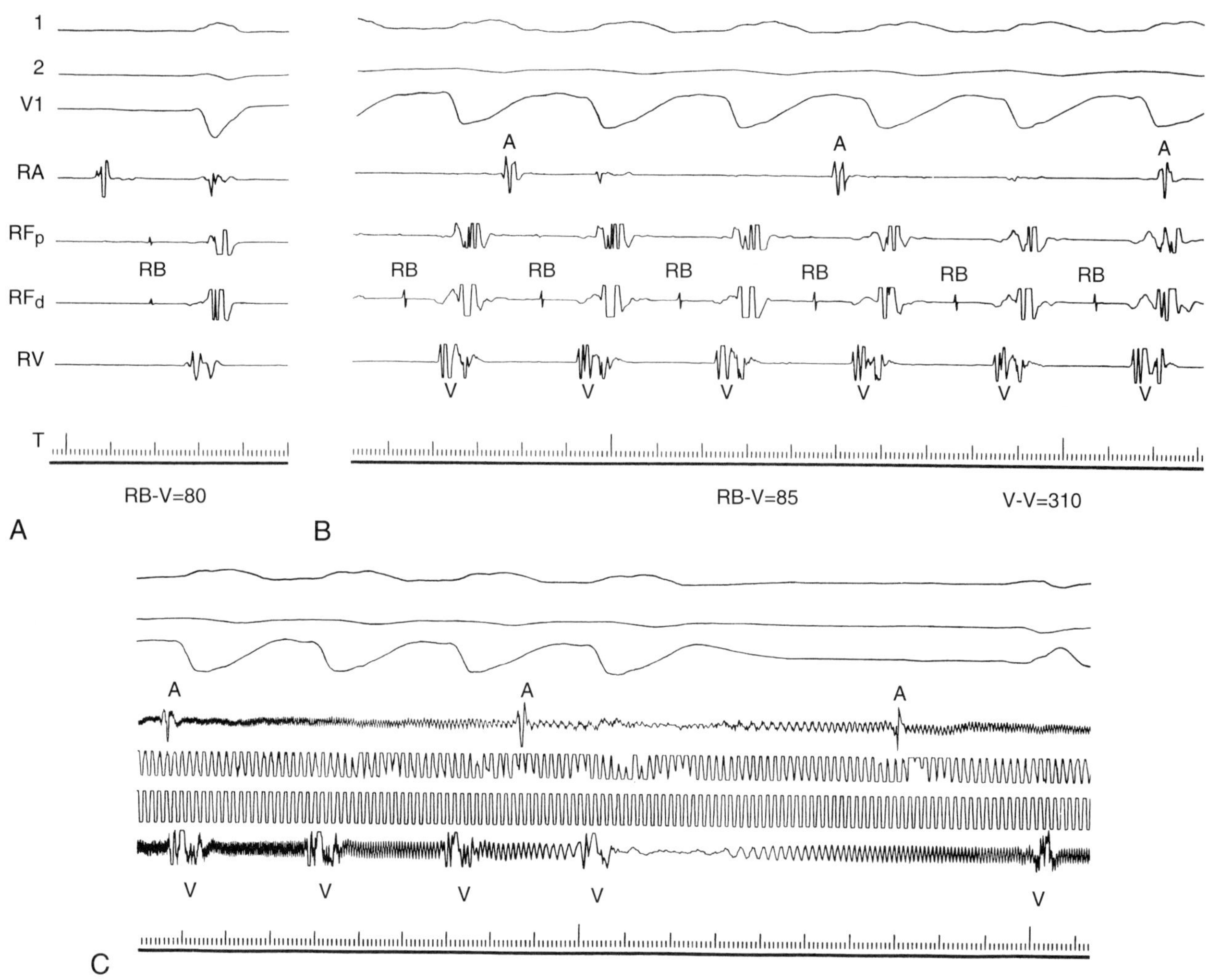

FIGURE 45-7 Termination of bundle branch reentrant ventricular tachycardia during radiofrequency catheter ablation of the right bundle. Shown (*top to bottom*) are surface ECG leads 1, 2, and V1, right atrium (RA), proximal and distal ablating catheters (RF_p, RF_d), right ventricular electrogram (RV), and time lines (T). All labeled intervals are in milliseconds. **A,** Activation of the right bundle (RB) is recorded in the proximal and distal bipole of the ablating catheter during sinus rhythm. **B,** Depicts bundle branch reentrant VT with a LBBB QRS morphology. Activation of the RB is recorded from the distal pole of the ablation catheter. **C,** Shows termination of tachycardia within six seconds of energy application. Note the expected right bundle branch block QRS morphology following termination of the tachycardia. (From Blanck Z, Dhala A, Deshpande S, et al: Bundle branch reentrant ventricular tachycardia: Cumulative experience in 48 patients. J Cardiovasc Electrophysiol 1993;4:253-262.)

Interfascicular ventricular tachycardia may be present along with bundle branch reentry, requiring, instead, ablation of the left bundle branch fascicles for treatment of both conditions.[83-85] Although ablation of the right bundle branch is usually well tolerated, in patients with abnormal left bundle branch conduction, excessive H–V interval prolongation may result from ablation, necessitating permanent pacemaker implantation.[82]

Notably, myocardial ventricular tachycardia may also be inducible in up to 43% of bundle branch reentry patients, potentially requiring adjunctive ICD or antiarrhythmic drug therapy.[82] Also, the mortality related to progressive congestive heart failure in this group of patients is considerable. Despite these limitations, catheter ablation of the right or left bundle remains an effective treatment for this highly symptomatic ventricular arrhythmia.

VENTRICULAR TACHYCARDIA ASSOCIATED WITH RIGHT VENTRICULAR DYSPLASIA

Arrhythmogenic right ventricular dysplasia (ARVD) is an idiopathic disorder characterized by fibrofatty atrophy of the right ventricle and ventricular arrhythmias.[86,87] In a small series of five patients with recurrent ventricular tachycardia, Ellison and colleagues recently reported on the expanded role of ablation in ARVD. Pacing entrainment was successfully used to map and ablate eight of 19 tachycardias in this report. Although four of five patients remained inducible following ablation, none had spontaneous recurrence in follow-up (mean 17 months).[88] Although no procedural complications were reported in this small series, the right ventricular thinning characteristic of ARVD may increase the risk of perforation. Additionally, although mapping and ablation appear feasible, the progressive nature of ARVD, with the potential for developing new arrhythmias, suggests that the role of ablation will remain palliative.[89]

VENTRICULAR TACHYCARDIA FOLLOWING SURGICAL CORRECTION OF CONGENITAL HEART DEFECTS

Ventricular arrhythmias occur commonly late after repair of congenital heart disease, particularly in patients with tetralogy of Fallot.[90,91] The frequent co-localization of the tachycardia origin and the site of surgical correction implicate the surgical procedure itself in the genesis of these ventricular arrhythmias.[92-94] In patients with unsuccessful medical management of symptomatic ventricular tachycardia, radiofrequency ablation represents a viable therapeutic alternative. In a series of 16 patients with refractory ventricular tachycardia late after surgical correction of tetralogy, ventricular septal defect (VSD), severe congenital pulmonary stenosis, or transposition with VSD, the long-term efficacy of radiofrequency ablation guided by conventional mapping techniques was 88%.[94] These results are comparable with those reported for patients with idiopathic ventricular tachycardia.

VENTRICULAR TACHYCARDIA ASSOCIATED WITH DILATED CARDIOMYOPATHIES

The dilated nonischemic cardiomyopathies represent a heterogeneous group of disorders, commonly associated with nonsustained ventricular arrhythmias. Induction of sustained monomorphic ventricular tachycardia at electrophysiological study in these individuals, however, is relatively uncommon. The mechanisms of ventricular tachycardia in these patients also appear diverse, including reentry within the His-Purkinje system, intramyocardial reentry, triggered activity, and enhanced automaticity.[95] Although the optimal approach to mapping and ablation of ventricular tachycardia associated with nonischemic cardiomyopathy remains ill-defined, recent reports indicate that ablation may play an important adjunctive role in these patients. Kottkamp and associates used conventional mapping strategies to guide radiofrequency ablation in eight patients with idiopathic dilated cardiomyopathy, rendering 67% of targeted ventricular tachycardias noninducible.[96]

More recently, Delacretaz and coworkers published their experience with radiofrequency ablation of recurrent sustained monomorphic ventricular tachycardia in 26 patients with nonischemic heart disease. They reported that 62% of the patients had scar-related reentry, 19% had bundle branch or interfascicular reentry, and 27% had focal automaticity as the underlying mechanism of ventricular tachycardia. Overall, catheter ablation prevented clinical recurrences in 77% of patients during a mean follow-up period of 15 ± 12 months.[95] These data indicate that it is essential to exclude His-Purkinje system reentry in the subset of patients with dilated cardiomyopathy and monomorphic ventricular tachycardia because this entity may account for a significant number of tachycardias, and is readily amenable to catheter ablation. Additionally, selected dilated cardiomyopathy patients with refractory or incessant myocardial ventricular tachycardia may benefit from catheter ablation as adjunctive therapy.

In summary, ventricular tachycardia ablations are often lengthy, technically difficult procedures. Success rates vary significantly according to the underlying cardiac disorder, the mechanism of tachycardia, and the operator experience. In the case of reentrant VT, extensive radiofrequency applications and repeat procedures may be necessary, raising concerns about potential complications, particularly in patients with depressed left ventricular function. In selected patients, however, ablation may improve quality of life and may be life saving.

REFERENCES

1. El-Sherif N, Scherlag BJ, Lazzara R, et al: Re-entrant ventricular arrhythmias in the late myocardial infarction period. Circ 1977; 55:686-702.
2. Downar E, Harris L, Mickleborough LL, et al: Endocardial mapping of ventricular tachycardia in the intact human ventricle: Evidence for reentrant mechanisms. J Am Coll Cardiol 1988;11: 783-791.
3. Stevenson WG, Friedman PL, Sager PT, et al: Exploring postinfarction reentrant ventricular tachycardia with entrainment mapping. J Am Coll Cardiol 1997;29:1180-1189.
4. El-Sherif N, Mehra R, Gough WB, et al: Reentrant ventricular arrhythmias in the late myocardial infarction period. Circ 1983; 68:644-656.
5. DeBakker JMT, Van Capelle FJL, Janse MJ, et al: Reentry as a cause of ventricular tachycardia in patients with chronic ischemic heart disease: Electrophysiologic and anatomic correlation. Circ 1988;77:589-606.
6. Wilber DJ, Kopp DE, Glascock DN, et al: Catheter ablation of the mitral isthmus for ventricular tachycardia associated with inferior infarction. Circulation 1995;92:3481-3489.
7. Downar E, Kimber S, Harris L, et al: Endocardial mapping of ventricular tachycardia in the intact human heart. Evidence for multiuse reentry in a functional sheet of surviving myocardium. J Am Coll Cardiol 1992;20:869-878.
8. Downar E, Saito J, Doig JC, et al: Endocardial mapping of ventricular tachycardia in the intact human ventricle. Evidence of multiuse reentry with spontaneous and induced block in portions of reentrant path complex. J Am Coll Cardiol 1995;25: 1591-1600.
9. Waldo A: The canine sterile pericarditis model of atrial flutter. In Waldo A, Touboul P (eds): Atrial Flutter: Advances in Mechanisms and Management, Armonk, NY, Futura, 1996, pp 173-192.
10. Spach MS, Dolber PC, Heidlage JF: Influence of the passive anisotropic properties on directional differences in propagation following modification of the sodium conductance in human atrial muscle: A model of reentry based on anisotropic discontinuous propagation. Circ Res 1988;62:811-832.
11. Fast VG, Darrow BJ, Saffitz JE, et al: Anisotropic activation spread in heart cell monolayers assessed by high-resolution optical mapping. Circ Res 1996;79:115-127.
12. Clerc L: Directional differences of impulse spread in trabecular muscle from mammalian heart. J Physiol (London) 1976;255: 335-346.
13. Spach MS, Dolber PC: Relating extracellular potentials and their derivatives to anisotropic propagation at a microscopic level in human cardiac muscle: Evidence for electrical uncoupling of side-to-side fiber connections with increasing age. Circ Res 1986; 58:356-371.
14. Morady F, Harvey M, Kalbfleisch SJ, et al: Radiofrequency catheter ablation of ventricular tachycardia in patients with coronary artery disease. Circulation 1993;87:363-372.
15. Stevenson WG, Sager PT, Natterson PD, et al: Relation of pace mapping QRS configuration and conduction delay to ventricular

tachycardia reentry circuits in human infarct scars. J Am Coll Cardiol 1995;26:481-488.
16. Coggins DL, Lee RJ, Sweeney J, et al: Radiofrequency catheter ablation as a cure for idiopathic tachycardia of both left and right ventricular origin. J Am Coll Cardiol 1994;23:1333-1341.
17. Tsai C-F, Chen S-A, Tsai C-T, et al: Idiopathic monomorphic ventricular tachycardia: Clinical outcome, electrophysiologic characteristics and long-term results of catheter ablation. Int J of Card 1997;62:143-150.
18. Rodriguez L-M, Smeets JLRM, Timmermans C, et al: Predictors for successful ablation of right- and left-sided idiopathic ventricular tachycardia. Am J Cardiol 1997;79:309-314.
19. Wen M-S, Taniguchi Y, Yeh S-J, et al: Determinants of tachycardia recurrences after radiofrequency ablation of idiopathic ventricular tachycardia. Am J Cardiol 1998;81:500-503.
20. Ben-Haim SA, Osadehy D, Schuster I, et al: Non-fluoroscopic, in vivo navigation, and mapping technology. Nat Med 1996;2: 1393-1395.
21. Gepstein L, Hayam G, Ben-Haim SA: A novel method for non-fluoroscopic catheter-based electroanatomical mapping of the heart: In vitro and in vivo accuracy results. Circulation 1997;95: 1611-1622.
22. Shpun S, Gepstein L, Hayam G, et al: Guidance of radiofrequency endocardial ablation with real-time three-dimensional magnetic navigation system. Circulation 1997;96:2016-2021.
23. Gepstein L, Evans SJ: Electroanatomic mapping of the heart: Basic concepts and implications for the treatment of cardiac arrhythmias. PACE 1998;21:1268-1278.
24. Nademanee K, Kosar EM: A nonfluoroscopic catheter-based mapping technique to ablate focal ventricular tachycardia. PACE 1998;21:1442-1447.
25. Varanasi S, Dhala A, Blanck Z, et al: Electroanatomic mapping for radiofrequency ablation of cardiac arrhythmias. J Cardiovasc Electrophysiol 1999;10:538-544.
26. Kottkamp H, Chen X, Hindricks G, et al: Radiofrequency catheter ablation of idiopathic left ventricular tachycardia: Further evidence for microreentry as the underlying mechanism. J Cardiovasc Electrophysiol 1994;5:268-273.
27. Ohe T, Shimomura K, Aihara N, et al: Idiopathic sustained left ventricular tachycardia: Clinical and electrophysiologic characteristics. Circulation 1988;77:560-568.
28. Okumura K, Matsuyama K, Miyagi H, et al: Entrainment of idiopathic ventricular tachycardia of left ventricular origin with evidence for reentry with an area of slow conduction and effect of verapamil. Am J Cardiol 1988;62:727-732.
29. Nishizaki M, Arita M, Sakurada H, et al: Demonstration of Purkinje potential during idiopathic left ventricular tachycardia: A marker for ablation site by transient entrainment. PACE 1997;20:3004-3007.
30. Nakagawa H, Beckman KJ, McClelland JH, et al: Radiofrequency catheter ablation of idiopathic left ventricular tachycardia guided by a Purkinje potential. Circulation 1993;88:2607-2617.
31. Gursoy S, Chiladakis I, Kuck K-H: First lessons from radiofrequency catheter ablation in patients with ventricular tachycardia. PACE 1993;16:687-691.
32. Stevenson WG, Weiss JN, Wiener I, et al: Fractionated endocardial electrograms are associated with slow conduction in humans: Evidence from pace-mapping. J Am Coll Cardiol 1989;13:369-376.
33. Harada T, Stevenson WG, Kocovic DZ, et al: Catheter ablation of ventricular tachycardia after myocardial infarction: Relation of endocardial sinus rhythm late potentials to the reentry circuit. J Am Coll Cardiol 1997;30:1015-1023.
34. Stevenson WG, Khan H, Sager P, et al: Identification of reentry circuit sites during catheter mapping and radiofrequency ablation of ventricular tachycardia late after myocardial infarction. Circulation 1993;88:1647-1670.
35. Bogun F, Bahu M, Knight BP, et al: Response to pacing at sites of isolated diastolic potentials during ventricular tachycardia in patients with previous myocardial infarction. J Am Coll Cardiol 1997;30:505-513.
36. Kocovic DZ, Harada T, Friedman PL, et al: Characteristics of electrograms recorded at reentry circuit sites and bystanders during ventricular tachycardia after myocardial infarction. J Am Coll Cardiol 1999;34:381-388.
37. Blanck Z, Akhtar M: Therapy of ventricular tachycardia in patients with nonischemic cardiomyopathies. J Cardiovasc Electrophysiol 1996;7:671-683.
38. Fitzgerald DM, Friday KJ, Wah JAYL, et al: Electrogram patterns predicting successful catheter albation of ventricular tachycardia. Circulation 1988;77:806-814.
39. Kottkamp H, Chen X, Hindricks G, et al: Idiopathic left ventricular tachycardia: New insights into electrophysiological characteristics and radiofrequency catheter ablation. PACE 1995;18: 1285-1297.
40. Tsuchiya T, Okumura K, Honda T, et al: Significance of late diastolic potential preceding Purkinje potential in verapamil-sensitive idiopathic left ventricular tachycardia. Circulation 1999;99:2408-2413.
41. Stevenson WG: Ventricular tachycardia after myocardial infarction: From arrhythmia surgery to catheter ablation. J Cardiovasc Electrophysiol 1995;6:942-950.
42. Bogun F, Bahu M, Knight BP, et al: Comparison of effective and ineffective target sites that demonstrate concealed entrainment in patients with coronary artery disease undergoing radiofrequency ablation of ventricular tachycardia. Circulation 1997;95:183-190.
43. El-Sherif N, Gough WB, Restivo M: Reentrant ventricular arrhythmias in the late myocardial infarction period: Mechanisms of resetting, entrainment, acceleration, or termination of reentrant tachycardia by programmed electrical stimulation. PACE 1987;10:341-371.
44. Waldecker B, Coromilas J, Saltman AE, et al: Overdrive stimulation of functional reentrant circuits causing ventricular tachycardia in the infarcted canine heart. Circulation 1993;87:1286-1305.
45. Friedman PL: Is the postpacing interval of any value during ablation of postinfarction ventricular tachycardia? J Cardiovasc Electrophysiol 1999;10:52-55.
46. Hadjis TA, Harada T, Stevenson WG, et al: Effect of recording site on postpacing interval measurement during catheter mapping and entrainment of postinfarction ventricular tachycardia. J Cardiovasc Electrophysiol 1997;8:398-404.
47. Bogun F, Knight B, Goyal R, et al: Clinical value of the postpacing interval for mapping of ventricular tachycardia in patients with prior myocardial infarction. J Cardiovasc Electrophysiol 1999;10:43-51.
48. El-Shalakany A, Hadjis T, Papageorgiou P, et al: Entrainment/mapping criteria for the prediction of termination of ventricular tachycardia by single radiofrequency lesion in patients with coronary artery disease. Circulation 1999;99:2283-2289.
49. Ellison KE, Stevenson WG, Sweeney MO, et al: Catheter ablation for hemyodynamically unstable monomorphic ventricular tachycardia. J Cardiovasc Electrophysiol 2000;11:41-44.
50. Kornowski R, Hong MK, Gepstein L, et al: Preliminary animal and clinical experiences using an electromechanical endocardial mapping procedure to distinguish infarcted from healthy myocardium. Circulation 1998;98:1116-1124.
51. Marchlinski FE, Callans DJ, Gottlieb CD, et al. Linear ablation lesions for control of unmappable ventricular tachycardia in patients with ischemic and nonischemic cardiomyopathy. Circulation 2000;101:1288-1296.
52. Schilling RJ, Peters NS, Davies DW: A non-contact catheter for simultaneous endocardial mapping in the human left ventricle: Comparison of contact and reconstructed electrograms during sinus rhythm. Circulation 1998;98:887-898.
53. Peters NS, Jackman WM, Schilling RJ, et al: Human left ventricular endocardial activation mapping using a novel noncontact catheter. Circulation 1997;95:1658-1660.
54. Schilling RJ, Peters NS, Davies DW: Feasibility of a noncontact catheter for endocardial mapping of human ventricular tachycardia. Circulation 1999;99:2543-2552.
55. Strickberger SA, Knight BP, Michaud GF, et al: Mapping and ablation of ventricular tachycardia guided by virtual electrograms using a noncontact, computerized mapping system. J Am Coll Cardiol 2000;35:414-421.
56. Sosa E, Scanavacca M, d'Avila A, et al: Endocardial and epicardial ablation guided by nonsurgical transthoracic epicardial mapping to treat recurrent ventricular tachycardia. J Cardiovasc Electrophysiol 1998;9:229-239.

57. Sosa E, Scanavacca M, d'Avila A, et al: Nonsurgical transthoracic epicardial catheter ablation to treat recurrent ventricular tachycardia occurring late after myocardial infarction. J Am Coll Cardiol 2000;35:1442-1449.
58. Arruda M, Chandrasekaran K, Reynolds D, et al: Idiopathic epicardial outflow tract ventricular tachycardia: Implications for RF catheter ablation. PACE 1996;19:183[Abstract].
59. Stellbrink C, Diem B, Schauerte P, et al: Transcoronary venous radiofrequency catheter ablation of ventricular tachycardia. J Cardiovasc Electrophysiol 1997;8:916-921.
60. Sosa E, Scanavacca M, d'Avila A, et al: A new technique to perform epicardial mapping in the electrophysiology laboratory. J Cardiovasc Electrophysiol 1996;7:531-536.
61. Sosa E, Scanavacca M, d'Avila A, et al: Radiofrequency catheter ablation of ventricular tachycardia guided by nonsurgical epicardial mapping in chronic chagasic heart disease. PACE 1999; 22:128-130.
62. Tomassoni G, Stanton M, Richey M, et al: Epicardial mapping and radiofrequency catheter ablation of ischemic ventricular tachycardia using a three-dimensional nonfluoroscopic mapping system. J Cardiovasc Electrophysiol 1999;10:1643-1648.
63. Zardini M, Thakur RK, Klein GJ, et al: Catheter ablation of idiopathic left ventricular tachycardia. PACE 1995;18:1255-1265.
64. Wilber DJ, Baerman J, Olshansky B, et al: Adenosine-sensitive ventricular tachycardia, clinical characteristics and response to catheter ablation. Circulation 1993;87:126-134.
65. Globits S, Kreiner G, Frank H, et al: Significance of morphological abnormalities detected by MRI in patients undergoing successful ablation of right ventricular outflow tract tachycardia. Circulation 1997;96:2633-2640.
66. Breithardt G, Borggrefe M, Wichter T: Catheter ablation of idiopathic right ventricular tachycardia. Circulation 1990;82: 2273-2276.
67. Bogun F, El-Atassi R, Daoud E, et al: Radiofrequency ablation of idiopathic left anterior fascicular tachycardia. J Cardiovasc Electrophysiol 1995;6:1113-1116.
68. Stevenson WG, Friedman PL, Kocovic D, et al: Radiofrequency catheter ablation of ventricular tachycardia after myocardial infarction. Circulation 1998;98:308-314.
69. Rothman SA, Hsia HH, Cossu SF, et al: Radiofrequency catheter ablation of postinfarction ventricular tachycardia, long-term success and the significance of inducible nonclinical arrhythmias. Circulation 1997;96:3499-3508.
70. Strickberger SA, Man KC, Daoud EG, et al: A prospective evaluation of catheter ablation of ventricular tachycardia as adjuvant therapy in patients with coronary artery disease and an implantable cardioverter-defibrillator. Circulation 1997;96:1525-1531.
71. Gonska B-D, Cao K, Schaumann A, et al: Catheter ablation of ventricular tachycardia in 136 patients with coronary artery disease: Results and long-term follow-up. J Am Coll Cardiol 1994;24: 1506-1514.
72. Kim YH, Sosa-Suarez G, Trouton TG, et al: Treatment of ventricular tachycardia by transcatheter radiofrequency ablation in patients with ischemic heart disease. Circulation 1994;89:1094-1102.
73. El-Shalakany A, Hadjis T, Papageorgiou P, et al: Entrainment/mapping criteria for the prediction of termination of ventricular tachycardia by single radiofrequency lesion in patients with coronary artery disease. Circulation 1999;99:2283-2289.
74. Callans DJ, Zado E, Sarter BH, et al: Efficacy of radiofrequency catheter ablation for ventricular tachycardia in healed myocardial infarction. Am J Cardiol 1998;82:429-432.
75. Calkins H, Bigger JT, Ackerman SJ, et al: Cost-effectiveness of catheter ablation in patients with ventricular tachycardia. Circulation 2000;101:280-288.
76. Cox JL: Patient selection criteria and results of surgery for refractory ischemic ventricular tachycardia. Circulation 1989;79: I-163-I-177.
77. Blanchard SM, Walcott GP, Wharton JM, et al: Why is catheter ablation less successful than surgery for treating ventricular tachycardia that results from coronary artery disease? PACE 1994;17:2315-2335.
78. Akhtar M, Gilbert C, Wolf FG, et al: Reentry within the His-Purkinje system, elucidation of reentrant circuit using right bundle branch and His bundle recordings. Circulation 1978;58: 295-304.
79. Caceres J, Jazayeri M, McKinnie J, et al: Sustained bundle branch reentry as a mechanism of clinical tachycardia. Circulation 1989;79:256-270.
80. Blanck Z, Dhala A, Deshpande S, et al: Bundle branch reentrant ventricular tachycardia: Cumulative experience in 48 patients. J Cardiovasc Electrophysiol 1993;4:253-262.
81. Blanck Z, Jazayeri M, Dhala A, et al: Bundle branch reentry: A mechanism of ventricular tachycardia in the absence of myocardial or valvular dysfunction. J Am Coll Cardiol 1993;22: 1718-1722.
82. Cohen TJ, Chien WW, Lurie KG, et al: Radiofrequency catheter ablation for treatment of bundle branch reentrant ventricular tachycardia: Results and long-term follow-up. J Am Coll Cardiol 1991;18:1767-1773.
83. Crijns HJGM, Smeets JLRM, Rodriguez LM, et al: Cure of interfascicular reentrant ventricular tachycardia by ablation of the anterior fascicle of the left bundle branch. J Cardiovasc Electrophysiol 1995;6:486-492.
84. Berger RD, Orias D, Kasper EK, et al: Catheter ablation of coexistent bundle branch and interfascicular reentrant ventricular tachycardias. J Cardiovasc Electrophysiol 1996;7:341-347.
85. Blanck Z, Deshpande S, Jazayeri MR, et al: Catheter ablation of the left bundle branch for the treatment of sustained bundle branch reentrant ventricular tachycardia. J Cardiovasc Electrophysiol 1995;6:40-43.
86. Marcus F, Fontaine G: Arrhythmogenic right ventricular dysplasia/cardiomyopathy: A review. PACE 1995;18:1298-1314.
87. Basso C, Thiene G, Corrado D, et al: Arrhythmogenic right ventricular cardiomyopathy, dysplasia, dystrophy, or myocarditis? Circulation 1996;94:983-991.
88. Ellison KE, Friedman PL, Ganz LI, et al: Entrainment mapping and radiofrequency catheter ablation of ventricular tachycardia in right ventricular dysplasia. J Am Coll Cardiol 1998;32:724-728.
89. Feld GK: Expanding indications for radiofrequency catheter ablation: Ventricular tachycardia in association with right ventricular dysplasia? J Am Coll Cardiol 1998;32:729-731.
90. Deanfield JE, McKenna WJ, Hallidie-Smith KA: Detection of late arrhythmia and conduction disturbances after correction of tetralogy of Fallot. Br Heart J 1980;44:248-253.
91. Webb Kavey RE, Blackman MS, Sondheimer HM: Incidence and severity of chronic ventricular dysrhythmias after repair of tetralogy of Fallot. Am Heart J 1982;103:342-350.
92. Biblo LA, Carlson MD: Transcatheter radiofrequency ablation of ventricular tachycardia following surgical correction of tetralogy of Fallot. PACE 1994;17:1556-1560.
93. Chinushi M, Aizawa Y, Kitazawa H, et al: Successful radiofrequency catheter ablation for macro-reentrant ventricular tachycardias in a patient with tetralogy of Fallot after corrective surgery. PACE 1995;18:1713-1716.
94. Gonska B-D, Cao K, Raab J, et al: Radiofrequency ablation of right ventricular tachycardia late after repair of congenital heart defects. Circulation 1996;94:1902-1908.
95. Delacretaz E, Stevenson WG, Ellison KE, et al: Mapping and radiofrequency catheter ablation of the three types of sustained monomorphic ventricular tachycardia in nonischemic heart disease. J Cardiovasc Electrophysiol 2000;11:11-17.
96. Kottkamp H, Hindricks G, Chen X, et al: Radiofrequency catheter ablation of sustained ventricular tachycardia in idiopathic dilated cardiomyopathy. Circulation 1995;92:1159-1168.

Chapter 46 Imaging Techniques in Interventional Electrophysiology

DAVID SCHWARTZMAN

The art of electrophysiological intervention has evolved decisively in the direction of the adage "anatomy is destiny." The development of new cardiac imaging technologies has been a key adjuvant. Herein, we use the term *imaging* in the most general sense, that is, any technique or technology that provides insight into an arrhythmia substrate so that the interventional electrophysiologist may interact with it more effectively. Several different technologies are now in routine clinical use. Their individual versatility, combined with the potential synergy of simultaneous use, provide a formidable arsenal. However, in targeting a specific substrate, the choice of imaging technologies may not be intuitive, and often more than one approach may be effective. In this chapter, our approach to the localization and ablation of left atrial foci, which are associated with triggering of atrial fibrillation is utilized as an illustrative vehicle, to provide a practical dimension to the use of current imaging technologies in the interventional electrophysiology laboratory. We chose this target because (1) from the perspective of the electrophysiologist, the left atrium is the most difficult cardiac chamber to image; and (2) the art of intervention in this area has evolved rapidly over the past few years, associated with adaptation of old imaging techniques and technologies and the introduction of new ones.

Our present approach has been cobbled together in trial-and-error fashion over the past several years, a process that has involved the evaluation of most of the presently available imaging technologies. It is our hope that this "target-centered" presentation will be of more use to the practicing electrophysiologist than a general review of individual technologies. Specifically, we want to demonstrate how current technologies are adaptable to a target (often altering the function originally envisioned by their inventor), where their shortcomings are, and the value of their simultaneous use. It is important to emphasize that this is just one approach and has no intrinsic merit. Where possible, we also review the reported experiences with other approaches using the same and/or other technologies. Comprehensive reviews of individual technologies have been published previously.[1-9]

Methodology

PREOPERATIVE CHARACTERIZATION OF LEFT ATRIAL ANATOMY

As will be discussed in more detail in subsequent sections, none of the modalities currently clinically available for intraoperative imaging provides a global, detailed rendering of the left atrial endocardial anatomy. Earlier pathologic studies have shown that the anatomy, particularly the posterior atrium in and about the pulmonary vein orifices, is quite complex. We thought that detailed knowledge of this anatomy would be an important step toward optimization of the safety and efficacy of the ablation procedure. We have used static imaging in this pursuit, employing multidetector computed tomography (CT).[10] Magnetic resonance (MR) imaging has also been used for this purpose.[11] We have compared these techniques and have found the image quality and information content to be roughly equivalent.[12] In our center, CT is the mainstay. Images are acquired using General Electric (Milwaukee, WI) Light Speed or Light Speed Plus 16-detector scanners. So the scan timing coincides with peak distal pulmonary vein/left atrial body opacification, a test dose of 20 mL of iodinated contrast (Optiray, Mallinckrodt, Inc., St. Paul, MN) is given via a peripheral vein at a rate of 4 mL/second using a power injector. Then, after an additional dose of 125 mL of contrast is administered at the same rate, images are obtained through the chest. Scanning is performed from the top of the aortic arch to the diaphragm during a single breath hold, in 1.25 mm increments. Image reconstruction is performed using General Electric software, available with the Advantage Windows workstation. By considering shape, continuity, and density criteria, sequential axial images are manually edited to remove all structures, except for left atrium

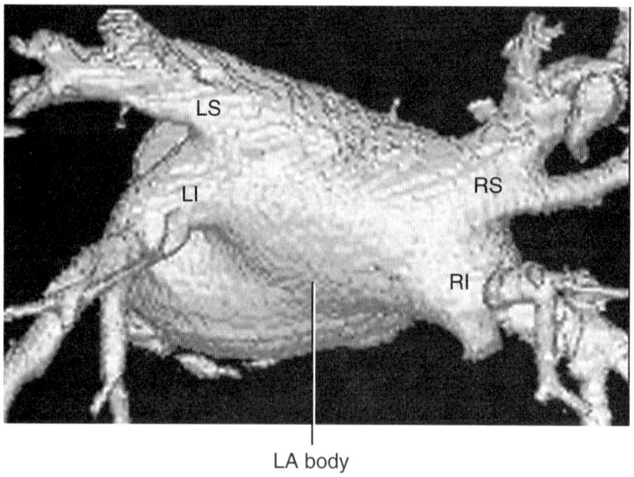

FIGURE 46-1 Extra-atrial view of the left atrium from an approximately posterior vantage. LI, Left inferior; LS, left superior pulmonary vein; RI, right inferior; RS, right superior.

and distal pulmonary veins. Three-dimensional reconstruction is then performed by coalescing these images, using two different volume-rendering techniques: (1) extra-atrial: to demonstrate global left atrial and pulmonary vein anatomy (Fig. 46-1); (2) intra-atrial: to demonstrate the posterior left atrial anatomy, specifically the pulmonary venous ostial regions (Fig. 46-2). In each individual, a multiplicity of intra- and extra-atrial viewing angles are examined on the workstation to achieve sufficient clarity of detail. As expected, anatomic detail is highly complex and interindividual variation is marked. These images have proved invaluable for

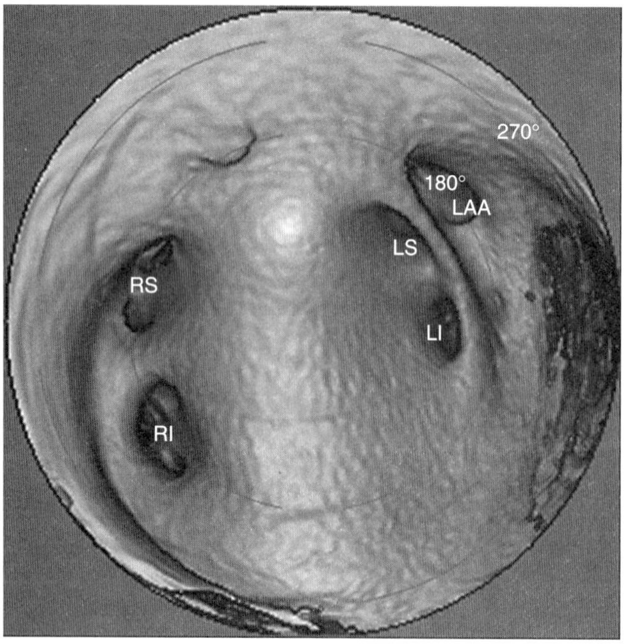

FIGURE 46-2 Intra-atrial views of the posterior left atrium, demonstrating pulmonary venoatrial junction regions. LAA, Left atrial appendage orifice region; LS, left superior pulmonary vein; RI, right inferior; RS, right superior.

planning intraoperative mapping and ablation procedures. Using intracardiac echocardiography, we have documented the accuracy of CT-derived atrial dimensions and anatomic detail.[10]

INTRAOPERATIVE LOCALIZATION OF SITES ASSOCIATED WITH AF INITIATION

Global Imaging

We have developed a technique for "global" simultaneous bi-atrial electrogram mapping as a modality to image the region from which atrial premature beats associated with atrial fibrillation (AF) onset originate (Fig. 46-3).[4] We have used this technique in patients with clinical syndromes of paroxysmal or persistent atrial fibrillation, with and without structural heart disease. Onset of atrial fibrillation was observed spontaneously and/or with drug, pacing, or shock stimulation.[4,13] In greater than 90% of onsets, the region of earliest activation during the initiating beat was the posterior left atrium, associated with one of the pulmonary vein microcatheter electrodes (Fig. 46-4). Occasionally, locally confined electrical events are detected serendipitously (Fig. 46-5); these are usually observed to initiate AF at other times.

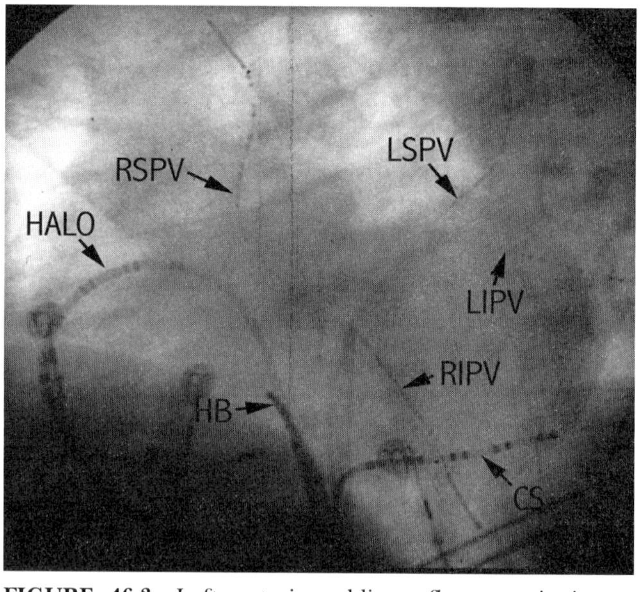

FIGURE 46-3 Left anterior oblique fluoroscopic image demonstrating electrode positions for "global" bi-atrial mapping. In each main pulmonary vein, a 16-electrode microcatheter (Pathfinder, Cardima Inc., Fremont, CA) is placed such that its proximal electrode pair straddles the venoatrial junction. A decapolar catheter is placed into the coronary sinus, a duodecapolar catheter (Halo, Biosense-Webster, Diamond Bar, CA) in the right atrium, a quadripolar catheter at the high septal right atrium, and a quadripolar catheter in the superior vena cava along its posterior wall to straddle the venoatrial junction (*not shown*). In some patients, an additional quadripolar microcatheter (Mini, Cardima, Inc.) is placed into the coronary sinus to cannulate the vein of Marshall (*not shown*). CS, Coronary sinus; HB, His bundle (high septal right atrium); LIPV, left inferior pulmonary vein; LSPV, left superior pulmonary vein; RIPV, right inferior pulmonary vein; RSPV, right superior pulmonary vein.

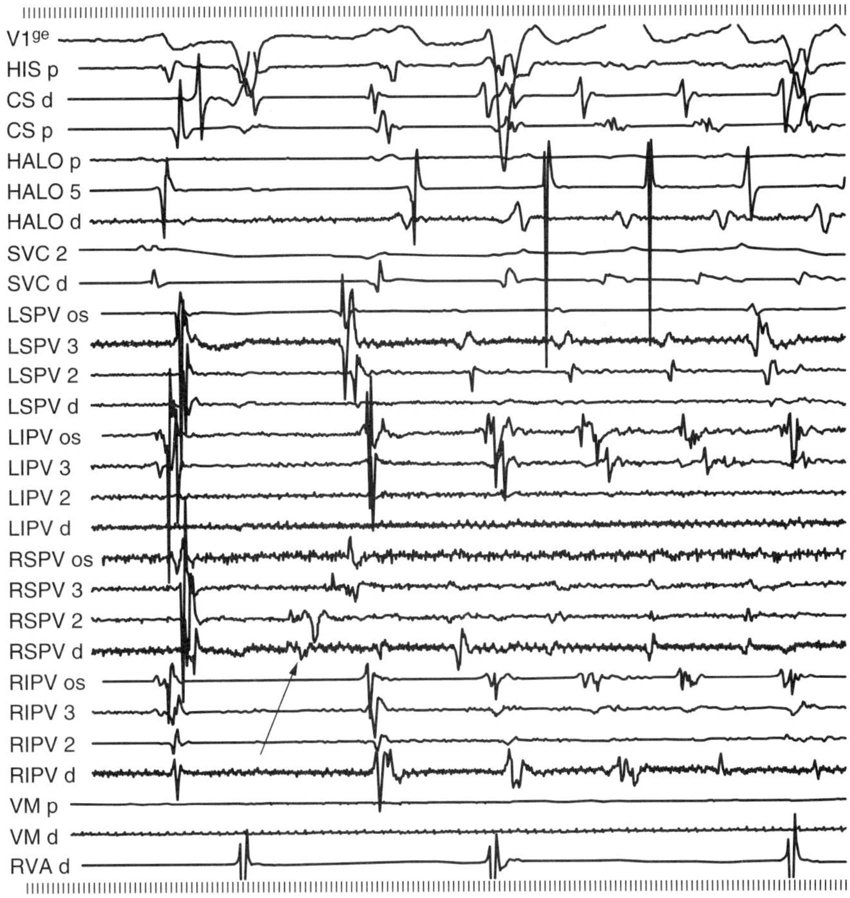

FIGURE 46-4 Sample recording demonstrating sinus beat (first beat), then atrial fibrillation-initiating atrial premature beat (second beat). The earliest electrical activity associated with the atrial premature beat is in the electrode pair deepest in the right superior pulmonary vein (*arrow*). This beat is associated with a distinct change in surface P wave morphology from sinus. The earliest intracardiac electrical activity begins over 75 ms before the earliest evidence of a surface P wave. V1, Precordial surface lead; HIS p, proximal recording from high septal right atrial quadripolar catheter; CS d, distal coronary sinus electrode pair; CS p, proximal coronary sinus electrode pair; HALO p, proximal Halo electrode pair; HALO 5, middle Halo electrode pair; HALO d, distal Halo electrode pair; SVC 2, proximal superior caval electrode pair; SVC d, distal superior caval electrode pair; LSPV os, pulmonary vein microcatheter pair straddling atriovenous junction. Remaining recordings are in spatial sequence deeper into the vein—the most distal site where sharp, reproducible potentials could be recorded was termed RSPV d. The same labeling was used for the other veins; V Mp, V Md, proximal and distal electrode pairs, respectively, from a quadripolar microcatheter in the vein of Marshall; RVA d, recording from the right ventricular apex.

In individual patients, multiple distinct regions from which initiating beats emanated were common. Occasionally, different regions were simultaneously active. Given the paucity of recordings, this methodology can only regionalize the origin of atrial fibrillation initiating ectopy. Other technologies have been similarly used, including multielectrode "basket" catheters (Boston Scientific, Inc., Sunnyvale, CA or Biosense-Webster, Inc., Diamond Bar, CA)[5] and noncontact mapping (Ensite, Endocardial Solutions, Inc., St Paul, MN).[14,15] These have varying capabilities regarding the number of electrograms (actual in the case of the

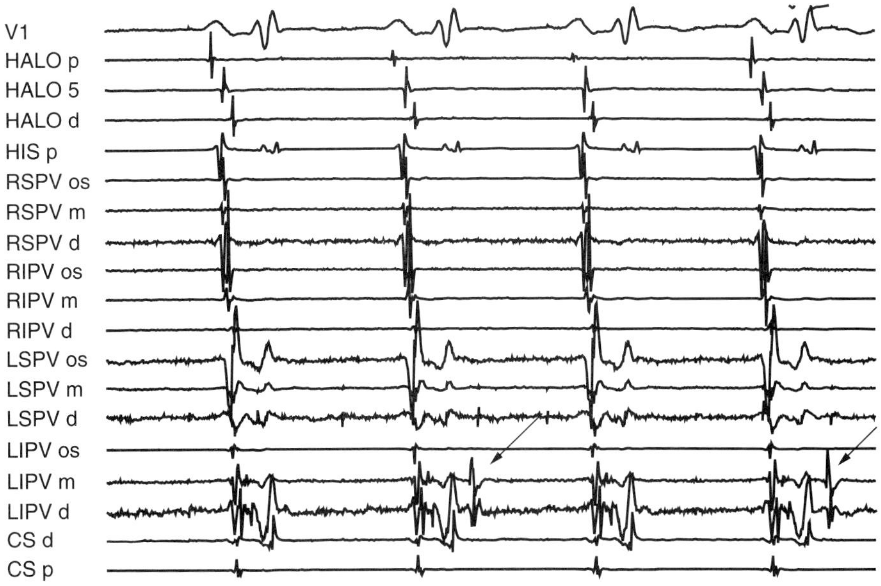

FIGURE 46-5 Electrogram recordings using global multielectrode technique shown in Figure 46-3. Labeling is similar to Figure 46-5. The ambient rhythm is sinus. There are sporadic extrasystoles (*arrows*) confined to the left inferior pulmonary vein. These were invisible to atrial body and surface electrodes. With pharmacologic manipulation (isoproterenol or adenosine), these extrasystoles were observed to depolarize the remainder of atrial myocardium, frequently initiating atrial fibrillation.

basket; virtual in the case of the noncontact system). In addition to electrographic imaging, we have also evaluated whether morphologic or mechanical imaging could be used to locate arrhythmogenic regions. Using CT or MRI (morphology) and intracardiac echocardiography (morphology, mechanics) we could find no differences between pulmonary veins associated with AF-initiating ectopy and those that were not.[16]

Local Imaging

A number of technologies have been used in an attempt to pinpoint rather than regionalize the site of origin of AF-initiating ectopy. Point-by-point electrographic imaging is possible, but only if the target of interest activates in a uniform and reproducible manner over a period of time sufficient to acquire the data. We, and others, have used linear array-type multielectrode catheters positioned longitudinally within the pulmonary veins.[4,17] However, the technique provides a highly simplified image of a very complex anatomy, and is plagued by irreproducibility. As such, it is inadequate for imaging sites of origin of AF-initiating ectopy to a degree of precision that would permit (optimally) single ablation lesion eradication. Other investigators have reported the use of other electrographic imaging technologies to improve precision, including circular,[18] basket,[5] and noncontact arrays.[14,15] The circular catheter may improve circumferential definition of segmental electrical communication between myocardium within the vein and that of the left atrial body. However, it still provides two-dimensional information, which does not move one much closer to imaging intravenous arrhythmogenic ectopy at its source. In addition, given that it generally finds stability 0.5 or more centimeters distal to the atriovenous junction, the circular array is a poor technology for imaging ectopy originating in the proximal vein, atriovenous junction, or contiguous left atrial body. Conceptually, a basket catheter placed within a pulmonary vein should provide more three-dimensional imaging information, and there are reports of its effective use in this regard.[5] Our experience with this technology has been disappointing, for two reasons. First, despite the fluoroscopic appearance of stable positioning of the basket splines (arms of the basket on which electrodes are mounted) around the vein circumference, imaging of actual position with intracardiac echocardiography revealed to us that splines were often floating in the lumen of the vein, leaving large portions of the vein wall without proximate electrode representation. In individual veins, experimentation with different basket sizes demonstrated a "best fit," but in most veins the noncontact problem was quite significant. Our impression was that the problem was due to (1) noncoaxial introduction of the basket into the vein, a common problem not easily overcome, and (2) a mismatch between the shape of the basket and that of the vein. This was particularly problematic near the atriovenous junction, where the vein circumference was expanding while the basket circumference was not. The motile and noncircular/planar features of the junction exacerbated the problem. Attempts to oversize the basket or force contact by putting pressure on the shaft of the catheter were limited by safety concerns or distal migration of the catheter. Interestingly, "near-field" appearing electrograms were commonly recorded from the floating basket electrodes. Second, as is demonstrated in Figures 46-1 and 46-2, pulmonary venous anatomy is not akin to a simple tube or trumpet-like shape. Incident to the main body of a vein are independent small veins or proximal branches, some of which may be invested with myocardium. Activity emanating from these branches enters the basket field in complex and often confusing ways. Noncontact mapping has also been evaluated in the quest for more precise electrical imaging of pulmonary venous myocardial investiture. We do not have experience with this technology, but reports from those who do suggest that its usefulness for this purpose is not good.[19,20]

ABLATION IN REGIONS ASSOCIATED WITH ATRIAL FIBRILLATION INITIATION

As originally reported by Haissaguerre and colleagues, the myocardial investiture of the left atriovenous junctions and pulmonary veins is disproportionately involved with the initiation of atrial fibrillation.[21] Following original case reports describing the "focal" nature of initiation, a profound deployment of focus and resources to this target has occurred. Based on the complexity of the target anatomy and the infrequent, nonrepetitive nature of the electrical event of interest (initiation of atrial fibrillation), more recent paradigms for ablation have moved away from traditional source activation mapping/ablation toward electrical isolation of arrhythmogenic regions.[13] The success of these paradigms are dependent on effective imaging and negotiation of the complex anatomy of these regions.

Our current paradigm involves electrical isolation of posterior atrial myocardium by circumferentially ablating the left atrial side of the atriovenous junction of individual, targeted pulmonary veins.[3] We begin by examining the CT scans and/or MR images of anatomy of the posterior left atrium, specifically the regions of the right and left atriovenous junctions. As demonstrated above, these regions demonstrate tremendous complexity and variability, including nonuniform motility, noncircular/nonplanar shape, supernumerary branches, and branches adjacent to the primary vein ostia (see Figs. 46-1 and 46-2). We have found "still" images (that can be obtained only preoperatively) to be invaluable because they provide a level of three-dimensional endocardial topographic detail that is not possible with available intraoperative imaging modalities. Based on these images, the intraoperative strategy can be planned, including specific sites of atrial transseptal puncture, catheter selection, likely vein targets, and design of ablation lesions to encompass these targets.

Intraoperatively, our imaging strategy does not involve fluoroscopy, except for gross catheter manipulation. For the purpose of ablation in the posterior left atrium, we have found this technology to be, at best, unhelpful and, at worst, misleading. Specifically (1) fluoroscopy provides no information regarding anatomic detail or location of catheters relative to the anatomy, only a scaffold on which to superimpose an arbitrary, mind's-eye view of these; (2) fluoroscopy provides misleading information as to the nature of catheter-endocardial contact (discussed later). Our strategy involves the synergistic use of three technologies (1) CARTO (Biosense-Webster, Inc.); (2) Rotating mechanical transducer intracardiac echocardiography (RMT-ICE; Boston Scientific, Inc); and (3) Phased-array intracardiac echocardiography (PA-ICE; Acuson Inc., Mountain View, CA).

The CARTO system uses synchronous extracorporeal magnetic fields to determine the location and orientation of a sensor mounted within a catheter near its distal electrode.[1] During endocardial mapping, the electrode is moved in point-by-point fashion along the endocardium. The system synthesizes the spatial locations of the summation of acquired points to present a three-dimensional shell representing the endocardial contour. Local information can be superimposed on this shell, including activation time and local electrogram amplitude. In theory, CARTO can provide anatomically accurate endocardial reconstruction, and can, therefore, be conceptualized as a "stand-alone" technology for paradigms of anatomy-defined ablation, such as ours. In fact, there are reports of its use for this purpose.[22] However, it is crucial to understand the limitations of CARTO that follow:

1. Although the endocardial contour that the CARTO system presents to the viewer does use the actual point data, it is the software that makes assumptions as to where the endocardium is between the points. Thus, in general, more points acquired is associated with a more accurate endocardial representation. However, the number of individual spatial points required to present a locally accurate representation of the endocardial anatomy is too high to be of practical utility intraoperatively from the standpoint of time commitment. Even with the time commitment, in my opinion, the system is not capable of adequate accuracy due to (A) the fact that the location sensor is at a finite distance from the distal catheter electrode, which is actually the element in contact with the endocardium. This distance permits a variable relationship between the sensor and the endocardium, depending on catheter-endocardial angulation and contact pressure; (B) respiratory motion, with heart position moving relative to the position of a reference sensor, generally located on the body surface; (C) unstable catheter electrode-endocardial contact (intermittent and/or sliding) result in a blurring of the image. Recent versions of the CARTO software have made the endocardial contour appear more "realistic" to the operator by smoothing an interpoint contour, which with earlier versions of the software was jagged, acting as a constant reminder that the image could not be representative of the actual anatomy.
2. The system software is unaware and cannot correct for anatomic complexities common to the atrial endocardium, including ridges and ostia. Spatially contiguous points are connected linearly, whether they are on the same or different side of a ridge or in and out of an ostium. Although the software permits designation and "reconstruction" of pulmonary veins as a way of demarcating them from the left atrial body, these are coarse constructs, not at all representative of the actual anatomy.
3. Given the lack of direct endocardial visualization, the system is incapable of designating discrete anatomic sites (e.g., mitral annulus).
4. The system acquires points as commanded by the operator. Except for those points that are operator-edited, the system assumes that all points were acquired when the distal mapping electrode was in stable contact with the endocardium. Therefore, points taken during a period of poor or noncontact will be incorporated into the map. As mentioned earlier, particularly in and about the pulmonary veins, contact cannot be judged using fluoroscopy or electrogram characteristics. Thus, CARTO map construction is left to the operator, who does not possess the proper tools to quality control the points that will compose the map.
5. The system does not provide real-time position data. Position information is updated once each cardiac cycle.

These limitations are not limited to the CARTO system. Other available "three-dimensional endocardial imaging systems," share some or all of these problems, including Non-Contact Mapping, LocaLisa (Medtronic, Inc., Minneapolis, MN), and the Realtime Position Management System (RPM, Cardiac Pathways/Boston Scientific, Sunnyvale, CA).

Intracardiac echocardiography is the only technology for intraoperative imaging that is real-time and that provides direct visualization of the endocardium. We have experience with each of the ICE technologies just listed:

1. RMT-ICE—This system uses a 9F catheter incorporating a rotating transducer operating at 9 MHz.[23] For electrical isolation of pulmonary venous myocardium, the transducer is placed into the left atrium contiguous to the ablation electrode/target, approximately 0.5 centimeters away from the atriovenous junction (atrial body side; Fig. 46-6). Although for this catheter the imaging plane is circumferential, orthogonal to the catheter axis, in this position the full circumference of the atriovenous junction is not imaged simultaneously. This is due to variation between the catheter axis and the major axis of the vein, and to the fact that the atriovenous junction is neither circular nor planar. It is important to emphasize that images obtained with the RMT-ICE device are two-dimensional. Although attempts can be made to mentally create a three-dimensional image, this inevitably fails due to changes in the angle of incidence between ultrasound beam and endocardium due to (A) changes in catheter axis during manipulation (often subtle and impossible to

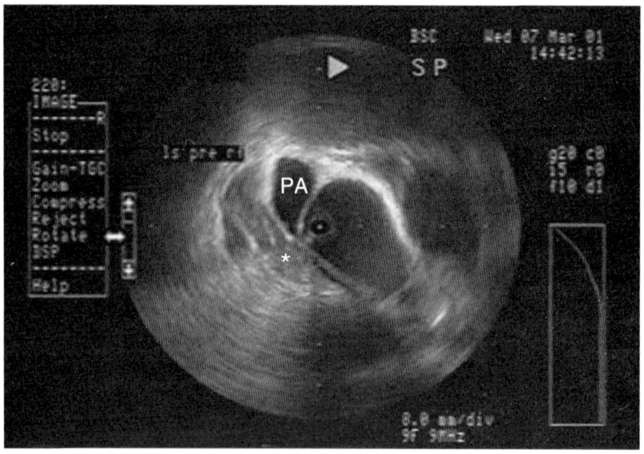

FIGURE 46-6 RMT-ICE image with transducer (*center of image*) at a pulmonary venoatrial junction. The contiguous chamber (PA) is a pulmonary artery. The shadow contiguous to the transducer is an ablation electrode (*).

discern), and (B) abrupt changes in endocardial topography. As such, ICE alone is unacceptable for three-dimensional imaging. We have performed several cases using a serial technique that can create accurate three-dimensional images using the RMT-ICE catheter (Fig. 46-7). However, these images require reconstruction and are, thus, not real time, eliminating the dominant value of ICE.[16]

We use RMT-ICE to characterize and monitor in continuous, real time the electrode-endocardial interface during catheter ablation procedures. Based on ICE, we manipulate the electrode to (A) ensure stable electrode-endocardial contact at every ablation site before, during, and after ablation energy application. In smooth atrial endocardial zones (all but the left atrial appendage), we have learned that using the standard electrode catheter's firm, stable contact can be very difficult to achieve. Intermittent local contact in the form of electrode sliding or lifting is common. Intermittent contact is often inapparent to operator "feel," fluoroscopic image, or electrogram criteria. Long sheath support is always necessary. Stable contact is critical to safe and effective ablation. Power titration (maximal power and duration) becomes relatively constant, thus avoiding problems with inaccuracy of power titration guided by electrode thermometry (underreporting of maximal myocardial temperature, resulting in overtitration of power). This phenomenon is particularly prevalent in smooth atrial endocardial areas with high-flow velocity, such as at an atriovenous junction.[24] (B) monitor the electrode-endocardial interface during radiofrequency applications. We have demonstrated the usefulness of ICE for avoiding complications of overtitration of ablation energy, as well as for imaging the physiologic impact of ablation lesions on cardiac structure and function.[16]

Although central to our ablation paradigm, the use of RMT-ICE is not without technical problems: (A) the catheter has no intrinsic deflection capability, limiting operator ability to perform fine spatial manipulation of the imaging plane. This is very important because angulation of the ablation electrode relative to the imaging plane markedly influences the quality of the electrode-endocardial interfacial image. We have documented the usefulness of a transponder mounted in the ablation electrode to ameliorate this influence.[25] This technology has not yet been released for clinical use; (B) at the 9-MHz operating frequency of the RMT-ICE catheter, image quality attenuates very rapidly with distance, even under blood. This requires the transducer to be physically located very close to the ablation electrode-endocardial interface, which can be difficult to achieve and maintain.

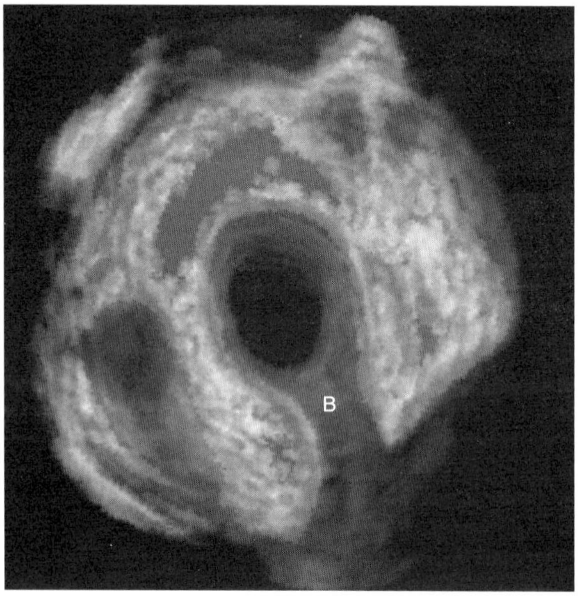

FIGURE 46-7 RMT-ICE image of a pulmonary vein with three-dimensional reconstruction. View of the vein is *en face*. A major branch of the vein (B) is seen.

2. PA-ICE—This system uses a 10F catheter incorporating an electronic, phased array transducer, which is capable of multiple operating frequencies ranging from 5.5 to 10 MHz. Unlike the RMT-ICE technology, the system is Doppler capable. In addition, the lower frequency permits high quality imaging at a distance, not only under blood but also between chambers.[9]

Our experience with the incremental value of this technology for assisting pulmonary venous myocardial electrical isolation has been disappointing. Although quality images of the veins can be obtained with the transducer positioned in the right atrium, equally often one or more of the veins cannot be well imaged. This has forced us to move the transducer into the left atrium, which has yielded reliable, excellent quality images. However, the sector-type imaging plane (e.g., parallel to long axis of the catheter) is poorly suited to imaging our current mode of circumferential ablation at the atriovenous junction. Unlike the RMT-ICE technology, which has a circumferential imaging plane and is self-stabilizing, to assist circumferential guidance with the PA-ICE technology, the catheter must be constantly manipulated and held in stable position to maintain the image. Orientation in the longitudinal "tomographic" images of the atriovenous junction is easily lost. We are aware of efforts to couple this technology with three-dimensional imaging software, but as for the RMT-ICE efforts in this regard, image reconstruction would not be real time, therefore eliminating one of the major benefits of ICE. We have found that PA-ICE is useful for imaging the physiologic impact of ablation on the atriovenous junction, a useful tool in avoiding unnecessary morbidity.[16]

Our technique for electrical isolation of posterior atrial myocardium by circumferentially ablating the left atrial side of the atriovenous junction of individual targeted pulmonary veins thus involves a synergy between different imaging technologies, minimizing their individual shortcomings (Table 46-1). The circumferential lesion is, then, the sum of a series of focal lesions applied adjacent to one another over a predetermined, anatomy-defined course. Using ICE, the appropriate anatomic zone in which to deploy the lesion is chosen, and firm/stable electrode contact is achieved and maintained during focal lesion applications at each site. Using CARTO, focal lesions are presented as spheric icons, each with a diameter of 4 mm. The ablation electrode is positioned such that the icon edges abut one another throughout the lesion course. Using this approach, we have had uniform success in achieving electrical isolation of the targeted region with no procedural morbidity and no short- or long-term evidence of pulmonary vein stenosis.

Several other approaches to electrical isolation of posterior left atrial myocardium have been reported. Pappone and coworkers have reported success using CARTO alone.[22] We are unaware of a published report of isolation performed guided by noncontact mapping. However, preliminary reports of "anatomy-based" ablation in other atrial areas guided by noncontact mapping have been favorable.[26] Similarly, we are unaware of reports of isolation guided solely by ICE. As detailed earlier, we would likely not be able to equal our present success rate using RMT-ICE alone. However, new ablation devices may change the efficacious use of ICE as an imaging adjunct. For example, Natale and associates reported a balloon catheter design for electrical isolation of myocardium distal to the atriovenous junction.[27]

TABLE 46-1 Evaluation of Imaging Technologies

	Intracardiac Echocardiography	Virtual Endocardial Imaging*
Direct endocardial visualization?	Yes	No
Real-time assessment of electrode-endocardial contact?	Yes	No
Real-time 3-D reconstruction	No	Yes

*CARTO, Ensite, LocaLisa, RPM.

In theory, this is a single-pass device; efficacy is dependent on occlusion of blood flow between the targeted vein and the left atrium during lesion application. In association with this ablation technology, PA-ICE has shown tremendous value in assessing proper balloon-vein "fit," including effective occlusion of blood flow.[9] Insights provided by ICE into the mechanisms of isolation failures have accelerated its development toward a safe and efficacious tool for use in humans.

SUMMARY

In this review, we have used a "target-specific" (posterior left atrium) approach as a vehicle to illustrate the practical utility of imaging technologies currently available to the interventional electrophysiology laboratory. The development of nonfluoroscopic imaging modalities has clearly been a key to the development of interventional procedures in this area. However, to optimize procedural safety and efficacy, it is important to recognize the advantages and limitations of individual modalities. As illustrated here, limitations can be overcome by coupling different technologies. It is also important to recognize that this chapter reflects a snapshot in time. The rapid pace of development of both imaging and ablative techniques and technologies will undoubtedly continue, and these will undoubtedly change the requirements and goals of imaging.

REFERENCES

1. Schwartzman D, Kuck KH: Anatomy-guided linear atrial lesions for radiofrequency catheter ablation of atrial fibrillation. PACE 1998;21:1959-1978.
2. Wittkampf FH, Wever EF, Derksen R, et al: LocaLisa: New technique for real-time 3-dimensional localization of regular intracardiac electrodes. Circulation 1999;99:1312-1317.
3. Schwartzman D: Catheter ablation to suppress atrial fibrillation: Evolution of technique at a single center. J Intervent Cardiac Electrophysiol 2003;9:295-300.
4. Schwartzman D: The common left pulmonary vein: A consistent source of arrhythmogenic atrial ectopy. J Cardiovasc Electrophysiol 2004;15:560-566.
5. Michael MJ, Haines DE, DiMarco JP, Paul MJ: Elimination of focal atrial fibrillation with a single radiofrequency ablation: Use of a basket catheter in a pulmonary vein for computerized activation sequence mapping. J Cardiovasc Electrophysiol 2000;11:1159-1164.
6. Schneider MA, Ndrepepa G, Zrenner B, et al: Noncontact mapping-guided catheter ablation of atrial fibrillation associated with left atrial ectopy. J Cardiovasc Electrophysiol 2000; 11:475-479.
7. deGroot N, Bootsma N, van der Velde ET, Schalij MJ: Three-dimensional catheter positioning during radiofrequency ablation in patients: First application of a real-time position management system. J Cardiovasc Electrophysiol 2000;11:1183-1192.
8. Ren J-F, Schwartzman D, Callans DJ, et al: Imaging technique and clinical utility for electrophysiologic procedures of lower frequency (9 MHz) intracardiac echocardiography. Am J Cardiol 1998;82:1557-1560.
9. Packer DL, Stevens CL, Curley MG, et al: Intracardiac phased-array imaging: Methods and initial clinical experience with high resolution, under blood visualization. J Am Coll Cardiol 2002; 39:509-516.
10. Schwartzman D, Lacomis JL, Wigginton WG: Characterization of left atrium and distal pulmonary vein morphology using multidimensional computed tomography. J Am Coll Cardiol 2003; 41:1349-1357.
11. Tsao H-M, Yu W-C, Cheng H-C, et al: Pulmonary vein dilation in patients with atrial fibrillation: Detection by magnetic resonance imaging. J Cardiovasc Electrophysiol 2001;12:809-813.

12. Schwartzman D, Lacomis JM: Three-dimensional imaging of the left atrium and pulmonary veins: CT or MRI? J Am Coll Cardiol 2003;41:132A.
13. Schwartzman D: Radiofrequency catheter ablation of atrial fibrillation. In Ganz LI (ed): Management of Cardiac Arrhythmias. 2002, Humana Press,Totowa, NJ, pp 145-162.
14. Asirvatham S, Packer DL: Validation of non-contact mapping to localize the site of simulated pulmonary vein ectopic foci. Circulation 2000;102:II-441.
15. Hindricks G, Kottkamp H: Simultaneous non-contact mapping of left atrium in patients with paroxysmal atrial fibrillation. Circulation 2001;104(3):297-303.
16. Schwartzman D, Kanzaki H, Bazaz R, Gorcsan J: Impact of catheter ablation on pulmonary vein morphology and mechanical function. J Cardiovasc Electrophysiol 2004;15:1-7.
17. Hsieh MH, Chen SA, Tai CT, et al: Double multielectrode mapping catheters facilitate radiofrequency catheter ablation of focal atrial fibrillation originating from pulmonary veins. J Cardiovasc Electrophysiol 1999;10(2):136-144.
18. Haissaguerre M, Shah DC, Jais P, et al: Electrophysiological breakthroughs from left atrium to the pulmonary veins. Circulation 2000;102:2463-2465.
19. Darbar D, Olgin JE, Miller JM, Friedman PA: Localization of the origin of arrhythmias for ablation: From electrocardiography to advanced endocardial mapping systems. J Cardiovasc Electrophysiol 2001;12:1309-1325.
20. Seidl K, Beatty G, Drogemuller A, et al: Catheter-based right and left atrial compartmentalization procedure in chronic atrial fibrillation using a novel non-contact mapping system: Feasibility and safety. J Am Coll Cardiol 2000;35:123.
21. Haissaguerre M, Jais P, Shah DC, et al: Spontaneous initiation of atrial fibrillation by ectopic beats originating in the pulmonary veins. N Engl J Med 1998;339:659-666.
22. Pappone C, Oreto G, Lamberti F: Catheter ablation of paroxysmal atrial fibrillation using a 3D mapping system. Circulation 1999;100:1203-1208.
23. Ren JF, Schwartzman D, Callans DJ, Marchlinski FE: Intracardiac echocardiograpy (9 MHz) in humans: Methods, imaging views and clinical utility. Ultrasound Med Biol 1999;25:1077-1086.
24. Bazaz R, Schwartzman D: Optimizing radiofrequency power titration during pulmonary vein ablation: Insights from intracardiac echocardiography. Circulation 2003;108:IV-473.
25. Menz V, Vilkomerson D, Ren J-F, et al: Echocardiographic transponder-guided catheter ablation: Feasibility and accuracy. J Intervent Cardiac Electrophysiol 2001;5:203-209.
26. Packer D, Asirvatham S, Munger T: Utility of non-contact mapping for identifying gaps in long linear lesions in patients with atrial fibrillation. PACE 2000;23:673.
27. Natale A, Pisano E, Shewchik J, et al: First human experience with pulmonary vein isolation using a through-the-balloon circumferential ultrasound ablation system for recurrent atrial fibrillation. Circulation 2000;102(16):1879-1882.

Chapter 47 Noninvasive Electrophysiology

IRINA SAVELIEVA AND A. JOHN CAMM

Modern noninvasive cardiac electrophysiology involves a comprehensive assessment of the processes governing the electrical activity of the heart and provides a basis for cardiac risk stratification. However, until very recently, identification of patients at increased risk of potentially fatal arrhythmic events was originally based on the evaluation of left ventricular function, estimation of the intensity of ventricular ectopic activity, and investigation of ventricular conduction abnormalities. Following the results of several recent multicenter studies, the importance of cardiac autonomic balance and ventricular repolarization has been appreciated. The practical implications of these methods include diagnostic evaluation, assessment of prognosis, risk stratification, and possibly guidance of treatment. Table 47-1 lists currently available noninvasive methods and indicates which have been useful for these purposes. In this chapter, routine and recently proposed techniques for noninvasive assessment of the electrical activity of the heart will be discussed.

Basic Electrocardiographic Tests

AMBULATORY ELECTROCARDIOGRAPHY

Technologic Advances in Holter Monitoring Systems

Since a radiotelemetry system for continuous ECG recording was first suggested by Norman Holter in 1961,[1] substantial advances in signal acquisition and signal analysis techniques have enabled further refinement of the ambulatory ECG monitoring technique from the traditional two- or three-lead analog tape recorders to the digital high-resolution devices supporting sample rates of 1000 samples per second in three or more channels without data compression. The major difference between the magnetic (tape) or compressed (solid state memory) data recording devices and the digital recorders is that in the latter, the signal is fed into an analog-to-digital converter immediately after it is amplified and filtered directly in the recorder using a crystal-controlled time base, which eliminates timing distortions and provides superior data quality. This significantly improves the quality of the recording, increases the accuracy of heart rate variability analysis, and allows the Holter ECG data to be used for a variety of signal analyses, such as the signal-averaged ECG, T wave alternans, and T wave morphology.

Recently, digital recorders have been introduced that enable a continuous 12-lead ECG recording and provide improved high-resolution capabilities ideal for research purposes and promising for clinical investigations. The diagnostic accuracy of the 12-lead systems was found to be superior to that of the traditional Holter systems in a variety of clinical conditions, such as supraventricular tachycardias (for the evaluation of P wave polarity and P wave axis and the P/R relationship); broad complex tachycardias and specific ventricular tachycardia morphologies (better identification of QRS axis and QRS morphology and, therefore, the site of origin of tachycardia); and ST segment shift, which may be identifiable only in leads V_{2-4}.[2] Figure 47-1 demonstrates a feasibility of a 12-lead Holter ECG recording to identify fascicular ventricular tachycardia based on the criteria of QRS complex morphology and QRS axis deviation, compared with the standard two-channel Holter ECG recording.

CLINICAL STUDIES

The ambulatory ECG enables both qualitative and quantitative assessment of the arrhythmia, which allows improved diagnosis in patients with suspected or proven arrhythmias, provides valuable prognostic information, especially in postmyocardial infarction patients, and may be useful for the assessment of the efficacy of drug therapy treatment, for example, for rate control in patients with permanent atrial fibrillation.

The presence of ventricular ectopic activity, frequent ventricular premature complexes (VPC) and/or nonsustained VT is an independent prognostic factor in postmyocardial infarction patients[3,4] and is associated with a two- to threefold increase in the mortality rate. Although higher in the first six months after myocardial infarction, the risk for mortality associated with nonsustained VT persists over a three-year period of follow-up and is significantly greater in the presence of left ventricular dysfunction. Furthermore, the MADIT (Multicenter Automatic Defibrillator Implantation Trial)

TABLE 47-1 Diagnostic and Prognostic Usefulness of Noninvasive Methods

Method	Diagnostic Value	Prognostic Value	Monitoring of Therapy
Holter ECG monitoring	Yes	Yes	Yes*
Exercise stress test	Yes	Yes, limited†	Yes, limited†
Tilt test	Yes	Yes	Yes
Signal-averaged ECG	Yes, limited‡	Yes	No
Heart rate variability	No	Yes	No
Baroreflex sensitivity	No	Yes	Unknown
Q–T interval and QT dispersion	Yes	Yes	Yes§
T wave alternans	No	Yes	Unknown
T wave morphology descriptors	Unknown	Yes	Unknown
Heart rate turbulence/dynamics/harmony	No	Yes	Unknown

*Routinely used but not supported by large-scale studies.

†Mainly related to the demonstration of myocardial ischemia or the induction of exercise-dependent arrhythmias; e.g., adrenergically mediated atrial fibrillation or catecholamine-sensitive ventricular tachycardia.

‡Useful for diagnosis of unexplained syncope.

§Important for dose adjustment of antiarrhythmic drugs prolonging repolarization to prevent proarrhythmias.

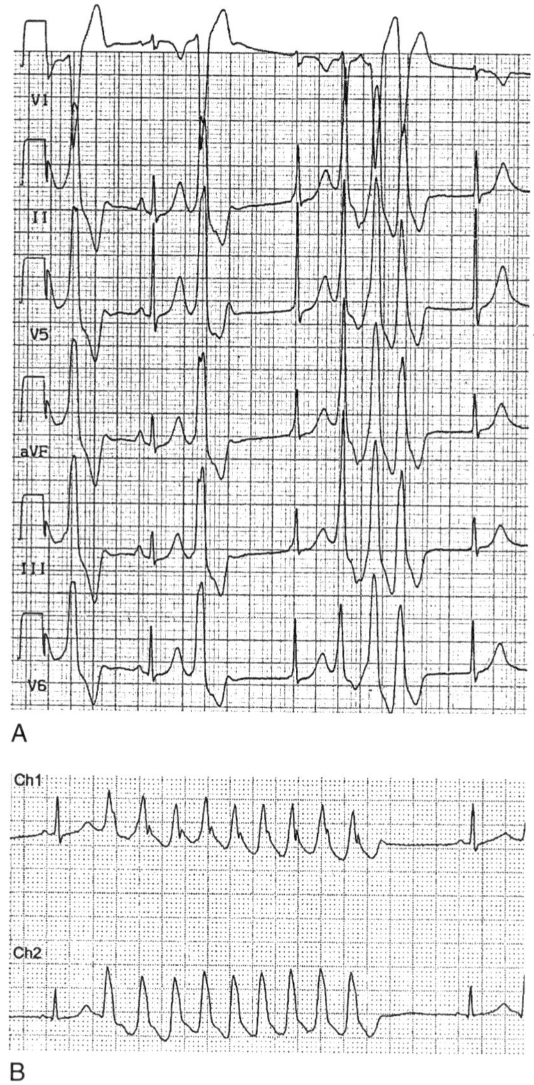

FIGURE 47-1 Twelve-lead Holter ECG strip (**A**) 6 leads are shown) allows the correct diagnosis of ventricular tachycardia originating from the right ventricular outflow tract compared with the standard 2-lead Holter ECG recording (**B**). The diagnosis is based on evaluation of QRS complex morphology of left bundle branch block and inferior QRS axis deviation.

and MUSTT (Multicenter Unsustained Tachycardia Trial) trials used nonsustained asymptomatic VT as an additional risk factor to support implantation of a cardioverter-defibrillator (ICD) in patients with a history of myocardial infarction, left ventricular ejection fraction below 35% to 40%, and inducible, nonsuppressible VT during electrophysiological study.[5,6] In the Autonomic Tone and Reflexes After Myocardial Infarction (ATRAMI) trial of 1284 patients with moderately impaired left ventricular function (mean ejection fraction 49%), nonsustained VT entailed a threefold increased risk of all-cause mortality and a 16-fold increased risk for arrhythmic death or sustained VT.[6a]

However, the analysis of pooled data has revealed a surprisingly low sensitivity and positive-predictive accuracy of frequent VPC (usually >10/hour) and nonsustained VT, for risk stratification in postmyocardial infarction patients and patients with congestive heart failure (Fig. 47-2).[7] The results of the ESVEM (Electrophysiological Study Versus Electrocardiographic Monitoring) study demonstrated that although univariate analysis suggested the association between the presenting arrhythmia and outcome, multivariate analysis failed to substantiate the predictive value of the presenting arrhythmia for sudden death and cardiac arrest.[8]

In 1080 patients with congestive heart failure studied in the PROMISE (Prospective Randomised Milrinone Survival Evaluation) trial, the frequency of VPC (>30/hour), presence of nonsustained VT, number of nonsustained VT runs, and the duration of the longest nonsustained VT episodes were each a significant independent predictor of all-cause mortality but failed to discriminate between modes of death.[9] Although in the GESICA (Gruppo de Estudio de la Sobrevida en la Insuficiencia Cardiaca en Argentina) study, patients with nonsustained VT had a higher incidence of sudden death compared with those without this arrhythmia (23.7% versus 8.7%, relative risk 2.77 in univariate analysis), the prognostic value of ventricular arrhythmias was not substantiated by multivariate analysis.[10]

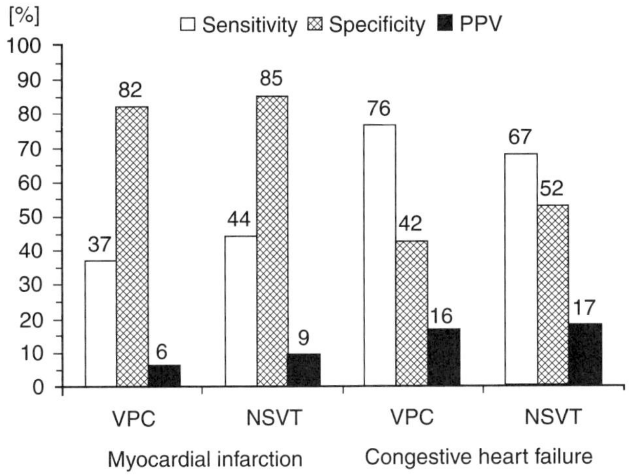

FIGURE 47-2 Pooled data on sensitivity, specificity, and positive-predictive value from 24 studies in postmyocardial infarction patients and four studies in patients with congestive heart failure. NSVT, nonsustained ventricular tachycardia; PPV, positive-predictive value; VPC, ventricular premature complex. (Modified from Hodges M, Bailey JJ: Ambulatory electrocardiography: Use in arrhythmia risk assessment. In Malik M (ed): Risk of Arrhythmia and Sudden Death. London, BMJ Books, 2001, pp 194-201.)

Similarly, in the CHF-STAT (Congestive Heart Failure-Survival Trial of Antiarrhythmic Therapy) trial, the presence of nonsustained VT on ambulatory ECG was not a predictor of sudden death in multivariate analysis.[11] The recent results of the MADIT II study have shown that in patients with a prior myocardial infarction and advanced left ventricular dysfunction, even in the absence of nonsustained VT, prophylactic implantation of an ICD has significantly improved survival, suggesting that the results of ambulatory Holter monitoring may not provide additional information to guide therapeutic interventions in this patient population.[12]

It is plausible to speculate that wide use of thrombolytic therapy and/or angioplasty to restore the patency of the infarct-related coronary artery has modified the arrhythmogenic substrate and, therefore, the incidence and characteristics of ventricular arrhythmias in patients with myocardial infarction. This may have resulted in a decrease of the prognostic significance of ventricular ectopic activity on Holter ECG after infarction. Indeed, Hohnloser and colleagues[13a] reported a low prevalence (9%) of nonsustained VT in 325 patients with acute myocardial infarction treated according to contemporary guidelines. Nonsustained VT was present in only 2.4% of patients with depressed left ventricular ejection fraction, and its prognostic value was inferior to that of an ejection fraction and impaired autonomic tone. The Gruppo Italiano per lo Studio della Sopravvivenza nell' Infarto miocardico 2 (GISSI-2) Investigators failed to demonstrate that nonsustained VT (6.8% of the total patient population) was an independent predictor of cardiac mortality in postinfarction patients who underwent thrombolysis.[13] In the ATRAMI study, although nonsustained VT conferred increased risk for mortality, only 13.4% had nonsustained VT during Holter monitoring.[6a] Furthermore, lack of predictive power of nonsustained VT in patients with congestive heart failure may be due to the multifactorial etiology of sudden cardiac death in this clinical setting.[14] There is no agreement whether reduction in the frequency of arrhythmias is associated with a decreased risk of arrhythmic events and sudden death.[15,16,17]

OTHER DEVICES FOR ELECTROCARDIOGRAPHIC MONITORING

External Event and Loop Recorders

The major limitation of Holter recorders is a relatively short monitoring period of 24 to 48 hours. Event recorders allow the capture of surface ECG at the time of cardiac events, usually during suspected arrhythmias. Less frequently, they are employed to detect ST segment changes, to monitor cardiac rhythm, and to identify asymptomatic arrhythmias in pharmacotherapy trials. Of note, in the recent PAFAC (Prevention of Atrial Fibrillation After Cardioversion) study involving more than 1000 patients with AF, 90% of arrhythmia recurrences were rendered asymptomatic by antiarrhythmic drug therapy and could be detected only by daily transtelephonic ECG monitoring.[18]

The duration of the recording varies but is typically in the range of 1 to 5 minutes. Event recorders can be nonlooping devices that are applied to the skin only at the time of symptoms or at prearranged intervals. External loop recorders are capable of storing from 30 seconds to 5 minutes of the preceding ECG continuously in a circular buffer. When the patient activates the event button, the loop is permanently stored and the device acquires a further period of ECG in real time. Looping event recordings of a 12-lead ECG signal are available but are constrained by the need to wear 10 electrodes for an extended period. Alternatively, the full 12-lead ECG can be very nearly reconstructed from limited channels by digital signal processing techniques.

The diagnostic yield of event recorders is higher than that of Holter monitors (24% to 54% versus 15% to 35%) but insufficient for patients with infrequent symptoms, especially for those with syncope.[19,20] Some technical characteristics of looping and nonlooping event recorders and Holter monitors are compared in Table 47-2.

Implantable Loop Recorders

The need for a loop recorder with a practical monitoring period of weeks to months led to the development of an implantable ECG monitor. The device is generally implanted when an arrhythmia is suspected and when the external loop recorder has not yielded a diagnosis within a month. The Reveal insertable loop recorder (Medtronic, Inc.) is implanted subcutaneously in prepectoral or submammary positions and enables the recording of an electrogram from electrodes at both ends of the device without the need for venous cannulation or

TABLE 47-2 Comparative Characteristics of Looping Event Recorders, Nonlooping Event Recorders, and Holter Recorders

	Looping Event Recorder	Nonlooping Event Recorder	Holter Recorder
Diagnostic yield for sustained palpitations	High	High	Moderate
Diagnostic yield for brief palpitations and syncope	Moderate	None	Low
ECG channels	Usually 2	Usually 1 or 2, but 12 available	Usually 2 or 3, but 12 available
ECG nature	Lead II or a Holter lead	Short bi-pole, clinically unfamiliar	Bipolar Holter lead
ECG quality	Poor if electrodes improperly applied	Poor signal if device not firmly applied during recording	Usually good quality

subcutaneous patches. The device is activated externally and records up to 40 minutes of current cardiac rhythm and rhythm preceding the activation as a continuous loop. It has a programmable memory bin selection for multiple events and real-time telemetry, and a shelf half-life of at least 14 months. Both programming and interrogation can be performed with a standard pacemaker programmer. Newer versions of this device are able to detect bradycardia or tachycardia and initiate recording automatically.

Krahn and colleagues,[21] using the Reveal device implanted in 85 patients with unexplained syncope, found that an arrhythmic mechanism was responsible for nearly one half the syncopal episodes that occurred in 58 patients. Figure 47-3 presents a rhythm strip from the Reveal device showing an episode of monomorphic VT in a patient with infrequent palpitations and in another patient, the Reveal captured an episode of sinus arrest associated with syncope.

Implantable loop recorders are believed to be superior to Holter monitoring systems and external looping devices in the diagnosis of very infrequent arrhythmic attacks, identification of asymptomatic arrhythmias, assessment of treatment efficacy, and detection of proarrhythmias. It has been suggested that using this device may improve the cost-effectiveness of the investigation in some groups of patients.

Implantable Therapeutic Devices

Modern implantable pacemakers and ICDs are equipped with relatively large memory capacities (usually 32 Kb for pacemakers and 256 Kb for ICDs). The monitoring data stored in these devices are used to evaluate the device performance and to confirm the efficacy and appropriateness of the device therapy. Episode data can be viewed in summary form to assess arrhythmia trends or as detailed interval and electrogram data. Some devices provide a long predetection buffer for a limited number of episodes to more carefully examine the events preceding the arrhythmia.

Recently, the ability of pacemakers and ICDs to identify previously unknown or unsuspected arrhythmias has been fully recognized. For example, new generation pacemakers have provided abundant data on the prevalence of atrial arrhythmias, modes of onset, duration of episodes, and correlation with symptoms in patients without a history of these arrhythmias.[22,23] In a large series of patients with an ICD implanted for ventricular tachyarrhythmias, 18% episodes of ventricular fibrillation were preceded by atrial arrhythmias.[24] Furthermore, there were 64 spontaneous ventricular episodes (tachycardia or fibrillation) in patients with atrial fibrillation without known ventricular arrhythmias.[25]

SUMMARY

Event recorder technology is advancing rapidly, and a wide range of additional features is now offered. These include increasing the amount of ECG data that can be stored, increased speed of transtelephonic transmission and transmission via the Internet, and multichannel ECG. These advances substantially increase the clinical applications of event recorders. Sensing T waves and ST segment changes, which is feasible in new implantable devices, will enable the automatic detection of myocardial ischemia and measurement of the Q–T interval. Combination of rhythm and hemodynamic monitoring will further improve the diagnostic performance of implantable devices.

EXERCISE STRESS TEST FOR THE ASSESSMENT OF ARRHYTHMIAS

Although an exercise stress test is more commonly used as a tool in the assessment of myocardial ischemia in patients with coronary heart disease, the physiologic effects of exercise may assist in the diagnosis of adrenergically mediated arrhythmias and may be useful for the assessment of the efficacy of treatment.

Supraventricular arrhythmias are fairly common during exercise. The most frequently encountered are atrial premature complexes that have been reported to occur in 5% of normal individuals and up to 40% of patients with organic heart disease.[26] The incidence of atrial tachycardias and atrial fibrillation varies between 0.1% and 2.8% and 0.3% and 1.1%, respectively.[27] In patients with permanent atrial fibrillation, the exercise test may be employed to assess control of the ventricular response during physical activity (Fig. 47-4).[28] Treatment with atrioventricular nodal blocking drugs may be titrated to the ventricular response seen on

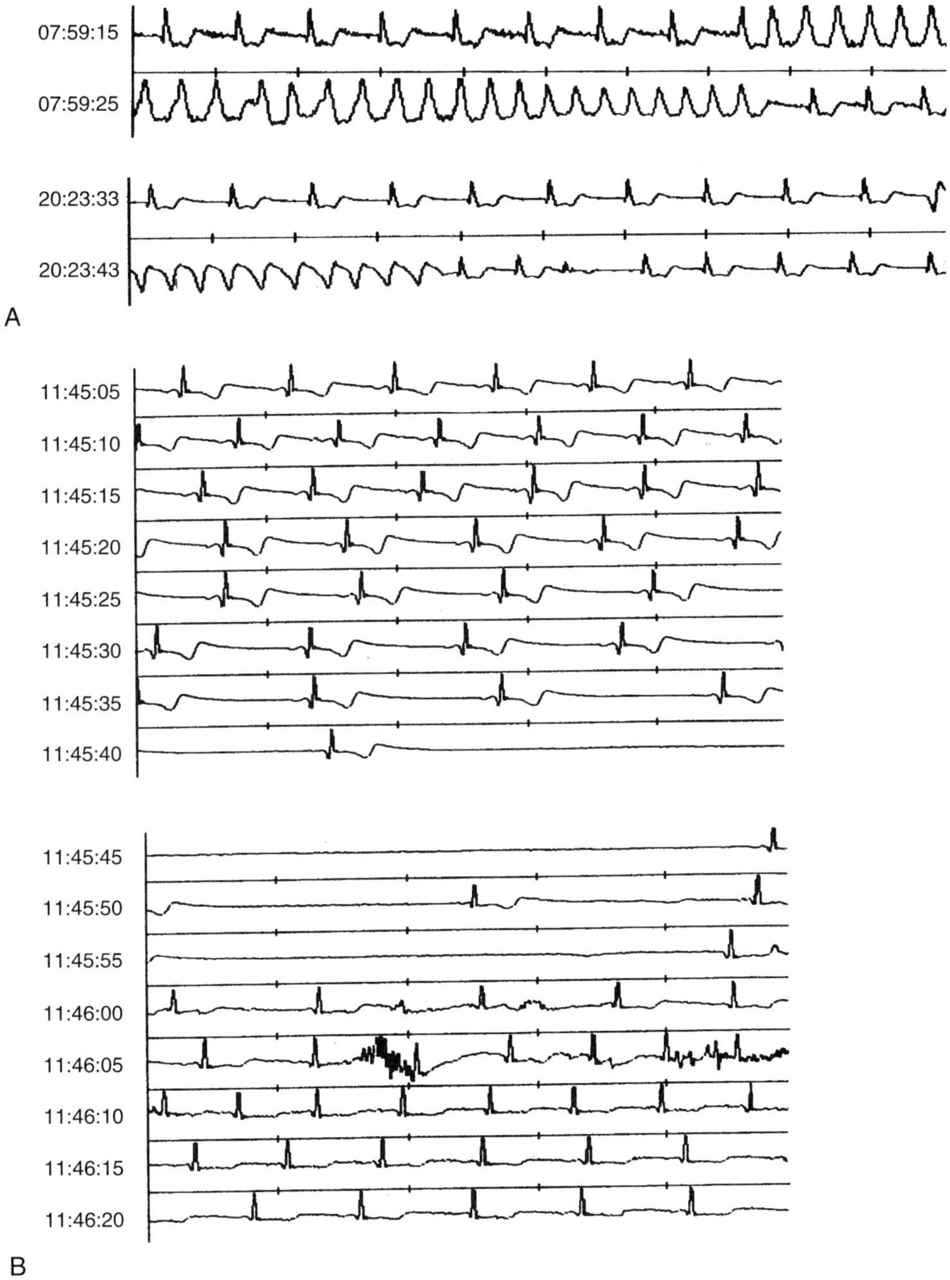

FIGURE 47-3 A rhythm strip from the Reveal device showing monomorphic ventricular tachycardia of two different morphologies in a patient with infrequent palpitations (**A**) and an episode of sinus arrest in a patient with unexplained syncope (**B**).

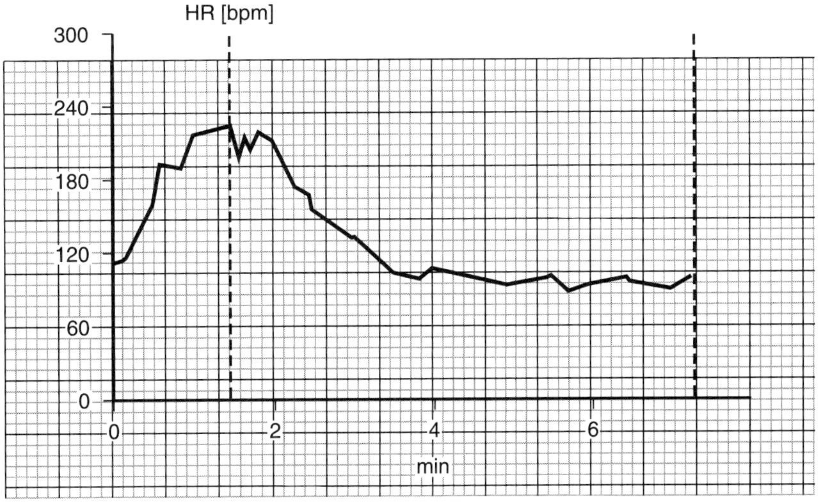

FIGURE 47-4 Tachogram showing inadequate ventricular rate control during exercise in a patient with permanent atrial fibrillation. After 90 seconds of exercise, ventricular rate are is 200 to 230 beats/minute.

exercise testing, thus allowing therapy to be individually optimized. Lack of increase in heart rate at peak exercise and an abrupt fall in heart rate following exercise may indicate chronotropic incompetence suggestive of sinus node or autonomic dysfunction.[29]

The incidence of complex or repetitive ventricular activity during exercise varies from 0% to 2.4% in apparently healthy individuals to 31% in patients over 50 years with structural heart disease.[26,30] Sustained VT is relatively uncommon (0.15% to 1.7%),[31,32] with the exception of right ventricular outflow tract VT or idiopathic nonischemic VT, which may be induced during exercise in as many as 62% of patients.[33,34]

In the prethrombolytic era, exercise-induced ventricular tachyarrhythmias at predischarge testing were strongly associated with increased one-year mortality: 15% versus 7% in those without arrhythmias.[35] However, this trend has changed since the widespread introduction of early revascularization,[36] rendering the exercise test less useful for identification of patients at high risk for arrhythmias and limiting its use for investigation of T wave alternans for risk stratification.

TILT TEST

Introduced in 1986,[37] the tilt test allows observation of the evolution of syncopal episodes resulting from orthostatic stress, such that vagovagal and dysautonomic responses may be distinguished. *The vasovagal response* occurs after some period of hemodynamic stability, indicating relatively normal autonomic nervous system function, followed by reflex-mediated vasodepression and bradycardia. The abrupt decrease in blood pressure is caused by the Bezold-Jarisch reflex and likely represents a hypersensitive response to an otherwise normal stimulus.[38] The triggering mechanism in vasovagal syncope is believed to be due to central hypovolemia resulting from blood pooling in the lower extremities. The afferent arc of this reflex may be mediated by activation of mechanoreceptors of the heart in response to vigorous contraction in the presence of low diastolic filling. As a result, the efferent sympathetic stimulation decreases and vagally mediated bradycardia occurs (Fig. 47-5).

The dysautonomic response has a multifactorial etiology and is characterized by a gradual decrease in blood pressure without significant changes in heart rate, in which case patients are unable to compensate for the acute reduction in venous return, resulting in true orthostatic hypotension.[39] This pathophysiologic response has multiple etiologies, including Parkinson's disease, multisystem atrophy, and pure autonomic failure.

Finally, in *postural orthostatic tachycardia syndrome*, a sudden and significant increase in heart rate is associated with only mildly reduced blood pressure.[40] The underlying mechanism is a failure of the peripheral vasculature to appropriately constrict in response to orthostatic stress, which is compensated by an excessive increase in heart rate. Differences in mechanisms of vasovagal and dysautonomic responses observed during the tilt test have important implications for treatment selection.

A tilt test generally is performed for 30 to 45 minutes at an angle 60° to 80°, with 70° being most common.[41] The preference for a 45-minute test is based on two standard deviations beyond the mean time to syncope, which has been shown to identify 95% of patients who are likely to develop syncope during testing. The appropriate end-point for a tilt test is induction of syncope or presyncope associated with hemodynamically significant hypotension and/or bradycardia that results in an inability to maintain postural tone.[42] The sensitivity of the tilt test has been reported to be 32% to 85% and the specificity, in the absence of pharmacologic provocation, 90%. The sensitivity of the test may be improved, at the cost of specificity, by using longer tilt durations, steeper angles, and provocative agents, such as nitroglycerin and isoproterenol. Nitroglycerin is given sublingually at a dose of 300 to 400 μg at the end of a 30- to 45-minute tilt test while the patient is still upright. The test is continued for another 20 minutes.[43] Isoproterenol is administered intravenously at a dose of 1 to 3 μg/minute at the end of a 30- to 45-minute tilt test with the patient returned to a supine position.[44] After achieving hemodynamic steady state, the patient is tilted head-up for 5 to 15 minutes and then returned to the supine position before the isoproterenol dose is further increased.

In 1996, the Expert Consensus on the use of tilt testing was published by the American College of Cardiology, stating that the test should be done in patients with recurrent syncope or in high-risk patients after a single syncopal episode.[41] Although the tilt test is used primarily in symptomatic patients without apparent structural heart disease, it may also be used in patients with organic heart disease, especially if other methods, including electrophysiological study, have failed to establish the diagnosis.

Assessment of Ventricular Depolarization

TIME DOMAIN ANALYSIS OF SIGNAL-AVERAGED ELECTROCARDIOGRAM

Methodology

In 1973, Berbari and coworkers[45] using the signal-averaging technique, first recorded the fragmented continuous electrical activity of ischemic myocardium from the body surface that was associated with reentrant VT. Eight years later, Simson[46] applied high-pass bidirectional filters to the signal-averaged ECG and suggested the criteria for ventricular late potentials that since then had been widely used in clinical practice. These include the duration of the filtered QRS complex greater than 120 ms, the duration of low amplitude (<40 μV) signals in the terminal part of the filtered QRS complex greater than 40 ms, and the mean root square voltage of the last 40 ms of the filtered QRS complex less than 25 μV for 25-Hz filter setting. The corresponding values for 40-Hz filter setting are 114 ms, 38 ms, and 20 μV, respectively.[47]

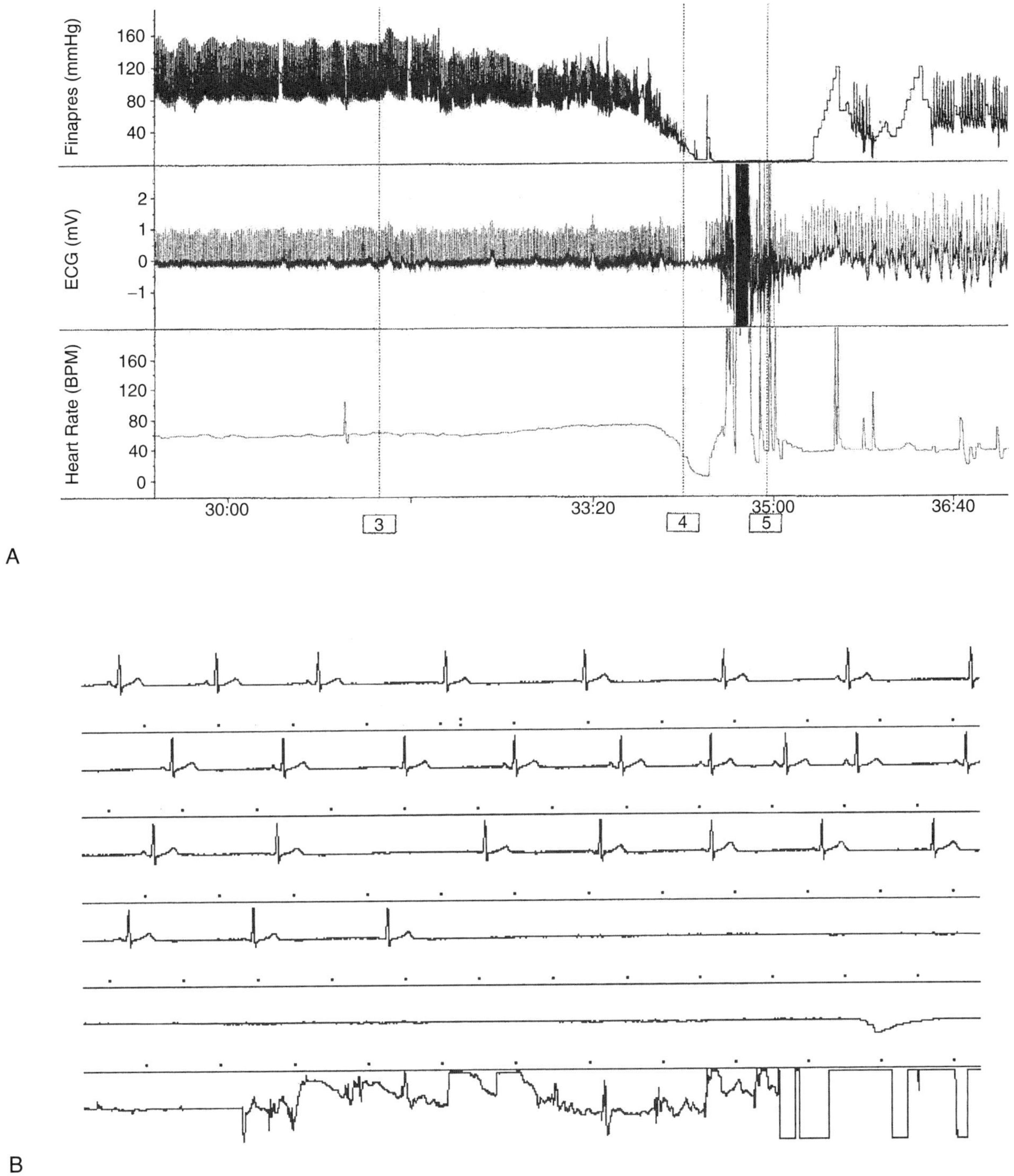

FIGURE 47-5 Vaso-vagal syncope during the tilt test. Blood pressure continuously recorded with a Finapress device and QRS amplitude and heart rate trends demonstrate the development of hypotension and bradycardia resulting in syncope (**A**). A rhythm strip shows a prolonged episode of asystole (**B**). Sinus rhythm and normal blood pressure restored after the patient was returned to a supine position.

The most often used technique for recording of late potentials is high gain amplification (usually × 10,000), complex bidirectional high frequency (25 to 250 Hz or 40 to 250 Hz) filtering, and averaging of approximately 200 to 300 identical beats to improve the signal-to-noise ratio by reducing the random noise to 0.3 μV or less. An orthogonal bipolar XYZ lead set is most commonly used. Consecutive ventricular complexes are compared with the template QRS complex and a predetermined value for coefficient of correlation with the template QRS (usually $r > 0.95$) is used to accept or to reject the QRS complex.

Although in 1991, the Task Force Committee of the European Society of Cardiology, the American Heart Association and the American College of Cardiology has made an attempt to standardize the signal-averaging

ECG techniques to achieve comparable results obtained in different laboratories,[47] there are still controversies as to normal and abnormal values in different clinical settings, filters, type of windows for ECG analysis, lead systems, and general approaches to the recording and analysis of the data.

Clinical Studies

In 1983, Breithardt and associates[48] published the first study that recognized the prognostic importance of the signal-averaged ECG in postmyocardial infarction patients. The prevalence of late potentials after myocardial infarction has been reported to be 25% to 30% in the absence of ventricular arrhythmias and 67% to 100% in the presence of sustained VT.[48-51] In patients with a history of polymorphic VT or ventricular fibrillation, late potentials are less frequent, which probably reflects the differences in the arrhythmogenic substrate and mechanisms of these arrhythmias.[52]

A meta-analysis of seven studies in more than 1000 postmyocardial infarction patients demonstrated a six- to eightfold increase in risk of arrhythmic events in the presence of the abnormal signal-averaged ECG.[53] Table 47-3 summarizes select studies showing the ability of the signal-averaged ECG to predict arrhythmic events and sudden death in postmyocardial infarction patients.[48-51,54-58,59-61]

The restoration of the patency of the infarct-related coronary artery with thrombolysis or primary angioplasty and the wide use of surgical revascularization have modified arrhythmogenic substrate in postmyocardial infarction patients, resulting in significant changes in the prevalence and characteristics of arrhythmic events which, in turn, have led to a noticeable reduction in the predictive power of the signal-averaged ECG.[60] It has been suggested that the presence of late potentials as a single marker is no longer useful for the identification of postmyocardial infarction patients at risk.[61-63] However, a high negative predictive value of 89% to 99% rendered the signal-averaged ECG a useful tool for differential diagnosis of unexplained syncope[64] and broad complex tachycardias.[65]

Of note, among the three signal-averaged ECG parameters, the duration of the filtered QRS complex has proved the best to identify patients with sustained ventricular tachyarrhythmias. In the PILP (Post Infarction Late Potential) study, the combination of the filtered QRS complex less than or equal to 103 ms and the root mean square of successive R–R interval (RR) differences greater than or equal to 36 ms identified patients without a risk for arrhythmic events with a 100% specificity.[55,56] Recent data from the MUSTT (Multicenter Unsustained Tachycardia Trial) signal-averaged ECG substudy suggest that the filtered QRS complex duration >114 ms is a powerful predictor of arrhythmic death and cardiac death (relative risks 1.90 and 1.73, respectively), although treatment assignment in MUSTT was not based on the results of the signal-averaged ECG.[58] The combination of abnormal signal-averaged ECG and low ejection fraction identified a very high-risk patient group that constituted nearly one quarter of the total study population and 36% of which succumbed to arrhythmic death.

The noninvasive electrocardiology substudy of the MADIT II study has been designed to determine

TABLE 47-3 Select Studies of Prognostic Significance of Signal-Averaged ECG in Patients with Myocardial Infarction

Study	Number of Patients	Follow-up, Months	Thrombolysis or PCA	Number of Events	Sensitivity, %	Specificity, %	PPV, %	NPV, %
Breihardt et al,[48] 1986	511	18	No	30	79	63	6	97
Kuchar et al,[49] 1987	210	14	No	15	93	65	17	99
Gomes et al,[50] 1987	102	12	No	15	87	63	29	96
Bloomfield et al,[51] 1996	177	14	No	16	69	62	15	95
Pedretti et al,[54] 1993	305	15	Yes	19	79	75	17	98
Makijavri et al,[55] 1995	778	6	Yes	33	76	63	8	98
Denes et al,[56] 1994	787	10	Yes	33	60	98	20	89
Zimmerman et al,[57] 1997	458	70	Yes	32	44	83	20	94
Gomes et al,[58] 1999	1268	39	Yes	236 AD/CA 341 CD	Not stated	Not stated	24 34	86 78
Steinbigler et al,[59] 1999	1120	30	Yes	88	56	68	14	89
Gold et al,[225] 2000	313	12	Yes	22	56	83	47	88
Ikeda et al,[226a] 2002	834	25	Yes	67	50	84	10	98

AD, arrhythmic death; CA, cardiac arrest; CD, cardiac death; NPV, negative-predictive value; PCA, primary coronary angioplasty; PPV, positive-predictive value. Note a progressive decrease in sensitivity of the signal averaged ECG for prediction of arrhythmic events in the thrombolytic era.

whether 12-lead ECG- and Holter-based risk factors can identify patients at high risk of death in the conventional treatment group and patients who are likely to benefit from ICD therapy. The results have not yet been published, but the preliminary data suggest that QRS duration greater than 120 ms on a standard or signal-averaged ECG conferred a nearly twofold increased risk for total mortality in the conventional arm and predicted appropriate ICD discharge in the ICD arm. In the total study population, ICD therapy on top of optimal conventional treatment reduced a risk of death by 31%, while patients with QRS greater than 120 ms benefited a remarkable 63% risk reduction.

A nearly linear relationship between the duration of unfiltered QRS complexes and annual mortality rates suggests that QRS duration measured on the 12-lead ECG may serve as a readily available clinical risk factor. Prolonged QRS duration greater than or equal to 130 ms, present in 33.6% of 915 patients with ventricular arrhythmias, was associated with a twofold increase in mortality.[58a] For every 10 ms increase in QRS duration, there was a 10% increase in mortality rate (Figure 47-6).

Finally, the results of the Munich and Berlin Infarct Study (MABIS) of 1120 patients with myocardial infarction have shown that the detection of *functional*, or dynamic, late potentials during the periods of reduced heart rate variability, increased heart rate or ST segment shift were associated with significantly improved sensitivity, specificity, and positive-predictive accuracy (77%, 92%, and 40%, respectively), with a negative-predictive value of 98%.[59]

The usefulness of late potentials in other clinical settings, including nonischemic dilated cardiomyopathy and hypertrophic cardiomyopathy has not yet been established. Although the specificity of this method to identify high-risk patients was generally satisfactory, the sensitivity and positive-predictive accuracy were too low to recommend the use of this technique as a noninvasive test for risk stratification.[66,66a,67]

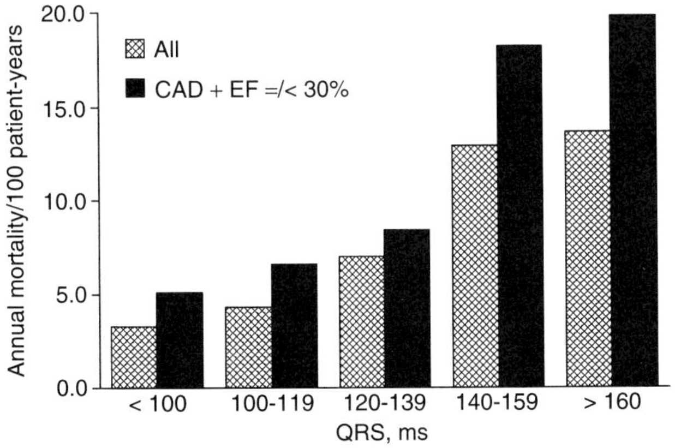

FIGURE 47-6 QRS duration and annual mortality rates in patients who underwent risk stratification for ventricular arrhythmias. The hatched bars represent all patients; the black bars represent patients with coronary artery disease (CAD) and an ejection fraction of less than or equal to 30%. (Modified from Kalahasti V, Nambi V, Martin DO, et al: QRS duration and prediction of mortality in patients undergoing risk stratification for ventricular arrhythmias. Am J Cardiol 2003;92:798-803.)

SPECTRAL ANALYSIS OF SIGNAL-AVERAGED ELECTROCARDIOGRAM

There is considerable contemporary interest in the development of new methods for the detection of the entire depolarization wavefront propagation abnormalities, because the electrophysiological arrhythmogenic substrate involved in VT circuits is mostly activated during the QRS complex and not in the early ST segment. Therefore, it is necessary to interrogate a larger portion of the QRS complex to identify more precisely arrhythmogenic areas of the myocardium.

To overcome the limitations of the time domain analysis of the signal-averaged ECG, various types of spectral analyses of signal-averaged ECG have been extensively investigated, among which are spectral temporal mapping,[68] spectral turbulence analysis,[69] analysis of intra-QRS potentials,[70] and combined spatial and spectral analyses of the entire cardiac cycle.[71] However, these methods did not yield the diagnostic and prognostic values superior to those of the conventional time domain analysis in large prospective studies.

Spectral temporal mapping, which is based on the assessment of the consistency of high frequency components within the terminal portion of the QRS complex, lacks the ability to identify the arrhythmogenic substrate in the entire QRS complex. Spectral turbulence analysis, which allows quantification of spectral changes in the entire high-gain QRS complex, has proved useful in the subset of patients with ventricular conduction defects, but in the absence of conduction abnormalities its predictive value is lower than that of the time domain analysis. Furthermore, it seems likely that beat-to-beat analysis of depolarization patterns is more appropriate than static signal averaging. Hence, new methods pertaining to the identification of the electrophysiological arrhythmogenic substrate should concentrate on a combination of signal averaging, allowing the best possible signal-to-noise ratio, with a beat-to-beat analysis of the QRS complexes.

WAVELET DECOMPOSITION ANALYSIS OF SIGNAL-AVERAGED ELECTROCARDIOGRAM

Wavelet decomposition analysis (WDA) enables the detection of low amplitude and transient perturbations within the signal-averaged QRS complex, which reflect the inhomogeneity of depolarization wave as it propagates through areas with slow conduction. First introduced in electrocardiography by Morlet in 1993,[72] WDA of the signal-averaged ECG may provide improved risk stratification for ventricular tachyarrhythmias.[73,74]

Methodology

The principle difference between WDA and fast Fourier transformation is a special window function using two parameters: a dilatation, which determines the width of the filter window, and a scale, which establishes the frequency score of the window. Their combination enables the design of a window with special quality: the length of the window is not fixed but varies according to the value of scale and dilatation, automatically adjusting to the frequency of components of the signal. The narrow window is used for the detection of high frequency components (small scales) and the extended window is used for the detection of low frequency components (large scales). The window function is called the *analyzing wavelet.* Computer implementation of WDA uses the Morlet analyzing wavelet with 54 scales that cover the range of 40 to 250 Hz.

Heterogeneity of depolarization wavefront propagation is described by four parameters:

1. Wavelet complex length (WCL), which corresponds to the duration of the wavelet decomposition of the QRS complex (Fig. 47-7)[75];
2. Wavelet maxima count (WMC), which corresponds to the total number of local maxima, or peaks, in wavelet curves of all scales within the WCL interval and detects conduction singularities within the QRS complex;
3. Wavelet surface area (WSA), which reflects the curvature and sheerness on the surface of the three-dimensional wavelet and reveals irregularities mainly in the middle portion of the QRS complex; and
4. Wavelet relative length (WRL), which represents the mean length of wavelet curves of all scales normalized to the height within the WCL and characterizes disturbances and irregularities in the initial and the terminal portions of the signal-averaged QRS complex.

Normal values of WDA parameters obtained from 104 healthy volunteers (54 men, mean age 50 ± 17 years) are presented in Table 47-4. Figure 47-8 shows examples of WDA of the signal-averaged ECG, accompanied by the results of the conventional time domain analysis.

The reproducibility of WDA parameters is at least as good as the conventional time domain parameters, with

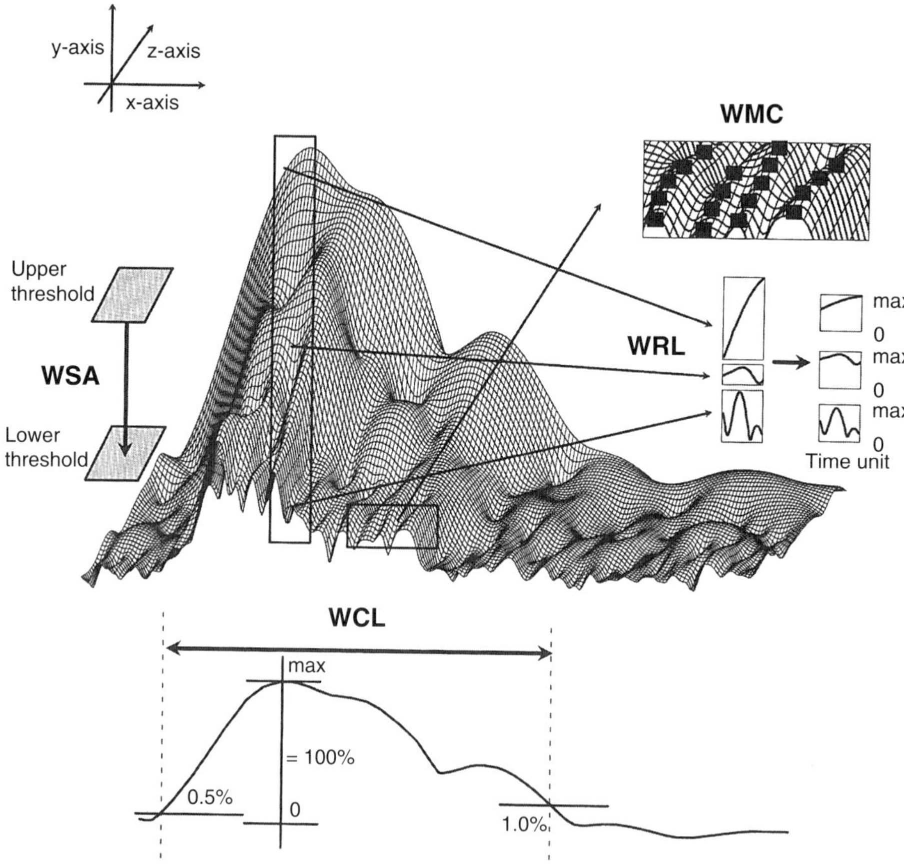

FIGURE 47-7 Derivation of wavelet decomposition parameters. The horizontal X axis shows time in ms, the vertical Y axis shows the wavelet vector magnitude in the absolute units, and the sagittal Z axis shows scale (frequency) in Hz. WCL is derived from the average of wavelet curves of individual scales, normalized within an interval of 0.1, where the highest value corresponds to the maximum. WCL is the duration of the interval between the start point, where the averaged wavelet curves fall below 0.5% of the maximum, and the end-point, where the averaged wavelet curves fall below 1% of the maximum. WCL measures the duration of wavelet decomposition of the QRS complex. A greater WCL indicates the more complex and delayed depolarization. WMC is the number of local wavelet maxima in wavelet curves of all scales within the WCL and detects conduction singularities within the decomposed QRS complex. WSA reflects the curvature and sheerness on the surface of the three-dimensional wavelet by counting its surface in a horizontal bound limited by fixed lower and upper thresholds, and reveals irregularities mainly in the middle portion of the QRS complex. Increased WMC and WSA are associated with greater conduction abnormalities. WRL represents the mean length of wavelet curves of all scales normalized to the height within the WCL and characterizes disturbances and irregularities in the initial and the terminal portions of the signal-averaged QRS complex. The more irregular and bumpy the wavelet curves, the lower the value of WRL. WCL, wavelet complex length; WMC, wavelet maxima count; WRL, wavelet relative length; WSA, wavelet surface area.

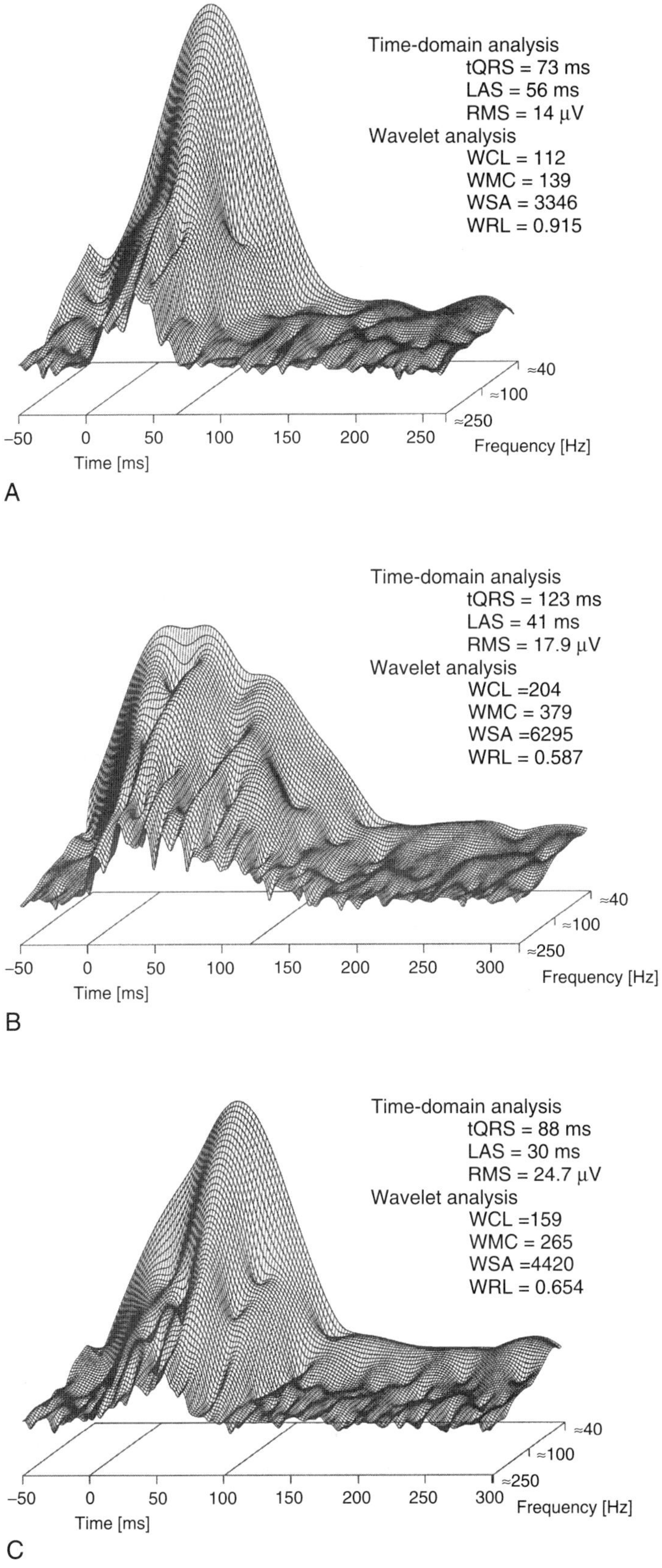

FIGURE 47-8 Examples of wavelet decomposition of the signal-averaged ECG, accompanied by the results of the conventional time domain analysis, in a healthy control with normal wavelet decomposition and time domain parameters (**A**), a patient with a remote myocardial infarction and sustained monomorphic ventricular tachycardia who presents with both wavelet decomposition and time domain parameters being substantially abnormal (**B**), and a patient with myocardial infarction and cardiac arrest in whom all time domain parameters are within normal values (**C**). Wavelet decomposition reveals conduction abnormalities hidden inside the QRS complex. Note a narrow and smooth surface of the wavelet decomposition in a healthy control compared with a highly irregular and bumpy surface in a patient with ventricular tachycardia. In a patient with cardiac arrest, the surface of wavelet decomposition contains some irregularities of the wavelet curves; WMC and WSA are increased, indicating depolarization inhomogeneity. TQRS, total filtered QRS complex; LAS, low amplitude signals in the terminal portion of the QRS complex; RMS, root mean square voltage of the terminal 40 ms of the QRS complex; WCL, wavelet complex length; WMC, wavelet maxima count; WRL, wavelet relative length; WSA, wavelet surface area. (Data from Hnatkova K, Malik M, Kulakowski P, Camm AJ: Wavelet analysis of signal-averaged electrocardiograms. I. Design, clinical assessment, and reproducibility of the method. Ann Noninvas Electrocardiol 2000;5:4-19.)

TABLE 47-4 Wavelet Decomposition Parameters in Normal Population

Parameter [absolute units]	Mean ± SD	Normal Limit
WCL	134.21 ± 11.65	153
WMC	193.23 ± 30.32	241
WSA	4031 ± 358.5	4610
WRL	0.7756 ± 0.0782	0.6543

WCL, wavelet complex length; WMC, wavelet maxima count; WRL, wavelet relative length; WSA, wavelet surface area. (Data from Hnatkova K, Malik M, Kulakowski P, Camm AJ: Wavelet analysis of signal-averaged electrocardiograms. I. Design, clinical assessment, and reproducibility of the method. Ann Noninvasive Electrocardiol 2000;5:4-19.)

the best reproducibility being for the total filtered QRS duration in the standard signal-averaged ECG and the wavelet surface area derived from WDA.

Clinical Studies

In 551 postmyocardial infarction patients from the St. George's Hospital Post Infarction Research Survey Programme, the WDA parameters were consistently different between patients who died from a cardiac cause or developed ventricular tachycardia or cardiac arrest and those in whom a two-year follow-up was uneventful.[75] Although WDA marginally improved the identification of patients at risk of VT compared with the conventional time domain analysis of the signal-averaged ECG, it acted as a much stronger predictor of sudden cardiac death and total cardiac mortality.

Very recently, the ability of WDA to identify patients at high risk of ventricular tachyarrhythmias and sudden cardiac death has been evaluated in 246 patients with hypertrophic cardiomyopathy who were followed for 68 months[76] and 82 patients with idiopathic dilated cardiomyopathy followed for 23 months.[77] In both subsets, WDA failed to predict risk of arrhythmic events but identified patients who died nonsuddenly and who developed progressive heart failure. A possible explanation is that, in these clinical settings, ventricular arrhythmias may occur due to mechanisms other than reentry, and that WDA measurements may merely reflect abnormalities, arising from fibrotic myocardium, which per se, may not form a substrate for ventricular tachyarrhythmias.

SUMMARY

Wavelet decomposition is a valid alternative approach to the analysis of the signal-averaged ECG that, in some aspects, is superior to the conventional time domain analysis. In particular, WDA enables better identification of patients at risk of sudden death. It has been recognized that postmyocardial infarction risk stratification should be multifactorial and should include a combination of time domain and wavelet decomposition analyses of the signal-averaged ECG with other established risk factors. However, it is plausible that the *depolarization-based* risk stratifiers are largely independent of other risk factors that are mainly based on the ventricular performance, autonomic cardiac status, triggered activity, and repolarization abnormalities.

P WAVE SIGNAL AVERAGING

An important consequence of unraveling the multiwavelet reentry as the most common mechanism of atrial fibrillation has been delineation of the role of the P wave signal-averaged ECG as a marker of the arrhythmia. Similar to the prolonged filtered QRS complex on ventricular signal-averaged ECG which is a recognized marker of global ventricular conduction delay and reentrant ventricular arrhythmias, a prolonged duration of the filtered P wave on atrial SAECG has been proposed for noninvasive identification of global atrial conduction delay and vulnerability to atrial tachycardias and fibrillation.[78-80]

In a prospective study of atrial fibrillation occurring after cardiac surgery, two factors were found to predict the arrhythmia: preoperative left ventricular ejection fraction and the filtered P wave duration of greater than 140 ms, providing a sensitivity of 77% and a positive-predictive accuracy of 37%.[79] The likelihood of the development of atrial fibrillation after cardiac surgery was increased almost fourfold if the P wave on the preoperative signal-averaged ECG was greater than 140 ms, irrespective of other clinical variables. An abnormal P wave signal-averaged ECG predicted the occurrence of atrial fibrillation in patients with congestive heart failure with a 90% sensitivity and a 69% specificity.[80] In this study, abnormal P wave signal-averaged ECG was associated with an 18-fold increase in risk of atrial fibrillation.

A potential usage of P wave signal-averaged ECG includes, but is not restricted to, the prediction of recurrence of the arrhythmia after cardioversion for atrial fibrillation,[81] the prediction of transition to a permanent form of the arrhythmia,[82] identification of sinus node dysfunction,[83] and assessment of stroke of unknown etiology,[84] and possibly, evaluation of the efficacy of antiarrhythmic drug therapy. However, lack of standardization of the technique and lack of uniformity in definitions of the abnormal P wave signal-averaged ECG,[85] low positive predictive value and a wide spread of cutoff points for discrimination between patients with and without atrial fibrillation (dichotomized values for the filtered P wave duration vary from 110 ms to 155 ms), substantial errors of automatic measurements,[86] and poor reproducibility[87,88] have led to hindering the applicability of the P wave signal-averaged ECG.

Recently, the frequency domain analysis has been proposed to identify patients with early recurrence of atrial fibrillation following cardioversion.[89] However, conflicting results yielded with the above-mentioned techniques and methodologic drawbacks have led to delineation of the importance of P wave morphology analysis.[90]

Assessment of Cardiac Autonomic Status

Since the 1980s, experimental and clinical studies have established the pervasive impact of the autonomic nervous system on cardiovascular morbidity and mortality and have provided strong evidence in favor of different autonomic tests for risk stratification for arrhythmia and sudden cardiac death.

Experimental Basis for Autonomic Tests

There are a number of ways in which disturbed autonomic activity can affect the cardiovascular system, including both direct and mediating factors. Experimental and clinical data have strongly implicated an increased sympathetic activity and loss of vagal influences in triggering life-threatening arrhythmias.[91-96] An increase in sympathetic activity following experimental acute myocardial ischemia is associated with a reduction in the ventricular fibrillation threshold and an augmented vulnerability to ventricular arrhythmias, an effect that can be realized through different direct electrophysiological mechanisms, including increased automaticity in Purkinje fibers, enhanced early afterdepolarizations, and increased dispersion of refractoriness, resulting in disturbances in impulse formation, conduction, or both.[97] Indirect effects of excessive sympathetic stimulation include increased cardiac metabolic activity, coronary vasoconstriction, and impaired oxygen demand-to-supply ratio.

Extensive experimental studies have demonstrated the cardioprotective effect of vagal stimulation, associated with an increase in the ventricular fibrillation threshold during acute myocardial ischemia superimposed on a healed myocardial infarction.[94,95] Antifibrillatory mechanism of vagal stimulation stems from the antagonism of detrimental effects of excessive sympathetic activity.[91,98] The molecular and cellular bases for this antagonism appear to be presynaptic inhibition of norepinephrine release and activation of muscarinic receptors that attenuate the response to catecholamines at the receptor site. During sinus tachycardia, a decrease in heart rate caused by vagal stimulation has been shown to inhibit the susceptibility to ventricular fibrillation by increasing diastolic perfusion time and reducing myocardial oxygen deficit.

Autonomic Imbalance After Myocardial Infarction

Experimental studies have delineated the arrhythmogenic nature of sympathetic stimulation and the cardioprotective effects of vagal activity. The likely importance of autonomic imbalance in postmyocardial infarction settings stimulated the development of autonomic markers for postinfarction risk stratification. Apart from profibrillatory sympathetic and antifibrillatory vagal effects in the acute phase of myocardial infarction, cardio-cardiac sympatho-vagal reflexes are likely to be involved at later stages. Changes in myocardial geometry due to the presence of necrotic and noncontracting segments may increase afferent sympathetic stimulation by mechanical distortion of afferent fibers. This may blunt the baroreceptor-mediated reflex, interacting with efferent vagal stimulation.[98] Noteworthy, postmyocardial infarction changes in cardiac autonomic status may also involve alterations of sympathetic reflexes, but to a lesser degree than those of vagal responses, resulting in autonomic imbalance with a shift to the sympathetic dominance.

Noninvasive Assessment of Cardiac Autonomic Nervous System Activity

Sympathetic and vagal activities and their roles in the genesis of ventricular tachyarrhythmias have mainly been investigated by stimulating or blocking the neural pathways to the heart or by administering various agents that are known to interfere with autonomic balance. Clinical assessment of cardiac autonomic status has been limited to the analysis of instantaneous heart rate and blood pressure changes in response to external stimuli that were thought to modify cardiac autonomic reflexes. These include the deep breath test, Valsalva's maneuver, cold pressor test, isometric handgrip, and orthostatic (tilt) test.[99]

Assessment of heart rate was the earliest and the easiest approach to the evaluation of sympathovagal balance. It has long been recognized as an indicator of health status. Resting heart rate appears to be a strong prognostic factor for all-cause mortality and sudden cardiac death in large population-based studies.[100-104] Mean heart rate measured in two hours of ambulatory ECG was significantly associated with all-cause mortality and cardiac events in an elderly cohort of the Framingham Heart Study.[101] In a study of 7735 British men, who were followed-up for eight years, a 4.8 age-adjusted relative risk has been found for a resting heart rate greater than beats per minute compared with a heart rate of less than beats per minute.[104] In postmyocardial infarction participants in St. George's Post Infarction Research Survey Programme, the mean R–R interval calculated from the 24-hour Holter ECG and dichotomized at 800 ms (which corresponds to a heart rate of 75 beats/minute) independently predicted total mortality in multivariate analysis.[105] These observations suggest that heart rate itself can be a target for risk prevention interventions.

Recently, *heart rate variability* (HRV) and *baroreflex sensitivity* (BRS) permitting the quantification of the cardiac autonomic status have been introduced. Both methods provide different facets of a comprehensive insight into sympathovagal cardiovascular interaction, the former representing the *tonic* and the latter assessing the *phasic* activities of the autonomic nervous system. HRV and BRS evaluate the integrity of cardiovascular autonomic innervation, the physiologic status of autonomic activity, and the vulnerability to cardiac arrhythmias. Considerable success in the identification of patients at high risk of life-threatening arrhythmias has been reported when employing these methods in large cohorts of postmyocardial infarction patients.[106,107]

HEART RATE VARIABILITY

Nearly 70 years ago, Samaan showed that section of vagal nerves abolished respiratory sinus arrhythmia and that autonomic nervous system determined not only the rate but also the rhythm of the heart.[108] In 1981, Akselrod and colleagues[109] demonstrated that muscarinic and/or β-adrenergic blockade suppressed different components of spectral heart rate fluctuations and provided a physiologic basis for investigations of HRV.

The first report by Wolf and coworkers[110] on the association of low HRV with poor prognosis in myocardial infarction, went almost unnoticed and was rediscovered only after the association of reduced HRV with poor survival was observed in the MPIP (Multicenter Post Infarction Program) study.[106] Published in 1996, the report of the European Society of Cardiology and North American Society of Pacing and Electrophysiology Task Force established standards for HRV measurements, physiologic interpretation, and clinical use.[111] Although substantial development has occurred since its publication, the report still provides a reasonable standard to follow.

Methodology

In the initial studies, HRV assessment was based on rather simple formulas, for example, the absolute or relative difference between the maximal and minimal R–R intervals during a variety of provocative tests. Contemporary methods of HRV measurements can be divided into time domain, which include statistic and geometric methods, frequency domain, spectral methods (Table 47-5), and nonlinear techniques.

The statistical methods quantify R–R interval variations usually over 24-hour ECG recording. A simple variable is the standard deviation of all normal R–R intervals (SDNN), which represents all cyclic components responsible for sinus variability during the recording and corresponds to the total power of spectral analysis. Other commonly used statistical indices include the standard deviation of the average of normal R–R intervals calculated over five-minute periods (SDANN), which expresses long-term variations with frequency below 0.0033 Hz. While SDNN and SDANN measure HRV in the sequence of R–R intervals irrespective of their order, other statistical indices (rMSSD and pNN50) investigate the beat-to-beat differences between adjacent R–R intervals and are believed to estimate high frequency variations in heart rate. The normal values for most commonly used HRV parameters are listed in Table 47-6.

Geometric methods are less affected by noise and artifacts in the computerized processing of long-term ECGs.[112] Using this approach, HRV is assessed from basic measurements of a geometric pattern, specifically from the R–R interval distribution. The simplest geometric method is the HRV triangular index that is determined by dividing the total number of all R–R intervals by the number of R–R intervals in the modal frequency.

Visual analysis of the patterns of R–R interval dynamics can also be based on classification of images created by plotting the length of each R–R interval against the length of the preceding R–R interval. These images are known as the Lorenz or Poincarè plots. Preserved physiologic HRV produces a relatively wide spread plot sometimes called the *fan-shaped* plot, whereas a decreased HRV results in a compact, or *narrow-shaped* plot.[113]

Spectral analysis of HRV aims at separating different frequency components of sinus rhythm modulations and at quantifying each of these components in numeric terms.[114] The nonparametric approach to the computation of spectral components is most frequently based on the fast Fourier transform, whereas the parametric approach uses an autoregressive model.

TABLE 47-5 Select Measures of Heart Rate Variability

Parameter	Definition	
	Statistical Measures (Time Domain)	
SDNN [ms]	Standard deviation of all sinus rhythm R–R intervals (N–N intervals)	
SDANN [ms]	Standard deviation of the mean of N–N intervals in each 5-min segment of the entire recording	
SDNN index [ms]	Mean of the standard deviations of all N–N intervals in each 5-min segment of the entire recording	
rMSSD [ms]	Mean root square of the successive differences between adjacent N–N intervals	
SDSD [ms]	Standard deviation of the differences between adjacent N–N intervals	
NN50 count	Total number of pairs of adjacent N–N intervals that differ by >50 ms	
pNN50 [%]	NN50 normalized by the total number of all N–N intervals	
AVGNN [ms]	Mean of all N–N intervals	
CV	Coefficient of variance = the mean (SDNN/AVGNN) in each 5-min segment of the entire recording	
	Geometric Measures (Time Domain)	
HRV triangular index	Total number of all N–N intervals divided by the height of the histogram of all N–N intervals	
TINN	Baseline width of the minimum square difference triangular interpolation of the highest peak of the histogram of all N–N intervals	
	Frequency Domain Measures	
	Cycle Length	**Frequency Range**
Total power [ms^2]	24 cycles/min–1 cycle/24 hr	1.15×10^{-5} – 0.4 Hz
Ultra-low frequency power [ms^2]	5 min–24 hr/cycle	1.15×10^{-5} – 0.033 Hz
Very low frequency power [ms^2]	25 sec–5 min/cycle	0.0033–0.04 Hz
Low frequency power (LF) [ms^2]	6.7 sec–25 sec/cycle	0.04–0.15 Hz
High frequency power (HF) [ms^2]	9–24 cycles/min	0.15–0.4 Hz
LF/HF ratio	—	—

TABLE 47-6 Normal Values of Measures of Heart Rate Variabilities Frequently Used in Clinical Studies

HRV Index	Normal Value
Statistical Measures (Time Domain)	
SDNN [ms]	141 ± 39
SDANN [ms]	127 ± 35
rMSSD [ms]	27 ± 12
Geometric Measures (Time Domain)	
HRV triangular index	27 ± 15
Frequency Domain Measures	
Total power [ms^2]	3466 ± 1018
Ultra-low frequency power [ms^2]	1170 ± 416
Very low frequency power [ms^2]	975 ± 203
Low frequency power (LF) [ms^2]	54 ± 4
High frequency power (HF) [ms^2]	29 ± 3
LF/HF ratio	1.5–2.0

HRV, heart rate variability.

Despite mathematic differences, both methods assume relative stationarity of R–R interval modulations that contain inherent periodicity during the entire analyzed period. The high-frequency spectral components mainly reflect vagal modulations of the cardiac rhythm; the low frequency components are associated with vasomotor modulations and sympathetic activity.

Recently, several mathematic approaches based on nonlinear mathematics and the chaos theory have been suggested for analysis of long-term fluctuations of the heart rhythm. These include a scaling index method,[115] nonlinear mathematics quantification of the fractal properties,[116] and entropy measures.[117] Loss of normal fractal dynamics and higher regularity have been shown to characterize heart rate dynamics in postmyocardial infarction patients and in patients with heart failure.[116,118]

Clinical Studies

In the population-based Rotterdam and Framingham studies,[102,119] and in the recent ARIC (Atherosclerosis Risk in Communities) study involving 14,672 men and women without coronary heart disease,[120] reduced HRV has been found to be associated with up to a fourfold increase in all-cause mortality and with a significant increase in morbidity from myocardial infarction, coronary artery disease, and congestive heart failure. Prognostic value of impaired HRV has been shown in patients with chronic heart failure,[121] dilated and hypertrophic cardiomyopathies,[122,123] and heart transplant recipients.[124] In the ARIC study, the annual mortality rate in patients with chronic heart failure and SDNN less than 50 ms was 51.4% compared with 5.5% in those with SDNN greater than 100 ms.[121]

Although encouraging results have been obtained regarding the prognostic value of HRV in different clinical settings, HRV-based risk stratification has mainly centered around the possibility of using it as a prognostic tool in postmyocardial infarction settings. The predictive power of HRV was recognized in 1987 in a landmark study by Kleiger and colleagues[106] who evaluated 808 postmyocardial infarction patients and reported a 5.3-fold mortality risk associated with SDNN less than 50 ms compared with preserved HRV (SDNN >100 ms) over a 31-month follow-up. Five years later, the prognostic value of the frequency-domain analysis of HRV was shown by Bigger and associates[125] in 715 patients with myocardial infarction who participated in the CAPS (Cardiac Arrhythmia Pilot Study) trial and were followed for three years. Data from the St. George's Post-Infarction Research Survey Programme confirmed these findings in a large cohort of postmyocardial infarction patients, using an HRV triangular index of less than 16 as a definition of the depressed HRV.[126] In the ALIVE (AzimiLide post Infarct surVival Evaluation) trial of 3717 postmyocardial infarction patients with low ejection fraction, placebo patients with depressed HRV (defined as a triangular index ≤20) had a significantly higher one year mortality than those with an HRV index greater than 20 U (15% versus 9.5%; hazard ratio, 1.64; 95% CI, 1.24 to 2.17 $P < .0005$) despite nearly identical ejection fractions.[126a] Although the HRV measurement results were not part of the eligibility requirements, they were used to prospectively stratify the at-risk trial population. Patients with baseline HRV of less than or equal to 20 U were assigned to the "high-risk" cohort. Patients with baseline HRV greater than 20 U are referred to as the "low-risk" group. However, low HRV did not predict arrhythmic mortality.

Although data from large postmyocardial infarction survival studies have repeatedly shown that reduced HRV is a strong independent predictor of poor prognosis, the predictive value of HRV taken alone often did not exceed 20% within the range of relevant levels of sensitivity. A combination of HRV with other risk factors yields better positive-predictive accuracy. For example, the combination of HRV with left ventricular ejection fraction or with frequent VPC during Holter monitoring usually doubled the positive-predictive accuracy.[125,126] Table 47-7 summarizes the representative studies of HRV in combination with other risk stratifiers in postmyocardial infarction patients.

Because it seems plausible that different risk factors may predict specific risks, information on combined risk factors can be used to identify patients with a high total mortality and a high proportion of either arrhythmic or nonarrhythmic death. By selecting the patients with a high risk of arrhythmic mortality (depressed HRV) and excluding those with a high risk of nonarrhythmic mortality (low left ventricular ejection fraction), Hartikainen and colleagues[127] identified a patient group in whom 75% of deaths were of arrhythmic nature. Similarly, a patient group having 75% of deaths due to nonarrhythmic mechanisms was defined by selecting the patients with the lowest left ventricular ejection fraction and excluding those with the lowest HRV. However, an analysis of the EMIAT (European Myocardial Infarct Amiodarone Trial) population has shown that the predictive power of HRV is preserved when patients with depressed left ventricular ejection fraction are considered.[128] The HRV triangular index

TABLE 47-7 Representative Studies of Heart Rate Variability and Baroreflex Sensitivity in Combination with Other Risk Factors for Postmyocardial Infarction Risk Stratification

Study	Number of Patients	Follow-up	Number of Events	Combined Risk Factors	Predictive Values
Kleiger et al[106]	808	4 yrs	127 deaths	SDNN, VPC, LVEF	Mortality: SDNN < 50 ms + AVGNN < 750 ms: 37% SDNN < 50 ms + LVEF < 30%: 49% SDNN < 50 ms + VPC >10/hr: 40% SDNN < 50 ms + VPC runs or couplets: 50%
Bigger et al[125]	715	4 yrs	119 deaths	Frequency domain HRV analysis, LVEF	Mortality: Decreased ULF or VLF + VPC > 3/hr: 40%–44% Decreased ULF or VLF + LVEF < 40%: 49% Decreased ULF and VLF + VPC > 3/hr: 53% Decreased ULF and VLF + LVEF < 40%: 56%
Farrel et al[126]	416	612 days	24 sudden deaths and arrhythmic events	HRV index + VPC + late potentials	Sensitivity 20%, specificity 99%, PPA 58%, NPA 95%
Pedretti et al[54]	303	15 mo	19 arrhythmic events	HRV index RR ≤ 750 ms Late potentials (filtered QRS) VPC, VT LVEF EPS	Relative risk for arrhythmic events: HRV index ≤ 29: 15, PPA 15% VT ≥ 2 runs: 6, PPA 25% Filtered QRS ≥ 106 ms: 9, PPA 17% LVEF < 40%: 16, PPA 26% Filtered QRS + VPC + LVEF: PPA 30% EPS (+): PPA 65%
Hartikainen et al[127]	575	>2 yrs	29 arrhythmic, 18 nonarrhythmic events	HRV index, VPC, VT runs, LVEF	HRV index < 20 + VT runs predicted arrhythmic death HRV index < 20 + frequent VPC + LVEF < 40% predicted arrhythmic death LVEF < 40% predicted nonarrhythmic death
La Rovere et al[107]	1284	21 mo	44 cardiac deaths, 5 nonfatal cardiac arrests	SDNN, BRS	SDNN < 70 ms + BRS < 3.0 ms/mm Hg (17% mortality) versus SDNN > 105 + BRS > 6.1 ms/mm Hg (2% mortality)
Farrel et al[139]	122	1 yr	10 arrhythmic events	BRS, HRV index, late potentials, VPC, LVEF, exercise stress test	Relative risk for arrhythmic events: BRS < 3 ms/mm Hg: 21.3 HRV index < 13: 10.1 Late potentials: 9.1 VPC > 10/hr: 20.6 LVEF< 40%: 10.4 Exercise stress test (+): 2.8
Hohnloser et al[142]	95	16 mo	41 ICD therapies	SDNN, VT runs, BRS, LVEF, TWA, late potentials, QTD	TWA ≥ 1.9 μV + LVEF ≤ 35% predicted VT recurrence (sensitivity 44%, specificity 72%, PPA 60%, NPA 58%) In coronary artery disease: BRS ≤ 3 ms/mm Hg + TWA ≥ 1.9 μV (77% VT recurrence) versus BRS > 3 ms/mm Hg + TWA < 1.9 μV (0% VT recurrence)
Camm et al[126a]	1690 in the placebo arm	1 yr	196 deaths	HRV index LVFF	Mortality: HRV index ≤20: 15% HRV index > 20: 9.5% Relative risk 1.64

AVGNN, the mean R–R interval over 24-hour ECG recording; EPS, electrophysiological study; HRV, heart rate variability; ICD, implantable cardioverter-defibrillator; LVEF, left ventricular ejection fraction; NPA, negative-predictive accuracy; PPA, positive-predictive accuracy; QTD, QT dispersion; SDNN, standard deviation of normal R–R interval during 24-hour ECG recording; TWA, T wave alternans; ULF, ultra-low frequency; VLF, very low frequency; VF, ventricular fibrillation; VPC, ventricular premature complex; VT, ventricular tachycardia.

of less than or equal to 20 allowed the selection of approximately 30% of patients with left ventricular dysfunction in the placebo arm, in whom a 70% increase in all-cause mortality and a 90% increase in arrhythmic mortality was observed, compared with all-cause and arrhythmic mortality rates in the total placebo arm.

Another important factor that can modify the predictive power of HRV is substantial changes in treatment of acute myocardial infarction, that is, revascularization therapy and more active use of β-blocking agents. Although in the GISSI-2 study impaired HRV continued to predict mortality, higher stratifying values had to

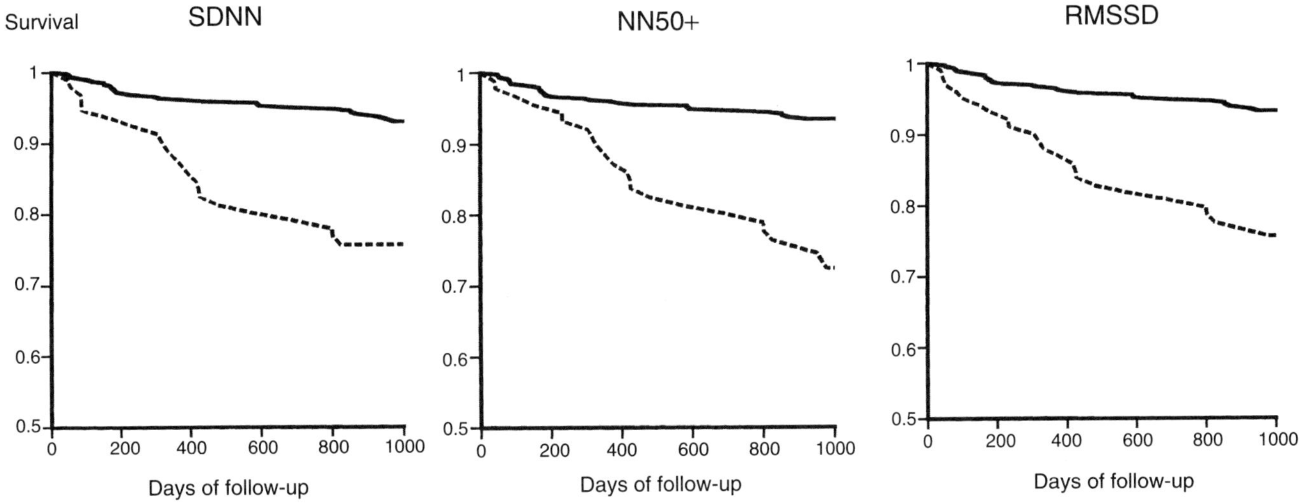

FIGURE 47-9 Survival curves for patients with low (*dashed lines*) and high (*solid lines*) HRV for the three indexes used. Cutoff values were SDNN, 70 ms; NN50+, 200 per 24 hours; and rMSSD, 17.5 ms. NN50+, total number of pairs of adjacent sinus rhythm R–R intervals in which the subsequent R–R interval is longer than the preceding by greater than 50 ms; rMSSD, mean root square of the successive differences between adjacent N–N intervals; SDNN, standard deviation of all normal R–R intervals over a 24-hour ECG recording. (Reproduced from Zuanetti G, Neilson James MM, et al: Prognostic significance of heart rate variability in post-myocardial infarction patients in the fibrinolytic era: The GISSI-2 results. Circulation 1996;94:432-484.)

be used.[129] SDNN of less than 70 ms was associated with a threefold increase in total mortality (Fig. 47-9), whereas the use of a 50-ms cutoff, suggested in the MPIP study, resulted in the reduction of patients identified at risk from 16% to 7.8%. Finally, modulating effects of β-blockade on sympathovagal balance, resulting in a decrease in sympathetic and an increase in parasympathetic activities, have been clearly demonstrated.[130]

BAROREFLEX SENSITIVITY

Methodology

Although HRV usually provides information on the continuous basal sympathovagal balance, the baroreflex sensitivity (BRS) method is aimed mainly at the quantitative assessment of the ability of the autonomic nervous system to react to acute stimulation that involves primarily vagal reflexes. BRS assessment was initially suggested in 1969 by Smyth and coworkers[131] for clinical studies related to hypertension. Methods used for BRS assessment include Valsalva's maneuver, neck cuff, negative pressure applied to the lower part of the body, carotid massage, and intravenous administration of vasoconstricting (phenylephrine) or vasodilating (nitroglycerin, nitroprusside) agents. BRS measurement obtained with different methods varies significantly, which is probably the result of assessment of different aspects of baroreflex function.[132]

Phenylephrine Test

Intravenous phenylephrine (α-adrenoreceptor agonist) administration (25 to 100 μg in healthy subjects, 150 to 350 μg in cardiac patients) over 30 seconds is the most standardized method for BRS assessment.[133] The dose of phenylephrine depends on the underlying heart disease and may vary from 150 to 350 μg (2 to 4 μg/kg) in postmyocardial infarction patients to 750 μg (10 μg/kg) in patients with congestive heart failure.[134,135] Under physiologic conditions, the administration of phenylephrine causes an increase in blood pressure by 20 to 30 mm Hg accompanied by an increase in R–R interval duration. The rate-pressure response is characterized by a linear relationship, the slope of which, measured by a linear regression, is used to express the baroreflex function numerically (Fig. 47-10). It was generally assumed that the slope reflects mainly vagal activity with little changes in the sympathetic tone. However, a flat slope, which is common in cardiac patients, may also indicate sympathetic hyperactivity. The normal values of BRS have been reported to be 14.8 to 16 ms/mm Hg, whereas in hypertensive and postmyocardial infarction patients, a significantly decreased BRS was observed (mean values of 3 and 7.5 ms/mm Hg, respectively).[133-136]

The disadvantage of the phenylephrine test is a simultaneous activation of multiple reflexogenic areas. Beyond its direct pressor action, phenylephrine may activate cardiac mechanoreceptors by increasing afterload and may produce direct effects on preganglionic and postganglionic cardiac parasympatheic nerves, altering the BRS index, which in this case is derived as a measure of the net autonomic balance to the heart.

Noninvasive Methods

The assessment of spontaneous changes in baroreflex function, using finger cuff pressure/pulse measurements, represents another attractive tool for quantification of cardiac autonomic status. Similar to HRV, it applies statistical (time domain) and spectral approaches and is

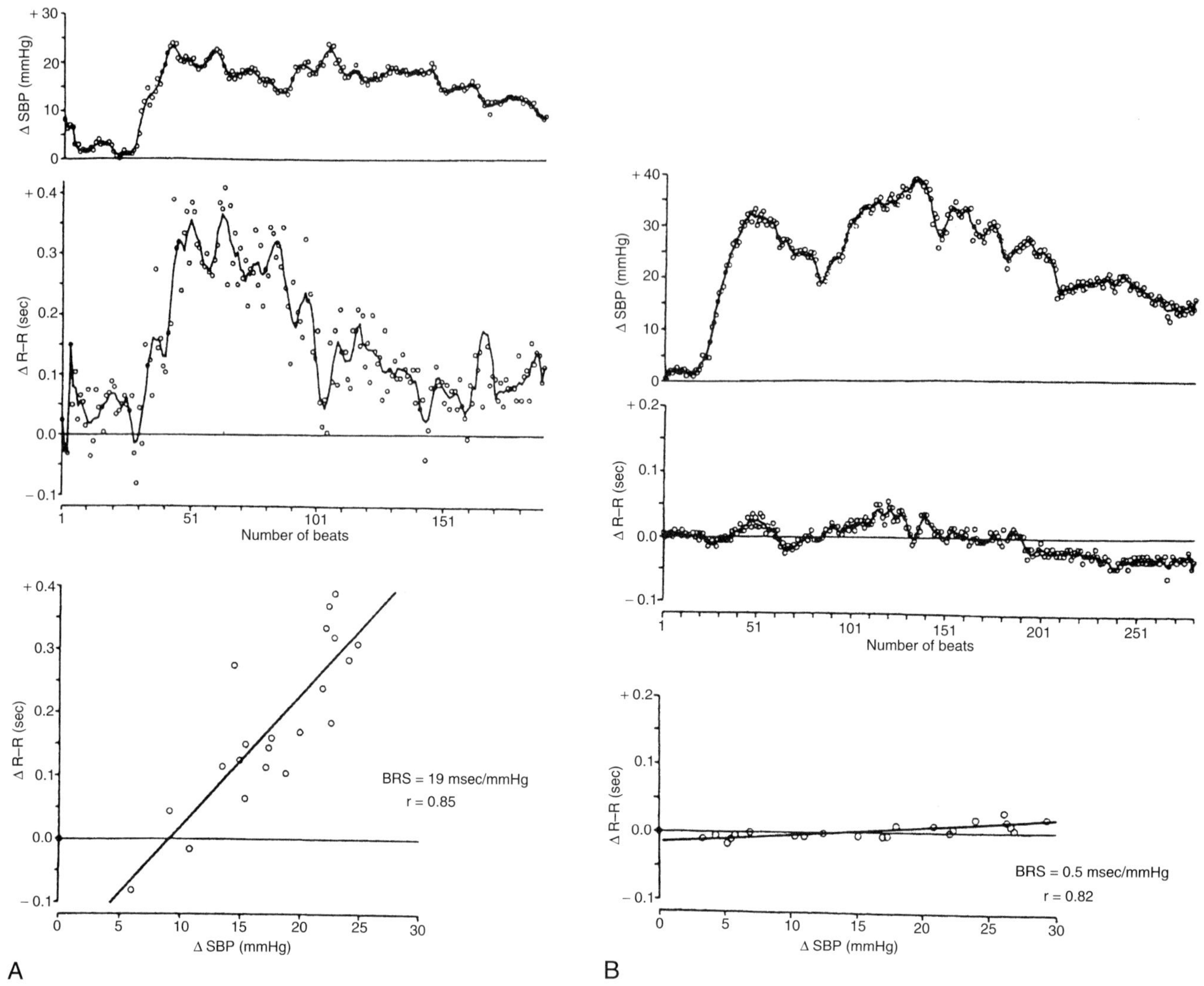

FIGURE 47-10 **A,** Beat-to-beat changes in systolic arterial blood pressure (Δ SBP; **top panel**) and R–R intervals (Δ RR; **middle panel**) after phenylephrine injection compared with baseline level. The regression line (**bottom panel**) is constructed using points from the first major increase in blood pressure and corresponding changes in heart rate. Note a 25-mm Hg increase in blood pressure and a marked acceleration of heart rate after phenylephrine showing a well-preserved baroreflex sensitivity (19 ms/mm Hg) expressed as a slope of the regression line. **B,** Despite an increase in systolic blood pressure, heart rate does not change resulting in a "flat" regression line demonstrating a remarkably reduced baroreflex sensitivity (–0.5 ms/mm Hg). BRS, Baroreflex sensitivity; SBP, systolic blood pressure. (Reproduced from La Rovere MT, Specchia G, Mortara A, Schwartz PJ: Baroreflex sensitivity, clinical correlates and cardiovascular mortality among patients with a first myocardial infarction: A prospective study. Circulation 1988;78:816-824.)

based on naturally occurring changes in arterial blood pressure that increase arterial baroreceptor firing and trigger vagal efferent stimulation.[133,137] The time domain approach identifies sequences of three or more consecutive beats characterized by a progressive increase in systolic blood pressure accompanied by a lengthening of R–R intervals, or a progressive decrease in blood pressure and a shortening of R–R intervals. Then, a slope of the regression line between changes in blood pressure and heart rate is computed. The threshold values are 1 mm Hg for blood pressure changes and 6 ms for R–R interval changes.

The concept of the spectral approach is that each spontaneous oscillation in blood pressure elicits an oscillation of a similar frequency in the R–R interval. Hence, the baroreflex gain can be obtained by dividing the amplitude of the R–R interval oscillation by the amplitude of the corresponding blood pressure oscillation. The function describing this gain is the modulus of the transfer function between systolic pressure and the R–R interval.

However, spontaneous oscillations of arterial blood pressure and heart rate, which are used by noninvasive techniques, are characterized by the interplay of multiple reflexes from the heart, lung, and vessels, and are probably affected by other purely hemodynamic or mechanical influences, not mediated via the baroreceptor reflex arch. Controversies exist regarding the agreement between the spectral analysis of spontaneous BRS and the results of the phenylephrine method.[137]

Prognostic Significance of Baroreflex Sensitivity

Although HRV and BRS have become established markers of autonomic imbalance and vulnerability to arrhythmia, it was not until 1998 when the ATRAMI (Autonomic Tone and Reflexes After Myocardial Infarction) study was completed that the prognostic values of combined cardiac autonomic status measures were shown in a prospective multicenter trial.[107] In the ATRAMI study, involving 1284 postmyocardial infarction patients, low values of HRV (SDNN < 70 ms) and BRS (<3 ms/mm Hg) carried a significant risk for cardiac mortality (3.2 and 2.8, respectively). The combination of low SDNN and BRS further increased mortality risk (Fig. 47-11). A two-year mortality rate was 17% with both variables below cutoff points and only 2% when both were well preserved (SDNN > 105 ms and BRS > 6.1 ms/mm Hg). When left ventricular ejection fraction of less than 35% was added to low HRV or BRS, the mortality risk increased to 6.7 and 8.7, respectively. In this study, HRV and BRS were not well correlated, confirming that the two methods evaluate different aspects of autonomic nervous system activity. In patients with moderate-to-severe heart failure, depressed BRS (<1.3 ms/mm Hg) paralleled the deterioration of clinical and hemodynamic status and was associated with poor survival, especially in ischemic cardiomyopathy: relative risk for mortality was 3.6 compared with 2.8 in idiopathic cardiomyopathy.[135]

In another series, blunted baroreceptor-mediated reflex was associated with a 23-fold increased risk for arrhythmic events and a 36-fold increased risk for VT inducibility during programmed stimulation.[138,139] Interestingly, it has been suggested that low BRS in patients with VT may be responsible for poor tolerability and hemodynamic collapse, probably due to inadequate baroreflex-mediated sympathetic nervous activity during paroxysms, whereas less reduced BRS may indicate more flexible cardiovascular reflexes to compensate for the adverse hemodynamic effects of tachycardia.[140,141] Finally, among 95 patients with an ICD for recurrent ventricular tachyarrhythmias, none of the ten patients with BRS greater than 3 ms/mm Hg and without T wave alternans developed VT or fibrillation during a 2.5-year follow-up, whereas 23 of 33 patients with abnormal BRS and positive T wave alternans, experienced one or more arrhythmic events.[142]

HEART RATE TURBULENCE

Very recently, the concept of heart rate (HR) turbulence was introduced by Schmidt and associates.[143] HR turbulence describes the physiologic biphasic response of the sinus node to a VPC. It has been noted that in low-risk postmyocardial infarction patients, modulation of the heart rate following each spontaneous VPC occurs in a rather systematic fashion, resulting in the early acceleration and later deceleration of sinus rhythm, whereas in patients with unfavorable outcomes this response is blunted (Fig. 47-12).

Mechanisms of Heart Rate Turbulence

Heart rate turbulence resembles ventriculophasic sinus arrhythmia, a phenomenon operating through baroreceptor activation caused by the ventricular contraction, which was first described by Erlanger and Blackman[144] in complete heart block nearly a hundred years ago and supported by further experimental and clinical studies.[145,146] HR turbulence also appears to be initiated by the baroreflex response.[147] When a VPC occurs early in the cardiac cycle, the subsequent stroke volume is reduced and, as a consequence, a shortening of the R–R interval occurs—perhaps compensating for the reduction in cardiac output. This is associated with the initial vagal withdrawal. Conversely, after the compensatory

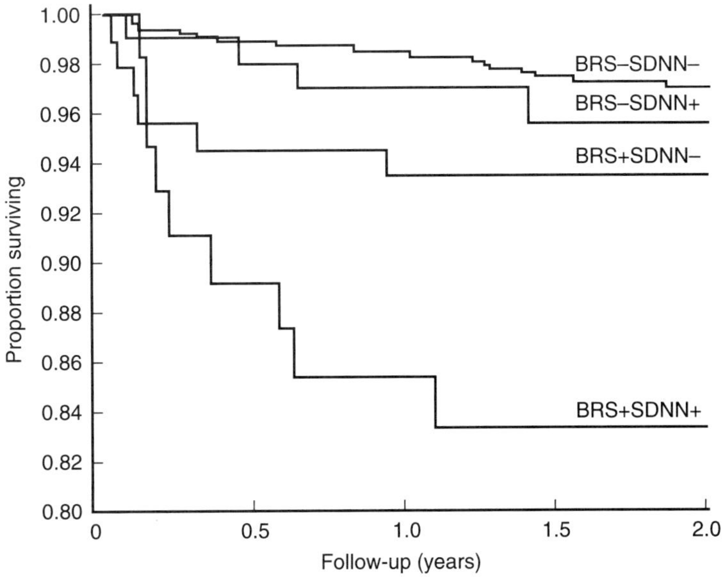

FIGURE 47-11 Kaplan-Meier survival curves for total cardiac mortality according to the presence of abnormal BRS and/or SDNN. BRS–, Baroreflex sensitivity greater than or equal to 3 ms/mm Hg; BRS+, baroreflex sensitivity less than 3 ms/mm Hg; SDNN–, standard deviation of all sinus R–R intervals over a 24-hour ECG recording greater than or equal to 70 ms; SDNN+, standard deviation of all sinus R–R intervals over a 24-hour ECG recording less than 70 ms. (Modified from La Rovere MT, Bigger JT Jr, Marcus FI, et al: The ATRAMI [Autonomic Tone and Reflexes After Myocardial Infarction] Investigators. Baroreflex sensitivity and heart rate variability in prediction of total cardiac mortality after myocardial infarction. Lancet 1998;351:478-484.)

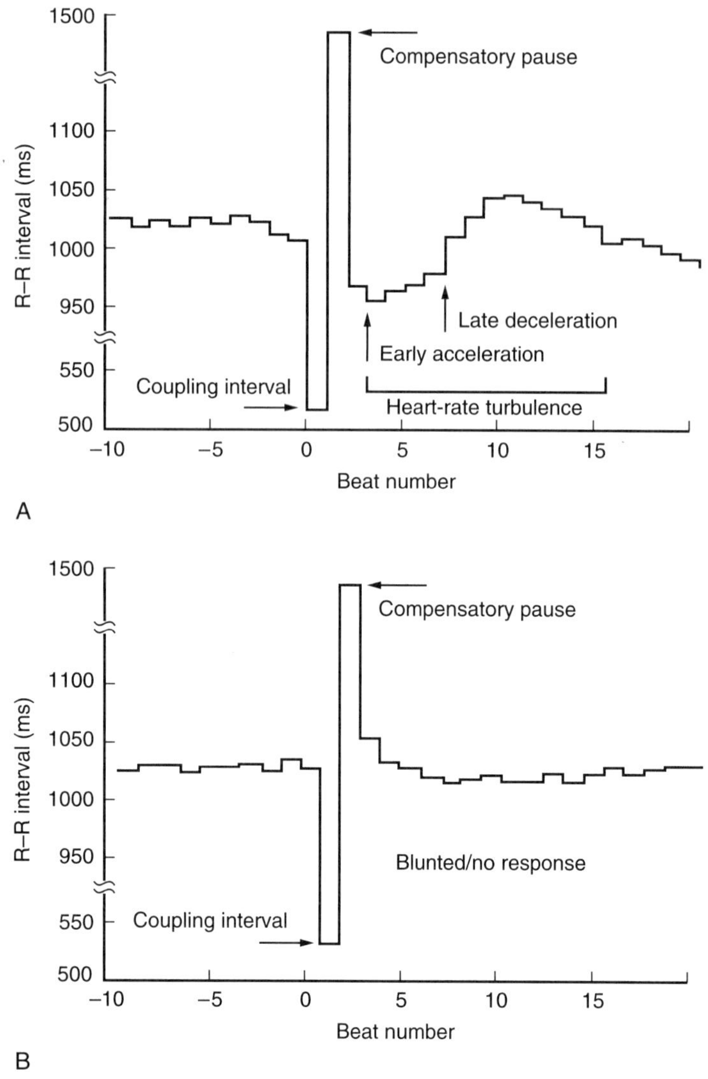

FIGURE 47-12 R–R interval tachogram demonstrates the characteristic acceleration and deceleration phases of sinus rhythm modulation following the ventricular premature complex indicating preserved heart rate turbulence (**A**). R–R interval tachogram illustrates a blunted pattern of sinus rhythm response to the ventricular premature complex indicating a loss of heart rate turbulence (**B**). (Data from Schmidt G, Malik M, Barthel P, et al: Heart rate turbulence after ventricular premature beats as a predictor of mortality after acute myocardial infarction. Lancet 1999;353:1390-1396.)

pause, there is an increase in stroke volume and blood pressure for the following sinus beats. The compensatory increase in vagal tone causes R–R intervals to lengthen. When autonomic control is preserved, transient perturbations in blood pressure caused by a VPC are immediately registered by a brisk response of the sinus node in the form of HR turbulence. If vagal control is impaired, such a response is reduced or absent, resulting in the absence of HR turbulence.

Physiologic Correlates of Heart Rate Turbulence

To date, the physiologic correlates of HR turbulence remain speculative. HR turbulence parameters are moderately correlated with BRS,[148] therefore, it seems very plausible to speculate that HR turbulence, reflects the phasic activity of the autonomic nervous system, yet represents intrinsic baroreflex which is somewhat different from the externally induced arterial reflexes used in BRS estimation. The physiology of HR turbulence, characterizing the almost instantaneous response of the sinus node pacemaker to a VPC, is also different from the long-term measurement of SDNN that is governed by various physiologic factors including autonomic, thermoregulatory, emotional, and environmental influences. On the contrary, HR turbulence appears to be triggered by endogenous stimuli that are likely to cause a more systematic response. Hence, although HR turbulence, BRS, and HRV all reflect autonomic nervous system activity, each characterizes different facets of autonomic modulation.

Methodology

HR turbulence is characterized by turbulence onset (TO) and turbulence slope (TS). Turbulence onset quantifies the acceleration phase of sinus rhythm modulations and represents the differences between the mean of the first two sinus rhythm R–R intervals after the compensatory pause and the mean of the last two sinus R–R intervals preceding the VPC, divided by the sum of the last two sinus R–R intervals preceding the VPC. It is expressed in percentage.

$$TO = ([RR_1+RR_2]-[RR_{-2}+RR_{-1}])/(RR_{-2}+RR_{-1})$$

given that RR_0 is a post-extrasystolic pause.

Consequently, a negative TO value indicates sinus rhythm acceleration and a positive TO value indicates sinus rhythm deceleration following a VPC. Turbulence slope expresses the subsequent deceleration of the heart rate and is quantified by the steepest regression line between the sinus R–R interval count and length and represents the maximum positive slope of regression lines constructed over any sequence of the five consecutive sinus R–R intervals within the first 20 sinus complexes after a VPC. Turbulence slope is expressed in milliseconds per R–R interval (ms/RRI).

In the original publication by Schmidt and colleagues,[149] the patterns of sinus R–R intervals following a single VPC were averaged over 24-hour Holter recordings and subsequently processed using the above expressions. Dichotomy limits for TO at 0% and TS at 2.5 5 ms/RRI were derived form the testing set of 100 patients with frequent VPCs.

TABLE 47-8 Composite Autonomic Index and Relative Risks for Fatal and Nonfatal Cardiac Arrests in the ATRAMI Substudy (Multivariate Analysis)

Model Using TO, TS, BRS, and SDNN Variables		
Variable	**Relative risk (95% CI)**	***p* Value**
1 factor abnormal	1.51 (0.66-3.44)	0.33
2 factors abnormal	2.09 (0.66-6.60)	0.21
3 factors abnormal	2.92 (0.94-9.06)	0.063
All 4 factors abnormal	8.67 (2.72-27.65)	0.0003
LVEF < 35%	2.86 (1.31-6.28)	0.0087

BRS, baroreflex sensitivity; LVEF, left ventricular ejection fraction; SDNN, standard deviation of sinus R–R intervals over 24-hours; TO, turbulence onset; TS, turbulence slope. Factors are considered abnormal if: (1) BRS < 3 ms/mm Hg; (2) SDNN < 70 ms; (3) TO ≥ 0%; TS ≤2.5 ms/R–R interval. (Data from Ghuran A, Reid F, La Rovere MT, et al: Heart rate turbulence-based predictors of fatal and nonfatal cardiac arrest (the Autonomic Tone and Reflexes After Myocardial Infarction Substudy). Am J Cardiol 2002;89:184-190.)

Clinical Studies

HR turbulence has been retrospectively evaluated in three independent patient populations from large-scale postmyocardial infarction trials: 577 patients from a prethrombolytic MPIP study,[106] 614 patients from the placebo arm of the EMIAT study,[150] and 1212 patients participated in the ATRAMI trial (data on HR turbulence were available in 981 participants).[107] In both the MPIP and EMIAT sub-studies, the absence of biphasic response of the sinus node to a VPC has been associated with a 3.2-fold increase in risk for total mortality (Fig. 47-13).[143] In the ATRAMI substudy, reduced HR turbulence predicted the composite end-point of fatal and nonfatal cardiac arrests.[148]

Because autonomic modulations appear to be involved, it is plausible that the combination of these factors may represent a comprehensive index for the assessment of cardiac autonomic status. Such an index, based on HR turbulence (TO, TS), HRV (SDNN), and BRS measurements, has been validated in the ATRAMI substudy and has been found to be associated with a nearly ninefold increase in fatal and nonfatal cardiac arrests, if all four factors were abnormal (Table 47-8).

The prospective Innovative Stratification of Arrhythmic Risk study enrolled 1455 patients with acute myocardial infarction, 90% of whom underwent primary coronary angioplasty, and followed them for an average of 22 months. Abnormal HR turbulence was the strongest predictor of all-cause mortality, with a hazard ratio of 5.9, followed by low ejection fraction (hazard ratio 4.5).[150a] The presence of nonsustained VT (9%), impaired HRV, and a mean 24-hour heart rate greater than 75 beats per minute were all associated with increased mortality in univariate analysis but were overpowered by HR turbulence in multivariate analysis. Furthermore, HR turbulence identified a high-risk subgroup (two-year mortality 15%) among patients with an ejection fraction greater than 30%.

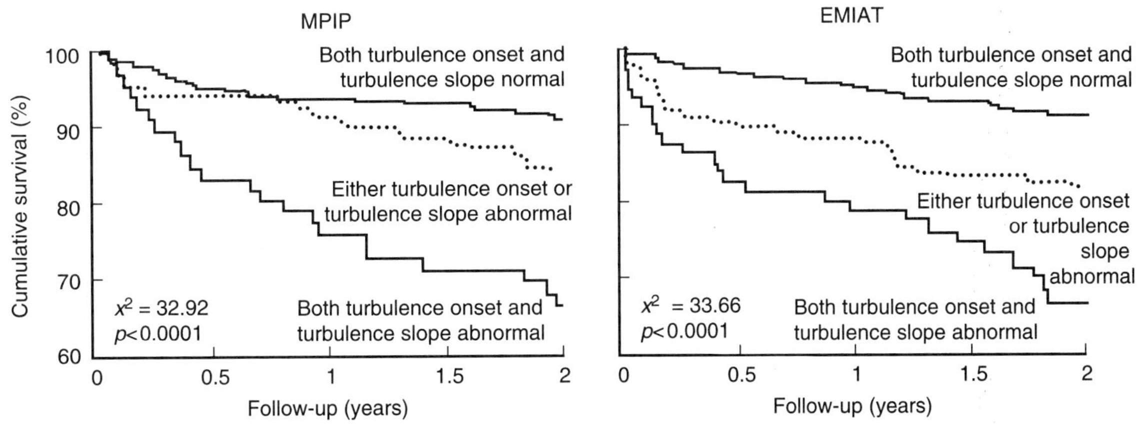

FIGURE 47-13 Kaplan-Meier cumulative survival curves for all-cause mortality in MPIP and EMIAT patients stratified according to TO and TS dichotomy values: TO less than 0% and TS greater than 2.5 ms/RRI (both factors are normal); either TO greater than or equal to 0% or TS less than or equal to 2.5 ms/RRI (one of the factors is abnormal); TO greater than or equal to 0% and TS less than or equal to 2.5 ms/RRI (both factors are abnormal). EMIAT, European Myocardial Infarct Amiodarone Trial; MPIP, Multicenter Post Infarction Program; RRI, R–R interval; TO, turbulence onset; TS, turbulence slope. (Data from Schmidt G, Malik M, Barthel P, et al: Heart rate turbulence after ventricular premature beats as a predictor of mortality after acute myocardial infarction. Lancet 1999;353:1390-1396.)

Although HR turbulence as a risk predictor has been investigated mainly in postmyocardial infarction settings, there are data suggesting that it may also be used as a prognostic marker in patients with heart failure.[151] When compared with HRV (relative risk 2.04, $P < 0.044$), BRS (relative risk 2.63, $P < 0.11$), and nonsustained VT (relative risk 3.3, $P = 0.001$), the combination of abnormal TO and TS was the strongest univariate predictor of cardiac mortality in this subset of patients.

Finally, the preliminary results from a small series of patients with VT referred for electrophysiological evaluation, have recently demonstrated the presence of HR turbulence after an atrial premature stimuli.[152] Although the response of the sinus node to an atrial premature beat was attenuated compared with that to a VPC, and the mechanism of this phenomenon remains unclear, this observation opens a possibility of the assessment of HR turbulence in patients with atrial arrhythmias.

SUMMARY

Compelling experimental and clinical evidence supports the value of the assessment of cardiac autonomic status and shows that methods used to quantify cardiac autonomic nervous system activity represent potent tools for arrhythmia risk stratification. Various methods, such as HRV, BRS, and HR turbulence, offer insight into different facets of cardiovascular autonomic status. Their inter-relationships require further study.

Ventricular Repolarization Assessment

Experimental and clinical evidence has strongly implicated prolonged and heterogeneous ventricular repolarization in the development of ventricular tachyarrhythmias and sudden death. Therefore, electrocardiographic patterns of ventricular repolarization may represent a potential tool for risk stratification. Different methods have been proposed for this purpose, including the assessment of the Q–T interval and its dispersion, QT/RR variability, T wave alternans, and principal component analysis of the T wave (a descriptor of the complexity of repolarization), only to emphasize the fact that no one ideal approach has yet been found. Most recently, a concept of T wave morphology dispersion based on the analysis of the T wave loop and characterizing the sequence of repolarization has been introduced with promising initial results regarding the prediction of all-cause mortality and arrhythmic death in various clinical settings.

Q–T INTERVAL

Q–T Interval and Mortality

A close link between delayed ventricular repolarization expressed as the prolonged Q–T interval on the surface ECG and arrhythmogenesis has long been recognized in the congenital long QT syndrome where the evidence for the occurrence of ventricular arrhythmias and sudden cardiac death is most cogent.[153-155] In 1978, Schwartz and Wolf[156] suggested that prolonged Q–T intervals in patients with myocardial infarction, associated with a twofold risk of sudden death, might reflect alterations in electrophysiological properties of myocardium favoring the occurrence of ventricular tachyarrhythmias. These initial observations have been later confirmed in the population-based prospective studies showing that a corrected Q–T interval typically greater than 440 ms conferred a 2.3 to 3-fold increased risk for sudden cardiac death (Table 47-9).[157-169,169a,192]

In the Dutch Civil Servants study and the Zutphen study, prolongation of Q–Tc interval greater than or equal to 420 ms was associated with an increased risk for sudden death, death from coronary heart disease, and nonfatal myocardial infarction, especially in men and in elderly subjects.[164,165] In more recent studies of 3455 Danish and 10,717 Finnish citizens,[166,167] the Q–T interval predicted cardiac death and nonfatal cardiovascular morbidity in a subgroup of patients with established coronary heart disease. Because the pathophysiologic link between the duration of the Q–T interval and death is an increased susceptibility to ventricular arrhythmias, in studies with multiple end-points, the prolonged Q–T interval is more likely to present the highest risk for sudden cardiac death and moderate risk for cardiac and all-cause mortality.[165-167]

However, in high-risk patients with left ventricular dysfunction, Q–T interval prolongation did not further increase the mortality risk.[157] The Q–T or J–T intervals were not significant independent markers of all-cause mortality, sudden cardiac death, or death from progressive heart failure in 554 patients with congestive heart failure, participating in the UK-HEART study.[170] Furthermore, the results of the Framingham Study have not confirmed the association between the Q–T interval prolongation and all-cause mortality, although a trend toward higher mortality rates has been noted for the two Q–T interval extremes, QTc equal to or greater than 440 ms and less than or equal to 360 ms.[171] It has been suggested that the absence of the association between Bazett's Q–Tc interval and mortality in the Framingham study may be due to a U-shaped relationship, conferring increased mortality in subjects with short or long Q–T intervals. However, De Bruyne and colleagues[168] did not find the risk associated with the heart rate-corrected Q–T interval to be influenced by the correction formula used. Of note, the prognostic value of the uncorrected Q–T interval was lower than that of the corrected Q–T interval, suggesting that heart rate may be a confounder that plays an important role in the prediction of mortality by Q–T interval.

Q–T INTERVAL VARIABILITY

Rate Correction of Q–T Interval

For more than 80 years, the Q–T interval has been known to be largely dependent on heart rate. In 1920, Fridericia[172] described this relationship by a cube root formula:

$$QTc = QT \times (1/RR)^{1/3}$$

TABLE 47-9 Q–T Interval and Mortality in Randomized and Population-based Studies

Study	Number of Patients	Follow-up	End-point	Dichotomy Values of Q–T Interval [ms]	Relative Risk (95% CI)
Schwartz and Wolf[156]	55 post MI	7 yrs	Sudden death	QTc ≥ 440	2.2 (1.3-3.6)
Ahnve et al[157]	214 post MI	1 yr	Cardiac death	QTc ≥ 440	13.3 (3.8-46.1)
Peters et al[158]	3692 post MI (BHAT study)	2.1 yrs	Sudden death	QTc ≥ 450	1.8 (1.2-2.7) placebo 1.9 (1.2-2.7) propranolol
Fioretti et al[159]	474 post MI	1 yr	All-cause death	QTc ≥ 440	1.8 (1.0-3.2)
Wheelan et al[160]	518 post MI	1.5 yrs	Sudden death	QTc ≥ 440	(–7-21)
Puddu and Bourassa[161]	1157 CAD	3.8 yrs	Sudden death	QTc ≥ 440	1.7 (1.0-2.9)
Algra et al[162]	5589 patients with cardiac complaints	2 yrs	Sudden death	QTc ≥ 440	2.3 (1.4-3.9) LVEF > 40% 1.0 (0.5-1.9) LVEF < 40%
Schouten et al[164]	3091 civil servants aged 40-65 yrs	15 yrs*	All-cause death Cardiac death Death from CAD	QTc > 440 versus QTc < 420	1.7 men, 1.6 women 1.8 men, 1.4 women 2.1 men, 0.9 women
Dekker et al[165]	877 men aged 40-65 yrs; 835 men aged 65-85 yrs	5-25 yrs	Nonfatal MI Death from CAD Sudden death	QTc ≥ 420 versus QTc < 385	1.3 (0.7-2.5) middle-aged, 2.4 (0.9-6.1) elderly 4.4 (1.2-16.4) middle-aged; 3.0 (1.2-7.3) elderly; 1.4 (0.3-5.7) middle-aged; 3.0 (1.0-8.9) elderly
Goldberg et al[170]	5125 subjects aged 30-62 yrs (the Framingham Study)	30 yrs	All-cause death Sudden death Death from CAD	QTc ≥ 440 versus QTc ≤ 360	1.02 (0.70-1.49) 1.31 (0.60-2.86) 0.85 (0.48-1.50)
Okin et al[163]	1839 subjects aged 45-74yrs (the Strong Heart Study)	3.7 yrs	All-cause death Cardiac death	QTc > 460	2.0 (1.4-3.0) 2.1 (1.0-4.4)
Elming et al[166]	1797 men, 1658 women aged 30-60 yrs	11-13 yrs	All-cause death Cardiac death Nonfatal cardiovascular morbidity All-cause death Cardiac death Nonfatal cardiovascular morbidity	QT ≥ 430 versus QT < 360 QTc ≥ 440 versus QTc < 380	1.87 (1.06-3.27) 2.93 (1.09-7.84) 2.88 (1.24-6.67) 1.89 (1.04-3.37) 3.31 (1.04-9.91) 1.91 (0.92-3.96)
Karjalainen et al[167]	5598 men, 5119 women	23 yrs	Sudden death Cardiac death All-cause death Sudden death Cardiac death All-cause death	QT > 430 (men) QT > 437 (women)	1.5 (0.78-2.89) 1.42 (1.06-1.96) 1.33 (1.08-1.63) — 1.32 (0.89-1.97) 0.98 (0.70-1.37)
De Bruyne et al[168]	2083 men, 3158 women aged ≥55 yrs (the Rotterdam Study)	4 yrs	All-cause death Cardiac death	QTc (Bazett) >437 (men) QTc (Bazett) >440 (women)	All: 1.8 (1.3-2.4) All: 1.7 (1.0-2.7)
Zabel et al[197]	280 post MI	32 mo	All-cause death (21) Arrhythmic events (19)	QTc >440	Not predictive
Brendorp et al[169]	1518 CHF; 703 (Placebo 402, Dofetilide 301) (DIAMOND-CHF QT substudy)	18 mo	All-cause mortality (285) Placebo (168) Dofetilide (117)	QTc < 479 versus QTc < 429 Placebo Dofetilide	0.7 (0.5-1.2); 0.9 (0.5-1.8)† 1.3 (0.8-1.9); 1.9 (1.0-3.6)†
Dekker et al[169a]	14548 men and women aged 45-64 yrs	2 yrs	All-cause death Cardiac death	QTc > 440 (men) QTc > 454 (women)	2.28 (1.73-3.00) 3.91 (2.40-6.37)†

*After 28 years, these associations were weaker.

†Patients with bundle branch block were excluded.

ARIC, Atherosclerosis Risk In Communities; BHAT, β-Blocker Heart Attack Trial; CAD, coronary artery disease; DIAMOND-CHF, Danish Investigation of Arrhythmia and Mortality on Dofetilide; MI, myocardial infarction; QTc, Q–T interval corrected for heart rate.

whereas Bazett[173] independently suggested that the square root formula:

$$QTc = QT \times (1/RR)^{1/2}$$

should be used to express the relation between the electric systole and heart rate. Other methods for rate correction of the Q–T interval that have been used in the population-based studies include, but are not restricted to, the linear regression formula (the Framingham formula):

$$QTc = QT + \beta \times (1 - RR)$$

with $\beta = 0.140$ in men and $\beta = 0.163$ in women[171] and the normogram method:

$$QTc = QT + \beta \times (1 - RR)$$

with $\beta = 0.116$ for heart rates of less than 60 beats/minute, $\beta = 0.156$ for heart rates from 60 to 90 beats/minute, and $\beta = 0.384$ for heart rates over 100 beats/minute.[174] However, none of the existing methods appears to describe accurately the complex relationship between the heart rate and the duration of ventricular repolarization under long-term physiologic conditions because they correlate the Q–T interval to the immediately preceding R–R interval and ignore any time delay between changes in heart rate and the subsequent changes in the Q–T interval. Furthermore, there is evidence based on a small series of healthy subjects, suggesting that the QT/RR relationship may vary greatly between the individuals and may not be described correctly by a single formula but rather should be assessed on the individual basis of continuous ECG recordings (Fig. 47-14).[175] It is possible that the intersubject variability of the QT/RR pattern may reflect inter-individual variations in the expression and/or autonomic control of different ionic repolarizing currents.

Physiologic Background of QT Dynamicity

Recently, it has been hypothesized that the adaptation of Q–T interval to changes in heart rate is likely to contain additional, information regarding ventricular repolarization to that of the Q–T interval alone.[176] Assessment of the dynamic aspect of ventricular repolarization may represent a more complex insight into interactions between the Q–T interval, heart rate variability, and autonomic tone modulations. Disturbances in circadian variation of the Q–T and R–R intervals may be a sign of autonomic imbalance with a loss of parasympathetic tone and/or sympathetic hyperactivity.

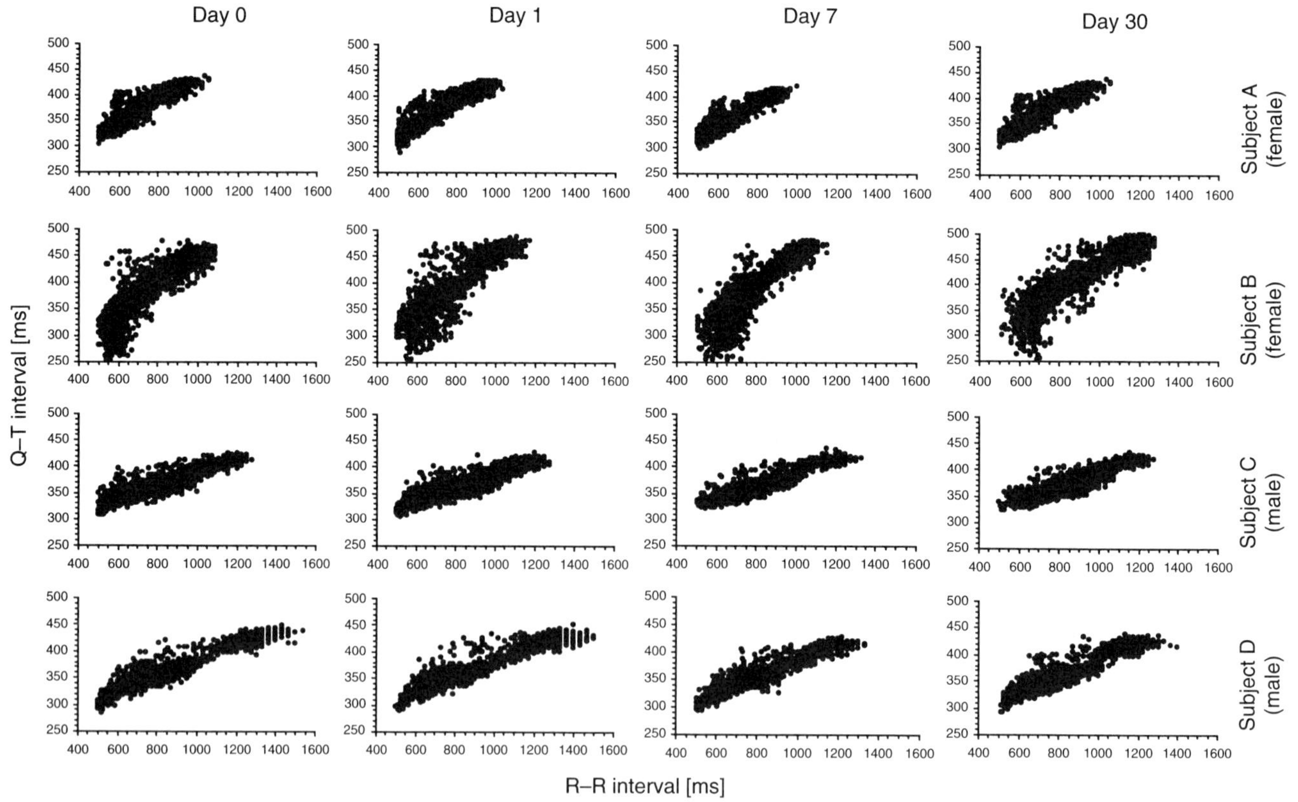

FIGURE 47-14 Representative examples of QT/RR patterns in two women and two men. Each row shows four consecutive recordings in one subject. Note that, in each individual, the QT/RR pattern remains stable over the time, but significant inter-individual differences in the QT/RR relationship are clearly shown. (Reproduced from Batchvarov VN, Ghuran A, Smetana P, et al: QT-RR relationship in healthy subjects exhibits substantial intersubject variability and high intersubject stability. Am J Physiol Heart Circ Physiol 2002;282:H2356-H2363.)

Under physiologic conditions, two kinds of the QT/RR relationship have been described (1) fast adaptation of the Q–T interval to changes in the heart rate leading to a strong dependence of the Q–T interval duration on the preceding R–R interval, and (2) a slower adaptation lasting for several minutes and termed *hysteresis* which is supposed to be due to cardiac memory mechanisms.[177,178] A typical pattern of R–R interval oscillations, the short-long sequence, immediately preceding ventricular tachyarrhythmias, including torsades de pointes, may be associated with exaggerated heterogeneity of repolarization favoring reentrant arrhythmias which may be initiated by early afterdepolarizations.

QT variability from Holter recording may be investigated in two ways, namely, by analyzing the changes in the Q–T interval corrected for heart rate[179] and by analyzing the QT variability on the beat-to-beat basis during a Holter ECG recording.[180]

Clinical Studies

Partial disintegration of the relation between the ventricular recovery time and the cycle length have first been noted in the long QT syndrome.[181] The circadian variation in the Q–T interval was similar in long QT gene carriers (*KCNQ1*) with long QT and normal QT phenotypes (a steeper QT/RR slope at night) but opposite to that observed in control subjects, including those from the same pedigree (a steeper QT/RR slope during the day) (Fig. 47-15).[182] In patients with myocardial infarction, a QT/RR slope greater than 0.18 during the daytime conferred a twofold increased risk for total mortality and a sixfold increased risk for sudden death.[183] An increased QT/RR slope, especially in the early morning hours and immediately after awakening, has been reported in patients with ischemic and idiopathic cardiomyopathy.[184] Although it has been suggested that compromised QT adaptation to heart rate influences may have a prognostic significance in these clinical settings, this marker has not yet been validated in the large prospective studies and the methodology for its assessment is controversial.

Studies of proarrhythmic effects of antiarrhythmic agents or noncardiac drugs, such as macrolide antibiotics, tricyclic antidepressants, and antihistamines with I_{Kr} potassium channel blocking effects, also implicated impaired QT dynamicity in the genesis of torsades de pointes.[185,186] In patients with the acquired long QT syndrome, a progressively greater increase in the Q–T interval at slow heart rates may contribute to bradycardia-dependent ventricular arrhythmias. These studies also support a concept that the pattern of the Q–T interval variability may have more clinical potential than the Q–T interval duration itself.

Other Methods for QT Dynamicity Assessment

Frequency analysis of the variability of the Q–T interval has revealed that it is characterized by rhythmic oscillations with low (0.04 to 0.15 Hz) and high (0.15 to 0.4 Hz) frequencies but with an almost 20-times lower

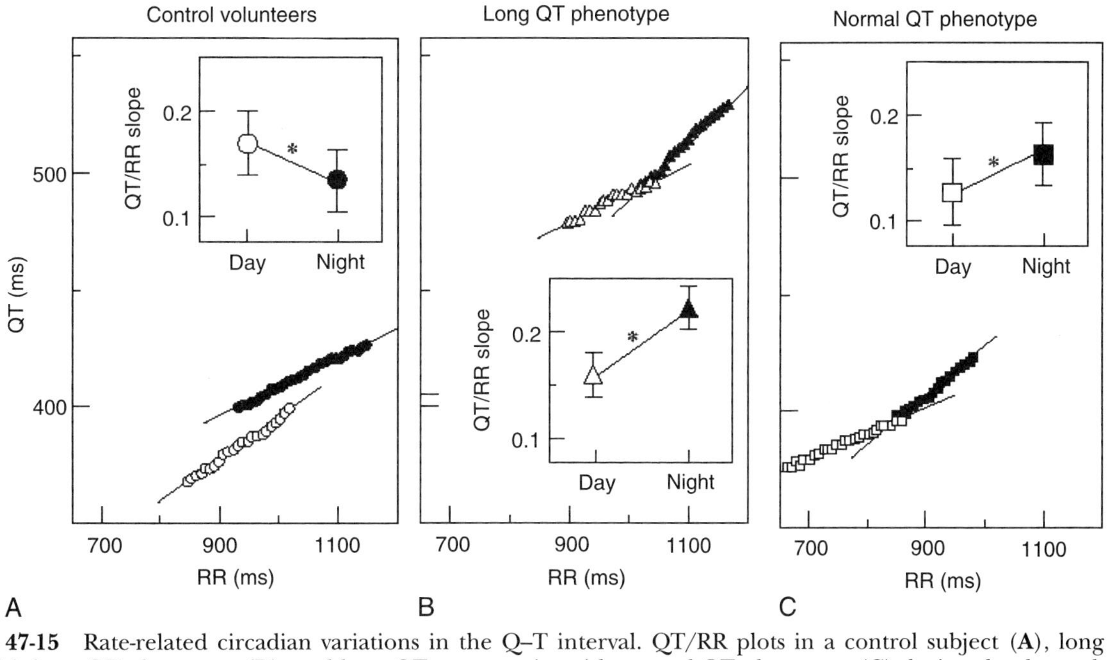

FIGURE 47-15 Rate-related circadian variations in the Q–T interval. QT/RR plots in a control subject (**A**), long QT gene carrier with long QT phenotype (**B**), and long QT gene carrier with normal QT phenotype (**C**) during the day and the night. The inset shows mean ± standard deviation of QT/RR slope values during the day and the night. Note that control subjects have a steeper QT/RR slope during the day, whereas long QT gene carriers show a reversed circadian QT/RR pattern. (Reproduced from Lande G, Kyndt F, Baró I, et al: Dynamic analysis of the QT interval in long QT1 syndrome patients with a normal phenotype. Eur Heart J 2001;22:410-422.)

amplitude than those of the R–R interval.[187] However, until further data are available, it can merely be concluded from the previous experience that the analysis of the QT variability from long-term recordings is of potential benefit.

QT DISPERSION

The prolonged Q–T interval may reflect either the uniform increase in action potential duration or a nonuniform recovery of myocardial excitability. When accompanied by a marked increase in dispersion of repolarization (and, thus, in refractoriness), action potential prolongation may result in early afterdepolarizations favoring reentry. An important early study identifying heterogeneity of myocardial repolarization as an arrhythmogenic factor was the one performed by Han and Moe,[188] in which refractory periods at several canine left ventricular sites were measured in a variety of arrhythmogenic conditions. It was emphasized that, regardless of changes in refractory periods, it was the inhomogeneity in recovery of excitability that was critical for the development of arrhythmias.

It has been noticed that the duration of the Q–T interval varies between individual leads in the surface ECG, and in 1990, Day and coworkers[189] suggested using the QT *inter-lead* variability (QT dispersion) for the assessment of repolarization inhomogeneity. Further experimental studies, showing a strong correlation between QT dispersion and the dispersion of 90% duration of monophasic action potential, seem to support the hypothesis that QT dispersion reflects regional variations in the ventricular recovery times and, therefore, may represent an attractive tool for noninvasive risk stratification.[190]

Clinical Studies

Over the past decade, the prognostic value of QT dispersion has been extensively studied in a variety of clinical conditions (Table 47-10).[163,166,170,191-201] Figure 47-16 presents pooled data from 79 studies with a total of 6883 patients with heart disease and 8455 normal subjects.[202] There is a clear trend toward an increase in QT dispersion in various heart diseases, with the highest mean values reported in patients with long QT syndrome. However, a significant overlap of values

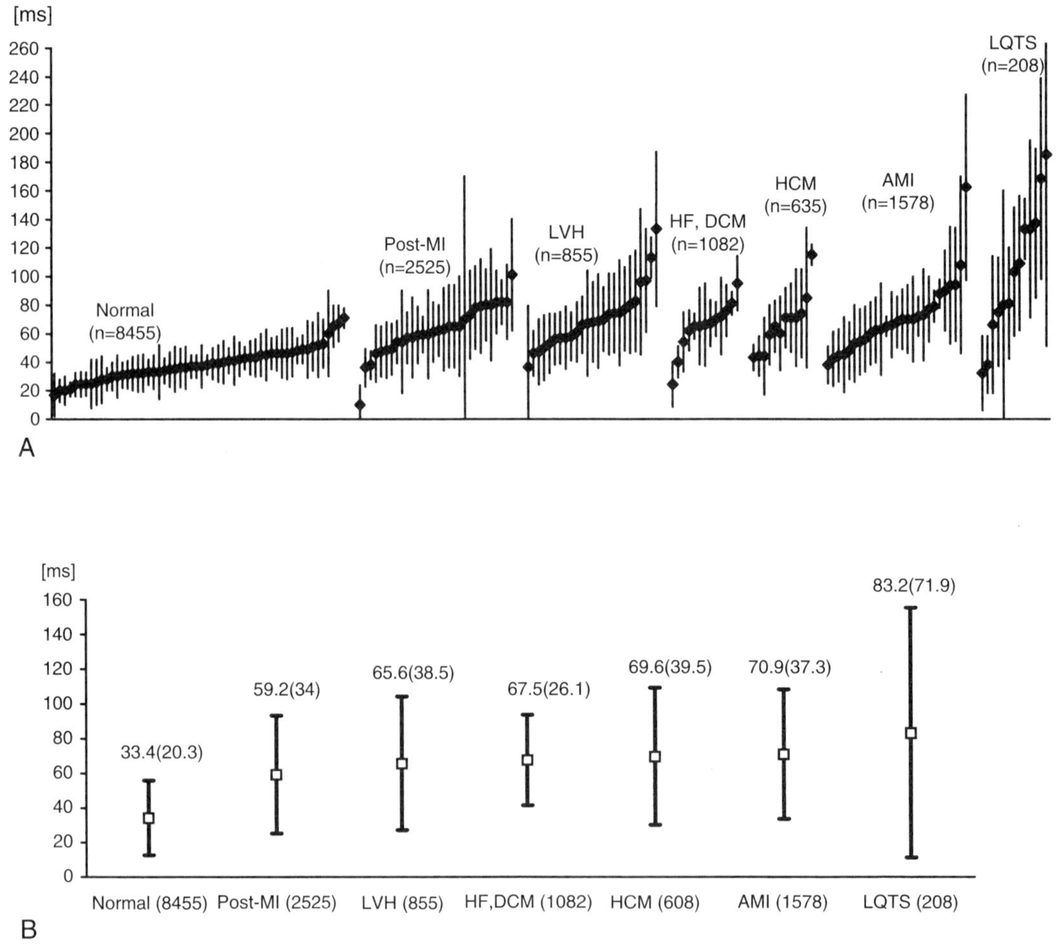

FIGURE 47-16 Weighted mean values of QT dispersion (**A**) in normal subjects and patients with remote myocardial infarction (post MI), acute myocardial infarction (AMI), left ventricular hypertrophy (LVH) excluding hypertrophic cardiomyopathy, hypertrophic cardiomyopathy (HCM), heart failure (HF) and dilated cardiomyopathy (DCM), and long QT syndrome (LQTS). Average mean of the weighted means of QT dispersion in the same clinical settings (**B**). Error bars represent standard deviation. (Data from Malik M (ed): Risk of Arrhythmia and Sudden Death. London, BMJ Books, 2001, pp.194-201.)

TABLE 47-10 QT Dispersion and Mortality in Randomized and Population-based Studies

Study	Number of Patients	Follow-up	End-point	Dichotomy Values of QT Dispersion (ms)	Relative Risk (95% CI)
Elming et al[166]	1797 men, 1658 women aged 30-60 yrs	11-13 yrs	All-cause death (331) Cardiac death (100) Nonfatal cardiovascular morbidity (172)	QTD ≥ 90 versus QTD < 40	1.45 (0.91-2.27) 3.87 (1.15-13.03) 2.28 (1.25-4.18)
Okin et al[163]	1839 subjects aged 45- 74 yrs (the Strong Heart Study)	3.7 yrs	All-cause death (188) Cardiac death (55) All-cause death (188) Cardiac death (55)	QTD >58	1.9 (1.1-3.3) 3.4 (1.5-7.5) Not predictive* 2.8 (1.1-6.6)*
De Bruyne et al[191]	2358 men, 3454 women aged ≥55 yrs (the Rotterdam Study)	4 yrs	All-cause death (568) Cardiac death (166) Sudden death (73) Nonfatal cardiac events (193)	QTD > 60 QTcD > 66	1.4 (1.1-1.8) 2.4 (1.5-3.8) 1.8 (0.9-3.5) 1.3 (0.9-1.8) 1.3 (1.0-1.6) 2.1 (1.4-3.3) 1.7 (0.9-3.1) 1.4 (1.0-1.9)
Macfarlane et al[192]	6595 men aged 45-64 yrs with moderate hyper-lipidemia; 1501 healthy subjects aged 18-78 yrs; 1784 children aged 0-17 yrs (WOSCOPS)		Death from CAD + Nonfatal MI	QTD > 44 10-ms increment	36% (2%-81%) 13% (4%-22%)
Zabel et al[197]	280 post MI	32 mo	All-cause death (21) Arrhythmic events (19)	QTD ≥ 61 Tp–Te interval ≥ 92	Not predictive Not predictive
Spargias et al[195]	603 Post MI+CHF (AIREX study); 501 QTD substudy	6 yrs	All-cause mortality	QTcD ≥ 108 versus < 78.2 10-ms increment	1.05 (1.01-1.09) 1.05 (1.02-1.09)
Brooksby et al[198]	554 CHF (495 QTD)	471 days	All-cause death Sudden death (24) Progressive CHF (30)	Not indicated QTD 89 ± 33 dead QTD 79 ± 32 alive	Not predictive Not predictive Not predictive
Brendorp et al[199]	1518 CHF; 704 QTD substudy (DIAMOND-CHF)	1 yr	All-cause mortality (285) Cardiac mortality (219) Arrhythmic mortality (131) Cardiac nonarrhythmic death (88)	QTD ≥ 70	Not predictive Not predictive Not predictive Not predictive
Padmanabhan et al[200]	2263 CHF; 1496 QTD substudy	7 yrs	All-cause mortality (46%)	QTD ≥ 35	Predictive in the elderly and in patients with severe left ventricular dysfunction
Fu et al[201]	163 CHF	26 mo	All-cause mortality (285) Cardiac mortality (219) Arrhythmic mortality (131) Cardiac nonarrhythmic death (88)	JTcD ≥ 85	Relative risk 8.4

* After adjustment to cardiovascular risk factors.

AIREX, Acute Infarction Efficacy Ramipril Extension study; CAD, coronary artery disease; CHF, congestive heart failure; DIAMOND-CHF, Danish Investigation of Arrhythmia and Mortality on Dofetilide in Congestive Heart Failure; JTcD, JT dispersion corrected for heart rate; MI, myocardial infarction; QTD, QT dispersion; QTcD, QT dispersion corrected for heart rate; WOSCOPS, West Of Scotland COronary Prevention Study.

between patient groups and between patients and healthy individuals, and significant variations within each clinical setting render any attempt at establishing distinct reference values impossible.

Several large prospective studies assessed the prognostic value of QT dispersion in the general population. In the Rotterdam and Danish studies, a QT dispersion greater than or equal to 90 ms was associated with a 1.5-fold risk for total mortality and nearly fourfold risk for cardiac death.[166,191] Similar to findings in the prolonged Q–T interval, the association between increased QT dispersion and all-cause mortality and cardiovascular death was stronger in patients with known cardiovascular disease. In the Strong Heart study, a QT dispersion greater than 58 ms (the upper 95th percentile in a separate population of normal subjects) was associated with a twofold increase in risk for total mortality and a 3.4-fold increase in risk for death due to cardiovascular causes.[163] However, in this study, QT dispersion lost its predictive power for total mortality after adjustment for other risk factors. The West Of Scotland COronary Prevention Study (WOSCOPS) investigators reported that in 6595 middle-aged men with moderate hyperlipidemia but no previous myocardial infarction, a 10-ms increment in QT dispersion increased the risk for death from coronary heart disease and risk for nonfatal myocardial infarction by 13% (95% CI 4% to 2%).[192] However, in this study, QT dispersion was dichotomized at 44 ms and predicted adverse outcome with a specificity of 93.8% but a sensitivity of only 8.8%.

Although many studies have found that postmyocardial infarction patients with ventricular arrhythmias have greater QT dispersion than those without rhythm disturbances,[193,194] for most studies with positive results, another study can be found that shows exactly the opposite.[195-197] Zabel and associates[197] showed that none of the 26 indices of repolarization dispersion had any predictive value for adverse outcome in 280 postmyocardial infarction patients who were followed for 32 months. Similarly, recent substudies of large-scale trials (UK-HEART, ELITE, DIAMOND-CHF) have not confirmed previous findings that QT dispersion predicts mortality and morbidity risks in heart failure.[170,198,199]

Several studies have provided serious arguments against the concept of QT dispersion as a measure of regional heterogeneity of myocardial repolarization.[202-205] First, it has been shown that a similar value of QT dispersion is measured in full 12-lead ECGs and in their reconstruction from the rectangular XYZ leads, in which the separate contribution by different myocardial regions has been filtered out. Second, QT dispersion has been found to correlate closely with the parameters of vectorcardiographic T-loop morphology. These observations suggest that variable projection of the T wave loop on separate leads can produce "QT dispersion" in the absence of regional heterogeneity of repolarization.

TRANSMURAL DISPERSION OF VENTRICULAR REPOLARIZATION

The cellular basis for the heterogeneity of ventricular repolarization is not yet completely understood, although regional differences in action potential duration and activation times are believed to contribute to its pathogenesis. Recent experimental studies by Antzelevitch and colleagues[206-208] have provided evidence that M cells found in the deep layers of ventricular myocardium are directly implicated in the inscription of the T wave and in the genesis of QT prolongation and related TU wave abnormalities. The unique feature of M cells is the ability of their action potential to prolong disproportionately compared with endocardial and epicardial layers at slower rates and in response to agents known to increase action potential duration (Fig. 47-17).

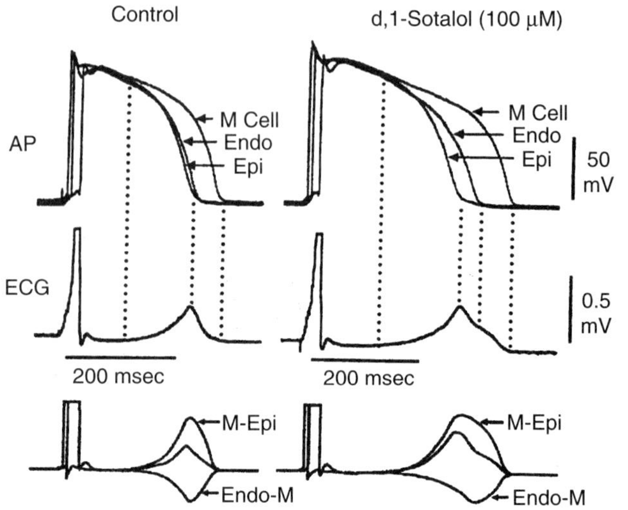

FIGURE 47-17 The effect of d,l-sotalol on action potentials recorded from endocardial (Endo), epicardial (Epi), and M region sites of the canine left ventricular wedge preparation **(top)** and the Q–T interval and T wave in the ECG recorded across the wedge **(middle)**. M region action potential is responsible for the T wave inscription at baseline. Note preferential prolongation of the M region action potential compared with endocardial and epicardial action potentials in response to d,l-sotalol accounting for QT prolongation. (Reproduced from Yan GX, Antzelevitch C: Cellular basis for the normal T wave and the electrocardiographic manifestation of the long QT-syndrome. Circulation 1998;98:1928-1936.)

The ionic basis for these properties of M cells is a weaker slowly activating delayed rectifier current (I_{Ks}) and a stronger late sodium current (late I_{Na}).[209] Indeed, measurements of the transmembrane action potential made from epicardial, mid-myocardial, and endocardial sites of perfused wedges of canine ventricles, have shown that the midmyocardial layer is the last to repolarize and marks the end of the T wave, both in normal conditions and in the presence of QT prolonging agents.[210] The initial portion of the ascending limb of the T wave is due to a rapid decline of the plateau, or phase 2, of the epicardial action potential resulting in the voltage gradient across the wall of the ventricle. The gradient increases as repolarization progresses, reaching the maximum when the epicardial layer is fully repolarized. This corresponds to the peak of the T wave. The next region to repolarize is endocardium, accounting for the initial portion of the descending limb of the T wave. The last region to repolarize is the M cell region, contributing to the terminal portion of the T wave. Thus, the duration of M cell potential determines the duration of the Q–T interval.

The interval between the peak and the end of the T wave on the surface ECG is thought to represent transmural dispersion of repolarization. In the long QT syndrome, differences in action potential duration between the three myocardial layers giving rise to transmural dispersion of repolarization are believed to account for reentrant ventricular tachyarrhythmias initiated by afterdepolarization-induced triggered activity in M cells. Of note, transmural dispersion of repolarization is expected to be smaller in the intact left ventricle than in the wedge preparation, mainly due to electrotonic coupling, which attenuates transmural voltage gradients between different layers of the ventricular wall.

REPOLARIZATION RESERVE

An elegant hypothesis giving insights into mechanisms underlying drug-induced torsades de pointes in the absence of noticeable Q–T interval prolongation under normal conditions is the concept of repolarization reserve.[211] The concept of repolarization reserve suggests that this phenomenon is genetically predetermined and that a decreased reserve is often associated with low-penetration mutations linked to long QT genes, especially those encoding rapidly activating delayed rectifier (I_{Kr}) channels. Therefore, mutations that produce only slight phenotypic effects may be present in the general population and, therefore, antiarrhythmic and noncardiac drugs with I_{Kr} channel blocking properties may pose a risk for the development of torsades de pointes in patients with normal baseline Q–T intervals or in apparently healthy individuals. Factors that are likely to be associated with a reduced repolarization reserve are female gender, slow heart rate, QT dispersion, hypokalemia, diuretic therapy, and the presence of organic heart decease.

Indeed, the proarrhythmic risk is two- to threefold higher in women than in men. Although the etiology of this difference is not yet identified, it may be related to different expression and function of ion channels and the densities of repolarizing currents, which may, in turn, be influenced by hormonal effects. There is experimental evidence for a decreased number of ion channels responsible for repolarization currents in female rabbit ventricular cardiomyocytes[212] and for the greater effects of I_{Kr} blocker dofetilide on the prolongation of the action potential duration, leading to a higher incidence of early afterdepolarizations in female compared with male rabbit hearts.[213] Consistent with these observations, analysis of T wave morphology has revealed substantial gender-specific differences in homogeneity of ventricular repolarization, with women having much greater regional repolarization heterogeneity (Fig. 47-18).[214]

T WAVE ALTERNANS

First noted nearly 100 years ago,[215] electrical alternans of the heart can occur in the QRS complex, ST segment, and the T wave.[216] Of these, the T wave alternans (TWA) has attracted significant attention and has been extensively studied in various pathologic conditions to

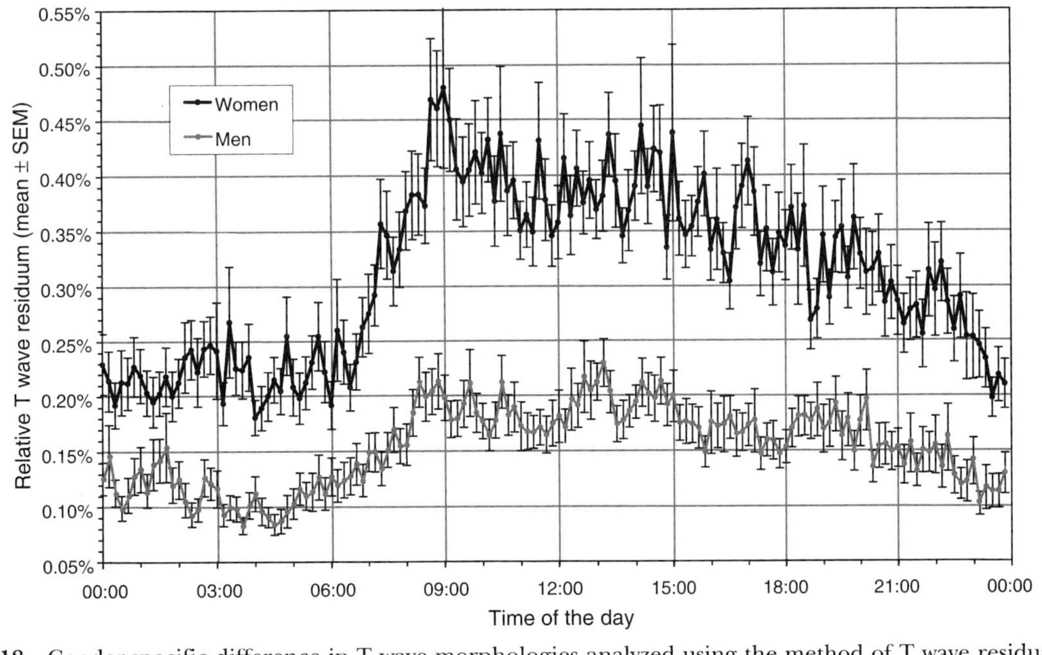

FIGURE 47-18 Gender-specific difference in T wave morphologies analyzed using the method of T wave residua, which represent the proportion of nondipolar components in the total energy of the repolarization signal and reflect local repolarization inhomogeneity. Note a significant increase in heterogeneity of ventricular repolarization in women compared with men during the day and, particularly, early morning hours. (Data from Smetana P, Batchvarov VN, Hnatkova K, et al: Sex differences in repolarisation homogeneity and its circadian pattern. Am J Physiol Heart Circ Physiol 2002;282:H1889-1897.)

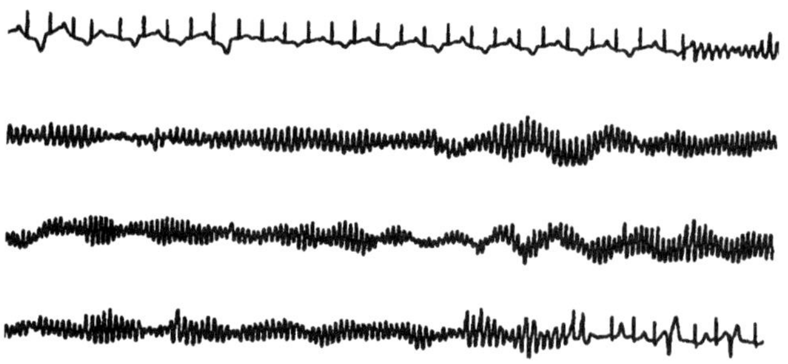

FIGURE 47-19 The two-minute ECG strip from a 24-hour Holter recording in a patient with long QT syndrome who died suddenly. Note a prominent T wave alternans preceding the episode of torsades de pointes that persisted for more than one minute. (Reproduced from Armoundas AA, Nanke T, Cohen RJ: T-wave alternans preceding torsades de pointes ventricular tachycardia. Circulation 2000;101:2550.)

best identify patients at high risk. TWA is defined as consistent beat-to-beat alternations in T wave morphology, amplitude, and polarity occurring during sinus rhythm or atrial pacing. Apparent TWA detected on the surface ECG in 2.5% of patients with long QT syndrome has long been recognized as a marker of an increased risk of sudden death.[217] Figure 47-19 shows a two-minute ECG strip from the Holter recording showing prominent TWA preceding the episode of torsades de pointes in a patient with long QT syndrome.[218]

Mechanisms of the T Wave Alternans

Two observations strongly support the concept that TWA plays an important role in the genesis of ventricular arrhythmias: (1) its association with dispersion of repolarization, and (2) concentration of TWA in the vulnerable phase of the cardiac cycle. In 1984, using finite element analysis of cardiac conduction, Smith and Cohen[219] suggested that the mechanism linking TWA and arrhythmias was regional dispersion of refractoriness, which led to wavefront fractionation and reentry by creating functional conduction block. They demonstrated the presence of 2:1 conduction in some regions of the myocardium resulting in electrical alternans in the simulated ECG.

Recent experimental studies have strongly implicated regional alterations in the action potential duration in the genesis of TWA in the surface ECG, suggesting that TWA is not only a marker of an increased risk but it is directly involved in the development of ventricular arrhythmias.[208,220] Furthermore, TWA observed at a rapid stimulation in the presence of the prolonged Q–T interval was largely due to alternation in the M cells action potential duration and increased transmural dispersion of repolarization resulting in unidirectional conduction block and reentry.[208]

Although ionic mechanisms responsible for TWA have not yet been identified, experimental observations of the ability of calcium channel blockers to suppress ischemia-induced TWA suggest the calcium current involvement in the genesis of TWA.[221] Recordings of the cell transmembrane action potential from different layers of myocardium have provided strong evidence that changes in the potassium current I_{Ks} may also account for TWA occurrence.[208,209]

Methodology

The conventional method based on visual inspection of changes in the amplitude, width, and shape of the T wave, is not useful for the detection of the subtle electrical alternans that occurs at a microvolt level. One of the computerized techniques devised for the quantitative assessment of TWA uses a type of harmonic analysis, or complex demodulation based on the assumption that the T wave represents a sinusoidal signal with a varying amplitude and phase at the alternans frequency.[222] The main advantage of complex demodulation analysis is its robustness against the nonstationary ECG signal and its potential usefulness for dynamic tracking of short-term changes in repolarization alternans. Because of significant analytic complexity, this method is still the prerogative of experimental studies.

Another spectral analysis, using fast Fourier transform, constructs a series of T wave amplitudes, usually for 128 beats, taken within the time window between the J point + 60 ms and the end of the T wave, always at the same time relative to the preceding QRS complex, to generate the power spectrum.[223] As the spectrum reflects measurements taken once per beat, its frequencies are expressed in cycles per beat. Alternans level is defined by a peak centered at exactly 0.5 cycles/beat, which represents oscillations occurring on an every-other-beat basis.

Two numeric indices, the alternans voltage, V_{alt}, and the alternans ratio k, are suggested for the quantification of TWA. V_{alt} is calculated as the square root of the amplitude of the alternans peak above the mean baseline noise level and represents the magnitude of the variation in T wave morphology compared with the mean T wave. The alternans ratio k is expressed as alternans power relative to the standard deviation of the baseline noise. TWA is considered clinically significant if V_{alt} is greater than or equal to 1.9 µV and k is greater than or equal to 3 at a heart rate of less than or equal to 110 beats per minute or less than or equal to 70% of predicted maximum heart rate, and at least one minute in duration. Figure 47-20 presents the positive (a) and negative (b) results of TWA analysis during an exercise test using commercial equipment (Cambridge Heart, Inc.).

Clinical Studies

The first experience with assessment of the prognostic value of TWA during atrial pacing at 100 beats per minute in 83 patients referred to an electrophysiological study showed that the presence of TWA predicted not only inducibility of ventricular arrhythmias but, more importantly, the occurrence of arrhythmic events during 20-month follow-up.[224] In this study, only 6% of patients who did not have TWA developed ventricular tachyarrhythmias compared with 81% of patients who presented with TWA during atrial pacing.

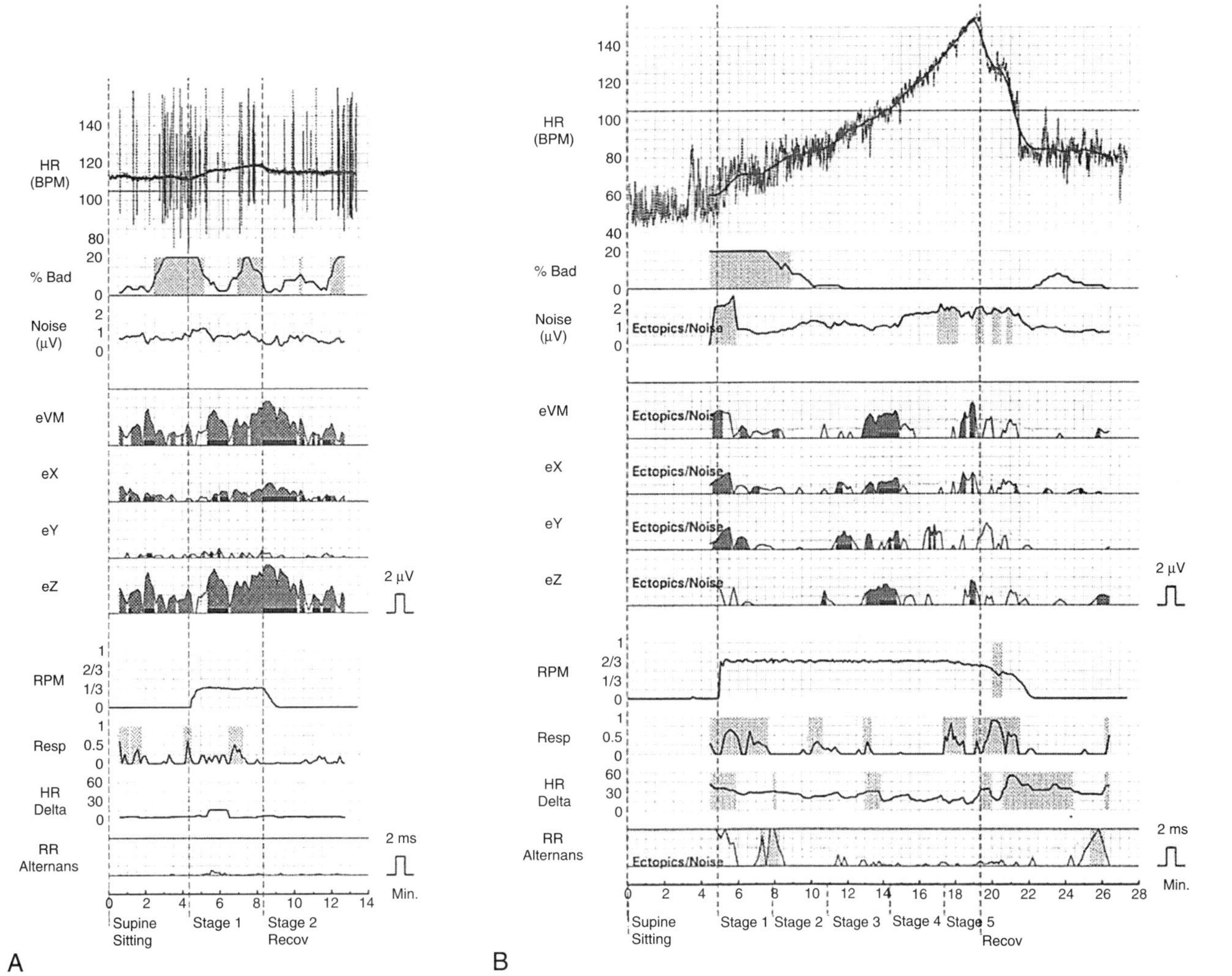

FIGURE 47-20 The results of T wave alternans analysis during exercise test using commercial equipment (Cambridge Heart, Inc., Bedford, MA). Tracings *from top to bottom*: heart rate (HR), percent of ectopic beats (% Bad), noise level, the alternans voltage (V_{alt}) in vector magnitude (eVM) and in orthogonal leads XYZ, pedaling rate (RPM), respiration (Resp), changes in heart rate (HR Delta), and RR alternans magnitude. Shaded area corresponds to alternans ratio κ greater than or equal to 3. V_{alt} greater than or equal to 1.9 μV and κ greater than or equal to 3 at a heart rate less than or equal to 110 beats/minute indicate the presence of T wave alternans in a patient with myocardial infarction and nonsustained ventricular tachycardia (**A**). Compare with a negative test in a normal subject (**B**).

The presence of TWA during an exercise-stress test was associated with a 4.3-fold increased risk for inducible VT during the electrophysiological study and a 10.6-fold increased risk for recurrent VT during a 1-year follow-up.[225] In 95 patients with ICD implanted for recurrent sustained ventricular tachyarrhythmias, TWA and left ventricular ejection fraction were the only predictors of an appropriate ICD discharge in univariate analysis, and only TWA was associated with an appropriate ICD discharge in multivariate analysis during 36 months of follow-up.[142] Of note, in the subgroup of patients with coronary artery disease, preserved BRS and the absence of TWA were associated with 100% event-free survival (Fig. 47-21). TWA at discharge was a stronger predictor of future sustained VT than low ejection fraction and the presence of late potentials in 102 patients with acute myocardial infarction.[226] The combination of abnormal TWA, low ejection fraction, late potentials, and nonsustained VT identified patients at very high (19-fold) risk of death with a 96% predictive accuracy.[226a] In patients with idiopathic dilated cardiomyopathy and congestive heart failure, TWA conferred a nearly 3.5-fold increased risk of ventricular tachyarrhythmias and sudden death and significantly overpowered left ventricular ejection fraction and other noninvasive risk factors.[226b,226c] None of the 107 patients with congestive heart failure who had normal TWA during an exercise stress test died or suffered an arrhythmic event during a 14-month follow-up period.[226c] These observations suggest that TWA probably is more closely involved in mechanisms of ventricular arrthythmias and, therefore, may better predict sudden death and arrhythmic events than other established risk stratifiers.

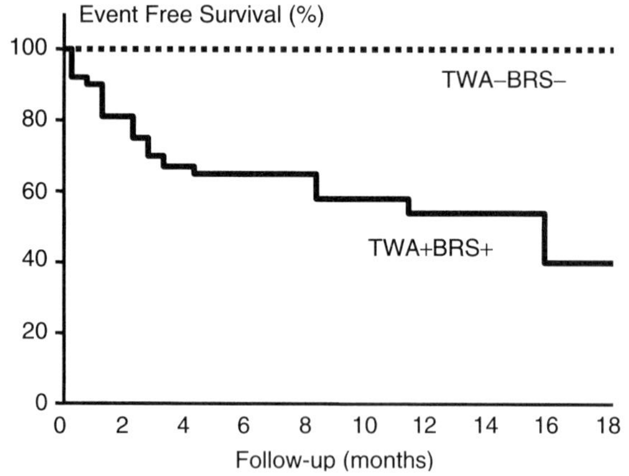

FIGURE 47-21 Event-free survival in patients with coronary artery disease and a cardioverter-defibrillator implanted for ventricular tachyarrhythmias stratified according to the presence or absence of T wave alternans and abnormal baroreflex sensitivity. TWA+, T wave alternans is present; TWA–, T wave alternans is absent; BRS+, baroreflex sensitivity less than or equal to 3 ms/mm Hg; BRS–, baroreflex sensitivity greater than 3 ms/mm Hg. (Modified from Hohnloser SH, Klingenheben T, Yi-Gang Li, et al: T wave alternans as a predictor of recurrent ventricular tachyarrhythmias in ICD recipients: Prospective comparison with conventional risk markers. J Cardiovasc Electrophysiol 1998;9:1258-1268.)

ASSESSMENT OF T WAVE MORPHOLOGY

Several methods have been suggested for the evaluation of the T wave shape, including principal component analysis of the T wave, algebraic decomposition of the T wave, cluster analysis, and the analysis of energy distribution in the T wave.

Principal component analysis (PCA) has been proposed for the assessment of the complexity of ventricular repolarization from the standard ECG and from the 12-lead digital Holter ECG recordings.[227] The method defines the principal, nonredundant spatial components into which the T wave is decomposed and which contribute to morphology of the T wave. The significance of each component is measured by eigenvalue. When repolarization is uniform, most of the information about morphology of the T wave is contained in the first, main component. When repolarization is heterogeneous and T wave morphology is complex, the relative value of the next, smaller component of the T wave increases. The ratio between the second and the first eigenvalues of the T wave vector (PCA ratio) characterizes T wave loop morphology. Recent data from the Strong Heart study have shown that a PCA greater than the 90th percentile (24.6% in men and 32% in women) was associated with a 2.77-fold increased risk for cardiac death in men and a 3.68-fold increased risk for cardiac death in women during a 4-year follow-up (Fig. 47-22).[228] Gender differences in the PCA ratio can be attributed to the lower T wave amplitudes and T wave areas found in women, despite similar or slightly longer durations of repolarization, which would, on average, result in higher ratios of the second to first eigenvalues in women than in men.

Recently, among various signal-processing methods, wavelet transformation analysis has been applied to the components of ventricular repolarization, including the T and U waves. This approach involves decomposition of the ECG signal in a set of coefficients. Each analyzing wavelet has its own time duration, time location, and frequency band. Wavelet coefficient corresponds to a measurement of the ECG components in given time segment and frequency bands. In families with long QT syndrome it has been shown that time-frequency characteristics of ventricular repolarization differ in gene carriers and noncarriers.[229] Wavelet transformation, which does not require the detection of the end of the T wave, may represent an alternative to the time-domain analysis of repolarization abnormalities.

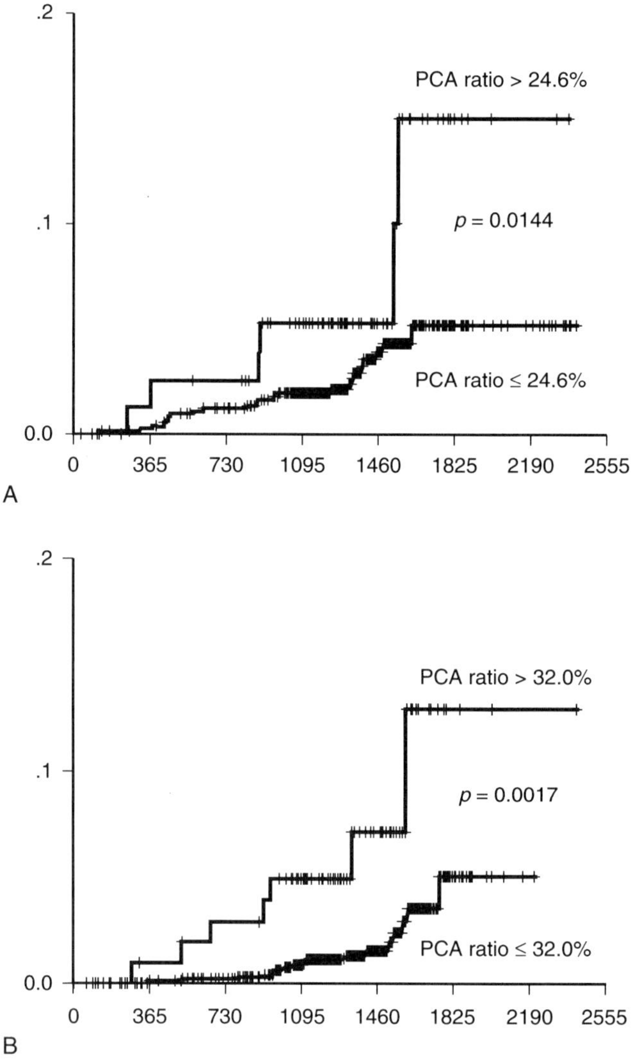

FIGURE 47-22 Kaplan-Meier curve plots of cumulative cardiovascular disease mortality in men (**A**) and women (**B**) grouped according to PCA ratios dichotomized at 24.6% and 32% (the upper 95th percentile value), respectively. PCA, Principal component analysis. (Modified from Okin PM, Devereux R, Fabsitz RR, et al: Principal component analysis of T wave and prediction of cardiovascular mortality in American Indians: The Strong Heart study. Circulation 2002;105:714-719.)

Most recently, a concept of measuring dispersion of T wave morphology has been suggested by Malik and colleagues.[230] Based on singular value decomposition of the 12-lead ECG and on the computation of eigenvalues of the signal, several indices have been proposed to characterize temporal and spatial variations in T wave morphology:

1. Dispersion of T wave morphology, which evaluates the morphologic dissimilarities between the T waves in the individual leads of the 12-lead electrocardiogram and reflects the spatial variations in the T wave pattern
2. Normalized T wave loop area, which describes the complexity of the three-dimensional T wave loop and characterizes the temporal variations in the T wave pattern
3. Dispersion of T wave loop, which expresses the temporal variations in ventricular activation sequence
4. Total cosine R to T (TCRT), which reflects the spatial angle between depolarization and repolarization main vectors

These descriptors have been validated in 280 post-myocardial infarction patients with prospectively defined end-points.[231] In a multivariate model including age, left ventricular ejection fraction, heart rate, QRS width, heart rate variability (SDNN), thrombolytic therapy, β-blocker therapy, and new repolarization descriptors, four variables—heart rate, ejection fraction, reperfusion therapy, and TCRT—yielded an independent predictive value for event-free survival. In 1047 patients with myocardial infarction enrolled in St. George's Hospital Post Infarction Research Survey Programme, a TCRT of less than or equal to –(–0.88) and left ventricular ejection fraction of less than 33% risk were independent predictors of 5-year cardiac mortality (relative risks 3.4 and 2.9, respectively).[232] Five-year cardiac mortality was 8% in patients with both a TCRT greater than –0.88 and an ejection fraction greater than 33%, increasing to 18% in patients with either TCRT ejection fraction, and to more than 50% in patients with both abnormal TCRT and significantly reduced ejection fraction (Fig. 47-23).

SUMMARY

The compelling experimental and clinical evidence supports the concept that the assessment of ventricular repolarization on the surface ECG represents a potential tool for arrhythmia risk stratification in a variety of clinical conditions. Although many attempts regarding the quantitative assessment of this phenomenon have been undertaken, ranging from the primitive measurements of QT dispersion to complex mathematical analyses, none of these technologies is able to evaluate reliably spatial and temporal heterogeneities of repolarization. Further development of practical approaches to the quantitative assessment of ventricular repolarization is essential for better understanding of physiologic and pathophysiologic processes governing this phenomenon, for clinical appraisal of its abnormalities, and for improving the arrhythmia risk stratification.

Although substantial advances have been made in noninvasive cardiac electrophysiology, in large-scale prospective clinical studies in patients with different underlying pathologies, none of the known noninvasive risk factors alone was able to select high-risk groups with sufficient sensitivity and positive predictive accuracy. Therefore, multiple combinations of several risk stratifiers have been suggested. The rationale for such combinations is often unclear, and practical methods for combining several factors are far from simple. Multivariate risk stratification strategies and stepwise strategies are becoming more popular. Currently, several prospective randomized intervention trials aimed at primary prevention of sudden death are being pursued based on stepwise risk stratification strategies.

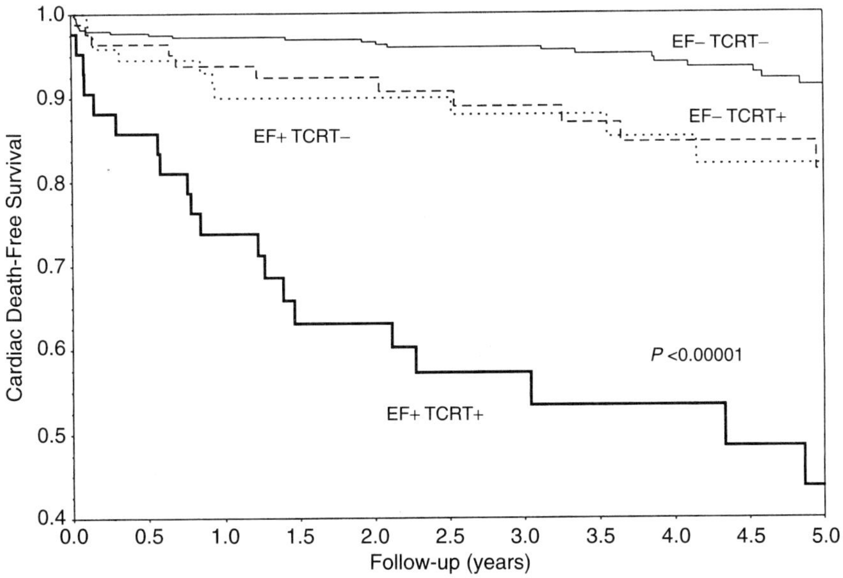

FIGURE 47-23 Cumulative five-year cardiac death-free survival. Note that in patients with depressed left ventricular ejection fraction (<33%, EF+), those with preserved and decreased TCRT (>–0.88 and ≤–0.88, TCRT– and TCRT+, respectively) have mortality rates of approximately 18% and 56%, respectively. (From Batchvarov VN, Hnatkova K, Ghuran A, et al: Ventricular gradient is an independent risk factor in survivors of acute myocardial infarction. J Am Coll Cardiol 2002;39:125A [Abstract].)

REFERENCES

1. Holter NJ: New method for heart studies: Continuous electrocardiography of active subjects over long periods is now practical. Science 1961;134:1214-1220.
2. Klootwijk P, Meij S, von Es GA, et al: Comparison of usefulness of computer assisted continuous 48-h 3-lead with 12-lead ECG ischaemia monitoring for detection and quantitation of ischaemia in patients with unstable angina. Eur Heart J 1997;18:931-940.
3. Bigger JT, Fleiss JL, Kleiger R, et al: The Multicenter Post-Infarction Research Group: The relationships among ventricular tachyarrhythmias, left ventricular dysfunction, and mortality in the 2 years after myocardial infarction. Circulation 1984;69:250-258.
4. Maggioni AP, Zuanetti G, Franzosi MG, et al: The GISSI-2 Investigators: Prevalence and prognostic significance of ventricular arrhythmias after acute myocardial infarction in the fibrinolytic era: The GISSI-2 results. Circulation 1993;87:312-322.
5. Moss AJ, Hall WJ, Cannom DS, et al: The Multicenter Automatic Defibrillator Implantation Trial Investigators: Improved survival with an implanted defibrillator in patients with coronary disease at high risk for ventricular arrhythmia. N Engl J Med 1996;335: 1933-1940.
6. Buxton AE, Lee KL, Fisher JD, et al: The Multicenter Unsustained Tachycardia Trial Investigators: A randomized study of the prevention of sudden death in patients with coronary artery disease. N Engl J Med 1999;341:1882-1890.

6a. La Rovere MT, Pinna GD, Hohnloser SH, et al: The ATRAMI (Autonomic Tone and Reflexes After Myocardial Infarction Investigators): Baroreflex sensitivity and heart rate variability in the identification of patients at risk for life-threatening arrhythmias: Implications for clinical trials. Circulation 2001;103:2072-2077.

7. Hodges M, Bailey JJ: Ambulatory electrocardiography: Use in arrhythmia risk assessment. In Malik M (ed): Risk of Arrhythmia and Sudden Death. London, BMJ Books, 2001, pp 194-201.
8. Caruso AC, Marcus FI, Hahn EA, et al: The ESVEM Investigators. Predictors of arrhythmic death and cardiac arrest in the ESVEM trial. Circulation 1997;96:1888-1892.
9. Teerlink JR, Jalaluddin M, Anderson S, et al: The PROMISE (Prospective Randomized Milrinone Survival Evaluation) Investigators: Ambulatory ventricular arrhythmias in patients with heart failure do not specifically predict an increased risk of sudden death. Circulation 2000;101:40-46.
10. Doval HC, Nul DR, Grancelli HO, et al: Nonsustained ventricular tachycardia in severe heart failure: Independent marker of increased mortality due to sudden death: GESICA-GEMA Investigators. Circulation 1996;94:3198-3203.
11. Singh SN, Fisher SG, Carson PE, Fletcher RD: Prevalence and significance of nonsustained ventricular tachycardia in patients with premature ventricular contractions and heart failure treated with vasodilator therapy: Department of Veterans Affairs CHF STAT Investigators. J Am Coll Cardiol 1998;32:942-947.
12. Moss AJ, Zareba W, Hall WJ, et al: The Multicenter Automatic Defibrillator Implantation Trial Investigators: Prophylactic implantation of a defibrillation in patients with myocardial infarction and reduced ejection fraction. N Engl J Med 2002; 346:877-883.
13. Franzosi MG, Santoro E, De Vita C, et al: GISSI Investigators. Ten year follow-up of the first mega-trial testing thrombolytic therapy in patients with acute myocardial infarction: Results of the Gruppo Italiano per lo Studio della Sopravvivenza nell'Infarcto-1 Study. Circulation 1998;2659-2665.

13a. Hohnloser SH, Klingenheben T, Zabel M, et al: Prevalence, characteristics and prognostic value during long-term follow-up of nonsustained ventricular tachycardia after myocardial infarction in the thrombolytic era. J Am Coll Cardiol 1999;33:1895-1902.

14. MERIT-HF Study Group. Effect of metoprolol CR/XL in chronic heart failure: Metoprolol CR/XL Randomised Intervention Trial in Congestive Heart Failure (MERIT-HF). Lancet 1999;353:2001-2007.
15. Mason JW: A comparison of electrophysiologic testing with Holter monitoring to predict antiarrhythmic drug efficacy for ventricular tachyarrhythmias: Electrophysiologic Study Versus Electrocardiographic Monitoring Investigators. N Engl J Med 1993;329: 445-451.
16. Echt DS, Liebson PR, Mitchell LB, et al: The CAST Investigators, et al: Mortality and morbidity in patients receiving encainide, flecainide, or placebo: The Cardiac Arrhythmia Suppression Trial. N Engl J Med 1991;324:781-788.
17. The Cardiac Arrhythmia Suppression Trial II Investigators. Effect of the antiarrhythmic agent moricizine on survival after myocardial infarction. N Engl J Med 1992;327:227-233.
18. Fetsch T, Breithardt G, Engberding R, et al: Can we believe in symptoms for detection of atrial fibrillation in clinical routine? The results of the PAFAC study. Circulation 2001;104:II-699 [Abstract].
19. Zimetbaum PJ, Josephson ME: The evolving role of ambulatory arrhythmia monitoring in clinical practice. Ann Intern Med 1999;130:848-856.
20. Waktare JEP: Current status and developments in cardiac event recorders. Card Electrophysiol Rev 1999;3:254-256.
21. Krahn AD, Klein GJ, Yee R, et al: The Reveal Investigators. Use of an extended monitoring strategy in patients with problematic syncope. Circulation 1999;99:406-410.
22. Defaye P, Dournaux F, Mouton E: The AIDA Multicenter Study Group. Prevalence of supraventricular arrhythmias from the automated analysis of data stored in the DDD pacemakers of 617 patients: The AIDA (Automatic Interpretation for Diagnostic Assistance) study. PACE 1998;21:250-255.
23. Hügl BJ, Sperzel J, Ziegenbalg K, et al: Incidence of atrial arrhythmias in a DDD-pacemaker population: Results of the European Pacemaker Kappa Registry database. Circulation 2001;23 [Abstract].
24. Stein KM, Hess M, Hannon CH, et al: Simultaneous atrial and ventricular tachyarrhythmias in defibrillator recipients: Does AF beget VF? J Am Coll Cardiol 1999;33:115A [Abstract].
25. Gold MR, Shorovsky SR, Rashba EJ, et al: Incidence of spontaneous ventricular arrhythmias in defibrillator patients with atrial fibrillation but no history of ventricular arrhythmias. Circulation 2000;102:II-395-II-396 [Abstract].
26. McHenry PL, Fisch C, Jordan JW, et al: Cardiac arrhythmias observed during maximal treadmill exercise testing in clinically normal men. Am J Cardiol 1972;29:331-336.
27. Gooch AS, McConnell D: Analysis of transient arrhythmia and conduction disturbances during submaximal treadmill exercise testing. Progr Cardiovasc Dis 1970;16:497-522.
28. Corbelli R, Masterson M, Wilkoff BL: Chronotropic response to exercise in patients with atrial fibrillation. PACE 1990;13:179-187.
29. Johnston FA, Robinson JF, Fyfe T: Exercise testing in the diagnosis of sick sinus syndrome in the elderly: Implications for treatment. PACE 1987;10:831-838.
30. Poblete PF, Kennedy HL, Caralis DG: Detection of ventricular ectopy in patients with coronary heart disease and normal subjects by exercise testing and ambulatory electrocardiography. Chest 1978;74:402-407.
31. Yang JC, Wesley RCJ, Froelicher VF: Ventricular tachycardia during routine treadmill testing: Risk and prognosis. Arch Intern Med 1991;151:349-353.
32. Codini MA, Sommerfeldt L, Eybel CE, Messer JV: Clinical significance and characteristics of exercise-induced ventricular tachycardia. Catheter Cardiovasc Diag 1981;7:227-234.
33. Buxton AE, Waksman HL, Marchlinski FE, et al: Right ventricular tachycardia; Clinical and electrophysiologic characteristics. Circulation 1983;68:917-927.
34. Mont L, Seixas T, Brugada P, et al: Clinical and electrophysiologic characteristics of exercise-related idiopathic ventricular tachycardia. Am J Cardiol 1991;68:897-900.
35. Fioretti P, Deckers J, Baardman T, et al: Incidence and prognostic implications of repetitive ventricular complexes during predischarge bicycle ergometry after myocardial infarction. Eur Heart J 1987;8[Suppl D]:51-54.
36. Newby KH, Thompson T, Stebbins A, et al: The GUSTO Investigators. Sustained ventricular arrhythmias in patients receiving thrombolytic therapy: Incidence and outcomes. Circulation 1998;98:2567-2573.
37. Kenny RA, Bayliss J, Ingram A, Sutton R: Head-up tilt: A useful test for investigating unexplained syncope. Lancet 1986;1:1352-1355.
38. Benditt DG: Neurally mediated syncopal syndromes: Pathophysiological concepts and clinical evaluation. PACE 1997;20:572-584.

39. Grubb PB: Pathophysiology and differential diagnosis of neurocardiogenic syncope: A brief review. Am J Cardiol 1999;84: 3Q-9Q.
40. Grubb BP, Kosinski D, Boehm K, Kip K: The postural orthostatic tachycardia syndrome: A neurocardiogenic variant identified during head up tilt table testing. PACE 1997;20:2205-2212.
41. Benditt DG, Ferguson DW, Grubb PB, et al: Tilt table testing for assessing syncope: The American College of Cardiology. J Am Coll Cardiol 1996;28:263-275.
42. Fitzpatrick A, Sutton R: Tilting towards a diagnosis in recurrent unexplained syncope. Lancet 1989;1:658-660.
43. Raviele A, Menozzi C, Brignole M, et al: Value of head-up tilt testing potentiated with sublingual nitroglycerin to assess the origin of unexplained syncope. Am J Cardiol 1995;76:267-272.
44. Almquist A, Goldenberg IF, Milstein S, et al: Provocation of bradycardia and hypotension by isoproterenol and upright posture in patients with unexplained syncope. N Engl J Med 1989;320: 346-351.
45. Berbari EJ, Scherlag BJ, Hope RR, Lazzara R: Recording from the body surface of arrhythmogenic ventricular activity during the S-T segment. Am J Cardiol 1978;41:697-702.
46. Simson MB: Use of signals in the terminal QRS complex to identify patients with ventricular tachycardia after myocardial infarction. Circulation 1981;64:235-242.
47. Breithardt G, Cain ME, El-Sherif N, et al: Standards for analysis of ventricular late potentials using high resolution or signal-averaged electrocardiography: A statement by a Task Force Committee between the European Society of Cardiology, the American Heart Association and the American College of Cardiology. Eur Heart J 1991;12:473-480.
48. Breithardt J, Schwarzmaier M, Borggrefe K, et al: Prognostic significance of late ventricular late potentials after acute myocardial infarction. Eur Heart J 1983;4:487-495.
49. Kuchar DL, Thorburn CW, Sammel NL: Prediction of serious arrhythmic events after myocardial infarction: Signal-averaged electrocardiogram, Holter monitoring, and radionuclide ventriculography. J Am Coll Cardiol 1987;9:531-538.
50. Gomes JA, Winters SL, Steward D, et al: A new non-invasive index to predict sustained ventricular tachycardia and sudden death in the first year after myocardial infarction: Based on signal-averaged electrocardiogram, radionuclide ejection fraction and Holter monitoring. J Am Coll Cardiol 1987;10:349-357.
51. Bloomfield DM, Snyder JE, Steinberg JS: A critical appraisal of quantitative spectro-temporal analysis of signal-averaged ECG: Predicting arrhythmic events after myocardial infarction. PACE 1996;19:768-777.
52. Dennis AR, Ross DL, Richards DA, et al: Differences between patients with ventricular tachycardia and ventricular fibrillation as assessed by signal-averaged electrocardiogram, radionuclide ventriculography and cardiac mapping. J Am Coll Cardiol 1988;11:276-283.
53. Steinberg JS, Berbari EJ: The signal-averaged electrocardiogram: Update on clinical applications. J Cardiovasc Electrophysiol 1996;7:972-988.
54. Pedretti R, Etro MD, Laporta A, et al: Prediction of late arrhythmic events after acute myocardial infarction from combined use of noninvasive prognostic variables and inducibility of sustained monomorphic ventricular tachycardia. Am J Cardiol 1993;71:1131-1141.
55. Makijavri M, Fetsch T, Reinhardt L, et al: The Postinfarction Late Potential (PILP) Study: Comparison and combination of late potentials and spectral turbulence analysis to predict arrhythmic events after myocardial infarction in the Postinfarction Late Potential (PILP) Study. Eur Heart J 1995;16:651-659.
56. Denes P, El-Sherif N, Katz R, et al: The Cardiac Arrhythmia Suppression Trial (CAST) SAECG Substudy Investigators: Prognostic significance of signal-averaged electrocardiogram after thrombolytic therapy and/or angioplasty during acute myocardial infarction (CAST Substudy). Am J Cardiol 1994;74: 216-220.
57. Zimmerman M, Sentici A, Adamec R, et al: Long-term prognostic significance of ventricular late potentials after a first acute myocardial infarction. Am Heart J 1997;134:1019-1028.
58. Gomes JA, Cain ME, Buxton AE, et al: Prediction of long-term outcomes by signal-averaged electrocardiography in patients with unsustained ventricular tachycardia, coronary artery disease, and left ventricular dysfunction. Circulation 2001;104: 436-441.
58a. Kalahasti V, Nambi V, Martin DO, et al: QRS duration and prediction of mortality in patients undergoing risk stratification for ventricular arrhythmias. Am J Cardiol 2003;92:798-803.
59. Steinbigler P, Haberl R, Brueggemann T, et al: The Munich and Berlin Infarct Study (MABIS): Functional late potential analysis in the Holter-ECG is predictive for malignant ventricular arrhythmia in 1120 patients. Circulation 1999;100:I-244 [Abstract].
60. Klingenheben T, Hohnloser SH: Usefulness of risk stratification for future cardiac events in infarct survivors with severely depressed versus near-normal left ventricular function: Results from a prospective long-term follow-up study. Ann Noninvasive Electrocardiol 2003;8:68-74.
61. Fetsch T, Reinhardt L, Borggrefe M, et al: The role of ventricular late potentials for risk stratification in a postinfarction population with reduced left ventricular function: Results from EMIAT. Circulation 1997;96:I-459 [Abstract].
62. Klingenheben T, Credner SC, Mauss O, et al: Comparison of measures of autonomic tone and the signal-averaged ECG for risk stratification after myocardial infarction: Results of a prospective long-term follow-up trial in 411 consecutive patients. Eur Heart J 1999;20:342 [Abstract].
63. Zareba W, Steinberg JS, Moss AJ, et al: The MADIT Investigators. Signal-averaged ECG in the Multicenter Automatic Defibrillator Implantation Trial (MADIT). J Am Coll Cardiol 1997;29:31A [Abstract].
64. Cook JR, Flack JE, Gregory GA, et al: The CABG Patch Trial: Influence of the perioperative signal-averaged electrocardiogram on left ventricular function after coronary bypass graft surgery in patients with left ventricular dysfunction. Am J Cardiol 1998;82:285-289.
65. Steinberg JS, Prystowsky E, Freedman RA, et al: Use of signal-averaged electrocardiogram for predicting inducible ventricular tachycardia in patients with unexplained syncope: Relation to clinical variables in a multivariate analysis. J Am Coll Cardiol 1994;23:99-106.
66. Griffith MJ, de Belder MA, Mehta D, et al: Signal averaging of the electrocardiogram in the remote differential diagnosis of broad complex tachycardia. Eur Heart J 1991;12:777-783.
66a. Grimm W, Hoffman J, Knop U, et al: Value of time- and frequency-domain analysis of signal-averaged electrocardiography for arrhythmia risk prediction in idiopathic dilated cardiomyopathy. PACE 1996;19:1923-1927.
67. Kulakowski P, Counihan PJ, Camm AJ, McKenna WJ: The value of time and frequency domain, and spectral temporal mapping analysis of the signal-averaged electrocardiogram in identification of patients with hypertrophic cardiomyopathy at increased risk of sudden death. Eur Heart J 1993;14:941-950.
68. Haberl R, Jilge G, Putler R, et al: Spectral mapping of the electrocardiogram with Fourier transform for identification of patients with sustained ventricular tachycardia and coronary artery disease. Eur Heart J 1989;10:316-322.
69. Kelen GJ, Henkin R, Starr AM, et al: Spectral turbulence analysis of the signal-averaged electrocardiogram and its predictive accuracy for inducible sustained monomorphic ventricular tachycardia. Am J Cardiol 1991;67:965-975.
70. Lander P, Gomis P, Goyal R, et al: Analysis of abnormal intra-QRS potentials: Improved predictive value for arrhythmic events with the signal-averaged electrocardiograms. Circulation 1997;95:1386-1393.
71. Kavesh NG, Cain ME, Ambos HD, Arthur RM: Enhanced detection of distinguishing features in signal-averaged electrocardiograms from patients with ventricular tachycardia by combined spatial and spectral analyses of entire cardiac cycle. Circulation 1994;90:254-263.
72. Morlet D, Peyrin F, Desseigne P, et al: Wavelet analysis of high-resolution signal-averaged ECGs in postinfarction patients. J Electrocardiol 1993;23:311-320.
73. Reinhardt L, Makijavri M, Fetsch T, et al: Predictive value of wavelet correlation functions of signal-averaged electrocardiogram in patients after anterior versus inferior myocardial infarction. J Am Coll Cardiol 1996;27:53-59.

74. Malik M, Hnatkova K, Staunton A, Camm AJ: Wavelet analysis of signal-averaged electrocardiograms. II. Risk stratification after acute myocardial infarction. Ann Noninvas Electrocardiol 2000;5:20-29.
75. Hnatkova K, Malik M, Kulakowski P, Camm AJ: Wavelet analysis of signal-averaged electrocardiograms. I. Design, clinical assessment, and reproducibility of the method. Ann Noninvasive Electrocardiol 2000;5:4-19.
76. Englund A, Hnatkova K, Kulakowski P, et al: Wavelet decomposition analysis of the signal averaged electrocardiogram used for risk stratification of patients with hypertrophic cardiomyopathy. Eur Heart J 1998;19:1383-1390.
77. Yi G, Hnatkova K, Mahon NG, et al: Predictive value of wavelet decomposition of the signal-averaged electrocardiogram in idiopathic dilated cardiomyopathy. Eur Heart J 2000;21:1015-1022.
78. Fukunami M, Yamada T, Ohmori M, et al: Detection of patients at risk for paroxysmal atrial fibrillation during sinus rhythm by P wave-triggered signal-averaged electrocardiogram. Circulation 1991;83:162-169.
79. Steinberg JS, Zelenkofske S, Wong SC, et al: Value of the P-wave signal-averaged ECG for predicting atrial fibrillation after cardiac surgery. Circulation 1993;88:2618-2622.
80. Yamada T, Fukunami M, Shimonagata T: Prediction of paroxysmal atrial fibrillation in patients with congestive heart failure: A prospective study. J Am Coll Cardiol 2000;35:405-413.
81. Raitt MH, Ingram KD, Thurman SM: Signal-averaged P wave duration predicts early recurrence of atrial fibrillation after cardioversion. PACE 2000;23:259-265.
82. Abe Y, Fukunami M, Yamada T, et al: Prediction of transition to chronic atrial fibrillation in patients with paroxysmal atrial fibrillation by signal-averaged electrocardiography: A prospective study. Circulation 1997;96:2612-2616.
83. Yamada T, Fukunami M, Shimonagata T, et al: Identification of the involvement of sinus node dysfunction in patients with paroxysmal atrial fibrillation by atrial early potentials. J Am Coll Cardiol 1999;33:143A [Abstract].
84. Gencel L, Poquet F, Gosse P, et al: Correlation of signal-averaged P wave with electrophysiological testing for atrial vulnerability in strokes of unexplained etiology. PACE 1994;17:2118-2124.
85. Ehlert FA, Korenstein D, Steinberg JS: Evaluation of P wave signal-averaged electrocardiographic filtering and analysis methods. Am Heart J 1997;134:985-993.
86. Savelieva I, Aytemir K, Hnatkova K, et al: Agreement between automatic and manual measurement of atrial and ventricular signal-averaged electrocardiograms in normal subjects. Ann Noninvas Electrocardiol 2000;5:133-138-127.
87. Savelieva I, Aytemir K, Hnatkova K, et al: Short-, mid-, and long-term reproducibility of the atrial signal averaged electrocardiogram in healthy subjects: Comparison with the conventional ventricular signal averaged electrocardiogram. PACE 2000;23: 122-127.
88. Stafford PJ, Cooper J, Fothergill J, et al: Reproducibility of the signal averaged P wave: Time and frequency domain analysis. Heart 1997;77:412-416.
89. Yamada T, Fukunami M, Shimonagata T, et al: Dispersion of signal-averaged P wave duration on precordial body surface in patients with paroxysmal atrial fibrillation. Eur Heart J 1999;20:211-220.
90. Platonov PG, Carlson J, Ingemansson MP, et al: Detection of concealed interatrial conduction defects with unfiltered signal-averaged P-wave ECG in patients with lone atrial fibrillation. Europace 2000;2:32-41.
91. Lown B, Verrier RL: Neural activity and ventricular fibrillation. N Engl J Med 1976;294:1165-1170.
92. Malliani A, Schwatz PJ, Zanchetti A: A sympathetic reflex elicited by experimental coronary occlusion. Am J Physiol 1969;217: 703-709.
93. Rosen M, Hordof AJ, Ilvento JP, et al: Effects of adrenergic amines on electrophysiological properties and automaticity of neonatal and adult canine Purkinje fibres: Evidence of alpha- and beta-adrenergic actions. Circ Res 1977;40:390-400.
94. Kolman BS, Verrier RL, Lown B. The effect of vagus nerve stimulation upon vulnerability of the canine ventricle: Role of the sympathetic-parasympathetic interactions. Circulation 1975;52: 578-585.
95. De Ferrari GM, Vanoli E, Stramba-Badiale M, et al: Vagal reflexes and survival during acute myocardial ischemia in conscious dogs with healed myocardial infarction. Am J Physiol 1991;261: H63-H69.
96. Cerati D, Schwartz PJ: Single cardiac fiber activity, acute myocardial ischemia, and risk for sudden death. Circ Res 1991;69: 1389-1401.
97. Raeder EA, Verrier RL, Lown B: Intrinsic sympathomimetic activity and the effects of beta-adrenergic blocking drugs on vulnerability to ventricular fibrillation. J Am Coll Cardiol 1983;1: 1442-1446.
98. Gnecchi Ruscone T, Lombardi F, Malfatto G, et al: Attenuation of baroreceptive mechanisms by cardiovascular sympathetic afferent fibers. Am J Physiol 1987;253:H787-H791.
99. Lombardi F: Basic autonomic tests. In Malik M (ed): Risk of Arrhythmia and Sudden Death. London, BMJ, 2001, pp 209-212.
100. Levine HJ: Rest heart rate and life expectancy. J Am Coll Cardiol 1997;30:1104-1106.
101. Kannel WB, Kannel C, Paffenbarger RS, et al: Heart rate and cardiovascular mortality: The Framingham study. Am Heart J 1987;113:1489-1494.
102. Algra A, Tijssen JGP, Roetlandt JRTC, et al: Heart rate variability from 24-hour electrocardiography and the 2-year risk of sudden death. Circulation 1993;88:180-185.
103. Benetos A, Rudnichi A, Thomas F, et al: Influence of heart rate on mortality in a French population: Role of age, gender, and blood pressure. Hypertension 1999;33:44-52.
104. Wannamethee G, Shaper AG, Macfarlane PW, et al: Risk factors for sudden cardiac death in middle-aged British men. Circulation 1995;91:1749-1756.
105. Copie X, Hnatkova K, Staunton A, et al: Predictive power of increased heart rate versus depressed left ventricular ejection fraction and heart rate variability for risk stratification after myocardial infarction: A two-year follow-up. J Am Coll 1996;27: 270-276.
106. Kleiger RE, Miller JP, Bigger JT Jr, Moss AJ and the Multicenter Postinfarction Research Group: Decreased heart rate variability and its association with increased mortality after acute myocardial infarction. Am J Cardiol 1987;59:256-262.
107. La Rovere MT, Bigger JT Jr, Marcus FI, et al: The ATRAMI (Autonomic Tone and Reflexes After Myocardial Infarction) Investigators: Baroreflex sensitivity and heart rate variability in prediction of total cardiac mortality after myocardial infarction. Lancet 1998;351:478-484.
108. Samaan A: The antagonistic cardiac nerves and heart rate. J Physiol 1935;83:332-340.
109. Akselrod S, Gordon D, Ubel FA, et al: Power spectrum analysis of heart rate fluctuations: A quantitative probe of beat-to-beat cardiovascular control. Science 1981;213:220-222.
110. Wolf MM, Varigos GA, Hunt D, Sloman JG: Sinus arrhythmia in acute myocardial infarction. Med J Aust 1978;2:52-53.
111. Malik M, Bigger JT Jr, Camm AJ, et al: Task Force of the European Society of Cardiology and the North American Society of Pacing and Electrophysiology. Heart rate variability: Standards of measurement, physiological interpretation, and clinical use. Circulation 1996;93:1043-1065.
112. Malik M: Geometrical methods for heart rate variability assessment. In Malik M, Camm AJ (eds): Heart Rate Variability. Armonk, NY, Futura, 1995, pp 47-61.
113. Huikuri HV, Seppänen T, Koistenen MJ, et al: Abnormalities in beat-to-beat dynamics of heart rate before the spontaneous onset of life-threatening ventricular tachyarrhythmias in patients with prior myocardial infarction. Circulation 1996;93: 1836-1844.
114. Cerutti S, Bianchi AM, Mainardi LT: Spectral analysis of heart rate variability signal. In Malik M, Camm AJ (eds): Heart Rate Variability. Armonk, NY, Futura, 1995, pp 63-74.
115. Schmidt G, Morfill GE: Nonlinear methods for heart rate variability assessment. In Malik M, Camm AJ (eds): Heart Rate Variability. Armonk, NY, Futura, 1995, pp 87-98.
116. Lombardi F, Sandrone G, Mortara A, et al: Linear and non-linear dynamics of heart rate variability after acute myocardial infarction and reduced left ventricular ejection fraction. Am J Cardiol 1996;77:1283-1288.
117. Signorini MG, Cerutti S, Guzzetti S, Parola R: Non-linear dynamics of cardiovascular variability signals. Meth Inform Med 1994;33:81-84.
118. Mäkikallio TH, Huikuri HV, Hintze U, et al: Fractal analysis and time- and frequency-domain measures of heart rate variability as predictors of mortality in patients with heart failure. J Am Coll Cardiol 2000;87:178-182.

119. Tsui H, Larson MG, Venditti FJ Jr, et al: Impact of reduced heart rate variability on risk for cardiac events: The Framingham Heart Study. Circulation 1996;97:2850-2855.
120. Dekker JM, Crow RS, Folson AR, et al: The ARIC (Atherosclerosis Risk In Communities) study. Low heart rate variability in a 2-minute rhythm strip predicts risk of coronary heart disease and mortality from several causes: Circulation 2000;102: 1239-1244.
121. Nolan J, Batin PD, Andrews R, et al: Prospective study of heart rate variability and mortality in chronic heart failure: Results of the United Kingdom Heart Failure Evaluation and Assessment of Risk Trial (UK-HEART). Circulation 1998;98:1510-1516.
122. Szabò BM, van Veldhuisen DJ, van der Veer N, et al: Prognostic value of heart rate variability in chronic congestive heart failure secondary to idiopathic or ischemic dilated cardiomyopathy. Am J Cardiol 1997;79:978-980.
123. Counihan PJ, Lü Fei, Bashir Y, et al: Assessment of heart rate variability in hypertrophic cardiomyopathy: Association with clinical and prognostic features. Circulation 1993;88:1682-1690.
124. Ramaekers D, Ector H, Vanhaecke J, et al: Heart rate variability after cardiac transplantation in humans. PACE 1996;2112-2119.
125. Bigger JT Jr, Fleiss JT, Steinman RC, et al: Frequency domain measures of heart rate variability and mortality after myocardial infarction. Circulation 1992;85:164-171.
126. Farrell TG, Bahir Y, Cripps T, et al: Risk stratification for arrhythmic events in postinfarction patients based on heart rate variability, ambulatory electrocardiographic variables and the signal-averaged electrocardiogram. J Am Coll Cardiol 1991;18:687-697.
126a. Camm AJ, Pratt CM, Schwartz PJ, et al: AzimiLide post Infarct surVival Evaluation (ALIVE) Investigators: Mortality in patients after a recent myocardial infarction: A randomized, placebo-controlled trial of azimilide using heart rate variability for risk stratification. Circulation 2004;109:990-996.
127. Hartikainen JEK, Malik M, Staunton A, et al: Distinction between arrhythmic and nonarrhythmic death after acute myocardial infarction based on heart rate variability, signal-averaged electrocardiogram, ventricular arrhythmias and left ventricular ejection fraction. J Am Coll Cardiol 1996;28: 296-304.
128. Malik M, Camm AJ, Janse MJ, et al: Depressed heart rate variability identifies postinfarction patients who might benefit from prophylactic treatment with amiodarone. J Am Coll Cardiol 2000;35:1263-1275.
129. Zuanetti G, Neilson James MM, Latini R, et al: Prognostic significance of heart rate variability in post-myocardial infarction patients in the fibrinolytic era: The GISSI-2 results. Circulation 1996;94:432-484.
130. Vanoli E, Cerati D, Pedretti RF: Autonomic control of heart rate: Pharmacological and nonpharmacological modulation. Basic Res Cardiol 1998;Suppl 1:133-142.
131. Smyth HS, Sleight P, Pickering GW: Reflex regulation of arterial pressure during sleep in man: A quantitative method of assessing of baroreflex sensitivity. Circ Res 1969;24:109-121.
132. Goldstein DS, Horwitz D, Keiser HR: Comparison of techniques for measuring baroreflex sensitivity in man. Circulation 1982;66:432-439.
133. La Rovere MT, Pinna GD, Mortara A: Assessment of baroreflex sensitivity. In Malik M (ed): Clinical Guide to Cardiac Autonomic Tests. Dordrech, Kluwer Academic Publishers 1998:257-281.
134. La Rovere MT, Specchia G, Mortara A, Schwartz PJ: Baroreflex sensitivity, clinical correlates and cardiovascular mortality among patients with a first myocardial infarction: A prospective study. Circulation 1988;78:816-824.
135. Mortara A, La Rovere MT, Pinna GD, et al: Arterial baroreflex modulation of heart rate in chronic heart failure: Clinical and hemodynamic correlates and prognostic implications. Circulation 1997;96:3450-3458.
136. Bristow JD, Honour AJ, Pickering JW, et al: Diminished baroreflex sensitivity in high blood pressure. Circulation 1969;39:48-54.
137. Pinna GD, La Rovere MT, Maestri R, et al: Comparison between invasive and non-invasive measurements of baroreflex sensitivity: Implications for studies on risk stratification after a myocardial infarction. Eur Heart J 2000;21:1522-1529.
138. Farrel TG, Paul V, Cripps TR, et al: Baroreflex sensitivity and electrophysiological correlates in patients after acute myocardial infarction. Circulation 1991;83:945-952.
139. Farrel TG, Odemuyiwa O, Bashir Y, et al: Prognostic value of baroreflex sensitivity testing after acute myocardial infarction. Br Heart J 1992;67:129-137.
140. Landolina M, Mantica M, Pessano P, et al: Impaired baroreflex sensitivity is correlated with hemodynamic deterioration of sustained ventricular tachycardia. J Am Coll Cardiol 1997;29: 568-575.
141. Hamdam MH, Joglar JA, Page RL, et al: Baroreflex gain predicts blood pressure recovery during simulated ventricular tachycardia in humans. Circulation 1999;100:381-386.
142. Hohnloser SH, Klingenheben T, Yi-Gang Li, et al: T wave alternans as a predictor of recurrent ventricular tachyarrhythmias in ICD recipients: Prospective comparison with conventional risk markers. J Cardiovasc Electrophysiol 1998;9:1258-1268.
143. Schmidt G, Malik M, Barthel P, et al: Heart rate turbulence after ventricular premature beats as a predictor of mortality after acute myocardial infarction. Lancet 1999;353:1390-1396.
144. Erlanger J, Blackman JR: Further studies in the physiology of heart block in mammals: Chronic auriculo-ventricular heart block in the dog. Heart 1909;1:177.
145. Parsonett AE, Miller R, Newark NJ: Heart block: The influence of ventricular systole upon the auricular rhythm in complete and incomplete heart block. Am Heart J 1944;27:676-687.
146. Scanes AC, Tang AS: Ventriculophasic modulation of atrioventricular nodal conduction in humans. Circulation 1998;97: 2245-2251.
147. Mrowka R, Persson PB, Theres H, et al: Blunted arterial baroreflex causes pathological heart rate turbulence. Am J Physiol Regulatory Integrative Comp Physiol 2000;279:R1171-1175.
148. Ghuran A, Reid F, La Rovere MT, et al: Heart rate turbulence-based predictors of fatal and nonfatal cardiac arrest (the Autonomic Tone and Reflexes After Myocardial Infarction Substudy). Am J Cardiol 2002;89:184-190.
149. Schmidt G, Morfill GE, Barthel P, et al: Variability of ventricular premature complexes and mortality risk. PACE 1996;19:976-980.
150. Julian DG, Camm AJ, Frangin G, et al: European Myocardial Infarct Amiodarone Trial Investigators: Randomised trial of effect of amiodarone on mortality in patients with left ventricular dysfunction after recent myocardial infarction: EMIAT. Lancet 1997;349:667-674.
150a. Barthel P, Schneider R, Bauer A, et al: Risk stratification after acute myocardial infarction by heart rate turbulence. Circulation 2003;108:1221-1226.
151. Morley-Davies A, Cobbe SM, Dargie HJ, et al: Heart rate turbulence and mortality in chronic heart failure. J Am Coll Cardiol 2000;35:106A [Abstract].
152. Savelieva I, Wichterle D, Harries M, et al: Heart rate turbulence after atrial and ventricular premature beats: Relation to left ventricular function and coupling intervals. PACE 2003;26:401-405.
153. Jervell A, Lange-Nielsen F: Congenital deaf-mutism, functional heart disease with prolongation of the QT interval and sudden death. Am Heart J 1957;54:59-68.
154. Romano C, Gemme G, Pongiglione R: Aritmie cardiache rare dell'eta' pediatrica. Clin Pediatr 1963;45:656-683.
155. Ward OC: A new familial cardiac syndrome in children. J Iri Med Assoc 1964;54:103-106.
156. Schwartz PJ, Wolf S: QT interval prolongation as predictor of sudden death in patients with myocardial infarction. Circulation 1978;57:1074-1077.
157. Ahnve A, Gilpin E, Madsen EB, et al: Prognostic importance of QTc interval at discharge after acute myocardial infarction: A multicenter study of 865 patients. Am Heart J 1984;108:395-400.
158. Peters RW, Barker A, Byington R: The BHAT study group: Prognostic value of QTc prolongation: The BHAT experience. Circulation 1984;70: II-6 [Abstract].
159. Fioretti P, Tijssen JGP, Azar AJ, et al: Prognostic value of predischarge 12 lead electrocardiogram after myocardial infarction compared with other routine clinical variables. Br Heart J 1987;57:306-312.
160. Wheelan K, Murkharji J, Rude RE, et al: The MILIS Study Group. Sudden death and its relation to QT-interval prolongation after myocardial infarction: Two-year follow-up. Am J Cardiol 1986;57:745-750.
161. Puddu PE, Bourassa MG: Prediction of sudden death from QTc interval prolongation in patients with chronic ischemic heart disease. J Electrocardiol 1986;19:203-212.

162. Algra A, Tijssen JGP, Roelandt JRTC, et al: QTc prolongation measured by standard 12-lead electrocardiography is an independent risk factor for sudden death due to cardiac arrest. Circulation 1991;83:1888-1894.
163. Okin PM, Devereux RB, Howard BV, et al: Assessment of QT interval and QT dispersion for prediction of all-cause and cardiovascular mortality in American Indians: The Strong Heart Study. Circulation 2000;101:61-66.
164. Schouten EG, Dekker JM, Meppelink P, et al: QT interval prolongation predicts cardiovascular mortality in an apparently healthy population. Circulation 1991;84:1516-1523.
165. Dekker JM, Schouten EG, Klootwijk P, et al: Association between QT interval and coronary heart disease in middle-aged and elderly men: The Zutphen study. Circulation 1994;90:779-785.
166. Elming H, Holm E, Holm E, et al: The prognostic value of the QT interval and QT interval dispersion in all-cause and cardiac mortality and morbidity in a population of Danish citizens. Eur Heart J 1998;19:1398-1400.
167. Karjalainen J, Reunanen A, Ristola P, Viitasalo M: QT interval as a cardiac risk factor in a middle aged population. Heart 1997;77:543-548.
168. De Bruyne MC, Hoes AW, Kors JA, et al: Prolonged QT interval predicts cardiac and all-cause mortality in the elderly: The Rotterdam study. Eur Heart J 1999;20:278-284.
169. Brendorp B, Elming H, Jun L, et al: The DIAMOND Study Group: QTc Interval as a guide to select those patients with congestive heart failure and reduced left ventricular systolic function who will benefit most from antiarrhythmic treatment with dofetilide. Circulation 2001;103:1422-1427.
169a. Dekker JM, Crow RS, Hannam PJ, et al: Heart rate-corrected QT interval prolongation predicts risk of coronary artery disease in black and white middle-aged men and women: The ARIC study. J Am Coll Cardiol 2004;43:565-571.
170. Brooksby P, Batin PD, Nolan J, et al: The relationship between QT intervals and mortality in ambulant patients with chronic heart failure: The United Kingdom Heart Failure Evaluation and Assessment of Risk Trial (UK-HEART). Eur Heart J 1999; 20:1335-1341.
171. Goldberg RJ, Bengtson J, Chen Z, et al: Duration of the QT interval and total and cardiovascular mortality in healthy persons (the Framingham Heart Study experience). Am J Cardiol 1991;67:55-58.
172. Fridericia LS: Die systolendauer im Elektrokardiogramm bei normalen Menschen und bei Herzkranken, Acta Med Scand 1920;53:469.
173. Bazett HC: An analysis of time relations of the electrocardiogram. Heart 1920;7:353-370.
174. Karjalainen J, Vitasalo M, Manttari M, Manninen V: Relation between QT intervals and heart rates from 40 to 120 beats/min in rest electrocardiograms of men and a simple method to adjust QT interval values. J Am Coll Cardiol 1994;23:1547-1553.
175. Batchvarov VN, Ghuran A, Smetana P, et al: QT-RR relationship in healthy subjects exhibits substantial intersubject variability and high intrasubject stability. Am J Physiol Heart Circ Physiol 2002;282:H2356-H2363.
176. Maison Blanche P, Coumel P: T wave dynamicity. In Malik M (ed): Risk of Arrhythmia and Sudden Death. London, BMJ Books, 2001, pp 249-255.
177. Franz MR, Swerdlow CD, Liem LB, Shaefer J: Cycle length dependence of human action potential duration in vivo: Effects of single extrastimuli, sudden sustained rate acceleration and deceleration, and different steady-state frequencies. J Clin Invest 1988;82:972-979.
178. Lau CP, Ward W: QT hysteresis: The effect of an abrupt change in pacing rate. In Butrous G, Schwartz PJ (eds): Clinical Aspects of Ventricular Repolarization. London, Farrand Press, 1989,pp 175-184.
179. Neyroud N, Maison Blanche P, Denjoy I, et al: Diagnostic performance of QT interval variables from 24-hour electrocardiography in the long QT syndrome. Eur Heart J 1998;19: 158-165.
180. Ahmed MW, Kadish AH, Goldberger JJ: Autonomic effects on the QT interval. Ann Noninvas Electrocardiol 1996;1:44-53.
181. Merri M, Moss AJ, Benhorin J, et al: Relation between ventricular repolarization duration and cardiac cycle length during 24-hour Holter recordings: Findings in normal patients and patients with long QT syndrome. Circulation 1992;85:1816-1821.
182. Lande G, Kyndt F, Baró I, et al: Dynamic analysis of the QT interval in long QT1 syndrome patients with a normal phenotype. Eur Heart J 2001;22:410-422.
183. Chevalier P, Burri H, Adeleine P, et al: The Groupe d'Etude du Pronostic de l'Infarctus du Myocarde (GREPI): QT dynamicity and sudden death after myocardial infarction: Results of a long-term follow-up study. J Cardiovasc Electrophysiol 2003;14: 227-233.
184. Berger RD, Kasper EK, Baughman KL, et al: Beat-to-beat QT interval variability: Novel evidence for repolarization lability in ischemic and nonischemic dilated cardiomyopathy. Circulation 1997;96:1557-1565.
185. Kadish AH, Weisman HF, Veltri EP, et al: Paradoxical effects of exercise on the QT interval in patients with polymorphic ventricular tachycardia receiving type Ia antiarrhythmic agents. Circulation 1990;81:14-19.
186. Buckingham TA, Bhutto ZR, Telfer EA, et al: Differences in corrected QT intervals at minimal and maximal heart rates may identify patients at risk for torsades de pointes during treatment with antiarrhythmic drugs. J Cardiovasc Electrophysiol 1994;5: 408-411.
187. Lombardi F, Sandrone G, Porta A, et al: Spectral analysis of short term $R\text{-}T_{apex}$ interval variability during sinus rhythm and fixed atrial rate. Eur Hear J 1996;17:769-778.
188. Han J, Moe GK: Nonuniform recovery of excitability in ventricular muscle. Circ Res 1964;14:44-60.
189. Day CP, McComb JM, Campbell RWF: QT dispersion: An indication of arrhythmia risk in patients with long QT intervals. Br Heart J 1990;63:342-344.
190. Zabel M, Portnoy S, Franz MR: Electrocardiographic indexes of dispersion of ventricular repolarization: An isolated heart validation study. J Am Coll Cardiol 1995;25:746-752.
191. De Bruyne MC, Hoes AW, Kors JA, et al: QTc dispersion predicts cardiac mortality in the elderly: The Rotterdam study. Circulation 1998;97:467-472.
192. Macfarlane PW, on behalf of the WOSCOP study group: QT dispersion—lack of discriminating power. Circulation 1998;98: I-81 [Abstract].
193. Oikarinen L, Viitasalo M, Toivonen L: Dispersions of the QT interval in postmyocardial infarction patients presenting with ventricular tachycardia or ventricular fibrillation. Am J Cardiol 1998;81:694-697.
194. Glancy JM, Garrat CJ, Woods KL, et al: QT dispersion and mortality after myocardial infarction. Lancet 1995;345:945-948.
195. Spargias KS, Lindsay SJ, Kawar GI, et al: QT dispersion as a predictor of long-term mortality in patients with acute myocardial infarction and clinical evidence of heart failure. Eur Heart J 1999;20:1158-1165.
196. Milletich A, Latini R, Garrido G, et al: The GISSI-ECG Collaborative Group. Lack of prognostic value of QT dispersion at discharge in patients recovering from acute myocardial infarction: A case control study from GISSI database. Eur Heart J 1996;17:30 [Abstract].
197. Zabel M, Klingenheben T, Franz MR, Hohnloser SH: Assessment of QT dispersion for prediction of mortality or arrhythmic events after myocardial infarction. Circulation 1998;97:2543-2550.
198. Brooksby P, Robinson PJ, Segal R, et al: The ELITE study group: Effects of losartan and captopril on QT dispersion in elderly patients with heart failure. Lancet 1999;334:395-396.
199. Brendorp B, Elming H, Jun L, et al: The DIAMOND study group: QT dispersion has no prognostic information for patients with advanced congestive heart failure and reduced left systolic ventricular function. Circulation 2001;103:831-835.
200. Padmanabhan S, Silvet H, Amin J, Pai RG: Prognostic value of QT interval and QT dispersion in patients with left ventricular systolic dysfunction: Results from a cohort of 2265 patients with an ejection fraction of < or =40%. Am Heart J 2003;145:132-138.
201. Fu GS, Meissner A, Simon R: Repolarization dispersion and sudden cardiac death in patients with impaired left ventricular function. Eur Heart J 1997;18:281-289.
202. Malik M, Batchvarov VN: Measurement, interpretation and clinical potential of QT dispersion. J Am Coll Cardiol 2000;36: 1749-1766.

203. Macfarlane PW, McLaughlin SC, Roger C: Influence of lead selection and population on automated measurement of QT dispersion. Circulation 1998;98:2160-2167.
204. Kors JA, Van Herpen G, Van Bemmel JH: QT dispersion as an attribute of T-loop morphology. Circulation 1999;99:1458-1463.
205. Malik M, Acar B, Yap YG, et al: QT dispersion does not represent electrocardiographic interlead heterogeneity of ventricular repolarization. J Cardiovasc Electrophysiol 2000;11:835-843.
206. Sicouri S, Antzelevitch C: A subpopulation of cells with unique electrophysiological properties in the deep subepicardium of canine ventricle. The M cell. Circ Res 1991;68:1729-1741.
207. Yan GX, Antzelevitch C: Cellular basis for the normal T wave and the electrocardiographic manifestation of the long QT-syndrome. Circulation 1998;98:1928-1936.
208. Shimuzi W, Antzelevitch C: Cellular and ionic basis for T wave alternans under long QT conditions. Circulation 1999;99: 1499-1507.
209. Liu DW, Antzelevitch C: Characteristics of the delayed rectifier current (I_{Kr} and I_{Ks}) in canine ventricular epicardial, midmyocardial, and endocardial myocytes: A weaker I_{Ks} contributes to the longer action potential of the M cell. Circ Res 1995;76:351-365.
210. Anyukhovsky EP, Sosunov EA, Rosen MR: Regional differences in electrophysiological properties of epicardium, midmyocardium, and endocardium. Circulation 1996;94:1981-1988.
211. Roden DM: Taking "idio" out of "idiosyncraticL:" Predicting torsades de pointes. PACE 1998;21:1029-1034.
212. Ebert SN, Liu XK, Woosley RL: Female gender as a risk factor for drug-induced cardiac arrhythmias: Evaluation of clinical and experimental evidence. J Womens Health 1998;7:547-557.
213. Pham TV, Sosunov EA, Gainullin RZ, et al: Impact of sex and gonadal steroids on prolongation of ventricular repolarization and arrhythmias induced by I_K-blocking drugs. Circulation 2001;103:2207-2212.
214. Smetana P, Batchvarov VN, Hnatkova K, et al: Sex differences in repolarisation homogeneity and its circadian pattern. Am J Physiol Heart Circ Physiol 2002;282:H1889-1897.
215. Lewis T: Notes upon alteration of the heart. Q J Med 1910;4: 141-144.
216. Surawitz B, Fish C: Cardiac alternans: Diverse mechanisms and clinical manifestations. J Am Coll Cardiol 1992;20:483-499.
217. Zareba W, Moss AJ: T wave alternans in idiopathic long QT syndrome. J Am Coll Cardiol 1994;23:1541-1546.
218. Armoundas AA, Nanke T, Cohen RJ: T-wave alternans preceding torsades de pointes ventricular tachycardia. Circulation 2000; 101:2550.
219. Smith JM, Cohen RJ: Simple finite element model accounts for wide range of cardiac dysrhythmias. Proc Natl Acad Sci USA 1984;81:233-237.
220. Chinushi M, Restivo M, Caref EB, El-Sherif N: Electrophysiological basis of arrhythmogenicity of QT/T alternans in the long QT syndrome. Circ Res 1998;83:614-628.
221. Hashimoto H, Suzuki K, Miyake S, Nakashima M: Effects of calcium antagonists on the electrical alternans of the ST segment and on associated mechanical alternans during acute coronary occlusion in dogs. Circulation 1983;68:667-672.
222. Nearing BD, Huang AH, Verrier RL: Dynamic tracking of cardiac vulnerability by complex demodulation of the T wave. Science 1991;252:437-440.
223. Smith JM, Clancy EA, Valeri CR, et al: Electrical alternans and cardiac electrical instability. Circulation 1988;77:110-121.
224. Rosenbaum DS, Jackson LE, Smith JM, et al: Electrical alternans and vulnerability to ventricular arrhythmias. N Engl J Med 1994; 330:235-241.
225. Gold MR, Bloomfield DM, Anderson KP, et al: A comparison of T wave alternans, signal-averaged electrocardiography and programmed ventricular stimulation for arrhythmia risk stratification. J Am Coll Cardiol 200;36:2247-2253.
226. Ikeda T, Takami M, Kondo N, et al: Combined assessment of T-wave alternans and late potentials used to predict arrhythmic events after myocardial infarction. J Am Coll Cardiol 2000;35: 722-730.
226a. Ikeda T, Saito H, Tanno K, et al: T-wave alternans as a predictor for sudden death after myocardial infarction. Am J Cardiol 2002;89:79-82
226b. Hohnloser SH, Klingenheben T, Bloomfield D, et al: Usefulness of microvolt T-wave alternans for prediction of ventricular tachyarrhythmic events in patients with dilated cardiomyopathy: Results from a prospective observational study. J Am Coll Cardiol 2003;41:2220-2224.
226c. Klingenheben T, Zabel M, D'Agostino RB, et al: Predictive value of T-wave alternans for arrhythmic events in patients with congestive heart failure. Lancet 2000;356:651-652.
227. Priori SG, Mortara DW, Napolitano C, et al: Evaluation of the spatial aspects of T-wave complexity in the long-QT syndrome. Circulation 1997;96:3006-3012.
228. Okin PM, Devereux R, Fabsitz RR, et al: Principal component analysis of T wave and prediction of cardiovascular mortality in American Indians: The Strong Heart study. Circulation 2002; 105:714-719.
229. Couderc JP, Zareba W, Burattini L, Moss AJ: Beat-to-beat repolarization variability in LQTS patients with the SCN5A sodium channel gene mutation. PACE 1999;22:1581-1592.
230. Acar B, Yi G, Hnatkova K, Malik M: Spatial, temporal and wavefront direction characteristics of 12-lead T wave morphology. Med Biol Eng Comput 1999;37:574-584.
231. Zabel M, Acar B, Klingenheben T, et al: Analysis of 12-lead T wave morphology for risk stratification after myocardial infarction. Circulation 2000;102:1252-1257.
232. Batchvarov VN, Hnatkova K, Ghuran A, et al: Ventricular gradient is an independent risk factor in survivors of acute myocardial infarction. J Am Coll Cardiol 2002;39:125A [Abstract].

Chapter 48 Antiarrhythmic Surgery

JOHN M. MILLER and YOUSUF MAHOMED

Cardiac tachyarrhythmias are sources of significant morbidity and mortality in the general population. Between 300,000 and 500,000 sudden cardiac deaths occur per year in the United States, the majority of which are due to ventricular tachyarrhythmias. Another 0.2% to 0.4% of the population has recurrent episodes of supraventricular tachycardia (SVT) due to Wolff-Parkinson-White (WPW) syndrome,[1,2] atrioventricular (AV) nodal reentry, or atrial tachycardia, and 1% to 2% of North Americans have episodes of atrial fibrillation (AF), the prevalence of which increases dramatically with age.[3] The concept of being able to cure a cardiac rhythm disturbance by excising or incising a portion of the heart has been a goal for decades. However, the ability to affect *only* the arrhythmogenic tissue and not inflict unwanted damage on residual, normal myocardium awaited the development of mapping tools and techniques that allowed this degree of discrimination between normally and abnormally functioning tissue. In this chapter, we will discuss the principles of cardiac surgical mapping and the application of surgical techniques to a wide variety of arrhythmias as well as the current status of surgical therapy of arrhythmias.

Surgery for Supraventricular Arrhythmias

HISTORY AND DEVELOPMENT

The first successful operation for SVT was for WPW syndrome in 1968.[4] Following this success, the number of patients undergoing surgery for this disorder rapidly increased until the development of catheter ablation techniques. AV nodal reentry was treated with surgery during the early 1980s, and in the late 1980s, the maze procedure was developed for treatment of AF. Its subsequent modifications and use of pulmonary vein isolation techniques for AF therapy have continued through the turn of the century.

ARRHYTHMIAS AND SURGICAL TECHNIQUES

A variety of different supraventricular arrhythmic substrates have been successfully addressed by surgical therapy. Intraoperative mapping has been used to guide the surgical procedure in cases in which focal ablation was required (such as WPW syndrome and atrial tachycardias), but has been less helpful in AF. The principles of mapping are generally the same, and stimulation and/or mapping are generally used to assess efficacy of therapy if not to guide it.

Wolff-Parkinson-White Syndrome and Concealed Accessory Pathways

WPW syndrome (and its variants) was the first, and until recently, the most common disorder for which surgery had been most widely applied. In this syndrome, a strand of myocardium spans the AV groove to connect atrium to ventricular myocardium extrinsic to the normal AV node and may participate in SVT in which the impulse propagates anterogradely over the AV node, through the ventricles, retrogradely over the accessory pathway, through the atria, and back to the AV node. The surgical approach consists of first determining the location of the pathway along the AV groove, then ablating or incising it. The localization process is carried out after exposure of the heart via median sternotomy and construction of a pericardial cradle. Atrial and ventricular reference and stimulation electrodes are attached and the pathway's location is sought (Fig. 48-1). This can be determined by mapping the pathway's atrial and/or ventricular insertion. The atrial insertion is located by mapping along the atrial aspect of the AV groove during SVT or ventricular pacing, seeking the site of earliest atrial activation (keeping in mind that retrograde conduction may also occur over the AV node and yield confusing results). The pathway's ventricular insertion can be discerned by mapping the ventricular aspect of the AV groove during sinus rhythm or atrial pacing to maximize preexcitation. Mapping can be performed using a roving electrode in a point-to-point fashion, or more quickly and accurately by using a multipolar strip of electrodes aligned along the AV groove. Once the pathway's location has been determined, it can be ablated or incised. Two general approaches have been used: endocardial and epicardial (Fig. 48-2). Using the endocardial approach,[5] cardiopulmonary bypass must be instituted

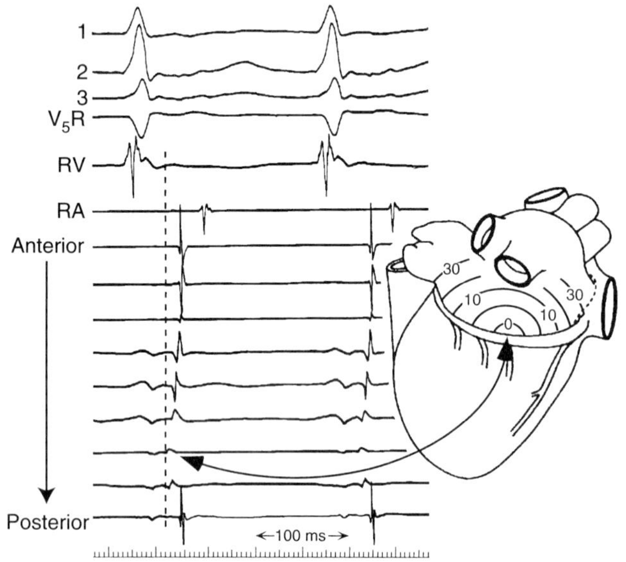

FIGURE 48-1 Surgical mapping during SVT in Wolff-Parkinson-White syndrome. Surface ECG leads and epicardial electrograms recorded during SVT with a multipolar strip electrode along the mitral annulus are shown. The *dashed line* shows the onset of atrial activity, which begins in the posterolateral left atrium (*arrow*). *Inset diagram* has heart viewed from left posterolateral aspect showing isochrones of left atrial activation derived from the map. SVT, Supraventricular tachycardia.

and the heart is arrested with cold cardioplegia and the atrium into which the pathway inserts is opened. An incision is made 2 to 3 mm above the AV valve annulus and is carried along most of the circumference of the annulus. The epicardial fat thereby exposed is gently pushed away until the epicardial surface of the ventricle has been cleaned. The supra-annular atriotomy is closed and cryolesions applied at each end to prevent escape of impulses around the incision, in the event that the ventricular cleaning did not interrupt the pathway. The entrance atriotomy is closed, the heart is rewarmed, and pacing is performed to ensure that the pathway has indeed been severed.

Using the epicardial approach,[6] the AV groove is cleaned with gentle blunt dissection and a nerve hook, and cryoablation is performed in the region indicated by mapping as additional insurance. Because this procedure can be done on the normothermic, beating heart, the surgical team can get instantaneous feedback as to when the pathway is actually interrupted (loss of delta wave). Because the epicardial approach can be performed without cardioplegia—and in some cases, without even being on cardiopulmonary bypass—and can provide confirmation of efficacy, it has become the preferred procedure. This procedure was successful in eliminating SVT episodes in more than 98% of cases, with less than a 1% mortality rate. However, because of the success of radiofrequency (RF) catheter ablation in WPW and concealed accessory pathways, these operations are very rarely performed now. When they are, it is usually after one or more failed attempts at catheter ablation and the anatomy of the AV groove can be distorted and scarred by previous damage from unsuccessful ablation efforts.

Atrioventricular Nodal Reentry

In this common arrhythmia, the reentrant circuit is composed of slowly conducting AV nodal tissue between the tricuspid annulus and coronary sinus ostium (the "slow pathway") and more rapidly conducting tissue closer to the apex of Koch's triangle (the "fast pathway"). Impulses conduct anterogradely to the His bundle and ventricles at almost the same time as the atria are being activated retrogradely, resulting in the characteristic ECG appearance (P wave "buried" in the QRS complex). Currently, catheter ablation of the slow pathway is curative in more than 95% of cases. Until this development, drug-resistant cases were sometimes treated with catheter ablation of the node or His bundle to effect complete heart block, followed by implantation of a permanent pacemaker. For a few years before the development of slow pathway ablation, surgical therapy for AV nodal reentry enjoyed a short period of popularity because it offered the possibility of arrhythmia cure without causing heart block. Although a number of approaches were employed, probably the most common procedure was perinodal cryoablation (Fig. 48-3).[7,8] This procedure consisted of making a wide right atriotomy during normothermic cardiopulmonary bypass, mapping of the location of the His bundle, and

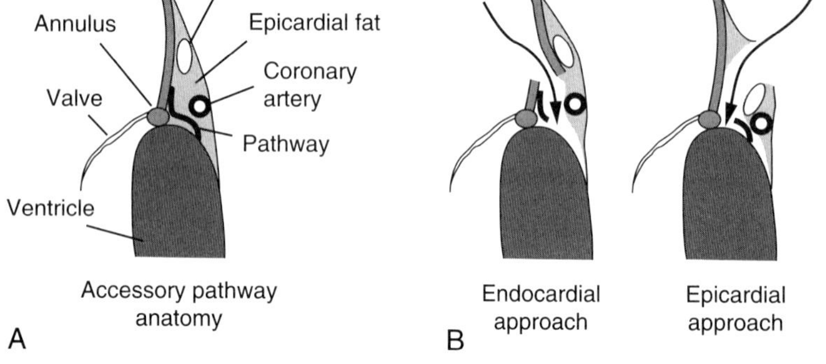

FIGURE 48-2 Surgical procedures for Wolff-Parkinson-White syndrome. **A,** A diagram of a cross-section of the left AV groove is shown. Note the accessory pathway connects atrium and ventricle extrinsic to the annulus. **B,** Surgical approaches; endocardial (*on left*) showing incision in atrium above annulus and dissection onto the ventricular surface, cleaning away epicardial fat, and epicardial approach (*on right*) showing cleaning of the AV groove from outside the heart after retracting epicardial vessels.

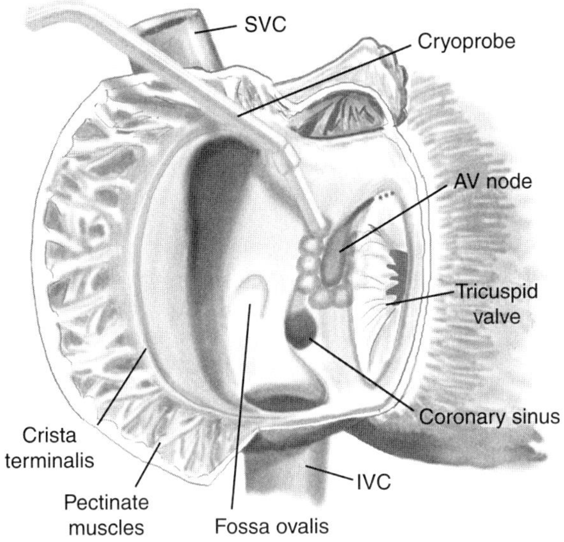

FIGURE 48-3 Perinodal cryoablation for AV nodal reentry. The right atrium is depicted as opened along the tricuspid annulus, viewing the septum. The AV nodal area has been nearly surrounded by a series of cryolesions, restricting input and output from the node. AV, Atrioventricular; IVC, inferior vena cava; SVC, superior vena cava.

application of a series of 5-mm cryolesions surrounding Koch's triangle. Mapping during the arrhythmia was generally not performed, though attempts at initiation of episodes before and after ablation were useful to determine efficacy. AV conduction was continuously monitored during cryoablation; if the P–R interval increased or AV block occurred, the cryoprobe was immediately warmed and removed. This technique resulted in a cure rate of greater than 95% with minimal surgical morbidity and almost no mortality. Because slow pathway ablation for cure of AV nodal reentry has been refined, surgical therapy is no longer needed for this disorder.

Atrial Flutter

As with other forms of SVT, catheter ablation is almost always successful in eliminating atrial flutter. In this disorder, a wavefront of electrical activation propagates around a large circuit in the right atrium, traveling between the crista terminalis and tricuspid annulus (TA) on the lateral wall, between the inferior vena caval (IVC) orifice and TA in the floor of the right atrium, along the atrial septum, and back to the crista terminalis–TA corridor. Most typically, the direction of wavefront propagation is in the order just described (counterclockwise when viewed from the left anterior oblique direction) but can be reversed (clockwise). The narrowest portion of this circuit is between the IVC and TA; catheter ablation that interrupts conduction in this zone is now standard therapy for atrial flutter. Before the refinement of this technique, some patients underwent surgical therapy guided by electrophysiological (EP) mapping. Cryolesions on the floor of the right atrium, accomplishing the same result as current catheter ablation, could be curative. Although surgical therapy provided some important insights into the nature of the flutter circuit, it is almost never needed.

Atrial Tachycardia

Atrial tachycardia (AT) can be categorized into two large groups: those that have a focal origin and others that are due to reentry within a relatively large circuit. There has been relatively little surgical experience with AT, in part because it is a less common arrhythmia than other types and, in part, because of significant technical difficulties. In cases of AT with a focal origin, the focus may become quiescent in the presence of general anesthetic agents (despite being nearly incessant while the patient is awake). Because the goal of surgery is to remove or destroy only the small portion of atrial tissue responsible for the arrhythmia, the location of which can only be determined by electrophysiological mapping, it is imperative that tachycardia be present during the surgical procedure. Fortunately, catecholamine infusion often facilitates resumption of automatic firing from the focus that allows its mapping and excision or cryoablation, but not in all cases.[9] Mapping of the focus is performed by moving a roving electrode over the atrial surface or acquiring data from multiple sites simultaneously, comparing local activation times with those of a reference electrode. The site of earliest activation (preceding the surface P wave onset by 20 to 50 ms) is the focus. This can be confirmed by cooling with the cryoprobe, which causes gradual slowing and termination of AT, before excising or freezing the area completely. Cryoablation alone has resulted in occasional recurrences; thus many recommend excision as well. There is far less surgical experience with macroreentrant AT; ironically, more cases of this type of arrhythmia are *caused* by surgery than are *cured* by it, because in most cases a long atriotomy forms the basis for reentry. An atriotomy incision can form a nonconductive barrier around which a reentrant wavefront can circulate because electrical connections are not reestablished along the atriotomy despite closing the incision.[10] An atriotomy is commonly used as endocardial access for closure of atrial or ventricular septal defects, for repair of tetralogy of Fallot, and for mitral valve repair.

Atrial Fibrillation

AF is the most common clinical arrhythmia. Until relatively recently, nonpharmacologic therapy consisted of AV nodal ablation and pacemaker insertion for control of rapid ventricular responses. In the late 1980s, as experience was growing with surgical treatment of other arrhythmias, several investigators sought ways to treat AF. Mapping studies during AF are difficult because of the apparent disorganization of atrial activity, but have generally shown a complex series of four to eight wavefronts processing across the atrial myocardium, with eddies and small reentrant circuits forming and extinguishing in seemingly random fashions.[11] There did not appear to be any one area responsible for initiation or perpetuation of the arrhythmia (such as with AT), nor a single reentrant pathway (as in WPW) that could be targeted for ablation. Instead, the entire atrial mass seemed to be involved. The "corridor" operation was developed to try to maintain physiologic rate response by isolating a band of atrial muscle connecting the sinus node and the AV node from the rest of the atrium (Fig. 48-4, right).[12] This preserved sinus rhythm and avoided a pacemaker, but because the rest of the atrial mass continued to fibrillate, true AV synchrony was absent and anticoagulation was still needed. In an effort to maintain sinus rhythm and effective atrial contraction, Cox and colleagues devised the so-called *maze procedure*, in which a series of incisions were made in both atria, including removal of both appendages (see Fig. 48-4, left).[13] The intent was to decrease the amount of contiguous atrial tissue below that required to support continued reentry, based on available mapping data. The initial results were very successful, but a high incidence of sinus node damage requiring pacemaker implantation led to modifications. The current version, the Maze-III, has a success rate of greater than 90% in preventing recurrent AF, less than a 1% mortality rate, and less than 5% of patients need a pacemaker.[14] Careful studies have shown that although the atrial mass does not operate as a coordinated unit, contractile function is preserved to an extent that atrial contraction provides a significant contribution to ventricular filling, and most patients are not maintained on chronic anticoagulation. Although the procedure can be performed as the sole indication for cardiac surgery, it is more commonly performed in conjunction with other cardiac surgical procedures (typically mitral valve repair or replacement).[15]

Pulmonary Vein Isolation

Recent data indicating that many episodes of AF are caused or maintained by rapid firing from extensions of atrial muscle in the pulmonary veins (PV)[16] have led to attempts to surgically isolate all four veins from the atrial myocardium. This is typically accomplished with RF energy using specially configured electrodes.[17] Early results are encouraging; however, there are reports of PV stenosis with resulting pulmonary hypertension following endocardial RF ablation within PVs.[18] Further studies are needed to determine if the surgical application of RF will be free of this negative outcome or not.

Although not directed at arrhythmia prophylaxis, some surgeons routinely amputate or oversew the left atrial appendage during any type of cardiac surgery to attempt to decrease the risk of systemic thromboembolic events in the event that AF occurs at some subsequent date. Although there are no data to suggest that this approach is beneficial, it is typically neither difficult nor time consuming to perform.

CURRENT STATUS

Surgical therapy for supraventricular arrhythmias is almost entirely confined to treatment of AF, with occasional cases of WPW syndrome that have failed repeated attempts at catheter ablation. Adaptation of minimally invasive cardiac surgical techniques in the treatment of AF may broaden the application of this mode of therapy.

Surgery for Ventricular Arrhythmias

HISTORY AND DEVELOPMENT

The first reported surgical cure of VT resulted from excision of an LV aneurysm.[19] This led to aneurysmectomy as a treatment for VT; however, it became apparent that this procedure had high mortality and low efficacy rates.[20,21] Similarly, bypass grafting was used to try to control VT with similarly poor results. From this

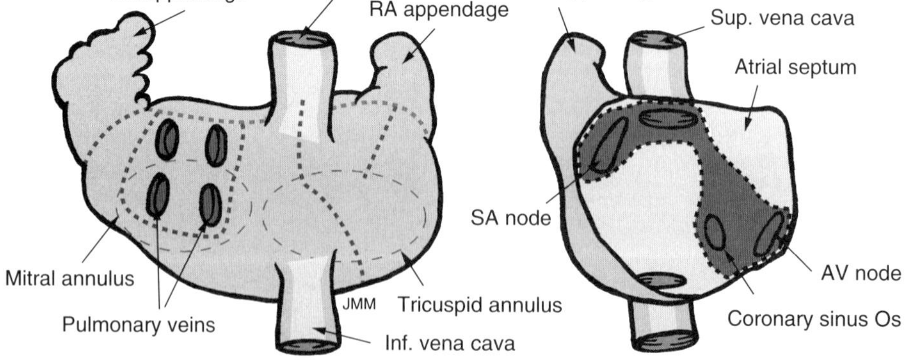

FIGURE 48-4 Surgical procedures for atrial fibrillation. The maze procedure is depicted (*left*) on both atria viewed from behind. *Thick dashed lines* indicate where incisions are made and closed to electrically subdivide the atrial mass into segments too small to support fibrillation. The corridor procedure (*right*) is shown on a frontal view of the opened right atrium. The shaded area bounded by thick dashes indicates the corridor of atrial muscle connecting the sinus and AV nodes, allowing normal sinus modulation or ventricular rate despite ongoing AF in the remainder of the atrial mass. AF, atrial fibrillation; AV, atrioventricular; LA, left atrium; RA, right atrium, SA, sinoatrial.

experience, it became apparent that VT episodes usually did not result from acute ischemia and that although aneurysms were frequently present in patients with VT, the VT did not originate in the aneurysm. Subsequent epicardial and endocardial mapping studies suggested that VT originated instead near the border between the aneurysm and more normal muscle. Several means of therapy targeting these regions were devised in the late 1970s, including excision (endocardial resection[22]), an isolating incision (encircling ventriculotomy[23]), and cryoablation.[24] Additional modalities have been used, including intraoperative laser photoablation.[25]

ARRHYTHMIAS AND SURGICAL TECHNIQUES

By far, the most experience with surgical therapy for ventricular arrhythmias is with uniform-morphology VT following myocardial infarction. The principles of mapping are similar to those for supraventricular arrhythmias, but the actual surgical means of arrhythmia cure are both different and more varied than with SVTs.

Ventricular Tachycardia

In the vast majority of cases, uniform-morphology VT (that is, identical QRS complexes during an episode) is due to reentry within damaged myocardium. The most extensive experience in surgical therapy of VT is in the setting of earlier myocardial infarction. In the typical case, damage from a large MI heals by replacing necrotic myocardium with dense collagenous scar tissue. In some areas (typically at the periphery of the infarct zone), some layers of surviving myocardial cells remain, interspersed with scar tissue.[26] If these layers connect at one or more points, they can form a potential circuit of conductive tissue, with an insulative layer of scar tissue protecting it from most outside influences. These surviving muscle bundles are most commonly situated near the endocardial surface, rather than in deeper myocardial layers. In patients with VT in the setting of dilated cardiomyopathy, the same pathophysiology is probably present although more diffusely within the myocardium.

Preoperative Studies

Because most patients who undergo surgery for VT have had an earlier infarction from coronary atherosclerosis, cardiac catheterization with coronary arteriography is an important part of the preoperative evaluation. Coronary stenoses that should undergo bypass grafting can thus be identified and any possible valvular lesions identified. Ventriculography (assessing overall LV function) can be helpful in estimating surgical risk. In the presence of coronary stenoses of uncertain flow-limiting significance, stress testing with some form of perfusion imaging can help determine whether bypass grafting of particular arteries would be beneficial. Preoperative EP study is almost always indicated (except in cases of severe left main stenosis or three vessel disease in which the rapid heart rates encountered during induced VT could possibly be detrimental). This study can provide information as to how readily inducible VT is as a guide to how vigorous intraoperative stimulation will need to be. Endocardial catheter mapping can also be performed at this time to help focus attention on certain areas that need to be ablated during surgery. However, if mapping is performed in this setting, it is almost always accompanied by ablation attempts; if the ablation successfully eliminates inducibility of one or more morphologies of VT, they no longer need surgical attention. If ablation is not successful, the validity of the mapping information may be called into question. Thus, preoperative mapping may have a more limited role than once thought.

Principles of Cardiac Mapping

If mapping is to be performed, several pieces of equipment are necessary beyond the usual requirements for open-heart surgery.[27-30] These include electrodes, a mapping system (consisting of amplifiers and a recording and analysis system), and a stimulator with which to initiate VT. Several commercial mapping systems are available, capable of recording from 64 to 256 electrodes simultaneously, on-line analysis of activation times, and generation of isochronal maps. Most systems record onto magnetic or optical media for data archiving.

At the time of surgery, the ECG leads from the patient are connected to the mapping system. The heart is exposed through a median sternotomy and a pericardial cradle is formed. Cannulae are placed for cardiopulmonary bypass and reference electrodes are inserted and tested. The heart is inspected for regions of infarct or aneurysm though which the ventricle may be entered without damaging viable myocardium, sometimes this area is only evident after the patient is on cardiopulmonary bypass and the LV has been vented. Cardiopulmonary bypass is then established maintaining a perfusate temperature of 37° to 38° C. This is necessary because cardiac electrical activity deteriorates at cooler temperatures and arrhythmias may not be inducible or mappable. Radiant heat loss of the heart surface during mapping may necessitate even slightly higher perfusate temperatures.

Epicardial mapping during sinus rhythm or VT can be performed with the heart closed, but this is usually omitted in the interest of time unless specific circumstances suggest its usefulness. In most cases, once the heart-lung machine is running well, the LV is then opened through the previously-identified infarct or aneurysm. The endocardial surface is inspected for the presence of adherent thrombus, which is then removed. An electrode or multipolar electrode array is then placed on the endocardial surface in an area of obvious scar tissue. At this time, sinus rhythm mapping can be performed to designate areas with abnormal electrograms indicating damaged myocardium; this step usually adds little and is omitted except in special situations. In most cases, VT is then initiated using previously-placed electrodes, and endocardial mapping is performed during VT. Some tissue is being activated at every instant during a reentrant arrhythmia, even during the diastolic interval between discrete QRS complexes; sites from which diastolic activation is recorded during VT are of particular interest because these regions are often in protected corridors that are critical for continued reentry.

Computerized mapping systems are able to quickly display these areas of diastolic activation (Fig. 48-5). Because the majority of patients with postinfarct VT have greater than 1 ECG morphology of VT that may arise from the same or different regions,[31] attempts are made to initiate and map other morphologies of VT. Because of time limitations, rigorous entrainment studies (overdrive pacing during VT) to validate the activation mapping data are often not performed.[32]

Recently, a technique called *return cycle mapping* has been proposed.[33] With this technique, entrainment is performed from several sites and isochronal lines are plotted connecting sites at which the electrogram on the beat following cessation of pacing occurs at an interval equal to the VT cycle length. Once a family of such isochronal lines has been made, the point of their convergence denotes the location of the protected diastolic corridor. This technique has yielded good results in animals but has not been tested in humans. Once all inducible morphologies of VT have been mapped, the electrode array is removed and the ablation portion of the procedure begins. Several different procedures have been used for this purpose (Fig. 48-6), including:

1. Endocardial resection,[22] in which a 1- to 2-mm thick layer of endocardial tissue is dissected from subjacent layers; this is designed to remove all or

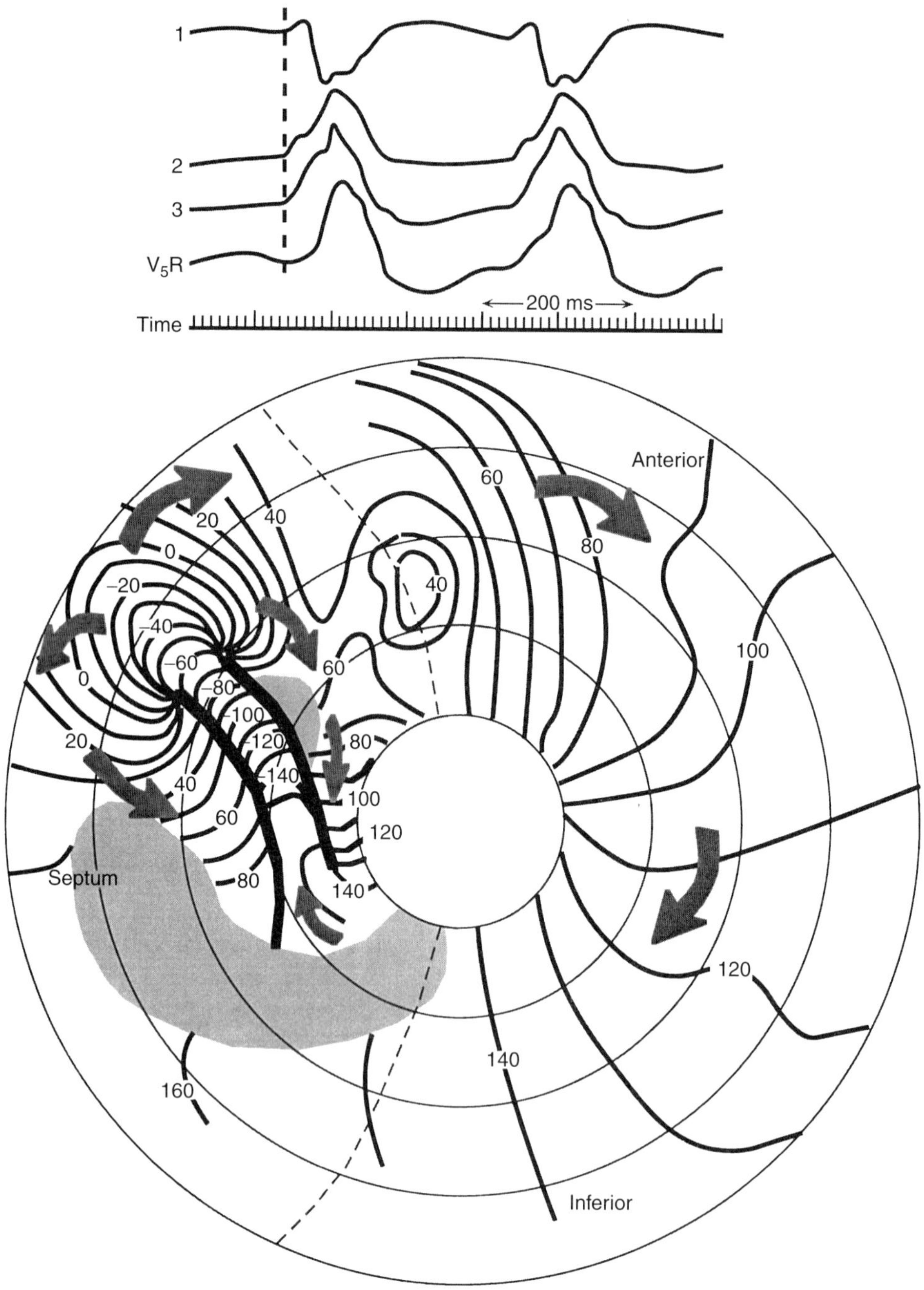

FIGURE 48-5 Isochronal maps during VT. The endocardial surface is shown in a polar projection with the apex at the center, septum at left, lateral wall at right. Lines of isochronal activation during VT are shown at 10-ms intervals; *thick dark lines* denote arcs of block; *shaded curved arrows* denote direction of wavefront propagation. Four ECG leads are shown above with a time line. A nearly complete figure-of-eight reentrant pattern is shown; mid-diastolic recordings are made from the narrow zone between arcs of block. No electrical activity could be recorded from *gray shaded areas.* VT, ventricular tachycardia.

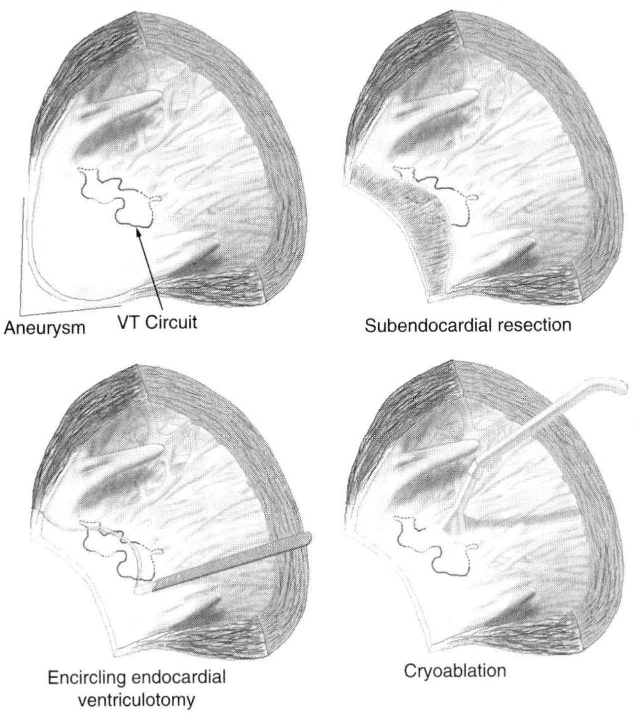

FIGURE 48-6 Surgical procedures for ventricular tachycardia (VT). In each panel, the left ventricle is depicted as viewing the septal endocardial surface showing papillary muscles and a thin, smooth aneurysm as opposed to the trabeculated normal endocardium. A VT circuit is shown, some of which is near the endocardial surface (*solid line*) and some in deeper layers (*dotted line*). At lower left, encircling endocardial ventriculotomy is shown incising near the perimeter of the endocardial scar and transecting a portion of the circuit. Subendocardial resection (*upper right*) removes a superficial portion of the circuit, while cryoablation destroys a portion by freezing. Of note, none of these procedures needs to destroy or remove the *entire* circuit in order to prevent VT, only a critical portion of it.

part of the region of diastolic activation required for reentry.

2. Encircling endocardial ventriculotomy,[23,34] in which an incision is made circumscribing the regions indicated by mapping, extending about halfway through the myocardial wall; this is intended to isolate the arrhythmogenic region or incise through critical portions of circuits.
3. Focal cryoablation using a 1.5-cm probe cooled to −70° C for 1 to 2 minutes[35]; this destroys myocardial cells without removing or disrupting the fibrous stroma.
4. Encircling cryoablation, combining focal cryoablation with encircling ventriculotomy.[36]
5. Laser photoablation, using Nd:YAG, argon or carbon dioxide sources[25]; this procedure has fundamentally the same effect as cryoablation.

Following the ablation portion, attempts are usually made to reinitiate VT with the same type of stimulation used during mapping. If VT can still be initiated, mapping is repeated; this usually reveals sites just outside the area ablated, or deep to it (intramural). Ablation is performed again (often cryoablation of deeper sites) and cycles of stimulation, mapping and ablation are repeated until VT can no longer be induced. This technique results in postoperative noninducibility of VT approaching 90%.[37] In inferior infarction, cryoablation of the isthmus of ventricular myocardium between the ventriculotomy and the mitral annulus has increased surgical success rates.[38] The ventriculotomy is closed and other procedures are performed, such as bypass grafting or valve surgery. If no other procedures are needed, the patient is weaned from cardiopulmonary bypass. Recent innovations in how the ventriculotomy is closed (aneurysmorrhaphy) have led to greater improvements in postoperative LV function.[39] Patients typically undergo EP study five to seven days after surgery to assess its antiarrhythmic efficacy.

In some cases, there is no obvious site for making a ventriculotomy to enter the ventricle (no infarct or aneurysm). This is especially true in inferior infarctions and cases in which early reperfusion of the infarct (pharmacologic or mechanical) yielded only a partial success. In these cases, an electrode-studded inflatable latex balloon can be inserted through the mitral orifice after a purse-string left atriotomy. Once the balloon is in the ventricular cavity, it is inflated such that its electrodes contact the endocardial surface.[40] Mapping can then be carried out as indicated earlier but without ventriculotomy, and ablation can be performed either with electrical energy (DC shock or RF energy) delivered through appropriately located electrodes, or using a cryoprobe advanced through the mitral orifice as the array had been.

Because of the presence of diseased myocardium, many areas have abnormal electrograms during sinus rhythm or VT, but do not actually participate in the reentrant wavefront. This low specificity, as well as the fact that up to 15% of VTs have been mapped to regions with relatively normal electrograms during sinus rhythm,[41] limits the usefulness of simply ablation of areas with abnormal electrograms. Furthermore, some sites have diastolic activation during VT but represent "side branches" that are not in the reentrant path. Some of these display 2:1 conduction of potentials during VT; obviously these sites are not in the reentrant circuit because VT continues despite their absence on alternate beats, but they are typically recorded from regions near the actual circuit.[42]

Mapping or No Mapping?

Although early map-guided surgical series reported good results, some investigators questioned the contribution of mapping to the results, noting that mapping equipment is expensive and electrophysiologists and surgeons skilled in these techniques are not universally available. They also argued that the same results might be obtained by ablation of regions of visible scarring, because these areas coincided with those designated by time-consuming mapping studies.[43] While this is largely true, critical circuit pathways are outside areas of visible scarring in about 15% of cases and in VT occurring within the first few weeks after infarction, the endocardial appearance of arrhythmogenic zones has not yet become distinct from normal areas. In addition, removal of all endocardial scarring would necessitate papillary muscle removal and mitral valve replacement in a large proportion of patients (which has been advocated by some[44]). There is evidence that the more thoroughly mapped a patient's VTs are, the higher the surgical success rate.[45]

Efficacy

The success rates for the various surgical ablation procedures vary according to the procedure as well as the definition of *success.* Most data indicate that elimination of all inducible VT at postoperative EP study is the most reliable endpoint of success, and that simply eliminating inducibility of certain VT morphologies (but not all VTs) will not prevent subsequent recurrences of other VTs.[46] Measured against the more stringent standard of eliminating all inducible VT, most of the procedures discussed earlier yield success rates from 70% to 90%.[39,47-51]

Surgical Risk

Due to the preexisting extensive myocardial damage, the surgical mortality rate of VT surgery is higher than other forms of elective cardiac surgery. Mortality rates range from 3% to 20% in different patient subsets. Because of this, several investigators have evaluated risk factors for VT surgery and have identified LV wall motion score, age more than 65 years, and emergency surgery as factors that increase the risk of surgical mortality.[52-54]

Patient Selection

Because of the above features and the availability of lower risk alternative therapies, surgery is rarely performed for VT. The benefits and risks of surgery, as well as availability of the necessary mapping equipment and electrophysiological and surgical expertise, must be carefully weighed before proceeding.[55]

Other Substrates of Ventricular Tachycardia

Reentrant VT can occur in other settings, such as idiopathic cardiomyopathy, arrhythmogenic right ventricular dysplasia (ARVD), Chagas' disease, or following right ventriculotomy for repair of congenital heart disease (ventricular septal defect and tetralogy of Fallot). There is relatively little surgical experience with *cardiomyopathic VT.* The arrhythmogenic substrate appears to be located on the endocardial aspect of the LV in most cases, as in postinfarct VT; unlike the latter, however, there is usually no readily available region of scarring or aneurysm through which the ventricle can be opened to allow endocardial access. It is generally undesirable to make a large incision through normal-appearing myocardium, the closure of which could itself conceivably become a source of reentrant arrhythmias postoperatively. Some cases of VT due to ARVD were managed surgically in the 1970s and 1980s; the substrate in these cases is relatively thin strands of surviving normal myocardium separated by large zones of adipose and fibrous tissue that have replaced portions of the RV free wall. Three major regions were identified in which this transformation was most commonly observed: the apex, the outflow tract, and the inferolateral basal RV near the tricuspid annulus.

The first surgical efforts in this disorder were directed at incising these visually identifiable zones of fatty replacement of the RV wall.[56] However, postoperative VT recurrences in some cases were determined to be due to reentry in other regions of the RV that had apparently not been sources of arrhythmia earlier. This led to the RV disarticulation procedure,[57] in which the entire RV free wall was incised from its septal attachments and myocardium that could not be completely incised (infundibulum, near the valve annuli) was frozen with a cryoprobe. The RV free wall was then reattached with a long suture line. In this way, VT could still occur in the isolated RV, but not spread to the LV (which remained in sinus rhythm). The RV then functioned as a conduit for blood flow through the right heart because it had no coordinated mechanical function. Although conceptually elegant, the results of this procedure were disappointing; arrhythmia "cure" rates were high, but poor hemodynamic performance, especially in the early postoperative period, caused significant morbidity and even mortality. As a result, this procedure is rarely performed; patients with ARVD most commonly receive ICD therapy.

In cases of VT following right ventriculotomy, mapping studies have usually shown a reentrant wavefront circulating around the incision that acts as a nonconducting barrier (as is the case following right atriotomy noted earlier). This can be treated by continuing the incision to another nonconducting barrier, such as the pulmonic valve annulus.[58] This is usually accomplished with extensive cryoablation. In some cases, VT appears to be due to reentry around the ventricular septal defect patch.

Rarely, VT occurs in the absence of structural heart disease. Several syndromes of VT in this setting have been relatively well characterized, including exercise-provoked right ventricular outflow tract (RVOT) VT of focal origin, and a reentrant arrhythmia involving a portion of the LV Purkinje system, right bundle branch block-left axis deviation (RBBB-LAD) VT. An uncommon variant of RVOT VT arises in the left ventricle, often near the aortic or mitral valve annulus. Although some of the earliest reports of surgical therapy of VT were for these arrhythmias,[24] surgery is rarely needed because of the very high success rates of catheter ablation (>90%). Several patients underwent surgery for RVOT VT before the development of catheter ablation techniques. As was noted for focal atrial tachycardias, general anesthesia can have a profound quieting effect on these types of VT; catecholamine infusion may restore firing of the focus, but can also simply increase the sinus rate over the intrinsic rate of the VT. We have had to pack the sinus node in ice to prevent this latter effect and allow the VT to become manifest to map its origin. There have been recent reports of minimally invasive surgical therapy for epicardial foci of VT that have not responded to catheter ablation.[59] Only rare cases of RBBB-LAD VT have undergone surgical therapy; in one fascinating case, the tachycardia terminated following incision of an LV false tendon.[60] Histologic examination revealed a large number of Purkinje cells, strengthening the association of the specialized conduction system with this unusual arrhythmia.

Current Status

Because of the high efficacy and low morbidity and mortality rates of ICD implantation and catheter ablation, surgical therapy for VT has largely been relegated to the second or even third tier of therapy, after these other methods have failed to effect adequate arrhythmia control. However, there are situations in which surgical therapy warrants earlier consideration:

1. In patients who are undergoing other indicated cardiac surgical procedures, such as coronary artery bypass or valve surgery, VT surgery can be performed at the same time. This additional procedure almost surely increases surgical risk to some degree, but there are no studies that have addressed this question directly.
2. In patients with incessant or very frequent, hemodynamically poorly tolerated VT episodes, in whom medical therapy has failed to control recurrences, VT surgery is often the only reasonable option. ICD therapy is contraindicated when it would result in frequent repeated shocks, and hemodynamic instability or presence of LV thrombus often precludes any consideration of catheter ablation. In these emergency situations, VT surgery can be lifesaving.
3. In patients who simply prefer the chance of "cure" of their arrhythmia rather than rescue from episodes (such as the ICD provides), in whom catheter ablation is not possible or is not preferred, VT surgery is a reasonable option. Careful counseling of the patient and family members is necessary for this unusual situation.

Ventricular Fibrillation

Following cardiac arrest, the rhythm recorded on arrival of the rescue personnel may be bradycardia or asystole, VT, or VF. When VF is the first observed rhythm, there are several possible explanations. VF may have indeed been the primary rhythm disturbance at the beginning of the episode. In this setting, possible causes include VF due to severe ischemia (with or without earlier myocardial damage), VF in the setting of prior damage (infarct or cardiomyopathy), long QT syndrome (congenital or acquired), idiopathic VF, Brugada syndrome, and other less common settings. Conversely, an episode of rapid ventricular tachycardia may have degenerated to VF that was present at the time of the first ECG recording. The patient's medical history, resting ECG, and assessment of LV function can help suggest one or another of these causes. Distinguishing between a solely ischemic cause and one that has a basis in prior infarct or cardiomyopathy (substrate) is of critical importance, because the treatments for these disorders are so drastically different. If severe ischemia is the cause, revascularization alone suffices and antiarrhythmic drug or ICD therapy is, at best, unhelpful and may even be proarrhythmic. If previously damaged myocardium provides the substrate for VT or VF, treatment of ischemia alone will not prevent arrhythmia recurrences. Although one can sometimes make this distinction on clinical grounds, it is often not possible to make a certain diagnosis. For instance, many patients with severe coronary stenoses that may be capable of producing VF also have an area of previous infarction. Because the consequences of choosing the wrong therapy are so serious, every effort should be made to determine the correct diagnosis. Patient evaluation should include clinical history, physical examination, and standard ECG; assessment of resting LV function; and assessment for the presence and extent of myocardial ischemia (functional assessment as well as coronary arteriography). The clinical history may strongly suggest severe ischemia as the provocation for VF (i.e., the patient was exercising or arguing vigorously and complained of increasing chest pain just before collapse), but often a detailed history of the event is not available. Exercise stress testing, at which severe ischemia at low workloads leads to nonsustained or sustained polymorphic VT or VF, may be diagnostic but should not be performed before knowing the patient's coronary anatomy. Other tests, such as T wave alternans, may be useful in the future but currently available data are not adequate to form the basis for clinical judgments.

Based on the results of these tests, patients may be characterized in several distinct subsets:

1. Severe coronary stenoses (typically >90%) with readily provocable ischemia, normal LV function, no scarring—the most likely cause is pure ischemically mediated VF.
2. Moderate-to-severe coronary stenoses with or without significant ischemia, reduced LV function, with myocardial scarring—these patients may have postinfarct myocardial substrate for rapid VT that degenerates to VF or ischemically mediated VF.
3. Moderate-to-severe coronary stenoses without significant ischemia, markedly reduced LV function, no scarring—the most likely diagnosis is cardiomyopathy with an ill-defined myocardial substrate for arrhythmias (VT or VF).
4. No significant coronary stenoses, normal LV function, no scarring—the most likely diagnoses are idiopathic VF, Brugada syndrome or its variants, or a form fruste of long QT syndrome.

Patients in the first group, with very severe coronary stenoses, normal LV function and wall motion, and no clinical, electrocardiographic or angiographic evidence of prior damage, appear to respond well to revascularization alone.[61] Although there is always a theoretical risk of recurrent VF if graft occlusion or restenosis occurs, this has not been a reported clinical problem (with limited numbers of cases). Patients in the second group (with significant coronary stenoses, reduced LV function, and prior scarring), for whom revascularization is planned, should undergo EP testing before and following revascularization to determine whether ventricular arrhythmias are inducible with programmed stimulation. If VT or VF can be initiated before surgical revascularization but not following the procedure, the chance of arrhythmia recurrence is significantly lower than if the VT or VF can still be initiated postoperatively.[62] Although data are limited, it appears that patients in this latter group (persistent arrhythmia inducibility following revascularization) should undergo ICD therapy as well because they have evidence of the ongoing presence of substrate for these arrhythmias.[62] To date, there are no comparable data for efficacy of percutaneous revascularization.

Some patients in the group with myocardial scarring and VF have frequent recurrences of arrhythmia in the absence of significant ischemia; some of these have responded well to endocardial resection or cryoablation guided by mapping of fractionated electrograms during sinus rhythm.[63] Patients in the third group (cardiomyopathy) cannot be expected to have antiarrhythmic benefit from revascularization and should undergo ICD therapy, as should patients in the fourth group (no evidence of structural disease) who have no other reliable means of preventing arrhythmic recurrences.

Patients presenting with VF comprise a very heterogeneous group with a wide variety of causes of arrhythmia for which optimal treatments vary markedly. These patients must undergo thorough evaluations because an improper therapy based on an incorrect diagnosis can have disastrous results.

SUMMARY

Current surgical therapy for supraventricular arrhythmias is almost exclusively for atrial fibrillation, with an occasional patient operated for catheter ablation-resistant WPW. Most often, a modified maze procedure or pulmonary vein isolation is performed in conjunction with other heart surgery (e.g., mitral valve repair or replacement). Surgical therapy for ventricular arrhythmias is much less commonly performed than it was a decade ago, primarily because of advances in implantable cardioverter-defibrillator technology, but it still has a role in selected cases of uniform-morphology ventricular tachycardia and ischemic ventricular fibrillation. Especially in the treatment of ventricular tachycardia, the combination of a surgeon and an electrophysiologist, who are experienced in these techniques and proper equipment for mapping and ablating the arrhythmia, plays a key role in the ultimate success of the procedure.

REFERENCES

1. Munger TM, Packer DL, Hammill SC, et al: A population study of the natural history of Wolff-Parkinson-White syndrome in Olmsted County, Minnesota, 1953-1989. Circulation 1993;87:866-873.
2. Sorbo MD, Buja GF, Miorelli M, et al: The prevalence of the Wolff-Parkinson-White syndrome in a population of 116,542 young males. G Ital Cardiol 1995;25:681-687.
3. Wolf PA, Benjamin EJ, Belanger AJ, et al: Secular trends in the prevalence of atrial fibrillation: The Framingham Study. Am Heart J 1996;131:790-795.
4. Cobb FR, Blumenschein SD, Sealy WC, et al: Successful surgical interruption of the bundle of Kent in a patient with Wolff-Parkinson-White syndrome. Circulation 1968;38:1018-1029.
5. Cox JL, Gallagher JJ, Cain ME: Experience with 118 consecutive patients undergoing operation for the Wolff-Parkinson-White syndrome. J Thorac Cardiovasc Surg 1985;90:490-501.
6. Guiraudon GM, Klein GJ, Yee R, et al: Surgical epicardial ablation of left ventricular pathway using sling exposure. Ann Thorac Surg 1990;50:968-971.
7. Cox JL, Ferguson TB, Lindsay BD, et al: Perinodal cryosurgery for atrioventricular node reentry tachycardia in 23 patients. J Thorac Cardiovasc Surg 1990;99:440-449.
8. Mahomed Y: Surgery for atrioventricular nodal reentrant tachycardia. In Zipes DP, Jalife J, (eds): Cardiac Electrophysiology: From Cell to Bedside. 2nd ed. WB Saunders, Philadelphia, 1994, pp 1577-1583.
9. Seals AA, Lawrie GM, Magro S, et al: Surgical treatment of right atrial focal tachycardia in adults. J Am Coll Cardiol 1988;11: 1111-1117.
10. Kalman JM, Van Hare GF, Olgin JE, et al: Ablation of "incisional" reentrant atrial tachycardia complicating surgery for congenital heart disease. Use of entrainment to define a critical isthmus of conduction. Circulation 1996;93:502-512.
11. Konings KT, Kirchhof CJ, Smeets JR, et al: High-density mapping of electrically induced atrial fibrillation in humans. Circulation 1994;89:1665-1680.
12. Guiraudon GM, Klein GJ, Yee R, et al: Surgery for atrial tachycardia. Pacing Clin Electrophysiol 1990;13:1996-1999.
13. Cox JL, Schuessler RB, Cain ME, et al: Surgery for atrial fibrillation. Sem Thoracic Cardiovasc Surg 1989;1:67-73.
14. Cox JL, Boineau JP, Schuessler RB, et al: Five-year experience with the maze procedure for atrial fibrillation. Ann Thorac Surg 1993;56:814-823.
15. Kosakai Y, Kawaguchi AT, Isobe F, et al: Cox maze procedure for chronic atrial fibrillation associated with mitral valve disease. J Thorac Cardiovasc Surg 1994;108:1049-1054.
16. Haïssaguerre M, Jaïs P, Shah DC, et al: Spontaneous initiation of atrial fibrillation by ectopic beats originating in the pulmonary veins. N Engl J Med 1998;339:659-666.
17. Melo J, Adragao PR, Neves J, et al: Electrosurgical treatment of atrial fibrillation with a new intraoperative radiofrequency ablation catheter. Thorac Cardiovasc Surg 1999;47 Suppl 3: 370-372.
18. Robbins IM, Colvin EV, Doyle TP, et al: Pulmonary vein stenosis after catheter ablation of atrial fibrillation. Circulation 1998;98: 1769-1775.
19. Couch OA: Cardiac aneurysm with ventricular tachycardia and subsequent excision of aneurysm. Circulation 1959;20: 251-253.
20. Buda AJ, Stinson EB, Harrison DC: Surgery for life-threatening ventricular tachyarrhythmias. Am J Cardiol 1979;44:1171-1177.
21. Mason JW, Stinson EB, Winkle RA, et al: Relative efficacy of blind left ventricular aneurysm resection for the treatment or recurrent ventricular tachycardia. Am J Cardiol 1982;49:241-248.
22. Josephson ME, Harken AH, Horowitz LN: Endocardial excision: A new surgical technique for the treatment of recurrent ventricular tachycardia. Circulation 1979;60:1430-1439.
23. Guiraudon G, Fontaine G, Frank R, et al: Encircling endocardial ventriculotomy: A new surgical treatment for life-threatening ventricular tachycardias resistant to medical treatment following myocardial infarction. Ann Thor Surg 1978;26:438-444.
24. Gallagher JJ, Anderson RW, Kasell J, et al: Cryoablation of drug-resistant ventricular tachycardia in a patient with a variant of scleroderma. Circulation 1978;57:190-197.
25. Selle JG, Svenson RH, Sealy WC, et al: Successful clinical laser ablation of ventricular tachycardia: A promising new therapeutic method. Ann Thorac Surg 1986;42:380-384.
26. Fenoglio JJ, Duc PT, Harken AH, et al: Recurrent sustained ventricular tachycardia: Structure and ultrastructure of subendocardial regions in which tachycardia originates. Circulation 1983; 68:518-533.
27. de Bakker JMT, Janse MJ, Van Capelle FJL, et al: Endocardial mapping by simultaneous recording of endocardial electrograms during cardiac surgery for ventricular aneurysm. J Am Coll Cardiol 1983;2:947-953.
28. Downar E, Parson ID, Mickleborough LL, et al: On-line epicardial mapping of intraoperative ventricular arrhythmias: Initial clinical experience. J Am Coll Cardiol 1984;4:703-714.

29. de Bakker JM, van Capelle FJ, Janse MJ, et al: Macroreentry in the infarcted human heart: The mechanism of ventricular tachycardias with a "focal" activation pattern. J Am Coll Cardiol 1991;18: 1005-1014.
30. Miller JM, A. RS, Hsia HH, et al: The role of electrophysiologic mapping for ventricular tachycardia ablation. In Singer I, (ed): Nonpharmacological Therapy of Arrhythmias for the 21st Century: The State of the Art. Futura, Armonk, NY, 1998, pp 607-630.
31. Miller JM, Kienzle MG, Harken AH, et al: Morphologically distinct sustained ventricular tachycardias in coronary artery disease: Significance and surgical results. J Am Coll Cardiol 1984;4:1073-1079.
32. Stevenson WG, Friedman PL, Sager PT, et al: Exploring postinfarction reentrant ventricular tachycardia with entrainment mapping. J Am Coll Cardiol 1997;29:1180-1189.
33. Nitta T, Schuessler RB, Mitsuno M, et al: Return cycle mapping after entrainment of ventricular tachycardia. Circulation 1998;97:1164-1175.
34. Cox JL, Gallagher JJ, Ungerleider RM: Encircling endocardial ventriculotomy for refractory ischemic ventricular tachycardia. IV. Clinical indications, surgical technique, mechanism of action, and results. J Thorac Cardiovasc Surg 1982;83:865-872.
35. Tweddell JS, Branham BH, Stone CM, et al: Focal cryoablation guided solely by intraoperative potential mapping ablates ventricular tachycardia of endocardial origin. Surg Forum 1989;40:216-218.
36. Guiraudon GM, Thakur RK, Klein GJ, et al: Encircling endocardial cryoablation for ventricular tachycardia after myocardial infarction: Experience with 33 patients. Am Heart J 1994;128: 982-989.
37. Haines DE, Lerman BB, Kron IL, et al: Surgical ablation of ventricular tachycardia with sequential map-guided subendocardial resection: Electrophysiologic assessment and long-term follow-up. Circulation 1988;77:131-141.
38. Hargrove WI, Miller JM, Vassallo JA, et al: Improved results in the operative management of ventricular tachycardia related to inferior wall infarction: Importance of the annular isthmus. J Thorac Cardiovasc Surg 1986;92:726-732.
39. Sosa E, Scanavacca M, d'Avila A, et al: Long-term results of visually guided left ventricular reconstruction as single therapy to treat ventricular tachycardia associated with postinfarction anteroseptal aneurysm. J Cardiovasc Electrophysiol 1998;9:1133-1143.
40. Downar E, Mickleborough L, Harris L, et al: Intraoperative electrical ablation of ventricular arrhythmias: A 'closed heart' procedure. J Am Coll Cardiol 1987;10:1048-1056.
41. Cassidy DM, Vassallo JA, Buxton AE, et al: The value of catheter mapping during sinus rhythm to localize site of origin of ventricular tachycardia. Circulation 1984;69:1103-1110.
42. Miller JM, Vassallo JA, Hargrove WC, et al: Intermittent failure of local conduction during ventricular tachycardia. Circulation 1985;72:1286-1292.
43. Moran JM, Kehoe RF, Loeb JM, et al: Extended endocardial resection for the treatment of ventricular tachycardia and ventricular fibrillation. Ann Thorac Surg 1982;34:538-552.
44. Moran JM, Kehoe RF, Loeb JM, et al: The role of papillary muscle resection and mitral valve replacement in the control of refractory ventricular arrhythmia. Circulation 1983;68:154-160.
45. Miller JM, Gottlieb CD, Marchlinski FE, et al: Does ventricular tachycardia mapping influence the success of antiarrhythmic surgery? J Am Coll Cardiol 1988;11:112A.
46. Miller JM, Josephson ME, Hargrove WC: Significance of "nonclinical" ventricular arrhythmias induced following surgery for ventricular tachyarrhythmias. In Breithardt G, Borggrefe M, Zipes DP (eds): Nonpharmacological Therapy of Tachyarrhythmias. Futura, Mount Kisco, NY, 1987, pp 133-141.
47. Miller JM, Kienzle MG, Harken AH, et al: Subendocardial resection for ventricular tachycardia: Predictors of surgical success. Circulation 1984;70:624-631.
48. Caceres J, Werner P, Jazayeri M, et al: Efficacy of cryosurgery alone for refractory monomorphic sustained ventricular tachycardia due to inferior wall infarction. J Am Coll Cardiol 1988;11:1254-1259.
49. Hargrove WC: Surgery for ischemic ventricular tachycardia—operative techniques and long-term results. Semin Thorac Cardiovasc Surg 1989;1:83-87.
50. Ostermeyer J, Kirklin JK, Borggrefe M, et al: Ten years of electrophysiologically guided direct operations for malignant ischemic ventricular tachycardia—Results. Thorac Cardiovasc Surg 1989;37:320-327.
51. Nath S, Haines DE, Kron IL, et al: The long-term outcome of visually directed subendocardial resection in patients without inducible or mappable ventricular tachycardia at the time of surgery. J Cardiovasc Electrophysiol 1994;5:399-407.
52. Miller JM, Gottlieb CD, Hargrove WC, et al: Factors influencing operative mortality in surgery for ventricular tachycardias. Circulation 1988;78 (II):44.
53. Van Hemel NM, Kingma JH, Defauw JAM, et al: Left ventricular segmental wall motion score as a criterion for selecting patients for direct surgery in the treatment of postinfarction ventricular tachycardia. Eur Heart J 1989;10:304-315.
54. Mittleman RS, Candinas R, Dahlberg S, et al: Predictors of surgical mortality and long-term results of endocardial resection for drug-refractory ventricular tachycardia. Am Heart J 1992;124:1226-1232.
55. Cox JL: Patient selection criteria and results of surgery for refractory ischemic ventricular tachycardia. Circulation 1989;79:163-177.
56. Guiraudon GM, Klein GJ, Sharma AD, et al: Surgical therapy for arrhythmogenic right ventricular adiposis. Eur Heart J 1989;10 (Suppl D):82-83.
57. Guiraudon GM, Klein GJ, Gulamhusein SS, et al: Total disconnection of the right ventricular free wall: Surgical treatment of right ventricular tachycardia associated with right ventricular dysplasia. Circulation 1983;67:463-470.
58. Downar E, Harris L, Kimber S, et al: Ventricular tachycardia after surgical repair of tetralogy of Fallot: Results of intraoperative mapping studies. J Am Coll Cardiol 1992;20:648-655.
59. Frey B, Kreiner G, Fritsch S, et al: Successful treatment of idiopathic left ventricular outflow tract tachycardia by catheter ablation or minimally invasive surgical cryoablation. Pacing Clin Electrophysiol 2000;23:870-876.
60. Suwa M, Yoneda Y, Nagao H, et al: Surgical correction of idiopathic paroxysmal ventricular tachycardia possibly related to left ventricular false tendon. Am J Cardiol 1989;64:1217-1220.
61. Kelly P, Ruskin JN, Vlahakes GJ, et al: Surgical coronary revascularization in survivors of prehospital cardiac arrest: Its effect on inducible ventricular arrhythmias and long-term survival. J Am Coll Cardiol 1990;15:267-273.
62. Wilber DJ, Garan H, Finkelstein D, et al: Out-of-hospital cardiac arrest: Use of electrophysiologic testing in the prediction of long-term outcome. N Engl J Med 1988;318:19-24.
63. Bourke JP, Campbell RWF, Renzulli A, et al: Surgery for ventricular tachyarrhythmias based on fragmentation mapping in sinus rhythm alone. Eur J Cardiothorac Surg 1989;3:401-406.

Chapter 49 Device Technology for Congestive Heart Failure

ANGELO AURICCHIO, HELMUT KLEIN, and
SANJEEV SAKSENA

Over the last decade, treatment of heart failure has markedly improved. Mortality due to pump failure and sudden death caused by ventricular tachyarrhythmic events has declined significantly.[1,2] Hospitalizations for severe symptoms of heart failure have decreased after increasing the use of ACE-inhibitors, β-blockers, diuretics, digoxin, and most recently, spironolactone. An important beneficial effect of medical therapy in heart failure is reduction in the abnormally increased sympathetic tone and the consequent interruption of the secondary activation of the renin-angiotensin-system. Heart transplantation is considered the therapy of choice in end-stage heart failure, but the limited availability of donor organs and the still unresolved issue of tissue rejection after transplantation have stimulated research in other nonpharmacologic approaches to symptomatic heart failure.

A common finding in the failing heart is a delay in the spread of ventricular activation caused by structural abnormalities of the myocardium, leading to asynchronous ventricular contraction and potentially life-threatening ventricular tachyarrhythmias.[3] Recent reports have demonstrated that ventricular conduction delay represents an independent risk marker for the development of heart failure and increased mortality.[4-6] It seems logical that devices, such as cardiac pacemakers that can potentially correct delayed cardiac electrical activation or defibrillators with pacing capabilities that can also prevent sudden arrhythmic death, may prove to be effective tools in the armamentarium of heart failure therapy. In the last few years cardiac pacing for heart failure has become a promising new therapeutic option for a selected group of patients with severe chronic heart failure. Twelve years after the first description of biventricular pacing for heart failure, more than 10,000 patients have been successfully treated with this new electrical therapy. In the interim, new terminology has developed for this therapy. Current understanding of stimulation therapy for heart failure indicates that discoordinate contraction pattern in the ventricles is corrected. For this reason, it is more appropriate to use the term *cardiac resynchronization therapy* (CRT).

Pathophysiologic Concept of Resynchronization Therapy

ELECTRICAL AND MECHANICAL ABNORMALITIES IN HEART FAILURE

Prolongation of the atrioventricular (A–V) interval is associated with impaired atrial contribution to ventricular contraction and reduced diastolic ventricular filling. Intra- as well as interventricular conduction delays prolong the pre-ejection time, reduce global and regional ventricular ejection fraction caused by an asynchronous contraction and relaxation pattern, and prolong mitral regurgitation.[7,8] One electrical indicator for delayed asynchronous ventricular contraction is the presence of left bundle branch block (LBBB). These contraction abnormalities can be assessed by echocardiography or magnetic resonance imaging. In addition, LBBB may be responsible for unequal regional distribution of ventricular work and wall stress.[9] At the beginning of ventricular systole, the region of earliest ventricular activation (usually the intraventricular septum) contracts against minimal workload because the remaining ventricular myocardium (usually the lateral and posterolateral left ventricular regions) is still in the relaxation or in a nonactivated phase. The regions of early ventricular activation waste contraction energy as no effective intraventricular pressure can develop. In contrast, the delayed depolarized regions; i.e., the lateral and posterolateral ventricular regions, have to contract against a preexisting stiffened portion of the ventricular wall (the septum) This generates increased wall stress with increased cardiac work. These changes in regional wall stress can contribute to myocyte damage, production of fibrous tissue, development of regional hypertrophy, and may induce regional apoptosis.

VENTRICULAR ASYNCHRONY

A broad QRS complex with LBBB causes delayed contraction of the left ventricular lateral wall, whereas the ventricular septum exhibits a paradoxical movement.[10]

Preexcitation of the left lateral ventricular wall with atrio-biventricular pacing in hearts with LBBB resynchronizes the ventricular contraction pattern by bypassing the conduction delay, thus restoring a normal contraction pattern.[11] Restoration of ventricular coordination also improves pump function of the heart. It is conceivable that other intraventricular delay patterns produce different regional wall motion abnormalities that can be similarly addressed.

Programming the AV delay during sequential pacing of the left ventricle is essential for the improvement of hemodynamics. A very long AV delay will not support ventricular resynchronization, because the atrial electrical impulse will follow the same route as during sinus rhythm without pacing. A very short AV delay, on the contrary, will cause early depolarization at the site of left ventricular stimulation, leaving the ventricle partially or totally refractory by the time the regularly conducted impulse reaches this region. An AV delay between these two extremes causes a collision of two activation wavefronts: one coming from the regular His-Purkinje system and the other from the preexcited ventricular activation. The region of collision depends on individual intraventricular and interventricular conduction properties (Fig. 49-1). Therefore, appropriate timing with respect to the AV delay, as well as the left lateral ventricular stimulus, is crucial for achieving the hemodynamic benefits of resynchronized ventricular contraction pattern in a failing heart.

ATRIOVENTRICULAR SYNCHRONY

Mechanical synchrony between the atrium and the ventricle disappears when AV conduction is pathologically prolonged. Prolongation of AV conduction reduces the active ventricular filling phase and shortens passive diastolic filling creating a ventriculoatrial gradient causing presystolic mitral regurgitation.[12,13] A prolonged mechanical A–V interval is frequently found in a significant number of patients with heart failure, even with an almost normal electrical A–V interval.[14] A prolonged electromechanical A–V interval can be assessed by measuring the onset of the P wave of the surface ECG to the aortic valve closure. This interval can be prolonged even in the absence of a prolonged P–R interval on the surface ECG. The mechanical A–V interval is always longer than the electrical A–V interval. An apparently *nonphysiologic* short electrical A–V interval has to be used with AV sequential biventricular pacing in order to *resynchronize* the AV asynchrony and reduce presystolic mitral regurgitation.

The discrepancy between electrical and mechanical AV sequences is most likely due to a prolonged intraventricular conduction time. In patients with LBBB, the onset of the electrical depolarization of the left ventricular free wall is significantly delayed.[15,16] This causes delayed mechanical onset of the left ventricular systole.[15] Consequently, AV sequential pacing with a shortened AV delay is able to restore an adequate mechanical AV synchrony.[17] Maximal hemodynamic benefit is achieved when the peak of the atrial pressure curve coincides with the onset of the mechanical ventricular systole (Fig. 49-2).

EFFECT OF RESYNCHRONIZATION THERAPY ON MITRAL REGURGITATION

Patients with symptoms of heart failure and a dilated left ventricle often demonstrate moderate or even severe mitral regurgitation. This is often seen in patients with LBBB. The slow ventricular activation caused by intercellular fibrotic tissue enhances mechanical asynchrony between different ventricular regions and can involve both papillary muscles.[18] The geometric distortion of the dilated left ventricle[19,20] and delayed left ventricular free wall activation further decrease the efficiency of mitral leaflet closure along with an increased tethering of the mitral apparatus.

Pacing from the left lateral wall, especially from the proximity of the posterior papillary muscle, diminishes conduction delay and decreases mitral regurgitation. Mean capillary wedge pressure drops significantly with

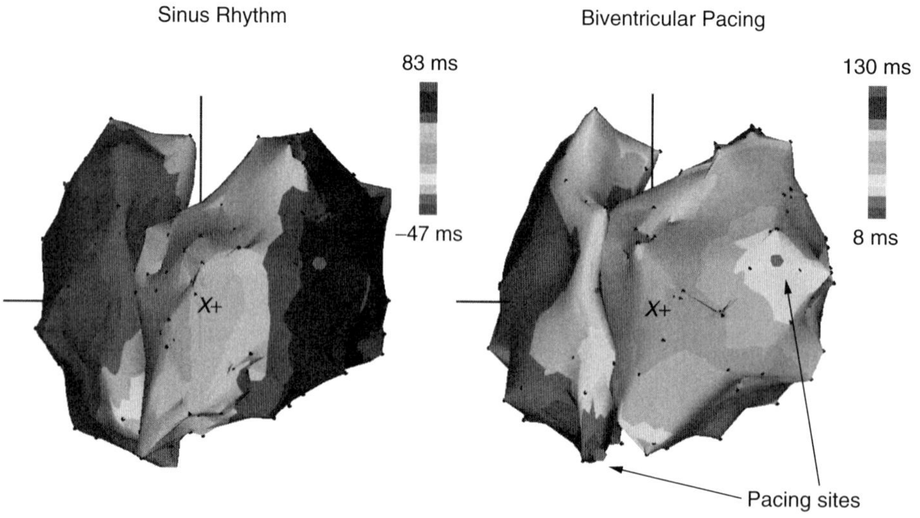

FIGURE 49-1 (See also Color Plate 49-1.) Three-dimensional electroanatomic, nonfluoroscopic mapping in a patient with dilated cardiomyopathy during sinus rhythm and during biventricular stimulation. In sinus rhythm **(left panel)** the earliest ventricular activation (*red*) is located at the anterolateral wall of the right ventricle. After about 60 ms, the activation breaks through into the left ventricle and slowly proceeds (cell-to-cell conduction) from the septum to the lateral and posterolateral wall. The simultaneous pacing from the apex of the right ventricle and lateral wall restored a more homogeneous electrical activation of both ventricles.

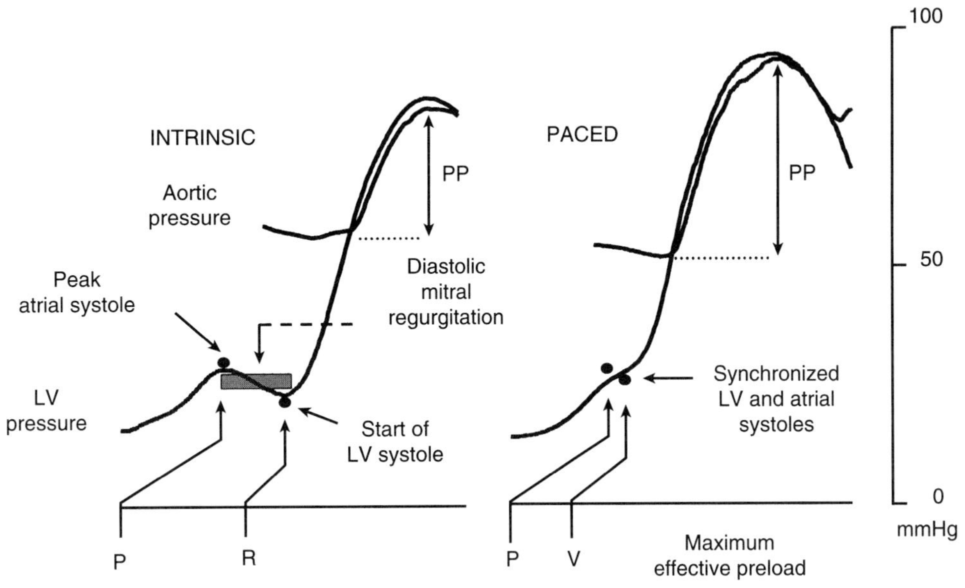

FIGURE 49-2 Schematic representation of the hemodynamic benefit obtained when the maximum of the atrial contraction curve is coincident with the onset of the mechanical ventricular systole.

left ventricular free wall pacing, whereas systolic blood pressure rises. The altered ventricular geometry of the dilated failing ventricle results in incomplete closure of the mitral valve at the onset of ventricular contraction causing early systolic mitral regurgitation. Therefore, shortening of the A–V interval by atrial sequential pacing along with left ventricular preexcitation by stimulation diminishes—or even abolishes—mitral regurgitation.[21]

Results of Resynchronization Therapy

ACUTE RESULTS

Acute hemodynamic testing has demonstrated that the type of intraventricular conduction block and the pacing site location are the primary determinants of hemodynamic benefits. In addition, acute data suggest a dichotomous behavior in patients who present with a QRS duration between 120 and 150 ms.[22] Patients with a QRS duration of greater than 150 ms showed the largest hemodynamic benefit. The Pacing Therapies for Congestive Heart Failure (PATH-CHF) results and data from Kass and colleagues[15] suggest that patients with RBBB and diffuse intraventricular conduction delay tend to benefit more from biventricular and right ventricular pacing. Atrial synchronous left or biventricular stimulation at a nominal AV delay is significantly more beneficial than is right ventricular pacing alone. Parameters of acute systolic function significantly improved during pacing of the left ventricular free wall alone or synchronously with the right ventricle, but not by pacing the right ventricular apex or septum alone.

Pressure-volume loops of the left ventricle show that in patients with LBBB, left ventricular pacing but not right ventricular pacing increased the stroke volume while minimally affecting the end-diastolic volume. Pulmonary capillary wedge pressure dropped significantly with left ventricular pacing and biventricular pacing but not with right ventricular pacing alone. Resynchronization of ventricular contraction improves mechanical efficiency with a net decrease in myocardial energy consumption.[16]

Data from the PATH-CHF study indicates that the acute benefits of resynchronization therapy are dependent on the A–V interval. The results showed that for each ventricular pacing site, the shortest and longest AV delays were suboptimal. In general, a range of AV delay around 100 ms produced the most beneficial hemodynamic effect. There was, however, a large variability in the optimal AV delay during sequential RV pacing, ranging from 50 to 120 ms, and during biventricular stimulation ranging from 100 to 150 ms.

LONG-TERM RESULTS

Acutely, hemodynamic results of biventricular or left ventricular pacing alone clearly demonstrate hemodynamic improvement but it is necessary to show that this translates into long-term symptomatic improvement and eventually survival benefit. Several prospective randomized studies on CRT have been completed, and two large patient registries have been created (Table 49-1). The four early prospective randomized controlled trials on biventricular pacing (PATH-CHF I),[17] the Multisite Stimulation in Cardiomyopathy study (MUSTIC),[18] the Contak CD trial, and the Multicenter InSync Randomized Chronic Evaluation (MIRACLE) have provided evidence that CRT increases exercise tolerance, improves quality of life, and reduces hospitalization in patients with CHF and ventricular conduction disturbances (Table 49-2). The vast majority of patients enrolled in these four randomized studies had severely impaired functional capacity (NYHA classification III to IV), LV systolic dysfunction (ejection fraction ≤35%), wide QRS complex (>120 ms), in most cases LBBB, and none of the patients had conventional indications for pacing therapy. The MUSTIC study, the Contak CD trial, and the MIRACLE study exclusively assessed the

TABLE 49-1 First Generation Trials and Registries of CRT

Trials	Year Concluded	No. of Patients Enrolled	No. of Randomized Patients (%)	Primary Endpoints	Secondary Endpoints
Prospective Controlled Randomized Trials					
PATH-CHF I	1999	42	41 (98)	VO_2, 6MW	QOL, NYHA, HF
MUSTIC (sinus rhythm)	2000	67	58 (87)	6MW	VO_2, QOL, NYHA, HF
MUSTIC (atrial fibrillation)	2000	65	48 (74)	6MW	VO_2, QOL, NYHA, HF
MIRACLE	2001	285	266 (93)	6MW	VO_2, QOL, NYHA, CE
Contak CD	2001	581	490 (84)	HF, VT/VF	VO_2
PATH-CHF II	2001	101	90 (89)	VO_2, 6MW	QOL, NYHA, HF
Registries					
InSync	1999		110	NYHA	
Contak	2000		1000	NYHA	

CE, composite endpoints; HF, hospitalization frequency; NYHA, New York Heart Association (functional classification); QOL, quality of life (Minnesota Living with Heart Failure Questionnaire); VO_2, oxygen consumption; 6MW, six-minute hall walk testing; VT/VF, frequency of ventricular tachycardia/fibrillation.

role of biventricular stimulation in patients with heart failure. The PATH-CHF I study was performed to address the question of whether acutely optimized atrial-synchronous ventricular stimulation (RV-, LV- or BV-stimulation based on acute hemodynamic evaluation) reduces heart failure in patients with intraventricular conduction defects.

Each of the four prospective studies showed that the oxygen consumption at maximal exercise capacity increased on an average from 11 to 12 mL/kg/minute prior to pacing to 15 to 16 mL/kg/minute after three to six months of pacing (see Table 49-2). The six-minute walk performance, a generally accepted parameter of physical exercise capacity, increased on average by 10% to 15%. Furthermore, quite consistently in the four studies, quality of life improved consistently. Nearly two thirds of patients who underwent biventricular therapy improved to NYHA class I or II from class III or IV. Patients in whom the CRT device was turned on were hospitalized less frequently and needed fewer days in the hospital for worsening of heart failure.

TABLE 49-2 Summary of North American Long-Term Clinical Studies Evaluating Cardiac Resynchronization

	Miracle	Mustic	Miracle ICD	Contak CD
QOL	+++	+++	+++	+++
NYHA Classification	+++	+++	+++	+++
Six-minute hall walk	+++	+++	+	+++
Peak VO_2	+++	+	+++	+++
Exercise time	+++		+++	
LVEDD	+++		+++	+++
Ejection fraction	+++	+	+	+
MR jet area	+++	+		
Freedom from hospitalization, IV inotropes	+++		+	+

+, Showed improvement with cardiac resynchronization without achieving clinical significance; +++, showed clinically significant improvement with cardiac resynchronization.; LVEDD, left ventricular end-diastolic diameter.

Consistent with the study hypothesis of the PATH-CHF, there were no clinical differences between hemodynamically optimized biventricular and univentricular (predominantly left ventricular) pacing.[19] Due to the small sample size it was difficult to predict if optimal hemodynamic benefit can be achieved with univentricular or biventricular pacing. The PATH-CHF I study showed that resting heart rate was significantly reduced after three months of pacing. Heart rate variability increased during CRT, whereas during the CRT off phase, an almost complete reversion to baseline values was observed. Changes in heart rate variability and resting heart rate reflect changes of the autonomic nervous system, confirming the positive effect of CRT on neurohumoral activation reported in the VIGOR-CHF study.[20] This study showed a significant reduction of the norepinephrine plasma level after 16 weeks of continuous biventricular stimulation.

The Contak CD study differed from all other studies of CRT as it included patients with class I indication for an ICD and is considered the first generation study in this category. There was a reduction of 21% in the overall combined endpoint, which was not statistically significant ($P < 0.17$). This may be due to the fact that the relatively large proportion of patients in NYHA class II enrolled in Contak CD did not benefit from CRT. Nevertheless, CRT in the Contak CD trial was associated with fewer deaths (23% relative risk reduction), a lower hospitalization rate (13% relative risk reduction), and a smaller proportion (26% relative risk reduction) of patients with worsening heart failure. The number of ventricular tachyarrhythmias was only modestly reduced (9% relative risk reduction) in patients receiving CRT. All patients showed a significant increase in peak oxygen consumption. Patients in an advanced functional class (NYHA class III or IV) showed double the average increase of oxygen consumption. This indicates that patients with more advanced heart failure benefit more than those with less severe symptoms. A recent analysis from our laboratory supports this hypothesis. We found that patients with more depressed oxygen consumption at peak exercise (<10 mL/kg/minute) showed a 50%

increase in peak oxygen consumption with CRT, which is much more impressive than the 2% increase found in patients with less reduced (>18 mL/kg/minute) pretreatment peak oxygen consumption.

SECOND GENERATION CRT TRIALS

More randomized studies on CRT are in progress. The PATH-CHF II study investigates the short- and midterm effect of univentricular CRT (mainly LV pacing alone). The patients enrolled in the PATH-CHF II study must have a QRS complex duration of equal to or greater than 120 ms, must be in an NYHA class equal to or greater than II, and be able to receive either a pacemaker or an implantable cardioverter-defibrillator (ICD), if indicated. Primary endpoints of the PATH-CHF II are changes in oxygen uptake at peak exercise and at anaerobic threshold as well as differences in walking distance during a six-minute hall walk. Secondary endpoints are changes in quality of life and LV ejection fraction.

Second generation prospective randomized controlled studies examined the effect of CRT (exclusively biventricular pacing) and defibrillation therapy on the outcome and morbidity. The PACMAN (Pacing for Cardiomyopathies) study investigated the effect of CRT on exercise performance and hospitalization frequency in patients with severe heart failure, wide QRS duration (>150 ms), including patients with classic ICD indications. The COMPANION (Comparison of Medical Therapy, Pacing and Defibrillation in Chronic Heart Failure) is a multicenter trial evaluating the effect of CRT on mortality, morbidity, and exercise performance in symptomatic heart failure patients without ICD indications. Similarly, the CARE-HF (Cardiac Resynchronization in Heart Failure) is a mortality and morbidity trial including patients with severe heart failure and major mechanical and electrical conduction delays but without an ICD indication.

Technical Aspects of Cardiac Resynchronization Therapy

The complexity of heart failure seen as an electromechanical disease in patients with intraventricular conduction defect needs unique and alternative technology for both leads and devices. New lead technologies, devices specifically designed for heart failure patients with integrated monitoring features, and new sensor technologies require considerable efforts for beneficial and optimal management of heart failure for patients with CRT. A short description of the principles of device system insertion and technology follows.

IMPLANTATION TECHNIQUE

The implantation technique for CRT devices does not differ from the currently used technique for standard pacemakers or ICDs. The right or left subclavian or cephalic vein approach is used for insertion of a conventional right atrial and right ventricular lead. An additional lead is required for pacing of the left ventricle. Initially, the left ventricular lead was fixated on the epicardial surface of the left ventricular wall via a small left lateral thoracotomy. In the meantime, alternative techniques were developed to insert the lead via the coronary sinus into the coronary veins. The left lateral wall is stimulated epicardially from the coronary vein. The anatomy of the coronary vein is, therefore, crucial for correct positioning of the left ventricular lead in order to achieve the most beneficial effect of resynchronization.

The sequence of insertion of the three leads is not standardized and is based on personal experience. We usually prefer to insert first the active fixation leads for the right atrium and the right ventricle. A preshaped long introducer or guiding catheter, which stabilizes the lead while introducing into the coronary sinus, is necessary.[21,22] Severe dilation of the atrial or ventricular chambers may modify the usual position of the coronary sinus os and its anatomic course, so that extensive and prolonged manipulation of the guiding catheter or different approaches may be required.

We preload the guiding catheter with a flexible 0.038 inch coated guidewire (e.g., Terumo guidewire), which is used for exploring the inferoseptal portion of the right atrium and facilitates the atraumatic insertion of a large (8 to 10 Fr) guiding catheter into the coronary sinus. To adapt to the anatomic variation of the coronary sinus, a series of guiding catheters[21] with different shapes is available (Fig. 49-3). Other investigators preload the guiding catheter with a steerable catheter used for electrophysiological studies. These catheters have deflecting tips to shape different curves. Major obstacles for cannulating the coronary sinus may be a thick lamina cribrosa or a large valve. We have rarely seen narrowing of the body of the coronary sinus. Such narrowing may occur in patients after previous open heart surgery.

Once the guiding catheter is inserted, for better evaluation of the coronary vein anatomy, occlusive angiography

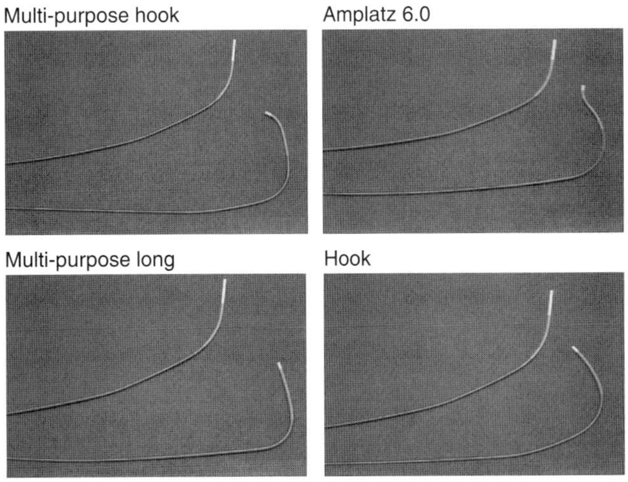

FIGURE 49-3 Guiding catheters specifically designed for insertion into the coronary sinus. A large variety of curves can be appreciated.

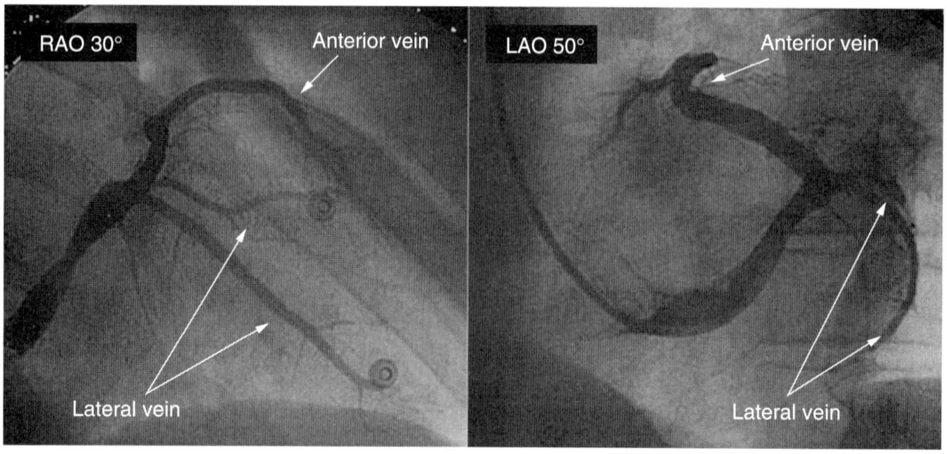

FIGURE 49-4 Selective occlusive angiography of the anterior and lateral coronary vein. LAO, Left anterior oblique; RAO, right anterior oblique.

by an inflated large balloon catheter is highly recommended (Fig. 49-4). Radiographic examination from at least two different plane views should be performed (RAO 30° and LAO 30°-40°). Additional radiographic examination (RAO 25°—caudal 25°, or an anteroposterior view) are suggested when a tortuous or small lateral coronary vein (<2.5 mm in diameter) is noted (Fig. 49-5).

Careful examination of the coronary vein anatomy often reveals one or more large lateral or posterolateral vein. These are the target veins for permanent lead implantation. Particular attention should be paid when performing balloon occlusion retrograde angiography of the coronary sinus and coronary vein. Inappropriate sizing of the balloon, or inflation of the balloon at the most proximal portion of a side vein can produce endothelial damage or extensive intimal lesions, or even rupture of the coronary vein. These dramatic complications are rare and in the range of 1% to 3%.

LEAD TECHNOLOGY

Early attempts to pace from the coronary veins were done by using either standard endocardial leads that were modified for coronary venous placement by removing the tines, or were used for other coronary sinus techniques.[21-23] It was soon realized that specifically designed pacing catheters were necessary to better navigate the coronary sinus, and chronically insert into the coronary veins. To achieve this, innovative coronary venous lead systems were designed that could incorporate components, accessories, and elements patterned after angioplasty devices (Fig. 49-6).

The most promising approach for chronic stimulation of the left ventricle is an over-the-wire pacing lead. This lead has an open lumen lead that tracks over a standard 0.014 inch or 0.010 inch guidewire to allow access to any coronary veins. Guidewire technology has evolved over the years to provide safe, atraumatic

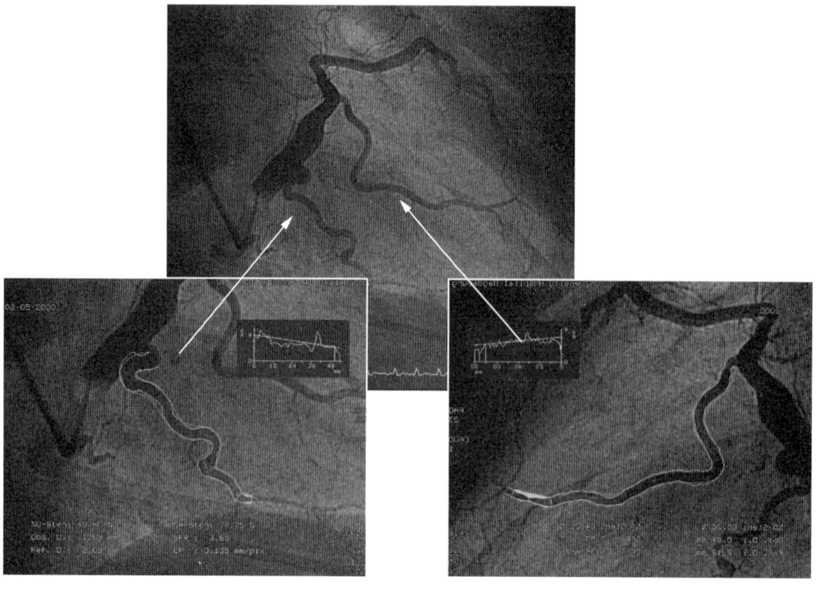

FIGURE 49-5 A tortuous coronary vein anatomy of two lateral veins (*arrows*) can here be appreciated. The sharp take-off of both veins prevented any possibility of inserting either a guidewire or a coronary vein lead.

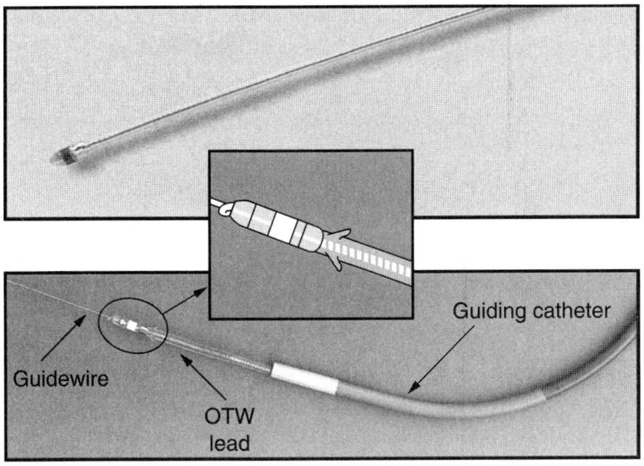

FIGURE 49-6 Two over-the-wire (OTW) permanent coronary vein leads.

navigation through sharp angles and changes in direction within coronary vessels. As in coronary angioplasty procedures, the implanting physician can select a host of available guidewires to obtain the ideal combination of stiffness, torque, stability, and flexibility required for the individual patient's coronary vein anatomy. The lead incorporates a number of characteristics that allow smooth and safe placement in select vein branches. Because of their particular design, over-the-wire leads usually require a deeper insertion—wedge position—into the coronary vein for a stable chronic implantation.

There is now extensive use with over-the-wire lead systems (Easytrak lead, Guidant, and Attain OTW—Medtronic model 4193 lead). Successful placement of the coronary vein lead is now possible in the majority of patients within a reasonable time. The over-the-wire design allows a 95% implantation success rate, including particularly challenging vein anatomy or when repositioning in a second vein is needed. For sensing, the mean R wave amplitude at the left free lateral wall is usually comparable with that achieved in the right ventricle. Pacing impedance and threshold are frequently higher than in the right ventricular apex. Sensing amplitude, pacing threshold, and impedance varies from lead to lead, but additional factors—such as myocardial viability, fat tissue around the coronary vein, and scar tissue after myocardial infarction—affect electrical stability.

MONITORING FEATURES OF CRT DEVICES

The monitoring features of market released pacemakers or ICDs are mainly focused on device-related parameters, such as pacing threshold, impedance, and R-wave amplitude. In addition, storage of intracardiac electrograms has been implemented. Careful hemodynamic control and electrical monitoring is now implemented into CRT devices.

Hemodynamic and Heart Rate Monitoring

Hemodynamic parameters can be either directly or indirectly monitored via the implanted devices. Specific sensors of the maximal pressure derivative are already implemented. The accuracy of monitoring chronic hemodynamic changes in heart failure patients has yet to be determined. Hemodynamic changes can be indirectly assessed by monitoring heart rate and heart rate variability (HRV). Heart rate at rest and during exercise, and heart rate variability, both depend on a variety of autonomic mechanisms, such as baro-receptor reflex, autonomic feedback, and contractility. Changes in heart rate or HRV can indirectly track spontaneous hemodynamic changes, as well as those induced by CRT.

Patients with heart failure present with a significantly reduced heart rate variability and high resting heart rate. Reduced standard deviation measured over normal intervals (SDNN index) has been shown to be associated with left ventricular (LV) dysfunction and increased mortality. Since pacemakers and ICDs can measure the intrinsic heart rate and perform a beat-to-beat analysis, one is able to automatically monitor the daily, weekly, or monthly changes of mean heart rate, minimal heart rate, and heart rate variability. A combination plot including heart rate, HRV, and frequency count more than each parameter alone. The automatic permanent chronic monitoring capability of one or all of these parameters is helpful for assessing the efficacy of certain medications, as well as for evaluating the therapeutic influence of changes in pacing device programming. It is conceivable in the future that these parameters will allow us to automatically titrate the parameters that determine the efficacy of CRT, such as AV delay and pacing site and mode.

A modern device for CRT incorporates three-dimensional plots of heart rate, heart rate variability, and frequency count. Initial experience showed significant changes during the first 12 weeks after the implantation compared with the preimplantation status (Fig. 49-7). Whether changes in heart rate variability really reflect improvement of heart failure and whether they are associated with lower overall mortality remains to be determined.

Evaluation of Physical Activity

Patients with heart failure have a significantly reduced exercise capacity because of limited pump efficacy, impaired systemic hemodynamics, altered muscular activity, and alteration of metabolic demand. The six-minute hall walk distance test has been used to assess the submaximal exercise capacity of patients with heart failure.[24] Although the walk distance in patients with heart failure has been shown to predict overall mortality and need for hospitalization,[25] the walking distance can be monitored only periodically and may be influenced by several factors, including patient motivation, coaching, familiarity with the test, and physical limitations. Variabilities in the walking distance between two subsequent tests are not unusual in patients with heart failure.[26] A more objective measurement of the functional capacity that is continuous, unbiased, and based on activities of daily life is necessary in the future. In this regard, continuous recording of accelerometer signals, reflecting the patient's physical activity, is already used in pacemakers and ICDs.

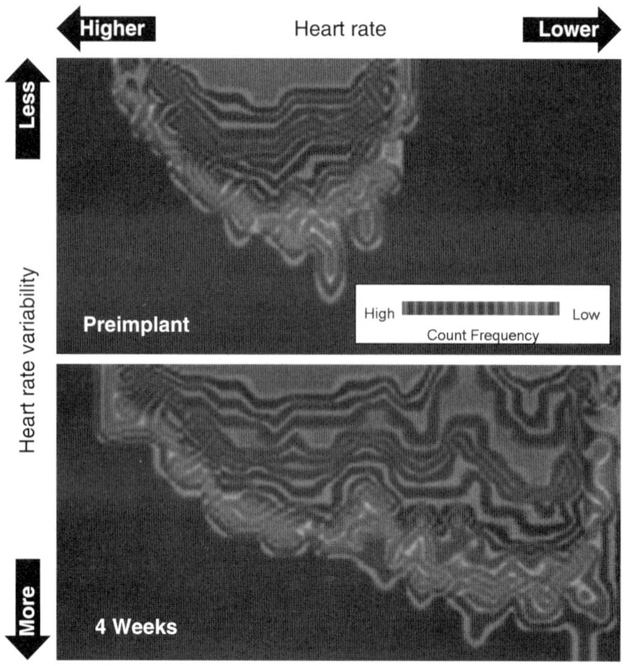

FIGURE 49-7 (See also Color Plate 49-7.) Three-dimensional plot of heart rate variability and heart rate before implantation of a CRT device and after 12 weeks of continuous pacing. At baseline, a markedly depressed heart rate variability was noted as well as a high resting heart rate. After CRT, the heart rate variability largely increased and the heart rate was greatly reduced, thus leading to a rightward shift of the plot. CRT, Cardiac resynchronization therapy.

This could be used for monitoring the patient's activity under heart failure treatment. Initial experience with recording of patient activity tracked by accelerometer has shown that activity monitored by accelerometers or activity log index (ALI) is highly sensitive and specific in detecting the patient's physical activity.[27] Studies have demonstrated that the ALI increased substantially when CRT was applied in a blinded fashion. The magnitudes of ALI change significantly and correlate with the change in the walking distance measured with six-minute hall walk testing (Fig. 49-8).

Devices

The introduction of sophisticated sensors, new monitoring features, and innovative lead technology required substantial changes in the software and hardware of devices specifically designed for providing CRT. Independent pacing outputs for each of the paced chambers with independently programmable AV delay for the right and left ventricle along with variable interventricular delay are currently under evaluation. Specific pacing modes for patients with heart failure and atrial fibrillation are currently being tested. Such new pacing algorithms enable continuous and uninterrupted pacing during atrial fibrillation. It could be shown that the "regularization" algorithm is able to modulate the AV node conduction leading to a more uniform distribution of the R–R interval.

FOLLOW-UP AND PROGRAMMING ISSUES

The follow-up of patients with a CRT device is more time consuming than a standard pacemaker or ICD check-up. The extensive follow-up visit of the heart failure patient needs testing of each of the three leads, careful evaluation of the AV delay, and assessment of the maximal tracking rate.

In the first generation of CRT devices (Contak TR, Guidant Corp., Minneapolis, MN and InSync, Medtronic, Inc. Minneapolis, MN) the two ventricular outputs were electrically connected (i.e., there was a *parallel pacing* of the right and left ventricle). All electrical parameters (i.e., the pacing threshold and R wave sensing) were the algebraic sum of each of the two ventricles. Pacing threshold was indirectly measured by monitoring the

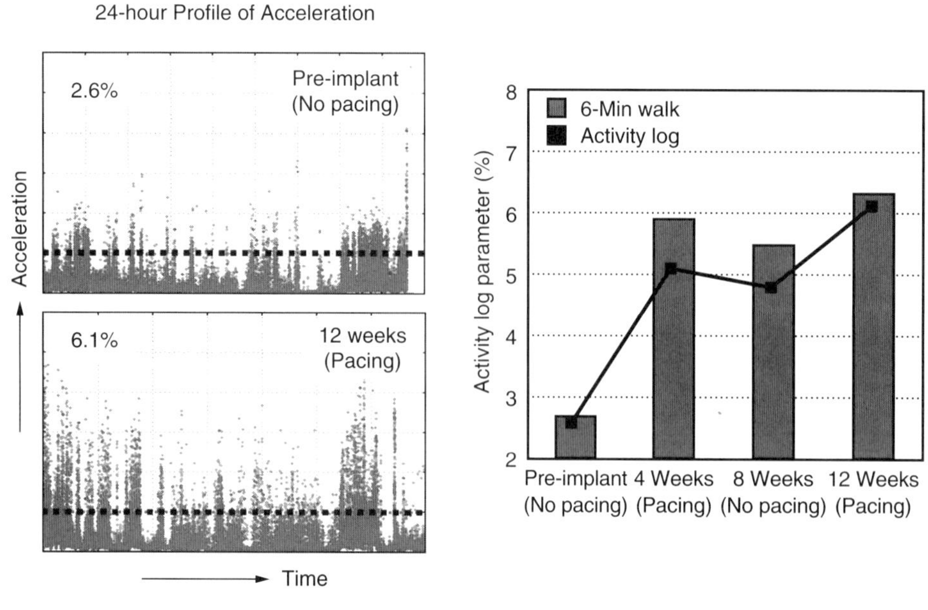

FIGURE 49-8 Changes in the activity log index in a patient with a CRT device. The changes in the activity log index (ALI) correlated well with distance walked during six minutes of testing. CRT, Cardiac resynchronization therapy.

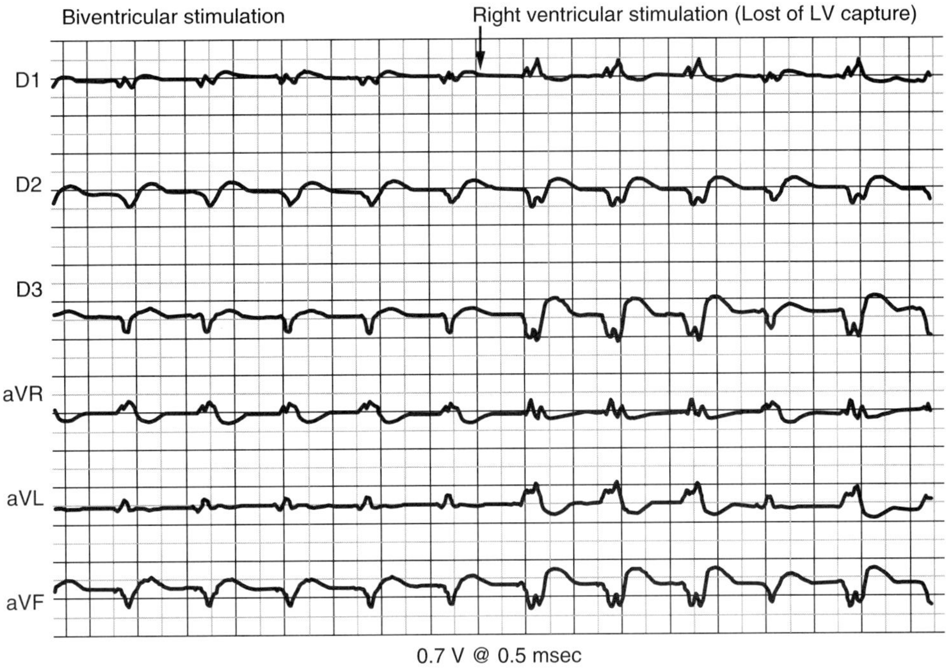

FIGURE 49-9 Evaluation of the pacing threshold in a patient with a first generation CRT device. In this device (Contak TR), the ventricular outputs are electrically connected in parallel in the header of the device. There is no other way for determining the pacing threshold of each chamber than to progressively reduce the output while recording the ECG. The sudden loss of left ventricular capture is characterized by a change in the QRS duration and axis orientation. CRT, Cardiac resynchronization therapy.

ECG signals and changes in the axis orientation as well as the QRS duration (Fig. 49-9). Because the left ventricle often presents with a higher pacing threshold, a sudden change of the axis orientation from a superior rightward orientation (range, −90 to −130°) to a superior leftward rotation (range −45 to −90°) with only right ventricular pacing can be noted (see Fig. 49-9). A change in the duration of the QRS complex is easy to recognize (see Fig. 49-9). Programming the pacing output to at least twice pacing threshold is strongly recommended, because a rise of left ventricular pacing threshold is not unusual during the first few weeks after implantation. New generation devices yield separate output programming for each of the three leads and also automatically performe continuous electrical measurements.

Atrioventricular Delay Optimization

Echocardiographic measurements of the most suitable AV delay are usually performed. Preload optimization methods, namely evaluation of the relationship of the E and A waves with respect to ventricular systolic contraction, are not necessarily as precise as once thought. Because improvement of systolic pump function in patients receiving a CRT device is both preload (AV optimization) and non-preload dependent (improvement of ventricular synchrony), the echocardiographic optimization of the AV delay may lead to the selection of an AV delay that is usually longer than the AV delay, which maximizes the pulse pressure measured during acute testing.[28]

Continuous adjustment of the AV delay is rarely required over the follow-up for patients with QRS duration of greater than 150 ms. This is due to the flat increase of pulse pressure while changing the AV delay within the range of 70 to 150 ms; minor adjustment, however, may not be relevant in this specific subgroup of patients, and A–V interval optimization may be performed by echocardiographic methods of preload (E and A waves). In contrast, patients with narrow QRS complex (120 to 150 ms), the hemodynamically determined optimal AV delay is very close to the intrinsic value. For this reason, optimization methods that use preload-dependent parameters are not suitable. It may be more suitable to optimize the degree of ventricular resynchronization by changes of left ventricular dP/dt_{max} or tissue Doppler imaging, rather than measuring preload.[29,30]

An important issue of AV delay optimization during follow-up visits is whether narrowing of the QRS complex during biventricular or left ventricular pacing alone should be pursued. Some reports indicated a relationship between narrowing QRS and clinical benefit. Acute hemodynamic data and limited electrophysiological evaluations, however, do not support this observation. A good example of this still unresolved controversy is given by the abnormal QRS duration, usually larger than the baseline, and the positive acute and chronic effects (comparable with biventricular pacing) during left ventricular pacing alone.

Evaluation of the Maximal Tracking Rate

To continuously provide resynchronization therapy at rest as well as during exercise, careful assessment of heart rate during physical activity above the anaerobic threshold is important during the follow-up of patients with CRT. We perform a symptom-limited exercise testing before and one month after the implantation of a CRT device in all patients. Almost all patients in severe heart failure demonstrated high resting heart rate and chronotropic incompetence, with a markedly reduced heart rate reserve during exercise. CRT increases both maximal and submaximal exercise capacities and

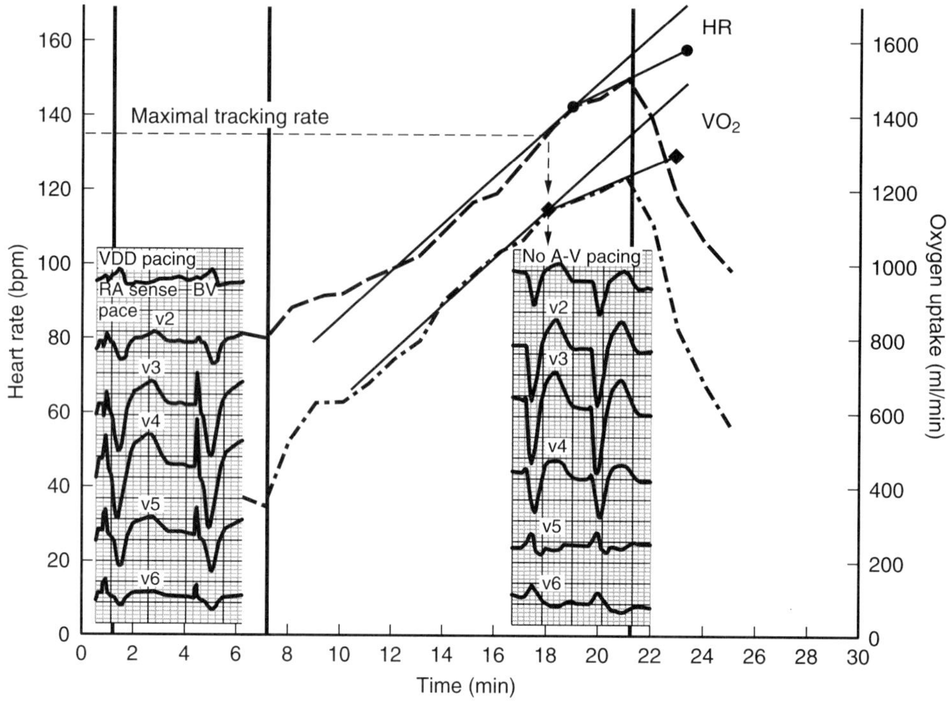

FIGURE 49-10 Loss of CRT at the maximal atrial tracking rate (135 bpm) of the device causes an immediate drop of the continuous increase of heart rate and oxygen pulse in a patient with dilated cardiomyopathy. A 1:1 atrioventricular biventricular stimulation (atrium sensed—both ventricles paced) is attained throughout the exercise testing. Please note the parallel increases of both heart rate and oxygen consumption. As soon as the sensed atrial rate reaches the maximal tracking rate of the device (*dashed line*: 135 bpm in this patient), no CRT is evident as demonstrated by absence of the ventricular spike with sudden prolongation of the P–R interval and QRS duration. The oxygen consumption immediately dropped (*continuous thicker line*). Also the increase of heart rate (*continuous thinner line*) was less pronounced than before the 1:1 synchrony. Drops of the systolic and diastolic blood pressure levels were also observed within several seconds. The exercise testing had been stopped within a few minutes after the loss of the 1:1 AV tracking sequence due to sudden onset of dyspnea and fatigue. AV, Atrioventricular; CRT, cardiac resynchronization therapy.

improves ventilation efficiency and heart rate adaptation during physical activity. When the patient's heart rate overrides the maximal tracking rate of the device, an immediate loss of preexcitation of the left ventricular lateral wall together with a drop in stroke can be measured (Fig. 49-10). A loss of CRT leads to dyspnea and fatigue. Thus, sudden loss of CRT during exercise may prevent the patient from otherwise beneficial physical activity.

Indications for CRT

Cardiac resynchronization is achieved with atrial synchronous biventricular or left ventricular pacing and leads to early hemodynamic and midterm functional improvement in patients who suffer from moderate-to-severe systolic heart failure. This has been demonstrated in several prospective studies (PATH-CHF, MUSTIC, MIRACLE). The long-term benefit of CRT on overall mortality has been demonstrated in some of the trials that have included a larger patient population, such as COMPANION.

The complete understanding of CRT and its effect on the diseased myocardial structure is still awaited. Preliminary results have shown that CRT can lead to reverse remodeling with a significant reduction of left ventricular diameters. A beneficial effect is also achieved with a decrease in mitral regurgitation. We are currently unable to demonstrate whether CRT is directly involved in the improvement of ventricular function or if we are dealing with an indirect effect of improvement by an achievement of better functional capacity with increased oxygen consumption and reduced heart rate. Together, it can be confirmed that CRT leads to significant reduction of heart failure symptoms, improves quality of life, and leads to less frequent hospitalization. Not all patients with heart failure can be considered candidates for CRT, and the current indications for CRT are limited to a group of patients who fulfill the disease and selection criteria used in the clinical trials.

The currently accepted ACC/AHA/NASPE indication for CRT is a class II indication for a device implant[31] and is reproduced below in italics:

Biventricular pacing is indicated in medically refractory, symptomatic NYHA class III or IV patients with idiopathic dilated or ischemic cardiomyopathy, prolonged QRS interval (greater than or equal to 130 ms), LV end-diastolic diameter greater than or equal to 55 mm, and ejection fraction less than or equal to 35%. (Level of Evidence: A)

TABLE 49-3 Patient Selection for Resynchronization Therapy in 2004

Drug—Refractory New York Heart Association class III and IV heart failure
Optimal medical treatment with ACE-I, β-blockers, diuretics, and aldosterone-antagonists
Ischemic and nonischemic cardiomyopathies
Intra- or interventricular conduction abnormality, including left bundle branch block with a QRS duration ≥130 msec
Left ventricular ejection fraction of less than 35%
Left ventricular end-diastolic diameter >55 mm
Stable sinus rhythm

For clinical practice purposes, patient selection criteria are enumerated in Table 49-3. Because the concept of CRT with biventricular pacing is based on morphologically induced severe conduction abnormality within the left or both ventricles, in general, only patients with delayed activation of the left ventricular wall will benefit.[32] Therefore, pacing of the left lateral wall is only indicated in patients who demonstrate a LBBB pattern with a QRS duration of more than 130 ms. Patients with a smaller QRS complex or even a normal QRS duration should not undergo CRT, even though they might have severe heart failure symptoms.

The type of LBBB may play an important role in predicting a beneficial effect of pacing. The standard ECG is, however, unable to predict the real spread of activation within both ventricles. Our own experience[33] has taught us that the exact spread of activation in patients with LBBB can only precisely assessed with electroanatomic mapping. This, however, is difficult to perform in daily practice. In some cases it may, indeed, help to select the optimal pacing site by accurately detecting the delayed region of activation of the left ventricle. Patients with a broad QRS complex, but demonstrating a right bundle branch block pattern, have rarely demonstrated benefit from pacing and should not be considered candidates for CRT unless future studies show that pacing site on the right ventricle will also achieve resynchronization of the failing right ventricle.

Recently published data[34-36] have shown that CRT is able to reduce abnormal myocardial strain distribution and induce reverse remodeling, demonstrating a significant decrease of left ventricular volume within the first six months after initiation of CRT. This may be due to a reduction of regional wall stress or a reduction of increased oxygen demand of the asynchronously contracting ventricles. There might be, however, a critical size of the left ventricular end-diastolic or end-systolic volume where reverse remodeling cannot be achieved. One recent study has demonstrated that CRT produced improvement of the New York Heart Association functional classification, left ventricular ejection fraction, left ventricular end-diastolic and end-systolic diameters, mitral regurgitation area, and interventricular delay, and deceleration time, in patients with intra- and interventricular dysynchronies, irrespective of normal or prolonged QRS duration. The six-minute walk test improved in both groups.[37]

The majority of patients in whom CRT was beneficially performed were in stable functional class III or IV, whereas in patients with less severe heart failure symptoms, benefits of CRT were less prominent or even absent. Therefore, patients in NYHA class II are poorer candidates for CRT. The value of prophylactic use of CRT to avoid the progression of left ventricular dysfunction is currently not confirmed by studies.[32]

Biventricular pacing is not a "stand-alone" therapy or a "replacement" therapy for medical therapy in heart failure patients. CRT should always be an additional step in therapy when drug therapy is unable to relieve symptoms or improve quality of life. Medical therapy of heart failure, as it is currently recommended in various guidelines, should be thoroughly tried before CRT is initiated. A careful titration—lasting over months—of ACE inhibitors, β-blocking agents, and diuretic compounds, including aldosterone antagonists, is mandatory and should be continued after initiating CRT.

To achieve the full benefit of CRT, it is necessary to improve left-sided atrioventricular synchronization, which can also reduce concomitant existing mitral regurgitation. Thus, it is preferable that the patient can maintain sinus or an atrial paced rhythm. Atrial pacing also may be necessary because of the use of high dose β-blocker therapy in these patients. However, programming the optimal AV delay during atrial pacing in patients with heart failure has not been fully investigated. The currently available results of CRT and atrial fibrillation are less promising and lack significant clinical improvement in several studies. However, one recent clinical study shows benefits of CRT in this population.[38] Therefore, the indication for CRT in patients with permanent atrial fibrillation is still being debated. Performing an ablation of the bundle of His may be a solution to achieve a permanent tracking of both ventricles, but data are lacking.

The role of the underlying disease causing heart failure and influencing the outcome of CRT has been debated. The use of CRT in coronary artery disease can be as effective as in patients with idiopathic dilated cardiomyopathy of nonischemic origin—where there is more consensus on CRT benefit. More importantly, patient outcome is dependent on inherent progression of the basic disease process and comorbidities, such as diabetes or renal failure. Whether or not reverse remodeling achieved by CRT will reduce the incidence of sudden arrhythmic death still remains unproved.

The guidelines for ICD implantation in patients with severely depressed ventricular function are also applicable to candidates for CRT.[39,40] The standard ICD indications (secondary or primary prevention of sudden death) can be applied once the patient has become a candidate for CRT. Devices that incorporate ICD therapy with biventricular pacing are now available for this group of patients. Two major issues have to be solved: (1) Should all patients in whom biventricular pacing promises additional benefit with CRT on top of the best medical therapy receive additional ICD back-up (because sudden cardiac death strikes more often in patients with moderate-to-severe heart failure, low ejection fraction, and broad QRS complex)? (2) Should all patients who need ICD therapy because of ventricular tachyarrhythmic events or because they meet a MADIT I—or now MADIT II—ICD indication receive additional CRT (because most patients with life-threatening VT/VF events have poor ventricular function and are most often in heart failure)?

IMPLANTABLE CARDIOVERTER DEFIBRILLATOR THERAPY IN HEART FAILURE POPULATIONS

Mortality in patients with heart failure (HF) results from a variety of causes. These can include ventricular tachyarrhythmias as well as ventricular fibrillation (VF), primary bradyarrhythmias, and conduction disturbances resulting in asystolic cardiac arrest, bradycardia or tachycardia-dependent polymorphic ventricular tachycardia (VT), and mechanical pump failure. Sudden cardiac death is a catastrophic event. The annual incidence increases from 2% to 6% per year in patients with NYHA class II symptoms and up to 24% per year for patients with class III or IV symptoms.[41] Dual chamber ICD devices have traditionally been employed in these patients but some pacing features may be deleterious. Data from the DAPPAF and DAVID studies suggest that adverse physiologic effects and outcomes result from chronic right ventricular apical pacing.[42,43] CRT can be considered when intraventricular conduction disturbances, as discussed earlier, are present. However, the use of CRT in ICD devices is primarily based on the assumption that sudden death prevention in heart failure populations will provide survival benefits not seen with pacing alone. In a follow-up of 153 patients receiving biventricular-pacing therapy alone, there was a 20% to 36% mortality rate, of which 33% to 47% was due to sudden death.[44]

This new hybrid approach (i.e., CRT and defibrillation therapy) has posed special challenges. In the early phase, CRT patients who needed ICD therapy in addition to CRT had to receive a biventricular pacing device and a separate ICD. This was associated with technical problems mostly related to interference between both devices. In the meantime, the technical advancement in the industry has developed devices that combine biventricular pacing and ICD therapy, thus solving many of the initial problems of interference. However, double counting of the sensed QRS complex due to simultaneous inputs from the right and left ventricle often appeared. The most recent generation of ICDs with CRT have independently programmable ventricular and atrial channels so that most of these technical issues have been solved (Fig. 49-11).

Early ICD trials did not systematically select patients with advanced heart failure patients despite observational data and clinical trial subgroup analyses that suggested major benefits in patients with severe left ventricular dysfunction. Several early analyses[45-48] reported improved survival rates and effective use of ICD therapy in patients with severe LV dysfunction. In the AVID study, survival of patients with ejection fraction (EF) below 35% was superior in the ICD arm.[48] The survival benefit of ICD patients in the MADIT I study was almost entirely confined to patients with an EF below the median value of 26%.[49] The Canadian Implantable Defibrillator Study (CIDS) showed the greatest benefit was derived by patients in the highest-risk quartiles, namely a low EF and a poorer NYHA functional classification.[50]

Table 49-4 summarizes the experience with biventricular pacing and ICD therapy in multicenter clinical trials in this population. Initial experience with an ICD incorporating ventricular resynchronization therapy was assessed in a prospective study using the Insync model 7272 ICD (Medtronic Inc., Minneapolis, MN).[51] Eighty-one patients with class I indications for ICD having symptomatic heart failure despite optimal medical therapy, LV ejection fraction of less than 35%, and QRS duration of more than 120 ms received this device.

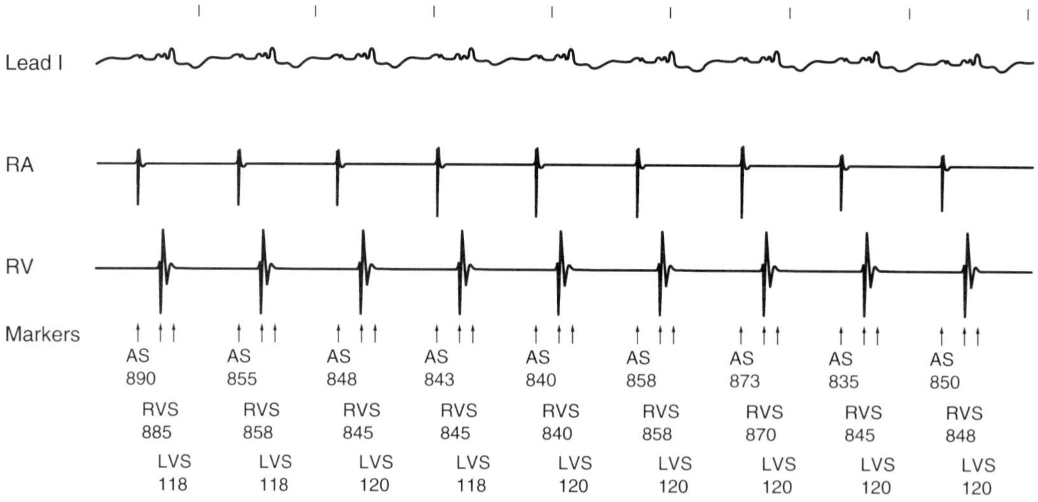

FIGURE 49-11 Printout of an ICD-CRT device (Renewal II, Guidant). Simultaneous intracardiac recording of the atrium and of both ventricles. Annotation markers help in correct classification of the device sensing. ICD-CRT, Implantable cardioverter defibrillator-cardiac resynchronization therapy; RA, right atrium; RV, right ventricle.

TABLE 49-4 ICD and CRT Device Trials in Heart Failure Populations

Trials	Device	Patient Profile	No. of Patients	Follow-Up (months)	Benefits
MADIT II	ICD	POST MI EF <30%	1232	20	31% decrease in the risk of death
PATH-CHF	BiV Pacing	Short-term and long-term LV/BiV pacing in patients with heart failure and increased IVCD	53	12	Improved six-minute walk distance, LV EF, and QOL
MUSTIC	BiV Pacing	NYHA III AND IVCD*	131	12	Increase in six-minute walk distance 20%, EF 5%, QOL 36%, NYHA 25%
INSYNC ICD	BiV ICD	EF <35%, QRS >130 ms	84	6	Improved six-minute walk distance, NYHA, and QOL
COMPANION	BiV ICD	NYHA III and IV, QRS >120 ms, PR >150 ms, EF <35%, LVEDD >60 mm	1634	12	19% decrease in all-cause mortality and 39.5% decrease in heart failure, hospitalizations, and mortality

Biv, biventricular; EF, ejection fraction; ICD, implantable cardioverter defibrillator; IVCD, intraventricular conduction delay; LV, left ventricular; MI, myocardial infarction; NYHA, New York Heart Association (classification); QOL, quality of life.

Reproduced from Rao BH, Saksena S: Implantable cardioverter-defibrillators in cardiovascular care: Technologic advances and new indications. Curr Opin Crit Care 2003;9:362-368.

There was significant improvement of heart failure symptoms and LV dimensions in these patients, particularly those in the NYHA classes III and IV. Patients significantly improved in the six-minute walk test at three and six months. All ventricular tachyarrhythmias were correctly identified and double counting of sensed QRS events did not occur.

The available results from the CONTAK-CD trial were also promising. Of the 4990 randomized patients enrolled in the study, all having class I indication for ICD therapy and being in heart failure NYHA class II or more, a 21% reduction of overall mortality was found. Although this did not reach statistical significance ($P < 0.17$), there seemed to be a benefit of biventricular pacing along with ICD therapy. A 23% reduction of death from all causes, a 13% lower need for hospitalization, a 26% lowering of worsening heart failure during the follow-up period, and an almost 10% reduction of tachyarrhythmic events occurred. There was a significant improvement of the peak oxygen consumption at peak exercise with even more benefit in patients with more advanced heart failure (functional NYHA class III or IV).

Long-term outcome data on clinically relevant endpoints, such as hospitalizations for heart failure, survival, and cardiac morbidity have been examined in the Comparison of Medical Therapy, Pacing and Defibrillation in Heart Failure (COMPANION) trial.[52] This is a randomized three-arm trial of patients in NYHA class III or IV with an LVEF of less than 35% and a prolonged QRS duration (>120 ms) and LV dilation (LV end-diastolic diameter >60 mm). The stated objectives of the study were to determine whether optimal pharmacologic therapy used with CRT alone or CRT in combination with cardioverter defibrillator was superior to optimal pharmacologic treatment alone in reducing combined all-cause mortality, hospitalizations, and cardiac morbidity, and in improving functional capacity cardiac performance and quality of life. In this study, CRT alone improved NYHA classification, quality of life, and reduced heart failure hospitalizations. However, significant mortality was achieved only in the CRT-ICD arm. These data are consistent with the original pacing trials in this population, where it appears that sudden death can limit the benefits achieved with CRT and challenges the notion that CRT per se reduced SCD. However, the delayed separation (after nine months) of mortality curves ($P = .12$) between the pacing and medical therapy arms in this study raises the possibility of ventricular remodeling trending to improving survival. The impact of CRT-ICD therapy is potentially huge. Initial estimates of ICD patients being suitable candidates for biventricular pacing have been revised from 7.3% to 25% to 40% in different analyses.[50,51]

The PACMAN trial is not a mortality study. The primary endpoint is functional improvement measured by six-minute walk testing of patients in functional NYHA class III, wide QRS complex (>150 ms) and left ventricular ejection fraction of 35% or less. The trial includes, in one arm, patients who have a history of cardiac arrest due to VT or VF, or are patients who have the criteria fulfilled in the MADIT I study. The second arm enrolls patients with biventricular pacing and optimal drug treatment of heart failure alone. The PACMAN uses patients as their own controls. After the first six months of randomized pacing mode (biventricular pacing on or off), all patients will have biventricular pacing activated.

OUTLOOK AND FUTURE DEVELOPMENTS

During the last decade of clinical pacing there has been a growing interest of the hemodynamic effect of the "abnormal" sequence of activation induced by ventricular pacing.[53] There is, indeed, increasing evidence that chronic pacing for bradycardia at the right ventricular apex is associated with structural changes of the heart and may be associated with the induction of heart failure. Animal studies showed that chronic changes during continuous right ventricular pacing consisted of asymmetrical hypertrophy, ventricular dilation, fiber disarray, increased myocardial catecholamine concentration, and impaired perfusion.[53] The potential worsening of pump function during ventricular pacing is of great practical importance, because the right ventricular apex is traditionally the most frequently used pacing site. If a left or right ventricular conduction delay is preexisting, it is not unlikely that abnormal electrical activation induced by right ventricular pacing may further increase the three-dimensional temporal and spatial derangement. The long-term effect of such a heterogeneous distribution of stretch and work load may be responsible for triggering maladaptation mechanisms that might lead to decreases in pump function, pump failure, or progression of preexisting heart failure.[54]

Appropriate stimulation of both ventricles can reduce, or perhaps normalize, the spatio-temporal distribution of activation and restore the mechanical synchrony of contraction. Recent animal studies showed that biventricular stimulation compared with right ventricular pacing reduced the temporal asynchrony and the spatio-temporal asynchrony of the left ventricular midwall contraction.[55] Proper stimulation in canines has demonstrated that ventricular function is restored to baseline levels after induction of LBBB by radiofrequency ablation of the left bundle.[56] These data, however, have been generated in healthy canine hearts, which might largely differ from a diseased human heart. The issue as to whether patients in need of standard pacemaker therapy should undergo conventional, biventricular, or left ventricular pacing remains to be determined and deserves further research.

There are important and challenging questions concerning hybrid CRT-ICD therapy. One of the most important issues is whether electrocardiographic evidence of left bundle branch block is an ideal selection criterion for patients with heart failure who are candidates for CRT.[57] Indeed, increasing evidence points to the presence of mechanical evidence of dyssynchrony as the major basis for CRT intervention. This will need serious study in the future. The optimal site of CRT electrodes will then become open to examination. CRT could, therefore, be applied in two or more sites in either ventricle. In fact, newer techniques with minimally invasive and robotic lead epicardial placement can make this a real possibility and are an important avenue of future research.[58]

The main goal of the hybrid CRT-ICD device technology is to achieve the most reliable sensing of biventricular signals, as well as atrial signals, and the lowest defibrillation threshold. It is unclear whether relief of heart failure symptoms and successful resynchronization of ventricular function will reduce the incidence of ICD activation and this is being actively examined. The question as to whether anti-tachycardia pacing or shock therapy should be delivered in the left ventricle or delivered through both ventricular leads needs further evaluation. The success of CRT combined with ICD backup will depend on a low incidence of inappropriate ICD discharges, few technologic failures, and appropriate lead position on the left ventricle. There is no doubt that inclusion of atrial fibrillation therapies or preventive pacing in patients with atrial fibrillation will further increase the technical challenges in this type of device.

SUMMARY

Treatment of heart failure by cardiac resynchronization therapy is one of the most fascinating new avenues of research in clinical cardiology. It links hemodynamics to electrophysiology. New electromechanical pathophysiologic concepts will need be to be presented, which will impact the various concepts of treatment for heart failure. There are several acute studies and promising data suggesting that combining hemodynamically optimized ventricular resynchronization therapy (atrial-synchronous biventricular or univentricular pre-excitation) with optimal medical therapy are most effective in treating heart failure. They improve ventricular function without stressing cardiac reserve or increasing cardiac metabolic demand. The resynchronization of ventricular contraction improves exercise capacity, functional classification, and quality of life in patients with moderate-to-severe heart failure, sinus rhythm, and ventricular conduction delay. CRT appears to achieve this goal by improving left ventricular systolic function through restoring synchronous ventricular contraction and optimizing atrioventricular synchrony.

Long-term outcome studies must confirm these preliminary observations. Better devices and leads specifically designed for heart failure treatment are mandatory and new algorithms for monitoring the immediate, short- and long-term effects of ventricular resynchronization therapy need to be developed in the future. If these expectations are achieved, new indications for implantation of both pacemakers and ICDs will emerge, and an impact on the important public health problem of heart failure can be anticipated.

REFERENCES

1. Cohn JN, Johnson G, Ziesche S, et al: A comparison of enalapril with hydralazine-isosorbide dinitrate in the treatment of chronic congestive heart failure. N Engl J Med 1991;325:303-310.
2. Captopril Multicenter Research Group: A placebo-controlled trial of captopril in refractory chronic congestive heart failure. J Am Coll Cardiol 1983;2:755-763.
3. Herman MV, Heinle RA, Klein MD, Gorlin R: Localized disorders in myocardial contraction: Asynergy and its role in congestive heart failure. N Engl J Med 1967;277:222-232.
4. Xiao HB, Roy C, Fujimoto S, Gibson DG: Natural history of abnormal conduction and its relation to prognosis in patients with dilated cardiomyopathy. Int J Cardiol 1996;53:163-170.
5. Aaronson KD, Schwartz S, Chen T-M, et al: Development and prospective validation of a clinical index to predict survival in ambulatory patients referred for cardiac transplant evaluation. Circulation 1997;95:2660-2667.
6. Shamim W, Francis DP, Yousufuddin M, et al: Intraventricular conduction delay: A prognostic marker in chronic heart failure. Int J Cardiol 1999;70:171-178.
7. Grines CL, Bashore TM et al: Functional abnormalities in isolated left bundle branch block. The effect of interventricular asynchrony. Circulation 1989;79:845-853.
8. Xiao HB, Brecker SJD, Gibson DG: Differing effects of right ventricular pacing and left bundle branch block on left ventricular function. Br Heart J 1993; 69:166-173.

9. Curry CW, Nelson GS, Wyman BT, et al: Mechanical dyssynchrony in dilated cardiomyopathy with intraventricular conduction delay as depicted by 3D tagged magnetic resonance imaging. Circulation 2000;101:E2.
10. Xiao HB, Roy C, Gibson DG: Nature of ventricular activation in patients with dilated cardiomyopathy: Evidence for bilateral bundle branch block. Br Heart J 1994;72:167-174.
11. Saxon LA, Kerwin WF, Cahalan MK, et al: Acute effects of intraoperative multisite ventricular pacing on left ventricular function and activation/contraction sequence in patients with depressed ventricular function. J Cardiovasc Electrophysiol 1998; 9:13-21.
12. Freedman RA, Yock PG, Echt DS, Popp RL: Effect of variation in PQ interval on pattern of atrioventricular valve motion and flow in patients with normal ventricular function. J Am Coll Cardiol 1986;7:595-602.
13. Ishikawa T, Sumita S, Kimura K, et al: Critical PQ interval for the appearance of diastolic mitral regurgitation and optimal PQ interval in patients implanted with DDD pacemakers. PACE Pacing Clin Electrophysiol 1994;17:1989-1994.
14. Auricchio A, Salo RW: Acute hemodynamic improvements by pacing in patients with severe congestive heart failure. Pacing Clin Electrophysiol 1997;20:313-324.
15. Kass DA, Chen CH, Curry C, et al: Improved left ventricular mechanics form acute VDD pacing in patients with dilated cardiomyopathy and ventricular conduction delay. Circulation 1999;99:1567-1573.
16. Nelson GS, Berger RD, Fetics BJ, et al: Left ventricular or biventricular pacing improves cardiac function at diminished energy cost in patients with dilated cardiomyopathy and left bundle-branch block. Circulation 2000;102:3053-3059.
17. Auricchio A, Stellbrink C, Sack S, et al: The PATH-CHF Study Investigators. The Pacing Therapies for Congestive Heart Failure (PATH-CHF) Study: Rationale, design and end-points of a prospective randomized multicenter study. Am J Cardiol 1999; 83:130D-135D.
18. Cazeau S, Leclercq C, Lavergne T, et al: Effects of multisite biventricular pacing in patients with heart failure and intraventricular conduction delay. N Engl J Med 2001;344:873-880.
19. Auricchio A, Stellbrink C, Sack S, et al: The Pacing Therapies in Congestive Heart Failure Study Group: Long-term clinical effect of hemodynamically optimized cardiac resynchronization therapy in patients with heart failure and ventricular conduction delay. J Am Coll Cardiol 2002;39:2026-2033.
20. Saxon LA, DeMarco T, Chatterjee K, Boehmer J: The VIGOR-CHF Investigators: The magnitude of sympathoneural activation in advanced heart failure is altered with chronic biventricular pacing [abstract]. PACE Pacing Clin Electrophysiol 1998;21:914.
21. Lau CP, Barold S, Tse HF, et al: Advances in devices for cardiac resynchronization in heart failure. J Interv Card Electrophysiol 2003;9:167-181.
22. Auricchio A, Klein H, Tockman B, et al: Transvenous biventricular pacing for heart failure: Can the obstacles be overcome? Am J Cardiol 1999;83:136D-142D.
23. Daubert CJ, Ritter P, LeBreton H, et al: Permanent left ventricular pacing with transvenous leads inserted into the coronary veins. PACE 1998;21:239-245.
24. Cahalin LP, Mathier MA, Semigran MJ, et al: The six-minute walk test predicts peak oxygen uptake and survival in patients with advanced heart failure. Chest 1996;110:325-332.
25. Bittner V, Weiner DH, Yusuf S, et al: Prediction of mortality and morbidity with a 6-minute walk test in patients with left ventricular dysfunction. JAMA 1993;270:1702-1707.
26. Pinna D, Opasich C, Mazza A, et al: Reproducibility of the six-minute walking test in chronic heart failure patients. Stat Med 2000;19:3087-3094.
27. Kadhiresan VA, Pastore J, Auricchio A, et al: The PATH-CHF Study Group: A novel method—the activity log index—for monitoring physical activity of patients with heart failure. Am J Cardiol 2002;89:1435-1437.
28. Auricchio A, Stellbrink C, Sack S, et al: The Pacing Therapies in Congestive Heart Failure Study Group: Long-term effect of hemodynamically optimized cardiac resynchronization therapy in patients with heart failure and ventricular conduction delay. J Am Coll Cardiol 2002;39:2026-2033.
29. Ansalone G, Giannantoni P, Ricci R, et al: Doppler myocardial imaging to evaluate the effectiveness of pacing sites in patients receiving biventricular pacing. J Am Coll Cardiol 2002;39: 489-499.
30. Penicka M, Bartunek J, De Bruyne B, et al: Improvement of left ventricular function after cardiac resynchronization therapy is predicted by tissue Doppler imaging echocardiography. Circulation 2004;109:978-983.
31. Gregoratos G, Abrams J, Epstein AE, et al: Guideline update for implantation of cardiac pacemakers and antiarrhythmia devices—summary article: A report of the American College of Cardiology/American Heart Association Task Force on Practice Guidelines (ACC/AHA/NASPE Committee to Update the 1998 Pacemaker Guidelines). J Am Coll Cardiol 2002;40: 1703-1719.
32. Saxon L, De Marco T: NASPE Expert Consensus Statement: Resynchronization therapy for heart failure. E pub. December 2003; www. naspe.org.
33. Auricchio A, Stellbrink C, Butter C, et al: Clinical efficacy of cardiac resynchronization therapy using left ventricular pacing in heart failure patients stratified by severity of ventricular conduction delay. J Am Coll Cardiol 2003;42:2109-2116.
34. Yu CM, Fung WH, Lin H, et al: Predictors of left ventricular reverse remodeling after cardiac resynchronization therapy for heart failure secondary to idiopathic dilated or ischemic cardiomyopathy. Am J Cardiol 2003;15;91:684-688.
35. St John Sutton MG, Plappert T, Abraham WT, et al: Multicenter InSync Randomized Clinical Evaluation (MIRACLE) Study Group: Effect of cardiac resynchronization therapy on left ventricular size and function in chronic heart failure. Circulation 2003;22:1985-1990.
36. Breithardt OA, Stellbrink C, Herbots L, et al: Cardiac resynchronization therapy can reverse abnormal myocardial strain distribution in patients with heart failure and left bundle branch block. J Am Coll Cardiol 2003;42:486-494.
37. Achilli A, Sassara M, Ficili S, et al: Long-term effectiveness of cardiac resynchronization therapy in patients with refractory heart failure and "narrow" QRS. J Am Coll Cardiol 2003;42: 2117-2124.
38. Leclercq C, Walker S, Linde C, et al: Comparative effects of permanent biventricular and right-univentricular pacing in heart failure patients with chronic atrial fibrillation. Eur Heart J 2002; 23:1780-1787.
39. Gregaratos G, Cheitlin MD, Conill A, et al: ACC/AHA guidelines for implantation of cardiac pacemakers and antiarrhythmia devices. J Am Coll Cardiol 1998:31:1175-1209.
40. Gregaratos G, Abrams J, Epstein AE, et al: ACC/AHA/NASPE 2002 guideline update for implantation of cardiac pacemakers and antiarrhythmia devices: Summary article. A report of the American College of Cardiology/American Heart Association Task Force on Practice Guidelines (ACC/AHA/NASPE Committee to Update the 1998 Pacemaker Guidelines). J Cardiovasc Electrophysiol 2002;13:1183-1199.
41. Goldman S, Johnson G, Cohn J, et al: Mechanisms of death in heart failure. Circulation 1993;87:V124-131.
42. Prakash A, Saksena S, Ziegler P, et al: DAPPAF Investigators. Dual site atrial pacing for prevention of atrial fibrillation (DAPPAF) trial: Echocardiographic evaluation of atrial and ventricular function during a randomized trial of support: High right atrial and dual site right atrial pacing. Pacing Clin Electrophysiol 2001;24:579.
43. Wilkoff BL, Cook JR, Epstein AE: Dual-chamber pacing or ventricular backup pacing in patients with an implantable defibrillator: The Dual Chamber and VVI Implantable Defibrillator (DAVID) Trial. JAMA 2002;288:3115-3131.
44. Gras D, Leclercq C, Anthony SL, et al: Cardiac Resynchronization therapy in advanced heart failure: The multicenter InSync Clinical Study. Eur J Heart Failure 2002;4:311-320.
45. Tchou PJ, Kadri N, Anderson J, et al: Automatic implantable cardioverter defibrillators and survival in patients with left ventricular dysfunction and malignant ventricular arrhythmias. Ann Intern Med 1988;109:529-534.
46. Axtell K, Tchou P, Akhtar M: Survival in patients with depressed left ventricular function treated by implantable cardioverter defibrillator. Pacing Clin Electrophysiol 1991;14:291-296.

47. Mehta D, Saksena S, Krol RB, et al: Device use patterns and clinical outcome of implantable cardioverter defibrillator patients with moderate and severe impairment of left ventricular function. Pacing Clin Electrophysiol 1993;16:179-185.
48. Domanski MJ, Saksena S, Epstein AE, et al: Relative effectiveness of the implantable cardioverter-defibrillator and antiarrhythmic drugs in patients with varying degrees of left ventricular dysfunction who have survived malignant ventricular arrhythmias. AVID investigators. J Am Coll Cardiol 1999;34:1090-1095.
49. Moss AJ: Implantable cardioverter defibrillator therapy: The sickest patients benefit the most. Circulation 2000;101:1638-1640.
50. Sheldon R, Connoly S, Krahn AA, et al: Identification of patients most likely to benefit from implantable cardioverter-defibrillator therapy: The Canadian Implantable Defibrillator Study. Circulation 2000;101:1660-1664.
51. Kuhlkamp V: Initial experience with an implantable cardioverter-defibrillator incorporating cardiac resynchronization therapy. J Am Coll Cardiol 2002;39:790-797.
52. Feldman A, Bristow MR: The Companion Investigators. Comparison of medical therapy, pacing, and defibrillation in chronic heart failure. Presented at the 52nd annual scientific sessions, 2003.
53. Badke FR, Boinay P, Covell JW: Effects of ventricular pacing on regional left ventricular performance in the dog. Heart Circ Physiol 1980;7:H858-867.
54. Leclercq C, Faris O, Halperin H, et al: Regional disparity of calcium handling and stress protein expression in failing hearts with dyssynchronous contraction. Circulation 2001;104:II-128-II.
55. Prinzen FW, Van Oosterhout MF, Vanagt WY, et al. Optimization of ventricular function by improving the activation sequence during ventricular pacing. Pacing Clin Electrophysiol 1998;21:2256-2260.
56. Prinzen FW, Hunter WC, Wyman BT, et al: Mapping of regional myocardial strain and work during ventricular pacing: Experimental study using magnetic resonance imaging tagging. J Am Coll Cardiol 1999;33:1735-1742.
57. Saksena S: Bundle branch block and cardiac resynchronization therapy: Do we need to look further before we leap? J Interv Card Electrophysiol 2003;8:163-164.
58. Jansens JL, Jottrand M, Preumont N, et al: Robotic-enhanced biventricular resynchronization: An alternative to endovenous cardiac resynchronization therapy in chronic heart failure. Ann Thorac Surg 2003;76:413-417.
59. Rao BH, Saksena S: Implantable cardioverter-defibrillators in cardiovascular care: Technologic advances and new indications. Curr Opin Crit Care 2003;9:362-368.

Index

Note: Page numbers followed by the letter f refer to figures; those followed by the letter t refer to tables.

B

C

D

E

F

G

H

I

J

K

L

M

Q

R

V

W

X

Z